COMPREHENSIVE MEDICINAL CHEMISTRY II

COMPREHENSIVE MEDICINAL CHEMISTRY II

Editors-in-Chief

Dr John B Taylor

Former Senior Vice-President for Drug Discovery, Rhône-Poulenc Rorer, Worldwide, UK

Professor David J Triggle

State University of New York, Buffalo, NY, USA

Volume 6

THERAPEUTIC AREAS I: CENTRAL NERVOUS SYSTEM, PAIN, METABOLIC SYNDROME, UROLOGY, GASTROINTESTINAL AND CARDIOVASCULAR

Volume Editor

Dr Michael Williams

Northwestern University, Chicago, IL, USA

ELSEVIER

AMSTERDAM BOSTON HEIDELBERG LONDON NEW YORK OXFORD
PARIS SAN DIEGO SAN FRANCISCO SINGAPORE SYDNEY TOKYO

Elsevier Ltd.
The Boulevard, Langford Lane, Kidlington, Oxford OX5 1GB, UK

First edition 2007

British Library Cataloguing in Publication Data
A catalogue record for this book is available from the British Library

Library of Congress Catalog Number: 2006936669

ISBN-13: 978-0-08-044513-7
ISBN-10: 0-08-044513-6

Printed and bound in Spain

06 07 08 09 10 10 9 8 7 6 5 4 3 2 1

Disclaimers

Both the Publisher and the Editors wish to make it clear that the views and opinions expressed in this book are strictly those of the Authors. To the extent permissible under applicable laws, neither the Publisher nor the Editors assume any responsibility for any loss or injury and/or damage to persons or property as a result of any actual or alleged libellous statements, infringement of intellectual property or privacy rights, whether resulting from negligence or otherwise.

Knowledge and best practice in this field are constantly changing. As new research and experience broaden our knowledge, changes in practice, treatment and drug therapy may become necessary or appropriate. Readers are advised to check the most current information provided (i) on procedures featured or (ii) by the manufacturer of each product to be administered, to verify the recommended dose or formula, the method and duration of administration, and contraindications. It is the responsibility of the practitioner, relying on their own experience and knowledge of the patient, to make diagnoses, to determine dosages and the best treatment for each individual patient, and to take all appropriate safety precautions. To the fullest extent of the law, neither the Publisher, nor Editors, nor Authors assume any liability for any injury and/or damage to persons or property arising out or related to any use of the material contained in this book.

Contents

Contents of all Volumes

Volume 7 Therapeutic Areas II: Cancer, Infectious Diseases, Inflammation & Immunology and Dermatology

Volume 8 Case Histories and Cumulative Subject Index

Preface

The first edition of *Comprehensive Medicinal Chemistry* was published in 1990 and was intended to present an integrated and comprehensive overview of the then rapidly developing science of medicinal chemistry from its origins in organic chemistry. In the last two decades, the field has grown to embrace not only all the sophisticated synthetic and technological advances in organic chemistry but also major advances in the biological sciences. The mapping of the human genome has resulted in the provision of a multitude of new biological targets for the medicinal chemist with the prospect of more rational drug design (CADD). In addition, the development of sophisticated in silico technologies for structure–property relationships (ADMET) enables a much better understanding of the fate of potential new drugs in the body with the subsequent development of better new medicines.

It was our ambitious aim for this second edition, published 16 years after the first edition, to provide both scientists and research managers in all relevant fields with a comprehensive treatise covering all aspects of current medicinal chemistry, a science that has been transformed in the twenty-first century. The second edition is a complete reference source, published in eight volumes, encompassing all aspects of modern drug discovery from its mechanistic basis, through the underlying general principles and exemplified with comprehensive therapeutic applications. The broad scope and coverage of *Comprehensive Medicinal Chemistry II* would not have been possible without our panel of authoritative Volume Editors whose international recognition in their respective fields has been of paramount importance in the enlistment of the world-class scientists who have provided their individual 'state of the science' contributions. Their collective contributions have been invaluable.

Volume 1 (edited by Peter D Kennewell) overviews the general socioeconomic and political factors influencing modern R&D in both the developed and developing worlds. Volume 2 (edited by Walter H Moos) addresses the various strategic and organizational aspects of modern R&D. Volume 3 (edited by Hugo Kubinyi) critically reviews the multitude of modern technologies that underpin current discovery and development activities. Volume 4 (edited by Jonathan S Mason) highlights the historical progress, current status, and future potential in the field of computer-assisted drug design (CADD). Volume 5 (edited by Bernard Testa and Han van de Waterbeemd) reviews the fate of drugs in the body (ADMET), including the most recent progress in the application of 'in silico' tools. Volume 6 (edited by Michael Williams) and Volume 7 (edited by Jacob J Plattner and Manoj C Desai) cover the pivotal roles undertaken by the medicinal chemist and pharmacologist in integrating all the preceding scientific input into the design and synthesis of viable new medicines. Volume 8 (edited by John B Taylor and David J Triggle) illustrates the evolution of modern medicinal chemistry with a selection of personal accounts by eminent scientists describing their lifetime experiences in the field, together with some illustrative case histories of successful drug discovery and development.

We believe that this major work will serve as the single most authoritative reference source for all aspects of medicinal chemistry for the next decade and it is intended to maintain its ongoing value by systematic electronic upgrades. We hope that the material provided here will serve to fulfill the words of Antoine de Saint-Exupery (1900–44) and allow future generations of medicinal chemists to discover the future.

'As for the future, your task is not to foresee it but to enable it'
Citadelle (1948)

John B Taylor and David J Triggle

Preface to Volume 6

Since the publication of the first edition of *Comprehensive Medicinal Chemistry* some 15 years ago, the fundamental processes involved in drug discovery have undergone many remarkable changes. These include the mapping of the human genome, with attendant expectations of new, safer drug targets, the promise of which has yet to be realized; numerous screening and lead optimization technologies; and advancements in the assessment of the pharmacokinetic (absorption, distribution, metabolism, and excretion (ADME)) properties of new chemical entities (NCEs) at a much earlier stage of the NCE selection and optimization process. Despite such changes, as of 2006, productivity in the form of new drug approvals (NDAs) has steadily declined. While being potentially attributed to: (1) a lack of investment by the pharmaceutical industry in R&D; (2) a consideration that all the easy problems (e.g., the proverbial 'low-hanging fruit') had been solved with only the difficult ones remaining; and (3) the need for additional time and investment for new technologies to bear fruit, late David Horrabin[1] suggested that current biomedical research is based on 'heavily funded and heavily hyped techniques' that have, in overly embracing reductionistic approaches to understanding disease pathology and conducting drug R&D, made 'no real contribution to real-world issues.' While certainly a controversial and polarizing viewpoint, Horrabin's view is to a major extent reflected in (1) the recent trend of excluding core technologies like pharmacology from the drug discovery process[2]; and (2) an overdependence on automated technologies rather than intellect to strategize and implement the drug discovery process, a good example of Kubinyi's 'turn on the computer, turn off the brain' mentality.[3]

In the present volume, an overarching theme is that of the key roles of medicinal chemistry and pharmacology, in vitro and in vivo, in reviving the fortunes of the drug discovery process. The present volume covers, in detail, the therapeutic areas of:

- *Central nervous system diseases* including schizophrenia, affective disorders, anxiety, attention deficit hyperactivity disorder, sleep, addiction, neurodegeneration (including Alzheimer's and Parkinson's diseases), autoimmune disorders, stroke, epilepsy, and agents for the treatment of ophthalmic disorders).
- *Pain* including acute and neuropathic pain, anesthesia, and migraine.
- *Metabolic syndrome* including obesity, diabetes, cholesterol metabolism, bone metabolism, and hormone replacement.
- *Urogenital* including sexual dysfunction, incontinence, benign prostatic hyperplasia (BPH), and renal dysfunction.
- *Gastrointestinal* including gastric and mucosal ulceration, irritable bowel syndrome (IBS), idiopathic inflammatory bowel disease (IDB), and emesis/prokinetic agents.
- *Cardiovascular* including hypertension, antiarrythmics, and thrombolytics.

I would like to thank the authors of this volume for their hard work, patience, dedication, and scholarship in the lengthy process of writing, editing, and revising their contributions, the Editors-in-Chief, John Taylor and David Triggle, for their support, insights and common sense and, finally, Holly and Heather for their understanding for the many nights, weekends, and holidays that it took to finally get this volume into production.

Michael Williams

References

1. Horrabin, D. F. *Nat. Rev. Drug Disc.* **2003**, *2*, 151–154.
2. Williams, M. *Biochem. Pharmacol.* **2005**, *70*, 1707–1716.
3. Kubinyi, H. *Nat. Rev. Drug Disc.* **2003**, *2*, 665–668.

Editors-in-Chief

John B Taylor, DSc, was formerly Senior Vice President for Drug Discovery at Rhône-Poulenc Rorer. He obtained his BSc in chemistry from the University of Nottingham in 1956 and his PhD in organic chemistry at the Imperial College of Science and Technology with Nobel Laureate Professor Sir Derek Barton in 1962. He subsequently undertook postdoctoral research fellowships at the Research Institute for Medicine and Chemistry in Cambridge (US) with Sir Derek and at the University of Liverpool (UK), before entering the pharmaceutical industry.

During his career in the pharmaceutical industry Dr Taylor spent more than 30 years covering all aspects of research and development in an international environment. From 1970 to 1985 he held a number of positions in the Hoechst Roussel organization, ultimately as research director for Roussel Uclaf (France). In 1985 he joined Rhône-Poulenc Rorer holding various management positions in the research groups worldwide before becoming Senior Vice President for Drug Discovery in Rhône-Poulenc Rorer.

Dr Taylor is the co-author of two books on medicinal chemistry and has more than 50 publications and patents in medicinal chemistry. He was joint executive editor for the first edition of Comprehensive Medicinal Chemistry, a visiting professor for medicinal chemistry at the City University (London) from 1974 to 1984 and was awarded a DSc in medicinal chemistry from the University of London in 1991.

David J Triggle, PhD, is the University Professor and a Distinguished Professor in the School of Pharmacy and Pharmaceutical Sciences at the State University of New York at Buffalo. Professor Triggle received his education in the UK with a BSc degree in chemistry at the University of Southampton and a PhD degree in chemistry at the University of Hull working with Professor Norman Chapman. Following postdoctoral fellowships at the University of Ottawa (Canada) with Bernard Belleau and the University of London (UK) with Peter de la Mare he assumed a position in the School of Pharmacy at the University at Buffalo. He served as Chairman of the Department of Biochemical Pharmacology from 1971 to 1985 and as Dean of the School of Pharmacy from 1985 to 1995. From 1996 to 2001 he served as Dean of the Graduate School and from 1999 to 2001 was also the University Provost. He is currently the University Professor, in which capacity he teaches bioethics and science policy, and is President of the Center for Inquiry Institute, a secular think tank located in Amherst, New York.

Professor Triggle is the author of three books dealing with the autonomic nervous system and drug–receptor interactions, the editor of a further dozen books, some 280 papers, some 150 chapters and reviews, and has presented over 1000 invited lectures worldwide. The Institute for Scientific Information lists him as one of the 100 most highly cited scientists in the field of pharmacology. His principal research interests have been in the areas of drug–receptor interactions, the chemical pharmacology of drugs active at ion channels, and issues of graduate education and scientific research policy.

Editor of Volume 6

Michael Williams, PhD, DSc, received his PhD (1974) from the Institute of Psychiatry and his Doctor of Science degree in Pharmacology (1987) from the University of London. Williams has worked in the US-based pharmaceutical industry for 30 years, at Merck, Sharp and Dohme Research Laboratories, Nova Pharmaceutical, CIBA-Geigy and Abbott Laboratories. He retired from the latter in 2000 and after serving as a consultant with various biotechnology/pharmaceutical companies in the US and Europe, joined Cephalon, Inc. in West Chester, in 2003; he is the Vice President of Worldwide Discovery Research. He has published some 300 articles, book chapters, and reviews; he is Adjunct Professor in the Department of Molecular Pharmacology and Biological Chemistry at the Feinberg School of Medicine, Northwestern University, Chicago, Illinois.

Contributors to Volume 6

S Abdelhadi
University of Medicine and Dentistry of New Jersey, Newark, NJ, USA

E R Bacon
Worldwide Discovery Research, Cephalon, Inc., West Chester, PA, USA

T Blackburn
Helicon Therapeutics Inc., Farmingdale, New York, USA

J M Blair
Anzac Research Institute, University of Sydney, NSW, Australia

T A Bowdle
University of Washington, Seattle, WA, USA

D Bozyczko-Coyne
Worldwide Discovery Research, Cephalon, Inc., West Chester, PA, USA

K E Browman
Abbott Laboratories, Abbott Park, IL, USA

S Chatterjee
Worldwide Discovery Research, Cephalon, Inc., West Chester, PA, USA

R E Davis
3-D Pharmaceutical Consultants, San Diego, CA, USA

C R Dunstan
Anzac Research Institute, University of Sydney, NSW, Australia

A El-Gengaihy
University of Medicine and Dentistry of New Jersey, Newark, NJ, USA

A A Elmarakby
Medical College of Georgia, Augusta, GA, USA

S Evangelista
Menarini Ricerche SpA, Firenze, Italy

R D Feldman
Robarts Research Institute, London, ON, Canada

A C Foster
Neurocrine Biosciences Inc., San Diego, CA, USA

G B Fox
Abbott Laboratories, Abbott Park, IL, USA

S Freedman
Elan Pharmaceuticals, South San Francisco, CA, USA

J D Gale
Pfizer Global Research and Development, Sandwich, UK

G F Gebhart
University of Pittsburgh, Pittsburgh, PA, USA

D R Gehlert
Eli Lilly and Company, Indianapolis, IN, USA

P J Goadsby
Institute of Neurology, The National Hospital for Neurology and Neurosurgery, London, UK

K R Gogas
Neurocrine Biosciences Inc., San Diego, CA, USA

D E Grigoriadis
Neurocrine Biosciences Inc., San Diego, CA, USA

E D Hall
Spinal Cord and Brain Injury Research Center, University of Kentucky Medical Center, Lexington, KY, USA

R A Hegele
Robarts Research Institute, London, ON, Canada

G A Hicks
Novartis Pharmaceuticals Corporation, East Hanover, NJ, USA

P Honore
Abbott Laboratories, Abbott Park, IL, USA

J D Imig
Medical College of Georgia, Augusta, GA, USA

M F Jarvis
Abbott Laboratories, Abbott Park, IL, USA

J F Kirmani
University of Medicine and Dentistry of New Jersey, Newark, NJ, USA

P Klimko
Alcon Research, Ltd., Fort Worth, TX, USA

L J S Knutsen
Worldwide Discovery Research, Cephalon Inc., West Chester, PA, USA

S M Lechner
Neurocrine Biosciences Inc., San Diego, CA, USA

I Lieberburg
Elan Pharmaceuticals, South San Francisco, CA, USA

M J Marino
Worldwide Discovery Research, Cephalon Inc., West Chester, PA, USA

S Markison
Neurocrine Biosciences Inc., San Diego, CA, USA

W McCarthy
Neurocrine Biosciences Inc., San Diego, CA, USA

A W Meikle
University of Utah School of Medicine, Salt Lake City, UT, USA

H Meltzer
Psychiatric Hospital Vanderbilt, Nashville, TN, USA

E Messersmith
Elan Pharmaceuticals, South San Francisco, CA, USA

B Metzger
Feinberg School of Medicine, Northwestern University, Chicago, IL, USA

I Mori
Pfizer Global Research and Development, Nagoya, Japan

D L Nelson
Eli Lilly and Company, Indianapolis, IN, USA

A H Newman
National Institutes of Health, Baltimore, MD, USA

S Phillips
Urodoc Ltd, Herne Bay, UK

D M Pollock
Medical College of Georgia, Augusta, GA, USA

T Prisinzano
University of Iowa, Iowa City, IA, USA

N Pullen
Pfizer Global Research and Development, Sandwich, UK

A I Qureshi
University of Medicine and Dentistry of New Jersey, Newark, NJ, USA

R B Rothman
National Institutes of Health, Baltimore, MD, USA

R Rydel
Elan Pharmaceuticals, South San Francisco, CA, USA

G Sachs
University of California at Los Angeles, CA, USA, and Veterans' Administration Greater Los Angeles Healthcare System, Los Angeles, CA, USA

L Schmeltz
Feinberg School of Medicine, Northwestern University, Chicago, IL, USA

A Scriabine
Yale University School of Medicine, New Haven, CT, USA

M J Seibel
Anzac Research Institute, University of Sydney, and Concord Hospital, Concord, NSW, Australia

N A Sharif
Alcon Research, Ltd., Fort Worth, TX, USA

J M Shin
University of California at Los Angeles, CA, USA, and Veterans' Administration Greater Los Angeles Healthcare System, Los Angeles, CA, USA

J A Sikorski
AtheroGenics, Alpharetta, GA, USA

P P S So
University of Toronto, Toronto, ON, Canada

D A Taylor
East Carolina University, Greenville, NC, USA

R Tuttle
El Cerrito, CA, USA

M J A Walker
University of British Columbia, Vancouver, BC, Canada

J Wasley
Simpharma LLC, Guilford, CT, USA

J P Williams
Neurocrine Biosciences Inc., San Diego, CA, USA

M Williams
Northwestern University, Chicago, IL, USA and Worldwide Discovery Research, Cephalon, Inc., West Chester, PA, USA

M G Wyllie
Urodoc Ltd, Herne Bay, UK

T Yednock
Elan Pharmaceuticals, South San Francisco, CA, USA

H Zhou
Anzac Research Institute, University of Sydney, NSW, Australia

6.01 Central Nervous System Drugs Overview

M J Marino and M Williams, Worldwide Discovery Research, Cephalon, Inc., West Chester, PA, USA

6.01.1 Introduction

Diseases of the central (brain; CNS) and peripheral nervous systems (spinal cord; PNS) represent a major healthcare burden that ranges from lifelong disorders like epilepsy and schizophrenia, to genetically based neurodegenerative disorders like multiple sclerosis and Huntington's disease, traumatic disorders, e.g., stroke and spinal cord injury, and diseases of aging like Alzheimer's (AD) and Parkinson's diseases and other neurodegenerative disorders.

Neurological and psychiatric disorders have a marked deleterious effect on quality of life, including the ability to work, maintain normal social interactions, and live independently, and can have a significantly negative impact on lifespan. While CNS disorders are highly debilitating on an individual basis, they also exact a significant toll on family members and on society as a whole. CNS disorders have been estimated to account for up to 15% of work days lost to disease, and approximately 20% of the nationwide cost of healthcare in the USA. Since age is a major risk factor for many CNS disorders, it is anticipated that these numbers will rise significantly as the population ages. Based on the size of the affected population and the impact on quality of life, this area as a whole continues to represent a major unmet medical need.

6.01.2 Social Cost and Market

The pervasiveness of psychiatric disorders, e.g., depression, substance abuse, anxiety, schizophrenia, etc., and the comorbidity of depression and anxiety with neurological disease has enormous costs for society, estimated in the trillions of dollars. Additionally, it has been estimated that over 60% of individuals with diagnosable mental disorders do not seek treatment.

Various projections indicate that sales of drugs to treat neurological diseases alone will approach $20 billion by 2007 reflecting an absence of effective treatments for AD and the use of generic L-dopa as first-line treatment for Parkinson's disease. Drugs for the treatment of psychiatric disorders represent a much larger market with sales of the selective 5HT reuptake inhibitor (SSRI) class of antidepressants currently in the $10 billion range. Together the current market for CNS drugs is in excess of $70 billion. Should effective drugs be identified for the treatment of AD, this will account for an additional $6–8 billion in sales given the incidence of these diseases and their long-term nature. At the decade beginning in 2001, the global CNS drug market (including pain) was approximately $50 billion[1] with estimates of $105 billion in 2005, $200 billion in 2010, and one estimate[2] approaching $1.2 trillion.

6.01.3 Diagnosis

The current system for the diagnosis of CNS disorders in the USA is DSM-IV-TR (Diagnostic and Statistical Manual of Mental Disorders, Fourth Edition, Text Revision).[3] This text, which has a long and chequered history,[4] 'facilitates the identification and management of mental disorders' to help provide a consensus view of these disorders and their treatment, and a better understanding of their etiology. The general classification of mental disorders that comprise DSM-IV-TR is shown in **Table 1** where the major headings are shown with illustrative data, where appropriate, on the syndromes within each heading. These diagnostic criteria lack a specific etiological conceptualization[3] and are thus organized principally around syndromes – a group or pattern of symptoms that appear in individuals in a temporal manner – reflecting, to a major extent, comorbidities. However DSM-IV-TR is confounded by ethnic, societal, and gender differences both in terms of diagnosis (physician) and disease (patient). Thus in the postgenomic era, it is anticipated that DSM-V, due to be published in 2011[5](or later) will incorporate disease genotypes as part of the diagnostic tree. Given that over 30 gene associations have been shown to date for schizophrenia, the challenge will be in incorporating these into a cohesive framework and avoid having 14 or more categories of schizophrenia based on gene association. One issue with genome-based disease association is that many of the diseased patient cohorts used for genotyping represent historical diagnoses from more than one physician or geographical center, a problem that can confound inclusion criteria.

6.01.4 Central Nervous System Disorders: Psychiatric and Neurodegenerative

Disorders of the CNS are broadly categorized as either psychiatric or neurodegenerative with a major degree of overlap in symptoms. Thus neurodegenerative disorders also have a high incidence of psychiatric comorbities including anxiety and depression. Psychiatric disease includes a variety of disorders such as schizophrenia, depression, obsessive compulsive disorder (OCD), attention deficit hyperactivity disorder (ADHD), and others. The underlying pathology is usually considered to be the result of synaptic dysfunction driven by: (1) a dysregulation of neurotransmitter availability or (2) signaling, the latter at the receptor and/or signal transduction levels. The net result is an alteration in neuronal circuitry involving multiple neurotransmitter/neuromodulator systems.

Neurodegenerative disease involves a defined degenerative process in which neurons are lost either by necrosis or apopotosis. This category is typified by slow chronic disorders that include Parkinson's disease and AD, but also includes more acute cell loss due to traumatic insults including stroke (brain attack) and spinal or brain damage. There are major issues with diagnosis in the absence of robust biomarkers.[6] For example, there are cases where the AD disease phenotype is predominant and where the diagnosis of AD has not been supported by autopsy.

6.01.4.1 Psychiatric Disorders

6.01.4.1.1 Schizophrenia

Schizophrenia (*see* 6.02 Schizophrenia) is a complex and debilitating neurodevelopmental psychiatric disorder that affects approximately 1% of the population. It is characterized by diminished drive and emotion during childhood followed by a deviation from reality with hallucinations, and appears to have both genetic and epigenetic causality. Schizophrenia presents with a spectrum of positive, negative, and cognitive symptoms. Positive symptoms include auditory and visual hallucinations, delusions, disorganized thought, and antisocial or violent behavior. Negative symptoms include dissociation, apathy, difficulty or absence of speech, and social withdrawal. Cognitive symptoms

Table 1 DSM-IV-TR major classifications

Disorders usually first diagnosed in infancy, childhood, or adolescence	Mental retardation
	Learning disorders
	Motor skills disorders
	Communication disorders
	Pervasive developmental disorders, e.g., autism
	Attention deficit and disruptive behavior disorders, e.g., ADHD
	Feeding and eating disorders of infancy or early childhood
	Tic disorders, e.g., Tourette's disorder
	Elimination disorders
Delirium, dementia and amnestic and other cognitive disorders	Delirium
	Dementia, e.g., dementia of the Alzheimer's type, vascular dementia
	Dementia due to human immnodeficiency virus (HIV) disease, head trauma, Parkinson's disease, Huntington's disease, etc.
	Amnestic disorders
Mental disorders due to a general medical condition not elsewhere classified	Catatonic disorder
Substance related disorders	Alcohol related disorders
	Alcohol induced disorders
	Amphetamine (or amphetamine-like) related disorders
	Amphetamine induced disorders
	Caffeine related disorders
	Cannabis related disorders
	Cannabis induced disorders
	Cocaine related disorders
	Cocaine induced disorders
	Hallucinogen related disorders
	Hallucinogen use disorders
	Hallucinogen induced disorders
	Inhalant use disorders
	Inhalant induced disorders
	Nicotine related disorders
	Nicotine use disorders
	Nicotine induced disorders
	Opioid related disorders
	Opioid use disorders
	Opioid induced disorders
	Phencyclidine-related disorders
	Phencyclidine use disorders
	Phencyclidine induced disorder
Sedative, hypnotic, or anxiolytic related disorders	Sedative, hypnotic, or anxiolytic use disorders
	Sedative, hypnotic, or anxiolytic induced disorders
	Polysubstance related disorder
	Other (or unknown) substance related disorders
Schizophrenia and other psychotic disorders	Paranoid, disorganized, catatonic, etc.
Mood disorders	
Depressive disorders	Major depressive disorder
Bipolar disorders	Bipolar I disorder
	Bipolar II disorder

continued

Table 1 Continued

Anxiety disorders	Panic disorder without agoraphobia
	Panic disorder with agoraphobia
	Social phobia
	Obsessive compulsive disorder
	Posttraumatic stress disorder
	Generalized anxiety disorder
Somatoform disorders	Somatization disorder
	Pain disorder
Factious disorders	Factious disorders
Dissociative amnesia	Dissociative amnesia
	Dissociative fugue
Sexual and gender identity differences	Sexual desire disorders
	Sexual arousal disorders
	Orgasmic disorders
	Sexual pain disorders
	Gender identity disorders
Sleep disorders	Dyssomnias – primary insomnia, narcolepsy
Primary	Parasomnias – nightmare disorder
Related to another mental disorder	
Other sleep disorders	
Impulse control disorders not elsewhere classified	Kleptomania, pathological gambling
Adjustment disorders	
Personality disorders	Paranoid personality disorder
Other conditions that may be the focus of clinical attention	

include disorganized thought, difficulty in attention or concentration, and poor memory. Symptoms usually begin in adolescence or early adulthood, but can occur at any stage of life including childhood. Current diagnostic criteria rely on the DSM-IV-TR classification of schizophrenia. While considerable research has been directed at the genetics of the disorder and significant advances have been made in the development of imaging tools and methods, there is currently no objective clinical test for diagnosis.

Dopamine (DA) and glutamate have been implicated in the molecular pathophysiology of schizophrenia. Thus stimulants that activate brain dopaminergic systems, e.g., cocaine or amphetamine, induce a paranoid psychosis similar to that seen with the positive symptom core of the disease suggesting that overactive DA transmission is a key facet of the disease. Similarly, based on the ability of the psychotomimetics phencyclidine and ketamine to block glutamate receptors (*N*-methyl-D-aspartate (NMDA)-subtype) the glutamate hypothesis suggests a hypoactivity of excitatory glutamatergic systems.[7]

Current treatment modalities include typical antipsychotics that act by blocking DA D2 receptors, e.g., haloperidol, thoridazine, and second-generation, atypical antipsychotics that include clozapine, risperidone, olanzepine, quetiapine, and ziprasidone and the partial DA agonist, aripiprazole. These latter agents are all modeled on molecular attributes of clozapine, the only antipsychotic with demonstrated superiority in efficacy to other antipsychotics in treating both positive and negative symptoms. All atypical antipsychotics block $5HT_2$ receptors in addition to D2 receptors. Their superiority to classical antipsychotics is controversial.[8] Thus, the gold standard therapy is clozapine, the use of which is limited due to a rare but fatal occurrence of agranulocytosis associated with treatment. Research efforts over the past 30 years have been focused on understanding the mechanism of action of clozapine, most of which have been highly empirical. Most recently, *N*-desmethylclozapine, a metabolite of clozapine with potent muscarinic activity, has been identified as another candidate for the mystery factor.[9] Novel approaches to therapies for schizophrenia are focused on addressing glutamatergic hypofunction.[7]

The overall prognosis for schizophrenia is poor. Only 60–70% of patients respond to currently available therapies, and the response is incomplete. In particular, the negative symptoms are usually refractory to treatment with atypical antipsychotics. In addition, there are significant adverse effects associated with prolonged use of antipsychotics including weight gain, increased production of prolactin, and tardive dyskinesia.

6.01.4.1.2 Affective disorders: depression and bipolar disease

Major depression (*see* 6.03 Affective Disorders: Depression and Bipolar Disorders) is a chronic disorder that affects 10–25% of females and 5–12% of males. Suicide in 15% of chronic depressives makes it the ninth leading cause of death in the USA. Presenting complaints for depression include: depressed or irritable mood, diminished interest or pleasure in daily activities, weight loss, insomnia or hypersomnia, fatigue, diminished concentration, and recurrent thoughts of death. The World Health Organization (WHO) has estimated that approximately 121 million individuals worldwide suffer from depression and that depression will become the primary disease burden worldwide by 2020. In the majority of individuals episodes of depression are acute and self-limiting. The genetics of major depression are not well understood and have focused on functional polymorphisms related to monoaminergic neurotransmission as the majority of effective antidepressants act by facilitating monoamine availability. Positive associations have been reported between the polymorphism in the serotonin transporter promoter region (5HT TLPR)[10] and bipolar disorder, suicidal behavior, and depression-related personality traits but not to major depression. There is preliminary data on the association of polymorphisms in brain-derived neurotrophic factor (BDNF) and depression. However, these are controversial.[22]

Other mood disorders include bipolar affective disorder (BPAD) and dysthymia. BPAD is diagnosed by discrete periods of abnormal mood and activity that define depressive and manic or hypomanic episodes and occurs in 10% of individuals with major depression. BPAD occurs as two types, I and II, the latter having a familial association and a higher incidence of hypomania. Dysthymia is a chronically depressed mood, usually present for 2 years or more, that does not warrant a diagnosis as major depression.

Depression has traditionally been treated by drugs and electroconvulsive therapy (ECT). The initial drugs to treat depression, the monoamine oxidase inhibitors (MAOIs), e.g., isoniazide, were found by serendipity,[11] and from these second-generation MAOIs, e.g., phenelzine, tranylcypromine, and the tricyclic antidepressants (TCAs) were developed and include impramine, desipramine, amitriptyline, etc. The latter are monoamine uptake inhibitors and while highly efficacious, have limiting side effects. Current treatment modalities for depression include more selective monoamine uptake blockers, and include: the SSRIs (selective 5HTuptake blockers), fluoxetine, sertraline, citalopram, paroxetine, etc., the SNRIs (serotonin norepinephrine reuptake inhibitors) that include venlfaxine, nefazodone, and mitrazepine. BPAD is typically treated with mood stabilizers including lithium and valproate, the mechanisms of action of which are currently becoming clearer.

The prognosis for patients diagnosed with major depressive disorders is relatively good with 70–80% of patients exhibiting significant favorable response to treatment. By comparison, the prognosis for bipolar disorder patients is poor. Only 50–60% of BPAD patients gain control of their symptoms with currently available therapies, and a mere 7% become symptom free.

6.01.4.1.3 Anxiety

Anxiety disorders (*see* 6.04 Anxiety) are characterized by an abnormal or inappropriate wariness. There are several disorders that fall under the heading of anxiety including panic disorders, phobias, generalized anxiety disorder (GAD), acute stress disorder, and posttraumatic stress disorder (PTSD). Panic disorder is characterized by rapid and unpredictable attacks of intense anxiety that are often without an obvious trigger. Phobias are examples of life-disrupting anxiety or fear associated with an object or situation, including social phobias. GAD develops over time and involves the generalization of fears and anxieties to other, usually inappropriate situations until they ultimately result in an overwhelming anxiety regarding life in general. Acute stress disorder involves the response to a threatened or actual injury or death and is characterized by dissociation, detachment, and depersonalization. Acute stress disorder usually resolves within a few weeks; however it can progress into more severe anxiety disorders such as PTSD. PTSD is characterized by intrusive, anxiety-provoking memories of a stress or trauma (death or injury, terrorist attack, combat experience) that recur and become disruptive to daily function. The symptoms of PTSD include nightmares, obsessive thoughts, flashbacks (re-experiencing the trauma), and avoidance of situations associated with the stressor and a generalization of anxiety.

All of these categories of anxiety can produce crippling and overwhelming emotional and physical symptoms. Anxiety disorders are common with as many as 25% of adults suffering from some form of clinical anxiety at some point in their life with 19 million individuals affected in the USA. Anxiety is estimated to cost worldwide approximately $40 billion annually.

Diagnosis of anxiety disorders is based upon interview with a psychologist or psychiatrist. While a considerable progress has been made in the understanding of the neuronal circuitry underlying fear responses, the biological basis of anxiety disorders is not well understood. Findings of a modest genetic influence including polymorphisms in COMT (catechol *O*-methyl transferase), the serotonin transporter gene *SLC6A4*, and 5HT receptors indicate the need for additional genetic linkage studies; however there is a marked environmental/epigenetic component, especially stress in early life, that appears to play a greater role in determining if an individual will develop an anxiety disorder.

Current treatments for anxiety disorders vary depending on the diagnosis. Acute stress disorder can be treated successfully with psychotherapy, and will often resolve itself. Phobias are also successfully treated through behavioral therapy and seldom require medication. The remaining disorders are treated with a combination of therapy and anxiolytic drugs. The earliest anxiolytics were barbiturates, which have now largely been replaced by the benzodiazepines (BZs), e.g., diazepam, clonazepam, etc. While BZs are highly efficacious and have a greater safety margin, there are still significant issues with sedation, withdrawal, dependence liability, and possible overdose. Antidepressants including SSRIs, SNRIs, and 5HT antagonists have also been used with mixed success.

While defined as an anxiety disorder, obsessive compulsive disorder (OCD) has several distinct features that set it apart. It is characterized by an overwhelming sense that negative consequences will arise from the failure to perform a specific ritual. The rituals are repetitive, taking a variety of forms including hand washing, checking to see if the stove is turned off, or the recitation of particular words or phrases. DSM-IV-TR criteria require that the symptoms include obsessions or compulsions that cause marked distress and occupy at least 1 h per day. OCD interferes with normal routine and produces a social impairment equivalent to schizophrenia. In addition to the disabling effect OCD has on the patient, it also produces a significant strain on family relations as relatives become caregivers and are drawn into a routine of providing reassurance and ritual maintenance. OCD becomes apparent at between 22 and 36 years, affecting males and females equally. OCD is highly prevalent, affecting 2–3% of the US population, yet in spite of its clear negative impact on quality of life, OCD is typically underdiagnosed. This is due to a combination of the lack of familiarity of a physician with OCD symptoms and patients tending to be embarrassed and hiding their symptoms.

In contrast to many other psychiatric disorders, OCD patients often have insight into the illogical and extreme nature of their behaviors. Unfortunately there is still reluctance to admit to the symptoms, and it is necessary that the physician probe carefully in order to make an appropriate diagnosis. Positron emission tomography (PET) and single positron emission computed tomography (SPECT) imaging studies suggest a role for increased activity in the orbitofrontal cortex, the cingulated cortex, and the caudate nucleus. Interestingly, these regions have been found to further activate when patients are confronted with a stimulus known to provoke symptoms, and show a decrease activity in response to therapy. While imaging methods are not routinely employed, these studies suggest that an objective diagnostic test is feasible.

Current treatment strategies for OCD include the use of the TCA, clomipramine or the SSRIs, fluoxetine or fluvoxamine at doses higher than those typically used in depressed patients. This approach is effective in 40–60% of patients; however, it only produces a modest 20–40% improvement in symptoms. Therefore, it is critical that patients also undergo behavioral therapy. The pharmacological approach has been found to greatly enhance the success rate of behavioral therapy, and the combined therapies have a much lower relapse rate than either when used alone. For refractory patients, SSRIs can be augmented by the addition of other drugs, e.g., dopamine antagonists. Neurosurgical approaches are available for severely affected patients that do not respond to combined pharmacological and behavioral therapy; however, this approach has not been extensively validated and the overall success rate is less than 50%.

While the prognosis for OCD has improved dramatically with the SSRIs, the overall outlook ranges from moderate to poor. At least 40% of patients are refractory to treatment, and those that respond often experience an incomplete decrease in symptoms and a significant propensity for relapse.

The prognosis for an individual with an anxiety disorder is dependent on the category and severity of the disorder. However, with combined pharmacological and behavioral interventions, the outlook is moderate to good even with the most severe cases of GAD. To satisfy current unmet medical need, efforts are being focused on new chemical entities (NCEs) that have the efficacy of BZs but lack the associated adverse effects. Such agents include γ-amino-butyric acid ($GABA_A$) receptor subunit selective 'BZ-like' NCEs[12] including neurosteroids (e.g., ganaxolone), direct acting $GABA_A$ agonists like gaboxadol, and newer approaches to anxiety including cannabinoids, metabotropic glutamate receptor modulators, nicotinic receptor agonists, and modulators of the corticotrophin-releasing factor (CRF) family of receptors.

6.01.4.1.4 **Attention deficit/hyperactivity disorder**

Attention deficit/hyperactivity disorder (ADHD (*see* 6.05 Attention Deficit Hyperactivity Disorder)) is one of the most common childhood psychiatric disorders. An estimated 5–10% of children worldwide are afflicted by ADHD. In addition, there is now increasing acceptance that ADHD occurs in adulthood in approximately 4% of the population. DSM-IV-TR diagnostic criteria classify ADHD symptoms under the headings of inattention or hyperactivity–impulsivity. Symptoms of inattention include lack of attention to detail, carelessness, difficulty in sustaining attention, difficulty in organizing and completing tasks, ease of distraction, and forgetfulness. Hyperactivity–impulsivity includes symptoms such as fidgeting or squirming, excessive and inappropriate running or climbing or a feeling of restlessness, excessive talking, difficulty in awaiting turn, and frequent interruption of others. The subjective nature of these criteria has led to the suggestion that ADHD is simply an extreme in normal behavioral variation that is possibly exacerbated by environmental circumstances. However, there is now a wealth of data clearly demonstrating a genetic inheritance of ADHD, albeit a polygenic disorder, with multiple genes each conferring a small risk. Genes associated with ADHD include: the dopamine transporter (DAT), dopamine receptors (D2, D4, and D5) and dopamine β-hydroxylase (DBH) gene, the α4 neuronal nicotinic receptor subunit, and the synaptosomal-associated protein, SNAP-25.

Current treatment strategies for ADHD involve the use of classical stimulants such as methylphenidate and amphetamine. These compounds act on DAT and norepinephrine transporters (NET) to enhance monoaminergic transmission in these systems. While effective, methylphenidate has a short half-life and must be administered every 4 h producing logistical and social stigma issues when dosing school-aged children, leading to low compliance rates. Long-acting formulations of methylphenidate and the development of atomoxetine, a selective NET inhibitor, have addressed this issue although the use of atomoxetine is confounded by potential hepatotoxicity issues. The atypical stimulant, modafinil, is also effective in treating ADHD without the abuse liability seen with classical stimulants.

While current therapies for ADHD have liabilities such as abuse potential, they produce robust effects on the core symptoms of the disorder. As newer generation formulations and novel targets are evaluated, e.g., histamine H_3 receptor antagonists, novel nonstimulant agents that would prove efficacious while avoiding the abuse potential of currently available therapeutics should be feasible.

6.01.4.1.5 Sleep disorders

Research into sleep disorders (*see* 6.06 Sleep) has intensified over the last decade given new generations of hypnotics and the success of the novel wake-promoting agent modafinil. The sleep spectrum involves insomnia, narcolepsy, and excessive daytime sleepiness (EDS). Dyssomnias are primary sleep disorders characterized by an abnormal amount, quality, or timing of sleep, and include primary insomnia, narcolepsy, and breathing-related sleep disorder. Primary insomnia is defined as a difficulty in initiating or maintaining sleep, or an inability to obtain restorative sleep. Insomnia is a highly prevalent sleep disorder estimated to affect 35% of the population during the course of a year, with 60% of those afflicted reporting chronic insomnia lasting longer than 1 month. Chronic insomnia is often accompanied by impairments in social and occupational function. Insomnia is roughly twice as common in females as in males, is more frequently observed in patients 60 years and older, and is comorbid with a number of other disorders including depression, Parkinson's disease, and AD. Narcolepsy is a chronic disorder characterized by repeated irresistible sleep attacks, cataplexy, and vivid hallucinations caused by an intrusion of rapid eye movement (REM) sleep into the transitional period between sleep and wake. These attacks typically last from a few seconds to several minutes, and occur at any time without warning. In addition, narcoleptics experience frequent wakening during the night resulting in an overall decrease in night-time sleep duration and quality. Narcolepsy afflicts approximately 200 000 people in the USA and is the third most diagnosed primary sleep disorder in sleep clinics. However, the disorder is believed to be underdiagnosed and actual prevalence may be higher. Of the breathing-related sleep disorders, obstructive sleep apnea (OSA) is the most common, being characterized by repeated obstruction of the upper airway during sleep. These apneic episodes typically last from 20 to 30 s and occur as many as 300 times in a single night. Episodes are typically noted as silent periods that interrupt loud snoring. This disruption in normal breathing leads to frequent awakenings, disruption in sleep architecture, and consequent nonrestorative sleep. In addition, OSA is associated with an increased risk of stroke, heart attack, and high blood pressure. OSA is very common, estimated to affect over 12 million individuals in the USA. The disorder is more common in males than females and is associated with risk factors including obesity and large neck size.

Diagnosis of sleep disorders is usually made by the primary care physician based on patient interview and clinical examination. It is unusual for a patient to be referred to a specialist or sleep clinic for follow-up polysomnography due to the subtle distinctions necessary for accurate diagnosis. For example, the three primary dyssomnias discussed have distinct etiologies and diverse treatments, yet for the most part all three present clinically as EDS. It is therefore not surprising that narcolepsy is not definitively diagnosed in most patients until 10–15 years after the first symptoms become apparent, or that only an estimated 30% of all insomniacs are diagnosed.

From a disease causality perspective, more than 90% of narcoleptic individuals carry the human leukocyte antigen (HLA) gene haplotype, *HLA-DR2/DQ1*, a reliable genetic marker for narcolepsy with a 100% association of the disease, but this is neither a necessary nor a sufficient causative factor for the disorder. A COMT polymorphism has been associated with the severity of narcolepsy although it may be noted that COMT polymorphisms have also been associated with gender differences in sensitivity to pain, breast cancer, and schizophrenia. Mutations in the orexin/hypoctein systems in both canines and humans have been associated with narcolepsy–cataplexy and numerous genes have been identified in *Drosophila melanogaster* that affect sleep/wake behaviors including circadian clock genes.

As mentioned above, the choice of treatment for dyssomnia depends upon the diagnosis. Insomnia is typically treated with BZ hypnotics (e.g., triazolam) or more commonly non-BZ hypnotics (e.g., zolpidem). Significant adverse effects are observed with these compounds including a next day 'hangover' effect that can significantly impair function, tolerance, rebound of REM sleep on withdrawal, and addiction. While the non-BZ hypnotics are better tolerated, the same liabilities exist. Patients are occasionally treated with sedating TCAs (e.g., amitriptyline) with mixed results.

Narcoleptic patients are typically treated with stimulant amphetamines or more recently with the novel wake-promoting drug, modafinil. The amphetamines (e.g., methylphenidate) are effective at reducing daytime sleepiness, but carry significant liabilities including irritability, night-time sleep disruption, tolerance, abuse potential, and rebound hypersomnolence. A characteristic of central nervous system (CNS) stimulants like amphetamine is that their positive effects on wake promotion are accompanied by a rebound hypersomnolence following withdrawal, a more profound sleepiness than that which initially led to the use of medication.

Modafinil can alleviate excessive daytime sleepiness without these adverse effects and does not induce tolerance, indicating a lower abuse potential. Both TCAs and SSRIs are effective in treating this aspect of narcolepsy. OSA is typically treated by lifestyle changes such as weight loss and decreased alcohol consumption. Patients may also be placed on a nasal continuous positive airway pressure (CPAP) device. CPAP acts to keep the airway open during the night thus avoiding the apneic episodes. Modafinil is also approved for use in OSA patients with EDS.

Prognosis for these sleep disorders is good for patients with narcolepsy or OSA, and relatively good for patients suffering from insomnia. As discussed above, the currently available treatments for insomnia produce significant adverse effects, so their use is somewhat limited, particularly in the elderly population. Future development of compounds that act outside of the GABA system and restore normal sleep architecture will represent a major breakthrough in the treatment of insomnia and include histamine H_1 receptor agonists, ligands interacting with the orexin/hypocretin system, $5HT_2$, and melatonin receptor modulators. For wake promotion, there is considerable focus on histamine H_3 inverse agonists, modulators of monoamine availability and of the orexin/hypocretin system. It is anticipated that the considerable work ongoing in *Drosophilia* will aid in identifying new targets for the pharmacotherapy of sleep disorders. An overriding issue however, is the relatively poor diagnosis rate. This is a significant unmet need since these disorders are not only debilitating but in many case life-threatening.

6.01.4.1.6 Addiction/substance abuse

DSM-IV-TR defines 11 classes of commonly abused substances (*see* 6.07 Addiction). These include alcohol, amphetamine and amphetamine-like compounds, caffeine, cannabis, hallucinogens, inhalants, nicotine, opioids, phencyclidine, and the class of drugs defined as sedatives, hypnotics, or anxiolytics. Substance dependence is defined as a pattern of repeated self-administration that can result in tolerance, withdrawal, or compulsive drug-taking behavior, and has as a basis an anhedonia, the inability to gain pleasure from normally pleasurable experiences. Tolerance is evident as either a need for increased amount of substance to produce a desired effect, or the diminished effect of the same dose of substance over time. All substances of abuse produce tolerance, but the actual degree of tolerance varies across classes. Withdrawal involves maladaptive physiological changes that occur with declining drug concentrations. These changes tend to be unpleasant and produce cognitive and behavioral consequences that lead the individual to seek to maintain a constant dosing regimen. This can lead to compulsive drug-taking behavior that can include the individual taking large amounts of drug over a long period of time, spending significant amounts of time seeking a supply of substance, a reduction in time spent in social, occupational, or recreational activities, and an inability to stop or decrease drug use despite frequent attempts. This leads a situation of hedonic dysregulation with associated hypofrontality, a decrease in prefrontal cortex function, and cell loss and remodeling.

Statistics compiled by the National Institute on Drug Abuse (NIDA) indicate that the prevalence of substance abuse and the cost of this disorder are considerable. In 1994, an estimated 9.4% of the US population was involved in substance abuse. Since most of the drugs of abuse are illegal, prevalence estimates often come from either treatment program data or an assessment of the incarcerated population. In 2003, 1.7 million people were admitted to publicly funded treatment programs. A survey of 14 major metropolitan areas between 2000 and 2002 found that 27–49% of male arrestees tested positive for cocaine. Substance abuse is also surprisingly common in school-aged children, with 2004 estimates of 8.4% of 8th-graders reporting illicit drug use during the last 30 days, a number that rises to 23% in the 12th grade population. The cost to society is estimated to be in excess of $320 billion annually with significant amounts going directly to healthcare costs and law enforcement. Additionally, a large proportion of healthcare costs are due to medical complications associated with the substance abuse. For example, intravenous drug use is the major vector for transmission of human immunodeficiency virus (HIV), accounting for one-third of all acquired immune deficiency syndrome (AIDS) cases, and is reaching epidemic proportions.[13]

Current approved treatments for substance abuse are directed at decreasing craving and preventing relapse. These include methadone and buprenorphine treatment for heroin addiction, and naltrexone for the treatment of alcohol abuse. Currently no drug is approved for the treatment of cocaine addiction, although several, including disulfiram and modafinil, have shown promise in randomized control trials. Notably, none of these compounds was developed for their potential to treat substance abuse. This stems in part from the persistent and ill-informed prejudice that addicts use compounds like methadone as a surrogate 'crutch.' It is clearly evident from the current statistics that substance abuse represents a major unmet medical need and a significant opportunity for future drug development.

The prognosis for addicted individuals varies significantly depending on the class of abuse substance and the degree of available psychological and social support. For example, nicotine addicts have several over-the-counter options that have been proven effective and the recent approval of the neuronal nicotinic receptor agonists. The availability of treatment combined with strong social pressure to stop smoking leads to a relatively good prognosis. On the other hand, the prognosis for cocaine addicts is poor due to the fact that they are less likely to seek medical attention and have relatively few pharmacological treatments available.

Ongoing areas of research that may yield new treatments for substance abuse include novel DAT inhibitors with different pharmacokinetic and pharmacodynamic properties to cocaine, new classes of κ-receptor opioid ligands and modulators of regulator of G protein signaling (RGS) protein function.

In addition to substance abuse there a number of addictions some of which are included in DSM-IV-TR under the classification of impulse control disorders which include kleptomania, pathological gambling, pyromania, and tricotillomania. Interestingly, recent reports have described an increase in compulsive gambling in Parkinson's disease patients receiving DA agonist treatment.

Addictions that may be added to these disorders include nymphomania, compulsive shopping, and overeating, all of which, in excess, lead to behaviors that are both illogical and harmful. Returning to the anhedonia context of addiction behaviors, the inability to gain pleasure from normally pleasurable experiences, it is debatable whether the milder forms of addiction are not in fact manifestations of depression. In this context, it is noteworthy that current medications for the treatment of obesity, the impulse dyscontrol related to food consumption, are antidepressants.

6.01.4.2 Neurodegenerative Diseases

Neurodegenerative diseases (*see* 6.08 Neurodegeneration) include: AD; Parkinson's disease; amyotrophic lateral sclerosis (ALS); demyelinating diseases, e.g., multiple sclerosis: neuropathies, e.g., diabetic, HIV, and chemotoxin-induced; Down's syndrome (DS); prion diseases, e.g., Creutzfeldt–Jakob disease; tauopathies, e.g., Pick's disease, frontal temporal dementia with Parkinsonism (FTDP); trinucleotide repeat or polyglutamine (polyQ) diseases, e.g., Huntington's disease (HD); spinocerebellar ataxias (SCA); dentatorubral-pallidolysian atrophy (DRPLA); Friedreich's ataxia; multiple systems atrophy (MSA); stroke and traumatic brain injury.

Current treatment strategies for all neurodegenerative disorders, where available, are palliative.

6.01.4.2.1 Alzheimer's disease

AD or dementia of the Alzheimer type (DAT) is a progressive and debilitating disorder characterized by dementia and a loss of cognitive function. Approximately 15 million individuals in developed countries are estimated to currently suffer from AD. Initial symptoms include confusion, loss of initiative, progressive decline in executive function, mood changes, and memory loss. However a decline in cognitive abilities ultimately results in a complete loss of mental function with disease progression. While approximately 10% of cases are early onset, AD predominately afflicts those over the age of 65. It is estimated that 5% of those over the age of 65% and 24% of those over the age of 85 suffer from AD. Diagnosis is made on an exclusionary basis using general medical and psychological evaluation, interview of caregivers, and assessment of family history of the disease. While recent advances in neuroimaging hold promise for future early detection, there is currently not a consensus that this or any of the currently available diagnostic tools are of value and it may be another decade before reliable diagnostics are available.[6]

DSM-IV-TR defines subtypes of AD: With Early Onset ($\leq$65 years); With Late Onset ($>$65 years); With Delusions; With Depressed Mood; Uncomplicated and/or With Behavioral Disturbance (e.g., wandering). In addition to memory impairment, diagnosis can include aphasia (deterioration in language function), agnosia (impaired object recognition), and apraxia (impairment in motor activity execution). AD is a disease of slow onset and gradual decline. Mild cognitive impairment (MCI), characterized by isolated episodes of long-term memory impairment, is thought to be the precusor of AD as 40–60% of MCI patients progress to AD within 3–4 years.

The initial hypothesis for AD causality was the 'cholinergic hypothesis'[14] based on a loss of forebrain cholinergic neurons in AD that led to the first generation of palliative therapy for AD, the cholinesterase inhibitors, e.g., tacrine, donzepil, galanthamine. The two pathophysiolgical hallmarks found at autopsy are amyloid deposits, principally Aβ42 and neurofibrillary tangles (NFTs). Some 20 missense mutations in Aβ42 producing proteins/enzymes have been identified in linkage studies in AD patients. Hyperphosphorylation of the low-molecular-weight microtubule associated protein tau has been implicated in the formation of paired helical filaments (PHF) which aggregate to form the neurofibrillary tangle (NFT). Reactive gliosis and microglial activation are also found in the AD brain with increased levels of inflammatory mediators supporting an inflammatory hypothesis of AD.

The only currently approved treatments for AD are the palliative acetylcholinesterase inhibitors donepezil and rivastigmine, which act to increase brain acetylcholine and offset aspects of the cognitive decline during early stages of the disease. The efficacy of these compounds is modest and short-lived as the disease progresses.

Based on the newer hypotheses of AD causality there are four active approaches to identifying new drugs to treat AD: (1) preventing or reducing Aβ42 formation; (2) reducing tau hyperphosphorylation; (3) inhibiting neuronal apoptosis; and (4) reducing brain inflammation. Progress in these initiatives has been slow. Inhibitors of the enzymes responsible for the formation of Aβ42 have been difficult to develop in the absence of true animal models of AD while a novel vaccine approach to aid in the clearance of Aβ42 encountered toxicity problems in clinical trials. Enzyme inhibitors to prevent tau hyperphosphorylation have also been difficult to identify given the multiple sites on tau amenable to phosphorylation and the diverse group of kinases that can act on these sites. Based on a positive retrospective analysis of the efficacy of the nonsteroidal anti-inflammatory drug (NSAID) indomethacin, reducing brain inflammation appeared to hold promise for AD treatment. However, prospective trials with both NSAIDs and COX-2 inhibitors have shown ambiguous outcomes. In all three instances, safe, efficacious, and brain bioavailable NCEs that are selective for discrete molecular targets will be necessary to provide proof of concept for disease etiology.

6.01.4.2.2 Non-Alzheimer's dementias

6.01.4.2.2.1 Vascular dementia

Vascular dementia (VD) represents a heterogeneous group of disorders, approximately eight in number, the common feature being a dementia associated with a disturbance in blood supply to the brain. Diagnostic criteria for VD are not considered robust but the National Institute of Neurological Disorders and Stroke–Association Internationale pour la Recherché et l' Enseignement en Neurosciences (NINDS–AIREN) criteria[15] describe VD in terms of an abrupt deterioration in cognitive function, gait disturbance or frequent falls, urinary frequency, focal neurological findings including sensory loss, lower facial weakness, depression, mood lability, and psychiatric symptoms. Treatment of VD is primary prevention with education regarding risk factors that include many of the same factors for stroke, e.g., smoking, coronary artery disease, alcohol abuse, age, etc.

6.01.4.2.2.2 Dementia with Lewy bodies

Dementia with Lewy bodies (DLB) is associated with a progressive decline in memory, language, praxis, and reasoning, the defining pathological characteristic being cortical Lewy bodies, the density of which is correlated with the severity of the dementia. DLB is often misdiagnosed as AD. Age of dementia in DLB is earlier than that of AD, with increased visual hallucinations with a presentation of parkinsonism that proceeds to dementia.

6.01.4.2.2.3 Acquired immune deficiency syndrome dementia

Dementia in AIDS is an exclusionary diagnosis characterized by cognitive and motor disturbances and behavioral changes including impaired short- and long-term memory, decreased concentration, and slowed thought processing. The pathophysiology involves neurovirulent strains of HIV, excitotoxicity, and inflammation. AIDS dementia is treated with the antiviral zidovidine, together with drugs for the treatment of the associated psychosis and depression.

6.01.4.2.3 Parkinson's disease

Parkinson's disease[16] afflicts 1.5 million Americans with an additional 60 000 cases diagnosed each year (*see* 6.08 Neurodegeneration). The cardinal symptoms of Parkinson's desease include tremor, bradykinesia, rigidity, and postural instability, the result of the loss of DA-containing cells in the substania nigra. Parkinson's disease patients also suffer from a variety of nonmotor symptoms including sleep disturbance, depression, and dementia. As with AD, Parkinson's disease is a disease that predominately affects those over the age of 65; however, an estimated 15% of patients develop symptoms before the age of 50. Diagnosis is made based on neurological examination with imaging methods used primarily to rule out other conditions. Response to the gold standard of Parkinson's disease treatment, L-dopa, is a diagnostic criterion for this disease.

Parkinson's disease is effectively treated early in the course of the disease with DA replacement therapies that include the DA precursor, L-dopa as well as direct-acting DA agonists, e.g., apomorphine, bromocriptine, pramipexole, and ropinirole, MAO-B inhibitors, e.g., rasagiline, and COMT inhibitors tolcapone and entacapone. While remarkably effective during the first few years of treatment, these approaches to enhancing DA availability loose efficacy as DA neurons die and ultimately lead to a variety of motor complications including disabling dyskinesias, motor freezing, and on–off fluctuations. Recent efforts to better control the temporal aspects of DA replacement have shown promise in reducing these adverse effects as evidenced in success of L-dopa/carbidopa/entacapone combination therapy.

Newer palliative approaches to Parkinson's disease include the indirect DA agonist, istradefylline, a selective adenosine A_{2A} receptor antagonist. Novel disease modifying approaches based on inhibition of neuronal apoptosis including the mixed lineage kinase inhibitor, CEP-1347, and the glyceraldehyde-3-phosphate (GAPDH) ligand, TCH346, have failed in advanced clinical trials.[16]

Overall, the prognosis for both AD and Parkinson's disease remains poor. While palliative treatments can provide some relief early on, the process of neurodegeneration will ultimately progress. The clear need for effective neuroprotective or neurorestorative approaches is hampered by an incomplete understanding of disease etiology. The fact that a number of genetic loci are associated with familial AD and Parkinson's disease has driven considerable efforts directed at understanding the discrete cellular mechanisms responsible for neuronal death in these disorders. Such mutations account for only 10% or fewer of diagnosed cases, the remainder being idiopathic. It is also becoming apparent that neurons do not die for any one reason but rather succumb to a complex array of insults including aberrant protein processing, proteasomal dysfunction, oxidative stress, neuroinflammation, and excitotoxicity. Therefore, while a mutation may predispose a neuron to die, targeting that mutation may have little or no effect in the established disease state. The numerous failures in the clinic indicate that major discrepancies in the ability to preclinically model the interrelated mechanisms responsible for cell death in these disorders may provide a currently insurmountable barrier to progress. Certainly, the utility of the various single, double, and triple transgenic animals that reflect aspects of the disease process show little utility in the human condition.

In addition to AD, VD, DLB, AIDS dementia, and Parkinson's disease, there are a large number of neurodegenerative conditions that can cause dementia. While these disorders can be collectively referred to as non-Alzheimer's dementias, they do not all share common features or underlying causes. These include dementia due to Huntington's disease, Pick's disease, Creutzfeldt-Jakob's disease, stroke, and head trauma.

6.01.4.2.4 Autoimmune/neuromuscular disorders

A number of autoimmune/neuroinflammatory disorders (*see* 6.09 Neuromuscular/Autoimmune Disorders) affect either the central or peripheral nervous system. Many of these disorders are exceptionally rare such as Moersch–Woltman syndrome (stiff-man), Lambert–Eaton myasthenic syndrome, and myasthenia gravis (MG). While uncommon, these disorders tend to be highly debilitating as they directly alter neuromuscular transmission. The most common of these disorders is MG which affects an estimated 60 000 people in the USA. The primary pathology underlying MG appears to be the production of autoantibodies directed against the alpha subunit of the neuromuscular nicotinic acetylcholine receptor. Through direct interference and complement-mediated lysis of the postsynaptic muscle membrane, the autoantibodies cause disruption in the motor endplate that leads to a weakness in skeletal muscle throughout the body. The autoimmune disorder systemic lupus erythematosus (SLE) and the neuroinflammatory disease multiple sclerosis are significantly more common. SLE is a chronic inflammatory autoimmune disease with diverse clinical manifestations. Common symptoms include fatigue, headache, joint pain, and photosensitivity. SLE afflicts approximately 240 000 people in the USA, and estimates suggest that 20–70% of SLE cases involve effects within the CNS, possibly due to a breakdown in the blood–brain barrier. These so-called neuropsychiatric SLE (NPSLE) patients develop a number of CNS-related syndromes including stroke, optic neuropathy, chorea, parkinsonism, epilepsy, dementia, psychosis, and depression.

Diagnosis of these disorders can be challenging. Fortunately, there are several diagnostic tests that allow for quantitative assessment of autoantibodies or inflammatory markers that greatly facilitate accurate diagnosis. Clinical interview, including a careful assessment of symptom history and temporal development of symptoms, is essential. MG patients will usually present with specific muscle dysfunction or weakness. Common initial symptoms include ptosis or diplopia. Definitive tests for MG include assessment of the presence of autoantibodies directed against the human nicotinic acetylcholine receptor, a decrement in muscle response with repetitive nerve stimulation measure through electromyography, and palliative response to cholinesterase inhibitors. SLE patients have a variety of symptoms including swelling of the joints, rash or scarring lesions of the skin, and numbness, weakness, or burning sensations. The most definitive test for SLE is an assessment of antinuclear autoantibodies. Additional diagnostic tests measure the presence of antibodies directed against double-stranded DNA, or Smith antigen.

In all three disorders, corticosteroids (prednisone) or immunosuppressants (cyclosporine, azathioprine) can provide effective therapies that promote remission and prevent further symptom flare-ups. Acetylcholinesterase inhibitors (pyridostigmine, neostigmine) can be used to enhance muscle strength in MG patients; NSAIDs are prescribed for pain and SSRIs for depression.

6.01.4.2.4.1 Multiple sclerosis

Multiple sclerosis is an inflammatory demylinating disease directed at the brain and spinal cord characterized by lesions that are due to the perivascular infiltration of monocytes and lymphocytes into the brain parenchyma, brain stem, optic nerve, and spinal cord. There is evidence that its etiology might be viral in origin but the data are inconclusive. The resulting demyelination leads to impaired nerve signaling and subsequent impairment in vision, sensation, movement, and cognition. Multiple sclerosis lesions tend to be focal, and therefore the symptoms are often discretely tied to a particular body region, and multiple symptoms may occur in a single patient. Multiple sclerosis primarily affects adults, with a typical age of onset between 20 and 50 years. An estimated 300 000 individuals in the USA are afflicted with multiple sclerosis. While a relationship between multiple sclerosis risk and latitude exists in North America, Australia, and New Zealand, this is not consistent. Israel, a country with a high immigration rate, has a higher incidence of multiple sclerosis than its latitude would predict. The highest rates of multiple sclerosis are found in the Orkney and Shetland Isles in the UK, and multiple sclerosis is rare in native Africans and Japanese, suggesting a genetic component, supported by association with certain HLA alleles.

Multiple sclerosis patients present with a broad array of symptoms including reduced or abnormal sensations, weakness, vision changes, clumsiness, and loss of bladder control. The diversity of initial symptoms is a reflection of the focal nature of the disease and makes accurate diagnosis a challenge. A number of signs can be assessed to help in making the diagnosis including abnormal eye movements or pupillary response, altered reflex responses, impaired coordination or sensation, and evidence of spasticity or weakness in the arms or legs. Definitive diagnosis is made by a number of tests including blood tests to rule out other possible diagnoses (e.g., Lyme disease), an examination of cerebrospinal fluid to assess the presence of elevated immunoglobulin G (IgG), and oligoclonal banding, a visual evoked potential test to determine if there is a slowing in signal conduction, and a magnetic resonance imaging (MRI) scan to assess the presence of periventricular lesions. Multiple sclerosis patients can go through remission periods that last for as long as 5 years in which the disease is relatively stable.

Glucocorticoid therapy and immunomodulatory drugs are used for the treatment of multiple sclerosis. The former include cyclophosphamide, cladribine, methotrexate, azathiopurine, cyclosporine, and interferon, the latter include Type I (IFN-α and IFN-β) and Type II (IFN-γ). Glatiramer acetate, a mixture of synthetic polypeptides composed of the acetate salts of the L-amino acids glutamic acid, alanine, glycine, and lysine that are found in high abundance in myelin basic protein (MBP), reduces the frequency of relapse in patients with relapsing–remitting multiple sclerosis. It is thought to either binding directly to major histocampatibility complex (MHC) class II complexes or induce MBP-specific suppressor cells.

Natalizumab is a humanized monoclonal antibody to α4 integrin that was approved for the treatment of relapsing forms of multiple sclerosis. It has robust effects on relapse rate and lesion activity in multiple sclerosis and may affect disease progression. Following three cases of progressive multifocal leukoencephalopathy (PML) in patients treated with natalizumab in combination with immunosuppressant therapy, the compound was put on a clinical hold which has now been removed following an additional assessment of the risk–benefit ratio for the use of this biological.

The overall prognosis for patients with autoimmune or neuroinflammatory disease appears good. While there is currently no drug available that will elicit a full remission or reversal of symptoms, there are a variety of choices for both palliative control of symptoms and modification of disease progression. There are significant adverse effects associated with corticosteroid, immunosuppression, and immunomodulation; however, the immunomodulators are relatively well tolerated.

6.01.4.2.5 **Stroke and nervous system trauma**

Stroke, an acute syndrome of brain ischemia, a 'brain attack,' traumatic brain injury (TBI), and spinal cord injury (SCI) are acute neurological insults that have devastating consequences on both survival and quality of life (*see* 6.10 Stroke/Traumatic Brain and Spinal Cord Injuries) The catastrophic injury that can occur in the nervous system with the resulting disability following such insults is not due to the primary ischemic infarct, but results from the secondary cascade of biochemical events, glutamate-mediated excitotoxicity, intracellular calcium overload, and reactive oxygen species (ROS)-induced oxidative damage, which take place within the first minutes, hours, and days after the traumatic event. In stroke, when a arterial occlusion is only temporary, e.g., less than 24 h, a transient ischemic attack, recovery can be rapid.

Unlike TBI and SCI where the trauma is typically immediately obvious, stroke is an exclusionary clinical diagnosis that often occurs when the reduction in blood flow in the ischemic core has resulted in an ischemic penumbra being formed. Immediate treatment, like that in cardiac ischemia, is to restore blood flow with thrombolytic therapy, e.g., t-PA (tissue plasminogen activator) accompanied by aspirin or the $P2Y_{12}$ receptor antagonists, e.g., ticlopidine or clopidogrel, both of which block platelet aggregation and prevent recurrent stroke.

The incidence of stroke in the USA approaches 700 000 per year, the majority occurring in the elderly. Of these, 90% are 'ischemic' involving a thromboembolic blockage of a brain artery impairing cerebral blood flow and oxygenation, causing infarction of the brain region. The remaining 10% of strokes are 'hemorrhagic' involving intracerebral hemorrhage where blood is released into the brain parenchyma producing brain damage by triggering brain edema leading to secondary ischemia and subarachnoid hemorrhage where the blood is released into the subarachnoid space. There are about 30 000 aneurysmal subarachnoid hemorrhages and 1.5 million cases of TBI each year in the USA, and 11 000 new cases of SCI each year with an overall prevalence of 250 000. Most cases of TBI and SCI occur in the second and third decades of life. Given the incidence, limited treatment options, and disabilities resulting from stroke, TCI, and SCI, there has been considerable effort over the past 20 years to identify agents that address the secondary biochemical cascade of events, including glutamate antagonists, specifically those for the NMDA receptor, calcium channel blockers, and NCEs that modulate ROS production. Without exception, these agents, that include the glutamate antagonists, selfotel, cerestat, lubeluzole, memantine, and citicholine, the GABA partial agonist, chlomethiazole, the calcium channel blocker, nimodipine, the ROS-scavengers, PEG-superoxide dismutase and the 21-amino steroid tirilazad, and the antioxidant/weak NMDA antagonist, dexabinol, failed to show consistent efficacy in controlled clinical trials. As a result while the unmet clinical need remains high, various caveats regarding the relevance of the animal models used to advance NCEs into clinical trials and the dosing regimens used have resulted in a need to restrategize approaches at the research level. Key have been the 3 h window for NCE treatment that is currently incompatible with an exclusionary diagnosis in stroke and the pre- or coadministration of NCE therapy with the traumatic insult in animal models when the human situation requires treatment post stroke, TBI, or SCI. Like trials in stroke, use of the above-mentioned agents where tested in TBI and SCI also gave negative results. The first generation of drug discovery and development efforts in neuroprotection has identified several deficiencies that must be avoided in future campaigns.

6.01.4.2.6 Epilepsy

Epilepsy describes a large class of seizure disorders in which normal patterns of neuronal activity become disturbed, leading to unusual emotions, behaviors, sensations, convulsions, muscle spasms, and loss of consciousness. Epileptic seizures can be divided into either partial (focal) or generalized seizures. Partial seizures (*see* 6.11 Epilepsy) occur in one part of the brain and can be either simple or complex. Simple partial seizures tend to involve sudden and unexplained sensations or emotions, while complex partial seizures typically involve a loss of, or alteration of, consciousness that may present as repetitive unproductive behaviors. Focal seizures are usually brief, lasting only a few seconds. Generalized seizures are more broadly expressed in the brain, typically involve both hemispheres, and may cause loss of consciousness, spasms, and falls. There are multiple classes of generalized seizure including absence, tonic, clonic, atonic, and tonic–clonic. Epilepsy currently afflicts over 2.7 million Americans with estimates of over 180 000 new cases each year. The disorder usually becomes apparent in childhood or adolescence; however, it can develop at any time in life. In some cases, the risk of seizure has a significant impact on daily life, in particular limiting activities such as driving. Furthermore, severe or treatment-resistant seizures have been associated with reduced life expectancy and cognitive impairment.

Diagnosis of epilepsy relies on obtaining a detailed medical history that includes description of symptoms and duration of seizures. Electroencephalography (EEG) is commonly employed in the diagnosis and can provide a powerful tool to rule out disorders such as narcolepsy. In addition to verifying the diagnosis of epilepsy, the EEG can identify specific epileptic syndromes and provide a more detailed assessment of disease prognosis. Rapid diagnosis is essential since many treatments that seem to work well after the first reported seizure are much less effective once the seizures are established.

Current treatments are derived from a broad class of compounds termed antiepileptics or anticonvulsants. While effective, many of these drugs have poorly defined or multiple mechanisms of action. These drugs can be loosely grouped into four categories: blockers of voltage-dependent sodium channels (e.g., phenytoin, carbamazepine), enhancers of GABAergic transmission (e.g., BZs, tiagabine), t-type calcium channel blockers (ethosuximide), and compounds that posses either multiple or unknown mechanisms of action (e.g., valproate, gabapentin, lamotrigine, topiramate). For the most part these compounds are prescribed on the basis of type of seizure and on previous data demonstrating efficacy. For example, partial seizures may be treated with lamotrigine or carbamazepine as a front-line therapy, with the addition of tiagabine or gabapentin as an adjunct if needed. A similar front-line approach is taken for generalized tonic–clonic seizure; however, second-line treatment would include phenytoin or clonazepam. Ethosuximide is a common choice for treating absence seizure with clonazepam or topiramate used as a second-line therapy.

The prognosis for a patient diagnosed with epilepsy is good. Available drugs are efficacious; however, the efficacy varies depending on the seizure type and the individual patient. While many of the older anticonvulsants have significant side effect liabilities, the newer antiepileptic drugs have demonstrated efficacy with much more acceptable therapeutic windows. In spite of this, there are still an estimated 30% of patients who receive no or modest benefit from available treatments. These patients may, as a last resort, turn to surgical resection or vagus nerve stimulation to prevent seizure. Therefore, new compounds that are effective in these resistant patients will be a useful addition to current therapies.

6.01.4.2.7 **Ophthalmic agents**

The eye is a key part of the CNS connecting to the brain via the second cranial or optic nerve. Many other receptors and signal transduction process in the eye recapitulate those present in the CNS proper although the cellular architecture of the eye and its function are unique (*see* 6.12 Ophthalmic Agents). The anterior pole of the eye, composed of the cornea, iris, and lens, serves to focus light onto the photoreceptors of the retina. The retina is a layered structure composed of retinal ganglion cells, amacrine cells, horizontal cells, and photoreceptors that transduces, processes, and integrates visual stimuli. The ganglion cells are the output neurons of the retina, sending axons to the lateral geniculate nucleus by way of the optic nerve. This highly evolved complex system allows for accurate processing of visual stimuli with an exceptional dynamic range. However, it is susceptible to multiple disorders including glaucomas and macular degeneration that ultimately can produce visual defect and blindness.

The glaucomas are a group of disease associated with elevated intraocular pressure. If left untreated, the increase in pressure produces irreversible vision loss by damaging the optic nerve. Glaucoma is the second most common cause of vision loss, and the leading cause of blindness in the USA, affecting over 3 million individuals. Glaucoma typically produces no overt symptoms until vision is irreparably lost. Fortunately, early detection is enabled by a number of facile and routine diagnostic tests such as air-puff tonometry. First-line treatment is usually with topical application of agents that reduce the secretion of aqueous humor such as β-adrenoceptor blockers (e.g., timolol), and carbonic anhydrase inhibitors (e.g., dorzolamide), or agents that increase the outflow of aqueous humor such as prostaglandin analogs (e.g., latanoprost) and parasympathomimetic drugs (e.g., pilocarpine).

Macular degeneration is a disorder in which the macula, the cone-rich central part of the retina responsible for fine detail and color vision, degenerates. The degeneration can be either a dry form which appears as an atrophy of the retinal pigmented epithelium, or a wet form in which degeneration is caused by the leaking of choroidal neovascularizations. The disorder develops gradually and with age, and is therefore often termed 'age-related macular degeneration.' It is the leading cause of vision loss affecting more than 1.6 million individuals in the USA. Symptoms include a loss of visual acuity, blurred or distorted vision, or a loss of vision in the center of the visual field. Diagnosis is typically made by visual examination, tests of central visual acuity, and fluorescein angiography to assess neovascularization.

Until recently, treatment options were only available for wet macular degeneration and were limited to surgical approaches such as photocoagulation. Several emerging therapies show promise including, intravitreal injections of the vascular endothelial growth factor (VEGF) antagonist, pegaptanib for wet macular degeneration, or the use of high dose combination antioxidant zinc therapy to slow the progression of dry macular degeneration.

6.01.5 **Future Prospects**

Over the past two decades, there has been a remarkable increase in: (1) efforts to identify novel, disease-related targets (albeit many of these are unvalidated); (2) the ability to perturb discrete neuronal circuits and their signaling pathways; and (3) a proficiency in designing highly target-selective NCEs. However, this has not translated into any significant increase in either the quality or quantity of NCEs available for disorders of the CNS.

Two areas of especial note in excellent and comprehensive science not translating to meaningful clinical progress are the neurokinen-1 (NK-1) receptor antagonists for pain and depression[17] and the various excitotoxic and ROS blockers that failed in stroke. Unfortunately, progress in the area of neurodegeneration is becoming reminiscent of what happened in these two areas: the inability to reduce fundamental research to practice.

CNS drug discovery has a long history of serendipity.[11] For example, the first antidepressant, the MAO-1 iproniazid, was originally developed for the treatment of tuberculosis while the anticonvulsant actions of a variety of NCEs were found to be due to the vehicle in which they were dissolved, valproic acid. The complexity of CNS diseases and the empirical nature of the animal models designed to show efficacy have led to many NCEs entering the clinic for one indication and being found useful in another. A key example in this regard is the antipsychotic clozapine.[9,18] Discovered in the prebiotech era of the 1970s based on empirical similarities to the dopamine antagonist haloperidol, clozapine was introduced in the early 1970s as a novel antipsychotic with a superior human efficacy profile, the mechanism for which was unknown. The superior attributes of clozapine were limited by the incidence of sometimes fatal agranulocytosis leading to a search for clozapine-like agents lacking this side effect. Clozapine has multiple receptor activities, including interactions with both the DA and 5HT receptor families and with each new receptor identified, clozapine has been found to have addition nuances in its target interactions both in terms of efficacy ($5HT_6$) and side effects (H_4). It has also been suggested that these various targets interactions might be additive or synergistic with one another.[18] Despite knowledge of the evolving molecular interactions of clozapine, numerous attempts to identify second-generation NCEs with similar efficacy to clozapine in effectively treating the entire spectrum of symptoms associated with schizophrenia have been unsuccessful and there is still no consensus on its mechanism of action.

The paucity of NCEs in the area of CNS disorders suggests that the current reductionist approach, driven largely by advances in molecular biology including a widespread use of transgenic animals that appear to present more questions than they answer, represents a major disadvantage when attempting to find drugs to restore normal function to a system as complex as the CNS. The lack of progress in the areas of pain and depression underlines the shortcoming of this approach and provides further fuel to concept of whether the search for a 'magic bullet,' an NCE active at a single target, has currency in treating diseases with a poly target, poly genomic causality like many of those in the CNS. It has further been argued[16,19,20] that current animal models of psychiatric disorders fail to account for basic aspects of synaptic plasticity that are often stress-related and that if these were more widely used both they and the NCEs tested in them might provide a more realistic assessment of the hurdles in advancing NCEs to clinical trials.

The high level of serendipity in the discovery of CNS drugs[11] and the extensive need to use clinical information in refining and redefining preclinical activities indicate that while the molecular basis of human CNS diseases is being conceptually defined at the basic preclinical and clinical levels, a process that may take several more decades, safe and effective NCEs should be rapidly advanced along a translational medicine path[21] to gain rapid proof of concept, a return in essence to what happened over the past 60 years when: (1) the antidepressant actions of ipronizazid were observed in tubercular humans, and (2) the phenothiazine chlopromazine, designed to treat surgical shock, became the first antipsychotic.

References

1. Williams, M.; Coyle, J. T.; Shaikh, S.; Decker, M. W. *Annu. Rep. Med. Chem.* **2001**, *35*, 1–10.
2. Uhl, G. R.; Grow, R. W. *Arch. Gen. Psychiatr.* **2004**, *61*, 223–229.
3. First, M. B.; Tasman, A. *DSM-IV-TR Diagnosis, Etiology and Treatment*; Wiley: New York, 2004.
4. Spiegel, A. *New Yorker*, Jan, 2005, pp 56–71.
5. Kupfer, D. J.; First, M. B.; Regier, D. A. *A Research Agenda for DSM-V*; American Psychiatric Publishing: Washington, DC, 2002.
6. Sunderland, T.; Linker, G.; Mirza, M.; Putnam, K. T.; Friedman, D. L.; Kimmel, L. H.; Bergeson, J.; Manetti, C. J.; Zimmermann, M.; Tang, B. *JAMA* **2003**, *289*, 2094–2103.
7. Coyle, J. T.; Tsai, G. *Int. Rev. Neurobiol.* **2004**, *59*, 491–515.
8. Lieberman, J. A.; Stroup, T. S. S.; McEvoy, J. P. et al. *N. Engl. J. Med.* **2004**, *353*, 1209–1223.
9. Weiner, D. M.; Meltzer, H. Y.; Veinbergs, I.; Donohue, E. M.; Spalding, T. A.; Smith, T. T.; Mohell, N.; Harvey, S. C.; Lameh, J. H.; Nash, N. et al. *Psychopharmacol.* **2004**, *177*, 207–216.
10. Caspi, A.; Sugden, K.; Moffitt, T. E.; Taylor, A.; Craig, I. W.; Harrington, H.; McClay, J.; Mill, J.; Martin, J.; Braithwaite, A.; Poulton, R. *Science* **2003**, *301*, 386–389.
11. Sneader, W. *Drug Discovery: The Evolution of Modern Medicines*; John Wiley: Chichester, UK, 1985.
12. Korpi, E. R.; Sinkkonen, S. T. *Pharmacol. Ther.* **2006**, *109*, 12–32.
13. NIDA Research Report – HIV/AIDS NIH Publication No. 06-5760, November, 2005.
14. Bartus, R. T. *Exp. Neurol.* **2000**, *163*, 495–529.
15. Roman, G. C.; Tatemichi, T. K.; Erkinjuntti, T. et al. *Neurology* **1993**, *43*, 250–260.
16. Waldmeier, P.; Bozyczko-Coyne, D. B.; Williams, M.; Vaught, J. *Biochem. Pharmacol.* **2006**, *72*, 1197–1206.
17. Williams, M. *Biochem. Pharmacol.* **2005**, *70*, 1707–1716.
18. Roth, B. L.; Sheffler, D. J.; Kroeze, W. K. *Nat. Rev. Drug Disc.* **2004**, *3*, 353–359.
19. Horrabin, D. F. *Nat. Rev. Drug Disc.* **2003**, *2*, 151–154.
20. Spedding, M.; Jay, T.; Costa e Silva, J.; Perret, L. *Nat. Rev. Drug Disc.* **2005**, *4*, 467–478.
21. FitzGerald, G. A. *Nat. Rev. Drug Disc.* **2005**, *4*, 815–818.
22. Surtees, P. G.; Wainwright, N. W. J.; Willis-Owen, S. A. G.; Luben, R.; Day, N. E.; Flint, J. *Biol. Psychiatr.* **2006**, *59*, 224–229.

Biographies

Michael J Marino received his PhD in 1995 from the University of Pittsburgh Department of Neuroscience where he employed behavioral and electrophysiological methods to the study of G-proteins and G protein-coupled receptors. As a postdoctoral fellow with Dr Jeff Conn at Emory University, he performed pioneering work on the role of metabotropic glutamate receptors in regulating the circuitry of the basal ganglia. This work led to the identification of potential targets for the treatment of Parkinson's disease. In 2001, he joined Merck Research Laboratories where he worked on novel drug discovery projects related on the modulatory control of brain circuitry underlying Parkinson's disease and related movement disorders. He is currently Associate Director of CNS biology at Cephalon, where he continues to focus on drug discovery for disorders of the CNS.

Michael Williams received his PhD (1974) from the Institute of Psychiatry and his Doctor of Science degree in pharmacology (1987) both from the University of London. Dr Williams has worked in the US-based pharmaceutical industry for 30 years at Merck, Sharp and Dohme Research Laboratories, Nova Pharmaceutical, CIBA-Geigy, and Abbott Laboratories. He retired from the latter in 2000 and served as a consultant with various biotechnology/pharmaceutical venture capital companies in the USA and Europe. In 2003 he joined Cephalon, Inc. in West Chester, where he is Vice President of Worldwide Discovery Research. He has published some 300 articles, book chapters, and reviews and is an Adjunct Professor in the Department of Molecular Pharmacology and Biological Chemistry at the Feinberg School of Medicine, Northwestern University, Chicago, IL.

Comprehensive Medicinal Chemistry II
ISBN (set): 0-08-044513-6

ISBN (Volume 6) 0-08-044519-5; pp. 1–16

6.02 Schizophrenia

M J Marino, Worldwide Discovery Research, Cephalon, Inc., West Chester, PA, USA
R E Davis, 3-D Pharmaceutical Consultants, San Diego, CA, USA
H Meltzer, Psychiatric Hospital Vanderbilt, Nashville, TN, USA
L J S Knutsen and M Williams, Worldwide Discovery Research, Cephalon, Inc., West Chester, PA, USA

6.02.1 Introduction

Schizophrenia is a chronic, debilitating mental disease, the etiology and pathophysiology of which is incompletely known.[1,80] It affects approximately 1% of the global population and has no defined ethnic or social boundaries. Schizophrenia is characterized by a well-defined set of symptoms, a plethora of pathological and neurochemical brain alterations, and, despite diverse therapeutic options, a limited ability to treat the core dimensions of the illness other than psychotic symptoms, e.g., auditory and visual delusions and hallucinations. Despite the use of drug or electroconvulsive treatment, approximately 15% of schizophrenic patients have persistent moderate to severe positive symptoms.

While new insights into the disorder have appeared over the past decade, psychotic symptoms have been recognized throughout human history. In most societies, such aberrations have been considered to be signs of severe mental illness, although at times they have been interpreted as a sign of religious exaltation or possession. The clinical features of psychosis were noted in early literature and have remained remarkably unchanged over time. In the latter half of the nineteenth century, features of psychosis were variously categorized as 'folie circulaire' (cyclical madness), 'hebephrenia' (a silly, vacuous state of mind), and ultimately, due to the clinical acumen of Emil Kraepelin at the turn of the twentieth century, dementia praecox, or 'dementia of early onset.'[2] Thus, Kraepelin deserves full credit for recognizing the relationship between psychosis and cognitive impairment. In 1911, Bleuler first used the term schizophrenia based upon an emphasis on nonpsychotic clinical symptomatology, specifically: blunted affect; loose associations; inability to experience pleasure; poverty of thought and a preoccupation with the self and one's own thoughts. The cognitive component of schizophrenia was all but forgotten from the time when chlorpromazine **1**, the first antipsychotic drug, was serendipitously discovered in 1951[3] until resurrected recently. Indeed, cognitive dysfunction has now emerged as the major unmet medical need in developing novel therapies for the treatment of schizophrenia.

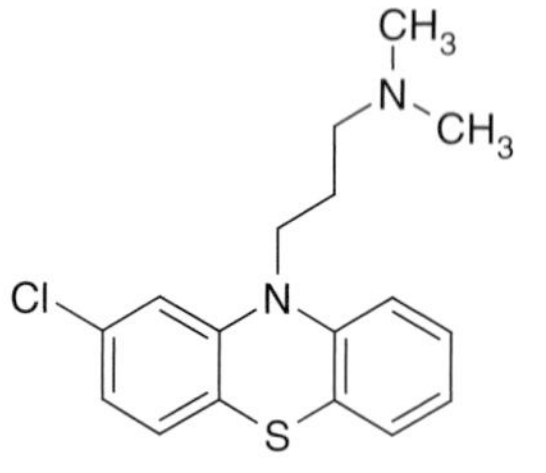

Chlorpromazine **1**

6.02.2 Disease State/Diagnosis

The *Diagnostic and Statistical Manual of Mental Disorders* (DSM-IV-TR) category for the schizophrenic disease spectrum is designated as 295.xx, 'schizophrenia and other psychotic disorders'[4] with the main subclasses being: paranoid type (295.30); disorganized type (295.10); catatonic type (295.10); undifferentiated type (295.90); and residual type (295.6) as well as schizophreniform disorder (295.40); schizoaffective disorder (295.70; including bipolar and depressive types); delusional disorder (297.1); brief psychotic disorder (298.8); and shared psychotic disorder (297.3).

The three main symptoms of schizophrenia are 'positive', 'negative' symptoms, and 'cognitive dysfunction.' Positive symptoms (although by no means positive to the patients) are defined as an excess or distortion of normal function and include: bizarre behavior, auditory, and, more rarely, visual hallucinations, paranoid and other types of delusions, and disorganized thought. Negative symptoms include a diminution or loss of normal function and include: affective flattening, anhedonia, social withdrawal, lack of motivation and spontaneity, and alogia and avolition – poverty of thought and speech. Cognitive impairment in schizophrenia begins before the onset of the psychosis and remains severe, with some worsening, throughout the illness. While the precise domains of cognitive dysfunction in patients with schizophrenia remain to be elucidated, schizophrenia is clearly associated with widespread, multifaceted impairments in cognitive function, including executive function, attention, processing, vigilance, verbal learning and memory, verbal and spatial working memory, semantic memory, and social cognition. Recent evidence suggests that cognitive impairment may be of equal or greater importance than positive or negative symptoms in predicting functional outcomes, such as work status, quality of life, and social problem solving.[5] For example, cognitive dysfunction along with disorganization symptoms discriminates schizophrenic patients who are able to work from those who are not.

There are six diagnostic criteria for schizophrenia[4]: (1) characteristic symptoms, that include (i) delusions, (ii) hallucinations, (iii) disorganized speech, (iv) grossly disorganized or catatonic behavior, and (v) negative symptoms (affective flattening, alogia (poverty or absence of speech), avolition (lack of interest and drive)) – two or more of which are present for a significant duration over a 1 month period; (2) social/occupational dysfunction; (3) duration – continuous signs of the disturbance for at least 6 months (unless successfully treated after early diagnosis); (4) schizoaffective and mood disorder exclusion; (5) substance/general medical condition exclusion; (6) relationship to a pervasive developmental disorder, e.g., autistic disorder. While considerable research has been directed at the genetics of the disorder and significant advances have been made in the development of imaging tools and methods, there is currently no objective clinical test for diagnosis.

The importance of cognitive dysfunction has recently been highlighted by an initiative to define the guidelines for drug approval for this indication. A consortium[6] headed by the National Institutes of Mental Health (NIMH) and including academic groups and representatives from the Food and Drug Administration (FDA) and industry is validating methods for evaluating drug effects on cognitive function in patients with schizophrenia, the Measurement and Treatment Research to Improve Cognition in Schizophrenia (MATRICS) initiative has been formed as an outgrowth of these interactions and this group has promulgated a cognitive testing battery that currently is undergoing validation studies in patients.

More recently, it has been argued that comorbid mood disorders are sufficiently common in schizophrenic patients to justify a fourth set of characteristic symptoms. Depression and bipolar disorder are highly comorbid in schizophrenia and are one of the key factors contributing to the increased risk for suicide in this disorder. Individuals with schizophrenia attempt suicide more often than people in the general population, and a high percentage, in particular younger adult males, succeeds in the attempt. Controversy remains over whether these mood disorders are in fact a manifestation of the disorder (i.e., share a common etiology), or an epiphenomenon associated with either the disease state or treatment. Regardless of the actual cause, comorbid mood disorders represent a clear risk in treating the schizophrenic patient population and are carefully considered along with the more traditional positive, negative, and cognitive symptoms. Schizophrenia is usually diagnosed in adolescence and the symptoms follow a characteristic pattern of development. Cognitive symptoms are first manifest during adolescence, accompanied by changes in social interactions. The first signs can include a change in peer group, declining academic performance, and increased irritability. Unfortunately, many of these symptoms are observed, albeit to a lesser degree, in normal adolescents. Therefore, diagnosis is not commonly made until the emergence of positive symptoms. Positive symptoms usually develop in men in their late teens and early twenties and in women in their mid-twenties to early thirties. While this describes the most common progression, rare cases of schizophrenia have been reported to emerge in children as young as 5 years, and adults past 45 years of age. The disorder is equally prevalent in both sexes and occurs at similar rates worldwide in all ethnic groups.

6.02.3 Disease Basis

Like the majority of central nervous system (CNS) disorders, the initial understanding of the factors causing schizophrenia was based on serendipity, in this instance, the finding that chlorpromazine, the first drug used for the treatment of the disease, was a dopamine (DA) receptor antagonist.[3] Since then it has been well established that schizophrenia is a multifactorial disease involving both genetic and epigenetic factors[7] that may also exist in several distinct subtypes. Identified risk factors for schizophrenia include: winter birth; low socioeconomic status; cannabis use; obstetric complications and intrauterine infection related to birth; immigration; living in a city (urbanicity) and the neighborhood cognitive social capital[8]; low intelligence quotient; and a family history of the disorder. There is increasing data[81] that schizophrenia can be associated with autoimmune diseases, e.g., celiac disease, acquired hemolytic anemia, thyrotoxicosis, interstitial cystitis, and Sjögren's syndrome, and that this association resulted in a 45% increase in risk for schizophrenia. The antipsychotics used to treat schizophrenia can be divided into two distinct classes, typical and atypical. The distinction between the two can be based on their time of introduction to market, typicals preceding atypicals; their receptor-binding profile, the atypicals antagonizing both D2 and $5HT_2$ receptors with additional binding to D3 and D4 receptors; but most importantly, the ability, albeit limited, of the atypical neuroleptics to address the negative symptoms of schizophrenia together with a lower risk of developing the tardive dyskinesia associated with the older, typical antipsychotic agents. A more controversial distinction is that typical antipsychotics are neurotoxic while atypical agents are metabolic poisons.[9]

6.02.3.1 The Dopamine (DA) Hypothesis

Until recently, the modal hypothesis on the pathophysiology of schizophrenia was that excessive dopaminergic transmission in the forebrain is a key causative factor. This DA hyperfunction hypothesis was primarily based on the observation that all clinically effective antipsychotic drugs have potent antagonist or inverse agonist activity at DA D2 receptors, and that the therapeutic efficacy of these compounds was highly correlated with their affinity for striatal D2 receptors. In addition, the psychotomimetic properties of indirect DA agonists like amphetamine and cocaine, and observed alterations in striatal DA release in schizophrenic patients, further supported the involvement of DA in the pathophysiology of schizophrenia.

The DA hypothesis has been useful in stimulating research on the neurochemical alterations underlying schizophrenia, placing the DA D2 receptor at the center of antipsychotic drug development, in essence recapitulating existing antipsychotic agents in a circular manner. For example, DA D2 antagonists are effective in schizophrenia; therefore, schizophrenia is a dysfunction of DA D2 receptor signaling. There are several important criticisms of the DA hypothesis that suggest DA hyperfunction is not the sole cause of schizophrenia. As noted, current antipsychotics, all of which act at least in part by blocking DA D2 receptors, are uniformly ineffective in treating either the negative and cognitive symptoms of schizophrenia. In fact, at least 15% of schizophrenic patients do not respond in any significant manner to DA antagonist therapy. Interestingly, the clinical efficacy of currently available drugs develops with a time course much slower than can be accounted for by a simple model in which the antipsychotic drug binds to the DA D2 receptor. This suggests that the clinical efficacy of D2 antagonists is not an immediate consequence of acute D2 receptor blockade but is dependent on additional effects that only occur with chronic treatment, e.g., neurogenesis.[10]

6.02.3.2 The Serotonin (5HT) Hypothesis

The serotonin (5HT) hypothesis of schizophrenia actually pre-dates that of DA. The ability of the hallucinogen lysergic acid diethylamide (LSD) to antagonize the effects of 5HT on smooth muscle led to the hypothesis that schizophrenia was caused by a decrease in central serotonergic function.[11] This theory, largely predicated on the similarities between schizophrenic psychosis and LSD-induced hallucination, was modified with the discovery that LSD could act as a 5HT agonist in some systems. These findings led to the search for endogenous psychotogens, an effort that never bore convincing fruit. With the discovery of chlorpromazine **1**, interest in DA D2 receptors rapidly supplanted interest in the 5HT system. The observation that $5HT_{2A}$ antagonism is a defining characteristic of the newer 'atypical' antipsychotics, together with recent evidence that all effective antipsychotics are $5HT_{2A}$ inverse agonists,[12] has reawakened interest in the role of 5HT in schizophrenia. To date there has been no compelling evidence to suggest a global increase in brain 5HT markers in schizophrenia. However, there are significant interactions between the 5HT and DA systems suggesting that a more subtle alteration in 5HT may be involved.

6.02.3.3 The Glutamate Hypofunction Hypothesis

Glutamate is the major excitatory neurotransmitter in the CNS, and antagonists of the NMDA (*N*-methyl-D-aspartate) subtype of glutamate receptor, the psychotomimetics, phencyclidine **2** (PCP) and ketamine **3**, mimic the positive, negative, and cognitive symptoms of schizophrenia.[13] In the clinic, NMDA receptor antagonists faithfully mimic the symptoms of schizophrenia to the extent that it is difficult to differentiate the two. Controlled human studies of psychosis induced by the NMDA receptor modulators, PCP or ketamine, as well as observations in recreational PCP abusers, have resulted in a convincing list of similarities between the psychosis induced by NMDA receptor antagonism and schizophrenia. In addition, NMDA receptor antagonists can exacerbate the symptoms in schizophrenics, and can trigger the re-emergence of symptoms in stable patients. Finally, the finding that NMDA receptor coagonists such as glycine **4**, D-cycloserine **5**, D-serine **6**, and milacemide **7**, produce modest benefits in schizophrenic patients implicates

NMDA receptor hypofunction in this disorder.[14] While these findings suggest a decrease in activation of NMDA receptors in schizophrenia, a simple model of decreased NMDA function in the forebrain is insufficient. In fact, acute systemic treatment with an NMDA antagonist in freely behaving rats induces a dramatic increase in cortical glutamate levels, leading to an increase in non-NMDA-mediated glutamatergic transmission and a subsequent increase in DA efflux.[15] The importance of this increase in cortical glutamate release is further supported by the finding that activation of group II metabotropic glutamate receptors by LY 354740 **8**, which decreases presynaptic glutamate release, blocks both the PCP-induced increase in glutamate release and the behavioral effects of the psychostimulant. This, combined with studies demonstrating a dissociation between cortical DA levels and the behavioral effects of PCP, suggests that NMDA antagonist-induced increases in cortical glutamate represent a key event underlying the psychotomimetic effects of PCP and ketamine.

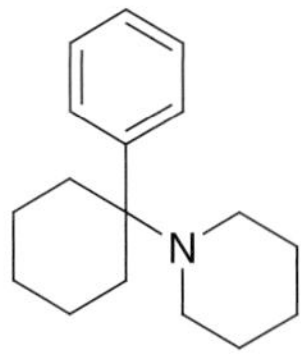

PCP **2** 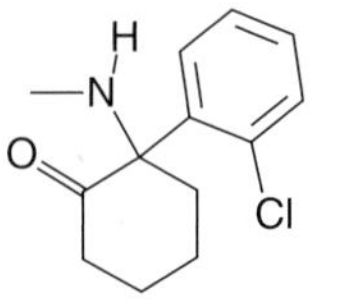 | Ketamine **3** 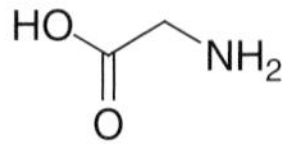 | Glycine **4** 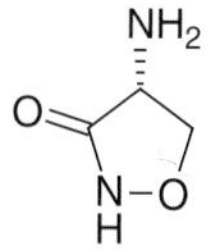 | D-Cycloserine **5**

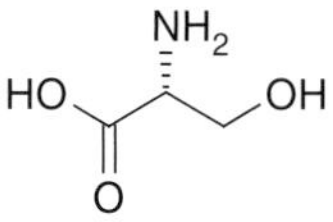

D-Serine **6** 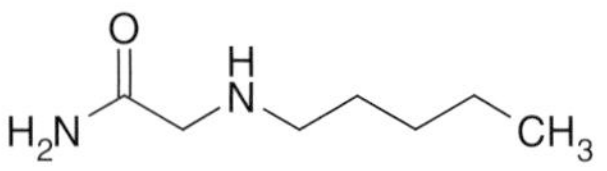 | Milacemide **7** 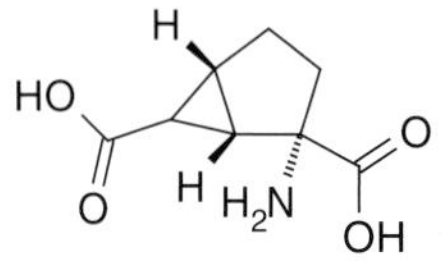 | LY-354740 **8**

One of the more attractive features of the glutamate/NMDA hypofunction hypothesis is that it does not contradict either the DA or 5HT hypotheses. Rather, glutamate, and in particular NMDA-mediated glutamatergic transmission, may provide a unifying link between the two systems. For example, the observation that cortical DA efflux increases after NMDA antagonist treatment[15] suggests that NMDA hypofunction could be a causative factor in inducing a hyperdopaminergic state and may, in part, explain the efficacy of typical and atypical antipsychotics in reversing the psychotomimetic actions of NMDA antagonists in both animal models and the clinic. Furthermore, studies on the interaction between glutamate and 5HT in the prefrontal cortex have led to a better understanding of the hallucinogenic actions of $5HT_{2A}$ agonists indicating that the 5HT hypothesis of schizophrenia may be compatible with the NMDA hypofunction model. Activation of $5HT_{2A}$ receptors increases the frequency of excitatory postsynaptic potentials at thalamocortical synapses in neocortical layer V pyramidal neurons. $5HT_2$ agonists also induce an asynchronous release of glutamate producing a slow, late excitatory postsynaptic potential in layer V neurons.[16] Therefore, both NMDA antagonists and $5HT_{2A}$ agonists may share a common path of psychotomimetic action through the induction of increased glutamate release in the cortex.

6.02.3.4 The γ-Amino-Butyric Acid (GABA) Hypothesis

γ-Amino-butyric acid (GABA) is the major inhibitory transmitter in the CNS, and has many effects that are opposite to those of glutamate, some of which involve GABAergic inhibition of glutamate function. The GABA uptake inhibitor, CI-966 **9**, has been associated with psychotic episodes in humans,[17] a similar phenotype to that seen with the psychotomimetics that block the effects of glutamate at the NMDA receptor. A role of GABA in the etiology of schizophrenia was first proposed in the early 1970s based on GABAergic regulation of DA neuronal function with a special focus on the role of GABA in working memory. GABA uptake sites are decreased in hippocampus, amygdala, and left temporal cortex in schizophrenics with some evidence of $GABA_A$ receptor upregulation[18] and reductions in GABA interneurons.[19] An extensive review of the use of benzodiazepines, the classical $GABA_A$ agonists, the $GABA_B$ agonist baclofen **10**, and the anticonvulsant valproic acid (VPA; **11**), the latter being used as a potential GABAergic agent, in combination with antipsychotics in clinical studies noted mixed results.[18]

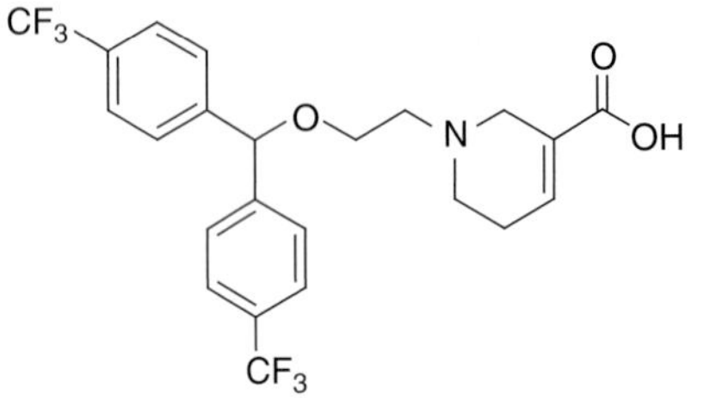

CI-966 **9** 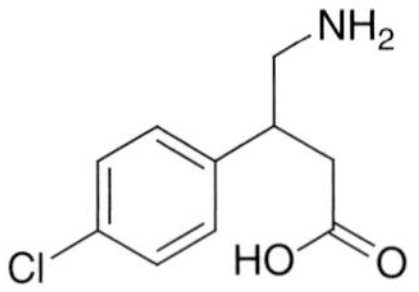 Baclofen **10** 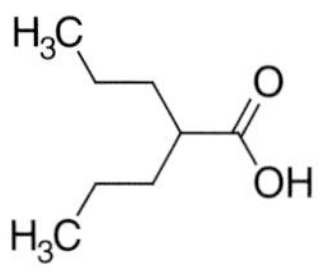 VPA **11**

6.02.3.5 The Clozapine Hypothesis

Clozapine **12** is the prototypic atypical antipsychotic, which has broad spectrum efficacy in schizophrenia, being efficacious in the treatment of refractory schizophrenics, with potential efficacy in treating cognitive deficits and having a lower extrapyramidal side effects (EPSs) liability.[20,21] These positive attributes are however limited by a high incidence of potentially fatal agranulocytosis that requires continuous monitoring in the clinical situation.

Based on the favorable therapeutic profile of clozapine, there has been an intense effort in the pharmaceutical industry over the past 30 years to identify clozapine-like new chemical entities (NCEs) that have the efficacy attributes of clozapine without the agranulocyotosis. This has led to a slew of second-generation atypical antipsychotics, discussed in detail below, none of which has achieved the efficacy seen with clozapine. This search has been largely based on the receptor binding profile of clozapine, in itself a challenge since, with each newly identified CNS receptor, the subsequent evaluation of clozapine tends to result in yet another potential pharmacological property being added to the profile of the compound.[22] Thus the chemist is challenged to synthesize compounds active at what are thought to be the primary molecular targets of clozapine, the DA D2 and $5HT_2$ receptors, anticipating that the binding profile will result in an approximation of the intrinsic activity and receptor ratios of clozapine at both the primary and the as-yet unknown ancillary targets, that lead to the unique antipsychotic profile of this molecule. A facile approach to the discovery of new antipsychotics has been to replicate the molecular properties of clozapine on a more or less trial-and-error basis. More recently, the *N*-desmethyl metabolite of clozapine **13**, which functions as a muscarinic receptor agonist, has been proposed as the key to understanding the unique properties of the parent drug.[23]

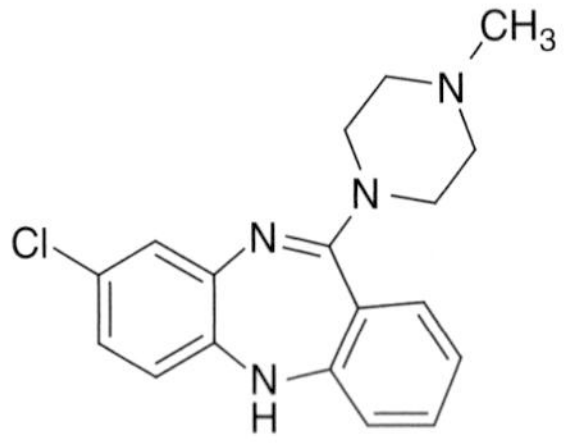

Clozapine **12** 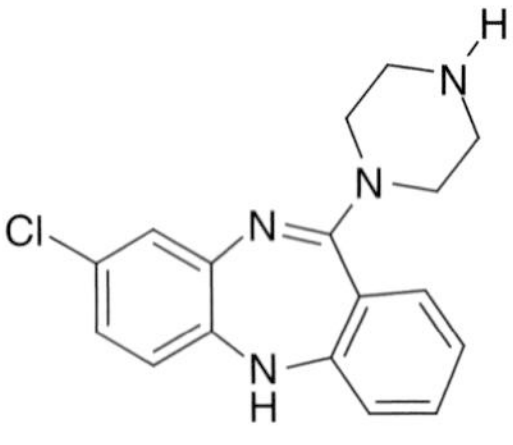 *N*-Desmethylclozapine **13**

6.02.3.6 Genetic Insights

Identifying the genetic causality of schizophrenia has become a major focus in CNS research. Multiple family, twin, and adoption studies have demonstrated that schizophrenia may be inherited. If a first-degree relative has schizophrenia there is a 10% chance of an individual developing schizophrenia, and if the first-degree relative is an identical twin the probability of developing schizophrenia rises to between 40% and 65%. However, no single specific genetic defect has yet been identified that explains this association nor has any genetic alteration been found that can account for more than a small proportion of the risk of inheriting schizophrenia. Thus schizophrenia is thought to be genetically heterogeneous with several discrete genes contributing to disease causality – the schizophrenia spectrum.[24] Twin studies have also indicated that there is a major epigenetic component of schizophrenia causality,[7,8] leading to the use of genome-wide scans to identify susceptibility loci.[25]

A large number of vulnerability genes have been identified that number in excess of 30.[26,27] Many of these genetic associations have failed to replicate when examined in different patient populations while allelic variations (e.g., Val108/158Met) in other genes, e.g., catechol *O*-methyltransferase (COMT) have been implicated in more than one disease state, in this instance gender-related pain sensitivity, temporomandibular joint disorder, breast cancer, and blood pressure.

6.02.3.6.1 Catechol *O*-methyltransferase

COMT is localized to chr22q11 where a microdeletion results in velocardiofacial syndrome (22qDS; DiGeorge or Shprintzen syndrome), a genetic subtype of schizophrenia. The COMT gene exists in two versions: Met158 and Val158, the former coding for a form of COMT that is less thermostable and thus has lower activity than the Val158. COMT is important for regulating DA but not norepinephrine (NE) levels in the prefrontal cortex.[28] Val158Met heterozygotic mice which have high COMT activity and, correspondingly, low prefrontal cortex DA levels show greater tyrosine hydroxylase expression in the midbrain, indicative of increased DA synthetic capability. In human Val158 carriers, neuroimaging studies showed greater midbrain F-DOPA uptake than Met158 carriers, consistent with increased DA biosynthesis. DA levels in prefrontal cortex play a key role in cognitive function and high-activity Val158 is associated with poorer performance and 'inefficient' prefrontal cortex function in some but not all studies. Despite several studies, the relationship of COMT dysfunction to schizophrenia is controversial with Val158 being considered a 'weak risk factor for schizophrenia'[28] that may reflect COMT variation providing 'a weak general predisposition to neuropsychiatric disease.'[29] In addition to the association with schizophrenia, the Val158Met mutation has also been associated with alterations in pain sensation[82] and breast cancer risk[83] adding further complexity to the psychiatric association. Nonetheless, in normal subjects and in Val158 carriers, COMT inhibitors like tolcapone **14** can improve aspects of working memory and executive function.

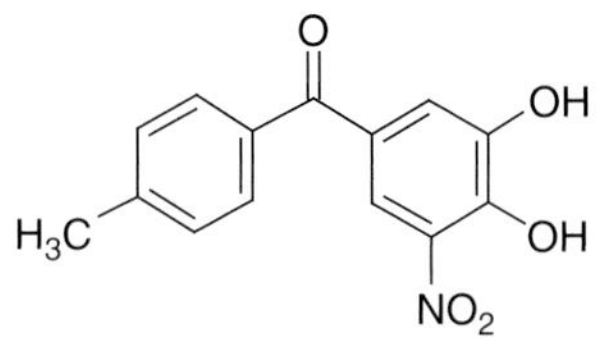

Tolcapone **14**

6.02.3.6.2 Neuregulin

Neuregulin (NRG1) on chr8p13 contains a core epidermal growth factor (EGF) domain and activates Erb-3 or -4 receptor tyrosine kinase.[30] It is involved in migration of GABAergic interneurons, signaling between axons and Schwann cells and the interactions between NMDA receptors and PSD-95. NRG-1 knockout mice show symptoms of schizophrenia. While 17 of 21 genetic association studies showed polymorphisms in NRG-1 associated with schizophrenia, the majority lie in noncoding regions of the genome. The NRG-1 genotype has been associated with the therapeutic response to clozapine in a Finnish population. The evidence for the involvement of NRG-1 in schizophrenia is typified as 'substantial but not incontrovertible'[30] and there have been studies with failure to replicate.[31]

6.02.3.6.3 Dysbindin-1

Dysbindin-1 (*DTNBP1*) at chr6p22.2 has been identified as a schizophrenia susceptibility gene in 10 separate patient cohorts[26] and reduced levels of expression of *DTNBP1* messenger RNA (mRNA) and protein have been observed in post-mortem brain samples from schizophrenics. Dysbindin is part of the dystrophin/dystrobrevin glycoprotein complex that is located in postsynaptic densities. Cultured neurons with reduced *DTNBP1* expression have a reduction in glutamate release.

6.02.3.6.4 Regulator of G protein signaling 4

Regulator of G protein signaling 4 (RGS4) at chr1q21-22 is present an area that has been associated by linkage analysis as a schizophrenia susceptibility locus.[32] The gene product of RGS4 downregulates signaling at both DA and 5HT receptors and its expression is modulated by stress.

6.02.3.6.5 *G72*

G72 at chr13q32-22 is a brain-expressed protein that has been associated with schizophrenia[33] that binds to DAAO (D-amino acid oxidase), an enzyme that oxidizes D-serine **6**, a potent activator of NMDA receptor, thus modulating glutamate function. While the association of *G72* with schizophrenia has been confirmed,[34] there is also evidence of an association with bipolar disorder, part of the schizophrenia spectrum.[24,35]

6.02.3.6.6 Glutamate receptors

Polymorphisms in GRM-3 (metabotropic glutamate receptor 3) at chr7q21.1-q21.2 and GRIN1 (NMDA subunit gene) at chr9q34 that encodes for the NMDA receptor subunit NR-1 have been associated with schizophrenia.[36] A reduction in the expression of GRIN1 in mice produces a schizophrenia-like phenotype.[37]

6.02.3.6.7 Disrupted-in-Schizophrenia 1 (*DISC1*)

Disrupted-in-schizophrenia 1 (DISC1) at chr1q is a component of the microtubule-associated dynein motor complex that is key to maintaining the centrosome complex and maintaining microtubular function. Depletion of endogenous DISC1 or mutated DISC1 causes neurite dysfunction in vitro and impairs cerebral cortex function in vivo, suggesting a neurodevelopmental role in schizophrenia.[38] More recently, DISC1 has been found to interact with the UCR2 domain of the phosphodiesterase PDE4B, suggesting a possible role in cAMP signaling processes that may involve CREB (cAMP response element binding protein) elements.

6.02.3.6.8 *CHRNA7*

Alternative phenotypes of *CHRNA7* (α7 neuronal nicotinic receptor, NNR) on chr15q14 have been associated with deficits in the P50 auditory response, a key phenotype of schizophrenia.[39] Agonists of the NNR like GTS-21 **15** and ABT-418 **16** have shown benefit in animal models of PPI, the rodent analog of P50.

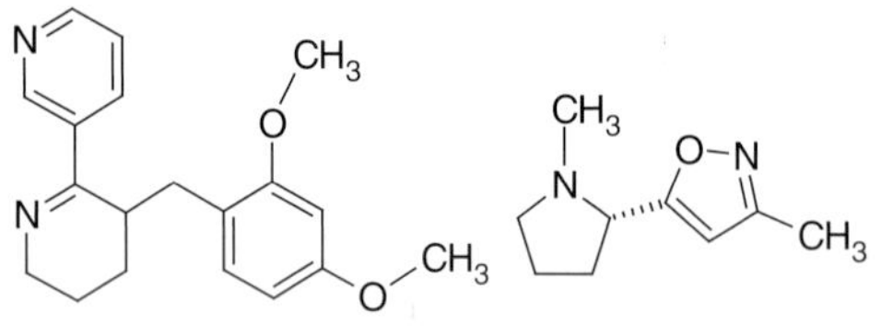

GTS-21 **15**

ABT-418 **16**

6.02.3.6.9 Miscellaneous associations

Additional schizophrenia associations include: GABBR1 ($GABA_B$ receptor) on chr6p21.3, a Ser9Gly polymorphism in the DA D3 receptor,[40] the $5HT_{2A}$ receptor, and CAPON (carboxy-terminal PDZ ligand of neuronal nitric oxide synthase). The reproducibility and uniqueness of these associations to schizophrenia is still under debate. The large number of studies focused on elucidating the genetic basis of schizophrenia and the multiple genetic foci thus far identified appear to be inversely proportional to the knowledge gained in understanding disease causality and treatment.[27] Thus it is likely that another decade will be needed for the many interesting gene targets and their contribution to the hypotheses of disease causality will be put in an appropriate context. Schizophrenia is a complex genetic disorder with multiple risk genes of small effect that are made more complex by possible allelic heterogenicity and epistatic influences.[7] In several instances (G72, GRM-3, and GRIN1) the genetic findings recapitulate (and are probably influenced by) existing hypotheses, in this instance, the glutamate hypothesis. In others, e.g., DISC1, the target for therapeutic intervention is unclear and in one, COMT, ignoring the multiple disease implications,[82,83] both the disease associations and the drugs, e.g., COMT inhibitors, are available to prove the value of genome-based disease causality and its treatment. This has not occurred, a possible reflection of the currently available COMT inhibitors lacking access to the brain. In this context, it is also of considerable interest that there is minimal agenetic evidence to support the concept that DA receptor antagonists, i.e., the current treatment modalities for schizophrenia, would be useful in the treatment of schizophrenia.

6.02.4 Experimental Disease Models

Schizophrenia is clearly a disorder with a primary impact on higher cognitive function. Therefore, the modeling of this disorder in less cognitively developed species than human represents a significant challenge. The validity of an animal model for any disorder can be rated on three scales: predictive, construct, and face validity.

'Predictive validity' focuses on how well results produced in the animal model are borne out in the clinic. More often, and particularly in the case of schizophrenia, animal models are back-validated using clinical benchmarks to provide a basis for arguing for future predictive validity. While this reasoning seems to hold for recent atypical antipsychotics in that they produce preclinical effects similar to that observed for older atypicals, the fact that these newer compounds are largely subtle variations on the clozapine theme discussed above raises questions regarding the usefulness of this back-validation. This is a significant caveat in that there is little confidence that a truly novel antipsychotic medication will appear as a positive in available preclinical models.

'Construct validity' has to do with the theoretical rationale underlying the model. A model with a high degree of construct validity would disrupt the same neurotransmitter systems and engage the same neuronal circuitry as the human disorder. Based on recent imaging studies and measurement of biological markers, there is an increased confidence that many of the pharmacological models of schizophrenia have a reasonable degree of construct validity. However, it is important to note that all of the pharmacological models currently employed are based primarily on the psychotomimetic properties of the compounds in humans. The inability to assess psychosis in rodents treated with these psychotomimetics thus raises an issue of face validity.

'Face validity' is a measure of how accurately the model reproduces the symptoms of the human disorder. This is particularly difficult in schizophrenia as many, if not all, of the symptoms are either uniquely human, or are impossible to probe in an animal mode.

6.02.4.1 Behavioral Assays

Before discussing the various means by which a schizophrenic state may be induced in animals, it is important to briefly consider the dominant behavioral paradigms employed to evaluate antipsychotic efficacy.

6.02.4.1.1 Conditioned avoidance

The ability of a compound to inhibit the conditioned avoidance response (CAR) to an aversive stimulus is one of the oldest predictors of antipsychotic efficacy. In this test, rats are trained to move from one side of shuttle box to the other on presentation of an audible cue (the conditioned stimulus) in order to avoid a footshock (the unconditioned stimulus). Once the animals have been trained, both typical and atypical antipsychotics are effective in decreasing the CAR to the conditioned stimulus without altering the escape response elicited by the unconditioned stimulus. This inhibition of the CAR is thought to be mediated by a reduction in dopaminergic function in the striatum and nucleus accumbens.[41] Therefore, inhibition of CAR is not an actual preclinical model of schizophrenia, but rather a facile in vivo method of detecting DA receptor blockade. The comparison between doses of antipsychotics that inhibit CAR and doses that induce catalepsy provides a convenient method to determine a therapeutic index for EPS.

6.02.4.1.2 Locomotor activity

Practically all antipsychotic agents decrease spontaneous locomotor activity and decrease locomotor activity that has been pharmacologically increased by amphetamine **17**, PCP **2**, or apomorphine **18**. As described for CAR, decreased locomotor activity can be interpreted as an in vivo readout of DA antagonism. However, the ability of nondopaminergic agents to induce hyperlocomotion that is sensitive to antipsychotics, and the ability of novel nondopaminergic compounds to reduce hyperlocomotion elicited by amphetamine suggest that this particular model involves a more complex circuit that may possibly have some relevance to the clinical state.

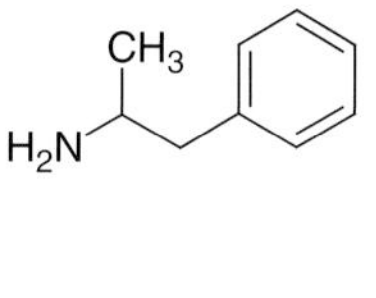

CH_3 H_2N HO OH H N CH_3

Amphetamine **17** 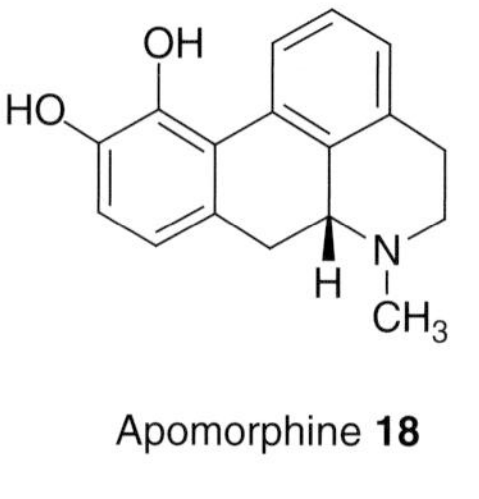 Apomorphine **18**

6.02.4.1.3 Latent inhibition

Latent inhibition is the ability of a pre-exposed nonreinforced stimulus to inhibit later stimulus-response learning.[42] This behavior can be disrupted by amphetamine in both rodents and humans. While often put forward as a model of positive symptoms with significant face validity, a careful review of the literature reveals significant disagreement on key facts, including the prevalence of disrupted latent inhibition in schizophrenic patients, the responsiveness of amphetamine-disrupted latent inhibition to atypical antipsychotics, and key differences between experimental paradigms used in human and animal studies. Results employing the latent inhibition assay must be interpreted with caution until these controversies are fully addressed.

6.02.4.1.4 Prepulse inhibition

A disruption in sensory and cognitive gating is hypothesized to be at the core of many of the symptoms of schizophrenia. Prepulse inhibition (PPI) refers to the ability of a low-intensity stimulus, or prepulse, to diminish the startle response elicited by a higher-intensity stimulus. This model has gained significant favor in recent years largely due to the findings that schizophrenic patients exhibit deficits in sensory and cognitive gating. This is particularly evident in studies of event-related potentials (ERPs) in the electroencephalogram of schizophrenic patients. The most common ERP paradigm has examined the latency and amplitude of P300 in response to an unpredictable change in a stimulus series (the 'oddball' paradigm). Reduced P300 amplitude in response to novelty in schizophrenic patients has been observed in a large number of studies.[43] The P50 ERP exhibits interesting behavior in response to pairs of brief auditory stimuli presented in rapid succession. In normal subjects, the P50 response to the second stimulus is attenuated relative to the first. Schizophrenic subjects, in contrast, often do not show this decreased P50 response to the second stimulus.[44] These differences in ERPs suggest that schizophrenic patients have a deficit in the gating or processing of sensory information. Consistent with this, several studies have examined PPI in schizophrenic patients[45,46] and found impairment relative to normal control subjects. This impaired sensorimotor gating may underlie the vulnerability in schizophrenia to sensory flooding, cognitive fragmentation, and conceptual disorganization. PPI is disrupted by a wide range of psychotomimetics and can be rescued by treatment with antipsychotic drugs.[47] Based on the high degree of face validity, apparent predictive validity, and the ability to strengthen construct validity by disrupting the behavior with multiple classes of psychotomimetics, PPI stands out as the current 'gold standard' assay for evaluating animal models of schizophrenia.

6.02.4.2 Pharmacological Models

Based largely on psychotomimetic properties in humans, a number of pharmacological tools have been used to induce a state in laboratory animals that is presumed to have some of the underlying pathophysiology of schizophrenia. These pharmacological models fall into three basic categories: DA releasing agents, $5HT_2$ agonists, and NMDA receptor antagonists.

According to the DA hyperfunction hypothesis of schizophrenia, agents that act to increase dopaminergic transmission should induce psychosis in normal individuals, and should precipitate or exacerbate psychosis in schizophrenics. Consistent with this, a large number of studies have demonstrated psychotomimetic effects of amphetamine **17**, cocaine **19**, and methylphenidate **20**, all compounds that act to increase DA release. Interestingly, these effects do not seem to be mimicked by direct-acting DA agonists, suggesting that a degree of circuit-based spatial selectivity may be important for the psychotomimetic effects of enhanced dopaminergic transmission. The 5HT hypothesis of schizophrenia is driven largely by the hallucinogenic effects of LSD **21**, and the finding that LSD interacts with 5HT receptors. Indolamines such as LSD **21** and psilocybin **22**, and phenethylamines including dimethoxymethylamphetamine **23** and mescaline **24**, constitute the two main classes of hallucinogens. Interestingly, these agents all have relatively high affinity for the $5HT_2$ subclass of 5HT receptors and in particular have been found to act as $5HT_{2A}$ agonists. The glutamatergic system has also been implicated in schizophrenia based on the findings that PCP **2** induces a psychotic state in humans that is very similar to that observed in schizophrenic patients, coupled with the finding that the main mode of action of PCP is that of a noncompetitive antagonist of the NMDA subtype of ionotropic glutamate receptor. In support of this NMDA receptor hypofunction model, other noncompetitive NMDA receptor antagonists including ketamine and dizocilpine (MK-801; **25**) induce a similar schizophrenia-like state.

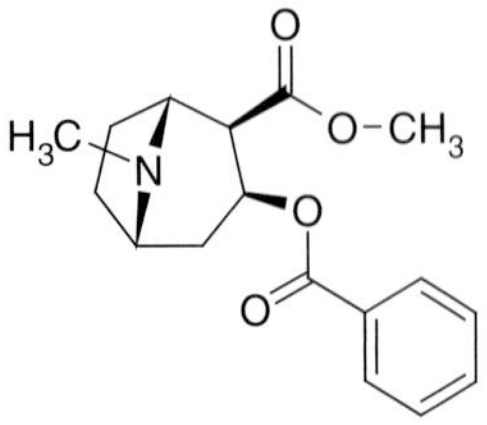

Cocaine **19** 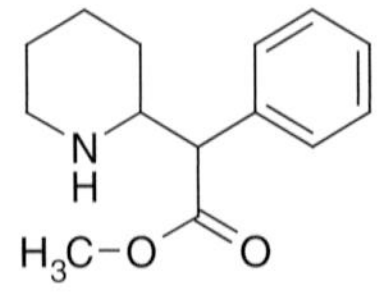 Methylphenidate **20** 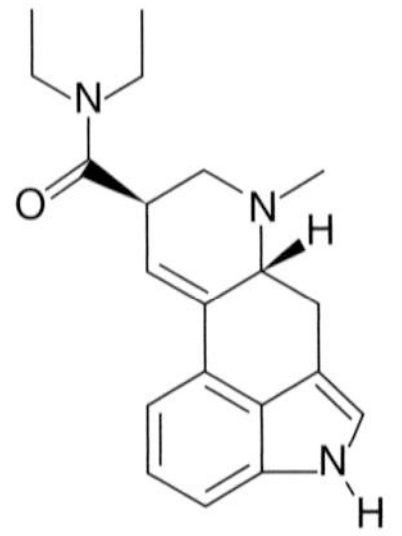 LSD **21** 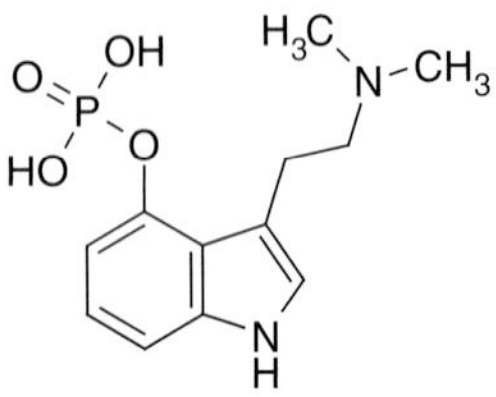 Psilocybin **22**

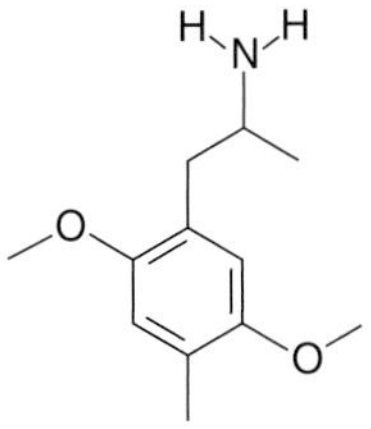

Dimethoxymethylamphetamine **23** 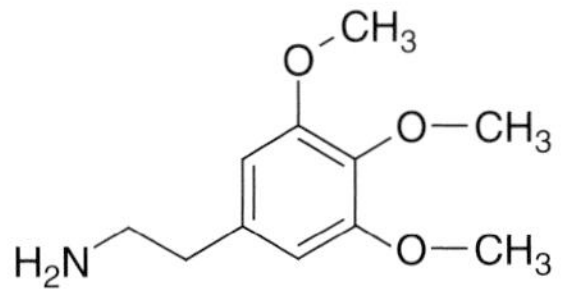 Mescaline **24** 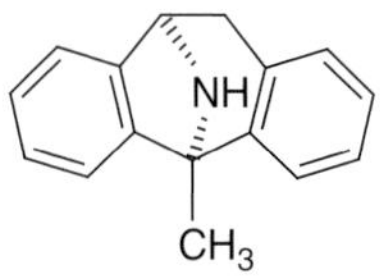 Dizocilpine **25**

DA-releasing agents, $5HT_2$ agonists, and NMDA receptor antagonists are all effective in a variety of rodent assays. Most notably, all three classes of compound disrupt PPI, and this disruption is responsive to atypical antipsychotics.[47] Since a large body of evidence suggests a dysregulation of each of these neurotransmitter systems in schizophrenia, the pharmacological disruption of PPI provides an animal model with considerable construct validity. This combined with the face validity of the PPI paradigm and the suggested predictive validity provided by back-validation studies with atypical antipsychotics provide support for this as a relevant model of positive symptomatology. However, an important caveat exists in the fact that the disruptions are pharmacological in nature. It is not unexpected that a DA antagonist reverses the effects of a DA-releasing agent, and given the incomplete understanding of the pathophysiology of schizophrenia a purely pharmacological disruption of the model cannot be ruled out, even with a truly novel NCE. It is therefore important to interpret these studies with care and to rely on more than one pharmacological tool to induce disruption of behavior.

6.02.4.3 Lesion Models

While the topic of much debate, there is some evidence that schizophrenia may have a neurodegenerative or neurodevelopmental basis.[48] For example, increases in the size of the ventricles, decrease in cortical and hippocampal volume, and selective decrease in subpopulations of GABAergic interneurons all suggest that neuronal loss may play a role in this disorder. Based on the lack of gross neurodevelopmental abnormalities that would occur from disruption of early stage neurogenesis, it has been proposed that the primary pathological changes that underlie schizophrenia may occur during the pre- or perinatal period. Based on this, a number of rodent models have been developed in which a lesion is produced during the pre-, peri-, or postnatal period. These include modeling of perinatal hypoxia and anoxia to mimic obstetric complications, and the aspiration, electrolytic, or excitotoxic lesioning of the prefrontal cortex or hippocampus. Hippocampal lesions in particular have been the focus of attention due to hippocampal projections to the nucleus accumbens and the ability of these projections to modulate the impact of prefrontal cortical inputs.

Interestingly, several neonatal lesion models exhibit a delayed onset of symptoms. The development of symptoms by the lesioned rodents in adolescence has been suggested to mirror the clinical presentation of symptoms in schizophrenic humans. For example, rats given a postnatal excitotoxic lesion of the ventral hippocampus are indistinguishable from nonlesioned littermates of postnatal day 35. However, by postnatal day 56, the animals begin to exhibit an increased sensitivity to amphetamine, a reduced sensitivity to haloperidol **26**, disrupted latent inhibition, and deficits in PPI that respond to atypical antipsychotics.[49] While these models appear to have some face validity, and the response to atypical antipsychotics suggests potential predictive validity, the extent and nature of the lesions are unlike anything observed in schizophrenic brains. Therefore, these models have questionable construct validity.

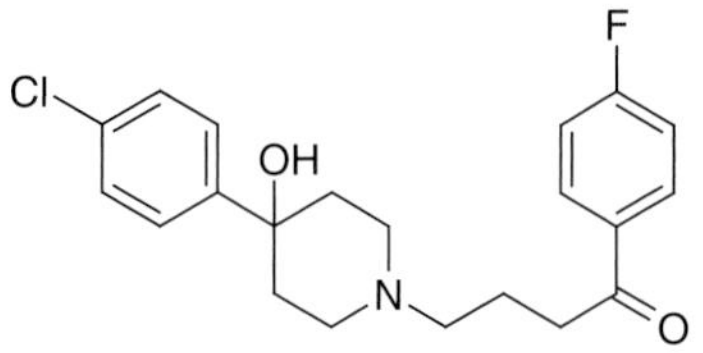

Haloperidol **26**

6.02.4.4 Social Isolation Model

The post-weaning social isolation of rats produces a model that exhibits behavioral abnormalities with some potential relevance to schizophrenia, including hyperactivity in response to novelty and amphetamine, disruption in PPI, and decreased social interactions. These abnormal behaviors are at least partially responsive to atypical antipsychotics.[50]

Social isolation is known to produce a variety of alterations in biochemical, electrophysiological, and anatomical measures. For example, stimulation of the ventral tegmental area in rats subjected to post-weaning social isolation produces typical evoked plateau depolarizations in prefrontal cortex that is accompanied by an abnormal firing or a short hyperpolarization that is not observed in control animals.[51] This suggests that social isolation may alter mesocortical dopaminergic modulation of the prefrontal cortex.

While social isolation produces an interesting behavioral model with some face validity, there are clear issues regarding construct validity. Since many of the systems disrupted by social isolation appear to be similar to changes observed in schizophrenic patients, this model may provide a fruitful path for basic research into potential developmental mechanisms of schizophrenia. However the incomplete response to atypical antipsychotics suggests that results from this model should be interpreted with caution.

6.02.4.5 Genetic Models

Attempts have been made to produce transgenic mouse models with a reasonable degree of construct validity. For example, neurotransmitter receptors believed to be relevant to schizophrenia including D1–D5 DA and NMDA receptors have been selectively knocked out or knocked down. One interesting example of this approach is the near complete knockdown of the obligatory NR1 subunit of the NMDA receptor to approximately 5% of normal expression levels.[37] This decrease in functional NMDA receptor expression produces increased spontaneous hyperlocomotion and social deficits that respond to antipsychotics. While this model appears to hold face validity, there is actually little evidence for decreased NMDA receptors in schizophrenia. Therefore, the overall validity of this model is still questionable. As discussed above, the current state of the human genetics of schizophrenia indicates that this is a nonhomogeneous multifactorial disorder. Therefore, there is not likely to be a 'schizophrenic mouse' model. While selective alteration of genes implicated in schizophrenia either on the basis of pathophysiology or genetic linkage may provide some insight into disease pathogenesis, this approach will likely result in incomplete models of the disorder. These models must be interpreted with great care as they carry all of the usual caveats of genetic models, including the potential for compensatory developmental changes in noninducible knockouts.

It is important to note that almost all of the currently available models and assays focus on the positive symptoms of schizophrenia. This has been a fruitful path for developing clozapine-like molecules with increase safety margins. However, this approach has led to the generation of therapies that do not alter the debilitating negative or cognitive symptoms of the disease. Currently there are no accepted models of negative or cognitive symptoms; however, as attention focuses on these more refractory symptoms of schizophrenia, it will be crucial to better characterize existing models and begin to develop new models to meet the obvious challenge of generating novel drugs that address these symptoms.

6.02.5 Clinical Trial Issues

Trials for novel antipsychotic drugs usually employ the randomized double blind placebo-controlled design and focus on a reduction in acute psychotic symptoms and the prevention of relapse as primary outcomes.[52] Trials usually average 50–60 participants, but large studies have been reported with sample sizes ranging from 200 up to 2000 subjects. Typical trial length is 6 weeks or less, but trials lasting for more than 6 months have provided valuable information on long-term treatment. Several issues are apparent in generalizing the results obtained from clinical trials to the general population of schizophrenics, and to clinical practice. Most large clinical trials for novel antipsychotic medications are carried out in a population of acute exacerbated schizophrenic patients, or in patients that are resistant to available treatments. The trials typically exclude individuals with comorbid psychiatric disorders or those that require continued use of antidepressants or mood stabilizers. In addition, patients with a history of substance abuse, violent behavior, or suicidal tendencies are usually excluded. These criteria are largely necessitated by the need to gather a large population of subjects with a relatively homogeneous disease state. While this approach aids in trial recruitment, it raises several important caveats. First, the selection process obviously favors a subgroup of patients that may not be representative of the whole population. Second, the individuals in the chosen population likely have a history of antipsychotic drug use that could alter response to an NCE. Lastly, the clinical trial conditions are not representative of the actual clinical situation in which comorbidity and polypharmacy are the rule rather than the exception.

In an attempt to improve the signal-to-noise ratio, many trials initiate with a short washout period during which the subject is treated with placebo. This serves several purposes beyond insuring a medication-free starting point. It allows for a more accurate identification of the actual response to the novel treatment and an enhanced ability to detect side effect liabilities. Furthermore, it provides an opportunity to screen out subjects that respond to placebo as well as noncompliant subjects. An alternative to the placebo washout is found in trials that selectively enroll first-episode schizophrenics. This design provides an opportunity to study efficacy in a drug-naive population. Unfortunately, this represents a significant challenge in subject recruitment resulting in relatively small sample size.

To date, most clinical trials for novel antipsychotic drugs have focused on amelioration of positive symptoms, or in some cases a combination of positive and negative symptoms. The tools typically employed are the Brief Psychiatric Rating Scale (BPRS), and the Positive and Negative Syndrome Scale (PANSS).[2,4] Additional scales are usually employed to assess extrapyramidal side effects, including the Simpson–Angus Extrapyramidal Symptom Scale, and the Abnormal Involuntary Movement Scale. This focus on a single aspect of core symptoms and potential adverse effects associated with known treatments in part explains the state of modern antipsychotic medication, which tends to primarily act to reduce positive symptoms with an improved therapeutic index for EPS. While claims have been made for improved efficacy in treating negative and cognitive symptoms, these claims are almost always based on secondary outcome measures that are by nature multiple comparisons and therefore lack the statistical rigor necessary to make firm statements that can be carried into clinical practice.

6.02.6 Current Treatment

None of the currently available drugs used for the treatment of schizophrenia can be considered cures, both because they have limited efficacy and because continuous therapy is needed. While representing major advances in positive symptom control, current drugs leave many patients with significant residual symptom burden, poor quality of life, lifelong functional impairment, and often undesirable side effects. Limited or noncompliance with treatment is very common because of limited efficacy and intolerable side effects.

6.02.6.1 First-Generation 'Typical' Antipsychotic Drugs

The serendipitous finding in 1951 that the major tranquilizer, chlorpromazine **1**, was effective in treating delusions and hallucinations associated with schizophrenia and other psychotic disorders marks the beginning of modern therapy for schizophrenia.[3] Unfortunately, treatment with chlorpromazine was accompanied by the development of EPS, some appearing even after the first dose (e.g., dystonias, akathisia). Other adverse effects were delayed for days or weeks such as parkinsonism, and the sometimes fatal neuroleptic malignant syndrome. Tardive dyskinesia, characterized by abnormal involuntary movements of the tongue, facial muscles, or limb muscles, develops in about 20% of patients and may be irreversible. Chlorpromazine also increased prolactin secretion leading to gynecomastia, galactorrhea, menstrual irregularities, sexual dysfunction, and possibly bone loss over the long term. Sedation, hypotension, and weight gain were also common with chlorpromazine. Despite these concerns, the discovery that the mechanism of action of chlorpromazine involved blockade of DA receptors led to an explosion of drugs with a similar mechanism. These included other phenothiazines (fluphenazine **27**, mesoridazine **28**, methotrimeprazine **29**, pericyazine **30**, perphenazine **31**, prochlorperazine **32**, thioproperazine **33**, thioridazine **34**, and trifluoperazine **35**); the thioxanthenes (e.g., chlorprothixene **36**, flupenthixol **37**, and thiothixene **38**); the butyrophenones (e.g., phenylbutylpiperadines including haloperidol **26**, melperone **39**, bromperidol **40**, and pimozide **41**); the substituted benzamides (e.g., sulpiride **42**, amisulpride **43**); and others (the dihydroindolone, molindone **44**, and the dibenzoxazepine, loxapine **45**). However, while all of these agents are effective in controlling the positive symptoms of schizophrenia, these drugs probably do not treat or may worsen the negative symptoms and cognitive dysfunction associated with this disease. Equally important, at therapeutic doses, these agents are associated with a similar set of side effects such as those seen after chlorpromazine.

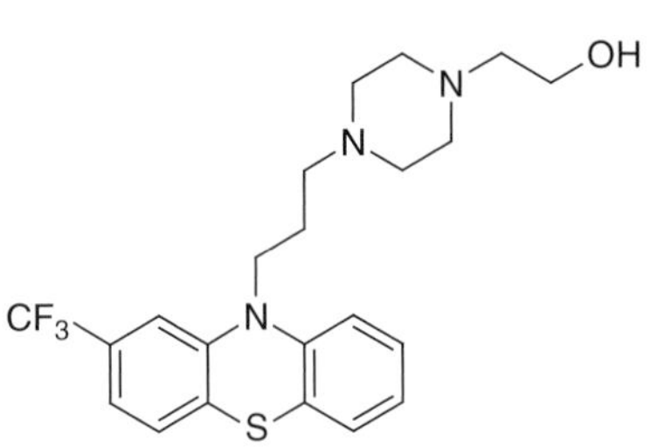

Fluphenazine **27**

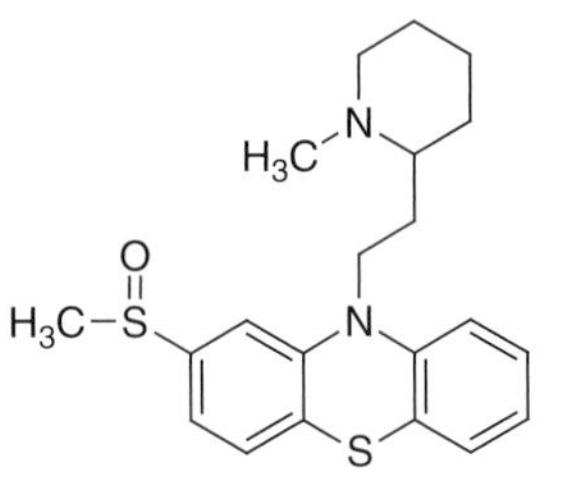

Mesoridazine **28**

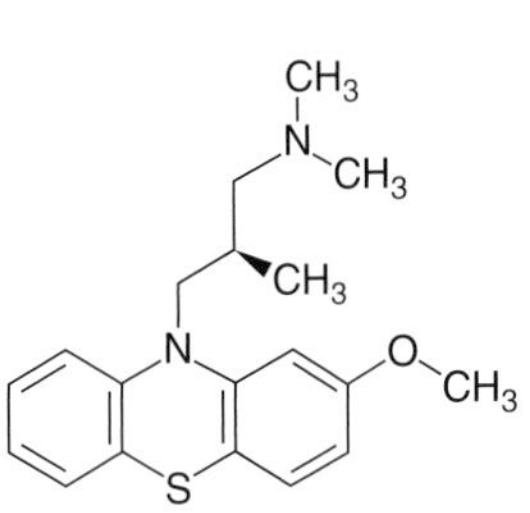

Methotrimeprazine **29**

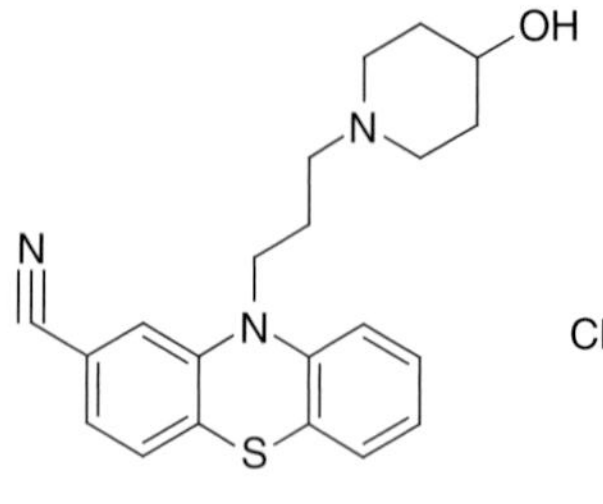

Pericyazine **30**

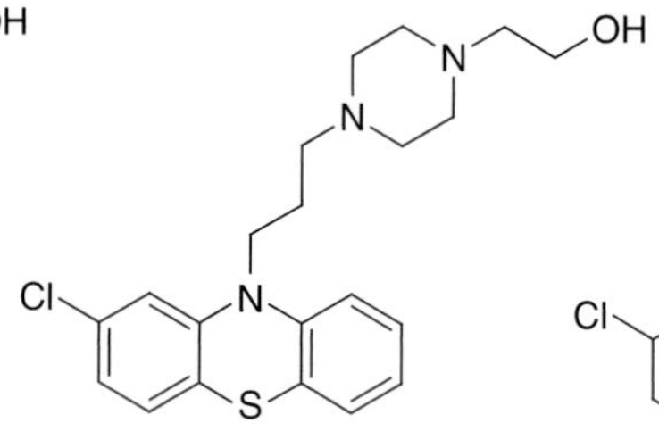

Perphenazine **31**

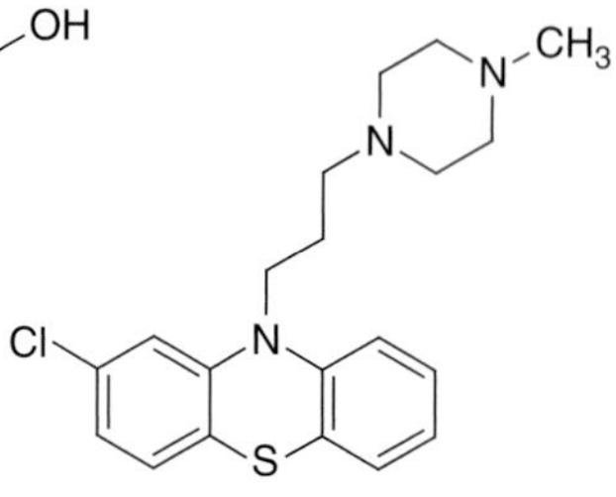

Prochlorperazine **32**

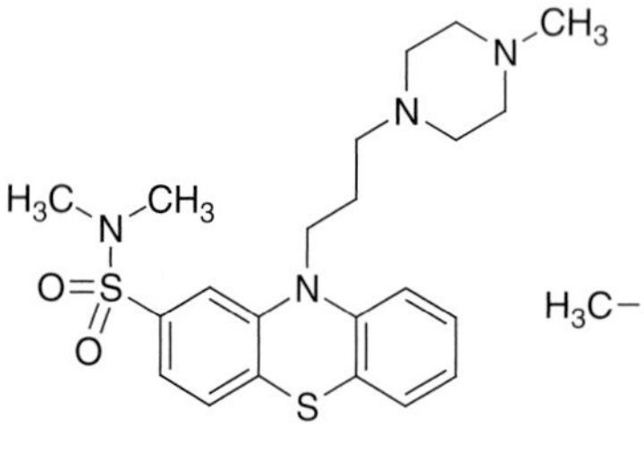

Thioproperazine **33**

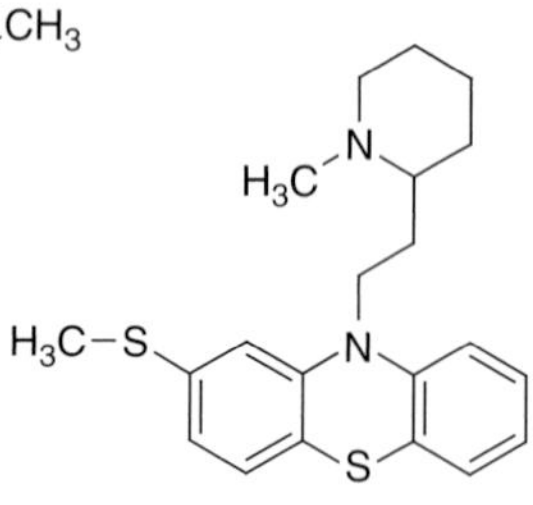

Thioridazine **34**

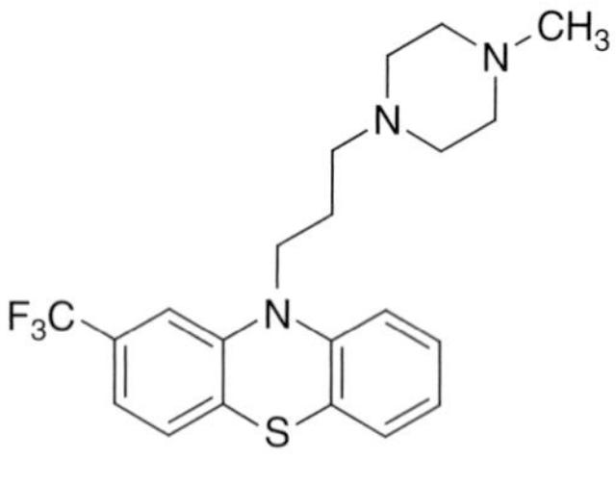

Trifluoperazine **35**

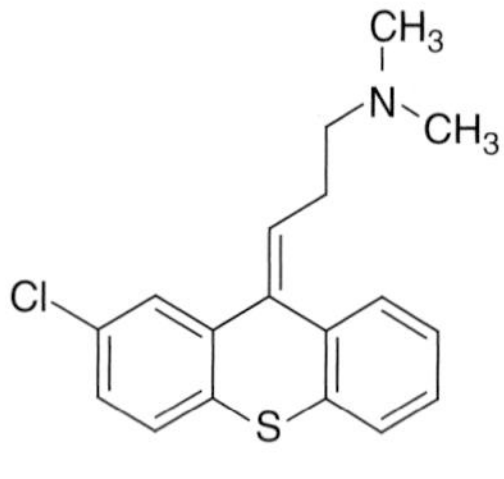

Chlorprothixene **36**

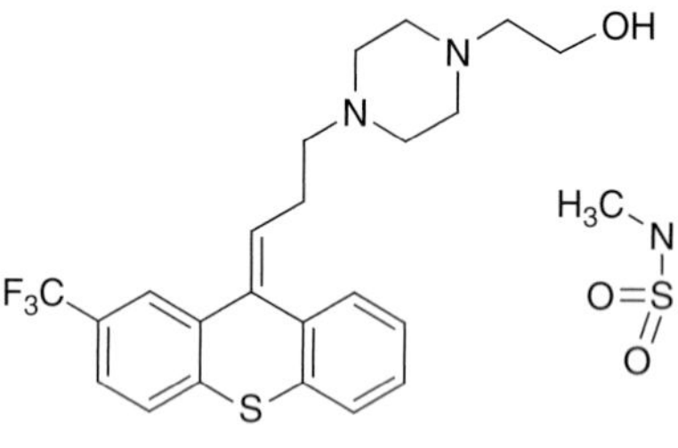

Flupenthixol **37**

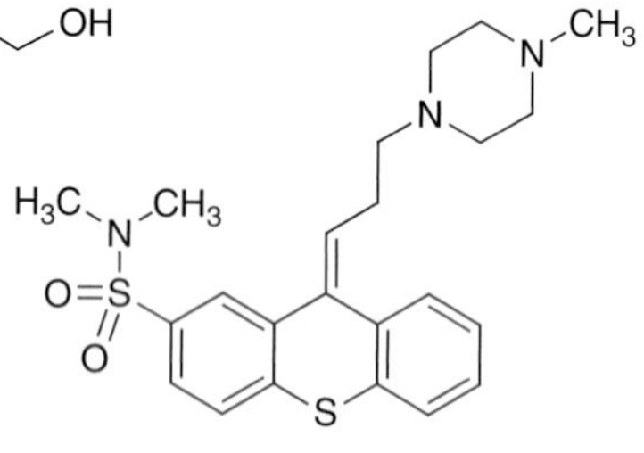

Thiothixene **38**

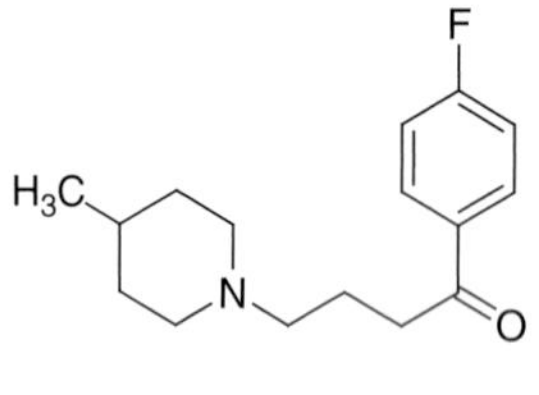

Melperone **39**

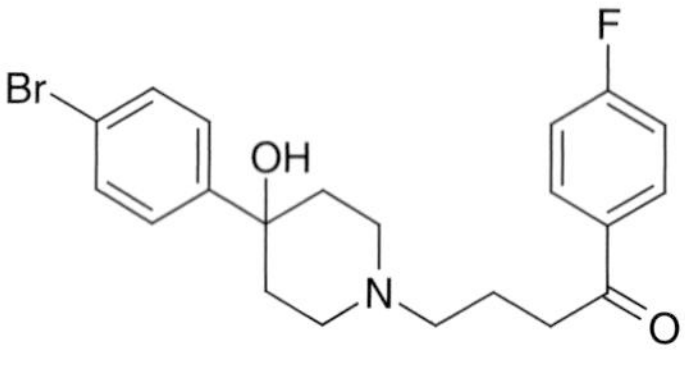

Bromperidol **40**

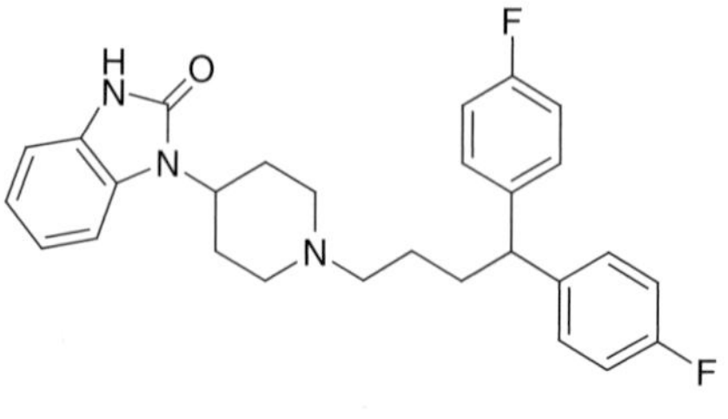

Pimozide **41**

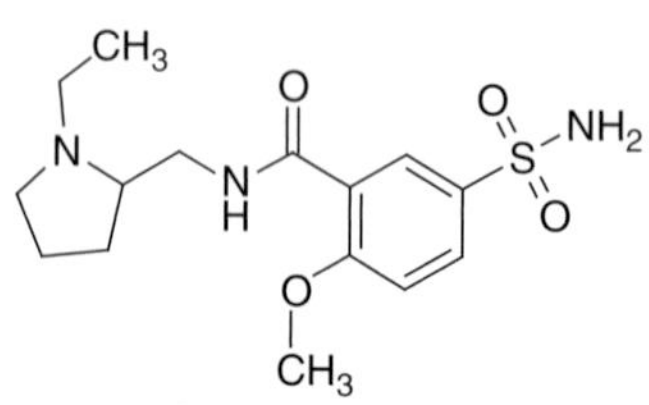

Sulpiride **42**

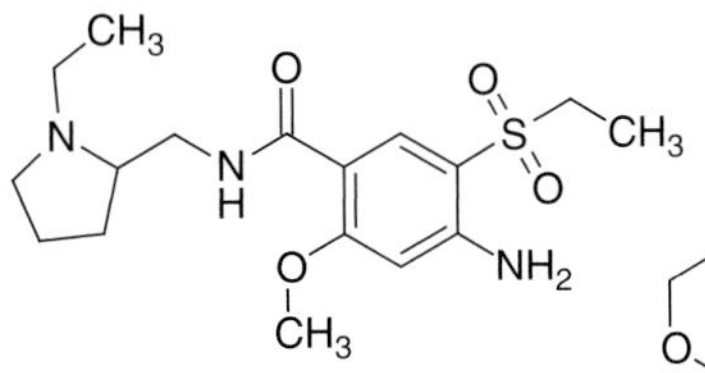

Amisulpride **43** 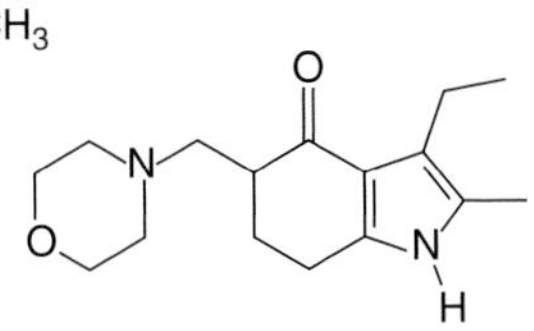 Molindone **44** 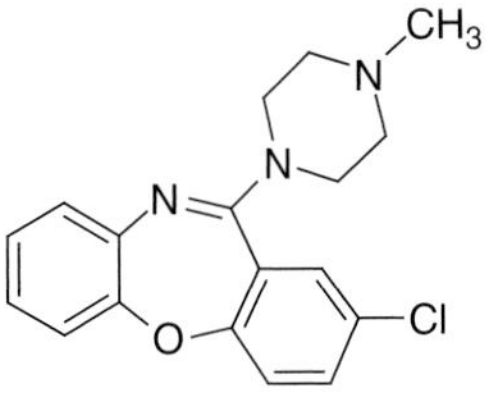 Loxapine **45**

Ultimately, it became clear that the antipsychotic action, EPS liabilities, and serum prolactin elevations of this class of antipsychotics resulted from the antagonism of the DA D2 receptors in the limbic system, dorsal striatum, and anterior pituitary gland, respectively. The positive correlation between in vitro DA receptor occupancy and average clinical dose suggested that for typical antipsychotic drugs such as haloperidol and chlorpromazine, antipsychotic activity, EPS liabilities, and prolactin elevations were closely linked to DA receptor blockade.[53,54] Positron emission tomography (PET) studies demonstrated that therapeutic response to most antipsychotic drugs occurs when ≥65–70% of striatal D2 receptors were blocked, even though this was mainly a measure of the region of the brain where EPSs are mediated. Because EPSs were shown to occur more commonly at a D2 receptor occupancy of ≥80%, a 'therapeutic window' of 60–80% D2 receptor blockade has been proposed as a means of striking a balance between the need for enough D2 receptor blockade to achieve clinical efficacy and the importance of avoiding EPS. None of these measures is particularly relevant to clozapine or quetiapine, two atypical antipsychotic drugs, which are effective as antipsychotic drugs at much lower occupancy of striatal D2 receptors than is the case for typical neuroleptic drugs, olanzapine, risperidone, and ziprasidone.

Because of their side effects and lack of efficacy against multiple symptom domains, and the development of more effective and tolerable agents, the above drugs no longer represent the preferred treatment for schizophrenia. Results from a recent NIMH-sponsored clinical trial (Clinical Antipsychotic Trials of Intervention Effectiveness; CATIE), however, have suggested that the use of first-generation antipsychotics be revisited. This study, which has only been partially reported,[55] suggested that perphenazine **31**, a first-generation drug, may be comparable in efficacy to quetiapine **46**, risperidone **47**, and ziprasidone **48**, and only moderately less effective than olanzapine **49**, which had more metabolic side effects than the other drugs. There are, however, numerous limitations with this study with regard to dosage, patient population inclusion criteria, prior antipsychotic medications, and primary endpoint, which preclude definitive conclusions regarding the preferred treatment for schizophrenia. Additionally, it has been noted that the CATIE is one of some 25 trials comparing antipsychotic agents. While the first generation of antipsychotic drugs is much cheaper than the newer drugs, at least at the present time, they cannot be recommended over the newer agents on the basis of cost effectiveness, spectrum of efficacy, or overall side effects. In particular, the risk of tardive dyskinesia is so much greater with first-generation antipsychotic drugs that they should not be substituted for the atypical antipsychotic drugs without clear warning to patients of the risk they are incurring by so doing.

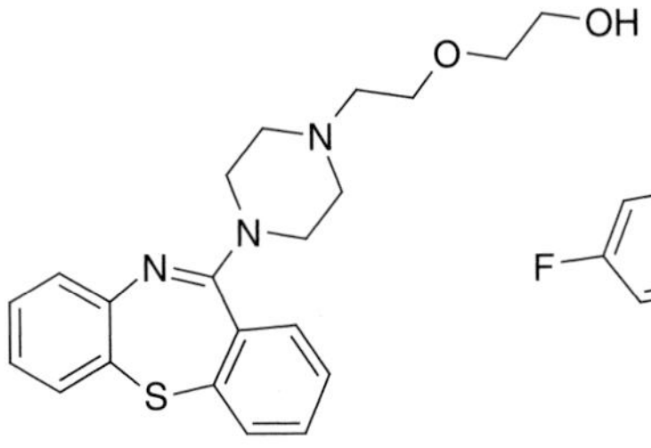

Quetiapine **46** 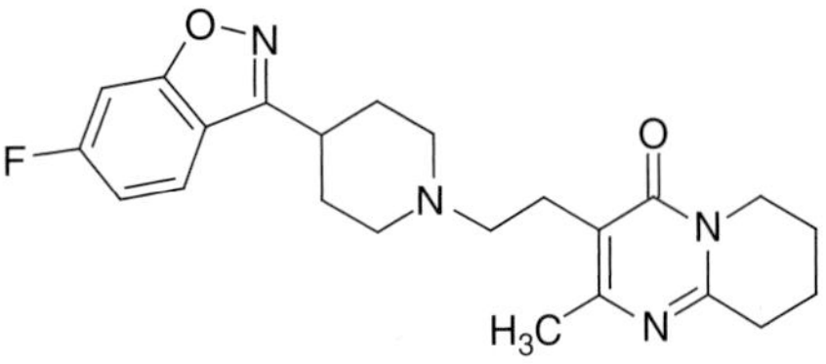 Risperidone **47**

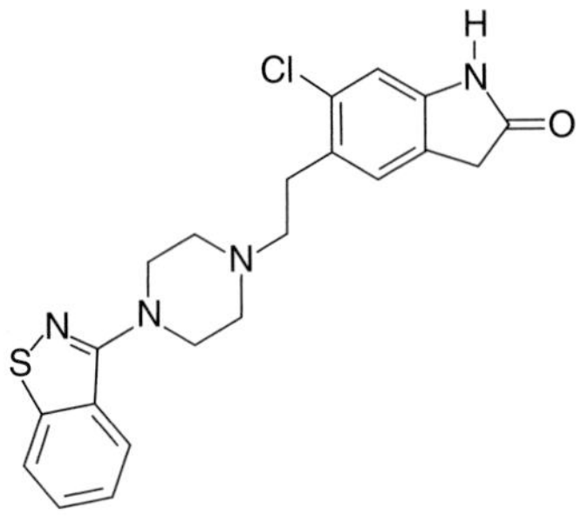

Ziprasidone **48** 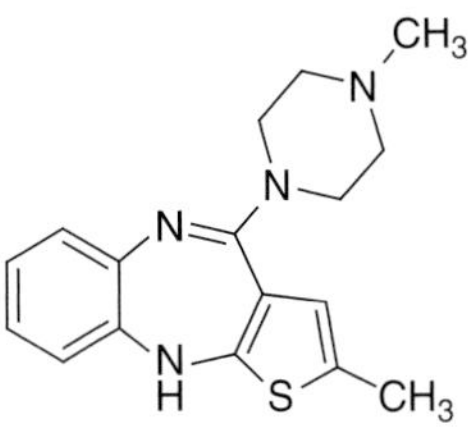 Olanzepine **49**

6.02.6.2 Second-Generation 'Atypical' Antipsychotic Drugs

The discovery of the benzodiazepine, clozapine **12**, in 1959 ushered in a new generation of potentially superior antipsychotic drugs. Clozapine was able to block DA-mediated behavior in animals and exerted antipsychotic effects in humans at doses that did not elicit EPS or produce sustained elevations in serum prolactin levels in humans. The motor symptom profile was sufficiently different from the first-generation antipsychotic drugs such that clozapine was labeled 'atypical' and clozapine became the blueprint for the development of other atypical antipsychotic drugs. Based on this blueprint, close clozapine analogs were developed that include loxapine **45**, olanzapine **49**, quetiapine **46**, and asenapine **50**, while the structurally dissimilar benzisoxidil group including risperidone **47**, iloperidone **51**, and ziprasidone **48**, and the phenylindole derivative, sertindole **52**.

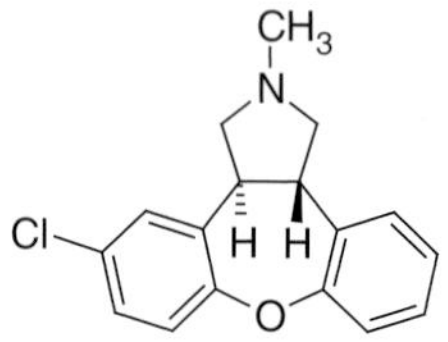

Asenapine **50** 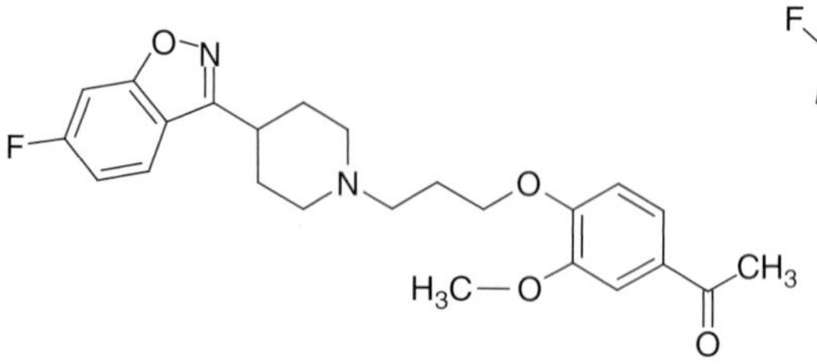 Iloperidone **51** 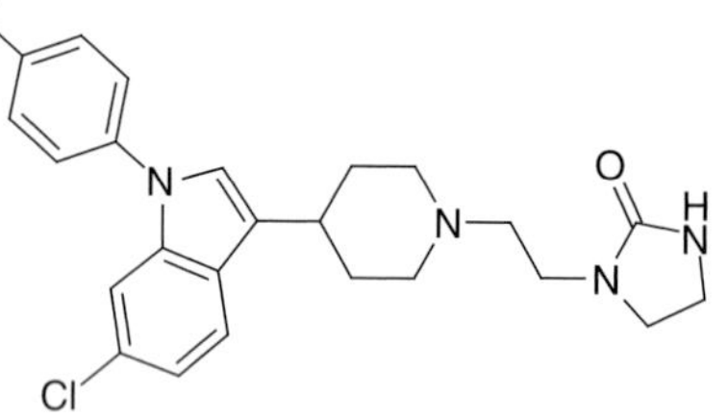 Sertindole **52**

The pharmacological mechanisms underlying the 'atypical' clinical properties of clozapine as well as the other newer antipsychotic agents have been richly debated. Since the major recognized benefit of these 'atypical' drugs appears to be a reduction in EPS liability and hyperprolactinemia, it was proposed that the differences in EPS liability of the older and newer antipsychotic drugs lie in the degree of DA receptor occupancy. According to this view, the older antipsychotic drugs are thought to more completely and persistently occupy the D2 receptor at efficacious concentrations than do atypical antipsychotic drugs. If so, a simple reduction in drug concentrations/dosage of these older drugs should, therefore, lead to emergence of the atypical profile. Thus drugs like haloperidol, thioridazine, or perphenazine should have atypical clinical profiles at low doses and typical clinical profiles at higher doses. However, this is not the case. Even at very low doses, drugs like haloperidol induce very different neurochemical effects in rodent, primate, and human brain than do drugs like clozapine. Further, in the patients with Parkinson's disease who experience psychotic episodes, and who have an already compromised DA system, even very low doses of thioridazine, the first-generation drug with the fewest motor side effects in schizophrenia, induces EPS. In addition, most atypical antipsychotics other than clozapine and quetiapine may cause EPS in Parkinson's patients at low doses and in non-Parkinson's patients in a dose-dependent manner. Rarely, cases of neuroleptic malignant syndrome and tardive dyskinesia have also been reported with these agents, suggesting that even agents like olanzapine and risperidone may cause too much and persistent D2 receptor occupancy in vulnerable individuals, e.g., those with inherently weak dopaminergic function. Clozapine and quetiapine, because of their low binding affinity, dissociate rapidly from the D2 receptor, which may account for their tolerability in Parkinson's disease patients. This property appears to occur independent of dose, and may thus preserve clinically 'atypical' properties of clozapine, quetiapine, and other atypical agents, even at high doses. However, under the 'fast dissociation' model, risperidone would be reclassified as not being atypical, and some first-generation agents (loxapine, molindone) would be considered pharmacologically atypical. The uniqueness of the atypical antipsychotic drugs must be driven by biological processes that add to or counteract some of the influences of strong D2 receptor blockade of the typical antipsychotic drugs. Higher antagonist affinity at the $5HT_{2A}$ receptor relative to the D2 receptor has been proposed to distinguish atypical from typical antipsychotic drugs.[12] More recently, nearly all compounds with antipsychotic activity have been characterized as potent and efficacious $5HT_{2A}$ inverse agonists (i.e., they turn off the intrinsic basal activity of the receptor). Furthermore, 'atypical' antipsychotic drugs that can occupy D2 receptors (e.g., clozapine, olanzapine, risperidone, and ziprasidone) are unique in having higher potency as $5HT_{2A}$ inverse agonists than as DA antagonists. This correlation strongly indicates that $5HT_{2A}$ inverse agonism can predict the 'atypical' nature of these kinds of antipsychotic drugs. These antipsychotic drugs occupy $>$70–80% of cortical $5HT_{2A}$ receptors, while occupancy of the striatal D2 receptors is much less. There also may be important differences between typical and atypical drugs in striatal D2 receptor occupancy relative to extrastriatal D2 receptor occupancy. Recent PET studies[56] have shown that these compounds also occupy fewer D2 receptors in the ventral tegmental/substantia nigra, thalamic and limbic areas of the brain that contain the cell bodies of the DA neurons which project to the limbic system and cortex, and caudate-putamen, respectively.

Typical antipsychotic drugs appear to preferentially interact with striatal D2 receptors while atypical antipsychotic drugs have a complex pattern of effects on extrastriatal D2 receptors, as noted above. Consistent with the above receptor binding interactions, atypical antipsychotic drugs appear to preferentially increase DA release in rat prefrontal cortex relative to the nucleus accumbens. The reverse is true for typical antipsychotic drugs. This differential neurochemistry largely is attributable to $5HT_{2A}$ antagonism, $5HT_{1A}$ agonism, and weak D2 antagonism. In the striatum, antagonism of $5HT_{2A}$ receptors located on dopaminergic neurons may release these cells from serotonergically mediated inhibition. The end result may be preservation of striatal dopaminergic tone and a reduction in EPS liability.

Thus, $5HT_{2A}$ and D2 receptor interactions are shared by all atypical antipsychotics and the combined interaction with these two receptors is sufficient to drive the antipsychotic efficacy against positive symptoms while maintaining a low risk for inducing EPS and hyperprolactinemia. Recent meta-analyses of randomized controlled trials have suggested that atypical antipsychotics are also superior to typical antipsychotics in improving negative symptoms, although the difference is slight. The superiority for improving cognitive deficits is larger and more robust. As alluded to earlier, animal studies have shown that $5HT_{2A}$ antagonism may increase DA efflux in the prefrontal cortex and hippocampus, key brain regions associated with cognitive functioning. They have lesser effect on DA efflux in the limbic system. Thus, 5HT–DA interactions may contribute to the relative success of atypical agents for improving cognitive and deficit symptoms that were beyond the reach of first-generation medications. Some thought leaders, however, believe this advantage does not reflect an absolute improvement in cognitive or negative symptoms as much as a lack of impairment in these symptoms due to treatment with first-generation antipsychotic drugs. What is certain is that most of these atypical agents do not normalize function in these domains and patients still have residual deficits that call for a more effective treatment.

It also is important to note that while atypical antipsychotic drugs are relatively free of EPS liability and hyperprolactinemia, treatment with this class of drugs results in other serious side effects associated with their use. These adverse events are thought to be unrelated to the interactions with the $5HT_{2A}$ and D2 receptors but more to the polypharmic or 'dirty' nature of these compounds[22] and their propensity to interact with many off-target receptors, including but not limited to the H_1 (sedation, obesity), $5HT_{2C}$ (obesity), muscarinic M_1 (blurry vision, dry mouth, constipation, urinary retention, cognitive dulling, and other mental status changes in susceptible individuals), muscarinic M_2 and M_3 (modulation of parasympathetic tone), and α_{1A} (orthostasis, reflex tachycardia) receptors. Adverse events observed in association with the use of atypical agents in clinical trials include, in addition to the above effects, hypersalivation (clozapine only), sweating, hypotension, myocarditis (clozapine), syncope, gastrointestinal complaints including constipation and nausea, and drug-induced fever. Olanzapine and clozapine are associated with increased weight gain, changes in adiposity, dyslipidemia, induction of insulin resistance (potentially progressing to type 2 diabetes mellitus), and, rarely, pancreatitis. Quetiapine and risperidone are considered to be more intermediate with regard to dysmetabolic risk, while ziprasidone and aripiperazole appear to be relatively free of these side effects. The most significant safety concerns with clozapine are the occurrence of seizures and agranulocytosis in some patients. Currently, the occurrence of clozapine-induced agranulocytosis is estimated to be 0.8% at 1 year and 0.91% at 18 months. While potentially fatal, morbidity and mortality due to agranulocytosis resulting from clozapine administration have been dramatically reduced by careful blood monitoring and the use of colony cell stimulating factors. Unlike the idiosyncratic occurrence of clozapine-associated blood dyscrasias, seizure risk appears to be dose related. Certain antipsychotic medications appear to affect other excitable tissues. For instance, the atyipal antipsychotics ziprasidone, iloperidone, and sertindole prolong the corrected QT interval (QTc); however, there is as yet no evidence that routine clinical use of ziprasidone is associated with the development of potentially fatal arrhythmias such as *torsade de pointes*. Several first-generation antipsychotic agents are also known to prolong the QTc interval, including thioridazine, mesoridazine, droperidol, pimozide, and haloperidol (in large intravenous doses).

What is the clinical impact of having $5HT_{2A}$ inverse agonism/antagonism in addition to D2 antagonism? Does this combination merely reduce the consequences of high D2 receptor occupancy or does this combination expand the therapeutic domains available to treatment? These questions can be answered by the use of combination strategies wherein relatively pure D2 antagonists are combined in various proportions with relative pure $5HT_{2A}$ antagonists. A low dose of mirtazapine, a $5HT_{2A}$ antagonist, significantly reduced the acute akathisia produced by high doses of typical antipsychotics.[57] Additionally, in normal healthy volunteers and rodents, $5HT_{2A}$ inverse agonists/antagonists have been shown to reduce high dose haloperidol-induced increases in prolactin secretion. The addition of mianserin, a combined $5HT_{2A/2C}$ and $\alpha 2$-adrenoceptor antagonist, to first-generation antipsychotic treatment in stabilized schizophrenic patients was associated with improvements in neurocognitive performance. What remains to be determined is whether $5HT_{2A}$ inverse agonism/antagonism can expand the treatment domains of low dose treatment with typical antipsychotics and atypical antipsychotics with 'too much' D2 antagonist activity. A number of hurdles face this combination approach, including determining the correct ratio of $5HT_{2A}$ to D2 receptor occupancy, pharmacokinetic and metabolism issues, etc. Solving these problems could provide a new approach to treating psychosis and a means to avoid the side effects caused by off-target activities of existing atypical antipsychotics. Flexibility in selecting the optimal $5HT_{2A}$ to D2 ratio would be an advantage of a combination approach that could not be achieved with a single molecule having a fixed intrinsic ratio.

6.02.6.3 Dopamine Receptor Modulators/Dopamine Partial Agonists

To the extent that DA overactivity and excessive stimulation of the D2 family of DA receptors are responsible for at least the positive symptom component of schizophrenia, therapeutic avenues other than direct antagonists may be available to modulate these DA neurochemical abnormalities. By exploiting the differential receptor reserve of various tissues and brain regions, it may be possible to use partial agonists of differing intrinsic activity to modulate DA activity. The full spectrum of pharmacology is available for the DA system from full inverse agonists to neutral antagonists to partial agonists to full agonists. Most existing antipsychotic agents are inverse agonists at the D2 receptor family. Regardless of the tissue receptor reserve, inverse agonists will act as antagonists. However, partial agonists have, by definition, lower intrinsic activity. Partial agonists, therefore, can behave as antagonists or agonists depending upon the tissue receptor reserve and their level of intrinsic activity. In tissues with high receptor reserve, these compounds may act as agonists but will act as antagonists in tissues with low receptor reserve.

It is thought that mesolimbic DA systems have low receptor reserve while the mesocortical dopamine system has high receptor reserve. Importantly, D2 autoreceptors are thought to have high receptor reserve. Depending upon the intrinsic activity of an NCE, it should be possible, therefore, for it to act as antagonist in the mesolimbic system in which overactivity is thought to elicit positive symptoms, a weak agonist in the mesocortical system in which reduced DA activity is thought to be associated with heightened negative symptoms and cognitive impairment, and a presynaptic agonist on DA neurons that should reduce dopaminergic neuronal activity and consequently postsynaptic dopamine tone. On theoretical grounds, a partial agonist should be an ideal antipsychotic agent and possess the ability to dynamically modulate dopamine neurotransmission differentially depending upon the state of the dopaminergic tone. The prototypical partial DA agonist, 3-PPP **53**, and in particular the *S*(−)-isomer, (−)-3-PPP, preclamol **54**, was the first partial DA agonist to be tested clinically. Patients with schizophrenia appeared to show initial improvements relative to placebo during the first week of therapy but efficacy waned with continued treatment. Other partial DA agonists, including terguride **55**, have shown antipsychotic efficacy albeit in small clinical trials.

At least part of the beneficial effects of aripiperazole **56**[58] may be related to its partial DA agonist properties. As expected of a partial agonist, this drug acts as an antagonist at the D2 receptors in a state of excessive dopaminergic neurotransmission, while it acts as an agonist at the D2 receptor in a state of low dopaminergic neurotransmission. Aripiperazole, therefore, appears to stabilize dopaminergic neurotransmission. However, aripiperazole also exhibits high affinity for serotonin $5HT_{1A}$ and $5HT_{2A}$ receptors, moderate affinity for DA D4, $5HT_{2c}$ and $5HT_7$, α_1-adrenergic and histamine H_1 receptors. Thus, its activity as a partial DA agonist must be viewed against the background of other receptor activities. Still, aripiperazole has modest intrinsic activity at DA D2 receptors and this intrinsic activity is less than that of (−)-3-PPP. Aripiperazole has been shown to be effective in treating the symptoms of acutely exacerbated patients with schizophrenia in a series of well-controlled clinical trials. This drug appears to improve both positive and negative symptoms with less risk of eliciting EPS and essentially no propensity for causing hyperprolactinemia (it actually reduces prolactin secretion). This compound also shows a lower risk for weight gain, QTc prolongation, and metabolic side effects.

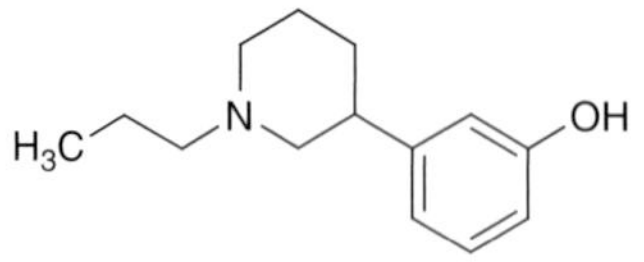

3-PPP **53**

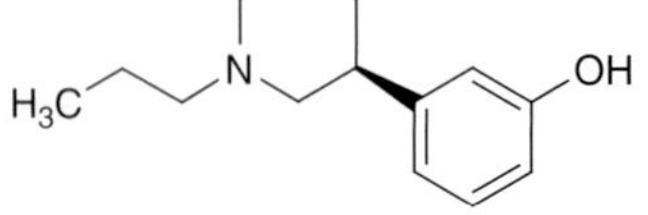

Preclamol **54**

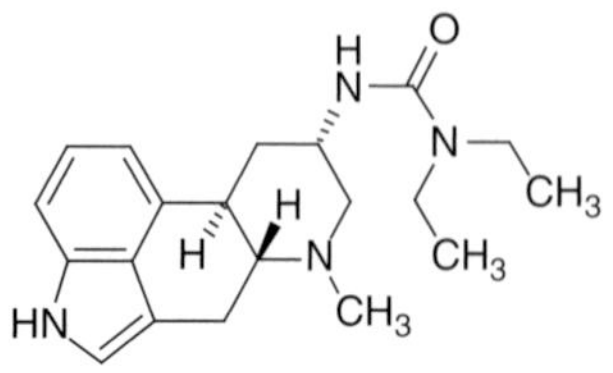

Terguride **55**

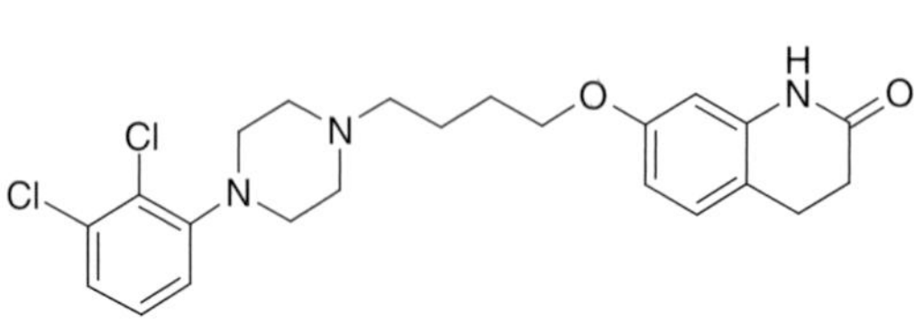

Aripiperazole **56**

Bifeprunox (DU-127090; **57**) also has partial D2 and $5HT_{1A}$ agonist activity with little activity at $5HT_{2A}$, $5HT_{2c}$, α_1-adrenergic, and histamine H_1 receptors. It is thus quite interesting in the context of the the hypothesis that $5HT_{1A}$ agonism can substitute for $5HT_{2A}$ antagonism.[12] Bifeprunox appears to act as a partial D2 agonist in rodents and to exhibit activity consistent with other antipsychotic drugs in animal models of psychosis. This compound is currently in phase III clinical trials but little clinical information is currently available on its efficacy.

Another partial DA D2 agonist with clinical potential is OSU 6162 (PNU-96391A; **58**), that has shown antidyskinetic and antipsychotic efficacy in small preliminary trials in patients suffering from Parkinson's disease and Huntington's disease. This NCE has also shown interesting activity as an add-on therapy in patients with refractory schizophrenia. Still, it remains to be demonstrated in larger studies that partial DA receptor agonism is sufficient to reliably elicit antipsychotic efficacy and whether these kinds of drugs are different enough to add to the antipsychotic pharmacopeia. This approach remains theoretical and needs empirical validation. Further, the optimum level of intrinsic partial agonism required for therapeutic benefit needs to be elucidated. In the end, the level of intrinsic activity may be quite different among individuals, making dose selection difficult. This has not proved to be the case for aripiperazole, where 15–20 mg day^{-1} has proved to be effective for almost all responsive patients.

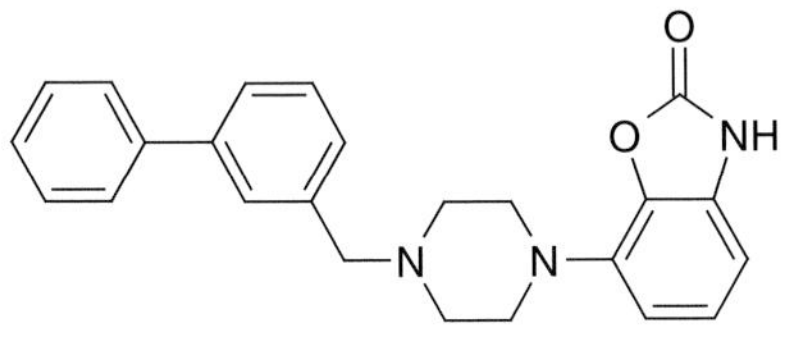

Bifeprunox **57**

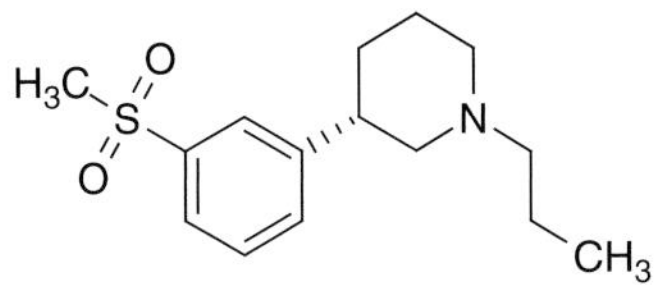

(–)-OSU 6162 **58**

6.02.7 Unmet Medical Needs

6.02.7.1 Reduced Side Effect Liability

Current pharmacological approaches to the treatment of schizophrenia suffer from two major issues; side effects and limited efficacy. The first major unmet medical need is improved side effect liability. Even the best of the modern atypical antipsychotics produce significant side effects with a low therapeutic index. Individuals with schizophrenia have an increased risk of death and, in general a 20% shorter life span[84,85] that, in part, may be attributable to the use of current antipsychotic medications. As discussed above, atypical antipsychotics have a clear lower risk of inducing EPS and hyperprolactinemia (with the exception of risperidone) when compared to typical antipsychotics, but the risk still exists. Furthermore, several atypicals, particularly clozapine and olanzapine, increase the risk of sedation, obesity, high blood sugar and diabetes, and dyslipidemia. Very rare cases of neuroleptic malignant syndrome, a rare but potentially fatal reaction characterized by fever, altered mental status, muscular rigidity, and autonomic dysfunction, have also been reported in patients treated with atypical antipsychotic drugs. It is hoped that drugs directed against novel targets will have a much larger therapeutic index with lower associated risk.

6.02.7.1.1 QTc liability

As mentioned earlier, several antipsychotics are associated with prolongation of the cardiac QTc, risk of developing *torsade de pointes*, and sudden cardiac death. Sudden death occurs nearly twice as often in antipsychotic treated patients as in the general population with 10–15 such events in 10 000 person-years of observation.[59] Pimozide, sertindole, and haloperidol elicit *torsade de pointes* and sudden death, with thioridazine having the greatest risk. While ziprasidone prolongs the QT interval, there is no evidence that this leads to increased incidence of *torsade de pointes* or sudden death. There appears to be no sudden death association with olanzapine, quetiapine, or risperidone. In a population-based case-control study of 554 cases of sudden cardiac death, the current use of antipsychotics was associated with a threefold increase in risk of sudden cardiac death.[60] QTc prolongation has become a major issue in the drug discovery process such that assessment of NCEs for this property occurs early in the lead optimization process while drugs like cisapride have been removed from the marketplace because of the QTc liability.

6.02.7.1.2 Metabolic syndrome

Many of the newer atypical antipsychotics have been associated with weight gain and in some instances diabetes,[61] the prevalence of which is twofold greater among schizophrenics than in the general population.[62] These are serious side effects for drugs that are used chronically. However, metabolic syndrome, including abdominal obesity, dyslipidemia, hypertension, and insulin resistance, have an increased prevalence in schizophrenics[63] which may be genetically related. Weight gain in schizophrenics has been associated with the histamine H_1 receptor affinity of antipyschotic medications[86] with clozapine and olanazepine being the most potent of the current medications ($K_i \sim 1$ nM) and haloperidol ($K_i = 1800$ nM, the least). The CATIE study[55] noted that olanzapine, of the four atypical antipsychotics studied, was associated with the greatest incidence of weight gain (0.9 kg month^{-1}) with greater increases in glycosylated hemoglobin, total cholesterol, and triglycerides, effects consistent with the development of metabolic syndrome.

6.02.7.2 Broader Efficacy

The second major unmet medical need is improved efficacy across all symptom domains. While all current drugs are effective at treating the positive symptoms of schizophrenia in the majority of patients, the negative and cognitive symptoms, along with comorbid mood disorders, are not sufficiently improved. Current inabilities to treat all of the core symptoms of schizophrenia are due in part to the primacy of the DA hypothesis, which has dominated treatment strategies for over 50 years. Necessarily, drug intervention based on this hypothesis has focused on efficacy as defined by a reduction in positive symptoms. No drug has been developed or approved for human use that does not reduce positive symptoms. Yet, the challenge of achieving optimal outcome in patients with schizophrenia has not been met. A large number of patients with schizophrenia in the USA are homeless or incarcerated. By no means do current therapies normalize function and allow the afflicted individual to re-enter society. This is particularly devastating given that first signs of schizophrenia appear in the late teens and early twenties and the disorder is lifelong. Residual impairments such as negative and cognitive symptoms which escape current treatment preclude normal socialization of the patient with schizophrenia, and are more closely correlated with functional outcome than positive symptoms. Typical antipsychotic drugs, like haloperidol, fail to improve and may further impair various domains of cognition in schizophrenic patients. Newer atypical antipsychotic drugs like clozapine, risperidone, quetiapine, and olanzapine appear not to impair cognitive function in patients with schizophrenia and may improve performance in some but not all domains of cognition. It is important to note that each drug may influence different aspects of cognitive function and no drug normalizes cognitive function in all patients.

6.02.8 New Research Areas

6.02.8.1 Muscarinic Agonists

Activation of muscarinic receptors may exert antipsychotic-like activity in animals and humans. The first indication that activation of muscarinic receptors might have antipsychotic activity in humans arose from the use of cholinesterase inhibitors to treat behavioral disturbances in demented individuals, particularly those with dementia with Lewy bodies (DLBs) (*see* 6.08 Neurodegeneration).[64] These patients exhibit visual hallucinations, delusions, apathy, agitation, dementia, and mild parkinsonism. Cholinergic deficits in this disorder are even more severe than in Alzheimer's disease (AD). Improvements in neuropsychiatric symptoms have been reported from small studies of DLB patients treated with the cholinesterase inhibitors tacrine **59**, rivastigmine **60**, donepezil **61**, and galantamine **62**. Because of the overwhelming peripheral cholinergic side effects of these drugs, it is unlikely that this class of agents will be used broadly but these studies have opened new avenues for future studies with safer drugs.

Because acetylcholinesterase inhibitors nonspecifically increase the release of ACh and lead to the activation of muscarinic and nicotinic receptors, it is difficult to ascribe the effects of these agents on psychotic behavior to either subtype. While selective muscarinic and nicotinic agonists have not been studied for antipsychotic activity in humans, xanomeline **63**, a broadly acting muscarinic partial agonist with limited selectivity for the M_1 and M_4 receptor subtypes, has been shown to reduce behavioral disturbances in patients with AD. Reductions in hallucinations, agitation, delusions, vocal outbursts, and suspiciousness were seen in these patients with AD after xanomeline treatment.[79] Xanomeline and other more selective muscarinic agonists have an antipsychotic-like profile in various animal models of psychosis. Muscarinic agonists have activity in these rodent models that is very similar to that seen with D2 antagonists, with the exception that muscarinic agonists do not cause catalepsy. These findings led to the suggestion that muscarinic agonists may have potential antipsychotic effects. A small study in patients with schizophrenia suggested xanomeline had antipsychotic activity in this population as well as AD. The utility of selective muscarinic agonists in patients with schizophrenia needs to be evaluated. An NCE that selectively activates the M_1 muscarinic receptor would be of interest as long as it did stimulate the M_3 and to a lesser extent the M_2 muscarinic receptors causing unwanted peripheral cholinergic side effects.

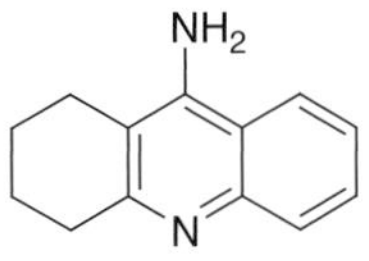

Tacrine **59**

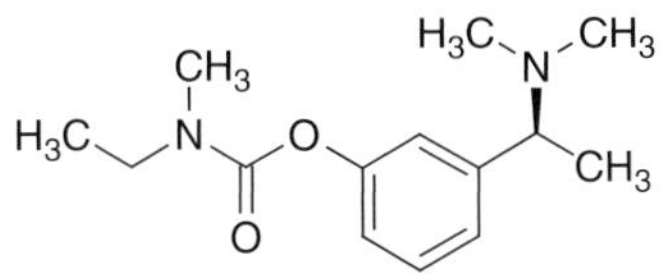

Rivastigmine **60**

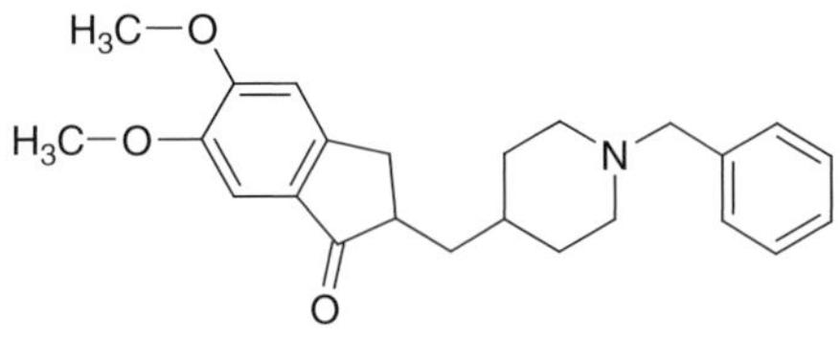

Donepezil **61**

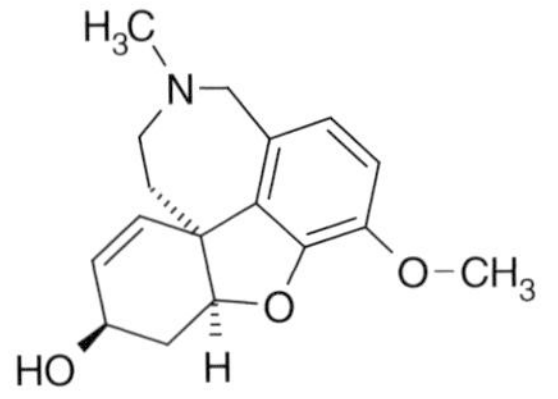

Galantamine **62**

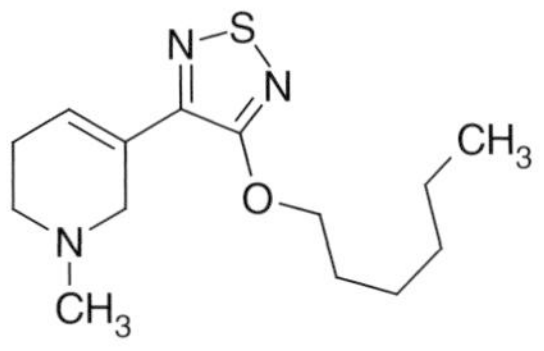

Xanomeline **63**

6.02.8.1.1 ***N*-Desmethyclozapine**

N-Desmethylclozapine (norclozapine, NDMC **13**) is one of the two major metabolites of the atypical antipsychotic, clozapine, formed by demethylation by cytochrome P450 enzymes in all species examined, including humans. NDMC is found at serum concentrations comparable to those of clozapine. Like other atypical antipsychotics, NDMC is a weak partial agonist of DA D2 receptors and a potent inverse agonist of $5HT_{2A}$ receptors. However unlike any other antipsychotic, NDMC is a potent and efficacious muscarinic receptor agonist.[23] Specifically, NDMC is a partial agonist of M_1 and M_5 receptors, and a competitive antagonist of M_3 muscarinic receptors. In addition, NDMC was shown to potentiate NMDA receptor currents in CA1 pyramidal cells through an activation of muscarinic receptors,[65] and is orally active in several animal models thought to be predictive of antipsychotic activity.

It is possible that the muscarinic agonist properties of NDMC may underlie some of the unique effects of treatment with clozapine. The multiple lines of evidence reported above support a procognitive effect of potentiating central cholinergic neurotransmission, including the clinical effects of acetylcholinesterase inhibitors and direct-acting muscarinic receptor agonists. High-dose clozapine therapy in treatment-refractory schizophrenics may serve to raise brain levels of NDMC to achieve central muscarinic receptor agonist activity, particularly M_1 receptor stimulation, rather than recruiting additional lower potency receptor interactions. Given the competing actions at M_1 receptors, NDMC/clozapine plasma ratios should be more predictive of therapeutic response, particularly for cognition, than absolute clozapine levels. The NDMC/clozapine ratio is superior to clozapine levels as a predictor of clinical response to clozapine,[66] particularly for negative symptoms and cognitive enhancement. Based principally on these data, NDMC is currently being evaluated as a potential stand-alone treatment in patients with schizophrenia.

6.02.8.2 Glutamate Modulators

The glutamate hypofunction hypothesis of schizophrenia has already been discussed. In rodents, NMDA antagonists increase locomotor activity and enhance amphetamine-induced DA release. However, $5HT_{2A}$ antagonists are more effective than D2 antagonists in blocking this increase in locomotor activity, suggesting a possible mode of action of drugs that are both $5HT_{2A}$ and D2 antagonists, e.g., clozapine and risperidone. This suggests that dysregulation of DA function associated with schizophrenia might be secondary to NMDA hypofunction. Thus the balance between D2 antagonism and NMDA receptor modulation may be pivotal for the improvement of both positive and negative symptoms in schizophrenia. Drugs that directly activate NMDA receptors, therefore, might be useful in treating patients with schizophrenia. Such agents, however, have been shown to excessively increase neuronal excitability, to cause seizures, and to be neurotoxic. Recently, many pharmacologically more subtle strategies involving activation of glutamate receptors have been suggested.

Glycine is an obligatory positive allosteric modulator of the NDMA receptor activity. Glycine site agonists and partial agonists (glycine, D-serine **6**, D-cycloserine) and the glycine transporter (GlyT) inhibitor sarcosine **64**, are effective in treating some of the symptoms, particularly negative symptoms and cognitive impairment, of schizophrenia when used adjunctively with existing antipsychotic drugs.[67] Of these agents, D-serine appears to be the most promising based on pharmaceutical characteristics and clinical outcomes. When combined with drugs like risperidone and olanzapine, D-serine improves positive and negative symptoms in treatment-resistant patients in some but not all studies. However, the efficacy of this compound appears to attenuate with repeated use, and it is not useful in patients who are receiving clozapine. Agonists of glutamate metabotropic receptors are also being explored as a potential treatment for schizophrenia.[68] Metabotropic glutamate receptors are G protein-coupled receptors (GPCRs) that have been classified into eight discrete subtypes. The group II mGluRs are often located presynaptically and typically modulate glutamate release. Reduction in glutamate release by group II agonists can block the behavioral activation caused by PCP and amphetamine in rats. Early stage group II mGluR ligands include LY 404039 **65**, LY 354740 **8**, and MGS 0039 **66**. Both orthotopic and allosteric modulators have been shown to have antipsychotic-like activity in animals but allosteric modulators appear not to suffer from rapid tachyphylaxis and, therefore, could have utility as a novel approach as a maintenance therapy for patients with schizophrenia. In addition, the group I mGluR, mGluR5, has been implicated as a potential novel target for schizophrenia. Both pharmacological blockade and genetic ablation of mGluR5 have been found to decrease prepulse inhibition in rodent studies, and CDPPB **67**, a selective positive allosteric modulator of mGluR5, reverses the psychotomimetic effects of amphetamine in rat models.[69] Since mGluR5 is linked to potentiation of NMDA receptor currents in a number of brain regions, it has been suggested that a selective activator of this receptor may be an effective drug for restoration of cognitive function.

Another class of agents that has been shown, at least in vitro, to allosterically enhance AMPA receptor activity is the ampakines. One of these agents, CX-516 **68**, was studied in combination with clozapine in a small clinical study and was shown to improve some symptoms associated with schizophrenia. As a sole agent, CX-516 does not appear to be effective in improving positive symptoms or cognition in patients with schizophrenia. Another ampakine, ORG 24448/ CX-619, is being tested as an adjunctive therapy as part of the NIMH effort to facilitate the development of medications to enhance cognition in patients with schizophrenia.

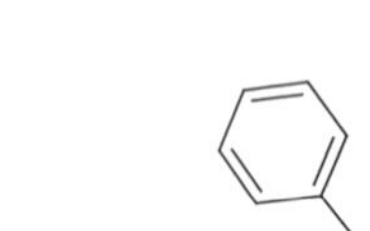

Sarcosine **64**

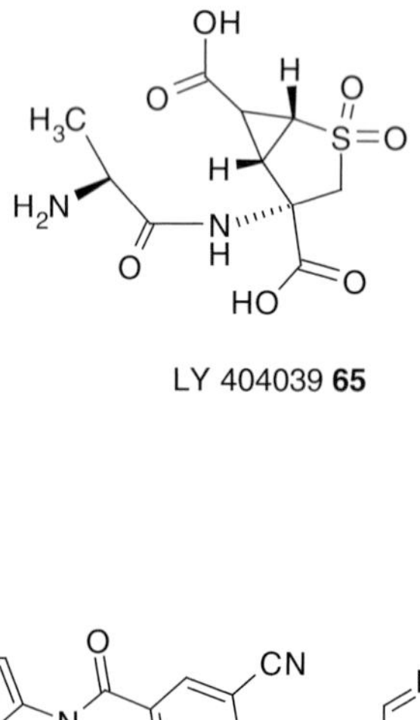

LY 404039 **65**

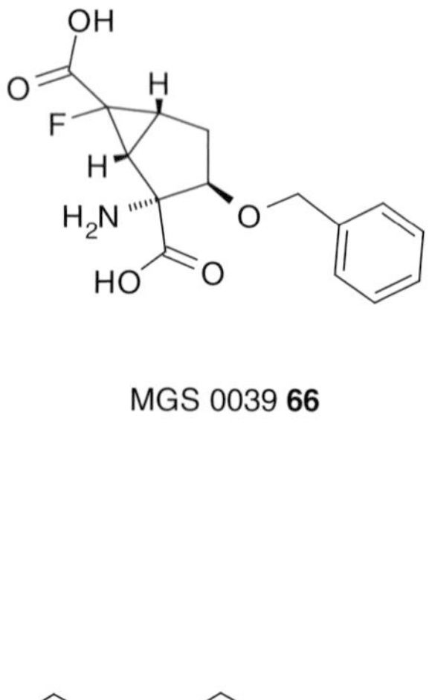

MGS 0039 **66**

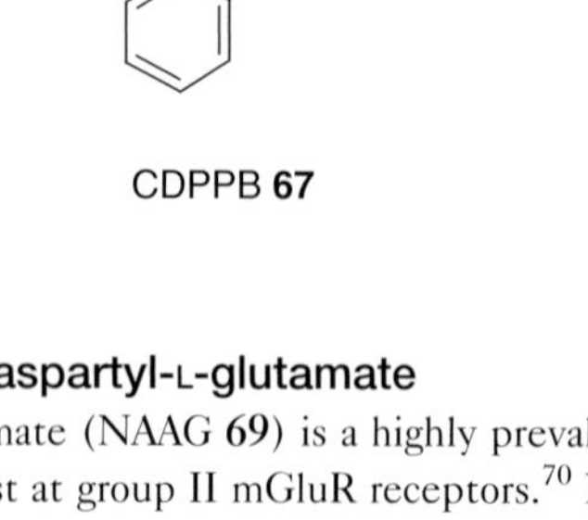

CDPPB **67**

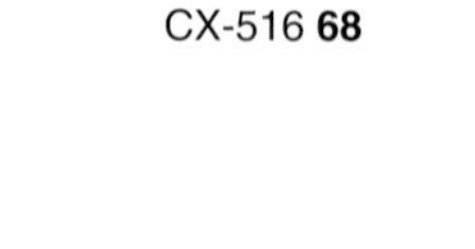

CX-516 **68**

6.02.8.2.1 *N*-Acetyl-L-aspartyl-L-glutamate

N-Acetyl-L-aspartyl-L-glutamate (NAAG **69**) is a highly prevalent small peptide neutotransmitter that is suggested to act as an endogenous agonist at group II mGluR receptors.[70] NAAG is catabolized to *N*-acetylaspartate and glutamate by NAAG peptidases glutamate carboxypeptidase II and III localized to the extracellular surface of astrocytes.

As discussed above, activators of the group II mGluRs hold promise as novel antipsychotics. Therefore, it has been suggested that inhibitors of NAAG peptidase could increase the levels of NAAG and provide antipsychotic efficacy through the activation of group II mGluRs.[71] A number of potent and selective NAAG peptidase inhibitors have been discovered, including 2-PMPA **70**, the more potent analog GPI-5693 **71**, and the structurally distinct ZJ38 **72**. The ability to selectively target these enzymes provides an opportunity to explore potential efficacy in animal models of schizophrenia.

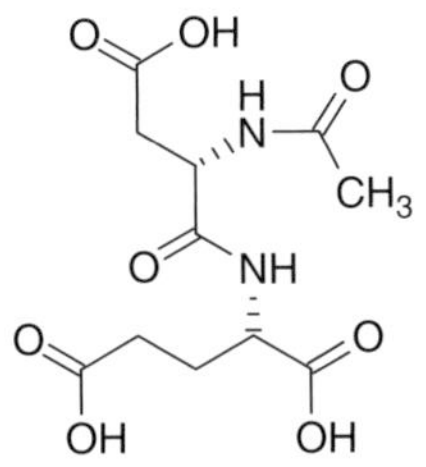

NAAG **69** 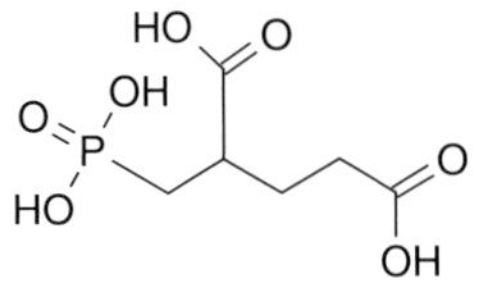 2-PMPA **70**

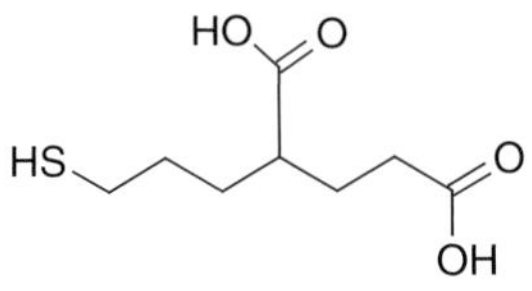

GPI-5693 **71** 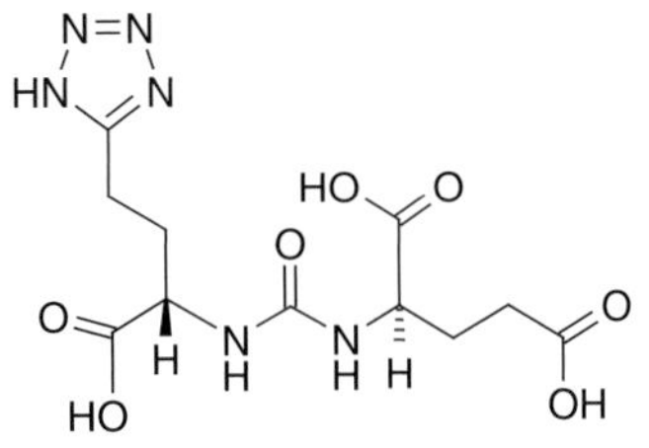 ZJ 38 **72**

6.02.8.3 Neuronal Nicotinic Receptor Agonists

The incidence of smoking is high in patients with schizophrenia, a rate at least three times higher than the general population. In fact, nicotine appears to produce a modest transient improvement in cognitive and sensory deficits in these patients. It has been suggested that smoking in schizophrenia represents an attempt to self-medicate.[39] However, these views must be interpreted with caution. Overall, schizophrenic patients have a high degree of comorbid abuse of a variety of substances including nicotine, alcohol, cannabis, cocaine, and amphetamine. Importantly, the rate of substance abuse is higher than in the general population for all of these substances in spite of the fact that such abuse is associated with poorer outcomes, exacerbation of positive symptoms, increased hospitalization, and increased frequency of homelessness. This increased propensity to abuse a variety of substances regardless of consequences suggests that there may be a disregulation of reward systems in schizophrenics. Alternatively, high levels of D2 receptor occupancy by atypical antipsychotics may blunt DA-mediated reward and lead to an enhanced abuse drive. In support of this, a correlation exists between D2 receptor occupancy by antipsychotics and the number of cigarettes smoked by schizophrenics.[72]

Genetic linkage studies have shown that polymorphisms in the NNR gene are associated with schizophrenia, being linked to deficiency in the normal inhibition of the P50 auditory-evoked response,[39] an electrographic index of auditory sensory gating that correlates with the PPI deficit observed in patients with schizophrenia. Deficits in sensory gating, therefore, may contribute to cognitive symptoms and perceptual disturbances and activation of nicotinic receptors may be a viable drug target for treating the sensory gating deficits of patients with schizophrenia.[73]

Clozapine but not haloperidol can improve sensory gating deficits in mice through a nicotinic mechanism. α7 NNR agonists, e.g., GTS-21 **15**, AZD-0328 **73**, SSR-180711 **74**, and W-56203 **75**, are currently being targeted for the treatment of schizophrenia. It remains to be determined whether these kinds of agents will broadly impact cognitive deficits in patients with schizophrenia, simply improve sensory gating, or be of no utility. Long-term use of nicotinic agonists also may be precluded because this class of receptors is susceptible to rapid tachyphylaxis.

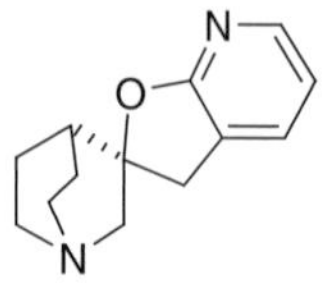

AZD-0328 **73** 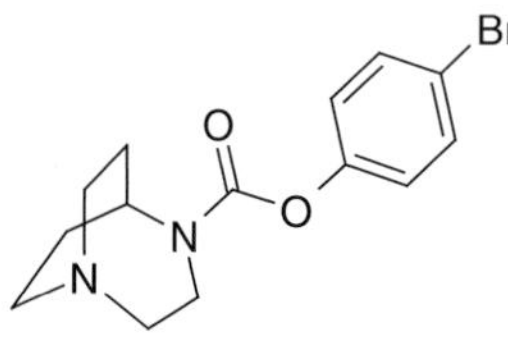 SSR-180711 **74** 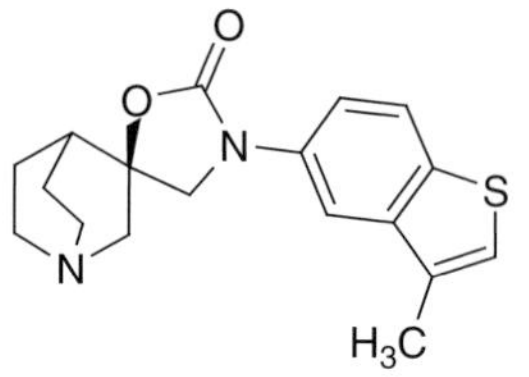 W-56203 **75**

6.02.8.4 Neurokinin$_3$ Antagonists

Two neurokinin$_3$ (NK_3) antagonists, osanetant **76** and talnetant **77**, which have selective interactions with the NK_3 receptor in the range of 1 nM, have been reported to have clinical efficacy in schizophrenia.[74] Osanetant had similar efficacy to haloperidol on positive symptoms with reduced EPS and weight gain liabilities. NK_3 receptors present on DA neurons in the A9 and A10 groups are thought to modulate DA release, agonists increasing DA release.

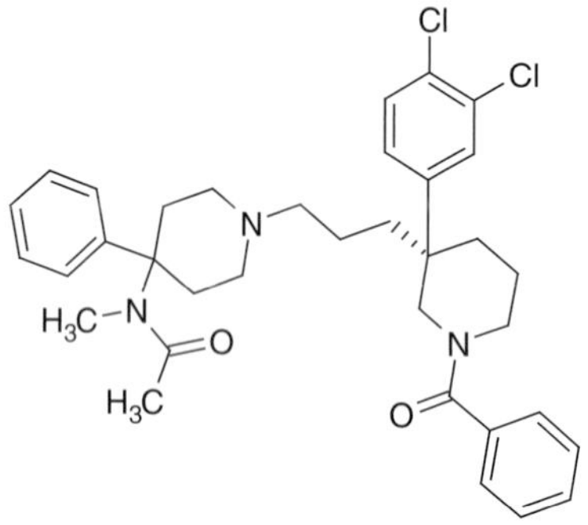

Osanetant **76** 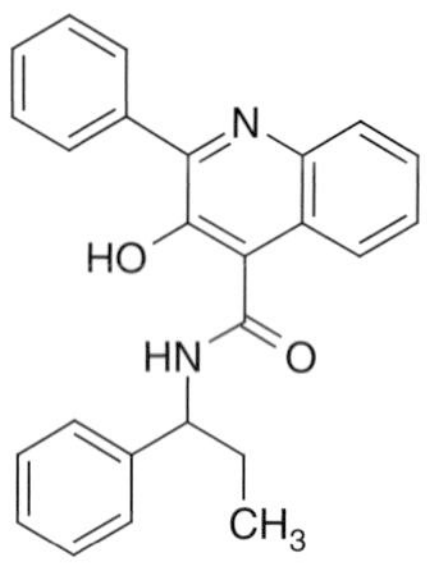 Talnetant **77**

6.02.9 Future Directions

Better drugs are clearly needed for the treatment of schizophrenia and these are most likely to come from a better understanding of disease origin and pathophysiology, and better novel molecular targets. Without these advances, it is likely that only incremental improvements on existing drugs will be possible. Indeed, all existing antipsychotic drugs and many drugs currently in development represent modest clinical and chemical improvements on earlier drugs; variations on a theme (olanzapine versus clozapine); metabolites of existing drugs (9-hydroxyrisperidone (paliperidone) versus risperidone); or attempts to mimic a limited set of models (aripiperazole (partial DA agonism) and asenapine ($5HT_{2A}$/D2 antagonism)).

Evolving information on the polygenic nature of schizophrenia and the fact that the atypical antipsychotics have polypharmic actions have led to the suggestion that 'magic bullets' for schizophrenia should be replaced by 'magic shotguns'[22] or 'selectively' nonselective agents.[75] In addition to considerations of the genetics and neurochemistry of schizophrenia, it has been proposed[10] that disrupted cortical circuitry, a consequence of neuronal apoptosis, may be a key event in the pathophysiology of schizophrenia. Changes in enzymes involved in the apopotic cascade, Bax/Bcl2, have been observed in the brains of patients with schizophrenia.[48] Such changes have been discussed in terms of proapoptoic stress in schizophrenia, a theme that Spedding *et al.* have developed regarding stress-related dysfunction of neuronal plasticity mechanism and neurogenesis in psychiatric disorders and that may reflect a 'failure to recover'.[76] Neurogenesis appears to be a key event in the delayed onset of antidepressants[77] and may also be involved in the delayed onset of action of antipsychotics.[78] The information revolution driven by genomics and understanding of the neurocircuitry and plasticity of the brain is just beginning to impact antipsychotic drug discovery efforts. Whether this information will be able to usefully provide new insights into the molecular targeting of the next generation of antipsychotic drugs remains to be seen.

References

1. Falkai, P.; Wobrock, T.; Lieberman, J.; Glenthoj, B.; Gattaz, W. F.; Moller, H. J. *World J. Biol. Psychiat.* **2005**, *6*, 132–191.
2. Carpenter, W. T.; Conley, R. R.; Buchanan, R. W. In *Pharmacological Management of Neurological and Psychiatric Disorders*; Enna, S. J., Coyle, J. T., Eds.; McGraw-Hill: New York, 1998, pp 27–51.
3. Deniker, P. In *Discoveries in Pharmacology: Vol. 1, Psycho- and Neuro-Pharmacology*; Parnham, M. J., Bruinvels, J., Eds.; Elsevier: Amsterdam, the Netherlands, 1983, pp 163–180.
4. First, M. B.; Tasman, A. In *DSM-IV-TR Mental Disorders: Diagnosis, Etiology and Treatment*; John Wiley: Chichester, UK, 2004, pp 639–701.
5. McGurk, S. R. *J. Psychiatr. Pract.* **2000**, *6*, 190–196.
6. Marder, S. R.; Fenton, W. *Schizophr. Res.* **2004**, *72*, 5–9.
7. Petronis, A. *Biol. Psychiatry* **2004**, *55*, 965–970.
8. Krabbendam, L.; Van Os, J. *Schizophr. Bull.* **2005**, *31*, 795–799.
9. Charlton, B. G. *Med. Hypotheses* **2005**, *65*, 1005–1009.
10. Lewis, D. A.; Lieberman, J. A. *Neuron* **2000**, *28*, 325–334.
11. Gaddum, J. H.; Hameed, K. A. *Br. J. Pharmacol. Chemother.* **1954**, *9*, 240–248.
12. Meltzer, H. Y. *Neuropsychopharmacology* **1999**, *21*, 106S–115S.
13. Coyle, J. T. *Harv. Rev. Psychiatry* **1996**, *3*, 241–253.
14. Millan, M. J. *Curr. Drug Targets. CNS Neurol. Disord.* **2002**, *1*, 191–213.
15. Moghaddam, B.; Adams, B.; Verma, A.; Daly, D. *J. Neurosci.* **1997**, *17*, 2921–2927.
16. Aghajanian, G. K.; Marek, G. J. *Brain Res.* **1999**, *825*, 161–171.
17. Taylor, C. P. *Drug Dev. Res.* **1990**, *21*, 151–160.
18. Wassef, A.; Baker, J.; Kochan, L. D. *J. Clin. Psychopharmacol.* **2003**, *23*, 601–640.
19. Benes, F. M. *Brain Res. Brain Res. Rev.* **2000**, *31*, 251–269.
20. Wahlbeck, K.; Cheine, M.; Essali, M. A. *Cochrane Database Syst. Rev.* **2000**, CD000059.
21. Bagnall, A. M.; Jones, L.; Ginnelly, L.; Lewis, R.; Glanville, J.; Gilbody, S.; Davies, L.; Torgerson, D.; Kleijnen, J. *Health Technol. Assess.* **2003**, 7, 1–193.
22. Roth, B. L.; Sheffler, D. J.; Kroeze, W. K. *Nat. Rev. Drug Disc.* **2004**, *3*, 353–359.
23. Davies, M. A.; Compton-Toth, B. A.; Hufeisen, S. J.; Meltzer, H. Y.; Roth, B. L. *Psychopharmacology* **2005**, *178*, 451–460.
24. Kendler, K. S. *Am. J. Psychiatry* **2003**, *160*, 1549–1553.
25. Takahashi, S.; Faraone, S. V.; Lasky-Su, J.; Tsuang, M. T. *Psychiatr. Res.* **2005**, *133*, 111–122.
26. Owen, M. J.; Craddock, N.; O'Donovan, M. C. *Trends Genet.* **2005**, *21*, 518–525.
27. Williams, M. *Curr. Opin. Investig. Drugs* **2003**, *4*, 31–36.
28. Tunbridge, E. M.; Harrison, P. J.; Weinberger, D. R. *Biol. Psychiatry* **2006**, *60*, 141–151.
29. Funke, B.; Malhotra, A. K.; Finn, C. T.; Plocik, A. M.; Lake, S. L.; Lencz, T.; DeRosse, P.; Kane, J. M.; Kucherlapati, R. *Behav. Brain Funct.* **2005**, *1*, 19.
30. Harrison, P. J.; Law, A. *J. Biol. Psychiatry* **2006**, *60*, 132–140.
31. Duan, J.; Martinez, M.; Sanders, A. R.; Hou, C.; Krasner, A. J.; Schwartz, D. B.; Gejman, P. V. *Psychol. Med.* **2005**, *35*, 1599–1610.
32. Talkowski, M. E.; Chowdari, K. V.; Lewis, D. A.; Nimgaonkar, V. L. *Schizophr. Bull.* **2006**, *32*, 203–208.
33. Chumakov, I.; Blumenfeld, M.; Guerassimenko, O.; Cavarec, L.; Palicio, M.; Abderrahim, H.; Bougueleret, L.; Barry, C.; Tanaka, H.; La Rosa P. et al. *Proc. Natl. Acad. Sci. USA* **2002**, *99*, 13675–13680.
34. Ma, J.; Qin, W.; Wang, X. Y.; Guo, T. W.; Bian, L.; Duan, S. W.; Li, X. W.; Zou, F. G.; Fang, Y. R.; Fang, J. X. et al. *Mol. Psychiatry* **2006**, *11*, 479–487.
35. Schumacher, J.; Jamra, R. A.; Freudenberg, J.; Becker, T.; Ohlraun, S.; Otte, A. C.; Tullius, M.; Kovalenko, S.; Bogaert, A. V.; Maier, W. et al. *Mol. Psychiatry* **2004**, *9*, 203–207.
36. Zhao, X.; Li, H.; Shi, Y.; Tang, R.; Chen, W.; Liu, J.; Feng, G.; Shi, J.; Yan, L.; Liu, H.; He, L. *Biol. Psychiatry* **2006**, *59*, 747–753.
37. Mohn, A. R.; Gainetdinov, R. R.; Caron, M. G.; Koller, B. H. *Cell* **1999**, *98*, 427–436.
38. Kamiya, A.; Kubo, K.; Tomoda, T.; Takaki, M.; Youn, R.; Ozeki, Y.; Sawamura, N.; Park, U.; Kudo, C.; Okawa, M. et al. *Nat. Cell Biol.* **2005**, 7, 1067–1078.
39. Freedman, R.; Adler, L. E.; Leonard, S. *Biol. Psychiatry* **1999**, *45*, 551–558.
40. Staddon, S.; Arranz, M. J.; Mancama, D.; Perez-Nievas, F.; Arrizabalaga, I.; Anney, R.; Buckland, P.; Elkin, A.; Osborne, S.; Munro, J. et al. *Schizophr. Res.* **2005**, *73*, 49–54.
41. Koob, G. F.; Simon, H.; Herman, J. P.; Le Moal, M. *Brain Res.* **1984**, *303*, 319–329.
42. Weiner, I. *Psychopharmacology* **2003**, *169*, 257–297.
43. Ford, J. M. *Psychophysiology* **1999**, *36*, 667–682.
44. Louchart-de la Chapelle, S.; Levillain, D.; Menard, J. F.; Van der Elst, A.; Allio, G.; Haouzir, S.; Dollfus, S.; Campion, D.; Thibaut, F. *Psychiatry Res.* **2005**, *136*, 27–34.
45. Braff, D. L.; Grillon, C.; Geyer, M. A. *Arch. Gen. Psychiatry* **1992**, *49*, 206–215.
46. Grillon, C.; Ameli, R.; Charney, D. S.; Krystal, J.; Braff, D. *Biol. Psychiatry* **1992**, *32*, 939–943.
47. Geyer, M. A.; Krebs-Thomson, K.; Braff, D. L.; Swerdlow, N. R. *Psychopharmacology* **2001**, *156*, 117–154.
48. Glantz, L. A.; Gilmore, J. H.; Lieberman, J. A.; Jarskog, L. F. *Schizophr. Res.* **2006**, *81*, 47–63.
49. Lipska, B. K.; Jaskiw, G. E.; Weinberger, D. R. *Neuropsychopharmacology* **1993**, *9*, 67–75.
50. Van den Buuse, M.; Garner, B.; Koch, M. *Curr. Mol. Med.* **2003**, *3*, 459–471.
51. Peters, Y. M.; O'Donnell, P. *Biol. Psychiatry* **2005**, *57*, 1205–1208.
52. Stroup, T. S.; Alves, W. M.; Hamer, R. M.; Lieberman, J. M. *Nat. Rev. Drug Disc.* **2005**, *5*, 107–114.
53. Creese, I.; Burt, D. R.; Snyder, S. H. *Science* **1976**, *192*, 481–483.
54. Seeman, P.; Lee, T. *Science* **1975**, *188*, 1217–1219.
55. Lieberman, J. A.; Stroup, T. S.; McEvoy, J. P.; Swartz, M. S.; Rosenheck, R. A.; Perkins, D. O.; Keefe, R. S.; Davis, S. M.; Davis, C. E.; Lebowitz, B. D. et al. *N. Engl. J. Med.* **2005**, *353*, 1209–1223.

56. Kessler, R. M.; Ansari, M. S.; Riccardi, P.; Li, R.; Jayathilake, K.; Dawant, B.; Meltzer, H. Y. *Neuropsychopharmacology* **2005**, *30*, 2283–2289.
57. Poyurovsky, M.; Epshtein, S.; Fuchs, C.; Schneidman, M.; Weizman, R.; Weizman, A. *J. Clin. Psychopharmacol.* **2003**, *23*, 305–308.
58. Hirose, T.; Kikuchi, T. *J. Med. Invest.* **2005**, *52*, 284–290.
59. Glassman, A. H.; Bigger, J. T., Jr. *Am. J. Psychiatry* **2001**, *158*, 1774–1782.
60. Straus, S. M.; Bleumink, G. S.; Dieleman, J. P.; van der Lei, J.; 't Jong, G. W.; Kingma, J. H.; Sturkenboom, M. C.; Stricker, B. H. *Arch. Intern. Med.* **2004**, *164*, 1293–1297.
61. Newcomer, J. W. *J. Clin. Psychiatry* **2004**, *65*, 36–46.
62. Meyer, J.; Koro, C. E.; L'Italien, G. J. *Int. Rev. Psychiatry* **2005**, *17*, 173–180.
63. Lindenmayer, J. P.; Czobor, P.; Volavka, J.; Citrome, L.; Sheitman, B.; McEvoy, J. P.; Cooper, T. B.; Chakos, M.; Lieberman, J. A. *Am. J. Psychiatry* **2003**, *160*, 290–296.
64. Simard, M.; van Reekum, R. *J. Neuropsychiat. Clin. Neurosci.* **2004**, *16*, 409–425.
65. Sur, C.; Mallorga, P. J.; Wittmann, M.; Jacobson, M. A.; Pascarella, D.; Williams, J. B.; Brandish, P. E.; Pettibone, D. J.; Scolnick, E. M.; Conn, P. J. *Proc. Natl. Acad. Sci. USA* **2003**, *100*, 13674–13679.
66. Weiner, D. M.; Meltzer, H. Y.; Veinbergs, I.; Donohue, E. M.; Spalding, T. A.; Smith, T. T.; Mohell, N.; Harvey, S. C.; Lameh, J.; Nash, N. et al. *Psychopharmacology* **2004**, *177*, 207–216.
67. Javitt, D. C. *Curr. Opin. Investig. Drugs* **2002**, *3*, 1067–1072.
68. Chavez-Noriega, L. E.; Schaffhauser, H.; Campbell, U. C. *Curr. Drug Targets CNS Neurol. Disord.* **2002**, *1*, 261–281.
69. Kinney, G. G.; O'Brien, J. A.; Lemaire, W.; Burno, M.; Bickel, D. J.; Clements, M. K.; Chen, T. B.; Wisnoski, D. D.; Lindsley, C. W.; Tiller, P. R. et al. *J. Pharmacol. Exp. Ther.* **2005**, *313*, 199–206.
70. Wroblewska, B.; Wroblewski, J. T.; Pshenichkin, S.; Surin, A.; Sullivan, S. E.; Neale, J. H. *J. Neurochem.* **1997**, *69*, 174–181.
71. Zhou, J.; Neale, J. H.; Pomper, M. G.; Kozikowski, A. P. *Nat. Rev. Drug Disc.* **2005**, *4*, 1015–1026.
72. deHaan, L.; Booij, J.; Lavalaye, J.; van Amelsvoort, T.; Linszen, D. *Psychopharmacology* **2006**, *183*, 500–505.
73. Martin, L. F.; Kem, W. R.; Freedman, R. *Psychopharmacology* **2004**, *174*, 54–64.
74. Spooren, W.; Riemer, C.; Meltzer, H. *Nat. Rev. Drug Disc.* **2005**, *4*, 967–975.
75. Spedding, M.; Jay, T.; Costa e Silva, J.; Perret, L. *Nat. Rev. Drug Disc.* **2005**, *4*, 467–476.
76. Insel, T. R.; Scolnick, E. M. *Mol. Psychiatry* **2006**, *11*, 11–17.
77. Santarelli, L.; Saxe, M.; Gross, C.; Surget, A.; Battaglia, F.; Dulawa, S.; Weisstaub, N.; Lee, J.; Duman, R.; Arancio, O. et al. *Science* **2003**, *301*, 805–809.
78. Kippin, T. E.; Kapur, S.; Van Der Kooy, D. *J. Neurosci.* **2005**, *25*, 5815–5823.
79. Bodick, N. C.; Offen, W. W.; Levey, A. I.; Cutler, N. R.; Gauthier, S. G.; Satlin, A.; Shannon, H. E.; Tollefson, G. D.; Rasmussen, K.; Bymaster, F. P. et al. *Arch. Neurol.* **1997**, *54*, 465–473.
80. Jones, P. B.; Buckley, P. F. *Schizophrenia*; Lundbeck Institute: Skodsborg, Denmark, 2006.
81. Eaton, W. W.; Byrne, M.; Ewald, H.; Mors, O.; Chen, C.-Y.; Agerbo, E.; Preben Bo Mortensen, P. B. *Am. J. Psychiatry* **2006**, *163*, 521–528.
82. Zubieta, J.-K.; Heitzeg, M. M.; Smith, Y. R.; Bueller, J. A.; Xu, K.; Xu, Y.; Koeppe, R. A.; Stohler, C. S.; Goldman, D. *Science* **2003**, *299*, 1240–1243.
83. Wu, A. H.; Tseng, C. C.; Van Den Berg, D.; Yu, M. C. *Cancer Res.* **2003**, *63*, 7526–7529.
84. Brown, S.; Birtwistle, J.; Roe, L.; Thompson, C. *Psychol. Med.* **1999**, *29*, 697–701.
85. Casey, D. E.; Hansen, T. E. In *Medical Illness and Schizophrenia*; Meyer, J. M., Nasrallah, H. A., Eds.; Am. Psychiatr. Publ.: Washington, DC, 2003, pp 13–34.
86. Kroeze, W. K.; Hufeisen, S. J.; Popadak, B. A.; Renock, S. M.; Steinberg, S.; Ernsberger, P.; Jayathilake, K.; Meltzer, H. Y.; Roth, B. L. *Neuropsychopharmacology* **2003**, *28*, 519–526.

Biographies

Michael J Marino, PhD, received his PhD in 1995 from the University of Pittsburgh Department of Neuroscience where he employed behavioral and electrophysiological methods to the study of G-proteins and G-protein coupled receptors. As a postdoctoral fellow with Dr Jeff Conn at Emory University, he performed pioneering work on the role of metabotropic glutamate receptors in regulating the circuitry of the basal ganglia. This work led to the identification of potential targets for the treatment of Parkinson's disease. In 2001, he joined Merck Research Laboratories where he worked on novel drug discovery projects related on the modulatory control of brain circuitry underlying Parkinson's disease and related movement disorders. He joined Cephalon, Inc. in 2005 and is currently Associate Director of CNS biology at Cephalon, where he continues to focus on drug discovery for disorders of the CNS. He has published over 40 scientific articles, book chapters, and reviews.

Robert E Davis, PhD, has spent almost 25 years in the discovery and development of small molecule drugs for human disease. Dr Davis has worked in large and start-up pharmaceutical companies. He currently serves as the President of 3D Pharmaceutical Consultants. In this role, he provides strategic and operational consulting services to small and large pharmaceutical companies. Before founding this firm he served as Executive Vice President of Drug Discovery and Development for ACADIA Pharmaceuticals Inc holding that position since February 2001.

He also was a founding member of ACADIA's Scientific Advisory Board from 1996 and served as a consultant to ACADIA from November 2000 until becoming an employee. From January 1994 until October 2000, Dr Davis held various positions at MitoKor, a development stage biotechnology company, serving at various times as its President, Chief Executive Officer and Chief Scientific Officer. This company was recently sold to another company.

Earlier in his career, Dr Davis held various positions at Parke-Davis Pharmaceutical Research, Warner-Lambert Company including Director of Neurodegenerative Diseases. While at Parke-Davis, Dr Davis has chaired or participated in research and development teams that advanced 12 new chemical entities into human clinical trials, including Cognex, the first drug approved by the FDA and other countries for Alzheimer's disease.

Herbert Y Meltzer received his AB with honors in Chemistry from Cornell, MA in Chemistry from Harvard, and MD from Yale. Dr Meltzer is Bixler/May/Johnson Professor of Psychiatry and Pharmacology and Director of the Psychobiology Program for Translational Research at the Vanderbilt University School of Medicine, Chairman of the International Psychopharmacology Algorithm Project, Chair of the Young Investigator Grant Review for NARSAD, and Director of the Schizophrenia Program of Centerstone Mental Health System. Dr Meltzer has been president of the Collegium Internationale Neuropsychopharmacologicum (CINP) and the American College of Neuropsychopharmacology (ACNP). He served as editor of 'Psychopharmacology: The Third Generation of Progress' and co-editor of the ACNP journal, 'Neuropsychopharmacology.' He has received the Noyes Prize of the Commonwealth of Pennsylvania, the Edward J Sachar Award from Columbia University, the Lieber Prize from NARSAD, the Stanley Dean Award of the American College of Psychiatry for his research on schizophrenia, the Gold Medal Award of the Society of Biological Psychiatry, the Earl Sutherland Prize for Lifetime Achievement in Research of Vanderbilt University (2004), and the Research Prize of the American Psychiatric Association (2005). Dr Meltzer is a member of the editorial board of numerous scientific journals. His research interests include: the psychopharmacology of schizophrenia, the mechanism of action of antipsychotic drugs, genetic factors in schizophrenia and pharmacogenomics, prevention of suicide in schizophrenia, and cognitive impairment in schizophrenia.

Lars J S Knutsen began his research career at Glaxo in Ware, Herts., UK having completed an MA in Chemistry at Christ Church, Oxford, in 1978. While at Glaxo he completed a PhD in Nucleoside Chemistry joining Novo Nordisk in Denmark in 1986. While there he led the project that identified tiagabine, a marketed anticonvulsant acting by blocking GABA uptake. In 1997, he joined Vernalis (Cerebrus) in the UK, initiating the adenosine A_{2A} antagonist project that led to V2006, currently in clinical trials with Biogen-IDEC for Parkinson's disease. He joined Ionix Pharmaceuticals Ltd., in Cambridge, UK in 2002 as Director of Chemistry. Dr Knutsen joined the CNS Medicinal Chemistry group at Cephalon Inc. in 2006. He has over 35 peer-reviewed publications and 18 issued US patents.

Michael Williams, PhD, DSc, received his PhD (1974) from the Institute of Psychiatry and his Doctor of Science degree in Pharmacology (1987) both from the University of London. Dr Williams has worked in the US-based pharmaceutical industry for 30 years at Merck, Sharp and Dohme Research Laboratories, Nova Pharmaceutical, CIBA-Geigy, and Abbott Laboratories. He retired from the latter in 2000 as Divisional Vice President, Neurological and Urological Diseases Research and after serving as a consultant with various biotechnology/pharmaceutical/venture capital companies in the US, Canada and Europe, joined Cephalon, Inc. in West Chester, in 2003 where he is Vice President of Worldwide Discovery Research. He has published some 300 articles, book chapters and reviews and is an Adjunct Professor in the Department of Molecular Pharmacology and Biological Chemistry at the Feinberg School of Medicine, Northwestern University, Chicago, Illinois.

Comprehensive Medicinal Chemistry II
ISBN (set): 0-08-044513-6

ISBN (Volume 6) 0-08-044519-5; pp. 17–44

6.03 Affective Disorders: Depression and Bipolar Disorders

T Blackburn, Helicon Therapeutics Inc., Farmingdale, New York, USA
J Wasley, Simpharma LLC, Guilford, CT, USA

6.03.1 Introduction

6.03.1.1 Disease State/Diagnosis

6.03.1.1.1 Affective disorders: depression and bipolar disorder

Depressive disorders involve all major bodily functions, mood, and thoughts, affecting the ways in which an individual eats and sleeps, feel about themselves, and think. Without treatment, depressive symptoms can last for weeks, months, or years. Fortunately, however, appropriate treatment can help most individuals suffering from depression. An increasing number of treatment options have become available over the past 20 years for individuals with major depression disorder (MDD), accompanied by a growing body of evidence-based medicine describing their effectiveness, efficacy, and safety that provides clinicians with multiple options to determine the most appropriate treatment for each patient.[1]

Bipolar affective disorder (BPAD or manic-depressive illness) is a common, recurrent, and severe psychiatric disorder characterized by episodes of mania, depression, or mixed states (simultaneously occurring manic and depressive symptoms). BPAD is frequently unrecognized and goes untreated for many years without clinical vigilance. Newer screening tools assist physicians in making the diagnosis and several drugs are now available to treat the acute mood episodes of BPAD and to prevent further episodes with maintenance treatment.

Two consensus reviews on depression and one on BPAD have been published by the American College of Neuropsychopharmacology (ACNP),[2] the other by the British Association of Psychopharmacology (BAP) on depression.[3] Consensus reviews on BPAD have also been published by BAP[4] and the American Psychiatric Association (APA).[5] Together, these encompass the major findings and lessons learned from antidepressant and BPAD research over the last decade, providing treatment guidelines for these conditions (see below). They also bring some consensus to disparities that exist in the clinical diagnosis of depression and mania.

6.03.1.1.1.1 Types of depression

Depressive disorders exhibit different phenotypes with variations in the number of symptoms, their severity, and persistence according to DSM-IV[6] and ICD-10[7] classifications.

6.03.1.1.1.1.1 Major depression disorder This manifests as a combination of symptoms (**Table 1**).

6.03.1.1.1.1.2 Dysthymia This is a less severe form of depression involving long-term, chronic symptoms that do not disable but keep individuals from functioning well or from feeling good. Many people with dysthymia also experience major depressive episodes at some time in their lives.

6.03.1.1.1.1.3 Bipolar affective disorder BPAD, also called manic-depressive illness, is not nearly as prevalent as other forms of depressive disorder. BPAD is characterized by cycling mood changes: severe highs (mania) and lows (depression). These mood switches can be dramatic and rapid, but more often they are gradual. In the depressed cycle, individuals can have any or all of the symptoms of a depressive disorder. In the manic cycle, they may be overactive, over

Table 1 Classification of depressive states

Classification used in guideline	*DSM-IV[a] (code)*	*ICD-10[b] (code)*
Major depression	Major depressive episode, single episode, or recurrent (296)	Depressive episode – severe (F32.2), moderate (F32.1), or mild with at least five symptoms[c] (F32.0)
		Recurrent depressive disorder current episode severe (F33.2), moderate (F33.1), or mild with at least five symptoms[c] (F33.0)
Milder depression	Depressive disorder not otherwise specified (311)	Depressive episode – mild with four symptoms[c] (F32.0)
		Recurrent depressive disorder current episode mild with four symptoms (F33.0)
		Mixed anxiety and depressive disorder (F41.2)
	Adjustment disorder with depressed mood/mixed anxiety and depressed mood (309)	Adjustment disorder – depressive reaction/mixed and depressive reaction (F43.2)
		Other mood (affective) disorders (F38)
Dysthymia	Dysthymia (300.4)	Dysthymia (F34.1)

Reprinted by permission of Sage Publications Ltd from Anderson, I. M.; Nutt, D. J.; Deakin, J. F. W. *J. Psychopharmacol.* **2000**, *14*, 3–20. Copyright (© British Association for Psychopharmacology, 2000).
[a] Fourth Revision of the American Psychiatric Association's Diagnostic and Statistics Manual.[6]
[b] Tenth Revision of the International Classification of Diseases.[7]
[c] For a list of symptoms, see the abridged DSM-IV criteria in **Table 2**. Must include at least two of (1) depressed mood, (2) loss of interest or pleasure, (3) decreased energy or increased fatigability.

talkative, and have a great deal of energy. Mania often affects thinking, judgment, and social behavior in ways that cause serious problems and embarrassment. For example, an individual in a manic phase may feel elated and full of grand schemes that might range from unwise business decisions to romantic sprees. Mania, left untreated, can worsen to a psychotic state.

6.03.1.1.1.2 Prevalence

In 1987, the selective serotonin (5HT) reuptake inhibitors (SSRIs; e.g., fluoxetine **1**) were introduced into clinical use. Their improved safety and side effect profile resulted in a move away from older antidepressant agents, the tricyclic antidepressants (TCAs) and monoamine oxidase inhibitors (MAOIs). Drugs for depression sold for an estimated $20 billion in the US in 2004, with growth predictions of 1.2% year-on-year. However, this is a mature market with a number of key drugs facing patent expiration in the next 6 years, and one that is awaiting the next paradigm shift in new chemical entity (NCE) treatment therapy.

Unipolar major depression is ranked fourth as a disease burden measured in disability adjusted for life years in 1990.[8] Despite available antidepressant medications, unipolar major depression is ranked second behind ischemic heart disease as a potential disease burden by 2020. The risk for unipolar major depression, especially for females in developed countries, is 1 in 10. There is considerable evidence that depression is associated with increased risk for cardiovascular and infectious diseases as well as immunological and endocrine changes. The World Health Organization has predicted that depression will become the leading cause of human disability by 2020.[1]

6.03.1.1.1.3 Global depression market

Although depression is a global phenomenon, the global market for antidepressants is unevenly distributed. In 2004, the US antidepressant market accounted for 71% of the global market compared to 24% in Europe and 5% in the rest of the world (mainly Japan). While depression is a mature market, there are several key drivers for its further growth including: (1) improvements in the efficacy; (2) speed of onset; (3) safety/tolerability of NCEs; and (4) a reduction in remission rates and relapse/recurrence.

A National Institutes of Mental Health (NIMH) National Comorbidity Survey of more than 9000 US adults in 2005 using DSM-IV-TR (Text Revision 2001) criteria found that 6% of those studied had a debilitating mental illness,[8] yet treatment was difficult to obtain, with only one-third of those in care receiving minimally adequate treatment, such as appropriate drugs or a few hours of therapy over a period of several months. In general, the investigators found that things had not changed much over the past decade. A Massachusetts Institute of technology (MIT) survey estimated the direct and indirect cost of mood disorders in the US to be $43 billion in 1990.[9] Depressed individuals incur almost twice the medical cost burden as nondepressed patients, the main part (80%) being for medical care rather than psychiatric or psychological services, with the bulk of antidepressant prescriptions (approximately 80% worldwide) being written by primary care physicians.

Effective treatment for moderate-to-severe depression includes a combination of somatic therapies (pharmacotherapy or electroconvulsive therapy (ECT)). ECT has been rejuvenated for the treatment for the most severe, melancholic depressions, particularly in the elderly (who are more prone to adverse effects of drugs) and in patients who do not respond to antidepressants.

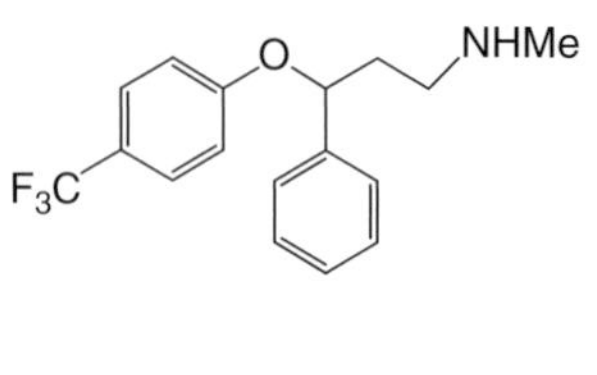

Fluoxetine **1** 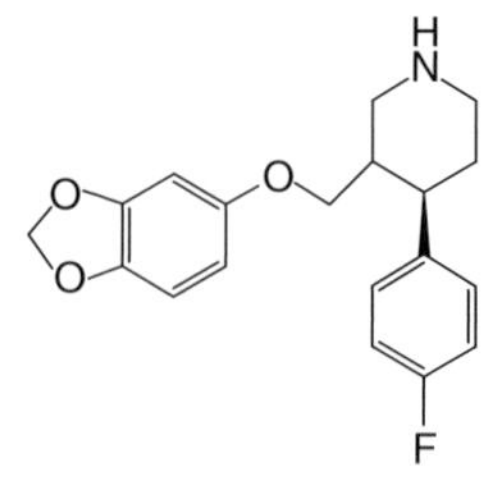 Paroxetine **2**

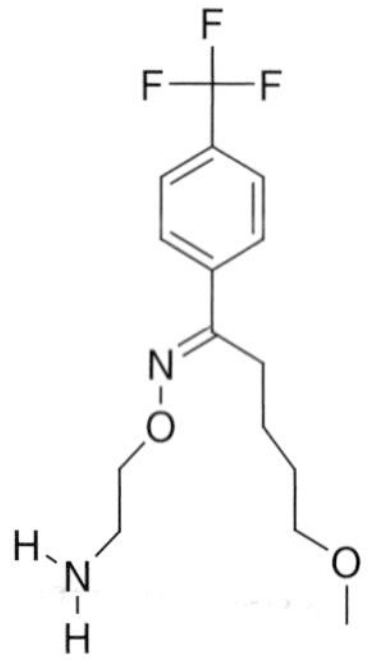

Fluvoxamine **3** 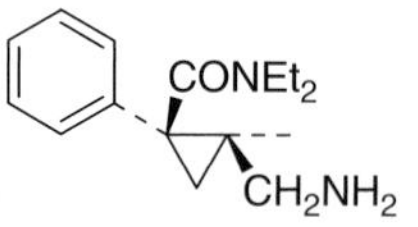 Milnacipran **4** 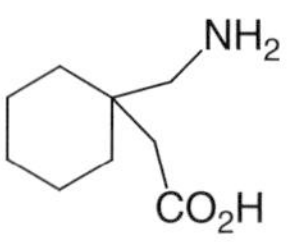 Gabapentin **5**

6.03.1.1.1.4 Prescription trends

Prescription trends of antidepressants and antimanic agents vary across countries, and recent research by Datamonitor (in *Strategic Perspective: Antidepressants*, published in 2001) argues that patients fail to meet treatment criteria by as much as 48% in Europe, 43% in the US, and 15% in Japan.[77] The latter can be accounted for by the relatively slow progression of the Japanese antidepressant market in comparison to other markets. Second-generation antidepressants were only launched in Japan in 1999, over a decade after other comparable markets. The fact that SSRIs and other second-generation antidepressants are superior to the older antidepressants, such as TCAs, has raised physicians' expectations of the newer agents. In Japan, the influence of paroxetine (**2**), fluvoxamine (**3**), and milinacipran (**4**) has yet to be fully felt, so physician expectation is modest resulting in physicians being comfortable with current prescribing habits.[77] Similar arguments exist for the treatment for BPAD, where the newer atypical antipsychotic agents, anticonvulsants, and gabapentin (**5**) have yet to establish themselves as first-line therapy over the gold standard, lithium.[77]

6.03.1.1.1.5 Classification and subclassification of major depression disorder and bipolar (affective) disorder

6.03.1.1.1.5.1 Major depressive disorder

The symptom criteria for major depression in DSM-IV-TR and ICD-10 guidelines are very similar (**Table 2**), although the coding systems are somewhat different.[6,7] One difference is that the ICD-10 has a separate, optional subdiagnosis for depression with and without somatic symptoms. The latter is not present in the DSM-IV system. Both sets of guidelines have depressive disorder subdiagnoses for the following:

- mild, moderate, and major severity;
- single and recurrent episodes;
- with and without psychotic symptoms; and
- partial and full remission.

Table 2 Abridged DSM-IV criteria for major depressive episode

A. Over the last 2 weeks, five of the following features should be present most of the day, or nearly every day (must include one or two):
  1 Depressed mood
  2 Loss of interest or pleasure in almost all activities
  3 Significant weight loss or gain (more than 5% change in 1 month) or an increase or decrease in appetite nearly every day
  4 Insomnia or hypersomnia
  5 Psychomotor agitation or retardation (observable by others)
  6 Fatigue or loss of energy
  7 Feelings of worthlessness or excessive or inappropriate guilt (not merely self-reproach about being sick)
  8 Diminished ability to think or concentrate, or indecisiveness (either by subjective account or observation of others)
  9 Recurrent thoughts of death (not just fear of dying), or suicidal ideation, or a suicide attempt, or a specific plan for committing suicide
B. The symptoms cause clinically significant distress or impairment in functioning
C. The symptoms are not due to a physical/organic factor or illness. The symptoms are not better explained by bereavement (although this can be complicated by major depression)

From the Fourth Revision of the American Psychiatric Association's Diagnostic and Statistics Manual.[6]
Reprinted by permission of Sage Publications Ltd from Anderson, I. M.; Nutt, D. J.; Deakin, J. F. W. *J. Psychopharmacol.* **2000**, *14*, 3–20. Copyright (© British Association for Psychopharmacology, 2000).[3]

Table 3 Types of BPAD

Bipolar I (BPI) (DSM-IV 296.00–296.06, 296.40–296.7)

- 1% prevalence
- at least one episode of full-blown mania
- episodes of hypomania, mixed states, and depression

Bipolar II (BPII) (DSM-IV 296.89)

- 0.5–3% prevalence
- at least one episode of depression
- at least one episode of hypomania

Bipolar spectrum disorder (DSM-IV includes all of above plus 301.13)

- up to 5% prevalence
- includes BPI, BPII, schizoaffective disorders, cyclothymia

DSM-IV comprises additional subcategories for catatonic, melancholic, and atypical features and for postpartum onset. Both ICD-10 and DSM-IV present affective disorders together in one section, distinguishing bipolar from unipolar disorder, including dysthymia, and using clear definitions. Operational problems with ICD-10 include complexity, use of different clinical and research definitions, emphasis on single versus recurrent episodes, and the lack of some clinically useful subtypes. DSM-IV is less complex but assigns separate unjustified categories of medical and substance-induced mood disorders, and fails to code its useful qualifiers.[10]

6.03.1.1.1.5.2 Bipolar disorder This describes a spectrum of disorders in which episodes of depression and mania occur, interspersed with periods of normal mood. It is also known as bipolar depression or manic depression. BPAD is characterized by cycles of mania and depression, which cause a person with bipolar disorder to experience severe mood swings. There are three types of BPAD according to DSM-IV (**Table 3**)[6]:

6.03.1.1.1.5.3 Biomarkers in depression and bipolar disorder The main uses of biomarkers in drug development are[11]:

- discovery and selection of lead NCEs;
- generation of pharmacokinetic (PK) and pharmacodynamic (PD) models;
- aid in clinical trial design and expedite drug development;

Table 4 Examples for biomarkers in psychotropic drug development

	Biomarker procedures
Brain imaging technique	Computed tomography (CT), regional cerebral blood flow (rCBF), magnetic resonance imaging (MRI), positron emission tomography (PET), single photon emission computed tomography (SPECT), magnetic resonance spectroscopy (MRS), magnetoencephalography (MEG)
Cell-based imaging	Fluorescent resonant energy tranfer
	Confocal imaging in brain slices
Electophysiological marker	Electroencephalogram (EEG), pupilometry, saccadic eye movements
Laboratory-based marker	Concentrations of catecholamines, hormones, enzymes, proteins, drugs, and drug metabolites
Psycho-immunological marker	Immunoglobulin, lymphocyte responses, lymphokine, cytokine, interleukin, interferon, viral serology, Alz-50, anticardiolipin antibodies (ACA)
Neuroendocrine marker	Dexamethasone suppression test (DST), thyrotropin releasing hormone stimulation teat (TRHST), growth hormone (GH) challenge test
Provocative anxiety tests	Lactate infusion, carbon dioxide (CO_2) challenge, cholecystokinin (CCK) challenge
Genetic markers	DNA banking, genotyping, restriction fragment length polymophesis (RFLPs)
Proteomic identification	Nuclear magnetic resonance (NMR), lipoprotein fractions and subfractions, matrix-assisted laser desorption/ionization–mass spectrometry (MALDI-MS)

Matrices mostly from plasma, urine, CSF, tissue, saliva, and hair.
Reprinted with permission from Bieck, P. R.; Potter, W. Z. *Annu. Rev. Pharmacol. Toxicol.* **2005**, *45*, 227–246, © 2005 by Annual Reviews www.annualreviews.org.

- serving as surrogates for clinical or mortality endpoints;
- optimizing drug therapy based on genotypic or phenotypic factors; and
- definition of patient enrollment in studies and help with stratification.

Biomarkers are currently being investigated in psychiatry and neurology through a wide variety of procedures (**Table 4**) and the knowledge gain in the use of biomarkers is slowly being integrated into databases for use by the scientific community, e.g., NIMH drug companies.

The advantages, disadvantages, and limitations of selected biotechnologies for assessing the access of an NCE to the brain are listed in **Table 5**. Brain imaging technology is a facile methodology for studying the interaction of an NCE with its target, making this a preferred technique. However, there is a limited number of targets for which validated positron emission tomography (PET) or single photon emission computed tomography (SPECT) ligands are available. Furthermore, recent PET studies with a neurokinin 1 (NK1) receptor antagonist showed that the NCE had greater than 90% occupancy of central NK1 receptors and was active in an early Phase II efficacy study of depression but not subsequent larger Phase II studies, making this a striking example of a novel target with an excellent PET ligand that fails to be supported by clinical data and thereby questions the entire clinical hypothesis and the NCE.[11]

According to Bieck and Potter, a single approach may not provide the answer to addressing the question of brain penetration and drug efficacy.[11] Instead, a multimodal approach to biomarkers in CNS disorders may well be the answer, using a combination of imaging technology (where PET ligands are feasible) and CSF studies.[11]

6.03.2 Disease Basis

6.03.2.1 Causes of Depression

The etiology of depression and BPAD is unknown. Depression is polygenic in nature with both genetic and epigenetic components, making the use of genetically engineered rodents as models for drug discovery precarious.[12,13] Moreover, emerging understanding of the biochemical mechanisms is compromised by the fact that most of the drugs used to treat depression and bipolar disorders (e.g., lithium and antidepressants in general) have complex and ill-defined pleiotypic mechanisms of action.[12]

Table 5 Advantages and limitations of selected biomarker technologies

Technology	*Advantages*	*Limitations*
CSF	• Measurement of drug PK in central compartment • Several surrogate markers available • Possible to combine PK and PD measures in one protocol	• Invasive • Sensitive assay required
EEG	• Unequivocal positive results provide evidence of CNS effects of intervention • Convenient for repeated measures within subjects • Noninvasive	• Observed effects generally not easily linked to specific mechanism of interaction • Prone to artifacts and not fully standardized
MRI, MRS, fMRI, sMRI	• Pharmacological doses of certain drugs can be accurately measured • Noninvasive	• Motion artifacts in agitated patients • More validation necessary for surrogate marker use • Expensive
PET/SPECT	• Highly sensitive detection of drugs that can be radiolabeled • Direct evidence of effect of drug at site of action • Noninvasive	• Exposure to ionizing radiation • Tracer not available for every application • Expensive

PK, pharmacokinetic(s); PD, pharmacodynamic(s); CSF, cerebrospinal fluid; EEG, electroencephalography; MRI, magnetic resonance imaging; MRS, magnetic resonance spectroscopy; fMRI, functional MRI; sMRI, structural MRI; PET, positron emission tomography; SPECT, single photon emission computed tomography.

6.03.2.1.1 The monoamine theory

The monoamine theory of depression and drugs acting on monoamine neurotransmission has dominated the treatment of depression for over 30 years. Indeed, the monoamine reuptake inhibitors and the MAOIs were shown to have antidepressant activity by chance, and the discoveries of their modes of action were instrumental in developing the monoamine theory. In the days when the monoamine theory of depression was evolving, the focus was more on norepinephrine (NE) than 5HT or dopamine (DA). The theory developed from observations that reserpine depleted monoamines and caused depression, whereas the MAOIs and monoamine reuptake inhibitors enhanced monoamine function and thereby relieved depression. More recent studies indicate that it may be the inhibition of glial uptake 2 sites by normetanephrine (NMN) or other inhibitors of uptake 2 that results in an enhanced accumulation of NE in the synapse. The hypothesis proposed by Schildkraut and Mooney suggests that drugs or other agents that increase levels of NMN or otherwise inhibit the extraneuronal monoamine transporter, uptake 2, in the brain will accelerate the clinical effects of NE reuptake inhibitor antidepressant drugs.[13] This hypothesis, as well as others discussed below, continue to drive pharmaceutical research for the Holy Grail, that is, a fast-acting antidepressant. However, Duman and others[14] would contend that this approach may not be possible based on their neurogenesis hypothesis of antidepressant efficacy (*see* Section 6.03.3.1.1).[19] To discover an antidepressant that has an effect within days rather than weeks has challenged researchers for decades to understand the reasons for the delay in onset of the antidepressant action. The focus has been mainly on the mechanism of action of SSRIs and one theory that inhibition of 5HT reuptake initially causes activation of the presynaptic $5HT_{1A}$ receptors on the cell bodies in the dorsal and median raphe nucleus.[15] This inhibits the firing of 5HT neurons, so reducing rather than increasing the release of 5HT from the terminals.[16]

According to the hypothesis first proposed by Blier *et al.* the primary mechanism of action of SSRIs is an increased activation of 5HT postsynaptic receptors in the forebrain, but this is not achieved until the raphe $5HT_{1A}$ receptors become desensitized.[16] The problem with putting this into practice clinically is that there are no selective $5HT_{1A}$ receptor antagonists available for clinical use. Artigas *et al.* pioneered the use of pindolol as a $5HT_{1A}$ receptor antagonist for proof of concept studies.[15] However, evidence from at least half of the trials failed to show that the combination provided a faster onset of action. A few cases showed an improved response rate, but there was no benefit in treating resistant depression. Despite the limited clinical efficacy observed with this combination, pharmaceutical companies are still developing compounds with this combined approach to achieve fast onset of action.

6.03.2.1.2 Vesicular neurotransmitter transporters (VMATs) and neurotransmitter transporters (SCDNTs)

Neurotransmitters are synthesized in neurons and concentrated in vesicles for their subsequent Ca^{2+}-dependent release into the synaptic cleft, where they are inactivated by either enzymatic degradation or active transport in neuronal and or glial cells by neurotransmitter transporters.[17] Molecular cloning has allowed the pharmacological and structural characterization of a large family of related genes encoding Na^{+}/Cl^{-}-dependent neurotransmitter transporters (SCDNTs). The identification of a superfamily of monoamine transporters (dopamine transporter (DAT), norepinephrine transporter (NET), and serotonin transporter (SERT)) has been the focus of recent research on the association of monoamine transporters with psychiatric illnesses, as discussed below.[18]

6.03.2.1.2.1 Norepinephrine transporter polymorphisms

At least 13 polymorphisms of NET have been identified,[20] the functional significance of which is unknown. Alterations in the concentration of NE in the CNS have been hypothesized to cause, or contribute to, the development of psychiatric illnesses such as major depression and BPAD. Many studies have reported altered levels of NE and its metabolites NMN and dihydroxyphenylglycol (DHPG) in the CSF, plasma, and urine of depressed patients as compared with normal controls. These variances could reflect different underlying phenotypes of depressive disorders with varying effects on NE activity. The melancholic subtype of depression (with positive vegetative features, agitation, and increased hypothalamic–pituitary–adrenal (HPA) axis activity) is most often associated with increased NE. Alternatively, so-called atypical depression is associated with decreased NE and HPA axis hypoactivation. In one study, urinary NE and its metabolites were found to be significantly higher in unipolar and bipolar depressed patients than in healthy volunteers, suggesting that unmedicated unipolar and bipolar depressed patients have a hyperresponsive noradrenergic system. Increased NE activity has also been observed to be a contributor to the borderline personality disorder traits of impulsive aggression and affective instability, high levels of risk taking, irritability, and verbal aggression. Furthermore, abnormal regulation or expression of the human NET has been reported in major depression. In post-mortem human brain, [^{3}H]-nisoxetine binding to NET was highest in dorsal *raphe nuclei* and *locus coeruleus*. Low levels of NET in the locus coeruleus in major depression may reflect a compensatory downregulation of this transporter protein in response to insufficient availability of its substrate (NE) at the synapse. These studies suggest that abnormalities that can cause impaired noradrenergic transmission could contribute to the pathophysiology of certain psychiatric disorders. However, results from other studies suggest that the investigated polymorphisms are not the main susceptibility factors in the etiology of major depression.[19] This was also the case for the NET DNA sequence variants identified in patients suffering from schizophrenia or BPAD. Subsequent case-control studies did not reveal any significant association between the variances and those diseases.[19]

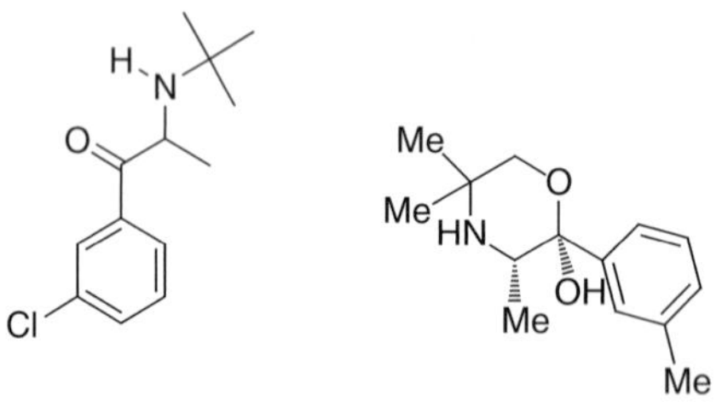

Bupropion **6**

Radafaxine **7**

6.03.2.1.2.2 Dopamine transporter polymorphisms

DAT terminates dopaminergic neurotransmission by reuptake of dopamine (DA) in presynaptic neurons and plays a key role in DA recycling. DAT can also provide reverse transport of DA under certain circumstances. Psychostimulants such as cocaine and amphetamines and drugs used for attention deficit hyperactivity disorder (ADHD) such as methylphenidate exert their actions via DAT. Altered DAT function or density has been implicated in various types of psychopathology, including depression, BPAD, suicide, anxiety, aggression, and schizophrenia. Altered transport properties associated with some of the coding variants of DAT suggest that individuals with these DAT variants could display an altered DA system.[17,20] Multiple human dopamine transporter (hDAT, *SLC6A3*) coding variants have been described, though to date they have been incompletely characterized. The antidepressant, bupropion (**6**) dose-dependently increases vesicular DA uptake; an effect also associated with VMAT-2 protein redistribution. Another purported antidepressant that weakly blocks DATs is radafaxine (**7**) ((2*S*,3*S*)-2-(3-chlorophenyl)-3,5,5,-trimethyl-2-morpholinol hydrochloride). NCEs that block DATs may have potential reinforcing effects and abuse liability, which needs to be addressed before clinical development.

6.03.2.1.2.3 Serotonin transporter polymorphisms

Reduced binding of imipramine and paroxetine to brain and platelet SERTs in patients with depression and suicide victims indicates that altered SERT function might contribute to aberrant behaviors. Two polymorphic regions have been identified in the SERT promoter and implicated in anxiety, mood disorders, alcohol abuse, and in various neuropsychiatric disorders.[21] Thus, studies are emerging to support the notion that impaired regulation might contribute to human disease conditions such as those seen in human variants of the SERT coding region.

6.03.2.2 Genetic and Environmental Origins of Depression and Bipolar Disorder

In the case of major depressive illness genetic factors account for about 30% of the variance and environmental factors play a major role in inducing the illness.[22,23] The first direct evidence of the importance of genetic variation in drug response was shown in depressed patients with a short form of the SERT promoter, who had a worse response to SSRIs than those with the long isoform.[21] Other genes have been associated with antidepressant treatment and undoubtedly the field of pharmacogenomics and its application to the pathophysiological mechanisms of affective disorders will continue to grow together with the technological advances resulting in the human genome mapping of psychiatric mood disorders.[22]

6.03.2.3 Epidemiological Status

It has been estimated that over a lifetime, the global prevalence of depression is 21.7% for females and 12.7% for males who suffer from depression at some point. The APA has estimated that 5–9% of women and 2–3% of men in the US suffer from depression at any given time.[9] A Norwegian study showed that 24% of women suffer major depression at some point in their lives and 13.3% suffer from dysthymia, while 10% of males suffer from major depression at some point and 6% suffer from dysthymia.[23] Depression in children and adolescents is a cause of substantial morbidity and mortality in this population, being a common disorder that affects 2% of children and up to 6% of adolescents. Although antidepressants are frequently used in the treatment of this disorder, there has been recent controversy about the efficacy and safety of these medications in this population. This led to the US Food and Drug Administration (FDA) publishing a list of recommendations from the Psychopharmacologic Drugs and Pediatric Advisory Committees in 2004.[26] A critical appraisal on the treatment of depression in children and adolescents has been published.[26] Some types of depression are familial, indicating that there is inherited vulnerability. Similarly, in studies of families in which members of each generation develop BPAD it has been found that those with the illness have a somewhat different genotype from those who do not become ill. However, the reverse is not true: not all individuals with a purported BPAD genotype will develop the illness. This suggests that additional factors, possibly stresses at home, work, or school or other coping factors, are involved in the onset of the disease.[27] In some families, major depression also seems to occur generation after generation, although it can also manifest in individuals who have no family history of depression. It is clear that a combination of genetic, psychological, and environmental factors are involved in the onset of a depressive disorder. However, it is now emerging, particularly in the case of MDD, that changes in brain structures (particularly the hippocampal regions) due to an impairment of neurogenesis may well be associated to some extent with the pathophysiology of depression and the mechanism whereby antidepressants exert their action.[13]

6.03.3 Experimental Disease Models

Drug discovery in depression has been hampered by the lack of a universally accepted animal model(s) that can be used to screen NCEs for antidepressant effects. Although there are several animal models that reproduce some features of depression in the context of stress and/or separation, it is questionable as to whether these are relevant to the human disorder MDD or BPAD. The advantages and disadvantages of animal models for depression are summarized in **Table 6**. In all cases, the behavioral features can be reversed by conventional antidepressant drug treatment. However, despite their intrinsic limitations, the full potential of these models has not yet been realized and they represent an under-explored opportunity. The heuristic value and the knowledge gain from behavioral animal models in psychopharmacology is, explicitly or implicitly, the central preoccupation of psychopharmacologists.[24] There are a number of compelling reasons to believe in the legitimacy of animal models in the development of

Table 6 Animal models of major depression

Model	*Rationale*	*Description*	*Phenotype reproduced*	*Predictive validity*
Part 1				
Uncontrollable stress models: • Learned helplessness model (LHM) • Behavioral despair/ forced swim test (BD/FST) • Tail suspension test (TST)	Exposure to uncontrollable stress produces performance deficits in learning tasks that are not subsequently seen in subjects exposed to identical stressors that are under the subjects' control	LHM: rats or mice exposed to uncontrollable stress, for example uncontrollable foot shock. BD is a variant of LHM: rats or mice are forced to swim in a confined environment. TST is theoretically similar to BD: mice are suspended by their tails for 6 min; the amount of time they spend immobile is recorded	Loss of appetite and weight, decreased locomotor activity and poor performance in both appetitively and aversively motivated tasks	Pharmacological treatments clinically effective in depression are effective in reducing the behavioral and 'physical' abnormalities seen in animals exposed to uncontrollable stress. These models have a high degree of predictive validity in terms of pharmacological isomorphism
Reward models: • Intracranial self-stimulation (ICSS) • Sucrose preference • Place preference	Stress induces abnormalities in reward processes. These paradigms are not considered models of an entire syndrome, but rather provide operational measures of anhedonia, a core feature of depression and a negative symptom of schizophrenia	ICSS: brief electrical self-stimulation of specific brain sites, which is very reinforcing. In reward models, the rate of responding and/or the psycho-physically defined threshold(s) can be used to measure the reward value of the stimulation. The effects of stress are used to change the animal response to reward models; for example, sucrose preference is decreased by chronic stress models, the decrement in sucrose ingestion being a measure of anhedonia. Amphetamine withdrawal seems to be a suitable substrate for inducing depression-like symptoms in rodents	Anhedonia is the markedly diminished interest or pleasure in all, or almost all, activities most of the day, nearly every day	Stress-induced alterations in ICSS behavior were reversed by antidepressant treatment. Anhedonic effect of stress was reversed by antidepressant treatment, but not by antipsychotic, anxiolytic, amphetamine or morphine treatment, indicating good predictive validity in terms of pharmacological isomorphism
Olfactory bulbectomy	Olfactory bulbs are extensions of the rostral telencephalon and constitute 4% of the total brain mass in adult rats. Extensive connections with the limbic and higher brain centres implicate this model in wide-ranging effects other than anosmia	Bilateral removal of olfactory bulbs of rodents (mainly rats)	Behavioral, neurochemical, neuroendocrine and neuroimmune alterations seem comparable to changes in depression	Reliable prediction of response to antidepressants in rats

Chronic mild stress (CMS)	Exposure to mild, unpredictable stressors induces long-term changes resembling those found in depressed patients	CMS has two major readouts: CMS depresses the consumption of sucrose solutions and it also decreases brain reward function (see above)	Behavioral, neurochemical, neuroendocrine and neuroimmune alterations that resemble those seen in depression	Anhedonia-like behaviors are reversed by chronic but not acute antidepressant treatment. Poor reliability of results in rats has limited the utilization of this model
Part 2				
Drug withdrawal	Withdrawal from drugs of abuse reduces brain reward function	Amphetamine withdrawal seems to be a substrate for inducing depression-like phenotypes. Readouts can be ICSS, sucrose preference, BD, TST or LHM	Anhedonia is the markedly diminished interest or pleasure in all, or almost all, activities most of the day, nearly every day	Anhedonia-like behaviors are reversed by chronic but not acute antidepressant treatment
Manual deprivation	Development of stress-responsive systems is maternally regulated. Separating mother and pups results in activation of stress systems	Litters are separated from their mother for the neonatal period and are subsequently tested	HPA axis alterations. Constellation of behavioural changes in the adult rat resembling features of human depressive psychopathology	Limited testing of antidepressants has been conducted
Neonatal clomipramine	Clomipramine increases monoaminergic availability at the synaptic level and suppresses REM sleep	Rat pups are treated from postnatal day 5 to 21 with clomipramine 15 mg per kg subcutaneously twice daily	Behavioural changes in adulthood, circadian disturbances and sexual behaviour alterations	Limited testing of antidepressants has been conducted
Drosophila melanogaster	Basis for further dissection of the genetic components of vesicular monoamine transporters (VMATs)	Genetic analysis of vesicular transporter function and regulation in the Drosophila homolog of the vesicular monoamine transporter (dVMAT)	The regulation of mammalian VMAT and the related vesicular acetylcholine transporter (VAChT) has been proposed to involve membrane trafficking, similar to mammalian systems	Similar to mammalian VMAT homologs and are inhibited by reserpine and the environmental toxins
Caenorhabditis elegans	*C. elegans* expresses both the vesicular acetylcholine transporter and the vesicular monoamine transporter	*C. elegans* contains monoamine and tyramine receptors, that individual isoforms may differ significantly in their sensitivity to other physiologically relevant biogenic amines	*C. elegans* monoaminergic and tyrpaminergic signaling may be important in understanding of the regulation of mammalian systems with regard to trace amines	The pharmacological profile of *C. elegans* VMAT is closer to mammalian VMAT2 than VMAT1
Zebrafish (*Danio rerio*)	Linking genes to brain, behavior and neurological diseases	Comparison of dopaminergic (DA) and serotonergic (5HT) neurons in larval and adult zebrafish, both wild type and mutants and their relationship to human neurons	Zebrafish offer a potentially important source of mutant neurons that will aid understanding of the development as well as the function of forebrain DA and 5HT neurons	Potential to provide important insights into the relationship between genes, neuronal circuits and behavior in normal as well as diseased states

continued

Table 6 Continued

Model	*Rationale*	*Description*	*Phenotype reproduced*	*Predictive validity*
Part 3				
Genetic models: selective animal breeding	Rats are selectively bred for hypo- or hypersensitivity or specific receptor subtypes whose altered function has been thought to be involved in aetiology of depression	Cholinergic-noradrenergic neurotransmitter imbalance (Flinders Sensitive Line rats); the 8-OH-DPAT line of rats Fawn-Hooded (FH/Wjd); Rouen 'depressed' mice	Animals are more susceptible to stress-induced behavioral disturbances	Increased immobility in the forced swim test and foot shock response to antidepressants
Transgenics	Transgenic mice (knockout or overexpressers) that exhibit depression or antidepressant-related behavior	• Serotonin (5HT) system • Norepinephrine (NE) system • Monoamine oxidase • Opioid • GABA • Glutamate substance P system • HPA axis • Galanin • Immunological intracellular signaling molecules and transcription factors • Other neurotransmitters or receptors	Phenotype is variable and each model has been accessed by one or more paradigm listed above (such as BD, TST, LHM, and CMS)	Phenotypes can be depression-like or antidepressant
Other useful tests	These tests may address specific psychiatric behavior/symptoms	EEG characterization energy expenditure, nesting behavior, social avoidance or withdrawal swimming, treadmill/running wheel	Fatigue or loss of energy can be assessed by these tests; social withdrawal can also be a measure of anxiety	These tests measure changes in signs that are not necessarily specific for depression
		Attention, spatial memory working memory	Decreased ability to think or concentrate	Same as above
		Prenatal stress	May reproduce stress or conditions that influence behavioral or physiological changes during adulthood	Same as above
		Novelty suppressed feeding	May reflect anxiety	Behavioral effects following chronic treatment
		Resident intruder	May reflect social stress or anxiety	Change in agonistic behavior during the course of antidepressant drug treatment
		LPS-induced immunological activation	Might reproduce neuroimmume or neuroendocrine changes that occur during stress	Sensitive to tricyclic antidepressants

BD, behavioral despair; CMS, chronic mild stress; EEG, electroencephalogram; GABA, γ-amino-butyric acid; HPA, hypothalamic–pituitary–adrenal; LHM, learned helplessness model; LPS, lipopolysaccharide; REM, rapid eye movement; TST, tail suspension test.
Modified from Wong and Licinio (2004).[25]

new improved drugs for the treatment of mental disorders; however, these models need to be based on the following criteria[24,25]:

- Predictive validity: the ability of a model to accurately predict clinical efficacy of a psychoactive pharmacological agent.
- Face validity: the similarity of the model to clinical manifestations of phenomenon/disorder in terms of major behavioral and/or physiological symptoms and etiology.
- Construct validity: the strength of the theoretical rationale upon which the model is based.

Animal models have been defined as experimental preparations developed in one species for the purpose of studying or understanding a phenomenon occurring in another species (e.g., the 5HT syndrome crosses species). In the case of animal models of human psychopathology, the aim is to develop syndromes that resemble those in humans in order to study selected aspects of psychopathology. The behavioral models are explicitly related to a broader body of theory, as they fulfill a valuable function in forcing the clinician and psychopharmacologists alike to critically examine their assumptions of the manifestations and pathophysiology of depression and bipolar disorders.[25]

To denigrate animal models of psychiatric disorders seems unwise when various examples can be clearly replicated in animals and in humans, which argues for their validity in creating phenotype models of mental disorders. It is clear that the etiology of psychiatric disorders is still in its infancy; however, a healthy skepticism provides a valuable service in pointing out the many shortcomings when animal models are measured against the complexities of human behavior.[29]

6.03.3.1 Animal Models of Depression

Animal models of depression[26–28] and BPAD have proved to be of considerable value in elucidating basic pathophysiological mechanisms and in developing novel treatments. However, the challenges faced by psychopharmacologists in modeling human affective disorders in experimental animals are fraught with difficulties. As new targets emerge through hypothesis-driven research or serendipity, the challenge is to link the mechanism to a clinical complex and heterogeneous disorder. Consequently, much of the animal research today is framed around physiological and neurobiological phenomena that may bear little resemblance to the disease state. However, Matthews *et al.*[29] argue that the poverty of reliable clinical science feedback needs to be addressed first, which would aid future model development. **Table 6** outlines the pros and cons of the classical models of depression (e.g., Porsolt forced swim test (FST), tail suspension test, olfactory bulbectomy, learned helplessness, chronic mild stress, and resident intruder) that have stood the test of time in the development of novel antidepressants.

6.03.3.1.1 Neurogenesis: creation of new neurons critical to antidepressant action

The seminal studies by Duman and co-workers[14] on neurogenesis may help to explain why antidepressants typically take a few weeks to have an effect and may indicate why a rapid-acting antidepressant may not be a viable propositition.[14] These workers created a strain of $5HT_{1A}$ 'knockout' mice that as adults show anxiety-related traits, such as a reluctance to begin eating in a novel environment. While unaffected by chronic treatment with the SSRI fluoxetine, the mice became less anxious after chronic treatment with TCAs that act via another neurotransmitter, NE, suggesting an independent molecular pathway. While chronic fluoxetine treatment doubled the number of new hippocampal neurons in normal mice, it had no effect in the knockout mice. The tricyclic imipramine boosted neurogenesis in both types of mice, indicating that the $5HT_{1A}$ receptor is required for neurogenesis induced by fluoxetine but not imipramine. Chronic treatment with a $5HT_{1A}$-selective drug confirmed that activating the $5HT_{1A}$ receptor is sufficient to spur cell proliferation. An extension of this work using the SSRI fluoxetine in a transgenic cell line from dentate gyrus showed that the SSRI does not affect division of stem-like cells but increases division of amplifying neuroprogenitor cells that results in new neurons in dentate gyrus. This effect was specific for dentate gyrus.[30] Although these findings strengthen the case that neurogenesis contributes to the effects of antidepressants, the authors caution that ultimate proof may require a 'cleaner' method of suppressing this process, such as transgenic techniques that will more precisely target toxins to the hippocampal circuits involved. Ultimately, these results suggest that strategies aimed at stimulating hippocampal neurogenesis provide novel avenues for the treatment of anxiety and depressive disorders. However, the Holy Grail of current treatment strategies is to develop antidepressants with a fast onset of action. In this light, the neurogenesis hypothesis, also reported in antipsychotic drug treatment (*see* 6.02 Schizophrenia), would therefore not support this approach.

6.03.3.1.2 Transgenic and knockdown and out mice: models of psychiatric disease

Transgenic animal technology (knockdown and out), from DNA microinjection to gene targeting and cloning, resulting in 'loss-of-function' mutants can be used to clarify the role of molecules thought to be involved in development and structural maintenance of the nervous system. Transgenic models of human disease are used extensively to assess the validity of therapeutic applications before clinical trials although there is an active debate as to the utility of transgenics in drug discovery.[25]

The most compelling evidence of a link between genetic variation and the role of the SERT in depression and anxiety led to SERT knockout mice that show increased anxiety-like behaviors, reduced aggression, and exaggerated stress responses. Appropriate functioning of SERT and monoamine oxidase A (MAO-A) during early life appear critical to the normal development of these systems. MAO-A and SERT knockout mice mimic in some respects the consequences of reduced genetic expression in humans. MAO-A knockout mice exhibit high levels of aggression, similar to the elevated impulsive aggression seen in humans lacking this gene. SERT knockout mice may thus represent a more exaggerated version of the reduced SERT expression found in certain subjects, and a partial model of the increased vulnerability to anxiety and affective disorders seen in human subjects with the low expressing allele. **Table 5** lists some of the genetically modified mice that have been reported to show depressive or antidepressant-like behavior in simple behavioral tests, e.g., Porsolt FST.

6.03.3.1.3 Ribonucleic acid (RNA) interference (small interfering RNA (siRNA))

RNA interference (RNAi) allows posttranscriptional gene silencing where double-stranded RNA induces degradation of the homologous endogenous transcripts, mimicking the effect of the reduction or loss of gene activity. This technique, therefore, holds promise but is still in its infancy. However, results from recent siRNA-mediated knockdown of the SERT in the adult mouse brain would support the concept.[31] Selectivity and side effects remain an issue.

6.03.4 Clinical Trial Issues

6.03.4.1 Symptoms of Depression and Bipolar Disorder

It is clear that not all individuals with depression or undergoing manic episodes experience every symptom, with severity varying between individuals and over time. With ever increasing numbers of treatment options available for patients with major depression and BPAD, and a growing body of evidence describing their efficacy and safety, clinicians often find it difficult to determine the best and most appropriate evidence-based treatment for each patient. Therefore, European and US consensus guidelines using statistical methods to synthesize and evaluate data from a number of studies (meta-analyses) have been published with recommendations for the treatment of major depression and bipolar disorder.[2–4] These guidelines cover the nature and detection of depressive disorders, acute treatment with antidepressant drugs, choice of drug versus alternative treatment, practical issues in prescribing, management when initial treatment fails, continuation treatment, maintenance treatment to prevent recurrence, and stopping treatment. Based on such guidelines the criteria for diagnosis of depression and BPAD are now better defined.

However, differences do exist between the European and North American guidelines; for example, the US guidelines differ on two main issues for the treatment of bipolar disorder[4,10]:

- antipsychotics are recommended as effective first-line antimanic agents; and
- the lack of emphasis on long-term treatments (mood stabilizers) in acute phase treatment of mania.

6.03.4.2 Clinical Efficacy

Clinical studies with antidepressants invariably involve self-reporting of symptoms using standardized questionnaires including the Hamilton Depression Rating (HAM-D) scale for depression (17 or 21 items of a 23-item scale). A positive NCE treatment usually registers at greater than 50% in the baseline HAM-D score on a 17- or 21-item scale. The Montgomery–Asberg Depression Rating Scale (MADRAS) is a five-item scale that is used to identify anxiolytic-like activity, and is gaining prominence in the US and Europe. Whichever rating scale is used, a wealth of clinical data exist to show that all current antidepressants seem to be effective in 20–70% of those treated, with placebo responses occurring in 30–50% of treated individuals.[32] Thus, the major issue with the current clinical instruments and trial design is that the overall efficacy of antidepressants may be less than 50%. Retrospective analysis of completed antidepressant trials has revealed that four out of six trials do not differentiate from placebo.[32]

The limitations and challenges of antidepressant clinical trials are well documented and relate to several inherent variables; these include the spontaneous remission observed in the length of the normal 6- to 8-week clinical trial and the power of placebo in these studies. The different phenotypes of depression, ranging from mild to severe forms of the illness, add to the 'noise' of the trial. To meet regulatory requirements and approval in the US, Europe, and Japan, large clinical trials are required with at least 2500 patients (at a cost of around $15 000/patient based on 2005 figures). As long as the Regulators' requirements are for a double-blind, placebo-controlled trial with a positive arm, the high cost to risk ratio of such studies is driving many pharmaceutical companies to seek alternative clinical assessment strategies, for example, the seminal work by Kahn and co-workers on 'fixed versus flexible' dose design and to engage in continuous phenotypic refinement of trial populations to determine patient subsets (stratification) that will improve efficacy scale ratings.[35]

Consensus documents agree on the use of DSM-IV-TR criteria (**Table 1**) providing guidelines to improve the management and outcome measures of antidepressant and bipolar depression trials in the future. Efficacy varies little between classes of antidepressants, and the advantages of the newer compounds such as SSRIs are based on their improved side-effect profile rather than their antidepressant efficacy (**Table 7**).[2–5]

6.03.4.3 Treatment Phases and Goals

Clinicians are usually consistent in their agreement with the APA or BAP recommended guidelines for the treatment of depression.[2,3] The goals of the three phases of treatment are defined in **Table 8**. Owing to the widespread use of antidepressants, clinicians now demand that an NCE has an acceptable safety profile (**Table 7**). The incidence of all transient side-effects, such as nausea, headache, dizziness, agitation, sexual dysfunction, and weight gain, should be measured.

6.03.4.4 Combining Data from Placebo- and Non-Placebo-Controlled Studies

The randomized, double-blind, placebo-controlled trial (RCT) remains the gold standard for treatment comparisons. However, some European countries do not permit inclusion of a placebo-controlled group. The decision of whether to combine data from both placebo- and non-placebo-controlled studies (i.e., active-comparator only trials) is debatable.[10]

6.03.4.4.1 Primary endpoint

For the HAM-D, the FDA usually requires a minimal meaningful improvement of a 50% decrease to a score of less than 10. Although a rating score reduction of 50% is accepted by the FDA, clinicians consider remission rate as a more meaningful endpoint. Remission rate is defined as a score of less than 12 according to the MADRAS scale at any point in time during the study. The MADRAS score decreases as depression symptoms improve. MADRAS measures the severity of a number of depressive symptoms including mood and sadness, tension, sleep, appetite, energy, concentration, suicidal ideation, and restlessness and is gaining in use in the European Union.

6.03.4.4.2 Secondary endpoint(s)

The HAM-A (Anxiety) scale and Clinical Global Impressions Scale (CGS-I) are included for hypothesis setting or to support the information obtained from primary endpoints.

6.03.4.5 Clinical Trial Design

6.03.4.5.1 The role of the placebo response and future clinical trial design

Less than 50% of the active treatment arms showed a significant difference from placebo and the magnitude of the change in the placebo group had a greater influence on the drug–placebo difference than the change in the drug group. The proportion of trials in which antidepressants were shown to give a significantly better HAM-D score than placebo ($P<0.01$) was 59.6% (34/57 trials) for flexible dose trials and 31.4% (11/35) for fixed dose trials. Several researchers have questioned such outcome measures and have noted that unidimensional subscales of the HAM-D are more sensitive to drug–placebo differences than is the total HAM-D score.[32]

Treatment effects are often evaluated by comparing change over time. However, valid analyses of longitudinal data can be problematic, particularly if some data are missing for reasons related to outcome measures. Last Observation Carried Forward (LOCF) protocols are a common method of handling missing data because of their simplicity and conservative nature. Recent advances in statistical theory and their implementation have made methods with far less restrictive assumptions than LOCF readily accessible resulting in the use of likelihood-based repeated measures

Table 7 Side effect profiles and lethality in overdose of commonly used antidepressant drugs

Drug	*Side effect*[a]										
	Action	*Anticholinergic*	*Sedation*	*Insomnia/ agitation*	*Postural hypotension*	*Nausea/gastro intestinal*	*Sexual dysfunction*	*Weight gain*	*Specific adverse effects*	*Inhibition of hepatic enzymes*	*Lethality in overdose*
Tricyclic antidepressants											
Amitriptyline, dothiepin	NRI > SRI	+ +	+ +	–	+ +	–	+	+ +		+ +	High
Clomipramine	SRI + NRI	+ +	+ +	+	+ +	+	+ +	+		+ +	Moderate
Imipramine	NRI > SRI	+ +	+	+	+ +	–	+	+		+ +	High
Desipramine, nortriptyline	NRI	+	+	+	+	–	+	–		+ +	High
Lofepramine	NRI	+	–	+	+	–	?	–		+	Low
Selective serotonin reuptake inhibitors											
Citalopram, sertraline	SRI	–	–	+	–	+ +	+ +	–		–	Low
Fluoxetine, fluvoxamine, paroxetine	SRI	–	–	+	–	+ +	+ +	–		+ +	Low
Escitalopram	SRI	–	–	+	–	+ +	+ +	–		–	Low
Other reuptake inhibitors											
Maprotiline	NRI	+ +	+ +	–	–	–	+	+ +	Increase seizure potential	?	High
Venlafaxine	SRI > NRI	–	–	+	–	+ +	+ +	–		–	Low
Reboxetine	NRI	+	–	–	–	–	+	–	Hypertension	–	Low

Receptor antagonists											
Trazodone	$5HT_2 + \alpha_1 > SRI$	–	+ +	–	+ +	–	–	+	Priapism	?	Low
Nefazodone	$5HT_2 > SRI$	+	+	–	+	+	–	+ +		+ +	Low
Miansetin	$5HT_2 + \alpha_1 + \alpha_2$	+	+ +	–	–	–	–	–	Blood dyscrasia	?	Low
Mirtazapine	$5HT_2 +$ $5HT_3 + \alpha_2$	–	+ +	–	–	–	–	+ +		–	Low
Monoamine oxidase inhibitors											
Phenelzine, tranylcypromine, isocarboxazide	Irreversible	+	+	+ +	+ +	+	+ +	+ +	Hypertensive crisis with sympatheto-mimetics	?	High
Moclobemide	RIMA	–	–	+	–	+	–	–		–	Low

NRI, noradrenaline reuptake inhibitor; SRI, serotonin reuptake inhibitor, $5HT_2$, $5HT_2$ antagonist; $5HT_3$, $5HT_3$, antagonist; α_1,/α_2, α_1 antagonist/α_2 antagonist; RIMA, reversible inhibitor of monoamine oxidase-A; + +, relatively common or strong; +, may occur or moderately strong; – absent or rare/weak; ?, unknown/insufficient information.

Reprinted by permission of Sage Publications Ltd from Anderson, I. M. ; Nutt, D. J.; Deakin, J. F. W. *J. Psychopharmacol.* **2000**, *14*, 3–20.

[a] These refer to symptoms commonly caused by muscarinic receptor blockade including dry mouth, sweating, blurred vision, constipation, and urinary retention; however, the occurrence of one or more of these symptoms may be caused by other mechanisms and does not necessarily imply that the drug binds to muscarinic receptors.The side-effect profiles given are not comprehensive and are for an approximate comparison only. Details of drugs used and potential cautions and interactions should be looked up in a reference book such as the latest US or British National Formulary.

Table 8 The three treatment phases for antidepressants

Phase	*Length*	*Treatment goal*
Acute	6–12 weeks	Achieve remission/stabilization
Countinuation	4–12 months	Prevent relapse
Maintenance	Varies	Protect against recurrence

Remission = a virtual elimination of symptoms; score of 7 or less on 17-item HAM-D scale. Relapse = re-emergence of significant depressive symptoms.

approaches, which have a number of theoretical and practical advantages for analysis of longitudinal data and dropouts. One method that is gaining acceptance is mixed model repeated measures (MMRM) and this has been extensively studied in the context of neuropsychiatry clinical trials.[33] Data from studies suggest that MMRM yields 75% empirical power compared with 50% for LOCF. MMRM is simple to use, easy to implement, and to specify *a priori*. It is also more likely than LOCF to give adequate control of type I (false-positive) and type II (false-negative) errors. In other words, the use of either MMRM or LOCF will lead to the same conclusions but MMRM is likely to yield fewer mis-steps along the way.[33]

6.03.4.6 Special Issues

To date, no published comparative study of newer antidepressants has enrolled a sufficiently large group of patients to have the power to reliably detect the differences between two effective treatments according to a recent critique.[32] One possible exception to this is the NIMH-sponsored Sequenced Treatment Alternatives to Relieve Depression (STAR*D) project, which will enroll 5000 patients in a comparative treatment trial.[34] Unfortunately, owing to the cost and resources required to conduct studies of sufficient size, the average RCT evaluating antidepressant effects is woefully underpowered. For example, in a recent review of 186 RCTs examining the efficacy and tolerability of amitriptyline in comparison with other antidepressants, the average number of patients per treatment group was 40.[35] In an analysis of pivotal studies (i.e., well-designed, well-controlled studies on which the FDA bases decisions about the efficacy of NCEs) for seven newer antidepressants, only 65–75 patients were included per study arm.[32] Thus, the average study comparing two effective antidepressants would have less than 20% power to find a real, albeit modest (i.e., 10%), difference in response rates. Put another way, the likelihood of a false-negative finding (i.e., a type II error) would be four times greater than the chance of observing a statistically significant difference.

It is apparent that specific treatment effects have declined in recent decades. This may be due to selection bias at work that differs from that of a generation ago. The sample size, the number of centers, treatment arms, dosing (e.g., flexible dosing versus fixed), and different expectation biases all potentially influence results. For example, in the 1960s, more trials evaluated hospitalized patients who are generally less responsive to placebo and who appear to have a more robust response to antidepressants.[32] Beyond the issue of inpatient/outpatient status, older studies were more likely to enroll patients with BPAD, psychosis, and recurrent melancholic subtypes of depression. In addition, the efficacy of antidepressant interventions was less well understood then (which may have lowered expectations of the patient or clinician) and fewer potential participants had ever received an effective course of pharmacotherapy.

Contemporary trials, on the other hand, may be enrolling a somewhat different population: highly selected ambulatory patients who are often contacted through the mass media. These subjects may be less severely depressed and are rarely treatment naive.[32,36] Attempts to lessen these problems by restricting enrollment to patients with relatively high levels of pretreatment severity have often, in fact, accentuated them by inadvertently causing an inflation of entry depression scores.[36] Many clinical trials use entry criteria based in part on a minimum score for the same instrument used to evaluate efficacy. Investigators may be motivated, consciously or not, to increase baseline scores slightly in order to enter subjects into the trial. Such scores may then decrease by that same amount once the subject is entered, thus contributing to what appears to be a placebo effect (if not analyzed appropriately).[32]

Another factor influencing the apparent effectiveness of antidepressants is the so-called 'file-drawer effect': the bias introduced by the tendency to publish positive but not negative studies. This bias is most evident when comparing reviews of published studies with reports that are based on data sets that have been submitted to the FDA for regulatory review.[32] For example, on the basis of studies conducted for the registration of new antidepressants from fluoxetine to citalopram the effects of antidepressants appear to be only about half the size (relative to placebo) once the unpublished studies are taken into account.

6.03.4.6.1 Suicidality in children and adolescents being treated with antidepressant medications

In 2004, the FDA directed manufacturers of all antidepressant drugs to revise the labeling for their products to include a boxed warning and expanded warning statements that alert healthcare providers to an increased risk of suicidality (suicidal thinking and behavior) in children and adolescents being treated with these agents, and to include additional information about the results of pediatric studies.[36] The FDA also informed manufacturers that it had determined that a Patient Medication Guide (MedGuide) to be given to patients receiving the drugs to advise them of the risk and precautions that can be taken, is appropriate for these drug products. These labeling changes are consistent with the recommendations made to the agency at a joint meeting of the Psychopharmacologic Drugs Advisory Committee and the Pediatric Drugs Advisory Committee in September 2004.[36] The drugs that are the focus of this new labeling language are all those included in the general class of antidepressants and are listed on the FDA website.

The suicidality risk associated with these agents was identified in a combined analysis of short-term (up to 4 months) placebo-controlled trials of nine antidepressant drugs, including the SSRIs among others, in children and adolescents with MDD, obsessive compulsive disorder (OCD), or other psychiatric disorders. A total of 24 trials involving over 4400 patients was included.[32] The analysis showed a greater risk of suicidality during the first few months of treatment in those receiving antidepressants. The average risk of such an event on drug was 4% – twice the risk on placebo (2%). No suicides occurred in these trials. Based on these data, the FDA has determined that the following points are appropriate for inclusion in the boxed warning:

- Antidepressants increase the risk of suicidal thinking and behavior (suicidality) in children and adolescents with MDD and other psychiatric disorders.
- Anyone considering the use of an antidepressant in a child or adolescent for any clinical use must balance the risk of increased suicidality with the clinical need.
- Patients who are started on therapy should be observed closely for clinical worsening, suicidality, or unusual changes in behavior.
- Families and caregivers should be advised to closely observe the patient and to communicate with the prescriber.
- A statement regarding whether the particular drug is approved for any pediatric indication(s) and, if so, which one(s).

Among the antidepressants, only fluoxetine is approved for use in treating MDD in pediatric patients. Pediatric patients being treated with antidepressants for any indication should be closely observed for clinical worsening, as well as agitation, irritability, suicidality, and unusual changes in behavior, especially during the initial months of a course of drug therapy or at times of dose changes, either increases or decreases. This monitoring should include daily observation by families and caregivers and frequent contact with the physician. It is also recommended that prescriptions for antidepressants be written for the smallest quantity of tablets consistent with good patient management, in order to reduce the risk of overdose.

In 2004, an American Medical Association (AMA) report noted that black box warnings on antidepressants should not be interpreted in such a way that would decrease their use in children and adolescents who would benefit from the drugs.[32] In January 2005, the FDA added black box warnings to the labels of antidepressants concerning the risk of suicidal thinking and behavior in children and adults. In addition to the boxed warning, FDA and European MedGuides on the risk of suicidality in children and adolescents are being prepared for all antidepressants to provide information directly to patients, their families, and caregivers.

6.03.4.7 New Methods of Conducting and Evaluating Clinical Trails

One particular initiative designed to evaluate psychiatric medicines is the New Clinical Drug Evaluation Unit Program (NCDEU) funded by the NIMH, which comprises over 1000 clinicians and industry and regulatory personnel. The NCDEU recently addressed the question of whether clinical trials of antidepressants reflect drug potential, and several groups involved in the initiative focused on different aspects of trial design, e.g., heightened placebo effect from such factors as a high dropout rate, poor site selection or poor protocol design, and their effect on masking the potential of active drugs. The NCDEU team reviewed 37 clinical trials, all of which had used the HAM-D. The HAM-D 'depressed mood' items according to DSM-IV-TR/ICD-10ACC are defined by the following core criteria[37]:

- persistent anxious or 'empty' mood;
- feelings of hopelessness, pessimism;
- feelings of guilt, worthlessness, helplessness;

- loss of interest or pleasure in hobbies and activities that were once enjoyed, including sex;
- decreased energy, fatigue, being 'slowed down;'
- difficulty concentrating, remembering, making decisions;
- insomnia, early morning awakening, or oversleeping; and
- appetite and/or weight loss or overeating and weight gain.

Also included in this evaluation were the CGS-I severity and CGS-I improvement scores to measure improvement from baseline at the end of week 6 and week 8 of the acute NCE treatment phase. In 13 (35%) of the trials, efficacy was not demonstrated on any of the four measures, and certain SSRIs achieved only 50% at best. The importance of controlling the confounding variables in the development of a new antidepressant compound is clear from the NCDEU findings.[37] In another study, the NCDEU team found that spontaneous improvements of depressive symptoms contribute to the placebo effect. Retrospective analysis showed that in placebo-controlled depression trials, the placebo effect is more prominent during the single-blind placebo run-in phase. The difference was unlikely to be due to different rates of spontaneously improved depression between the two trial phases. In addition to validating an NCE, comparisons were made between fixed-dose clinical studies to establish a minimal effective dose of the new agent and discourage subsequent use of excessively high doses with associated heightened side effects. It is argued that variable dose studies are more cost effective when attempting to demonstrate efficacy, whereas fixed-dose studies require larger sample size and subject the patients to either too low or too high a dose of a novel drug.

Another variable that demands attention is the adequacy of the statistical methods used to account for subject dropouts in clinical studies. Some forms of statistical analysis can make untenable assumptions, which are confounded by trial design. The introduction of computerized survival analysis at any single time point helps to distinguish between patients who remit at different times during a trial and those who remit by a particular time point. Survival analysis is now being used as a research tool that enables the statistical power of a study to be planned with more precision and the size of the treatment effect to be determined more accurately.[34]

6.03.4.7.1 Rating scales

One of the fundamental issues related to gauging the effectiveness of antidepressants is the choice of a validated rating scale, which is often arbitrary, and that depression symptom scales vary in their psychometric properties, conceptual focus, response burden, and discriminating power. In a study reported as part of the NCDEU sponsored program at Duke University, a community control comparison of four scales in 688 patients was conducted (559 patients with major depression and 129 normal volunteers). The Duke study employed different assessment scales including the MADRAS and the Center for Epidemiologic Studies Depression Scale (CESID), as well as the Carroll Depression Scale (CDS, 52-item scale) and Brief Carroll Depression Scale (BCDS, 12-item scale). The findings from this study using meta-analysis found that the four scales intercorrelated highly significantly, diagnostic specificity was high, and mean depression scores for patients were significantly greater than those of the normal subjects.[39] The NCDEU concluded that while validated depression scales effectively separated clinically depressed patients from the community control subjects, they vary in the cognitive burden placed on the patient and the effectiveness measure appears less robust when used in primary care.

Assessing the speed of onset of a new medication is being addressed on several fronts, both from the pharmacological standpoint and in terms of techniques and clinical instruments needed to record the speed of onset of novel antidepressant agents. One enabling technology that is becoming commonplace in clinical studies is the interactive voice response (IVR) technology over the phone line. IVR technology enables remote evaluation of treatment response 24 h a day from any touch phone.[39] Such accessibility allows a more frequent monitoring of speed of onset of medication in antidepressant studies. IVR represents a new era of clinical instruments that are currently under investigation to aid the design and outcome measures of clinical trials in depression.[34]

6.03.4.7.2 Bridging studies

In a 'bridging study,' dosage is optimized early in development by determining the maximum tolerated dose of a compound in patients. Consecutive panels of patients each receive higher doses of study drug until a minimum intolerated dose is reached. The dose immediately below this one is then considered the maximum tolerated dose. Careful subject selection, adequate facilities, and highly qualified, experienced personnel are critical to the successful implementation of a bridging study. Correctly done, bridging can streamline the overall drug development process, while making the Phase II and III trials safer for patients. The International Conference on Harmonization (ICH) E5 (1998) guidelines were developed to provide a general framework for evaluating the potential impact of ethnic factors on the acceptability of foreign clinical data to facilitate global drug development and registration of NCEs, and to reduce the number of clinical trials for international approval. An essential prerequisite for the acceptance of foreign data is a Complete Clinical Data Package including foreign data that have to meet all regional regulatory requirements. Furthermore, sponsors were requested to show whether the 'foreign clinical data' could be appropriately extrapolated to the new geographical region. The assessment of medicines sensitivity to ethnic factors has to be done according to given intrinsic as well as extrinsic ethnic factors. Supplemental 'bridging studies' may become necessary to provide clinical or PD as well as PK data allowing an extrapolation to the population of the new region. Extrapolation via a bridging study can avoid the need to conduct additional, expensive clinical trials in the new region, and can facilitate access to superior treatment to patients in a timely fashion. The technical challenge of a bridging study is to demonstrate 'similarity' of profile (statistics, sample size, etc.) of an NCE.

While the requirements of major regions such as North America and Europe are met largely by a global clinical development program within multinational Phase II/III trials, the acceptability of foreign data for registration in Japan and other Asian countries is still a substantial challenge. During the 5-year period following the adoption of ICH E5 by the Prescription Drug Marketing Act (PDMA), it became clear that Complete Clinical Data Packages were not accepted without a supplemental bridging study allowing sufficient extrapolation to the Japanese patient population. In the majority of cases, extrinsic ethnic factors, such as different medical practice in the new region, were the source of concern in terms of affecting the efficacy and safety of new medicines. Data on clinical efficacy addressing the dose–response relationship in the Japanese population are essential for sufficient extrapolation. Stand-alone PK studies were not considered valid for the bridging of foreign data. In the case of extrapolation of foreign data from a clinical Phase III program to the Japanese population, a Phase II dose–response study accompanied by a set of domestic single and repeated dose Phase I trials and/or an early Phase II study is requested to provide the dose–response data in the Japanese population. Sponsors should consult with the Japanese authorities early on in the development process to define the necessity, topic, and design of bridging studies.

6.03.4.7.3 Study trial length

Although there is much controversy in the clinical community, a review of the six most commonly prescribed antidepressants found that efficacy studies were virtually always short-term, rarely exceeding 12 weeks of treatment.[3] Hence, it can be argued that in the real world, the clinical value derived from such studies remains relatively restricted. Thus, only 18% of the observed changes during these short-term studies in patients with nonsevere forms of depression could be attributed to the active effects of medication. Active medication and placebo both shared 82% of the maximum clinical changes observed, leading to the conclusion that, assuming the effects of the antidepressant medication and placebo are additives, the effects of the medication, even in this clinically favorable group (very few or no psychotic or suicidal participants, very little comorbidity), is extremely modest, perhaps negligible, and potentially of little clinical significance.

In summary, such studies strike at the core of the understanding of neuropsychopharmacological drug development. Indeed, there has been considerable debate as to the nature of these analyses and how they can be best interpreted.[33] However, the single most comforting suggestion for psychopharmacology is that the powerful antidepressant effects of these drugs are actually masked by current clinical trial designs and that the research strategy for the evaluation of novel psychotropic agents, according to Matthews and colleagues, needs re-evaluation.[29]

6.03.5 Current Treatment

6.03.5.1 Drug Classification

As reviewed in Section 6.03.2, the majority of antidepressants in clinical use today act by enhancing the neurotransmission of monoamines, serotonin (5HT), NE, DA, or all three, either directly or indirectly.[38] This is done by either blocking reuptake via monoamine transporters (**Table 9**) or blocking the metabolism of monoamines by inhibiting the major catalytic degradation enzymes, monoamine oxidase and catechol-*O*-methyltransferase (COMT). Other modes of action include direct or indirect modulation of receptors or signal transduction mechanism. Antidepressants that modulate neurotransmission of monoamine uptake inhibition are divided into those that are nonselective (e.g., TCAs, with dual action), SSRIs, and selective norepinephrine reuptake inhibitors (SNRIs). An additional class of antidepressant is the polypharmic heterocyclics acting at both reuptake sites and receptors. Other neuropeptide agents and transcription factors with potentially novel mechanisms of action, albeit acting through a final common monoaminergic pathway have been investigated.

Table 9 Antidepressants: inhibition of human serotonin (SERT), norepinephrine (NET) and dopamine (DAT) transporters[a]

Generic name	*Human SERT, K_d (nM)*	*Human NET, K_a (nM)*	*Human DAT, K_d (nM)*	*Selectivity SERT versus NET*
Amitriptyline	4.3	35	3250	8
Amoxepine	58	16	4310	0.3
Bupropion	9100	52000	520	5.7
Citalopram	1.2	4070	28 100	3500
Escitalopram[b]	1.1[b]	7841[b]	27 410[b]	7128[b]
R-citalopram[b]	36.5[b]	12 270[b]	18 720[b]	336.1[b]
Clomipramine	0.3	38	2190	130
Desipramine	17.6	0.5	3190	0.05
Dothiepin	8.6	46	5310	5.3
Daxepine	68	29.5	12 100	0.4
Fluoxetine	0.8	240	3600	300
Fluvoxamine	2.2	1300	9200	580
Imipramine	1.4	37	8500	27
Lofepramine	70	5.4	18 000	0.05
Maprotiline	5800	11.1	1000	0.002
Mirtazapine	$>$100 000	4600	$>$100 000	–
Nefazodone	200	360	360	1.8
Nortriptyline	18	4.4	1140	0.24
Paroxetine	0.13	40	490	300
Protriptyline	19.6	1.4	2100	0.07
Reboxetine	129	1.1	$>$10 000	0.002
Sertraline	0.29	420	25	1400
Trazodone	160	8500	7400	53
Trimipramine	149	2450	780	16
Venlafaxine	8.9	1060	9300	120

The results are expressed as equilibrium dissociation constants (K_d) in nM, using ^{3} H-imipramine binding to human serotonin transporter, [^{3}H]-nisoxetine binding to human norepinephrine transporter, and ^{3} H-WIN or [^{125}]RTI-55[a] binding to human dopamine transporter, or for reboxetine, [^{3}H]-citalopram binding to human serotonin transporter and ^{3}H-nisoxetine binding to the human norepinephrine transporter.

[a] Data is reproduced and modified with permission from Burger's *Medicinal Chemistry and Drug Discovery*, 6th ed.: Vol. 6, *Nervous System Agents*; Abraham, D. J., Ed. J. Wiley & Sons, Inc.: New York, **2003**, p 500.[39]

[b] Data from Owens, M. J.; Knight, D. L.; Nemeroff, C. B. *Biol. Psychiatry* **2001**, *50*, 345–350.[40]

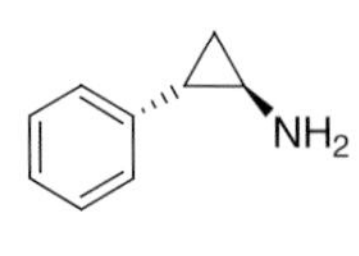

Tranylcypromine **8** 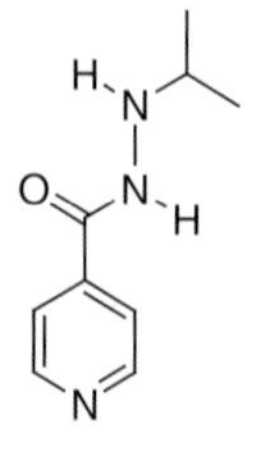 Iproniazid **9** 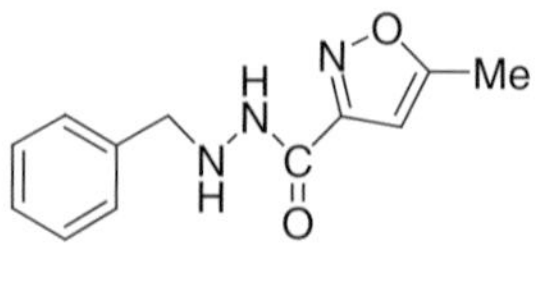 Isocarboxazide **10** 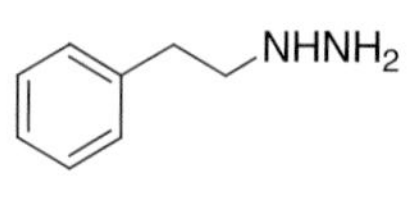 Phenelzine **11**

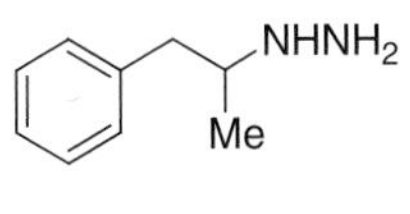

Pheniprazine **12** 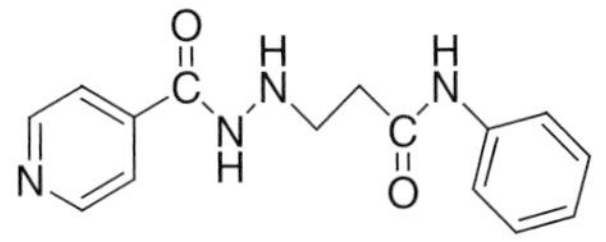 Nialamide **13** 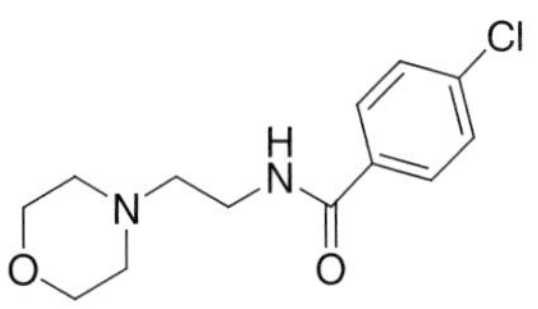 Moclobemide **14**

6.03.5.2 Antidepressants

6.03.5.2.1 Monoamine oxidase inhibitors

First-generation MAOIs like tranylcypromine (**8**), iproniazid (**9**), isocarboxazide (**10**), phenelzine (**11**), pheniprazine (**12**), and nialamide (**13**) revolutionized the treatment of depression.[22] These are generally less utilized in current clinical practice due to their poor side-effect profile and potentially dangerous interactions with other drugs and foods, the latter reflecting the cheese, or tyramine, effect. The current leading MAOIs are tranylcypromine (**8**), phenelzine (**11**), and moclobemide (**14**).

The first generations of MAOI antidepressants were hydrazine derivatives, e.g., phenelzine and isocarboxazide, which are probably converted into hydrazine to produce long-lasting inhibition of MAO. Tranylcypromine is essentially a cyclized amphetamine without the covalent bond. Selegiline (**15**), a propargylamine MAOI, contains a reactive acetylenic bond that interacts irreversibly with the flavin cofactor of MAO resulting in prolonged MAOI activity. Selegiline is still used in clinical practice today, mainly in Parkinson's disease. However, a new patch delivery formulation of selegiline that is proposed to overcome the adverse events associated with MAOIs is in Phase III studies for major depression. Rasagiline (**16**) is currently marketed for Parkinson's disease in Europe.

The major challenge over the last 60 years has been to synthesize short-acting, reversible MAOIs, of which moclobemide (**14**), rasagiline (**16**), toloxatone (**17**), befloxatone (**18**), brofaromine (**19**), secloramine (**20**), and cimoxatone (**21**) are representatives of third-generation MAOIs.

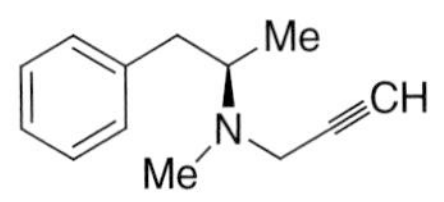

Selegiline **15** 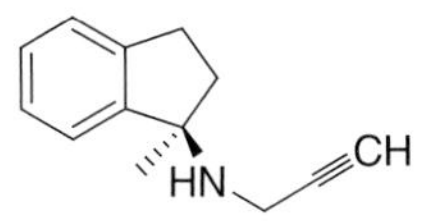 Rasagiline **16** 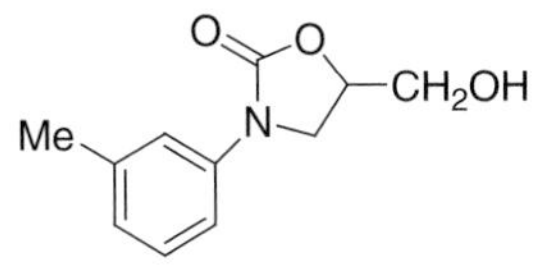 Toloxatone **17**

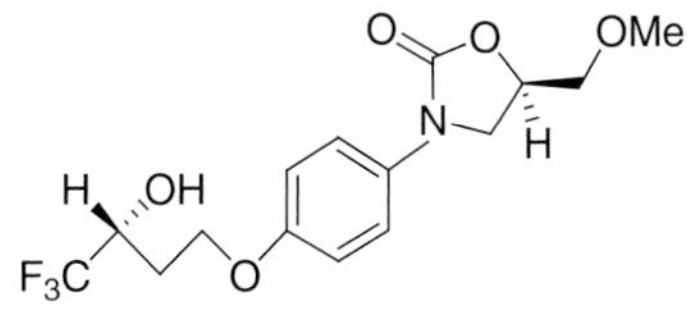

Befloxatone **18** 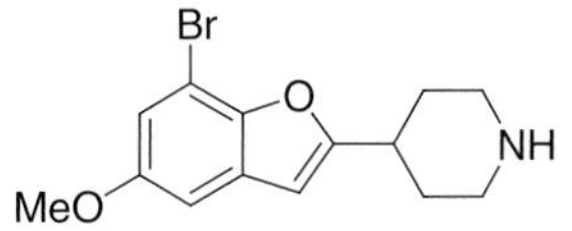 Brofaromine **19** 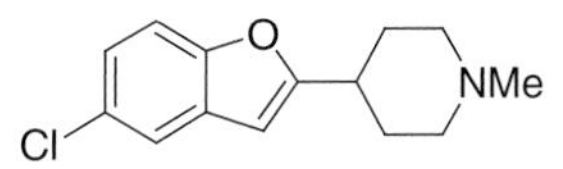 Sercloramine **20**

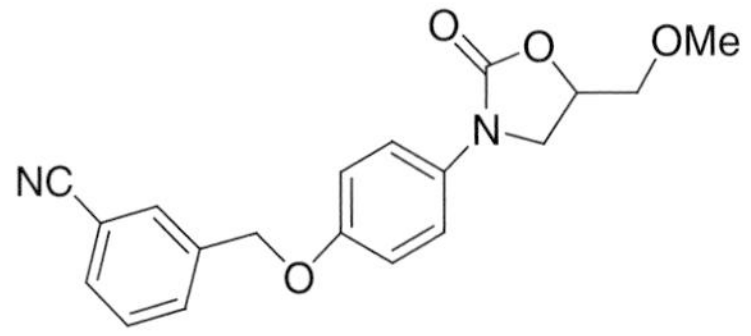

Cimoxatone **21** 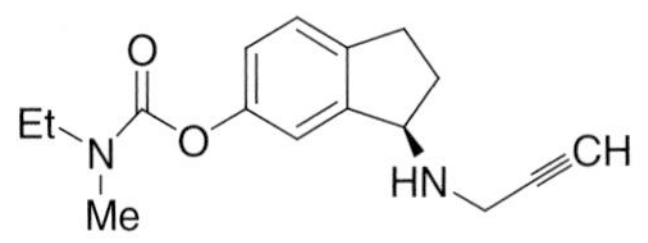 Ladostigil **22**

Isozyme-selective and partly reversible MAOIs have been developed over the last two decades and are described as reversible inhibitors of monoamine oxidase type A (RIMAs). In contrast to the irreversible nonselective inhibitors, selective RIMAs are isozyme selective – MAO-A or MAO-B. MAO-A inhibition results in increases in NE, 5HT, and DA in the synaptic cleft, while selective MAO-B inhibitors increase only DA. The molecular action mechanism of the

RIMAs is the same as in the nonselective NCEs: increase of monoamines, near to the receptor, leads, after a number of intermediate steps, to activation of functional proteins in the cell. The control of therapy with MAO-A inhibitors is easier, because of their reversibility. Of the current RIMAs, only moclobemide has been shown to have mild to moderate effects in depression.[41] The development of RIMAs, as antidepressants and with possible anti-Parkinson activity, has had limited success to date, and is constrained by the tyramine effect, since the amine can displace the inhibitor from its binding site on the enzyme. Therefore, the Holy Grail of RIMAs would be to inhibit both forms of the enzymes to get the full functional activities of the amine neurotransmitters, without inducing a tyramine effect or 'cheese reaction.' This was not possible until recently, with the development of the novel cholinesterase-brain-selective MAO-A/B inhibitor ladostigil (**22**), a carbamate derivative of the irreversible MAO-B inhibitor rasagiline. Ladostigil (**22**) is a brain-selective MAO-A and MAO-B inhibitor; even after 2 months of daily administration it has little or no effect on the enzyme in the intestinal tract and liver.[43] Pharmacologically, it has a limited tyramine potentiation effect, similar to moclobemide, but it has the antidepressant, anti-Parkinson, and anti-Alzheimer activities of an MAO-A/B inhibitor in the respective animal models used to develop such NCEs. No clinical studies have been reported with this novel compound.

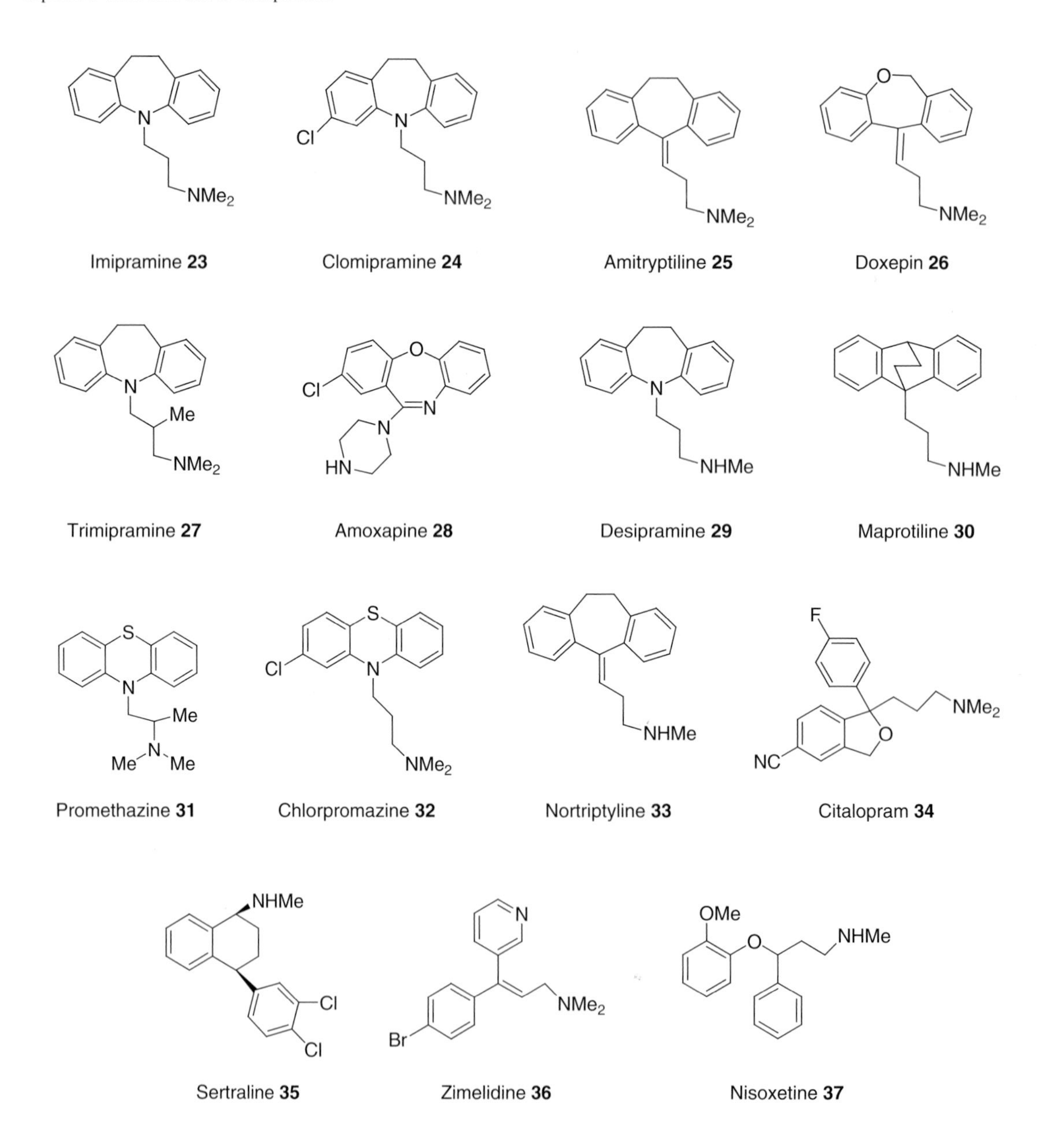

6.03.5.2.2 Tricyclic antidepressants

TCAs were introduced in the 1950s and became the gold standard treatment for depression before the launch of the first SSRI, fluoxetine, in 1987.[42] All current antidepressants rely upon the principle of enhancing monoamine neurotransmission interfering with the presynaptic transporter that reimports the neurotransmitter from the synaptic cleft once released from presynaptic nerve terminals. The classic tricyclic compounds were much less specific, representing a shotgun approach to several neurotransmitter receptors and transporters. Now they are increasingly used as a third-line therapy as the side-effect profile of second-generation SSRI antidepressants is far superior. The leading TCAs are imipramine (**23**), clomipramine (**24**), and amitriptyline (**25**), along with a number of other marketed products.

The TCAs such as imipramine exert their therapeutic actions by inhibition of both 5HT and NE reuptake/transporter sites (**Table 9**). Unfortunately, their polypharmic target profiles, which extend to antagonism of α_1-adrenoceptors, muscarinic receptors, and histamine receptors, as well as cardiac ion channels, NMDA and SK1, underlies their poor tolerance. Based on their complex pharmacology this class exhibits a variety of adverse side effects (see **Table 7**). Although some of these side effects are not of great concern, others are life threatening and only have a narrow therapeutic safety index (**Table 7**). The most serious side effects of the classical tertiary (e.g., imipramine (**23**), clomipramine (**24**), amitriptyline (**25**), doxepin (**26**), trimipramine (**27**)) and secondary amine (e.g., amoxapine (**28**), desipramine (**29**), maprotiline (**30**)) tricyclics are attributed to direct quinidine-like actions on the heart, interfering with normal conduction and prolongation of the QRS or QT interval with associated cardiac arrhythmia and arrest. Other toxic effects include respiratory depression, delirium, seizures, shock, and coma. Anticholinergic side effects including dry mouth, blurred vision, urinary retention, and sinus tachycardia are perhaps the most commonly reported (**Table 7**). The adverse event profile of tricyclics is also associated with a greater risk due to their interaction with cytochrome P450 (CYP) isoenzymes, in particular CYP2D6, which is a highly polymorphic gene. A considerable minority of the population has CYP2D6 polymorphic variations, gene deletion, or gene duplication. CYP2D6 polymorphism results in ultrarapid or ultraslow metabolism, leading to varying degrees of drug bioavailability according to genotype, which can lead to increased bioavailability and severe toxicity in some cases.

All the 'classical' TCAs have a basic three-ring pharmacophore. The therapeutic and commercial success of *N*-aminoalkylphenothiazines such as promethazine (**31**), promazine, and chlorpromazine (**32**) initiated an enormous effort in the molecular modification of the polycyclic phenothiazine ring structure and its *N*-aminoalkyl side chain. The substitution of the sulfur bridge of the phenothiazine ring of promethazine with an ethylene bridge resulted in imipramine, the first clinically useful TCA. It did not take long for medicinal chemists focusing on the diamine structure to substitute the additional $N{-}CH_2$ group in imipramine for a $C{=}CH$ group in amitriptyline. Following earlier dimethylamine work, monomethyl amine derivatives were then synthesized, from which desipramine (**29**) and nortriptyline (**33**) evolved.

6.03.5.2.3 Selective serotonin reuptake inhibitors (SSRIs)

While the primary mode of action of TCAs was thought to be inhibition of NE reuptake, a reassessment of the actions of the diverse antihistamines on the reuptake of various biogenic amines, especially 5HT, led to the hypothesis that an increase in brain noradrenergic function caused the energizing and motor stimulating effects of the TCAs, but that an increase in 5HT function was responsible for their mood-elevating effects.[42] Structural analogs of diphenhydramine were sought as novel antidepressants. The phenoxyphenylpropylamine pharmacophore was used to identify fluoxetine (**1**), the first SSRI.[33] The phenomenal success of fluoxetine (**1**) as an antidepressant led to the identification of other SSRIs, e.g., paroxetine (**2**), citalopram (**34**), fluvoxamine (**3**), and sertraline (**35**).

Most SSRIs are aryl or aryloxalkylamines. Within this chemical genus, fluoxetine (**1**), citalopram (**34**), and zimelidine (**36**) are racemates, and sertraline and paroxetine are separate enantiomers. The (*S*)-enantiomers of citalopram (escitalopram) and of fluoxetine (norfluoxetine) are relatively more potent at SERT (**Table 9**). Structure–activity relationships (SARs) are not well established for SSRIs, although the *para*-CF_3 substituent of fluoxetine is critical for potency as removal and substitution at the *ortho*-position of a methoxy group yields nisoxetine (**37**), a highly selective NE uptake (NET) inhibitor.

The potency for SERT inhibition varies amongst this group, as does the selectivity for 5HT relative to NE and DA reuptake inhibition. The relative potency of sertraline for DA reuptake inhibition differentiates it from other SSRIs (**Table 9**). Affinity for neurotransmitter receptors such as $sigma_1$, muscarinic, and $5HT_{2C}$ also differs widely (**Table 7**). Furthermore, the inhibition of nitric oxide synthetase by paroxetine, and possibly other SSRIs, may have significant pharmacodynamic effects. Fluoxetine has a long-acting and pharmacologically active metabolite, with high affinity for the

$5HT_{2C}$ receptor. Other important clinical differences among the SSRIs include differences in their half-lives, linear versus nonlinear PK, effect of patient age on their clearance, and their potential to inhibit drug metabolizing CYP isoenzymes. These differences underly the increasingly apparent important clinical differences among the SSRIs (**Table 8**).

Of the very limited comparative clinical data available, a meta-analysis of 20 short-term comparative studies of five SSRIs (citalopram, fluoxetine, fluvoxamine, paroxetine, and sertraline) showed no difference in efficacy between individual compounds but a slower onset of action of fluoxetine. The most common adverse reactions to the SSRIs, according to FDA and MHRA websites, were gastrointestinal (especially nausea) and neuropsychiatric (particularly headache, tremor, discontinuation reactions, and sexual dysfunction; **Table 7**).

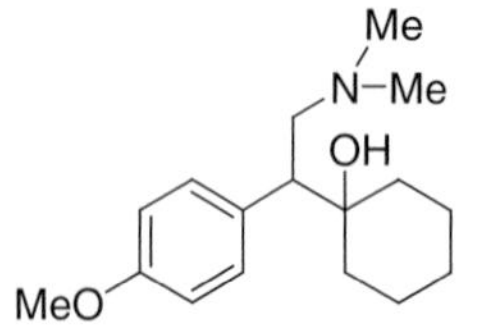

Venlafaxine **38**

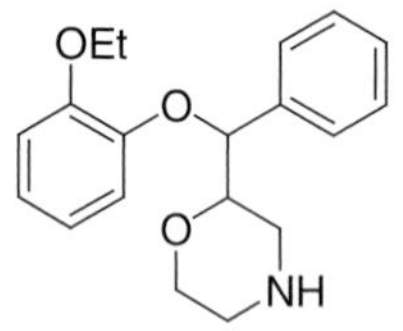

Reboxetine **39**

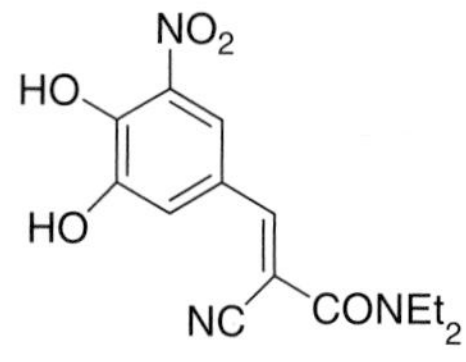

Entacapone **40**

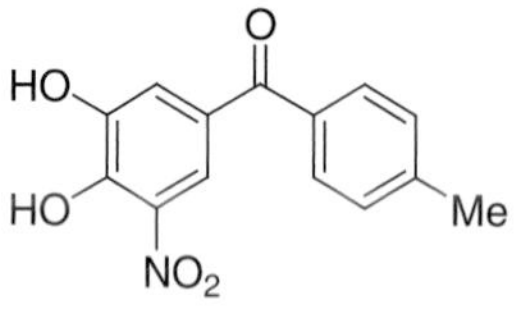

Tolcapone **41**

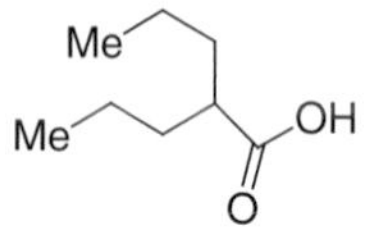

Valproate **42**

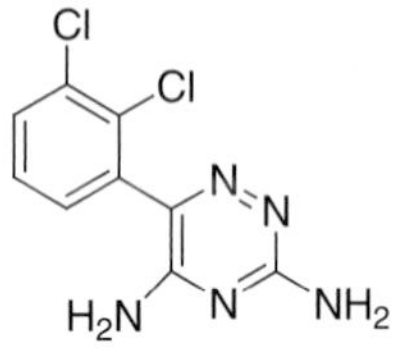

Lamotrigine **43**

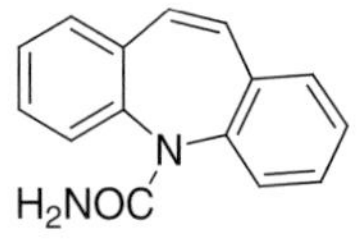

Carbamazepine **44**

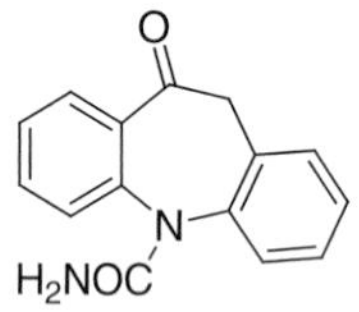

Oxcarbazepine **45**

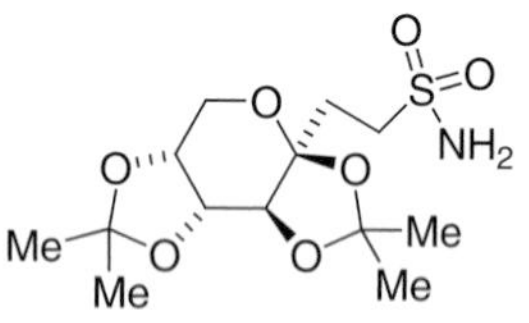

Topiramate **46**

6.03.5.2.4 Selective norepinephrine reuptake inhibitors

SNRIs are a class of antidepressants characterized by a mixed action on both major monoamines of depression: NE and serotonin. In essence, SNRIs are improved TCAs with less off-target activity, e.g., muscarinic, histaminic and α_1-adrenergic receptors, and MAOI. The combination of inhibition of 5HT and NE uptake confers a profile of effectiveness comparable to TCAs and is reported to be higher than SSRIs, especially in severe depression. SNRIs are purported to be better tolerated than TCAs and more similar to SSRIs without the associated sexual dysfunction seen with the latter. Venlafaxine (**38**) and milnacipran (**4**) have been approved so far, and several others are in development. They are active on depressive symptoms, as well as on certain comorbid symptoms (anxiety, sleep disorders) frequently associated with depression. SNRIs appear to have an improved rate of response and a significant rate of remission, decreasing the risk of relapse and recurrence in the medium and long term and address two of the current goals of antidepressant treatment. Reboxetine (**39**) is another SNRI, which was approved in Europe for the treatment of major depression but is not available in the USA.

6.03.5.2.5 Catechol-*O*-methyltransferase inhibitors

An increase in the functional monoamines NE, DA, and 5HT can precipitate mania or rapid-cycling in an estimated 20–30% of affectively ill patients. A strong association between velo-cardio-facial syndrome (VCFS) patients diagnosed with rapid-cycling bipolar disorder, and an allele encoding the low enzyme activity catechol-*O*-methyltransferase variant (COMT L) has been identified.[43] Between 85% and 90% of VCFS patients are hemizygous for COMT. There is nearly an equal distribution of L and H alleles in Caucasians. Individuals with low-activity allele (COMT LL) would be expected to have higher levels of transynaptic catecholamines due to a reduced COMT degradation of NE and DA. COMT inhibitors (entacapone (**40**) and tolcapone (**41**)) could therefore be beneficial as adjuncts to L-dopa not only in Parkinson's disease but also in the coincident depressive illness associated with rapid cycling.

6.03.5.3 Bipolar Disorder

6.03.5.3.1 Treatment of bipolar disorder

6.03.5.3.1.1 Acute mania

Lithium is generally the drug of choice to stabilize the person and is usually very effective in controlling mania and preventing new episodes. Response to lithium treatment may take several days initially. If the individual is experiencing psychotic symptoms, antipsychotic medications, e.g., clozapine, olanzapine, and other atypical antipsychotic agents, may be prescribed.

Mood-stabilizing anticonvulsant drugs, such as valproate (**42**), lamotrigine (**43**), carbamazepine (**44**), oxcarbazepine (**45**), and topiramate (**46**), may also be used. Often these medications are combined with lithium for maximum effect with mixed benefit to the patient.

6.03.5.3.1.2 Acute depression

Lithium can be a very effective treatment for the depression that occurs in bipolar disorder. Antidepressants, including SSRIs, may also be prescribed. Antidepressant medications used to treat the depressive symptoms of bipolar disorder, when taken without a mood-stabilizing medication, can increase the risk of switching into mania or hypomania, or developing rapid cycling, in people with bipolar disorder. Therefore, mood-stabilizing medications are generally required, alone or in combination with antidepressants, to protect patients with bipolar disorder from this switch. Lithium and valproate are the most commonly used mood-stabilizing drugs today.

6.03.5.3.2 Current treatments of bipolar disorder

Current medications used for the treatment of bipolar disorder are summarized in **Table 10**.

6.03.5.3.2.1 Lithium

Since the early 1800s, lithium has been a first-line medication in the treatment of bipolar disorder (manic depression). The mechanism of action by which lithium produces efficacy is unknown. Lithium inhibits production of inositol monophosphate, which plays a role in gene expression.[44] Lithium is a well-established treatment for bipolar disorder, being effective in the treatment of both the manic and depressive phases.[45,46] It has an approximate response rate of 79%.[4,5,47]

6.03.5.3.2.2 Anticonvulsants

The treatment spectrum for bipolar disorder has broadened since the use of anticonvulsants, such as valproate, carbamazepine, lamotrigine, and gabapentin. Patients with rapid cycling or mixed episode are more likely to benefit from treatment with anticonvulsants than patients with other types of bipolar disorder.[48]

6.03.5.3.2.2.1 Valproate The mechanisms by which valproate exerts its therapeutic effects have not been established. It has been suggested that its activity in epilepsy is related to increased brain concentrations of GABA and

Table 10 Current medications commonly used for the treatment of bipolar disorder

Acute mania/mixed episodes	• Lithium • Valproate • Carbamazepine • Atypical antipsychotics (e.g., olanzapine, quetiapine, and risperidone) • Typical antipsychotics (e.g., haloperidol)
Maintenance treatment	• Lithium • Carbamazapine • Antidepressants
Investigational agents	Atypical antipsychotics (e.g., clozapine, olanzapine, ziprasidone, and aripiprazole)

Adapted from Keck, P. E., Jr.; McElroy, S. L.; Arnold, L. M. *Med. Clin. North Am.* **2001**, *85*, 645–661.

also its sodium channel antagonist properties. Divalproex sodium dissociates into the valproate ion in the gastrointestinal tract. Valproate is the most frequently prescribed mood stabilizer in the US, where it is approved for the treatment of acute mania, but only when lithium and carbamazepine have failed or are not well tolerated. While it is currently not widely used in Europe or Japan, the incidence of its use is increasing.[46,48] The clinical utility of valproic acid (VPA) is controversial.[44] Sodium valproate is better tolerated than valproate (valproic acid (**42**)), with less gastrointestinal effects and therefore less discontinuation liability.

6.03.5.3.2.2.2 Carbamazepine Carbamazepine (**44**) has antiepileptic psychotropic and neurotropic actions. It is believed that its primary mechanism of action is blockade of voltage-sensitive sodium and calcium channels, although the molecular mechanisms underlying the actions of carbamazepine and other mood-stabilizing drugs used in the treatment of bipolar disorders are still largely unknown.

Compared to lithium and placebo, carbamazepine is effective for the long-term treatment of bipolar disorder, but is not approved for this indication worldwide.[46,48] It has antimanic and antidepressive properties, both as a monotherapy and in combination with lithium or antidepressants. In one study, 53% of depressed patients responded rapidly to the blind addition of lithium to carbamazepine.[44] Lithium therapy is superior to carbamazepine therapy, and combination therapy is better than monotherapy, especially in rapid cycles. This may be due, in part, to the ability of carbamazepine to induce its own metabolism by the CYP microsomal enzyme system.[44]

6.03.5.3.2.2.3 Lamotrigine The effects of lamotrigine (**43**) are thought to include inhibition of excitatory amino acid and voltage-dependent sodium channels and blockade of $5HT_3$ receptors.[49] A number of studies have shown lamotrigine to be effective for the treatment of the depressive phase of bipolar disorder and rapid cycling bipolar disorder. The adverse effects of lamotrigine are similar to other anticonvulsants, with a slightly higher rate of headache.[49] Lamotrigine is not practical for the treatment of the manic phase of bipolar disorder; this is partly due to the need for slow dose escalation.[44,49] Lamotrigine has been reported to enhance the effectiveness of valproate in bipolar disorder; however, there is a risk of rash with this dosing regimen. To reduce the risk of rash a slow dosage titration is recommended. In contrast, when coadministering lamotrigine with carbamazepine a more rapid dose increase is recommended.[48] Lamotrigine is emerging as a potentially useful agent for the treatment of bipolar disorder.[48]

6.03.5.3.2.2.4 Gabapentin Gabapentin (**5**) is an antiepileptic drug that is structurally related to GABA. Its mechanism of action is unknown, but its distinct profile of anticonvulsant activity in animal seizure models and its lack of activity at many drug binding sites associated with other antiepileptic drugs indicate that its mechanism of action is novel. Gabapentin was formed by the addition of a cyclohexyl group to GABA, which allowed this form of GABA to cross the blood–brain barrier. Despite its structural similarity to GABA, gabapentin does interact with GABA receptors in the CNS. Its mechanism of action is unknown, but may involve enhanced neuronal GABA synthesis.

Gabapentin differs from other mood-stabilizing drugs in two ways: (1) it is sometimes effective for patients who have failed to respond to antidepressants or mood stabilizers; and (2) it has a relatively benign side-effect profile.[50] Gabapentin has been successful in controlling rapid cycling and mixed bipolar states in a few people who did not receive adequate relief from carbamazepine and/or valproate. It also appears that gabapentin has significantly more antianxiety and antiagitation potency than either carbamazepine or valproate. Gabapentin also may be useful as a treatment for people with antipsychotic-induced tardive dyskinesia.[50] It is possible that gabapentin will prove to be a useful treatment for individuals with other mood disorders. The final dose of gabapentin is usually between 900 and 2000 $mg\,day^{-1}$ when used as an antidepressant or as a mood-stabilizing agent. Some individuals require doses as high as 4800 $mg\,day^{-1}$ to achieve efficacy.

6.03.5.3.2.3 Antidepressants

Standard antidepressants are effective for the treatment of BPAD-1 in combination with a mood stabilizer. The most commonly recommended antidepressants are SSRIs, TCAs, MAOIs, and bupropion (**6**).[45]

6.03.6 Unmet Medical Needs

Onset of action, drug efficacy, and side-effect profile are the three key unmet needs with current mood disorder therapies according to physicians interviewed in a recent Datamonitor survey (2001).[77] An improvement in general efficacy of the drug was cited as the single most important factor physicians require in new antidepressant medications. It is widely acknowledged that the characteristics of any new entrant into the marketplace should target the areas listed below if they are to maximize their potential uptake.[77] The 'ideal' antidepressant drug should: (1) have efficacy in all depressed patients, irrespective of the various subcategories of this mental disorder; (2) have no adverse side effects; (3) have rapid onset of clinical efficacy; (4) be nontoxic when taken in overdose; and (5) have no drug dependence or rebound withdrawal effects.

Approximately 20–35% of patients do not respond to any of the currently available agents and discontinuation rates are high (up to 50% within the first 4 months of treatment), primarily due to side effects related to the sexual dysfunction associated with the SSRIs. ECT is useful, particularly for individuals whose depression is severe or life threatening or who cannot take antidepressant medication. It is often effective in cases where antidepressant medications do not provide sufficient relief of symptoms.[51]

A variety of antidepressants are often tried before finding the most effective medication or combination of medications, and often the dosage must be increased to be effective. Although some improvement may be seen in the first few weeks, antidepressant medications must be taken regularly for 3–5 weeks (in some instances 8 weeks) before the full therapeutic effect occurs.[51] Patient compliance is often the most cited reason to stop medication due to lack of efficacy or adverse effects (**Table 7**).

6.03.6.1 Unmet Needs for Treating Bipolar Disorder

More than 2 million adults in the United States are living with BPAD. Less than half of these patients achieve a response to existing therapy that allows them to lead active lives, while many continue to suffer frequent recurrences of mania and depression. Lithium has been the treatment of choice for BPAD, as it can be effective in smoothing out the mood swings common to this disorder. Its use must be carefully monitored, as the range between an effective dose and a toxic one is limited. Valproate is approved for first-line treatment of acute mania. Other anticonvulsants used for BPAD include lamotrigine and gabapentin together with second-generation antipsychotics.

6.03.7 New Research Areas

The monoamine hypothesis of depression has been the cornerstone of antidepressant treatment for several decades.[52] However, many questions remain unanswered as to the underlying pathophysiology of affective disorders and if monoamines themselves are responsible for regulating unipolar and bipolar depressives states. It is clear, as stated earlier, that the etiology of depression and bipolar disorder is still unknown. Arguably, however, the clinical and preclinical data supporting the monoamine hypothesis are beyond question. With this in mind, the fact remains that the clinical response is delayed several weeks following administration of monoaminergic antidepressant agents, suggesting that other mechanisms may well be involved in the efficacy of these agents. It has long been suggested that alterations in gene expression may be a contributing factor for the delayed clinical response, thereby resulting in changes in signal transduction mechanisms (**Figure 1**).[35,44,53] Several purported mechanisms may account for the delay in the clinical response: (1) receptor downregulation; (2) other components of cellular signaling that are regulated by cyclic AMP, which are prominent transcription factors in the brain (phosphorylated cAMP response element-binding protein, CREB); and (3) factors controlling cellular plasticity such as brain-derived neurotrophic factor (BDNF).

At present, the number one challenge is to develop novel antidepressants with greater efficacy and rapid onset of action. To this end, several pharmaceutical companies continue to bet on the tried and tested monoamine reuptake inhibitors approach, the latest entrant in the US depression market being duloxetine (**47**). This was launched in 2004, and is reported to have a faster onset of action and efficacy in treating the physical pain associated with depression. Escitalopram, the active enantiomer of citalopram, was launched in 2003 and is marketed as having a faster onset of action and lower rates of discontinuation. This perceived low-risk (risk aversive) approach has resulted in several other mixed-monoaminergic reuptake inhibitors, which are currently under development (Phase II/III), including: desvenlafaxine (**48**), a metabolite of venlafaxine (*see* Section 6.03.5.2.4); GW353162 (**49**), a metabolite of bupropion; GW372475, SEP-225289, ORG 4420 (**50**), DOV 216,303 (**51**), DOV 21,947, and DOV 102,677.

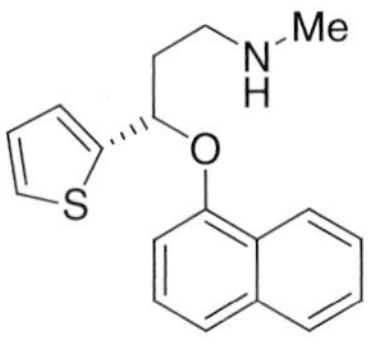

Duloxetine **47** 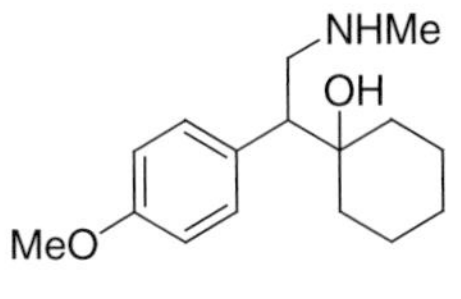 | Desvenlafaxine **48** 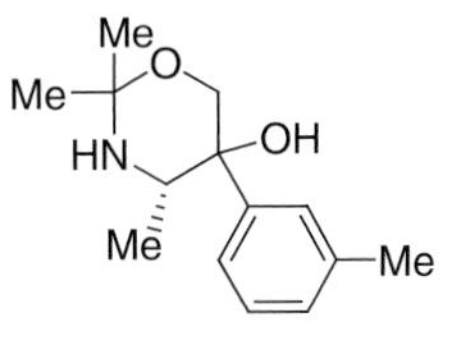 | GW353162 **49** 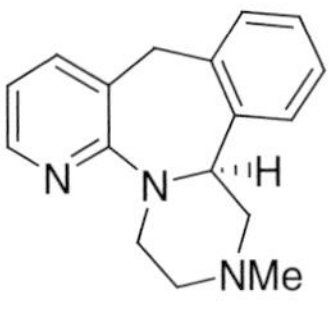 | ORG 4420 **50**

DOV 216,303 **51**

Gepirone **52**

Ipsapirone **53**

Zalospirone **54**

GR127935 **55**

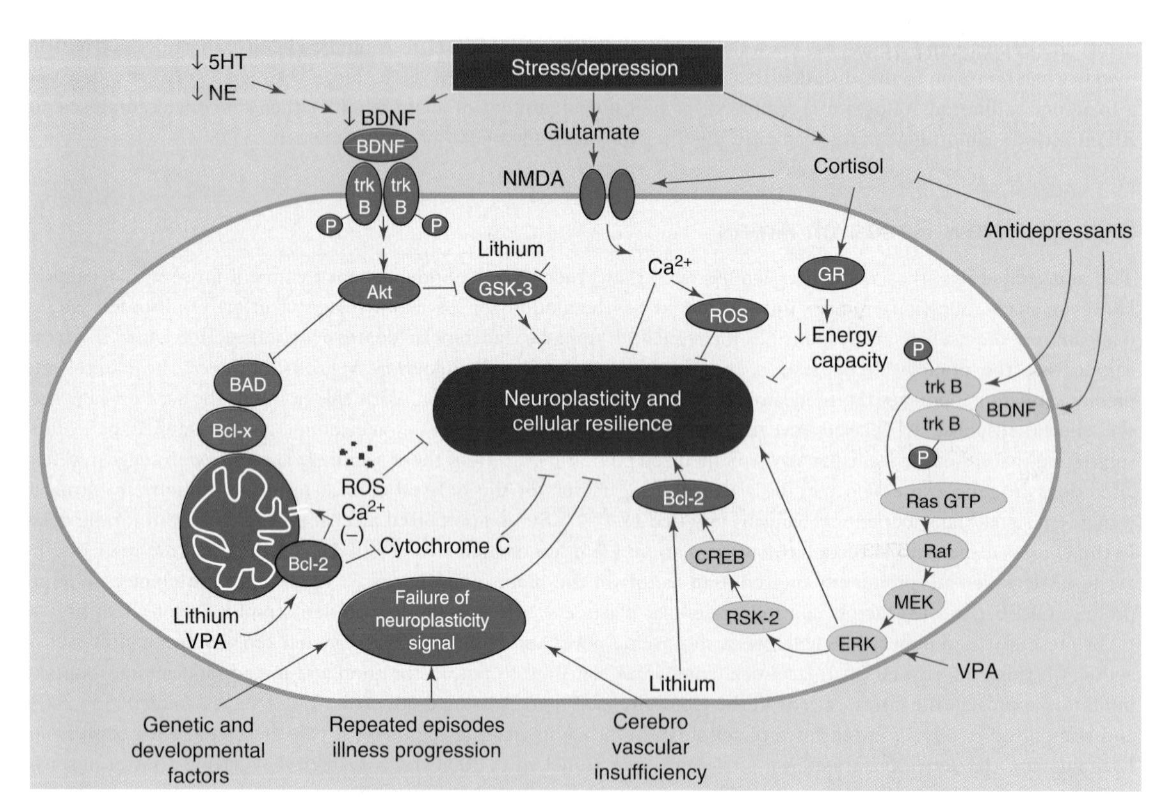

Figure 1 Hypothetical signal transduction pathways in depression. (Reproduced with permission from Manji, H. *Am. J. Psychiatry* **2003**, *160*, 24.)[53]

6.03.7.1 Novel Strategies in Depression and Bipolar Disorder: Ongoing Proof of Efficacy Studies

6.03.7.1.1 Serotonin agents

The fact that repeated SSRI treatment can downregulate a number of pre- and postsynaptic 5HT receptors in both healthy subjects and depressed patients has led to intensive research to unravel the role of 5HT receptor subtypes in the pathophysiology of depression and BPAD.[51]

The $5HT_{1A}$ area has been studied in Section 6.03.5.2.3; however, interest in this area still persists based on the ability of azapirone partial agonists (e.g., gepirone (**52**), isapirone (**53**), and zalospirone (**54**)) to modify 5HT function by an action on the 5HT autoreceptor. Other $5HT_{1A}$ agonist compounds in this class include GR127935 (**55**) and WAY163426, which are reported to raise serotonin levels more rapidly than an SSRI, and exhibit activity in chronic models of depression that are consistent with a more rapid onset. VPI-013 (**56**) and PRX-00023 (**57**) have mixed $5HT_{1A}$ receptor antagonist and sigma agonist properties and are currently in Phase II studies. $5HT_{1B/D}$ ligands mediate their effects at autoreceptors involved in the local inhibitory control of 5HT release and may play a role in the pathogenesis of MDD and the antidepressant effects of the SSRIs in patients. A number of compounds in this class include GR125743 (**58**), SB-224289 (**59**), and L-694247 (**60**). AZD8129 (**61**), a $5HT_{1B}$ antagonist, and CP-448187 (**62**), a $5HT_{1B/1D}$ receptor antagonist, are in Phase II studies for depression.

Although SSRIs lower the sensitivity of postsynaptic 5HT receptors, their primary action is to facilitate 5HT neurotransmission by increasing synaptic 5HT levels.[54] A potential for $5HT_{2A}$ receptors is based on the antidepressant efficacy of nefazadone (**63**), a combined SSRI and $5HT_{2A}$ receptor antagonist that has been 'black boxed' by the FDA and taken off the market in Europe due to risk–benefit analysis, including consideration of the risk of hepatic failure associated with nefazadone treatment. Other 5HT receptor targets under examination include $5HT_{2C}$ selective agonists (e.g., WAY163909 (**64**), Ro-60-0175 (**65**), Ro-60-0332, Org-12962 (**66**)). Paradoxically, $5HT_{2C}$ receptor inverse agonists (e.g., SB-243213 (**67**) and WAY163909) are also believed to be in development for the treatment of depression.[55] The fact that both compounds show a similar pharmacology in several animal models of depression reflects the ability of the $5HT_{2C}$ receptor agonists and antagonists to desensitize the $5HT_{2C}$ receptor as observed following chronic SSRI treatment.

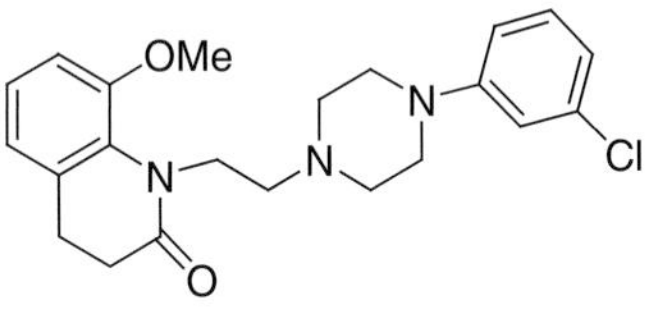

VPI-013 **56**

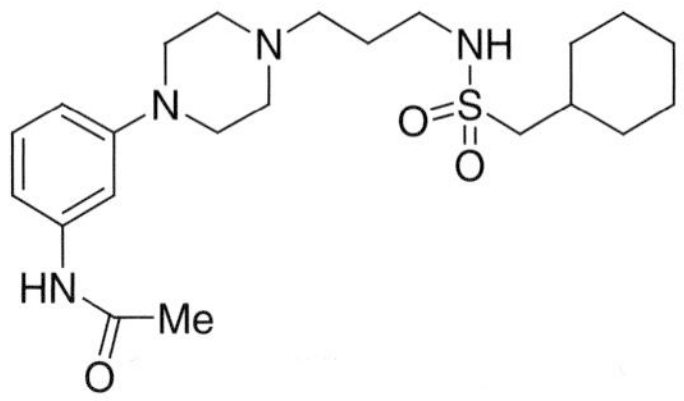

PRX-00023 **57**

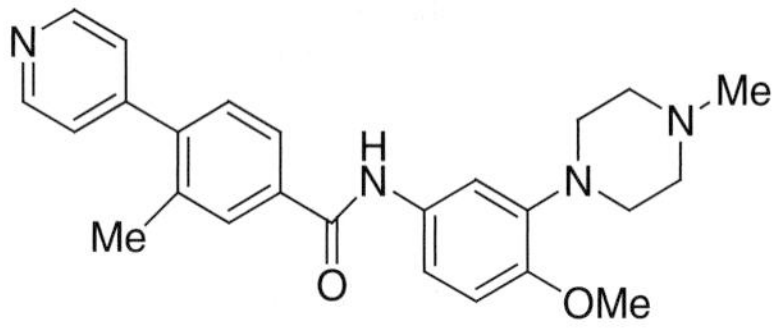

GR125743 **58**

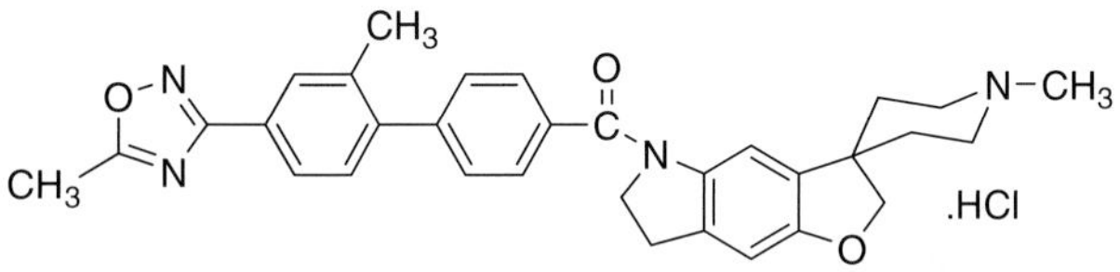

SB-224289 **59**

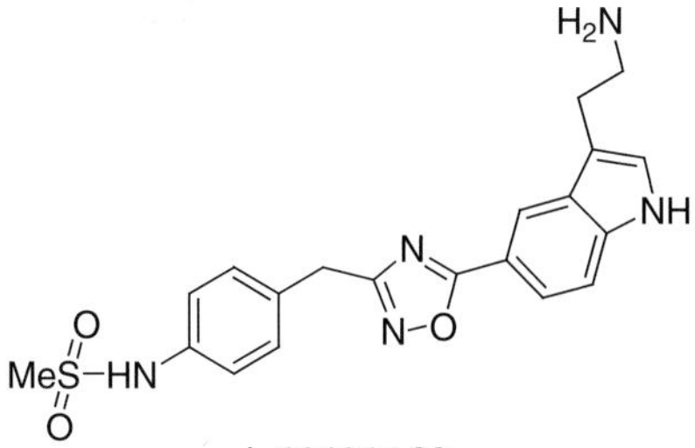

L-694247 **60**

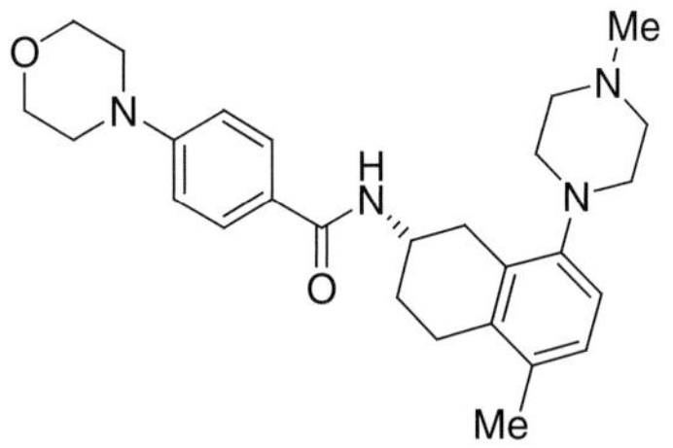

AZD8129 **61**

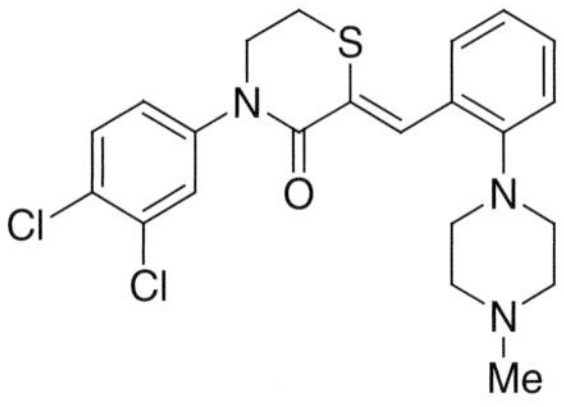

CP-448187 **62**

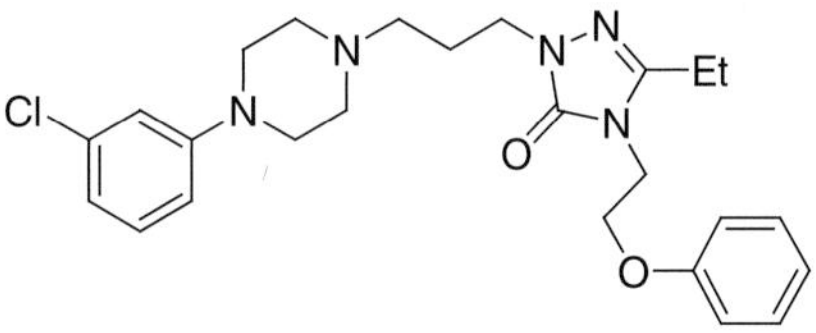

Nefazadone **63**

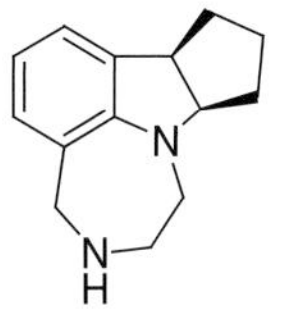

WAY163909 **64**

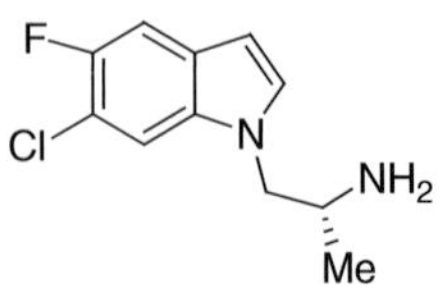

Ro-60-0175 **65**

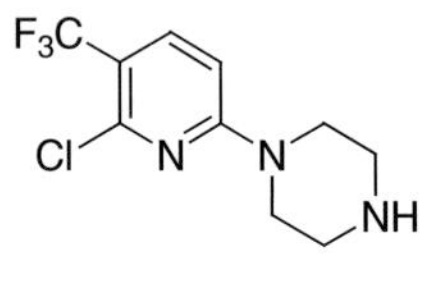

Org-12962 **66**

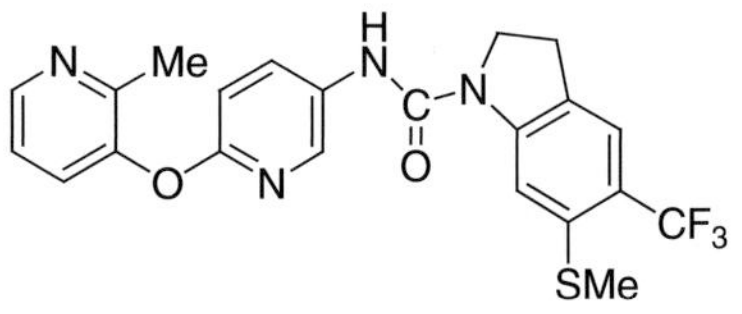

SB-243213 **67**

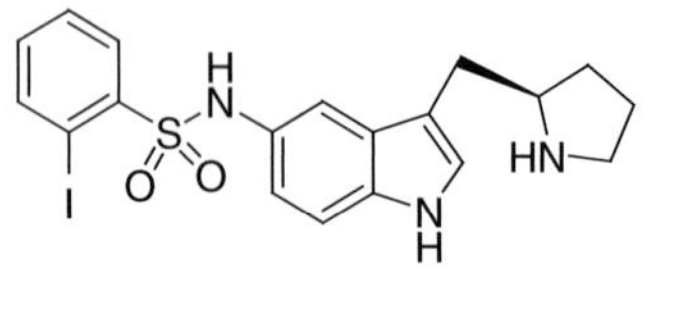

WAY466 **68**

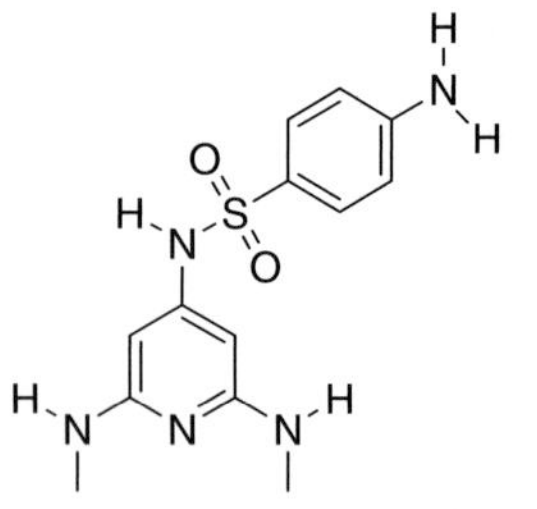

Ro-63-0563 **69**

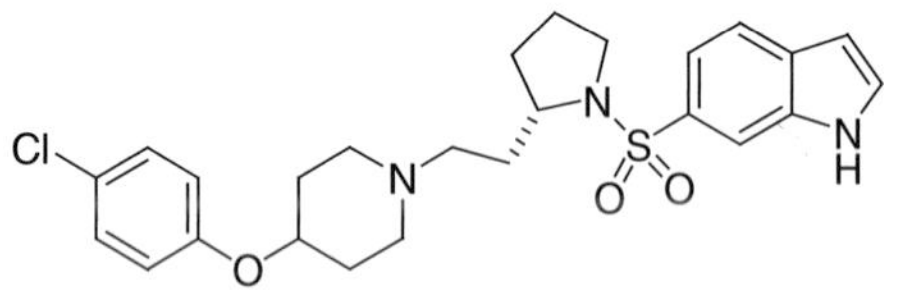

SB-656104-A **70**

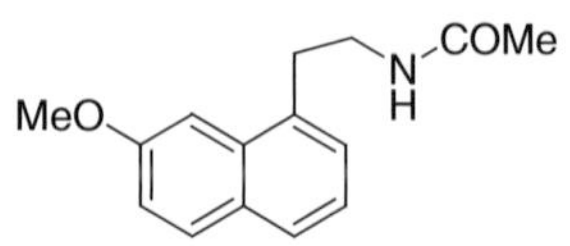

Agomelatine **71**

Selective $5HT_6$ agonists and antagonists (e.g., WAY466 (**68**), Ro-63–0563 (**69**), SB-171046, and GW 742457) may show therapeutic promise while a selective $5HT_7$ receptor antagonist (SB-656104-A (**70**)) has recently been reported to modulate REM sleep, part of the sleep architecture that has long been associated with depression.

Agomelatine (**71**) is the first melatonergic (MT1 and MT2 receptor) agonist antidepressant, with several advantages over existing antidepressant treatments.[56] Besides being an effective antidepressant, agomelatine improves the often disrupted sleep patterns of depressed patients, without affecting daytime vigilance. It has no effects on sexual function, tolerability problems, or discontinuation symptoms.[61]

6.03.7.1.2 Neuropeptide approaches

6.03.7.1.2.1 NK1 receptor antagonists

SP is an 11 amino acid peptide belonging to the tachykinin family; it mediates its biological actions through G tachykinin (NK1) receptors. Evidence to support a major role of the NK1 receptor system in stress-related behaviors has guided the clinical development of several NK1 receptor antagonists, including aprepitant (MK-869; **72**), lanepitant (**73**), dapitant (**74**), vestiptant (**75**), PD-174424 (**76**), and NBI 127914 (**77**). The antidepressant efficacy of the first NK1 receptor antagonist MK-0869 (Aprepitant; **72**) was demonstrated in patients with major depression and high anxiety, and has recently been replicated with a second compound, L759274. Aprepitant improved depression and anxiety symptoms in a quantitatively similar manner to SSRIs. However, it failed to show efficacy in Phase III clinical trials for depression.[11,57] NK2 receptor ligands (e.g., NKP 608 (**78**), GR159897 (**79**)) are also under investigation for their potential role in depression and anxiety disorders.

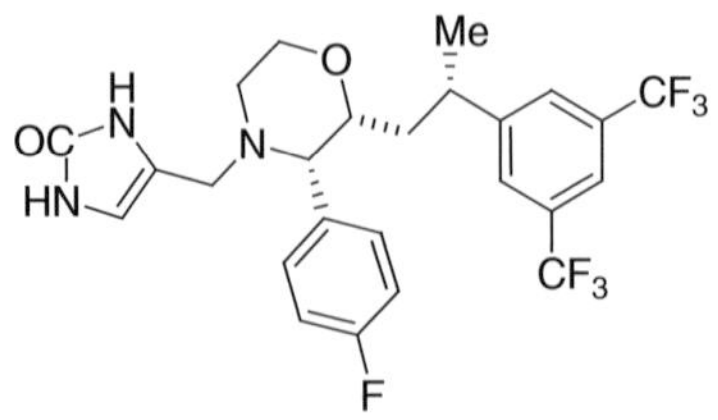

Aprepitant (MK-869) **72**

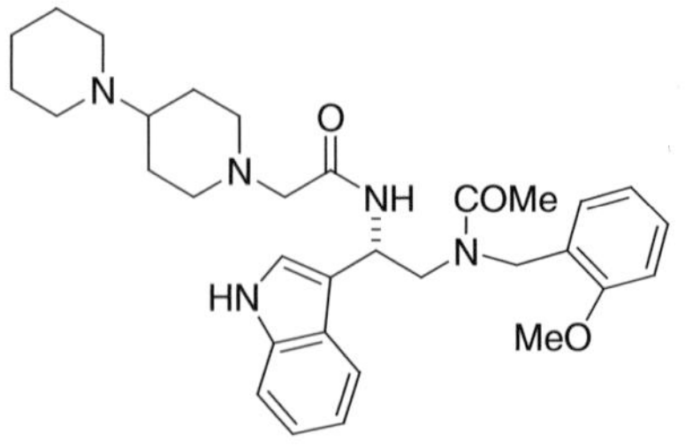

Lanepitant **73**

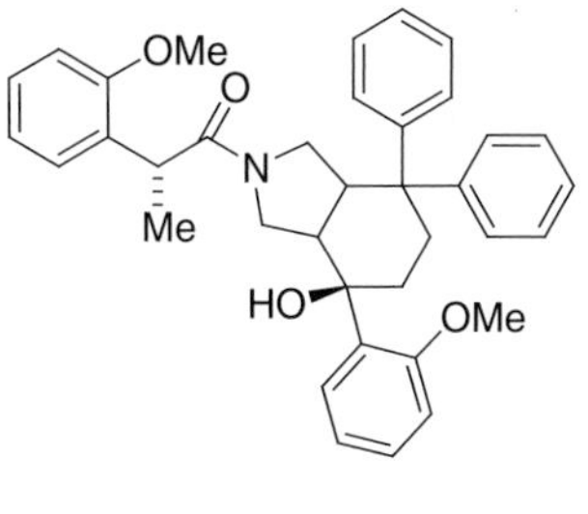

Dapitant **74**

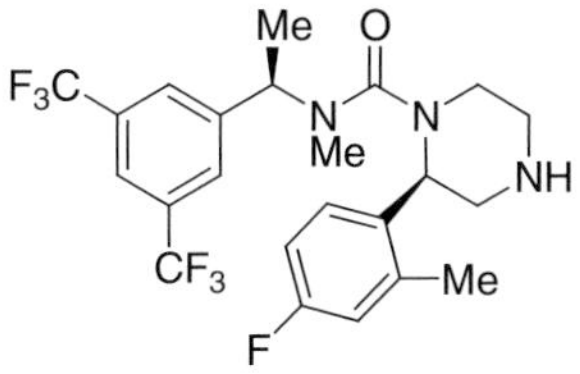

Vestiptant **75**

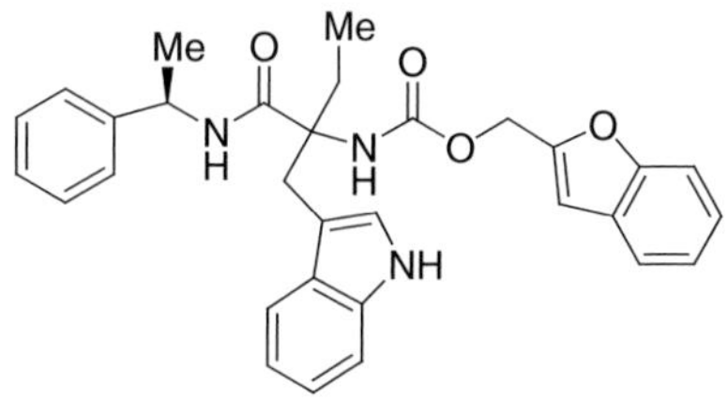

PD-174424 **76**

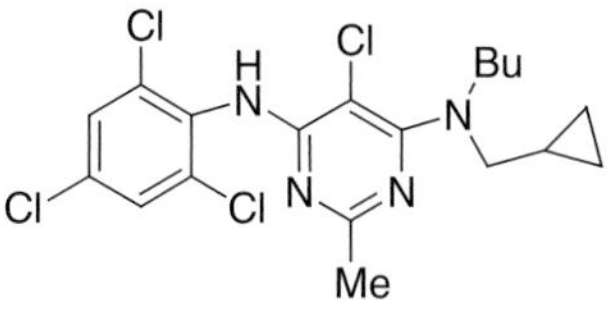

NBI 127914 **77**

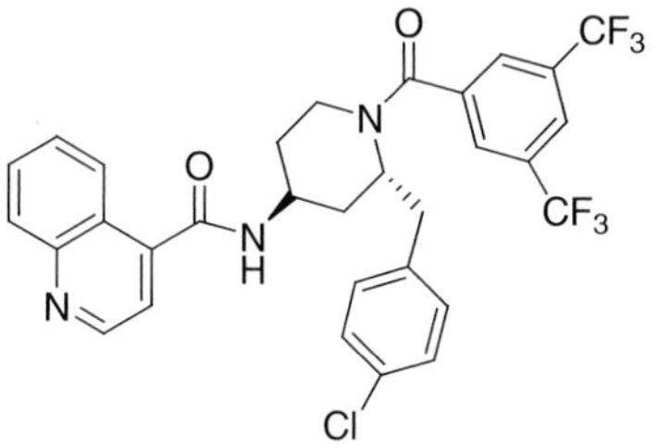

NKP 608 **78**

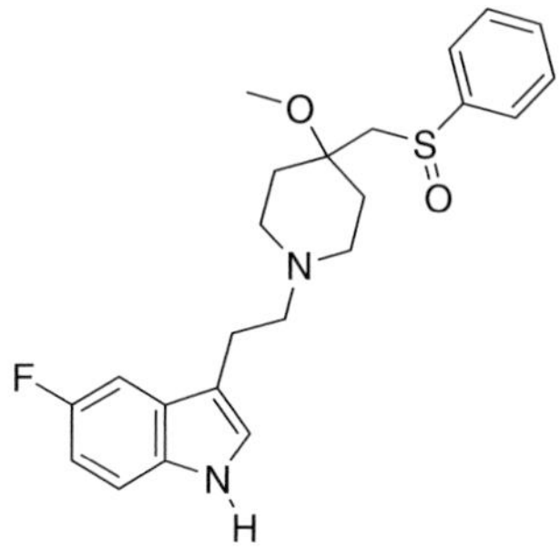

GR159897 **79**

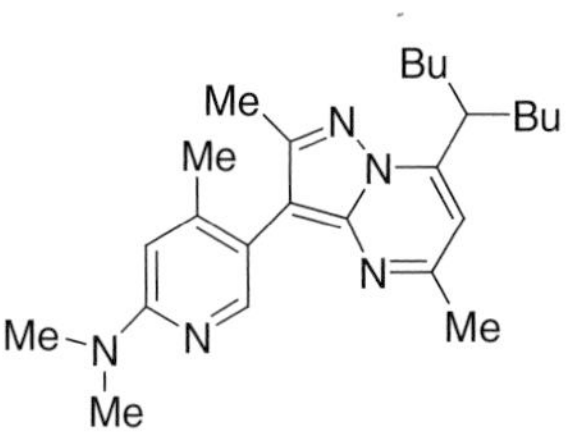

R-12919/NBI-30775 **80**

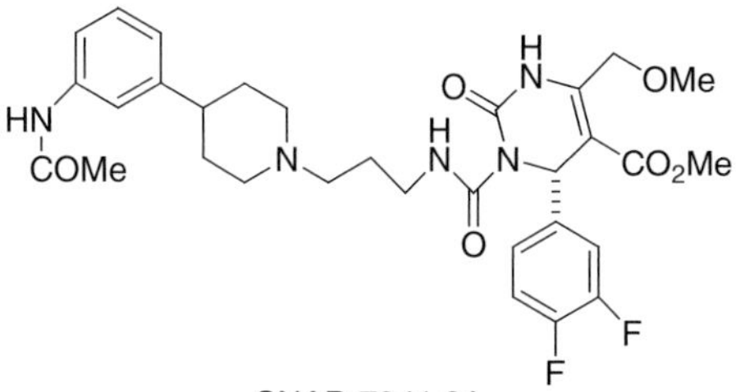

SNAP-7941 **81**

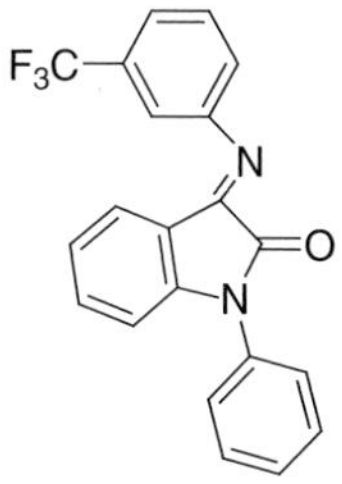

SNAP-37889 **82**

6.03.7.1.2.2 Corticotropin-releasing factor (CRF)

The 41 amino acid neuropeptide CRF initiates the HPA axis response to stress, and may have utility in the treatment of depression and anxiety.[58] In a small, open-label clinical trial, symptoms of anxiety and depression in patients with major depression were reduced during treatment with the CRF1 receptor antagonist R-12919 (**80**).

6.03.7.1.2.3 Melanin concentrating hormone (MCH)

MCH is an orexigenic hypothalamic neuropeptide that plays an important role in the complex regulation of energy balance and body weight. SNAP-7941 (**81**) is a selective, high-affinity MCH1 receptor (MCH1-R) antagonist that has been shown to inhibit food intake stimulated by central administration of MCH and reduce consumption of palatable food; chronic administration of SNAP-7941 to rats with diet-induced obesity resulted in a marked, sustained decrease in body weight.[59] In addition to the orexigenic effects, SNAP-7941 produced effects similar to clinically used antidepressants and anxiolytics in animal models of depression/anxiety: the rat FST, rat social interaction, and guinea pig maternal separation vocalization tests. Given these observations, it has been suggested that MCH1-R antagonists may be useful not only in the management of obesity but also as a treatment for depression and/or anxiety.

6.03.7.1.2.4 Galanin

Galanin is a biologically active neuropeptide that is widely distributed in the central and peripheral nervous systems and the endocrine system. The amino acid sequence of galanin is very conserved (almost 90% among species), indicating the importance of the molecule. Galanin has multiple biological effects.[60] In the CNS, galanin alters the release of several neurotransmitters. In particular, the ability of galanin to inhibit acetylcholine release together with the observation of hyperinervation of galanin fibers in the human basal forebrain of Alzheimer's disease patients suggests a possible role for galanin in this neurodegenerative disease. Galanin may also be involved in other neuronal functions, such as learning and memory, epileptic activity, nociception, spinal reflexes, and feeding. Three human and rodent galanin receptor subtypes have been cloned. The GAL_3 receptor antagonist SNAP-37889 (**82**) may have utility as a novel antidepressant.[61,62]

6.03.7.1.2.5 Neuropeptide Y

Neuropeptide Y (NPY) is abundantly expressed in numerous brain areas, including the locus coeruleus, hypothalamus, amygdala, hippocampus, nucleus accumbens, and neocortex.[59] Central NPY colocalizes with NE, GABA, somatostatin, and agouti-related protein. Actions of NPY are mediated through the Y1, Y2, and Y5 receptor. In rodent models, NPY is expressed and released following stress, and attenuates the behavioral consequences of stress. Rodent studies have demonstrated antidepressant-like effects of centrally administered Y1 receptor agonists.[59]

6.03.7.1.2.6 Vasopressin VP1b receptor antagonists

The nonpeptide vasopressin, which is synthesized in the PVN and supraoptic nucleus, acts via vasopressin receptors (V1a and V1b) expressed mainly in limbic areas and in the hypothalamus.[57] Abnormalities in vasopressin expression or receptor activity occur in both clinical depression and rodent genetic models of depression, whereas vasopressin release predicts anxiety reactions to stress provocation in healthy volunteers. The nonpeptide V1b receptor antagonist SSR-149415 (**83**) exerts marked anxiolytic- and antidepressant-like effects in rodents.[57]

6.03.7.1.3 Glutamate receptors

6.03.7.1.3.1 NMDA receptors

Several types of antidepressants can alter the expression of mRNA for the NMDA receptor while glutamate levels are elevated in the occipital cortex of medication-free subjects with unipolar major depression.[63] Direct NMDA receptor antagonists have antidepressant effects in animal models including behavioral despair and chronic mild stress paradigms. Two drugs with putative glutamatergic properties, lamotrigine (**43**) and riluzole, have been reported to have antidepressant activity in the clinic trials while a preliminary study using the noncompetitive NMDA receptor antagonist, ketamine (**84**), showed rapid and short-lived antidepressant actions. A more recent study in 18 subjects[64] showed that ketamine ($0.5\,mg\,kg^{-1}$ i.v.) significantly improved depressive symptoms (HAMD-21 score) within 2 h of administration to the same extent as seen with bupropion, SSRIs, and venlafaxine at 8 weeks. This effect remained significant through the following week after treatment. These data suggest that NMDA antagonists, like ketamine which acts at the PCP site on the receptor, may be beneficial in the treatment of depression.

6.03.7.1.3.2 Group 1 glutamate receptors (mGlu 1–8)

Metabotropic glutamate (mGlu) receptors (mGlu 1–8) are a heterogeneous family of GPCRs that modulate brain excitability via presynaptic, postsynaptic, and glial mechanisms. Glutamate receptors regulate brain glutamate activity and may play an important role in the pathophysiology of depression. NMDA receptor antagonists, mGluR1 and mGluR5 antagonists, and positive modulators of AMPA receptors have antidepressant-like activity in a variety of preclinical models. A single intravenous dose of an NMDA receptor antagonist is sufficient to produce sustained relief from depressive symptoms.[65]

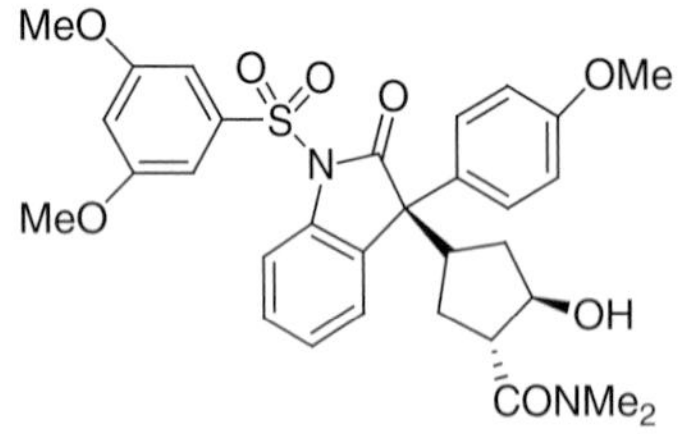

SSR-149415 **83** 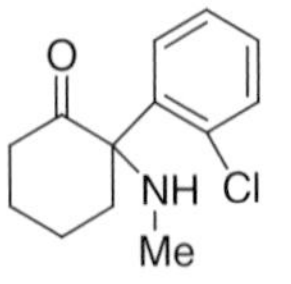 Ketamine **84** 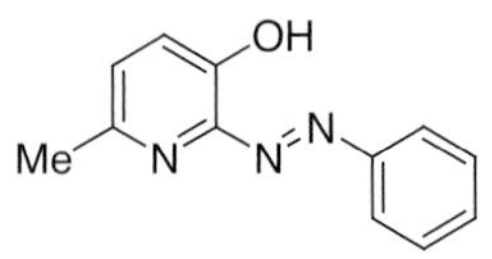 SIB-1757 **85**

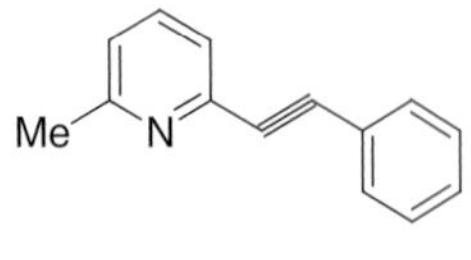

MPEP **86** 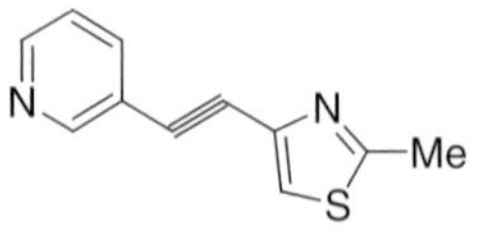 MTEP **87** 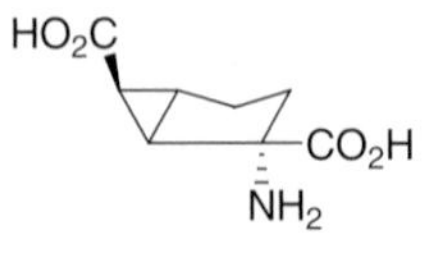 LY-354740 **88** 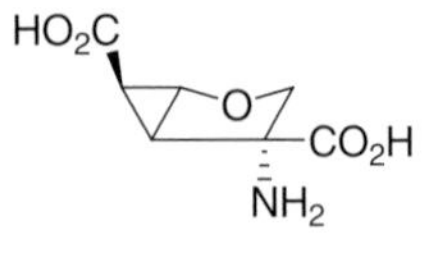 LY-379268 **89**

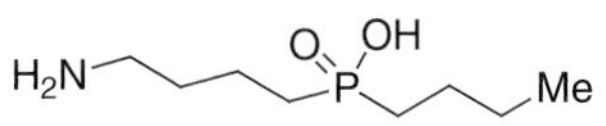

SGS742 **90** 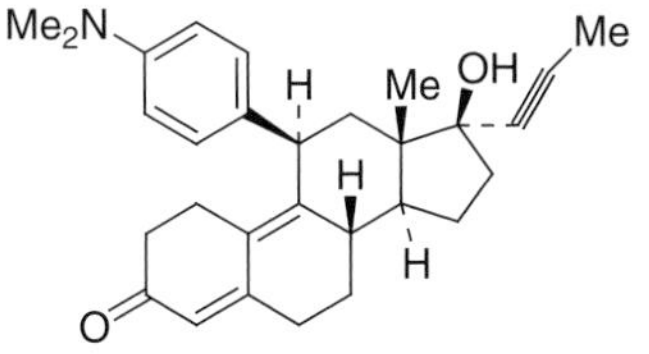 | Mifepristone (RU-486) **91** 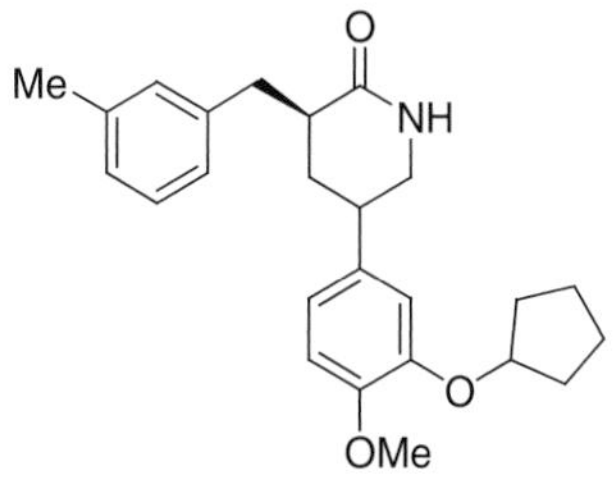 | HT-0712/IPL-455903 **92**

6.03.7.1.3.2.1 mGlu5 receptor The first, potent, selective, and structurally novel mGlu5 receptor antagonists identified were SIB-1757 (**85**) and SIB 1893, which led to the noncompetitive antagonists MPEP (**86**) and MTEP (**87**).[62] mGlu1/5 receptors are involved in long-term potentiation (LTP), chronic depression, and memory formation.[66]

6.03.7.1.3.2.2 mGlu2/3 The mGlu2/3 agonists LY-354740 (**88**) and LY-379268 (**89**) are active in animal models of anxiety[65] acting as functional $5HT_{2A}$ antagonists with therapeutic implications in the treatment of mood disorders.[67–69]

6.03.7.1.3.3 $GABA_B$ receptor antagonist

Dysfunction of GABAergic systems are implicated in the pathophysiology of anxiety and depression. Recent evidence points to a role of $GABA_B$ receptors in anxiety and depression. Metabotropic $GABA_B$ receptors predominantly function as heterodimers of $GABA_{B1}$ and $GABA_{B2}$ subunits, but $GABA_{B1}$ can also form functional receptors in the absence of $GABA_{B2}$. Mice lacking the $GABA_{B1}$ subunit have altered behavioral responses in tests for anxiety and depression. Both $GABA_{B1}$- and $GABA_{B2}$-deficient mice were found to be more anxious than wild type in the light–dark box paradigm.[70] In contrast, these mice exhibited an antidepressant-like behavior in the FST. Taken together, these data suggest that heterodimeric $GABA_{B1,2}$ receptors are required for the normal regulation of emotional behavior and offer a novel approach for the treatment of mood disorders.[70] $GABA_B$ receptor antagonists in preclinical and clinical development for depression and cognitive disorders include AVE 1876 and SGS742 (**90**).

6.03.7.2 Glucocorticoids

Glucocorticoids are important in the pathogenesis of depression, but this potentially serious psychological side effect is often overlooked in clinical practice. The unwanted behavioral effects of anabolic steroids are widely known, but those of glucocorticoid therapy, though recognized for over 45 years, receive less attention. Placebo-controlled studies have revealed that a third of patients taking glucocorticoids experience significant mood disturbance and sleep disruption. More importantly, up to 20% of patients on high-dose glucocorticoids report psychiatric disorders including depression, mania, psychosis, or a mixed affective state. A recent double-blind, placebo-controlled trial of corticosteroid administration in healthy individuals showed that 75% of subjects developed disturbances in mood and cognition, which reversed when steroids were stopped. Dysregulation of the HPA axis in depression is one of the oldest and most consistent findings in biological psychiatry. A large-scale meta-analysis of over 140 studies using the low-dose dexamethasone suppression test illustrated that persistent adrenocortical hyperactivity is a robust indicator of poor prognosis and a weaker predictor of suicide in depression. Overall, data from conditions of both exogenous and endogenous steroid excess provide support for a glucocorticoid theory of depression.

HPA axis hyperactivity is found in bipolar disorder related to depression and mixed states. Patients with bipolar disorder also have cognitive difficulties and endocrine disturbances may contribute to such dysfunction. Antiglucorticoid therapies are novel treatments of mood disorder. Preliminary data in psychotic depression suggest that mifepristone (RU-486), a glucocorticoid receptor antagonist (**91**), has antidepressant and salutary cognitive effects in a matter of days. The positive effects of mifepristone in severe bipolar depression in a parallel, double-blind, placebo-controlled experiment were recently reported with improvement in two-thirds of patients in the medium- and high-dose groups within 7 days.[71] The other major treatment for psychotic major depression is a combination of antidepressants and antipsychotics, which improve symptoms in roughly 60% of cases.[72] However, side effects from mifepristone are very low compared to these combinations of drugs.[73] The glucocorticoid ORG 34517/34850 is also in Phase II clinical studies of BPAD.

6.03.7.3 Transcription Factors

6.03.7.3.1 cAMP response element-binding protein (CREB)

Antidepressants usually take weeks to exert significant therapeutic effects. This lag phase is suggested to be due to neural plasticity, which may be mediated by the coupling of receptors to their respective intracellular signal transduction pathways. Phosphorylated CREB, a downstream target of the cAMP signaling pathway, is a molecular state marker for the response to antidepressant treatment in patients with MDD. The transcriptional activity of CREB is upregulated by antidepressant treatment. Therefore, it has been hypothesized that antidepressant treatment exerts its therapeutic effect by this mechanism.[74] NCEs under clinical development in this area and other CNS disorders include ND1251, MEM 1917, and HT-0712/IPL-455903 (**92**).

6.03.7.4 Trace Amines

In addition to the classical monoamines NE, 5HT, and DA, there exists a class of 'trace amines' that are found in very low levels in mammalian tissues, and include tyramine, β-phenylethylamine (β-PEA), tryptamine, and octopamine.[74] The rapid turnover of trace amines, as evidenced by the dramatic increases in their levels following treatment with MAOIs or deletion of the MAO genes, suggests that the levels of trace amines at neuronal synapses may be considerably higher than predicted by steady-state measures. Although there is clinical data in the literature that supports a role for trace amines in depression as well as other psychiatric disorders, the role of trace amines as neurotransmitters in mammalian systems has not been thoroughly examined.[75] Because they share common structures with the classical amines and can displace other amines from their storage vesicles, trace amines have been referred to as 'false transmitters.' Thus, many of the effects of trace amines are indirect and are caused by the release of endogenous classical amines. However, there is a growing body of evidence suggesting that trace amines function independently of classical amine transmitters and mediate some of their effects via specific receptors. Saturable, high-affinity binding sites for [^{3}H]-tryptamine, *p*-[^{3}H]-tyramine, and β-[^{3}H]PEA have been reported in rat brain, and both the pharmacology and localization of these sites suggest that they are distinct from the amine transporters. However, although binding sites have been reported no specific receptors for these trace amines have yet been identified conclusively. A family of 15 GPCRs has been described, two members of which are activated by trace amines. TA_1 is activated most potently by tyramine and β-PEA, and TA_2 by β-PEA. At least 15 distinct receptors have been described to date, along with the orphan receptor (putative neurotransmitter receptor (PNR)) and the pseudogenes GPR58, GPR57 and the $5HT_4$ pseudogene share a high degree of sequence homology and together form a subfamily of rhodopsin GPCRs distinct from but related to 5HT, NE, and DA receptors. The localization of the TA_1 receptor in human and rodent tissues, as well as the chromosomal localization of the human members of this family, has been well characterized. The identification of this family of receptors should facilitate the understanding of the roles of trace amines in the mammalian nervous system and their role in affective disorders.

Clinical studies have examined the levels of PEA in many conditions and discovered significant associations. Patients with depression have decreased PEA levels while levels are increased in schizophrenic and psychopathic subjects. The administration of PEA has been found to reduce depression. Likewise, the administration of its precursor, phenylalanine, has been found to improve depression on its own and the therapy outcome when combined with some antidepressants.[76] Trace amines may play a role in the pathophysiology of depression.[76] YKP-10A (R228060) is reported to be a phenylalanine-like molecule.

6.03.8 Summary

The future trends in the development of drugs for the treatment affective disorders have followed the high degree of anatomical overlap in the distribution and coexistence of monoamine and neuropeptide neurons in limbic regions of the human brain. The intimacy of the relationship between monoamines and neuropeptides suggests a common downstream effect on neural systems that mediate stress. The relationship between stress-related interactions among glutamate, CRF, galanin, NPY, SP, and vasopressin V1B and monoamines in the brain is well documented. However, the role of neuropeptides in depression is still in its infancy and it is premature to speculate on whether and how these systems might exert common downstream effects on brain pathways that mediate stress and emotion. Thus, although SP, CRF, vasopressin, NPY, MCH, and galanin demonstrate important functional interactions with monoamines implicated in the etiology and treatment of stress-related disorders, their effects almost certainly go beyond modulation of these neurotransmitters. Understanding these effects is a central goal of future research in this field. There are certain characteristics common to neuropeptides that might make them attractive targets for novel therapeutics.

Because neuropeptides possess a more discrete neuroanatomical localization than monoamines and GABA, neuropeptide receptor ligands might be expected to produce relatively little disruption of normal physiology. Thus, pharmacological alteration of neuropeptide function might normalize pathological activity in circuits mediating stress, such as the HPA axis, without producing unwanted side effects. Moreover, antagonists might be less likely than agonists to produce tolerance and dependence. Indeed, drugs that are antagonists at CRF, vasopressin, NPY, and galanin receptors might have a particularly low side-effect burden because such compounds would not be expected to disrupt normal physiology in the absence of neuropeptide release. Preliminary clinical data appear to be encouraging in this regard.

Several ligands that target neuropeptide receptors are currently undergoing clinical evaluation to determine whether they provide efficacious alternatives to existing drug treatments for depression and anxiety disorders. Establishing the safety, therapeutic efficacy, and an acceptable tolerability profile of drugs that target neuropeptides in depression and anxiety disorders would represent a major advance in the treatment of these diseases. The validation of future targets will be facilitated by the generation of mutant rodents to elucidate neuropeptide function, particularly where a paucity of selective, brain-penetrant ligands limits conventional psychopharmacological approaches. The effects of constitutive neuropeptide mutations can be skewed by developmental alterations that compensate for the mutated neuropeptide or cause changes in other systems that confound interpretation of stress-related phenotypes. Therefore, engineering neuropeptide mutations that are limited to specific developmental stages and brain regions, or improving other molecular manipulations (e.g., RNA interference, antisense, and viral vector delivery techniques), will be valuable. Once promising targets are identified, a further challenge is the generation of small-molecule neuropeptide receptor ligands that are soluble, bioavailable, brain penetrant, and have a low potential for tachyphylaxis. Although there are important obstacles to surmount, neuropeptide-based therapeutic strategies for depression and affective disorders represent a highly promising approach to treating these debilitating conditions.

References

1. *World Health Report 2001 – Mental Health: New Understanding, New Hope.* World Health Organization: Geneva, Switzerland, 2001.
2. Lieberman, J. F.; Greenhouse, J.; Robert, J.; Hamer, M.; Ranga Krishnan, K.; Nemeroff, C. B.; Sheehan, D. V.; Thase, M. E.; Martin, B.; Keller, M. B. *Neuropsychopharmacology* **2005**, *30*, 445–460.
3. Anderson, I. M.; Nutt, D. J.; Deakin, J. F. W. *J. Psychopharmacol.* **2000**, *14*, 3–20.
4. Goodwin, G. M. J. *Psychopharmacol* **2003**, *17*, 149–173.
5. Duffy, F. F.; Narrow, W.; West, J. C.; Fochtmann, L. J.; Kahn, D. A.; Suppes, T.; Oldham, J. M.; McIntyre, J. S.; Manderscheid, R. W.; Sirovatka, P.; Regier, D. et al. *Psychiatr. Q.* **2005**, *76*, 213–230.
6. American Psychiatric Association Diagnostic and Statistical Manual of Mental Disorders DSM-IV; APA: Washington, DC, 1994.
7. World Health Organization. *The ICD-10 Classification of Mental and Behavioural Disorders: Diagnostic Criteria for Research*; World Health Organization: Geneva, Switzerland, 1993.
8. Murray, C. J.; Lopez, A. D. *Lancet* **1997**, *349*, 1498–1504.
9. Greenberg, P. E.; Kessler, R. C.; Birnbaum, H. G.; Leong, S. A.; Lowe, S. W.; Berglund, P. A.; Corey-Lisle, P. K. *J. Clin. Psychiatr.* **2003**, *64*, 1465–1475.
10. Paykel, E. S. *Psychopathology* **2002**, *35*, 94–99.
11. Bieck, P. R.; Potter, W. Z. *Annu. Rev. Pharmacol. Toxicol.* **2005**, *45*, 227–246.
12. Roth, B. L.; Sheffler, D. J.; Kroeze, W. K. *Nat. Rev. Drug Disc.* **2004**, *3*, 353–359.
13. Schildkraut, J. J.; Mooney, J. J. *Am. J. Psychiatry* **2004**, *161*, 909–911.
14. Santarelli, L.; Saxe, M.; Gross, C.; Surget, A.; Battaglia, F.; Dulawa, S.; Weisstaub, N.; Lee, J.; Duman, R.; Arancio, O. et al. *Science* **2003**, *301*, 805–809.
15. Artigas, F.; Celada, P.; Laruelle, M.; Adell, A. *Trends Pharmacol. Sci.* **2001**, *22*, 224–228.
16. Blier, P. *J. Clin. Psychiatry* **2001**, *62*, 12–17.
17. Tellioglu, T.; Robertson, D. *Expert Rev. Mol. Med.* **2001**, *19*, 1–10.
18. Zill, P.; Engel, R.; Baghai, T. C.; Juckel, G.; Frodl, T.; Muller-Siecheneder, F.; Zwanzger, P.; Schule, C.; Minov, C.; Behrens, S. et al. *Neuropsychopharmacology* **2002**, *26*, 489–493.
19. Hadley, D.; Hoff, M.; Holik, J.; Reimherr, F.; Wender, P.; Coon, H.; Byerley, W *Hum. Hered.* **1995**, *45*, 165–168.
20. Lin, Z.; Uhl, G. R. *Pharmacogenomics J.* **2003**, *3*, 159–168.
21. McCauley, J. L.; Olson, L. M.; Dowd, M.; Amin, T.; Steele, A.; Blakely, R. D.; Folstein, S. E.; Haines, J. L.; Sutcliffe, J. S. *Am. J. Med. Genet. Part B (Neuropsychiatric Genetics)* **2004**, *127*, 104–112.
22. NIMH. *Bipolar Disorder Research at the National Institute of Mental Health.* NIH publication number 00–4502, 2000.
23. Malt, U. F.; Robak, O. H.; Madsbu, H. P.; Loeb, B. M. *Br. Med. J.* **1999**, *318*, 1180–1184.
24. Willner, P. *Behavioural Models in Psychopharmacology – Theoretical, Industrial and Clinical Perspectives.* Oxford University Press: Oxford, UK, 1991.
25. Wong, M.-L.; Licino, *J. Nat. Rev. Drug Disc.* **2005**, *3*, 136–151.
26. Anisman, H.; Matheson, K *Neurosci. Biobehav. Rev.* **2005**, *29*, 525–546.
27. Frazer, A.; Morilak, D. A. *Neurosci. Biobehav. Rev.* **2005**, *29*, 515–523.
28. Dunn, D. A.; Pinkert, C. A.; Kooyman, D. L. *Drug Disc. Today* **2005**, *10*, 757–767.
29. Matthews, K.; Christmas, D.; Swan, J.; Sorrell, E *Neurosci. Biobehav. Rev.* **2005**, *29*, 503–513.
30. Encinas, J. M.; Vaahtokari, A.; Enikolopov, G. *Proc. Nat. Acad. Sci. USA* **2006**, *103*, 8233–8238.
31. Thakker, D. R.; Natt, F.; Hüsken, D.; van der Putten, H.; Maier, R.; Hoyer, D.; Cryan, J. F. *Mol. Psychiatry* **2005**, *10*, 714.

32. Khan, A.; Khan, S. R.; Walens, G.; Kolts, R.; Giller, E. L. *Neuropsychopharmacology* **2003**, *28*, 552–557.
33. Thase, M. E. J. *Clin. Psychiatry* **1999**, *60*, 3–9, discussion 31–35.
34. Fava, M.; Rush, A. J.; Trivedi, M. H.; Nierenberg, A. A.; Thase, M. E.; Sackeim, H. A.; Quitkin, F. M.; Wisniewski, S.; Lavori, P. W.; Rosenbaum, J. F.; Kupfer, D. *J. Psychiatr. Clin. North Am.* **2003**, *26*, 457–494.
35. Barbui, C.; Hotopf, M. *Acta Psychiatr. Scand.* **2001**, *104*, 92–95.
36. Licinio, J.; Wong, M.-L. *Rev. Drug Disc.* **2005**, *4*, 165–171.
37. Bender, K. J. *Psychiatric Times* **1998**, *15*, 10.
38. Millan, M. J. *Eur. J. Pharmacol.* **2004**, *500*, 371–384.
39. Burger's Medicinal Chemistry and Drug Discovery. In: *Nervous System Agents*, 6th Ed.; Abraham, D. J., Ed.; J. Wiley & Sons, Inc., 2003; Vol 6, p. 500.
40. Owens, M. J.; Knight, D. L.; Nemeroff, C. B. *Biol. Psychiatry* **2001**, *50*, 345–350.
41. Youdim, M. B. H.; Bakhle, Y.S. *British. J. Pharmacol.* **2006**, *S287–S296*.
42. Wong, D. T.; Perry, K. W.; Bymaster, F. P. *Nat. Rev. Drug Disc.* **2005**, *4*, 764–774.
43. Shifman, S.; Bronstein, M.; Sternfeld, M.; Pisante A.; Weizman, A.; Reznik, I.; Spivak, B.; Grisaru, N.; Karp, L.; Schiffer, R.; Kotler, M.; Strous, R. D.; Swartz-Vanetik, M.; Knobler H. Y.; Shinar, E. *et al. Am. J. Med. Genet.* **2004**, *128B*, 61–64.
44. Manji, H. K.; Moore, G. J.; Rajkowska, G.; Chen, G. *Mol. Psychiatry* **2000**, *5*, 578–593.
45. Keck, P. E., Jr.; McElroy, S. L.; Arnold, L. M. *Med. Clin. North Am.* **2001**, *85*, 645–661.
46. Brandish, P. E.; Su, M.; Holder, D. J.; Hodor, P.; Szumiloski, J.; Kleinhanz, R. R.; Forbes, J. E.; McWhorter, M. E.; Duenwald, S. J.; Parrish, M. L. et al. *Neuron* **2005**, *25*, 861–872.
47. Compton, M. T.; Nemeroff, C. B. Clin. *J Psychiatry* **2000**, *61*, 57–67.
48. Licht, R. W. *Acta Psychiatr. Scand.* **1998**, *97*, 387–397.
49. Bowden, C. L.; Asnis, G. M.; Ginsberg, L. D.; Bentley, B.; Leadbetter, R.; White, R. *Drug Safety* **2004**, *27*, 173–184.
50. Berk, M.; Segal, J.; Janet, L.; Vorster, M. *Drugs* **2001**, *61*, 1407–1414.
51. Delgado, P. L.; Miller, H. L.; Salomon, R. M.; Licinio, J.; Heninger, G. R.; Gelenberg, A. J.; Charney, D. S. *Psychopharmacol. Bull.* **1993**, *29*, 389–396.
52. Preskorn, S. H. *J. Clin. Psychiatry* **1995**, *56*, 12–21.
53. Manji, H. *Am. J. Psychiatry* **2003**, *160*, 24.
54. Whale, R.; Clifford, E. M.; Bhagwagar, Z.; Cowan, P. J. *Br. J. Psychiatr* **2001**, *178*, 454–457.
55. Wood, M. D.; Reavill, C.; Trail, B.; Wilson, A.; Stean, T.; Kennett, G. A.; Lightowler, S.; Blackburn, T. P.; Thomas, D.; Gager, T. L. et al. *Neuropharmacology* **2001**, *41*, 186–199.
56. Millan, M. J.; Gobert, A.; Lejeune, F.; Dekeyne, A.; Newman-Tancredi, A.; Pasteau, V.; Rivet, J.-M.; Cussac, D. *JPET* **2003**, *306*, 954–964.
57. Holmes, A.; Heilig, M.; Rupniak, N. M. J.; Steckler, T.; Griebel, G. *Trends Pharmacol. Sci.* **2003**, *24*, 580–595.
58. Todorovic, C.; Jahn, O.; Tezval, H.; Hippel, C.; Spiess, J. *J. Neurosci. Biobehav. Rev.* **2005**, *29*, 1323–1333.
59. Borowsky, B.; Durkin, M. M.; Ogozalek, K.; Marzabadi, M. R.; DeLeon, J.; Lagu, B.; Heurich, R.; Lichtblau, H.; Shaposhnik, Z.; Daniewska, I. et al. *Nat. Med.* **2002**, *8*, 779–781.
60. Brewer, A.; Echevarria, D. J.; Langel, U.; Robinson, J. K. *Neuropeptides* **2005**, *39*, 323–326.
61. Blackburn, T. P.; Swanson, C. J.; Wolinsky, T. D.; Konkel, M. J.; Walker, M. M.; Durkin, M.; Artymyshyn, R. P.; Smith, K. E.; Jones, K. J.; Craig, D. A. et al. *Neuropsychopharmacology* **2005**, *30*, S99.
62. Cosford, N. D.; Tehrani, L.; Roppe, J.; Schweiger, E.; Smith, N. D.; Anderson, J.; Bristow, L.; Brodkin, J.; Jiang, X.; McDonald, I. et al. *J. Med. Chem.* **2003**, *46*, 204–206.
63. Skolnick, P. *Eur. J. Pharmacol.* **1999**, *375*, 31–40.
64. Zarate, C. A.; Singh, J. B.; Carlson, P. J.; Brutsche, M. E.; Ameli, R.; Luckenbaugh, D. A.; Charney, D. S.; Manji, H. K. *Arch. Gen. Psychiatry* **2006**, *63*, 856–864.
65. Schoepp, D. D.; Wright, R. A.; Levine, L. R.; Gaydos, B.; Potter, W. Z. *Stress* **2003**, *6*, 189–197.
66. Spooren, W.; Gasparini, F *Drug News Perspect.* **2004**, *17*, 251–257.
67. Grillon, C.; Cordova, J.; Levine, L. R.; Morgan, C. A. *Psychopharmacology (Berl.)* **2003**, *168*, 446–454.
68. Marek, G. J.; Wright, R. A.; Gewirtz, J. C.; Schoepp, D. D. *Neuroscience* **2001**, *105*, 379–392.
69. Moldrich, R. X.; Beart, P. M. *Curr. Drug Targets CNS Neurol. Disord.* **2003**, *2*, 109–122.
70. Cryan, J. F.; Kaupmann, K. *Trends Pharmacol. Sci.* **2005**, *26*, 36–43.
71. Flores, B. H.; Kenna, H.; Keller, J.; Solvason, H. B.; Schatzberg, A. F. *Neuropsychopharmacology* **2006**, *3*, 628–636.
72. Grillon, C.; Cordova, J.; Levine, L. R.; Morgan, C. A. *Psychopharmacology (Berl.)* **2003**, *168*, 446–454.
73. Young, A. H.; Gallagher, P.; Watson, S.; Del-Estal, D.; Owen, B. M. and Ferrier, I. N. *Neuropsychopharmacology* **2004**, *29*, 1538–1545.
74. O'Donnell, J. M.; Zhang, H.-T. *Trends Pharmacol. Sci.* **2004**, *25*, 158–163.
75. Branchek, T. A.; Blackburn, T. P. *Curr. Opin. Pharmacol.* **2003**, *3*, 90–97.
76. Lindermann, L.; Hoener, M. *Trends Pharmacol. Sci.* **2005**, *26*, 5, 274–280.
77. Datamonitor Strategic Perspectives 2001: Antidepressants DMHC1695. Datamonitor: New York, 2001.

Biographies

Thomas P Blackburn, DSc, FBPhS, is an internationally recognized neuroscientist whose industry experience embraces both major pharmaceutical and biotech companies. He received his graduate degree from Manchester University, England. His expertise lies in neuropharmacology and clinical drug development, and he was responsible for the preclinical and clinical development of several compounds, including Paxil. He has published over 100 papers in the field of neuroscience and he is a co-inventor on several patents. He is a Past President of the British Pharmacological Society and a nonexecutive board member for Motac Neuroscience Ltd.

Jan Wasley has a scientific background in organic and medical chemistry. He received his BSc and PhD at the University of Nottingham, England. Dr Wasley held positions of increasing responsibility within the Research Department of Ciba-Geigy in drug discovery and drug development activities. He joined Neurogen Corporation in 1993 as Director of Chemistry and became involved in the development of Neurogen's Accelerated Intelligent Drug Design (AIDD) technologies. He retired from Neurogen in 2001.

Comprehensive Medicinal Chemistry II
ISBN (set): 0-08-044513-6

ISBN (Volume 6) 0-08-044519-5; pp. 45–83

6.04 Anxiety

K R Gogas, S M Lechner, S Markison, J P Williams, W McCarthy, D E Grigoriadis, and A C Foster, Neurocrine Biosciences Inc., San Diego, CA, USA

6.04.1 Overview

In the early 1980s, anxiety was subdivided into distinct entities including social phobia (aka social anxiety disorder, SAD), posttraumatic stress disorder (PTSD), obsessive compulsive disorder (OCD), generalized anxiety disorder (GAD), and panic in the *Diagnostic and Statistical Manual of Mental Disorders IV/IV-TR* (DSM IV/IV-TR).[1,2] Anxiety is usually associated with fear, nervousness, apprehension, and panic.

6.04.1.1 Generalized Anxiety Disorder

GAD has a lifetime prevalence of approximately 5%[3] with more than 90% of individuals having at least one other comorbid psychiatric disorder including depression or another anxiety disorder.[3] GAD is second to only depression as the most frequent psychiatric disorder in the primary care setting.[4] DSM IV-TR[2] characterizes GAD as excessive anxiety and apprehension occurring more days than not, over at least a 6 month period. The anxiety and apprehension are accompanied by somatic symptoms such as restlessness, disturbed sleep, difficulty concentrating, and muscle tension. These symptoms, taken together, are also associated with a significant subjective distress and impairment in both social and work functioning. GAD tends to be a chronic and recurrent disease with less than one-half of cases remitting without medical treatment. Approximately 48% of GAD patients seek professional help with 25% of these patients receiving medication for the disorder.

6.04.1.2 Social Anxiety Disorder

SAD or social phobia is a common anxiety disorder often associated with serious role impairment. The 12-month prevalence rate for all types of SAD was estimated to be 8% with a lifetime prevalence of 13%. SAD is a chronic disease with a slightly higher prevalence in females than males (15% versus 11%, respectively), with retrospective studies showing an average duration of 25 years. Overall, females with poor baseline functioning at the time of diagnosis have the greatest risk of disease chronicity. DSM IV-TR defines SAD as a marked and persistent fear of social or performance situations in which embarrassment may occur. The diagnosis of SAD is only made when the fear, avoidance, or anxious anticipation of the event persists for at least 6 months and the phobia directly interferes with daily function or when the individual is distressed about having the phobia. Although SAD can relate to a specific set of circumstances (i.e., public speaking), there are cases in which there can be a broad array of fears that include both performance and interactional factors with the latter being considered more severe and disabling than the former.[5] Individuals with comorbid GAD demonstrated more severe psychopathology and exacerbation of the clinical course of SAD over those individuals without comorbid GAD.[6]

6.04.1.3 Posttraumatic Stress Disorder

DSM-IV-TR classifies PTSD as an anxiety disorder with the major criteria of an extreme precipitating stressor, intrusive recollections, emotional numbing, and hyperarousal. Individuals at risk for PTSD include, but are not limited to, soldiers and victims of motor accidents, sexual abuse, violent crime, accidents, terrorist attacks, or natural disasters such as floods, earthquakes or hurricanes.[7] PTSD has acute and chronic forms. In the general population, the lifetime prevalence of PTSD ranges from 1% to 12%[7] and is frequently comorbid with anxiety disorders, major depressive disorder, and substance abuse disorders with a lifetime prevalence of comorbid disease ranging from 5% to 75%.[7] PTSD is often a persistent and chronic disorder and a longitudinal study of adolescents and youth with PTSD showed that more than one-half of individuals with full DSM-IV-TR PTSD criteria at baseline remained symptomatic for more than 3 years and 50% of those individuals with subthreshold PTSD at baseline remained symptomatic at the 34–50 month follow-up.[8]

6.04.1.4 Obsessive Compulsive Disorder

OCD is a chronic and often disabling disorder that affects 2–3% of the US population. OCD has been labeled a 'hidden epidemic' and is ranked 20th in the Global Burden of Disease studies among all diseases as a cause of disability-adjusted life years lost in developed countries. OCD is often associated with substantial quality of life impairment especially in individuals with more severe symptoms. The disease usually begins in adolescence or early adulthood with 31% of first episodes occurring at 10–15 years of age and 75% by the age of 30. The essential features of OCD are recurrent obsessions or compulsions that are severe enough to be time consuming (i.e., take more than 1 h per day) and/or cause significant levels of distress or interference with normal daily activities. OCD can have comorbidity with major depression and social phobia, as well as other mental disorders such as eating disorders and schizophrenia.[9]

6.04.2 Neural and Genetic Basis of Anxiety

6.04.2.1 Neurocircuitry

The anatomical basis of fear and anxiety in humans has not been definitively established; however, through the use of animal models of stress and conditioned fear, particularly those of Pavlovian fear conditioning and fear potentiated startle,[10] a basic model of normal fear responding has been developed and refined. While anxiety is not identical to fear, it is closely linked to fear responding. From these preclinical studies, a 'fear neurocircuitry' has been proposed, which involves an extended anatomical network that centers on the critical involvement of the amygdala (AMYG), located in the anterior part of the medial temporal lobe.[11]

The AMYG is comprised of numerous subnuclei that include the basolateral complex (BLC; lateral, basal, and accessory basal nuclei) and the central nucleus (CeA; anterior, medial, and lateral nuclei). Together, these nuclei receive and process sensory input necessary for assessment of a potentially threatening stimulus, incorporate memory and prior experience into this appraisal, and coordinate the autonomic and behavioral components of fear responding. In addition to extensive 'cross-talk' and reciprocity between nuclei within the AMYG many additional structures function with the AMYG during fear learning. Key among these are the sensory thalamus and cortices, additional mesiotemporal nuclei, the orbital and medial prefrontal cortex, the anterior insula, the hypothalamus, and multiple brainstem nuclei. A schematic of this circuitry can be found in **Figure 1**.[12]

Sensory information that is vital to threat assessment is received by the AMYG via two pathways, a *monosynaptic* pathway from the anterior thalamus and a *polysynaptic* pathway from sensory cortical areas. Projections from sensory thalamus to lateral AMYG are thought to be involved in rapid conditioning to simple auditory and visual features of a stimulus, while projections from primary sensory and sensory association cortices to lateral AMYG appear to be involved in conditioned responses to more complex sensory stimuli. The most extensive extra-AMYG projections to the sensory thalamus are reciprocal to the basal and accessory basal amygdalar nuclei and these nuclei are thought to be involved in the development of long-lasting memory traces for fear conditioning.[13]

In addition to the reception of sensory information, the AMYG facilitates the acquisition of additional information regarding specific threats via reciprocal projections to subcortical and limbic cortical regions. The rostral perirhinal cortex, the anterior insula, and the hippocampus are particularly involved in modulation of contextual fear, providing information to the AMYG about the context of a potentially threatening stimulus or situation. In addition to a role in spatial contextual conditioning, projections from the hippocampus to the AMYG provide information about the environment that is retrieved from specific memory stores. The medial frontal cortex (mFC), including infralimbic,

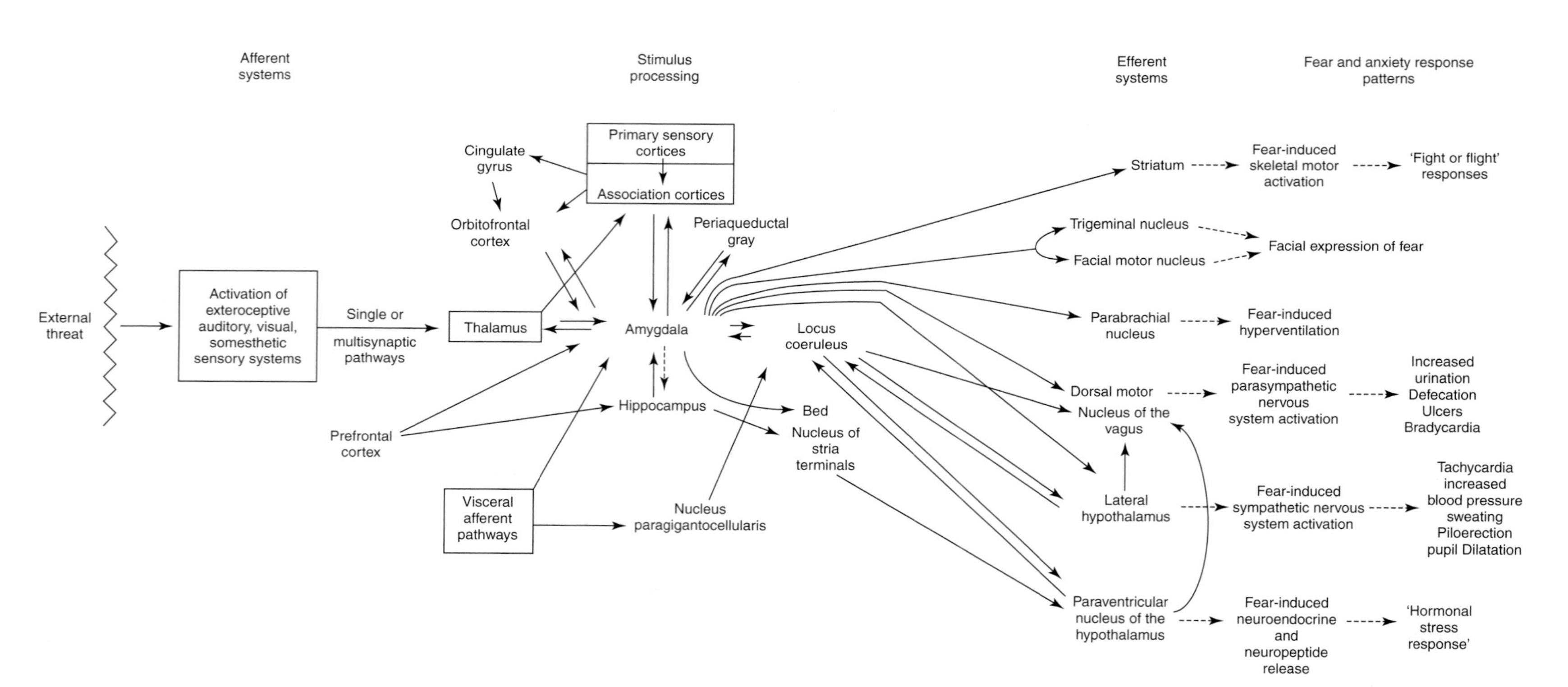

Figure 1 Key anatomical pathways comprising the 'fear neurocircuitry' underlying fear- and anxiety-induced responses. This circuitry centers on the amygdala and includes afferent input from sensory cortex and thalamus in response to external threat, activation of numerous reciprocal connections to the amygdala that are involved in processing this afferent information and incorporating information from memory stores, and the efferent systems ultimately responsible for eliciting the physiological and behavioral responses characteristic of fear and anxiety states. (Reproduced with permission from Charney, D. S.; Deutch, A. *Crit. Rev. Neurobiol.* **1996**, *10*, 419–446.)

anterior cingulate, and medial and orbitofrontal cortex, is involved in attenuating fear responses and extinguishing behavioral responses to fear-conditioned stimuli that are no longer reinforced. In this way, the mFC ameliorates the fear response when the threat has been removed or its import altered. In addition, the flow of information reaching the AMYG is regulated at the level of the thalamus, via cortico-striato-thalamic projections.

The AMYG has considerable reciprocal connections with the bed nucleus of the stria terminalis (BNST) that is proposed to be part of an 'extended amygdala,' based on its chemoarchitecture and parallel neurocircuitry. The BNST, as well as the CeA, project to the hypothalamus and other brainstem nuclei that play a role in regulation of the autonomic and neuroendocrine aspects of fear and anxiety. Stimulation of the lateral hypothalamus via direct projections from the CeA, as well as through projections via the locus coeruleus, results in activation of the sympathetic nervous system and leads to increases in blood pressure, heart rate, sweating, piloerection, and pupil dilation, whereas projections to the dorsal motor nucleus of the vagus are involved in activation of the parasympathetic nervous system and lead to increases in urination, defecation, and decreased heart rate. Similarly, projections from the CeA to the parabrachial nucleus are involved in activation of respiration, or fear-induced hyperventilation. Projections from the CeA and the locus coeruleus to the paraventricular nucleus of the hypothalamus are important in regulating the hypothalamic-pituitary-adrenal (HPA) axis response to stressful or fearful stimuli.

The behavioral components of fear responding are also coordinated to a large extent by the AMYG. Direct projections from the CeA to the ventral lateral periaqueductal gray (PAG) are thought to mediate freezing behavior in response to threatening stimuli.[14] In contrast, efferent projections from the CeA to lateral PAG, as well as from the basal nucleus to the nucleus accumbens and striatum are thought to mediate active avoidance behaviors to fear.[14] It is a complex interaction of pathways that are involved in eliciting motor responses to threat, including activation of the musculoskeletal system for 'fight or flight.' These include cortical association areas and motor cortex and multiple areas within the striatum, including the nucleus accumbens, the olfactory tubercle and parts of the caudate and putamen. The AMYG densely innervates both the prefrontal cortex and the striatum and is positioned to play a role in the regulation of both of these systems in response to threatening stimuli.

While the neuroanatomical correlates underlying fear and anxiety are quite complex, the brain areas described above are thought to be key components of this neurocircuitry and provide a basis for the exploration into the human neurocircuitry underlying specific anxiety disorders.

6.04.2.2 Neuroimaging

Structural and functional neuroimaging studies have been used to investigate the human neuroanatomy and neurocircuitry of specific anxiety disorders.[11,15] Structural neuroimaging techniques include morphometric magnetic resonance imaging (mMRI), which allows for accurate assessment of brain structure volumes, and magnetic resonance diffusion tensor imaging, which allows for determination of white matter tract orientations. Functional imaging techniques focus on acquisition of data that reflect regional brain activity. These include positron emission tomography (PET) with tracers that measure blood flow (^{15}O-carbon dioxide or water) or glucose metabolism (^{18}F-fluorodeoxyglucose), single photon emission computed tomography (SPECT) with tracers that measure correlates of blood flow, and functional magnetic resonance imaging (fMRI), which measures blood oxygenation level-dependent (BOLD) signal changes. Functional imaging studies can be performed using several paradigms.[11] Neutral state paradigms image subjects in a resting state while performing a nonspecific continuous task, symptom provocation paradigms image subjects in a neutral state and then again when an anxious state is induced behaviorally or pharmacologically, and cognitive activation paradigms image subjects while performing cognitive tasks.

6.04.2.2.1 Generalized anxiety disorder

Little information utilizing neuroimaging is available on the neurocircuitry underlying GAD, although some associations have been made between GAD and major depressive disorder (MDD). These studies are difficult to interpret due to the high degree of comorbidity between the two disorders.

6.04.2.2.2 Social anxiety disorder

Neurocircuitry models of SAD are still being developed and implicate dysfunctional connections between the AMYG and areas involved in processing social cues. Structural studies, as well as neutral state functional studies, have shown few differences between SAD patients and healthy volunteers in any brain regions. In symptom provocation paradigms contrasting public versus private speaking conditions and in cognitive activation studies where subjects were exposed to human face stimuli or to these stimuli paired with an aversive stimulus, increased activity was observed in the AMYG and related medial temporal lobe areas (orbitofrontal, cingulate, insular cortices).[16]

6.04.2.2.3 Posttraumatic stress disorder

Neurocircuitry models of PTSD have focused on the interactions between the AMYG and the medial prefrontal cortex and hippocampus. Hypersensitivity and lack of habituation to repetitive nonthreatening stimuli characteristic of PTSD may be a result of decreased regulation of AMYG activity by the anterior cingulate cortex suggesting an impairment of processes mediating extinction to trauma-related stimuli, as well as decreased hippocampal function, implying an impaired ability to form appropriate contextual associations and dysfunctional memory processing with respect to trauma-related stimuli. mMRI studies of PSTD have focused primarily on the hippocampus. Reduced hippocampal volumes in Vietnam combat veterans have been reported and studies in adults with either childhood or adult traumatic experiences also support an association between reduced hippocampal volume and PTSD. Studies in pediatric populations, however, have failed to show reductions in hippocampal volumes, suggesting that this decrease may develop over time. It is also possible that smaller hippocampal volumes may be a risk factor for developing PTSD, rather than a consequence of the trauma.[17] Functional imaging studies using a neutral state paradigm have shown increased regional cerebral blood flow in the caudate and putamen, as well as the orbitofrontal and anterior cingulate cortices. Several studies of PTSD using symptom provocation paradigms have shown increased activity in the AMYG and a lack of involvement of the anterior cingulate cortex.[18] Studies examining responses in a working memory task suggest patients with PTSD have a distinct pattern of brain activation in response to the organization and processing of verbal working memory, relying more heavily on nonverbal working memory and shifting away from verbal processing.

6.04.2.2.4 Obsessive compulsive disorder

Neurocircuitry models of OCD suggest that OCD symptoms are mediated by hyperactivity in orbitofrontal-subcortical circuits, with a primary pathology centered in the striatum.[15] Functional imaging studies have demonstrated increased activity in orbitofrontal and anterior cingulate cortices in neutral state paradigms, which is accentuated in symptom provocation and cognitive activation paradigms, and attenuates with successful treatment.[19] Increased metabolism in the striatum has also been observed in untreated OCD patients and correlates well with the degree of symptom severity.

6.04.2.3 Neurochemical Basis of Anxiety

Neurochemical modulators exist at every level within the complex neurocircuitry described above, providing a large number of potential therapeutic targets for the treatment of anxiety. Despite the variations in clinical presentation of the various anxiety disorders, extensive preclinical studies have provided significant evidence for the role of multiple neurochemical systems in the etiology of anxiety, in general, with a significant amount of overlap between the individual disorders themselves. The major neuromodulators and neuromodulatory systems that have been implicated both through preclinical and clinical studies include the monoamines (5-hydroxytryptamine (5HT), norepinephrine (NE), and dopamine), corticotrophin-releasing factor (CRF), γ-amino-butyric acid (GABA), glutamate, neuropeptides (substance P, neuropeptide Y, galanin), cholecystokinin, and neurotrophic factors.[20]

6.04.2.4 Genetic Basis of Anxiety

Both human and animal studies have provided evidence for a genetic influence in the etiology of anxiety. A growing number of genetically altered mice have been produced in an effort to identify specific genes conferring enhanced or reduced risk for anxiety disorders. A meta-analysis of family studies, as well as twin studies and prospective studies in children of anxious parents, have demonstrated a significant familial aggregation for all anxiety disorders. From these studies, the range of reported genetic effect has been 25–65%. It has also been postulated that anxiety disorders may represent a complex genetic trait in which genetic risk factors are probabilistic rather than deterministic in nature.[21]

The development of knockout and transgenic mice has facilitated investigations into the contribution of several classes of neurotransmitters/receptors to anxiety.

6.04.2.4.1 GABAergic

Endogenous GABA levels were altered in mice by targeting the two isoforms of glutamic acid dehydrogenase (GAD) responsible for the synthesis of GABA. Mice lacking GAD67, which is thought to be responsible for maintenance of basal GABA levels, produced a prenatal lethal phenotype. Deletion of the GAD65 gene, thought to be activated in response to high endogenous GABA demand, reduced GABA levels, was associated with a lower rate of lethality than the GAD67 mutant mice, and exhibited an increase in anxiety-like behavior in the elevated zero maze and open field tests.[22]

Deletion of the $\beta 2$ subunit gene of the $GABA_A$ receptor decreased the total number of $GABA_A$ receptors in the brain, but did not produce any clear anxiety-related phenotype.[23] Mice deficient in both splice variants of the $\gamma 2$ subunit ($\gamma 2S$, $\gamma 2L$) of the $GABA_A$ receptor were insensitive to either the behavioral or electrophysiological effects of benzodiazepines.[24] Homozygous mutants failed to survive past postnatal day 18, and behavioral testing with heterozygote $\gamma 2$ mutants showed anxiogenic-like behavior in the open field, elevated plus maze (EPM), and light/dark box tests.[24] Although the $\gamma 2$ subunit knockout mice showed some increases in anxiety-like behavior, transgenic mice overexpressing either the $\gamma 2S$ or $\gamma 2L$ subunits of the $GABA_A$ receptor failed to exhibit any anxiety-like phenotype when compared to the responses of their wild-type littermate controls.[25]

6.04.2.4.2 **Noradrenergic**

The major mammalian enzymes involved in catecholamine metabolism are monoamine oxidase (MAO) and catechol *O*-methyl transferase (COMT). Mice lacking MAO-A have increased levels of NE and 5HT in the brain.[26] Behavioral evaluation of these animals showed a reduction in anxiety-related behaviors in the open field, but it remains unclear whether this effect is anxiety specific or related to a general alteration in sensory ability in the animals.[26] MAO-B mutants exhibit an increase in brain and urinary β-phenylethylamine (the preferential MAO-B substrate), with no alterations in brain levels of NE, 5HT, or DA.[27] The MAO-B mutants did not exhibit any behavioral differences in anxiety tests such as the open field or EPM.[27] Deletion of the gene for COMT produced an increase in dopamine in the frontal cortex of male, but not female, mutant mice, and no change in NE or 5HT levels in any region.[28] When tested behaviorally, only female COMT mutants exhibited anxiogenic (increased anxiety) behavior in the light/dark box test.

Reuptake of NE via the norepinephrine transporter (NET) is a major means of inactivation of the released transmitter in the brain and peripheral nervous system.[29] Deletion of the NET gene results in a sixfold decrease in the clearance rate for NET and a twofold increase in the extracellular NE levels.[30] NET-deficient mice exhibited reduced anxiety-like behavior in the open field test, but this is difficult to interpret given the overall difference in basal activity levels in the mutant versus wild-type mice.[30]

Deletion of the α_2-adrenoceptor resulted in an anxiogenic phenotype in the EPM and open field tests.[31] This is thought to be related to the increased NE turnover and synaptic concentrations of NE found in the mutant mice.

6.04.2.4.3 **Serotonergic**

Deletion of the serotonin transporter (SERT) in mice produced an alteration in brain 5HT homeostasis, decreased tissue concentrations of 5HT, and an increase in anxiety-related behaviors (EPM, light/dark box test) that was more pronounced in females than males.[32]

Deletion of the gene for the $5HT_{1A}$ receptor produced an 'anxious' phenotype in a variety of anxiety models. In addition to this anxiogenic profile, $5HT_{1A}$ knockout mice also showed increased autonomic activity (i.e., elevated body temperature, tachycardia) in response to stress, changes that are similar to those seen in humans with anxiety or stress disorders.

Deletion of the $5HT_{1B}$ receptor gene has been associated with increased aggression. The presence of an anxiety-related phenotype in these mice is equivocal as increased, decreased, or unchanged anxiety levels have been reported in the EPM and light/dark box. $5HT_{1B}$ knockouts exhibited reduced anxiety in the EPM and the novel exploration and novelty suppressed feeding tasks but showed no difference in the light/dark box test.[33]

Mutant $5HT_{2C}$ receptor mice have a reduced anxiety phenotype, perhaps consistent with the findings that show an anxiolytic effect of $5HT_{2C}$ antagonists in animals.[34] Experiments with mice lacking the gene for the $5HT_{5A}$ receptor suggest that this receptor may modulate exploratory behavior rather than anxiety per se.[35]

6.04.2.5 **Human Genetics**

The search for susceptibility genes for anxiety is a major priority. However, several caveats are noteworthy in attempting to interpret the data generated to date. While many investigators define patient populations based on DSM-IV-TR criteria, controversy exists over whether anxiety phenotypes should be defined more broadly or more narrowly. Thus, discrepancies among similar studies may be due to diagnosis selection criteria. In addition, factors such as comorbidity and heterogeneity within specific anxiety disorder populations must be taken into consideration. Molecular genetic strategies themselves provide differing levels of reproducibility. Linkage studies involve genotyping unknown DNA markers across the genome in large pedigrees or affected relative pairs to determine the approximate chromosomal location of susceptibility genes. In these studies the gene locus does not need to be known, however genes involved in small effects remain undetected. Genetic association studies involve the selection of candidate genes based on location determined from linkage studies or on disease pathology. However, due to both population variability and the choice of polymorphisms, they also show the greatest susceptibility for discrepancies.

Association studies in PTSD have implicated the *DRD₂* dopamine receptor and the DAT gene as susceptibility gene candidates. Studies in SAD have provided a potential linkage to chromosome 16, a region containing the gene for NET, *SLC6A2*. Studies in OCD have linked the disorder to 9p24, an area where the glutamate transporter gene, *SLC1A1*, is located, although no evidence for biased transmission of this transporter was obtained. Association studies have shown varying results but strong susceptibility gene candidates include genes for: $5HT_{1B}$, $5HT_{2A}$ receptors, *5HTTLPR* (SERT), and the DRD4 dopamine receptor. Novel candidates awaiting further analysis in OCD include genes for: BDNF (brainderived neurotrophic factor), the *GRIN2B* (glutamate receptor, ionotropic *N*-methyl-D-aspartate subunit 2B), and *MOG* (myelin oligodendrocyte glycoprotein). Association studies in GAD have demonstrated a positive association with the MAO-A gene (catecholamine degradation enzyme). For a review of the genetics of anxiety disorders, see [36].

6.04.3 Animal Models of Anxiety

There are several classical animal models used to evaluate the anxiolytic potential of novel and existing compounds. It is important to remember that when using animal models of CNS disorders, there is a risk of both false positives and false negatives. Thus, it is important to evaluate a novel entity in several anxiety models to determine the 'signature' of the compound to compare it with clinically effective anxiolytics.

6.04.3.1 Elevated Plus Maze/T-maze

The EPM is a rodent model of anxiety that has been used extensively in the characterization of both established drugs and new chemical entities for anxiolytic activity.[37] It is an ethologically based test that uses nonpainful, nonaversive stimuli to induce fear and anxiety thus reducing the possible confounds of motivational and perceptual states. The test is a modification of the Y-maze test that relies on the propensity of a rodent to spend less time in the open areas of the maze than the closed areas. A typical apparatus is the shape of a 'plus' sign and has two elevated arms, one open and one closed. The center of the maze is an open area, and the animal is placed on this center area at the start of the test. Over the course of a 5 min period, the time the animal spends in either the open or closed areas is recorded. Since rodents have an innate fear of height and openness untreated/naive animals spend less time on the open versus closed arms, and anxiolytics like diazepam will increase the amount of time spent in the open arm reflecting a decrease in the anxiety of the animal.

The elevated T-maze is a derivation of the EPM and consists of three arms, with one being enclosed by lateral walls. Unlike the EPM where the animal is placed on the open center area at the start of the test, there are two options when running the T-maze test. The animal can be placed in the enclosed arm and since the rodent does not see the open arms until it emerges from the enclosed arm, it can be trained in an inhibitory avoidance response if placed in the enclosed arm multiple times. When placed on the end of the open arms, the animal can move towards the enclosed arm thus performing an escape response. The model has also been pharmacologically validated[38] using benzodiazepines (BZs; diazepam), azapirones (buspirone, ipsapirone), and the $5HT_2$ antagonist, ritanserin. All three drug classes produced anxiolytic effects in the inhibitory avoidance but not the escape component suggesting that the former rather than the latter is related to GAD.

6.04.3.2 Light–Dark Box

Like the EPM, the light–dark test is an ethological model that relies on both the innate aversion of rodents to brightly illuminated areas and their tendency to exhibit spontaneous locomotor behavior in response to novel environments.[39] The typical apparatus has two connected compartments, one is lit (aversive area) and the other is darkened (safe area), with an opening between the two compartments. This test was developed using male mice and the C57BL/6J and SW-NIH strains are the best strains in which to evaluate the potential anxiolytic effect of new chemical entities (NCEs). Mice are placed in the light area and then begin moving along the periphery of the compartment until they locate the opening to the dark compartment. The time the animal spends moving, rearing, and transitioning from the dark to light area is recorded over a 5–10 min period. Anxiolytics increase locomotion and time spent in the light area while anxiogenic compounds have opposite effects. While BZs are reliably detected in this model as being anxiolytic, the sedative effects of this class of compounds at high doses can confound the results. Inverse agonists at the $GABA_A$ site also produce anxiogenic effects in the light-dark test. There is controversy concerning the effectiveness of drugs that modulate 5HT neurotransmission in this test. Drugs acting on $5HT_1$ receptors (e.g., buspirone) work in this test, but tend to do better in conflict-based models (see below). $5HT_2$ agonists and $5HT_3$ antagonists show anxiolytic effects in this model, while SSRIs tend to be ineffective.[40]

6.04.3.3 Conflict Models

There are two types of conflict tests, both of which involve conditioned and punished responses. In the Geller–Seifter (GS) test, food-deprived rats are trained to press a lever for food and when the animals reach a stable state of lever pressing for the food, a shock component is introduced such that responding for food also elicits a mild electric shock (conflict component). NCEs with anxiolytic activity release the suppression of the punished response.[41] An important parameter of the test is that it measures punished as well as nonpunished responding so that an assessment of the potential motor effects of the NCE can be evaluated within the test session itself. This test has proven invaluable for evaluating the potential anxiolytic properties of NCEs, but requires a lengthy daily training period before an NCE can be evaluated. An alternative conflict model is the Vogel conflict test (VCT), in which water-deprived rats receive a mild shock from the spout of the test bottle during portions of the test session. This model circumvents the long training process required for the GS test but because it does not measure nonpunished as well as punished responses, there should be a thorough examination of whether the NCE under evaluation has motor or analgesic properties since these effects could confound the test results. Both models have good predictive validity with $GABA_A$, serotonergic, glutamatergic, and adrenergic compounds producing reversal of the punished response in both paradigms.[42]

The four-plate test is an animal model of anxiety in which novel exploratory behavior in novel surroundings is suppressed by delivering mild foot shock to plates that are organized into four quadrants.[43] Similar to the VCT and GS tests described above, administration of standard anxiolytic drugs such as diazepam and alprazolam, as well as the SSRIs (e.g., sertraline, citalopram) and serotonin-norepinephrine reuptake inhibitors (SNRIs; e.g., milnacipran, venlafaxine) produce a significant increase in punished responding.[44]

6.04.3.4 Fear Potentiated Startle

The startle response is an autonomic, reflexive response that is seen across species.[45] It consists of a rapid and sequential muscle contraction that is thought to protect the body and facilitate the flight reaction to avoid a sudden attack. The fear potentiated startle (FPS) test can be used to assess differences in startle reactivity and in both animals and humans is thought to represent a form of anticipatory anxiety. The amplitude of the startle reflex can be augmented in the presence of a cue that has previously been paired with foot shock to provide a conditioned stimulus. A light or visual stimulus can be used as the conditioned stimulus and either sound or air puff to induce startle. The basic training paradigm for the FPS test consists of two phases, conditioning and testing. The conditioning phase includes two sets of rats: one set receives light-shock pairings at a fixed interval; and the second set receives lights and shock in random pairings. During the testing phase, startle is elicited by an auditory stimulus (e.g., 100 dB burst of white noise) in the presence or absence of the conditioned stimulus. The difference in the amplitude of the startle response in the presence or absence of the conditioned stimulus indicates the fear potentiated component of the response. A similar paradigm has been used on human subjects using an air puff as the startle stimulus and a wristband fitted with an electrode as the conditioned stimulus. A variety of drugs are effective in the rat FPS model, including the α_2 agonist, clonidine, the opioid agonist morphine, $GABA_A$ positive allosteric modulators such as diazepam, and the $5HT_{1A}$ partial agonist, buspirone.[46] It is interesting to note that the data in the human FPS model are equivocal with respect to the effects of benzodiazepines, although this class of compounds has been shown to reduce the startle response in humans.[47] This difference is perhaps consistent with the finding that this class of compounds does not work well in the treatment of fear-related disorders such as phobias.

6.04.3.5 Defensive Burying

Defensive burying[48] refers to the rodent behavior of displacing bedding material with treading and shoveling movements of their heads and forepaws towards a stimulus that they perceive to be a near and/or imminent threat (e.g., introduction of a wall-mounted electrified probe into the cage). The animal is placed into the cage with the probe exposed, and following the initial contact with the probe behavior is observed for 10–15 min. Although initial studies recorded only probe directed burying time, later studies included measurement of several concurrent behavioral indices of fear/anxiety, reactivity, and exploration (for review see [49]). In addition to defensive burying of the probe, animals also display increases in freezing postures away from the probe indicating both active and passive avoidance components of the test. The predictive validity of this test is better than other tests of anxiety, with $GABA_A$ and serotonergic compounds showing dose-related suppression of defensive burying and freezing behavior. Further, the relative potency of clinically used compounds in this test is roughly comparable to those used in the therapeutic setting.[49] Suppression of burying has also been reported for putative anxiolytics with novel mechanisms of action including CRF antagonists,[50] neurosteroids,[51] and angiotensin II antagonists.[52]

In addition to the defensive burying of electrified probes, mice also bury harmless objects like glass marbles. This latter test is relatively simple to set up and requires placing mice into a cage containing a layer of marbles evenly spaced on top of the bedding. Thirty minutes later, the animal is removed and the number of buried marbles recorded. Inhibition of marble-burying behavior in mice is considered to be a correlative, rather than a direct, measure of anxiety behavior since the marbles themselves are not a fear-provoking stimulus as in the case of the electrified probe. SSRIs, buspirone, and neurokinin (NK_1) receptor antagonists are effective in this test.[53,54] Given this pharmacological profile, it has been suggested that the marble burying behavior by mice may be more reflective of OCD rather than anxiety although the validity of this test as a model for OCD remains to be determined.

6.04.3.6 Open Field

The open field test is a widely used model of anxiety-like behavior developed to evaluate emotionality in animals and is based on subjecting an animal to an unknown environment whose escape is prevented by surrounding walls. The animal is placed into the center or close to the walls of an apparatus (circular, square, or rectangular) and s everal behavioral parameters (usually horizontal/vertical activity and grooming) are recorded over a set time, e.g., 5 min. Rodents typically prefer not to be in the center, lit area of the apparatus and tend to walk close to the walls (thigmotaxis) since this is a novel, and presumably, stressful, environment to the animal. NCEs with anxiolytic activity increase time spent in the open center area and also decrease the latency to enter the center area after being placed into the apparatus. The predictive validity of this test is generally good with chlordiazepoxide, diazepam, and $5HT_{1A}$ agonists, which consistently produce anxiolytic effects; however, SSRIs are inactive in this test.[55]

6.04.3.7 Defensive Test Battery

This test battery is based on rodent unconditioned behaviors such as freezing, hiding, taking flight, defensive threat/attack, and risk assessment[56,57] with the flight component being thought to relate to panic. NCEs that are clinically effective in treating panic disorder as well as anxiety are effective in this model. The fear/defense test battery developed for rats uses a long (6 m) oval runway apparatus with a human experimenter representing the threat stimulus. When the runway is in the oval configuration, rats exhibit flight and escape behaviors; changing the runway to a straight alleyway elicits a freezing response. This test is primarily conducted using wild trapped *Rattus norvegicus* and *R. rattus* since laboratory bred rats have blunted escape, freezing, and defensive behaviors.[58] BZs selectively reduce defensive threat vocalizations while $5HT_{1A}$ agonists reduce both defensive threat vocalizations and defensive attack while having no effect on flight avoidance or freezing.[58–60] The anxiety/defense test battery also uses a threatening stimulus, usually a cat or cat odor, but the stimulus does not approach the animal during the test.[57] Thus, this task measures the inhibition of normal behaviors, freezing and avoidance of the area containing the threatening stimulus. It is also used to measure risk assessment: the orientation of the animal to openings/corners in a stretched posture followed by episodes of head poking to scan and sniff the area.[60] Diazepam and chlordiazepoxide reduce avoidance of the threat stimulus as well as general behavioral inhibition and also reduce risk assessment when cat odor is present, while $5HT_{1A}$ agonists like gepirone and 8-OH-DPAT reduce the avoidance behavior and behavioral inhibition but have no effect on risk assessment or freezing behavior.[57]

The mouse defensive battery is a parallel test to the rat fear/defense and anxiety/defense test battery.[57] Overall, the behaviors are similar to those of the rat with one exception being the observation of putative risk assessment behaviors in the fear/defense portion of the assessment. These behaviors include flight, flight speed, freezing, defensive vocalizations and attacks, risk assessment, and contextual defense that involve attempts to escape after the predator (anesthetized rat) is no longer near. In the mouse battery, both nonselective $GABA_A$ positive allosteric modulators and compounds with functional selectivity for $\alpha 2$ subunit-containing $GABA_A$ receptors (SL-651498), are active,[61] as are $5HT_{1A}$ receptor antagonists.[62] Selective and mixed SNRI compounds as well as MAO inhibitors have clear effects on the flight response when administered chronically but not acutely.[63,64] This is perhaps consistent with the use of fluoxetine (SSRI), imipramine (SNRI), and phenelzine (MAO inhibitor) for the treatment of panic disorder in the clinic. CRF antagonists (reduce effective components of defensive behaviors), NK_2 receptor antagonists (decrease flight, risk assessment, and contextual anxiety), neurotensin antagonists (reduce risk assessment), and vasopressin antagonists (reduce defensive aggression) also have activity in this battery of tests suggesting that it can be used to evaluate anxiolytic activity of compounds with novel mechanisms of action.[57]

6.04.3.8 Social Interaction (SI) Test

The SI test[65] was the first ethologically based anxiety model that used natural behavior as a dependent variable. In general, rats are placed together in pairs in a test arena and the time spent interacting with each other (sniffing, grooming, following) is recorded. Environmental manipulations can increase or decrease the amount of time that the rats interact thus allowing for assessment of either anxiolytic (increased interaction) or anxiogenic (decreased interaction) effects of the NCE. The four test conditions are: low light, familiar arena (LF lowest level of anxiety); high light, familiar and low light, unfamiliar arena (HU, LU moderate anxiety); and high light, unfamiliar arena (HU highest level of anxiety). To evaluate the anxiolytic effect of compounds, the HU environment is selected while anxiogenic effects can be evaluated using the LF condition. Although this test is relatively easy to set up and run, there are a number of behavioral (single versus grouped housing) and environmental (noise, stress) factors that must be addressed prior to using the model to evaluate the anxiolytic/anxiogenic effects of drugs (for review see [65]). In addition to the rat SI test, this test has also been developed for use in mice and gerbils. A range of $GABA_A$ positive allosteric modulators are effective anxiolytics in the SI test after subchronic (5 days) but not acute administration since the sedative effects seen after acute dosage interferes with the measurement of SI.[65] Buspirone has anxiolytic effects at low doses and anxiogenic effects at higher doses in the SI test.

6.04.3.9 Stress-Induced Hyperthermia

Stress-induced hyperthermia is an autonomic response that occurs prior to and during stress and/or stress-related events.[66] The first paradigms used group housed mice and evaluated the change in rectal temperature twice at 10 min intervals. The procedure produced reliable elevations in animal core temperatures with anxiolytics reducing the stress-induced hyperthermia response. Later studies showed that similarly robust effects could be produced using singly housed mice, allowing for an average of 10% fewer mice per study. Diazepam and chlordiazepoxide had anxiolytic activity in this model,[67] although subunit selective $GABA_A$ compounds like zolpidem exhibited only marginal activity.[67] $5HT_{1A}$ receptor agonists (e.g., flesinoxan) have a dose-related inhibition of the hyperthermia response, while partial agonists like buspirone produce a lesser effect.[66] In general, antidepressants (e.g., imipramine, chlomipramine, and mianserin) are ineffective at reversing stress-induced hyperthermia, although fluoxetine has shown partial reversal of the response.[66]

6.04.3.10 Ultrasonic Vocalization (USV) in Rat Pups

USV is an ethologically relevant anxiety model since it relies on the innate response of rat pups between 9 and 11 days postnatal to emit 35–45 kHz ultrasonic vocalization in response to separation from their mother and littermates.[68] Basically, on the day of the study, the pups are separated from their mothers and kept in a warm environment (either home cage or test apparatus) until the NCE is administered. The USV response is assessed via a high-frequency microphone connected to a US signal detection device with data being analyzed to determine the occurrence, frequency, and amplitude of the response.[69] Changes in body temperature, motor activity, and respiratory rate can confound USV, so there must be clear evidence that the compound being evaluated does not affect any of these parameters prior to using the USV to evaluate anxiolytic activity.[69] BZs (e.g., diazepam), $5HT_{1A}$ receptor agonists (e.g., flesinoxan, buspirone), and SSRIs (fluvoxamine, chlomipramine) effectively reduce rat pup USV.[69]

6.04.3.11 Marmoset Human Threat Test

Several models of fear/anxiety can be used in primates,[70] one being the human threat test in marmosets. When confronted by a human observer, marmosets exhibit body movements towards the threatening stimulus including tail postures (elevation of tail to expose genital region), slit stares (flattened ear tufts and partial eye closure), scent marking of cage surfaces, and arching of the back with associated piloerection. They also tend to spend less time at the front of their home cage. Diazepam antagonizes these behaviors and increases the amount of time spent in the front of the home cage in the presence of the human observer. $5HT_{1A}$ agonists (e.g., buspirone) and nonpeptide NK2 tachykinin receptor antagonists also antagonize these responses and cause an increase in the time spent in the front of the home cage.[70]

6.04.4 Clinical Trials in Anxiety

6.04.4.1 Generalized Anxiety Disorder

A consensus meeting of The European College of Neuropsychopharmacology (ECNP)[71] defined the parameters and considerations to be used to evaluate NCEs for the treatment of GAD. The conventional parallel group, double-blind, placebo-controlled trial design is typically used to evaluate the efficacy of NCEs in GAD. To compensate for a potential exacerbation of anxiety resulting from discontinuation of previous treatment, a placebo run-in period is often used prior to randomization. Cross-over studies are not typically used for GAD given the difficulty in assessing carry-over effects in GAD patients. At least two positive, double-blind, placebo-controlled trials using the intent to treat population are needed to demonstrate efficacy in GAD. A minimum severity entry criteria based on a score on a pivotal scale such as the Hamilton anxiety scale (HAM-A) is typically used to select patients for GAD trials, with scores of 18–22 being the typical range for patients at trial initiation. While the HAM-A scale is the most widely used for GAD trials, it overemphasizes somatic anxiety symptoms rather than the anxious worrying and tension that are of more relevance to GAD. Thus, anxious mood and psychic tension are often used together as a pivotal subscale in controlled GAD trials. Other anxiety scales used in GAD trials include the psychopathological rating scale and the hospital anxiety and depression scale but experience with their use as pivotal scales is limited. The recommended duration of acute trials is 8 weeks and that for long-term studies is at least 6 months.

6.04.4.2 Social Anxiety Disorder

The ECNP consensus meeting on SAD[72] recommended the conventional double-blind, placebo-controlled, randomized group comparison study to demonstrate efficacy in SAD. Since the latter can be either generalized or nongeneralized, studies investigating efficacy in this disease concentrate on generalized SAD with symptoms of at least four distinct social situations. As in other anxiety disorders, there is a larger drug versus placebo effect seen in patients in the severe to moderate SAD subgroups, as based on scores of 50–70 in the Liebowitz social anxiety scale (LSAS). A potential confounding issue in SAD trials is the presence or absence of comorbid disease. For example, patients suffering from major depressive disorder over the previous 3–6 months should be excluded from the SAD trial if the study will include a potential antidepressant so the results can be deemed relevant for the general population with SAD rather than those with comorbid MDD. The LSAS is the current gold standard scale used to establish efficacy in SAD trials.[73] The brief social phobia scale (BSPS) has also been widely used to establish efficacy.[74] The clinical global scales (clinical global impression (CGI) improvement and CGI severity) have been used in SAD trials, but these global scales are not recommended as primary scales and are used as secondary scales to help judge the clinical relevance of the trial finding. Trials of 8–12 weeks duration are used to establish efficacy in SAD.[72,73]

6.04.4.3 Posttraumatic Stress Disorder

Double-blind, placebo-controlled studies are required to establish efficacy in PTSD[75] with patient selection requiring a minimum prior duration of the disorder of at least 1 month for acute and 3 months for chronic forms of the disorder. This is required to differentiate PTSD from an acute stress response that could resolve on its own with no therapeutic intervention thus contributing to a high placebo response rate. Comorbid depression is common in the PTSD population, thus complicating the interpretation of studies where the NCE is, or has the potential to be, an antidepressant. Additionally, a thorough investigation is required in the clinical trial population to ensure that there are no previous diagnoses of MDD, GAD, or OCD that could confound the planned efficacy study. Several scales are used to measure response in placebo-controlled PTSD trials, including the clinician-rated treatment outcome PTSD rating scale (TOP-8) and the clinician-administered PTSD scale (CAPS-2). There are also a number of self-rating scales including the impact of events scale[76] and the Davidson trauma scale (DTS).[74] Recent PTSD studies have shown efficacy by 8–10 weeks of treatment (for review see [77]), although it is recommended that placebo-controlled studies of short-term duration for this indication have a duration of 10–12 weeks with at least 6 months being required to show long-term efficacy.[78]

6.04.4.4 Obsessive Compulsive Disorder

Double-blind, placebo-controlled multicenter trials are required to establish efficacy in OCD and require exclusion of patients with comorbid conditions of marked severity, including anxiety, depression, mixed schizoid, and schizotypal disorders. The Y-BOCS (Yale–Brown obsessive compulsive scale), a 10 item, clinician-administered scale, is the most widely used rating scale for assessing symptom severity in OCD.[79] In controlled clinical trials, a decrease of greater than 35% is widely accepted as clinically meaningful with a global improvement rating of much or very improved.[80] Controlled clinical trials with SSRIs for OCD treatment have ranged from 10 to 12 weeks in length with 40–60% of patients achieving improvement on the Y-BOCS. Demonstration of efficacy maintenance (i.e., more than 6 months) is important as chlomipramine and SSRIs generally remain effective in long-term treatment.

6.04.5 Current Treatments

The current pharmacological treatment of anxiety disorders is dominated by two drug classes, the BZs and the monoamine uptake inhibitors (including SSRIs, SNRIs), and, to a lesser extent, the older tricyclic antidepressants (TCAs). $5HT_{1A}$ agonists like buspirone, beta-adrenoceptor blockers, atypical antipsychotics, and anticonvulsants are also used. The dominance of the BZs in the treatment of anxiety disorders has decreased due to concerns regarding the potential for abuse, relapse, and withdrawal symptoms once treatment has ceased. In addition, BZs produce side effects such as sedation, fatigue, decreased psychomotor performance, and cognitive impairment. Since comorbidity exists between depression and anxiety, it is not surprising that antidepressant drugs were tested for their effectiveness in anxiety disorders. Despite a slower onset of action than the BZs, SSRIs and SNRIs currently dominate the pharmacotherapy of anxiety due to their similar efficacy, safety and reduced side effect liability, lack of abuse potential, and ability to treat comorbid depression.

6.04.5.1 Benzodiazepines

The origin of BZs stemmed from the 'diversity oriented,'[81] search for a new tranquilizer. Chlordiazepoxide was synthesized in 1955 but was not evaluated in vivo until 3 years later when it was found to be a potent CNS depressant. The reason for not subjecting chlordiazepoxide to biological evaluation earlier was the fact that the compound was thought to be quinazoline-*N*-oxide **1**, which resulted from the failed synthesis of heptoxdiazine **2** (**Figure 2a**). Several analogs of **1** were subjected to in vivo evaluation and all failed to show the desired effect. The attempted synthesis of **2** was performed simply because the compounds would be diverse relative to the known tranquilizers of the day

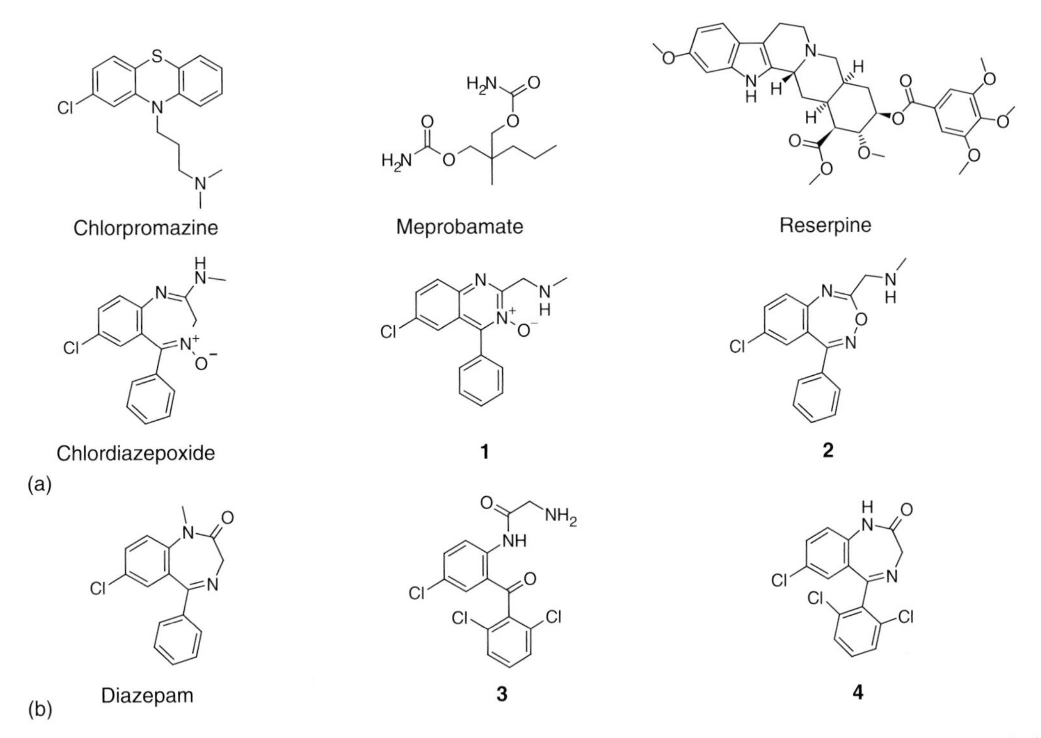

Figure 2 (a) Structures of early tranquilizers, chlordiazepoxide and related compounds. (b) Structures of diazepam and related compounds.

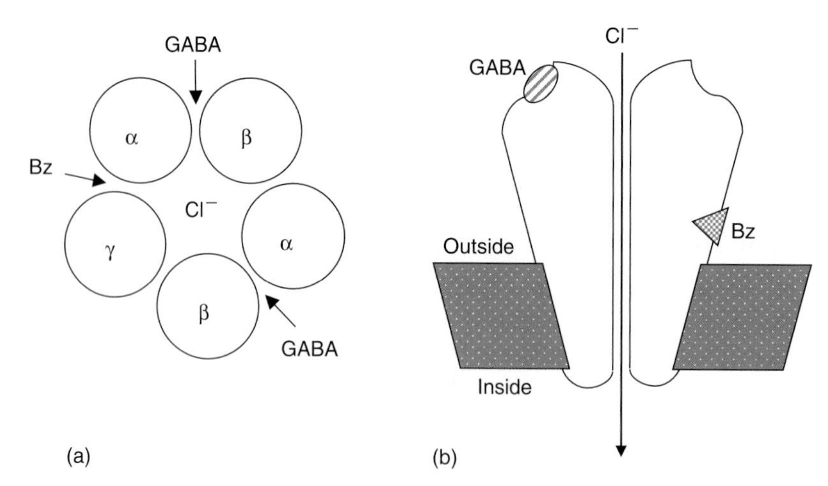

Figure 3 $GABA_A$ receptor. (a) Horizontal cross-section through the $GABA_A$ receptor showing the pentameric arrangement of two α subunits, one β subunit, and one γ subunit to form the central chloride permeable channel. Arrows indicate the two recognition sites for GABA at the α/β interface and the benzodiazepine (Bz) binding site at the α/γ interface. (b) Vertical cross-section.

(chlorpromazine, meprobamate, and reserpine). Chlordiazepoxide eventually was tested as a result of a laboratory clean up and an interest in publishing the chemistry. The compound ultimately proved active in animal models used to test for tranquilizers and its structure was accurately determined. Ultimately, chlordiazepoxide was evaluated successfully in the clinic and two and a half years elapsed between the first in vivo test and its commercial launch.

BZs interact with the $GABA_A$ ligand-gated ion channel.[82] GABA is the primary inhibitory neurotransmitter in the mammalian CNS. The effect of synaptically released GABA on receptive neurons is to inhibit neuronal activity through activation of the $GABA_A$ receptor to increase the postsynaptic chloride ion permeability of the cell leading to hyperpolarization of the membrane potential. The $GABA_A$ receptor is a ligand-gated ion channel (**Figure 3**) formed by the hetero-oligomeric arrangement of five transmembrane spanning subunits from 16 different genes, α (1–6), β (1–3), γ (1–3), δ, ε, π, and θ.[82–84] In most neurons, two α subunits, two β subunits, and one γ subunit form the typical $GABA_A$ receptor. The δ, ε, π, and θ subunits have some reported selective functions but are not yet fully understood. The five subunits assemble to form an internal pore that is selectively permeable for anions. GABA binding to a specific site on each α subunit produces a conformational change in the complex that allows chloride ions to pass through the pore down a concentration gradient. BZs enhance the effect of GABA by binding to a distinct site at the interface of the α and γ subunits increasing the affinity of GABA for its recognition site, acting as allosteric modulators of $GABA_A$ receptor function. Ligands that bind to the BZ site can enhance the effect of GABA (positive allosteric modulators, aka BZ agonists), reduce the effect of GABA (negative allosteric modulators, also known as BZ inverse agonists), or bind with no consequence for GABA, but compete with both positive and negative allosteric modulators (neutral allosteric modulators, also known as BZ antagonists). A continuum of allosteric effects is thus possible for BZ ligands. The degree of efficacy of BZ ligands (full or partial) is an important determinant of the pharmacological effect. The historical BZ anxiolytics, which also have anticonvulsant and muscle relaxant activities, e.g., diazepam (**Figure 2b**), are high efficacy positive allosteric modulators that produce a high degree of $GABA_A$ receptor enhancement, and the high efficacy negative allosteric modulators, e.g., the beta-carbolines, have the opposite effect, producing anxiolytic and convulsant effects by reducing $GABA_A$ receptor function.

Two moieties that were initially presumed to determine the activity of chlordiazepoxide, the methyl amine and *N*-oxide, could be removed without loss of biological activity. This structure–activity relationship (SAR) work led to the identification of the more potent BZ, diazepam. Acyclic analog **3** (amino ketone benzodiazepine precursor) is less active in $[^3H]$-diazepam binding with IC_{50} values of greater than 1000 nM versus the cyclized analog **4**, $IC_{50} = 5.5$ nM, indicating that the 7-membered ring maintains the key features in an optimal 3D arrangement (see **Figure 2b**).

Analogs have also been prepared where the lactam carbonyl has been incorporated into a 1,2,4-triazolyl heterocycle as seen in alprazolam and triazolam.[85] Complete removal of the lactam carbonyl results in a significant loss (by two orders of magnitude) of biological activity. Beyond the two aromatic moieties there appears to be much flexibility with respect to substitution around the ring. Electron withdrawing groups on the aryl group are generally observed but polar substituents attached to the lactam amide group, illustrated in flurazepam, and polar hydroxyl or carboxy groups at the 3-position, present in oxazepam, temazepam and chlorazepate, hint at a significant flexibility in those regions (see **Figure 4**).

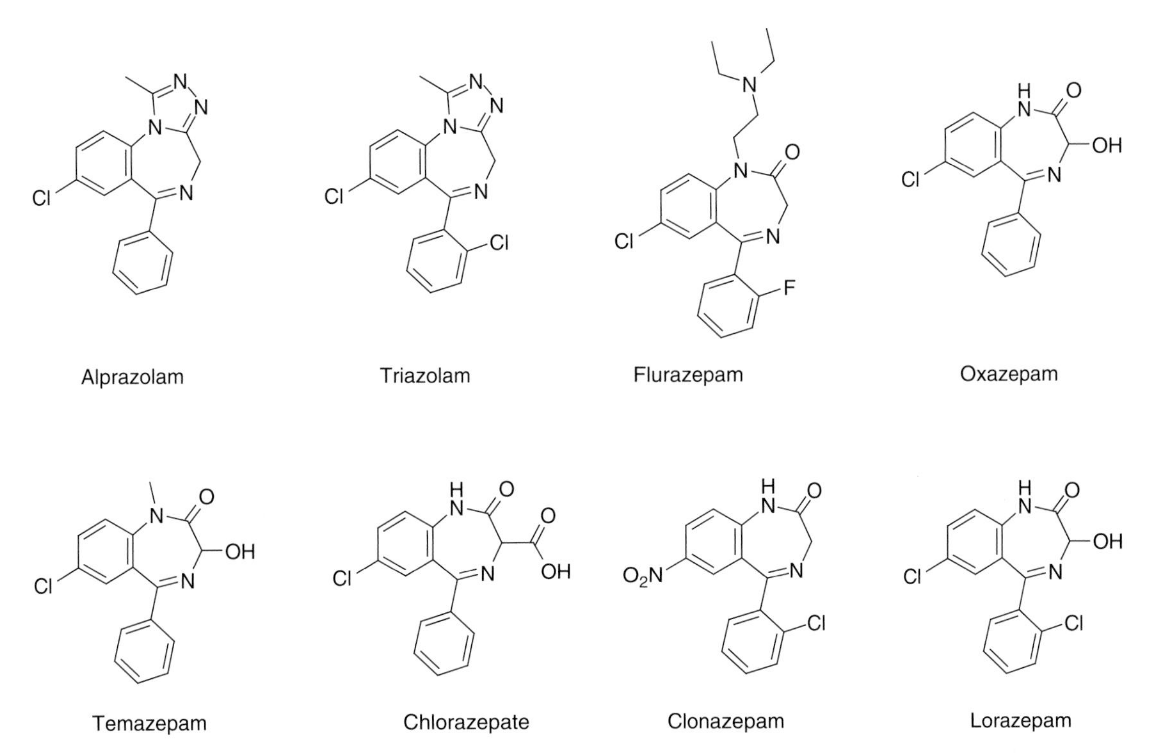

Figure 4 Benzodiazepine anxiolytics.

Clinically, BZs are used in the treatment of panic disorder, PTSD, and GAD when rapid onset is required. Selection of the specific BZ generally depends on the desired pharmacokinetic outcome. For instance, clonazepam has a $t_{1/2}$ of 20–50 h, which can be used once or twice a day. Lorazepam ($t_{1/2} = 10$–20 h) is used if drug accumulation is an issue and alprazolam offers a short half-life option (see **Figure 4**).

Use of the BZs has been curtailed due to potential for abuse, relapse, and significant withdrawal symptoms once treatment has ceased.[86] The observation that anxiety and depression are comorbid conditions has led current therapy away from BZ therapy alone[1,86] and introduced the use of drugs typically used for 'mood elevation' in anxiety disorders.

6.04.5.2 Monoamine Transport Inhibitors

The tricyclic antidepressants (TCAs) were originally designed to improve upon the efficacy and side effect profile of the phenothiazine class of antipsychotics. Their pharmacological spectrum was quite well understood in that these compounds interact with multiple brain neurotransmitter systems. The TCAs inhibit reuptake of monoamine neurotransmitters (dopamine (DA), 5HT, and NE) increasing their levels and function in the brain. TCAs include imipramine, desipramine, nortriptyline, amitriptyline, clomipramine, and doxepin (**Figure 5**). These compounds also interact with a variety of biological targets like muscarinic receptors, complicating their pharmacology and contributing to side effects such as orthostasis, dry mouth, and constipation. Clomipramine is the most effective TCA for panic disorder, OCD, and SAD[87] but more selective reuptake inhibitors have displaced the use of the tricyclics due to their improved side effect profile.

The origin of the first SSRIs, fluoxetine, resides in the early observation that diphenhydramine blocked monoamine uptake.[88] Preparation and testing of analogs of diphenhydramine led to the identification of nisoxetine, which was over 500-fold more selective for NET. Additional SAR studies around the phenoxy substituent ultimately led to the discovery of the 4-trifluoromethyl derivative, fluoxetine. The subtle change in phenoxy substitution appears to be independent of the electronic nature of the substituent since the 4-chloro analog is approximately 10-fold less active and marginally selective for 5HT uptake, whereas the 4-methoxy derivative is only fourfold less active and almost 20-fold more selective for SERT (**Figure 6**).

Other SSRIs in clinical use are paroxetine, sertraline, citalopram, and fluvoxamine[89] (**Figure 6**). Like fluoxetine, all these compounds have phenyl rings flanking a basic amine moiety except for fluvoxamine, which has only a single 4-trifluoromethyl substituted phenyl ring. The SSRIs emerged as the drugs of choice for the treatment of depression

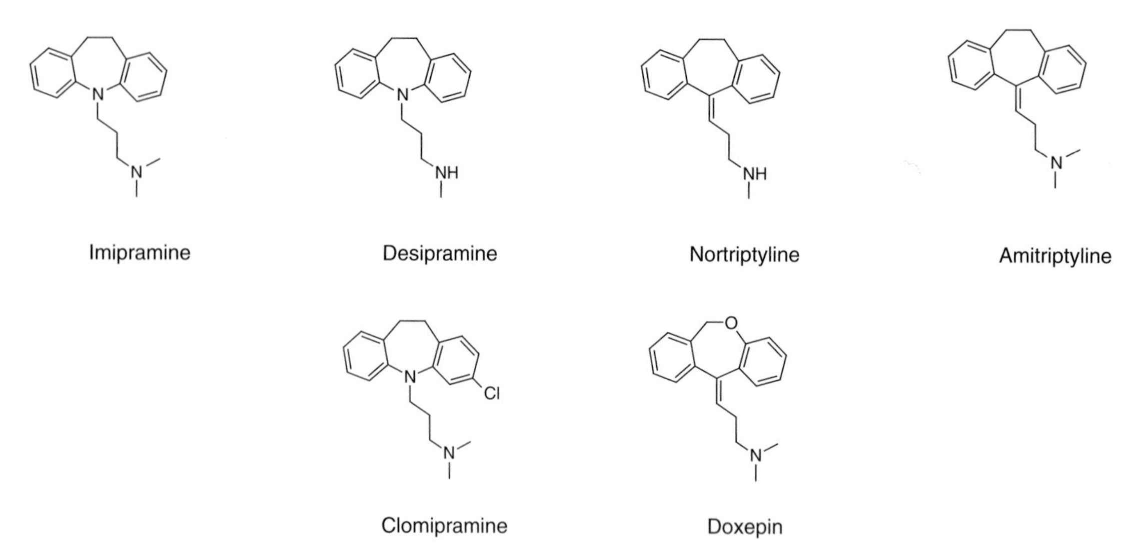

Figure 5 Tricyclic antidepressants (TCAs).

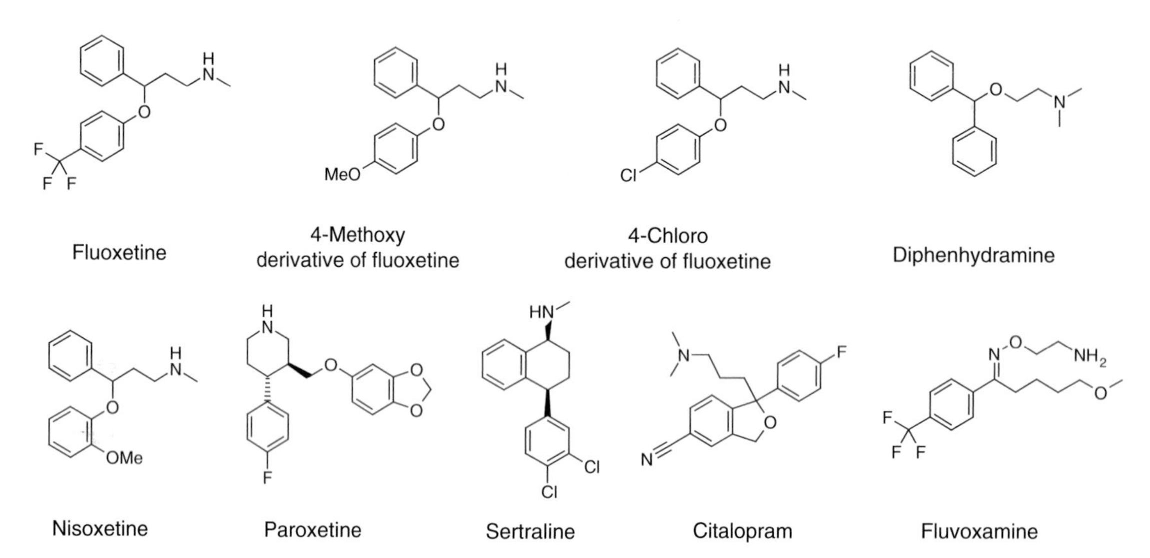

Figure 6 Selective serotonin reuptake inhibitors (SSRIs) and the selective norepinephrine reuptake inhibitor, nisoxetine.

due to their improved tolerability and safety on overdose as compared to the BZs. Clinical studies have expanded the indications for the SSRIs and they are currently being used to treat a variety of anxiety disorders including SAD, panic disorder, OCD, and social phobia.[87] For panic disorder, SSRIs are first line therapy but if an accelerated response is necessary (SSRIs require at least 2–3 weeks to achieve efficacy) the combination of an SSRI (sertraline) and a fast-acting BZ (clonazepam) has proven effective with clonazepam being given for the first 4 weeks and then tapered over a 3 week period to minimize side effects.

Like their use in depression, use of antidepressants for anxiety requires 4–6 weeks to achieve desired efficacy. From a disease mechanism point of view, the slow onset of efficacy observed across the spectrum of monoamine reuptake inhibitors is inconsistent with their primary mechanism of action, the blockade of reuptake of neurotransmitters that occurs within minutes/hours following compound administration. A plausible hypothesis for this inconsistency could be that as the levels of monoamines increase in the synapse, activation of presynaptic autoreceptors (by the monoamine) acts against the activity of the reuptake inhibition, thus counteracting the initial increase in synaptic neurotransmitter concentration. Only following long-term use of the reuptake inhibitor do these presynaptic autoreceptors (for example

$5HT_{1A}$ receptors) desensitize, allowing the increased neurotransmitter levels to have their effect. Recent attempts to visualize this mechanism have demonstrated that using PET techniques, there can be a 30–40% decrease in the surface dendritic expression of $5HT_{1A}$ receptors following fluoxetine treatment.[90] Thus, use of specific PET probes may provide a viable and rapid means of assessing the potential of antidepressant therapy in any given individual and may serve as a surrogate marker for efficacy. Recent work has provided evidence that the delayed onset of action of antidepressants is dependent on hippocampal neurogenesis.[91] Treatment of mice with SSRIs or SNRIs increased the differentiation of hippocampal progenitor cells into neurons over a time course consistent with the onset of anxiolytic activity; focal hippocampal irradiation to deplete dividing cells prevented the anxiolytic-like effects. Evidence suggests that $5HT_{1A}$ receptors may mediate these effects along with brain-derived growth factor.

Amongst the SNRIs that have also been developed for depression and other disorders, venlafaxine and nefazedone are used in anxiety disorders, and an extended release formulation of venlafaxine was recently approved for use in GAD (see **Figure 7**).

6.04.5.3 $5HT_{1A}$ Agonists

The azapirone class of drugs includes the $5HT_{1A}$ agonist, buspirone (**Figure 8**), which also has weak dopaminergic activity.[92] $5HT_{1A}$ receptors have opposing effects on the release of 5HT depending on whether they are located on pre- or postsynaptic sites, and it is hypothesized that the activation of the autoreceptors on the raphe nuclei contributes to the delayed onset of SSRIs in the clinic. Thus, agonism of the $5HT_{1A}$ receptor (pre- and post-synaptically) could produce changes in brain 5HT systems that are similar to those seen following SSRI administration, but only in brain areas expressing the $5HT_{1A}$ receptor. Buspirone was efficacious in several animal models of anxiety providing an impetus for its evaluation in a series of double-blind, placebo-controlled anxiety trials. These trials established the efficacy of buspirone, without accompanying neuroleptic side effects or abuse potential. Buspirone was the first non-BZ

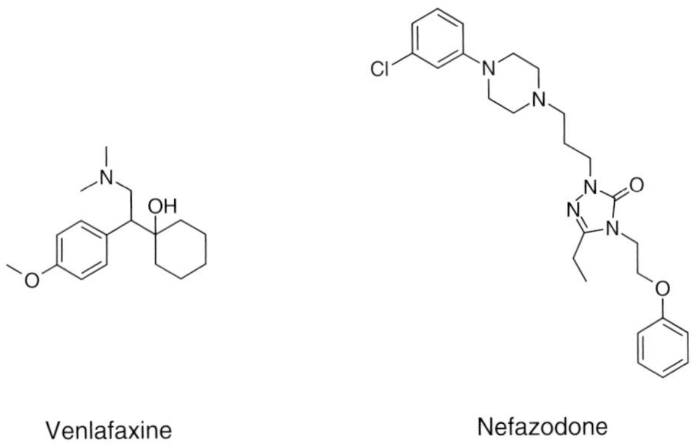

Venlafaxine

Nefazodone

Figure 7 Structures of SNRIs.

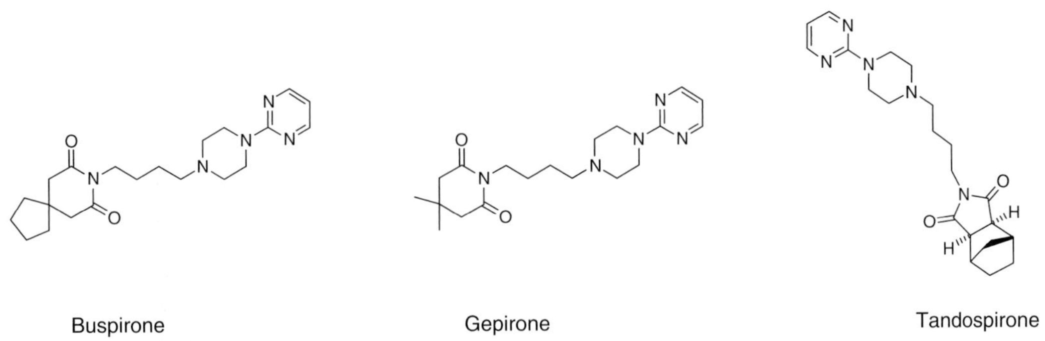

Buspirone

Gepirone

Tandospirone

Figure 8 $5HT_{1A}$ agonist anxiolytics.

nonsedating anxiolytic approved for the treatment of GAD.[93] Despite its clinical efficacy buspirone is not widely prescribed mainly because of a delay in onset of action compared to BZs and also because it is ineffective in patients who have previously taken BZs for their anxiety.[93]

Gepirone (**Figure 8**) is an azapirone that is structurally similar to buspirone, but is more selective for the $5HT_{1A}$ receptor. Initial studies of gepirone in depressed patients using an immediate release formulation resulted in good antidepressant efficacy but this was limited by tolerability issues at higher doses.[93] An extended release formulation of gepirone (Gepirone ER) is under development for treatment of depression. In addition to major depression, gepirone may have utility in patients with anxious depression[94] as well as GAD.[95] Another second-generation azapirone, tandospirone, was launched in Japan (1996) and China (2004) for treatment of GAD (**Figure 8**).

6.04.6 New Research Areas

6.04.6.1 Corticotropin-Releasing Factor

In terms of a common physiological pathway that may underlie many of the symptoms of anxiety and depression, the stress axis has both a strong rationale and mechanism-based evidence both preclinically and clinically.

6.04.6.1.1 The CRF hypothesis and stress

CRF is synthesized in the hypothalamus and elicits the release of adrenocorticotropic hormone (ACTH) from the pituitary. CRF was isolated from sheep hypothalamus and its structure as a 41-amino-acid peptide determined.[96] The hypothalamic paraventricular nucleus (PVN) is the major region in the brain of CRF-containing cell bodies and through axonal projections to the capillaries of the median eminence can secrete CRF directly into the portal system where it acts at the pituitary to regulate ACTH secretion into the circulation. The principal role of ACTH is to stimulate the release of cortisol from the adrenal gland, thus completing the HPA axis, a primary component of the neuroendocrine response to stress. Similarly, projections from the PVN to the lower brainstem and spinal cord have been demonstrated to regulate autonomic function and help to further mediate the behavioral responses to stress. High densities of CRF-containing neurons are localized in particular to prefrontal, cingulate, and insular areas and appear to regulate behavioral actions of the peptide. Above and beyond its direct role in mediating HPA activity, the extrahypothalamic anatomical distribution of CRF suggests that this neuroactive peptide is localized in discrete brain regions that are thought to regulate mood and stress-related behaviors.[89] Administration of CRF into the *locus coeruleus* initiates stress responses accompanied by increases in catecholamine turnover and activity in the frontal cortex suggesting that the peptide produces anxiety-related behavior by increasing the activity of the noradrenergic system.[89] Such studies implicate CRF and related peptides in the mechanisms through which various stressors can alter behavior.[89]

6.04.6.1.2 A role for CRF in anxiety-related disorders

Whilst there is considerable data on the role of CRF in depression,[89] CRF systems may also play a role in anxiety states. CRF has central, behavioral arousal properties characteristic of other anxiogenic compounds that relate directly to the hyperarousal that defines anxiety disorders. CRF has been implicated as a mediator in panic disorder based on the observation that patients with this disorder exhibit a blunted ACTH response to CRF as compared to normal individuals. This blunted response likely reflects processes occurring above the level of the hypothalamus related to hypersecretion of CRF. In rats, CRF disrupts PPI in a manner similar to that observed in patients with panic disorder where the CRF system is thought to be overactive.[97] These effects could be reversed by CRF receptor antagonists.[89]

6.04.6.1.3 Potential CRF receptor antagonists for anxiety

As CRF appears to plays a major role in the regulation of stress responses and the subsequent anxiety a concerted effort has focused on the search for orally active, nonpeptide antagonists. Examples of the different compound classes with high affinity for the CRF_1 receptor are shown in **Figure 9**.[98] Interestingly, while these molecules are *chemically* distinct, they appear to fit a common pharmacophore indicating that the receptor has strict requirements to allow a high-affinity interaction.[98]

6.04.6.1.4 Human studies

To date, only one CRF antagonist, the pyrazolopyrimidine R121919 (NBI 30775, **Figure 9**), has shown clinical effects in a population of patients with MDD. R121919 reduced measures of depression and anxiety on the HAM-D and

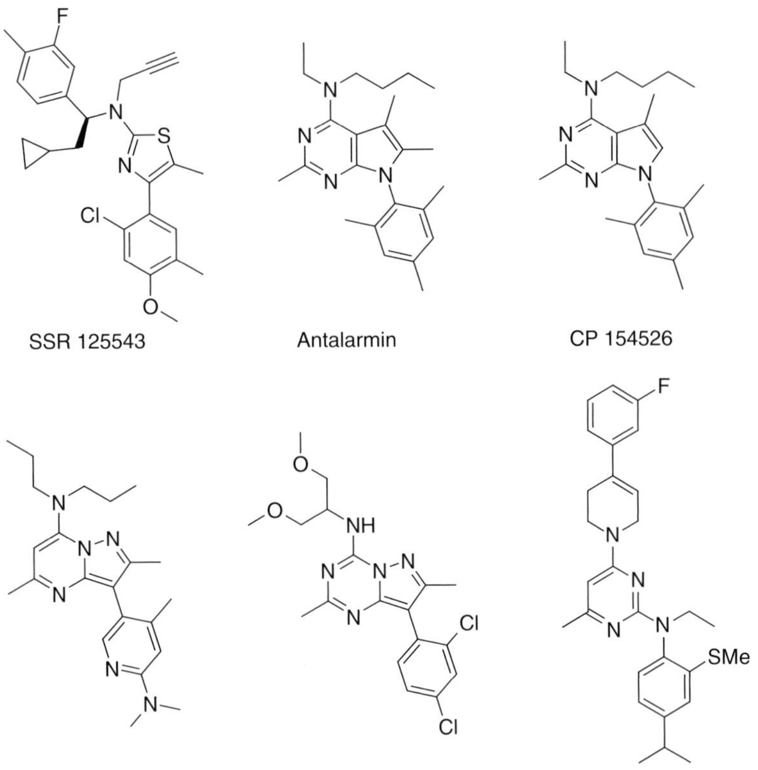

SSR 125543 Antalarmin CP 154526

NBI 30775 DMP 696 CRA 1001

Figure 9 Small molecule CRF_1 receptor antagonists.

HAM-A rating scales without affecting either basal HPA activity or producing any significant blunting of an exogenously administered CRF-induced ACTH response[99] in a Phase IIA open label, nonplacebo-controlled study. The compound also improved sleep-EEG patterns in humans and rats,[100,101] but caused reversible liver enzyme elevations that precluded further clinical development.

6.04.6.2 Subtype-Selective $GABA_A$ Receptor Modulators

As noted, BZs have been the primary treatment for anxiety for nearly half a century. Despite their proven efficacy and rapid onset of action, their use is limited by side effects, e.g., sedation and amnesia, the development of tolerance, and concerns about dependence and withdrawal.[82,102] These side effects are a natural extension of their mechanism of action. Considerable effort has been expended over the past three decades to discover and develop novel, anxioselective BZ ligands that have improved side effect profiles.

6.04.6.2.1 Benzodiazepine site ligands

Since the allosteric potentiation of $GABA_A$ receptor function was first elucidated using BZs, this site is commonly referred to as the 'BZ site'; however, compounds of different chemical classes, 'non-BZ' ligands, have been discovered that also bind to the BZ site and produce similar therapeutic effects. These compounds provided the first evidence for heterogeneity of BZ sites and for subtypes of $GABA_A$ receptors that mediate the different pharmacological actions of BZ site ligands. Other allosteric sites present on the $GABA_A$ receptor complex include those for barbiturates, ethanol, etomidate, neurosteroids, and zinc, compounds that also modulate $GABA_A$ receptor function.[103]

Molecular genetics in combination with classical pharmacology have greatly facilitated understanding of how BZ site ligands interact with the $GABA_A$ receptor to produce their biological effects. The existence of 16 different $GABA_A$ subunits theoretically implies that there are many thousands of possible subunit combinations and therefore thousands of possible $GABA_A$ subtypes. However, only a limited number of subtype combinations have been found in native systems with $\alpha1\beta2\gamma2$, $\alpha2\beta3\gamma2$, and $\alpha3\beta3\gamma2$ being the most abundant.[103] The assembly of α, β, and γ subunits is required to produce functional $GABA_A$ receptors that exhibit all the pharmacological properties of native $GABA_A$ receptors. The diversity of subunits and their heterogeneous distribution between brain regions raises the question of whether $GABA_A$ receptors composed of different subunit combinations play different functional roles in the brain. $GABA_A$ receptors containing α1, α2, α3, or α5 subunits in combination with β,γ subunits bind to and are potentiated by BZs. In contrast, α4- or α6-subunit-containing receptors are insensitive to classical BZs like diazepam (exceptions being the BZ antagonist, Ro15-1788 and the inverse agonist, Ro 15-4513, which show weaker activity at α4 and α6). Differences in BZ ligand affinity between α subunits results from a histidine (His 101 in rat α1) at the homologous position in α1, α2, α3, and α5, while an arginine is present in the homologous position in the α4 and α6 subunits. Transgenic mouse lines with a single histidine (H) to arginine (R) mutation at this position for individual α1, α2, α3, or α5 subunits results in an insensitivity to diazepam and other BZ ligands, while $GABA_A$ receptor expression and GABA-evoked responses remain normal.[84] $GABA_A$ receptors containing α1 subunits mediate the sedative properties of diazepam as the α1 (H101R) mutant mice are resistant to the depressant effects of diazepam as measured by locomotor activity assays.[104–106] $GABA_A$ receptors containing α2 subunits are proposed to play a dominant role in anxiolytic effects since α2 (H101R) mutant mice are resistant to the antianxiety effects of diazepam.[105] The role of the α3-subunit-containing $GABA_A$ receptors is unclear since α3 (H126R) mutant mice respond normally to diazepam,[105] although a role in anxiolytic effects has been proposed on the basis of pharmacological evidence (*see* Section 6.04.6.2.4). $GABA_A$ receptors containing α5 subunits are of considerable interest given their involvement in memory processes, due to their localization and proposed physiological role in the hippocampus,[107] and because knockout of α5[108] or H105R mutation of α5[109] attenuates the effects of diazepam on memory tasks. Tolerance to the sedative effects of diazepam is absent in α5 (H105R) mice, suggesting that activation of α5-subunit-containing $GABA_A$ receptors may be important for the development of tolerance.[110]

6.04.6.2.2 Approaches to anxioselective $GABA_A$ receptor modulators

Two approaches have been taken in the pursuit of improved anxiolytic compounds acting via the BZ receptor. Compounds with partial efficacy as positive allosteric modulators could elicit anxiolytic effects with a reduced propensity for sedation and amnesia. This approach assumes that only a limited degree of $GABA_A$ receptor potentiation is required to produce anxiolysis, whereas a greater potentiation is necessary to produce side effects. The second approach is focused on developing a positive allosteric modulator selective for α2- (and/or α3-) subunit-containing $GABA_A$ receptors that would be anxiolytic lacking the sedative, amnestic, and tolerance properties of classical BZ ligands. These two approaches are not mutually exclusive and a compound with combined subtype selectivity and partial efficacy may be a superior anxiolytic. The challenge of finding such compounds has been such that no anxioselective BZs have advanced to the market.[82,102,111]

6.04.6.2.3 Benzodiazepine site ligands with partial efficacy

The imidazobenzodiazepines bretazenil, imidazenil, and FG-8205 (**Figure 10a**) are high-affinity partial agonists for $GABA_A$ receptor subtypes that contain α1, α2, α3, and α5 subunits. These compounds are anxiolytic in animal models, with a greater separation between doses required for anxiolytic versus sedative effects than that observed for diazepam. Where studied, these compounds also show little evidence for tolerance to the anxiolytic effects, do not display withdrawal, and have a lower abuse potential than full BZ agonists in nonhuman primates. Bretazenil (**Figure 10a**) entered clinical studies in the mid-1980s and was efficacious in GAD and panic disorder,[112] with a lower abuse potential than diazepam.[113] However, bretazenil was sedative, with little evidence for separation between anxiolytic and sedative effects,[114] and its clinical development was discontinued.

The β-carboline abecarnil (**Figure 10a**), has high affinity for α1-, α2-, α3-, and α5-subunit-containing $GABA_A$ receptors, and has shown partial agonist activity in some, but not all, studies. Some reported pronounced anxiolytic versus sedative and muscle relaxant activity,[115] while others reported marked sedation.[116] Abecarnil was evaluated in the clinic and shown to be superior to placebo, equivalent to a BZ comparator although these effects were not always significant. The compound produced drowsiness and was discontinued in the 1990s.

The pyrazolopyrimidine, ocinaplon (**Figure 10a**), has relatively low affinity for rat brain BZ binding sites, being a positive allosteric modulator of $GABA_A$ receptors exhibiting partial agonist activity at all subtypes, with EC_{50} values in the 3–10 μM range.[117] In animal models, it produces anxiolytic activity equivalent to diazepam with reduced sedative and muscle relaxant effects. Ocinaplon had efficacy in GAD patients with 2 weeks of dosing, with a side effect profile equivalent to placebo[117] and is currently in phase III development, although it is currently on hold due to liver toxicity.

The pyridobenzimidazole, RWJ-51204 (**Figure 10a**), has low nanomolar affinity for BZ binding sites with the characteristics of a partial agonist,[116] although no data are available for its actions on discrete α-subunit-containing

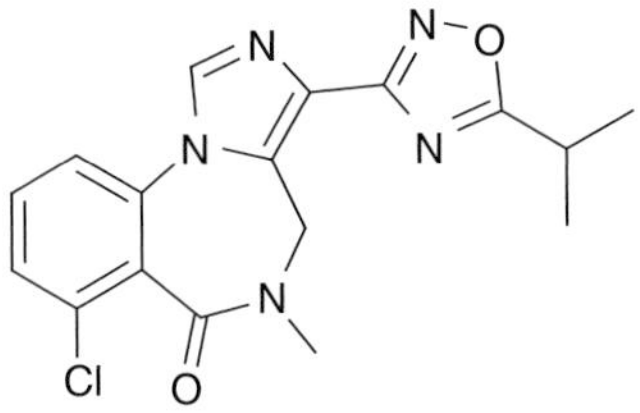

FG-8205

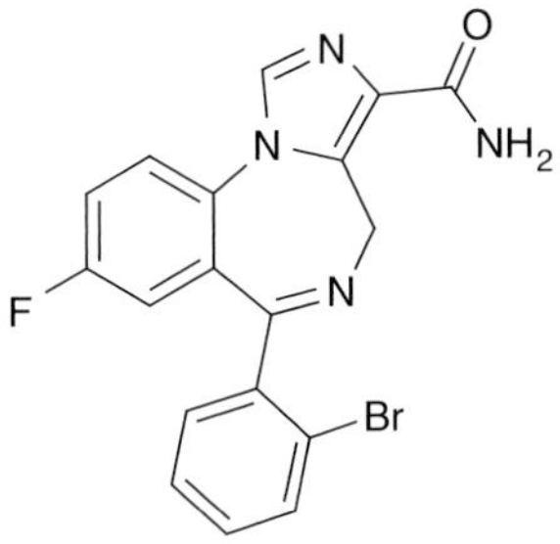

Imidazenil

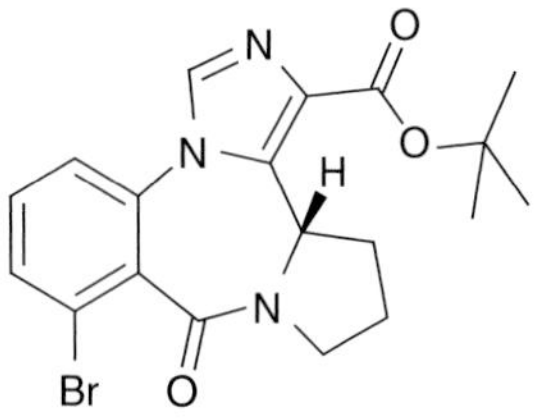

Bretazenil (Ro-16-6028)

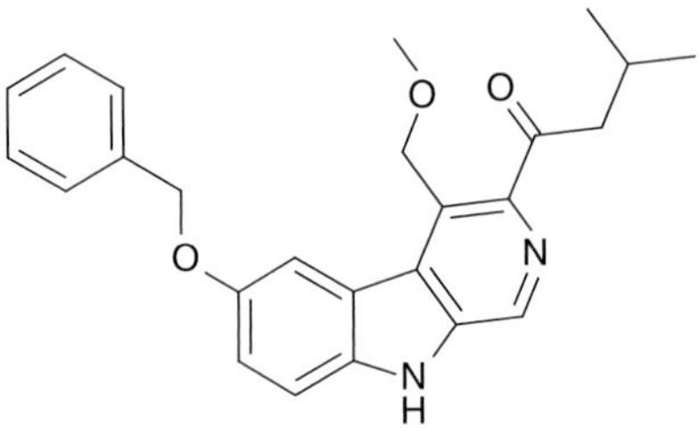

Abecarnil

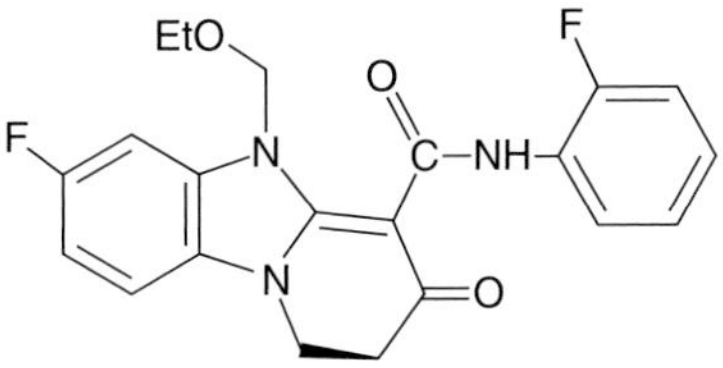

RWJ-51204

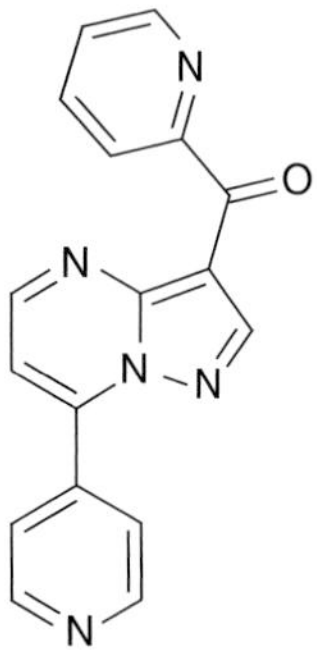

Ocinaplon (CL 273,547)

(a)

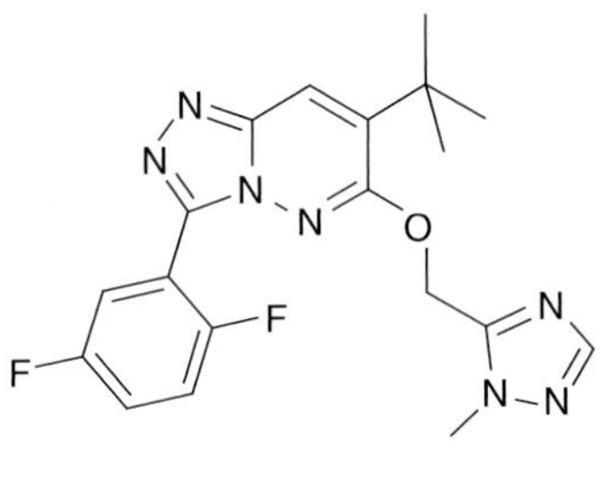

L-838417

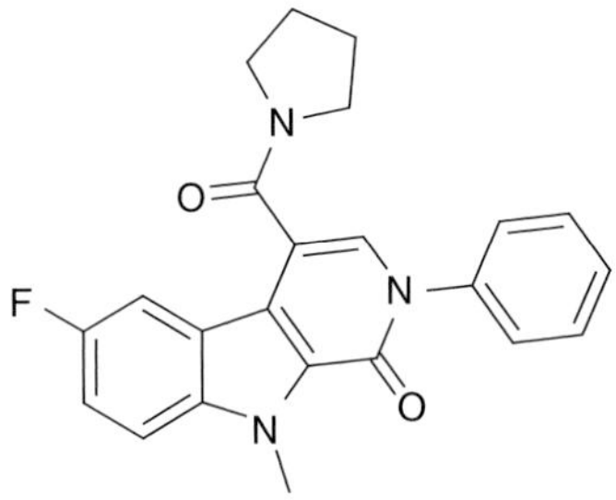

SL 651498

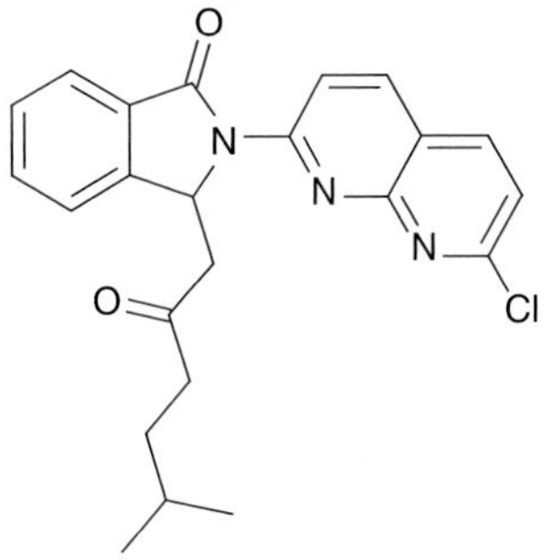

Pagoclone

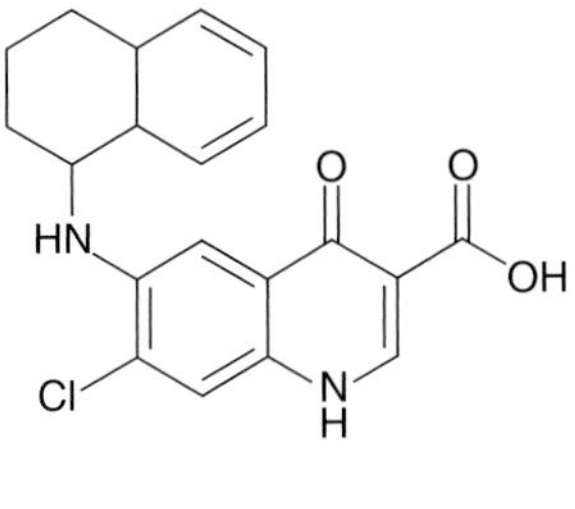

Quinolone ‘compound 4’

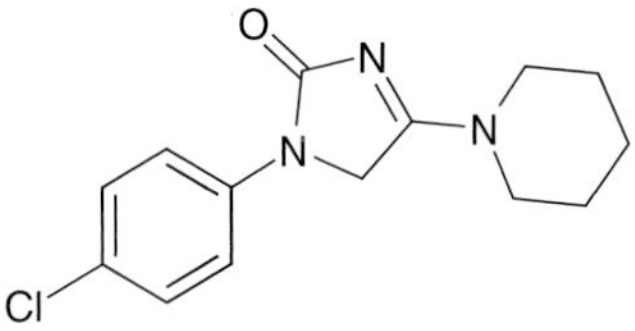

ELB-139

(b)

Figure 10 Anxioselective $GABA_A$ receptor-positive allosteric modulators. (a) Compounds with partial efficacy. (b) Compounds with subtype selectivity.

$GABA_A$ receptors. In mice and monkeys, RWJ-51204 showed anxiolytic effects that were separated from doses causing sedation, although in the rat, sedative and anxiolytic effects overlapped. In monkeys the anxiolytic effects of the compound showed a lower maximum compared with benzodiazepine full agonists and were similar to other partial agonist comparators. In 1999, the compound was reported to reduce CCK-4-induced panic in normal volunteers,[116] but further clinical development activities have not been reported.

6.04.6.2.4 $GABA_A$ subtype-selective benzodiazepine site ligands

No NCEs have yet been reported with a high degree of selective affinity for $GABA_A$ receptor subtypes believed to be important for anxiolysis. However, compounds have been identified that, despite having similar affinities, show a degree of functional selectivity for different α-subunit-containing $GABA_A$ receptors (i.e., they behave as full or partial modulators of certain subtypes and neutral modulators of others). A compound preferentially potentiating α2-subunit-containing $GABA_A$ receptor responses could provide anxiolytic effects with a separation from sedation and memory impairment. A pyridine-2-one derivative, which behaves as a negative allosteric modulator with functional selectivity for α3-subunit-containing $GABA_A$ receptors, increased anxiety responses in rodents[82,118] suggesting that potentiation of α3-subunit-containing $GABA_A$ receptor responses may also provide anxiolysis.

The triazalopyradizine, L-838417 (**Figure 10b**), had high affinity for α1-, α2-, α3-, and α5-subunit-containing $GABA_A$ receptors, but failed to affect GABA responses mediated by α1-subunit-containing $GABA_A$ receptors, and was a high affinity positive allosteric modulator at the remaining subtypes.[106] In rodents, this compound had anxiolytic effects comparable to chlordiazepoxide in the EPM and fear-potentiated startle assays with little evidence for sedation. Anxiolytic activity was also observed in primates without sedation, with a low propensity for self-administration that might suggest lower abuse potential. [119] Additional compounds with functional selectivity for α2/3- or α3-subunit-containing $GABA_A$ receptors have been described[118,120] with one reaching phase II clinical trials for GAD, but was discontinued.[111]

The pyridoindole derivative, SL-651498 (**Figure 10b**), had full efficacy as a positive allosteric modulator at α2- and α3-subunit-containing $GABA_A$ receptors, and was a partial agonist at α1- and α5-subunit-containing $GABA_A$ receptors. The compound has anxiolytic efficacy in a variety of tests, with higher doses required for sedative and ataxic effects.[61] It reached phase II clinical trials, but no outcome or further studies have been reported.

The cyclopyrrolone, pagoclone (**Figure 10b**), had low nanomolar affinity for α1-, α2-, α3-, and α5-subunit-containing $GABA_A$ receptors and was a full agonist at $GABA_A$-receptor-containing α3 subunits, with partial agonist activity at those containing α1, α2, and α5 subunits.[82] In animals, pagoclone was anxiolytic with a lower propensity for sedative and muscle relaxant effects. Initial trials showed the pagoclone was effective in panic attack and GAD patients without sedative effects. Additional studies failed to replicate these initial findings.

Other compounds with some degree of functional selectivity towards α2- and/or α3-subunit-containing $GABA_A$ receptors and anxiolytic effects include: the quinolone 'compound 4,'[121] NGD 91-3, and ELB-139[82,122]; few details are available on these compounds.

There is clear evidence from clinical studies that positive allosteric modulators with partial efficacy (bretazenil, ocinaplon) are anxiolytic with efficacy comparable to standard BZs, but why bretazenil produces sedation at anxiolytic doses, whereas this appears not to be the case with ocinaplon is difficult to reconcile with the receptor profiles of these compounds. Clinical studies with compounds showing functional selectivity for α2- and α3-subunit-containing $GABA_A$ receptors have been inconclusive (pagoclone) and further studies are clearly needed to validate the potential benefit of this profile. It may also be that the $GABA_A$ receptor subtype story is more complicated. Evidence exists that a substantial proportion of individual native $GABA_A$ receptors contain two different α subunits, e.g., α1α2βγ α1α3βγ.[123] The cellular location and functional role of these 'mixed α subunit' $GABA_A$ receptor subtypes, as well as their functional response to BZ site ligands remains to be seen, but may explain some of the anomalies observed to date.

6.04.6.3 Metabotropic Glutamate Receptors

G protein-coupled receptors for glutamate, metabotropic glutamate receptors (mGlu), play a diverse modulatory role in neurotransmission and are potential targets for novel anxiolytic drugs.[124] The eight known mGlu receptors are divided into three families, group I ($mGlu_{1,5}$), group II ($mGlu_{2,3}$) and group III ($mGlu_{4,6,7,8}$), based on structure, pharmacology, and signal transduction. The focus for anxiolytics has been on group I and II mGlus, although data from $mGlu_7$ and $mGlu_8$ knockout mice suggest that group III receptors may also be relevant for altered anxiety and stress responses.

6.04.6.3.1 Group II mGlu agonists

$mGlu_2$ and $mGlu_3$ are expressed in brain areas important for anxiety disorders, e.g., amygdala, hippocampus, and prefrontal cortex. $mGlu_2$ receptors are located on glutamate-releasing nerve terminals where they suppress glutamate release, whereas $mGlu_3$ is found both pre- and postsynaptically, and on glia. Activation of group II mGlus also suppresses the release of GABA, monoamines, and neuropeptides.[125] The constrained glutamate analogs LY354740 and MGS 0028 (**Figure 11a**) are potent, selective agonists of $mGlu_2$ and $mGlu_3$ receptors.[124,126] Systemic administration of LY354740 produces anxiolytic effects in a range of animal models, including fear potentiated startle, EPM, and conflict tests. Where compared, the anxiolytic effects of these compounds are similar to those for standard BZs, but occur in the absence of sedation or ataxia. LY354740 may, however, disrupt memory processes in animals,[127] although the compound reduced ketamine-induced deficits in working memory in humans.[128] LY354740 demonstrated efficacy in two clinical studies: carbon dioxide-induced anxiety in panic attack patients and a phase II trial in GAD patients.[124] LY544344, a prodrug of LY354740, is in phase III trials for anxiety. It is unclear whether agonist activity at both $mGlu_2$ and $mGlu_3$ is required for the efficacy of LY354740, since anxiolytic efficacy is lost in both $mGlu_2$ or $mGlu_3$ knockout mice.[129] However, compounds with $mGlu_2$ selectivity have also shown anxiolytic activity in animal models (see below).

Allosteric modulators of mGlus have been described that bind to sites on the receptor distinct from the orthotopic glutamate recognition site, and act as positive, negative, or neutral allosteric modulators of glutamate-induced receptor function. Several classes of positive allosteric modulators with high affinity and selectivity for $mGlu_2$ have been reported (**Figure 11b**). These have better 'drug-like' properties than glutamate site agonists and show greater selectivity towards mGlu subtypes. In addition, they have the theoretical advantage over glutamate site agonists in that their action is to potentiate mGlu receptor function when it is activated physiologically by the endogenous ligand, rather than indiscriminate activation by an exogenous agonist. Positive allosteric modulators of $mGlu_2$ are anxiolytic in animal models.[130]

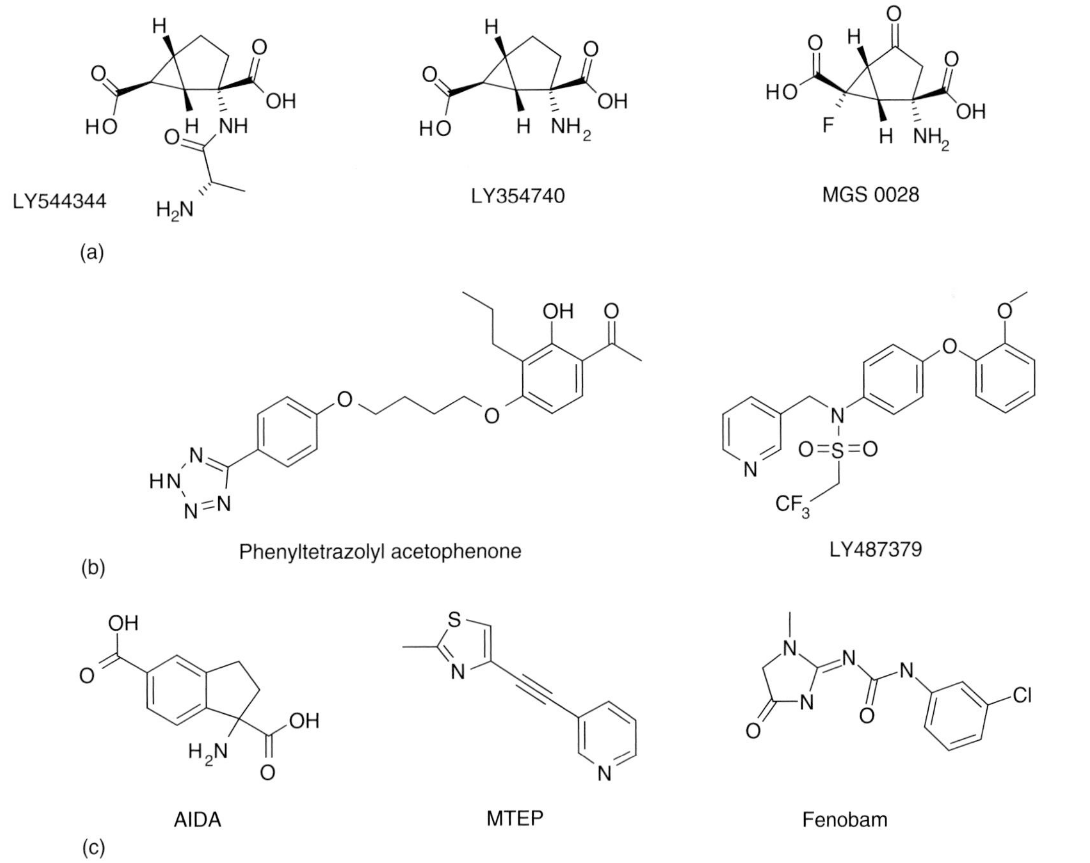

Figure 11 Anxiolytic mGlu receptor ligands: (a) $mGlu_{2/3}$ agonists; (b) $mGlu_2$-positive allosteric modulators; and (c) $mGlu_{1/5}$ antagonists.

6.04.6.3.2 Group I mGlu antagonists

$mGlu_1$ and $mGlu_5$ have reciprocal distributions in the brain and are located predominantly postsynaptically where they are thought to augment neurotransmission. Antagonists of both $mGlu_1$ and $mGlu_5$ have shown anxiolytic activity in animal models.[124] For $mGlu_1$ antagonists, the glutamate analog AIDA (**Figure 11c**) and the allosteric antagonist JNJ16259685 are both effective after systemic administration in the Vogel test, and positive results were also reported for AIDA in the EPM and open field, but not the four-plate test.[131,132] The $mGlu_5$ allosteric antagonist MPEP was efficacious in fear-potentiated startle, EPM, conflict tests, and stress-induced hypothermia with systemic administration,[124,133] as was the analog MTEP (**Figure 11c**).[134] In general, anxiolytic effects were observed without sedation; however, group I mGlu receptors are involved in learning and memory processes, consequently blockade of $mGlu_1$ and $mGlu_5$ may be problematic in this respect. Fenobam (**Figure 11c**), a drug efficacious in patients with anxiety, is a potent negative allosteric modulator of $mGlu_5$[135] representing the first validation of the potential anxiolytic efficacy of a group I mGlu antagonist in humans.

6.04.6.4 Gabapentinoids

Gabapentin and its backup pregabalin (**Figure 12a**) were initially developed as anticonvulsants, but are clinically effective in a range of CNS indications, including neuropathic pain and anxiety. Gabapentinoids have no direct effect on

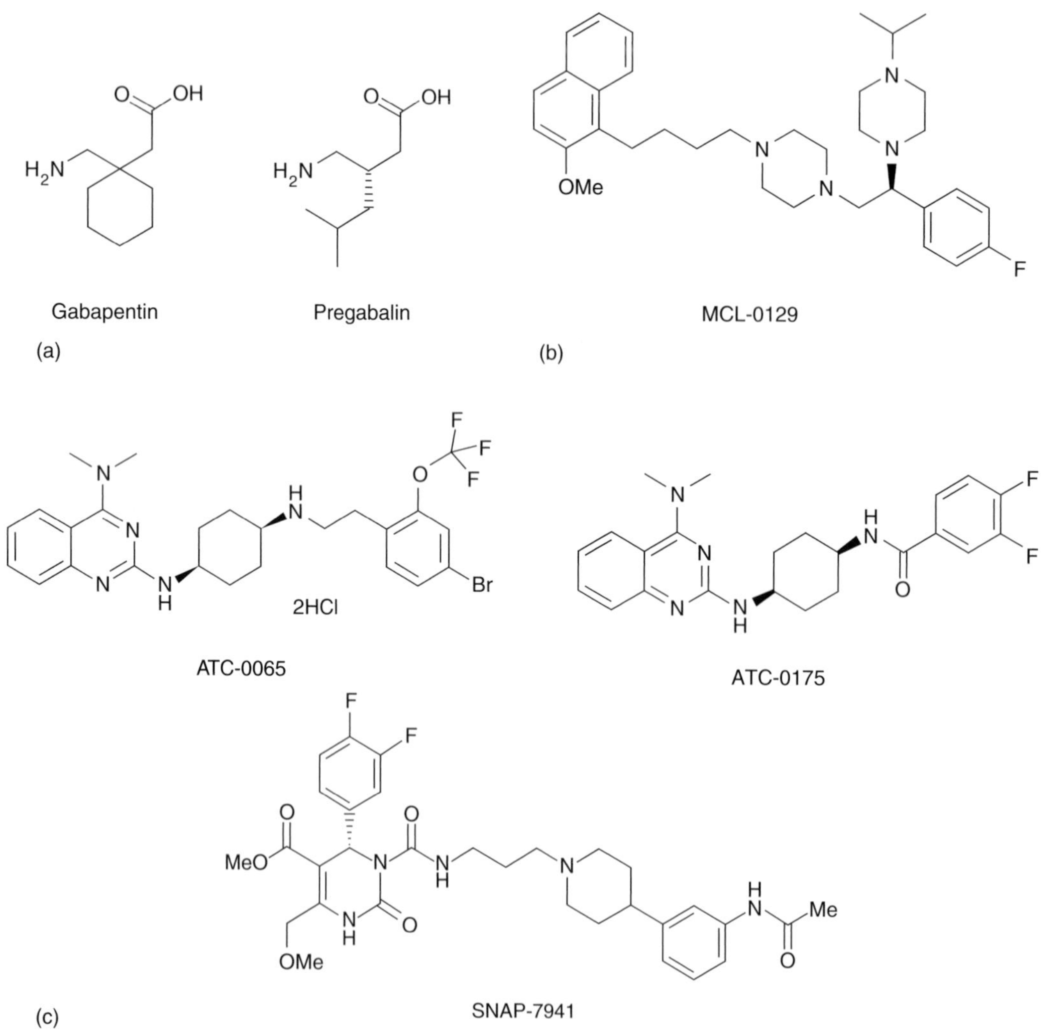

Gabapentin Pregabalin (a)

MCL-0129 (b)

ATC-0065 ATC-0175

SNAP-7941 (c)

Figure 12 (a) Gabapentin and pregabalin; (b) a small-molecule MC_4 receptor antagonist; and (c) small-molecule MCHR1 receptor antagonists.

GABA systems, and their precise mechanism of action remains undefined and controversial.[136,137] Some reports indicate that gabapentin and pregabalin bind with high affinity to the α2δ subunit of voltage-dependent calcium channels but no robust effects have been demonstrated on calcium channel function. Additionally, they are required to be substrates for an amino acid transporter to produce in vivo efficacy, presumably to allow their efficient transit as polar molecules across the gut and blood–brain barriers.[138] Both gabapentin and pregabalin have anxiolytic effects in rodent models including conflict, elevated X-maze, and light/dark box tests.[139] Gabapentin is also effective in the marmoset human threat test.[140]

Pregabalin has anxiolytic activity comparable to concurrently studied BZ or SNRI comparators, after 1 week of dosing.[141,142] Both gabapentin and pregabalin were effective in patients with social phobia[143,144] with gabapentin having efficacy in panic disorder subjects.[145] Compared to established anxiolytic drugs, pregabalin shows less sedative effects, although somnolence, ataxia, and dizziness can be present, but these effects are mild to moderate, transient, and tolerate over a 6 week period.[141] No tolerance to the anxiolytic effects was noted, and withdrawal symptoms were not clinically significant. Pregabalin is approved for use for epilepsy and neuropathic pain but not GAD. The compound showed euphoric effects and is now being scheduled.

6.04.6.5 Melanocortins

The melanocortin (MC) receptor family comprises five known subtypes. MC_3 and MC_4 receptors and their natural peptide ligands, α-MSH and AgRP, are expressed within the CNS and are involved in feeding behavior and energy regulation.[146,147] The localization of these receptors to amygdala[148] suggests they may be involved in anxiety-related behaviors. α-MSH, the endogenous nonselective $MC_{3/4}$ receptor agonist, given i.c.v. produces anxiogenic effects in various behavioral models including: increased separation distress vocalizations in the chick[149,150]; decreased exploratory behavior in the hole-board test in the rat[151]; reduced time spent in the open arms of the rat EPM[151]; and decreased licking in the Vogel conflict model in rat.[152] Similarly, the synthetic nonselective $MC_{3/4}$ agonist MTII is active in the Vogel conflict model in the rat[152] and decreases the amount of time rats engage in social interaction.[153] The MC_4 receptor antagonist, HS014,[154] and the more selective antagonist, MCL-0020,[152] attenuate anorexia in the rat produced by immobilization/restraint stress, a possible measure of anxiety. MCL-0020 and MCL-0129 prevent swim stress-induced decreases in light area exploration in the light/dark exploration test in mice.[152,155] Furthermore, MCL-0129, a selective MC_4 receptor antagonist (**Figure 12b**), increased the amount of time rats spend in social interaction after a week of administration.[153] Taken together, these preclinical findings suggest that MC_4 antagonists may be useful in the treatment of anxiety disorders. The true utility of this target awaits proof of concept studies from human clinical trials.

6.04.6.6 Melanin-Concentrating Hormone

The expression pattern of melanin-concentrating hormone (MCH) neurons and receptors in the CNS support a potential role for MCH in a variety of physiological functions including stress, regulation of neuroendocrine processes, and feeding. Of these functions, MCH effects on feeding behavior and energy homeostasis have been most studied,[156] but $MCHR_1$ antagonists may be a novel approach to the treatment of anxiety. MCH cell bodies are located in the lateral hypothalamus and *zona incerta* with widespread projections to limbic structures such as hippocampus, septum, amygdala, *locus coeruleus*, dorsal raphe nucleus, and the shell of the *nucleus accumbens*, areas that are involved in the regulation of emotion.[157] The $MCHR_1$ nonpeptide antagonists ATC-0065 and ATC-0175 (**Figure 12c**) reverse swim stress-induced anxiety in the EPM in rats and decrease temperature change in the stress-induced hyperthermia model in mice.[158] ATC-0175 also increased social interaction between unfamiliar rats and decreased distress vocalizations in guinea pig pups. The compounds had no effect on marble burying in mice, a model of OCD. The $MCHR_1$ antagonist SNAP-7941 (**Figure 12c**) decreased distress vocalizations in guinea pig pups and increased social interaction time compared with vehicle in rats.[159]

While the few small molecule antagonists that have been tested seem to produce anxiolytic effects in animal models, studies using the endogenous peptide agonist MCH have produced mixed results. Consistent with the antagonist results, MCH injected bilaterally into the medial preoptic area decreases the amount of time rats spend in the open arms of the elevated plus maze.[151] In contrast, when MCH is injected into the cerebral ventricles, it increases licking in the rat Vogel conflict test[160] and increases time spent in the open arms of the elevated plus maze[161] supporting an anxiogenic effect. It is unclear why MCH seems to produce opposite effects in some cases.

6.04.6.7 Neurokinins

The neurokinin (NK) system is one of the most studied peptide systems in the CNS. Three NK receptors exist, each with a preferred ligand: substance P (SP) binds with greatest affinity to NK_1, neurokinin A prefers NK_2, and neurokinin B has preferential binding to NK_3. The NK system has been implicated in a wide range of functions including nociception, emetic response/vomiting, inflammatory conditions, asthma, irritable bowel syndrome, migraine, anxiety, and depression. SP and the NK_1 receptor are broadly distributed in the amygdala, *locus ceruleus*, hypothalamus, substantia nigra, peduncular nuclei, caudate putamen, *nucleus accumbens*, raphe nuclei, and lamina I of the spinal cord. The distribution of the receptor in limbic regions (amygdala, hypothalamus) overlaps with neurotransmitter systems known to be involved in the regulation of mood (e.g., 5HT, NE). Thus, the SP-NK_1 receptor system has been hypothesized to play an important role in affective behaviors (i.e., anxiety and depression).[162]

Preclinical studies showed that SP delivered to the lateral septal nucleus, the medial amygdala, or into the intracerebral ventricles increased time spent in the closed arms of the EPM, an anxiogenic response. Furthermore, noxious stimuli and acute and chronic stress produce SP release in the brain. Specifically, maternal separation in the guinea pig and immobilization stress elevate SP levels in amygdala. Given these data NK_1 receptors would be anticipated to have anxiolytic activity. Thus, the NK_1 receptor knockout mouse has a phenotype that resembles mice treated with anxiolytics. Compared to wild-type controls, NK_1 receptor knockout mice display fewer vocalizations in the maternal separation test, spend more time in the open arms of the EPM, and show increased exploratory behavior in the open field.

A variety of NK_1 receptor antagonists (**Figure 13**) have been developed, primarily as analgesics, and produce anxiolytic effects in a variety of animal models[162]: CP-96345 is effective in the mouse light-dark shuttle box; NKP-608 increases social interaction in rat and in gerbil; L-733060, L-760735, MK-869, CP-99994, and GR-205171 attenuate guinea pig vocalizations in the maternal separation test; and GR-205171 and RP-67580 block marble burying in mice. In the EPM RP-67580 increased open arm activity in mouse with FK-888 causing similar effects in rat and mice and MK-869, L-742694, L-7330360, and CP-99994 are efficacious in the gerbil. The anxiolytic actions of these compounds occur in the absence of motor impairment or sedation.

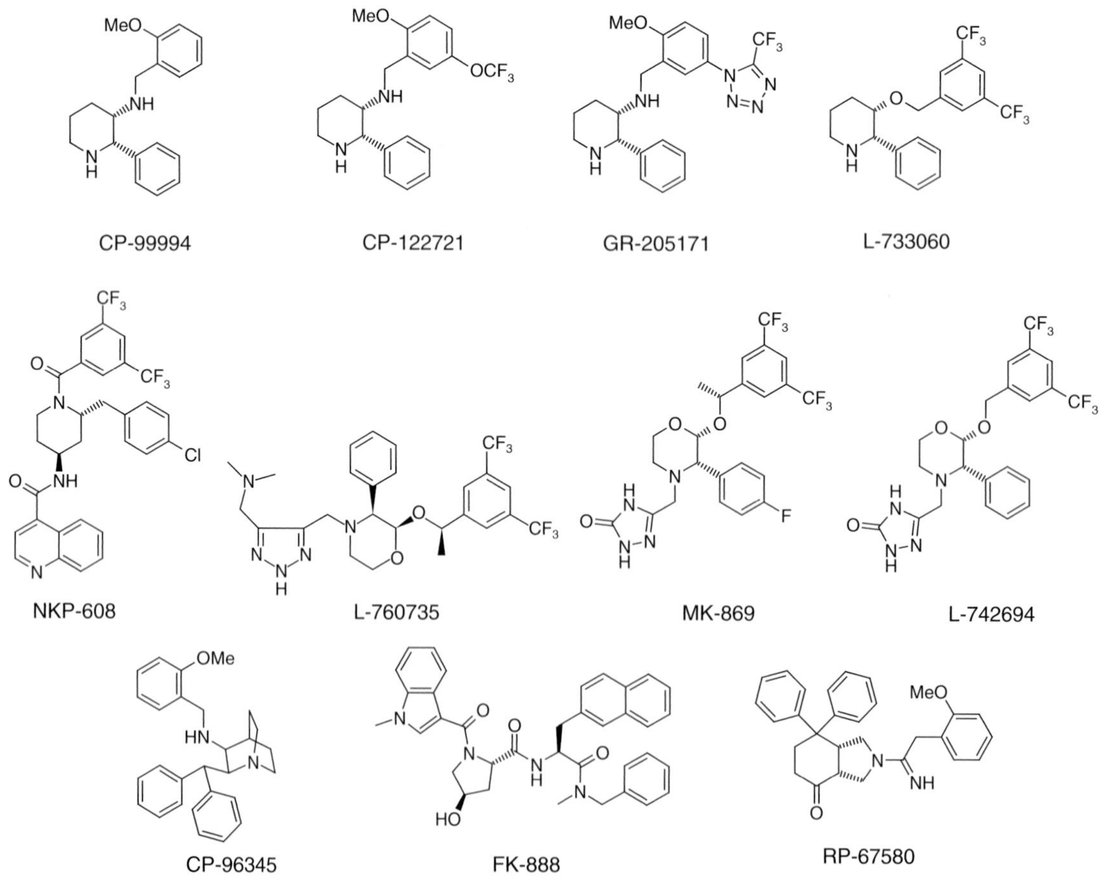

CP-99994 CP-122721 GR-205171 L-733060

NKP-608 L-760735 MK-869 L-742694

CP-96345 FK-888 RP-67580

Figure 13 Small-molecule NK_1 receptor antagonists.

NK_1 receptor antagonists have shown mixed effects in clinical trials.[162] The initial target for NK_1 antagonists was pain. Despite robust activity in numerous preclinical models, none of the NK_1 antagonists advanced to the clinic showed efficacy.[163] Based on the efficacy of MK-869 in the guinea pig vocalization model it was advanced to clinical trials in anxiety and at 300 mg showed initial improvements in depression and anxiety symptoms comparable to the SSRI paroxetine (20 mg) in a 6 week randomized, double-blind, placebo-controlled multicenter study in patients with major depression.[164] MK-869 failed to show efficacy in subsequent phase III trials and was discontinued. A backup compound, L-759274, significantly improved HAM-A symptoms and indices of depression. CP-122721 has shown efficacy similar to fluoxetine in depressed patients in a 6 week double-blind, placebo-controlled trial.

Mixed $NK_{1/2}$ or $NK_{1/2/3}$ receptor antagonists have also been targeted as anxiolytics,[165] the rationale being that blockade of more than one receptor subtype might be a more effective strategy to antagonize the effects of substance P and more efficacious in the treatment of anxiety and depression.

References

1. Kessler, R. C.; McGonagle, K. A.; Zhao, S.; Nelson, C. B.; Hughes, M.; Eshleman, S.; Wittchen, H. U.; Kendler, K. S. *Arch. Gen. Psychiatry* **1994**, *51*, 8–19.
2. American Psychiatric Association *DSM-IV-TR: Diagnostic and Statistical Manual of Mental Disorders*, IV-TR ed.; American Psychiatric Association: Washington, DC, 2000.
3. Nutt, D. J.; Ballenger, J. C.; Sheehan, D.; Wittchen, H. U. *Int. J. Neuropsychopharmacol.* **2002**, *5*, 315–325.
4. Sartorius, N.; Ustun, T. B.; Lecrubier, Y.; Wittchen, H. U. *Br. J. Psychiatry Suppl.* **1996**, 38–43.
5. Katzelnick, D. J.; Greist, J. H. *J. Clin. Psychiatry* **2001**, *62*, 11–15, discussion 15–16.
6. Bruce, S. E.; Yonkers, K. A.; Otto, M. W.; Eisen, J. L.; Weisberg, R. B.; Pagano, M.; Shea, M. T.; Keller, M. B. *Am. J. Psychiatry* **2005**, *162*, 1179–1187.
7. Katzman, M. A.; Struzik, L.; Vivian, L. L.; Vermani, M.; McBride, J. C. *Expert Rev. Neurotherapeut.* **2005**, *5*, 129–139.
8. Perkonigg, A.; Pfister, H.; Stein, M. B.; Hofler, M.; Lieb, R.; Maercker, A.; Wittchen, H. U. *Am. J. Psychiatry* **2005**, *162*, 1320–1327.
9. Fireman, B.; Koran, L. M.; Leventhal, J. L.; Jacobson, A. *Am. J. Psychiatry* **2001**, *158*, 1904–1910.
10. Davis, M. *Biol. Psychiatry* **1998**, *44*, 1239–1247.
11. Kent, J. M.; Rauch, S. L. *Curr. Psychiatry Rep.* **2003**, *5*, 266–273.
12. Charney, D. S.; Deutch, A. *Crit. Rev. Neurobiol.* **1996**, *10*, 419–446.
13. Maren, S. *Trends Neurosci.* **1999**, *22*, 561–567.
14. Amorapanth, P.; LeDoux, J. E.; Nader, K. *Nat. Neurosci.* **2000**, *3*, 74–79.
15. Talbot, P. S. *Curr. Psychiatry Rep.* **2004**, *6*, 274–279.
16. Tillfors, M.; Furmark, T.; Marteinsdottir, I.; Fischer, H.; Pissiota, A.; Langstrom, B.; Fredrikson, M. *Am. J. Psychiatry* **2001**, *158*, 1220–1226.
17. Gilbertson, M. W.; Shenton, M. E.; Ciszewski, A.; Kasai, K.; Lasko, N. B.; Orr, S. P.; Pitman, R. K. *Nat. Neurosci.* **2002**, *5*, 1242–1247.
18. Pissiota, A.; Frans, O.; Fernandez, M.; von Knorring, L.; Fischer, H.; Fredrikson, M. *Eur. Arch. Psychiatry Clin. Neurosci.* **2002**, *252*, 68–75.
19. Saxena, S.; Brody, A. L.; Maidment, K. M.; Dunkin, J. J.; Colgan, M.; Alborzian, S.; Phelps, M. E.; Baxter, L. R., Jr. *Neuropsychopharmacology* **1999**, *21*, 683–693.
20. Kent, J. M.; Mathew, S. J.; Gorman, J. M. *Biol. Psychiatry* **2002**, *52*, 1008–1030.
21. Arnold, P. D.; Zai, G.; Richter, M. A. *Curr. Psychiatry Rep.* **2004**, *6*, 243–254.
22. Kash, S. F.; Tecott, L. H.; Hodge, C.; Baekkeskov, S. *Proc. Natl. Acad. Sci. USA* **1999**, *96*, 1698–1703.
23. Sur, C.; Wafford, K. A.; Reynolds, D. S.; Hadingham, K. L.; Bromidge, F.; Macaulay, A.; Collinson, N.; O'Meara, G.; Howell, O.; Newman, R. et al. *J. Neurosci.* **2001**, *21*, 3409–3418.
24. Gunther, U.; Benson, J.; Benke, D.; Fritschy, J. M.; Reyes, G.; Knoflach, F.; Crestani, F.; Aguzzi, A.; Arigoni, M.; Lang, Y. et al. *Proc. Natl. Acad. Sci. USA* **1995**, *92*, 7749–7753.
25. Wick, M. J.; Radcliffe, R. A.; Bowers, B. J.; Mascia, M. P.; Luscher, B.; Harris, R. A.; Wehner, J. M. *Eur. J. Neurosci.* **2000**, *12*, 2634–2638.
26. Kim, J. J.; Shih, J. C.; Chen, K.; Chen, L.; Bao, S.; Maren, S.; Anagnostaras, S. G.; Fanselow, M. S.; De Maeyer, E.; Seif, I. et al. *Proc. Natl. Acad. Sci. USA* **1997**, *94*, 5929–5933.
27. Grimsby, J.; Toth, M.; Chen, K.; Kumazawa, T.; Klaidman, L.; Adams, J. D.; Karoum, F.; Gal, J.; Shih, J. C. *Nat. Genet.* **1997**, *17*, 206–210.
28. Gogos, J. A.; Morgan, M.; Luine, V.; Santha, M.; Ogawa, S.; Pfaff, D.; Karayiorgou, M. *Proc. Natl. Acad. Sci. USA* **1998**, *95*, 9991–9996.
29. Amara, S. G.; Kuhar, M. J. *Annu. Rev. Neurosci.* **1993**, *16*, 73–93.
30. Xu, F.; Gainetdinov, R. R.; Wetsel, W. C.; Jones, S. R.; Bohn, L. M.; Miller, G. W.; Wang, Y. M.; Caron, M. G. *Nat. Neurosci.* **2000**, *3*, 465–471.
31. Lahdesmaki, J.; Sallinen, J.; MacDonald, E.; Kobilka, B. K.; Fagerholm, V. *Neuroscience* **2002**, *113*, 289–299.
32. Murphy, D. L.; Li, Q.; Engel, S.; Wichems, C.; Andrews, A.; Lesch, K. P.; Uhl, G. *Brain Res. Bull.* **2001**, *56*, 487–494.
33. Brunner, D.; Buhot, M. C.; Hen, R.; Hofer, M. *Behav. Neurosci.* **1999**, *113*, 587–601.
34. Eison, A. S.; Eison, M. S. *Prog. Neuropsychopharmacol. Biol. Psychiatry* **1994**, *18*, 47–62.
35. Grailhe, R.; Waeber, C.; Dulawa, S. C.; Hornung, J. P.; Zhuang, X.; Brunner, D.; Geyer, M. A.; Hen, R. *Neuron.* **1999**, *22*, 581–591.
36. Leonardo, E. D.; Hen, R. *Annu. Rev. Psychol.* **2006**, *57*, 117–137.
37. Dawson, G. R.; Tricklebank, M. D. *Trends Pharmacol. Sci.* **1995**, *16*, 33–36.
38. Graeff, F. G.; Netto, C. F.; Zangrossi, H., Jr. *Neurosci. Biobehav. Rev.* **1998**, *23*, 237–246.
39. Crawley, J.; Goodwin, F. K. *Pharmacol. Biochem. Behav.* **1980**, *13*, 167–170.
40. Bourin, M.; Hascoet, M. *Eur. J. Pharmacol.* **2003**, *463*, 55–65.
41. Sanger, D. J. *J. Pharmacol. Exp. Ther.* **1990**, *254*, 420–426.
42. Millan, M. J.; Brocco, M. *Eur. J. Pharmacol.* **2003**, *463*, 67–96.
43. Hascoet, M.; Bourin, M.; Colombel, M. C.; Fiocco, A. J.; Baker, G. B. *Pharmacol. Biochem. Behav.* **2000**, *65*, 339–344.
44. Ripoll, N.; Nic Dhonnchadha, B. A.; Sebille, V.; Bourin, M.; Hascoet, M. *Psychopharmacology (Berlin)* **2005**, *180*, 73–83.
45. Koch, M. *Prog. Neurobiol.* **1999**, *59*, 107–128.

46. Davis, M. *Braz. J. Med. Biol. Res.* **1993**, *26*, 235–260.
47. Baas, J. M.; Grillon, C.; Bocker, K. B.; Brack, A. A.; Morgan, C. A., III; Kenemans, J. L.; Verbaten, M. N. *Psychopharmacology (Berlin)* **2002**, *161*, 233–247.
48. Pinel, J. P. J.; Triet, D. *J. Comp. Physiol. Psychol.* **1978**, *92*, 708–712.
49. De Boer, S. F.; Koolhaas, J. M. *Eur. J. Pharmacol.* **2003**, *463*, 145–161.
50. Basso, A. M.; Spina, M.; Rivier, J.; Vale, W.; Koob, G. F. *Psychopharmacology (Berlin)* **1999**, *145*, 21–30.
51. Picazo, O.; Fernandez-Guasti, A. *Brain Res.* **1995**, *680*, 135–141.
52. Tsuda, A.; Tanaka, M.; Georgiev, V.; Emoto, H. *Pharmacol. Biochem. Behav.* **1992**, *43*, 729–732.
53. Ichimaru, Y.; Egawa, T.; Sawa, A. *Jpn. J. Pharmacol.* **1995**, *68*, 65–70.
54. Millan, M. J.; Girardon, S.; Mullot, J.; Brocco, M.; Dekeyne, A. *Neuropharmacology* **2002**, *42*, 677–684.
55. Prut, L.; Belzung, C. *Eur. J. Pharmacol.* **2003**, *463*, 3–33.
56. Blanchard, D. C.; Griebel, G.; Blanchard, R. J. *Neurosci. Biobehav. Rev.* **2001**, *25*, 205–218.
57. Blanchard, D. C.; Griebel, G.; Blanchard, R. J. *Eur. J. Pharmacol.* **2003**, *463*, 97–116.
58. Blanchard, D. C.; Rodgers, R. J.; Hendrie, C. A.; Hori, K. *Pharmacol. Biochem. Behav.* **1988**, *31*, 269–278.
59. Blanchard, D. C.; Hori, K.; Rodgers, R. J.; Hendrie, C. A.; Blanchard, R. J. *Psychopharmacology (Berlin)* **1989**, *97*, 392–401.
60. Blanchard, R. J.; Blanchard, D. C. *J. Comp. Psychol.* **1989**, *103*, 70–82.
61. Griebel, G.; Perrault, G.; Simiand, J.; Cohen, C.; Granger, P.; Decobert, M.; Francon, D.; Avenet, P.; Depoortere, H.; Tan, S. et al. *J. Pharmacol. Exp. Ther.* **2001**, *298*, 753–768.
62. Griebel, G.; Rodgers, R. J.; Perrault, G.; Sanger, D. J. *Psychopharmacology (Berlin)* **1999**, *144*, 121–130.
63. Griebel, G. *Pharmacol. Ther.* **1995**, *65*, 319–395.
64. Griebel, G.; Perrault, G.; Sanger, D. J. *Psychopharmacology (Berlin)* **1997**, *131*, 180–186.
65. File, S. E.; Seth, P. *Eur. J. Pharmacol.* **2003**, *463*, 35–53.
66. Olivier, B.; Zethof, T.; Pattij, T.; van Boogaert, M.; van Oorschot, R.; Leahy, C.; Oosting, R.; Bouwknecht, A.; Veening, J.; van der Gugten, J. et al. *Eur. J. Pharmacol.* **2003**, *463*, 117–132.
67. Olivier, B.; Bouwknecht, J. A.; Pattij, T.; Leahy, C.; van Oorschot, R.; Zethof, T. J. *Pharmacol. Biochem. Behav.* **2002**, *72*, 179–188.
68. Gardner, C. R. *J. Pharmacol. Methods* **1985**, *14*, 181–187.
69. Insel, T. R., Winslow, J. T. Rat Pup Ultrasonic Vocalizations: an Ethologically Relevant Behaviour Responsive to Anxiolytics. In *Animal Models in Psychopharmacology*; Birkhauser Verlag: Basel, Switzerland, 1991, pp 15–36.
70. Barros, M.; Tomaz, C. *Neurosci. Biobehav. Rev.* **2002**, *26*, 187–201.
71. Montgomery, S.; van Zwieten-Boot, B.; Angst, J.; Baldwin, D.; Bourin, M.; Buller, R.; Hackett, D.; Kasper, S.; Kern, U.; Lader, M. et al. *Eur. Neuropsychopharmacol.* **2002**, *12*, 81–87.
72. Montgomery, S. A.; Lecrubier, Y.; Baldwin, D. S.; Kasper, S.; Lader, M.; Nil, R.; Stein, D.; Van Ree, J. M. *Eur. Neuropsychopharmacol.* **2004**, *14*, 425–433.
73. Versiani, M. *World J. Biol. Psychiatry* **2000**, *1*, 27–33.
74. Davidson, J. R.; Miner, C. M.; De Veaugh-Geiss, J.; Tupler, L. A.; Colket, J. T.; Potts, N. L. *Psychol. Med.* **1997**, *27*, 161–166.
75. Montgomery, S.; Bech, P.; Angst, J.; Davidson, J.; Delini-Stula, A.; van Ree, J.; van Zwieten-Boot, B.; Zohar, J.; Dunbar, G. *Eur. Neuropsychopharmacol.* **2000**, *10*, 297–303.
76. Horowitz, M.; Wilner, N.; Alvarez, W. *Psychosomatic Med* **1997**, *41*, 209–218.
77. Schoenfeld, F. B.; Marmar, C. R.; Neylan, T. C. *Psychiatr. Serv.* **2004**, *55*, 519–531.
78. Davidson, J. R.; Connor, K. M.; Hertzberg, M. A.; Weisler, R. H.; Wilson, W. H.; Payne, V. M. *J. Clin. Psychopharmacol.* **2005**, *25*, 166–169.
79. Goodman, W. K.; Price, L. H.; Rasmussen, S. A.; Mazure, C.; Fleischmann, R. L.; Hill, C. L.; Heninger, G. R.; Charney, D. S. *Arch. Gen. Psychiatry* **1989**, *46*, 1006–1011.
80. Koran, L. *Obsessive Compulsive and Related Disorders in Adults*; University Press: Cambridge, UK, 1999.
81. Schreiber, S. L. *Science* **2000**, *287*, 1964–1969.
82. Atack, J. R. *Expert Opin. Invest. Drugs* **2005**, *14*, 601–618.
83. Barnard, E. A.; Skolnick, P.; Olsen, R. W.; Mohler, H.; Sieghart, W.; Biggio, G.; Braestrup, C.; Bateson, A. N.; Langer, S. Z. *Pharmacol. Rev.* **1998**, *50*, 291–313.
84. Mohler, H.; Fritschy, J. M.; Rudolph, U. *J. Pharmacol. Exp. Ther.* **2002**, *300*, 2–8.
85. Hester, J. B., Jr.; Rudzik, A. D.; Kamdar, B. V. *J. Med. Chem.* **1971**, *14*, 1078–1081.
86. Ballenger, J. C.; Davidson, J. R.; Lecrubier, Y.; Nutt, D. J.; Baldwin, D. S.; den Boer, J. A.; Kasper, S.; Shear, M. K. *J. Clin. Psychiatry* **1998**, *59*, 47–54.
87. Schatzberg, A. F. *J. Clin. Psychiatry* **2000**, *61*, 9–17.
88. Wong, D. T.; Perry, K. W.; Nat. Bymaster, F. P. *Rev. Drug Discov.* **2005**, *4*, 764–774.
89. Grigoriadis, D. E. *Expert Opin. Ther. Targets* **2005**, *9*, 651–684.
90. Riad, M.; Zimmer, L.; Rbah, L.; Watkins, K. C.; Hamon, M.; Descarries, L. *J. Neurosci.* **2004**, *24*, 5420–5426.
91. Santarelli, L.; Saxe, M.; Gross, C.; Surget, A.; Battaglia, F.; Dulawa, S.; Weisstaub, N.; Lee, J.; Duman, R.; Arancio, O. et al. *Science* **2003**, *301*, 805–809.
92. Brunton, L.; Lazo, J.; Parker, K. *Goodman & Gilman's The Pharmacological Basis of Therapeutics*, 11th ed; McGraw-Hill: New York, NY, 2005.
93. Ninan, P.; Muntasser, S. Buspirone and Gepirone. *Textbook of Psychopharmacology*; American Psychiatric Publishing, Inc: Washington, DC, 2004.
94. Alpert, J. E.; Franznick, D. A.; Hollander, S. B.; Fava, M. *J. Clin. Psychiatry* **2004**, *65*, 1069–1075.
95. Csanalosi, I.; Schweizer, E.; Case, W. G.; Rickels, K. *J. Clin. Psychopharmacol.* **1987**, *7*, 31–33.
96. Vale, W.; Spiess, J.; Rivier, C.; Rivier, J. *Science* **1981**, *213*, 1394–1397.
97. Risbrough, V. B.; Hauger, R. L.; Roberts, A. L.; Vale, W. W.; Geyer, M. A. *J. Neurosci.* **2004**, *24*, 6545–6552.
98. Saunders, J.; Williams, J. *Prog. Med. Chem.* **2003**, *41*, 195–247.
99. Zobel, A. W.; Nickel, T.; Kunzel, H. E.; Ackl, N.; Sonntag, A.; Ising, M.; Holsboer, F. *J. Psychiatr. Res.* **2000**, *34*, 171–181.
100. Lancel, M.; Muller-Preuss, P.; Wigger, A.; Landgraf, R.; Holsboer, F. *J. Psychiatr. Res.* **2002**, *36*, 197.
101. Held, K.; Kunzel, H.; Ising, M.; Schmid, D. A.; Zobel, A.; Murck, H.; Holsboer, F.; Steiger, A. *J. Psychiatr. Res.* **2004**, *38*, 129–136.
102. Basile, A. S.; Lippa, A. S.; Skolnick, P. *Eur. J. Pharmacol.* **2004**, *500*, 441–451.
103. McKernan, R. M.; Whiting, P. J. *Trends Neurosci.* **1996**, *19*, 139–143.

104. Rudolph, U.; Crestani, F.; Benke, D.; Brunig, I.; Benson, J. A.; Fritschy, J. M.; Martin, J. R.; Bluethmann, H.; Mohler, H. *Nature* **1999**, *401*, 796–800.
105. Low, K.; Crestani, F.; Keist, R.; Benke, D.; Brunig, I.; Benson, J. A.; Fritschy, J. M.; Rulicke, T.; Bluethmann, H.; Mohler, H. et al. *Science* **2000**, *290*, 131–134.
106. McKernan, R. M.; Rosahl, T. W.; Reynolds, D. S.; Sur, C.; Wafford, K. A.; Atack, J. R.; Farrar, S.; Myers, J.; Cook, G.; Ferris, P. et al. *Nat. Neurosci.* **2000**, *3*, 587–592.
107. Caraiscos, V. B.; Elliott, E. M.; You-Ten, K. E.; Cheng, V. Y.; Belelli, D.; Newell, J. G.; Jackson, M. F.; Lambert, J. J.; Rosahl, T. W.; Wafford, K. A. et al. *Proc. Natl. Acad. Sci. USA* **2004**, *101*, 3662–3667.
108. Collinson, N.; Kuenzi, F. M.; Jarolimek, W.; Maubach, K. A.; Cothliff, R.; Sur, C.; Smith, A.; Otu, F. M.; Howell, O.; Atack, J. R. et al. *J. Neurosci.* **2002**, *22*, 5572–5580.
109. Crestani, F.; Martin, J. R.; Mohler, H.; Rudolph, U. *Br. J. Pharmacol.* **2000**, *131*, 1251–1254.
110. van Rijnsoever, C.; Tauber, M.; Choulli, M. K.; Keist, R.; Rudolph, U.; Mohler, H.; Fritschy, J. M.; Crestani, F. *J. Neurosci.* **2004**, *24*, 6785–6790.
111. Atack, J. R. *Curr. Drug Targets CNS Neurol. Disord.* **2003**, *2*, 213–232.
112. Pieri, L.; Hunkeler, W.; Jauch, R.; Merz, W. A.; Roncari, G.; Timm, U. *Drugs Fut.* **1988**, *13*, 730–735.
113. Busto, U.; Kaplan, H. L.; Zawertailo, L.; Sellers, E. M. *Clin. Pharmacol. Ther.* **1994**, *55*, 451–463.
114. van Steveninck, A. L.; Gieschke, R.; Schoemaker, R. C.; Roncari, G.; Tuk, B.; Pieters, M. S.; Breimer, D. D.; Cohen, A. F. *Br. J. Clin. Pharmacol.* **1996**, *41*, 565–573.
115. Stephens, D. N.; Schneider, H. H.; Kehr, W.; Andrews, J. S.; Rettig, K. J.; Turski, L.; Schmiechen, R.; Turner, J. D.; Jensen, L. H.; Petersen, E. N. et al. *J. Pharmacol. Exp. Ther.* **1990**, *253*, 334–343.
116. Dubinsky, B.; Vaidya, A. H.; Rosenthal, D. I.; Hochman, C.; Crooke, J. J.; DeLuca, S.; DeVine, A.; Cheo-Isaacs, C. T.; Carter, A. R.; Jordan, A. D. et al. *J. Pharmacol. Exp. Ther.* **2002**, *303*, 777–790.
117. Lippa, A.; Czobor, P.; Stark, J.; Beer, B.; Kostakis, E.; Gravielle, M.; Bandyopadhyay, S.; Russek, S. J.; Gibbs, T. T.; Farb, D. H. et al. *Proc. Natl. Acad. Sci. USA* **2005**, *102*, 7380–7385.
118. Collins, I.; Moyes, C.; Davey, W. B.; Rowley, M.; Bromidge, F. A.; Quirk, K.; Atack, J. R.; McKernan, R. M.; Thompson, S. A.; Wafford, K. et al. *J. Med. Chem.* **2002**, *45*, 1887–1900.
119. Rowlett, J. K.; Platt, D. M.; Lelas, S.; Atack, J. R.; Dawson, G. R. *Proc. Natl. Acad. Sci. USA* **2005**, *102*, 915–920.
120. Russell, M. G.; Carling, R. W.; Atack, J. R.; Bromidge, F. A.; Cook, S. M.; Hunt, P.; Isted, C.; Lucas, M.; McKernan, R. M.; Mitchinson, A. et al. *J. Med. Chem.* **2005**, *48*, 1367–1383.
121. Johnstone, T. B.; Hogenkamp, D. J.; Coyne, L.; Su, J.; Halliwell, R. F.; Tran, M. B.; Yoshimura, R. F.; Li, W. Y.; Wang, J.; Gee, K. W. *Nat. Med.* **2004**, *10*, 31–32.
122. Langen, B.; Egerland, U.; Bernoester, K.; Dost, R.; Unverferth, K.; Rundfeldt, C. *J. Pharmacol. Exp. Ther.* **2005**, *314*, 717–724.
123. Benke, D.; Fakitsas, P.; Roggenmoser, C.; Michel, C.; Rudolph, U.; Mohler, H. *J. Biol. Chem.* **2004**, *279*, 43654–43660.
124. Swanson, C. J.; Bures, M.; Johnson, M. P.; Linden, A. M.; Monn, J. A.; Schoepp, D. D. *Nat. Rev. Drug Discov.* **2005**, *4*, 131–144.
125. Cartmell, J.; Schoepp, D. D. *J. Neurochem.* **2000**, *75*, 889–907.
126. Shimazaki, T.; Iijima, M.; Chaki, S. *Eur. J. Pharmacol.* **2004**, *501*, 121–125.
127. Spinelli, S.; Ballard, T.; Gatti-McArthur, S.; Richards, G. J.; Kapps, M.; Woltering, T.; Wichmann, J.; Stadler, H.; Feldon, J.; Pryce, C. R. *Psychopharmacology (Berlin)* **2005**, *179*, 292–302.
128. Krystal, J. H.; Abi-Saab, W.; Perry, E.; D'Souza, D. C.; Liu, N.; Gueorguieva, R.; McDougall, L.; Hunsberger, T.; Belger, A.; Levine, L. et al. *Psychopharmacology (Berlin)* **2005**, *179*, 303–309.
129. Linden, A. M.; Shannon, H.; Baez, M.; Yu, J. L.; Koester, A.; Schoepp, D. D. *Psychopharmacology (Berlin)* **2005**, *179*, 284–291.
130. Johnson, M. P.; Barda, D.; Britton, T. C.; Emkey, R.; Hornback, W. J.; Jagdmann, G. E.; McKinzie, D. L.; Nisenbaum, E. S.; Tizzano, J. P.; Schoepp, D. D. *Psychopharmacology (Berlin)* **2005**, *179*, 271–283.
131. Klodzinska, A.; Tatarczynska, E.; Stachowicz, K.; Chojnacka-Wojcik, E. *J. Physiol. Pharmacol.* **2004**, *55*, 113–126.
132. Steckler, T.; Lavreysen, H.; Oliveira, A. M.; Aerts, N.; Van Craenendonck, H.; Prickaerts, J.; Megens, A.; Lesage, A. S. *Psychopharmacology (Berlin)* **2004**.
133. Spooren, W.; Gasparini, F. *Drug News Perspect.* **2004**, *17*, 251–257.
134. Klodzinska, A.; Tatarczynska, E.; Chojnacka-Wojcik, E.; Nowak, G.; Cosford, N. D.; Pilc, A. *Neuropharmacology* **2004**, *47*, 342–350.
135. Porter, R. H.; Jaeschke, G.; Spooren, W.; Ballard, T.; Buettelmann, B.; Kolczewski, S.; Peters, J. U.; Prinssen, E.; Wichmann, J.; Vieira, E. et al. *J. Pharmacol. Exp. Ther.* **2005**, *315*, 711–721.
136. Taylor, C. P. *Rev. Neurol. (Paris)* **1997**, *153*, S39–S45.
137. Urban, M. O.; Ren, K.; Park, K. T.; Campbell, B.; Anker, N.; Stearns, B.; Aiyar, J.; Belley, M.; Cohen, C.; Bristow, L. *J. Pharmacol. Exp. Ther.* **2005**, *313*, 1209–1216.
138. Belliotti, T. R.; Capiris, T.; Ekhato, I. V.; Kinsora, J. J.; Field, M. J.; Heffner, T. G.; Meltzer, L. T.; Schwarz, J. B.; Taylor, C. P.; Thorpe, A. J. et al. *J. Med. Chem.* **2005**, *48*, 2294–2307.
139. Field, M. J.; Oles, R. J.; Singh, L. *Br. J. Pharmacol.* **2001**, *132*, 1–4.
140. Singh, L.; Field, M. J.; Ferris, P.; Hunter, J. C.; Oles, R. J.; Williams, R. G.; Woodruff, G. N. *Psychopharmacology (Berlin)* **1996**, *127*, 1–9.
141. Lauria-Horner, B. A.; Pohl, R. B. *Expert Opin. Investig. Drugs* **2003**, *12*, 663–672.
142. Pohl, R. B.; Feltner, D. E.; Fieve, R. R.; Pande, A. C. *J. Clin. Psychopharmacol.* **2005**, *25*, 151–158.
143. Pande, A. C.; Davidson, J. R.; Jefferson, J. W.; Janney, C. A.; Katzelnick, D. J.; Weisler, R. H.; Greist, J. H.; Sutherland, S. M. *J. Clin. Psychopharmacol.* **1999**, *19*, 341–348.
144. Pande, A. C.; Feltner, D. E.; Jefferson, J. W.; Davidson, J. R.; Pollack, M.; Stein, M. B.; Lydiard, R. B.; Futterer, R.; Robinson, P.; Slomkowski, M. et al. *J. Clin. Psychopharmacol.* **2004**, *24*, 141–149.
145. Pande, A. C.; Pollack, M. H.; Crockatt, J.; Greiner, M.; Chouinard, G.; Lydiard, R. B.; Taylor, C. B.; Dager, S. R.; Shiovitz, T. *J. Clin. Psychopharmacol.* **2000**, *20*, 467–471.
146. Cone, R. D. *Nat. Neurosci.* **2005**, *8*, 571–578.
147. Farooqi, I. S.; O'Rahilly, S. *Annu. Rev. Med.* **2005**, *56*, 443–458.
148. Kishi, T.; Aschkenasi, C. J.; Lee, C. E.; Mountjoy, K. G.; Saper, C. B.; Elmquist, J. K. *J. Comp. Neurol.* **2003**, *457*, 213–235.
149. Vilberg, T. R.; Panksepp, J.; Kastin, A. J.; Coy, D. H. *Peptides* **1984**, *5*, 823–827.
150. Panksepp, J.; Abbott, B. B. *Peptides* **1990**, *11*, 647–653.
151. Gonzalez, M. I.; Vaziri, S.; Wilson, C. A. *Peptides* **1996**, *17*, 171–177.

152. Chaki, S.; Ogawa, S.; Toda, Y.; Funakoshi, T.; Okuyama, S. *Eur. J. Pharmacol.* **2003**, *474*, 95–101.
153. Shimazaki, T.; Chaki, S. *Pharmacol. Biochem. Behav.* **2005**, *80*, 395–400.
154. Vergoni, A. V.; Bertolini, A.; Wikberg, J. E.; Schioth, H. B. *Eur. J. Pharmacol.* **1999**, *369*, 11–15.
155. Chaki, S.; Hirota, S.; Funakoshi, T.; Suzuki, Y.; Suetake, S.; Okubo, T.; Ishii, T.; Nakazato, A.; Okuyama, S. *J. Pharmacol. Exp. Ther.* **2003**, *304*, 818–826.
156. Kowalski, T. J.; McBriar, M. D. *Expert Opin. Investig. Drugs* **2004**, *13*, 1113–1122.
157. Saito, Y.; Cheng, M.; Leslie, F. M.; Civelli, O. *J. Comp. Neurol.* **2001**, *435*, 26–40.
158. Chaki, S.; Funakoshi, T.; Hirota-Okuno, S.; Nishiguchi, M.; Shimazaki, T.; Iijima, M.; Grottick, A. J.; Kanuma, K.; Omodera, K.; Sekiguchi, Y. et al. *J. Pharmacol. Exp. Ther.* **2005**, *313*, 831–839.
159. Borowsky, B.; Durkin, M. M.; Ogozalek, K.; Marzabadi, M. R.; DeLeon, J.; Lagu, B.; Heurich, R.; Lichtblau, H.; Shaposhnik, Z.; Daniewska, I. et al. *Nat. Med.* **2002**, *8*, 825–830.
160. Kela, J.; Salmi, P.; Rimondini-Giorgini, R.; Heilig, M.; Wahlestedt, C. *Regul. Pept.* **2003**, *114*, 109–114.
161. Monzon, M. E.; De Barioglio, S. R. *Physiol. Behav.* **1999**, *67*, 813–817.
162. McLean, S. *Curr. Pharm. Des.* **2005**, *11*, 1529–1547.
163. Hill, R. *Trends Pharmacol. Sci.* **2000**, *21*, 244–246.
164. Kramer, M. S.; Cutler, N.; Feighner, J.; Shrivastava, R.; Carman, J.; Sramek, J. J.; Reines, S. A.; Liu, G.; Snavely, D.; Wyatt-Knowles, E. et al. *Science* **1998**, *281*, 1640–1645.
165. Giardina, G. A.; Gagliardi, S.; Martinelli, M. *IDrugs* **2003**, *6*, 758–772.

Biographies

Kathleen R Gogas, PhD, is a director in the Neuroscience group at Neurocrine Biosciences, a bio-pharmaceutical company located in San Diego, California, USA. She joined Neurocrine Biosciences in 1996 as associate director in neuroscience following $5\frac{1}{2}$ years at Roche Biosciences where she held positions of increasing responsibility that included Head of the Neurobehavior Group and Deputy Head of the CNS Therapeutic Area at Roche Bioscience in Palo Alto. Dr Gogas earned her PhD in pharmacology from Albany Medical College and conducted postdoctoral studies in neuroanatomy, pharmacology, and molecular biology of endogenous pain systems at University of California at San Francisco and Harvard University.

Sandra M Lechner, PhD, is a director in the Neuroscience group at Neurocrine Biosciences, a bio-pharmaceutical company located in San Diego, California, USA. She joined Neurocrine Biosciences in 2004 following 3 years at Merck Research Laboratories where she established a core histology/imaging group for the analysis of CNS targets at MRL in San Diego. Prior to this, she spent 3 years at Allelix Neuroscience, as group leader, Behavioral Pharmacology. Dr Lechner completed a postdoctoral fellowship at Hahnemann University where she studied CRF and the neurobiology of stress, after earning her PhD in physiology/pharmacology from University of California, San Diego.

Stacy Markison, PhD, is a principal scientist in the Department of Neuroscience at Neurocrine Biosciences. Dr Markison received her PhD from the University of Florida in 1998 and was a postdoctoral fellow at the University of Pennsylvania until she joined Neurocrine Biosciences in 2002. At Neurocrine, she is responsible for developing in vivo models to screen and characterize compounds being developed for the treatment of central nervous system disorders including anxiety, depression, Parkinson's disease, and metabolic diseases such as cachexia, diabetes, and obesity.

John P Williams, PhD, is a senior director in the Medicinal Chemistry group at Neurocrine Biosciences. He joined Neurocrine Biosciences in 1999 as associate director following 4 years at Combichem where he was involved in many aspects of combinatorial chemistry and its subsequent application to drug discovery and development, and 7 years at Gilead Sciences where he was involved in the optimization of the pharmaceutical properties of GS-522, an oligonucleotide that binds and inhibits human thrombin. Dr Williams received his PhD in organic chemistry from the University of Michigan in 1990 and conducted postdoctoral studies at Ohio State University, where he was an American Cancer Society postdoctoral fellow working in the area of alkaloid total synthesis.

William McCarthy is an associate director in the Commercial group at Neurocrine Biosciences, a bio-pharmaceutical company located in San Diego, California, USA. He joined Neurocrine Biosciences in 2005 following 2 years at IMS Health, where he was an engagement director in the firm's Pricing and Reimbursement group. Prior to IMS, he spent 2 years in Deloitte Consulting's health care practice, and 4 years at Cambridge Pharma Consultancy. Throughout his career, Mr McCarthy has focused on developing strategies for pharmaceutical product commercialization and market access. Mr McCarthy holds an MBA from London Business School and an MS in Economics from Exeter University, UK.

Dimitri E Grigoriadis, PhD, is a Neurocrine Fellow at Neurocrine Biosciences, and is head of the Pharmacology and Lead Discovery department, responsible for the initial lead discovery efforts and subsequent structure activity lead optimization assays for a variety of GPCR, transporter and ion channel targets in the Neurocrine portfolio. He joined Neurocrine in 1993 following 4 years in the CNS department at the DuPont Merck Pharmaceutical Company. Dr Grigoriadis is a neuropharmacologist interested in the discovery and development of drugs targeting CNS diseases and has authored over 100 peer-reviewed publications including many in his primary area of interest in depression and anxiety-related disorders.

Alan C Foster, PhD, is a Neurocrine Fellow at Neurocrine Biosciences, a bio-pharmaceutical company located in San Diego, California, USA. He joined Neurocrine in 1996 as head of the Neuroscience group following 4 years as head of CNS at Gensia, and 8 years in the CNS research group at Merck & Co. A neuropharmacologist by training, Dr Foster is interested in the discovery and development of drugs targeted at G protein-coupled receptors, ion channels, transporters, and enzymes for the treatment of nervous system diseases. Dr Foster has authored over 150 peer-reviewed publications on basic research and drug discovery aspects of neuroscience.

Comprehensive Medicinal Chemistry II
ISBN (set): 0-08-044513-6

ISBN (Volume 6) 0-08-044519-5; pp. 85–115

6.05 Attention Deficit Hyperactivity Disorder

K E Browman and G B Fox, Abbott Laboratories, Abbott Park, IL, USA

6.05.1 Introduction

Attention defect hyperactivity disorder (ADHD) affects 3–7% of school-aged children, and reports suggest similar percentages in adults. Children and adults manifest the symptoms of hyperactivity, impulsivity, and inattention differently, however, and diagnostic criteria are typically oriented around the children's core symptoms. While diagnostic criteria have developed significantly since the 1970s, underdiagnosis is a key issue with regard to adult ADHD. In addition, many of the controlled clinical studies have largely ignored the use of therapeutics in populations other than school-aged Caucasian males. On the positive side, it is clear that, once diagnosed, medical treatment is often very effective in ameliorating the effects of ADHD, conferring strong benefits to those who have struggled with ADHD-related problems at work, with family, and with social issues.

There are no objective laboratory measures of ADHD, and there is no one definitive animal model. There are a number of evolving animal models of aspects of ADHD, which have greatly furthered research directed toward the development of novel therapeutics for treating symptoms of ADHD. Perhaps one of the most relevant animal assays is the five-choice serial reaction time test (5-CSRTT), originally designed as a preclinical correlate of the continuous performance test (CPT) of attention used in clinical studies. In addition, while imaging measures currently lack specificity for use in diagnosing ADHD, it is clear that recent imaging studies are shedding light on the etiology of the disorder.

The biological underpinnings of ADHD focus largely on monoamine neurotransmitter systems, including dopamine and norepinephrine. Imaging studies show decreased activation in frontal brain regions (areas involved in attention and working memory). Genetic studies indicate heritability for ADHD of about 70%. Genetic variants related to genes coding for dopamine and norepinephrine neurotransmission may confer susceptibility to ADHD. Stimulant medications, which are still among the most effective treatments for ADHD, act on the dopaminergic system.

Of the available treatments, the stimulants (a class that includes methylphenidate) are first-line agents for both children and adults diagnosed with ADHD. About 70–80% of children respond to methylphenidate, and, with appropriate dosing regimens, adults often respond as well as children to treatment. Long-acting formulations are a key new area in the treatment of ADHD, as these permit once-a-day dosing that eliminates interruptions during the school day for dosing. Other treatment options include the newly approved drug atomoxetine, which is the first nonscheduled drug approved for this disorder. Off-label use of antidepressants and antihypertensive medications may be effective in patients who do not respond to(or cannot tolerate the side effects of) stimulant medications.

Future research will need to consider associated comorbidities with ADHD, which, if properly identified, may influence successful treatment options. New therapeutics with the efficacy of stimulants but without the scheduling concerns would clearly be advantageous in the treatment of ADHD. Finally, emphasis should be given to appropriate diagnostic criteria for children, adolescents, and adults.

6.05.2 Disease State

ADHD is characterized by: (1) the inattentive; (2) the hyperactive/impulsive (hyperactivity without inattention); or (3) the combined type.[1] The condition arises before the age of 7 and frequently persists through adolescence and into adulthood, although the behavioral features of the condition vary at different ages. Children with the hyperactive/impulsive subtype usually develop symptoms of ADHD by 4 years of age, with significant difficulties contributing to academic challenges by the age of 8. In contrast, children with the inattentive subtype tend to develop difficulties later, with a typical age of presenting with noticeable difficulties around 9–10 years of age. In adolescence, the hyperactive and impulsive symptoms may become less evident, but the problems with inattention often persist. Generally ADHD is assumed to have a ratio of 10:1 (boys to girls) based on clinical estimates, while community estimates are 3:1.[2] In adult samples, the ratios of men to women are approximately equal.

ADHD, defined by the American Psychiatric Association *Diagnostic and Statiscal Manual-IV* (DSM-IV), is diagnosed using behavioral criteria. Behavioral heterogeneity among ADHD sufferers complicates the diagnosis and makes it difficult to establish its prevalence. ADHD prevalence is typically estimated at 4–12% of school-aged children.[3] One of the difficulties in estimating the prevalence of ADHD is that, unlike for other psychiatric disorders, physicians rely on recognition of various types of behaviors in different combinations (somewhat subjective judgment) for diagnosis. More than 4.5 million individuals have been diagnosed with ADHD in the US, the vast majority of whom are males between 5 and 19 years of age. Studies have suggested that 30–50% of children with ADHD continue to have symptoms of the disorder into adulthood, although a smaller proportion of 10% have clinically significant symptoms. This suggests that 1–3% of adults may suffer ADHD symptoms, and are likely deserving treatment. The negative social and occupational impact of ADHD in adults is now recognized, with prevalence in adulthood estimated at 1–3%. The recognition of distinct subtypes of ADHD, particularly the inattentive subtype, in more recent definitions of ADHD has resulted in increased prevalence estimates. Moreover, inattentive behavior is hypothesized to be more prevalent in girls, who are generally viewed as being underdiagnosed for ADHD.

It can be difficult to estimate worldwide prevalence rates, as not all countries use the DSM-IV ADHD criteria. In Europe, DSM-IV criteria are not routinely used for diagnosis (the *International Classification of Disease* manual (ICD-10) is used), leading to a much lower rate of diagnosis and a more severe population being treated. In Japan, ADHD is virtually nonexistent and this is not expected to change in the midterm. When operational definitions of ADHD are used, studies of children aged 4–16 years old were found to have the following frequencies: New Zealand 7%, US 8%, Canada 6%, Puerto Rico 9%, UK 5%, and Hong Kong 9%. When ICD-10 psychiatric diagnosis criteria are applied, the following frequencies were found: Sweden, 2%; Germany 2–4%; UK 2%; and Hong Kong 1%.[4] Studies in Brazil have estimated an ADHD prevalence of 5.8%.[5]

In the US, ADHD is the commonest neurobehavioral disorder of childhood and is among the most prevalent chronic health conditions affecting school-aged children. ADHD accounts for 30–50% of all referrals for child mental health services in the US,[1] and comprises the majority of the economic cost of childhood mental disorders. Recent projections from the US Census Bureau indicate that the school age population is growing by approximately 300 000 per year. By 2010, the number of children under 19 years of age will have grown from 78.5 million in 2000 to 81 million. Assuming the mean prevalence of 6%, the incidence of pediatric ADHD will be approximately 4.8 million. In addition, most physicians do not consider hyperactivity to be a stand-alone disorder. As many as one-third of children diagnosed with ADHD have at least one coexisting condition – oppositional defiant disorder (35.2%), conduct disorder (25.7%), anxiety disorder (25.8%), and/or depression (18.2%).[2] Although it is generally accepted that ADHD persists into adulthood, there is a noticeable lack of a consensus as to diagnostic criteria suitable for adults with ADHD. As children approach adulthood, a number of developmental changes occur, and diagnostic criteria are not currently identifying how the subtypes change in this transition period. Furthermore, adult diagnostic criteria currently rely on symptomology identified in younger populations, and likely do not adequately capture all of the adults with ADHD. In general, there are three approaches for diagnosing adult ADHD: (1) the Wender Utah criteria; (2) the DSM-IV criteria; and (3) laboratory assessments.[6]

In the 1970s–80s evidence emerged suggesting that ADHD persisted into adulthood. At the same time the diagnostic framework for the DSM-III was developed, focusing on childhood hyperactivity. The Wender Utah diagnostic criteria were developed based on the premise that diagnostic criteria appropriate for children were not appropriate for adults and include a retrospective childhood diagnosis, ongoing difficulties with inattentiveness and hyperactivity, and the inclusion of two other core symptoms of adult ADHD. As detailed by McGough and Barkley in 2004,[6] while the Utah approach to adult ADHD underscores the importance of retrospective childhood diagnosis and evaluation of current symptoms, the Wender Utah diagnostic criteria do not identify patients with predominantly inattentive symptoms, exclude some patients with comorbid psychopathology, and differ from the DSM diagnostic framework.[6]

The following general categories are used for ADHD diagnosis using DSM-IV criteria:

- the presence of six or more symptoms for at least 6 months
- demonstration of clinically significant impairment in social, academic, or occupational functioning
- observation of symptoms in two or more settings
- onset of symptoms before the age of 7 years
- the exclusion of other disorders.

Although the potential areas of impairment include occupational functioning, the DSM-IV focuses largely on children. As an example, in discussing possible symptoms, these tend to focus on examples such as 'cannot play quietly' or 'runs and climbs excessively' – items clearly missing the areas of dysfunction pertinent to adults and lacking face validity for adults. One of the difficulties with both the Wender Utah scale and the DSM-IV criteria are that they require a diagnosis of ADHD in childhood. If the physician does not examine the patient until adulthood, this requires anecdotal recall, which is often unreliable. Although the DSM-IV criteria have been used (successfully) in diagnosing adult ADHD, the scale is likely not the most appropriate for diagnosing adults. Other scales that have been validated include the Adult Self-Report Scale (ASRS), developed in conjunction with the World Health Organization,[95] and the Conners Adult ADHD Rating Scale (CAARS), comprised of self-reports and observer ratings, and providing normative data for comparison.

Other diagnostic criteria that have been proposed include laboratory-based diagnostic paradigms such as tests of executive functioning and working memory,[7] ecological measures of academic performance and classroom behavior,[8] the CPT,[9] electroencephalography,[10] and neuroimaging.[11] Obviously there is interest in the development of laboratory tests for ADHD, as research as well as diagnostic criteria. The other advantage of laboratory tests is the correlation with animal models. For example, the CPT is hypothesized to have similarities with the 5-CSRTT discussed under animal models. To date, however, there is a lack of data supporting the use of these measures as diagnostic tools.[6]

While not yet appropriate for diagnosing ADHD, the development of laboratory tests is positively influencing other areas for future research, including efforts directed toward understanding the heterogeneity of ADHD. Studies investigating the underlying neurobiology should lead to new understanding in this area. Imaging techniques are giving important insight, and this should ultimately contribute to a greater knowledge of the disorder. Pharmacogenomic studies are in their infancy, although the National Human Genome Research Institute is currently sponsoring a study on the genetic analysis of ADHD.

Future research will also need to consider a full range of medications, disease manifestations, and pharmacokinetic criteria. Associated comorbidities with ADHD are poorly understood, and need to be clearly identified. A recently published practice parameter discussed the use of stimulant medications in the treatment of children, adolescents, and adults, and summarized limited studies investigating the combination of stimulants and other psychotropic agents useful for treating comorbid conditions,[12] suggesting this as a key area for future research.

6.05.3 Disease Basis

Until relatively recently, controversy surrounded the status of ADHD as a genuine medical condition. It was commonly believed that ADHD was a childhood reaction to poor parenting or family stress. However, recent findings that ADHD is found in a variety of cultures and that there is a strong genetic component bolster the validity of designating ADHD as a medical disorder.[11] ADHD is likely a polygenic disorder, meaning that multiple genes contribute, each conferring a small risk.

Candidate gene searches have focused on the dopaminergic system, in part due to the effective medications acting primarily on this neurotransmitter system (for a review of pharmacogenomic ADHD studies, see [13]). Genes that have been associated with ADHD include the dopamine transporter (*DAT1*), dopamine D2 (*DRD2*) and D4 (*DRD4*) and D5 receptor (*DRD5*) subtypes, and the dopa-β-hydroxylase gene (*DBH*).[13]

Recently, an intron mutation in the gene coding for the α4 subunit of neuronal nicotinic receptors (NNRs) was associated with ADHD characterized by severe inattention. The functional consequence of this mutation is unknown but its location is suggestive of effects on pre-mRNA stability or splicing.[14] Interestingly, adolescents with ADHD have at least a twofold higher rate of smoking cigarettes than the average smoking population (excluding individuals with other neurological disorders) and this finding continues into adulthood.

Polymorphisms within the gene encoding synaptosomal-associated protein of 25 kDa (SNAP-25), a presynaptic plasma membrane protein with an integral role in synaptic transmission, have also been implicated in ADHD. SNAP-25 forms a complex with syntaxin 1a and synaptic vesicle proteins such as VAMP-2 (synaptobrevin 2) and synaptotagmin, which mediates calcium-dependent exocytosis of neurotransmitter from the synaptic vesicle into the synaptic cleft.[15] SNAP-25 is differentially expressed in the brain, with high levels found in regions such as the hippocampus, neocortex, thalamus, substantia nigra, and cerebellum. In a recent study of 93 ADHD nuclear families in Ireland, significant increased preferential transmission of the SNAP-25 polymorphism DdeI allele 1 was found.[16] In a separate study of Canadian families with ADHD, significant increased preferential transmission of the SNAP-25 polymorphism DdeI allele 2 was found.[17] Additional haplotype analysis of SNAP-25 implicates SNAP-25 in the etiology of ADHD.

Imaging studies have also supported the characterization of ADHD as a medical condition, suggesting that ADHD appears to be associated with decreased activity of the prefrontal cortex, which is consistent with the impaired attention and executive function characteristic of the disease.[11]

6.05.4 Experimental Disease Models

6.05.4.1 General Considerations for Modeling Attention Defect Hyperactivity Disorder

6.05.4.1.1 Diverse clinical symptoms with unknown etiology

ADHD is a clinically heterogeneous neuropsychiatric disorder with symptomatic components of hyperactivity, inattention, and impulsivity that usually present during childhood but can remain into adulthood. There are no objective laboratory measures for the diagnosis of ADHD and, similarly, there is no one definitive animal model of ADHD. This is largely due to a reliance on assessment of behavioral phenotypes that likely result from one or more genetic or neurodevelopmental disturbances across interacting neuronal networks, as well as an incomplete understanding of neurotransmitter systems that subserve these fundamental behavioral functions. Nonetheless, individual behavioral symptoms are readily assessed in laboratory animals,[18] and recent molecular biological[19] and functional imaging studies[20] are shedding light on the etiology of the disorder.

There are many advantages to developing appropriate animal models of any disorder: a simpler system may be easier to interpret than the complex clinical syndrome, potential treatment groups are genetically homogeneous, and testing environments can be tightly controlled. Thus, the researcher can avoid complications associated with many clinical studies such as comorbidities, previous drug exposure, and heterogeneous environmental conditions. Ideally, animal models of ADHD should closely resemble the clinical disorder in as many ways as possible, including etiology, pathophysiology, behavioral phenotype, and response to pharmacotherapies that are clinically effective. Thus, to be considered as a valid ADHD animal model, one must be able to show that: (1) the model is based on a valid etiological theory such as a proposed pathophysiology or genetic mutation (construct validity); (2) behavioral deficits in the model closely resemble those commonly observed in the clinic (face validity); and (3) the model can selectively predict efficacy of known and unknown therapeutics or underlying aspects of the disorder (predictive validity).

For many years, rodents have been used to model ADHD. Manipulations to produce a behavioral phenotype similar to the disorder are numerous, including exposure to neurotoxins/environmental pollutants during development, neonatal anoxia, selective lesions of neurotransmitter systems, x-irradiation of selected brain regions, and genetic manipulations. Several of the more popular animal models are reviewed here.

6.05.4.1.2 Choosing behavioral assays

Testing paradigms utilized by different laboratories can vary but tend to revolve around using so-called standard assays, e.g., open-field/automated locomotor activity in a novel environment and general tests of cognitive function, e.g., the water maze, radial arm maze, or passive avoidance paradigm. While versions of the open field are sensitive to showing the hyperactivity component of ADHD and are useful for demonstrating face validity, standard tests of cognitive function are arguably less demonstrative in that behavioral domains affected in ADHD (e.g., decreased response inhibition or impulsivity, inattention) are not specifically addressed. Therefore, more sophisticated tests are desirable. Three such tests of varying degrees of complexity are used routinely by others or us. These include: (1) fixed-interval/extinction schedule operant responding, where rodents are required to withhold responding (usually pressing on a lever) for a defined time period (up to several minutes) for a food or drink reward – motor-based impulsiveness is operationalized as bursts of responses with short interresponse times prior to the end of the fixed interval, i.e., poor response inhibition[21]; (2) five-trial inhibitory avoidance, in which juvenile rats are required to withhold transferring from a brightly illuminated 'safe' compartment of a shuttle-like apparatus to a darkened compartment, where they receive a very mild (0.1 mA) footshock – the test is usually conducted sequentially for five trials (maximum 180 s long) spaced 1 min apart and impulsiveness is operationalized in this instance as low latencies to transfer to the darkened compartment, i.e., poor response inhibition[22]; and (3) 5-CSRTT, in which selective attention, impulsivity, motivation, and motor function can all be evaluated individually.[23] The 5-CSRTT is perhaps the single most relevant assay since sustained attention as well as the other afore-mentioned behaviors can be assessed. Originally designed as a preclinical correlate of the CPT of attention used in clinical studies, the 5-CSRTT relies on visual cues that predict a food reward that is only delivered when the rodent correctly chooses the location of a short-duration (typically less than 1 s during testing) light stimulus. Sustained attention is operationalized as percent correct or incorrect (errors of commission) choices or the number of missed responses (errors of omission) over a test session (usually 30 min or 100 trials long), while impulsivity is measured by assessing the number of responses between trials (termed the 'limited hold' period). Latencies to respond as well as number of missed responses provide additional information on motor function. All of these measures are assessed over the same time period in a rat that is on a food-restricted diet for motivational purposes. The 5-CSRTT can also be conducted in nonhuman primates,[24] although research using monkeys in this regard lags considerably behind that conducted in rodents. A downside of the 5-CSRTT, as well as fixed-interval tests, is that training schedules of 8 weeks or more are typical, although three or four compounds can usually be sequentially evaluated in the same animals using a cross-over design[25] typical of similar studies in the clinic. In contrast, the five-trial inhibitory avoidance is usually conducted acutely over a period of 15 min, although the test is labor-intensive and can take 2–4 weeks to generate a full dose–response curve when evaluating the effects of a novel drug.[26]

6.05.4.2 Animal Models of Attention Deficit Hyperactivity Disorder

6.05.4.2.1 The spontaneously hypertensive rat (SHR)

The most widely accepted animal model of ADHD is the SHR, a strain that was originally developed from Wistar Kyoto (WKY) rats in Japan more than 40 years ago. Unexpectedly, when selecting for hypertension, hyperactivity was also observed. Since then, the SHR has been studied extensively from face, predictive, and construct validity perspectives.

6.05.4.2.1.1 Face validity

SHRs exhibit many behavioral features characteristic of ADHD. Hyperactivity in the SHR strain is observed across multiple behavioral paradigms, including single- and multiple-compartment open-field as well as automated spontaneous locomotor tests. The nature of the hyperactivity is dependent on the test environment, with maximal disturbances compared to control rat strains observed in unfamiliar environments and more modest hyperactivity observed in the home cage.[27]

Excessive responding is also seen using fixed-interval/extinction schedules.[21] Fixed-interval schedules (pressing a lever for a food pellet that is only rewarded after a fixed interval of time has expired) are useful for studying rate-dependent learning, also a measure of reward, which is impaired in ADHD. A typical 'scallop' pattern is usually observed during baseline responding in control rat strains such that responses are typically low during the early part of the interval and higher in the later part of the interval, closer to the time for the food reward. The SHR 'scallop' is shifted considerably to the left compared to other strains such as the WKY rat, demonstrating a different reactivity to the food reward, as well as motor impulsiveness. When extinction periods are added (effectively a time-out period), the SHR shows component differences in sensitivity to the stimulus change as well as decreased sustained attention.[21]

Response inhibition and cognitive function are also impaired in the SHR in five-trial inhibitory avoidance[22] (**Figure 1**) and 5-CSRTT[28] when compared with age- and sex-matched controls from WKY and other rat strains. Interestingly, these components of behavioral dysfunction correspond to the clinically observed symptoms of hyperactivity, impulsivity, and inattention/cognitive impairment, making the SHR a good model for studying multiple facets of ADHD. For example, in

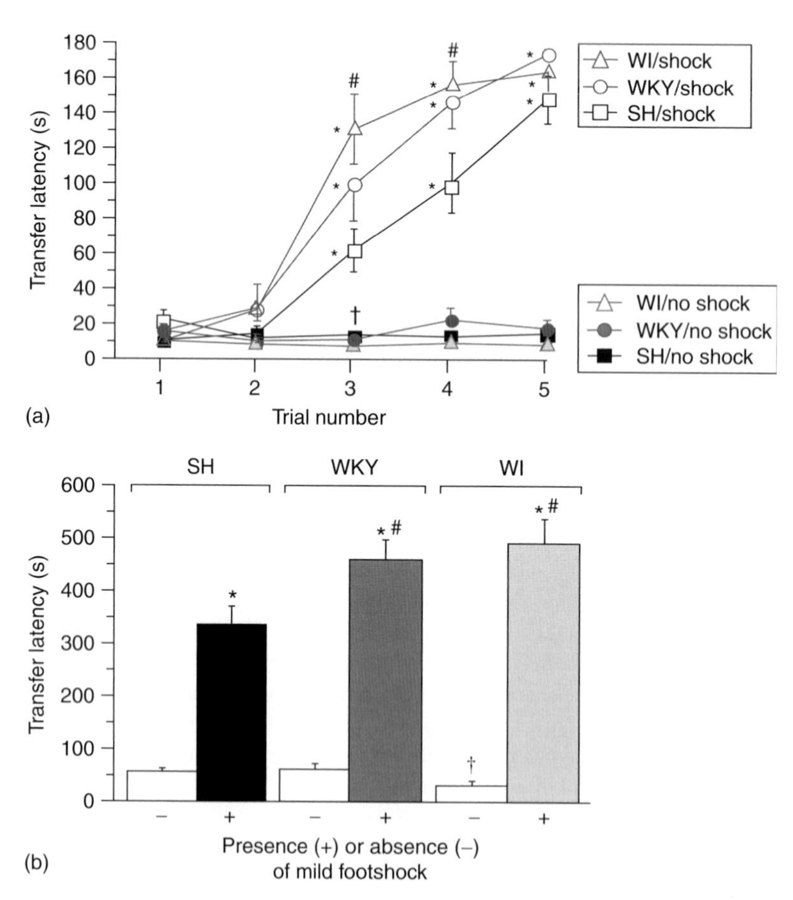

Figure 1 Acquisition of a five-trial inhibitory avoidance test in rat pups from three different strains: spontaneously hypertensive (SH), Wistar-Kyoto (WKY) and Wistar (WI). In the absence of the mild footshock (closed symbols), there is no difference in transfer times between the illuminated and darkened compartments. When the precision shock generator is active, however, pups from all three strains acquired the test after five trials. However, SHR pups exhibited a shallower learning curve compared with WKY or WI pups. This effect is readily apparent when data from trials 2–5 inclusive are summed and the mean taken. (Modified from Fox, G. B.; Pan, J. B.; Esbenshade, T. A.; Bennani, Y. L.; Black, L. A.; Faghih, R.; Hancock, A. A.; Decker, M. W. *Behav. Brain Res.* **2002**, *131*, 151–161, reprinted with permission of Elsevier.)

five-trial inhibitory avoidance, SHR pups continue to transfer to a darkened chamber (which they prefer) from a brightly illuminated chamber (which is aversive to a rodent) over five trials spaced only 1 min apart, despite the presence of a mild footshock upon crossing. This is measured as relatively low transfer latencies, that are higher in pups from other strains.[22] Similarly in the 5-CSRTT, SHR adults exhibit impaired sustained visual attention, although this aspect has been attributed to the hypertension that is evident in older adult SHRs.[28] Of course, SHRs also have impaired cognitive function in a number of other tests, including the water maze,[28–30] a measure of spatial learning and memory, the radial arm[31,32] and Y-maze,[33] measures of spatial working memory, as well as other operant-based tasks such as the delayed nonmatching to position test,[28] a measure of short-term or working memory, and the lateralized reaction time task, a behavioral measure of visuospatial divided attention.[34]

6.05.4.2.1.2 Predictive validity

Juvenile SHRs, that do not exhibit the potential confound of hypertension that develops later in the adult SHR, respond to stimulant drugs that are clinically efficacious for treating ADHD, such as methylphenidate (Ritalin). Locomotor hyperactivity and decreased delayed reinforcement in SHRs are reportedly attenuated in a dose-related manner by low doses of methylphenidate,[33,35] while impaired spontaneous alternation behavior in the Y-maze, which may be gender-specific (worse in male rats) was restored by methylphenidate treatment.[33] Similarly, increased impulsivity assessed in an elevated plus maze was also lowered by methylphenidate,[33] although it is difficult to dissociate this from potential anxiety-related measures. Impaired response inhibition/attention in five-trial inhibitory avoidance is effectively reversed by doses of methylphenidate,[22] producing plasma levels (approximately 4–17 ng mL^{-1}) similar to those efficacious in the clinic (approximately 8–10 ng mL^{-1}). Similarly, non-stimulants with different mechanisms, such as the NNR agonist, ABT-418, that are effective in clinical trials for ADHD[36] also reverse impairments in response inhibition/attention in five-trial inhibitory avoidance[22] at efficacious plasma levels consistent between SHRs (3.5 ng mL^{-1}) and humans (8–30 ng mL^{-1}). Similar data were also obtained very recently with the selective $\alpha 4\beta 2$ NNR agonist, ABT-089, currently in development for treating cognitive dysfunction in neurological disorders such as ADHD.[37] Further, in adults with ADHD, nicotine administration improves cognitive function and behavioral inhibition to a level at least as comparable with methylphenidate.[38] Nicotine also reverses behavioral impairments in SHRs tested in the Y-maze, an effect that is blocked by the $\alpha 4\beta 2$ NNR antagonists mecamylamine and dihydro-β-erythroidine.[39] SHRs are also reportedly sensitive to D-amphetamine, demonstrating reduced motor activity and impulsiveness.[27]

6.05.4.2.1.3 Construct validity

Diagnosis of ADHD is based on behavioral symptoms, following DSM IV criteria. As with ADHD, similar behavioral impairments observed in SHRs are believed to have a genetics-based etiology. In recent years in humans diagnosed with ADHD, several genetic studies have identified candidate genes that are consistent with the widely held belief that hypofunction of the mesocortical dopaminergic system contributes to the main symptoms of ADHD. In three genomewide scans of three different samples from US, Dutch, and German populations, the *DAT1* candidate gene was linked to ADHD symptomatology.[19] In adult SHRs, the DAT is significantly overexpressed in the striatum when compared with control WKY rats, while dopamine levels in the striatum and frontal cortex are decreased. Increases in D1 receptors occur in some brain regions in SHRs, which is presumably linked to deficient activation of the dopaminergic system.[27] This is consistent with stimulant efficacy in both ADHD patients and SHRs, given that mechanisms of action proposed for drugs such as methylphenidate and amphetamine include increased release of dopamine. Increased neuronal uptake of norepinephrine, which is also evident in SHRs,[40] has been implicated in mediating hypertension rather than ADHD-related abnormal behavior. SHRs also appear to have impaired vesicular storage of dopamine, causing leakage of the neurotransmitter into the cytoplasm. Striatal dopamine release is also impaired.

Cholinergic function is also disrupted in SHRs. NNRs containing the $\alpha 4$ subunit are decreased in multiple brain regions, especially the frontal cortex in prehypertensive juvenile SHRs, worsening with age.[30,32,41] This finding appears unrelated to the hypertension that also develops with age, since treatment of SHRs with the antihypertensive agent hydralazine prevented the development of hypertension in adult SHRs but did not affect the reduced expression of nicotinic receptors.[41]

6.05.4.2.2 **Dopamine transporter (DAT) knockout mouse**

A recently developed animal model that may be relevant to ADHD is the DAT knockout mouse, which shows about a 300-fold decrease in the rate of clearance of extracellular dopamine[42] due to the lack of the gene that encodes DAT-1. These knockout mice also show evidence of behavioral abnormalities similar to those observed in ADHD, although there remain several important issues regarding dopamine autoreceptor downregulation, serotonergic tone, and activation of trace amine receptors in these animals that need to be addressed with regard to relevance to ADHD.[43]

6.05.4.2.2.1 Face validity

DAT knockout mice demonstrate a behavioral phenotype that, on the surface, appears to mirror symptomatology associated with ADHD.[43] For example, hyperactivity was one of the earliest observations in these mice, which was subsequently revealed to be particularly sensitive to a novel environment, where locomotor activity was determined to be 12-fold higher in the knockouts compared with wild-type controls. Further, while locomotor habituation to the environment (decreased activity following prolonged exposure to the environment) was evident in wild types, DAT knockout mice remained hyperactive even after 4 h. In addition, repetitive exposure of the knockout animals to the activity environment only resulted in augmentation of the locomotor response. Lack of habituation to novel stimuli was also observed in a version of the novel object test as well as a modified Y-maze. Taken together, these data suggest that the DAT knockouts are not only hyperactive, but are less able to adapt to novel stimuli and may exhibit reward-like behavior. These data are also in agreement with siRNA knockdown[44] or pharmacological inhibition of DAT[45] in mice, which also produce pronounced hyperactivity.

Impaired cognitive function is also evident in DAT knockout mice. In a spatial working memory test using a radial arm maze, knockout mice were essentially unable to acquire the test conducted over 21 sessions. Knockout mice also had significantly higher preservation errors compared to wild-type mice and these errors remained elevated for the duration of the study, suggesting that the knockouts had difficulty suppressing inappropriate responses.[43]

6.05.4.2.2.2 Predictive validity

Psychostimulants such as methylphenidate, amphetamine, and cocaine (all at relatively high doses) paradoxically robustly attenuated hyperactivity in DAT knockout animals in a novel environment.[43] Interestingly, these 'calming-like' effects of the stimulants were delayed, but long-lasting, particularly for methylphenidate (up to 4 h). Further, the effects of methylphenidate were dose-dependent in the DAT knockouts, but methylphenidate dosing over the same range in wild-type controls produced the more usual enhancement of activity, following an inverted U-shaped dose response. While these data were exciting, subsequent work described in the same article provided some puzzling information: extracellular dopamine concentrations in the striatum of DAT knockouts, measured by in vivo microdialysis in freely moving mice after the administration of methylphenidate, did not change, in contrast to wild-type controls, which were significantly elevated. The most likely reason for this is that dopamine levels in the DAT knockouts are already highly elevated and were probably already at ceiling. This posed a dilemma: if dopaminergic stimulants can act to decrease hyperactivity dramatically in a novel environment in DAT knockout mice, and dopamine levels are not affected, what is the mechanism of action of psychostimulants? Further experiments with the selective norepinephrine transporter (NET) inhibitor nisoxetine demonstrated no effect. However, the serotonin transporter (SERT) inhibitor (selective serotonin reuptake inhibitor (SSRI)), fluoxetine, as well as the nonselective serotonin receptor agonist quipazine and 5-HT precursor substrates, 5-hydroxytryptophan and L-tryptophan, also dramatically decreased hyperactivity in the DAT knockout mice.[43] Striatal dopamine levels were not affected by these various 5-HT treatments, however. Additional experiments with fluoxetine and the dopamine receptor agonist apomorphine in dopamine-depleted DAT knockout mice suggested that the serotonergic effects on decreasing hyperactivity were mediated downstream of dopaminergic neurotransmission, with later findings indicating that limbic brain regions were of especial importance. The relevance of these findings to ADHD is currently unclear, as SSRIs are not effective in treating ADHD and one of the side effects of these drugs in the clinic is motor stimulation.[46] Similarly, the norepinephrine reuptake (NET) inhibitor, atomoxetine (Strattera), is somewhat effective in treating ADHD, while nisoxetine was without effect on hyperactivity in the DAT knockout mice.[43] Despite these limitations, DAT knockouts may be useful in elucidating the nondopaminergic neural mechanisms underlying ADHD.

6.05.4.2.2.3 Construct validity

Disruption of the *DAT* gene in mice significantly increases extracellular concentrations of dopamine and leads to a reduction in mRNA levels for the DRD1- and DRD2-like receptors in the striatum,[47–49] findings that run somewhat counter to those found in ADHD patients. In clinical ADHD, however, patients with the DAT1 polymorphism show mutations within the 3′ untranslated portion of the mRNA, i.e., outside the protein-coding region. Thus, it remains unclear how this might affect dopamine reuptake in ADHD compared to the complete lack of DAT function in DAT1 knockout mice. Further, there is evidence to suggest that the DAT1 is actually overexpressed in ADHD patients.

According to a single photon emission computed tomography (SPECT) study, the DAT is elevated by approximately 70% in adults with ADHD.[50] Other SPECT studies have also demonstrated that the efficacy of the stimulant methylphenidate is attributed to an approximate 50% block of DAT: maximal clinical efficacy is also observed at a time (approximately 60 min after oral dosing) when 50% DAT blockade is achieved.[51] In contrast, evidence exists for dopamine hypofunctional prefrontal cortex in both DAT knockouts and in ADHD, which may be responsible for mediating the attentional/cognitive impairments observed in both cases. Thus, while the behavioral abnormalities observed in DAT knockouts is intriguing, the direct relevance of DAT knockout to ADHD remains to be determined.

6.05.4.2.3 **Synaptosome-associated protein of 25 kDa (SNAP-25, coloboma mutant)**

The coloboma mutation arose from neutron irradiation mutagenesis studies, producing a deletion on chromosome 2 that disrupted coding of four known genes for the proteins phospholipase β1 and β4, jagged 1, and SNAP-25.[52,53] The mutation is homozygous lethal. Adult heterozygote mice are viable, although a distinctive ocular dysmorphology leading to 'sunken' eyes in some mice limits behavioral testing.[54]

6.05.4.2.3.1 Face validity

SNAP-25 mutants are hyperactive when spontaneous locomotor activity is assessed in novel environments, reaching two- to fourfold above basal activity (**Figure 2**) of control littermates.[55] However, this hyperactivity is very variable and is hypothesized to reflect a loss of control of activity rather than a simple increase in basal activity. This can be confounded by head-bobbing and other stereotypies such as repeated jumping against the wall on one side of the activity arena (**Figure 2**). The hyperactivity is attenuated by administration of relatively high doses of amphetamine (4 mg kg^{-1}), without producing further stereotypies.[55] Unfortunately, given the ocular dysmorphology evident in many SNAP-25 mutants, the effects of SNAP-25 deletion on attention/cognitive function in vivo is not known. However, SNAP-25 constitutes an integral protein for synaptic vesicle fusion and neurotransmitter release studies suggest that impaired neurotransmitter (and perhaps neuropeptide) release is regionally specific.[56] SNAP-25 has also been shown to be important for hippocampal-based memory consolidation in several studies in rats utilizing long-term potentiation[54,57,58]; intracerebroventricular antisense oligonucleotide also induced deficits in several cognitive tests such as long-term inhibitory avoidance and fear conditioning,[59] although these tests may hold little relevance for cognitive deficits observed in ADHD.

6.05.4.2.3.2 Predictive validity

The psychostimulant amphetamine attenuated hyperactivity in coloboma mice, while it increased activity, as expected, in littermate controls.[55] Thus, by reversing the dopamine plasma membrane pump, increased dopamine is released in regions such as the dorsal striatum where it is low due to impaired vesicular release caused by deficient SNAP-25. In contrast, methylphenidate does not reverse hyperactivity, but instead increased activity in both controls and SNAP-25 mutants.[55] Little other pharmacological work has been performed and thus, the predictive validity of SNAP-25 mutants is unclear.

6.05.4.2.3.3 Construct validity

Behavioral changes in the coloboma mouse are associated with a mutation of the gene encoding the important synaptosome-associated protein, SNAP-25. Polymorphism analysis of the gene encoding SNAP-25 in humans identified biased transmission of the haplotypes of the alleles of two polymorphisms, implicating SNAP-25 in the etiology of ADHD.[16,60,61] However, neurobiological deficits associated with mutation of SNAP-25 in mice, such as decreased neuronal number in neocortex, hippocampal dysfunction, and ocular dysmorphology, are not consistent with ADHD in the clinic.

6.05.4.2.4 **6-Hydroxydopamine (6-OHDA) lesions**

Exposure of rat pups to 6-OHDA, usually via intracerebroventricular or intracisternal injection, selectively lesions dopamine projections to the frontal cortex, resulting in an age-dependent increase in spontaneous locomotor activity.[27] While rats lesioned in this manner also show evidence of cognitive impairment, these are also age-dependent. Impulsivity is not present in 6-OHDA-lesioned rats and, since impulsivity and cognitive impairment are evident in adults with ADHD, these rats may be a useful model regarding elucidation of mechanisms that cause hyperactivity in ADHD.

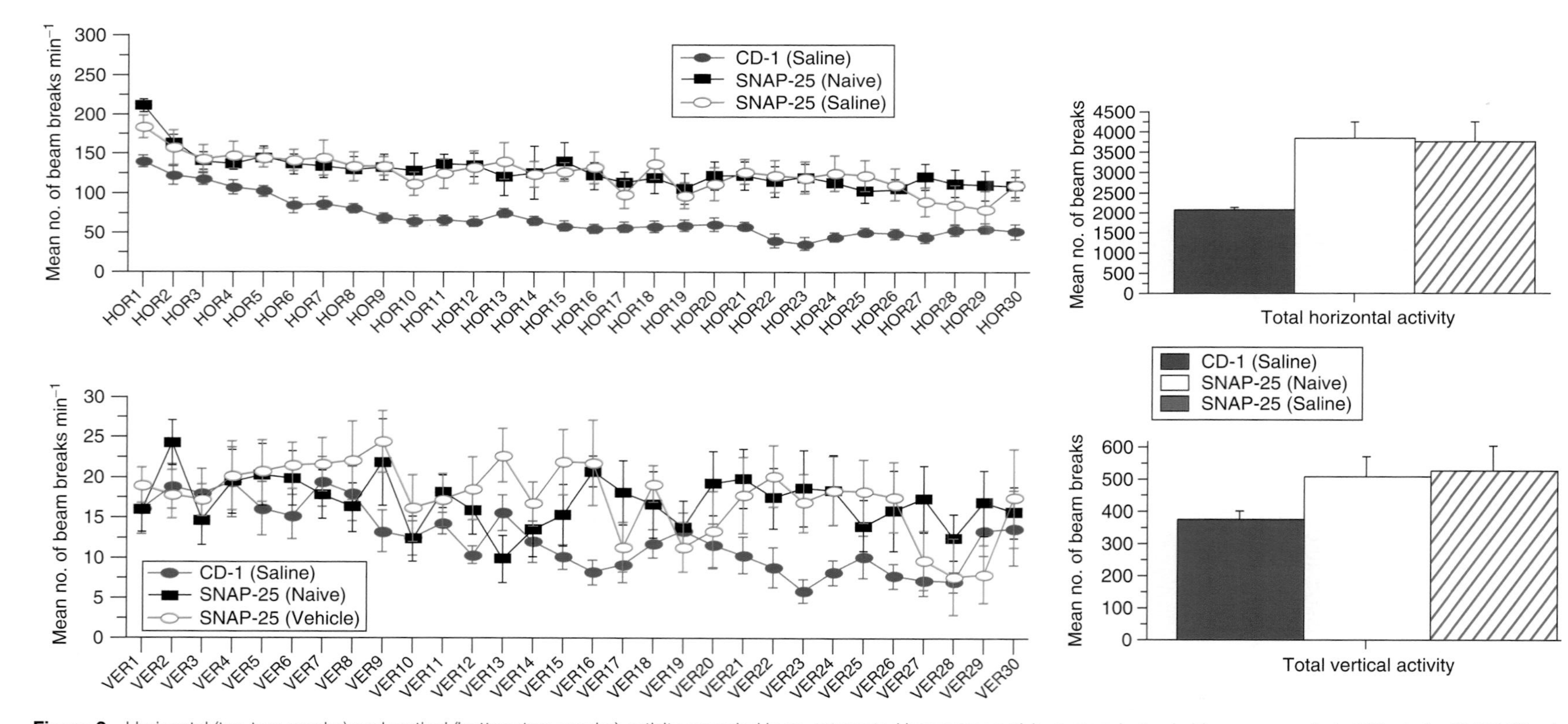

Figure 2 Horizontal (top two graphs) and vertical (bottom two graphs) activity recorded in an automated locomotor activity system in 1-min bins over a period of 30 min for SNAP-25 mice compared to CD-1 mice. SNAP-25 mice, whether naive or saline-treated, were significantly more active than CD-1 mice.

6.05.4.2.4.1 Face validity

When neonatal rats are injected intracisternally with 100 μg 6-OHDA, on postnatal day 5, pronounced spontaneous hyperactivity is observed in a novel environment during a periadolescent development stage at postnatal days 21–30,[62] but not at later stages of development at postnatal days 36 or 59.[63] The nature of the hyperactivity suggests a lack of habituation to the novel environment of the activity test arena, since significant differences in spontaneous activity compared to sham control groups only become apparent during the later time bins (15–90 min) and not at earlier times (0–15 min).[64] These findings are consistent with clinical ADHD since hyperactivity is generally only found in adolescence and usually resolves in adulthood (although the cognitive deficits do not). Similarly, ADHD patients exhibit difficulty habituating to novel situations and acting appropriately.[65]

6.05.4.2.4.2 Predictive validity

Psychostimulants effective in treating ADHD, such as methylphenidate and amphetamine, are also effective at alleviating the hyperactivity produced by 6-OHDA lesions.[62,64] While the effects of both stimulants are dose-related, the best effect is seen with relatively high doses (e.g., 10 mg kg^{-1} intraperitoneally for D-methylphenidate and 3 mg kg^{-1} intraperitoneally for amphetamine). Interestingly, a recent report described efficacy of atomoxetine against hyperactivity in 6-OHDA-lesioned rats.[66] These effects of atomoxetine, a NET inhibitor currently labeled as a nonstimulant for the treatment of ADHD, were observed at a more reasonable dose of 1 mg kg^{-1} i.p.

6.05.4.2.4.3 Construct validity

Consistent with clinical imaging data of dopamine hypofunction in the prefrontal cortex of ADHD patients, 6-OHDA lesioning in neonatal rats caused a loss of developing dopamine projections in the rat forebrain, as measured by decreased DAT binding.[64] (DAT binding does not appear to correlate with the increased spontaneous motor activity in this model.) However, while motor hyperactivity induced by neonatal 6-OHDA lesions is mainly associated with destruction of mesolimbic dopamine pathways, behavioral responses of individual rats do not appear to be correlated with dopamine concentrations in any particular brain region.

Neonatal 6-OHDA lesions result in a remarkable resistance to the motor-impairing effect of alcohol (0.5–1.0 g kg^{-1}) as well as to the motor-suppressive effect of diazepam (5.0 mg kg^{-1}), which may have clinical relevance since ADHD is a risk factor for alcohol abuse and antisocial behavior.[65] However, learning deficits that are apparent soon after neonatal lesioning are maximal at weaning but disappear in adulthood.[49] This is in contrast to the clinical situation, where learning impairments persist in many ADHD patients into adulthood, while the hyperactivity component generally remits.

Neonatal dopamine lesioning also induces neuronal adaptations that differ from responses to the same lesions in adult rats. For example, hyperinnervation of the striatum by serotonergic neurons during the periadolescent period is observed following early neonatal lesions but not following lesions in the adult.[63,67] The functional significance of this is unclear, however, since serotonergic agents are generally not effective in treating symptoms of ADHD. In contrast to the adaptive changes seen in the serotonergic system, no major effects on the norepinephrine system are apparent following 6-OHDA lesioning.[49] Since the calming effects of stimulants can be mimicked by NET inhibitors in lesioned rats, these data may be relevant to clinical ADHD, where such inhibitors also show efficacy, e.g., atomoxetine.

DA4 receptor polymorphisms have been repeatedly linked to ADHD so it is interesting to note that DA4 receptor upregulation correlated with 6-OHDA lesion-induced hyperactivity and that DA4 receptor antagonists ameliorated this hyperactivity in a dose-related manner, whereas DA2/DA3 receptor antagonists were without effect.[62,68,69] Interestingly, mice lacking DA4 receptor that were lesioned with 6-OHDA on postnatal day 2 did not develop hyperactivity when tested between postnatal days 25 and 29; this was in contrast to lesioned wild-type controls, which did exhibit hyperactivity.[70] Thus, the DA4 receptor appears to be important for hyperactivity and impaired behavioral inhibition in this model.

6.05.4.3 Other Models

6.05.4.3.1 Dorsal prefrontal cortex (dPFC) lesions

Lesions of the prefrontal cortex produce cognitive deficits such as impaired regulation of attention and impulsivity as well as disrupted working memory; these deficits are thought to be mediated by hypofunctioning catecholamine neurotransmitter systems.[71] In a recent report describing pronounced deficits in attention in rats with lesions of the dPFC using a new combined attention memory task, the psychostimulant amphetamine reversed these deficits in a manner that was specific for attention.[72] Importantly, the effective dose of amphetamine in this model was low, 0.2 mg kg^{-1} intraperitoneally, which is more in line with clinical expectations compared to efficacy with the same stimulant in 6-OHDA-lesioned rats. While this new model is quite laborious to conduct – taking 4–6 months to train rats to baseline performance – data generated may have important implications for the cognitive deficits seen in ADHD as well as in schizophrenia.

6.05.4.3.2 Environmental toxins

Chronic exposure to lead or polychlorinated biphenyls can cause motor hyperactivity as well as cognitive impairment and impulsivity in rodents and may contribute to some instances of human ADHD.[27] There are some older data indicating that psychostimulants such as methylphenidate and amphetamine may ameliorate some of these symptoms in mice, but these models have not been well characterized and the number of cases of ADHD caused by exposure to environmental toxins is likely small.

6.05.4.3.3 Neonatal hypoxia

Anoxia in perinatal humans is a risk factor for developing ADHD. In rats, transient neonatal hypoxia produces behavioral symptoms similar to those observed in clinical ADHD, such as age-related hyperactivity, most pronounced in the periadolescent period, as well as longer-term deficits in cognitive functioning, possibly as a result of hippocampal damage. Complex age-related changes in neurotransmitter levels in the cortex and cerebellum as well as neurodegeneration in the hippocampus as a result of anoxia are not readily interpretable with respect to clinical relevancy.[27]

6.05.4.3.4 Animals selected for inattention and impulsivity

Using a 5-CSRTT similar to that described earlier, rats that are poor performers can be selected and preferentially bred for deficient sustained attention and increased impulsiveness.[73] Methylphenidate treatment can improve accuracy in these otherwise poor performers; however, since these rats are not hyperactive to begin with, this model may only serve as a measure of the cognitive components of ADHD.[73] Interestingly, local injection of a DA1 receptor agonist into the frontal cortex of poor performers (<75% accuracy) in the 5-CSRTT enhanced attention[74] (high-performers were impaired), which may be pertinent to the dopamine hypofunctioning evident in clinical ADHD.

6.05.4.4 Summary

There is no one definitive animal model of ADHD, but rather several models that address different aspects of this complex behavioral syndrome. Interestingly, a factor common to many of the most frequently used models is a hypofunctioning dopamine system, especially in frontal cortex and striatal brain regions and associated pathways. Perhaps the most studied and best characterized is the SHR, which exhibits many of the behavioral and genetic aspects of ADHD and is responsive to drugs selective for different molecular targets that have proven efficacious in the clinic for treatment of ADHD. Other more recent genetically based (DAT knockout, SNAP-25 mutant) and cortical lesion (dPFC) models may also shed new light on the underlying pathology that causes ADHD. Finally, an important point to make at this juncture is the necessity to study the effects of new drugs using these models, which respond differentially to treatment due to altered neurotransmitter systems, compared to naive animals.

6.05.5 Clinical Trial Issues

Despite the use of stimulant medications in the treatment of hyperkinetic children over 60 years ago, and the common observation that stimulant treatment for children with ADHD is arguably the best treatment, there are relatively few studies systematically investigating efficacy in patients with ADHD. Other issues in evaluating the efficacy of treatments are a lack of understanding as to how improvement on laboratory measures reflects the real-life impact of medication.[8] In fact, a study suggests that there is a minimal agreement between laboratory and natural settings, especially when investigating drug effects.[75] Finally, there is evidence to suggest that the laboratory paradigms do not truly reflect the school setting in which ADHD patients are involved, which includes test-taking and note-taking.[8]

Among the difficulties in clinical ADHD research arethe diagnostic criteria, and concerns about the validity of existing diagnostic criteria.[6] Much of the focus on adult diagnostic criteria stems from longitudinal studies of ADHD children, illustrating the perseverance of symptoms into adulthood. As an understanding of the disorder has advanced, it is clear that many of these studies excluded children that would today meet the DSM-IV criteria for ADHD, and in particular have ignored the inattentive subtype. Thus, there are really no validation studies of ADHD diagnostic criteria in adults.[6] In conducting clinical trials diagnostic criteria are key for patient selection, and if the criteria are inaccurate it is difficult to assess the efficacy of novel therapeutics accurately.

The American Academy of Pediatrics (AAP) has developed a practice guideline for the diagnosis of ADHD among children from 6 to 12 years of age who are evaluated by primary care clinicians.[3] The AAP has also developed a treatment guideline for school-aged children with ADHD.[76] The significant components of the diagnostic guideline include: (1) the use of explicit criteria for the diagnosis, using DSM-IV criteria; (2) the importance of obtaining information about the child's symptoms in more than one setting (especially from schools); and (3) the search for coexisting conditions that may make the diagnosis more difficult or complicate treatment planning. While this guideline is intended to provide a framework for primary care clinicians, rather than the sole diagnostic criteria, it is clear that this type of approach is needed in the field. It is further evident that a similar approach is needed toward the diagnosis of adolescents and adults with ADHD, as symptoms are likely different between children, adolescents, and adults.

6.05.6 Overview and Comparison of Drug Classes

Historically the treatment of ADHD relied on agents affecting monoaminergic neurotransmission, and largely consisted of the stimulants, antidepressants, and antihypertensive agents (**Table 1**).[1,65] The primary pharmacological treatment for ADHD continues to be the use of stimulants, particularly methylphenidate and amphetamines. Methylphenidate has long been the leading treatment for ADHD; however it has a short duration of action and a midday dose is required. In schoolchildren, this necessitates dosing during the school day and contributes to poor compliance and social stigma. The importance of avoiding this midday dose is underscored by the success of Concerta (J & J), a novel formulation of methylphenidate designed to provide both rapid and sustained release, and Adderall XR (Shire), a QD formulation of mixed amphetamine salts.

The first nonstimulant therapy designed for ADHD, Strattera (atomoxetine, Lilly), was introduced in January 2003 and has rapidly gained acceptance. Atomoxetine is not a scheduled drug, has a low risk for abuse and dependency, has a nonstimulant side-effect profile (although sleep and growth disturbances are shared with stimulants), and is the first indicated product for adults with ADHD. Unlike stimulants, for example, atomoxetine lacks methylphenidate-like drug reinforcement properties in monkeys,[77] leading to the conclusion of reduced likelihood for abuse potential in human patients. On the other hand, there is no consensus that patients with ADHD abuse prescribed stimulants.[46] The efficacy of atomoxetine is not better, and perhaps less, than methylphenidate. In one clinical report,[78] atomoxetine was reported to have better effects on inattentive symptoms compared to the hyperactive/impulsive symptoms consistent with the proposed role for norepinephrine in measures of distractibility. Atomoxetine use is associated with a number of adverse events that are an extension of its pharmacology (e.g., elevated blood pressure, urinary dysfunction), or to those of the primary metabolite (4-OH-atomoxetine) that shows modest affinity[79] for several opioid receptors (e.g., constipation or other gastrointestinal disturbances).

Table 1 Comparison of the main agents currently used for the treatment of ADHD

Product	*Company*	*Mechanism of action*	*Comments*
Strattera	Lilly	Norepinephrine reuptake inhibitor	First non-controlled agent. Indicated for adults and children. Once or twice a day dosing allowed. Nausea and appetite suppression as major limitations. Sexual dysfunction is problematic for adults
Adderall XR	Shire	Amphetamine salts	Provides more sustained effects compared to unformulated MPH
Concerta	ALZA/J&J	MPH controlled release	1st QD MPH formulation. Clinical data demonstrate efficacy for ~12 h equivalent to MPH
Ritalin	Novartis	MPH immediate release	Has 1–4-h duration of action. Multiple generics exist
Medadate CD	Celltech/Medava	MPH controlled release	Combines immediate and sustained release like Concerta
Focalin	Novartis	Dexmethylphenidate MPH (d-isomer)	Lower dosing, longer duration and better side effect profile than MPH
Dexedrine SR Dexedrine	GlaxoSmithKline	Dextroamphetamine Slow release Immediate release	9-h duration. Subject to greater abuse than MPH. Generics exist

The third line of therapy for ADHD, antidepressants and antihypertensive therapies, is used less frequently than stimulants, but can be effective. These drugs are sometimes used in combination with stimulants in patients with comorbid symptoms. Commonly used agents are tricyclic antidepressants (TCAs), bupropion, or alpha-adrenoceptor agonists (clonidine, guanfacine, etc.). The therapeutic limitations of these diverse compounds include weight loss, sleep disturbances, abuse liability, and social stigma of stimulants, and the cardiovascular side effects and less well-defined efficacy of atomoxetine, TCAs, or alpha-adrenergic agents.

6.05.6.1 Stimulants

The primary pharmacological treatment for ADHD continues to be the use of stimulants, particularly methylphenidate (Ritalin, Ritalin LA, Ritalin SR, Concerta, Focalin, Methylin, Methylin ER, Metadate ER, and Metadate CD) and amphetamines (Adderall, Adderall XR, Dexedrine, Dexedrine Spansule, and DesxtroStat).[1,46,65] Stimulants are estimated to be effective in 70% of adolescents and appear to improve both cognitive deficits and general behavior.[8,46] The beneficial effects of stimulants are similar for male and female adolescents, and for younger children. With lower doses of therapies, adult response rates are in the range of 50%. When higher doses were used, adults responded as well to treatment as do children.[80]

6.05.6.1.1 Sites and mechanism of action

Stimulant medications increase the synaptic availability of dopamine, and this mechanism is hypothesized to underlie the therapeutic effect. However, it is important to recognize that these compounds also have direct or indirect effects on other neurotransmitters, e.g., acetylcholine, norepinephrine, and 5HT. Based on preclinical data, for example, it has been suggested that the stimulant-induced decreases in hyperactivity may be due to increased 5HT[81,82] (but see the discussion of SSRIs below); both norepinephrine and acetylcholine play important roles in attentional processing. Stimulant activity per se is not the critical determinant of efficacy in ADHD, however, since caffeine, a stimulant with a different mechanism of action, appears to be ineffective.[83]

While the stimulants act to increase the availability of neurotransmitters such as dopamine and norepinephrine, they do so via different mechanisms. Methylphenidate blocks reuptake transporters of dopamine, norepinephrine, and 5HT. Methylphenidate acts to block the reuptake of these monoaminergic neurotransmitters via action on the external surface of the neuron, but does not act as a substrate at these receptor sites. Methylphenidate binds to a site that has high affinity for cocaine. In contrast, amphetamine acts with the transporter substrate recognition site on DAT, but also crosses the cell membrane through transport and diffusion. In this way, amphetamine is able to displace dopamine and activate the release of newly synthesized (nonvesicular) dopamine by reverse transport, ultimately resulting in more synaptic dopamine. The increased abuse liability observed following amphetamine (compared to methylphenidate) may be due to these differences in mechanism of action.

6.05.6.1.2 Structure–activity relationship

Methylphenidate (methyl 2-phenyl-2-(2′-piperidyl)-acetate) can exist as four possible stereoisomers, although Ritalin is a racemic mixture of threo diastereomers.[84] The erythoro isomers have little efficacy in the treatment of ADHD symptomology, but contribute to the hypertensive effects of Ritalin. In an attempt to decrease side effects while maintaining efficacy, a single isomer entity dexmethylphenidate (Focalin) was introduced. **Figure 3** shows the chemical structures for both Ritalin and Focalin. Focalin is a chirally pure oral dexmethylphenidate (D-MPH); the single stereoisomer version of methylphenidate (DL-MPH). In studies comparing D-MPH to DL-MPH, the treatment effects of D-MPH were obtained at half the dose of DL-MPH. While both agents showed a reduction of ADHD on primary and secondary endpoints, patients treated with D-MPH showed significant improvements in math test scores at 6 h postdosing. In phase III studies, D-MPH was superior to the racemic DL-MPH in all parameter measures, with a longer duration of action and a low incidence of side effects (**Figure 3**).

6.05.6.1.3 Comparison within drug class

While methylphenidate has long been the leading treatment for ADHD, the compound has a short duration of action and a midday dose is required. In schoolchildren, this necessitates dosing during the school day and contributes to poor compliance and social stigma. Extended-release formulations of methylphenidate (Concerta), a novel formulation of methylphenidate designed to provide both rapid and sustained release, and amphetamine (Adderall XR), a formulation of mixed amphetamine salts, provide durations of action up to 12 h. The longer-acting compounds not only result in

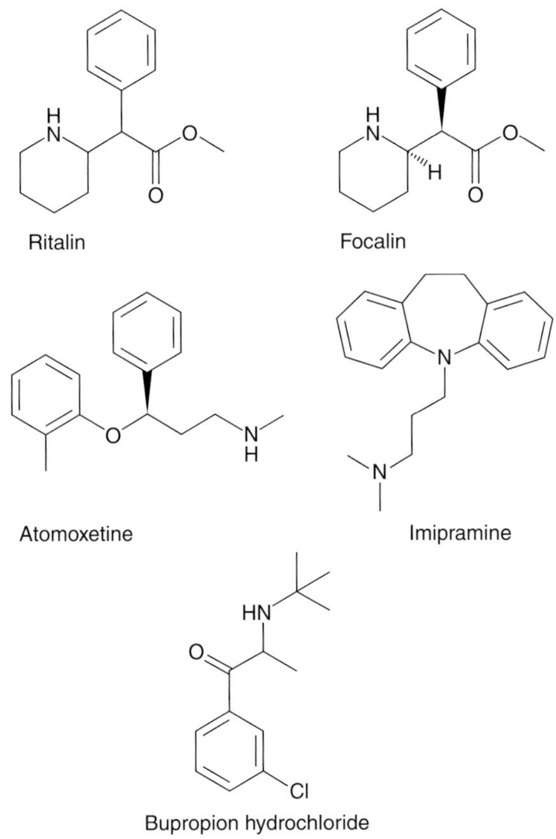

Ritalin

Focalin

Atomoxetine

Imipramine

Bupropion hydrochloride

Figure 3 Chemical structures of compounds used in the treatment of ADHD.

fewer compliance issues arising from the need for multiple daily dosing, but also provide less potential for diversion and abuse. Shire has transferred two-thirds of the Adderall business to the once-a-day formulation, in spite of the availability of generic forms of Adderall, underlining the importance of once-a-day dosing in this condition. Concerta (J & J) is a novel formulation of methylphenidate designed to provide an initial rapid-release component followed by sustained release. Clinical data demonstrate sustained levels in the efficacious range for approximately 12 h and equivalent or superior efficacy to methylphenidate (4 h) on behavioral rating scales.

6.05.6.1.4 **Limitations**

Common short-term adverse effects, such as appetite suppression, sleep disturbances, and abdominal pain,[8,65] are reported with the stimulant class. Longer-term adverse effects are still debated, with reports suggesting motor tic development as well as some height/weight decreases among adolescents with ADHD.[12,65] While the use of stimulants in children has always been somewhat controversial, recent safety warnings may have a significant impact on future market development. Adderall XR was removed from the Canadian market in 2005 due to an unexpected rate of sudden death and stroke in pediatric patients. The US Food and Drug Administration is currently investigating prior to any final decision on label changes or marketing status. In 2005, warnings of Ritalin links to increased risk of cancer were reported in the US media.

Diversion of stimulant medications continues to be an issue. The medications may be taken orally, or ground into powder and snorted. There are reports of patients selling their stimulant medication to peers for profit, and reports of college students (without ADHD) using methylphenidate or amphetamine recreationally. In addition, the stimulants are contraindicated in patients with a history of illicit use or abuse of stimulants. As new drugs enter the market devoid of abuse liability and not scheduled by the US Drug Enforcement Administration (DEA; *see* Section 6.05.6.2), it is likely that, if efficacious, these treatments will become the market leaders.

6.05.6.1.5 Future directions

Up to 30% of patients do not respond well to an initial trial of stimulants,[83] although approximately 70% of children with ADHD respond to either methylphenidate or dexamphetamine.[12] One of the areas where stimulants are most effective is reducing the hyperactivity in ADHD. However, the stimulants are less effective in addressing some of the working memory, organization, and planning (i.e., 'executive function') deficits characteristic of ADHD. Thus, new therapies addressing these other cognitive aspects are likely to be successful in the marketplace. Modafinil (Provigil) is available as Sparlon for the treatment of ADHD in 2006. While modafinil is able to increase extracellular dopamine, it dose not release dopamine in the way that classic stimulants such as amphetamine do. In addition, while there is evidence in nonhuman primates that modafinil is self-administered, modafinil is not DEA-scheduled. Modafinil has been used off-label for the treatment of ADHD, but it is not clear how the approval of this product will change the market.

6.05.6.2 Norepinephrine Uptake Inhibitors

The first nonstimulant therapy for ADHD, Strattera, rapidly gained acceptance, largely due to its lack of scheduling, low risk for abuse and dependency, nonstimulant side-effect profile (although sleep and growth disturbances are shared with stimulants), and the fact that it was the first indicated product for adults with ADHD. Atomoxetine is a highly specific NET inhibitor, consistent with the likely mode of action described below for the antidepressants and antihypertensive drugs. In addition to efficacy for the treatment of ADHD, atomoxetine has beneficial effects for co-occurring disorders such as anxiety, tics, and depression.

Atomoxetine has similar efficacy and tolerability among adolescents, relative to school-aged children with ADHD. Short-term side effects may include sedation, appetite suppression, nausea, vomiting, and headaches. Atomoxetine also has some cardiovascular liabilities, is slowly metabolized in a subset of patients (because of metabolism via CYP2D6), and has known drug–drug interactions with paroxetine, fluoxetine, albuterol, and other common drugs. In 2005 a warning in bold type of the potential for severe liver injury was added to the label. In contrast to the stimulants, some emerging longer-term side-effect data suggest that normal growth in height and weight is found with atomoxetine treatment.[85]

Atomoxetine is not scheduled, affording an advantage over stimulant treatment for ADHD. Atomoxetine can be dosed once or twice a day, although the efficacy is not as immediate as found with stimulants, with peak efficacy developing over 2–6 weeks. Physicians may consider Strattera as a second-line therapy behind stimulants because of inferior efficacy, but the lack of scheduling is a key benefit.

6.05.6.2.1 Structure–activity relationships

Atomoxetine exists as two enantiomers, (*R*)-(LY-139603) and (*S*)-(LY-139602). It is selective, does not bind to monoamine receptors, and is devoid of activity for acetylcholine, histamine H_1, α_1- and α_2-adrenergic, and dopamine receptors, allowing a safer profile of action than typical antidepressants. The NET selectivity prompted the testing of atomoxetine in ADHD, since norepinephrine is hypothesized to play a role in the control of sensory processing, with a facilitating action on the dopaminergic system, thought to be impaired in ADHD. The therapeutic utility of atomoxetine underscores the concept that ADHD is not only a disorder of dopamine hypofunction, but has a noradrenergic component, especially for aspects of attention and arousal that are disrupted in ADHD.[86]

6.05.6.2.2 Sites and mechanism of action

The lack of binding at the monoamine receptors suggests that atomoxetine should have few side effects. In addition, atomoxetine has little effect on the reuptake of dopamine and 5HT, although it binds to a very-low-affinity site, from which it is displaced by citalopram, a selective 5HT reuptake inhibitor SERT.[87] In synaptosomal preparations of rat brain, the order of functional potency (K_i value) was 4.5, 152, and 657 nM for norepinephrine, 5HT, and dopamine, respectively. The specificity of atomoxetine for the norepinephrine uptake site is very high, with a K_i for NET inhibition of 0.7–1.9 nM.[88]

Atomoxetine inhibits norepinephrine uptake in nerve terminals in both the central and peripheral nervous systems. In both pigeons and monkeys, atomoxetine (1–10 $mg\,kg^{-1}$) fully substitutes for the psychostimulants imipramine, cocaine, and methamphetamine, effects blocked by α_1-antagonists, further supporting activity of atomoxetine at the norepinephrine receptor. A study using cocaine as the discriminative stimulus in rhesus monkeys, however, failed to demonstrate substitutive properties for atomoxetine.

Another in vivo model that provides evidence of the actions of atomoxetine on the norepinephrine system is the antagonism of cataplexy in the dog, an orexin-deficient canine model of human narcolepsy. Current treatment for narcolepsy includes modafinil, which is the first-line therapy, and combined TCAs and amphetamine-like compounds. Atomoxetine ($0.001–1\ mg\,kg^{-1}$) is very effective at suppressing cataplexy in the dog – an effect shared by NET inhibitors.[89]

6.05.6.2.3 Limitations

While the nonstimulant component to atomoxetine has permitted differentiation from the stimulant drugs, and initially led to rapid share gains, sales have since reched a plateau, with physicians considering atomoxetine a second-line therapy behind stimulants, primarily because of inferior efficacy and a delayed onset of action. Atomoxetine also has cardiovascular liabilities, is slowly metabolized in a subset of patients (because of metabolism via CYP2D6), and has known drug–drug interactions with paroxetine, fluoxetine, albuterol, and other common drugs. A warning in bold type of the potential for severe liver injury has been added to the label.

6.05.6.2.4 Future directions

Despite these liabilities, atomoxetine is well received in the marketplace, which underscores the importance of lack of scheduling for future treatments of ADHD.

6.05.6.3 Antidepressants and Antihypertensives

A third line of therapies for ADHD, antidepressants and antihypertensive medications, is not generally approved for use in ADHD, but can be effective off-label. These drugs are sometimes used in combination with stimulants in patients with comorbid symptoms. The antidepressants with demonstrated utility for the treatment of ADHD include the TCAs, bupropion, and monoamine oxidase inhibitors with the antihypertensives, including α_2-adrenoceptor agonists.

6.05.6.3.1 Structure–activity relationships

Imipramine (**Figure 3**) is a TCA. Imipramine is comprised of a tricyclic nucleus (including two phenyl rings and a central cycloheptadien ring). A short side chain and a terminal amine group are additional structural features characterizing imipramine and related members of the same chemical series.[90] The short side chain and the terminal amine group appear to be the conditions that are important for antidepressant activity. The tertiary amines, including imipramine, are more selective for SERT than NET, while the secondary amines, including nortriptyline, are more NET-selective.[86]

Bupropion hydrochloride (**Figure 3**) is an aminoketone antidepressant related to the phenylisopropylamines.[86] Bupropion appears to act as an indirect dopamine agonist as well as having specific noradrenergic effects, being a functional NET inhibitor.

6.05.6.3.2 Comparison within drug class

There is substantial clinical evidence that TCAs have efficacy in ADHD.[86] TCAs, including imipramine, desipramine, and nortriptyline, block neurotransmitter reuptake, including norepinephrine. TCAs, while less effective than stimulants for treating cognitive impairments, are effective in controlling behavioral problems and do improve cognitive impairments. Desipramine and nortriptyline have short- and long-term efficacy in adolescents.

It is assumed that the activity of TCAs in the treatment of ADHD stems from actions on reuptake, and in particular norepinephrine reuptake. Advantages of TCAs include their relatively long half-life, up to 12 h, which, as discussed in the stimulant section above, is an important aspect of treatment. Of the studies available investigating the efficacy of TCAs, 91% reported beneficial effects. Interestingly, desipramine has efficacy in placebo-controlled clinical trials in children as well as adults with ADHD.

Antidepressants with dopaminergic modes of action, such as bupropion are also effective and well tolerated in the treatment of ADHD.[86] In contrast, SSRIs, which have an improved safety profile, appear to be less effective in treating symptoms of ADHD, although they may be useful in treating common comorbid disorders. Venlefaxine, likely due to its NET inhibition, may have some modest efficacy for ADHD.[91] The limited efficacy of SSRIs in ADHD suggests that effects on 5HT may be of less importance than effects on other neurotransmitters, a view supported by lack of efficacy seen with fenfluramine.[83] Monoamine oxidase inhibitors are also efficacious in the treatment of ADHD.[46] While a limited number of studies suggest that monoamine oxidase inhibitors may be effective in juvenile and adult ADHD, the side effects (see below) limit their utility.

Although α_2-adrenoceptor agonists (clonidine is the most frequently used) have been used in children diagnosed with ADHD, there are few controlled studies supporting efficacy. Clonidine appears to have the greatest benefit in hyperactive and aggressive juveniles.[92] A metaanalysis suggests a moderate effect size of clonidine for some co-occurring symptoms of ADHD, including aggression or conduct disorder.[92]

6.05.6.3.3 Limitations

Concerns regarding the safety of these medications, particularly in children, has limited their use. Common side effects of TCAs among adolescents include sedation, weight gain, dry mouth, constipation, and headache. Three deaths among children with ADHD treatment with desipramine have been reported,[46] although the link may not be causal. Heart rate increases are often observed following TCA administration, and thus electrocardiographic monitoring is suggested.

The potential for hypertensive crises associated with tyramine-containing foods (e.g., cheese) and interactions with prescribed, illicit, and over-the-counter drugs limit the usefulness of monoamine oxidase inhibitors in the treatment of ADHD.

The sedative properties of clonidine limit its utility. Sedation is not the basis for the efficacy of clonidine in ADHD, since guanfacine, a more selective α_2-adrenergic agonist that produces less sedation, is also effective.[46] Although the antihypertensive effects of clonidine appear to be the result of reduced norepinephrine activity (probably through stimulation of presynaptic α_2-adrenergic autoreceptors), this is not the likely mechanism of action in ADHD. Instead, studies in monkeys suggest that the effects of α_2-adreneceptor agonists on attention are related to stimulation of postsynaptic receptors, particularly in the prefrontal cortex, an area believed to be involved in the pathophysiology of ADHD.[93]

6.05.6.3.4 Future directions

While side effects limit the utility of many of the antidepressants and antihypertensive agents, Venlafaxine, due to its noradrenergic reuptake inhibition, may have some modest efficacy for ADHD,[65] but side effects, including increased hyperactivity, need to be better understood. Several other compounds and medication classes are in development for ADHD (**Table 2**). Compounds include longer-acting extended-release formulations, such as transdermal methylphenidate (MethyPatch), $GABA_B$ antagonists, and selective partial agonists at $\alpha 4\beta 2$ NNRs.

6.05.7 Unmet Medical Need

Adult ADHD represents a relatively untreated condition. In adults, the hyperactive component is less pronounced and the attention deficit a larger treatment issue than in adolescents. ADHD is not just a lifestyle condition in that inattention and impulsivity result in significant functional impairment and have negative occupational and interpersonal consequences. Impulsivity and the resulting impaired risk assessment also contribute to a higher incidence of illicit drug use among ADHD patients.

From a research perspective, there is a need for objective diagnostic tools and indicators to predict which drug will be the most effective therapy for a given patient. Approximately 70% of children with ADHD respond positively to stimulants as first-line therapy. Additionally, roughly two-thirds of children who do not respond to the first stimulant usually respond to another type (i.e., Adderall or methylphenidate, or vice versa). Hence, the total response rate appears to be about 90%. Overall, studies indicate that multiple unrelated pharmacological agents are efficacious in treating ADHD across the lifespan, with efficacious agents sharing noradrenergic and dopaminergic mechanisms of action. Stimulants are most effective in reducing the hyperactivity in ADHD but less effective in addressing some of the working memory, organization, and planning (i.e., 'executive function') deficits characteristic of ADHD.[94] Research also suggests that the impulsivity and/or aggression associated with this condition are often suboptimally treated with existing therapies.

Key additional areas of unmet need for the treatment of ADHD include stimulant-like efficacy on hyperactivity in a nonscheduled agent, improved efficacy on impulsivity/aggression, alternative therapies for stimulant nonresponders, therapeutics that do not affect growth, sleep, and appetite, nonstimulants without a risk for abuse, and once-daily formulations.

Table 2 Summary of some drugs in development for the treatment of ADHD

Name	*Company*	*Stage*	*Mechanism of action*	*Comments*
Desmethylsibutramine	Sepracor	Phase I	DA, NE, 5-HT reuptake inhibitor	Active metabolite of sibutramine.
DOV-102677	DOV Pharma	Phase I	DA uptake inhibitor	
Aricept (donepezil)	Eisai	Phase II	AChEI	May improve cognitive function more than behavioral symptoms (adjunctive therapy).
SGS-742	Novartis	Phase II	GABA B antagonist, Nootropic agent, anticonvulsant agents	
TC-1734	Targacept	Phase II	Selective partial agonist at nicotinic receptor subtypes α4β2	Efficacy in multiple Ph II cognition trials, especially memory and attention components
Provigil (Sparlon)	Cephalon	Phase II/III	Stimulant (Schedule IV)	Marketed for narcolepsy. Positive Phase III trial in children, not significant in adults.
MethyPatch (transdermal MPH)	Noven	Phase III	MPH	24-h delivery system requiring lower MPH dosage.
SPD 503 (guanfacine)	Shire Pharma	Phase III	Non-scheduled, a2-adrenergic	New use for ADHD in adults and children; on market as antihypertensive.
Adderall XR 2 (SPD 465)	Shire	Phase III	Longer acting than Adderall XR	
NRP-104	Shire/New River Pharma	Phase III	Amphetamine conjugate with amino acid	Prodrug of amphetamine postulated to have lower abuse liability but recommended as Schedule II by FDA

6.05.8 New Research Areas

In conclusion, it is clear that we have come a long way in the more than 60 years since serendipity led to the observation that amphetamine can reduce symptoms in hyperkinetic children. We now have a clear understanding of ADHD as a bona fide disorder, have interesting animal paradigms to model aspects of ADHD, have diagnostic criteria for adolescents, and have therapies that successfully treat the majority of patients, albeit with some limiting side effects. Ongoing research and future directions are clearly going to enhance our ability to understand, diagnose, and treat ADHD successfully in all populations. One promising new molecular target is the histamine H_3 receptor, where inverse agonists have shown some efficacy in animal models of ADHD.[96]

References

1. Chung, B.; Suzuki, A. R.; McGough, J. J. *Exp. Opin. Emerg. Drugs* **2002**, *7*, 269–276.
2. Biederman, J. *Biol. Psychiatry* **2005**, *57*, 1215–1220.
3. AAP Clinical practice guideline. *Pediatrics* **2000**, *105*, 1158–1170.
4. Swanson, J. M.; Sergeant, J. A.; Taylor, E.; Sonuga-Barke, E. J.; Jensen, P. S.; Cantwell, D. P. *Lancet* **1998**, *351*, 429–433.
5. Rohde, L. A.; Biederman, J.; Busnello, E. A.; Zimmermann, H.; Schmitz, M.; Martins, S.; Tramontina, S. *J. Am. Acad. Child Adolesc. Psychiatry* **1999**, *38*, 716–722.
6. McGough, J. J.; Barkley, R. A. *Am. J. Psychiatry* **2004**, *161*, 1948–1956.
7. Gallagher, R.; Blader, J. *Ann. NY Acad. Sci.* **2001**, *931*, 148–171.
8. Evans, S. W.; Pelham, W. E.; Smith, B. H.; Bukstein, O.; Gnagy, E. M.; Greiner, A. R.; Altenderfer, L.; Baron-Myak, C. *Exp. Clin. Psychopharmacol.* **2001**, *9*, 163–175.
9. Epstein, J. N.; Johnson, D. E.; Varia, I. M.; Conners, C. K. *J. Clin. Exp. Neuropsychol.* **2001**, *23*, 362–371.

10. Monastra, V. J.; Lubar, J. F.; Linden, M. *Neuropsychology* **2001**, *15*, 136–144.
11. Faraone, S. V.; Biederman, J. *Biol. Psychiatry* **1998**, *44*, 951–958.
12. Greenhill, L. L.; Pliszka, S.; Dulcan, M. K.; Bernet, W.; Arnold, V.; Beitchman, J.; Benson, R. S.; Bukstein, O.; Kinlan, J.; Mcclellan, J. et al. *J. Am. Acad. Child Adolesc. Psychiatry* **2002**, *41*, 26S–49S.
13. McGough, J. J. *Biol. Psychiatry* **2005**, *57*, 1367–1373.
14. Todd, R. D.; Lobos, E. A.; Sun, L. W.; Neuman, R. J. *Mol. Psychiatry* **2003**, *8*, 103–108.
15. Bark, C.; Bellinger, F. P.; Kaushal, A.; Mathews, J. R.; Partridge, L. D.; Wilson, M. C. *J. Neurosci.* **2004**, *24*, 8796–8805.
16. Brophy, K.; Hawi, Z.; Kirley, A.; Fitzgerald, M.; Gill, M. *Mol. Psychiatry* **2002**, 7, 913–917.
17. Barr, C. L.; Feng, Y.; Wigg, K.; Bloom, S.; Roberts, W.; Malone, M.; Schachar, R.; Tannock, R.; Kennedy, J. L. *Mol. Psychiatry* **2000**, *5*, 405–409.
18. Russell, V. A.; Sagvolden, T.; Johansen, E. B. *Behav. Brain Funct.* **2005**, *1*, 9.
19. Heiser, P.; Friedel, S.; Dempfle, A.; Konrad, K.; Smidt, J.; Grabarkiewicz, J.; Herpertz-Dahlmann, B.; Remschmidt, H.; Hebebrand, J. *Neurosci. Biobehav. Rev.* **2004**, *28*, 625–641.
20. Bush, G.; Valera, E. M.; Seidman, L. J. *Biol. Psychiatry* **2005**, *57*, 1273–1284.
21. Sagvolden, T. *Neurosci. Biobehav. Rev.* **2000**, *24*, 31–39.
22. Fox, G. B.; Pan, J. B.; Esbenshade, T. A.; Bennani, Y. L.; Black, L. A.; Faghih, R.; Hancock, A. A.; Decker, M. W. *Behav. Brain Res.* **2002**, *131*, 151–161.
23. Robbins, T. W. *Psychopharmacology (Berl.)* **2002**, *163*, 362–380.
24. Taffe, M. A.; Davis, S. A.; Yuan, J.; Schroeder, R.; Hatzidimitriou, G.; Parsons, L. H.; Ricaurte, G. A.; Gold, L. H. *Neuropsychopharmacology* **2002**, *27*, 993–1005.
25. Bizarro, L.; Stolerman, I. P. *Psychopharmacology (Berl.)* **2003**, *170*, 271–277.
26. Fox, G. B.; Esbenshade, T. A.; Pan, J. B.; Radek, R. J.; Krueger, K. M.; Yao, B. B.; Browman, K. E.; Buckley, M. J.; Ballard, M. E.; Komater, V. A. et al. *J. Pharmacol. Exp. Ther.* **2005**, *313*, 176–190.
27. Sagvolden, T.; Russell, V. A.; Aase, H.; Johansen, E. B.; Farshbaf, M. *Biol. Psychiatry* **2005**, *57*, 1239–1247.
28. De Bruin, N. M.; Kiliaan, A. J.; De Wilde, M. C.; Broersen, L. M. *Neurobiol. Learn. Mem.* **2003**, *80*, 63–79.
29. Prediger, R. D.; Pamplona, F. A.; Fernandes, D.; Takahashi, R. N. *Int. J. Neuropsychopharmacol.* **2005**, 1–12.
30. Terry, A. V., Jr.; Hernandez, C. M.; Buccafusco, J. J.; Gattu, M. *Neuroscience* **2000**, *101*, 357–368.
31. Nakamura-Palacios, E. M.; Caldas, C. K.; Fiorini, A.; Chagas, K. D.; Chagas, K. N.; Vasquez, E. C. *Behav. Brain Res.* **1996**, *74*, 217–227.
32. Hernandez, C. M.; Hoifodt, H.; Terry, A. V., Jr. *J. Psychiatry Neurosci.* **2003**, *28*, 197–209.
33. Ueno, K. I.; Togashi, H.; Mori, K.; Matsumoto, M.; Ohashi, S.; Hoshino, A.; Fujita, T.; Saito, H.; Minami, M.; Yoshioka, M. *Behav. Pharmacol.* **2002**, *13*, 1–13.
34. Jentsch, J. D. *Behav. Brain Res.* **2005**, *157*, 323–330.
35. Sagvolden, T.; Metzger, M. A.; Schiorbeck, H. K.; Rugland, A. L.; Spinnangr, I.; Sagvolden, G. *Behav. Neural Biol.* **1992**, *58*, 103–112.
36. Wilens, T. E.; Biederman, J.; Spencer, T. J.; Bostic, J.; Prince, J.; Monuteaux, M. C.; Soriano, J.; Fine, C.; Abrams, A.; Rater, M. et al. *Am. J. Psychiatry* **1999**, *156*, 1931–1937.
37. Pataki, C. S.; Feinberg, D. T.; McGough, J. J. *Exp. Opin. Emerg. Drugs* **2004**, *9*, 293–302.
38. Levin, E. D.; Conners, C. K.; Silva, D.; Canu, W.; March, J. *Exp. Clin. Psychopharmacol.* **2001**, *9*, 83–90.
39. Ueno, K.; Togashi, H.; Matsumoto, M.; Ohashi, S.; Saito, H.; Yoshioka, M. *J. Pharmacol. Exp. Ther.* **2002**, *302*, 95–100.
40. Myers, M. M.; Whittemore, S. R.; Hendley, E. D. *Brain Res.* **1981**, *220*, 325–338.
41. Gattu, M.; Terry, A. V., Jr.; Pauly, J. R.; Buccafusco, J. J. *Brain Res.* **1997**, *771*, 104–114.
42. Jones, S. R.; Gainetdinov, R. R.; Jaber, M.; Giros, B.; Wightman, R. M. et al. *Proc. Natl. Acad. Sci. USA* **1998**, *95*, 4029–4034.
43. Gainetdinov, R. R.; Wetsel, W. C.; Jones, S. R.; Levin, E. D.; Jaber, M.; Caron, M. G. *Science* **1999**, *283*, 397–401.
44. Thakker, D. R.; Natt, F.; Husken, D.; Maier, R.; Muller, M.; Van Der Putten, H.; Hoyer, D.; Cryan, J. F. *Proc. Natl. Acad. Sci. USA* **2004**, *101*, 17270–17275.
45. Giros, B.; Jaber, M.; Jones, S. R.; Wightman, R. M.; Caron, M. G. *Nature* **1996**, *379*, 606–612.
46. Biederman, J.; Spencer, T.; Wilens, T. *Int. J. Neuropsychopharmacol.* **2004**, 7, 77–97.
47. Gainetdinov, R. R.; Jones, S. R.; Fumagalli, F.; Wightman, R. M.; Caron, M. G. *Brain Res. Brain Res. Rev.* **1998**, *26*, 148–153.
48. Jaber, M.; Dumartin, B.; Sagne, C.; Haycock, J. W.; Roubert, C. et al. *Eur. J. Neurosci.* **1999**, *11*, 3499–3511.
49. Davids, E.; Zhang, K.; Tarazi, F. I.; Baldessarini, R. J. *Brain Res. Brain Res. Rev.* **2003**, *42*, 1–21.
50. Krause, K. H.; Dresel, S. H.; Krause, J.; Kung, H. F.; Tatsch, K. *Neurosci. Lett.* **2000**, *285*, 107–110.
51. Volkow, N. D.; Wang, G. J.; Fowler, J. S.; Logan, J.; Franceschi, D.; Maynard, L.; Ding, Y. S.; Gatley, S. J.; Gifford, A.; Zhu, W. et al. *Synapse* **2002**, *43*, 181–187.
52. Hess, E. J.; Jinnah, H. A.; Kozak, C. A.; Wilson, M. C. *J. Neurosci.* **1992**, *12*, 2865–2874.
53. Hess, E. J.; Collins, K. A.; Copeland, N. G.; Jenkins, N. A.; Wilson, M. C. *Genomics* **1994**, *21*, 257–261.
54. Wilson, M. C. *Neurosci. Biobehav. Rev.* **2000**, *24*, 51–57.
55. Hess, E. J.; Collins, K. A.; Wilson, M. C. *J. Neurosci.* **1996**, *16*, 3104–3111.
56. Raber, J.; Mehta, P. P.; Kreifeldt, M.; Parsons, L. H.; Weiss, F.; Bloom, F. E.; Wilson, M. C. *J. Neurochem.* **1997**, *68*, 176–186.
57. Steffensen, S. C.; Wilson, M. C.; Henriksen, S. J. *Synapse* **1996**, *22*, 281–289.
58. Roberts, L. A.; Morris, B. J.; O'Shaughnessy, C. T. *Neuroreport.* **1998**, *9*, 33–36.
59. Hou, Q.; Gao, X.; Zhang, X.; Kong, L.; Wang, X.; Bian, W.; Tu, Y.; Jin, M.; Zhao, G.; Li, B. et al. *Eur. J. Neurosci.* **2004**, *20*, 1593–1603.
60. Mill, J.; Curran, S.; Kent, L.; Gould, A.; Huckett, L.; Richards, S.; Taylor, E.; Asherson, P. *Am. J. Med. Genet.* **2002**, *114*, 269–271.
61. Mill, J.; Richards, S.; Knight, J.; Curran, S.; Taylor, E.; Asherson, P. *Mol. Psychiatry* **2004**, *9*, 801–810.
62. Zhang, K.; Tarazi, F. I.; Baldessarini, R. J. *Neuropsychopharmacology* **2001**, *25*, 624–632.
63. Zhang, K.; Davids, E.; Tarazi, F. I.; Baldessarini, R. J. *Brain Res. Dev. Brain Res.* **2002**, *137*, 135–138.
64. Davids, E.; Zhang, K.; Tarazi, F. I.; Baldessarini, R. J. *Psychopharmacology (Berl.)* **2002**, *160*, 92–98.
65. Wolraich, M. L.; Wibbelsman, C. J.; Brown, T. E.; Evans, S. W.; Gotlieb, E. M.; Knight, J. R.; Ross, E. C.; Shubiner, H. H.; Wender, E. H.; Wilens, T. *Pediatrics* **2005**, *115*, 1734–1746.
66. Moran-Gates, T.; Zhang, K.; Baldessarini, R. J.; Tarazi, F. I. *Int. J. Neuropsychopharmacol.* **2005**, *8*, 439–444.
67. Joyce, J. N.; Frohna, P. A.; Neal-Beliveau, B. S. *Neurosci. Biobehav. Rev.* **1996**, *20*, 453–486.
68. Zhang, K.; Tarazi, F. I.; Davids, E.; Baldessarini, R. J. *Neuropsychopharmacology* **2002**, *26*, 625–633.
69. Zhang, K.; Davids, E.; Tarazi, F. I.; Baldessarini, R. J. *Psychopharmacology (Berl.)* **2002**, *161*, 100–106.

70. Avale, M. E.; Falzone, T. L.; Gelman, D. M.; Low, M. J.; Grandy, D. K.; Rubinstein, M. *Mol. Psychiatry* **2004**, *9*, 718–726.
71. Dalley, J. W.; Cardinal, R. N.; Robbins, T. W. *Neurosci. Biobehav. Rev.* **2004**, *28*, 771–784.
72. Chudasama, Y.; Nathwani, F.; Robbins, T. W. *Behav. Brain Res.* **2005**, *158*, 97–107.
73. Puumala, T.; Ruotsalainen, S.; Jakala, P.; Koivisto, E.; Riekkinen, P., Jr.; Sirvio, J. *Neurobiol. Learn. Mem.* **1996**, *66*, 198–211.
74. Granon, S.; Passetti, F.; Thomas, K. L.; Dalley, J. W.; Everitt, B. J.; Robbins, T. W. *J. Neurosci.* **2000**, *20*, 1208–1215.
75. Pelham, W.; Milich, R. Individual Differences in Response to Ritalin in Classwork. In *Ritalin: Theory and Patient Management*; Greenhill, L., Osman, B. P., Eds.; Mary Ann Liebert: New York, 1991; pp 203–221.
76. AAP Clinical practice guideline. *Pediatrics* **2001**, *108*, 1033–1044.
77. Wee, S.; Woolverton, W. L. *Drug Alcohol Depend* **2004**, *75*, 271–276.
78. Michelson, D.; Adler, L.; Spencer, T.; Reimherr, F. W.; West, S. A.; Allen, A. J.; Kelsey, D.; Wernicke, J.; Dietrich, A.; Milton, D. *Biol. Psychiatry* **2003**, *53*, 112–120.
79. Creighton, C. J.; Ramabadran, K.; Ciccone, P. E.; Liu, J.; Orsini, M. J.; Reitz, A. B. *Bioorg. Med. Chem. Lett.* **2004**, *14*, 4083–4085.
80. Biederman, J.; Spencer, T. *J. Attent. Disordt* **2002**, *6*, S101–S107.
81. Garland, E. J. *J. Psychopharmacol.* **1998**, *12*, 385–395.
82. Patrick, K. S.; Markowitz, J. D. *Hum. Psychopharm.* **1997**, *12*, 527–546.
83. Biederman, J.; Spencer, T. Pharmacotherapy of Attention Deficit Hyperactivity Disorder: Nonstimulant Treatments. In *NIH Consensus Development Conference on Diagnosis and Treatment of Attention Deficit Hyperactivity Disorder*; Jensen, P. S.; Cooper, J. R.; Elliot, J. M., Eds.; National Institutes of Health: Bethesda, MD, 1998, pp 97–104.
84. Thai, D. L.; Sapko, M. T.; Reiter, C. T.; Bierer, D. E.; Perel, J. M. *J. Med. Chem.* **1998**, *41*, 591–601.
85. Michelson, D.; Allen, A. J.; Busner, J.; Casat, C.; Dunn, D.; Kratochvil, C.; Newcorn, J.; Sallee, F. R.; Sangal, R. B.; Saylor, K. et al. *Am. J. Psychiatry* **2002**, *159*, 1896–1901.
86. Biederman, J.; Spencer, T. *Biol. Psychiatry* **1999**, *46*, 1234–1242.
87. Gehlert, D. R.; Schober, D. A.; Gackenheimer, S. L. *J. Neurochem.* **1995**, *64*, 2792–2800.
88. Shaw, J.; Threlkeld, P.; Wong, D. T. *Soc. Neurosci* **1999**, *25*, 1963.
89. Mignot, E.; Renaud, A.; Nishino, S.; Arrigoni, J.; Guilleminault, C.; Dement, W. C. *Psychopharmacology (Berl.)* **1993**, *113*, 76–82.
90. Labrid, C.; Moleyre, J.; Poignant, J. C.; Malen, C.; Mocaer, E.; Kamoun, A. *Clin. Neuropharmacol.* **1988**, *11*, S21–S31.
91. Hedges, D.; Reimherr, F. W.; Rogers, A.; Strong, R.; Wender, P. H. *Psychopharmacol. Bull.* **1995**, *31*, 779–783.
92. Connor, D. F.; Fletcher, K. E.; Swanson, J. M. *J. Am. Acad. Child Adolesc. Psychiatry* **1999**, *38*, 1551–1559.
93. Arnsten, A. F.; Steere, J. C.; Hunt, R. D. *Arch. Gen. Psychiatry* **1996**, *53*, 448–455.
94. Zametkin, J.; Ernst, M. *N. Engl. J. Med.* **1999**, *340*, 40–46.
95. Adult Self-Report Scale (ASRS), developed in conjunction with the World Health Organization. http://www.med.nyu.edu/psych/assets/adhdscreener.pdf.
96. Esbenshade, T. A.; Fox, G. B.; Cowart, M. D. *Mol. Interv.* **2006**, *6*, 77–88.

Biographies

Kaitlin E Browman received her BA (with honors) from the Department of Psychology at the University of California, Santa Cruz (US) in 1992. She then pursued her doctoral degree at the University of Michigan (US) in the Department of Biopsychology under the tutelage of Dr Terry Robinson. While pursuing her PhD, Kaitlin focused on environmental factors influencing the development of locomotor sensitization to the psychomotor stimulants amphetamine and cocaine. After receiving her doctoral degree in 1998, Kaitlin completed her postdoctoral work in the laboratory of Dr John Crabbe at Oregon Health Sciences University in Portland, OR (US). As a postdoctoral fellow, Kaitlin expanded her scientific knowledge by studying the behavioral genetics of alcoholism. From Oregon Health Sciences University Kaitlin joined discovery efforts at Bristol-Myers Squibb in Wallingford, CT, in 2000, where she worked on the development of animal models of affective disorders, novel therapeutics, and genomic approaches to drug discovery. Currently Kaitlin is at Abbott Laboratories in North Chicago, IL, working on animal models of cognition, and supporting discovery projects focused on ADHD, Alzheimer's disease, or cognitive deficits associated with schizophrenia.

Gerard B Fox received his BSc (Hons) from the Department of Pharmacology, University College, Dublin (Ireland) in 1989. He then joined the Discovery Biology Team at Pfizer Central Research, Sandwich, Kent (UK) where he spent almost 4 years developing animal models of stroke and contributing to the drug discovery process. Gerard returned to Dublin to assume dual responsibilities as a neuroscientist with the biotechnology company, American Biogenetic Sciences, Inc., while completing his PhD in neuropharmacology at University College. During this time, Gerard also served as special lecturer in Advanced CNS Pharmacology. Subsequently, a postdoctoral fellowship at Georgetown University Medical Center, Washington, DC (USA) allowed Gerard to extend his expertise in behavioral neuroscience, neuroplasticity, as well as biomedical imaging, before joining the Global Pharmaceutical R&D division at Abbott Laboratories, Chicago, IL (US) in 1998. He is currently focused on discovering new drug treatments for CNS disorders such as attention deficit hyperactivity disorder (ADHD), Alzheimer's disease, and cognitive dysfunction associated with schizophrenia or following brain injury. Gerard is author of more than 55 peer-reviewed articles and book chapters.

Comprehensive Medicinal Chemistry II
ISBN (set): 0-08-044513-6

ISBN (Volume 6) 0-08-044519-5; pp. 117–138

6.06 Sleep

E R Bacon, S Chatterjee, and M Williams, Worldwide Discovery Research, Cephalon, Inc., West Chester, PA, USA

6.06.1 Introduction

Sleep is a physiological state of consciousness, the functional role of which is the subject of considerable debate.[1] Humans optimally spend nearly a third of each day sleeping during which time the brain is actually in a dynamic and complex physiological state of activity. Whether to maintain brain plasticity to enhance learning and memory or to allow the brain to rest, the recuperative value of normal sleep and its positive impact on individual performance is without doubt critically important for personal well-being and quality of life, the more so given the impact of societal demands in terms of current day work and social schedules and travel across time zones, all of which promote sleep deficits.

6.06.1.1 Incidence and Prevalence of Sleep Disorders

Sleep disorders are highly prevalent in the developed world with approximately one-third of the adult population experiencing some difficulty in falling asleep that leads to insomnia. Adolescents may also experience sleep disorders. Transient sleep loss or its converse, sleepiness, occurs as the result of both external and internal events including travel, work scheduling, stress, anxiety, pain, and other disease states.

Sleep disorders are divided into two categories: (1) conditions of abnormal somnolence or excessive sleepiness (hypersomnia); (2) insomnia, a condition involving difficulties in initiating or maintaining sleep, or more frequently, poor quality of sleep. Despite its prevalence, insomnia often remains undiagnosed and untreated. Short-term insomnia can result from stressful, albeit self-limiting situations while chronic insomnia can be provoked by a variety of medical and psychiatric conditions.[2] There is also comorbidity of mood disorders with insomnia, the latter being common in depressed patients. Mood disorder association is not limited to insomnia but can also involve hypersomnia.

6.06.1.2 Socioeconomic Impact of Sleep Disorders

Alterations in sleep homeostasis, deviations from the 'early to bed, early to rise' adage, the result of altered work (shift work scheduling, military duty, jet lag, extended commuting, etc.) and social (late night TV, early rising to avoid traffic, family schedules, etc.) patterns can lead to a state of perpetual tiredness with associated fatigue, impaired judgement, cognitive dysfunction, irritability, mood disorders, and anxiety that can exacerbate existing conditions and compromise any sense of well-being. For instance, it is unclear whether the loss of sleep associated with pain states is a cause-and-effect phenomenon; is the pain causing impaired sleep or is a lack of rest leading to hyperalgesia?

Economically, therefore, the impact associated with dysfunctional 'sleep need' is considerable. Direct costs of insomnia have been estimated to be in excess of $10 billion[3] with indirect costs being three times this based on a 10% prevalence in the USA alone.[4] At the opposite end of the sleep–wake spectrum are disorders involving excessive somnolence, e.g., excessive daytime sleepiness (EDS). Narcolepsy, a disabling disease characterized by more severe and chronic EDS, is frequently accompanied by attacks of muscular weakness or paralysis known as cataplexy. Although these syndromes may be less prevalent that those associated with insomnia, the consequences of excessive somnolence have profound risks for individual health and safety. In 1988 it was estimated that the cost of accidents related to all sleep-related disorders (insomnia and excessive sleepiness) was $50 billion.[5] Hypersomnia and insomnia-related syndromes are also associated with psychiatric, neurodegenerative, and pain-related diseases. Obstructive sleep apnea (OSA), a disorder reflecting increased upper airway resistance that results in difficulty in breathing and consequently disturbed sleep (sleep fragmentation), is estimated to have an incidence of 8–27% in the general pediatric population.

6.06.1.3 Measures of Somnolence and Sleepiness Testing

Sleep medicine is a relatively new discipline, based on advances in the fundamental understanding of the physiology and pharmacology of sleep and its accompanying pathophysiology.[6] While sleep disorders are common in the general population, their diagnosis, treatment, and in-depth study has only emerged in the last three decades. The study of sleep disorders, in particular, excessive sleepiness, led to the development of a number of subjective measures of sleep behavior based on questionnaires and scales of condition related responses.[7] The Stanford Sleepiness Scale a self-rating scale assessing sleepiness on a 1–7 scale (1 being the most alert) has largely been replaced by the Epworth Sleepiness Scale (ESS), a self-rated questionnaire evaluating the propensity to fall asleep under a variety of normal circumstances, e.g., reading and watching television, during the normal course of the day. In this system, a total score of 10 indicates a high degree of sleepiness. While these subjective measures are useful in the diagnosis of sleepiness syndromes, objective tests have also been developed to assess the severity of insomnia. The Hamilton Rating Scale for Depression (HAM-D) is used to assess anergy and fatigue and also provides information regarding difficulties in falling asleep, waking from sleep during the wake period, and early waking from sleep. The Pittsburgh Sleep Quality Index is a

19-item questionnaire developed specifically for psychiatric patients.[7] Sleep logs and patient interviews can also be used to identify abnormal sleep patterns.

Since many of the actual changes in brain and body function associated with sleep are outwardly unobservable, specific physiological techniques are employed for their analysis including: polysomnography (PSG), a technique that measures various electrophysiological parameters during the sleep phase including encephalogram wave oscillations (EEG), muscle tonicity as measured by the electromyogram (EMG), and eye movements recorded via the electrooculogram (EOG). Additional physiological variables that can be monitored include oral/nasal airflow, respiration, blood oxygen saturation, and carbon dioxide levels during exhalation, all of which are important diagnostic indicators for sleep apnea. Peripheral EMG activity is also helpful in detecting movements during the sleep cycle. PSG recordings are most frequently conducted the normal sleep time of an individual but may also be evaluated during the day when excessive sleepiness is suspected. The Multiple Sleep Latency Test (MSLT) is utilized when EDS is under investigation, sleep latency being defined as the time required to fall asleep. A variation of this test, the Maintenance of Wakefulness Test (MWT) evaluates the ability to maintain wakefulness under conditions of low light and quiet background.

Functional neuroimaging is also being used to study sleep–wake syndromes.[8] Positron emission tomography (PET) and functional magnetic resonance imaging (fMRI) have shown that in normal subjects, brain activity declines from waking to sleep stages in several key anatomical regions including the prefrontal cortex and thalamus. Imaging studies have shown cerebral hypoperfusion in the basal ganglia during non-rapid eye movement (NREM) sleep compared to normal subjects. Elevated function, as determined by PET imaging, in the ventromedial prefrontal cortex, an area linked to the prefrontal cortex, may contribute to arousal dysfunction. Functional imaging can also study changes in metabolism in deep brain structures.

An important measure in assessing quality of sleep is WASO (wake after sleep onset), which is a measure of the time between which an individual falls asleep and wakes up. A normal physiological WASO score will be in the 6–8 h range, with shorter time periods being reflective of sleep pathophysiology or sleep disorders secondary to anxiety, depression, Alzheimer's disease, etc.

6.06.1.4 Sleep Stages

Two distinct physiological states occur during the sleep process (**Figure 1**): non-REM (NREM or non-rapid eye movement) and REM (rapid eye movement). During nocturnal sleep, individuals cycle between NREM and REM approximately every 1.5–2 h with the alternation typically occurring several times over an 8 h period. NREM is a synchronized sleep stage and is divided into four stages: Stage 1 – a drowsiness phase that is brief, lasting less than 5–7 min in which the eyes are closed, a phase that may be readily interrupted by waking; Stage 2 – a period of light sleep (10–25 min) following the transition from wakefulness. During this phase, there are spontaneous periods of muscle tone and relaxation while the body is entering deep sleep; Stages 3 and 4 represent increasingly intense sleep

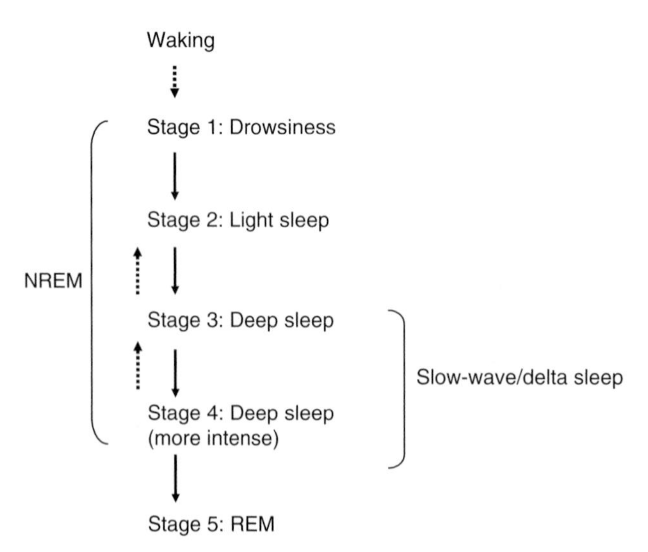

Figure 1 Normal nocturnal sleep cycle. NREM, non-rapid eye movement; REM, rapid eye movement.

stages with a combined duration of 20–40 min that is referred to as slow wave sleep (SWS) or delta sleep. The early stages of NREM repeat prior to the onset of REM. As sleep progresses, each successive cycle differs from the preceding one in terms of duration and NREM–REM organization. The last cycle is usually the longest with SWS tending to diminish significantly in the last cycle. Stage 5, the final stage, is the REM phase (dream, paradoxical, or desynchronized sleep). Sleep architecture refers to the amount and distribution of the individual sleep stages, e.g., minutes of REM and NREM sleep.

6.06.1.5 Overview of Sleep Physiology and Pharmacology

Sleep and wake states are evaluated using three physiological variables: EEG, EOG, and EMG. Functional neuroimaging can also be used to probe sleep–wake physiology and metabolic processes, particularly those in deep brain structures. EEG expression of sleep–wake activity results from the synchronization/desynchronization of the thalamocortical circuitry.

Sleep is normally tightly regulated, playing a critical role in mammalian homeostasis.[9] Over the last two decades, considerable information has accumulated regarding the neuroanatomical structures and neurotransmitter systems involved in sleep–wake physiology.[10,11] Sleep is controlled via the ventrolateral preoptic nucleus (VLPO) with wake being controlled via the hypothalamus (**Figure 2**).[12] While the brainstem (ascending reticular activating system) and thalamocortical system have been historically viewed as being dominant in sleep physiology, the hypothalamus is now recognized as the key anatomical center for sleep regulation. Hypothalamic sleep regulating systems interact with the endogenous circadian pacemaker, the superchiasmatic nuclei (SCN) lying just above the optic chiasm (**Figure 2**). The SCN is sensitive to photic stimuli relayed from the retina via the retinohypothalamic tract that is distinct from the visual transduction pathway.

The SCN modulates sleep–wake cycles with a 24 h circadian rhythm synchronized by a variety of stimuli including light and melatonin and regulated by specific clock genes.[13,14] The SCN also generates signals that: (1) oppose or (2) initiate sleep drive, the latter a function of the time an individual remains awake, and thus, the greater that the circadian signal exceeds sleep drive, the greater will be the wake propensity.[15]

6.06.1.5.1 Wake promotion

The wake/sleep systems broadly represent counterelements of the 'sleep switch,' a hypothalamic circuit sustaining either a state of wake or sleep, with either system inhibiting the other when active. Wake is sustained through excitatory activity in wake promoting structures in the hypothalamus and brainstem. These include the locus ceruleus, dorsal raphe, tuberomammillary nucleus (TMN) and the laterodorsal tegmental and pedunculopontine tegmental nuclei (PPT/LDT) in the pons.

6.06.1.5.2 Sleep promotion

Multiple inhibitory neuronal projections from the VLPO innervate wake-promoting systems to directly suppress wake-promoting nuclei via γ-amino-butyric acid (GABA) and galanin systems. The sleep switch functions as an on/off switch with negligible intermediate physiological states between sleep and wake. Sleep–wake control represents a reciprocal interaction between brainstem monoaminergic and cholinergic systems, with the paracrine neurotransmitter adenosine mediating sleep homeostasis/need, with levels in the basal forebrain increasing during the wake period, as compared to sleep-inhibiting cholinergic neurons activated during wake thus promoting the sleep state.[12,16]

6.06.1.6 Neurotransmitter Systems

Many neurotransmitter candidates are involved in regulating sleep and wake/arousal, the primary of which include: norepinephrine (NE), serotonin (5HT), histamine (HIS), and dopamine (DA).[10,11] During wake, neurons sensitive to these neurotransmitters are activated inhibiting the sleep-promoting neurons of the VLPO. Additional neurotransmitter systems involved in sleep–wake include: orexin/hypocretin (Ox/Hcrt), acetylcholine (ACh), adenosine (Ado), melatonin (MT), glutamate (Glu), and GABA.[10] During wake and arousal, ACh, NE, HIS, and orexin neurons are active decreasing during the SWS phase, being quiescent during REM sleep except for ACh neurons which discharge at a high rate during the REM stage. GABAergic neurons act in an opposite manner to those involved in arousal. Interactions between these various neurotransmitters are complex and perhaps redundant, reflecting the importance of sleep homeostasis for survival. It is only in the past decade that research tools and systems have been available to begin to dissect the pharmacology and pathophysiology of sleep. The only unambiguous data is for adenosine as the endogenous sleep-inducing agent.[12]

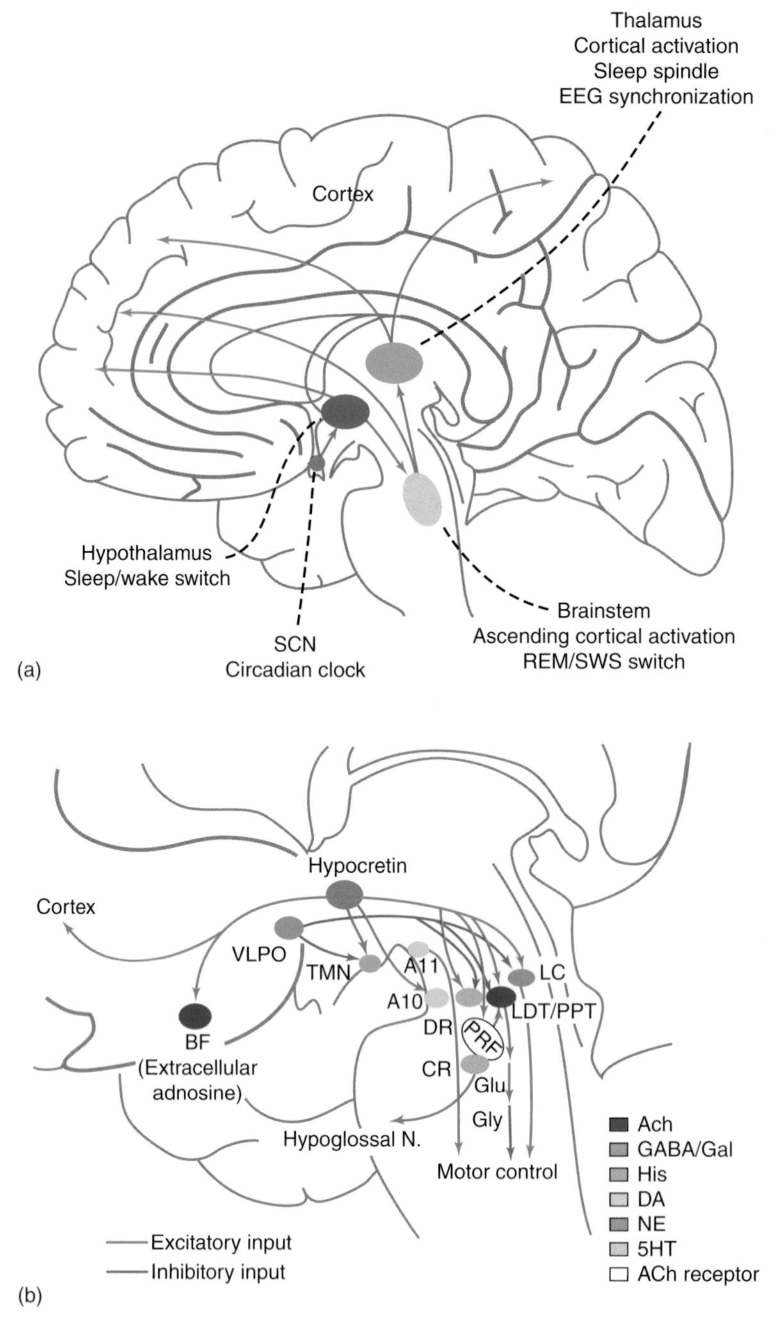

Figure 2 Systems regulating sleep–wake biology. Upper panel (a): Sleep-wake states are controlled through a delicate balance of activities between the thalamus (cortical activation and EEG synchronization), the hypothalamus (sleep/wake switch), the superchiasmatic nucleus (SCN or circadian clock) and the brainstem (ascending cortical activation, REM/SWS switch). Lower panel (b): Multiple neurotransmitters are involved in sleep-wake regulation. The ventrolateral preoptic area (VLPO) contains sleep promoting GABA/Galanin (Gal) neurons whereas wake-promoting orexin (hypocretin) neurons reside in the hypothalamus. The VLPO and orexin systems innervate key areas of the ascending arousal system: locus coeruleus (LC; adrenergic), dorsal raphe (DR; serotonergic) and tuberomammillary nucleus (TMN; histaminergic). Other key regions include the dopaminergic ventral tegmental center (VTA or A10 and A11 projections) which plays a role in alertness and may be important in cataplexy and restless leg syndrome. Serotonin plays a role in upper airway resistance control, thus potentially important in obstructive sleep apnea. Histamine is also involved in wakefulness. Other components of the arousal system are: BF (basal forebrain; cholinergic), LDT/PPT (laterodorsal tegmental nuclei/pediculopontine tegmental nuclei), CR (caudal raphe), PRF (pontine reticular formation); NE (norepinephrine); ACh (acetylcholine); GLY (glycine) and GLU (glutamate). (Reprinted with permission from Mignot, E.; Taheri, S.; Nishino, S. *Nat. Neurosci.* **2002**, *5*, 1071–1075 © Nature Publishing Group.)

6.06.1.6.1 Adenosine

Adenosine (Ado), a paracrine neuromodulator produced in response to changes in cellular energy metabolism and tissue trauma, has both direct and indirect effects on neuronal function. Directly, Ado activates four receptors, A_1, A_{2A}, A_{2B}, and A_3, which comprise the P1 receptor family. Indirectly, via presynaptic A_1 heteroreceptors, Ado inhibits the release of neurotransmitters including ACh, glutamate, and GABA. Ado is released during wake and induces sleep onset, particularly after prolonged wakefulness.[17] Adenosine effects on sleep are also thought to involve adenosine A_{2A} receptor activation.[98] The alerting effects of caffeine, **1** (*see* Section 6.06.6), occur via antagonism of P1 receptors. Ado may be the common final pathway for several sleep factors including endogenous somnolytic mediators, e.g., prostaglandin D_2 (PGD_2), tumor necrosis factor α (TNF-α), and interleukin-1 (IL1).

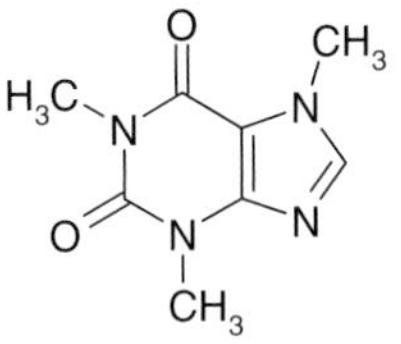

1 Caffeine 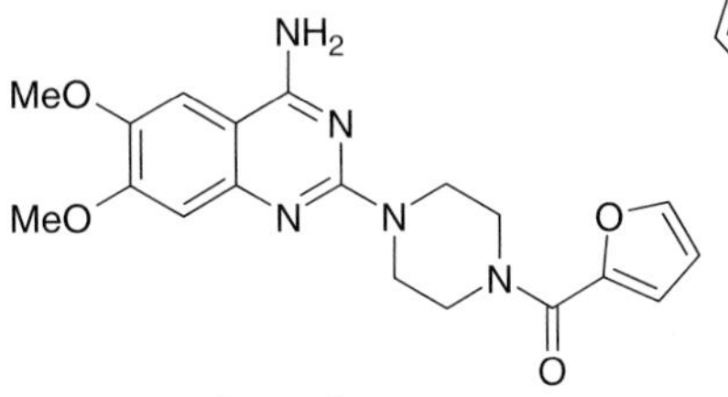 **2** Prazosin 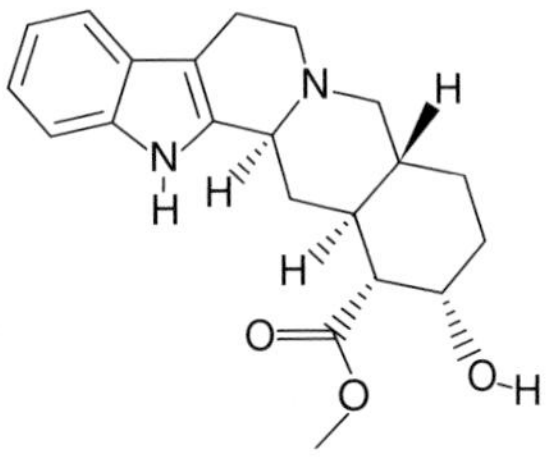 **3** Yohimbine

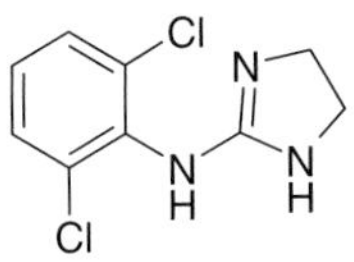

4 Clonidine 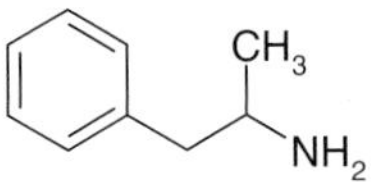 **5** Amphetamine 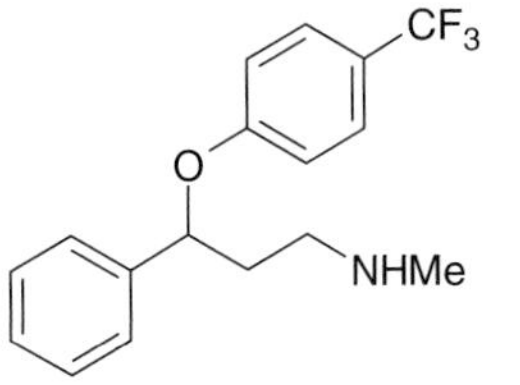 **6** Fluoxetine

6.06.1.6.2 Glutamate

Glutamate (Glu) is the major excitatory neurotransmitter in the central nervous system (CNS) and is present in high concentrations in neuronal projections important for maintenance of cortical activation and arousal, e.g., the brainstem reticular formation (RF) and thalamocortical projections to the cerebral cortex.[11] Glu produces its effects via three classes of postsynaptic receptor that include AMPA (kainate), NMDA (*N*-methyl-D-aspartate), and metabotropic/*t*-ACPD (*trans*-aminocyclopentane-1,3-dicarboxylic acid).

6.06.1.6.3 Norepinephrine

The effects of norepinephrine (NE) may be excitatory or inhibitory depending on the type of adrenoceptor involved. The α_1-antagonist prazosin, **2**, blocks postsynaptic NE receptors and facilitates sleep onset while α_2-antagonists, e.g., yohimbine, **3**, delay sleep latency via blockade of inhibitory α_2-adenoceptors. The α_2-receptor agonist, clonidine, **4**, has hypnotic properties via attenuation of arousal, an effect that may be due to autoreceptor activation and inhibition of NE release. Stimulants like amphetamine, **5**, that block the uptake and stimulate release of NE enhance wake and are used in the treatment of hypersomnolence.

6.06.1.6.4 Dopamine

Dopamine (DA)-containing neurons located in the mesencephalic tegmentum of the substantia nigra and ventral tegmental area that innervate the hypothalamus, nucleus accumbens, and frontal cortex play a key role in wake maintenance. DA neuron firing occurs with arousal during both wake and REM stages. Inhibitors of DA uptake, e.g., cocaine, elicit arousal as do stimulants like amphetamine that facilitate monoamine release leading to cortical activation and waking. These can be used to treat narcolepsy–cataplexy although they have marked dependence liabilities. Antipsychotic DA antagonists result in drowsiness which limits their use.

6.06.1.6.5 Serotonin

Serotonin (5HT) neurons are located in key anatomical locations (e.g., dorsal raphe) linked to wake/arousal. 5HT neuronal firing increases during waking, decreases during SWS, and ceases during REM. The large number of distinct 5HT receptor subtypes offers considerable complexity in regard to their role of in sleep homeostasis. Selective 5HT reuptake inhibitors (SSRIs) like fluoxetine, **6**, have hypnotic/sedative effects.

6.06.1.6.6 Acetylcholine

Acetylcholine (ACh) is a key mediator of cortical activation and vigilance during waking acting via two major groups of cholinergic neurons: the pontine mesencephalic reticular formation giving rise to projections in thalamic nuclei in the forebrain and the lateral hypothalamus and the basal forebrain (e.g., nucleus basalis) projections. ACh produces its effects via muscarinic 7TM receptors and multiple nicotinic (nACh) ion channels. ACh neurons fire with cortical activation but not with arousal. Thalamocortical neuronal excitation is directly mediated by muscarinic M_1 receptors and nAChRs, and indirectly by inhibitory M_2 ACh receptors on GABA-containing neurons. In the brainstem RF, ACh produces either cortical activation or motor inhibition depending on the receptor activated. Acetylcholinesterase inhibitors produce cortical activation with waking during the daytime. At night, however, when other excitatory neurotransmitter systems are inactive, they induce REM sleep.

6.06.1.6.7 Histamine

Histamine (HIS)-containing neurons project to key sleep–wake regulating centers in the TMN and posterior hypothalamus and are active during wake being inhibited by VLPO GABAergic input during sleep. Antihistamines, acting via H_1 receptors, are used to treat allergic rhinitis and other allergic conditions and have pronounced sedative effects. Direct administration of HIS into the brain produces a waking effect via complex interactions with one of four 7TM receptors, H_1, H_2, H_3, and H_4.[18] At H_1 receptors, HIS promotes waking with antagonists eliciting sedation, sleepiness, and decreased alertness. H_3 receptors are a family of presynaptic auto- and heteroreceptors that inhibit the release of HIS, NE, DA, GABA, Glu, and ACh, all of which are involved in sleep–wake regulation. H_3 antagonists may therefore be useful wake-promoting agents.

6.06.1.6.8 GABA

GABA is the major inhibitory neurotransmitter in the CNS, having a key role in regulation of sleep–wake. GABA produces its effects via two receptor families: the $GABA_A$ (which includes the retinal $GABA_C$ subtype) ligand-gated ion channels and the $GABA_B$, 7TM/G protein-coupled receptors (GPCRs). The $GABA_A$ receptor is the site of action of the benzodiazepines (BZs) and other hypnotics, e.g., pentobarbital, which allosterically enhance GABA-mediated neurotransmission. The $GABA_A$–BZ receptor complex consists of at least five transmembrane subunits (two α_1-subunits, two β_1-subunits, and one γ_2-subunit).[19] The binding site for BZs appears to be at the junction of the α- and γ-subunits. Several receptor isoforms are known: ($\alpha_1\beta_2\gamma_2$ and $\alpha_2\beta_2\gamma_2$). Most traditional BZs appear to bind to both subtypes. Current evidence suggests that the sedating and memory properties of the BZs are both mediated by the α_1-subunit. GABAergic neurons are distributed in the brainstem, forebrain, and throughout the anterior hypothalamus–preoptic area where the wake/activating systems are. GABA release in the posterior hypothalamus is higher during SWS than during wake or paradoxical sleep while GABAergic neurons in the brainstem RF and locus ceruleus are active during sleep and inhibit wake-promoting neurons. GABAergic neurons in the thalamus inhibit thalamocortical neurons attenuating cortical activation. GABAergic transmission is thus intimately involved in the induction and maintenance of SWS. The direct $GABA_A$ agonists muscimol and gaboxadol promote SWS. $GABA_B$ receptor agonists like baclofen enhance cortical slow-wave activity and sleep. The therapeutic potential of compounds acting on $GABA_{A/C}$ receptors in treating sleep and memory disorders has been recently reviewed.[20]

6.06.1.6.9 Orexin/Hypocretin

The hypocretin (orexin) signaling pathway is involved in sleep–wake homeostasis and plays a central role in maintaining a stable, basal 'waking tone.'[21] Two novel hypothalamic peptides, hypocretin-1 (Hcrt-1) and -2 (Hcrt-2), also known as orexin-A and orexin-B, produce their effects via OX_1R and OX_2R receptors.[22] The latter are expressed in regions with dense orexin neuronal projections. OX_2R receptors predominate in the tuberomammillary nucleus (TMN). OX_1R are abundantly expressed in the locus ceruleus. Both the TMN and locus ceruleus are anatomical regions critical for maintenance of arousal.

6.06.1.6.10 Melatonin

Melatonin (MT) is an endogenous hormone, normally secreted from the pineal gland at night. Three subtypes of mammalian receptors have now been cloned, the MT1, MT2 and MT3 receptors, respectively. MT1 and MT2 are members of the 7TM GPCR family and share approximately 60% homology. MT1 messenger RNA (mRNA) is present in the human SCN and in most mammalian species. MT2 mRNA is less abundant than MT1 but has also been found in the human SCN and retina. The role of these receptors in the regulation of sleep and circadian rhythms is well established. A third receptor subtype, MT3, may be a form of quinone reductase, an enzyme widely distributed in different tissues across species but for which the relevance of melatonin binding is unclear. Melatonin and melatonin agonists are useful in the treatment of insomnia and other sleep–wake syndromes.[23]

6.06.1.6.11 Miscellaneous neuromodulators

Other neuromodulators implicated in sleep homeostasis include vasoactive intestinal peptide (VIP), substance P, neurotensin, corticotrophin releasing factors (CRFs), thyrotrophin releasing factors (TRFs), galanin, growth hormone-releasing factor (GHRF), and PGD_2.[10]

6.06.2 Disease State

6.06.2.1 Classification of Primary Sleep Disorders

Sleep disorders encompass a wide variety of clinical syndromes that prior to 1979 lacked a formal classification system. For instance, narcolepsy was considered a neurological disorder while insomnia was classified as a psychiatric disease. The International Classification of Sleep Disorders (ICSD) classified primary sleep disorders into two categories, the dyssomnias (insomnia or excessive sleepiness disorders) and parasomnias (abnormal behaviors or psychological events occurring during sleep) which are distinct from psychiatric sleep disorders.[24] The Diagnostic and Statistical Manual of Mental Disorders, Fourth Edition, Text Revision (DSM-IV-TR) (**Table 1**) divides sleep disorders into four major categories by etiology: (1) primary sleep disorders; (2) sleep disorders related to other mental disorders; (3) sleep disorders due to general medical conditions; and (4) substance-induced sleep disorders.[25]

Primary sleep disorders are divided into the same two main categories as ICSD, dyssomnias and parasomnias. These are further subclassified: within dyssomnias primary insomnia, primary hypersomnia, narcolepsy, breathing-related sleep disorder, circadian rhythm sleep disorder, and dyssomnias not otherwise specified; parasomnia subdivisions include syndromes such as nightmare disorder, sleep terror disorder, sleepwalking disorder, and similarly parasomnias not otherwise specified (those falling outside of specific DSM-IV criteria). The three additional DSM-IV-TR sleep disorder categories are outside the scope of this review and are only briefly described. Sleep disorders related to other mental conditions (2), are viewed as integral to the pathophysiology of the specific mental disorder that affects sleep–wake activity indirectly. Sleep disregulation due to general medical conditions (3) may be the direct effect of a disease on sleep–wake regulation. Substance-induced sleep disorders (4) result from substance use/abuse or withdrawal.

Table 1 DSM-IV classification of primary sleep disorders

Dyssomnias
Primary insomnia 307.42
Primary hypersomnia 307.44
Narcolepsy 347
Breathing-related sleep disorder 780.59
Circadian rhythm sleep disorder 307.45
Dyssomnias not otherwise specified 307.47
Parasomnias
Nightmare disorder 307.47
Sleep terror disorder 307.46
Sleepwalking disorder 307.46
Parasomnia, not otherwise specified 307.47

6.06.2.2 Narcolepsy and Primary Hypersomnia

Hypersomnia manifests as a strong desire to sleep during the day or prolonged night-time sleep with difficulty awakening. Narcolepsy is the best known of the excessive somnolence disorders in which hypersomnia is a primary symptom. The essential feature of primary hypersomnia is a period ($\geq$1 month) of excessive sleepiness characterized by daily episodes of sleepiness or prolonged sleep episodes. Sleep associated with primary hypersomnia although continuous, is not a recuperative process, therefore affected individuals do not easily recover from this unproductive sleep surplus. Nocturnal PSG evaluation of primary hypersomnia patients shows normal to prolonged sleep duration, reduced sleep latency, and generally normal REM/NREM sleep cycles (sleep architecture). While the actual prevalence of primary hypersomnia is unknown, it is less common than narcolepsy.[26] Some 5–10% of patients presenting to sleep clinics with EDS symptoms are ultimately diagnosed as having primary hypersomnia with lifetime prevalence being as high as 16% with onset occurring at between 15–30 years and gradually progressing to a chronic and stable state without drug treatment.[25] Recurrent hypersomnia is described by periods of excessive sleepiness lasting at least 3 days, several times over the course of a year and for a period of at least 2 years. The worldwide prevalence of narcolepsy is 1 in 2000 individuals, making it a common disorder.[26]

6.06.2.2.1 Excessive daytime sleepiness

Excessive daytime sleepiness (EDS) is frequently associated with nervous system disorders although excessive sleepiness may dominate the clinical presentation, despite a primary, underlying CNS pathology. Viral infections, structural lesions in the brain, head trauma, tumors, and a wide variety of other causative factors produce EDS. Infectious agents also cause sleepiness, the best known being trypanosomiasis (sleeping sickness). Excessive sleepiness is also associated with neurodegenerative diseases like Parkinson's and Alzheimer's disease. Narcolepsy syndrome is the best-known hypersomnolence condition with hallmark symptoms of hypersomnia and muscle weakness.[26] A disabling neurological disorder, it is characterized by EDS, cataplexy, sleep paralysis, and hallucinations. Excessive, irresistible sleepiness must occur daily for a minimum period of 3 months to qualify the diagnosis of narcolepsy.

6.06.2.2.2 Cataplexy

Cataplexy is characterized by reversible episodes of bilateral loss of muscle tone without loss of consciousness. These episodes are sudden, last from seconds to minutes, and are precipitated by incidents of intense emotion or excitement, e.g., laughing, anger, fear, and surprise. Hypersomnia and cataplexy are two main features of narcolepsy.

6.06.2.2.3 Sleep paralysis

Sleep paralysis is a condition where an individual becomes immobilized during sleep and lacks the ability to move the head and limbs and may have difficulty breathing. Hallucinations can occur at sleep onset or at the onset of wakening.

The diagnosis of narcolepsy in 50–80% of patients is based on a clear association between hypersomnia and well-defined cataplexy. Nocturnal polysomnography coupled with the Multiple Sleep Latency Test (MSLT) indicates that narcoleptic patients display poorly organized sleep architecture, disordered sleep, short REM latency, and significant sleep-onset REM intrusions. The prevalence of narcolepsy–cataplexy is both societal and gender dependent.[27]

6.06.2.3 Primary Insomnia

Insomnia is difficulty in falling asleep, remaining asleep, early morning awakening, and/or sleep that is nonrestorative, all of which may lead to daytime consequences including fatigue, impaired cognition, irritability, mood disorders, and anxiety. It is among the most common of clinical complaints and can either be persistent or transient. A diagnosis of persistent insomnia should be aggressively managed due to significant attendant health concerns, e.g., depression and suicide, and precipitation of manic episodes in bipolar disorder.[28] Insomnia is a cardinal symptom in depression and anxiety with treatment of the underlying psychiatric disorder frequently relieving the sleep disorder.

6.06.2.4 Circadian Rhythm Sleep Disorders

Shift work is classified according to different factors including continuity (continuous, semicontinuous, or discontinuous) and shift type (number of days spent during shift period, clockwise or counterclockwise rotation, and rotation cycle which describes the shift alternation). Syndromes falling under the general category of circadian rhythm sleep disorders include jet lag (time zone change) syndrome, shift work sleep disorder (SWSD), delayed sleep phase syndrome (DSPS), advanced sleep phase syndrome (ASPS), irregular sleep–wake pattern (ISWP), and non-24 hour sleep–wake disorder (hypernycthemeral syndrome).[7,29] These sleep disorders have a common basis through a mismatch in an individuals endogenous circadian cycle and the actual sleep demands (timing and duration), rather than a dysfunction of the specific mechanisms generating sleep and wakefulness. Jet lag syndrome and shift work sleep disorder are the most well recognized of the circadian rhythm syndromes. In jet lag syndrome, the traveler experiences difficulty sleeping due to changing time zones, particularly within a short period of time. Shift work sleep disorder is an important type of circadian rhythm disorder where an individual must change the normal bedtime, most commonly due to a change in work schedule (daytime to night-time and then reverse). In these circumstances the body does not readily adapt to these changes and individuals can suffer from either periods of insomnia or hypersomnia. Nearly 23% of subjects involved in shift work experienced a circadian shift disorder relative to individuals on a normal (daytime) work schedule.[7] Delayed sleep phase type syndrome is characterized by delays in the bedtime whereby the patient cannot fall asleep until early in the morning resulting in waking much later in the day. This syndrome primarily affects adolescents. In advanced sleep phase type syndrome, the process is advanced compared to delayed sleep syndrome, and thus the bedtime is moved forward thereby causing difficulty in staying awake later and may be due to age-related changes in the circadian rhythm of an individual. Non-24 hour sleep–wake syndrome is a recently recognized somewhat rare sleep disorder. Individuals presenting with this disorder are either blind or have specific personality disorders, e.g., introversion with schizoid behavior. Irregular sleep–wake pattern is a sleep disorder most commonly found in nursing homes and in patients suffering from dementia.

6.06.2.5 Breathing-Related Sleep Disorders

Breathing-related sleep disorder syndrome is associated with sleep disruption with excessive sleepiness being a prominent symptom. Within the context of breathing-related sleep disorder, respiratory abnormalities fall into three distinct categories: apneas (breathing cessation), hypopneas (slow or shallow respiration), and hypoventilation (with abnormal oxygen and carbon dioxide levels). Breathing-related sleep disorder can also be subdivided into three types: obstructive sleep apnea syndrome (OSAS), central sleep apnea (CSA), and central alveolar hypoventilation syndrome. OSAS, the most common type of this sleep disorder, is characterized by episodes of apnea and hypopnea that repeat during the sleep phase, is more common in overweight individuals, and is associated with high morbidity and mortality.[30] Given the close clinical correspondence between sleep hypopnea syndrome and sleep apnea syndrome, the syndrome is generally referred to as obstructive sleep apnea–hypopnea syndrome (OSAHS). Upper airway resistance syndrome (UARS) is a variant in which changes in ventilation are small and difficult to detect but for which the modest increase in airway resistance initiated a cascade of events producing symptoms similar to OSAHS.

Central sleep apnea syndrome has a similar presentation to OSAHS with periodic cessation of breathing followed by respiration, but without airway obstruction, and is more commonly observed in elderly patients where ventilation regulation may be compromised due to neurological or cardiac conditions. In central alveolar hypoventilation syndrome, control of ventilation is impaired resulting in lower arterial oxygen saturation, which may be further exacerbated by apneas and hypopneas. Individuals with breathing-related sleep disorder present with a variety of symptomatology including chest discomfort at night, choking or suffocation, and anxiety associated with apneas and difficult breathing. Patients complain of restless sleep with marginal recuperative value and feel worse upon waking than before falling asleep.

6.06.3 Disease Basis

6.06.3.1 Narcolepsy and Primary Hypersomnia

More than 90% of narcoleptic patients carry the human leukocyte antigen (HLA) gene HLA-DR2/DQ1 (HLA-DR15-DQ6) located on chromosome 6 and is considered to be a reliable genetic marker for narcolepsy. Although there is a 100% association of the disease with this HLA haplotype, the genetic link is not a necessary or sufficient causative factor.[31] A catechol-*O*-methyl transferase (COMT) polymorphism has also been associated with the severity of narcolepsy.[32]

A narcoleptic phenotype was identified in a dachshund in 1973 and extended to other canine breeds which then served as a natural model of the disorder.[31] This phenotype was subsequently shown to be associated with a deletion mutation in OX-2 receptors.[31] A narcolepsy-like phenotype occurs in orexin knockout mice, with a phenotype of disrupted sleep–wake cycles and periods of behavioral inactivity resembling cataplectic episodes. The phenotype of OX_1R knockout mice is that of fragmented wake and sleep states while OX_2R knockouts demonstrate sleep disruption and cataplectic-like activity.[33] *Drosophila melanogaster* is also a useful model for studying sleep biology and circadian rhythms.[34]

A total of eight mammalian clock genes – *Clock, Arntl, Cnkle, Cry1, Cry2, Per1–3* – have now been identified creating a feedback loop leading to the 24 h circadian period.[12] Circadian rhythm regulation in *Drosophila* has a high degree of mammalian homology, thus providing a relevant and potentially facile model for sleep research. Newer clock genes regulating circadian activity identified in *Drosophila* include *period* (*per*) and *timeless* (*tim*).[34] Compounds active in humans also alter sleep–wake patterns in *Drosophila* including caffeine, cyclohexyladenosine, hydroxyzine, and modafinil.[35]

6.06.3.2 Primary Insomnia

Primary insomnia is a state of 'hyperarousal' that is poorly understood.[36] Genetic factors that can be linked to insomnia are not well known although a recent study suggested the existence of a familial predisposition.[37] A rare form of insomnia known as fatal familial insomnia (FFI) is a prion disease in the same family as Creutzfeld–Jakob disease (CJD).[38]

6.06.3.3 Circadian Rhythm Sleep Disorder

The circadian rhythm in humans is generated by a pacemaker located in the SCN of the anterior hypothalamus. In rodents, but not in primates, lesioning eradicates all circadian rhythms.[39] Circadian timing has three primary components: an oscillator with a 24 h rhythm; inputs for light stimuli; and outputs for pacemaker regulation. Melatonin is well established as a chronobiotic in a several species including human. In addition to modulating the circadian clock, melatonin has sleep-promoting properties in humans. Daytime administration of melatonin (when endogenous levels would be low) induces sleep in normal subjects and when given during the wake maintenance zone (just before bedtime).[40]

6.06.3.3.1 Shift work sleep disorder

Given different permutations shift work, several factors contribute to the ability of an individual to adjust to shift work: chronobiological, sleep, and domestic factors. When the phase shift period is on the order of 6 h, individual circadian rhythms do not automatically and rapidly readjust to the new time phase. Wake–sleep cycles generally adjust the most rapidly (several days), while temperature and hormonal rhythms can take a week or more to reset. With these differential rates in adjustment to time shifts, normal circadian rhythms become desynchronized. Adjusting to night work is also confounded by the loss of two primary circadian inputs, social stimulation and light, which are powerful factors for the circadian synchronization that facilitates the shift back to a normal day cycle.

Chronobiological factors are also important; clockwise time shifts are more easily tolerated than those in the opposite direction. Individuals referred to as 'night-persons' rather than 'morning-persons' tend to adapt more readily as do younger individuals. In shift work, sleep quality is an important tolerance factor and this is naturally disrupted with an altered cycle. NREM sleep occurs primarily at the beginning of the night-time in a normal diurnal cycle. When this schedule is altered there is a profound effect on the quality and amount of the NREM sleep stage which impacts REM sleep making it more difficult to obtain during the off-cycle. Social, domestic, and environmental factors also contribute to difficulties in sleep during the daytime for an individual on a shift work cycle. As compared to daytime workers, sleep time and quality tend to be shortened and perturbed, respectively, and are common complaints of shift workers.

6.06.3.3.2 Jet lag syndrome

Jet lag syndrome is caused by the desynchronization between the internal clock of an individual and the external clock. The duration of this process depends on the speed at which the internal pacemaker shifts phase, which can be directly related to light exposure (duration and intensity) during travel. The severity and duration of desynchronization varies widely on an individual basis and is linked to the number of time zones crossed, direction of travel, time of departure and arrival, and age, with a phenotype of sleep duration reduction, fragmented sleep, and excessive sleepiness. The symptoms of jet lag can extend to mood disorders, increased anxiety, and decreased cognitive performance. The amount of sleep prior to travel is a significant factor but inconsistent with symptomatology, especially for disruptive long-distance travel when time zones are not traversed (e.g., north–south routes). The inability to more readily adapt to westbound travel (compared to eastbound) is the 'asymmetrical effect.' Jet lag syndrome is more pronounced in the elderly population than in younger individuals.

6.06.3.3.3 Delayed sleep phase syndrome

Delayed sleep phase syndrome (DSPS) is a common circadian rhythm disorder with an estimated prevalence of 7–10% of patients suffering from insomnia but less than 1% overall in the general population.[41] This form of sleep onset insomnia is more common in individuals less than 30 years of age with the greatest prevalence in high school and college students. Students with DSPS generally went to bed two or more hours later than normal on weekends describing symptoms of sleep-onset insomnia and inadequate sleep during the week.

While the underlying cause(s) for DSPS has yet to be elucidated, it may result from: (1) delayed core body temperature rhythms which would subsequently align what is known as the wake maintenance zone with bedtime. Delayed rhythms of melatonin and core temperature occur in patients with DSPS; however, it is unclear whether these changes are causal or secondary to delayed sleep[42]; (2) different phase relationships between the sleep–wake cycle and the intrinsic circadian pacemaker[43] with a 25 h periodicity of the endogenous pacemaker that resets each day; and (3) a stronger wake drive in patients with DSPS that facilitates a delay in sleep onset, pushing the cycle to later in the evening. Individuals with DSPS appear to be at greater risk for developing the non-24 h sleep–wake syndrome, as they tend to sleep at later circadian phases. Differences in light sensitivity, an established circadian synchronizer, may also occur in patients with DSPS and either a decreased light response or an increased sensitivity to evening light could impact the sleep phase, with the latter, heightened light response directly suppressing melatonin secretion until later in the evening. As nearly 75% of individuals who are diagnosed with DSPS present with depression or have a history of depressive illness the most compelling causal factors may be psychological and social.[41]

6.06.3.3.4 Advanced sleep phase syndrome

Advanced sleep phase syndrome (ASPS) is a sleep disorder originally identified in older populations that is rare in children, in the absence of depression. It is characterized by evening sleepiness with an atypically early time of sleep onset. Prevalence for this syndrome has been estimated between 15% and 50%.[44] In contrast to younger individuals, older subjects tend to go to bed earlier, sleep less overall, and awaken earlier, changes that are apparent around middle age and may result from a reduced circadian period. A familial form of this syndrome has been linked to a phosphorylation defect in the human homolog of the period gene, *hPer2*.[45]

6.06.3.3.5 Non-24 hour sleep–wake syndrome

This syndrome (hypernycthemeral syndrome) is quite rare in individuals without ocular pathology and may be a more extreme form of delayed sleep phase syndrome (DSPS).[7] Animal studies suggest that this may result from a lengthening of the rest–activity cycle that would increase the endogenous circadian period, altering the normal 24 h cycle. In blind individuals, fluctuating rhythms (e.g., melatonin, sleep, body temperature) result from the inability to respond to light stimulus as a circadian synchronizer. Without retinal input (photons), the internal clock operates at a period greater than 24 h and the individual generally responds to environmental cues for sleep phase initiation. This syndrome is also seen in patients who have no visual impairment who live at very high latitudes and are unable entrain a normal circadian cycle through the protracted period of winter darkness.

6.06.3.4 Breathing-Related Sleep Disorder

6.06.3.4.1 Obstructive sleep apnea–hypopnea syndrome

Of the three primary variants of breathing-related sleep disorder, obstructive sleep apnea–hypopnea syndrome (OSAHS) is the best defined. Mechanistically, apnea derives from loss of control of ventilation brought on by airway muscle collapse. Many factors may contribute to upper airway occlusion including anatomical abnormalities and airway patency but the latter of is largely dependent on muscle tonicity, specifically, the action of the oropharyngeal dilator and abductor muscles. The pharynx can collapse when the negative upper airway pressure generated by inspiration exceeds the force produced the muscles. Patients with OSAHS have diminished control of pharyngeal muscle activity during sleep, as evidenced by lack of compensatory responses to experimental airway occlusion during sleep. A narrowing of the upper airway is common in OSAHS. Therefore, mechanical and functional factors appear to play a central role and it has been suggested that in apnea episodes during wakefulness, upper airway muscle activity compensates for reduced airway caliber, and that this compensatory response is significantly attenuated during sleep. Obesity has also been linked to OSAHS and familial cases have been reported suggesting a genetic basis of the disorder.[30] A hormonal basis for OSAHS may also be possible based upon a male predominance and the rarity of the OSAHS in women before menopause. Growth hormone secretion generally occurs during deep NREM stages and a growth hormone deficiency has been reported in OSAHS. Most recently, a causal relationship between obstructive sleep apnea and metabolic syndrome ('Syndrome X'; *see* 6.19 Diabetes/Syndrome X) has been proposed.[46]

6.06.3.4.2 Central sleep apnea

In central sleep apnea (CSA) syndrome, the respiratory dysfunction results from aberrant control of respiratory function outside of normal wakefulness where respiratory function tends to be normal. As early as the 1950s, it was recognized that certain medullary brain lesions could negatively affect respiration during the sleep phase but not during the wake period. CSA appears to be a disorder of metabolic control of respiration that can be caused by a wide variety of factors including metabolic, neurologic, pulmonary, and cardiac parameters.

6.06.4 Experimental Disease Models

6.06.4.1 Narcolepsy and Primary Hypersomnia

To date, no true surrogate animal model for primary hypersomnia has been reported. Orexin knockout mice evidence a narcolepsy-like phenotype and have disrupted sleep–wake cycles with periods of behavioral inactivity resembling cataplectic episodes relative to wild-type mice.[33] Despite the gross similarity of the narcoleptic phenotype in knockout mice to the human syndrome, they do not truly mimic human pathology as occurs in humans orexin-producing cells degenerate rather than completely cease to function as in knockout animals. Mouse and rat models have been developed where animals begin to lose hypocretin producing cells at a phase of cellular development when hypocretin expression is high. In this model, a prepro-*Hcrt* promoter drives expression of the neurotoxin, ataxin-3, producing cellular ablation. The effects on sleep are similar to the normal knockout in which wake activity is interrupted by periods of sleep and behavioral inactivity during the active phase (lights-out) and sleep is also disrupted during the inactive (lights-on) phase.[47] Evolving animal models to test the effects of NCEs for their potential as wake promoting agents involve sleep deprivation induced by handling animals or placing them on a rotating platform. The typical time period for inducing deprivation is 6 h, termed the SD6 model. The effects of stimulants in the SD6 model can differ from those in the regular sleep-deprivation models.

6.06.4.2 Primary Insomnia

Animal models are not a true surrogate for insomnia although there are models of sleep impairment induced by housing rats under conditions of novel light/dark cycle or individual housing.[48] The sleep–wake cycle of the guinea pig may be a close physiological approximation of insomnia.[49]

6.06.4.3 Circadian Rhythm Sleep Disorder

Few defined animal models of circadian rhythm disorders exist. *Drosophila* can be used as a surrogate for the study of mammalian sleep disorders while circadian clock mutations occur in the mouse.[34] For example, mice with visual impairment cannot be entrained to light dark cycles and those with an immature SCN display disorganized circadian rhythms.[50] In rats, delayed sleep phase syndrome can be mimicked by maintaining animals in complete darkness for a period of months followed by return to a light dark cycle of 12 h for each phase.[51]

6.06.4.4 Breathing-Related Sleep Disorder

Models of sleep apnea use rats and monkeys chronically implanted for polysomnography using EEG and EMG electrodes to simultaneously measure sleep and respiration.[52] More recently, a monkey model has been established where liquid collagen injections are used to create sleep-disordered breathing and abnormal sleep.[53] The anatomy of the tongue and uvula in monkey closely approximates that of human.

6.06.4.5 The SCORE System

The SCORE system is a microcomputer-based real-time sleep scoring system for rodents that uses a 48-dimension EEG vector related to four arousal states: wakefulness, theta-dominated wakefulness, REM sleep, and NREM sleep[54] that is capable of measuring sleep signature profiles.[55]

6.06.4.6 Transgenic Models and Sleep

An obvious extension of the use of animal models is to characterize sleep–wake profiles in transgenic models that potentially recapitulate human disease states. The Alzheimer Tg2576 mouse model that overexpresses Aβ had longer circadian wheel-running rhythms than wild-type mice with a blunting of the delta (1–4 Hz) power that normally occurs during NREM sleep. Donexepil, a cholinesterase inhibitor that has wake-promoting efficacy, was also lower in Tg2576 mice. At 22 months, female Tg2576 mice showed a REM deficit that could be abolished by an N-terminal antibody to Aβ.[56] SERT knockout mouse exhibit more REM sleep than wild-type littermates (11% versus 7%) with more frequent and longer-lasting REM sleep bouts, an effect similar to that seen with chronic treatment with SSRIs.[57]

6.06.5 Clinical Trial Issues

6.06.5.1 Narcolepsy and Primary Hypersomnia

A basic challenge in clinical trials is establishing sleep disorder as the primary syndrome and dissociating it from other conditions where sleep dysfunction is a side effect. Sleep analyses are complicated by the different methods used to measure sleep dysfunction and quality with more than 20 different algorithms not one of which is fully encompassing.[58] Studies may also be complicated by the fact that measures of sleepiness and alertness are not always correlated when tests such as the MSLT, MWT, and ESS are utilized. In some studies, individuals with sleepiness as determined by MSLT results can report normal alertness while other individuals with apparent normal sleep function have reduced latency on the MLST.[59]

6.06.5.2 Primary Insomnia

The complex nature of insomnia contributes to the challenges associated with it clinical evaluation and patient treatment. PSG is used to study patients with certain sleep abnormalities but it is now evident that objective sleep measures do not always provide a high correlation to the patients' actual symptomatic experiences. Thus evaluating new chemical entity (NCE) effects using PSG may not provide a clear clinical picture. As mentioned above, frank insomnia and comorbid psychiatric illness are frequently difficult to differentiate and a key question is whether treatment of insomnia provides a positive outcome for coexisting conditions. While treating secondary insomnia provides patient benefit, little actual evidence exists supporting benefit in situations of comorbid illness.[2]

6.06.5.3 Circadian Rhythm Sleep Disorder

The correlation of endogenous levels of melatonin in patients to the effectiveness of exogenously administered melatonin in the treatment of sleep disorders remains under investigation[60] suggesting that patient selection will likely be an important consideration. Interpatient melatonin receptor sensitivity and dosing regimens may also play a role in the clinical outcome will also be an important considerations dosing. Surprisingly, the effects on sleep of exogenous melatonin may extend beyond the dosing period for up to weeks, suggesting that the sleep-promoting mechanism is different from that of the widely used benzodiazepine hypnotics.[61] Melatonin and melatonin agonists thus have potential in the treatment of chronic insomnia.

6.06.6 Current Treatment

6.06.6.1 Narcolepsy and Primary Hypersomnia

Treatment of primary hypersomnia parallels pharmacological approaches to treat narcolepsy with classical stimulants including amphetamine, **5**, methamphetamine, **7**, methylphenidate, **8**, pemoline, **9**, and selegilene, **10**, having historically been the first line approach for the treatment of fatigue and hypersomnolence syndromes, e.g., EDS and narcolepsy.

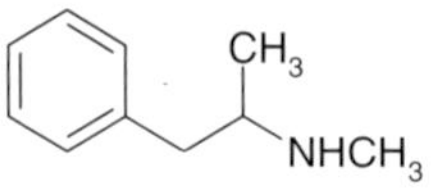

CH_3 $NHCH_3$

7 Methamphetamine

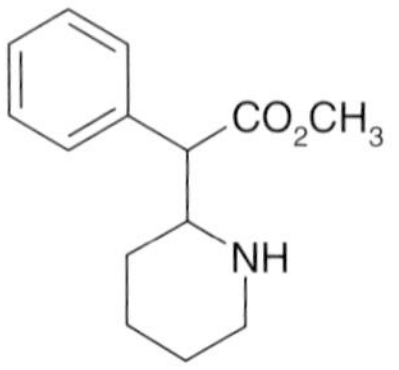

CO_2CH_3 NH

8 Methyphenidate

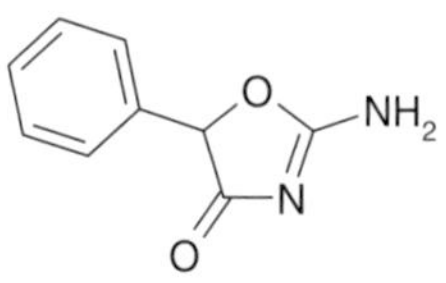

O NH_2 N O

9 Pemoline

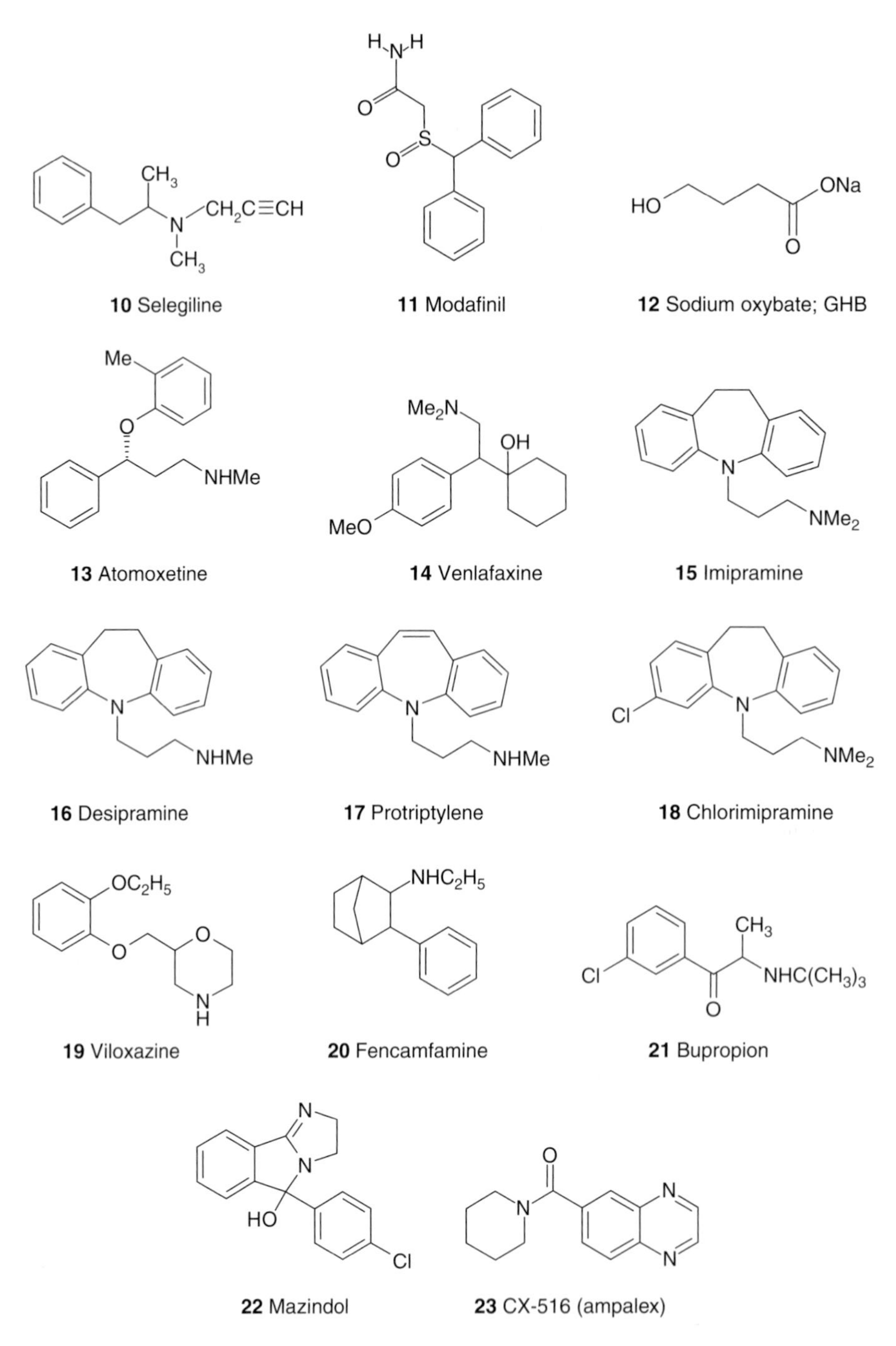

10 Selegiline

11 Modafinil

12 Sodium oxybate; GHB

13 Atomoxetine

14 Venlafaxine

15 Imipramine

16 Desipramine

17 Protriptylene

18 Chlorimipramine

19 Viloxazine

20 Fencamfamine

21 Bupropion

22 Mazindol

23 CX-516 (ampalex)

Caffeine, **1**, the most widely used natural psychostimulant can combat drowsiness, fatigue and impaired alertness but has limited efficacy and tolerates with excessive use which also causes withdrawal headache. Concerns regarding abuse liability and cardiovascular side effects limit the use of stimulants.[62] Agents currently approved for use in the treatment of narcolepsy and hypersomnia are shown in **Table 2**. Two new drugs, modafinil, **11**, and sodium oxybate (γ-hydroxybutyric acid, GHB, **12**) have been approved for the treatment of narcolepsy. Sodium oxybate has also been filed for the treatment of cataplexy. The efficacy of tricyclic antidepressants (TCAs) in treating cataplectic symptoms was reported in the 1960s and monoamine oxidase inhibitors (MAOIs) were also reported effective but with serious side effect potential. These compounds were effective in suppressing REM sleep with a rapid onset of action in stark contrast to delayed time (weeks) for antidepressant efficacy. TCAs typically show anticataplectic efficacy below their antidepressant doses.[63] Drugs used to treat cataplexy are: fluoxetine, **6**, sodium oxybate, **12**, atomoxetine, **13**, venlafaxine, **14**, imipramine, **15**, desipramine, **16**, protriptylene, **17**, chlorimipramine, **18**, and viloxazine, **19**.

Table 2 Agents marketed/in development for the treatment of narcolepsy and hypersomnia

Compound	*Structure*	*Company*	*Development phase*	*Mode of action*
Modafinil		Cephalon, Inc.	Launched	Precise mechanism of action is not known
Armodafinil	*R*-isomer of racemic modafinil	Cephalon, Inc.	NDA submitted	Precise mechanism of action is not known
Sodium oxybate		Orphan Medical/Jazz Pharmaceuticals	Launched	$GABA_B$ agonist
Methylphenidate		Novartis	Launched	Enhancement of monoaminergic transmission (particularly dopamine and norepinephrine)
Dexmethylphenidate		Celgene/Novartis	Launched	
Amphetamine		GlaxoSmithKline	Launched	CNS stimulant
Dexamphetamine			Launched	
CX-516 (ampalex)		Cortex Pharmaceuticals	Phase I	AMPA modulator
CX-717	Unknown	Cortex Pharmaceuticals	Phase I	AMPA modulator
JNJ-17216498	Unknown	Johnson and Johnson	Phase I	Histamine H_3 antagonist

6.06.6.2 Amphetamine-Like Stimulants

Amphetamines including mixed salts have long been first line therapy of hypersomnolence and narcolepsy.[64] Amphetamine is a central stimulant that promotes the release of DA, NE, and 5HT. Their primary effects are mediated through reverse efflux of DA via inhibition of the DAT. At higher doses, inhibition of vesicular monoamine transporter (VMAT) which is involved in amine storage enhances stimulant pharmacology, increasing amine quantities available for efflux. Dexamphetamine, the D-isomer, is more active than the racemate. Methamphetamine, the *N*-methyl analog of amphetamine, is also a potent CNS stimulant showing higher brain permeability due to increased lipophilicity. Various slow-release formulations of these drugs are available with extended efficacy. Amphetamine-like compounds are potent CNS stimulants with high abuse potential and are scheduled substances. Typical side effects for the amphetamines include headache, nausea, anxiety, and appetite suppression in addition to cardiovascular and pulmonary side effects. Stimulant efficacy is dependent on the route of administration, an effect related to pharmacokinetics.[65] The effects of amphetamine are dependent on an intact DAT since amphetamine is without wake promoting activity in DAT knockout mice.[66] A characteristic of CNS stimulants like amphetamine is that their positive effects on wake promotion are accompanied by rebound hypersomnolence following withdrawal, a more profound sleepiness than that which led to the use of stimulants.

6.06.6.2.1 Methylphenidate and dexmethylphenidate

Methylphenidate, **8**, and its D-isomer dexmethylphenidate, are stimulants structurally different from amphetamine, that are used for the treatment of ADHD. Mechanistically, the methylphenidates block monoamine uptake (DA > NE ≫ 5HT) but have less effect on reverse amine efflux or VMAT function than amphetamine[64] and the primary pharmacological stimulant effect is dopaminergic.[65] Methylphenidate was introduced for the treatment of narcolepsy in 1959 and is the most widely prescribed psychostimulant.[64] The more potent D-isomer has a half-life of 6 h contributing to its greater usage compared to amphetamine. Various slow-release formulations of these drugs are also available. Side effects for methylphenidate include increased blood pressure, palpitations, appetite suppression (less than observed with amphetamine), tremor, and insomnia.

6.06.6.2.2 Pemoline and selegiline

Pemoline, **9**, is similar in its pharmacology to amphetamine but is a milder stimulant with less potent sympathomimetic properties, a slower onset of action, and a long half-life (12–16 h). Its primary mechanism of action is DAT inhibition. Pemoline was approved for the treatment of attention deficit/hyperactivity disorder (ADHD) but it and related generic products have been withdrawn (2005) from the market due to the risk of fatal hepatotoxicity. The MAO-B inhibitor selegiline (L-deprenyl, **10**), which prolongs extracellular monoamine effects, has been largely replaced by other first-line agents for the treatment of hypersomnolence. It is metabolized to L-amphetamine and L-methamphetamine which likely accounts for its efficacy.[64]

6.06.6.2.3 Miscellaneous stimulants

Fencamfamine, **20**, a phenethylamine derived stimulant, is not currently available for use in the USA. Bupropion, **21**, originally developed as an antidepressant, is a selective DA uptake blocker with weak activity for inhibiting NE uptake ($>$60-fold less potent than imiprimine) and 5HT that may have utility in treating EDS.[64] Insomnia has been observed in some patients in depression trials. Mazindol, **22**, also blocks uptake of NE and is a weak DA releaser. Despite high affinity for DAT, it has low abuse potential. Mazindol has shown efficacy in treating cataplexy and EDS in humans but is rarely used in the USA and appears to have been recently discontinued.

6.06.6.3 Atypical Stimulants

6.06.6.3.1 Modafinil and armodafinil

Modafinil, **11**, a novel, racemic sulfoxide derivative, is the only drug approved for the treatment of EDS associated with narcolepsy, and is first-line therapy for this indication. It is also approved for the treatment of EDS associated with obstructive sleep apnea/hypopnea syndrome and shift work sleep disorder. It is without effect against cataplexy. The mechanism of action of modafinil is controversial,[64] its weak inhibition (IC_{50} $\sim$3.7 μM) of DAT[67] being the only reported pharmacological effect of the compound. Despite the low affinity for DAT, there is considerable focus on an enhancement of dopaminergic transmission in its mechanism of action as modafinil lacks wake-promoting activity in DAT knockout mice.[6,66] The wake-promoting activity of modafinil has also been attributed to: (1) stimulation of central α_1-adrenoceptors; (2) inhibition of the transporter for NE (NET); (3) a selective activation of the hypothalamocortical pathways involved in sleep–wake regulation; (4) effects on GABAergic systems; and (5) effects on histamine release.[64,68–70] Despite the DAT inhibiting effects of modafinil, the compound differs from the central stimulants amphetamine and methylphenidate in that it does not produce rebound hypersomnolence.[71] Armodafinil, the longer-acting *R*-enantiomer of modafinil, like the racemate does not produce rebound hypersomnolence.[72]

6.06.6.3.2 Sodium oxybate

Sodium oxybate (γ-hydroxybutyric acid, GHB, **12**) is approved for the treatment of cataplexy associated with narcolepsy.[73] GHB is a weak agonist of the $GABA_B$ receptor and also effects DA transmission.[10] At therapeutically effective doses for cataplexy narcolepsy, GHB is well tolerated; however, like amphetamine and other central stimulants, it is abusable. The CNS effects of GHB are sedating at lower doses and it has also been investigated as an anesthetic agent. No structural analogs of this drug have been disclosed.

6.06.6.3.3 Ampakines

Ampakines like CX-516, **23**, and CX-717 are modulators of the AMPA receptor. CX-516 is in phase I trials for Alzheimer's disease and mild cognitive impairment (MCI) and in a recent clinical study showed positive effects in wake promotion. A clinical study in shift work with the CX-717 is planned.[74]

6.06.6.3.4 H_3 antagonists

Modulation of the histamine (H_3) receptor represents another approach to treating sleep–wake disorders and H_3 receptor antagonists have been shown to enhance wakefulness in rats and cats.[18] The first H_3 antagonist advanced to the clinic was GT-2331, **24**, an imidazole derivative that reached phase II trials for ADHD but was subsequently withdrawn.[74,75] The benzofuran, ABT-239, **25**, was rumored to be in trials for cognitive dysfunction but was not actually advanced. ABT-834 (structure unknown) is in phase I trials for ADHD. The benzazepine, GSK-189254, **26**, is in phase I trials for the treatment of dementias.[75] JNJ-17216498 is an orally active and selective H_3 antagonist of undisclosed structure that reduces cataplectic episodes in preclinical narcolepsy models and reported to be in phase I trials for narcolepsy.[74] JNJ-5207852, **27**, is another NCE antagonizing the H_3 receptor.[75]

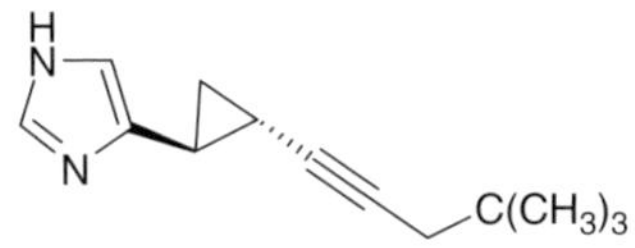

24 GT-2331

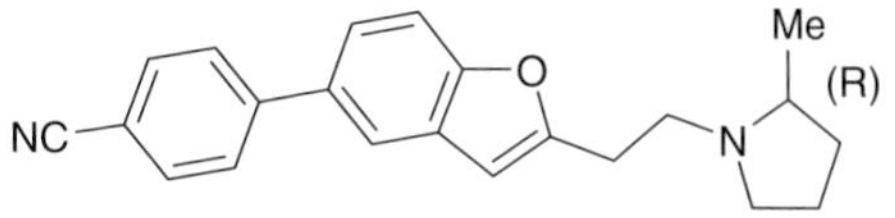

25 ABT-239

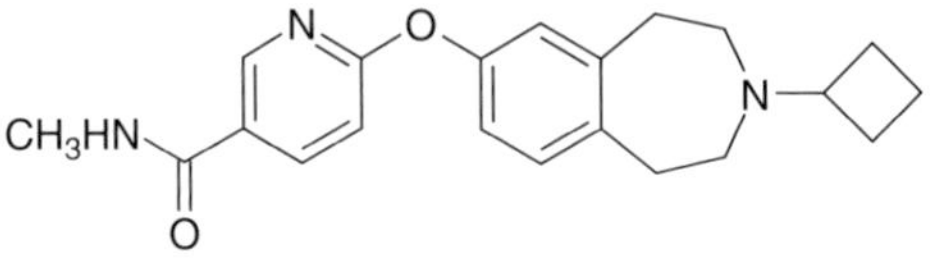

26 GSK-189254

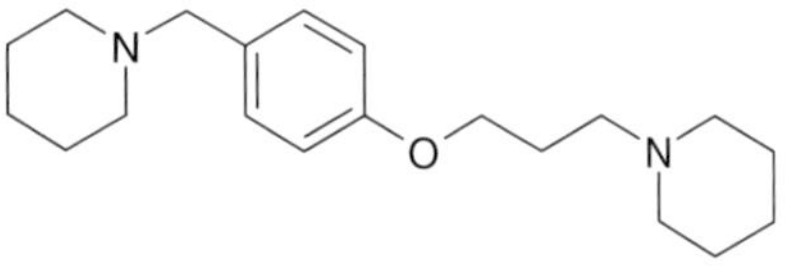

27 JNJ-5207852

6.06.6.4 Primary Insomnia

6.06.6.4.1 Antidepressants

Tricyclic antidepressants (TCAs) and other antidepressants including amitriptyline, **28**, trimipramine, **29**, doxepine, **30**, mirtazapine, **31**, trazodone, **32**, and nefazodone, **33**, are used to treat insomnia,[23] but their therapeutic hypnotic benefit derives from their sedative side effect profiles.

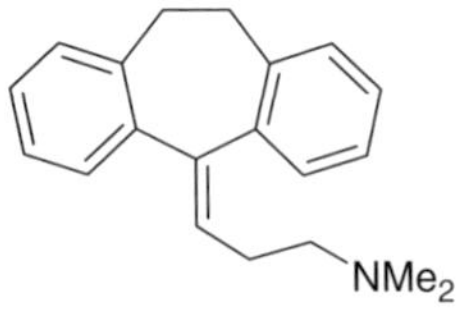

28 Amitriptyline

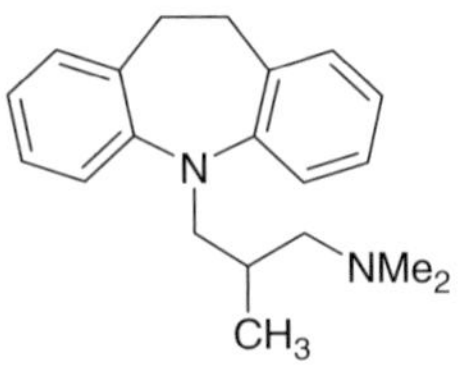

29 Trimipramine

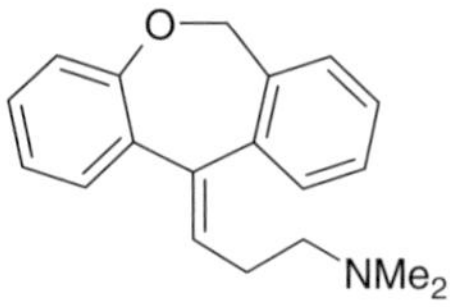

30 Doxepine

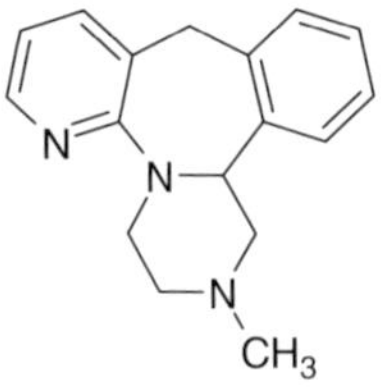

31 Mirtazapine

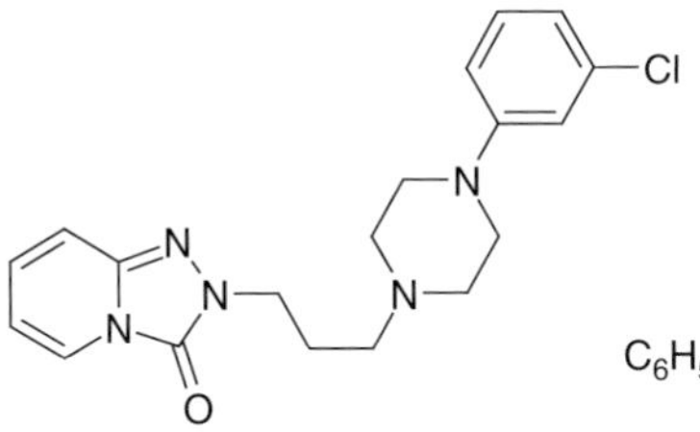

32 Trazodone

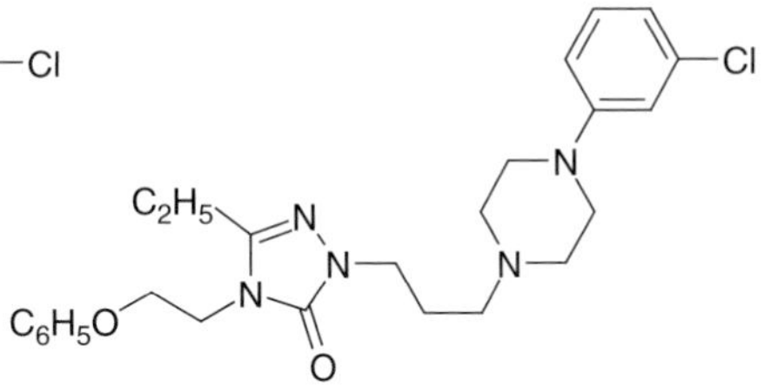

33 Nefazodone

6.06.6.4.2 Antihistamines (H_1-antagonists)

The early CNS-permeable H_1 antagonists have long been known for their sedating side effects. Such compounds include diphenhydramine, **34**, doxylamine, **35**, chlorpheniramine, **36**, and hydroxyzine, **37**. Diphenhydramine or doxylamine are common components of over-the-counter antihistamines used for insomnia.

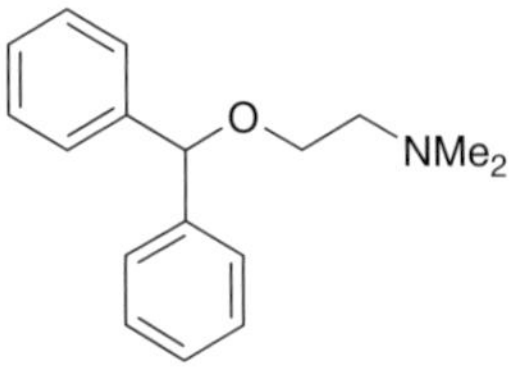

34 Diphenydramine

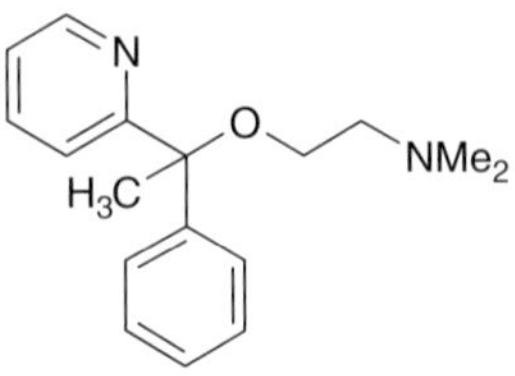

35 Doxylamine

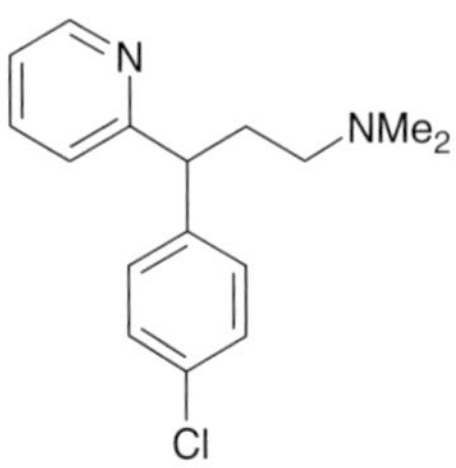

36 Chlorpheniramine

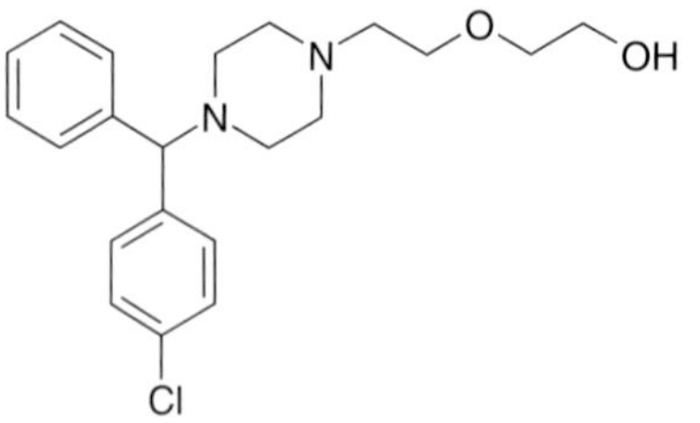

37 Hydroxyzine

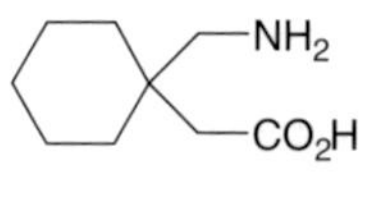

38 Gabapentin

6.06.6.4.3 Classical benzodiazepines

Classical benzodiazepines (BZs) such as diazepam and triazolam have been mainstay hypnotic therapy for decades and target GABA, the major inhibitory neurotransmitter in the CNS. The classical BZs are allosteric modulators of the benzodiazepine site on $GABA_A$ receptors (*see* 6.04 Anxiety)[19] and will not be discussed further in this chapter. Advances in sleep biology have also led to the development of other mechanistic approaches for the treatment of insomnia and new drugs are starting to appear. These agents (launched or in clinical evaluation) are described below and listed in **Table 3**.

6.06.6.4.4 $GABA_A$ benzodiazepine receptor (ω_1) antagonists

Despite their widespread use, the side effect profiles of classical BZs, which include altered sleep architecture, rebound insomnia, tolerance, respiratory depression, and abuse potential, have led to a search for more selective NCEs.[76] A variety of novel selective, BZ receptor ligands that were not structurally BZs (the non-BZs) were identified, e.g., zolpidem, that had hypnotic efficacy similar to that of the traditional BZs with a reduced incidence of side effects have been identified (**Table 3**). These ligands specifically act at the ω_1 site on the $GABA_A$ receptor while classical BZs interact at both ω_1 and ω_2 sites, the latter site being implicated in their side effect profiles. The ω_1 $GABA_A$ receptor subtype is located in regions of the brain linked to sleep–wake regulation, mood, and cognition.

6.06.6.4.4.1 Zopiclone, zolpidem, and eszopiclone

Zopiclone and zolpidem (**Table 3**) were the first two non-BZs ('Z'-drugs based on their names) interacting at the ω_1 binding site although this was not known when they advanced to the clinic. These drugs have the efficacy without the major liabilities of the classical BZs.[77] Eszopiclone, the active *S*-(+)-isomer of racemic zopiclone was launched in the USA. It is approximately 50-fold more potent than the (−)-enantiomer in binding studies and is responsible for the hypnotic effects of the racemate. Eszopiclone is efficacious in insomnia and well tolerated.[78] Eszopiclone produces euphoria to a similar to diazepam, in patients with histories of BZ abuse. No serious withdrawal symptoms were noted in clinical studies. Like the BZs these compounds are scheduled.

6.06.6.4.4.2 Zaleplon

Zaleplon is a pyrazolopyrimidine non-BZ hypnotic that has sedative, anxiolytic, muscle relaxant, and anticonvulsant activities.[76] Zaleplon is equieffiacious to the BZs and zolpidem. Zaleplon does not appear to cause significant morning memory impairment following night-time administration (which is sometimes observed with hypnotic therapy) nor rebound insomnia and it did not appear to cause withdrawal symptoms following discontinuation. The abuse liability is similar that that of other BZ hypnotics and the drug is classified schedule IV.

6.06.6.4.4.3 Indiplon

Indiplon is a triazolopyrimidine non-BZ that is a selective $GABA_A$ ω_1 agonist.[79] In preclinical studies it was more potent than both zaleplon and zolpidem.[74] In clinical trials, indiplon was well tolerated even in elderly patients, a major population among chronic insomniacs. Due to its short half-life (1–1.5 h) with no residual impairment, this NCE should be beneficial to insomniacs who find it difficult to resume sleep once awakened in the middle of the night.

6.06.6.4.5 Melatonin receptor agonists

6.06.6.4.5.1 Ramelteon

Melatonin and melatonin agonists are useful in the treatment of insomnia and other sleep–wake syndromes and ramelteon (TAK-375) is the first of a new class of hypnotics selectively interacting with the MT1 receptor (**Table 3**).

Table 3 Agents marketed/in development for the treatment of insomnia

Compound	*Structure*	*Company*	*Development phase*	*Mode of action*
Zolpidem	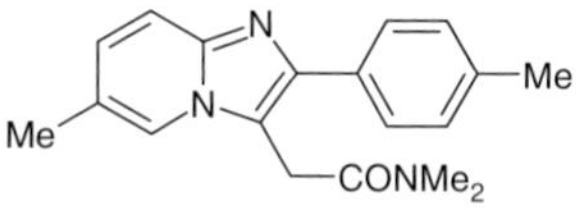	Sanofi-Synthelabo	Launched	$GABA_A$ benzodiazepine receptor (ω_1) agonist
Zopiclone	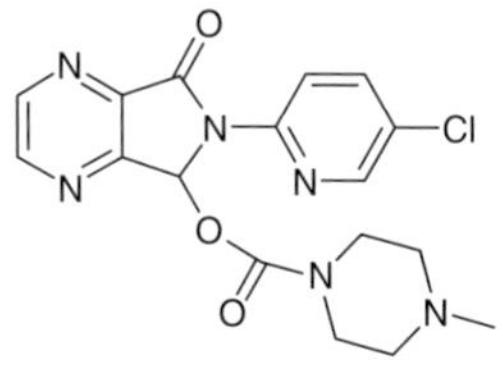	Rhone-Poulenc Rorer (Aventis)	Launched	$GABA_A$ benzodiazepine receptor (ω_1) agonist
Eszopiclone	Single isomer (*S*-) of zopiclone	Sepracor/Aventis	Launched	$GABA_A$ benzodiazepine receptor (ω_1) agonist
Zaleplon	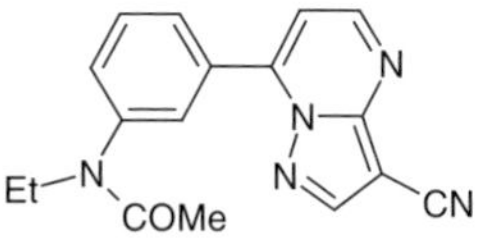	Wyeth (American Home Products)	Launched	$GABA_A$ benzodiazepine receptor (ω_1) agonist
Indiplon	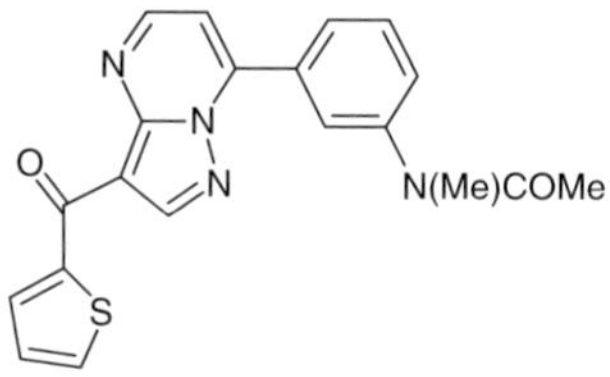	Neurocrine Biosciences/Pfizer	NDA submitted	$GABA_A$ benzodiazepine receptor (ω_1) agonist
Mirtazapine	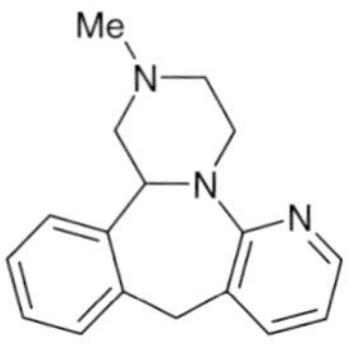	Organon	Launched	$5HT_2/5HT_3$ antagonist, α_2-adrenoreceptor antagonist
TAK-375 Ramelteon	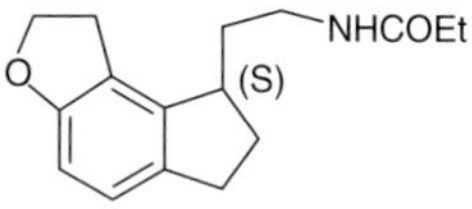	Takeda	Launched	Melatonin (MT1) receptor agonist
Gaboxadol	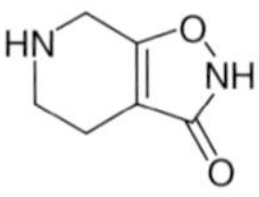	H Lundbeck A/S/ Merck	Phase III	$GABA_A$ agonist
Tiagabine	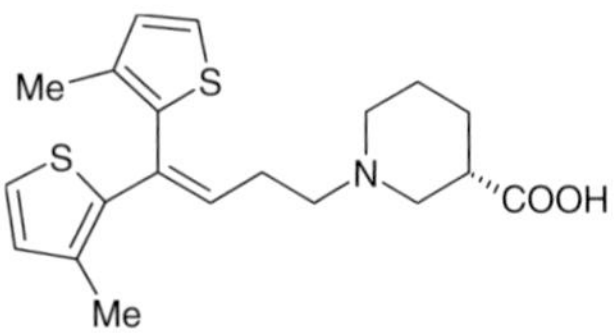	Cephalon, Inc.	Phase II	GABA reuptake inhibitor

continued

Table 3 Continued

Compound	*Structure*	*Company*	*Development phase*	*Mode of action*
Eplivanserin	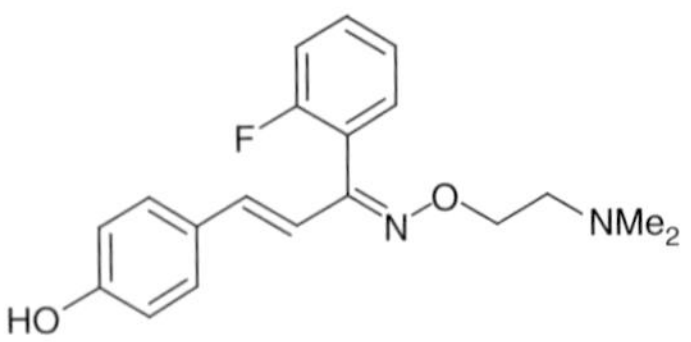 F; HO; N; O; NMe_2	Sanofi-Aventis	Phase II	$5HT_2$ antagonist
PD-6735	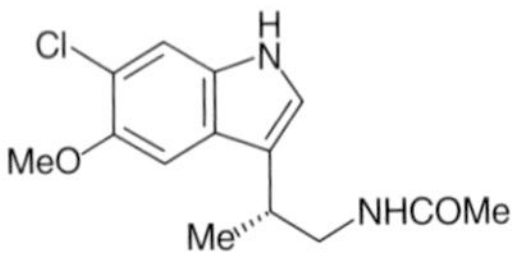 Cl; H N; MeO; Me; NHCOMe	Phase 2 Discovery Inc.	Phase II	Melatonin (MT1) receptor agonist
EMD-28104 (EMR-62218)	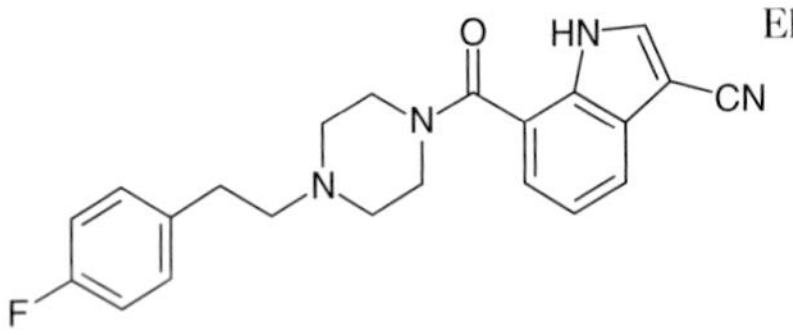 O; HN; CN; N; N; F	Eli Lilly & Co., Merck KGaA	Phase I	$5HT_{2A}$ antagonist
APD-125	Not disclosed	Arena Pharmaceuticals Inc.	Phase I	$5HT_{2A}$ inverse agonist
NDG-96-3	Not disclosed	Neurogen-Pfizer	Phase I	GABA agonist
ORG-50081	Not disclosed	Organon	Phase II	$5HT_{2A}$ antagonist
M-100907		Sanofi-Aventis	Phase IIb	$5HT_{2A}$ antagonist

It is a melatonin analog in which the indole nucleus has been replaced by an indane ring.[80] TAK-375 has 15-fold greater affinity for the MT1 receptor than melatonin[23] and in clinical studies was efficacious in reducing latency to persistent sleep and sleep onset as well as sleep maintenance.[81] TAK-375 is the first nonscheduled drug in this area.

6.06.6.4.5.2 PD-6735 (LY 156735)

PD-6735 (**Table 3**) is an MT1 agonist and an analog of melatonin being developed for the treatment of sleep disorders, especially insomnia. It was granted orphan status in 2004 for the treatment of circadian sleep disorders in the blind and was efficacious in phase II trials.[82]

6.06.6.4.5.3 Miscellaneous NCEs acting at melatonin receptors

Agomelatine (S-20098), an NCE with MT1 agonist and $5HT_{2C}$ antagonist activities, is reportedly in clinical evaluation for depression.[23] BMS-214778 is a melatonin agonist of unknown structure.[23]

6.06.6.4.6 **$GABA_A$ receptor agonists/modulators**

6.06.6.4.6.1 Gaboxadol (THIP)

Gaboxadol is a direct-acting GABA receptor agonist and GABA mimetic, being developed for insomnia.[82] It is a conformationally restricted analog of the GABA mimetic muscimol, where the hydroxyisoxazole moiety mimics the carboxyl group of GABA. Gaboxadol is relatively nonselective for GABA receptor subtypes and is less potent than muscimol in binding studies. Preclinical studies indicate that tolerance to gaboxadol does not develop quickly and that sleep patterns do not change dramatically following drug withdrawal. In the clinic, gaboxadol did not produce a hangover effect, a side effect characteristic of other hypnotic drugs.

6.06.6.4.6.2 Additional GABA agonists/modulators

NG-2-73 (structure undisclosed) is a $GABA_A$ receptor partial agonist reported to be in phase II development.[83]

6.06.6.4.7 Selective GABA reuptake inhibitors

6.06.6.4.7.1 Tiagabine

Tiagabine (**Table 3**) was originally developed as a novel anticonvulsant drug. It is a blocker of GABA reuptake selectively interacting with GAT-1, reducing GABA uptake, and potentiating the inhibitory effects of GABA. Tiagabine is a nipecotic acid analog. Tiagabine promoted slow wave sleep and delta EEG activity in normal elderly subjects[84] supporting a role as a hypnotic agent.

6.06.6.4.8 Miscellaneous agents working through GABA

Like tiagabine, gabapentin, **38**, was developed as an anticonvulsant although the mechanism of action is not completely defined despite its structural similarity to GABA. It is widely used for the treatment of neuropathic pain, bipolar disorder, and insomnia. Gabapentin does block GABA uptake but may interact with ligand gated ion channel subunits.[23]

6.06.6.4.9 $5HT_{2A/2C}$ receptor antagonists

6.06.6.4.9.1 Eplivanserin

The $5HT_{2A/2C}$ receptor has been targeted as a new therapeutic approach for the treatment of insomnia.[12] Eplivanserin, a potent $5HT_{2A}$ receptor antagonist with 20-fold selectivity over the $5HT_{2C}$ receptor, was originally developed as an antipsychotic but is currently in phase II trials for chronic insomnia and sleep apnea syndrome. Its sleep-inducing effects may be the result of its anxiogenic effects. This NCE may have reduced abuse potential.

6.06.6.4.9.2 *S*-(+)-Mirtazapine

S-(+)-Mirtazapine is the optically active isomer of the antidepressant mirtazapine. Mirtazapine is an antagonist at $5HT_{2C}$ receptors as well as $5HT_1$, $5HT_2$, H_1, α_1, and α_2 receptors. It is reported to be in phase II for insomnia.[23,74]

6.06.6.4.9.3 EMD-28104

EMD-28104 (**Table 3**) is an indole derivative that is a potent and selective $5HT_{2A}$ receptor antagonist.[85] It advanced to phase I trials as an antipsychotic but is now being investigated for use in the treatment of sleep disorders.[74,83]

6.06.6.4.9.4 Miscellaneous 5HT receptor antagonists for insomnia

ORG-50081 is a $5HT_{2A}$ receptor in phase II trials for insomnia. APD-125 is a $5HT_{2A}$ inverse agonist in phase I trials for sleep disorders. M-100907 was an antipsychotic NCE that reached phase III trials but is now in phase II for sleep indications (**Table 3**).[83]

6.06.6.4.10 Miscellaneous drugs for insomnia

6.06.6.4.10.1 SO-101

SO-101 is a low dose formulation of doxepine, **30**, that has recently advanced to phase III trials.[74] The mechanism of action of doxepine like other TCAs is complex with interactions at NE, 5HT, H_1, and H_2 receptors and sodium channels[23]; thus the use of a low dose formulation should improve the side effect profile. Unlike other insomnia drugs, doxepine is not a scheduled substance and if shown to have efficacy as a hypnotic agent will represent the first compound its class. Other H_1 receptor-based hypnotics are in development.

6.06.6.4.10.2 TH-9507

TH-9507 is a growth hormone-releasing hormone (GHRH) agonist.[83] GHRH is located in hypothalamic and ventromedial neurons and promotes sleep in mice, rats, rabbits, and humans and is thought to be one of the key components of the sleep-promoting system.[86]

6.06.6.5 Circadian Rhythm Sleep Disorder

6.06.6.5.1 Shift work sleep disorder

The primary approach to treating shift work sleep disorder is the adoption of a schedule to allow for 1 week of consecutive sleep at a consistent time, either daytime or night-time, in concert with adjunctive light treatment; however, this is often difficult to accommodate in practice due to the shift work schedule. Stimulation by bright light results in inhibition of melatonin secretion via the retinohypothalamic pathways. Melatonin, which is released in a circadian manner with peak levels at night, has been used to resynchronize circadian rhythms. Hypnotics like zolpidem and zopiclone (**Table 3**) have been used to treat sleep disorders related to shift work as has modafinil.[87]

6.06.6.5.2 Jet lag syndrome

Managing jet lag syndrome is generally accomplished via nonpharmacological measures, with the approach depending on how many time zones have been traveled. The duration of symptoms can be reduced by light exposure but the effectiveness of this approach is highly dependent on body temperature cycling. The maximum effect of bright light on phase-shifting generally occurs several hours prior to or following the endogenous temperature minimum, which occurs approximately during the midpoint of the usual sleep cycle. Light intensity is also an important factor as ambient indoor lighting is generally not of sufficient intensity to promote proper phase shifting. Short-acting hypnotics are used for the insomnia resulting from jet lag syndrome but tend to exacerbate insomnia on subsequent nights.[88] Melatonin reduces the symptoms of jet lag syndrome but has phase cyclic effects opposite to those produced by bright light exposure.[89]

6.06.6.5.3 Delayed sleep phase syndrome

Several treatment options exist to manage delayed sleep phase syndrome. Generally, treatment is predicated upon simple phase-shifting of the sleep cycle. This can be accomplished by chronotherapy–resetting of the clock through progressively delayed sleeping times, timed exposure to bright light, and pharmacological intervention with BZ hypnotics or melatonin.[29]

6.06.6.5.4 Advanced sleep phase syndrome

Treatment for advanced sleep phase syndrome is similar to that for delayed sleep phase syndrome and includes chronotherapy, light exposure, and pharmacotherapy. The most effective treatment has been shown to be 2 h of bright light therapy in the evenings and a shift in the sleep schedule by short (15–30 min) daily intervals. Melatonin therapy has limited utility.[29]

6.06.6.5.5 Non 24-hour sleep–wake syndrome

This syndrome is particularly challenging from the perspective of treatment, and rigorous scheduling of the individual's activities is critical. In patients with ocular disorder but where retinal signaling pathways are intact, treatment with high-intensity light can be used to reset the circadian rhythm.

6.06.6.5.6 Irregular sleep–wake pattern

Irregular sleep–wake pattern is treated through management of patient activities and increased exposure to light. Hypnotics, stimulants, and neuroleptics are of little value and are contraindicated.

6.06.6.6 Breathing-Related Sleep Disorders

6.06.6.6.1 Obstructive sleep apnea–hypopnea syndrome

In moderate to severe cases obstructive sleep apnea–hypopnea syndrome (OSAHS) is treated with continuous positive airway pressure (CPAP) therapy, a highly effective technique in the control of apnea. Automatic CPAP devices that continually adjust airway pressure have been developed. Patient compliance can be a limiting factor in treatment outcomes, and initial or early patient rejection rates as high as 50% occur.[90] Treatment approaches include agents that reduce REM sleep which improve symptomatology, such as antidepressants, e.g., fluoxetine[91] and mirtazapine[92] and adenosine A_1 agonists, e.g., GR79236.[93]

6.06.6.6.2 Central sleep apnea

Therapeutic treatment of central sleep apnea (CSA) is based on the presumed causative factor, i.e., if cardiac insufficiency or congestive heart failure is suspected treatment of the underlying pathology would be the primary course. Nasal CPAP, oxygen, and carbon dioxide therapy may be useful based on the underlying etiology. The carbonic anhydrase inhibitor acetazolamide, **39**, is useful for treating CSA in some patients by facilitating bicarbonate diuresis, reducing respiratory acidosis and hypocapnea which may precipitate episodes of apnea. Acetazolamide is used prophylactically by high-altitude (>5000 m) climbers prior to ascending, to prevent apnea brought on by acute mountain sickness and associated sleep disruption and high-altitude respiratory sequelae such as pulmonary edema.[94]

6.06.7 Unmet Medical Needs

6.06.7.1 Narcolepsy and Primary Hypersomnia

There is a key need for nonstimulant and, by extrapolation, nonscheduled NCEs to replace amphetamine-like agents. The success of modafinil, which has been described as producing 'a sense of "being awake," compared with the experience being "stimulated on sympathomimetcs,"[10] has prompted a resurgence of activity in the search for novel NCEs that is only limited by a lack of knowledge of the mechanism of action of this novel wake-promoting agent. The search for novel therapeutics that address fundamental disease biology beyond targeting symptomatology are starting to yield novel agents. Medicinal chemistry approaches for the identification of selective orexin antagonists are still in the early stages but compounds of potential interest are starting to emerge that can be used to probe the orexin-1 receptor SB-334867, **40**, and SB-408124, **41**, show high selectivity (50–100-fold) for OX_1 relative to OX_2 and other GPCR targets, and with relatively high antagonist potency.[95]

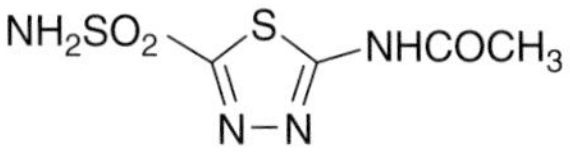

39 Acetazolamide

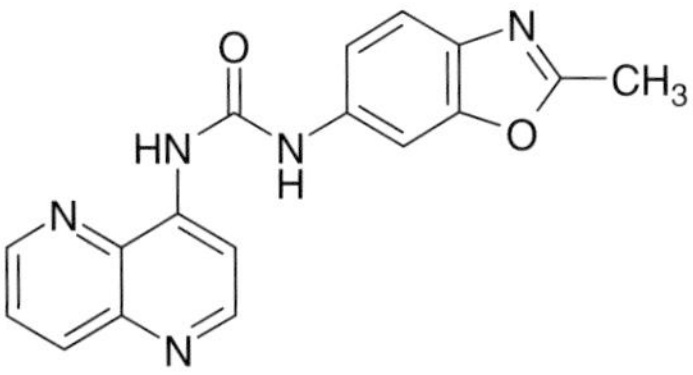

40 SB-334867

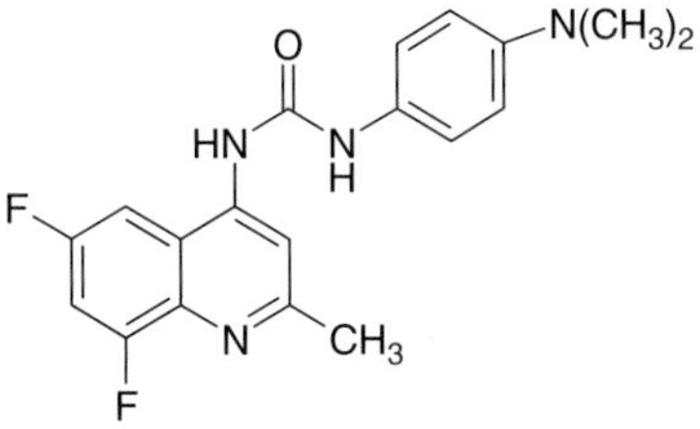

41 SB-408124

6.06.7.2 Primary Insomnia

Based on the unequivocal role of Ado in eliciting sleep, an obvious target for the development of new hypnotics would appear to be the A_1 receptor agonists. However, while adenosine is a highly effective paracrine neuromodulator, its widespread distribution has confounded efforts in NCE discovery given the sedative effects of agonists in the CNS and the hypotensive actions of the purine nucleoside. Despite 30 years of research, the only Ado agonist approved for human use is Ado itself for the treatment of supraventricular tachycardia (*see* 6.33 Antiarrythmics) where the short half-life (10 s) of the natural agonist limits its side effect potential. Nonetheless, A_1 receptor agonists like GR79236 are effective in reducing sleep apnea in rats.

The non-BZ based GABA ω_1 agonist replacements for traditional BZs represented a major step in the treatment of management of insomnia. Despite the properties of the 'Z-drugs' (zolpidem, zopiclone, and zaleplon), concerns for a residual/next-day effect, tolerance, and rebound insomnia still exist. These agents are also scheduled drugs and carry the stigma of 'sleeping pills.' The ideal hypnotic would be nonscheduled, nonabusable, rapidly inducing, and capable of maintaining sleep for 6–8 h without awakenings, preserving sleep architecture with a high safety margin. Preservation of sleep architecture is a characteristic that may enable drugs to provide a more recuperative sleep compared to those therapies that reduce the time that a patient spends in delta and REM stages. For the elderly patient population, where polypharmacy is typical, hypnotics must also work effectively in those settings with few side effects. There is increasing evidence that a combination of pharmacologic treatment and behavioral modifications (cognitive behavioral therapy) may produce clinically sustainable positive benefits.

6.06.7.3 Circadian Rhythm Sleep Disorder

Advances in the biology and genetics of circadian clock rhythm have led a greater understanding of circadian sleep disorders but few drug targets have been identified. Therapeutic interventions such as melatonin treatment have demonstrated some utility but are not approved. Light therapy to reset circadian rhythms is still relatively empirical in its implementation.

6.06.7.4 Breathing-Related Sleep Disorder

Sleep apnea is a relatively common clinical disorder, second only to asthma in prevalence among the chronic respiratory diseases, and can develop in children as well as adults. While CPAP and related mechanical devices can be effective in the management of moderate to severe sleep apnea, patient compliance can be a limiting factor in their effectiveness. Thus, the area of breathing-related sleep disorders would appear to be a logical target for one or more effective pharmacologic treatment approaches that would obviate the issue of patient compliance regarding mechanical treatment devices.

6.06.8 New Research Areas

6.06.8.1 Narcolepsy and Primary Hypersomnia

The multiple receptor systems involved in sleep–wake regulation[11] could provide avenues for future research as the biology becomes more fully elucidated. Histamine receptor (H_3) antagonists are an active area of research where NCEs have reached clinical evaluation but proof of concept is still required. The orexin pathway has a clear link to the pathophysiology of narcolepsy, and selective OX_2 antagonists are being targeted for clinical proof of concept. Drugs that can target the circadian clock mechanisms and related regulatory systems may offer new therapeutic approaches. Drugs targeting receptors for transforming growth factor α (TGFα) and prokineticin are two potential avenues for circadian rhythm modulation.[12] Recent studies suggest a role for urotensin II receptors in the regulation of REM sleep through direct activation of cholinergic neurons in the brainstem but it is too early to know whether agents could be identified that could directly target the CNS receptors avoiding the peripheral vasoconstrictive effects mediated by urotensin II.[96]

6.06.8.2 Primary Insomnia

Agents that directly target VLPO neuronal projections may represent novel hypnotics. Gaboxadol is a GABAergic agent that affects slow wave sleep and may be beneficial in narcolepsy. Melatonin receptor agonists offer a new perspective on sleep-inducing agents as being more sleep enhancing by virtue of effects on the circadian pacemaker.

6.06.8.3 Circadian Rhythm and Breathing-Related Sleep Disorders

In sleep apnea syndromes, e.g., OSAHS, upper airway restriction leads to deteriorated sleep, ultimately resulting in daytime sleepiness and impaired cognition. Treatments with classical stimulants like amphetamine are limited due to the side effect profile and potential for abuse. Modafinil has been evaluated as adjunctive therapy to CPAP treatment in OSAHS. In a randomized double blind placebo controlled study, modafinil in conjunction with CPAP therapy significantly improved both the subjective and objective measures of daytime sleepiness compared to CPAP alone as assessed on the ESS testing findings supported in a larger study.[97] Advances in the treatment of circadian rhythm disorders are still at an early stage as the basic mechanisms and potential drug targets are better understood. Behavioral interventions and light therapy are still the mainstays of circadian disorder therapy.

References

1. Siegel, J. M. *Nature* **2005**, *437*, 1264–1271.
2. Benca, R. M. *J. Clin. Psychiatry* **2001**, *62*, 33–38.
3. Brunello, N.; Armitage, R.; Feinberg, I.; Holsboer-Trachsler, E.; Leger, D.; Linkowski, P.; Mendelson, W. B.; Racagni, G.; Saletu, B.; Sharpley, A. L. et al. *Neuropsychobiology* **2000**, *42*, 107–119.
4. Martin, S. A.; Aikens, J. E.; Chervin, R. D. *Sleep Med. Rev.* **2004**, *8*, 63–72.
5. Balter, M. B.; Uhlenhuth, E. H.; *J. Clin. Psychiatry* **1992**, *53*, 34–39; discussion 40–42.
6. Saper, C. B.; Scammell, T. E.; Lu, J. *Nature* **2005**, *437*, 1257–1263.
7. Doghramji, K. *J. Clin. Psychiatry* **2004**, *65*, 17–22.
8. Nofzinger, E. A. *Sleep Med. Rev.* **2005**, *9*, 157–172.
9. Hobson, J. A. *Nature* **2005**, *437*, 1254–1256.
10. Mignot, E.; Nishino, S. *Sleep* **2005**, *28*, 754–763.
11. Jones, B. E. *Trends Pharmacol. Sci.* **2005**, *26*, 578–586.
12. Mignot, E.; Taheri, S.; Nishino, S. *Nat. Neurosci.* **2002**, *5*, 1071–1075.
13. Gooley, J. C.; Saper, C. B. Anatomy of the Mammalian Circadian System. In *Principles and Practice of Sleep Medicine*; Kryger, M. H., Roth, T., Dement, C., Eds.; Elsevier Saunders: Philadelphia, PA, 2005, pp 335–350.
14. Rosenwasser, A. M.; Turek, F. Physiology of the Mammalian Circadian System. In *Principles and Practice of Sleep Medicine*; Kryger, M. H., Roth, T., Dement, C., Eds.; Elsevier Saunders: Philadelphia, PA, 2005, pp 351–374.
15. Dijk, D. J.; Lockley, S. W. *J. Appl. Physiol.* **2002**, *92*, 852–862.

16. Retey, J. V.; Adam, M.; Honegger, E.; Khatami, R.; Luhmann, U. F.; Jung, H. H.; Berger, W.; Landolt, H. P. *Proc. Natl. Acad. Sci. USA* **2005**, *102*, 15676–15681.
17. Porkka-Heiskanen, T.; Alanko, L.; Kalinchuk, A.; Stenberg, D. *Sleep Med. Rev.* **2002**, *6*, 321–332.
18. Leurs, R.; Bakker, R. A.; Timmerman, H.; de Esch, I. J. *Nat. Rev. Drug Disc.* **2005**, *4*, 107–120.
19. Mendelson, W. B. Hypnotic Medications: Mechanisms of Action and Pharmacologic Effects. In *Principles and Practice of Sleep Medicine*; Kryger, M. H., Roth, T., Dement, C., Eds.; Elsevier Saunders: Philadelphia, PA, 2005, pp 444–451.
20. Chebib, M.; Hanrahan, J. R.; Mewett, K. N.; Duke, R. K.; Johnston, G. A. R. *Annu. Rep. Med. Chem.* **2004**, *39*, 13–23.
21. Saper, C. B.; Chou, T. C.; Scammell, T. E. *Trends Neurosci.* **2001**, *24*, 726–731.
22. Sakurai, T. *Sleep Med. Rev.* **2005**, *9*, 231–241.
23. Buysse, D. J.; Schweitzer, P. K.; Moul, D. E. Clinical Pharmacology of Other Drugs Used as Hypnotics. In *Principles and Practice of Sleep Medicine*; Kryger, M. H., Roth, T., Dement, C., Eds.; Elsevier Saunders: Philadelphia, PA, 2005, pp 452–467.
24. Thorpy, T. Classification of Sleep Disorders. In *Principles and Practice of Sleep Medicine*; Kryger, M. H., Roth, T., Dement, C., Eds.; Elsevier Saunders: Philadelphia, PA, 2005, pp 615–623.
25. American Psychiatric Association. *Diagnostic and Statistical Manual of Mental Disorders*, 4th ed., Text Revision; American Psychiatric Association: Washington, DC, 2000.
26. Black, J. E.; Brooks, S. N.; Nishino, S. *Semin. Neurol.* **2004**, *24*, 271–282.
27. Billiard, M.; Dauvilliers, Y. Narcolepsy. In *Sleep, Physiology, Investigations and Medicine*; Billiard, M., Ed.; Kluwer Academic/Plenum Publishers: New York, 2003, pp 403–428.
28. Benca, R. M. *Psychiatr. Serv.* **2005**, *56*, 332–343.
29. Reid, K. J.; Zee, P. C. Circadian Disorders of the Sleep–Wake Cycle. In *Principles and Practice of Sleep Medicine*; Kryger, M. H., Roth, T., Dement, C., Eds.; Elsevier Saunders: Philadelphia, PA, 2005, pp 691–701.
30. Redline, S.; Tishler, P. V. *Sleep Med. Rev.* **2000**, *4*, 583–602.
31. Nishino, S.; Okura, M.; Mignot, E. *Sleep Med. Rev.* **2000**, *4*, 57–99.
32. Dauvilliers, Y.; Neidhart, E.; Lecendreux, M.; Billiard, M.; Tafti, M. *Mol. Psychiatry* **2001**, *6*, 367–372.
33. Willie, J. T.; Chemelli, R. M.; Sinton, C. M.; Tokita, S.; Williams, S. C.; Kisanuki, Y. Y.; Marcus, J. N.; Lee, C.; Elmquist, J. K.; Kohlmeier, K. A. et al. *Neuron* **2003**, *38*, 715–730.
34. Hendricks, J. C.; Sehgal, A. *Sleep* **2004**, *27*, 334–342.
35. Hendricks, J. C.; Kirk, D.; Panckeri, K.; Miller, M. S.; Pack, A. I. *Sleep* **2003**, *26*, 139–146.
36. Bonnet, M. H.; Arand, D. L. *Sleep Med. Rev.* **1997**, *1*, 97–108.
37. Bastien, C. H.; Morin, C. M. *J. Sleep Res.* **2000**, *9*, 49–54.
38. Montagna, P. *Sleep Med. Rev.* **2005**, *9*, 339–353.
39. Edgar, D. M.; Dement, W. C.; Fuller, C. A. *J. Neurosci.* **1993**, *13*, 1065–1079.
40. Zhdanova, I. V. *Sleep Med. Rev.* **2005**, *9*, 51–65.
41. Regestein, Q. R.; Monk, T. H. *Am. J. Psychiatry* **1995**, *152*, 602–608.
42. Oren, D. A.; Turner, E. H.; Wehr, T. A. *J. Neurol. Neurosurg. Psychiatry* **1995**, *58*, 379.
43. Duffy, J. F.; Dijk, D. J.; Hall, E. F.; Czeisler, C. A. *J. Investig. Med.* **1999**, *47*, 141–150.
44. Buysse, D. J.; Reynolds, C. F., III; Monk, T. H.; Hoch, C. C.; Yeager, A. L.; Kupfer, D. J. *Sleep* **1991**, *14*, 331–338.
45. Toh, K. L.; Jones, C. R.; He, Y.; Eide, E. J.; Hinz, W. A.; Virshup, D. M.; Ptacek, L. J.; Fu, Y. H. *Science* **2001**, *291*, 1040–1043.
46. Vgontzas, A. N.; Bixler, E. O.; Chrousos, G. P. *Sleep Med. Rev.* **2005**, *9*, 211–224.
47. Beuckmann, C. T.; Sinton, C. M.; Williams, S. C.; Richardson, J. A.; Hammer, R. E.; Sakurai, T.; Yanagisawa, M. *J. Neurosci.* **2004**, *24*, 4469–4477.
48. Michaud, J. C.; Muyard, J. P.; Capdevielle, G.; Ferran, E.; Giordano-Orsini, J. P.; Veyrun, J.; Roncucci, R.; Mouret, J. *Arch. Int. Pharmacodyn. Ther.* **1982**, *259*, 93–105.
49. Gvilia, I.; Darchia, N.; Oniani, T. *J. Sleep Res.* **2000**, *9*, 76.
50. Schwartz, W. J.; Zimmerman, P. *J. Neurosci.* **1990**, *10*, 3685–3694.
51. Armstrong, S. M.; McNulty, O. M.; Guardiola-Lemaitre, B.; Redman, J. R. *Pharmacol. Biochem. Behav.* **1993**, *46*, 45–49.
52. Radulovacki, M.; Pavlovic, S.; Carley, D. W. *Sleep* **2004**, *27*, 383–387.
53. Philip, P.; Gross, C. E.; Taillard, J.; Bioulac, B.; Guilleminault, C. *Neurobiol. Dis.* **2005**, *20*, 428–431.
54. Van Gelder, R. N.; Edgar, D. M.; Dement, W. C. *Sleep* **1991**, *14*, 48–55.
55. Edgar, D. M. *Methods Find. Exp. Clin. Pharmacol.* **2002**, *24*, 71–72.
56. Wisor, J. P.; Edgar, D. M.; Yesavage, J.; Ryan, H. S.; McCormick, C. M.; Lapustea, N.; Murphy, G. M., Jr. *Neuroscience* **2005**, *131*, 375–385.
57. Wisor, J. P.; Wurts, S. W.; Hall, F. S.; Lesch, K. P.; Murphy, D. L.; Uhl, G. R.; Edgar, D. M. *Neuroreport* **2003**, *14*, 233–238.
58. Devine, E. B.; Hakim, Z.; Green, J. *Pharmacoeconomics* **2005**, *23*, 889–912.
59. Arand, D.; Bonnet, M.; Hurwitz, T.; Mitler, M.; Rosa, R.; Sangal, R. B. *Sleep* **2005**, *28*, 123–144.
60. Zhdanova, I. V.; Wurtman, R. J.; Regan, M. M.; Taylor, J. A.; Shi, J. P.; Leclair, O. U. *J. Clin. Endocrinol. Metab.* **2001**, *86*, 4727–4730.
61. Kunz, D.; Mahlberg, R.; Muller, C.; Tilmann, A.; Bes, F. *J. Clin. Endocrinol. Metab.* **2004**, *89*, 128–134.
62. Auger, R. R.; Goodman, S. H.; Silber, M. H.; Krahn, L. E.; Pankratz, S.; Slocumb, N. L. *Sleep* **2005**, *28*, 667–672.
63. Mignot, E. *Sleep Med. Rev.* **2004**, *8*, 333–338.
64. Nishino, S.; Mignot, E. Wake Promoting Medications: Basic Mechanisms and Pharmacology. In *Principles and Practice of Sleep Medicine*; Kryger, M. H., Roth, T., Dement, C., Eds.; Elsevier Saunders: Philadelphia, PA, 2005, pp 468–483.
65. Boutrel, B.; Koob, G. F. *Sleep* **2004**, *27*, 1181–1194.
66. Wisor, J. P.; Nishino, S.; Sora, I.; Uhl, G. H.; Mignot, E.; Edgar, D. M. *J. Neurosci.* **2001**, *21*, 1787–1794.
67. Mignot, E.; Nishino, S.; Guilleminault, C.; Dement, W. C. *Sleep* **1994**, *17*, 436–437.
68. Duteil, J.; Rambert, F. A.; Pessonnier, J.; Hermant, J. F.; Gombert, R.; Assous, E. *Eur. J. Pharmacol.* **1990**, *180*, 49–58.
69. Scammell, T. E.; Estabrooke, I. V.; McCarthy, M. T.; Chemelli, R. M.; Yanagisawa, M.; Miller, M. S.; Saper, C. B. *J. Neurosci.* **2000**, *20*, 8620–8628.
70. Gallopin, T.; Luppi, P. H.; Rambert, F. A.; Frydman, A.; Fort, P. *Sleep* **2004**, *27*, 19–25.
71. Edgar, D. M.; Seidel, W. F. *J. Pharmacol. Exp. Ther.* **1997**, *283*, 757–769.
72. Wisor, J. P.; Dement, W. C.; Aimone, L.; Williams, M.; Bozyczko-Coyne, D. *Sleep* **2006**, submitted for publication.
73. Arnulf, I.; Mignot, E. *Sleep* **2004**, *27*, 1242–1243.

74. Thomson. http://www.IDdb3.com (accessed June 2006).
75. Celanire, S.; Wijtmans, M.; Talaga, P.; Leurs, R.; de Esch, I. J. P. *Drug Disc. Today* **2005**, *10*, 1613–1627.
76. Wagner, J.; Wagner, M. L. *Sleep Med. Rev.* **2000**, *4*, 551–581.
77. Terzano, M. G.; Rossi, M.; Palomba, V.; Smerieri, A.; Parrino, L. *Drug Saf.* **2003**, *26*, 261–282.
78. Rosenberg, R.; Caron, J.; Roth, T.; Amato, D. *Sleep Med.* **2005**, *6*, 15–22.
79. Walsh, J. K.; Roehrs, T.; Roth, T. Pharmacologic Treatment of Primary Insomnia. In *Principles and Practice of Sleep Medicine*; Kryger, M. H., Roth, T., Dement, C., Eds.; Elsevier Saunders: Philadelphia, PA, 2005, pp 749–760.
80. Uchikawa, O.; Fukatsu, K.; Tokunoh, R.; Kawada, M.; Matsumoto, K.; Imai, Y.; Hinuma, S.; Kato, K.; Nishikawa, H.; Hirai, K. et al. *J. Med. Chem.* **2002**, *45*, 4222–4239.
81. Roth, T.; Stubbs, C.; Walsh, J. K. *Sleep* **2005**, *28*, 303–307.
82. Huckle, R. *Curr. Opin. Investig. Drugs* **2004**, *5*, 766–773.
83. Holzinger, E. *IDrugs* **2005**, *8*, 410–415.
84. Mathias, S.; Wetter, T. C.; Steiger, A.; Lancel, M. *Neurobiol. Aging* **2001**, *22*, 247–253.
85. Bartoszyk, G. D.; van Amsterdam, C.; Bottcher, H.; Seyfried, C. A. *Eur. J. Pharmacol.* **2003**, *473*, 229–230.
86. McGinty, D.; Szmusiak, R. Sleep-Promoting Mechanisms in Mammals. In *Principles and Practice of Sleep Medicine*; Kryger, M. H., Roth, T., Dement, C., Eds.; Elsevier Saunders: Philadelphia, PA, 2005, pp 169–184.
87. Czeisler, C. A.; Walsh, J. K.; Roth, T.; Hughes, R. J.; Wright, K. P.; Kingsbury, L.; Arora, S.; Schwartz, J. R.; Niebler, G. E.; Dinges, D. F. *N. Engl. J. Med.* **2005**, *353*, 476–486.
88. Morris, H. H., III; Estes, M. L. *J. Am. Med. Assoc.* **1987**, *258*, 945–946.
89. Petrie, K.; Conaglen, J. V.; Thompson, L.; Chamberlain, K. *Br. Med. J.* **1989**, *298*, 705–707.
90. Engleman, H. M.; Wild, M. R. *Sleep Med. Rev.* **2003**, 7, 81–99.
91. Hanzel, D. A.; Proia, N. G.; Hudgel, D. W. *Chest* **1991**, *100*, 416–421.
92. Carley, D. W.; Radulovacki, M. *Am. J. Respir. Crit. Care Med.* **1999**, *160*, 1824–1829.
93. Carley, D. W.; Radulovacki, M. *Exp. Neurol.* **1999**, *159*, 545–550.
94. Weil, J. V. Respiratory Physiology: Sleep at High Altitudes. In *Principles and Practice of Sleep Medicine*; Kryger, M. H., Roth, T., Dement, C., Eds.; Elsevier Saunders: Philadelphia, PA, 2005, pp 245–255.
95. Selbach, O.; Eriksson, K. S.; Haas, H. L. *Drug News Perspect.* **2003**, *16*, 669–681.
96. Huitron-Resendiz, S.; Kristensen, M. P.; Sanchez-Alavez, M.; Clark, S. D.; Grupke, S. L.; Tyler, C.; Suzuki, C.; Nothacker, H. P.; Civelli, O.; Criado, J. R. et al. *J. Neurosci.* **2005**, *25*, 5465–5474.
97. Schwartz, J. R.; Hirshkowitz, M.; Erman, M. K.; Schmidt-Nowara, W. *Chest* **2003**, *124*, 2192–2199.
98. Huang, Z.-L.; Qu, W.-M.; Eguchi, N.; Chen, J.-F.; Schwarzschild, M. A.; Fredholm, B. B.; Urade, Y.; Hayaishi, O. *Nat. Neurosci.* **2005**, *8*, 858–859.

Biographies

Edward R Bacon obtained his PhD degree in organic chemistry in 1980 from the University of Washington (Seattle, WA) under the direction of Prof Niels H Andersen working in the area of sesquiterpene total synthesis. He then moved to the department of Medicinal Chemistry at Sterling Winthrop Research Institute where he worked in a variety of areas including pain, depression, and predominantly in the cardiovascular arena, focusing on congestive heart failure. During this period, he also conducted research in the area of diagnostic contrast media for x-ray/computed tomography (CT) and magnetic resonance imaging (MRI) utilizing nanoparticle drug delivery technology. In 1994 he moved to Nycomed-Amersham where he continued contrast media research including synthesis of targeted radioconjugates for nuclear medicine applications and development of novel devices for cancer brachtherapy. In 2000 he moved to Cephalon, Inc. (West Chester, PA), where currently he is Senior Director, CNS Medicinal Chemistry, and has been engaged in efforts to discover novel agents for the treatment of sleep disorders, neurodegenerative diseases, and oncology.

Sankar Chatterjee obtained his PhD degree in organic chemistry from the Pennsylvania State University (University Park) under the supervision of Prof Maurice Shamma in 1984 working on total synthesis of natural products and study of reaction mechanisms. He then did his postdoctoral research with Prof Steven M Weinreb at the same university in the area of total synthesis of natural product and with Prof Franklin A Davis at Drexel University (Philadelphia, PA) in the area of development of novel synthetic reagents, respectively. In 1987, he joined Franklin Research Center (Philadelphia, PA) to work in the area of parasitic diseases. In 1990, he joined Cephalon, Inc. (West Chester, PA), where he has been involved in the discovery of novel medicinal agents in the areas of neuroscience (wake and cognition promoting agents, stroke, and Alzheimer's disease) and oncology. Currently he holds the title of Associate Director, CNS Medicinal Chemistry.

Michael Williams received his PhD (1974) from the Institute of Psychiatry and his Doctor of Science degree in pharmacology (1987), both from the University of London. Dr Williams has worked in the US-based pharmaceutical industry for 30 years at Merck, Sharp and Dohme Research Laboratories, Nova Pharmaceutical, CIBA-Geigy, and Abbott Laboratories. He retired from the latter in 2000 and after serving as a consultant with various biotechnology/ pharmaceutical companies in the USA and Europe, joined Cephalon, Inc. in West Chester, PA, in 2003 where he is Vice President of Worldwide Discovery Research. He has published some 300 articles, book chapters, and reviews and is Adjunct Professor in the Department of Molecular Pharmacology and Biological Chemistry at the Feinberg School of Medicine, Northwestern University, Chicago, IL.

Comprehensive Medicinal Chemistry II
ISBN (set): 0-08-044513-6

ISBN (Volume 6) 0-08-044519-5; pp. 139–167

6.07 Addiction

A H Newman and R B Rothman, National Institutes of Health, Baltimore, MD, USA

Published by Elsevier Ltd.

6.07.1 Disease State

Drug and alcohol addictions affect more than 30 million people in the USA and Europe.[1] The overall cost of drug abuse and addiction in the USA in 1998 was estimated to be greater than $143 billion; estimates include costs associated with loss of productivity, healthcare costs, and other related expenses including those related to the criminal justice system. Based on these data the projected overall cost of addiction for 2006 may exceed $180 billion.

Addictive behaviors are also associated with gambling, shopping, sexual behavior, and eating, although terming these behaviors 'addictions' remains controversial. Hence these will not be discussed further in this chapter but rather the interested reader is referred to relevant reviews.[2–4] Nevertheless, the biochemical bases of all addictive disorders appear to have at least some overlap, as do the hypotheses of addiction. Indeed, the questions asked, and the discoveries made, in drug addiction research may be applied toward a better understanding of these other addictive disorders and may be implemented in treatment strategies to prevent secondary medical consequences of these behaviors, such as obesity in overeating disorder.[4,5] A list of terms and definitions used in this chapter can be found in **Table 1**.[6]

The definition of addiction has permutated over time, and hypotheses related to manifestation are described in Section 6.07.2. However, one guide for clinicians is the DSM-IV Criteria of Addictions (**Table 2**). As noted, the

Table 1 Definitions of terms used in drug addiction

Craving (formerly called psychological dependence)	An intense desire to reexperience the effects of a psychoactive substance. Craving is a cause of relapse after long periods of abstinence.
Physical or physiological dependence	Physical tolerance and the withdrawal syndrome.
Priming	New exposure to a formerly abused substance. This exposure can precipitate rapid resumption of abuse at previous levels or at higher levels.
Relapse	Resumption of drug-seeking or drug-taking behavior after a period of abstinence. Priming, environmental cues (people, places, or things associated with past drug use), and stress can trigger intense craving and cause a relapse.
Reward	A stimulus that the brain interprets as intrinsically positive or as something to be attained.
Sensitization	The increase in the expected effect of a drug after repeated administration (e.g., increased locomotor activation after the administration of psychostimulants). Sensitization also refers to persistent hypersensitivity to the effect of a drug in a person with a history of exposure to that drug (or to stress). Sensitization may be one of the neurobiologic mechanisms involved in craving and relapse.
Substance abuse	Behavior characterized by recurrent and clinically significant adverse consequences related to the repeated use of substances, such as failing to fulfill major role obligations, use of drugs in situations in which it is physically hazardous, occurrence of substance-related legal problems, and continued drug use despite the presence of persistent or recurrent social or interpersonal problems.
Substance dependence	As defined by The Diagnostic and Statistical Manual of Mental Disorders-Fourth Edition (DSM-IV), a cluster of cognitive, behavioral, and physiological symptoms indicating that a person is continuing to use a substance despite having clinically significant substance-related problems. For substance dependence to be diagnosed, at least three of the following must be present: symptoms of tolerance; symptoms of withdrawal; the use of a substance in larger amounts or for longer periods than intended; persistent desire or unsuccessful attempts to reduce or control use; the spending of considerable time in efforts to obtain the substance; a reduction in important social, occupational, or recreational activities because of drug use; and continued use of a substance despite attendant health, social, or economic problems.
Withdrawal syndrome	A constellation of signs and symptoms that follows the abrupt discontinuation or reduction in the use of a substance or after blockage of the actions of a substance with antagonists (e.g., naloxone in heroin addiction). The syndrome can also be produced by cues associated with substance use (conditioned withdrawal). Symptoms tend to be the opposite of those produced after short-term exposure to a drug. Withdrawal is one of the causes of compulsive drug-taking behavior and short-term relapse.

Modified from Table 1 in Cami, J.; Farre, M. *N. Engl. J. Med.* **2003**, *349*, 975–986.

Table 2 DSM-IV definition of substance dependence (addiction)

A maladaptive pattern of substance use, leading to clinically significant impairment or distress, as manifested by three (or more) of the following, occurring at any time in the same 12-month period:

1. Tolerance, as defined by either of the following:
   (a) A need for markedly increased amounts of the substance to achieve intoxication or desired effect.
   (b) Significantly diminished effect with continued use of the same amount of the substance.
2. Withdrawal, as manifested by either of the following:
   (a) The characteristic withdrawal syndrome for the substance.
   (b) The same (or a closely related) substance is taken to relieve or avoid withdrawal symptoms.
3. The substance is often taken in larger amounts or over a longer period than was intended.
4. There is a persistent desire or unsuccessful efforts to cut down or control substance use.
5. A great deal of time is spent in activities necessary to obtain the substance, use the substance, or recover from its effects.
6. Important social, occupational, or recreational activities are neglected or reduced because of substance use.
7. The substance use is continued despite knowledge of having a persistent or recurrent physical or psychological problem that is likely to have been caused or exacerbated by the substance.

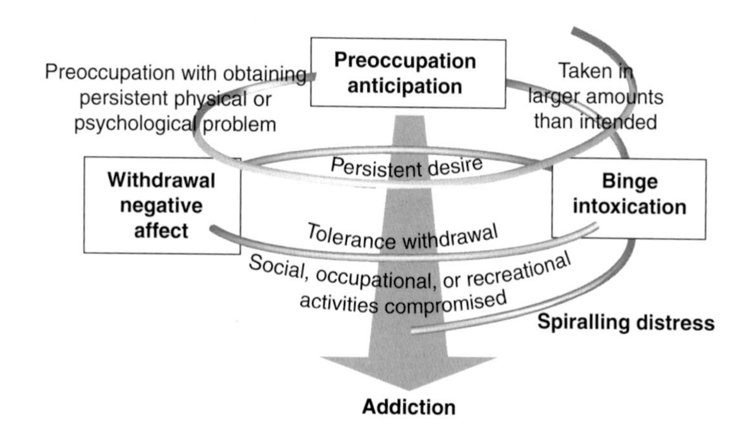

Figure 1 The addiction cycle. (Reprinted from Koob, G. F.; Le Moal, M. *Neuropsychopharmacology* **2001**, *24*, 97–129, with permission from Elsevier.)

occurrence of both tolerance and/or dependence are not described as being crucial to the development of addiction, although they are two of several criteria that are regarded as components of the disorder. Rather, this definition emphasizes the loss of control that develops, such that a person will often ingest larger quantities of drug then intended, persistently seek out the drug, be unsuccessful in reducing drug use, and continue using drugs despite adverse physical, social, occupational, and psychological consequences. In other words, drug addicts can exhibit all these behaviors and not demonstrate measurable tolerance or withdrawal symptoms. **Figure 1** illustrates the vicious cycle of addiction, where intoxication leads to withdrawal and negative affective states, which then lead to greater drug use in an attempt to ameliorate the spiraling psychological, and in some cases, physical distress.

Addiction to drugs does not typically develop following a single episode of drug use. Rather, addiction develops over a period of time, evolving from sporadic or intermittent use to regular use, and finally, in vulnerable individuals, to addiction.[7] Numerous factors influence vulnerability, including genetic factors (40–60%), environmental factors, such as drug availability, socioeconomic status, poor parental support, and stress, as well as comorbid mental illness.[8] Although many of the genetic factors that determine vulnerability to addictions remain to be defined, major advances have been made in understanding the brain mechanisms underlying many of the effects of addictive drugs. In particular, many studies indicate that drugs of abuse, despite the differences in their molecular targets, share in common the ability to activate the mesolimbic dopamine system. With cocaine and amphetamine, this results directly from the release of dopamine (DA) from the nerve dopaminergic terminals in the nucleus accumbens (NAc). Other drugs, such as the opioids, increase the firing rate of DA cell bodies located in the ventral tegmentum area (VTA), which project to the NAc via inhibiting GABAergic interneurons that themselves tonically exert inhibitory effects on DA cell bodies.[7] The ability of drugs of abuse to release DA in the NAc underlies the ability of these drugs to support self-administration behavior, which is one way to measure the reinforcing nature of these agents. Natural reinforcers, such as food and sex, also release DA in the NAc, but these do not generally lead to addiction perhaps because natural reinforcers produce elevations in DA that are longer in duration and substantially lower in magnitude, and more discrete in terms of anatomical distribution, than drugs of abuse.[8] Chronic treatment of rodents with drugs of abuse, because of their common effect on dopaminergic mechanisms, also produce similar long-term changes in brain function via changes in gene transcription.[9] These long-term changes in gene expression are thought to contribute to the progression of occasional drug use to addiction.

The central role of DA and the NAc in mediating the rewarding effects of drugs of abuse should not obscure the fact that nondopaminergic brain neurons and circuits contribute to the development and continuance of addictive behavior. For example, chronic stimulant use produces neuronal deficits in serotonergic function that resemble those observed in major depression,[10] supporting the hypothesis that nondopaminergic mechanisms substantially contribute to the development of addiction.[11] Moreover, Childress (and others) reports that cocaine craving, triggered by cocaine-related cues, is associated with differential activation of limbic structures of the brain.[12] Other studies implicate cocaine-associated changes in the frontal orbital cortex and the cingulate cortex as important contributors to cocaine addiction.[13] Indeed, **Figure 2** emphasizes the fact that many brain circuits besides the well-studied reward circuitry of rodents contribute to addiction in humans. For example, the loss of frontal lobe volume observed in addicts may contribute to their poor judgment and reduced impulse control.

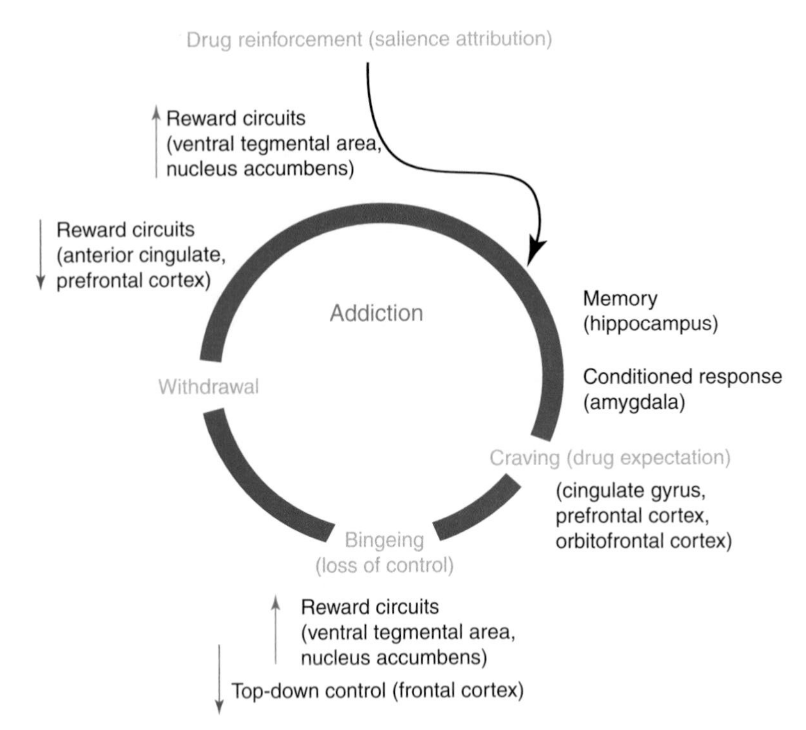

Figure 2 Integrative model of brain and behavior.[13] (Reprinted from Goldstein, R. Z.; Volkow, N. D. *Am. J. Psychiat.* **2002**, *159*, 1642–1652, with permission from the American Journal of Psychiatry (Copyright 2002) American Psychiatric Association.)

Addiction to alcohol, opioids, cocaine, and amphetamines, once considered as primarily a legal, social, and moral problem, has finally been recognized as a chronic brain disorder/disease with potentially treatable solutions.[1] The combination of pharmacotherapy and psychosocial intervention provides the basis of addiction recovery. However, a clearer understanding of biochemical bases of drug addiction and relapse is required in order to identify targeted and effective treatment strategies. The severe medical consequences of drug abuse and addiction result from neurobehavioral modifications due to chronic drug use. In addition, secondary diseases such as human immunodeficiency virus–acquired immunodeficiency syndrome (HIV–AIDS) or hepatitis B from needle sharing, chronic liver disease from alcohol abuse, and lung cancer from smoking are just a few examples of long-term health problems associated with addiction. As addiction is a compulsive behavioral disorder that is not curbed by negative consequences, the demise of family connections, and loss of employment and income ensures that an addict will continue seeking and taking drugs. Current treatment modalities are discussed in Section 6.07.5. However, the hypotheses addressing the transition from casual drug use to addiction are instructive in order to understand the biochemical processes involved in this disorder and then to further target mechanistically based pharmacotherapies. A list of the primary drugs of abuse that can lead to addiction can be found in **Table 3** and their chemical structures are depicted in **Figure 3**.

6.07.2 Biochemical Basis for Addiction and Relapse: Current Hypotheses

The development of addiction requires the transition from casual drug use to compulsive patterns of excessive drug use that lead to neuropsychological changes to produce chronic drug seeking/taking behavior. The hypothetical explanations of this transition to addiction are varied and include: (1) the traditional view that drug pleasure followed by unpleasant, anhedonic symptoms of drug abstinence/withdrawal lead to addiction; (2) addiction is due to drug-induced abnormal or aberrant learning that affects stimulus–response habits; (3) sensitization of the neural system causes compulsive motivation to take addictive drugs despite consequences; and (4) dysfunction of the cortical systems in the brain leads to impaired judgment and decision-making leading to impulsive drug taking.[14,15] For each of these hypotheses, a neurochemical explanation has been proposed, based on experimental models, neuroanatomical changes induced by chronic drug taking, and brain imaging experiments. The interesting notion that despite varying pharmacological mechanisms, all drugs of abuse may share common biochemical mechanisms underlying their addiction has been proposed and supported with convergent data that particularly focuses on the brain reward circuitry involving

Table 3 Drugs of abuse and their molecular targets

Drug	*Target*
Cocaine	Indirect dopamine agonist via inhibition of the DA transporter and promotion of DA release
Amphetamine/methamphetamine	Indirect dopamine agonists via release of DA mediated via the DA transporter
Opioids (e.g., heroin)	Activation of μ opioid receptors
Tetrahydrocannabinol (e.g., marijuana)	Activation of CB_1 and CB_2 receptors
Ethanol	Facilitates GABAergic transmission via $GABA_A$ receptors and other actions
Nicotine	Agonist at nicotinic acetylcholine receptors

Adapted from Nestler, E. J. *Nat. Rev. Neurosci.* **2001**, *2*, 119–128.

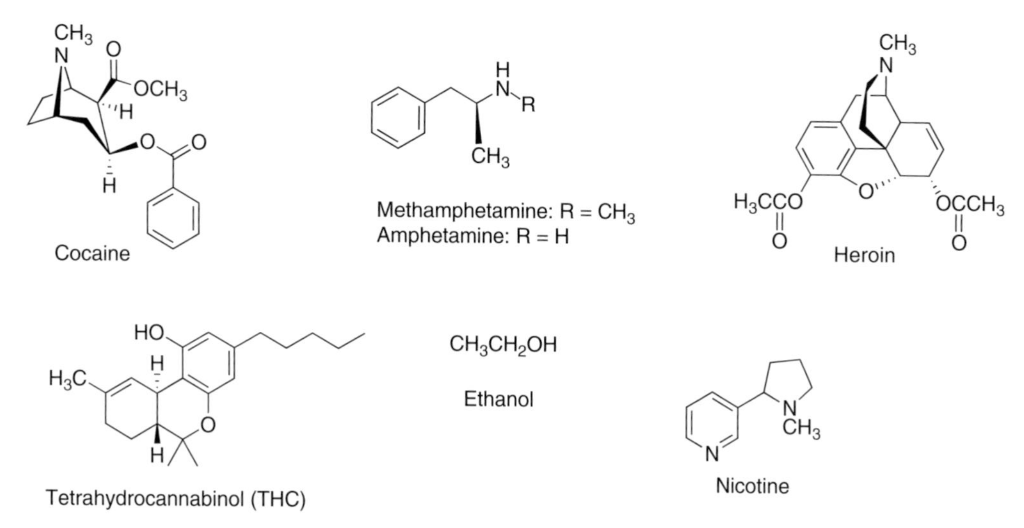

Figure 3 Chemical structures of commonly abused addictive drugs.

DA-containing neurons in the VTA that target areas in the limbic region of the brain, especially the NAc and frontal region of the cerebral cortex (**Figure 4**).[16,17] In addition, upon chronic drug use leading to addiction, the neurotransmitters involved in regulation of the VTA DA neurons are also affected. As such, glutamatergic receptors, both ionotropic and metabotropic, GABAergic, cholinergic, serotonergic, and noradrenergic input can all be altered.[18] Another factor that impacts addictive behavior in humans is drug-induced damage to cortical regions, which normally control executive functions such as decision-making and judgment.[19,20]

6.07.2.1 Opponent Motivational Process Hypothesis

The most intuitive explanation for addiction comes from the concept that drugs of abuse are initially taken for pleasure and that continued seeking of that pleasant effect leads to repeated exposure, which ultimately leads to neuroadaptations that cause tolerance, dependence, and unpleasant symptoms when the drug is absent. One prevailing view is that the mesolimbic dopaminergic system provides a 'reward circuit' which is stimulated through enhancement of intracellular DA levels in the NAc, either directly (e.g., cocaine and methamphetamine through indirect agonist actions), or indirectly through secondary stimulation of DA-release via opioidergic and cholinergic receptor agonist actions (e.g., opioids and nicotine).[21] Chronic use of these drugs leads to disruption of homeostatic levels of many neurotransmitters including but not exclusively DA, and withdrawal produces a state of dysphoria[18] leading to the cycle of addiction. Koob and colleagues have proposed that drug addiction is a result of dysregulation of the reward mechanism whereby a shift in positive reinforcement by the substance is replaced by negative reinforcement driving the behavior to

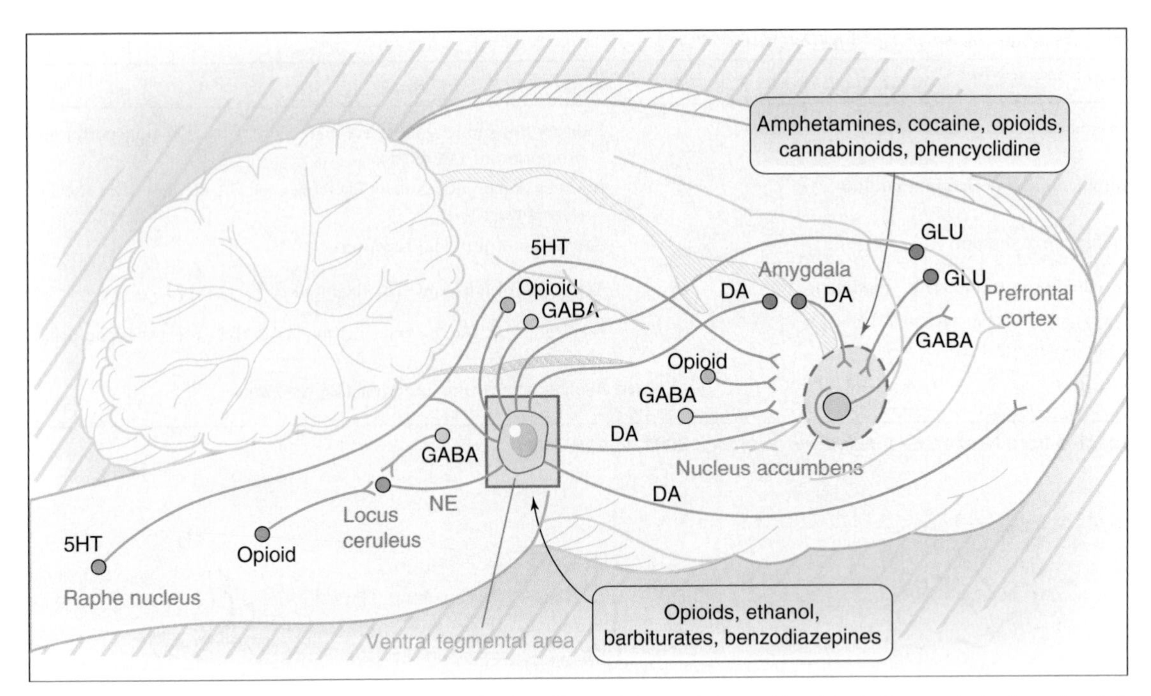

Figure 4 Neural reward circuits involved in the reinforcing effects of drugs of abuse. (Reprinted with permission from Cami, J.; Farre, M. *N. Engl. J. Med.* **2003**, *349*, 975–986, copyright © 2003 Massachusetts Medical Society. All rights reserved.)

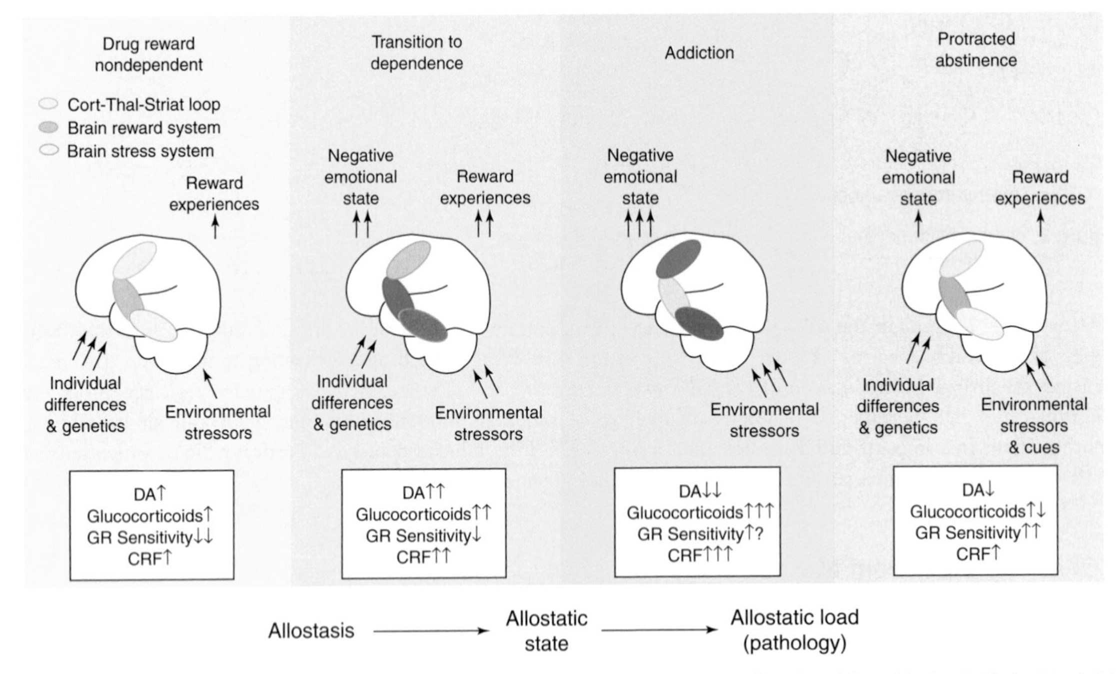

Figure 5 The transition from drug taking to addiction and protracted abstinence. (Reprinted from Koob, G. F.; Le Moal, M. *Neuropsychopharmacology* **2001**, *24*, 97–129, with permission from Elsevier.)

take drugs (**Figure 5**).[22] This is known as the 'opponent motivational process.' According to this model, the emergence of dysphoria, anxiety, and a general negative emotional state promotes the transition from drug use to addiction. The most recent concept in this hypothesis is that of 'allostasis' which is a state that is proposed to develop from chronic drug use that sets a new and higher 'hedonic' threshold and has been linked to dysregulation of molecular systems.[23]

The primary criticism of this hypothesis is that in the absence of withdrawal symptoms and despite long periods of drug abstinence, relapse still occurs.[14] Nevertheless, long-term neurochemical and cellular changes may account for relapse, despite the absence of overt symptoms.

6.07.2.2 Aberrant Learning Hypothesis

The role of learning in addiction has been explored, in part based on the fact that repeated exposure to drugs of abuse causes neuroadaptations in this reward learning circuitry that have been termed 'aberrant learning' as it differs from learning associated with natural rewards.[14] The neuroadaptations or neuroplasticity within the brain networks that involve motivation and emotion form the basis for the hypothesis that aberrant learning can lead to habit and automatic behaviors associated with addiction.[24] Critics of this hypothesis suggest that it is too simplistic to think that knowledge can, by itself, cause the behavior of drug taking that leads to addiction. On the other hand, recent data that cue-induced drug seeking behavior can be attenuated by the administration of pharmacological agents strongly suggests that this research area may provide new therapeutic agents for treating addiction.[25,26]

6.07.2.3 Incentive Sensitization Hypothesis

Repeated administration of psychostimulants produces greater increases in motor activity with each administration, a phenomenon termed locomotor sensitization. The incentive-sensitization theory of addiction and relapse, proposed by Robinson and Berridge,[14] extends this observation to psychostimulant reward, positing that sensitization of a neural system that attributes incentive salience causes compulsive motivation or 'wanting' to take addictive drugs. According to this hypothesis, as reviewed by Hyman,[27] drug-induced increases in DA do not produce drug-liking (i.e., the 'high') but rather shape behavioral responses so as to maximize the attainment of future rewards. Repeated exposure to the drug of abuse sensitizes this process, increasing the motivational significance of cues that predict the delivery of drug. Counterindications to this theory are the limited data that support its occurrence in human addicts and other data, reviewed by Di Chiara,[28] that psychomotor-induced increases in brain DA in humans correlates with the subjective experience of a 'high.'

6.07.2.4 Craving, Relapse, and the Role of Stress

The phenomenon of relapse is part of the definition of addiction and understanding its neurobiological basis has been the focus of recent investigation. Two major precipitating factors for relapse have been identified as craving and stress. These risk factors are further exacerbated by prolonged withdrawal symptoms, comorbidity of other psychiatric disorders, socioeconomic status, and perceived drug availability.[29] Understanding the long-term brain changes that serve to maintain craving and promote relapse and resumption of chronic drug use, even after years of drug abstinence, is at the forefront of addiction research. For example, signaling factors activated by brain-derived neurotrophic factor (BDNF) may be one type of cellular and molecular neuroadaptation that is associated with craving and relapse.[29,30] Differential activation of BDNF-induced signaling is suggested to contribute to the differences between behaviors associated with natural versus drug reward.[31] Another area of increasing interest is the involvement of stress on craving and relapse, which has been related to increases in central noradrenergic and corticotrophin releasing factor.[32] Thus, there is considerable interest in developing a corticotrophin releasing hormone (CRH) receptor antagonist as a potential treatment agent for drug dependence.

6.07.3 Experimental Models of Addiction

In order to identify potential medications for the treatment of drug addiction, animal models have been created that allow the elucidation of the underlying mechanisms of drug-induced behaviors. By its definition, addiction is a unique and complicated human behavior and a single animal model simply cannot predict medication efficacy in humans. Thus most investigators use an arsenal of in vitro and in vivo tests to study neurochemical mechanisms underlying the pharmacological actions and abuse liability of various drugs of abuse, as well as for discovering potential medications. In vitro binding and functional assays are generally used to determine mechanisms of action of test compounds (new compound entities, NCEs) and then those with the desired in vitro profile are further investigated in animal models. Biochemical assessment using in vivo microdialysis is often employed to further delineate NCE effects on neurotransmitter levels in various brain regions. Ultimately, behavioral models, initially in rodents, but ultimately nonhuman primates, have been developed to study drug seeking, taking, reward, reinstatement, and the molecular mechanisms associated with these behaviors that lead to addiction.

There are numerous animal models of the pharmacological actions of drugs of abuse including locomotor activity for psychostimulants and tail flick or the hotplate test for assessing analgesic effects of opioids. In addition, drug discrimination has emerged as a very important paradigm for ascertaining like- or attenuation of discriminative stimulant effects of the training drug.[33] However, in this chapter the discussion is limited to animal models that are most commonly used to assess the addictive liability of a drug or the ability of a potential medication to attenuate drug reward or relapse.

6.07.3.1 Self-Administration/Reinstatement Model

Self-administration in animals requires an operant behavior (e.g., lever press) for the intravenous injection of a drug. A drug or NCE is considered to be reinforcing when either the rate of responding on the drug-correct lever exceeds that of the control or vehicle-paired lever, or if the response rate of the drug-receiving animal is greater than that of a yoked-control animal. Once self-administration behavior is established, the experimenter can increase the number of lever presses required for a drug injection and thereby measure the reinforcing properties of the drug. When the number of lever presses required exceeds the 'work' that the animal is willing to do to receive an injection, extinction of lever-pressing/self-administration will ensue. Several schedules of reinforcement can be employed. The fixed ratio (FR) schedule requires the animal to press the lever a fixed number of times before receiving a drug injection. The fixed interval (FI) schedule is used to determine how many presses an animal will make in a fixed interval of time. The most commonly used schedule is the progressive ratio (PR) in which a progressively larger number of lever presses are required to receive the same drug injection. The highest response requirement a drug will sustain is called the 'break point' and this has been used to measure drug reinforcement properties. Furthermore, the attenuation of the break point has been used to evaluate potential medications or other conditions that might decrease the reinforcing effects of the drug.[33] Most drugs abused by humans are generally self-administered by animals and laboratory-based self-administration in humans has also been used to study drug addiction.[34]

In addition, this model can be extended to a reinstatement model of relapse.[35] In the reinstatement model, the animal is trained to self-administer the drug of abuse. Once this behavior is established, the drug injections are replaced with saline injections. In time, the animal will no longer press the lever for a drug injection and this is called 'extinction.' Several investigators have demonstrated that drug- or cue-induced reinstatement of drug taking behavior (e.g., lever pressing) can be induced by a single priming injection of the drug of abuse or a cue that had previously been paired with the drug injection, during training. In addition, aversive conditions used as a model of stress, such as foot shock, can also reinstate lever pressing behavior. This test has face validity as a model for relapse to drug use by humans and has been suggested to further model the effects of stress on drug taking behavior.[35] Further, the attenuating effects by certain medication candidates in this model have prompted intense interest in particular mechanisms for drug development, such as DA D3 and mGluR5 receptors (*see* Section 6.07.7).[36,37]

6.07.3.2 Intracranial Self-Stimulation (ICSS)

This model has also been used to investigate rewarding and addictive properties of drugs.[38] Rats will press a lever to self-stimulate due to a concomitant passage of a small electrical current into regions in the brain that involve the reward circuitry (e.g., VTA, prefrontal cortex, NAc). All addictive drugs, when acutely administered, will lower the threshold of stimulation required to maintain ICSS. This is interpreted as the ability of these drugs to enhance the stimulating effects of the ICSS. Conversely, acute withdrawal from drugs of abuse will elevate ICSS reward thresholds and is interpreted to reflect a precipitation of a deficit in brain reward function that can be alleviated by increasing ICSS. This observation suggests that ICSS may be a relevant behavioral model of the aversive emotional state associated with drug withdrawal[39] further substantiating VTA DA cells as projections that play a key role in perpetuating the addiction cycle.[22]

6.07.3.3 Conditioned Place Preference (CPP)

This paradigm is a Pavlovian conditioning procedure in which an animal learns to prefer an environment that is paired with a drug of abuse, and has been promoted as modeling the environmental cues associated with drug addiction in humans.[40] Indeed drugs of abuse are able to engender conditioned place preference, whereby the animal remains in the drug-paired side of the chamber for a longer period of time than in the other side (**Figure 6**).[6] Thus, in CPP the degree of drug reinforcement and addictive liability is inferred from the degree of preference or

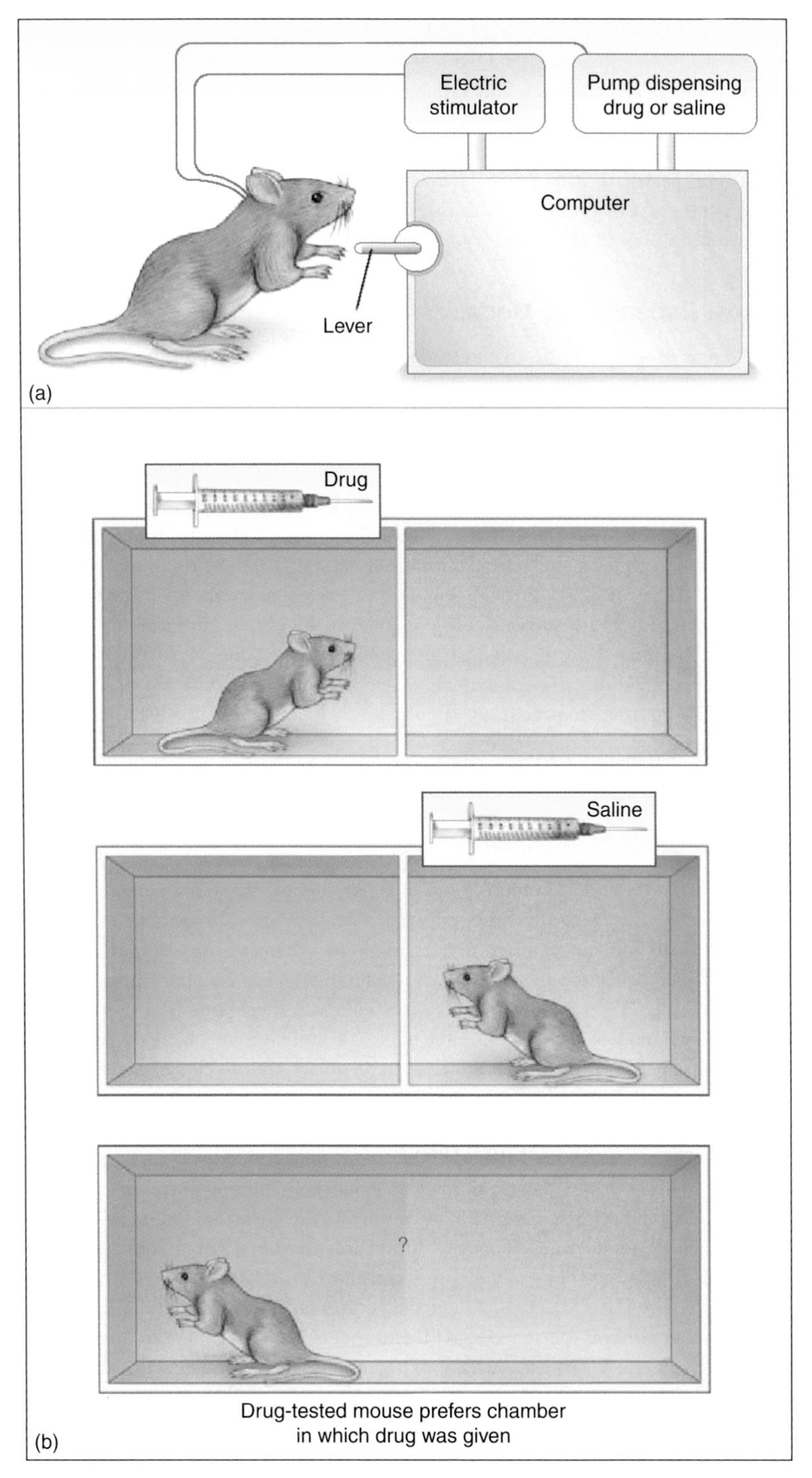

Figure 6 Conditioned place preference model in mice used to study the positive reinforcing effects of drugs. (Reprinted with permission from Cami, J.; Farre, M. *N. Engl. J. Med*. **2003**, *349*, 975–986, copyright © 2003 Massachusetts Medical Society. All rights reserved.)

time spent in the drug-paired environment, as compared to the vehicle-paired environment. Conversely, aversion to a drug or condition can be also modeled in this paradigm, and, depending on dose, certain drugs can cause an animal to stay in the nondrug-paired side of the chamber. Motivational (affective) symptoms of drug withdrawal can be measured using this paradigm, even in the absence of overt symptoms of withdrawal. Indeed, studies using CPP for studying nicotine dependence have demonstrated that the reinforcing effects of nicotine can be controlled by environmental stimuli. This can be correlated to the high rates of relapse to cigarette smoking in humans, which in part may be due to the pairing of environments with the reinforcing effects of nicotine, such as home and workplace that are impossible to avoid.[41]

6.07.3.4 Physical Dependence Models

As described above, chronic drug use leads to neuroadaptations. Cessation of drug use can lead to biochemical, physiological, and behavioral changes that together constitute a withdrawal or abstinence syndrome. This is also described as a state of physical dependence. When an addict discontinues drug use, symptoms of withdrawal can range from dysphoria and anxiety, to paranoia, physical illness, seizures, and death. The withdrawal symptoms depend on the drug of abuse, its pharmacology, pharmacokinetics, and the nature of the adaptations that have occurred due to chronic drug intake. Physical dependence is especially prominent with opioids, barbiturates, and alcohol. For these drugs, withdrawal symptoms are highly aversive, dangerous, and at times, life-threatening, and are believed to play a major role in relapse.[33] In animal studies, subjects are chronically dosed, often continuously via intragastric infusion for a prolonged period of time (e.g., 1 month). Behavioral and observational assessments during this phase of the testing provide documentation of physical and behavioral changes induced by chronic drug taking. Withdrawal can then be assessed either by abruptly stopping the drug administration or by precipitating withdrawal via administering a known antagonist (e.g., naltrexone for opioid-dependent animals). Comparisons of physical and behavioral assessments during this period to before and during drug treatment are then made.[33] These tests provide the means to determine the potential of a drug to produce physical dependence, as well as the means to test the ability of NCEs to interrupt the development of physical dependence or to ameliorate the symptoms and signs of withdrawal.

6.07.3.5 Genetic Models of Addiction

Different strains of rats behave differently to various drugs of abuse, and this has been demonstrated, for example, using CPP.[40] More recently the use of genetically mutated strains or transgenic mice that have specific receptors or transporters deleted has provided another approach to determining mechanistic correlates to drug abuse and addiction. Successful manipulation of gene expression by homologous recombination has enabled researchers to knockout specific receptors or transporters (e.g., DAT knockout) to ultimately compare these genetically mutated mice to wild-type littermates in various behavioral models of drug abuse and addiction.[42] These knockout (ko) mice can provide valuable information, especially when selective drugs are not available to pharmacologically 'knockout' the receptor or transporter. One of the seminal studies using DAT knockout mice in cocaine abuse research demonstrated that these animals were extremely hyperactive and did not demonstrate further increases in locomotor activity when cocaine or amphetamine was administered.[43] Subsequently work showed that DAT knockout mice indeed self-administer cocaine, and that only the dual DAT/SERT (serotonin transporter) knockout mice no longer self-administer cocaine, suggesting that additional mechanisms may be involved in the reinforcing effects of cocaine.[44] Nevertheless, strain, as well as compensatory and developmental differences, in these ko mice requires caution in interpretation of these results in the absence of substantive pharmacological data. Other transgenic mouse strains have also been created for drug dependence research, including, but not limited to, DA D1, D2, D3, D4, and D5 receptors, as well as $5HT_{1B}$, and nicotine-acetylcholine receptor kos. The production of subtype-specific DA, 5HT, and acetylcholine receptor ko mice as well as dual kos (e.g., DAT/SERT) have provided genetic tools with which to further delineate molecular mechanisms underlying the addictive properties of drugs. When conditional mutation of these genes is achieved concerns about compensatory and development modification that occur upon gene deletion procedures will be reduced.[42]

6.07.4 Clinical Trial Challenges

Drug dependence is a chronic relapsing medical illness similar in many respects to other chronic medical diseases such as diabetes and hypertension.[45] Clinical trials in the drug dependence arena face special challenges. These include high cost, difficulties recruiting patients, the high rate of comorbidities, high dropout rates, poor patient compliance, ambiguous and difficult-to-measure clinical trial endpoints, and a lack of validated biomarkers.[46]

The high level of comorbidity between psychiatric illness, e.g., depression and bipolar disorder, and drug dependence not only complicates the task of recruiting homogeneous patient populations for clinical trials, but also suggests that any medication that might be found to have some efficacy in treating an 'uncomplicated' group of patients would not work in a more typical community sample of addiction patients with a higher level of psychiatric comorbidity. These considerations indicate the importance of analyzing data from clinical trials to determine the efficacy of the medication on subgroups of patients defined on the basis of their psychiatric history.[47] A significant factor that can complicate the design of clinical trials for pharmacotherapeutic agents for cocaine addiction is the fact that chronic cocaine use alters many aspects of brain neurochemistry and circuitry. Thus, testing a medication that works via a single well-defined mechanism of action might be doomed to failure, since a single mechanism agent would 'normalize' just one of many brain mechanisms dysregulated by chronic cocaine. Viewed from this perspective, perhaps it is not surprising that most controlled trials of medications for cocaine addiction have failed to demonstrate efficacy.[48] One way of addressing this problem is to conduct clinical trials with two (or more) medications that act via different mechanisms, or alternatively, as suggested for developing medications for schizophrenia and mood disorders, the ideal medications for substance dependence disorder may well be "selectively nonselective drugs."[49] Finally, given the intrinsic difficulties for developing pharmacotherapeutics for substance abuse disorders, some argue that, rather than defining a successful medication as one that 'cures' drug addiction, which seems unlikely due to its chronic and relapsing nature, a medication that reduces drug use or mitigates the harm produced by a drug of abuse,[50] would also be a useful pharmacotherapeutic.

A significant factor that complicates medication development efforts in the drug abuse field is that the most successful therapeutics to date apply the principle of agonist substitution therapy: e.g., nicotine, delivered by various routes, and methadone, a μ opioid receptor agonist, for heroin addiction. Despite being clinical practice since the 1960s,[7] the use of a potential drug of abuse to treat an addiction to a similar substance remains troubling to many. Not surprisingly, the possible use of stimulants, such as D-amphetamine, to treat stimulant dependence, although promising, is controversial.[51] Recent research shows, however, that it is possible to design medications that act in a manner similar to amphetamine or cocaine but which lack their stimulant and reinforcing properties.[52,53]

Decades of research led to the development of US Food and Drug Administration (FDA) approved medications for nicotine, alcohol, and heroin dependence (**Figure 7**). For a variety of reasons, although progress is being made, the effort to develop medications for cocaine and methamphetamine dependence has been slow.

Another significant factor that impacts the development of medications for cocaine dependence, and to a lesser extent for alcohol dependence, is that large pharmaceutical companies do not invest resources into this area. As described by Gorelick *et al.*,[54] market factors that disinterest the large pharmaceutical companies in this endeavor include: the small and uncertain market for cocaine dependence medication, a substance abuse treatment system with limited physician involvement and antimedication attitudes among nonphysician clinicians, poor insurance coverage, and the expectation of low prices by the government entities that fund most substance-abuse treatment programs. Thus, the development of medications for substance dependence indications, and addiction research in general, falls to the public sector.

6.07.5 Current and Targeted Medications

A current list of medications approved for the treatment of drug addiction is shown in **Table 4**. A brief description of these medications for treatment of specific drugs of abuse is described below. In all cases, new medications are under clinical and preclinical investigation. Examples of potential medications currently under clinical investigation are listed in **Table 5** with their chemical structures being shown in **Figure 8**. As the mechanisms of actions of these agents are widely varied, no attempt to describe structure–activity relationships (SARs) within these classes of molecules has been made. However, the interested reader is referred to relevant review articles throughout the text that describe in more detail the drug design and synthesis of these agents.

6.07.5.1 Alcohol Addiction

The basic treatments available for alcohol dependence and alcohol withdrawal are well known and the reader is referred to textbooks or review articles on the subject for an in-depth presentation.[55,56] Pharmacotherapy of alcoholism works best when used in conjunction with specific relapse prevention counseling methods and attendance at self-help groups such as Alcoholics Anonymous (AA). For over 40 years, the aversive agent disulfiram was the only FDA approved medication for the treatment of alcoholism. Interestingly, a large double-blind placebo-controlled clinical trial conducted by the Veterans Administration showed that disulfiram was no better than placebo in helping patients

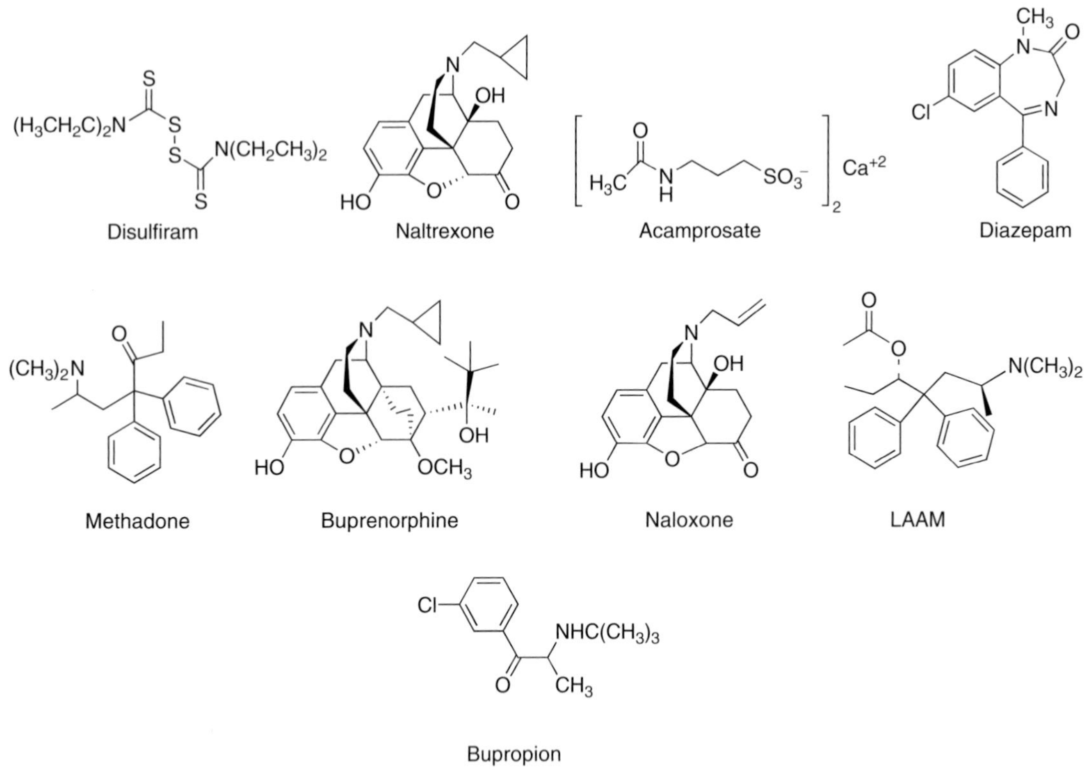

Disulfiram Naltrexone Acamprosate Diazepam

Methadone Buprenorphine Naloxone LAAM

Bupropion

Figure 7 Current FDA approved medications for treatment of drug addiction.

Table 4 Medications approved by the FDA for substance abuse disorders

Indication	*Medication*	*Mechanism of action and/or target*
Alcohol	Disulfiram (Antabuse)	Disulfiram interferes with alcohol metabolism producing unpleasant side effects due to the accumulation of acetaldehyde
	Naltrexone (Revia)	Relapse prevention. Naltrexone is a μ opioid antagonist
	Acamprosate	Relapse prevention; the mechanism of action remains to be determined, but may involve blockade of NMDA receptors and actions
	Diazepam or lorazepam	Facilitate GABAergic transmission; Management of alcohol withdrawal
Heroin	Methadone	Long-acting μ opioid agonist
	Buprenorphine Buprenorphine + naloxone	Long-acting μ opioid agonist Naloxone, a narcotic antagonist, is inactive after oral administration but will prevent the effects of buprenorphine if it were injected intravenously
	Levo-α-acetyl-methadol (LAAM)	Long-acting μ opioid agonist
	Naltrexone (Revia, Trexan)	μ opioid antagonist
Nicotine	Nicotine delivered via a transdermal patch, gum, lozenge, oral inhaler or intranasal spray	Agonist at the nicotinic cholinergic receptor
	Sustained release bupropion	Antidepressant with actions on the dopaminergic and noradrenergic systems
Cocaine and methamphetamine	None	

Table 5 Medications that have shown promising results in double-blind placebo-controlled efficacy studies

Drug	*Presumed mechanism*
Disulfiram	Under investigation
Baclofen	$GABA_B$ agonist
Tiagabine	GABA uptake blocker
Topiramate	Unknown mechanism; GABA or glutamate modulator?
Modafinil	Unknown mechanism; glutamate or DA release?
Anti-cocaine vaccine	Immunotherapy

For additional information see [54,76,93,94] and references therein.

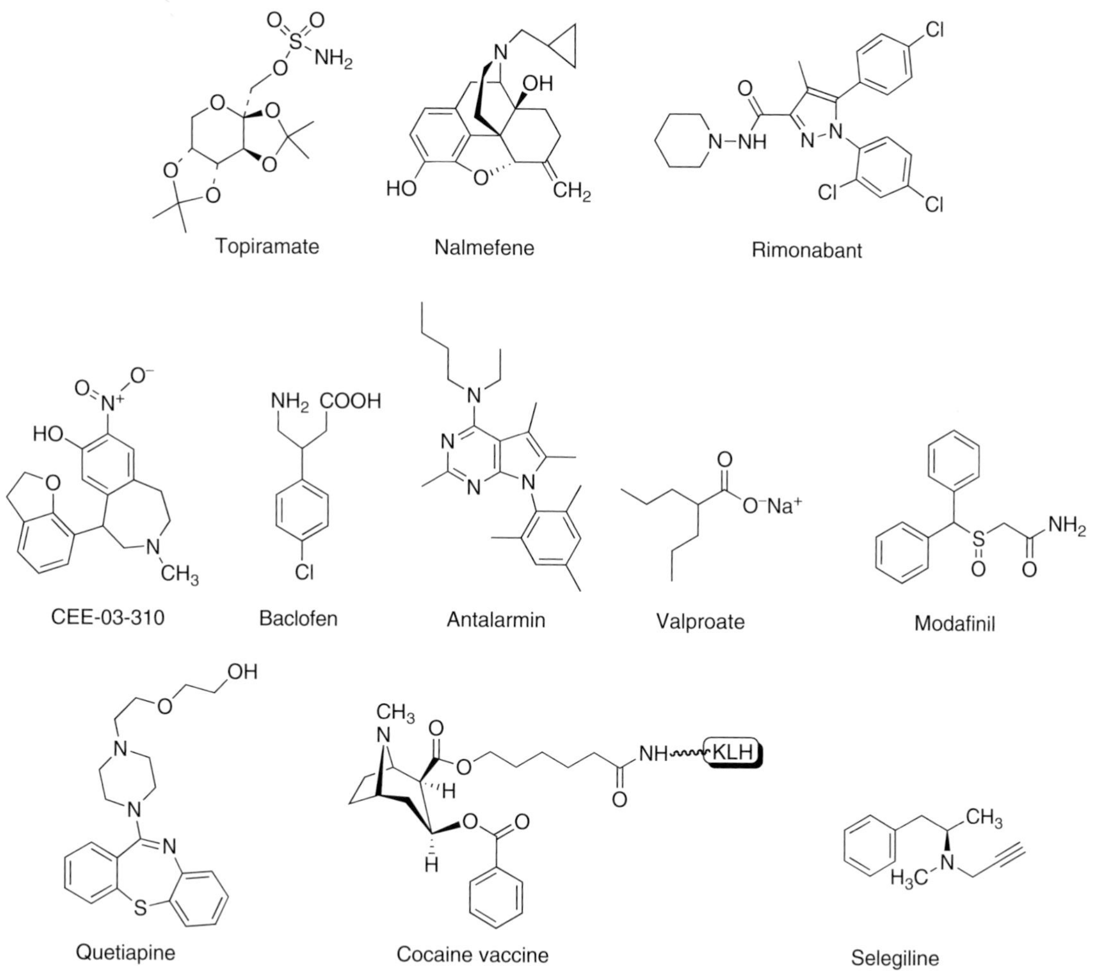

Topiramate Nalmefene Rimonabant

CEE-03-310 Baclofen Antalarmin Valproate Modafinil

Quetiapine Cocaine vaccine Selegiline

Figure 8 Chemical structures of compounds under clinical investigation as potential medications for the treatment of drug addiction.

achieve abstinence or delaying relapse,[55] although this negative result might have been due to poor compliance.[57] This drug interferes with alcohol metabolism by inhibiting alcohol dehydrogenase, leading to an accumulation of acetaldehyde, making the patient feel quite ill. According to the drug monograph for disulfiram:

> Symptoms of the disulfiram–alcohol reaction include flushing, throbbing in the head and neck, throbbing headache, dypsnea, nausea, copious vomiting, sweating, thirst, chest pain, palpitation, hyperventilation, tachycardia,

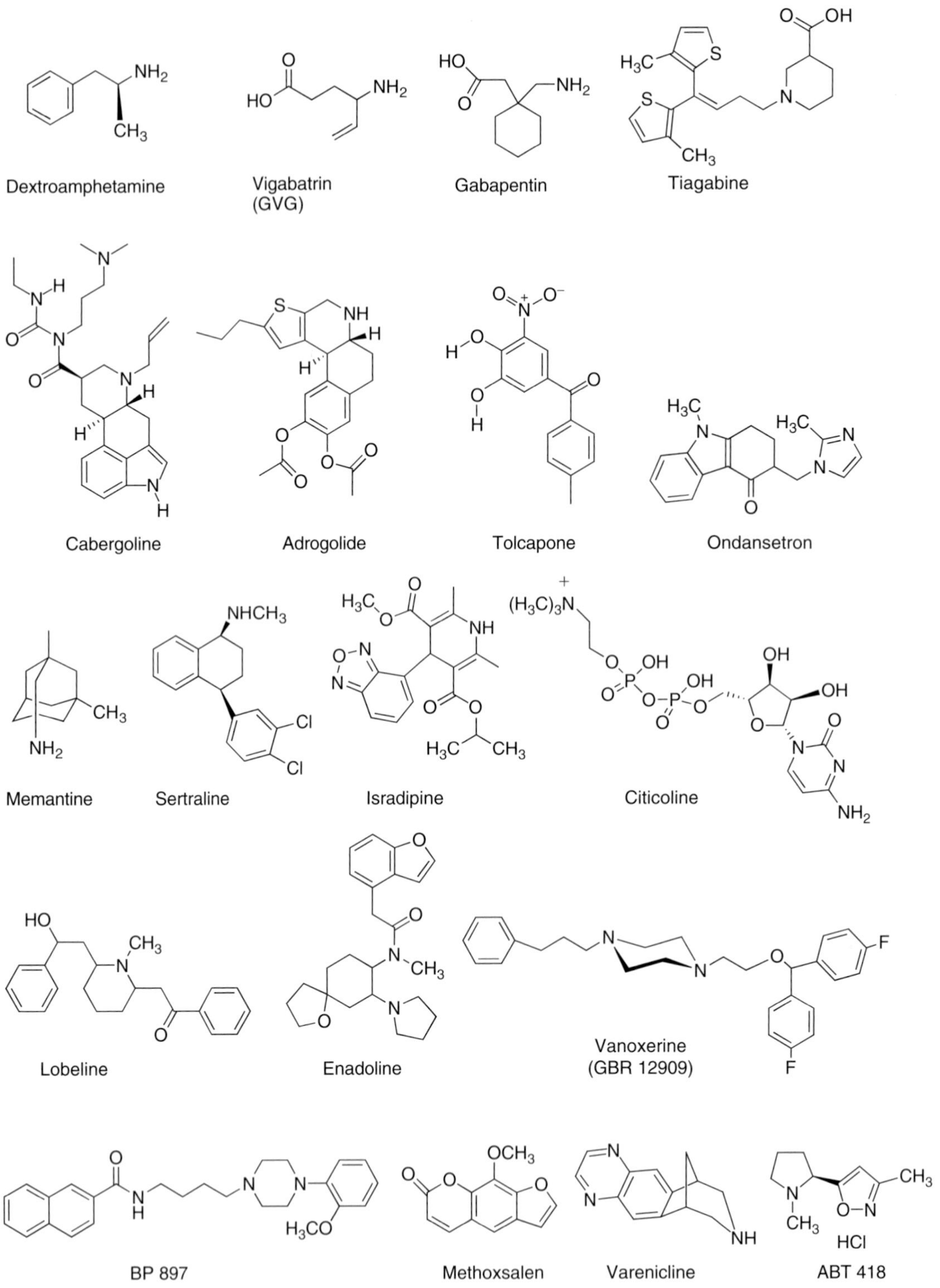

Dextroamphetamine

Vigabatrin (GVG)

Gabapentin

Tiagabine

Cabergoline

Adrogolide

Tolcapone

Ondansetron

Memantine

Sertraline

Isradipine

Citicoline

Lobeline

Enadoline

Vanoxerine (GBR 12909)

BP 897

Methoxsalen

Varenicline

ABT 418

Figure 8 Continued

hypotension, syncope, anxiety, weakness, vertigo, blurred vision, and confusion. Mild reactions are usually followed by sound sleep and complete recovery. However, symptoms may progress to respiratory depression, cardiovascular collapse, arrhythmias, myocardial infarction, acute congestive heart failure, unconsciousness, seizures, and death.

Disulfiram is a medication with a significant risk for serious toxicity, both in the absence and presence of alcohol, and its usefulness is also limited by the high degree of patient compliance that is required.

In recent years, other medications for alcohol dependence have been approved by the FDA. Naltrexone is a μ opioid receptor antagonist that partially attenuates the rewarding effects of alcohol. Initial clinical trials indicated that naltrexone prevents relapse to heavy drinking, and reduces alcohol craving, days of drinking per week, and the relapse rate among those who drank. Thus, naltrexone reduces, but does not generally eliminate, drinking. The fact that several clinical trials did not report positive findings and that this medications is not widely used by clinicians may reflect its modest effectiveness.[55]

Acamprosate (calcium acetylhomotaurinate) was recently approved by the FDA for the maintenance of abstinence from alcohol in patients with alcohol dependence that are abstinent at treatment initiation. The mechanism of action is still under investigation. However, some evidence indicates that acamprosate acts to normalize the glutamate system, and possibly the GABA system as well, whose normal function is disrupted by chronic alcoholism. Treatment with acamprosate is moderately effective, helping to significantly reduce relapse.[55] A number of other medications are under investigation for alcohol dependence (see **Table 6**).

6.07.5.2 Nicotine Addiction

Nicotine replacement (patches, gum) and sustained release bupropion are the only FDA-approved treatments for nicotine dependence. Nicotine, administered in a manner to produce steady and sustained nicotine blood levels, helps decrease withdrawal symptoms and doubles the initial success rate for quitting smoking. Unfortunately, a substantial number of patients ultimately relapse,[58] illustrating the chronic relapsing nature of nicotine dependence and other addictive disorders. FDA-approved formulations of nicotine include: the transdermal nicotine patch, and the nicotine polacrilex gum and lozenge that are available over the counter. The nicotine nasal spray and inhaler are available via prescription. The rationale and use of nicotine replacement therapy is reviewed by Henningfield.[58]

Bupropion is an antidepressant that also helps to reduce nicotine withdrawal symptoms. The usual course of treatment is for 6 months, resulting in about a 30% quit rate at the 1-year follow-up. Bupropion and its metabolite hydroxybupropion[59] are active as inhibitors of DA and NE uptake, as well as being weak antagonists of the $\alpha_4\beta_2$ neuronal nicotinic receptor. These actions probably explain its ability to help patients stop smoking. Other medications that are considered effective for smoking cessation include the α_2 agonist clonidine, and the tricyclic antidepressant nortriptyline.[55] The relatively low success rate of current treatments for nicotine dependence illustrates the need for additional pharmacological adjuncts to smoking cessation treatment programs. **Table 6** provides a list of some additional medications that are currently under investigation for this purpose.

6.07.5.3 Heroin (Opioid) Addiction

The mainstay treatment for opioid dependence is agonist substitution therapy with long-duration opioid agonists such as methadone, levomethadyl acetate (LAAM), or buprenorphine. Without methadone treatment, most patients relapse to drug use again, supporting the hypothesis that heroin addicts suffer from a dysregulated endorphin system, which is normalized by treatment with a μ receptor agonist, and that long-term treatment is necessary. Methadone treatment reduces criminal behavior and diseases such as hepatitis that occur as a result of intravenous drug use.[7,60]

LAAM is a long-acting methadone analog that can be dosed at 48- or 72-h intervals, allowing take-home doses. Unfortunately, LAAM can increase the cardiac QT interval, which can cause a number of serious or fatal cardiac adverse effects, such as *torsades de pointes* and cardiac arrest. When this became known from postmarketing reports, LAAM was removed from the EU market and is considered to be a second-line medication in the USA.

Buprenorphine is a partial μ opioid receptor agonist and a potent κ receptor antagonist.[61] For take-home doses, buprenorphine is combined with naloxone. Since naloxone is not bioactive after oral administration, but is after intravenous administration, this combination serves to prevent diversion of the medication to intravenous use. When dispensed in a clinic, buprenorphine does not have to be coadministered with naloxone. Since buprenorphine is a partial μ agonist, it is less useful than methadone in heroin addicts who require higher doses of methadone ($80–150\,\mathrm{mg\,day^{-1}}$).[7]

Naltrexone has been available since the 1970s as a treatment for heroin addiction. The rationale for its use is to block the effects of heroin by virtue of its long-lasting antagonist action at μ opioid receptors. Naltrexone is quite effective in this regard; however, compliance with this treatment among heroin addicts is quite poor. The underlying

Table 6 Medications for addictive disorders under clinical investigation

Clinical indication	*Medication*	*Biological targets*
Alcohol dependence	Topiramate	GABA/glutamate
	Nalmefene	μ Opioid antagonist
	Rimonabant	CB1 receptor antagonist
	Neramexane	NMDA receptor antagonist
	CEE-03-310	D1 receptor antagonist
	Baclofen	$GABA_B$ receptor agonist
	Antalarmin	CRF_1 Receptor antagonist
	Valproate	Unknown mechanism
	Various naltrexone formulations	μ Opioid antagonist
Cocaine dependence	Disulfiram	Unknown
	Topiramate	GABA/glutamate
	Modafinil	Unknown
	Cocaine vaccine (TA-C)	Cocaine
	Quetiapine	Dopamine $D_2/5HT_2$ antagonist
	Selegiline	MAO-B inhibitor
	γ-Vinyl-GABA	GABA transaminase inhibitor
	Gabapentin	Multiple mechanisms
	Tiagabine	GABA transport inhibitor
	Baclofen	$GABA_B$ receptor agonist
	Cabergoline	Dopamine D_2 receptor agonist
	Adrogolide (DAS-431)	Dopamine D_1-like receptor agonist
	Tolcapone	COMT inhibitor
	Vigabatrin (GVG)	Irreversible GABA transaminase inhibitor
	Ondansetron	$5HT_3$ receptor antagonist
	Memantine	NMDA receptor antagonist
	Sertraline	Serotonin transporter inhibitor
	Isradipine	Calcium channel antagonist
	Citicoline	Enhanced phospholipid synthesis
	Dextroamphetamine	Dopamine releaser
	Lobeline	Nicotinic receptor antagonist, inhibitor at the $VMAT_2$
	BP 897	D3 receptor partial agonist
	Vanoxerine (GBR 12909)	Dopamine transporter inhibitor
	Enadoline (CI 977)	Opioid κ receptor agonist
	Selegiline	MAO-B inhibitor
Nicotine dependence	Rimonabant	CB1 receptor antagonist
	Methoxsalen	CYP2A6 inhibitor that inhibits nicotine metabolism
	Nicotine conjugate vaccine (NicVax)	Nicotine
	Varenicline, SSR-591813, ABT 418	Nicotinic α4β2 agonists
	Various formulations of nicotine	

Adapted from [8,51,54,75,76].

reason for the poor compliance may result from naltrexone-induced dysphoric effects. Compliance with naltrexone treatment is higher among healthcare workers and parolees who by law can not participate in methadone clinics, and who are required to participate in treatment as a condition of their licensure.[7]

6.07.5.4 Stimulant Addiction

There are no FDA-approved treatments for cocaine and methamphetamine dependence at the present time. However, there are a number of promising medications that are under clinical investigation presently (**Table 5**). As reviewed elsewhere, these include but are not limited to disulfiram, and the GABA uptake inhibitor tiagabine, as well as topirimate, modafinil, baclofen,[62] and anti-cocaine vaccines.[63]

The anti-cocaine vaccine induces the production of anti-cocaine antibodies, which are hypothesized to bind peripherally circulating cocaine, preventing it from penetrating the blood–brain barrier. Thus this immunogenic response serves as a 'sponge' mechanism to prevent cocaine from getting into the brain.[64] Cocaine itself cannot induce this immune response, but when it is covalently linked to a carrier protein, it can become a haptenic molecule with immunogenic properties. The carrier protein both stimulates T cell-mediated antibody production but also provides a scaffold upon which cocaine is able to cross-link immunoglobulin on the surface of B cells.[64] Cocaine, or a cocaine derivative, is linked to a carrier protein such as keyhole limpet hemocyanin (KLH) or bovine serum albumin (BSA). One such conjugate has been shown to achieve levels of anti-cocaine antibody that suppressed cocaine-induced locomotor activity and stereotypical behaviors.[64] By replacing the 2-position ester with an amide linkage, the resulting vaccine could suppress cocaine-induced locomotor activity for up to 12 days, which was a dramatic improvement and suggested long-term protection against CNS effects of cocaine. Currently, safety and efficacy studies are being conducted in humans to assess medication potential in humans.

The development of medications to treat stimulant addiction remains a challenging area. It is likely that no single agent will provide an entirely satisfactory treatment, but that combining agents that work via different mechanisms to optimize therapeutic response will be the pathway that cocaine therapeutic development follows. Section 6.07.7.1 describes current preclinical targets with candidates that appear to show promise for treating cocaine abuse in animal models.

6.07.5.5 Marijuana (THC) Addiction

Addiction to tetrahydrocannabinol (THC), the psychoactive component in marijuana, has only recently been defined in animal models and compared to human marijuana abuse.[65] Hence, although marijuana use in humans is quite prevalent, and this drug has been coined the 'gateway' drug to other drugs of abuse, efforts to develop medications to specifically treat marijuana addiction are in early stages. As such, this remains an area for future investigation.

6.07.5.6 Behavioral/Environmental Modification

Although medication development for the treatment of drug addiction has been highlighted in this chapter, a brief discussion of the importance and role of behavioral therapy must be included for completeness. Indeed, drug addiction results from human behavior gone awry, and in most cases addicts who seek recovery from their addiction are more successful if they also engage in some form of behavioral/environmental modification. The literature is vast on the topic of conditioned behavior and no attempt to further describe it will be made herein. However, as mentioned in the Introduction of this chapter, environmental factors and associations play a large role in relapse, and clinical intervention in the form of therapy and behavioral modification are proving successful.[66] Recent investigations have shown that concepts of rewarding and aversive mechanisms can be applied to therapy, which, if available, can further be paired with pharmacotherapy.[67] In contingency management, reinforcement, in the form of vouchers for services, is given for drug abstinence, and is withheld when the drug addict fails to remain drug-free. This form of behavioral modification is based on the reinforcement/aversive effects associated with drug taking and drug abstinence. Thus far, high rates of increased cocaine abstinence have been reported using this approach.[67] In general, treatment, in whatever form available, is beneficial in reducing substance use, alleviating social, employment, and legal barriers to recovery, and identifying and treating medical and psychiatric problems that can potentiate relapse.[66]

6.07.6 Unmet Medical Needs

Substance dependence is complicated by psychiatric[68] and by medical comorbidity,[69] and has a significant and costly impact on individual and public health.[70] Alcohol- and drug-dependent persons without primary medical care have a

Table 7 Major medical comorbidities associated with drugs of abuse

Drug	*Major medical comorbidities*
Nicotine (tobacco)	Lung cancer, certain nonrespiratory cancers, cardiovascular disease, cerebrovascular disease, chronic respiratory disease, gastric ulcers, adverse pregnancy outcomes
Alcohol	Heavy use increases risk for diabetes, hypertension, and cardiomyopathy
	Lower amounts of alcohol use reduces risk of coronary heart disease, but heavier use increases risk; cirrhosis of the liver
	Overdoses
Cocaine/methamphetamine	Myocardial infarction, left ventricular dysfunction, stroke, transmission of HIV and hepatitis B and C via intravenous use
	Overdoses
Heroin	HIV, hepatitis B and C, endocarditis
	Overdoses

From Wadland, W. C.; Ferenchick, G. S. *Psychiat. Chin. N. Am.* **2004**, *27*, 675–687.

substantial burden of medical illness compared to age- and gender-matched US population controls.[70] Drug-related complications occur not only in public sector patients, but also in a community sample of patients enrolled in a health maintenance organization.[71] Some of the widely accepted medical complications are summarized in **Table 7**. Drug abuse is a leading contributor to the spread of AIDS, HIV, and hepatitis C, and drug abuse treatment can reduce the spread of these illnesses.[8] Not unexpectedly, intravenous drug use causes various surgical complications.[72]

The medical and psychiatric needs of patients with substance dependence are far from met. There are a variety of reasons for this, as described in detail in the 1999 report of the US Surgeon General.[73] Unfortunately, the various infrastructure and financial factors that affect the delivery of mental health and drug abuse treatments adversely impact the ability to develop new medications for substance dependence treatment.[74]

6.07.7 New Research Areas

Intensive investigation toward novel clinical candidates for treatment of drug addiction has ensued and many candidates with a diversity of mechanisms of action have been identified and are in various stages of preclinical and clinical development. A list of these potential medications is given in **Table 6** with their chemical structures shown in **Figure 8**. These compounds have been identified based on their current clinical use and/or their mechanism of action which has been deemed pertinent to drug addiction, as discussed in Section 6.07.2. Their efficacy in treating drug addiction remains to be established but in all cases, a proof of concept in animal models has been provided.[8,51,54,75,76]

6.07.7.1 Preclinical Targets for Stimulant Addiction

Since there are no FDA-approved medications for the treatment of cocaine and amphetamine addiction, intensive efforts have been directed toward discovering medication candidates. In this regard, over the past decade, a focus on elucidating mechanisms underlying the reinforcing effects and addictive liability of cocaine, as well as those underlying relapse, has prompted the identification of receptor/transporter candidates for drug discovery. Although many targets have been proposed, the following receptors/transporters and examples of promising drug candidates that have high affinity and selectivity for these targets have been widely studied and currently hold the most potential for medication development.

The monoamine neurotransmitter transporters are the principle sites of action of the psychostimulants cocaine and methamphetamine. These psychostimulants bind to the monoamine transporters and inhibit the reuptake of their respective neurotransmitters. Methamphetamine is further transported into the neuron and binds to the vesicular monoamine transporter (VMAT), resulting in a release of neurotransmitter into the synapse by disruption of vesicular storage and reverse transport. The psychostimulant actions of these drugs are primarily due to the release of DA, which stimulates postsynaptic DA receptors. However, especially with methamphetamine, there is also significant release of serotonin. A widespread concern with methamphetamine is its increasing use and its neurotoxic potential. Thus, identifying medications that prevent these illicit drugs from producing their actions is of primary importance in drug abuse research. Thus, a large investment in discovering and developing novel DA uptake inhibitors, both selective for DAT and also nonselective monoamine transport inhibitors has transpired over the past 15 years, resulting in an arsenal of compounds, based on cocaine, GBR 12909, benztropine, mazindol, methylphenidate, and rimcazole.[53,77,78] Many of these compounds demonstrate a cocaine-like behavioral profile in animal models, and could be potential 'substitute' therapy candidates. Several compounds have been identified that are potent and selective DA uptake inhibitors but do not produce cocaine-like behavioral effects. Importantly, some of these compounds attenuate cocaine-induced behaviors, such as cocaine-induced locomotor activity and the cocaine discriminative stimulus in animals.[79]

Examples of DAT inhibitors that show promise in this regard and are in various stages of preclinical development can shown in **Figure 9**. The novel DAT inhibitor RTI 336 represents the enormous class of 3-phenyltropane analogs reported over the years as high-affinity DAT inhibitors.[77] In general these drug candidates have longer durations of action than cocaine, but are cocaine-like in most animal models of cocaine abuse. Although GBR 12909 (**Figure 8**)

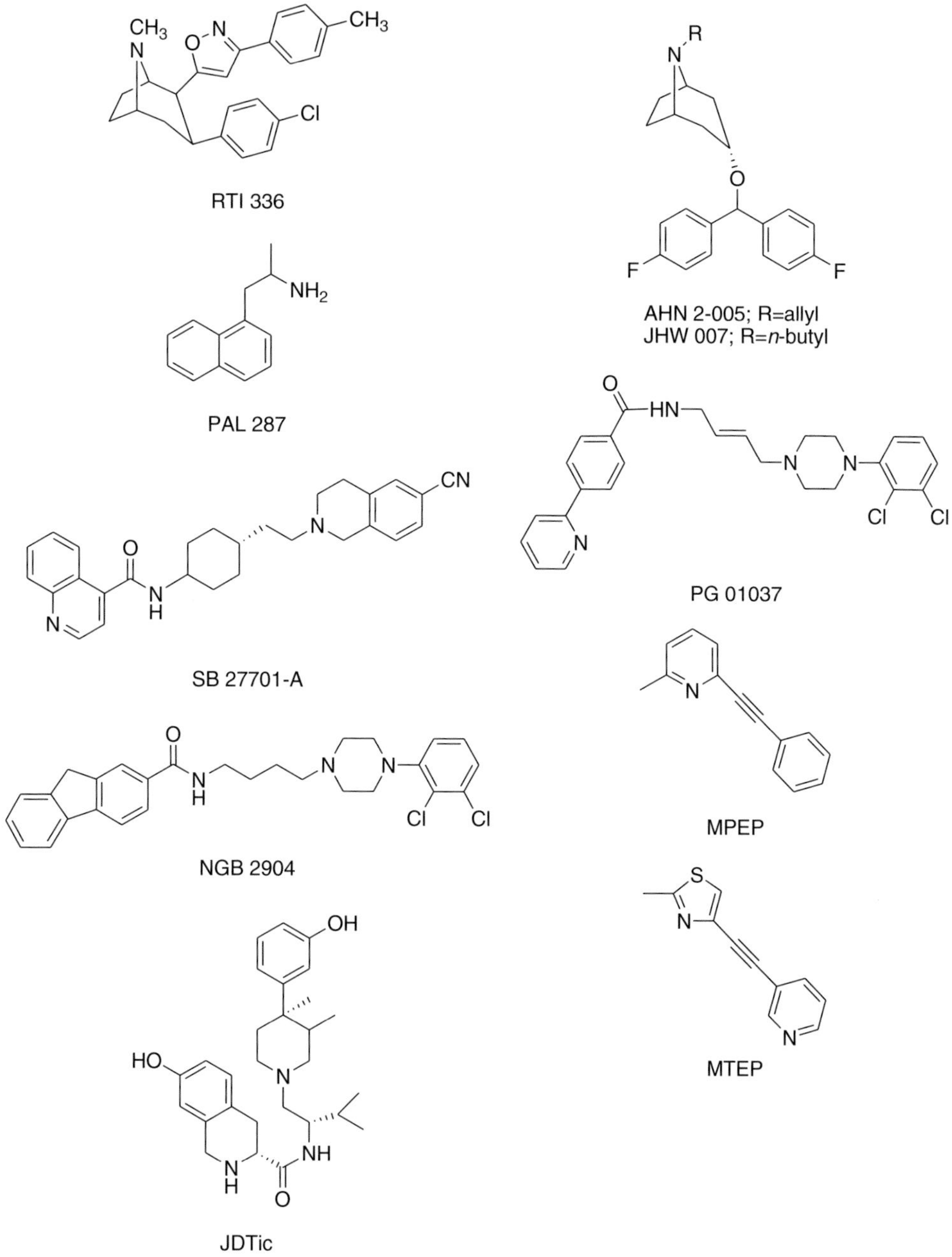

Figure 9 Representative structures of preclinical candidates for cocaine addiction medications.

showed cardiotoxicity (increases in QTc interval) in phase I clinical trials[80] and will not move forward in the clinic, structurally similar analogs of benztropine such as JHW 007 and AHN 2-005 show promise preclinically as potent and selective DAT inhibitors that are not cocaine-like in numerous animal models and attenuate cocaine-induced behaviors.[79,81] This unique behavioral profile is proposed to be due to their slow association onto the DAT and long duration of action, and is currently being further evaluated.

Agents that release both DA and 5HT via a DAT and SERT carrier-mediated exchange mechanism (an amphetamine-like action), were proposed to be potential treatment agents for cocaine addiction that would have low abuse liability yet still be able to decrease drug seeking behavior.[82] In support of this hypothesis, recent work demonstrated that the dual DA/5HT releaser PAL 287, a naphthylisopropylamine, is not self-administered yet suppresses cocaine self-administration in rhesus monkeys.[52] The compounds discussed above target the monoamine transporters, since these are primary proteins through which cocaine and methamphetamine produce their psychostimulant and reinforcing actions. However, additional targets that may be more associated with long-term effects of drugs of abuse on the reward circuitry, and which may provide new targets for treatment of drug relapse have been identified. These are briefly described below.

DA receptor subtypes (D1-like and D2-like) have been targeted for many years as yielding potential medications for the treatment of cocaine abuse.[83] Unfortunately, for the most part, these compounds have provided important insight into the roles the D1- and D2-like receptors play in the pharmacology of substances of abuse, but untoward side effects and a lack of efficacy that have limited their potential as medications.[84] Recent investigations have demonstrated that selective and high-affinity DA D3 receptor ligands attenuate cocaine seeking behaviors and show promise in animal models of relapse. Furthermore, these compounds do not demonstrate behavioral toxicity associated with other nonselective D2 receptor ligands.[85] Promising candidates from the D3 class of compounds are primarily antagonists or partial agonists that fall into a rather narrow SAR profile as recently reviewed.[85] Compounds BP 897 (**Figure 8**), SB 277011, NGB 2904, and PG 01037 (**Figure 9**) are the 'drugs of choice' for studying the role of the DA D3 receptor in drug abuse, and BP 897 is currently in clinical trials for treatment of psychiatric disorders, including drug abuse.[75] Although SB 277011 has a metabolic profile that is unsuitable for clinical use, future analogues are currently under investigation.[85] PG 01037 is an analog of NGB 2904 that demonstrates high affinity ($K_i = 0.7$ nM) for hD3 receptors and greater than 130-fold selectivity over hD2 receptors in vitro, and is also water soluble. Like the other D3 antagonists, PG 01037 is a D3 receptor antagonist in several in vivo models, and attenuates cocaine-induced reinstatement in monkeys without behavioral toxicity (R. Spealman, D. Platt, and A. Newman, unpublished observations). Further investigation into the D3 receptor as a target for psychostimulant abuse medications will determine whether these or other compounds should be developed further.

Both depression and stress promote relapse in drug abusers. The endogenous κ-opioid agonist dynorphin has been implicated in modulating stress via the κ-opioid receptor system. Thus, κ-opioid receptor antagonists have been under development for some time as potential treatments for drug addiction. Recently, JDTic (**Figure 9**) emerged as a lead compound to test this hypothesis because of its antidepressant effect and efficacy in animal models of drug relapse. Other preclinical studies support the clinical development of this agent and future studies will determine its potential as a cocaine abuse medication in humans.[86] Similarly, the ability of corticotropin receptor 1 (CRH1) receptor antagonists to reduce stress and anxiety has led investigators to explore the effect of CRH1 receptor antagonists on cocaine-related behaviors. As reviewed by Gurkovskaya,[87] CRH1 receptor antagonists can reduce cocaine self-administration as well as cue-induced cocaine seeking behavior.

Several other nondopaminergic targets are under active investigation as potential targets for drug addiction medication development. One such target is the metabotropic glutamate receptor subtype 5 (mGluR5). Initial studies in rats showed that the selective mGluR5 antagonist MPEP (**Figure 9**) attenuated cocaine seeking behavior, and that a mGluR5 ko mouse would also not respond for cocaine.[88] These initial studies suggested a role for this receptor subtype in the addictive effects of cocaine. Subsequently, efforts in several industrial and academic laboratories have focused on discovering novel high-affinity and selective mGluR5 noncompetitive antagonists/allosteric modulators that have improved bioavailability over MPEP. This has been a formidable challenge, since even small changes in the MPEP structure lead to a loss of activity. One successful modification was the replacement of the 2-methylpyridyl ring of MPEP with a 2-methylthiazole and the minor modification substituting the phenyl ring with the isosteric pyridyl ring system to give the close analog MTEP (**Figure 9**),[89] which has been used for several recent in vivo studies. These and additional mGluR5 antagonists are providing the tools to further elucidate the role of the mGluR5 in drug addiction.[90] In addition the mGluR2/3 receptors are also under intensive investigation toward further understanding the roles of these receptors in drug addiction and neuropsychiatric disorders, and the therapeutic potential of allosteric modulation of these receptors has been recently disclosed.[91] RGS proteins are another potential target,[92] but much research remains to investigate the utility of this family of proteins for developing drug abuse treatment medications.

6.07.8 Summary

Addiction is a disorder that has a biological basis. Addiction to drugs and/or alcohol result in enormous costs to society in terms of healthcare, family destruction, loss of productivity, and direct and indirect effects on our criminal justice system. The biochemical bases of addiction and relapse have been investigated and numerous hypotheses of how drug seeking for pleasure evolves into compulsive drug taking, in the face of personal destruction, have ensued. These studies have identified central targets to which potential medications can be designed. In addition, medications that have been discovered to successfully treat related mental disorders such as anxiety and depression have also been identified as potential medication candidates and some of these are clinically prescribed. Moreover, clinical evaluation of medications that are successfully used in the treatment of one addiction (e.g., nicotine) may also be useful for treatment of another (e.g., cocaine). Other NCEs are in various stages of clinical development and their success or failure will dictate future trends in research toward discovering treatment candidates, especially for psychostimulant abuse. It is clear that addiction is a complex disorder that will likely not be cured with a 'magic bullet' but rather an array of treatment strategies that may need to be individualized depending on the patient, the drug or drugs to which he is addicted, and any comorbid disorders that he may also suffer from. As such, a multidisciplinary approach to drug abuse research will be required and consideration of many relevant biological targets as well as combination therapies involving medication and behavioral treatments will likely prove to have the greatest success.

Acknowledgments

This work was supported by the National Institute on Drug Abuse – Intramural Research Program, National Institutes of Health.

References

1. Pouletty, P. *Nat. Rev. Drug Disc.* **2002**, *1*, 731–736.
2. Bullock, K.; Koran, L. *Drugs Today* **2003**, *39*, 695–700.
3. Goudriaan, A. E.; Oosterlaan, J.; de Beurs, E.; Van den Brink, W. *Neurosci. Biobehav. Rev.* **2004**, *28*, 123–141.
4. Wang, G. J.; Volkow, N. D.; Fowler, J. S. *Exp. Opin. Ther. Targets* **2002**, *6*, 601–609.
5. Volkow, N. D.; Wise, R. A. *Nat. Neurosci.* **2005**, *8*, 555–560.
6. Cami, J.; Farre, M. *N. Engl. J. Med.* **2003**, *349*, 975–986.
7. Kreek, M. J.; LaForge, K. S.; Butelman, E. *Nat. Rev. Drug Disc.* **2002**, *1*, 710–726.
8. Volkow, N. D.; Li, T. K. *Nat. Rev. Neurosci.* **2004**, *5*, 963–970.
9. Nestler, E. J. *Nat. Rev. Neurosci.* **2001**, *2*, 119–128.
10. Baumann, M. H.; Rothman, R. B. *Biol. Psychiat.* **1998**, *44*, 578–591.
11. Rothman, R. B.; Glowa, J. R. *Mol. Neurobiol.* **1995**, *11*, 1–19.
12. Childress, A. R.; Mozley, P. D.; McElgin, W.; Fitzgerald, J.; Reivich, M.; O'Brien, C. P. *Am. J. Psychiat.* **1999**, *156*, 11–18.
13. Goldstein, R. Z.; Volkow, N. D. *Am. J. Psychiat.* **2002**, *159*, 1642–1652.
14. Robinson, T. E.; Berridge, K. C. *Annu. Rev. Psychol.* **2003**, *54*, 25–53.
15. Chao, J.; Nestler, E. J. *Annu. Rev. Med.* **2004**, *55*, 113–132.
16. Nestler, E. J. *Neuropharmacology* **2004**, *47*, 24–32.
17. Betz, C.; Mihalic, D.; Pinto, M. E.; Raffa, R. B. *J. Clin. Pharm. Ther.* **2000**, *25*, 11–20.
18. Melis, M.; Spiga, S.; Diana, M. *Int. Rev. Neurobiol.* **2005**, *63*, 101–154.
19. Bolla, K. I.; Cadet, J. L.; London, E. D. *J. Neuropsychiat. Clin. Neurosci.* **1998**, *10*, 280–289.
20. Jentsch, J. D.; Taylor, J. R. *Psychopharmacology* **1999**, *146*, 373–390.
21. Wise, R. A. *Curr. Opin. Neurobiol.* **1996**, *6*, 243–251.
22. Koob, G. F.; Le Moal, M. *Neuropsychopharmacology* **2001**, *24*, 97–129.
23. Koob, G. F.; Ahmed, S. H.; Boutrel, B.; Chen, S. A.; Kenny, P. J.; Markou, A.; O'Dell, L. E.; Parsons, L. H.; Sanna, P. P. *Neurosci. Biobehav. Rev.* **2004**, *27*, 739–749.
24. Kelley, A. E. *Neuron* **2004**, *44*, 161–179.
25. Backstrom, P.; Hyytia, P. *Neuropsychopharmacology* **2005**, *31*, 778–786.
26. Miller, C. A.; Marshall, J. F. *Neuron* **2005**, *47*, 873–884.
27. Hyman, S. E. *Am. J. Psychiat.* **2005**, *162*, 1414–1422.
28. Di Chiara, G. *Behav. Pharmacol.* **2002**, *13*, 371–377.
29. Weiss, F. *Curr. Opin. Pharmacol.* **2005**, *5*, 9–19.
30. Bolanos, C. A.; Nestler, E. J. *Neuromol. Med.* **2004**, *5*, 69–83.
31. Grimm, J. W.; Lu, L.; Hayashi, T.; Hope, B. T.; Su, T. P.; Shaham, Y. *J. Neurosci.* **2003**, *23*, 742–747.
32. Vocci, F. J.; Acri, J.; Elkashef, A. *Am. J. Psychiat.* **2005**, *162*, 1432–1440.
33. Ator, N. A.; Griffiths, R. R. *Drug Alcohol Depend.* **2003**, *70*, S55–S72.
34. Fischman, M. W.; Foltin, R. W. *Ciba Found. Symp.* **1992**, *166*, 165–173.
35. Shaham, Y.; Shalev, U.; Lu, L.; De Wit, H.; Stewart, J. *Psychopharmacology* **2003**, *168*, 3–20.
36. Lee, B.; Platt, D. M.; Rowlett, J. K.; Adewale, A. S.; Spealman, R. D. *J. Pharmacol. Exp. Ther.* **2005**, *312*, 1232–1240.
37. Vorel, S. R.; Ashby, C. R., Jr.; Paul, M.; Liu, X.; Hayes, R.; Hagan, J. J.; Middlemiss, D. N.; Stemp, G.; Gardner, E. L. *J. Neurosci.* **2002**, *22*, 9595–9603.

38. Kornetsky, C.; Esposito, R. U.; McLean, S.; Jacobson, J. O. *Arch. Gen. Psychiat.* **1979**, *36*, 289–292.
39. Markou, A.; Koob, G. F. *Neuropsychopharmacology* **1991**, *4*, 17–26.
40. Tzschentke, T. M. *Prog. Neurobiol.* **1998**, *56*, 613–672.
41. Le Foll, B.; Goldberg, S. R. *Trends Pharmacol. Sci.* **2005**, *26*, 287–293.
42. Pich, E. M.; Epping-Jordan, M. P. *Ann. Med.* **1998**, *30*, 390–396.
43. Giros, B.; Jaber, M.; Jones, S. R.; Wightman, R. M.; Caron, M. G. *Nature* **1996**, *379*, 606–612.
44. Sora, I.; Hall, F. S.; Andrews, A. M.; Itokawa, M.; Li, X. F.; Wei, H. B.; Wichems, C.; Lesch, K. P.; Murphy, D. L.; Uhl, G. R. *Proc. Natl. Acad. Sci. USA* **2001**, *98*, 5300–5305.
45. McLellan, A. T.; Lewis, D. C.; O'Brien, C. P.; Kleber, H. D. *JAMA* **2000**, *284*, 1689–1695.
46. Tankosic, T. *Drug Market Devel. Newsl.* **2002**. http://pharmalicensing.com/features/disp/1039693784_3df877d80dede (accessed June 2006).
47. Carroll, K. M.; Nich, C.; Rounsaville, B. J. *NIDA Res. Monogr.* **1997**, *175*, 137–157.
48. de Lima, M. S.; de Oliveira Soares, B. G.; Reisser, A. A.; Farrell, M. *Addiction* **2002**, *97*, 931–949.
49. Roth, B. L.; Sheffler, D. J.; Kroeze, W. K. *Nat. Rev. Drug Disc.* **2004**, *3*, 353–359.
50. Gray, N. J.; Henningfield, J. E. *Nicotine Tob. Res.* **2004**, *6*, 759–764.
51. Grabowski, J.; Shearer, J.; Merrill, J.; Negus, S. S. *Addict Behav.* **2004**, *29*, 1439–1464.
52. Rothman, R. B.; Blough, B. E.; Woolverton, W. L.; Anderson, K. G.; Negus, S. S.; Mello, N. K.; Roth, B. L.; Baumann, M. H. *J. Pharmacol. Exp. Ther.* **2005**, *313*, 1361–1369.
53. Newman, A. H.; Kulkarni, S. *Med. Res. Rev.* **2002**, *22*, 429–464.
54. Gorelick, D. A.; Gardner, E. L.; Xi, Z. X. *Drugs* **2004**, *64*, 1547–1573.
55. Sofuoglu, M.; Kosten, T. R. *Psychiat. Clin. N. Am.* **2004**, *27*, 627–648.
56. Anton, R. F. In *Conn's Current Therapy*, 57th ed.; Rakel, R. E., Bope, E. T., Eds.; Elsevier Saunders: Philadelphia, PA, 2005, pp 1253–1256.
57. O'Brien, C. P. *Am. J. Psychiat.* **2005**, *162*, 1423–1431.
58. Henningfield, J. E. *N. Engl. J. Med.* **1995**, *333*, 1196–1203.
59. Damaj, M. I.; Carroll, F. I.; Eaton, J. B.; Navarro, H. A.; Blough, B. E.; Mirza, S.; Lukas, R. J.; Martin, B. R. *Mol. Pharmacol.* **2004**, *66*, 675–682.
60. Anonymous. *JAMA* **1998**, *280*, 1936–1943.
61. Romero, D. V.; Partilla, J. S.; Heyliger, S. O.; Ni, Q.; Rice, K. C.; Rothman, R. B. *Synapse* **1999**, *34*, 83–94.
62. Vocci, F.; Ling, W. *Pharmacol. Ther.* **2005**, *108*, 94–108.
63. Kosten, T.; Owens, S. M. *Pharmacol. Ther.* **2005**, *108*, 76–85.
64. Carrera, M. R.; Meijler, M. M.; Janda, K. D. *Bioorg. Med. Chem.* **2004**, *12*, 5019–5030.
65. Gonzalez, S.; Cebeira, M.; Fernandez-Ruiz, J. *Pharmacol. Biochem. Behav.* **2005**, *81*, 300–318.
66. McGovern, M. P.; Carroll, K. M. *Psychiat. Clin. N. Am.* **2003**, *26*, 991–1010.
67. Higgins, S. T.; Heil, S. H.; Lussier, J. P. *Annu. Rev. Psychol.* **2004**, *55*, 431–461.
68. Kessler, R. C.; McGonagle, K. A.; Zhao, S.; Nelson, C. B.; Hughes, M.; Eshleman, S.; Wittchen, H. U.; Kendler, K. S. *Arch. Gen. Psychiat.* **1994**, *51*, 8–19.
69. Wadland, W. C.; Ferenchick, G. S. *Psychiat. Clin. N. Am.* **2004**, *27*, 675–687.
70. De Alba, I.; Samet, J. H.; Saitz, R. *Am. J. Addict.* **2004**, *13*, 33–45.
71. Mertens, J. R.; Lu, Y. W.; Parthasarathy, S.; Moore, C.; Weisner, C. M. *Arch. Intern. Med.* **2003**, *163*, 2511–2517.
72. Warner, R. M.; Srinivasan, J. R. *Surgeon* **2004**, *2*, 137–140.
73. *Mental Health: A Report of the Surgeon General*; US Department of Health and Human Services, Substance Abuse and Mental Health Services Administration, Center for Mental Health Services, National Institutes of Health, National Institute of Mental Health: Bethesda, MD, 1999.
74. Committee on Immunotherapies and Sustained-Release Formulations for Treating Drug Addiction, N. R. C. *New Treatments for Addiction: Behavioral, Ethical, Legal, and Social Questions*; National Academies Press: Washington, DC, 2004.
75. Heidbreder, C. A.; Hagan, J. J. *Curr. Opin. Pharmacol.* **2005**, *5*, 107–118.
76. Vocci, F. J.; Elkashef, A. *Curr. Opin. Psychiat.* **2005**, *18*, 265–270.
77. Carroll, F. I. *J. Med. Chem.* **2003**, *46*, 1775–1794.
78. Prisinzano, T.; Rice, K. C.; Baumann, M. H.; Rothman, R. B. *Curr. Med. Chem. (CNSA)* **2004**, *4*, 47–59.
79. Desai, R. I.; Kopajtic, T. A.; Koffarnus, M.; Newman, A. H.; Katz, J. L. *J. Neurosci.* **2005**, *25*, 1889–1893.
80. Vocci, F.; Ling, W. *Pharmacol. Ther.* **2005**, *108*, 94–108.
81. Desai, R. I.; Kopajtic, T. A.; French, D.; Newman, A. H.; Katz, J. L. *J. Pharmacol. Exp. Ther.* **2005**, *315*, 397–404.
82. Rothman, R. B.; Baumann, M. *Pharmacol. Ther.* **2002**, *95*, 73–88.
83. Khroyan, T. V.; Barrett-Larimore, R. L.; Rowlett, J. K.; Spealman, R. D. *J. Pharmacol. Exp. Ther.* **2000**, *294*, 680–687.
84. Kosten, T. R.; George, T. P.; Kosten, T. A. *Exp. Opin. Invest. Drugs* **2002**, *11*, 491–499.
85. Newman, A. H.; Grundt, P.; Nader, M. A. *J. Med. Chem.* **2005**, *48*, 3663–3679.
86. Beardsley, P. M.; Howard, J. L.; Shelton, K. L.; Carroll, F. I. *Psychopharmacology* **2005**, *183*, 118–126.
87. Gurkovskaya, O. V.; Palamarchouk, V.; Smagin, G.; Goeders, N. E. *Synapse* **2005**, *57*, 202–212.
88. Chiamulera, C.; Epping-Jordan, M. P.; Zocchi, A.; Marcon, C.; Cottiny, C.; Tacconi, S.; Corsi, M.; Orzi, F.; Conquet, F. *Nat. Neurosci.* **2001**, *4*, 873–874.
89. Cosford, N. D.; Tehrani, L.; Roppe, J.; Schweiger, E.; Smith, N. D.; Anderson, J.; Bristow, L.; Brodkin, J.; Jiang, X.; McDonald, I. et al. *J. Med. Chem.* **2003**, *46*, 204–206.
90. Kenny, P. J.; Markou, A. *Trends Pharmacol. Sci.* **2004**, *25*, 265–272.
91. Niswender, C. M.; Jones, C. K.; Conn, P. J. *Curr. Topics Med. Chem.* **2005**, *5*, 847–857.
92. Traynor, J. R.; Neubig, R. R. *Mol. Interv.* **2005**, *5*, 30–41.
93. Dackis, C. A. *Curr. Psychiat. Rep.* **2004**, *6*, 323–331.
94. Dackis, C. A.; Kampman, K. M.; Lynch, K. G.; Pettinati, H. M.; O'Brien, C. P. *Neuropsychopharmacology* **2005**, *30*, 205–211.

Biographies

Amy Hauck Newman received her PhD in medicinal chemistry from the Medical College of Virginia, Virginia Commonwealth University, Richmond, VA, in 1985. After postdoctoral studies at the National Institute on Diabetes, Digestive and Kidney Diseases, National Institutes of Health (NIH), she joined the Intramural Research Program of the National Institute on Drug Abuse (NIDA-IRP), NIH. She is currently a Senior Investigator and Chief of the Medicinal Chemistry Section, Medications Discovery Research Branch. She has coauthored more than 130 original articles and reviews on the design, synthesis, and pharmacological evaluation of CNS active agents and is an inventor on several US patents. The emphasis of Dr Newman's research is the design and synthesis of selective ligands for the dopaminergic and glutamatergic systems, as molecular probes and as leads toward potential treatment medications for cocaine abuse.

Richard B Rothman received an MD and a PhD degree in pharmacology from the University of Virginia, Charlottesville, VA, in 1982. After postdoctoral studies at the National Institute of Mental Health (NIMH) (1982–1984), Dr Rothman served a residency in psychiatry at St Elizabeths Hospital, Washington, DC (1984–1987) and later a 2-year fellowship in psychiatry at the NIMH. In 1991, Dr Rothman joined the Intramural Research Program of the National Institute on Drug Abuse (NIDA-IRP) in 1991. He has coauthored more than 320 original articles focused mostly on the pharmacology of opioids and stimulants, and is an inventor on several US patents. Dr Rothman is board certified in psychiatry, and currently serves as Chief of the Clinical Psychopharmacology Section, Medications Discovery Research Branch.

Comprehensive Medicinal Chemistry II
ISBN (set): 0-08-044513-6

ISBN (Volume 6) 0-08-044519-5; pp. 169–191

6.08 Neurodegeneration

D Bozyczko-Coyne and M Williams, Worldwide Discovery Research, Cephalon, Inc., West Chester, PA, USA

6.08.1 Overview

The underlying pathophysiology of all neurodegenerative diseases is the dysfunction and ultimate death of neurons. Neurodegenerative diseases are progressive in nature with a worsening of clinical symptoms over time. The neuronal loss in different neurodegenerative diseases is asynchronous with symptoms progressing as a result of a cumulative loss of the neurons that synthesize the neurotransmitters essential to signal propagation through the particular brain circuitry associated with a given disease. Different disease phenotypes occur with neuron loss depending on the neuronal population affected, the insult initiating the cell death cascade, and the genetic makeup of the individual. For example, memory or coordinated movement is not the function of a single specific neuronal population or brain region but involves several, such that the loss of one key element can perturb neurotransmitter homeostasis (excitatory versus inhibitory), thus affecting the final integrated output of the system.

Alzheimer's disease (AD), Parkinson's disease (PD), Huntington's disease (HD), and amyotrophic lateral sclerosis (ALS) are the best known neurodegenerative diseases, but also included in the category of neurodegeneration are demyelinating diseases (e.g., multiple sclerosis, Charcot–Marie–Tooth, neuromyelitis optica), neuropathies (e.g., diabetic, human immunodeficiency virus (HIV), chemotoxic), Down's syndrome, prion diseases (e.g., Creutzfeldt–Jakob disease), tauopathies (e.g., Pick's disease, frontal temporal dementia with parkinsonism (FTDP)), additional trinucleotide repeat or polyglutamine (polyQ) diseases (e.g., spinocerebellar ataxias, dentatorubral-pallidolysian atrophy, Freidreich's ataxia), multiple systems atrophy, stroke, and traumatic brain injury. The current chapter focuses on AD, PD, HD, and ALS: multiple sclerosis and stroke are covered elsewhere in this volume (*see* 6.09 Neuromuscular/Autoimmune; 6.10 Stroke/Traumatic Brain and Spinal Cord Injuries).

AD affects cognitive function at its onset, while in PD, HD, and ALS, motor function is initially impaired. In the later stages of PD and HD, and within a specific variant of ALS, progressive cognitive dysfunction and dementia may occur.

Neurodegenerative diseases can be classified into two predominant types based upon their nature of inheritance. If there is no definable pattern of inheritance, the neurodegenerative disease is referred to as *idiopathic* or *sporadic*, whereas diseases with a definable inheritance pattern and genetic linkage are referred to as *familial*. The two categories may represent different entities. AD, PD, and ALS exist in both idiopathic and familial disease forms while HD is a familial disease, with 100% penetration with inheritance of the identified gene. A number of genetic risk factors (susceptibility genes) as well as putative environmental exposure factors (epigenetics) are associated with the development of idiopathic forms of AD, PD, and ALS. Thus, as with other disease states, genetics and environment (nature and nurture) play key roles in the development of neurodegenerative diseases. Since multiple genetic risk factors are associated with central nervous system (CNS) degenerative diseases, it is anticipated that treatment will need to be prophylactic, based on individual genetic profiles. However, this will require the development of drugs with relatively benign safety profiles to allow for long-term treatment and biomarkers to diagnose the disease and to assess disease progression and drug responses.

In each neurodegenerative disease, there is selective neuronal vulnerability such that the primary neuronal population affected differs. Common factors that are potential mediators of the disease process include: increased oxidative burden (reactive oxygen species); impaired energy metabolism; lysosomal dysfunction; protein aggregation/inclusion-body formation; inflammation; excitoxicity; necrosis; and/or apoptosis. Since many of these features underlie multiple CNS disease states, new chemical entities (NCEs) targeting these mechanisms may have general utility. As evidence of this, evaluation of antioxidants, antiinflammatory agents, glutamate receptor antagonists, and antiapoptotic agents for effects on neuronal survival has been conducted in animal models of neurodegeneration with some having been clinically evaluated in multiple neurodegenerative disease states (e.g., minocycline, riluzole).

Drugs approved over the last 10–15 years for the treatment of neurodegenerative diseases, mainly cholinesterase inhibitors, have provided modest symptomatic improvement. However, evidence is emerging that some of these drugs may alter the course of disease (e.g., dopamine (DA) agonists for treatment of PD). In contrast, the cholinesterase inhibitors approved for AD have modest efficacy.

Current research is focused on finding NCEs to halt disease progression – the 'holy grail' for neurodegenerative disease. However, in the absence of a clear understanding of the mechanism(s) causing cell death, this research is almost exclusively dependent on identifying potent, selective, and drug-like NCEs, active at molecular targets thought to underlie the process of neurodegeneration, to build confidence for these targets and derive proof of concept. The theoretical effects of symptomatic treatment versus neuroprotective treatment on disease progression are shown in **Figure 1**. Basically, with symptomatic treatment, a relatively immediate improvement in functional response occurs but on withdrawal the natural rate of disease progression resumes. Neuroprotectants are anticipated to alter the rate (e.g., change the slope) of disease progression but not necessarily improve functional responses in the short term. This is an important issue in that many of the animal models used to evaluate NCEs have been developed using short-term measures, e.g., Morris water maze.

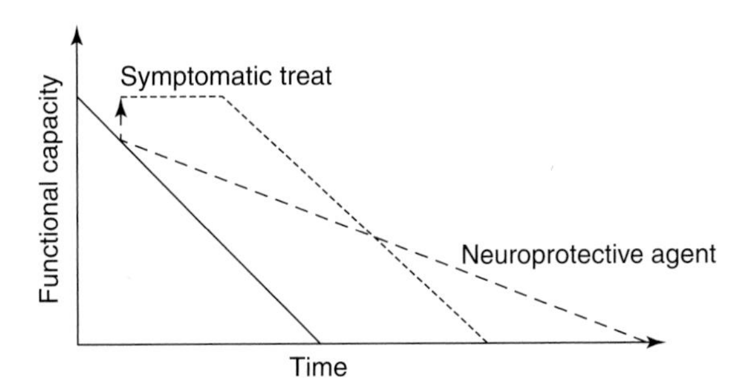

Figure 1 Time course of Alzheimer's disease (AD) progression. The solid line indicates normal disease progression – functional capability of the AD patient decreases in a linear manner with time. Symptomatic treatment, e.g., cholinesterase inhibitors, can produce an acute stabilization of the disease, resulting in a time shift in disease progression. It is debatable whether the acute stabilization is a drug-related or placebo effect. A neuroprotective agent, none of which has yet progressed to proof of concept in humans, is shown in the figure as markedly delaying disease progression such that normal aging may precede serious cognitive decline. An 'ideal' neuroprotective agent may even reverse disease progression, resulting in a progression, line approaching the horizontal.

As the average human lifespan has increased, a concomitant increase in the incidence of deaths attributed to neurodegenerative disease has occurred.[1] According to the National Vital Statistics Report of the US Centers for Disease Control, life expectancy from birth in 2003 was 77.6 years.[1] This represents a 0.3-year increase from 2002, which was partially driven by increases in the number of deaths listed as due to AD, the eighth leading cause, and PD, the 14th leading cause. In comparison, the average life expectancy in 1900 was 47.3 years. Thus, with time, the pharmacoeconomic impact of neurodegenerative diseases on society is increasing. At the same time, new drug approvals have decreased, especially in the CNS area, suggesting that as technology has advanced, productivity has fallen. An interesting debate is ongoing that cancer may represent a risk factor in dementia,[2] reflecting the yin and yang of cell death therapeutics. In cancer, the task is to enhance cell death, while in neurodegeneration the converse is true.

New technologies in genetics (transgenic or gene-null vertebrates and invertebrates) have been used as novel preclinical models of disease and for testing NCEs. For neurodegenerative diseases, rodent, fly (*Drosophilia*), and worm (*Caenorhabditis elegans*) models exist for familial forms of AD, PD, HD, and ALS.[3] Clinical back-validation of these models has been limited since, for most of these engineered models, the human disease process is not fully recapitulated. However, invertebrate models provide facile systems for cost-effective screening of NCEs for their effects on the basic biological process that may underlie disease pathophysiology. Genetically engineered nonhuman primate models are also being developed, although the availability of animals and costs will limit general applicability and widespread utility.

A major discourse in drug discovery as productivity has decreased is that the 'low-hanging fruit' has been harvested. In the area of neurodegeneration, the drugs launched over the past two decades target the same mechanism, e.g., cholinesterase inhibitors for AD and DA agonists for PD. While improvements in pharmacokinetics or reduced side effect liabilities have been made in second-generation compounds and have aided treatment management, these have had limited impact in advancing the understanding of disease causality. Additionally, a number of promising compounds acting at newer targets based on elegant preclinical hypotheses have singularly failed in the clinic, e.g., estrogen and anti-inflammatory agents in AD.

Conceptually, drug therapies for neurodegenerative disease can be divided into three treatment categories: (1) symptomatic; (2) protective; and (3) curative. Strategies for symptomatic and protective treatments will be discussed in this chapter. Curative treatment, yet to be achieved, is typically envisaged in terms of tissue (stem cell) transplantation and gene replacement strategies that are beyond the scope of this chapter.

6.08.2 Alzheimer's Disease

6.08.2.1 Disease State/Diagnosis

AD or dementia of the Alzheimer's type (DAT) is named for Alois Alzheimer, who first described the cognitive impairment and later the neuropathological hallmarks of the disease in 1907: amyloid plaques, neurofibrillary tangles (NFTs), and inflammation marked by astrocytic gliosis and reactive microglia. AD is a mentally debilitating disease with profound socio- and pharmaco-economic impact. Over 4 million individuals in the US and 15–20 million individuals worldwide have AD. AD onset typically occurs at 60–65 years, with the risk of disease doubling every 5 years above the age of 65. AD prevalence is 3% at ages 65–74 and ~50% at 85 and older. Early memory impairment, specifically episodic memory, and progressive decline in executive function are overt clinical signs of AD accompanied by drastic alterations in personality, including aggressive behavior, which make management of AD patients challenging. The mean survival time postdiagnosis ranges from 5 to 10 years and can be as long as 20 years.

AD is the most common form of dementia diagnosed, but it is an exclusionary diagnosis of other probable causes of dementia, including amnestic disorders, dementia due to multiple etiologies or other medical conditions, vascular dementia, substance intoxication, or substance withdrawal. The *Diagnostic and Statistical Manual of Mental Disorders* (DSM-IV-TR)[4] and the World Health Organization ICD-10 disease classification systems define AD subtypes: with early onset (<65 years), with late onset (>65 years); with delusions; with depressed mood; uncomplicated; and/or with behavioral disturbance (e.g., wandering). The primary diagnostic of any dementia is memory impairment but can include aphasia (deterioration of language function), agnosia (impairment in object recognition), apraxia (impairment in executing motor activities), and disturbances in executive function. In AD, these cognitive deficits are typically observed with disease progression, which is usually of slow onset and gradual decline. Mild cognitive impairment (MCI), characterized by isolated episodes of long-term memory impairment, may be the precusor of AD. Longitudinal studies indicate that up to 80% of MCI patients will progress to dementia within 6 years.[5]

The definitive diagnosis of AD is postmortem and is based on clinical documentation of dementia and neuropathological criteria. The pathological classification of AD is based on criteria established by Braak and Braak,[6] the Consortium to Establish a Registry for Alzheimer Disease (CERAD),[7] and/or the National Institute for Aging and Ronald and Nancy Reagan Institute of the Alzheimer's Association (NIA–RIA).[8] Braak and Braak's criteria are based on an evaluation of the distribution of NFTs and neuropil threads in the brains of 83 nondemented and demented individuals at autopsy. Six stages of neuropathology were defined and correlated to cognitive function. A general progression of NFT distribution from the transentorhinal cortical area (stage I) to the entorhinal cortex (stage II), extending to the hippocampus (stage III) and increasing in number (stage IV), followed by involvement of the neocortex (stage V) and ultimately the primary cortex (stage VI), was related to cognitively normal (stage I/II), cognitively impaired (stage III/IV), and demented (stage V/VI) individuals. The CERAD criteria are based on the histopathological evaluation of three brain regions: the second frontal gyrus, the first temporal gyrus, and the supramarginal gyrus at the levels of the caudate putamen, amygdala, and parietooccipital sulcus, respectively, and the age of the patient. Neuritic plaque density is quantified in these regions and classified into three stages: stage A (<2), stage B (~ 6), and stage C (>30). AD is diagnosed at autopsy if patients younger than 50 years were classified as stage A; patients at 50–75 years were classified as stage B; or if patients older than 75 years were classified as stage C and had a clinical history of dementia. According to the NIA–RIA criteria, AD is categorized as low, intermediate, and of high likelihood based on both the Braak and CERAD classifications: stages I/II and A; stages III/IV and B, and stages V/VI and C, respectively. The utility of definitive disease diagnosis postmortem is however of little consequence in treating the patient, especially when the accuracy of clinical diagnosis of probable AD is in the 75–92% range. The ability to accurately diagnose and assess AD progression is critical to understanding the disease in order to develop and evaluate NCEs at a time when there is sufficient brain function remaining for restorative approaches to be efficacious.

Newer diagnostic tools for AD in living patients are being developed but require correlation with both the clinical dementia rating (CDR) and the histopathological evaluation of NFTs at autopsy in longitudinal studies for validation. Brain neuroimaging using computed tomography (CT) or magnetic resonance image (MRI) scanning to measure brain volume has been evaluated as a potential surrogate marker for disease progression.[9] While reduced volume in specific brain regions (e.g., entorhinal cortex, hippocampus) can be correlated with cognitive dysfunction, it is not usually evident until significant brain atrophy has occurred and thus cannot serve as an early marker of the disease process prior to obvious clinical symptoms.[9] Functional magnetic resonance imaging (fMRI), proton magnetic resonance spectroscopy, and positron emission tomography scanning are used to evaluate brain activity and its correlation to different cognitive states to aid in clinical diagnosis. Cortical hypometabolism is associated with mild cognitive disability (e.g., misplacement of objects, inability to recognize familiar faces) and can be identified in individuals having specific genetic risk factors associated with AD, e.g., ApoE4 allele, but are as of yet symptomatic.[9]

Blood–brain barrier-permeable molecules with affinity for the β-amyloid peptide (Aβ) are also being developed as imaging agents. Most of these are derivatives of Congo Red or thioflavin that stain brain amyloid plaques. Methoxy-X04, a derivative of Congo Red, can detect plaque formation in animals, and the thioflavin T derivative, PIB, has been successfully evaluated in AD patients.[9]

Biomarkers are also necessitated for improved diagnosis and medical management of AD. Aβ, tau, and phosphorylated tau (ptau) proteins are potential biomarkers, and clinical studies over the course of the past decade have shown that monitoring the levels of Aβ42 and ptau (pthr231, pthr181, pser199) in cerebrospinal fluid (CSF) can potentially serve to discriminate AD from normal aging and other neurological disorders and can predict conversion of patients from MCI to AD, with low Aβ42 levels associated with both high total and ptau levels being a distinguishing feature.[10] These biomarkers are not robust in being able to distinguish AD from other forms of dementias, and alternate biomarkers are being sought using quantitative proteomic analysis to identify a panel of proteins, a molecular fingerprint that can differentiate AD and be used diagnostically – a major challenge. The National Institute of Health's BIOCARD study (Biomarkers in Older Controls at Risk for Dementia) reported 'tremendous' variability in CSF peptide markers (Aβ42, tau, and ptau) between controls and patients, making the current data rate-limiting in developing an accurate biomarker.[11] However, combining MRI and CSF biomarkers may improve sensitivity and specificity for AD diagnosis.

6.08.2.2 Disease Basis

A number of hypotheses as to the causes of AD have led to therapeutic strategies to ameliorate symptoms as well as modify disease.

6.08.2.2.1 The cholinergic hypothesis

The cholinergic hypothesis of AD, now some 40 years old, resulted from the discovery of reduced choline acetyltransferase activity, the enzyme that synthesizes acetylcholine (ACh), and basal forebrain cholinergic neurons in postmortem AD brain.[12] These findings led to the hypothesis that dementia resulted from cholinergic neuron dysfunction and/or loss. This hypothesis was supported by animal studies in which cholinergic neuron loss or dysfunction impaired tasks requiring memory or cognitive abilities. The exact relationship between cholinergic dysfunction and dementia is unclear; however, it is generally accepted that cholinergic systems are involved in both cognitive (e.g., attention and memory) and noncognitive (e.g., apathy, depression, pychosis, aggression, and sleep disturbances) behaviors that manifest during AD progression. Patients with MCI have increased choline acetyltransferase activity in the frontal cortex and hippocampus at autopsy, suggesting that upregulation of this enzyme may be a compensatory change occurring early in disease to counteract progression to AD.[13]

6.08.2.2.2 The amyloid hypothesis

Amyloid is a proteinaceous material that deposits both within neurons and as extracellular 'plaques' in brain tissue and is a hallmark (tombstone) neuropathological feature of AD.[14] Two peptides, amyloid beta (Aβ) peptide 1–40 (Aβ40) and 1–42 (Aβ42), derived from proteolytic cleavage of amyloid precursor protein (APP), are major constituents of plaques. Genetic linkage studies in AD patients have identified mutations in APP (20 missense) and in presenilin proteins -1 (140 known) and -2 (10 known) that result in increased production of APP cleavage products. Aβ peptides are hydrophobic, being derived from the transmembrane domain sequence of APP, and thus have a propensity to aggregate in aqueous environments. Aβ peptides associate in a different conformations of increasing molecular weight and decreasing solubility: oligomers, Aβ-derived diffusible ligands, protofibrils, and fibrils, all of which can interfere with neuronal function, elicit toxicity, and have the potential to elicit inflammatory cytokine production.[15] Aβ oligomers block long-term potentiation underlying memory consolidation.[16] Different forms of Aβ can activate cell surface receptors (e.g., α7 neuronal nicotinic receptors, p75 neurotrophin receptor, receptor for advanced glycosylation end-products; RAGE)[17] and intracellular signaling pathways that culminate in neuronal cell death.[18]

6.08.2.2.3 The tau hypothesis

The low-molecular-weight microtubule-associated protein, tau, has been implicated in the pathophysiology of AD and other CNS neurodegenerative diseases for which dementia and NFT formation is a common feature (e.g., FTDP linked to chromosome 17 (FTDP-17); Niemann–Pick disease type C).[19] Tau plays an important role in axonal transport, stabilizing microtubules, a process regulated by phosphorylation at one or more of the 30 identified serine and threonine sites on the protein. Kinases known to phosphorylate tau include microtubule affinity-regulating kinase, cyclin-dependent kinase-5 (CDK-5), glycogen synthase kinase-3 (GSK-3), c-jun N-terminal kinase (JNK), protein kinase A, protein kinase C, and calcium/calmodulin-dependent protein kinase II. Dephosphorylation of tau occurs via protein phosphatase-1, -2A, and -2B activity as well as the prolyl isomerase, Pin 1. In pathological states, tau is hyperphosphorylated, which causes its dissociation from microtubules and aggregation into a fibrillar form, paired helical filaments, which further aggregate to constitute the NFT. In AD, dementia is more closely correlated with NFT formation than amyloid plaques; thus, mechanisms involved in NFT formation have been of interest for drug development, with selective kinase inhibitors and inhibitors of tau protein aggregation being potential targets.

6.08.2.2.4 The inflammatory cascade hypothesis

Reactive gliosis and microglial activation are consistent findings in AD brain with a host of inflammatory mediators (e.g., complement proteins, cytokines, and chemokines) being present at higher levels than in normal brain. These observations support an inflammatory hypothesis of AD that postulates that either neurodegeneration in AD brain is secondary to an inflammatory response to senile plaques and NFTs (cause) or inflammation triggers formation of NFTs and senile plaques, which in turn activate immune reactions that drive a cycle of neuronal destruction.[20]

6.08.2.2.5 Genetic risk factors

Numerous genetic risk factors, genes with multiple alleles, one of which may predispose an individual to disease, have been identified, which increase the likelihood of development of AD.[21] A few recently identified AD-associated genetic risk factors are listed in **Table 1**. The first described with the highest risk association with disease to date is the E4 allele of the apolipoprotein E (APOE). This allele has been shown to influence Aβ oligomerization and its

Table 1 Neurodegenerative disease characteristics

Disease	*Alzheimer's*	*Parkinson's*	*Huntington's*	*ALS*
Phenotype	Episodic memory loss Cognitive decline	Tremor Rigidity Bradykinesis	Psychiatric symptoms Chorea Dementia	Muscle weakness/ fatigue Spasticity
Neuronal loss	Cortical/hippocampal	Dopaminergic	Striatal interneuron	Motor neuron
Number of affected individuals	4 million (US)	1.5 million (US)	30 000 (US)	30 000
Mean age of onset (years)	Early (EOAD) $\leqslant 65 \geqslant$ Late (LOAD)	60	Midlife	53–57
Life expectancy postdiagnosis (years)	5–10	10–20	15–20	3–5
Familial disease gene linkages	Amyloid precursor protein (APP) Presenilin 1 Presenilin 2	α-synuclein Parkin UCH-L1 PINK1 DJ-1 NR4A2 Synphilin-1 LRRK2	IT15 gene; expanded GAG (poly Q) repeat length	Cu/Zn superoxide dismutase (SOD-1)
Percent familial disease	5–10	5–10	100	5–10
Risk factors	Apolipoprotein E genotype VEGF mutation -2578A/A GAPDH SNPs (LOAD) D10S1423	Environmental exposure to toxins (e.g., pesticides; herbicides)	Mutant gene inheritance	Environmental exposure to toxins

EOAD, early-onset Alzheimer's disease; LOAD, late-onset Alzheimer's disease.

export from the brain and is associated with hypercholesterolemia, an additional factor associated with AD. D10S1423 on chromosome 10 that codes for a low-density lipoprotein-like receptor has shown a 97–98% association with AD diagnosis at autopsy.[22]

6.08.2.3 Experimental Disease Models

Recapitulating human disease pathophysiology in animals is a challenge, with many examples where activity in animal models has not been predictive of activity in human disease. This is especially true in the area of cognitive function where the actual tasks performed by animals differ substantially in terms of sophistication to those in humans. The complexity of the brain circuitry and the dominating brain regions differ widely, i.e., those distinguishing features separating humans from animals. Nonetheless, several models have been developed, which capture some, but not all, pathological features of AD and serve to provide systems in which to test mechanistic hypotheses.

6.08.2.3.1 Pharmacological and lesion models

Putative AD-like cognitive deficits are modeled in animals by producing electrolytic, mechanical, neurochemical, or immunotoxin lesions of the brain, predominantly the basal forebrain,[12] a strategy based on the cholinergic hypothesis of AD. Cholinergic denervation produces attention and memory dysfunction in multiple preclinical test species. Scopolamine, a muscarinic ACh receptor (mAChR) antagonist, also produces cognitive dysfunction. Lesion of the basal forebrain and in particular the *nucleus basalis magnocellularis* (NBM) in the rat with the excitotoxic glutamate receptor agonists, ibotenate, quinolate, kainate, quisqualate, or AMPA, can model the cognitive deficits associated with AD. These excitotoxins are predominately axon sparing and cause neuronal death by an excessive influx of Ca^{2+}. Reactive gliosis is also observed in these models.

The specific cholinergic toxin, AF64 **1**, binds to the high-affinity choline uptake system, and also produces lesions in the brain. High-dose intracerebroventricular AF64 produces widespread damage across the brain, while low dose produces more damage selective to the NBM. Immunotoxins can also produce lesions using cell surface receptor expression of cholinergic neuron markers, e.g., the low-affinity nerve growth factor receptor p75NGF. Saporin-conjugated antibodies injected intracerebroventricularly or into specific brain regions produce dose-dependent cholinergic neuron loss. To mimic AD pathophysiology more closely, infusion of fibrillar β-amyloid (Aβ42) into the NBM has also been employed, causing transient hypofunction of cholinergic neurons, i.e., function is restored when fibrils are cleared.

A major advantage of lesion models is that they produce a deficit in a short time span in readily accessible animals, i.e., rodents, providing the opportunity to evaluate NCEs rapidly. Deficits in attention and short-term memory resulting from cholinergic lesions can be measured postlesion using a variety of behavioral assays, e.g., delayed alternation in a T-maze or the Morris water maze. However, such cognitive deficits are of short duration and do not mimic the longer-term progressive dysfunction observed in AD. Also, the AD tombstones, plaques, and tangles are not observed, although an inflammatory response may occur.

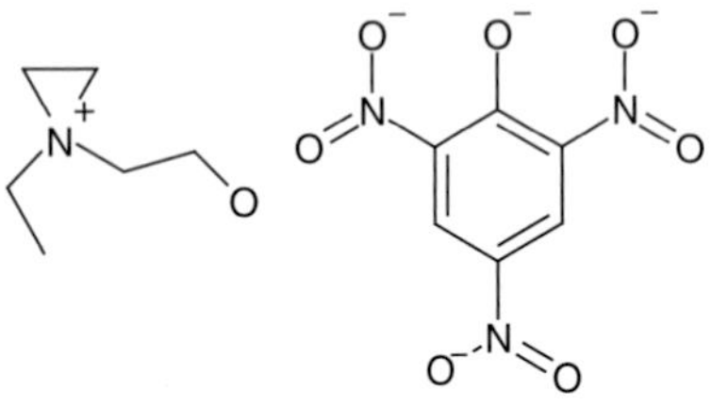

AF64 (**1**)

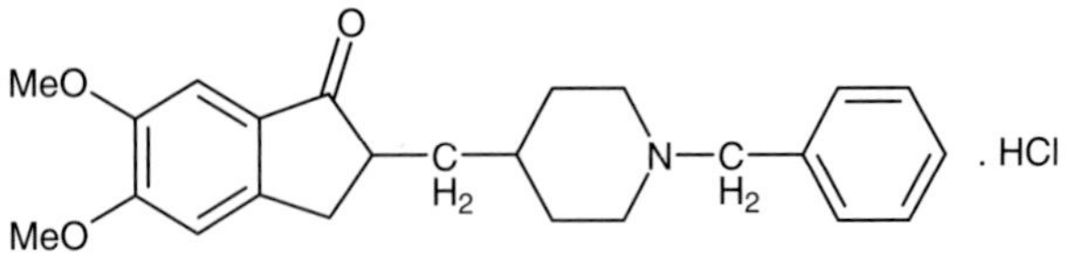

Donepezil (**2**)

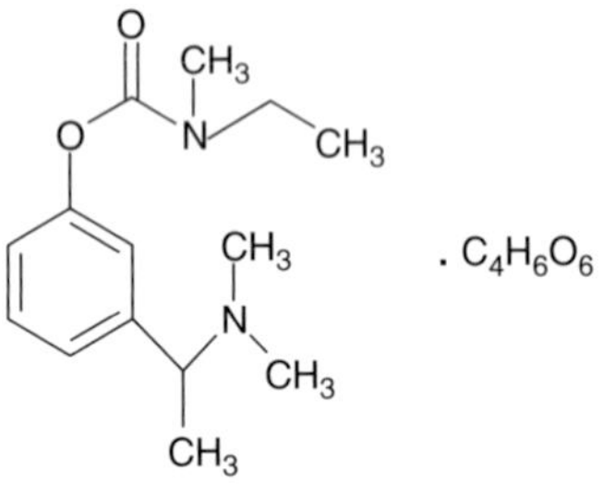

Rivastigmine (**3**)

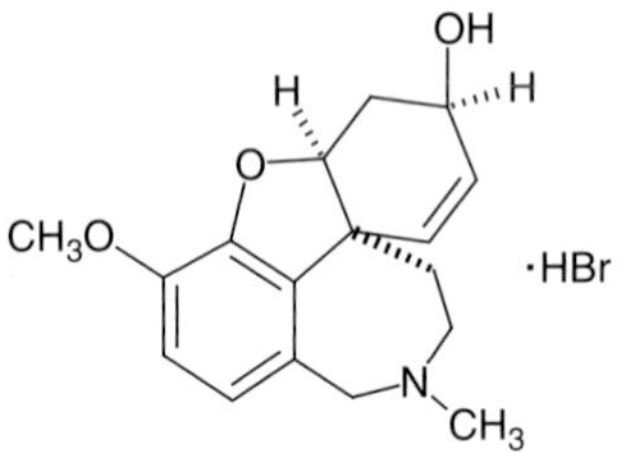

Galantamine (**4**)

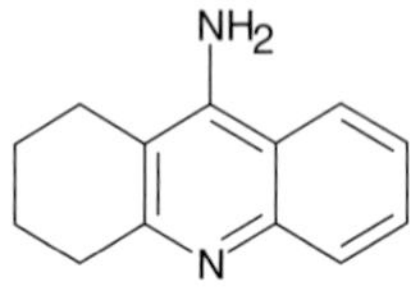

Tacrine (**5**)

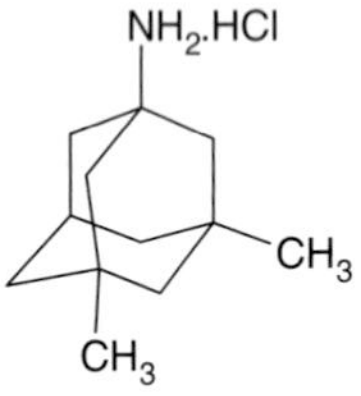

Memantine (**6**)

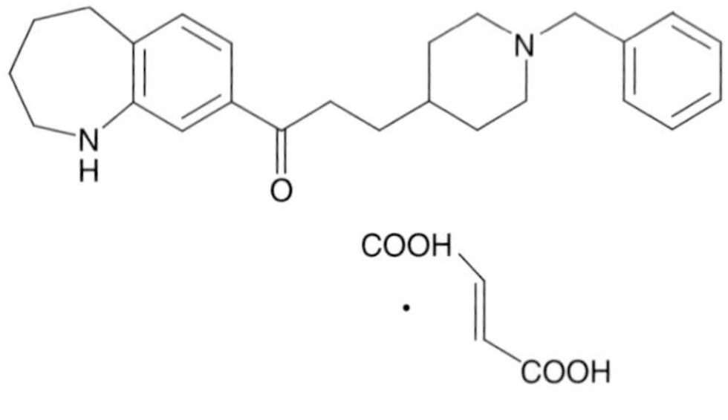

TAK147 (**7**)

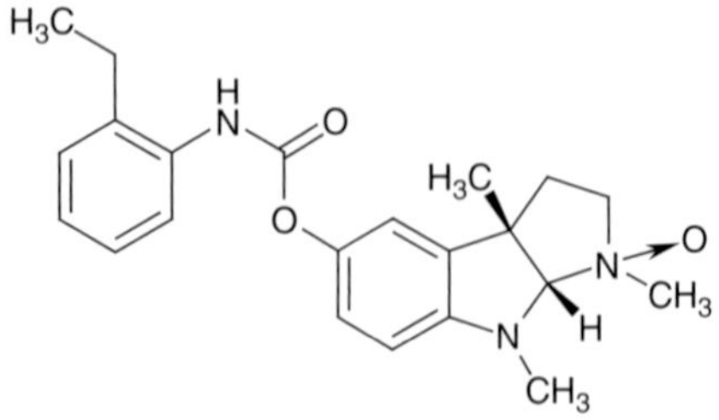

CHF2819 (**8**)

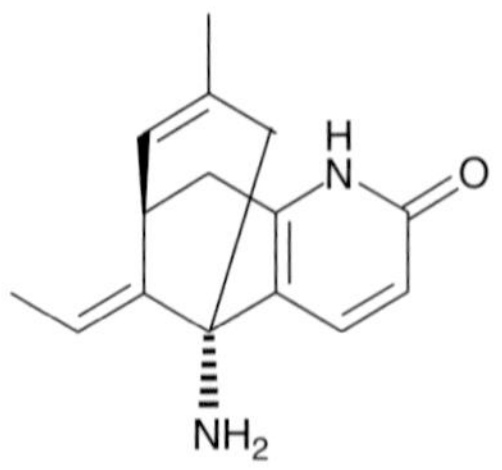

Huperazine A (**9**)

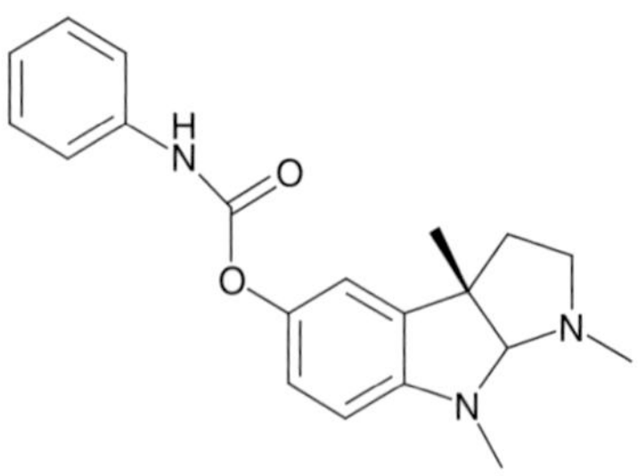

Phenserine (**10**)

6.08.2.3.2 Natural models

Aged animals, in particular aged rats, dogs, and nonhuman primates, can also be used as natural models of age-related cognitive decline. Each differs in its lifespan, so the term 'aged' is relative. For rats, adults are designated in the range of 5–6 months and aged animals are typically 20–26 months. In cognitive tests, e.g., delayed match-to-sample or Morris water maze, aged rats perform worse than their adults. However, no amyloid or neurofibrillary pathology is evident in aged rats; thus, the cognitive decline is attributed to age-related cholinergic hypofunctionality. The aged canine (7–14 years) has gained face validity as a model of human aging and dementia, i.e., AD, as dogs accumulate Aβ neuropathology in their brain and exhibit impairments in learning and memory with progressing age.[23] Like humans, the more toxic and less soluble peptide Aβ42 accumulates prior to deposition of Aβ40 in dogs and, importantly, the amyloid sequence is identical. Based upon performance in visuospatial and object recognition memory tasks, dogs can be classified as successful agers, impaired or severely impaired, paralleling the human classifications of successful aging, MCI, and dementia. Nonhuman aged (>20 years) primates (rhesus, macaque, and cynomolgus monkeys) that more closely model human cognition can be used to assess task impairment and reversal by NCEs. Like the dog, aged monkeys show a propensity to accumulate Aβ42 and develop senile plaques,[24] thus representing a model with greater face validity to AD than the rat.

6.08.2.3.3 Genetic models

The identification of gene mutations with linkage to early-onset AD (EOAD) provided an approach to studying the pathophysiology of autosomal dominant familial disease. Transgenic mice and rats have been developed that express the APP and/or the presenilins (PS1, PS2) with familial AD mutations. An array of pathology is observed in these animals, including amyloid deposition in diffuse and core plaques, dystrophic neuritis, and inflammatory gliosis. Widespread neurodegeneration is however not prevalent and NFT formation is absent, although immunohistological and biochemical analyses reveal phosphorylated and high-molecular-weight insoluble tau.[25] The Tg2576 model, the Hsaio mouse, introduced in 1996, was cross-bred to animals transgenic for PS1-containing FAD mutations. The mutant PS1 gene on the Tg2576 mouse accelerated the time course of amyloid deposition. Transgenic animals harboring mutations (P301L) in the tau protein gene responsible for FTDP exhibit motor and behavioral deficits and show age and gene dose-dependent development of NFTs in the spinal cord and in some (amygdala, septal nuclei, hypothalamus, midbrain, pons), but not all, brain regions (cortex, hippocampus, basal ganglia). Mice expressing both tau and APP familial AD mutations show enhanced development of NFTs in different brain regions, depending on whether there are single or multiple mutations in the tau gene. Reduced neuron numbers in both the cortex and hippocampus occur in double transgenic mice with multiple tau mutations. In triple transgenics with mutant APP, PS1, and tau proteins, synaptic dysfunction is evident early on and hallmark lesions (senile plaques and NFTs) follow a similar pattern of progression and expression in brain regions predominantly affected in AD. A more recent study using the Tg2576 AD mouse model showed decreases in dendritic spine density, impairment of long-term potentiation (LTP) and behavior that occurred several months before plaque deposition.[26] Decreases in spine density in the outer molecular layer of the dentate gyrus occurred at 4 months of age while plaque deposition did not occur until 12–18 months, re-emphasizing the need to study new treatment modalities designed to arrest disease progression at markers distinct from plaque formation when their efficacy may be limited. Despite the genetic sophistication of these mouse models, there are no NCE back-validated models of AD in human.

6.08.2.4 Clinical Trial Issues

Various clinical instruments have been used in the diagnosis of AD and can be used to monitor drug efficacy. These include the AD Assessment Scale (ADAS), noncognitive versus cognitive (cog), the Blessed Dementia Scale (BDS), the Blessed Information Memory Concentration (BIMC), the Behavior Rating Scale for Dementia (BRSD), the Clinical Dementia Rating (six categories; CDR), the Clinician's Interview-based Impression of Change (CIBIC), the Sum of Boxes (Global CDR, CDR–SB), the Dementia Rating Scale (DRS), the Extended Scale for Dementia (ESD), Global Deterioration Scale (GDS), the Mini-Mental State Examination (MMSE), the Progressive Deterioration Scale (PDS), and the Severe Impairment Battery (SIB).[27] Each scale measures both cognitive ability and activity of daily living but they are difficult to compare as their absolute ranges are very different, making comparison of results across clinical trials using different scales difficult. Thus, the only basis to compare data is on the percentage change from a baseline value. The ADAS–cog, CIBIC, and MMSE are most frequently used as primary endpoints in AD clinical trials. An unusual feature of AD clinical trials is that informed consent is typically consent given by the caregiver and not by the patient.

6.08.2.5 Current Treatment

6.08.2.5.1 Cholinesterase inhibitors

The cholinergic hypothesis of AD has led to the development of NCEs that increase ACh, which is lost when neurons of the basal forebrain degenerate. ACh is degraded by cholinesterase activity. Two major forms of cholinesterase exist, acetylcholinesterase (AChE) and butyrylcholinesterase (BuChE), with differential localization. Both enzymes are found in the brain, although AChE predominates ($\sim$10-fold greater) and is in synaptic clefts. BuChE is also found in brain, mainly in glia, but is preferentially localized to the periphery (intestine, liver, heart, lungs). AChE is selective for ACh hydrolysis, thus for terminating cholinergic neurotransmission, while BuChE hydrolyzes multiple choline esters. Its function in normal physiology is unknown. Two isoforms of AChE exist in the brain, a tetrameric form (G4) and a monomeric form (G1).[28] AChE levels decline with AD progression with preferential loss of the G4 isoform, while BuChE levels increase in the plasma and CSF of AD patients. This enzyme associates with neuritic plaques and tangles. With disease progression, there is an apparent shift in the enzyme dependence of ACh metabolism. Compounds that reversibly inhibit ACh hydrolysis would conceptually facilitate cholinergic neurotransmission by maintaining ACh in the synaptic cleft and maintaining cholinergic tone at muscarinic (mAChR) or nicotinic (nAChR) receptors (**Figure 2**).

Cholinesterase inhibitors approved for the treatment of mild to moderate AD include: donepezil **2**, rivastigmine **3**, and galantamine **4**. Each drug is structurally dissimilar and all are improvements over the first cholinesterase inhibitor, the aminoacridine tacrine **5**, which is no longer used due to its hepatotoxicity and the need for q.i.d. administration.

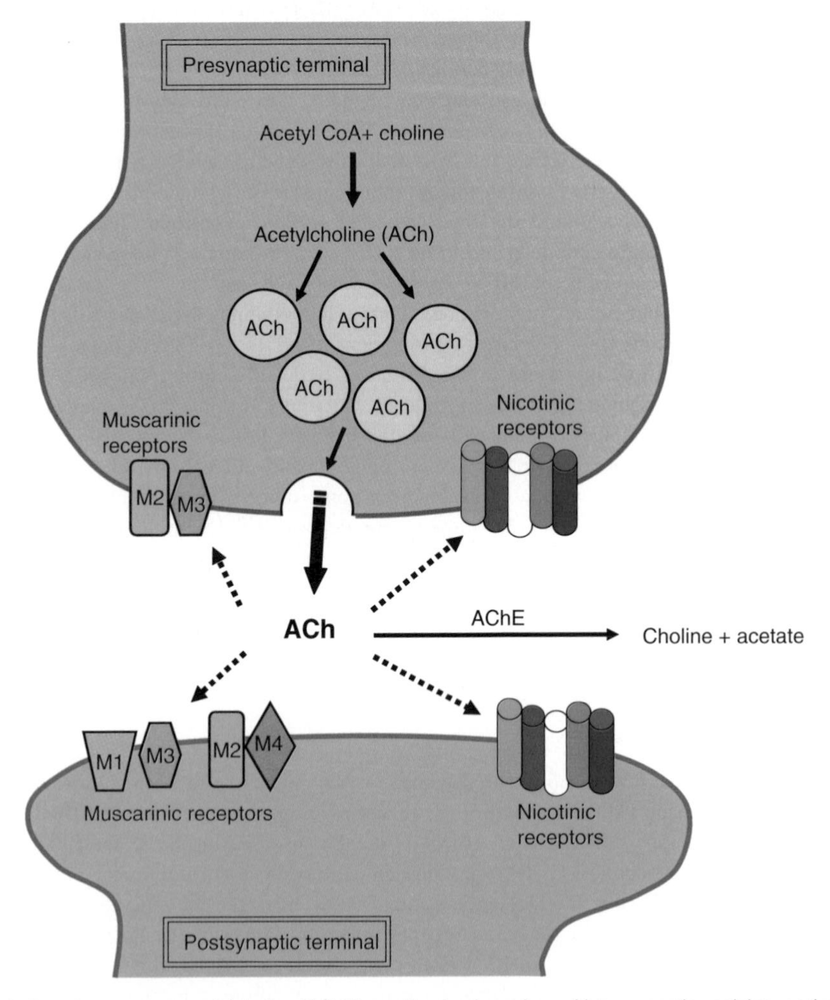

Figure 2 The cholinergic synapse: acetylcholine (ACh) is synthesized, packaged into synaptic vesicles, and released into the synaptic cleft. Termination of cholinergic transmission is accomplished by acethylcholinesterase (AChE) which degrades ACh. Blockade of AChE thus will sustain ACh in the synaptic cleft and facilitate cholinergic transmission through interaction with both muscarinic and nicotinic receptors. (Reprinted with permission from Wilkinson, D. G.; Francis, P. T.; Schwam, E.; Payne-Parrish, *J. Drugs Aging* **2004**, *21* 453–478.)

6.08.2.5.1.1 Donepezil

Donepezil **2** was approved for the treatment of AD in 1996, and is an indanone derivative of tacrine and physostigmine selectivity for AChE relative to BuChE (1:405). It has a very long elimination half-life, ranging from approximately 60 h in young adults to approximately 100 h in the elderly,[29] consistent with q.d. use. Donepezil showed significant improvement versus placebo in ADAS–cog measures (1.5–2.5 units at 5 mg; 2.88–3.1 at 10 mg dose). For the CIBIC, a higher percentage of patients were rated as improved, that is, 26–52% (5 mg) and 25–38% (10 mg) versus 11–22% (placebo).

6.08.2.5.1.2 Rivastigmine

Rivastigmine **3** targets AChE and BuChE, and exhibits prolonged inhibition of both enzymes, being termed a 'pseudo-irreversible' as well as 'brain region'-selective cholinesterase inhibitor.[28] The pharmacodynamic effect of rivastigmine (10 h) exceeds its short plasma elimination half-life (1 h) because of the nature of its pseudo-irreversible interaction with AChE. Rivastigmine displays greater potency ($\sim$4–6 times) for the G1 versus the G4 isoform of AChE, that is reduced in AD, and may thus afford a more prolonged treatment window opportunity relative to other cholinesterase inhibitors without this isoform selectivity profile.

6.08.2.5.1.3 Galantamine

Galantamine **4**, a naturally occurring plant tertiary alkaloid, is a reversible AChE inhibitor that is a weak positive allosteric modular of nAChRs.[30] Thus galantamine has been proposed both to increase ACh levels and to potentiate the interaction of ACh presynaptically to enhance further ACh release and postsynaptically to facilitate downstream signaling. The proposed presynaptic actions of galantamine may also lead to the modulation of glutamate, serotonin, and norepinephrine release, all of which are affected in AD. Galantamine is approximately 50-fold selective for AChE over BuChE; however, selectivity for erythrocyte AChE versus brain AChE only differs by 10-fold.

Unlike rivastigmine, galantamine is metabolized by cytochrome P450 enzymes, CYP2D6 and CYP3A4, with doses of this drug being lowered if it is co-administered with other drugs that inhibit these enzymes.[30] Phase III trials of 3–6 months' duration, with one study extending to 12 months, showed improvements in the ADAS–cog, the disability assessment for dementia (DAD), and the AD cooperative study/activities of daily living (ADCS/ADL) inventory following galantamine treatment.[30,31]

6.08.2.5.2 *N*-Methyl-D-aspartate (NMDA) receptor antagonists

Memantine **6** is an orally active, noncompetitive NMDA receptor antagonist approved for treatment of moderate-to-severe AD. Memantine also has efficacy in mild-to-moderate AD and is pending labeling approval in the US.

The NMDA receptor plays a key role in long-term potentiation, a process underlying memory and learning consolidation. Pathological activation of NMDA receptors via excess glutamate and calcium drives excitotoxicity associated with AD. A limitation in targeting this receptor is that compounds interacting at the glutamate or glycine sites on this receptor, e.g., phencyclidine (PCP), are known psychomimetics.

Memantine binds weakly to the ion channel-binding site on the NMDA receptor when it is in an open state and thus blocks the tonic pathological activation, induced by micromolar glutamate concentrations. It also normalizes NMDA receptor activation in AD patients, allowing physiological responses to occur, preventing excitotoxic events.[32] In nine randomized, double-blind, placebo-controlled clinical trials in dementia and AD patients over a 12-year period, memantine showed improvement in at least one, if not multiple, primary endpoints (1992–2004).

6.08.2.6 Unmet Medical Needs

The major unmet medical need for AD, as with all neurodegenerative diseases, is for disease-modifying therapies that can be used with reliable early diagnostic markers. There are currently no approved drugs that modify overall disease progression, although recent data suggest that donepezil may provide some protective benefit.[33] Neuropsychiatric symptoms, including depression, pychosis, agitation (aggressive or disruptive behavior), and wandering associated with AD, are currently addressed with drugs used offlabel for AD (e.g., citalopram, clomipramine, sertraline, risperidone, olanzapine).

6.08.2.7 New Research Areas

Multiple strategies for AD treatment have emerged from an evolving understanding over the past 10–20 years regarding the pathological basis of the disease. The limited time window for effective symptomatic treatment with AChE inhibitors observed in clinical practice in patients with mild to moderate AD highlights the need for additional treatment strategies. Memantine was a partial solution to this need in treating later AD stages (i.e., moderate to severe). Other AChE inhibitors (e.g., TAK147 **7**, CHF2819 **8**, huperzine A **9**) that have improved CNS selectively are in clinical trials with slow progress. AChE also appears to have alternate activities beyond catabolism of ACh, including accelerating the rate of formation of Aβ fibrils, and inhibitors that affect both activities are under development,[34] including analogs of phenserine **10**.[35]

Since cholinergic afferent neurons projecting to the basal forebrain and hippocampus are lost in AD, agonists that stimulate ACh release by interaction with presynaptic receptors and also mimic ACh to stimulate postsynaptic receptors (i.e., mAChRs and nAChRs) and facilitate neurotransmission have been investigated as therapeutic treatments. NCEs that target these postsynaptic ACh receptors in theory would serve as add-ons to AChE inhibitors. Strategies to halt disease progression include targeting production, aggregation, and clearance of Aβ, hyperphosphorylation of tau, and neuroinflammation, all of which are key elements leading to neuronal dysfunction and cell death. Multiple strategies to block neuronal death via trophic support or by interfering with death signaling pathways are also being explored and a host of targets that facilitate improvements in cognition are under investigation.

6.08.2.7.1 Muscarinic acetylcholine receptors (mAChRs)

The mAChR receptor is a 7TM/G protein-coupled receptor (GPCR) family with five members (M1–M5). Muscarinic M1 and M3 receptor (mAChR) activation can stimulate the α-secretory pathway for processing of APP to yield greater amounts of sAPPα and, theoretically, less Aβ peptide.[36] M2 receptor activation may also inhibit sAPPα secretion. M1 agonists also decrease the tau protein phosphorylation and hyperphosphorylation resulting from ApoE deficiency and improve learning and memory in animal models with cholinergic deficits. Muscarinic agonists advanced to the clinic that include civemeline **11**, alvameline **12**, and xanomeline **13** have failed due to a lack of receptor selectivity, poor bioavailability, a lack of robust efficacy, and/or a narrow therapeutic index, usually gastrointestinal side effects.

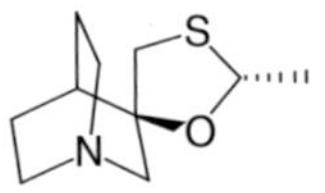

Civemiline (**11**) 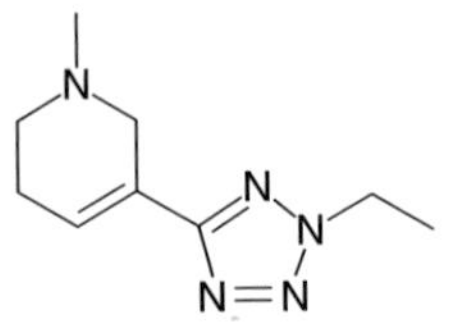   Alvameline (**12**) 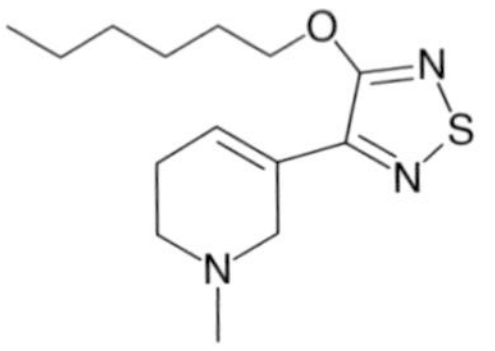   Xanomeline (**13**)

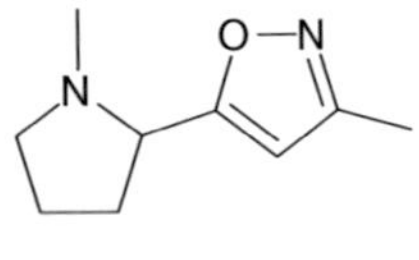

ABT-418 (**14**) 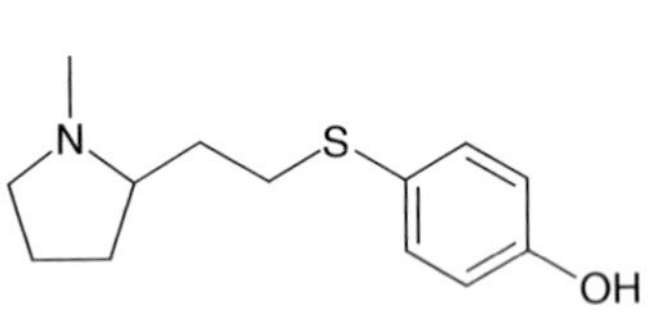   SIB-1553A (**15**) 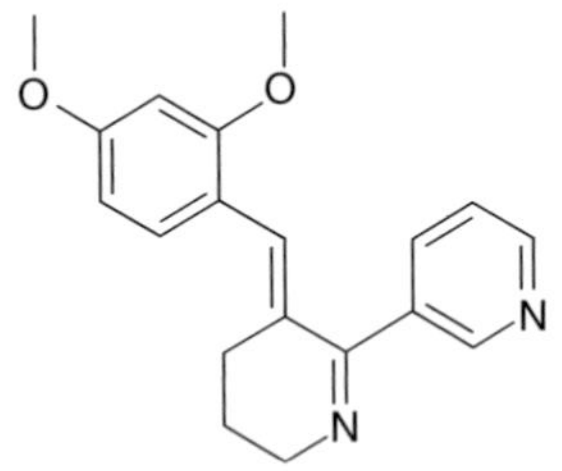   GTS-21 (**16**)

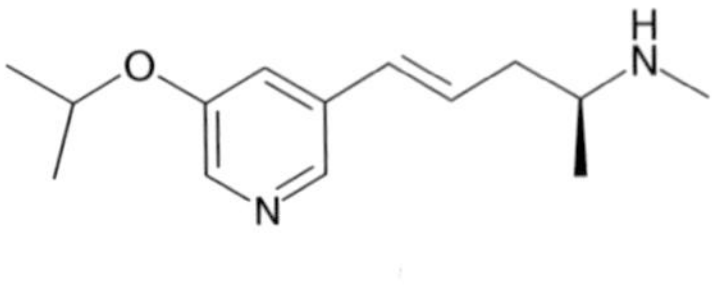

TC-1734 (**17**) 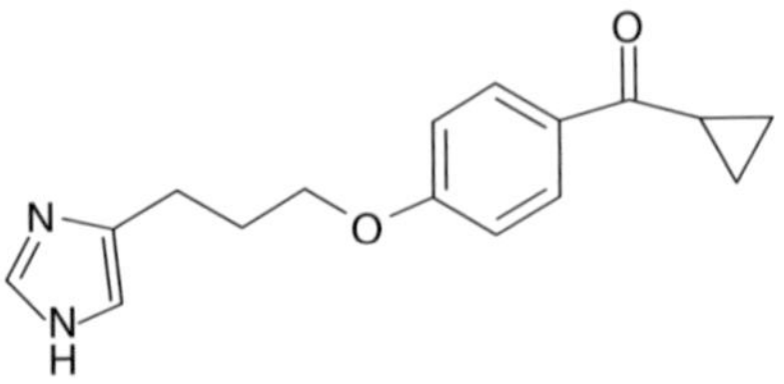   Ciproxifan (**18**) 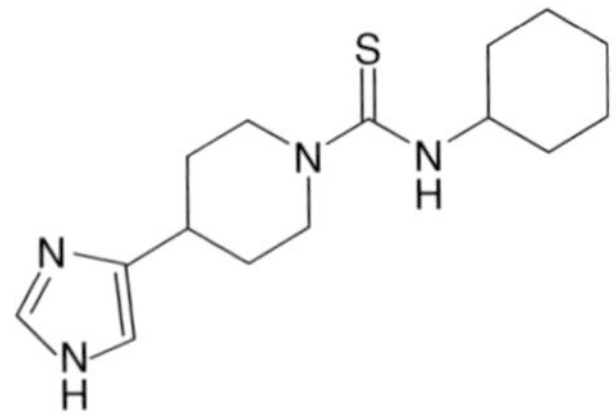   Thioperamide (**19**)

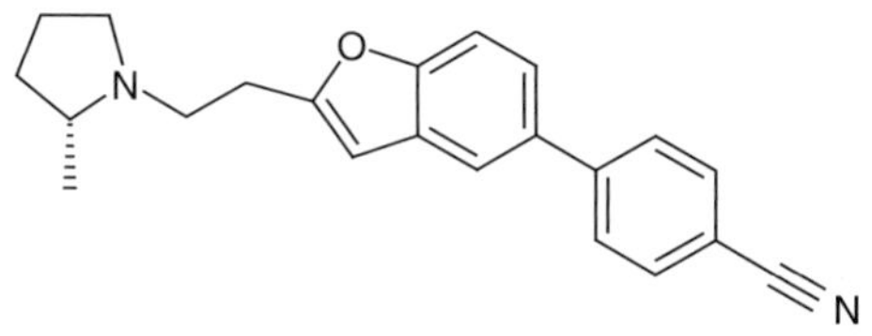

ABT-239 (**20**)

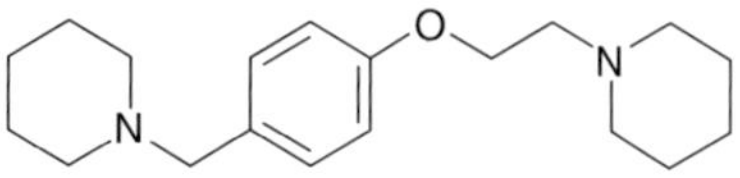

JNJ-5207852 (**21**)

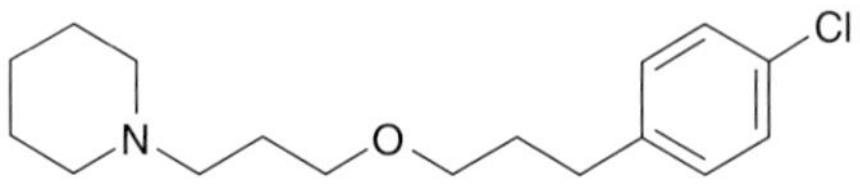

BF 2649 (**22**)

6.08.2.7.2 Nicotinic acetylcholine receptors (nAChRs)

nAChRs are homo- or heteropentameric ligand-gated ion channels present in the CNS, peripheral nervous system, and neuromuscular junction. Various subunits can combine to provide a diversity of receptor subtypes with unique brain and neuron-specific distributions.[37] Activation of nAChRs mediates calcium influx and neurotransmitter release, again specific to the neuronal subtype (i.e., cortical, hippocampal). Interest in nAChRs comes from observations that they are reduced in AD and that nicotine improves attention in AD patients. Also Aβ42 can bind to α7 nAChRs and antagonists of this receptor promote neuron survival. Several nAChR agonists have entered the clinic (e.g., ABT-418 **14**, SIB-1553A **15**, GTS-21 **16**, TC-1734 **17**). The progress of these agents has been slow due to efficacy issues and side effects, including emesis, motor dysfunction, and hallucinations, although these are reduced in comparison to nicotine due to improved receptor subtype selectivity. The first α4β2 agonist to undergo clinical trials, ABT-418, showed efficacy in acute studies in AD patients, but failed to show efficacy, as did nicotine, in a double-blind placebo-controlled trial. TC-1734 is a second-generation, orally active α4β2 nAChR agonist with enhanced receptor subtype selectivity.[37]

6.08.2.7.3 Histamine receptor antagonists

Of the four members of the histamine receptor superfamily, the H_3 is predominately expressed in the brain, localizing to cerebral cortex, amygdala, hippocampus, striatum, thalamus, and hypothalamus. H_3 receptors are localized presynaptically on histaminergic nerve terminals and act as inhibitory autoreceptors; thus, when activated by histamine, histamine release and biosynthesis are blocked.[38] Of relevance to AD, H_3 receptors expressed on nonhistaminergic nerve terminals can modulate the release of ACh, DA, γ-amino-butyric acid (GABA), glutamate, and serotonin. Thus, H_3-receptor antagonists or inverse agonists, by blocking the inhibitory effects of histamine, will facilitate the release of multiple neurotransmitters, reminiscent of the effects of nAChR activation. H_3-receptor antagonists enhance vigilance, promoting wakefulness in rats, mice, and cats, and improve cognitive function in a variety of preclinical models. H_3-receptor inverse agonists are thus being targeted as therapeutics for AD and attention-deficit hyperactivity disorder. Prototypic H_3-receptor antagonists include thioperamide **18** and ciproxifan **19**. Newer H_3-receptor antagonists include ABT-239 **20**, JNJ-5207852 **21**, and BF 2649 **22**.

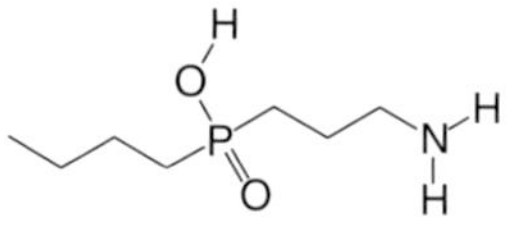

SGS-742 (**23**)

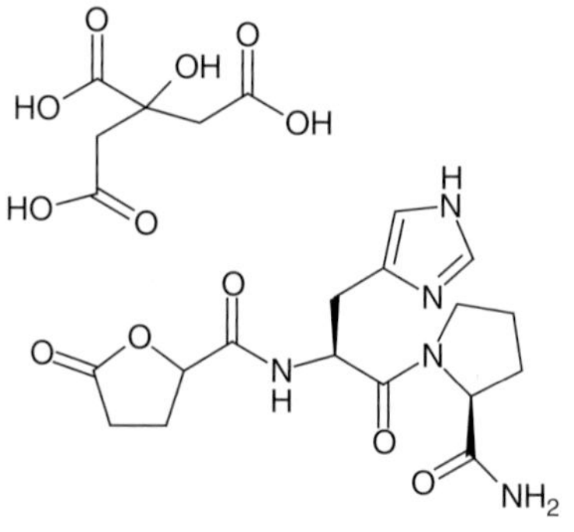

DN-1417 (**24**)

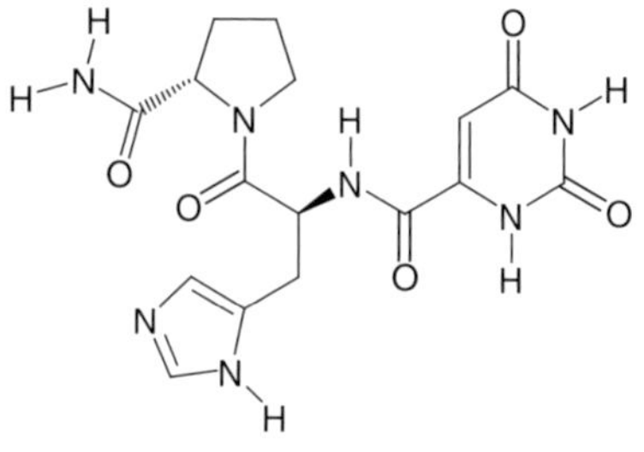

Orotirelin (**25**)

RX-77368 (**26**)

MK-771 (**27**)

Posatirelin (**28**)

Taltirelin (**29**)

Leuprolide (**30**)

ONO-1603 (**31**)

6.08.2.7.4 γ-Amino-butyric acid receptor antagonists

GABA, the major inhibitory neurotransmitter in the CNS, acts via $GABA_A$, $GABA_B$, and putative $GABA_C$ receptors. $GABA_A$ and $GABA_C$ receptors are ligand-gated ion channels, while $GABA_B$ receptors are 7TM-GPCRs. The $GABA_B$

receptor is of most interest as a target for AD. Both autoreceptors and heteroreceptors exist for the $GABA_B$ receptor family that modulate the presynaptic release of GABA as well as glutamate, various monoamines, and neuropeptides. GABA receptor activation inhibits learning; thus receptor antagonists may be useful in reversing this deficit. SGS-742 (CGP-36742: **23**), is a low-potency $GABA_B/GABA_C$ antagonist (IC_{50} values = 36 and 63 μM, respectively) currently in the clinic.[39] Chronic administration of SGS-742 upregulated $GABA_B$ receptor expression, increased protein levels of nerve growth factor (NGF) and brain-derived neuronal factor (BDNF) in rat cortex and hippocampus, and facilitated the release of glutamate, aspartate, glycine, and somatostatin. In mice and rats, SGS-742 improved learning and memory tasks at doses of 3–30 $mg\,kg^{-1}$ (i.p. or p.o.).

6.08.2.7.5 5-Hydroxytryptamine (5HT)-receptor agonists

The 5HT-receptor 7TM/GPCR receptor family comprises at least 16 members: the $5HT_4$ subtype is of interest in AD. $5HT_4$-receptor activation reverses deficits in synaptic transmission in hippocampal slice preparations from transgenic mice overexpressing Aβ.[40]

6.08.2.7.6 Thyrotropin-releasing hormone agonists

The neuropeptide, thyrotropin-releasing hormone (TRH), modulates CNS function, including activation of the cholinergic system. TRH analogs that overcome the short plasma half-life (4–5 min) include DN-1417 **24**, orotirelin **25**, montirelin, RX-77368 **26**, MK-771 **27**, azetirelin, and posatirelin **28**.[41] Taltirelin (TA-0910: **29**) is a small-molecule TRH agonist, with improved potency (up to 300-fold, depending on the biological assay) and pharmacokinetics (plasma $t_{1/2}$ ~3 h) introduced in Japan in 2000 for the treatment of spinocerebellar degeneration. In animals, it dose-dependently increased extracellular ACh, potentiated ACh-induced neuronal excitation in cortical neurons, and inhibited ACh-induced desensitization. Taltirelin improved MMSE scores in demented patients and is reportedly in trials for AD.

6.08.2.7.7 Hormone replacement therapy (HRT) – gonadatropin-releasing hormone (GnRH) agonists (estrogen)

Epidemiological studies showing an increased prevalence of AD in postmenopausal females focused interest on the regulation of the hypothalamic–pituitary–gonadal (HPG) axis in AD. Additionally, females with high levels of endogenous estrogen were less prone to develop AD.[42] The possibility that estrogen might serve as an AD preventive led to several preclinical studies examining its protective effects against amyloid toxicity and on cognitive performance. HRT was found not only to be ineffective in preventing AD in postmenopausal women over the age of 65 years in the National Institute of Health-sponsored Women's Health Initiative Memory Study, but increased the risk of dementia. Despite this finding, regulation of the HPG axis is still thought to be key in the development of AD, since AD patients show a twofold increase in gonadotropins, specifically luteinizing hormone in AD patients. Luteinizing hormone is thought to cause reactivation of the cell cycle in neurons leading to cell death, as well as processing of APP via the amyloidogenic pathway.[43] Leuprolide (leuprolide acetate: **30**), a GnRH agonist that suppresses luteinizing hormone production by downregulating GnRH receptors, is currently in phase III trials.

6.08.2.7.8 Prolyl endopeptidase inhibitors

Prolyl endopeptidases hydrolyze proline-containing peptide hormones, several of which are implicated in learning and memory. ONO-1603 (**31**) is a potent prolyl endopeptidase inhibitor (K_i = 12 nM) that reverses scopolamine-induced deficits in learning and memory in rats, restores decreased levels of choline and 5HT in aged rats and improves cognitive performance, and is also neuroprotective to CNS neurons undergoing apoptosis.[44] While ONO-1603 was advanced to phase II trials for senile dementia in Japan, it was discontinued in 1995.

6.08.2.7.9 Secretase inhibitors

Secretases are enzymes that process APP into nonamyloidogenic and amyloidogenic fragments and exist in three subtypes: α, β, and γ. Cleavage of APP by α-secretase followed by γ-secretase produces peptide fragments in the nonamyloidogenic pathway, whereas cleavage by β-secretase followed by γ-secretase produces the Aβ peptide fragments that aggregate and deposit as plaques (**Figure 3**). Thus, both β- and γ-secretases are potential targets for AD therapy, although γ-secretases have shown toxic liabilities. Two zinc-dependent metalloproteinases, ADAM10 and ADAM 17 (TACE), are candidate genes for α-secretase. Since the latter enzyme produces cleavage products with neurotrophic properties, inhibition of this enzyme may be counterproductive.

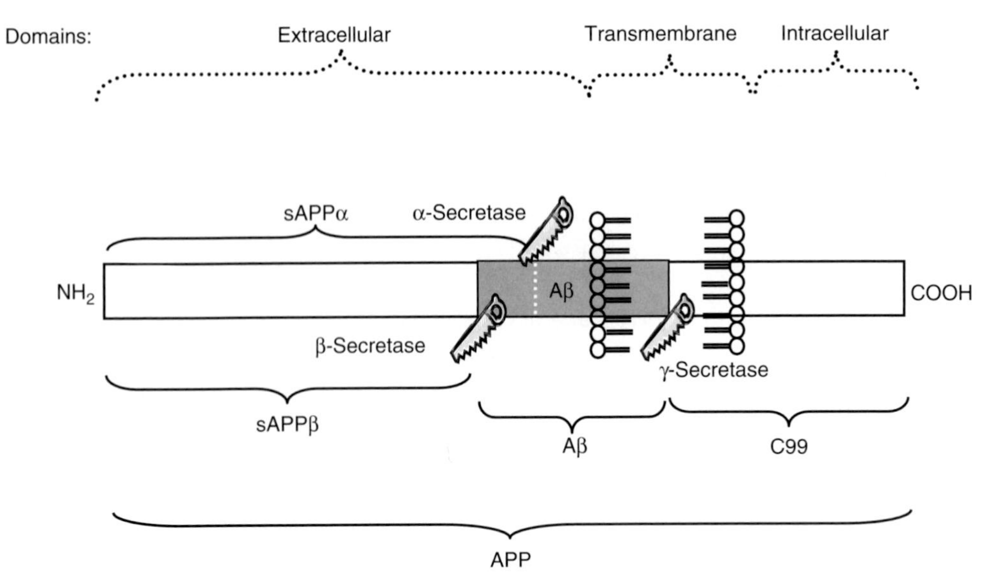

Figure 3 Amyloid precursor protein (APP) processing. APP is enzymatically cleaved by secretases (= α, β, and γ) to liberate multiple peptide fragments. Processing by α-secretase alone produces a neurotrophic soluble(s) APPα peptide; simultaneous cleavage by β- and γ-secretases produces the pathogenic Aβ, C99, and sAPPβ peptides.

Pepstatin (**32**)

LY450139 (**33**)

Clioquinol (**34**)

(**35**)

(**36**)

Azakenpaullone (**37**)

Indirubin (**38**)

Pyrazolopyrimidines (**39**)

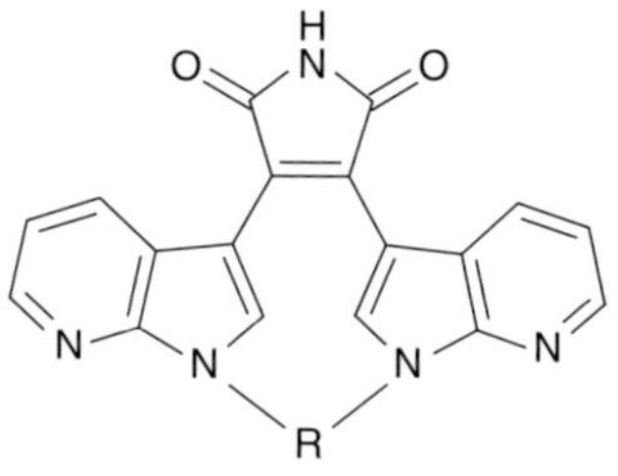

(40)

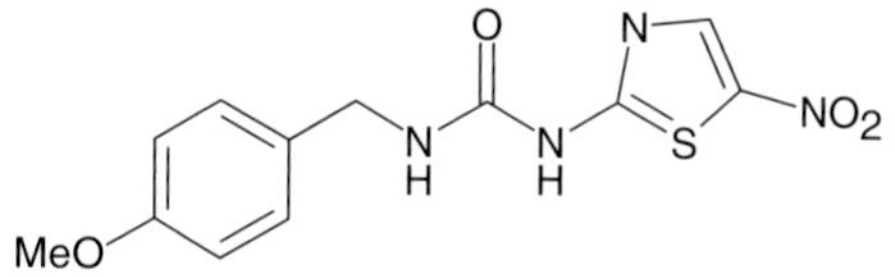

AR-A014418 (41)

β-Secretase (BACE-1, β-site APP cleavage enzyme-1; Asp2; Memapsin) is a membrane-bound aspartyl proteinase. It is a zymogen, the activation of which requires the cleavage of its prodomain by a Furin-like proteinase. BACE-1 knockout mice develop normally and lack overt abnormal behaviors. Inhibitors of BACE-1 are drug targets to limit Aβ formation without toxicity. The prototypic BACE-1 inhibitor is pepstatin **32**. Newer BACE-1 inhibitors have potent activity against the isolated enzyme ($IC_{50} = 1$–400 nM) but limited cellular potency ($IC_{50} \sim 1$–50 μM).

γ-Secretase is an apartyl protease catalyzing the intramembrane cleavage of APP together with β-secretase to produce the Aβ40 and Aβ42 peptides. The proteins linked to EOAD, the presenilins (mainly PS-1), constitute a complex with γ-secretase together with the membrane proteins: nicastrin, Aph-1, and Pen-2. The challenge in finding inhibitors of γ-secretase is that the enzyme has many substrates in addition to APP, some of which, e.g., Notch proteins, have a role in development.[45] The γ-secretase inhibitor, LY-450139 (**33**: $K_i = 0.9$ nM), inhibits Aβ formation in cells, transgenic animals, and beagle dogs. LY-450139 (30 mg day^{-1} for 1 week followed by 40 mg day^{-1} for 5 weeks) decreased plasma levels of Aβ40 ~40% in AD patients on stable doses of an AChE inhibitor. No differences in measures of cognition were observed, perhaps reflecting the need for longer trials in order to reveal added benefit of γ-secretase inhibition.[46]

6.08.2.7.10 β-Amyloid aggregation blockers

Alternate strategies to reduce the Aβ plaque load include: (1) inhibition of the formation of Aβ dimers and their aggregation into β-pleated sheets; and (2) clearing of Aβ deposits by immune surveillance. Only small molecule-based strategies will be reviewed here; however, both passive and active immunization strategies to facilitate brain Aβ clearance, e.g., AN-1792, have shown success in animal models. Clinical exposure with AN-1792 resulted in significant adverse events with second-generation vaccines, e.g., AAB001, under evaluation.

Aβ polymerization is dependent on pH, temperature, and time, with a number of cofactors, e.g., heparin sulfates, glycosaminoglycans, and metals (e.g., Cu^{2+}, Zn^{2+}, Fe^{3+}, Al^{3+}), modulating the process. Linear peptides, 5–7 amino acids in length, corresponding to the central hydrophobic region of the Aβ sequence inhibit Aβ fibril formation. Specifically, peptides with a proline residue preclude β-sheet formation as well as those that contain Phe in the second and Leu in the third position.[47] No peptide-based inhibitors of Aβ aggregation are in clinical trials.

Metal chelators have also been targeted, as metal ions form intermolecular bridges with the histidine residues of adjacent Aβ peptides and can stabilize aggregates. Clioquinol (PBT-1: **34**), an antimalarial agent, selectively chelates Zn^{2+} and Cu^{2+}, avoiding nonspecific depletion of tissue metal ions.[48] It reduced brain Aβ deposition in genetically engineered mice and increased the fraction of solubilized Aβ, thus reducing the total brain Aβ burden. Whether increased soluble Aβ is a positive event is controversial as soluble Aβ has toxic effects. Additionally, the safety profile of clioquinol is questionable as it reduces vitamin B_{12}, an action that may be related to subacute myelooptic neuropathy seen in patients taking clioquinol as an antimalarial. Nonetheless, clioquinol advanced to clinical trials, with patients being supplemented with vitamin B_{12} and folic acid. In a 36-week phase II dose escalation study in AD patients, clioquinol lowered plasma Aβ42 levels, improving performance on the ADAS–cog in severely affected patients, although the behavioral improvement was not apparent at the 6-month time point.

6.08.2.7.11 Tau kinase inhibitors

Prevention of tau phosphorylation is an additional strategy for AD therapy. The multiple epitopes on tau appear essential to AD pathology, e.g., hyperphosphorylation promotes aggregation and formation of paired helical filaments. Inhibition of the phosphorylation of sites close to the microtubule-binding domain of tau, e.g., pthr231 and pser396, is thought to be critical for preventing tau aggregation, and their hyperphosphorylation has been associated with the incidence of AD. Two kinases identified in the phosphorylation of these sites are CDK-5 and GSK-3β.

CDK-5 (also known as neuronal CDC2-like kinase or NCLK) is a proline-directed serine/threonine kinase present in all tissues; however, its activity is dependent on noncyclin regulatory proteins that are highly expressed in neurons.[49] Excessive CDK-5 activity results in neurotoxicity and a pathological event that facilitates increased enzyme activity is proteolysis of its regulatory protein, p35, to a smaller protein p25 by the Ca^{2+}-dependent enzyme, calpain. Several lines of evidence support a role of CDK-5 in AD pathogenesis: calpain activity is increased in AD brain together with the p25:p35 ratio; overexpression of CDK-5 in primary neurons increases tau phosphorylation at AD-relevant epitopes, and induces cytoskeletal disruption and apoptotic cell death. However, in models of p25 overexpression, NFT formation only occurs with coexpression of mutant tau, suggesting that CDK-5 activity alone is not sufficient for inducing this pathological event. There is no clinical data on small molecule CDK-5 inhibitors. Prototypic compounds include 3′-substituted indolones[50] (**35**) and [1,3,5]triazine-pyridine biheteroaryls[51] (**36**).

GSK-3β (tau protein kinase I, TPK1) is a serine/threonine kinase that constitutively inactivates glycogen synthase.[52] GSK-3β is linked to AD pathogenesis via its ability to hyperphosphorylate tau and colocalization with NFTs. Aβ exposure of hippocampal neurons activates GSK-3β enhancing tau phosphorylation at multiple sites (Ser181, Ser199, Ser202, Ser396, Ser404, and Ser413). Mice overexpressing GSK-3β show tau hyperphosphorylation in hippocampal and cortical neurons with somatodendritic localization that is accompanied by neuronal death and reactive gliosis and spatial leaning deficits in the Morris water maze. Inhibition of GSK-3β may be beneficial in AD. However, this kinase is also involved in the *wnt* signaling pathway and β-catenin expression, which is linked to colon cancer and may thus have unacceptable side effects.

Prototype GSK-3β inhibitors include kenpaullones **37**, indirubins **38**, pyrazolopyrimidines **39**, macrocyclic maleimides **40**, and thiazoles (e.g., AR-A014418; **41**)[53].

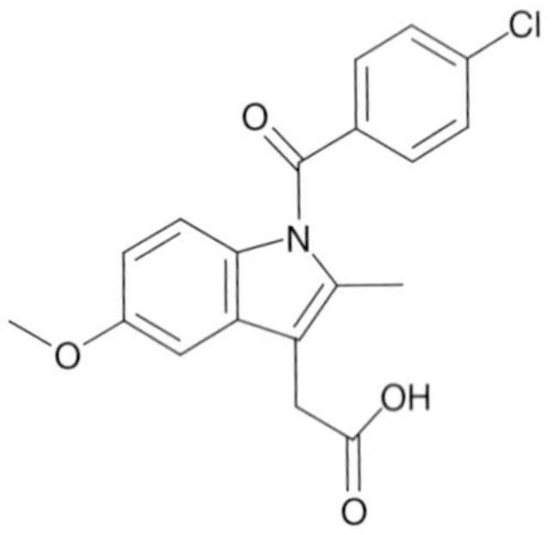

Indomethacin (**42**)

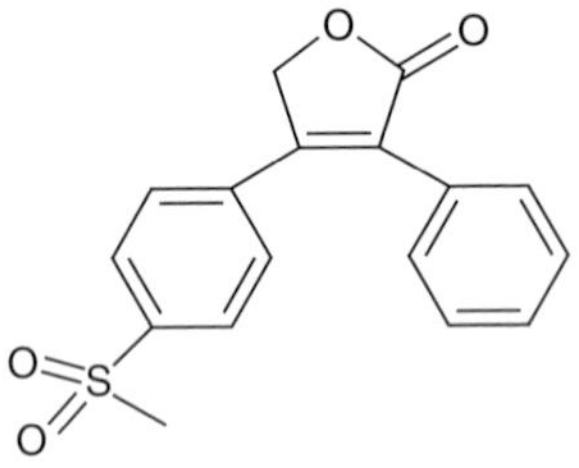

Rofecoxib (**43**)

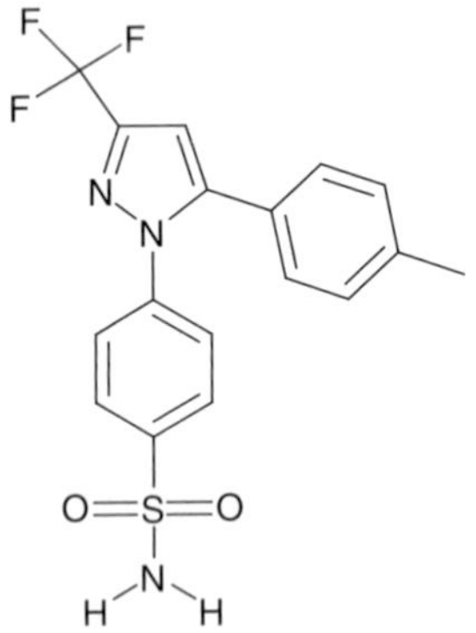

Celecoxib (**44**)

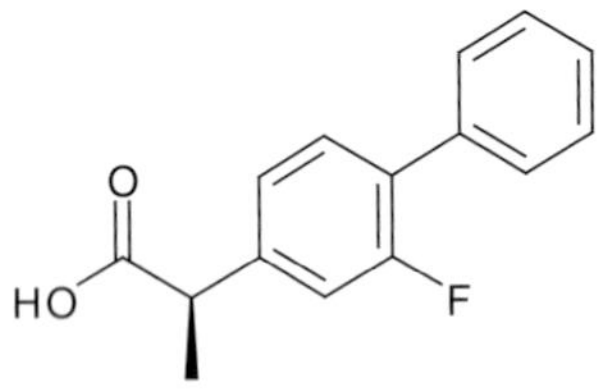

(R) Flurbiprofen (**45**)

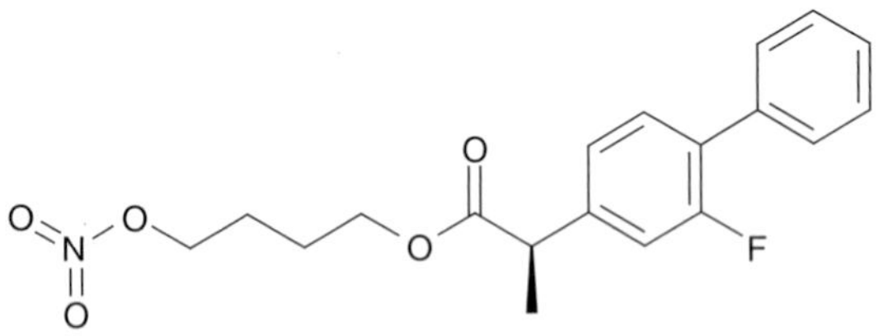

Nitroflurbiprofen (**46**)

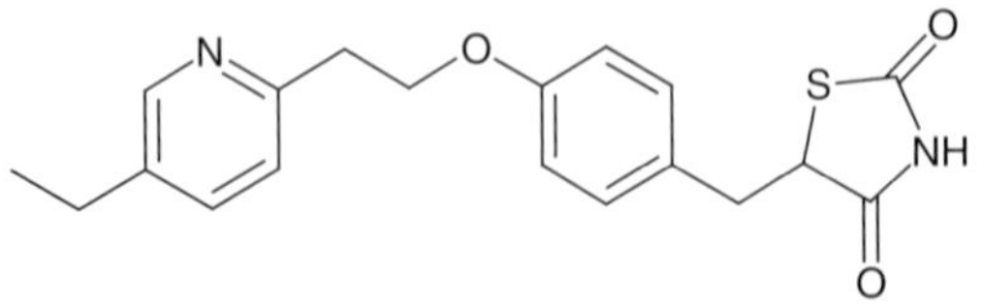

Pioglitazone (**47**)

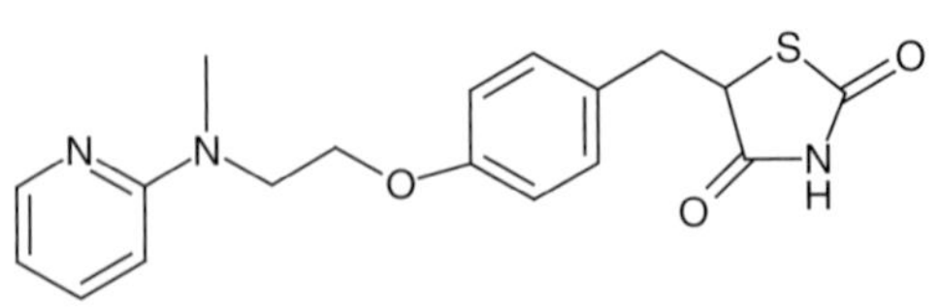

Rosiglitazone (**48**)

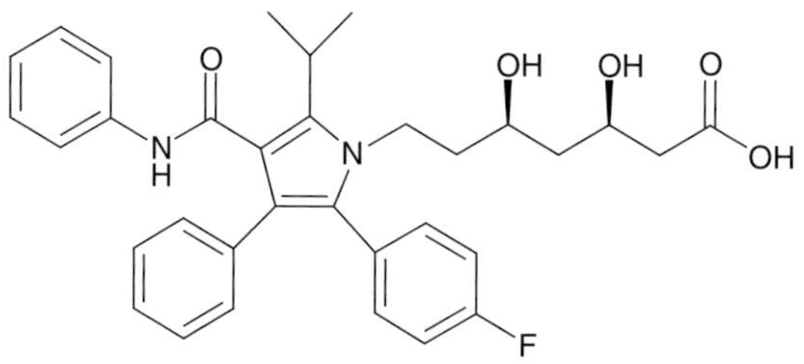

Atorvastatin (**49**)

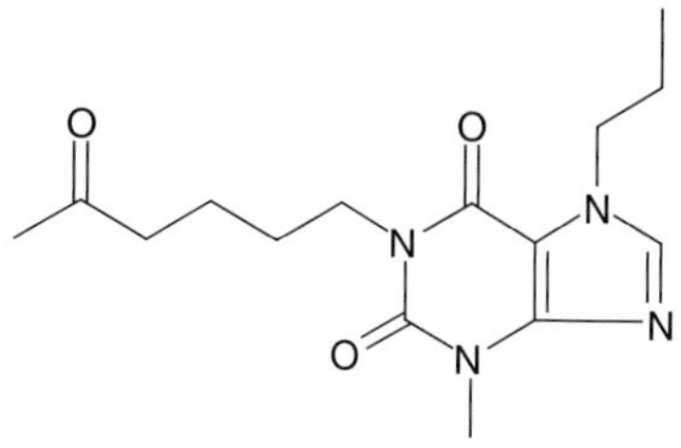

Propentofylline (**50**)

6.08.2.7.12 **Anti-inflammatory compounds**

Brain inflammation and activation of microglia are early pathological events in AD, and epidemiological studies have suggested that nonsteroidal anti-inflammatory drugs (NSAIDs: cyclooxygenase (COX)-1 and -2 inhibitors) may slow disease onset. Initial retrospective studies of patients with AD using indomethacin **42** showed a positive effect in reducing AD progress.[54] However, prospective trials with NSAIDs, e.g., rofecoxib **43**, celecoxib **44**, and naproxen, failed to demonstrate efficacy in AD. Nonetheless, the COX-1 inhibitor, (*R*)-flurbiprofen **45**, is in Phase III AD trials, despite limited benefit in a Phase II study. Nitroflurbiprofen (NCX-2216; **46**), an NO-donating flurbiprofen derivative, with reduced gastrointestinal side effects, is also in Phase II trials in AD. In preclinical studies, nitroflurbiprofen reduced lipopolysaccharide-induced microglial activation and decreased both the microglial activation and the amount of Aβ deposits in the brains of transgenic mice with familial AD mutations in the APP gene.[55]

In addition to their anti-inflammatory effects, NSAIDs may modulate amyloidogenic processing of APP via interactions with the peroxisome proliferator-activated receptor-γ (PPAR-γ).[20] NSAIDs, e.g., indomethacin **42**, activate PPAR-γ, hence the interest in evaluating this target in AD. Pioglitazone **47**, a PPAR-γ agonist used in the treatment of type 2 diabetes, showed equivocal effects on Aβ pathology in AD mouse models but clinical improvement was demonstrated with another PPAR-γ agonist, rosiglitazone **48**, in patients with mild AD or amnestic MCI.[56]

6.08.2.7.13 **Statins**

Dietary cholesterol affects the formation of Aβ and accelerates the appearance of AD pathology in mouse models; high serum cholesterol and low high-density lipoprotein cholesterol levels are AD risk factors.[57] Cholesterol-lowering agents, e.g., statins, are of interest as treatments for AD. These agents inhibit β-hydroxy-β-methyl glutaryl-CoA reductase, a key regulatory enzyme in the cholesterol biosynthetic pathway. Atorvastatin **49**, evaluated in an intention-to-treat clinical trial, showed positive benefit in ADAS–cog performance at 6 months.[58] An observational study of AD patients treated with lipid-lowering agents, 47% of which were statins, showed a positive benefit (i.e., slower decline) on the MMSE.[59] Thus, with further study statins may become standard therapy.

6.08.2.7.14 **Neuroprotectants**

Enhancing neuronal survival via the use of trophic agents, the majority of which are proteins, has been challenging as the latter do not readily cross the blood–brain barrier. The identification of small molecule neutrophin receptor agonists has also been challenging. Alternate strategies to target neurotrophic pathways involve the identification of: (1) NCEs that enhance neurotrophic protein production, e.g., propentofylline and SR-57746; and (2) small molecule peptide agonists that can be delivered intranasally, e.g., AL-108. Despite their preclinical efficacy, these neuroprotective agents have had limited clinical success.

6.08.2.7.14.1 Propentofylline

Propentofylline (**50**: HWA 285) is a xanthine phosphodiesterase inhibitor with weak adenosine uptake inhibitory properties that can potently stimulate astrocyte NGF synthesis. It partially restores age-associated decreases in cortical NGF in vivo and attenuates bilateral forebrain-induced impairments in animal behavioral tasks, and can also inhibit advanced glycation end-product-induced TNF-α release.

Propentofylline reduces Aβ plaque burden and attenuated tau hyperphosphorylation in transgenic mice overexpressing APP with familial AD mutations.[60] Two European phase III trials showed positive outcomes of propentofylline in AD but a longer-term 72-week trial failed to show benefit, and development was discontinued.

6.08.2.7.14.2 AL-108

AL-108 is an 8-mer (NAPVSIPQ) of activity-dependent neuroprotective protein that protects neurons from a variety of insults and protects APOE-deficient mice against developmental retardation and learning impairments.[61] It also improved rat performance in the water maze following cholinergic lesioning AL-108 in phase I trials.

6.08.3 Parkinson's Disease

6.08.3.1 Disease State/Diagnosis

PD is primarily a movement disorder resulting from the loss of dopaminergic neurons of the substantia nigra pars compacta (SNpc). It is diagnosed by the presence of resting tremor, bradykinesia (slowness of voluntary movement), rigidity (stiff muscles, expressionless face), and postural instability (poor recovery of balance). The gait of individuals with PD is also distinctive, with a tendency to lean unnaturally forwards or backwards, a diminished arm swing, and a shuffling step. Although loss of DA-containing neurons is a hallmark pathological feature of the disease, other neuronal populations are also affected.[62] Structures that are initially affected in stages 1–2 of PD include the vagal nerve, the dorsal motor nucleus and the olfactory bulb, and the anterior olfactory nucleus. At stage 3, involvement of basal portions of the midbrain and forebrain and a number of cholinergic nuclei and the tuberomamillary nucleus is apparent. At stage 4, portions of the cerebral cortex are affected. Stages 3–4 are where individuals begin the transition from presymptomatic to symptomatic phases of PD and progressively deteriorate through stages 5–6, with complete loss of midbrain neurons. PD symptoms are not manifest until 50–80% of dopaminergic neurons are lost, making maintenance of the survival of the remaining 20–50% of these neurons critical.

It is estimated that 1.5 million individuals in the US are affected by PD. The disease usually presents after the age of 60 with a disease course that can last 10–20 years. The most commonly used scales for PD diagnosis and clinical trial evaluation are the Hoehn and Yahr staging of PD, the unified PD rating scale (UPDRS), and the Swab and England activities of daily living scales.[63] However, like AD, PD is a diagnosis based on exclusion, and definitive diagnosis is made at autopsy.

There are currently no biomarkers approved for PD diagnosis to determine disease status or clinical outcomes. Radiotracer imaging with probes such as [^{18}F]fluorodopa, [^{123}I]β-CIT, and [^{11}C]dihydrotetrabenazine, to measure neuronal viability based on the presence of the DA transporter (DAT), the vesicular monoamine transporter (VMAT), or dopaminergic or monoaminergic nerve terminals, is being used in clinical trials, although its utility as a surrogate marker is unclear since the loss of marker can reflect either overt neuron loss or a downregulation of the radiotracer target, DAT.[64] For instance, kinase inhibitors can downregulate DAT,[65] making data interpretation complex and questioning the degree of validation of this biomarker.

6.08.3.2 Disease Basis

Although the neuropathology of PD is well established, the molecular mechanisms underlying neurodegeneration have not been elucidated. Biochemical abnormalities in PD patients include: dysfunction in complex I of the mitochondrial respiratory chain with consequent energy (ATP) depletion, increased free radical damage, and oxidative stress. Environmental factors, e.g., exposure to pesticides and herbicides like rotenone and paraquat, inhibitors of complex I, may be causal in PD. Reduction of ATP reduces proteasomal protein degradation. Defects in protein handling occur in PD and appear to underlie the formation of Lewy bodies, an intracytoplasmic aggregate of several proteins, including α-synuclein and ubiquitin that is a hallmark feature of the disease.

Genetic linkage studies of familial PD have revealed abnormal expression or mutation in at least eight proteins (**Table 1**), some of which are associated with mitochondrial function (e.g., DJ-1, PINK-1).[66]

6.08.3.3 Experimental Disease Models

6.08.3.3.1 Neurotoxin models

Several neurotoxins have been used to produce selective dopaminergic neuron loss to model sporadic disease, and, in some species, e.g., the nonhuman primate, parkinsonism-like symptoms are evident following exposures.[67] The three major toxins used to lesion the brain are 1-methyl-4-phenyl-1,2,3,6-tetrahydropyridine (MPTP: **52**), 6-hydroxydopamine (6-OHDA: **53**), and rotenone **54**. Unilateral lesion of the substantia nigra, medial forebrain bundle, or striatum with 6-OHDA produces dopaminergic neuron loss, albeit at different rates, leading to a 'hemiparkinson' model where the injected side of the brain exhibits degeneration. Administration of DA agonists like apomorphine or the stimulant amphetamine elicits rotation or circling toward the contralateral side in animals, unmasking the DA loss. Although 6-OHDA causes dopaminergic neuron loss, other neurons affected in PD (e.g., lower brainstem nuclei and locus coeruleus) are unaffected and Lewy bodies do not occur.

The MPTP model of PD is based on the idiopathic parkinsonism seen in individuals abusing synthetic heroin contaminated with MPTP. MPTP is converted by glial cells to a toxic metabolite methylpyridium (MPP +), a complex I inhibitor, which selectively affects DA neurons as it is a ligand for DAT. MPTP neurotoxicity varies across species (the rat is not susceptible), with mouse, particularly the C57Bl6 strain, and nonhuman primates being the most extensively studied. Both acute and chronic administration of MPTP is used to model PD and to evaluate NCEs. In acute models, a high dose of MPTP causes a rapid degenerative response while in chronic models multiple applications of low doses of MPTP cause a more protracted degenerative response, more closely resembling human PD progression. Shortcomings of the MPTP model include the lack of Lewy body formation (although cytoplasmic inclusions occur) and the lack of degeneration in other monoaminergic structures (e.g., the *locus coeruleus*) that are also affected in PD.

Based on epidemiological studies, rotenone, a naturally occurring complex ketone and high-affinity inhibitor of complex I used as a herbicide, has been used to develop a model of PD. Systemic administration of rotenone to rats results in a selective loss of dopaminergic neurons, despite widespread depression of complex I activity throughout the brain, suggesting that dopaminergic neurons are uniquely susceptible to this toxin. Like other PD toxin models, the rotenone model displays PD-like behavioral disturbances, including postural instability, gait unsteadiness, and paw tremor. However, unlike other models, the neuropathology of PD is closely mimicked, e.g., α-synuclein and ubiquitin-containing inclusions develop with ultrastructural similarities in fibril formation to those of Lewy bodies. Microglial activation also occurs prior to overt dopaminergic neurodegeneration. The enhanced toxicity of rotenone may be associated with an inflammatory response and to NADPH oxidase-derived O_2 release from microglial cells.

6.08.3.3.2 Genetic models

Several genetic manipulations have been used to study the pathophysiology, etiology, and pathogenesis of PD.[68] Mice with altered DA function can be created by knocking out enzymes involved in DA formation, including tyrosine hydroxylase (TH) catabolic enzymes (monoamine oxidase (MAO), catechol *O*-methyltransferase (COMT)), DAT, VMAT, or altering DA receptor expression. The findings from these mice are that, unless a marked alteration occurs in striatal DA levels or postsynaptic neurotransmission, no gross abnormalities can be observed.

The identification of genes linked to PD spurred the development of both vertebrate and invertebrate animal models of the disease via either knockout or overexpression strategies. α-Synuclein knockouts show no DA neuron loss or abnormal motor behavior, but do exhibit a subtle alteration of DA receptors in the striatum and altered responses to amphetamine. Transgenic wild-type and mutant (A53T, A30P) α-synuclein mice display neuropathological and behavioral features of PD; however, no one vertebrate model encompasses all disease features.[68] Some models show striatal DA deficits accompanied by motor abnormality while others show motor abnormality without deficits in DA content but Lewy body-like pathology. The only model that displays all features of PD is in *Drosophila*, where there is a loss of DA neurons, cytoplasmic inclusion formation, a DA deficit, and associated motor dysfunction. The only confound is that these features occur with overexpression of either wild-type or mutant α-synuclein, although polymorphisms in the α-synuclein gene promoter as well as genomic multiplication of the α-synuclein locus have been linked in PD patients; thus there is a disease basis for overexpression of the normal protein as perhaps causing the associated dysfunction.

6.08.3.4 Clinical Trial Issues

Like AD, a major clinical trial issues for all CNS neurodegenerative diseases is in the design of studies to differentiate symptomatic from neuroprotective benefits of NCEs.

For PD, two different clinical study designs (drug wash-out and randomized delayed starts) have been used in attempts to distinguish this difference for NCEs.[69] The first uses a drug wash-out period at the end of the treatment course. Patients are followed for different time periods following drug withdrawal to determine if a benefit persists in the absence of continued drug use. While this is a logical design, there are inherent problems. One issue is that the pharmacodynamic effects of treatment may outlast the duration of the physical presence of the drug. Another is that patients with more severe symptoms cannot tolerate discontinuation of a symptomatic treatment; thus patients with less severe symptoms may be preferentially retained in a study. The second design is a randomized delayed-start trial in which patients are randomized to treatment groups following receipt of placebo for various durations. This study design also has challenges: (1) treatment earlier in the course of disease may provide greater symptomatic benefit; and (2) patients may drop out of the study prior to receiving treatment.

6.08.3.5 Current Treatment

While non-dopaminergic agents used for adjunct PD treatment include the weak NMDA antagonist amantadine **55** and the anti-cholinergics, trihexylphenidyl **56** and benztropine **57**, the most effective drugs used to treat PD all modulate the dopaminergic system, either via regulation of DA synthesis, blockade of synaptic reuptake or via direct activation of DA receptors. Drugs used for PD treatment provide symptomatic improvement, either as primary treatment or adjunct therapy. The gold standard is L-dopa **51** introduced in the mid-1970s. Direct DA receptor agonists, e.g., apomorphine **58**, bromocriptine **59**, pramipexole **60**, ropinirole **61**, and pergolide **62**, have been increasingly used as firstline therapy to delay onset of motor fluctuations and dyskinesias that typically occur with long-term L-dopa use.[70] In patients with advanced PD, DA receptor agonists are used in combination with L-dopa to 'spare' L-dopa therapy. There is currently no approved treatment for slowing PD progression.

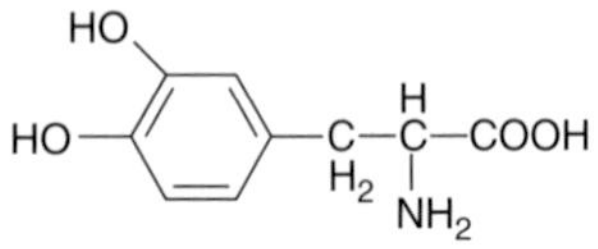

L-Dopa (**51**)

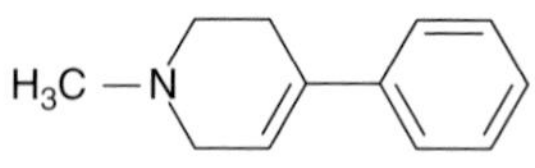

MPTP (**52**)

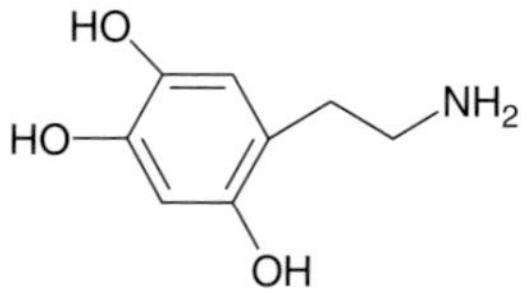

6-OHDA (**53**)

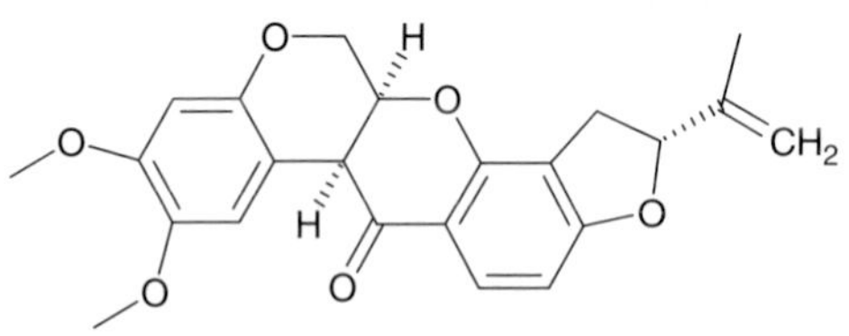

Rotenone (**54**)

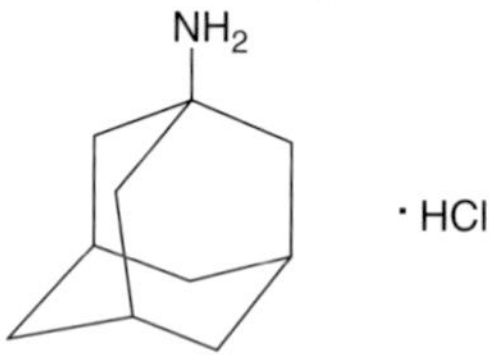

Amantadine (**55**)

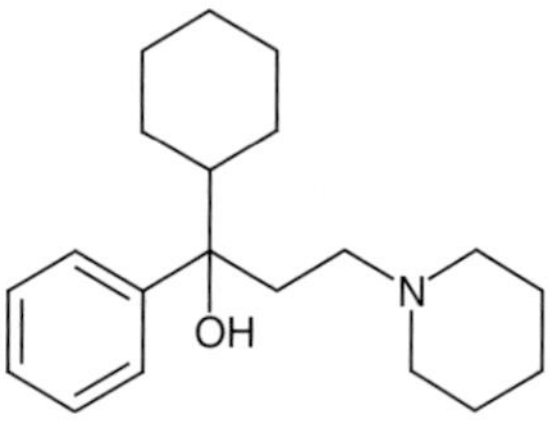

Trihexyphenidyl (**56**)

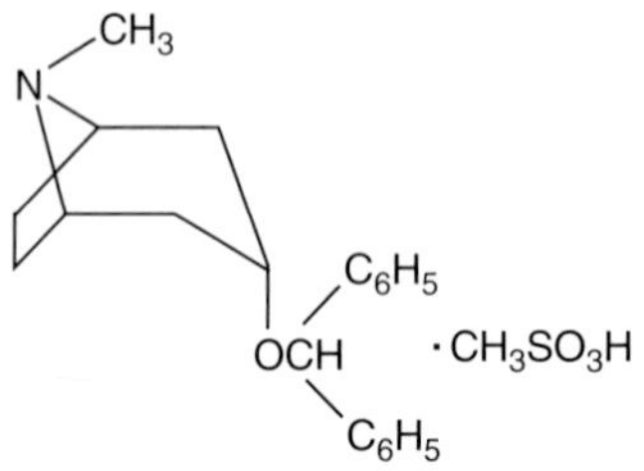

Benztropine (**57**)

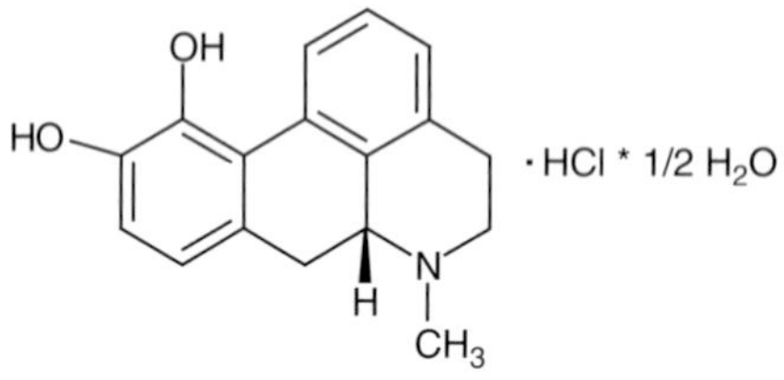

Apomorphine (**58**)

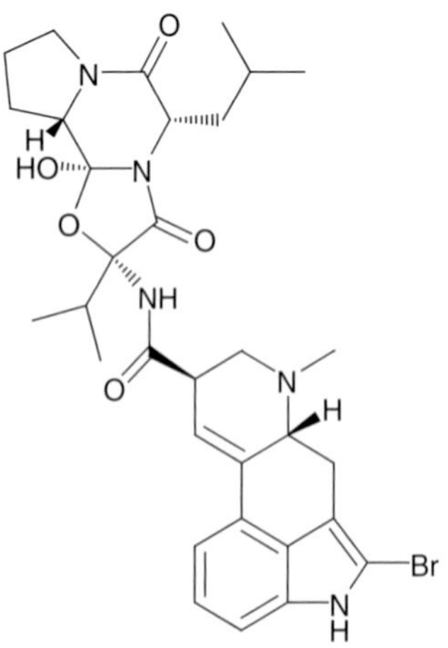

Bromocriptine (**59**)

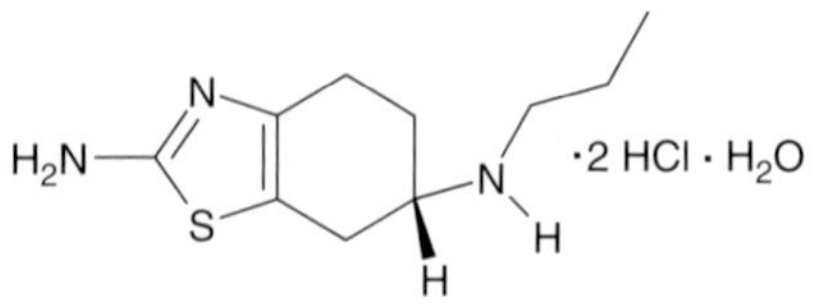

Pramipexole (**60**)

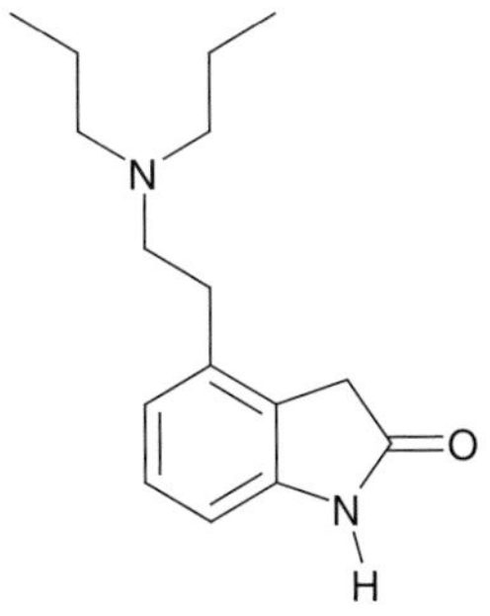

Ropinirole (**61**)

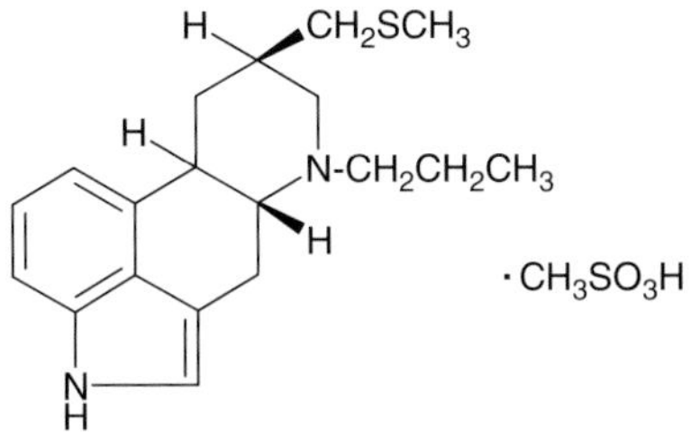

Pergolide (**62**)

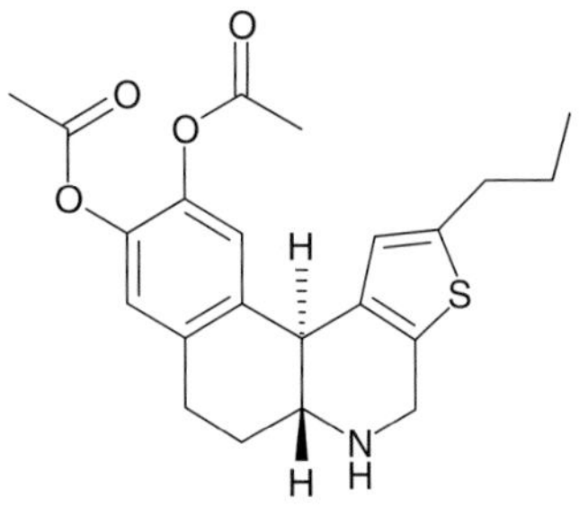

ABT-431 (adrogolide) (**63**)

6.08.3.5.1 Dopamine replacement

L-dopa was the first drug approved for treatment of PD, acting to restore decreased DA levels as a result of the loss of dopaminergic neurons. This DA precursor acts locally using remaining DA neurons to synthesize DA. Peripheral metabolism of L-dopa is prevented by peripheral decarboxylase inhibitors like carbidopa. L-Dopa treatment has a number of limitations: (1) efficacy is dependent on DA neurons; as these die as a result of disease progression, L-dopa becomes less effective; and (2) L-dopa may potentiate DA neurotoxicity. DA and/or its metabolites are toxic to neurons both in vitro and in vivo, an effect that occurs via inhibition of oxidative phosphorylation, generation of reactive oxygen species, or via direct DA receptor interactions. A randomized, placebo-controlled trial ELLDOPA (earlier versus later levodopa therapy in PD) examined both the neurotoxic and the disease-modifying potential of L-dopa.[71] A dose-dependent reduction in the worsening of symptoms for patients on L-dopa with no evidence of any acceleration in the rate of clinical decline was observed, suggesting a neuroprotective effect.

6.08.3.5.2 Dopamine receptor agonists

Of the two major classes of DA receptors, D_1 and D_2, it is generally accepted that D_1 receptors function in the direct pathway in the basal ganglia whereas the D_2 receptors function in a parallel indirect pathway. Thus, D_1 receptor activation facilitates movement whereas D_2 receptor activation inhibits movement. DA receptor agonists used in the clinic comprise two broad chemical classes: the non-ergolines, including apomorphine **58**, pramipexole **60**, and ropinirole **61**; and ergolines, bromocriptine **59** and pergolide **62**. Pramipexole is a D_2 selective, full agonist with ropinirole having a similar profile being also active at D_3 receptors. Apomorphine binds with highest affinity to the D_4 receptor but also interacts with D_1 and D_2 receptors. While newer D_1 agonists like ABT-431 **63** failed to show improved benefit in Phase II trials, additional D_2 agonists are in clinical trials and include the partial D_2 agonist, SLV-308, and the D_2 agonist, rotigotine.

6.08.3.5.3 Dopamine metabolism inhibitors

Inhibition of DA-metabolizing enzymes to enhance brain DA levels is an alternative strategy to L-dopa. Drugs approved for PD treatment target MAO-B or COMT. Inhibition of either of these enzymes prevents the catabolism of DA to its major metabolites, 3,4-dihydroxyphenyl-acetic acid and homovanillic acid.

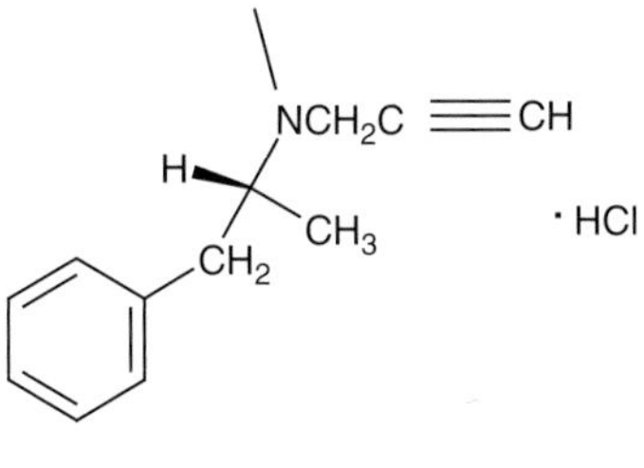

Selegeline (**64**)

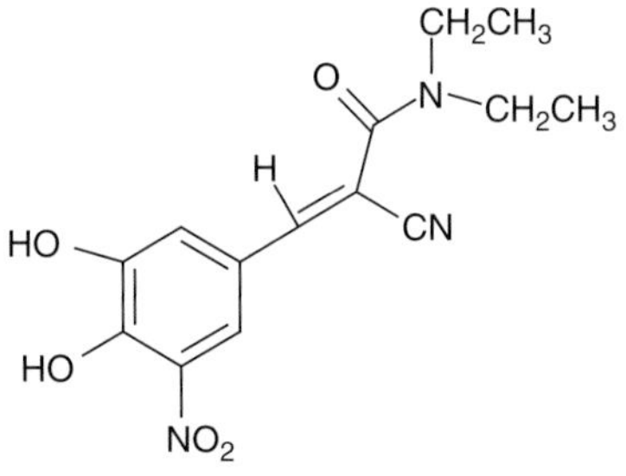

Entacapone (**65**)

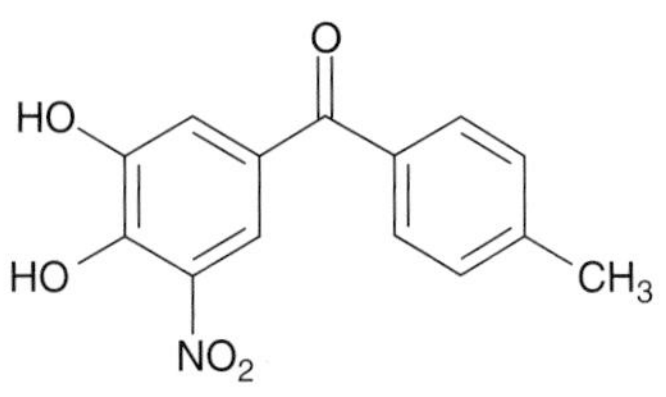

Tolcapone (**66**)

Rasagiline (**67**)

SPD743 (**68**)

DU-127090 (bifeprunox) (**69**)

Talipexole (**70**)

GPI1046 (**71**)

V10367 (**72**)

CEP-1347 (**73**)

TCH346 (**74**)

ONO-2506 (**75**)

E-2007 (**76**)

Istradefylline (**77**)

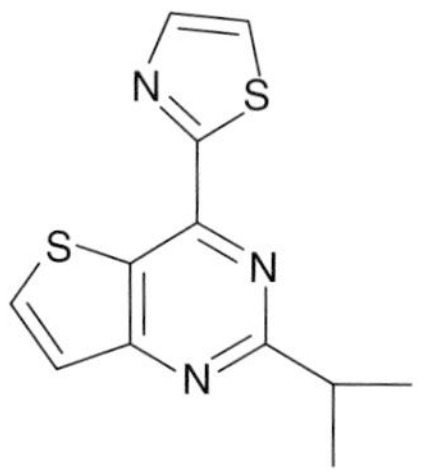

V-2006 (**78**)

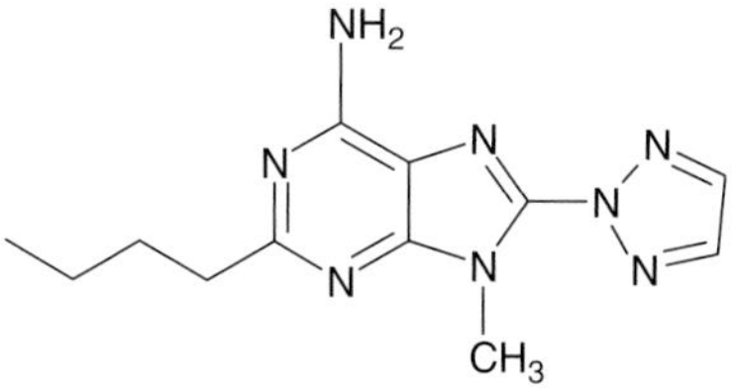

ST-1535 (**79**)

6.08.3.5.3.1 Monoamine oxidase-B inhibitors

Selegiline **64** is an irreversible MAO-B inhibitor approved as adjunct therapy in PD patients being treated with L-dopa. Rasagaline (**67**) is a newer irreversible MAO inhibitor highly selective for MAO-B: it is 5–10 times more active than selegeline.[72] It cannot be metabolized to amphetamine and thus does not produce amphetamine-like behavioral effects.

6.08.3.5.3.2 Catechol *O*-methyltransferase inhibitors

Inhibition of COMT prevents metabolism of L-dopa. Entacopone **65** and tolcapone **66** are COMT inhibitors approved for adjunct treatment to L-dopa in PD. Entacopone is an improvement over tolcapone in terms of liver toxicity issues.

6.08.3.6 Unmet Medical Needs

As with AD, a large unmet medical need in PD is the development of drugs that slow disease progression. A paucity of established biomarkers for early disease diagnosis confounds indicators of treatment efficacy. This is particularly evident for drugs in which there is no symptomatic benefit demonstrated, since clinical trials of long duration ($\sim$18–24 months) need to be carried out with large patient numbers in order to power the study appropriately to see perhaps small but significant effects. For drugs that do have symptomatic effects, neuroprotective events of the compounds can be established using a drug wash-out study design. This strategy has been used for selegine and pramipexole, but as of yet no drug is approved and has US FDA labeling indicative of a disease-modifying effect.

6.08.3.7 New Research Areas

A number of drugs in development for PD have been discontinued. These include melevodopa; irreversible MAO-B inhibitors (e.g., rasagaline **67**); newer COMT inhibitors (eBIA-3-202); DA (brasofensine) and monoamine reuptake inhibitors (SPD-743, **68**); DA receptor agonists, in particular D_2 full (e.g., rotigotine; sumanirole, now discontinued) or partial agonists, in combination with $5HT_{1A}$ agonist activity (e.g., SLV-473; DU-127090, **69**) or α_2-adrenoceptor agonist activity (e.g., talipexole, **70**); and a variety of neuroprotectants, including the immunophilins (e.g., GPI1046, **71**; V10367, **72**), mixed-lineage kinase (MLK) inhibitors (e.g., CEP-1347, **73**), glyceraldehyde phosphate dehydrogenase (GAPDH) inhibitors (e.g., TCH-346, **74**), growth factor agonists (e.g., leteprenim, an NGF agonist), modulators of astrocytic glutamate release (ONO-2506, **75**), and AMPA antagonists (e.g., E-2007, **76**).[70] Adenosine A_{2A} receptor antagonists, e.g., istradefylline **77**,[73] appear to be the most robust of the replacement therapies beyond DA replacement.

6.08.3.7.1 Adenosine A_{2A} receptor antagonists

The adenosine A_{2A} receptor is a member of the P1 GPCR family that modulates striatal output in the indirect basal ganglia pathway. Inhibition of adenosine A_{2A} receptors produces motor stimulant effects, and in both the MPTP and 6-OHDA models of PD, adenosine A_{2A} receptor antagonists have positive effects on motor impairment, potentiating DA agonist-stimulated behavior and thus acting as indirect DA agonists.

Istradefylline (KW-6002: **77**) is an orally available, xanthine-based adenosine A_{2A} receptor antagonist ($K_i = 2.2$ nM) that is in Phase III clinical trials for PD.[73] It has shown positive benefit in reducing the tremor duration and general slowness/stiffness in advanced PD patients. V-2006 (**78**) is a more potent and selective, nonxanthine adenosine A_{2A} receptor antagonist that has recently entered Phase II clinical studies. ST1535 **79** is an earlier stage A_{2A} receptor antagonist.

6.08.3.7.2 Neuroimmunophilins

Neuroimmunophilin ligands were derived from the immunosuppressant, rapamycin (FK-506), which had potent neurotrophic properties.[74] The first of these NCEs, GPI-1046 **71**, demonstrated a 'sprouting' effect by increasing tyrosine hydroxylase immunoreactivity in the striatum of MPTP-treated mice and reduced amphetamine-induced ispilateral rotations in unilateral 6-OHDA-lesioned rats. GPI-1046 lacked efficacy in MPTP-treated nonhuman primates. A second-generation NCE, GPI-1485, was advanced to Phase II clinical trials to investigate its potential neuroprotective effects on dopaminergic neurons but was discontinued.

6.08.3.7.3 Glyceraldehyde phosphate dehydrogenase inhibitors

GAPDH is a glycolytic enzyme with key functions in apoptosis. Interest in GAPDH as a target for PD evolved from the finding that the in vitro neuroprotective effects of selegeline were independent of MAO-B inhibition. Selegeline **64** bound weakly to GAPDH and analogs were developed that were selective for GAPDH over MAO-B and had robust neuroprotective activity.[75] One of these, TCH346 **74**, was advanced to clinical trails and discontinued due to lack of efficacy.

6.08.3.7.4 Mixed-lineage kinase inhibitors

The c-jun JNK pathway leading to c-jun phosphorylation is involved in neuronal death in response to a wide variety of stimuli, including trophic factor withdrawal, Aβ, MPTP glutamate, and NMDA toxicity. Although JNK is a target for preventing neuronal death, upstream kinases in this pathway, in particular, the MLKs are of interest.

CEP-1347 **73** is an MLK inhibitor that blocks JNK signaling and has broad neuroprotective activity in a variety of in vitro and in vivo models of neuronal cell death.[76] Of relevance to PD, CEP-1347 promotes survival of embryonic dopaminergic neurons[77] and blocks MPP+ induced cell death and MPTP-dependent increases in nigral JNK pathway activation and loss of SNpc DA neurons and striatal nerve terminals. Evaluation of CEP-1347 in a phase II/III trial for its ability to modify the course of PD showed no benefit of the NCE, and its development for PD was discontinued.[75]

6.08.4 Huntington's Disease

6.08.4.1 Disease State/Diagnosis

HD, first described in 1872, is a rare disease affecting both motor and cognitive abilities. It is one of the nine polyQ diseases that affect the CNS. HD onset usually occurs around age 40 but cases have been reported in infants as young as 2 years and in elderly patients up to 80 years of age. The mean duration of disease from time of diagnosis is 15–20 years.[78] Approximately 30 000 people in the US have the disease, with the worldwide prevalence being approximately 1 in 20 000. Symptoms include motor dysfunction, with patients displaying writhing or jerking movements of the upper limbs and trunk that are continuous and involuntary (termed choreatic movements, chorea, or choreaoathetosis) in addition to impairment in voluntary movements, e.g., gait, speech, and swallowing.[79] Psychiatric symptoms and cognitive decline also occur. Postmortem assessment shows selective neuronal loss in the striatum, particularly GABAergic medium spiny neurons. Loss of GABA inhibitory neurotransmission disrupts basal ganglia signaling; involuntary movements become excessive and voluntary movements become impaired. HD is predominantly a familial disease with an autosomal dominant pattern of inheritance; thus the presence of a single mutant gene dictates disease occurrence.

6.08.4.2 Disease Basis

The abnormal expansion of a polyglutamine (polyQ, GAG repeat) region in the amino terminus of the *IT15* gene that encodes the normally cytoplasmic ~350 kDa huntingtin (Htt) protein, is the molecular basis of HD.[79] In normal individuals, the polyQ region of the htt gene is a GAG trinucleotide repeat length of 35. HD manifests when the GAG trinucleotide repeat length is larger than 40, with the age of onset of disease being inversely related to the size of the polyQ expansion such that the higher the number of repeats, the earlier the age of disease onset.[78] However, the GAG trinucleotide repeat length is not predictive of the severity of symptoms or the rate of disease progression. At least one potential genetic modifier of disease is the glutamate receptor 6 (GluR6) subunit of the kainate receptor, suggesting that excitotoxicity may be involved in disease pathogenesis.[80]

The normal function of htt is unknown, but it may have a role in vesicle transport, clathrin-mediated endocytosis, postsynaptic signaling, and neuronal survival.[81] HD results from a 'gain of function' in the mutant htt protein coupled with a loss of function of the wild-type protein. The processing of the htt protein appears central to disease pathophysiology. Mutant htt is enzymatically cleaved by caspases and calpains to generate small N-terminal fragments that exacerbate the development of intranuclear and cytoplasmic protein aggregates, also termed inclusions.[82] The proteosome also appears to play a role in inclusion formation as an inhibitor of this enzyme complex. Lactacystin increases aggregate formation while overexpression of chaperone proteins, e.g., HSP70, involved in refolding of abnormal proteins, can rescue cells from htt-induced death. It is controversial whether aggregate formation is a toxic event. Aggregates sequester the protein mammalian target of rapamycin (mTOR) that is involved in the inhibition of autophagy that serves to clear htt fragments.[81] Thus, aggregate formation could serve to stimulate enhanced clearance of htt fragments and act in a protective manner. The reason for selective vulnerability of GABAergic medium spiny interneurons in HD is unknown. Research has focused on the distribution and function of the 20 or so htt-associated proteins, e.g., HAP1, HIP 1 and 2.

6.08.4.3 Experimental Disease Models

Experimental models of HD include both neurotoxin- and genetic-based models. Peripheral administration of 3-nitropropionic acid causes selective degeneration of striatal neurons and has been used to study mechanisms of neuronal death thought to be associated with HD. Genetic models exist in *C. elegans*, *Drosophila,* and mice, all of which are based on the overexpression of various forms of the mutant htt protein containing various lengths of CAG repeats. The first HD model, the Bates or R6/2 model, was developed in the mouse via transgenic expression of the first exon of the protein containing 141–157 GAG repeats. This model has hindlimb behavioral anomalies (e.g., grasping, grooming stereotypies) that occur shortly (5–6 weeks of age) after intraneuronal inclusions are observed (3–4 weeks of age) that precede neuronal degeneration. Other, less well characterized mouse models include a model with an N-terminal htt fragment of 171 amino acids with 82 CAG repeats (N171) and another with yeast artificial chromosomes (YACs) containing human genomic DNA spanning the full length of the gene and having 72 repeats (YAC72).

6.08.4.4 Clinical Trial Issues

The major clinical trial issue is the relatively small population of individuals affected by HD that make recruitment difficult and limit statistical power. This also makes Huntington's an orphan indication.

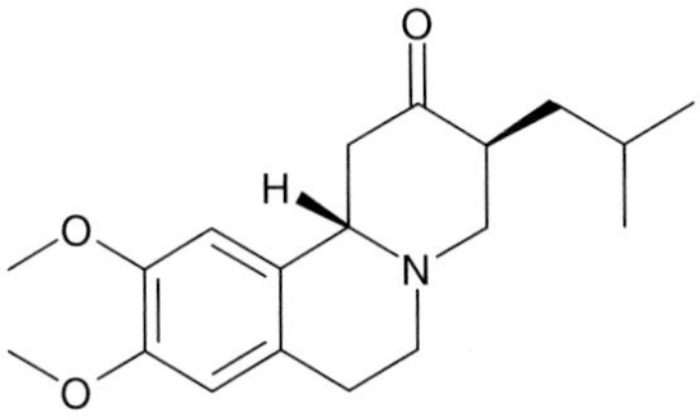

Tetrabenazine (**80**)

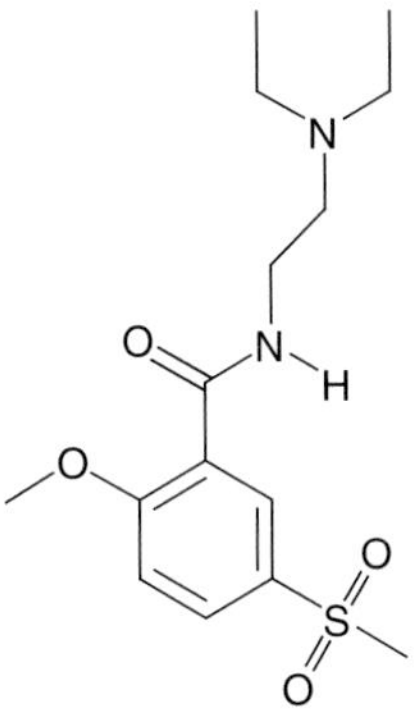

Tiapride (**81**)

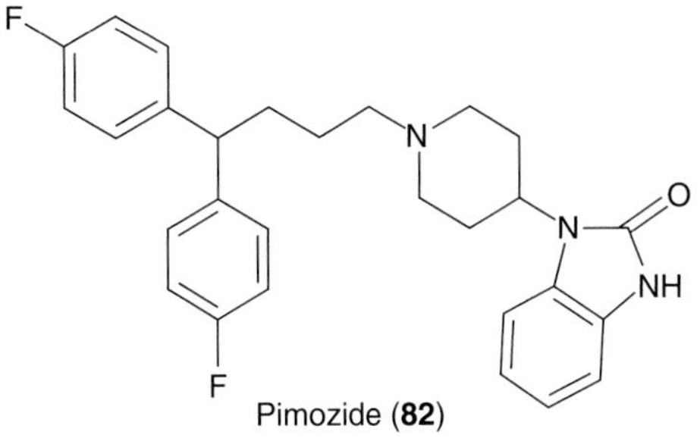

Pimozide (**82**)

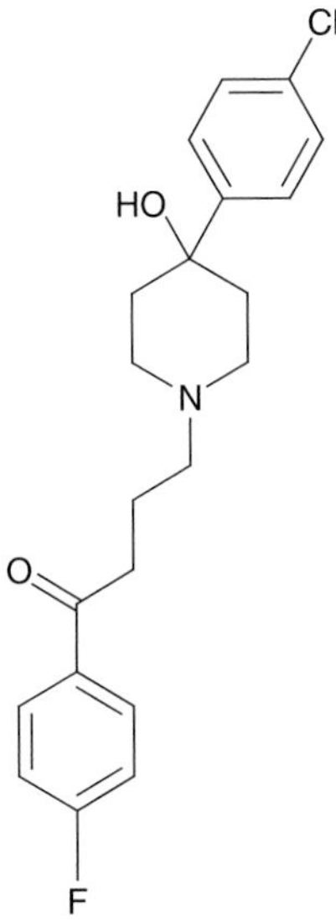

Haloperidol (**83**)

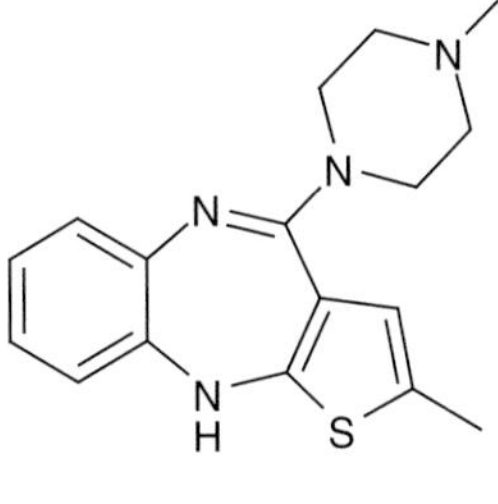

Olanzapine (**84**)

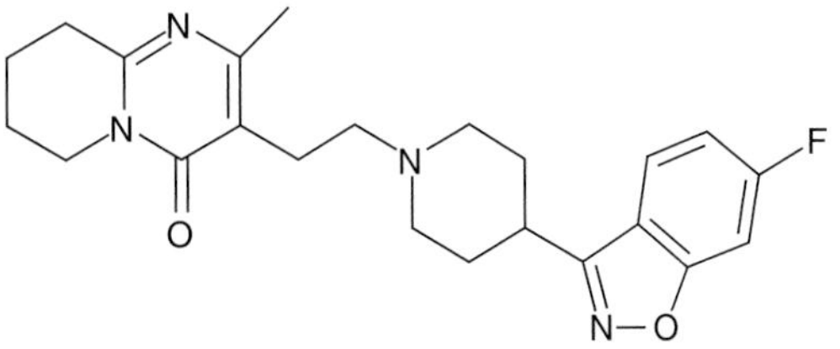

Risperidone (**85**)

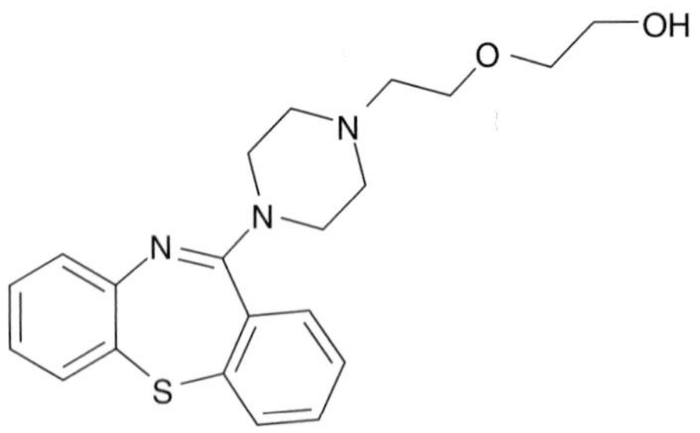

Quetiapine (**86**)

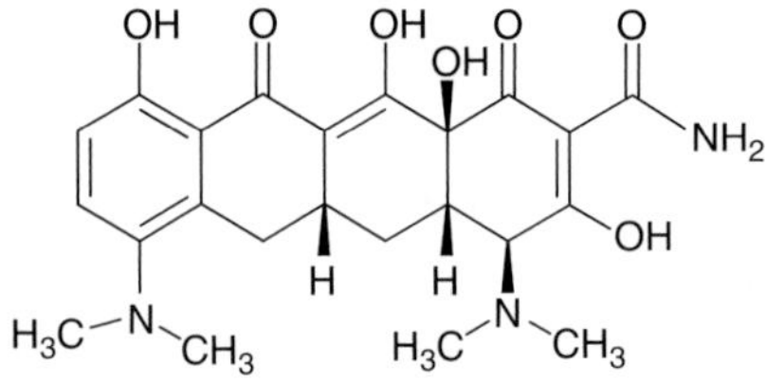

Minocycline (**87**)

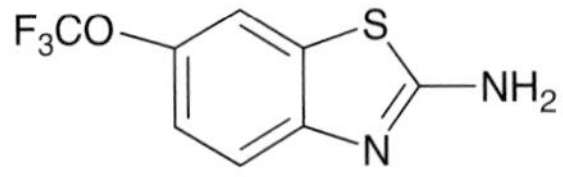

Riluzole (**88**)

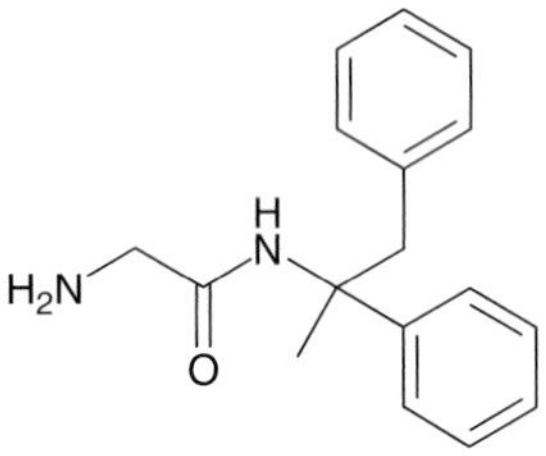

Remacemide (**89**)

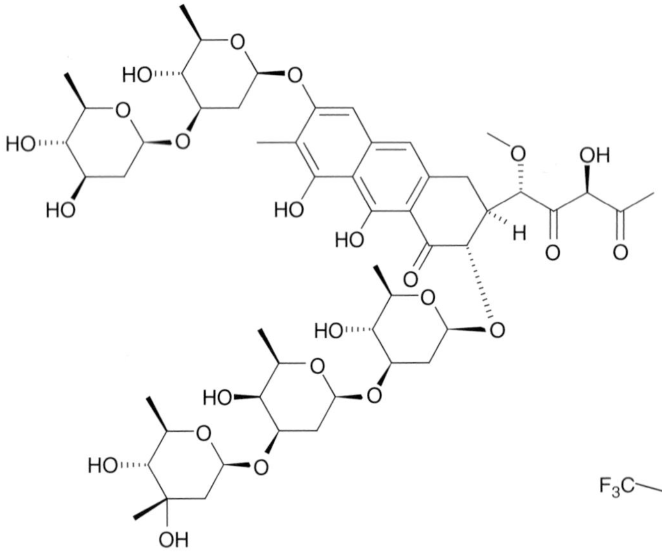

Mithramycin (**90**)

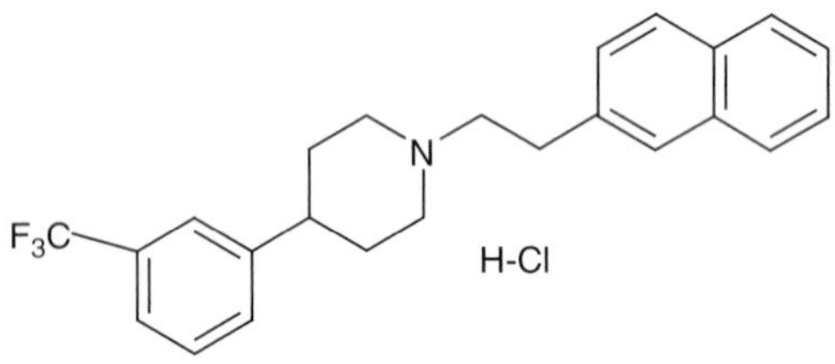

Xaliproden (**91**)

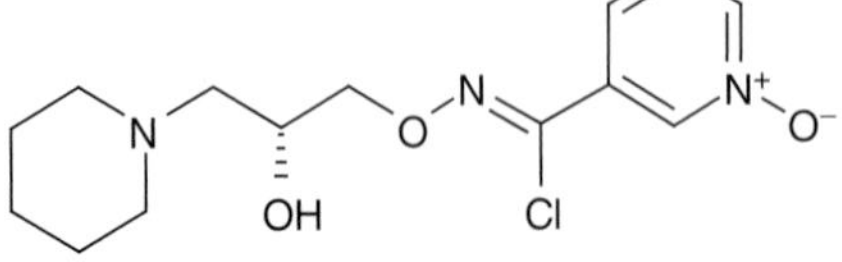

Arimoclomol (**92**)

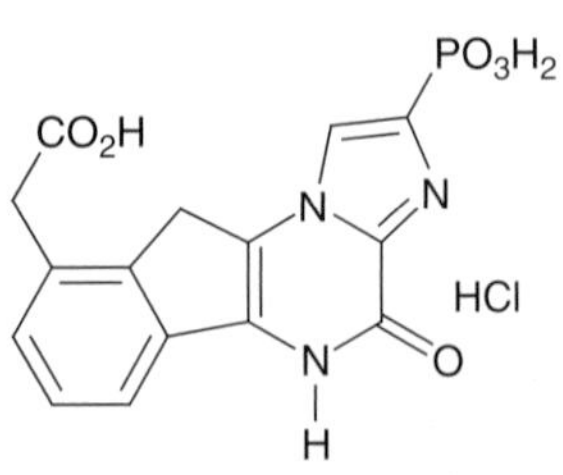

RPR-119990 (**93**)

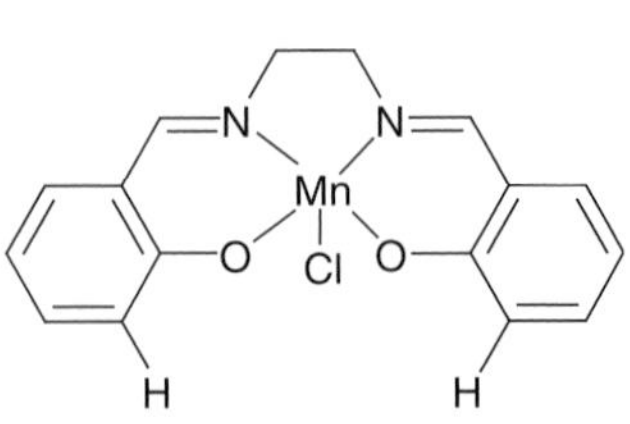

EUK-8 (**94**)

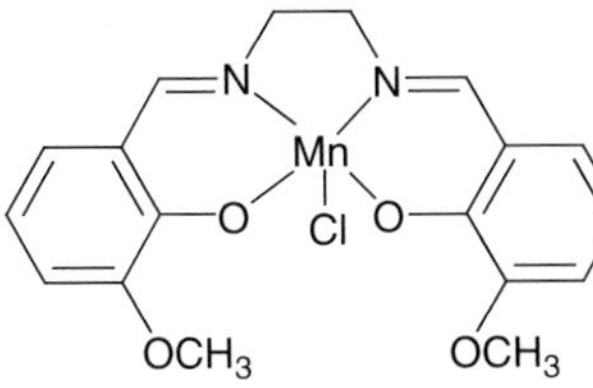

EUK-134 (**95**)

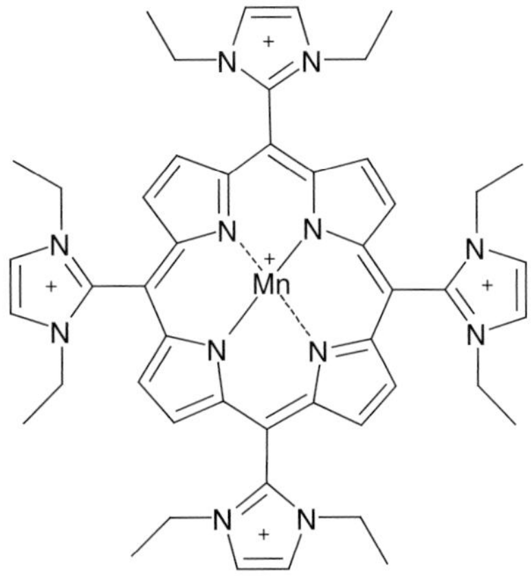

AEOL-10150 (**96**)

6.08.4.5 Current Treatment

There are currently no drugs approved for use in the treatment of HD. The major clinical focus has been on management of the motor problems. However, in HD, there are a host of associated psychiatric conditions that include aggression, and irritability (~40% of cases), depression and suicide (~30% of cases), mania (~10% of cases), apathy (~60% of cases), anxiety (~ 30% of cases, with a greater prevalence in men), obsessive-compulsive disorder (50%), and dementia. Anti-dopaminergic agents such as tetrabenazine **80** and DA receptor antagonists, including tiapride **81**, pimozide **82**, haloperidol **83**, olanzapine **84**, risperidone **85**, and quetiapine **86**,[79,83] are used to treat chorea. Psychiatric conditions associated with HD are treated with amantadine **55**, fluphenazine, SSRIs, and mirtazapine.[79]

6.08.4.6 Unmet Medical Needs

HD is a disease that affects a small number of patients and could be prevented through genetic counseling. However, there are many ethical considerations to this approach of disease management, and thus efforts continue to discern new disease targets and therapeutic approaches focused toward neuroprotection.

6.08.4.7 New Research Areas

Antiapoptotic agents (e.g., caspase inhibitors), anti-inflammatory agents (e.g., minocycline **87**), bioenergetic enhancers/antioxidants (e.g., creatine, coenzyme Q, lipoic acid, dicholoroacetate, cystamine), excitotoxic anatgonists (e.g., riluzole **88**, remacemide **89**), histone deacetylase inhibitors (sodium butyrate and suberoylanilide hydroxamic acid), and transcriptional inhibitors (e.g., mithramycin **90**) have all been examined in the R6/2 mouse model of HD and have shown some benefit.[84] Some of these agents have been advanced to clinical trials (**Table 2**). Aggregation inhibitors are also a target.[85]

6.08.5 Amyotrophic Lateral Sclerosis

6.08.5.1 Disease State/Diagnosis

ALS, known as 'Lou Gehrig's disease,' was originally described in 1869 and is a rapidly progressing neurodegenerative disease affecting both upper and lower motor neurons of the ventral horn of the spinal cord, motor cortex, and brainstem.[86] Large axonal caliber neurons are predominantly affected, and a pathological feature is the presence of axonal inclusions of protein aggregates. As with all other neurodegenerative diseases, an inflammatory response is also a pathological feature of the disease.[87]

Disease onset occurs typically in midlife and, upon diagnosis, life expectancy is in the range of 3–5 years. Males appear to have a higher incidence of ALS and, overall, 1–7 people in 100 000 develop the disease.[86–88] In comparison to AD and PD, the incidence of ALS is low.

ALS diagnosis is based upon clinical presentation and electrophysiological and neuropathological evidence of motor neuron degeneration. ALS is classified into three variants/types: (1) sporadic (no genetic component); (2) familial

Table 2 NCEs evaluated in genetic models of Huntington's disease

Compound	*% Survival increase*	*Mechanism*	*Clinically evaluated*
Mithramycin	29.1	Transcription inhibition	
z-VAD-fmk	24.8	Caspase inhibition	
Sodium butyrate	21.7	HDAC inhibition	
Cystamine	19.5	Htt inhibition/antioxidant	
Creatine	17.4	Bioenergetic/antioxidant	✓
Congo Red	16.4	HDAC inhibition	
Remacemide	15.5	Antiexcitotoxic	
BN82451	15.0	Antioxidant	
Coenzyme Q	14.5	Bioenergetic/antioxidant	✓
Minocycline	13.5	Antiapoptotic/antiinflammatory	✓
Riluzole	10.2	Antiexcitotoxic	✓
Lipoic acid	7.1	Bioenergetic/antioxidant	
Dicholoracetate	6.8	Bioenergetic	

Reproduced with permission from Hersch, S. M.; Ferrante, R. J. *NeuroRx* **2004**, *1*, 298–306.[84]

(fALS: inherited); and (3) Guamanian (of dietary origin). The majority of ALS cases are sporadic. Only 5–10% of cases are of the familial type. Accurate diagnosis of ALS rules out other diseases in which upper motor neuron dysfunction is absent (e.g., progressive muscular atrophy, spinal muscular atrophy) or is the only feature present (e.g., primary lateral sclerosis). Clinical features of ALS include areflexia (absence of spinal reflexes), hypotonia (loss of muscle tone), fasciculations (muscle twitching), and generalized muscle weakness that typically first occurs in the arms and legs. Involvement of upper motor neurons is evident if excessive salivation, dysphagia (impaired swallowing), and dysarthria (impaired speech) occur. Motor neuron death innervating the diaphragm muscle leads to the need for trachoestomy and at the latest stage ventilator-assisted respiration.

6.08.5.2 Disease Basis

The cause of sporadic ALS is unknown but it is generally accepted that susceptibility to disease is based upon genetic risk factors and environmental exposure. The disease represents a heterogeneous set of disorders caused by different initiating factors that manifest in a common event, i.e., motor neuron loss. Included among the genetic risk factors for sporadic disease are mutations or polymorphisms in the genes for dynactin (DCTN1, p150 subunit),[89] peripherin (PRPH),[90] neurofilament heavy chain (NF-H),[91] glutamate transporter-2 (EAAT-2), and vascular endothelial growth factor (VEGF).[92] Mutations in VEGF have also been associated with multiple neurodegenerative diseases. Familial disease dominantly inherited falls into two classes: disease associated with superoxide dismutase-1 (SOD-1) mutations, for which over 90 are documented, and disease not associated with SOD-1 mutations. For dominantly inherited fALS not associated with SOD-1 mutations, linkages have been made on chromosomes 16, 18, and 20.[93] Juvenile fALS forms also exist. For these, genes within chromosomal locations 15q15-22 and 2q33 are linked and inherited in a recessive manner. Although the identity of the gene on chromosome 15 is yet to be determined, ALS2 is the gene localized to chromosome 2. ALS2 encodes a protein of 184 kDa (alsin) that contains three putative guanine-nucleotide exchange factor domains. Alsin is suspected to be involved in protein trafficking since it acts, at least in vitro, as an exchange factor for Rab5a, which is known to be involved in endosome function.

6.08.5.3 Experimental Disease Models

The most common models of ALS are based upon the gene mutations associated with fALS linked to SOD-1 mutations. SOD-1–3 convert superoxide radicals to molecular oxygen and hydrogen peroxide. Copper/zinc (Cu/Zn)

SOD-1 is predominantly localized to the cytosol, but may also be associated with the inner mitochondrial membrane; manganese (Mn) SOD-2 is localized to the mitochondria, and SOD-3 is localized extracellularly. Identified fALS-associated mutations in SOD-1 do not cluster around the catalytic site but are distributed throughout the protein. Originally, it was thought that mutations might decrease the enzyme activity of SOD-1; thus, genetically engineered mice that lacked SOD-1 were produced but failed to recapitulate the phenotype of fALS. Transgenic rodents carrying human SOD-1 with fALS mutations G86R, G93A, G37R, or H46R recapitulate essential features of the disease (motor neuron degeneration, muscle atrophy, paralysis) with involvement of the lower and upper motor neurons as well as lifespan being dependent on gene dosage.

6.08.5.4 Clinical Trial Issues

Several issues have arisen in the conduct of clinical trials in ALS, including the application of diagnostic criteria for patient enrollment, administration of clinical tests, and study design. The El Escorial World Federation of Neurology established diagnostic criteria in 1994[94] that were revised in 2000 to define more accurately the patient population selected for ALS clinical trials. Primary endpoint measures often used in clinical trials include: (1) muscle strength; (2) pulmonary function; (3) activities of daily living; and (4) survival. At least two different clinical instruments are being used to measure muscle strength, including maximum voluntary isometric contraction and manual muscle testing. Respiratory function is measured by maximum inspiratory pressure and forced vital capacity, both of which measure the strength of diaphragm muscle contractions.[95] Two of the earliest rating scales for ALS were the Appel and Norris scales. The ALS functional rating scale (ALSFRS) is a favored primary endpoint owing to its short time for administration (5 min), its correlation with physiological measures of disease progression, and relevance to the ability of a patient to function and perform tasks associated with daily living.

6.08.5.5 Current Treatment

Riluzole **88** was approved in 1995 and remains the only drug available for the treatment of ALS. Its mechanism of action is unknown. Preclinically, it has been characterized as a noncompetitive inhibitor of blockade of glutamate release, presumably by reducing Ca^{2+} influx and voltage-gated sodium channel activity, albeit at micromolar concentrations.[96]

Two large randomized, double-blind, placebo-controlled phase III clinical trials established the efficacy of riluzole in the treatment of ALS. Overall riluzole extended patient survival by $\sim$3 months but lacked any measurable effects on muscle strength or respiratory function. Riluzole is well absorbed but extensively metabolized by the liver. N-Hydroxylation by CYP1A2 and subsequent formation of O- and N-glucuronides appear to be responsible for its elimination in the urine. Elevation of serum aminotransferase is a likely event within the first months of treat ment with riluzole, and monitoring of liver enzymes while on this drug is important to avoid or reduce hepato-toxicity.

6.08.5.6 Unmet Medical Needs

Unmet medical needs for ALS fall into two categories: (1) improved means/methods for early diagnosis; and (2) improved medications. Diagnosis of ALS, especially nonfamilial disease, is difficult, and definitive diagnosis is often delayed for up to 1.5 years after symptom onset. This typically results from general practitioners not recognizing early signs of disease owing to the limited number of the patients that they encounter.

The paucity of approved treatments for ALS underscores the critical need for further research and drug development. Moreover, the limited clinical efficacy of riluzole warrants improved therapeutic treatments. Unfortunately, the lack of knowledge regarding the mechanism of action of riluzole has limited the potential for second-generation compounds within this drug class.

Since the approval of riluzole, at least 10 therapeutic approaches which showed promise in preclinical models have been tested in the clinic but have failed or shown equivocal results (such as for insulin-like growth factor-1). These include neurotrophic factor replacement strategies (ciliary neurotrophic factor, BDNF, glial-derived neurotrophic factor, and insulin-like growth factor-1), glutamate/NMDA receptor antagonists (topiramate, dextromethorphan) or putative glutamate modulators (gabapentin, lamotrogine, N-acetylcysteine), antioxidants (vitamin E), MAO/glyceraldehyde phosphate dehydrogenase inhibitors (selegeline), and mitochondrial transition pore blockers (creatine).[86,87,95] Insulin-like growth factor-1 has shown improvement in quality of life and benefit in ALS patients.[97]

6.08.5.7 New Research Areas

6.08.5.7.1 Neurotrophic/Neuroprotectant small molecules

Neurotrophic factor replacement strategies for ALS treatment have met with limited success to date; whether this is a result of short half-lives of proteins or an inability to cross the blood–brain barrier is not fully understood. However, preclinical data strongly support a role of these proteins in the maintenance of survival as well as rescue from a variety of toxic insults to motor neurons. To circumvent the potential issues of protein therapeutics in treating CNS disease, several alternate strategies have been taken either to enhance endogenous levels of neurotrophic proteins or to identify small molecules with neurotrophic-like properties. Xaliproden (SR-57746A **91**) is a small-molecule, $5HT_{1A}$ agonist that stimulates BDNF production in vitro and promotes motor neuron survival both in vitro and in vivo.[98] In a double-blind Phase II study, patients taking xaliproden, as compared to those on placebo, who completed the 32-week study showed a slower rate of deterioration in vital capacity.[99] However, two phase III trials failed to reach significance.[100]

Arimoclomol **92** amplifies the heat shock response by increasing the levels of heat shock proteins that have antiapoptotic properties.[101] In the SOD-1 G93A ALS model, arimoclomol improved hindlimb muscle function and motor neuron survival and resulted in a 22% increase in lifespan. This NCE is proceeding to Phase II clinical trials.

6.08.5.7.2 Anti-inflammatory agents

Minocycline **87**, a tetracycline antibiotic with anti-inflammatory activity, appears to have multiple effects on cell function, including inhibition of the mitochondrial permeability transition-mediated release of cytochrome *c*, an event key to the initiation of the apoptotic cascade. Additional activities of minocycline include: inhibition of reactive microgliosis, caspase-1, caspase-3, and nitric oxide synthase transcriptional upregulation, and of p38 MAPK activation. It protects against motor neuron loss in a mouse ALS model (SOD-1 G93A) alone or in combination with creatine.[87]

The COX-2 inhibitor, celecoxib **44**, preclinically slowed motor neuron deterioration in an organotypic model of motor neuron disease and improved survival up to 30% following oral administration in an ALS transgenic model.[87] COX-2 is upregulated in both the transgenic SOD-1 G93A mouse model and in the CNS of ALS patients.[86] COX-2 catalyzes the conversion of arachidonic acid to prostaglandin E_2, a proinflammatory mediator, which is also elevated in the CSF of ALS patients. One possible mechanism for the neuroprotective properties of celecoxib is a reduction in astrocyte glutamate release stimulated by prostaglandin synthesis.[87] Prostaglandin E_2 receptor activation also rescues motor neuron loss induced by glutamate toxicity.[102] Given the cardiovascular problems with COX-2 inhibitors, it is unlikely that this approach will be pursued.

6.08.5.7.3 AMPA antagonists

The role of glutamate in ALS pathophysiology remains speculative and the exact mechanism underlying selective motor neuron loss in the disease is unknown. However, a role of AMPA receptors in selective motor neuron vulnerability is emerging. The AMPA receptor is a tetrameric complex of four subunits (GluR1–GluR4). Its Ca^{2+} permeability is regulated by posttranscriptional RNA editing of the GluR2 subunit, such that, normally, permeability is low. Failure in the RNA-editing process can lead to a receptor with altered permeability properties. Reduced editing efficiency of GluR2 mRNA has been reported in spinal motor neurons from human sporadic ALS patients.[103] Transgenic animals harboring modified GluR2 subunits with increased Ca^{2+} permeability display late-onset degeneration of spinal motor neurons and decline of motor function.[104] The increased Ca^{2+} influx via modified GluR2 AMPA receptors promotes misfolding of SOD-1[105] and crossing modified GluR2 transgenic animals to mutant SOD-1 (G93A) animals accelerates disease progression and decline in motor function and exacerbates death.[104]

RPR-119990 **93** is a selective, potent ($K_i = 107$ nM) AMPA receptor antagonist that improved muscle strength in SOD-1 mutant (G93A) mice and prolonged survival relative to vehicle-treated animals.[106] FP-0011 an antiglutamatergic agent of unknown mechanism is in phase I trials.

6.08.5.7.4 Superoxide dismutase mimetics

Increased oxidative burden contributes to the pathophysiology of ALS with advanced glycation end-products, markers of lipid peroxidation (4-hydroxy-2-nonenal-histidine, crotonaldehyde-lysine), and markers of protein glycooxidation localized to motor neurons of ALS patients. A number of free radicals, including superoxide and nitric oxide, likely cause damage to motor neurons via the formation of peroxynitrite and subsequent nitration of tyrosine residues on proteins (e.g., neurofilaments), which are central to neuronal function. Spinal cord tissue from ALS patients has increased levels of 3-nitrotyrosine and 3-nitro-4-hydroxyphenylacetic acid, both products formed through the action of peroxynitrite.

EUK-8 **94** and EUK-134 **95** demonstrated a benefit in the ALS mouse model. Given prophylactically, they delayed disease onset.[107] AEOL-10150 **96** is a manganese porphyrin catalytic antioxidant that decomposes biological oxidants such as peroxinitrite via its ability to cycle between Mn(III) and Mn(IV) states, and may reduce the oxidative burden. In a transgenic mouse model of ALS (SOD-1 (G93A)), AEOL-10150 improved motor activity of mice and resulted in an approximately 30-day increase in the mean survival age.[108]

6.08.6 Future Aspects

With the aging of the population, the incidence of neurodegeneration is increasing dramatically in the absence of either effective therapeutic interventions or a clear understanding of the discrete pathophysiology of neurogenerative disease states. In the AD area, following the identification of Aβ and tau/NFTs as potential targets for disease amelioration, there has been a wealth of research around these two targets, using both biochemical and genetic information with multiple approaches to their modulation. Despite this, there has been little practical outcome as successive mechanistic targets, some with substantive retrospective data sets, e.g., indometacin, estrogen, have failed to show efficacy in prospective controlled clinical trials or resulted in the identification of unacceptable side effects. Similarly, in the PD area, while new palliative approaches are showing promise, e.g., istradefylline, treatments for disease amelioration, despite robust preclinical data, have failed to show efficacy, e.g., CEP1347, TCH346. There are a number of challenges with current approaches to both finding and testing novel treatments for AD and PD:

1. The numerous hypotheses related to disease causality have yet to be validated, and require a safe and efficacious NCE that can be demonstrated to be effective in the human disease state. Thus, current approaches require the testing of both the hypothesis (which may be multifactorial) and the NCE (the effects of which may be redundant in the context of the complexity of the human disease state).
2. The majority of the biological systems in which NCEs are being evaluated are highly sophisticated, synthetically engineered models of hypothetical concepts of the disease causality rather than true reflections of the human disease state, with many, if not all, transgenic mouse models, recapitulating the molecular hypothesis of disease causality rather than the disease. Thus, NCEs are tested in a highly reductionistic mode, with both the NCEs and the models representing a self-contained loop that lacks real relevance to the actual disease, thus achieving Glass Bead Game status.[109]
3. Many of the genetic findings that have been applied to understanding causality have been controversial in that they have not been replicable and have usually focused on familial rather than the more typical idiopathic forms of the disease.
4. The anticipated endpoints to measure the effects of NCEs that modify disease outcomes in AD and PD are complex, require large patient cohorts, and are lengthy. This necessitates biomarker approaches for both disease diagnosis and assessment of disease progression that are as predictively robust and noninvasive as possible. To date, such biomarkers have been difficult to validate.[11]
5. In the absence of biomarkers to diagnose neurodegenerative diseases appropriately, when the signs and symptoms of PD and AD appear, it is usually far too late to treat the disease effectively. Thus, promising approaches that could have major benefit if used early enough in the disease process usually prove ineffective at advanced disease stages, further elaborating on the urgency for biomarkers.

Nonetheless, there is considerable optimism[110] that the past 20 years of research have provided the basis for true advances in neurodegenerative disease modification therapy. This will however require a less reductionistic approach to preclinical research, with less dependence on transgenic models, in vivo test tubes, and a concerted effort in the area of translational medicine to develop robust biomarkers of disease progression.

References

1. Hoyert, D. L.; Kung, H. C.; Smith, B. L. *Natl. Vital Stat. Rep.* **2005**, *53*, 1–48.
2. Heflin, L. H.; Meyerowitz, B. E.; Hall, P.; Lichtenstein, P.; Johansson, B.; Pedersen, N. L.; Gatz, M. *J. Natl. Cancer Inst.* **2005**, *97*, 854–856.
3. Westlund, B.; Stilwell, G.; Sluder, A. *Curr. Opin. Drug Disc. Dev.* **2004**, *7*, 169–178.
4. American Psychiatric Association. *DSM-IV-TR: Diagnostic and Statistical Manual*, 4th ed., test revision; American Psychiatric Association: Washington, DC, 2000.
5. Chong, M. S.; Sahadevan, S. *Lancet Neurol.* **2005**, *4*, 576–579.
6. Braak, H.; Braak, E. *Acta Neuropathol. (Berl.)* **1991**, *82*, 239–259.

7. Mirra, S. S.; Heyman, A.; McKeel, D.; Sumi, S. M.; Crain, B. J.; Brownlee, L. M.; Vogel, F. S.; Hughes, J. P.; van Belle, G.; Berg, L. *Neurology* **1991**, *41*, 479–486.
8. National Institute on Aging and Reagen Institute Working Group on Diagnosis Criteria for the Neruopathological Assessment of Alzheimer Disease. *Neurobiol. Aging* **1997**, *18*, S1–S3.
9. Villemagne, V. L.; Rowe, C. C.; Macfarlane, S.; Novakovic, K. E.; Masters, C. L. *J. Clin. Neurosci.* **2005**, *12*, 221–230.
10. Andreasen, N.; Blennow, K. *Clin. Neurol. Neurosurg.* **2005**, *107*, 165–173.
11. Sunderland, T.; Linker, G.; Mirza, N.; Putnam, K. T.; Friedman, D. L.; Kimmel, L. H.; Bergeson, J.; Manetti, G. J.; Zimmermann, M.; Tang, B. et al. *JAMA* **2003**, *289*, 2094–2103.
12. Bartus, R. T. *Exp. Neurol.* **2000**, *163*, 495–529.
13. DeKosky, S. T.; Ikonomovic, M. D.; Styren, S. D.; Beckett, L.; Wisniewski, S.; Bennett, D. A.; Cochran, E. J.; Kordower, J. H.; Mufson, E. J. *Ann. Neurol.* **2002**, *51*, 145–155.
14. Wirths, O.; Multhaup, G.; Bayer, T. A. *J. Neurochem.* **2004**, *91*, 513–520.
15. Walsh, D. M.; Selkoe, D. J. *Protein Pept. Lett.* **2004**, *11*, 213–228.
16. Dickey, C. A.; Gordon, M. N.; Mason, J. E.; Wilson, N. J.; Diamond, J. M.; Guzowski, J. F.; Morgan, D. *J. Neurochem.* **2004**, *88*, 434–442.
17. Verdier, Y.; Zarandi, M.; Penke, B. *J. Pept. Sci.* **2004**, *10*, 229–248.
18. Bozyczko-Coyne, D.; Saporito, M. S.; Hudkins, R. L. *Curr. Drug Targets CNS Neurol. Disord.* **2002**, *1*, 31–49.
19. Froelich-Fabre, S.; Bhat, R. V. *Drug Disc. Today: Dis. Mechanisms* **2004**, *1*, 391–398.
20. Townsend, K. P.; Pratico, D. *FASEB J.* **2005**, *19*, 1592–1601.
21. Dekosky, S. T. *Clin. Cornerstone* **2001**, *3*, 15–26.
22. Zubenko, G. S.; Hughes, H. B., III; Stiffler, J. S. *Mol. Psychiatry* **2001**, *6*, 413–419.
23. Head, E.; Torp, R. *Neurobiol. Dis.* **2002**, *9*, 1–10.
24. Kimura, N.; Yanagisawa, K.; Terao, K.; Ono, F.; Sakakibara, I.; Ishii, Y.; Kyuwa, S.; Yoshikawa, Y. *Neuropathol. Appl. Neurobiol.* **2005**, *31*, 170–180.
25. Spires, T. L.; Hyman, B. T. *NeuroRx* **2005**, *2*, 423–437.
26. Jacobsen, J. S.; Wu, C-C.; Redwine, J. M.; Comery, T. A.; Bowlby, M.; Martone, R.; Morrison, J. H.; Pangalos, M. N.; Reinhart, P. H.; Bloom, F. E. *Proc. Natl. Acad. Sci. USA* **2006**, *103*, 5161–5166.
27. Chan, P. L.; Holford, N. H. *Annu. Rev. Pharmacol. Toxicol.* **2001**, *41*, 625–659.
28. Gottwald, M. D.; Rozanski, R. I. *Exp. Opin. Investig. Drugs* **1999**, *8*, 1673–1682.
29. Shigeta, M.; Homma, A. *CNS Drug Rev.* **2001**, 7, 353–368.
30. Lilienfeld, S. *CNS Drug Rev.* **2002**, *8*, 159–176.
31. Zarotsky, V.; Sramek, J. J.; Cutler, N. R. *Am. J. Health Syst. Pharm.* **2003**, *60*, 446–452.
32. Molinuevo, J. L.; Llado, A.; Rami, L. *Am. J. Alzheimers Dis. Other Demen.* **2005**, *20*, 77–85.
33. Hashimoto, M.; Kazui, H.; Matsumoto, K.; Nakano, Y.; Yasuda, M.; Mori, E. *Am. J. Psychiatry* **2005**, *162*, 676–682.
34. Rosini, M.; Andrisano, V.; Bartolini, M.; Bolognesi, M. L.; Hrelia, P.; Minarini, A.; Tarozzi, A.; Melchiorre, C. *J. Med. Chem.* **2005**, *48*, 360–363.
35. Utsuki, T.; Yu, Q.-S.; Davidson, D.; Chen, D.; Holloway, H. W.; Brossi, A.; Sambamurti, K.; Lahiri, D. K.; Greig, N. H.; Giordano, T. *J. Pharmacol. Exp. Ther.* **2006**, *318*, 855–862.
36. Beach, T. G. *Curr. Opin. Investig. Drugs* **2002**, *3*, 1633–1636.
37. Breining, S. R.; Mazurov, A. A.; Miller, C. H. *Annu. Rep. Med. Chem.* **2005**, *40*, 3–16.
38. Leurs, R.; Bakker, R. A.; Timmerman, H.; de Esch, I. J. *Nat. Rev. Drug Disc.* **2005**, *4*, 107–120.
39. Bullock, R. *Curr. Opin. Investig. Drugs* **2005**, *6*, 108–113.
40. Spencer, J. P.; Brown, J. T.; Richardson, J. C.; Medhurst, A. D.; Sehmi, S. S.; Calver, A. R.; Randall, A. D. *Neuroscience* **2004**, *129*, 49–54.
41. Brown, W. M. *IDrugs* **2001**, *4*, 1389–1400.
42. Marlatt, M. W.; Webber, K. M.; Moreira, P. I.; Lee, H. G.; Casadesus, G.; Honda, K.; Zhu, X.; Perry, G.; Smith, M. A. *Curr. Med. Chem.* **2005**, *12*, 1137–1147.
43. Casadesus, G.; Atwood, C. S.; Zhu, X.; Hartzler, A. W.; Webber, K. M.; Perry, G.; Bowen, R. L.; Smith, M. A. *Cell Mol. Life Sci.* **2005**, *62*, 293–298.
44. Katsube, N.; Sunaga, K.; Aishita, H.; Chuang, D. M.; Ishitani, R. *J. Pharmacol. Exp. Ther.* **1999**, *288*, 6–13.
45. Pollack, S. J.; Lewis, H. *Curr. Opin. Investig. Drugs* **2005**, *6*, 35–47.
46. Siemers, E.; Skinner, M.; Dean, R. A.; Gonzales, C.; Satterwhite, J.; Farlow, M.; Ness, D.; May, P. C. *Clin. Neuropharmacol.* **2005**, *28*, 126–132.
47. Cuello, A. C.; Bell, K. F. S. *Curr. Med. Chem.* **2005**, *5*, 15–28.
48. Huckle, R. *Curr. Opin. Investig. Drugs* **2005**, *6*, 99–107.
49. Monaco, E. A., III *Curr. Alzheimer Res.* **2004**, *1*, 33–38.
50. Johnson, K.; Liu, L.; Majdzadeh, N.; Chavez, C.; Chin, P. C.; Morrison, B.; Wang, L.; Park, J.; Chugh, P.; Chen, H. M. et al. *J. Neurochem.* **2005**, *93*, 538–548.
51. Kuo, G. H.; Deangelis, A.; Emanuel, S.; Wang, A.; Zhang, Y.; Connolly, P. J.; Chen, X.; Gruninger, R. H.; Rugg, C.; Fuentes-Pesquera, A. et al. *J. Med. Chem.* **2005**, *48*, 4535–4546.
52. Woodgett, J. R. *Curr. Drug Targets Immune Endocrinol. Metab. Disord.* **2003**, *3*, 281–290.
53. Benbow, J. W.; Helal, C. J.; Kung, D. W.; Wager, T. T. *Annu. Rep. Med. Chem.* **2005**, *40*, 135–147.
54. Rogers, J.; Kirby, L. C.; Hempelman, S. R.; Berry, D. L.; McGeer, P. L.; Kaszniak, A. W.; Zalinski, J.; Cofield, M.; Mansukhani, L.; Willson, P. et al. *Neurology* **1993**, *43*, 1609–1611.
55. Scatena, R. *Curr. Opin. Investig. Drugs* **2004**, *5*, 551–556.
56. Watson, G. S.; Cholerton, B. A.; Reger, M. A.; Baker, L. D.; Plymate, S. R.; Asthana, S.; Fishel, M. A.; Kulstad, J. J.; Green, P. S.; Cook, D. G. et al. *Am. J. Geriatr. Psychiatry* **2005**, *13*, 950–958.
57. Shobab, L. A.; Hsiung, G.-Y. R.; Feldman, H. H. *Lancet Neurol.* **2005**, *4*, 841–852.
58. Sparks, D. L.; Sabbagh, M. N.; Connor, D. J.; Lopez, J.; Launer, L. J.; Browne, P.; Wasser, D.; Johnson-Traver, S.; Lochhead, J. et al. *Arch. Neurol.* **2005**, *62*, 753–757.
59. Masse, I.; Bordet, R.; Deplanque, D.; Khedr, A. A.; Richard, F.; Libersa, C.; Pasquier, J. *J. Neurol. Neurosurg. Psychiatry* **2005**, *76*, 1624–1629.
60. Chauhan, N. B.; Siegel, G. J.; Feinstein, D. L. *Neuropharmacology* **2005**, *48*, 93–104.
61. Gozes, I.; Brenneman, D. E. *J. Mol. Neurosci.* **2000**, *14*, 61–68.
62. Braak, H.; Ghebremedhin, E.; Rub, U.; Bratzke, H.; Del Tredici, K. *Cell Tissue Res.* **2004**, *318*, 121–134.
63. Swab and England Activities of Daily Living scale. http://neurosurgery.mgh.harvard.edu/pdstages.htm (accessed Aug 2006).

64. Ravina, B.; Eidelberg, D.; Ahlskog, J. E.; Albin, R. L.; Brooks, D. J.; Carbon, M.; Dhawan, V.; Feigin, A.; Fahn, S.; Guttman, M. et al. *Neurology* **2005**, *64*, 208–215.
65. Moron, J. A.; Zakharova, I.; Ferrer, J. V.; Merrill, G. A.; Hope, B.; Lafer, E. M.; Lin, Z. C.; Wang, J. B.; Javitch, J. A.; Galli, A. et al. *J. Neurosci.* **2003**, *23*, 8480–8488.
66. Vila, M.; Przedborski, S. *Nat. Med* **2004**, *10S*, S58–S62.
67. Orth, M.; Tabrizi, S. J. *Mov. Disord.* **2003**, *18*, 729–737.
68. Fernagut, P. O.; Chesselet, M. F. *Neurobiol. Dis.* **2004**, *17*, 123–130.
69. PSG Study Group. *Arch. Neurol.* **2004**, *61*, 561–566.
70. Johnston, T. H.; Brotchie, J. M. *Curr. Opin. Investig. Drugs* **2004**, *5*, 720–726.
71. Parkinson's Study Group. *Neurology* **2003**, *60*, A80–A81.
72. Finberg, J. P. M.; Youdim, M. B. H. *Neuropharmacology* **2002**, *43*, 1110–1118.
73. Chase, T. N.; Bibbiani, F.; Bara-Jiminez, W.; Dimitrova, T.; Oh-Lee, J. D. *Neurology* **2003**, *61*, S107–S111.
74. Marshall, V. L.; Grosset, D. G. *Curr. Opin. Investig. Drugs* **2004**, *5*, 107–112.
75. Waldmeier, P.; Williams, M.; Bozyczko-Coyne, D.; Vaught, J. *Biochem. Pharmacol.* **2006**, *72*, 1197–1206.
76. Mucke, H. A. CEP-1347 (Cephalon). *Drugs* **2003**, *6*, 377–383.
77. Boll, J. B.; Geist, M. A.; Kaminski Schierle, G. S.; Petersen, K.; Leist, M.; Vaudano, E. *J. Neurochem.* **2004**, *88*, 698–707.
78. Young, A. B. *J. Clin. Invest.* **2003**, *111*, 299–302.
79. Bonelli, R. M.; Hofmann, P. *Exp. Opin. Pharmacother.* **2004**, *5*, 767–776.
80. Hannan, A. J. *Acta Biochim. Pol.* **2004**, *51*, 415–430.
81. Landels, C.; Bates, G. P. *EMBO Rep.* **2004**, *5*, 958–963.
82. Qin, Z. H.; Gu, Z. L. *Acta Pharmacol. Sin.* **2004**, *25*, 1234–1239.
83. Grimergen, Y. A. M.; Roos, R. A. C. *Curr. Opin. Investig. Drugs* **2003**, *4*, 51–54.
84. Hersch, S. M.; Ferrante, R. J. *NeuroRx* **2004**, *1*, 298–306.
85. Zhang, X.; Smith, D. L.; Meriin, A. B.; Engemann, S.; Russel, D. E.; Roark, M.; Washington, S. L.; Maxwell, M. M.; Marsh, J. L.; Thompson, L. M. et al. *Proc. Natl. Acad. Sci. USA* **2005**, *102*, 892–897.
86. Carter, G. T.; Krivickas, L. S.; Weydt, P.; Weiss, M. D.; Miller, R. G. *IDrugs* **2003**, *6*, 147–153.
87. Weiss, M. D.; Weydt, P.; Carter, G. T. *Exp. Opin. Pharmacother.* **2004**, *5*, 735–746.
88. McGeer, E. G.; McGeer, P. L. *BioDrugs* **2005**, *19*, 31–37.
89. Munch, C.; Sedlmeier, R.; Meyer, T.; Homberg, V.; Sperfeld, A. D.; Kurt, A.; Prudlo, J.; Peraus, G.; Hanemann, C. O.; Stumm, G. et al. *Neurology* **2004**, *63*, 724–726.
90. Leung, C. L.; He, C. Z.; Kaufmann, P.; Chin, S. S.; Naini, A.; Liem, R. K.; Mitsumoto, H.; Hays, A. P. *Brain Pathol.* **2004**, *14*, 290–296.
91. Robberecht, W. *J. Neurol.* **2000**, *247*, 2–6.
92. Terry, P. D.; Kamel, F.; Umbach, D. M.; Lehman, T. A.; Hu, H.; Sandler, D. P.; Taylor, J. A. *J. Neurogenet.* **2004**, *18*, 429–434.
93. Bruijn, L. I.; Miller, T. M.; Cleveland, D. W. *Annu. Rev. Neurosci.* **2004**, *27*, 723–749.
94. Brooks, B. R.; Miller, R. G.; Swash, M.; Munsat, T. L. *Amyotroph. Lateral Scler. Other Motor Neuron Disord.* **2000**, *1*, 293–299.
95. Gordon, P. H. *Curr. Neurol. Neurosci. Rep.* **2005**, *5*, 48–54.
96. Wang, S. J.; Wang, K. Y.; Wang, W. C. *Neuroscience* **2004**, *125*, 191–201.
97. Nagano, I.; Shiote, M.; Murakami, T.; Kamada, H.; Hamakawa, Y.; Matsubara, E.; Yokoyama, M.; Moritaz, K.; Shoji, M.; Abe, K. *Neurol. Res.* **2005**, *27*, 768–772.
98. Duong, F. H.; Warter, J. M.; Poindron, P.; Passilly, P. *Br. J. Pharmacol.* **1999**, *128*, 1385–1392.
99. Lacomblez, L.; Bensimon, G.; Douillet, P.; Doppler, V.; Salachas, F.; Meininger, V. *Amyotroph. Lateral Scler. Other Motor Neuron Disord.* **2004**, *5*, 99–106.
100. Meininger, V.; Bensimon, G.; Bradley, W. R.; Brooks, B.; Douillet, P.; Eisen, A. A.; Lacomblez, L.; Leigh, P. N.; Robberecht, W. *Amyotroph. Lateral Scler. Other Motor Neuron Disord.* **2004**, *5*, 107–117.
101. Kieran, D.; Kalmar, B.; Dick, J. R.; Riddoch-Contreras, J.; Burnstock, G.; Greensmith, L. *Nat. Med.* **2004**, *10*, 402–405.
102. Bilak, M.; Wu, L.; Wang, Q.; Haughey, N.; Conant, K.; St Hillaire, C.; Andreasson, K. *Ann. Neurol.* **2004**, *56*, 240–248.
103. Kawahara, Y.; Ito, K.; Sun, H.; Aizawa, H.; Kanazawa, I.; Kwak, S. *Nature* **2004**, *427*, 801.
104. Kuner, R.; Groom, A. J.; Bresink, I.; Kornau, H. C.; Stefovska, V.; Muller, G.; Hartmann, B.; Tschauner, K.; Waibel, S.; Ludolph, A. C. et al. *Proc. Natl. Acad. Sci. USA* **2005**, *102*, 5826–5831.
105. Tateno, M.; Sadakata, H.; Tanaka, M.; Itohara, S.; Shin, R. M.; Miura, M.; Masuda, M.; Aosaki, T.; Urushitani, M.; Misawa, H. et al. *Hum. Mol. Genet.* **2004**, *13*, 2183–2196.
106. Canton, T.; Bohme, G. A.; Boireau, A.; Bordier, F.; Mignani, S.; Jimonet, P.; Jahn, G.; Alavijeh, M.; Stygall, J.; Roberts, S. et al. *J. Pharmacol. Exp. Ther.* **2001**, *299*, 314–322.
107. Jung, C.; Rong, Y.; Doctrow, S.; Baudry, M.; Malfroy, B.; Xu, Z. *Neurosci. Lett.* **2001**, *304*, 157–160.
108. Crow, J. P.; Calingasan, N. Y.; Chen, J.; Hill, J. L.; Beal, M. F. *Ann. Neurol.* **2005**, *58*, 258–265.
109. Horrabin, D. F. *Nat. Rev. Drug. Disc.* **2003**, *2*, 151–154.
110. Forman, M. S.; Trojanowski, J. Q.; M-Y. Lee, V. *Nat. Med.* **2004**, *10*, 1055–1063.

Biographies

Donna Bozyczko-Coyne, PhD, received her doctorate in Biochemistry (1988) from the University of Pennsylvania, School of Medicine, with a focus on neurosciences. Under the direction of Dr Alan F Horwitz, she studied the role of cell adhesion molecules (integrins) in the development of the neuromuscular junction and in neurite outgrowth. Continuing an interest in developmental neurosciences, she conducted postdoctoral studies with Dr Art Morris at the Wistar Institute, Philadelphia, PA, studying the role of IGF-1 in the development of the oligodendrocyte lineage. In 1990, Dr Bozyczko-Coyne accepted a Research Scientist position at Cephalon, Inc., where she has conducted drug discovery research in the therapeutic areas of stroke/ischemic neuronal damage, Parkinson's and Alzheimer's diseases. In 2000, Dr Bozyczko-Coyne became Director of Neurobiology and is, today, Director of Research and Development Strategic Coordination at Cephalon, Inc.

Michael Williams, PhD DSc, received his PhD (1974) from the Institute of Psychiatry and his Doctor of Science degree in Pharmacology (1987), both from the University of London. Dr Williams has worked in the US-based pharmaceutical industry for 30 years at Merck, Sharp, and Dohme Research Laboratories, Nova Pharmaceutical, CIBA-Geigy, and Abbott Laboratories. He retired from Abbott Laboratories in 2000 and, after serving as a consultant with various biotechnology/pharmaceutical companies in the US and Europe, joined Cephalon, Inc. in West Chester in 2003, where he is Vice President of Worldwide Discovery Research. He has published some 300 articles, book chapters, and reviews, and is Adjunct Professor in the Department of Molecular Pharmacology and Biological Chemistry at the Feinberg School of Medicine, Northwestern University, Chicago, IL.

Comprehensive Medicinal Chemistry II
ISBN (set): 0-08-044513-6

ISBN (Volume 6) 0-08-044519-5; pp. 193–228

6.09 Neuromuscular/Autoimmune Disorders

I Lieberburg, T Yednock, S Freedman, R Rydel, and E Messersmith, Elan Pharmaceuticals, South San Francisco, CA, USA
R Tuttle, El Cerrito, CA, USA

6.09.1 Introduction

The immune system responds to invading foreign antigens. Once the target foreign antigen has been eliminated, the reduction in the immune response is rapid, and the affected host tissue is able to recover from the damage inflicted by the inflammatory response. In autoimmune diseases, the immune system responds to self-antigens; as a result, the target antigen is usually never eliminated, thus the response is often sustained and the tissue damage may not completely subside. Periods of reduced damage are, however, common – thus the relapsing–remitting course of all of the chronic autoimmune diseases discussed in this review. Autoimmunity is now believed to be a naturally occurring phenomenon gone awry in autoimmune disease.

Many of the diseases discussed in this review are presumed to be autoimmune (**Table 1**). In theory, certain criteria should be met if a disease is to be considered to have an autoimmune etiology: autoantibodies or tissue-directed T cells should be detectable in serum or the affected tissue; they should be present before the onset of symptoms; and there should be evidence of an immune-mediated pathology in the affected tissue. Examples of experimental evidence are if isolated autoantibodies, activated T cells, or the presumptive autoantigen recreate disease when injected into healthy animals. This type of direct evidence is perhaps strongest for Myasthenia Gravis (MG) and Gullain–Barre syndrome (GBS), where neonates sometimes mimic the autoimmune disease of their mother. Further, symptoms should show improvement with the use of immunosuppressive or immunomodulatory drugs. Lastly, in some autoimmune diseases, there may be evidence of pre-existing infections, associated with disease onset or relapse; this is often the case for GBS, and perhaps for chronic inflammatory demyelinating polyradiculoneuropathy (CIDP) and multiple sclerosis (MS) as well.

Each autoimmune disease or syndrome described in this review represents a heterogeneous set of disorders with common features (see **Table 1**). This heterogeneity can result from a common disease manifesting itself in distinct genetic and epigenetic circumstances – seen, for example, in animal models of MS, where the disease course varies with the genetic background. Alternatively, heterogeneity can arise from distinct pathogenic mechanisms manifesting as observed common clinical features. The establishment of useful classification schemes is a dynamic, ongoing process influenced by one or more factors – including clinical features, prognosis, therapeutic response, underlying pathogenic mechanisms, and biomarker data. These classification systems are valuable for diagnostic purposes – and so inclusion in laboratory and clinical research studies, for prognostic purposes and for determining individualized treatment strategies (i.e., predicting treatment response or risk of adverse effects).

Autoimmune diseases of the nervous system can be grouped according to whether they affect tissue of the central nervous system (CNS), peripheral nervous system (PNS), or both (see **Table 1**). As with all classification systems, this one has its limitations. CNS lesions are sometimes present in patients with PNS autoimmune disease, such as CIDP; however, these CNS features are usually clinically silent. Conversely, clinically silent PNS lesions are sometimes present in patients with MS. The general rule is that patients with demyelination in both the CNS and PNS are usually symptomatic at one site or the other, but not both. Systemic lupus erythematosus (SLE) is one of the exceptions to this rule. In addition, there are poorly understood CNS–PNS overlap syndromes. Further, some diseases have their etiology outside the nervous system altogether, but end up affecting the nervous system as the disease progresses; the development of CIDP in patients with diabetes provides one example.

Autoimmune diseases tend to be predominantly (auto)antibody mediated or Th1 cell mediated. Examples of the former are MG, GBS, and SLE; MS is a typical T cell-mediated disease. The situation with CIDP is less clear. There is diagnostic and therapeutic advantage to identifying the autoantigens and, among them, the primary target. In the case of antibody-mediated autoimmune disease, characterization of patient autoantibodies can reveal the identity of the autoantigen(s), although not always the primary one. In T cell-mediated autoimmune diseases, on the other hand, the autoantigen recognized by the T cells is often unknown.

One might expect autoimmune diseases to be characterized by autoantigens that are exposed to immune attack – that is, either secreted or intramembrane molecules. However, in some diseases, patients exhibit autoantibodies to intracellular antigens. In MG, this results from a process known as intermolecular epitope spreading, whereby cytoplasmic epitopes are exposed upon destruction of the cell membrane. In SLE, many of the autoantigens are located in the nucleus.

Autoimmune diseases can present asymmetrically, symmetrically, or unilaterally. It is perhaps curious that a disease with a systemic component should present asymmetrically or even unilaterally, but this is the case in MS (particularly ON), MG, in some forms of CIDP (MMN and MMSN, see **Table 1**), and is often the case in amyotrophic lateral sclerosis (ALS) or in brachial neuritis. Perhaps use or disuse, as well as trauma, can predispose one not necessarily to disease itself but to a particular asymmetric manifestation of disease.

Sexual dimorphisms are apparent in the prevalence of some autoimmune diseases. For example, MS, early onset MG, and SLE disproportionately affect females; for SLE, predominance can be as high as 5:1 in childbearing years. Further, the clinical onset of SLE often coincides with menarche, pregnancy, postpartum, or menopause. MS patients

Table 1 Autoimmune diseases of the nervous system

Disease	*US Incidence (annual, per 100 000)*	*Subtypes*
Central nervous system		
Multiple sclerosis (MS)	3	Clinically isolated syndrome (CIS)
		Relapsing–remitting MS (RRMS)
		Secondary progressive MS (SPMS)
		Primary progressive MS (PPMS)
		Relapsing progressive MS (RPMS)
		Benign MS
		Optic neuritis (ON)
		Neuromyelitis optica (NMO) or Devic's syndrome
Central nervous system/peripheral nervous system		
Systemic lupus erythematosus (SLE)	7.3	Juvenile SLE
		Lupus nephritis (LN)
		Neuropsychiatric SLE (NPSLE)
		Non-neuropsychiatric SLE (non-NPSLE)
		Anti-phospholipid antibody syndrome (APS)
Peripheral nervous system		
Guillain–Barré syndrome (GBS)	1.5–3	Acute inflammatory demyelinating polyradiculoneuropathy (AIDP)
		Miller–Fisher syndrome
		Acute motor axonal neuropathy (AMAN)
		Acute motor sensory axonal neuropathy (AMSAN)
		Pure sensory neuropathy
		Pure autonomic neuropathy (acute pandysautonomia)
		Atypical presentations of AIDP
		Regional presentations of AIDP
Myasthenia gravis (MG)	0.2–2	Generalized MG:
		Acetylcholine receptor (AChR) Ab-positive (85% of generalized MG patients)
		Early onset MG (<50): higher AChR Ab titer; thymic hyperplasia; no other Abs
		Late onset MG: lower AChR Ab titer; thymic atrophy; titin, RyR Abs in 50%
		Thymoma MG; lower AChR Ab titer; thymoma (usually cortical); titin, RyR Abs in 95%
		AChR Ab-negative (15% of generalized MG patients)
		Early onset MG: MuSK-positive and -negative subtypes
		Late onset MG: MuSK-positive and -negative forms
		Ocular MG:
		AChR Ab-positive (60% of ocular MG patients)
Chronic inflammatory demyelinating polyradiculoneuropathy (CIDP)	ND	Sensory ataxic neuropathy (SAN)
		Subacute motor sensory demyelinating neuropathy (SMSDN)
		Chronic motor sensory demyelinating neuropathy (CMSDN)
		Multifocal motor sensory neuropathy (MMSN, Lewis–Summer neuropathy)
		Pure motor demyelinating neuropathy (PMDN)
		Multifocal motor neuropathy (MMN)
		Diabetes-associated CIDP
		CIDP-associated with monoclonal gammopathies

have lower relapse rates during pregnancy, worsening of MS during menstruation, and correlation of high estradiol and low progesterone with magnetic resonance imaging (MRI) monitoring of disease activity. These findings and the gender asymmetries in prevalence, as well as clinical trial data, suggest a possible role for estrogens and androgens in the pathogenesis of some autoimmune diseases.[1] In fact, current clinical trials are testing the therapeutic safety and efficacy of sex hormones in the treatment of MS.

In addition to possible hormonal factors, the development of these autoimmune diseases appears, in general, to be influenced by both genetic and environmental factors.[2] In addition, the disproportionate prevalence of some autoimmune disease in older patients (e.g., late onset MG and GBS) suggests that the effects of aging on the immune system may also play a role.

In this chapter, the general features of several autoimmune diseases and syndromes affecting the nervous system are examined with consideration of the therapies currently in use and in development to treat them.

6.09.2 Autoimmune Diseases

6.09.2.1 Diseases Affecting the Central Nervous System

6.09.2.1.1 Multiple sclerosis

6.09.2.1.1.1 Overview

MS is a chronic disease of the CNS, characterized by cell-mediated inflammation, demyelination, and variable degrees of axonal loss.[3,4] MS is also characterized by lesions in the white matter of brain, brainstem, optic nerve, or spinal cord that are visible by MRI, and formed by infiltration of CNS parenchyma by monocytes and lymphocytes. Degeneration of postmitotic oligodendrocytes and oligodendrocyte progenitors is also a hallmark of MS lesions and an impediment to remyelination. The lesions are heterogeneous in nature, and have been classified into four histopathological subtypes. MS is the most prevalent inflammatory disease of the CNS, and is far more common in females (1.6–2:1). It can occur at any age, but is most commonly diagnosed between the third and fourth decades.

It is unknown if MS is a single disease or a syndrome. There are, however, multiple subtypes, the most common being relapsing–remitting MS (RRMS), which occurs in 85% of new patients (see **Table 1**). Relapses are defined by a clinical episode or by subclinical activity detected by MRI. More than half of RRMS patients progress to secondary progressive MS (SPMS). Six additional subtypes have been described: PPMS and RPMS – both progressive diseases from onset, CIS, benign MS, ON, and NMO (Devic's syndrome). Axonal loss occurs early in the disease course, suggesting that all forms of MS – except benign – are progressive.

PPMS patients can present with weakness in one limb and progress without remission to total paralysis; spinal cord involvement is pronounced. Benign MS can only be diagnosed retrospectively, and is seen in a small fraction of MS patients who have no signs of physical disability even decades after diagnosis. NMO, which is generally relapsing, is distinct in that it affects only the optic nerves and spinal cord, does not progress to SPMS, exhibits distinct demographics, and exacerbations are usually severe with lasting disability.

There are no MS-specific biomarkers. However, elevated levels of immunoglobulin in cerebrospinal fluid (CSF), mostly immunoglobulin G (IgG), are sometimes used to support the diagnosis of MS. NfH, a neurofilament heavy chain phosphoform monitored in plasma or CSF, is indicative of neurodegeneration and a poor prognostic sign; it is elevated in the plasma of RRMS patients versus healthy controls. In addition, nitric oxide metabolites are elevated in CSF, but not serum, of MS patients – particularly those with milder disability – suggesting a role for nitric oxide in the early phase of MS. Finally, S100B, a marker for astrocytic activation, is often elevated in plasma of RRMS patients who respond to interferon (IFN)-β therapy compared with nonresponders or healthy controls.

6.09.2.1.1.2 Pathogenesis

The two currently favored hypotheses for the pathogenesis of MS are immunopathological and apoptotic. Histopathological data as well as the ability of immunosuppressive therapies, such as natalizumab, to reduce or stop clinical relapses and suppress MRI activity support the idea that inflammation is an important pathogenic mechanism, and thus a therapeutic target for controlling relapses in RRMS. Further, the natalizumab trials support the role of α_4 integrin, and the immune cells that express it, in MS pathogenesis. The presence of autoantibodies in some patients lends additional support to the autoimmune or immunopathological hypothesis, as does the association of MS with genes of the major histocompatibility complex. On the other hand, the apoptotic theory is favored by a pathological report from a very early case of MS, and by the general failure of immunosuppressive therapies to slow disease progression. Indeed, progression seems to be independent of relapses – it occurs regardless of the presence of superimposed relapses. Poor recovery from relapses does, however, account for the acquisition of disability.

Both polygenetic and environmental factors are thought to contribute to MS etiology. Concordance rates in monozygotic twins are relatively modest (about 25%), supporting a role for environmental and polygenetic factors. One or more genes in the area of the major histocompatibility complex are thought to account for some of the genetic risk. MS is more common in Caucasian populations. Prevalence data also show a north–south gradient in the northern hemisphere that reverses in the southern hemisphere; moving to high-risk areas before the age of 15 years increases the risk of developing MS. These data suggest that a decrease in sunlight exposure or a temporal zone pathogen (such as virus) may be causative in MS. Prevalence also correlates positively with the level of socioeconomic development of a country, which can be explained by many factors, including decreased exposure to sunlight and a reduction in childhood infections.[4]

MS is a cell-mediated immune disorder, with strong evidence for both CD4+ and CD8+ T cells playing a role in disease pathogenesis.[4] Epidemiological studies suggest that autoreactive CD4+ Th1 cells may be established for 10–20 years before they become active and contribute to disease.

While MS pathogenesis is primarily mediated by T cells, a role for B cells is also well described in MS patients. While the autoantigens in MS are unknown, most putative ones are myelin-associated antigens: myelin basic protein (MBP), proteolipoprotein, myelin-associated glycoprotein, and myelin oligodendrocyte glycoprotein (MOG); other putative autoantigens are transaldolase, 2′,3′-cyclic nucleotide 3′-phosphodiesterases, and α/β-crystallin. A pathogenic role for autoantibodies has not been clearly demonstrated. However, antibodies to MOG are pathogenic in animal models, have been detected in brain parenchyma from MS patients, including lesions, and are present early in the disease. MOG is a type I membrane protein expressed exclusively in CNS. Interestingly, while antibodies from RRMS and SPMS patients bind to oligodendrocyte precursors, binding to a neuronal cell line is increased in SPMS. Many of these studies are plagued by problems and shortcomings common to antigen–antibody assays.

6.09.2.2 Diseases Affecting Both the Central Nervous System and Peripheral Nervous System

6.09.2.2.1 Systemic lupus erythematosus

6.09.2.2.1.1 Overview

SLE is a chronic autoimmune disorder affecting nearly all organs – including the CNS, PNS, and muscle. Clinical disease follows a relapsing–remitting course, and manifestations are diverse – ranging from mild arthritis or fatigue and musculoskeletal complaints to life-threatening renal or cerebral disease. Atherosclerosis and nephritis are major causes of morbidity and mortality. Psychiatric and neurologic disorders are manifested in a large fraction of SLE patients. The most common neurologic manifestations of SLE are encephalopathies; these diffuse syndromes correlate with patchy areas of dysfunction seen with functional MRI, but not conventional MRI. Peripheral manifestations of SLE include peripheral neuropathy and myopathy – including disturbances of the neuromuscular junction, which may clinically duplicate MG.

The current classification scheme for SLE, which is under revision,[5] utilizes multiple criteria – four dermatological, but only one for each of the other affected organ systems: neurologic disorder, arthritis, serositis, renal disorder, hematologic, and immunologic. In fact, it is from the dermatological manifestations that the Latin name, *lupus* – wolf, derives; a thirteenth-century physician used it to describe erosive facial lesions reminiscent of the bite of a wolf. Only four out of 14 criteria need to be validated for an SLE classification. The current classification system has several flaws. Among them are that some of the criteria may not be present early in the disease course and the neurological criteria, psychosis and seizure, are relatively infrequent. The subclasses presented for SLE in **Table 1** simply represent some commonly discussed subtypes – not a classification scheme per se.

Elevated anti-double-stranded DNA (dsDNA) antibodies are highly specific to and, therefore, diagnostic for SLE[6]; a majority of patients with active SLE have a positive anti-DNA test. Also diagnostic for SLE are antibodies to Sm, a ribonucleoprotein; anti-Sm antibodies are found in about a third of SLE patients, and are not found in any other disease. Most individuals with SLE have either anti-DNA or anti-Sm antibodies. Antiribosomal P protein antibodies, which are highly specific for SLE, are associated with CNS involvement and nephritis. Clinical data suggest that increases in dsDNA antibodies and decreases in C3a, a fragment of the complement protein C3, are predictive of flare-ups; their monitoring, therefore, could permit earlier, preventive intervention in the management of SLE.[1] Clinical trials are underway to further test this hypothesis.

SLE is found in all ethnic groups and races; susceptibility genes, which may increase the risk of SLE, vary across ethnic populations. The prevalence of various organ manifestations and autoantibodies also differs by ethnicity, and perhaps nation. All age groups are affected; however, peak incidence is in young adulthood.

6.09.2.2.1.2 Pathogenesis

Over 100 different autoantibodies have been identified in SLE patients.[7] However, it is not known if these autoantibodies are pathogenic. Although a single patient might have a large number of autoantibodies, most are found in only a minority of patients. Nearly all SLE patients have antibodies to nuclear antigens, and, as mentioned earlier, the vast majority have anti-dsDNA antibodies, which correlate positively with disease activity; current clinical trials are testing whether lowering levels has any therapeutic benefit. It is unclear how these intracellular antigens become exposed to immune system components.[8]

The presence of these autoantibodies in SLE patients, and the finding that they are present before clinical symptoms are manifested, lends strong support for an autoimmune etiology of SLE. Interestingly, many of these antibodies (e.g., anti-dsDNA) are present years before disease onset, whereas others (e.g., anti-Sm) are present months before diagnosis. Also supportive of an autoimmune etiology is the presence of activated complement proteins and localized inflammation in kidney or other tissues in lupus patients with active nephritis.

Environmental factors are thought to play a role in triggering SLE, although genetic factors may influence susceptibility (most monozygotic twins are discordant for SLE). The large number of autoantibodies suggests that SLE results from a polygenetic defect that results in polyclonal B cell overactivation – loss of B cell tolerance; indeed, removal of B cells prevents the formation of SLE in mice. Further, SLE patients commonly overexpress B lymphocyte stimulator (BLyS), which induces B cell proliferation and immunoglobulin secretion.

A role for sex hormones in SLE pathogenesis is suggested by the fact that clinical onset often coincides with menarche, pregnancy, postpartum, or menopause. Further support comes from the fact that SLE shows a strong female predominance.[1]

6.09.2.3 Diseases Affecting the Peripheral Nervous System

6.09.2.3.1 Guillain–Barré syndrome

6.09.2.3.1.1 Overview

GBS is a collection of acute or sometimes subacute, self-limiting postinfectious autoimmune polyneuropathies of the PNS. GBS involves both myelin and axons, and results in flaccid paralysis of at least two limbs. Other symptoms can include generalized weakness, areflexia, and usually numbness, a varying degree of sensory disturbances, and involvement of cranial nerves. Weakness of respiratory muscles renders about 25% of patients respirator-dependent. Symptoms begin 1–3 weeks after various infections, and peak from a few days up to 4 weeks, with patients recovering spontaneously. Recovery can take weeks or years. Many patients have persistent fatigue, and some are unable to walk 1 year after onset; the younger the patient, the better the prospect for complete recovery. Mortality rates range from 3% to 12%, with patients dying of complications during the acute stage.

GBS patients can be subtyped on the basis of antecedent infection, anti-ganglioside antibodies, and neurological deficits (see **Table 1**). GBS subtypes include acute inflammatory demyelinating neuropathy (AIDP), which accounts for about 90% of GBS cases, atypical presentations of AIDP, Miller–Fisher syndrome, in which weakness is restricted to extraocular and other craniobulbar muscles, and the axon loss variants – including acute motor axonal neuropathy (AMAN), which occurs in up to 10% of patients. Some patients exhibit clinical features typical of more than one subtype – so-called overlap syndromes. Making a rapid diagnosis is often difficult as there are many mimics of GBS,[9] and CSF may not be diagnostic for up to 2 weeks.

The AIDP subtype is characterized by an infiltration of endoneural capillaries by lymphocytes and a macrophage-mediated demyelination of both sensory and motor nerves in a segmental fashion with deposition of complement activation products on Schwann cell surface membranes. AMAN, in which motor axons appear to be the target of immune attack, is characterized by Wallerian-like degeneration without significant demyelination and with little or no lymphocytic infiltration.

While GBS occurs throughout the world, the axonal forms of the disease are much less common in Europe and North American, and more common in China, Japan, India, and Central America. Contrary to many autoimmune diseases, males are more commonly affected than females.[10] Age distribution is bimodal – with peaks in young adults and the elderly. There are no reliable serological markers.

6.09.2.3.1.2 Pathogenesis

GBS etiology has a clear environmental component: about two-thirds of GBS patients have antecedent gastrointestinal or respiratory infections by *Campylobacter jejuni*, Epstein–Barr virus, cytomegalovirus, or *Mycoplasma* pneumonia. The most frequently identified cause of GBS is *C. jejuni* infection – identified in up to 41% of patients. Patients with antecedent *C. jejuni* infection are more likely to require ventilation and have prolonged severe disability.[10]

Clinical data support the hypothesis that autoantibodies are involved in GBS pathogenesis: the efficacy of plasma exchange (PE) and intravenous immunoglobulin (IVIg) therapies and the case of flaccid paralysis in a newborn of a mother with GBS. There is also strong support for a specific role for ganglioside antibodies: the identification of GBS-like symptoms in some patients treated with gangliosides; approximately 40% of GBS patients have anti-ganglioside antibodies in their plasma; disease subtypes or symptoms correlate with the ganglioside antibody subtype, as well as with the differential expression of those subtypes in the PNS; and rabbits sensitized with *C. jejuni* lipooligosaccharide developed anti-GM1 IgG antibody, flaccid limb weakness, and peripheral nerve pathology identical to GBS.[11] The last study provides the strongest and most direct evidence that at least some types of GBS are the result of carbohydrate mimicry between an infectious agent and peripheral nerve ganglioside epitopes.[12]

Although gangliosides are present in both CNS and PNS, GBS appears to affect only the PNS. Thin-layer chromatogram immunostaining data suggest that antibodies from some patients may recognize ganglioside complexes, but have little or no reactivity for individual gangliosides; patients with these antibodies tend to have severe disabilities and cranial nerve deficits.[13]

Other data suggest possible functions of ganglioside antibody in GBS pathogenesis: IgG ganglioside antibody from GBS patients recognizes antigens at the node of Ranvier and the motor endplate presynaptic terminal, and in vitro experiments with rodent tissues demonstrate reversible conduction blockade with GM1 ganglioside monoclonal antibody.[11] In addition, anti-GM1- or GD1a-specific IgG containing sera from GBS patients induce leukocyte inflammatory functions such as degranulation and phagocytosis via FcγR.[14]

Based on these findings, a pathogenetic sequence has been proposed for GBS with antecedent *C. jejuni* infection: infection by *C. jejuni* carrying a complex ganglioside-like lipooligosaccharide induces – in some patients – high production of anti-ganglioside IgG that binds to complex ganglioside on motor axons, leading, in turn, to macrophage-mediated and complement-dependent axonal degeneration.[11,15]

Estimates of the number of cases of *C. jejuni* infection complicated by the occurrence of GBS are one in 1000 or less, suggesting a role for host genetic factors in GBS etiology. The search for genetic factors revealed polymorphisms in the promoter region of *Fas* and in the gene for FcγRIII in a West European population that may represent mild disease-modifying factors in GBS.[16,17] While these *Fas* and *FcγRIII* polymorphisms do not differ between GBS patients and controls, and thus do not represent general susceptibility factors for GBS, there does appear to be a weak association with the presence of anti-ganglioside antibodies and disease severity, respectively.

6.09.2.3.2 Chronic inflammatory demyelinating polyradiculoneuropathy

6.09.2.3.2.1 Overview

Chronic inflammatory demyelinating polyradiculoneuropathy (CIDP) is a chronic demyelinating disease of the PNS that can involve sensory components, motor components, or both. The duration of the onset phase is generally more than 8 weeks; however, some CIDP patients show acute onset, as with GBS. It is characterized by progressive or relapsing–remitting weakness and numbness. There is a high incidence of fatigue that may persist for years. Electrophysiological features of CIDP include nerve conduction block and slowed conduction velocities resulting, most likely, from demyelination.

CIDP is a heterogeneous disorder with disagreement on which diseases to include under the CIDP umbrella (see **Table 1**), and a lack of clarity with regard to diagnostic criteria. In most subtypes, the symptoms are symmetrical. However, the Lewis–Summer motor sensory subtype is an asymmetric form with persistent conduction block at individual sites over long periods. Multifocal motor neuropathy (MMN), a motor form, is also asymmetric, and can show some sensory loss. Conversely, the sensory form of CIDP shows evidence of motor involvement. One of the subtypes, diabetes-associated CIDP, arises secondarily in up to 17% of patients with either type 1 or type 2 diabetes mellitus (*see* 6.19 Diabetes/Syndrome X).

CIDP is most common in the fifth and sixth decades, but may occur at any age. The MMN subtype is more common in men than women (2.6:1). A better outcome is reported to be related to younger age at onset, a relapsing–remitting course, and the absence of axonal damage.

6.09.2.3.2.2 Pathogenesis

Like MS, CIDP has a relapsing or progressive course, focal demyelination, axon degeneration that can occur early or late in the disease course, and an immune-mediated pathophysiology.

CIDP is thought to be an autoimmune disorder caused by either humoral or cell-mediated immunity to axonal, myelin, or other Schwann cell antigens. The autoantigens associated with CIDP are generally unknown. However, MMN is associated with anti-GM1 IgM antibodies, the levels of which decrease with clinical improvement.

Evidence for an autoimmune basis for CIDP is centered mainly on its general responsiveness to immunotherapy, and the fact that an inflammatory response is observed at the site of the disease. Unlike GBS, CIDP is rarely preceded by infection.

6.09.2.3.3 Myasthenia Gravis

6.09.2.3.3.1 Overview

MG is a chronic autoimmune disease of the neuromuscular junction characterized by fluctuating weakness and fatigability of skeletal and extraocular muscles. The natural history is characterized by exacerbations and remissions. Symptoms increase during the day, and may be masked by rest. Patients may have respiratory (and swallowing) difficulties during myasthenic crisis, requiring artificial ventilation and airway protection; respiratory compromise is the most common cause of death in MG patients. In 1997, the death rate was reported to be less than 10%.

In adults, there are two forms – ocular and generalized. In children, there are three forms: autoimmune or juvenile MG, genetic or congenital MG and transient neonatal. In addition, the adult forms can be subtyped by autoantibody profile, the presence or absence of thymic abnormalities, age of onset, and severity[18,19] (see **Table 1**). Approximately half of MG patients present with the ocular form and symptoms of ptosis and diplopia; most of these progress to generalized MG, with involvement of the bulborpharyngeal and skeletal muscles. Eventually, almost all MG patients have some eye muscle involvement.

MG is rare during childhood, and almost never occurs before 1 year of age; in North America, onset before the age of 20 years accounts for about 10–15% of all patients with MG. The demographic data are complex, varying by subtype.[18] Generally, however, the disease shows two peaks for age of onset: one, between 20 and 40 years, is dominated by women; the other, between 60 and 80 years, is shared equally by men and women.

6.09.2.3.3.2 Pathogenesis

While the etiology of MG is not known, it is perhaps the best understood of the autoimmune diseases.[18] Unlike some types of GBS, microbial infections have not been shown to cause any of the chronic autoimmune diseases. However, onset and exacerbation of symptoms of MG may follow febrile illness and insect bites, and some data suggest a possible role for molecular mimicry. A genetic basis is suggested by the finding that 30% of patients have one maternal relative with MG or another autoimmune disorder.

Thymic pathology occurs in 80–90% of MG patients: about 15% of MG patients have a thymoma and 50–70% have thymic hyperplasia with proliferation of germinal centers. There is evidence to suggest that thymoma-associated MG is a paraneoplastic disease. In normal thymus, muscle epitopes can be found at low levels and confined to the medulla. In thymoma MG, however, muscle epitopes (AChR-like and others) and co-stimulatory molecules are overexpressed on neoplastic antigen-presenting epithelial cells found throughout the thymoma; these, in turn, are surrounded by large numbers of T cells.[18]

Although both T helper and B cells are involved in the autoimmune response, the attack on the neuromuscular junction is carried out exclusively by antibodies – most commonly against a portion of the extracellular domain of the AChRα subunit, the main immunogenic region. AChR antibodies impair neuromuscular transmission by three basic mechanisms, all involving the actual or functional loss of AChRs. The first and primary mechanism is complement-mediated destruction of the postsynaptic membrane. The second mechanism is blockade of ACh binding (the ACh-binding site and the main immunogenic region are both located on the α subunit), although such blocking antibodies are present in small amounts and probably play a minor role in most cases. The third mechanism is cross-linking of AChRs by bivalent antibody molecules, leading to accelerated AChR internalization; this mechanism appears to play a relatively minor role as well. Interestingly, the titer of AChR antibody does not appear to correlate with disease severity.

Some of the autoantibodies found in MG patients are against proteins localized to the cytoplasm, where they should be protected from immune attack. For example, 15% of MG patients with AChR antibodies also have antibodies against the muscle cytoplasmic protein rapsyn. Some of these antibodies to intracellular antigens appear to play a pathogenic role. They may be formed as a result of intermolecular epitope spreading.

6.09.2.4 Candidate Autoimmune Diseases

6.09.2.4.1 Amyotrophic lateral sclerosis

Amyotrophic lateral sclerosis (ALS; also *see* 6.08 Neurodegeneration) ALS is a neurodegenerative disease characterized by selective loss of spinal cord and cranial motor neurons. Patients are devastated by progressive muscle weakness and atrophy, and die of respiratory failure. Over 90% of ALS cases are sporadic; of the familial cases of ALS, a fraction is caused by mutations in the gene for the free-radical scavenger superoxide dismutase (SOD)-1. It has been hypothesized that sporadic ALS is an autoimmune disease.

The evidence for this hypothesis comes from animal and human studies. In SOD-1 mutant mice, the onset of disease is preceded by an inflammatory response, involving microglial activation and astrocytosis. In addition, minocycline, a tetracycline antibiotic with anti-inflammatory effects (**Figure 1** and **Table 2**), extends survival in mouse models of ALS. In ALS patients, spinal cord fluid shows elevated levels of proteins involved in inflammation. Also intriguing is a controlled study noting morphological evidence of motor neuron degeneration in mice injected with IgG purified from ALS patients.[20]

The strongest evidence against the ALS autoimmune hypothesis derives from the fact that immunosuppression and PE fail to alter disease progression, the most recent failure being with the cyclooxygenase 2 inhibitor, rofecoxib, an anti-inflammatory drug commonly used to treat arthritis, including rheumatoid arthritis (RA). While rofecoxib improved the life span of mutant SOD-1 mice, in clinical trials it had no effect on ALS progression.

There are currently three clinical trials addressing the ALS autoimmune hypothesis: two Phase II trials evaluating thalidomide, which has anticytokine, tumor necrosis factor (TNF)-α-inhibitory activity – as well as antiangiogenic properties and a third trial, Phase III, examining minocycline (see **Figure 1** and **Table 2**).

6.09.3 Experimental Animal Models

6.09.3.1 Overview

Animal models – used both in vivo and in vitro – are indispensable yet imperfect tools that are infrequently predictive of response in humans. Animal models, almost exclusively in rats and mice, can provide only proof-of-concept. In fact, the US Food and Drug Administration (FDA) does not require that efficacy studies be carried out in animal models – only that drug developers provide a reasonable biological rationale for why they think a particular therapy might be effective. On the other hand, testing – in normal animals – is generally required to provide risk assessment data; beyond this, sponsors use animal models to identify therapeutic targets, and to convince themselves of a drugs potential efficacy before proceeding with its development. Ultimately, however, a potential therapeutic must be studied in humans, since beyond problems in predicting efficacy, many side effects (e.g., certain infections, nausea, and headache) cannot be detected in animal studies.

A flaw with many animal models is that they focus on replication of symptoms rather than pathogenic mechanisms. The fact that the fundamental differences between chronic and acute response in autoimmune disease are not understood also limits the development of appropriate animal models. In addition, there is often a mismatch between the observed pathology of the human and animal diseases – perhaps because the former tend to be from late stage, post-mortem analysis, while the animal data are often from the earliest stages of disease; indeed, because of the short life span of many animals, they may never develop the pathology seen in late-stage humans.

Drosophila melanogaster and other invertebrates have proven to be excellent model systems for many neuro-degenerative diseases that faithfully replicate key neuropathological features of the human disorders. Invertebrates have many advantages, particularly the possibility for sophisticated and rapid genetic analysis, the short life span, and modest cost. However, there are, as yet, no invertebrate models for autoimmune disease – although there are ongoing efforts to develop one for ALS. This is, in large part, due to the fact that *Drosophila* lacks an adaptive immune response, although it appears to have an innate immune response (humans have both).

Rodents are the most commonly utilized animal model in preclinical research. However, only a small fraction of the many treatments that proved effective in rodents with experimental autoimmune disease have reached the clinic. The reasons for these failures are numerous and diverse. The diseases are often chronic while the animal models are acute. Both rats and mice have the disadvantage of small brain and body size, which presents problems for imaging studies and longitudinal blood sampling; rabbits are sometimes substituted for just this reason. In addition, their relatively short life span, while offering obvious advantages, does not provide a good model for chronic therapies. Further, the diversity characteristic of all autoimmune diseases and syndromes may be lacking in rodent models utilizing inbred strains. Lastly, there are many immunological differences between rodents and humans, and laboratory rodents are generally pathogen-free, while patients have a long history of infections. With all that said, however, the advantages of rodents are real: genetically susceptible strains, transgenic animals (knockouts, knock-ins), numerous reagents developed for use in rodents, and cells and tissues that can be easily established in vitro.

Nonhuman primates have many advantages over rodents as models of human disease (cost is obviously not one of them). Foremost is their close immunological and genetic proximity to humans and their outbred nature. In addition,

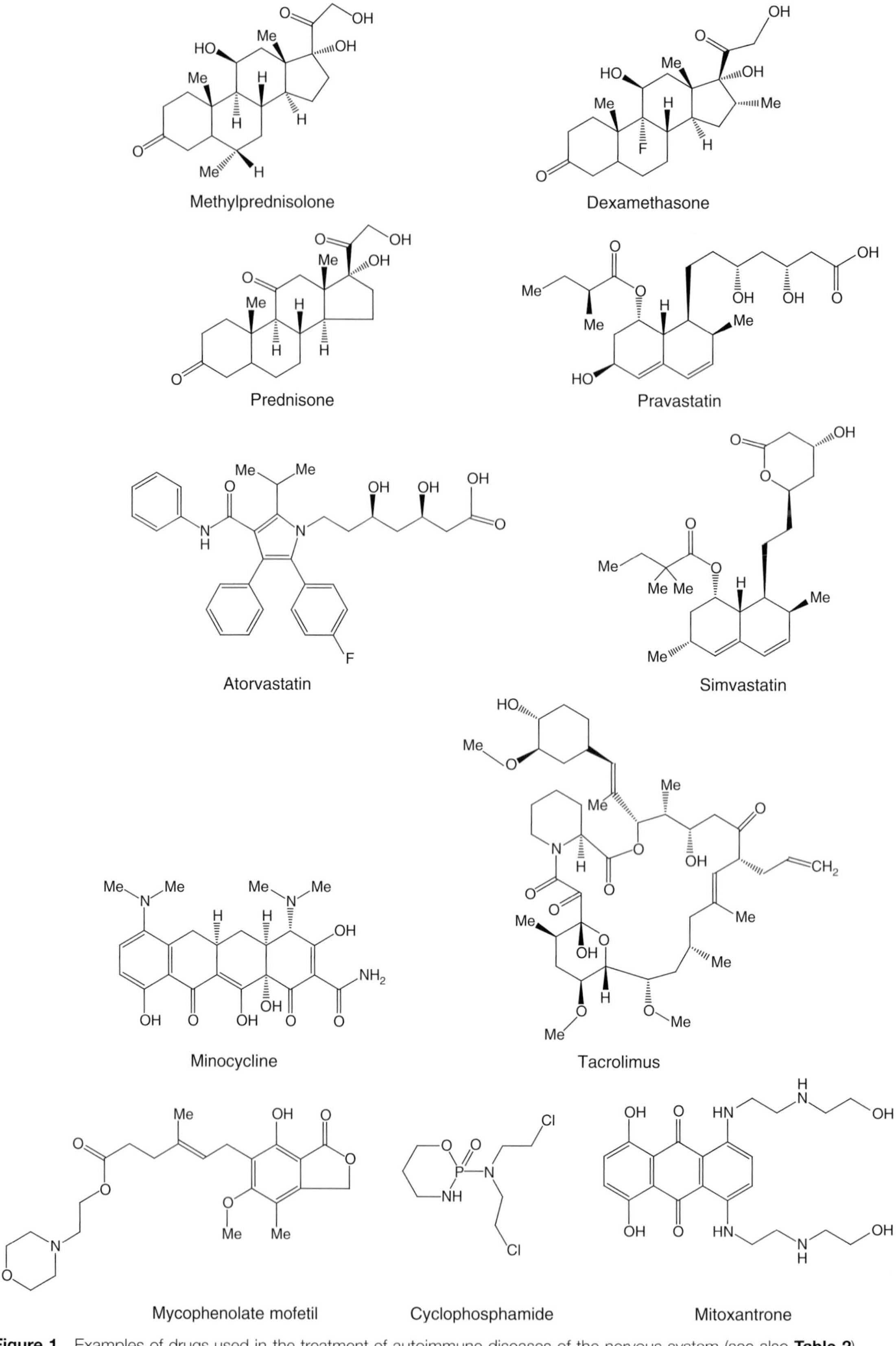

Figure 1 Examples of drugs used in the treatment of autoimmune diseases of the nervous system (see also **Table 2**).

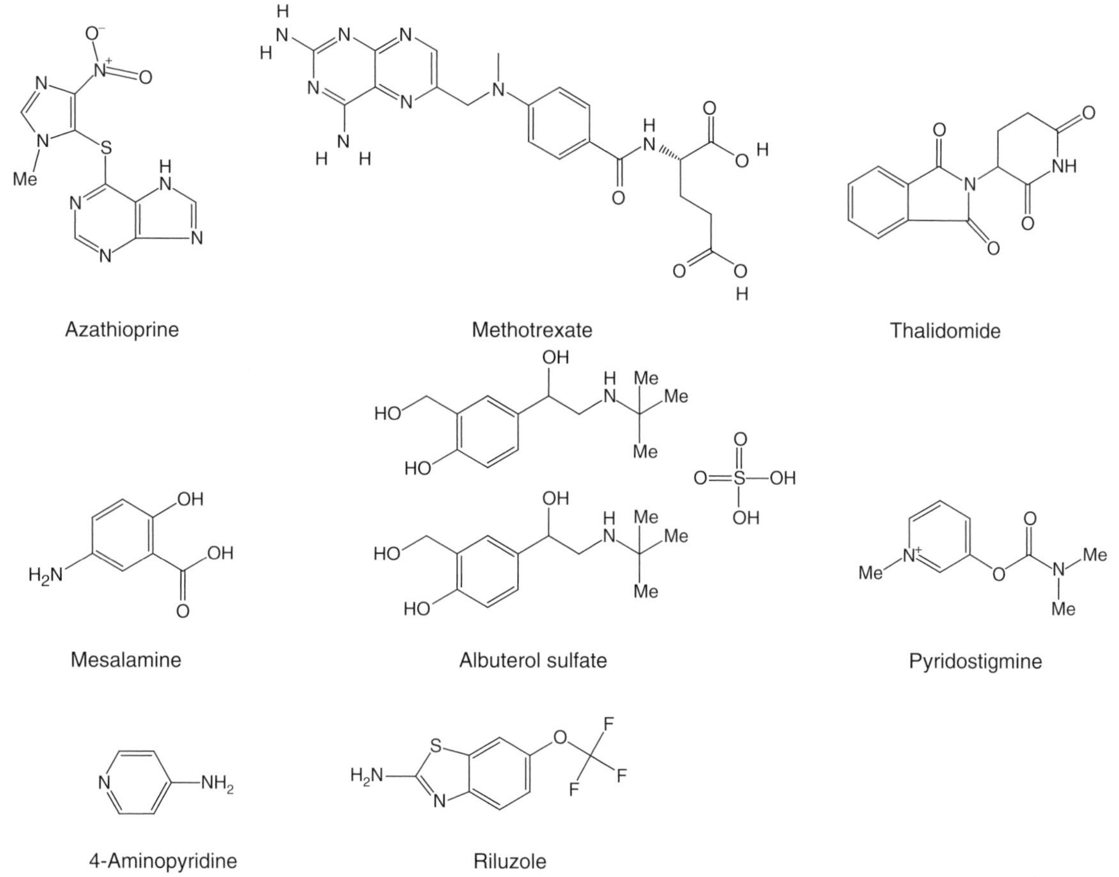

Azathioprine

Methotrexate

Thalidomide

Mesalamine

Albuterol sulfate

Pyridostigmine

4-Aminopyridine

Riluzole

Figure 1 continued

their large size permits imaging studies and longitudinal blood analysis. In the end, however, there are no good nonhuman primate models, and so there is no regulatory requirement for proof-of-concept studies in them. Instead, nonhuman primates tend to be used only for safety studies, and then only when such studies are required by regulatory bodies such as the US FDA.

6.09.3.2 Disease-Specific Animal Models

Experimental autoimmune encephalomyelitis (EAE) is the most common animal model for MS.[21] Multiple means have been utilized to generate EAE in both rats and mice (also in nonhuman primates). Usually an immune response is evoked by injecting myelin-derived protein plus adjuvant. In one EAE variant, activated T cell clone specific for MBP is injected into rats. EAE was used very successfully for the development of the MS drug natalizumab. In this case, the EAE model was predictive not only for efficacy but also for the effective dose range for humans; not surprisingly, however, it proved to be an incomplete predictor of safety. In another case, the MS therapy glatiramer acetate was developed to exacerbate EAE, but instead ameliorated it – demonstrating how little is really known about EAE. Further, IFN-β is only modestly efficacious in the EAE model, but much more so in RRMS – illustrating discord between EAE and RRMS. Indeed, several immunological manipulations that reduced or abrogated EAE were inert in MS or caused worsening.

For SLE, there are several murine models, including an SLE-prone mouse (NZB/WF1), as well as a lupus-like mouse that overexpresses BLyS. Conversely, there is a BLyS knockout mouse that ameliorates the disease.

Experimental autoimmune neuritis (EAN) is an animal model for both acute and chronic inflammatory demyelinating polyneuropathy (i.e., GBS and CIDP, respectively). EAN can be generated in rats, mice, or rabbits, and is typically induced by injection of myelin proteins (P0, P2, PMP22, and MAG). Two other vaccination protocols have been used to develop models for GBS in Japanese white rabbits: a ganglioside vaccination protocol that elicits

Table 2 Current treatment options for autoimmune diseases of the nervous system

Pharmacotherapy	*Action*	*Nature*	*Route*	*Indications*		
				Approved by the US Food and Drug Administration	*Off-label*	*Clinical trials*
Biopharmaceuticals (biologics)						
Interferon β-1a (Avonex)	IM	Cytokine	INJ	RRMS		CIDP, CIS, SPMS with relapses
Interferon β-1b (β seron)	IM	Cytokine	INJ	RRMS, SPMS		CIS
Interferon β-1a (Rebif)	IM	Cytokine	INJ	RRMS		
Natalizumab (Tysabri)	Selective adhesion molecule inhibitor; blocks lymphocyte migration, IS, AI	MCAB to $\alpha_4\beta_1$ integrin, H	IV	Relapsing MS		Crohn's disease, RRMS
Belimumab (LymphoStat-B)	B cell depletion, IS	MCAB to BLyS, H	IV			SLE
Rituximab (Rituxan)	B cell depletion, AN	MCAB to CD20 R, chimeric	INJ	Refractory RA	CIDP, MG	PPMS, RRMS, SLE
Epratuzumab	B cells, depletion?	MCAB to CD22, H	IV			SLE
Alemtuzumab (Campath)	B/T cell depletion, AN	MCAB to CD52, H	IV			RRMS
CNTO 1275	IL-12 inhibitor, AI, IS	MCAB to IL-12/IL-23, H	INJ			Crohn's disease, MS
J695	IL-12 inhibitor, AI, IS	MCAB to IL-12, H	INJ			Crohn's disease
Daclizumab (Zenapex)	Inhibits T cell growth, IS	MCAB to IL-2 R α, H	IV			RRMS
MRA	IL-6 blocker, cytokine modulation	MCAB to IL-6 R, H	IV			SLE
Infliximab (Remicade)	TNF-α inhibitor	MCAB to TNF-α, chimeric	IV	Crohn's disease		Crohn's disease
Adalimumab (D2E7)	TNF-α inhibitor	MCAB to TNF-α, H	INJ			Crohn's disease
RG2077	Co-stimulatory molecule inhibitor, block CD28:B7, IM	Soluble fusion P of CTLA4	IV			RRMS, SLE
Abatacept (Orencia)	Co-stimulatory molecule inhibitor, block CD28:B7, IM	Soluble fusion P of CTLA4	IV	RA		SLE
Glatiramer acetate	Vaccine, IM	Random, synthetic peptide mix	INJ	RRMS		RRMS
Abetimus sodium	Toleragen, IM	dsDNA-based bioconjugate	IV			SLE
EN101	Antisense, cholinesterase inhibitor	Oligonucleotide	Oral, IV			MG (Israel)
Pharmaceuticals (drugs)						
Methylprednisolone	AI, IS	Corticosteroid	IV, oral		CIDP, MG, relapsing MS, SLE	CIDP

Dexamethasone	AI, IS	Corticosteroid	IV		CIDP, MG, relapsing MS, SLE	Crohn's disease, RRMS, MMMS
Prednisone	AI, IS	Corticosteroid	Oral	SLE	CIDP, MG, relapsing MS	SLE
Pravastatin	Hyolipidemic agent, IM	HMG-CoA reductase inhibitor (statin)	Oral			RRMS, SLE
Atorvastatin	Hyolipidemic agent, IM	HMG-CoA reductase inhibitor (statin)	Oral			CIS, MS, SLE
Simvastatin	Hyolipidemic agent, IM	HMG-CoA reductase inhibitor (statin)	Oral			CIS
Minocycline	AI, antibacterial, neuroprotective	Antibiotic (tetracycline)	IV, oral			ALS, RRMS
Tacrolimus	IS	Antibiotic (macrolide)	Oral, INJ		MG	LN
Mycophenolate mofetil	Guanosine nucleotide inhibitor, IS	2-Morpholinoethyl ester of mycophenolic acid; prodrug	Oral		LN, MG, SLE	Early MS, LN, MG
Cyclophosphamide	Alkylating agent, AN, IS	2-[Bis(2-chloroethyl)amino] tetrahydro-2*H*-13,2-oxazaphosphorine 2-oxide monohydrate; prodrug	Oral, IV		CIDP, MG, MS, SLE	Crohn's disease, LN, SLE, SPMS
Mitoxantrone	AN, IS	Anthracenedione compound	IV	RPMS, SPMS, RRMS		
Azathioprine	IS, antimetabolite	Imidazolyl derivative of 6-mercaptopurine	Oral		CIDP, MG, RRMS	Crohn's disease, LN, SLE
Methotrexate	Inhibits dihydrofolate reductase, antimetabolite, AN, IS	*N* -[4-[[(2,4-Diamino-6-pteridinyl) methyl]methyl-amino]benzoyl]-L-glutamic acid	Oral, INJ	RA		Crohn's disease, RRMS
Thalidomide	IM	α-(*N*-Phthalimido)glutarimide	Oral			ALS
Mesalamine	AI	5-Amino-2-hydroxybenzoic acid	Oral			Crohn's disease
Albuterol sulfate (salbutamol sulfate)	β_2-adrenergic bronchodilator	α_1[(*t*-Butylamino)methyl]-4-hydroxy-*m*-xylene-α,α'-diol sulfate	Oral, inhaled			RRMS
Pyridostigmine	Cholinesterase inhibitor	3-Hydroxy-1-methylpyridinium bromide dimethylcarbamate	Oral	MG		
4-Aminopyridine	K^+ channel blocker	4-Aminopyridine, sustained release	Oral			GBS, MS
Riluzole	Inhibits glutamatergic transmission	2-Amino-6-(trifluoromethoxy) benzothiazole	Oral	ALS		

AI, antiinflammatory; AN, antineoplastic; H, humanized; IM, immunomodulatory; INJ, injected; IS, immunosuppressive; MCAB, monoclonal antibody; P, protein; R, receptor; RA, rheumatoid arthritis; TNF-α, tumor necrosis factor α.

specific IgM and IgG responses and GBS-like symptoms,[22] and a protocol that injects *C. jejuni* lipooligosaccharide generating anti-GM1 antibodies, flaccid muscle weakness, and a pathology identical to GBS.[11] The latter model requires multiple injections and 133–329 days to develop paresis or tetraparesis, thereby reducing its efficacy for experimental studies, and suggesting an imperfect match with the human pathogenesis.

An immunization protocol is also utilized to generate experimental autoimmune MG; the immunogen is a peptide unique to the α subunit of the AChR.

6.09.4 Clinical Trial Issues

The most common problem with clinical trials is the lack of them – particularly good, randomized, double-blinded, controlled studies. There are often few data for dose refinement, or even basic recommendations, in children, for example. In too many instances, the efficacy of many current – and often very expensive – therapeutic options has simply not been demonstrated in controlled, randomized trials. Sadly, there are only a handful of clinical trials currently for GBS, MG, and CIDP. With a couple of hundred current trials, the treatment horizon is more hopeful for MS and SLE patients.

Other problems arise from design flaws – sometimes unavoidable ones. For ethical reasons, placebo control treatment arms are often lacking when some other treatment has become the mainstay of therapy (e.g., corticosteroids in MG). Another flaw is that although the therapeutic aim is to improve both short- and long-term prognoses, clinical trials often examine only short-term time points.

Further, scales used to measure disease severity or progression are imperfect. One needs a standardized scale that can be used to provide a quantitative measure of a patient's clinical status and course or stage of progression. Scales currently in use have limitations – the Kurtzke Expanded Disability Status Scale (EDSS) and the Multiple Sclerosis Functional Composite score used in MS, for example.

Several patient recruitment issues can negatively impact trials. Disease classification criteria are imperfect, and diagnoses are often flawed – leading to inappropriate inclusion in or exclusion from clinical trials. Racial, ethnic, and socioeconomic issues can influence outcome in clinical trials,[1] yet are often overlooked at either the patient recruitment or analysis stages. Inclusion of patients with mild or stable disease does not allow for an effect size sufficient to show statistical differences in treatment arms. Often, only patient subsets are included in trials – for example, when, for ethical reasons, new therapies are tested on patients who have failed conventional therapies; this may not pose a problem for safety studies, but it can skew early results for efficacy trials. Together, these problems diminish the power of the conclusions that can be drawn.

Statistical power is also negatively impacted by low patient numbers, which are mandated in Phase I trials. Even when patient numbers are statistically sufficient at the outset, patient attrition inevitably occurs as a result of side effects, an inability to follow a particular treatment regimen, patient expectations not being met, or patients being allowed to switch therapies. Some measures can be taken to minimize dropout rates by, for example, permitting patients to take the highest tolerated dose instead of fixed doses, and making sure that patients are well informed at the outset and that their expectations are reasonable.

Interpretation of data is complicated in progressive and relapsing diseases by the fact that deteriorations can be due either to ongoing disease progression or the effect of the treatment. In diseases with a relapsing–remitting course, or with drugs that have a slow onset of action, conclusions are particularly difficult to draw from anecdotal observations. Finally, even in an optimal clinical trial, clinicians are faced with the question of the applicability of the overall results to a particular clinical subset or patient.

6.09.5 Treatment Challenges

One obvious target for any chronic autoimmune disease is to simply turn off the autoimmune response; however, little is known of what drives the chronicity of autoimmune disease. Instead, therapies tend to target points in the autoimmune/inflammatory cascade: factors involved in lymphocyte homing to target tissues, enzymes that are critical for the penetration of blood vessels and extracellular matrix (ECM) by immune cells, cytokines that mediate pathology within the tissues, various cell types that mediate damage at the site of the disease, as well as the antigen-specific adaptive receptors of these cells – including the T cell receptor, and immunoglobulin and other toxic mediators – such as complement and nitric oxide. In each of these instances, the therapeutic target is involved in both host defense and autoimmune pathology, and so the overall risk:benefit ratio needs to be evaluated; natalizumab provides a timely example.

The dynamic character of chronic progressive autoimmune diseases presents special challenges for drug development. For instance, autoantigens can differ at various stages of a disease; this appears to be the case in MS, SLE, and

MG. Also, in the case of MS, disease most often progresses from a relapsing–remitting form to a progressive form distinguished by progressive axonal degeneration and little inflammation. Therefore, a therapy that works in one phase may be ineffectual in another; for example, most MS therapies target only relapsing forms, with little, if any, effect on progressive forms. Early intervention is always best to minimize the accumulation of cellular damage and any resulting functional deficit, as well as epitope spreading. However, early diagnosis with accurate prognostic assessment – in CIS, for example – is not always possible, and so treatments are often initiated years after the onset of the pathogenic process. The presence of both inflammatory and degenerative features presents a greater therapeutic challenge than a disease with only one of these features – another argument for early intervention when symptoms are likely to be less complex.

Molecular accessibility is an issue in the development of therapeutics. Barriers exist at the cell and tissue level. For this reason, molecular targets, particularly for biologics, are often extracellular domains of cell surface receptors or secreted molecules.

Excluding corticosteroids, the only approved therapies for the autoimmune diseases listed in **Table 1** are for MS, and these are all parenteral – requiring injection or intravenous infusion – and almost all are protein based (see **Table 2**). Patients prefer oral therapies; in some instances, injections are not tolerated because of concomitant disease or injection site side effects, which are the focus of multiple current clinical trials. Unfortunately, protein-based therapies administered orally appear to have little or no therapeutic effect. The development of oral small-molecule agents for autoimmune disease would impact patient acceptance and adherence issues. Small-molecule agents are also favored by the fact that protein-based therapies have a much greater potential for immunogenicity (although small-molecule drugs can also elicit an autoimmune response). This potential for autoimmune response to the therapeutic itself adds a further complication to drug development, as this risk and its clinical consequences – separate from the efficacy of the therapy – need to be assessed; yet, as discussed earlier, there are no good, predictive animal models for doing so. In addition to presenting a potential clinical risk, any resulting neutralizing antibodies in humans can negatively impact efficacy. This problem of neutralization has been minimized for antibody therapeutics by the development of fully humanized monoclonals. However, it remains a concern with recombinant proteins and other antigens – for example, IFN-β therapies. On the other hand, protein-based therapies are natural body constituents, and so any toxicity from protein-based therapies is likely to be mechanism related, or a result of improper manufacturing procedures.

6.09.6 Biological Treatments

6.09.6.1 Selective Adhesion Molecule Inhibitors

Natalizumab is a humanized monoclonal antibody to α_4 integrin that was approved by the US FDA in 2004 for the treatment of relapsing forms of MS.[23–25] The integrins are a large family of adhesion molecules mediating cell–cell and cell–ECM interactions. Expression of α_4 integrin is predominantly on lymphocytes, monocytes, eosinophils, and basophils, but is usually undetectable on neutrophils. It forms heterodimers with either β_1 or β_7. One ligand for $\alpha_4\beta_1$ integrin is VCAM-1 (vascular cell adhesion molecule), which is expressed on vascular endothelial cells and, perhaps, glial cells. Other ligands of $\alpha_4\beta_1$ integrin include the ECM proteins, osteopontin, thrombospondin, and fibronectin. Osteopontin has adhesion and cytokine activity, and its expression is increased in MS lesions and spinal cord of EAE rats; mice deficient in osteopontin are relatively resistant to EAE.

Leukocytes are concentrated in lymphatic tissues, and present in other tissues, including brain, in low numbers. In MS, and other inflammatory diseases, their migration across endothelium into tissues (in the case of MS, across the blood–brain barrier to sites of new or active brain lesions) is enhanced. Binding of $\alpha_4\beta_1$ integrin, in the case of lymphocytes, leads to phenotypic transformation to promote migration, activation, and proliferation. Indeed, translocation of CD4+ T cells to brain parenchyma is dependent on the expression of $\alpha_4\beta_1$ integrin.

In Crohn's disease, an organ-specific autoimmune disease in which the immune system attacks the intestinal mucosa, the therapeutic target of natalizumab is $\alpha_4\beta_7$ integrin, which, like $\alpha_4\beta_1$ integrin, is recognized by α_4 integrin-specific antibodies. This integrin has multiple functions, including mediating migration of gut-homing T cells via its counter-receptor in the gut, MadCAM.

Early in 2005, all dosing of natalizumab was voluntarily suspended by the sponsor after learning of two cases of progressive multifocal leukoencephalopathy (PML) in MS patients receiving natalizumab and IFN-β combination therapy; a third case of PML was subsequently discovered in a patient with Crohn's disease receiving natalizumab following extensive therapy with various immunosuppressive medications. One of the two MS patients recovered; however, the other two cases proved fatal. Following an exhaustive safety review, the US FDA approved the use of natalizumab in open-label monotherapy clinical trials. The FDA subsequently granted approval for reintroduction based on the review of natalizumab clinical trial data; revised labeling with enhanced safety warnings; and a risk management plan (TOUCH Prescribing Program) designed to inform physicians and patients of the benefits and risks of natalizumab treatment and minimize potential risk of progressive multifocal leukoencephalopathy (PML). Because of the increased risk of PML, natalizumab monotherapy is generally recommended for patients who have had an inadequate response to, or are unable to tolerate, alternate MS therapies.

In a monotherapy clinical trial, natalizumab seemed to significantly reduce the progression of disability, the occurrence of relapse, and the formation of lesions visualized by MRI in patients with relapsing MS.[26] A second trial suggested that natalizumab–IFN-β combination therapy was significantly more effective than IFN-β alone in a population of RRMS patients who experienced breakthrough disease during IFN-β monotherapy.[27] Prior to the PML reports, side effects were modest, and there were no reports of adverse effects, suggesting the most promising risk:benefit ratio for an MS therapy to date. However, no head-to-head studies with IFN-β or glatiramer acetate therapies have been completed.

6.09.6.2 Anti-Inflammatory Cytokines

There are currently three US FDA approved cytokine treatments for RRMS (see **Table 2**) – preparations of two recombinant forms of the cytokine IFN-β (the interferons were named for their ability to interfere with viral infection). While one form, IFN-β-1b is approved for SPMS, IFN therapy, in general, is not recommended for patients with progressive MS unless there is clear clinical or subclinical (MRI) evidence for inflammatory disease activity.

IFN-β therapy appears to positively impact the early course of MS – increasing the time to a second clinical attack, reducing the relative risk of relapse, and reducing progression of disease activity on MRI. Side effects are mainly flu-like symptoms, liver abnormalities, and injection site reactions, which all decrease with time. Although the overall efficacy of IFN-β in MS is relatively modest, the efficacy in individuals who remain neutralizing antibody-negative is considerably better than in those who become persistently positive for these antibodies.[28] Therefore, when evaluating the clinical efficacy of IFN-β therapy, comparisons should perhaps be limited to cohorts that remain neutralizing antibody-negative. Current clinical trials are testing the antigenicity of new formulations.

For IFN-β, early initiation of treatment was recommended. Immediate initiation of IFN-β therapy following a first clinical demyelinating event – in patients defined by MRI criteria as high risk – delayed the development of clinical definite MS and decreased the development of new or enlarging lesions. On the other hand, while there are no evidence-based recommendations for treatment duration, if a patient is demonstrating a positive risk:benefit analysis, IFN therapy is generally continued.

Multiple mechanisms have been postulated to underly the therapeutic efficacy of IFN-β: modulation of cytokine production toward anti-inflammatory cytokines; inhibitory effects on proliferation of leukocytes and antigen presentation; and inhibition of T cell migration across the blood–brain barrier by downregulating the expression of adhesion molecules, and inhibiting the activity of T cell matrix metalloproteinases.

Current IFN-β clinical trials are testing different IFN-β preparations and doses, single versus combination therapies, the impact of early treatment, IFN-β treatment for CIDP, IFN-β compared against other US FDA-approved MS therapeutiucs, as well as examining the long-term clinical course of IFN-β patients.

6.09.6.3 Proinflammatory Cytokines and Their Receptors

Proinflammatory cytokines, such as TNF-α and interleukin (IL)-12, are produced in increasing quantities in peripheral blood cells of MS patients prior to relapse, and protein and mRNA of proinflammatory cytokines are found in MS plaques. Increased levels of TNF-α have also been found in the CSF of MS patients, and IL-2 receptors are expressed on the surface of activated lymphocytes.

Anti-TNF-α therapies are approved for the treatment of RA and Crohn's disease. However, monoclonal antibodies against TNF-α induce SLE, and lead to increased and prolonged exacerbations in MS patients. Further, there is some clinical trial evidence that etanercept, a soluble recombinant TNF receptor Fc protein, designed as a TNF-α inhibitor, may actually lead to the upregulation of TNF-α expression in some MG patients – leading to worsening of the disease.

Monoclonal antibodies against IL-12 are currently being tested in clinical trials for MS and Crohn's disease. A monoclonal antibody to the IL-2 receptor daclizumab functions as an IL-2 antagonist. Daclizumab inhibits IL-2-mediated stimulation of lymphocytes, is approved for use in kidney transplant patients, and is being tested in clinical trials as a therapy for RRMS.

The monoclonal antibody antagonist of the cytokine BLyS can also be included in this category (see below).

6.09.6.4 B and T Cell Targets

Chronic T and B cell responses occur in many autoimmune diseases, and, therefore, represent a logical therapeutic target. Therapeutics currently in development aim to reduce T cell migration (see natalizumab above), and reduce overall T and B cell numbers or control their activation.

6.09.6.4.1 B and T Cell depletion mechanisms

BLyS is a member of the TNF family of proinflammatory cytokines that is expressed on monocytes and induces B cell proliferation and immunoglobulin secretion. Constitutive overexpression of BLyS protein can result in SLE-like disease in mice, and treatment with a BLyS protein antagonist or knockout models ameliorates disease progression and enhances survival. In a large subset of SLE patients, circulating levels of BLyS protein are elevated. Belimumab is a humanized monoclonal antibody to BLyS – a BLyS antagonist that induces a reduction in B cells and appears to be well tolerated. Two ongoing Phase II trials are testing its efficacy and safety in SLE and RA, respectively.

Rituximab is another monoclonal antibody therapy that leads to B cell depletion. This monoclonal antibody is directed against the CD20 receptor protein present on B lymphocytes. CD20 is expressed on B cells in intermediate stages of development – not on plasma cells – and plays a role in the differentiation of B cells into plasma cells. Rituximab appears to have no significant adverse events. It is licensed for a type of non-Hodgkins lymphoma, and is currently being tested in clinical trials for progressive and relapsing forms of MS as well as for SLE.

Epratuzumab is a monoclonal antibody to CD22, a transmembrane sialoglycoprotein expressed on the surface of mature B cells. Epratuzumab is being tested as a therapy for SLE. The precise mechanism of action of epratuzumab has not been defined but may relate to effects on B cell signaling, and may render cells more prone to apoptosis.

Alemtuzumab is a monoclonal antibody directed against the glycoprotein CD52, which is expressed on the surface of essentially all B and T lymphocytes and a majority of monocytes and macrophages as well as natural killer cells. It is used as a neoplastic drug that induces profound lymphopenia. It has US FDA approval as a second-line therapy for B cell chronic lymphocytic leukemia, but is associated with numerous and common adverse effects. It is currently being tested in clinical trials for RRMS.

Mitoxantrone is a synthetic anthracenedione (see **Figure 1** and **Table 2**) with both antineoplastic and immunosuppressive activities.[29] Unlike many small drugs, mitoxantrone is administered intravenously every 3 months, since it is poorly absorbed orally. It easily crosses the blood–brain barrier and intercalates into DNA, inhibiting DNA replication and DNA-dependent RNA synthesis; it is also a potent inhibitor of topoisomerase II, an enzyme involved in DNA repair. Mitoxantrone is thought to have a broad range of action, based mainly on preclinical studies. It is cytotoxic to both proliferating and nonproliferating human cells in vitro. It inhibits B cell, T cell, and macrophage proliferation, leading to apoptosis (patients taking mitoxantrone can have severe leukopenia). Mitoxantrone also impairs antigen presentation (by causing apoptosis of the cells involved) and decreases secretion of proinflammatory cytokines – including IL-2 and TNF-α. In MS patients, mitoxantrone has a positive impact on progression (thus the FDA approval for SPMS), relapse rate as well as MRI benefits. Mitoxantrone therapy is associated with many adverse effects, including treatment-related leukemia; however, the major limitation relates to potential cardiotoxic effects, which limits the cumulative lifetime dose.

Mycophenolate mofetil is a prodrug that is metabolized to mycophenolic acid (MPA; see **Figure 1** and **Table 2**), an immunosuppressant that acts by reversible inhibition of IMP dehydrogenase, the rate-limiting enzyme in purine synthesis. MPA has a selective effect on T and B cells, since their proliferation is dependent on de novo synthesis of purines (other cell types can utilize salvage pathways). MPA has a broad spectrum of action, mainly as a consequence of its suppression of T and B cell proliferation (leukopenia is a prominent adverse reaction). Mycophenolate mofetil is approved for the prevention of allograft rejection; clinical trials are underway testing mycophenolate mofetil for MS, MG, and SLE.

6.09.6.4.2 Co-stimulatory molecule inhibitors

In the adaptive immune response, nonspecific signals are generated from the interaction of pairs of receptors, so-called co-stimulatory molecules.[30] Inhibition of the co-stimulatory signal results in suppression of the immune response. The CD28:B7, CTLA4:B7, and CD40:CD40L receptor:ligand pairs of co-stimulatory molecules have been the targets of drug development for autoimmune disease. There are two conceptually different approaches to blocking co-stimulation: by inhibiting either the induced expression of co-stimulatory molecules or the transmission of their specific intracytoplasmic signal. The latter approach is described here.

CTLA4 (cytotoxic T lympocyte antigen 4) is a co-stimulatory molecule expressed cytoplasmically in activated T cells, but found on the surface of mainly antigen-experienced T cells; this finding has therapeutic relevance, since established T cell responses driven by antigen-experienced T cells often underly chronic immunopathology. CTLA4 is homologous to CD28 – both are members of the immunoglobulin superfamily, and both bind the ligand B7, which is expressed almost exclusively on antigen-presenting cells. However, CTLA4 transmits an inhibitory signal to T cells (a negative regulator of T cell activation), whereas CD28 transmits a stimulatory signal. Indeed, in experimental models of SLE, CTLA4 inhibited the autoimmune response. While CTLA4 is a transmembrane protein, two companies have developed soluble fusion protein versions (CTLA4-Ig), which appear to mimic the effects of endogenous CTLA4. One of these products (abatacept) has been recommended to receive US FDA approval for the treatment of RA. In addition, clinical trials of CTLA4-Ig are underway for SLE and RRMS.

Two therapies targeting the CD40:CD40L interaction failed at the clinical trial stage. Both were monoclonal antibodies to CD40L – one was associated with life-threatening prothrombotic activity, and the other failed to demonstrate efficacy.

6.09.6.5 Therapeutic Vaccines and Toleragens

While vaccines have traditionally been used as prophylactics, they are increasingly being used as therapies for already established chronic disease.[31] For autoimmune disease therapy, a peptide similar to that causing the disease is administered to the patient; in theory, the vaccine then stimulates an immune response to the T cells reacting in the disease. This presumes that the autoantigen(s) is known, and that the disease is caused by one or a few antigens. An advantage to this hypothetical mechanism is that the anti-T cell response should, in theory, be specific for the T cells contributing to disease pathogenesis.

Glatiramer acetate is the first vaccine that has been used to treat an autoimmune disease – in this case RRMS[32]; it was approved by the US FDA in 1996. It is a mixture of many synthetic peptides – random polymers – that mimic the antigenic portion of MBP. It reduces the relapse rate in RRMS, impacts MRI markers of disease activity, is well tolerated, and does not appear to induce the formation of neutralizing antibodies. Several mechanisms of action have been proposed for glatiramer acetate, none of which reflect the theoretical mechanism proposed above for a prophylactic vaccine. Glatiramer acetate is a strong inducer of Th2 cells, and thus anti-inflammatory cytokine production; one proposed mechanism is that it acts by inducing a Th1 to Th2 shift. In addition to glatiramer acetate, a vaccine for MG is in preclinical development.[31]

Another vaccine-like therapeutic approach is to use a toleragen, which acts to induce tolerance to the disease-causing immunogen, without otherwise impacting immune system function. A promising therapy in this realm is abetimus sodium, which is being tested in Phase III trials for the treatment of SLE.[33] Abetimus sodium is composed of four dsDNA molecules fused with a nonimmunogenic polyethylene glycol. It was specifically designed to decrease the severity and number of renal flares in SLE by inducing tolerance in B cells directed against dsDNA, and thereby selectively reducing antibodies to dsDNA (since anti-dsDNA levels are associated with a risk of renal flare, reducing its levels might represent a therapeutic objective in SLE patients with LN). Abetimus sodium reduces anti-dsDNA antibody levels in patients, and appears to be well tolerated and have a quality of life impact. It remains to be seen whether it has a therapeutic impact on the incidence of either renal or major SLE flares in patients with high-affinity antibodies to abetimus at the baseline.

6.09.6.6 Statins

HMG-CoA reductase inhibitors, the so-called statins, are currently being tested in clinical trials for their efficacy as MS therapeutics,[34] and for their ability to reduce cholesterol in adolescents with SLE (see **Figure 1** and **Table 2**). Beyond their cholesterol-reducing properties, statins have pleiotropic effects – including anti-inflammatory and neuroprotective properties. In preclinical studies, statins inhibited the onset and progression of disease in EAE by inducing a shift from Th1 (proinflammatory) toward Th2 (anti-inflammatory) cytokine production. In vitro studies suggest that statins inhibit lymphocyte migration across the blood–brain barrier. Further, statins may promote remyelination. In an open-label study of 30 RRMS patients, a simvastatin regimen decreased lesional activity assessed by MRI, and was well tolerated; however, it did not affect EDSS scores or yearly relapse rates.[35] Statins have many advantages: beside being well tolerated, they are administered orally, have been used for many years, and have a low cost.

6.09.6.7 Corticosteroids

The dominant effect of corticosteroids is anti-inflammatory – reducing the expression of inflammatory cytokines and adhesion molecules, and reducing the trafficking of inflammatory cells. At high doses, they may induce apoptosis in immune cells. Their effects are often transient. Further, frequent use of corticosteroids is associated with long-term adverse effects, such as osteoporosis, and their use has to be carefully managed. On the other hand, individual pulses of corticosteroids are relatively free from adverse effects.

6.09.7 Other Treatment Strategies

6.09.7.1 Intravenous Immunoglobulin and Plasma Exchange

PE is a process in which blood is removed from a patient, the plasma and cellular components separated mechanically, and the cellular components then reintroduced to the patient. PE works by simply removing circulating autoantibodies.

IVIg is the infusion of human immunoglobulin pooled from 3000 to 10 000 donors.[36] IVIg is thought to act by several different mechanisms, varying according to the disease. It is thought to negatively impact autoantibody levels by supplying anti-idiotypic antibodies that can potentially neutralize pathogenic autoantibodies, and bind to and negatively regulate antibody production by B cells. In addition, IVIg could theoretically accelerate the catabolism of autoantibodies by saturating the protective transport receptors (FcRn) in endocytotic vesicles. IVIg also appears to act by inhibition of the complement pathway – by inhibiting complement uptake and formation and deposition of membranolytic attack complex. IVIg modulates Fc receptors on macrophages – either by blockade of the receptors or by increasing the ratio of receptors to favor inhibition over the activation of phagocytosis. IVIg also neutralizes superantigens, and could prevent the activation and clonal expansion of superantigen-triggered cytotoxic T cells; this effect may be relevant in controlling relapses triggered by infections – as seen in MG and CIDP, for example. IVIg might also inhibit T cell function; indeed, IVIg induces transient lymphopenia.

The onset for both IVIg and PE can vary greatly; for CIDP, it is generally 24–48 h, but can take up to 4 weeks, or a patient may see no effect. The duration of action is short – generally 1–2 months. Further, both IVIg and PE are associated with adverse events, although complications seem to be less with IVIg. Further, both take several hours to administer. IVIg is expensive and sometimes associated with recalls and shortages; in addition, there are many products on the market, although there is no evidence of differences in biological action. While all indications for IVIg are off-label, IVIg or PE are considered first-line therapy for GBS and CIDP, including MMN, and as a second-line therapy for MG, SLE, and MS. Currently, clinical trials for IVIg or PE for MG, CIDP, GBS, and SLE are ongoing.

6.09.7.2 Stem Cell Therapy

Safety and efficacy data in humans for stem cell therapies in autoimmune disease is lacking.[37] However, autologous peripheral blood stem cell transplantion and allogeneic hematopoietic stem cell transplantation are currently being tested in several clinical trials as treatments for MS (including RRMS and SPMS) and Crohn's disease.

6.09.7.3 Gene Therapy

Gene therapy as a treatment for autoimmune disease of the nervous system is currently in the preclinical stages of development with multiple issues to overcome, including the choice of vector and the delivery method.[38,39] As a general therapeutic approach, gene therapy offers the potential to overcome half-life issues for some protein therapies as well as overcoming the requirement for frequent administration. In addition, gene therapy can provide focal administration of a therapeutic; this is a drawback for multifocal diseases such as MS, but can, theoretically, be addressed by employing ex vivo engineered mobile cells. Since expression is to be long-term for chronic disease, the vector and transgene must be nonimmunogenic.

Gene therapy also can deliver a steady dose, as opposed to many systemic therapies that are often inefficiently administer at a high dose in order to achieve a therapeutic dose in the target tissue – the inverted V character for dose versus time typical of many systemic therapies. However, a constant dose is a drawback for acute autoimmune disease or chronic disease with acute phases; this issue can, theoretically, be overcome by having the gene expression under the control of a promoter that is responsive to the dynamics of the disease or by using a pharmacologically regulated expression system.

Direct injection is required because most viral and nonviral vectors cannot cross endothelial cell barriers. In the CNS, this is a potential problem for in vivo therapies, since the vector is likely to be transported anterogradely and retrogradely by any axon passing through the injection site – leading to broad multifocal expression.

6.09.8 Disease-Specific Treatment Strategies

6.09.8.1 Multiple Sclerosis

Currently available therapies for MS have limited efficacy, are not well tolerated, and have a very high cost.[40–42]

Treatment goals for relapsing MS are, in the short term, to achieve a remission of symptoms, and, in the long term, to minimize structural damage. It is recommended that treatment begins immediately, since patients who receive early treatment have slower progression. The true onset of MS may be long before the first clinical symptoms become evident, and axonal damage can already be detected at early stages of the disease. Thus, early treatment may impact progression by minimizing damage accrual. It is generally recommended that immunomodulatory treatments be continued indefinitely, unless a better therapy becomes available, there is no detectable benefit, or side effects are intolerable.

With natalizumab currently available only in clinical trials, IFN-β or glatiramer acetate are the first line of therapy (both discussed above). There are currently multiple clinical trials comparing IFN-β and glatiramer acetate in isolation and combination. The results from these studies will help refine currently available treatment strategies. Short-term pulse administration corticosteroids can be used alone or in combination with these therapies to hasten recovery from acute relapse.

IVIg or PE is indicated as a second-line therapy for treating severe disabling relapses, when first-line therapies are not effective, tolerated, or appropriate (e.g., women contemplating becoming pregnant). Mitoxantrone is a second-line therapy for patients with worsening MS despite first-line treatment, or for those with inflammatory episodes not responsive to corticosteroids, IVIg, or PE.

Symptomatic therapies are the cornerstone of therapy for patients with SPMS – to reduce neurological impairment and decrease disability and handicap.[43] For relapsing progressive forms of MS, IFN-β or mitoxantrone are indicated. For both relapsing and progressive forms of MS, cyclophosphamide has questionable efficacy combined with serious adverse effects and, as such, is only considered an option for patients who do not respond to other treatments (see **Figure 1** and **Table 2**). Another antineoplastic agent, methotrexate (see **Figure 1** and **Table 2**), appears to have modest efficacy for patients with relapsing forms of MS.

6.09.8.2 Systemic Lupus Erythematosus

The only current US FDA-approved therapy for SLE is prednisone (see **Figure 1** and **Table 2**); however, many biologics and drugs, some new, are being tested in clinical trials.[1,44] Corticosteroids (described above) have long been used to treat SLE. Prednisone given for a short period seems to prevent severe flares and exposure to high-dose steroids.

Cyclophosphamide has long been the standard treatment for severe SLE, particularly LN. A low-dose intravenous cyclophosphamide regimen followed by maintenance azathioprine (see **Figure 1** and **Table 2**) may be as, or more, effective, and less toxic than the standard high-dose cyclophosphamide regimen; nonetheless, a significant proportion of patients fail to achieve remission, or experience a relapse of active LN during maintenance therapy. Socioeconomic and ethnic factors have been shown to influence cyclophosphamide efficacy in patients with proliferative LN.

It is generally agreed that B cell dysfunction, specifically loss of tolerance, plays a central role in SLE pathogenesis. Thus, many of the pharmacotherapies currently being tested in clinical trials for SLE (mycophenolate mofetil, rituximab, belimumab, epratuzumab, and abetimus sodium – described above) exert their effects on some aspect of B cell proliferation or function. Mycophenolate mofetil appears to have an equal or superior efficacy to intravenous cyclophosphamide, with fewer side effects. This along with its oral administration argue for its consideration as an alternative to intravenous cyclophosphamide – at least for LN. Rituximab shows promise as a therapy for various SLE manifestations. Abetimus sodium holds promise as a new therapeutic option developed specifically to reduce the number and severity of renal flares in SLE.

IL-6 levels are elevated in both human and murine SLE. Blocking the action of IL-6 ameliorates disease activity in murine models of SLE; this approach is being tested in an early stage clinical trial using a monoclonal antibody to the IL-6 receptor.

Some patients on IFN-α develop autoimmune conditions, such as SLE. In addition, serum levels of IFN-α are elevated in SLE patients. This and animal studies suggest the potential therapeutic value of IFN-α inhibition for the treatment of SLE.

6.09.8.3 Guillain–Barré Syndrome

Both IVIg and PE speed recovery; while they appear to be therapeutically equivalent,[45] reports in the literature suggest that some subsets of patients may respond better to one or the other. Surprisingly, perhaps, there is no evidence that corticosteroids provide any benefit to GBS patients.

One therapeutic approach is to synthesize oligosaccharide ligands – truncated ganglioside epitopes – to neutralize or remove the autoantibodies either by systemic administration or extracorporeal immunoadsorption.

The only potential new drug in clinical stage development for GBS is 4-aminopyridine, a potassium channel blocker (see **Figure 1** and **Table 2**); it is also being tested for MS (and spinal cord injury). By blocking potassium channels, 4-aminopyridine facilitates the generation and conduction of action potentials – both of which are hampered in demyelinated axons.

6.09.8.4 Chronic Inflammatory Demyelinating Polyradiculoneuropathy

CIDP is generally treated with corticosteroids, PE, and IVIg – sometimes as a combination therapy.[46] While clinical trial data demonstrate efficacy for all three treatment options, there is heterogeneity in the treatment response that is not well understood. For example, MMN is not responsive to either corticosteroids or PE, but responds very well to IVIg. Many immunosuppressive agents have been investigated as potential therapies for CIDP: azathioprine (probably the most commonly used in CIDP), cyclophosphamide, cyclosporin A, mycophenolate mofetil, and etanercept. It is unclear if any of these agents has a positive impact on CIDP. While clinical trial data suggest that both cyclophosphamide and cyclosporin A are beneficial, their serious side effects discourage their use. The immunomodulatory therapies, IFN-β and IFN-α, have been used to treat CIDP, however the benefit of either is unclear, and, paradoxically, IFN-α has been reported to cause CIDP. A handful of clinical trials are currently in progress; one of these is examining IFN-β, while the remainder are investigating IVIg therapy.

6.09.8.5 Myasthenia Gravis

The primary goals for MG therapy are to induce and then maintain remission, and to make use of immunosuppressive treatments as early as possible to reduce the likelihood of epitope spreading. Secondary goals are to manage disease symptoms. Most current MG treatments are immunosuppressive, with either of two types of treatment effects: rapid but short lived and long term.[18,19,47]

Acetylcholinesterase inhibitors (e.g., pyridostigmine) are the first-line therapy for recently diagnosed MG patients (see **Figure 1** and **Table 2**); they are palliative treatments, bringing symptomatic relief, and are effective early in the disease course or in patients with mild disease, when the number of AChRs is sufficient to elicit a response. However, they have side effects (particularly at high doses), erratic absorption profiles, and short-lived effects (a few hours), thus requiring frequent administration. A greater worry perhaps is that their use can mask disease progression. Anti-sense technology is being used to develop a novel cholinesterase inhibitor, Monarsen (EN101), which is being tested in a Phase IIb clinical trial in Israel. Monarsen is reported to have a longer duration of efficacy than conventional inhibitors, with fewer side effects and equal or greater efficacy.

One immunosuppressive strategy for MG is thymectomy, the classic long-term treatment. Thymectomy appears to increase the likelihood of remission or remission maintenance and reduce or preclude the use of steroids, and is most effective when performed early in the disease. It is indicated for patients with neoplastic thymoma (thymomatous MG), but also for patients with thymic hyperplasia or no thymic abnormalities. There are, however, adverse effects, and little controlled randomized trial data to support its use for nonthymomatous MG; to address this issue, a clinical trial, testing the efficacy of thymectomy for nonthymomatous MG will begin shortly.

Corticosteroids, usually prednisone, are a short-term treatment that have been used for decades to treat MG, often in combination with thymectomy; numerous observational studies and expert opinion support their efficacy.[48] However, there is little or no evidence from randomized controlled clinical trials to clearly demonstrate their efficacy or suggest the optimal dosage or route of administration. The onset is slow, with improvement noted by about 6 weeks, and remission by 3 months. Once remission occurs, the steroid administration is slowly and gradually tapered until the patient shows signs of relapse; some patients can be completely weaned from steroids. During tapering, azathioprine, which has fewer side effects than corticosteroids, can be used to increase the success of this phase. Cyclophosphamide tends to only be used for patients with severe MG who do not tolerate or respond well to high-dose corticosteroids. Other short-term treatments include IVIg and PE, which tend to be used for patients with an actual or impending myasthenic crisis; they are often administered in response to an exacerbation during the tapering phase of corticosteroids.

Mycophenolate mofetil is indicated for patients with poorly controlled disease in combination with the therapeutics described above. Randomized controlled clinical trial data are lacking for nearly all MG therapies, and there are currently only a handful of clinical trials for MG – one testing IVIg and the others mycophenolate mofetil.

6.09.9 The Future

A major goal for the future is to better define and understand the heterogeneity of these diseases, and the heterogeneity of treatment response. Outcome data from a small clinical trial of MG with the drug etanercept – with some patients improving, others worsening, and a third category showing no effect – provide a good example of the need for better patient profiling prior to making individual therapeutic recommendations.

Some of the tools that can be used to achieve these goals are DNA microarray technology – including the polymorphism (SNP) arrays and gene expression or transcription profiling arrays.[49] Proteomic approaches – including autoantigen microarrays – also hold great promise.[6] Beyond helping to define disease and treatment response heterogeneity, these arrays can be used to monitor disease and to develop disease- and patient-specific tolerizing DNA vaccines. All of these array technologies will also yield greater insight into the pathogenic mechanisms of autoimmune disease.

MRI technology can also be used to define a patient's response to a drug. Newer MRI techniques – magnetization transfer, fluid-attenuated inversion recovery, MR spectroscopy – promise to yield important information regarding MS and SLE heterogeneity, prognosis, and treatment effects.

Finally, data analysis of the hundreds of current clinical trials for autoimmune disease should provide insights into pathological mechanisms and risk factors for progression, as well as characteristics of responsive and nonresponsive patients. More directly, a current clinical trial aims to identify predictive factors of therapeutic response in MS patients treated with high-dose IFN-β.

Once signatures are defined and validated, an individual patient's signature, or pharmacogenomic profile, could be accessed to select the optimal treatment strategy and the optimal dose and minimize the risk of drug-related adverse effects. This information would also be included in covariate analyses for clinical trials.

In terms of new therapeutic options for patients with autoimmune disease of the nervous system, the greatest, most immediate promise is likely to come from combination therapies. The lesson from RA (and cancer) is that more clinical efficacy and less toxicity result from the use of combination therapies. The aim in combination therapy is to partially block two or more signaling pathways rather than trying to completely block a single pathway. However, certain combinations may be risky and should probably be avoided. As the underlying disease pathogenesis and the mechanism of action for therapies are better understood, more effective combination therapies can be devised that maximize efficacy and minimize side effects.

Exploiting existing drugs, which have already passed safety hurdles for other indications, is an approach that also promises faster results than a conventional drug development path. Statins for autoimmune disease are just one such example. Even if they prove to be nonoptimal therapies on their own, these 'repurposed' therapies can point toward key alternative pathways, or may prove to be effective when used in synergy with other approaches. This approach is a reflection of a welcome trend to take – rather than an organ-specific approach, a more cross-disciplinary approach to exploit the commonalities of the various autoimmune diseases.

Finally, it is important to develop new therapeutic agents that are capable of targeting specific disease effector pathways without the risks associated with blanket immunosuppression. This will happen, no doubt, as our understanding of disease heterogeneity and pathogenesis becomes more sophisticated.

References

1. Ginzler, E. M.; Moldovan, I. *Curr. Opin. Rheumatol.* **2004**, *16*, 499–504.
2. Cooper, G. S.; Stroehla, B. C. *Autoimmun. Rev.* **2003**, *2*, 119–125.
3. Hafler, D. A.; Slavik, J. M.; Anderson, D. E.; O'Connor, K. C.; De Jager, P.; Baecher-Allan, C. *Immunol. Rev.* **2005**, *204*, 208–231.
4. Sospedra, M.; Martin, R. *Annu. Rev. Immunol.* **2005**, *23*, 683–747.
5. Petri, M. *Rheum. Dis. Clin. North Am.* **2005**, *31*, 245–254.
6. Graham, K. L.; Robinson, W. H.; Steinman, L.; Utz, P. J. *Autoimmunity* **2004**, *37*, 269–272.
7. Sherer, Y.; Gorstein, A.; Fritzler, M. J.; Shoenfeld, Y. *Semin. Arthritis. Rheum.* **2004**, *34*, 501–537.
8. Datta, S. K.; Zhang, L.; Xu, L. *J. Mol. Med.* **2005**, *83*, 267–278.
9. Levin, K. H. *Neurologist* **2004**, *10*, 61–74.
10. Hughes, R. A.; Rees, J. H. *J. Infect. Dis.* **1997**, *176*, S92–S98.
11. Yuki, N.; Susuki, K.; Koga, M.; Nishimoto, Y.; Odaka, M.; Hirata, K.; Taguchi, K.; Miyatake, T.; Furukawa, K.; Kobata, T.; Yamada, M. *Proc. Natl. Acad. Sci. USA* **2004**, *101*, 11404–11409.
12. Ang, C. W.; Jacobs, B. C.; Laman, J. D. *Trends Immunol.* **2004**, *25*, 61–66.
13. Kaida, K.; Morita, D.; Kanzaki, M.; Kamakura, K.; Motoyoshi, K.; Hirakawa, M.; Kusunoki, S. *Ann. Neurol.* **2004**, *56*, 567–571.
14. van Sorge, N. M.; van der Pol, W. L.; Jansen, M. D.; van den Berg, L. H. *Autoimmun. Rev.* **2004**, *3*, 61–68.
15. O'Hanlon, G. M.; Bullens, R. W.; Plomp, J. J.; Willison, H. J. *Neurochem. Res.* **2002**, *27*, 697–709.
16. van Sorge, N. M.; van der Pol, W. L.; Jansen, M. D.; Geleijns, K. P.; Kalmijn, S.; Hughes, R. A.; Rees, J. H.; Pritchard, J.; Vedeler, C. A.; Myhr, K. M. et al. *J. Neuroimmunol.* **2005**, *162*, 157–164.
17. Geleijns, K.; Laman, J. D.; van Rijs, W.; Tio-Gillen, A. P.; Hintzen, R. Q.; van Doorn, P. A.; Jacobs, B. C. *J. Neuroimmunol.* **2005**, *161*, 183–189.

18. Romi, F.; Gilhus, N. E.; Aarli, J. A. *Acta Neurol. Scand.* **2005**, *111*, 134–141.
19. Richman, D. P.; Agius, M. A. *Neurology* **2003**, *61*, 1652–1661.
20. Pullen, A. H.; Demestre, M.; Howard, R. S.; Orrell, R. W. *Acta Neuropathol. (Berl.)* **2004**, *107*, 35–46.
21. Steinman, L.; Zamvil, S. S. *Trends Immunol.* **2005**, *26*, 565–571.
22. Yuki, N.; Yamada, M.; Koga, M.; Odaka, M.; Susuki, K.; Tagawa, Y.; Ueda, S.; Kasama, T.; Ohnishi, A.; Hayashi, S. et al. *Ann. Neurol.* **2001**, *49*, 712–720.
23. Berger, J. R.; Koralnik, I. J. *N. Engl. J. Med.* **2005**, *353*, 414–416.
24. Rice, G. P.; Hartung, H. P.; Calabresi, P. A. *Neurology* **2005**, *64*, 1336–1342.
25. Steinman, L. *Nat. Rev. Drug Disc.* **2005**, *4*, 510–518.
26. Polman, C. H.; O'Connor, P. W.; Havrdova, E.; Hutchinson, M.; Kappos, L.; Miller, D. H.; Phillips, J. T.; Lublin, F. D.; Giovannoni, G.; Wajgt, A. et al. *N. Engl. J. Med.* **2006**, *354*, 899–910.
27. Rudick, R. A.; Stuart, W. H.; Calabresi, P. A.; Confavreux, C.; Galetta, S. L.; Radue, E. W.; Lublin, F. D.; Weinstock-Guttman, B.; Wynn, D. R.; Lynn, F. et al. *N. Engl. J. Med.* **2006**, *354*, 911–923.
28. Bertolotto, A. *Curr. Opin. Neurol.* **2004**, *17*, 241–246.
29. Fox, E. J. *Neurology* **2004**, *63*, S15–S18.
30. Brunner-Weinzierl, M. C.; Hoff, H.; Burmester, G. R. *Arthritis Res. Ther.* **2004**, *6*, 45–54.
31. Sela, M.; Mozes, E. *Proc. Natl. Acad. Sci. USA* **2004**, *101*, 14586–14592.
32. Miller, A. E. *Neurol. Clin.* **2005**, *23*, 215–231.
33. Cardiel, M. H. *Expert Opin. Investig. Drugs* **2005**, *14*, 77–88.
34. Neuhaus, O.; Stuve, O.; Archelos, J. J.; Hartung, H. P. *J. Neurol. Sci.* **2005**, *233*, 173–177.
35. Vollmer, T.; Key, L.; Durkalski, V.; Tyor, W.; Corboy, J.; Markovic-Plese, S.; Preiningerova, J.; Rizzo, M.; Singh, I. *Lancet* **2004**, *363*, 1607–1608.
36. Dalakas, M. C. *Pharmacol. Ther.* **2004**, *102*, 177–193.
37. Pluchino, S.; Martino, G. *J. Neurol. Sci.* **2005**, *233*, 117–119.
38. Chernajovsky, Y.; Gould, D. J.; Podhajcer, O. L. *Nat. Rev. Immunol.* **2004**, *4*, 800–811.
39. Baker, D.; Hankey, D. J. *Gene Ther.* **2003**, *10*, 844–853.
40. Rieckmann, P.; Toyka, K. V.; Bassetti, C.; Beer, K.; Beer, S.; Buettner, U.; Chofflon, M.; Gotschi-Fuchs, M.; Hess, K.; Kappos, L. et al. *J. Neurol.* **2004**, *251*, 1329–1339.
41. Giovannoni, G. *CNS Drugs* **2004**, *18*, 653–669.
42. Rizvi, S. A.; Agius, M. A. *Neurology* **2004**, *63*, S8–S14.
43. Frohman, E. M.; Stuve, O.; Havrdova, E.; Corboy, J.; Achiron, A.; Zivadinov, R.; Sorensen, P. S.; Phillips, J. T.; Weinshenker, B.; Hawker, K. et al. *Arch. Neurol.* **2005**, *62*, 1519–1530.
44. Vasoo, S.; Hughes, G. R. *Lupus* **2005**, *14*, 181–188.
45. Hughes, R. A.; Raphael, J. C.; Swan, A. V.; Doorn, P. A. *Cochrane Database Syst. Rev.* **2004**, CD002063.
46. Leger, J. M. *Expert Opin. Pharmacother.* **2005**, *6*, 569–582.
47. Sieb, J. P. *Curr. Opin. Pharmacol.* **2005**, *5*, 303–307.
48. Schneider-Gold, C.; Gajdos, P.; Toyka, K. V.; Hohlfeld, R. R. *Cochrane Database Syst. Rev.* **2005**, CD002828.
49. Fathman, C. G.; Soares, L.; Chan, S. M.; Utz, P. J. *Nature* **2005**, *435*, 605–611.

Comprehensive Medicinal Chemistry II
ISBN (set): 0-08-044513-6

ISBN (Volume 6) 0-08-044519-5; pp. 229–251

6.10 Stroke/Traumatic Brain and Spinal Cord Injuries

E D Hall, Spinal Cord and Brain Injury Research Center, University of Kentucky Medical Center, Lexington, KY, USA

6.10.1 Disease State/Diagnosis

Stroke, traumatic brain injury (TBI) and spinal cord injury (SCI) represent three of the most catastrophic occurrences that human beings can suffer. There are approximately 700 000 strokes per year in the USA, most, but not all, affecting the elderly. Ninety per cent of strokes are 'ischemic' in nature, involving a thromboembolic blockage of a brain artery that impairs cerebral blood flow and oxygenation to the point of causing infarction of the brain region that is dependent upon the blocked vessel for most of its blood supply. The remaining 10% of strokes are 'hemorrhagic'. Within this category, there are two types of insult. The first is intracerebral hemorrhage, where blood is released into the brain parenchyma and produces brain damage by triggering brain edema (swelling) and mass effects, which result in secondary ischemia within the tissue surrounding the intracerebral mass of blood (hematoma). The second type of hemorrhagic stroke is subarachnoid hemorrhage (SAH), where the blood is released into the subarachnoid space (between the subarachnoid and pial membranes covering the brain) from a burst congenital berry aneurysm ballooning out from one of the major arteries at the base of the brain. The pathophysiology of this type of hemorrhagic stroke also involves the triggering of a secondary ischemic insult due to the induction of delayed cerebral vasospasm that peaks in incidence at 4–7 days post-SAH. There are about 30 000 aneurysmal SAHs per year in the USA, mainly occurring in those in their early fifties and with a 2:1 female:male preponderance.

With regard to TBI, it is estimated that there are 1.5 million cases per annum in the USA, ranging from mild to severe. Of these, 1.2 million seek medical care. Of these most are 'mild' in severity, but about 58 000 are severe (Glasgow Coma Score 3–8) and 64 000 moderate (Glasgow Coma Score 9–12), and often require intensive medical treatment and extended recovery periods. In the case of SCI, there are about 11 000 new injuries each year in the USA, and the overall prevalence of SCI is approximately 250 000. Although TBI and SCI can victimize active individuals at any age, most occur in young adults in the second and third decades of life. Moreover, the majority of stroke, TBI, and SCI patients now survive their neurological insults due to improvements in emergency, neurological intensive care and surgical treatments. Nevertheless, the need for intensive rehabilitation and the reality of prolonged disability exacts a significant toll on the individual, his or her family, and society. Effective ways of maintaining or recovering function could markedly improve the outlook for those with stroke, TBI, or SCI by enabling higher levels of independence and productivity.

6.10.2 Disease Basis

6.10.2.1 Potential for Neuroprotective Drug Discovery

Much of the opportunity for pharmacological intervention to preserve neurological function after acute central nervous system (CNS) injuries such as stroke, TBI, and SCI is based on the fact that most of the vascular and neurodegeneration that follows these injuries is not due to the primary ischemic, hemorrhagic, or mechanical (i.e., shearing of blood vessels, and nerve cells) insults, but rather to secondary injury events set in motion by the primary injury. For example, most SCIs do not involve actual physical transection of the cord, but rather the spinal cord is damaged as a result of a contusive, compressive, or stretch injury. Some residual white matter, containing portions of the ascending sensory and descending motor tracts, remains intact, which allows for the possibility of neurological recovery. However, during the first minutes and hours following injury a secondary degenerative process, which is proportional to the magnitude of the initial insult, is initiated by the primary mechanical injury. Nevertheless, the initial anatomical continuity of the injured spinal cord in the majority of cases, together with our present knowledge of many of the factors involved in the secondary injury process, has led to the notion that pharmacological treatments that interrupt the secondary cascade, if applied early, could improve spinal cord tissue survival, and thus preserve the necessary anatomic substrates for functional recovery to take place. Similarly, the extent of neurological damage and the potential for a good outcome after stroke or TBI is mainly determined by the extent of the potentially treatable secondary pathophysiology and neurodegeneration. The goal of interfering with this process is referred to as 'neuroprotection'.

6.10.2.2 Overview of Secondary Central Nervous System Injury

The key players and the complex interrelationships involved in the secondary cascade of events occurring during the first minutes, hours, and days after traumatic CNS injury are shown in **Figure 1**.[1–5] Most of the players involved in ischemic and hemorrhagic CNS insults are the same as for traumatic injury. For TBI and SCI the most immediate event is mechanically induced depolarization and the consequent opening of voltage-dependent ion channels (i.e., Na^+, K^+, Ca^{2+}). Similarly, the onset of ischemia is quickly followed by loss of ionic homeostasis in the affected tissue. In the case of intracerebral hemorrhage or SAH, this loss of normal ion distribution is more insidious, requiring time for the secondary ischemic events to manifest themselves. The depolarization leads to massive release of a variety of neurotransmitters, including glutamate, which can cause the opening of glutamate-receptor-operated ion channels (e.g., NMDA and AMPA). The most important consequence of these rapidly evolving ionic disturbances is the accumulation of intracellular Ca^{2+} (i.e., Ca^{2+} overload), which initiates several damaging effects. The first is mitochondrial dysfunction, leading to a failure of aerobic energy metabolism, shift to glycolytic (i.e., anaerobic)

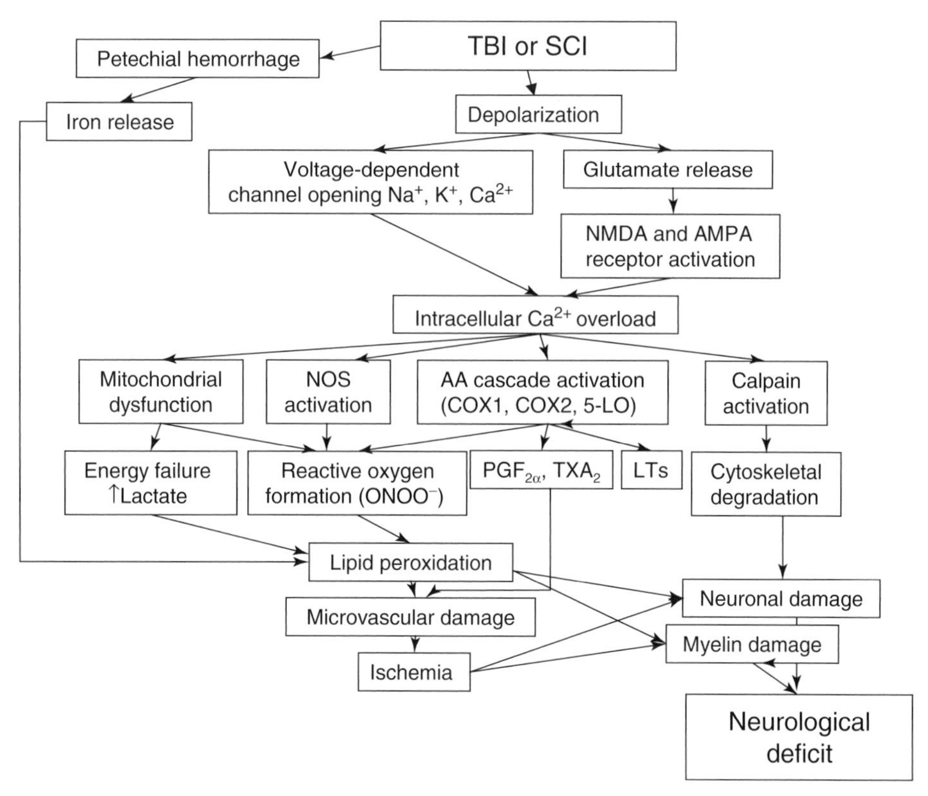

Figure 1 Schematic diagram illustrating the key factors involved in microvascular and neuronal secondary injury after TBI or SCI.

metabolism, and the accumulation of lactate. The second is activation of mitochondrial and cytoplasmic nitric oxide synthase (NOS) and nitric oxide (NO) production. The third is activation of phospholipase A_2 liberating arachidonic acid (AA), which is then converted by cyclooxygenases (COX-1 and COX-2) to a variety of prostanoids with deleterious actions. These include the potent vasoconstrictor prostaglandin $F_{2\alpha}$ ($PGF_{2\alpha}$) and the vasoconstrictor/platelet aggregation promoter thromboxane A_2 (TXA_2). In addition, activated lipoxygenases lead to an increase in tissue leukotrienes (LTs), some of which are chemoattractants for inflammatory polymorphonuclear (PMN) leukocyte and macrophage influx that can invade the injured or ischemic tissue and amplify the secondary injury process. The fourth consequence of intracellular Ca^{2+} overload is the activation of the calcium-activated protease calpain, which degrades a variety of cellular substrates, including cytoskeletal proteins.

A byproduct of mitochondrial dysfunction, COX and lipoxygenase activity, and NOS activation is the formation of reactive oxygen species (ROS), including peroxynitrite ($ONOO^-$). Peroxynitrite is a product of the reaction of superoxide radical with NO. Although peroxynitrite can trigger cellular damage by a variety of mechanisms, cell membrane (plasma and organellar) lipid peroxidation (LP) has been conclusively demonstrated to be a key mechanism.[2,6,7] However, iron is a powerful catalyst that accelerates the propagation of LP reactions. Glycolytically derived lactate promotes LP by stimulating the release of iron from storage sites, (e.g., ferritin). In addition, primary and secondary petechial hemorrhages supply hemoglobin-bound iron. Lipid peroxidation occurs in neurons and blood vessels, directly impairing neuronal and axonal membrane function and integrity, and causing microvascular damage and secondary ischemia that indirectly contributes to the secondary neuronal injury.

For SCI the secondary events occur initially in the central gray matter, spreading to the surrounding white matter. As implied above, the key issue in predicting recovery of function is the degree of preservation of the ascending and descending white matter tracts. Many of the axons that do survive, however, do not conduct impulses due to posttraumatic demyelination. Therefore, the goal of neuroprotective pharmacotherapy in the context of SCI is to preserve as many of the white matter axons and as much of their investing myelin as possible. In TBI a key determinant in neurological recovery is also the loss of axons. Based upon the often widespread loss of axons in the injured brain, this phenomenon is referred to as 'diffuse axonal injury'. However, it should be realized that a significant factor in influencing the extent of neural injury both in TBI and SCI is a decrease in brain or spinal cord microvascular perfusion (i.e., secondary ischemia). When this occurs, the result is an exacerbation of the injury process due to superimposed tissue ischemic hypoxia. Moreover, deficiencies in CNS hypoperfusion can be aggravated by systemic hypotension and/or hypoxia. Thus, it is important to note that secondary injury involves both neuronal and microvascular events.

For ischemic stroke, the goal is to limit the extent of the infarction by preventing secondary injury in the partially perfused penumbral region surrounding the core of the infarct. In SAH the main focus of attention has been on finding ways to prevent the delayed vasospasm phenomenon, which leads to secondary ischemic infarction. For intracerebral hemorrhage the hope is that the deleterious effects of the hematoma on the surrounding tissue can be prevented, and edema and ischemic damage limited.

6.10.3 Experimental Disease Models

A list of the in vivo models of acute neurological injury (i.e., ischemia, hemorrhage, TBI, and SCI) that have been, or are currently being, utilized for discovery evaluation of neuroprotective or neurorestorative agents is given in **Table 1**.

6.10.3.1 Focal Ischemic Stroke Models

Various stroke models have been developed in the past 20 years.[8] However, those in use today include the unilateral middle cerebral artery occlusion (MCAO) model used in rats and mice. Since these models were first developed in the 1980s, the MCAO has been variably induced by surgical ligation or cauterization via a small craniotomy over the middle cerebral artery, passage of a intraluminal nylon suture up into the ipsilateral cerebral circulation via the external carotid in the neck, or injection of a small autologous thrombus into the common carotid artery. The latter two methods are the most commonly employed today, and the thromboembolic paradigm is the most clinically relevant, since the majority of human focal ischemic strokes involve a thromboembolic occlusion of the middle cerebral artery. The MCAO models come in two varieties: temporary and permanent. The temporary MCAO involves removal of the vascular occlusion at varying times (30, 60, 90, 120, 180 min) after the onset in order to allow reperfusion of the ischemic tissue to take place. This experimental scenario, which is accomplished by surgical removal of the extraluminal or intraluminal occlusion device, mimics either the instance where spontaneous thrombus dissolution may take place during the first 3 h after the beginning of the stroke due to the activation of endogenous thrombolytic processes (believed to be a fairly rare occurrence), or the situation where the stroke victim is treated with the thrombolytic agent tissue plasminogen activator (TPA) for the purpose of dissolving the clot and restoring recirculation. Although removal of the vascular occlusion and re-establishment of the normal cerebral circulation is an obviously desirable therapeutic goal, it is a double-edged sword that can lead to 'reperfusion injury', which is caused by a burst of ROS in the previously ischemic brain tissue. Thus, there is a need in this situation for a neuroprotective agent, to reduce both the pathophysiological events set in motion by the ischemic insult and the subsequent deleterious side effects of pharmacological recirculation. Accordingly, the temporary MCAO models are most useful for evaluating neuroprotective strategies that may be used in conjunction with TPA and other thrombolytic agents. However, since pharmacological thrombolysis can lead to secondary cerebral hemorrhage if used beyond the first few hours, it can only be safely employed in patients who are available for emergency treatment during the first few hours after the onset of their strokes. This is only feasible in a small fraction of ischemic stroke patients. Furthermore, MCAO animal studies have shown that reperfusion beyond the first 3 h cannot lessen the extent of ischemic damage. Thus, the temporary MCAO models, although widely used in stroke research, actually have limited relevance to the majority of middle cerebral artery territory strokes.

The second variety of focal ischemic stroke model, the permanent MCAO, where the occlusion is permanently left in place, is therefore a better model of the vast majority of strokes, where recirculation has not been re-established either spontaneously or pharmacologically during the critical first few hours after stroke onset. In this instance, the therapeutic goal is simply to try to reduce the expansion of the ischemic damage from the severely ischemic 'core' area, which is doomed to infarction if reperfusion does not occur during the first 3 h, into the surrounding 'penumbral' area. The 'ischemic penumbra' is potentially salvageable for several hours due to its partial circulation from collateral blood vessels. In the author's opinion, the permanent MCAO version is the best option for preclinical evaluation of potential neuroprotective agents, since it is more relevant to the overall focal ischemic stroke population where early reperfusion is not that common. However, the testing of new compound entities in the temporary MCAO paradigm is also recommended, but not as the only model. Historically, there has been an unfortunate affinity for the temporary models among stroke therapeutic investigators due to the observation that it is generally easier to demonstrate infarct reductions with new compound entities.

Most stroke research with either the temporary or permanent MCAO models is carried out in mice or rats. The primary endpoints are histological demonstration of a reduction in brain infarct size and improvement in motor

Table 1 In vivo models employed for discovery of neuroprotective and neurorestorative agents

Stroke (focal ischemia)
Rat or mouse temporary middle cerebral artery occlusion (MCAO): microclip or intraluminal suture for 30 min to 2 h
Rat or mouse permanent MCAO: electrocoagulation or intraluminal suture
Cardiac arrest/resuscitation (transient global ischemia)
Rat two-vessel (bilateral carotid) occlusion plus hypotension for 5–30 min
Rat four-vessel occlusion for 5–30 min (permanent bilateral vertebral artery electrocoagulation followed 24 h later by transient bilateral carotid occlusion)
Gerbil bilateral carotid occlusion (BCO) for 5–15 min
Hemorrhagic stroke models (subarachnoid hemorrhage or intracerebral hemorrhage)
Rabbit, cat, or dog intracisterna magna injection of autologous blood
Rat intracranial injection of autologous blood via dorsolateral cranial burr hole
Monkey SAH via surgical placement of autologous blood clot around base of middle cerebral artery
Rat striatal intracerebral hemorrhage
Traumatic brain injury
Diffuse
Rat or mouse fluid percussion: can be combined with hypotension or hypoxia
Rat impact–acceleration: can be combined with hypotension or hypoxia
Mouse weight drop
Pig or primate rotational acceleration (nonimpact)
Focal
Rat or mouse controlled cortical impact (CCI)
Axonal
Mouse optic nerve stretch
Subdural hematoma
Rat intracranial injection of autologous blood via dorsolateral cranial burr hole (same as SAH model)
Spinal cord injury
Weight drop contusion
Wrathall device and model
Rat NYU (MASCIS) device and model
Rat OSU (ESCID) device and model
Rat or mouse UK (Infinite Horizons) device and model
Compression
Rat aneurysm clip compression (Fehlings and Tator model)
Cat weight compression (Anderson model)
Combination contusion and compression
Rat contusion followed by placement of Teflon wedges underneath vertebrae
Ischemic injury
Rabbit, balloon in descending aorta inflated transiently above level of lumbar spinal arteries
Rat, laser photoablation (Rose Bengal dye intravenously)
Excitotoxic injury
Kainic, quisqualic, or ibotinic acid, or dynorphin spinal cord microinjection
Regeneration models
Spinal cord transection, resection, or hemisection
Dorsal rhizotomy

function, typically determined between 24 and 72 h after stroke onset. However, MCAO models have been developed and are occasionally used in higher species, including the cat, monkey, and baboon. Obviously, the use of these for neuroprotective new compound entity evaluation carries considerable expense. Some investigators (a minority) believe that it is important to replicate pharmacological neuroprotective actions in these gyrencephalic species prior to movement of the compound into human clinical trials. In actuality, there is no solid comparative evidence that supports the notion that neuroprotective effects seen in rodent stroke models may not be predictive of human efficacy. Furthermore, there is presently no firm data that support the commonly held idea that the therapeutic time window for a particular neuroprotective mechanism in a rat stroke model (e.g., 1 h) may be longer (e.g., 6 h) in nonhuman primates or humans. On the contrary, the fact that various neuroprotective new compound entities that demonstrated a rather limited (1–2 h) therapeutic window for reduction of infarct size in rat MCAO models subsequently failed to improve outcome of stroke patients in clinical trials where the treatment initiation time varied from 6 to 24 h is consistent with the concept that the therapeutic window for neuroprotective effects may not be all that different between rodents and primates. At least no difference has been firmly demonstrated.

6.10.3.2 Hemorrhagic Stroke Models

As noted above, the two basic types of hemorrhagic strokes are the intracerebral hemorrhage and the SAH. For the former the approach is simply to inject a volume of the animal's own blood directly into the brain parenchyma, and then to analyze the volume of damage to the surrounding brain tissue. With regard to the latter, most of the available models involve the injection of a volume of the autologous blood, withdrawn immediately prior to SAH induction from the systemic circulation of the animal (e.g., rat, cat, rabbit, and dog), into the subarachnoid space via injection into the cisterna magna, or over one of the cerebral hemispheres via a small burr hole and puncture of the dura mater covering the brain. The most common endpoints for new compound entity evaluation involve measurement of blood–brain barrier compromise or decreases in cerebral blood flow during the first several post-SAH hours, or the assessment of cerebral vasospasm by histological or arteriographic methods between 2 and 7 days. The most sophisticated SAH model available involves the neurosurgical placement of an autologous blood clot around the base of the middle cerebral artery in monkeys, followed by measurement of arteriographic and histological ischemic damage at 7 days. However, the cost of evaluating a single-dose level of new compound entity for its ability to inhibit delayed cerebral vasospasm in that model runs into the hundreds of thousands of dollars.

6.10.3.3 Traumatic Brain Injury Models

In vivo TBI models include three basic types: diffuse, focal, and axonal injury (**Table 1**).[9] There are three diffuse injury models. The first of these is the rat fluid percussion TBI paradigm, which involves the application of a transient hydraulic pressure pulse onto the exposed dura mater, either over the midline of the brain or laterally over one of the hemispheres. The second is the rat impact–acceleration injury model, which involves a 0.5 or 1.0 kg weight drop onto a steel helmet cemented onto the exposed skull. The third is the mouse weight–drop concussion paradigm, and the last are the pig or primate rotational acceleration models, which are useful for studying the phenomenon of diffuse axonal injury.

For the induction of focal TBIs there is the widely employed controlled cortical impact (CCI) model, used in either rats or mice, that involves the infliction of a contusion injury through a small craniotomy. The magnitude of the injury is generally varied by the depth of the cortical indentation (usually 0.5–1.0 mm in mice and 1.0–2.0 mm in rats). The CCI model is a model of TBI-induced brain contusions, although a recent study has shown that the subsequent neurodegeneration is not as focal as is generally thought.[10] A relatively new in vivo model has been developed to examine the effects of stretch injury on axons, and involves the induction of a controlled stretch of the optic nerve in mice.

6.10.3.4 Subdural Hematoma Model

The rat lacks an arachnoid membrane, and thus the rat version of the SAH model involving injection of blood through the dura mater can also be thought of, and employed, as a subdural hematoma model. When the autologous, nonheparinized blood is injected through the dorsal burr hole through the dura mater, it will typically form a clot over the dorsal surface of the brain, mimicking a posttraumatic subdural hematoma. Just as is done in the management of human subdural hematomas, the experimental protocol involves surgical removal of the clot at a specified time followed by histological measurement of ischemic damage caused by the hematoma.[11]

6.10.3.5 Spinal Cord Injury Models

A large number of SCI paradigms have been developed over the past 100 years. As shown in **Table 1**, the current rodent models that are used to evaluate neuroprotective agents involve contusion, compression, ischemic, and excitotoxic injury mechanisms. By far, the contusion models predominate in the experimental acute SCI field, and in particular the New York University (NYU)[12] and University of Kentucky (UK)[13] controlled contusion devices dominate acute SCI research. For strategies that may influence axonal regeneration in the injured spinal cord, the usual models involve either complete transaction or hemisection of the spinal cord or dorsal roots (rhizotomy) followed by histological assessments of axonal growth across the lesion site. Assessment of neurological recovery in rat SCI models most commonly employs the Basso/Beattie/Breshahan locomotor recovery assessment tool (BBB Score).[14] However, other motor recovery assessment tools have been developed and are employed along with the BBB scoring system.

6.10.4 Clinical Trial Issues

In the early 1980s several pharmaceutical companies became attracted to the idea of discovering neuroprotective drugs for the acute treatment of stroke and CNS injury. As a result, several new compound entities were discovered that were entered into development, with some making their way into large double-blind, multicenter, phase III clinical trials for stroke (ischemic and SAH), TBI, and/or SCI. These efforts, which dominated neuroprotective clinical trials in the 1980s and 1990s, were primarily directed at three general pharmacological mechanistic strategies to interrupt secondary injury processes:

1. inhibition of glutamate-mediated excitotoxicity (i.e., glutamate receptor antagonists; GABA agonists)
2. reduction of intracellular calcium overload (L-type calcium channel blockers)
3. interruption of reactive-oxygen-mediated damage (i.e., free radical scavengers/antioxidants).

6.10.4.1 Glutamate Antagonists

Multiple glutamate receptor antagonists were taken into phase II and III trials, including: the competitive NMDA receptor antagonists selfotel (CGS 19755) and aptiganel (CNS 1102), which block the binding of glutamate to its receptor complex recognition site; eliprodil, which blocks the polyamine site; and CP-101,606, which blocks the NR2B subunit on the NMDA receptor complex. None of these produced a statistically significant improvement in neurological recovery in TBI or ischemic stroke trials.[15,16]

An alternative mechanism for countering glutamate excitotoxicity is to increase γ-amino-butyric acid (GABA) mediated inhibitory transmission with the administration of GABA receptor agonists. This approach resulted in the clinical evaluation of the GABA partial agonist new compound entity chlomethiazole in a phase III stroke trial. However, no significant beneficial effect was demonstrated.[15,16]

6.10.4.2 Calcium Channel Blockers

As noted previously, accumulation of intracellular calcium is a major player in secondary injury after CNS injury or stroke. One of the mechanisms for the postinsult calcium overload involves depolarization-induced entry via voltage-dependent L-type channels. Accordingly, the first neuroprotective approach to be tested in phase III clinical trials in TBI or stroke was the competitive L-type calcium channel blocker nimodipine, which entered clinical trials in the late 1970s. In two separate phase III, multicenter TBI (moderate and severe) trials[17] and a single stroke trial,[16] no overall benefit was revealed with nimodipine treatment. However, retrospective analysis of the TBI trials revealed that nimodipine may improve outcome in patients with traumatic SAH (tSAH).[17] This was not an insignificant finding, as about half of all patients with severe TBI have tSAH as part of the pathophysiology. Furthermore, nimodipine produces a slight, but significant, increase in survival in aneurysmal SAH patients, and has been approved in most countries for the treatment of that condition. Indeed, nimodipine represents the first agent to be approved for neuroprotective use, even though much of its effect is probably mediated via protection of the microvasculature and vasodilatation-mediated improvements in cerebral blood flow. Due to a manifestation of its microvascular vasodilatation, the compound must be used with care, because it can lower arterial and cerebral perfusion pressures that can exacerbate posttraumatic, postiischemic, or post-SAH secondary brain injury.

6.10.4.3 Free Radical Scavengers/Antioxidants

In the case of efforts to interrupt reactive oxygen damage, the polyethylene conjugated form of the superoxide radical scavenger Cu/Zn superoxide dismutase (PEG-SOD) was evaluated in trials conducted in moderate and severe TBI patients. Although it showed a positive trend in an initial small phase II trial,[18] subsequent phase III trials failed to show any enhancement of neurological recovery.[19] A larger development program was undertaken with the 21-aminosteroid LP inhibitor tirilazad (U-74006F; 'lazaroid'; **Figure 2**). Tirilazad was extensively evaluated in animal models of SCI, TBI, ischemic stroke, and SAH, and shown to exert a variety of neuro- and vasoprotective effects.[20,21] Based on these preclinical studies, clinical trials of tirilazad were conducted in TBI,[19,22] SAH,[23] ischemic stroke,[15,16]

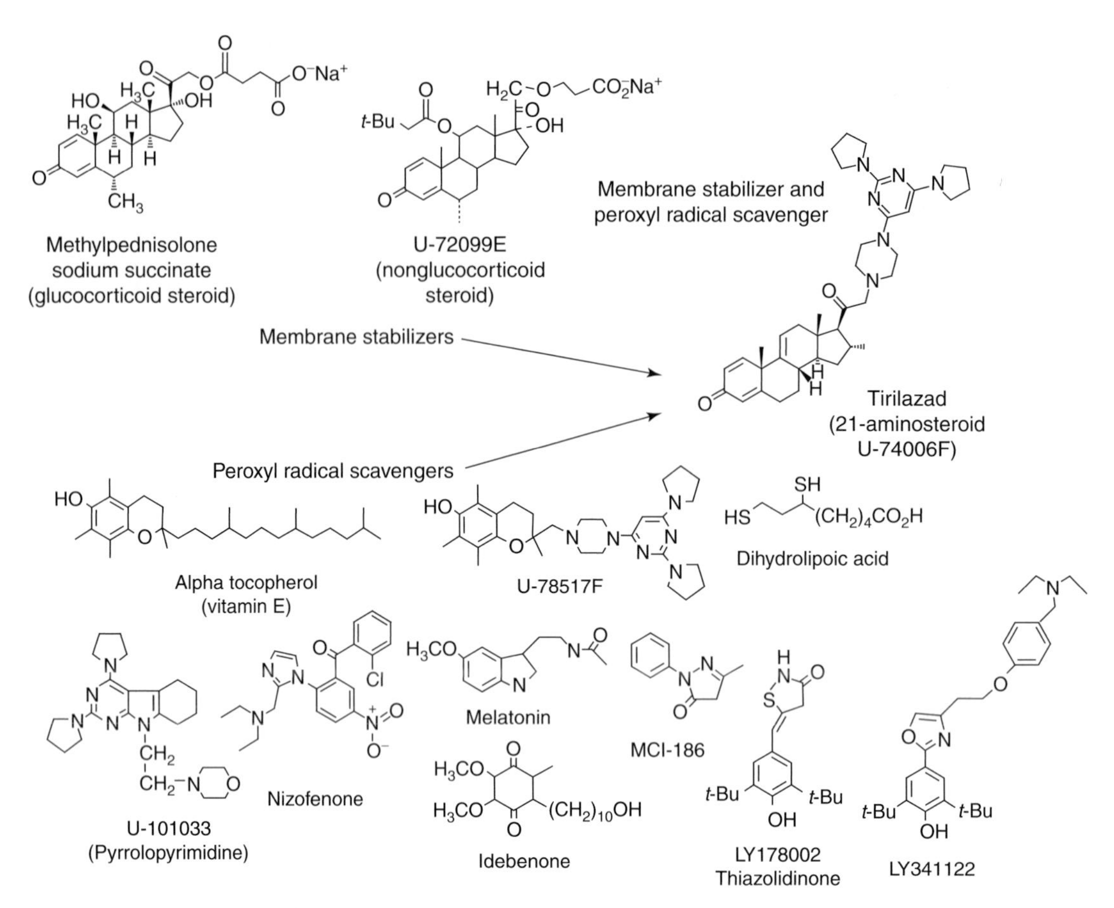

Figure 2 Chemical structures of lipid peroxidation inhibiting drugs that have been shown to be neuroprotective in models of CNS injury and/or ischemic stroke. Two classes are noted. The first class includes compounds that scavenge lipid peroxyl radicals, either by donation of an electron to form a lipid hydroperoxide or by binding the peroxyl radical followed by breakage of the peroxide bond, resulting a hydroxylated scavenger and a lipid alcohol. The second class are steroid-based compounds that inhibit peroxidation by 'stabilizing' cell membranes. Such compounds are believed to act by limiting the fluidity of membrane phospholipids and, as a result, the peroxidized lipids are not as free to interact with neighboring phospholipids. Two examples of this type of compound are the glucocorticoid steroid methylprednisolone and its nonglucocorticoid analog U-72099E. The 21-aminosteroid tirilazad is a dual mechanism compound that works by both membrane stabilization and by peroxyl radical scavenging.[20]

and SCI.[24] An initial North American trial of 1100 TBI patients compared tirilazad treatment with placebo for 5 days, both initiated with 4 h postinjury. The trial ended with such a confounding randomization imbalance that no meaningful efficacy analysis could be extracted. In contrast, a European phase III trial was successfully completed, but failed to show an overall effect in moderate and severely injured patients. However, post hoc analysis revealed that the new compound entity significantly improved survival in both moderately and severely injured male patients with tSAH.[22] This beneficial effect in the tSAH subgroup, representing about half of the severe TBIs, was not surprising, in that the new compound entity had previously been shown to improve recovery and survival in a phase III trial in aneurysmal SAH patients.[23] Interestingly, this effect in tSAH and aneurysmal SAH was mainly apparent in males. This gender difference was partially due to a faster rate of metabolism of the drug in females. Nevertheless, subsequent female-only trials with higher tirilazad doses that were calculated to duplicate the exposure levels in males failed to demonstrate the same level of efficacy as seen in male patients, although beneficial effects were apparent in females with the more severe SAH.[25,26] Thus the issue of gender differences in neuroprotective drug responsiveness clouds the interpretation of the neuroprotective efficacy of tirilazad.

Tirilazad was also extensively evaluated in four different phase III stroke trials.[15,16] The first two (TESS I in Europe, and RANTTAS I in the USA) evaluated the effects of 6 $mg\,kg^{-1}$ intravenous (i.v.) per day for 3 days, with treatment beginning within 6 h of stroke onset. No effect was seen on 3- or 6-month outcomes. Two subsequent higher dose trials (10 $mg\,kg^{-1}$ per day in males; 15 $mg\,kg^{-1}$ per day in females) were conducted. The first, the European

TESS II, which included patients enrolled within the first 6 h of the stroke occurring, was stopped prematurely due to a significant increase in morbidity and mortality in the high-dose tirilazad group. Prudence dictated the simultaneous cessation of the parallel US high-dose RANTTAS II trial. However, subsequent analysis of the 3-month recovery scores of the approximately 100 patients enrolled in RANTTAS II revealed a nearly significant improvement in neurological recovery. The only difference between the two trials was that in TESS II the enrollment window was 6 h, whereas in RANTTAS II treatment began within 4 h. The contrasting results of TESS II and RANTTAS II indicate that tirilazad may be effective in stroke patients if given in the first 4 h, but may in fact be harmful if delayed until 6 h. Another issue besides therapeutic window is the issue of how long to maintain treatment. The decision to treat stroke patients in all of the tirilazad trials for 72 h was based on the limits of safety rather than on a demonstration of the benefits of such lengthy treatment in preclinical stroke models.[20] The toxicity of the drug in TESS II indicates that it is possible to overtreat with tirilazad. Thus, the possibility exists that a shorter treatment duration may have yielded more positive results. The fact that neither the optimum therapeutic window nor the optimal treatment duration were ever determined for tirilazad or any other neuroprotective drug prior to their being advanced into clinical trials for TBI or stroke may have played a role in the failures of NMDA antagonists, the calcium channel blocker nimodipine, and the antioxidants PEG-SOD and tirilazad in achieving an overall beneficial effect.

Currently, however, the nitroxide free radical 'spin trapping' agent NXY-059 is being tested in phase III stroke trials.[27]

6.10.4.4 Dexanabinol (HU-211)

The cannabinoid analog dexanabinol, which demonstrated protective efficacy in preclinical TBI models,[28] has been tested in a phase II clinical trial in severe TBI patients. The new compound entity was found to be safe and showed that patients treated within 6 h tended to have better control of intracranial and cerebral perfusion pressures.[29] This new compound entity is unique in that it combines antioxidant and weak NMDA receptor antagonist properties along with selected anti-inflammatory properties. However, a recently completed multicenter, multinational, phase III trial has reportedly failed to show neuroprotective efficacy.

6.10.4.5 Lessons Learnt from Past Clinical Trials of Neuroprotective Agents

The brief history of neuroprotective drug discovery and development over the past 20–25 years outlined above could be fairly characterized as a series of often high-profile and expensive failures. Although these have largely dampened the enthusiasm of the pharmaceutical industry for this therapeutic area, much has been learned that could, and should, serve as a roadmap for future efforts aimed at pharmacological neuroprotection and improved neurological recovery after stroke, TBI, and SCI. Postmortem analyses of mistakes made in stroke[15,16] and TBI[19] drug development have been published, and a careful reading of these reveals a host of shortcomings in past preclinical testing of candidate neuroprotective agents and in clinical trial design and conduct that need to be addressed in the future. A summary of these issues is given in **Table 2**.

Firstly, the discovery of the first generation of neuroprotective agents, which included glutamate receptor antagonists, calcium channel blockers, and antioxidants, occurred prior to the elucidation of an adequate understanding of the intricacies of the targeted secondary injury mechanisms. In each case, knowledge of the time course and interrelationships of these events and their therapeutic windows for effective treatment intervention was lacking. In the case of reactive oxygen mechanisms, knowledge of the key ROS species and their sources and cellular targets was inadequate to guide the design of optimum antioxidant neuroprotective new compound entities. Secondly, the preclinical efficacy testing of new compound entities was, in some cases, woefully inadequate, and even naive. In contrast, it is now realized that several issues need to be addressed in preclinical evaluations in order to guide the design of clinical trials. These include:

1. the demonstration of the time course of the target pathophysiological mechanism in relevant animal models – this is needed to determine when treatment needs to begin and for how long it must be maintained
2. rigorous dose–response analysis with regard to effects on the target mechanism, and ability to reduce posttraumatic neurodegeneration and improve behavioral recovery
3. correlation of neuroprotective action with plasma and CNS tissue pharmacokinetics
4. correlation of plasma and CNS pharmacokinetics with a plasma or CNS biomarker (e.g., LP products)
5. comparison of single versus multiple dose regimens in order to establish the optimum treatment regimen (i.v. bolus plus infusion make the most sense)

Table 2 Reasons for past failures in stroke and TBI drug discovery and development

Inadequate understanding of secondary injury mechanisms
Lack of definition of time course of NMDA receptor functional changes
Lack of definition of the sources and spatial and temporal characteristics of reactive oxygen generation → inability to rationally determine therapeutic window and optimum treatment duration
Lack of understanding of the integration of secondary injury mechanisms
Focus on secondary injury mechanisms with short therapeutic windows → need to identify and target injury mechanisms with longer therapeutic windows
Lack of understanding of the relative therapeutic windows in animal models and humans. Is the time course of secondary injury in mice, rats, and humans similar?
Inadequate preclinical testing
Lack of testing in multiple models
Failure to compare efficacy in male and female animals
Incomplete dose–response and definition of therapeutic plasma levels
Incomplete definition of therapeutic window
Lack of definition of pharmacokinetics, timing of needed maintenance dosing, and optimum treatment duration
Poor clinical trial design
Gross mismatch between preclinical and clinical testing
Imprecise endpoints (e.g., Glasgow Coma Scale, Glasgow Outcome Scale, and ASIA scale)
Lumping together of all kinds of ischemic strokes or moderate and severe TBIs
Lack of identification of an a priori plan to analyze subgroups (e.g., tSAH)
Lack of a biomarker to follow the progression of the pathophysiology and to monitor mechanistic drug effects
Lack of standardization of neurorehabilitation protocols

6. determination of the therapeutic window to determine how early treatment must begin (this may vary between TBI, ischemic stroke, SAH, and SCI)
7. comparison of neuroprotective pharmacology in multiple injury models in order to determine whether the agent in question only works in certain types of injuries
8. determination of pharmacodynamic and pharmacokinetic interactions with other commonly used ancillary treatments (e.g., anticonvulsants).

Thirdly, in retrospect, there were several flaws in the early design of neuroprotective clinical trials in terms of the use of imprecise and insensitive endpoints, lumping together of all types of patients, failure to stratify according to subgroups, lack of a suitable biomarker (with which to monitor the progression of the pathophysiology and drug effects thereon) for use in early phase dose–response trials, and the failure to include and standardize neurorehabilitative techniques across all patients in the trials. The detailed preclinical evaluation of neuroprotective agents in stroke, TBI, or SCI models called for above should then be followed by a clinical trial design that is consistent with the results of preclinical trials. One of the most glaring examples of how this has not been done in the past concerns the fact that, even with agents in which the therapeutic efficacy window has been determined, the trials typically involved an enrollment window that far exceeded the postinsult time at which a particular agent can be expected to retain neuroprotective potential. It has been argued that even if a particular agent only has a 1-h window in a rat stroke, TBI, or SCI model, the window in humans with the corresponding condition is likely to be much longer. However, there is little or no evidence to support this assumption. Consequently, in the author's opinion, clinical trial design should take seriously the preclinical therapeutic window definition for a particular agent with regard to how soon the compound may need to be given to patients. With this in mind, a failure to demonstrate a clinically practical therapeutic window for a particular agent may mean that this agent and its corresponding secondary injury mechanism may be too short to be effectively addressed in real-world therapeutics.

6.10.4.6 Biomarkers for Neuroprotective Drug Evaluation

A key need to improve the translation of preclinical neuroprotective effects into successful mechanism-based clinical trials is the identification of measurable biomarkers that allow the pathophysiology and drug effects thereon to be determined. Three examples of promising CNS injury biomarkers are the LP-related isoprostanes[30] and neuroprostanes,[31] calpain and caspase-3-mediated degradation products of the cytoskeletal protein α-spectrin,[32] and the astrocytic injury marker S100β.[33,34] The validation of these for the monitoring of postischemic or posttraumatic injury is ongoing in multiple centers and should lead to a quantifiable means for the more efficient clinical evaluation of neuroprotective pharmacotherapies with regard to neuroprotective efficacy, dose–response, therapeutic window, and optimal dosing regimen determinations.

6.10.5 Current Treatment

6.10.5.1 Stroke

The only pharmacological agent available for the treatment of acute ischemic stroke in Western countries, including the USA, are the thrombolytic agents, which are used to dissolve acute thromboembolic blood clots. The first of these approved was recombinant tissue plasminogen activator (rTPA).[35] Although hailed as a breakthrough for stroke treatment when first approved, its use is severely limited, since it is only approved for use during the first 3 h after stroke onset. Its use beyond that therapeutic window can lead to secondary hemorrhagic stroke. Thus, only the handful of stroke patients who present for emergency treatment within the first 3 h can be treated. This relatively short window of opportunity is further complicated by the fact that all patients first undergo a computed tomography (CT) scan to rule out the possibility of the patient having suffered a hemorrhagic, rather than a thromboembolic, stroke. There is some indication that newer thrombolytic agents may have less risk than rTPA in terms of possible hemorrhagic conversion.[36] There are also experimental data which show that the combined use of a neuroprotective agent along with the thrombolytic agent may reduce the risk of secondary hemorrhage and thus extend the therapeutic window for safe thrombolysis. However, no neuroprotective agents are currently approved in the USA.

6.10.5.2 Aneurysmal Subarachnoid Hemorrhage

As noted earlier, the L-type calcium channel blockers nimodipine and nicardipine are available in most countries for the neuroprotective treatment of aneurysmal SAH. The approval of nimodipine specifically was achieved in the early 1990s on the basis of weakly positive data regarding an improvement in survival in SAH patients. The rationale for its use is the expectation that it will inhibit delayed cerebral vasospasm. However, it is more likely that the compound improves microvascular perfusion and/or exerts a direct protective effect on the ischemic brain parenchyma. Nimodipine's usage is accompanied by a great deal of clinical caution due to the legitimate concern about the risk of systemic vasodilatation and hypotension, which can compromise rather than improve cerebral blood flow. Nevertheless, the prior approval of nimodipine for aneurysmal SAH resulted in the subsequent tirilazad SAH clinical development being carried out on top of concomitant nimodipine usage. As noted above, in comparison to nimodipine alone, nimodipine plus tirilazad significantly improved neurological recovery and survival in male patients with aneurysmal SAH.[23] Consequently, in the mid-1990s tirilazad was approved for use in males with SAH in Canada, several European countries, Australia, and New Zealand. It has not been approved in the USA, and Pfizer, the current sponsor, is unlikely to rekindle its development for SAH or other neuroprotective indications. Thus, nimodipine and the orally active analog nicardipine are the only agents currently approved for use in SAH in the USA and many other countries.

6.10.5.3 Traumatic Brain Injury

There are currently no neuroprotective agents for the acute treatment of TBI. The only agents available for TBI management are the osmotic diuretic mannitol and the short-acting barbiturates, both of which are used to counteract posttraumatic brain edema. Therefore, acute TBI remains a totally unmet medical need in relation to the discovery and successful development of pharmacological therapies that will improve neurological recovery and survival.

6.10.5.4 Spinal Cord Injury

The glucocorticoid steroids (e.g., dexamethasone and methylprednisolone (MP); **Figure 2**) were extensively employed in the clinical treatment of SCI beginning in the mid-1960s and continuing throughout the 1970s. The mechanistic rationale for their use in that era was centered on the expectation that they would reduce posttraumatic spinal cord edema. This notion was based on the rather remarkable reduction of peritumoral brain edema induced by glucocorticoids in brain tumor patients. Furthermore, steroid pretreatment became a standard of care prior to neurosurgical procedures to prevent intra- and postoperative brain swelling.

In the mid-1970s a randomized, multicenter clinical trial, the National Acute Spinal Cord Injury Study (NASCIS I), was initiated to determine if steroid dosing was beneficial in improving neurological recovery in humans after SCI. It compared the efficacy of 'low-dose' MP (100 mg i.v. bolus per day for 10 days) and 'high-dose' MP (1000 mg i.v. bolus per day for 10 days) in affecting outcome after SCI.[37] The trial, which began in 1979, did not include a placebo group due to the prevailing belief that glucocorticoid dosing probably was beneficial and could not be ethically withheld. However, the results failed to show any difference between the low- and high-dose groups at either 6 months[37] or 1 year,[38] suggesting to the investigators that steroid dosing was of little benefit. Additionally, there was a suggestion that the 10-day high-dose regimen increased the risk of infections, a predictable side effect of sustained glucocorticoid dosing. Based on the negative results of NASCIS I, as well as waning neurosurgical enthusiasm for steroid treatment of CNS injury in general, the majority of neurosurgeons concluded after NASCIS that the conventional use of steroids in the acute management of spinal trauma was not beneficial, while at the same time being fraught with the potential for serious side effects.

Increasing knowledge of the posttraumatic LP mechanism in the 1970s and early 1980s prompted the search for a neuroprotective pharmacological strategy aimed at antagonizing oxygen-radical-induced LP in a safe and effective manner. Attention was focused on the hypothetical possibility that glucocorticoid steroids might be effective inhibitors of posttraumatic LP, based upon their high lipid solubility and known ability to intercalate into artificial membranes between the hydrophobic polyunsaturated fatty acids of the membrane phospholipids and to thereby limit the propagation of LP chain reactions throughout the phospholipid bilayer.[39]

Interest in the LP hypothesis of secondary SCI evolved during parallel investigations of the effects of high-dose MP (15–90 $mg\,kg^{-1}$ i.v.) on spinal cord electrophysiology in the context of improving impulse conduction and recovery of function in the injured spinal cord.[39] A similar high dose of MP, which enhanced spinal neuronal excitability and impulse transmission, was tested for its ability to inhibit posttraumatic spinal cord LP. In an initial study in cats, an i.v. bolus of MP inhibited posttraumatic LP in spinal cord tissue, but the doses required were much higher (30 $mg\,kg^{-1}$) than those previously hypothesized, or those empirically employed in the clinical treatment of acute CNS injury or tested in the NASCIS trial. Additional studies in cat SCI models showed that at a dose of 30 $mg\,kg^{-1}$ MP not only prevented LP but, in parallel, inhibited posttraumatic spinal cord ischemia, supported aerobic energy metabolism (i.e., reduced lactate and improved ATP and energy charge), improved recovery of extracellular calcium (i.e., reduced intracellular overload), and attenuated calpain-mediated neurofilament loss. However, the central effect in this protective scenario was the inhibition of posttraumatic LP.[39] The antioxidant neuroprotective action of MP is closely linked to the tissue pharmacokinetics of the drug. This prompted the hypothesis that prolonged MP therapy might better suppress the secondary injury process and lead to better outcomes compared with the effects of a single large i.v. dose. Indeed, subsequent experiments in a cat spinal injury model demonstrated that animals treated with MP using a 48-h antioxidant dosing regimen had improved recovery of motor function over a 4-week period.[39]

The early empirical treatment of peritumoral edema and acute SCI with glucocorticoid steroids was heavily weighted toward the use of dexamethasone, based upon the fact that it was, and is, the most potent synthetic glucocorticoid steroid available for parenteral use. Dexamethasone is about five times more potent than MP in terms of glucocorticoid receptor affinity and anti-inflammatory potency.[39] However, the antioxidant efficacy of MP is unrelated to its glucocorticoid steroid receptor activity. Indeed, a careful concentration–response study compared the ability of different glucocorticoid steroids to inhibit oxygen-radical-induced LP damage in rat-brain synaptosomal preparations, and confirmed that LP-inhibiting potencies and anti-inflammatory potencies do not correlate. Although dexamethasone is five times more potent than MP as a glucocorticoid, it is only slightly more potent than MP as an inhibitor of LP. Furthermore, the maximal antioxidant activity of MP appeared to be superior to that for dexamethasone. Hydrocortisone, the prototype glucocorticoid, completely lacks the ability to inhibit oxygen-radical damage in CNS tissue. Thus, the choice of a steroid for its potential antioxidant neuroprotective activity should not be predicated on glucocorticoid-receptor-mediated anti-inflammatory actions. In addition, selection of the most potent glucocorticoid would logically carry the greatest potential for concomitant steroid-related side effects.

The above-described studies with high-dose MP inspired the second National Acute Spinal Cord Injury Study (NASCIS II), even though the earlier NASCIS trial, which came to be known as NASCIS I, had failed to show any efficacy of lower MP doses even when administered over a 10-day period.[40] The NASCIS II trial compared 24-h dosing of MP versus placebo for the treatment of acute SCI. A priori trial hypotheses included the prediction that SCI patients treated within the first 8 h postinjury would respond better to pharmacotherapy than patients treated after 8 h. Indeed, the results demonstrated the effectiveness of 24 h of intensive MP dosing (30 $mg\,kg^{-1}$ i.v. bolus plus a 23-h infusion at 5.4 $mg^{-1}\,kg^{-1}\,h^{-1}$) when treatment was initiated within 8 h of injury. Significant benefit was observed in individuals with both neurologically complete (i.e., plegic) and incomplete (i.e., paretic) injuries. Although predictable side effects of steroid therapy were noted, including gastrointestinal bleeding, wound infections, and delayed healing, these were not significantly more frequent than those recorded in placebo-treated patients. Another finding was that the delay in the initiation of MP treatment until after 8 h is actually associated with decreased neurological recovery. Thus, treatment within the 8-h window is beneficial, whereas dosing after 8 h can be detrimental. Nevertheless, the NASCIS II results demonstrating the ability of high-dose MP treatment led to this treatment becoming the standard of care for acute SCI in most countries, including the USA. Furthermore, several regulatory agencies in western and Asian countries approved the use of high-dose MP (24-h dosing begun within first 8 h) for use in the treatment of acute SCI. Although MP had long been approved in the USA for various anti-inflammatory uses, the FDA did not approve the drug for SCI on the basis of NASCIS II alone due to the statutory requirement for the completion of a second phase III trial that verified the efficacy shown in the first trial.

The demonstrated efficacy of a 24-h dosing regimen of MP in human SCI in NASCIS II,[40] and the discovery of tirilazad,[20] led to the organization and conduct of NASCIS III.[24] In the NASCIS III trial, three patient groups were evaluated. The first (active control) group was treated with the 24-h MP dosing regimen shown to be effective in NASCIS II. The second group was also treated with MP, except that the duration of MP infusion was extended to 48 h. The purpose was to determine whether extension of the MP infusion from 24 to 48 h resulted in greater improvement in neurological recovery in acute SCI patients. The third group of patients was treated with a single $30\,mg\,kg^{-1}$ i.v. bolus of MP followed by 48-h administration of tirilazad. No placebo group was included because it was deemed ethically inappropriate to withhold at least the initial large bolus of MP. Another objective of the study was to ascertain whether treatment initiation within 3 h following injury was more effective than when therapy was delayed until 3 to -8 h post-SCI.

Completion of the NASCIS III trial showed that all three treatment arms produced comparable degrees of recovery when treatment was begun within the shorter 3-h window. When the 24-h dosing of MP was begun more than 3 h post-SCI recovery was poorer in comparison to the cohort treated within 3 h following SCI. However, in the 3 to 8 h post-SCI cohort, when MP dosing was extended to 48 h significantly better recovery was observed than with the 24-h dosing regimen. In the comparable tirilazad cohort (3–8 h post-SCI) recovery was slightly, but not significantly, better than in the 24-h MP group, and poorer than in the 48-h MP group. These results showed that:

1. initiation of high-dose MP treatment within the first 3 h was optimal
2. the nonglucocorticoid tirilazad was as effective as 24-h MP therapy
3. if treatment is initiated later than 3 h post-SCI, extension of the MP dosing regimen from 24 to 48 h is indicated.

However, in comparison with the 24-h dosing regimen, significantly more glucocorticoid-related immuno-suppression-based side effects were seen with more prolonged dosing (i.e., the incidence of severe sepsis and pneumonia increased). In contrast, tirilazad showed no evidence of steroid-related side effects, suggesting that this nonglucocorticoid 21-aminosteroid would be safer for extension of dosing beyond the 48-h limit used in NASCIS III.[24,41] The ability of tirilazad to improve neurological recovery in NASCIS III at least as well as MP (i.e., treatment within 3 h post-injury) while producing fewer side effects[24,41] strongly suggests that it is worthy of additional trials in acute SCI, which could ultimately show greater efficacy and safety in comparison to high-dose MP.

For a more complete discussion of the neuroprotective pharmacology of high-dose MP therapy in SCI, regulatory status, and limitations of the NASCIS II and III clinical trials (see Hall and Springer).[42]

6.10.6 Unmet Medical Need

All the acute neurological disorders discussed above still represent either largely or completely unmet medical needs. Experimental studies have revealed various targets for neuroprotective therapeutic intervention, and various prototype compounds have been shown to be effective in models of stroke, SAH, TBI, and SCI, supporting the feasibility of achieving successful pharmacological neuroprotection and improved recovery and survival in stroke, TBI, and SCI patients. However, a reasonably convincing demonstration of protective efficacy in the context of clinical trials has for the most part been limited to the development of thrombolytic agents for thromboembolic stroke, the calcium channel blocker nimodipine and the LP inhibitor tirilazad for aneurysmal SAH, and the use of high-dose MP for SCI.

6.10.7 New Research Areas

Despite the disappointments of previous neuroprotective drug development efforts for stroke and TBI, there is still a widely held belief that it is possible to treat these acute neurological conditions. This belief is based on three accomplishments:

1. the identification of past avoidable mistakes in preclinical and clinical trials that should be avoidable in future studies
2. an improved understanding of the intricacies of previously targeted postischemic or posttraumatic secondary injury mechanisms, their time course, and interrelationships
3. the discovery of better and more therapeutically practical neuroprotective drug targets.

The following is a discussion of some of the more promising neuroprotective strategies, some of which are not all that new, and some that are relatively novel.

6.10.7.1 New Concept Concerning Glutamate-Mediated Excitotoxicity

As noted above, a major target of past neuroprotective drug discovery and development has been the glutamate–NMDA receptor complex. The approaches taken have been to block NMDA receptor activation and the opening of its associated calcium channel with various types of compounds that either reduce glutamate release (e.g., riluzole), competitively inhibit glutamate binding to its recognition site (e.g., selfotel), one of its modulatory polyamine (e.g., ifenprodil, eliprodil, CP-101,606) or glycine (e.g., HA-966?) positive regulatory sites, or to noncompetitively block the associated calcium channel (e.g., $MgCl_2$, MK-801, aptiganel). Although these approaches resulted in neuroprotective effects in stroke, TBI, and even SCI models, those that have been the subject of stroke or TBI clinical trials failed to produce a significant effect on recovery or survival. A recent report has shed light on why this may be the case. Biegon *et al.*[43] found in a mouse TBI model using MK-801 autoradiography that, while the glutamate receptor and channel are hyperactivated immediately after injury, this is short lived (<1 h), and is followed by a long-lasting (>7 days) loss of function. Thus, attempts to pharmacologically block glutamate receptor activation beyond the first hour after injury are ineffective due to loss of receptor function and the concomitant loss of their neuroprotective activity. In contrast, administration of NMDA can actually promote posttraumatic neurological recovery if given between 24 h and 2 weeks postinjury, whereas NMDA antagonists actually blunt recovery in mice compared with no treatment. This study, together with the lack of any positive effects in past clinical trials, severely limits enthusiasm for the discovery of new glutamate antagonist compounds.

6.10.7.2 Newer Antioxidant Approaches

The role of ROS generation and the resulting oxidative damage mechanisms in secondary CNS injury is arguably one of the more established aspects of secondary CNS injury, based on findings that a multitude of free radical scavengers and LP-inhibiting antioxidants can reduce postischemic and/or posttraumatic damage in preclinical stroke, TBI, and SCI models (**Figure 2**). Following the discovery and attempted development of PEG-SOD and tirilazad, discussed above, considerable interest shifted to a variety of very potent nitrone-based 'spin-trapping' agents (**Figure 3**). The prototype of this antioxidant class was α-phenyl-*N*-tert-butylnitrone (PBN), which was protective in animal models of stroke and TBI. However, limitations to its solubility led to the incorporation of two sulfonyl groups to improve aqueous solubility, forming NXY-059, which was extensively tested in ischemic and hemorrhagic stroke models followed by entry into clinical trials in ischemic stroke patients. NXY-059 has at least a therapeutic window in rodent and primate MCAO stroke models of at least 4 h, and is currently undergoing a phase III trial.[44] Other nitrones (**Figure 3**), including tempol, the cyclic compound MDL 101,202, and the dual nitrone compound stilbazulenyl nitrone (STAZN), also reduce post-ischemic brain damage in rodent stroke models.[44,45]

During the years since the initiation of neuroprotective clinical trials with agents such as PEG-SOD and tirilazad, a new concept of ROS formation has been identified that appears to offer a more practical antioxidant target. Beginning in 1991, Beckman and co-workers[46] introduced the theory that the principal ROS involved in producing tissue injury in a variety of neurological disorders was peroxynitrite ($ONOO^-$), which is formed by the combination of NOS-generated $^\bullet NO$ radical and superoxide radical.[46] Since that time, the biochemistry of peroxynitrite, which is often referred to as a reactive nitrogen species, has been further clarified. Peroxynitrite-mediated oxidative damage is actually caused by peroxynitrite decomposition products that possess potent free-radical characteristics. These are formed in one of two ways. The first involves the protonation of peroxynitrite to form peroxynitrous acid (ONOOH), which can undergo

PBN

NXY-059

Tempol

MDL 101, 202

Stilbazulenyl nitrone

Figure 3 Chemical structures of nitrone-containing 'spin traps' that have been shown to be neuroprotective in models of CNS injury and/or stroke.

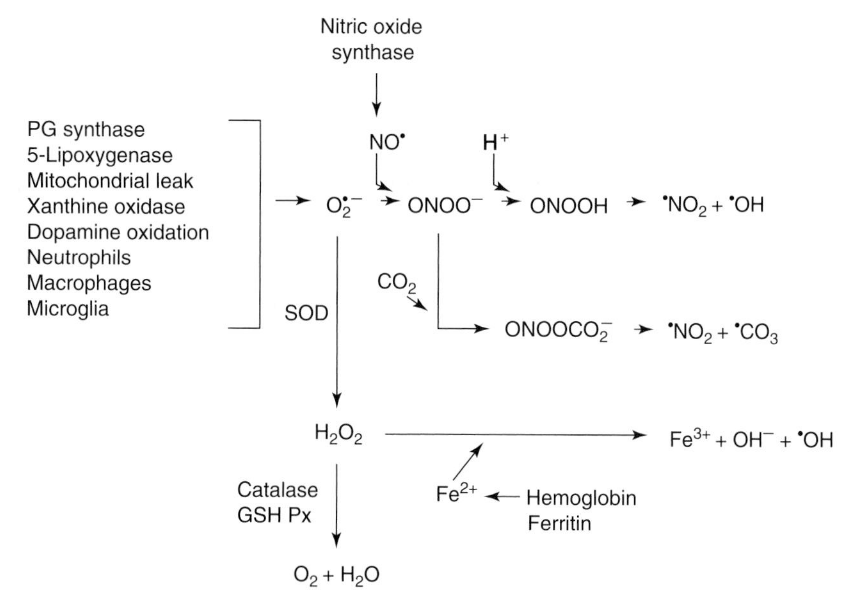

Figure 4 Biochemistry of oxygen radical formation via the peroxynitrite pathway and the dismutation/Fenton reaction pathway. $O_2{-}^{\bullet}$, superoxide radical; $NO^{\bullet}$, nitric oxide or nitrogen monoxide; $ONOO^-$, peroxynitrite anion; ONOOH, peroxynitrous acid; $^{\bullet}NO_2$, nitrogen dioxide; $^{\bullet}OH$, hydroxyl radical; $ONOOCO_2^-$, nitrosoperoxocarbonate; $^{\bullet}CO_3$, carbonate radical; SOD, superoxide dismutase; H_2O_2, hydrogen peroxide; GSH PX, glutathione peroxidase.

hemolytic decomposition to form the highly reactive nitrogen dioxide radical ($^{\bullet}NO_2$) and hydroxyl radical ($^{\bullet}OH$). Probably more important physiologically, peroxynitrite will react with carbon dioxide (CO_2) to form nitrosoperoxocarbonate ($ONOOCO_2$), which decomposes into $^{\bullet}NO_2$ and carbonate radical ($^{\bullet}CO_3$). **Figure 4** summarizes the biochemistry of peroxynitrite generation as well as the fact that dismutation of superoxide radical to hydrogen peroxide via superoxide dismutase (SOD) can also lead to the formation of $^{\bullet}OH$ via the iron-catalyzed Fenton reaction. This latter process undoubtedly plays a role along with peroxynitrite in posttraumatic and postischemic oxidative damage.

Each of the peroxynitrite-derived radicals ($^{\bullet}OH$, $^{\bullet}NO_2$, and $^{\bullet}CO_3$) can initiate lipid peroxidative damage (LP) cellular damage by abstraction of an electron from a hydrogen atom bound to an allylic carbon in polyunsaturated fatty acids, or cause protein carbonylation by reaction with susceptible amino acids (e.g., lysine, cysteine, or arginine). Moreover, the aldehydic LP products malondialdehyde (MDA) and 4-hydroxynonenal (4-HNE) can bind to cellular proteins, compromising their structural and functional integrity. 4-HNE is the more interesting of the two aldehydes in that it is itself neurotoxic. [47] In addition, $^{\bullet}NO_2$ can nitrate the 3 position of tyrosine residues in proteins; 3-NT is a specific footprint of peroxynitrite-induced cellular damage. Increased 3-NT has been found in injured CNS tissue in models of stroke,[48] TBI,[49] and SCI,[50] which is indicative of a role of peroxynitrite in each of these acute insults.

Compared with other short-lived ROS (e.g., H_2O_2) and free-radical species (e.g., superoxide and $^{\bullet}OH$), peroxynitrite possesses a much longer half-life (approximately 1 s), which enables it to diffuse across intracellular and cellular membranes. In addition, the relatively long lifespan of peroxynitrite makes it a more tractable target for successful scavenging. Indeed, several new compound entities have been identified as having the ability to react directly with peroxynitrite, decomposing it into nonreactive products. The chemical structures of several new compound entities that scavenge peroxynitrite, many of which have been shown to have neuroprotective properties in CNS injury animal models, are shown in **Figure 5** and include uric acid,[51] dimethylthiourea (DMTU),[52] the indoleamine melatonin,[53] the pyrrolpyrimidine antioxidant U-101033E,[54] and penicillamine.[55] Perhaps the best characterized with regard to chemical scavenging is penicillamine, which is a stoichiometric scavenger with one molecule of penicillamine decomposing one molecule of peroxynitrite; but penicillamine is also destroyed in the process (see the scavenging mechanism in **Figure 5**). Although such an inefficient scavenging mechanism may be adequate to produce neuroprotective effects, a catalytic scavenging mechanism wherein one molecule of the antioxidant can decompose many molecules of the oxidant without being destroyed would represent a more desirable peroxynitrite scavenging pharmacophore, but one that remains to be identified. Recent studies have demonstrated that the nitroxide new compound entity tempol can catalytically scavenge the peroxynitrite-derived radical species NO_2 and CO_3 (**Figure 6**).[56] Tempol has beneficial effects in rodent TBI, SCI, and stroke models that may be a manifestation of its ability to efficiently scavenge highly reactive peroxynitrite-derived radicals. Other nitroxides may share this property, although the structure–activity relationship for this potent and efficient radical scavenging property remains to be established.

Perhaps an even better antioxidant approach would be to combine two mechanistically additive or complementary antioxidant pharmacophores into a single new compound entity, such as a dual LP inhibitor or a LP inhibitor coupled

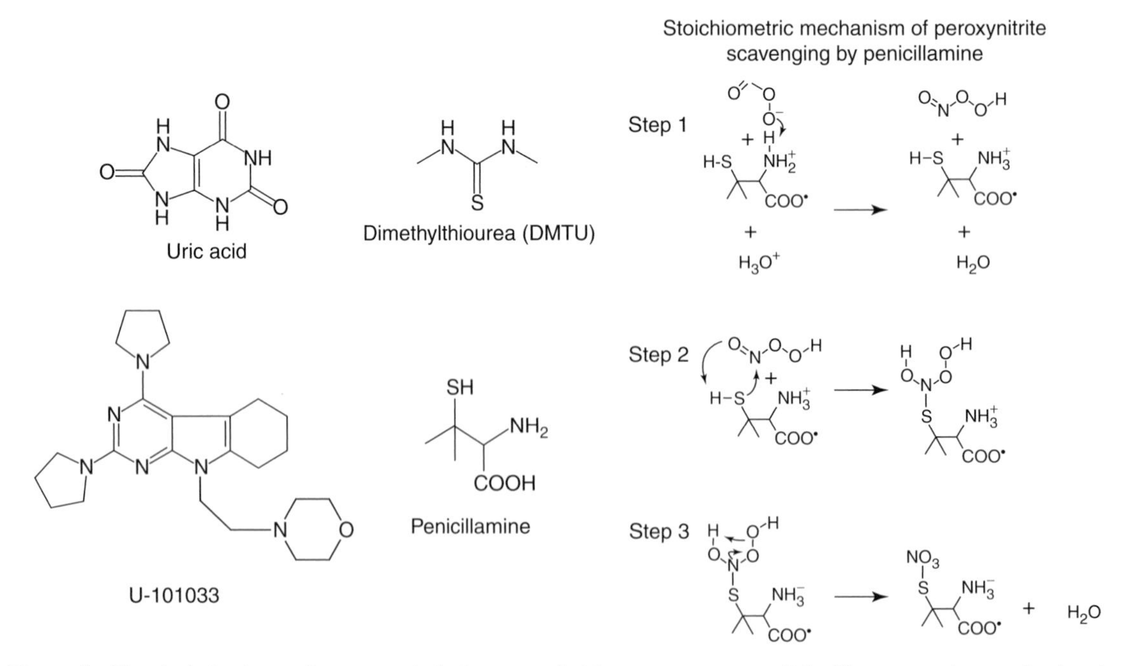

Figure 5 Chemical structures of compounds that are reported to scavenge peroxynitrite. The scavenging mechanism for penicillamine is shown.

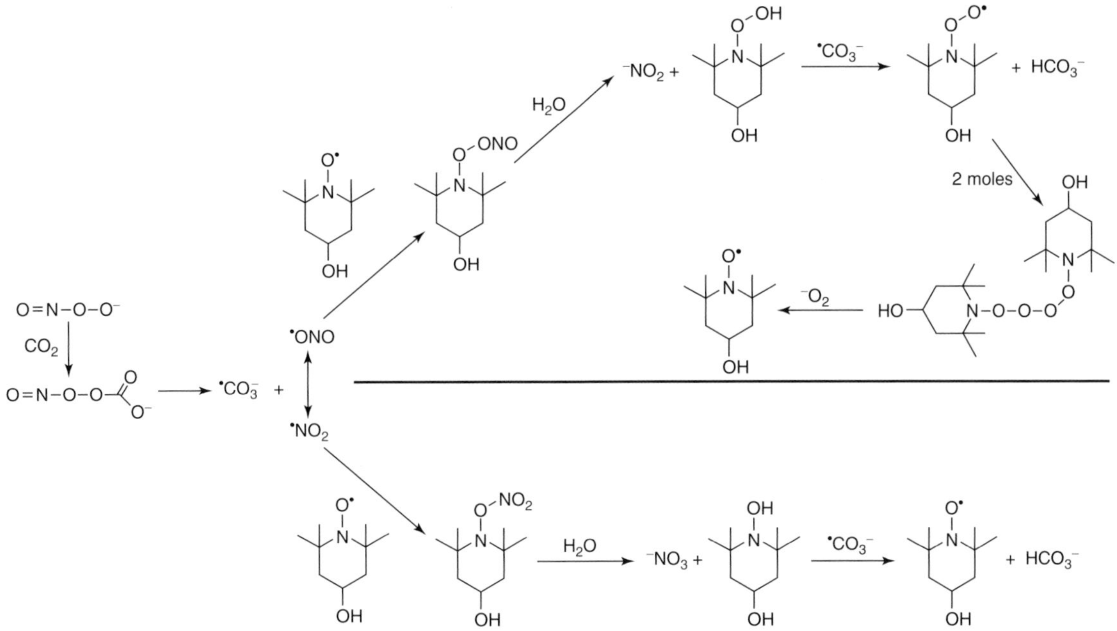

Figure 6 Mechanism by which the nitroxide-containing tempol can catalytically scavenge $^{\bullet}NO_2$ and $^{\bullet}CO_3$. (Modified from Carroll, R. T.; Galatsis, P.; Borosky, S.; Kopec, K. K.; Kumar, V.; Althaus, J. S.; Hall, E. D. *Chem. Res. Toxicol.* **2000**, *13*, 294–300.)

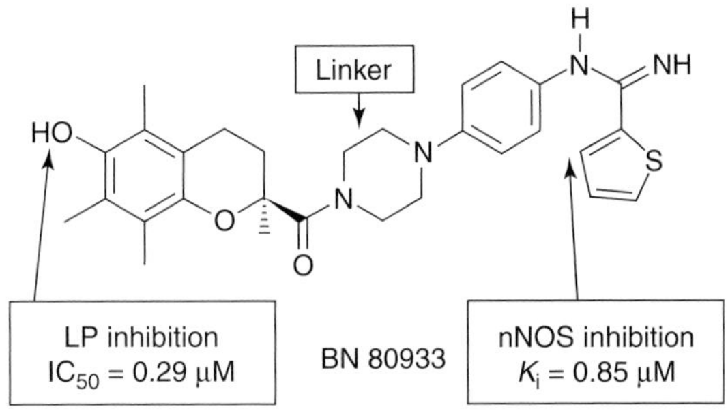

Figure 7 Chemical structure of BN 80933, a dual inhibitor of lipid peroxidation and neuronal nitric oxide synthase (nNOS).[58] The peroxidation-inhibiting chroman moiety works by donating an electron from the hydroxyl group to a lipid peroxyl radical.

with a moiety that will either inhibit $^{\bullet}NO$ production or scavenger peroxynitrite or peroxynitrite-derived radicals. Two examples of this approach are the 2-methylaminochroman U-78517F[57] (**Figure 2**) and BN-80933[58] (**Figure 7**). These new compound entities link the peroxyl radical scavenging chroman ring structure of vitamin E (**Figure 2**) via a piperazine bridge to the peroxyl radical scavenging amine moiety of tirilazad in the case of U-78517F, or with the an inhibitor nitric oxide synthase (NOS) in the case of BN-80933, which would decrease production of $^{\bullet}NO$ and consequently reduce peroxynitrite formation. U-78517F is more effective than either vitamin E or the vitamin E chroman in protecting culture spinal cord neurons from oxidative damage.[57] Similarly, BN-80933 has greater neuroprotective activity than other single antioxidant mechanism new compound entities in rodent stroke and TBI models than either the trolox or NOS-inhibiting moieties alone.[58]

6.10.7.3 Prevention of Mitochondrial Dysfunction

Mitochondrial dysfunction has been increasingly recognized as a key player in the postischemic and posttraumatic secondary injury.[59,60] It is clear that this is directly related to Ca^{2+} ions, which alter mitochondrial function and increase ROS production. For example, loss of mitochondrial homeostasis, increased mitochondrial ROS production, and disruption of synaptic homeostasis occur after TBI, implicating a pivotal role for mitochondrial dysfunction in the neuropathological sequalae that follow the mechanical trauma. This theory has been solidified by the demonstration that therapeutic intervention with the immunosuppressant, cyclosporine A (CsA) following experimental TBI reduces mitochondrial dysfunction and cortical damage, as well as cytoskeletal changes and axonal dysfunction.[61–63] These neuroprotective effects of CsA result from the ability of the drug to bind to cyclophilin D, thus preventing binding of the latter to the adenine nucleotide translocator protein (ANT), blocking the interaction of ANT with the mitochondrial permeability transition (MPT) pore. In the absence of cyclophilin D/ANT binding to the MPT pore, the MPT pore cannot open and MPT cannot occur; the mitochondrion is protected from a catastrophic loss of its membrane potential ($\Delta\psi$) and metabolic failure and the neuronal energy metabolism (ATP generation) is preserved. Cyclosporine A is currently in phase II clinical trials in TBI. However, its neuroprotective efficacy, based on its mitochondrial protective mechanism, is potentially compromised by its immunosuppressive properties and the well-known ability to elicit renal and/or CNS toxicity, effects related to the ability of CsA to bind calcineurin. To circumvent these properties, the *N*-methylleucine residue in position 3 of CsA has been converted in the recently described new compound entity (NIM811) to an *N*-methylisoleucine, which eliminates calcineurin binding and its associated immunosuppressive and toxic properties.[64] Recent studies have demonstrated that NIM811 can replicate the ability of CsA to protect brain and spinal cord mitochondrial function in TBI and SCI models, respectively.[65]

Evidence has begun to accumulate that the particular ROS formed by mitochondria is peroxynitrite. NO is present in mitochondria, and a mitochondrial NOS isoform (mtNOS) has been isolated, Although probably playing a key physiological role in mitochondria, dysregulation of mitochondrial $^{\bullet}$NO generation and the aberrant production of the toxic metabolite peroxynitrite appear to play a role in many, if not all, the major acute and chronic neurodegenerative conditions.[66] Exposure of mitochondria to Ca^{2+}, which renders them dysfunctional, results in peroxynitrite generation, which in turn triggers mitochondrial Ca^{2+} release (i.e., limits their Ca^{2+} uptake or buffering capacity).[67] Both peroxynitrite forms, $ONOO^-$ and $ONOOCO_2$, deplete mitochondrial antioxidant stores and cause protein nitration. The relatively long half-life of peroxynitrite in comparison to other short-lived ROS also allows for mtNOS-derived peroxynitrite to diffuse between cells.

Figure 8 illustrates the formation of peroxynitrite in mitochondria during ischemic or traumatic insults. As shown, O_2^- radical production is a byproduct of the mitochondrial electron transport chain during ATP generation. Electrons escape from the chain and reduce O_2 to O_2^-. Normally, cells convert O_2^- to H_2O_2, utilizing manganese superoxide dismutase (MnSOD), which is also localized to the mitochondria. However, if pathophysiological insults (e.g., mechanical trauma) trigger an increase in intracellular Ca^{2+}, causing an increase in mtNOS activity and $^{\bullet}$NO

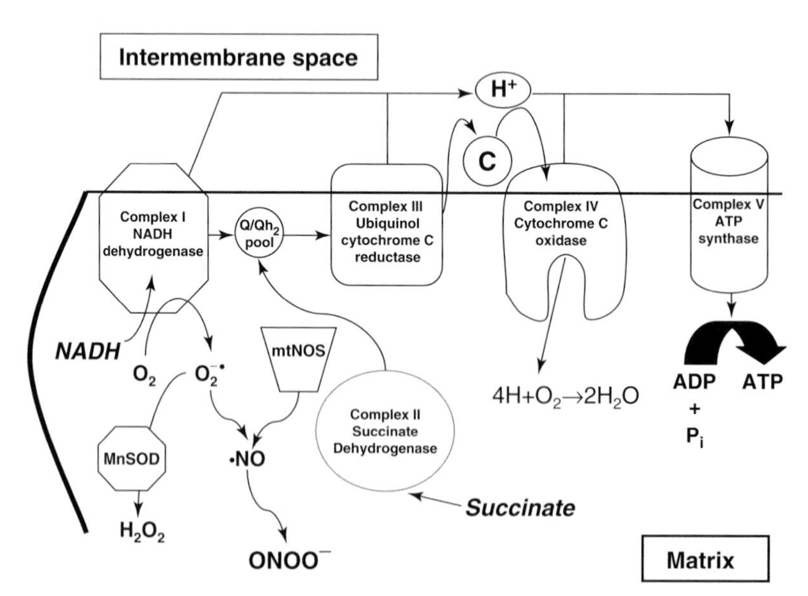

Figure 8 Schematic diagram of the mitochondrial electron-transport chain, showing the site of formation of superoxide (O_2^-) and nitric oxide ($^{\bullet}$NO). These two radicals interact to form peroxynitrite anion ($ONOO^-$). The rate constant for this reaction is reported to be $1.6 \times 10^{10} M^{-1} s^{-1}$, as reported by Nauser and Koppenol.[68] This is faster than the frequently reported rate constant for the dismutation of superoxide by the mitochondrial manganese superoxide dismutase (MnSOD), which is only $1.4 \times 10^9 M^{-1} sc^{-1}$. Thus the formation of peroxynitrite is favored.

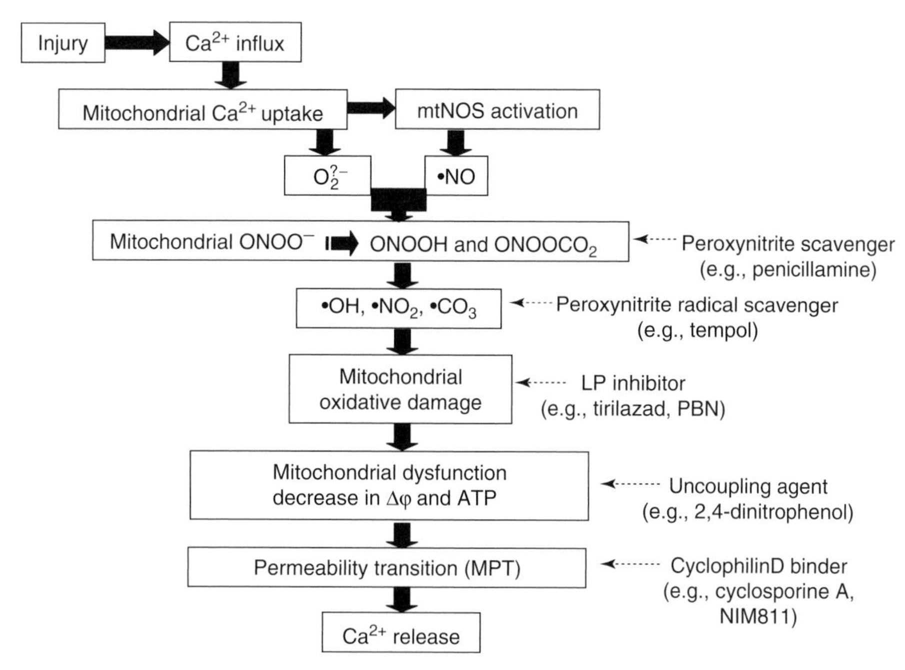

Figure 9 Schematic diagram of the sequence of events occurring in posttraumatic or postischemic mitochondrial damage and possible intervention points for neuroprotective therapy.

liberation, peroxynitrite formation is a certainty, since the rate constant for the reaction of $^{\bullet}NO$ with O_2^{-} greatly exceeds the rate constant for dismutation of O_2^{-} by MnSOD.[68] Peroxynitrite then damages mitochondria by tyrosine nitration and by causing LP and the production of 4-HNE that conjugates to mitochondrial membrane proteins, impairing their function. An efficient scavenger of peroxynitrite, or its derived radical species, would be expected to protect mitochondrial function from oxidative damage.

Another approach to mitochondrial protection involves the use of pharmacological 'uncoupling' agents that facilitate the proton movement from the mitochondrial inner membrane space into the mitochondrial matrix. This results in the pumping of protons out of the matrix via the electron-transport system, being 'uncoupled' from proton flow into the matrix during the operation of the ATP synthase. Consequently, the membrane potential $\Delta\psi$ is reduced, which in turn decreases mitochondrial calcium uptake and ROS production, which are linked to $\Delta\psi$. In support of this strategy, the well-known mitochondrial uncoupler 2,4-dinitrophenol protects mitochondrial function and exerts a neuroprotective effect in rodent models of stroke, TBI, and SCI.[60] However, the goal of this approach is to achieve a modest level of uncoupling and only slightly decreasing $\Delta\psi$ from a normal level of 160 mV down to 140 mV. Any greater decrease in membrane potential would decrease ATP synthesis and lead to mitochondrial failure. The currently available uncouplers, such as 2,4-DNP, although sufficient for proof of concept, have a sharp, U-shaped neuroprotective dose–response curve due to their ability to move readily from the desired 'modest uncoupling' mode to a level of severe uncoupling (decrease in $\Delta\psi$ to within the range 140–100 mV or lower). Whether it is possible to discover safer uncoupling compounds that will only be capable of decreasing $\Delta\psi$ down to a neuroprotective level, and not beyond, remains to be demonstrated. The search, however, appears worthwhile, since the magnitude of mild uncoupling-induced neuroprotection in rodent CNS injury models is truly impressive, and this mechanistic strategy is equally applicable to stroke, TBI, and SCI. **Figure 9** summarizes the different steps involved in postischemic or posttraumatic mitochondrial failure and possible points of pharmacological intervention.

6.10.7.4 Inhibition of Calpain-Mediated Proteolytic Damage

As noted earlier in this chapter, disruption of intracellular Ca^{2+} homeostasis is a critical issue in the secondary neurodegenerative pathophysiology of stroke, TBI, and SCI. Following an injurious event, there is a massive posttraumatic Ca^{2+} influx, initially caused by depolarization-induced glutamate release and the opening of glutamate receptor-operated and voltage-dependent Ca^{2+} channels.[3,4] Increased intracellular Ca^{2+} concentrations in TBI models have been demonstrated in each of the mainstream rodent TBI models.[5] Excessive intracellular Ca^{2+}

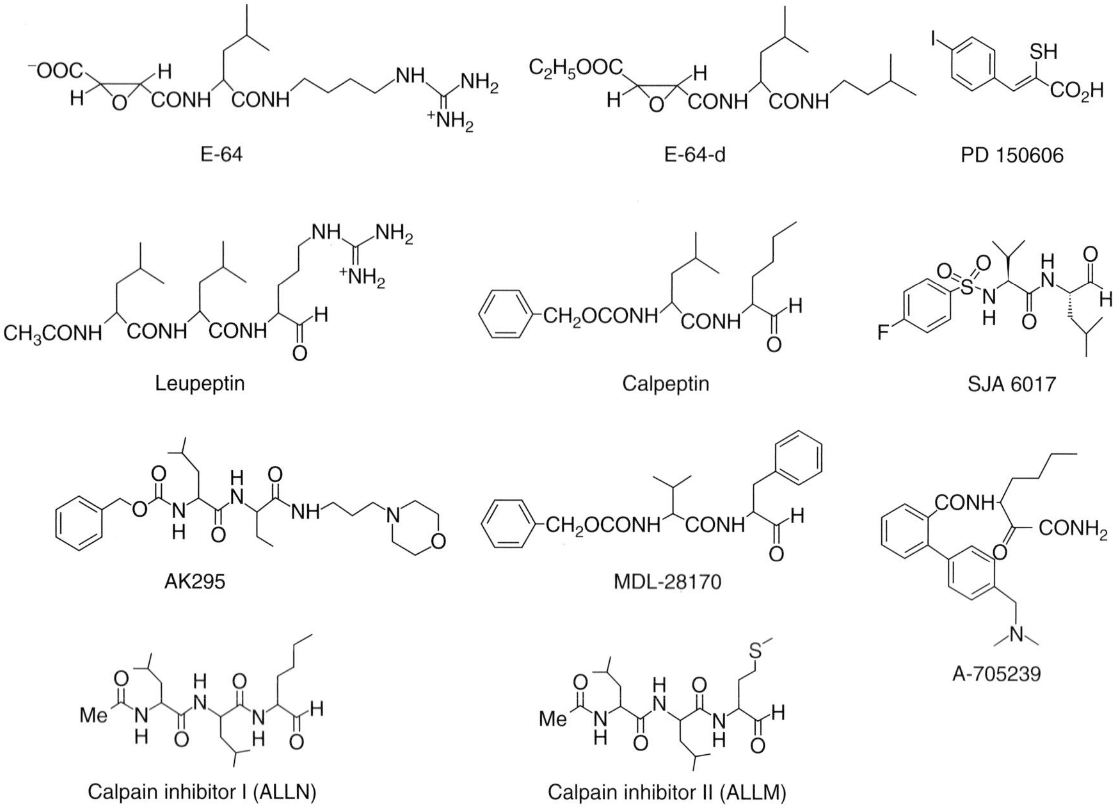

Figure 10 Chemical structures of a variety of calpain inhibitors, most of which have been shown to be neuroprotective in in vitro or in vivo models of neural injury.

accumulation then leads to neuronal degeneration by enzyme activation, including proteases, kinases, phosphatases, and phospholipases, and induction of additional ROS release.[5]

The principal mechanism of posttraumatic Ca^{2+}-mediated secondary neuronal injury involves the activation of the neutral proteases known as calpains.[69,70] When activated, calpains degrade cytoskeletal proteins, receptor proteins, signal transduction enzymes, and transcription factors. In the case of cytoskeleton proteins, α-spectrin (a 280 kDa protein that provides structural support to membranes) can be cleaved by calpain at tyrosine 1176 to yield a 150 kDa fragment (SBDP150), or at glycine 1230 to yield a 145 kDa fragment (SBDP145).[71] This is consistent with the notion that cytoskeletal damage is the final common mechanism of posttraumatic neurodegeneration. Therefore, its inhibition is a meaningful biochemical measure of the efficacy of upstream neuroprotective strategies. A key role of calpain activation in mediating posttraumatic axonal damage in acute SCI models has been demonstrated using prototypical calpain inhibitors, including leupeptin, antipain, calpain inhibitor I and II, calpeptin, E64, AK295, MDL28170, SJA6017, and PD150606, which have neuroprotective effects (**Figure 10**).[70] However, the potential translation of calpain inhibition into neuroprotective clinical trials has been limited by a lack of small-molecule inhibitors with sufficient CNS penetration and appropriate pharmaceutical and pharmacokinetic properties for use as drugs. Nevertheless, newer ketoamide calpain inhibitors with improved membrane penetration and water solubility have been discovered, and these need to be evaluated in stroke and CNS injury models (**Figure 10**).[72,73]

6.10.7.5 Anti-Inflammatory Strategies

As indicated in **Figure 11**, multiple inflammatory processes play a role in secondary injury and perhaps repair in stroke, TBI, and SCI. First of all, the arachidonic acid cascade is activated, resulting in the formation of various deleterious prostanoids and leukotrienes. COX-1 and -2 inhibitors can decrease the formation of $PGF_{2\alpha}$ and TXA_2, and have neuroprotective effects.[42]

In addition to this acute inflammatory mechanism initiated in the first minutes of the insult, a more delayed microglial and astrocyte activation and influx of neutrophils, monocytes, and T lymphocytes has been documented to

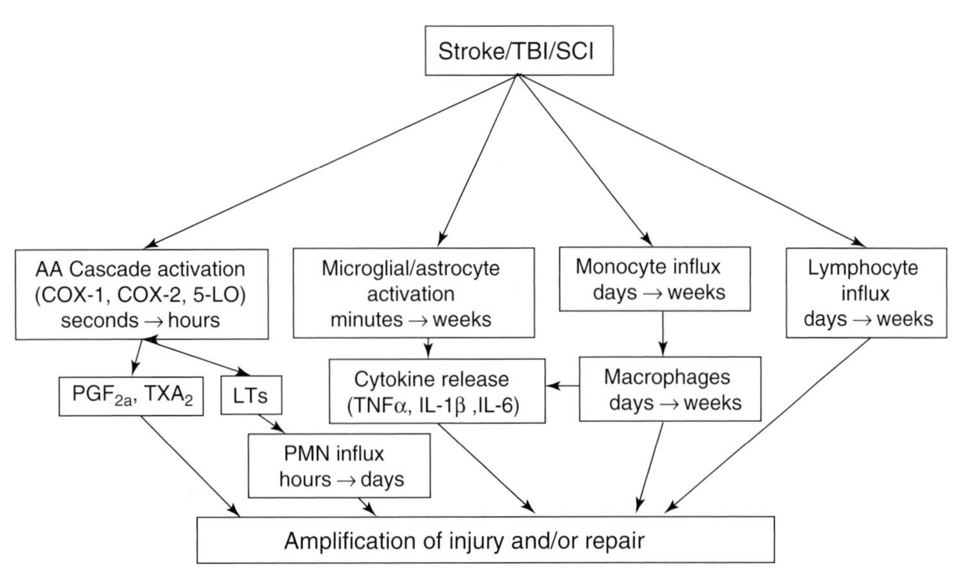

Figure 11 Schematic diagram showing the key inflammatory events that take place in the injured CNS that either amplify secondary injury or are involved in repair mechanisms.

begin within the first hour. Some of this cellular inflammatory response appears to continue for weeks and perhaps months, and may play a role in both secondary damage and repair processes.[74] Considerable evidence exists to support the idea that much of this cellular inflammation can amplify the secondary injury process, and therefore should be inhibited via anti-inflammatory, immunosuppressive, or antiproliferative agents. With regard to the latter mechanistic possibility, a recent study in a rodent TBI model has shown that there is a considerable posttraumatic increase in cell cycle proteins, many of which have kinase activities.[75] In the same study, the administration of the cell cycle inhibitor flavoperidol decreased posttraumatic elevations in neuronal and astrocyte cyclin D1 expression, reduced neuronal damage and astrocyte proliferation, and improved neurological recovery.

In the context of SCI models, the second-generation tetracycline derivative minocycline is neuroprotective and improves neurological recovery, together with suppressing microglial activation.[76,77] However, there is ongoing controversy as to whether the cellular inflammation is more bad than good for the injured spinal cord, or vice versa. Some believe that enhancement of cellular inflammation begun after the first 24 h may act to stimulate repair processes.[78]

6.10.7.6 Inhibition of Apoptotic Neurodegeneration

Most of the therapeutic strategies discussed thus far are mainly associated with inhibition of 'necrotic' or 'oncotic' CNS neuronal, glial, and microvascular damage after ischemic or traumatic insults. On the other hand, a potentially promising area of neuroprotective drug discovery for treating ischemic and traumatic CNS injury is the development of agents targeting apoptotic cell death. Although apoptosis probably plays some role in the acute phase (first 24–48 h) after ischemic or traumatic insults, a number of studies have documented that apoptosis is perhaps the dominant mechanism of ongoing cell loss over the days, weeks, and possibly months following CNS injury. For example, apoptotic cell death can be detected hours to several weeks following SCI, and occurs in numerous cell types, including neurons, oligodendroglia, and inflammatory cells such as neutrophils, microglia, and macrophages.[79]

A major factor and potential neuroprotective drug target involved in the intracellular apoptotic cascade is the activation of the cysteine protease caspase-3, which appears to be a common endpoint regardless of the initiating stimulus for apoptosis.[80] Indeed, caspase-3 activation has been demonstrated in all types of CNS injury models, including those for TBI, SCI, and stroke.[81] The biochemical pathway(s) leading to caspase-3 activation have been extensively studied using in vitro models of apoptotic cell death, and it has been well documented that the mitochondria are thought to play a critical, if not essential, role in caspase-3 activation. Apoptotic signals reaching mitochondria result in the release of cytochrome c and Smac/DIABLO[82]; the former then interacts with Apaf-1 (apoptosis protease activating factor-1) to promote the caspase-9 activation, an upstream activator of caspase-3. Smac/DIABLO promotes capsase-9 and caspase-3 activation by removing the inhibitory influence of members of the IAPs (inhibitor of apoptotic proteins) family.[83] In addition, the mitochondria can also release pro-apoptotic molecules, such as endonuclease g and apoptosis-inducing factor (AIF), which are thought to function independently of caspase activation.[84] Therefore, a number of signaling events associated with mitochondria regulate activation of a caspase-dependent and -independent apoptotic cell-death programs. Targeting these signaling events at the mitochondrial level, in addition to upstream apoptotic signaling, will be essential in determining the contribution of this cell-death process to neurological dysfunction in stroke, TBI, and SCI.

To date, it is evident that a clear understanding of the extracellular events leading to transduction of an apoptotic signal in the injured CNS is lacking. This is an important research area, as inhibiting the extracellular signals as far upstream as possible should limit postischemic or posttraumatic apoptosis to a greater degree than targeting the intracellular pathway(s). Several studies have implicated potential candidate molecules as extracellular apoptotic signals in the injured CNS, including pro-inflammatory cytokines, certain growth factors, Fas ligand, and glutamate. There is additional evidence that ROS (e.g., superoxide radical and peroxynitrite) can also lead to apoptosis. What is clear from this limited list is that, although caspase-3 can be considered as the common endpoint in the apoptotic biochemical cascade, there are a number of potentially interruptable extracellular signals that could contribute to

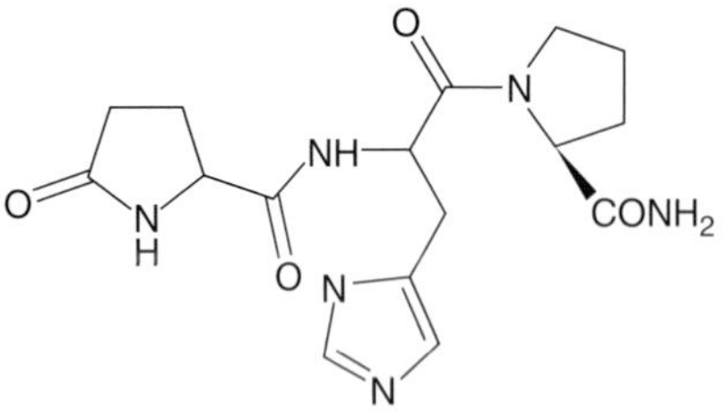

TRH – endocrine, analeptic, autonomic, and neuroprotection

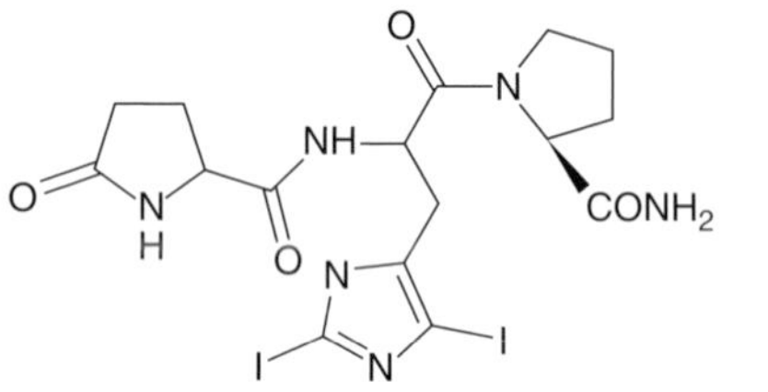

2,4-diiodo-TRH - neuroprotection only

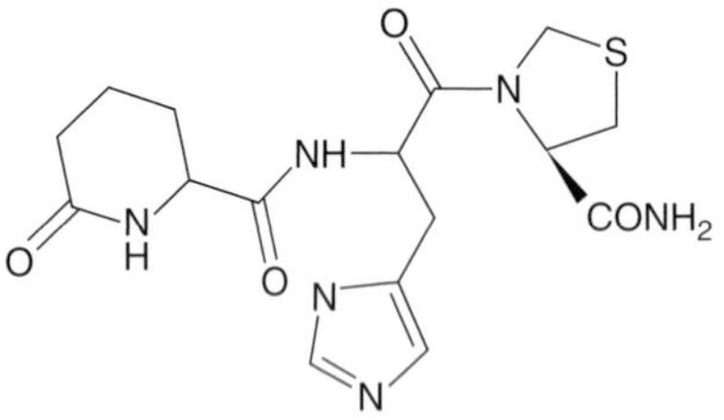

MK-771 – endocrine, analeptic, autonomic, but no neuroprotection

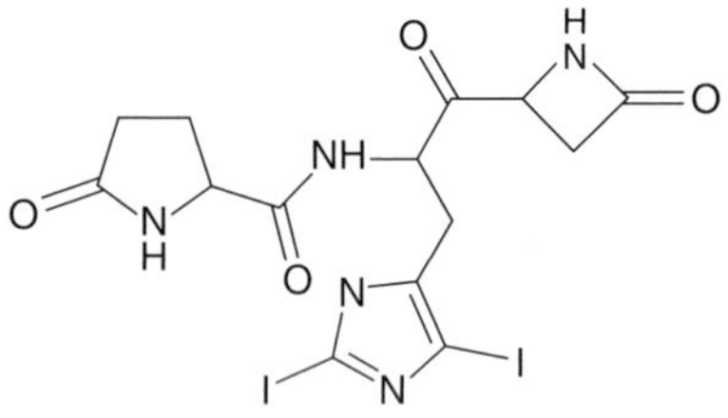

Compound 53a - neuroprotection only

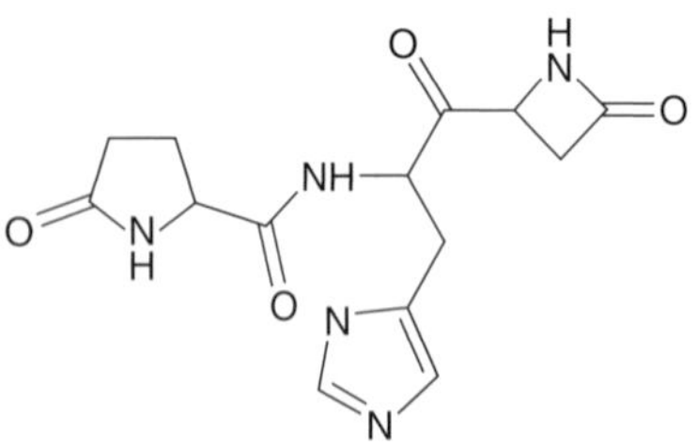

YM-14673 – endocrine, analeptic, autonomic, and neuroprotection

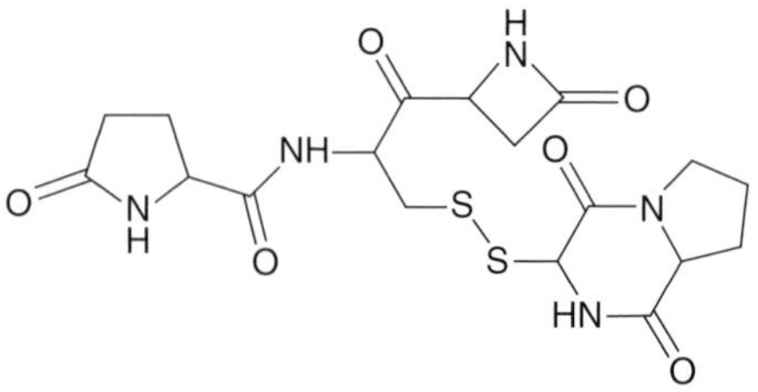

Compound 57a - neuroprotection only

Figure 12 Chemical structures of the tripeptide thyrotropin-releasing hormone (TRH) and analog. TRH and the synthetic analog YM-14673 possesses endocrine, analeptic, autonomic, and neuroprotective effects. The neuroprotective effects of TRH are lost in the case of MK-771, which retains the endocrine, analeptic, and autonomic properties. In contrast, the diiodoimidazle analogs of TRH (2,4-diiodo-TRH) and YM-14673 (Compound 53a), and diketopiperazine-containing Compound 57A are selective for neuroprotective effects of TRH only.[85]

transduction of the apoptotic signal. Therefore, it will be important to determine the contribution of each of these different extracellular events to the overall apoptotic cell loss that occurs in stroke and trauma models.

Exploration of the potential of blocking the apoptotic cascade as a neuroprotective strategy has been limited by a lack of small-molecule inhibitors of selected points in the apoptotic cascade that have drug-like pharmaceutical properties and good CNS penetrability. Current exploratory strategies involve the use of competitive caspase inhibitors containing tri- and tetrapeptide amino acid sequences that are preferred by certain caspases. While these inhibitors have been modified to enhance cell permeability and stability, the delivery of sufficient levels to the cells of interest over a broad time window and to areas distant to the site of injury may be very limited. Thus, evidence of their efficacy has been limited primarily to in vitro studies. In addition, the short half-life of these peptide-based compounds (e.g., the caspase-3 inhibitor *z*-DEVD-fmk) may reduce their effective concentration. For example, the presence of numerous proteases released in the injured brain or spinal cord could certainly limit the amounts of intact peptide that ultimately cross the blood–brain barrier. Nevertheless, these compounds are effective in models of CNS injury.[81] There are several unanswered concerns regarding the pharmacological inhibition of apoptosis in the injured CNS. Firstly, will inhibition of caspase-dependent apoptosis shift the mode of death to caspase-*in*dependent apoptosis of necrosis? Secondly, is caspase activation a normal physiological process that plays a role in cellular functions such as cytoskeletal rearrangement? Thirdly, might caspase inhibitors lead to the survival of dysfunctional cells? Further study with improved anti-apoptotic new compound entities will be required to address these questions.

6.10.7.7 Thyrotropin-Releasing Hormone Analogs

It has long been known that tripeptide thyrotropin-releasing hormone (TRH) and certain of its analogs (**Figure 12**) have neuroprotective properties in a variety of CNS injury models. However, their use has been hindered by the fact that TRH and many of related compounds have analeptic, autonomic, and endocrine effects that preclude their full exploration. More recently a series of diiodoimidazole analogs (e.g., 2,4-diiodo-TRH and Compound 53a) and a series of cyclized dipeptides (diketopiperazines; Compound 57a), both of which are selective for neuroprotective effects and have less potential for the unwanted properties of the TRH, have been identified.[85]

6.10.7.8 Neurorestorative Drug Discovery

Until a decade ago, it was firmly believed that once the brain or spinal cord was damaged by the acute and subacute secondary injury process there was little, if any, capability for regeneration of axons or the formation of new synapses. However, over the last several years it has been discovered that CNS neurons are indeed capable of significant structural and functional regeneration and repair. These processes might be pharmacologically enhanced, either by reawakening the growth potential of the surviving neurons, or by antagonizing the multiple inhibitory factors that have been discovered, the activity of which is aimed at inhibiting axonal growth and synaptogenesis. Several pharmacological mechanisms have been identified that can be targeted to try to enhance the function and/or structural plasticity of neuronal pathways that survive the ravages of postischemic or posttraumatic secondary injury.[86–89]

References

1. Anderson, D. K.; Hall, E. D. *Ann. Emerg. Med.* **1993**, *22*, 987–992.
2. Hall, E. D.; Braughler, J. M. *Res. Publ. Assoc. Res. Nerv. Ment. Dis.* **1993**, *71*, 81–105.
3. Hall, E. In *Neurosurgery 96: Manual of Neurosurgery*; Palmer, J., Ed.; Churchill-Livingstone: New York, 1995, pp 505–510.
4. McIntosh, T. K.; Smith, D. H.; Meaney, D. F.; Kotapka, M. J.; Gennarelli, T. A.; Graham, D. I. *Lab. Invest.* **1996**, *74*, 315–342.
5. McIntosh, T.; Saatman, K. E.; Raghupathi, R. *Neuroscientist* **1997**, *3*, 169–175.
6. Braughler, J. M.; Hall, E. D. *Free Rad. Biol. Med.* **1989**, *6*, 289–301.
7. Hall, E. D.; Braughler, J. M. *Free Rad. Biol. Med.* **1989**, *6*, 303–313.
8. Ginsberg, M. D.; Busto, R. In *Cerebrovascular Disease: Pathophysiology, Diagnosis and Management*; Ginsberg, M. D., Bogousslavsky, J., Eds.; Blackwell Science: Malden, MA, 1998; Vol. 1, pp 14–35.
9. Kline, A.; Dixon, C. In *Head Trauma: Basic, Preclinical and Clinical Directions*; Miller, L., Hayes, R., Eds.; Wiley-Liss: New York, 2001, pp 65–84.
10. Hall, E. D.; Sullivan, P. G.; Gibson, T. R.; Pavel, K. M.; Thompson, B. M.; Scheff, S. W. *J. Neurotrauma* **2005**, *22*, 252–265.
11. Miller, J. D.; Bullock, R.; Graham, D. I.; Chen, M. H.; Teasdale, G. M. *Neurosurgery* **1990**, *27*, 433–439.
12. Basso, D. M.; Beattie, M. S.; Bresnahan, J. C. *Exp. Neurol.* **1996**, *139*, 244–256.
13. Scheff, S. W.; Rabchevsky, A. G.; Fugaccia, I.; Main, J. A.; Lumpp, J. E. *J. Neurotrauma* **2003**, *20*, 179–193.
14. Basso, D. M.; Beattie, M. S.; Bresnahan, J. C. *J. Neurotrauma* **1995**, *12*, 1–21.
15. Cheng, Y. D.; Al-Khoury, L.; Zivin, J. A. *NeuroRx* **2004**, *1*, 36–45.
16. Labiche, L. A.; Grotta, J. C. *NeuroRx* **2004**, *1*, 46–70.
17. Murray, G. D.; Teasdale, G. M.; Schmitz, H. *Acta Neurochir. (Wien)* **1996**, *138*, 1163–1167.
18. Muizelaar, J. P.; Kupiec, J. W.; Rapp, L. A. *J. Neurosurg.* **1995**, *83*, 942.

19. Narayan, R. K.; Michel, M. E.; Ansell, B.; Baethmann, A.; Biegon, A.; Bracken, M. B.; Bullock, M. R.; Choi, S. C.; Clifton, G. L.; Contant, C. F. et al. *J. Neurotrauma* **2002**, *19*, 503–557.
20. Hall, E. D.; McCall, J. M.; Means, E. D. *Adv. Pharmacol.* **1994**, *28*, 221–268.
21. Hall, E. D. *Neuroscientist* **1997**, *3*, 42–51.
22. Marshall, L. F.; Maas, A. I.; Marshall, S. B.; Bricolo, A.; Fearnside, M.; Iannotti, F.; Klauber, M. R.; Lagarrigue, J.; Lobato, R.; Persson, L. et al. *J. Neurosurg.* **1998**, *89*, 519–525.
23. Kassell, N. F.; Haley, E. C.; Apperson-Hansen, C.; Alves, W. M. *J. Neurosurg.* **1996**, *84*, 221–228.
24. Bracken, M. B.; Shepard, M. J.; Holford, T. R.; Leo-Summers, L.; Aldrich, E. F.; Fazl, M.; Fehlings, M.; Herr, D. L.; Hitchon, P. W.; Marshall, L. F. et al. *JAMA* **1997**, *277*, 1597–1604.
25. Lanzino, G.; Kassell, N. F. *J. Neurosurg.* **1999**, *90*, 1018–1024.
26. Lanzino, G.; Kassell, N. F.; Dorsch, N. W.; Pasqualin, A.; Brandt, L.; Schmiedek, P.; Truskowski, L. L.; Alves, W. M. *J. Neurosurg.* **1999**, *90*, 1011–1017.
27. Wang, C. X.; Shuaib, A. *Int. J. Clin. Pract.* **2004**, *58*, 964–969.
28. Biegon, A. *Curr. Pharm. Des.* **2004**, *10*, 2177–2183.
29. Knoller, N.; Levi, L.; Shoshan, I.; Reichenthal, E.; Razon, N.; Rappaport, Z. H.; Biegon, A. *Crit. Care Med.* **2002**, *30*, 548–554.
30. Pratico, D.; Reiss, P.; Tang, L. X.; Sung, S.; Rokach, J.; McIntosh, T. K. *J. Neurochem.* **2002**, *80*, 894–898.
31. Morrow, J. D.; Roberts, L. J. *Am. J. Respir. Crit. Care Med.* **2002**, *166*, S25–S30.
32. Pike, B. R.; Flint, J.; Dutta, S.; Johnson, E.; Wang, K. K.; Hayes, R. L. *J. Neurochem.* **2001**, *78*, 1297–1306.
33. Wunderlich, M. T. *J. Neurol. Neurosurg. Psychiatry* **2003**, *74*, 827–828, author reply 828.
34. Pelinka, L. E.; Toegel, E.; Mauritz, W.; Redl, H. *Shock* **2003**, *19*, 195–200.
35. Kahn, J. H.; Viereck, J.; Kase, C.; Jeerakathil, T.; Romero, R.; Mehta, S. D.; Kociol, R.; Babikian, V. *J. Emerg. Med.* **2005**, *29*, 273–277.
36. Molina, C. A.; Saver, J. L. *Stroke* **2005**, *36*, 2311–2320.
37. Bracken, M. B.; Collins, W. F.; Freeman, D. F.; Shepard, M. J.; Wagner, F. W.; Silten, R. M.; Hellenbrand, K. G.; Ransohoff, J.; Hunt, W. E.; Perot, P. L., Jr. et al. *JAMA* **1984**, *251*, 45–52.
38. Bracken, M. B.; Shepard, M. J.; Hellenbrand, K. G.; Collins, W. F.; Leo, L. S.; Freeman, D. F.; Wagner, F. C.; Flamm, E. S.; Eisenberg, H. M.; Goodman, J. H. et al. *J. Neurosurg.* **1985**, *63*, 704–713.
39. Hall, E. D. *J. Neurosurg.* **1992**, *76*, 13–22.
40. Bracken, M. B.; Shepard, M. J.; Collins, W. F.; Holford, T. R.; Young, W.; Baskin, D. S.; Eisenberg, H. M.; Flamm, E.; Leo-Summers, L.; Maroon, J. et al. *N. Engl. J. Med.* **1990**, *322*, 1405–1411.
41. Bracken, M. B.; Shepard, M. J.; Holford, T. R.; Leo-Summers, L.; Aldrich, E. F.; Fazl, M.; Fehlings, M. G.; Herr, D. L.; Hitchon, P. W.; Marshall, L. F. et al. *J. Neurosurg.* **1998**, *89*, 699–706.
42. Hall, E.; Springer, J. E. *NeuroRx* **2004**, *1*, 80–100.
43. Biegon, A.; Fry, P. A.; Paden, C. M.; Alexandrovich, A.; Tsenter, J.; Shohami, E. *Proc. Natl. Acad. Sci. USA* **2004**, *101*, 5117–5122.
44. Green, A. R.; Ashwood, T.; Odergren, T.; Jackson, D. M. *Pharmacol. Ther.* **2003**, *100*, 195–214.
45. Ginsberg, M. D.; Becker, D. A.; Busto, R.; Belayev, A.; Zhang, Y.; Khoutorova, L.; Ley, J. J.; Zhao, W.; Belayev, L. *Ann. Neurol.* **2003**, *54*, 330–342.
46. Beckman, J. S. *J. Dev. Physiol.* **1991**, *15*, 53–59.
47. Kruman, I.; Bruce-Keller, A. J.; Bredesen, D.; Waeg, G.; Mattson, M. P. *J. Neurosci.* **1997**, *17*, 5089–5100.
48. Eliasson, M. J.; Huang, Z.; Ferrante, R. J.; Sasamata, M.; Molliver, M. E.; Snyder, S. H.; Moskowitz, M. A. *J. Neurosci.* **1999**, *19*, 5910–5918.
49. Hall, E. D.; Detloff, M. R.; Johnson, K.; Kupina, N. C. *J. Neurotrauma* **2004**, *21*, 9–20.
50. Bao, F.; Liu, D. *Neuroscience* **2003**, *116*, 59–70.
51. Scott, G. S.; Cuzzocrea, S.; Genovese, T.; Koprowski, H.; Hooper, D. C. *Proc. Natl. Acad. Sci. USA* **2005**, *102*, 3483–3488.
52. Whiteman, M.; Halliwell, B. *Free Rad. Biol. Med.* **1997**, *22*, 1309–1312.
53. Blanchard, B.; Pompon, D.; Ducrocq, C. *J. Pineal Res.* **2000**, *29*, 184–192.
54. Rohn, T. T.; Quinn, M. T. *Eur. J. Pharmacol.* **1998**, *353*, 329–336.
55. Hall, E. D.; Kupina, N. C.; Althaus, J. S. *Ann. NY Acad. Sci.* **1999**, *890*, 462–468.
56. Carroll, R. T.; Galatsis, P.; Borosky, S.; Kopec, K. K.; Kumar, V.; Althaus, J. S.; Hall, E. D. *Chem. Res. Toxicol.* **2000**, *13*, 294–300.
57. Hall, E. D.; Braughler, J. M.; Yonkers, P. A.; Smith, S. L.; Linseman, K. L.; Means, E. D.; Scherch, H. M.; Von Voigtlander, P. F.; Lahti, R. A.; Jacobsen, E. J. *J. Pharmacol. Exp. Ther.* **1991**, *258*, 688–694.
58. Chabrier, P. E.; Auguet, M.; Spinnewyn, B.; Auvin, S.; Cornet, S.; Demerle-Pallardy, C.; Guilmard-Favre, C.; Marin, J. G.; Pignol, B.; Gillard-Roubert, V. et al. *Proc. Natl. Acad. Sci. USA* **1999**, *96*, 10824–10829.
59. Sullivan, P. G.; Keller, J. N.; Mattson, M. P.; Scheff, S. W. *J. Neurotrauma* **1998**, *15*, 789–798.
60. Sullivan, P. G.; Springer, J. E.; Hall, E. D.; Scheff, S. W. *J. Bioenerg. Biomembr.* **2004**, *36*, 353–356.
61. Okonkwo, D. O.; Buki, A.; Siman, R.; Povlishock, J. T. *Neuroreport* **1999**, *10*, 353–358.
62. Scheff, S. W.; Sullivan, P. G. *J. Neurotrauma* **1999**, *16*, 783–792.
63. Sullivan, P. G.; Thompson, M. B.; Scheff, S. W. *Exp. Neurol.* **1999**, *160*, 226–234.
64. Hansson, M. J.; Mattiasson, G.; Mansson, R.; Karlsson, J.; Keep, M. F.; Waldmeier, P.; Ruegg, U. T.; Dumont, J. M.; Besseghir, K.; Elmer, E. *J. Bioenerg. Biomembr.* **2004**, *36*, 407–413.
65. Sullivan, P. G.; Rabchevsky, A. G.; Waldmeier, P. C.; Springer, J. E. *J. Neurosci. Res.* **2005**, *79*, 231–239.
66. Heales, S. J.; Bolanos, J. P.; Stewart, V. C.; Brookes, P. S.; Land, J. M.; Clark, J. B. *Biochim. Biophys. Acta* **1999**, *1410*, 215–228.
67. Bringold, U.; Ghafourifar, P.; Richter, C. *Free Rad. Biol. Med.* **2000**, *29*, 343–348.
68. Nauser, T.; Koppenol, W. H. *J. Phys. Chem.* **2002**, *106*, 4084–4086.
69. Bartus, R. *Neuroscientist* **1997**, *3*, 314–327.
70. Ray, S. K.; Hogan, E. L.; Banik, N. L. *Brain Res. Brain Res. Rev.* **2003**, *42*, 169–185.
71. Wang, K. K. *Trends Neurosci.* **2000**, *23*, 59.
72. Lubisch, W.; Beckenbach, E.; Bopp, S.; Hofmann, H. P.; Kartal, A.; Kastel, C.; Lindner, T.; Metz-Garrecht, M.; Reeb, J.; Regner, F. et al. *J. Med. Chem.* **2003**, *46*, 2404–2412.
73. Neuhof, C.; Gotte, O.; Trumbeckaite, S.; Attenberger, M.; Kuzkaya, N.; Gellerich, F.; Moller, A.; Lubisch, W.; Speth, M.; Tillmanns, H. et al. *Biol. Chem.* **2003**, *384*, 1597–1603.
74. Popovich, P. G.; Jones, T. B. *Trends Pharmacol. Sci.* **2003**, *24*, 13–17.

75. Di Giovanni, S.; Movsesyan, V.; Ahmed, F.; Cernak, I.; Schinelli, S.; Stoica, B.; Faden, A. I. *Proc. Natl. Acad. Sci. USA* **2005**, *102*, 8333–8338.
76. Stirling, D. P.; Koochesfahani, K. M.; Steeves, J. D.; Tetzlaff, W. *Neuroscientist* **2005**, *11*, 308–322.
77. Wells, J. E.; Hurlbert, R. J.; Fehlings, M. G.; Yong, V. W. *Brain* **2003**, *126*, 1628–1637.
78. Schwartz, M.; Kipnis, J. *Ann. NY Acad. Sci.* **2005**, *1051*, 701–708.
79. Beattie, M. S.; Farooqui, A. A.; Bresnahan, J. C. *J. Neurotrauma* **2000**, *17*, 915–925.
80. Springer, J. E.; Nottingham, S. A.; McEwen, M. L.; Azbill, R. D.; Jin, Y. *Clin. Chem. Lab. Med.* **2001**, *39*, 299–307.
81. Eldadah, B. A.; Faden, A. I. *J. Neurotrauma* **2000**, *17*, 811–829.
82. Verhagen, A. M.; Ekert, P. G.; Pakusch, M.; Silke, J.; Connolly, L. M.; Reid, G. E.; Moritz, R. L.; Simpson, R. J.; Vaux, D. L. *Cell* **2000**, *102*, 43–53.
83. Holcik, M.; Gibson, H.; Korneluk, R. G. *Apoptosis* **2001**, *6*, 253–261.
84. Cande, C.; Cecconi, F.; Dessen, P.; Kroemer, G. *J. Cell Sci.* **2002**, *115*, 4727–4734.
85. Faden, A. I.; Knoblach, S. M.; Movsesyan, V. A.; Lea, P. M.; Cernak, I. *Ann. NY Acad. Sci.* **2005**, *1053*, 472–481.
86. Fournier, A. E.; Strittmatter, S. M. *Curr. Opin. Neurobiol.* **2001**, *11*, 89–94.
87. Gallo, G.; Letourneau, P. C. *J. Neurobiol.* **2004**, *58*, 92–102.
88. McKerracher, L. *Neurobiol. Dis.* **2001**, *8*, 11–18.
89. Ren, J. M.; Finklestein, S. P. *Curr. Drug Targets CNS Neurol. Disord.* **2005**, *4*, 121–125.

Biography

Edward D Hall received his PhD in neuropharmacology from the Cornell University Graduate School of Medical Sciences in 1976. After completing a postdoctoral fellowship at Cornell University Medical College, he joined the Northeastern Ohio Universities College of Medicine, where he rose to the rank of associate professor of pharmacology. In 1982, he moved to The Upjohn Company, where he initiated and led an effort over many years to discover and develop agents for the treatment of traumatic brain and spinal cord injury and stroke. He played a leading role in the development of high-dose methylprednisolone therapy for acute spinal cord injury, demonstrating that its neuroprotective action involves inhibition of posttraumatic lipid peroxidation. For this work he received the Upjohn Achievement in Science and Medicine Award in 1991. In addition, he was co-discoverer of the 21-aminosteroids (lazaroids), including tirilazad mesylate. In 1997, Dr Hall left Upjohn and joined Parke-Davis Pharmaceutical Research, which is now part of Pfizer Global Research and Development (PGRD). In 2001, he was appointed senior director of CNS Pharmacology at PGRD-Ann Arbor. On July 1, 2002, he joined the University of Kentucky Medical Center, where he is director of the Spinal Cord and Brain Injury Research Center (SCoBIRC) and professor of Anatomy and Neurobiology, Neurosurgery and Neurology. His ongoing research is funded by the NIH and by the Kentucky Spinal Cord & Head Injury Research Trust.

Comprehensive Medicinal Chemistry II
ISBN (set): 0-08-044513-6
ISBN (Volume 6) 0-08-044519-5; pp. 253–277

6.11 Epilepsy

L J S Knutsen and M Williams, Worldwide Discovery Research, Cephalon Inc., West Chester, PA, USA

6.11.1 Introduction

Epilepsy is a chronic neurological disorder that is manifested in the form of recurrent, spontaneous seizure episodes or convulsions, the latter reflecting sudden, stereotyped episodes with accompanying changes in motor activity, sensation, and behavior.[1,2] Convulsions can also be accompanied by muscle spasms and a loss of consciousness depending on the type of epilepsy. These are thought to result from an imbalance, occurring in discrete anatomical pathways in the brain, between the major excitatory and inhibitory systems, glutamate and γ-amino-buytric acid (GABA), which leads to abnormal electrical discharges. Convulsions induce alterations in neurons, glia, and neuronal circuits that include alterations in membrane receptors and neurotransmitter uptake sites, both neurogenesis and apoptosis, astrocyte proliferation, and axonal sprouting. These phenomena increase the susceptibility to additional convulsive episodes and cognitive dysfunction.

Drugs to treat epilepsy (**Table 1**), antiepileptic drugs (AEDs) protect against seizures via actions at a diversity of targets, primarily ion channels.[3,4] In most instances, AEDs represent chronic therapy for individuals with epilepsy.

Epilepsy is the most common neurological disorder in humans, affecting 1–2% of the global population. According to World Health Organization data, approximately 50 million people worldwide, a prevalence of 50 per 100 000 of the general population, suffer from epilepsy. In the USA approximately 180 000 new cases of epilepsy are reported each year with 2.5 million patients having experienced active epileptic episodes within the past 5 years.[5] Between 0.8% and 1.2% of children and between 0.3% and 0.7% of adults experience recurrent seizures with 3% of the population having at least one seizure during their life. For the majority of people, lifelong treatment is necessary in the form of medications although vagal nerve stimulation[6] and surgery at defined resectable seizure foci[7–9] are also options to treat epilepsy. Approximately 30–40% of patients with epilepsy fail to achieve freedom from seizures, even with current

Table 1 Postulated mechanisms of action of anticonvulsant drugs[4]

Drug	*Voltage-sensitive ion channel blockade*		*GABA$_A$ receptor modulator*	*Increased GABA levels*	*Kainate/AMPA receptor blockade*	*Carbonic anhydrase inhibition*
	Sodium	*Calcium*				
Gabapentin	?	√	–	√	–	–
Lamotrigine	√	√	–	–	–	–
Tiagabine	–	–	–	√	–	–
Topiramate	√	√	√	√	√	√
Oxcarbazepine	√	√	–	–	–	–
Zonisamide	√	√	√	–	–	√

medications.[10] Thus while the majority of patients are seizure-free with the use of one or two drugs, a large number of patients remain refractory to drug treatment. An individual who fails to achieve freedom from seizures with adequate doses of two anticonvulsants is thus at high risk for failure on other anticonvulsant medications. Additionally, approximately 10–20% of patients have resistant seizures, despite the use of drug combinations. The major problem for the majority of patients, however, is that of side effects. For instance, the two most commonly prescribed anticonvulsants, sodium valproate (valproic acid; VPA) and carbamazepine (**Figure 1**), while highly effective in controlling seizures, produce fatigue, weight gain, and dizziness in addition to the latter having 'black box' warnings for hepatoxicity and teratogenicity.

In contrast to many other disease states, epilepsy receives little attention[2,10] such that there are numerous economic and social issues for individuals with epilepsy that result in a poor quality of life. Left uncontrolled, epilepsy results in significant morbidity, mortality, and financial burden to the healthcare system.[10] Health system costs due to epilepsy in the USA have been estimated at approximately $12.5 billion annually.[4] Despite significant genericization, the global epilepsy market grew fivefold in terms of sales over the decade beginning in 1994, with a historical compound annual growth rate (CAGR) of 21–26%. Combined sales of AEDs in the developed markets were approached $9 billion in 2005. Of these, approximately 80% of sales for gabapentin were as an adjuvant for neuropathic pain states and in the treatment of migraine and bipolar disorder.[10,12] It has been projected that the CAGR for AEDs over the next 10 years will be 6.8% with sales expecting to double. Market growth is however anticipated to be affected by generic versions of gabapentin, topiramate, and lamotrigine (**Figure 2**), will capture 70% of brand volume in the US. Unmet needs in AED therapy include the need for new chemical entities (NCEs) that can be used as monotherapy and that also have improved side effect profiles, more potent add-on drugs to treat patients with refractory epilepsy, and NCEs that can more effectively control the postictal state which can often be more problematic than the seizure itself,[7] causing drowsiness and confusion with the cognitive and behavioral comorbidities in children being viewed as 'catastrophic.'[13]

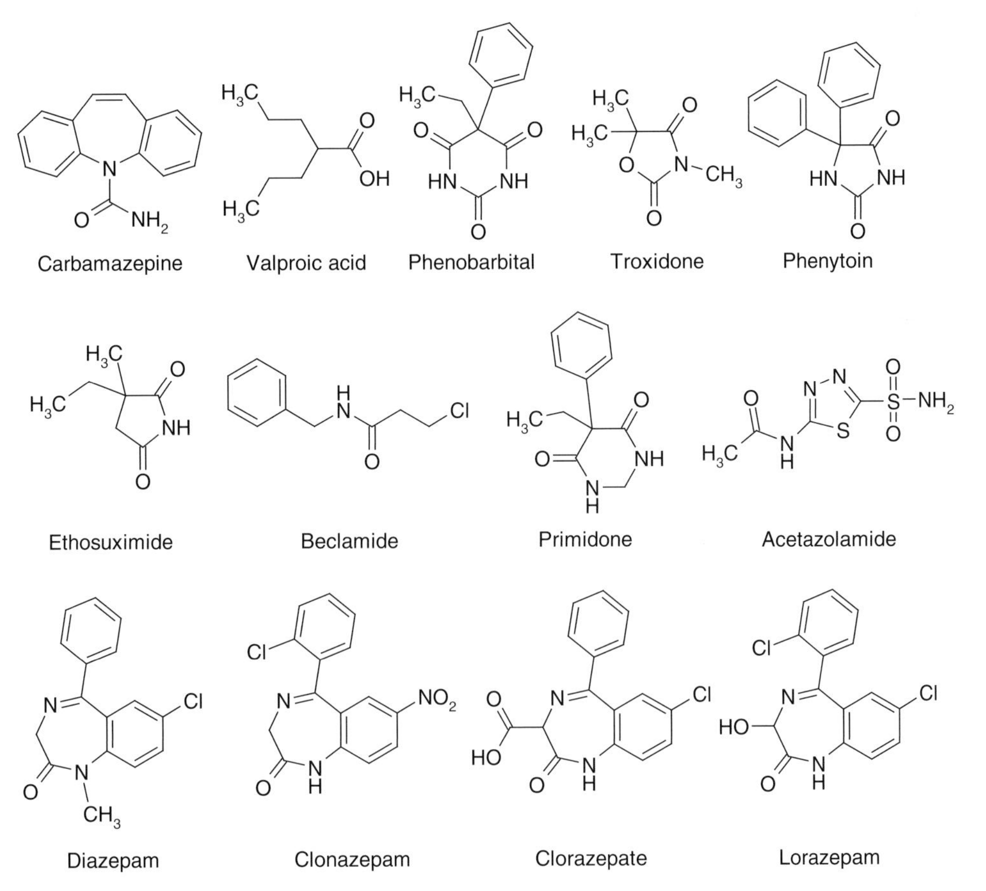

Figure 1 First-generation antiepileptic drugs.

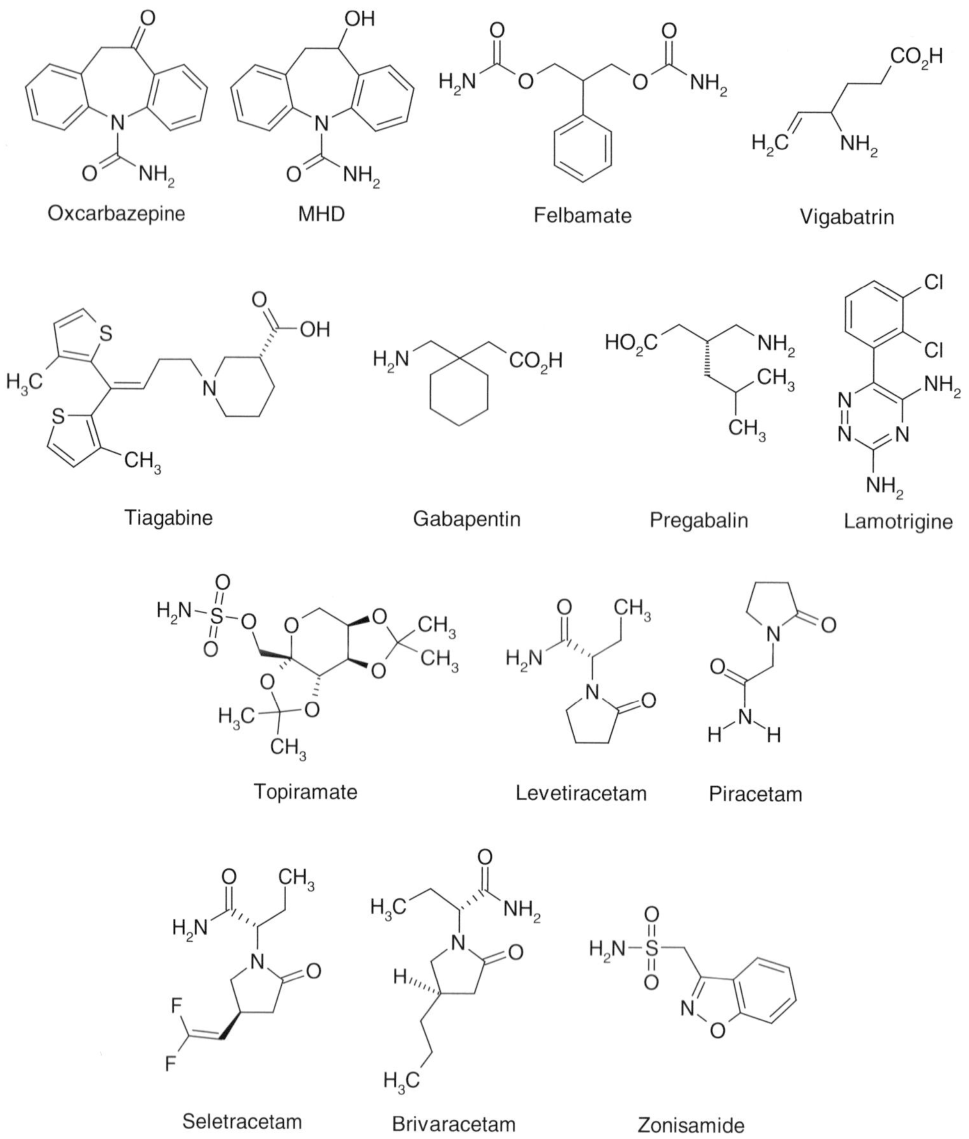

Oxcarbazepine MHD Felbamate Vigabatrin

Tiagabine Gabapentin Pregabalin Lamotrigine

Topiramate Levetiracetam Piracetam

Seletracetam Brivaracetam Zonisamide

Figure 2 Second-generation antiepileptic drugs.

6.11.2 Disease State

Epilepsy can be classified as *idiopathic* in origin when no etiology other than a genetic predisposition has been identified; *symptomatic* when epilepsy can be associated with trauma, e.g., poisoning, stroke, head injury or a brain lesion, Lennox–Gastaut syndrome, hippocampal sclerosis, and cerebral palsy; or *cryptogenic* where a symptomatic cause is suspected but unproven. Advances in the understanding of the genetics of epilepsy together with neuroimaging technologies have reduced the diagnosis of cryptogenic epilepsy. Approximately 70% of epilepsies can be attributed to a specific brain pathology. The many forms of epilepsy suggest that it is not a single disorder with a single molecular lesion but rather a spectrum of diseases.

6.11.2.1 Seizure Classification

The International Classification of Epileptic Seizures (ICES) established guidelines in 1981 that were refined by the Commission on Classification and Terminology of the International League Against Epilepsy (ILAE) in 1989.[14] These developed concise definitions of the various types of epilepsy and are widely used for diagnosis.[8] These classification systems however only describe disease phenotypes and provide little information regarding the causality or severity of the condition or prognosis for the patient.

6.11.2.1.1 Partial seizures

Partial seizures can be divided into: (1) *simple*, where only a part of the brain is involved and consciousness is not impaired, and (2) *complex*, which is differentiated from simple by consciousness also being impaired. Complex partial seizures are frequently preceded by a simple partial seizure or an aura (**Table 2**).

6.11.2.1.2 Generalized seizures

Generalized seizures (**Table 2**) involve both sides of the brain and result in tonic and clonic movements (primary or secondary generalized) or another type of primary generalized epilepsy (e.g., absence or atonic seizure).

6.11.2.1.2.1 Tonic–clonic, *grand mal*, or major motor seizures

Tonic–clonic, *grand mal* or major motor seizures are characterized by a loss of consciousness, falling, muscle rigidity and jerking, and an electrical discharge that involves all or most of the brain.

6.11.2.1.2.2 Absence or *petit mal* seizures

Absence or *petit mal* seizures reflect a primary generalized epileptic seizure that usually last for less than 20 s and are characterized by a stare sometimes associated with blinking or brief automatic movements of the mouth or hands. These usually begin in childhood, are well controlled with drugs, and are outgrown in approximately 75% of children.

6.11.2.1.2.3 Atypical absence seizures

Atypical absence seizures are characterized by a staring spell characterized by partial impairment of consciousness that often occurs in children with Lennox–Gastaut syndrome, a rare disorder begining in childhood that is characterized by mental retardation, multiple multifocal seizures that do not respond well to therapy and an electroencephalograph (EEG) that shows slow (less than 3 s^{-1}) spike-and-wave discharges.

6.11.2.1.2.4 Atonic or drop seizure

Atonic or drop seizure is an epileptic seizure characterized by a sudden loss of muscle tone that is usually not associated with loss of consciousness.

Table 2 Types of seizure

Types of seizure
Partial seizures
Simple (no loss of memory)
Sensory
Motor
Sensory–motor
Psychic
Autonomic
Complex (memory and consciousness impaired)
With aura
Without aura
With automatisms
Without automatisms
Secondarily generalized
Generalized seizures
Absence – *petit mal*
Tonic–clonic – *grand mal*
Atonic – drop seizures
Myoclonic

6.11.2.1.3 Febrile seizures

Febrile seizures are tonic–clonic convulsions that occur in children aged between 6 months and 5 years, that are provoked by fever and are in most cases benign. Some children can go on to develop a higher incidence of hippocampal sclerosis, a primary cause of idiopathic temporal lobe epilepsy. Febrile seizures are typically treated with phenobarbital. This produces hyperactivity and behavioral and learning problems, leading pediatric neurologists to believe that treatment of febrile seizures is worse than the seizure itself.

6.11.2.1.4 *Status epilepticus*

Status epilepticus describes a prolonged seizure.[15] Originally described as a series of repeated seizures or seizures lasting longer than 30 min that occur in almost any seizure type, more recent guidelines[7] indicate that treatment should begin after 5 min of convulsive seizures. *Status epilepticus* is associated with a high rate of mortality, 3% in children and 30% in adults, and is defined as a medical emergency, requiring rapid diagnosis and treatment. The probability of developing epilepsy after a episode of acutely precipitated *status epilepticus* is 41% in a 2-year period after the episide supporting a relationship between the prolonged seizures of *status epilepticus* and eliptogenesis.[9]

6.11.2.2 Antiepileptic Drug Mechanisms

Despite considerable research, the precise mechanism(s) of action of the majority of AEDs is essentially unknown, with the exception of the GABA transaminase inhibitor, vigabatrin, and the selective GABA GAT1 uptake inhibitor, tiagabine, both of which were designed as AEDs based on a defined mechanism of action (**Figure 2**). While a multiplicity of targets for AEDs have been identified, primarily ion channels, including voltage-gated sodium and calcium channels, $GABA_A$ receptors, and, more recently, *N*-methyl-D-aspartate (NMDA)- and α-amino-3-hydroxy-5-methylisoxazole-4-propionic acid (AMPA)-type glutamate receptors[3,4] as well as HCN (hyperpolarization-activated, cyclic nucleotide-gated cation) or pacemaker channels,[16] many of these represent only a part of a broader spectrum of the activity of different AEDs with the majority of these interactions occurring in the micromolar concentration range. For instance, while sodium valproate can block ion channels and increase GABA levels in brain, its use in the treatment of both epilepsy and bipolar disorder is predicated on its polypharmic role as a modulator of ion channel function and phosphatdiyl inositol signaling pathways, as an inhibitor of histone deacetylase (HDAC) and an activator of mitogen-activated protein kinases (MAPKs).[17]

6.11.3 Disease Basis

Once an individual has established an epileptic phenotype, EEG recording in normal and sleep-deprived conditions and brain magnetic resonance imaging (MRI) can be used to both diagnose the type of epilepsy and to decide the treatment course. EEG is not however a sensitive diagnostic test.[7] Epilepsies have symptomatic and genetic origins.[4]

6.11.3.1 Symptomatic Epilepsy

Symptomatic epilepsy can arise from: (1) head trauma, involving a penetrating injury to the brain; (2) brain tumors, both benign and malignant, that include meningiomas, astrocytomas, glioblastomas, oligidendromas, gangliogliomas, etc.; (3) bacterial, fungal, parasitic, and viral infections, especially *Herpes simplex* encephalitis and human immunodeficiency virus (HIV); (4) strokes; (5) dysplasias, where normal brain cells occupy an abnormal place in the brain; (6) chemical imbalances that can result from renal or hepatic failure, hypoxia, complications of pregnancy, treatment with aminophylline, phenothiazine antipsychotics, antidepressants, alcohol, cocaine, and other stimulants, and hormonal changes, e.g., estrogen. The penumbra resulting from a stroke can act as a seizure focus as can tumors.

6.11.3.2 Genetics of Epilepsy

Inheritance in epilepsy can be categorized according to the mechanism of inheritance: (1) mendelian disorders in which epilepsy forms part of the phenotype; (2) idiopathic epilepsy with mendelian inheritance; (3) epilepsy with complex inheritance; (4) idiopathic epilepsy associated with cytogenetic abnormality. Relatives of individuals with generalized epilepsy have a four-times greater risk of epilepsy than the general population. This risk is about two-times higher than in the general population for relatives of individuals with partial or focal epilepsy. In the majority of cases, the pattern of risk does not fit a simple genetic model (dominant, recessive, etc.), making it difficult to predict which family members will develop epilepsy. Genetic risk factors decrease with age. Specific genes for juvenile myoclonic epilepsy (JME), one of the more common epilepsy syndromes that accounts for 7% of all cases of epilepsy, include: the chloride channel, *CLCN2*, *BRD2*, a putative developmental transcription regulator, *EFHC1* which interacts with the calcium channel $Ca_v2.3$, and *ME2* which encodes for the mitochondrial enzyme malic enzyme 2, which is involved in GABA synthesis.[18] Mutations associated with progressive myoclonus epilepsy, a syndrome involving more than a dozen different diseases related to epilepsy, have been mapped to chromosome 21. As with the majority of central nervous system (CNS) diseases, epilepsy appears to be a polygenic disorder[19] with genetic associations being easier to obtain with familial forms of the disease than those occurring in the general population.

6.11.3.2.1 Neonatal epilepsy syndromes

These familial epilepsies include: benign familial neonatal convulsions (BFNC); generalized epilepsy with febrile seizures plus (GEFS+), and server myoclonic epilepsy of infancy (SMEI), and have been associated with mutations in ion channel genes. These mutations include voltage-gated sodium channels (*SCN1A*–chromosome 2q24, *SCN2A*–chr2q23-q24, *SCN1B*–chr19q13), voltage-gated potassium channels (*KCNQ2*–chr20q13.3, *KCNQ3*–chr8q24) and the GABA-gated receptor gene, *GABRG2* chr5q33-q34.[4,19,20] There is extensive locus heterogeneity with mutations in different genes causing similar epileptic phenotypes, e.g., GEFS+ can be caused by mutations in *SCN1A*, *SCN1B*, and *GABRG2*. Conversely, different mutations in the same gene can cause clinically distinct phenotypes (allelic heterogeneity). Thus different mutations in *SCN1A* can result in both GEFS+ and the more severe SMEI.

6.11.3.2.2 Genetic focal epilepsies

Genetic focal epilepsies are also familial in origin and include: autosomal-dominant nocturnal frontal lobe epilepsy (ADNFLE); familial mesial temporal lobe epilepsy (FMTLE); familial lateral temporal lobe epilepsy (FLTLE); and familial partial epilepsy with variable foci (FPEVF).[21] For ADNFLE, three loci and two genes for coding for the subunits of the neuronal nicotinic receptor (NNR) have been identified and produce the same phenotype. *ENFL1* (chr20q13.2) has four different mutations in the α4 NNR subunit coding gene, *CHRNA4*. *ENFL2* (chr15q24) is an unknown gene while *ENFL3* is associated with mutations in the NNR β2 subunit gene, *CHRNB2*. For FLTLE, mutations in the *LGI1* (leucine rich glioma-inactivated, epitempin) gene, which codes for a membrane protein of unknown function, have been reported.[18] For the remainder of the familial genetic focal epilepsies, either none, or multiple (and as yet unresolved) genetic foci have been reported.

6.11.3.2.3 Cavernous angiomas or malformations

Cavernous angiomas or malformations (CMs) are collections of large, closely clustered blood vessels separated from each other by a single layer of endothelium; they are weak and prone to leak blood or hemorrhage, and are inherited in an autosomal dominant manner. Cerebral CMs (CCMs) cause epilepsy in 50–85% of symptomatic patients. Three CCM loci (*CCM1*, *CCM2*, and *CCM3*) have been mapped to chr7q21–q22, chr7p13–p15 and chr3q25.2–q27, respectively.

6.11.3.2.4 Familial febrile convulsions

Familial febrile convulsions (FCs) have a significant genetic component, the presence of which results in a threefold greater risk for experiencing FCs. Genes for the loci *FEB1* (chr8q); *FEB2* (chr19p) have not yet been identified. *FEB3* (chr19q) has been identified in GEFS+ and is a mutation in the voltage-gated sodium channel β1 subunit gene, *SCN1B*.[19,22] Studies on the genetics of epilepsy in increasing numbers of patient cohorts has begun to establish that

> "there is no discrete biological boundary delineating what were once considered distinct clinical entities. Instead, mutations in ion channels appear to be able to produce an essentially continuous range of phenotypes from the mild to the severe end of the spectrum"[20]

is a comment that applies to CNS diseases in general. Additionally, these familial associations indicate that the same epileptic phenotype can be produced by mutations in different ion channel genes and that a single mutation can produce different phenotypes, the latter no doubt a reflection of epigenetic causality factors.

6.11.4 Experimental Disease Models

Models of epilepsy for the evaluation of NCEs include both electrophysiological models using transfected cell lines and brain slices,[23,24] *Drosophila*,[25] and animal models, both rodent and nonhuman primate.[9,26–29] Convulsions can be induced chemically, electrically, or audiogenically and can be extended to mouse strains with a genetic predisposition to convulse. While a number of animal models of epilepsy exist that reflect different aspects of the epileptic phenotypes seen in humans,[26,28] their utility has been debated both in terms of the inevitable relationship to the human form of the disease, e.g., differences in latency,[9] ictal, interictal, and postictal behaviors, and also limitations in being only capable of identifying 'me-too' AEDs.[28] Animal models can be categorized into models of seizures and those of epilepsy. Thus acute seizure activity without chronic epileptiform behavior, e.g., spontaneous seizures, is not a model of epilepsy.[26]

A 'good' animal model of epilepsy has been defined in terms of six criteria[26]: (1) the animal should exhibit similar electrophysiological correlates/patterns as the human condition; (2) the etiologies should be similar, reflecting an underlying genetic predisposition, injury, or neuronal migration disorder; (3) the temporal onset of epilepsy should reflect the human condition; (4) the focal lesions and cortical dysplasia seen in the human condition should be mimicked in the animal model; (5) the animal model should respond to the same AEDs shown to be effective in the human situation; and (6) seizure-induced behavioral manifestations and short- or long-term behavioral deficits should be evident. However, most models fall short of achieving these criteria. Nonetheless, a variety of mouse, rat, chicken (Fepi), rabbit, and primate models can be used to characterize NCEs as potential AEDs.[26]

6.11.4.1 In Vitro Models

Since epilepsy is a disorder associated with the sudden synchronous and repetitive firing of neurons, in vitro electrophysiological studies are especially useful and can be conducted in both normal and transfected cell lines, transfected oocytes, and brain slices to study the effects of NCEs on nerve conduction parameters and induced-seizure-like activity.[19,20] Generalized tonic–clonic seizures can be induced in animals by the periodic application of subconvulsive seizures[9] or by using alumina gel, zinc, cobalt, or Fe^{3+} implantation in the brain.[26] This leads to epileptiform activity in the whole animal and can be studied in ex vivo electrophysiological slice preparations that can be challenged with application of NCEs.

6.11.4.2 In Vivo Models

6.11.4.2.1 Induced models

6.11.4.2.1.1 Chemically induced seizures

Seizures can be induced in rodents using a variety of chemicals. These include pentylenetetrazole (PTZ), strychnine, picrotoxin, the $GABA_A$ antagonist bicuculline, the $GABA_A$ receptor inverse agonist methyl 6,7-dimethoxy-4-ethyl-β-carboline-3-carbonylate, glutamate antagonists, e.g. kainiate, domoic acid, NMDA, quisqualate, the potassium channel blockers 4-aminopyridine and dendrotoxin (DTX), the latter a snake venom given i.c.v., and glutamic acid decarboxylase (GAD) inhibitors.

6.11.4.2.1.2 Maximal electroshock (MES)

Maximal electroshock (MES) in mouse or rat is a widely used model of epilepsy. A 60 Hz alternating current is applied by corneal electrodes and is one of the more widely used models.

6.11.4.2.1.3 Kindled mice

Kindling, a seizure-induced plasticity of the nervous system, can be evoked in rats by repeated administration of a subconvulsive electrical stimulus via bipolar electrodes implanted in the amygdala, hippocampus, or entorhinal cortex which leads, after a few days, to secondarily generalized seizures that eventually become spontaneous reflecting a model of mesial temporal lobe epilepsy (MTLE).[9] Mice can similarly be kindled with kainate to provide a model of MTLE. The latter is characterized by frequent seizures initiated from the temporal lobe and is associated with hippocampal sclerosis, neuronal loss, and gliosis. Kindling can also be achieved using cobalt or Fe^{3+} implants. Kindled mice show neuronal loss in the hippocampus, e.g., CA1, CA3 regions, with spontaneous and reccurrent hippocampal discharge. Like MTLE in humans, the kainiate kindled mouse model is resistant to most AEDs.

6.11.4.2.1.4 *Status epilepticus*

Status epilepticus can be induced in mice with pilocarpine, lithium-pilocarpine, kainiate, or electrical stimulation of amygdala, perforant pathway, or hippocampus. Such animals develop spontaneous seizures after a latent period.

6.11.4.3 Traumatic Brain Injury

Traumatic brain injury models of epilepsy can be replicated in three models of posttraumatic hyperexcitability: (1) chronic isolated cortex (undercutting areas of the cortex while maintaining pial blood supply); (2) focal iron-induced epilepsy; and (3) fluid percussion, the latter involving injury to the dura that leads to hippocampal hilar interneuron loss.[9]

6.11.4.4 Genetic Models of Epilepsy

A number of genetic models of absence epilepsy have been identified and include: GAERS (Genetic absence epilepsy rats from Strasbourg); the WAG/Rij rat; the spontaneous epileptic rat; the *entla* mouse; and the DBA-2 and Frings AGS seizure-susceptible mouse models[26] in the DBA-2 strain due to a variant of the inwardly rectifying potassium ion channel *Kcnj10*. Genetic models of generalized seizures (tonic, tonic–clonic) include the $Otx^{-/-}$ mouse, the transgenic 'jerky' mouse, and the Ihara rat.

6.11.4.5 The NINDS Anticonvulsant Screening Program

The National Institute of Neurological Disorders and Stroke (NINDS) Anticonvulsant Screening Program (ASP) is a US government-sponsored program initiated over 30 years ago, to identify NCEs for the treatment of epilepsy.[27] Some 25 000 NCEs from nearly 400 partnerships, many with pharmaceutical companies, have been evaluated for their potential as AEDs in animal models. Five of these have become new AEDs approved for use in the USA. In addition to establishing efficacy in a variety of animal models of epilepsy, the ASP also measures acute toxicity in terms of muscle tone, rotarod performance, tolerance and subchronic effects on liver function.[27]

6.11.5 Clinical Trial Issues

NCEs with anticonvulsant potential are typically tested as add-on therapy with existing AEDs to test the efficacy of the NCEs as AEDs.[30] As efficacy is established in longer-term studies, monotherapy studies are initiated to establish the efficacy of the NCE on its own and the potential for side effects. Individuals with epilepsy require considerable data to switch AEDs as they are fearful of being exposed to agents that do not provided them coverage from episodes of epilepsy. For this reason, it takes 5 years or more for a new AED to achieve significant market share requiring a considerable investment on the part of the company advancing the new AED. Optimal treatment of epilepsy in adults requires a tailored approach that weighs the efficacy of individual AEDs in the specific epilepsy diagnosis against the risk for adverse events. Partial seizures can be effectively controlled by all the standard and newer AEDs. For the generalized epilepsies, valproate has been the drug of choice, but overall, many of the newer AEDs may offer a better tolerability than the standard agents because of more favorable pharmacokinetic (PK) characteristics and lack of interactions with drugs other than AEDs.[31] Monotherapy is the goal when AED treatment is instituted for the adult with epilepsy. In monotherapy-resistant patients, polytherapy will generally be necessary. A combination of two AEDs may produce antagonistic, additive, or synergistic anticonvulsant effects. However, when the supra-additive anticonvulsant efficacy is also associated with a distinct increase in toxicity, the therapeutic index may be not affected or even lowered. There is evidence of a synergistic interaction between combinations of valproate/phenytoin/ethosuximide, topiramate/carbamazepine/phenobarbital, and felbamate/all major conventional antiepileptics.[32]

6.11.6 Current Treatment

6.11.6.1 Historical Perspective

Pharmacotherapy currently represents the primary treatment modality for epilepsy. Historically, treatment of epilepsy included a variety of plant, animal, and mineral products ('metallotherapy'), in addition to drinking human blood, exorcism, sacrifice, strangulation, trepanation, castration, clitoridectomy, and circumcision.[33] The oldest bona fide therapeutic used for the treatment of epilepsy is bromine which was initially prescribed based on hypotheses that epilepsy resulted from masturbation in males and menstruation in females.[34] The antiepileptic effects of phenobarbital (**Figure 1**), introduced in 1912, were discovered serendipitously when it was used as a hypnotic to keep epileptic patients quiet during the night. Phenytoin was identified in 1938 in the electroshock seizure test established by Merritt and Putnam[35] and was less sedating than phenobarbital. Other early AEDs included troxidone, ethosuximide, beclamide, primidone, and acetazolamide (**Figure 1**). There are currently are approximately 20 drugs approved for the treatment of epilepsy including the benzodiazepines, diazepam, clonazepam, clorazepate, and lorazepam (**Figure 1**).

6.11.6.2 Approved Anticonvulsant Drugs

6.11.6.2.1 Phenobarbital

Phenobarbital (**Figure 1**) is a classical barbiturate and has been used as an AED for nearly 100 years. It produces its effects by modulating $GABA_A$ receptor function. It is superior to other barbiturates in being able to elicit anticonvulsant effects at doses below those that produce sleep although it is still sedating. Mephobarbital is the *N*-methyl analog of phenobarbital which functions as prodrug of the latter due to *N*-demethylation in the hepatic endoplasmic recticulum.

6.11.6.2.2 Phenytoin

Phenytoin is an orally active hydantoin (**Figure 1**) that is effective against partial and tonic–clonic seizures but not absence seizures. It is not a CNS depressant and produces its effects primarily via interactions with voltage-sensitive sodium (Na_v) channels. It is available as a prodrug, fosphenytoin, which can be infused more rapidly than phenytoin. Phenytoin can elicit cardiac toxicity, e.g., arrhythmias, gingival hyperplasia, and endocrine effects, the latter including inhibition of ADH and insulin secretion.

6.11.6.2.3 Ethosuximide

Ethosuximide, a succinimide (**Figure 1**), is currently the AED of choice for the treatment of absence seizures, being inactive against tonic–clonic seizures. It is thought to produce its anticonvulsant effect by inhibiting thalamic T-type calcium currents. Among its side effects are nausea, vomiting, anorexia, dizziness, and headache.

6.11.6.2.4 Valproic acid

Valproic acid (VPA), a branched chain dicarboxylic acid (**Figure 1**), was discovered to be an AED serendipitously when it was used as a vehicle for other NCEs.[34] VPA has broad-based anticonvulsant activity, being effective in the treatment of absence, myoclonic, partial, and tonic–clonic seizures. Side effects associated with VPA include anorexia, vomiting, nausea, sedation, ataxia, and tremor. It causes hepatotoxicity and teratogenicity, the latter including neural tube defects, which are of major concern and have led to the search for second-generation VPA-like AEDs that have the efficacy of VPA but reduced side effect liabilities.[36] As the mechanism of action of VPA is unclear, despite the identification of a multiplicity of targets including ion channels, kinases, and cell signaling pathways, the search for second-generation NCEs (*see* Section 6.11.7) has been focused on structurally related NCEs with similar anticonvulsant phenotypes.

6.11.6.2.5 Carbamazepine

Carbamazepine is an iminostilbene (**Figure 1**) related to the tricyclic antidepressants that was originally developed for the treatment of trigeminal neuralgia but has been in use since the 1970s as an AED. It is the primary AED for the treatment of both simple and complex partial and tonic–clonic seizures. Carbamazepine is thought to produce its anticonvulsant effects by slowing the rate of recovery of Na_v channels. It can produce nausea, vomiting, drowsiness, ataxia, vertigo, and blurred vision, and can result in convulsions and respiratory depression with acute intoxication and long-term use at high doses. Aplastic anemia occurs in approximately 1 in 195 000 patients using this drug.

6.11.6.2.6 Oxcarbazepine

Oxcarbazepine is structurally related to carbamazepine (**Figure 2**) but with different metabolism, producing its effects via the pharmacologically active 10-monohydroxy derivative (MHD; **Figure 2**). Thus oxcarbazepine is the prodrug of MHD and was approved in 2000 as both adjunctive and monotherapy for the treatment of partial seizures in adults with epilepsy and for the adjunctive treatment of partial seizures in children aged 4–16. Oxcarbazepine and MHD inhibit voltage-gated ion channels, with some use-dependence observed at sodium channels, and also antagonize the A_1 adenosine receptor, increase dopaminergic transmission, and potentiate voltage-gated potassium channels. Interactions of MHD with peripheral-type benzodiazepine receptors may also be relevant in the anticonvulsant action of MHD.[37]

6.11.6.2.7 Felbamate

Felbamate (**Figure 2**) is a dicarbamate, structurally related to the anxiolytic meprobamate. It is efficacious in patients with poorly controlled partial and secondarily generalized seizures and in Lennox–Gastuat syndrome. Felbamate was approved in 1993 for both adjunctive and monotherapy but was its use has been associated with analastic anemia leading to major restrictions on its use. Felbamate is thought to produce its anticonvulsant effects by inhibiting voltage-dependent sodium channels (Na_v) and blockade of NR_{1-2B} subunits of NMDA receptors, resulting in decreased excitatory amino acid neurotransmission. This effect may be use-dependent with the AED reducing seizure discharges but not normal neuronal activity.

6.11.6.2.8 Vigabatrin

Vigabatrin (**Figure 2**) is an irreversible inhibitor of the enzyme GABA transaminase. It is structurally related to GABA and dose-dependently increases GABA levels, the major inhibitory neurotransmitter, in brain and cerebrospinal fluid. It is effective in the treatment of partial seizures, Lennox–Gastaut syndrome, and infantile spasms. Its use is limited by the development of a visual field loss in 14–92% of patients.

6.11.6.2.9 Tiagabine

Tiagabine, a nipecotic acid derivative (**Figure 2**), functions as a GABA uptake inhibitor selectively inhibiting the GAT-1 subtype of the GABA transporter.[38] It is lipid soluble and readily crosses the blood–brain barrier. Like vigabatrin, tiagabine increases extracellular levels of GABA. It is effective as an AED as add-on for all subtypes of partial seizures including refractory partial seizures as well as infantile spasms. Side effects seen with tiagabine include dizziness, asthenia, somnolence, infection, headache, nausea, and nervousness.[38]

6.11.6.2.10 Gabapentin

Gabapentin is structurally related to GABA (**Figure 2**) and is indicated for use as adjunctive therapy for refractory partial seizures in adults. It had modest efficacy as monotherapy. There is considerable interest in its mechanism of action as the majority of prescriptions for gabapentin are for off-label use including neuropathic pain, migraine, and bipolar disorder, markets that are larger than epilepsy (*see* 6.14 Acute and Neuropathic Pain). Thus second-generation NCEs, like pregbalin (see below) are being aggressively sought. Gabapentin, despite its structural similarity to GABA, has no effect on GABAergic systems despite early reports that it modulated $GABA_B$ receptor function.[39] Findings that it selectively interacts with the $\alpha 2\delta$ subunit of the $Ca_v2.2$ channel remain the subject of debate.[40,41] Side effects of gabapentin include transient somnolence, dizziness, fatigue, and modest weight gain, but the compound is generally considered to have a wide margin of human safety. Pregabalin (**Figure 2**), is a second-generation agent to gabapentin, which like the latter is thought to produce its anticonvulsant effects via interactions with the $\alpha 2\delta$ subunit of the $Ca_v2.2$ channel.[40] In animal models, pregabalin is 3–10 times more potent than gabapentin and is effective in patients with partial seizures. It is generally well tolerated, with dizziness, somnolence, and vomiting as the major side effects. Like gabapentin, pregabalin has analgesic activity with additional anxiolytic activity. During phase III trials, pregabalin was found to produce euphoria and was listed as Schedule V, e.g., 'abuse of the drug or other substance may lead to limited physical dependence or psychological dependence.'

6.11.6.2.11 Lamotrigine

Lamotrigine (**Figure 2**) is a phenyltriazine derivative used as monotherapy and add-on-therapy for partial and secondarily generalized tonic–clonic seizures and Lennox–Gastaut syndrome. Its development was as an antifolate agent based on the premise that reducing folate would reduce convulsive episodes. Its effectiveness as an AED is unrelated to its antifolate activity. Side effects of lamotrigine include dizziness, ataxia, distorted vision, nausea, and vomiting.

6.11.6.2.12 Topiramate

Topiramate (**Figure 2**) is effective in treating refractory chronic partial seizures and has been used as monotherapy in adolescents and adults. It is a polypharmic AED with effects on carbonic anhydrase activity, AMPA and $GABA_A$ receptors as well as Ca_v and Na_v channels. Topiramate use is associated with cognitive dysfunction and the precipitation of psychosis and depression.

6.11.6.2.13 Levetiracetam

Levetiracetam is an analog of the nootropic, piracetam, and is thus structurally dissimilar to other AEDs (**Figure 2**). It is used as an adjunct in the treatment of partial drug-resistant epilepsy and refractory patients and can be used as monotherapy. The synaptic vesicle protein SV2A, a protein that modulates vesicle exocytosis, is a target for levetiracetam, although how this contributes to its anticonvulsant profile is unknown. Seletracetam (UCB-44212) (**Figure 2**) and brivaracetam (UCB-34714) (**Figure 2**) are second-generation AEDs, the latter being approximately 10 times more potent than levetiracetam as an anticonvulsant in audiogenic seizure-prone mice. Levetiracetam is well tolerated, the most common reported side effects being asthenia, somnolence, headache, and dizziness. It does not impair cognitive function.

6.11.6.2.14 Zonisamide

Zonisamide is a sulfonamide derivative that is chemically and structurally unrelated to other AEDs (**Figure 2**). It is used as add-on therapy for the treatment of partial seizures with or without secondary generalization in adults. It is polypharmic in its anticonvulsant activity with actions in blocking voltage-dependent sodium and T-type calcium channels showing a use-dependent inhibitory action on the release of excitatory neurotransmitters. Additional effects include a biphasic, dose-dependent modulation of dopaminergic activity, accelerated release of GABA from hippocampal slices derived from epileptic mice, inhibition of lipid peroxidation and free radical scavenging, and interactions of carbonic anhydrase. Zonisamide is a broad-spectrum anticonvulsant in animal models with efficacy equivalent to that of VPA, phenytoin, and carbamazepine.

6.11.6.3 Newer Anticonvulsant Drugs

6.11.6.3.1 Rufinamide

Rufinamide (**Figure 3**) is an orally active $GABA_B$ receptor antagonist targeted for used as adjunct therapy for Lennox–Gastaut syndrome and for partial-onset seizures with and without secondary generalization in adults and adolescents.[42]

6.11.6.3.2 Becampanel

Becampanel (**Figure 3**) is an aminomethylquinoxalinedione AMPA receptor antagonist ($IC_{50} = 11$ nM). It is under development for the potential treatment of *status epilepticus* and other types of seizures. It is also being evaluated for use in the treatment of neuropathic pain and cerebrovascular ischemia.

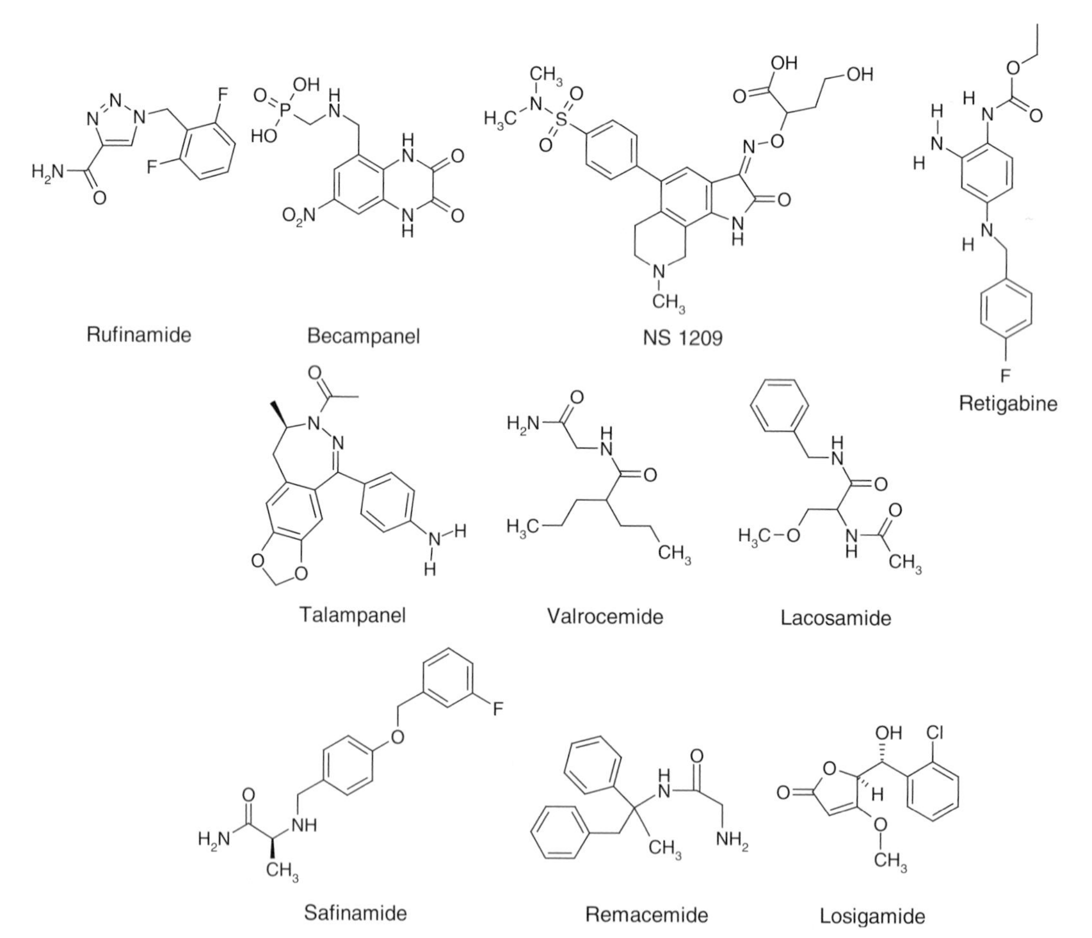

Figure 3 Newer antiepileptic drugs.

6.11.6.3.3 NS 1209

NS 1209 (SPD-502) (**Figure 3**) is an antagonist of AMPA and glutamate receptor 5 (GluR5) receptors. It is under development for the potential treatment of *status epilepticus* and other types of seizures.

6.11.6.3.4 Retigabine

Retigabine (**Figure 3**) is a carbamic acid ethyl ester that is thought to act as a selective potassium channel opener, for the potential treatment of complex partial seizures.

6.11.6.3.5 Talampanel

Talampanel (**Figure 3**) reduces seizure frequency in humans and is effective against PTZ, aminophylline, and MES induced seizures in rodents. It is a selective noncompetitive AMPA receptor antagonist.

6.11.6.3.6 E 2007

E 2007 (structure unavailable) is a selective, noncompetitive AMPA receptor antagonist which in addition to its anticonvulsant activity is being examined for its potential utility in the treatment of migraine, Parkinson's disease, and multiple sclerosis.

6.11.6.3.7 Lacosamide

Lacosamide (also known as harkoseride and erlosamide) (**Figure 3**) is a propioamide under development as an oral agent for the potential treatment of epilepsy. While it is rumored to act through a distinct site in the CNS, like other AEDs, it has polypharmic actions with inhibition of NMDA receptor function and potentiation of GABAergic transmission. It is predicted, based on activity in animal models, to be useful in the treatment of partial onset and generalized tonic-clonic seizures. It is also being evaluated for its utility in the treatment of neuropathic pain.

6.11.6.3.8 Safinamide

Safinamide (NW-1015) (**Figure 3**) had its origins in the weak AED, milacemide. Safinamide is a broad-spectrum AED, effective in the treatment of partial and generalized seizures in humans, producing its effects via a combination of ion channel-related activities including sodium channel blockade ($IC_{50} = 8\,\mu M$), calcium channel modulation, and glutamate release inhibition. It is also a potent MAO-B inhibitor ($IC_{50} = 100\,nM$), with some effects of a dopamine uptake inhibitor. In this latter context, it is in phase III trials for the treatment of Parkinson's disease (*see* 6.08 Neurodegeneration).

6.11.6.3.9 Remacemide

Remacemide (**Figure 3**) has been used as add-on therapy in refractory epilepsy. Mechanistically, it is a low-affinity NMDA receptor antagonist. In animals it blocks MES, NMDA, cocaine, 4-aminopyridine and kainic acid induced seizures but has no effect on picrotoxin-, bicuculline-, or strychnine-induced seizures.

6.11.6.3.10 Losigamone

Losigamone (**Figure 3**) is a novel AED with modest anticonvulsant activity. It is more potent than phenytoin or VPA in the MES test in mice. The mechanism of action is unknown but involves indirect activation of GABA-dependent chloride channels.

6.11.6.3.11 Second-generation valproic acid analogs

With the successful use of VPA not only as an broad-spectrum AED but also in the treatment of bipolar affective disorder (BAPD), a number of VPA analogs (**Figure 4**) have been under investigation as NCEs. These include: ABS-103, NPS-1776, TVP-1901 (valrocemide) (**Figure 4**), and the VPA prodrug, DP-VPA (**Figure 4**). While all have been phenotypically evaluated in animal models of epilepsy, given the predictivity of these models for the human situation, these agents may have considerable potential in BAPD and migraine, both larger and less genericized markets than epilepsy. To date, these NCEs, all closely related in structure to VPA, have moved slowly through the development process, with questionable if any improvements over efficacy and side effect liability profiles to VPA. ABT-769 (**Figure 4**)[36] is a newly synthesized VPA analog that has broad-spectrum anticonvulsant activity with a similar efficacy profile to VPA but an improved safety profile in terms of neither inducing neural tube defects nor affecting mitochondrial fatty acid β-oxidation, the latter related to the hepatotoxic effects of the 4-ene and 2,4-diene metabolites of VPA.

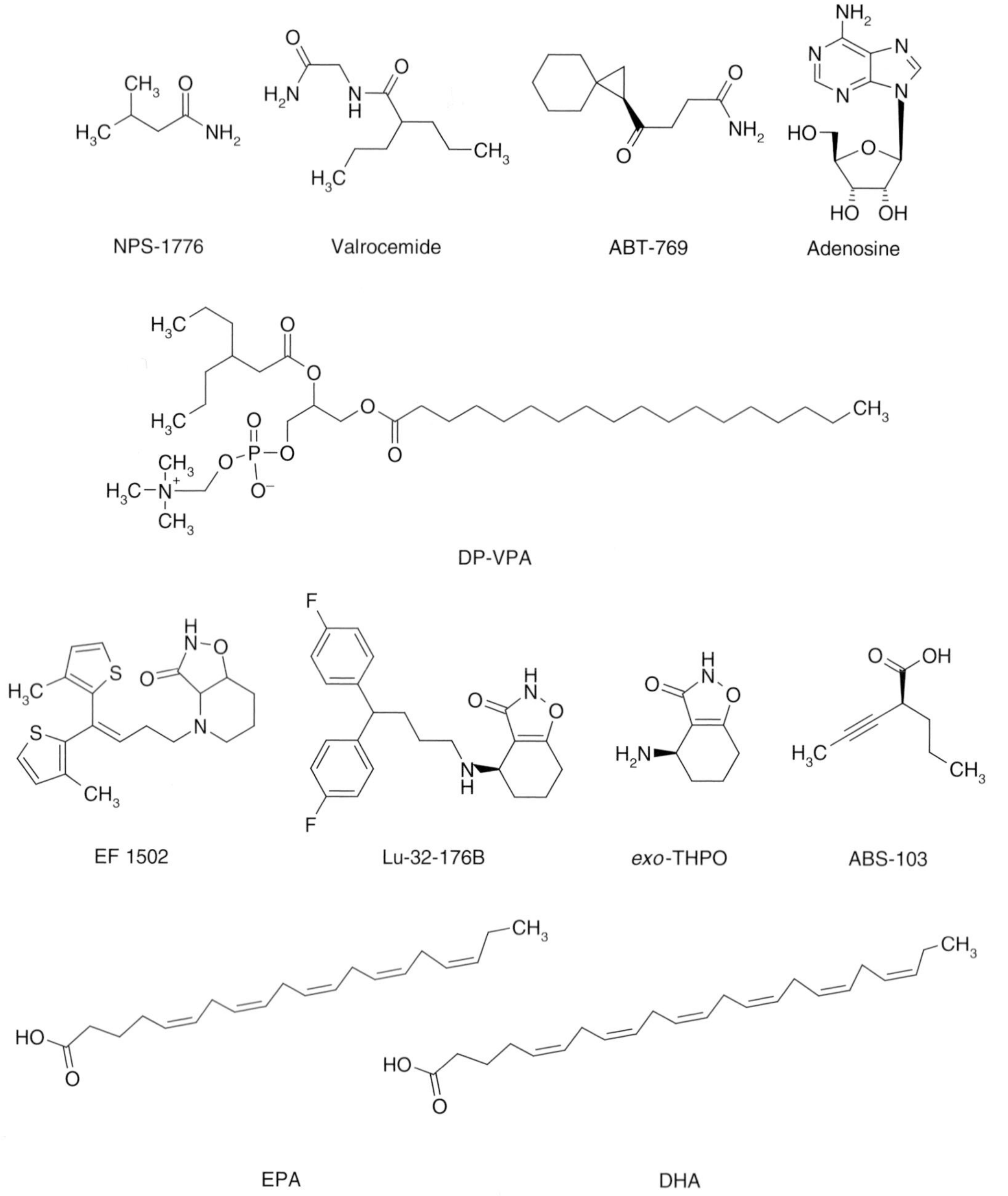

NPS-1776 Valrocemide ABT-769 Adenosine

DP-VPA

EF 1502 Lu-32-176B *exo*-THPO ABS-103

EPA DHA

Figure 4 Valproic acid analogs, noval GABA uptake inhibitors, and fatty acids antiepileptic drugs.

6.11.7 Unmet Medical Needs

There are a number of unmet needs in the field of anticonvulsant drug therapy, not the least of which is a clearer understanding of the discrete mechanisms by which the various AEDs produce their anticonvulsant effects. As noted, with the exception of vigabatrin and tiagabine which were synthesized in the context of a defined molecular hypothesis to enhance GABAergic transmission, the majority of clinically used AEDs have among other properties multiple effects on ion channel function, none sufficiently potent to highlight any one of these channels as a clear target for de novo synthetic efforts.

Additional unmet needs include the spectrum of anticonvulsant in different types of AED and their therapeutic index. In many instances, side effect liabilities are an extension of the AED's, albeit unknown, anticonvulsant pharmacology. In other instances, side effects may be due to totally unknown properties of the AED. The generally micromolar molecular effects of these compounds would, in a classical target-based drug discovery approach, make them potentially early lead NCEs requiring additional medicinal chemistry efforts to improve their potency rather than finished IND candidates.

Concerns have been raised in regard to current screening paradigms used in AED discovery, e.g., the NINDS in vivo panel, that are viewed as generating AEDs of similar efficacy (and limitations) to those already in use.[29,43] In the 10 years encompassing the 1990 Decade of the Brain, eight new AEDs were introduced, none of which appears to have had any impact on the treatment of intractable epileptic patients.[43] This same period also saw a doubling of the finding for epilepsy research from $40 million to $80 million and, in 1999, a major White House initiative, 'Curing Epilepsy – Focus on the Future,' focused on translational research initiatives to use the evolving knowledge of basic brain function at the genomic and proteomic levels to develop new models that would lead to new treatments for epilepsy, new AEDs as well as possible cures and prevention of the disorder(s). Like many of the debatably successful outcomes from the Decade of the Brain,[44] the transition of research findings to improved healthcare has been slow. Part of this has been due to the inherent complexity of the disease – which is viewed as providing a 'low impact' career in which to recruit research scientists – and also issues with the return on investment in commercially developing AEDs,[43] a tangible issue given that the federally funded NINDS Anticonvulsant Screening Program[27] represents the major path through which to identify NCEs in the epilepsy area. Additionally, the social stigma associated with epilepsy has resulted in it being a hidden, chronic condition with little focus in public awareness or concern despite the efforts of advocacy groups. Funding for epilepsy R&D in 2003 on a per person basis was a modest $38. Comparable numbers for diseases like autism, multiple sclerosis, and Parkinson's disease were in the range of $220–$248, with Alzheimer's disease R&D being funded at the level of $165 per affected individual.

An alternative to this somewhat pessimistic situation is the considerable potential, discovered in the clinic by serendipity, for the use of AEDs in neuropathic pain (*see* 6.14 Acute and Neuropathic Pain) and BAPD (*see* 6.03 Affective Disorders: Depression and Bipolar Disorders), which has increased interest in advancing AEDs to the clinic as multifactorial therapeutic agents and a renewed focus on understanding the mechanism(s) of action of these agents as anticonvulsants in order to understand the role of aberrant and spontaneous neuronal firing via epileptogenic-like foci in neuropathic pain (neuromas) and BAPD. Like chronic convulsive episodes, outcomes from chronic pain states include cell death, aberrant neuronal sprouting, and neuronal pathway remodeling (*see* 6.14 Acute and Neuropathic Pain).

6.11.8 Emerging Research Areas

6.11.8.1 Adenosine Producing Stem Cell Therapy

Adenosine (**Figure 4**) is an endogenous neuromodulatory agent that has anticonvulsant activity in a variety of animal models.[45] Using hippocampal microdialysis probes, adenosine levels were found to be increased 6–31-fold in patients with intractable complex partial epilepsy during seizures,[46] suggesting that compounds that mimic adenosine effects, e.g., synthetic adenosine analogs, or facilitate its actions, e.g., adenosine kinase (AK) inhibitors, may be potent and effect AEDs.[47] However, as in many other therapeutic areas where modulation of adenosine function has been viewed as a therapeutic option, e.g., neuropathic pain, stroke, asthma, chronic obstructive pulmonary disease (COPD), sleep promotion (*see* 6.06 Sleep), etc., the efficacy of adenosine and its analogs have been accompanied by unmanageable side effects including sedation and hypotension.[48] A novel approach to circumventing the side effects of adenosine has been an 'ex vivo gene therapy' approach, tailoring the local delivery of adenosine to an epileptic focus by implanting encapsulated fibroblasts.[49] An extension of this approach has been to engineer mouse myoblasts to release adenosine by genetic inactivation of AK.[50] In this latter instance, kindled mice grafted with adenosine-releasing implants had complete protection from convulsive seizures with a corresponding reduction in EEG after discharges for periods up to and in excess of 3 weeks without evidence of desensitization or sedation. Efforts are now ongoing to engineer adult stem cells from skin or bone marrow that can produce adenosine to generate autologous therapeutic grafts with a potential lifespan of up to 500 days.[50]

6.11.8.2 Novel GABA Transporter (GAT) Inhibitors

The clinical use of tiagabine as an AED has established GAT-1 as a bona fide anticonvulsant drug target.[38,51] An extensive structure–activity relationship study has been established around this AED[52] leading to newer GAT-1 inhibitors including *exo*-4,5,6,7-tetrahydroisoxazolo[4,5-c]pyridin-3-ol (*exo*-THPO) and Lu-32-176B (**Figure 4**).[53] EF1502 (**Figure 4**) is a *N*-substituted analog of THPO that inhibits both mouse GAT-1 and GAT-2 in transfected cell lines had synergistic rather than additive anticonvulsant effects in combination with *exo*-THPO in the rat PTZ-seizure model and the Frings audiogenic seizure-susceptible mouse. This effect occurred without synergistically impairing rotorod performance suggesting that the human form of mouse GAT-2 might be a novel AED target. GAT can also facilitate GABA release by reversing in response to depolarization, an effect that is enhanced by gabapentin and vigabatrin. Modulation of GAT reverse transport may thus be a physiological source of GABA during seizures.[54]

6.11.8.3 ω Fatty Acids

Omega fatty acids like (eicosapentaenoic acid) EPA and docosahexenoic acid (DHA) (**Figure 4**) have a variety of effects in CNS disease states with reports of efficacy in depression, schizophrenia, and Huntington's disease. EPA and DHA can raise seizure threshold in rats and can also reduce the proinflammatory mediator production seen in epilepsy models and in epileptic patients by inhibiting PLA_2 and COX-2 activity.[55] In a 12-week, double blind, placebo controlled trial, patients with refractory epilepsy treated with an ω-3 fatty acid supplement (1 g EPA; 0.7 g DHA) showed a reduction in seizure frequency over the first 6-weeks, an effect that was not sustained over the remaining trial period. These promising data are being extended to other ω-3 fatty acids, combinations, and formulations.[55]

References

1. Chang, B. S.; Lowenstein, D. H. *N. Engl. J. Med.* **2003**, *349*, 1257–1266.
2. McAuley, J. W.; Biederman, T. S.; Smith, J. C.; Moore, J. L. *Ann. Pharmacother.* **2002**, *36*, 119–129.
3. White, H. S. *J. Clin. Psychiat.* **2003**, *64*, 5–8.
4. Rogawski, M. A.; Loscher, W. *Nat. Rev. Neurosci.* **2004**, *5*, 553–564.
5. Annegers, J. F. *Am. J. Managed Care* **1998**, *4*, S453–S462.
6. Helmers, S. L.; Griesemer, D. A.; Dean, J. C.; Sanchez, J. D.; Labar, D.; Murphy, J. V.; Betti, D.; Park, Y. D.; Shuman, R. M.; Morris, G. L., III. *Neurologist* **2003**, *9*, 160–164.
7. Fisher, R. S. In *Pharmacological Management of Neurological and Psychiatric Disorders*; Enna, S. J., Coyle, J. T., Eds.; McGraw-Hill: New York, 1998, pp 459–503.
8. Benbadis, S. R.; Tatum, W. O., IV. *Am. Fam. Physician* **2001**, *64*, 91–98.
9. White, H. S. *Neurology* **2002**, *59*, S7–S14.
10. LaRoche, S. M.; Helmers, S. L. *JAMA* **2004**, *291*, 605–614.
12. Rogawski, M. A.; Loscher, W. *Nat. Med.* **2004**, *10*, 685–692.
13. Wirrell, E.; Farrell, K.; Whiting, S. *Can. J. Neurol. Sci.* **2005**, *32*, 409–418.
14. Commission on Classification and Terminology of the International League against Epilepsy. *Epilepsia* **1989**, *30*, 389–399.
15. Sirven, J. I.; Waterhouse, E. *Am. Fam. Physician* **2003**, *68*, 469–476.
16. Poolos, N. P. *Epilepsy Behav.* **2005**, 7, 51–56.
17. Yuan, P. X.; Huang, L. D.; Jiang, Y. M.; Gutkind, J. S.; Manji, H. K.; Chen, G. *J. Biol. Chem.* **2001**, *276*, 31674–31683.
18. Turnbill, J.; Lohi, H.; Kearney, J. A.; Rouleau, G. A.; Delgado-Esutea, A. V.; Meisler, M. H.; Cossette, P.; Minassian, B. A. *Hum. Mol. Genet.* **2005**, *14*, 2491–2500.
19. Mulley, J. C.; Scheffer, I. E.; Harkin, L. A.; Berkovic, S. F.; Dibbens, L. M. *Hum. Mol. Genet.* **2005**, *14*, R243–R249.
20. Burgess, D. L. *Epilepsia* **2005**, *46*, 51–58.
21. Andermann, F.; Kobyashi, E.; Andermann, E. *Epilepsia* **2005**, *46*, 61–67.
22. Daoud, A. J. *Pediat. Neurol.* **2004**, *2*, 9–14.
23. Hoffman, W. H.; Haberly, L. B. *J. Neurophysiol.* **1996**, *76*, 1430–1438.
24. Holopainen, I. E. *Neurochem. Res.* **2005**, *30*, 1521–1528.
25. Timpe, L. C.; Jan, L. Y. *J. Neurosci.* **1987**, 7, 1307–1317.
26. Sarkisian, M. R. *Epilepsy Behav.* **2001**, *2*, 201–216.
27. White, H. S.; Woodhead, J. H.; Wilcox, K. S.; Stables, J. P.; Kupferberg, H. S.; Wolf, H. H. In *Antiepileptic Drugs*, 5th ed.; Levy, R. H., Mattson, R. H., Meldrum, B. S., Perruca, E., Eds.; Lippincott Williams and Wilkins: Philadelphia, PA, 2002, pp 36–48.
28. Heinrichs, S. C.; Seyfried, T. N. *Epilepsy Behav.* **2005**, *8*, 5–38.
29. Rogawski, M. A. *Epilepsy Res.* **2006**, *68*, 22–28.
30. Porter, R. J. In *Antiepileptic Drugs*, 5th ed.; Levy, R. H., Mattson, R. H., Meldrum, B. S., Perruca, E., Eds.; Lippincott Williams and Wilkins: Philadelphia, PA, 2002, pp 58–65.
31. Matson, R. H. *Neurology* **1998**, *51*, S15–S20.
32. Czuczwar, S. J.; Borowicz, K. K. *Epilepsy Res.* **2002**, *52*, 15–23.
33. Meijer, J. W. A.; Meinardi, H.; Binnie, C. D. In *Discoveries in Pharmacology: Vol. 1, Pyscho- and Neuro-pharmacology*; Parnham, M. J., Bruinvels, J., Eds.; 1983; Elsevier: Amsterdam, the Netherlands, pp 447–488.
34. Sneader, W. *Drug Discovery. The Evolution of Modern Medicines*; Wiley: Chichester, UK, 1985, pp 41–47.
35. McNamara, J. In *Goodman and Gilman's The Pharmacological Basis of Therapeutics*, 11th ed.; Brunton, L. L., Lazo, J. S., Parker, K. L., Eds.; McGraw-Hill: New York, 2006, pp 501–525.
36. Giardina, W. J.; Dart, M. J.; Harris, R. R.; Bitner, R. S.; Radek, R. J.; Fox, G. B.; Chemburkar, S. R.; Marsh, K. C.; Waring, J. F.; Hui, J. Y. et al. *Epilepsia* **2005**, *46*, 1349–1361.
37. Ambrosio, A. F.; Soares-Da-Silva, P.; Carvalho, C. M.; Carvalho, A. P. *Neurochem. Res.* **2002**, *27*, 121–130.
38. Knutsen, L. J. S.; Andersen, K. E.; Lau, J.; Lundt, B. F.; Henry, R. F.; Morton, H. E.; Nærum, L.; Petersen, H.; Stephensen, H.; Suzdak, P. D. et al. *J. Med. Chem.* **1999**, *42*, 3447–3462.

39. Betrand, S.; Ng, G. Y. K.; Puriasi, M. G.; Wolfe, S. E.; Severidt, M. W.; Nouel, D.; Robitaille, R.; Low, M. J.; O'Neill, G. P.; Metters, K. et al. *J. Pharmacol. Exp. Ther.* **2001**, *298*, 15–24.
40. Sills, G. J. *Curr. Opin. Pharmacol.* **2006**, *6*, 108–113.
41. Taylor, C. P. *CNS Drug Rev.* **2004**, *10*, 183–188.
42. Stepien, K.; Tomaszewski, M.; Czuczwar, S. J. *Pharmacol. Rep.* **2005**, *57*, 719–733.
43. Sutula, T. P. *Epilepsy Behav.* **2005**, *6*, 296–302.
44. Williams, M.; Coyle, J. T.; Shaikh, S.; Decker, M. W. *Annu. Rep. Med. Chem.* **2001**, *36*, 1–10.
45. McGaraughty, S.; Cowart, M.; Jarvis, M. F.; Berman, R. F. *Curr. Topics Med. Chem.* **2005**, *5*, 43–58.
46. During, M. J.; Spencer, D. D. *Ann. Neurol.* **1992**, *32*, 618–624.
47. Boisson, D. *Neuroscientist* **2005**, *11*, 25–36.
48. Williams, M. In *Handbook of Experimental Pharmacology-Purinergic and Pyrimidinergic Neurotransmission Vol. 151, Part 2*; Abbracchio, M. P., Williams, M., Eds.; Springer-Verlag: Berlin, Germany, 2001, pp 407–434.
49. Huber, A.; Padrun, V.; Deglon, N.; Aebischer, P.; Mohler, H.; Boison, D. *Proc. Natl. Acad. Sci. USA* **2001**, *98*, 7611–7616.
50. Guttinger, M.; Padrun, V.; Pralong, W. F.; Boison, D. *Exp. Neurol.* **2005**, *193*, 53–65.
51. Schousboe, A.; Sarup, A.; Larsson, O. M.; White, H. S. *Biochem. Pharmacol.* **2004**, *68*, 1557–1563.
52. Andersen, K. E.; Braestrup, C.; Grønwald, F. C.; Jørgensen, A. S.; Nielsen, E. B.; Sonnewald, U.; Sørensen, P. O.; Suzdak, P. D.; Knutsen, L. J. S. *J. Med. Chem.* **1993**, *36*, 1716–1725.
53. White, H. S.; Watson, W. P.; Hansen, S. L.; Slough, S.; Perregaard, J.; Sarup, A.; Bolvig, T.; Petersen, G.; Larsson, O. M.; Clausen, R. P. et al. *J. Pharmacol. Exp. Ther.* **2005**, *312*, 866–874.
54. Richerson, G. B.; Wu, Y. *Adv. Exp. Med. Biol.* **2004**, *548*, 76–91.
55. Yeun, A. W. C.; Sander, J. W.; Fluegel, D.; Patsalos, P. N.; Bell, G. S.; Johnson, T.; Koepp, M. J. *Epilepsy Behav.* **2005**, 7, 253–258.

Biographies

Lars J S Knutsen began his research career at Glaxo in Ware, Herts., UK having completed an MA in Chemistry at Christ Church, Oxford, in 1978. While at Glaxo he completed a PhD in Nucleoside Chemistry joining Novo Nordisk in Denmark in 1986. While there he led the project that identified tiagabine, a marketed anticonvulsant acting by blocking GABA uptake. In 1997, he joined Vernalis (Cerebrus) in the UK, initiating the adenosine A_{2A} antagonist project that led to V2006, currently in clinical trials with Biogen-IDEC for Parkinson's disease. He joined Ionix Pharmaceuticals Ltd., in Cambridge, UK in 2002 as Director of Chemistry. Dr Knutsen joined the CNS Medicinal Chemistry group at Cephalon Inc. in 2006. He has over 35 peer-reviewed publications and 18 issued US patents.

Michael Williams received his PhD (1974) from the Institute of Psychiatry and his Doctor of Science degree in Pharmacology (1987) both from the University of London. Dr Williams has worked in the US-based pharmaceutical industry for 30 years at Merck, Sharp and Dohme Research Laboratories, Nova Pharmaceutical, CIBA-Geigy and Abbott Laboratories. He retired from the latter in 2000 and after serving as a consultant with various biotechnology/pharmaceutical/venture capital companies in the US and Europe, joined Cephalon, Inc. in West Chester, in 2003 where he is vice president of Worldwide Discovery Research. He has published some 300 articles, book chapters and reviews and is an adjunct professor in the Department of Molecular Pharmacology and Biological Chemistry at the Feinberg School of Medicine, Northwestern University, Chicago, IL.

Comprehensive Medicinal Chemistry II
ISBN (set): 0-08-044513-6

ISBN (Volume 6) 0-08-044519-5; pp. 279–296

6.12 Ophthalmic Agents

N A Sharif and P Klimko, Alcon Research, Ltd., Fort Worth, TX, USA

6.12.1 Introduction: Ocular Anatomy and Diseases

Sight is a very precious sense. Most information about our surroundings is gathered by the eye, which is literally a 'window' for the brain. With an aging population, the prevalence of sight-threatening ocular diseases continues to increase. Thus, for instance, more than 70 million people suffer from glaucoma worldwide.[1–3] Visual impairment caused by diabetes affects up to 90% of diabetics over 10 years of age. Likewise, pathological dry eye and ocular allergic conditions afflict >100 million patients worldwide, and age-related macular degeneration (AMD) is the leading cause of blindness among the elderly, affecting up to 28% of patients after the seventh decade of life.[2] Consequently, the discovery and development of therapeutic products for the treatment of these various ocular diseases is of paramount importance, and is being actively pursued within the pharmaceutical industry.[1–3]

The eye is a somewhat immune-privileged organ, composed of the cornea and conjunctiva on the ocular surface (**Figure 1**). The anterior chamber of the eye, between the lens and cornea, contains the aqueous humor (AH), a fluid that is continuously produced by the nonpigmented ciliary epithelial cells of the ciliary body process. The AH drains from the eye via the trabecular meshwork into the canal of Schlemm and then into the venous circulation. Under normal circumstances the rate of influx and efflux of AH is constant, maintaining a certain intraocular pressure (IOP), that keeps the eye rigid and thus permits light to be transmitted from the cornea through the lens to the retina at the back of the eye. The AH also provides vital nutrients to the avascular corneal and lens tissues. The iris lies in front of the lens and forms the pupil, which controls the amount of light transmitted through the lens to the retina. The retina relays the images to the brain via the optic nerve at the posterior end of the eye. Behind the lens is a jelly-like material (vitreous humor) that keeps the posterior chamber of the eye filled and rigid, and thereby provides a cushion and support for the retina lining the posterior wall of the eye. The retina is nourished by a vascular bed of capillaries of the choroid plexus that also remove the waste products from the retinal tissues. Nerve impulses from the retina are transmitted to the brain via the optic nerve. The various diseases of the eye, their etiologies, and treatments are discussed herein, with priority given to the most prevalent and sight-threatening diseases.

6.12.2 Disease State

6.12.2.1 Glaucoma

Glaucoma is a major cause of vision loss throughout the world.[1–3] It represents a heterogeneous group of slow but progressive optic neuropathies that culminate in blindness. Glaucoma involves a triad of tissues comprising

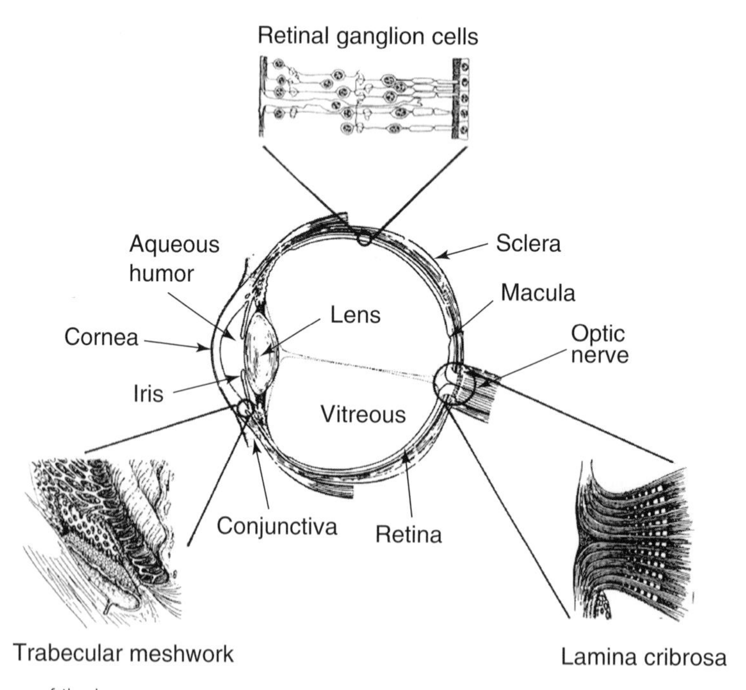

Figure 1 The anatomy of the human eye.

the trabecular meshwork (TM) in the anterior chamber; the optic nerve head (ONH); and the retinal ganglion cells (RGCs) in the posterior chamber of the eye. The exact etiology of glaucoma is unclear, but the eventual death of RGCs through apoptotic mechanisms causes gradual loss of peripheral vision and eventually leads to blindness.[1–3]

The major risk factor of glaucoma is elevated IOP in the anterior chamber of the eye. Over time the pressure is transmitted to the back of the eye, leading to structural deformation of the posterior chamber tissues (notably the retina and ONH), constriction of retinal blood vessels leading to ischemia, and apoptotic death of RGCs, followed by axonal loss, leading to optic disc cupping/constriction and gradual loss of vision. Primary open-angle glaucoma (POAG), which is characterized by ocular hypertension, is the most common form. Normal-tension glaucoma, where the IOP is not elevated, probably has a different set of causative factors and is less-well understood, prompting the belief that vascular abnormalities around the retinal architecture causing ischemia may be more important in this disease. Genetic linkage analyses of heritable forms of glaucoma have identified 17 glaucoma loci, including six different loci for POAG (i.e., *GLC1A* to *GLC1F*), where the first glaucoma gene (*MYOC*; myocilin) is mapped to the *GLC1A*.[2] *PITX2, FOXC1, CYP1B1*, and *OPTN* are additional genes associated with glaucoma but, as with the other genes mentioned, the exact linkage and mechanism(s) responsible for causing the disease are unknown.[2] The largest genetic contributor to POAG identified to date is the *GABRB3* locus, GLC11 on chromosome 15 (15q11-13 locus).[4] However, since a curative approach to glaucoma would involve disease-modifying therapeutic agents, it is hoped that research in the ocular genetic arena would eventually lead to the discovery of such drugs in the future. POAG afflicts 1–4% of the population over 45 years, amounting to >70 million people worldwide.[2] The most prominent risk factors for POAG are ocular hypertension, age, race (African-American > Caucasian) and a family history of glaucoma.[1,2] Additional risk factors are diabetes, high blood pressure, and other chronic systemic diseases.[2]

One major limitation is the inability to diagnose glaucoma, and currently the detection of ocular hypertension in patients is the only accepted phenotypic marker. However, since ocular hypertension itself is a ‘quiet’ disorder with no major measurable symptoms, patients are generally oblivious to this condition until they start to lose their peripheral vision, by which time the damage to the RGCs and retinal architecture has already progressed and is continuing to worsen. Therapeutic intervention with ocular hypotensive agents is aimed at slowing progression of visual field loss by reducing IOP.

6.12.2.2 Age-Related Macular Degeneration

In AMD, central visual acuity is lost due to death of macula photoreceptors located in the central portion of the retina. This pattern of vision loss is different than that in glaucoma, where peripheral vision is progressively lost. AMD can be broadly divided into two categories: *nonexudative*, or ‘dry,’ AMD; and *exudative*, or ‘wet,’ AMD. Dry AMD is characterized by the loss of photoreceptor cells in the macula following the death of supporting retinal pigmented epithelium (RPE) cells. Vision decline is usually gradual. Wet AMD is characterized by neovascularization of the choroidal capillaries (choroidal neovascularization (CNV)), which supply oxygen and nutrients to and remove waste products from photoreceptors and RPE cells. Wet AMD can be further subcategorized by the appearance upon examination by angiography of newly formed capillaries: (1) *predominantly classic*, in which most of the newly formed blood vessels are well formed; (2) *minimally classic*, in which most of the capillaries are poorly defined; and (3) *occult*, in which all of the capillaries are poorly defined. Progression from the occult to the minimally classic to the predominantly classic subtype correlates with increasing disease severity. Inappropriate blood vessel growth leads to retinal fibrosis, scar formation, and detachment, with resultant photoreceptor death and loss of visual acuity.[5]

Disease development and vision loss occur much more rapidly for wet than for dry AMD, frequently occurring on a timescale of months for the former as opposed to years for the latter. Thus progression of AMD from the dry to the wet form usually precipitates an accelerated decline in visual function. AMD is the most common cause of blindness in the western world in people over the age of 55 years. In the USA the number of individuals with features of AMD has been variously estimated between 2 million and 8 million, depending on the criteria used for disease classification.[6,7] Although about 90% of AMD patients have the dry form and 10% have the wet form, most cases of AMD-related blindness are due to the latter. Advancing age is the strongest demographic risk factor associated with AMD; for example, it has been reported that the approximately 1.5% incidence of AMD in the ≥40 year age group increases to about 15% in white female over the age of 80 years.[8] Race is also a notable risk factor, with the incidence of AMD being higher in caucasians than in African-Americans.[9]

6.12.3 Disease Basis

6.12.3.1 Glaucoma

Elevation of IOP results from a deposition of extracellular matrix (ECM) in the anterior chamber, and thus blockage of the AH drainage pathway, the TM and Schlemm's canal. The ECM is composed of numerous proteins, modified glycoproteins, and glycosaminoglycans, including hyaluronic acid, chondroitin sulfate, dermatan sulfate, keratin sulfate, and heparin sulfate. The abnormal deposition and/or clearance of ECM in glaucomatous patients may also result from TM cell death and loss of phagocytic activity during the aging process. Coupled with these phenomena is perhaps the decreased ability of TM cells to liberate local matrix metalloproteinases (MMPs), that can digest the ECM and re-model and maintain the anterior chamber architecture and function. There is increasing evidence that increased concentration of tissue growth factor-β2 (TGF-β2) observed in ocular hypertensive glaucoma patients contributes to the elevation of ECM in the TM. The loss of MMP activity could also be due to TGF-β2.[10] In addition, a soluble glycoprotein (sCD44) derived from the hyaluronic acid receptor (CD44 H) found in the AH of glaucoma patients appears to be toxic to TM cells,[11] and may be responsible for the decreased TM cellularity in glaucoma patients' eyes. Perhaps deposition of other neurotoxic apolipoproteins, including serum amyloid-A (SAA)[12] and/or amyloid-β and related proteins[13] in the anterior chamber of the eye and/or in the retina, could also be involved in the etiology of ocular hypertension and glaucoma. A recent study strongly correlated RGC apoptosis with elevated IOP and ECM deposition,[14] thus strengthening the link between ocular hypertension and loss. Consequently, the mainstay pharmaceutical treatment for POAG has been to treat ocular hypertension either by limiting the generation of AH by the ciliary body or by promoting the efflux of the AH from the anterior chamber. Since the AH can leave the eye via the TM/Schlemm's canal (conventional outflow) and via the spaces between the ciliary muscle bundles and into venous blood vessels (uveoscleral outflow), therapeutic approaches have focused recently on reducing IOP by the latter pathways rather than reducing AH production. Agents that reduce AH production can actually deprive the anterior chamber tissues of much needed nutrients and oxygen, and reduce the elimination of waste products. However, patients refractory to outflow agents have to rely on drugs that reduce inflow and/or undergo surgical treatments such as laser trabeculoplasty and/or filtration surgery (see below).

6.12.3.2 Age-Related Macular Degeneration

When considering potential causes and treatments of macular degeneration, it is informative to consider the normal vision process from a functional and morphological standpoint. Light that is transmitted through the surface and the anterior chamber of the eye is absorbed in the outer retina by opsin-bound (as a Schiff base) retinaldehyde, which is present in specialized light-gathering cells called photoreceptors. There are two types of photoreceptors: *rods*, which are used primarily for low light vision, and *cones*, which are responsible for color perception and visual acuity. Light-induced isomerization of the retinaldehyde Schiff base position 11 olefin from *cis* to *trans* geometry generates a signal, which is converted to a nerve impulse for eventual relay to RGCs in the inner retina (**Figure 2**). RGCs transmit this signal through the optic nerve head (ONH) and into the brain, where the signal is integrated.

Photoreceptor function is only possible with the support of specialized cells in the retinal pigmented epithelium (RPE). These RPE cells take up released 11-*trans*-retinaldehyde (in the form of the reduced retinol) and isomerize the olefin geometry back to the photoactive 11-*cis* form. RPE cells also phagocytize photoreceptor outer membrane segments that are continuously shed and replaced. Choroidal capillaries provide nutritional support (oxygen, proteins, hormones, etc.) to and remove waste products from photoreceptors and RPE cells, and are separated from them by

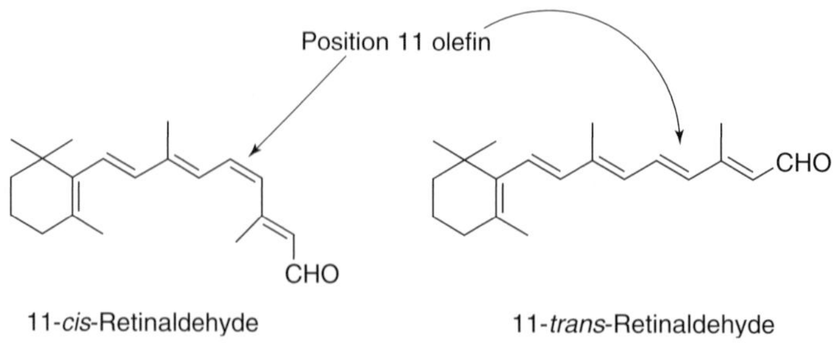

Position 11 olefin

CHO

CHO

11-*cis*-Retinaldehyde

11-*trans*-Retinaldehyde

Figure 2 11-*cis*- and 11-*trans*-retinaldehyde.

Bruch's membrane. A normally functioning Bruch's membrane is sufficiently permeable to allow diffusional exchange of nutrition and waste products between the choroidal capillaries and RPE cells.

AMD is a chronic disease resulting from a malfunctioning of this system at several loci. Although there are many gaps in understanding of the disease pathology and progression, a general hypothesis with several variations has emerged, based on consistent clinical and epidemiologic observations in AMD patients.[2,15] Dry AMD is characterized by increased deposition of a material called '*drusen*', which can be observed clinically using an ophthalmoscope, between the RPE and Bruch's membrane. Drusen likely originate from material discharged from RPE cells. Although this process probably occurs throughout life due to the heavy phagocytosis burden of RPE cells, the composition of the material expelled from dying/dysfunctional cells may be more heavily weighted toward injurious components such as oxidized lipids, proinflammatory cytokines, and acute phase inflammatory proteins.[16] Intriguingly, it has recently been reported that drusen commonly contain amyloid beta, a proinflammatory protein associated with protein misfolding and neurodegenerative pathology in Alzheimer's disease (*see* 6.08 Neurodegeneration).[17] The more highly cytotoxic composition of drusen derived from dying RPE cells may itself induce inflammation-induced oxidative stress, increased RPE cell death, and the formation of yet more drusen in a positive feedback loop.[18] Consequently, photoreceptor cell death occurs, since photoreceptors cannot survive without their supporting RPE cells. The loss of macular photoreceptors is termed 'geographic atrophy'. Without functioning macular photoreceptors, central visual acuity is gradually lost.

Although it is widely accepted that RPE cell dysfunction plays a central role in AMD disease progression, the complex interaction of the RPE with photoreceptors, Bruch's membrane, and the choroidal blood supply have made it difficult to unentangle the cause and effect of RPE cell death. Cell dysfunction likely begins with intralysosomal accumulation of a fluorescent material called lipofuscin. Lipofuscin is a complex mixture rich in polyunsaturates, and is probably derived from phagocytosed photoreceptor outer segments that are undigestible. Over time this material inactivates lysosomal enzymes and raises the pH level, causing lysosomal membrane breach and cell death. This process may be accelerated by efficient lipofuscin photon absorption and fluorescence to generate reactive oxygen species, which damages surrounding proteins and membranes.

A likely functionally important component of lipofuscin is the amphiphilic pyridinium ion A2E[19] (**Figure 3**), that is believed to be formed in vivo from the condensation of one equivalent of phosphatidylethanolamine with two equivalents of 11-*trans*-retinaldehyde, followed by phospholipase-D-catalyzed dephosphorylation. A2E might be the major component of lipofuscin, producing reactive oxygen species (ROS).[20] A2E is thought to be the major component of lipofucsin producing ROS in the presence of light and oxygen.[20] Enhanced production of A2E can occur as the result of the dysfunctional transport of photobleached 11-*trans* retinal/phosphatidyethanolamine Schiff base (NRPE) through the photoreceptor disk membrane resulting in the accumulation of RPE cell lysosomes. BNRPE transport ossurs via an ABCR (ATP-binding cassette transporter) mechanism. Individuals null for the ABCR gene suffer from Stardgardt's macular dystrophy, an early-onset macular degeneration-like disease.[21] Individuals with one allele of the ABCR gene may be at higher risk for developing AMD,[22] but this is not a major risk factor.

What are the characteristics of dry AMD that cause progression to the wet form in susceptible patients? The answer to this question may lie in the consequences of RPE cell dysfunction. As mentioned above, continuous RPE cell discharge of intracellular material, such as oxidized lipids and proinflammatory proteins, leads to accumulation of this material in drusen between the RPE and Bruch's membrane, and deposition of this material within Bruch's membrane.

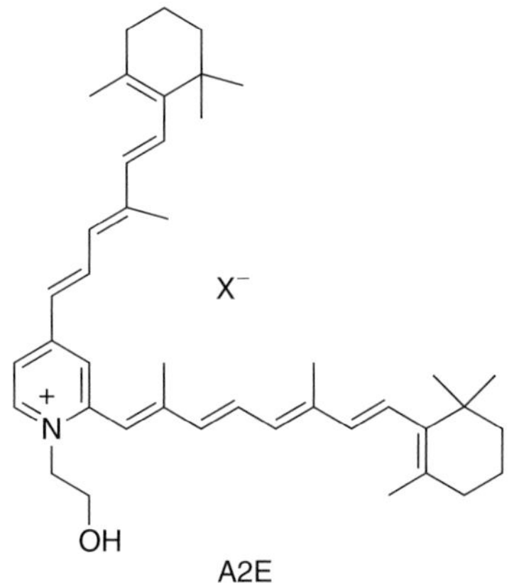

Figure 3 The structure of the amphiphilic pyridinium ion A2E.

Over time, this leads to thickening, increased protein cross-linking, and increased hydrophobicity of the membrane. Consequently, the membrane has decreased permeability both to plasma-borne nutrients (especially oxygen) from the choroidal capillaries and to waste products from remaining RPE cells. Ostensibly the resultant hypoxic condition induces stabilization and nuclear import of hypoxia-inducible transcription factors such as HIF-1α, which upregulate the production of pro-angiogenic proteins in an attempt to re-establish adequate blood flow. One of the most important of these proteins is probably vascular endothelial growth factor (VEGF), that promotes the proliferation of new capillaries from existing ones, and these breach Bruch's membrane. This leads to macular accumulation of fluid and blood from the leaky new vessels. VEGF is a potent factor in increasing blood vessel permeability and the formation of fibrous deposits and scar tissue in the retina, which rapidly cause retinal detachment and therefore loss of visual function.

Besides the morphological observations mentioned above, there is a considerable body of epidemiological evidence that suggests the importance of excessive inflammation and oxidative stress in the pathology of AMD.[15] For example, it is widely accepted that cigarette smoke, that is known to significantly increase oxidative stress in the body, is likely the most important environmental factor in increasing AMD risk.[6,23] An increased blood plasma concentration of the proinflammatory proteins, C-reactive protein (CRP) and interleukin-6 (IL-6) has also been positively correlated with AMD progression.[24]

Recent genetic and clinical studies have suggested that hyperactivation of the complement immune system is the ultimate source of the inflammation–AMD link. In a landmark group of publications, four independent laboratories have provided strong evidence that individuals harboring a specific mutation in the gene coding for the complement factor H protein (CFH) (i.e., the substitution of histidine for tyrosine – the Y402 H mutation) have upto a sevenfold increased risk of developing AMD when both alleles are affected.[25,26] It is estimated that as many as 50% of all AMD cases are due to this mutation. In addition, CFH polymorphisms were found that were associated with a reduced risk of AMD progression. CFH is an endogenous inhibitor of the innate immune system that binds to and inhibits/prevents the activation of inflammatory proteins like CRP. The Y402 H mutation probably generates a CFH variant that binds more weakly to its cognate proinflammatory protein(s), leading to a reduced endogenous anti-inflammatory response. The resultant chronic inflammatory stimulus leads to generation of ROS, increased tissue damage, and excessive remodeling that replaces functional with scar tissue.

Hyperactivation of the immune system may be initiated in susceptible individuals by infection. Exposure to *Chlamydia pneumoniae*[27] and cytomegalovirus[28] have been correlated with AMD disease progression.

A disturbance in lipoprotein/cholesterol metabolism has also been suggested to be a contributing factor to AMD. In particular, the $\varepsilon 4$ allele of the apolipoprotein E (apoE) gene has been suggested to be protective, with the $\varepsilon 2$ allele being detrimental, with respect to AMD development.[29,30] However, other studies have not found a statistically significant association between apoE ε allele status and AMD,[31] while an apoE $\varepsilon 4$ transgenic mouse develops an AMD-like disease faster than $\varepsilon 2$ and $\varepsilon 3$ variants.[32] These conflicting studies are likely to be due in part to confounding factors such as the different ethnic groups studied and a possibly insufficient consideration of blood cholesterol concentration independent of apoE ε allele status. Thus the influence of the magnitude and direction of the apoE ε gene influence on AMD progression is unclear. Nevertheless, a potential effect of lipid metabolism on disease development seems plausible, given the aforementioned hypothesized role of Bruch's membrane dysfunction in disease progression.

In summary, while there appear to be a variety of risk factors that influence AMD development current evidence suggests that hyperactivation of the complement component of the immune system with ensuing chronic inflammation and oxidative stress is the most important one. This hyperactivation likely results from a combination of genetic (e.g., CFH haplotype) and environmental (e.g., cigarette smoke, infection) components.

6.12.4 Experimental Disease Models

6.12.4.1 Glaucoma

The simplest assay system is represented by cells grown in vitro. Isolated cells of the TM, ciliary muscle, and ciliary process, and various retinal neurons in culture have proven useful for validating disease targets, for studying the mechanisms of action of ocularly indicated drugs, for studying the etiologies of ocular hypertension and glaucoma, and for discovering new drugs for treating ocular hypertension and glaucoma. For example, human CM and TM cells[33] have mRNAs and functionally active receptors for FP-class PGs that potently and efficaciously lower IOP in nonhuman primates and humans. These cells release MMPs[4,34] in response to FP-receptor activation and are involved in digesting the ECM in the anterior chamber, thereby enhancing the AH efflux and lowering IOP. Similar human TM and CM cell-based studies have demonstrated the existence of the cellular machinery for $5HT_2$ receptor agonists, which also induce ocular hypotension.[35] However, one major limitation of primary cultures of ocular cells is their inability to propagate through many passages, thereby limiting the number of studies. Therefore, immortalization of cells from the anterior chamber (e.g., TM3 cells) and retina (e.g., RGC-5 cells) has proven useful in the development of more robust in vitro models, as these engineered cells appear to represent the parent cells well in terms of their phenotypic responses and behaviors. Nevertheless, results from the immortalized cell should be interpreted with caution and should be replicated in normal primary cells.

An ex vivo model that has proven quite useful for directly measuring IOP are isolated perfused bovine, porcine, and human anterior eye segments in culture. Since the medium flows out of the TM this model represents a conventional outflow model that can actually remain viable for several days. Agents like dexamethasone have been shown to raise IOP, while other compounds, such as PGE_2, epinephrine, ethacrynic acid, ticrynafen, cytochalasin D, and bumetanide, have been shown to increase outflow.[36]

Clearly, the best models to study glaucoma are animals and human subjects. However, spontaneous glaucoma rarely occurs in mammals, although the beagle dog and the DBA/2J mouse exhibit secondary glaucoma due to anterior chamber synechiae or iridial pigment dispersion.[2] Consequently, ocular hypertensive animal models have been artificially created using a variety of techniques, such as laser-induced ablation of the outflow TM pathway in mice, rats, rabbits, and monkeys.[37,38,81] Other experimental animal models include measures of IOP in normal rat eyes, eyes that have received intracameral injections of hyaluronic acid,[39] or models where hypertonic saline is injected into the corneal veins[40] or where the episcleral veins are photocoagulated.[41] IOP can also been studied in normal and knockout mouse eyes.[3]

Rat models of ocular hypertension demonstrate IOP elevation that induces RGC cell death by apoptosis with subsequent axonal loss, and similar ONH damage as that observed in human glaucomatous neuropathy.[42] Rat models of partial, graded, or complete optic nerve transection are used to study mechanisms of axonal and RGC loss, and to assess the retinoprotective effects of various agents.[43] Cynomolgus and rhesus monkey eye responses to ocular hypotensive agents are usually quite predictive of human responses,[38,81] although the dose–response and duration of action parameters do not often match.

Other animal models being utilized for studying RGC death and for identifying agents that could attenuate RGC loss involve an ischemia–reperfusion insult to the retina following occlusion of the posterior ciliary artery supplying blood to the ONH.[2] Other more chronic models have been developed using rats, rabbits, and monkeys, where a vasoconstrictor such as endothelin is delivered to the ONH blood vessels via implanted delivery devices to cause a protracted but constant ischemic condition of the optic nerve. Furthermore, a model that reduces axonal transport within the optic nerve and thus kills the RGCs involves direct intravitreal injection of endothelin.[44]

6.12.4.2 Age-Related Macular Degeneration

Experimental disease models for AMD attempt to mimic what are thought to be important disease-contributing conditions in the retina, but they suffer from one major limitation: AMD is a chronic disease that takes years to develop, while the existing cellular and animal models produce morphological and functional characteristics on a weeks to months timescale. Largely this is out of necessity for timely evaluation of new chemical entities (NCEs), but partly this is due to the difficulty of efficient disease phenotype generation with a weak but long-lasting stimulus, which probably more accurately reflects pathology in man. As with most diseases, in vitro AMD models provide convenient screening tools and frequently use RPE cells, while being limited by the lack of heterogeneous, dynamically interacting components (e.g., RPE, photoreceptors, and Bruch's membrane) that an animal model provides. Nonhuman primates would constitute the preferred in vivo model; however, rodent models are mostly used due to ethical and cost issues and the availability of mouse genetic manipulation. Anatomic differences that may lessen the translatability of results from rodent models include a different surface volume/area ratio of eyes between man and rodents and the fact that rodents do not have a macula and thus technically cannot develop AMD.[2] Also, because current disease models focus on a single disease-inducing stimulus, they do not accurately reflect the RPE dysfunction–photoreceptor dropout–choridal capillary neovascular breach of Bruch's membrane–retinal detachment disease etiology, and thus might be expected to overestimate the anti-AMD efficacy of agents that are effective in those models.

6.12.4.2.1 Dry age-related macular degeneration

In vitro models of dry AMD typically subject an anatomically relevant cell type, such a photoreceptor or an RPE cell, to an AMD risk factor-linked stressor, e.g., high light flux or oxidative insult. The ability to prevent cytotoxicity by genetic manipulation or intervention with a protein or small molecule can then be measured.[45] Although the aforementioned lack of a complex, dynamically interacting matrix requires caution to interpretation of cell culture results, these models provide a useful filter for selection of candidates for evaluation in an animal model.

Most current in vivo models of dry AMD can be classified into two categories: genetic models and light-damage models, sometimes in combination.[46] Light-damage models typically expose rodents to bright white or blue light for a set period of time.[47] Popular genetic models used for drug candidate screening include:

- The Royal College of Surgeons (RCS) rat, which contains a recessive genetic defect that prevents RPE phagocytosis of photoreceptor shed outer segments, leading to photoreceptor and RPE cell death (the specific genetic defect causing retinal degeneration having been determined[48]).
- A rat with a proline-to-histidine mutation at amino acid position 23 of rhodopsin (P23H rat), which, although is more appropriate as a model for the degenerative disease, *retinitis pigmentosa*, is used as an AMD model.[49]
- The *abcr*$^{-/-}$ mouse in which the ABCR protein, involved in transport of 11-*trans*-retinaldehyde to the RPE cell for recycling back to the photoactive 11–*cis* isomer, is knocked out.[50]

Other genetic animal models of retinal diseases, have been reviewed.[51] One mouse model of the highly significant complement dysregulation–AMD link described above, where monocyte chemoattractant protein-1 (MCP-1) or its cognate receptor chemokine receptor-2 (CCR-2) is knocked out.[52] Appealingly, unlike other models it reproduces much of the human disease pathology, from drusen accumulation under the RPE to photoreceptor dropout to progression to choroidal neovascularization (i.e., wet AMD).

6.12.4.2.2 Wet age-related macular degeneration

The two most useful models of wet AMD are the oxygen-induced retinopathy (OIR) model and the laser-induced CNV model.[53] In the OIR model, neonatal animals (usually mice or rats) are initially exposed to a high oxygen concentration for 1–2 weeks, after which time room air is introduced. During the early hyperbaric period the disappearance of many capillaries is observed. In contrast, the lower oxygen concentration of room air is perceived as a hypoxic condition by the retina, evoking sprouting of new capillaries from existing ones. The neovascular response is believed to be due to secretion of pro-angiogenic proteins, e.g., VEGF. The OIR model simulates retinopathy of prematurity, a blinding disease sometimes observed in premature infants that are typically on oxygen support therapy at the beginning of life due to insufficient lung development. The development pattern of neovascularization is also somewhat reminiscent of that in proliferative diabetic retinopathy. However, since the OIR model does not involve the CNV response with breach of Bruch's membrane that is a hallmark of wet AMD, there are questions as to whether it is an appropriate stand-in for the disease.

The laser-induced CNV model involves rupture of Bruch's membrane, leading to an inflammatory/wound-healing response and concomitant CNV. Presumably the neovascularization response is due to upregulation of VEGF and other pro-angiogenic factors. This perhaps mimics wet AMD disease pathology better than the OIR model, in that the choroidal capillaries are explicitly involved in the neovascular response, and this model produces a similar angiographic appearance to the disease. This model, however, provides a more powerful angiogenic stimulus than likely occurs in wet AMD due to the more extensive injury to Bruch's membrane.

6.12.5 Clinical Trial Issues

6.12.5.1 Glaucoma

Clinical trials for testing ocular hypotensive agents in human subjects have all the attendant problems associated with any NCE, although the accessibility of the eye makes direct exposure to the target tissues much easier. Nevertheless, safety is of prime concern, since even topical ocular dosing can produce a high systemic exposure due to the ocular drainage from the tear ducts into the nasal passages, followed by exposure to the back of the throat, tongue, and ultimately the stomach. Since NCE absorption from the nasal epithelium, throat, and tongue is quite robust, topical ocular dosing needs to limit the amount that will ultimately enter the systemic circulation. Thus, for instance, topical ocular dosing with beta blockers, α_2-adrenergic agonists, and other systemically active drugs can have profound effects on the blood pressure, heart rate, and pulmonary circulation, as well as respiratory activity and CNS function. Other side-effects of topical ocular dosing of drugs (see below), including burning, stinging, foreign body sensation, and hyperemia, can severely affect patient compliance and thus the net result of the trial.

From an efficacy perspective, clinical trials of ocular hypotensive agents can be relatively short with well-defined endpoints of IOP reduction following once or twice daily dosing. However, recruitment of glaucoma and ocular hypertensive patients and the provision of the NCE in the correct formulation with reasonable shelf-life and stability are formidable challenges fraught with logistics issues and, ultimately, patient compliance.

6.12.5.2 Age-Related Macular Degeneration

6.12.5.2.1 Dry age-related macular degeneration

Perhaps the most significant hurdle for dry AMD clinical trials is that disease onset occurs late in life as a cumulative result of chronic damage to the RPE cells and their dependent photoreceptors. Most dry AMD patients are likely unaware of their disease until visual acuity begins to decline, at which point substantial morphological damage and dysfunction have occurred. It is thus unclear that intervening in a recognized AMD risk-increasing pathway (e.g., by administering an anti-inflammatory drug or by cessation of smoking) will work at this late stage. Since the RPE, photoreceptors, and Bruch's membrane target tissue are in the posterior segment of the eye, delivering an efficacious NCE concentration is usually ineffective using a topical eyedrop and instead requires locally invasive (e.g., intraocular injection) or systemic methods. On the positive side, since the most important disease risk factor is aging, a therapy that slows but does not arrest or reverse progression of dry AMD and its transition to wet AMD could still effectively prevent disease-induced blindness for many people due to life expectancy considerations (i.e., patients die of old age before they become blind). In addition, clinical determination of functional improvement using visual acuity testing is straightforward.

Although clinical trials for several different approaches to treat dry AMD could in theory be justified based on clinical observations and preclinical research (e.g., statins, apoptosis inhibitors, and anti-inflammatory drugs), there are only a few methods currently being investigated in clinical trials. One is the use of the carotenoid pigment, lutein (**Figure 4**) as an oral supplement. In theory the anti-oxidant and preferential retinal accumulation properties of lutein would be expected to have a salutary effect on dry AMD progression, but the unexpected negative effect of β-carotene consumption on lung cancer development in smokers counsels caution.

Another method being studied is the use of a blood filtration procedure called rheopheresis. In rheopheresis blood is removed from the patient, the platelets and plasma are separated, the plasma is filtered through a membrane designed to remove high molecular mass proteins, including low-density lipoprotein (LDL), and the platelets and plasma are reintroduced into the patient. The rationale behind this treatment can be summarized as follows: removal of high molecular mass proteins from blood leads to decreased blood viscosity and increased leaching of the same high molecular mass, permeability-decreasing proteins from Bruch's membrane. This in turn leads to improved diffusional exchange of oxygen/nutrients and waste products between the choroidal capillaries and the RPE through Bruch's membrane, and finally disease arrest. In a 43-patient study where eight rheopheresis filtrations were performed over 10 weeks, the treatment group demonstrated enhanced visual acuity compared with controls after 1 year.[54] Note that the functional benefit of treatment versus placebo was evident even though the rheopheresis group received no treatment for more than four-fifths of the time (42/52 weeks). Although the procedure has several drawbacks (e.g., a several hour filtration and reinfusion timeframe, intravenous intervention, and contraindication in those patients without adequate venous access), if the clinical benefit is confirmed in larger phase III clinical trials this could represent a major advance in dry AMD treatment.

6.12.5.2.2 Wet age-related macular degeneration

Due to the more rapidly progressing nature of wet than dry AMD, the most important clinical goal is to slow significantly CNV and the associated loss of visual acuity. As is the case for dry AMD, drug delivery remains a difficult issue, with intravenous and intraocular injection methods predominating in current clinical trials. An advantage for clinical trials for wet as opposed to dry AMD treatment is the more detailed knowledge of the immediate causes for the

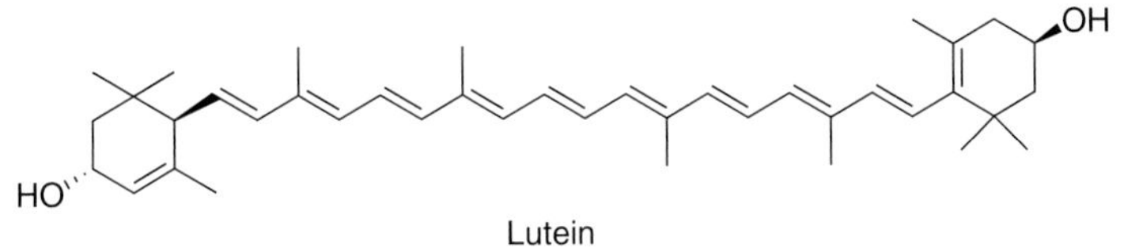

OH

HO

Lutein

Figure 4 The structure of the carotenoid pigment lutein.

former, largely due to research on cancer angiogenesis. Antiangiogenic agents to treat cancer also provide a pool of potential anti-wet-AMD therapeutics, e.g., squalamine, pegaptanib sodium, and ranibizumab.[55]

Table 1 summarizes ongoing and recently completed FDA-approved clinical trials for the treatment of wet AMD as of mid-2005 with structures shown in **Figure 5**.

Two important goals of potential anti-wet-AMD therapeutics in clinical trials are the ability to improve visual acuity and the use of a noninvasive, preferably local delivery method. With respect to the first objective, recently released phase III clinical data for the anti-VEGF antibody, ranibizumab[55] demonstrated that patients had improved, while the untreated control group had worse, visual acuity scores after receiving therapy. This is the first anti-wet-AMD therapy to demonstrate functional improvement in patients instead of only slowing loss and may represent an important therapeutic advance.

With respect to the second aim, there are a variety of drug-delivery methods being investigated for wet AMD treatment, including: topical ocular application (none yet in humans, but see ciliary neurotrophic factor (CNTF), nerve growth factor (NGF), and nepafenac information below), noninvasive posterior segment delivery as in a juxtascleral depot (e.g., anecortave acetate), oral delivery (e.g., celecoxib), intravitreal injection (e.g., ranibizumab), intravenous infusion (e.g., squalamine), and the use of a drug-impregnated intravitreal insert device (e.g., fluocinolone). While each of these methods has associated benefits and risks, in general the ideal is topical ocular application for reasons of convenience and minimization of local and systemic side effects. Although this is widely perceived to be a daunting problem in the art, recent publications indicate that this is surmountable.[56] Topical ocular application of recombinant CNTF provided almost complete functional and morphological retinal protection in streptozotocin-induced diabetic rats,[57] and topical ocular application of NGF to adult rats efficiently delivered the protein to the retina and ONH.[58] With respect to small molecules, topical ocular delivery of the nonsteroidal anti-inflammatory drug nepafenac,

Table 1 Therapeutic agents in clinical trials for wet AMD

Agent	*Mechanism*	*Delivery*	*Clinical trial phase*
Adenoviral-delivered PEDF gene	Endogenous anti-angiogenic and neuroprotective protein	Intravitreal injection	I
AG-13958	Small molecule VEGF kinase inhibitor	Sub-tenon injection	I
Anecortave acetate	Angiostatic cortisene	Posterior juxtascleral depot	III
cand5	siRNA against VEGF	Intravitreal injection	I
Celecoxib + verterporfin	COX-2 inhibitor + PDT	Oral celecoxib + intravenous verteporfin	II
Combretastatin A_4 phosphate	Microtubule disruption leading to neovascular capillary occlusion	Intravenous injection	I
Fluocinolone	Angiostatic steroid	Intravitreal implant	II
Ranibizumab	Antibody against three isoforms of VEGF	Intravitreal injection	III
Rostaporfin	PDT	Intravenous	III
siRNA-027	siRNA against VEGF	Intravitreal injection	I
Squalamine lactate	Angiostatic aminosterol	Intravenous	III
Talaporfin sodium	PDT	Intravenous	I
Triamcinolone acetonide + verteporfin	Anti-inflammatory?/angiostatic steroid + PDT	Triamcinolone intravitreal injection + intravenous verteporfin	III
VEGF trap	Antibody–VEGF receptor ligand-binding domain chimera that binds to VEGF	Intravenous	I

COX-2, cyclooxygenase-2 enzyme; PDT, photodynamic therapy (see description in Section 6.12.6.2.2.5) PEDF, pigment epithelium-derived factor; siRNA, small interfering ribonucleic acid.

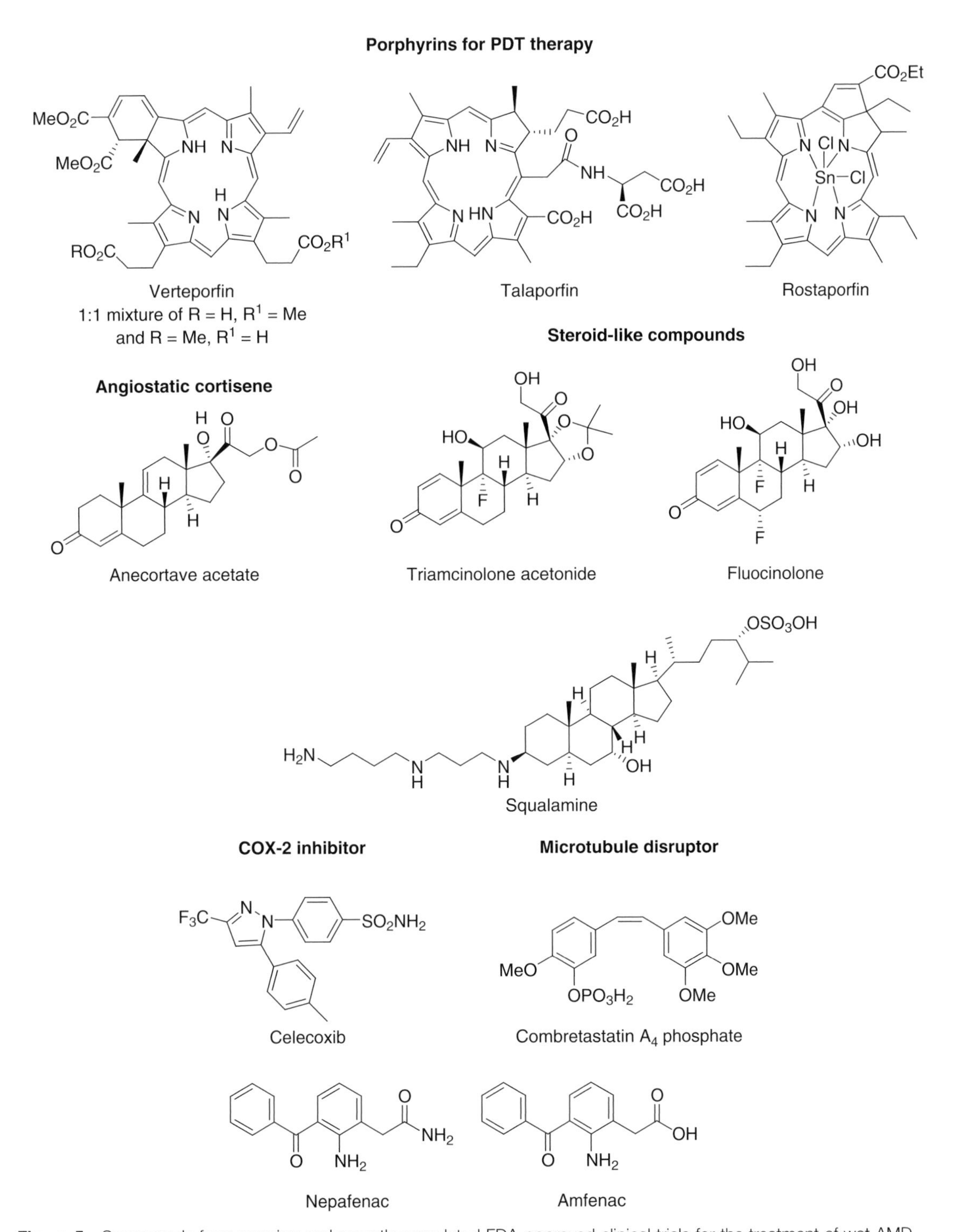

Figure 5 Compounds from ongoing and recently completed FDA-approved clinical trials for the treatment of wet AMD.

a pro-drug of the mixed COX-1/COX-2 inhibitor amfenac (**Figure 5**) inhibited retinal neovascularization in both the mouse OIR and laser-induced CNV models.[59] It is tempting to speculate that efficient diffusion of the uncharged pro-drug to the posterior segment is followed by retinal enzymatic hydrolysis to the active amfenac, which as an obligately charged species (an amino acid) inefficiently diffuses out of the retina.[60]

As a final note, the use of gene therapy to treat wet AMD, e.g., adenoviral vector-induced PEDF gene upregulation (**Table 1**) may offer advantages over other therapies in terms of reducing off-target activity, harnessing endogenous mechanisms for controlling angiogenesis, and decreasing the frequency of agent administration. However, usual gene therapy challenges apply, e.g., avoiding immune response and providing a method to quickly and efficiently switch off gene expression if necessary.

6.12.6 Current Treatments

6.12.6.1 Glaucoma

As mentioned above, the major risk factor for POAG is elevated IOP, and thus drugs that reduce IOP represent the mainstay treatment for glaucoma at present (**Figure 6**). Drugs that reduce AH production include β-adrenoceptor antagonists (e.g., timolol, betaxolol, and levobunolol), α_2-adrenoceptor agonists (e.g., brimonidine and apraclonidine), carbonic anhydrase inhibitors (e.g., dorzolamide and brinzolamide), and the rarely used Na/K-ATPase inhibitors (e.g., ouabain). The practice of medicine continues to limit the use of such inflow reducing drugs, as AH is important for the nourishment of the anterior chamber tissues. However, β blockers and carbonic anhydrase inhibitors are often prescribed first, partly because of their relatively lower cost, to begin treatment for ocular hypertension with a subsequent switching to use of drugs that enhance AH efflux. Conventional outflow-promoting drugs include muscarinic cholinergic agonists (e.g., pilocarpine), while uveoscleral outflow promoting drugs are represented by FP-class prostaglandins (e.g., latanoprost, travoprost, bimatoprost, and unoprostone isopropyl ester). Presently, the FP-class prostaglandins are the primary drugs of choice for treating ocular hypertension and glaucoma when cost is not an issue. The latter drugs work by liberating MMPs from the ciliary muscle cells (and perhaps from TM cells) that digest the ECM and help the AH egress via the uveoscleral pathway and, to a small degree, via the TM conventional outflow pathway.

Attempts have been made to combine both inflow- and outflow-enhancing drugs, primarily using timolol and an FP prostaglandin analog, thereby maximizing IOP reduction. However, the FDA has not approved such a combination use of such drugs, even though a combination of latanoprost and timolol is approved for glaucoma treatment in Europe, Canada, and Australia.

The mechanism of action of bimatoprost, the ethyl amide of 17–phenyl-$PGF_{2\alpha}$, despite the fact that it is structurally related to $PGF_{2\alpha}$ (**Figure 6**). An uncharacterized so-called 'prostamide receptor',[61] through which bimatoprost exerts its IOP-lowering action, is thought to exist.[62] Other investigators contend that bimatoprost is simply a pro-drug that is hydrolyzed in the eye to liberate the potent FP-receptor agonist, bimatoprost free acid (17–phenyl-$PGF_{2\alpha}$), which then activates the classic FP PG receptor to lower IOP.[62]

Those patients whose IOP is not controlled by drugs become candidates for surgical treatments in order to reduce the IOP and thus slow down their vision loss. Laser trabeculoplasty involves creating holes in the TM to promote AH drainage to lower the IOP. A recent in vitro study demonstrated that femtosecond laser photodisruption of human TM could create holes in the TM without collateral damage.[63] Glaucoma filtration surgery is usually the last resort and involves creating a fistula of the scleral tissue to the conjunctiva (the 'bleb') to promote drainage of the AH from the anterior chamber. While such surgery is very effective in lowering IOP, the patency of the bleb is often compromised and limited by the body's healing mechanism, which tends to cover up the drainage hole and eventually leads to elevation of the IOP again. Since the IOP cannot be regulated there is often a danger of anterior segment collapse. Drugs to prevent or slow down the local healing process in the bleb, such as mitomycin C or 5–fluoruracil (**Figure 7**), and thus keep the drainage pathway open show promise but have a narrow therapeutic index. However, agents that would release MMPs locally to keep the bleb open would be useful adjuncts to the surgical procedures in the future for reducing and maintaining the IOP at an acceptable level.

6.12.6.2 Age-Related Macular Degeneration

6.12.6.2.1 Dry age-related macular degeneration

There is currently no accepted pharmacological treatment of dry AMD. A prospective clinical trial designed to measure the effect of vitamin and mineral supplementation on the development of AMD reported that for the subgroup of persons at highest risk for developing advanced AMD, supplementation with a combination of high-dose vitamins E and C, β-carotene, and zinc afforded a 25% risk reduction for progression of high-risk patients to wet AMD.[64] The interpretation of the results of this study has been controversial.[65]

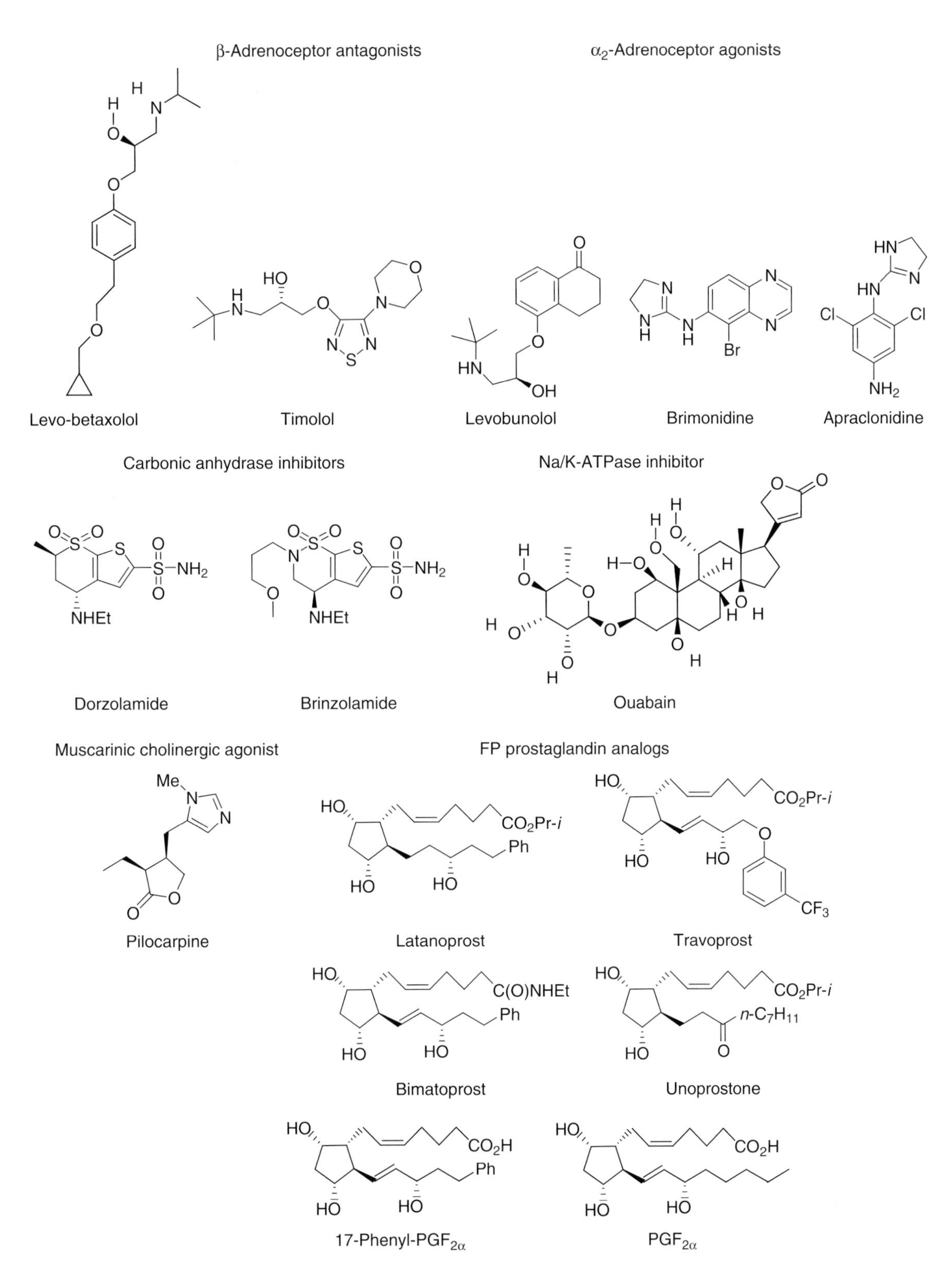

Figure 6 Drugs that reduce IOP.

6.12.6.2.2 Wet age-related macular degeneration

With respect to wet AMD, there are two approved drug treatments as of mid-2005. The first is photodynamic therapy (PDT), which was introduced in the 1999–2000 timeframe. The currently approved PDT therapy uses a solution of

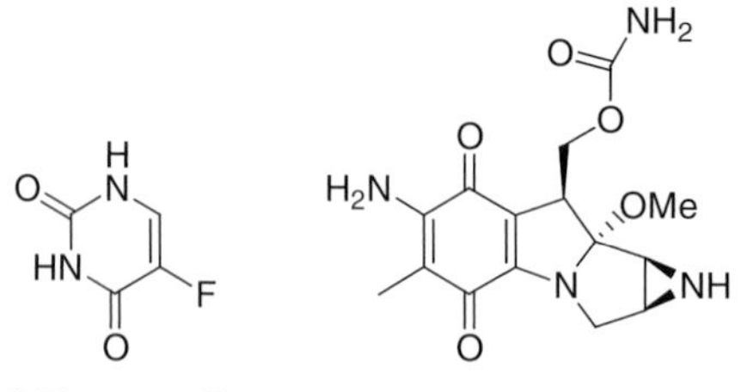

5-Fluorouracil

Mitomycin C

Figure 7 The structures of 5-fluoruracil and mitomycin C.

verteporfin (**Figure 5**) as the active agent. In PDT, a verteporfin-containing solution is administered intravenously to the patient, and the verteporfin apparently preferentially collects in the endothelium of neoangiogenic capillaries. About 15 min later a red laser is shone into the AMD-affected eye(s), and resultant photon absorption by the porphyrin produces an electronically excited state that transfers energy to oxygen to produce reactive oxygen species. Oxidation of molecules in the endothelium ensues, leading to vessel obstruction and collapse of the affected capillaries. The dosing frequency is approximately once every 3 months.

PDT is different from and offers several distinct advantages over laser retinal photocoagulation, which was previously the only treatment option for wet AMD. The latter case involves thermal destruction of the neovascular lesion with a laser, which because of the vagaries of laser targeting and thermal energy transfer leads to collateral destruction of some surrounding tissue. The destroyed tissue permanently loses visual functionality, and patients frequently experience an immediate drop in visual acuity. PDT, however, involves the use of a 'cold' laser that only directly transfers energy to the porphyrin. The reactive oxygen species produced as a downstream event is likely to react only with the closest surrounding tissue, leading to less collateral damage. The apparent preferential accumulation of verteporfin in newly formed capillaries also helps limit unwanted tissue destruction.

There are several limitations of PDT however. First, relative to placebo this therapy only slows the rate of loss, but does not improve visual acuity. Second, the dosing regimen using an intravenous infusion of verteporfin over a 10–15 minute period can be inconvenient. Third, after treatment the patient is somewhat photosensitive and is advised to avoid sun exposure for several days. Fourth, like laser photocoagulation, PDT works not by interrupting on the molecular level a fundamental mechanism of angiogenesis (e.g., by inhibiting an overexpressed receptor) but instead uses a massive chemical insult to destroy the lesion. As such it seems unlikely that the disease pathology/progression can be reversed or even significantly slowed. Fifth, the FDA currently approves PDT using verteporfin for the treatment of only the predominantly classic subtype of wet AMD, although there are clinical trials evaluating its use for prevention of disease progression in the minimally classic and occult sub-types.[66]

The second approved therapy uses a solution of the VEGF-binding aptamer pegaptanib sodium. Pegaptanib, a 28-residue, RNA-based oligonucleotide that is capped on both ends with a residue containing polyethylene glycol, has a molecular mass of approximately 50 kDa. The pegaptanib solution is injected into the vitreous humor of the patient every 6 weeks, and is approved for all three subtypes of wet AMD (classic, minimally classic, and occult). In phase III clinical trials AMD patients treated with 0.3 mg of pegaptanib solution over either a 1- or 2-year period experienced less loss of visual acuity than did patients receiving placebo.[67]

Pegaptanib is thought to work by binding with high affinity to the $VEGF_{165}$ isoform, which is believed to be the most pathologically important isoform.[68] Therefore $VEGF_{165}$ is prevented from binding to and activating the VEGF receptor. With the angiogenic signal thus blocked, new capillaries stop forming.

The main advantage of interfering with VEGF–VEGF receptor binding in general and pegaptanib in particular is that a fundamental pathological pathway is intercepted. Since pegaptanib is an aptamer, in theory there should be fewer problems with adverse immune system effects as compared with an antibody. The highly selective action of the drug suggests a favorable side-effect profile, at least with respect to off-target effects.

The disadvantages to pegaptanib therapy arise from drug specific and drug class considerations. Pegaptanib is administered by intravitreal injection, which is inconvenient, requires highly skilled delivery by an ophthalmologist, and in clinical trials has demonstrated a higher rate of intraocular infection (endophthalmitis) and retinal detachment than in controls.[67] It is not known if pegaptanib's lack of binding to other VEGF isoforms lessens its effectiveness as compared with pan-isoform binders like the anti-VEGF antibody ranibizumab (*see* Section 6.12.5.2.2). Pegaptanib does not improve visual acuity or arrest its degradation, which would be ultimately desirable.

Efficient blockade of the VEGF axis in general may hold pitfalls. VEGF functions as a protective agent in animal models of several neurodegenerative diseases, e.g., stroke (*see* 6.10 Stroke/Traumatic Brain and Spinal Cord Injuries) and amyotrophic lateral sclerosis (*see* 6.09 Neuromuscular/Autoimmune Disorders).[69] Genetically induced enhancement of VEGF production is being investigated as a treatment for coronary vascular disease.[70] Thus, direct interruption of the VEGF axis could inhibit the body's endogenous repair response to ocular neuronal and vascular deficiencies, which may occur with heightened frequency in AMD. In a related fashion, if excess VEGF production occurs as a response to perceived hypoxia in the retina, it is not clear that the underlying conditions causing the hypoxia are resolved when VEGF activity is inhibited.

6.12.7 Unmet Medical Needs

6.12.7.1 Glaucoma

Many drugs that are being used to lower and control IOP and thus slow down the progression of glaucomatous damage and vision loss. As with most drugs the side effects need serious consideration prior to prescription. Topical ocular use of FP-class PG analogs causes mild but protracted hyperemia; burning and stinging; iridial hyperpigmentation; orbital skin color pigmentation; and eyelash thickening and growth. The major side effect of topical ocular use of β blockers is transient stinging, burning, and some foreign body sensation. The systemic effects of these drugs are more pronounced and may include hypotension, bradycardia, palpitations, arrhythmias, and bronchospasms and other related pulmonary side-effects, and are thus contraindicated for asthmatic patients. Topical ocular β blockers may also cause CNS effects such as sleep disturbance, loss of memory and libido, depression, anxiety, and confusion. Pilocarpine and other muscarinic agonists, although much less prescribed nowadays, cause miosis, brow ache, and accommodative problems. Topical administration of carbonic anhydrase inhibitors can cause temporary but significant ocular surface discomfort, blurred vision, and a bitter taste, which sometimes limits their utility for treating ocular hypertension. Likewise, first-generation adrenoceptor agonists such as epinephrine cause brow ache, eye pain, headache, conjunctival hyperemia, and numerous systemic side effects, such as systemic hypertension and arrhythmias. Newer α_2-agonists, e.g., brimonidine and apraclonidine, can cause an ocular allergy-type reaction, manifested as hyperemia, itching, tearing, foreign body sensation, and conjunctival edema. Other side effects of these latter drugs are taste aversion, dry mouth/nose, headache, fatigue, lethargy, and sedation. Topical ocular use of hypotensive drugs and their side effects are discussed in detail elsewhere.[1,2]

Since some ocular hypertensive patients become refractory to treatment to some or all of the currently available drugs, there is a continuing need to discover new potent and efficacious drugs with reduced side effects. In addition, better surgical procedures and drugs that can render the latter more successful are still being sought.

Diagnosis of glaucoma has relied on short-wavelength automated perimetry (SWAP), frequency doubling perimetry (FDP),[2] and multifocal visual evoked potential (MVEP) measurement, which represent psychophysical measurements of vision and visual fields, coupled with scanning laser polarimetry (SLP)[2] and optical coherence tomography,[2] which measure retinal nerve fiber layer thickness and optic nerve cupping. However, further advances in these techniques and simpler and more predictive genetic tests are still warranted to aid the diagnosis of glaucoma patients at earlier stages of the disease process.

6.12.7.2 Age-Related Macular Degeneration

6.12.7.2.1 Dry age-related macular degeneration

The main unmet medical need is that there is currently no accepted medical therapy to treat dry AMD. Based on epidemiological studies many cases of dry AMD could be prevented by reducing cigarette smoking; for example, almost 30 000 cases of AMD in the UK could be attributed to smoking.[23] The role of oxidative stress in disease pathology suggests the advisability of antioxidant, carotenoid, and vitamin supplementation. One difficulty with dry AMD is that, as a largely asymptomatic and slow-developing disease, it is frequently the case that significant morphological damage (RPE cell and photoreceptor death) has occurred before the patient experiences visual deficit and thus is aware of the disease. On the positive side, a therapy that even just significantly slowed AMD progression in an elderly population might be sufficient to mostly preserve visual acuity for the rest of a patient's life.

6.12.7.2.2 Wet age-related macular degeneration

With the existing and pending approval of several treatment options, such as PDT, VEGF inhibitors, and angiostatic steroids, the prospects for wet-AMD treatment look bright. This is especially encouraging given the frequently aggressive course of the disease. Still, the two approved therapies as of mid-2005 (pegaptanib, PDT with verteporfin) can only slow the loss of visual acuity, instead of the ideal case of improving it or at least arresting its decline. However, recent exciting phase III clinical trial data for ranibizumab (see above) suggests the attainability of improving visual function.

The delivery of most of these agents is less than desirable: intravenous delivery, as used for example for verteporfin, is inconvenient and gives systemic exposure to the drug, while the intravitreal injection route, as used for example for pegaptanib, ranibizumab, and triamcinolone acetonide, carries the risk of enhanced intraocular infection. Less invasive delivery methods, as in the case of anecortave acetate, where a drug depot is created in the back of the eye by injection of the compound onto the posterior sclera without penetration of the eyeball, can offer a distinct benefit if a therapeutically relevant concentration of the drug can diffuse to the choroidal capillaries. In addition, the use of intravitreally placed drug devices impregnated with a drug may come to represent an important delivery method, especially if the frequency of implant removal and reinstallation is significantly less frequent than for other methods (e.g., once a year).

The ideal treatment for either wet or dry AMD would involve topical ocular or locally noninvasive delivery of the therapeutic agent, which would have little if any local or systemic side effects, could be administered infrequently (once a day or less often), and would significantly impede disease regression and improve visual acuity. Although such a treatment may require some time to be discovered, these criteria provide a useful benchmark with which to evaluate new therapies.

6.12.8 New Research Areas

6.12.8.1 Glaucoma

While FP prostaglandin analog esters currently dominate the therapeutic scene for treatment of ocular hypertension and glaucoma, investigators are keen to discover new drugs that may supplant the prostaglandins and/or be combined with the prostaglandins in the future. On the horizon are some new ocular hypotensive agents with a variety of mechanisms of action. These include protein kinase inhibitors such as *rho* kinase inhibitors (e.g., Y-27632)[71]; myosin-II ATPase inhibitor (blebbistatin)[72]; $5HT_2$ receptor agonists (α-methyl-5HT; AL-34662)[44]; adenosine agonists (2-alkynyladenosine); diadenosine polyphosphates[73]; calcium antagonists (nivaldepine and flunarizine)[74]; an angiotensin AT_1 receptor antagonist (CS-088)[75]; $5HT_{1A}$ agonists[76]; melatonin agonists[74]; dopamine agonists (PD128907, 3–PPP, CHF1035, and CHF1024)[77]; agents that degrade glyocosaminoglycans (AL-3037A)[78]; cannabinoid agonists[79]; natriuretic peptides[80]; marine macrolides (e.g., latrunculins and bumetamide)[2]; chloride transport inhibitors (ethacrynic acid)[82]; and various prostanoids of DP- and TP-class. Additional compounds of undefined mechanisms that appear to lower IOP in various animal models are listed in **Table 2**. However, due to major species differences in the ocular hypotensive activities of these agents, it is unclear which if any will become bona fide drugs to treat ocular hypertension and glaucoma in the future.

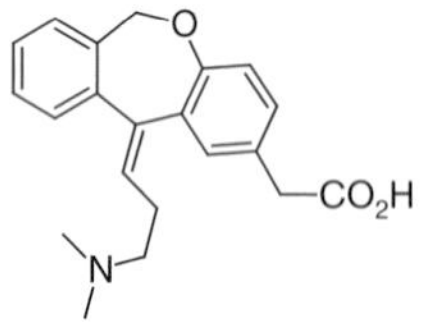

Olapatadine

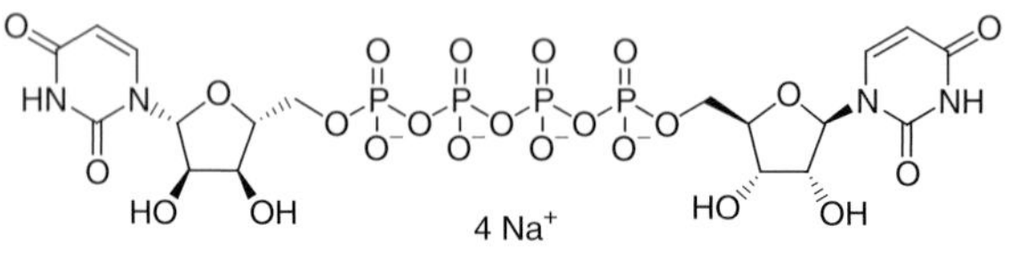

INS-365

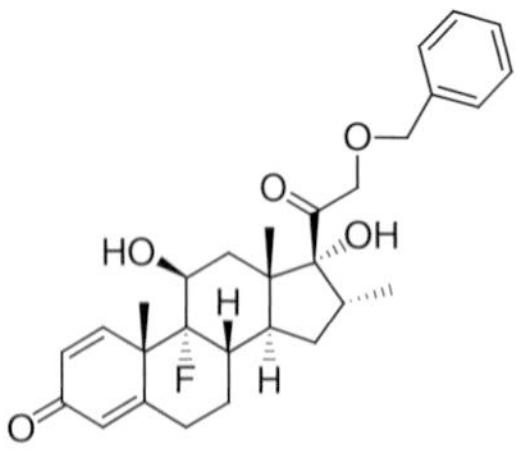

AL-2512

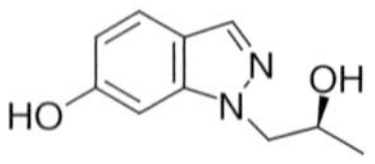

AL-34662

Since reduction of IOP in some patients fails to blunt the damage to OHN and/or reduce the loss of RGCs and the patients continue to lose visual acuity, it has been recognized that an important goal for treating glaucoma is to reduce or attenuate the progression of RGC death in addition to reducing ocular hypertension. The etiology of optic neuropathy has many routes spanning many different mechanisms that result in RGC demise and optic nerve damage, including mechanical constriction and vasoconstriction of retinal blood vessels, and increased sensitivity of RGCs

Table 2 Ocular hypotensive anti-glaucoma agents

Class of agent	*Specific drug or compound*	*Mechanism of action*
Inflow-inhibition agents		
β-Adrenergic antagonists	Timolol maleate, Betaxolol, Levobunolol, Metipranolol, Timolol hemihydrate	Block β-adrenergic receptors in the ciliary process, decrease AH formation
Carbonic anhydrase inhibitors	Dorzolamide, Brinzolamide	Inhibit ciliary process CA-II and CA-IV to reduce bicarbonate AH production
α_2-Adrenergic agents	Apraclonidine, Brimonidine, Clonidine	Stimulate α_2-adrenergic receptors in the ciliary process and thus decrease AH production.
Dopamine agonists	SDZ GLC-756, (S)-(−)-3-Hydroxyphenyl)-*N*-n-propylpiperidine (3-PPP) PD128907, CHF1035, CHF1024	Probably inhibit release of norepinephrine and prevent AH formation; may release natriuretic peptides
Na^+-K^+-ATPase inhibitors	Ouabain, Endothelin, Nitric oxide inhibitors	Inhibition of ciliary process Na^+-K^+-ATPase leading to direct inhibition of AH formation
Chloride channel inhibitors	5-Nitro-2-(3-phenylpropylamino)benzoate (NPPB)	Prevent ion fluxes of nonpigmented ciliary epithelial cell of the ciliary process to decrease AH formation
Conventional outflow agents		
Muscarinic agonists	Pilocarpine, Pilocarpine, Carbachol, Acecledine	Contract ciliary muscle to promote outflow of AH via the TM
Chloride transport inhibitors	Ethacrynic acid, Indacrinone, Ticrynafen	Inhibition of Na-K-Cl-transporter activity in the TM changes cell shape and volume increases AH efflux
Protein kinase inhibitors	Y-27632, Blebbistatin, Chelerythrine	Inhibition of rho kinase, myosin-II ATPase, and other kinases modifies actomyosin contractility, leading to changes in the actin cytoskeleton of TM and AH efflux
Marine macrolides	Latrunculins A and B, Bumetanide Swinholide	Sequestration of actin monomers and dimers in TM causes cell-shape change and AH efflux
Guanylate cylase activators	ANP, CNP, SNP	Type-A and type-B ANP receptor activation leads to cGMP production, TM relaxation, and AH efflux
κ-Opioid agonists	Bremazocine, Dynorphin	Release natriuretic peptides increases thus raise cGMP in TM, leading to relaxation and AH efflux
Cannabinoids	SR141716A	Cannabinoid receptor activation opens BK_{Ca} channels and relaxes TM, promotes AH efflux
Serotonin-2 agonists	α-Methyl-5HT, AL-34662	Probably contraction/relaxation of ciliary muscle and TM (may also promote uveoscleral outflow via release of MMPs)
MMP activators and AP-1 activators	t-Butylhydroquinone (t-BHQ), β-Naphthoflavone, 3-Methylcholanthrene, IL-1α	Agents that stimulate local production of MMPs cause ECM degradation around TM and open pathways for AH efflux
Uveoslceral outflow agents		
FP-class PGs	Travoprost, Latanoprost, Bimatoprost, Unoprostone isopropyl ester	Activation of FP receptors in CM and TM releases MMPs that break down ECM around CM and promote AH efflux
$5HT_2$ agonists	α-Methyl-5HT AL-34662	Contraction/relaxation of CM and TM by activation of $5HT_2$ receptors may release MMPs and/or PGs or other local mediators that stimulate CM remodeling and thus promote AH efflux

continued

Table 2 Continued

Class of agent	*Specific drug or compound*	*Mechanism of action*
Combination products		
	Timolol & Latanoprost, Timolol Dorzolamide, Timolol Travoprost	Inhibit AH formation and promote AH efflux via TM and CM-associated uveoscleral outflow pathway
Other IOP-lowering agents		
Melatonin agonists	Melatonin and analogs	Inflow inhibition
Purines	Agonists and antagonists, diadenosine polyphosphates	Alteration in signal transducer
Phosphatase Inhibitors	Calyculin	Outflow via multiple mechanisms
$5HT_{1A}$ agonists	Flesinoxan	Inflow and outflow promotion; confusing literature
Ca^{2+}-channel blockers	Verapamil, Nivaldipine, Brovincamine, Nifedipine	Most increase retinal blood flow to provide protection in normal-tension glaucoma patients
α-Adrenoceptor antagonists	Oxymetazoline, Ketanserin 5-Methylurapidil	Mostly via outflow, but mechanism needs to be defined
Endothelin agonists	Sarafotoxin-S6c	Undefined mechanism
Calcitonin gene-related peptide (CGRP)	CGRP	Apparently promotes outflow, especially when combined with $PGF_{2\alpha}$
MMPs and other enzymes or activators of enzymes	MMPs, Hyaluronidases, Chondroitinases AL-3037A, t-BHQ	Degrade ECM to promote outflow. May upregulate ECM-degrading enzymes such as MMPs
Prostanoids	AL-6598, BW245C, Sulprostone AGN192093	Multiple mechanisms of action involving cAMP production; Ca^{2+} mobilization leading to relaxation/contraction of ciliary muscles/TM

to locally produced neurotoxic agents (e.g., glutamate, oxidants, nitric oxide (NO), cytokines, and vasogenic/neurotoxic peptides like endothelin). Consequently, there is now a heightened search for drugs that are neuroprotective. Obviously, agents that possess both ocular hypotensive and neuroprotective activities will be extremely useful in the future. Even though much preclinical data support such dual pharmacophoric claims for brimonidine[83] and betaxolol,[84] the clinical endpoints for neuroprotection afforded by these drugs are not yet fully realized. In the meantime, numerous classes of drugs have been studied as potential neuroprotectants for the treatment of glaucoma (**Figure 8**), including caspase inhibitors, nitric oxide inhibitors (e.g., aminoguanidine), calcium channel blockers (e.g., nimodipine and betaxolol),[82] endothelin, and tumor necrosis factor receptor antagonists. Furthermore, glutamate receptor subtype antagonists (e.g., eliprodil, memantine, and CP101-606), flupirtine, HU-211, anti-apoptotic agents, and certain neurotrophins/growth factors have exhibited neuroprotective activity in experimental models of glaucoma.[2,3] Memantine is currently in clinical trials as a retinoprotectant,[85] but its clinical efficacy and utility remains undefined.

6.12.8.2 Age-Related Macular Degeneration

New research areas for the treatment of AMD will be guided by a combination of biochemical reasoning, results from in vitro and animal models, and retrospective and prospective epidemiological studies of disease-associated risk factors in humans. This last method represents an especially relevant source that will likely increasingly inform therapeutic research, as evidenced for example by the CFH mutation–AMD genetic linkage discovery.

Table 3 summarizes some potential new AMD research areas and therapeutic targets.

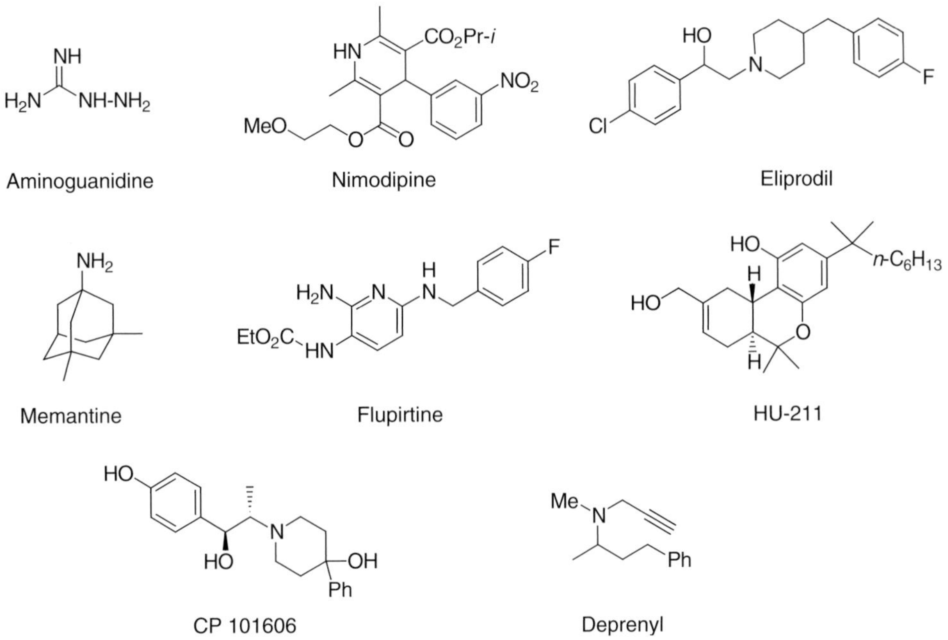

Aminoguanidine Nimodipine Eliprodil

Memantine Flupirtine HU-211

CP 101606 Deprenyl

Figure 8 Compounds that have been studied as potential neuroprotectants for the treatment of glaucoma.

Table 3 Possible new AMD therapeutic agents and areas

Therapy	*AMD type*	*Example*	*Potential advantages*	*Potential disadvantages*
Statins[7]	Wet and dry	Lovastatin	Extensive clinical experience in heart disease patients; co-morbidity of heart disease and AMD	Oral delivery leading to systemic side effects
ACE inhibitors/AT2 antagonists[7]	Wet and dry	Losartan	Extensive clinical experience in heart disease patients; co-morbidity of heart disease and AMD	Oral delivery leading to systemic side effects
Neurotrophic factors[89]	Dry	NGF	Possibility of small-molecule-driven upregulation; rescue of RPE cells harnessing endogenous mechanism	Neuronal survival may not persist if the basis of the chronic insult is not addressed
Apoptosis inhibitors[51]	Dry	Bcl-2	Possibility of small-molecule-driven upregulation; possible effectiveness in face of multiple insults	Rescue of distressed cells may not improve visual function; potential cancer risk?
ω-3 Fatty acids[90]	Dry and wet?	DHA	Photoreceptor cell membranes are rich in DHA; certain oxidative metabolites may be endogenous neuroprotectants	Overwhelming oxidative stress may generate electrophilic/ toxic fatty acid degradants
TNF-α antagonists[91]	Wet	Infliximab	Many existing agents already marketed	Side effect tolerance for some agents (cf. thalidomide)
Embryonic stem cell implantation[92]	Dry and wet	Embryonic stem cells	Replacement of dead photoreceptors could restore lost sight	Replacement cells may not have full function; cancer risk?

ACE, angiotensin-converting enzyme; AT_2, angiotensin II; DHA, (4*Z*,7*Z*,10*Z*,13*Z*,16*Z*,19*Z*)-4,7,10,13,16,19-docosahexaenoic acid; NGF, nerve growth factor; TNF-α, tumor necrosis factor α.

6.12.9 Other Diseases of the Retina

Vision is totally dependent on the ability of the retina to receive images from outside the eye, convert the information through photochemical and biochemical reactions into electrical signals, and transmit these signals to the visual cortex in the brain via the optic nerve. Consequently, the retina is an extremely complex tissue composed of ten different layers of cells that process the visual information and send it on to the brain. Defects and/or malfunctions in the physiology of these cells have a significant effect on vision. The major retinal diseases stem from either genetic defects or diabetes (and other systemic diseases) or are related to the normal or abnormal aging process such as AMD.

6.12.9.1 Drugs to Treat Diabetic Retinopathy and Angiogenesis

Diabetic retinopathy (DR) occurs in 27% of diabetics after 5–10 years of contracting diabetes and up to 90% have DR after more than 10 years of diabetes. DR essentially stems from retinal ischemia due to thickening of retinal capillary basement membranes when capillary pericytes start to die off. The ensuing ischemia causes the overexpression of VEGF, which in turn causes capillaries to leak and cause edema. VEGF also stimulates the growth of new blood vessels. This neovascularization could be perceived as a compensatory mechanism, but it is actually detrimental, since the new blood vessels disrupt retinal function. Deterioration of vision in DR is due to macular edema and proliferation of fibrovascular membranes that can lead to retinal detachment. Current treatments for DR are limited to laser photocoagulation of the leaky blood vessels, but this unfortunately destroys the underlying and surrounding retinal tissue. A much more specific approach is the use of photodynamic therapy, where a photosensitizing dye is given intravenously followed by ocular laser irradiation. Pharmacotherapy is not currently possible for DR, although a number of agents are in various stages of development.[86] These include the protein kinase C-β inhibitor (LY333531; ruboxistaurin mesylate), which blocks the VEGF signaling pathway[87]; intravitrially administered glucocorticoids (e.g., dexamethasone, trancinolone, and fluocinolone], which are expected to reduce retinal edema and perhaps also curb neovascularization; and nonsteroidal anti-inflammatory agents, such as celecoxib, which may reduce the local inflammation and edema in DR.[86] It is possible that the $P2Y_2$ agonist INS37217 may find future utility in DR treatment by promoting fluid absorption away from the edematous sites.[73] Angiostatic cortisines, e.g., anecortave acetate, may also prove useful for treating the neovascularization found in DR.[88]

6.12.10 Ocular Surface Diseases

The cornea and conjunctiva are exposed to the environment and are thus susceptible to overexposure to light, radiation, chemicals, and airborne pathogens and allergens. Consequently, ocular surface infections, allergies, dryness, and ulcerations represent the major disorders of the cornea and conjunctiva. While these conditions are not blinding by themselves, they severely limit the quality of life for millions of people.

6.12.10.1 Antiallergic Drugs

Allergic conjunctivitis and perennial conjunctivitis are fairly acute ocular surface disorders. However, vernal conjunctivitis, atopic keratonjunctivitis, and giant papillary conjunctivitis are chronic disorders.[93] The hallmarks of allergic conjunctivitis are itching, redness, swelling, tearing, and temporary acute photophobia caused by various mast cell mediators released from mast cells when an allergen contacts the conjunctiva. The acute allergic conjunctivitis may develop into a chronic disease if left untreated, and this causes corneal and conjunctival remodeling and ulceration, sometimes accompanied by bacterial infection. Since mast-cell-derived histamine is the major culprit in allergic conjunctivitis, topical ocular antihistamines, such as emedastine, levocabastine, azelastine, and ketotifen, were considered the drugs of choice, often supplemented with vasconstrictors (e.g., oxymetazoline) and edema reducers (nonsteroidal anti-inflammatory agents and corticosteroids). However, with the discovery of a dual pharmacophoric drug, olopatadine,[94] it became clear that agents with mast-cell stabilizing and antihistaminic activities are the much preferred drugs of choice in the treatment of allergic conjunctivitis. Olopatadine offers both these activities and has a long duration of action, as its clearance from the histamine-1 receptor is very slow. A single drop of olopatadine was effective against ocular itching for up to 8 hours and was shown to be more efficacious than a 2-week load with nedocromil, a mast-cell stabilizer, in a model of allergic conjunctivitis.

Suitable in vitro models for testing ocular antiallergic and/or antihistaminic drug candidates include the use of isolated human conjunctival mast cells,[96] human conjunctival and corneal epithelial cells, and corneal and conjunctival fibroblasts.[94]

6.12.10.2 Drugs for Treating Dry Eye

Another series of disorders of the ocular surface is 'dry eye.' Dry eye is characterized by deficits in tear production/secretion and deficiencies in the quality of tears, thereby causing ocular discomfort, itching, and a foreign body sensation on the ocular surface.[95] It has recently been recognized that several aspects of dry eye involve an inflammatory cascade. Sjögren's syndrome (keratoconjunctivis sicca) is a common (affects 0.5% of adult women) autoimmune disorder of the lacrimal and salivary glands that causes ocular dryness. Treatments for dry eye have traditionally sought to reduce the symptoms of dryness by hydrating and lubricating the ocular surface with artificial tears. However, this is rarely sufficient and pharmacotherapy is necessary to prevent corneal and/or conjunctival damage. Therapeutic approaches include use of immune suppressive/anti-inflammatory agents, such as the fungal-derived peptide, cyclosporin, and topical corticosteroids, such as methylprednisolone, loteprednol etabonate, and fluorometholone. However, the relatively low efficacy of cyclosporin and the IOP-raising and cataract forming side effects of steroids limit their utility. Perhaps corticosteroids with reduced liabilities, such as AL-2512, may be more suitable for such treatment. New-generation anti-inflammatory agents, such as nepafenac,[66] may also find utility in dry eye treatment in the future. More recent approaches include the use of $P2Y_2$ purinergic agonists, such as INS-365, to increase ocular moisture by promoting water transport across the conjunctiva[73] and thereby alleviating the dryness of the ocular surface. Perhaps the most promising therapeutic agents for treating dry eye involve increasing the local release of ocular surface mucins by 15–S-HETE,[97] thereby lubricating the ocular surface and retaining moisture.

6.12.10.3 Anti-Inflammatory Drugs

Trauma to the eye, including bacterial infections, causes local inflammation. Thus antibiotics are often prescribed with anti-inflammatory drugs. The most common inflammatory ocular conditions are conjunctivitis, uveitis, and corneal keratitis, which require the use of glucocorticoids and nonsteroidal anti-inflammatory agents that prevent generation of prostaglandins and other autocoids. Rimexolone is an effective ocular inflammatory steroid with a low potential for raising IOP. Loteprednol etabonate is claimed as a 'soft,' easily metabolized steroid with minimal IOP effects. Chronic inflammation of the ocular surface can lead to corneal ulceration and neovascularization, which impair vision. Treatment with angiostatic steroids, such as anecortave acetate, can prevent corneal neovascularization,[98] although some other agents may also be useful.[88] Nonsteroidal anti-inflammatory agents, such as nepafenac, ketorolac, bromfenac, flubiprofen, and diclofenac, lead the way as drugs of choice for treating ocular inflammation. However, corneal defects are a major cause of concern related to the use of these agents.

6.12.10.4 Anti-Infective Drugs

The greatest hurdle for the treatment of ocular infections such as bacterial keratitis, as with other types of infections, is the emergence of bacterial resistance. Ocular bacterial infections are currently treated with fourth-generation fluoroquinolones, such as moxifloxacin and gatifloxicin.[99] Levofloxacin, ciprofloxacin, and ofloxacin are being less frequently prescribed due to the development of bacterial resistance to these somewhat older antibiotics. Future anti-infective products must be broad spectrum, efficacious, prophylactically useful, comfortable, convenient to use, and nontoxic, in particular being free of benzalkonium chloride.[99]

6.12.11 Conclusion

The major sight-threatening diseases are glaucoma, AMD, and diabetic retinopathy. Intense research is ongoing to find more potent and efficacious drugs with reduced side-effect potential to treat these diseases. Traditional research centered around receptor-based pharmaceutics, but now more emphasis seems to be directed toward discovering and developing enzyme inhibitors. Furthermore, agents that impart multiple therapeutically useful activities, for instance IOP-lowering and retinoprotective agents, are being actively pursued to add value to the treatment of these ocular diseases. A better ability to noninvasively diagnose and treat retinal diseases and dysfunctions at an earlier stage would greatly enhance the overall success rate of treatment. The generation of novel techniques (e.g., proteomics and gene therapy) and reagents such as small interfering RNA (siRNA)[100] may represent tomorrow's 'drugs' to treat ocular diseases such as AMD, diabetic retinopathy, and glaucoma. We look forward to such advances in the drug discovery/development and disease diagnosis processes.

References

1. Sugrue, M. F. *J. Med. Chem.* **1997**, *40*, 2793–2809.
2. Clark, A. C.; Yorio, T. *Nat. Rev. Drug Disc.* **2003**, *2*, 448–459.
3. Pang, I.-H.; Wang, W.-H.; Clark, A. C. *Exp. Eye Res.* **2005**, *80*, 207–214.
4. Allingham, R. R.; Wiggs, J. L.; Hauser, E. R.; Larocque-Abramson, K. R.; Santiago-Turla, C.; Broomer, B.; Del Bono, E. A.; Graham, F. L.; Haines, J. L.; Pericak-Vance, M. A. et al. *Invest. Ophthalmol. Vis. Sci.* **2005**, *46*, 2002–2005.
5. Kincaid, M. C. In *Age-Related Macular Degeneration: Principles and Practice*; Hampton, G. R., Nelson, P. T., Eds.; Raven Press: New York, 1992, Chapter 2, pp 37–61.
6. Clemons, T. E.; Milton, R. C.; Klein, R.; Seddon, J. M.; Ferris, F. L., III. *Ophthalmology* **2005**, *112*, 533–539.
7. Friedman, E. *Br. J. Ophthalmol.* **2004**, *88*, 161–163.
8. Friedman, D. S.; O'Colmain, B. J.; Munoz, B.; Tomany, S. C.; McCarty, C.; de Jong, P. T.; Nemesure, B.; Mitchell, P.; Kempen, J. *Arch. Ophthalmol.* **2004**, *4*, 564–572.
9. Congdon, N.; O'Colmain, B.; Klaver, C. C.; Klein, R.; Munoz, B.; Friedman, D. S.; Kempen, J.; Taylor, H. R.; Mitchell, P. *Arch. Ophthalmol.* **2004**, *122*, 477–485.
10. Fuchshofer, R.; Welge-Lussen, U.; Luten-Drecoll, E. *Exp. Eye Res.* **2003**, 77, 757–765.
11. Choi, J.; Miller, A. M.; Nolan, M. J.; Yue, B. Y. J. T.; Thotz, S. T.; Clark, A. C.; Agarwal, N.; Knepper, P. A. *Invest. Ophthalmol. Vis. Sci.* **2005**, *46*, 214–222.
12. Urieli-Shoval, S.; Cohen, P.; Eisenberg, S.; Matzner, Y. *J. Histochem. Cytochem.* **1998**, *46*, 1377–1384.
13. McKinnon, S. J.; Lehman, D. M.; Kerrigan-Baumrind, L. A.; Merges, C. A.; Pease, M. E.; Kerrigan, D. F.; Ransom, N. L.; Tahzib, N. G.; Reitsamer, H. A.; Levkovitch-Verbin, H. et al. *Invest. Ophthalmol. Vis. Sci.* **2002**, *43*, 1077–1087.
14. Guo, Li.; Moss, S. E.; Alexander, R. A.; Ali, R. R.; Fitzke, F. W.; Cordeiro, M. F. *Invest. Ophthalmol. Vis. Sci.* **2004**, *46*, 175–182.
15. Zarbin, M. A. *Arch. Ophthalmol.* **2004**, *122*, 598–614.
16. Hageman, G. S.; Luthert, P. J.; Chong, N. H. V.; Johnson, L. V.; Anderson, D. H.; Mullins, R. F. *Prog. Ret. Eye Res.* **2001**, *20*, 705–732.
17. Anderson, D. H.; Talaga, K. C.; Rivest, A. J.; Barron, E.; Hageman, G. S.; Johnson, L. V. *Exp. Eye Res.* **2004**, *78*, 243–256.
18. Anderson, D. H.; Mullins, R. F.; Hageman, G. S.; Johnson, L. V. *Am. J. Ophthalmol.* **2002**, *134*, 411–431.
19. Ben-Shabat, S.; Parish, C. A.; Hashimoto, M.; Liu, J.; Nakanishi, K.; Sparrow, J. R. *Bioorg. Med. Chem. Lett.* **2001**, *11*, 1533–1540.
20. Pawlak, A.; Wrona, M. *Photochem. Photobiol.* **2003**, 77, 253–258.
21. Azarian, S. M.; Travis, G. H. *FEBS Lett.* **1997**, *409*, 247–252.
22. Allikmets, R.; Shroyer, N. F.; Singh, N.; Seddon, J. M.; Lewis, R. A.; Bernstein, P. S.; Peiffer, A.; Zabriskie, N. A.; Li, Y.; Hutchinson, A. et al. *Science* **1997**, *277*, 1805–1807.
23. Evans, J. R.; Fletcher, A. E.; Wormald, R. P. *Br. J. Ophthalmol.* **2005**, *89*, 550–553.
24. Seddon, J. M.; George, S.; Rosner, B.; Rifai, N. *Arch. Ophthalmol.* **2005**, *123*, 774–782.
25. Klein, R. J.; Zeiss, C.; Chew, E. Y.; Tsai, J. Y.; Sackler, R. S.; Haynes, C.; Henning, A. K.; SanGiovanni, J. P.; Mane, S. M.; Mayne, S. T. et al. *Science* **2005**, *308*, 385–389.
26. Haines, J. L.; Hauser, M. A.; Schmidt, S.; Scott, W. K.; Olson, L. M.; Gallins, P.; Spencer, K. L.; Kwan, S. Y.; Noureddine, M.; Gilbert, J. R. et al. *Science* **2005**, *308*, 419–421.
27. Kijlstra, A.; La Heij, E.; Hendrikse, F. *Ocul. Immunol. Inflamm.* **2005**, *13*, 3–11.
28. Miller, D. M.; Espinosa-Heidmann, D. G.; Legra, J.; Dubovy, S. R.; Suner, I. J.; Sedmak, D. D.; Dix, R. D.; Cousins, S. W. *Am. J. Ophthalmol.* **2004**, *138*, 323–328.
29. Zareparsi, S.; Reddick, A. C.; Branham, K. E.; Moore, K. B.; Jessup, L.; Thoms, S.; Smith-Wheelock, M.; Yashar, B. M.; Swaroop, A. *Invest. Ophthalmol. Vis. Sci.* **2004**, *45*, 1306–1310.
30. Schmidt, S.; Klaver, C.; Saunders, A.; Postel, E.; De La Paz, M.; Agarwal, A.; Small, K.; Udar, N.; Ong, J.; Chalukya, M. et al. *Ophthalmic Genet.* **2002**, *23*, 209–223.
31. van Leeuwen, R.; Klaver, C. C.; Vingerling, J. R.; Hofman, A.; van Duijn, C. M.; Stricker, B. H.; de Jong, P. T. *Am. J. Ophthalmol.* **2004**, *137*, 750–752.
32. Malek, G.; Johnson, L. V.; Mace, B. E.; Saloupis, P.; Schmechel, D. E.; Rickman, D. W.; Toth, C. A.; Sullivan, P. M.; Rickman, C. B. *Proc. Natl. Acad. Sci. USA* **2005**, *102*, 11900–11905.
33. Sharif, N. A.; Williams, G. W.; Crider, J. Y.; Xu, S. X.; Davis, T. L. *J. Ocular Pharmacol. Ther.* **2004**, *20*, 489–508.
34. Weinreb, R. N.; Lindsey, J. D. *Invest. Ophthalmol. Vis. Res.* **2002**, *43*, 716–722.
35. Sharif, N. A.; Kelly, C. R.; Crider, J. Y.; Senchyna, M. *Invest. Ophthalmol. Vis. Sci.* **2005**, Suppl. Abst. 3688.
36. Pang, I.-H.; McCartney, M. D.; Steeley, H. T.; Clark, A. C. *J. Glaucoma* **2000**, *9*, 468–479.
37. Anderson, M. G.; Smith, R. S.; Hawes, N. L.; Zabaleta, A.; Chang, B.; Wiggs, J. L.; John, S. W. *Nature Genet.* **2002**, *30*, 81–85.
38. May, J. M.; McLaughlin, M. A.; Sharif, N. A.; Hellberg, M. R.; Dean, T. R. *J. Pharmacol. Expt. Ther.* **2003**, *306*, 301–309.
39. Pang, I.-H.; Wang, W.-H.; Millar, J. C.; Clark, A. C. *Exp. Eye Res.* **2005**, *81*, 359–360.
40. Moreno, M. C.; Marcos, H. J. A.; Croxatto, J. O.; Sande, P. H.; Campanelli, J.; Jaliffa, C. O.; Benozzi, J.; Rosentein, R. E. *Exp. Eye Res.* **2005**, *81*, 71–80.
41. Shareef, S. R.; Garcia-Valenzuela, E.; Salierno, A.; Walsh, J.; Sharma, S. C. *Exp. Eye Res.* **1995**, *61*, 379–382.
42. Reitsamer, H. A.; Kiel, J. W.; Harrison, J. M.; Ransom, N. L.; McKinnon, S. J. *Exp. Eye Res.* **2004**, *78*, 799–804.
43. Avila, M. Y.; Stone, R. A.; Civan, M. M. *Invest. Ophthalmol. Vis. Res.* **2002**, *43*, 3021–3026.
44. Stokely, M. E.; Brady, S. T.; Yorio, T. *Invest Ophthalmol. Vis. Sci.* **2002**, *43*, 3223–3230.
45. Crawford, M. J.; Krishnamoorthy, R. R.; Rudick, L. V.; Collier, R. J.; Kapin, M.; Agarwal, B. B.; Al-Ubaidi, M. R.; Agarwal, N. *Biochem. Biophys. Res. Commun.* **2001**, *281*, 1304–1312.
46. Chader, G. J. *Vision Res.* **2002**, *42*, 393–399.
47. Verkman, A. S. *Exp. Eye Res.* **2003**, *76*, 137–143.
48. Agarwal, N.; Martin, E.; Krishnamoorthy, R. R.; Landers, R.; Wen, R.; Krueger, S.; Kapin, M. A.; Collier, R. J. *Exp. Eye Res.* **2002**, *74*, 445–453.
49. D'Cruz, P. M.; Yasumura, D.; Weir, J.; Matthes, M. T.; Abderrahim, H.; LaVail, M. M.; Vollrath, D. *Hum. Mol. Genet.* **2000**, *9*, 645–651.

50. Machida, S.; Kondo, M.; Jamison, J. A.; Khan, N. W.; Kononen, L. T.; Sugawara, T.; Bush, R. A.; Sieving, P. A. *Invest. Ophthal. Vis. Sci.* **2000**, *41*, 3200–3209.
51. Fauser, S.; Luberichs, J.; Schuttauf, F. *Surv. Ophthalmol.* **2002**, *47*, 357–367.
52. Ambati, J.; Anand, A.; Fernandez, S.; Sakurai, E.; Lynn, B. C.; Kuziel, W. A.; Rollins, B. J.; Ambati, B. K. *Nat. Med.* **2003**, *9*, 1390–1397.
53. Ambati, J.; Ambati, B. K.; Yoo, S. H.; Ianchulev, S.; Adamis, A. P. *Surv. Ophthalmol.* **2003**, *48*, 257–293.
54. Pulido, J. S.; Sanders, D.; Klingel, R. *Can. J. Ophthalmol.* **2005**, *40*, 332–340.
55. Melnikova, I. *Nat. Rev. Drug Disc.* **2005**, *4*, 711–712.
56. Koevary, S. B. *Curr. Drug Metab.* **2003**, *4*, 213–222.
57. Aizu, Y.; Katayama, H.; Takahama, S.; Hu, J.; Nakagawa, H.; Oyanagi, K. *Neuroreport* **2003**, *14*, 2067–2071.
58. Lambiase, A.; Tirassa, P.; Micera, A.; Aloe, L.; Bonini, S. *Invest. Ophthalmol. Vis. Sci.* **2005**, *46*, 3800–3806.
59. Takahashi, K.; Saishin, Y.; Mori, K.; Ando, A.; Yamamoto, S.; Oshima, Y.; Nambu, H.; Melia, M. B.; Bingaman, D. P.; Campochiaro, P. A. *Invest. Ophthalmol. Vis. Sci.* **2003**, *44*, 409–415.
60. Ke, T. L.; Graff, G.; Spellman, J. M.; Yanni, J. M. *Inflammation* **2000**, *24*, 371–384.
61. Woodward, D. F.; Krauss, A. H.-P.; Chen, J.; Lai, R. K.; Spada, C. S.; Burke, R. M.; Andrews, S. W.; Shi, L.; Liang, Y.; Kedzie, K. M. et al. *Surv. Ophthalmol.* **2001**, *45*, S337–S345.
62. Hellberg, M. R.; Ke, T.-L.; Haggard, K.; Klimko, P. G.; Dean, T. R.; Graff, G. *J. Ocular Pharmacol. Ther.* **2003**, *19*, 97–103.
63. Toyran, S.; Liu, Y.; Sigha, S.; Shan, S.; Cho, M. R.; Gordon, R. J.; Edward, D. P. *Exp. Eye Res.* **2005**, *81*, 298–305.
64. Age-Related Eye Disease Study Research Group Report 8. *Arch. Ophthalmol.* **2001**, *119*, 1417–1436.
65. Gaynes, B. I. *Arch. Ophthalmol.* **2003**, *121*, 416–417.
66. Azab, M.; Boyer, D. S.; Bressler, N. M.; Bressler, S. B.; Cihelkova, I.; Hao, Y.; Immonen, I.; Lim, J. I.; Menchini, U.; Naor, J. et al. *Arch. Opthalmol.* **2005**, *123*, 448–457.
67. Gragoudas, E. S.; Adamis, A. P.; Cunningham, E. T., Jr.; Feinsod, M.; Guyer, D. R. *N. Engl. J. Med.* **2004**, *351*, 2805–2816.
68. Adamis, A. P.; Shima, D. T. *Retina* **2005**, *25*, 111–118.
69. Storkebaum, E.; Lambrechts, D.; Carmeliet, P. *Bioessays* **2004**, *26*, 943–954.
70. Syed, I. S.; Sanborn, T. A.; Rosengart, T. K. *Cardiology* **2004**, *101*, 131–143.
71. Tian, B.; Kaufman, P. L. *Exp. Eye Res.* **2005**, *80*, 215–225.
72. Zhang, M.; Rao, P. V. *Invest. Ophthalmol. Vis. Sci.* **2005**, *46*, E-Abst. 1351.
73. Pintor, J. *Curr. Opin. Invest. Drugs* **2005**, *6*, 76–80.
74. Campana, G.; Bucolo, C.; Murari, G.; Spampinato, S. *J. Pharmacol. Expt. Ther.* **2002**, *303*, 1086–1094.
75. Wang, R.-F.; Podos, S. M.; Mittag, T. W.; Yokoyoma, T. *Exp. Eye Res.* **2005**, *80*, 629–632.
76. Chidlow, G.; Cupido, A.; Melena, J.; Osborne, N. N. *Curr. Eye Res.* **2001**, *23*, 144–153.
77. Chu, E.; Socci, R.; Chu, T. C. *J. Ocular Pharmacol. Ther.* **2004**, *20*, 15–23.
78. Pang, I.-H.; Moll, H.; McLaughlin, M. A.; Knepper, P. A.; DeSantis, L.; Epstein, D. L.; Clark, A. C. *Exp. Eye Res.* **2001**, *73*, 815–825.
79. Chein, F. Y.; Wang, R.-F.; Mittag, T. W.; Podos, S. M. *Arch. Ophthalmol.* **2003**, *121*, 87–90.
80. Millar, J. C.; Shahidullah, M.; Wilson, W. S. *J. Ocular Pharmacol. Ther.* **1997**, *13*, 1–11.
81. Hellberg, M. R.; McLaughlin, M. A.; Sharif, N. A.; DeSantis, L.; Dean, T. R.; Kyba, E. P.; Bishop, J. E.; Klimko, P. G.; Zinke, P. W.; Selliah, R. D. et al. *Surv. Ophthalmol.* **2002**, *47*, S13–S33.
82. Epstein, D. L.; Roberts, B. C.; Skinner, L. L. *Invest. Ophthalmol. Vis. Res.* **1997**, *38*, 1526–1534.
83. Woldemussie, E.; Ruiz, G.; Wijono, M.; Wheeler, L. A. *Invest. Ophthalmol. Vis. Sci.* **2001**, *42*, 2849–2855.
84. Osborne, N. N.; Cazevieille, C.; Carvalho, A. L.; Larsen, A. K.; DeSantis, L. *Brain Res.* **1997**, *751*, 113–123.
85. Hare, W.; Woldemussie, E.; Lai, R. K.; Ton, H.; Ruiz, G.; Chun, T.; Wheeler, L. *Invest. Ophthalmol. Vis. Sci.* **2004**, *45*, 2625–2639.
86. Speicher, M. A.; Danis, R. P.; Criswell, M.; Pratt, L. *Expert Opin. Emerg. Drug* **2003**, *8*, 239–250.
87. Aiello, L. P. *Surv. Ophthalmol.* **2002**, *47*, S263–S269.
88. Clark, A. C.; Bingaman, D. P.; Kapin, M. A. *Expert Opin. Ther. Patents* **2000**, *10*, 427–448.
89. Thanos, C.; Emerich, D. *Expert Opin. Biol. Ther.* **2005**, *5*, 1443–1452.
90. Mukherjee, P. K.; Marcheselli, V. L.; Serhan, C. N.; Bazan, N. G. *Proc. Natl. Acad. Sci. USA* **2004**, *101*, 8491–8496.
91. Markomichelakis, N. N.; Theodossiadis, P. G.; Sfikakis, P. P. *Am. J. Ophthalmol.* **2005**, *139*, 537–540.
92. Arnhold, S.; Klein, H.; Semkova, I.; Addicks, K.; Schraermeyer, U. *Invest. Ophthalmol. Vis. Sci.* **2004**, *45*, 4251–4255.
93. Bielroy, L.; Lien, K. W.; Bigelsen, S. *Drugs* **2005**, *65*, 215–228.
94. Yanni, J. M.; Sharif, N. A.; Gamache, D. A.; Miller, S. T.; Weimer, L. K.; Spellman, J. M. *Acta Ophthalmol. Scand.* **1999**, 77, 33–37.
95. Stern, M. E.; Pflugfelder, S. C. *Ocular Surface* **2004**, *2*, 124–130.
96. McNatt, L. G.; Weimer, L.; Yanni, J.; Clark, A. C. *J. Ocular Pharmacol. Ther.* **1999**, *15*, 413–423.
97. Gamache, D. A.; Wei, Z.; Weimer, L. K.; Miller, S. T.; Spellman, J. M.; Yanni, J. M. *J. Ocular Pharmacol. Ther.* **2002**, *18*, 349–361.
98. BenEzra, D. P.; Griffin, B. W.; Maftzir, G.; Sharif, N. A.; Clark, A. F. *Invest. Ophthalmol. Vis. Sci.* **1997**, *38*, 1954–1962.
99. Schlech, B. A.; Blondeau, J. *Surv. Ophthalmol.* **2005**, *50*, S64–S67.
100. Oshitari, T.; Brown, D.; Roy, S. *Exp. Eye Res.* **2005**, *81*, 32–37.

Biographies

Naj A Sharif was educated and trained in England. He received his PhD in neuropharmacology (1982) from the University of Southampton (UK) and conducted postdoctoral work at the University of Maryland and at University of Nottingham (UK). Dr Sharif was a Staff Scientist at Parke-Davis Research Unit (Cambridge, UK) and at Syntex Research (Palo Alto, CA), and a Group Leader at Synaptic Pharmaceutical Corp (NJ). He joined Alcon Laboratories in 1992 and is a Director and Head of Pharmacology. Dr Sharif has been involved in pharmacological drug discovery research for the last 20 years, publishing over 150 papers/book chapters and editing two scientific books. Dr Sharif has filed numerous Memoranda of Invention and has authored numerous patent applications. Dr Sharif is on the Editorial Boards of several journals and reviews manuscripts for leading pharmacology and ocular journals. Dr Sharif is an Adjunct Professor at the University of North Texas (UNT) and at the University of Houston.

Peter Klimko was born in Detroit, Michigan, in 1965. He obtained his BSc degree in chemistry from Ohio State University in Columbus, Ohio, in 1987 and his PhD in synthetic organic chemistry from Texas A&M University in 1992 under the direction of Professor Daniel Singleton. After postdoctoral studies with Professor Charles Swindell at Bryn Mawr College, investigating the development of an efficient synthetic route to taxane diterpenes, Dr Klimko started at Alcon Laboratories in 1993 as a medicinal chemist. His research interests have included the exploration of prostaglandin SAR with respect to lowering intraocular pressure and the effect of arachidonic acid metabolites on dry eye symptoms.

Comprehensive Medicinal Chemistry II
ISBN (set): 0-08-044513-6

ISBN (Volume 6) 0-08-044519-5; pp. 297–320

6.13 Pain Overview

T Prisinzano, University of Iowa, Iowa City, IA, USA
G F Gebhart, University of Pittsburgh, Pittsburgh, PA, USA

6.13.1 Introduction

Pain is the most common reason for medical appointments, resulting in an estimated 40 million visits annually and costing an estimated $100 billion each year in healthcare and lost productivity. According to the American Pain Foundation, one in three Americans lose more than 20 h of sleep each month due to pain. In addition, over 26 million people between the ages of 20 and 64 experience frequent back pain and two-thirds of Americans will have back pain during their lifetime. Furthermore, only an estimated one in four people with pain receive proper treatment.

Drugs for the treatment of pain, or analgesics, have been historically placed into one of two general categories: (1) opioids (narcotics) such as morphine, codeine, and fentanyl; and (2) non-opioid inhibitors of cyclooxygenase (COX), like aspirin, ibuprofen, and celecoxib. Both opioid and non-opioid analgesics are efficacious and widely used (e.g., sales exceeded $8 billion in the US alone in 2004), although unwanted effects can limit their use. Accordingly, development of analgesics with novel mechanisms and improved side-effect profiles remains a high priority.

Acute pain (pricks, cuts, and burns), postoperative pain, and many chronic pains are initiated by activation of a nociceptor. Nociceptors are the sensory receptors in the periphery that respond to stimuli termed 'noxious' and that commonly give rise to pain. Nociceptors are present in all tissues. Nociceptors in skin respond to stimuli that damage or threaten damage to skin (cutting, crushing, burning, etc.), but the adequate noxious stimuli that activate nociceptors in muscle, joints, and the viscera differ from those in skin. For example, chemical mediators in muscle (e.g., lactic acid) and distension of hollow organs are noxious in those tissues.

Technically, nociceptors are the end-organs that transduce noxious stimulus energy (mechanical, thermal, and chemical) into electrical signals that travel along an axon to pass on information to a second-order neuron, typically in the spinal cord. Conventionally, however, the term 'nociceptor' is commonly applied to the sensory neuron in which the nociceptor is embedded. The peripheral terminals of nociceptors in all tissues are very fine and typically without a defining structure (i.e., there is no encapsulated end-organ and nociceptors are often characterized as 'free' nerve endings).

6.13.2 New Analgesic Targets

The mechanisms by which opioid and non-opioid analgesics work are well understood, but neither class of drugs is ideal as either an analgesic or antihyperalgesic. Accordingly, considerable effort continues to be directed at improved understanding of nociceptor function and development of selective analgesics that do not have unwanted effects, such as associated with opioid and non-opioid analgesics. Recent focus has turned to a variety of ion channels or receptors that respond to changes in the chemical milieu in which the nociceptor resides. These include, but are not limited to:

- the transient receptor potential vanilloid (TRPV1), or capsaicin, receptor;
- acid-sensing ion channels (ASICs);

- purinergic P2X and P2Y receptors;
- ionotropic and metabotropic glutamate receptors; and
- tetrodotoxin-resistant voltage-gated sodium channels (Na_V) and some voltage-gated calcium and potassium channels.

The putative roles of most of these molecular entities have been examined by either knocking out (genetically null mice), knocking down (antisense oligonucleotide or siRNA strategies), or blocking (antagonists) the receptors to reveal altered behaviors to application of noxious stimuli.

6.13.2.1 Transient Receptor Potential (TRP) Ion Channels

Members of the transient receptor potential (TRP) family are nonselective cation channels that are involved in several physiological conditions, including pain.[1,2] TRPV1 is of particular interest because it responds to thermal stimuli in the normal noxious range for skin (42–52 °C), to protons and a reduction in tissue pH (as commonly occurs during inflammation), and to chemicals, notably capsaicin, but also a variety of endogenous lipids produced when tissue is insulted. Other members of the TRP family of receptors respond principally to very high (>52 °C) or very low (<18 °C) temperatures in the tissue-damaging range, and are thus noxious.

TRPV1 is a serpentine protein that has six transmembrane (TM) segments flanked by specific ankyrin domains in the cystolic N-terminal portion.[3] The channel pore (P-loop) is located between TM5 and TM6. It is believed to exist in oligomeric form, and heterogeneous assembly of TRPV1 with other channels has been speculated to account for observed discrepancies in cellular assays of vanilloid activity in native and cloned receptors.[3] TRPV1 is the prototypical member of the TRPV subfamily which comprises six members, TRPV1–TRPV6.[4] Among TRPV channels, TRPV1 is unique in its interaction with two potent natural products, capsaicin and resiniferatoxin, which appear to gate the channel by decreasing the threshold of temperature activation.[2]

Three different classes of lipids, derived from the metabolism of arachidonic acid, have been characterized as activators of TRPV1. These are the endocannabinoid anandamide, several lipoxygenase products of arachidonic acid, and *N*-arachidonyldopamine.[5,6] Characteristically, these endogenous lipids have the ability to modulate TRPV1, functionally increasing or enhancing its activation. TRPV1 is expressed in neuronal and non-neuronal tissues, including nociceptors and central nervous system neurons. Wherever located, activation of TRPV1 results in rapid increase in intracellular Ca^{2+} and consequent activation of intracellular mechanisms.[6]

Studies have shown that capsaicin-sensitive afferent neurons participate in the regulation of normal urinary bladder function, gastrointestinal circulation, secretion, mucosal homeostasis, motility, and both somatic and visceral nociception.[4,7,8] For example, TRPV1 null mice exhibit reduced hyperalgesia in inflammatory[9] and neuropathic[8] pain models when compared to wild-type mice, reduced bladder function,[7] and colon mechanosensitivity.[10] These (and other) reports emphasize why the TRPV1 receptor is an attractive target for the development of novel molecules for the treatment of both inflammatory somatic and visceral pain.

6.13.2.2 Acid-Sensing Ion Channels

ASICs are members of the NaC/DEG superfamily. This family includes epithelial Na^+ channels (ENaC), degenerins (DEG), and a peptide-gated Na^+ channel, Phe-Met-Arg-Phe-NH_2 (FaNaC).[11] The major structural features of ASICs, based on the cloning of ASIC1a, are two TM domains (TM1 and TM2), a large extracellular loop, and the C- and N-terminus are intracellular.[12] Evidence suggests that the ASIC1a assembles as a tetramer. ASICs are mainly expressed in nervous system tissue, but not in glia, and in some non-neuronal sites. In the periphery, ASICs are present in all sensory ganglia and are regulated by the autonomic nervous system in different organs.

At present, few pharmacological approaches are effective or selective for ASICs. These channels are blocked by amiloride, a nonselective antagonist; a tarantula psalmotoxin specifically blocks the ASIC1a subunit. Phenylalanine amide-related peptides (FRMF amides) are a family of peptides that slow down desensitization of ASIC3 homomers and heteromers. High concentrations of nonsteroidal anti-inflammatory drugs (NSAIDs) reportedly inhibit ASICs and their inflammation-induced expression,[13] suggesting that ASICs may be a new target for the development of novel NSAIDs.[14]

6.13.2.3 P2 Receptors

Neurons in the central nervous system are endowed with adenosine triphosphate (ATP)-sensitive receptors belonging to the P2X and P2Y types.[15] P2X receptors are multimeric ligand-gated ion channels whereas P2Y receptors are

G protein-coupled receptors. ATP is ubiquitous and is the principal agonist for P2X receptors. ATP itself, when administered exogenously, has long been appreciated as 'algogenic' (pain-producing). When released from epithelium in the skin or internal organs (e.g., bladder urothelium), ATP activates P2X receptors on nerve terminals, including nociceptors, and is believed to be an important contributor to pain.

P2X receptors are purine nucleotide-gated cation channels that mediate fast cell–cell signaling in excitable tissues.[16] Seven P2X receptors have been idenitifed and are designated $P2X_{1-7}$. Their molecular structure consists of two membrane-spanning regions, an intracellular N- and C-terminal, and a cysteine-rich extracellular loop.[16] P2X receptors are oligomeric and composed of more than one subunit. A functional receptor appears to be a trimer. These receptors form cation-selective channels with almost equal permeability to Na^+ and K^+. P2X receptor antagonists selective for $P2X_{1/3}$ (TNP-ATP) and $P2X_{2/3}$ (A-317491) reveal a role for P2X channels in inflammatory and other pain states,[17,18] which is supported by results obtained in $P2X_3$ and $P2X_{2/3}$ null mutant mice.[19,20]

P2Y receptors are G protein-coupled and are also activated by ATP, but adenosine diphosphate and related analogs have greater potency.[15] There are 10 cloned and functionally defined P2Ysubtypes. Eight receptors ($P2Y_1$, $P2Y_2$, $P2Y_4$, $P2Y_6$, $P2Y_{11}$, $P2Y_{12}$, $P2Y_{13}$, and $P2Y_{14}$) are present in human tissue and several ($P2Y_1$, $P2Y_6$, $P2Y_{11}$, $P2Y_{12}$, $P2Y_{13}$, and $P2Y_{14}$) occur in the central nervous system. At present less information is available on the potential role of P2Y receptors in nociceptive processes compared to P2X receptors. However, $P2Y_1$ and $P2Y_2$ receptors have been implicated in the ATP-mediated sensitization of TRPV1 receptors.[16] In addition, there is additional evidence that P2Y receptors may be involved in chronic pain.[16]

6.13.2.4 Excitatory Amino Acid Receptors

The excitatory amino acid glutamate plays a major role in nociceptive processing.[21] Glutamate released from sensory neurons, including nociceptors, exerts its actions on two major types of receptors: ionotropic glutamate receptors (iGluRs) and metabotropic glutamate receptors (mGluRs). iGluRs are ligand-gated ion channels that include *N*-methyl-D-aspartate (NMDA), α-amino-3-hydroxy-5-methyl-4-isoxazolepropionic acid (AMPA), and kainate receptors. mGluRs are G protein-coupled receptors that modulate the actions of glutamate through various second-messenger systems.

The metabotropic glutamate receptor family consists of at least eight members (mGluR1–mGluR8) which can be divided into three main groups based on sequence similarity.[21] Group 1 consists of mGluR1 and mGluR5, which couple most commonly to phospholipase C. Group 2 includes mGluR2 and mGluR3, whereas group 3 includes mGluR4, mGluR6, mGluR7, and mGluR8. Groups 2 and 3 inhibit adenyl cyclase and modulate G protein-coupled inwardly rectifying potassium channels (GIRKs) or voltage-gated calcium channels.

NMDA receptors have been studied in great detail as potential targets for the development of novel analgesics. While efficacious, NMDA receptor antagonists suffer from side effects such as fatigue, dizziness, pyschosis, and at higher concentrations, amnesia and neurotoxicity.[22,23] Thus, there has been movement away from the development of nonselective NMDA receptor antagonists[24] to modulatory site (e.g., $glycine_B$ site) and NMDA receptor subunit site (e.g., NR2B) antagonists, which have shown potential for reduced side effects while maintaining efficacy. Agents selective for the various subtypes of mGluRs are also a target for development. However, validation of mGluRs as a drug target is incomplete at present.

6.13.2.5 Voltage-Sensitive (Na_V) Channels

The voltage-sensitive Na^+ channel is a TM protein which generates action potentials in excitable cells. A variety of different isoforms of this channel have now been identified.[25] Mammalian Na^+ channels consist of an α subunit (around 260 kDa) and a β subunit (30–40 kDa).[25] Different isoforms of the α subunit have been identified. Nine isoforms have been grouped into a single family, Na_V1, and some of these isoforms have a role in nociceptive processing.

Genetic and pharmacological studies indicate that Na_V 1.8 (SNS/PN3), a tetrodotoxin (TTX)-resistant channel, is likely important in nociception.[24,26] Intrathecal administration of specific antisense oligodeoxynucleotides (ODNs) to knockdown this TTX-resistant sodium channel reversed tactile allodynia and thermal hyperalgesia in models of neuropathic pain in rats.[27,28] In addition, Na_V $1.8^{-/-}$ mice showed blunted responses to intracolonic capsaicin, but normal nociceptive behavior to acute noxious stimuli.[29] Furthermore, nonselective Na_V 1.8 channel blockers have been shown to produce antinociceptive effects in rodent models of neuropathic pain.[30] Currently, there are no selective Na_V 1.8 TTX-resistant channel blockers.

6.13.2.6 Voltage-Gated Calcium Channels (VGCCs)

Voltage-gated calcium channels (VGCCs) have also been shown to play an important role in the modulation of nociceptive processing.[31] Calcium ions are a universal second messenger for intracellular signaling in many cell types.

Several individual subtypes of calcium channels are under active investigation as targets for the treatment of chronic pain. VGCCs are found in the plasma membrane of all excitable cells, including peripheral and central neurons. They are localized throughout the neuron and trigger the release of neurotransmitters.[31] Ten VGCCs have been identified, of which nine are expressed in the nervous system.[32] L-type channels were originally classified as channels with a large single channel conductance and long open time. T-type channels were channels with a tiny single channel conductance and transient open time. Further work identified N-type (found in neurons) and P-type (found in cerebellar Purkinje neurons) channels.[31] A numerical numbering system has been adopted to classifiy VGCCs into three families (Ca_V1–Ca_V3) according to the sequence homology of their α_1 subunits.[33,34]

Among all VGCCs, N-type calcium channels encoded by the $Ca_V2.2$ gene have attracted the most attention as a target for the treatment of chronic pain.[31] N-type channel blockers stop the release of neuropeptides such as substance P.[35,36] Studies have shown that N-type channel null mice are less sensitive to neuropathic and inflammatory pain when compared to wild-type mice.[37,38] Furthermore, ziconotide, a potent N-type channel blocker, has been approved for the treatment of neuropathic pain.[32]

Ziconotide is a synthetic form of the peptide ω-conotoxin MVIIA, originally isolated from the venom of the cone snail *Conus magnus*.[39] Toxins from other cone snails have also been identified as potent VGCC inhibitors.[40] Currently, ziconotide is only available as an intrathecal infusion[41] and there is an effort to identify other small-molecule N-type calcium-channel blockers.

6.13.3 Future Considerations

Pain is a complex sensory experience, contributed to by cognitive factors, environment (setting, society, and culture), experience, and gender, and significantly modulated by the central nervous system. We have focused here on the most tractable aspect of the experience of pain, namely the nociceptor, because most, but not all, pain is associated with activation of a nociceptor in peripheral tissue. The drug receptors/ion channels briefly discussed above are all present on nociceptors and thus are targets for drug development. These same receptors are present in the central nervous system as well, and effects on central P2, ASIC, or TRP receptors may also contribute to desirable analgesic or antihyperalgesic effects, or to unanticipated and unwanted effects.

References

1. Calixto, J. B.; Kassuya, C. A. L.; Andre, E.; Ferreira, J. *Pharmacol. Ther.* **2005**, *106*, 179–208.
2. Ferrer-Montiel, A.; Garcia-Martinez, C.; Morenilla-Palao, C.; Garcia-Sanz, N.; Fernandez-Carvajal, A.; Fernandez-Ballester, G.; Planells-Cases, R. *Eur. J. Biochem.* **2004**, *271*, 1820–1826.
3. Szallasi, A.; Appendino, G. *J. Med. Chem.* **2004**, *47*, 2717–2723.
4. Holzer, P. *Eur. J. Pharmacol.* **2004**, *500*, 231–241.
5. van der Stelt, M.; Di Marzo, V. *Eur. J. Biochem.* **2004**, *271*, 1827–1834.
6. Cortright, D. N.; Szallasi, A. *Eur. J. Biochem.* **2004**, *271*, 1814–1819.
7. Birder, L. A.; Nakamura, Y.; Kiss, S.; Nealen, M. L.; Barrick, S.; Kanai, A. J.; Wang, E.; Ruiz, G.; De Groat, W. C.; Apodaca, G. et al. *Nat. Neurosci.* **2002**, *5*, 856–860.
8. Honore, P.; Wismer, C. T.; Mikusa, J.; Zhu, C. Z.; Zhong, C.; Gauvin, D. M.; Gomtsyan, A.; El Kouhen, R.; Lee, C. H.; Marsh, K. et al. *J. Pharmacol. Exp. Ther.* **2005**, *314*, 410–421.
9. Davis, J. B.; Gray, J.; Gunthorpe, M. J.; Hatcher, J. P.; Davey, P. T.; Overend, P.; Harries, M. H.; Latcham, J.; Clapham, C.; Atkinson, K. et al. *Nature* **2000**, *405*, 183–187.
10. Jones, R. C. W., III.; Xu, L.; Gebhart, G. F. *J. Neurosci.* **2005**, *25*, 10981–10989.
11. Waldmann, R.; Champigny, G.; Lingueglia, E.; De Weille, J. R.; Heurteaux, C.; Lazdunski, M. *Ann. N. Y. Acad. Sci.* **1999**, *868*, 67–76.
12. Krishtal, O. *Trends Neurosci.* **2003**, *26*, 477–483.
13. Voilley, N.; de Weille, J.; Mamet, J.; Lazdunski, M. *J. Neurosci.* **2001**, *21*, 8026–8033.
14. Voilley, N. *Curr. Drug Targets Inflamm. Allergy* **2004**, *3*, 71–79.
15. Illes, P.; Ribeiro, J. A. *Curr. Top. Med. Chem.* **2004**, *4*, 831–838.
16. Liu, X. J.; Salter, M. W. *Curr. Opin. Investig. Drugs* **2005**, *6*, 65–75.
17. Jarvis, M. F.; Burgard, E. C.; McGaraughty, S.; Honore, P.; Lynch, K.; Brennan, T. J.; Subieta, A.; Van Biesen, T.; Cartmell, J.; Bianchi, B. et al. *Proc. Natl. Acad. Sci. USA* **2002**, *99*, 17179–17184.
18. McGaraughty, S.; Wismer, C. T.; Zhu, C. Z.; Mikusa, J.; Honore, P.; Chu, K. L.; Lee, C. H.; Faltynek, C. R.; Jarvis, M. F. *Br. J. Pharmacol.* **2003**, *140*, 1381–1388.
19. Cockayne, D. A.; Dunn, P. M.; Zhong, Y.; Rong, W.; Hamilton, S. G.; Knight, G. E.; Ruan, H. Z.; Ma, B.; Yip, P.; Nunn, P. et al. *J. Physiol.* **2005**, *567*, 621–639.
20. Vlaskovska, M.; Kasakov, L.; Rong, W.; Bodin, P.; Bardini, M.; Cockayne, D. A.; Ford, A. P.; Burnstock, G. *J. Neurosci.* **2001**, *21*, 5670–5677.
21. Varney, M. A.; Gereau, R. W. *Curr. Drug Targets CNS Neurol. Disord.* **2002**, *1*, 283–296.
22. Eide, K.; Stubhaug, A.; Oye, I.; Breivik, H. *Pain* **1995**, *61*, 221–228.
23. Jevtovic-Todorovic, V.; Todorovic, S. M.; Mennerick, S.; Powell, S.; Dikranian, K.; Benshoff, N.; Zorumski, C. F.; Olney, J. W. *Nat. Med.* **1998**, *4*, 460–463.

24. LoGrasso, P.; McKelvy, J. *Curr. Opin. Chem. Biol.* **2003**, *7*, 452–456.
25. Ogata, N.; Ohishi, Y. *Jpn. J. Pharmacol.* **2002**, *88*, 365–377.
26. Lai, J.; Porreca, F.; Hunter, J. C.; Gold, M. S. *Annu. Rev. Pharmacol. Toxicol.* **2004**, *44*, 371–397.
27. Lai, J.; Gold, M. S.; Kim, C. S.; Bian, D.; Ossipov, M. H.; Hunter, J. C.; Porreca, F. *Pain* **2002**, *95*, 143–152.
28. Porreca, F.; Lai, J.; Bian, D.; Wegert, S.; Ossipov, M. H.; Eglen, R. M.; Kassotakis, L.; Novakovic, S.; Rabert, D. K.; Sangameswaran, L. et al. *Proc. Natl. Acad. Sci. USA* **1999**, *96*, 7640–7644.
29. Laird, J. M.; Souslova, V.; Wood, J. N.; Cervero, F. *J. Neurosci.* **2002**, *22*, 8352–8356.
30. Erichsen, H. K.; Hao, J. X.; Xu, X. J.; Blackburn-Munro, G. *Eur. J. Pharmacol.* **2003**, *458*, 275–282.
31. McGivern, J. G.; McDonough, S. I. *Curr. Drug Targets CNS Neurol. Disord.* **2004**, *3*, 457–478.
32. Doering, C. J.; Zamponi, G. W. *Curr. Pharm. Des.* **2005**, *11*, 1887–1898.
33. Ertel, E. A.; Campbell, K. P.; Harpold, M. M.; Hofmann, F.; Mori, Y.; Perez-Reyes, E.; Schwartz, A.; Snutch, T. P.; Tanabe, T.; Birnbaumer, L. et al. *Neuron* **2000**, *25*, 533–535.
34. Catterall, W. A.; Striessnig, J.; Snutch, T. P.; Perez-Reyes, E. *Pharmacol. Rev.* **2003**, *55*, 579–581.
35. Altier, C.; Zamponi, G. W. *Trends Pharmacol. Sci.* **2004**, *25*, 465–470.
36. Smith, M. T.; Cabot, P. J.; Ross, F. B.; Robertson, A. D.; Lewis, R. J. *Pain* **2002**, *96*, 119–127.
37. Hatakeyama, S.; Wakamori, M.; Ino, M.; Miyamoto, N.; Takahashi, E.; Yoshinaga, T.; Sawada, K.; Imoto, K.; Tanaka, I.; Yoshizawa, T. et al. *Neuroreport* **2001**, *12*, 2423–2427.
38. Saegusa, H.; Kurihara, T.; Zong, S.; Kazuno, A.; Matsuda, Y.; Nonaka, T.; Han, W.; Toriyama, H.; Tanabe, T. *EMBO J.* **2001**, *20*, 2349–2356.
39. McIntosh, M.; Cruz, L. J.; Hunkapiller, M. W.; Gray, W. R.; Olivera, B. M. *Arch. Biochem. Biophys.* **1982**, *218*, 329–334.
40. Jones, R. M.; Bulaj, G. *Curr. Pharm. Des.* **2000**, *6*, 1249–1285.
41. Wermeling, D. P. *Pharmacotherapy* **2005**, *25*, 1084–1094.

Biographies

Thomas E Prisinzano was born in New York City, and studied at the University of Delaware, where he obtained a BS in 1995, and Virginia Commonwealth University, where he completed his PhD in 2000 under the direction of Professor RA Glennon. He was then awarded an Intramural Research Training Award Fellowship to study in the laboratory of Dr Kenner C Rice at the National Institute of Diabetes and Digestive and Kidney Diseases (NIDDK). While at NIDDK, he worked on drugs of abuse and treatment agents. Subsequently, he took up his present position as an Assistant Professor in Division of Medicinal and Natural Products Chemistry at the University of Iowa in May 2003. His scientific interests include the development of novel analgesics, in particular, the development of novel compounds to study the neurochemical mechanisms of drug dependence and tolerance.

G F (Jerry) Gebhart was born in Chicago, and earned a BS (1967, Pharmacy) at the University of Illinois (Chicago) and MS and PhD degrees (in 1969 and 1971, respectively, Pharmacology) at the University of Iowa. After 2 years of postdoctoral training with Herbert H Jasper in Montreal, he returned to Iowa where he rose to the rank of Professor in 1981 and was Head of the Department of Pharmacology from 1996 to 2006. He is currently Director of the Pittsburgh Center for Pain Research at the University of Pittsburgh. He spent a sabbatical year with Manfred Zimmermann in Heidelberg, Germany, in the early 1980s. His scientific interests include descending modulation of pain and mechanisms of nociception, particularly visceral nociception and visceral hypersensitivity, which have been supported by NIH funding continuously since the early 1970s.

Comprehensive Medicinal Chemistry II
ISBN (set): 0-08-044513-6

ISBN (Volume 6) 0-08-044519-5; pp. 321–326

6.14 Acute and Neuropathic Pain

P Honore and M F Jarvis, Abbott Laboratories, Abbott Park, IL, USA

6.14.1 Introduction: Pain States

The International Association for the Study of Pain (IASP) defines pain as an "unpleasant sensory and emotional experience associated with actual or potential tissue damage, or described in terms of such damage."[1] This definition clearly indicates that pain is a multidimensional experience. Pain is not homogenous and can be classified temporally as acute or chronic. Acute or physiological pain is an early warning against potential injury, a vital defense mechanism, whereas chronic pain does not play any useful role. As such, chronic pain can be very detrimental to the quality of life of an individual, disrupting sleep and normal living, and degrading health and functional capability.[2]

Chronic pain is one of the most common complaints for which individuals solicit medical attention. It can affect general and psychological health (as is evident by the high degree of comorbidity of chronic pain and emotional disorders such as depression and anxiety), and can also have deleterious socioeconomic consequences since it is often associated with the loss of work time and an increased use of healthcare resources. Published estimates of the prevalence of chronic pain typically range from 2% to 45% with 50% of respondents reportedly suffering from chronic pain.[3] Interestingly, there was no significant difference between genders and, as might be expected, the proportion and the degree of pain significantly increased with age. Furthermore, 33–40% of chronic pain patients included in their study were unhappy with medical examinations, medical tests, and treatments related to their chronic pain state.[4] This lack of satisfactory pain relief in chronic pain patients was also identified in a 2004 'Americans Living with Pain Survey' conducted on behalf of the American Chronic Pain Association.[5] In this study, 50% of the chronic pain patients surveyed felt that their pain was not under control. In the face of this growing unmet medical need is an increasing awareness of undertreated pain resulting in a more aggressive use of analgesics, especially opioids. As the population of elderly people continues to increase the demand for therapies to treat arthritis, pain associated with osteoporosis, and other painful diseases of the aged will also increase. These results clearly illustrate the need for better and more efficacious pain management medications, programs, and therapies. One of the greatest challenges in creating more efficacious medications for pain control has been the heterogeneity of the condition itself, including: the causes and underlying pathologies; the redundancies in pain perception; and the usefulness of current pharmacological therapies.

6.14.2 Neuropathophysiology of Pain

The conceptualization of the neurobiology of pain has undergone continuous refinement with increasing knowledge of multiple nociceptive targets and pathways.[6,7] The psychophysical parameters used to describe nociceptive processing have thus been refined to differentiate acute withdrawal behaviors in response to dangerous (e.g., sharp or hot stimuli) stimuli in the environment (acute nociception) from increased sensitivity to mildly painful stimuli (hyperalgesia) or to otherwise innocuous stimuli (allodynia) (**Figure 1**).[8] An increase in stimulus intensity in any sensory modality will eventually become noxious (**Figure 1**). Obviously, the sensation of noxious environmental stimuli (acute pain) is physiologically protective. However, following injury, this psychophysical function shifts such that previous noxious stimuli are now perceived as exceedingly painful (*hyperalgesia*). Additionally, tissue injury results in ongoing or spontaneous pain and the perception that normally nonnoxious stimuli are pain generating (*allodynia*). It is now

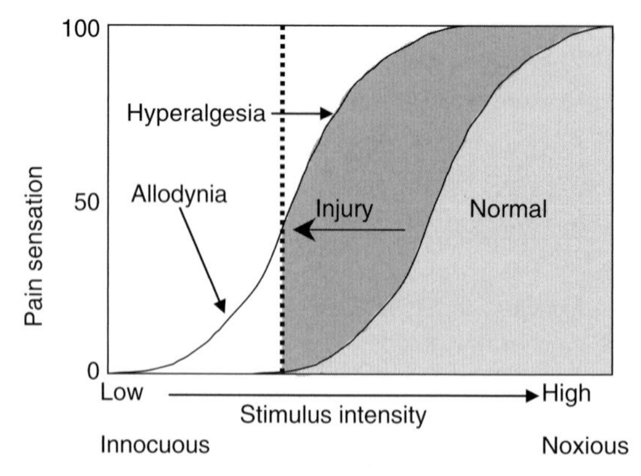

Figure 1 Psychophysical representation of hyperalgesia and allodynia sensory sensitivity. Following injury, being tissue or nerve injury, pain transmission and perception are changed so that a normally painful stimulus is going to be felt more painful (hyperalgesia) and a normally nonpainful stimulus is going to be felt as painful (allodynia). (Adapted from Cervero, F.; Laird, J. M., *Pain* **1996**, *68*, 13–23.)

well appreciated that distinct sensory mechanisms contribute to physiological pain, to pain arising from tissue damage (inflammatory or nociceptive pain), and to pain arising from injury to the nervous system (neuropathic pain).[9]

Nociceptive pain is caused by the ongoing activation of A-δ and C-nociceptors in response to a noxious stimulus (injury, disease, inflammation) (**Figure 2**). It can be further classified into visceral pain (deep cramping sensation associated with referred pain), superficial somatic pain (skin; well-localized sharp, pricking, or burning sensation), and deep somatic pain (muscle, joint capsules, and bone; diffuse dull or aching sensation). Under normal physiological conditions, there is a close correspondence between pain perception and stimulus intensity, and the sensation of pain is indicative of real or potential tissue damage. As the nervous system becomes sensitized (responding more strongly than normal to peripheral stimuli), in addition to spontaneous pain, nociceptive pain is also associated with evoked hyperalgesic and allodynic conditions.[10–12] In general, nociceptive pain abates completely upon the resolution of injury if the disease process is controlled. Because of this, the use of disease-modifying therapies is being emphasized in the treatment of nociceptive chronic pain as illustrated in the treatment of rheumatoid arthritis not only by antiinflammatory agents but also by biological therapies such as tumor necrosis factor-α (TNF-α) antagonists.[13]

Unlike nociceptive pain, neuropathic pain can persist long after the initiating injurious event has been removed and any damage has healed. This then leads to abnormal processing of sensory information by the nervous system. Neuropathic pain can be classified as peripheral (painful peripheral mononeuropathy and polyneuropathy) or central (post stroke, following spinal cord injury) and can originate from nerve injury following a wide array of conditions/events, e.g., direct trauma to nerves, inflammation/neuritis/nerve compression, diabetes, infections (herpes zoster, human immunodeficiency virus (HIV)), tumors (nerve compression/infiltration), toxins (chemotherapy), and primary neurological diseases.[14,15] Following nerve injury, changes occur in the central nervous system (CNS) that can persist indefinitely. Under these conditions of sensitization, pain can occur without a specific stimulus or can be disproportionate to the stimulus intensity. The sensation of neuropathic pain may also be constant or intermittent and is felt in many different ways (e.g., allodynia or hyperalgesia associated with mechanical or thermal stimuli but also spontaneous sensations such as burning, tingling, prickling, shooting, deep aching, and spasm).[14,15]

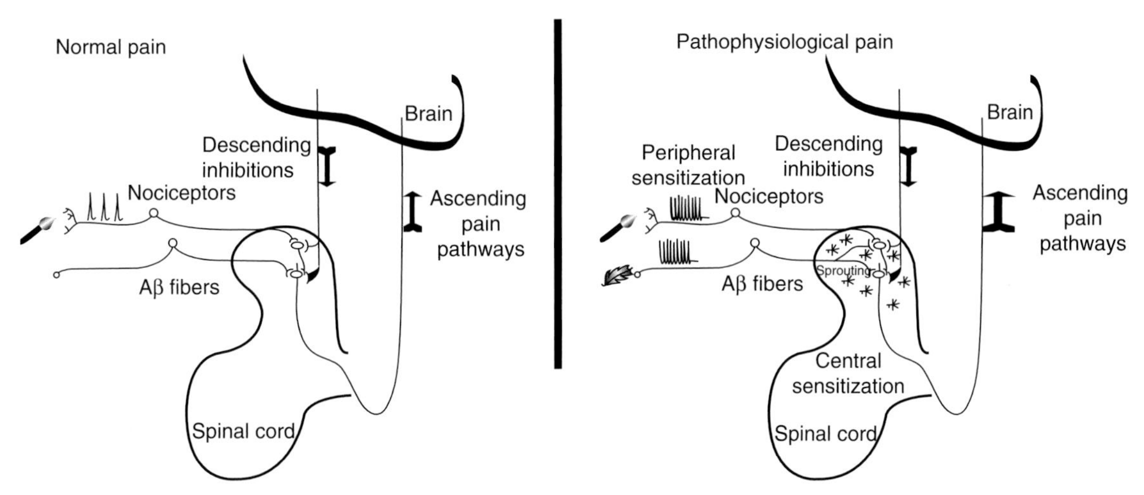

Figure 2 Normal and pathophysiological transmission of pain. Under normal conditions, pain stimuli such as noxious heat are transmitted from the peripheral site (e.g., skin or joints) through the nociceptive primary afferent fibers to the spinal cord and the brain. Pathophysiological conditions are associated with peripheral and central sensitization. Peripheral sensitization can result from the sensitization of nociceptors by inflammatory mediators, neurotrophic factors released during tissue damage or by inflammatory cells. Peripheral sensitization is also associated with intense, repeated, or prolonged action potential generation in primary sensory afferents that is mediated by altered expression and activity of voltage-gated sodium and calcium channels. Consequences of peripheral sensitization are a lowering of the activation threshold of nociceptors and an increase in their firing rate. These changes result in the production of hyperalgesia and allodynia associated with nociceptive chronic pain. Central sensitization (long-lasting increases in dorsal horn neuron excitability and responsiveness) is associated with spontaneous dorsal horn neuron activity, responses from neurons that normally only respond to low intensity stimuli (altered neural connections following sprouting of Aβ fibers to superficial laminae) and reduction in central inhibition. Central sensitization is associated with persistent pain, hyperalgesia, allodynia, and the spread of pain to uninjured tissue. In addition, it reflects a complex series of changes occurring in the spinal cord that may promote long-lasting increases in dorsal horn neuron excitability including the involvement of astrocytes and microglia activation.

Tissue injury results in the release of pronociceptive mediators that sensitize peripheral nerve terminals (*peripheral sensitization*), leading to phenotypic alterations of sensory neurons and increased excitability of spinal cord dorsal horn neurons (*central sensitization*).[7,9] Nerve injury may be associated with abnormal firing of the injured neurons, leading also to central sensitization and phenotypic changes in spinal cord neurons.[16,17] In addition, descending supraspinal systems modulate nociceptive responses.[18] A multitude of receptors, transmitters, second messenger systems, transcription factors, and other signaling molecules are now appreciated to be involved in pain pathways (**Figures 2** and **3**).[6]

As noted above, two mechanisms play a key role in the development and maintenance of chronic pain, namely peripheral and central sensitization.[19] Peripheral sensitization can result from the sensitization of nociceptors by inflammatory mediators (e.g., prostaglandin E2 (PGE2), serotonin (5HT), bradykinin, epinephrine, adenosine), by neurotrophic factors released during tissue damage (e.g., nerve growth factor (NGF)) or by inflammatory cells (proinflammatory cytokines including interleukin-1 (IL1)). Peripheral sensitization is also associated with intense, repeated, or prolonged action potential generation in primary sensory afferents that is mediated by altered expression and activity of voltage-gated sodium and calcium channels.[6] Consequences of peripheral sensitization are a lowering of the activation threshold of nociceptors and an increase in their firing rate. These changes result in the production of hyperalgesia and allodynia associated with nociceptive chronic pain. In addition, peripheral sensitization plays also an important role in the development and maintenance of central sensitization.[6,17]

Central sensitization (long-lasting increases in dorsal horn neuron excitability and responsiveness) is associated with spontaneous dorsal horn neuron activity, responses from neurons that normally only respond to low-intensity stimuli (altered neural connections following sprouting of Aβ fibers to superficial laminae), expansion of dorsal horn neuron receptive fields, and reduction in central inhibition.[20–22] Central sensitization is associated with persistent pain, hyperalgesia, allodynia, and the spread of pain to uninjured tissue, i.e., secondary hyperalgesia due to increased receptor

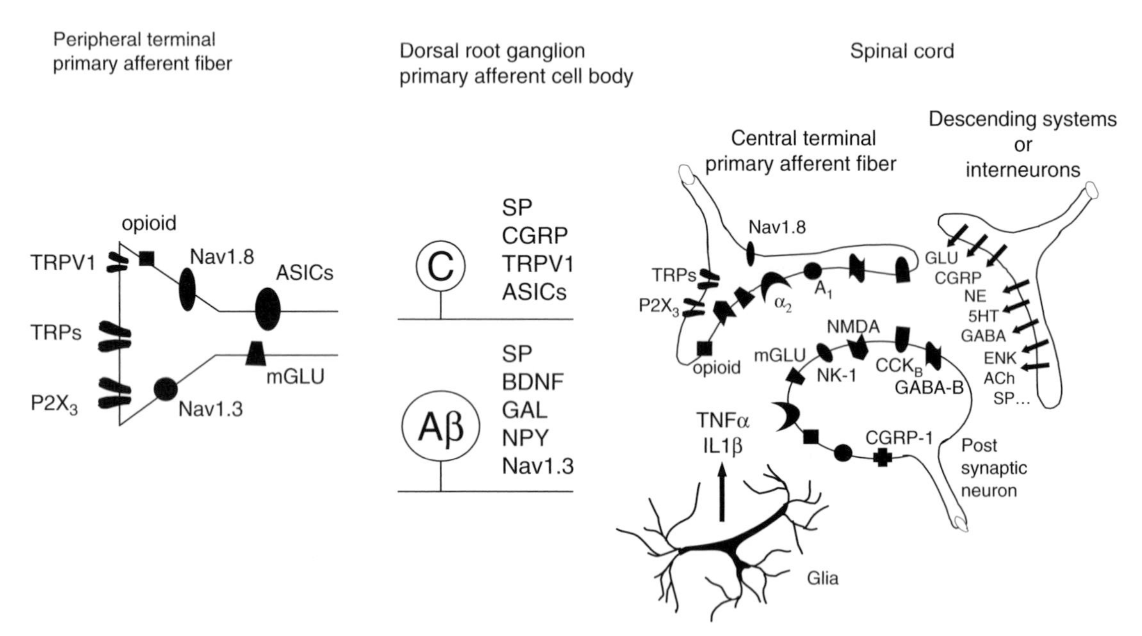

Figure 3 Pain transmission sites can be simply divided into two: the peripheral compartment (e.g., skin, muscle, organs) that encompasses primary afferent fibers and dorsal root ganglions and the central compartment that includes the spinal cord and brain. A multitude of receptors, transmitters, second messenger systems, transcription factors, and other signaling molecules located all along pain transmission pathways are now appreciated to be involved in pain signaling. Tissue injury results in the release of pronociceptive mediators that activate and sensitize peripheral nerve terminals (*peripheral sensitization*) through various receptors/channels, leading to phenotypic alterations of sensory neurons with changes of receptor expression in DRG cell bodies and changes in neurotransmitter contents and increased excitability of spinal cord dorsal horn neurons (*central sensitization*) due to a variety of changes in receptor expression patterns and neurotransmitter release. In addition, interneurons and descending supraspinal systems modulate nociceptive responses either through excitatory or inhibitory effects. 5HT (serotonin), A_1 (adenosine 1), α_2 (alpha 2 adrenergic), ACH (acetylcholine), ASICs (acid sensing ion channels), BDNF (brain derived nerve growth factor), CCK_B (cholecystokinin), CGRP (calcitonin gene related peptide), ENK (enkephalines), GABA, GAL (galanin), GLU (glutamate), mGlu (metabotropic glutamate receptors), Nav1.3 (sodium channels), Nav1.8 (sodium channels), NE (norepinephrine), NK-1 (neurokinin-1 receptor), NMDA (*N*-methyl-D-aspartate receptor), NPY (neuropeptide Y), opioid, $P2X_3$ (ATP receptor), SP (substance P), TRP (transient receptor potential), and TRPV1 (vanilloid receptor 1).

field of spinal neurons. In addition, it reflects a complex series of changes occurring in the spinal cord that may promote long-lasting increases in dorsal horn neuron excitability. This process is also know as 'wind-up' in that the response of sensitized dorsal horn neurons is exaggerated relative to the normal situation.[20–22] While both peripheral and central sensitization play a role in nociceptive chronic pain, central sensitization clearly plays a key role in neuropathic pain. Thus, central sensitization also explains the observation that established pain is more difficult to suppress than acute pain because of the maladaptive changes that have taken place in the CNS.[6,17] Interestingly, not only neurons, but also glia, e.g. astrocytes and microglia, as well as infiltrating mast cells are involved in the generation and maintenance of central sensitization.[23,24]

In addition to the activation of pronociceptive inflammatory and/or neurotrophic messengers noted above, the sensitization of the nervous system in response to chronic pain involves the alteration and/or activation of many neurotransmitter systems that have been extensively reviewed elsewhere.[6,15] Chronic pain is mediated by altered neuronal excitability involving activation of sodium and calcium channels in both peripheral and spinal neurons. Additionally, there is enhanced glutamatergic activity and a concomitant decrease in GABAergic inhibitory neuromodulation at the level of the dorsal spinal horn.[6,15] This altered neurochemical profile contributes to the heightened state of neuronal excitability (e.g., wind-up) and can be viewed as a shift in the balance of excitatory and inhibitory systems that also incorporates activation of intracellular signaling cascades (e.g., ras-mitogen-activated protein kinase (MAPK) pathway) and recruitment of neurotrophic neuropeptides including substance P, neuropeptide Y, and brain-derived neurotrophic factor (BDNF). These changes in spinal neuron neurochemistry are also accompanied by upregulation of specific excitatory amino acid receptors (e.g., α-amino-3-hydroxy-5-methylisoxazole-4-propionic acid (AMPA) and *N*-methyl-D-aspartate (NMDA)), as well as increased calcium and potassium ion channel activity. Taken together, chronic pain is associated with a large variety of deranged patterns of neurotransmission at multiple levels of the neuroaxis with considerable target and pathway redundancy. Thus, in the absence of ongoing injury, chronic pain can be viewed as a disease in itself. The enhanced appreciation of the many neurochemical and neurophysiological alterations in neuronal function associated with chronic pain has led to the development of both new preclinical models of pain and a variety of potentially useful therapeutic interventions.

6.14.3 Experimental Pain Models

To facilitate the study of pain transmission and the characterization of novel analgesic compounds, an array of experimental animal pain models has been developed mainly in rodents, reflecting all types of pain, from acute to chronic, somatic to visceral, and nociceptive to neuropathic and cancer-related pain. Depending on the model, pain measurements can encompass spontaneous pain behaviors as well as pain evoked by various sensory modalities. It is important to note that in rodents, measuring spontaneous pain is very difficult and is generally limited to the observation of quantifiable nocifensive (pain-escape) behaviors such as hind paw lifting or altered grooming. However, experimental measures of evoked pain are well characterized and are analogous to clinical diagnostic methods. In the following overview, acute pain refers to pain that lasts from seconds to a day while chronic pain typically refers to experimental pain manipulations that persist for at least several days. In addition, this section will focus on the most-widely used preclinical pain models.

The majority of these animal models of pain were originally developed in rats. Except as noted below, essentially all have also been successfully carried out using various mouse strains including gene-disrupted (knockout) mice.[25] However, significant differences in the basal nociceptive sensitivity and analgesic response have been noted for different mouse strains serving to further complicate the interpretation of the knockout phenotype.[26]

6.14.3.1 Models of Acute Pain

Animal models of acute pain allow the evaluation of the effects of potential analgesics on pain sensation/transmission in an otherwise normal animal. In addition, the same tests may be used to measure stimulus-evoked pain in animals with chronic inflammation or nerve injury. Usually, these tests rely on an escape behavior/withdrawal reflex or vocalization as an index of pain. The animals have control over the duration of the pain, that is, their behavioral response leads to termination of the noxious stimulus.

6.14.3.1.1 Acute thermal pain

Models have been developed to interrogate acute thermal pain sensitivity, using various means of applying a noxious heat stimulus to the paw or the tail of rodents. These models have been widely used in the characterization of opioid analgesics. Usually, 'latency to behavioral response' is recorded and a cutoff is set to avoid any tissue damage to the animal. The tail-flick test involves the application of a focused heat (usually light) source on the tail until a tail-flick (rapid removal of the tail) reflex occurs. This test has an advantage in that it does not involve repeated assessments of animal behavior, i.e., animals learning with time when the stimulus is going to be applied and anticipating the test.

The hot plate assay uses a hot plate set at a fixed temperature, usually 50–55 °C. Latency to licking, shaking of hind limbs or fore limbs, in addition to latency to jump can be recorded and statistically analyzed for groups of animals. This assay can be difficult to standardize since the heat stimulus is not delivered in a controlled fashion. Possible sources of variability include differential exposure to the heated plat depending on how much weight the animal puts on each limb.

Another approach to assess acute thermal pain is the use of a radiant heat source.[27] Using this methodology, the temperature of the heat source applied to the hind paw increases over time until it reaches a painful threshold. Latency to hind paw withdraw is recorded and analyzed. In each test session, each animal is tested in three to four sequential trials at approximately 5 min intervals to avoid sensitization of the response. One of the advantages of this method versus the tail-flick assay is that both paws can be tested. This important control has proven a useful behavioral assessment in models of unilateral inflammation or nerve injury, the contralateral paw serving as control for the injured paw. In addition, in this assay, rats are confined in plastic chambers but not manually restrained as in the tail-flick assay or in the immersion tests (see below), decreasing the stress level of the test subjects. This method also uses a heated (30 °C) glass test surface to prevent paw cooling and to minimize sensitization artifacts.

Acute thermal pain can also be evaluated using a fixed temperature (45–50 °C) water bath and assessment of latency to withdraw of a hind limb or tail from the hot water. One of the advantages of this method is that the water bath can be set at various temperatures and it can be less sensitive to environmental conditions. However, this assay requires handling of the animals when testing for nociceptive behavior, making this measure highly dependent on experimenter experience/comfort handling/restraining animals by hand.

This method can also be used to test for reactivity to cold, using a 4 or 10 °C water bath and recording latency to withdraw as an index of pain. Another method uses a cold plate cooled by cold water circulating under it. Latency to nociceptive behavior or duration of guarding behavior can be recorded. As for the hot plate assay, the cold plate test has the advantage of not requiring animal restraint. However, depending on the position of the animal paw on the plate (or just above it), the cold stimulation can be very variable. Another widely used method is application of a drop of cold acetone on the plantar skin of animals resting on an elevated mesh floor. Acetone produces a distinct cooling sensation as it evaporates. Normal rats will not respond to this stimulus or with a very small response (in amplitude and duration) while nerve-injured rats will almost always respond with an exaggerated response.

6.14.3.1.2 Acute mechanical pain

Models have been developed to interrogate acute mechanical pain/sensitivity, using various means of stimulating the paw or the tail of a mouse or a rat. A common method for the assessment of acute mechanical pain is determination of withdrawal thresholds to paw/tail pressure using the Randall Selitto test.[28] This apparatus allows for the application of a steadily increasing pressure to the dorsal surface of the hind paw/tail of a rat via a dome-shaped plastic tip. The threshold (in g) for either paw/tail withdrawal or vocalization is recorded. Usually, two or three measurements are conducted on each paw or tail. This apparatus was designed originally for measuring mechanical sensitivity of inflamed paws and its use in normal noninflamed paws can produce great variability in the response, depending on the location of the stylus (soft tissue between the metatarsal/bone/joint). It is worth noting that training helps generate a more stable response with this assay.

Another approach for assessment of acute mechanical pain is to use a pinprick, applying painful pressure to the plantar surface of the hind paw. This is similar to the pricking pain test done during the neurological examination in patients. The behavior can be measured by the duration of paw lifting following the pinprick application or recorded as a frequency of withdrawal (percentage of response to the pinprick in 10 trials).

Finally, mechanical hypersensitivity can also be tested with von Frey monofilaments. These are a series of hairs/nylon monofilaments of various thicknesses that exert various degrees of force when applied to the planter surface of the hind paw. Responses can be quantified as percentage response or duration of response to a given monofilament force applied several times, or mechanical threshold can be determined using the up–down method.[29]

6.14.3.1.3 Acute chemical pain

Usually, when studying acute chemical pain, behaviors such as flinching, biting, or licking the injected paw are recorded at various time points following the injection of a chemical irritant (capsaicin, formalin, PGE2, mustard oil, αβ-methylene ATP). The duration of the nociceptive behavior as well as the number of behaviors can be quantified and analyzed. Two models are mostly used to study acute chemical pain: nocifensive behaviors following injection of capsaicin or formalin into the hind paw.[30] Doses of capsaicin can vary from 1 to 10 μg per 10 μL injected to the dorsal surface of the rat hind paw. The injection of capsaicin is immediately followed by an intense period of nocifensive behaviors that are usually recorded for 5 min following capsaicin injection. Following formalin injection (usually 5% per 50 μL) into the dorsal or sometimes plantar surface of the rat hind paw, a biphasic behavioral response can be observed. Phase I of the formalin response is defined as the period of time immediately following injection of formalin until 10 min after the formalin injection and corresponds to acute thermal pain by direct activation of nociceptors by formalin. Following a 'quiet' period of little or no nocifensive behavior, the second phase of the formalin response can be observed (20–60 min post formalin injection) that corresponds to a more persistent inflammatory state.

6.14.3.2 Models of Nociceptive Pain

Models of nociceptive pain are defined as models of pain following tissue injury induced by trauma, surgery, inflammation, and cancer. As stated above, spontaneous pain in these models is difficult to measure. However, evoked pain behaviors have been well characterized and can be induced by the methods described above in the 'acute pain' section. The focus of this section will be on models of nociceptive pain, mimicking as closely as possible rheumatoid arthritis and osteoarthritis clinical conditions since they have been the most studied and widely used. Models of postoperative pain or cancer pain will not be described, as they are recent and still under validation.

6.14.3.2.1 Adjuvant-induced arthritis

Experimental arthritis is generated by an intravenous injection of complete Freund's adjuvant (CFA) at the base of the tail. The development of the joint inflammation is progressive and dramatic, leading to a multijoint arthritis with dramatic swelling and permanent joint tissue destruction.[31] In this model, it is clear that the animals are in chronic pain, all their joints are swollen, they have decreased appetite, they limp, and have lower threshold for limb withdrawal or vocalization to paw pressure/joint manipulation. This model is rarely used today as the polyarthritic rat has significant systemic disease with abnormal hunchback posture and piloerection.

6.14.3.2.2 Unilateral inflammation

To further study inflammatory pain, various models have been developed to induce a localized inflammatory reaction by injecting various substances, e.g., formalin, carrageenan, or CFA into the paw or the joint. Following the initial injection, pain can be measured minutes to days later, at the site of inflammation or away from the primary site of injury. Usually, the inflamed paw/joint becomes very sensitive to both thermal and mechanical stimuli while the contralateral paw remains 'normal.' Sometimes, secondary mechanical hypersensitivity can also develop on the contralateral side as observed 2 weeks following carrageenan injection into the knee joint when testing on the contralateral paw. These models of more localized inflammation/inflammatory pain have been widely used in pain research to test the effects of potential analgesic compounds but also in electrophysiological and gene expression studies to determine the plastic changes that initiate/maintain chronic inflammatory pain.

6.14.3.2.3 Models of osteoarthritic pain

More recently, models have been developed to mimic osteoarthritic (OA) pain observed in the clinic. Contrary to rheumatoid arthritis (RA) and the models of inflammatory pain, OA in the clinic and in animal models is not associated with a large amount of inflammation. In addition, to mimic more closely the clinical situation, pain evaluation in OA pain models relies on functional measures such as weight bearing or grip force of the affected limb rather than evaluation of withdrawal latencies to thermal or mechanical stimuli. Two models have been widely used, intraarticular administration of sodium monoiodoacetate (MIA) into the knee and partial meniscectomy.[32] Contrary to what is observed in the polyarthritic rat, no changes in body weight were observed over a 4-week period after either iodoacetate injection or partial medial meniscectomy. In addition, the general health of the animals is good with no signs of spontaneous nociceptive behavior, impaired locomotion, or distress. Furthermore, both iodoacetate injection and partial medial meniscectomy in the knee joint of the rat induced histological changes and pain-related behaviors characteristic of clinical OA. Although the behavioral changes and histology both worsened over time, the majority of the pain responses were apparent within one week of surgery or iodoacetate injection. It is important to note that the pain behaviors are less pronounced in the surgery model than in the MIA model and that these findings agree with the clinical situation. Indeed, magnetic resonance imaging (MRI) studies have shown that although meniscal lesions in humans are common, they are also rarely associated with pain.

6.14.3.3 Models of Neuropathic Pain

6.14.3.3.1 Direct trauma to nerves

To mimic nerve injury observed in the clinic, a number of different animals models have been developed. One of the most studied models is the L5–L6 spinal nerve ligation (SNL, Chung model) (**Figure 4**) model.[33] In this model, following sterilization procedures, a 1.5 cm incision is made dorsal to the lumbosacral plexus, the paraspinal muscles are separated from the spinous processes, the L5 and L6 spinal nerves are isolated, and tightly ligated with 3–0 silk thread. Usually the animals are allowed to recover from surgery for 7 days before being tested for mechanical allodynia using von Frey monofilaments (up–down method or percentage response to 10 applications of innocuous or noxious von Frey monofilament). While the spinal nerve injured rats also develop cold allodynia and thermal hyperalgesia, they have a greater degree of mechanical allodynia and most pharmacological studies with these animals have involved mechanical allodynia endpoints.

Another widely used model of direct nerve injury is a partial nerve ligation model (PNL) (**Figure 4**).[33] The sciatic nerve is exposed unilaterally, just distal to the descendence of the posterior biceps semitendinosus nerve from the sciatic. The dorsal 1/3–1/2 of the nerve thickness is then tightly ligated with an 8–0 silk suture. Following injury, these animals develop guarding behavior of the injured hind limb suggesting the possibility of spontaneous pain. In addition, the animals develop mechanical allodynia as well as thermal hyperalgesia and bilateral mechanical hyperalgesia.

6.14.3.3.2 Inflammation/neuritis/nerve compression

Neuropathic pain can also result from inflammation around peripheral nerves and peripheral nerve compression. Two preclinical models have been developed to attempt to mimic this phenomenon. The first model is the chronic constriction injury (CCI) (Bennett model) (**Figure 4**) of the sciatic nerve model.[33] In this model, a 1.5 cm incision is made 0.5 cm below the pelvis. The biceps femoris and the gluteus superficialis are separated and the sciatic nerve exposed, isolated, and four loose ligatures (5–0 chromic catgut) with 1 mm spacing are placed around it. CCI animals develop mechanical allodynia, cold allodynia, and thermal hyperalgesia. When compared to the 'SNL injured animals,' CCI animals do develop thermal hyperalgesia and cold allodynia to a greater extent.

The second model, developed more recently, is the SIN model or zymosan-induced sciatic inflammatory neuritis.[34] In this model, a chronic indwelling perisciatic catheter is used to inject zymosan around the sciatic nerve. After aseptic exposure of the sciatic nerve at midthigh level, the gelfoam is threaded around the nerve so as to minimize nerve displacement. Suturing and insertion of a sterile 'dummy' injection tube during implantation maintained catheter patency and ensured replicable drug delivery close to the nerve. After anchoring to the muscle, the external end is tunneled subcutaneously to exit 1 cm rostral to the tail base. After removal of the 'dummy' injector, the external end of the silastic tube is protected. Usually, catheter placement can be verified at sacrifice by visual inspection. The catheter

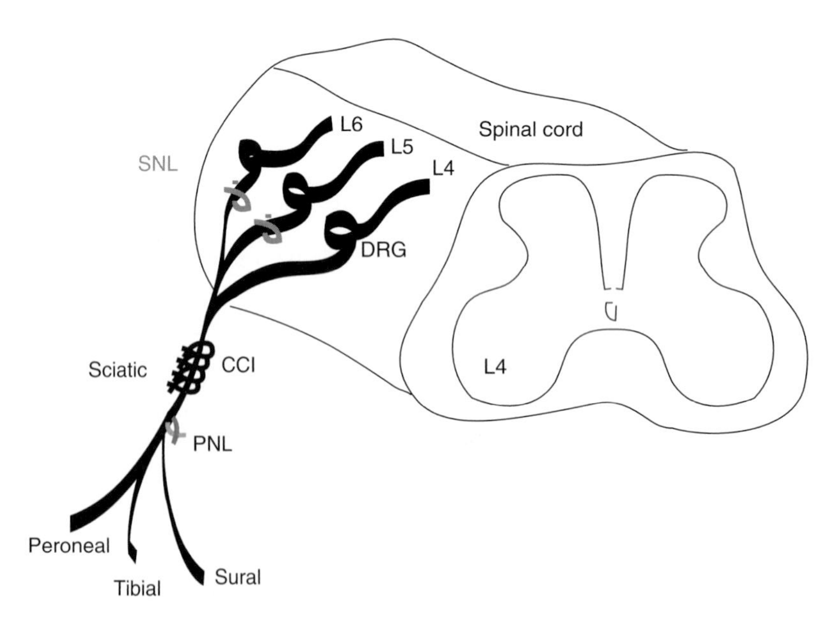

Figure 4 Animal models of neuropathic pain. This schematic illustrates the three main rodent models of neuropathic pain associated with direct nerve injury; the L5–L6 spinal nerve ligation (SNL) model (Chung model), the chronic constriction injury (CCI) of the sciatic nerve model (Bennett model), and the partial nerve ligation (PNL) model (Seltzer model).

is used for a single injection 4–5 days after surgery conducted in freely moving rats. In this model, perisciatic zymosan injection induces unilateral mechanical allodynia at low dose and bilateral mechanical allodynia at high dose. Interestingly, the same high dose injected into gelfoam in neighboring muscles does not induce mechanical allodynia, suggesting that immune activation must occur in close proximity to peripheral nerves to create allodynia and that zymosan spread to systemic circulation cannot explain allodynia created by perisciatic zymosan. Interestingly, no thermal hyperalgesia is observed in this model.

6.14.3.3.3 Diabetes

Another major cause of neuropathic pain in the clinic is neuropathic pain observed in diabetic patients. In rodents, this is mimicked by streptozotocin (STZ) injection to induce diabetes and subsequent neuropathic pain symptoms.[33] Usually, diabetes is induced by a single injection of STZ (75 mg kg^{-1} intraperitoneal). Diabetes is confirmed by testing for blood glucose levels. Not all animals show signs of neuropathic pain immediately following STZ administration. Generally it takes usually between 4 and 8 weeks to observed neuropathic pain symptoms, mostly mechanical allodynia assessment with von Frey monofilaments, in a group of streptozotocin-treated rats.

6.14.3.3.4 Chemotherapy-induced neuropathic pain (vincristine/paclitaxel/platine)

The last 'type' of neuropathic pain models are chemotherapy-induced neuropathic pain models. Cancer-related pain is a significant clinical problem that will likely increase in its extent as the average lifespan continues to rise and cancer therapies continue to improve. The two main sources of cancer-related pain are that from the malignancy itself and from the treatments utilized to alleviate the cancer (surgery, radiation, and chemotherapy). Peripheral neuropathy and subsequent neuropathic pain related to chemotherapeutic treatment can be dose limiting, and the pain is often resistant to standard analgesics. To date, no one drug or drug class is considered to be both a 'safe and effective analgesic' in the treatment of chemotherapy-induced pain, and three preclinical models of chemotherapy-induced neuropathic pain have been recently developed to further our understanding of the pathophysiology of such neuropathic pain states. Chemotherapy-induced neuropathic pain can be induced by the injection of either vincristine, platine, or paclitaxel.[33] Depending on the experimental protocol, they can be injected as a bolus, for several days or weeks or as a continuous intravenous infusion using osmotic pump. Interestingly, as observed in the clinic, thermal hyperalgesia is not observed in these animals. However, both mechanical allodynia and cold allodynia are observed.

The differential efficacy of analgesic medications for different types of pain that is seen in the clinic is also observed in animal pain models. For example, while opioid analgesics like morphine (**Figure 5**) are potent and efficacious in all animal pain models, anti-inflammatory agents such as ibuprofen and celecoxib (**Figure 5**) are most potent and effective in animal models associated with inflammation, and anti-epileptics like lamotrigine and gabapentin (**Figure 6**) are most potent and efficacious in animal models of neuropathic pain (**Table 1**). As preclinical models of the various forms of pain appear to have selective and differential predictive validity for efficacy in the clinical setting, they should be useful in determining if new chemical entities (NCEs) with a novel molecular mechanism have the promise to be broad-spectrum analgesics.

6.14.4 Clinical Trial Issues

Traditionally, the assessment of novel analgesics has been based on methods and models based on the clinical utility of opioid analgesics.[14] Many of the endpoints measured involve the use of self-report methodologies including the classical visual analog scale (VAS) with which patients rate their pain from a score of 0 (no noticeable pain) to 10 (worst pain imaginable). For the specific assessment of neuropathic pain, clinical studies have used tools like the McGill Pain Questionnaire, Neuropathic Pain Scale, and the Neuropathic Pain Symptoms Inventory. While the use of many of these analgesic endpoints has been validated in the clinical setting, the use of specific combinations of these scales may be useful to enhance the sensitivity of clinical outcomes for new analgesic compounds.

Clinical trial designs often employ parallel placebo-controlled and randomized withdrawal types of experimental manipulations.[14,35] The majority of these designs have been well validated using opioids; however, the relative utility in assessing novel analgesics that target specific aspects of chronic pain (e.g., neuropathic allodynia) await further evaluation (e.g., non-NSAID (nonsteroidal anti-inflammatory drug) mediated analgesia in the third molar extraction model).

A number of nociceptive tests have also been used in experimental clinical trials including acute heat sensitivity, topical and intradermal capsaicin, heat/capsaicin combinations, and quantitative sensory testing using both mechanical and thermal stimuli.[36,37] Some of these tests can also be coupled to other functional readouts such as functional MRI

Morphine Oxycodone Fentanyl Tramadol

Ibuprofen Indomethacin Aspirin

Celecoxib Rofecoxib

Figure 5 Opioid analgesics, nonsteroidal anti-inflammatory analgesics, and cyclooxygenase-2 (COX-2) inhibitors.

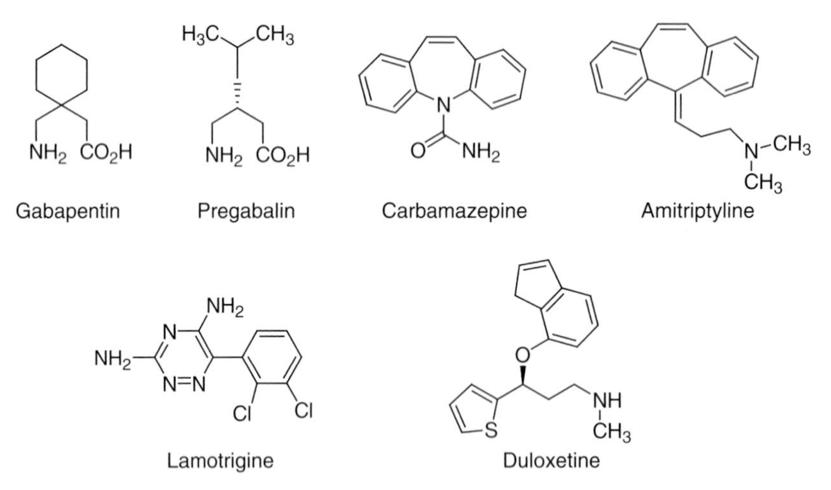

Figure 6 Analgesic adjuvant agents.

(fMRI) and nerve conduction velocity assessments to further enhance measurements of drug action.[38] For example, in assessing the acute antinociceptive effects of remifentanyl, the apparent sensitivity of fMRI analysis of human cortical oxygen utilization was significantly greater than patient's subjective ratings using traditional VAS scales. The use of fMRI coupled with experimental pain models like the capsaicin-evoked secondary hyperalgesia model may provide a reliable early assessment of novel analgesic efficacy. Many of these techniques have received clinical validation using opioids and are now being used to characterize mechanistically novel therapeutics.

Table 1 Effects of clinically used analgesics in preclinical models of acute, nociceptive, and neuropathic pain; relative analgesic efficacy of clinically useful analgesics in experimental pain models in the absence of psychomotor side effects. Data derived from both in house and literature values (see text)

	Acute pain		*Nociceptive pain*					*Neuropathic pain*		
	T	*C*	*F*	*Car*	*CFA*	*OA*	*RA*	*SNL*	*CCI*	*Vinc*
Morphine	2	2	3	4	5	4	5	2	3	3
Ibuprofen	0	1	2	5	5	2	NT	2	3	0
Celecoxib	0	1	2	5	5	5	1	2	3	0
Gabapentin	0	1	3	2	3	2	NT	5	5	3
Lamotrigine	2	2	3	2	3	2	NT	5	4	3

Scale: 0, no effect; 1, <20% effect; 2, 20–40% effect; 3, 40–60% effect; 4, 60–80% effect; 5, 80–100% effect, NT, Not tested. Abbreviations: T, tail flick; C, abdominal constriction; F, formalin; Car, carrageenan; CFA, complete Freund's adjuvant; OA, osteoarthritis; RA, rheumatoid arthritis; SNL, spinal nerve ligation; CCI, chronic constriction injury; Vinc, vincristine.

6.14.4.1 Translational Medicine in Pain

A major issue in all aspects of drug discovery is the extent of predictivity of the preclinical animal models used to characterize NCEs to the clinical situation. Of all areas of research, based on validation with opioids and nonsteroidal anti-inflammatory drugs (NSAIDs), pain models were always considered highly robust. However, the recent failure of multiple NCEs targeted for the neurokinin-1 (NK-1) or substance P receptor has raised serious concerns about the translation of preclinical analgesic data to analgesic efficacy in patients.[39,40] Structurally diverse NK-1 antagonists from several companies were evaluated clinically for acute pain (third molar extraction, migraine, or diabetic neuropathy).[40] While these NCEs attenuated nociceptive responses sensitized by either inflammation or nerve damage, they exhibited little effect on baseline (acute) nociception in preclinical models of pain. However, these same NCEs failed to demonstrate significant analgesic efficacy in early clinical trials. This situation has led to the re-evaluation of both the predictability of the preclinical animal models as well as the utility of clinical trial designs that were customized to assess analgesic by opiate and/or NSAID mechanisms. Is should also be noted that with the exception of the formalin model described above, animal pain models largely depend on stimulus-evoked pain behaviors as experimental endpoints. Consequently, assessments of ongoing nociception that are the foundation of most clinical studies are not modeled very well preclinically. It remains unclear whether the disconnect between the preclinical and clinical data for NK-1 antagonists reflects species-dependent roles of substance P in chronic pain or an imprecision in relating the nociceptive state of experimental animals to that of humans. Clearly, additional research is needed to adequately resolve these translational research issues for NK-1 antagonists. This issue highlights the practical need for the development of accurate and cost-effective translational medicine approaches to assess analgesic efficacy such as the coupling of fMRI studies with pain modeling in clinical studies described above.[38]

6.14.5 Current Treatments

Currently available analgesic agents can be broadly categorized as nonopioid analgesics (acetaminophen and NSAIDs), opioid analgesics (morphine and fentanyl), and adjuvant analgesics or coanalgesics (e.g., antiepileptics, anesthetics, and antidepressants). Nonopioid analgesics are mostly used to relieve mild to moderate nociceptive pain, adjuvant analgesics are used to relieve neuropathic pain, and opioid analgesics are used to treat severe pain of all origins, depending on the dose prescribed. Based on a 'number needed to treat' analysis, a variety of opioid medications including fentanyl and oxycodone produce equivalent analgesic efficacy in neuropathic pain as compared to gabapentin, tramadol, and analgesic adjuvants like tricyclic antidepressants and anticonvulsants.[41] The primary adverse event associated with chronic opioid therapy is decreased gastrointestinal (GI) motility. Respiratory depression and opioid dependence, which are routinely cited as a major issue in the use of opioids for pain therapy, are significantly less prevalent in chronic pain patients. This finding has led to ethical concerns related to patients not being given sufficient pain medication due to legal issues with drug scheduling.[35,42]

While currently available analgesics have therapeutic utility in different pain states, all suffer from drawbacks in clinical use. The opioids can produce tolerance and dependence, constipation, respiratory depression, and sedation.

The NSAIDs are associated with GI side effects and increased bleeding time, and do not effectively ameliorate severe pain. It has been estimated conservatively that 16 500 NSAID-related deaths occur among patients with rheumatoid arthritis or osteoarthritis every year in the USA.[43] The development of selective COX-2 inhibitors (e.g., celecoxib and rofecoxib (**Figure 5**)) offered NSAID-like analgesia with the potential for diminished GI side effects. However, in light of the withdrawal of rofecoxib from the market, the cardiovascular safety of this class of analgesics is questionable.[44] Adjuvant agents, e.g., gabapentin, lamotrigine, and amitryptiline, whose mechanism of action may be primarily mediated via a nonselective block of sodium channels, are associated with CNS and cardiovascular side effects. Currently available analgesics also have limited utility in the treatment of neuropathic pain. The anticonvulsant adjuvant gabapentin (**Figure 6**), which also has a poorly defined mechanism of action, has demonstrated clinical utility in the treatment of some forms of neuropathic pain.[45] Thus, there is a significant unmet medical need for safer and more effective analgesic agents.

6.14.5.1 Opioids

Opioid analgesics produce their effects by binding and activating the opioid receptor subtypes (e.g., μ, δ, κ receptors) in the CNS.[6] The cloning and characterization of the major opioid receptor subtypes (μ, OP1; δ, OP2; and κ, OP3) has stimulated significant basic and clinical research to discover new opioids with improved target selectivity, safety, and efficacy. Moreover, each of these opioid subtypes has been further subdivided into putative subtypes, and an 'orphan' member of this family, ORL-1 (OP4), has also been described.[46]

Given the anatomical distribution of the opioid receptors in the CNS, opioid analgesics can block pain transmission from the periphery to the spinal cord by blocking neurotransmitter release from the primary afferent fibers and by directly decreasing activation of postsynaptic dorsal horn neurons. The broad spectrum analgesic efficacy of the opioids, like morphine, fentanyl, and oxycodone (**Figure 5**), coupled with the fact that these agents do not show analgesic ceiling effects makes opioid compounds the mainstay in the control of moderate to severe pain.[46] The analgesic actions of opioid drugs are mediated at multiple sites of action including primary sensory afferent neurons, the dorsal horn of the spinal cord, and sites within the brain such as the brainstem and midbrain.

This multitude of opioid interactions also contributes to the side effects associated with opioid analgesic therapy including dependence, tolerance, immunosuppression, respiratory depression, and constipation.[41] Opioid dose titration can be achieved to manage some nociceptive conditions; however, this strategy does not provide full efficacy in all chronic pain syndromes such as cancer and neuropathic pain.[35,41] In addition, acting at higher brain centers, opioid analgesics can decrease pain transmission from the spinal cord to the brain, alter the limbic system, and increase descending inhibitory pathways to modulate pain transmission at the spinal level. While nociceptive pain is generally more responsive to opioid analgesics than neuropathic pain, nearly all types of pain respond to the right dose of opioid analgesics. Opioid analgesics are usually recommended to treat moderate to severe pain that does not respond to nonopioid analgesics alone and are often prescribed in combination with nonopioid analgesics. They do play a key role in the treatment of acute pain (postoperative pain), breakthrough pain, cancer pain, and some types of chronic noncancer pain.[35] However, the use of opioid analgesics is associated with side effects such as sedation, confusion, respiratory depression, constipation, nausea, and vomiting that can limit their utility. More recently, a number of controlled release or extended release formulations of opioids have been developed in an effort both to enhance the analgesic coverage of these medications and to reduce the severity of the adverse events.

The discovery of the opioid receptors in 1973 and their subsequent cloning has conceptually provided the tools necessary to develop receptor subtype selective NCEs that may improved efficacy and/or reduced side effect liabilities that would make them more useful analgesic agents. However, since 1975, more than $3 billion has been spent on research to find such improved NCEs but none has lived up to its preclinical promise in the clinical situation.

6.14.5.2 Anti-Inflammatory Drugs

6.14.5.2.1 Nonsteroidal anti-inflammatory drugs

The primary mechanism of action of NSAIDs is the blockade of prostaglandin synthesis via inhibition of the cyclo-oxygenase enzymes (COX-1 and COX-2). NSAIDs like aspirin, diclofenac, and ketorolac (**Figure 5**) have anti-inflammatory, antipyretic, and analgesic effects, and their anti-inflammatory effects can indirectly relieve nociceptive pain by reducing inflammation and tissue swelling, although potency across these effects can be compound dependent. These characteristics make them particularly efficacious is nociceptive pain conditions associated with peripheral inflammation.

6.14.5.2.2 Cyclooxygenase-2 inhibitors

Recently, selective inhibitors of COX-2 (the COX enzyme isoform inducible by inflammation), e.g., celecoxib and rofecoxib (**Figure 5**), have demonstrated similar efficacy to classical NSAIDs with reportedly fewer GI side effects at therapeutic doses.[47] However, this is controversial and these agents have also recently been associated with a significant increase in the risk of cardiovascular disease that has brought the entire class under intense scrutiny with rofecoxib having been withdrawn from the market.[44,48] As stated above, NSAIDs relieve mild to moderate pain associated with trauma, surgery, cancer, and arthritis. They are especially effective for certain types of somatic pain such as muscle and joint pain, sprains, bone/dental pain (tooth extraction), inflammatory pain, osteoarthritic pain, and in combination with disease-modifying therapies in rheumatoid arthritic pain.[49]

6.14.5.3 Analgesic Adjuvants

Adjuvant analgesics have a primary indication other than pain but have demonstrated analgesic effects in particular pain conditions. Antiepileptic drugs such as gabapentin, amitriptyline, carbamazepine, and lamotrigine (**Figure 6**) are one type of adjuvant analgesic and are used primarily in the treatment of neuropathic pain.[45] While it is hypothesized that their analgesic effects are due to their ability to reduce membrane excitability and tone down a hyperexcitable nervous system, their exact mechanism(s) of action remains unclear and not all antiepileptic drugs are good analgesics. As a class, they are most efficacious at treating peripheral neuropathic pain, postherpetic neuralgia, trigeminal neuralgia, and diabetic neuropathic pain.[50]

6.14.5.3.1 Amitriptyline

Amitriptyline is a tricyclic compound that has been approved for treatment of major affective disorders (e.g., depression) since the 1950s.[51] The antidepressant actions of amitriptyline generally are associated with blockade of the uptake of serotonin and norepinephrine in the CNS; however, amitriptyline possesses a multiplicity of other distinct pharmacological activities, e.g., antagonist actions at histamine, muscarinic, α1-adrenergic, and serotonin receptors at namomolar concentrations and at a number of ion channels (e.g., sodium, calcium, and potassium) at micromolar concentrations.[52,53] Amitriptyline also has micromolar affinity for blocking the uptake of the antinociceptive and anti-inflammatory purine adenosine.[52] In addition, amitriptyline has peripheral anti-inflammatory/analgesic actions in several in vivo models that are associated with acute, local delivery of low concentrations of amitriptyline. Many of these pharmacological activities are likely to contribute to its analgesic and anti-inflammatory actions.

6.14.5.3.2 Antiepileptics

Recent data suggest that newer antiepileptic drugs such as gabapentin or lamotrigine (**Figure 6**) are better alternatives to older agents of this class. Lamotrigine has been shown to be effective in patients with trigeminal neuralgia, complex regional pain syndrome, and neuropathic pain associated with multiple sclerosis and HIV infection.[54,55] In addition, it has been shown that gabapentin provides pain relief in diabetic neuropathic pain conditions and postherpetic neuralgia; it also has a more favorable side effect profile compared to other neuropathic pain agents.[55] Open-label studies suggest that gabapentin also may be useful in the management of trigeminal neuralgia, central pain, phantom limb pain, and neuropathy associated with HIV infection. Common side effects of antiepileptic drugs as a class include sedation, mental clouding, dizziness, nausea, or unsteadiness, and patients need to start at low doses and go through a slow titration in order to diminish the risk of side effects.

6.14.5.3.3 Pregabalin

Pregabalin (**Figure 6**), an alkylated analog of GABA, is a more potent anticonvulsant than gabapentin and has shown enhanced analgesic potency as well.[56] Pregabalin is clinically effective in pain associated with diabetic neuropathy and postherpetic neuralgia. The enhanced potency of pregabalin relative to gabapentin appears to be related to pharmacokinetic properties rather the mechanism of action. In this regard, the analgesic mechanism(s) for these anticonvulsant compounds has not been definitively determined. Recent data indicate that these GABA analogs bind the $\alpha 2\delta$ subunit of voltage-gated calcium channels with high affinity.[57] However, it should be noted that binding to the $\alpha 2\delta$ subunit is not likely to fully account for the analgesic properties of gabapentin or pregabalin.[58]

6.14.5.3.4 Antidepressants

In addition to antiepileptic drugs, tricyclic antidepressant drugs like amitriptyline are also used as adjunct analgesics in the treatment of neuropathic pain.[14,45] While other antidepressant medications such as selective serotonin reuptake inhibitors (SSRIs) have not proven to be particularly efficacious in treating neuropathic pain, some of the newer combination serotonin–norepinephrine reuptake inhibitors such as duloxetine and venlafaxine significantly attenuate neuropathic pain. Indeed, duloxetine (**Figure 6**) is the newest approved medications for the treatment of the pain associated with diabetic neuropathy.[59]

Interestingly, in parallel to their preferred efficacy in the clinic, clinically used analgesics also have a differential analgesic profile in preclinical pain models (see **Table 1**). While opioid analgesics like morphine are potent and efficacious in all animal models of pain, including acute pain, pain associated with inflammation in which they are particularly potent, and pain associated with neuropathy, NSAIDs such as ibuprofen and celecoxib are most potent and effective in animal models associated with inflammation, with near to no efficacy in models of acute pain and neuropathic pain, and antiepileptics like lamotrigine and gabapentin are most potent and efficacious in animal models of neuropathic pain.

6.14.6 Unmet Medical Needs

Due to the relatively poor efficacy to tolerability ratio for opioid analgesics in treating neuropathic pain, this area represents a major unmet medical need. While gabapentin and other adjuvant analgesics have been reported to be clinically effective in treating neuropathic pain, the efficacy rates are relatively small and are often accompanied by side effects of sufficient magnitude to limit compliance.[7] This situation is not surprising in that essentially all clinically used analgesic adjuvants were originally developed to treat other indications. Their clinical utility in treating pain has been based largely on clinical serendipity.

In the case of inflammatory or nociceptive pain, opioids provide significant analgesic efficacy; however, long-term use of these analgesics is limited by both opioid-mediated side effects including constipation, and regulatory concern of opioid dependence and abuse liability. NSAIDs also provide moderate pain relief in these pain states, but are associated with GI disturbances. The COX-2 inhibitors represent an analgesic advance due to their enhanced GI safety profile; however, the long-term cardiovascular safety of these agents is controversial.

6.14.7 New Research Areas

The identification clinically useful analgesic targeting nonopioid mechanisms has been challenging as demonstrated by the failure of several novel tachykinin NK-1 receptor antagonists in clinical trials.[39] Other novel analgesic agents have either not been advanced or are used in only limited conditions due to mechanism related toxicities. For example, the analgesic efficacy of ziconotide, a selective neuronal calcium channel (N-type, $Ca_v2.2$) blocker (discussed in more detail below) is limited to intrathecal administration in order to minimize severe cardiovascular adverse effects. Additionally, the development of several classes of adenosine kinase inhibitors as analgesics was halted due to the occurrence of vascular microhemorrhages in brain.[60] Despite these difficulties in translating advances in pain neurobiology into clinical useful analgesics, a number of novel analgesic mechanisms and compounds have been identified and validated in preclinical models. Some of these are described below.

6.14.7.1 Neuronal Nicotinic Receptor Agonists

Activation of neuronal nicotinic receptors (NNRs) represents a novel approach to pain management supported by the observation that epibatidine (isolated from *Epipedobates tricolor)* had significantly greater analgesic potency than morphine in assays of acute thermal pain.[61,62] While the mechanism of action of epibatidine was unknown, it was subsequently found to be a picomolar agonists at NNRs. NNR agonists with higher affinity for the $\alpha4\beta2$ subunit (the predominant subtype in the CNS) relative to the $\alpha1\beta1\delta\gamma$ nicotinic acetylcholine receptor subunit (located at the neuromuscular junction) had analgesic efficacy with a larger therapeutic window from severe side effects than did epibatidine.[63–67] These observations facilitated the discovery of ABT-594 (**Figure 7**), an $\alpha4\beta2$-preferring NNR agonist that was synthesized independently of the identification of the mechanism of action of epibatidine. ABT-594 has broad-spectrum analgesic activity in both acute (hot plate, tail flick, formalin) inflammatory (CFA), and neuropathic pain models. Importantly, ABT-594 showed less potential for analgesic tolerance than morphine in animal models and did not produce pharmacologic dependence.[63–65]

6.14.7.2 Vanilloid Receptor Modulators

The analgesic actions of topically applied capsaicin, the active ingredient in hot chillies, has been known for many years; however, the clinical utility of vanilloid-derived analgesics has been limited by the initial burning sensation these

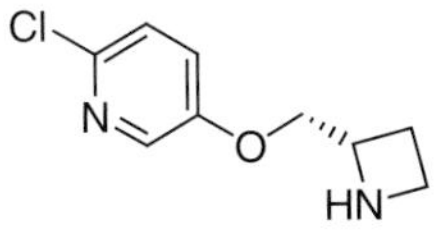

ABT-594

Figure 7 ABT-594, a neuronal nicotinic receptor (NNR) agonist.

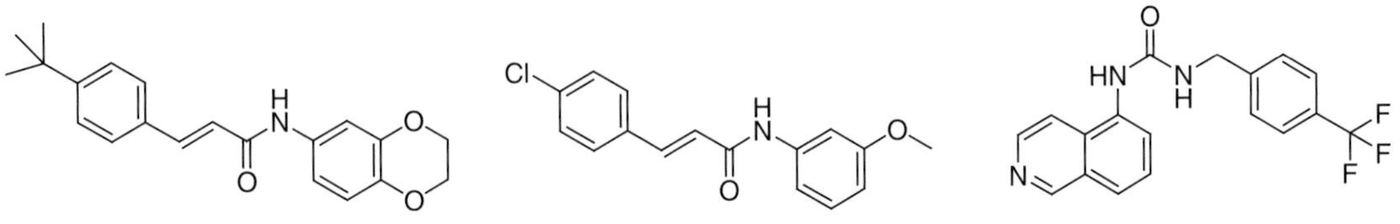

AMG 9810 SB-366791 A-425619

Figure 8 Novel nonvanilloid TRPV1 receptor antagonists.

compounds elicit. The cloning and characterization of the capsaicin-sensitive vanilloid receptor (TRPV1) has greatly enhanced understanding of the mechanism by which vanilloids, acid, and heat may alter nociceptor sensitivity.[68] TRPV1 receptors are one member for a larger family of transient receptor potential (TRP) proteins, several member of which have been associated with sensory function including cold (TRPA1, TRPM8) and mechanical (TRPV3, TRPV4) sensitivity.[68] These findings have also led to the discovery of several distinct classes of TRPV1 antagonists (**Figure 8**).[69,70] These compounds potently and selectively block capsaicin activation of TRPV1 in a competitive manner. Interestingly, these antagonists also effectively block TRPV1 activation by acid and heat, indicating that these agents may exert channel modulating activity as compared to direct channel activation. These latter effects may be species dependent with some capsaicin-competitive antagonists being ineffective in blocking acid activation of rat TRPV1 receptors. TRPV1 antagonists have demonstrated analgesic efficacy in animal models of inflammatory pain, thus illustrating their potential to be clinically effective analgesics.

6.14.7.3 Excitatory Amino Acid Receptor Antagonists

The excitatory amino acid (EAA) glutamate functions as a primary excitatory neurotransmitter in the CNS, and activation of EAA-specific ionotropic and metabotropic receptor superfamilies in the spinal cord underlies the process of central sensitization involved in chronic pain.[16] Activation of the heteromultimeric NMDA receptors (NR1/NR2B/NR2D subunits) expressed in spinal cord contributes to the expression of tactile and thermal hyperalgesia. A number of competitive and noncompetitive NMDA receptor antagonists including (±)-CPP, MK-801, ketamine, and dextromethorphan (**Figure 9**) block hyperalgesia in animal models and attenuate the process of central sensitization.[71,72] A problematic issue associated with NMDA receptor antagonists is their psychotomimetic effects that include both dysphoria and cognitive impairment. This has led to the search for ligands that are selective for specific NMDA channel subunits (e.g., NR2B) that may alter nociceptive processing with an improved therapeutic window relative to previous NMDA antagonists. CP-101,606 (**Figure 9**), a selective NR2B antagonist, effectively reduced pain in low back pain and spinal cord injury patients.[73] Adverse events associated with CP-101,606 treatment were dizziness, and hypoesthesia, but were tolerated by the patients.

Memantine (**Figure 9**) is a low-affinity (NR1/NR2B $IC_{50} = 820\,nM$) noncompetitive NMDA antagonist that has analgesic efficacy in humans. In early clinical studies, memantine attenuated ongoing neuropathic pain symptoms in both diabetic and postherpetic neuralgia patients that memantine failed in Phase III pain indication trials.[74] In addition to these agents, other NMDA antagonists are being investigated preclinically as potential analgesic. For example, MRZ-2/579 (**Figure 9**), a low-affinity noncompetitive NMDA antagonist, attenuates carrageenan-induced thermal hyperalgesia at doses that do not affect sensory-motor function.[75] GV 196771A (**Figure 9**) modulates NMDA receptor function by blocking the glycine binding site of the NMDA receptor complex. Like memantine and dextromethorphan, GV 196771A produces antihyperalgesia in animal models at doses that do not elicit CNS side effects.[76] Additionally, GV 196771A is only weakly active in cerebral stroke models suggesting differences in the physiological substrates of nociception and neuroprotection. NMDA receptor antagonists can provide opioid sparing effects and may prevent the

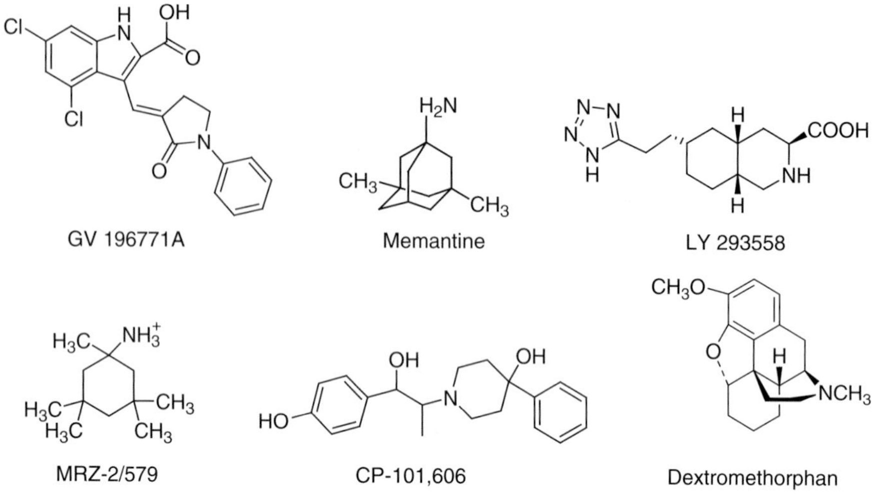

GV 196771A

Memantine

LY 293558

MRZ-2/579

CP-101,606

Dextromethorphan

Figure 9 Antagonists for excitatory amino acid (glutamate) receptors.

tolerance related to prolonged opioid use. As noted above, late stage clinical trials of a dextromethorphan/morphine combination (MorphiDex) are ongoing. Antagonists of the kainic acid subtype of the glutamate receptor include LY 293558 (**Figure 9**) which was active in acute migraine.[77]

6.14.7.4 Calcium Channel Modulators

Modulation of N-type ($Ca_v2.2$) calcium channels has been shown to provide an avenue for development of novel analgesics, as exemplified by ziconotide, a 25 amino acid polycationic peptide originally isolated from the venom of a cone snail. Its delivery is limited to the epidural and intrathecal routes, as systemic administration has led to a risk for orthostatic hypotension.[78] Ziconotide is approved for treating intractable cancer pain and chronic neuropathic pain. The clinical efficacy of ziconotide provides important validation of this novel analgesic mechanism. In addition to N-type calcium channels, T-type channels ($Ca_v3.1$ and $Ca_v3.2$) have also been implicated in nociceptive processing. T-type calcium channels are expressed on dorsal root ganglion neurons and intrathecal antisense treatment targeting the $Ca_v3.2$ subtype of T-type calcium channels effectively blocked all low-voltage calcium currents in dorsal root ganglion neurons and significantly attenuated both acute and inflammatory pain.[79]

6.14.7.5 Cannabinoids

Marijuana (cannabis) has been used to relieve pain for centuries.[80] However, clinical evaluation of the major active cannabinoid, Δ9-tetrahydrocannabinol (Δ9-THC), has produced equivocal results in chronic cancer pain patients. Furthermore, the analgesic actions of Δ9-THC could not be clearly separated from the other well-described psychotropic actions of Δ9-THC. Investigation of the pharmacological actions of the cannabinoids has been greatly aided by the recent discovery of specific cannabinoid receptor subtypes (CB_1 and CB_2), elucidation of their signal transduction pathways, and the identification of putative endogenous ligands (e.g., anandamide).[80] High densities of CB_1 receptors are found in the CNS, while CB_2 receptors are localized primarily to immune cells and peripheral nerve terminals. These advances in cannabinoid pharmacology suggest the possibility of identifying receptor subtype selective ligands.

Cannabimimetics have been shown to produce antinociception in animal pain models via spinal and supraspinal actions on CB_1 receptors, and by peripheral actions at CB_2 receptors on sensory afferents and, indirectly, on immune cells. Recent compounds in preclinical development include agonists with improved oral bioavailability and/or enhanced receptor subtypes selectivity. CT-3 (**Figure 10**) is an orally active and nonselective analog of THC that dose-dependently reduces acute nociception in the rat. Recently, CT-3 has been tested in the clinic in a phase II trial in chronic neuropathic pain patients and the results suggested that CT-3 could be useful in treating this condition.[80] O-1057 (**Figure 10**) is a potent and moderately CB_1 receptor selective analog of CT-3 that has improved water solubility and acute antinociceptive actions.[67] HU-308 (**Figure 10**) is a novel, highly CB_2 receptor selective agonist

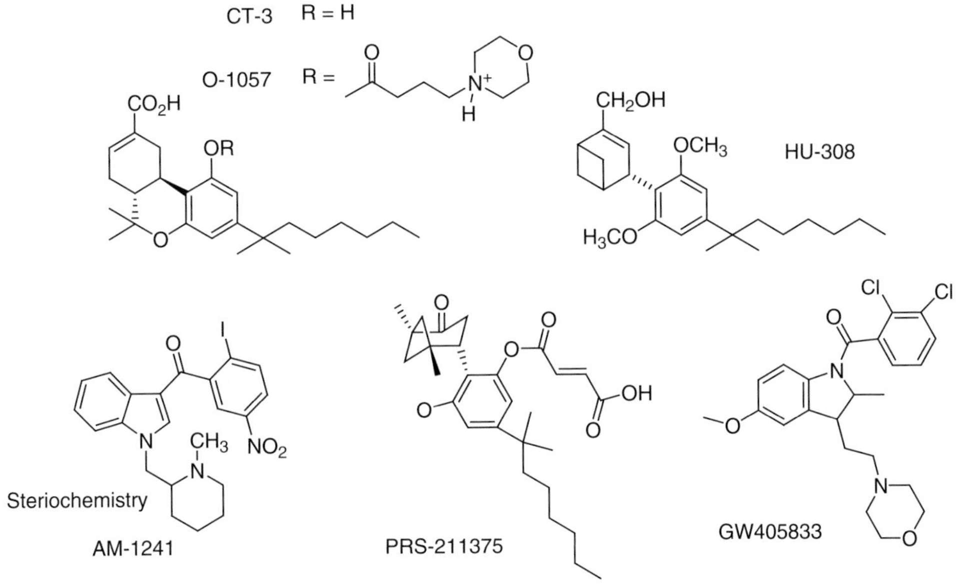

Figure 10 Agonists for cannabinoid receptors.

(K_i $CB_1 > 10$ mM, $CB_2 = 23$ nM) that has antinociceptive effects in the persistent phase of the mouse formalin test, but was inactive in the acute phase of the formalin test.[81] While no effects of HU-308 were observed on motor function, antinociceptive doses of the compound also reduced GI motility and blood pressure.[82]

More recently, the potential analgesic profile of CB_2 agonists has been extensively characterized in in vivo pain models in rodents. PRS-211375 (**Figure 10**) has been shown to be CB_2 selective (CB_2 $K_i = 9$ nM; CB_1 $K_i = 300$ nM) and a full agonist in adenylyl cyclase and GTPγS binding assays with similar potencies as observed in binding assays. PRS-211375 has good CNS penetration and has shown efficacy in various animal models including formalin, neuropathic pain models, acute thermal pain, visceral pain, and CFA-induced arthritis.[83] PRS-211375 recently completed phase I clinical trial. AM-1241 (**Figure 10**) is another selective CB_2 agonist used in preclinical models to identify the potential analgesic profile of CB_2 agonists. This compound has been shown to decrease acute, inflammatory, and neuropathic pain, its effects mediated through the release of β-endorphin and effects on opioid receptors.[84] GW405833 (**Figure 10**), another CB_2 agonist, has also been shown to produce analgesia in animal models of chronic pain including postoperative pain. However, contrary to the results obtained with AM-1241, the analgesic effects of GW405833 are not mediated through activity at opioid receptors.[85]

6.14.7.6 Sodium Channel Modulators

The activation of voltage-gated sodium channels is necessary for the generation of neuronal action potentials. A feature common to the local anesthetics and most analgesic adjuvants (e.g., carbamazepine, lamotrigine, and amitriptyline) is their ability to block sodium channels and this property may underlie the clinical utility of these agents in reducing pain. However, all of these agents possess other pharmacologically relevant activities that results in limits their analgesic effectiveness in the clinic. Notably, these include CNS sedation and/or untoward cardiovascular effects.

The cloning and characterization of several sensory nerve-specific sodium channel subtypes has raised interest in the possibility of developing subtype-specific inhibitors which might overcome the cardiovascular and proconvulsant liabilities of nonselective agents. The voltage-gated sodium channel gene family consists of multiple members, termed $Na_v1.1$ through $Na_v1.9$.[86] At least six of these channels are found in the peripheral nervous system.[87] Structurally, the family has a high overall degree of similarity (around 50% identity), with subfamilies being very closely related (up to 90% identity). Susceptibility to blockade by natural toxins, particularly tetrodotoxin (TTX), has been typically used to classify sodium channel currents. Two TTX-resistant channels are present in the periphery: $Na_v1.8$ (also called PN3 or SNS) and $Na_v1.9$ (also called NaN). $Na_v1.8$ is likely to be the more important sodium channel in regulating nociceptive signaling since in vivo antisense experiments targeting $Na_v1.9$ did not reduce chronic neuropathic pain.[88] $Na_v1.8$ immunoreactivity is increased in the carrageenan inflammatory pain model, and increased proximal to the site of nerve injury in rats and humans. Antisense oligonucleotides against $Na_v1.8$ prevented thermal hyperalgesia or mechanical allodynia from developing in animal models of neuropathic pain, and were also effective at reducing prostaglandin-induced hyperalgesia. $Na_v1.8$ knockout mice demonstrated a diminished response to noxious mechanical stimuli and delayed inflammatory hyperalgesia. While the PN3/SNS subtype has been a major focus of research, other sodium channels may also be appealing targets for pharmaceutical intervention of pain.[89,90] These data indicate that the development of antagonist that are selective for specific sodium channel subunits, like $Na_v1.8$ may provide novel analgesic agents. The feasibility of this approach is likely to be difficult based on the poor selectivity of currently available sodium channel blockers, however, two NCEs have been recently described that show increased potency for tetrodotoxin-resistant sodium channels as compared to the typical analgesic adjuvants. Ralfinamide (NW-1029) (**Figure 11**) inhibits TTX-resistant currents in rat dorsal root ganglion neurons with an IC_{50} value of 10 μM and dose-dependently reduced allodynia in neuropathic pain models.[91] CDA-54 (**Figure 11**) is a potent (IC_{50} ~200 nM) nonselective sodium channel blocker that reduces pain in inflammation pain models.[92]

6.14.7.7 Purines

6.14.7.7.1 P1 receptor agonists

The systemic or spinal administration of adenosine (ADO) or ADO A_1 receptor selective agonists has been shown to provide clinically effective analgesia.[93] However, analgesic efficacy of systemically administered ADO in neuropathic pain patients has been variable.[94] In contrast, intrathecal injection of ADO or an A_1 receptor agonist appears to consistently reduce pathological pain in neuropathic subjects or in healthy volunteers given intradermal capsaicin to induce central sensitization.[95–97] Despite these promising results, their profound effects on cardiovascular function have hampered the development of ADO receptors agonists as analgesics.

Inhibition of the primary metabolic enzyme, adenosine kinase (AK), which regulates ADO availability enhances the beneficial actions of adenosine at sites of tissue injury or trauma and AK inhibitors are potent antinociceptive and anti-inflammatory agents with an potentially improved therapeutic window over direct acting ADO receptor agonists.[98] ABT-702 (**Figure 12**), a non-nucleoside AK inhibitor, has both analgesic (e.g., effective in acute nociception) and antihyperalgesic activity (e.g., effective in blocking both hyperalgesia and allodynia).[98] The analgesic actions of ABT-702 are blocked by selectivity ADO A_1 receptor subtype antagonists and the locus of analgesic activity is in the spinal cord.[98] ABT-702 also shows significantly greater separation between its antihyperalgesic activity and its effects on CNS or cardiovascular function as compared to direct-acting ADO receptor agonists.

6.14.7.7.2 P2 receptor antagonists

The cloning and characterization of the family of ATP-sensitive ligand-gated ion channels (P2X receptors) provided mechanistic insights for the role of ATP as an extracellular signaling molecule.[99] The $P2X_3$ channel is localized

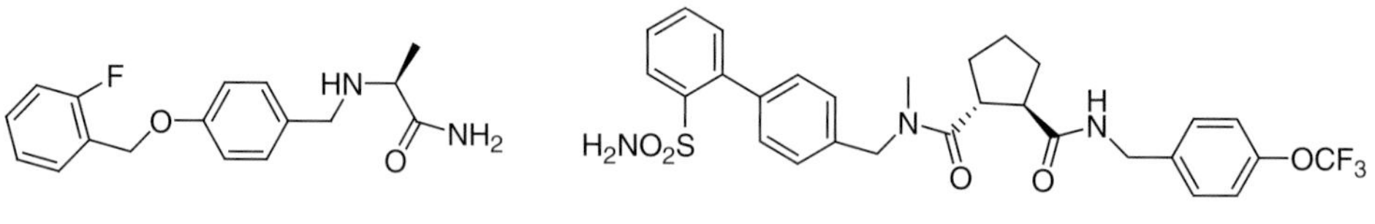

Ralfinamide

CDA-54

Figure 11 Novel voltage-gated sodium channel blockers.

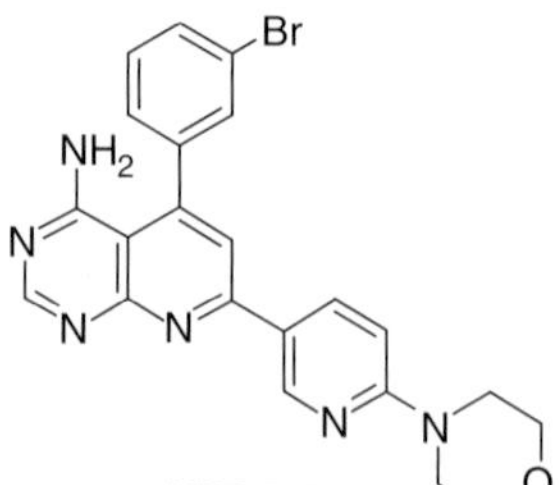

ABT-702

Figure 12 ABT-702, a nonnucleoside adenosine kinase (AK) inhibitor.

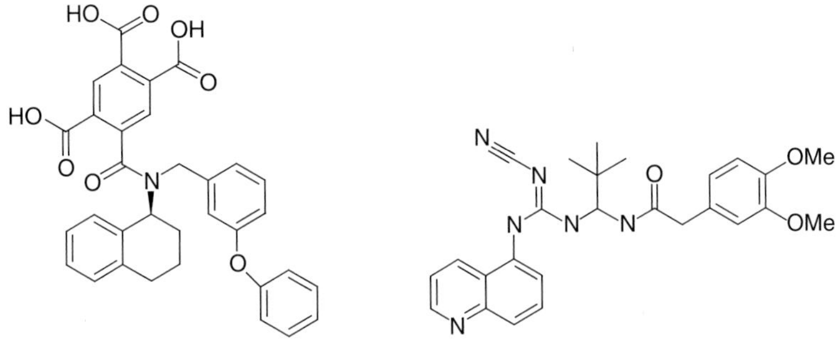

A-317491 A-740003

Figure 13 A-317491 and A-740003, selective $P2X_3$ and $P2X_7$ ligand-gated ion channel antagonists, respectively.

primarily to sensory neurons, suggesting a role in pain transmission. The $P2X_3$ messenger RNA (mRNA) occurs only in the trigeminal, dorsal root, and nodose ganglia, and the receptor is selectively expressed in sensory C-fiber neurons that project to the periphery and spinal cord, and which are predominantly nociceptors. In addition, $P2X_3$ receptors located presynaptically at the central terminals of primary afferent neurons may have a facilitatory role to enhance neurotransmission, leading to a further increase in pain sensation.[100]

A role of ATP in pain transmission is consistent with the observed induction of pain by ATP upon application to human skin, and with reports that intradermal and intrathecal application of ATP and ATP analogs (e.g., αβ-methylene ATP (αβmetATP)) into the rat hind paw evokes acute nociceptive behavioral responses. Transgenic disruption of $P2X_3$ receptors in rodents via knockout, antisense, or short interfering RNA (siRNA) manipulations leads to decreased nociceptive sensitivity.[101,102] A-317491 (**Figure 13**), a potent and selective antagonist of homomeric $P2X_3$ and heteromeric $P2X_{2/3}$ receptors that when given systemically and intrathecally dose-dependently reduces nociception in animal models of inflammatory and neuropathic pain indicating that blockade of spinal $P2X_3$ receptors may provide broad-spectrum analgesic effects in chronic pain states.[100]

$P2Y_2$ receptors signal through protein kinase C (PKC) and, in turn, modulates the activation of TRPV1 receptors.[103] Intrathecal antisense studies have also shown that $P2X_4$ receptors acting via a spinal neural–microglial interaction alter nociceptive processing in neuropathic rodents.[104] More recently, $P2X_7$ receptor knockout mice have been characterized and show reduced inflammation and neuropathic pain as compared to wild-type mice.[105] $P2X_7$ receptors are not localized on small-diameter neurons in the periphery, but are found on gial cells in the dorsal root ganglion.[106] Blockade of $P2X_7$ receptors leads to inhibition of IL-1β release from macrophages and glial cells, while antagonists like A-740003 are active in pain models.[110] In addition, $P2X_7$ receptors mediate the ability of ATP to stimulate glutamate release from glial cells, thus providing an additional mechanism for ATP-mediated fast neurotransmission.[107]

6.14.7.8 Emerging Pain Targets

As indicated above, a developing concept in pain research is the appreciation that neuroimmune interactions participate in nociceptive signaling in chronic pain states.[25] This has greatly added to the potential list of candidate mechanisms that may offer an avenue for analgesic intervention. In addition to traditional mechanisms associated with neuron–neuron communication, cytokines, chemokines, and inflammatory acute phase proteins are now known to contribute to nociceptive signaling. This collection of potential analgesic mechanisms is further complemented by findings that extracellular acid may also play a neuromodulatory role in the transmission of pain signals. Members of the TRP channel family are sensitive to both endogenous lipids as well as acidic pH.[68,69] The activation of specific acid-sensing ion channels (ASICs) also contributes to the encoding of noxious stimulation.[108] The development of selective antagonists of ASIC3 antagonist A-317567 (**Figure 14**) provided the demonstration that this acid-sensing channel contributes to both inflammatory hyperalgesia and pain associated with skin incision.[108]

As noted above, neurotrophic factors play important roles in the remodeling of the peripheral and central nervous systems in response to pain. Specifically, nerve growth factor (NGF) is a neurotrophin that is an important survival factor for sensory neurons.[109] However, NGF also has pronociceptive actions, producing pain and enhancing hyperalgesia in both experimental animals and in human clinical studies. The recent development of small-molecule antagonists like ALE0540 and PD90780 (**Figure 15**) as well as anti-NGF antibodies has enabled studies demonstrating that these agents effectively reduce chronic arthritic pain, skin incision pain, and tactile allodynia following peripheral nerve injury.[109]

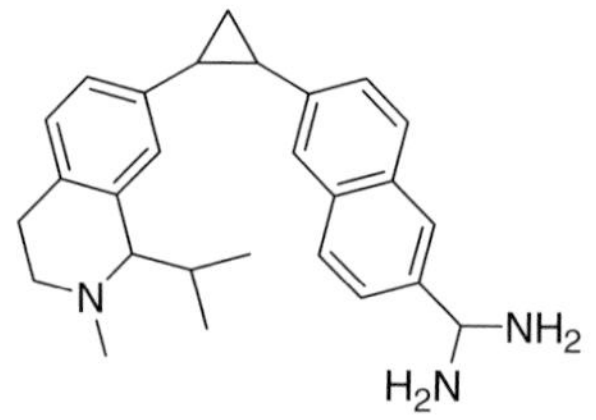

A-317567

Figure 14 A-317567, a selective antagonist of acid-sensing ion channels.

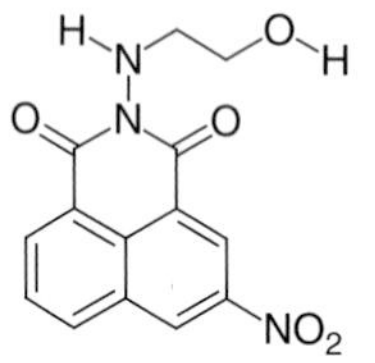

ALE 0540

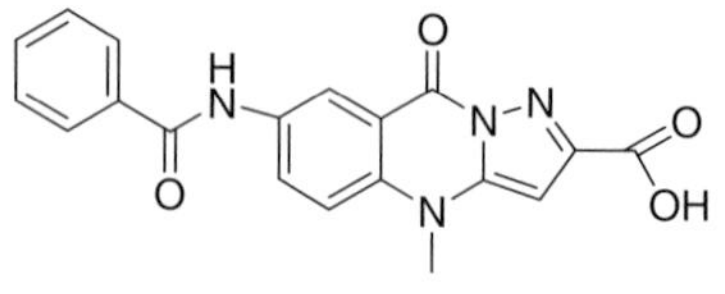

PD 90780

Figure 15 Antagonists of nerve growth factor (NGF) binding to trkA and p75^{NTR}.

6.14.8 Conclusions

The adverse physiological, psychological, and economic effects of inadequate pain management have become increasingly recognized in recent years. This has been accompanied by a growing awareness on the part of patients and caregivers that pain need not be tolerated, and an increased emphasis amongst physicians on the proactive treatment of pain. The unmet need for new analgesics remains substantial. Recent advances in the neurobiology of pain, together with the development of new preclinical and clinical pain paradigms, have revealed new opportunities for the development of analgesics, and raised the exciting possibility of entirely novel classes of analgesics in the future.

References

1. Merskey, H.; Bugduk, N. *Classification of Chronic Pain,* 2nd ed.; IASP Press: Seattle, WA, 1994.
2. Harden, N.; Cohen, M. *J. Pain Sympt. Mgmt.* **2003**, *25*, S12–S17.
3. Elliott, A. M.; Smith, B. H.; Penny, K. I.; Smith, W. C.; Chambers, W. A. *Lancet* **1999**, *354*, 1248–1252.
4. Ericksen, J.; Jensen, M. K.; Sjøgren, P.; Ekholm, O.; Rasmussen, N. K. *Pain* **2003**, *106*, 221–228.
5. *American living with pain survey*; American Chronic Pain Association, Roper Public Affairs and Media: Rocklin, CA, 2004.
6. Millan, M. J. *Prog. Neurobiol.* **1999**, *57*, 1–164.
7. Scholz, J.; Woolf, C. J. *Nat. Neurosci.* **2002**, *5*, 1062–1067.
8. Cervero, F.; Laird, J. M. *Pain* **1996**, *68*, 13–23.
9. Woolf, C. J.; Costigan, M. *Proc. Natl. Acad. Sci. USA* **1999**, *96*, 7723–7730.
10. Coda, B. A.; Bonica, J. J. *In Bonica's Management of Pain,* 3rd ed.; Lippincott: Baltimore, MD, 2001, pp 222–240.
11. Byers, M.; Bonica, J. J. *In Bonica's Management of Pain,* 3rd ed.; Lippincott: Baltimore, MD, 2001, pp 26–72.
12. Meyer, R. A.; Campbell, J. N.; Raja, S. N. *Advances in Pain Research and Therapy*; Raven: New York, 1985; Vol. 9, pp 53–71.
13. Abbott, J. D.; Moreland, L. W. *Exp. Opin. Invest. Drugs* **2004**, *13*, 1007–1018.
14. Dworkin, R. H.; Backonja, M.; Rowbotham, M. C.; Allen, R. R.; Argoff, C. R.; Bennett, G. J.; Bushnell, M. C.; Farrar, J. T.; Galer, B. S.; Haythornthwaite, J. A. *Arch. Neurol.* **2003**, *60*, 1524–1534.
15. Smith, P. A. *Drug News Perspect.* **2004**, *17*, 5–17.
16. Woolf, C. J.; Mannion, R. J. *Lancet* **1999**, *353*, 1959–1964.
17. Woolf, C. J.; Salter, M. W. *Science* **2000**, *288*, 1765–1768.
18. Urban, M. O.; Gebhart, G. F. *Proc. Natl. Acad. Sci. USA* **1999**, *96*, 7687–7692.
19. Treede, R. D.; Meyer, R. A.; Raja, S. N.; Campbell, J. N. *Prog. Neurobiol.* **1992**, *38*, 397–421.
20. McMahon, S. B.; Wall, P. D. *Pain* **1984**, *19*, 235–247.
21. Woolf, C. J. *Nature* **1983**, *306*, 686–688.
22. Woolf, C. J.; Thompson, S. W. *Pain* **1991**, *44*, 293–299.
23. Watkins, L. R.; Milligan, E. D.; Maier, S. F. *Pain* **2001**, *93*, 201–205.
24. Tsuda, M.; Inoue, K.; Salter, M. W. *Trends Neurosci.* **2005**, *25*, 101–107.
25. Honore, P.; Wade, C. L.; Zhong, C.; Harris, R. R.; Wu, C.; Ghayur, T.; Iwakura, Y.; Decker, M. W.; Sullivan, J. P.; Faltynek, C.; Jarvis, M. F. *Behav. Brain Res.* **2006**, *167*, 355–364.

26. Mogil, J. S.; Yu, L.; Basbaum, A. I. *Annu. Rev. Neurosci.* **2000**, *23*, 777–811.
27. Hargreaves, K.; Dubner, R.; Brown, F.; Flores, C.; Joris, J. *Pain* **1988**, *32*, 77–88.
28. Randall, L. O.; Selitto, J. J. *Arch. Int. Pharmacodyn. Ther.* **1957**, *111*, 409–419.
29. Chaplan, S. R.; Bach, F. W.; Pogrel, J. W.; Chung, J. M.; Yaksh, T. L. *J. Neurosci. Methods* **1994**, *53*, 55–63.
30. Dubuisson, D.; Dennis, S. G. *Pain* **1977**, *4*, 161–174.
31. Besson, J.-M.; Guilbaud, G. In *The Arthritic Rat as a Model of Clinical Pain*, Proceedings of the International Symposium, Saint-Paul de Vence, France, June 6–8, 1988; Besson, J.-M., Guilbaud, G., Eds.; Elsevier: Amsterdam, The Netherlands, 1988.
32. Fernihough, J.; Gentry, C.; Malcangio, M.; Fox, A.; Rediske, J.; Pellas, T.; Kidd, B.; Bevan, S.; Winter, J. *Pain* **2004**, *112*, 83–93.
33. Ueda, H. *Pharmacol. Ther.* **2006**, *109*, 57–77.
34. Chacur, M.; Milligan, E. D.; Gazda, L. S.; Armstrong, C.; Wang, H.; Tracey, K. J.; Maier, S. F.; Watkins, L. R. *Pain* **2001**, *94*, 231–244.
35. Portenoy, R. K.; Lesage, P. *Lancet* **1999**, *353*, 1695–1700.
36. Barden, J.; Edwards, J. E.; Mason, L.; McQuay, H. J.; Moore, R. A. *Pain* **2004**, *109*, 351–356.
37. Schuler, P. *Pain Therapeutics*, London, UK, 2005.
38. Bostock, H.; Campero, M.; Serra, J.; Ochoa, J. L. *Brain* **2005**, *128*, 2154–2163.
39. Boyce, S.; Hill, R. G. In *Progress in Pain Research and Management*; IASP Press: Seattle, WA, 1999; Vol. 16, pp 313–324.
40. Hill, R. *Trends Pharmacol. Sci.* **2000**, *21*, 244–246.
41. McQuay, H. *Lancet* **1999**, *353*, 2229–2232.
42. Portenoy, R. K. *Clin. Adv. Hematol. Oncol.* **2005**, *3*, 30–32.
43. Wolfe, M. M.; Lichtenstein, D. R.; Singh, G. *N. Engl. J. Med.* **1999**, *340*, 1888–1899.
44. Maxwell, S. R. J.; Webb, D. J. *Lancet* **2005**, *365*, 449–451.
45. McCleane, G. *Exp. Opin. Pharmacother.* **2004**, *5*, 1299–1312.
46. Pasternak, G. W.; Letchworth, S. R. *Curr. Opin. CPNS Invest. Drug.* **1999**, *1*, 54–64.
47. Davis, M. P.; Walsh, D.; Lagman, R.; LeGrand, S. B. *Lancet Oncol.* **2005**, *6*, 696–704.
48. Davies, N. M.; Jamali, F. *J. Pharmacol. Pharmaceut. Sci.* **2004**, 7, 332–336.
49. Hochberg, M. *Am. J. Managed Care* **2002**, *8*, S502–S517.
50. Hansson, P. T.; Dickenson, A. H. *Pain* **2005**, *113*, 251–254.
51. Williams, M.; Kowaluk, E. A.; Arneric, S. P. *J. Med. Chem.* **1999**, *42*, 1481–1500.
52. Sawynok, J.; Reid, A. R.; Esser, M. J. *Pain* **1999**, *80*, 45–55.
53. Eisenach, J. C.; Gebhart, G. F. *Anesthesiology* **1995**, *83*, 1046–1054.
54. Backonja, M.-M. *Neurology* **2002**, *59*, S14–S17.
55. Pappagallo, M. *Clin. Ther.* **2003**, *25*, 2506–2538.
56. Dworkin, R. H.; Kirkpatrick, P. *Nat. Rev. Drug Disc.* **2005**, *4*, 455–456.
57. Belliotti, T. R.; Capiris, T.; Ekhato, I. V.; Kinsora, J. J.; Field, M. J.; Heffner, T. G.; Meltzer, L. T.; Schwarz, J. B.; Taylor, C. P.; Thorpe, A. J. et al. *J. Med. Chem.* **2005**, *48*, 2294–2307.
58. Mortell, K. H.; Anderson, D. J.; Lynch, J. J.; Nelson, S. L.; Sarris, K.; McDonald, H.; Sabet, R.; Baker, S.; Honore, P.; Lee, C.-H. et al. *Bioorg. Med. Chem. Lett.* **2006**, *16*, 1138–1141.
59. Waitekus, A. B.; Kirkpatrick, P. *Nat. Rev. Drug Disc.* **2004**, *3*, 907–908.
60. McGaraughty, S.; Cowart, M.; Jarvis, M. F.; Berman, R. *Curr. Topics Med. Chem.* **2005**, *5*, 43–58.
61. Qian, C.; Li, T.; Shen, T. Y.; Libertine-Garahan, L.; Eckman, J.; Biftu, T.; Ip, S. *Eur. J. Pharmacol.* **1993**, *250*, R13–R14.
62. Spande, T. F.; Garraffo, H. M.; Edwards, M. W.; Yeh, H. J. C.; Pannell, L.; Daly, J. W. *J. Nat. Prod.* **1992**, *55*, 707–722.
63. Bannon, A. W.; Decker, M. W.; Holladay, M. W.; Curzon, P.; Donnelly-Roberts, D.; Puttfarcken, P. S.; Bitner, R. S.; Diaz, A.; Dickenson, A. H.; Porsolt, R. D. et al. *Science* **1998**, *279*, 77–81.
64. Bannon, A. W.; Decker, M. W.; Kim, D. J. B.; Campbell, J. E.; Arneric, S. P. *Brain Res.* **1998**, *801*, 158–163.
65. Bannon, A. W.; Decker, M. W.; Curzon, P.; Buckley, M. J.; Kim, D. J. B.; Radek, R. J.; Lynch, J. K.; Wasicak, J. T.; Lin, N. H.; Arnold, W. H. et al. *J. Pharmacol. Exp. Ther.* **1998**, *285*, 787–794.
66. Barlocco, D.; Cignarella, G.; Tondi, D.; Vianello, P.; Villa, S.; Bartolini, A.; Ghelardini, C.; Galeotti, N.; Anderson, D. J.; Kuntzweiler, T. A. et al. *J. Med. Chem.* **1998**, *41*, 674–681.
67. Kesingland, A. C.; Gentry, C. T.; Panesar, M. S.; Bowes, M. A.; Vernier, J. M.; Cube, R.; Walker, K.; Urban, L. *Pain* **2000**, *86*, 113–118.
68. Caterina, M. J. *Annu. Rev. Neurosci.* **2001**, *24*, 487–517.
69. Honore, P.; Wismer, C. T.; Mikusa, J.; Zhu, C. Z.; Zhong, C.; Gauvin, D.; Gomtsyan, A.; El-Kouhen, R.; Lee, C.-H.; Marsh, K. et al. *J. Pharmacol. Exp. Ther.* **2005**, *314*, 410–421.
70. Swanson, D. M.; Dubin, A. E.; Shah, C.; Nasser, N.; Chang, L.; Dax, S. L.; Jetter, M.; Breitenbucher, J. G.; Liu, C.; Mazur, C. et al. *J. Med. Chem.* **2005**, *48*, 1857–1872.
71. Herman, B. H.; Vocci, F.; Bridge, P. *Neuropsychopharmacology* **1995**, *13*, 269–293.
72. Kristensen, J. D.; Svensson, B.; Gordh, T. *Pain* **1992**, *51*, 249–253.
73. Sang, C. N.; Weaver, J. J.; Jinga, L.; Wouden, J.; Saltarelli, M. D. *2003 Abstract Viewer/Itinerary Planner*, Program No. 814.9; Society for Neuroscience: Washington, DC, 2003.
74. Sang, C. N.; Booher, S.; Gilron, I.; Parada, S.; Max, M. B. *Anesthesiology* **2002**, *96*, 1053–1061.
75. Parsons, C. In *New Developments in Glutamate Pharmacology*; Orlando, FL, Mar 4–5, 1999.
76. Quartaroli, M.; Fasdelli, N.; Bettelini, L.; Maraia, G.; Corsi, M. *Eur. J. Pharmacol.* **2001**, *430*, 219–227.
77. Sang, C. N.; Ramadan, N. M.; Wallihan, R. G.; Chappell, A. S.; Freitag, F. G.; Smith, T. R.; Silberstein, S. D.; Johnson, K. W.; Phebus, L. A.; Bleakman, D. et al. *Cephalgia* **2004**, *24*, 596–602.
78. Bowersox, S. S.; Gadbois, T.; Singh, T.; Pettus, M.; Wang, X. Y.; Luther, R. R. *J. Pharmacol. Exp. Ther.* **1996**, *279*, 1243–1249.
79. Bourinet, E.; Alloui, A.; Monteil, A.; Barrère, C.; Couette, B.; Poirot, P.; Pages, A.; McRory, J.; Snutch, T. P.; Eschalier, A. et al. *EMBO J.* **2005**, *24*, 315–324.
80. Grotenhermen, F. *Curr. Drug Targets CNS Neurol. Disorders* **2005**, *4*, 507–530.
81. Pertwee, R. G.; Gibson, T. M.; Stevenson, L. A.; Ross, R. A.; Banner, W. K.; Saha, B.; Razdan, R. K.; Martin, B. R. *Br. J. Pharmacol.* **2000**, *129*, 1577–1584.
82. Hanus, L.; Breuer, A.; Tchilibon, S.; Shiloah, S.; Goldenberg, D.; Horowitz, M.; Pertwee, R. G.; Ross, R. A.; Mechoulam, R.; Fride, E. *Proc. Natl. Acad. Sci. USA* **1999**, *96*, 14228–14233.

83. Meilin, S. B.; Bar-Joseph, A.; Reichstein, A.; Weksler, A.; Berkovich, Y.; Azulai, M.; Yacovan, A.; Amselem, S.; Nimrod, R.; David, P. *2005 Abstract Viewer/Itinerary Planner*, Program No. 172.4; Society for Neuroscience: Washington, DC, 2005.
84. Ibrahim, M. M.; Porreca, F.; Lai, J.; Albrechy, P. J.; Rice, F.; Khodorova, A.; Davar, G.; Makriannis, A.; Vanderah, T. W.; Mata, H. P. et al. *Proc. Natl. Acad. Sci. USA* **2005**, *102*, 3093–3098.
85. Whiteside, G. T.; Gottshall, S. L.; Boulet, J. M.; Chaffer, S. M.; Harrison, J. E.; Pearson, M. S.; Turchin, P. L.; Mark, P. L.; Garrison, A. E.; Valenzano, K. J. *Eur. J. Pharmacol.* **2005**, *258*, 65–72.
86. Goldin, A. L.; Barchi, R. L.; Caldwell, J. H.; Hofmann, F.; Howe, J. R.; Hunter, J. C.; Kallen, R. G.; Mandel, G.; Meisler, M. H.; Netter, Y. B. et al. *Neuron* **2000**, *28*, 365–368.
87. Novakovic, S. D.; Eglen, R. M.; Hunter, J. C. *Trends Neurosci.* **2001**, *24*, 473–478.
88. Porreca, F.; Lai, J.; Bian, D.; Wegert, S.; Ossipov, M. H.; Eglen, R. M.; Kassotakis, L.; Novakovic, S.; Rabert, D. K.; Sangameswaran, L.; Hunter, J. C. *Proc. Natl. Acad. Sci. USA* **1999**, *96*, 7640–7644.
89. Lai, J.; Porreca, F.; Hunter, J. C.; Gold, M. S. *Annu. Rev. Pharmacol. Toxicol.* **2004**, *44*, 371–397.
90. Wood, J. N.; Boorman, J. *Curr. Topics Med. Chem.* **2005**, *5*, 529–537.
91. Stummann, T. C.; Salvati, P.; Fariello, R. G.; Faravelli, L. *Eur. J. Pharmacol.* **2005**, *510*, 197–208.
92. Brochu, R. M.; Dick, I. E.; Tarpley, J. W.; McGowan, E.; Gunner, D.; Herrington, J.; Shao, P. P.; Ok, D.; Li, C.; Parsons, W. H. et al. *Mol. Pharmacol.* **2005**, *69*, 823–832.
93. Segerdahl, M.; Irestedt, L.; Sollevi, A. *Acta. Anaesth. Scand.* **1999**, *40*, 792–797.
94. Eisenach, J. C.; Rauck, R. L.; Curry, R. *Pain* **2003**, *105*, 65–70.
95. Karlsten, R.; Gordh, T., Jr. *Anesth. Analg.* **1995**, *80*, 844–847.
96. Eisenach, J. C.; Curry, R.; Hood, D. D. *Anesthesiology* **2002**, *97*, 938–942.
97. Eisenach, J. C.; Hood, D. D.; Curry, R. *Anesthesiology* **2002**, *96*, 29–34.
98. Jarvis, M. F. *Exp. Opin. Ther. Targets* **2003**, 7, 513–522.
99. North, R. A. *Physiol. Rev.* **2002**, *82*, 1013–1067.
100. Jarvis, M. F.; Burgard, E. C.; McGaraughty, S.; Honore, P.; Lynch, K.; Brennan, T. J.; Subieta, A.; van Biesen, T.; Cartmell, J.; Bianchi, B. et al. *Proc. Natl. Acad. Sci. USA* **2002**, *99*, 17179–17184.
101. Dorn, G.; Patel, S.; Wotherspoon, G.; Hemmings-Mieszczak, M.; Barclay, J.; Natt-François, J. C.; Martin, P.; Bevan, S.; Fox, A.; Ganju, P. et al. *Nucleic Acids Res.* **2004**, *32*, e49.
102. Honore, P.; Kage, K.; Mikusa, J.; Watt, A. T.; Johnston, J. F.; Wyatt, J. R.; Faltynek, C. R.; Jarvis, M. F.; Lynch, K. *Pain* **2002**, *99*, 11–19.
103. Moriyama, T.; Iida, T.; Kobayashi, K.; Higashi, T.; Fukuoka, T.; Tsumura, H.; Leon, C.; Suzuki, N.; Inoue, K.; Gachet, C. et al. *J. Neurosci.* **2003**, *23*, 6058–6062.
104. Tsuda, M.; Shigemoto-Mogami, Y.; Koizumi, S.; Mizokoshi, A.; Kohsaka, S.; Salter, M. W.; Inoue, K. *Nature* **2003**, *424*, 778–783.
105. Chessell, I. P.; Hatcher, J. P.; Bountra, C.; Michel, A. D.; Hughes, J. P.; Green, P.; Egerton, J.; Murfin, M.; Richardson, J.; Peck, W. L. et al. *Pain* **2005**, *114*, 386–396.
106. Zhang, X.-F.; Han, P.; Faltynek, C. R.; Jarvis, M. F.; Shieh, C.-C. *Brain Res.* **2005**, *1052*, 63–70.
107. Donnelly Roberts, D. L.; Namovic, M. T.; Faltynek, C.; Jarvis, M. F. *J. Pharmacol. Exp. Ther.* **2004**, *308*, 1053–1061.
108. Dube, G. R.; Lehto, S. G.; Breese, N. M.; Baker, S. J.; Wang, X.; Matulenko, M. A.; Honore, P.; Stewart, A. O.; Moreland, R. B.; Brioni, J. D. *Pain* **2005**, *117*, 88–96.
109. Hefti, F. F.; Rosenthal, A.; Walicke, P. A.; Wyatt, S.; Vergara, G.; Shelton, D. L.; Davies, A. M. *Trends Pharmacol. Sci.* **2006**, *27*, 85–91.
110. Honore, P.; Donnelly-Roberts, D.; Namovic, M. T.; Hsieh, G.; Zhu, C. Z.; Mikusa, J. P.; Hernadez, G.; Zhong, C.; Gauvin, D. M.; Chandran, P. et al. *J. Pharmacol. Exp. Ther.* **2006**, doi:10.1124/jpet.106.111559.

Biographies

Prisca Honore, PhD, a native of France, received a doctoral degree in molecular pharmacochemistry and experimental pharmacology from the University of Paris. Her postdoctoral training was completed at the University of Minnesota in Minneapolis, where she studied the role of substance P receptors expressing spinal cord neurons in pain transmission using an SP-Saporin toxin, in addition to investigating the mechanisms underlying bone cancer pain. Prisca joined Abbott Laboratories in 2000 and is a senior group leader in the Neuroscience Research division, at Abbott Park, IL. In 2004, she received the John C. Liebeskind Early Career Scholar Award from the American Pain Society.

Michael F Jarvis, PhD, received his doctoral training in behavioral neuroscience at Rutgers University. He is currently a project leader and associate research fellow in Neuroscience Research, at Abbott Laboratories. A continuing research interest has been the study of the role of purines in modulating nociceptive signaling. His current research is focused on the identification and characterization of novel analgesic drug targets and ligands.

Comprehensive Medicinal Chemistry II
ISBN (set): 0-08-044513-6

ISBN (Volume 6) 0-08-044519-5; pp. 327–349

6.15 Local and Adjunct Anesthesia

T A Bowdle, University of Washington, Seattle, WA, USA
L J S Knutsen and M Williams, Worldwide Discovery Research, Cephalon, Inc., West Chester, PA, USA

6.15.1 Introduction

Anesthetics are used to depress the peripheral and central nervous systems (CNS) by blocking nerve conduction in order to facilitate surgical and other noxious procedures.[1] Anesthetics can be divided into general (inhalation and parenteral) and local types, the former inducing a loss of waking consciousness in humans similar in many respects to sleep while the latter block local nerve conduction. Parenteral anesthetics are also termed intravenous hypnotics or intravenous induction agents and can be used alone or in combination with other agents.

General anesthesia can be defined in terms of a functional deafferentation reflecting a global loss of response to, and perception of, all external stimuli, e.g., a surgical incision. Additional facets of general anesthesia include: blunting of hemodynamic and endocrine responses, respiratory paralysis, and a lack of awareness and memory.[2] It involves sites in both the brain and spinal cord and has been defined as artificially (drug)-induced sleep. In the mid twentieth century, inhalation or volatile anesthetics were defined on the basis of their ability to induce a loss of righting reflex (motor tone) that was frequently referred to as 'sleep time,' and the amount of time required to induce an anesthetic state. Motor atonia can be dissociated from loss of wakefulness while controlled sedation can be equated with 'light sleep.'[3] The intertwining of anesthesia with sleep-state physiology is controversial[4] and reflects the complexity of the state of consciousness. While both conditions involve a common hypothalamic locus,[4] they can be distinguished by the fact that a sleeping individual can be aroused rapidly while an anesthetized individual can only be aroused when the drug is cleared from the brain.

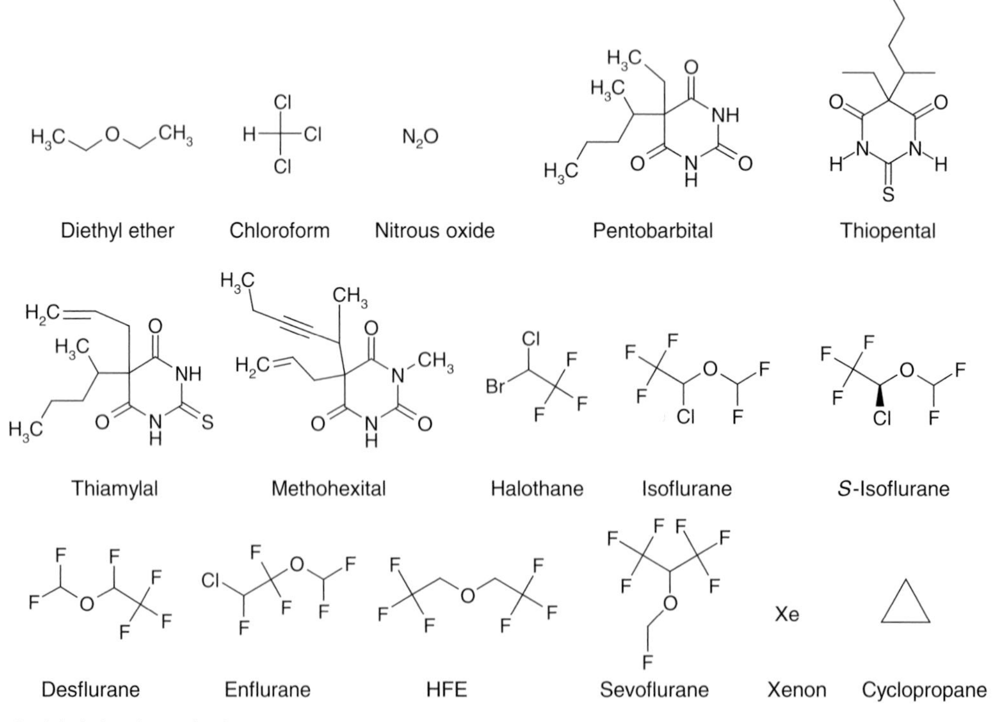

Figure 1 Inhalational anesthetics.

The first description of anesthetic use was in 1846 when diethyl ether (**Figure 1**) was used to induce unconsciousness in the operating room setting.[1] Despite the widespread and increasing use of anesthetics in human medicine for over 150 years, their mechanisms of action have yet to be clearly elucidated, reflecting: (1) the diverse structures of general anesthetics (**Figure 1**)[3,5]; (2) the high, generally millimolar, concentrations of anesthetics used to induce anesthesia[1]; (3) the lack of anesthetic antagonists; and (4) the complex nature of general anesthesia that involves amnesia, analgesia, unconsciousness, and immobility. Additionally, the proposal in the 1860s by the French physiologist, Claude Bernard, based on studies on the effects of anesthetic agents on plants and animals, that anesthetics like ether and chloroform acted by coagulating 'albuminoid' cell contents, focused attention on a relatively nonspecific interaction of these agents with the cell membrane. Subsequently, the seminal studies of Meyer and Overton in 1895 showed that anesthetic potency was correlated with the oil/gas partition coefficient, the Meyer–Overton rule, that the number of molecules dissolved in the lipid cell membrane and not the type of inhalation agent produces anesthesia, focusing attention on the lipid bilayer as the site of action of general anesthetics. The latter led to the unitary theory of anesthesia, namely that all anesthetics, both general and local, produce their effects by a simple perturbation of the cell membrane. In many respects, the lipid perturbation theory confounded both the search for discrete mechanisms of action for anesthetics and, consequently, the search for improved anesthetic agents. This theory has been extended to the lipophilic intrusion protection organization-integrative design (LIPOID) hypothesis of analgesia[8] that postulates that the multiple lipophilic properties of anesthetic molecules contribute to many simultaneous actions that result in anesthesia.

Continuing research[6–10] has focused on the interaction of anesthetics with membrane proteins, e.g., ion channels, as a primary mechanism of action. The observation[11] that several anesthetics, e.g., isoflurane, etomidate (**Figure 2**), as well as some steroids, have modest enantioselective effects, while further supporting the concept that specific membrane-binding sites or receptors are responsible for the actions of anesthetics, are at odds with Pfiffer's rule,[12] given the low affinities measured to date with general anesthetics. It has also been noted, however, that "each anaesthetic agent has a unique action, the end result of which is to produce a state of anaesthesia."[7]

Inhalation anesthetics include a variety of gases and volatile liquids, e.g., halothane, desflurane, nitrous oxide (**Figure 1**). These agents have a low safety margin with therapeutic indices in the range of 2–4, and with each inhalation anesthetic having a distinct side-effect profile,[1] a finding that reinforced the concept of each anesthetic having a unique mechanism of action.[7] Their differing side-effect profiles and evidence for gender- and anatomically specific differences in the actions of general anesthetics additionally confused the logic of a search for a common site of action.

The parenteral general anesthetics, administered by the intravenous route, include barbiturates like thiopental and other agents, including propofol, ketamine, and etomidate (**Figure 2**), as well as certain opioids, e.g., morphine, fentanyl, alfentanil, sufentanil (**Figure 3**), that can be used for the maintenance of general anesthesia.

Local anesthetics include cocaine, lidocaine, and bupivacaine (**Figure 3**), that act by blocking voltage-sensitive sodium channels (VSSCs) or Na_v channels.[13]

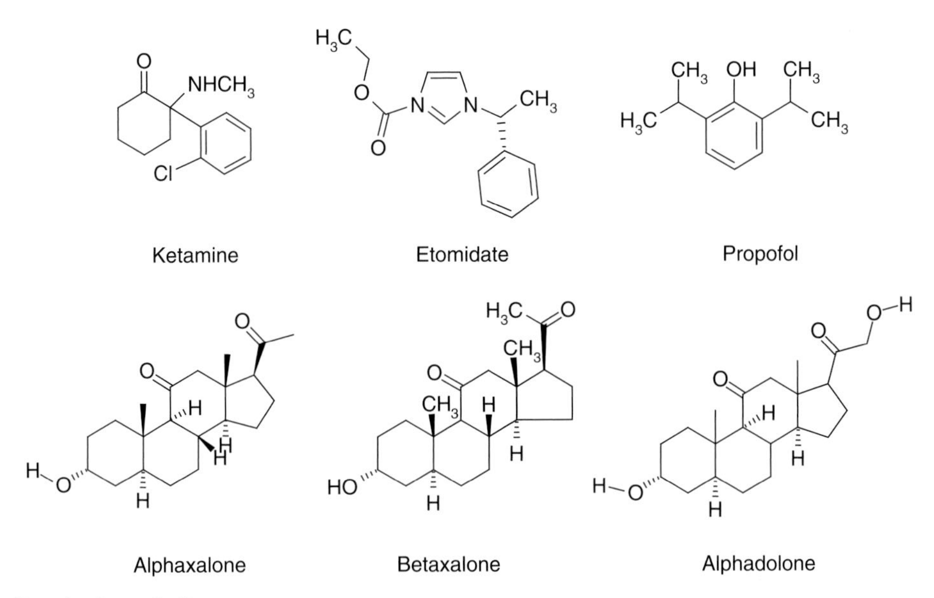

Figure 2 Parenteral anesthetics.

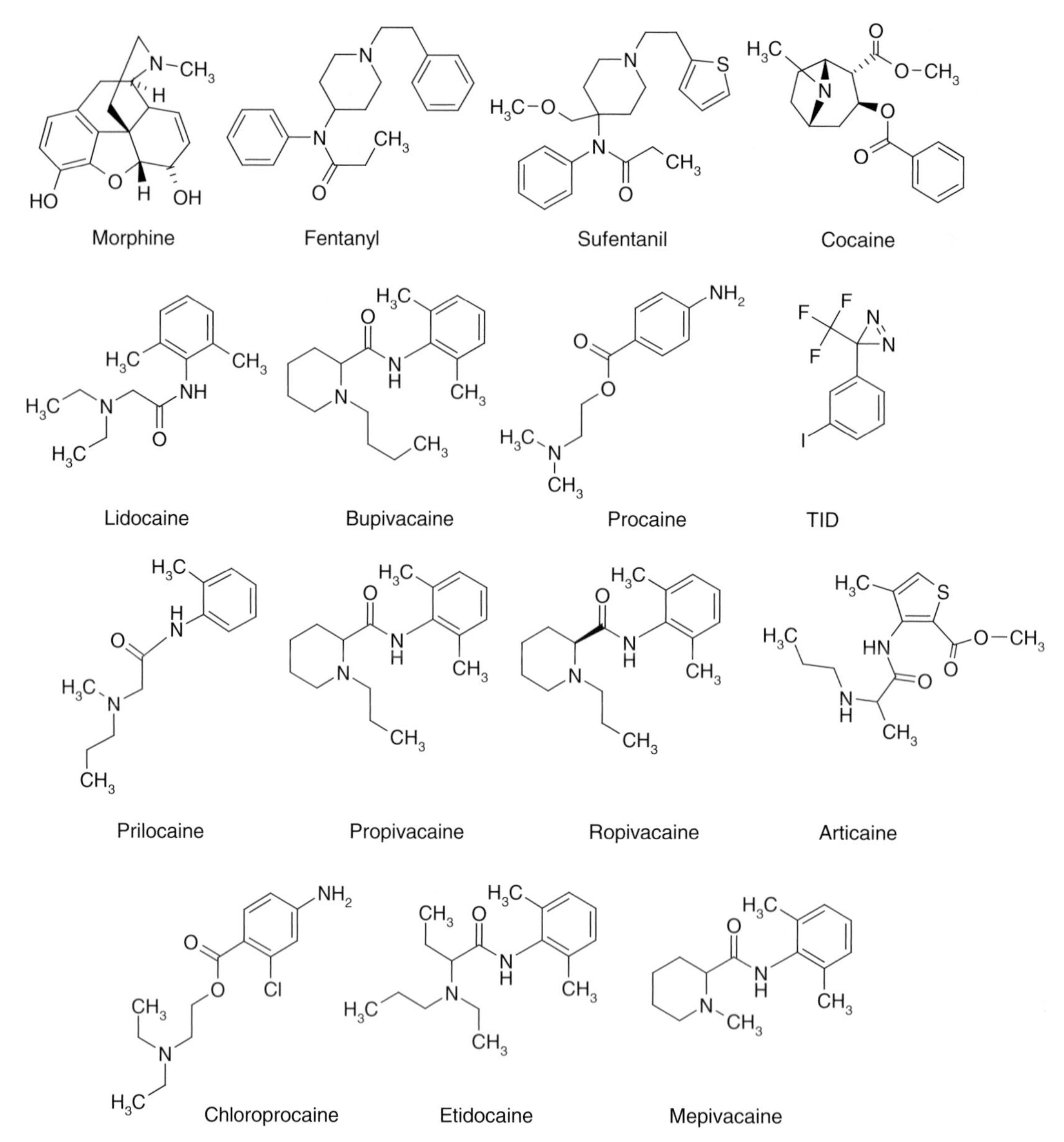

Figure 3 Opioids and local anesthetics.

6.15.2 Anesthetic Classes

The use of anesthetic agents is dictated in large part by the nature of the surgical procedure. In some cases there is a requirement for the patient to be fully 'asleep' in order to avoid discomfort and interference with the procedure, whereas in other cases, local or regional anesthesia (e.g., pudenal block in pregnancy, dental surgery) may suffice and the patient may remain awake during the procedure.

6.15.2.1 Inhalation or General Anesthetics

Inhalation or general anesthetics are administered via the lungs and allow the depth of anesthesia to be rapidly altered by controlling the amount of inhaled anesthetic. This can be advantageous over the use of intravenous anesthetics that require the agent to be gradually removed from the body via urinary or fecal excretion. The rapid elimination of inhalation anesthetics also provides for a shorter duration of postoperative respiratory depression; however, inhalational anesthetics provide no postoperative analgesia. Inhalation anesthetics used in surgical procedures require the use of a general anesthesia machine for administration. This comprises a ventilator to maintain breathing during deeper stages of anesthesia, vaporizers to deliver the anesthetic into the breathing circuit, as well as monitors and alarms to monitor patient safety. The use of inhalation anesthetics is driven by several considerations[1]: (1) minimizing the side effects of the anesthetic, e.g., renal and hepatic toxicity, undesirable cardiovascular side effects, e.g., cardiac depression, cerebral vasodilation, and the techniques used for its administration; (2) optimizing pharmacokinetics, e.g., rapid onset and offset; (3) maintaining physiological homeostasis; and (4) improving postoperative outcomes.

6.15.2.1.1 General anesthetic potency

The potency of volatile anesthetics is measured as the minimum alveolar concentration (MAC) – the alveolar concentration (measured as end-tidal concentration) of the anesthetic (at one atmosphere) that prevents movement in 50% of patients in response to pain like a surgical skin incision. Measuring the MAC for skin incision does not predict the concentration of inhalation anesthetics necessary to avoid the motor responses to other painful stimuli, e.g., endotracheal intubation. Modifications of MAC include MAC_{awake}, the ability to respond to verbal commands. Following a brief equilibration period, the alveolar concentration of the anesthetic equals the blood concentration and subsequently reflects the brain concentration, e.g., the partial pressure of the anesthetic in the CNS. The MAC is age-dependent, being highest in infants and decreasing as a function of age. MAC values for inhaled anesthetics are additive and can be modulated by opioid treatment. When given alone, inhalation anesthetics frequently do not suppress hemodynamic responses to painful stimuli.

6.15.2.2 Parenteral Anesthetics

Parenteral anesthetics are small-molecular-weight compounds, the hydrophobicity of which dictates their pharmacokinetics and efficacy.[1] These agents preferentially partition into the lipophilic tissues in the central and peripheral nervous systems within a single circulation cycle. As blood levels fall, the anesthetic depot in nervous tissue re-equilibrates, resulting in a distribution of the agent to other lipophilic tissues. Patients treated with a single bolus of thiopental emerge from anesthesia in 10 min while those receiving a prolonged infusion may take a day or more to recover.[1]

6.15.2.3 Local Anesthetics

Local anesthetics are used to reduce sensation in a part of the body, e.g., skin, oral cavity, without the loss of either consciousness or control of vital functions.[14] In contrast to general anesthesia, local anesthesia avoids the major physiological consequences of the former, e.g., amnesia and impairment of cardiac or respiratory function, and can be used to involve the patient in assessing pain responses.

6.15.2.3.1 Infiltration anesthesia

Infiltration anesthesia involves the induction of local anesthesia by injecting a local anesthetic, e.g., lidocaine, procaine, or bupivacaine (**Figure 3**), directly into tissue to avoid impacting other body functions. Epinephrine can be used to enhance the duration of infiltration anesthesia.

6.15.2.3.2 Nerve block

Nerve block involves the injection of a local anesthetic proximal to discrete peripheral nerves to produce large areas of anesthesia. Injection directly into the nerve is avoided to prevent pain. Examples of nerve block include: (1) trigeminal for dental surgery; (2) brachial for surgical procedures in the shoulder; (3) cervical plexus for neck surgery; and (4) sciatic and femoral for surgery distal to the knee. Local anesthetics used to produce nerve block can be divided into three classes based on duration of action: (1) short (20–30 min: procaine); (2) intermediate (1–2 h; lidocaine, mepivicane); and (3) long (6–7 h; bupivicaine, ropivicaine).

6.15.2.3.3 Field block anesthesia

Field block anesthesia involves the subcutaneous injection of local anesthetic to produce anesthesia distal to the injection site and involves similar paradigms to those used for infiltration anesthesia.

6.15.2.3.4 Topical anesthesia

Topical anesthesia involves the application of aqueous solutions of a local anesthetic, such as cocaine or lidocaine, to mucous membranes in the nose, throat, mouth, esophagus, tracheobrachial tree, and the urogenital system. Peak anesthetic effect occurs within 2–10 min as local anesthetics are rapidly absorbed into the mucous membrane. Topical anesthesia can be used as a prelude to nerve block, especially in the dental setting.

6.15.2.3.5 Intravenous regional anesthesia

Intravenous regional anesthesia, also known as Bier's block, uses the vascular system to introduce local anesthetics to the nerve trunks and endings. A double tourniquet is used in the affected area, typically a limb, and local anesthetic, such as lidocaine (0.5% solution; 40–50 mL), is injected into a peripheral vein in the same area, resulting in anesthesia of the area from the level of the tourniquet. The tourniquet prevents toxic amounts of anesthesic from entering the general circulation.

6.15.2.3.6 Spinal or subarachnoid anesthesia

Spinal or subarachnoid anesthesia involves the injection of local anesthetic, lidocaine or bupivicaine, into the cerebrospinal fluid of the lumbar space to block the fibers in the spinal nerve roots. This type of anesthesia is typically used in hernia repair, gynecological or urological surgery, and lower-extremity orthopedic procedures. It is widely used in older patients and individuals with chronic respiratory disease, hepatic and renal impairment, and diabetes. The duration of the anesthesia can be increased by the addition of epinephrine.

6.15.2.3.7 Epidural anesthesia

Epidural anesthesia involves the introduction of local anesthetic into the epidural space, usually via catheterization, and can provide repeat administration or infusion of the anesthetic. The key difference between spinal and epidural anesthesia is that a much larger amount of anesthetic is required to produce epidural block because the anesthetic must diffuse into the spinal cord, nerve roots, and cerebrospinal fluid from the epidural space. Consequently, the onset of anesthesia is significantly slower with epidural as compared to spinal block. Blood levels of local anesthetic will be significantly higher with epidural administration, although this is seldom a clinically important problem. The ease of placing a catheter in the epidural space allows for continuous drug administration, which is advantageous for control of pain during labor and postoperatively. Altering the concentration of local anesthetic in the epidural space can be used to advantage to produce a differential block of sympathetic, sensory, and motor pathways, while obtaining these gradations of effect with subarachnoid administration is more difficult. Opioids can also be used to provide spinal and epidural analgesia.[15]

6.15.3 Anesthetic Mechanisms of Action

Over the past decade, considerable evidence has been gathered to indicate that anesthetics may produce their effects via discrete molecular mechanisms.[9] This contrasts with the considerable historical data on nonspecific effects of anesthetics, especially general anesthetics, on cell membrane lipids and proteins.

6.15.3.1 Protein Interactions

Considerable efforts have been focused on studying the interaction of general anesthetics with artificial proteins.[16,17] These entities are distinct from receptors and ion channels. Millimolar binding constants have been determined for the interactions of isoflurane and halothane with bovine serum albumin using radioligand binding, ^{19}F nuclear magnetic resonance, and tryptophan fluoresence quenching.[15] Such binding involves hydrophobic, electrostatic, and van der Waals interaction as well as hydrogen bonding. General anesthetics are thought to bind to cavities within the transmembrane four-helix bundles of receptors, perhaps acting as allosteric modulators. Binding of halothane and isoflurane to ferritin, a 24mer of four-helix bundles, identified a motif at which micromolar binding was observed, higher than that previously reported for general anesthetic–protein interactions, that may reflect a common anesthetic binding pocket within the interhelical dimerization interface. This high-affinity anesthetic–apoferritin complex suggests a greater selectivity than had previously been thought and that direct protein actions, provided these interactions result in alterations in membrane function, may explain the physiological effects of anesthetics at concentrations lower than surgical levels of anesthetic, including loss of awareness.[17] The ω-3 fatty acids, e.g., eicosapentanoic acid, represent another class of drug-like agent that may produce their effects by changing membrane fluidity.[18,19]

6.15.3.2 Lipid Interactions

With the focus of early studies of the effects of anesthetics at the membrane level, considerable attention has been paid to the lipophilicity and hydrophobicity of anesthetics.[3,16] Coupled with their structural diversity and the high concentrations required to elicit anesthesia, Bernard's unitary theory of anesthesia, stating that all anesthetics produce their effects by a simple perturbation of the cell membrane, had a certain logic. The lack of pharmacological antagonists of general anesthetics and the debate on the modest stereoselectivity of anesthesics[3,16] has made the unitary theory "both elegant and attractive."[3] The increased anesthetic efficacy observed with increases in methylene chain length, until the anesthetic cut-off (C = 12 and 13), is also consistent with this role of hydrophobicity and is further supported by the reversal in anesthesia when the hydrostatic pressure is increased, a phenomenon known as pressure reversal.

The search for hydrophobic pockets with which anesthetics interact has been limited by the modest changes in membrane lipid-ordering produced by clinical doses of anesthetics. Nonetheless, at high concentrations, general anesthetics can shift the conformational equilibrium of the neuronal nicotinic receptor (nAChR) from a resting to desensitized state,[20] while halothane can photolabel nAChR subunits. 3-Trifluoromethyl-3-(3-iodophenyl)diazirine (TID: **Figure 3**), which has undergone preliminary characterization as a novel general anesthetic in tadpole with an EC_{50} value of approximately 600 nmol L^{-1},[21] can label the lipid–protein interface of nAChR subunits. However TID is better known as a hydrophobic photoreactive probe than a general anesthetic.[22,23] Point mutations of the nAChR have resulted in the identification of a discrete binding site for isoflurane in the pore-forming M2 domain.[24] Using large-scale molecular dynamic simulations in a gramicidin A (gA) channel model, halothane interactions were localized to the channel–lipid–water interface, reinforcing a global, as opposed to local, change in channel dynamics.[25] Studies with hexafluoroethane (HFE: **Figure 1**), an analog of halothane that is devoid of anesthetic activity in the gA channel model, showed that this inactive control, in contrast to halothane, had no significant effect on gA channel dynamics, further supporting a role for changes in global membrane dynamics in the action of general anesthetics.[26] One caveat to these studies, however, is that the gA channel is a highly reductionistic model, not a physiological drug target.

6.15.3.3 Ion Channels

Several ion channels have been proposed as potential molecular targets for anesthetic actions.[5–10,14,27–30] These include voltage-gated calcium,[27] voltage-gated and two-pore-domain potassium,[6] sodium, NMDA and hyperpolarization-activated, cyclic nucleotide-gated (HCN) channels,[5,9,14] nAChRs,[3,10] and $GABA_A$ and glycine receptors.[5,10,28]

6.15.3.3.1 γ-Aminobutyric $acid_A$ ($GABA_A$) receptors

$GABA_A$ receptors are a site of action of the major inhibitory transmitter in the CNS, GABA. The benzodiazepines (BZs) are a class of drugs with anxiolytic, hypnotic, muscle relaxant, and alcohol-potentiating actions that also interact with the $GABA_A$ receptor via an allosteric site, the BZ receptor. Similarly, the barbiturates can also interact with a discrete allosteric site on the $GABA_A$ receptor with pentobarbital (**Figure 1**), facilitating GABA currents and inhibitory neurotransmission in the brain and spinal cord. Inhalation anesthetics like halothane, enflurane, and isoflurane also interact with $GABA_A$ receptors.[27] Halothane, for instance, has an EC_{50} value of 230 μmol L^{-1}, close to its EC_{50} value for general anesthesia.[31] The S-isomer of isoflurane (**Figure 1**) is twice as potent as R-isoflurane in prolonging GABA-evoked inhibitory postsynaptic potentials (i.p.s.ps) in cultured rat hippocampal neurons.[32] Similarly, the steroid anesthetic, alphaxalone (**Figure 2**), but not its 3-β-hydroxy-isomer, betaxalone (**Figure 2**), which is devoid of anesthetic activity, can enhance GABA currents.[33] Point mutation/chimeric studies of the $GABA_A$ receptor have identified two amino acid residues in TM2 and TM3 that are critical for enflurane and isoflurane, but not propofol, interactions with the $GABA_A$ receptor.[29] The sedative effects of propofol and pentobarbital are mediated via $GABA_A$ receptors in the tuberomammillary nucleus, as shown by their blockade with the $GABA_A$ receptor antagonist, gabazine. Those of ketamine (**Figure 2**), which produces its anesthetic actions via the NMDA receptor, were unaffected by gabazine.[5] Propofol delays desensitization of $GABA_A$ receptors and it, and a series of analogs, shows a distinct structure–activity relationship for their anesthetic potency that does not correlate with lipid solubility.[34]

6.15.3.3.2 Voltage-gated potassium channels

Voltage-gated potassium channels (VKCs or K_v channels), a family that comprises 12 distinct members,[35] have been implicated as the target of anesthetics, with considerable interest in the *Shaw*-related family, K_{v4}, as a potential target for anesthetics.[36] Isoflurane can also inhibit 'leak' or two-pore-domain potassium channels,[37] including the weakly inward-rectifying K^+ channel (TWIK), the acid-sensitive K^+ channel (TASK), the TWIK-related K^+ channel (TREK), and the TWIK-related arachidonic acid-stimulated K^+ channel (TRAAK).

6.15.3.3.3 Voltage-sensitive calcium channels

Voltage-sensitive calcium channels (Ca_v) are comprised of three major families, Ca_v1–3, of which there are a total of 10 members.[38] While these have been implicated in anesthetic interactions, this only occurs at high concentrations in the millimolar range[27] and may not be of physiological significance.

6.15.3.3.4 Voltage-gated sodium channels

Voltage-gated sodium channels (Na_v) are a family of nine distinct receptors, Na_v 1.1–1.9, with distinct pharmacology.[13] While sodium currents in giant axons of the squid and crayfish are insensitive to inhalation anesthetics, those in small unmyelinated axons in the hippocampus are depressed. $Na_v1.2$ is inhibited via a voltage-independent block of peak current and a hyperpolarizing shift in voltage dependence of the steady state.[39] Volatile anesthetics also interact with $Na_v1.4$, $Na_v1.5$, and $Na_v1.6$ channels,[9] leading to the potential to modulate neurotransmitter release.

6.15.3.3.5 *N*-Methyl-D-aspartate (NMDA) and glycine receptors

The best-known anesthetic interacting with NMDA receptors is ketamine (**Figure 2**), which, as has already been noted, is an atypical dissociative anesthetic with hallucinogenic side effects. The NMDA receptor may also be the site at which ethanol, xenon, nitrous oxide, and cyclopropane produce their anesthetic effects.[9] Studies on the glycine receptor, a component allosteric modulatory site on the NMDA receptor, have shown that propofol can enhance activation of the glycine receptor and, by extrapolation, NMDA receptor function.

6.15.3.3.6 Neuronal nicotinic receptors

nAChRs were the first ion channels to be available in sufficient quantities to study interactions with pharmacological agents. As such, there is a considerable body of data to support an interaction with nAChRs in the context of the unitary hypothesis of anesthetic action. The isomers of isoflurane do however show a twofold stereoselectivity in its ability to inhibit nAChR-mediated currents,[40] while the N-terminal domain of the $\alpha7$ subunit is involved in the effects of inhalation anesthetics.[41]

6.15.3.3.7 Hyperpolarization-activated, cyclic nucleotide gated (HCN) or pacemaker channels

HCN or pacemaker channels can be inhibited by enflurane and halothane that decrease I_h conductance (which is involved in resting membrane potential) in brainstem motor neurons[42]and thalamic neurons.[43] Propofol slows HCN1 channel activation with an EC_{50} value of $6\,\mu mol\,L^{-1}$ and can also interact with HCN2 and HCN4 channels.[44] While HCN neurons may represent "an under-appreciated anesthetic target site,"[9] propofol interactions with HCN channels may contribute to bradyarrhythmias.[44]

6.15.4 Animal Models of Anesthetic Action

Evaluation of the potential for new chemical entities as potential anesthetics typically involves a behavioral assessment of the loss of righting reflex[45] or, more recently, electro encephalogram (EEG) recording.[46] Loss of righting can be measured by administering anesthetic either in a barometric chamber or in emulsion form via a cannula in the tail vein. Immediately after anesthetic administration, the animal is placed on its back and considered anesthetized if it fails to right itself within 30 s of the injection. An animal that rights itself is deemed as being awake. Alternatively anesthesia can be assessed by the ability of an induced animal to stand on all four limbs when placed in a rolling cylinder. Dose–response curves can be used to determine the MAC. Using techniques similar to those used in the study of sleep, rodents implanted with miniaturized transmitters can be used to record EEG and electromyogram (EMG) waveforms to determine doses that elicit classical changes in the EEG signal. Arterial blood pressure and temperature can also be recorded. EEG-based approaches can also be used in conjunction with nociceptive stimuli, e.g., unprovoked and provoked righting reflexes.

6.15.5 Currently used Anesthetics

6.15.5.1 Historically used Agents

6.15.5.1.1 Diethyl ether

Diethyl ether (**Figure 1**) is a potent anesthetic whose actions are accompanied by analgesic and muscle relaxant activity. It has a long induction period and is highly flammable and explosive.

6.15.5.1.2 Chloroform

Chloroform (**Figure 1**) is a halogenated hydrocarbon that, unlike ether, is not flammable but has significant toxicity, including carcinogenicity, hepato-, and nephrotoxicity.

6.15.5.1.3 Barbiturate anesthetics

Barbiturate anesthetics include sodium thiopental, thiamylal, and methohexital (**Figure 1**). While these agents have demonstrated enantioselective effects, they are typically used as racemates. Thiopental (3–5 mg kg^{-1}) induces anesthesia within 30 s, with a duration of 5–8 min. Barbiturates reduce cerebral metabolism leading to decreases in cerebral blood flow and intracranial pressure. They also produce respiratory depression and can elicit dose-dependent decreases in blood pressure.

6.15.5.1.4 Cyclopropane

Cyclopropane (**Figure 1**) was discovered as a contaminant of propylene that was being investigated for its anesthetic potential in 1928. Like ether, it is an explosive gas and has been largely replaced by newer generations of inhaled anesthetic.

6.15.5.2 Currently used Anesthetics

6.15.5.2.1 Inhalation anesthetics

6.15.5.2.1.1 Nitrous oxide

Nitrous oxide (**Figure 1**) is an inorganic, nonflammable gas. It boils at −88 °C with low solubility and results in rapid induction or awakening. It is a weak anesthetic, producing reliable surgical anesthesia only in hyperbaric conditions, and is used to supplement other inhalation agents. High concentrations of nitrous oxide increase the alveolar concentration of simultaneously administered gases – the second-gas effect. At 70% nitrous oxide, the MAC for other inhalation anesthetics can be reduced by 60%. Nitrous oxide does not inhibit the hypoxic pulmonary vasoconstriction response in the lungs nor produce skeletal muscle relaxation. It is a sympathomimetic and increases systemic vascular resistance. Nitrous oxide diffuses into air-containing cavities more than 30 times faster than nitrogen leaves, causing dangerous accumulation of volume and increase in pressure in closed spaces, e.g., bowel, middle ear, and pneumothorax. In patients with ileus, the volume of air in the bowel can double within 4 h of nitrous oxide administration. The maximum dose of nitrous oxide should not exceed 70%, to avoid the occurrence of hypoxemia. It may produce its effects via interactions with the NMDA receptor.[9] Nitrous oxide has significant analgesic properties at subanaesthetic concentrations, hence its application in the treatment of dental and labor pain while the patients remain conscious.

6.15.5.2.1.2 Halothane

Halothane (**Figure 1**) is a nonflammable halogenated alkene with a blood/gas coefficient of 2.3 and an MAC in 100% oxygen of 0.74 and 0.29 in 70% nitrous oxide. Halothane has superior hypnotic activity but lacks analgesic properties. Induction of anesthesia is rapid and can be achieved by using 1–3% halothane in air or oxygen, or by using 0.8% halothane in 65% nitrous oxide. Halothane produces a dose-dependent reduction (20–25%) in arterial blood pressure at its MAC and increases cerebral blood flow, raising intracranial pressure. It has no effect on systemic vascular resistance but causes myocardial depression and has negative inotropic actions. It sensitizes the myocardium to the arrhythmogenic effects of epinephrine. Halothane can elicit malignant hyperthermia, a hypermetabolic, condition that is potentially lethal. It does not cause coronary artery vasodilatation and therefore does not lead to coronary artery steal syndrome. Halothane hepatitis and hepatic necrosis are complications of halothane anesthesia that occur in 1 in 6,000–35,000 cases and can be fatal.

6.15.5.2.1.3 Isoflurane

Isoflurane (**Figure 1**) is a nonflammable halogenated methyl ethyl ether with a blood/gas coefficient of 1.4 and the MAC in 100% oxygen of 1.15 and in 70% nitrous oxide of 0.50. Induction of anesthesia is achieved with 3–4% isoflurane in air or in oxygen, or 1.5–3% in 65% nitrous oxide. It is administered following induction due to its pungent odor. The use of isoflurane alone can lead to coughing and apneic periods and it is therefore used in combination with intravenous anesthetics. Emergence from anesthesia with isoflurane is faster than with halothane or enflurane. Isoflurane produces a dose-dependent reduction in arterial blood pressure due to peripheral vasodilatation but has no effect on cardiac output. It increases cerebral blood flow, and may raise intracranial pressure, but this effect is much less potent than with halothane. Isoflurane does not sensitize the myocardium to produce arrhythmias. Coronary artery vasodilatation can lead to coronary artery steal; however, this does not appear to be a significant clinical problem. Like halothane, isoflurane can trigger malignant hyperthermia.

6.15.5.2.1.4 Enflurane

Enflurane (**Figure 1**) is a nonflammable fluorinated ethyl methyl ether with a blood/gas coefficient of 1.8. The MAC in 100% oxygen is 1.68 and that in 70% nitrous oxide is 0.57. Induction of anesthesia is achieved by using 3–4% enflurane in air or in oxygen or 1.5–3% in 65% nitrous oxide. It is used as a maintenance, rather than an induction, agent. Enflurane enhances the action of paralyzing agents more than other inhalation anesthetics and produces a dose-dependent reduction in arterial blood pressure due to its negative inotropic effects. Enflurane does not sensitize the heart and does not elicit coronary artery steal. Enflurane can increase intracranial pressure and, especially in combination with hyperventilation, increases the risk of seizure activity. Like other inhalation anesthetics, it can also trigger malignant hyperthermia. Emergence from anesthesia with enflurane is somewhat slower than with isoflurane.

6.15.5.2.1.5 Sevoflurane

Sevoflurane (**Figure 1**) is a nonflammable fluorinated isopropyl ether with a blood/gas coefficient of 0.59. The MAC in 100% oxygen is 1.71 and that in 63% nitrous oxide is 0.66. Sevoflurane reacts with carbon dioxide absorbents to form the haloalkene, compound A, that has been linked to renal injury in some studies, but not in others. Sevoflurane dose-dependently decreases arterial blood pressure due to peripheral vasodilatation. It is considered superior to isoflurane and halothane for inhalant anesthetic induction as its low blood solubility results in rapid induction of anesthesia without intravenous anesthetic use, with a concomitant rapid recovery that favors its use in the outpatient situation. Sevoflurane causes less airway irritation, resulting in less coughing and excitation during induction. It can be used without intravenous anesthetics.

6.15.5.2.1.6 Desflurane

Desflurane (**Figure 1**) is a nonflammable fluorinated ether derivative with a blood/gas coefficient of 0.42. The MAC in 100% oxygen is 6.0 and in 60% nitrous oxide is 2.8. Unlike other inhalation anesthetics, desflurane requires the use of electrically heated rather than standard vaporizers. Its low tissue solubility results in rapid elimination and awakening. Desflurane evokes a dose-dependent reduction in arterial blood pressure due to peripheral vasodilatation. Desflurane can cause coughing and is not used without intravenous anesthetics.

6.15.5.2.1.7 Xenon

Xenon is an inert gas that produces rapid induction and recovery due to its insolubility in blood and tissues and has minimal side effects.[47,48] Its status as an inert gas limits its availability and it is expensive. It is not approved for use in the US.

6.15.5.2.2 Parenteral anesthetics

6.15.5.2.2.1 Propofol

Propofol (**Figure 2**) has a similar anesthetic profile to that of thiopental; however, it is distinguished by having a much larger clearance (propofol clearance is greater than liver blood flow, while thiopental clearance is much smaller than liver blood flow), and is therefore much shorter-acting when given in repetitive boluses or by constant infusion. It is the most commonly used parenteral anesthetic in the US. Propofol infusion (25–300 $\mu g\,kg^{-1}\,min^{-1}$) is commonly used for procedural sedation and for the maintenance of general anesthesia. Propofol acts at least in part by blocking Na_v channels.

6.15.5.2.2.2 Ketamine

Ketamine (**Figure 2**) is an arylcyclohexylamine congener of phencyclidine. While used as a racemate, its S-isomer is a more potent anesthetic with fewer side effects. Ketamine is a dissociative anesthetic, producing a hypnotic state distinct from that of other parenteral anesthetics that is reminiscent of cataplexy, in addition to being an efficacious analgesic. It also increases blood pressure, cerebral blood flow, and intracranial pressure, effects that can be attenuated by coadministration of BZs or thiopental. Individuals treated with ketamine do not respond to verbal commands and have amnesia. At the same time they are able to breathe spontaneously and keep their eyes open, a situation that is accompanied by nystagmus, salivation, and lacrimation. Ketamine can be given via the oral, rectal, and intramuscular as well as intravenous routes. Emergence from ketamine anestheia is frequently accompanied by hallucinations and symptoms of delirium. Mechanistically, ketamine acts as a modulator of the NMDA receptor and can also block monoamine uptake, leading to indirect sympathomimetic activity.

6.15.5.2.2.3 Etomidate

Etomidate (**Figure 2**) is a substituted imidazole that is used in the form of its active D-isomer. Etomidate has the fewest cardiovascular side effects of any of the intravenous anesthetics, producing no cardiac depression and having little or no direct effect on peripheral vascular resistance. It is frequently considered the intravenous induction agent of choice for patients with hypovolemia, myocardial depression, and other forms of cardiovascular instability. Myoclonic jerking may be seen following administration, more commonly than with thiopental or propofol. A single induction dose of etomidate inhibits the biosynthesis of cortisol, thereby preventing the adrenocortical stress response. Cortisol plasma levels remain normal, but do not rise in response to surgery as would ordinarily occur. Inhibition of cortisol biosynthesis by etomidate has not been shown to be harmful, except when etomidate was infused (e.g., for sedation in the intensive care unit) continuously for days, resulting in Addisonian crisis in some patients.

6.15.5.2.2.4 Alphaxalone

Alphaxalone (**Figure 2**) is a steroid anesthetic, a neurosteroid, used for short-term anesthesia that is similar in its effects to barbiturates but has a wider margin of safety than barbiturates and produces less apnea on induction. It produces its anesthetic effects via interactions with a neurosteroid binding site on the $GABA_A$ receptor. It is primarily used in for veterinary surgery being given together with alphadolone (**Figure 2**). The latter was initially used to improve the solubility of alphaxalone in the excipient, Cremophor, but was subsequently found to have anesthetic activity on its own. The alphaxalone/alphadolone was withdrawn from human use in the 1980s because of major anaphylactic reactions to Cremophor.

6.15.5.2.3 Local anesthetics

Local anesthetics (**Figure 3**) produce their effects by blocking Na_V channels in the neuronal membrane, thus preventing nerve impulse generation and conduction.[13]

6.15.5.2.3.1 Cocaine

Cocaine is an effective blocker of nerve impulse conduction and, as such, an efficacious local anesthetic. However, it is a potent and highly addictive stimulant due to its ability both to stimulate dopamine release and block its reuptake.

6.15.5.2.3.2 Procaine

Procaine is the first synthetic local anesthetic and was introduced into use in 1905. It has been replaced by newer generations of local anesthetics, its use being reserved for infiltration anesthesia and diagnostic nerve blocks.

6.15.5.2.3.3 Lidocaine

Lidocaine is an aminoethylamide that is superior to procaine in its speed to onset, intensity, extensiveness, and duration of effect. It is used in a patch formulation for the treatment of the pain associated with postherpetic neuralgia and is also used as an antiarrhythmic agent. Lidocaine has been extensively used for spinal anesthesia when relatively short duration is required. However, spinal administration of lidocaine has been associated with nerve root irritation (transient radicular irritation), the mechanism of which is not well understood, resulting in the reluctance of many anesthesiologists to continue using lidocaine for spinal anesthesia. Lidocaine continues to be widely used for peripheral nerve block, infiltration, and epidural block.

6.15.5.2.3.4 Bupivacaine

Bupivacaine is a butyl piperidine amide anesthetic that has a long duration of action that is widely used during labor and postoperative pain. It is much more cardiotoxic than lidocaine. Overdosage of bupivacaine can produce intractable cardiac arrest from ventricular tachycardia or fibrillation due to tight binding to cardiac sodium channels. Ropivacaine and levobupivacaine (**Figure 3**) were developed in an attempt to find a local anesthetic with long duration similar to bupivacaine but with less cardiotoxicity. All of the amide local anesthetics except for lidocaine contain an asymmetric carbon, and the (S)-enantiomer of the cardiotoxic amide local anesthetics, such as bupivacaine, appears to be somewhat less cardiotoxic than the R-enantiomer. Bupivacaine is typically used as the racemate. The (S)-enantiomer of bupivacaine (levobupivacaine; **Figure 3**) and the (S)-enantiomer of propivacaine (ropivacaine) appear to be somewhat less cardiotoxic than bupivacaine.

6.15.5.2.3.5 Other local anesthetics

Other local anesthetics include articaine, chloroprocaine, etidocaine, mepivacaine, prilocaine, and ropivicaine (**Figure 3**). Their use is mainly dictated by differences in their pharmacokinetic profiles, potency of motor blockade, and cardiac toxicity.

6.15.5.3 Anesthetic Adjuvants

Anesthetic adjuvants or adjuncts are given together with general anesthetics to reduce the doses of the latter and result in reduced side effects.

6.15.5.3.1 Benzodiazepines

BZs (e.g., midazolam, diazepam, lorazepam; **Figure 4**) produce anesthesia similar to the barbiturates but are limited in this use due to the potential for prolonged sedation and amnesia, unless this is desired. They are used as adjuncts prior to induction with their anxiolytic, amnestic, and sedative effects providing patient benefit. Midazolam has a pharmacokinetic advantage over other BZs with sedative doses (0.10–0.07 mg kg^{-1} intravenously) that have a peak effect in 2 min that is sustained for approximately 30 min. The incidence of thrombophlebitis following intravenous administration of midazolam is much lower compared to diazepam because of the pH-dependent ring opening of the diazepine ring of midazolam. At low pH, the diazepine ring is in the open form, resulting in greater water solubility and suitable for formulation in an aqueous vehicle. By contrast, highly lipid-soluble diazepam is commonly formulated in propylene glycol, an irritating vehicle that results in a high rate of thrombophlebitis. Once midazolam is injected, physiologic pH results in predominance of the closed form of the diazepine ring, and greatly increased lipid solubility, suitable for crossing the blood–brain barrier.

6.15.5.3.2 Analgesics

Ketamine (**Figure 2**) and nitrous oxide (**Figure 1**) are the only general anesthetics with significant analgesic activity. As a result, analgesics are typically coadministered with anesthetic agents to reduce anesthetic dose and to minimize hemodynamic and nociceptive responses to painful stimuli. The latter is termed preemptive analgesia. In surgical procedures, pain often persists long after the procedure. The introduction of an analgesic regimen, e.g., opioid or nonsteroidal anti-inflammatory drug (NSAID: ketoprofen; **Figure 4**), before the surgical procedure has been shown in some instances to prevent the sensitization or 'wind-up' of the nervous system to noxious stimuli, reducing the pain cascade. Intravenous opioids, novel analgesics like adenosine, H_1 antagonists like doxepine (**Figure 4**), local anesthetic infiltration, NSAIDs and nerve, subarachnoid, and epidural block can reduce the pain following surgery, thus decreasing the recovery period. Parenteral opioids used as anesthetic adjuncts include the μ-opioid agonists fentanyl, sufentanil, alfentanil, remifentanil, meperidine, and morphine (**Figure 3**), with sufentanil being the most potent. The choice between these opiods is generally based on their duration of action. The duration of action of remifentanil is ultrashort, being approximately 10 min, and rapid clearance takes place by ester hydrolysis. Remifentanil is an example of successful rational drug design. Modeled after several other ester drugs with extremely rapid clearance due to ester hydrolysis (e.g., succinylcholine, esmolol), a series of fentanyl ester analogs were synthesized with the intent of discovering a fentanyl ester analog with potent opioid activity that would be rapidly cleared to an inactive metabolite. Remifentanil was chosen for development from the series, and has proven to be an extremely valuable opioid in anesthetic practice.

6.15.5.3.3 α_2-Adrenoceptor agonists

Dexmedetomidine (**Figure 4**) is an imidazole-based, short-acting α_2-adrenoceptor agonist modestly selective for the α_{2A} adrenoceptor that is used in the intensive care setting to sedate critically ill adults. It has associated analgesic and anxiolytic effects.[49,50] Dexmedetomidine does not produce respiratory depression; its main side effects are hypotension, bradycardia, and nausea.[49]

6.15.5.3.4 Neuromuscular blockers

Muscle relaxants, e.g., succinylcholine (**Figure 4**), are used during anesthetic induction to relax the muscles of the jaw, neck, and airways to facilitate laryngoscopy and endotracheal intubation, or for relaxation of thoracic, abdominal, hip, or shoulder muscles to facilitate surgical exposure. Mivacurium and rocuronium (**Figure 4**) are newer myorelaxants in development; the former has a short half-life due to its esterase susceptibility.[51]

6.15.6 Unmet Medical Needs

While major strides have been made in understanding the mechanisms through which general, parenteral, and local anesthetics produce their beneficial effects, there still remains an unmet medical need in terms of the differing side effects of the various agents in current use which have been outlined above. The therapeutic index of general or

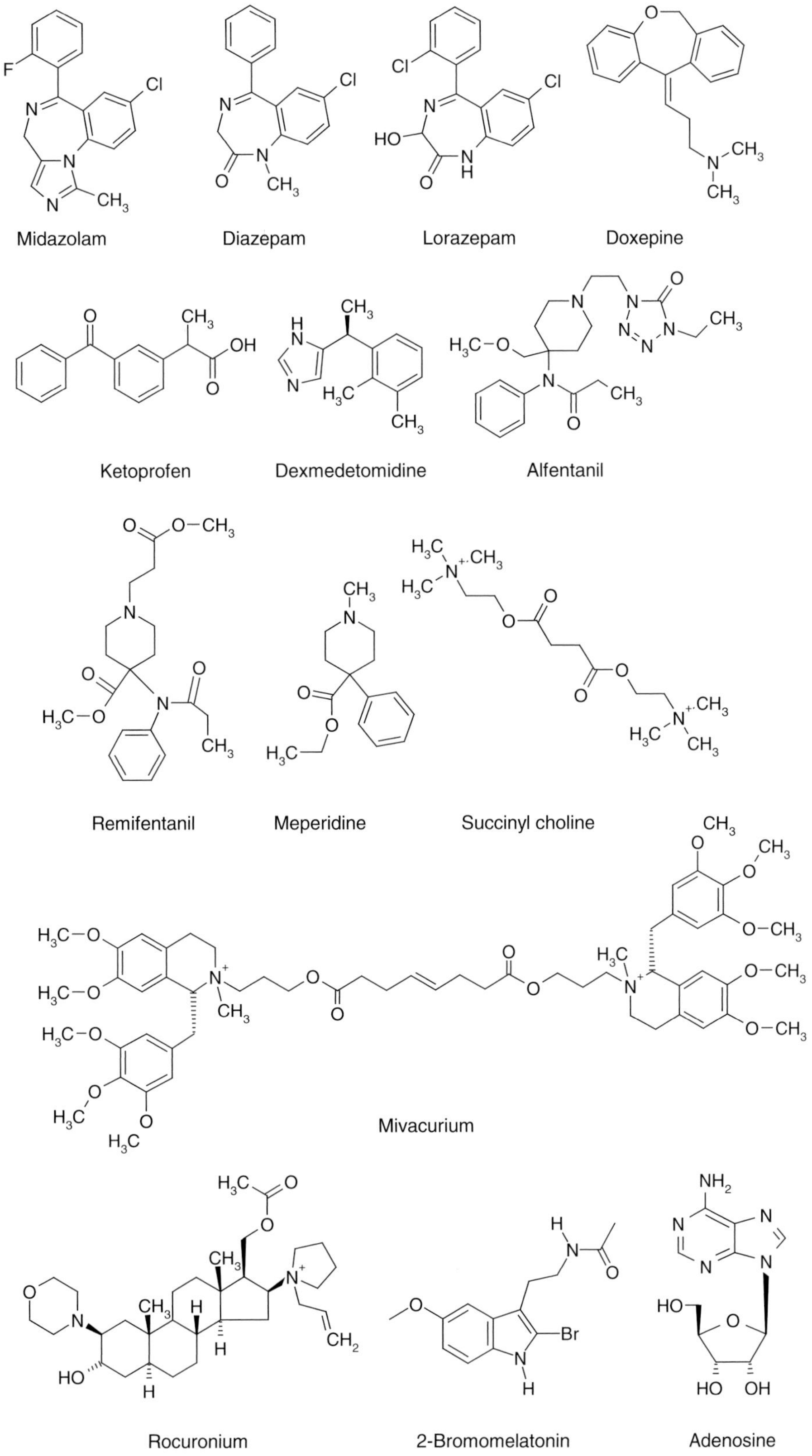

Midazolam

Diazepam

Lorazepam

Doxepine

Ketoprofen

Dexmedetomidine

Alfentanil

Remifentanil

Meperidine

Succinyl choline

Mivacurium

Rocuronium

2-Bromomelatonin

Adenosine

Figure 4 Anesthetic adjuvants and novel anesthetics.

inhalational anesthetics is low, providing an obvious challenge for improvement, although the conflicting data on specific versus global anesthetic effects will make it difficult to plan de novo new chemical entitity approaches. An issue in the mechanistic studies is that much of the data has been derived in artificial membrane systems and as such is sufficiently reductionistic to question its value. Similarly, the stereospecific effects that have been reported for various anesthetics are, at best, modest. A recent review[51] has concluded that anesthesiology has evolved in two phases over the past 60 years. The first phase, from 1954 to 1978, was characterized by the introduction of new anesthetics and new surgical techniques; the second, from 1979 to 2004, had a greater emphasis on clinical excellence, outcome, and quality of patient care both in the operating room and elsewhere in the hospital, and research.

6.15.6.1 Analgesic Side Effects

Opioid drugs are the current mainstay of perioperative pain control, and are widely used during general anesthesia to suppress responses to surgical stimuli. Opioids have serious side effects; respiratory depression is the most dangerous. Side effects commonly limit the effective use of analgesic drugs. Development of opioid or nonopioid analgesics with superior side-effect profiles could dramatically alter the treatment of acute and chronic pain.

6.15.6.2 Local Anesthetic Side Effects

Local anesthetics are relatively safe drugs in clinical practice; however there are unmet needs for less cardiotoxic, long-acting local anesthetics and for a local anesthetic with onset and duration similar to lidocaine for spinal anesthesia but without the risk for transient radicular irritation.

6.15.6.3 Rapid-Offset Intravenous Anesthetics

Improved pharmacokinetic properties can result in significant improvement in the utility of anesthetic drugs, even in the absence of significant changes in pharmacodynamic properties. Propofol (**Figure 2**) and remifentanil (**Figure 4**) are examples of intravenous anesthetic drugs (a hypnotic drug and an opioid, respectively) that are mainly distinguished by being more rapidly cleared than similar, alternative drugs. Generally, there is a considerable advantage to anesthetic drugs with rapid offset that promote rapid recovery from anesthesia. This has been a significant theme in the development of inhaled and intravenous anesthetics for a number of years. Development of an intravenous hypnotic drug that would have a significantly faster offset than propofol, comparable to the offset of remifentanil, would be a significant advance.

6.15.6.4 Long-Lasting Side Effects of Inhaled Anesthetics

Pharmacologic dogma suggests that the effects of inhaled anesthetics are fully reversible. However, decrements in mental performance have been noticed following anesthesia, especially in the elderly, that can persist for long periods of time. This has not been well understood. Recently, in vitro and animal studies have suggested that inhalational agents may produce lasting changes in brain chemistry, raising concerns about the safety of inhaled anesthetics. Intravenous anesthetics may not produce similar effects, raising the possibility that total intravenous anesthesia (usually consisting of continuous intravenous infusion of a hypnotic and an opioid drug, e.g., propofol and remifentanil) may be superior to inhaled anesthetics for general anesthesia. Much more investigative work is needed in this area. If inhaled anesthetics prove to have significant long-lasting side effects, and if intravenous anesthetic drugs do not have these effects, there will be a profound effect on the direction of development of new anesthetic agents.

6.15.7 New Research Areas

Clearly the tremendous advances, however controversial, in the understanding of targets for anesthetics have enhanced the possibility of discovering novel agents. A key question is what precisely can be improved in new agents over currently used agents. It seems highly likely that this question will have to be initially addressed at the molecular level in vitro using electrophysiological approaches to identify NCEs that can be evaluated in behavioral and telemeterized rodent models.

6.15.7.1 Anesthetic-Specific Benzodiazepines (BZs)

Like advances in subunit-selective $GABA_A$ receptor ligands for use in the treatment of anxiety,[52] it is to be anticipated that, with additional knowledge of the precise regions/subunits of the $GABA_A$ receptor involved in anesthetic induction, improved parenteral anesthetics may be feasible. One approach may be the development of improved neurosteroids that are more soluble than alphaxalone (*see* 6.04 Anxiety).

6.15.7.2 Melatonin

2-Bromomelatonin (**Figure 4**) is comparable to propofol in its rapid induction and short duration of activity in rodent models of anesthetic activity while being 6–10 times less active.[53] Unlike propofol, the antinocifensive activity of 2-bromomelatonin persisted after the anesthetic effect, as measured by the fact that the righting reflex had ceased. 2-Bromomelatonin analogs may represent another novel approach to parenteral anesthesia.[54]

6.15.8 Future Directions

It has been argued[8] that general anesthesia cannot be adequately explained until there is a clearer understanding of the phenomenon of consciousness and the associated states of awareness and sleep. Examination of individual components of anesthesia, immobility, antinociception, and amnesia underlines the need for an integrated approach to such understanding since, for instance, pain can lead to increased awareness. The search for discrete mechanisms for anesthetics of all three classes – inhalation, parenteral and local – is gradually advancing but has yet to be used for the de novo discovery of new anesthetic agents. The ability to reconcile the unitary hypothesis of anesthesia with multiple molecular targets will be an interesting research challenge and will require that the sophisticated molecular modeling of anesthetic interactions with membranes extends beyond artificial, reductionistic constructs,[3,10,16,17,25,26] that, although eminently interesting, add little to advancing research in the area without a consideration of the effects of anesthetics in more complex native membrane systems.

References

1. Evers, A. S.; Crowder, M.; Balser, J. R. General Anesthetics. In *Goodman and Gilman's The Pharmacological Basis of Therapeutics*, 11th ed.; Brunton, L. L., Lazo, J. S., Parker, K. L., Eds.; McGraw Hill: New York, 2006, pp 341–368.
2. Saidman, L. J. *JAMA* **1995**, *273*, 1661–1662.
3. Forman, S. A.; Raines, D. E.; Miller, K. W. The Interactions of General Anesthetics with Membranes. In *Anesthesia: Biologic Foundations*; Yaksh, T. L., Maze, M., Lynch, C., Biebuyck, J. F., Zapol, W. M., Saidman, L. J., Eds.; Lippincott-Raven: Philadelphia, PA, 1997, pp 5–18.
4. Lydic, R.; Baghdoyan, H. Cholinergic Contributions to the Control of Consciousness. In *Anesthesia: Biologic Foundations*; Yaksh, T. L., Maze, M., Lynch, C., Biebuyck, J. F., Zapol, W. M., Saidman, L. J., Eds.; Lippincott-Raven: Philadelphia, PA, 1997, pp 433–450.
5. Nelson, L. E.; Guo, T. Z.; Lu, J.; Saper, C. B.; Franks, N. P.; Maze, M. *Nat. Neurosci.* **2002**, *5*, 979–984.
6. Århem, P.; Klement, G; Nilsson, J. *Neuropsychopharmacology* **2003**, *28*, S40–S47.
7. Angel, A. *Br. J. Anaesth.* **1993**, *71*, 148–163.
8. Urban, B. W. *Br. J. Anaesth.* **2002**, *89*, 167–183.
9. Hemmings, H. C., Jr.; Akabas, M. H.; Goldstein, P. A.; Trudell, J. R.; Orser, B. A.; Harrison, N. L. *Trends Pharmacol. Sci.* **2005**, *26*, 503–510.
10. Yamakura, T.; Bertaccini, E.; Trudell, J. R.; Harris, R. A. *Annu. Rev. Pharmacol. Toxicol.* **2001**, *41*, 23–51.
11. Tomlin, S. L.; Jenkins, A.; Lieb, W. R.; Franks, N. P. *Anesthesiology* **1998**, *88*, 708–717.
12. Pfiffer, C. C. *Science* **1956**, *124*, 29–31.
13. Catterall, W. A.; Goldin, A. L.; Waxman, S. G. *Pharmacol. Rev.* **2005**, *57*, 397–409.
14. Catterall, W. A.; Mackie, K. Local Anesthetics. In *Goodman and Gilman's The Pharmacological Basis of Therapeutics*, 11th ed.; Brunton, L. L., Lazo, J. S., Parker, K. L., Eds.; McGraw Hill: New York, 2006, pp 369–386.
15. Yaksh, T. L.; Rudy, T. A. *Science* **1976**, *192*, 1357–1358.
16. Eckenhoff, R. G.; Johnasson, J. S. Inhalation Anesthetic Interactions with Proteins. In *Anesthesia: Biologic Foundations*; Yaksh, T. L., Maze, M., Lynch, C., Biebuyck, J. F., Zapol, W. M., Saidman, L. J., Eds.; Lippincott-Raven: Philadelphia, PA, 1997, pp 19–32.
17. Liu, R.; Loll, P. J. *Eckenhoff, R. G. FASEB J.* **2005**, *19*, 567–576.
18. Peet, M. *Prostaglandin, Leukotrienes, Essential Fatty Acids* **2004**, *70*, 417–422.
19. Kemperman, R. F. J.; Veurink, M.; van der Wal, T.; Knegtering, H.; Bruggeman, R.; Fokkema, M. R.; Kema, I. P.; Korf, J.; Muskiet, F. A. J. *Prostaglandin, Leukotrienes, Essential Fatty Acids* **2006**, *74*, 75–85.
20. Firestone, L. L.; Alifimoff, J. K.; Miller, K. W. *Mol. Pharmacol.* **1994**, *46*, 508–515.
21. Leal, S. M.; Evers, A. S. *Anesthesiology* **1994**, *81*, A897.
22. Taylor, J. M.; Jacob-Mosier, G. G.; Lawton, R. G.; VanDort, M.; Neubig, R. R. *J. Biol. Chem.* **1996**, *271*, 3336–3339.
23. Leite, J. F.; Blanton, M. P.; Shahgholi, M.; Dougherty, D. A.; Lester, H. A. *Proc. Natl Acad. Sci. USA* **2003**, *100*, 13054–13059.
24. Forman, S. A.; Miller, K. W.; Yellen, G. *Mol. Pharmacol.* **1995**, *48*, 574–581.
25. Tang, P.; Xu, Y. *Proc. Natl Acad. Sci. USA* **2002**, *99*, 16035–16040.
26. Liu, Z.; Xu, Y.; Tang, P. *Biophys. J.* **2005**, *88*, 3784–3791.
27. Franks, N. P.; Lieb, W. R. *Nature* **1994**, *367*, 607–614.

28. Rudolph, U.; Antkowiak, B. *Nat. Rev. Neurosci.* **2004**, *5*, 709–720.
29. Mihic, S. J.; Ye, Q.; Wick, M. J.; Koltchine, V. V.; Krasowski, M. D.; Finn, S. E.; Mascia, M. P.; Valenzuala, C. F.; Hanson, K. K.; Greenblatt, E. P. et al. *Nature* **1997**, *389*, 385–389.
30. Franks, N. P. *Br. J. Pharmacol.* **2006**, *147*, S72–S81.
31. Wakamori, M.; Ikemoto, Y.; Akaike, N. *J. Neurophysiol.* **1991**, *66*, 2014–2021.
32. Jones, M. V.; Harrison, N. L. *J. Neurophysiol.* **1993**, *70*, 1339–1349.
33. Cottrell, G. A.; Lambert, J. J.; Peters, J. A. *J. Neurosci.* **1987**, *90*, 491–500.
34. Krasowski, M. D.; Jenkins, A.; Flood, P.; Kung, A. Y.; Hopfinger, A. J.; Harrison, N. L. *J. Pharmacol. Exp. Ther.* **2001**, *297*, 338–351.
35. Gutman, G. A.; Chandy, K. G.; Grissmer, S.; Lazdunski, M.; McKinnon, D.; Pardo, L. A.; Robertson, G. A.; Rudy, B.; Sanguinetti, M. C.; Stühmer, W. et al. *Pharmacol. Rev.* **2005**, *57*, 473–508.
36. Harris, T.; Graber, A. R.; Covarrubias, M. *Am. J. Physiol. Cell Physiol.* **2003**, *285*, C788–C796.
37. Patel, A. J.; Honore, E. *Anesthesiology* **2001**, *95*, 1013–1021.
38. Catterall, W. A.; Perez-Reyes, E.; Snutch, T. P.; Striessnig, J. *Pharmacol. Rev.* **2005**, *57*, 411–425.
39. Rehberg, B.; Xiao, Y-H.; Duch, D. S. *Anesthesiology* **1996**, *84*, 1223–1233.
40. Franks, N. P.; Lieb, W. R. *Science* **1991**, *254*, 427–430.
41. Zhang, L.; Oz, M.; Stewart, R. R.; Peoples, R. W.; Weight, F. F. *Br. J. Pharmacol.* **1997**, *120*, 574–581.
42. Strois, J. E.; Lynch, C., III; Bayliss, D. A. *J. Physiol.* **2002**, *541*, 717–729.
43. Wan, X.; Mathers, D. A.; Puil, E. *Neuroscience* **2005**, *121*, 947–958.
44. Cacheaux, L. P.; Topf, N.; Tibbs, G. R.; Schaefer, U. R.; Levi, R.; Harrison, N. L.; Abbott, G. W.; Goldstein, P. A. *J. Pharmacol. Exp. Ther.* **2005**, *315*, 517–525.
45. Dickenson, R.; White, I.; Lieb, W. R.; Franks, N. P. *Anesthesiology* **2000**, *93*, 837–843.
46. Gustafsson, L. L.; Ebling, W. F.; Osaki, E.; Stanski, D. R. *Anesthesiology* **1996**, *84*, 415–442.
47. Gottschalk, A.; Smith, D. S. *Am. Fam. Phys.* **2001**, *63*, 1979–1986.
48. Goto, T.; Nakata, Y.; Morita, S. *Anesthesiology* **2003**, *98*, 1–2.
49. Bhana, N.; Goa, K. L.; McClellan, H. J. *Drugs* **2000**, *59*, 263–268.
50. Jalonen, J.; Hynynen, M.; Kuitunen, A.; Heikkila, H.; Perttila, J.; Salmenpera, M.; Valtonen, M.; Aantaa, R.; Kallio, A. *Anesthesiology* **1997**, *86*, 331–345.
51. Flood, P. *Curr. Opin. Pharmacol.* **2005**, *5*, 322–327.
52. Shephard, D. A. E. *Can. J. Anesth.* **2005**, *52*, 238–248.
53. Nadeson, R.; Goodchild, C. S. *Br. J. Anaesth.* **2001**, *86*, 704–708.
54. Naguib, M.; Baker, M. T.; Spadoni, G.; Grgegerson, M. *Anesth. Analg.* **2003**, *97*, 763–768.

Biographies

T A Bowdle, MD, PhD, obtained his undergraduate education at University of California at Davis and BS degree in 1974. He obtained his graduate education at University of Washington; Medical Scientist Training Program, MD, and PhD (Pharmacology) degrees in 1980; his internship at Virginia Mason Medical Center, 1980–81 and Anesthesiology residency at University of Washington, 1980–83. He is currently a professor of anesthesiology and pharmaceutics (adjunct), Chief of the Division of Cardiothoracic Anesthesiology at University of Washington. Major interests are local anesthetic and opioid pharmacology; pharmacokinetics; clinical outcomes (drug administration errors, complications of central venous catheter placement); depth of anesthesia monitoring.

L J S Knutsen began his research career at Glaxo in Ware, Herts, UK having completed an MA in chemistry at Christ Church, Oxford, in 1978. While at Glaxo, he completed a PhD in nucleoside chemistry joining Novo Nordisk in Denmark in 1986. There he led the project that identified tiagabine, a marketed anticonvulsant acting by blocking GABA uptake. In 1997, he joined Vernalis (Cerebrus) in the UK, initiating the adenosine A_{2A} antagonist project that led to V2006, currently in clinical trials with Biogen-IDEC for Parkinson's disease. He joined Ionix Pharmaceuticals Ltd., in Cambridge, UK in 2002 as Director of Chemistry. Dr Knutsen joined the CNS Medicinal Chemistry group at Cephalon Inc., in 2006. He has over 35 peer-reviewed publications and 18 issued US patents.

Michael Williams, PhD, DSc, received his PhD (1974) from the Institute of Psychiatry and his DSc degree in pharmacology (1987) both from the University of London. Dr Williams has worked in the US-based pharmaceutical industry for 30 years at Merck, Sharp and Dohme Research Laboratories, Nova Pharmaceutical, CIBA-Geigy and Abbott Laboratories. He retired from the latter in 2000 and after serving as a consultant with various biotechnology/pharmaceutical/venture capital companies in the US and Europe, joined Cephalon, Inc. in West Chester, in 2003 where he is vice president of Worldwide Discovery Research. He has published some 300 articles, book chapters, and reviews, and is an adjunct professor in the Department of Molecular Pharmacology and Biological Chemistry at the Feinberg School of Medicine, Northwestern University, Chicago, IL.

Comprehensive Medicinal Chemistry II
ISBN (set): 0-08-044513-6

ISBN (Volume 6) 0-08-044519-5; pp. 351–367

6.16 Migraine

P J Goadsby, Institute of Neurology, The National Hospital for Neurology and Neurosurgery, London, UK

6.16.1 Introduction

Headache is a common affliction[1] that can be highly disabling.[2] There are two basic forms of headache – *primary* headaches, where the headache is the disease, such as migraine and tension-type headache, and *secondary* headaches, where another disease process produces headache as part of its pathophysiology, such as meningitis or the headache associated with brain tumors.[3] Monographs are available covering the range of headache problems.[4] Migraine is a particular, common, and disabling form of primary headache[5] that has been well served by developments in serotonin pharmacology over the last 20 years.

Migraine is generally an episodic headache with certain associated features, such as sensitivity to light, sound, or movement, and often with nausea or vomiting accompanying the headache (**Table 1**). None of the features is compulsory,[3] and indeed given that the migraine aura, visual disturbances with flashing lights, or zigzag lines moving across the fields or other neurological symptoms are reported in only about 30% of patients,[6] a high index of suspicion is required to diagnose migraine. A headache diary can often be helpful in making the diagnosis,[7] in assessing disability, or recording how often patients use acute attack treatments. It is relevant to mention in the context of preventive treatments that migraine can be daily, as a subset of the broad group of problems known as chronic daily headache.[8] In differentiating the two main primary headache syndromes seen in clinical practice, migraine at its most simple level is headache with associated features, and tension-type headache is headache that is featureless, i.e., headache alone.

Table 1 Simplified diagnostic criteria for migraine

Repeated attacks of headache lasting 4–72 h that have these features, normal physical examination, and no other reasonable cause for the headache:	
At least two of:	At least one of:
• Unilateral pain	• Nausea/vomiting
• Throbbing pain	• Photophobia and phonophobia
• Aggravation by movement	
• Moderate or severe intensity	

Adapted from the International Headache Society Classification.[86]

6.16.2 Disease Basis

Understanding migraine can be facilitated by grasping the basic neurobiology of head pain, its processing in the brain, and modulation by centers that facilitate or inhibit sensory transmission. In experimental animals the detailed anatomy of the connections of the pain-producing intracranial extracerebral vessels and the dura mater has built on the classical human observations of Wolff[9] and others. It is crucial to remember that, while head pain can be an important manifestation of these disorders, for migraine in particular it is more accurate to think of it as a pansensory abnormality of processing, not simply a disorder generating pain signals.

The key structures whose input must be modulated when considering intracranial nociceptive processing are[4]:

- the large intracranial vessels and dura mater
- the peripheral terminals of the trigeminal nerve that innervate these structures
- the central terminals and second-order neurons of the caudal trigeminal nucleus and dorsal horns of C_1 and C_2, trigeminocervical complex.

6.16.2.1 Trigeminovascular System

The innervation of the large intracranial vessels and dura mater by the trigeminal nerve is known as the trigeminovascular system.[10] The cranial parasympathetic autonomic innervation provides the basis for symptoms, such as lacrimation and nasal stuffiness, which are prominent in cluster headache and paroxysmal hemicrania, although they may also be seen in migraine. It is clear from human functional imaging studies that vascular changes in migraine and cluster headache are driven by these neural vasodilator systems, so that these headaches should be regarded as neurovascular.[10] The concept of a primary vascular headache should be consigned to the dustbin of history since it neither explains the pathogenesis of what are complex central nervous system disorders, nor does it necessarily predict treatment outcomes. The term 'vascular headache' has no place in modern practice when referring to primary headache.[5]

6.16.2.2 Migraine Genetics

Migraine is an episodic syndrome of headache with sensory sensitivity, such as to light, sound, and head movement, probably due to dysfunction of aminergic brainstem/diencephalic sensory control systems.[11] The first of the migraine genes identified was for familial hemiplegic migraine,[12] in which about 50% of families[13] have mutations in the gene for the $Ca_V2.1$ (α_{1A}) subunit of the neuronal P/Q voltage-gated calcium channel. This finding and the clinical features of migraine suggest it might be part of the spectrum of diseases known as channelopathies,[14] disorders involving dysfunction of voltage-gated channels. Subsequently identified migraine genes *ATP1A2* (FHM2; chromosome 1q23), encoding a transmembrane transporter (α_2-subunit of the Na^+, K^+-ATPase),[15] and *SCN1A,* encoding the neuronal voltage-gated sodium channel $Na_V1.1$ (FHM3; 2q24-),[16] suggest that migraine aura at least may be better described as an ionopathy.

6.16.2.3 Migraine Aura

Migraine aura is defined as a focal neurological disturbance manifest as visual, sensory, or motor symptoms.[3] It is seen in about 30% of patients,[6] and it is neurally driven.[17] The case for the aura being the human equivalent of the cortical spreading depression of Leao[18] has been made.[19] In humans visual aura has been described as affecting the visual field, suggesting the visual cortex, and it starts at the center of the visual field and propagates to the periphery at a speed of $3\,mm\,min^{-1}$.[20]

6.16.2.4 Migraine Disorder Modulation of Afferent Traffic

Stimulation of nociceptive afferents in the superior sagittal sinus of the cat activates neurons in the ventrolateral periaqueductal gray (PAG) matter.[21] PAG activation in turn feeds back to the trigeminocervical complex with an inhibitory influence. The PAG is clearly included in the area of activation seen in positron emission tomography (PET) studies in migraineurs,[22] although the activation may be more generically antinociceptive.[23] This typical negative-feedback system will be further considered below as a possible mechanism for the symptomatic manifestations of migraine.

Another potentially modulatory region activated by stimulation of nociceptive trigeminovascular input is the posterior hypothalamic gray.[24] This area is crucially involved in several primary headaches, notably cluster headache,[25] short-lasting unilateral neuralgiform headache attacks with conjunctival injection and tearing (SUNCT),[26] paroxysmal hemicrania,[27] and hemicrania continua.[28] Moreover, the clinical features of the premonitory phase,[29] and other features of the disorder, suggest dopamine neuron involvement. Orexinergic neurons in the posterior hypothalamus can be both pro- and antinociceptive,[30] offering a further possible region whose dysfunction might involve the perception of head pain.

6.16.2.4.1 Brain imaging in humans

Functional brain imaging with PET has demonstrated activation of the dorsal midbrain, including the PAG, and in the dorsal pons, that would include the *nucleus locus coeruleus*, in studies during migraine without aura.[22] Dorsolateral pontine activation is seen with PET in spontaneous episodic[31] and chronic migraine,[32] and with nitrogylcerin-triggered attacks.[33] These areas are active immediately after successful treatment of the headache but are not active interictally. The activation corresponds to the brain region causing migraine-like headache when stimulated in patients with electrodes implanted for pain control.[34] Similarly, it had been noted[35] that excess iron in the PAG of patients with episodic and chronic migraine, and chronic migraine, can develop after a bleed into a cavernoma in the region of the PAG[35] or with a lesion of the pons.[36]

6.16.2.4.2 Animal experimental studies of sensory modulation

It has been shown in the experimental animal that stimulation of *nucleus locus coeruleus*, the main central noradrenergic nucleus, reduces cerebral blood flow in a frequency-dependent manner through an α_2-adrenoceptor-linked mechanism.[37] This reduction is maximal in the occipital cortex.[38] In addition the main serotonin-containing nucleus in the brainstem, the midbrain *dorsal raphe nucleus*, can increase cerebral blood flow when activated.[11] Furthermore, stimulation of PAG will inhibit sagittal sinus-evoked trigeminal neuronal activity in cat while blockade of P/Q-type voltage-gated Ca^{2+} channels in the PAG facilitates trigeminovascular nociceptive processing[39] with the local GABAergic system in the PAG still intact.[21]

6.16.3 Experimental Disease Models

There is no model of migraine. However, there are model of parts of the pathophysiology in animals and in humans.[40] The essential issue is: what is one trying to do? If the question is to screen or evaluate new chemical entities (NCEs) for migraine, then basic animal studies are used. For exploration of identified questions in humans, there are now models of trigeminal function in humans that can allow, for example, dose selection for further clinical trials. A list of models is shown in **Table 2**. A case history example would be the development to proof-of-principle of the potential for adenosine A_1 receptor agonists in the acute treatment of migraine (*see* Section 6.16.7.4).

6.16.4 Clinical Trial Issues

The methods for doing clinical trials in migraine have been largely set out. The International Headache Society (IHS) has provided guidelines[41] that evolved with the sumatriptan development program.[42] In general terms, for acute attack treatments, since patients seek rapid, quick, and prolonged relief of pain, the primary endpoint of the 2 h painfree

Table 2 Models of aspects of migraine pathophysiology

Model	*Method*	*Comment*
Vascular in vitro Vascular in vivo	Isolated arteries and veins Carotid arterial bed, arteriovenous anastomoses, pial arteries	Allows classic pharmacology and receptor characterization but will not address unmet needs in primary headache: such needs include nonvascular treatments
Plasma protein extravasation (PPE)	Evaluation of protein plasma extravasation with dyes or radiolabeled tracers	Screen for pharmacology of peripheral trigeminal transmission with a high false-positive rate[48]
Activation of nucleus trigeminalis caudalis	Meningeal stimulation or electrical/ mechanical stimulation of the superior sagittal sinus	Useful for identifying neuronal targets for inhibiting trigeminal nociceptive transmission[49]
Cortical spreading depression	Cortical application of KCl or electrical stimulation with electrophysiology or cerebral blood flow measurements	Excellent for understanding aura with unproven benefit in identifying either preventive or acute attack treatments[50]
Effects of nitric oxide donors	Nitric oxide donors used to activate experimental animal or human headache	Useful particularly in the human setting[51]
Blink reflex studies	Measurement of nociceptive-specific blink reflex	Useful in determining in humans dose at which a new compound can inhibit trigeminal nociceptive processing[52]

response, or the 24 h sustained painfree response, seem good choices. The latter is a composite of an initial painfree response by 2 h followed by 24 h of pain freedom and no need to take a rescue medication. Parallel groups are most often used, and placebo is very often required. Adaptive designs have proved useful in both identifying new therapies and demonstrating a lack of effectiveness of a putative therapy.[43] A particular issue in clinical trial design is the very high placebo rates that are reported in adolescent studies.[44]

Preventive therapies have recently been studied, again with the program that evaluated topiramate.[45] The IHS recently revised its guidelines in this area.[46] The botulinum toxin A[47] studies recently reported with electronic diary information will no doubt have an influence on trial design as data capture was more complete and clearly contemporaneous. A fundamental issue is the primary end-point, such as reduction in number of migraine attacks, or reduction in migraine periods, a 24 h time epoch in which the patient experiences migraine symptoms for at appears 2 h or takes a migraine-specific medication. The latter has sensitivity to change and allows for varying lengths of attack that clearly impact on disease burden. This is an interesting area of clinical research.

6.16.5 Current Treatment

The management of migraine may be divided into three parts. First, the diagnosis, which rests on a careful clinical history and neurological examination.[4] Second, after an explanation of what can and cannot be done, and particularly advice about lifestyle in the context of migraine as an episodic, probably ionopathic disorder. Third, physicians can offer preventive treatment to reduce the attack frequency and severity, or acute attack treatment for individual attacks.

6.16.5.1 Preventive Treatments for Migraine

The basis of considering preventive treatment from a medical viewpoint is a combination of acute attack frequency and attack tractability. Attacks that are unresponsive to abortive medications are easily considered for prevention, while simply treated attacks may be less obviously candidates for prevention. The other part of the equation relates to what is happening with time. If a patient diary shows a clear trend of an increasing frequency of attacks it is better to get in early with prevention than wait for the problem to become chronic.

A simple rule for frequency might be that for 1–2 headaches a month there is usually no need to start a preventive, for 3–4 it may be needed but not necessarily, and for 5 or more a month, prevention should definitely be on the agenda for discussion. Options available for treatment[4] vary somewhat by country in the European Union and again compared to North America. The largest problem with preventives is not that there are none, but that they have fallen into migraine from other indications. They have disparate mechanisms of action and thus it is difficult to make too many

Table 3 Oral acute migraine treatments

Nonspecific treatments	Specific treatments
(often used with antiemetic/prokinetics, such as domperidone (10 mg) or metaclopramide (10 mg))	
Aspirin (900 mg)	Ergot derivatives
Acetaminophen (1000 mg)	• Ergotamine (1–2 mg)
NSAIDs	Triptans
• Naproxen (500–1000 mg)	• Sumatriptan (50 or 100 mg)
• Ibuprofen (400–800 mg)	• Naratriptan (2.5 mg)
• Tolfenamic acid (200 mg)	• Rizatriptan (10 mg)
	• Zolmitriptan (2.5 or 5 mg)
	• Eletriptan (40 or 80 mg)
	• Almotriptan (12.5 mg)
	• Frovatriptan (2.5 mg)

generic conclusions about migraine from its preventive treatments. Often the doses required to reduce headache frequency produce marked and intolerable side effects. While it is not absolutely clear how preventives work, it seems likely that they modify the brain sensitivity that underlies migraine.[53]

6.16.5.2 Acute Attack Therapies for Migraine

Acute attack treatments for migraine can be usefully divided into disease nonspecific treatments – analgesics and NSAIDs – and relatively disease-specific treatments – ergot-related compounds and triptans (**Table 3**). It must be said at the outset that most acute attack medications seem to have a propensity to aggravate headache frequency and induce a state of refractory daily or near-daily headache – medication overuse headache. This propensity appears related to migraine biology. Codeine-containing compound analgesics are a particularly pernicious problem when available in over-the-counter preparations.

6.16.5.2.1 Treatment strategies

Given the array of options to control an acute attack of migraine, how does one start? The simplest approach to treatment has been described as stepped care. In this model all patients are treated, assuming no contraindications, with the simplest treatment, such as aspirin **1** 900 mg or acetaminophen **2** 1000 mg with an antiemetic. Aspirin is an effective strategy, has been proven so in double-blind controlled clinical trials, and is best used in its most soluble formulations. The alternative would be a strategy known as stratified care, by which the physician determines, or stratifies, treatment at the start based on likelihood of response to levels of care.[54] An intermediate option may be described as stratified care by attack. The latter is what many headache authorities suggest and what patients often do when they have the options. Patients use simpler options for their less severe attacks, relying on more potent options when their attacks or circumstances demand them.

6.16.5.2.2 Nonspecific acute migraine attack treatments

Since simple drugs, such as aspirin **1** and acetaminophen **2** are inexpensive and can be very effective, they can be employed in many patients. Dosages should be adequate and the addition of domperidone **3** (10 mg p.o.)[55] or metaclopramide **4** (10 mg p.o.)[56] can be very helpful. NSAIDs can very useful when tolerated. Their success is often limited by inappropriate dosing, and adequate doses of naproxen **5** (500–1000 mg p.o. or p.r., with an antiemetic),[57] ibuprofen **6** (400–800 mg p.o.),[58] or tolfenamic acid (200 mg p.o.)[59] can be extremely effective.

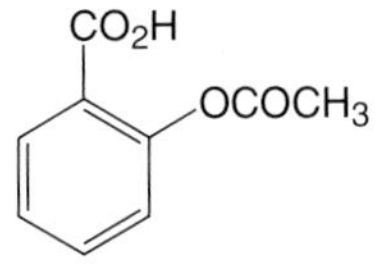

1 Aspirin

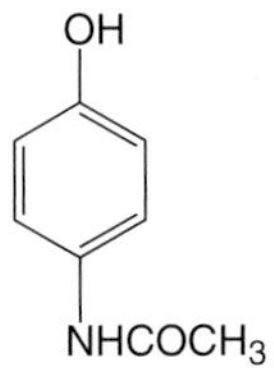

2 Acetaminophen

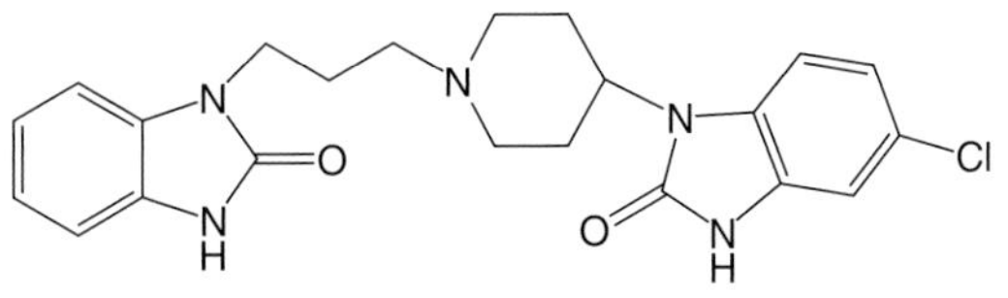

3 Domperidone

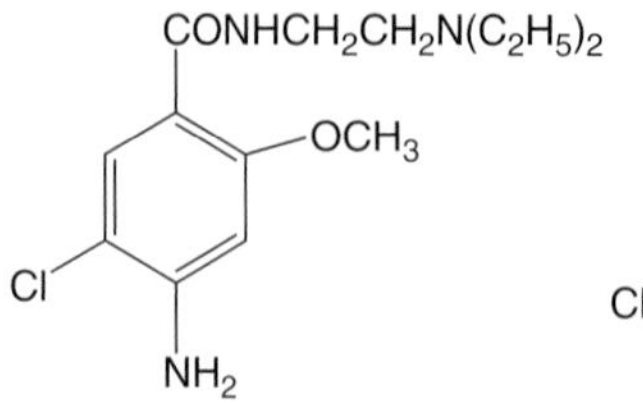

4 Metoclopramide

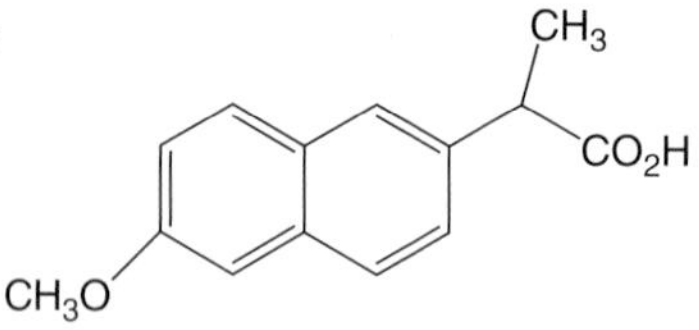

5 Naproxen

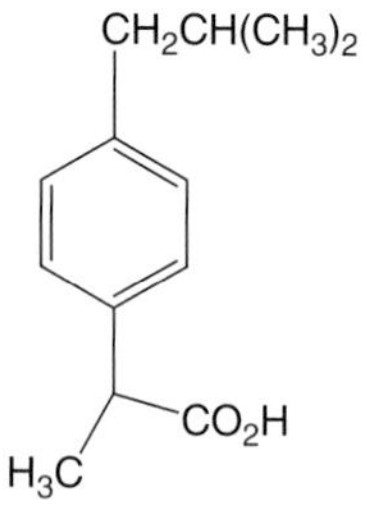

6 Ibuprofen

6.16.5.2.3 Specific acute migraine attack treatments

When simple measures fail or more aggressive treatment is required, the specific treatments are required. While ergotamine remains a useful antimigraine compound, it can no longer be considered the treatment of choice in acute migraine.[60] There are particular situations in which ergotamine is very useful, but its use must be strictly controlled as ergotamine overuse produces dreadful headache, in addition to a host of vascular problems. The triptans have revolutionized the life of many patients with migraine and are clearly the most powerful option available to stop a migraine attack.

6.16.5.2.3.1 Triptans

Triptans, $5HT_{1B/1D}$ receptor agonists, e.g., sumatriptan **7**, etc. deserve special mention since their well-described pharmacology allows one to begin to piece together both their antimigraine effect and, by inference, some aspects of migraine pathophysiology. Triptans are extremely effective acute attack treatments for migraine.[61] These compounds can potentially reverse each of the components of trigeminovascular activation, viz., constrict large cranial vessels, inhibit peripheral trigeminal terminals, and inhibit activity in the trigeminocervical complex. This effect is reflected in inhibition of release of calcitonin gene-related peptide (CGRP), that is elevated in migraine,[62] although as a generic biomarker for clinical trials it may not be useful. This mechanism of action led to the development of highly potent CGRP receptor antagonists,[63] and the demonstration in a recent clinical trial that the CGRP receptor antagonist BIBN-4096 **8** is effective in acute antimigraine. At the primary end-point of 2 h, 66% of patients had a headache response while 27% of patients had responded to placebo.[64]

Although the pharmacology of the triptans is very similar, individual patient responses can be remarkably different. Clinical areas where particular triptans are useful are listed in **Table 4**.

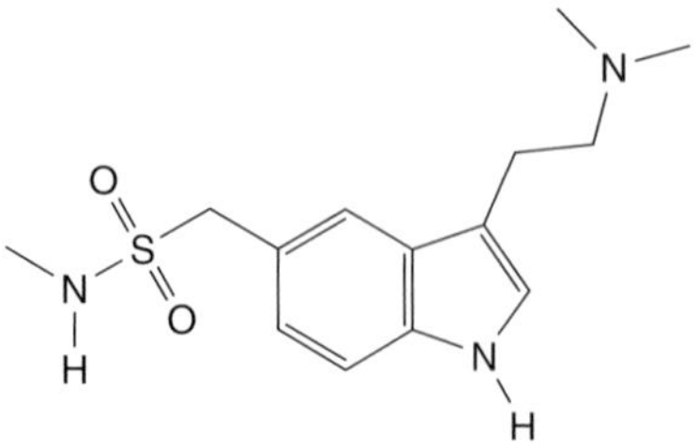

7 Sumatriptan

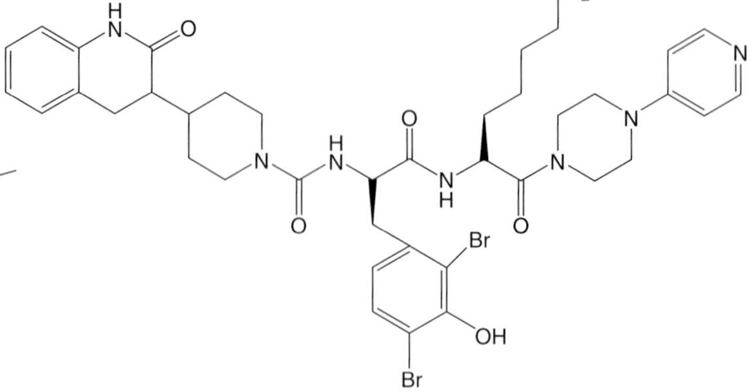

BIBN-4096
8 Olcegepant

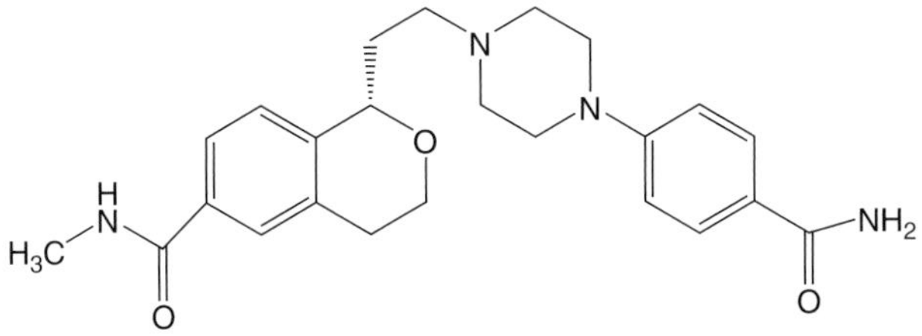

9 PNU-142633

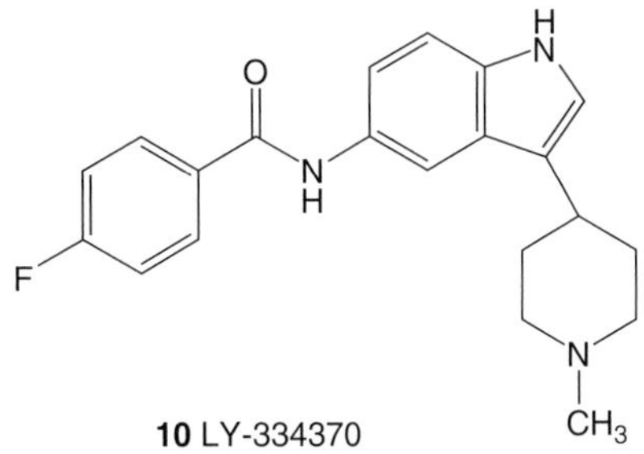

10 LY-334370

Table 4 Clinical stratification of acute specific migraine treatments

Clinical situation	*Treatment options*
Failed analgesics/NSAIDs	First tier • Sumatriptan 50 mg or 100 mg p.o. • Rizatriptan 10 mg p.o. • Almotriptan 12.5 mg • Eletriptan 40 mg p.o. • Zolmitriptan 2.5mg p.o. Slower effect/better tolerability • Naratriptan 2.5 mg • Frovatriptan 2.5 mg Infrequent headache • Ergotamine 1–2 mg p.o. • Dihydroergotamine nasal spray 2 mg
Early nausea or difficulties taking tablets	Sumatriptan 20 mg nasal spray Zolmitriptan 5 mg nasal spray Rizatriptan 10 mg MLT wafer Zolmitriptan 2.5 mg dispersible
Headache recurrence	Ergotamine 2 mg (perhaps most effective p.r./usually with caffeine) Naratriptan 2.5 mg p.o. Eletriptan 80 mg
Tolerating acute treatments poorly	Naratriptan 2.5 mg Frovatriptan 2.5 mg
Early vomiting	Sumatriptan 25 mg p,r. Sumatriptan 6 mg s.c.
Menstrually related headache	Prevention • Ergotamine p.o. nocte • Estrogen patches Treatment • Triptans • Dihydroergotamine nasal spray
Very rapidly developing symptoms	Sumatriptan 6 mg s.c. Dihydroergotamine 1 mg i.m.

6.16.6 Unmet Medical Needs

The greatest unmet need in migraine is a broader recognition of the biological basis of the problem. It remains true after the decade of the triptans, 1995–2005, that the majority of migraine sufferers in the western world, and certainly the developing world, still do not access best-practice care. If diagnostic rates in most of the world reflect those of the US, where nearly 50% of migraine sufferers do not have the diagnosis made,[65] and only a minority are treated with migraine-specific treatments, the opportunity to do good is huge. Any medicine development needs to bear this in mind. Having said that, the obvious need in acute treatment is a purely neuronally acting treatment that has no vascular side effects; this would be a major opportunity given the perceived cardiac safety issues with triptans.[66] A drug with less recurrence, i.e., a greater 24-h sustained painfree response, would also be most welcome by the fully one-third of migraineurs who suffer headache recurrence: the attack is settled by treatment but reappears some hours later. For preventive medicines, a well-tolerated treatment would be a major, immediate, and welcome bonus, even if it did not help all sufferers, since the choice of side effects is daunting to patients. Lastly, the holy grail might be a medicine that would work as an acute therapy, preemptively if the patient feels an attack, and in short-term prophylaxis, for example when there will be a known trigger or an important event over some days. There is considerable opportunity.

6.16.7 New Research Areas

Serotonin $5HT_{1B/1D}$ receptor agonists, the triptans **7**, and related 5HT receptor agonists, **9**, **10**, provided the most important advance in migraine therapeutics in the four millennia that the condition has been recognized. Simultaneously the development of triptans, with their vasoconstrictor action, produced a small clinical penalty in terms of coronary vasoconstriction and an enormous intellectual question: the extent to which migraine is a vascular problem. Functional neuroimaging and neurophysiological studies have consistently developed the theme of migraine as a brain disorder and thus demand that the search for neurally acting antimigraine drugs should be undertaken. CGRP receptor blockade is an effective acute antimigraine strategy and is nonvasconstrictor terms of the mechanism of action. It is likely that direct blockade of CGRP release by inhibition of trigeminal nerves would be similarly effective. Options for such an action based on preclinical work include: serotonin $5HT_{1F}$ and $5HT_{1D}$ receptor agonists, glutamate excitatory amino acid receptor antagonists, adenosine A_1 receptor agonists, nociceptin, vanilloid TRPV1 receptors, cannabinoid CB_1 receptors, orexin receptors, and nitric oxide (NO) mechanisms.[49] Those mechanisms tested in patients are discussed below. Adenosine A_1 receptor agonists were discussed above.

6.16.7.1 $5HT_{1F}$ Receptor Agonists

The potent specific $5HT_{1F}$ agonist LY-334370 blocks neurogenic plasma protein extravasation.[67] LY-334370 **10** is effective in acute migraine, albeit at doses with some central nervous system side effects and no cardiovascular problems.[68] Unfortunately, development was stopped because of a nonhuman toxicity problem. $5HT_{1F}$ receptor activation is inhibitory in the trigeminal nucleus in rat and cat, albeit in cat seeming less potent than $5HT_{1B}$ or $5HT_{1D}$ receptor activation.[68] There is a good expectation that $5HT_{1F}$ receptor agonists would be both nonvascular and probably useful in migraine and cluster headache.

6.16.7.2 $5HT_{1D}$ Receptors

$5HT_{1D}$ receptor agonists are potent inhibitors of neurogenic dural plasma protein extravasation[70] and have no vascular effects. Peptidergic nociceptors express these receptors in a manner that is activation-dependent.[71] Specific potent $5HT_{1D}$ agonists have been developed by taking advantage of similarities between human and nonhuman primate $5HT_{1B}$ and $5HT_{1D}$ receptors. PNU-142633 **9** was clinically ineffective,[72] although it was a relatively weak agonist when compared to sumatriptan in in vitro studies,[73] and was poorly brain-penetrant. Since this NCE was developed using gorilla receptors, there are questions as to whether this was the correct compound to test the $5HT_{1D}$ hypothesis. Preclinical studies are able to dissect out a potent $5HT_{1D}$ receptor-mediated inhibition of the trigeminocervical complex,[69] so that this mechanism remains both plausible and not fully tested.

6.16.7.3 Nitric Oxide Synthase Inhibitors

Much has been written of NO and migraine.[51] Nitroglycerin triggers migraine by a necessary dilation of cranial vessels. However, three recent observations suggest that dilation is an epiphenomenon. First, nitroglycerin triggers premonitory symptoms in many patients.[74] These were no different to those reported in spontaneous attacks[52] and occurred well after any vascular change would have been present. Secondly, downstream activation of the cyclic guanosine monophosphate pathway by sildenafil can induce migraine without any change in middle cerebral artery diameter.[75] Thirdly, dilation of the internal carotid artery after nitroglycerin administration in cluster headache patients is dissociated in time from the onset of the attack.[76] Together these observations suggest that, while NO mechanisms may play a role in some part of the pathophysiology of these disorders, it need not be a vascular effect. A role for inducible NO synthase has been suggested,[77] and inhibition of trigeminocervical complex *fos* expression occurs with NO synthase blockade.[78] Both examples provide a nonvascular approach, although potentially with rather different NO synthase subtype targets. The available data, therefore, suggest that NO-based developments may find clinical utility in migraine, and indeed the initial clinical study was positive.[79]

The prospect of a nonvasconstrictor acute migraine therapy offers a real opportunity to patients, and perhaps more importantly, provides a therapeutic rationale to plant migraine firmly in the brain as a neurological problem, where it undoubtedly belongs.

HN OH N N N N HO O HO OH

11 GR-79236

6.16.7.4 Adenosine

Adenosine antagonists decreased nociceptive thresholds in rats.[80] The antinociceptive effects of adenosine are mediated via the A_1 receptor.[81,82] This has been localized in human trigeminal ganglia,[83] suggesting a potential ability of adenosine A_1 receptor agonists to inhibit the trigeminal nerve. Selective adenosine A_1 receptor agonists, e.g., GR-79236, **11**, can inhibit trigeminovascular activation, both in the trigeminal nucleus and by inhibition of release of CGRP in the cranial circulation.[84] The effect within the trigeminal nucleus reflects a central action, whilst inhibition of CGRP release is likely to be attributable to an action at adenosine A_1 receptors on peripheral terminals of the trigeminal nerve. Both effects are in keeping with the concept of adenosine A_1 receptors being located prejunctionally on primary afferent neurons and causing inhibition of transmitter release. Adenosine A_1 receptor agonists, such as GR-79236, have no effect on resting meningeal artery diameter in rats.[85] Moreover, GR-79236 can inhibit the nociceptive trigeminal blink reflex at doses in humans[52] that are both trigeminally inhibitory and without vascular effects in experimental animals.

References

1. Rasmussen, B. K.; Olesen, J. *Neurology* **1992**, *42*, 1225–1231.
2. Menken, M.; Munsat, T. L.; Toole, J. F. *Arch. Neurol.* **2000**, *57*, 418–420.
3. Headache Classification Committee of The International Headache Society. *Cephalalgia* **2004**, *24*, 1–160.
4. Lance, J. W.; Goadsby, P. J. *Mechanism and Management of Headache*; Elsevier: New York, 2005.
5. Goadsby, P. J.; Lipton, R. B.; Ferrari, M. D. *N. Engl. J. Med.* **2002**, *346*, 257–270.
6. Rasmussen, B. K.; Olesen, J. *Cephalalgia* **1992**, *12*, 221–228.
7. Russell, M. B.; Rassmussen, B. K.; Brennum, J. et al. *Cephalalgia* **1992**, *12*, 369–374.
8. Welch, K. M. A.; Goadsby, P. J. *Curr. Opin. Neurol.* **2002**, *15*, 287–295.
9. Wolff, H. G. *Headache and Other Head Pain*; Oxford University Press: New York, 1948.
10. May, A.; Goadsby, P. J. *J. Cereb. Blood Flow Metab.* **1999**, *19*, 115–127.
11. Goadsby, P. J.; Zagami, A. S.; Lambert, G. A. *Headache* **1991**, *31*, 365–371.
12. Ophoff, R. A.; Terwindt, G. M.; Vergouwe, M. N. et al. *Cell* **1996**, *87*, 543–552.

13. Ducros, A.; Denier, C.; Joutel, A. et al. *N. Engl. J. Med.* **2001**, *345*, 17–24.
14. Kullmann, D. M. *Brain* **2002**, *125*, 1177–1195.
15. De Fusco, M.; Marconi, R.; Silvestri, L. et al. *Nat. Genet.* **2003**, *33*, 192–196.
16. Goadsby, P. J.; Kullmann, D. K. *Lancet* **2005**, *366*, 345–346.
17. Cutrer, F. M.; Sorensen, A. G.; Weisskoff, R. M. et al. *Ann. Neurol.* **1998**, *43*, 25–31.
18. Leao, A. A. P. *J. Neurophysiol.* **1944**, *7*, 391–396.
19. Lauritzen, M. *Brain* **1994**, *117*, 199–210.
20. Lashley, K. S. *Arch. Neurol. Psychiatry* **1941**, *46*, 331–339.
21. Knight, Y. E.; Bartsch, T.; Goadsby, P. *J. Neurosci. Lett.* **2003**, *336*, 113–116.
22. Weiller, C.; May, A.; Limmroth, V. et al. *Nat. Med.* **1995**, *1*, 658–660.
23. Tracey, I.; Ploghaus, A.; Gati, J. S. et al. *J. Neurosci.* **2002**, *22*, 2748–2752.
24. Benjamin, L.; Levy, M. J.; Lasalandra, M. P. et al. *Neurobiol. Dis.* **2004**, *16*, 500–505.
25. Goadsby, P. J. *Lancet Neurol.* **2002**, *1*, 37–43.
26. May, A.; Bahra, A.; Buchel, C.; Turner, R.; Goadsby, P. J. *Ann. Neurol.* **1999**, *46*, 791–793.
27. Matharu, M. S.; Cohen, A. S.; Frackowiak, R. S. J.; Goadsby, P. J. *Cephalalgia* **2005**, *25*, 859.
28. Matharu, M. S.; Cohen, A. S.; McGonigle, D. J. et al. *Headache* **2004**, *44*, 462–463.
29. Giffin, N. J.; Ruggiero, L.; Lipton, R. B. et al. *Neurology* **2003**, *60*, 935–940.
30. Bartsch, T.; Levy, M. J.; Knight, Y. E.; Goadsby, P. J. *Pain* **2004**, *109*, 367–378.
31. Afridi, S.; Giffin, N. J.; Kaube, H. et al. *Brain* **2005**, *62*, 1270–1275.
32. Matharu, M. S.; Bartsch, T.; Ward, N. et al. *Brain* **2004**, *127*, 220–230.
33. Afridi, S.; Matharu, M. S.; Lee, L. et al. *Brain* **2005**, *128*, 932–939.
34. Veloso, F.; Kumar, K.; Toth, C. *Headache* **1998**, *38*, 507–515.
35. Welch, K. M.; Nagesh, V.; Aurora, S.; Gelman, N. *Headache* **2001**, *41*, 629–637.
36. Goadsby, P. J. *Cephalalgia* **2002**, *22*, 107–111.
37. Afridi, S.; Goadsby, P. J. *J. Neurol. Neurosurg. Psychiatry* **2003**, *74*, 680–682.
38. Goadsby, P. J.; Duckworth, J. W. *Brain Res.* **1989**, *476*, 71–77.
39. Knight, Y. E.; Bartsch, T.; Kaube, H.; Goadsby, P. J. *J. Neurosci.* **2002**, *22*, 1–6.
40. Edvinsson, L. *Experimental Headache Models in Animals and Man*; Martin Dunitz: London, 1999.
41. International Headache Society Committee on Clinical Trials in Migraine. *Cephalalgia* **1991**, *11*, 1–12.
42. Pilgrim, A. J. *Eur. Neurol.* **1991**, *31*, 295–299.
43. Roon, K. I.; Olesen, J.; Diener, H. C. et al. *Ann. Neurol.* **2000**, *47*, 238–241.
44. Winner, P.; Lewis, D.; Visser, W. H. et al. *Headache* **2002**, *42*, 49–55.
45. Brandes, J. L.; Saper, J. R.; Diamond, M. et al. *JAMA* **2004**, *291*, 965–973.
46. Tfelt-Hansen, P.; Block, G.; Dahlof, C. et al. *Cephalalgia* **2000**, *20*, 765–786.
47. Silberstein, S.; Mathew, N.; Saper, J.; Jenkins, S. *Headache* **2000**, *40*, 445–450.
48. May, A.; Goadsby, P. J. *Exp. Opin. Investig. Drugs* **2001**, *10*, 1–6.
49. Goadsby, P. J. *Nat. Rev. Drug Disc.* **2005**, *4*, 741–750.
50. Goadsby, P. *J. Ann. Neurol.* **2001**, *49*, 4–6.
51. Thomsen, L. L.; Olesen, J. *Curr. Opin. Neurol.* **2001**, *14*, 315–321.
52. Giffin, N. J.; Kowacs, F.; Libri, V. et al. *Cephalalgia* **2003**, *23*, 287–292.
53. Goadsby, P. J. *Cephalalgia* **1997**, *17*, 85–92.
54. Lipton, R. B.; Stewart, W. F.; Stone, A. M.; Lainez, M. J. A.; Sawyer, J. P. C. *JAMA* **2000**, *284*, 2599–2605.
55. Cottrell, J.; Mann, S. G.; Hole, J. *Cephalalgia* **2000**, *20*, 269.
56. Oral Sumatriptan and Aspirin plus Metaclopramide Comparative Study Group. *Eur. Neurol.* **1992**, *32*, 177–184.
57. Welch, K. M. A. *Cephalalgia* **1986**, *6*, 85–92.
58. Kellstein, D. E.; Lipton, R. B.; Geetha, R. et al. *Cephalalgia* **2000**, *20*, 233–243.
59. Codispoti, J. R.; Prior, M. J.; Fu, M.; Harte, C. M.; Nelson, E. B. *Headache* **2001**, *41*, 665–679.
60. Tfelt-Hansen, P.; Saxena, P. R.; Dahlof, C. et al. *Brain* **2000**, *123*, 9–18.
61. Ferrari, M. D.; Goadsby, P. J.; Roon, K. I.; Lipton, R. B. *Cephalalgia* **2002**, *22*, 633–658.
62. Gallai, V.; Sarchielli, P.; Floridi, A. et al. *Cephalalgia* **1995**, *15*, 384–390.
63. Moreno, M. J.; Abounader, R.; Hebert, E.; Doods, H.; Hamel, E. *Neuropharmacology* **2002**, *42*, 568–576.
64. Olesen, J.; Diener, H.-C.; Husstedt, I.-W. et al. *N. Engl. J. Med.* **2004**, *350*, 1104–1110.
65. Lipton, R. B.; Diamond, S.; Reed, M.; Diamond, M. L.; Stewart, W. F. *Headache* **2001**, *41*, 638–645.
66. Dodick, D.; Lipton, R. B.; Martin, V. et al. **2004**, *44*, 414–425.
67. Johnson, K. W.; Schaus, J. M.; Durkin, M. M. et al. *NeuroReport* **1997**, *8*, 2237–2240.
68. Goldstein, D. J.; Roon, K. I.; Offen, W. W. et al. *Lancet* **2001**, *358*, 1230–1234.
69. Goadsby, P. J.; Classey, J. D. *Neuroscience* **2003**, *122*, 491–498.
70. Maneesi, S.; Akerman, S.; Lasalandra, M. P.; Classey, J. D.; Goadsby, P. J. *Cephalalgia* **2004**, *24*, 148.
71. Ahn, A. H.; Fields, H. L.; Basbaum, A. I. *Neurology* **2004**, *62*, A440–A441.
72. Gomez-Mancilla, B.; Cutler, N. R.; Leibowitz, M. T. et al. *Cephalalgia* **2001**, *21*, 727–732.
73. Pregenzer, J. F.; Alberts, G. L.; Im, W. B. et al. *Br. J. Pharmacol.* **1999**, *127*, 468–472.
74. Afridi, S.; Kaube, H.; Goadsby, P. J. *Pain* **2004**, *110*, 675–680.
75. Kruuse, C.; Thomsen, L. L.; Birk, S.; Olesen, J. *Brain* **2003**, *126*, 241–247.
76. May, A.; Bahra, A.; Buchel, C.; Frackowiak, R. S. J.; Goadsby, P. J. *Neurology* **2000**, *55*, 1328–1335.
77. Reuter, U.; Bolay, H.; Jansen-Olesen, I. et al. *Brain* **2001**, *124*, 2490–2502.
78. Hoskin, K. L.; Bulmer, D. C. E.; Goadsby, P. J. *Neurosci. Lett.* **1999**, *266*, 173–176.
79. Lassen, L. H.; Ashina, M.; Christiansen, I.; Ulrich, V.; Olesen, J. *Lancet* **1997**, *349*, 401–402.
80. Paalzow, G.; Paalzow, L. *Acta Pharmacol. Toxicol.* **1973**, *32*, 22–32.
81. Sjolund, K.-F.; Sollevi, A.; Segerdahl, M.; Hansson, P.; Lundeberg, T. *NeuroReport* **1996**, 7, 1856–1860.
82. Salter, M. W.; Sollevi, A. *Hanb. Exp. Pharmacol.* **2001**, *151*, 371–401.

83. Schindler, M.; Harris, C. A.; Hayes, B.; Papotti, M.; Humphrey, P. P. A. *Neurosci. Lett.* **2001**, *297*, 211–215.
84. Goadsby, P. J.; Hoskin, K. L.; Storer, R. J.; Edvinsson, L.; Connor, H. E. *Brain* **2002**, *125*, 1392–1401.
85. Honey, A. C.; Bland-Ward, P. A.; Connor, H. E.; Feniuk, W.; Humphrey, P. P. A. *Cephalalgia* **2000**, *22*, 260–264.
86. Headache Classification Committee of The International Headache Society. *Cephalalgia* **1988**, *8*, 1–96.

Biography

Peter J Goadsby obtained his basic medical degree and training at the University of New South Wales, Australia. His neurology training was done under the supervision of Prof James Lance in Sydney. After postdoctoral work in New York with Don Reis at Cornell, with Jacques Seylaz in Paris, and postgraduate neurology training at Queen Square in London, he returned to the University of New South Wales, and the Prince of Wales Hospital, Sydney as a consultant neurologist and, in 1994 as associate professor of neurology.

He was appointed a Wellcome Senior Research Fellow at the Institute of Neurology, University College London in 1995. He is currently professor of clinical neurology and consultant neurologist at the National Hospital for Neurology and Neurosurgery, Queen Square, and the Hospital for Sick Children, Great Ormond St., London. His major research interests are in the neural control of the cerebral circulation and the basic mechanisms of head pain in both experimental settings and in the clinical context of headache. The work of the Headache Group involves human imaging and electrophysiological studies in primary headache, as well as experimental studies of trigeminovascular nociception. We aim to understand what parts of the brain drive and modulate headache syndromes, and how those might be modified by treatment.

Comprehensive Medicinal Chemistry II
ISBN (set): 0-08-044513-6

ISBN (Volume 6) 0-08-044519-5; pp. 369–379

6.17 Obesity/Metabolic Syndrome Overview

R D Feldman and R A Hegele, Robarts Research Institute, London, ON, Canada

6.17.1 Introduction

Of the next wave of chronic diseases that are expected to reach epidemic proportions in developed nations, the metabolic syndrome (MetS) must be viewed as among the most important. The rise of MetS to its current preeminence as a clinical problem is even more remarkable considering that this disease entity was essentially unknown – or at least invisible – before 1980. The heightened awareness of the importance of MetS as a worldwide major health concern is based on: (1) the shear breadth of its prevalence in westernized and westernizing countries; (2) its potential impact upon public health; and (3) its clinical importance as a target of new therapies. One index of the attention given to MetS is the citation of ~1000 articles in PubMed in the first 6 months of 2005 using the search term 'metabolic syndrome'; moreover, this pace of scientific citation and clinical awareness has been constantly accelerating since its initial description.

Although the clustering of the component metabolic disturbances that are characteristic of diabetes and hypertension had been well recognized for more than 40 years, the concept of MetS was first popularized in the mid-1980s by several scientists, especially by Reaven, who initially termed it 'syndrome X.'[1] The significance of defining MetS as a clinical entity relates to its importance as a risk factor that can predict the subsequent development of diabetes and cardiovascular disease. The growing prevalence of MetS over the next several decades will feed the looming epidemic of diabetes and the resulting resurgence of atherosclerotic disease, which has been declining in relative terms over the past two decades.

6.17.2 Definition of Metabolic Syndrome

MetS is a commonly occurring cluster of clinical phenotypes that are individually and collectively strongly related to cardiovascular disease.[2] MetS is characterized by disturbed carbohydrate and insulin metabolism, and is clinically defined by threshold values applied to indices of central obesity, dysglycemia, dyslipidemia, and/or elevated blood pressure, which must be present concurrently in any one of a variety of combinations.[2,3] The cardinal feature of MetS is abdominal obesity, as quantified most directly by increased waist circumference.[4,5] Biochemically, MetS is characterized by insulin resistance – hyperinsulinemia – and by dyslipidemia – most typically raised triglycerides and/or reduced HDL cholesterol. Additionally, a range of biochemical abnormalities have been secondarily associated, including increased serum concentrations of apolipoprotein B, fibrinogen, free fatty acids, C-reactive protein (CRP), tumor necrosis factor (TNF)-α, interleukin-6, and plasminogen activator inhibitor-1, and with depressed concentrations of adiponectin. Hemodynamically, MetS is characterized by elevated blood pressure – at least in the high normal or 'prehypertensive' range (*see* 6.32 Hypertension). The International Diabetes Federation (IDF) has recently updated the consensus worldwide definition of MetS.[6,7] It has identified both a definition for clinical practice (**Table 1**) and a 'platinum standard' definition that includes additional metabolic criteria that are currently used investigationally (**Table 2**).

Table 1 IDF metabolic syndrome definition: 2005

A person to be defined as having the metabolic syndrome must have: central obesity (defined as waist circumference >94 cm for Europid men and >80 cm for Europid women, with ethnicity-specific values for other groups) plus any two of the following four factors:

1. Raised triglyceride level: >150 mg dL^{-1} (1.7 mmol L^{-1}), or specific treatment for this lipid abnormality
2. Reduced HDL cholesterol: <40 mg dL^{-1} (1.03 mmol L^{-1}*) in males and <50 mg dL^{-1} (1.29 mmol L^{-1}*) in females, or specific treatment for this lipid abnormality
3. Raised blood pressure (BP): systolic BP >130 or diastolic BP >85 mmHg, or treatment of previously diagnosed hypertension
4. Raised fasting plasma glucose (FPG) >100 mg dL^{-1} (5.6 mmol L^{-1}), or previously diagnosed type 2 diabetes. (If above 5.6 mmol L^{-1} or 100 mg dL^{-1}, oral glucose tolerance test (OGTT) is strongly recommended but is not necessary to define the presence of the syndrome)

Source: International Diabetes Federation.

Table 2 Additional metabolic syndrome criteria, for research

Abnormal body fat distribution	General body fat distribution (DXA)
	Central fat distribution (CT/MRI)
	Adipose tissue biomarkers: leptin, adiponectin
	Liver fat content (MRS)
Atherogenic dyslipidemia	(beyond apoB (or non-HDL-c) elevated triglyceride and low HDL)
	Small LDL particles
Dysglycemia	OGTT
Insulin resistance (other than elevated fasting glucose)	Fasting insulin/proinsulin levels
	HOMA-IR
	Insulin resistance by Bergman minimal model
	Elevated free fatty acids (fasting and during OGTT)
	M value from euglycemic hyperinsulinemic clamp
Vascular dysregulation (beyond elevated blood pressure)	Measurement of endothelial dysfunction
	Microalbuminuria
Proinflammatory state	Elevated high-sensitivity C-reactive protein (SAA)
	Elevated inflammatory cytokines (e.g., TNF-α, IL-6)
	Decrease in adiponectin plasma levels
Prothrombotic state	Fibrinolytic factors (PAI-1 etc.)
	Clotting factors (fibrinogen etc.)
Hormonal factors	Pituitary–adrenal axis

DXA, dual-energy x-ray absorptiometry; CT, computed tomography; MRI, magnetic resonance imaging; MRS, magnetic resonance spectroscopy; LDL, low-density lipoprotein; HOMA-IR, homeostatic model assessment for insulin resistance; SAA, serum amyloid A; PAI-1, plasminogen activator inhibitor-1; OGGT, oral glucose tolerance test.
Source: International Diabetes Federation.

6.17.3 Scope of the Metabolic Syndrome Epidemic: Prevalence and Risk

Estimates of MetS prevalence vary according to age, country, and ethnic group. Interpopulation comparisons have been complicated by the fact that it is probably inappropriate to use single threshold values for the defining quantitative traits. Abdominal obesity appears to be the most frequently found MetS component observed in a number of population-based studies, irrespective of ethnicity.[8–10] There has been recent consideration of the appropriateness of uniform cut-off points for abdominal obesity, drawing attention to the need for ethnic-specific guidelines.[11] Some investigators have already employed modified MetS criteria in order to identify better Asian individuals with MetS,[12] yet clearly more investigation is required in this respect. Interestingly, other components, such as blood pressure, show distinctive variability between ethnic groups, being dominant features in African-American populations[10] and Korean males.[10]

In several countries, MetS prevalence by any definition has been reported to rise from greater than 10% in adults aged 20–40 years up to 40–70% in adults aged over 60.[13,14] Among patients with hypertension, between a third and a half will concurrently have MetS, at least if hyperinsulinemia is considered as the defining biochemical feature.[15–17] The risk of having MetS is most tightly linked to obesity: MetS prevalence increases to greater than 50% in both obese adults[16] and in children with severe obesity.[18] More specifically, MetS is linked with visceral – also called abdominal or central – obesity. Visceral obesity seems to precede the development of insulin resistance and elevated blood pressure, and likely also temporally precedes the onset of most serum biochemical disturbances in MetS. Despite the availability of numerous advanced technologies to quantify the extent of visceral obesity, determination of waist circumference has the advantage of simplicity, familiarity, and standardization against more resource-intensive measurements.[19]

MetS is associated with increased risk of cardiovascular disease and also increased risk for the development of its associated risk factors – diabetes and hypertension. The presence of MetS according to the National Cholesterol Education Program (NCEP) definition was prospectively associated with the development of type 2 diabetes[20] (*see* 6.19 Diabetes/Syndrome X) and both all-cause and cardiovascular mortality.[21] Cardiovascular disease risk has been reported to be increased two- to fivefold in both men and women with MetS (greater in men than in women) when adjusted for age, cholesterol, and tobacco use.[22] Some national guidelines for risk factor management include MetS as a factor that can be used to restratify patients into a higher level of risk than would be assumed by simple tallying of traditional risk factors.[23]

6.17.4 Etiology of Metabolic Syndrome: The Search for the Single Causal Mechanism

The recognition of the cluster of disorders characterizing MetS has sparked an ongoing examination of the potential for a single causal mechanism underlying the syndrome. This search for etiological mechanisms has largely focused on either the central role of insulin resistance and/or the hormonal/biochemical consequences of visceral obesity.

With the appreciation of the similarity of MetS complex to the pattern of vascular and lipid metabolism changes seen in type 2 diabetes, the central role of insulin resistance as the etiology of the disorders of MetS has been intensively studied. This has especially focused on the vascular effects of insulin/insulin resistance, especially with respect of endothelial function.

Initial interest focused on the direct effect of insulin on vascular function and the impact of insulin resistance. Insulin mediates a direct vasodilator effect on a range of vascular beds, both arterial and venous.[24,25] The effect has been predominantly related to activation of nitric oxide synthase activity,[26] – although other endothelial mechanisms have also been suggested.[26] Further, insulin's vasodilator effect has been suggested to be an important determinant of the extent of glucose uptake in skeletal muscle.[27] Further studies indicated that resistance to the vascular effects of insulin paralleled systemic insulin resistance in diabetes and in hypertension as well as with obesity.[28] Also, modalities that improve systemic insulin resistance (e.g., angiotensin-converting enzyme inhibitors, salt loading) also improve vascular insulin sensitivity.[29,30]

However, vascular insulin resistance ultimately could not be viewed as a specific cause of the vascular manifestation of MetS. In all of the component disorders in which vascular insulin resistance – including hyperinsulinemia, dyslipidemia, hypertension, diabetes, and obesity – has been reported, a more global defect in endothelial-mediated vasodilation has also been demonstrated.[31] Further, the maneuvers that improve systemic and/or vascular insulin resistance more generally remediate the defect in 'global' endothelial function, including therapy with fibrates,[32] glitazones,[33] and metformin.[34] Beyond that, a range of other hemodynamic disturbances, including activation of the sympathetic nervous system, and disordered renal salt/water handling, could not be directly related to vascular insulin resistance.[35] Thus the current view is that vascular insulin resistance, although a contributing mechanism to both the defects in peripheral resistance and skeletal muscle glucose uptake characteristic of MetS, cannot be viewed as the single causal mechanism underlying this condition.

The other major focus in the search for the single causal mechanism underlying MetS has been the cause/consequences of visceral obesity. These relate to both hormonal and metabolic consequences of obesity. The impact of several adipocytokines has been examined for their roles as etiological factors in MetS. These include leptin, adiponectin, and resistin as well as inflammatory hormones (TNF-α) and additionally nonesterified free fatty acids.[36] The pattern of endothelial dysfunction and sympathoadrenal activation has been linked to these mediators as well as the inflammatory phenotype characteristic of patients with MetS.

6.17.5 Genes and the Metabolic Syndrome: Complex Trait Genetics

The genetic basis of MetS has increasingly been appreciated, especially regarding the development of obesity. In this respect, MetS can be considered to result from the interaction of environmental factors, such as caloric excess and physical inactivity, with genetic susceptibility factors.[37–39] While the increased prevalence in MetS in the broader population is undoubtedly related to lifestyle – primarily an imbalance between caloric intake and expenditure – it is likely that genetic factors will also be proven to be important. Given the complexity of MetS phenotype, one approach to parse the relative roles of genetic and nongenetic factors is to study MetS in genetically isolated human subpopulations. For instance, in certain North American aboriginal communities, such as the Oji-Cree of Ontario, the combined prevalence of impaired glucose tolerance and type 2 diabetes is approximately 40%.[40] This development has been inextricably linked with the recent doubling of hospitalizations for coronary heart disease among Oji-Cree, despite declining rates in the general Canadian population.[41] Among Oji-Cree adults aged 35 and older, 43% had MetS.[42] Furthermore, 8.7% of female Oji-Cree adolescents had MetS, as defined by the NCEP Adult Treatment Panel (ATP) III[47] criteria. Increased waist girth and depressed HDL cholesterol were the most prevalent individual components in subjects with MetS. Common functional polymorphisms in genes encoding proteins involved in the renin angiotensin system, the G-protein family (GNB3) and components of triglyceride-rich lipoproteins were each significantly associated with MetS in Oji-Cree adults, especially women.[43] Such studies suggest that the component of genetic susceptibility in MetS is possibly related to genetic basis of the defining intermediate quantitative traits of obesity, dyslipidemia, dysglycemia, and hypertension.

6.17.6 Genes and the Metabolic Syndrome: Monogenic Disorders

In addition to identifying genes for 'garden-variety' MetS, careful characterization of patients with rare monogenic disorders that recapitulate features of MetS have led to important insights that can be translated to the more common complex form. For instance, familial partial lipodystrophy (FPLD) syndromes are autosomal dominant disorders that occur with a frequency of perhaps 1:100 000 individuals in the general population.[37–39] There are three distinct forms: FPLD1, FPLD2, and FPLD3, of which the latter two have been found to be due to mutations, respectively, in the *LMNA* gene, encoding nuclear lamin A/C and the PPAR-γ gene, encoding peroxisome proliferator-γ activated receptor (PPAR-γ). Both FPLD2 and FPLD3 feature a loss of fat tissue in peripheral depots such as the extremities and gluteal region, with preservation of central and visceral fat stores. The infinite ratio of visceral to peripheral subcutaneous fat creates an extreme form of common visceral obesity, which leads to the development of several key features of the common MetS, including increased risk of atherosclerosis endpoints.[44] FPLD2 (the *LMNA* form) implicates structural abnormalities of the nuclear envelope as a cause of insulin resistance, diabetes, and ultimately atherosclerosis. FPLD3 (the PPAR-γ form) proved that inherited partial lipodystrophy was genetically heterogeneous, clarified the metabolic phenotype of PPAR-γ-deficiency due to mutant PPARG, and confirmed the key role of PPAR-γ in adipogenesis and metabolism. Comparison with FPLD2 indicated that FPLD3 was associated with less severe lipodystrophy and more severe insulin resistance,[37] suggesting additional mechanisms underlying insulin resistance and metabolic changes beyond those attributed solely to adipose tissue redistribution. This suggests that intervening on the PPAR-γ pathway could have multiple beneficial effects upon metabolic phenotypes and downstream complications. It is no coincidence that pharmacological agonists of PPAR-γ have been shown to improve insulin resistance, lessen the intermediate metabolic disturbances, and likely reduce vascular endpoints.

6.17.7 Insights into Metabolic Syndrome Progression from the Monogenic Disorders

Nondiabetic FPLD2 subjects have high plasma insulin, triglycerides, free fatty acids, and CRP, together with low plasma HDL cholesterol and adiponectin long before diabetes developed.[43] Thus, the characteristic cluster of biochemical abnormalities precedes the relatively late decompensation of glycemic control among carriers predisposed to lipodystrophy. A similar pattern of progression – i.e., early hyperinsulinemia with dyslipidemia and altered adipocytokines followed by hypertension and finally diabetes – may be important in 'garden-variety' MetS. In both cases, spillover of FFA from adipose tissue and uptake and storage viscerally and in ectopic sites, such as muscle and liver, may be a key inciting pathophysiological event. In FPLD, spillover of free fatty acids occurs because there is an anatomical absence of the peripheral subcutaneous fat buffer. In 'garden-variety' MetS, this spillover occurs because peripheral fat stores become saturated. If this pattern of metabolic progression is pathophysiologically important for most patients, then it would suggest that there may be several points for intervention, such as early intervention focused on preventing maldistributed adipose tissue (especially visceral body fat), followed by management of plasma lipoprotein and adipocytokines. Successful intervention at the early milestones of metabolic progression may delay the need to control hypertension and impaired glucose tolerance, which occur at a relatively later stage of disease evolution.

6.17.8 Management of the Metabolic Syndrome

The current management of MetS reflects two major considerations: (1) promotion of healthy lifestyle changes to ameliorate the root environmental causes; and (2) the therapeutic approaches to manage the individual components of MetS.

Healthy lifestyle promotion includes: (1) dietary changes in terms of moderating calorie intake to achieve a 5–10% loss of body weight in the first year and changing dietary composition; (2) increasing physical activity; and (3) changes in dietary composition to lower intake of saturated fats, *trans* fats and cholesterol, and include carbohydrates with a low glycemic index and high content of soluble fiber.

Finally, because of the heightened risk of vascular disease, MetS patients should receive other broad-spectrum preventive treatments such as aspirin. The targets for lipids and blood pressure should follow guidelines and recommendations for higher-risk patients. The treatment of the atherogenic dyslipidemia in MetS is aimed at correcting the fundamental disturbances – namely reducing plasma triglycerides and raising plasma HDL cholesterol. Among currently available agents, the fibrates, which are PPAR-γ agonists, are best at normalizing MetS dyslipidemia. In contrast, statins target LDL cholesterol, but their role in vascular disease prevention is supported by a much wider evidence base. Many clinicians will first choose a statin as dyslipidemia therapy for MetS patients, especially when triglycerides are only mildly elevated. Other treatment alternatives include the judicious use of niacin preparations and also combinations of a fibrate, statin, and/or cholesterol absorption inhibitor, such as ezetimibe.

The management of associated hypertension follows conventional guidelines for therapy (with a preference toward angiotensin-converting enzyme inhibitors and angiotensin receptor blockers in patients who progress to type 2 diabetes).[45] The role of specific therapy for the insulin resistance associated with MetS (in the absence of frank diabetes mellitus) remains problematic. Specifically, the use of biguanides, thiazolidinediones, acarbose, and orlistat has been proposed to delay the onset of frank diabetes.[46] Their long-term effectiveness in reducing the atherosclerotic complications of MetS has yet to be established. There is an urgent need for prospective randomized studies for treatment strategies in patients with MetS.

References

1. Reaven, G. M. *Annu. Rev. Med.* **1993**, *44*, 121–131.
2. National Cholesterol Education Program (NCEP) Expert Panel. *JAMA* **2001**, *285*, 2486–2497.
3. Eckel, R. H.; Grundy, S. M.; Zimmet, P. Z. *Lancet* **2005**, *365*, 1415–1428.
4. Wahrenberg, H.; Hertel, K.; Leijonhufvud, B. M.; Persson, L. G.; Toft, E.; Arner, P. *Br. Med. J.* **2005**, *330*, 1363–1364.
5. Lofgren, I.; Herron, K.; Zern, T.; West, K.; Patalay, M.; Shachter, N. S.; Noo, S. I.; Fernandez, M. L. *J. Nutr.* **2004**, *134*, 1071–1076.
6. International Diabetes Federation. http://www. idf. org/webdata/docs/Metabolic_syndrome_rationale.pdf (accessed July 2006).
7. International Diabetes Federation. http://www. idf. org/webdata/docs/IDF_Metasyndrome_definition.pdf (accessed July 2006).
8. Bonora, E.; Kiechl, S.; Willeit, J.; Oberhollenzer, F.; Egger, G.; Bonadonna, R. C.; Muggeo, M. *Int. J. Obes. Relat. Metab. Disord.* **2003**, *27*, 1283–1289.
9. Ford, E. S.; Giles, W. H.; Dietz, W. H. *JAMA* **2002**, *287*, 356–359.
10. Kim, M. H.; Kim, M. K.; Choi, B. Y.; Shin, Y. J. *J. Korean Med. Sci.* **2004**, *19*, 195–201.
11. Razak, F.; Anand, S.; Vuksan, V.; Davis, B.; Jacobs, R.; Teo, K. K.; Yusuf, S. SHARE Investigators. *Int. J. Obes. Relat. Metab. Disord.* **2005**, *29*, 656–667.
12. Tan, C. E.; Ma, S.; Wai, D.; Chew, S. K.; Tai, E. S. *Diabetes Care* **2004**, *27*, 1182–1186.
13. Ford, E. S.; Giles, W. H.; Dietz, W. H. *JAMA* **2002**, *287*, 356–359.
14. Azizi, F.; Salehi, P.; Etemadi, A.; Zahedi-Asl, S. *Diabetes Res. Clin. Pract.* **2003**, *61*, 29–37.
15. Cuspidi, C.; Meani, S.; Fusi, V.; Severgnini, B.; Valerio, C.; Catini, E.; Leonetti, G.; Magrini, F.; Zanchetti, A. *J. Hypertens.* **2004**, *22*, 1991–1998.
16. Kelishadi, R.; Derakhshan, R.; Sabet, B.; Sarraf-Zadegan, N.; Kahbazi, M.; Sadri, G. H.; Tavasoli, A. A.; Heidari, S.; Khosravi, A.; Amani, A. et al. *Ann. Acad. Med. Singapore* **2005**, *34*, 243–249.
17. Rantala, A. O.; Kauma, H.; Lilja, M.; Savolainen, M. J.; Reunanen, A.; Kesaniemi, Y. A. *J. Intern. Med.* **1999**, *245*, 163–174.
18. Park, Y. W.; Zhu, S.; Palaniappan, L.; Heshka, S.; Carnethon, M. R.; Heymsfield, S. B. *Arch. Intern. Med.* **2003**, *163*, 395–397.
19. Weiss, R.; Dziura, J.; Burgert, T. S.; Tamberlane, W. V.; Taksali, S. E.; Yeckel, C. W.; Allen, K.; Lopes, M.; Savoye, M.; Morrison, J. et al. *N. Engl. J. Med.* **2004**, *350*, 2362–2374.
20. Laaksonen, D. E.; Lakka, H. M.; Niskanen, L. K.; Kaplan, G. A.; Salonen, J. T.; Lakka, T. A. *Am. J. Epidemiol.* **2002**, *156*, 1070–1077.
21. Lakka, H. M.; Laaksonen, D. E.; Lakka, T. A.; Niskanen, L. K.; Kumpusalo, E.; Tuomilehto, J.; Salonen, J. T. *JAMA* **2002**, *288*, 2709–2716.
22. Poirier, P.; Lemieux, I.; Mauriège, P.; Dewailly, E.; Blanchet, C.; Bergeron, J.; Després, J. P. *Hypertension* **2005**, *45*, 363–367.
23. Genest, J.; Frohlich, J.; Fodor, G.; McPherson, R. *Can. Med. Assoc. J.* **2003**, *169*, 921–924.
24. Feldman, R. D.; Bierbrier, G. S. *Lancet* **1993**, *342*, 707–709.
25. Baron, A. D.; Brechtel, G. *Am. J. Physiol.* **1993**, *265*, E61–E67.
26. Zeng, G.; Quon, M. J. *J. Clin. Invest.* **1996**, *98*, 894–898.
27. Baron, A. D.; Steinberg, H. O.; Chaker, H.; Leaming, R.; Johnson, A.; Brechtel, G. J. *Clin. Invest.* **1995**, *96*, 786–792.
28. Baron, A. D. *J. Investig. Med.* **1996**, *44*, 406–412.
29. Feldman, R. D.; Logan, A. G.; Schmidt, N. D. *Clin. Pharmacol. Ther.* **1996**, *60*, 444–451.
30. Feldman, R. D.; Schmidt, N. D. *J. Hypertens.* **2001**, *19*, 113–118.
31. Steinberg, H. O.; Baron, A. D. *Diabetologia* **2002**, *45*, 623–634.
32. Avogaro, A.; Miola, M.; Favaro, A.; Gottardo, L.; Pacini, G.; Manzato, E.; Zambon, S.; Sacerdoti, D.; de Kreutzenberg, S.; Piliego, T. et al. *Eur. J. Clin. Invest.* **2001**, *31*, 603–609.
33. Watanabe, Y.; Sunayama, S.; Shimada, K.; Sawano, M.; Hoshi, S.; Iwama, Y.; Mokuno, H.; Daida, H.; Yamaguchi, H. *J. Atheroscler. Thromb.* **2000**, *7*, 159–163.
34. Mather, K. J.; Verma, S.; Anderson, T. J. *J. Am. Coll. Cardiol.* **2001**, *37*, 1344–1345.

35. Hall, J. *Curr. Hypertens. Rep.* **2000**, *2*, 139–147.
36. Fornoni, A.; Raij, L. *Curr. Hypertens. Rep.* **2005**, 7, 88–95.
37. Hegele, R. A. *Int. J. Obes. Relat. Metab. Disord.* **2005**, *29*, S31–S35.
38. Hegele, R. A. *Trends Endocrinol. Metab.* **2003**, *14*, 371–377.
39. Hegele, R. A. *Trends Cardiovasc. Med.* **2004**, *14*, 133–137.
40. Harris, S. B.; Gittelsohn, J.; Hanley, A.; Barnie, A.; Wolever, T. M.; Gao, J.; Logan, A.; Zinman, B. *Diabetes Care* **1997**, *20*, 185–187.
41. Harris, S. B.; Zinman, B.; Hanley, A.; Gittelsohn, J; Hegele, R.; Connelly, P. W.; Shah, B.; Hux, J. E. *Diabetes Res. Clin. Pract.* **2002**, *55*, 165–173.
42. Pollex, R. L.; Hanley, A. J.; Zinman, B.; Harris, S. B.; Khan, H. M.; Hegele, R. A. *Atherosclerosis* **2005**, *184*, 121–129.
43. Hegele, R. A.; Kraw, M. E.; Ban, M. R.; Miskie, B. A.; Huff, M. W.; Cao, H. *Arterioscler. Thromb. Vasc. Biol.* **2003**, *23*, 111–116.
44. Hegele, R. A. *Circulation* **2001**, *103*, 2225–2229.
45. Isomaa, B.; Almgren, P.; Tuomi, T.; Forsén, B.; Lahti, K.; Nissén, M.; Taskinen, M. R.; Groop, L. *Diabetes Care* **2001**, *24*, 683–689.
46. Khan, N. A.; McAlister, F. A.; Lewanczuk, R. Z.; Touyz, R. M.; Padwal, R.; Rabkin, S. W.; Leiter, L. A.; Lebel, M.; Herbert, C.; Schiffrin, E. L. et al. *Can. J. Cardiol.* **2005**, *21*, 657–692.
47. Adult Treatment Panel III Guidelines. At-a-glance Quick Desk Reference US Department of Health and Human Services, Public Health Service, Bethesda, MD, 2001, NIH Publication No. 01-3305.

Biographies

Ross D Feldman, MD, FACP, FRCPC, is the RW Gunton Professor of Therapeutics, Departments of Medicine (and of Pharmacology and Physiology) at the University of Western Ontario. Since 2001, he has been the deputy scientific director of the Robarts Research Institute. He received his medical degree from Queen's University in Kingston, Ontario in 1977 and training in Internal Medicine (University of Toronto) and Clinical Pharmacology (Vanderbilt University). Following his postgraduate training, he has held teaching positions at the University of Iowa College of Medicine in Iowa City and then at the University of Western Ontario (since 1989). Among a number of awards and scholarships, he was the recipient of a Career Investigator Award (Heart and Stroke Foundation of Ontario), the George Morris Piersol Teaching and Research Scholarship (American College of Physicians), and the Burroughs-Wellcome Clinical Pharmacology Award. He is a member of editorial boards for several journals, including *American Journal of Physiology- (Cellular and Endocrine sections)*, *Clinical Pharmacology and Therapeutics*, and *Pharmacological Reviews*. He has been vice president of the American Society of Clinical Pharmacology and Therapeutics, Chair of the Clinical Pharmacology Division of the American Society of Pharmacology and Experimental Therapeutics, as well as President of the Canadian Hypertension Society. He just completed a 3-year term as Chair of the Steering Committee of the Canadian Hypertension Education Program which, since 1999 has produced and disseminated yearly updates of the *Canadian Recommendations for the Management of Hypertension*.

Robert A Hegele, MD, FACP, FRCPC, is the Jacob J Wolfe Chair in Functional Genomics and the Edith Schulich Vinet Canada Research Chair in Human Genetics in the Faculty of Medicine and Dentistry at the University of Western Ontario. He received his MD (Honours) from the University of Toronto in 1981, followed by an Internal Medicine residency and an Endocrinology fellowship at the University of Toronto. His postdoctoral fellowships in metabolism and human genetics were at the Rockefeller University in New York and the Howard Hughes Medical Institute in Salt Lake City. In 1989, he joined the University of Toronto faculty and was staff endocrinologist at St Michael's Hospital. In 1997, he joined the Robarts Research Institute and the University of Western Ontario, both in London, Ontario, and is currently Professor of Medicine and Biochemistry. He is the director of the Blackburn Cardiovascular Genetics Laboratory and the London Regional Genomics Centre. He directs a tertiary referral clinic for lipoprotein and metabolic disorders at the London Health Sciences Centre. He is a consulting editor for *Arteriosclerosis, Thrombosis and Vascular Biology* and is a member of the editorial boards of the *Journal of Lipid Research* and the *Journal of Human Genetics*. He previously served on the Advisory Board for the Institute for Aboriginal Peoples' Health of the Canadian Institutes for Health Research.

Comprehensive Medicinal Chemistry II
ISBN (set): 0-08-044513-6

ISBN (Volume 6) 0-08-044519-5; pp. 381–387

6.18 Obesity/Disorders of Energy

D L Nelson and D R Gehlert, Eli Lilly and Company, Indianapolis, IN, USA

6.18.1 Disease State

Obesity is a multifactorial, chronic disorder that has reached epidemic proportions in most industrial countries and is now threatening to become a global epidemic.[1] In addition, the number of overweight or obese children and adolescents has doubled in the past two to three decades in the US. It is likely that the trend will continue for the foreseeable future as overweight children and adolescents are more likely to become overweight or obese adults. While considered a cosmetic problem by a majority of the public, the association between obesity and increased morbidity and mortality has elicited concern from various sectors of the healthcare community. Obese patients are at higher risk for a number of chronic diseases, including coronary artery disease, hypertension, hyperlipidemia, diabetes mellitus, cancers, cerebrovascular accidents, osteoarthritis, restrictive pulmonary disease, and sleep apnea (**Table 1**).

In the US alone, it is estimated that obesity is responsible for approximately 300 000 deaths per year, and the direct and indirect costs of obesity exceed $100 billion per year. Obesity is commonly assessed using a weight parameter known as body mass index (BMI) and is calculated by dividing body weight in kg by height in m^2 (**Table 2**). Individuals are defined as being overweight when they have a BMI of 25.0–29.9 $kg\,m^{-2}$, while obese is defined as a BMI of 30 $kg\,m^{-2}$

Table 1 Medical complications associated with obesity

Cardiovascular	Hypertension, coronary heart disease, congestive heart failure, dysrhythmias, pulmonary hypertension, ischemic stroke, venous stasis, deep-vein thrombosis, pulmonary embolus
Endocrine/metabolic	Metabolic syndrome, insulin resistance, impaired glucose tolerance, type 2 diabetes mellitus, dyslipidemia, polycystic ovary syndrome
Gastrointestinal	Gallstones, pancreatitis, abdominal hernia, nonalcoholic fatty liver disease (NAFLD) (steatosis, steatohepatitis, cirrhosis), and possibly gastroesophageal reflux disease
Respiratory	Abnormal pulmonary function, obstructive sleep apnea, obesity, hypoventilation syndrome
Musculoskeletal	Osteoarthritis, gout, lower-back pain
Gynecologic	Abnormal menses, infertility
Genitourinary	Urinary stress incontinence
Ophthalmologic	Cataracts
Neurologic	Idiopathic intracranial hypertension (pseudotumor cerebri)
Cancer	Esophagus, colon, gallbladder, prostate, breast, uterus, cervix, kidney
Postoperative events	Atelectasis, pneumonia, deep-vein thrombosis, pulmonary embolus

Adapted from Klein, S.; Wadden, T.; Sugerman, H. J. *Gastroenterology* **2002**, *123*, 882–932.

Table 2 Relationship between BMI and comorbid disease risk

	Obesity class	*BMI ($kg\,m^{-2}$)*	*Comorbid disease risk*
Underweight		<18.5	Increased
Normal		18.5–24.9	Normal
Overweight		25.0–29.9	Increased
Obesity	I	30.0–34.9	High
	II	35.0–39.9	Very high
Extreme obesity	III	>40.0	Extremely high

Adapted from Klein, S.; Wadden, T.; Sugerman, H. J. *Gastroenterology* **2002**, *123*, 882–932.

or more. Of the estimated 97 million adults who have a BMI of $>25\,\mathrm{kg\,m^{-2}}$, 44.3 million are obese. In general, the higher the BMI, the greater the risk of adiposity-related diseases and premature mortality (**Table 2**). However, the BMI does not capture the complete story. Other factors, such as physical fitness, fat distribution, and the time of life that the weight gain occurred can affect the overall risk profile. In general, higher levels of physical fitness decrease the risk profile while excess upper-body fat increases the risk of cardiovascular and diabetic diseases. Because of the importance of fat distribution in determining risk, this is often captured as waist circumference or waist-to-hip ratio. Weight gain during adulthood (greater than 18 years old) also increases the cardiovascular and diabetic risks.

Obesity is also associated with the metabolic syndrome, which consists of insulin resistance or type 2 diabetes, hypertension, high triglycerides, high uremic acid levels, and low levels of high-density lipoprotein cholesterol. A recent report[35] outlines that the prevalence of the metabolic syndrome increases with age, affecting more than 40% of those older than 60 years. Based on age-adjusted estimates, about a quarter of the US population has the metabolic syndrome, representing approximately 47 million people. In the recently released third report[36] of the National Cholesterol Education Program Expert Panel on detection, evaluation, and treatment of high blood cholesterol in adults (adult treatment panel III), the metabolic syndrome is defined for the first time and its importance in the morbidity and mortality of cardiovascular disease is established. This report also highlights the importance of this syndrome as a new target of risk reduction therapy.

The root cause of obesity is the ingestion of more energy than is expended. Very small but chronic differences between energy intake and expenditure over long periods of time can lead to large increases in body fat. Ingestion of only 8 kcal of excess energy per day over a period of 30 years may lead to an increase of 10 kg in body weight. This is the average amount of weight American adults gain during the 30-year period from 25 to 55 years of age. The prevalence of obesity in the US increases progressively from 20 to 50 years of age, but begins to decline after 60–70. In all, approximately 61% (110 million) of adults in the US (10–74 years of age) are overweight or obese. Since 1960, the prevalence of overweight has increased slightly while the prevalence of obesity has more than doubled, from 12.8% to 27%. The prevalence of obesity is particularly high in many ethnic-minority women such as African-American, Mexican-American, Native American, Pacific Islander American, Puerto Rican, and Cuban-American women. In the UK and Europe, approximately 15% of men and 20% of women are obese. Obesity has also increased in South-East Asia, Japan, and China. In the adult population worldwide, it is estimated that there are more than 500 million overweight and greater than 250 million obese people. It is also likely that these numbers will continue to increase.

6.18.2 Genetics

Body size is determined by both genetic and environmental factors. Genetic background may explain 40% or more of the variance in body mass in humans. The genetics of obesity are complex and likely involve the interaction between multiple genes. Through a number of studies, over 250 genes, markers, and chromosomal regions have been identified and associated with human obesity. The clinical importance of these associations is still under investigation. In rare cases, monogenic causes of obesity have been identified in humans. These mutations include the genes for leptin, leptin receptor, prohormone convertase, proopiomelanocortin, melanocortin-4 receptor, and SIM1. While these rare mutations have not provided a solution to the obesity epidemic, they have been critical to enabling our understanding the etiology of the disorder.

It is highly unlikely that the increase in prevalence of obesity in the past 20 years can be attributed to genetic changes. It is much more likely that this is the result of environmental influences. The most likely cause of the obesity epidemic is a combination of the increase in energy intake coupled with a decline in physical activity. In recent years, there has been a greater availability of highly palatable and convenient food. In addition, more meals are eaten outside the home and serving sizes are generally larger. At the same time, a large portion of the population has decreased daily physical activity due to sedentary work and social activities as well as a proliferation of mass and individual motorized transportation. When placed in a modern lifestyle, individuals with susceptible genetic backgrounds are particularly predisposed to increased adiposity. Perhaps the best studied example is the Pima Indians. Those Pima Indians, who live in the Sierra Madre mountains of northern Mexico, eat a traditional diet with 15% of their intake as fat. They also labor at physically demanding occupations such as farming or working in a sawmill. On the other hand, the Pima Indians living in Arizona, US, are much more sedentary and eat a high-fat diet, consisting of 50% of their calories as fat that is provided by government surplus commodities. The latter environment has led to a dramatic increase in obesity and diabetes within this Indian population. Similar examples have been described in the aboriginal population of northern Australia and in natives of Papua New Guinea. Urbanization of these populations has also led to a dramatic increase in obesity, type 2 diabetes, and hypertriglyceridemia.

6.18.3 Disease Basis

For body weight to remain stable, food or energy intake must remain equivalent to energy expenditure. Daily total energy expenditure (TEE) is comprised of resting energy expenditure (REE) that constitutes approximately 70% of total energy expenditure. The remainder of TEE consists of the energy expenditure associated with the ingestion and absorption of food (approximately 10% of TEE) and the energy expended during physical activity that constitutes approximately 20% of TEE. The hypothalamus is generally considered to integrate signals that influence energy intake and energy expenditure. This brain region processes signals that arise during meals that indicate caloric quantity and quality to the brain. Collectively, these are known as satiety signals. In general, gastrointestinal peptides provide the signals that help coordinate digestion and signal satiety. Many of them also act on receptors found on sensory neurons innervating the gut so their signal can be conveyed to the brainstem and subsequently relayed to the hypothalamus. Some of these peptides are also believed to cross the blood–brain barrier and interact directly with brain centers that control feeding and metabolism. A second set of signals regulate the amount of fat stored throughout the body. These consist of polypeptide hormones that are released in direct proportion to body fat and interact with neurons located in the hypothalamus. The best-known of these adiposity signals are insulin and leptin. When these hormones are administered directly into the brain, animals eat less food and expend more energy. A third set of signals include a variety of neurotransmitters and modulators that are synthesized and released within the brain itself. These signals broadly regulate anabolic versus catabolic activity and, as such, constitute attractive drug targets. Examples of these are the monoamine neurotransmitters and neuropeptides such as neuropeptide Y (NPY), Agouti-related peptide (AGRP), and melanin-concentrating hormone (MCH) (*see* Section 6.18.7).

The hypothesis that obesity results from defects in energy metabolism has been pursued by a number of investigators. However, there are very few data that support this theory. In obesity, REE is typically greater than observed in lean individuals of the same height. This is believed to be primarily due to the energy requirements needed to support the increased lean and adipose tissues found in obese persons. Small, but potentially important, differences in the energy required for the digestion of food have been reported between obese and lean volunteers. In load-bearing activities, obese individuals expend more energy than lean because of the increased work required to carry the additional body weight. Overall, there is very little evidence that a defect in TEE is the root cause of obesity. However, it is still possible that inherent differences in energy expenditure contribute to the pathogenesis of human obesity. Several longitudinal studies have been performed, indicating that those with the lowest REE or TEE exhibited the greatest weight gain during subsequent years. On the other hand, the Baltimore Longitudinal Study did not find a relationship between initial REE and weight gain during a 10-year average follow-up of 775 men.[1]

Although weight gain always occurs when energy intake exceeds energy expenditure, the amount of weight that is gained after overfeeding may be genetically determined and certain individuals may be more resistant to weight gain than others. In a study of 12 monozygotic twin pairs, chronic overfeeding of 1000 kcal per day caused a variable increase in body weight gain.[1] However, the weight gain by one twin was similar to the weight gain by the other. Some of this difference in weight gain may be attributed to differences in the thermogenic properties in response to overfeeding. Another study examined 8 weeks of overfeeding and attributed body fat gain differences to differences in nonvolitional energy expenditure (fidgeting, etc.). Therefore, voluntary or nonvoluntary physical activities appear to be important components in determining weight gain induced by excess ingested energy.

One of the difficulties in controlling weight gain by reducing energy consumption is that this approach results in a decrease in REE that then contributes to weight regain. This is part of the normal metabolic adaptation to reduced energy consumption. The decline in REE only appears during dietary restriction and does not persist during weight maintenance. This phenomenon has led to the theory that body weight is predetermined so that weight loss reduces REE in an effort to return body weight to a predetermined setpoint. Since the changes in metabolic rate do not appear to persist into the maintenance phase after dietary restriction, it is unlikely that this contributes directly to the obesity epidemic but rather contributes to the difficulty to lose weight by dietary restriction.

6.18.4 Experimental Disease Models

There are a vast number of animal models that have been used in an effort to attempt to mimic food consumption and obesity in humans. Historically, rats and mice have been the predominant models of human energy homeostasis and obesity. Ideally, an animal model should closely mimic the behavior and physical traits observed in human disorder. In general, rats and mice share similar hormones and neurotransmitters that regulate feeding and metabolism, thus making them a suitable species for early preclinical research. Their relatively small size and body weight also make optimal use of precious peptides and small molecules that are being evaluated for their antiobesity effects. While there are many similarities in how animals control food intake, energy expenditure, adipose tissue physiology, and gastrointestinal function, there are many differences that limit the predictive validity of these models. For instance, rats have no gallbladder and are unable to store bile and therefore this changes the digestive process compared to humans. Also, there are many notable differences in hormonal regulation that need to be considered. For example, in rodents, leptin produced impressive reductions in food intake, increased metabolism, and reduced adiposity (*see* Section 6.18.7). In most human obese patients, leptin had very little effect on any of these parameters. With this in mind, these types of studies become important to facilitate understanding of the connectivity of animal research to human obesity.[2]

Many of the measures used to evaluate human obesity, such as BMI or waist-to-hip ratio, are not useful to evaluate small-animal obesity. However, recent advances have allowed the measurements of body composition in small animals using dual-energy x-ray absorptiometry (DEXA) scanning or nuclear magnetic resonance (NMR), allowing a sensitive measure of adiposity and lean mass in living animals. Using rats and mice, a number of laboratories have sought to mimic the western diet by using chow consisting of highly palatable, energy-dense foods which can also induce obesity in these species. In addition, the sedentary lifestyles that are also believed to be an important contribution to the obesity epidemic can be easily attained in rodents. What is not accurately mimicked are the heterogeneous genetics, lifestyles, and schedules of human obesity. Rats and mice are typically from controlled strains and specific breeders and are individually housed in small cages that preclude social interactions and robust physical activity. The cages are maintained in rooms controlled for temperature, humidity, environmental stress, and noise level and have a fixed light–dark cycle. A single type of food is usually freely available and the animals consume a large portion of their calories during the dark cycle. While, these conditions are optimal for experimental control, they cannot mimic many of the aspects of human feeding and energy expenditure.

Historically, genetic models of obesity were used that consisted of spontaneous mutations that resulted in an obese phenotype. While these animal models have provided interesting insight into the processes of feeding and metabolism that contribute to adiposity, recent developments have rendered many of these animal models less useful. For instance, the *ob/ob* mouse, *db/db* mouse, and the Zucker rat were all commonly used for proof of concept studies for antiobesity compounds. With the cloning of the leptin gene, these animals were found to have defects in either the leptin gene or the gene encoding the leptin receptors. Once leptin was found to have limited utility in human clinical trials (see below), the predictive validity of these animals came into question. More recently, investigators have turned to dietary-induced obesity in an effort to mimic the Western diet and its consequences.[3] Some strains of mice and rats are particularly susceptible to developing obesity when fed high-fat or combined high-fat and high-carbohydrate diets, such as the A/J and C57/BL mice and the Osborne–Mendel rats. Other strains of mice (SWR) and rats (S5B/Pl) are resistant to developing obesity when placed on similar diets. Based on dietary fat dose–response curves conducted in rodents, dietary fat needs to exceed 25% before obesity will develop. Rodents fed a high-fat diet develop moderate obesity with an increase in energy intake and insulin resistance. Therefore, there is considerable face validity to utilizing dietary-induced obesity as a model of the human disease.

Another aspect of the high-fat, high-carbohydrate diet contributes to its ability to induce obesity. Unlike protein and carbohydrate, fat stimulates excess energy intake by its high palatability and its inability to induce satiety. In human studies, periodic exposure to high-fat meals was sufficient to lead to excess energy consumption, resulting in increased adiposity. In overfeeding studies, overconsumption of a high-carbohydrate diet led to rapid adjustments in metabolism to compensate for the excess energy consumption. By comparison, metabolic compensation to overconsumption of a high-fat diet was relatively slow. This is particularly the case when the feed consists of both high-fat and high-sugar content. The reasons for this are likely related to the high palatability of high-carbohydrate, high-fat foods, resulting in overconsumption. In rats maintained on a high-fat diet for prolonged periods of time, switching to a lower-fat diet did not result in weight loss to return the animals to control levels. These results indicate that increased fat intake may be particularly important in inducing obesity while a reduction in dietary fat has a limited effect on weight loss.

Looking forward, it will continue to be important to study experimental therapies in rodent models of obesity. Given the many similarities in consumption of highly palatable foodstuffs in these animals, there is considerable face validity to studying feeding and obesity in these species. In addition, many of the aspects of hormonal regulation and metabolism in humans are seen in rodents. Only continued testing of compounds in these experimental models and, subsequently, in humans will allow us to understand the predictive validity. In addition, given the complexities of human feeding, it will become increasingly important to understand primate feeding, metabolism, and obesity for end-stage testing of promising new therapies.

6.18.5 Clinical Trial Issues

The path forward for the clinical development of antiobesity drugs is guided by the recommendations of the various governmental agencies involved in the approval of new medicinal agents. In the US the Food and Drug Administration (FDA) issued a draft guidance in 1996 to aid in constructing appropriate clinical trials to meet the requirements for the drug approval process.[30] The European Agency for the Evaluation of Medicinal Products, through its Committee for Proprietary Medicinal Products (CPMP), issued a similar guidance in 1997.[31] A primary feature underlying the guidance in these two documents is the concern for an adequate risk-to-benefit ratio for drugs used to treat obesity. This is a prime consideration because obesity is not usually considered a life-threatening condition (at least at the time of initiation of therapy), and because pharmacotherapy will likely involve long-term drug administration. Also, of concern to the FDA is data suggesting that in the US the majority of prescriptions written for obesity drugs are for women 18–44 years of age. **Table 3** summarizes some of the major recommendations from the FDA and CPMP documents. The FDA document provides some guidance as to the potential size of the efficacy trials while the CPMP does not. In reality, the number of subjects needed for a convincing phase III trial for registration may be significantly greater than that noted in the current guidance document. For the phase III studies (started in 2001) for the CB-1 receptor antagonist rimonabant, over 6600 patients were involved in the obesity trials according to information provided by Sanofi-Aventis, Inc.[32]

While BMI is the most commonly used measure to assess the degree of overweight, there are concerns about how well it defines adiposity among different body types, between the genders, and among different racial and age groups. Since BMI does not really measure body composition, obesity trials may include other measures. For example, accumulation of central or visceral fat might be a better index of the overall association of obesity with morbidity, and therefore, a measure of waist-to-hip ratio may be appropriate in addition to calculating BMI. Noninvasive methods of measuring body composition in humans are available, e.g., DEXA and hydrostatic weighing. However, these can add substantially to trial costs and may not be readily available at most clinical trial sites.

6.18.6 Current Treatment

6.18.6.1 History

In spite of the large literature on obesity (a PubMed search of the term in 2005 yielded over 78 000 references), the history of the pharmacologic treatment of obesity is relatively brief, and the types of agents used have been relatively few. In fact, in the US, as safety concerns grow, the list of compounds approved by the FDA for the treatment of obesity has actually been shrinking.

Perhaps the earliest pharmacologic treatment of obesity that had a scientific basis was the use of thyroid extract, which dates from the late 1800s.[4] Also of historical interest was the use of dinitrophenol, which came to be used after the observation in the early 1900s that textile workers lost weight after on-the-job exposure.[5] Needless to say, this uncoupler of oxidative phosphorylation was discontinued due to serious side effects.

Table 3 Summary of the FDA and CPMP clinical trials recommendations

	FDA	*CPMP*
Basic trial design to show efficacy	Randomized, double-blind, and placebo-controlled. Study subjects (both placebo and drug) should be given similar instructions in diet, exercise, behavior modification, and other lifestyle changes	Randomized, double-blind, and placebo-controlled
Study subjects	Moderately to markedly obese with a body mass index (BMI) of ≥30 for otherwise healthy individuals and BMI ≥27 for those with comorbid conditions	BMI of ≥30 for otherwise healthy males and females; BMI ≥27 for those with comorbid conditions
Definition of weight loss	1. Demonstration that the drug effect is significantly greater than the placebo effect and the mean drug-associated weight loss exceeds the mean placebo weight loss by at least 5% or: 2. Demonstration that the proportion of subjects who reach and maintain a loss of at least 5% of their initial body weight is significantly greater in subjects on drug than in those on placebo	'Demonstration of a significant degree of weight loss of at least 10% of baseline which is also statistically greater than that associated with placebo'

The majority of drugs that have been approved over the years for the treatment of obesity have been the so-called anorexiants (also called anorectics or anorexigenics), i.e., compounds that decrease food intake. Most of these drugs have been structural variants of phenylethylamine and produce their anorexic effects by modulating one or more of the monoamine neurotransmitter systems, serotonin (5-HT), norepinephrine (NE), and/or dopamine (DA). The earliest of these compounds to be associated with weight loss was amphetamine sulfate (Benzedrine).[4,5]

6.18.6.2 Antiobesity Drugs that Act through Central Nervous System (CNS) Mechanisms

6.18.6.2.1 Currently marketed compounds

Safety concerns that have arisen for antiobesity drugs in recent years have resulted in a steadily diminishing number of approved compounds. Currently, in the US seven anorexiant drugs are approved/marketed for the treatment of obesity. These are summarized in **Table 4**. The structure of the prototypical compound phenylethylamine is included for comparison.

Table 4 Chemical structures of approved anorectic drugs

Generic name	*Structure*	*Controlled substance class*	*US trade names*	*Year approved in US*
Phenylethylamine	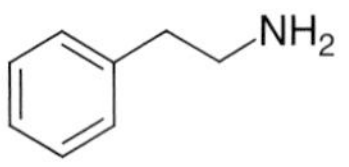 NH_2			
Amphetamine	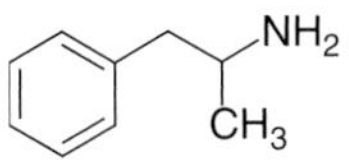 NH_2, CH_3	Schedule 2	Adderall	1939[a]
Methamphetamine	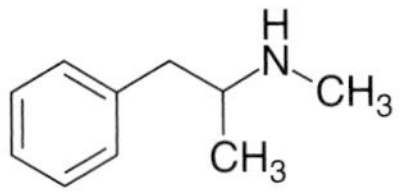 H, N, CH_3, CH_3	Schedule 2	Desoxyn	1943[a]
Phentermine	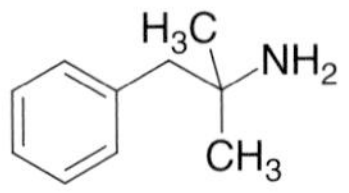 H_3C, NH_2, CH_3	Schedule 4	Pro-Fast Ionamin Adipex-P	1959
Benzphetamine	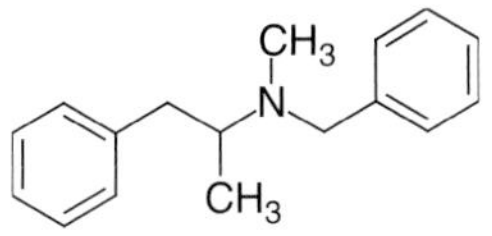 CH_3, N, CH_3	Schedule 3	Didrex	1960
Diethylpropion	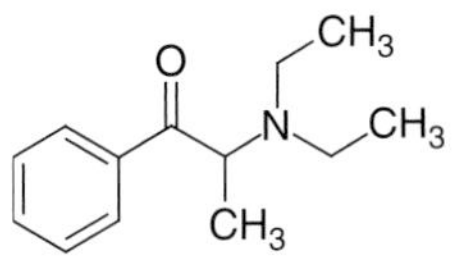 O, CH_3, N, CH_3, CH_3	Schedule 4	Tenuate	1959
Phendimetrazine	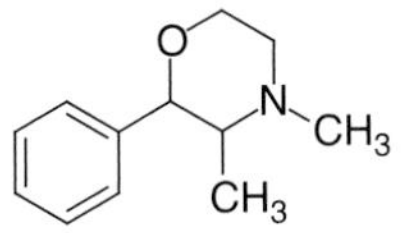 O, N, CH_3, CH_3	Schedule 3	Bontril Melfiat Prelu-2	1959
Sibutramine	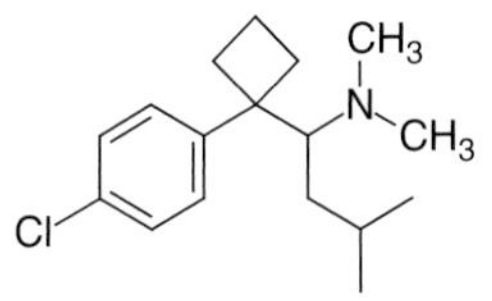 CH_3, N, CH_3, Cl	Schedule 4	Meridia	1997

[a] These compounds were originally approved for indications other than obesity.

6.18.6.2.2 Amphetamine and methamphetamine

6.18.6.2.2.1 Mechanism of action

The amphetamines are generally classified as CNS stimulants or sympathomimetics. This classification is consistent with their primary pharmacologic effects, i.e., the release of the catecholamines NE and DA and the inhibition of uptake of the catecholamines through the NE and DA plasma membrane transporters.[6] The effect of the amphetamines on appetite suppression is thought to be mediated in the CNS, with the lateral hypothalamus considered a likely target. However, other CNS areas may contribute to the overall effect.

6.18.6.2.2.2 Indications

Exogenous Obesity

Amphetamine and methamphetamine are approved as short-term adjuncts in a regimen of weight reduction based on caloric restriction for patients refractory to alternative therapy (e.g., repeated diets, group programs, other drugs).

Other

Amphetamine and methamphetamine are also approved for the treatment of narcolepsy and attention-deficit disorder with hyperactivity.

The history of amphetamines and effects on eating and weight loss go back to at least 1938 when Lesses and Myerson[4] showed that Benzedrine (amphetamine sulfate) was effective in the management of obesity.

6.18.6.2.2.3 Issues, side effects, and contraindications

From early on in the use of amphetamines it was recognized that the CNS-stimulant properties and the peripheral cardiovascular effects (e.g., tachycardia and increased blood pressure) were liabilities for the long-term treatment of disease. Also, tolerance to their anorectic effects can occur. Of even greater concern with the amphetamines is their abuse potential. Amphetamine and especially methamphetamine have a high propensity for abuse. There have been periods when large-scale abuse of the amphetamines has occurred. As such, these are now controlled substances with a class II labeling.

6.18.6.2.3 Phentermine

6.18.6.2.3.1 Mechanism of action

Like the amphetamines, phentermine is classed as a sympathomimetic and CNS stimulant. While phentermine has been reported to produce weight loss similar in magnitude to amphetamine,[4] it appears to have significantly less propensity for abuse than amphetamine. In vitro studies show that phentermine has lower potency than amphetamine to affect catecholamine release or to interact with catecholamine transporters.[6]

6.18.6.2.3.2 Indications

Phentermine is approved as a short-term (a few weeks) adjunct in a regimen of weight reduction based on exercise, behavioral modification, and caloric restriction in the management of exogenous obesity for patients with an initial BMI $\geq 30\,\mathrm{kg\,m^{-2}}$, or $\geq 27\,\mathrm{kg\,m^{-2}}$ in the presence of other risk factors (e.g., hypertension, diabetes, hyperlipidemia).

Phentermine received a great deal of attention as the result of a comprehensive study published in 1992 that was supported by the National Heart, Lung and Blood Institute of the National Institutes of Health.[7] This study, which combined phentermine with the drug fenfluramine, showed the clear benefit of this drug combination on significant long-term weight loss, and the drug combination became known as phen-fen. This spurred a tremendous increase in the prescribing of phentermine to be used in combination with fenfluramine until fenfluramine's removal from the market in 1997.

Clinical data on the potential long-term effectiveness of phentermine alone are sparse. Yet, in studies ranging from 6 to 9 months, phentermine appears to retain its efficacy and may actually have better efficacy than fenfluramine[9] over the treatment periods studied.

6.18.6.2.3.3 Issues, side effects, and contraindications

Generally, phentermine appears to be relatively well tolerated. It can produce side effects consistent with its catecholamine-releasing properties, e.g., tachycardia, increased heart rate, and increased alertness, but the incidence and magnitude of these appear to be less than with the amphetamines. There is no indication that phentermine has the abuse potential of the amphetamines.

6.18.6.2.4 **Benzphetamine**

6.18.6.2.4.1 Mechanism of action

As with related compounds, benzphetamine is classed as a sympathomimetic and CNS stimulant. Presumably, benzphetamine produces its effects through mechanisms similar to amphetamine, but a review of the literature failed to reveal studies that have actually demonstrated the mechanism of action of benzphetamine. The side-effect profile has been reported to be similar to phenmetrazine.[4]

6.18.6.2.4.2 Indications

Benzphetamine is approved as an adjunct in the short-term (a few weeks) treatment of obesity.

6.18.6.2.4.3 Issues, side effects, and contraindications

The side-effect profile has been reported to be similar to phenmetrazine,[4] and the general side effects and precautions to use are similar to those for amphetamine.[33] There is surprisingly little quantitative information about benzphetamine in the literature. Because of the typical concerns about sympathomimetic anorexiant drugs, e.g., rapid development of tolerance to the anorectic effects, CNS stimulation, and increased blood pressure, benzphetamine use appears to be very limited.

6.18.6.2.5 **Diethylpropion**

6.18.6.2.5.1 Mechanism of action

Diethylpropion is classed as a sympathomimetic and CNS stimulant. Over the years there has been some debate about the mechanism(s) by which it produces its effects. Yu *et al.*[10] have provided some insights into this by characterizing diethylproipion and its major metabolites for effects on monoamine neurotransmitter release and uptake. In humans diethylpropion is rapidly and extensively metabolized to produce two major metabolites (**Figure 1**).

Evaluation of diethylpropion in vitro on either the uptake or release of DA, NE, or 5-HT revealed no significant effect, prompting Yu *et al.*[10] to conclude that it might be acting as a prodrug. In contrast the N-dealkylated metabolite (M1 in **Figure 1**) was relatively potent at causing the release of NE, with weaker effects on DA. Yu *et al.*[10] concluded that M1 was about 10 times more potent at the NE transporter than at the DA transporter. Like the parent molecule, the reduced metabolite (M2) also had essentially no effect on either the uptake or release of the monoamine neurotransmitters. The relatively greater effect on NE systems than DA may explain the inability of diethylpropion to substitute for cocaine in clinical trials and its apparent lower abuse liability than the amphetamines.

6.18.6.2.5.2 Indications

Diethylpropion is approved for the short-term (a few weeks) management of obesity in patients with an initial BMI of $\geq 30\,\mathrm{kg\,m^{-2}}$ combined with a weight reduction regimen based on caloric restriction. These should be patients who have not responded to a regimen of diet and/or exercise alone.

6.18.6.2.5.3 Issues, side effects, and contraindications

Constraints in the use of diethylpropion should be consistent with the sympathomimetic class of compounds, e.g., avoiding use in patients with hypertension, relatively rapid tolerance to the anorectic effects, and CNS stimulation.

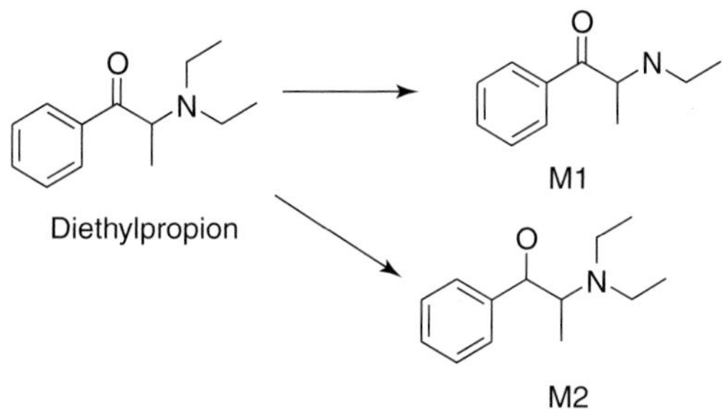

Figure 1 Dethylpropion and its major metabolites.

6.18.6.2.6 **Phendimetrazine**

6.18.6.2.6.1 Mechanism of action

Phendimetrazine is yet another sympathomimetic, CNS-stimulant type of anorexiant. After oral administration it shows significant metabolism to phenmetrazine (**Figure 2**). Differences in the in vivo effects of phendimetrazine and phenmetrazine, which depend on the route of administration, have led to the speculation that phendimetrazine might actually be a prodrug and that one of its metabolites is the active compound.[11]

Examination of phendimetrazine for effects on the uptake or release of DA, NE, and 5-HT revealed no interaction of the parent compound with either uptake or release of these monoamines.[11] Phendimetrazine is a racemic mixture of the *trans* configuration, producing both *trans* isomers of phenmetrazine when metabolized in vivo. Unlike the parent compound, both of these isomers affected both NE and DA release (**Table 5**).[11] In this regard the (+) isomer appeared to be the more potent isomer. Thus, it appears likely that biologic effects of phendimetrazine are due primarily to the effects of the metabolite phenmetrazine on central NE and DA release.

6.18.6.2.6.2 Indications

Phendimetrazine is approved as a short-term (a few weeks) adjunct for the treatment of obesity in a weight reduction plan based on caloric restriction.

6.18.6.2.6.3 Issues, side effects, and contraindications

The principal issues with phendimetrazine are the same as for the other CNS stimulants, e.g., tolerance to the anorectic effects, CNS stimulation/excitation, and increased blood pressure.

6.18.6.2.7 **Sibutramine**

6.18.6.2.7.1 Mechanism of action

Sibutramine is believed to produce its therapeutic effects through the inhibition of the reuptake of NE, 5-HT, and DA. Unlike the other anorectic agents discussed above, it does not cause the release of any of these monoamines. In vivo sibutramine is N-demethylated to form both desmethylsibutramine and didesmethylsibutramine (**Figure 3**).

While sibutramine has some effect on the reuptake of the monoamine neurotransmitters, most of the activity appears to reside in the metabolites. Sibutramine is a racemic mixture that produces enantiomers of both demethylated metabolites in vivo. Glick *et al.*[12] have examined the parent compound as well as these metabolites for their effects on monoamine uptake (**Table 6**). It appears likely that the pharmacologic effects of sibutramine in vivo are due to its metabolites, and that there is a clear selectivity between the isomers, with the *R*-isomer being the more potent.

Phendimetrazine → Phenmetrazine

*denotes chiral centers

Figure 2 Metabolism of phendimetrazine to phenmetrazine.

Table 5 Comparison of uptake inhibition and release by phendimetrazine and phenmetrazine

	Norepinephrine		*Dopamine*		*Serotonin*	
	Reuptake IC_{50} (nmol L^{-1})	*Release EC_{50} (nmol L^{-1})*	*Reuptake IC_{50} (nmol L^{-1})*	*Release EC_{50} (nmol L^{-1})*	*Reuptake IC_{50} (nmol L^{-1})*	*Release EC_{50} (nmol L^{-1})*
Phendimetrazine	8300±445	>10 000	19 000±537	>10 000	>100 000	>100 000
(+)-Phenmetrazine	240±24	37.5±4.3	359±23	87.4±7.8	>10 000	3246±263
(−)-Phenmetrazine	388±54	62.9±9.5	1669±189	415±45	>10 000	>10 000

Values are given as the mean±SD.

Reproduced from Rothman, R. B.; Katsnelson, M.; Vu, N.; Partilla, J. S.; Dersch, C. M.; Blough, B. E.; Baumann, M. H. *Eur. J. Pharmacol.* **2002**, *447*, 51–57.

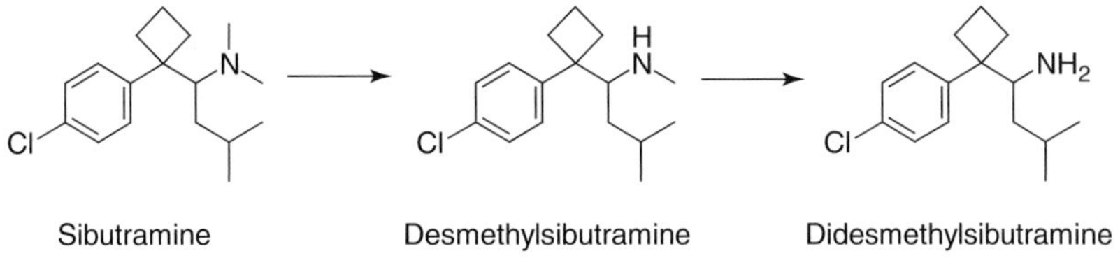

Sibutramine Desmethylsibutramine Didesmethylsibutramine

Figure 3 Sibutramine and its metabolites.

Table 6 Comparison of sibutramine and its metabolites on monoamine uptake in vitro

	Inhibition of uptake, IC_{50} ($nmol\,L^{-1}$)		
	Norepinephrine	*Dopamine*	*Serotonin*
(*RS*)-Sibutramine	350	1200	2800
(*R*)-Desmethylsibutramine	4	12	44
(*S*)-Desmethylsibutramine	870	180	9200
(*R*)-Didesmethylsibutramine	13	8.9	140
(*S*)-Didesmethylsibutramine	62	12	4300

Reproduced from Glick, S. D.; Haskew, R. E.; Maisonneuve, I. M.; Carlson, J. N.; Jerussi, T. P. *Eur. J. Pharmacol.* **2000**, *397*, 93–102.

Given that sibutramine is a monoamine uptake inhibitor, there has been much debate about which type of uptake inhibition is responsible for the effects on weight loss. The extensive clinical use of selective 5-HT uptake inhibitors, e.g., fluoxetine, paroxetine, and sertraline, has shown that this property alone is not sufficient to cause significant long-term weight loss.[13] The clinical use of selective NE uptake inhibitors, e.g., reboxetine and atomoxetine, has also not revealed significant long-term weight loss. Even dual 5-HT/NE uptake inhibition, e.g., duloxetine, has not been reported to produce significant weight reduction in humans. This leads to the conclusion that the DA uptake inhibition by the sibutramine metabolites, alone or in combination with 5-HT and/or NE uptake inhibition, may be a key contributor to the weight-reducing effects of sibutramine.

6.18.6.2.7.2 Indications

Sibutramine is approved for the management of obesity, including weight loss and maintenance of weight loss, and should be used in conjunction with a reduced-calorie diet. It is recommended for obese patients with an initial BMI ≥ 30 or $\geq 27\,kg\,m^{-2}$ in the presence of other risk factors (e.g., diabetes, dyslipidemia, controlled hypertension).

6.18.6.2.7.3 Issues, side effects, and contraindications

Sibutramine appears to be relatively well tolerated, with the most common side effects being dry mouth, insomnia, and constipation. The most significant concern with sibutramine is its ability to increase blood pressure and heart rate. While these changes are on average small, they can be quite significant in some patients, and thus, it is recommended that blood pressure and pulse should be measured before starting therapy and should be monitored at regular intervals thereafter. It is also recommended that sibutramine should not be used in patients with a history of coronary artery disease, congestive heart failure, arrhythmias, or stroke. Given that many obese patients have coexisting cardiovascular morbidities, these warnings preclude the use of sibutramine in such obese subjects.

6.18.6.2.8 Other anorexiant drugs

6.18.6.2.8.1 Fenfluramine

Although it is no longer on the market, fenfluramine had such an impact on the pharmacotherapy of obesity that it deserves mentioning when discussing the other anorexiants. Fenfluramine has been around since the 1960s, and D,L-fenfluramine was approved by the FDA for the short-term treatment of obesity in 1973. The use of fenfluramine got a significant boost with the publication of a series of papers in 1992[7] that demonstrated the magnitude and sustainability of weight loss that could be seen when fenfluramine was combined with phentermine (the combination that was known popularly as phen-fen), and in 1996, the FDA approved the D-enantiomer of fenfluramine for the long-term treatment of obesity. However, fenfluramine was withdrawn from the market in 1997 due to accumulating evidence that chronic use of fenfluramine could result in a significant incidence of valvulopathy.[13]

Fenfluramine is a substituted amphetamine (**Figure 4**), and like amphetamine, fenfluramine causes release of monoamines. Specifically, fenfluramine and its metabolite norfenfluramine cause release of 5-HT and NE (**Table 7**).[8] In addition, high-dose fenfluramine produces a long-lasting depletion of brain 5-HT in animal models.[14]

Fenfluramine is metabolized in vivo (in several different species, including humans) to norfenfluramine. Like the parent compound, norfenfluramine is also a monoamine neurotransmitter-releasing agent (**Table 7**). In addition to monoamine neurotransmitter release, norfenfluramine is a potent agonist at the 5-HT_2 family of receptors. The current literature is consistent with the anorectic effect of fenfluramine primarily due to norfenfluramine stimulation of central 5-HT_{2C} receptors.[15]

The combination of fenfluramine plus phentermine appeared to produce greater efficacy in treating human obesity, suggesting that using a combination of pharmacologies may be a better overall approach to treating obesity than the use of single agents. Given that the principal pharmacology of fenfluramine is 5-HT release and that of phentermine is NE release (**Table 7**), it is tempting to suggest that it is the combination of these two pharmacologies that gave phen-fen such good clinical results. However, as noted above, the metabolite of fenfluramine, norfenfluramine, is a potent agonist of the 5-HT_{2C} receptor and appears to be responsible for much of the anorectic effect of fenfluramine through this mechanism. Also, norfenfluramine can also cause release of NE, although it is less potent than phentermine in doing this. So, the exact mechanisms by which fenfluramine and phentermine produce their combined efficacy remain to be determined.

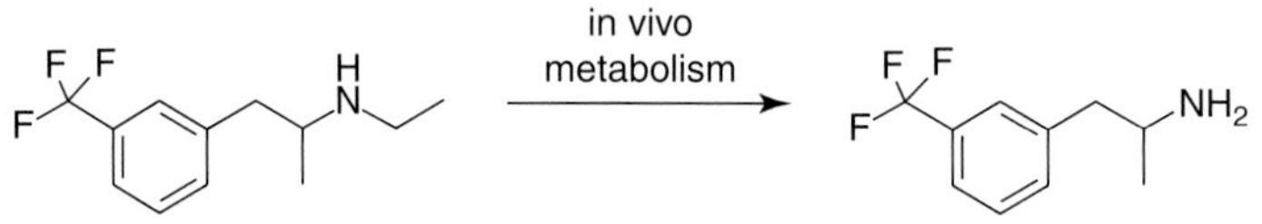

in vivo metabolism

Fenfluramine

Norfenfluramine

Figure 4 Metabolism of fenfluramine to norfenfluramine.

Table 7 In vitro monoamine-releasing effects of simple amphetamines

	NE release, EC_{50} (nmol L^{-1})	*DA release, EC_{50} (nmol L^{-1})*	*5-HT release, EC_{50} (nmol L^{-1})*
(+)Amphetamine	7.07±0.95	24.8±3.5	1765±94
(+)Methamphetamine	12.3±0.7	24.5±2.1	736±45
(+)Fenfluramine	302±20	>10 000	51.7±6.1
(–)Fenfluramine	>10 000	>10 000	147±19
(+)Norfenfluramine	72.7±5.4	924±112	59.3±2.4
(–)Norfenfluramine	474±40	>10 000	287±14
Phentermine	39.4±6.6	262±21	3511±253
Diethylpropion	>10 000	>10 000	>10 000
N-dealkylated diethylpropion	99.3±6.6	>1000	2118±98
Phendimetrazine	>10 000	>10 000	>100 000
(+)Phenmetrazine	37.5±4.3	87.4±7.8	3246±263
(–)Phenmetrazine	62.9±9.5	415±45	>10 000

Data taken from [6,8,10,11].

6.18.6.2.8.2 Ephedrine and ephedrine/caffeine combinations

Another pharmacologic approach to inducing weight loss has been through the use of the sympathomimetic ephedrine alone or in combination with caffeine. This was done primarily through the use of dietary supplements, i.e., the use of plant materials that naturally contain ephedrine (and related compounds), so that the product could be marketed without the need of a prescription. This was a very popular over-the-counter medication, and the plant-derived ephedrine/caffeine combination was often referred to as herbal phen-fen. While much has been written about the effects of ephedrine and the ephedrine/caffeine combinations on obesity, it is difficult to dissect the mechanisms of the effects from the lore surrounding the use of these agents.

Ephedrine and related compounds are natural constituents of a number of different plant species, especially of the family Ephedraceae (e.g., *Ephedra sinica, E. equisetina*, and *E. gerardiana*). The plants of this group are often collectively referred to as ephedra or ma huang (mahuang). Ephedrine can also be made synthetically. Since it contains two chiral centers, four stereoisomers are possible (**Figure 5**). The literature has described ephedrine as both a direct-acting (i.e., directly interacting with adrenoceptor) and an indirect-acting (i.e., NE-releasing) sympathomimetic. However, a comprehensive evaluation of ephedrine and its enantiomers suggests that there is little, if any, direct effect on adrenergic receptors and that the most likely mechanism of action is through promoting the release of NE.[16]

A number of small studies looking at the effects of ephedrine/caffeine combinations or herbal combinations containing ephedrine and caffeine have concluded that they are efficacious in reducing weight.[13] However, it is not clear that the magnitude of this effect is different than that described for the other anorectic agents that act by releasing NE.

A milestone in the use of ephedrine to promote weight loss was the final ruling of the FDA in 2004 prohibiting the sale of dietary supplements containing ephedrine alkaloids (ephedra). In issuing this ruling the FDA concluded that "dietary supplements containing ephedrine alkaloids pose a risk of serious adverse events, including heart attack, stroke, and death, and that these risks are unreasonable in light of any benefits that may result from the use of these products."[34]

While the sale of dietary supplements containing ephedrine alkaloids was banned, ephedrine itself is still approved as a prescription medication for certain indications. Approved indications include use as a vasopressor in shock, as a bronchodilator to treat asthma/bronchospasm, and as a decongestant for the relief of nasal congestion due to cold, hayfever, rhinitis, or sinusitis. It should be noted that ephedrine is not approved for the treatment of obesity.

Caffeine (**Figure 6**) is a plant-derived material, and it is generally categorized as a CNS stimulant. A broad segment of society is exposed to caffeine via its natural occurrence in tea and coffee and through its addition to many other beverages. Caffeine appears to produce its effects primarily by acting as an antagonist at both adenosine A_1 and A_{2A} receptors. By itself, caffeine seems to have little effect on body weight. It is claimed that caffeine facilitates the weight-reducing effects of ephedrine; however, the rigor with which this has been demonstrated can be questioned.

Ephedrine, either alone or in combination with caffeine, has often been described as thermogenic, i.e., having the ability to increase energy expenditure, and this has often been used to explain the weight-reducing effects of ephedrine/caffeine. While it is clear that this combination can produce small increases in energy expenditure, it is not clear that this is significantly different from other sympathomimetic compounds or that this contributes significantly to the overall effects of these compounds on weight reduction.

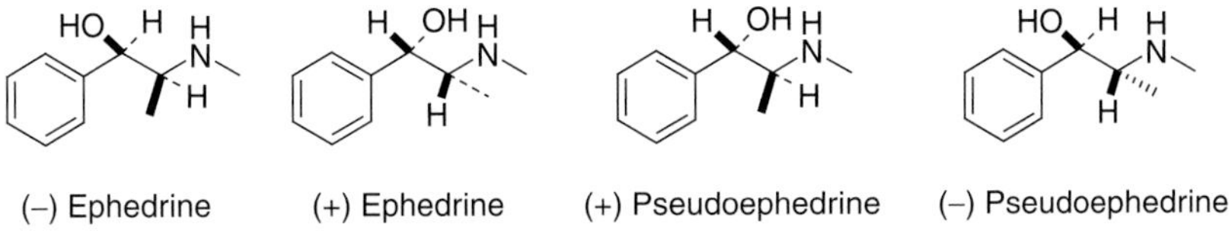

(−) Ephedrine (+) Ephedrine (+) Pseudoephedrine (−) Pseudoephedrine

Figure 5 The stereoisomers of ephedrine.

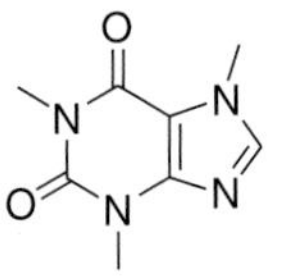

Figure 6 Structure of caffeine.

6.18.6.3 Antiobesity Drugs that Act through Peripheral Mechanisms

6.18.6.3.1 Orlistat

6.18.6.3.1.1 Mechanism of action

Orlistat (Xenical: **Figure 7**) has a unique mechanism of action among agents approved for the treatment of obesity and was approved in the US in 1999 and in the European Union in 2003. Most dietary fat is in the form of triglycerides, i.e., three fatty acids covalently bound to a glycerol backbone. To be absorbed into the body the triglycerides are broken down into monoglycerides and free fatty acids by intestinal lipases (**Figure 8**). Orlistat is a reversible inhibitor of these lipases, and the unmetabolized triglycerides are passed on through the bowel and eliminated. About 30% of dietary fat absorption is inhibited by a dose of 120 mg given three times daily.

6.18.6.3.1.2 Approved indications

Orlistat is indicated for obesity management, including weight loss and weight maintenance when used in conjuction with a reduced-calorie diet. It is also indicated to reduce the risk of weight regain after prior weight loss. Orlistat is indicated for obese patients with an initial BMI of ≥ 30 or $\geq 27\,\mathrm{kg\,m^{-2}}$ in the presence of other risk factors (e.g., hypertension, diabetes, dyslipidemia).

6.18.6.3.1.3 Side effects, warnings, and contraindications

Since orlistat is poorly absorbed from the gastrointestinal tract, systemic exposure is minimal. Thus, side effects are primarily confined to actions in the gut, and are primarily related to the mechanism of action, i.e., prevention of absorption of fats so that they are passed out of the body through the intestine. The nature, frequency, and extent of these side effects may be influenced by the fat content of the patient's diet. The commonly observed side effects seen with orlistat include oily spotting from the rectum, flatus with discharge, fecal urgency, fatty/oily stool, oily evacuation, increased defecation, and fecal incontinence.

6.18.6.4 Concluding Remarks Regarding Currently Available Pharmacotherapies for Obesity

The overall performance of the currently approved medicines for treating obesity is much less than desired, both from side-effect profiles and efficacy measures.[5,13] Because of the side-effect issues, only two agents are currently approved by the FDA for long-term treatment. Sibutramine's use is limited because of patient, physician, and regulatory

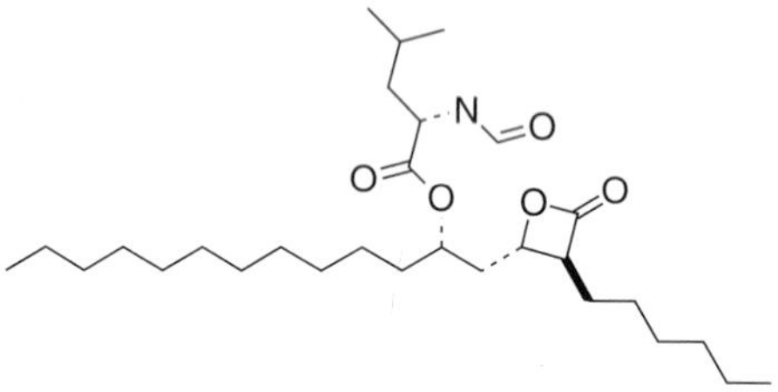

Figure 7 The chemical orlistat.

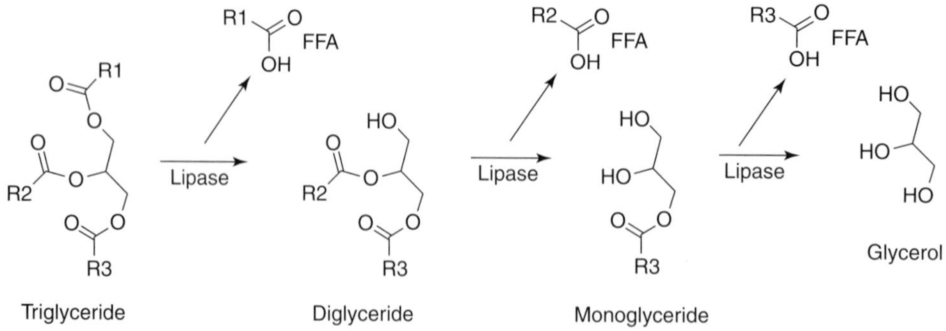

Figure 8 Conversion of triglyceride to free fatty acids.

concerns about potential cardiovascular side effects and orlistat's use is limited because of patient unhappiness with the gastrointestinal side effects. Evaluation of overall weight loss with these agents is somewhat difficult to assess because of differences in study design, in placebo response, and in study length. In general, it appears that sustained weight loss of 5–10% of starting weight is about the average that can be seen with these drugs. It also appears that better results can be achieved when drug therapy is accompanied by a complete weight management program.

6.18.7 Unmet Medical Needs

As noted in Section 6.18.1, obesity has reached epidemic proportions in the developed nations of the world. From what is known about the genesis of obesity it is clear that a combination of dietary caloric restriction and exercise is the safest and surest way to achieve and maintain a healthy body weight. It is also clear that most overweight individuals cannot maintain a diet and exercise regimen that will allow them to achieve a healthy body weight in the face of a calorie-rich environment and a society that promotes a sedentary lifestyle.

Thus, pharmacotherapy would seem to be in an excellent position to aid in the battle to achieve a healthy weight. However, as pointed out in previous sections of this chapter, currently available pharmacotherapies for obesity leave large unmet medical needs. The following summarizes some of the key issues or needs regarding the unmet needs for the current pharmacotherapy for obesity.

6.18.7.1 Safety/Tolerability

Safety is a paramount concern for the pharmacologic treatment of obesity for a variety of reasons. First, it is generally agreed that obesity is typically not acutely life-threatening, that the health concerns are cumulative over time. Second, a large proportion of the treated population will be women of child-bearing age. Third, there is the potential of many patients taking pharmacotherapy for cosmetic obesity treatment rather than for medically necessary treatment. Fourth, given the large percentage of the population that is obese or overweight, the numbers of individuals that might take a compound is potentially very large, so that even a low incidence of a serious side effect will result in a large number of affected individuals. Given these concerns, an antiobesity agent has to be very safe to provide an acceptable benefit-to-risk ratio. Tolerability for antiobesity agents has to be high in order to achieve good patient compliance. If patients cannot maintain an appropriate diet and level of exercise to manage weight, it is unrealistic to expect them to comply with long-term pharmacotherapy that has any significant level of unpleasant side effects. A current example of this can be found with the lipase inhibitor orlistat. Patients do not like having to watch the amount of fat in their diets to minimize the unpleasant side effects of having unabsorbed fat pass through the intestines. A key component of safety and tolerability is that these have to be maintained over very long periods of time, i.e., periods measured in years of treatment.

6.18.7.2 Efficacy

There is a clear need for pharmacotherapies that have better overall efficacy than currently available agents. While there is continuing debate about what the minimal efficacy should be, there seems be a somewhat general consensus that pharmacotherapy should produce at least a 10% reduction in body weight in excess of that produced by placebo. Ideally, this weight loss should occur at a rate that keeps the patient motivated to continue therapy and this minimal (or better) effect should occur in a large proportion of patients taking the compound. Coupled to this is the need for durability or sustainability of effect. Once a stable weight has been achieved, it is desirable that this plateau be maintained for at least a year and preferably longer. One of the issues with the current therapies targeting central NE systems is that with time the antiobesity effect often begins to wane.

6.18.7.3 Multiple Mechanisms

As noted previously, the currently available agents that focus on neuronal systems are less than ideal because the magnitude of weight loss is generally small and the effect often diminishes with chronic treatment. Humans are programmed to eat, since survival of the species is dependent on an adequate nutritional state that supports individual survival and procreation. As such, it is not surprising that we would have multiple redundant systems that keep us eating. From an evolutionary standpoint, it would be undesirable if interfering with any one system could prevent maintenance of that adequate nutritional state, and it might be expected that chronic interference with one system would result in other components gradually adapting to minimize that effect. Thus, it is likely that adequate pharmacotherapy of obesity will require either individual molecules that have multiple pharmacologies or molecules with a single mechanism that can be combined with other molecules having different mechanisms of antiobesity action. This is analogous to other successful therapies that target neuronal systems. For example, the most efficacious treatments for schizophrenia or depression are those that target multiple different neuronal systems or processes. The success of the combination of phentermine with fenfluramine also argues for the concept of the polypharmacology approach.

6.18.7.4 Patient Selection

This is an area of unmet medical need that is certainly not unique to the treatment of obesity, and it affects each of the other areas of need described above. It is clear that genetic differences among individuals can affect the safety, tolerability, and efficacy of all drug therapies. Because of the strict requirements for a large benefit-to-risk ratio for antiobesity drugs, it will be especially important to understand genetic differences that affect antiobesity efficacy so that compounds (or mechanisms) can be selected for individual patients that have the highest likelihood of producing successful weight loss.

6.18.8 New Research Areas

6.18.8.1 Leptin

The discovery of the peptide hormone leptin was a major advance in obesity research.[17] Leptin is a 148-amino-acid protein that is primarily secreted from adipose tissue in proportion to fat mass. Structurally, it belongs to the type 1 cytokine superfamily and is characterized by a long-chain four-helical bundle structure similar to growth hormone (GH), prolactin, and interleukin-3. The leptin receptor (Ob-R) was originally cloned from mouse choroid plexus and is alternatively spliced, giving rise to six different forms of the receptor, known as Ob-Ra, Ob-Rb, Ob-Rc, Ob-Rd, Ob-Re, and Ob-Rf. Ob-R is a member of the class 1 cytokine receptor superfamily. Ob-Rb is expressed at high levels in the hypothalamus and is believed to mediate the central effects of peripherally secreted leptin. Natural mutation of the leptin gene is found in genetically obese (*ob/ob*) mice, while the gene for the receptor is mutated in fatty (*fa/fa*) rats and in diabetic (*db/db*) mice. Chronic administration of leptin to rodents produced impressive reductions in food consumption, body weight, and adiposity. Given these data, the obesity research community had high expectations for leptin as a therapeutic agent for obesity.

In clinical studies, it was found that obese patients had raised plasma levels of leptin when compared to normal-weight controls. Furthermore, increased plasma levels of leptin were associated with increases in adiposity. In phase II clinical studies, administration of leptin or leptin analogs did not reduce food consumption or body weight in a broad segment of the population.[18] However, in select individuals with mutations in the leptin sequence, leptin did reduce food consumption and body weight. While Amgen has discontinued the commercial development of leptin, this is still a very active area of research and a number of alternative approaches are being attempted to understand regulation of body weight by the leptin pathway.

6.18.8.2 Cannabinoid Receptor Antagonists

The ability of marijuana to stimulate appetite in humans has been well known for centuries.[19] Administration of cannabinoids, the active ingredients in marijuana, has been established to stimulate food intake in animal models of feeding. Central and peripheral administration of anandamide (**Figure 9**), one of the major endocannabinoids, also increases food intake in preclinical models. The cannabinoid receptor 1 (CB1) is present in brain regions known to control food intake. Cannabinoid-induced feeding can be antagonized by the use of CB1 antagonists and starvation-induced feeding can be reduced by the administration of CB1 antagonists. Interestingly, CB1 knockout mice are leaner than wild-type controls primarily due to decreased food intake and lipogenesis in white fat during early postnatal development. In older animals, increased peripheral energy expenditure appears to be the predominant defense against increased adiposity in the knockout animals. The CB1 receptor is expressed within the key hypothalamic peptidergic systems that regulate appetite, including neurons containing corticotropin-releasing factor (CRF) in the paraventricular nucleus, cocaine- and amphetamine-related transcript (CART) in the arcuate nucleus, and MCH and orexin in the lateral hypothalamus and perifornical regions. Therefore, it is likely that endocannabinoids, through the CB1 receptor, influence these peptidergic systems to regulate both food intake and peripheral storage of energy.

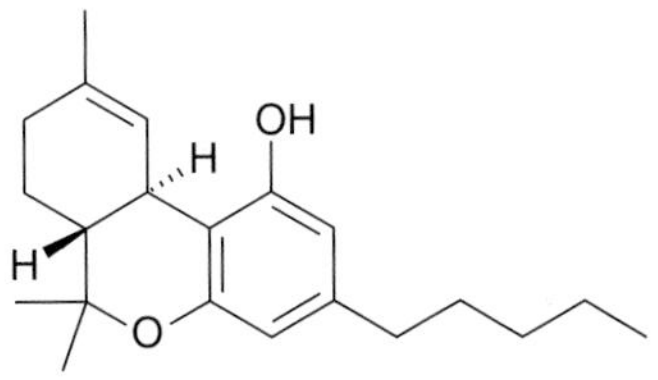

Figure 9 Chemical Δ-9-tetrahydrocannabidiol, anandamide, and the CB-1 antagonist, rimonabant.

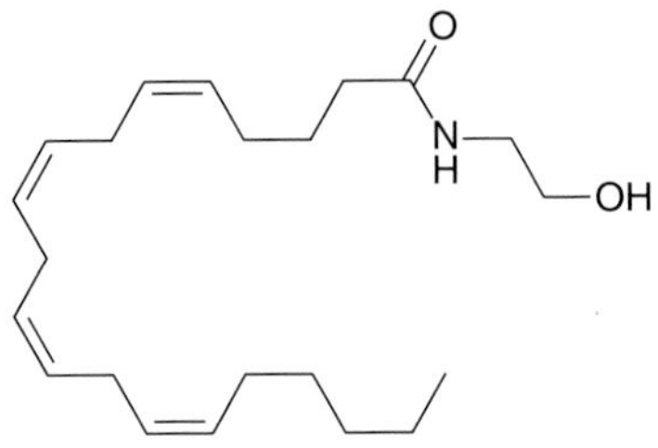

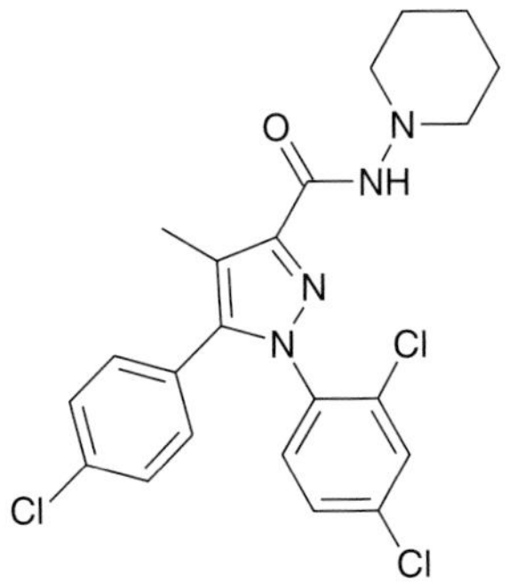

Rimonabant (SR141716, **Figure 9**) is a potent and selective CB1 antagonist that is centrally active. Rimonabant has been used by a number of research groups to investigate the role of CB1 receptors in food consumption and body weight maintenance. This compound antagonizes the increase in food consumption produced by both anadamide and Δ-9-tetrahycannabinol (**Figure 9**), the major active component of marijuana. In addition to antagonizing the orexigenic effects of cannabinoids, rimonabant produces changes in ingestive behaviors when administered alone. It primarily reduces the consumption of sweet, palatable food without much of an effect on a normal bland diet. This suggests that CB1 receptors may also play an important role in the reward aspects of feeding. During a 5-week study of rimonabant in diet-induced obese mice, there was a transient decrease in food intake during week 1 but a sustained reduction in body weight and adiposity. At the end of the 5-week treatment period, the insulin resistance observed initially in these animals was corrected and decreases were observed in plasma leptin, insulin, and free fatty acid levels. Rimonabant also produces a greater weight loss in 24 h fasting, indicating that it affects metabolic rate as well.[19]

Recently, this CB1 antagonist has entered late-stage clinical trials for the treatment of obesity. In a multicenter clinical trial, rimonabant was tested in 287 male and female subjects aged 18–65 years old with a BMI of 29–41. Subjects received either placebo or 5, 10, or 20 mg rimonabant while on a slightly hypocaloric diet. After 16 weeks of treatment, patients receiving the drug lost 2–4 times more weight (2–4 kg or 5–8 lb) than patients on placebo. In addition, the waist circumference among patients taking rimonabant decreased more than twice as much than it did for those on placebo. While there were a number of improvements in surrogate markers, no serious side effects were reported in the study. Rimonabant has been tested in phase III clinical trials involving more than 6000 obese subjects in the US and Europe. Rimonabant has been approved for use in Europe and is awaiting approval in the US.

6.18.8.3 β_3-Adrenoceptor Agonists

β_3-Adrenoreceptors were cloned in 1989 and found to be responsible for the thermogenic activities of β-agonists in brown adipose tissue.[18] In obese rodents, β_3-agonists (**Figure 10**) produce weight loss due to increased energy expenditure. Body composition improved due to fat loss, with some increase in lean mass. Lean animals have less of a response to the compound due to the decreased fat mass in these animals. At doses lower than required to elicit the antiobesity effects, β_3-agonists also improved insulin sensitivity in obese and insulin-resistant rodents. Initial clinical results with this class of compounds have produced mixed effects. A number of the initial compounds that were explored clinically produced significant tremors, limiting their clinical utility. CL-316243 increased fat oxidation and improved insulin-mediated glucose uptake in healthy lean subjects. However, the development of the compound was limited by poor bioavailability.

One of the reasons evoked for the poor efficacy seen in many of the early trials was that, in humans, the β_3-receptor is primarily expressed in brown adipose tissue. Human brown adipose tissue is primarily expressed during infancy and adult humans exhibit very little. Only low levels of β_3-receptor expression have been detected in human skeletal muscle and white adipose tissue. In addition, species differences in the pharmacological selectivity of β_3-receptor agonists have been observed between humans and rodents. Therefore, compounds that looked to be highly selective in rodent studies had substantial affinity and efficacy for human β_1- and β_2-adrenoceptors. Using the human receptors, a number of pharmaceutical companies are working on improved versions of the first-generation compounds. Several of these compounds are shown in **Figure 10**. LY377604 was reported to produce an 18% increase in metabolic rate at the highest dose tested in normal and obese subjects.[18] These data indicate that this class of compounds may produce meaningful increases in energy expenditure in humans. Future human studies will need to address the long-term effects of these compounds and whether energy intake will compensate for the increase in metabolic rate.

6.18.8.4 Neuropeptide Y Antagonists and Agonists

NPY is one of most abundant peptides in brain and is particularly concentrated in the hypothalamus.[20] It is a member of a peptide family that includes the endocrine peptides peptide YY (PYY) and pancreatic polypeptide (PP). NPY is found in neuronal cell bodies within the arcuate nucleus of the hypothalamus, where it is coexpressed with AGRP and regulated by leptin. When NPY, PYY, and certain analogs were administered centrally to rodents, a strong feeding response was observed. Repeated central administration of this peptide produces sustained increases in feeding, increased adiposity, and decreased metabolism. Using either antisense oligonucleotides or immunoneutralization,

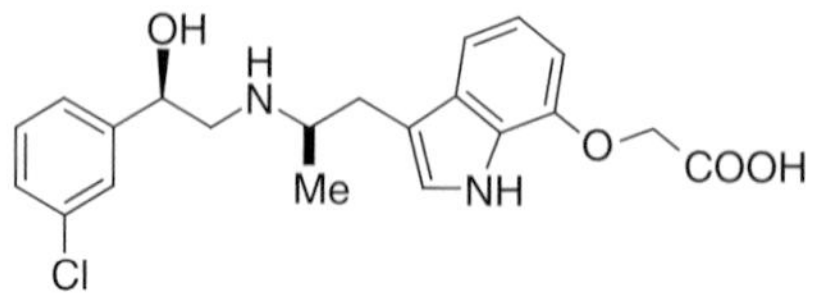

AD-9677

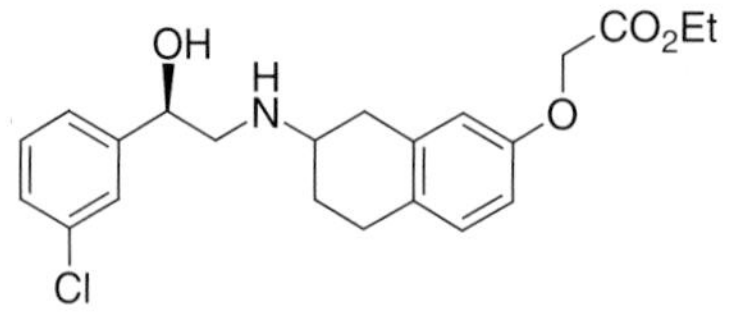

SR-58611

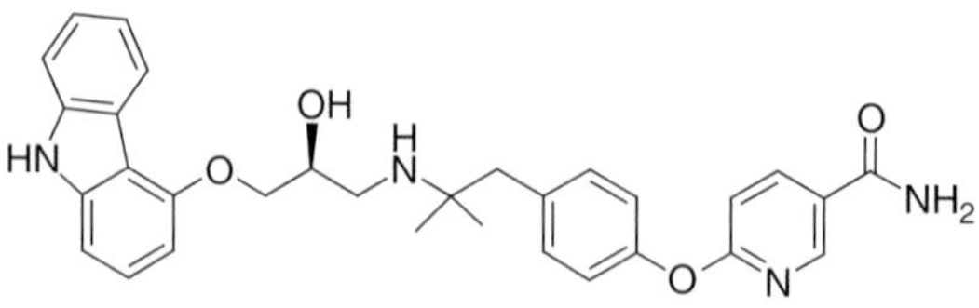

LY377604

Figure 10 β_3-Adrenoceptor agonists.

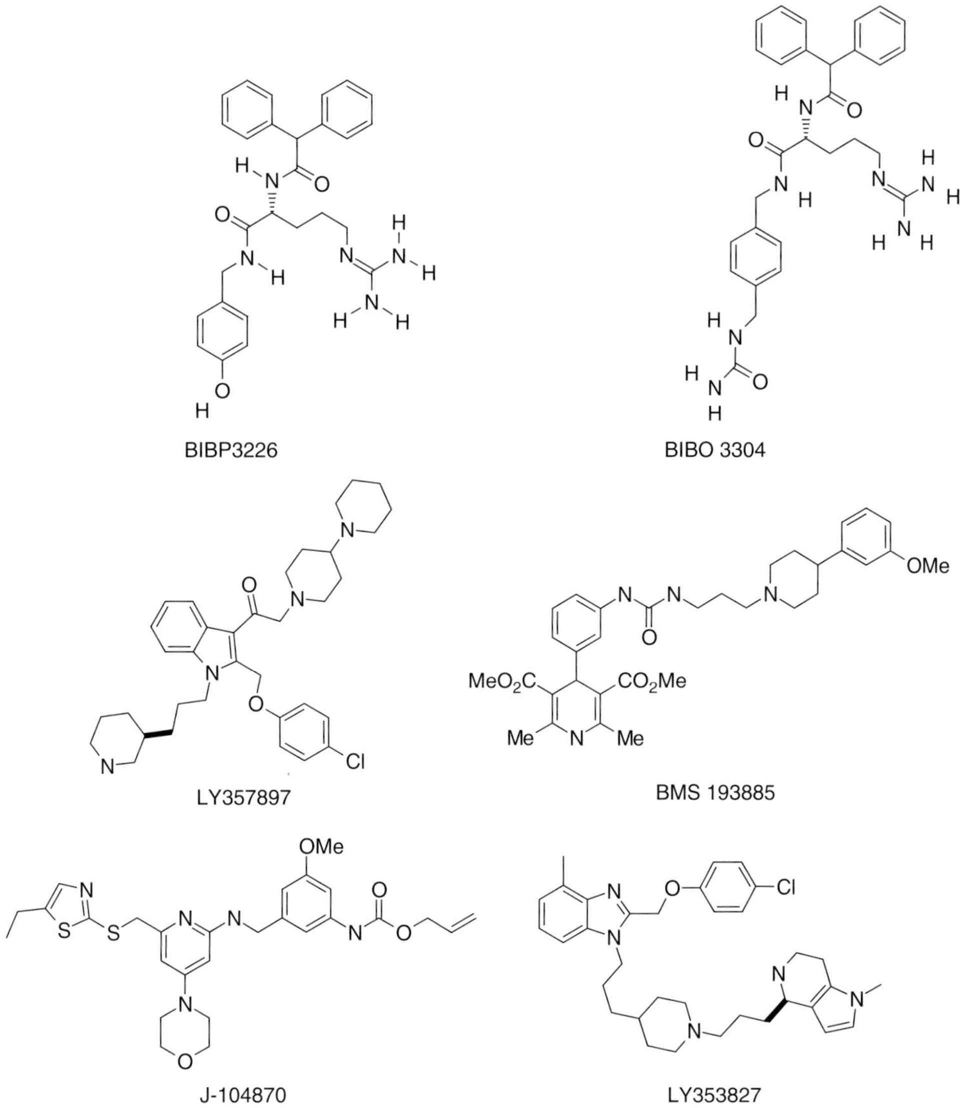

BIBP3226

BIBO 3304

LY357897

BMS 193885

J-104870

LY353827

Figure 11 NPY Y1 antagonists.

central NPY levels can be reduced, leading to a decrease in food intake. Four G protein-coupled receptor (GPCRs: Y1, Y2, Y4, and Y5) have been identified in humans that bind NPY, PYY, and PP with varying affinities. Strangely, mice that lack the Y1, Y2, or Y5 receptors all develop a mild, late-onset obesity. Y1 and Y5 knockout mice exhibited a reduced orexigenic response to centrally administered NPY. The paradoxical effects observed in the knockout mice have confounded target validation for this area of research. However, several major pharmaceutical companies are still pursuing the discovery and development of selective Y-receptor antagonists for the treatment of obesity. Most efforts have targeted the Y1 or Y5 receptors and representative chemical structures are shown in **Figures 11** and **12**.

A fragment of PYY has also been proposed to regulate feeding.[20] The major circulating form of this peptide is PYY3–36, which is a preferential ligand for the presynaptic Y2 receptor. PYY3–36 is released from the gastrointestinal tract following a meal in proportion to the energy content of the meal. In human obese subjects, the post-prandial PYY3–36 release is diminished when compared to lean subjects, even though the obese subjects consumed a greater number of calories. Recently, it was demonstrated that peripheral administration of PYY3–36 inhibits food intake and body weight gain in rats and mice via interaction with hypothalamic Y2 receptors, though this finding has been disputed by a number of other laboratories. Since PYY3–36 decreased hypothalamic NPY mRNA expression, it is presumed that presynaptic Y2 receptor stimulation reduced the synthesis and release of

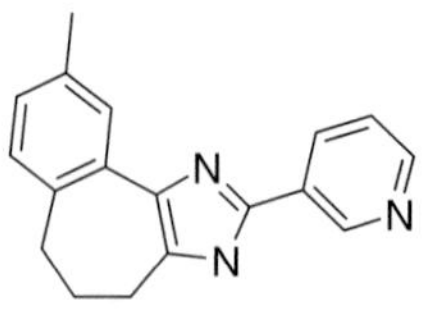

FR252384

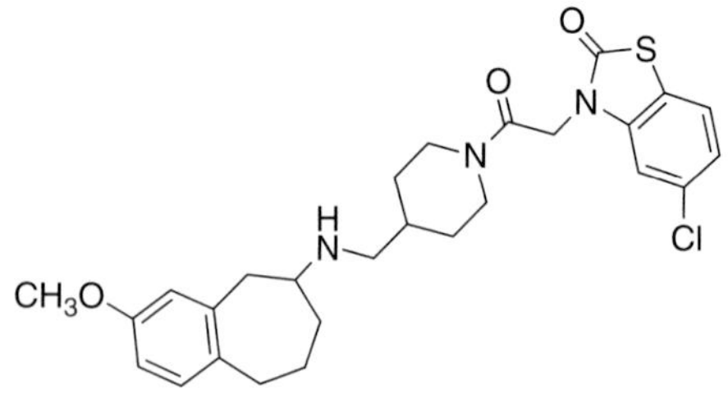

FR233118

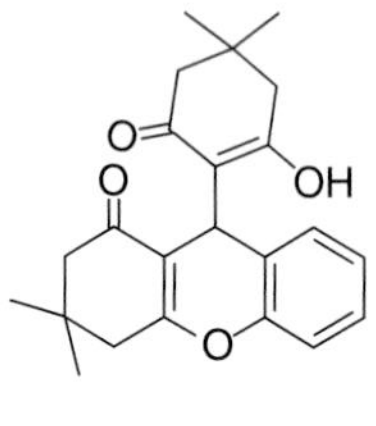

L-152,804

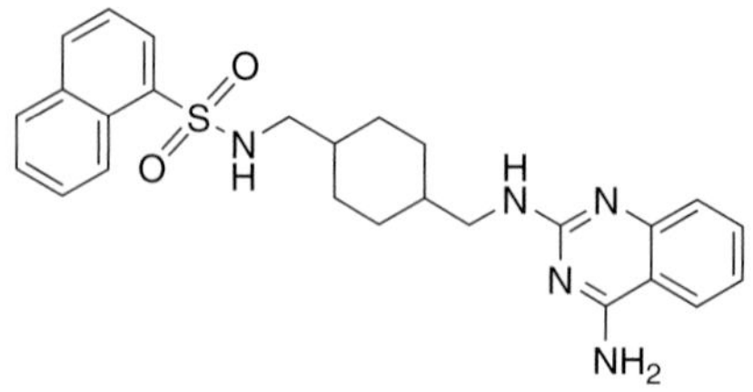

CGP71683A

Figure 12 NPY Y5 antagonists.

NPY on the postsynaptic Y1 and Y5 receptors to produce the effects. In human trials, PYY3–36 infusion 2 h before an 'all you can eat' buffet reduced food consumption by approximately 30% compared to saline-infused control treatment.[21] Intranasal administration of PYY3–36 is being pursued by several companies as a potential treatment for obesity.

6.18.8.5 Ghrelin Antagonists

Ghrelin is a stomach-derived 28-amino-acid peptide with an *n*-octanoyl modification of Ser^3.[21–23] It acts as the endogenous ligand for the GH secretagogue receptor (GHS-R). Ghrelin and GHS-R are also expressed in the hypothalamus and pituitary, where they modulate the secretion of GH from the pituitary as well as the secretion of CRF and vasopressin. Prominent among the biological effects of ghrelin is a pronounced orexigenic effect that, with repeated administration, can lead to obesity in preclinical models. In rodents, the feeding response to ghrelin appears to be dependent on NPY and AGRP, since immunoneutralization or administration of a receptor antagonist to these peptides abolished the acute orexigenic effects of ghrelin. Based on these experimental findings, an antagonist or inverse agonist to the GHS-R will be an interesting approach for the treatment of obesity. However, impairment of GH secretion and the consequent reduction in metabolism will warrant further study in the development of these agents.

6.18.8.6 Melanin-Concentrating Hormone Antagonists

The peptide MCH was originally identified in fish as a neurohypophyseal hormone that affects skin pigmentation in teleost fishes.[21] In humans, MCH is a cyclic nonadecapeptide with high homology to the salmon MCH amino acid sequence. Subsequently, MCH was found to be an orexigenic peptide after central administration to rodents. Central administration of MCH into normal rats produces a marked increase in food intake, while fasting will increase the expression of mRNA encoding MCH in mice. In the hypothalamus of obese *ob/ob* mice, MCH mRNA is upregulated and fasting will further increase the expression. MCH produces its biological effects through two GPCRs called MCH1 and MCH2. Since MCH2 is not expressed by rodents, the MCH1 receptor has been studied more extensively. Peripheral administration of the small-molecule MCH1 receptor antagonist (SNAP 7941, **Figure 13**) has been reported to block food consumption induced by central MCH administration. Chronic administration to rats with dietary-induced obesity results in a significant and sustained decrease in body weight. In addition, this antagonist exhibited anxiolytic and antidepressant-like activity in preclinical models. Therefore MCH1 receptor antagonists may have broad therapeutic utility for the treatment of obesity, eating disorders, anxiety, and depression.

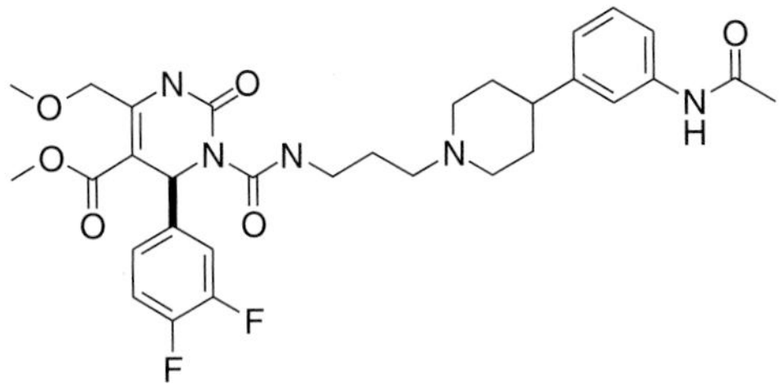

Figure 13 MCH1r antagonist SNAP-7941.

6.18.8.7 MC4 Receptor Agonists

The MC4 receptor is a GPCR for the endogenous hormone α-melanocyte stimulating hormone (α-MSH).[18] The MC4 receptor is localized primarily in the brain and is highly concentrated in hypothalamic regions. Several lines of evidence implicate the system in the control of energy homeostasis. MC4 knockout mice demonstrate increased food consumption and increased adiposity, indicating that the tonic stimulation of the MC4 receptor is important in the regulation of food intake and body weight. Heterozygotes for the MC4 deletion also become obese but to a lesser degree compared to the homozygote. Subsequently, the gene for AGRP was identified based on homology to the Agouti protein, a protein that determines coat color in mice. This peptide was found to be an endogenous antagonist of MC3 and MC4 receptors. AGRP was subsequently localized to NPY-containing neurons within the arcuate nucleus of the hypothalamus and AGRP was found to be upregulated in this region by fasting and leptin deficiency. When centrally administered, AGRP increases appetite for up to a week. Finally, synthetic agonists of MC4 decreased food intake while synthetic antagonists of this receptor increase food intake. Therefore, it is likely that interplay between AGRP and α-MSH is important in the maintenance of body weight and composition. As such, the discovery and development of specific, centrally active MC4 agonists has become a high priority within the pharmaceutical industry.

6.18.8.8 Neuromedin U (NmU)

Neuromedin U (NmU) is a highly conserved neuropeptide and was originally isolated from porcine spinal cord.[24] It is broadly distributed throughout the body, with particularly high levels found in the gastrointestinal tract and pituitary. Administration of NmU intracerebroventricularly or directly into the periventricular nucleus of the hypothalamus decreases feeding and increases gross locomotor activity, body temperature, heat production, and oxygen consumption in rats. NmU interacts with two receptor subtypes and it is not clear which receptor subtype mediates these effects, although expression of the NmU-R1 is found in the paraventricular nucleus. More recently, mice deficient in NmU were found to be obese. In addition peripheral administration of NmU has been shown to decrease food intake and body weight. Taken collectively, a NmU agonist may provide an innovative antiobesity agent that would decrease consumption and increase the metabolic rate.

6.18.8.9 Corticotropin-Releasing Factor, Urocortin I, II, and III

CRF was originally discovered as a hypothalamic factor involved in the regulation of the hypothalamic–pituitary–adrenal axis.[25] This peptide is a 42mer and is highly concentrated in the hypothalamus and pituitary. Two GPCR subtypes (CRF_1 and CRF_2) were identified that mediated increases in intracellular cyclic adenosine monophosphate. The CRF_1 receptor was found to mediate the stresslike endocrine and behavioral responses to centrally administered CRF. Subsequently, centrally administered CRF was found to decrease feeding and increase metabolism by interaction with the CRF_2 receptor. However, deletion of neither the CRF_1 nor CRF_2 receptors in mice had any prominent effect on body weight. Homology screening using CRF and urotensin allowed the identification of three new related endogenous peptides, urocortin (Ucn) I, Ucn II, and Ucn III. Given the relatively low affinity of CRF for the CRF_2 receptor, the high affinity that these three peptides display for the CRF_2 receptor makes it likely that one or more of these peptides is the endogenous ligand for this receptor. Therefore, it will be interesting to evaluate these and more selective and stable CRF_2 agonists in the regulation of feeding and metabolism.

6.18.8.10 Cocaine- and Amphetamine-Related Transcript

CART was originally discovered as an mRNA encoding an 89mer that was regulated by drugs of abuse in certain brain regions.[26] Two C-terminal CART-derived peptides, CART 42–89 and CART 49–89, have been isolated from the rat hypothalamus and pituitary, although these fragments have not yet been observed in human samples. Central administration of CART and CART fragments decreases food intake, although the actual brain region responsible for this action is somewhat debated. CART also has effects on the control of normal movement which may complicate preclinical food intake studies. Following fasting, rats show a decrease in CART mRNA in the arcuate nucleus of the hypothalamus, while decreased CART expression was found in obese animals. However, the CART knockout mouse does not become obese on a normal diet and the receptor for this peptide is currently unknown. Further study will be required to appreciate fully the potential of this target for the treatment of obesity.

6.18.8.11 Bombesin and Related Peptides

Bombesin is a 14-amino-acid peptide originally isolated from the skin of an amphibian, *Bombina bombina*.[18] In subsequent studies, large quantities of bombesin-like immunoreactivity were found in the CNS and gastrointestinal tract in various species. This immunoreactivity was not due to bombesin, since it is not expressed in mammals, but rather to the highly related mammalian peptides gastrin-releasing peptide (GRP) and neuromedins B and C. These peptides are collectively known as the bombesins and have a variety of actions on feeding, plasma levels of metabolic hormones, blood pressure, and hypothalamic–pituitary–adrenal function. Both central and peripheral administration of bombesin and GRP decrease food intake, suggesting these peptides may function as gut–brain-signaling molecules. The anorectic effects of bombesin and GRP have also been observed in humans, though the effect seems to be absent in obese individuals. At present, at least four different bombesin receptors (BB1–4) have been identified to date and these receptors can be subclassified into neuromedin B-preferring (BB1) or GRP-preferring (BB2). BB3 and BB4 exhibit low affinity for bombesin and the endogenous ligand for these receptors is currently unknown. Genetic deletion of the bombesin receptors in mice has led to some insight as to the potential molecular entities involved in feeding. Deletion of the BB1 and BB2 receptors does not result in any overt phenotype related to obesity. In the BB1 knockout mice, neither bombesin nor GRP suppresses glucose intake, suggesting this is the receptor involved in feeding. However, the lack of an overt phenotype suggests that GRP performs only a short-term role in satiety. Interestingly, the BB3 knockout exhibits an age-associated mild obesity with hypertension, glucose intolerance, and elevated insulin levels. The increased adiposity appears to be due to a reduction in energy expenditure without a change in food consumption or locomotor activity. These results suggest that the BB3 receptor may be a promising target for the discovery of novel antiobesity agents.

6.18.8.12 Enterostatin

Enterostatins are pentapeptides derived from N-terminal cleavage of pancreatic procolipase in the small intestine.[18] Colipase serves as a protein cofactor for pancreatic lipase and is necessary for intestinal fat digestion. Enterostatin is absorbed from the gut and functions as a potent anorectic peptide to reduce fat consumption in rodents selectively. Three enterostatin amino acid sequences have been identified in rats and humans and are described by their single letter amino acid abbreviations: VPDPR (rat and human), APGPR (human), and VPGPR (rat). While all three of these peptides are capable of reducing food intake, little is known about the receptor(s) mediating these effects. The anorectic effect of enterostatin in preclinical models appears to require prior chronic ingestion of a high-fat diet. In addition, plasma concentrations of enterostatin have been correlated to fat preference in different strains of rats. The relevance of the preclinical findings to human obesity may be limited since intravenous enterostatin produced no effects on feelings of hunger, satiety, or food preference in a phase II, double-blind randomized crossover placebo-controlled study of 18 obese patients. Further advances in this area will be dependent on the identification of the receptor for enterostatins as well as a better understanding of the peptide's ability to penetrate the blood–brain barrier.

6.18.8.13 Orexins

Orexin-A and orexin-B are 33 and 28mers, respectively.[18] These peptide sequences are the results of posttranslational processing of the same preproorexin peptide. These peptides were first identified in extracts of rat brain and bovine hypothalamus and were distinguished by their ability to evoke transient elevations of intracellular calcium concentrations in cells expressing an orphan GPCR now known as the orexin-1 or OX1 receptor. A second receptor (OX2) was identified based on sequence homology to the OX1 receptor. The OX1 receptor has higher affinity for orexin-A than orexin-B, while the OX2 receptor exhibits similar affinity for the two peptides. In general, orexin-A exhibits a greater orexigenic effect than orexin-B, though orexin-B appears to be more rapidly metabolized. While the absolute increase in food intake is not as great as observed with NPY or a GRP, increases in food intake are similar to those observed with galanin or MCH. Food deprivation has been demonstrated to increase orexin mRNA levels. Intracerebroventricular injection of an orexin-A antibody reduces food intake in fasted rats while a selective OX1 receptor antagonist (SB-334867, **Figure 14**) was reported to reduce food intake.

While the orexins were named for their effects on food intake, an important role for this peptide has been identified in the sleep–wake cycle. OX2 receptor knockout mice exhibit spontaneous sleep episodes similar to those observed in human narcolepsy. In addition, canine narcolepsy was associated with a mutation in the OX2 receptor gene. Based on these data, a compelling argument could be made for targeting the OX2 receptor for the development of therapeutics for sleep disorders and narcolepsy. While the OX1 receptor appears to be more involved in food intake, a role for the OX2 receptor cannot be excluded at this time. This is a very active area of research by a number of pharmaceutical companies and, hopefully, significant advances will be made in the discovery and development of selective antagonists in the coming years.

6.18.8.14 Cholecystokinin Agonists

Cholecystokinin (CCK) and CCK peptide fragments are synthesized and secreted from cells in the duodenum in response to nutrients in the lumen.[22] The specific nutrients that are most effective in stimulating CCK secretion vary according to the species studied. CCK also functions to stimulate the exocrine pancreas and the gallbladder to facilitate ingestion. Administration of CCK before a meal will dose-dependently reduce meal size in a variety of species. Plasma CCK has a very short half-life and, therefore, is considered to be a short-term regulatory factor. In addition, the satiety-inducing effect of continuous intravenously administered CCK or CCK analogs rapidly desensitizes. CCK is believed to produce its biological effects through two GPCRs, CCK1 and CCK2. Rats with a spontaneous mutation of the CCK1 receptor (Otsuka Long–Evans Tokushima fatty rats) show increased meal size and increased body weight. In contrast, CCK1 receptor knockout mice have a normal body weight. Glaxo-Smith-Kline (GSK) has had an active program in developing nonpeptide CCK1 agonists for the treatment of obesity and for gallstone prophylaxis. However, the lead compound, GI181771, was recently discontinued following phase II trials that showed disappointing clinical efficacy for the treatment of obesity. Development of an earlier molecule discovered at Abbott laboratories, A-71623, has also been discontinued (**Figure 15**).

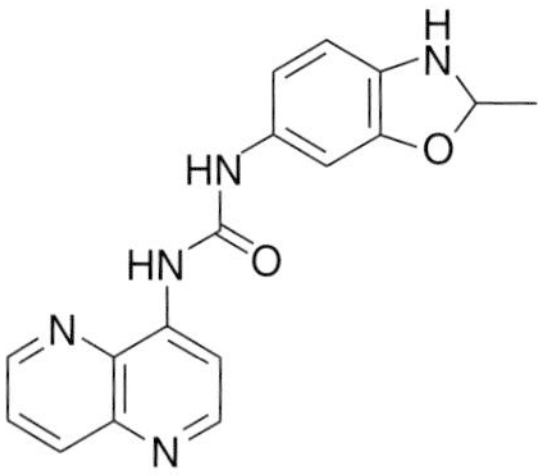

Figure 14 Orexin-1 receptor antagonist SB-334867.

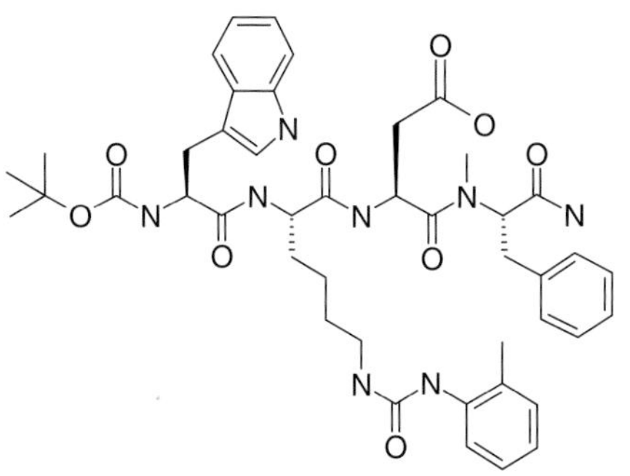

A-71623

Figure 15 CCK_A agonist, A-71623.

6.18.8.15 Opioid Antagonists

The regulation of food intake by opioids is complex, involving five major peptide classes (β-endorphin, dynorphin, enkephalins, endomorphans, and nociceptin (NOP: Orphanin FQ)) and four GPCRs μ opioid (MOP), delta opioid (DOP), κ opioid (KOP), and NOP receptors.[27] The interest in the opioid system and feeding began with the observation that the mixed MOP, DOP, and KOP antagonist, naloxone, decreased deprivation-induced feeding in rats. Subsequent studies documented the efficacy of mixed and subtype-selective opioid agonists to induce feeding in preclinical studies. The mixed opioid antagonist LY255582 (**Figure 16**) reduced food intake in both normal-weight and obese Zucker rats and long-term administration of this compound resulted in reduced food intake and body weight.

The opioid peptide NOP (Orphanin FQ) also stimulates feeding after intracerebroventricular administration. The increased feeding is antagonized by naloxone. Central administration of NOP stimulated both carbohydrate and fat intake in fat-preferring rats, but did not stimulate food intake in sucrose-preferring animals or those without a dietary preference. NOP and the nonpeptide agonist Ro 64–6198 (**Figure 16**) reversed the anorexic effects of restraint stress or central administration of CRF. These effects were antagonized by prior treatment with a NOP antagonist. Very little has been done to understand the effects of chronic agonists and antagonists on food consumption, body weight gain, and body composition. These data would be very helpful to understand the potential utility of these compounds for the treatment of obesity. NOP agonists have also shown promise as antistress/anxiolytic compounds, while NOP antagonists such as J-113397 (**Figure 16**) have been reported to have efficacy in pain models.

6.18.8.16 Uncoupling Proteins

Uncoupling protein 1 (UCP1) is found in brown adipose tissue that is abundant in small animals and human infants.[28] UCP1 functions to uncouple mitochondrial fatty acid oxidation from the production of ATP. During activation of UCP1 (by cold or diet), energy derived from the oxidation of fat is not available to drive the phosphorylation of ATP and is consequently dissipated as heat. UCP1 is of limited clinical interest because brown adipose tissue is barely detectable in adult humans. Furthermore, mice, which have abundant brown adipose tissue, lacking UCP1 do not become obese even when placed on a high-fat diet. Therefore UCP1 probably plays an important role in murine thermal regulation that may not be an important regulator of body composition. Molecular cloning efforts identified additional protein sequences that have high homology to UCP1. In contrast to UCP1, UCP2 is expressed in all tissues examined so far, while UCP3 is expressed predominantly in skeletal muscle and brown adipose tissue with lower levels in white adipose tissue and the heart. Additional homology cloning efforts identified brain mitochondrial carrier protein 1 (BMCP1 or

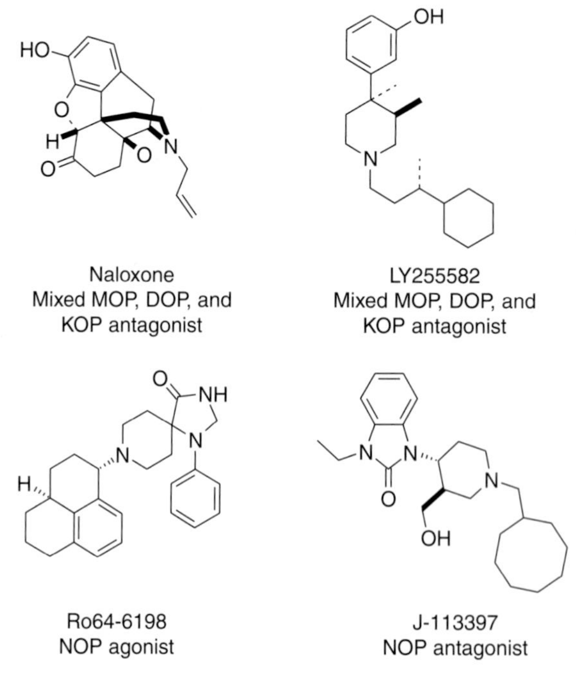

Figure 16 Mixed opioid receptor antagonists and a selective NOP antagonist.

UCP5). UCP5 is primarily expressed in brain and other neural tissues, with lower levels in white and brown adipose tissue, liver, skeletal muscle, gut, kidney, heart, and testis. While the UCP1 homologs have been the subject of intensive research activities, there are sparse data available to associate these proteins with thermogenesis and body weight regulation. Mice lacking UCP1, UCP2, or UCP3 or mice lacking both UCP1 and UCP3 do not become obese and have a phenotype similar to control mice. In addition, these mice do not show major differences in resting metabolic rate thermogenesis or energy expenditure. Further study will be required to validate one or more of these proteins as potential targets for obesity therapies.

6.18.8.17 Glucagon-Like Peptide 1 and Oxyntomodulin

Glucagon-like peptide 1 (GLP-1) is a gastrointestinal peptide isolated from the L cells of the intestine.[18,29] GLP-1 and the related peptide oxyntomodulin (Oxm) are produced by posttranslational processing of the preproglucagon gene. Subsequently, both these peptides have been found in the CNS, intestine, and the colon. In the brain, GLP-1-containing neurons are found in the nucleus of the solitary tract and the hypothalamus, with high densities of immunoreactive fibers within the hypothalamus. GLP-1 receptors are found in a variety of brain regions but are particularly high in regions of the hypothalamus. In the periphery, these peptides are released into the circulation from the intestine following food intake. They then act synergistically with glucose to stimulate insulin secretion from the pancreatic beta cells, reduce gastric motility and secretion, and induce satiety.

Central administration of GLP-1 to rats inhibits food intake and activates hypothalamic neurons. Peripheral administration of GLP-1 to rats and humans has also been reported to induce satiety. Oxm produces similar effects when administered centrally or peripherally to rats. These findings open the possibility that GLP-1 and Oxm interact directly with the brain after peripheral administration. Interestingly, high levels of GLP-1 receptors are localized in areas of the brain outside the blood–brain barrier, such as the subfornical organ, area postrema, and choroid plexus, indicating that these peptides may interact with these sites to regulate central appetite control. The effects of GLP-1 and Oxm can be blocked by the GLP-1 receptor antagonist, exendin (9–39), suggesting that Oxm signals through the GLP-1 receptor. Chronic central administration of GLP-1 has been reported to decrease food intake and body weight in normal and obese Zucker rats. In addition to decreasing food intake, GLP-1 increases oxygen consumption and energy expenditure, with a preferential increase in fat oxidation. Chronic central or peripheral administration of Oxm has been shown to cause a reduction in weight gain that is greater than that observed in pair-fed controls, indicating that Oxm also increases energy expenditure.

In humans, intravenous infusion of GLP-1 increases satiety, lowers energy intake, and increases REE. Oxm has also been demonstrated to reduce calorie intake, though the effects on energy expenditure have not yet been explored. Oxm infusion has also been reported to reduce plasma ghrelin levels in humans, a mechanism that may contribute to its efficacy by the peripheral route. Given the promise these peptides have shown in both preclinical and clinical studies, a number of pharmaceutical companies are pursuing this target for drug discovery. The peptide, exendin 4 (exenatide, Bayetta), a GLP-1 agonist, is approved for the treatment of type 2 diabetes. In clinical trials, this peptide has been shown to improve glycemic control as well as reduce body weight.

6.18.8.18 5-HT_{2c} Receptor Agonists

Central 5-HT-containing systems are well known to be involved in regulating feeding behavior.[15] As noted in Section 6.18.6.2.2.1, the anorectic agent fenfluramine was originally described in the 1970s, and its actions as a 5-HT-releasing agent helped reinforce a role for serotonergic activation in decreasing feeding behavior, as well as producing clinically relevant decreases in body weight. The discovery of multiple 5-HT receptors (approximately 14, depending how one counts species-specific receptors and splice variants) stimulated much interest in finding specific 5-HT receptor subtypes that were involved in the regulation of feeding. Early work with nonselective agonists, such as *m*-chlorophenylpiperazine (mCPP) and trifluoromethylphenypiperzine (**Figure 17**) suggested that the 5-HT_{2C} receptor might be involved in the decreased feeding produced by these compounds. Work with nonselective antagonists, such as ritanserin, and selective 5-HT_{2C} receptor antagonists, such as SB-242084 (**Figure 17**), reinforced the idea of the involvement of the 5-HT_{2C} receptor. The role of the 5-HT_{2C} receptor in feeding behavior was solidified with the development of a transgenic mouse having the 5-HT_{2C} receptor knocked out. These mice were observed to begin gaining weight during mouse middle age, and this weight gain was due to an increase in fat mass. The anorectic effects of mCPP and low-dose fenfluramine were lost in the 5-HT_{2C} knockout animals. Additional studies have shown that norfenfluramine, a metabolite fenfluramine (**Figure 4**), has high affinity for the 5-HT_{2C} receptor, and that it is the likely mediator of most of the anorectic effects of fenfluramine.

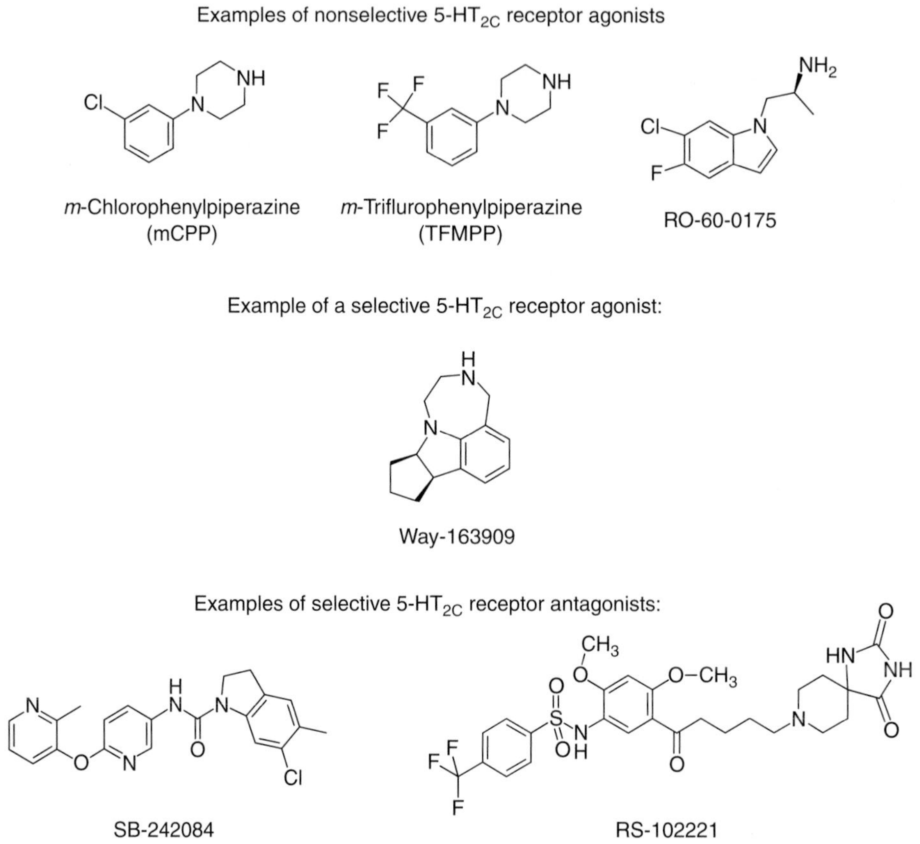

Examples of nonselective 5-HT_{2C} receptor agonists

m-Chlorophenylpiperazine (mCPP)

m-Triflurophenylpiperazine (TFMPP)

RO-60-0175

Example of a selective 5-HT_{2C} receptor agonist:

Way-163909

Examples of selective 5-HT_{2C} receptor antagonists:

SB-242084

RS-102221

Figure 17 5-HT_{2C} receptor agonists and antagonists.

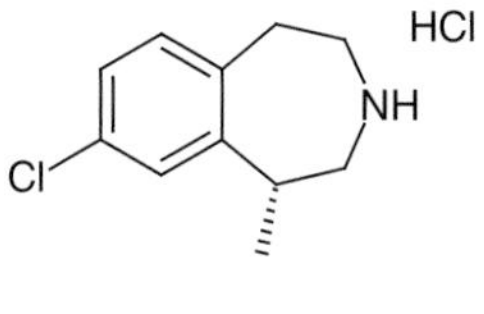

HCl

Figure 18 APD356 (lorcaserin hydrochloride).

Based on the strength of the evidence implicating the 5-HT_{2C} receptor in the control of feeding behavior, as well as other CNS pathways, there has been much interest in developing selective, druggable 5-HT_{2C} receptor agonists. The patent literature suggests that a number of companies have been involved in this effort, including Bayer, Biovitrum, Bristol-Myers Squibb, Eli Lilly, Pfizer, Vernalis, Wyeth, Yamanouchi, and Arena.

At least two relatively selective 5-HT_{2C} receptor agonists have been reported in clinical trials. Biovitrum reported that BVT-933 produced a significant and clinically relevant weight reduction compared to placebo in a phase IIa trial. However, the phase IIb trial was discontinued in favor of developing a more selective compound. Currently, the structure of BVT-933 has not been publicly revealed, so it is not known what molecular features allow this molecule to have selectivity for the 5-HT_{2C} receptor relative to other 5-HT receptors, especially the closely related 5-HT_{2A} and 5-HT_{2B} receptors. Arena has completed both a 28-day phase IIa trial and a 12-week phase IIb trial with the 5-HT_{2C} receptor agonist APD356 (lorcaserin; **Figure 18**). There are no peer-reviewed published data on APD356; however, the Arena website (http://www.arenapharm.com/wf/page/apd356) indicates that APD356 is approximately 15-fold more selective for the 5-HT_{2C} receptor compared to 5-HT_{2A} and about 100-fold more selective for the 5-HT_{2C} receptor relative to 5-HT_{2B}. In the phase IIb clinical trial, three doses of ADP356 were compared to placebo over 12 weeks with the following results: placebo, 0.7 lb decrease; 10 mg, 4.0 lb decrease; 15 mg, 5.7 lb decrease; 20 mg (10 mg twice daily), 7.9 lb decrease.

References

1. Klein, S.; Wadden, T.; Sugerman, H. J. *Gastroenterology* **2002**, *123*, 882–932.
2. Thibault, L.; Woods, S. C.; Westerp-Plantenga, M. S. *Br. J. Nutr.* **2004**, *92*, S41–S45.
3. Bray, G. A.; Paeratakul, S.; Popkin, B. M. *Physiol. Behav.* **2004**, *83*, 549–555.
4. Samuel, P. D.; Burland, W. A Review of the Medicinal Treatment of Obesity. In *Obesity Symposium: Proceedings of a Servier Research Institute symposium held in December 1973*; Burland, W. L., Samuel, P. D., Yudkin, Y., Eds.; Churchill Livingstone: London, 1974, pp 293–315.
5. Wangsness, M. *Minn. Med.* **2000**, *83*, 21–26.
6. Rothman, R. B.; Baumann, M. H.; Dersch, C. M.; Romero, D. V.; Rice, K. C.; Carroll, F. I.; Partilla, J. S. *Synapse* **2001**, *39*, 32–41.
7. Weintraub, M. *Clin. Pharmacol. Ther.* **1992**, *51*, 642–646.
8. Rothman, R. B.; Clark, R. D.; Partilla, J. S.; Baumann, M. H. *J. Pharmacol. Exp. Ther.* **2003**, *305*, 1191–1199.
9. Glazer, G. *Arch. Intern. Med.* **2001**, *161*, 1814–1824.
10. Yu, H.; Rothman, R. B.; Dersch, C. M.; Partilla, J. S.; Rice, K. C. *Bioorg. Med. Chem.* **2000**, *8*, 2689–2692.
11. Rothman, R. B.; Katsnelson, M.; Vu, N.; Partilla, J. S.; Dersch, C. M.; Blough, B. E.; Baumann, M. H. *Eur. J. Pharmacol.* **2002**, *447*, 51–57.
12. Glick, S. D.; Haskew, R. E.; Maisonneuve, I. M.; Carlson, J. N.; Jerussi, T. P. *Eur. J. Pharmacol.* **2000**, *397*, 93–102.
13. Bray, G. A.; Greenway, F. L. *Endocrinol. Rev.* **1999**, *20*, 805–875.
14. Rothman, R. B.; Jayanthi, S.; Cadet, J. L.; Wang, X.; Dersch, C. M.; Baumann, M. H. *Ann. NY Acad. Sci.* **2004**, *1025*, 151–161.
15. Vickers, S. P.; Dourish, C. T. *Curr. Opin. Invest. Drugs* **2004**, *5*, 377–388.
16. Rothman, R. B.; Vu, N.; Partilla, J. S.; Roth, B. L.; Hufeisen, S. J.; Compton-Toth, B. A.; Birkes, J.; Young, R.; Glennon, R. A. *J. Pharmacol. Exp. Ther.* **2003**, *307*, 138–145.
17. Otero, M. L. R.; Lago, F.; Casanueva, F. F.; Dieguez, C.; Gomez-Reino, J. J.; Gualillo, O. *FEBS Lett.* **2005**, *579*, 295–301.
18. Clapham, J. C.; Arch, J. R. S.; Tadayyon, M. *Pharmacol. Ther.* **2001**, *89*, 81–121.
19. Cooper, S. J. *Eur. J. Pharmacol.* **2004**, *500*, 37–49.
20. Berglund, M. M.; Hipskind, P. A.; Gehlert, D. R. *Exp. Biol. Med.* **2003**, *228*, 1–28.
21. Nisoli, E.; Carruba, M. O. *Pharmacol. Res.* **2004**, *50*, 453–469.
22. Woods, S. C. *Am. J. Physiol. Gastrointest. Liver Physiol.* **2004**, *286*, G7–G13.
23. Woods, S. C.; Benoit, S. C.; Clegg, D. J.; Seeley, R. J. *Best Pract. Clin. Endocrinol. Metab.* **2004**, *18*, 497–515.
24. Brighton, P. J.; Szekeres, P. G.; Willars, G. B. *Pharmacol. Rev.* **2004**, *56*, 231–248.
25. Richard, D.; Lin, Q.; Timofeeva, E. *Eur. J. Pharmacol.* **2002**, *440*, 189–197.
26. Kuhar, M. J.; Adams, L. D.; Hunter, R. G.; Vechia, S. D.; Smith, Y. *Regul. Peptides* **2000**, *89*, 1–6.
27. Bodnar, R. J. *Peptides* **2004**, *25*, 697–725.
28. Dulloo, A. G.; Seydoux, J.; Jacquest, J. *Physiol. Behav.* **2004**, *83*, 587–602.
29. Murphy, K. G.; Bloom, S. R. *Exp. Physiol.* **2004**, *89*, 507–516.
30. Food and Drug Administration. Draft guidelines (1996). Available online at: http://www.fda.gov/cder/guidance/obesity.pdf (accessed April 2006).
31. European Agency for the Evaluation of Medicinal Products Committee for Proprietary Medicinal Products. Guidance (1997). Available online at: http://www.emea.eu.int/pdfs/human/ewp/028196en.pdf (accessed April 2006).
32. Data on file. Sanofi-Synthelabo, Inc. Available online at: http://www.sanofi-Aventis.us/index.html (accessed April 2006).
33. Pfizer. Product insert. Available online at: http://www.pfizer.com/pfizer/download/uspi_didrex.pdf (accessed April 2006).
34. Food and Drug Administration. Ruling (2004). Available online at: http://www.fda.gov/oc/initiatives/ephedra/february2004/finalsummary.html (accessed April 2006).
35. Meigs, J. B. *Am. J. Manag. Care* **2002**, *8*, S283–S292; Quiz S293–S296.
36. Carr, D. B.; Utzschneider, K. M.; Hull, R. L.; Kodama, K.; Retzlaff, B. M.; Brunzell, J. D.; Shofer, J. B.; Fish, B. E.; Knopp, R. H.; Kahn, S. E. *Diabetes* **2004**, *53*, 2087–2094.

Biographies

David L Nelson, PhD, is a research fellow at Lilly Research Laboratories, a division of Eli Lilly and Company. With a BS in pharmacy, Dr Nelson received his PhD in pharmacology from the Department of Pharmacology, University of Colorado Medical Center. His postdoctoral training was in neuropharmacology at the College de France, Paris. Before joining Lilly, Dr Nelson was an associate professor in the Department of Pharmacology and Toxicology, School of Pharmacy, University of Arizona. Dr Nelson's research career has focused on monaminergic systems, with emphasis on the pharmacology and function of serotonergic receptors. His work, which includes two citation classics and the original description and characterization of the 5-HT_{1A} and 5-HT_{1B} receptor subtypes, is expressed in over 100 publications and 12 issued patents. Since joining Lilly in 1990, Dr Nelson has led and participated in a number of different drug development projects focused on nervous system targets, including obesity, migraine, schizophrenia, depression, and cognition.

Donald R Gehlert, PhD, is a research fellow at Lilly Research Laboratories, a division of Eli Lilly and Company. In addition, Dr Gehlert is an adjunct associate professor in Neuroscience and Psychiatry at the Indiana University Medical School. During his 16 years at Lilly, he has led a number of drug discovery efforts, including atomoxetine (Strattera) for the treatment of attention deficit disorder. More recently, Dr Gehlert has focused his efforts toward understanding the role of neuropeptides in psychiatric and metabolic disorders. His research efforts have resulted in over 150 published papers on neuropharmacology and neurochemistry and he is a co-inventor on 13 patents. He is active in a number of scientific societies and serves on the editorial boards of a number of scientific journals. Dr Gehlert received his PhD in pharmacology from the University of Utah School of Medicine and later was awarded a Pharmacology Research Associate Training (PRAT) fellowship from the National Institute of General Medical Sciences (NIGMS) to conduct his postdoctoral research at the National Institutes of Health campus in Bethesda. Afterwards, he held a senior staff fellow position at the National Institute of Neurological Disorders and Stroke.

Comprehensive Medicinal Chemistry II
ISBN (set): 0-08-044513-6

ISBN (Volume 6) 0-08-044519-5; pp. 389–416

6.19 Diabetes/Syndrome X

L Schmeltz and B Metzger, Feinberg School of Medicine, Northwestern University, Chicago, IL, USA

6.19.1 Introduction

Diabetes mellitus is a heterogeneous group of disorders characterized by hyperglycemia due to a relative or absolute insulin deficiency. Metabolic abnormalities in carbohydrate, fat, and protein metabolism contribute to chronic hyperglycemia that leads to microvascular and macrovascular complications.

Type 1 diabetes mellitus (T1DM), previously called insulin-dependent diabetes or juvenile diabetes, is caused by immune-mediated pancreatic β-cell destruction that leads to a decrease and eventual cessation of insulin secretion. In the absence of insulin, individuals are ketosis-prone and can develop life-threatening metabolic abnormalities. Most common in children and adolescents, T1DM accounts for 5–10% of all cases of diabetes. The incidence of T1DM varies geographically, ranging from areas of high incidence, such as Scandinavia (35/100 000 per year) and the USA (8–17/100 000 per year), to areas of low incidence, such as Japan and China (1–3/100 000 per year).

Type 2 diabetes mellitus (T2DM), formerly known as non-insulin dependent diabetes mellitus (NIDDM), is the most prevalent abnormality of glucose homeostasis, accounting for approximately 90% of all cases of diabetes. Individuals with T2DM have relative insulin deficiency due to insulin resistance or abnormalities in the insulin receptor–signaling pathway combined with impairment in insulin secretion. At disease onset, hyperglycemia elicits elevated levels of circulating insulin in most individuals, but they have decreased peripheral glucose utilization and impaired suppression of endogenous glucose production as a result of insulin resistance. Decline of β-cell function tends to be progressive, and can ultimately result in an insulin-dependent state. T2DM is associated with obesity, older age, physical inactivity, certain racial and ethnicity populations, a family history of T2DM, a history of gestational diabetes, or impaired glucose metabolism (impaired glucose tolerance (IGT), or pre-diabetes).

Gestational diabetes mellitus (GDM), the new onset or recognition of glucose intolerance during pregnancy, affects 5–8% of all pregnancies and requires treatment to normalize maternal blood glucose levels to avoid neonatal complications. Women with GDM have a high lifetime risk of developing T2DM.

Other types of diabetes can result from genetic defects in insulin secretion or insulin action, metabolic abnormalities that impair insulin secretion, mitochondrial abnormalities, and other disease states that cause abnormal glucose utilization. Together, these disorders account for 1–5% of all diagnosed cases of diabetes. *Mature onset diabetes of the young* (MODY) results from genetic mutations that lead to primary defects in insulin secretion and is inherited in an autosomal-dominant inheritance pattern. Pancreatic parenchymal disorders can destroy the islet population and result in insulin deficiency. Endocrinopathies, such as Cushing's disease and acromegaly, can alter glucose metabolism as well.

Obesity, particularly visceral or central obesity, develops as a consequence of a high-fat, high-energy diet in conjunction with a sedentary lifestyle. Obesity is a known risk factor for the development of insulin resistance, diabetes, coronary artery disease, cerebrovascular disease, hypertension, cholelithiasis, osteoarthritis, obstructive sleep apnea, and multiple types of cancer (uterine, breast, colorectal, kidney, and gall bladder). Obesity is also associated with the development of hypercholesterolemia, hyperandrogenism leading to hirsutism and menstrual irregularities, complications of pregnancy and surgical procedures, and psychological disorders, specifically depression.

The co-occurrence of abdominal obesity, impaired glucose metabolism, dyslipidemia, and hypertension define the metabolic syndrome, also referred to as the insulin resistance syndrome or syndrome X. The metabolic syndrome is associated with an increased risk for cardiovascular mortality, development of T2DM, and presence of the polycystic ovary syndrome (PCOS).

6.19.1.1 Epidemics of Obesity, Diabetes, and the Metabolic Syndrome

Obesity, diabetes mellitus, and the metabolic syndrome are increasing concurrently at epidemic levels in the USA and worldwide. The prevalence of overweight and obesity (a body mass index (BMI) greater than $25\,\mathrm{kg\,m^{-2}}$ and $30\,\mathrm{kg\,m^{-2}}$, respectively) has steadily increased across all ages (including children), racial and ethic groups, socioeconomic levels, and both genders. Results from the 1999–2002 National Health and Nutrition Examination Survey (NHANES) indicate that an estimated 65% of US adults are either overweight or obese, a 16% increase in the age-adjusted overweight estimates obtained from NHANES III (1988–94), and a 32% increase in the age-adjusted overweight estimates obtained from NHANES II (1976–80)[1] (**Figure 1**).

Diabetes affects 20.8 million individuals or 9.6% of the US adult population. Since 1991, the rise in the incidence of diabetes in the US population correlates with the increasing obesity trend.[2] In individuals age 20 years or older, T2DM affects 8.7% of Caucasians, 9.5% of Hispanic Americans, 13.3% of African-Americans, and as many as 35% of certain Native American groups such as the Pima Indians.[3] The Center for Disease Control (CDC) estimates that 1 in 3 individuals born within the USA in 2000 have a lifetime risk of developing diabetes. This epidemic is not limited to the boundaries of the USA. A decade ago, King and Rewers,[4] in conjunction with the World Health Organization (WHO), estimated that 150 million people had diabetes mellitus worldwide and projected that the number of persons with T2DM will double to 300 million by 2025.

Epidemiologic studies illustrate the association between weight gain, increase in BMI, and diabetes risk. It has been estimated that for every 1-kg increase in weight over a 10-year span, the risk of developing T2DM increases by 4.5%. Every one-unit increase in BMI raises the risk of developing T2DM by 12.1%.[5]

Results from the third National Health and Nutrition Examination Survey (NHANES III) indicate that 22% of Americans fulfill criteria for the metabolic syndrome (*see* Section 6.19.2.6). The metabolic syndrome was present in 5% of normal weight, 22% of overweight, and 60% of obese individuals, signifying obesity as a major risk factor. Prospective studies have shown a strong association between the presence of the metabolic syndrome and the subsequent development of T2DM. In the Beaver Dam Study[6] it was found that subjects without T2DM, but having at least four criteria for the metabolic syndrome had a 9-to 34-fold increased risk of developing T2DM within 5 years compared with those with one or fewer criteria. Similarly, in the West of Scotland Coronary Prevention Study (WOSCOPS),[7] men with four or five features of the metabolic syndrome had a 7-to 24-fold increased risk of developing T2DM during the 5-year follow-up compared with men with no features.

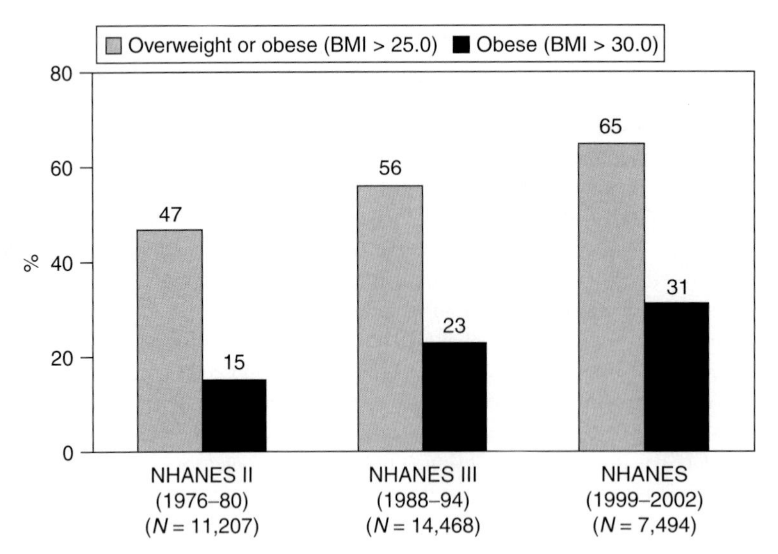

Figure 1 The increasing prevalence of overweight and obesity. Age-adjusted prevalence of overweight and obesity among US adults, age 20–74 years. Age-adjusted by the direct method to the year 2000 US Bureau of the Census estimates using the age groups 20–39, 40–59, and 60–74 years. (Adapted from National Center for Health Statistics Website. Prevalence of Overweight and Obesity among Adults: United States, 1999–2002, with permission from the Center for Disease Control.)

6.19.1.2 Impact of High Prevalence of Diabetes, Obesity, and the Metabolic Syndrome

6.19.1.2.1 Cardiovascular disease

Obese individuals have a two-fold increase in all-cause mortality compared with normal-weight individuals, with the largest increased risk being in cardiovascular mortality. Similarly, individuals with diabetes have a twofold increase in all-cause mortality compared with non-diabetics[8] and life expectancy is reduced by approximately 10 years. Cardiovascular disease (CVD) is the reported cause of death in 65% of individuals with diabetes in the USA. The risk of ischemic heart disease among individuals with diabetes is increased in all age groups, and in both men and women. Epidemiological evidence suggests the risk for macrovascular disease starts before the development of diabetes. Several studies have reported an increased risk of CVD in individuals with impaired glucose tolerance (adjusting for traditional cardiovascular risk factors such as hypertension, dyslipidemia, family history, and smoking status). In the WOSCOPS Study, men with four or five features of the metabolic syndrome had a 3.7-fold increased risk of CVD compared with men without these features.

6.19.1.2.2 Diabetic complications – retinopathy, nephropathy, and neuropathy

Chronic hyperglycemia is associated with vascular damage in the smallest blood vessels, or microvasculature, with potentially devastating effects on the retina, kidneys, and peripheral nerves. The risk of diabetic microvascular complications increases in proportion to the severity and duration of hyperglycemia. Many individuals with T2DM have asymptomatic hyperglycemia for a prolonged period prior to their clinical presentation. With an average delay of 4–7 years between disease onset and initial clinical presentation of T2DM,[9] approximately 20% of individuals with T2DM have evidence of diabetic microvascular complications at the time of diagnosis.

Diabetic complications are associated with significant morbidity that impacts quality of life. Diabetic retinopathy, the leading cause of blindness in those aged 20–74 years, accounts for 12 000–24 000 new cases of blindness each year. Diabetic nephropathy is the leading cause of end-stage renal disease requiring either dialysis or renal transplantation. Almost half of all adults with diabetes have at least one lower-extremity condition that increases their lifetime risk for lower-extremity amputations.

6.19.1.3 Financial Impact on the Health Care System

The overall impact of diabetes on direct medical and indirect expenditures was estimated to be $132 billion in 2002.[2] Direct medical expenditures included $23.2 billion for diabetes care, $24.6 billion for management of chronic diabetic complications, and $44.1 billion for other co-morbid medical conditions. The average person with diabetes incurred $13 243 in healthcare expenditures in 2002 compared with $2560 in non-diabetic individuals. Currently, $1 of every $5 healthcare dollars in the USA is spent on individuals with diabetes. The indirect economic impact of diabetes is also large.

6.19.1.4 Reasons for the High Prevalence of Diabetes and Obesity

Overweight and obesity result from the complex interactions of genetic factors, chronic energy imbalance, environmental influences, and concurrent disease or medications. Excessive caloric consumption and inadequate physical activity lead to weight gain, and both these factors are rampant in US society. In 2002, the US Department of Health and Human Services reported that less than one-third of US adults have regular leisure-time physical activity and 10% of adults reported none at all. Physical activity is associated with improved weight control, decreased coronary artery disease and associated cardiac mortality, and decreased risk of developing T2DM, hypertension, and colon cancer.

6.19.1.5 Diabetes Prevention Trials

Risk factors for the development of diabetes have been known for many years. In the last two decades, efforts have been made to decrease the incidence or delay the appearance of new cases of diabetes. The DaQing Study in China[10] and the Finnish Diabetes Prevention Study[11] showed that lifestyle interventions, including diet, exercise, and weight reduction, can decrease the risk of developing overt T2DM in individuals with IGT.

The Diabetes Prevention Program (DPP), the largest study to date, recruited a racially/ethnically, socio-economically, heterogeneous cohort of US adult men and women (ages 25–82 years) to participate in a randomized trial of lifestyle modification or metformin for prevention of T2DM.[12] This prospective, multicenter clinical trial enrolled 3,234 people with IGT. The goal of the intensive lifestyle intervention was to reduce weight by $>7\%$ using low fat, lower calorie diets and 150 min or more of exercise per week. During the 3-year follow-up period, 29% of the control group developed T2DM compared with 14% of the intensive lifestyle change group and 22% of the metformin arm. Individuals in the lifestyle intervention and metformin groups had a 58% and 31% relative risk reduction for developing T2DM, respectively.

Additional studies have since shown that other classes of oral hypoglycemic agents can prevent or delay T2DM. The STOP-NIDDM study determined that the α-glucosidase inhibitor, acarbose, reduced the risk of conversion from IGT to T2DM by 36%.[13] The TRIPOD study (Troglitazone in Prevention of Diabetes) showed efficacy of the thiazolidedione, troglitazone, in preventing T2DM in Hispanic women with a history of GDM with a relative risk reduction of 55%.[14]

6.19.2 Classification of Diabetes Mellitus and the Metabolic Syndrome

In 2003, the American Diabetes Association (ADA) revised the etiologic classification of diabetes mellitus, removing the distinction between primary and secondary causes of diabetes[15] (**Table 1**). The nomenclature now uses Arabic rather than Roman numerals to designate T1DM and T2DM. Terms such as insulin-dependent, non-insulin-dependent, juvenile-onset, maturity-onset, and adult-onset diabetes are eliminated. Thus, diabetes is now classified according to etiology and pathophysiology, without distinction as to age of onset or type of treatment.

6.19.2.1 Type 1 Diabetes Mellitus

6.19.2.1.1 Immune-mediated diabetes mellitus (Type 1A)

Immune-mediated diabetes, previously referred to as insulin-dependent diabetes, type I diabetes, and juvenile-onset diabetes, accounts for 5–10% of all cases of diabetes. Immune-mediated diabetes typically develops in childhood and adolescence, but has a variable age of onset ranging from infancy to the eighth and ninth decades of life. Abnormalities in glucose homeostasis result from severe insulin deficiency due to cell-mediated autoimmune inflammation (insulitis) and subsequent destruction of the β-cells of the pancreas. In T1DM, daily exogenous insulin administration is a life-sustaining intervention, and the absence of insulin can result in a state of acute metabolic decompensation known as diabetic ketoacidosis (DKA). Individuals with T1DM are at increased risk for other autoimmune disorders such as Graves' disease, Hashimoto's thyroiditis, Addison's disease, vitiligo, celiac sprue, autoimmune hepatitis, myasthenia gravis, and pernicious anemia.

Table 1 Etiologic classification of diabetes mellitus

Type 1 Diabetes Mellitus (β-cell destruction, usually leading to absolute insulin deficiency)	
Immune mediated	
Idiopathic	
Type 2 Diabetes Mellitus (insulin resistance and impairment of β-cell insulin secretion)	
Gestational Diabetes Mellitus (GDM)	
Other specific types	
Genetic defects of β-cell function	
Chromosome 12, HNF-1α (MODY3)	Chromosome 7, glucokinase (MODY2)
Chromosome 20, HNF-4α (MODY1)	Chromosome 13, IPF-1 (MODY4)
Chromosome 17, HNF-1β (MODY5)	Chromosome 2, *NeuroD1* (MODY6)
Mitochondrial DNA	Others
Genetic defects in insulin action	
Type A insulin resistance	Leprechaunism
Rabson–Mendenhall syndrome	Lipoatrophic diabetes
Others	
Diseases of the exocrine pancreas	
Pancreatitis	Trauma/pancreatectomy
Neoplasia	Cystic fibrosis
Hemochromatosis	Fibrocalculous pancreatopathy
Others	
Endocrinopathies	
Acromegaly	Cushing's syndrome
Glucagonoma	Pheochromocytoma
Hyperthyroidism	Somatostatinoma
Aldosteronoma	Others
Drug- or chemical-induced	
Vacor	Pentamidine
Nicotinic acid	Glucocorticoids
Thyroid hormone	Diazoxide
β-Adrenergic agonists	Thiazides
Dilantin	α-Interferon
Others	
Infections	
Congenital rubella	Cytomegalovirus
Others	
Uncommon forms of immune-mediated diabetes	
'Stiff-man' syndrome	Anti-insulin receptor antibodies
Others	
Other genetic syndromes sometimes associated with diabetes	
Down's syndrome	Klinefelter's syndrome
Turner's syndrome	Wolfram's syndrome
Friedreich's ataxia	Huntington's chorea
Laurence–Moon–Biedl syndrome	Myotonic dystrophy
Porphyria	Prader–Willi syndrome
Others	

6.19.2.1.2 Idiopathic diabetes mellitus (Type 1B)

Individuals who present with a phenotype suggestive of T1DM (insulinopenia and ketosis prone) without evidence of autoimmunity are classified as idiopathic or type 1B diabetes. This form of diabetes is rare, strongly inherited, and most commonly found in persons with African or Asian ancestry.

6.19.2.2 Type 2 Diabetes Mellitus

T2DM accounts for approximately 90% of all cases of diabetes. Historically it has been referred to as adult-onset diabetes, non-insulin-dependent diabetes and type II diabetes. T2DM is characterized by a combination of insulin resistance (in skeletal muscle, adipose tissue, and liver) and impaired insulin secretion. As reviewed below (*see* Section 6.19.3.3.3.1) individuals with similar levels of glucose intolerance vary widely in their severity of insulin resistance and deficiency of insulin secretion. The prevalence of T2DM increases with age, sedentary lifestyle, obesity, and family history of diabetes, including GDM. The genetic predisposition to T2DM is strong and suspected to be polygenic. Since insulin deficiency in T2DM is relative, individuals are not usually ketosis prone, but ketoacidosis can occur in the setting of severe stress or illness. However, as β-cell function deteriorates, many people with T2DM require exogenous insulin to achieve satisfactory glycemic control.

6.19.2.3 Other Specific Types of Diabetes

Diabetes with pathophysiology not congruent with that of T1DM and T2DM is classified into a broad category of 'other specified types.'

6.19.2.3.1 Genetic β-cell defects

Monogenic defects of pancreatic β-cells are characterized by impaired insulin secretion without obligatory abnormalities in insulin action. The most common defects involve the glucose-sensing enzyme glucokinase and various transcription factors such as hepatocyte nuclear factors (HNF-4α, HNF-1α, and HNF-1β) and insulin promoter factor (IPF-1). Collectively, these autosomal-dominantly inherited mutations are known as maturity onset diabetes of the young (MODY), and affected individuals typically develop hyperglycemia before the age of 25 years. Abnormalities in insulin sensitivity can still occur in these individuals from environmental factors, altering their clinical course and clouding their underlying pathophysiology and classification.

6.19.2.3.2 Genetic defects in insulin action

Rarely, mutations of the insulin receptor or the postreceptor signal transduction pathways cause impairment of insulin action, resulting in metabolic abnormalities ranging from hyperinsulinemia with mild hyperglycemia to severe diabetes. Clinical manifestations usually develop within the first year of life, and each mutation commonly has a disease-specific phenotype that accompanies the metabolic abnormalities.

6.19.2.3.3 Diseases of the exocrine pancreas

Any disease process involving the pancreatic parenchyma can affect glucose homeostasis. Acute and chronic pancreatitis, pancreatic carcinoma, trauma, infection, cystic fibrosis, and infiltrating diseases such as hemochromatosis can all affect islet cell function and insulin secretion. To elicit endocrine dysfunction, a significant loss of β-cell mass is required.

6.19.2.3.4 Endocrinopathies

Many hormones are involved in glucose homeostasis, and excess secretion of any of them can affect insulin resistance. When present in excess, counterregulatory hormones (growth hormone, cortisol, glucagon, and epinephrine) that normally protect against hypoglycemia antagonize insulin action. Acromegaly, Cushing's syndrome, glucagonoma, and pheochromocytoma are associated with overt diabetes in individuals with an underlying genetic predisposition. Other endocrinopathies such as hyperthyroidism (which increases β-adrenergic activity) and primary hyperaldosteronism (which impairs insulin secretion due to hypokalemia) are infrequently associated with T2DM. Neuroendocrine tumors of the pancreas, specifically those secreting glucagon, vasoactive intestinal peptide, and somatostatin can also affect glucose homeostasis. In all these disorders, glycemic abnormalities may be reversible with treatment of the underlying condition.

6.19.2.3.5 Drug or chemical-induced diabetes

Medications do not usually cause overt diabetes independently. Drugs that impair insulin secretion, alter insulin sensitivity, or lead to β-cell destruction may precipitate diabetes in an individual with underlying glucose intolerance. Antihypertensive agents including β-adrenoceptor antagonists, calcium channel antagonists, diuretics, some antipsychotics, and calcineurin inhibitors may reversibly inhibit insulin secretion. Pentamidine and toxins such as Vacor can result in permanent β-cell destruction. Use of glucocorticoids increases insulin resistance and may be associated with the development of diabetes. Several chemotherapeutic agents (mithramycin, asparaginase, and α-interferon) are associated with the development of diabetes, although the precise mechanism is unknown.

6.19.2.3.6 Infections

Certain viruses, including congenital rubella, Coxsackie virus B, cytomegalovirus, adenovirus, and mumps, have been associated with β-cell destruction and have been implicated as an environmental trigger for the immune response in T1DM. However, no specific virus appears to be responsible for the majority of cases.

6.19.2.3.7 Uncommon forms of immune-mediated diabetes

Autoantibodies to insulin or the insulin receptor are associated with rare forms of diabetes. Stiff-man syndrome, an autoimmune disease of the central nervous system resulting in axial muscle stiffness and spasms, is commonly associated with high titers of the autoantibodies to glutamic acid decarboxylase (GAD) and a predilection to develop diabetes. Individuals with other autoimmune diseases, such as systemic lupus erythematosus or dermatomyositis, can develop autoantibodies to the insulin receptor that act as competitive inhibitors of insulin or, in some cases, activators of the insulin receptor resulting in hypoglycemia.

6.19.2.3.8 Genetic syndromes sometimes associated with diabetes

Impaired glucose tolerance or diabetes appear with increased frequency in a number of congenital disorders, including Down's syndrome, Klinefelter's syndrome, Turner's syndrome, Wolfram's syndrome, Friedreich's ataxia, Huntington's chorea, Laurence–Moon–Biedl syndrome, myotonic dystrophy, porphyria, and Prader–Willi syndrome.

6.19.2.4 Gestational Diabetes Mellitus

GDM is defined as glucose intolerance with onset or recognition during pregnancy. Individuals with pre-existing T1DM or T2DM are excluded from this classification. Individuals are at increased risk of developing GDM if they have a family history of diabetes or GDM, are a member of a racial/ethnic group with a high prevalence of T2DM (e.g., Native American, African American, Hispanic American, or Asian American), are obese, or of high maternal age, (e.g., >38–40 years). GDM is associated with increased risk of pregnancy-induced hypertension, macrosomia, cesarean delivery, and other obstetric complications. Women are treated with lifestyle modifications and may require insulin for optimal glycemic control. GDM identifies women at very high risk for GDM in future pregnancies and T2DM later in life.

6.19.2.5 Diagnostic Criteria for Diabetes

In 1979, the National Diabetes Data Group (NDDG) established criteria that included diagnostic levels for fasting plasma glucose (FPG) and oral glucose tolerance test (OGTT). FPG $\geq 140\,\mathrm{mg\,dL^{-1}}$ and/or 2-h plasma glucose value (2hPG) $\geq 200\,\mathrm{mg\,dL^{-1}}$ during an OGTT were chosen as the diagnostic threshold because individuals with average glucose levels above these values had been found to develop microvascular complications. Those with OGTT 2hPG values $> 140\,\mathrm{mg\,dL^{-1}}$, but below the threshold for diabetes ($200\,\mathrm{mg\,dL^{-1}}$) were classified as having impaired glucose tolerance (IGT).

In 1997, an ADA 'Expert Panel' recommended a reduction in the FPG threshold for the diagnosis of diabetes to $\geq 126\,\mathrm{mg\,dL^{-1}}$ ($7.0\,\mathrm{mmol\,L^{-1}}$) as it correlated better with a 75 g OGTT 2-h value $\geq 200\,\mathrm{mg\,dL^{-1}}$. By reducing the FPG criteria, some individuals previously diagnosed as IGT were reclassified as having diabetes, resulting in a modest increase in the prevalence of diabetes. A new category, called impaired fasting glucose (IFG), was added for those individuals with a FPG ≥ 110 and $< 126\,\mathrm{mg\,dL^{-1}}$. In 2003, the ADA lowered the threshold for IFG to $100\,\mathrm{mg\,dL^{-1}}$ and reinstated an old term, 'Pre-diabetes,' to describe individuals with either IGT or IFG. As a result, risk stratification and lifestyle modifications can be initiated earlier to prevent the development of overt diabetes and other associated vascular complications.

The current ADA diagnostic criteria[16] (**Table 2**) identify three independent parameters by which the diagnosis of diabetes can be made: FPG, 75 g 2-h OGTT, and random plasma glucose in the setting of symptoms suggestive of diabetes. Any combination of two abnormal tests performed on different days confirms a diagnosis of diabetes. The OGTT remains the diagnostic test for GDM and is preferred by many experts for postpartum evaluation for persistent abnormalities in glucose homeostasis.

6.19.2.6 Diagnostic Criteria for the Metabolic Syndrome

The metabolic syndrome (MetS), also known as the insulin-resistance syndrome or syndrome X, is characterized by insulin resistance, abdominal obesity, elevated blood pressure, and lipid abnormalities consisting of elevated levels of triglycerides and low levels of high-density lipoprotein (HDL) cholesterol (*see* 6.17 Obesity/Metabolic Syndrome

Table 2 Criteria for the diagnosis of diabetes mellitus[a]

	Normoglycemia	*Pre-diabetes*	*Diabetes*
		Impaired fasting glucose	
Fasting plasma glucose	$<100\,\mathrm{mg\,dL^{-1}}$	≥ 100 and $<126\,\mathrm{mg\,dL^{-1}}$	$\geq 126\,\mathrm{mg\,dL^{-1}}$
	$<5.6\,\mathrm{mmol\,L^{-1}}$	≥ 5.6 and $<7.0\,\mathrm{mmol\,L^{-1}}$	$\geq 7.0\,\mathrm{mmol\,L^{-1}}$
		Impaired glucose tolerance	
75 g OGTT	$<140\,\mathrm{mg\,dL^{-1}}$	≥ 140 and $<200\,\mathrm{mg\,dL^{-1}}$	$\geq 200\,\mathrm{mg\,dL^{-1}}$
2-h post-load glucose	$<7.8\,\mathrm{mmol\,L^{-1}}$	≥ 7.8 and $<11.1\,\mathrm{mmol\,L^{-1}}$	$\geq 11.1\,\mathrm{mmol\,L^{-1}}$
Random Glucose			$\geq 200\,\mathrm{mg\,dL^{-1}}$
			$\geq 11.1\,\mathrm{mmol\,L^{-1}}$
			+ symptoms of diabetes

OGTT, oral glucose tolerance test.

[a] Any combination of two abnormal tests performed at different points in time is diagnostic for diabetes. Fasting is defined as no caloric intake for at least 8 h. Random glucose is defined as any time of day without regard to time since last meal. Symptoms of marked hyperglycemia include polyuria, polydipsia, weight loss, sometimes with polyphagia, and blurred vision.

Table 3 Diagnostic criteria for metabolic syndrome

Component	*WHO diagnostic criteria (insulin resistance + ≥2 of the following)*	*ATP III diagnostic criteria (≥3 of the following)*
Abdominal/central obesity	Waist-to-hip ratio:	Waist circumference:
	>0.90 for men	>102 cm (40 in.) in men
	>0.85 for women	>88 cm (35 in.) in women
	Body mass index (BMI) $>30\,\mathrm{kg\,m^{-2}}$	
Triglycerides	$\geq 150\,\mathrm{mg\,dL^{-1}}$ ($\geq 1.7\,\mathrm{mmol\,L^{-1}}$)	$\geq 150\,\mathrm{mg\,dL^{-1}}$ ($\geq 1.7\,\mathrm{mmol\,L^{-1}}$)
HDL cholesterol	$<35\,\mathrm{mg\,dL^{-1}}$ ($<0.9\,\mathrm{mmol\,L^{-1}}$) for men	$<40\,\mathrm{mg\,dL^{-1}}$ ($<1.036\,\mathrm{mmol\,L^{-1}}$) for men
	$<39\,\mathrm{mg\,dL^{-1}}$ ($<1.0\,\mathrm{mmol\,L^{-1}}$) for women	$<50\,\mathrm{mg\,dL^{-1}}$ ($<1.295\,\mathrm{mmol\,L^{-1}}$) for women
Blood pressure	≥140/90 mmHg or documented use of antihypertensive therapy	≥130/85 mmHg or documented use of antihypertensive therapy
Fasting glucose	Impaired glucose tolerance	$\geq 110\,\mathrm{mg\,dL^{-1}}$ ($\geq 6.1\,\mathrm{mmol\,L^{-1}}$)[a]
	Impaired fasting glucose	
	Diabetes	
Urinary microalbumin	Urinary albumin/creatinine ratio: $30\,\mathrm{mg\,g^{-1}}$	
	Albumin excretion rate: $20\,\mathrm{\mu g\,min^{-1}}$	

ATP, Adult Treatment Panel; HDL, high-density lipoprotein; WHO, World Health Organization.

Adapted from Alberti, K. G.; Zimmet, P. Z. *Diabet. Med.* **1998**, *15*, 539–553 and Executive Summary of the Third Report of the National Cholesterol Education Program (NCEP) Expert Panel on Detection, Evaluation, and Treatment of High Blood Cholesterol in Adults (Adult Treatment Panel III). *JAMA* **2001**, *285*, 2486, with permission from the American Medical Association.

[a] The American Diabetes Association and National Heart Lung and Blood Institute have suggested lowering this threshold to $100\,\mathrm{mg\,dL^{-1}}$.

Overview). This combination of risk factors is responsible for a majority of cardiovascular morbidity among overweight and obese individuals and those with T2DM. Each component of the metabolic syndrome has been independently associated with an increased risk of CVD.[17]

The WHO initially defined criteria for the diagnosis of the metabolic syndrome[18] in 1998 (**Table 3**), requiring the presence of insulin resistance and two or more of the following metabolic abnormalities: waist-to-hip ratio (>0.90 for men; >0.85 for women), triglycerides ($\geq 150\,\mathrm{mg\,dL^{-1}}$), HDL cholesterol ($<35\,\mathrm{mg\,dL^{-1}}$ for men; $<39\,\mathrm{mg\,dL^{-1}}$ for women), blood pressure (≥140/90 mmHg or documented use of antihypertensive therapy), or microalbumiuria (Urinary albumin/creatinine ratio $>30\,\mathrm{mg\,g^{-1}}$).

In its third report, the National Cholesterol Education Program Expert Panel on Detection, Evaluation, and Treatment of High Blood Cholesterol in Adults (NCEP-ATP III) developed an alternative classification for metabolic syndrome,[19] requiring the presence of at least three metabolic abnormalities to make the diagnosis. The NCEP-ATP III criteria do not make insulin resistance an absolute requirement to define the metabolic syndrome, has lower threshold values for blood pressure ($\geq$130/85 mmHg or documented use of antihypertensive therapy) and HDL cholesterol ($<$40 mg dL^{-1} for men; $<$50 mg dL^{-1} for women), uses waist circumference ($>$102 cm in men; $>$88 cm in women) instead of waist-to-hip ratio, and does not include microalbuminuria.

6.19.3 Disease Basis

6.19.3.1 Normal Glucose Homeostasis

Glucose homeostasis is maintained as a closed feedback loop involving the pancreatic islet cells, liver, and peripheral tissues including the brain, muscle, and adipose. The islets are interspersed throughout the pancreas and make up 1–2% of pancreatic mass. Islets are composed of 70–80% β-cells (insulin producing), 20–25% α-cells (glucagon producing), and 5–10% somatostatin and pancreatic polypeptide secreting cells.[20]

The islet architecture is quite vascular. Insulin, synthesized in the β-cells, enters the islet vasculature, is sensed by α-cells to regulate glucagon secretion, and flows into the portal venous circulation. Hyperglycemia increases insulin and reduces glucagon secretion. Conversely, hypoglycemia is characterized by enhanced glucagon secretion and diminished insulin secretion. By flowing into the portal circulation, the liver is exposed to insulin and glucagon concentrations 5- to 10-fold higher than in peripheral tissues, facilitating tight regulation of hepatic glucose production. Peripheral glucose uptake, both insulin dependent and insulin independent, is modulated in part by the upregulation and downregulation of GLUT transport proteins in target cell membranes.

In the basal state, after an overnight fast, hepatic glucose production (via glycogen breakdown and gluconeogenesis) and renal glucose production (via gluconeogenesis) are responsible for supplying the basal glucose requirement to peripheral tissue. Approximately 75–80% of glucose utilization in the fasting state occurs in insulin-independent cells including the brain, gut, and red blood cells. Insulin-dependent cells such as muscle and adipose cells, therefore, have minimal glucose requirements in the fasting state.

In the fed state, a glucose load increases the plasma glucose level and stimulates glucose-dependent insulin release from the β-cells. Insulin stimulates peripheral glucose uptake in muscle and adipose tissue, and suppresses endogenous hepatic and renal glucose production. A concurrent decrease in glucagon suppresses hepatic glucose production. Contrary to the fasting state, glucose utilization in the fed state is dominated by muscle and adipose tissue accounting for 85% and 5% of glucose uptake, respectively. A rise in peripheral insulin concentrations also results in decreased plasma free fatty acid (FFA) via inhibition of lipolysis. Decreased FFA levels contribute to suppression of endogenous glucose production and increased hepatic glucose utilization in muscle.

Physiologic insulin secretion is also modulated by non-glucose nutrients, hormones, and neural influences. Amino acids, specifically leucine, arginine, and lysine, stimulate insulin secretion in the absence of glucose and augment insulin secretion in its presence. Glucose absorption in the gastrointestinal tract stimulates release of gastrointestinal peptide hormones, known as incretins, including glucagon-like peptide-1 (GLP-1) and glucose-dependent insulinotropic polypeptide (previously called gastric inhibitory polypeptide (GIP)), and cholecystokinin.[21] Incretins do not independently stimulate insulin secretion,[22] but augment β-cell response to the glucose load in a glucose concentration-dependent manner. At equivalent plasma glucose concentrations, insulin secretion after an oral glucose load is higher than with an intravenous glucose load, illustrating the incretin effect.[23] Neural influences on islet cells are mediated via cholinergic and adrenergic innervation. Insulin secretion is increased with parasympathetic or vagal stimulation and suppressed by sympathetic activation. However, both sympathetic and parasympathetic stimulation have been associated with increased glucagon secretion.[24] β-Adrenoceptor stimulatory and α-adrenoceptor inhibitory influences affect insulin secretion and are altered by pharmacological blockade.

6.19.3.1.1 Insulin biosynthesis

Insulin consists of 51 amino acids arranged in two peptides (an A chain composed of 21 amino acids and a 30-residue B chain) that are linked by two disulfide bonds. The A chain contains species-specific sites whereas the B chain contains the core of biological activity. Synthesis of insulin begins with genetic transcription forming preproinsulin (a single-chain, 86 amino acid polypeptide) in the ribosomes of the rough endoplasmic reticulum[25] (**Figure 2**). Removal of the amino-terminal signal peptide of preproinsulin results in the insulin precursor proinsulin, which is structurally similar to the

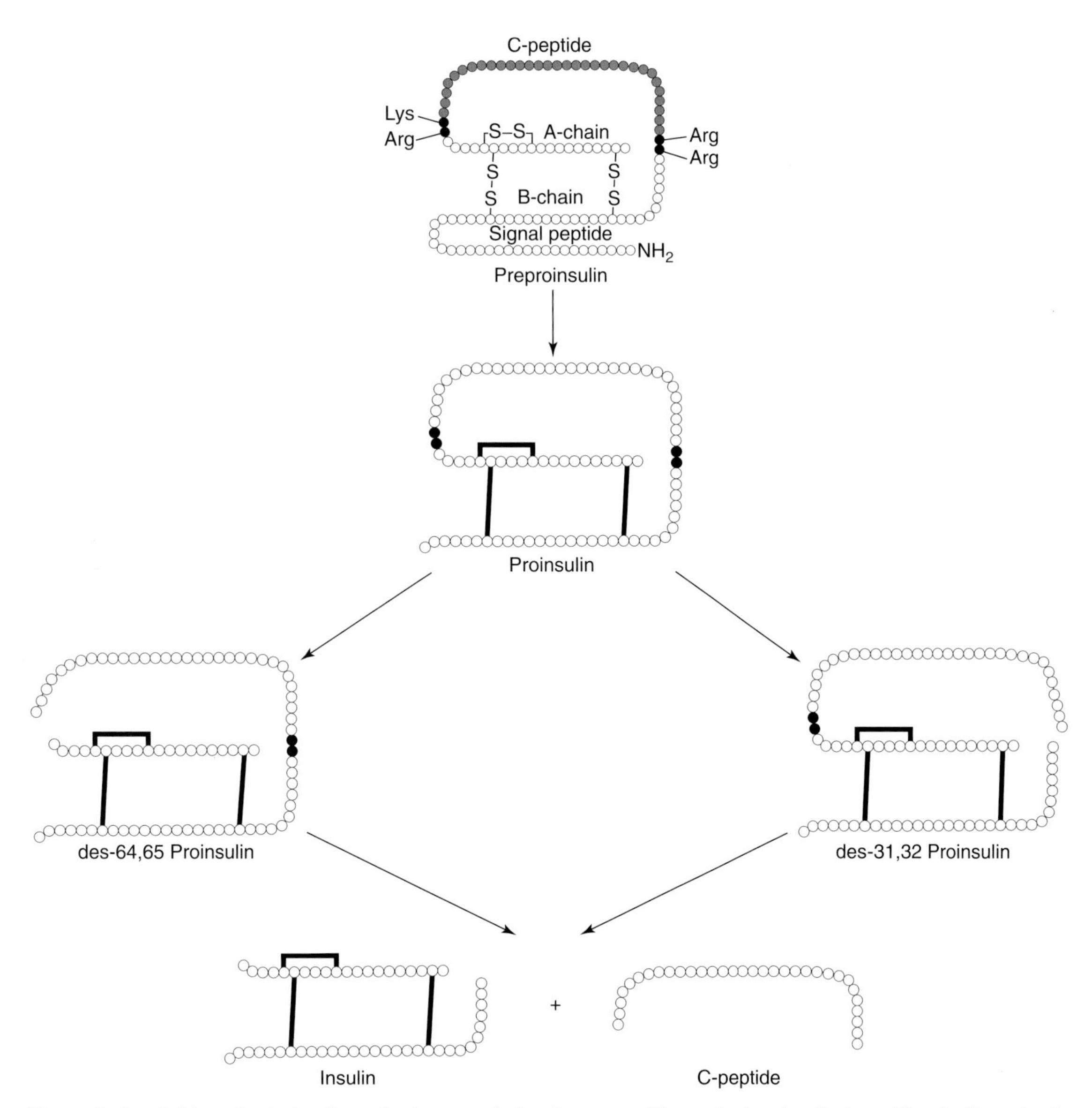

Figure 2 Insulin biosynthesis. Insulin synthesis occurs in the ribosomes of the endoplasmic reticulum of β-cells. Preproinsulin consists of four distinct domains. The signal peptide represents the first 24 residues and is cleaved to produce proinsulin. Conversion of proinsulin to insulin occurs via two endopeptidases, which cleave either arginine–arginine (B-chain/C-peptide junction) or arginine–lysine (A-chain/C-peptide junction) resulting in two proinsulin conversion intermediates des-31,32 proinsulin and des-64,65 proinsulin. A second step of endoproteolytic cleavage generates insulin and C-peptide. (Adapted from Rhodes, C. J.; Shoelson, S.; Halban, P. A. In *Joslin's Diabetes Mellitus*, 14th ed.; Baltimore, OH: Lippincott Williams & Wilkins, 2004, Chapter 5, pp 66, 69, with permission from Lippincott Williams & Wilkins.)

insulin growth factors (IGF-1 and IGF-2) and has a low affinity for the insulin receptor. Formation of disulfide linkages between the A and B chains precedes cleavage of an internal 31 amino C-peptide acid fragment. The Golgi apparatus, located near the cell membrane, packages insulin and C-peptide into secretory granules in equimolar amounts. C-peptide has no physiologic function, but is a clinically useful marker of β-cell function and endogenous insulin secretion.

6.19.3.1.2 Insulin secretion

Glucose stimulated insulin secretion is linked to its metabolism. Glucose enters the β-cell via the GLUT 2 glucose transporter and is subsequently phosphorylated by glucokinase, the glucose sensor of the β-cell, forming glucose-6-phosphate[26] (**Figure 3**). This is the rate-limiting step of insulin secretion. Metabolism of glucose-6-phosphate via glycolysis leads to mitochondrial production of ATP. Increasing the intracellular ATP to ADP ratio closes ATP-sensitive

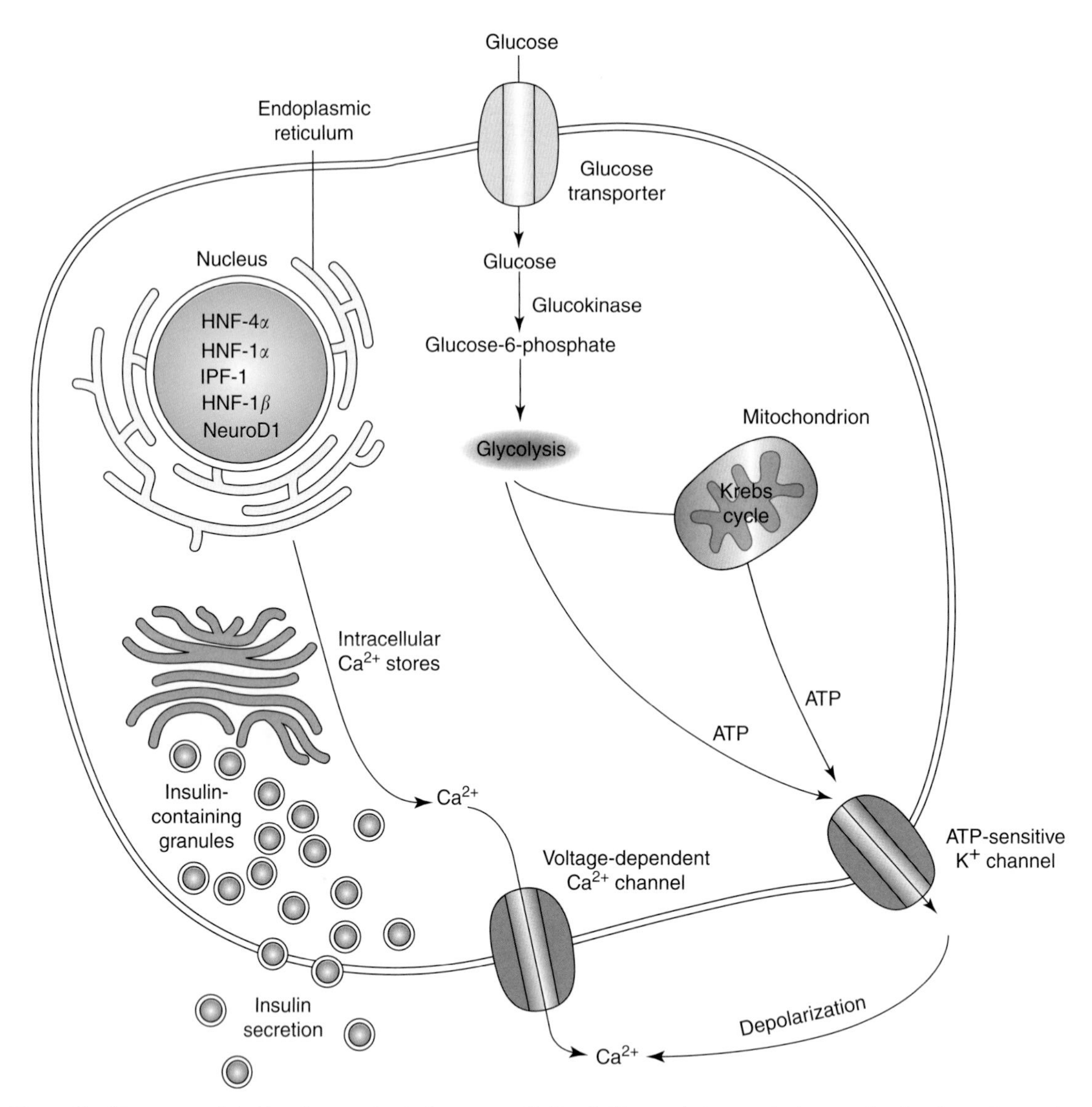

Figure 3 Glucose-mediated insulin secretion in the pancreatic β-cell. Glucose is transported across the β-cell membrane via a glucose transporter GLUT-2. Glucokinase, the glucose sensor of the β-cell and rate-limiting step of glucose-mediated insulin secretion, catalyzes the phosphorylation of glucose to form glucose-6-phosphate. ATP synthesis via glycolysis and the Krebs cycle leads to closure of the ATP-sensitive potassium (K^+) channels in the β-cell membrane, depolarization of the cell membrane, opening of voltage-dependent calcium channels, and influx of calcium ions that stimulates fusion of insulin-containing secretory granules with the plasma membrane releasing insulin into the circulation. (Reprinted from Fajans, S. S.; Bell, G.; Polonsky, K. S. *N. Engl. J. Med.* **2001**, *345*, 971–980. Copyright © 2006 Massachusetts Medical Society. All rights reserved.)

potassium (K^+) channels in the β-cell membrane. The K^+-ATP channel is composed of two subunits: a cytoplasmic binding site for sulfonylureas and ATP, designated the sulfonylurea receptor type 1 (SUR1); and a K^+ channel that acts as the pore-forming unit. Inhibition of this K^+ channel induces β-cell membrane depolarization, opens voltage-dependent calcium channels and permits influx of calcium ions. The increase in intracellular calcium concentration leads to the fusion of insulin-containing secretory granules with the plasma membrane and the release of insulin into the portal circulation. In addition to stimulating insulin secretion, glucose stimulates insulin biosynthesis by increasing the transcription rate of the insulin gene.

At plasma glucose levels $<70\,\mathrm{mg\,dL^{-1}}$, insulin is secreted at a low basal rate. Glucose-mediated insulin secretion increases with glucose levels $<70\,\mathrm{mg\,dL^{-1}}$ and follows an S-shaped response curve. The β-cell response to glucose concentration plateaus at glucose levels $>450\,\mathrm{mg\,dL^{-1}}$.[27]

There are two phases of insulin release in response to an acute increase in glucose concentration. First-phase insulin secretion occurs rapidly (within 2 min), resulting in a sharp peak in plasma insulin concentrations; it then returns to basal levels within 10 min. If exposure to glucose is prolonged, a slower, second-phase insulin secretion is observed. Continuous exposure to a high concentration of glucose results in a relative reduction in insulin secretion, due to the desensitization of β-cells to glucose. First-phase insulin secretion represents the release of mature secretory granules close to the β-cell plasma membrane; second-phase insulin secretion corresponds both to release of stored secretory granules and de novo synthesis of insulin. Abnormalities in the insulin secretory patterns constitute the earliest sign of β-cell dysfunction in most types of diabetes.

6.19.3.1.3 Insulin action

The liver degrades up to 50% of insulin secreted into the portal venous circulation ('first-pass effect'). The remaining insulin enters the systemic circulation and travels to target tissues. After insulin binds to an insulin receptor on a target cell, the activated receptor tyrosine kinase catalyses autophosphorylation of the insulin receptor, which itself acts as a tyrosine kinase, phosphorylating insulin receptor substrate (IRS) proteins. Through a complex cascade of phosphorylation and dephosphorylation reactions, the metabolic effects of insulin are initiated and include glucose transport, glycogen synthesis, protein synthesis, lipogenesis, and regulation of various genes in insulin-responsive cells. In muscle and adipose tissue, translocation of the glucose transport protein GLUT4 to the cell membrane is facilitated by activation of the phosphatidylinositol-3-kinase (PI-3-K) pathway[28] (**Figure 4**).

Insulin has profound effects on carbohydrate, lipid, and protein metabolism, and significant influences on mineral metabolism. Consequently, derangements in insulin signaling have widespread and devastating effects on the function of many organs and tissues. Most importantly, insulin facilitates entry of glucose into muscle, adipose, and other tissues

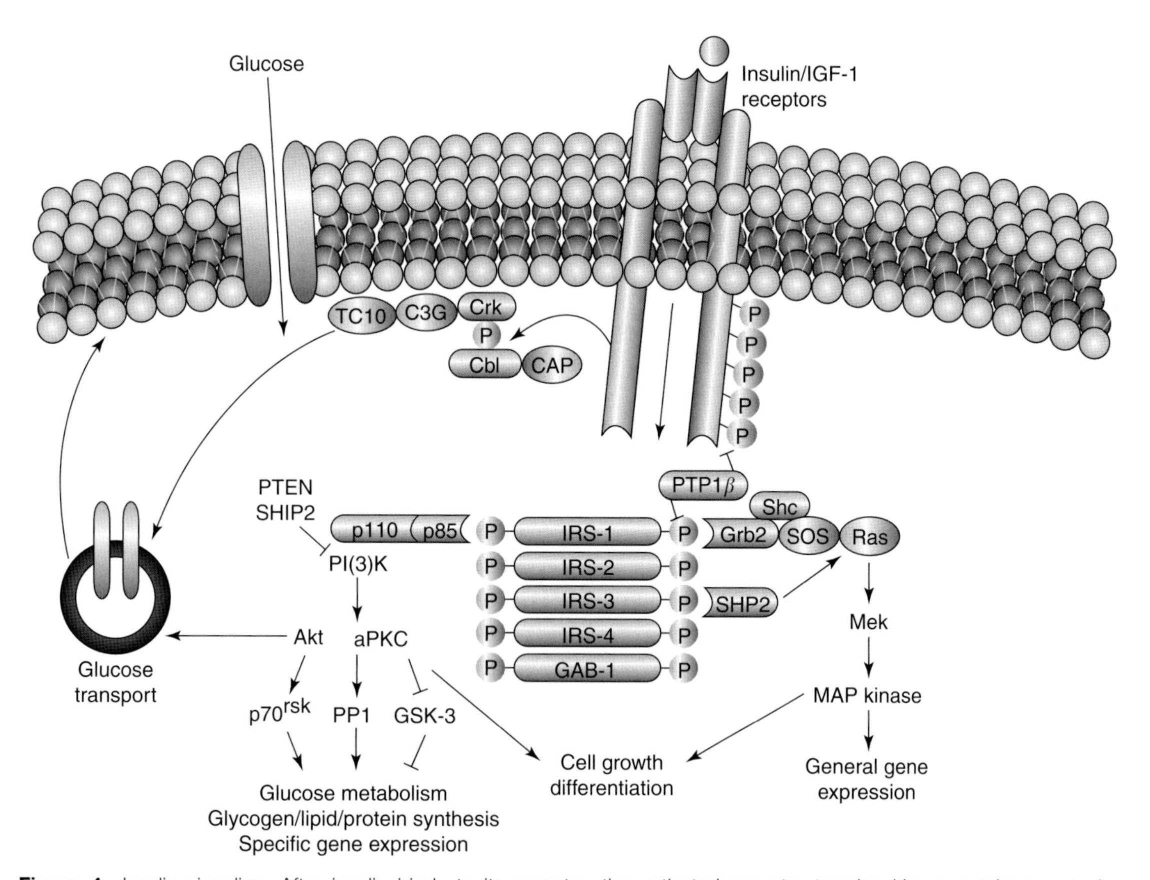

Figure 4 Insulin signaling. After insulin binds to its receptor, the activated receptor tyrosine kinase catalyses autophosphorylation of the receptor and subsequent phosphorylation of intracellular substrates, such as the IRS proteins. Phosphorylated tyrosine residues serve as docking and activation platforms for multiple signaling molecules, such as phosphatidylinositol (PI) 3-kinase, that mediate many metabolic actions of insulin. (Reprinted from Saltiel, A. R.; Kahn, C. R. *Nature* **2001**, *414*, 799–806, with permission from Nature.)

with insulin-dependent glucose transporters, as described previously. In the liver, insulin inhibits the activity of glucose-6-phosphatase and stimulates glycogen synthase and phosphofructokinase, enzymes that are directly involved in glycogen synthesis (glucose storage). With regard to lipid metabolism, insulin promotes fatty acid synthesis in the liver and inhibits lipolysis of stored triglyceride. In addition to increasing peripheral glucose uptake into target cells, insulin stimulates the uptake of amino acids, potassium, magnesium, and phosphate ions, which are critical in glycogen and protein synthesis.

6.19.3.2 Pathogenesis of Type 1 Diabetes Mellitus

The development of T1DM involves the interaction of genetic, environmental, and autoimmune components that ultimately result in severe insulin deficiency due to cell-mediated autoimmune inflammation and destruction of the pancreatic β-cells. The preclinical phase involves the presence of autoantibodies and reactive T cells with β-cell antigens in the setting of normoglycemia. Lymphocytic infiltration of the pancreatic islet cells ensues and results in β-cell destruction. As β-cell destruction occurs, insulin secretory capacity progressively decreases, but hyperglycemia usually does not develop until almost 90% of β-cell mass has been lost. The interval during which β-cell destruction occurs is variable, and ranges from months to years. The end result is absolute insulin deficiency. Islet cell injury is reflected by the presence of autoantibodies to islet cells and specific autoantibodies to insulin, to glutamic acid decarboxylase (GAD), and to the tyrosine phosphatases IA-2 and IA-2β.

Hyperglycemia clinically manifests as a syndrome of symptoms including polyuria, polydipsia, polyphagia, blurry vision, weight loss, and dehydration. Commonly, T1DM initially presents with acute metabolic decompensation known as diabetic ketoacidosis (DKA). DKA is frequently triggered by a physical stress, such as puberty or an infection, during which insulin demand escalates without the capacity to concurrently increase insulin secretion.

Although individuals are rarely obese when they develop T1DM, the presence of obesity is not incompatible with the diagnosis. Individuals with T1DM can still develop insulin resistance, particularly if they develop abdominal obesity.

6.19.3.2.1 Genetic predisposition

Genetic predisposition for T1DM is polygenic in origin. In monozygotic twins, the concordance for T1DM is approximately 50%[29] and the risk to a first-degree relative is only 5%,[30] implying that additional factors are involved in disease development. The familial aggregation of T1DM is partially related to genes in the major histocompatibility locus on the short arm of chromosome 6. This region encodes human leukocyte antigen (HLA) class II molecules responsible for antigen presentation to helper T cells. The most important HLA molecules, DQ and DR, code for antigens expressed on macrophage and B lymphocyte cell surfaces. Alterations in these proteins, even a single amino acid substitution, may change the binding affinity of antigens for HLA molecules and the subsequent immune response. Individuals expressing a DR3 and/or DR4 haplotype are at highest risk for T1DM, whereas the presence of HLA-DR2 is associated with protection from T1DM. One non-HLA gene has been clearly associated with an increased risk of T1DM. The IDDM-2 gene, a promoter region of the insulin gene, is located on chromosome 11 and accounts for up to 10% of familial aggregation of T1DM. However, most individuals with predisposing haplotypes do not develop diabetes.

6.19.3.2.2 Presence of autoimmunity

In individuals in the preclinical and early clinical phase of T1DM, pancreatic β-cell destruction correlates with the presence of autoantibodies toward the antigens noted above. Other less clearly defined autoantigens include an islet ganglioside, carboxypeptidase H, and phogrin (insulin secretory granule protein). Serological evaluation for autoantibodies can be used clinically in classifying individuals with T1DM and to identify at-risk individuals. At the time of diagnosis, 85–90% of individuals with T1DM have detectable levels of one or more of these autoantibodies. Other types of islet cells are spared from this autoimmune process.

Pathologically, lymphocytic infiltration of the pancreatic islet cells ensues and results in β-cell destruction. β-cells are susceptible to the toxic effects of cytokines including tumor necrosis factor α (TNF-α), interferon γ (INF-γ), and interleukin-1 (IL1). The mechanism of β-cell death is unknown, but putative mechanisms implicate nitric oxide metabolites, apoptosis, and T cell cytotoxicity. After complete β-cell destruction, the inflammatory process ceases and titers of circulating autoantibodies decline, often to undetectable levels.

6.19.3.2.3 Environmental factors

Demographic differences illustrate the important influence that environmental factors play in the development of T1DM. Suspected environmental triggers of the autoimmune process in genetically susceptible individuals include viruses (mumps, rubella, and coxsackie virus B4), vaccinations, bovine milk ingestion in early childhood, toxic chemicals, and other destructive cytotoxins. Congenital rubella infection is the only clearly identifiable environmental factor associated with an increased risk of developing T1DM. Linkage between environmental factors and the development of T1DM is difficult to establish, as the environmental trigger can precede the clinical onset of diabetes by many years. The mechanisms by which genetic and environmental factors trigger the autoimmune process remain incompletely understood. One hypothesis invokes molecular mimicry, meaning the immune system mistakenly targets β-cell proteins that share homologies with certain viral or other foreign peptides and elicits an immune response.

6.19.3.3 Pathogenesis of Type 2 Diabetes Mellitus

T2DM is a heterogeneous syndrome characterized by impaired insulin secretion and insulin resistance, resulting in increased hepatic glucose production and decreased peripheral glucose uptake in target tissues. Both defects in glucose homeostasis are present prior to the onset of hyperglycemia, but after hyperglycemia ensues, the predominant factor in disease progression is deterioration of β-cell function and impaired insulin secretion, with some secondary increase in insulin resistance[31] (**Figure 5**).

6.19.3.3.1 Abnormalities in glucose homeostasis

Fasting hyperglycemia in T2DM develops primarily as a result of increased endogenous glucose production due to relative insulin deficiency and hepatic insulin resistance. Elevated circulating concentrations of glucagon, FFAs, and precursors of gluconeogenesis potentiate an increase in endogenous glucose production. In the fasting state, peripheral glucose uptake and metabolism are relatively insulin-independent processes. Postprandially, glucose enters the systemic circulation via intestinal absorption. Normally, glucose concentration reaches its peak 30–60 min after beginning a meal and returns to basal levels within 2–3 h, primarily due to suppression of hepatic glucose production and the stimulation of peripheral glucose uptake. The elevated insulin levels suppress glucagon and FFA and reduce hepatic glucose production. In T2DM, postprandial glucose excursions are exaggerated and prolonged because of dual defects: decreased insulin sensitivity and impaired insulin secretion. The result is subnormal suppression of endogenous glucose production and reduced glucose uptake in muscle and adipose tissue.

6.19.3.3.2 Genetic predisposition

T2DM has strong genetic predisposition. In monozygotic twins, the concordance of T2DM is nearly 90%.[32] If both parents have T2DM, the lifetime prevalence for their prodigy is around 40%. T2DM is assumed to be polygenic in

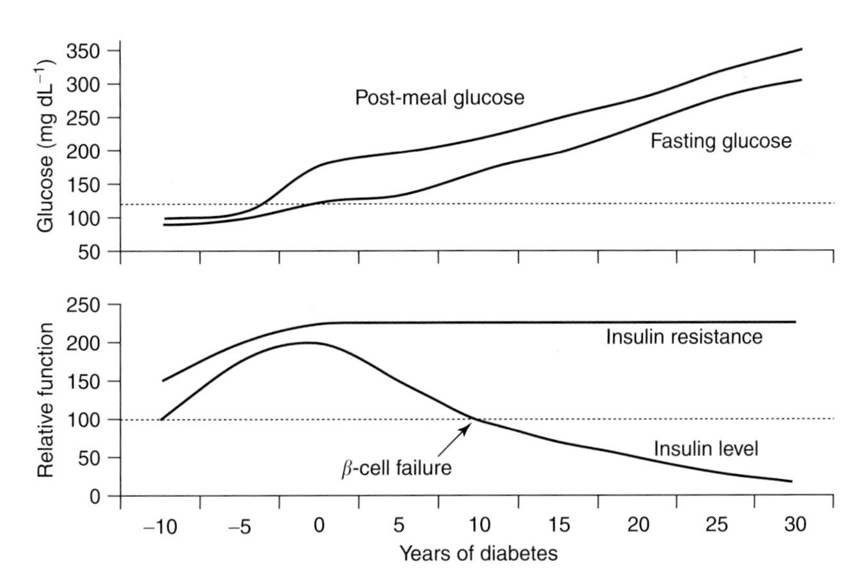

Figure 5 Natural history of type 2 diabetes mellitus. Insulin resistance requires an increase in insulin synthesis and secretion to maintain normoglycemia. When β-cells are unable to compensate for the level of insulin resistance, hyperglycemia develops. Once hyperglycemia ensues, the predominant factor in disease progression is further deterioration of β-cell function and impaired insulin secretion. (Adapted from Bergenstal, R. M.; Kendal, D. M.; Franz, M. J.; Rubenstein, A. H. In *Endocrinology*, 4th ed.; DeGroot, L. J., Jameson, J. L., Eds.; WB Saunders: Philadelphia, PA, 2001, Chapter 58, p 822, with permission from Elsevier.)

origin, but to date only susceptibility loci, not causative genes, have been identified. In genetically predisposed individuals, the risk of developing T2DM is influenced by environmental factors such as obesity, level of physical activity, and dietary fat consumption. Studying and identifying the genetic component of T2DM is difficult as the phenotype is dependent on these environmental factors or concurrent genetic defects affecting insulin secretion or action. Monogenic alterations of the insulin receptor, insulin-signaling cascade, or other enzymes involved in glucose homeostasis account for only a small subset of individuals with T2DM.

6.19.3.3.3 Metabolic abnormalities in type 2 diabetes

6.19.3.3.3.1 Insulin resistance

A decrease in insulin sensitivity, commonly referred to as 'insulin resistance', results from reduced activity in the insulin receptor or the insulin-signaling cascade, impeding insulin action. Insulin resistance is relative and can be overcome at least in part by higher insulin concentrations. In the presence of insulin resistance, in vivo and in vitro dose–response curves of insulin action display a rightward shift, indicating that a higher insulin level is needed to attain the same level of insulin action. In muscle, insulin resistance results in decreased glucose transport across the cell membrane (impaired glucose uptake) and decreased glycogen formation, thereby contributing to postprandial hyperglycemia. In the liver, insulin resistance impairs suppression of glucose production in the fasting and postprandial states, resulting in a high basal plasma glucose concentration. In adipose tissue, insulin resistance decreases inhibition of lipolysis, resulting in an increase in plasma FFA concentrations.

Many components contribute to the degree of insulin resistance manifested by an individual, including genetic factors, age, physical fitness, dietary nutrients, medications, and body fat distribution. Potentially modifiable factors, in particular obesity, account for the majority of insulin resistance that predisposes individuals to develop the metabolic syndrome and T2DM. Specifically, the amount of visceral adiposity is strongly correlated with insulin resistance. Adipocytes, initially thought to function solely in energy storage, function as an active endocrine organ, synthesizing and secreting various adipocytokines that play an active role in glucose homeostasis. For example, decreased insulin sensitivity is associated with elevations of TNF-α, interleukin-6, FFAs, adipson, macrophage and monoctye chemoattractant protein-1 (MCP-1), plasminogen activator inhibitor-1 (PAI-1), and resistin.[33] Conversely, leptin, adiponectin, and visfatin are linked to greater insulin sensitivity, and a reduction or deficiency in these adipocytokines may, in fact, worsen insulin resistance. Genetic factors, the mediation of which is unknown, can also affect an individual's visceral fat composition. Relatively lean people may have reduced insulin sensitivity, and in some cases, a higher than expected proportion of visceral fat content may be a contributing factor.

6.19.3.3.3.2 Impaired insulin secretion

In the presence of insulin resistance, β-cells increase insulin synthesis and secretion to maintain normoglycemia, and insulin levels can rise 2–3 times above those in individuals with high insulin sensitivity. Despite this need for compensatory hyperinsulinemia, the majority of obese individuals do not develop overt diabetes because, to do so, a concomitant defect in insulin secretion must be present. One of the earliest manifestations of impaired β-cell function in persons at risk for T2DM is reduced first-phase insulin response to an acute rise in glucose concentration. This defect, and others, may already be evident in those with IGT and in first-degree relatives of individuals with T2DM who have normal glucose tolerance.[34] As T2DM progresses, basal insulin secretion and second-phase insulin responses to glucose challenges also become impaired and hyperglycemia worsens. β-cell function progressively deteriorates and can result in severe β-cell failure with absolute insulin deficiency.

The quantitative relationship between insulin sensitivity and insulin secretion can be used as a measure of β-cell function. With normal β-cell function, the sensitivity–secretion relationship is most efficiently expressed as a rectangular hyperbola[35,36] (**Figure 6**). The product of insulin sensitivity and insulin secretory response equals a constant referred to as the *disposition index* (DI). The DI measures the ability of the β-cells to compensate for insulin resistance such that shifts in insulin sensitivity are accompanied by compensatory alterations in β-cell sensitivity to glucose. Thus, an increase in insulin resistance should upregulate β-cell secretion of insulin, whereas a decrease in insulin resistance would downregulate the β-cell secretory response. Loss of β-cell function is associated with a lower DI, represented by a curve shifted closer to the origin. A reduced DI may be the earliest phenotypic sign of a β-cell defect in an otherwise glucose-tolerant individual.

The underlying mechanism(s) leading to progressive β-cell dysfunction in T2DM is not fully defined. Time-dependent decreases in both mass and function of β-cells have been documented. Some factors hypothesized to play a role include: a genetic β-cell defect superimposed on insulin resistance; increased apoptosis and decreased regeneration

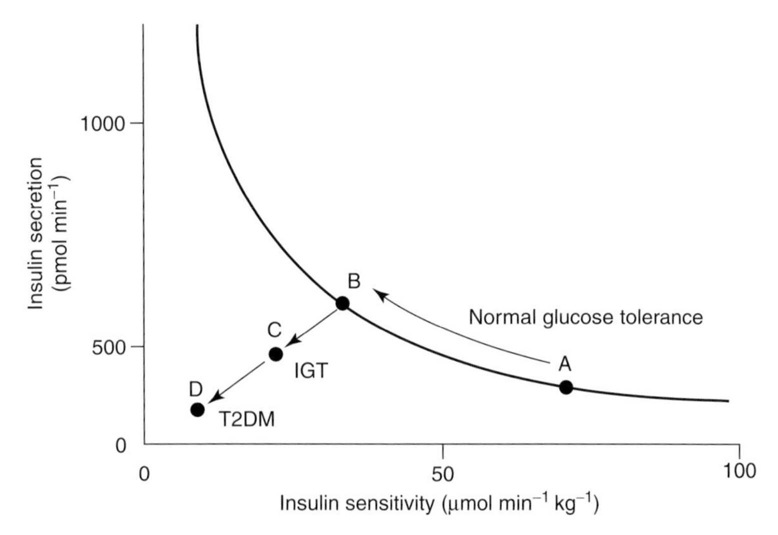

Figure 6 The hyperbolic insulin sensitivity/secretion curve. In normal individuals, insulin secretion and insulin sensitivity are related and follow a hyperbolic curve. As insulin sensitivity decreases (or insulin resistance increases) by moving up the curve from point A to point B, insulin secretion increases. In individuals with abnormal glucose homeostasis, insulin secretion cannot compensate appropriately for a change in insulin resistance, leading to the development of impaired glucose tolerance (IGT; point C) or type 2 diabetes (T2DM; point D). (Adapted with permission from Kahn, S. E. *J. Clin. Endocrinol. Metab.* **2001**, *86*, 4047. Copyright 2001, The Endocrine Society.)

of β-cells; 'exhaustion' of β-cell function secondary to long-standing insulin resistance; β-cell desensitization from glucose or lipid toxicity as a result of chronic hyperglycemia; and elevated FFA levels, and amyloid deposition, causing a reduced β-cell mass.

6.19.4 Animal Models of Diabetes

Animal models have been essential to diabetes research since before the discovery of insulin in 1922. For more than 50 years, the insulin used clinically for treatment was isolated from bovine and swine pancreases obtained at packing houses throughout the world. Animal models have been very informative for studies of regulation of nutrient metabolism and understanding the mechanisms of insulin resistance. Models of T1DM consist of insulinopenic animals whose β-cells have been destroyed chemically by autoimmunity or are genetically absent. Animal models of T2DM have been developed by the selective breeding of naturally diabetic animals and genetic alterations inducing diabetes, obesity, or insulin resistance. This section is not intended to be a comprehensive review of animal models of diabetes, but to provide a sample of current research models. **Table 4** summarizes the animal models that are currently used in diabetes research.[37]

6.19.4.1 Animal Models of Type 1 Diabetes

6.19.4.1.1 Animals with β-cell destroyed by chemical cytotoxins

Alloxan, a pyrimidine structurally similar to glucose and uric acid, directly disrupts β-cell membrane permeability and produces irreversible β-cell damage within 12 h in most animals. Intracellularly, alloxan is reduced to dialuric acid, which then undergoes autoreoxidation to alloxan. As a byproduct of this cycle, superoxide radicals are produced and cause damage to various cellular constituents.[38] Furthermore, inhibition of the tricarboxylic acid cycle and calcium (Ca^{2+})-dependent dehydrogenases in the mitochondria causes ATP deficiency, cessation of insulin production, and eventual β-cell death.

Streptozotocin is an equally destructive agent for β-cells. The molecular structure of streptozotocin includes a glucose-like moiety that facilitates transportation across the cell membrane, and a nitrosourea moiety that causes β-cell toxicity. Streptozotocin induces DNA strand breaks and reduces the intracellular level of NAD, a substrate for poly (ADP-ribose) synthase that is involved in DNA repair.[39] Streptozotocin-treated animals may retain some insulin-secreting capacity due to its dose-dependent effects, thereby avoiding a ketotic state and dependence on exogenous insulin for survival. Streptozotocin-treated animals, however, have a high frequency of pancreatic islet tumors.

Table 4 Animal models in diabetes research

Animals with β-cells destroyed by chemical cytokines	Alloxan Streptozocin
Animals with autoimmune diabetes with spontaneous onset causing β-cell loss	BB rats NOD mice LETL rats Torri rat LEW.1AR1/ZTM-iddm rat
Genetically altered animals with various forms of diabetes	Multiple transgenic and gene-disrupted animals (mainly mice)
Insulin-resistant mutant rodents with diabetes and obesity	C57BKs *db* mice (*lepr^db^*) C57BL6J *ob* mice (*lep^ob^*) Yellow A^{v} and A^{vy} mice KK mice NZO mice Zucker *fa* rats (*lepr^fa^*) and BBZ/Wor rats Zdf/Drt-*fa* rats Wistar–Kyoto diabetic/fatty rat group Corpulent rat group including SHR/N-*cp*, LA/N-*cp*, SHHF/Mcc-*cp*
Rodents with spontaneous diabetes of varying etiology	NON mice WBN/Kob rats eSS rats BHE/Cdb rats OLETF rats NSY mice Koletzky (SHROB) rats (*fa^k^*) Hypertriglyceridemic (HTG) rats
Rodents with overnutrition-evoked diabetes and obesity	*Psammomys obesus* (sand rats) *Acomys cahirinus* (spiny mice) C57BL/6J mice
Diabetic rodents isolated by selective breeding from normal pools	GK (Goto–Kakizaki) rats Cohen sucrose-induced rats
Diabetic non-rodents	Primates Dogs and cats

Reprinted from Shafrir, E. In *Ellenberg & Rifkin's Diabetes Mellitus,* 6th ed.; Porte, D., Jr., Sherwin, R. S., Baron, A., Eds.; McGraw-Hill: New York, 2003, Chapter 16, p 231, with permission from McGraw-Hill.

6.19.4.1.2 Animals with autoimmune diabetes

Animal models of T1DM (autoimmune diabetes in BB rats and NOD mice) have been highly instrumental in the rapid development of our current understanding of the basic mechanisms of autoimmune disorders in humans, especially T1DM.

BB rats abruptly develop classic features of decompensated diabetes (weight loss, polyuria, polydipsia, glucosuria, and ketoacidosis) between 60 and 120 days of age. They have marked hyperglycemia, hypoinsulinemia, insulitis, and loss of β-cells. Over the course of 1–5 days, endogenous insulin levels diminish and these diabetic animals become totally insulin dependent for survival. The autoimmune process[40] in BB rats is not isolated to the islet cells, but frequently results in a polyendocrine syndrome associated with lymphocytic thyroiditis, and autoantibodies against smooth and skeletal muscle, gastric parietal cells, thyroglobulin, and thyroid cells.

The non-obese diabetic (NOD) mouse has provided insight into the complex interaction between a genetic predisposition for T1DM and the role of environmental, nutritional, and hormonal influences on disease penetrance. NOD mice have a polygenic mutation that causes the absence of a histocompatibility molecule and affects immune function. NOD mice develop insulitis around 4–5 weeks of age, characterized by lymphocytic islet infiltration and β-cell destruction.[41] Progression to overt diabetes occurs between 13 and 30 weeks. Classically, the mice develop severe hyperglycemia, insulin deficiency, and β-cell loss.

6.19.4.2 Animal Models of Type 2 Diabetes

6.19.4.2.1 Genetically altered animals

Genetic manipulation has been widely used to delete specific genes in the insulin signal transduction pathway, providing significant insights into molecular mechanism and biochemical pathways of human metabolism. Homologous recombination targeted gene knockouts in mice have become a powerful strategy for the study of monogenic and polygenic disorders.

Monogenic defects in insulin action have been demonstrated through deletion of whole-body or tissue-specific components of the insulin-signaling pathway (**Figure 4**). Various degrees of insulin resistance can be created in mice, depending on the specific protein and on its localization or importance in the insulin-signaling cascade. Insulin receptor heterozygous knockout mice exhibit a 50% reduction in insulin receptor expression and develop overt diabetes. IRS-deficient mice manifest significant hepatic insulin resistance and lack compensatory β-cell hyperplasia.[42] GLUT-4 heterozygotic knockout mice show a 50% decrease in GLUT-4 expression in adipose tissue and skeletal muscle and have a phenotype with hypertension, insulin resistance, and T2DM. However, GLUT-4 homozygotic knockout mice only develop moderate insulin resistance, suggesting a compensatory mechanism for glucose transport that has not been identified. Polygenic models support the hypothesis that multiple minor defects in insulin secretion and insulin action leads to T2DM and emphasize the importance of interactions of different genetic loci in the production of diabetes. Knockout animal models, however, provide insight only into the function of specific gene products and do not necessarily invoke meaning into interspecies physiology.

6.19.4.2.2 Insulin-resistant mutant rodents with diabetes and obesity

Homozygous mice for Lep^{ob} mutation fail to produce leptin, resulting in a markedly obese, hyperphagic mouse with insulin resistance and elevated insulin levels. Additional hypothalamic abnormalities of leptin deficiency contribute to obesity by causing a hypometabolic state. Mice with homozygosity for the leptin receptor mutation (Lepr^{db}) have a more severe diabetic phenotype, becoming mildly hyperglycemic at 6–8 weeks and overtly diabetic within 4–6 months due to β-cell dysfunction.[43] The New Zealand obese (NZO) mouse[44] has a defect in the glycolytic pathway in pancreatic β-cells resulting in defective glucose-dependent insulin secretion and increased adiposity. Subsequent hepatic and peripheral insulin resistance with glucose intolerance is seen in obese NZO males. Unfortunately, NZO mice are extremely susceptible to autoimmune disorders affecting both connective tissue and the insulin receptor, which make analysis of this model complicated. The Zucker Diabetic Fatty (ZDF) rat is a useful model for T2DM in that it develops impaired glucose tolerance in the presence of an inherited obesity gene mutation that results in a shortened leptin receptor protein and leptin insensitivity. The ZDF rat phenotype includes obesity, elevated leptin levels, hyperglycemia, insulin resistance, T2DM, hypertriglyceridemia, and hypercholesterolemia. Female ZDF rats do not have the same profile as their male counterparts, possibly due to a slower β-cell depletion rate than that of the male.

6.19.4.2.3 Rodents with spontaneous diabetes

Breeding stock for the Bureau of Home Economics (BHE/Cdb) rat[45] was selected for the presence of hyperlipidemia and hyperglycemia in the absence of obesity and renal dysfunction. Maternally inherited mitochondrial DNA mutations are responsible for hepatic defects in the coupling of mitochondrial respiration to ATP synthesis and results in elevated rates of gluconeogenesis. Initially, the increase in hepatic glucose production is compensated by increased peripheral tissue glucose uptake, oxidation, and lipogenesis. Diminished glucose-stimulated insulin secretion follows and leads to diabetes.

Otsuka Long–Evans Tokushima Fatty (OLETF) rats[46] have a selective deletion mutation in a cholecystokinin receptor and are associated with a sex-linked gene *Odb-1* that leads to a phenotype comparable to T2DM. The characteristic features of OLETF rats are mild obesity, hyperlipidemia, a late onset of hyperglycemia (after 18 weeks of age) followed by development of overt DM.

6.19.4.2.4 Rodents with overnutrition-evoked diabetes and obesity

C57BL/6J mice develop severe obesity, hyperglycemia, hyperinsulinemia, and hyperlipidemia if exposed to a high-fat, high-sucrose diet after weaning. Insulin resistance and a blunted glucose-dependent insulin secretion have been observed in this animal model. The gerbil, *Psammomys obesus* (sand rat),[47] is characterized by primary insulin resistance and is a well-defined model for dietary-induced T2DM. Feeding *Psammomys* a high caloric diet results in a reversible metabolic syndrome consisting of hyperglycemia, hyperinsulinemia, hypertriglyceridemia, and a moderate elevation in body weight. Reducing caloric intake restores normoglycemia in 90% of the animals with recovery of insulin secretion and β-cell granulation. The metabolism of *Psammomys* is well adapted toward life in a low-energy environment as found in the desert. The capacity to constantly accumulate adipose tissue allows for sustenance and breeding in periods of scarcity.

6.19.5 Clinical Trial Issues

Large-scale clinical trials have been very instrumental in the major advances in diabetes treatment and outcomes that have been achieved in the last 15–20 years. The first large clinical study, the University Group Diabetes Program (UGDP), which ended in 1970, reached the very controversial conclusion that use of the oral sulfonylurea, tolbutamide for the treatment of T2DM was associated with an increase in CVD mortality.[48] Despite the controversy, the successful conduct of the UGDP was a major achievement. It illustrated that clinical trials of complex disease entities, such as diabetes, could be successfully conducted and paved the way for landmark intervention trials including the Diabetes Control and Complications Trial (DCCT), the United Kingdom Prospective Diabetes Study (UKPDS), and the Diabetes Prevention Program (DPP) that in the last two decades have generated the results that guide clinical care today.

From a current perspective, it is difficult to realize that until the DCCT (*see* Section 6.19.6.1) was successfully completed, there was much uncertainty about the direct relationships between metabolic control and the development of microvascular complications. Indeed, it was commonly thought that glucose control was less important than genetic factors in placing individuals with diabetes at risk for complications. This uncertainty about the role of glycemic control was greatly confounded by the fact that, prior to the time that HbA_{1C} and capillary blood glucose testing were available, efforts to derive integrated indices of metabolic control relied heavily on semiquantitative measures of urinary glucose.

Randomized controlled clinical trials have also been highly instrumental in demonstrating the effectiveness of new therapeutic options for the treatment or prevention of diabetic retinopathy (laser photocoagulation) and nephropathy (angiotensin-converting enzyme inhibitors (ACE-I) and angiotensin receptor blocking (ARB) drugs). Cardiovascular diseases account for two-thirds of the mortality of persons with diabetes. Compared with the general population, women with diabetes have an even greater relative risk for development of CVD than men. In the last three decades there have been major improvements in treatment of and reductions in mortality from CVD in the population at large. It has been more difficult to determine the effectiveness of various approaches to CVD prevention and treatment in persons with diabetes. In part, this has stemmed from early tendencies to preclude persons with diabetes from intervention trials because it was known that CVD was so highly prevalent and at younger age in individuals with diabetes. This trend has been reversed and individuals with diabetes (including women) are specifically included in ongoing CVD clinical trials.

Clinical trials in diabetes are difficult to conduct, expensive, labor intensive on both investigators and participants and, thus, require a significant commitment of resources. Often, long intervals of time are necessary to verify observational data or to measure the effects of therapeutic interventions on glycemic control or diabetic complications. Nevertheless, it is very likely that well-designed clinical trials will be necessary in the future as in the past.

6.19.6 Current Treatment

6.19.6.1 Benefits of Glycemic Control

Glycemic control is critical to the prevention of diabetic complications. The duration and severity of hyperglycemia correlates with both the development and rate of progression of diabetic retinopathy, nephropathy, and neuropathy. The DCCT and the UKPDS established the effectiveness of glycemic control in the prevention of diabetic complications in T1DM and T2DM, respectively, during the 1990s.

The DCCT studied the effect of an intensive insulin regimen on the primary and secondary prevention of diabetic microvascular complications in individuals with T1DM.[49] This prospective study included 1441 individuals with T1DM who were randomized to receive intensive glycemic management or conventional insulin therapy. Those assigned to intensive therapy attempted to achieve and maintain near-normal glucose levels by using ≥ 3 daily insulin injections or an insulin pump, self-monitoring blood glucose levels ≥ 4 times/day, and monthly visits to a multidisciplinary healthcare team. The conventional treatment group continued to use 1–2 injections of daily insulin, daily self-monitoring of blood sugars, and had a routine follow-up visit every 3 months. At study initiation, both groups had a mean HbA_{1C} around 9%. Within 3 months, the intensive group achieved and maintained an average HbA_{1C} of 7%, whereas the mean HbA_{1C} remained at 9% throughout the study in the conventional treatment group.

Intensive therapy reduced the risk for developing retinopathy by 76% (primary prevention) and slowed progression of pre-existing retinopathy by 54%, the incidence of severe retinopathy by 47%, and the need for laser therapy by 56% (secondary prevention). In the primary prevention cohort, the appearance of neuropathy was reduced by 69% at 5 years in subjects on intensive therapy compared with subjects on conventional therapy. In the secondary prevention cohort, intensive therapy reduced the appearance of clinical neuropathy at 5 years by 57%. Furthermore, intensive therapy prevented the development and slowed the progression of diabetic kidney disease by 50%. Intensive therapy, however, was associated with greater weight gain (about 1 kg $year^{-1}$) and a threefold greater risk of severe hypoglycemia (loss of consciousness or need of second person assistance) compared with conventional therapy.[50]

The Epidemiology of Diabetes Interventions and Complications (EDIC) study is examining the long-term effects of conventional versus intensive diabetes treatment received during the DCCT on the subsequent development and progression of microvascular, neuropathic, and cardiovascular complications. The previous intensive and conservative treatment arms are combined and have a mean HbA_{1C} of 8%. Results from EDIC indicate that individuals with T1DM have a metabolic memory, where the benefits of a period of good glucose control continue even after HbA_{1C} levels have risen.[51] Specifically, DCCT subjects who received intensive therapy have shown a 57% reduction in cardiovascular endpoints (myocardial infarctions and strokes)[52] and similar benefits with regards to the development of nephropathy,[53] hypertension, and retinopathy.

The UKPDS, conducted from 1977 to 1997 in 23 centers throughout England, Northern Ireland, and Scotland, verified that optimal glycemic control has similar benefits in T2DM. This prospective, randomized intervention trial of 5102 newly diagnosed individuals with T2DM examined the effects of glucose control on cardiovascular and diabetic complications via diet therapy or pharmacologic intervention with sulfonylureas, metformin, or insulin therapy.[54] After a 3-month dietary run-in period, subjects with persistent hyperglycemia were randomized to conventional therapy with diet alone (goal FPG $<270\ mg\,dL^{-1}$ ($15\ mmol\,L^{-1}$)) or intensive therapy (goal FPG $<110\ mg\,dL^{-1}$ ($6\ mmol\,L^{-1}$)) with pharmacological monotherapy. Subjects receiving pharmacologic intervention were further randomized to receive sulfonylurea, metformin, or insulin therapy. Concurrently, a cohort of the subjects was tested to determine if blood pressure control reduced morbidity and mortality in T2DM. The antihypertensive agents used primarily were the ACE inhibitor, captopril, and the beta blocker, atenolol.

Similarly to the DCCT, the UKPDS found that a reduction in HbA_{1C} resulted in decreased microvascular complications. Over the duration of the study, the intensive group had a mean HbA_{1C} of 7.0% versus 7.9% in the conventional group. Diabetic retinopathy was reduced by 25% and early diabetic nephropathy (defined by the presence of microalbuminuria) by 33%. Overall, a 1% decrement in HbA_{1C} translated to a 35% reduction in risk of diabetic complications. Similar to the DCCT, more frequent hypoglycemia and greater weight gain was observed in the intensive therapy cohort, with the exception of the metformin arm. Independently, improvements in blood pressure control reduced the risk of diabetes-related endpoints by 24%, diabetes-related deaths by 32%, and microvascular endpoints by 37%.

6.19.6.2 Goals of Diabetic Management

The goal of diabetic management is to achieve and maintain normal or near-normal fasting, pre-meal and postprandial blood glucose concentrations. The ADA has defined goals as follows: HbA_{1C} $<7\%$, FPG 90–130 $mg\,dL^{-1}$ (5.0–7.2 $mmol\,L^{-1}$), and a 2hPG $<180\ mg\,dL^{-1}$ ($10\ mmol\,L^{-1}$).

The DCCT and UKPDS demonstrated the benefits of glycemic control on the prevention of diabetic microvascular complications and potential benefit in diabetes-associated CVD. Independently, diabetes is a risk factor for CVD morbidity and mortality equivalent to having pre-existing coronary artery disease. Therefore, the treatment goals for modifiable cardiac risk factors, specifically blood pressure and abnormal lipid levels, are more aggressive in individuals with diabetes[55] (**Table 5**). The ADA recommends a goal blood pressure of $<130/80$ mmHg, LDL cholesterol of $<100\ mg\,dL^{-1}$ ($<70\ mg\,dL^{-1}$ in individuals with known coronary artery disease), HDL cholesterol of $>40\ mg\,dL^{-1}$ ($>50\ mg\,dL^{-1}$ in women), and triglycerides $<150\ mg\,dL^{-1}$.

6.19.6.3 Lifestyle Modifications

The UKPDS study examined the glycemic response to the 3-month dietary run-in subjects completed prior to randomization.[56] Only 16% of individuals achieved the glycemic target (FPG $<108\ mg\,dL^{-1}$ ($6.0\ mmol\,L^{-1}$)) with

Table 5 Goals for the treatment of type 2 diabetes

Measure	*Value*
Glucose	
HbA_{1C}[a]	< 7%
Fasting plasma glucose	90–130 mg dL^{-1} (5.0–7.2 mmol L^{-1})
Peak postprandial glucose	<180 mg dL^{-1} (<10.0 mmol L^{-1})
Blood pressure	
Systolic	<130 mmHg
Diastolic	<80 mmHg
Lipids	
Low-density lipoprotein (LDL) cholesterol[b]	<100 mg dL^{-1} (<2.6 mmol L^{-1})
Diabetes + coronary heart disease	<70 mg dL^{-1}
High density lipoprotein (HDL) cholesterol[c]	>40 mg dL^{-1} (>1.1 mmol L^{-1})
Triglycerides	<150 mg dL^{-1} (<1.7 mmol L^{-1})

[a] HbA_{1C} is the primary target for glycemic control. Glycemic goals should be individualized, with certain populations (children, pregnant women, and elderly) receiving special considerations, and less intensive glycemic goals in patients with severe or frequent hypoglycemia. Normoglycemia (HbA_{1C} <6%) is optimal but increases the risk of hypoglycemic events.
[b] LDL cholesterol goals are lower in patients with overt coronary heart disease.
[c] HDL cholesterol goals are increased by 10 mg dL^{-1} in women.

diet alone. The cohort that achieved the glycemic target lost an average of 15–20 pounds in the 3-month period. At 1-year follow-up only about half (9% of the starting group) maintained the target glycemic control on diet alone.

Sustained lifestyle modification is an essential element of all successful treatment regimens for T2DM. As primary or combined therapy, effective diet and exercise allow individuals with diabetes to utilize endogenous insulin more efficiently, thereby decreasing the amount of medication or supplemental insulin required to achieve euglycemia. The Diabetes Prevention Program (DPP) found that lifestyle interventions alone (goal for weight reduction ≥7%) decreased the risk of progression from IGT to T2DM by 58%, this being nearly twice as effective as metformin in preventing new cases of diabetes.[13]

Weight management follows a simple theory. If caloric intake exceeds caloric usage, then weight is accrued. If caloric usage exceeds caloric intake, weight is lost. Once a summative accrual or loss of 3500 calories occurs, 1 pound is gained or lost, respectively. Modest, but consistent, reduction in caloric intake is the key component to sustained weight loss. Increased physical activity plays an important ancillary role in preventing rebound weight gain. Improved glycemic control associated with a hypocaloric diet can be seen before appreciable weight loss occurs. Unfortunately, most efforts to change lifestyle are not sustained. Maintaining a hypocaloric diet and increased physical activity proves to be difficult for most individuals. Bariatric surgery is an option for some severely obese individuals, commonly resulting in dramatic and sustained weight loss.

6.19.6.4 Overview and Summary of Oral Hypoglycemic Agents

The four classes of oral hypoglycemic agents currently used to treat T2DM facilitate glycemic control via separate mechanisms (**Table 6** and **Figure 7**). The insulin secretagogues (sulfonylureas and meglitinides) act on pancreatic β-cells to increase insulin secretion and bioavailability. Biguanides suppress excessive hepatic glucose production and improve hepatic insulin action. Thiazolinediones improve peripheral insulin sensitivity, especially in muscle and adipose tissues. α-Glucosidase inhibitors delay gastrointestinal absorption of dietary glucose, decreasing postprandial glucose excursions. Each class can be used as monotherapy or in combination.

6.19.6.4.1 Sulfonylureas

For nearly 50 years, sulfonylureas, derived from sulfonic acid and urea, have had a central role in oral hypoglycemic therapy of T2DM. Sulfonylureas increase endogenous insulin secretion and can only be used in individuals that have retained significant β-cell function. The combination of efficacy, low incidence of adverse events, and low cost has contributed to their success and continued use. Sulfonylureas are generally safe and are relatively inexpensive. Hypoglycemia is the most common adverse event that is encountered with their use. First-generation

Table 6 Summary of oral antidiabetic therapies

	Diet and exercise	*Sulfonylureas and glitinides*	*Metformin*	*α-Glucosidase Inhibitors*	*Thiazolidinediones*
Primary mechanism	↓ Insulin resistance	↑ Insulin secretion	↓ Hepatic glucose output	↓ GI absorption of carbohydrates	↑ Insulin sensitivity
Typical improvement in HbA_{1C}	0.5–2.0%	1.0–2.0%	1.0–2.0%	0.5–1.0%	0.5–1.0%
Typical improvement in FPG		60–75 mg dL^{-1} 3.3–4.2 mmol L^{-1}	50–70 mg dL^{-1} 2.8–3.9 mmol L^{-1}	25–30 mg dL^{-1} 1.9–2.2 mmol L^{-1}	60–80 mg dL^{-1} 3.3–4.3 mmol L^{-1}
Recommended starting dose	Caloric restriction to reduce weight by 1–2 kg $month^{-1}$	Glyburide 1.25 mg day^{-1} Glipizide 2.5 mg day^{-1} Glimepiride 1 mg day^{-1} Nateglinide 60 mg before meals Repaglinide 0.5 mg before meals	Metformin 500 mg before breakfast and dinner	Acarbose 25 mg with meals Miglitol 50 mg with meals	Rosiglitazone 4 mg day^{-1} Pioglitazone 7.5 mg day^{-1}
Maximal daily dose	Can use meal substitutes or add orlistat or sibutramine	Glyburide 20 mg day^{-1} Glipizide 40 mg day^{-1} Glimepiride 8 mg day^{-1} Nateglinide 120 mg before meals Repaglinide 16 mg day^{-1} (4 mg before meals)	Metformin 2550 mg day^{-1} (850 mg with each meal)	Acarbose 300 mg day^{-1} (100 mg with each meal) Miglitol 300 mg day^{-1}	Rosiglitazone 8 mg day^{-1} Pioglitazone 45 mg day^{-1} (100 mg with each meal)
Effect on weight	Decrease	Increase	Decrease	No effect	Increase
Effect on lipids	↑ HDL ↓ LDL ↓ Triglycerides	↔ HDL ↔ LDL ↔ Triglycerides	↑ HDL ↓ LDL ↓ Triglycerides	↔ HDL ↔ LDL ↔ Triglycerides	↑ HDL ↔ or ↑↑ LDL ↔ or ↓ Triglycerides
Adverse effects	Injury	Hypoglycemia Weight gain	GI symptoms Lactic acidosis	Weight gain Flatulence GI discomfort	Weight gain Edema Hepatotoxicity
Major contraindications	None	Impaired renal function	Impaired renal function Impaired hepatic function Congestive heart failure	Intestinal Disease	Impaired hepatic function Congestive heart failure
Agents used in combination	Sulfonylureas Glitinides Metformin α-Glucosidase inhibitors Thiazolidinediones Insulin	Metformin α-Glucosidase inhibitors Thiazolidinediones	Sulfonylureas Glitinides α-Glucosidase inhibitors Thiazolidinediones Insulin	Sulfonylureas Glitinides Metformin Thiazolidinediones Insulin	Sulfonylureas Glitinides Metformin α-Glucosidase inhibitors Insulin

sulfonylureas possess a lower binding affinity for the ATP-sensitive potassium channel and thus require higher doses to achieve efficacy. Second-generation sulfonylureas are much more potent compounds (~100-fold) with a rapid onset of action, shorter plasma half-lives, and longer duration of action compared with the first-generation agents[57] (**Table 7**).

6.19.6.4.1.1 Pharmacokinetics

Both first-generation sulfonylureas (tolbutamide, chlorpropamide, tolazamide, and acetohexamide) and second-generation sulfonylureas (glyburide, glipizide, and glimepiride) are rapidly absorbed after oral administration. The half-lives are between 4 and 10 h with the duration of glycemic effect ranging from 6 to 24 h. The exception is chlorpropamide, which has a half-life of 25–60 h and a duration of glycemic effect of 24–72 h. Most sulfonylureas are hepatically metabolized and renally cleared. Active circulating metabolites may prolong the hypoglycemic effect, especially in individuals with acute or chronic renal impairment.

6.19.6.4.1.2 Mechanisms of action

The sulfonylurea receptor on the cell membrane of pancreatic β-cells is a component of the ATP-sensitive potassium (K^+) channel (**Figure 3**). The K^+-ATP channel is composed of two subunits: a cytoplasmic binding site for sulfonylureas

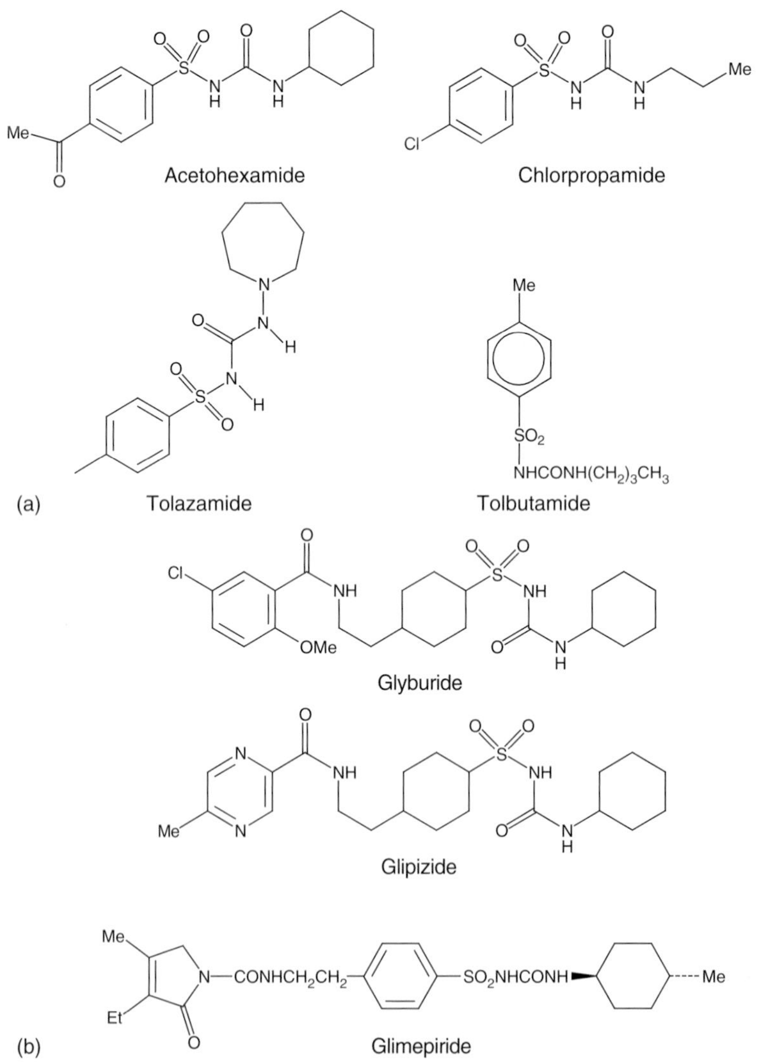

Figure 7 Chemical structures of current oral pharmacological therapies used to treat type 2 diabetes: (a) Sulfonylureas (first generation), (b) Sulfonylureas (second generation), (c) glitinides, (d) biguanides, (e) thiazolidinediones, (f) α-glucosidase inhibitors, and (g) other potential therapeutic agents.

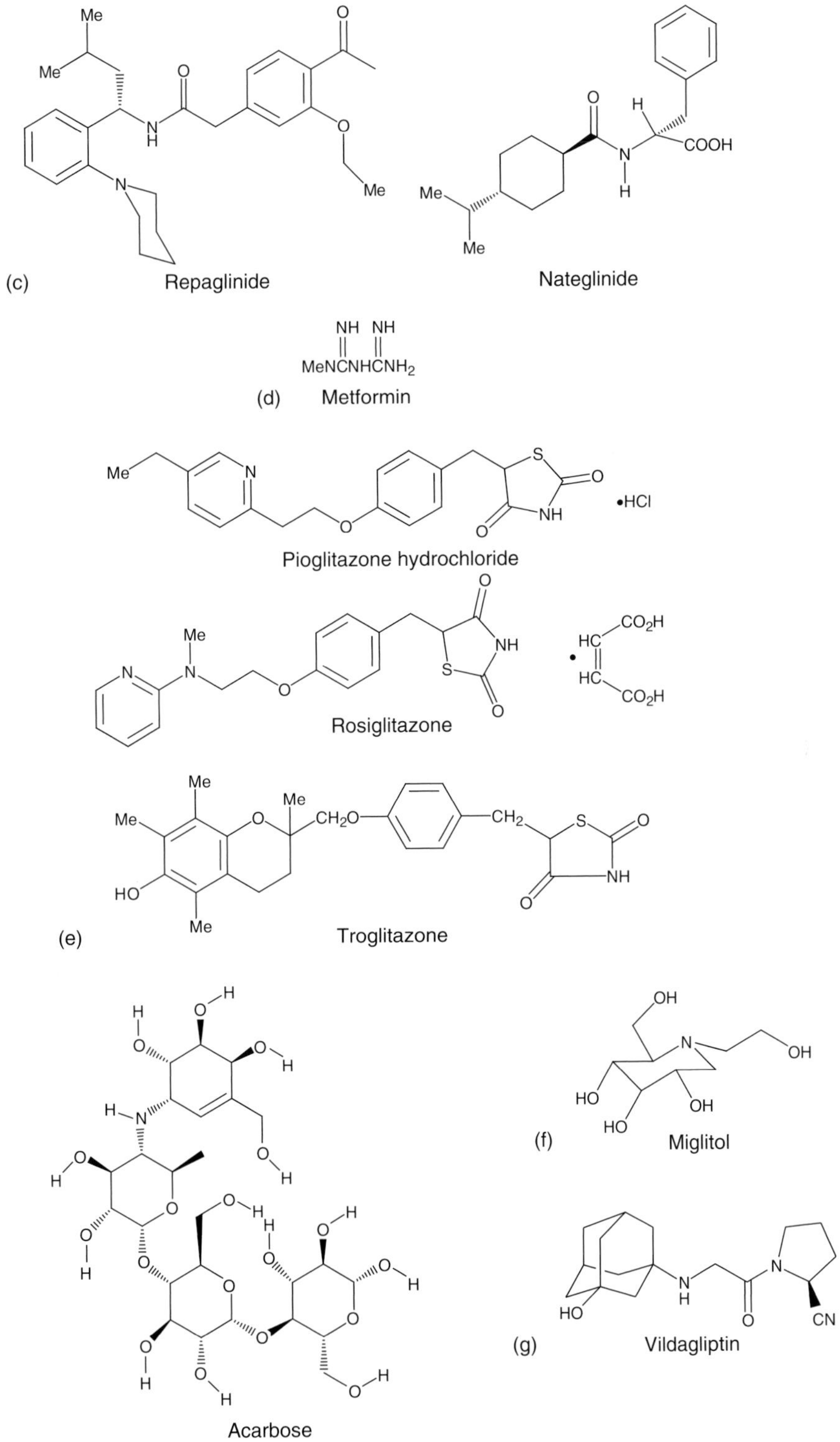

(c) Repaglinide Nateglinide

(d) Metformin

Pioglitazone hydrochloride

Rosiglitazone

(e) Troglitazone

Acarbose

(f) Miglitol

(g) Vildagliptin

Figure 7 Continued

and ATP (sulfonylurea receptor type 1 (SUR1)), and a K^+ channel that acts as the pore-forming unit. Sulfonylurea binding and subsequent receptor activation leads to inhibition and closure of these K^+-ATP channels. Inhibition of K^+ channels induces β-cell membrane depolarization, opening voltage-dependent calcium channels and allowing calcium ion influx. The increase in intracellular calcium concentration leads to the fusion of insulin-containing secretory granules with

Table 7 Overview of first- and second-generation sulfonylureas

Name	*Initial dose ($mg\ day^{-1}$)*	*Recommended maximal dose ($mg\ day^{-1}$)*	*Doses/day*	*Plasma half-life (hours)*	*Duration of hypoglycemic action (hours)*	*Mode of metabolism*	*Activity of Metabolites*	*Excretion in urine (%)*
First generation								
Tolbutamide	500–1500	3000	2–3	4–6.5	6–10	Hepatic carboxylation	Inactive	100
Chlorpropamide	100–250	500	1	36	60	Hepatic hydroxylation or side-chain cleavage	Active	80–90
Tolazamide	100–250	1000	1–2	7	12–14	Hepatic metabolism	3 inactive; 3 weakly active	85
Acetohexamide	250–500	1500	1–2	4–6	12–18	Hepatic reduction to 1-hydroxyhexamide	2.5 times original	60
Second generation								
Glyburide (glibenclamide)	1.25–2.50	20	1–2	4–11	24	Hepatic metabolites	Mostly inactive	50
(micronized formulation)	0.75–1.50	12	1	Micronized in 4 h	24	Hepatic metabolites	Mostly inactive	50
Glipizide	2.5–5	20	1–2	2.5–4.7	Up to 24	Hepatic metabolites	Inactive	50
Glipizide XL	5	20	1	2.5–4.7	Up to 24	Hepatic metabolites	Inactive	50
Gliclazide	40	320	1–2	8–11	Up to 24	Hepatic metabolites	Probably inactive	60–70
Glimepiride	1–2	8	1	5–9	24	Hepatic metabolites	Mildly active	60

Adapted from Mudaliar, S.; Henry, R. R. In *Ellenberg & Rifkin's Diabetes Mellitus*, 6th ed.; Porte, D., Jr., Sherwin, R. S., Baron, A., Eds.; McGraw-Hill: New York, 2003, Chapter 32, pp 532–533, with permission from McGraw-Hill.

the plasma membrane and release of insulin into the portal circulation. The net effect is increased responsiveness of β-cells to stimulation by both glucose and non-glucose secretagogues (such as amino acids), resulting in more insulin release at all blood glucose concentrations, and thus increasing the risk of hypoglycemia between meals.

6.19.6.4.1.3 Clinical use

Sulfonylureas increase endogenous insulin secretion. Their efficacy, measured by the plasma glucose lowering effect, is greatest in individuals with newly diagnosed T2DM. Clinical studies have shown that sulfonylureas reduce mean FPG to 54–72 mg dL^{-1} and HbA_{1C} levels by 1.5–2%. The benefit of sulfonylurea therapy depends on the initial degree of hyperglycemia, duration of diabetes (more effective in T2DM of shorter duration), and previous use of other oral hypoglycemic agents. Because many of the sulfonylurea metabolites are active, more conservative dosing is advised in any patient who may be at high risk for decreased hepatic metabolism or renal clearance of the active drug or metabolites.

6.19.6.4.1.4 Side effects/contraindications

Sulfonylurea-induced hypoglycemia is the most common side effect and usually occurs due to excessive dosage, impaired metabolism, impaired clearance, improper diet, or excessive physical activity. Mild hypoglycemic events occur in approximately 2–4% of individuals, and severe hypoglycemic reactions that require hospitalization occur at a frequency of 0.2–0.4 cases per 1000 patient-years of treatment.

6.19.6.4.1.5 Advantages

Because they enhance insulin secretion, sulfonylureas are often effective when used in combination with metformin or thiazolidinediones. Concurrent use with insulin therapy is of little benefit. The improvement in dyslipidemia seen in sulfonylurea therapy is probably due to the improvement in glycemic control rather than to a direct effect on lipid metabolism.

6.19.6.4.1.6 Disadvantages

As discussed above, the primary disadvantage of sulfonylurea use is a risk of hypoglycemia, especially in individuals with hepatic or renal dysfunction. Drug interactions between sulfonylureas and other pharmacologic agents (salicylates, sulfonamides, fibric acid derivatives, and warfarin) can prolong the activity of either medication, and requires both close glucose monitoring and measurement of serum drug levels when possible. The activity of metabolites prolongs the drug effect, which can be beneficial in improving glycemic control, but dangerous if buildup of the metabolites occurs. Finally, weight gain is common in persons on sulfonylurea therapy.

6.19.6.4.2 **Glitinides (non-sulfonylurea secretagogues)**

Glitinides, like sulfonylureas, act at the K^+-ATP channels to augment glucose-stimulated insulin secretion. Glitinides differ from sulfonylureas by having a more rapid onset and shorter duration of action, primarily lowering postprandial glucose levels. They are only effective in individuals with viable pancreatic β-cells and are therefore not suitable for individuals with T1DM. Repaglinide and nateglinide (**Figure 7**) are the two glitinides approved by the FDA for use in the USA.

6.19.6.4.2.1 Pharmacokinetics

Repaglinide is rapidly absorbed from the gastrointestinal tract, with a mean absolute bioavailability of 56%. Peak plasma concentrations are reached within 1 h of administration. Nateglinide has a mean absolute bioavailability of 75% and also achieves peak plasma concentrations within 1 h. When nateglinide is administered postprandially there is a significantly diminished rate of absorption, with a prolonged time to peak plasma concentration.

6.19.6.4.2.2 Mechanisms of action

Repaglinide is a carbamoylmethyl benzoic acid (CMBA) derivative that binds to the ATP-sensitive potassium channels in the pancreatic β-cell, resulting in increased insulin secretion. Pharmacologic activity is both dose-dependent and glucose-dependent. In vitro studies using islet cells of mice showed that repaglinide causes a greater insulin secretion than glyburide in the presence of moderate glucose concentration. Conversely, glyburide caused greater insulin secretion than repaglinide at the extremes of glucose concentrations, either in the absence of glucose or in high concentrations of glucose.[58] Nateglinide, a D-phenylalanine derivative, has a similar mechanism of action as repaglinide but a lower binding affinity for the sulfonylurea receptor.

6.19.6.4.2.3 Clinical use

The clinical efficacy of glitinides is similar to that of the sulfonylureas. In short-term, placebo-controlled trials, repaglinide monotherapy reduced HbA_{1C} levels from 8.5% at baseline to 7.8% after only 12 weeks. In individuals with T2DM not previously treated with other oral hypoglycemic agents, repaglinide resulted in a 30% decrease in HbA_{1C}, from 6.9% to 4.8%, with fasting and postprandial blood glucose levels decreased by 70 and 112 $mg\,dL^{-1}$, respectively. Repaglinide[59] and nateglinide[60] as combination therapy with metformin are more effective than either agent alone. Glitinides are taken before every meal but are not taken if a meal is skipped.

6.19.6.4.2.4 Side effects/contraindications

As with other insulin secretagogues, hypoglycemia can occur with administration of repaglinide and nateglinide. Due to the short half-life of these drugs hypoglycemia is usually postprandial and of short duration. Repaglinide and nateglinide are both metabolized in the liver and serum drug concentrations may rise in individuals with hepatic dysfunction. Repaglinide is metabolized by the cytochrome P450 enzyme CYP3A4 and concurrent use with other drugs that induce or suppress this enzyme may alter its glycemic effects.

6.19.6.4.2.5 Advantages

Due to their rapid absorption and short half-life, use of glitinides allows more flexibility in the frequency and timing of meals. Those who are prone to delay or miss meals do not need to be concerned about hypoglycemia as they would on sulfonylurea or long-acting insulin therapy. Doses of repaglinide and nateglinide do not need to be reduced in individuals with renal dysfunction.

6.19.6.4.2.6 Disadvantages

Compared with sulfonylureas, glitinides are considerably more expensive as they are still under patent and have no generic equivalent.

6.19.6.4.3 **Biguanides**

Biguanides have been used in the treatment of T2DM since the1950s. Derived from the French lilac, *Galega officinalis*, phenformin was the initial biguanide available for clinical use. It was taken off the market in the 1970s after numerous cases of lactic acidosis were found to be associated with its use. The etiology of the lactic acidosis was related to impaired hydroxylation of phenformin in a subset of individuals. In contrast, individuals taking metformin are at very low risk of developing this complication. Over time, metformin has proven to be safe and remains the only biguanide approved by the FDA for use in the USA.

6.19.6.4.3.1 Pharmacokinetics

Metformin (**Figure 7**) is absorbed via the small intestine, with 50–60% bioavailability. The maximal plasma concentration of 1–2 $\mu g\,mL^{-1}$ occurs 1–2 h after an oral dose of metformin, with minimal binding to plasma proteins. The plasma half-life is estimated at 1.5–4.9 h. Tissue distribution is proportional to the circulating plasma concentration, with higher concentrations found in the liver and kidney. The highest concentrations are found in salivary glands and intestinal mucosa. Metformin is not metabolized in vivo and is renally cleared, with approximately 90% elimination within 12 h. Impaired glomerular filtration or tubular secretion can cause high circulating levels of metformin.

6.19.6.4.3.2 Mechanisms of action

The specific mechanisms of action of metformin have not been definitively demonstrated. In the liver, metformin increases insulin-dependent suppression of gluconeogenesis and decreases the glucagon-dependent stimulation of gluconeogenesis, resulting in an overall decrease in hepatic glucose production. Animal models suggest additional mechanisms of action for metformin, including insulin-dependent glucose uptake by muscle[61] and adipose tissue,[62] with resultant increases in glycogen formation, glucose oxidation, and lipogenesis. De Fronzo *et al.*[63] demonstrated that, in humans, improvement in fasting blood glucose on metformin results from reduction in basal hepatic glucose production. In studies using the glucose/insulin clamp techniques metformin did not improve whole-body insulin sensitivity in individuals with T2DM. Since the glucose-lowering effect of metformin occurs without stimulation of insulin secretion, metformin is not associated with hypoglycemia when used as monotherapy.

6.19.6.4.3.3 Clinical use

Metformin can be used as monotherapy or in combination with any other oral hypoglycemic agent. Metformin typically decreases HbA_{1C} by 1–2%. The US Multicenter Metformin Study Group randomized individuals with T2DM who were inadequately controlled on diet alone to either metformin or placebo.[64] After 6 months, the mean FPG was $189\,mg\,dL^{-1}$ ($10.6\,mmol\,L^{-1}$) in the metformin group as compared with $244\,mg\,dL^{-1}$ ($13.7\,mmol\,L^{-1}$) in the placebo group.

6.19.6.4.3.4 Side effects/contraindications

Metformin is generally well tolerated and safe. Gastrointestinal side effects (diarrhea, flatulence, abdominal discomfort, anorexia, nausea, and metallic taste) predominate and are commonly dose-related. Dose reduction, administration with food, or an extended-release formulation can improve drug tolerance. Over 50% of individuals tolerate the maximum daily dose of metformin, but 5–10% of people need to discontinue the drug altogether.[65]

Lactic acidosis is a rare, but potentially fatal, adverse complication, with an estimated incidence of 0.03 cases per 1000 patient-years. Most cases of lactic acidosis occur in a setting in which metformin therapy is continued inappropriately: during an acute or chronic medical condition that may alter the pharmacokinetics or renal clearance of metformin (renal impairment (plasma creatinine $\geq 1.5\,mg\,dL^{-1}$ in men; $\geq 1.4\,mg\,dL^{-1}$ in women)); or, in circumstances when hypoxia or reduced peripheral perfusion may occur (liver disease, cardiac or respiratory conditions, or a history of previous lactic acidosis).

6.19.6.4.3.5 Advantages

The risk of hypoglycemia on metformin monotherapy is extremely low. Combination therapy with other oral hypoglycemic agents increases this risk. Metformin therapy, unlike other oral hypoglycemic agents, is not associated with weight gain but a modest weight loss of 2–3 kg. Long-term therapy with metformin is also associated with decreases in serum concentrations of plasma triglycerides (10–20% reduction), total cholesterol (5–10% reduction), and LDL cholesterol, with some studies noting a slight increase in HDL cholesterol.

6.19.6.4.3.6 Disadvantages

Besides the gastrointestinal side effects and low risk of lactic acidosis, long-term therapy with metformin can reversibly reduce intestinal absorption of vitamin B_{12} and folate, but rarely leads to macrocytic anemia. Metformin therapy should be discontinued in any situation that may cause a rapid change in renal clearance, such as acute illness requiring hospitalization or prior to intravenous radiologic contrast studies. Metformin can be restarted once renal function has stabilized.

6.19.6.4.4 α-Glucosidase inhibitors

Two α-glucosidase inhibitors, acarbose and miglitol (**Figure 7**), are currently approved by the FDA. Acarbose, the first α-glucosidase inhibitor, is a nitrogen-containing pseudo-tetrasaccharide, and miglitol is a synthetic analog of 1-deoxynojirimycin.

6.19.6.4.4.1 Pharmacokinetics

Acarbose acts locally in the gastrointestinal tract on the surface of enterocytes. Only 2% of an oral acarbose dose is systemically absorbed, with a plasma half-life of 2 h. Acarbose is metabolized by the intestinal microbial flora and digestive enzymes and excreted in the stool.

Miglitol requires systemic absorption to achieve its therapeutic effect. Oral absorption is rapid and nearly complete, but saturable; nearly all of a 25-mg dose is absorbed systemically, whereas only 50–70% of a 100-mg dose is absorbed. Miglitol is not metabolized in vivo and is renally excreted. In individuals with severe renal dysfunction, plasma levels of both acarbose and miglitol may rise.

6.19.6.4.4.2 Mechanisms of action

α-Glucosidase inhibitors are competitive, reversible inhibitors of pancreatic α-amylase and membrane-bound intestinal α-glucosidase hydrolase enzymes that line the brush border of the enterocytes in the proximal portion of the small intestine. These enzymes normally convert dietary polysaccharides into absorbable monosaccharide. Enzyme inhibition delays hydrolysis of polysaccharides, reducing the rate of absorption of monosaccharides, and thereby decreasing the peak of postprandial blood glucose and blunting the plasma insulin response.

6.19.6.4.4.3 Clinical use

α-Glucosidase inhibitors reduce postprandial glucose excursions and have little effect on FPG concentrations. To exert enzymatic inhibition in the small intestine effectively, acarbose and miglitol are administered immediately prior to or within 15 min after the start of each meal.

When used as monotherapy, acarbose and miglitol lower mean postprandial glucose levels by 40–60 mg dL^{-1} and mean FPG levels by 10–20 mg dL^{-1}. Mean HbA_{1C} levels are reduced by 0.5–1.0%.[66] Acarbose, as combination therapy with sulfonylureas, metformin, and insulin, further reduces HbA_{1C} by 0.3–0.5% and mean postprandial glucose by 25–30 mg dL^{-1}. Miglitol, in combination with sulfonylureas, reduces mean HbA_{1C} by 0.7% and mean postprandial glucose by 60–70 mg dL^{-1}. In individuals with T1DM, acarbose therapy decreases the amplitude of postprandial glycemic excursions and HbA_{1C} values.

6.19.6.4.4.4 Side effects/contraindications

α-Glucosidase inhibitors are associated with significant gastrointestinal side effects that affect over 50% of individuals. Increased colonic gas production due to fermentation of unabsorbed carbohydrate cause abdominal bloating, cramping, increased flatulence, or diarrhea. Individuals with acute or chronic diseases involving the gastrointestinal tract should not be prescribed α-glucosidase inhibitors. Renal impairment (serum creatinine >2.0 mg dL^{-1}) and hepatic dysfunction are contraindications to α-glucosidase inhibitor therapy.

Hypoglycemia in acarbose- or miglitol-treated subjects must be treated with simple carbohydrates found in milk, juices, or glucose tablets. Disaccharides or polysaccharides (sucrose (table sugar), candy, and soft drinks) cannot be used because the α-glucosidase inhibitory effects delay their hydrolysis and absorption.

6.19.6.4.4.5 Advantages

When used as monotherapy, acarbose and miglitol are not associated with hypoglycemia or significant weight changes. Blocking the absorption of complex carbohydrates decreases the caloric uptake of the small intestine, but the large intestine compensates to assure that adequate caloric goals are met. α-Glucosidase inhibitors do not significantly affect LDL or HDL cholesterol concentrations, but triglyceride levels decline. These agents may prove to be useful in the management of severe hypertriglyceridemia in both the diabetic and non-diabetic population.

6.19.6.4.4.6 Disadvantages

Due to the enteral action of α-glucosidase inhibitors, pharmacological agents that alter the normal gastrointestinal flora may cause increased gastrointestinal side effects, decrease distal colonic absorption of glucose, and decrease overall caloric uptake. α-Glucosidase inhibitors may also affect pharmacologic agents absorbed in the proximal small intestine, such as digoxin and warfarin. Combined use of acarbose, alcohol, and acetaminophen can be fatal and should be avoided. Both acarbose and alcohol enhance the activity of the hepatic isoenzyme CYP2E1, which is responsible for acetaminophen metabolism to a toxic reactive metabolite. Buildup of this metabolite can put persons at risk for acute hepatic injury.

6.19.6.4.5 **Thiazolidinediones**

Thiazolidinediones, or glitazones, are insulin-sensitizing agents that are selective ligands of the nuclear transcription factor peroxisome proliferator-activated receptor γ (PPAR-γ). An insulin sensitizer has the potential to target insulin resistance directly, a key underlying factor in the pathogenesis of T2DM. Insulin resistance also contributes to dyslipidemia, abnormal coagulation, altered fibrinolysis, and hypertension.

Rosiglitazone and pioglitazone (**Figure 7**) are the two thiazolidinediones approved by the FDA for use in the USA. In January 1997, troglitazone was the first thiazolidinedione to receive FDA approval. However, it was withdrawn from use in March 2000 due to hepatotoxicity. Rosiglitazone and pioglitazone were approved for use in 1999 and do not appear to have the same hepatic profile as their predecessor.

6.19.6.4.5.1 Mechanisms of action

The peroxisome proliferator-activated receptors (PPARs) are a subset of the 48-member nuclear-receptor superfamily and directly regulate gene expression. To date, three PPARs have been identified, PPAR-α, PPAR-δ (also known as PPAR-β), and PPAR-γ.

PPAR-α expression[67] is primarily found in the liver, heart, skeletal muscle, and smooth muscle cells of the vascular wall (**Table 8**). PPAR-α activation regulates gene transcription directly, influencing lipoprotein metabolism, fatty acid uptake and oxidation, and some anti-inflammatory products. PPAR-α agonists have also been shown to prevent or slow the progression of atherosclerosis in animal models[68] and in humans.[69] Common ligands for PPAR-α are the fibrates, including fenofibrate, bezafibrate, ciprofibrate, and gemfibrozil.

Table 8 Target organs of PPAR action

Liver (PPAR-α)	*Skeletal muscle (PPAR-α and PPAR-β)*	*Adipose tissue (PPAR-γ)*	*Vascular wall (PPAR-α, PPAR-β, PPAR-γ)*
Lipoprotein metabolism • Decreased apolipoprotein C-III • Increased apolipoprotein A-I, II **Fatty acid uptake** • Increased fatty acid transport protein-1 • Increased fatty acid translocase/CD36 **Fatty acid catabolism** • Increased CPT I, II **Decreased inflammation** • Decreased C-reactive protein (αγ)	**Fatty acid catabolism (PPAR-α)** • Increased CPT I, II **Glucose uptake (PPAR-γ)** • Increased GLUT4 • Increased phosphatidyl 3-kinase • Decreased PDK-4	**Adipocyte differentiation** **Fatty acid uptake and storage** • Increased fatty acid transport protein-1 • Increased acyl-co-enzyme A synthetase **Intravascular lipolysis** • Increased lipoprotein lipase **Glucose uptake** • Increased GLUT4 • Increased phosphatidyl 3-kinase • Increased insulin-receptor substrates IRS-1 and IRS-2 • Increased CAP • Increased glycerol kinase **Other effects** • Increased adiponectin • Decreased	**Adhesion molecules** • Decreased intercellular adhesion molecule-1 (γ) • Decreased vascular-cell adhesion molecule-1 (αγ) **Inflammation** • Increased nuclear factor κB (α) • Decreased cyclooxygenase-2 (α) **Decreased endothelin (αγ)** **Cholesterol efflux** • Increased ABCA1 (αβγ) • Increased scavenger receptor-B1 (αγ) **Other** • **Decreased iNOS (γ)** • Decreased TNF-α (α) • Decreased interleukin-6 (αγ) • Decreased MMP-9 (γ) • Decreased MCP-1 (γ) • Decreased tissue factor (α)

ABC, ATP-binding cassette; CAP, Cbl-associated protein; CPT, carnitine palmitoyl transferase; HSD, hydroxysteroid dehydrogenase; iNOS, inducible nitric oxide synthase; MCP, monocyte chemoattractant protein; MMP, matrix metalloproteinase; PDK, pyruvate dehydrogenase kinase.
Adapted from Yki-Järvinen, H. *N. Engl. J. Med.* **2004**, *351*, 1106–1118.

PPAR-δ (also known as PPAR-β) expression is highest in adipose tissue, brain, and skin cells. Although not targeted pharmacologically yet, PPAR-δ has been shown to delay wound closure and alter myelination.

PPAR-γ expression is primarily found in adipose tissue, pancreatic β-cells, vascular endothelial cells, and macrophages. Tissues with high PPAR-γ expression generally have lower expression of PPAR-α, and vice versa. PPAR-γ activation regulates adipocyte differentiation, fatty acid uptake and storage, intravascular lipolysis, and glucose uptake.

6.19.6.4.5.2 Pharmacokinetics

Thiazolidinediones are rapidly absorbed, with peak serum concentrations of rosiglitazone occurring within 1 h and pioglitazone within 2 h of oral administration. Administration with food does not prevent drug absorption, but may delay the time to peak serum concentration. Both rosiglitazone and pioglitazone are avidly protein-bound, primarily to serum albumin. The plasma half-life for rosiglitazone is 3–4 h and it is extensively metabolized. Pioglitazone has a half-life of 3–7 h and it is metabolized by hydroxylation and oxidation. The active, but less potent, metabolites have half-lives of 16–24 h.

6.19.6.4.5.3 Clinical use

The thiazolidinediones are approved for use as monotherapy or in combination with secretagogues, metformin, or insulin therapy. Unlike most antidiabetic medications, which have a rapid onset of activity, pioglitazone and rosiglitazone require 8–12 weeks or more to achieve maximal clinical benefit. This is probably related to their mode of action, i.e., the regulation of gene expression. To date, there have been no head-to-head studies comparing rosiglitazone and pioglitazone, and no studies on patient characteristics that may predict a good treatment response with these agents.

When used as monotherapy or adjunctive therapy, thiazolidinediones typically decrease HbA_{1C} by 1–1.5%.[70] Clinical studies with rosiglitazone monotherapy lowered FPG by up to 55 mg dL^{-1}. Similarly, pioglitazone (15, 30, and 45 mg doses) reduced FPG (by 30, 32, and 56 mg dL^{-1}, respectively) and HbA_{1C} (by 0.3, 0.3, and 0.9%, respectively).[71] Dividing the daily dose (twice daily) was more effective in reducing both FPG levels and HbA_{1C} levels compared with single daily dosing.[72]

6.19.6.4.5.4 Side effects/contraindications

The most common side effects associated with thiazolidinedione use are fluid retention, weight gain, and anemia. The weight increase may be attributed to an expansion of the subcutaneous fat depot and the development of peripheral edema. Plasma volume increases 6–7% on glitazone therapy and results in clinically notable edema in 4–6% of individuals, increasing the risk of symptomatic heart failure in predisposed individuals. The anemia associated with thiazolidinedione use is mild. Hepatotoxicity, as seen with the first thiazolidinedione (troglitazone), does not appear to be a drug-class effect. In clinical trials with rosiglitazone and pioglitazone, the incidence of hepatotoxicity and liver enzyme elevations is similar to placebo.

6.19.6.4.5.5 Comparisons of rosiglitazone and pioglitazone

In vitro studies reveal that pioglitazone acts as a PPAR-γ and partial PPAR-α agonist, whereas rosiglitazone is only a PPAR-γ agonist. This difference may explain their different effects on lipid metabolism.[73] Both rosiglitazone and pioglitazone improve HDL cholesterol levels by approximately 10%. Rosiglitazone is associated with an increase in LDL cholesterol concentration of 8–16%, which has not been seen with pioglitazone therapy. On the other hand, pioglitazone is associated with a decrease in triglyceride levels that has not been seen with rosiglitazone. Both agents decrease serum FFA concentrations by approximately 20–30%.

6.19.6.4.5.6 Advantages

Thiazolidinediones, when used as monotherapy, are not associated with hypoglycemia, as they do not increase insulin secretion. Thiazolidinediones decrease the secretory demands and the strain on the β-cell caused by chronic insulin resistance and may preserve pancreatic β-cell function. Individuals using rosiglitazone or pioglitazone in combination with secretagogues or insulin therapy are at risk for hypoglycemic events.

In addition to their contribution as insulin sensitizers for glycemic control, the thiazolidinediones potentially have beneficial effects on endothelial function, atherogenesis, fibrinolysis, and ovarian steroidogenesis. Thiazolidinediones exhibit potent anti-inflammatory properties, and in recent clinical studies have been shown to retard the progression and reverse carotid intimal medial thickening, and decrease coronary intimal hyperplasia, in individuals with and without coronary stents. It remains unclear if these anti-inflammatory properties are unique to thiazolidinediones specifically or result from improvements in insulin resistance.

6.19.6.4.5.7 Disadvantages

Pioglitazone shares a hepatic cytochrome metabolism pathway (CYP3A4 and CYP2C8) with ethinyl estradiol and norethindrone (CYP3A4), and reduces the bioavailability of certain oral contraceptives and their effectiveness. Higher dose oral contraceptive agents are required to overcome this effect in order to provide effective contraception. Rosiglitazone is metabolized by a separate cytochrome pathway (CYP2C8) that does not interfere with any oral contraceptive agents. The pharmacokinetics of rosiglitazone and pioglitazone are not affected by renal dysfunction and, therefore, do not require dose adjustments.

6.19.6.5 Insulin

The essential role of insulin in the treatment of T1DM is universally accepted, whereas its use in T2DM is often considered the last resort. Insulin should be viewed as a valuable therapeutic tool for early intervention, to attain and maintain target levels of glycemic control.

The evolution of insulin preparations, from those 'purified' from animal pancreases, to human-insulin produced with recombinant DNA technology, and the present use of insulin analogs, represents more than 80 years of collaboration among industry, clinical, and basic research, and millions of people with diabetes.

6.19.6.5.1 Sources of insulin

With the availability of human insulin, use of animal insulin declined dramatically. Currently, the biosynthesis of human insulin involves insertion of the human proinsulin gene into either *Saccharomyces cerevisiae* (baker's yeast) or a non-pathogenic laboratory strain of *Escherichia coli*, which serve as the production organism. Human insulin is then isolated and purified.

Sophisticated knowledge of the structure–function aspects of insulin has led to the development and production of analogs to human insulin. The molecular structure of insulin has been modified slightly to alter the pharmacokinetic properties of insulin, primarily affecting absorption from subcutaneous tissue. The B26–B30 region of the insulin molecule is not critical for insulin receptor recognition and it is in this region that amino acids have been substituted[74] (**Figure 8**). Thus, the insulin analogs are recognized by the insulin receptor and bind to the receptor with similar affinity as native insulin.

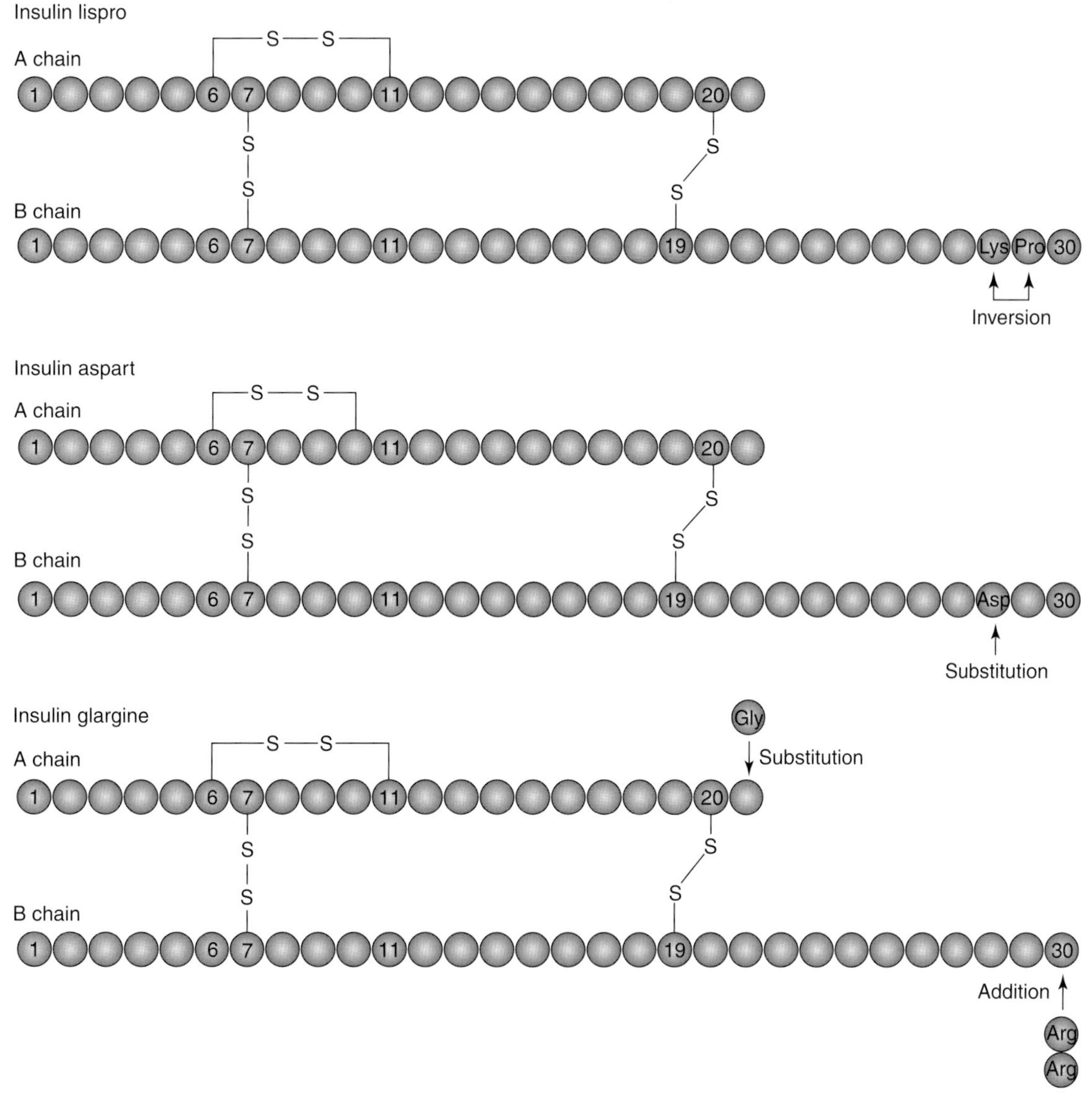

Figure 8 The structure of three insulin analogs. The human insulin molecule, derived from its proinsulin precursor, consists of an A chain and a B chain. The insulin lispro molecule has a proline residue at position B28 that has been transposed with the lysine residue at position B29 of the B chain of the human insulin molecule. Insulin aspart has a B28 amino acid proline substitution with aspartic acid. In the insulin glargine molecule, glycine has replaced the asparagine residue at position A21 and two arginine residues have been added to the carboxyl terminus of the B chain. (Reprinted from Hirsch, I. B. *N. Engl. J. Med.* **2005**, *352*, 174–183.)

6.19.6.5.2 Pharmacodynamics and pharmacokinetics of insulin

Insulin was initially available as a soluble, clear solution. The relatively short duration of action meant that individuals had to take multiple daily injections. Combining insulin with protamine or high concentrations of zinc led to the formation of suspensions that are slowly absorbed with a longer and variable duration of action. Insulin analogs were designed with more predictable characteristics of rapid or attenuated absorption and action. The pharmacodynamic profiles of the different types of insulin and insulin analogs are summarized in **Table 9**.

In solution, and in the β-cell where its concentration is high, insulin tends to self-associate, forming dimers, hexamers, and larger aggregates. In the β-cell, this self-association facilitates the transportation, conversion, and intracellular storage of insulin crystals. However, self-association retards absorption of soluble insulin after subcutaneous administration because hexamers must dissociate to monomers before entering the bloodstream. The rapid-acting analogs differ from regular insulin by virtue of their capacity to dissociate rapidly into monomers in subcutaneous tissue. The rate of absorption from a subcutaneous site also varies day to day and from one injection site to another, and is also influenced by ambient temperature and physical activity.

Subcutaneous administration of insulin does not truly duplicate endogenous insulin physiology. In normal individuals the pancreas secretes insulin with a rapid first-phase secretion followed by a prolonged second-phase release directly into the portal circulation. High insulin concentrations reach the liver, regulate endogenous glucose production, and undergo first-pass hepatic metabolism, resulting in lower circulating insulin concentrations to peripheral tissues. Subcutaneous insulin, on the other hand, is absorbed peripherally, allowing the highest plasma insulin concentrations to reach the peripheral tissues before arriving at the liver.

The liver and kidneys account for the majority of insulin degradation. Normally, the liver degrades 60% of endogenous insulin (delivered via portal vein blood flow) and the kidneys 30–40%.[75] When insulin is injected exogenously, the degradation profile is altered, as insulin is not delivered directly to the portal vein. The kidney plays a greater role in insulin degradation with subcutaneous insulin (approximately 60%), with the liver degrading 30–40%.[76] Severe renal dysfunction reduces the clearance of insulin and prolongs its effect.[77]

6.19.6.5.3 Types of insulin

Human insulin is available in rapid, short, intermediate, and long-acting forms (regular, NPH, lente, and ultralente). In addition, five insulin analogs are currently available for clinical use (insulins aspart, lispro, glulisine, detemir, and glargine).

6.19.6.5.3.1 Rapid-acting insulin analogs (lispro, aspart, and glulisine)

Changes in the amino acid sequence of the insulin analogs lispro, aspart, and glulisine reduce the tendency to self-associate into hexamers, resulting in more rapid onset and a shorter duration of action compared to regular human insulin. Insulin lispro (Humalog) has a reversal of amino acid sequence at the B28 (proline) and B29 (lysine) positions, resulting in insulin lysine-proline. Insulin aspart (Novolog) has a B28 amino acid proline substitution with aspartic acid. Insulin glulisine (Apidra) has two amino acid substitutions and differs from human insulin in that B3 asparagine is replaced by lysine, and B29 lysine is replaced by glutamic acid.[78]

Table 9 Pharmacodynamics of insulin and insulin analogs

Type of insulin	*Time to onset of activity (hours)*	*Peak concentration (hours)*	*Duration of action (hours)*
Rapid acting			
Insulin lispro	Within 15 min	30–90 min	3–5
Insulin aspart	Within 15 min	30–90 min	3–5
Insulin glulisine	Within 15 min	30–90 min	3–5
Short acting			
Regular	30–60 min	2–4	5–8
Intermediate acting			
NPH	2–4	4–10	12–18
Lente	2–4	6–14	12–20
Insulin detemir	1–2	4–10	16–20
Long acting			
Ultralente	6–10	14–18	18–24
Insulin glargine	2–4	Flat	24

Compared to regular insulin, insulins lispro, aspart, and glulisine have more rapid onset, earlier peak effect, and shorter duration of action. Optimally, these analogs are injected immediately prior to or after starting a meal, due to their onset of action within 5–15 min. Regular insulin, with a longer onset of action, is injected 30–60 min prior to a meal, and can be difficult to time accurately. With a pharmacokinetic profile closer to physiologic insulin, the rapid-acting analogs are potentially more effective in lowering postprandial blood glucose levels and reducing risk of hypoglycemia compared to regular insulin.

6.19.6.5.3.2 Short-acting insulin (regular insulin)

Human recombinant DNA produced regular insulin has an onset of action between 30 and 60 min after and a peak effect 2–4 h after injection, with a usual duration of action of 6–8 h. Regular insulin was the shortest acting insulin available prior to the availability of the rapid-acting analogs. Its duration of action extends beyond the duration of digestion and absorption of most meals, thereby increasing the risk of hypoglycemia.

6.19.6.5.3.3 Intermediate-acting insulins Neutral Protamine Hagedorn ((NPH), lente, and insulin detemir)

Neutral Protamine Hagedorn (NPH) and insulin zinc (lente) are different types of intermediate-acting insulins. NPH is a suspension of medium-sized crystals, which include zinc and protamine. Lente is a suspension of large, zinc-containing crystals that have been precipitated in an acetate buffer. These crystals dissolve slowly after subcutaneous injection. NPH and lente have similar pharmacodynamic profiles, with an onset of action approximately 2 h, a peak effect 6–14 h, and duration of action up to 24 h following subcutaneous injection. Intermediate-acting insulins can provide basal and/or prandial insulin, depending on dose and time of administration.

Insulin detemir is an acylated insulin analog with threonine removed at position B30, and lysine at position B29 is acylated with a 14-carbon myristoyl fatty acid acylation, resulting in delayed action due to increased binding to albumin. It has a longer duration of action (20 h) than NPH or lente insulins, and less intradose pharmacokinetic variability.

6.19.6.5.3.4 Long-acting insulins (ultralente and insulin glargine)

Long-acting insulins are used to provide a basal level of insulin. Ultralente insulin reaches a peak gradually (14–18 h after injection), with a duration of action of up to 24 h. Similar to lente insulin, ultralente is a suspension of large, zinc-containing crystals that have been precipitated in an acetate buffer.

Insulin glargine (Lantus) contains two modifications of the human insulin molecule that change both the onset and duration of action. Two arginine residues are added to the carboxyl terminal end of the insulin B-chain, and glycine is substituted for asparagine at the end of the A-chain (position A21). The latter modification prevents deamidation and dimerization. Overall, these changes result in a stable molecule that is soluble at an acidic pH but insoluble at the neutral pH of subcutaneous tissues. When insulin glargine is injected subcutaneously, the acidic solution is neutralized. Microprecipitates of insulin glargine form in the subcutaneous tissue and are slowly absorbed over a period of up to 24 h, resulting in a nearly constant level of insulin throughout the day. Insulin glargine may be given at any time of day and has been shown to cause less nocturnal hypoglycemia when used at bedtime than NPH insulin. Insulin glargine cannot be mixed prior to injection with any other insulin or solution because this will alter its pH and affect its absorption profile.

6.19.6.5.4 **General principles of insulin therapy**

The goal of insulin replacement therapy is to achieve a blood insulin profile that mimics physiologic insulin secretion. Basal insulin suppresses endogenous glucose production between meals and overnight and boluses of insulin are necessary with meals to promote postprandial glucose utilization. The type, frequency, and timing of insulin injections necessary to achieve this is greatly influenced by the amount of residual β-cell function and concurrent use of oral hypoglycemic agents.

6.19.6.5.4.1 Type 1 diabetes

As β-cell function fails, individuals with T1DM require exogenous insulin for survival. Daily insulin requirements are usually between 0.5 and 1.0 units kg^{-1}. During the early stages of T1DM, residual levels of endogenous insulin can reduce the daily insulin requirements below this range. Intensive insulin therapy, defined as ≥ 3 insulin injections daily, provides more flexibility of lifestyle and often better glycemic control in individuals with T1DM than is achieved with 1–2 daily insulin injections.

6.19.6.5.4.2 Type 2 diabetes

Because of progressively declining β-cell function, many individuals with T2DM eventually require insulin therapy. In the UKPDS less than 25% of individuals treated with a sulfonylurea as monotherapy were maintaining a target HbA_{1C} <7.0% after 9 years of treatment. Since insulin resistance is a major component in T2DM, daily insulin requirements often exceed 1 unit kg^{-1}.

6.19.6.5.4.3 Combination therapy with insulin and oral agents

Insulin has been used in combination with every other type of pharmacological therapy for glycemic control in T2DM. Combination therapy with a sulfonylurea has been extensively studied, with overall improved glycemic control in individuals on a bedtime dose of intermediate or long-acting insulin.[79,80] The concurrent use of insulin and metformin can modestly decrease HbA_{1C} values and may also decrease the amount of weight gain associated with insulin alone. Thiazolidinediones maintain or slightly lower HbA_{1C} levels when used with insulin, and further decrease the required insulin dose, but may result in additional weight gain. A clinical study investigating the use of acarbose with insulin therapy showed a 0.5% decrease in HbA_{1C} values, which is likely due to an improvement in postprandial hyperglycemia.

It remains unclear, however, if the benefits of combination therapy outweigh the added expense and only modest improvements in HbA_{1C}. Combination therapy also has the potential for added toxicity, drug interactions, and decreased compliance.

6.19.6.5.5 Side effects of insulin

The most significant adverse effect of insulin therapy is hypoglycemia. This is especially the case for treatment of T1DM, but is also true for T2DM. Insulin allergy and lipoatrophy were commonly seen with the use of animal insulin before 'pure' and biosynthetic preparations became available. Both reactions are now rare, but can be seen, probably because there is some degradation during storage and/or with depot injection into tissues that can induce an immune response. Weight gain commonly occurs following improved glycemic control with insulin therapy. In the UKPDS, individuals receiving insulin therapy had an average weight gain of 4.0 kg over the course of the study.[81]

6.19.6.6 Amylin Analogs (Pramlintide)

Amylin is a 37 amino acid peptide that is formed almost exclusively within pancreatic β-cells and co-secreted with insulin. Amylin acts as a neuroendocrine signal that complements the actions of insulin in postprandial glucose homeostasis via suppression of postprandial glucagon secretion and via inhibition of gastric emptying. Individuals with T1DM have an absolute deficiency of both insulin and amylin, whereas individuals with T2DM have a relative deficiency of both hormones.

Pramlintide is the first amylin analog commercially available and received FDA approval in March 2005 for therapy in both T1DM and T2DM. Pramlintide, studied as an adjunctive therapy to insulin, has been shown to improve postprandial and overall glycemic control in individuals with both T1DM and T2DM (improvements in HbA_{1C} of 0.67%[82] and HbA_{1C} of 0.62%,[83] respectively) without increasing the incidence of hypoglycemia or weight gain. The glycemic improvements with pramlintide had no significant effects on lipid concentrations or blood pressure and showed no evidence of cardiac, hepatic, or renal toxicity. The most frequent adverse side effects associated with pramlintide therapy include transient mild to moderate nausea and anorexia. In its current formulation, pramlintide is administered via subcutaneous injection separately from insulin.

6.19.6.7 Glucagon-Like Peptide-1 (GLP-1) Agonists (Exenatide)

Glucagon-like peptide-1 (GLP-1) is a 30/31 amino acid peptide released from the distal small bowel and colon and undergoes rapid inactivation by dipeptidyl peptidase-IV (DPP-IV). GLP-1, with a half-life of approximately 90 s, augments glucose-mediated β-cell insulin secretion, inhibits glucagon secretion, promotes β-cell proliferation (in animal models), and slows gastric emptying. Augmentation of insulin secretion by GLP-1 is dependent on glucose concentration, and therefore rarely contributes to hypoglycemia when used as monotherapy. Clinically, endogenous levels of GLP-1 are significantly reduced in individuals with T2DM.

Exenatide, the first GLP-1 agonist to receive FDA approval, is much more potent than native GLP-1, largely due to its resistance to DPP-IV-mediated inactivation. Structurally, exenatide has a glycine for alanine substitution at position 2 of the GLP-1 peptide, rendering it a poor substrate for DPP-IV inactivation and degradation.

In clinical trials, exenatide improved postprandial and overall glycemic control in patients with T2DM on metformin monotherapy (decrease in HbA_{1C} of 0.78%) and was associated with a modest weight loss (mean 2.8 kg) without increasing the incidence of hypoglycemia.[84] Similar improvements in glycemic control and sustained weight reduction with adjunctive exenatide therapy were found in individuals with T2DM on sulfonylurea monotherapy[85] and sulfonylurea and metformin combination therapy.[86]

Adverse events of exenatide include mild or moderate nausea, diarrhea, and a slightly increased risk of hypoglycemia with concurrent sulfonylurea therapy. Disadvantages of the GLP-1 analog are that it needs to be injected and cannot be mixed with insulin.

6.19.7 Unmet Medical Needs

6.19.7.1 Limitations of Existing Therapies

6.19.7.1.1 Obesity

Lifestyle modifications are essential for successful treatment of obesity and T2DM. When weight loss is achieved and sustained, complications associated with obesity are reduced and T2DM can be prevented in high-risk individuals. The lack of a cost-effective, long-term means of achieving this goal is at the root of the current epidemics of these two diseases. Development of either non-pharmacological or pharmacological means of preventing or reducing obesity that can be maintained long term would have a profound impact on healthcare costs and the use of healthcare resources.

6.19.7.1.2 Type 2 diabetes mellitus

Although the diagnosis of T2DM is straightforward, it is extremely difficult to monitor and to treat successfully long-term. Therapy and the monitoring of its effectiveness are complex, invasive, and expensive. Lack of patient education and resources compounds the problem and contributes to non-compliance and suboptimal glycemic control. The asymptomatic nature of chronic hyperglycemia does not allow the patient to truly understand the risk of diabetic complications until irreversible damage has developed. Therapies that have been added to available options in recent years have new and different mechanisms of action. However, when used as monotherapy, none of them (save insulin and its analogs) sustain optimal control for long intervals of time in the majority of patients.

6.19.7.1.3 Diabetic complications

Diabetic retinopathy, neuropathy, and nephropathy have significant morbidity and mortality that impact the quality of life in individual with diabetes. To date, good glycemic control is the primary means of preventing complications. Retinopathy and nephropathy can be treated and often arrested; however, therapy of neuropathy is symptomatic. As of 2006, techniques to identify persons at highest risk for complications are also very limited.

6.19.7.2 Arresting Loss of β-Cells

Progressive loss of β-cell function over time in individuals with T2DM progressively changes the therapeutic options available to maintain satisfactory glycemic control. Efforts to develop methods to arrest the loss and/or restore the function of β-cells are needed urgently. If successful, enormous benefits will follow.

6.19.8 New Areas of Research

Ongoing work in several arenas holds promise for the development of new agents or new approaches to treatment, or the prevention of obesity, the metabolic syndrome, and T2DM.

6.19.8.1 Incretin Augmentation of Insulin Secretion

The observation that food ingestion or enteral glucose administration provoked a greater stimulation of insulin release than similar amounts of glucose infused intravenously led to the recognition of gastrointestinal hormones known as incretins. Although a number of neurotransmitters and gut hormones possess incretin-like activity, several lines of evidence (immunoneutralization, administration of antagonists, and knockout studies) suggest that glucose-dependent insulinotropic polypeptide (GIP) and glucagon-like peptide-1 (GLP-1) represent the dominant peptides responsible for nutrient-augmented stimulation of insulin secretion. Exenatide, the first GLP-1 agonist, has recently received FDA approval for adjuvant therapy in T2DM (*see* Section 6.19.6.7)

Dipeptidyl peptidase-IV (DPP-IV) is a membrane-associated peptidase that is widely distributed in tissues and also exists as a soluble circulating form. Several DPP-IV inhibitors have been characterized and shown to lower blood glucose via prolongation of circulating GLP-1 and GIP action.[87] Animal studies with DPP-IV inhibitors show promising results. Progressive improvement in glycemic control, enhanced insulin secretory response, increased insulin-stimulated muscle glucose uptake, and improved hepatic and peripheral insulin sensitivity have been noted. Vildagliptin (LAF-237), a DPP-IV inhibitor, is in phase III clinical trials for T2DM. It markedly reduces DPP-IV activity within 30 min and continues to show activity for at least 10 h. In a recently reported trial that compared metformin plus LAF-237 50 mg day^{-1} with metformin plus placebo, LAF-237 reduced HbA_{1C} levels by 0.7% in 12 weeks.[88] The LAF-237 group also showed significant reductions in FPG levels and improvements in mean prandial glucose. The dipeptidyl peptidase-IV (DPP-IV) inhibitor, sitagliptin was approved as monotherapy for type 2 diabetes in October 2006.

Glucose-dependent insulinotropic polypeptide (GIP), also known as gastric inhibitory polypeptide, is a 42 amino acid peptide released from intestinal cells in response to nutrient absorption. Similarly to GLP-1, GIP undergoes rapid metabolism by DPP-IV via amino-terminal inactivation. GIP promotes β-cell proliferation[89] and stimulates glucose-dependent β-cell insulin secretion,[90] but has no effect on glucagon secretion or gastric emptying. GIP infusion in normal human subjects did not affect glucose, insulin, or C-peptide at normoglycemia. Although there are no GIP analogs currently in phase III clinical trials, there remains interest in the possibility that GIP analogs resistant to DPP-IV may exhibit therapeutic potential.

6.19.8.2 Other Potential Therapeutic Targets

Our understanding about the pathways of insulin action downstream from the insulin receptor is rapidly expanding (see **Figure 4**). Every step of the insulin receptor pathway has the potential to serve as a target for pharmacological intervention in the treatment of T2DM and obesity. Two such examples are summarized here in GSK-3 and PTP1B.

Glycogen synthase kinase 3 (GSK-3) is a protein kinase the activity of which is inhibited by insulin.[91] GSK-3 serves a regulatory function to phosphorylate glycogen synthase and thereby inactivate it. Insulin action stimulates the PI-3 kinase pathway, resulting in Akt activation, which phosphorylates and inactivates GSK-3. Elevated levels of GSK-3 have been observed in the skeletal muscle of individuals with T2DM. Thus, compounds that inhibit GSK-3 suggest that specific inhibitors of GSK-3 could mimic some of the actions of insulin and hold the potential as novel therapeutics for diabetes. The identification of a phosphorylation site unique to GSK-3 in the insulin-signaling pathway allows for the development of non-ATP competitive inhibitors that would selectively inhibit some functions of GSK-3 but not others, diminishing the possibility of unwanted side effects. These drugs are predicted to be of particular therapeutic relevance to the treatment of diabetes in the future.

Protein tyrosine phosphatase-1B (PTP1B) dephosphorylates the insulin receptor, thus blunting its ability to initiate the signal transduction cascade upon insulin binding.[92] Genetically modified mice that lack PTP1B protein expression and animals treated with a specific PTP1B antisense oligonucleotide inhibit PTP1B and thereby restore activity to the insulin receptor, resulting in increased insulin sensitivity, improved glycemic control, and resistance to diet-induced obesity. PTP1B inhibition also reduces adipose tissue storage of triglyceride under conditions of overnutrition, and was not associated with any obvious toxicity. The effects of the loss of PTP1B in vivo were also remarkably specific for components of the insulin action cascade, in spite of cell studies suggesting that PTP1B may exert a regulatory influence on a variety of other signaling pathways. Overall, these studies have paved the way for the commercial development of PTP1B inhibitors, which may serve as a novel type of 'insulin sensitizer' in the management of T2DM and the metabolic syndrome.

6.19.8.3 Alternative Methods of Insulin Delivery

Delivery of insulin via the pulmonary route can potentially provide the benefits of bolus insulin therapy without injections. Pulmonary delivery of insulin uses the well-vascularized, highly permeable alveoli of the lungs as the port of entry for macromolecules. Several formulations of inhaled insulin are in clinical trials or awaiting regulatory approval. Skyler *et al.*[93] provided the proof-of-concept study that illustrates the efficacy of mealtime use of inhaled insulin in individuals with T1DM. Glycemic control was similar in subjects receiving preprandial inhaled insulin plus subcutaneous ultralente insulin at bedtime (HbA_{1C} 7.9% at 12 weeks) or a usual insulin regimen of two to three injections per day (HbA_{1C} 7.7%). Similarly, Cefalu *et al.*[94] showed similar efficacy of inhaled insulin in individuals with T2DM whose treatment with combination oral agents had failed. In both these studies inhaled insulin was well tolerated without adverse pulmonary effects. However, due to its inefficient absorption, approximately 10-fold higher doses of insulin must be administered to achieve a therapeutic response. Data concerning long-term efficacy and safety are currently being collected.

Another alternative insulin delivery approach uses a metered-dose inhaler to direct fine aerosolized droplets of liquid insulin into the mouth for transmucosal absorption. One such formulation (Oral-lyn) has already received regulatory approval in Ecuador and is currently in phase III clinical trials in the Canada and Europe.

6.19.8.4 The Artificial Pancreas

The development of a 'closed-loop' insulin delivery system (artificial pancreas) has been intensively pursued for more than two decades. In the interval, insulin pump delivery systems have been improved greatly and reduced dramatically is size. Sophisticated delivery algorithms and options for delivery of insulin (and even other hormones such as glucagon) directly into the portal circulation have been developed. The lack of a sensing system that is implantable, accurate, and stable for a minimum of 6–12 months continues to be the limiting barrier to this approach. While success is plausible, it appears that such a system will not be available for clinical use in the near future.

6.19.8.5 β-Cell Replacement Therapy

Reinstitution or regeneration of pancreatic islet cells to achieve insulin independence shows promise in reversing the underlying insulin deficiency in both T1DM and T2DM. Pancreatic and islet cell transplantation are currently available, and exciting research in stem cell transplantation and gene therapy is ongoing.

Pancreas transplantation has become a widely accepted treatment for individuals with T1DM who have undergone a previous or simultaneous kidney transplant. The success rate, defined as the maintenance of normoglycemia and insulin independence, is approximately 80% at 3 years.[95] However, solitary solid-organ pancreas transplantation is not widely accepted because of the associated surgical complications and the need for vigorous immunosuppression, both of which contribute considerably to the overall morbidity and costs of this procedure.

Islet cell transplantation has the promise of β-cell replacement without complete pancreatic transplantation, allowing a less invasive alternative to solid-organ transplantation, lower degrees of immunosuppression, fewer complications, and a decreased financial burden. Shapiro *et al.*[96] reported 100% success in seven individuals undergoing islet transplantation in 2000, raising excitement for a clinically practical breakthrough. The largest study to date, published in 2002, reported on 54 islet infusions in 30 individuals with T1DM with separate infusions from two donor organs given approximately 1 month apart.[97] The follow-up data on 17 consecutive individuals shows achievement of insulin independence, with mean HbA_{1C} levels of 6.1%. To date, islet cell transplantation has been limited by constraints in islet cell harvesting, loss of islet cells during the purification and isolation processes, and a fall-off of patients who remain insulin independent after the first year.

The availability of human islets for transplantation falls far short of the potential need. Thus, alternative sources of insulin-producing cells are considered as an attractive solution for long-term β-cell replacement therapies. Potential alternatives include adult/embryonic stem cells or genetic engineering of non β-cells to produce insulin. Differentiated cell types derived from embryonic stem cells possess many of the physiologic and functional capacities of normal cells.

Under certain conditions, with selected differentiation factors, the embryonic stem cells can be differentiated into insulin-producing β-cells.[98] More importantly, the long-term function or expansion of such cells when they are transplanted into diabetic animals has been demonstrated.[99]

Induction of synthesis and regulated release of human insulin by non β-cells utilizes in vivo gene transfer technology to create 'surrogate β-cells.' Autologous cells escape rejection without immunosuppressive therapy and, therefore, are suitable surrogates. Hepatocytes possess the key molecules (GLUT-2 and glucokinase) used by β-cells to sense glucose variation and are extremely good surrogate candidates. An alternative possibility is that multipotent progenitor cells reside within the adult pancreas. Regeneration of β-cells after tissue injury has been observed in several model systems, indicating that certain cells within the mature pancreas retain the ability to partially restore β-cell mass after injury.[100] Recent studies indicate that differentiated exocrine acinar and/or ductal cells may potentially transdifferentiate into β-cells.[101]

References

1. National Center for Health Statistics Website. Prevalence of Overweight and Obesity Among Adults: United States, 1999–2002. http://www.cdc.gov/nchs/products/pubs/pubd/hestats/obese/obse99.htm (accessed on June 2005).
2. Centers for Disease Control and Prevention. *National Diabetes Fact Sheet: General Information and National Estimates on Diabetes in the United States, 2005.* US Department of Health and Human Services, Centers for Disease Control and Prevention: Atlanta, GA, 2005.
3. Mokdad, A. H.; Ford, E. S.; Bowman, B. A.; Dietz, W. H.; Vinicor, F.; Bales, V. S.; Marks, J. S. *JAMA* **2003**, *289*, 76–79.
4. King, H.; Rewers, M. *Diabetes Care* **1993**, *16*, 157–177.
5. Ford, E. S.; Williamson, D. F.; Liu, S. *Am. J. Epidemiol.* **1997**, *146*, 214–222.
6. Klein, B. E. K.; Klein, R.; Lee, K. E. *Diabetes Care* **2002**, *25*, 1790–1794.
7. Sattar, N.; Gaw, A.; Scherbakova, O. *Circulation* **2003**, *108*, 414–419.
8. Kleinman, J. C.; Donahue, R. P.; Harris, M. I. *Am. J. Epidemiol.* **1999**, *128*, 389.
9. Harris, M. I.; Klein, R.; Welbom, J. A.; Knuiman, M. W. Onset of NIDDM occurs at least 4–7 yr before clinical diagnosis. *Diabetes Care* **1992**, *15*, 815–819.

10. Pan, X. R.; Li, G. W.; Hu, Y. H.; Wang, J. X.; Yang, W. Y.; An, Z. X.; Hu, Z. X.; Lin, J.; Xiao, J. Z.; Cao, H. B. et al. *Diabetes Care* **1997**, *20*, 537–544.
11. Tuomilehto, J.; Lindstrom, J.; Eriksson, J. G.; Valle, T. T.; Hamalainen, H.; Ilanne-Parikka, P.; Keinanen-Kiukaanniemi, S.; Laakso, M.; Louheranta, A.; Rastas, M. et al. *N. Engl. J. Med.* **2001**, *344*, 1343–1350.
12. The Diabetes Prevention Program (DPP) Research Group. *N. Engl. J. Med.* **2002**, *346*, 393–403.
13. Chiasson, J. L.; Josse, R. G.; Gomis, R.; Hanefeld, M.; Karasik, A.; Laakso, M. *JAMA* **2003**, *290*, 486–494.
14. Buchanan, T. A.; Xiang, A. H.; Peters, R. K.; Kjos, S. L.; Marroquin, A.; Goico, J.; Ochoa, C.; Tan, S.; Berkowitz, K.; Hodis, H. N.; Azen, S. P. *Diabetes* **2002**, *51*, 2796–2803.
15. 2005 ADA Clinical Practice Guidelines. *Diabetes Care* **2005**, *28*, S40.
16. 2005 ADA Clinical Practice Guidelines. *Diabetes Care* **2005**, *28*, S41.
17. Alexander, C. M.; Landsman, P. B.; Teutsch, S. M.; Haffner, S. M. *Diabetes* **2003**, *52*, 1210–1214.
18. Alberti, K. G.; Zimmet, P. Z. *Diabet. Med.* **1998**, *15*, 539–553.
19. Executive Summary of the Third Report of the National Cholesterol Education Program (NCEP) Expert Panel on Detection, Evaluation, and Treatment of High Blood Cholesterol in Adults (Adult Treatment Panel III). *JAMA* **2001**, *285*, 2486.
20. Seaquist, E. In *Endocrine Physiology*, 1st ed.; Niewoehner, C. B., Ed.; Fence CreekH New Haven, CT, 1998, Chapter 8, p 118.
21. Andersen, D.; Elahi, D.; Brown, J. C. *J. Clin. Invest.* **1978**, *62*, 152–161.
22. Dupre, J.; Ross, S. A.; Watcon, D.; Brown, J. C. *J. Clin. Endocrinol. Metab.* **1973**, *37*, 826–828.
23. Creutzfeldt, W.; Nauck, M. *Diabetes Metab. Rev.* **1992**, *8*, 149.
24. Nishi, S.; Seino, Y.; Ishida, H. J. *Clin. Invest.* **1987**, *79*, 1191–1196.
25. Rhodes, C. J.; Shoelson, S.; Halban, P. A. In *Joslin's Diabetes Mellitus*, 14th ed.; Baltimore, OH: Lippincott Williams & Wilkins, 2004, Chapter 5, pp 66, 69.
26. Fajans, S. S.; Bell, G.; Polonsky, K. S. *N. Engl. J. Med.* **2001**, *345*, 971–980.
27. Ward, W. K.; Bolgiano, D. C.; McKnight, B.; Halter, J. B.; Porte, D., Jr. *J. Clin. Invest.* **1984**, *74*, 1318–1328.
28. Saltiel, A. R.; Kahn, C. R. *Nature* **2001**, *414*, 799–806.
29. Redondo, M. J.; Yu, L.; Hawa, M.; Mackenzie, T.; Pyke, D. A.; Eisenbarth, G. S. *Diabetologia* **2001**, *44*, 927.
30. Tillil, H.; Kobberline, J. *Diabetes* **1987**, *36*, 93–99.
31. Bergenstal, R. M.; Kendal, D. M.; Franz, M. J.; Rubenstein, A. H. In *Endocrinology*, 4th ed.; DeGroot, L. J., Jameson, J. L., Eds.; WB Saunders: Philadelphia, PA, 2001, Chapter 58, p 822.
32. Medici, F.; Hawa, M.; Ianari, A. *Diabetologia* **1999**, *42*, 146–150.
33. Lazar, M. *Science* **2005**, *307*, 373–375.
34. Pimenta, W.; Kortytkowski, M.; Mitrakou, A. *JAMA* **1995**, *273*, 1855–1861.
35. Bergman, R. N.; Phillips, L. S.; Cobelli, C. *J. Clin. Invest.* **1981**, *68*, 1456–1467.
36. Kahn, S. E. *J. Clin. Endocrinol. Metab.* **2001**, *86*, 4047.
37. Shafrir, E. In *Ellenberg & Rifkin's Diabetes Mellitus*, 6th ed.; Porte, D., Jr., Sherwin, R. S., Baron, A., Eds.; McGraw-Hill: New York, 2003, Chapter 16, p 231.
38. Malaisse, W. J. *Biochem. Pharmacol.* **1982**, *31*, 3527.
39. Okamoto, H. In *Molecular Biology of the Islets of Langerhans*; Okamoto, H., Ed.; Cambridge University Press: Cambridge, 1990.
40. Crisa, L.; Mordes, J. P.; Rossini, A. A. *Diabetes Metab. Rev.* **1992**, *8*, 9.
41. Harada, M.; Makino, S. *Ann. Rep. Shionogi Res. Lab.* **1992**, *42*, 70.
42. Araki, E.; Lipes, M. A.; Patti, M. E. *Nature* **1994**, *372*, 182.
43. Coleman, D. L. *Metabolism* **1983**, *32*, 162.
44. Andrikopoulos, S.; Thorburn, A. W. E.; Proietto, J. In *Primer on Animal Models of Diabetes*; Sima, A. A. F., Shafrir, E., Eds.; Harwood Academic Press: Reading, 2000, p 171.
45. Berdanier, D. D. In *Lessons from Animal Diabetes*; Shafrir, E., Ed.; Smith-Gordon: London, 1995, p 231.
46. Kawano, K.; Hirashima, T.; Mori, S. In *Lessons from Animal Diabetes*; Shafrir, E., Ed.; Birkhauser: New York, 1996, p 227.
47. Ziv, E.; Shafrir, E. In *Lessons from Animal Diabetes*; Shafrir, E., Ed.; Smith-Gordon: London, 1995, p 231.
48. University Group Diabetes Program. *Diabetes* **1970**, *19*, 747–830.
49. The DCCT Research Group. *N. Engl. J. Med.* **1993**, *329*, 977–986.
50. The DCCT Research Group. *Diabetes Care* **1995**, *18*, 1415–1427.
51. Diabetes Control and Complications Trial/Epidemiology of Diabetes Interventions and Complications Research Group. *JAMA* **2002**, *287*, 2563–2569.
52. Diabetes Control and Complications Trial/Epidemiology of Diabetes Interventions and Complications (DCCT/EDIC) Study Research Group. *N. Engl. J. Med.* **2005**, *353*, 2643.
53. Diabetes Control and Complications Trial/Epidemiology of Diabetes Interventions and Complications Research Group. *JAMA* **2003**, *290*, 2159–2167.
54. UK Prospective Diabetes Study Group. *Lancet* **1998**, *352*, 837–853.
55. 2006 ADA Clinical Practice Guidelines. *Diabetes Care* **2006**, *29*, S10.
56. UK Prospective Diabetes Study Group. *Metabolism* **1990**, *39*, 905–912.
57. Mudaliar, S.; Henry, R. R. In *Ellenberg & Rifkin's Diabetes Mellitus*, 6th ed.; Porte, D., Jr., Sherwin, R. S., Baron, A., Eds.; McGraw-Hill: New York, 2003, Chapter 32, pp 532–533.
58. Fuhlendorff, J.; Rorsman, P.; Kofod, H.; Brand, C. L.; Rolin, B.; MacKay, P.; Shymko, R.; Carr, R. D. *Diabetes* **1998**, *47*, 345–351.
59. Moses, R.; Slobodniuk, R.; Boyages, S.; Colagiuri, S.; Kidson, W.; Carter, J.; Donnelly, T.; Moffitt, P.; Hopkins, H. *Diabetes Care* **1999**, *22*, 119–124.
60. Horton, E. S.; Clinkingbeard, C.; Gatlin, M.; Foley, J.; Mallows, S.; Shen, S. *Diabetes Care* **2000**, *23*, 1660–1665.
61. Rossetti, L.; DeFronzo, R. A.; Gherzi, R. *Metabolism* **1990**, *39*, 425–435.
62. Cigolini, M.; Bosello, O.; Zancanaro, C.; Orlandi, P. G.; Fezzi, O.; Smith, U. *Diabetes Metab.* **1984**, *10*, 311–315.
63. DeFronzo, R. A.; Barzilai, N.; Simonson, D. C. *J. Clin. Endocrinol. Metab.* **1991**, *73*, 1294–1301.
64. DeFronzo, R. A.; Goodman, A. M. *N. Engl. J. Med.* **1995**, *333*, 541–549.
65. Bailey, C. J. Biguanides and NIDDM. *Diabetes Care* **1992**, *15*, 755–772.
66. Campbell, L. K.; White, J. R. *Ann. Pharmacother.* **1996**, *30*, 1255.
67. Yki-Järvinen, H. *N. Engl. J. Med.* **2004**, *351*, 1106–1118.

68. Duez, H.; Chao, Y. S.; Hernandez, M.; Torpier, G.; Poulain, P.; Mundt, S.; Mallat, Z.; Teissier, E.; Burton, C. A.; Tedgui, A. et al. *J. Biol. Chem.* **2002**, *277*, 48051–48057.
69. Diabetes Atherosclerosis Intervention Study Investigators. *Lancet* **2001**, *357*, 905–910.
70. Lebovitz, H. E.; Dole, J. F.; Patwardhan, R.; Rappaport, E. B.; Freed, M. I. *J. Clin. Endocrinol. Metab.* **2001**, *86*, 280–288.
71. The Pioglitazone 001 Study Group. *Diabetes Care* **2000**, *23*, 1605–1611.
72. Phillips, L. S.; Grunberger, G.; Miller, E.; Patwardhan, R.; Rappaport, E. B.; Salzman, A. *Diabetes Care* **2001**, *24*, 308–315.
73. Sakamoto, J.; Kimura, H.; Moriyama, S. *Biochem. Biophys. Res. Commun.* **2000**, *278*, 704–711.
74. Hirsch, I. B. *N. Engl. J. Med.* **2005**, *352*, 174–183.
75. Ter Braak, E. W.; Woodworth, J. R.; Bianchi, R.; Cerimele, B.; Erkelens, D. W.; Thijssen, J. H. H.; Kurtz, D. *Diabetes Care* **1996**, *19*, 1437–1440.
76. Nolte, M. A.; Karam, J. H. In *Basic and Clinical Pharmacolog*, 8th ed; Katzung, B. G., Ed.; Appleton and Lang: Stamford, CT, 2001, p 711.
77. Rabkin, R.; Ryan, M. P.; Duckworth, W. C. *Diabetologia* **1984**, *27*, 351–357.
78. Barlocco, D. *Curr. Opin. Invest. Drugs* **2003**, *4*, 1240–1244.
79. Wright, A.; Burden, A. C.; Paisey, R. B.; Cull, C. A.; Holman, R. R. *Diabetes Care* **2002**, *25*, 330–336.
80. Pugh, J. A.; Wagner, M. L.; Sawyer, J.; Ramirez, G.; Tuley, M.; Friedberg, S. J. *Diabetes Care* **1992**, *15*, 953–959.
81. UK Prospective Diabetes Study (UKPDS) Group. *Lancet* **1998**, *352*, 837–853.
82. Whitehouse, F.; Kruger, D. F.; Fineman, M.; Shen, L.; Ruggles, J. A.; Maggs, D. G.; Weyer, C.; Kolterman, O. G. *Diabetes Care* **2002**, *25*, 724–730.
83. Hollander, P. A.; Levy, P.; Fineman, M.; Shen, L.; Maggs, D. G.; Strobel, S. A.; Weyer, C.; Kolterman, O. G. *Diabetes Care* **2003**, *26*, 784–790.
84. DeFronzo, R. A.; Ratner, R. E.; Han, J.; Kim, D. D.; Fineman, M. S.; Baron, A. D. *Diabetes Care* **2005**, *28*, 1092–1100.
85. Exenatide-113 Clinical Study Group. *Diabetes Care* **2004**, *27*, 2628–2635.
86. Kendall, D. M.; Riddle, M. C.; Rosenstock, J.; Zhuang, D.; Kim, D. D.; Fineman, M. S.; Baron, A. D. *Diabetes Care* **2005**, *28*, 1083–1091.
87. Pederson, R. A.; White, H. A.; Schlenzig, D.; Pauly, R. P.; McIntosh, C. H.; Demuth, H. U. *Diabetes* **1998**, *47*, 1253–1258.
88. Ahren, B.; Gomis, R.; Standl, E.; Mills, D.; Schweizer, A. *Diabetes Care* **2004**, *27*, 2874–2880.
89. Trumper, A.; Trumper, K.; Horsch, D. *J. Endocrinol.* **2002**, *174*, 233–246.
90. Nauck, M. A.; Heimesaat, M. M.; Orskov, C.; Holst, J. J.; Ebert, R.; Creutzfeldt, W. *J. Clin. Invest.* **1993**, *91*, 301–307.
91. Frame, S.; Cohen, P. *Biochem. J.* **2001**, *359*, 1–16.
92. Zhang, Z. Y.; Lee, S. Y. *Expert Opin. Invest. Drugs* **2003**, *12*, 223.
93. Skyler, J. S.; Cefalu, W. T.; Kourides, I. A.; Landschulz, W. H.; Balagtas, C. C.; Cheng, S. L.; Gelfand, R. A. *Lancet* **2001**, *357*, 331–335.
94. Cefalu, W. T.; Skyler, J. S.; Kourides, I. A.; Landschulz, W. H.; Balagtas, C. C.; Cheng, S.; Gelfand, R. A. *Ann. Intern. Med.* **2001**, *134*, 203–207.
95. Gruessner, A. C.; Sutherland, D. E.; Dunn, D. L. *J. Am. Soc. Nephrol.* **2001**, *12*, 2490–2499.
96. Shapiro, A. M.; Lakey, J. R.; Ryan, E. A.; Kobutt, G. S.; Toth, E.; Warnock, G. L.; Kneteman, N. M.; Rajotte, R. V. *N. Engl. J. Med.* **2000**, *343*, 230–238.
97. Ryan, E. A.; Lakey, J. R. T.; Paty, B. W.; Imes, S.; Korbutt, G. S.; Kneteman, N. M.; Bigam, D.; Rajotte, R. V.; Shapiro, A. M. J. *Diabetes* **2002**, *51*, 2148–2157.
98. Hori, Y.; Rulifson, I. C.; Tsai, B. C. *Proc. Natl. Acad. Sci.* **2002**, *99*, 16105–16110.
99. Lumelsky, N.; Blondel, O.; Laeng, P. *Science* **2001**, *292*, 1389–1394.
100. Hayashi, K. Y.; Tamaki, H.; Handa, K. *Arch. Histol. Cytol.* **2003**, *66*, 163–174.
101. Bonner-Weir, S. *J. Mol. Endocrinol.* **2000**, *24*, 297–302.

Biographies

Lowell Schmeltz received his undergraduate degree from the University of Michigan and medical degree from the University of Pittsburgh School of Medicine. He completed his internship and residency in Internal Medicine at the Feinberg School of Medicine of Northwestern University in Chicago, IL. Dr Schmeltz is currently an endocrinology fellow at Northwestern University, pursuing clinical research in diabetes management, and has published online updates of various endocrinology topics in collaboration with Dr J Larry Jameson for *Harrison's Principles of Internal Medicine*.

Boyd Metzger was an undergraduate at the University of South Dakota and received his MD from the University of Iowa. He was an Intern, Medical Resident, and Resident in Endocrinology & Metabolism at Michael Reese Hospital in Chicago, and a Research Fellow in the Department of Biological Chemistry at Washington University in St Louis. In 1967, he joined the faculty of Northwestern University School of Medicine in the Department of Medicine and the Center for Endocrinology Metabolism and Nutrition. Currently, he is Tom D Spies Professor of Metabolism & Nutrition. His research interests have included the pathophysiology and prevention of type 2 diabetes mellitus, regulation of intermediary metabolism in normal and diabetic pregnancy, and the intrauterine, perinatal, and life-long impact of the metabolic environment of maternal diabetes. Because gestational diabetes is a common precursor of type 2 diabetes, he has used it as a model for studies of insulin resistance and β-cell function in subjects at high risk for progression to overt diabetes. Currently, his is Principal Investigator of the Hyperglycemia & Adverse Pregnancy Outcome (HAPO) Study, an international, multicenter epidemiological study aimed at determining the level of maternal hyperglycemia associated with clinically meaningful risk of adverse pregnancy outcome. The data should permit investigators and clinicians to establish 'outcome-based' criteria for the diagnosis of gestational diabetes.

Comprehensive Medicinal Chemistry II
ISBN (set): 0-08-044513-6

ISBN (Volume 6) 0-08-044519-5; pp. 417–458

6.20 Atherosclerosis/Lipoprotein/Cholesterol Metabolism

J A Sikorski, AtheroGenics, Alpharetta, GA, USA

6.20.1 Disease State

6.20.1.1 Atherosclerosis

Atherosclerosis refers to the underlying progression in arterial dysfunction and remodeling that restricts blood flow to vessels in the peripheral vasculature.[1] Taken from the Greek, *athero* (gruel or paste) and *sclerosis* (hardness), atherosclerosis describes the buildup of lipid-laden fatty deposits within the vessel wall that is often called atheroma or atherosclerotic plaques. As illustrated in **Figure 1**, temporal morphological changes accompany the progression of disease. In its early stages, atherosclerosis causes significant changes to the vessel wall with the accumulation of cholesterol and scar tissue. The lesion is preceded by a fatty streak in the intima below the endothelium, composed of lipid-enriched macrophages and T cells. These fatty streaks are usually symptomatically silent, are frequently observed in teenagers and young adults, and may either disappear or progress to more advanced disease. As the lesion develops, it is enriched in cholesterol-laden foam cells, becomes calcified, and hardens. In later stages, the developing lesion induces a thickening of the vessel wall due to smooth-muscle cell (SMC) migration and proliferation above the foam cells, accompanied by the formation of a tough, fibrous cap over the lipid-laden core. The resulting plaque may be stable for long periods or may be unstable and prone to rupture.

Two types of vessel wall remodeling have been observed using intravascular ultrasound (IVUS) imaging techniques: (1) positive remodeling that largely maintains the lumen diameter as the lesion grows inward through the vessel wall and the external elastic membrane area expands; and (2) negative remodeling, where the external elastic membrane area contracts and lesion growth narrows the luminal diameter and restricts blood flow.[2] Historically, lesions that impeded blood flow were thought to be the most dangerous as the progressive narrowing restricted blood flow and oxygen supply to the heart, eventually leading to heart attack and myocardial infarct (MI). However, recent angiographic studies indicate that in many cases the most dangerous lesions appear at sites where blood flow is not severely restricted. The latest studies indicate that plaque activation and rupture rather than stenosis are considered to be the primary underlying events preceding ischemia and infarct.

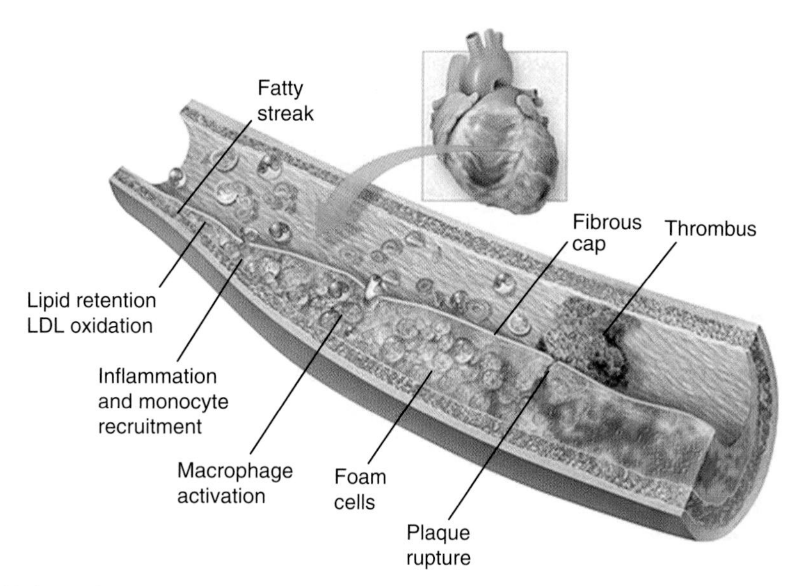

Figure 1 The initiation and progression of atherosclerosis in arterial vessels starting from LDL modification and oxidation leading to lipid retention, monocyte recruitment and inflammation, then macrophage activation and foam cell development ending in plaque rupture and thrombus formation. (Reprinted with permission by Keith Kasnot.)

Atheromas that either impede or do not impede blood flow are subject to rupture. Typically, the fibrous cap becomes weakened as inflammatory mediators from within the lesion are secreted and degrade the overlying extracellular matrix. The cap then develops cracks that rupture the lesion, exposing tissue factor from below the vessel wall that induces the extrinsic coagulation cascade locally at the site and quickly leads to thrombus formation. The resulting loss of oxygen supply to the heart produces cellular and tissue injury and ultimately culminates in chest pain (angina), heart attack, and/or MI. The disrupted flow to the carotid artery or in microvessels within the brain produces stroke (*see* 6.10 Stroke/Traumatic Brain and Spinal Cord Injuries). The underlying atherosclerotic process develops slowly and asymptomatically in early adulthood, where clinical signs and symptoms are typically not observed until after age 45 in males and later after menopause in females. Both MI and stroke frequently occur without any prior warning, require emergency treatment and hospitalization, and have immediate life-threatening consequences.

6.20.1.2 Heart Disease Imaging Methods

Since disease onset is unpredictable, early diagnosis is important. Several noninvasive testing procedures are frequently used in heart disease diagnosis, including resting electrocardiogram (ECG), echocardiogram, exercise stress test (with or without radiolabel tracer), IVUS, computed tomography (CT), and magnetic resonance imaging (MRI).[2] All have their advantages and limitations. Unfortunately, no one test provides a complete, reliable assessment as to whether the observed atherosclerotic plaques are vulnerable to rupture. A more invasive procedure, coronary angiography, delivers imaging dyes through small tubes inserted in the artery. This technique has been especially helpful over the last two decades in identifying vessels with severe blockage. This procedure is often followed by either balloon angioplasty to widen the artery or by the placement of a small, expandable, hollow, wire mesh tube (stent) to widen the artery and restore blood flow. In the last few years, drug-eluting stents have risen in popularity; such stents markedly reduce the incidence of restenosis. More than a million of these revascularization procedures are now performed annually in the US alone.

6.20.1.3 Heart Disease Mortality

Cardiovascular disease (CVD) is the leading cause of death worldwide. According to statistics compiled by the World Health Organization (WHO), more than 13 million people die each year from coronary heart disease (CHD) and stroke.[3] WHO statistics also show that more people have died worldwide since 1990 from CHD than from any other disease. Approximately 40% of the annual deaths in the US are attributed to CHD for both men and women. Thus, US death rates from CHD greatly exceed those from all types of cancers. The incidence of this disease in the US is expected to grow as a larger percentage of the population develops obesity and diabetes. As a result of lifestyle changes, primarily in reduced smoking, improved diet and exercise, more aggressive preventive medicine, as well as modern improvements in detection, diagnosis, and treatment, the mortality rates from CHD in both men and women are significantly decreasing in western Europe and the US. Nevertheless, despite these improvements, nearly 25–30% of patients in the US still die within a year of their first heart attack.[3] In contrast, from 1990 to 2002 the most dramatic and rapid increases in both CHD incidence and mortality have occurred in developing countries such as Russia, China, India, and the countries of Eastern Europe. Proportionate increases of smaller magnitude have also been observed in South America and Africa.[3] While previously considered a disease attributed to the sedentary lifestyle and fatty diets of modern American and western cultures, more than 75% of deaths resulting from CHD now occur in the poorer countries in the world. Based on these trends, WHO estimates that more than 20 million deaths from CHD and stroke will occur globally by 2020.[3]

CHD has an enormous global economic impact in terms of both the costs to society for treatments and hospitalizations as well as in decreased longevity. WHO estimates that in 2002 the 'healthy years of life lost' due to heart disease approached 10% in low and middle-income countries and rose to approximately 18% in higher-income countries.[3] In the US alone, the economic costs due to CHD have been estimated at $400 billion annually.

6.20.2 Disease Basis

6.20.2.1 Dyslipidemia as a Risk Factor for Coronary Heart Disease

While smoking, hypertension, age, obesity, diet, and family history all contribute to CHD, dyslipidemia is one of the most prominent risk factors for this disease. Historically, CHD has often been considered a disease primarily associated with hyperlipidemia, i.e., high plasma cholesterol levels, particularly cholesterol associated with low-density lipoprotein (LDLc).[4] Indeed, atherosclerotic plaques taken from heart transplant patients contained significantly higher percentages of cholesterol (19–26% $\pm$ 10%) than nonatherosclerotic coronary tissue taken from the same donors (4% $\pm$ 3%).[5]

Recent studies have more clearly defined the underlying pathobiology and biochemical mechanisms contributing to disease initiation and progression. The accumulation and retention of lipids from modified LDL particles activate endothelial cells, initiating an inflammatory response within the arterial wall. As discussed in more detail below, the developing atherosclerotic lesion mechanistically involves the same monocyte and leukocyte recruitment process and shares many of the common inflammatory mediators typically observed in severe chronic inflammatory diseases. As a result, atherosclerosis is now considered to be a chronic inflammatory disease, and modified lipoproteins play a central proinflammatory role in initiating this disease.[6–8]

6.20.2.2 Lipoproteins: Composition, Structure, Function, and Lipid Transport

A proper balance between phospholipids and free cholesterol (FC) is required to maintain optimal cell membrane structure and fluidity.[9] Many cells maintain their FC membrane requirements through endogenous biosynthesis. Other cells, including both the macrophages and underlying SMC involved in atherosclerotic lesion formation, acquire cholesterol by internalization of FC from lipoproteins. Because of their poor water-solubility, neutral lipids such as triglycerides (TGs), FC, and cholesteryl ester generally are not freely circulating in plasma, but instead are packaged together and assembled into larger lipoprotein particles that have amphipathic lipids and proteins as surface components. These submicroscopic spherical particles contain phospholipid (PL) and FC in the outer layer surrounding a core of neutral lipids, primarily cholesteryl ester and TG, held together by noncovalent forces. The cholesteryl ester found in the lipoprotein core is typically derivatized by saturated fatty acids, such as myristate, while the triglycerides contain mixed esters of saturated, unsaturated, and polyunsaturated fatty acids. The proteins associated with the particle surface are known as apolipoproteins (apo), and they control the structural integrity, functionality, and scavenger receptor binding of lipoprotein particles. These apolipoproteins also act as cofactors or inhibitors for remodeling lipases and other enzymes.

Lipoprotein particles have traditionally been classified by their relative density following isolation by ultracentrifugation and by the type of accompanying primary associated apolipoproteins. As summarized in **Table 1**, there are six major subtypes of lipoproteins: very low-density lipoproteins (VLDL) and chylomicrons are the largest and least dense of the lipoproteins whose cores are enriched in TG; LDL and HDL are the smallest, densest particles whose cores are enriched in cholesteryl ester; remodeling of VLDL by lipoprotein lipase (LpL) produces intermediate-density lipoprotein (IDL) and then LDL; and lastly, remodeling of LDL produces an LDL-like particle, lipoprotein(a) (Lp(a)). Each of these classes represents a heterogeneous population of particles that vary by their relative size, shape, charge, density, lipid composition, and associated secondary proteins (e.g., apoC-(I, II, III), apoD, apoE).

Table 1 Size and density characteristics of the major classes of lipoproteins and their associated apolipoproteins

Lipoprotein	*Origin*	*Structural apolipoprotein*	*Secondary apolipoproteins*	*Average diameter (nm)*	*Average density* ($g\,mL^{-1}$)
Chylomicrons	Intestine	apoB-48	apoA-I, apoA-II, apoC-I, apoC-II, apoC-III, apoE, apoH	>100	<0.95
VLDL[10]	Liver	apoB-100	apoC-I, apoC-II, apoC-III, apoE	~60	0.95–1.01
IDL	Remodeling of VLDL	apoB-100	apoC-III, apoE	~30–50	1.01–1.02
LDL[12,13]	Liver	apoB-100	apoE	20	1.02-1.06
HDL[15]	Liver, intestine, plasma	apoA-I, apoA-II	apoA-I, apoA-II, apoC-I, apoC-II, apoC-III, apoD, apoE	~10	1.06–1.20
Lp(a)	Liver remodeling of LDL	apo(a)-apoB-100		~25	1.05–1.09

Lipid transport and transfer are also mediated by plasma proteins. For example, cholesteryl ester transfer protein (CETP) mediates cholesteryl ester and TG transfer between HDL and either VLDL or LDL particles. For every cholesteryl ester acquired from HDL and transferred to LDL or VLDL, CETP acquires one TG molecule from LDL or VLDL and transfers it to HDL. Thus, CETP plays a critical role in both cholesteryl ester and TG transfer among lipoproteins. Similarly, PL transfer protein (PLTP) mediates the transfer of phospholipids between the various lipoprotein particles. The entire lipid transport process is highly regulated both in terms of lipoprotein particle and apolipoprotein production as well as the transfer of lipids to and from the particles.

6.20.2.2.1 Very low-density lipoprotein, chylomicrons, and triglyceride transport

Chylomicrons and VLDL particles each contain surface apolipoprotein-B (apoB). Chylomicrons are assembled primarily in the intestine and contain a smaller version, apoB-48, whereas VLDL particles contain the larger apoB-100 surface protein and are primarily assembled in the liver. The functional role for VLDL and chylomicron particles is to deliver TG to peripheral tissue. TG accounts for nearly half (48%) of the mass in VLDL particles analyzed from healthy human control groups (**Table 2**).[10] VLDL particles have commensurately lower levels of FC and cholesteryl ester. Human TG plasma levels in healthy subjects are considered normal if they fall below 200 $mg\,dL^{-1}$.[10] Plasma TG levels in hypertriglyceridemic subjects exceed 200 $mg\,dL^{-1}$ due to significantly increased plasma VLDL.

In the periphery, LpL acts on these particles to release the TG to adipose tissue for fat storage or to muscle tissue where they can be used for energy. The resulting TG-depleted and cholesteryl ester-enriched VLDL remnant particles can be recycled to the liver by the interaction of a specific hepatic LDL receptor (LDLr) with apoB and apolipoprotein-E (apoE). Alternatively, further remodeling of VLDL remnants produces cholesteryl ester-enriched particles: first IDL and then LDL. In contrast to VLDL and chylomicrons, the smaller and denser LDL and HDL particles help move cholesterol to and from the periphery, respectively.

6.20.2.2.2 Low-density lipoprotein: structure and composition

Like VLDL, LDL particles contain apoB-100 as surface proteins and are primarily assembled in the liver. With a molecular weight >550 kDa and over 4500 amino acid residues, apoB-100 is one of the largest and most insoluble human proteins.[11,12] The apoB-100 is so tightly associated with the surface of these particles that it is unexchangeable without disrupting the integrity of the particle. LDL particles have an average diameter of about 20 nm and are about three times smaller than VLDL particles. The remodeling of VLDL by LpL reduces its overall TG core content, increases its cholesteryl ester content, and shrinks the diameter of the particle, with the net effect of producing an

Table 2 Chemical compositions of LDL and HDL subclasses

Lipoprotein	*Apolipoprotein*	*Percentage of total lipoprotein particle mass*				
		% Apolipoprotein	*% CE[a]*	*% FC[b]*	*% PL[c]*	*% TG[d]*
VLDL[10]	apoB-100	4	12	4	19	48
LDL[12,13]	apoB-100	21	38	9	25	7
Pre-β-HDL[15]	apoA-I	>90	2	0.3	7	Negligible
HDL_3[15]	apoA-I, apoA-II	55	15	3	23	3
HDL_2[15]	apoA-I, apoA-II	40	18	6	31	4

[a] CE, cholesteryl ester.
[b] FC, free cholesterol.
[c] PL, phospholipid.
[d] TG, triglyceride.

LDL particle via IDL. Thus, the apoB-100 in LDL can also originate from VLDL. The apoB-100 comprises over 20% of the mass of an LDL particle (**Table 2**), and nearly half of the total mass is composed of FC (9%) and cholesteryl ester (38%).

Lipases, including phospholipase A_2 (PLA_2), also metabolize LDL. PLA_2 is primarily produced by macrophages, and the majority of circulating plasma PLA_2 is associated with LDL. The remodeling of LDL by PLA_2 in effect creates a smaller, denser, modified LDL particle with a higher apoB content that can more easily penetrate peripheral tissue and deliver cholesterol by interaction with cellular scavenger receptors. The resulting intracellular FC is transformed to cholesteryl ester by the action of the enzyme, acyl-CoA:cholesterol acyltransferase (ACAT). These small, dense, modified LDL particles have been associated with an accompanying increased coronary risk. Increased plasma concentrations of PLA_2 have also been linked to a higher CHD risk.

The heterogeneity of LDL particle populations has so far prevented any detailed, high-resolution, x-ray structural information from being elucidated by protein crystallography, although an informative spherical model has been developed from low-resolution data.[11,12] As depicted in **Figure 2**, modeling indicates that apoB-100 surrounds the LDL particle with a thick, wide ribbon composed primarily of β-sheet that covers approximately 30% of the PL surface and occasionally penetrates below PL to interact with the neutral lipid core. Approximately 20% of the particle surface is covered by more helical components of apoB-100 that also form a protruding cap that acts as a ligand for the hepatic LDLr.[11]

Nearly 60% of the circulating cholesterol in human plasma is present as LDLc, and the majority of this cholesterol is removed from plasma through the liver via uptake by the LDLr.[12,13] Indeed, functional mutations in the LDLr gene significantly decrease cholesterol clearance by this pathway and have been linked to the genetic disease, familial hypercholesterolemia (FH).[14] The average human plasma concentration of LDLc in normal healthy subjects from one study was approximately 106 $mg\,dL^{-1}$, accounting for nearly 60% of the total average plasma cholesterol levels of 170 $mg\,dL^{-1}$ (**Table 3**).[13] In CHD patients with higher overall total plasma cholesterol levels, most of the excess cholesterol is found associated with LDLc.

6.20.2.2.3 High-density lipoprotein: structure and composition

In contrast to LDL, HDL particles (**Figure 3**) remove excess cholesterol from the periphery and return it to the liver either by the direct uptake of HDL particles via hepatic HDL scavenger receptor type B-1 (SR-B1) or indirectly by CETP-mediated transfer of cholesteryl ester from HDL to LDL or VLDL and hepatic uptake of these apoB particles via the LDLr. The process of collecting excess peripheral cholesterol by HDL for disposal via the liver is the reverse cholesterol transport (RCT) pathway. This is an extremely important process to maintain cholesterol homeostasis since most cells do not have the capability to metabolize excess cholesterol. The key role that HDL plays in mediating cholesterol efflux from cholesterol-rich macrophages and foam cells in the atherosclerotic lesion is regarded as the primary mechanism by which HDL reduces or reverses atherosclerosis.[16]

With an average diameter of approximately 10 nm, HDL particles have a smaller overall surface size and higher density than LDL particles. The major HDL structural protein is apolipoprotein-AI (apoA-I) which is occasionally accompanied by apoA-II. In contrast to apoB-100 in LDL, apoA-I is much smaller (243 amino acids, ~28 kDa), primarily α-helical, and found in both lipid-free and lipid-associated states. However, circulating, lipid-free apoA-I

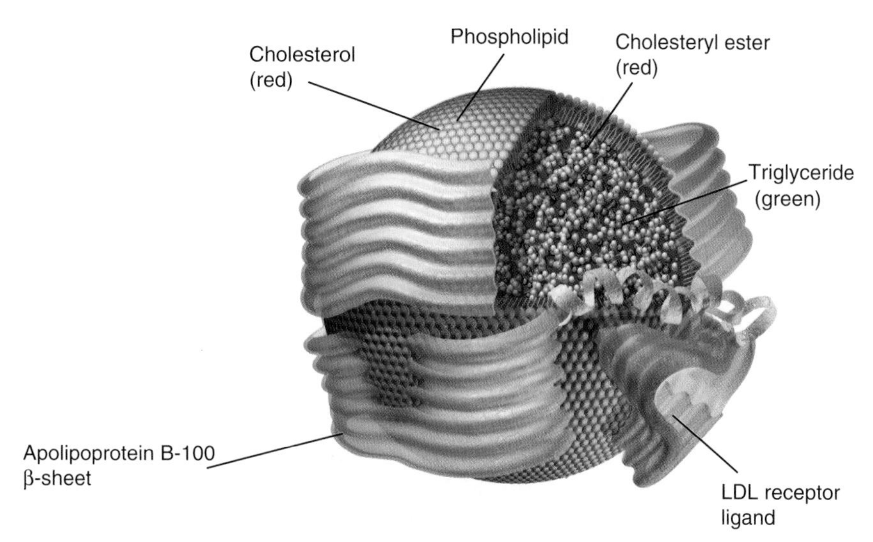

Figure 2 An illustrated model of a low-density lipoprotein (LDL) particle depicting the spherical surface composed primarily of phospholipids (PLs) with their polar head groups held together by non-covalent interactions on the surface and their associated hydrophobic fatty acid ester tails pointing inward toward the interior of the particle. Free cholesterol (FC) molecules are occasionally interspersed among the PLs on the surface. The interior core of the particle is composed of neutral lipids, primarily cholesteryl ester (CE), with smaller amounts of triglycerides (TGs). Nearly 50% of the surface of the LDL particle is covered with the tightly associated, large (MW >550 kDa) apolipoprotein, apoB-100, either as a wide ribbon with β-sheet secondary structure or defined α-helical loops. ApoB-100 occasionally dips below the PL surface to penetrate the interior core. α-Helical regions of apoB-100 also form ligand recognition loops for various receptors, including the hepatic LDL receptor. (Reprinted with permission by Keith Kasnot.)

Table 3 Average human plasma concentrations of Lp(a), LDL, and HDL components in normal subjects

Lipoprotein	*Apolipoprotein*	*Average human plasma concentration in normal subjects*	
		Apolipoprotein ($mg\,dL^{-1}$)	*Lipoprotein cholesterol* ($mg\,dL^{-1}$)
LDL[12,13]	apoB-100	80 ± 11	106 ± 13
Pre-β-HDL[15]	apoA-I	6–8	Negligible
HDL_3[15]	apoA-I	34	31
HDL_2[15]	apoA-I	14	18
Lp(a)	apo(a)-apoB-100	Highly variable	

represents an extremely small percentage (<5%) of apoA-I found in human plasma where discoidal and spherical forms of HDL predominate.

As a result, mature HDL particle assembly (**Figure 4**) is a complex process involving several intermediates that occurs throughout the circulation. As shown in **Table 2**, several subclasses of HDL particles have been observed which can be separated by gel filtration or size exclusion chromatography, including a discoidal pre-β-HDL, and two spherical subclasses, HDL_2 and HDL_3. HDL_3 particles have higher densities and smaller diameters than HDL_2 particles.[15] Pre-β-HDL particles contain mostly apoA (>90%) and have only a small relative percentage of cholesteryl ester and FC. In contrast to LDL particles, the spherical HDL particle subclasses also have a higher overall apolipoprotein content (40–55%) and correspondingly lower cholesteryl ester and FC mass percentages.

As illustrated in **Figure 4**, mature HDL assembly begins when lipid-free apoA-I accepts FC at sites in the liver, intestine, endothelial cells, or macrophages by a receptor-mediated process involving adenosine triphosphate-binding cassette (ABC) transporters, primarily ABCA1.[17] The resulting FC associated with apoA-I is esterified to cholesteryl ester by the action of lecithyl cholesterol acyl transferase (LCAT) which utilizes apoA-I as a cofactor. This LCAT-assisted fatty acid esterification process drives the formation of the discoidal pre-β-HDL particles. The pre-β-HDL

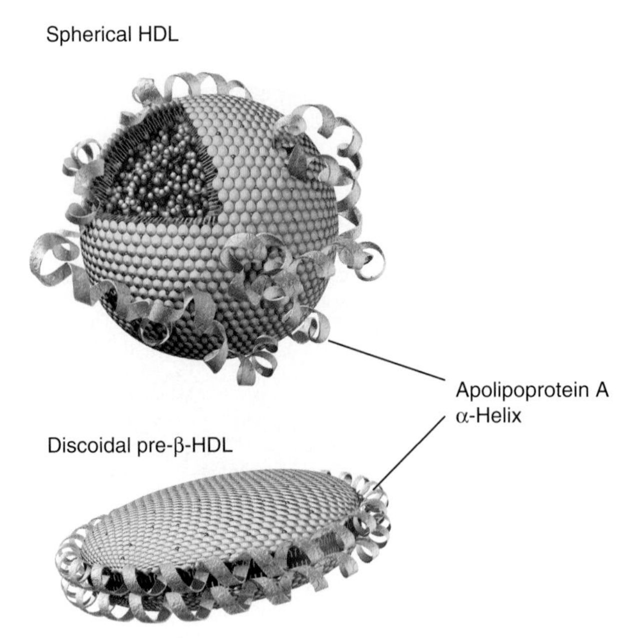

Figure 3 An illustrated model of mature spherical high-density lipoprotein (HDL) and discoidal pre-β-HDL. Discoidal pre-β-HDL forms as an intermediate in mature HDL particle assembly (**Figure 4**), as the small, helical apolipoprotein AI (apoA-I) accumulates phospholipid (PL) and free cholesterol (FC) which is further esterified to cholesteryl ester (CE) following the action of lecithyl cholesterol acyl transferase (LCAT). Like LDL, mature, spherical HDL is composed primarily of PLs with their polar head groups held together by non-covalent interactions on the surface and their associated hydrophobic fatty acid ester tails pointing inward toward the interior of the particle. FC molecules are occasionally interspersed among the PLs on the spherical surface. In contrast to LDL, the neutral lipids in the interior core of the HDL particle are composed primarily of triglycerides (TGs) with smaller amounts of CE. The surface of the HDL particle is covered with the loosely associated braids of small (MW = 28 kDa), α-helical apoA-I. The flexibility of this helical protein provides multiple opportunities for ligand recognition loops for various receptors, including scavenger receptors and adenosine triphosphate-binding cassette (ABC) transporters. (Reprinted with permission by Keith Kasnot.)

particles also accept more FC from ABCA1 and develop further into spherical HDL_2 and HDL_3 particles, following the action of LCAT. Spherical HDL_2 and HDL_3 particles and mature HDL do not interact well with ABCA1, but can access additional FC through apoA-I-mediated interaction with related specific macrophage ABCG1 transporters or the SR-B1 scavenger receptors found in peripheral cells. Intact spherical HDL_2 and HDL_3 particles and mature HDL can deliver FC and cholesteryl ester to the liver by a similar apoA-I-mediated interaction with the hepatic SR-B1 receptor. While ABCA1 and ABCG1 promote cholesterol efflux from cells to the various forms of HDL, the SR-B1 receptors mediate bidirectional transport of cholesterol in both the liver and the periphery.

While the heterogeneity of the HDL particle population precludes detailed structural studies by protein crystallography, a spherical model for mature HDL has been developed.[15] As illustrated in **Figure 3**, the smaller helical apoA-I surrounds the sphere in a loosely held braid. The flexibility of the apoA-I helix affords multiple opportunities for the formation of exposed loops that can act as ligand sites for ABC transporters or SR-BI receptors.

As shown in **Table 3**, much lower plasma levels of apoA-I are found in healthy humans than apoB-100. Similarly, HDL cholesterol (HDLc) represents a smaller overall percentage of total cholesterol. Comparing HDLc levels of 45–50 mg dL^{-1} with a total cholesterol concentration of 170–180 mg dL^{-1} indicates that HDLc represents ~30% of total cholesterol in normal subjects.[15]

6.20.2.3 The Role of Modified Low-Density Lipoprotein, Inflammatory Mediators, and Reactive Oxygen Species (ROS) in Atherosclerosis

While the accumulation and retention of cholesterol in macrophages and other leukocytes contribute to atherosclerosis, recent studies have uncovered a number of key inflammatory mediators that participate in the activation of the endothelium. This endothelial activation leads to the recruitment and activation of both monocytes and T cells at the site of the developing lesion.[6–8] The resulting activated cells release ROS (e.g., hydrogen peroxide, hydroxyl radical,

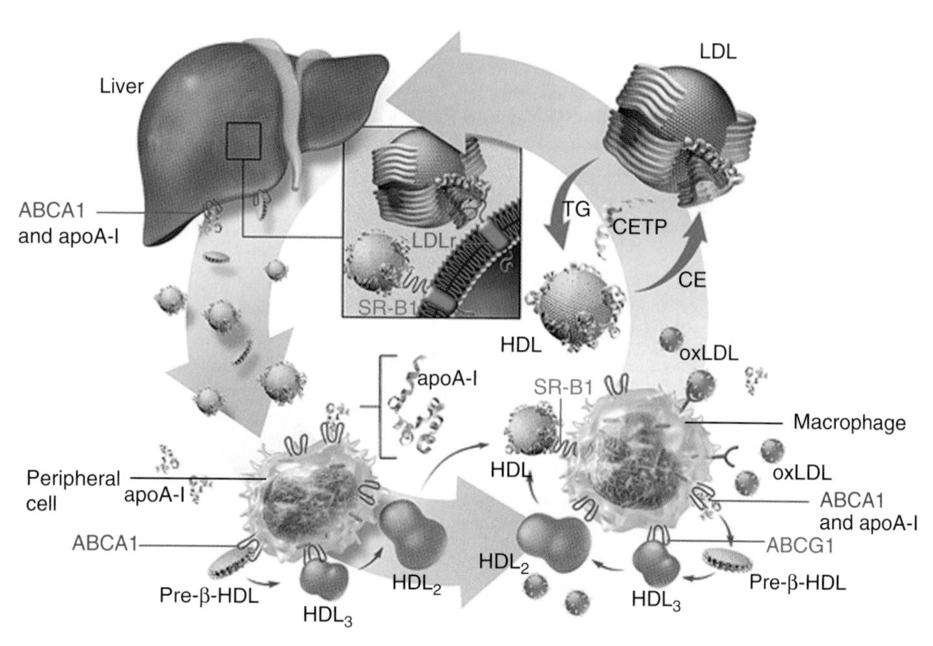

Figure 4 An illustrated model of high-density lipoprotein (HDL) particle assembly and its role in reverse cholesterol transport (RCT). The process of collecting excess peripheral cholesterol by HDL for disposal via the liver is called the RCT pathway. HDL plays a key role in removing cholesterol from peripheral tissue. Lipid-free apolipoprotein AI (apoA-I) is secreted into the general circulation from the liver. ApoA-I accepts free cholesterol (FC) from cellular sites in the periphery including endothelial cells and macrophages by receptor-mediated processes including adenosine triphosphate-binding cassette (ABC) transporters, primarily ABCA1. Intermediate discoidal pre-β-HDL forms as the small, helical apoA-I accumulates phospholipid (PL) and free cholesterol (FC) which is further esterified to cholesteryl ester (CE) following the action of lecithyl cholesterol acyl transferase (LCAT). The pre-β-HDL particles also accept more FC from ABCA1 and develop further into spherical HDL_2 and HDL_3 particles, following the action of LCAT. Spherical HDL_2 and HDL_3 particles and mature HDL do not interact well with ABCA1, but can access additional FC through apo-AI-mediated interaction with related specific macrophage ABCG1 transporters or the SR-B1 scavenger receptors found in peripheral cells. Intact spherical HDL_2 and HDL_3 particles and mature HDL can deliver FC and CE directly to the liver by a similar apo-AI-mediated interaction with the hepatic SR-B1 receptor. Alternatively, CE is transferred from HDL to LDL or VLDL following the action of cholesteryl ester transfer protein (CETP) and hepatic uptake of these apo-B particles removes excess cholesteryl esters via the LDL receptor. (Reprinted with permission by Keith Kasnot.)

superoxide anion) and reactive nitrogen species (RNS, e.g., nitric oxide, peroxynitrite) that alter the surrounding redox environment and stimulate the expression and secretion of several proinflammatory proteins that produce an elevated inflammatory response.

6.20.2.3.1 Low-density lipoprotein oxidation, modification, and retention

As illustrated in **Figure 5**, remodeled and modified LDL particles accumulate in the intima below the vascular endothelium. Oxidized LDL (oxLDL) forms as minimally modified LDL (mmLDL) is subject to oxidation by a variety of chemical oxidants and enzymes. Both oxLDL and mmLDL are cytotoxic to cells and inhibit the migration of the macrophage out of the extravascular space. Several proinflammatory oxidative enzymes have been implicated in this process, including: 5-lipoxygenase (5-LO), 15-lipoxygenase (15-LO), and myeloperoxidase (MPO). Hypochlorous acid (HOCl) and RNS generated from MPO cause chemical modifications to the apoB protein by introducing specific changes in tyrosine and cysteine residues.[18] MPO-mediated HOCl also oxidizes the polyunsaturated fatty acids contained in the LDL lipid core. Elevated plasma levels of MPO have been positively correlated with an increased cardiovascular risk, while individuals with MPO deficiency display a significantly reduced risk.[18]

Similarly, 5-LO and 15-LO also oxidize unsaturated fatty acids bound to surface PL, generating oxidized PL (oxPL) or oxidize polyunsaturated arachidonic acids found within or released from the mmLDL particle by the action of PLA_2, creating a number of reactive fatty acid hydroperoxide species. These hydroperoxides act as a source of damaging free radicals by propagating radical chain reactions within the tissue. OxPL species are found in both early- and late-stage atheroma tissue, suggesting that they contribute to all phases of disease progression.[19] The transformation of arachidonic acid by 5-LO initiates the leukotriene biochemical cascade (**Figure 6**) that produces the powerful

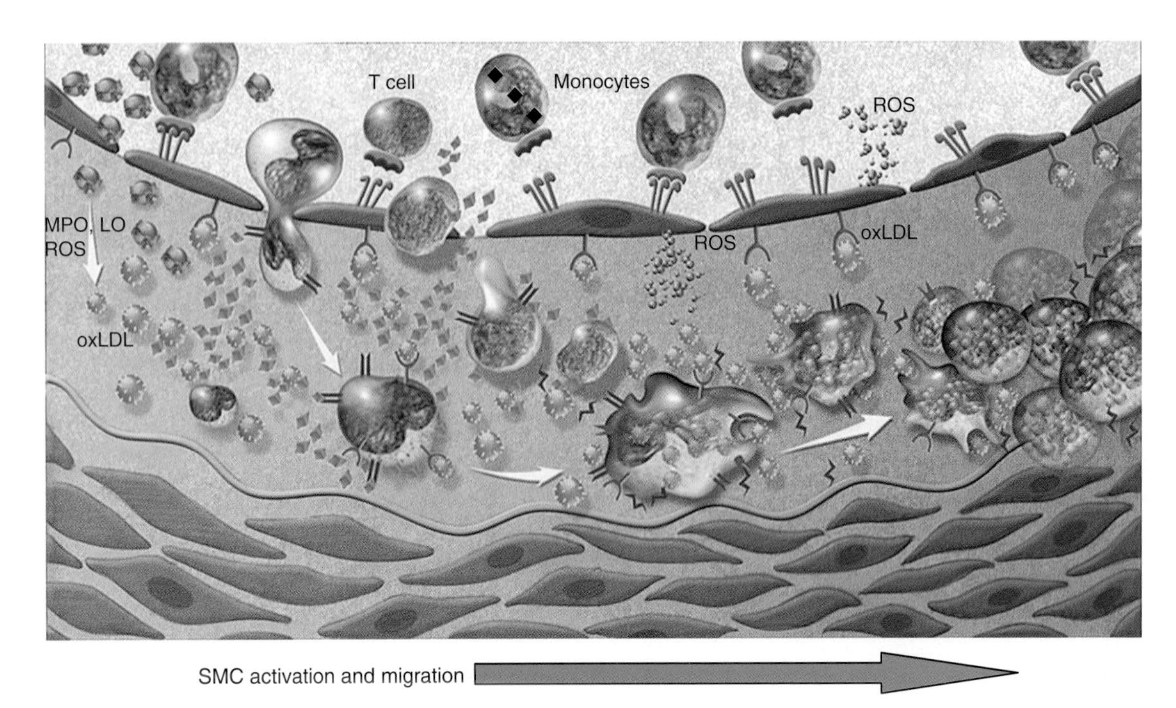

Figure 5 An illustrated model of atherosclerosis as an inflammatory disease. Modified LDL is oxidized by MPO and ROS (red circles), activating endothelial cells to express cell adhesion molecules and recruit monocytes and T cells. Oxidized LDL induces cytokine and chemokine (blue diamonds) expression to magnify the inflammatory response, leading to macrophage activation and more ROS production. The activated macrophages secrete other inflammatory mediators, accumulate additional lipid, and develop into foam cells. ROS and inflammatory mediators induce SMC activation and migration. (Reprinted with permission by Keith Kasnot.)

chemoattractant, LTB_4, which plays a key role in SMC migration. Both oxPL and oxLDL are injurious to the endothelial lining, increase the secretion of superoxide anion, and activate the endothelial cells to express surface cell adhesion molecules, notably vascular cell adhesion molecule-1 (VCAM-1). Cell adhesion molecules act as coreceptors for VLA-4 integrins found on the surfaces of monocytes and T cells. The resulting specific interaction between cell adhesion molecules and integrins captures the rolling leukocytes and facilitates their localization and recruitment through the endothelial lining. Similarly, activated monocytes, macrophages, and SMC secrete ROS and RNS in response to oxLDL and oxPL, thus perturbing the delicate redox balance of the surrounding tissue.

Activated endothelial cells and monocytes secrete both chemokines[20] (e.g., monocyte chemoattractant protein-1 (MCP-1)) and several common proinflammatory cytokines (e.g., tumor necrosis factor alpha, (TNF-α)). Whereas MCP-1 is secreted into the bloodstream, fractalkine (CX3CL1) is a membrane-bound chemokine and adhesion molecule, expressed on the activated endothelium. Both MCP-1 and fractalkine act as powerful chemotactic agents that recruit more monocytes/macrophages and T cells to the atherosclerotic site. TNF-α activates VCAM-1 expression and contributes to the maintenance of the inflammatory cascade.

As more cholesterol-rich mmLDL particles are taken up by phagocytosis or endocytosis within the monocytes, the intracellular cholesterol concentration increases as FC is converted to cholesteryl ester through the action of ACAT-2 within the macrophage. The resulting activated macrophages, which are transformed into developing foam cells, secrete proteases, inflammatory cytokines, and ROS to promote the inflammatory response. Many of these same inflammatory mediators also induce profound changes to the fibrous cap covering the mature lesion. These mediators induce the expression of matrix-degrading metalloproteinases that weaken the overlying extracellular matrix and create a lesion more susceptible to rupture. Both plasma MPO[18] levels and the expression of 5-LO[21] within atheroma have been positively correlated with increased vulnerability to plaque instability and rupture.

The altered redox state due to inflammation at the atherosclerotic site also changes HDL functionality. For example, apoA-I is a specific target for MPO-catalyzed oxidation of HDL, producing chemical modifications of key tyrosine residues. These changes significantly reduce the capability of the oxidized HDL particle to participate in ABCA1-mediated cholesterol efflux from macrophages.[16,18] MPO dramatically alters the delicate balance between

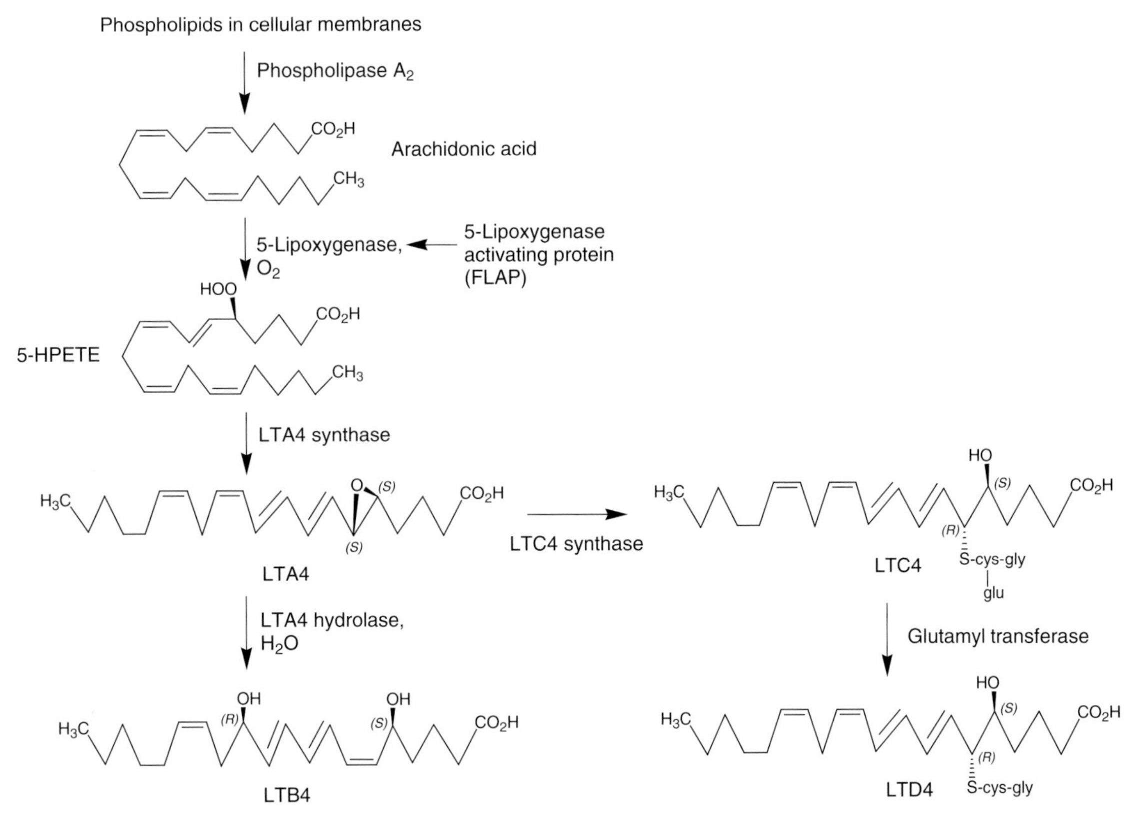

Figure 6 Selected steps in leukotriene biosynthesis. The action of phospholipase A_2 (PLA_2) on phospholipids in cellular membranes releases free arachidonic acid which can be oxidized by the enzyme, 5-lipoxygenase (5-LO), to form 5-hydroperoxy-eicosa-6,8,11,14-tetraenoic acid (5-HPETE). Subsequent conversion of 5-HPETE into leukotriene A4 (LTA4, 4-(3-tetradeca-1(*E*),3(*E*),5(*Z*),8(*Z*)-tetraenyl-2(*S*),3(*S*)-oxiranyl)-butyric acid) is catalyzed by LTA4 synthase. LTA4 can be hydrolyzed by LTA4 hydrolase to form leukotriene B4 (LTB4, 5(*S*),12(*R*)-dihydroxy-6(*Z*),8(*E*),10(*E*),14(*Z*)-eicosatetraenoic acid). Alternatively, LTA4 undergoes enzyme-catalyzed ring opening of the epoxide with the tripeptide, gly-cys-gly, and LTC4 synthase to form the tripeptide adduct, leukotriene C4 (LTC4). LTC4 is converted to the dipeptide adduct, leukotriene D4 (LTD4), by the action of glutamyl transferase.

LDL-mediated cholesterol uptake and HDL-mediated cholesterol efflux in the activated macrophage, and thus plays a key role in the underlying lipid accumulation and activation process.

Activated immune cells such as T cells and mast cells also participate in the inflammatory response within the lesion. The proinflammatory cytokines secreted within the atheroma induce activated T cells to differentiate into Th1 effector cells, producing the cytokine interferon-γ (IFN-γ). IFN-γ is a potent activator of macrophages and increases synthesis and secretion of the proinflammatory cytokines TNF-α, interleukin-1 (IL1), and IL6.[6] As a result of these combined responses, the underlying pathology and biochemical changes accompanying plaque initiation and progression closely resemble the inflammatory cascade more commonly associated with joint or peripheral tissue injury. However, the accumulation and retention of cholesterol and modified LDL in atherosclerosis are unique characteristics of lesion formation that differentiates atherosclerosis from other inflammatory diseases.

6.20.2.3.2 The antioxidant paradox

While the oxidation of LDL by several oxidative enzymes is believed to be a key component driving the retention of mmLDL, foam cell development, and the progression of plaque, the levels of extracellular antioxidants have been found to be quite high within atheroma and approach those of normal human plasma. LDL contains significant quantities of α-tocopherol (vitamin E), a phenolic antioxidant that helps protect LDL from one-electron radical oxidation processes, and does not appear to be depleted in isolated oxLDL. Thus, LDL lipid peroxides and apoB protein oxidation products form even in the presence of α-tocopherol. These and other data challenge the conventional thinking about the importance of modified and oxidized LDL particles in the lipid retention process.[22] Alternatively, many of the key oxidation processes that contribute to LDL oxidation may be derived from two-electron oxidations, for which radical scavengers such as α-tocopherol offer no protection.

6.20.2.3.3 The atheroprotective effects of high-density lipoprotein cholesterol

While excessive mmLDL plays an important proinflammatory role in atherosclerosis, HDL and its associated apoA-I exert a corresponding cardiovascular benefit through the RCT pathway and several other important protective mechanisms.[16] HDL also has beneficial antioxidant and anti-inflammatory properties and improves endothelial dysfunction. HDL contains two antioxidant enzymes, paraoxonase and platelet-activating factor (PAF) acetylhydrolase (PAFAH), that not only inhibit the oxidation of LDL but also help degrade oxidized phospholipids within oxLDL. HDL and its associated apoA-I protect endothelial cells from damage induced by oxLDL. HDL also displays anti-inflammatory properties, since pretreatment of human endothelial cells with HDL inhibits the cytokine-induced expression of cell adhesion molecules.

Alternatively, overall HDL functionality may be more important than absolute HDLc levels, since recent data indicate that HDL has both proinflammatory and anti-inflammatory properties. HDL isolated and characterized from the plasma of a small group of CHD patients exhibited a more proinflammatory profile in vitro based on its lipid peroxide content and ability to alter LDL-induced monocyte chemotaxis than HDL obtained from age- and sex-matched controls.[23] These preliminary results suggest that the proinflammatory characteristics of HDL from CHD patients may be a better predictor of overall risk than absolute HDLc levels. However, no large trials have yet been conducted to confirm this hypothesis.

6.20.2.4 Low-Density Lipoprotein Cholesterol and Coronary Heart Disease Risk

The link between elevated plasma total cholesterol levels, high LDLc, and increased CHD risk has been known from epidemiological studies since the early 1980s. The Framingham Study followed over 5000 males and females, monitoring plasma lipid levels and the incidence of MI. In this cohort, subjects with elevated serum cholesterol ($>275\,\mathrm{mg\,dL^{-1}}$) had more adverse events whether they were healthy or already had CHD. The prevalence of plasma cholesterol levels above $240\,\mathrm{mg\,dL^{-1}}$ in subjects who experienced an MI was 35–52% in males and 66% in females.[24] LDLc levels in this CHD subpopulation were well above $100\,\mathrm{mg\,dL^{-1}}$ and were most prevalent at $160\,\mathrm{mg\,dL^{-1}}$ Subjects who experienced an MI and had high plasma cholesterol levels were at increased risk for another MI or death from either CHD or other causes.

Similarly, the Multiple Risk Factor Intervention Trial (MRFIT) monitored 300 000 middle-aged males. Using a comparative baseline plasma cholesterol of $200\,\mathrm{mg\,dL^{-1}}$, subjects in this study having total cholesterol of $250\,\mathrm{mg\,dL^{-1}}$ had a twofold increased risk of death from CHD, while those with values of $300\,\mathrm{mg\,dL^{-1}}$ had a threefold higher risk. The Atherosclerosis Risk in Communities (ARIC) Study followed over 12 000 males and females who were free of CHD and monitored their incidence of coronary events over a 10-year period. The lowest incidence of events occurred in those subjects with the lowest LDLc levels ($<100\,\mathrm{mg\,dL^{-1}}$) in both males and females. An associated 40% increased CHD risk occurred in both males and females for every $\sim 40\,\mathrm{mg\,dL^{-1}}$ increase in LDLc. Based upon these and other studies, it appeared that cardiovascular risk increased significantly above a certain threshold value of total plasma cholesterol and LDLc.[4]

Total cholesterol levels in FH patients exceed twice the normal range, and can reach higher than $500\,\mathrm{mg\,dL^{-1}}$ in homozygous populations. LDL catabolism is dramatically lowered in these FH subjects due to functional impairment in their LDLr.[14] Mutations in the LDLr of FH patients impedes LDL uptake by the liver, and excess LDL accumulates. While only one in a 1 million homozygous FH subjects occur in the general population, they develop severe atherosclerosis in early childhood and have one of the highest incidences of childhood CHD mortality.[4]

Based upon these and other studies, cholesterol reduction treatment strategies were initiated that sought to lower total plasma cholesterol below the threshold values of $200\,\mathrm{mg\,dL^{-1}}$ and LDLc below $130\,\mathrm{mg\,dL^{-1}}$, particularly in patients having other risk factors. Recently, a more aggressive approach has been recommended by the National Cholesterol Education Program Adult Treatment Panel following multiple successful outcome trials with lipid-lowering therapies.[25,26]

6.20.2.5 Lipoprotein(a) and Coronary Heart Disease Risk

While Lp(a) particles have been identified in atheroma tissue, their natural biological function is unknown.[4] Lp(a) has proinflammatory and chemoattractant properties since it stimulates the expression of cell adhesion molecules. While

Lp(a) and apo(a) plasma levels vary widely throughout the general population, high Lp(a) plasma levels have been associated with an increased risk of CHD, stroke, and early MI in both middle-aged and elderly males. In these subjects, there was also a close correlation between increased plasma oxPL levels and higher Lp(a) concentrations. Although elderly females had overall higher average Lp(a) plasma concentrations than males, these females did not display an associated higher CHD risk.[27,28]

6.20.2.6 Markers of Inflammation and Oxidant Stress in Coronary Heart Disease

Currently, there are no validated biomarkers of either inflammation or oxidant stress that can be used predictably for drug intervention in CHD patients. C-reactive protein (CRP), whose biological function is undetermined, has been proposed as a potential marker of inflammation, particularly in patients with acute coronary syndromes (ACS).[6] Whether CRP is produced in response to inflammation or contributes directly to an inflammatory response is still unknown. However, elevated plasma CRP levels may represent an independent risk factor for CHD in the general population, even in subjects with near-normal cholesterol levels.[1,6] Subjects with the lowest quintile of plasma cholesterol had a twofold higher risk of CHD when their plasma CRP levels fell in the highest CRP quintile. Similarly, subjects having plasma cholesterol levels in the highest quintile doubled their CHD risk as their CRP levels increased from the lowest to the highest quintile. Subjects with near-normal cholesterol levels and the highest CRP quintile were as much at risk as subjects with the highest quintile of plasma cholesterol and the lowest quintile of CRP. Plasma CRP levels may account in part for the majority of the CHD population who are at risk even though they have near-normal levels of plasma cholesterol. In elderly males and females, subjects with plasma CRP levels above $3\,\mathrm{mg\,L^{-1}}$ had an increased cumulative 10-year risk of developing CHD.

6.20.2.7 High-Density Lipoprotein Cholesterol and Coronary Heart Disease Risk

In contrast to the elevated CHD risk associated with higher LDLc levels, several epidemiological studies have demonstrated an inverse relationship between serum HDLc levels and the incidence of ischemic heart disease in both middle-aged males and the elderly.[16] In the Framingham cohort,[24] a higher prevalence (19%) of heart disease occurred in subjects due solely to low ($<35\,\mathrm{mg\,dL^{-1}}$) HDLc compared to the prevalence of disease (12%) in those who only had elevated ($>130\,\mathrm{mg\,dL^{-1}}$) LDLc. Thus, low levels of HDLc represent a significant independent risk factor in CHD irrespective of whether patients have elevated LDLc.[26] A risk prediction model developed from the Framingham cohort showed that a nearly 50% higher cardiovascular risk was associated with male patients having HDLc levels below $35\,\mathrm{mg\,dL^{-1}}$ compared with those in the normal range (35–59 $\mathrm{mg\,dL^{-1}}$). An even higher (twofold) associated risk was observed in females with HDLc below $35\,\mathrm{mg\,dL^{-1}}$. In contrast, a significantly reduced risk (0.61 versus 1.00) was present in both males and females having HDLc levels above $60\,\mathrm{mg\,dL^{-1}}$.[29] Elevated HDLc levels ($>60\,\mathrm{mg\,dL^{-1}}$) in both males and females reduced plaque growth in patients with preexisting lesions, and were shown to be protective in atherosclerotic plaque progression.

These atheroprotective effects of HDL have been primarily attributed to apoA-I. Humans with genetic deficiencies in apoA-I have very low levels of plasma HDL and exhibit premature CHD.[16] Infusion or overexpression of apoA-I in animals produced profound reductions in atherosclerotic lesions.[16] Human subjects have been identified with hyperalphalipoproteinemia, i.e., markedly elevated HDLc, as much as three- to sixfold higher than normal. The elevated HDLc in these subjects has been attributed to CETP deficiency or genetic CETP defects that produce lower CETP plasma levels and/or impaired CETP activity.[16] The Honolulu Heart Study followed more than 3000 elderly Japanese-American males with CETP deficiency arising from two different CETP gene mutations and found that the relationship between CETP deficiency and CHD risk was complex and possibly dependent upon the accompanying HDLc and TG levels. The prevalence of disease was higher in males with CETP gene mutations (21%) than in subjects without mutations (16%), even though subjects with mutations had reduced CETP levels and an overall higher average of HDLc. However, those subjects with HDLc levels greater than $60\,\mathrm{mg\,dL^{-1}}$ had a significantly reduced and near-normal cardiovascular risk.[30]

HDLc is now recognized as one of the best predictors of cardiovascular risk in females of all ages and in males after middle age.[31] The National Cholesterol Education Panel (NCEP) has recently raised the threshold for treatment intervention in those coronary artery disease patients at risk because of low HDLc levels to include subjects having HDLc below $40\,\mathrm{mg\,dL^{-1}}$.[26] In these patients, it is still uncertain to what extent HDLc must be raised to produce a cardiovascular benefit, but elevations of 50% or more may be required.

6.20.3 Experimental Disease Models

6.20.3.1 Murine Models of Atherosclerosis

The induction of diet-induced atherosclerosis is generally difficult in rodent species that have low LDL and VLDL plasma concentrations. Mice and rats carry most of their cholesterol in atheroprotective HDL particles and are naturally deficient in CETP.[16] Low VLDL and LDL plasma concentrations in mice are maintained by an avid uptake and catabolism via the LDLr. Two strains of genetic knockout mice can overcome this limitation. Specific murine knockout strains have been developed for either the LDLr (LDLr$^{-/-}$) or apoE (apoE$^{-/-}$) that impede uptake by the LDLr and induce higher LDLc and lower HDLc plasma concentrations in mice closer to human levels. However, higher plasma concentrations of VLDLc and LDLc are observed in apoE$^{-/-}$ mice than in apoE-deficient humans. As a result, these knockouts are susceptible to atherosclerosis induced by high-fat, cholesterol-enriched diets and provide a reliable short-term animal model for this disease. While drug-induced changes in LDLc plasma concentrations can be detected almost immediately in these models, alterations in foam cell and plaque progression generally require multiple (4–12) weeks of dosing.[32] Similar effects on HDLc lowering, LDLc increases, and cholesterol-induced atherosclerosis progression have been observed in transgenic mice overexpressing human CETP either alone in wild-type mice or on a LDLr$^{-/-}$ or apoE$^{-/-}$ genetic background.[16]

Both LDLr$^{-/-}$ and apoE$^{-/-}$ models have been further exploited by backcrossing with other knockouts or transgenic mice overexpressing specific proteins to create new transgenic models that have helped define many of the mechanistic components underlying this disease. As a result, the specific effects of both human proinflammatory, anti-inflammatory, oxidative, antioxidant, and redox-regulated proteins on the atherosclerotic process have been confirmed in animal studies.

6.20.3.2 Chronic Models in Higher Species

In contrast to rodents, animals such as rabbits, who naturally express high CETP activity, are spontaneously susceptible to diet-induced atherosclerosis. Rabbit models of atherosclerosis have been used to define the long-term quantitative efficacy of experimental drug products on lesion progression under more chronic conditions. These models typically require dosing up to 6 months or longer. Transgenic rabbit models have also confirmed the contributions that various gene products make to disease progression in rabbits, including apolipoproteins (apo(a), apoA-I, apoB, apoE), hepatic lipase (HL), LCAT, LpL, and 15-LO.[33] Primate models of atherosclerosis have also been employed to define the long-term efficacy of experimental drugs for up to 1 year of dosing.

6.20.4 Clinical Trial Issues

6.20.4.1 Low-Density Lipoprotein Cholesterol-Lowering Agents

Plasma cholesterol levels are determined by the balance between dietary intake, de novo biosynthesis, and reabsorption processes in the gut as well as by biliary clearance and excretion. While the majority of plasma cholesterol concentrations are derived from cellular biosynthesis in the body, primarily in the liver, about one-third is accounted for from absorption after dietary intake. Prior to the late 1980s, few alternatives existed to lower plasma cholesterol levels. While changes in lifestyle and diet were recommended, polymeric resins that sequestered bile acids in the gut, such as cholesteryamine and colestipol, were used to lower LDLc. While these agents were extremely safe due to their lack of systemic absorption, the high doses required for clinical efficacy limited patient compliance. Even at maximal doses of grams per day, these resins only achieved LDLc reductions of no more than 25%. Over the last 25 years, tremendous progress has been made in providing alternatives for lipid-lowering therapies. Two complementary, pharmacological intervention approaches have been successfully utilized in clinical settings to reduce coronary events by lowering total plasma cholesterol and LDLc levels: cholesterol biosynthesis inhibitors (statins) and cholesterol absorption inhibitors.

As discussed below, the results generated from controlled trials with these new agents have been an important contributor to the changes in drug treatment guidelines proposed by the NCEP Adult Treatment Panel based on patient lipid levels and relative CHD risk (*see* 6.19 Diabetes/Syndrome X). For moderately high-risk patients (two or more risk factors), reducing LDLc level below 130 mg dL^{-1} was recommended, with more aggressive treatment to lower plasma LDLc below 100 mg dL^{-1} being considered. For very-high-risk patients, particularly those diagnosed with CHD, reducing LDLc levels below 100 mg dL^{-1} was recommended, and more aggressive treatment to lower plasma LDLc below 70 mg dL^{-1} was strongly advised.[26]

6.20.4.1.1 Clinical efficacy of hydroxymethylglutaryl-coenzyme A reductase inhibitors (statins) as low-density lipoprotein cholesterol-lowering agents in patients with coronary heart disease

Statins represent a class of drugs that specifically inhibit hydroxymethylglutaryl-coenzyme A (HMG-CoA) reductase, an enzyme that catalyzes the rate-controlling step in de novo cholesterol biosynthesis.[34] Two general classes of statin have been identified: the earlier natural product-based statins (mevastatin **1**, lovastatin **2**, simvastatin **3**, pravastatin **4**; **Figure 7**) and the more recent new generation of synthetic statins (atorvastatin **6**, fluvastatin **5**, cerivastatin **7**, rosuvastatin **8**, pitavastatin **9**; **Figure 7**). Since their initial discovery in the mid-1970s, multiple statins have been

The reaction catalyzed by HMG-CoA reductase:

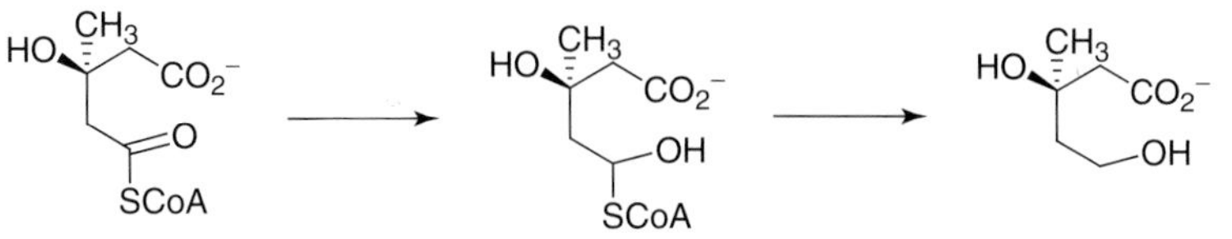

HMG-CoA → Intermediate → Mevalonate

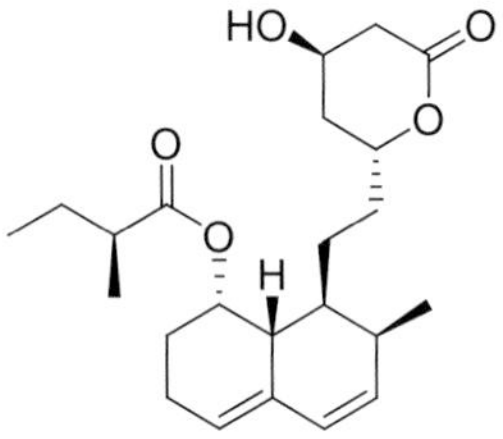

1 Mevastatin

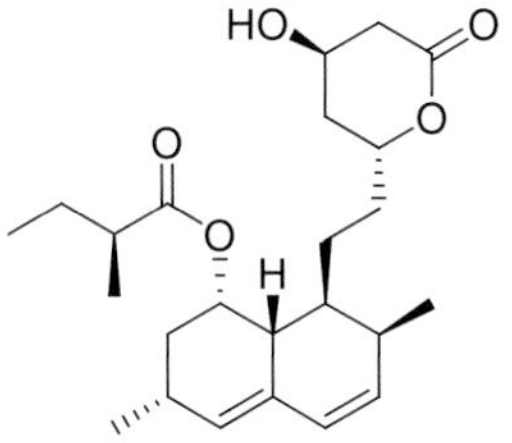

2 Lovastatin

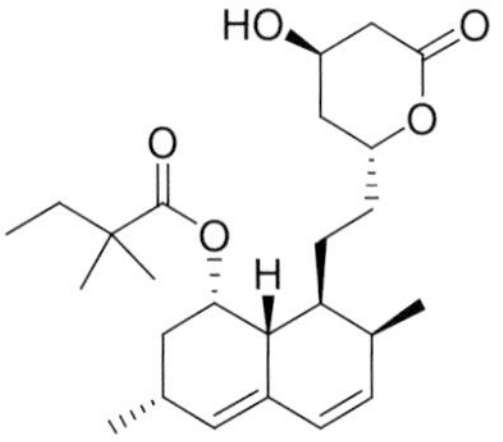

3 Simvastatin

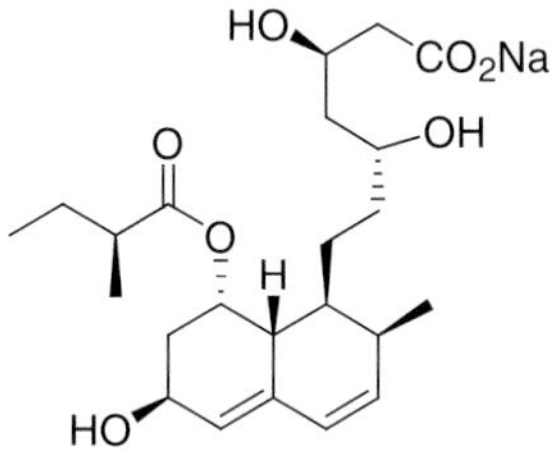

4 Pravastatin sodium

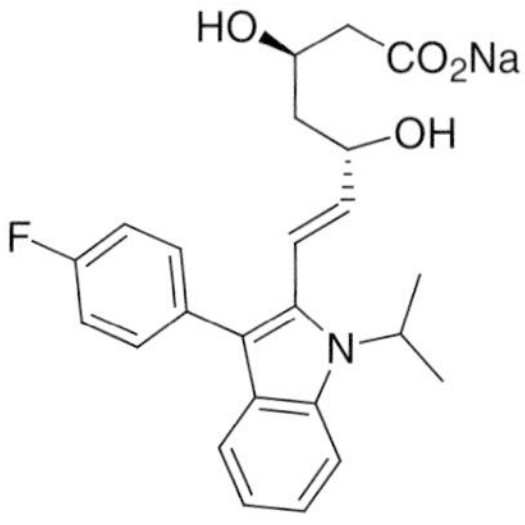

5 Fluvastatin sodium

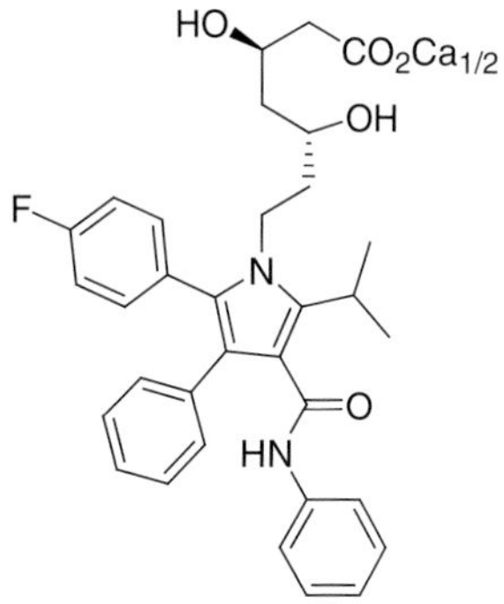

6 Atorvastatin calcium

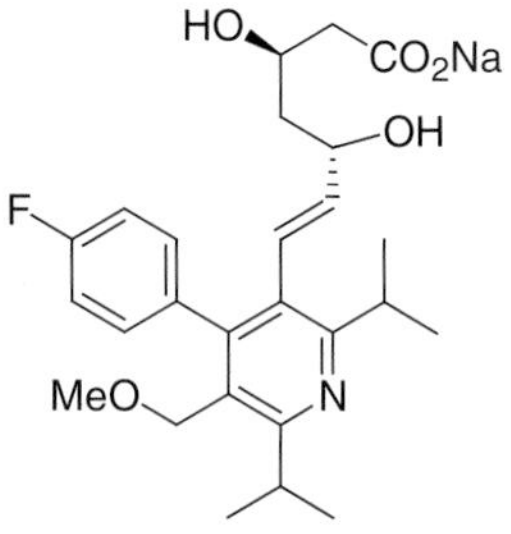

7 Cerivastatin sodium

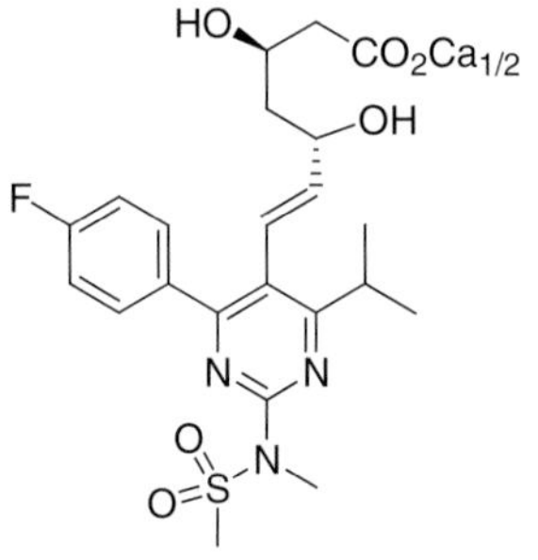

8 Rosuvastatin calcium

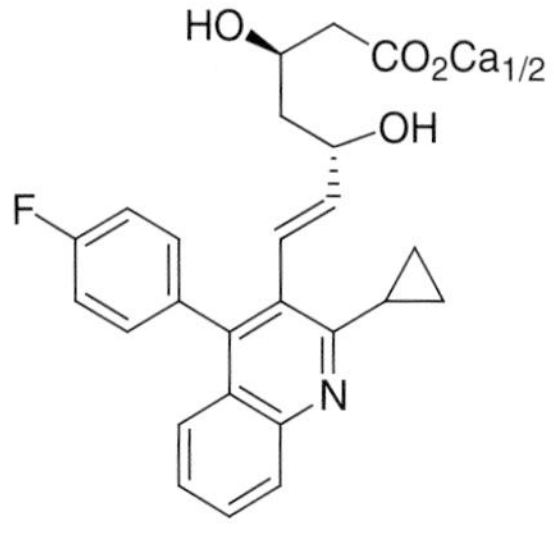

9 Pitavastatin calcium

Figure 7 Structures of clinically tested HMG-CoA reductase inhibitors (statins).

Table 4 Select clinical data obtained with HMG-CoA reductase inhibitors (statins)

Trial	*Statin treatment*	*Daily dose (mg)*	*Patient size*	*Baseline mean LDLc ($mg\,dL^{-1}$)*	*Final mean LDLc ($mg\,dL^{-1}$)*	*% Mean change in LDLc*	*% Change in major coronary events*
4S[a]: sp	Simvistatin	20–40	>4000 men and women	188	122	−35	−34
CARE[b]: sp	Pravastatin	40	>4000 men and women	139	98	−29	−24
WOSCOPS[c]: pp	Pravastatin	40	>6500 men	197	142	−28	−31
ASCOT[d]	Atorvastatin		>10 000 men and women	131	85	−35	−29
REVERSAL[e]	Atorvastatin	80	>500 men and women	150	79	−46	ND
REVERSAL[e]	Pravastatin	40	–	150	110	−25	ND
TNT[f]	Atorvastatin	80	>10 000 men and women	97	77	−21	−22% versus 10 mg event rate
TNT[f]	Atorvastatin	10	–	98	101	NS	
LIPS[g]	Fluvastatin	80	>1600 men and women	–	–	−27	−22
STELLAR[h]	Rosuvastatin	10	>2400 men and women	–	–	−46	ND
STELLAR[h]	Rosuvastatin	20	–	–	–	−52	ND
STELLAR[h]	Rosuvastatin	40	–	–	–	−55	ND

NS, no significant change from baseline; ND, not determined.
[a] 4S, Scandinavian Simvistatin Survival Study,[36] sp, secondary prevention trial.
[b] CARE, Cholesterol and Recurrent Events.[37]
[c] WOSCOPS, West of Scotland Coronary Prevention Study,[38] pp, primary prevention trial.
[d] ASCOT, AngloScandinavian Coronary Outcomes Study.[39]
[e] REVERSAL, Reversal of Atherosclerosis with Aggressive Lipid Lowering.[40,41]
[f] TNT, Treating to New Targets.[42]
[g] LIPS, Lescol Intervention Prevention Study.[43]
[h] STELLAR, Statin Therapies for Elevated Lipid Levels Compared across Doses.[49]

successfully tested in clinical trials and introduced as breakthrough therapies that effectively reduced both mortality and morbidity in CHD patients.

Several benchmark clinical trials successfully demonstrated with representatives from each of the statin subclasses that pharmacological intervention safely and effectively lowered total and LDL cholesterol and produced a concomitant, significant reduction in the onset of both primary and secondary events in patients with CHD (**Table 4**). Recent subgroup analyses from two secondary prevention trials using two different statins indicated a common therapeutic response and the benefits of lowering LDLc below 130 mg dL^{-1}.[35–37] Notably, the approximate 29–35% reduction in plasma LDLc levels observed in patients with established CHD treated with simvastatin in the Scandinavian Simvastatin Survival Study (4S) trial or with pravastatin in the Cholesterol and Recurrent Events (CARE) trial produced a commensurate 24–34% reduction in major coronary events. Thus, a nearly one-for-one benefit in reduced risk was observed in patients with high cholesterol and established CHD: for every 1% lowering in LDLc attained, a commensurate 1% reduction in coronary events was observed.[35] Similar results have also been reported from multiple clinical studies with atorvastatin, including the Angloscandinavian Coronary Outcomes Study (ASCOT) trial, where the benefit of lowering LDLc below 100 mg dL^{-1} was also shown to be beneficial.[39] Based on these and other clinical data, statins effectively reduce coronary risk in patients with established CHD and LDLc above 130 mg dL^{-1}. As a result, statin therapy has become the standard of care in CHD patients. The therapeutic benefit achieved by the statin class in CHD patients has been well recognized by regulatory authorities, and LDLc-lowering in hyperlipidemic patients has been used as a surrogate marker to progress newer analogs through the initial approval process.

Similar beneficial effects were also observed in the West of Scotland Coronary Prevention Study (WOSCOPS), the first reported statin coronary primary prevention study.[38] Hypercholesterolemic subjects with no prior evidence of CHD were enrolled in this study and, following daily treatment with pravastatin, achieved an average 28% reduction in plasma LDLc that was accompanied by a 31% reduction in major coronary events (**Table 4**). As a result, statins are recommended as a first-line therapy for patients with diagnosed CHD and as primary prevention therapy in those at high risk of developing CHD. Interestingly, in this study, all patient subgroups attained a similar reduction in coronary events, regardless of their initial baseline LDLc levels. Thus, the WOSCOPS trial was unable to define clearly the optimal LDLc target to reach for high-risk patients. Since then, multiple studies have been undertaken to determine if more aggressive LDLc lowering would achieve improved therapeutic benefits.

Three landmark studies have recently demonstrated the overall effectiveness of more aggressive lipid lowering in CHD patients. The Reversal of Atherosclerosis with Aggressive Lipid Lowering (REVERSAL) trial compared the overall benefits of moderate lipid lowering, administering 40 mg pravastatin with more aggressive LDLc reductions using 80 mg atorvastatin, on the progression of atherosclerotic lesions in CHD patients as measured by IVUS.[40,41] While a 25% reduction in LDLc to a mean value of 110 mg dL^{-1} was achieved over baseline levels in the pravastatin group, the atorvastatin-treated arm attained an average 46% reduction over baseline LDLc, reaching a mean value of 79 mg dL^{-1}. IVUS measurements indicated that the primary endpoint, progression of atherosclerotic lesion development, continued unchecked in the pravastatin group, but was slightly reduced in the atorvastatin group. Thus, the overall change in atheroma volume as measured by IVUS increased by 2.7% in the pravastatin group, but decreased (-0.4%) in the atorvastatin arm. Those patients in the atorvastatin-treated arm exhibited no change in atheroma burden. This was the first report where aggressive lipid lowering was shown to have such a significant direct effect on disease progression.

CRP plasma levels were also reduced by 36% in the atorvastatin-treated group, whereas those treated with pravastatin achieved only a 5% reduction in plasma CRP. These results suggest that higher doses of atorvastatin may have beneficial anti-inflammatory properties beyond those of low-dose pravastatin. The overall reduction in plasma CRP levels observed with atorvastatin was independently correlated with atheroma progression. Those patients who achieved reductions in both LDLc and plasma CRP below the median from baseline had significantly less disease progression than patients who achieved reductions in either marker alone.[40,41]

In the Treating to New Targets (TNT) study,[42] the beneficial effects of 80 mg versus 10 mg doses of atorvastatin were compared in more than 10 000 patients with stable CHD. Patients were monitored for nearly 5 years, and the primary endpoint for this study was the occurrence of a major coronary event, including death, MI, and fatal or nonfatal stroke. Patients entered the study with mean baseline LDLc levels of 97–98 mg dL^{-1}. The 10 mg atorvastatin group showed virtually no change from baseline in LDLc lowering, whereas the 80 mg group attained nearly a 21% reduction in plasma LDLc accompanied by a nearly 22% improvement in the overall risk of a coronary event compared with those in the 10 mg group. Both doses were well tolerated, but there was a higher incidence of liver enzyme elevations (1.2%) in the 80 mg arm than in the 10 mg group (0.2%).[42] The tradeoffs between safety and tolerability versus clinical efficacy are likely to become more pronounced and a higher incidence of significant side effects may be observed as more studies are conducted that push the statin dose to achieve maximal reductions in LDLc.

The combined benefits of aggressive lipid lowering with reduction in CRP levels on coronary clinical outcomes were also reported from the Pravastatin or Atorvastatin Evaluation and Infection Therapy Thrombolysis in Myocardial Infarction (PROVE-IT TIMI-22) trial that compared a 40 mg dose of pravastatin with 80 mg atorvastatin and followed nearly 4000 patients with ACS for 2 years.[46,47] Each statin was effective in lowering plasma LDLc or CRP levels. However, atorvastatin was approximately four times more likely to reduce both LDLc and CRP levels than pravastatin. In this study, there was no clear correlation between LDLc lowering and reduction in plasma CRP levels. Nevertheless, a linear correlation was observed between reductions in plasma LDLc and the concomitant reduced risk of MI or death from coronary events. A more pronounced ~32% reduction in coronary events was observed in those patients who achieved LDLc levels below 70 mg dL^{-1} compared with those whose plasma LDLc levels stayed above 70 mg dL^{-1}. A similar ~28% reduction in the relative risk of event recurrence was observed when statin therapy reduced CRP levels below 2 mg L^{-1}, compared with patients whose CRP plasma levels stayed above 2 mg L^{-1}, independent of the change in LDLc achieved. The ~27% of the study population who achieved the dual targets of LDLc of <70 mg dL^{-1} and plasma CRP levels below 2 mg L^{-1} had the lowest rate of MI recurrence and achieved a 28% reduction in coronary events. Approximately 80% of the patients in this subgroup were treated with atorvastatin. Like the results observed in the REVERSAL trial, atorvastatin again demonstrated an apparent superiority over pravastatin and is now considered first among the statin class.

Thus, the combined reductions in LDLc and plasma CRP levels observed in PROVE-IT led to a significant reduction in coronary events in ACS patients. These results strongly suggest that the therapeutic benefits of at least some statins extend beyond simple LDLc lowering and include an anti-inflammatory component. However, in PROVE-IT, neither drug in this study was successful in bringing either LDLc or CRP levels into the desired range required for optimal therapeutic benefit in the majority of treated patients. The overall impact of CRP reductions on coronary risk therefore appears to be beneficial, but requires additional study, particularly to define its benefit in other target patient populations or by administering alternative therapies.

Fluvastatin **5** followed atorvastatin **6** as another member of the synthetic statins with improved efficacy over the natural product analogs. The Lescol Intervention Prevention Study (LIPS) trial monitored lipid lowering in over 800 patients for nearly 4 years after having undergone angioplasty by percutaneous transluminal coronary intervention (PTCI).[43] In this trial, an 80 mg (2×40 mg b.i.d.) daily dose of fluvastatin effectively lowered LDLc by 27%, and the corresponding reduction in the risk of major coronary events was 22%. This benefit was observed even in those patients with normal cholesterol levels. Fluvastatin was approved by the US Food and Drug Administration (FDA) in 2003 to reduce the risk of coronary revascularization procedures in CHD patients.

Cerivastatin **7** (**Figure 7**) is one of the most potent HMG-CoA reductase inhibitors identified ($K_i = 10$ pM), being approximately 100 times more potent in vitro than other members of the class. As a result, lower daily doses of cerivastain have been used in clinical settings. In phase IIb/III studies, low daily doses of 0.025–0.4 mg cerivastatin reduced plasma LDLc levels dose-dependently. The 0.4 mg daily dose reduced LDLc by $>40\%$.[48] Unfortunately, these low doses of cerivastatin were linked to severe side effects, and it was withdrawn from the market in 2001 after more than 50 fatalities from renal failure due to severe rhabdomyolysis.

Rosuvastatin **8** was approved by the FDA in 2003 for lipid lowering in hypercholesterolemic patients as the seventh member of the statin class. The Statin Therapies for Elevated Lipid Levels Compared Across Doses (STELLAR) trial demonstrated that, after 6 weeks, rosuvastatin reduced LDLc and plasma apoB levels in hypercholesterolemic subjects across the daily 5–40 mg dosing range. Daily doses of 10 mg rosuvastatin reduced LDLc by 46% and were superior to the LDLc-lowering effects achieved with comparable daily doses of atorvastatin (-37%), pravastatin (-20%), and simvastatin (-28%). At the highest 40 mg dose, rosuvastatin reduced LDLc by 55%, achieved comparable efficacy to 80 mg doses of atorvastatin (-51%), and was superior to 80 mg of simvastatin (-46%).[49] Rosuvastatin (80 mg day^{-1}) also lowered plasma LDLc, but with a much higher incidence of severe myopathy and consequently this dose was not approved.

Pitavastatin **9** is another extremely potent HMG-CoA reductase inhibitor that can be administered in very low doses. Pitavastatin (2 mg day^{-1}) in hypercholesterolemic subjects for 12 weeks reduced LDLc by 37% and was more effective than the comparable 10 mg dose of pravastatin that lowered LDLc by 18%.[50] Pitavastatin was approved in Japan in 2003 for treating hyperlipidemia, but has not yet received approval in the US or Europe.

6.20.4.1.2 Safety concerns with statins

Data from animal studies have indicated that pronounced inhibition of cholesterol biosynthesis with high doses of statins induced multiple toxic side effects, including liver enzyme elevations, cataracts, skeletal muscle changes, central nervous system lesions, and certain tumors. However, overall, good safety and tolerability have been observed for the first- and second-generation statins in multiyear clinical trials, and millions of patients have been safely treated with approved doses of these agents. Withdrawals due to adverse events from statins in several multiyear trials were similar to placebo, and the overall incidence of clinically significant adverse events has been quite low. Initial concerns about cataract formation are no longer considered a significant safety issue since several clinical studies monitoring optical lens opacity with the early statins showed that these occurred at the same rate in both statin-treated and placebo groups. Since statins target cholesterol biosynthesis in the liver, recommendations have been made to monitor liver enzyme functionality for several months as patients are newly treated with statins, increase their dose, or change statin therapy. As higher numbers of patients have been treated with statins, the incidence of hepatic transaminase elevations more than three-times above the upper limit of normal has been quite low, generally 1% or less, depending on the dose. Most of the observed liver enzyme elevations were usually asymptomatic, and no significant long-term hepatic toxicity has been observed.[51]

The most important adverse effect has been the incidence of muscle pain or weakness (myopathy) and, in its severest manifestation, rhabdomyolysis, observed with all statins. Rhabdomyolysis can lead to hospitalizations and, in rare cases, renal failure. While these skeletal muscle effects occur fairly rarely ($<0.1\%$), sizable numbers of patients have experienced these effects because of the large treated patient population. The incidence of skeletal muscle side effects is more pronounced at higher doses, particularly with higher potency inhibitors, which raises additional concerns as more aggressive lipid lowering is utilized to improve clinical outcomes. Patients who complain of muscle soreness are monitored for excessive plasma levels of creatine phosphokinase (CPK), and elevations in CPK 10 times above the upper limit of normal are a potential indication of myopathy. The mechanistic basis for this statin side effect is not well-understood, despite being observed for over a decade. The most effective treatment involves statin withdrawal followed by the addition of mevalonate to supplement the biochemical intermediate suppressed by inhibition of HMG-CoA reductase.[51]

Cerivastatin **7**, an extremely potent third-generation statin, was withdrawn from patients postapproval in 2001 after a higher incidence of severe myopathy and renal failure was observed in cerivastatin-treated patients even though they were taking doses up to 100 times lower than those treated with other statins to achieve their LDLc-lowering target. The mechanistic basis for this increased risk with cerivastatin alone has not been determined. However, more cases of severe myopathy and rhabdomyolysis were observed in clinical trials using cerivastatin in combination with gemfibrozil **13** than with cerivastatin alone. These results have been attributed to drug–drug interactions that occurred as a result of potent inhibition of cytochrome P450-2C8 (CYP2C8) by gemfibrozil **13**, one of the key metabolizing enzymes of cerivastatin, resulting in elevated plasma concentrations of cerivastatin.

Similar concerns have recently been raised with rosuvastatin postapproval, where more than a twofold greater incidence of adverse events was observed in patients from rosuvastatin phase IV trials than in those treated with atorvastatin, lovastatin, or pravasatin.[44,45] In addition to higher rates of muscle damage from myopathy and rhabdomyolysis, rosuvastatin treatment was associated with kidney complications, including proteinuria, and, in rare cases, kidney failure. The incidence of these adverse events was low, as only about 145 cases of muscle or kidney damage were observed in over 5 million prescribed patients. Regulatory authorities are in the process of reviewing these results.[44,45]

6.20.4.1.3 Clinical efficacy of cholesterol absorption inhibitors as low-density lipoprotein cholesterol-lowering agents

Since dietary intake accounts for up to a third of plasma cholesterol levels and a redundant system in the enterohepatic recirculation is used to reabsorb and conserve cholesterol-derived bile acids and steroidal intermediates, various inhibitors of cholesterol absorption have been explored to identify alternative lipid-lowering agents. Such agents should provide an advantage over a simple low-fat diet and have particular benefit for patients who either do not respond or are unable to tolerate statin therapy. Ezetimibe **10** represents the first cholesterol absorption inhibitor approved for LDLc lowering. In clinical studies, daily doses of 0.25–10 mg of ezetimibe as monotherapy were safe and well-tolerated, and did not alter the plasma concentrations of lipid-soluble vitamins. In this study, ezetimibe dose-dependently lowered plasma LDLc, and the 10 mg dose lowered LDLc by 17–18% in hypercholesterolemic patients after 12 weeks of dosing.[52] There was no statistically significant change in HDLc after 12 weeks. No long-term safety studies have yet been reported. Ezetimibe represents the first drug approved for lipid lowering that acts by an alternate mechanism since the statins were discovered.

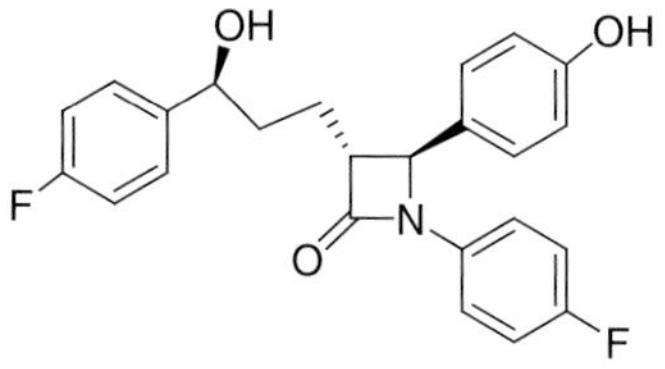

10

6.20.4.1.4 Combinations of statins and cholesterol absorption inhibitors

The complementary nature of the two inhibitory mechanisms targeting HMG-CoA reductase and cholesterol absorption suggests that combinations of these agents may have additive or perhaps even synergistic clinical benefits. The first clinical evidence supporting this hypothesis was observed from the additional lipid lowering that could be achieved by using ezetimibe as an add-on therapy to patients who are already taking statins. In this study, 10 mg ezetimibe was given to patients on stable simvastatin therapy. The addition of 10 mg ezetimibe produced an additional 25% reduction in LDLc, indicating a significant synergistic effect.[53] In another trial, the addition of 10 mg of ezetimibe to 10 mg atorvastatin achieved $>50\%$ reduction in LDLc, comparable to that seen with an 80 mg dose of atorvastatin alone. Thus, combinations of statins with ezetimibe offer the opportunity to lower significantly the statin dose needed to achieve lipid targets, perhaps with potentially increased safety.

Based on these and other studies, a fixed combination of ezetimibe (10 mg) and various simvastatin doses (20–80 mg) has been developed, marketed as Vytorin. In clinical trials, the 10/20 fixed combination of ezetimibe and simvastatin lowered plasma LDLc levels by 52% in hypercholesterolemic patients after 12 weeks of dosing.[54] Vytorin was approved by the FDA for LDLc lowering in 2004.

6.20.4.2 Antioxidants as Monotherapy in Coronary Heart Disease

The clinical success of antioxidant-based therapies has been extremely limited.[22] Several studies have explored the potential contribution that supplementation with antioxidant-based natural vitamins, such as vitamin C or vitamin E, might play in reducing cardiovascular risk, with inconsistent results. In primary prevention studies, the Antioxidant Supplementation in Atherosclerosis Prevention (ASAP) trial followed subjects on supplemental vitamin E or vitamin C for 3 years. Supplementation with vitamin C or vitamin E did not reduce the progression of atherosclerosis compared to placebo. Instead, vitamin E appeared to produce a small disease-promoting effect. A recent examination of the combined data from secondary prevention trials treating over 81 000 patients with vitamin E concluded that vitamin E therapy produced no significant reduction in coronary events. Similar results were observed using vitamin E in combination with other antioxidants where the combination of antioxidant therapies had no significant effect in reducing coronary events or in some cases actually worsened outcomes.[22]

Probucol **11** is a synthetic, highly lipophilic, phenolic antioxidant capable of interfering with both one- and two-electron oxidations.[22] Probucol demonstrated potent antiatherogenic properties in animal studies and was effective in lowering plasma LDLc levels in animals and in hypercholesterolemic patients. Probucol reduces atherogenesis at least in part by reducing LDL oxidation. In clinical trials, patients with mean baseline LDLc levels of 166 $mg\,dL^{-1}$ treated with probucol at 500 mg b.i.d. for up to 24 months, attained a modest 24–26% reduction in LDLc but their HDLc was also lowered by 21%. This response was accompanied by an approximate 14% decrease in intima media thickness (IMT) and a corresponding 14% reduction in coronary events.[55] These results confirm the potential of novel antioxidant-based therapies as CHD treatments. Unfortunately, serious concerns remain about the safety of chronic dosing with high doses of probucol since both the parent drug and its metabolites have been associated with QTc prolongation and fatal arrhythmia.

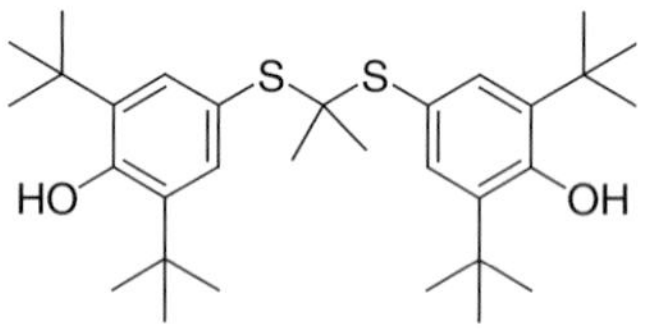

S S HO OH

11

6.20.4.2.1 Antioxidants in combination therapy with statins

Several clinical studies have examined antioxidants in combination with various statins. However, no significant beneficial effects in coronary event reduction were achieved by adding vitamin E to simvastatin, pravastatin, or atorvastatin. Notably, in the secondary prevention HDL Atherosclerosis Treatment (HATS) trial, treatments with antioxidants such as vitamin E, vitamin C, and beta-carotene were compared alone or in joint therapy with a combination of simvastatin **3** and niacin **17**. In contrast to the lipid-lowering effects and reduction in coronary events observed with the simvastatin/niacin combination alone, antioxidants by themselves demonstrated no significant benefit in disease progression and coronary outcomes. However, the beneficial effects of simvastatin/niacin on lipid lowering and disease progression were essentially negated when this combination was administered together with antioxidants. Thus, supplementation with antioxidants produced a significant negative outcome for patients, and consequently their use in combination with statins has been actively discouraged.[22]

6.20.4.3 High-Density Lipoprotein Cholesterol-Elevating Agents

Epidemiological studies clearly indicate that low HDLc ($<40\,mg\,dL^{-1}$) represents a significant independent risk factor for CHD. While statin therapy effectively lowered LDLc by up to 45%, statins generally produce only very modest increases in HDLc levels, usually less than 10%.[34] Currently, only two alternatives are available for treating low HDLc: fibrates and niacin-based (**Figure 8**) therapies. Fibrates increase HDLc by ~10–15%, but have a more pronounced effect on TG lowering. Fibrates have been available since the early 1970s. The fibrate mechanism of action

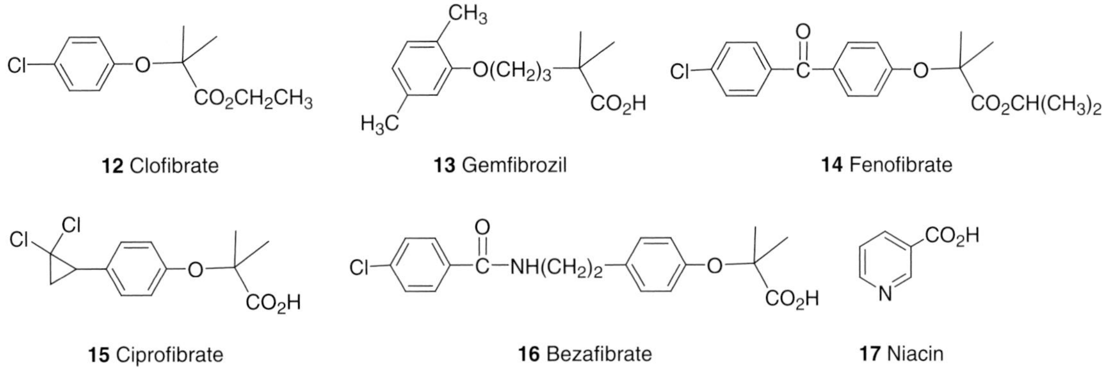

12 Clofibrate **13** Gemfibrozil **14** Fenofibrate

15 Ciprofibrate **16** Bezafibrate **17** Niacin

Figure 8 Structures of approved fibrates and niacin as HDLc-elevating agents.

is complex and includes activation of peroxisome proliferator-activated receptor (PPAR), a family of nuclear receptors (PPAR-α, PPAR-β/δ, PPAR-γ) that act as transcription factors and control genes involved in both lipid and lipoprotein metabolism as well as glucose homeostasis. Fibrates are selective PPAR-α agonists.[56] Niacin **17** is currently the best available therapy for raising HDLc, attaining increases of 30% or more.

6.20.4.3.1 Fibrates as high-density lipoprotein cholesterol-elevating agents

Several clinical studies have demonstrated that a reduction in CHD events is positively correlated with treatments that raised HDLc. The controlled Helsinki Heart Trial monitored coronary event rates in over 4000 asymptomatic middle-aged males with dyslipidemia for 5 years and showed that correcting low levels of HDLc with gemfibrozil led to a 34% decrease in coronary events.[4,57] For each $1\,\mathrm{mg\,dL^{-1}}$ rise in HDLc among the study subjects, the average response was a 3% decrease in CHD risk.

Similarly, in the Veterans Affairs HDL Intervention Trial (VA-HIT), 2531 males with CHD having low HDLc ($\leqslant 40\,\mathrm{mg\,dL^{-1}}$), near-normal LDLc ($\leqslant 140\,\mathrm{mg\,dL^{-1}}$), and elevated TG ($\leqslant 300\,\mathrm{mg\,dL^{-1}}$) were treated for an average of 5 years with gemfibrozil.[58] In this study, LDLc was not significantly altered, TG was lowered 31%, HDLc was raised 6%, and the combined number of CHD events was lowered 24% (a 2–3% decrease in CHD for every 1% rise in HDLc). In comparison, prospective statin trials showed only about a 1% reduction in CHD risk for every 1% decrease in LDLc (**Table 4**).[34] However, attributing the event reductions observed in the VA-HIT trial exclusively to HDL elevations is confounded by the concomitant reductions in triglycerides. In contrast, no significant reductions in coronary events were observed in patients treated with bezafibrate **16** in the Bezafibrate Infarction Program (BIP) trial.

6.20.4.3.2 Niacin as a high-density lipoprotein cholesterol-elevating agent

Niacin **17** (nicotinic acid, pyridine-3-carboxylic acid, **Figure 8**) in doses up to $1000\,\mathrm{mg\,day^{-1}}$ raises HDLc and lowers LDLc and TG. Niacin elevated HDLc by up to 30%, but side effects such as flushing were common and limited patient compliance. Similar increases in total HDLc of 26–28% were achieved with fewer side effects following long-term treatment with Niaspan, an extended-release form of niacin, either alone or in combination with simvastatin. In other studies, prolonged treatment of patients with low HDLc with immediate-release niacin alone had some clinical benefit in reducing total mortality, but side effects were common and limited patient compliance. Niacin has multiple and complex mechanisms of action in dyslipidemia, but its ability to raise HDLc has been attributed in part to the inhibition of apoA-I-mediated hepatic uptake of HDL.

6.20.4.3.3 Combinations of statins with high-density lipoprotein cholesterol-elevating agents

While statin therapy offers a significant therapeutic benefit to the subset of patients that respond to these agents, typically, more than 60% of the statin-treated patients in controlled trials continued to develop cardiovascular disease and failed to experience a therapeutic benefit.[34] Most of these nonresponders also had low HDLc levels. Since statins produce only modest increase of HDLc ($<10\%$),[34] several studies have been conducted to define the potential benefit using statins in combinations with either fibrates or niacin. For example, in the HATS secondary prevention trial, CHD patients with low HDLc ($<31\,\mathrm{mg\,dL^{-1}}$) and normal LDLc levels were treated with a combination of simvastatin and niacin. This combination produced an elevation of 26% in HDLc and a surprising 42% reduction in LDLc. These combined effects on lipids were accompanied by a remarkable 60–90% reduction in CHD events. Since this was a small trial, it would be useful to have these results confirmed in a larger patient population.[34] In the Arterial Biology for the Investigation of the Treatment Effects of Reducing Cholesterol (ARBITER-2) trial, daily doses of 1000 mg Niaspan were added to CHD patients with low HDLc on background statin therapy, and changes in IMT were monitored after 12 months of therapy.[59] In this study, HDLc increased by 21% after 12 months of niacin treatment. IMT increased significantly in the statin plus placebo group, but was unchanged in the statin plus niacin-treated group. Thus, extended-release niacin plus statin therapy slowed atherosclerosis progression in diagnosed CHD patients with low HDLc. A fixed combination of lovastatin with extended-release niacin is available as Advicor.

Similar beneficial effects were expected with statins in combination with fibrates. However, a heightened awareness of the potential for severe side effects with this combination therapy has occurred as more cases of severe myopathy and rhabdomyolysis were observed in clinical trials administering cerivastatin in combination with gemfibrozil than with cerivastatin alone. These results have been attributed to drug–drug interactions occurring as a result of the potent inhibition of CYP2C8, a key metabolizing enzyme of cerivastatin, by gemfibrozil, resulting in elevated plasma concentrations of cerivastatin. These adverse effects contributed to the withdrawal of cerivastatin from the marketplace.

6.20.5 Current Treatments

6.20.5.1 Low-Density Lipoprotein Cholesterol-Lowering Agents: Hydroxymethylglutaryl-Coenzyme A Reductase Inhibitors (Statins)

Statins specifically target and potently inhibit HMG-CoA reductase, the enzyme that catalyzes mevalonate formation and represents the rate-controlling step in de novo cholesterol biosynthesis (**Figure 7**).[34] No toxic metabolites accumulate as a result of the inhibition of HMG-CoA reductase. HMG is water-soluble and has a number of metabolic pathways available to preclude buildup of this intermediate. However, in vivo, the resulting decrease in mevalonate that occurs from inhibition of HMG-CoA reductase reduces the overall concentration of the steroid pool. In response to this depletion, HMG-CoA reductase expression increases and the hepatic LDLr is upregulated.

Approved statins (**Figure 7**) consist of two subclasses: the early first-generation natural product-derived or their related semisynthetic analogs (mevastatin **1**, lovastatin **2**, simvastatin **3**, pravastatin **4**), and the newer second- and third-generation synthetic analogs containing a central core aromatic heterocycle (atorvastatin **6**, cerivastatin **7**, fluvastatin **5**, rosuvastatin **8**, pitavastatin **9**). All of the statin structures share a common substituted 3,5-dihydroxypentanoic acid moiety similar to that found in the reduced enzymatic product, mevalonate (**Figure 7**). Six of these are currently approved for patient treatment in the US (lovastatin **2**, simvastatin **3**, pravastatin **4**, atorvastatin **6**, fluvastatin **5**, and rosuvastatin **8**). Cerivastatin was withdrawn from the market postapproval due to safety issues, and pitavastatin has not been approved in the US.

All statins are potent, competitive inhibitors of HMG-CoA reductase with picomolar to low nanomolar activity. The three-dimensional x-ray structures of six different statins bound to HMG-CoA reductase have been reported. The statin binding site overlaps with the HMG-CoA substrate site and thus confirms their observed competitive inhibition properties. However, some rearrangement of the substrate site is required to accommodate statin binding, and subtle differences were noted between the natural product-derived decalin analogs and the planar synthetic statins in their interactions with the protein. Several common binding features were also noted. For example, the common 4-F-phenyl moieties in the synthetic statins share the same binding space as the butyrate ester groups in the fermentation products, while the *iso*-propyl groups of the synthetic statins occupy part of the space covered by the decalin ring.[60]

6.20.5.1.1 Mevastatin and lovastatin

The statin class was first identified in the early 1970s by investigators at Sankyo who discovered that the natural product, compactin (mevastatin **1**), isolated from microbial fermentations of *Penicillium citrinum* in a search for new antimicrobial agents, potently inhibited HMG-CoA reductase and lowered serum cholesterol levels in animals.[61] Compactin (**Figure 7**) also effectively lowered serum total cholesterol and LDLc in heterozygous FH patients. However, development was stopped in 1980 for unknown reasons that may have been related to toxicity issues that were uncovered in longer term animal studies.[51]

Following the compactin lead, lovastatin **2** (**Figure 7**) was identified from fermentation broths from *Aspergillus terreus*. Lovastatin and compactin are very close structural homologs, with lovastatin containing an additional methyl group in the bicyclic decalin ring. Both compounds are administered as inactive lactone prodrugs that must be metabolized in vivo to the corresponding open hydroxy-acid forms before HMG-CoA reductase inhibition can occur.[54,62] Like compactin, lovastatin effectively lowered plasma LDLc in animals and produced profound reductions in serum LDLc in heterozygous FH patients. The association between lipid lowering and HMG-CoA reductase inhibition was later shown with lovastatin to be due to an upregulation of the LDLr and the resulting enhanced clearance from plasma of excess LDLc.[34] The development of lovastatin was slowed by safety considerations surrounding compactin and the perceived underlying uncertainties associated with the potential consequences of long-term cholesterol biosynthesis inhibition. Lovastatin **2** was approved by the FDA in 1987.[51]

6.20.5.1.2 Simvastatin

The protracted development associated with lovastatin prompted the search for another statin-derived fermentation product. The intact lovastatin lactone was limited by its relatively poor oral bioavailability in patients (12–20%), so alternatives were needed. Simvastatin **3** (**Figure 7**) was the second entrant to the statin market and contains two additional methyl groups retaining the methyl group found in the decalin ring of lovastatin while having an extra methyl moiety in the butyrate ester group. The intact simvastatin lactone has a much higher (>80%) overall oral bioavailability than lovastatin. Both lovastatin and simvastatin were developed as hydrophobic lactones as they offered the initial advantages of passive transport across the hepatocyte membranes, facile metabolic conversion to the biochemically active hydroxy-acids, and subsequent selective accumulation in the liver, while the corresponding hydroxy-acids provided a significantly reduced oral pharmacokinetic profile.[62] Simvastatin **3** was approved in 1988 and was the first generic statin to be approved as an over-the-counter medication in the UK. Over-the-counter use of statins in the US has not been approved.

6.20.5.1.3 Pravastatin

The successful introduction of HMG-CoA reductase inhibitors for lipid lowering sparked intense competition in the field. pravastatin **4** (**Figure 7**) is made in a two-step fermentation process that first generates compactin with *Penicillium citrinum*, and after hydrolysis of the lactone, employs a biological hydroxylation with *Streptomyces carbophilus* to introduce the allylic 6-alcohol group regioselectively. Pravastatin has higher water solubility and provided several potential advantages over the more hydrophobic analogs, lovastatin and simvastatin. Unlike lovastatin and simvastatin, pravastatin sodium is administered as the sodium salt of the corresponding open-chain hydroxy-acid. The polar nature of pravastatin, however, limited its overall human absolute bioavailability to ~17%, presumably due to incomplete absorption and first-pass metabolism. However, pravastatin has been used to demonstrate the facilitated cellular uptake of compounds through organic anion transporter proteins. Compared to lovastatin and simvastatin, pravastatin has the added benefit of not being metabolized by CYP3A4 enzymes, thus reducing its potential for drug–drug interactions.[62] Because of its polarity, much less pravastatin is protein-bound (~45%) compared to the high protein binding (>95%) observed for lovastatin and simvastatin. Pravastatin **4** was approved for human use in 1991.

6.20.5.1.4 Atorvastatin

Atorvastatin **6** (**Figure 7**) is a synthetic statin that contains an unusual penta-substituted pyrrole ring in place of the hexahydronaphthalene found in the early statins. The rationale for optimization employed a combination of insights from molecular modeling and hypothesis-based chemistry to define key structure–activity relationships.[63] The genesis of the atorvastatin pyrrole core developed (**Figure 9**) from early observations that the complex hexahydronaphthalene system could be replaced with a simpler, achiral *ortho*-biphenyl moiety as in **18** and still retain HMG-CoA reductase inhibitory activity similar to that of the fungal metabolites. From this observation, it was proposed that suitably substituted pyrrole analogs, e.g., **19**, might exhibit similar activities. Initial work focused on incorporating small aliphatic substituents at R_5 and led to the first early leads **20a** and **20b** (**Figure 9**) with submicromolar potency. The addition of more bulky groups such as cyclic aliphatic rings or aromatic groups at R_5 reduced activity. However, both **20a** and **20b** were about 20-fold less potent than mevastatin ($IC_{50} = 0.03\,\mu M$) after hydrolysis of the lactone to the corresponding hydroxy-acids.

Modeling comparisons of **20b** versus **18** suggested that additional hydrophobic groups at R_3 and R_4 may be required for increased potency. The symmetrical dichloro (**21a**) and dibromo (**21b**) derivatives achieved the desired potency, and lowered lipid levels in animals comparable to mevastatin. However, these compounds were toxic in preclinical testing. In contrast, the unsymmetrical 3-phenyl-4-carboethoxy derivative **21c** had greater potency than **20a**. Further optimization of **21c** with a short, focused series identified the anilide **21d** with low nanomolar potency, about fivefold better than mevastatin, after hydrolysis to the corresponding hydroxy-acids. Chiral syntheses of both enantiomers demonstrated that all of the activity resided in the one (*R*,*R*,+)-stereoisomer shown. In several animal species, **21d** was found to be more effective at LDLc lowering than lovastatin. Atorvastatin **6** (**Figure 7**) was eventually selected for development as the calcium salt of the open hydroxy-acid analog of **21d**, and atorvastatin became the first totally synthetic statin to enter clinical development.[63]

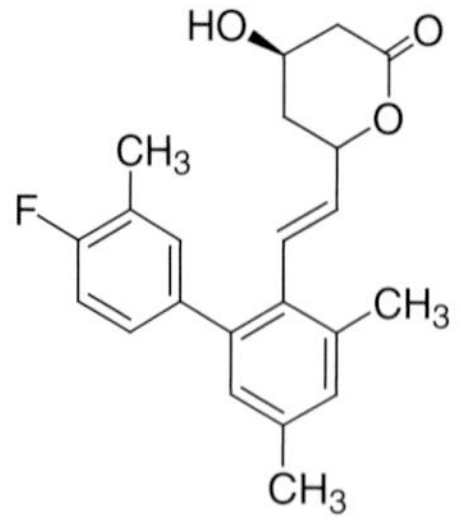

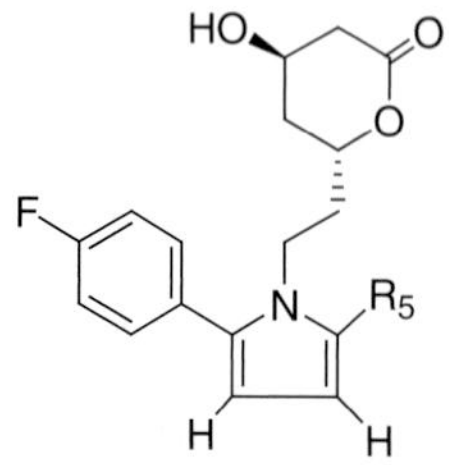

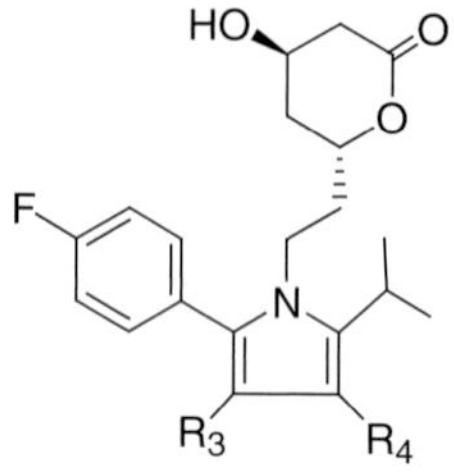

20a R_5=CH_3: IC_{50}=0.6 μM
20b R_5=$(CH_3)_2CH$: IC_{50}=0.4 μM
20c R_5=$(CH_3)_3C$: IC_{50}=1.6 μM
20d R_5=cyclo-C_3H_5: IC_{50}=2.2 μM

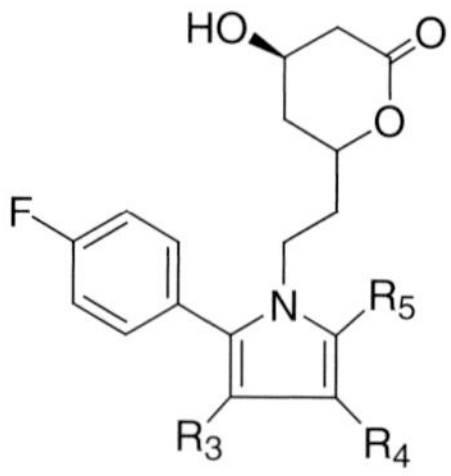

21a R_3,R_4=Cl: IC_{50}=0.03 μM
21b R_3,R_4=Br: IC_{50}=0.03 μM
21c R_3=Ph, R_4=CO_2Et: IC_{50}=0.17 μM
21d R_3=Ph, R_4=CONHPh: IC_{50}=0.007 μM

Figure 9 Structures of key compounds and structure–activity relationships leading to the discovery of atorvastatin.

The greater potency of oral atorvastatin in animals was confirmed in clinical trials where nearly 60% reductions of LDLc were achieved using an 80 mg dose of atorvastatin, substantially greater reductions than could be attained with the highest dose of other statins. Like the natural product-derived statins, orally administered atorvastatin has a relatively low overall bioavailability (~12%) in humans. Atorvastatin resembles lovastatin and simvastatin in being highly protein-bound (>95%), and is likewise metabolized by CYP3A4. Atorvastatin was approved for human use in 1997.

6.20.5.1.5 Fluvastatin

Fluvastatin **5** (**Figure 7**) was discovered at about the same time as atorvastatin **6**[64,65] and contains a disubstituted indole core in place of the hexahydronaphthalene found in the fungal fermentation products. In contrast to atorvastatin, fluvastatin has an unsaturated *E*-heptenoic acid side chain. Some detailed structure–activity data for the optimization process used to identify fluvastatin have been reported.[64]

Also, attracted by the reported efficacy of **18** (**Figure 9**), several potential replacements for the decalin ring of compactin, including poly-substituted imidazoles, naphthalenes, indenes, and indoles, were explored, leading to the identification of the *N*-methyl indole **22a** with activity comparable to compactin (**Figure 10**). The introduction of small bulky indole *N*-substituents identified the *N-iso*-propyl group of fluvastatin **22c** with optimal potency (IC_{50} = 0.007 μM), whereas the incorporation of larger groups in **22d–22f** gave significantly less activity.

Structural variations of the 4-fluorophenyl group in **22c**, as shown in **23a–23i**, demonstrated that one or more *ortho* or *meta* methyl groups in addition to or in place of the 4-fluoro substituent gave analogs **23a–23c**, with slightly better or nearly comparable activity as **22c**. For example, the combination of 2-methyl and 4-fluoro substituents in **23a** resulted in twice the potency of the singular 4-fluoro substituent in **22c** and fourfold more potency than the 2-methyl group alone in **23g**. Incorporating bulkier electron-withdrawing groups (**23f**) or more electron-rich or polar *para* substituents (**23h** and **23i**) generally produced weaker activity. Similarly, a few aliphatic and alkoxy groups were explored in the benzo-fused ring of the indole, and none was more active than the unsubstituted indole in **22c**.

Two structure–activity relationship results were generated from compounds **24** and **25** (**Figure 10**). The reduced dihydro derivative **24** (IC_{50} = 0.114 μM) was nearly 20-fold less active than **22c**, demonstrating the importance that the unsaturated *E*-heptenoic acid contributed to potency. Conversely, compound **25** (IC_{50} = 0.002 μM), that switched the relative orientations of the *iso*-propyl and 4-fluorophenyl groups, was three times more potent than **22c**, suggesting an unexpected symmetry to the enzymatic binding sites of these indole derivatives.

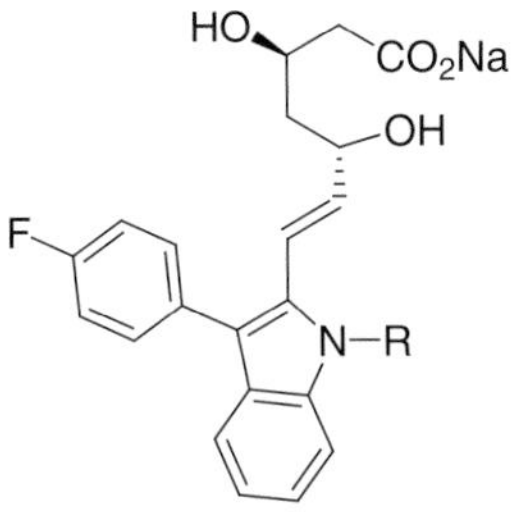

22a R = CH_3: IC_{50} = 0.6 μM
22b R = CH_3CH_2: IC_{50} = 0.1 μM
22c R = $(CH_3)_2CH$: IC_{50} = 0.007 μM
22d R = $(CH_3)_2CHCH_2$: IC_{50} = 0.24 μM
22e R = *cyclo*-C_6H_{11}: IC_{50} = 50 μM
22f R = $C_6H_5CH_2CH_2$: IC_{50} = 49 μM

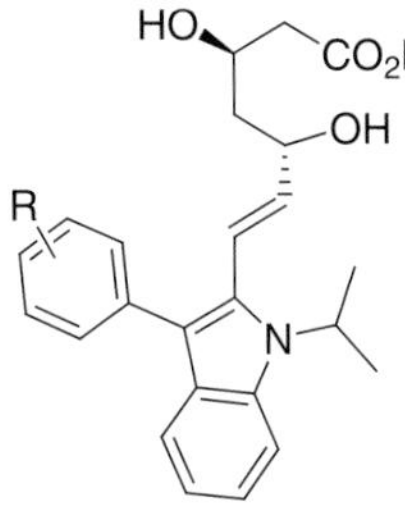

23a R = 2-CH_3, 4-F: IC_{50} = 0.004 μM
23b R = 3,5-$(CH_3)_2$: IC_{50} = 0.005 μM
23c R = 3-CH_3, 4-F: IC_{50} = 0.009 μM
23d R = H: IC_{50} = 0.017 μM
23e R = 3,5-$(CH_3)_2$, 4-F: IC_{50} = 0.02 μM
23f R = 4-CF_3: IC_{50} = 0.09 μM
23g R = 2-CH_3: IC_{50} = 0.14 μM
23h R = 4-SCH_3: IC_{50} = 1.1 μM
23i R = 4-CO_2Na: IC_{50} >10 μM

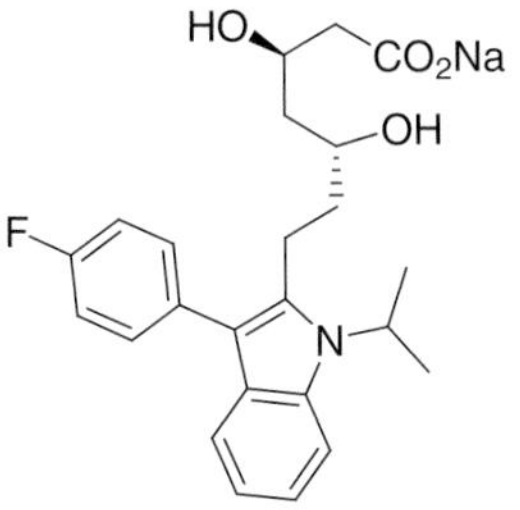

24 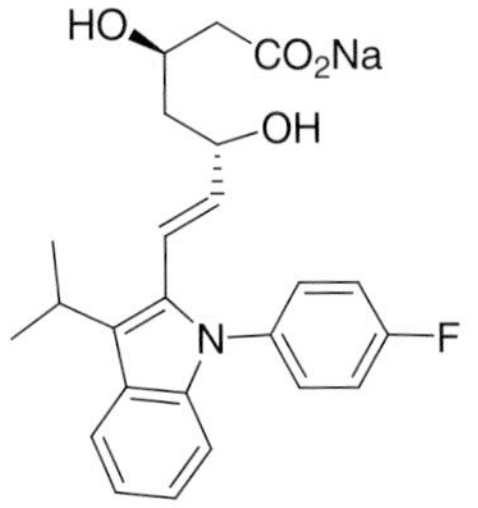 **25**

Figure 10 Structures of key compounds and structure–activity relationships leading to the discovery of fluvastatin.

In its racemic form, fluvastatin is approximately four times more potent in vitro than lovastatin, while the pure (3*R*,5*S*) enantiomer is approximately 10-fold more potent than lovastatin in biochemical assays. As a hydroxy-acid analog, fluvastatin has intermediate physiochemical properties, being about twice as hydrophilic as lovastatin and about 40 times more lipophilic than pravastatin.[62] In contrast to other statins, orally administered fluvastatin is nearly completely absorbed. However, the high protein binding (99%) of fluvastatin dictates that the overall plasma-free drug concentration is very low and thus limits its exposure to peripheral tissue. The metabolism of fluvastatin is mediated by CYP2C9 and not by CYP3A4, and as a result has a lower potential for drug–drug interactions with CYP3A4 inhibitors. Oral administration of fluvastatin dose-dependently reduced plasma VLDLc and LDLc concentrations in a variety of animal species at daily doses between 2 and 50 mg kg^{-1}.[64,65] Fluvastatin sodium **5** was approved for human use in Europe in 1995 and was launched in the US in 2000.

6.20.5.1.6 Cerivastatin

Cerivastatin **7** (**Figure 7**), another entirely synthetic statin, is one of the most potent inhibitors of HMG-CoA reductase (K_i = 1.3 pM) identified.[66] Cerivastatin contains a unique penta-substituted pyridine core, retains the unsaturated *E*-heptenoic acid side chain with the identical (3*R*, 5*S*) chirality found in fluvastatin, and is about 100-fold more potent than lovastatin. Cerivastatin was orally active in animal models in the low μg kg^{-1} range. For example, in normal chow-fed dogs, oral cerivastatin at 0.01–0.1 mg kg^{-1} day^{-1} dose-dependently reduced LDLc plasma levels after 18 days, and this reduction reached a maximum of 75% with the 0.1 mg kg^{-1} day^{-1} dose. Similarly, low oral daily doses of 0.1 mg kg^{-1} of cerivastatin reduced plasma LDLc and the progression of atherosclerosis in rabbit models after 9 weeks of dosing.

The lower doses of cerivastatin administered in animal studies were also effective in humans where, at daily doses of 0.025–0.4 mg, cerivastatin reduced LDLc levels dose-dependently and by greater than 50% using the 0.4 mg daily dose. Cerivastatin has a much higher overall oral bioavailability (~60%) than other statins and is metabolized in humans by both CYP3A4 and CYP2C8. Like many other statins, it is highly protein-bound (>95%), in spite of its fairly low *C*logP (1.5). Cerivastatin **7** was first approved for human use in the UK in 1997. It was voluntarily withdrawn in 2001 after higher incidences of severe myopathy and renal failure were observed in cerivastatin-treated patients, particularly in patients co-treated with gemfibrozil.

6.20.5.1.7 Rosuvastatin

Rosuvastatin **8** (**Figure 7**) contains an unusual *N*-methyl *N*-linked sulfonamide moiety in addition to its fully substituted pyrimidine core,[67] while retaining the unsaturated *E*-heptenoic acid side chain with the identical (3*R*, 5*S*) chirality found in fluvastatin. Rosuvastatin **26a** was more potent in vitro ($IC_{50} = 0.16\,nM$) than any of the related pyrimidine **26b–26g** or pyrrole **27a** and **27b** analogs (**Figure 11**) tested, being approximately 10-fold more potent in vitro than atorvastatin. The x-ray structure of enzyme-bound rosuvastatin revealed that it generated the most ligand–protein interactions of any of the statins studied crystallographically.[60] The combination of the pyrimidine ring and the sulfonamide moiety serves to lower dramatically the *c*log*P* of rosuvastatin to −0.3, near that of pravastatin. As such, rosuvastatin is the most hydrophilic synthetic statin reported.

Rosuvastatin reduced total plasma cholesterol in beagle dogs by 26% after 14 days of oral dosing at $3\,mg\,kg^{-1}\,day^{-1}$ compared to an 18% reduction obtained with $3\,mg\,kg^{-1}\,day^{-1}$ of pravastatin. In cynomolgus monkeys, rosuvastatin reduced plasma cholesterol by 22% at a daily oral dose of $12.5\,mg\,kg^{-1}$, comparable to the effects obtained with a $50\,mg\,kg^{-1}\,day^{-1}$ oral dose of pravastatin that reduced cholesterol by 19%. The overall bioavailability of rosuvastatin is ~20%, exceeding that of atorvastatin. The greater polarity associated with rosuvastatin results in reduced protein binding (88%) compared with other synthetic statins. Interestingly, rosuvastatin, like pravastatin, is not metabolized significantly by CYP enzymes, including CYP3A4, so the drug–drug interaction potential is reduced. Rosuvastatin is slowly metabolized by CYP2C9 and CYP2C19 to generate small quantities of the corresponding *N*-desmethyl derivative. Rosuvastatin has a prolonged duration of action and a longer plasma half-life ($t_{1/2} = 20\,h$) in humans than any other statin. Rosuvastatin **8** was first approved for human use in the US and Europe in 2003.

6.20.5.1.8 Pitavastatin

Pitavastatin **9** (**Figure 7**) contains a trisubstituted quinoline core and again retains the unsaturated *E*-heptenoic acid side chain with the identical (3*R*,5*S*) chirality found in fluvastatin.[68,69] It differs from other synthetic statins by incorporating a *cyclo*-propyl moiety in place of the more typical *iso*-propyl ring substituent. Pitavastatin is approximately fourfold more potent than pravastatin ($K_i = 1.7\,nM$). While detailed structure–activity data used to identify pitavastatin

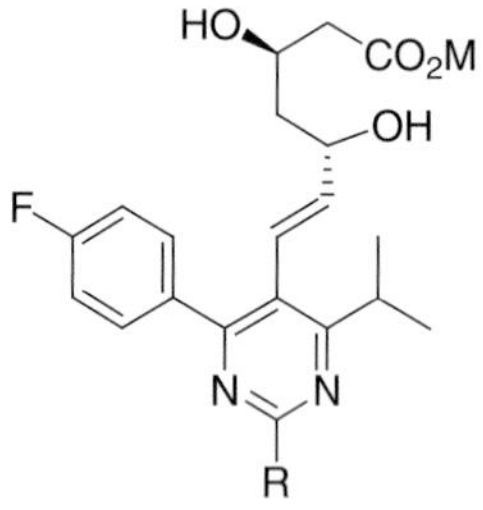

26a R = $CH_3SO_2N(CH_3)$: M = Ca
26b R = CH_3SO_2: M = Na
26c R = CH_3S: M = Na
26d R = $CH_3SO_2NHN(CH_3)$: M = Na
26e R = $(CH_3)_2NSO_2N(CH_3)$: M = Na
26f R = CH_3O: M = Na
26g R = $CH_3CON(CH_3)$: M = Na

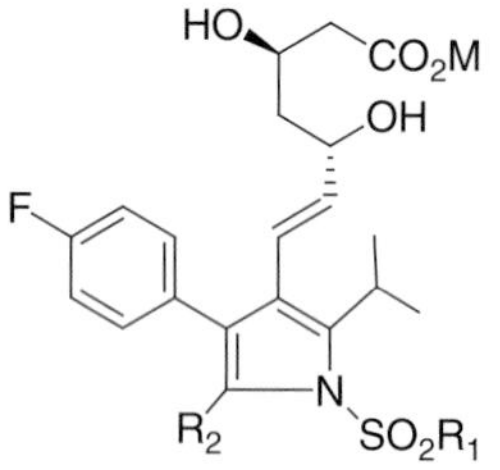

27a R_1, $R_2 = CH_3$: M = Na
27b $R_2 = CH_3$: R_1 = H: M = Na

Figure 11 Structures of key compounds leading to the discovery of rosuvastatin.

have not been reported, synthetic details have been described.[70] In animal studies, pitavastatin also dose-dependently reduces total plasma cholesterol and LDLc, with concomitant reductions in serum TG.

Oral administration of pitavastatin ($0.5\,mg\,kg^{-1}\,day^{-1}$) to hyperlipidemic rabbits for 26 weeks reduced total plasma cholesterol by 7–20% and also reduced the atherosclerotic lesion area. Pitavastatin is rapidly absorbed and has the highest overall bioavailability of any statin ($\sim$80%). Pitavastatin is also highly protein-bound ($>$95%). Like rosuvastatin and pravastatin, pitavastatin is not significantly metabolized by CYP3A4 enzymes, so the potential for drug–drug interactions is reduced. Pitavastatin is metabolized by CYP2C9 and CYP2C8. Pitavastatin has a reasonable duration of action and plasma half-life ($t_{1/2}$ = 4–5 h). Pitavastatin **9** was launched in Japan in 2003 and is being tested clinically in the US.

6.20.5.2 Low-Density Lipoprotein Cholesterol-Lowering Agents: Cholesterol Absorption Inhibitors

Ezetimibe **10** was discovered as part of a program directed at identifying novel ACAT inhibitors for lipid lowering. These azetidinone leads, although weak as ACAT inhibitors, were efficacious in lowering cholesterol in animals, suggesting that they might work by a different mechanism.[71–73] This discovery represents a tour de force in lead optimization, driven primarily by increasing in vivo efficacy by minimizing metabolism, and is in sharp contrast to the more common target-directed discovery approaches employed today. The biochemical target for ezetimibe was unknown at the time that discovery efforts began and has only recently been elucidated.[74]

The separation between ACAT potency and in vivo efficacy was pronounced for azetidinone **28** (**Figure 12**). It was a relatively weak ACAT inhibitor (IC_{50} = 2.6 μM), yet when given orally to hamsters at $10\,mg\,kg^{-1}\,day^{-1}$, lowered serum cholesterol by 28% (ED_{50} = $2\,mg\,kg^{-1}$) and also lowered serum cholesteryl ester by 93%. Compound **28** was even more potent in dogs and monkeys. The results suggested that **28** blocked cholesterol absorption near the intestinal wall by an unknown mechanism, with the in vivo efficacy of **28** being attributable to its metabolite, **29**. The introduction of a chiral hydroxyl group on the aliphatic side chain improved the in vivo efficacy of **30** (ED_{50} = $0.9\,mg\,kg^{-1}$) twofold compared with **28**. Since metabolic degradation occurred by hydroxylation of the unsubstituted phenyl ring in **30**, *p*-fluorophenyl groups were introduced at two of the rings to give **31**, and the related phenol **10** minimized systemic exposure to localize the compound in the intestines. The resulting compound, ezetimibe **10**, displayed excellent in vivo oral efficacy in lowering plasma cholesterol across a variety of animal species, including hamsters (ED_{50} = $0.04\,mg\,kg^{-1}$), rat (ED_{50} = $0.03\,mg\,kg^{-1}$), monkey (ED_{50} = $0.0005\,mg\,kg^{-1}$), and dog (ED_{50} = $0.007\,mg\,kg^{-1}$), with minimal systemic exposure. Thus, ezetimibe is more than 50-fold more effective in animals than the original lead, **28**.

Ezetimibe **10** has a long half-life, is suitable for once-a-day oral dosing, is rapidly absorbed, and becomes conjugated as its glucoronide metabolite that is excreted in the bile. This mechanism efficiently delivers ezetimibe to its site of action in the intestine, and recycling in the enterohepatic recirculation further insures that a high concentration of drug is maintained in the intestine. In contrast to statins that are preferentially prescribed for evening dosing, ezetimibe can

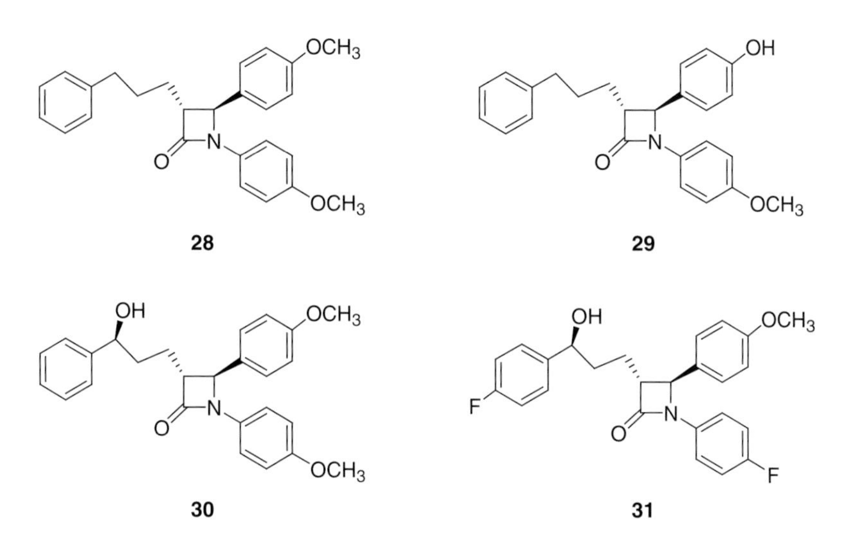

Figure 12 Structures of key compounds leading to the discovery of ezetimibe.

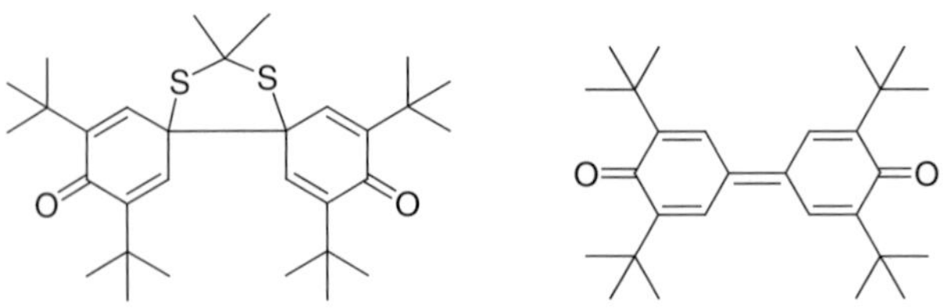

Figure 13 Structures of key probucol metabolites.

be taken at any time of day or night. Ezetimibe has no effect on the major drug-metabolizing enzymes and consequently its potential to cause drug–drug interactions is negligible. The molecular target for ezetimibe is the Niemann–Pick C1-Like1 protein that mediates cholesterol absorption within the brush-border membranes of intestinal enterocytes.[74] This discovery should facilitate the search for new-generation cholesterol absorption inhibitors. Ezetimibe **10** was approved in 2002 in the US as monotherapy, and in 2004 as a fixed combination with simvastatin, known as Vytorin.

6.20.5.3 Low-Density Lipoprotein Cholesterol-Lowering Agents: Synthetic Antioxidants

The synthetic antioxidant probucol **11** was identified in the early 1960s as a modest lipid-lowering agent. Probucol lowers plasma LDLc in both animals and humans, but it also profoundly reduces HDLc. In clinical testing, probucol lowers LDLc by up to 20% but can lower HDLc by as much as 20–30%. The precise mechanism by which probucol lowers LDLc and HDLc is unknown; however, probucol is known to prevent the oxidation of LDLc both in vitro and in vivo. Probucol has shown favorable benefits in reducing restenosis in patients, but its effects on atherosclerosis have been variable. Probucol undergoes extensive metabolism and oxidative degradation[75] to produce a variety of potentially toxic metabolites, including the spiroquinone **32** and the *bis*-quinone **33** (**Figure 13**), that have been implicated in causing QTc prolongation and fatal arrhythmia.[76] Probucol **11** was approved in the US in the early 1980s, but, because of these safety concerns, it was voluntarily withdrawn from the US market in 1995, although it is still prescribed in certain parts of the world.

6.20.6 Unmet Medical Needs

6.20.6.1 Limitations of Low-Density Lipoprotein Cholesterol Reduction and Statin Therapy

A clear overall therapeutic benefit has been achieved by lowering LDLc with multiple statins in a subset of patients with established CHD and in primary prevention for high-risk patients. Based upon these positive benefits, statin use has increased such that recent population studies now indicate that in 2002 over 9% of the US adult population were taking some form of statin.[77] However, even in controlled clinical settings, the majority of patients fail to reach their projected lipid levels and significant numbers of patients fail to achieve a therapeutic benefit as they continue to experience major coronary events while continuing statin treatment. While coronary event reductions of 25–30% are observed with statins as monotherapy, in secondary prevention trials, nearly 70% of the patients have coronary events that are not avoided. Thus, the search has continued for alternative strategies to lower LDLc or to increase the clinical effectiveness of statins using combination therapies. In those patients who have not tolerated statins, the need for safer alternative therapies is particularly acute.

6.20.6.2 High-Density Lipoprotein Cholesterol-Elevating Agents

While population studies indicate that increased statin usage has produced a progressive decrease in average LDLc levels between 1999 and 2002 compared to the LDLc levels observed in 1988–94, average HDLc levels remained unchanged over the same time period.[77] These data likely reflect a lack of comparable safe therapies for raising HDLc. Low HDLc represents a significant independent risk factor for CHD. In those patients at risk for CHD because of low HDLc, increases of 50% or more in HDLc may be required. The maximal response observed with fibrates or niacin led to much smaller percentage elevations, and severe side effects limit their use. To date, no large populations have been examined in secondary prevention trials to correlate HDLc increases greater than 30% through direct pharmacological intervention with an improved cardiovascular benefit. The choices in drug intervention for safely elevating HDLc to restore a cardioprotective level are currently quite limited.

6.20.7 New Research Areas

6.20.7.1 Low-Density Lipoprotein Cholesterol-Lowering Agents

With seven approved statins already available for patient treatment, some of which are reaching or approaching generic status, it seems unlikely that additional HMG-CoA reductase inhibitors beyond pitavastatin will be developed. However, several other approaches are currently being explored to offer patients alternatives for LDLc lowering. The recent discovery of the Niemann–Pick C1-Like1 protein as the molecular target for ezetimibe.[74] should stimulate new research for second-generation cholesterol absorption inhibitors.

6.20.7.1.1 Ileal bile acid transporter (IBAT) and apical sodium-co-dependent bile acid transporter (ASBT) inhibitors

As an alternative to cholesterol absorption inhibitors, specific inhibitors of the IBAT, also known as the apical co-dependent bile acid transporter are being sought. The IBAT system facilitates the reuptake of bile acids from the intestine, thus conserving the sterol pool. Functionally, an IBAT inhibitor should produce the same physiological effects of LDLc lowering achieved by anionic resins, such as cholestyramine, which sequester bile acids in the gut, without the high grams per day doses that dramatically limit patient compliance with sequesterants. The first entrant in this field, **34** (S-8921; **Figure 14**), has demonstrated oral efficacy in lowering LDLc with no accompanying change in HDLc at doses as low as 1 $mg\,kg^{-1}\,day^{-1}$.[78] Development of **34** has been discontinued as it showed no clinical benefit over existing products.

The benzothiazepine IBAT inhibitor, 264W94 **35** ($IC_{50}=0.25\,\mu M$),[79] dose-dependently (0.03–1.0 $mg\,kg^{-1}\,day^{-1}$ p.o. b.i.d.) reduced VLDLc and LDLc by up to 60% in diet-induced hypercholesterolemic rats, in only 3.5 days. Although it entered the clinic, its development has been discontinued. Despite these initial discouraging results, the opportunity exists to provide significant lipid lowering with IBAT inhibitors, either alone or in combination with a statin. In addition, a nonabsorbed version of an apical co-dependent bile acid transporter inhibitor based on a chiral poly-substituted benzothiepine core would have the advantage of delivering the inhibitor to the site of action with fewer accompanying side effects due to systemic exposure. Compounds **36** ($IC_{50}=0.28\,nM$) and **37** ($IC_{50}=0.75\,nM$) are potent apical co-dependent bile acid transporter inhibitors, which had no significant systemic exposure in rats but lowered LDLc upon oral dosing. The quaternary ammonium groups in **36** and **37** (**Figure 14**) also satisfied important solid-state criteria of nonhygroscopicity and crystallinity to advance in preclinical testing.[80]

6.20.7.1.2 Acyl-CoA:cholesterol acyltransferase inhibitors

Since the accumulation of cholesteryl ester within the macrophages and SMC of the developing lesion are important for atherosclerosis progression, inhibitors of ACAT, which esterifies intracellular free cholesterol, may have antiatherogenic properties. Selective inhibitors of ACAT-2, which is only found within peripheral macrophages, may be particularly effective. The acyl sulfamic ester, avasimibe **38** (**Figure 14**) is a selective ACAT inhibitor[81,82] that was orally effective as a lipid-lowering agent in multiple animal studies. Avasimibe (10 or 25 $mg\,kg^{-1}\,day^{-1}$ p.o.) reduced both VLDLc and LDLc by up to 40% in miniature pigs. It was safe and well tolerated and, in phase II trials, daily oral dosing of avasimibe at 50–500 mg decreased VLDLc by up to 30%, but no clear dose response was observed. While avasimibe was the first ACAT inhibitor to reach clinical proof of concept, it has been discontinued. Research is however continuing on other orally active ACAT inhibitors.

Pactimibe **39** (**Figure 14**) is a dual ACAT1 and ACAT2 inhibitor that has progressed to Phase II testing. When given orally, **39** reduced the progression of atherosclerosis in animal models by preventing cholesteryl ester accumulation in macrophages. Development of **39** was recently halted when it failed to meet its primary endpoint in Phase II trials.[83]

SMP-797 is an ACAT inhibitor for hyperlipidemia in phase II testing. A related analog, SMP-500 **40** (**Figure 14**), is a potent hepatic ACAT inhibitor ($IC_{50}=0.07$–$0.08\,\mu M$) that reduced serum cholesterol as well as serum cholesteryl ester levels in several animal models after oral dosing.[84]

6.20.7.2 High-Density Lipoprotein Cholesterol-Elevating Agents

In CHD patients at risk because of low HDLc, it is still unclear to what extent HDLc must be raised to produce a cardiovascular benefit, but elevations of 50% or more may be required.[16] Current therapies offer only modest improvements and are often limited by side effects.[16,56] A clear need exists to identify safer and more effective HDLc-raising drugs for use as monotherapy or in conjunction with other lipid-lowering agents. Several new strategies are under investigation that might lead to more pronounced HDLc elevations.[16,85] These include targeting enzymes associated with lipoprotein metabolism (e.g., hepatic triglyceride lipase, LpL, LCAT, or PLTP), modulating SR-B1 and

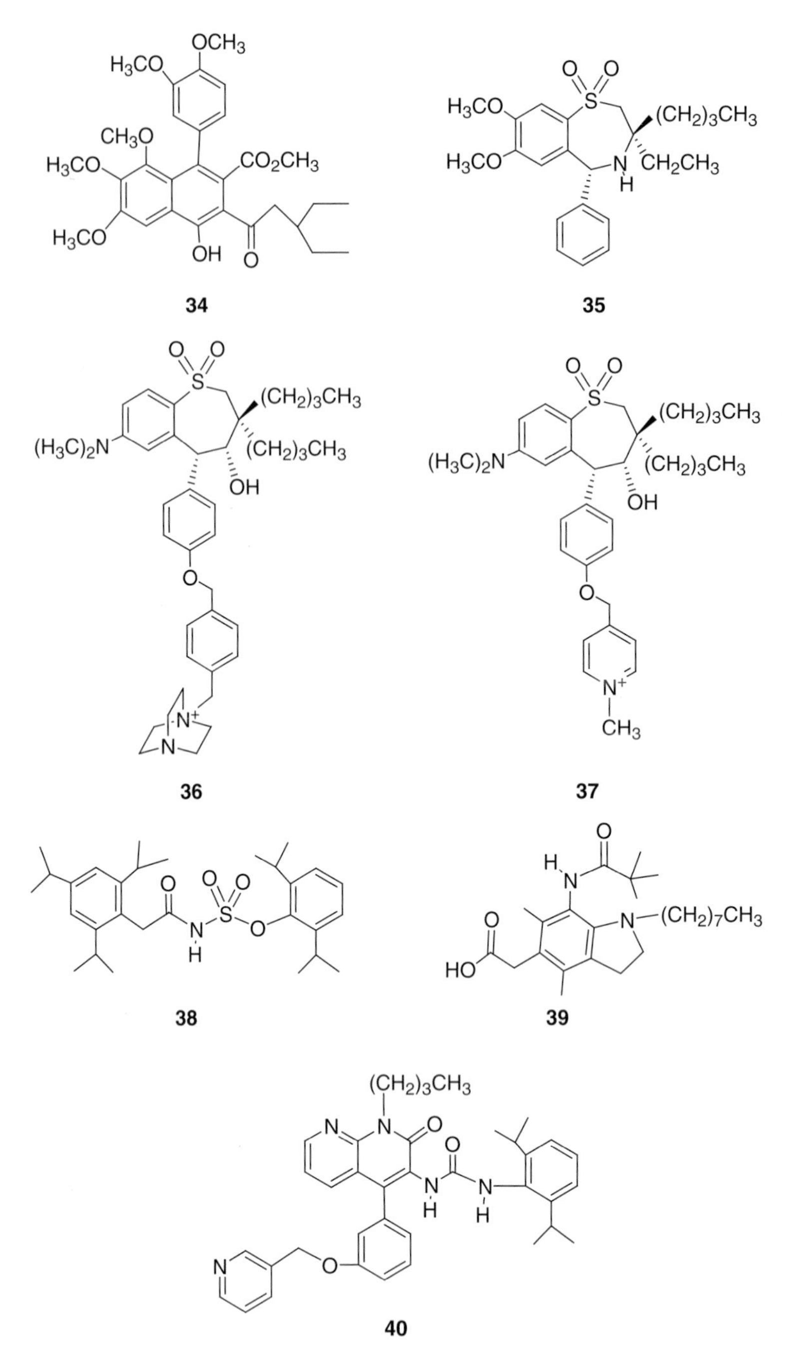

Figure 14 Structures of new IBAT, ASBT, and ACAT inhibitors.

ABC transporters, administering apoA-I or other HDL apolipoproteins like apoA-I_{Milano}, activating agonists of nuclear receptors such as PPARs, or inhibiting CETP. Of these, oral CETP inhibitors and oral PPAR agonists are the most advanced and have achieved clinical proof of concept.

6.20.7.2.1 Cholesteryl ester transfer protein inhibitors

Whether CETP contributes to atherosclerosis or plays an antiatherogenic role is a controversial topic.[16] In human plasma, CETP plays a potentially proatherogenic role by moving cholesteryl ester from HDL into VLDL and LDL particles, thereby lowering atheroprotective HDLc and raising proatherogenic VLDLc and LDLc. This equilibrium should be driven by the overall plasma concentration of CETP, and higher plasma CETP protein levels should therefore induce greater amounts of proatherogenic LDLc. On the other hand, CETP also plays a potentially key antiatherogenic role in the RCT process (**Figure 4**). By transferring cholesteryl ester from HDL to LDL and VLDL, CETP helps remove excess cholesterol from peripheral tissues (including atherosclerotic plaque) and increases the amount of cholesterol in these apoB lipoproteins that is taken up by the liver through the hepatic LDLr. Higher CETP plasma levels should therefore facilitate cholesterol removal by the RCT process.

A CETP inhibitor might be expected to raise plasma HDLc levels, lower LDLc, and provide a potential therapeutic benefit for the large number of patients at risk for CHD, because of low plasma HDLc levels. Recently, several new CETP inhibitors[86,87] have entered clinical trials, and two have now independently demonstrated phase II proof of concept by dramatically increasing HDLc levels by 50% or more in healthy volunteers and patients with low HDLc.[88,89] Thus, for the first time, pharmacological agents may be available that have the potential to elevate HDLc in patients with low HDLc to the same extent that the current standard-of-care drugs reduce LDLc in hyperlipidemic patients.

6.20.7.2.1.1 JTT-705

The thioester, JTT-705 **41a** (**Figure 15**), is a prodrug for the thiol-based, covalent modifier, **41b**, that irreversibly inhibits CETP. While **41b** effectively inhibited CETP in human plasma ($IC_{50} = 3\,\mu M$), it was too unstable for oral dosing and necessitated the identification of a suitable prodrug moiety. The *iso*-butyryl ester derivative **41a** ($IC_{50} = 6\,\mu M$) was nearly equally as active in human plasma and provided the optimal combination of chemical stability and oral absorption required to move into preclinical testing.

JTT-705 (30–300 mg kg^{-1} day^{-1} p.o.) dose-dependently reduced plasma CETP activity, raised HDLc, and lowered the overall atherogenic index (i.e., the non-HDLc/HDLc ratio) after 4 weeks in animal models. In a phase II study with nearly 200 healthy subjects with mild hyperlipidemia, daily oral doses of 300, 600, and 900 mg of JTT-705 for 4 weeks were well tolerated, with 900 mg producing a maximal 34% increase in HDLc levels. The 600 mg (26%) and 300 mg (15%)

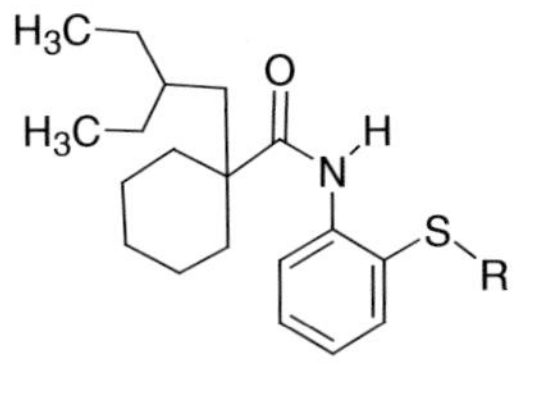

41a: R=$(CH_3)_2CHC(O)$–
41b: R=H

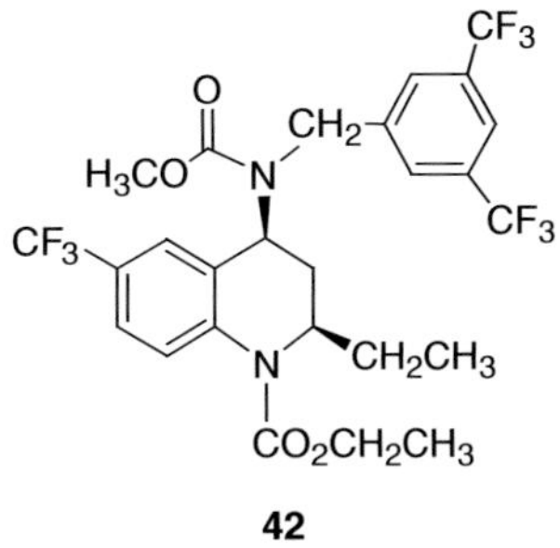

42

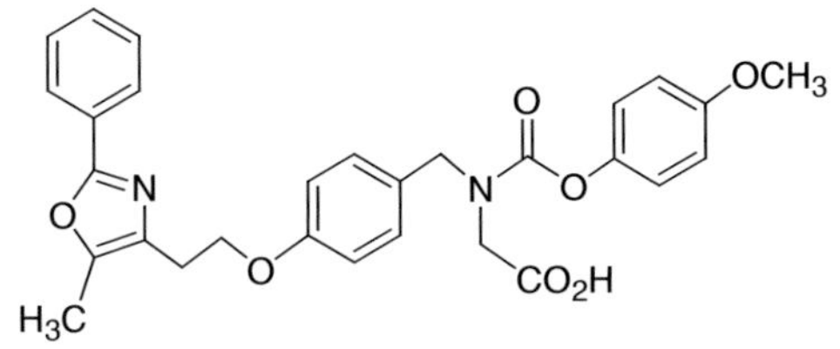

43

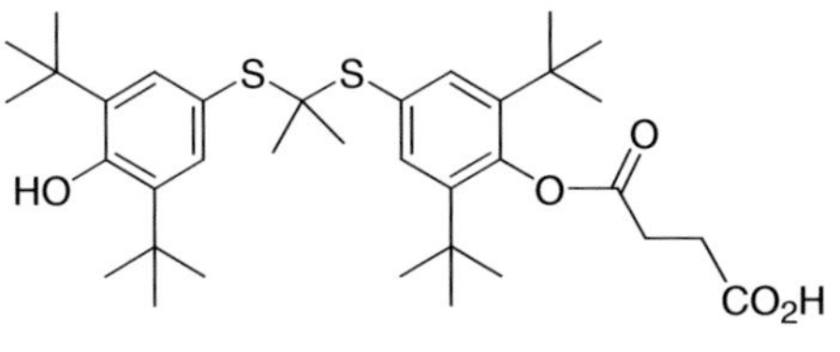

44

Figure 15 Structures of new CETP inhibitors, PPAR agonists, and anti-inflammatory antioxidants in clinical testing.

groups achieved smaller increases in HDLc. Accompanying these HDLc changes, a small dose-dependent decrease in LDLc was also observed that reached statistical significance with a maximal 7.4% decrease in LDLc in the 900 mg group.[88] The 34% increase in HDLc observed with the 900 mg group compared very favorably with the 25–30% increases frequently observed with multigram daily doses of niacin. Based upon these encouraging results, JTT-705 has entered phase III trials.

6.20.7.2.1.2 Torcetrapib

From a series of 4-amino-substituted 1,2,3,4-tetrahydroquinoline carbamates, the potent ($IC_{50} = 0.05\,\mu M$, human plasma) CETP inhibitor torcetrapib **42** (CP-529,414; **Figure 15**), was identified as a lead for clinical development. In contrast to JTT-705, which modifies a key cysteine residue within the CETP protein, torcetrapib is presumably a reversible inhibitor, since it contains no obviously reactive functionalities.

Torcetrapib **42** effectively inhibited CETP-mediated transfer activity in animals and dose-dependently increased HDLc and reduced LDLc. In rabbits, oral dosing with torcetrapib inhibited CETP-mediated cholesteryl ester transfer immediately by 70–80%, produced a concomitant nearly fourfold elevation in HDLc levels, and reduced the aortic atherosclerotic lesion area by nearly 60% in the torcetrapib-treated group versus vehicle controls.

Interestingly, the LDLc lowering of torcetrapib was additive when atorvastatin was dosed simultaneously in animals. The effects observed on LDLc lowering using the torcetrapib/atorvastatin combination were greater than those observed with atorvastatin alone. These data suggest that combinations of torcetrapib and atorvastatin may have utility, and this hypothesis is being tested clinically. These initial results are the first time that additive preclinical efficacy for lowering non-HDLc has been demonstrated by combining a potent CETP inhibitor with a statin.

In clinical trials, daily oral dosing with torcetrapib at doses up to 120 mg were safe and well tolerated. A small phase II clinical proof-of-concept study compared 120 mg torcetrapib q.d. or b.i.d. for 4 weeks to placebo in subjects having low HDLc ($<40\,mg\,dL^{-1}$) either as monotherapy or with a statin.[89] One arm of this study included patients co-treated once daily with both 120 mg torcetrapib and 20 mg atorvastatin. Treatment with 120 mg torcetrapib alone inhibited plasma CETP activity by 28% given q.d. and by 65% b.i.d. After 4 weeks, 120 mg torcetrapib alone produced increases in HDLc of 46% (q.d.) and 106% (b.i.d.), compared to placebo. However, the LDLc reductions observed in this study were small and not statistically significant, even though total cholesterol levels remained unchanged. Treatment with 120 mg torcetrapib alone produced nonsignificant reductions in LDLc of 8% (q.d.) and 17% (b.i.d.). Even though these LDLc changes were not statistically significant, the LDLc reductions observed with torcetrapib alone exceeded the 7% lowering in LDLc observed clinically with JTT-705 at $900\,mg^{-1}$ day.[88] After 4 weeks, subjects co-treated daily with both 120 mg torcetrapib and 20 mg atorvastatin attained inhibition of plasma CETP activity by 38%, accompanied by a 61% increase in HDLc and a statistically significant 17% reduction in LDLc beyond that achieved with atorvastatin alone.[89] Based upon these results, a torcetrapib/atorvastatin fixed combination has been selected for extensive testing in phase III trials.

6.20.7.2.2 Peroxisome proliferator-activated receptor agonists

The discovery that fibrates act as selective PPAR-α agonists has prompted the search for selective chemical entities as well as dual-acting PPAR agonists.[56,90] Since PPARs are activated by naturally occurring polyunsaturated fatty acids, it is perhaps not surprising that fibrates would act as ligands for these receptors. PPARs are intimately involved in several of the early monocyte recruitment steps in atherosclerosis.[90] PPAR-α agonists modulated chemokine (MCP-1) expression, whereas both PPAR-α and PPAR-γ inhibited VCAM-1 expression in activated endothelial cells. PPAR-α agonists, and not PPAR-γ, inhibited expression of the oxLDL receptor in endothelial cells. Both PPAR-α and PPAR-γ agonists stimulated cholesterol efflux from foam cells, whereas dual PPAR-βδ agonists promoted cholesterol accumulation in foam cells. PPAR-α agonists and dual PPAR-αγ agonists are being targeted as potential new treatments for dyslipidemia and atherosclerosis.

Muraglitazar **43** (**Figure 15**), a dual PPAR-αγ agonist, is the most advanced of these and is awaiting regulatory approval. In clinical trials, oral treatment of diabetic patients with daily doses of either 2.5 or 5 mg of muraglitazar as monotherapy lowered plasma glucose and TG levels (18–27%) dose-dependently and also increased plasma HDLc levels by 10–16%. Serum LDLc levels were modestly reduced by 3–5%.[91]

6.20.7.3 Anti-Inflammatory Based Approaches

Given the contribution that inflammatory mediators make to atherosclerosis, a number of companies are now seeking to identify suitable clinical candidates based on these underlying mechanisms. Two approaches have reached early proof of concept in phase II trials and are currently in phase III testing.

6.20.7.3.1 Oral 5-lipoxygenase-activating protein inhibitors

An oral small molecule inhibitor of 5-lipoxygenase-activating protein (FLAP), DG-031, of unknown structure, is in clinical trials with CHD patients. FLAP inhibitors are anticipated to reduce proinflammatory leukotriene mediators such as the potent chemotactic agent, LTB_4 (**Figure 6**). In a 4-week phase II trial in patients with a previous history of MI, DG-031 (250, 500, and 750 mg day^{-1} p.o.) at the highest dose reduced serum LTB_4 (26%) and MPO (12%) levels as well as soluble intercellular adhesion molecule. As discussed previously, MPO and LTB_4 are inflammatory markers associated with a higher risk of CHD. Plasma CRP levels were also reduced at the highest dose, but the response did not reach statistical significance. Accompanying these changes, LDLc levels increased slightly, by 8% with the highest dose.[92] These results represent the first demonstration that two key biomarkers of inflammation were reduced following oral dosing with an anti-inflammatory agent in a CHD patient population. It remains to be seen whether these biomarker results translate to a corresponding benefit in actual clinical events, especially given the absence of any profound changes in LDLc, in phase III trials.

6.20.7.3.2 New oral anti-inflammatory antioxidants

AGI-1067 (succinobucol **44**, **Figure 15**), a metabolically stable derivative of probucol, where the introduction of a monosuccinate ester group alters its overall in vitro and in vivo pharmacology, is a new oral antioxidant being tested in atherosclerosis. AGI-1067, like probucol, is a potent extracellular antioxidant but has an enhanced intracellular uptake capability compared to probucol. AGI-1067 inhibited intracellular ROS production whereas probucol had no effect and also inhibited proinflammatory gene expression in stimulated endothelial cells, including the expression of several proteins associated with atherosclerosis: VCAM-1 and MCP-1.[93] In LPS-challenged mice, oral dosing with AGI-1067 reduced VCAM-1 and MCP-1 mRNA levels. In $LDLr^{-/-}$ mice, AGI-1067 reduced aortic atherosclerosis by 49% in the absence of a lipid-lowering effect.[94] Like probucol, oral dosing with AGI-1067 inhibited restenosis in phase II trials, but an important antiatherosclerotic effect was observed in the nonintervened arterial segments. This effect was not observed with probucol. The observed reductions in atheroma volume were sufficiently robust for AGI-1067 to enter phase III testing in a secondary prevention trial targeting atherosclerosis in 2004. If successful, this trial should demonstrate the potential benefit of novel antioxidant-based anti-inflammatory agents as potential new therapies for CHD patients.[95,96]

6.20.8 Conclusions

Statins have demonstrated the clinical benefits associated with LDLc lowering and have dramatically improved the treatment options for CHD patients. Cholesterol absorption inhibitors now represent a viable alternative for lipid lowering, particularly in patients who are either unresponsive to or intolerant of statins. A variety of newer therapies are in clinical testing which may make alternative options available for LDLc lowering in patients. **Figure 16** summarizes some of these newer therapies and how they fit in mechanistically versus established drugs in providing new options to address disease initiation, progression, and/or regression.

Based upon the proof-of-concept clinical results observed with both JTT-705 and torcetrapib, viable new options may be on the horizon for increasing HDLc levels in patients at risk of CHD because of low HDLc. The percentage

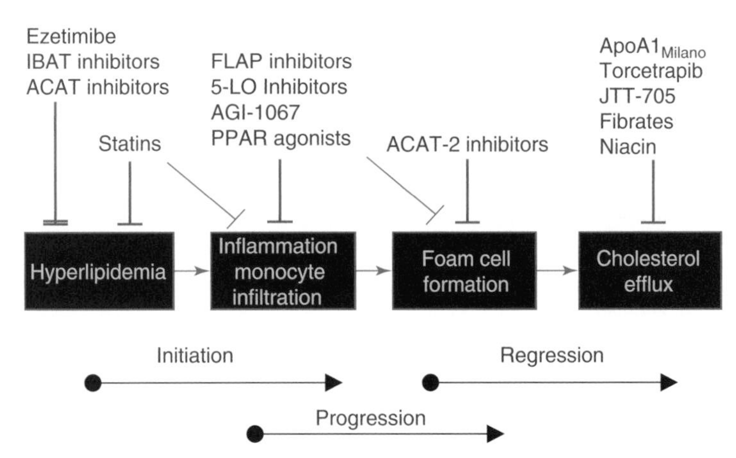

Figure 16 Current treatment strategies and new approaches for atherosclerosis.

increases in HDLc observed with torcetrapib and JTT-705 are higher than those observed clinically following statin, fibrate, or niacin treatment. It will however be important to determine whether these agents perform similarly across a broader, more diverse patient population without disrupting normal lipoprotein balance and function. At the same time, there is still a clear need to demonstrate their direct benefit on reducing morbidity and mortality in phase III trials with larger CHD patient populations.

While the inflammatory aspects underlying atherosclerosis initiation and progression are now better appreciated, the first potential new therapies based upon these hypotheses are undergoing key phase III testing. Oral FLAP inhibitors and VCAM-1 expression inhibitors have the potential to alter the treatment landscape for CHD patients dramatically. Beyond these, a number of new exploratory anti-inflammatory targets are also being pursued. It remains to be seen whether any of these new approaches for treating CHD patients succeed in the clinic and come to fruition as viable treatment options for the majority of patients who are still unable to avoid major coronary events with statin monotherapy.

References

1. Libby, P. *Sci. Am.* **2002**, *286*, 46–55.
2. Schoenhagen, P.; Ziada, K. M.; Vince, G.; Nissen, S. E.; Tuzcu, E. M. *J. Am. Coll. Cardiol.* **2001**, *38*, 297–306.
3. Mackay, J.; Mensah, G. *Atlas of Heart Disease and Stroke*; WHO Press: Geneva, Switzerland, 2004.
4. Loscalzo, J., Ed. *Molecular Mechanisms of Atherosclerosis*; Taylor and Francis: London, UK, 2005.
5. Romer, T. J.; Brennan, J. F.; Fitzmaurice, M.; Feldstein, M. L.; Deinum, G.; Myles, J. L.; Kramer, J. R.; Lees, R. S.; Feld, M. S. *Circulation* **1998**, *97*, 878–885.
6. Hansson, G. K *N. Engl. J. Med.* **2005**, *352*, 1685–1695.
7. Libby, P.; Ridker, P. M.; Maseri, A. *Circulation* **2002**, *105*, 1135–1143.
8. Ross, R. *N. Engl. J. Med.* **1999**, *340*, 115–126.
9. Tabas, I. *J. Clin. Invest.* **2002**, *110*, 905–911.
10. Brites, F. D.; Bonavita, C. D.; Cloes, M.; Yael, M. J.; Fruchart, J.-C.; Castro, G. R.; Wikinski, R. W. *Clin. Chim. Acta* **1998**, *269*, 107–124.
11. Segrest, J. P.; Jones, M. K.; De Loof, H.; Dashti, N. *J. Lipid Res.* **2001**, *42*, 1346–1367.
12. Teerlink, T.; Scheffer, P. G.; Bakker, S. J. L.; Heine, R. J. *J. Lipid Res.* **2004**, *45*, 954–966.
13. Teng, B.; Thompson, G. R.; Sniderman, A. D.; Forte, T. M.; Krauss, R. M.; Kwiterovich, P. O. *Proc. Natl. Acad. Sci. USA* **1983**, *80*, 6662–6666.
14. Goldstein, J. L.; Brown, M. S. *Science* **2001**, *292*, 1310–1312.
15. Lund-Katz, S.; Liu, L.; Thuahnai, S. T.; Phillips, M. C. *Frontiers Biosci.* **2003**, *8*, 1044–1054.
16. Linsel-Nitschke, P.; Tall, A. R. *Nat. Rev. Drug Disc.* **2005**, *4*, 193–205.
17. Schmitz, G.; Langmann, T. *Curr. Opin. Investig. Drugs* **2005**, *6*, 907–919.
18. Nichols, S. J.; Hazen, S. L. *Arterioscler. Thromb. Vasc. Biol.* **2005**, *25*, 1102–1111.
19. Berliner, J. A.; Watson, A. D. *N. Engl. J. Med.* **2005**, *353*, 9–11.
20. Quehenberger, O. *J. Lipid Res.* **2005**, *46*, 1582–1590.
21. Cipollone, F.; Mezzetti, A.; Fazia, M. L.; Cuccurullo, C.; Iezzi, A.; Ucchino, S.; Spigonardo, F.; Bucci, M.; Cuccurullo, F.; Prescott, S. M. et al. *Arterioscler. Thromb. Vasc. Biol.* **2005**, *25*, 1665–1670.
22. Stocker, R.; Keaney, J. F. *Physiol. Rev.* **2003**, *84*, 1381–1478.
23. Ansell, B. J.; Navab, M.; Hama, S.; Kamranpour, N.; Fonarow, G.; Hough, G.; Rahmani, S.; Mottahedeh, R.; Dave, R.; Reddy, S. T. et al. *Circulation* **2003**, *108*, 2751–2756.
24. Kannel, W. B. *Am. J. Cardiol.* **1995**, *76*, 69c–77c.
25. Grundy, S. M.; Cleeman, J. I.; Merz, N. B.; Brewer, H. B.; Clark, L. T.; Hunninghake, D. B.; Pasternak, R. C.; Smith, S. C.; Stone, N. J. *Arterioscler. Thromb. Vasc. Biol.* **2004**, *24*, 1329–1330.
26. Executive Summary of the Third Report of the NCEP Expert Panel on Detection, Evaluation and Treatment of High Blood Cholesterol in Adults. *JAMA* **2001**, *285*, 2486–2497.
27. Tsimikas, S.; Brilakis, E. S.; Miller, E. R.; McConnell, J. P.; Lennon, R. J.; Kornman, K. S.; Witztum, J. L.; Berger, P. B. *N. Engl. J. Med.* **2005**, *353*, 46–57.
28. Ariyo, A. A.; Thach, C.; Tracy, R. *N. Engl. J. Med.* **2003**, *349*, 2108–2115.
29. Wilson, P. W. F.; D'Agostino, R. B.; Levy, D.; Belanger, A. M.; Silbershatz, H.; Kannel, W. B. *Circulation* **1998**, *97*, 1837–1847.
30. Burchfiel, C. M.; Abbott, R. D.; Sharp, D. S.; Curb, J. D.; Rodriguez, B. L.; Yano, K. *Arterioscler. Thromb. Vasc. Biol.* **1996**, *16*, 1356–1364.
31. Brewer, H. B. B. *Am. Heart J.* **2004**, *148*, S14–S18.
32. Hofker, M. H.; van Vlijmen, B. J. M.; Havekes, L. M. *Atherosclerosis* **1998**, *137*, 1–11.
33. Moghadasian, M. H. *Life Sci.* **2002**, *70*, 855–865.
34. Mehta, J. L., Ed. *Statins: Understanding Clinical Use*; Saunders: Philadelphia, PA, 2004.
35. Grundy, S. M. *Circulation* **1998**, *97*, 1436–1439.
36. Pedersen, T. R.; Olsson, A. G.; Faergeman, O.; Kjekshus, J.; Wedel, H.; Berg, K.; Wilhelmsen, L.; Haghfelt, T.; Thorgeirsson, G.; Pyorala, K. et al. *Circulation* **1998**, *97*, 1453–1460.
37. Sacks, F. M.; Moye, L. A.; Davis, B. R.; Cole, T. G.; Rouleau, J. L.; Nash, D. T.; Pfeffer, M. A.; Braunwald, E. *Circulation* **1998**, *97*, 1446–1452.
38. West of Scotland Coronary Prevention Study Group. *Circulation* **1998**, *97*, 1440–1445.
39. Sever, P. S.; Dahlof, B.; Poulter, N. R. *Lancet* **2003**, *361*, 1149–1158.
40. Nissen, S. E.; Tuzcu, E. M.; Schoenhagen, P.; Brown, B. G.; Ganz, P.; Vogel, R. A.; Crowe, T.; Howard, G.; Cooper, C. J.; Brodie, B. et al. *JAMA* **2004**, *291*, 1071–1080.
41. Nissen, S. E.; Tuzcu, E. M.; Schoenhagen, P.; Crowe, T.; Sasiela, W. J.; Tsai, J.; Orazem, J.; Magorien, R. D.; O'Shaughnessy, C.; Ganz, P. *N. Engl. J. Med.* **2005**, *352*, 29–38.

42. LaRosa, J. C.; Grundy, S. M.; Waters, D. D.; Shear, C.; Barter, P.; Fruchart, J.-C.; Gotto, A. M.; Greten, H.; Kastelein, J. J. P.; Shepherd, J. et al. *N. Engl. J. Med.* **2005**, *352*, 1425–1435.
43. Serruys, P. W. J. C.; de Feyter, P.; Macaya, C.; Kokott, N.; Puel, J.; Vrolix, M.; Branzi, A.; Bertolami, M. C.; Jackson, G.; Strauss, B. et al. *JAMA* **2002**, *287*, 3215–3222.
44. Grundy, S. *Circulation* **2005**, *111*, 3016–3019.
45. Alsheikh-Ali, A. A.; Ambrose, M. S.; Kuvin, J. T.; Karas, R. H. *Circulation* **2005**, *111*, 3051–3057.
46. Ridker, P. M.; Cannon, C. P.; Morrow, D.; Rifai, N.; Rose, L. M.; McCabe, C. H.; Pfeffer, M. A.; Braunwald, E. *N. Engl. J. Med.* **2005**, *352*, 20–28.
47. Ridker, P. M.; Morrow, D. A.; Rose, L. M.; Rifai, N.; Cannon, C. P.; Braunwald, E. *J. Am. Coll. Cardiol.* **2005**, *45*, 1644–1648.
48. Davignon, J.; Hanefeld, M.; Nakaya, N.; Hunninghake, D. B.; Insull, W.; Ose, L. *Am. J. Cardiol.* **1998**, *82*, 32J–39J.
49. Jones, P. H.; Davidson, M. H.; Stein, E. A.; Bays, H. E.; McKenney, J. M.; Miller, E.; Cain, V. A.; Blasetto, J. W. *Am. J. Cardiol.* **2003**, *92*, 152–160.
50. Saito, Y.; Yamada, N.; Teramoto, T.; Itakura, H.; Hata, Y.; Nakaya, N.; Mabuchi, H.; Tushima, M.; Sasaki, J.; Ogawa, N. et al. *Atherosclerosis* **2002**, *162*, 373–379.
51. Tolbert, J. A. *Nat. Rev. Drug Disc.* **2003**, *2*, 517–526.
52. Knopp, R. H.; Gitter, H.; Truitt, T.; Bays, H.; Manion, C. V.; Lipka, L. J.; LeBeaut, A. P.; Suresh, R.; Yang, B.; Veltri, E. P. *Eur. Heart J.* **2003**, *24*, 729–741.
53. Davidson, M. H.; McGarry, T.; Bettis, R.; Melani, L.; Lipka, L. I.; LeBeaut, A. P.; Suresh, R.; Sun, S.; Veltri, E. P. *J. Am. Coll. Cardiol.* **2002**, *40*, 2125–2134.
54. Bays, H. E.; Ose, L.; Fraser, N.; Tribble, D. L.; Quinto, K.; Reyes, R.; Johnson-Levonas, A. O.; Sapre, A.; Donahue, S. R. *Clin. Ther.* **2004**, *26*, 1758–1773.
55. Sawayama, Y.; Shimizu, C.; Maeda, N.; Tatsukawa, M.; Kinukawa, N.; Koyanagi, S.; Kashiwagi, S.; Hayashi, J. *J. Am. Coll. Cardiol.* **2002**, *39*, 610–616.
56. Steiner, G. *Atherosclerosis* **2005**, *182*, 199–207.
57. Manninen, V.; Tenkanen, L.; Koskinen, P.; Huttunen, J. K.; Manttari, M.; Heinonen, O. P.; Frick, H. M. *Circulation* **1992**, *85*, 37–45.
58. Rubins, H. B.; Robins, S. J.; Collins, D.; Fye, C. L.; Anderson, J. W.; Elam, M. B.; Faas, F. H.; Linares, E.; Schaefer, E. J.; Schectman, G. et al. *N. Engl. J. Med.* **1999**, *341*, 410–418.
59. Taylor, A. J.; Sullenberger, L. E.; Lee, H. J.; Lee, J. K.; Grace, K. A. *Circulation* **2004**, *110*, 3512–3517.
60. Istvan, E. S.; Deisenhofer, J. *Science* **2001**, *292*, 1160–1164.
61. Endo, A. *Atherosclerosis Suppl.* **2004**, *5*, 67–80.
62. Hamelin, B. A.; Turgeon, J. *Trends Pharmacol. Sci.* **1998**, *19*, 26–37.
63. Roth, B. D. *Prog. Med. Chem.* **2002**, *40*, 1–22.
64. Kathawala, F. G. *Med. Res. Rev.* **1991**, *11*, 121–146.
65. Corsini, A.; Fumagilli, R.; Paoletti, R.; Bernini, F. *Drugs Today* **1996**, *32*, 13–35.
66. Bischoff, H.; Angerbauer, R.; Bender, J.; Bischoff, E.; Faggiotto, A.; Petzinna, D.; Pfitzner, J.; Porter, M. C.; Schmidt, D.; Thomas, G. *Atherosclerosis* **1997**, *135*, 119–130.
67. Watanabe, M.; Koike, H.; Ishiba, T.; Okada, T.; Seo, S.; Hirai, K. *Bioorg. Med. Chem.* **1997**, *5*, 437–444.
68. Isley, W. L. *Drugs Today* **2001**, *37*, 587–594.
69. Aoki, T.; Yamazaki, H.; Suzuki, H.; Tamaki, T.; Sato, F.; Kitahara, M.; Saito, Y. *Arzneim.-Forsch.* **2001**, *51*, 197–203.
70. Miyachi, N.; Yanagawa, Y.; Iwasaki, H.; Ohara, Y.; Hiyama, T. *Tetrahedron Lett.* **1993**, *34*, 8267–8270.
71. Clader, J. W. *J. Med. Chem.* **2004**, *47*, 1–9.
72. Wu, G.; Wong, Y.; Chen, X.; Ding, Z. *J. Org. Chem.* **1999**, *64*, 3714–3718.
73. Burnett, D. A. *Curr. Med. Chem.* **2004**, *11*, 1873–1887.
74. Garcia-Calvo, M.; Lisnock, J.; Bull, H. G.; Hawes, B. E.; Burnett, D. A.; Braun, M. P.; Crona, J. H.; Davis, H. R.; Dean, D. C.; Detmers, P. A. et al. *Proc. Natl. Acad. Sci. USA* **2005**, *102*, 8132–8137.
75. Witting, P. K.; Wu, B. J.; Raftery, M.; Southwell-Keely, P.; Stocker, R. *J. Biol. Chem.* **2005**, *280*, 15612–15618.
76. Namiki, A.; Yajima, H.; Yamamoto, T. *J. Med. Soc. Jpn.* **1992**, *39*, 284–289.
77. Carroll, M. D.; Lacher, D. A.; Sorlie, P. D.; Cleeman, J. I.; Gordon, D. J.; Wolz, M.; Grundy, S. M.; Johnson, C. L. *JAMA* **2005**, *294*, 1773–1781.
78. Hara, S.; Higaki, J.; Higashino, K. I.; Iwai, M.; Takasu, N.; Miyata, K.; Tonda, K.; Nagata, K.; Goh, Y.; Mizui, T. *Life Sci.* **1997**, *60*, 365–370.
79. Root, C.; Smith, C. D.; Sundseth, S. S.; Pink, H. M.; Wilson, J. G.; Lewis, M. C. *J. Lipid Res.* **2002**, *43*, 1320–1330.
80. Huang, H. C.; Tremont, S. J.; Lee, L. F.; Keller, B. T.; Carpenter, A. J.; Wang, C. C.; Banerjee, S. C.; Both, S. R.; Fletcher, T.; Garland, D. J. et al. *J. Med. Chem.* **2005**, *48*, 5853–5868.
81. Lee, H. T.; Sliskovic, D. R.; Picard, J. A.; Roth, B. D.; Wierenga, W.; Hicks, J. L.; Bousley, R. F.; Hamelehle, K. L.; Homan, R.; Speyer, C. et al. *J. Med. Chem.* **1996**, *39*, 5031–5034.
82. Burnett, J. R.; Wilcox, L. J.; Telford, D. E.; Kleinstiver, S. J.; Barrett, P. H.; Newton, R. S.; Huff, M. W. *J. Lipid Res.* **1999**, *40*, 1317–1327.
83. Daiichi Sankyo Co. Ltd., Press Release, October 26, 2005.
84. Ioriya, K.; Noguchi, T.; Muraoka, M.; Fujita, K.; Shimizu, H.; Ohashi, N. *Pharmacology* **2002**, *65*, 18–25.
85. Sviridov, D. *Curr. Med. Chem. Immun. Endoc. Metab. Agents* **2005**, *5*, 299–307.
86. Sikorski, J. A. *J. Med. Chem.* **2006**, *49*, 1–22.
87. Clark, R. W.; Chang, G.; Didiuk, M. T. *Curr. Med. Chem. Immun. Endoc. Metab. Agents* **2005**, *5*, 339–360.
88. de Grooth, G. J.; Kuivenhoven, J. A.; Stalenhoef, A. F. H.; de Graaf, J.; Zwinderman, A. H.; Posma, J. L.; van Tol, A.; Kastelein, J. J. P. *Circulation* **2002**, *105*, 2159–2165.
89. Brousseau, M. E.; Schaefer, E. J.; Wolfe, M. L.; Bloedon, L. T.; Digenio, A. G.; Clark, R. W.; Mancuso, J. P.; Rader, D. J. *N. Engl. J. Med.* **2004**, *350*, 1505–1515.
90. Duval, C.; Chinetti, G.; Trottein, F.; Fruchart, J.-C.; Staels, B. *Trends Mol. Med.* **2002**, *8*, 422–430.
91. Buse, J. B.; Rubin, C. J.; Frederich, R.; Viraswami-Appanna, K.; Lin, K. C.; Montoro, R.; Shockey, G.; Davidson, J. A. *Clin. Ther.* **2005**, *27*, 1181–1195.
92. Hakonarson, H.; Thorvaldsson, S.; Helgadottir, A.; Gudbjartsson, D.; Zink, F.; Andresdottir, M.; Manolescu, A.; Arnar, D. O.; Andersen, K.; Sigurdsson, A. et al. *JAMA* **2005**, *293*, 2245–2256.

93. Kunsch, C.; Luchoomun, J.; Grey, J. Y.; Olliff, L. K.; Saint, L. B.; Arrendale, R. F.; Wasserman, M. A.; Saxena, U.; Medford, R. M. *J. Pharmacol. Exp. Ther.* **2004**, *308*, 820–829.
94. Sundell, C. L.; Somers, P. K.; Meng, C. Q.; Hoong, L. K.; Suen, K. L.; Hill, R. R.; Landers, L. K.; Chapman, A.; Butteiger, D.; Jones, M. et al. *J. Pharmacol. Exp. Ther.* **2003**, *305*, 1116–1123.
95. Tardif, J.-C. *Curr. Atheroscler. Rep.* **2005**, *7*, 71–77.
96. Franks, A. M.; Gardner, S. F. *Ann. Pharmacother.* **2006**, *40*, 66–73.

Biography

James A Sikorski received a BS degree with honors in chemistry at Northeast Louisiana State University in 1970. He attended Purdue University and received MS and PhD degrees in organic chemistry, under the direction of Nobel Laureate, Prof Herbert C Brown. He joined Monsanto Agricultural Company in 1976, advancing to Science Fellow. In 1988, he moved to drug discovery research, first as Science Fellow in Medicinal Chemistry at Monsanto Corporate Research, then at Searle and Pharmacia, where he contributed to several anti-infective, anti-inflammatory, and cardiovascular projects. Dr Sikorski has received both the 1994 St Louis ACS Award and 1999 Kenneth A Spencer Award. In 2001 he joined AtheroGenics, Inc., and is currently Senior Director of Medicinal Chemistry, where his research is focused on the discovery of novel agents to treat chronic inflammatory diseases, including atherosclerosis, arthritis, and asthma.

Comprehensive Medicinal Chemistry II
ISBN (set): 0-08-044513-6

ISBN (Volume 6) 0-08-044519-5; pp. 459–494

6.21 Bone, Mineral, Connective Tissue Metabolism

C R Dunstan, J M Blair, and H Zhou, Anzac Research Institute, University of Sydney, NSW, Australia
M J Seibel, Anzac Research Institute, University of Sydney, and Concord Hospital, Concord, NSW, Australia

6.21.1 Background

Bone serves multiple functions in the body. It provides a strong and rigid frame to enable movement and to support and protect the soft tissues, and it provides a readily accessible store of calcium and phosphorus. To fulfill these complex and at times conflicting roles, bone is a dynamic structure undergoing continual remodeling in response to mechanical stress, damage, and physiological requirements for calcium and phosphorus. Metabolic bone diseases, through their impact on remodeling processes, can impair skeletal function, resulting in bone fractures and disturbances in calcium and phosphorus homeostasis. Significant advances have been made in developing therapies that address underlying disease processes affecting bone, but considerable unmet medical need remains due to the enormous size of affected populations and the limited efficacy and patient acceptance of current treatments.

6.21.2 Introduction

6.21.2.1 Bone Structure

Bone is a complex structure consisting of diverse cell types, extracellular matrix, and mineral. It is a metabolically active tissue in which the coupled processes of formation and resorption are continuous throughout life. Bone matrix is composed predominantly of extensively cross-linked collagen type I but also incorporates many other proteins, including osteocalcin, osteopontin, proteoglycans, and growth factors. The mineral phase consists of poorly crystalline carbonate apatite, containing calcium phosphate in the form of apatite and with extensive carbonate substitution of hydroxyl groups.[1] This mineral component provides mechanical strength and rigidity to the skeleton and is sufficiently labile to be easily removed, enabling it to act as an ion store for supporting calcium homeostasis.

6.21.2.2 Bone Cells

Bone cells (**Figure 1**) consist of two main groups or families of cells: osteoblasts, responsible for bone formation, and osteoclasts, responsible for bone resorption. The two cell types are derived from different progenitor cells. Osteoclasts

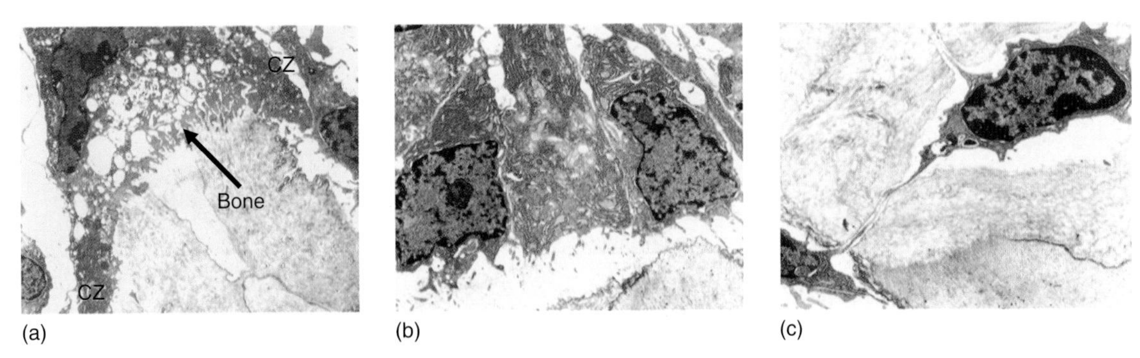

Figure 1 (a) Bone cell osteoclasts are bone-resorbing cells. They have the distinctive features of a clear zone (CZ), which seals an extracellular space between the cell and the bone, and the ruffled border (arrow) which enables the secretion of acid and proteolytic enzymes out of the cell and phagocytosis of bone fragments. (b) Osteoblasts are large cuboidal cells rich in endoplasmic reticulum. They lay down the collagenous matrix of bone and regulate its mineralization. (c) Osteocytes are small quiescent cells embedded in bone. They are derived from osteoblasts that become embedded in bone matrix. They form a dense network of canaliculi containing cell processes that interconnect with adjacent osteocytes and bone-lining cells.

are derived from the hemopoietic stem cell system,[2] whereas osteoblasts are derived from mesenchymal precursors, which progress through periosteal and marrow stromal stem cell stages.[3] Intercellular communication, particularly between osteoblasts and osteoclasts, is thought to be important in determining net effects of bone formation and resorption.

6.21.2.2.1 Osteoclasts

Osteoclasts resorb both the mineral and the organic phases of bone. They are giant, multinucleated cells containing from two to several hundred nuclei. Transmission electron micrographs show osteoclasts to have several distinctive features. One of the distinguishing features is the ruffled border, which is located at the interface between the resorbing bone surface and the cell surface within resorption lacunae. Secondly, the cell surface next to bone is characterized by close apposition of the plasma membrane to bone and an adjacent, organelle-free area, rich in actin filaments, called the clear zone.[4] The clear zone encircles the ruffled border completely so that the site of resorption is isolated and localized. Osteoclasts are richly endowed with lysosomal enzymes with one such enzyme, tartrate-resistant acid phosphatase (TRAP), being relatively specific to the osteoclast. Osteoclasts also contain carbonic anhydrase, which is utilized for the production of H^+ ions secreted at the ruffled border.

6.21.2.2.2 Osteoblasts

Osteoblasts are highly differentiated cells, responsible for laying down the organic matrix of new bone. They are able to synthesize type I collagen and various noncollagenous extracellular matrix proteins; they are rich in alkaline phosphatase; they possess receptors for, and respond to, many hormones and local factors. As the apposition of matrix and its mineralization progresses, the cells become thin lining cells on the bone surface. Bone-lining cells have lost the capacity to synthesize matrix. They are flattened cells with slender nuclei and contain few organelles. The function of lining cells remains uncertain, although they probably play a key role in the localization and initiation of remodeling.[5]

6.21.2.2.3 Osteocytes

Osteocytes represent the most mature stage of the osteoblastic line, as they are derived from osteoblasts that have become embedded in bone; they communicate with each other and with surface osteoblasts or lining cells by forming extensive networks of cell processes, which lie in canaliculi in the matrix.[6] This network of osteocytes provides the cellular structure that allows the detection of bone microdamage and enables bone as an organ to determine local needs for bone augmentation or reduction in response to mechanical demands.

6.21.2.3 Bone Remodeling

Bone remodeling (**Figure 2**) continues throughout adult life and is necessary for the maintenance of normal bone structure and bone quality. Maintenance of bone mass requires that bone formation and resorption should be balanced. Bone remodeling occurs in focal or discrete packets by groups of cells known as bone multicellular units (BMUs).[7] In this process, both bone formation and resorption occur at the same place so that there is no change in the shape of the

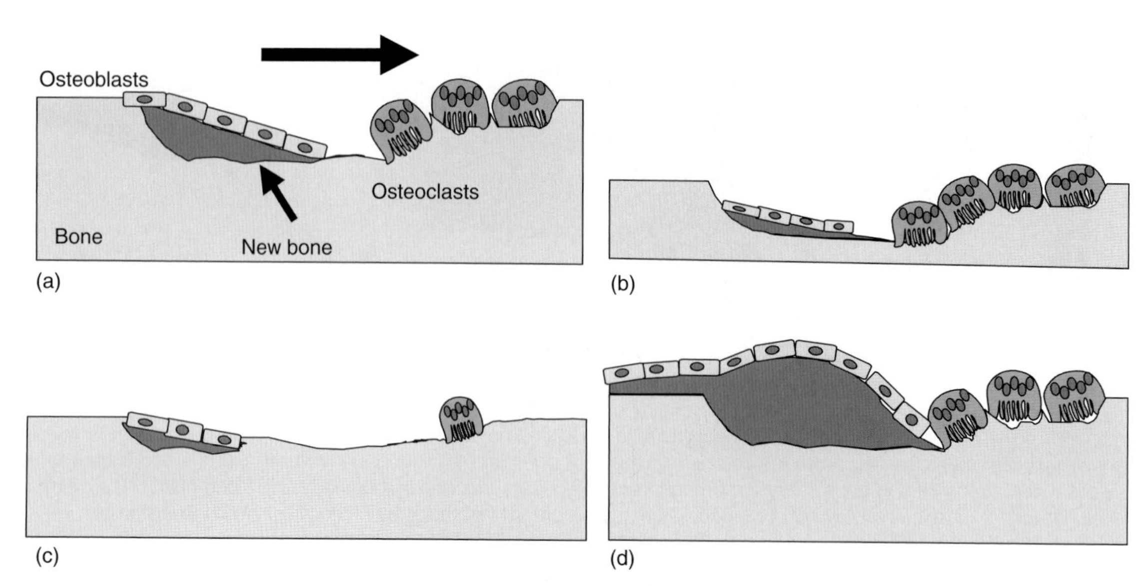

Figure 2 Cancellous bone remodeling. (a) Normal bone is remodeled by teams of osteoclasts and osteoblasts, forming a bone multicellular unit that progresses across the bone surface with osteoclasts eroding a pit, which is subsequently refilled by osteoblasts. (b) In osteoporosis, bone removed by each cycle is greater than that replaced, so that the trabeculae become thinner and can be perforated and lost. (c) Antiresorptive treatment inhibits osteoclasts so that numbers and size of pits are reduced, resulting in a coupled reduction in bone formation. Bone volume is increased as existing holes are filled and then maintained through the reduction in remodeling cycles. (d) Anabolic factors like teriparatide induce osteoblasts to form more bone during each remodeling cycle. Bone volume increases until osteoclast activity increases, apparently to compensate for the increased bone formation. Teriparatide action depends on the presence of remodeling cycles, as concurrent treatment with an antiresorptive reduces its effects.

bone. After a certain amount of bone is removed as a result of osteoclastic resorption and the osteoclasts have moved away from the site, a reversal phase takes place, in which a cement line is laid down. Osteoblasts then synthesize matrix, which becomes mineralized. The BMU remodeling sequence normally takes about 3 months to produce a bone structural unit (BSU), the unit of bone tissue formed by one BMU. The constant remodeling is important in three ways. Firstly, bone, like other structural materials that undergo repetitive cyclical loadings, is subject to fatigue. After a number of loading cycles, tiny cracks may form, which, if not repaired in a timely fashion, will accumulate and eventually lead to structural failures. Bone remodeling will replace bone containing cracks and prevent structural failure. Secondly, bone remodeling is required to adapt bone material properties to the mechanical demands that are placed on the skeleton. Finally, bone remodeling plays a critical role in regulating calcium homeostasis, which is critical for life. Bone is a major reservoir for calcium. When serum calcium becomes low and there is insufficient dietary calcium, calcium will be released from bone by osteoclasts to meet the demand.

6.21.2.3.1 **Hormonal regulation of bone remodeling**

Bone is a metabolically remarkably active tissue that throughout life undergoes constant remodeling by osteoclasts and osteoblasts, as described above. These cellular events are regulated by systemic and local modulators, namely parathyroid hormone (PTH), 1,25-dihydroxyvitamin D, sex and other steroid hormones, calcitonin, prostaglandins, growth factors, and cytokines.

Bone is the most important reservoir for body calcium. Ionized plasma calcium, which represents ~50% of the total plasma calcium pool, is essential for countless metabolic functions and is therefore tightly controlled through three major circulating hormones: PTH, 1,25-dihydroxyvitamin D (calcitriol), and, to a much lesser extent, calcitonin. To keep serum ionized calcium levels stable, these regulators act on three major tissues: (1) bone; (2) the intestine; and (3) the kidney.[8]

6.21.2.3.1.1 PTH, vitamin D, and calcitonin

The secretion of PTH by parathyroid glands is directly stimulated through a decrease in plasma levels of ionized calcium. Within minutes, PTH increases plasma calcium by activating osteoclasts and thus mobilizing calcium from the mineral store. It also increases renal tubular reabsorption of calcium (and the excretion of phosphate). Through the activation of 1α-hydroxylase and thereby the generation of 1,25-dihydroxyvitamin D_3, PTH also helps to increase intestinal calcium absorption. In both women and men, serum levels of PTH increase with advancing age, possibly due to an age-related decrease in calcium absorption, a decline in renal 1α-hydroxylase (CYP27B) activity, lower serum calcitriol levels, and age-associated vitamin D deficiency.[8]

1,25-Dihydroxyvitamin D (calcitriol) (**Figure 3**) is a potent secosteroid hormone that controls calcium and phosphate homeostasis by stimulating intestinal calcium absorption and bone resorption. In addition, calcitriol has pronounced effects on cell proliferation and differentiation (e.g., bone cells, bone marrow stem cells, skin basal cells, pancreatic islet cells) and on immunoregulatory functions.

The synthesis of calcitriol is usually induced through low serum calcium or inorganic phosphate levels. However, changes in active vitamin D levels occur slower than those in PTH as the 'storage form' of vitamin D, 25-hydroxyvitamin D (calcidiol), needs to be converted into the biologically active form of the hormone, i.e., 1, 25-dihydroxyvitamin D. This conversion/activation is achieved through the hydroxylation of 25-hydroxyvitamin D in its C1 position by CYP27B (α-hydroxylase). The activity of the latter enzyme, which is predominantly found in kidney parenchymal cells, is upregulated by PTH, and downregulated by plasma HPO_4 levels. Conversely, calcitriol inhibits the secretion of PTH from the parathyroid glands.[9] Measurement of serum 25 and 1,25 vitamin D levels may be useful in patients with osteoporosis, vitamin D deficiency, renal disease, hypercalcemia, and sarcoidosis.

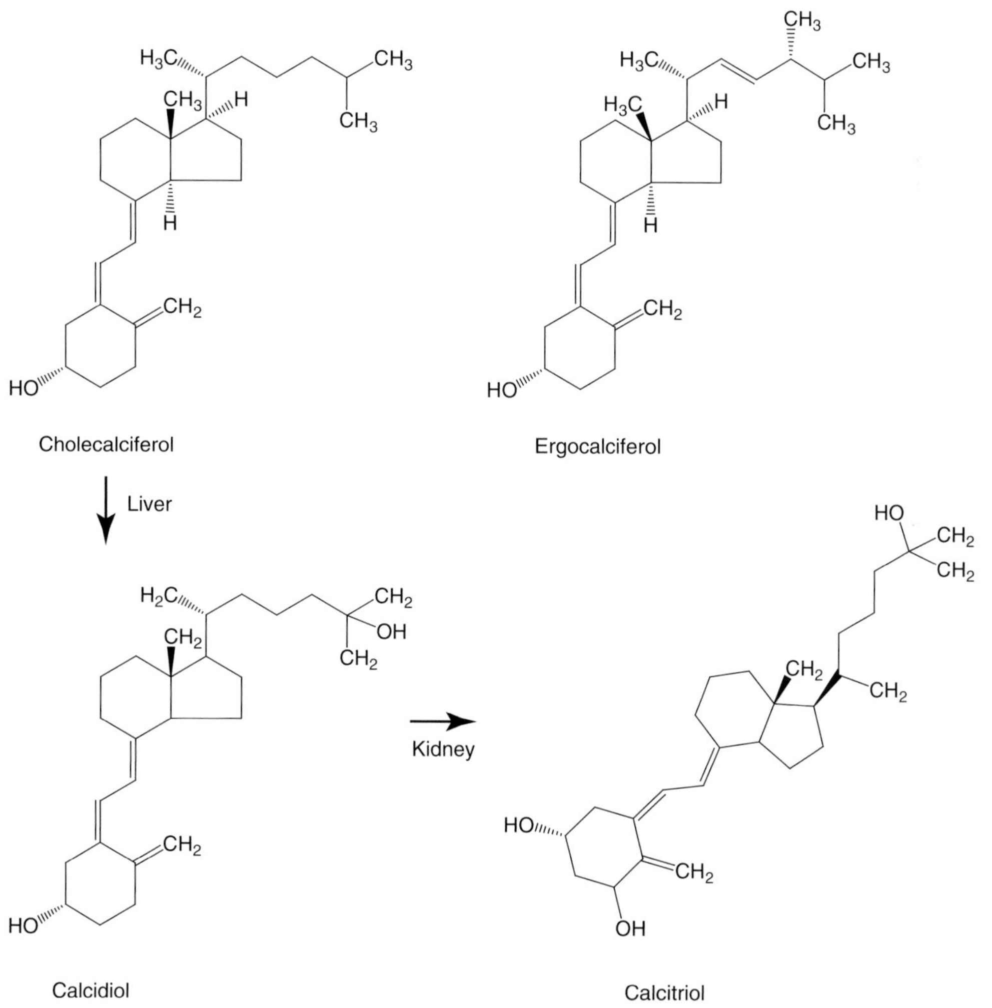

Figure 3 Chemical structure of vitamin D and its metabolites. Cholecalcifol (vitamin D_3) and ergocalciferol (vitamin D_2) are obtained from sunlight exposure and dietary sources, respectively. These molecules are hydroxylated at carbon 25 and carbon 1 in the liver and kidney, respectively, to produce the active metabolite/hormone as shown for calcidiol (25-hydroxy-vitamin D_3) and calcitriol (1,25-dihydroxyvitamin D_3).

Vitamin D deficiency is common amongst the elderly, mainly due to decreased cutaneous vitamin D production, and an inverse relationship between serum PTH and vitamin D levels has been observed. In addition, secondary hyperparathyroidism may also be induced by renal impairment. Serum PTH levels tend to increase with age and even in people with normal or near normal glomerular filtration rate (GFR). However, the contribution of renal insufficiency to the age-related rise in serum PTH becomes increasingly important as the GFR falls. Interestingly, serum PTH has been shown to be a predictor of falls and mortality in elderly men and women independent of vitamin D status, bone mass, and measures of general health.[10]

Calcitonin is a product of the thyroidal C cells. Its secretion is stimulated by high serum calcium levels. The peptide has been shown to inhibit osteoclast activity rapidly and thus decreases serum calcium levels. While calcitonin is an important clinical marker of medullary thyroid cancer, its relevance in human calcium homeostasis and bone metabolism has not been well established.[11]

Glucocorticoids seem to have different effects on bone, depending on the dose and duration of exposure. Excessive glucocorticoid levels result in rapid and profound reductions in bone mineral density (BMD) with greatly increased fracture risk. However, the precise mechanisms of how glucocorticoids affect bone are still obscure. In vivo, markers of bone formation are clearly suppressed during glucocorticoid treatment, reflecting the profound inhibitory effects of glucocorticoids on osteoblast activity, and markers of bone resorption are often increased. Histologically, mean wall thickness is decreased, reflecting the reduced amount of bone replaced in each remodeling unit. Thus, while glucocorticoids appear to exert their effects on the skeleton through a number of pathways (e.g., suppression of sex steroids, muscle wasting, changes in renal/intestinal calcium handling), it appears that the most important catabolic skeletal actions of glucocorticoids directly target the osteoblast.[12]

In vitro, osteoblasts and their precursors are highly responsive to glucocorticoids. Here, the predominant effect is to promote osteoprogenitor proliferation, lineage commitment, and osteoblast differentiation, resulting in the formation of bone nodules of increased size and number. Glucocorticoids stimulate osteoblast differentiation and osteogenesis in developmental animal models. However, glucocorticoids have also been shown to inhibit type I collagen expression in rat calvarial organ cultures and primary osteoblast cell cultures, and to decrease preosteoblastic replication. Also, glucocorticoids can promote apoptosis of osteoblasts and osteocytes. The inhibitory effects of glucocorticoids on bone formation may be in part due to downregulation of insulin-like growth factor 1 (IGF-1) expression in osteoblasts. Moreover, glucocorticoids modulate factors, including receptor activator of nuclear factor kappa B (NFκB) ligand (RANKL) and osteoprotegerin (OPG), and transcription factors, NFκB and activator protein-1 (AP-1), that mediate signaling in bone, and osteoclast function and lifespan – effects that may explain the action of glucocorticoids on osteoclasts.[13]

6.21.2.3.1.2 Sex hormones

Sex hormones are major regulators of bone turnover and remodeling in both genders. Estrogens reduce bone loss by inhibiting the generation of new osteoclasts, reducing the activation frequency of the BMU and promoting apoptosis of mature osteoclasts via mechanisms that are not well understood. Some of the effects of estrogen seem to be mediated via the modulation of growth factors and cytokines, while others are associated with binding to at least two different estrogen receptors (ERa, ERb). A reduction in circulating endogenous estrogen levels, as occurs during and after menopause, has been shown to prolong osteoclast survival and stimulate the recruitment and hence generation of osteoclasts. The result is an increase in the activation frequency of the BMU, reflected in a high bone turnover state.[14]

While there is no doubt that androgens (i.e., testosterone, dihydrotestosterone) play a dominant role in male bone health, it also appears that circulating estradiol levels are important for bone development in younger men, and for bone turnover in elderly men. However, it remains contentious just how important blood estradiol relative to blood testosterone levels are for the maintenance of bone in (older) men. Recent findings suggest that aromatization of testosterone to estradiol plays a significant role in the regulation of bone metabolism, and, ultimately, influences age-related bone loss in elderly males.[15]

6.21.2.4 Local Regulatory Factors

6.21.2.4.1 Regulation of osteoclasts

Osteoclasts are the cells that resorb both the mineral and matrix components of bone (*see* Section 6.21.2.2.1). Osteoclasts are derived from hematopoietic cells of the monocyte–macrophage lineage but their recruitment, differentiation, activity, and survival are regulated locally by mesenchymal cells of the osteoblast lineage during normal bone remodeling. The mesenchymal cells provide colony stimulating factor-1 (CSF-1), which is a permissive factor

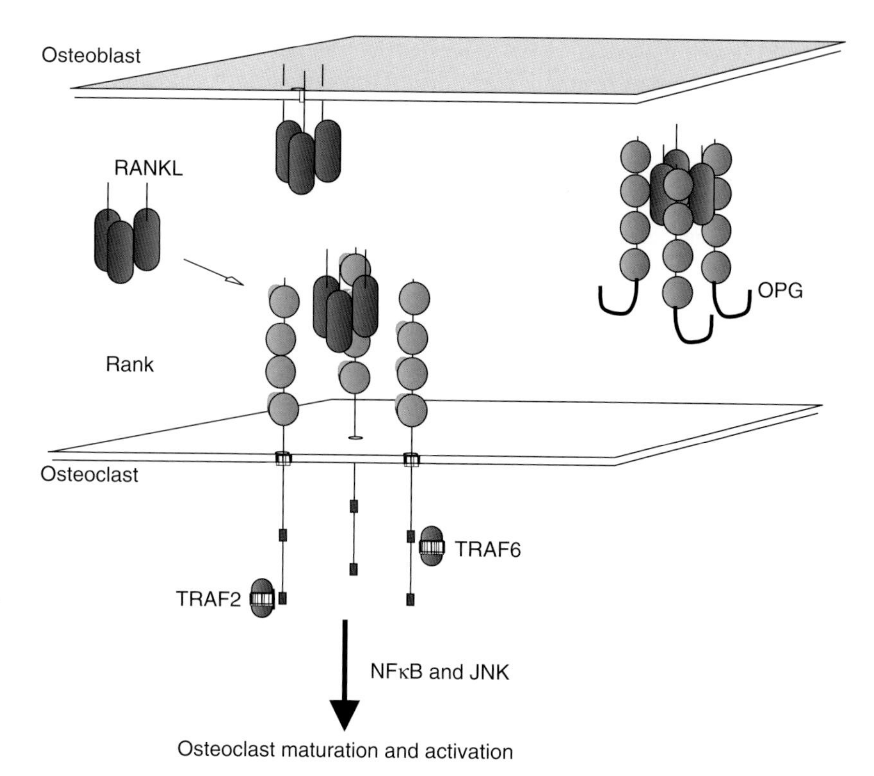

Figure 4 Osteoclast regulation. Osteoblast lineage cells integrate proresorptive hormonal and paracrine signals and induce bone resorption by producing the TNF family cytokine RANKL, as either a membrane-bound or soluble ligand for its receptor RANK on osteoclasts or their precursors. Ligand binding induces osteoclast differentiation and activation. Conversely, osteoblast lineage cells can secrete the decoy receptor OPG that binds to RANKL and prevents its interaction with RANK, thus inhibiting bone resorption.

essential for the survival in bone of osteoclast precursors. The importance of colony stimulating factor-1 is shown by the osteopetrotic phenotype of mice lacking this cytokine.[16]

In addition, osteoblast lineage cells regulate bone resorption by activating another cytokine pathway. This pathway consists of three players: (1) the receptor activator of NFκB (RANK), which is expressed on osteoclasts and their precursors; (2) its ligand RANKL, which is a member of the tumor necrosis factor (TNF) superfamily and expressed by osteoblast precursors; and (3) the decoy receptor for RANKL, osteoprotegerin (OPG), which is also expressed by osteoblast lineage cells (**Figure 4**). Osteoblast lineage cells integrate multiple systemic and local proresorptive signals and translate net impetus for bone resorption into changes in relative RANKL/OPG expression. The number and activity of osteoclasts are determined by the relative local levels of RANKL and OPG, which promote and inhibit osteoclast differentiation and activity, respectively. Other cells, such as activated T cells, are also able to express RANKL and may contribute to pathologic bone resorption during inflammation.[17]

6.21.2.4.2 Other local factors regulating bone resorption

In addition to RANKL and OPG, which are the primary regulators of bone resorption, other factors, such as inflammatory cytokines and prostaglandins, act locally to modulate osteoclast function to increase or decrease bone resorption. TNF-α and interleukin-1 beta (IL1β) are two important inflammatory cytokines that have profound proresorptive action through both RANKL-mediated and direct effects on osteoclasts. Both of these cytokines increase RANKL expression by osteoblasts. In addition, TNF-α activates NFκB and other pathways in osteoclasts that are common to RANKL signaling, and augments RANKL proresorptive effects. IL1β increases osteoclast activity and survival directly and thus also directly increases bone resorption during inflammatory diseases.[18]

6.21.2.4.3 Regulation of osteoblasts

The local regulation of osteoblast differentiation and function is not well understood. There is close coupling of bone resorption and bone formation both temporally and spatially but the signals that link these two processes are not identified. Many growth factors and cytokines are expressed in bone and have profound effects on osteoblast differentiation and function in vitro and in vivo; however, their relative importance and sequence of action in normal bone physiology are also unknown. Some factors are potent proliferative reagents for osteoblast precursors while others appear more important for lineage commitment and differentiation. It appears to be a regulatory system with high levels of redundancy and overlapping function. Many of these factors are made by osteoblasts and are present in significant amounts sequestered in bone matrix and so have the potential to be released by osteoclasts during bone resorption. However, while gaining an understanding of physiology is very important, our knowledge is sufficient to identify a large number of specific factors with proven and unproven therapeutic potential.[19]

6.21.2.4.4 Proliferative growth factors

Among the strongly proliferative factors are fibroblast growth factor 1 and 2 (FGF-1 and -2), transforming growth factor β (TGFβ) and platelet-derived growth factor (PDGF). In vitro, these factors potently promote osteoblast precursor proliferation but inhibit osteoblast differentiation. In vivo, they promote new bone formation but bone formed is disorganized woven bone and mineralization can be inhibited. PDGF in particular may have an important role in fracture repair where rapid formation of a fracture callus is required.[20] The systemic use of these factors is limited by their pleiotropic proliferative effects on other tissues.

IGF-1 and -2 are weaker mitogens for osteoblast precursors and also promote initial differentiation. Their physiologic role is in part to mediate the anabolic actions of growth hormone. They have modest anabolic effects on bone when given systemically, though dosing tends to be limited by agonist actions on the insulin receptor at higher circulating concentrations.

In contrast, bone morphogenetic proteins (BMPs) have weak proliferative effects in vitro but are potent differentiation agents for osteoblast precursors and can induce the formation of ectopic bone in vivo. BMP2 and BMP7 are used clinically to promote fracture repair and closure of bone defects. BMPs act physiologically as autocrine and paracrine factors that promote osteoblast lineage commitment and regulate osteoblast differentiation.[19]

Wnt proteins signal through β-catenin and are important in bone metabolism, as absence of *Wnt* signaling induces osteopenia and constitutive activation of the *Wnt* pathway produces a high-bone-mass phenotype in human patients. *Wnts* appear to have an important role in osteoblast lineage commitment and in providing autocrine signals vital for full osteoblast differentiation. The human high-bone-mass phenotype in particular has provided strong validation for the concept that a *Wnt* pathway agonist could have bone anabolic actions.[21]

6.21.3 Diseases of Bone – Epidemiology and Diagnosis

6.21.3.1 Metabolic Bone Diseases

6.21.3.1.1 Osteoporosis

Osteoporosis is defined as a chronic skeletal disorder characterized by compromised bone strength. The latter is caused by, or associated with, low bone mass and changes in bone size and geometry, bone turnover, and microarchitecture. Compromised bone strength leads to enhanced bone fragility, which predisposes to an increased risk of fracture. 'Minimal-trauma fractures' are the hallmark and major complication of osteoporosis, causing substantial morbidity, excess mortality, and high cost.

Amongst the metabolic bone diseases, osteoporosis is by far the most frequent one. The World Health Organization defines osteoporosis as a BMD of 2.5 standard deviations (SD) or more below the mean for young healthy individuals. According to this definition, approximately 30% of all postmenopausal women and 20% of all men older than 60 years of age have osteoporosis. The incidence of osteoporosis and of osteoporotic fractures increases with age: while only 4–5% of all women 60–70 years of age are found to have low BMD, this proportion rises to 50% in women aged 80 years or older. Similarly, according to the Dubbo Osteoporosis Epidemiological Study (DOES), the incidence of osteoporotic fractures is approximately 2000 per 100 000 person-years in the age group 60–70 years, but rises to almost 8000 per 100 000 person-years in women aged 80 years or more. One-third of women and one-sixth of men at age 90 years are estimated to have suffered a hip fracture.[22]

In the US, more than 1.3 million osteoporotic fractures occur each year. In 2001, the annual incidence of osteoporotic fractures in Australia was 64 500, which means that, every 8 min, one patient was being hospitalized for an osteoporotic fracture, resulting in a prevalence of ~1.9 million (present population: 20 million). Fifty percent of these fractures occur in the spine, while 25% are wrist and another 25% are hip fractures.[23] Clearly, the latter are by far the most significant complications of the disease, with a 12-month mortality of 25–30%, and lasting morbidity requiring long-term care for another 50% of patients who have suffered a hip fracture. Fewer than one-third of all hip fracture patients will return to their home, and even these individuals have been shown to experience lasting disabilities and a loss in quality of life.[24]

Apart from fractures, low BMD is one of the major features of osteoporosis. There is a relatively stringent relationship between BMD and fracture risk, where statistically 1 SD decrease in BMD is associated with a twofold increase in fracture risk. Absolute fracture risk, however, is determined by a host of other factors, including age, gender, nutritional status, body composition, fracture status, and others. Not surprisingly, the majority of women who experience a minimal-trauma fracture have normal (*T*-score > -1.0 SD) or osteopenosteopenic (*T*-score between -1.0 and -2.5 SD) peripheral bone density, and only 7% have osteoporosis as defined by bone densitometry measurement.[25]

The pathogenesis of osteoporosis is complex, even in those cases where a single cause is considered to be responsible for the bone damage (secondary osteoporosis). However, an imbalance in bone turnover and, therefore, in remodeling is currently considered the central process that ultimately may lead to a reduction in bone strength. The classical example is postmenopausal osteoporosis, where both bone resorption and bone formation are accelerated (high-turnover). A relative predominance of bone resorption over bone formation, maintained over many years, results in ongoing net loss not only of bone mass, but, more importantly, of mechanically viable bone (micro)structure. This will eventually result in mechanical failure of the bone.

The change in bone turnover, as seen in postmenopausal women, is most likely due to the rapid decline in endogenous estrogen levels around and shortly after menopause. Premenopausal estrogen levels inhibit osteoclast activity, reduce the activation frequency of the BMU, and promote apoptosis of mature osteoclasts via a mechanism that is not well understood. Low levels of circulating estrogen prolong osteoclast survival and stimulate the recruitment and differentiation/activation of osteoclasts. Some of the actions of estrogens are mediated through growth factors and cytokines (*see* Section 6.21.2.3.1.2).[14]

The situation is somewhat different in glucocorticoid-induced osteoporosis. Chronic treatment with glucocorticoids results in rapid and profound bone loss, with most of the change occurring during the first 6–12 months of treatment. Fracture risk is greatly increased amongst long-term hypercortisolemic patients. While the precise mechanisms of how glucocorticoids affect bone are still obscure, it is obvious that treatment with exogenous or excess of endogenous glucocorticoids results in a profound suppression of osteoblasts (bone formation) and a transient increase in bone resorption (osteoclasts). These catabolic effects of glucocorticoids on bone are in direct contrast to their known anabolic actions, where glucocorticoids promote the differentiation, maturation, and function of osteoblasts. Thus, while glucocorticoids appear to exert their effects on the skeleton through a number of pathways (e.g., suppression of sex steroids, muscle wasting, changes in renal/intestinal calcium handling), it appears that the most important catabolic skeletal actions directly target the osteoblast.[12]

Osteoporosis is a multifactorial, complex disease with a strong genetic determinant[26] (*see* Section 6.21.3.1.1). Identifying the patient at high risk for osteoporosis (and preventing the disease) is one of the major clinical aims. Existing (prevalent) osteoporotic fractures are amongst the strongest risk factors for future fractures. Compared to patients without prevalent fractures, subjects with one osteoporotic fracture have a three- to fivefold increased risk for further fractures. This relative risk increases to 10 in patients with three or more prevalent fractures.[27]

Other risk factors for osteoporosis include: advanced age, female gender, ethnic background, a family history of fracture, a low body mass index, early menopause/hypogonadism, low calcium intake, vitamin D deficiency, relative immobilization/sedentary lifestyle, previous hyperthyroidism, low BMD, high bone turnover, smoking, systemic inflammatory disease (e.g., rheumatoid arthritis, inflammatory bowel disease), celiac disease, prior or current therapy with glucocorticoids, anticonvulsant or neuroleptic drugs.[28]

Importantly, some of the above risk factors also affect falls risk. Particularly in the elderly, falls constitute a specific problem in that they predispose to fracture. Most peripheral fractures and, in particular, hip and forearm fractures, result from falls, even though the minority of falls actually results in a broken bone. Again, the association between falls and fracture is complex, but important factors include fall mechanics, bone geometry, neuromuscular coordination and body height, as well as fat mass.

6.21.3.1.2 Primary hyperparathyroidism

Primary hyperparathyroidism (pHPT) is a frequent disorder characterized by hypercalcemia and inappropriately elevated serum levels of intact PTH. In 80% of cases with pHPT, a solitary parathyroid adenoma is found. However, in approximately 15% of patients presenting with typical pHPT, hyperplasia of all four parathyroid glands is found and these patients often exhibit signs and symptoms of hereditary syndromes, such as multiple endocrine neoplasia (types 1 and 2a).

Before the introduction of automated multichannel chemistry autoanalyzers, the diagnosis of pHPT was difficult to make and patients often progressed to advanced disease stages, which regularly comprised symptoms such as nephrocalcinosis, fractures, osteitis fibrosa cystica, and muscle weakness or neuromuscular dysfunction ('stones, bones, and groans'). Today, patients with pHPT are often asymptomatic or very mildly symptomatic, and the diagnosis often follows the incidental finding of hypercalcemia during routine biochemistry checks. Subsequently, the correct diagnosis of pHPT is made by demonstrating inappropriately elevated serum PTH levels. The term 'inappropriately elevated' is clinically relevant, as up to 20% of patients with pHPT may present with normal or high normal serum PTH concentration in the presence of abnormally high serum calcium values. As hypercalcemia physiologically results in suppressed serum PTH levels, any patient with nonsuppressed serum PTH levels and hypercalcemia should be considered as having pHPT until proven otherwise. The only differential diagnosis in this situation is that of familial (benign) hypocalciuric hypercalcemia, although patients on long-term thiazides or lithium may present with similar constellations.

Due to the phosphaturic action of PTH, patients with pHPT often exhibit low normal or frankly reduced serum phosphorus concentrations. The urinary excretion of calcium (24-h urine) is high in about 40–50% of cases; low values are often observed in mild cases, as the primary renal effect of PTH is to increase calcium reabsorption. Serum 25-hydroxy vitamin D levels are usually low or low normal, while those of 1,25-dihydroxyvitamin D are often high normal or frankly elevated due to the stimulatory action of PTH on 1α-hydroxylase activity. Patients presenting with vitamin D deficiency and pHPT are considered to suffer from more severe skeletal disease, such as osteitis fibrosa cystica.

While most patients are oligo- or asymptomatic, a history of kidney stones is not uncommon and, together with osteopenia, nephrolithiasis is one of the more frequent manifestations of symptomatic pHPT. In long-standing pHPT, bone mass and density are typically reduced at predominantly cortical sites (e.g., distal radius), while at sites of cancellous bone (e.g., lumbar spine), bone density may be normal or only slightly reduced. These observations are attributed to the fact that pHPT has different effects on cortical and cancellous bone. However, approximately 15% of patients with pHPT present with osteopenia at the lumbar spine, and some patients may have osteoporosis. Of note, severe pHPT always results in global bone loss, and in these patients, fractures are not uncommon. Careful analysis of the patient with pHPT will reveal other symptoms, such as nonspecific neurological symptoms, fatigue, or depression.[29]

6.21.3.1.3 Paget's disease of bone

Paget's disease of bone is a chronic disease of unknown etiology, characterized by accelerated bone remodeling, abnormal bone structure, and hypertrophy of the affected bones. A number of factors, such as geographic distribution, secular changes, and familial occurrence, suggest that both genetic susceptibility and additional environmental factors play a role in the pathogenesis of the disease, Susceptibility loci for Paget's disease have been identified on chromosomes 18q21 (*PDB2*), 5q35 (*PDB3*), 5q31 (*PDB4*), and 2p361 (*PDB5*), and this list is still expanding. Furthermore, gene mutations involving sequestosome, a component of the ubiquitin signaling pathways, have been described in a common familial form of Paget's disease. The viral etiology of Paget's disease is still controversial, but there is good evidence to support the concept of viral protein cellular modulators.

The prevalence of the disease in the adult population (>40 years) is around 1–3%, increases with age, and varies by geographic location (common in the UK, Australia, New Zealand, and South Africa, but rare in Asia, the Middle East, Africa, and, surprisingly, in Scandinavia). Recent investigations indicate that the incidence of Paget's disease of bone is in decline, particularly in countries with traditionally high prevalence.

The disease may occur in any part of the skeleton and may affect only a single (monostotic) or several bones (polyostotic). Clinically, patients present with skeletal pain, nerve compression syndromes, deformities (often resulting in severe osteoarthritis), and fractures. Rarely, neoplastic transformation into osteosarcoma-like tumors may occur.

The diagnosis of Paget's disease is based upon the clinical presentation, increased skeletal uptake on whole-body bone scans, and typical radiographic changes in the corresponding bones. The latter include bulging of affected bones, cortical thickening, and a mix of osteolytic and sclerotic bone changes. Distinctive abnormalities in bone turnover (mainly serum total and/or bone specific alkaline phosphatase) are used to define the activity of the disease.[30]

Today, almost all patients with Paget's disease of bone are treated with oral or intravenous bisphosphonates. The treatment aim is to mitigate pain and to prevent the typical complications of the disease. Whether this is best achieved by reducing bone turnover, or, as suggested lately, by completely suppressing bone remodeling for extended periods of time remains unclear. The treatment response is usually assessed by changes in clinical presentation, radionucleotide/radiographic imaging, and biochemistry, although biochemical markers of bone turnover have proven to be of most use in the follow-up of these patients.[31]

6.21.3.2 Cancer-Induced Bone Disease

6.21.3.2.1 Bone tumors

Skeletal tumors are histologically classified according to the cell or tissue type from which they seemingly originate, such as from bone (osteosarcomas) and from cartilage (chondrosarcomas). They are broadly classified as benign or malignant but can exhibit characteristics of both (transitional), and their aggressive and metastatic behaviors also range widely. In addition to a physical examination and complete history, determining tumor type and staging of the tumor are assisted by imaging techniques such as radiography, computed tomography, and magnetic resonance imaging, and by histological analysis of biopsied material (reviewed by Whyte[32]).

6.21.3.2.1.1 Giant cell tumor of bone

Giant cell tumors (GCTs) of bone, also known as osteoclastomas, are benign locally lytic bone lesions that occur most commonly in men aged 20–40 years of age. The tumor mass is comprised of large multinucleated osteoclast-like cells and mononuclear macrophages within a mononuclear cell population of stromal cell appearance. The latter cells are the proliferative component of these tumors. By the time of presentation, GCTs have often produced chronic pain and pathological fracture. GCTs are mainly treated either by curettage or by excision followed by reconstructive surgery.[33]

6.21.3.2.1.2 Osteogenic sarcoma

Primary malignant bone tumors vary considerably in their clinical, pathological, and histological presentation. The most common primary bone tumor, typically occurring in male patients younger than 30 years of age, is osteosarcoma, or osteogenic sarcoma. Most are classic osteosarcomas, which present in the metaphysis of long bones, usually in the distal femur or proximal tibia, although they can form in any bone. They are aggressive and exhibit a wide phenotypic spectrum, producing osteolytic, mixed, or osteoblastic lesions. They are characterized by their rapid growth and metastasis to other areas of the bone or other organs, often causing pathological fracture and intractable pain. Treatment usually involves chemotherapy followed by limb-salvage surgery.[34]

6.21.3.2.1.3 Skeletal metastasis

The vast majority of bone tumors are metastasized solid tumors. The skeleton is a common site for the metastasis of solid tumors, including cancers of the prostate, breast, lung, kidney, and thyroid: other cancers, including bladder cancer, melanoma, and neuroblastoma, can also metastasize to the bone, but this is comparatively rare. Most metastases to the skeleton are lytic in nature, such as the majority of breast cancer bone metastases; however, some are mixed or osteosclerotic, as in the majority of prostate cancer bone metastases. Both the lytic and sclerotic effects of these cancers result from the dysregulation of normal bone remodeling, with relative excess in osteoclastic bone resorption mediating osteolysis and a relative excess of osteoblastic bone formation mediating osteosclerosis. However, even osteosclerotic metastases typically show increased bone resorption.[35]

Cancer cells induce bone resorption and formation by expressing bone active factors that act locally or systemically. Parathryroid hormone related protein (PTHrP) is a primary factor inducing bone resorption. The factors increasing bone formation are not clearly identified but candidates are endothelin-1, prostate-specific antigen, BMP6, TGFβ and IGF-1 and -2. It is likely that the increased bone resorption supports further growth of the tumor cells through the release of active proliferative factors from the bone matrix, thus establishing a vicious cycle, accelerating both bone destruction and tumor growth.[36]

Cancer bone metastases, regardless of phenotype, can cause insurmountable pain, pathological fractures, vertebral compressions, and hypercalcemia or hypocalcemia. Detectable bone metastasis, particularly in the case of prostate cancer bone metastasis, is correlated with significantly decreased survival. Treatment options are mainly palliative, although therapeutic strategies combining an antiresorptive approach, chemotherapy, and radiotherapy are showing promise.

6.21.3.2.1.4 Multiple myeloma

Multiple myeloma is a disease in which B cells are transformed and proliferate to fill bone marrow spaces. The majority of multiple myeloma patients show predominantly lytic lesions, though sclerotic or mixed lesions are occasionally seen. Multiple myeloma cells are able to induce osteoclastic bone resorption through cell–cell interactions and through macrophage inflammatory protein-1 alpha (MIP1α) secretion, and also can inhibit bone formation through secretion of dickkopf 1. These dual actions lead to both local and systemic bone loss. Multiple myeloma is associated with a high risk of skeletal complications, including fracture, bone pain, and hypercalcemia. Treatment of patients with bisphosphonates to block bone resorption has been found to be of benefit in reducing these complications.[37]

6.21.3.3 Inflammation

6.21.3.3.1 Rheumatic disease

This category encompasses a wide range of diseases that affect joint structures, including osteoarthritis, rheumatoid arthritis, systemic lupus erythematosus, and spondyloarthropathies. Osteoarthritis is typically limited to the articular structures, whereas the other disorders can also manifest systemic disease. Rheumatoid arthritis and other inflammatory diseases of joints are associated with local bone erosions, periarticular osteoporosis and systemic bone loss. These patterns of bone loss can contribute significantly to the progressive joint destruction that leads to permanent impairment of joint function and the loss of motility and independence associated with severe arthritis. The effects of inflammation on bone remodeling have been most studied in rheumatoid arthritis. Infiltration of the bone by proliferating synovial lining cells and inflammatory cells stimulates osteoclast formation, which stimulates osteolysis and produces radiographically apparent focal bone erosions. Osteopenic effects, both adjacent to the inflamed joint and systemic, are mainly mediated by stimulation of bone resorption brought about by increased numbers of osteoclasts.[38]

6.21.3.3.2 Periodontal disease

Periodontitis is diagnosed by erosion of the alveolar bone and a loss of soft-tissue attachment to the tooth, which can eventually result in abscesses and tooth loss. Some evidence of loss of attachment has been found in 90% of juveniles and adults, although only 15% of people exhibited clinically severe disease. Plaque bacteria are thought to initiate periodontitis and the presence of inflammatory factors, such as prostaglandins and cytokines, is correlated with alveolar bone loss, most likely by stimulating osteoclastogenesis and osteoclast activity.[39]

6.21.3.3.3 Wear particle foreign-body reactions (prostheses)

Joint replacement, particularly hip replacement, can fail in the long term due to aseptic loosening, requiring complex revision operations. This can occur due to the release from the articulating surfaces of wear particles from the prosthesis: these particles activate macrophages, which in turn induce inflammation at the implant and concomitantly stimulate osteolysis at the bone–implant interface. Anti-inflammatory and antiresorptive therapeutics and novel prosthesis materials are different approaches for the treatment or prevention of wear particle-induced bone loss.[40]

6.21.3.4 Genetic Diseases of Bone Metabolism

There are many congenital disorders affecting the skeleton either through direct impairment of cells involved in skeletal development or bone remodeling or through indirect effects mediated by endocrine or other metabolic abnormalities. These can result in bone of increased or decreased bone density, poorly mineralized bone, or disorganized bone undergoing excessive remodeling.

6.21.3.4.1 Genetic basis of bone diseases

Genetic predisposition is a significant factor in the development of primary bone diseases, including osteoporosis and Paget's disease of bone. Of greater interest is the genetic basis of osteoporosis. A family history of osteoporosis is a significant risk factor for developing osteoporosis. Twin studies indicate a strong inherited component in peak BMD. For this reason, there is considerable interest in identifying polymorphisms or other genetic markers that could predict an individual's risk of developing osteoporosis. However, inheritance of peak bone mass, bone architecture, and fracture risk has so far proved to be complex, with many genes and chromosomal loci currently implicated. Studies in two inbred strains of mice revealed multiple genetic associations with bone mass, with different bone sites and indices showing different associations.[56]

Interest in polymorphisms as predictors of risk of developing osteoporosis was initiated by findings of association between polymorphisms in the vitamin D receptor gene and BMD.[57] Large numbers of candidate genes have now been assessed for their association with various skeletal attributes and many positive associations reported. Unfortunately, many of these findings are not reproducible and no polymorphism is sufficiently predictive on its own to be clinically useful in an individual. Two genes with polymorphisms showing more robust association with BMD and fracture are the genes coding collagen 1α and the vitamin D receptor, but each of these exhibits inadequate predictive power to be useful in individual patients.[58,59] If useful diagnostic tools are to be developed for osteoporosis based on genetic testing, it is apparent that multiple genes will need to be assessed and interpreted based on extensive epidemiological data on how these genes interact to determine bone quality.

6.21.3.4.2 Congential osteoporosis

Osteogenesis imperfecta is the most common form of congenital childhood osteoporosis. As with forms of osteoporosis, it is marked by decreased BMD and increased bone fragility. Osteogenesis imperfecta can be of varying severity and of either dominant or recessive inheritance. Osteogenesis imperfecta results from mutations in collagen type 1 genes that cause insufficient or disrupted collagen content in bone matrix. The most severe forms are postnatally lethal and less severe forms can produce multiple minimal-trauma fractures, growth impairment, and inability to walk.[41] Recently, treatment with repeated infusions of pamidronate has been shown to be of benefit in these patients.[42] Osteoporosis can result from mutations in genes related to osteoblast regulation, e.g., osteoporosis pseudoglioma syndrome, which results from an inactivating mutation in LRP5 that inhibits *Wnt* signaling in osteoblasts.[43]

6.21.3.4.3 Osteopetrosis

Osteopetrosis is a congenital condition resulting from a failure of bone resorption due to defects in the generation or function (more commonly) of osteoclasts. It is marked clinically by dense bones and increased bone fragility. In severe cases, there is also impaired growth and tooth eruption, club-shaped long bones, deafness, nerve compression, immune deficiency, and anemia. The marrow spaces are occluded by dense trabecular structures containing mineralized cartilage rests that result from failure to remove mineralized cartilage and bone formed during endochondral bone formation. Mutations in genes encoding carbonic anhydrase II, the chloride channel *CLCN7*, and the $\alpha 3$ subunit of vaculolar H^+ pump have been identified as causes of osteopetrosis and would all impair the ability of osteoclasts to generate acid required for bone demineralization. Treatment is by marrow transplant or with interferon-γ. Pycnodysostosis is a similar bone disease with osteopetrotic phenotype, in this case due to mutations in cathepsin K, an osteoclast enzyme required for proteolytic breakdown of bone collagen.[44]

6.21.3.4.4 Osteosclerosis

Excessive bone formation can also produce bones of increased bone density. Mutations activating *Wnt* signaling have been found in the gene encoding LRP6 and in the gene encoding SOST; each produces a high-bone-mass phenotype.[45,46] The increased bone strength in these subjects has led to active research into *Wnt* signaling as a therapeutic target for the development of anabolic factors.

6.21.3.4.5 Skeletal dysplasias

Some genetic bone diseases are marked by high bone remodeling rate, disorganized bone structure, and, in some cases, fibrous replacement of bone marrow. This can be due to congenital hyperparathyroidism, mutations affecting osteoblast regulation (Gs-alpha subunit gene, *GNAS1*, in fibrous dysplasia), or in osteoclast regulation (RANK in familial expansile osteolysis, or OPG in familial hyperphosphatasia). Some of these conditions can be treated by reducing the impetus for high bone remodeling by parathyroidectomy or with antiresorptive treatments, e.g., calcitonin or bisphosphonates.[47]

6.21.3.4.6 Congenital rickets

This condition is marked skeletally by the failure of bone mineralization. Bone matrix remains as unmineralized osteoid and bone strength and rigidity are impaired. Congenital rickets can be induced by mutations affecting phosphate or calcium homeostasis, and by mutations in alkaline phosphatase, which are required for osteoblasts to induce matrix mineralization.

6.21.4 Animal Models of Bone Diseases

6.21.4.1 Animal Models of Osteoporosis

Postmenopausal osteoporosis has been studied in several animal models, including rats, sheep, ferrets, baboons, and monkeys.

6.21.4.1.1 Rat acute oophorectomy model of menopausal bone loss

Oophorectomy in rats results in development of a high turnover osteopenia, with most bone loss occurring over 4–6 weeks.[48] Treatment with estradiol is effective in inhibiting the bone loss. Thus, the rat has been found to provide a very useful model of the acute bone loss occurring in women following natural or surgical menopause due to estrogen withdrawal. Acute bone loss occurs in both young growing rats and mature rats following oophorectomy; however, it is preferable to use rats of at least 12 weeks of age that have a much reduced growth rate so that the results are not significantly influenced by growth-related effects of treatments. This model is particularly valuable in assessing treatments targeting bone resorption as inhibition of bone resorption effectively preserves normal bone mass. In contrast, oophorectomized mice are not a good model, as bone loss is variable and estradiol treatment is strongly anabolic, producing a profound osteosclerosis at high doses and indicating that hormonal regulation of bone mass is fundamentally different from that in rats and humans.

6.21.4.1.2 Rat aged oophorectomy model of established osteoporosis

Rats that are oophorectomized at 3 months after birth, and allowed to age for 3–6 months, develop severe osteopenia.[49] These animals provide a useful model of established osteoporosis. This model is particularly useful to evaluate anabolic therapies. Large increases in bone density, bone architecture, and bone strength can be produced in these rats with PTH treatment. This model can be used to assess antiresorptive treatments; however, large numbers of animals are required as treatment effects are small, as is also seen in osteoporotic human patients.

Aged mice can also be used as a model of established osteoporosis; however, there is almost a complete absence of trabeculae in aged mice, so response to treatment is limited by an absence of bone surfaces for agents to act on.

6.21.4.1.3 Large animal models of osteoporosis

The limitation in rodent models of osteoporosis is twofold. Epiphyses do not fuse in rodents, such that there are always possible confounding growth effects. Secondly, cortical bone in rodents does not undergo Haversian remodeling and so the role of this process in osteoporosis cannot be evaluated. Various larger animals have been evaluated as models of osteoporosis. Lower primates, such as baboons, cynomolgus, or rhesus monkeys, provide the best large-animal models of human osteoporosis as these animals respond similarly to oophorectomy and hormone treatments. Use of animals from dedicated facilities is preferable as wild-caught animals frequently have significant increases in bone density following capture due to improved nutrition. Bone structure and remodeling are also most similar to humans.

Sheep, pigs, dogs, and ferrets have also been used to model osteoporosis. Aged oophorectomized animals of these species provide a cheaper and ethically more acceptable alternative to primates. Problems encountered are that hormone cycling tends to be seasonal, bone loss with oophorectomy is variable, and, in the ungulates, cortical bone has a plexiform structure with only limited Haversian remodeling.

With all the large-animal models, bone changes occur slowly relative to those seen in rats, and thus the duration of studies is much longer. To assess treatment effects on bone strength, for example, would likely require duration of studies of 12–16 months.

6.21.4.1.4 Genetically modified mice

The function of specific genes in maintaining bone turnover and BMD has been determined by generating transgenic mice in which the gene of interest has either been introduced or knocked out. As an example of this approach, the overexpression or knockout of OPG, RANKL, and RANK in transgenic and knockout mice has shown the importance of this triad of molecular regulators in determining bone resorption level.[17]

6.21.4.2 Animal Models of Arthritis and Inflammatory Bone Diseases

The pathology of arthritis is related to changes in periarticular bone in addition to cartilage and ligament damage. Protection of the bony skeleton would be advantageous in the treatment of these disorders to limit progressive structural degeneration of joints. There are a number of animal models of the inflammatory joint disease that can be used to assess mechanism of disease and therapeutic responses. Joint inflammation can be induced in genetically susceptible rats and mice by the injection of collagen or strong adjuvants. In these models, severe inflammation is induced in the joints of the feet, which results in soft-tissue swelling, pannus formation and cartilage loss, local bone erosion, and systemic bone loss. These models tend to produce inflammation lasting 1–3 weeks, which is an advantage in allowing rapid assessment of treatments, but may not accurately reproduce all the pathology of human chronic inflammatory disease. The KRN mouse model develops spontaneous arthritis in mature mice. In mice of nonsusceptible background, such as genetically modified mouse strains, joint inflammation can be induced by the injection of serum from arthritic KRN mice or by using a combination of monoclonal antibodies targeting collagen type 2.[50]

6.21.4.3 Animal Models of Cancer Metastasis to Bone

Bone metastasis has been most commonly studied in immunocompromised mice bearing human cancer xenografts implanted directly into mouse bone, which allows the study of the interaction between cancer cells and the bone microenvironment. Tumor cells have also been introduced directly into the circulation by injection into the left ventricle of the heart or into arteries, e.g., the femoral artery: this approach facilitates study of the ability of the cancer cells to migrate towards bone. The bone metastatic behavior of cancers, most particularly those of prostate and breast cancers, has also been studied in mouse models bearing pieces of transplanted human bone: this is useful for studying the species-specific trafficking of cancer cells to human bone.[51]

6.21.5 Issues in Preclinical and Clinical Development for Drugs to Treat Osteoporosis

The clinical development of drugs for the treatment of osteoporosis is required to satisfy specific regulatory guidelines prior to approval for marketing. In 1994, the US FDA released draft guidelines covering preclinical and clinical development of osteoporosis drugs of postmenopausal osteoporosis.[52] The World Health Organization has developed similar guidelines[53] and these are available on its website. The FDA guidelines are under current review and variations may be expected. In addition, as these guidelines have continuing draft status, particular requirements could be open to negotiation, as shown by the recent approval of teriparatide (PTH 1-34), with phase III data of approximately 2 years' duration rather than the 3 years described in the guidelines.

6.21.5.1 Preclinical Study Requirements

In addition to the normal toxicological and pharmacological studies required, there is a specific requirement for pharmacology studies in two different recognized animal models of osteoporosis. One of these studies should be conducted in an oophorectomized rat model and be of 12 months' duration. A second study should be in a large-animal model that shows full skeletal remodeling (sheep, pigs, and primates are given as good examples, but the validity of dog models is questioned). The large-animal study should be of 16 months' duration. The purpose of these studies is to demonstrate that the new chemical entity (NCE) tested does not have deleterious effects on bone quality, despite positive effects on bone mass. This requirement reflects a concern initially raised about fluoride, which increases BMD but can reduce bone strength. Reflecting the concerns regarding bone quality, these two pharmacology studies must include evaluation of bone quality as endpoints. This can be achieved, showing bone matrix structure (lamellar or woven), trabecular architecture, and bone mechanical strength are maintained or improved, as well as bone mass increased. It may be possible to argue exception from these requirements if the drug being developed cannot be used in rats and/or larger species, as could be the case for a monoclonal antibody specific for a human protein. Use of surrogate markers of bone remodeling is required to confirm pharmacologic actions of NCEs throughout the duration of these studies.

6.21.5.2 Clinical Study Requirements

6.21.5.2.1 Phase I studies

These should be conducted using standard protocols to determine drug safety, tolerability, and pharmacokinetics. The existence of assays for various surrogate markers of bone formation and resorption makes it possible to leverage phase I studies to obtain valuable information on NCE pharmacology in human subjects. These surrogate markers provide a sensitive and dynamic indication of the nature, size, and duration of the biologic response to the tested drug, as demonstrated in the study of an OPG construct.[54] Measurement of telopeptides of collagen type I in serum or urine provides a clear measurement of antiresorptive effects of an NCE and measurement of bone-specific alkaline phosphatase or procollagen peptides gives a clear indication of changes in bone formation.

6.21.5.2.2 Phase II studies

Phase II studies extend safety data, confirm biological activity of the NCE in humans, and define dose and dosing frequency for phase III trials. For osteoporosis NCEs, phase II trials are typically 1 year in duration and utilize BMD measurement as the primary clinical endpoint. Surrogate markers of bone remodeling can also be used as these give excellent information to assess dose responses, maintenance of pharmacological action, and mechanism of action. Bone biopsy and histomorphometry provide information regarding NCE mechanism of action and maintenance or improvement in bone quality.

6.21.5.2.3 Phase III studies

FDA guidelines recommend a 3-year phase III trial with reduction in fracture incidence as the primary endpoint. There is an option for a BMD endpoint if the preclinical studies support maintenance or improvement of bone quality.

However, given the competitive market for osteoporosis drugs that already show fracture benefit, commercial considerations likely make the generation of data demonstrating fracture benefit an imperative. It is possible that shorter trials showing fracture benefit will be accepted, as was the case with teriparatide. All subjects should receive calcium and vitamin D supplementation, with a total calcium intake of 1500 mg being maintained. Trials are double-blinded placebo or active-controlled randomized trials. Currently osteoporosis NCE trials are typically placebo-controlled. However, the ethics of conducting a placebo rather than active NCE as comparator is coming under increased discussion. Very large and expensive trials would be required to gain acceptable confidence intervals for noninferiority in fracture benefit in an active-controlled trial. Alternatives are to continue with placebo-controlled studies but to exclude patients at high fracture risk and/or to allow shorter trials demonstrating fracture benefit after 1–2 years.

The simplest fractures in which to show benefit are vertebral fractures. These are commonly defined radiologically as deformations leading to a 20% reduction in the height of lumbar vertebrae. These occur with sufficient frequency (5/100 patient-years) in patients with low BMD and at least one prevalent fracture to allow trials of 1000–2000 subjects. Hip fractures are much more clinically important as they are associated with much greater morbidity and mortality. Reduction in hip fracture incidence was demonstrated for alendronate and risedronate, though the number of trial participants required was around 5000, as the frequency is about 0.8/100 patient-years in patients with one prevalent fracture.[55]

Secondary endpoints for the phase III osteoporosis trials are BMD, effects on surrogate markers of bone remodeling, and bone quality as assessed by histomorphometry. An ideal profile of a successful osteoporosis NCE will show increases in BMD in the lumbar vertebrae and in the femoral neck and reduction in both vertebral and hip fracture incidence. Due to the excellent safety profiles of approved osteoporosis treatments, a high standard for safety and tolerability will be applied to new drug applications in this area.

6.21.5.3 Clinical Issues in Developing Drugs to Treat Skeletal Effects of Metastatic Bone Disease

Metastatic bone disease is associated with significant morbidity: bone pain, fracture, and hypercalcemia are common. Bisphosphonates (**Figures 5** and **6**) have demonstrated that treatment of these patients with antiresorptives is of clinical benefit. As in the development of NCEs for osteoporosis, use of surrogate markers of bone resorption allows the leveraging of phase I and II trials to determine optimum dosing and dosing frequency accurately. More aggressive disease needs higher NCE doses for the effective suppression of bone resorption and clinical benefit in inhibiting skeletal events.

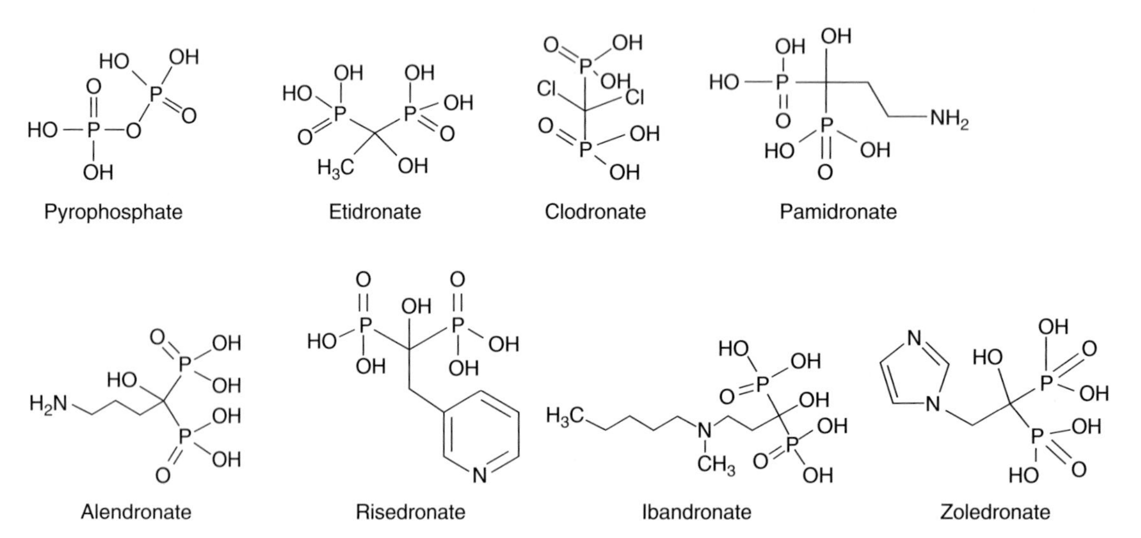

Figure 5 Chemical structure of pyrophosphate and the bisphosphonates. The bisphosphonates are analogs of pyrophosphate. The P-C-P group is a stable analog of the labile P-O-P group of pyrophosphate and provides high affinity for bone mineral. The side chain determines the potency of antiresorptive activity. The nitrogen-containing bisphosphonates, which are highly potent, act by inhibiting enzymes in the mevalonate pathway.

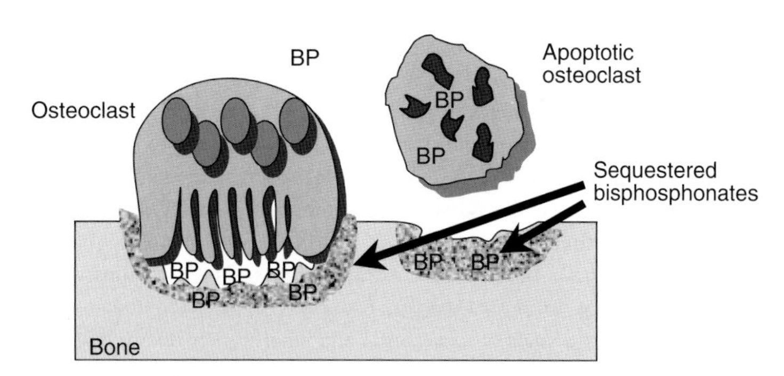

Figure 6 Bisphosphonate (BP) antiosteoclast actions. Administered BPs, such as pamidronate, are rapidly sequestered in bone. Osteoclasts release BP into the resorption cavity to produce a high local concentration and endocytose sufficient to induce osteoclast apoptosis (nitrogen-containing BPs) or inactivity (clodronate or etidronate) and arrest of bone resorption.[61] Other cells in the body are only exposed to a transient and low level of BPs. Thus, BPs have exquisite specificity for osteoclasts.

For approval, it is necessary to show that an NCE reduces skeletal complications of malignancy, usually through the evaluation of new 'skeletally related events,' that can include interventions for bone pain, fracture, surgery for nerve compression, and hypercalcemia. Trials are usually of 1–2 years' duration.

6.21.6 Current Treatments

6.21.6.1 Antiresorptives

6.21.6.1.1 Bisphosphonates

Bisphosphonates (**Figure 5**) are a drug class of small molecules, sharing a common P-C-P structure, that are potent and long-acting inhibitors of bone resorption. The molecular structure has similarities to the P-O-P structure of pyrophosphate. The P-C-P and P-O-P groups share similar high affinity with calcium phosphate crystal surfaces. The P-C-P group differs from P-O-P, however, in its high chemical stability and resistance to chemical and enzymatic hydrolysis. The presence and potency of antiresorptive activity and the cell pathways targeted are dependent on the side chain. Bisphosphonates are targeted to bone through P-C-P-mediated high affinity to hydroxyapatite crystals. Once sequestered in bone, negligible bisphosphonate is released, with the half-life in the bone compartment being approximately 8 years in adult humans.[60]

Bisphosphonates have high negative charge and are likely dependent on endocytosis to enter cells. The specific antiresorptive action of bisphosphonates is due to their being concentrated in bone and to their selective release from bone and endocytosis by osteoclasts during bone resorption (**Figure 6**) via the ruffled border, which forms on the cell side of the osteoclast-resorbing space (**Figure 1**).[61] The bisphosphonates that are currently most potent are those containing nitrogen in the side chain (e.g., pamidronate, alendronate, ibandronate, risedronate, and zoledronate). These bisphosphonates impair osteoclast function and induce osteoclast apoptosis and act to inhibit the mevalonate pathway enzymes squalene synthase, farnesyl diphosphate synthase, and isopentenyl diphosphate isomerase to block farnesylation and geranylgeranylation of proteins. As only geranylgeraniol can rescue cells from the apoptotic effects of nitrogen-containing bisphosphonates, the latter pathway is likely to be critical and is essential for the production of the small guanosine triphosphatases, such as Rho and Rab. In contrast, clodronate and etidronate appear to inhibit adenosine triphosphate processing and thus disrupt cell energy trafficking (reviewed by Rogers *et al.*[60]).

Thus, bisphosphonates are targeted cytotoxic drugs with an extraordinary specificity for osteoclasts. Bisphosphonates have toxic effects on other cell types, implying some uptake, but this occurs at much higher concentrations than those required to inhibit bone resorption.

Bisphosphonates can be given orally, though their bioavailability is very low ($<1\%$ for alendronate). To achieve adequate drug absorption, the patient is required to fast several hours before, and for 30 min following, drug ingestion. Risk of esophageal damage due to exposure to high concentrations of bisphosphonate is avoided by requiring the patient to remain upright after drug ingestion. As these drugs accumulate in bone, adequate levels can be absorbed over a few weeks to inhibit bone resorption. However, amounts absorbed can be inadequate to inhibit strong proresorptive stimuli, such as in lytic bone metastases and Paget's disease of bone. For these conditions, intravenous administration can be used to ensure high doses of bisphosphonate reach the bone, with infusion rather than bolus administration used to limit high circulating levels that can produce renal toxicity. Infusion of bisphosphonates produces rapid antiresorptive action (24–48 h) and high doses can be effective, even in high bone resorptive states.

Bisphosphonates have utility in all bone disorders characterized by a relative excess of bone resorption. Clinical utility has been demonstrated in a very wide range of bone disorders, including osteoporosis of various etiologies, high-turnover bone disease in Paget's disease of bone, lytic bone diseases related to primary or secondary cancers in bone, local bone loss associated with inflammatory disease such as rheumatoid arthritis, and in congenital diseases of bone metabolism such as osteogenesis imperfecta.

In particular, the bisphosphonates have been widely used in the treatment of postmenopausal and age-related osteoporosis. Bisphosphonate treatment in postmenopausal women with osteoporosis was effective in modestly increasing BMD (approximately 8% over 3 years of treatment) and significantly decreasing vertebral and femoral neck fracture risk by approximately 50%. The effect on fracture appears disproportionate to the increase in bone density and it is thought that, by inhibiting bone resorption, other bone changes contribute to increased bone strength. These include removal of resorption lacunae that may act as stress risers for crack initiation on bone surfaces, increased mineral content of bone matrix, reduced cortical porosity, and improvements in trabecular architecture.

Bisphosphonates are highly effective in the treatment of hypercalcemia of malignancy and moderately effective in the prevention of fracture, bone pain, and palliative surgery and other skeletal-related events in patients with bone metastases.

In osteogenesis imperfecta, there is a genetic abnormality in the collagen genes that results in impaired formation of bone matrix. For reasons that are not well understood, this change is associated with a high bone turnover state and children with this disorder have been found to respond well to treatment with regular pamidronate infusions that increase BMD, reduce bone pain and fracture, and improve mobility and skeletal growth.

The selection of which bisphosphonate to use and the mode of administration depends on the clinical application. The potency of the bisphosphonate appears more important in determining dose level and dose frequency, rather than clinical outcome. Oral weekly alendronate and risedronate are the most commonly used treatments in osteoporosis. Treatments using infrequent infusions of ibandronate or zoledronate are in late development for osteoporosis. Higher oral doses of alendronate and risedronate, or infusions of the more potent bisphosphonates, are effective in Paget's disease. Treatment of metastatic bone disease requires bisphosphonate treatment delivered by systemic infusion.

6.21.6.1.2 Estrogenic hormones

6.21.6.1.2.1 Estrogen

Estrogen deficiency, due to menopause, amenorrhea, chemotherapy, or excessive exercise, is associated with precipitate trabecular bone loss. This is due at least in part to increased osteoclastic activity, which occurs via increases in proresorptive cytokines, e.g., IL1 and IL6, and RANKL, and decreases in antiresorptive factors, such as OPG.

Estrogen therapy (ET) or estrogen plus progesterone therapy, commonly referred to as hormone replacement therapy (HRT), are effective antiresorptive strategies in preventing menopause-associated bone loss and in reducing osteoporotic fractures: reversal of bone resorption can occur within a month of treatment initiation. Whilst increases in the BMD of the spine, forearm, and hip were found in women treated with high-dose estrogen when compared with the placebo groups, lower-dose estrogen treatment is also bone-protective and may reduce side effects. Response to estrogen replacement depends upon patient age, the number of years since menopause, dose and type of estrogen, and whether or not progestins are included, length of treatment, supplementation with calcium and vitamin D, bone resorption rate at the start of treatment, baseline BMD, and other factors.[62]

6.21.6.2.1.2 Selective estrogen receptor modulators (SERMs)

Whilst HRT/ET results in increases in BMD and decreases in fracture incidence, long-term treatment is associated with several side effects, including cardiovascular events, uterine bleeding, and increased risk of breast cancer. Tamoxifen (**Figure 7**), the prototypical member of the SERM family, exhibits estrogen-like effects on the skeleton, reducing bone loss and fracture incidence but, in contrast to ET, has antiestrogen effects on breast tissue, leading to its therapeutic use in treating breast cancer. Tamoxifen treatment is also associated with decreases in serum low-density lipoprotein cholesterol. Whilst it is an improvement over using ET, there are still limitations; for example, tamoxifen is associated with increased incidences of vaginal discharge, hot flushes, thickening of the endometrial wall, and an increased risk of endometrial cancer, and with the development of cataracts.

The safety profile of raloxifene (**Figure 7**), a second-generation, more selective SERM, is improved over that with tamoxifen, with a single serious side effect, venous thromboembolism, although this occurred at a similar or lower rate in raloxifene-treated women than in women treated with HRT or with tamoxifen. In common with tamoxifen, raloxifene prevented bone loss, decreased bone turnover to premenopausal levels, and reduced pathological fracture in

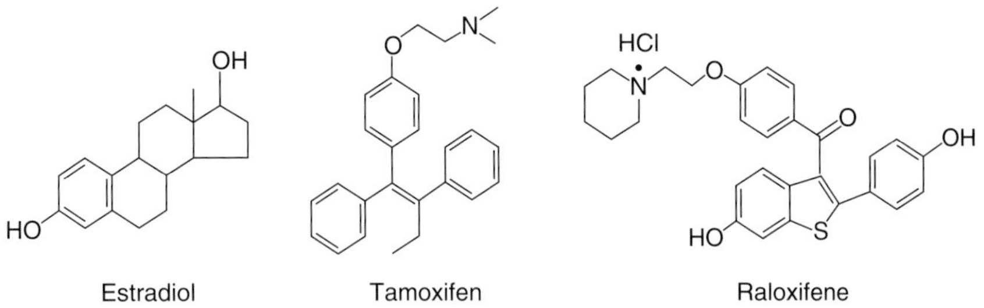

Estradiol Tamoxifen Raloxifene

Figure 7 Chemical structure of estradiol, tamoxifen, and raloxifene. Tamoxifen and raloxiphene are selective estrogen receptor modulators (SERMs) approved for clinical use. Tamoxifen and raloxifene bind to the estrogen receptors, but are selective in the tissue responses induced.

the spines of osteoporotic postmenopausal women and also profoundly decreased the incidence of breast cancer. Added benefits of raloxifene over tamoxifen include the findings that raloxifene treatment elicited no changes in endometrial thickness, proliferation, or hyperplasia and there was no difference between raloxifene-treated women and placebo in incidences of vaginal discharge and spotting. There was also a trend towards a reduction in cognitive decline in elderly women treated with raloxifene as compared to placebo.[63]

6.21.6.2 Calcium and Vitamin D

6.21.6.2.1 Calcium

Calcium has long been used for the treatment of osteoporosis, both in the form of dietary and pharmacological supplements. In patients with calcium deficiency, oral calcium at doses of 500–1500 mg per day corrects a negative calcium balance and suppresses PTH secretion. Sufficient calcium intake is important for the accrual of peak bone mass in the young but is also considered the basis of most antiosteoporotic regimens. In the elderly, supplementation with oral calcium and vitamin D reduces the risk of hip fracture by about 30–40%. However, the majority of controlled studies have failed to show an effect of calcium supplementation alone on fracture risk in postmenopausal women.

6.21.6.2.2 Vitamin D

Vitamin D and its metabolites are widely recommended for the treatment of osteoporosis. Natural vitamin D_3 (cholecalciferol) is synthesized in the skin through ultraviolet radiation (sun exposure) (**Figure 3**), while small amounts of vitamin D_2 (ergocalciferol) are contained in the diet. Both components are metabolized, in the liver, to 25-hydroxyvitamin D (calcidiol) (**Figure 3**). The latter compound is considered the 'storage form' of vitamin D, and although it probably has little biological activity, calcidiol is measured in the clinical setting to determine whether a patient is deficient of this steroid hormone. The biologically active form of vitamin D is generated in the kidney through hydroxylation in position C1, leading to the formation of 1,25-dihydroxyvitamin D or calcitriol (**Figure 3**). The activity of the hydroxylating enzyme, 1α-hydroxylase, is tightly controlled by a number of regulators, including PTH or serum phosphate levels. 1,25-dihydroxyvitamin D is a potent steroid hormone with almost countless effects throughout the body, most concerning the differentiation of immature cells. As regards bone, 1,25-dihydroxyvitamin D has differentiation-inducing as well as activating effects on both osteoblasts and osteoclasts. It also increases the calcium absorption from the gut.

Data from some controlled clinical trials suggest that daily supplementation with 500–1200 mg calcium and 700–1000 IU of oral calciferol (not calcitriol!) reduces the rate of bone loss in postmenopausal women. Of note, institutionalized elderly persons with a high prevalence of secondary hyperparathyroidism due to calcium and vitamin D deficiency appear to benefit greatly from calcium and vitamin D supplementation: the landmark study by Chapuy *et al.*[64] demonstrated that, in this elderly population, 1200 mg of calcium and 800 IU of vitamin D daily reduced the number of hip and nonvertebral fractures by 30%. In contrast, the recently published Randomised Evaluation of Calcium OR Vitamin D (RECORD) study failed to demonstrate a significant effect of oral calcium and vitamin D on osteoporotic fracture risk in men and women with prevalent low-trauma fractures ($n = 5292$ aged 70 years or older; 85% female).[65]

So-called 'active' vitamin D metabolites are still controversial with regard to their therapeutic use in postmenopausal and/or age-related osteoporosis. There is still no good evidence supporting the claim that, in postmenopausal or age-related osteoporosis, vitamin D metabolites such as 1α-hydroxy vitamin D or calcitriol are more efficacious with regard to bone loss or fracture reduction than either cholecalciferol or ergocalciferol. There is, however, evidence that calcitriol might be beneficial in certain secondary forms of osteoporosis, e.g., glucocorticoid-induced osteoporosis. While cholecalciferol and ergocalciferol have a broad therapeutic window with very few, if any, adverse effects, 'active' vitamin D metabolites are characterized by a narrow therapeutic window with hypercalciuria (kidney stones) and hypercalcemia being the most serious problems.

It should be noted that all of the recent drug trials used calcium and vitamin D as part of the therapeutic regimen (with the placebo group receiving calcium and vitamin D as active treatment) and that neither anabolic nor antiresorptive drugs should be given without calcium supplements.

Replacement of calcium and native vitamin D is critical when treating the most common form of osteomalacia, i.e., hypocalcemic osteomalacia, in the setting of severe vitamin D deficiency.

6.21.6.3 Anabolic Treatments

PTH has, for reasons that are not well defined, a dual effect on bone cells: given intermittently, PTH stimulates osteoblast activity, leading to substantial increases in bone density. In contrast, when secreted continuously at relatively high doses (as seen in patients with pHPT), PTH stimulates osteoclast-mediated bone resorption and suppresses osteoblast activity. Further to its direct effects on bone cells, PTH also enhances renal calcium reabsorption and phosphate clearance, as well as renal synthesis of 1,25-dihydroxyvitamin D. Both PTH and 1,25-dihydroxyvitamin D act synergistically on bone to increase serum calcium levels and are closely involved in the regulation of the calcium/phosphate balance. The anabolic effects of PTH on osteoblasts are probably both direct and indirect via growth factors, such as IGF-1 and TGFβ Multiple signaling pathways mediating PTH effects on bone cells include activation of adenylyl cyclase, phospholipase C, protein kinase C, tyrosine kinase c-src, alterations in intracellular protein phosphorylation, and generation of inositol 1,4,5-triphosphate.

Recombinant human (rh)PTH(1-34) (teriparatide) increases bone mass, trabecular connectivity, and bone strength, improving BMD in females with postmenopausal osteoporosis, males with idiopathic osteoporosis, and in individuals with glucocorticoid-induced osteoporosis. There is currently only a single randomized controlled trial on rhPTH(1-34) in postmenopausal osteoporosis. In this 19-month-long study of 1637 women with prior osteoporotic vertebral fractures, 20 and 40 μg of rhPTH(1-34) subcutaneously reduced the risk of new vertebral fractures by 65–69% and the risk of nonvertebral fragility fractures by 53–54%. Concomitantly, lumbar spine BMD increased by 10–14% and femoral neck BMD by 3–5%.[66] A comparative trial of teriparatide 40 μg subcutaneously versus alendronate 10 mg in women with postmenopausal osteoporosis showed that teriparatide was more potent than alendronate in increasing BMD and reducing fracture risk.[67] While osteosarcomas have been observed in certain rat strains treated with high doses of rhPTH(1-34), these had no clinical relevance in humans. rhPTH 1-34 has recently been approved for the treatment of severe osteoporosis.

6.21.6.4 Strontium Salts

Strontium salts have long been investigated as anabolic agents for bone. In animals, strontium stimulates bone formation and substitutes for calcium in hydroxyapatite crystals. In humans, studies have shown increased bone mass (after correction of BMD values for strontium content) and a reduction in vertebral and, to a lesser degree, nonvertebral fractures.[68] The mechanism of action of strontium salts is unknown and may involve modulation of calcium sensors or calcium channels.[69]

6.21.6.5 Fluoride

Fluoride ions stimulate bone formation by a direct mitogenic effect on osteoblasts altering hydroxyapatite crystals in the bone matrix. At low doses, fluorides induce lamellar bone, and at higher doses abnormal woven bone of inferior quality. The effect of fluorides on normal and abnormal (e.g., osteoporotic) bone is therefore dose-dependent. Fluoride has been used as a therapy for osteoporosis; however, clear benefit on fracture incidence has not been observed. Indeed, some high-dose studies have demonstrated increased fracture incidence.[70] If fluoride is of benefit, it is within a very narrow therapeutic range.

6.21.7 Unmet Medical Needs and New Research Areas

6.21.7.1 Improved Evaluation of Fracture Risk

Accurate evaluation of fracture risk is currently an area of significant medical need. While BMD can be measured accurately, it defines only a proportion of fracture risk for an individual. Further risk can be identified from medical history, including previous fractures, by measuring surrogate markers of bone formation and resorption,[71] identifying vitamin D deficiency, measuring muscle strength and balance skills, and by genetic analyses. There is a need for epidemiological studies to integrate these approaches to determine a global fracture risk to assist in treatment decisions for patients.[72]

6.21.7.2 Alternative Antiresorptive Treatments

While bisphosphonates are highly effective antiresorptive treatments, they are not always well tolerated orally. This is particularly an issue for nonambulant patients or those with existing gastrointestinal disease, indicating the need for alternative approaches or formulations. Parenteral administration of bisphosphonates is a viable alternative, although the risk of osteonecrosis of the jaw has been identified with this approach. New therapies targeting RANKL signaling, cathepsin K, other osteoclast enzymes, and integrins are currently being investigated. RANKL is being targeted through the development of a neutralizing, fully human monoclonal antibody designated denosumab.[54] Phase I results indicate that this antibody could be used in osteoporosis by subcutaneous injection at a dosing frequency of once every 6 months.

Cathepsin K is a proteolytic enzyme expressed with high specificity in osteoclasts and that is required for normal bone resorption. Extensive research activity by many research groups has identified potent inhibitors for this enzyme which show excellent in vivo activity in rat models of osteoporosis and are now in clinical trial. An example of these inhibitors is SB331750[73] (**Figure 8**).

Osteoclasts also need to adhere to bone to activate bone resorption and the integrin $\alpha_v\beta_3$ is an important mediator of adhesion and the intracellular signaling related to adhesion to bone. Small-molecule antagonists of $\alpha_v\beta_3$ (see **Figure 8**) with its ligand vitronectin are effective inhibitors of bone resorption in vivo and many of these are being investigated as therapeutic agents.[74]

In cancer metastasis patients, it is possible that even more potent antiresorptive action is required to improve skeletal protection, as clinical benefits from bisphosphonate treatment are only modest. Reductions in skeletal-related events of only 20–30% are seen and, while this is of significant benefit, in many patients skeletal complications of malignancy are not prevented. Studies with bisphosphonates have shown that incomplete control of bone resorption, as determined by the measurement of bone resorption surrogate markers, is associated with a higher risk of skeletal complications, providing support for this concept.

6.21.7.3 Anabolic Treatments

There is an enormous need for treatments that actually rebuild bone. Many patients with osteoporosis have reductions in BMD of more than 50%. Treatment with bisphosphonates results in only minor recovery (10% increase from current

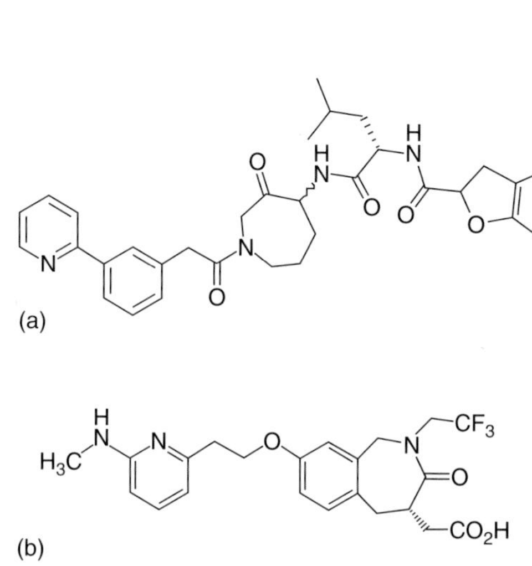

(a)

(b)

Figure 8 Chemical structure of a cathepsin K inhibitor and an αvβ3 integrin inhibitor. Structures are shown for (a) the cathepsin K inhibitor SB331750,[73] and (b) the $\alpha_v\beta_3$ integrin inhibitor (*S*)-3-oxo-8-[2-[6-(methylamino)-pyridin-2-yl]-1-ethoxy]-2-(2,2,2-trifluoroethyl)-2,3,4,5-tetrahydro-1H-2-benzazepine-4-acetic acid.[74] These molecules reduce bone resorption by respectively inhibiting osteoclast-mediated bone matrix degradation by cathepsin K or by impairing osteoclast adhesion to bone matrix mediated by $\alpha_v\beta_3$ integrin.

BMD or less than 5% of original peak BMD). Teriparatide (PTH 1-34) treatment is the only currently approved anabolic treatment. While it is perhaps twice as effective[67] as alendronate, teriparatide requires daily injections and does not return BMD to anywhere near original levels. New, more potent anabolic therapies are required and targets currently being investigated include *Wnt* pathway modulators, strontium (weak anabolic), IGF/IGFBP (IGF binding protein) combinations and other growth factors.[75] The concern with anabolics is that treatment produces structurally competent bone, as loss of bone architecture can result in increased fragility despite increased bone density (as seen with fluoride). Thus, demonstration of fracture efficacy is particularly important for anabolic treatments.

An activating mutation in low density lipoprotein receptor-related protein 5 (LRP5), a component of the *Wnt* signaling pathway, produces strong, dense bones, and this observation has demonstrated the therapeutic potential of this pathway for anabolic treatment of bone diseases. One way the *Wnt* pathway is being targeted is by the antagonism of the SOST gene product, sclerostin, with a neutralizing monoclonal antibody. This protein normally acts as an inhibitor of *Wnt* signaling through its binding to LRP5 or LRP6. Its absence results in an upregulation of *Wnt* signaling and thus an increase in bone density in mice.[76]

Strontium ranelate has recently been approved for the treatment of osteoporosis. It appears to increase bone formation modestly and decrease bone resorption and this uncoupling of these processes produces an increase in BMD and a decrease in vertebral fracture incidence by 30–45%.[75]

IGF-1 and growth hormone have also been investigated and do increase bone formation and thus increase BMD. IGF-1, however, is able to act as a ligand to the insulin receptor and this has limited dosing. Use of concurrent IGF-binding proteins has been proposed as a way to minimize this effect. Growth hormone has the theoretical ability to cause diabetes. Both of these treatments require controlled fracture trials before they can be accepted.

FGF-1 and -2 are profoundly anabolic to bone in rodents and have the unique ability to increase trabecular number.[77] However, FGF-1 and FGF-2 systemic treatment is associated with acute suppression of blood pressure and these growth factors have potential pleiotropic proliferative effects on many cell types. Thus the use of these agents is probably limited to local defect repair and systemic therapeutic use would likely require specific targeting to bone.

For local bone defects and nonunion fractures, considerable progress has been made with various combinations of natural and synthetic matrices and growth factors. Use of BMPs has been a significant advance. Two BMPs are currently approved for local bone repair: BMP-7 (also known as osteogenic protein-1) and BMP-2. These are both delivered in collagen-based matrices to provide a scaffold for new bone formation, which is enhanced by the presence of the BMP.[78,79] However, in this area, products with improved efficacy and/or ease of delivery would be of clinical value.

6.21.7.4 Gene Therapies and Stem Cell Therapies

Gene and stem cell therapies provide the long-term promise of sustained clinical improvement, particularly in patients with congenital disorders of bone metabolism. Treatments for children with osteogenesis imperfecta and osteopetrosis are particularly needed, given the continuing poor outcomes with current treatments. The use of a patient's own stem cells to fill defects or produce systemic improvements in bone formation in aged or severely osteoporotic patients is also an attractive concept. Significant challenges lie in producing sufficient genetically modified osteoprogenitors or early stem cells, delivering these cells to the whole skeleton, and producing consistent and sustained function.[80,81]

References

1. Rey, C.; Miquel, J. L.; Facchini, L.; Legrand, A. P.; Glimcher, M. J. *Bone* **1995**, *16*, 583–586.
2. Marks, S. C., Jr.; Walker, D. G. *Am. J. Anat.* **1981**, *161*, 1–10.
3. Owen, M. *Arthr. Rheumatism* **1980**, *23*, 1073–1080.
4. Holtrop, M. E.; King, G. J. *Clin. Orthop. Relat. Res.* **1977**, *123*, 177–196.
5. Matthews, J. L.; Talmage, R. V. *Clin. Orthop. Relat. Res.* **1981**, *156*, 27–38.
6. Weinger, J. M.; Holtrop, M. E. *Calcif. Tissue Res.* **1974**, *14*, 15–29.
7. Parfitt, A. M. *J. Cell. Biochem.* **1994**, *55*, 273–286.
8. Boden, S. D.; Kaplan, F. S. *Orthoped. Clin. North Am.* **1990**, *21*, 31–42.
9. Holick, M. F. *J. Cell. Biochem.* **2003**, *88*, 296–307.
10. Sambrook, P. N.; Chen, J. S.; March, L. M.; Cameron, I. D.; Cumming, R. G.; Lord, S. R.; Zochling, J.; Sitoh, Y. Y.; Lau, T. C.; Schwarz, J. et al. *J. Clin. Endocrinol. Metab.* **2004**, *89*, 1572–1576.
11. Inzerillo, A. M.; Zaidi, M.; Huang, C. L. *Thyroid* **2002**, *12*, 791–798.
12. Canalis, E.; Delany, A. M. *Ann. NY Acad. Sci.* **2002**, *966*, 73–81.
13. Hofbauer, L. C.; Heufelder, A. E. *J. Mol. Med.* **2001**, *79*, 243–253.
14. Rosen, C. J. *N. Engl. J. Med.* **2005**, *353*, 595–603.
15. Khosla, S.; Melton, L. J., III; Riggs, B. L. *J. Clin. Endocrinol. Metab.* **2002**, *87*, 1443–1450.

16. Wiktor-Jedrzejczak, W.; Gordon, S. *Physiol. Rev.* **1996**, *76*, 927–947.
17. Hofbauer, L. C.; Khosla, S.; Dunstan, C. R.; Lacey, D. L.; Boyle, W. J.; Riggs, B. L. *J. Bone Miner. Res.* **2000**, *15*, 2–12.
18. Goldring, S. R.; Gravallese, E. M. *Arthr. Res.* **2000**, *2*, 33–37.
19. Aubin, J. E. *Rev. Endocrinol. Metab. Dis.* **2001**, *2*, 81–94.
20. Fujii, H.; Kitazawa, R.; Maeda, S.; Mizuno, K.; Kitazawa, S. *Histochem. Cell Biol.* **1999**, *112*, 131–138.
21. Levasseur, R.; Lacombe, D.; de Vernejoul, M. C. *Joint, Bone, Spine: Rev. Rhumatisme* **2005**, *72*, 207–214.
22. Chang, K. P.; Center, J. R.; Nguyen, T. V.; Eisman, J. A. *J. Bone Miner. Res.* **2004**, *19*, 532–536.
23. Sambrook, P. N.; Seeman, E.; Phillips, S. R.; Ebeling, P. R. *Med. J. Aust.* **2002**, *15*, S1–S16.
24. Melton, L. J., III *J. Bone Miner. Res.* **2003**, *18*, 1139–1141.
25. Miller, P. D.; Siris, E. S.; Barrett-Connor, E.; Faulkner, K. G.; Wehren, L. E.; Abbott, T. A.; Chen, Y. T.; Berger, M. L.; Santora, A. C.; Sherwood, L. M. *J. Bone Miner. Res.* **2002**, *17*, 2222–2230.
26. Ralston, S. H. *Curr. Opin. Rheumatol.* **2005**, *17*, 475–479.
27. Black, D. M.; Arden, N. K.; Palermo, L.; Pearson, J.; Cummings, S. R. *J. Bone Miner. Res.* **1999**, *14*, 821–828.
28. Kanis, J. A.; Borgstrom, F.; De Laet, C.; Johansson, H.; Johnell, O.; Jonsson, B.; Oden, A.; Zethraeus, N.; Pfleger, B.; Khaltaev, N. *Osteoporos Int.* **2005**, *16*, 581–589.
29. Bilezikian, J. P.; Brandi, M. L.; Rubin, M.; Silverberg, S. J. *J. Int. Med.* **2005**, *257*, 6–17.
30. Reddy, S. V. *J. Cell. Biochem.* **2004**, *93*, 688–696.
31. Langston, A. L.; Ralston, S. H. *Rheumatology* **2004**, *43*, 955–959.
32. Whyte, M. P. In *Primer on the Metabolic Bone Disease and Disorders of Bone Metabolism*, 5th ed.; Favus, M. J., Ed.; American Society for Bone and Mineral Research: Washington DC, 2003, pp 479–487.
33. Wulling, M.; Engels, C.; Jesse, N.; Werner, M.; Delling, G.; Kaiser, E. *J. Cancer Res. Clin. Oncol.* **2001**, *127*, 467–474.
34. Sim, F. H.; Frassica, F. J.; Unni, K. K. *Orthopedics* **1995**, *18*, 19–23.
35. Blair, J. M.; Zhou, H.; Seibel, M. J.; Dunstan, C. R. *Nat. Clin. Pract. Oncol.* **2006**, *3*, 41–49.
36. Mundy, G. R. *Cancer* **1997**, *80*, 1546–1556.
37. Barille-Nion, S.; Barlogie, B.; Bataille, R.; Bergsagel, P. L.; Epstein, J.; Fenton, R. G.; Jacobson, J.; Kuehl, W. M.; Shaughnessy, J.; Tricot, G. *Hematology* **2003**, *2003*, 248–278.
38. Goldring, S. R.; Gravallese, E. M. *Arthr. Res.* **2000**, *2*, 33–37.
39. Takayanagi, H. *J. Periodont. Res.* **2005**, *40*, 287–293.
40. Haynes, D. R.; Crotti, T. N.; Potter, A. E.; Loric, M.; Atkins, G. J.; Howie, D. W.; Findlay, D. M. *J. Bone Joint Surg. Br.* **2001**, *83*, 902–911.
41. Fedarko, N. S.; Vetter, U.; Robey, P. G. *Connect. Tissue Res.* **1995**, *31*, 269–273.
42. Munns, C. F.; Rauch, F.; Travers, R.; Glorieux, F. H. *J. Bone Miner. Res.* **2005**, *20*, 1235–1243.
43. Gong, Y.; Slee, R. B.; Fukai, N.; Rawadi, G.; Heegers, S.; Sabatakos, G. et al. *Cell* **2001**, *107*, 513–523.
44. Whyte, M. P. Skeletal Disorders Characterized by Osteosclerosis or Hyperostosis. In *Metabolic Bone Disease*, 2nd ed.; Alvioli, L. V., Krane, S. M., Eds.; Academic Press: San Diego, CA, USA, 1997, pp 697–738.
45. Van Wesenbeeck, L.; Cleiren, E.; Gram, J.; Beals, R. K.; Benichou, O.; Scopelliti, D.; Key, L.; Renton, T.; Bartels, C.; Gong, Y. et al. *Am. J. Hum. Genet.* **2003**, *72*, 763–771.
46. Li, X.; Zhang, Y.; Kang, H.; Liu, W.; Liu, P.; Zhang, J.; Harris, S. E.; Wu, D. J. *Biol. Chem.* **2005**, *280*, 19883–19887.
47. Collins, M. T.; Bianco, P. Fibrous Dysplasia. In *Primer on the Metabolic Bone Disease and Disorders of Bone Metabolism*, 5th ed.; Favus, M. J., Ed.; American Society for Bone and Mineral Research: Washington, DC, 2003, pp 47–487.
48. Kalu, D. N. *Endocrinology* **1984**, *115*, 507–512.
49. Kalu, D. N.; Liu, C. C.; Hardin, R. R.; Hollis, B. W. *Endocrinology* **1989**, *124*, 7–16.
50. Kyburz, D.; Corr, M. *Springer Semin. Immunopathol.* **2003**, *25*, 79–90.
51. Sasaki, A.; Boyce, B. F.; Story, B.; Wright, K. R.; Chapman, M.; Boyce, R.; Mundy, G. R.; Yoneda, T. *Cancer Res.* **1995**, *55*, 3551–3557.
52. *Guidelines for the Preclinical and Clinical Evaluation of Agents Used in the Prevention or Treatment of Postmenopausal Osteoporosis.* United States Federal Drug Adminstration: Rockville, MD, 1994.
53. *Guidelines for Preclinical Evaluation and Clinical Trials in Osteoporosis.* World Health Organization: Geneva, 1998 (www.who.int/en/).
54. Bekker, P. J.; Holloway, D. L.; Rasmussen, A. S.; Murphy, R.; Leese, P. T.; Holmes, G. B.; Dunstan, C. R.; DePaoli, A. M. *J. Bone Miner. Res.* **2004**, *19*, 1059–1066.
55. Black, D. M.; Thompson, D. E.; Bauer, D. C.; Ensrud, K.; Musliner, T.; Hochberg, M. C.; Nevitt, M. C.; Suryawanshi, S.; Cummings, S. R. *J. Clin. Endocrinol. Metab.* **2000**, *85*, 4118–4124.
56. Bouxsein, M. L.; Uchiyama, T.; Rosen, C. J.; Shultz, K. L.; Donahue, L. R.; Turner, C. H.; Sen, S.; Churchill, G. A.; Muller, R.; Beamer, W. G. *J. Bone Miner. Res.* **2004**, *19*, 587–599.
57. Morrison, N. A.; Qi, J. C.; Tokita, A.; Kelly, P. J.; Crofts, L.; Nguyen, T. V.; Sambrook, P. N.; Eisman, J. A. *Nature* **1994**, *367*, 284–287.
58. Ralston, S. H. *J. Clin. Endocrinol. Metab.* **2002**, *87*, 2460–2466.
59. Brown, M. A.; Haughton, M. A.; Grant, S. F.; Gunnell, A. S.; Henderson, N. K.; Eisman, J. A. *J. Bone Miner. Res.* **2001**, *16*, 758–764.
60. Rogers, M. J.; Gordon, S.; Benford, H. L.; Coxon, F. P.; Luckman, S. P.; Monkkonen, J.; Frith, J. C. *Cancer* **2000**, *88*, 2961–2978.
61. Sato, M.; Grasser, W.; Endo, N.; Akins, R.; Simmons, H.; Thompson, D. D.; Golub, E.; Rodan, G. A. *J. Clin. Invest.* **1991**, *88*, 2095–2105.
62. Gallagher, J. C. *Rheum. Dis. Clin. North Am.* **2001**, *27*, 143–162.
63. Fontana, A.; Delmas, P. D. *Curr. Opin. Rheumatol.* **2001**, *13*, 333–339.
64. Chapuy, M. C.; Pamphile, R.; Paris, E.; Kempf, C.; Schlichting, M.; Arnaud, S.; Garnero, P.; Meunier, P. J. *Osteoporosis Int.* **2002**, *13*, 257–264.
65. Grant, A. M.; Avenell, A.; Campbell, M. K.; McDonald, A. M.; MacLennan, G. S.; McPherson, G. C.; Anderson, F. H.; Cooper, C.; Francis, R. M.; Donaldson, C. et al. *Lancet* **2005**, *365*, 1621–1628.
66. Neer, R. M.; Arnaud, C. D.; Zanchetta, J. R.; Prince, R.; Gaich, G. A.; Reginster, J. Y.; Hodsman, A. B.; Eriksen, E. F.; Ish-Shalom, S.; Genant, H. K. *N. Engl. J. Med.* **2001**, *344*, 1434–1441.
67. Body, J. J.; Gaich, G. A.; Scheele, W. H.; Kulkarni, P. M.; Miller, P. D.; Peretz, A.; Dore, R. K.; Correa-Rotter, R.; Papaioannou, A.; Cumming, D. C. et al. *J. Clin. Endocrinol. Metab.* **2002**, *87*, 4528–4535.
68. Meunier, P. J.; Roux, C.; Seeman, E.; Ortolani, S.; Badurski, J. E.; Spector, T. D.; Cannata, J.; Balogh, A.; Lemmel, E. M.; Pors-Nielsen, S. et al. *N. Engl. J. Med.* **2004**, *350*, 459–468.
69. Pi, M.; Quarles, L. D. *J. Bone Miner. Res.* **2004**, *19*, 862–869.

70. Gutteridge, D. H.; Stewart, G. O.; Prince, R. L.; Price, R. I.; Retallack, R. W.; Dhaliwal, S. S.; Stuckey, B. G.; Drury, P.; Jones, C. E.; Faulkner, D. L. et al. *Osteoporosis Int.* **2002**, *13*, 158–170.
71. Woitge, H. W.; Seibel, M. J. *Clin. Lab. Med.* **2000**, *20*, 503–526.
72. NIH Consensus Development Panel on Osteoporosis Prevention, Diagnosis, and Therapy. *JAMA* **2001**, *285*, 785–795.
73. Lark, M. W.; Stroup, G. B.; James, I. E.; Dodds, R. A.; Hwang, S. M.; Blake, S. M.; Lechowska, B. A.; Hoffman, S. J.; Smith, B. R.; Kapadia, R. et al. *Bone* **2002**, *30*, 746–753.
74. Lark, M. W.; Stroup, G. B.; Dodds, R. A.; Kapadia, R.; Hoffman, S. J.; Hwang, S. M.; James, I. E.; Lechowska, B.; Liang, X.; Rieman, D. J. et al. *J. Bone Miner. Res.* **2001**, *16*, 319–327.
75. Lane, N. E.; Kelman, A. *Arthr. Res. Ther.* **2003**, *5*, 214–222.
76. Warmington, K.; Ominsky, M.; Bolon, B.; Cattley, R.; Stephens, P.; Lawson, A.; Lightwood, D.; Perkins, V.; Kirby, H.; Moore, A. et al. *J. Bone Miner. Res.* **2005**, *20*, S22.
77. Dunstan, C. R.; Boyce, R.; Boyce, B. F.; Garrett, I. R.; Izbicka, E.; Burgess, W. H.; Mundy, G. R. *J. Bone Miner. Res.* **1999**, *14*, 953–959.
78. Geesink, R. G.; Hoefnagels, N. H.; Bulstra, S. K. *J. Bone Joint Surg. Br.* **1999**, *81*, 710–718.
79. Riedel, G. E.; Valentin-Opran, A. *Orthopedics* **1999**, *22*, 663–665.
80. Millington-Ward, S.; McMahon, H. P.; Farrar, G. J. *Trends Mol. Med.* **2005**, *11*, 299–305.
81. Pelled, G. G T.; Aslan, H.; Gazit, Z.; Gazit, D. *Curr. Pharm. Design* **2002**, *8*, 1917–1928.

Biographies

Colin R Dunstan was born in Sydney, Australia, and studied at the University of Sydney, where he obtained a BSc (Hons) in 1977 and a PhD in 1991 under the supervision of Dr RA Evans. After completing his PhD he worked for 3 years in the laboratory of Dr GR Mundy in San Antonio, Texas. Dr Dunstan then worked for 6 years in the Pathology Department of Amgen Inc. in Thousand Oaks, California, as a Senior Research Scientist. Subsequently, in 2002 he was awarded a New South Wales Government BioFirst Award and returned to Sydney to take up his present position as Principal Research Fellow in the Bone Research Program at the ANZAC Research Institute. His scientific interests include all aspects of bone metabolism, with particular interest in osteoclast regulation and processes promoting cancer metastasis to bone.

Julie M Blair was born in Londonderry, Northern Ireland and studied at the University of Sheffield, UK, where she obtained a BSc (Hons) in 1991 and an MPhil in 1994. She also studied at the University of Bath, where she obtained a

PhD in 1998 under the direction of Dr J N Beresford. After a 3-year senior postdoctoral fellowship in the laboratories of Prof R L Vessella at the University of Washington, US, she moved to Prince of Wales Hospital, Sydney, Australia, as a senior hospital scientist. Subsequently, she took up the position of senior hospital scientist in the Molecular Bone Biology Laboratory at the ANZAC Research Institute in 2005. Her scientific interests include all aspects of bone metastasis, with a particular focus on prostate and breast cancer bone tumors.

Hong Zhou was born in Shanghai, China, and studied at Ningxia Medical Collage, where she obtained an MD in 1983. She obtained her PhD in 1992 under the supervision of Associate Professor KW Ng at the University of Melbourne, Australia. She then worked for 11 years in the laboratory of Prof T J Martin in St Vincent's Institute of Medical Research, Melbourne. In 2004, she took up the position as Senior Research Fellow at ANZAC Research Institute, the University of Sydney. Her scientific interests include bone cell biology and molecular biology, in particular, effects of steroid hormones on osteoblast differentiation, osteoblast–osteoclast interactions, and the biology of bone cancers.

Markus J Seibel holds the position of Professor and Chair of Endocrinology at the University of Sydney, Australia. He is also the Head of the Department of Endocrinology & Metabolism at Concord General Hospital, Sydney, and the Director of the Bone Research Program at ANZAC Research Institute. The focus of his clinical and research activities is the pathophysiology of bone metabolism, especially in osteoporosis and metastatic bone disease.

He completed his medical training in Germany (University of Heidelberg), Switzerland (University of Basel), and in the US (Columbia University New York), and until 2001 was the Vice-Director at the Department of Endocrinology at the University of Heidelberg, Germany. In November 2001, he moved to Sydney, Australia.

Markus was the Past-President of the German Academy of Bone and Joint Sciences. He is a member of the Professional Practice Committee of the American Society of Bone and Mineral Research (ASBMR) and a member of the

International Osteoporosis Foundation's (IOF) Scientific Advisory Board. He is on the Editorial Boards of the *Journal of Clinical Endocrinology and Metabolism, Calcified Tissue International, Journal of Clinical Densitometry, Clinical Endocrinology, Clinical Laboratory*, and others. He has written over 200 scientific articles, reviews, editorials, and book chapters, and is the editor of five books.

Comprehensive Medicinal Chemistry II
ISBN (set): 0-08-044513-6

ISBN (Volume 6) 0-08-044519-5; pp. 495–520

6.22 Hormone Replacement

A W Meikle, University of Utah School of Medicine, Salt Lake City, UT, USA

6.22.1 Disease State/Diagnosis

Testosterone (**Table 1**) is the principal androgen in human circulation. Androgens stimulate and maintain masculine sexual characteristics of the genital tract, secondary sexual characteristics, fertility, and anabolic effects of somatic tissues.[1,2] Testosterone may be used clinically at physiological doses for androgen replacement therapy, but pharmacological androgen therapy is often abused. The goal of testosterone replacement therapy is to attempt to duplicate the normal physiological pattern of androgen exposure to the body.[3] This is currently best accomplished with testosterone, rather than synthetic derivatives of testosterone. Thus, an understanding of the normal and pathophysiology of testosterone secretion and action is required as a basis for replacement therapy.

Testosterone deficiency and infertility are major manifestations of male hypogonadism. Hypogonadism is often used to denote testosterone deficiency. Some clinical disorders in males present with infertility, but without testosterone

Table 1 Structure of testosterone and its derivatives

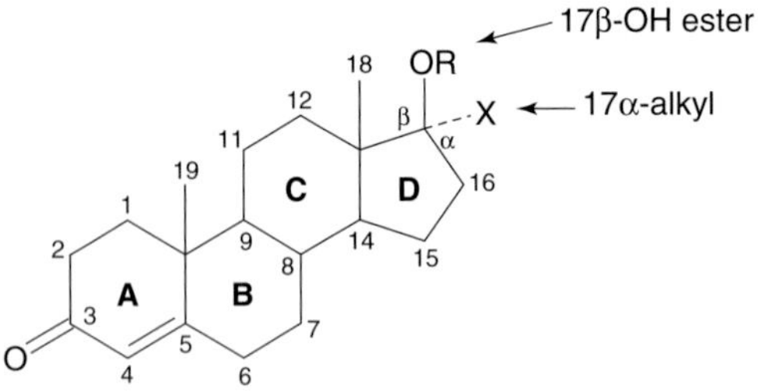

Generic name	*R*	*X*	*Other modifications*
Natural androgens			
Testosterone	H	H	
5α-Dihydrotestosterone	H	H	4,5-ane
Unmodified 17α esters			
Testosterone propionate	$COCH_2CH_3$	H	
Testosterone enanthate	$CO(CH_2)_5CH_3$	H	
Testosterone cypionate	$COCH_2CH_2C_5H_9$	H	
Testosterone undecanoate	$CO(CH_2)_9CH_3$	H	
Modified 17α esters			
Methenolone acetate	$COCH_3$	H	1-CH_3; 1,2-ene; 4,5-ane
Nandrolone phenylpropionate	$COCH_2CH_2C_6H_5$	H	19-nor CH_3
Nandrolone decanoate	$CO(CH_2)_8CH_3$	H	19-nor CH_3
17α-Alkylation			
Methyltestosterone	H	CH_3	
Fluoxymesterone	H	CH_3	9-F; 11-OH
Methandrostenolone	H	CH_3	1,2-ene
Oxandrolone	H	CH_3	C_2 replaced by O; 4,5-ane
Oxymethelone	H	CH_3	2 = CHOH; 4,5-ane
Stanozolol	H	CH_3	4,5-ane; {3,2-c}pyrazole
Danazole	H	CCH	{2,3-d}isoxazole
Norethandrolone	H	CH_2CH_3	19-nor CH_3
Ethylestrenole	H	CH_2CH_3	19-nor CH_3; 3-H
Modified androgen			
Mesterolone	H	H	1-CH_3; 4,5-ane

deficiency, and in many both testosterone deficiency and infertility coexist. Male hypogonadism is a relatively common disorder (**Tables 2** and **3**), particularly in aging males, and in those with diabetes, the metabolic syndrome, and Klinefelter's syndrome. It affects patients' fertility, sexual function, and general health.[1,2] The clinical features of male hypogonadism depend on the age and stage of sexual development.[1,2] If testicular function is deficient or if there is androgen unresponsiveness during fetal life, the phenotype is that of a female, and the classic presentation is testicular feminization syndrome. Both testosterone deficiency and infertility are observed with almost all patients with secondary or tertiary hypogonadism. In primary hypogonadism, infertility may present without testosterone deficiency, as observed in some males with Klinefelter's syndrome.

In adolescent males with prepubertal hypogonadism, androgen deficiency is seldom recognized before the typical age for onset of puberty except in adolescents with associated growth retardation or other anatomic and endocrine abnormalities. During the expected pubertal years, hypogonadism is suspected if the testes fail to enlarge normally or sexual maturation fails to occur.[1,2]

Pubertal failure is well characterized by several clinical features, as shown in **Table 4**. In adults, regression of secondary sexual characteristics and loss of libido and impotence suggest the onset of hypogonadism, as summarized in **Table 4**. Infertility and androgen deficiency are manifestations of postpubertal hypogonadism. The clinical symptoms and signs may evolve slowly and subtly, and may be incorrectly attributed to aging in older males (**Table 2**). Male-pattern body hair growth often slows, while changes of the voice and size of the phallus, testes, and prostate may be undetectable. In younger adult males, a delay in temporal hair recession and balding may go unnoticed.

6.22.2 Disease Basis

An understanding of normal physiology helps in using laboratory testing to define the cause of gonadal dysfunction. Pulsatile gonadotropin-releasing hormone (GnRH) secretion by the hypothalamus stimulates the release of gonadotropins (luteinizing hormone (LH) and follicle-stimulating hormone (FSH)).[1,2] Leydig cells synthesize and secrete testosterone in response to LH, and testosterone and its metabolites dihydrotestosterone (DHT) and estradiol have feedback effects on LH secretion. The seminiferous tubule compartment comprises about 85% of the mass of the testicular mass and contains Sertoli cells that are responsive to FSH, thereby increasing production of an androgen-binding protein and enabling the testes to concentrate testosterone manifold above the serum levels. Inhibin secreted by Sertoli cells feeds back on GnRH and FSH secretion.[4]

Circulating testosterone levels demonstrate distinct episodic and diurnal rhythms.[1,2] Diurnal patterns show morning peak testosterone levels and nadir levels in the afternoon, which are more evident in younger than older males. Consequently, reference ranges for testosterone measurements are established for morning blood samples on at least two different days.

Pinpointing the cause and extent of hypogonadism makes it possible to tailor successful replacement therapy (**Table 3**). Primary hypogonadism may be associated with small testes, azoospermia, or oligospermia and elevated FSH but with normal testosterone and LH. Both diminished sperm and testosterone with elevation of LH and FSH also occur. Gynecomastia is observed more often in males and boys with primary than in secondary or tertiary hypogonadism because estradiol production relative to testosterone is excessive in response to FSH stimulation of aromatase of the testes. Secondary and tertiary hypogonadism is more likely than primary hypogonadism to have concurrent reductions of both sperm and testosterone production, and LH and FSH are inappropriately low.

Table 2 lists the disorders with congenital and acquired hypogonadotropic hypogonadism and hypergonadotropic hypogonadism. The causes may be congenital, hereditary, or acquired. The most common cause of primary hypogonadism is Klinefelter's syndrome (1 in 500–1000 males).[1,2] Aging,[5–7] obesity,[8] diabetes,[9] and the metabolic syndrome[10] are the most common secondary/tertiary causes and affect 20–30% of males with these disorders. Patients with Kallmann's syndrome have a deficiency of GnRH as a hereditary disorder.[11]

6.22.3 Genetic Disease Models

As summarized in **Table 2**, disturbance of GnRH causes hypogonadism in Kallmann's syndrome by affecting the *Kal-X* gene, but GnRH is also affected in the Prader–Lahart–Willi syndrome and idiopathic hypogonadotrophic hypogonadism. Tumors, infiltrations, trauma, irradiation, ischemia, and surgery may cause hypothalamic or pituitary dysfunction.[1,2] Isolated FSH deficiency is found with the Pasqualini syndrome.

Defects in primary testicular function are also summarized in **Table 2**. The causes range from congenital and chromosomal to acquired. There are also disorders of testosterone synthesis, metabolism, and action. Androgen-resistant syndromes range from complete androgen resistance to male infertility.

Table 2 Differential diagnosis of hypogonadotropic and hypergonadotropic hypogonadism

Hypogonadotropic hypogonadism (HH)	Hypergonadotropic hypogonadism
Congenital HH	*Congenital*
Idiopathic hypogonadotropic hypogonadism (IHH)	Klinefelter's sydrome
Kallmann syndrome	47, XYY syndrome
Adult-onset IHH	Dysgenetic testes
Fertile eunuch syndrome	Androgen receptor defects
Adrenal hypoplasia congenita	Testicular feminization
Genetic defects of the gonadotropin subunits	Reifenstein's syndrome
HH associated with other pituitary hormone deficiencies	5-α-reductase deficiency
HH associated with obesity	Androgen synthesis defects
Prader–Willi syndrome	
Laurence–Moon–Biedl syndrome	*Acquired*
	Myotonic dystrophy
Acquired HH	Cryptorchidism
Structural	Vanishing testes syndrome
Tumors	Hemochromatosis
Craniopharyngiomas	Trauma
Pituitary adenomas (e.g., prolactinoma, nonfunctioning tumor)	Mumps orchitis
Germinoma, glioma, meningioma	Radiation
Infiltrative disorders	Chemotherapy
Sarcoidosis, hemochromatosis, histiocytosis X	Autoimmune
Head trauma	Sertoli cell only syndrome
Radiation therapy	Human immunodeficiency virus (HIV)
Pituitary apoplexy	Hepatic cirrhosis
Primary hypothyroidism	Chronic renal failure
Functional	Idiopathic
Exercise	Aging
Dieting	Diabetes
Anabolic steroids	Metabolic
Glucocorticoid therapy	
Narcotics	
Critical illness (cancer)	
Diabetes	
Metabolic syndrome	
Obesity	
Aging (mixed with Leydig cell dysfunction)	

Table 3 Laboratory testing of hypogonadism

Hypothalamic	*Primary hypogonadism*	*Seminiferous tubule disease*	*Leydig cell failure*	*Pituitary disease*	*Hypothalamic disease*
Testosterone	Low	Normal	Low	Low	Low
LH	High	Normal	High	Low	Low
FSH	High	High	Normal	Low	Low
Sperm count	Low	Low	Low	Low	Low
LH and FSH response to GnRH	Normal	Not done	Not done	Low	Normal

Results summary and further testing recommendations.

1. Testosterone low, LH and FSH elevated → primary hypogonadism: order karyotype.
2. Testosterone low, LH and FSH normal or low → secondary hypogonadism, obtain PRL and CT scan of head to screen for mass lesion; remaining pituitary hormones must be tested for deficiency.
3. Testosterone and LH normal, FSH high → abnormal seminiferous tubule compartment; order semen analysis.
4. Testosterone, LH and FSH high → androgen resistance syndrome.

Table 4 Clinical presentation of peripubertal and postpubertal hypogonadism

Peripheral hypogonadism

- Small testes, phallus, and prostate (prepubertal testes are between 3 and 4 mL in volume and less than 3 cm long by 2 cm wide; peripubertal testes are between 4 and 15 mL in volume and 3–4 cm long by 2–3 cm wide)
- Lack of male-pattern hair growth
- Scant pubic and axillary hair, if adrenal androgens are also deficient
- Disproportionately long arms and legs (from delayed epiphyseal closure, eunuchoidism; with the crown-to-pubis ratio < 1 and an arm span more than 6 cm greater than height)
- No pubertal growth spurt
- No increase in libido or potency
- Reduced male musculature
- Gynecomastia
- Persistently high-pitched voice

Postpubertal hypogonadism

- Progressive decrease in muscle mass
- Loss of libido
- Impotence
- Infertility with oligospermia or azoospermia
- Hot flashes (with acute onset of hypogonadism)
- Osteoporosis
- Anemia
- Adult testes are usually 15–30 mL and 4.5–5.5 cm long by 2.8–3.3 cm wide
- Mild depression
- Reduced energy

A well-characterized disorder relates to congenital 5-α-reductase deficiency. In genetically affected males, congenital 5-α-reductase deficiency due to mutation of the type 2 enzyme protein[12] causes genital ambiguity and undermasculinization. At puberty these males become virilized, including phallic growth and, occasionally, masculine gender reorientation, but the prostatic development remains rudimentary. This disorder establishes the dependence of the urogenital sinus-derivative tissues on strong expression of 5-α-reductase as a local amplification mechanism. Azasteroid 5-α-reductase inhibitors[13] were developed that inhibit type 2, 5-α-reductase (finasteride; 4-azaandrost-1-ene-17-carboxamine, *N*-(1,1-dimethyl)-3-oxo-, (5α, 17β)) and both type 1 and 2,5-α-reductases (dutasteride; (5α, 17β)-*N*-(2,5-bis(trifluoromethyl)phenyl)-3-oxo-4-azaandrost-1-ene-17-carboxamide). These agents block more than 95% of testosterone entering the prostate from being converted to DHT and have been used clinically to treat benign and malignant prostate growth and to reverse male-pattern hair loss in males. They must be used with great caution in females because of the potential sexual differentiation disorders that may occur in male fetuses.

Estrogen resistance has also been characterized in males. The biological importance of estradiol in male physiology is important, as illustrated by developmental defects in bone and other tissues of a man[14] and mouse line[15] with genetic mutations inactivating the estrogen receptor ERα. Inactivating genetic mutations of ERβ has little effect on male mouse phenotype,[16] but mutations in humans have yet to be reported. Males with aromatase deficiency not only have the same phenotype as in estrogen resistance but estrogen therapy produces significant bone maturation. These observations suggest that, at least in bone, estradiol is critical in males, but androgen action also contributes substantially to bone health. Bone mass in males is greater than in females, although circulating estradiol levels are lower in males than in premenopausal normal female. Animals and humans with androgen resistance but responsiveness to estrogen have normal bone mass, and androgens incapable of conversion to estrogen also increase bone mass in females. Further studies are thus needed to clarify the contributions of androgens and estrogens in bone physiology.

6.22.4 Current Treatment

6.22.4.1 Overview of Physiology of Androgens

Testosterone is synthesized by Leydig cells (interstitial) of the testes by an enzymatic sequence of steps beginning with cholesterol, which can be synthesized de novo, supplied from intracellular cholesterol ester stores or from circulating low-density lipoproteins.[1,2] LH mainly drives testicular testosterone secretion through its regulation of the rate-limiting conversion of cholesterol to pregnenolone. In males, testosterone is secreted transiently during the first trimester of intrauterine life and again during neonatal life and then continually after puberty to sustain virilization. After 30–40 years of age, circulating total and free testosterone levels decline gradually as gonadotrophin and sex hormone-binding globulin (SHBG) levels increase – the so-called 'andropause.' Andropause is attributed to impaired hypothalamic regulation of testicular function, Leydig cell attrition, and dysfunction.

Endogenous adrenal androgens contribute negligibly to direct virilization of males,[1,2] but in individuals with congenital adrenal hyperplasia they may be converted in sufficient amounts to testosterone to cause virilization. In children and females, adrenal androgens make a proportionately larger contribution to the much lower circulating testosterone concentrations, which are ~5–10% of values in males.

Testosterone circulates in blood avidly bound to SHBG.[1,2] Endogenous sex steroids and parenteral administration of testosterone that maintains physiological hormone concentrations (transdermal, injections, depot implants), have minimal effects on altering blood SHBG concentrations. Circulating SHBG levels may be elevated by acute or chronic liver disease, estrogens, thyroxine, and androgen deficiency. In contrast, obesity, glucocorticoids, androgens, protein-losing states, and genetic SHBG deficiency markedly lower SHBG. Under physiological conditions, 60–70% of circulating testosterone is SHBG-bound, with the remainder bound to albumin at lower affinity, and 1–2% remaining nonprotein-bound or free. The free (nonprotein-bound) fraction is the most biologically active while the albumin-bound plus the free testosterone is considered the bioavailable fraction of circulating testosterone. Free testosterone levels can be estimated by tracer equilibrium dialysis or ultrafiltration methods or calculated by a variety of nomograms based on immunoassays or mass spectrometry total testosterone and SHBG. The direct analog assay[17] and the free testosterone index are clearly invalid.

Testosterone is inactivated in the liver, kidney, gut, muscle, and adipose tissue but predominantly by hepatic oxidases, and in the liver predominantly by the cytochrome P450 3A family and by hepatic conjugation to glucuronides (phase II metabolism), which are excreted by the kidney.[1,2]

Testosterone undergoes metabolism to DHT predominantly in the skin and prostate and to estradiol. Approximately 4% of circulating testosterone is metabolized to the more potent androgen, DHT,[1,2,18] which has a three- to 10-fold greater molar potency than testosterone in binding to androgen receptors. Testosterone is converted to DHT by the enzyme 5-α-reductase that originates from two distinct genes (I and II). Type 1 5-α-reductase (EC 1.3.1.30) is expressed in liver, kidney, skin, and brain, and type 2, 5-α-reductase (EC 1.3.1.30) is characteristically robustly expressed in the prostate but also at lower levels in skin (hair follicles) and liver.

About 0.2% of testosterone is also converted by the enzyme aromatase (EC 1.14.14.1) to estradiol, which activates ERs. In normal males, about 80% of circulating estradiol is derived from extratesticular aromatization. The metabolic clearance rate of testosterone is reduced by the elevation of circulating SHBG levels, aging, liver dysfunction, and reduced hepatic blood flow. Rapid hepatic metabolic inactivation of testosterone reduces oral bioavailability[1,2] and duration of action of parenterally administered testosterone. Body mass index also correlates positively with the clearance of testosterone from the circulation. In androgen replacement, these limitations must be considered for parenteral depot testosterone formulations (e.g., injectable testosterone esters, testosterone implants, or transdermal testosterone) and oral delivery systems that involve portal bypass (buccal, sublingual, gut, lymphatic, or synthetic androgens).

Sex steroids initiate their biological actions by binding to their respective receptors located in most tissues and cause activation of the receptor and the biological response.[18] Local metabolism of testosterone can modulate and amplify the biological response by converting testosterone to DHT or estradiol.[18] The magnitude of the direct effects on the androgen receptor relative to indirect effects via active metabolites varies between androgens and target tissues. In the prostate, the magnification of testosterone to DHT is profound.

6.22.5 Clinical Trial Issues

Most clinical trials with testosterone replacement treatment have been done in postpubertal males with hypogonadism from various causes. With few exceptions, they have been open-label; double-blind studies would be a challenge, although preferable. Total testosterone concentrations have generally been used for the diagnosis of androgen deficiency and for adequacy of replacement testosterone therapy. Free or bioavailable testosterone concentrations may provide a better basis for diagnosis of deficiency and for adequate replacement therapy. Intrinsic potency, bioavailability, and rate of clearance from the circulation are determinants of the biological actions of androgens.

6.22.5.1 Goal of Androgen Therapy

A safe general principle in androgen replacement therapy is to approach normal physiology with concentrations of testosterone (350–1050 ng dL^{-1}) and its active metabolites[3] and this should not have untoward health hazards on the prostate, serum lipids, and cardiovascular, liver, and lung function. Thus, unphysiologically high or low serum testosterone concentrations should be avoided. Normal physiological responses to androgen replacement therapy allow virilization in prepubertal males and restoration or preservation of virilization in postpubertal males. Self-administration that is convenient, causes minimal discomfort, and results in reproducible daily pharmacokinetics at a reasonable cost is preferred. None of the currently available androgen replacement therapies achieves the ideal, but improvements have been made in achieving these goals.

6.22.5.2 Historical Aspects

Oral pure testosterone was ineffective, but the methyl derivative (methyltestostereone) was effective orally, as were implanted subcutaneous testosterone pellets. Testosterone ester injections gained widespread acceptance in the 1950s.[19] Other testosterone derivatives were made for their anabolic properties. Testosterone undecanoate (TU) administered orally became available for clinical use in some countries in the 1970s. Developers of androgens have avoided derivatives of testosterone with hepatic toxicity and have focused on delivering pure testosterone by oral, injectable, or transdermal preparations. Testosterone is metabolized to potent metabolites by steroid 5-α-reductase to form DHT and aromatase to form estradiol and chemical modification of testosterone results in substances that are poor substrates for these enzymes.

6.22.5.3 Adults

A 5 mg delivery dose of a patch or gel system or 200 mg of either testosterone enanthate or cypionate intramuscularly (IM) every 2 weeks is administered for androgen replacement therapy in males with hypogonadism. If IM testosterone enanthate or cypionate is used, an injection of 100 mg produces a better pattern of testosterone levels, but higher doses at less frequent intervals deviate much more from the physiologic normal testosterone range.[3]

The efficacy of androgen replacement therapy is best assessed by monitoring the patient's serum testosterone responses,[1,2] because variability in response to testosterone therapy in hypogonadal males in libido, potency, sexual activity, feeling of well-being, motivation, energy level, aggressiveness, stamina, and hematocrit is considerable. Increases in body hair, muscle mass and strength, and bone mass may require months to years of therapy. In sexually immature, eunuchoidal males, androgen replacement therapy may also cause secondary sexual characteristics and long bone growth.

When injectable forms of testosterone are administered, testosterone levels during therapy should be in the mid to normal range 1 week after an injection. Some hypogonadal males treated with testosterone esters experience fluctuations in sexual function, energy level, and mood, which are associated with fluctuations in serum testosterone concentrations between injections.

With patch and gel systems the recommended beginning dose is 5 mg testosterone for adults; smaller doses are recommended for some elderly males. Measurement of serum testosterone concentrations about 12 h after application after daily treatment for 7–14 days will provide information about adequate dosing. Then dosing adjustments can be made. Counseling of patients and their partners before beginning androgen replacement is recommended to help reduce or alleviate adjustment problems of increased sexual interest and performance.

Males with prepubertal hypogonadotropic hypogonadism require the combined treatment with human chorionic gonadotropin (hCG) plus human menopausal gonadotropins to initiate sperm production and fertility. In those with a selective deficiency of GnRH, such as Kallmann's syndrome, pulsatile GnRH therapy has been shown to stimulate testosterone production and spermatogenesis.

6.22.5.4 Prepubertal

Initiation of androgen replacement therapy in prepubertal hypogonadism is usually begun at about 14 years of age, and earlier if clinically indicated.[1,2] Growth, virilization, and psychological adjustment should be monitored. Distinguishing between simple delayed puberty and hypogonadism is often a challenge, and transient androgen therapy can be given until permanent hypogonadism is established.[1,2]

In boys gradual replacement therapy with testosterone less than an adult replacement dose is justified (50 or 100 testosterone enanthate or cypionate IM monthly[1,2] or 2.5 mg daily of transdermal testosterone), and the goal is to duplicate the changes in testosterone occurring naturally in normal boys and resulting in gradual virilization and progression of secondary sexual development. Then, the dosage of testosterone ester may be 50–100 mg every 2 weeks or 2.5 mg Androderm nightly for 12 h for approximately 6 months.[3] Stopping replacement for 3–6 months allows an assessment of the spontaneous onset and progression of puberty. If spontaneous puberty is not in evidence, androgen therapy is reinstituted for another 6 months with 100 mg testosterone ester every 2 weeks or 2.5 mg Androderm daily.

6.22.5.5 Testosterone Preparations

6.22.5.5.1 Intramuscular preparations

6.22.5.5.1.1 Testosterone esters: overview

Testosterone fatty acid esters at the 17 position prolongs IM retention and duration of activity of testosterone commensurate with fatty acid length (**Figure 1**).[49] When administered IM,[3] the androgen ester is slowly absorbed into the circulation where it is then rapidly metabolized to active unesterified testosterone. Testosterone propionate, a short fatty acid, releases testosterone for only 2–3 days and is not suitable for long-term replacement therapy. Intermediate-acting preparations, including testosterone enanthate, cypionate, and cyclohexane carboxylate are suitable for clinical use and have similar steroid release profiles when injected IM. Common regimens are administration of 200 mg of either testosterone enanthate or cypionate once every 2 weeks IM or 100 mg weekly.[3] T-*trans*-4-n-butylcyclohexyl-carboxylate and TU are even longer-acting preparations with sustained therapeutic blood levels for about 6–12 weeks depending on the dose.[20] As summarized in **Figure 1**, injections of testosterone enanthate (200–250 mg injected every 2 weeks) result in a maximal supraphysiological testosterone serum concentration as high as 51 nmol L^{-1} shortly after injection and then decline to the lower range of normal testosterone serum concentration (12 nmol L^{-1}) before 2 weeks.[3]

6.22.5.5.1.2 Nandrolone phenylpropionate and decanoate

Nandrolone phenylpropionate and decanoate are 17β-hydroxyl esters of 19-norT that have prolonged action following injection, and are used primarily to treat refractory anemias rather than for androgen replacement therapy.[3]

6.22.5.5.1.3 Testosterone buccilate

A dose of 600 mg testosterone buccilate injected IM in hypogonadal males produced serum testosterone concentrations within the normal range for about 8 weeks with a half-life of 29.5 days.[20] Serum DHT and SHBG concentrations were within the normal range; estradiol was only slightly increased and gonadotropins were significantly suppressed. No adverse biochemical and prostate effects were reported. This preparation may be used for male contraceptive therapy or replacement for hypogonadism.

6.22.5.5.1.4 Testosterone undecanoate

Testosterone undecanoate (TU) injections of 500 and 1000 mg in hypogonadal males resulted in increased mean serum testosterone levels from less than 10 nmol L^{-1} to 47.8 ± 10.1 and 54.2 ± 4.8 nmol L^{-1}, respectively, after about 1 week. By 50–60 days, serum testosterone levels decreased and reached the lower normal limit for adult males with a terminal

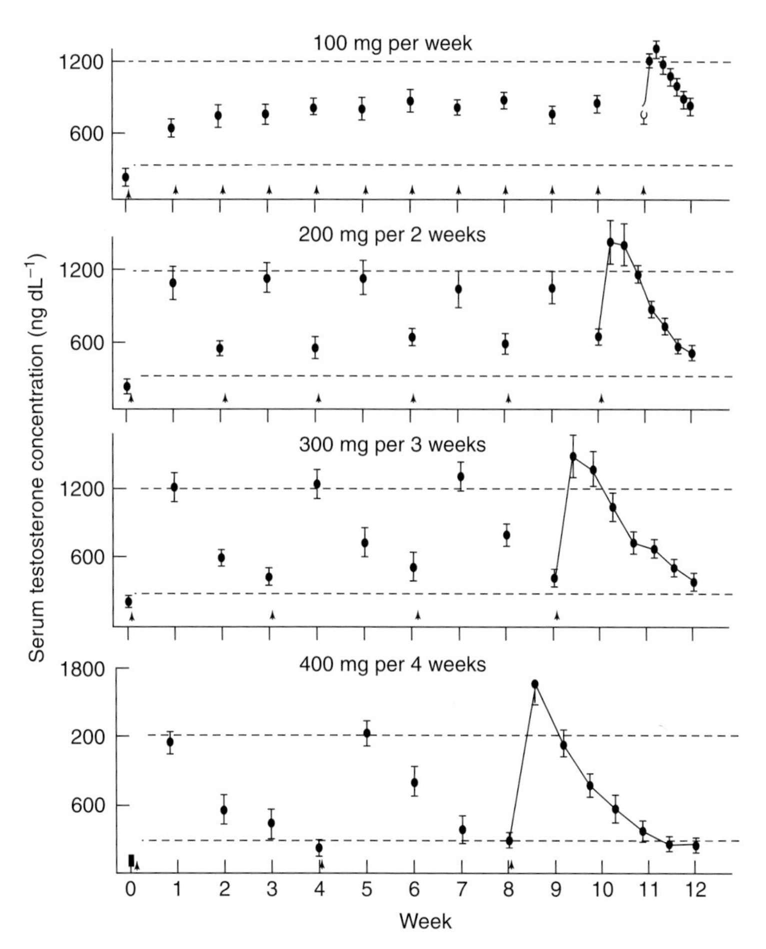

Figure 1 Serum testosterone concentrations during testosterone replacement therapy in adult primary hypogonadal men. Testosterone enanthate was administered by IM injection (arrows) for 12 weeks in four dosage regimens: 100 mg weekly; 200 mg every 2 weeks; 300 mg every 3 weeks; and 4000 mg every 4 weeks. Blood was sampled weekly until the last dose and more frequently thereafter. (Reproduced with permission from Snyder, P. J.; Lawrence, D. A. *J. Clin. Endocrinol. Metab.* **1980**, *51*, 1335–1339. Copyright 1980, The Endocrine Society.)

elimination half-life of 18.3 ± 2.3 and 23.7 ± 2.7 days.[21] Estradiol and DHT paralleled testosterone and remained within normal limits. In these short-term studies, no serious side effects were reported. IM TU has promise for long-term substitution therapy in hypogonadism and hormonal male contraception.[21]

6.22.5.5.1.5 Subcutaneous testosterone implants

Pure fused pellets or Silastic capsules implanted subcutaneously release testosterone in amounts to maintain physiologic concentrations of testosterone for between 4 and 6 months.[22]

Jockenhovel *et al.*[23] implanted 6–200 mg fused crystalline testosterone pellets in the subdermal fat tissue of the abdomen in hypogonadal males, and after an initial peak on the first day of administration a stable plateau lasted for 63 days. Testosterone values fell below the normal range by 180 days but took about 300 days to reach baseline. However, serum estradiol and DHT were elevated from day 21 to 105, and SHBG was decreased from day 21 to 168. Implants of testosterone pellets could be used for testosterone replacement therapy as well as for reversible male contraception.

6.22.5.5.2 **Transdermal testosterone**

Transdermal testosterone creams have been used in the treatment of microphallus in children and hypogonadism in adults.[3]

6.22.5.5.2.1 Percutaneous Dihydrotestosterone in hypogonadal males

A 125 mg dose of hydroalcoholic gel of DHT applied twice daily to the skin can result in sustained concentrations of DHT.[24] As expected, the ratio of DHT to testosterone increased to around 5 (normal ranges between 0.1 and 0.2); serum testosterone, estradiol, and SHBG concentrations did not rise and gonadotropins were unchanged. Treatment of testosterone-deficient hypogonadal males reportedly improved virilization and sexual function and decreased plasma low- and high-density lipoprotein (HDL) cholesterol levels moderately without causing enlargement of the prostate as determined by ultrasound study.[24]

6.22.5.5.3 Transscrotal testosterone

Scrotal skin is at least five times more permeable to testosterone than other skin sites. Testoderm and Testoderm with adhesive applied to the scrotum of hypogonadal males delivers 4 and 6 mg of testosterone daily.[25] Peak testosterone concentrations occurred at 3 weeks and then remained stable. DHT concentrations were elevated, and normal-range androgen concentrations (testosterone plus DHT) and estradiol were achieved in 80% of hypogonadal males.[3]

6.22.5.5.3.1 Nonscrotal testosterone patch therapy (Androderm)

After a 5 mg Androderm system was applied daily to nonscrotal skin (back, abdomen, thighs, and upper arms) at about 22:00 h, testosterone was continuously absorbed during the 24-h dosing period, and the serum testosterone concentration profile for more than 90% of males resembled the normal circadian variation in healthy young males (**Figure 2**).[3] In addition, bioavailable testosterone (BT), DHT and estradiol serum testosterone concentrations paralleled the serum testosterone profile (**Figure 2**) after patch application and remained within the normal reference range.

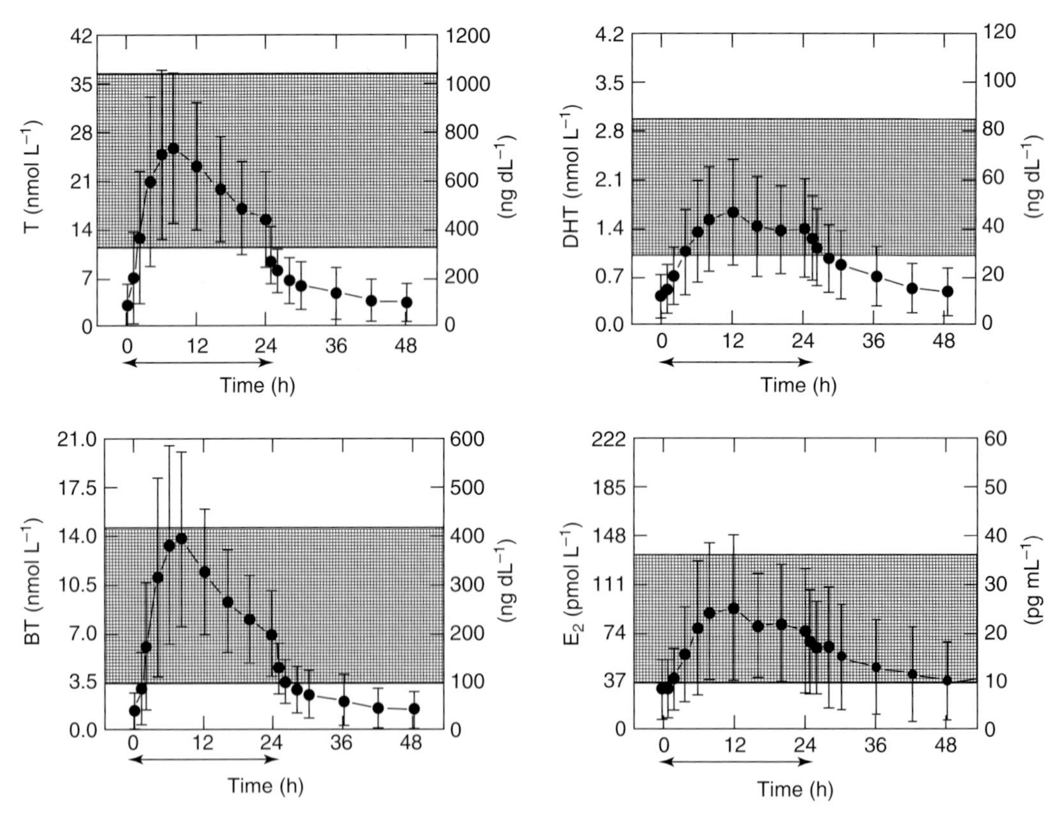

Figure 2 Serum concentration profiles of testosterone (T), BT, DHT, and 17β-estradiol (E_2) during and after the nighttime application of two testosterone transdermal systems (TTDs) to the back of 34 hypogonadal men (mean ± SD). The shaded areas represent 95% confidence intervals for morning hormone levels in normal men between the ages of 20 and 65 years. The arrow denotes the 24-h duration of TTD application.[50] (Reproduced with permission from Meikle, A. W.; Arver, S.; Dobs, A. S.; Sanders, S. W.; Rajaram, L.; Mazer, N. A. *J. Clin. Endocrinol. Metab.* **1996**, *81*, 1832–1840. Copyright 1996, The Endocrine Society.)

Hypogonadal symptoms, such as fatigue, mood, and sexual function, as determined from questionnaires and nocturnal penile tumescence, improved with Androderm therapy.[3]

6.22.5.6 Management of Skin Irritation

Chronic contact dermatitis (mainly from the alcohol component) occurs in about 10% of males following several weeks of use of Androderm. Two drops of 0.1% triamcinolone acetonide cream applied to the skin under the central drug reservoir greatly reduces contact dermatitis and itching without significantly affecting testosterone delivery or pituitary adrenal function.[3]

6.22.5.6.1 Comparison with intramuscular testosterone

Sixty-six patients were randomized to receive either Androderm or IM testosterone enanthate (200 mg every 2 weeks) treatment for 6 months.[3] In the Androderm-treated group the percentage of normal-range serum concentrations of testosterone, bioavailable testosterone, DHT, and estradiol was 82, 87, 76, and 81%, respectively, compared with 72, 39, 70, and 35%, respectively, for IM testosterone injections. Sexual function assessment and lipid profiles were similar between groups.

6.22.5.6.1.1 AndroGel

AndroGel 5, 7.5, or 10 g contain packets of 50, 75, or 100 mg testosterone, respectively; a pump dispenser is also available for dosing. On average approximately 10% of the applied testosterone dose is absorbed across skin of average permeability during a 24-h period, producing circulating concentrations of testosterone observed in normal males.[26] The 24-h pharmacokinetic profiles of testosterone for patients (**Figure 3**) maintained on 5 or 10 g AndroGel (to deliver 50 or 100 mg of testosterone, respectively) for 30 days show testosterone values in the normal range. The mean (±SD) daily testosterone concentration produced by AndroGel 10 g on day 30 was 792 (±294) ng dL^{-1} and by AndroGel 5 g 566 (±262) ng dL^{-1}.[3]

After skin application, AndroGel dries quickly, and release of testosterone into the systemic circulation[26] is sustained. After the first 10 g dose testosterone increases in serum testosterone within 30 min, and by 4 h of application most patients have a serum testosterone concentration within the normal range for the 24-h dosing interval. Steady-state levels are achieved by the second or third day of dosing.

Serum testosterone levels decrease after stopping application of the gel, but serum testosterone concentrations remain in the normal range for 24–48 h after the last application and take 5 days to return to baseline. By 180 days of treatment, 87% achieved an average serum testosterone level within the normal range.

Following AndroGel doses of 5 and 10 g day^{-1}, DHT and estradiol concentrations increased in parallel with testosterone concentrations, and the DHT-to-testosterone ratio and estradiol levels remained within the normal range. SHBG concentrations decreased modestly (1–11%) and serum levels of LH and FSH in males with

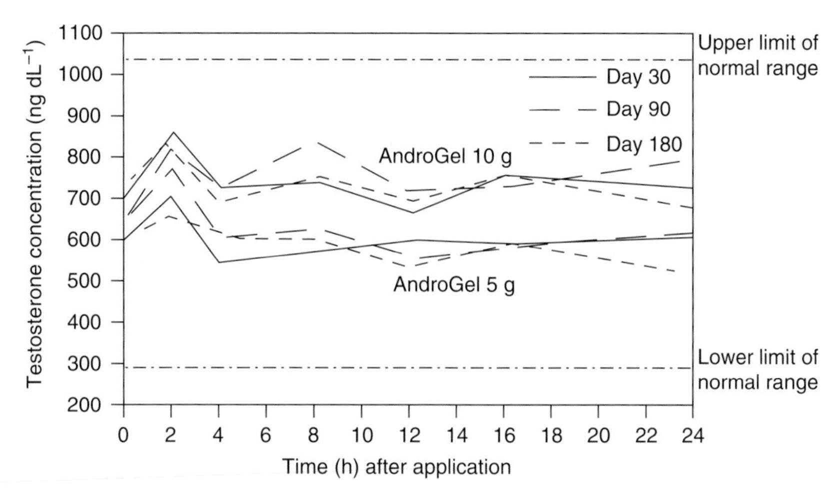

Figure 3 The mean steady-state serum testosterone concentrations in patients applying 5 g or 10 g Androgel once daily. (Data on file, Unimed Pharmaceuticals.)

hypergonadotropic hypogonadism fell in a dose- and time-dependent way during treatment with AndroGel. Although skin reactions were reported in 3–5% of patients using AndroGel for up to 6 months, none necessitated discontinuation of drug.

6.22.5.6.1.2 Potential partner testosterone transfer

Dermal testosterone transfer between males dosed with AndroGel and their untreated female partners following AndroGel use indicated that unprotected female partners had a serum testosterone concentration more than twice the baseline value at some time during the study. Using a barrier, such as a shirt, to cover the application site(s) abolished this transfer.[3]

A 5 g tube of Testim applied each morning to the upper arm or shoulder delivers physiologic amounts of testosterone. Following skin application, testosterone concentrations peak at 2–4 h, and average normal concentrations of testosterone are achieved in the 24-h period. In clinical trials, 74% of patients on Testim treatment had serum testosterone concentrations within the normal range at 90 days.[27,28] If the testosterone concentrations do not reach the normal range, the dose should be increased to 10 g. Precautions to avoid skin transfer to another person should be followed. Testim is easy to apply and dries quickly and is then invisible.

A new testosterone (T) gel (Cellegy Pharmacueticals, Inc.) has undergone clinical testing in about 200 hypogonadal males. The initial dose applied to the skin daily raises the serum concentrations of testosterone, and steady state is reached by 14 days. A metered canister allows the dose to be adjusted in 10 mg increments. After adjustment of the dose based on the testosterone concentrations on day 14, over 90% of males achieved a 24-h concentration average within the normal physiologic range; the ratio of estradiol to testosterone and DHT to testosterone was also normal. Its safety profile in terms of chemistries, lipid profiles, polycythemia, and prostate-specific antigen (PSA) were comparable to other transdermal testosterone preparations. Six months of therapy also resulted in improvement of hip and spine bone mineral density (BMD) of about 4%.[29]

6.22.5.6.1.3 Buccal

Striant is a buccal tablet preparation that is applied to the depression between the gum above the upper incisors and releases testosterone across the buccal mucosa into the peripheral circulation. Twice-daily application may be needed to sustain testosterone concentrations in the therapeutic range.[30,31]

6.22.5.6.1.4 Oral testosterone

The first orally active, synthesized derivative of testosterone was 17α-methyltestosterone (**Table 5**). After oral ingestion, peak blood levels occurred between 1.5 and 2 h, and its serum half-life was about 150 min, indicating several daily doses would be required to maintain a therapeutic level of the steroid. Hepatic toxicity, characterized as cholestasis, peliosis, and elevation of liver enzymes and reduction of HDL cholesterol has limited its use.[3]

Fluoxymesterone is a 17α-methyltestosterone steroid with fluorine in the 9 position and has a longer half-life in serum than the parent steroid, but risk of hepatoxicity also limits its clinical use.

6.22.5.6.1.5 Mesterolone

Mesterolone is derived from 5-α-dihydrotestosterone with a methyl group in the 1 position, is not hepatoxic, and is not metabolized to estrogen. Dosing is difficult to monitor.[3]

6.22.5.6.1.6 Testosterone undecanoate

When given with a meal, a 17-β long aliphatic side-chain ester to testosterone produces good absorption from the gut via the lymphatics rather than delivery to the hepatic portal system. Testosterone enters the systemic circulation (63%) without substantial hepatic transformation, and testosterone levels within the normal range are achieved over the first few hours of administration.[3] This preparation is not currently available in the US; an improved formulation being tested in the US may overcome some of these deficiencies, making it more suitable for androgen replacement therapy.

6.22.5.6.1.7 Human chorionic gonadotropin

hCG produced by the human placenta has a specific β subunit that binds to the LH receptors on Leydig cells and stimulates endogenous testosterone production from the testes.[3,32] In prepubertal boys between the ages of 4 and 9 years with cryptorchidism not caused by anatomical obstruction, hCG treatment is given to produce testicular descent. The following regimen is given: 4000 United States Pharmacoepia (USP) units of hCG three times weekly for

Table 5 Pharmacokinetics and safety of androgens

Preparation	*Peak*	*Trough*	*Testosterone monitoring*	*DHT*	*Estradiol*	*Liver dysfunction*	*HDL-cholesterol*	*Skin irritation*
Testosterone propionate	1 day	2–3 days	None	Dose-dependent	Dose-dependent	None	Dose-dependent	None
Testosterone enanthate	1–2 days	10–14 days	1 week	Dose-dependent	Dose-dependent	None	Dose-dependent	None
Testosterone cypionate	1–2 days	10–14 days	1 week	Dose-dependent	Dose-dependent	None	Dose-dependent	None
Testosterone buccilate	2–4 weeks	12–14 weeks	4–6 weeks	Dose-dependent	Dose-dependent	None	Dose-dependent	None
Testosterone pellets	1 month	6 months	3–4 months	Dose-dependent	Dose-dependent	None	Dose-dependent	None
Testosterone undeconoate	2–6 h	2–6 h		Increased	Normal	None	Dose-dependent	None
Testosterone cyclodextrin	1 h	6 h	2–4 h	Normal	Normal	None	Modest	None
Methyltestosterone	1.5–2 h	4–5 h	None	Low	Low	Yes	30% decrease	None
Fluoxymesterone				Low	Low	Yes		None
Mesterolone				Low	Low	None		None
Testosterone buccal	3–5 h	12 h	24 h	Normal	Normal	None	Modest	None
Testosterone gel	20–24 h	Days	24 h	Normal	Normal	None	Modest	Minimal
Testosterone scrotal	3–5 h	20–24 h	12 h	Elevated	Normal	None	Modest	Minimal
Testosterone nonscrotal	6–8 h	24 h	12 h	Normal	Normal	None	Modest	Yes
DHT topical	4–8 h	20–24 h	DHT 12 h	Elevated	Low	None	Modest	Minimal

DHT, dihydrotestosterone; HDL, high-density lipoprotein.

Table 6 Treatment of male hypogonadism

Group	*Goal of therapy*	*Plasma testosterone*	*Preparation*	*Usual dose*
Delayed adolescence	Short-term maintenance, initial	100–300 ng dL^{-1}	hCG	500 IU IM 1–2 times per week
			Androderm	2.5 mg patch
				12 h at night
			TE or TC	50–100 mg q 3–4 weeks
	Subsequent	300–400 ng dL^{-1}	Androderm	2.5 mg daily
			TE or TC	100 q 2 weeks
Adult	Long-term maintenance	400–1000 ng dL^{-1}		
Hypogonadotropic hypogonadism			GnRH[a]	5–30 μg SC q 2 h
Hypogonadism			hCG	1000–4000 IU IM 1–3 times per week
			Androderm, Testoderm, Androgel Striant	5 mg per day 30 mg b.i.d.
			Testoderm, scrotal	4 or 6 mg per day
			TE or TC	200 mg q 2 weeks or 100 q 1 week IM
	Subreplacement		Fluoxymesterone	5–10 mg per day p.o.
			Methyltestosterone	5–25 mg daily
			Testosterone	200 p.o. q 2 weeks
			Undecanoate[b]	

hCG, human chorionic gonadotropin; GnRH, gonadotropin-releasing hormone; TE, testosterone enanthate; TC, testosterone cypionate.
[a] Experimental, requires programmed pump.
[b] Not available in the US.

3 weeks; then 5000 units every 2 days for four injections, followed by 15 injections of 500–1000 units for 6 weeks; and then 500 units three times a week for 4–6 weeks.

In boys and men with gonadotropin deficiency, hCG is an alternative to testosterone therapy in inducing pubertal development in boys and treating androgen deficiency in men. A dosage of 1000–2000 international units (IU) IM 2–3 times weekly is used. The dose is then slowly increased during the next 1–2 years to the adult dosage of 2000–4000 IU IM, 2–3 times weekly (**Table 6**). The dosage needs to be adjusted to normalize serum testosterone concentrations. Therapy in prepubertal boys can be assessed by the clinical response to treatment and following the progression of virilization and growth and the serum testosterone concentration.[33]

hCG therapy has two major advantages over testosterone replacement therapy. It stimulates growth of the testes, which may be important to boys with delayed puberty, and may stimulate sufficient intratesticular testosterone production to facilitate initiation of spermatogenesis.[3,32]

The disadvantages of hCG treatment include frequent injections, expense, and elevation of estradiol relative to testosterone, which may cause gynecomastia. If blocking antibodies to hCG are induced, its efficacy may be diminished.[1,2]

6.22.5.6.2 Summary of androgen replacement therapy

6.22.5.6.2.1 Shorten and revise or delete

The hormone replacement goals for the management of male hypogonadism depend upon both the cause and the stage of sexual development in which gonadal failure occurs (**Table 6**). Androgen replacement therapy is indicated to stimulate and sustain normal secondary sexual characteristics, sexual function, and behavior in prepubertal boys and males with either primary or secondary hypogonadism. Several options for replacement therapy are available in various countries, and the availability of those preparations should be taken into consideration by the clinician before therapy is instituted. The goal is to normalize physiology as closely as possible and at the lowest cost. All of these factors influence the decision, which is shared with the physician and patient.

The available testosterone esters for IM injection (testosterone propionate, testosterone enanthate, testosterone cypionate, and testosterone cyclohexane carboxylate) do not achieve physiologic serum testosterone profiles for the treatment of male hypogonadism (**Table 6**).[3] Testosterone propionate has a duration of action of 2–4 days. The other esters have duration of actions of 2–3 weeks depending on the dose administered. However, they do achieve therapeutic responses when administered in appropriate doses and intervals. Doses and injection intervals frequently prescribed in the clinic result in initial supraphysiologic androgen levels and subnormal levels prior to the next injection. Injections of 100 mg testosterone enanthate or cypionate IM at weekly intervals would more closely approximate normal physiology than 200–250 mg every 2–3 weeks. However, an injection frequency more often than 2 weeks may be unacceptable to many patients. Further, since both the short-acting and intermediate-acting esters show maximal serum concentrations shortly after injection, there is no advantage in combining short-acting testosterone esters (i.e., testosterone propionate) and longer-acting esters (i.e., testosterone enanthate) for testosterone replacement therapy.

Of the clinically available injectable androgen esters, 19-nortestosterone hexoxyphenyipropionate shows the best pharmacokinetic profile. However, as a derivative of the naturally occurring testosterone, 19-nortestosterone might not possess its full pharmacodynamic spectrum, and therefore, is not an ideal drug for the treatment of male hypogonadism.[3]

Oral TU is easy to administer, and a new formulation may reduce the need for multiple daily doses. The most favorable pharmacokinetic profiles of testosterone are observed using either the transdermal patch or gel systems. The scrotal system has the disadvantage of supraphysiologic DHT concentrations and is not easy to use for many patients. Daily evening application of Androderm (a transdermal testosterone patch) results in serum testosterone concentrations in the normal range, mimicking the normal circadian rhythm. The nontransscrotal system has the disadvantage of local skin reactions, which can often be successfully managed with topical glucocorticoid administration. Testoderm transdermal testosterone system (TTS) also produces a circadian variation of testosterone delivery and satisfactory replacement therapy. Patch adherence is a problem for many patients, which reduces its therapeutic efficacy. Skin reactions are much less with Testoderm TTS than with Androderm. Androgel also produces testosterone levels within the normal range but does not have a distinct diurnal delivery pattern. Skin reactions are much less frequent with gels than with patch systems, and since the gels are transparent after application, they are more discreet. Person-to-person transfer is a potential problem with gels, but this problem can be avoided with appropriate precautions. Currently, several satisfactory options for testosterone replacement therapy are available to clinicians in various countries. Of course, the treatment must be tailored for each patient, and important considerations include ease of use, physiologic replacement, side effects, and cost. All of these should be discussed with the patient before making the selection. Although testosterone esters, testosterone enanthate and cypionate are effective, safe, and the least expensive androgen preparations available (particularly if self-administered), they require administration by injection into a large muscle. Testosterone enanthate and cypionate are considered equally effective and have been popular in the past for the treatment of hypogonadal males (**Table 6**).

6.22.5.6.3 Potential benefits of androgen therapy

6.22.5.6.3.1 Testosterone replacement therapy in andropause (aging males)

An age-related decrease in testosterone begins during the fourth decade of life, and by age 60 about 20% of males have serum testosterone concentrations of less than 300–350 ng dL^{-1}.[1,7,34] Free and bioavailable testosterone decline more profoundly than total testosterone because SHBG concentrations rise in aging males. Thus, measurement of free or bioavailable testosterone is recommended when the total testosterone concentration is between 200 and 400 ng dL^{-1}.

6.22.5.6.3.2 Testosterone effects on body composition and muscle strength

Testosterone replacement in older or younger hypogonadal males has generally confirmed that body fat and visceral fat mass declines and lean body mass and strength increase. In addition, some aspects of muscle strength, fat-free mass, and visceral fat improved.[35–37] Testosterone replacement may improve insulin sensitivity in hypogonadal, overweight males with type 2 diabetes by altering body composition, but studies are conflicting and further investigation is needed.

6.22.5.6.3.3 Mood

Within 3 months of testosterone therapy in elderly and younger hypogonadal males improvement is observed in spatial cognition, sense of well-being, libido, irritability, anger, fatigue, self-reported sense of energy ($49 \pm 19\%$ to $66 + 24\%$; $P = 0.01$), and sexual function ($24 \pm 20\%$ to $66 + 24\%$; $P < 0.001$) within the first 3 months of initiation of therapy.[36,38]

6.22.5.6.3.4 Impotence

Many males have androgen deficiency, develop impotence, and treating other than testosterone deficiency and their impotence may not be possible with testosterone therapy.[3]

6.22.5.6.4 Leptin

Leptin is an appetite suppressant and there are some studies on the influence of testosterone on serum leptin. Serum testosterone is negatively associated with leptin in males, but testosterone does not appear to be the major determinant of serum leptin in males. Testosterone therapy will reduce leptin and the relationship between serum leptin and adiposity is maintained.[3,39]

6.22.5.6.5 Osteoporosis

6.22.5.6.5.1 Androgen deficiency and osteoporosis

Androgens contribute to peak cortical and trabecular bone mass achieved during pubertal development and the subsequent amount of bone lost with hypogonadism.[3]

Surgical or medical castration in males causes a reduction in BMD, elevated biochemical markers of bone turnover, osteoporosis, and risk of fractures.[3] Hypogonadism with aging and with delayed puberty may contribute to osteoporosis. These findings establish that testosterone replacement in young or elderly hypogonadal males improves BMD.[3,36] Behre *et al.*[40] reported that, during the first year of testosterone therapy of hypogonadal males BMD increased by 26%. Testosterone replacement therapy improves BMD, markers of bone formation and resorption, and can also treats osteoporosis in hypogonadal males.[36,41,42]

6.22.5.6.5.2 Mechanism of action of androgens on bone

The mechanism of androgen effects on bone density suggests that DHT is not important but that conversion to estrogen has a major influence, as evidenced by males who have advanced osteoporosis and have estrogen resistance or aromatase deficiency. However, cortical BMD is greater in males than in females,[3,43] for unknown reasons.

6.22.5.6.5.3 Therapy of androgen-deficiency bone loss

Androgen replacement therapy has been shown to increase in BMD in younger and older hypogonadal males, and particularly those with skeletal immaturity.[44] Whether long-term testosterone therapy will prevent osteoporosis without causing undue risks awaits appropriate study. Androgens are contraindicated in some males with hypogonadism and prostate cancer; in such males, designer estrogens, such as raloxifene or bisphosphonates, may prevent bone loss.

6.22.6 Unmet Medical Needs

6.22.6.1 Precautions of Androgen Therapy in Aging Males

Androgen replacement therapy has been reported to cause water retention, polycythemia, hepatotoxicity, sleep apnea, prostate enlargement, and to reduce HDL cholesterol.[3] Hepatic toxicity is associated with derivatives of testosterone rather than pure testosterone. Polycythemia is observed more commonly in males receiving injectable testosterone. Thus, hematocrit should be measured periodically to minimize the risk of polycythemia. HDL cholesterol lowering is more profound with oral methyltestosterone than transdermal or injectable testosterone, but this effect is dose dependent, with higher doses causing more profound lowering of HDL cholesterol. Although testosterone is associated with prostate enlargement and prostate cancer, testosterone replacement therapy is rarely associated with an increase in urinary tract voiding symptoms, leading to cessation of therapy. In addition, prostate cancer surveillance can be done by measurement of PSA during the first 6-month interval after initiating testosterone replacement therapy and at yearly intervals thereafter. There is no compelling evidence to support a causal relationship between prostate cancer induction and testosterone replacement therapy, but further study is needed to exclude such a relationship. Testosterone replacement therapy in hypogonadal males delays time to ischemia, improves mood, and is associated with potentially beneficial reductions of total cholesterol and serum tumor necrosis factor-α (TNF-α).[45]

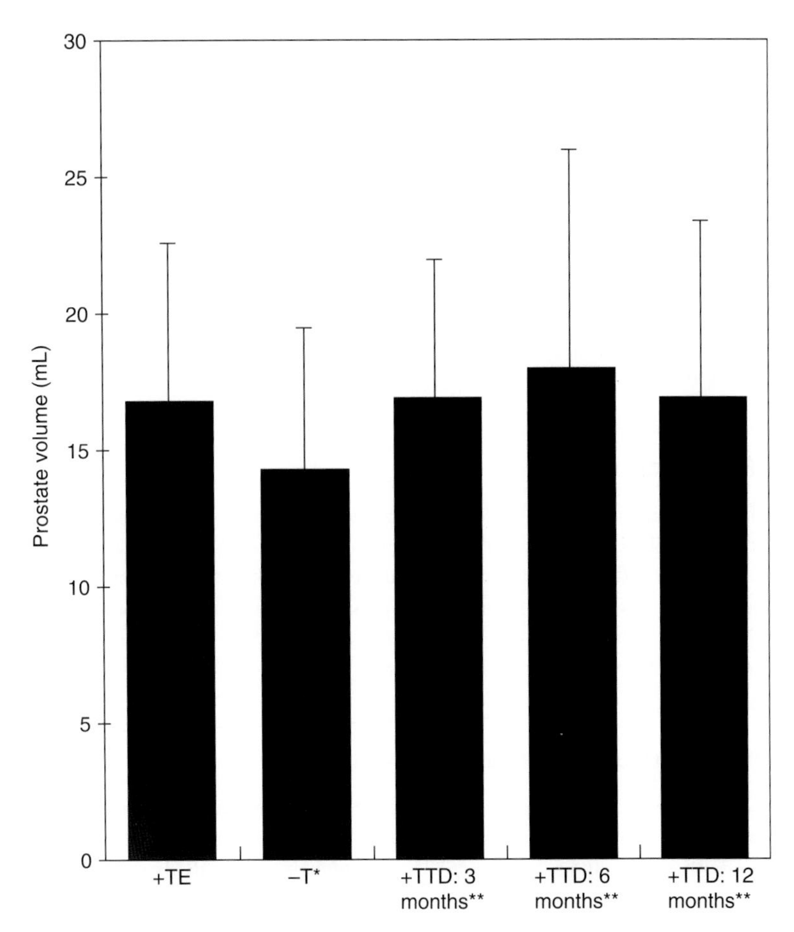

Figure 4 Prostate volumes measured by transrectal ultrasound (TRUS). T, testosterone; TE, testosterone enanthate; TTD, testosterone transdermal system. (Reproduced from Meikle, A. W.; Arver, S.; Dobs, A. S.; Adolfsson, J.; Sanders, S. W.; Middleton, R. G.; Stephenson, R. A.; Hoover, D. R.; Rajaram, L.; Mazer, N. A. *Urology* **1997**, *49*, 191–196, with permission from Elsevier.)

6.22.6.2 Prostate Disease

6.22.6.2.1 Benign prostate enlargement

Both the transition and peripheral zones of the prostate enlarge as males age in response to testosterone.[3,46]

Androgen withdrawal or deficiency results in a significant reduction in prostate volume of both zones (**Figure 4**),[47] but the age relationship to prostate size is maintained. Testosterone replacement therapy does not elevate serum concentrations of PSA above those expected in males of comparable age. Lower urinary tract symptoms (*see* 6.24 Incontinence (Benign Prostatic Hyperplasia/Prostate Dysfunction)) are strongly influenced by hereditary factors and prostate volume[48]; it follows that androgen replacement therapy is associated with symptomatic prostate symptoms.

6.22.6.2.2 Prostate cancer

Prostate cancer is an androgen-responsive cancer, but evidence that testosterone replacement therapy results in prostate cancer is lacking.[3] An undiagnosed prostate cancer may grow in response to testosterone replacement therapy. Therefore, a PSA and digital rectal examination are recommended in males aged 40 or over before initiation of androgen replacement therapy for hypogonadism with surveillance at 6-month intervals initially and then based on age.

6.22.6.3 Other Considerations of Treatment of Androgen Deficiency

The main use of androgen replacement therapy is in the management of males with testosterone deficiency.

Reversible causes of the hypogonadism should be established and therapy directed toward the underlying cause.[3] Hyperprolactinemia from a pituitary tumor can be treated with bromocriptine, which often corrects the testosterone deficiency. Other tumors of the pituitary or hypothalamus may cause other pituitary deficiencies and require surgical or irradiation therapy. Systemic acute and chronic illness and glucocorticoid therapy will cause hypogonadism; androgen replacement therapy may be considered in such patients. Androgen replacement therapy in aging males requires monitoring for diseases incident to age but may offer benefits in bone preservation, lean body mass, mood, and intellectual and sexual function.

For males desiring fertility, hormonal therapy to enhance spermatogenesis may be successful.

Gynecomastia, lack of secondary sexual characteristics, or small testes might contribute to psychological problems or social embarrassment for some hypogonadal adolescent and mature males. Gynecomastia generally does not regress and may worsen during hormonal replacement therapy of hypogonadism, which can be treated with plastic surgery (reductive mammoplasty). Some patients with secondary or tertiary hypogonadism may experience testicular enlargement in response to gonadotropin therapy. Implantation of testicular prostheses is an option for males who have psychological concerns about testicular size.

6.22.7 New Research Areas

Large, long-term clinical trials are needed in males over the age of 50 years to determine the benefits and risks of androgen replacement therapy. These studies could determine if cardiovascular risk, prostate cancer, frailty, fractures, osteoporosis, cognitive function, and life expectancy are influenced by androgen replacement therapy. A critical area of uncertainty is what testosterone concentration is needed to provide adequate androgenic effects. This is an important question because it relates to the concentration of testosterone where benefits might or might not be expected. Should free, bioavailable, or total testosterone concentrations be used?

6.22.8 Selective Androgen Receptor Modulators

The development of selective androgen receptor modulators (SARMs) that have anabolic effects on the muscle but that do not have adverse effects on the prostate and cardiovascular system has been of considerable interest for the treatment of older men with testosterone deficiency.[51–54] The nonsteroidal SARMs differ from testosterone in that they are not converted to active metabolites, such as estradiol and DHT, but they act as agonists in muscle and bone and only partial agonists in prostate and seminal vesicles. More favorable pharmacokinetics and androgen receptor specificity of nonsteroidal SARMs compared to testosterone provide promise for unique pharmacological interactions with the androgen receptor and actions that may allow more specific indications for their clinical use. The proposed mechanisms to explain the tissue selectivity of SARMs as nuclear receptor modulators include ligand binding specific conformational changes in the ligand-binding domain and modulation of surface topology, inducing protein–protein interactions between the androgen receptor and other co-regulators. If the ligand-specific receptor conformation and protein–protein interactions differ in tissues, this could result in tissue-specific gene regulation from interactions with androgen response elements, co-regulators, or transcription factors.

SARMs that promote both muscle strength and bone mechanical strength might have a considerable advantage over other treatments for osteoporosis that only affects bone by increasing bone density.[51–54]

SARMs that are prostate antagonists, but muscle agonists, might find a use in the treatment of benign prostatic hyperplasia or prostate cancer. Other SARMs with gonadotropin suppression but without prostate stimulation might find application for male contraception. Most of the SARMs research has been conducted in experimental animals. Therefore, the above-mentioned potential applications of SARMs require extensive clinical trials before beneficial use will be a reality.

Whether the absence of both DHT and estrogen action is good or bad is unknown.

6.22.8.1 Evidence-Based Approaches

Regardless of the study, evidence-based results must be part of the outcome analysis before the results will be widely accepted. This will be difficult because endpoints in humans are often suggestive but may not be definitive.

6.22.8.1.1 Stimulation of lueteinizing hormone and follicle-stimulating hormone

In males seeking fertility because of deficits in secretion of gonadotropins, much research is needed to enhance sperm count and fertility in a reliable and consistent way.

Acknowledgments

This work was supported by grants DK-45760, DK-43344, and RR-00064 from the National Institute of Health USPHS and Department of Internal Medicine and ARUP Institute for Experimental Pathology.

References

1. Santen, R. J. Testis: Function and Dysfunction. In *Reproductive Endocrinology*, 4th ed.; Yen, S. S. C., Jaffe, R. B., Barbieri, R. L., Eds.; Saunders: Philadelphia, PA, 1999, pp 632–668.
2. Griffin, J. E.; Wilson, J. D. Disorders of the Testes and Male Reproductive Tract. In *William's Textbook of Endocrinology*, 9th ed.; Wilson, J. D., Foster, D. W., Eds.; Saunders: Philadelphia, 1998, pp 819–875.
3. Meikle, A. W. Androgen Replacement Therapy of Male Hypogonadism. In *Endocrine Replacement Therapy in Clinical Practice*; Meikle, A. W., Ed.; Humana Press: Totowa, NJ, 2003, pp 333–368.
4. Anawalt, B. D.; Bebb, R. A.; Matsumoto, A. M.; Groome, N. P.; Illingworth, P. J.; McNeilly, A. S.; Bremner, W. J. *J. Clin. Endocrinol. Metab.* **1996**, *81*, 3341–3345.
5. Morley, J. E.; Kaiser, F. E.; Perry, H. M., III; Patrick, P.; Morley, P. M.; Stauber, P. M.; Vellas, B.; Baumgartner, R. N.; Garry, P. J. *Metabolism* **1997**, *46*, 410–413.
6. Tenover, J. L. *J. Androl.* **1997**, *18*, 103–106.
7. Harman, S. M.; Metter, E. J.; Tobin, J. D.; Pearson, J.; Blackman, M. R. *J. Clin. Endocrinol. Metab.* **2001**, *86*, 724–731.
8. Glass, A. R.; Swerdloff, R. S.; Bray, G. A.; Dahms, W. T.; Atkinson, R. L. *J. Clin. Endocrinol. Metab.* **1977**, *45*, 1211–1219.
9. Dhindsa, S.; Prabhakar, S.; Sethi, M.; Bandyopadhyay, A.; Chaudhuri, A.; Dandona, P. *J. Clin. Endocrinol. Metab.* **2004**, *89*, 5462–5468.
10. Laaksonen, D. E.; Niskanen, L.; Punnonen, K.; Nyyssonen, K.; Tuomainen, T. P.; Valkonen, V. P.; Salonen, J. T. *J. Clin. Endocrinol. Metab.* **2005**, *90*, 712–719.
11. Pitteloud, N.; Hayes, F. J.; Boepple, P. A.; DeCruz, S.; Seminara, S. B.; MacLaughlin, D. T.; Crowley, W. F., Jr. *J. Clin. Endocrinol. Metab.* **2002**, *87*, 152–160.
12. Imperato McGinley, J.; Gautier, T.; Pichardo, M.; Shackleton, C. *J. Clin. Endocrinol. Metab.* **1986**, *63*, 1313–1318.
13. Steers, W. D. *Urology* **2001**, *58*, 17–24.
14. Smith, E. P.; Boyd, J.; Frank, G. R.; Takahashi, H.; Cohen, R. M.; Specker, B.; Williams, T. C.; Lubahn, D. B.; Korach, K. S. *N. Engl. J. Med.* **1994**, *331*, 1056–1061.
15. Lubahn, D. B.; Moyer, J. S.; Golding, T. S.; Couse, J. F.; Korach, K. S.; Smithies, O. *Proc. Natl. Acad. Sci. USA* **1993**, *90*, 11162–11166.
16. Couse, J. F.; Mahato, D.; Eddy, E. M.; Korach, K. S. *Reprod. Fertil. Dev.* **2001**, *13*, 211–219.
17. Winters, S. J.; Kelley, D. E.; Goodpaster, B. *Clin. Chem.* **1998**, *44*, 2178–2182.
18. Wilson, J. D.; George, F. W.; Griffin, J. E. *Science* **1981**, *211*, 1278–1284.
19. Nieschlag, E.; Behre, H. M. Pharmacology and Clinical Uses of Testosterone. In *Testosterone: Action, Deficiency, Substitution*; Nieschlag, E., Behre, H. M., Eds.; Springer-Verlag: Berlin, 1990, pp 92–108.
20. Behre, H. M.; Nieschlag, E. *J. Clin. Endocrinol. Metab.* **1992**, *75*, 1204–1210.
21. Nieschlag, E.; Buchter, D.; Von Eckardstein, S.; Abshagen, K.; Simoni, M.; Behre, H. M. *Clin. Endocrinol. (Oxf.)* **1999**, *51*, 757–763.
22. Handelsman, D.; Conway, A.; Boylan, L. *J. Clin. Endocrinol. Metab.* **1992**, *75*, 1326–1332.
23. Jockenhovel, F.; Vogel, E.; Kreutzer, M.; Reinhardt, W.; Lederbogen, S.; Reinwein, D. *Clin. Endocrinol. (Oxf.)* **1996**, *45*, 61–71.
24. De Lignieres, B. *Ann. Med.* **1993**, *25*, 235–241.
25. Place, V. A.; Atkinson, L.; Prather, D. A.; Trunell, N.; Yates, F. E. Transdermal Testosterone Replacement Through Genital Skin. In *Testosterone: Action, Deficiency, Substitution*; Nieschlag, E., Behre, H. M., Eds.; Springer-Verlag: Berlin, 1990, pp 165–180.
26. Wang, C.; Berman, N.; Longstreth, J. A.; Chuapoco, B.; Hull, L.; Steiner, B.; Faulkner, S.; Dudley, R. E.; Swerdloff, R. S. *J. Clin. Endocrinol. Metab.* **2000**, *85*, 964–969.
27. Bouloux, P. *Clin. Ther.* **2005**, *27*, 286–298.
28. Marbury, T.; Hamill, E.; Bachand, R.; Sebree, T.; Smith, T. *Biopharm. Drug Dispos.* **2003**, *24*, 115–120.
29. Meikle, A. W.; Matthias, D.; Hoffman, A. R. *BJU Int.* **2004**, *93*, 789–795.
30. Wang, C.; Swerdloff, R.; Kipnes, M.; Matsumoto, A. M.; Dobs, A. S.; Cunningham, G.; Katznelson, L.; Weber, T. J.; Friedman, T. C.; Snyder, P. et al. *J. Clin. Endocrinol. Metab.* **2004**, *89*, 3821–3829.
31. Korbonits, M.; Slawik, M.; Cullen, D.; Ross, R. J.; Stalla, G.; Schneider, H.; Reincke, M.; Bouloux, P. M.; Grossman, A. B. *J. Clin. Endocrinol. Metab.* **2004**, *89*, 2039–2043.
32. Liu, P.; Handelsman, D. J. Hormonal Therapy of the Infertile Man. In *Endocrine Replacement Therapy in Clinical Practice*; Meikle, A. W., Ed.; Humana Press: Totowa, NJ, 2003, pp 453–470.
33. Finkel, D.; Phillips, J.; Snyder, P. *N. Engl. J. Med.* **1985**, *313*, 651–655.
34. Tenover, J. Testosterone and the Older Man. In *Endocrine Replacement Therapy in Clinical Practice*; Meikle, A. W., Ed.; Humana Press: Totowa, NJ, 2003, pp 397–414.
35. Wang, C.; Swedloff, R. S.; Iranmanesh, A.; Dobs, A.; Snyder, P. J.; Cunningham, G.; Matsumoto, A. M.; Weber, T.; Berman, N. *J. Clin. Endocrinol. Metab.* **2000**, *85*, 2839–2853.
36. Snyder, P. J.; Peachey, H.; Berlin, J. A.; Hannoush, P.; Haddad, G.; Dlewati, A.; Santanna, J.; Loh, L.; Lenrow, D. A.; Holmes, J. H. et al. *J. Clin. Endocrinol. Metab.* **2000**, *85*, 2670–2677.
37. Bhasin, S.; Woodhouse, L.; Casaburi, R.; Singh, A. B.; Mac, R. P.; Lee, M.; Yarasheski, K. E.; Sinha-Hikim, I.; Dzekov, C.; Dzekov, J. et al. *J. Clin. Endocrinol. Metab.* **2005**, *90*, 678–688.
38. Hirshkowitz, M.; Orengo, C.; Cunningham, G. Androgen Replacement: Sexual Behavior, Affect and Cognition. In *Endocrine Replacement Therapy in Clinical Practice*; Meikle, A. W., Ed.; Humana Press: Totowa, NJ, 2003, pp 368–369.
39. Soderberg, S.; Olsson, T.; Eliasson, M.; Johnson, O.; Brismar, K.; Carlstrom, K.; Ahren, B. *Int. J. Obes. Relat. Metab. Disord.* **2001**, *25*, 98–105.
40. Behre, H. M.; Kliesch, S.; Leifke, E.; Link, T. M.; Nieschlag, E. *J. Clin. Endocrinol. Metab.* **1997**, *82*, 2386–2390.
41. Wang, C.; Swerdloff, R. S.; Iranmanesh, A.; Dobs, A.; Snyder, P. J.; Cunningham, G.; Matsumoto, A. M.; Weber, T.; Berman, N. *Clin. Endocrinol. (Oxf.)* **2001**, *54*, 739–750.

42. Katznelson, L.; Finkelstein, J. S.; Schoenfeld, D. A.; Rosenthal, D. I.; Anderson, E. J.; Klibanski, A. *J. Clin. Endocrinol. Metab.* **1996**, *81*, 4358–4365.
43. Bonjour, J.; Theintz, G.; Buchs, B.; Slosman, D.; Rizzoli, R. *J. Clin. Endocrinol. Metab.* **1991**, *73*, 555–563.
44. Finkelstein, J. S.; Neer, R. M.; Biller, B. M.; Crawford, J. D.; Klibanski, A. *N. Engl. J. Med.* **1992**, *326*, 600–604.
45. Malkin, C. J.; Pugh, P. J.; Morris, P. D.; Kerry, K. E.; Jones, R. D.; Jones, T. H.; Channer, K. S. *Heart* **2004**, *90*, 871–876.
46. Tenover, J. Androgen Therapy in Aging Men. In *Pharmacology, Biology, and Clinical Applications of Androgens*; Bhasin, S., Galelnick, H., Spieler, J., Swerdloff, R., Wang, C., Eds.; Wiley-Liss: New York, 1996, pp 309–318.
47. Meikle, A. W.; Arver, S.; Dobs, A. S.; Adolfsson, J.; Sanders, S. W.; Middleton, R. G.; Stephenson, R. A.; Hoover, D. R.; Rajaram, L.; Mazer, N. A. *Urology* **1997**, *49*, 191–196.
48. Meikle, A. W.; Bansal, A.; Murray, D. K.; Stephenson, R. A.; Middleton, R. G. *Urology* **1999**, *53*, 701–706.
49. Snyder, P. J.; Lawrence, D. A. *J. Clin. Endocrinol. Metab.* **1980**, *51*, 1335–1339.
50. Meikle, A. W.; Arver, S.; Dobs, A. S.; Sanders, S. W.; Rajaram, L.; Mazer, N. A. *J. Clin. Endocrinol. Metab.* **1996**, *81*, 1832–1840.
51. Kearbey, J. D.; Wu, D.; Gao, W.; Miller, D. D.; Dalton, J. T. *Xenobiotica* **2004**, *34*, 273–280.
52. Gao, W.; Kearbey, J. D.; Nair, V. A.; Chung, K.; Parlow, A. F.; Miller, D. D.; Dalton, J. T. *Endocrinology* **2004**, *145*, 5420–5428.
53. Yin, D.; He, Y.; Perera, M. A.; Hong, S. S.; Marhefka, C.; Stourman, N.; Kirkovsky, L.; Miller, D. D.; Dalton, J. T. *Mol. Pharmacol.* **2003**, *63*, 211–223.
54. Gao, W.; Reiser, P. J.; Coss, C. C.; Phelps, M. A.; Kearbey, J. D.; Miller, D. D.; Dalton, J. T. *Endocrinology* **2005**, *146*, 4887–4897.

Biography

A Wayne Meikle was born in Smithfield, Utah, and studied at Utah State University, where he obtained a BSc in 1962, and at Vanderbilt University School of Medicine, where he obtained his MD in 1965. His residency in Internal Medicine was conducted at Vanderbilt University and the University of Utah, where he also completed a fellowship in Endocrinology and Metabolism in 1969 under Frank H Tyler. He returned to the University of Utah in 1972 after 3 years at Fitzsimons US Army Hospital in the Medical Nutrition Laboratory. He is professor of Endocrinology and Metabolism and Pathology and is director of the Endocrine Testing Laboratory at ARUP laboratories, which is affiliated with the University of Utah School of Medicine. His scientific interests include male reproductive endocrinology, the regulation of prostate cancer growth, and methodology, particularly for quantitation of sex steroids, including their biological activity.

Comprehensive Medicinal Chemistry II
ISBN (set): 0-08-044513-6

ISBN (Volume 6) 0-08-044519-5; pp. 521–540

6.23 Urogenital Diseases/Disorders, Sexual Dysfunction and Reproductive Medicine: Overview

M Williams, Northwestern University, Chicago, IL, USA

6.23.1 Introduction

Diseases, disorders, and treatments of urogenenital tract function (that also include the kidney: *see* 6.25 Renal Dysfunction in Hypertension and Obesity) fall into two distinct camps, those that involve a distinct disease state, e.g., bladder dysfunction related to age, cancers of the kidney, bladder and prostrate, infections, etc. and more life-style-related disorders/uses, the majority related to sexual function, the most well known of which is erectile dysfunction (ED).[1]

Current treatment of defined urogenital disorders reflects a mixture of; surgical procedures, e.g., transurethral resection of the prostate (TURP); devices, e.g., stents, that can be inserted into the urethra to help improve urine flow; and drug therapy. The latter is growing in importance due to its inherent convenience and improved side effect profile. For instance, TURP has as a potential side effect impotence due to tissue and nerve damage resulting from the surgery.

The markets for the various disorders of the urogenenital tract that also include ED, female sexual dysfunction, and reproductive function (and dysfunction) are large and growing, especially with the aging of the population and a focus on a continued active lifestyle, both in terms of sexual function and exercise. The market for drugs acting on the urinary tract was $4.7 billion in 2003 and is anticipated to double by 2010.

6.23.2 Bladder Dysfunction

Disorders involving bladder dysfunction are captured under the acronym LUTS: lower urinary tract symptoms. These disorders include benign prostatic hyperplasia or hypertrophy (hyperplasia being an increase in the number of the prostate cells and hypertrophy an increase in cell size), incontinence or overactive bladder, and bladder outlet obstruction.[1,2]

6.23.2.1 Benign Prostatic Hyperplasia

Benign prostatic hyperplasia (BPH) is a progressive disease characterized by prostate stromal and epithelial cell 'thickening' or enlargement that may lead in time to prostate cancer, although this remains a controversial viewpoint. BPH can be asymptomatic or be associated with LUTS, the latter including urinary frequency and urgency, nocturia, decreased and/or intermittent force of stream, and incomplete bladder voiding (*see* 6.24 Incontinence (Benign Prostatic Hyperplasia/Prostate Dysfunction)).

BPH is an age-related condition with the prevalence of symptomatic BPH being 2.7% in males aged 45–49 years and 24% in males over 80 years of age. BPH is thought to be initiated by hormone-related changes (primarily testosterone) associated with aging, coupled with a redifferentiation of prostate mesenchymal cells. This leads to excessive cell proliferation, inflammatory cell infiltration and inhibition of apoptosis, leading to a loss of normal prostate function. Prostate hyperplasia can result in bladder outlet obstruction (BOO), the latter an increase in urethral resistance to urine flow, acute bladder distension, and the development of LUTS.

Treatment options for BPH that lead to improvements in urinary flow include lifestyle changes, 'watchful waiting,' drug therapy, nonsurgical procedures, and major surgery. If the BPH symptoms are mild and do not affect quality of life, 'watchful waiting' is often recommended. Individuals undergo regular checkups; when the symptoms cause discomfort, affect activities of daily living, or endanger the health, drug treatment is recommended. Drugs can relieve the common urinary symptoms associated with BPH by either reducing the size of the prostate gland or slowing prostate growth.[1,2] Drug classes used to treat BPH include: the nonselective α-adenoceptor blockers, e.g., terazosin (**2**), doxazosin (**1**), and tamsulosin (**3**), which relax both prostate smooth muscle and the urethra, leading to an increase in urinary flow; 5-α-reductase inhibitors, e.g., finasteride (**5**) and dutasteride (**6**), which block testosterone production; or a combination of the two.[1,3] Alfuzosin (**4**) is a more uroselective α_1-adrenoceptor antagonist. As a general rule, α_1-adrenoceptor antagonists fail to fully relieve BPH symptoms.

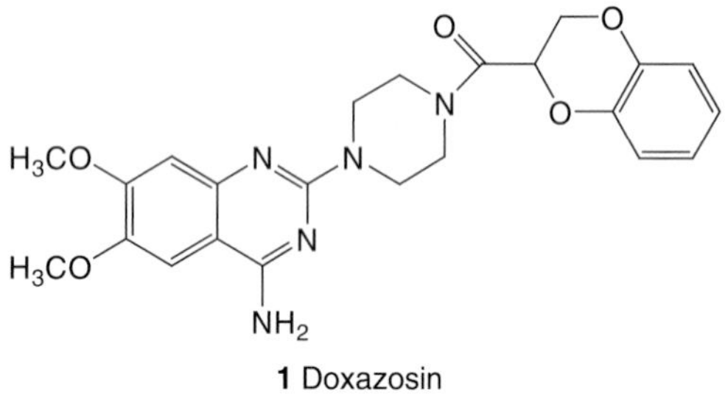

1 Doxazosin

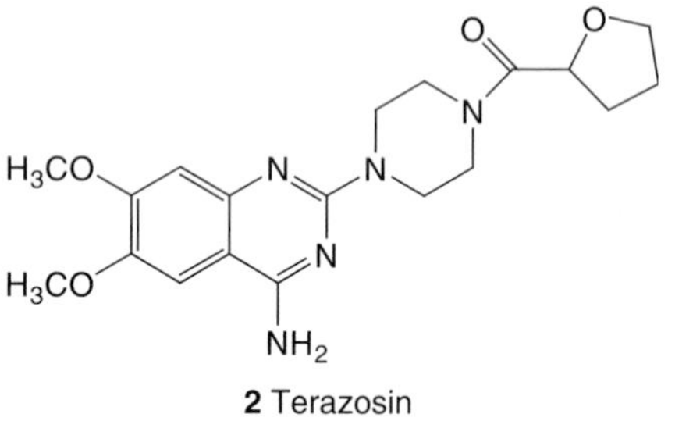

2 Terazosin

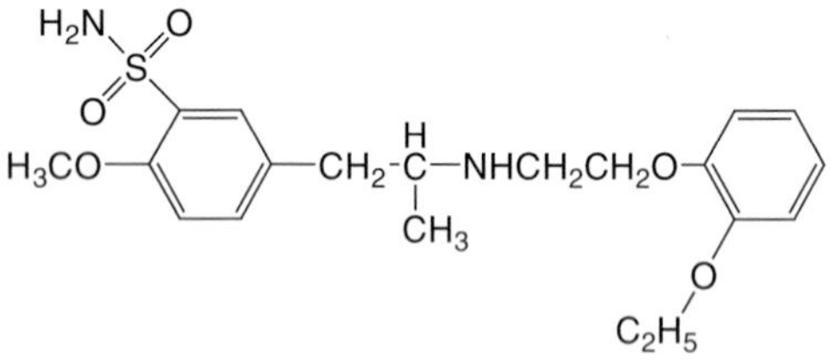

3 Tamsulosin

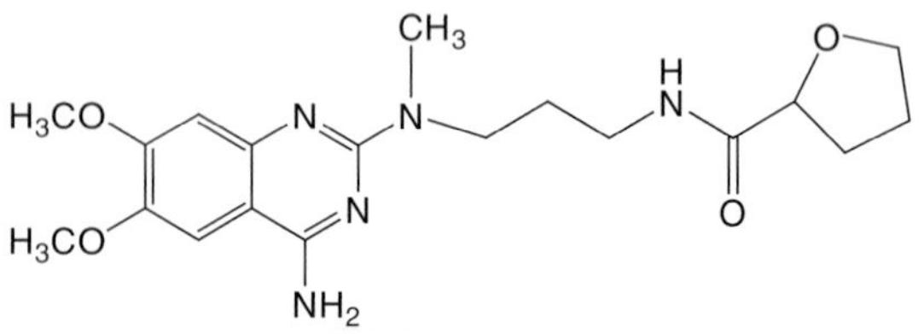

4 Alfuzosin

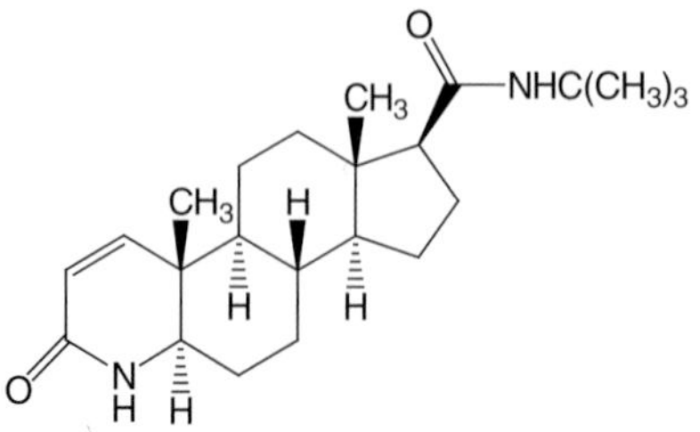

5 Finasteride

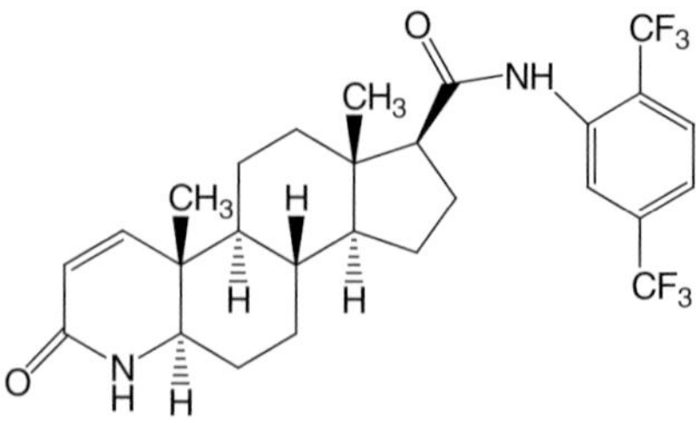

6 Dutasteride

A major concern with drug therapy for BPH (as well as incontinence) is the modest therapeutic index of the drugs used to treat these disorders, many of which are used chronically for disorders that are generally not life-threatening. In the field of α-adenoceptor antagonist new chemical entities (NCEs), the ability to design functionally selective (>100-fold) ligands among the six different types of receptor (α_{1A}, α_{1B}, and α_{1D}; α_{2A}, α_{2B}, and α_{2C}) has been challenging. The presence of synaptic and extrasynaptic receptors in the cardiovascular system additionally limits functional differences in separately modulating the urogential and vascular systems. Alternatives to prescription drugs in the treatment of BPH are herbal remedies or 'alternative medicines' that are thought to have fewer side effects than prescription medications and are hence 'safer.' One such agent is saw palmetto, a herbal medicine with an unknown mechanism of action, that is widely used to treat BPH, especially in Europe. While small clinical studies reported saw palmetto to be efficacious in BPH, a recent double-blind trial showed no difference between saw palmetto and placebo in terms of efficacy.[4]

If drug therapy fails to provide adequate symptom relief surgery is an option to help correct prostate hypertrophy. Surgical options for BPH include TURP and transurethral incision of the prostate (TUIP). TURP involves removal of the inner portion of the prostate while TUIP involves making one to two incisions into the prostate to reduce pressure. While TUIP is less invasive than TURP, the latter is the procedure of choice because of its effectiveness. Less invasive procedures for the treatment of BPH include transurethral needle ablation (TUNA), transurethral microwave hyperthermia (TUMT), and high-intensity focused ultrasound (HIFU). These procedures use heat in enlarged areas of the prostate to remove excessive growth. TUNA, TUMT, and HIFU are not as effective as TURP and only used in individuals who are unable to undergo surgery.

6.23.2.2 Bladder Outlet Obstruction (BOO)

Bladder outlet obstruction (BOO) is a disorder of the pelvic floor in females that is reflected as hesitancy in voiding, poor urine stream, and stop–start voiding.[5] BOO can be caused by prolapse of the large-anterior vaginal wall or by pelvic-floor dysfunction, and secondarily, may be due to surgical procedures for stress urinary incontinence (SUI). It is usually treated by catheterization.

6.23.2.3 Prostatitis

Prostatitis is an inflammation of the prostate gland that can be divided into: acute bacterial prostatitis, an acute bacterial infection of the prostate gland; chronic bacterial prostatitis, a recurrent infection of the prostate that is associated with chronic urinary infection; and nonbacterial prostatitis, inflammation in the absence of infection. Prostatitis is treated with quinolone antibiotics, anti-inflammatory agents, and α-adenoceptor blockers.

6.23.2.4 Urinary Incontinence/Overactive Bladder

Urinary incontinence (UI) is an involuntary loss of urine due to weakened bladder control (detrussor muscle function). This may be the result of: a loss of bladder tone; a lack of neural control over bladder function (as in Alzheimer's or Parkinson's diseases); or involuntary bladder muscle spasms, the latter being known as overactive bladder (OAB). Incontinence affects men and women equally with approximately 13 million individuals in the USA suffering from incontinence. Symptoms of incontinence include: frequency – urinating eight or more times in a 24-h period; night-time frequency or nocturia – urinating two or more times a night; urgency – sudden urges to urinate; urge incontinence – an urgent need to urinate followed by leakage or wetting incidents; and stress incontinence – loss of urine due to increased pressure on the bladder, caused by coughing, sneezing, laughing, or physical activity.[6]

Incontinence can be treated with antimuscarinics like oxybutynin (**7**), tolterodine, and darifenacin (**8**), which block contraction of the bladder by relaxing the bladder muscles. While oxybutynin is efficacious, its side effects, especially dry mouth, result in 80–85% of individuals discontinuing treatment. The latter is to some extent overcome by once a day, controlled-release oxybutynin (Ditropan XL). Tolterodine is another antimuscarinic developed for OAB has a lesser incidence of dry mouth. Darifenacin is a muscarinic M_3-receptor-selective antagonist that is also selective for bladder smooth muscle and may avoid dry mouth.

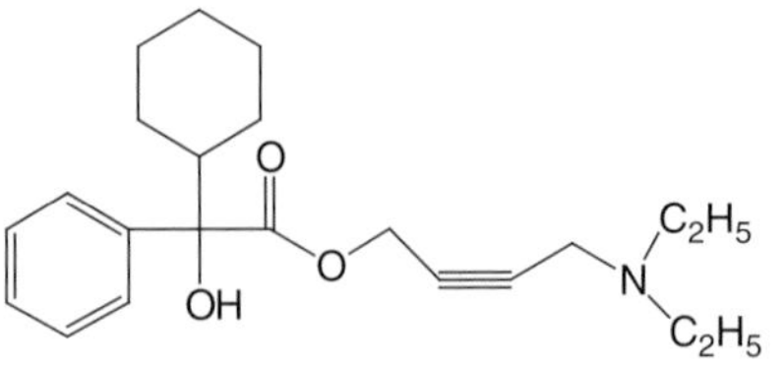

7 Oxybutynin

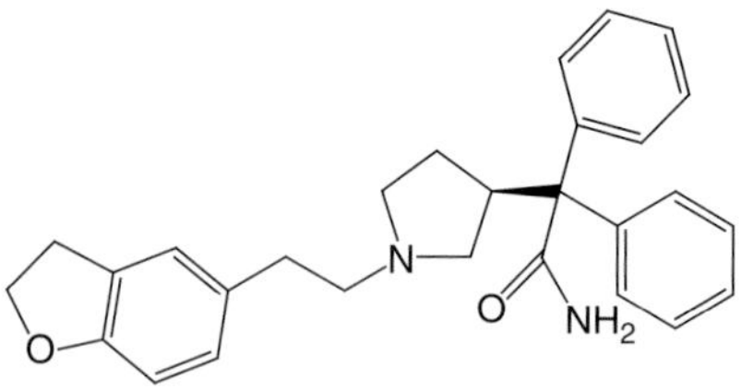

8 Darifenacin

Tricyclic antidepressants, particularly imipramine, can also be used for treatment, facilitating urine storage, by both decreasing bladder contractility and by increasing outlet resistance. The 5HT/NE uptake inhibitor (SNRI) duloxetine (**9**) increases extracellular monoamine levels leading to contraction of the urethral sphincter to prevent urine leakage. Bethanechol (**11**) is another cholinergic agent used to treat bladder (detrusor) underactivity in individuals with incomplete bladder emptying.

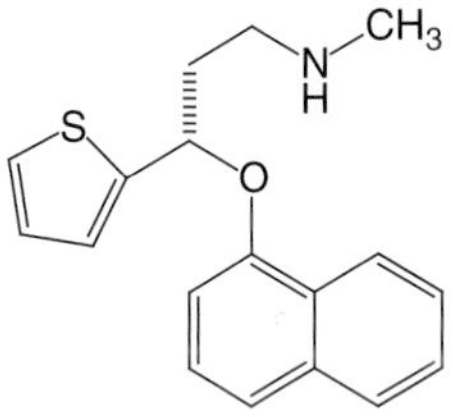

9 Duloxetine

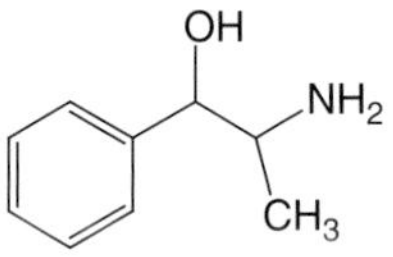

10 Phenylpropanolamine

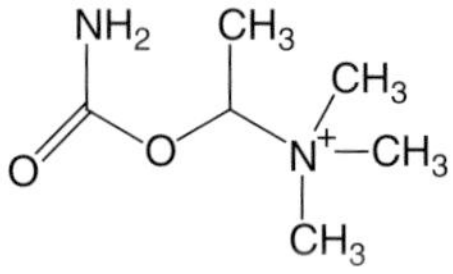

11 Bethanechol

Drug therapy in individuals with SUI involves the use of α-adrenergic agonists that increase bladder outlet resistance via actions on the bladder neck and base, and proximal urethra. Nonselective α-adrenoceptor agonists like phenylpropanolamine (**10**) and midodrine (**12**) have limited clinical efficacy in stress incontinence due to effects on blood pressure and heart rate that occur at doses comparable to those improving urethral function. α-Adrenoceptor agonists selective for the α_{1A} subtype versus the α_{1B} subtype, e.g., ABT-866 (**13**), represented one conceptual approach to urethral selectivity but proved to be only modestly uroselective with blood pressure increases being observed at doses only 3–10-fold higher than those eliciting effects on urethral pressure.[1]

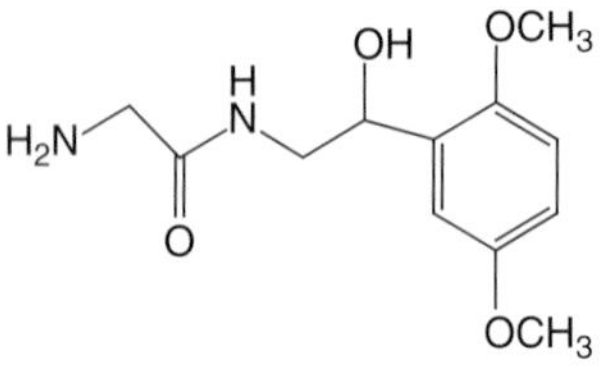

12 Midodrine

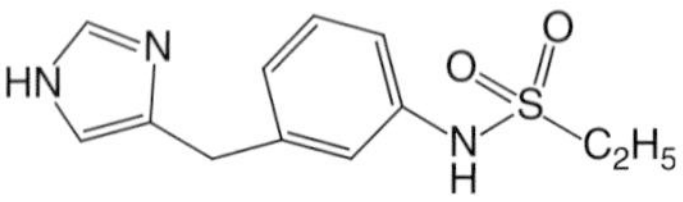

13 ABT-866 (NS-49)

In females, estrogen therapy is used to treat urge and mixed UI and OAB symptomatology, particularly in postmenopausal women. Estrogens, which are used both systemically or topically, restore the functional integrity of the urethral mucosa, increasing resistance to outflow. Newer drugs approved for UI and OAB are the anticholinergic, trospium chloride and duloxetine.

6.23.2.5 Nocturnal Enuresis

Nocturnal enuresis or bedwetting affects some 8–9 million children in the USA. This disorder exists in two types: *primary* when a child has never developed complete night-time bladder control, and *secondary* when a child has accidental wetting having had bladder control for 6 or more months. Causes include consuming fluids before bedtime; a deep sleep pattern; and laziness, e.g., not getting out of bed to void. Secondary nocturnal enuresis may involve neurogenic bladder and associated spinal cord abnormalities, urinary tract infections, etc. Treatments include enuresis alarms, that are triggered by moisture; used for 3–5 months these are effective in 70% of children. However, in 10–15% of children bedwetting returns. Drug treatments for nocturnal enuresis, while short-term solutions, work more rapidly than alarms and include tricyclic antidepressants and the synthetic analog of arginine vasopressin (antidiuretic hormone) desmopressin (DDAVP) given as an nasal spray or orally.

6.23.3 Prostate Cancer

More than 70% of all prostate cancers are diagnosed in males over age 65. Data on first-degree relatives indicates a 2–11-fold increase in the risk of prostate cancer in males who have a familial history. Prostate cancer is the most common noncutaneous malignancy in US males with an estimated lifetime risk of disease of 17–18%. In 2003, an estimated 220 000 new cases of prostate cancer were diagnosed in the USA.

6.23.4 Renal Dysfunction

In addition to a role in producing the products stored in, and voided from, the bladder, the renal system is a key system in the regulation of blood pressure via its role in regulating blood volume. Diseases of the kidney encompass renal dysfunction and end-stage renal disease (ESRD), the result of the metabolic syndrome, diabetes, obesity, and hypertension (*see* 6.25 Renal Dysfunction in Hypertension and Obesity). Treatments to deal with these issues initiate with lifestyle changes, diet, and exercise, are progressively followed with drug regimens to treat the diabetes, obesity, and hypertension, kidney dialysis to restore normal function, and kidney transplant.

6.23.5 Sexual Dysfunction

Sexuality, both male and female, is a complex and controversial subject, involving the central nervous system, vascular, and endocrine systems.

6.23.5.1 Erectile Dysfunction

ED is a repeated inability to achieve or sustain a penile erection of sufficient rigidity for sexual intercourse.[1,7,8] While ED frequently is referred to as impotence, the latter term extends to a multitude of other problems that interfere with sexual intercourse and reproduction, e.g., lack of sexual desire driven by psychological issues, problems with ejaculation or orgasm, and dysfunctional organ systems. ED can reflect a total inability to achieve erection, an inconsistent ability to do so, or a tendency to sustain only brief erections. In 1985, approximately 8 of 1000 men in the US made physician office visits for ED. In 1999, the rate had increased to 22 in 1000 due, in part, to the availability of vacuum devices to induce erections, in part due to injectable drugs for the penis but more the result of the introduction of the PDE5 inhibitor sildenafil (**14**), the orally active 'little blue pill' and a massive direct-to-consumer advertising campaign for sildenafil and other PDE5 inhibitors.

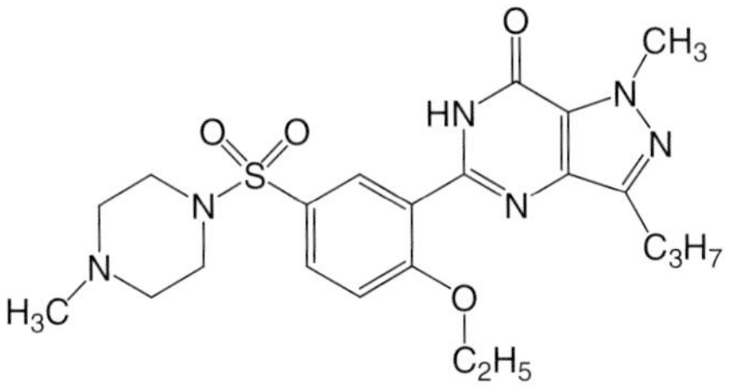

14 Sildenafil

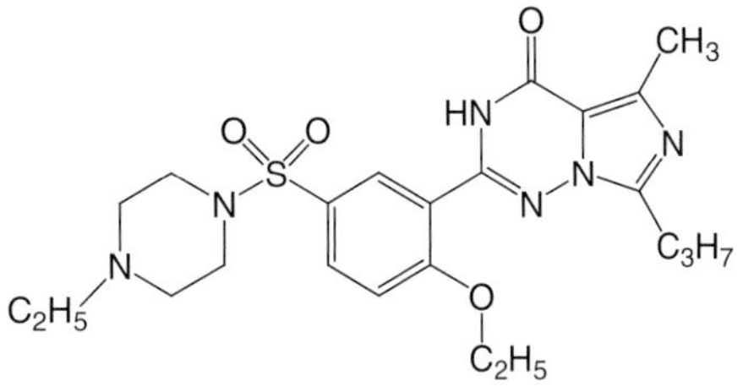

15 Vardenafil

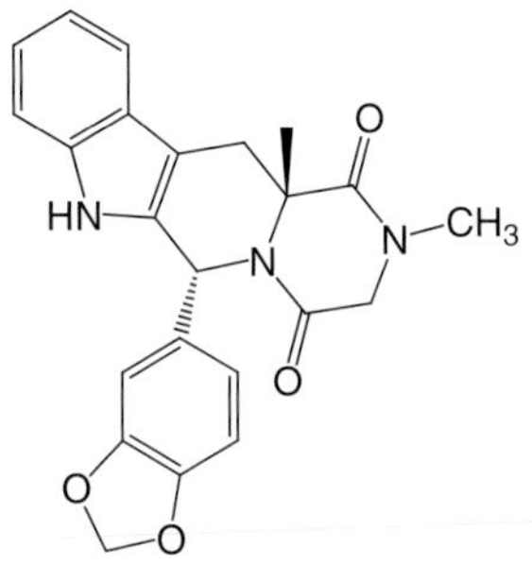

16 Tadalafil

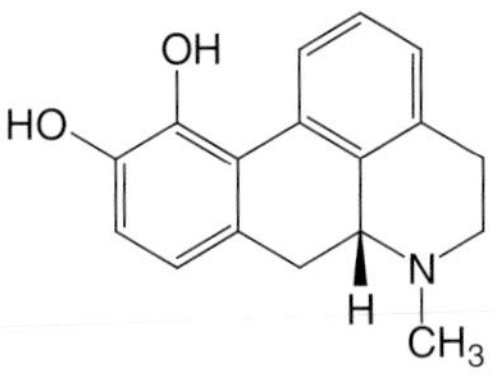

17 Apomorphine

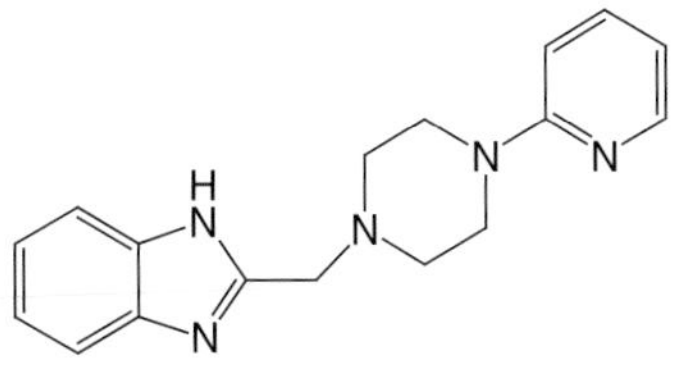

18 ABT-724

The worldwide prevalence of ED has been estimated at over 152 million males with projections for 2025 being in excess of 320 million. In older males, ED may have a physical cause, such as disease, injury, drug side effects, injury to nerves, arteries, smooth muscles, and fibrous tissues, or impaired blood flow in the penis. Other common causes of organic ED include the metabolic syndrome, diabetes, kidney disease, chronic alcoholism, multiple sclerosis, atherosclerosis, vascular disease, and neurologically related causes.[9] The incidence of ED increases with age with approximately 5% of 40-year-old males and 15–25% of 65-year-old males experiencing aspects of ED. Surgery, including radical prostatectomy and bladder cancer surgery, can injure nerves and arteries near the penis, causing ED. Antihypertensives, antihistamines, antidepressants, and hypnotics can produce ED as a side effect. Psychological factors including stress, anxiety, guilt, depression, low self-esteem, and fear of sexual underperformance failure can contribute to 10–20% of ED cases. Quality of life assessments suggest that ED can significantly contribute to the incidence of depression.

Penile erection is a spinal reflex under CNS (supraspinal) inhibitory control. Visual, tactile, olfactory, and imaginative stimuli from the higher cortical areas of the brain are integrated in the medial preoptical area of the hypothalamus via dopaminergic and oxytocinergic neuronal systems. These elicit the release of NO from 'nitrergic neurons' in the *corpus cavernosum* of the penis that then activates soluble guanylyl cyclase (sGC) in smooth muscle producing vasodilation and venoocclusion with engorging of the penis resulting in rigidity and erection.

Drug targets for the treatment of ED are present in both the brain and periphery. CNS targets include dopamine (DA) and melanocortin (MCH) while peripheral targets are focused on drugs that enhance smooth muscle vasodilation or block the adrenergic or endothelin (ET)-mediated vasoconstriction associated with penile flaccidity.

6.23.5.1.1 Centrally acting treatments for erectile dysfunction

Injection of DA D2 agonists into the rat medial preoptic area of the hypothalamus can induce penile erections. Apomorphine (**17**), a centrally acting agent, is a nonselective DA D2 agonist approved for use in Europe and Japan for the treatment of ED. The lack of DA receptor selectivity is associated with an emetic response, which to some patients has represented an unacceptable side effect, to others, a reinforcing stimulus. The DA D4 receptor was recently identified as being responsible for the erectile actions of apomorphine leading to the discovery and characterization of the selective DA D4 agonist, ABT-724 (**18**).[10] α-MSH (α-melanocyte-stimulating hormone) and oxytocin in the CNS increase penile erection via supraspinal α-MSH and supraspinal and spinal oxytocin receptors. An α-MSH agonist, PT-141, which produces its actions via melanocortin receptors, is in clinical trials being dosed via the intranasal route.[11] THIQ is a small-molecule, nonpeptide melanocortin MC-4 receptor agonist that elicits penile erection.

6.23.5.1.2 Peripherally acting treatments for erectile dysfunction

6.23.5.1.2.1 PDE5 inhibitors

Sildenafil was the first PDE5 inhibitor found to be effective in promoting spontaneous penile erections. While the mechanistic basis for this effect has become obvious in retrospect, it was initially noted as a frequent side effect of an NCE that originally entered the clinic for the treatment of congestive heart failure. Thus the utility of sildenafil for the treatment of ED was discovered by serendipity.[12] PDE5 inhibitors block the hydrolysis of the cGMP produced by sGC in response to NO, prolonging cGMP levels in corpus cavernosum smooth muscle. Sildenafil blocks sGC with an IC_{50} value of 5.3 nM. The second-generation PDE5 inhibitors tadalafil (**16**) and vardenafil (**15**) are also potent PDE5 inhibitors (IC_{50} = 3.6 and 0.4 nM, respectively) but have different pharmacokinetic profiles and/or selectivity profiles for other members of the PDE enzyme family.[13] Allsosteric activators of sGC like A-350619 (**19**) and BAY 41-2272 (**20**) can elicit penile erection in animals synergistically enhancing the effects of NO by stimulating cGMP synthesis without altering degradation of the cyclic nucleotide.

6.23.5.1.2.2 Prostaglandin E_1

Prostaglandin E_1 (PGE_1) is also used to treat ED, acting via specific PGE receptors EP_2 and EP_4 on smooth muscle to increase intracellular cAMP synthesis and potentiate smooth muscle relaxation. PGE_1 is an useful treatment for ED despite its delivery via penile injection and intraurethral routes. Injecting drugs like alprostadil into the penis can lead to stronger erections by engorging the penis with blood, but can lead to priapism. Blockade of endogenous vasoconstrictor-induced tone in corpus cavernosum smooth muscle can also induce penile erection. However, the nonselective α-adrenoceptor antagonist phentolamine showed only modest efficacy in mild to moderate ED. Endothelin (ET) antagonists were also considered to be a useful treatment option for ED based on the potent contractile activity of ETs in vitro in the penis. However, an ET_A-selective receptor antagonist, BMS-193884 (**21**), that was effective on rabbit and human corpus cavernosa, failed to show efficacy in the clinic.

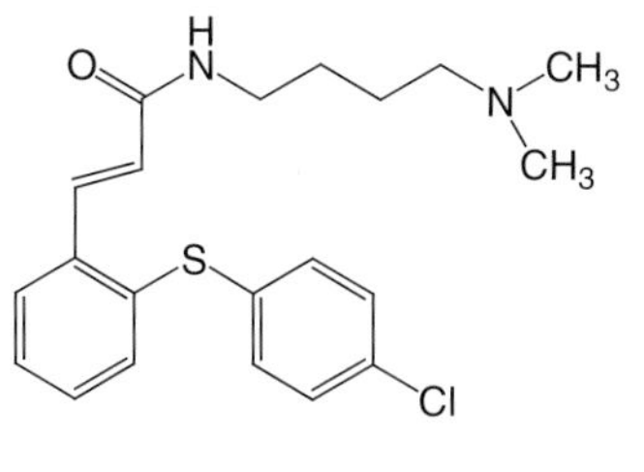

19 A-350,619 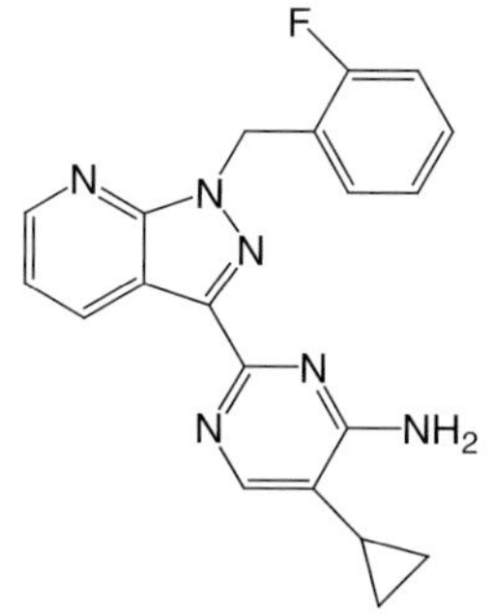 **20** Bay 41-2272 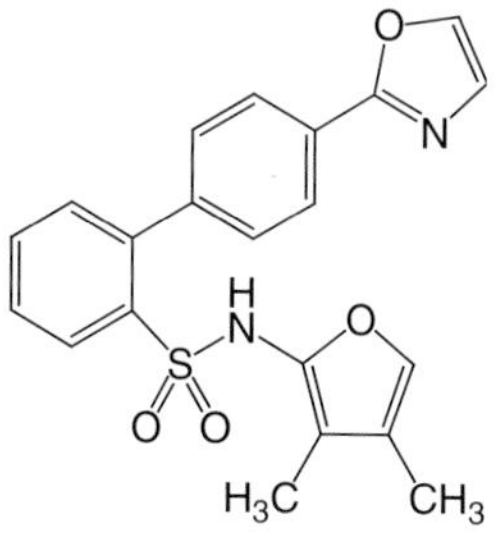 **21** BMS-193884

6.23.5.2 Female Sexual Dysfunction

Female sexual dysfunction (FSD) is a complex and controversial disorder that includes components of desire and arousal and orgasmic and sex pain disorders (dyspareunia and vaginismus).[14,15] FSD consists of four recognized disorders: *hypoactive sexual desire disorder* (HSDD) associated with decreased arousal and sexual aversion; *female sexual desire disorder* (FSAD) associated with decreased arousal; *orgasmic disorder*, a difficulty or inability to achieve orgasm; and *sexual pain disorder*. HSDD can be treated with a transdermal testosterone patch, although a Food and Drug Administration (FDA) advisory panel declined to recommend approval for the Intrinsa transdermal testosterone in 2005 due to concerns with long-term safety and clinical trial endpoint issues.[16] While there is a major psychological component in decreased sexual desire, as discussed above, the success of the PDE5 inhibitors in promoting erections in males has led to a hypothesis that FSAD is caused by a decreased blood flow to pelvic tissues and a decrease in muscle relaxation in the tissues that comprise the vaginal tube and clitoris. This hypothesis is based on the concept that the clitoris, given a common embryological origin, is similar to the penis. Thus as ED can be associated with decreased penile blood flow, female sexual dysfunction can be considered to have a similar causality. Decreased blood flow results in a decrease in vaginal lubrication with the vaginal tube fails to dilate adequately in preparation for penetration, resulting in painful intercourse, diminished vaginal sensation, and an inability to achieve an orgasm. While there has been considerable interest in the use of PDE5 inhibitors to treat female sexual dysfunction, this has been challenging, as there is a major challenge in measuring subtle changes in vaginal lubrication and penile rigidity. Blood flow to the sexual organs can however be measured. A noninvasive Doppler ultrasound probe can be used to measure blood flow in the penis, vagina, and clitoris. Vaginal photoplethysmography can also be used to measure changes in vaginal blood flow. Sildenafil has however, shown efficacy in FSAD trials.

The discovery of sildenafil and related PDE5 inhibitors has driven public awareness of sexual dysfunction as a drug-treatable disorder in both males and females and also introduced a richer than usual vocabulary into the staid halls of science, e.g., 'stuffers,' males who take ED medication, fail to achieve an erection but enthusiastically engage in attempting penetration nonetheless. However, after an initial explosion in sales, broad nonprescription use by recreational users, and extensive direct-to-consumer advertising, the ED drug market has failed to achieve projections. One reason for this has been a side effect profile involving 'blue vision' due to inhibition of PDE activity in the eye, another with 500 deaths being attributed to the use of sildenafil[12] and yet another the confounding issue of the psychology of sexual performance where 'mood' and fear of failure can color outcomes. In many instances in the clinical trial setting, NCEs have been ineffective when used in a monogamous relationship but extremely effective when a new partner was part of the relationship, reflecting the psychological novelty of the latter as an important component. Similarly, placebo is highly effective in that situation. In other instances, none of the treatment options for ED were effective due to insurmountable psychological or organic dysfunction, while in other situations a single use of an ED drug restored 'normal' sexual performance, acting as the catalyst for addressing sexual problems within a relationship. Once whatever hurdles have been overcome, 'little blue pills' become unnecessary.

6.23.5.3 Premature Ejaculation

Premature ejaculation (PE) affects 21–33% of males aged 18–19.[16] It is defined in DSM-IV-TR as 'the persistent or recurrent onset of orgasm and ejaculation with minimal sexual stimulation, before, on, or shortly after penetration and before the person wishes it.'[17] Mechanistically, the ejaculatory response represents a complex interaction between serotonergic and dopaminergic neuronal sytems with additional influences from cholinergic, oxytocinergic, adrenergic, and GABAergic systems. DA, acting via D2 receptors promotes ejaculation while 5HT inhibits ejaculation. The selective serotonin reuptake inhibitor (SSRI) antidepressants, e.g., fluoxetine, have been used for the treatment of PE and this is one of their side effects as a drug class. PDE5 inhibitors and local anesthetic formulations, e.g., NM-100061 and PSD-502, have also been used to treat PE. Dapoxetine is a newer SSRI developed for the treatment of PE as an on-demand oral agent. The pharmacokinetic profile of dapoxetine differs from that of other SSRIs in that peak levels are reached within 1 h with an initial half-life of 14 h. Other NCEs being evaluated for use in the treatment of PE include: UK 390957, a short-acting SSRI, VI-0162, a $5HT_3$ receptor antagonist, and LI-301, an SSRI/μ opioid agonist combination.[16]

A major issue involved in finding novel drugs for the treatment of human sexual dysfunction disorders is the benefit-to-risk ratio for drug therapy in what is considered both a non-life-threatening and lifestyle disorder.

6.23.6 Reproductive Medicine

Reproductive medicine encompasses all aspects of fertility including birth control and the reproductive results of aging of the reproductive organs. The utility of drugs for use as birth control agents and in the treatment of postmenopausal disorders like endometriosis is an underesearched area.

6.23.6.1 Birth Control

Birth control involves the use of practices, devices, and drug treatments to reduce the possibility of becoming pregnant and involves the use of both devices and drugs that act as contraceptives. These can be divided into agents that prevent: (1) fertilization and (2) implantation of the embryo after fertilization occurs. The latter agents are also characterized as abortifacients. Birth control, while somewhat less controversial than abortion, is a controversial issue both politically and ethically. Devices that prevent conception as an alternative to abstinence, e.g. male and female condoms, female diaphragms, and sponges, date back to the dawn of history, but were only formally approved by the FDA in 1989. They are used on their own or in combination with a spermicide like nonoxynol-9.

6.23.6.1.1 The birth control pill

Oral contraception in the form of the birth control pill or 'The Pill'[18] suppresses ovulation via the combined actions of steroid hormones, principally estrogen and progestin, to control fertility. It was approved in 1960. The Pill is considered by many to be one of the major achievements of the twentieth century 'enabling individuals, families and nations to control child-bearing in a responsible manner.'[19] Others feel it is a major factor in the decline of Western civilization and its values. Various iterations on estrogen and progestin for birth control include patches, vaginal contraceptive rings, implants, and injections.

6.23.6.1.2 RU-486

RU-486 or mifepristone is an antiprogestagenic abortifacient, the effect of which is to cause sloughing of the endometrium and termination of pregnancy 7–8 weeks after the last normal menstrual period. The use and approval of RU-486 have been the subject of heated political and religious debates regarding 'right-to-life' issues and women's rights groups.

6.23.6.2 Endometriosis

Endometriosis, the growth of endometrial tissue outside the uterus into 'nodules,' 'tumors,' 'lesions,' 'implants,' or 'growths,' that cause pain, infertility, and other problems, affects nearly 10% of all postmenopausal females. These endometrial growths are chiefly abdominal, being found in the ovaries, fallopian tubes, uterine ligaments supporting the area between the vagina and the rectum, the outer surface of the uterus, and the lining of the pelvic cavity. The etiology of endometriosis is unknown, but various theories exist including: (1) retrograde menstruation where menstrual tissue backs up through the fallopian tubes, implants in the abdomen, and grows; (2) distribution of endometrial tissue via the lymphatic and blood systems from the uterus to other parts of the body; (3) the development of quiescent embryonic tissue into endometrial tissue; (4) dedifferentiation of adult tissues to a reproductive tissue phenotype; or (5) a genetic component.[20] Endometrial growths are generally neither malignant nor cancerous although an increased frequency of malignancy in conjunction with endometriosis has been observed. While hysterectomy and removal of the ovaries are viewed as a 'definitive' cure for endometriosis, hormonal treatment including birth control pills, a progestin alone, e.g., danazol, gonadotropin releasing hormone GnRH agonists, and aromatase inhibitors, e.g., letrozole or anastrozole, are also used.

6.23.7 Future Directions

The incidence of disorders and dysfunctions of the urogenital tract including the kidney are increasing with the aging of the population. Drug treatments for both bona fide organic disorders (endometriosis, prostatitis, BPH, incontinence, BOO) and those more psychologically related (sexual dysfunction) are limited in their efficacy and side effect liabilities and perhaps with new insights into potential genetic causes of these disorders, may be treated via newer targets that have a lesser impact on cardiovascular system function.

The more effective treatment of the spectrum of disorders related to women's health has spurred initiatives both within the industry and government. The latter, the National Institutes of Health's Women's Health Initiative, now 15 years old has spent $725 million in studying 161 000 females. It is reportedly plagued with flawed clinical trial designs, patient enrollment criteria, and data misinterpretation, and has raised issues regarding the effectiveness of hormone therapy in the treatment of postmenopausal orders,[21] confounding the search for improvements on existing medications. Similarly, the evolution of drug treatments for the lifestyle disorders of ED and FSD, like that of nonmorbid obesity, are confounded by the major psychological component of each making it challenging to prioritize the targeting of new drugs for peripheral versus central targets.

References

1. Moreland, R. B.; Brioni, J. D.; Sullivan, J. P. *J. Pharmacol. Exp. Ther.* **2004**, *308*, 797–804.
2. de Groat, W. C.; Yoshimura, N. *Annu. Rev. Pharmacol. Toxicol.* **2001**, *41*, 691–721.
3. Speakman, M.; Batista, J.; Berges, R.; Chartier-Kastler, E.; Conti, G.; Desgrandchamps, F.; Dreikorn, K.; Lowe, F.; O'Leary, M.; Perez, M. et al. *Prostate Cancer Prostatic Dis.* **2005**, *8*, 369–374.
4. Bent, S.; Kane, C.; Shinohara, K.; Neuhaus, J.; Hudes, E. S.; Goldberg, H.; Avins, A. L. *N. Engl. J. Med.* **2006**, *354*, 557–566.
5. Lemack, G. E. *Nat. Clin. Pract. Urol.* **2006**, *3*, 38–44.
6. Wein, A. J. *Exp. Opin. Invest. Drugs* **2001**, *10*, 65–83.
7. Klein, E. A.; Thompson, I. M. *Curr. Opin. Urol.* **2004**, *14*, 143–149.
8. Moreland, R. B.; Hsieh, G.; Nakane, M.; Brioni, J. D. *J. Pharmacol. Exp. Ther.* **2001**, *296*, 225–234.
9. Rosen, E. C.; Friedman, M.; Kostis, J. B. *Am. J. Cardiol.* **2005**, *96*, 76M–79M.
10. Brioni, J. D.; Moreland, R. B.; Cowart, M.; Hsieh, G. C.; Stewart, A. O.; Hedlund, P.; Donnelly-Roberts, D. L.; Nakane, M.; Lynch, J. J., III; Kolasa, T. et al. *Proc. Natl. Acad. Sci. USA* **2004**, *101*, 6758–6763.
11. Martin, W. J.; MacIntyre, D. E. *Eur. Urol.* **2004**, *45*, 706–713.
12. Campbell, S. F. *Clin. Sci.* **2000**, *99*, 255–260.
13. Steif, C. G.; Uckert, S.; Jonas, U. *J. Men's Health Gender* **2005**, *2*, 87–94.
14. Phillips, N. A. *Am. Fam. Physician* **2000**, *62*, 127–136.
15. Kovalevsky, G. *Semin. Reprod. Med.* **2005**, *23*, 180–187.
16. Azam, U. *Br. J. Pharmacol.* **2006**, *147*, S153–S159.
17. First, M. B.; Tasman, A. *DSM-IV-TR Mental Disorders: Diagnosis, Etiology and Treatment*; John Wiley: Chichester, UK, 2004, pp 1069–1071.
18. Djerassi, C. In *Discoveries in Pharmacology, Vol. 2, Hemodynamics, Hormones and Inflammation*; Parnham, M. J., Bruinvels, J., Eds.; Elsevier: Amsterdam, The Netherlands, 2004, pp 339–361.
19. Greep, R. O. In *Discoveries in Pharmacology, Vol. 2, Hemodynamics, Hormones and Inflammation*; Parnham, M. J., Bruinvels, J., Eds.; Elsevier: Amsterdam, The Netherlands, 2004, pp 321–337.
20. Barlow, T. H.; Kennedy, S. *Annu. Rev. Med.* **2005**, *56*, 345–356.
21. Parker-Pope, T. *Wall Street Journal*, Feb 28, 2006, pp A1/A13.

Biography

Michael Williams received his PhD in 1974 from the Institute of Psychiatry and his Doctor of Science degree in Pharmacology (1987) both from the University of London. Dr Williams has worked for 30 years in the US-based pharmaceutical industry at Merck, Sharp, and Dohme Research Laboratories, Nova Pharmaceutical, CIBA-Geigy, and Abbott Laboratories. He retired from the latter in 2000 and served as a consultant with various biotechnology/pharmaceutical companies in the US and Europe. In 2003 he joined Cephalon, Inc. in West Chester, where he is vice president of Worldwide Discovery Research. He has published some 300 articles, book chapters, and reviews and is adjunct professor in the Department of Molecular Pharmacology and Biological Chemistry at the Feinberg School of Medicine, Northwestern University, Chicago, IL from which vantage point he publishes his personal viewpoints on the drug discovery process.

Comprehensive Medicinal Chemistry II
ISBN (set): 0-08-044513-6

ISBN (Volume 6) 0-08-044519-5; pp. 541–550

6.24 Incontinence (Benign Prostatic Hyperplasia/Prostate Dysfunction)

M G Wyllie and S Phillips, Urodoc Ltd, Herne Bay, UK

6.24.1 Introduction to the Prostate

The prostate is a fibromuscular glandular organ involved in the male reproductive system. It is located between the bladder and pelvic floor, and surrounds the prostatic urethra. It is comprised of a large, nonglandular region and a smaller, glandular region that consists of epithelium and stroma, and can be further subdivided into three main zones: the peripheral zone, which constitutes about 70% of the gland; the central zone, comprising 25% of the gland; and the transitional zone, which represents about 5–10% of the gland (see **Figure 1**). These zones have histological differences, mainly relating to secretory cell structure and the ratio of epithelium-to-stroma, which indicates that they have different functions.[1]

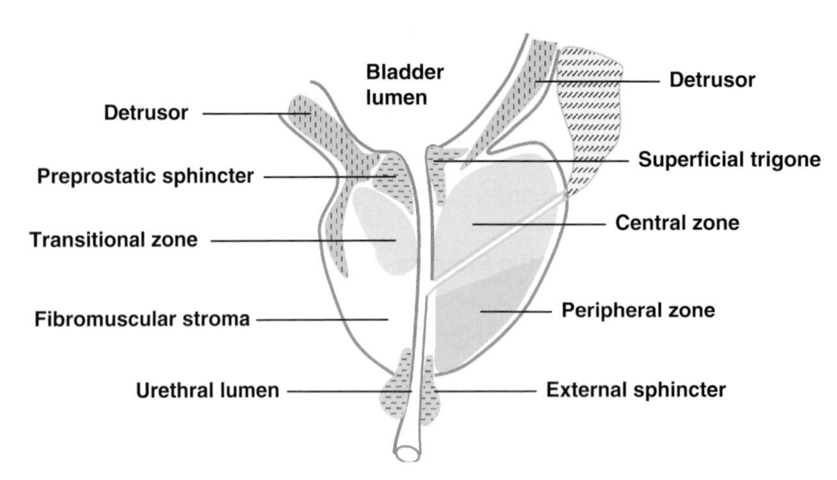

Figure 1 A sagittal section of the prostate, which shows anatomic subdivisions. (Reproduced with kind permission from Dixon, J. S.; Gosling, J. A. *Textbook of Benign Prostatic Hyperplasia*, 2nd ed.; Taylor & Francis: London, 2005, pp 3–10 © Taylor & Francis.)

The prostate is androgen dependent, and relies primarily on the androgen testosterone produced by the Leydig cells in the testes and adrenal glands for normal functioning. In the context of benign prostatic hyperplasia (BPH), testosterone is converted to its potent active metabolite dihydrotestosterone (DHT) by 5α-reductase isoenzymes, largely situated on the prostate cell membrane. Once formed, DHT binds to androgen receptors (adrenoceptors), forming a DHT–androgen receptor complex inside the prostate cell, where it produces a number of regulatory effects relating to prostate growth, gene activity, and initiation of other hormone response elements, including estrogen effects.[2]

A number of other mediators also play an important role in the maintenance of normal prostate functioning, including the 'balancing act' between cell proliferation and apoptosis (programmed cell death). Under the influence of DHT, a key growth factor known as fibroblast growth factor-2 (FGF-2), which is produced by stromal and epithelial cells, promotes stromal fibroblast growth. Fibroblasts produce keratinocyte growth factor (KGF), an important growth-promoting factor that exerts its effects in epithelial cells. In contrast, transforming growth factor beta (TGFβ), produced via stromal muscle cells, is a growth inhibitory factor that inhibits the epithelial actions of KGF. The predominant function exerted by FGF-2, KGF, and TGFβ varies according to their location in the stroma and/or epithelium: those located in the distal region are largely involved in proliferation, those in the intermediate region are involved in cell differentiation, and those in the proximal region handle apoptosis (see **Figure 2**). Maintaining a normal balance between these key mediators is essential for regulating prostate cell growth, cell functioning, and death.[2]

Additional regulatory factors, which are mainly secreted by neuroendocrine cells in the epithelium, also affect prostate function, growth and development, and neuroendocrine, endocrine, and exocrine secretion. These factors include vascular endothelial growth factor (VEGF), transforming growth factor alpha (TGFα), serotonin (5HT), bombesin (a gastrin-releasing peptide), the chromogranin family of polypeptides, the calcitonin family of peptides, somotostatin, parathyroid hormone-releasing protein (PTHrP), thyroid-stimulating hormone-like peptide, and the human chorionic gonadotropin-like peptide. In addition, insulin-like growth factors (IGFs) and interleukins (ILs) are key cell survival factors and Bcl-2 (a 26-kDa protein) and nitric oxide (NO)-synthase influence apoptosis.

It can be seen therefore that the prostate is under the influence of a complex set of pathways and mediators present in both the stroma and epithelium. This process is summarized in **Figure 3**, which represents a cross-section of a 'typical' working prostate cell.

6.24.1.1 The Aging Prostate

It is well established that testosterone production declines as the human male ages; longitudinal studies have shown that total testosterone falls by approximately 1% per year after the age of 50 years,[3] and unbound (free and bioavailable) testosterone falls more steeply, which may be due to an age-related increase in the testosterone binding protein known as sex hormone-binding globulin (SHBG). Epidemiology studies from the USA estimate that 30% of men aged 60–70 years, 70% of men aged 70–80 years, and 80% of men aged over 80 years have low unbound testosterone.[4] This age-related decline in testosterone is illustrated in **Figure 4**.

Multiple etiologies have been reported to play a role in the age-related decrease in testosterone; however, it appears that the Leydig cell, which is responsible for 90–95% of testosterone production, also ages, resulting in lower

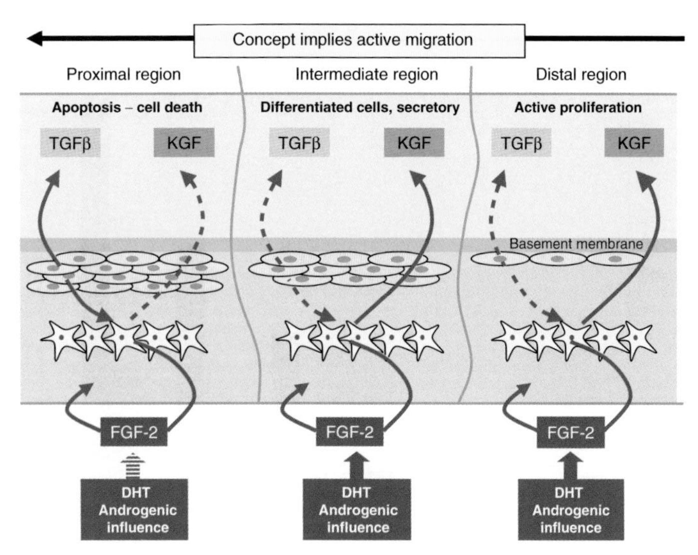

Figure 2 The process of prostate cell proliferation and apoptosis DHT, dihydrotestosterone; FGF-2, fibroblast growth factor-2; KGF, keratinocyte growth factor; TGFβ, transforming growth factor beta. (Reproduced with kind permission from Turkes, A.; Griffiths, K. *Textbook of Benign Prostatic Hyperplasia*, 2nd ed.; Taylor & Francis: London, 2005, pp 27–68 © Taylor & Francis.)

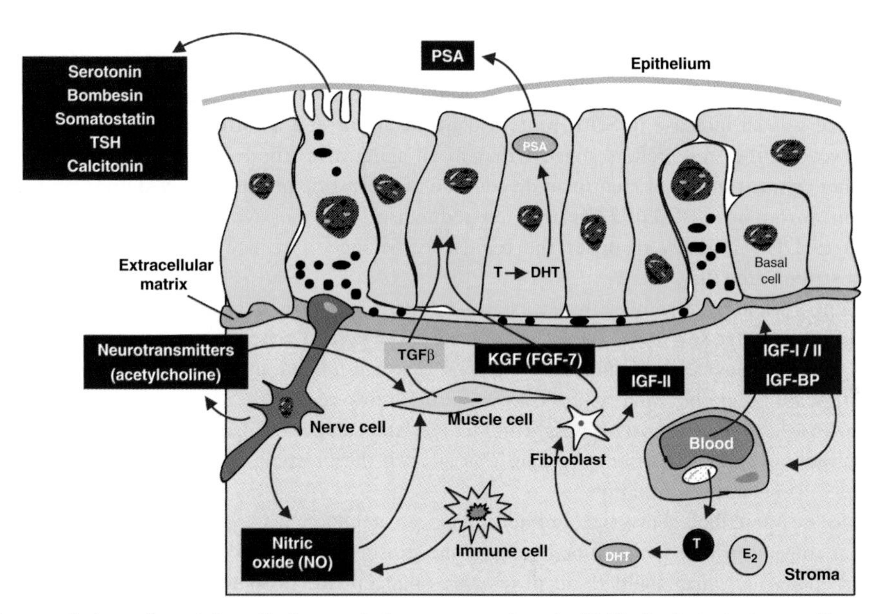

Figure 3 The morphology of prostate epithelium and stroma compartments. DHT, dihydrotestosterone; E_2, estradiol; FGF-7, fibroblast growth factor-7; IGF, insulin-like growth factor; IGF-BP, insulin-like growth factor-binding protein; KGF, keratinocyte growth factor; PSA, prostate-specific antigen; T, testosterone; TGFβ, transforming growth factor beta; TSH, thyroid-stimulating hormone. (Reproduced with kind permission from Turkes, A.; Griffiths, K. *Textbook of Benign Prostatic Hyperplasia*, 2nd ed.; Taylor & Francis: London, 2005, pp 27–68 © Taylor & Francis.)

testosterone production. More specifically, steroidogenic (i.e., testosterone) biosynthesis, which is initiated when the substrate cholesterol is transferred to the mitochondrial inner membrane, becomes 'rate-limited' as the steroidogenic acute regulatory (StAR) protein, which facilitates this initiation process, also declines. At the same time, cyclooxygenase-2 (COX-2) expression is upregulated, and this COX isoform may inhibit StAR gene expression and steroidogenesis in Leydig cells.[5]

In a recent animal (rodent) model study,[6] total testosterone decreased from 41% to 33% as the animal aged from 3 to 30 months, respectively. This decrease was accompanied by a 346% increase in COX-2 in Leydig cells and a 33% decrease in StAR protein expression (over that of young cells). When the COX-2 inhibitor

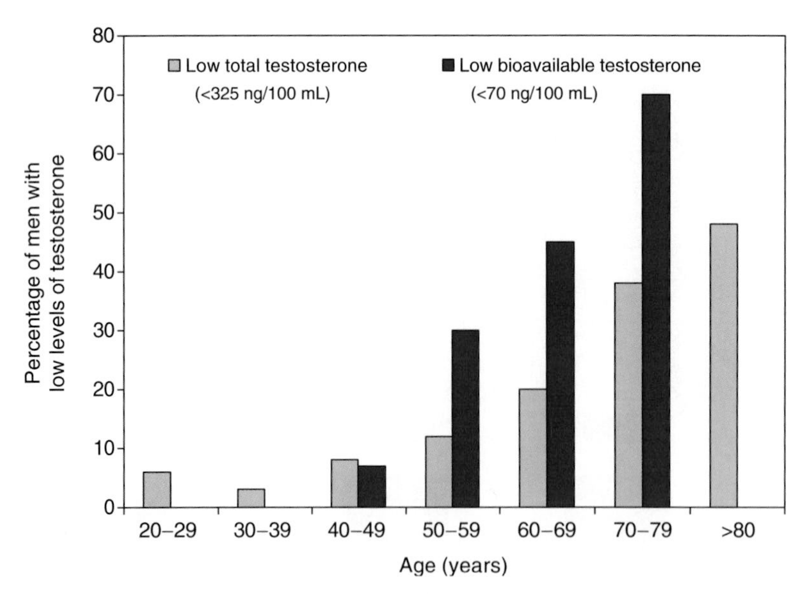

Figure 4 The age-related decline in total and bioavailable testosterone. Bioavailable testosterone was not measured in some groups. (Reproduced with kind permission from Rhoden, E. L. *N. Eng J. Med.* **2004**, *350*, 482–492 © 2004 Massachusetts Medical Society. All rights reserved.)

5,5-dimethyl-3-(3-fluorophenyl)-4-(4-methylfsulfonyl)phenyl-2-(5*H*)-furanone (DFU) was used as a dietary supplement, there was an increase in StAR protein concentrations and a corresponding increase in testosterone production. However, other researchers suggest that in an aging male there is an increase in lipid peroxidation resulting in higher concentrations of nicotinamide adenine dinucleotide phosphate (NADPH), and this in turn will increase the concentration of NADPH-sensitive 5α-reductase. An excessive conversion of testosterone (albeit reduced levels) to DHT is likely to upset the regulatory balance, and may lead to excessive DHT-mediated epithelium and stroma growth.

Prostate volume increases steadily with age, starting after the fourth decade of life. This process can occur independently of hormonal status, and is more pronounced in the central zone of the prostate (this is the area where BPH is usually more obvious). Some studies have shown that age-related annual growth ranges from 3.5% for the central zone to 1.5% for the periurethral zone and 0.8% for total prostate volume.[7] A more recent study, which assessed the relationship between age and prostate weight in 962 volunteers, showed that age-related volume increases were significant in epithelium and stroma from the transition zone of the prostate, and this was coupled with calculi and lymphocyte (i.e., inflammatory) infiltration.[8]

All of the processes described above (i.e., hormonal changes, nonhormonal growth, and inflammation) are likely to have a significant impact on the normal functioning of the prostate, and particularly the homeostasis between cell proliferation and apoptosis, which could result in prostate enlargement. In some men, progressive enlargement will result in the development of BPH (or benign prostatic enlargement), which can result in bladder obstruction and overt urinary symptoms (see **Figure 5**).

6.24.2 The Definition of Benign Prostatic Hyperplasia

BPH is a progressive disease characterized by prostate stromal and epithelial cell hyperplasia, which usually begins in the periurethral zone of the prostate. This process can be asymptomatic or associated with lower urinary tract symptoms (LUTS), typified by urinary frequency and urgency, nocturia, decreased and/or intermittent force of stream, and incomplete bladder emptying.[9] BPH is extremely common; one European survey of 80 774 subjects estimated that the prevalence of symptomatic BPH was 2.7% in men aged 45–49 years and 24% in men aged 80 years and over. Similar age-related trends have been observed in the USA, with an incidence of 34/100 000 in 20- to 39-year-olds compared with a peak incidence of 1803/100 000 in 60- to 69-year-olds[10] (see **Table 1**). Histopathologic evidence of BPH places the burden much higher, with a prevalence around 50% in men aged 60 years and 85% in men aged 90 years and over.[9,11]

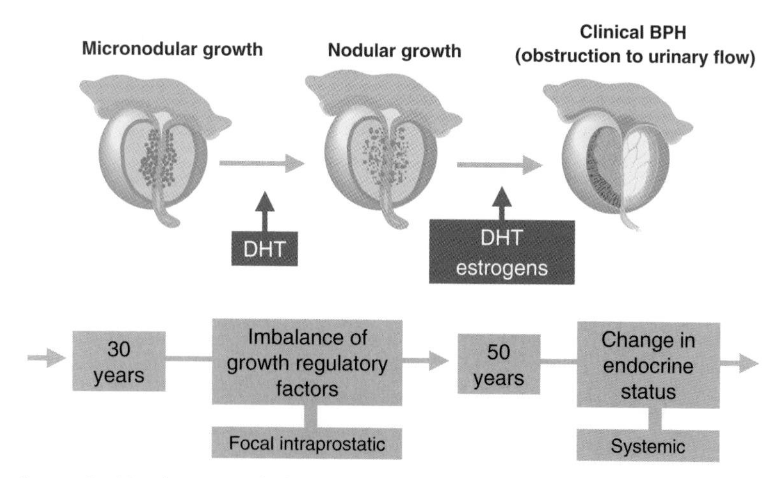

Figure 5 The development of benign prostatic hyperplasia (BPH) and associated urinary flow obstruction (bladder outlet obstruction) DHT, dihydrotestosterone. (Reproduced with kind permission from Turkes, A.; Griffith, K. *Textbook of Benign Prostatic Hyperplasia*, 2nd ed.; Taylor & Francis: London, 2005, pp 27–68 © Taylor & Francis.)

Table 1 Age-specific incidence of BPH in Europe and the USA

Age range (years)	*European incidence of BPH*[a]	*US incidence of BPH*[b]
20–39	–	65
40–49	–	415
45–49	91	–
50–59	495	951
60–69	686	643
70–79	675	205
>80	234	51
Total	2181	2330

From Sarma, A. V.; Jacobson, D. J.; McGree, M. E.; Roberts, R. O.; Lieber, M. M.; Jacobsen, S. J. *J. Urol.* **2005**, *173*, 2048–2053 and Verhamme, K. M.; Dieleman, J. P.; Bleumink, G. S.; van der Lei, J.; Sturkenboom, M. C.; Artibani, W.; Begaud, B; Berges, R.; Borkowski. A.; Chappel, C. R. *et al. Eur. Urol.* **2002**, *42*, 323–328.
[a] $n = 80\,774$ males (the Netherlands) between 1995 and 2000.
[b] $n = 392\,970$ men (Olmsted County, Minnesota) between 1987 and 1997.

6.24.3 The Pathophysiology of Benign Prostatic Hyperplasia

The prostate of aging males can be stimulated to undergo excessive growth. This is characterized by a number of cellular and molecular alterations leading to increased cell proliferation and reduced apoptosis in the prostate epithelium and stroma. The 'remodeling' that occurs as a result of these processes can permanently alter the appearance of the prostate, and may result in symptoms, long-term damage, and prostate cancer. It is not clear why these changes occur in some males, although it has been suggested that they are linked to a number of risk factors such as smoking, racial differences, obesity, liver cirrhosis, cardiovascular risks, and genetic predisposition. The pathogenesis of BPH is discussed in this section.

6.24.3.1 Genetics

A genetic involvement in the development and progression of BPH has been established. Molecular profiling using real-time quantitative reverse transcriptase-polymerase chain reaction on prostate tissue samples from BPH patients and normal controls has shown that at least 23 genes are upregulated and seven genes are downregulated in the BPH samples[12]; these altered gene expressions are shown in **Table 2**.

Table 2 Altered gene expression in BPH patients

Gene	*Mean change*[a]	*Gene*	*Mean change*[a]
Insulin-like growth factor 2	21.02	Endothelin receptor β	3.21
Kit ligand	15.46	Retinoblastoma	3.11
Fibroblast growth factor 7	6.56	Osteonectin	3.10
Fibroblast growth factor receptor 2IIIb	6.36	Platelet-derived growth factor receptor α	3.03
α2 Macroglobulin 2	4.85	α1 Integrin	2.95
Angiopoietin 2	4.55	T-cell lymphoma invasion plus metastasis	2.78
Homeobox 13b	4.50	Transmembrane protein A15	2.75
Neural crest transcription factor	4.16	C kit	2.72
Angiopoietin 1	4.02	KISS receptor	−9.52
Progesterone receptor	4.00	Hepsin	−8.33
Kinase interacting stathmin	3.88	Metastasis suppressor gene KISS1	−6.67
Bone morphogenic protein 4	3.85	Plasminogen	−6.67
Peroxisome proliferator activated receptor γ	3.70	Vascular endothelial growth factor 165	−6.45
Chemokine stromal-derived factor 1	3.52	Vascular endothelial growth factor A	−4.65
Relaxin	3.36	Ret protooncogene	−4.00

Adapted with kind permission from Fromont, G.; Chene, L.; Latil, A.; Bieche, I. Vidaud, M.; Vallancien, G.; Mangin, P.; Fournier, G.; Validire, P.; Cussenot, O. *J. Urol.* **2004**, *172*, 1382–1385.
[a] Mean normal value was 1.

In addition to those listed by Fromont *et al.*,[12] other studies have shown that polymorphism of the TGFβ1 gene, which plays a significant role in cell proliferation and apoptosis, is associated with BPH. Using a sample of 221 BPH patients and 303 male controls, those with the codon10 polymorphism had a 1.51-fold increased risk of BPH.[13] CYR61, a protein involved in BPH remodeling (*see* Section 6.24.3.4), also shows increased gene expression in patients with BPH. **Figure 6** shows that the expression of CYR61 was higher in those patients with symptomatic BPH or comorbid prostate cancer than in normal controls.[14] These findings suggest that abnormal expression of CYR61 may be a predisposing factor in the development of cancer. Other studies show that mutation of the SRD5A2 gene, which encodes for 5α-reductase, may be involved in the development of BPH. One longitudinal study of 510 men with BPH showed that those patients with polymorphism of this gene (referred to as LL/VL, AT/TT, and TA(0)/TA(0) genotypes) were considered high risk for BPH. These genotypes were also associated with larger prostate volumes.[15]

6.24.3.2 Cellular Changes

At the cellular level, BPH is characterized by a 're-awakening' of embryonic mesenchyme in the stroma and epithelium of the prostate periurethral zone; this induces proliferation of stromal smooth muscle cells, coupled with an increase in nonmuscle myosin heavy chain. This is followed by an increase in fibrous elements that can lead to the appearance of 'stromal BPH nodules.'[16] The epithelium also changes; obstruction of ducts (possibly due to stromal growth and BPH nodules) may cause the luminal secretory epithelium to flatten, resulting in a dramatic increase in the intraluminal space. The basal epithelium becomes attenuated, which may be due to a loss of cellular adhesion molecule, a key marker involved in the maintenance of epithelial cell differentiation, and the density of neuroendocrine cells is reduced.[16] This loss of normal prostate cell 'architecture' is characteristic of the histopathologic changes associated with BPH.

6.24.3.3 Changes in Proliferation, Apoptosis, and Senescence

BPH is associated with a change in the proliferation-to-apoptosis ratio, i.e., there is an increase in cell growth and a decline in cell death. The process by which cell proliferation increases during BPH development is largely unknown; however, there is evidence to suggest that local hypoxia in BPH tissue, caused by an excessive demand on oxygen supply from growing cells, may trigger an increase in growth factors such as FGF-7, IGF, and TGF, and interleukins such

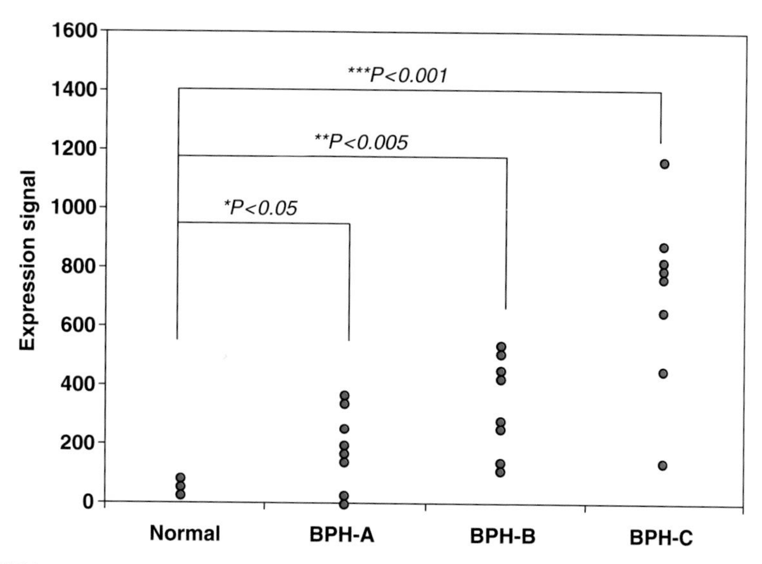

Figure 6 Mean CYR61 gene expression in the transitional zone of the prostate taken from normal controls, patients with asymptomatic benign prostatic hyperplasia (BPH) (A), symptomatic BPH (B), and BPH with prostate cancer (C): $*P<0.05$; $**P<0.005$; $***P<0.001$ versus normal controls. (Reproduced with kind permission from Sakamoto, S.; Yokoyama, M.; Zhang, X.; Prakash, K.; Nagao, K.; Hatanaka, T.; Getzenberg, R. H.; Kakehi, Y. *Endocrinology* **2004**, *145*, 2929–2940, copyright © 2004 The Endocrine Society.)

as IL1β, which results in more growth, more demand on oxygen and increased production of growth factors. An upregulation of hypoxia inducible factor-1 (HIF-1) is thought to be the key mediator responsible for excess production of these growth factors. It is possible that HIF-1 is a prostate' response to excessive cell proliferation, i.e., the promotion of a hypoxic state, which is likely to destroy cells by starving them of oxygen, may re-address the loss of apoptosis in BPH tissue. Alternatively, the prostate may be fooled into thinking that it is dying and HIF-1 could be increased as a way of promoting growth. A number of other growth-promoting factors have also been shown to increase during the development of BPH, including activated protein kinases and VEGF.[17] In addition, polyamines are important stimulators of proliferation, and it has been suggested that an increase in facilitators of the polyamine-biosynthesis pathway (such as ornithine decarboxylase) could be responsible for the resultant increase in cell growth.[18] With regard to apoptosis, there is evidence to suggest that abnormal expression of apoptotic regulators and growth suppression factors, such as Bcl-2, BAX and p27/Kip1, may be the trigger for reduced apoptosis in hyperplastic cells.[19]

Cells in BPH tissue can also lose their ability to proliferate or undergo apoptosis, these cells remain in a metabolically active state and can even accumulate; this process is known as senescence. Senescent cells are a common feature of aging tissue, and can be caused by deoxyribonucleic acid (DNA) damage, oxidative stress, and inappropriate expression of oncogenes. Senescent cells are highly prevalent in BPH tissue, particularly in the epithelium, and a key senescence biomarker known as senescence-associated β-galactosidase (SA-β-gal) has been found to be present in more than 80% of BPH tissues from large prostates, i.e., those with a weight greater than 55 g.[20] These cells can have altered function, characterized by increased expression of cytokines, growth factors and PTHrP, and reduced expression of protease inhibitors; and they may have an active role in the development of BPH as they drive the proliferation of neighboring nonsenescent cells. SA-β-gal strongly correlates with IL1α and IL8 production, leading researchers to suggest that ILs are actively involved in the development of senescence. IL8 also correlates with prostate weight. As senescence is thought to be a proliferation-limiting factor, it follows that prostate volume may be a trigger for increased production of ILs as a way of preventing further growth; however, the impact of senescent cells on their neighboring nonsenescent cells does not fit neatly into this theory.

6.24.3.4 Remodeling

Remodeling has recently been proposed as a model of BPH development and progression. It is characterized by hypertrophic basal cells, calcification, clogged ducts, inflammation and elevated proinflammatory markers, increased reactive oxygen species (ROS), and TGFβ1.[21] There is evidence to suggest that an increase in TGFβ1, which is possibly due to increased secretion from proliferating or senescent cells, converts fibroblasts into 'reactive' myofibroblasts that have increased secretory activities and contribute to extracellular matrix production.[22] This 'reactive state' and the increased expression of extracellular matrix components may contribute to the development of angiogenesis (the formation and differentiation of blood vessels), cell invasion and tumorigenesis, which can permanently alter the prostate 'architecture.'

Increased expression of CYR61, an extracellular matrix signaling protein, is evident in BPH tissue and is believed to play a key role in cell adhesion, proliferation, angiogenesis, anti-apoptosis, and prostate enlargement. CYR61 messenger ribonucleic acid (mRNA), which is regulated by lysophosphatidic acid (an endogenous lipid growth factor), is increased in both the epithelium and stroma of BPH tissue; this in turn leads to an increased production of CYR61 protein. CYR61 exerts its hyperplastic and proliferation effects mainly via an effect on growth factors such as platelet-derived growth factor.[14] This process may contribute to the development of prostate cancer.

6.24.3.5 Androgen-Related Changes

The age-related changes in testosterone and DHT, and the implications for BPH development, have been discussed in Section 6.24.1.1. However, a number of other androgen-related changes have also been shown to occur in BPH tissue; androgen receptor co-activators such as ARA54, ARA55, and SRCI are upregulated in BPH stroma whereas androgen receptor repressors such as DAX-1 are downregulated.[16] This may, in part, explain why there is an increased conversion to DHT in BPH tissue, and is in addition to an increase in NADPH-sensitive 5α-reductase, which occurs as a consequence of age-related lipid peroxidation.

Further studies have shown that two isoenzymes of 5α-reductase exist in the prostate, known as type 1 and type 2; the type 1 isoenzyme predominates in epithelial cells and the type 2 isoenzyme is present in both epithelium and stroma. Both have a key role to play in prostatic enlargement. The expression of type 1 mRNA (and consequently type 1 isoenzyme activity) has been reported to be much higher than that of type 2; however, the clinical significance of this has yet to be established.[24]

In animal models, androgen-induced prostatic enlargement has been shown to have a direct effect on bladder obstruction. In a 28-day study, canines were given DHT 75 mg day^{-1} and 17β-estradiol 0.75 mg day^{-1} (i.e., a realistic expression of BPH androgenic activity) via an implanted pump. Compared with controls, canines supplied with DHT and 17β-estradiol displayed obstructive micturition, increased detrusor pressure, lower urine-flow rate and a larger wet prostate weight. Furthermore, these changes were comparable to those observed in canines with natural BPH symptoms as shown in **Figure 7**.[25] These researchers proposed that DHT induces obstruction by overstimulating

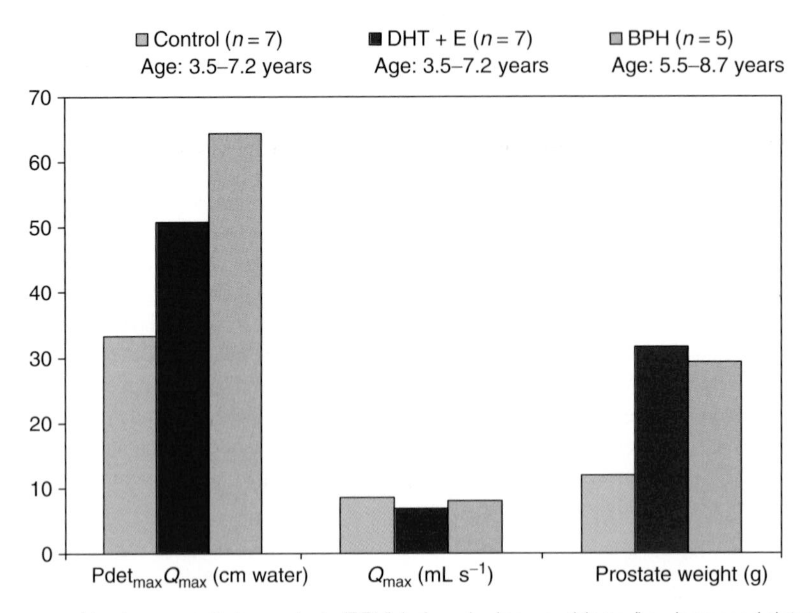

Figure 7 Androgen- and benign prostatic hyperplasia (BPH)-induced urinary problems (i.e., increased detrusor pressure and prostate volume and lower urine flow) compared with controls in canines. DHT, dihydrotestosterone; E, 17β-estradiol; $Pdet_{max}Q_{max}$, maximum detrusor pressure at maximum urine flow rate; Q_{max}, maximum urine flow rate. (Reproduced from Yokota, T.; Honda, K.; Tsuruya, Y.; Nomiya, M.; Yamaguchi, O.; Gotanda, K.; Constantinou, C. E. *Prostate* **2004**, *58*, 156–163, copyright © 2004 Wiley-Liss, Inc., A Wiley Company, with permission of John Wiley & Sons, Inc.)

muscarinic cholinergic receptors leading to a supersensitive effect on the detrusor muscle, and the magnitude of urethral constriction observed (and consequently low urine-flow rate) may depend on the duration of excessive DHT exposure. Although these findings are not directly comparable to the age-related effects of DHT in human males, they are still a useful hormonal-urodynamic model of BPH.

6.24.3.6 Inflammation

A number of inflammatory markers are associated with the cell proliferation and differentiation changes observed in BPH. It has been suggested that elevated numbers of T lymphocytes and macrophages leads to an increased secretion of various proinflammatory factors such as IL1, IL6, tumor necrosis factor alpha (TNF-α), and VEGF, resulting in chronic inflammation, clogged epithelial ducts, and disruption of glandular epithelium in BPH tissue.[21] Many of these factors are also responsible for the upregulation of COX-2, a key marker involved in the age-related decline in testosterone biosynthesis (*see* Section 6.24.1.1). Not surprisingly, COX-2 concentrations increase in epithelial areas where there are high levels of T lymphocyte and macrophage infiltration (as shown in **Figure 8**). COX-2 overexpression may also be linked to increased proliferation, cell death inhibition, prevention of oxy-radical defense systems in BPH tissue, and the development of prostate cancer.[26] Macrophages are also a source of ROS such as hydrogen peroxide, mainly via an effect on NADPH oxidase. Macrophage-induced ROS can lead to oxidative stress and tissue injury in areas where these cells have accumulated.[21]

6.24.3.7 Impact on Bladder Pathophysiology

Although BPH is largely confined to prostate abnormalities, a number of degenerative changes can take place in the bladder. It is thought that prostate hyperplasia can result in partial bladder outlet obstruction (PBOO), which increases urethral resistance to urine flow and causes acute bladder distension. The bladder responds to distension by inducing bladder wall growth (i.e., bladder hypertrophy) and angiogenesis, a compensation process that is designed to improve the bladder's ability to void. This is rapidly followed by a period of decompensation, where these growth mechanisms are switched off, possibly via genetic signals. Nevertheless, voiding-induced ischemia and tissue hypoxia are still features of the stabilized, hypertrophic bladder, and can have an effect on urinary retention. Marked remodeling of the bladder wall due to denervation, loss of smooth muscle and compliance, and increased fibrosis also remains, and is likely to have a significant long-term effect on bladder emptying and the development and expression of LUTS. Obstructed bladder dysfunction, secondary to BPH, is a chronic and progressive aspect of the disease, which may eventually result in renal damage.[27] The development of bladder dysfunction as a consequence of unrelieved PBOO is outlined in **Figure 9**.

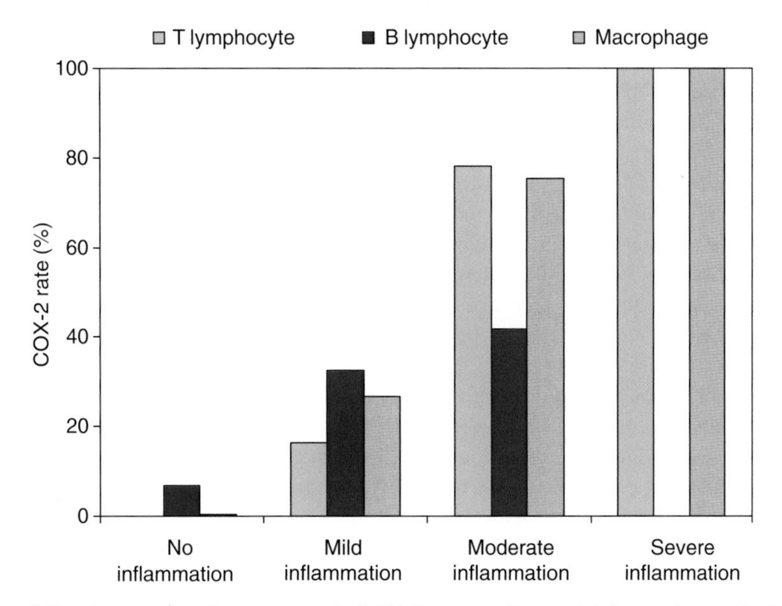

Figure 8 The correlation between cyclooxygenase-2 (COX-2) expression and inflammatory cells in benign prostatic hyperplasia (BPH) tissue. (Reproduced from Wang, W.; Bergh, A.; Damber, J.-E. *Prostate* **2004**, *61*, 60–72, copyright © 2004 Wiley-Liss, Inc., A Wiley Company, with permission of John Wiley & Sons, Inc.)

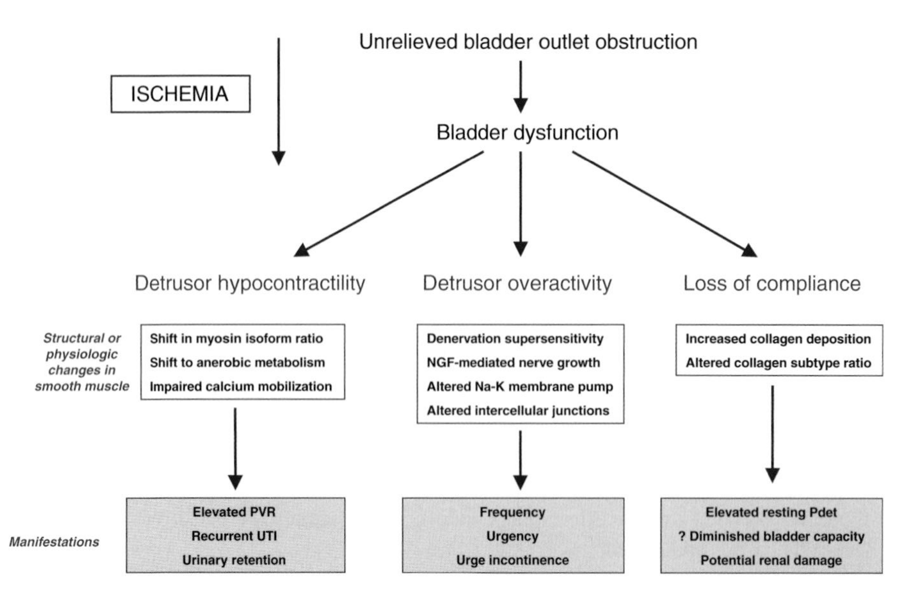

Figure 9 The effect of partial bladder outlet obstruction on bladder dysfunction: K, potassium; Na, sodium; NGF, nerve growth factor; Pdet, detrusor pressure; PVR, postvoid residue; UTI, urinary tract infection. (Reproduced with kind permission from Lemack, G. E.; McConnell, J. D. *Textbook of Benign Prostatic Hyperplasia*, 2nd ed.; Taylor & Francis: London, 2005, pp 113–115 © Taylor & Francis.)

6.24.3.8 Impact on Lower Urinary Tract Symptoms

It has been established that LUTS usually occur as a result of BPH-related bladder outlet obstruction (BOO). More recently, there is evidence to suggest that once present, BOO can induce overexpression of epithelial sodium channels (ENaCs) in the bladder (as shown by immunofluorescent staining and real-time quantitative reverse transcriptase-polymerase chain reaction), which has been shown to be associated with urinary storage symptoms. It is also thought that overexpression of ENaCs by BOO could be associated with the detrusor instability observed in BPH possibly via an effect on the bladder afferent activity.[28]

6.24.4 The Diagnosis of Benign Prostatic Hyperplasia

Several guidelines have been published to aid in the diagnosis of BPH; the most prominent ones are those published by the American Urological Association (AUA)[9] and by the European Association of Urology (EAU).[29] Both of these guidelines recommend that the diagnosis of BPH is based on a medical history, symptomatic assessment using a validated instrument such as the International Prostate Symptom Score (IPSS), which uses eight items to categorize BPH 'urinary' symptoms as mild (IPSS 0–7), moderate (IPSS 8–19) and severe (IPSS 20–30), and a physical examination consisting of a digital rectal examination (DRE) to evaluate the prostate size. The AUA guidelines have produced a diagnostic algorithm, which is illustrated in **Figure 10**.

The prostate-specific antigen (PSA) test is used to distinguish between BPH and prostate cancer. PSA, an endogenous serine protease secreted by the epithelium, is a key malignant tumor marker. At present, a number of methods are used to assess PSA (e.g., PSA velocity, PSA density, age-specific PSA, and free/total PSA ratio), which have been developed as a way of enhancing specificity of the measure. More recently, glycosylation of PSA is being examined as a way of further refining the distinction between BPH PSA and prostate cancer PSA. In one study, urinary PSA isoforms were characterized by ion-exchange chromatography and lectin affinity chromatography using immobilized plant lectins in patients with either prostate cancer or BPH. Although PSA recovery from urine was low (up to 60%) the researchers identified two main prostate cancer isoforms and two BPH isoforms, which had different binding affinities to *Ulex europaeus* agglutinin, *Aleuria aurantia* agglutinin, *Phaseolus vulgaris* erythroagglutinin, and *P. vulgaris* leukoagglutinin.[30] Other tumour markers are also being established; for example, telomerase activity in epithelium cells, which is determined by a telomeric repeat amplification protocol assay, was detected in 90% of prostate cancer cases compared with 13% in BPH cases. This measure had a specificity of 76% and a positive predictive value (of cancer) of 87%.[31]

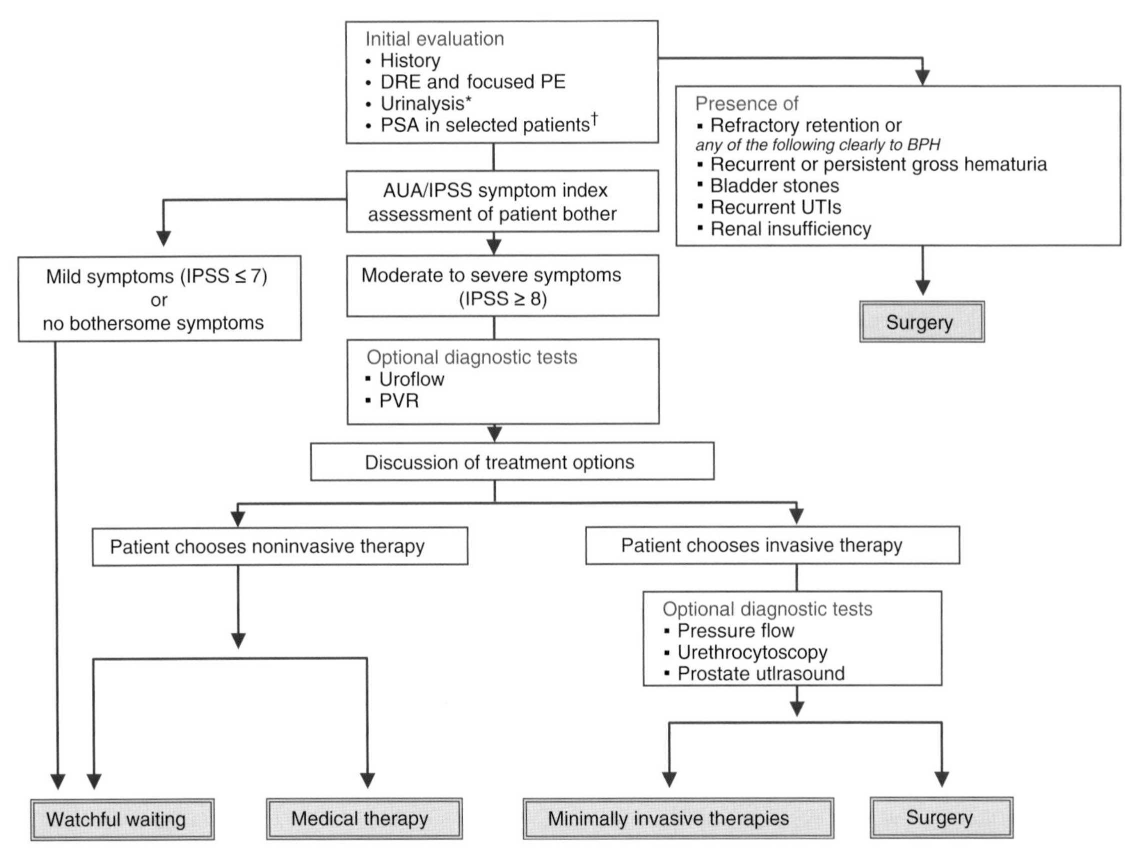

Figure 10 The AUA diagnostic and treatment algorithm for benign prostatic hyperplasia (BPH): *In patients with clinically significant prostatic bleeding; †patients with at least a 10-year life expectancy for whom knowledge of the presence of prostate cancer would change management or patients for whom the PSA measurement may change the management of voiding symptoms; DRE, digital rectal examination; IPSS, International Prostate Symptom Score; PE, physical examination; PSA, prostate-specific antigen; PVR, postvoid residue; UTI, urinary tract infection. (Reproduced with kind permission from AUA Practice Guidelines Committee. *J. Urol.* **2003**, *170*, 530–547.)

A number of optional diagnostic tests are also recommended by the AUA and EAU guidelines. Urinalysis (dipstick testing or microscopic examination of urine sediment) is recommended if hematuria or urinary tract infections (UTIs) are present. To determine the presence of obstruction (i.e., PBOO or BOO) in BPH patients with severe or bothersome LUTS, a number of urine flow and urine pressure assessments can be undertaken; in general, a low urine flow rate and high urine pressure are indicative of obstruction. Urine flow is usually assessed via uroflowmetry, a noninvasive measure that can reveal abnormal voiding. A number of flow-type measures have been developed as a way of improving this measure; for example, the maximum flow rate (Q_{max}) is a commonly used endpoint in clinical trials, which has a specificity of 78%.[32] Pressure-flow studies are used to categorize the severity of obstruction, and usually involve the measurement of both intravesical and intra-abdominal pressure. Postvoid residue (PVR) is a noninvasive measure obtained via transabdominal ultrasonography. Studies have shown that the normal PVR in healthy males is usually less than 5 mL and no more than 12 mL, whereas PVR volumes exceeding 200 mL may indicate that bladder dysfunction or detrusor instability are present.[33] Some studies have also reported a correlation between the increase in PVR and a decrease in detrusor contractility. Other diagnostic measures include endoscopy, which can provide additional information on the type and degree of obstruction, and imaging (renal ultrasonography) of the lower urinary tract.

Although these optional diagnostic measures are used to evaluate patients with LUTS suggestive of BOO, there is still few data validating their reliability. In a recent study, 152 men with LUTS and suspected BOO were examined using a number of procedures such as PVR, uroflowmetry, and pressure-flow studies. The researchers found that 45.8% had some form of obstruction, and the degree of obstruction was related to prostate volume, PVR, and Q_{max}; the severity of BOO was also strongly related to detrusor overactivity, which increased from 16% in the mildly obstructed group to 53.4% in the severely obstructed group.[34]

6.24.5 Benign Prostatic Hyperplasia and Lower Urinary Tract Symptoms

Clinically, BPH is characterized by LUTS, the incidence of which increases with age. Findings from the UK show that the prevalence increased from 3.5% for men aged 45–49 years to more than 30% for men aged 80 years and over. Findings from Norway also showed that LUTS increased with age, with approximately 5% of men under 40 years reporting LUTS compared with more than 30% in men aged 70 years or over; a number of factors such as body mass index, waist–hip ratio, alcohol consumption, smoking, and the presence of comorbid conditions such as diabetes, stroke, and osteoarthritis were also linked to moderate-to-severe LUTS.[35] A survey of 39 928 men from Sweden also showed that LUTS was age related, with frequent urination being the most common symptom in men younger than 70 years and nocturia being more common in men aged over 70 years, as shown in **Figure 11**.[36] Ethnic differences have also been reported. A survey of 2480 men from the USA (Olmsted County) showed that the prevalence of moderate-to-severe LUTS was significantly higher in black than Caucasian men (41% versus 34%, respectively; $P<0.001$).[37]

6.24.5.1 Lower Urinary Tract Symptoms and Sexual Dysfunction

There is increasing evidence to suggest that moderate-to-severe LUTS are linked with erectile dysfunction. A recent large-scale study known as the Multinational Survey of the Aging Male (MSAM-7), which surveyed 12 815 men in the USA and Europe (France, Germany, Italy, Netherlands, Spain, and the UK), showed that the prevalence of sexual dysfunction increased with LUTS severity, independently of age (see **Figure 12**).[38] The link between LUTS and sexual dysfunction is poorly understood, although a number of common components have been identified such as upregulation of α_1-adrenoceptor activity, alteration in α_1-adrenoceptor subtypes, decreased NO bioactivity, and sex hormone imbalance. α_1-Adrenoceptors are involved in the maintenance of smooth muscle contraction and relaxation in penile tissue. It has been suggested that any impairment of these receptors and their regulators such as the Rho/Rho-kinase pathway will contribute to smooth muscle dysfunction, which is evident in both LUTS and erectile dysfunction. Endothelium dysfunction as a result of ROS-induced NO breakdown can also contribute to LUTS and erectile dysfunction. Testosterone alterations are well established as a cause of BPH and LUTS, and it is likely that any androgen imbalances may also lead to erectile dysfunction. For more information on sexual dysfunction, see Chapter 6.23.

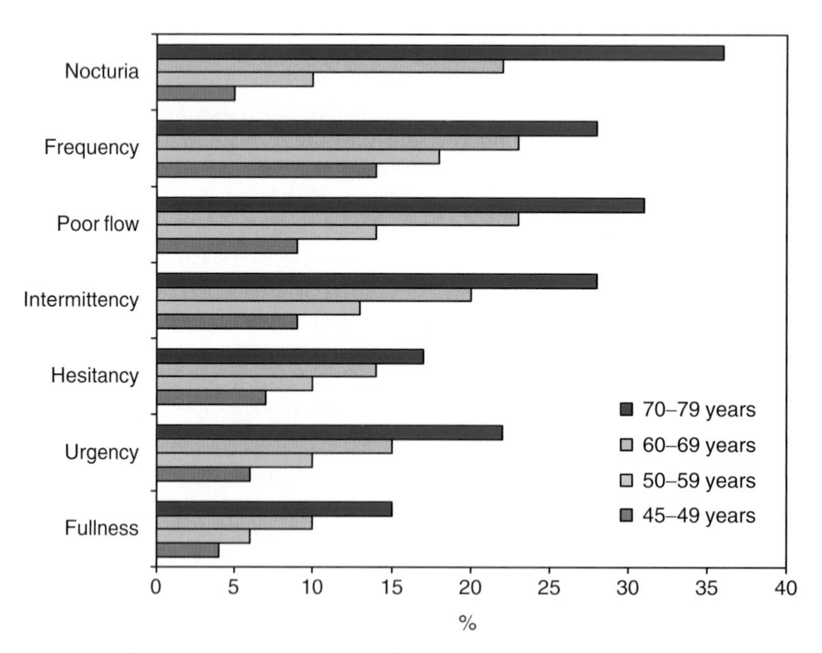

Figure 11 The percentage of lower urinary tract symptoms (LUTS) occurring 'more than half the time' in a population of men from Sweden ($n=39928$) stratified by age. (Reproduced with kind permission from Andersson, S.-O.; Rashidkhani, B.; Karlberg, L.; Wolk, A.; Johansson, J.-E. *BJU Int.* **2004**, *94*, 327–331 © 2004 Blackwell Publishing.)

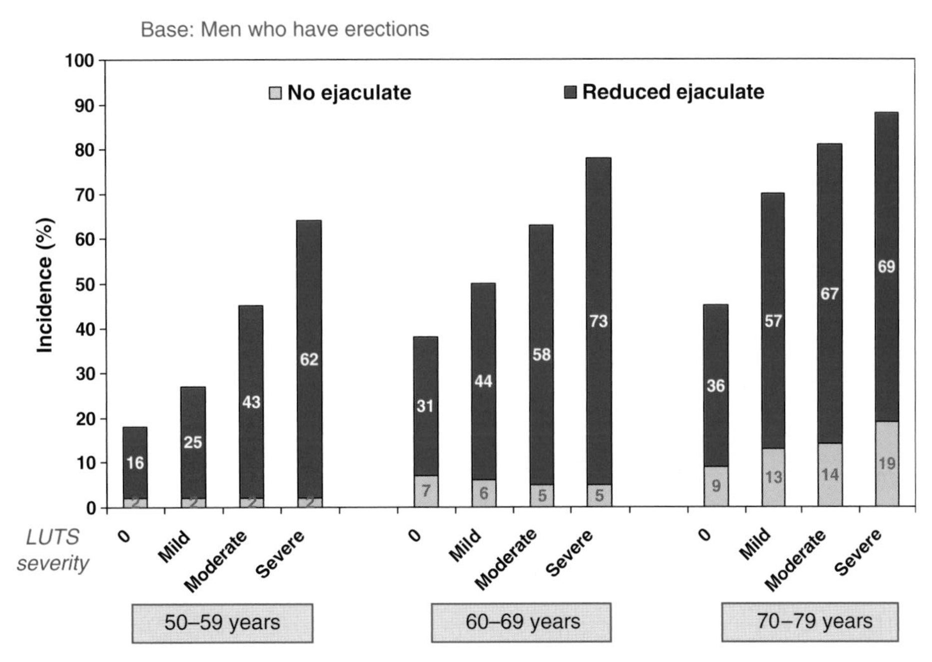

Figure 12 The relationship between lower urinary tract symptom (LUTS) severity and ejaculatory dysfunction (assessed via DAN-PSS) in relation to age. (Reproduced with kind permission from Rosen, R. C.; Altwein, J.; Boyle, P. *et al. Eur. Urol.* **2003**, *44*, 637–649.)

6.24.6 Benign Prostatic Hyperplasia and Renal Disease

The relationship between BPH and chronic kidney disease (CKD) has yet to be fully established. At present, there is some evidence to suggest that BPH is a weak risk factor for CKD. In a recent USA study of 476 men, 29 (6%) had CKD, characterized by a serum creatinine $\geqslant 133\,\mu M$. These men also had a significantly higher number of BPH symptoms than men without CKD including slow urinary stream (Q_{max} was $<15\,mL\,s^{-1}$ in 62% of men with CKD versus 35% of men without CKD; $P<0.01$), IPSS>7 (59% versus 33%; $P<0.01$), and chronic urinary retention (PVR> 100 mL in 17% versus 6%; $P<0.01$). The presence of these 'classic' BPH symptoms doubled or even trebled the risk of CKD (diagnosed by serum creatinine). A higher number of CKD patients also had BOO (48% versus 20%; $P<0.001$) although there was no association between CKD and prostatic enlargement.[39] For more information on renal disease see Chapter 6.25.

6.24.7 Treatment Targets for Benign Prostatic Hyperplasia

6.24.7.1 Receptor Targets

It is well established that α_1-adrenergic receptors (adrenoceptors) play a key role in the development of BPH and the manifestation of LUTS. At the molecular level these are designated as a, b, and c, whereas at the pharmacological level the nomenclature is A, B, and D. For the remainder of this chapter the molecular descriptor will in general be used. Adrenoceptors are 7-transmembrane G-protein-linked receptors, and three subtypes of α_1-adrenoceptors have been identified in prostate stroma (α_{1A}, α_{1B}, and α_{1D}) (**Figure 13**). These receptors, which are activated by epinephrine and norepinephrine, mediate smooth muscle contraction via phospholipase C activation, and it is thought that abnormal activity of these receptors may account (in part) for smooth muscle abnormalities (mainly contraction) leading to BOO, which contributes to LUTS.[40] α_{1A}-Adrenoceptors are the most common subtype in prostate stroma accounting for 70% of the receptor population, and this subtype may be the most important in the mediation of smooth muscle contraction, making it a key target for treatments. α_{1D}-Adrenoceptors predominate in the detrusor muscle of the bladder and the bladder neck, and enhanced expression of this subtype has been observed in BOO-related bladder hypertrophy; this subtype may also be important in the expression of BPH irritative symptoms such as frequency, urgency, and nocturia whereas the α_{1B}-adrenoceptor, which is widely distributed outside the urogenital tract, may be the major locus of the treatment-related cardiovascular side effects, such as orthostasis, dizziness, and fatigue.[41]

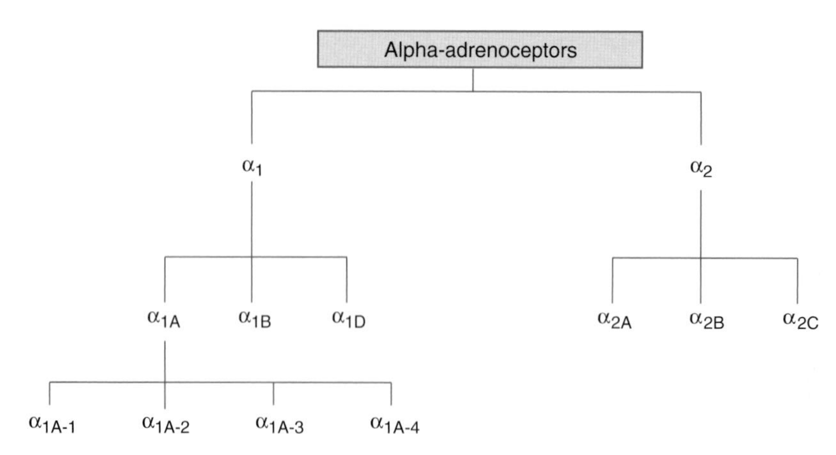

Figure 13 The α-adrenoceptor family tree. (Reproduced with kind permission from Roehrborn, C. G.; Schwinn, D. A. *J. Urol.* **2004**, *171*, 1029–1035.)

Table 3 Receptor binding affinity and selectivity of α_1-adrenoceptor antagonists

Compound	*Receptor affinity (pK$_i$)*			*Selectivity ratio*	
	α_{1A}	α_{1B}	α_{1D}	α_{1A}/α_{1B}	α_{1A}/α_{1D}
Prazosin	9.70	9.60	9.50	1.2	1.3
Tamsulosin	9.70	8.90	9.80	6.3	0.2
Doxazosin	8.56	8.98	8.78	0.4	1.6
Alfuzosin	8.20	8.53	8.40	0.5	1.4
Terazosin	8.16	8.71	8.48	0.3	1.8

Reproduced with kind permission from Lowe, F. C. *Clin. Ther.* **2004**, *26*, 1701–1713.

As a result of the key role that α_1-adrenoceptors play in BPH and LUTS, a number of agents that target these receptors (e.g., α_1-adrenoceptor antagonists) have been developed as treatments (*see* Section 6.24.7.2). Theoretically, an ideal agent (i.e., one with an optimal clinical profile) would be 'uroselective,' having high binding affinity for α_{1A}- and α_{1D}-adrenoceptors and low affinity for α_{1B}-adrenoceptors. α_1-Adrenoceptor antagonists (or α_1,-blockers) can be classified based on their receptor specificity (see **Table 3**).

A number of other receptors are also being examined for their role in BPH and as potential treatment targets. Of clinical importance is the recent finding that estradiol and androgen receptors have different subcellular distribution in the prostate of patients with BPH and prostate cancer. Vitamin D receptors, which are expressed by the human bladder, are also being investigated as a treatment target,[42] and bradykinin receptors, part of the kallikrein–kinin system, which are expressed in normal human prostate and have a role to play in cell proliferation via activation of extracellular signal-related kinases (ERK)-1/2 pathways, are also being explored.[43] For more information on choosing targets see Chapter 2.19.

6.24.7.2 Pharmacotherapy

6.24.7.2.1 α_1-Adrenoceptor antagonists

α_1-Adrenoceptor antagonists 'block' the adrenoceptors in the prostate and bladder neck, causing the smooth muscle to relax, which results in an improvement in urine flow and a decrease in residual bladder volume. The four α_1-adrenoceptor antagonists recommended by both the EAU and AUA guidelines are alfuzosin, doxazosin, terazosin (quinazoline derivatives), and tamsulosin (sulfonamide derivative) (see **Figure 14** for structures). Indirect comparisons of placebo-controlled studies using the α_1-adrenoceptor antagonists alfuzosin, doxazosin, terazosin, and tamsulosin have shown that they have comparable efficacy; symptom scores are improved by 30–45% and Q_{max} by 15–30% versus baseline.[44] More recently, there is evidence to suggest that doxazosin and terazosin also have proapoptotic effects in the prostate epithelium and stroma.[45]

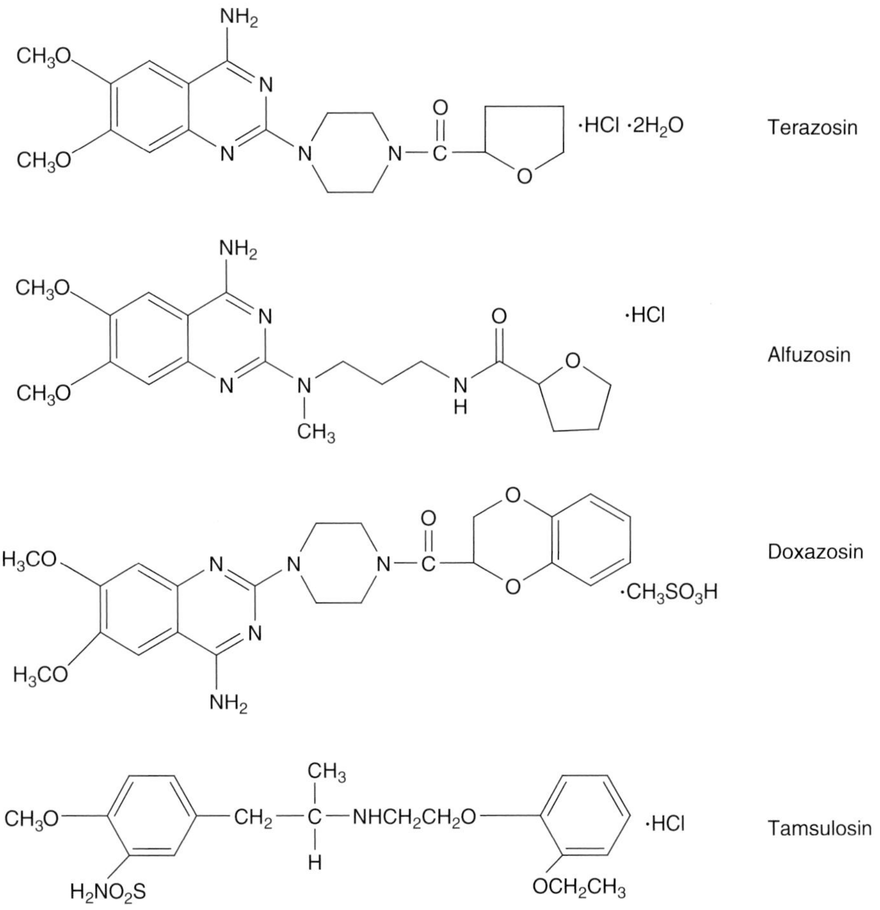

Figure 14 α_1-Adrenoceptor antagonists.

6.24.7.2.2 5α-Reductase inhibitors

5α-Reductase converts testosterone to DHT, and inhibition of this enzyme reduces the high levels of DHT evident in BPH. Two 5α-reductase inhibitors are recommended by guidelines (finasteride and dutasteride) (see **Figure 15** for structures). Finasteride is a selective inhibitor of type 2 5α-reductase isoenzyme, which has been shown to decrease plasma DHT by 60–70% and prostatic DHT by 85%. In comparison, dutasteride inhibits both isoenzymes and decreases DHT concentrations by more than 90%.[46] Data from placebo-controlled trials shows that both agents decrease prostate volume (−18% reduction for finasteride versus −26% reduction for dutasteride), decrease IPSS (−3.3 versus −4.5, respectively), and improve urine flow rate.[47]

6.24.7.2.3 Side effects

α_1-Adrenoceptor antagonists can be differentiated, to a certain extent, by their side effect profiles. At the maximal doses used in the clinic, with regard to cardiovascular adverse events, doxazosin and terazosin have higher dizziness rates (0–20%, corrected for placebo) than alfuzosin and tamsulosin (0–8%, corrected for placebo). Only a few direct comparative studies exist, which show that alfuzosin 2.5 mg once daily significantly decreases both diastolic and systolic blood pressure compared with tamsulosin 0.4 mg once daily. In addition, abnormal ejaculation, including retrograde ejaculation and reduced ejaculate volume, have been associated with one of these agents, tamsulosin, where the incidence of abnormal ejaculation has been shown to range from 4 to 30%.[48] One hypothesis is that blockade of α_{1A}- and α_{1D}-adrenoceptors in the bladder neck, seminal vesicles, and vas deferens might induce these effects, and greater bladder neck closure and seminal vesicle contraction, which has been observed with tamsulosin, may account for these differences. However, the unusually high degree of sexual dysfunction observed with tamsulosin may involve interactions with receptor subtypes other than α_1.[48]

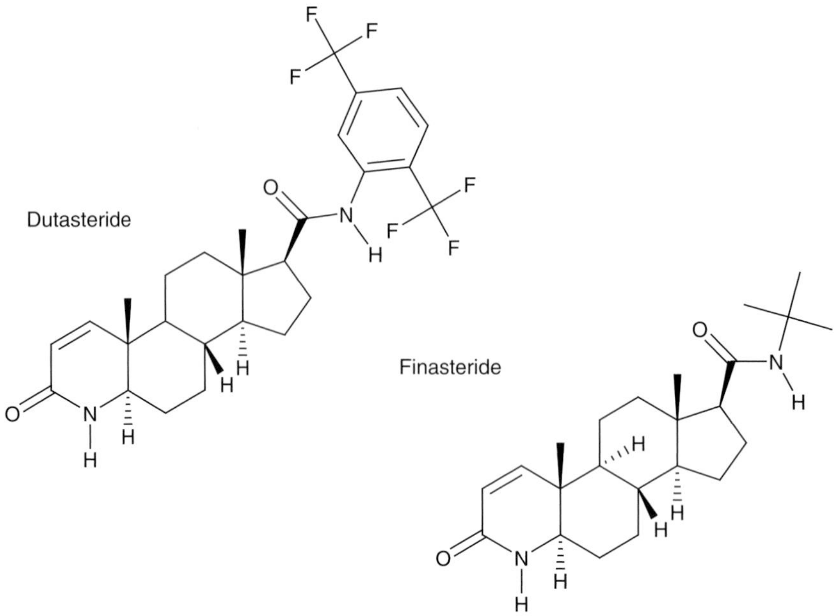

Dutasteride

Finasteride

Figure 15 5α-Reductase inhibitors.

5α-Reductase inhibitors have also been associated with sexual dysfunction side effects; findings from a large-scale trial known as the Medical Therapy Of Prostatic Symptoms (MTOPS) study showed that finasteride was significantly associated with decreased libido, erectile dysfunction, and ejaculatory disorders in comparison with placebo and doxazosin.[49]

At present, there is a 'trend' toward using combination therapies to target BPH and comorbid conditions such as erectile dysfunction, which has important consequences for safety considerations. In a recent placebo-controlled study of healthy volunteers, the combination of tadalafil (a phosphodiesterase type 5 (PDE5)-inhibitor used in the treatment of erectile dysfunction) with doxazosin significantly reduced the blood pressure compared with tadalafil and placebo (the mean difference in systolic blood pressure was 9.8 mmHg); however, this effect was not observed with tadalafil and tamsulosin, supporting the conservative use of this latter combination in the treatment of BPH patients with erectile dysfunction.[50] Whether, the PDE5-inhibitor can prevent the high incidence of abnormal ejaculation observed with tamsulosin remains a key topic for future studies assessing combination therapies. However, combination strategies have now become recommended as treatment approaches in guidelines (as discussed below).

6.24.8 Management Approaches

The overall aim of therapy is to improve both symptoms and quality of life. Both the EAU and AUA guidelines[9,29] list a number of recommended treatments for LUTS associated with BPH. Those listed by the EAU are outlined in **Table 4**; watchful waiting (i.e., no intervention) is advised if patients have minimal symptoms, 5α-reductase inhibitors or α_1-adrenoceptor antagonists are recommended in patients with moderate-to-severe LUTS, with combination therapy being an option for those with large prostate volumes, at risk of developing complications or progressing, and surgery is advised for those who do not improve after medical therapy, do not want medication, or have complications such as acute urinary retention (AUR), bladder stones, recurrent infections, or recurrent hematuria refractory to medication. The type of surgery undertaken is usually based on prostate size, surgeon's judgment, and the presence of comorbidities.

The most recent addition to these guidelines has been the inclusion of combination therapy with a 5α-reductase inhibitor (finasteride) and an α_1-adrenoceptor antagonist (doxazosin). This recommendation was based on findings from a recent large-scale trial known as MTOPS, which had a mean follow-up of 4–5 years and included a total of 3047 BPH patients. The study showed that combination of finasteride and doxazosin reduced the risk of symptomatic progression by 67% compared with 39% (doxazosin) and 34% (finasteride); the risk of AUR was reduced by 79% compared with 31% (doxazosin) and 67% (finasteride) and the risk of BPH-related surgery was reduced by 67%.[51]

Table 4 Treatment recommendations for LUTS associated with BPH

Treatment	*EAU 2004 recommendation*
Watchful waiting	Recommended
Medical therapy	
α1-Blocker	
Alfuzosin	Recommended
Doxazosin	Recommended
Tamsulosin	Recommended
Terazosin	Recommended
5ARI	
Dutasteride	Recommended
Finasteride	Recommended
Combination therapy	
α1-Blocker plus 5ARI	Recommended
Plant extracts	Not recommended
Minimally invasive therapies	
High-energy TUMT	Recommended
TUNA[a]	Recommended
Prostatic stents[b]	Recommended
Surgical therapies	
TUIP	Recommended
TURP	Recommended
Open prostatectomy	Recommended
Transurethral holmium laser enucleation	Recommended
Transurethral laser vaporization[a]	Recommended
Interstitial laser coagulation[a]	Recommended
Transurethral laser coagulation[a]	Recommended
Emerging therapies	
Ethanol injections	
High-intensity focused ultrasound	
Water-induced thermotherapy	
PlasmaKinetic tissue management	

5ARI, 5α-reductase inhibitors; TUIP, transurethral incision of the prostate; TUMT, transurethral microwave thermotherapy; TUNA, transurethral resection of the prostate; TURP, transurethral incision of the prostate.
Reproduced with kind permission from Madersbacher, S.; Alivizatos, G.; Nordling, J.; Sanz, C. R.; Emberton, M.; de la Rosette, J. J. M. C. H. *Eur. Urol.* **2004**, *46*, 547–554.
[a] Not as first-line treatment.
[b] Only to high-risk patients as an alternative to permanent catheterization.

6.24.8.1 New Avenues

Although not recommended by the EAU and AUA guidelines, phytotherapy has become a popular BPH treatment in parts of Europe. Recommendations from the 5th International Consultation on BPH regarding the use of plant extracts in the treatment of BPH suggest that every brand should be fully evaluated, and only extracts with proven clinical efficacy should be used.[52] A lipido-sterolic extract of *Serenoa repens* (Saw palmetto) is one such option, and there is a growing database supporting the use of one form (LSESr Permixon), which is a complex mixture containing 90% free fatty acids and 7% esterified long-chain fatty acids, which mainly consists of oleic, lauric, myristic, and palmitic acid.[53] In a recent meta-analysis of 14 randomized clinical trials involving over 4000 patients, with a maximum duration of 720 days, Permixon was associated with a mean decrease in IPSS of 4.78 and a mean improvement in Q_{max} of 1.02 mL s^{-1} above placebo.[54] Although the exact mechanism of action is unknown, there is evidence to suggest that Permixon is a 5α-reductase inhibitor, which also inhibits DHT binding, and has anti-inflammatory effects relating to leukotriene production.[53] Saw palmetto is also being explored in combination with nettle root, another herbal option, in a preparation known as Prostagutt forte. Other phytotherapeutic preparations also being assessed include red clover extracts, which contain high concentrations of isoflavones that have been shown to have symptomatic benefits.

In addition, luteinizing hormone-releasing hormone antagonists that inhibit the pituitary–gonadal axis, and vitamin D_3 analogs that target receptors in the bladder are also in development. Botulinum A toxin (Botox) has also been explored in patients who were poor candidates for surgery. In a recent study of 10 patients with BPH and AUR, all patients given 200 U Botox injections into the transition zone of the prostate showed improvements in PVR, Q_{max}, and prostate volume, which were maintained during the 12-month follow-up period.[55] Other areas being explored include de-sensitization of C-fibers, which are involved in nociception (pain perception), using intravesical resiniferatoxin solution[56] and the effects of beta-radiation on key growth factors, including TGFβ1 and bFGF. It is thought that beta-rays may shrink the hyperplastic cells and reawaken apoptosis.[57]

For existing products there is a trend toward new formulations such as extended-release doxazosin, which has the same clinical benefits as the standard preparation but with less need for dose-titration, and the use of these medications in comorbid conditions. In a recent study of BPH patients with comorbid sexual dysfunction, doxazosin extended-release 4 or 8 mg or doxazosin standard 1–8 mg were given to patients for 13 weeks. The study showed that sexual function improved after treatment with both formulations as shown by the International Index of Erectile Function questionnaire.[58]

6.24.9 Treatment Complications and Progression

BPH is considered a progressive condition. Longitudinal data from a large-scale trial known as the Proscar Long-Term Efficacy and Safety Study (PLESS) in which 3040 men were randomized to treatment with finasteride (Proscar) or placebo for 4 years, with an additional 2-year follow-up period showed that finasteride prevented progression.[59] However, baseline PSA and prostate volume could predict those patients at greater risk of symptom and quality of life deterioration, and those at risk for the development of AUR or BPH-related surgery. Those patients with a prostate size greater than 40 cm^3 and PSA of 1.4 ng mL or above were at the greatest risk for developing AUR, and serum PSA was a strong predictor of prostate growth; the annual growth ranged from 0.7 mL $year^{-1}$ for patients with a PSA 0.2–1.3 ng mL^{-1} to 3.3 mL $year^{-1}$ for PSA range of 3.3 to 9.9 ng mL^{-1}.[60] In the MTOPS study, which also assessed the impact of medical treatment of BPH progression, baseline PSA, Q_{max}, PVR, and prostate volume were all significantly associated with clinical progression.[60] **Figure 16** shows the effect of prostate volume on several progression endpoints, including AUR and BPH-related treatment.

Given that a number of baseline variables can predict outcomes in BPH, they are often used to chart management progress in these patients. In a recent survey of 1859 BPH patients, PSA was found to reflect prostatic enlargement with sufficient accuracy to be of clinical benefit. From the data, these researchers produced a nomogram that could predict prostate volume at 5-year intervals based on serum PSA values.[61] PSA has also been shown to be a predictor of BOO In a study of 302 men with LUTS and a PSA less than 10 mg nL^{-1}, where patients had a PSA greater than 4 ng mL^{-1}, mild or definite BOO was likely in 89% of cases whereas if the PSA was less than 2 ng mL^{-1} then the chance of having BOO was much lower (approximately 33%).[62] In another study, the appropriateness of choosing a particular treatment in relation to the patient risk profile was considered against a reference profile; these findings are shown in **Table 5**.[63]

6.24.10 Other Causes of Incontinence

In addition to BPH, a number of other conditions have been associated with the development of LUTS. In one survey of 119 men with LUTS, less than 50% had BPH; other causes included urethral stricture, bladder neck sclerosis,

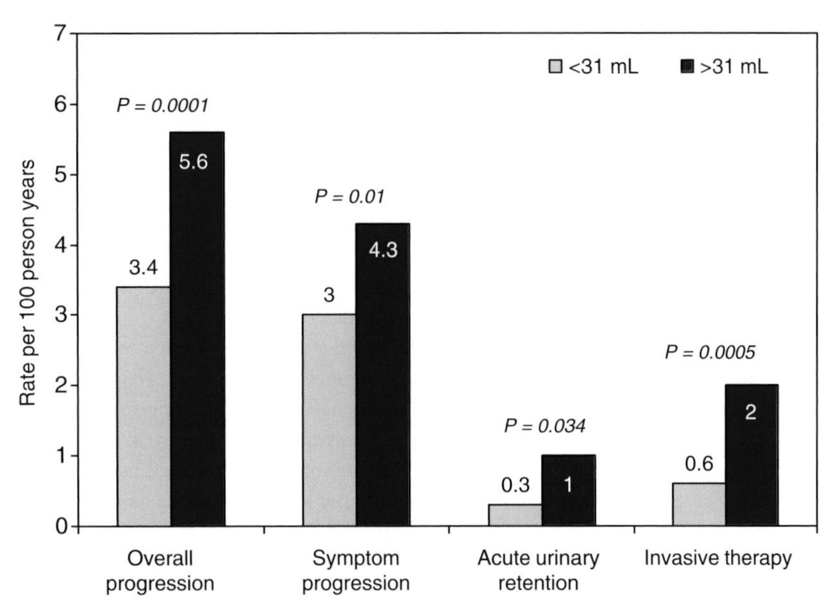

Figure 16 The risk of clinical progression (overall, symptomatic, acute urinary retention and invasive therapy) in relation to prostate volume (<31 mL or >31 mL). (Reproduced with kind permission from Kim, E. D. *Curr. Urol. Rep.* **2004**, *5*, 267–273.)

Table 5 Treatment choice as a function of clinical risks; values are calculated against a reference profile without any unfavourable risks (i.e., total IPSS 0–7; prostate volume <30 mL; PSA <1.4 ng mL^{-1}; Q_{max} >15 mL s^{-1} and PVR <50 mL)

Variable	*α-Blocker*	*5α-Reductase inhibitor*	*Combined therapy*	*Surgery*
Total IPSS 8–19	+ + + + +		+ +	+ + +
Total IPSS 20–35	+ + + +		+ +	+ + + + +
Prostate volume 30–59 mL		+ + + + +	+ + + + +	
Prostate volume >60 mL		+ + + + +	+ + + + +	+
Q_{max} 10–15 mL s^{-1}				+ +
Q_{max} <10 mL s^{-1}			+	+ + +
PVR 50–150 mL				+ + +
PVR > 150 mL		+	+	+ + + + +

IPSS, International Prostate Symptom Score; PVR, post-void residue; Q_{max}, maximum urinary flow rate.
Adapted with kind permission from Trachtenberg, J. *BJU Int.* **2005**, *95*, 6–11.

prostate or bladder cancer, hypocontractive bladder, overactive bladder, abacterial prostatitis, or a UTI.[64] Prostate surgery can also be a cause of refractory incontinence. There are also a number of nonurological causes including neurological diseases such as Parkinson's disease and back surgery. Some of the prostate-related causes of LUTS are discussed in this section.

6.24.10.1 Prostate Surgery and Incontinence

Prostate surgery can contribute to transient, short- or long-term incontinence. Radical retropubic prostatectomy is a surgical procedure used in prostate cancer, which involves the removal of adenomatous tissue via an incision in the surgical capsule of the prostate. However, this technique can result in refractory incontinence. One study of 146 prostate cancer patients showed that stress urinary incontinence was evident in the majority of patients (95%) after surgery, with the main cause of incontinence being intrinsic sphincter deficiency.[65] Another study in 120 prostate cancer patients showed that radical prostatectomy had a significant impact on nocturia and voiding frequency in a small proportion of patients.[66]

6.24.10.2 Prostatitis

Prostatitis is an extremely common condition, which is classified into several subgroups. Acute bacterial prostatitis is the result of an acute pathogenic infection and can lead to chronic bacterial prostatitis. Chronic nonbacterial prostatitis, or chronic pelvic pain syndrome (CPPS), is a multifactorial condition that may be linked to other conditions such as bladder neck obstruction, urethral stricture, detrusor sphincter, dyssynergia, or dysfunctional voiding. Nonculturable organisms and sexually transmitted infections (e.g., *Chlamydia trachomonas*) are also possible causes.

CPPS is the most common form affecting up to 14% of men, and approximately 50% will have this condition at some time in their life. CPPS is typically characterized by chronic discomfort or pain, which can occur in the lower back, tip of the penis, suprapubic area, and perineal area. Urinary symptoms include urinary frequency, dysuria, weak stream, incomplete emptying, and painful ejaculation; inflammation can also be present.[67]

Although the underlying causes of CPPS are not fully understood, there is evidence to suggest that immunologic, neurologic, endocrine, and psychologic factors all contribute as shown in **Figure 17**. It is thought that an initiation event such as an infection or trauma leads to inflammation, often characterized by abnormal levels of proinflammatory markers (cytokines), which have been shown to correlate with symptoms and play a role in nociception.[68] This inflammatory reaction is also linked to a number of other contributory factors; for example, increased levels of the inflammatory markers TNF-α and interferon-γ (IFN-γ) can alter the surface of epithelial cells resulting in increased inflammatory infiltration and decreased steroid hormone production. There is also evidence to suggest that inflammation increases the activation of nerve growth factor, an important neurotrophin involved in the regulation of nociceptive nerves and C-fibers, and this may lead to 'sensitization' to pain. A role for testosterone has also been

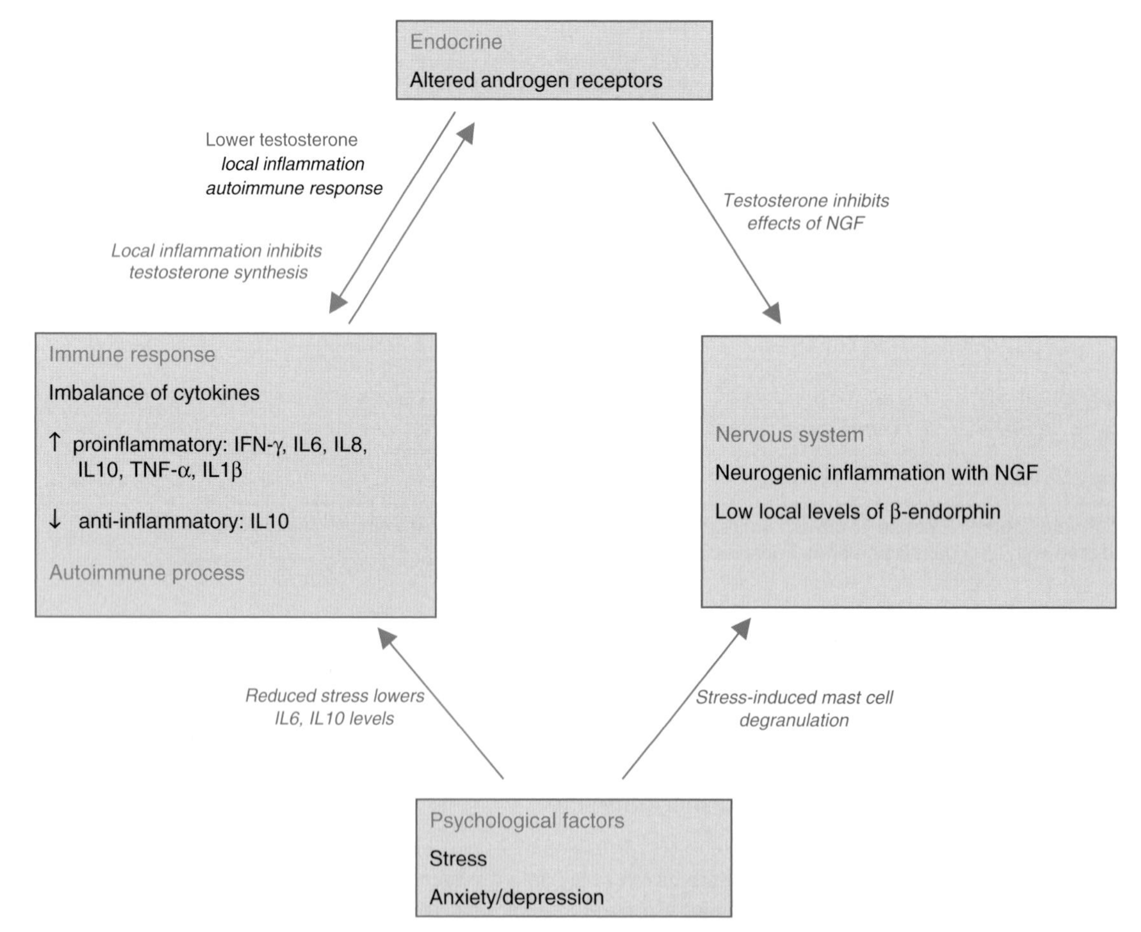

Figure 17 The role of immunologic, endocrine, neurologic, and psychologic factors in the development of chronic pelvic pain syndrome (CPPS) INF, interferon; IGF, insulin-like growth factor; IL, interleukins; NGF, nerve growth factor; TNF, tumor necrosis factor. (Reproduced with kind permission from Pontari, M. A.; Ruggieri, M. R. *J. Urol.* **2004**, *172*, 839–845.)

implicated. The inflammatory process can also inhibit the production of endogenous opioids such as β-endorphins, further enhancing the pain evident in CPPS.[69]

Given the multiple etiologies, prostatitis can be extremely difficult to diagnose and treat. Acute bacterial prostatitis is usually diagnosed via dipstick testing of midstream urine, together with bacterial cultures. If bacteria are present, then it is typically treated by hydration and intravenous antibiotics such as cephalosporins followed by oral antibiotics such as quinolones or co-trimoxazole. For those patients with recurrent (chronic) bacterial prostatitis, symptoms should be present for at least 3 months, and the management approach is both curative and suppressive, i.e., a long-term course of antibiotics with follow-up to check for relapse.[70] When leukocytes are present in prostatic secretions and ejaculate then inflammatory CPPS is suspected; other diagnostic measures include symptomatic questionnaires such as the Chronic Prostatitis Symptom Index, transrectal ultrasound, and the presence of obstruction assessed via urodynamics or endoscopic investigations.

However, due to the absence of approved pharmacotherapy in this area, the management of CPPS is often a 'trial-and-error' approach. At present, the three most commonly used medications are antibiotics, anti-inflammatories, and α_1-adrenoceptor antagonists (α-blockers). A recent review of randomized placebo-controlled trials using antibiotics, α_1-adrenoceptor antagonists, and anti-inflammatories has shown that the greatest response rates in CPPS have been observed with alfuzosin 5 mg (65% versus 24% for placebo after 6 months), rofecoxib 50 mg (63% after 6 weeks), and quercetin 500 mg (67% versus 20% for placebo after 4 weeks).[71]

6.24.10.3 Other Prostatitic Conditions Causing Incontinence

There is evidence to suggest that prostatitis may be pathologically linked to interstitial cystitis in men as both of these conditions share common symptoms. In a survey of 92 men with interstitial cystitis confirmed by the National Institute for Diabetes and Digestive and Kidney Diseases criteria, the most commonly reported initial symptoms were mild suprapubic discomfort (33%), nocturia (15%), urgency (15%), dysuria (11%), and frequency (11%); however, after 2.5 years these symptoms had become more severe and 89% reported dysuria, 85% had urinary frequency, 82% had severe suprapubic discomfort and/or urgency, and 56% had sexual dysfunction.[67] The overlap between prostatitis and interstitial cystitis has led researchers to consider that they have underlying causes. In a recent study of 50 patients with prostatitis, the majority of patients were also symptomatic on an interstitial cystitis questionnaire known as the pelvic pain and urgency/frequency questionnaire, and 77% with a score more than 7 also tested positive for the potassium-sensitivity test. It has been established that bladder epithelial dysfunction can develop in interstitial cystitis, allowing irritative substances such as potassium from the urine to penetrate the epithelium and provoke symptoms. These findings suggest that bladder dysfunction, commonly associated with interstitial cystitis, may be a key source of symptoms in prostatitis.[72]

Another classic form of incontinence known as urinary urge incontinence (UUI), which is a complaint of involuntary leakage of urine accompanied by or immediately preceded by urgency, is more common in males than females, accounting for 40–80% of male cases.[73] UUI is usually caused by detrusor overactivity in men. It is thought that obstruction caused by BPH can affect the local or ventral detrusor control, which results in overactivity, and this explains the higher prevalence in males. The presence of detrusor overactivity can also affect bladder contraction strength, and greater bladder contractions can lead to higher urge severity. It is possible that both UUI and overactive bladder can have similar underlying mechanisms to those observed in BPH patients with detrusor overactivity, which leads some authors to question whether BPH is part of a larger syndrome involving prostatitis (inflammatory), intersititial cystitis, UUI, and overactive bladder.

Other abnormalities of the prostate such those associated with the anterior fibromuscular stroma (AFMS) can also contribute to urinary disturbances. There is evidence to suggest that age-related urinary problems can be associated with poor movement of the AFMS in patients without evidence of BPH or bladder neck obstruction.[74] For more information on urological diseases, see Chapter 6.23.

References

1. Dixon, J. S.; Gosling, J. A. *Textbook of Benign Prostatic Hyperplasia*, 2nd ed.; Taylor & Francis: London, 2005, pp 3–10.
2. Turkes, A.; Griffiths, K. *Textbook of Benign Prostatic Hyperplasia*, 2nd ed.; Taylor & Francis: London, 2005, pp 27–68.
3. Morales, A.; Heaton, J. P.; Carson, C. C., III *J. Urol.* **2000**, *163*, 705–712.
4. Rhoden, E. L. *N. Engl. J. Med.* **2004**, *350*, 482–492.
5. Wang, X.; Stocco, D. M. *Mol. Cell. Endocrinol.* **2005**, *238*, 1–7.
6. Wang, X.; Shen, C.-L.; Dyson, M. T.; Eimerl, S.; Orly, J.; Hutson, J. C.; Stocco, D. *Endocrinology* **2005**, *146*, 4202–4208.
7. Bosch, J. L.; Hop, W. C.; Niemer, A. Q.; Bangma, C. H.; Kirkels, W. J.; Schroeder, F. H. *J. Urol.* **1994**, *152*, 1501–1505.

8. Fujikawa, S.; Matsuura, H.; Kanai, M.; Fumino, M.; Ishii, K.; Arima, K.; Shiraishi, T.; Sugimura, Y. *Prostate* **2005**, *65*, 355–364.
9. AUA Practice Guidelines Committee. *J. Urol.* **2003**, *170*, 530–547.
10. Sarma, A. V.; Jacobson, D. J.; McGree, M. E.; Roberts, R. O.; Lieber, M. M.; Jacobsen, S. J. *J. Urol.* **2005**, *173*, 2048–2053.
11. Verhamme, K. M.; Dieleman, J. P.; Bleumink, G. S.; van der Lei, J.; Sturkenboom, M. C.; Artibani, W.; Begaud, B.; Berges, R.; Borkowski, A.; Chappel, C. R. et al. *Eur. Urol.* **2002**, *42*, 323–328.
12. Fromont, G.; Chene, L.; Latil, A.; Bieche, I.; Vidaud, M.; Vallancien, G.; Mangin, P.; Fournier, G.; Validire, P.; Cussenot, O. *J. Urol.* **2004**, *172*, 1382–1385.
13. Li, Z.; Habuchi, T.; Tsuchiya, N.; Mitsumori, K.; Wang, L.; Ohyama, C.; Sato, K.; Kamoto, T.; Ogawa, O.; Kato, T. *Carcinogenesis* **2004**, *25*, 237–240.
14. Sakamoto, S.; Yokoyama, M.; Zhang, X.; Prakash, K.; Nagao, K.; Hatanaka, T.; Getzenberg, R. H.; Kakehi, Y. *Endocrinology* **2004**, *145*, 2929–2940.
15. Roberts, R. O.; Bergstrath, E. J.; Farmer, S. A.; Jacobson, D. J.; McGree, M. E.; Hebbring, S. J.; Cunningham, J. M.; Anderson, S. A.; Thibodeau, S. N.; Lieber, M. M. et al. *Prostate* **2005**, *62*, 380–387.
16. Lee, K. L.; Peehl, D. M. *J. Urol.* **2004**, *172*, 1784–1791.
17. Cordon-Cardo, C.; Koff, A.; Drobnjak, M.; Capodieci, P.; Osman, I.; Millard, S. S. *J. Natl. Cancer. Inst.* **1998**, *90*, 1284–1291.
18. Liu, X.; Wang, L.; Lin, Y.; Teng, Q.; Zhao, C.; Hu, H.; Chi, W. *Prostate* **2000**, *43*, 83–87.
19. Gandour-Edwards, R.; Mack, P. C.; deVere-White, R. W.; Gumerlock, P. H. *Prostate Cancer Prostatic Dis.* **2004**, 7, 321–326.
20. Castro, P.; Giri, D.; Lamb, D.; Ittmann, M. *Prostate* **2003**, *55*, 30.
21. Untergasser, G.; Madersbacher, S.; Berger, P. *Exp. Gerontol.* **2005**, *40*, 121–128.
22. Untergasser, G.; Gander, R.; Lilg, C.; Lepperdinger, G.; Plas, E.; Berger, P. *Mech. Aging Dev.* **2005**, *126*, 59–69.
24. Shirakawa, T.; Okada, H.; Acharya, B.; Zhang, Z.; Hinata, N.; Wada, Y. *Prostate* **2004**, *58*, 33–40.
25. Yokota, T.; Honda, K.; Tsuruya, Y.; Nomiya, M.; Yamaguchi, O.; Gotanda, K.; Constantinou, C. E. *Prostate* **2004**, *58*, 156–163.
26. Wang, W.; Bergh, A.; Damber, J.-E. *Prostate* **2004**, *61*, 60–72.
27. Lemack, G. E.; McConnell, J. D. *Textbook of Benign Prostatic Hyperplasia*; 2nd ed.; Taylor & Francis: London, 2005, pp 113–115.
28. Araki, I.; Du, S.; Kamiyama, M.; Mikami, Y.; Matsushita, K.; Komuro, M.; Furuya, Y.; Takeda, M. *Urology* **2004**, *64*, 1255–1260.
29. Madersbacher, S.; Alivizatos, G.; Nordling, J.; Sanz, C. R.; Emberton, M.; de la Rosette, J. J. M. C. H. *Eur. Urol.* **2004**, *46*, 547–554.
30. Jankovic, M. M.; Kosanovic, M. M. *Clin. Biochem.* **2005**, *38*, 58–65.
31. Vincentini, C.; Gravina, G. L.; Angelucci, A.; Pascale, E.; D'Ambrosio, E.; Muzi, P.; Di Leonardo, G.; Fileni, A.; Tubaro, A.; Festuccia, C. et al. *J. Cancer Res. Clin. Oncol.* **2004**, *130*, 217–221.
32. Abrams, P. *J. Urol.* **1977**, *117*, 71–74.
33. Bosch, R. *World J. Urol.* **1995**, *13*, 17–18.
34. Vesely, S.; Knutson, T.; Fall, M.; Damber, J.-E.; Dahlstrand, C. *Neurourol. Urodyn.* **2003**, *22*, 301–305.
35. Seim, A.; Hoyo, C.; Ostbye, T.; Vatten, L. *BJU Int.* **2005**, *96*, 88–92.
36. Andersson, S.-O.; Rashidkhani, B.; Karlberg, L.; Wolk, A.; Johansson, J.-E. *BJU Int.* **2004**, *94*, 327–331.
37. Sarma, A. V.; Wei, J. T.; Jacobson, D. J.; Dunn, R. L.; Roberts, R. O.; Girman, C. J.; Lieber, M. M.; Cooney, K. A.; Schottenfeld, D.; Montie, J. E. et al. *Urology* **2003**, *61*, 1086–1091.
38. Rosen, R. C.; Altwein, J.; Boyle, P. et al. *Eur. Urol.* **2003**, *44*, 637–649.
39. Rule, A. D.; Jacobson, D. J.; Roberts, R. O.; German, C. J.; McGree, M. E.; Lieber, M. M.; Jacobsen, S. J. *Kid. Int.* **2005**, *67*, 2376–2382.
40. Lowe, F. C. *Clin. Ther.* **2004**, *26*, 1701–1713.
41. Roehrborn, C. G.; Schwinn, D. A. *J. Urol.* **2004**, *171*, 1029–1035.
42. Crescioli, C.; Morelli, A.; Adorini, L.; Ferruzzi, P.; Luconi, M.; Vannelli, G. B.; Marini, M.; Gelmini, S.; Fibbi, B.; Donati, S. *J. Clin. Endocrinol. Metab.* **2005**, *90*, 962–972.
43. Srinivasan, D.; Kosaka, A. H.; Daniels, D. V.; Ford, A. P.; Bhattacharya, A. *Eur. J. Pharmacol.* **2004**, *504*, 155–167.
44. Djavan, B.; Chapple, C. R.; Milani, S.; Marberger, M. *Urology* **2004**, *64*, 1081–1088.
45. Chapple, C. R. *BJU Int.* **2004**, *94*, 738–744.
46. Roehrborn, C. G.; Boyle, P.; Nickel, J. C.; Hoefner, K.; Andriole, G.; the ARIA3001, ARIA3002 and ARIA3003 Study Investigators. *Urology* **2002**, *60*, 434–441.
47. Andriole, G.; Bruchovsky, N.; Chung, L. W. K.; Matsumoto, A. M.; Rittmaster, R.; Roehrborn, C.; Russell, D.; Tindall, D. *J. Urol.* **2004**, *172*, 1399–1403.
48. Andersson, K. E.; Wyllie, M. G. *BJU Int.* **2003**, *92*, 876–877.
49. Lowe, F. C. *BJU Int.* **2005**, *95*, 12–18.
50. Kloner, R. A.; Jackson, G.; Emmick, J. T.; Mitchell, M. I.; Bedding, A.; Warner, M. R.; Pereira, A. *J. Urol.* **2004**, *172*, 1935–1940.
51. Desgrandchamps, F. *Curr. Opin. Urol.* **2004**, *14*, 17–20.
52. Habib, F. K.; Wyllie, M. G. *Prostate Cancer Prostatic Dis.* **2004**, 7, 195–200.
53. Buck, A. C. *J. Urol.* **2004**, *172*, 1792–1799.
54. Boyle, P.; Robertson, C.; Lowe, F.; Roehrborn, C. *BJU Int.* **2004**, *93*, 751–755.
55. Kuo, H. C. *Urology* **2005**, *65*, 670–674.
56. Dinis, P.; Silva, J.; Ribeiro, M. J.; Avelino, A.; Reis, M.; Cruz, F. *Eur. Urol.* **2004**, *46*, 88–94.
57. Ma, Q.-J.; Gu, X.-Q.; Cao, X.; Zhao, J.; Kong, X.-B.; Li, Y.-X.; Cai, S.-Y. *Asian J. Androl.* **2005**, 7, 49–54.
58. Kirby, R. S.; O'Leary, M. P.; Carson, C. *BJU Int.* **2005**, *95*, 103–109.
59. Fong, Y. K.; Milani, S.; Djavan, B. *Curr. Opin. Urol.* **2005**, *15*, 35–38.
60. Kim, E. D. *Curr. Urol. Rep.* **2004**, *5*, 267–273.
61. Mochtar, C. A.; Kiemeney, L. A. L. M.; van Riemsdijk, M. M.; Barnett, G. S.; Laguna, M. P.; Debruyne, F. M. J.; de la Rosette, J. J. M. C. H. *Eur. Urol.* **2003**, *44*, 695–700.
62. Laniado, M. E.; Ockrim, J. L.; Marrionaro, A.; Tubaro, A.; Carter, S. S. *BJU Int.* **2004**, *94*, 1283–1286.
63. Trachtenberg, J. *BJU Int.* **2005**, *95*, 6–11.
64. Hedelin, H.; Johansson, N.; Stroberg, P. *Scand. J. Urol. Nephrol.* **2005**, *39*, 154–159.
65. Kielb, S. J.; Clemens, J. Q. *Urology* **2005**, *66*, 392–396.
66. Namiki, S.; Saito, S.; Ishidoya, S.; Tochigi, T.; Ioritani, N.; Yoshimura, K.; Terai, A.; Arai, Y. *Urology* **2005**, *66*, 147–151.
67. Forrest, J. B.; Schmidt, S. *J. Urol.* **2004**, *172*, 2561–2562.

68. Opree, A.; Kress, M. *J. Neurosci.* **2000**, *20*, 6289–6293.
69. Pontari, M. A.; Ruggieri, M. R. *J. Urol.* **2004**, *172*, 839–845.
70. Clinical Effectiveness Group (Association for Genitourinary Medicine and the Medical Society for the Study of Venereal Diseases). 2002 National Guideline for the Management of Prostatitis. www.guideline.gov (accessed Aug 2006).
71. Nickel, J. C. *BJU Int.* **2004**, *94*, 1230–1233.
72. Parsons, C. L.; Rosenberg, M. T.; Sassani, P.; Ebrahimi, K.; Koziol, J. A.; Zupkas, P. *BJU Int.* **2005**, *95*, 86–90.
73. Hunskaar, S.; Lose, G.; Sykes, D.; Voss, S. *BJU Int.* **2004**, *93*, 324–330.
74. Ukimura, O.; Iwata, T.; Ushijima, S.; Suzuki, K.; Honjo, H.; Okihara, K.; Mizutani, Y.; Kawauchi, A.; Miki, T. *Ultrasound Med. Biol.* **2004**, *30*, 575–581.

Biography

Mike G Wyllie, BSc, PhD After spells as a lecturer in Edinburgh and at Wyeth Laboratories, Dr Wyllie joined Pfizer in 1984. During the next decade as Director of Biology he was responsible for building up the Pfizer urology and sexual health groups that resulted in the development of doxazosin, sertraline, sildenafil, and darifenacin. Over the period 1994–98 during which both Cardura and Viagra were launched he was Director of Urology at Pfizer, NY. Subsequently, Dr Wyllie created Urodoc Ltd, a strategic planning and medical education company (http://www.urodoc.co.uk), and a specialist urology biopharmaceutics company specializing in urological product development, Plethora Solution PLC (http://www.plethorasolutions.co.uk). He is also chief scientific officer for the medical technology company Zi Medical PLC (http://www.zi-medical.com).

Dr Wyllie has written in excess of 100 articles and is the inventor of almost 90 patents. He is an Assistant Editor (Sexual Medicine Section) of the *British Journal of Urology* (*BJU International*) and a member of the European Association Of Urology (EAU), American Urological Association (AUA), and the British Association of Urological Surgeons (BAUS).

Comprehensive Medicinal Chemistry II
ISBN (set): 0-08-044513-6

ISBN (Volume 6) 0-08-044519-5; pp. 551–573

6.25 Renal Dysfunction in Hypertension and Obesity

A A Elmarakby, D M Pollock, and J D Imig, Medical College of Georgia, Augusta, GA, USA

6.25.1 Introduction

Renal dysfunction is defined as failure of the kidney to adequately retain essential nutrients and clear out toxic substances from the blood. It has become an increasing epidemic problem in the world, especially in the elderly population. An estimated 10–20 million Americans have impaired renal function and they have a high risk for

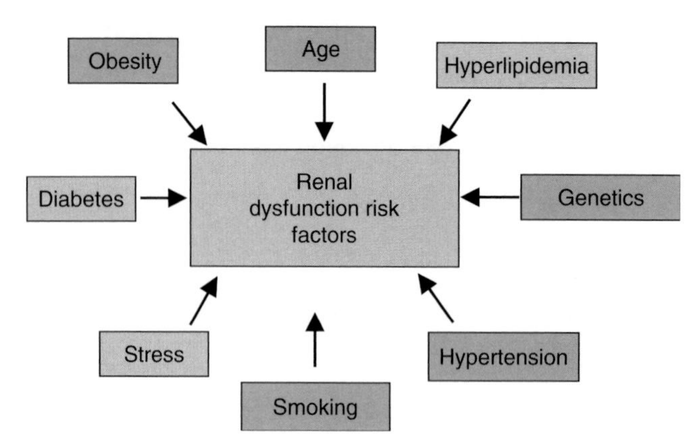

Figure 1 Risk factors that predispose renal dysfunction.

cardiovascular disease progression and sudden death.[1] The number of people diagnosed with renal dysfunction has doubled since 1990 and is expected to nearly double again by 2010.[1] Renal dysfunction often progresses to kidney failure and end-stage renal disease, requiring treatment with dialysis and kidney transplantation. End-stage renal disease has also progressively increased in developed countries. From 1997 to 2000, the incidence of end-stage renal disease among patients aged 65–74 and those greater than 75 increased to 7.8% and 22.3%, respectively.[1]

End-stage renal disease occurs when the kidney is no longer able to function normally to maintain fluid homeostasis. At this point, kidney function is so compromised that without dialysis or kidney transplantation, severe end-organ damage and death may occur from accumulation of fluids and waste products in the body. In the USA, nearly 300 000 people are on chronic dialysis and more than 20 000 have a functioning transplanted kidney.[2] The cost of end-stage renal disease has exponentially increased to $18 billion for healthcare alone with $2–4 billion lost income for patients.

The incidence of renal dysfunction and end-stage renal disease is increased in racial minorities including African Americans, Hispanics, and Native American. For example, African American and Native American are four times more likely to develop renal dysfunction than Caucasians. The factors underlying kidney disease progression are not well understood; however, genetics, environment, and nutrition may all contribute to the development of renal dysfunction. Diabetes and hypertension are the leading causes of renal dysfunction nowadays.[2] On the other hand, obesity is often associated with insulin resistance and may lead to diabetes and associated renal dysfunction. Taken together, hypertension, diabetes, and obesity account for over 70% of all the chronic kidney disease and end-stage renal disease cases (**Figure 1**).

6.25.2 Causes of Renal Dysfunction

6.25.2.1 Hypertension

Hypertension (*see* 6.32 Hypertension) in older subjects is defined as an increase in systolic blood pressure greater than 140 mm Hg and diastolic blood pressure greater than 90 mm Hg. One-quarter of the population worldwide has hypertension and this disease affects approximately 1 billion individuals. The incidence of hypertension is extremely elevated in patients with diabetes and chronic kidney disease. Systemic hypertension may increase glomerular pressure leading to glomerular hyperfiltration and vascular endothelial damage.[3] Consequently, albuminuria and proteinuria often result from endothelial dysfunction, increased glomerular pressure, and release of inflammatory factors. Hypertension also produces many changes in the vascular smooth muscle cells and the extracellular matrix of blood vessels. These changes lead to maladaptive blood vessels that exacerbate renal damage and contribute to the development of renal dysfunction. Additionally, hypertension impairs renal blood flow autoregulatory responses, alters responses to hormonal and paracrine vasoconstrictors, damages the endothelium, and induces pathologic changes in vascular structure. Collectively, these changes in renal structure and function may increase the degree of renal damage resulting from chronic exposure to hypertensive conditions. Hypertension-induced renal vascular injury can be attenuated by lowering systolic blood pressure to less than 140 mm Hg in patients without chronic kidney disease. Thus, antihypertensive therapies used in patients with chronic kidney disease should be evaluated for their ability to achieve blood pressure levels of less than 130/80 mm Hg and reduce associated albuminuria and proteinuria.

6.25.2.2 Obesity

Obesity is one of the top global health problems. Studies have shown that obesity is the leading cause of diabetes and its associated cardiovascular and renal complications (*see* 6.18 Obesity/Disorders of Energy). Obesity as a disease is defined as the voluntary consumption of more calories of food than is burned by the body. Therefore, regulation of food intake and burning any extra calories is considered the main approach for the treatment of obesity. Unfortunately, this approach has not worked and the prevalence of obesity has increased dramatically in the last two decades. In 1999–2000, the incidence of obesity, defined as a body mass index (BMI) of greater than 30, was 30.5% compared with 22.9% in the 1980s and early 1990s. The prevalence of overweight individuals, defined as a BMI greater than 25.0, has also increased to 64.5% of the population.[4] Western cultures have the greatest percentage of obese adults, and at least 50% of all individuals in the USA are 20% above their ideal weight.[5] Additionally, the number of obese children has increased dramatically in the last 20 years. Although obesity is one of the main causes of end-stage renal disease, the etiology of obesity remains unclear. Environmental factors play a role in the increase in food consumption and the decrease in physical activity. In this decade, modern technology and changing lifestyles make people less apt to exercise and decrease their chance of burning extra calories. The danger of obesity over time is that the increase in body weight will cause insulin resistance, dyslipidemia, atherosclerosis, and hypertension, what is also referred to as metabolic syndrome.

Metabolic syndrome (*see* 6.19 Diabetes/Syndrome X) was first described in 1988 as 'syndrome X' and refers to a number of cardiovascular risk factors.[6] These include obesity, hypertension, type 2 diabetes, insulin resistance, low high-density lipoproteins (HDL), elevated cholesterol, elevated triglycerides, and atherosclerosis (**Figure 2**). Visceral obesity is the driving force for metabolic syndrome as it can lead to insulin resistance and glucose intolerance, which can progress to type 2 diabetes. Type 2 diabetes, inflammation, and atherosclerosis are common complications associated with the increase in visceral fat that can lead to renal dysfunction.

The relationship between obesity and metabolic syndrome is well recognized but mechanisms remain unclear. Adipocytes play an important role in the pathogenesis of metabolic syndrome by secreting a variety of hormones, cytokines, and growth factors collectively named adipokines. These include prostaglandins, angiotensin, tumor necrosis factor-α (TNF-α), leptin, and interleukin-6 (IL6).[7] Leptin is an adipocyte hormone that has received considerable attention because of the way it acts in the hypothalamus to regulate appetite and energy expenditure and decrease food consumption. Malfunctions in leptin signaling result in increased food intake independent of body mass.[8] Other hormones and factors released by the gastrointestinal tract may also influence feeding behavior. For example, the stomach-derived circulating hormone ghrelin stimulates food intake whereas the intestinal factors cholecystokinin, glucagon-like peptide-1, and peptide YY stimulate satiety and decrease food intake.[9] Adiponetin is another protein produced by adipose tissue where its levels are inversely proportional to the amount of adipose tissue in the body. Low levels of adiponectin are a marker for obesity and diabetes and exogenous administration of adiponectin lowers blood glucose, enhances glucose uptake and utilization, and increases insulin sensitivity.[10] Collectively, adipocyte-secreting hormones play a major role in the pathogenesis of obesity and metabolic syndrome.

The relationship between obesity, hypertension, and renal dysfunction is now widely studied. Experimental studies demonstrate that weight gain can raise blood pressure and that losing weight is the most effective method to lower blood pressure in many hypertensive patients.[11] The mechanisms by which excessive weight gain alters renal function

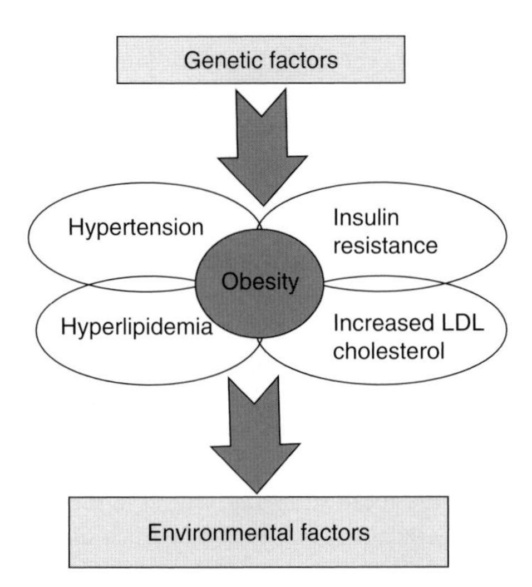

Figure 2 Signs and factors predisposing metabolic syndrome.

and raises blood pressure require further elucidation. Many studies have focused on the mechanism of obesity-induced hypertension and its subsequent physiological changes. These changes include increases in arterial pressure, cardiac output, heart rate, activation of the renin–angiotensin and sympathetic nervous systems, sodium and water retention, expansion of extracellular fluid volume, and increases in glomerular filtration rate (GFR).[11]

6.25.2.3 Diabetes

Diabetes is defined as failure of the β-cells of the pancreas to secrete sufficient insulin to satisfy body requirements and is one of the most common complications of obesity (*see* 6.19 Diabetes/Syndrome X). Diabetes can be classified into two main types: type 1 and 2. Type 1 generally shows a marked hyperglycemia with diabetic ketoacidosis. In contrast to type 1, type 2 diabetes shows less hyperglycemia and ketoacidosis. The incidence of diabetes has increased enormously in this decade. Diabetes can be diagnosed by determining blood glucose level where 8-h fasting blood glucose is $\geqslant 126\,\mathrm{mg\,dL^{-1}}$ or 2-h postprandial blood glucose $\geqslant 200\,\mathrm{mg\,dL^{-1}}$ following an oral glucose tolerance test. Additionally, symptoms of marked hyperglycemia include polyuria and polydipsia.[12]

6.25.2.3.1 Type 1 diabetes

Type 1 diabetes is defined as the failure of pancreatic β-cells to secret insulin and can be subdivided into type 1A and 1B diabetes. Type 1A results from chronic, progressive T cell-mediated autoimmune destruction of the β-cells of the pancreas, eventually leading to severe insulin deficiency. The rate of β-cell destruction is quite variable, being rapid in some individuals, especially infants and young children, and slower in adolescents and adults. The disease occurs throughout childhood and adolescence in genetically predisposed individuals. Type 1A diabetes predominantly affects European Caucasians and is less frequent in African Americans, Asians, and Native North Americans. Type 1B diabetes has a low prevalence rate with unknown etiology and it is strongly inherited. There is no evidence of β-cell autoimmunity associated with type 1B diabetes. Patients with type 1 diabetes should receive insulin frequently and do not respond well to oral hypoglycemic drugs.[13]

6.25.2.3.2 Type 2 diabetes

Type 2 diabetes occurs as a result of a decrease in insulin secretion and/or insulin resistance due to lack of exercise and increased body weight gain. The increased incidence of type 2 diabetes parallels the worldwide obesity epidemic. A family history of type 2 diabetes has repeatedly been reported in patients with this disease and patients often have other independent risk factors for cardiovascular disease, including hypertension, dyslipidemia, and microalbuminuria. Until recently, children virtually always had type 1 diabetes mellitus caused by absolute deficiency of insulin secretion, whereas type 2 diabetes was predominantly a disease of middle-aged and elderly. Nowadays, type 2 diabetes accounts for 45% of the new diabetic cases in children and adolescents in the USA. There is a greater occurrence of type 2 diabetes in African American, Hispanic, and Native American adolescents.[12] Children and adolescents with newly diagnosed type 2 diabetes are virtually always overweight or obese. Therefore, children and adolescents with type 2 diabetes may develop both renal micro- and macrovascular complications at a young age due to inadequate treatment. Patients with type 2 diabetes can be treated with oral hypoglycemic drugs and/or insulin.

There is a strong correlation between diabetes and the incidence of renal dysfunction. Diabetes mellitus is the most common cause of end-stage renal disease. Hyperglycemia itself can cause glomerular hyperfiltration, endothelial dysfunction, and albuminuria. Albumin constitutes about 40% of total urinary protein excreted in most patients with kidney disease and diabetes. Both albuminuria and proteinuria are early markers of kidney damage, particularly in diabetic kidney disease.[14] The high prevalence of abnormal albuminuria in individuals with diabetes suggests that hyperglycemia is often associated with renal abnormalities in some subjects and that these abnormalities precede the onset of diabetes. Microalbuminuria is also an independent risk factor for renal disease in both diabetic and nondiabetic individuals. Many diabetic patients will progress from renal dysfunction that is characterized by glomerular hyperfiltration and albuminuria to diabetic nephropathy. Diabetic nephropathy is one of the most common complications of diabetes, occurring in 57% of diabetic subjects and may increase morbidity rate due to severe organ damage. Diabetic nephropathy shows functional and structural changes in the kidney including decrease in glomerular filtration rate and renal fibrosis. Taken together, it is well established that diabetes can lead to renal dysfunction and without pharmacological intervention, 20% to 40% of patients with type 2 diabetes progress to severe nephropathy and about 20% of these patients will eventually develop end-stage renal disease.[15]

6.25.2.4 Prognosis of Renal Dysfunction

Kidney impairment usually progresses through several well-defined stages: microalbuminuria, macroalbuminuria, chronic renal insufficiency, chronic renal failure, and end-stage renal disease. This progression to end-stage renal disease collectively is called chronic kidney disease.[15] Chronic kidney disease is defined as structural or functional abnormalities of the kidney and is demonstrated most often by persistent albuminuria with or without decreased GFR. Microalbuminuria, a slight elevation in urinary albumin excretion, is the earliest clinical sign of nephropathy and it is considered an independent risk factor for morbidity. Microalbuminuria now is recognized as an important marker of renal and cardiovascular diseases. In the Third National Health and Nutrition Examination Survey (NHANES III), the prevalence of microalbuminuria was 28.1% in diabetic patients, 12.8% in nondiabetic patients with hypertension, and 4.8% in persons with neither diabetes nor hypertension. Patients with microalbuminuria may progress from renal dysfunction to chronic kidney disease manifested by macroalbuminuria or proteinuria, increased serum creatinine, and decreased GFR. Single-nephron glomerular pressure also increases with the progression of chronic kidney disease leading to glomerulosclerosis.[16] Higher blood pressure levels are often associated with a decline in GFR, and control of blood pressure slows the progression of chronic kidney disease.

Renal dysfunction is a challenging area for medical treatment. Currently, there is no effective treatment other than kidney transplantation to restore renal function once end-stage renal disease develops. One of the important complications of renal dysfunction is the development of cardiovascular disease such as coronary atherosclerosis, heart failure, angina pectoris, myocardial infarction, and aortic and arterial stiffening, which are the leading causes of death in patients with renal failure.[17] Cardiovascular disease mortality in patients with end-stage renal disease is 10–20 times greater than in the general population. The excessive risk for cardiovascular disease associated with nephropathy is due to a greater prevalence of risk factors such as age, hypertension, high lipid levels, diabetes mellitus, and physical inactivity in patients with renal disease.

6.25.2.5 Mechanisms of Renal Dysfunction

Studies have demonstrated significant renal hemodynamic changes in hypertension and diabetes, e.g., decreased renal blood flow, increased renal vascular resistance, elevated glomerular hydrostatic pressure, and proteinuria. The exact mechanisms for renal dysfunction require further investigation and are thought to involve many factors such as activation of the renin–angiotensin system (RAS), enhanced sympathetic activity, and impaired nitric oxide signaling. All these factors lead to endothelial dysfunction and the progression of end-stage renal disease (**Figure 3**).

6.25.2.5.1 Enhanced sympathetic activity

The sympathetic nervous system is an important regulator of cardiovascular function. Its activity is determined by psychological, neuronal, and humoral factors. Activation of neurohumoral systems as well as impairment of local

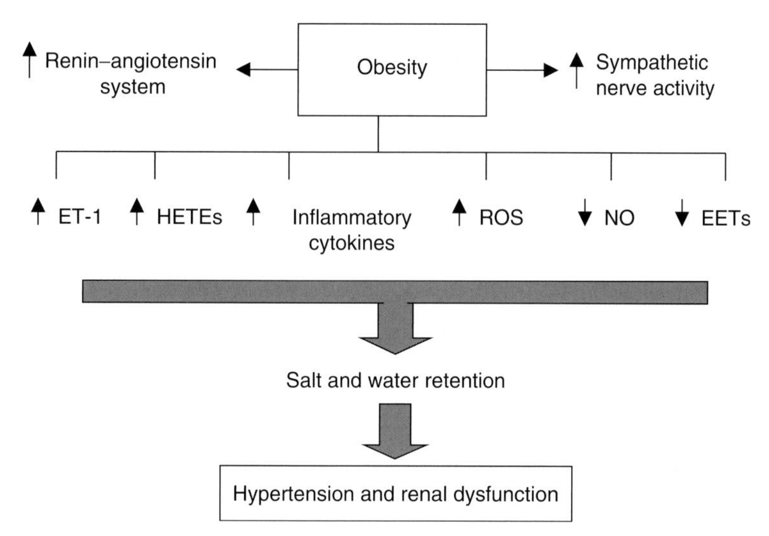

Figure 3 Mechanisms of obesity-induced hypertension.

regulatory mechanisms plays a significant role in the pathogenesis and prognosis of hypertension and cardiovascular diseases. The sympathetic nervous system activity increases with age, independent of disease state.[18] Elevated sympathetic activity correlates with hypertension, insulin resistance, and risk of coronary heart diseases.[19] The sympathetic nervous system activity contributes to the development of hypertension in early stages of the disease. Essential hypertension is thought to be associated with an enhanced sympathetic activity triggered at the level of the central nervous system in a complex manner.[20] Sympathetic activation leads to vasoconstriction and increase in blood pressure mediated by α-adrenoceptors on smooth muscle cells, whereas effects on the heart are mediated by β-adrenoceptors. The sympathetic nervous system also interacts with the RAS and the vascular endothelium. Stimulation of the β_1-adrenoreceptor of the juxtaglomerular apparatus leads to activation of the RAS via elevation of renin release; this mechanism increases blood pressure as well as sodium and water retention. Overall, stimulation of sympathetic outflow increases blood pressure, and impairs renal pressure natriuresis. Therefore, it is likely that interference with neuronal pathways involved in the regulation of sympathetic activation at the level of the central nervous system may reduce blood pressure and cardiovascular risk.

Obesity is one of the main causes of renal dysfunction being characterized by increased sympathetic activity. Increased blood pressure associated with obesity is also accompanied by impaired natriuresis which could be attributed to the increase in renal sodium reabsorption.[11] In chronic obesity, the increase in arterial pressure and glomerular hyperfiltration can lead to glomerular injury, gradual loss of renal function, and further impairment of renal pressure natriuresis. The increase in sodium reabsorption associated with weight gain could be the consequence of elevated renal sympathetic activity, activation of the RAS, and altered intrarenal physical forces. Overall, weight gain is associated with increased sympathetic activity and combined α- and β-adrenoceptor blockade markedly attenuates the elevation in blood pressure during the development of diet-induced obesity in animals as well as in obese hypertensive individuals.[21]

Recent studies have focused on exploring the mechanisms by which obesity increases sympathetic outflow. One of these is hyperleptinemia. Leptin can regulate energy balance by decreasing appetite and stimulating thermogenesis via sympathetic stimulation. Acute infusion of leptin increases sympathetic activity and the hypertensive effect of leptin was completely abolished by combined α- and β-adrenoceptor blockade.[22] Hyperinsulinemia also plays a role in the activation of the sympathetic nervous system associated with obesity. In rats, insulin causes an enhancement of sympathetic activity in different tissues such as the kidney.[23] High circulating levels of free fatty acids in obese subjects may participate in the activation of the sympathetic nervous system. Collectively, these data suggest that leptin, hyperisulinemia, and increased plasma free fatty acids could contribute to the activation of sympathetic system in obese subjects.

6.25.2.5.2 Endothelial dysfunction

Endothelial dysfunction commonly occurs in obesity, type 2 diabetes, and hypertension. The endothelium acts to regulate vascular homeostasis by maintaining a balance between vasodilation and vasoconstriction, inhibition and stimulation of smooth muscle cell proliferation and migration, and inhibition of platelet activation, adhesion, and aggregation.[24] Essential hypertension was first recognized to cause endothelial dysfunction early in the last decade where the increase in blood pressure has a direct influence on vascular function independent of other cardiovascular risk factors. Dysfunction of the endothelium could be due to decreased vasodilatory mediators and/or increased vasoconstrictor mediators. Factors that lead to reduction of vasodilation and endothelial dysfunction include a reduction in nitric oxide (NO) production, increased oxidative stress, a decrease in NO bioavailability; decreased prostacyclin levels, and a reduction of hyperpolarizing factors. Inflammatory responses also play a role in endothelial dysfunction as evidenced by the upregulation of adhesion molecules, generation of chemokines and production of plasminogen activator inhibitor-1. Vasoconstrictor peptides such as angiotensin and endothelin-1, hypercholesterolemia, and hyperglycemia contribute to endothelial dysfunction. Damage to the endothelium is an important risk factor for cardiovascular and renal diseases because it leads to structural changes such as thickening of the intima and media of the vessel wall. Because endothelial dysfunction is a complex process that results in hypertension and involves many factors, studies have focused on new approaches to improve endothelial dysfunction and slow he progression of hypertension.

6.25.2.5.3 Nitric oxide

Vascular tone is maintained by release of numerous dilator and constrictor substances where NO is the major vasodilator. The hallmark of endothelial dysfunction is impaired endothelium-dependent vasodilation. NO is formed by endothelial cells from L-arginine via the enzymatic action of endothelial NO synthase (eNOS), which is located in cell membrane caveolae. The protein caveolin-1 binds to calmodulin to inhibit activity of eNOS. The binding of calcium to calmodulin displaces caveolin-1, activating eNOS and leading to production of NO, which diffuses to vascular smooth muscle and causes relaxation by activating guanylate cyclase, to increase cyclic guanosine monophosphate (cGMP) levels which in turn produce a vasodilatory response.[25]

NO signaling is impaired in diabetic and hypertensive animal models with renal dysfunction such as stroke-prone spontaneously hypertensive rats and deoxycorticosterone acetate (DOCA)-salt hypertensive rats.[26] Many factors contribute to the impairment of NO signaling in these models including decreased L-arginine bioavailability, decreases in the cofactors required for NO synthesis such as tetrahydropiopterin, and/or increased production of superoxide, which scavenges NO. NO production could also be regulated by posttranslational phosphorylation of serine-threonine residues on eNOS in experimental models of hypertension, diabetes, and obesity.[27]

6.25.2.5.4 **Oxidative stress**

Oxidative stress is defined as an imbalance between prooxidants and antioxidants. Reactive oxygen species (ROS) are intermediary metabolites that are normally produced in the course of oxygen metabolism. There are many reactive oxygen species that are produced by all cell types and can have profound effects on the vascular system to impact blood pressure regulation. Oxidative stress increases during hypertension due to increased production of ROS such as superoxide, hydroxy radical, and hydrogen peroxide and/or decreased superoxide dismutase (SOD), which scavenges ROS. Most recent attention has been given to the role of superoxide. There are many enzymatic sources of superoxide including nicotinamide adenine dinucleotide phosphate (NADPH) oxidase, xanthine oxidase, nitric oxide synthase, and cytochrome P450. ROS can react with and denature proteins, lipids, nucleic acids, carbohydrates, and other molecules leading to inflammation, apoptosis, fibrosis, and cell proliferation. However, under normal conditions, ROS play a critical role as signaling molecules, and ROS produced by activated leukocytes and macrophages are essential for defense against invading microorganisms. The excess production of ROS and/or impaired antioxidant defense capacity leads to oxidative stress, which predisposes tissue damage and endothelial dysfunction.

Oxidative stress is a common manifestation of cardiovascular and renal complications. For example, it is involved in the pathogenesis of endothelial dysfunction and atherosclerosis.[28] This could be attributed to the ability of superoxide to scavenge NO and reduce its vasodilatory response. Oxidative stress is present in patients with mild to moderate renal insufficiency, as well as those with end-stage renal disease. Increased ROS production has also been shown in patients with essential, malignant, and renovascular hypertension. Agarwal *et al.*[29] reported an increase in plasma levels of malondialdehyde, a marker of oxidative stress, in patients with chronic renal failure compared with those with essential hypertension despite similar blood pressure suggesting that inflammation and altered cellular redox state could be the reasons for the increase in oxidative stress. NADPH oxidase has emerged as the main source of ROS in the cardiovascular system. It is well established that oxidative stress can cause inflammation and inflammation can cause oxidative stress. For example, oxidative stress activates the transcription factor, nuclear factor kappa B (NFκB), leading to the generation of proinflammatory cytokines and activation of ROS generation by leukocytes and macrophages. Overall, increases in ROS result in endothelial dysfunction via decreased NO levels, increased inflammatory cytokines production, and/or direct oxidative and nitrosative damage to the cell.

6.25.2.5.5 **Role of inflammation cytokines**

Inflammation is an important contributor to the renal injury and endothelial dysfunction observed in hypertension and obesity. Elevated circulating levels of IL6 and TNF-α are observed in obesity and metabolic syndrome patients. IL6 stimulates the central and the sympathetic nervous system, which may result in hypertension.[30] IL6 induces increases in hepatic triglyceride secretion in rats. IL6 also stimulates the production of C-reactive protein in liver and plasma levels of this protein are a good predictor of vascular inflammation. Another cytokine linked to obesity is TNF-α. TNF-α is overexpressed in the adipose tissue of obese patients, as compared with tissues from lean individuals. A positive correlation has been found between serum TNF-α concentration and both systolic blood pressure and insulin resistance in subjects with a wide range of adiposity.[31] TNF-α acutely raises serum triglyceride levels in vivo by stimulating very low-density lipoprotein (VLDL) production and hence it can play role in the increased incidence of obesity. Upregulation of TNF-α secretion occurs in peripheral blood monocytes from hypertensive patients. TNF-α is important as it activates the transcription factor NFκB, resulting in increased expression of adhesion molecules and chemokines, e.g., monocyte chemoattractant protein-1 (MCP-1). Increased expression of MCP-1 and adhesion molecules leads to vascular inflammation and dysfunction, which in turn participates in the elevation in blood pressure and renal injury in hypertension and diabetes. As inflammatory cytokines play a key role in hypertension, insulin resistance, and obesity, future therapeutic efforts should focus on the possibility of using anti-inflammatory therapy for the treatment of nephropathy associated with obesity and hypertension.

6.25.2.5.6 20-Hydroxyeicosatetraenoic acid and epoxyeicosatrienoic acids

Arachidonic acid is metabolized by cytochrome P450 (CYP450) enzymes in the kidney, liver, heart, brain, and peripheral vasculature to epoxyeicosatrienoic acids (EETs) and the hydroxyeicosatetraenoic acids (HETEs) 19-HETE and 20-HETE. Enzymes of CYP450 4A and 4F families catalyze the formation of the potent vasoconstrictor metabolite 20-HETE, and enzymes of CYP450 2C and 2J families catalyze the formation of the potent vasodilator metabolites EETs that possess antihypertensive activity. EETs and 20-HETE influence both renal function and peripheral vascular tone and are also involved in the long-term control of blood pressure (**Figure 4**).[32]

20-HETE is a potent vasoconstrictor metabolite enhancing the vasoconstrictor actions of several hormones that regulate blood pressure including angiotensin II, endothelin, and 5HT. 20-HETE depolarizes vascular smooth muscle cells by inhibiting calcium-activated potassium channels and increases the conductance of L-type calcium channels, both effects leading to increased calcium entry and increased blood pressure. Additional effects of 20-HETE include mediation of the myogenic response of small cerebral and renal arteries to elevations in transmural pressure and the autoregulation of cerebral and renal blood flow and GFR. 20-HETE contributes to the elevation in peripheral vascular tone and vascular reactivity in angiotensin II hypertension. Chronic administration of angiotensin II increases 20-HETE production and this coincides with the development of hypertension. Additionally, blocking formation of 20-HETE attenuates the development of hypertension.[32] In renal tubules 20-HETE produces natriuretic and antihypertensive effects by inhibiting sodium-potassium-ATPase, leading to diuresis and a fall in blood pressure. Induction of renal 20-HETE by fibrates lowers blood pressure and improves renal function in Dahl-salt sensitive rats while inhibition of 20-HETE promotes development of salt-sensitive hypertension in normotensive rats.[32]

EETs are produced by the endothelium and are potent vasodilators. In the kidney, EETs are produced in the proximal tubule and collecting duct where they inhibit sodium and water reabsorption.[33] EETs act as endothelium-derived hyperpolarizing factor (EDHF) independent of NO and prostaglandins where they hyperpolarize vascular smooth muscle cells by activating calcium-activated potassium channels.[34] The vasodilatory response to bradykinin in afferent arterioles of the rat juxtamedullary vascular preparation depends on cyclooxygenase metabolites and NO with a significant contribution by EETs.[35] Decreased EET levels may play a role in the induction of endothelial dysfunction in experimental models of hypertension. In the kidney microvascular expression of epoxygenase enzymes decreases in angiotensin II hypertension.[36] Additionally, increased expression of soluble epoxide hydrolase (sEH), the enzyme responsible for EETs degradation has been reported in angiotensin II-infused rats.[33] Thus, the deficiency in the renal formation of EETs may contribute to the development of angiotensin II-dependent forms of hypertension and its associated renal dysfunction. EET and 20-HETE production are altered in experimental and genetic models of hypertension, diabetes, uremia, and pregnancy toxemia, suggesting that the CYP450 metabolites of arachidonic acid contribute to the changes in renal function and vascular tone associated with hypertension and diabetes. However, the contribution of these CYP450 metabolites to renal dysfunction and end-stage renal disease remains largely unexplored.

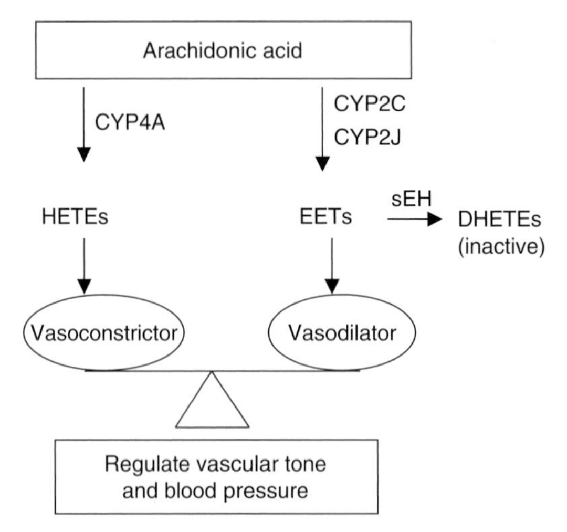

Figure 4 Pathway and effects of cytochrome P450 metabolites.

6.25.2.5.7 Endothelin in hypertension

Endothelin (ET-1) is a potent vasoconstrictor and mitogen that is thought to play a role in the development of hypertension. The vascular endothelium is a major source of ET-1 production. ET-1 is believed to act in a paracrine manner on ET_A and ET_B receptors on smooth muscle that mediate contraction, cell proliferation, and hypertrophy. Activation of ET_B receptors on endothelial cells stimulates the production of prostacyclin and NO to induce vasorelaxation. ET_B receptor activation also inhibits sodium transport in renal tubules. Renal ET-1 is increased in salt-sensitive hypertension and this effect is attenuated with ET_A receptor antagonists.[37] Interestingly, plasma ET-1 increases in hypertensive African American populations.[37] ET-1 levels are also increased in streptozotocin-induced diabetic rats where the diabetes-induced vascular hypertrophy and remodeling is ET-dependent. Patients with diabetes or hypertension have elevated ET-1 levels, but do not exhibit positive correlations between ET-1 levels and blood pressure.[38] These studies suggest ET-1 is involved in the pathogenesis of hypertension.

Aside from its potent vasoconstrictor effect, studies suggest that ET-1-induced vasoconstriction may be dependent, in part, on the production of superoxide anion. In vitro and in vivo studies have shown that ET-1 can stimulate superoxide anion formation in aortic rings.[39] In cultured pulmonary artery smooth muscle cells, ET-1 increases superoxide production via stimulation of ET_A receptor.[40] Furthermore, ET_A receptor blockade attenuates hypertension and reduced oxidative stress in salt-sensitive ET_B receptor-deficient rats as an example of a high ET-1 model of hypertension.[41] ET-1 also increases transforming growth factor-β (TGF-β), which plays a role in the vascular inflammation and fibrosis associated with hypertension and diabetes.[42] Thus, beside its potent vasoconstrictor effect, ET-1 may produce further elevation in blood pressure via stimulation of superoxide and cytokines production.

6.25.2.5.8 Renin–Angiotensin system

The RAS contributes to the development and maintenance of hypertension and mediates renal injury by inducing systemic and glomerular hypertension. Synthesis of angiotensin II (AT-II) depends on the release of renin, primarily by juxtaglomerular cells in the kidney (**Figure 5**). The release of renin is regulated by the hydrostatic pressure sensed at the glomerular afferent arterioles, AT-II levels, and the quantity of sodium delivered to the macula densa. Plasma potassium, atrial natriuretic peptide, and endothelin levels also affect renin synthesis and release. Renin acts to cleave the liver angiotensinogen to angiotensin I (AT-I). ACE then converts AT-I to AT-II. This proteolytic enzyme is found in the endothelial cells of the lung, vascular endothelium, and cell membranes of the kidneys, heart, and brain. ACE also degrades vasodilator bradykinin to inactive fragments. Non-renin and non-ACE pathways also exist in the body allowing the production of AT-II either directly from angiotensinogen or from angiotensin I.[43] The biologic actions of AT-II in the kidney are mediated by two well-characterized receptors: AT-II type 1 (AT-1) and AT-II type 2 (AT-2) receptors. In adult tissues, the AT-1 receptor is distributed in the vasculature, kidney, adrenal gland, heart, liver, and brain. In healthy adults, the AT-2 receptor is present only in the adrenal medulla, uterus, ovary, vascular endothelium, and

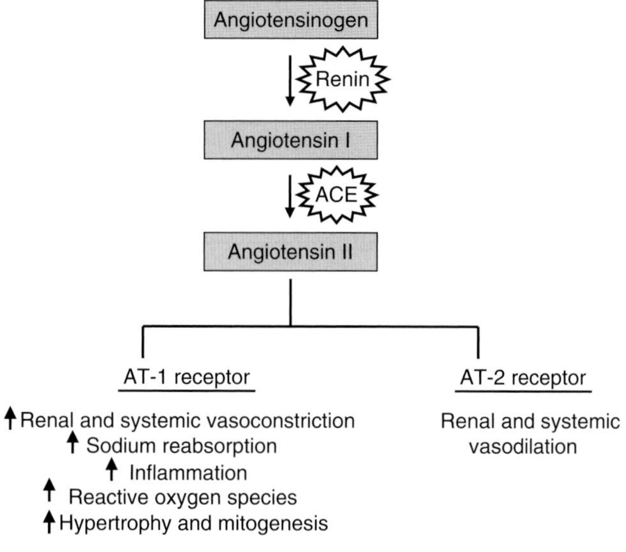

Figure 5 Pathway of the renin–angiotensin system (RAS) and its receptors.

distinct brain areas.[43] The AT-1 receptor mediates the hemodynamic actions, endocrine functions, and mitogenic effects of angiotensin II in the kidney whereas the AT-2 receptor possesses vasodilatory effects and can modulate the action of AT-1 receptor.

Stimulation of the RAS increases sodium and water reabsorption through direct actions on renal tubular transport function 44.[44] This effect is mainly due to the stimulation of AT-1 receptor by AT-II. Activation of angiotensin receptors by AT-II also stimulates synthesis of aldosterone in the zona glomerulosa of the adrenal gland. Aldosterone binds mineralocorticoid receptors expressed in the kidney and other organs leading to more salt and water retention. Aldosterone is involved in the development of obesity-induced hypertension. Plasma aldosterone levels are elevated in hypertensive patients with visceral obesity,[45] and blocking mineralocorticoid receptors with the specific antagonist, eplerenone, inhibits development of high blood pressure in dogs fed a high-fat diet.[46]

Beside its water-retaining ability, AT-II is also a potent vasoconstrictor with growth-promoting properties. AT-II regulates blood pressure and fluid and electrolyte homeostasis. In the kidney, AT-II infusion increases renal microvascular resistance and enhances preglomerular reactivity during the early and established phases of hypertension. The increase in renal vascular resistance contributes to the increase in glomerular capillary pressure and inability of the kidney to properly excrete sodium and water. AT-II also reduces medullary blood flow and diminishes renal interstitial pressure. Collectively, these changes promote movement of sodium and fluid from the proximal tubule into the interstitium and systemic circulation and result in progressive increase in arterial blood pressure.

AT-II not only increases blood pressure via its vasoconstrictor effects, but also via the stimulation of ET-1 synthesis[47] and increased ROS.[48] Indeed, a potential mediator mechanism for the proinflammatory effects of AT-II may be due to increased superoxide generation via NADPH oxidase stimulation. The AT-II-stimulated increase in the activity of NADPH oxidase appears to be via the AT-1 receptor, as this effect can be blocked by preincubation of endothelial cells with the AT receptor blocker losartan.[49]

Aside from its hemodynamic effects, AT-II stimulates cell proliferation, inflammation, and tissue remodeling by enhancing the synthesis of profibrotic cytokines and growth factors. Infusion of AT-II causes vascular inflammation and endothelial dysfunction where these effects could be mediated by the activation of NFκB signaling and transforming growth factor-β (TGF-β).[50] TGF-β stimulates production of other growth factors, causes proliferation of fibroblasts, and increases extracellular matrix protein synthesis. Both the production and activation of TGF-β are involved in glomerulosclerosis and interstitial fibrosis of the kidney.[51] Collagen deposition is also enhanced through inhibition of proteases that normally function to degrade abnormal tissue proteins. In total, AT-II plays a critical role in the pathogenesis of hypertension and renal dysfunction not only via its direct vasoconstrictor and sodium retaining effects but also via nonhemodynamic effects such as increased ET-1, ROS, and inflammatory cytokine production.

The RAS plays a role in obesity-induced hypertension and its associated renal dysfunction. All components of the RAS are expressed by adipose tissue, and are upregulated in obesity.[11] Obese persons have been shown to have high plasma renin activity, plasma angiotensinogen, ACE activity, and plasma AT-II levels. AT-II may also contribute to the increase glomerular injury and nephron loss associated with obesity. Angiotensin receptor blockers or ACE inhibitors blunt sodium retention and volume expansion in obesity. Studies in overweight patients with type 2 diabetes indicate that ACE and AT receptor blockers slow progression of renal disease.[44] Given this information, as well as the fact that obesity is closely associated with the two main causes of end-stage renal disease, hypertension and diabetes, obesity-mediated angiotensin activation may greatly increase the risk for end-stage renal disease.

6.25.3 Experimental Disease Models

Although studies suggest that renal dysfunction underlies the development of all forms of hypertension in human and experimental animals, most investigators believe that impaired renal function is a consequence of the hypertension rather than the primary basis of the disease. The induction of experimental hypertension and the subsequent renal dysfunction involves a reduction in the ability of kidney to excrete sodium and water at normal level of arterial pressure. This includes numerous animal models such as Goldblatt hypertension (renal artery stenosis), coarctation of the aorta, mineralocorticoid models of hypertension through the administration of aldosterone or DOCA, surgical reduction of renal mass, perinephritic models of hypertension, and chronic infusion of vasoconstrictor substance such as ET-1, AT-II, vasopressin, and epinephrine.[52] Impairment of endogenous vasodilatory substances that play an important role in the control of renal function and arterial pressure such as NO, prostaglandin, atrial natriuretic peptide (ANP), and EETs may induce hypertension and renal dysfunction. Additionally, genetic rat models of hypertension including spontaneously hypertensive rats (SHR), Dahl salt-sensitive rats, Lyon hypertensive rats, and transgenic renin gene rats can be used as good models of renal dysfunction where the pressure natriuresis response in these models is blunted and reset toward higher pressures.[52]

6.25.4 Clinical Trial Issues

Clinical studies have focused on improving the efficiency of current antihypertensive classes to reduce the incidence of renal dysfunction. For example, β-blockers are effective agents for the treatment of hypertension in both diabetic and nondiabetic patients with chronic renal disease. In the UK prospective diabetes study of patients with type 2 diabetes, the β-blocker, atenolol, was as effective as captopril in term of lowering blood pressure and protection against microvascular disease.[53] In a recent randomized controlled trial in patients with hypertension and chronic renal failure, microproteinuria diminished during short-term β-adrenoceptor blocker therapy, but prolonged therapy failed to reduce microproteinuria compared with ACE inhibitors or calcium channel blockers, despite equal antihypertensive efficacy. In contrast to conventional β-adrenoceptor blockers, several trials have documented a favorable effect of carvedilol on renal hemodynamics. Renal plasma flow has been shown to increase significantly with carvedilol, and microalbuminuria decreases.[54] Recent clinical trials mainly focus on the use of the new antihypertensive drugs such as ACE inhibitors, angiotensin receptor blocker, and calcium channel blockers for improving renal dysfunction. For example, ACE inhibition was more effective than amlodipine or β-adrenoceptor blockade in decreasing the progression of nondiabetic renal disease in African Americans with hypertension.[55] Losartan, an AT-1 receptor blocker, reduced the incidence of a doubling of the serum creatinine concentration as a risk factor of end-stage renal disease. Furthermore, the level of proteinuria also declined upon losartan treatment. Inhibition of the RAS with losartan in the Losartan Intervention for Endpoint Reduction (LIFE) trial was also associated with a reduction in the risk of new-onset diabetes.[56] Treatment with losartan was also associated with lower risk of development of diabetes. AT receptor blockers have proved to slow the progress of renal disease in clinical trials of patients with diabetic nephropathy.[57] Dual blockade of ACE and AT-1 receptor may provide a better effect than either treatment alone. Dual inhibition has been evaluated in diabetic patients with both microalbuminuria and macroalbuminuria. The Candesartan and Lisinopril Microalbuminuria (CALM) study showed that combined administration of candesartan and lisinopril provides a better effect in reducing blood pressure and microalbuminuria compared with either drug alone. Dual blockade of the RAS is associated with a better renoprotective effect in patients with nondiabetic proteinuric nephropathies than either treatment alone. Combined administration of losartan and enalapril decreased urinary protein excretion in patients with renal disease, whereas doubling the dose of either medication had no effect on proteinuria.[58] The combination of valsartan and benazepril in patients with nondiabetic kidney disease also reduced proteinuria more effectively than AT receptor blocker alone.[59]

6.25.5 Management of Renal Dysfunction

Management of hypertension in patients with kidney disease or diabetes is challenging, generally requiring 2-3 years of therapy to achieve the recommended blood pressure level. Studies have shown that some antihypertensive drugs can lower blood pressure without improving endothelial dysfunction and renal damage. For example, the traditional use of diuretics and α- and β-adrenoceptor blockers can lower blood pressure with no direct effect on endothelial dysfunction.[53] ACE inhibitors, AT receptor blockers, and calcium channel blockers are the most effectively used treatment for hypertension as they lower blood pressure and improve endothelial dysfunction in hypertensive patients (**Figure 6**).

Aside from the classic treatment of hypertension, an emerging approach to renal dysfunction is the treatment of the components that trigger endothelial dysfunction, e.g., NO bioavailability and oxidative stress. For example, oral treatment with L-arginine, the precursor of NO, reduces blood pressure and improves endothelial dysfunction in hypertensive patients.[60] Statins and lipid-lowering drugs improve endothelial dysfunction in hypertensive animal models by enhancing NO levels.[61] Antioxidants also can improve endothelial dysfunction in hypertensive animal models. For example, the SOD mimetic tempol decreases hypertension and oxidation stress and improves endothelium-dependent relaxation and kidney damage in hypertensive animal models such as AT-II-infused mice and Dahl salt-sensitive rats.[62]

6.25.5.1 Diuretics

Diuretics are the most common initial treatment for mild hypertension (**Figure 7**). Based on their site of action in the kidneys, diuretics can be classified into loop, thiazide, and potassium-sparing diuretics. Loop or high-ceiling diuretics,

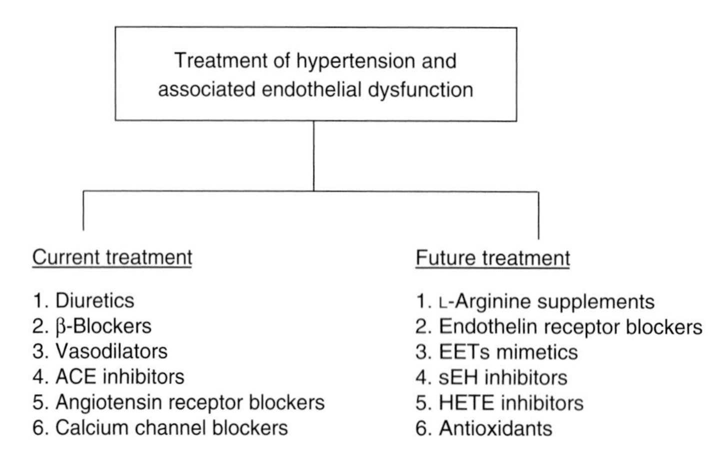

Figure 6 Current and future treatment of renal dysfunction and hypertension.

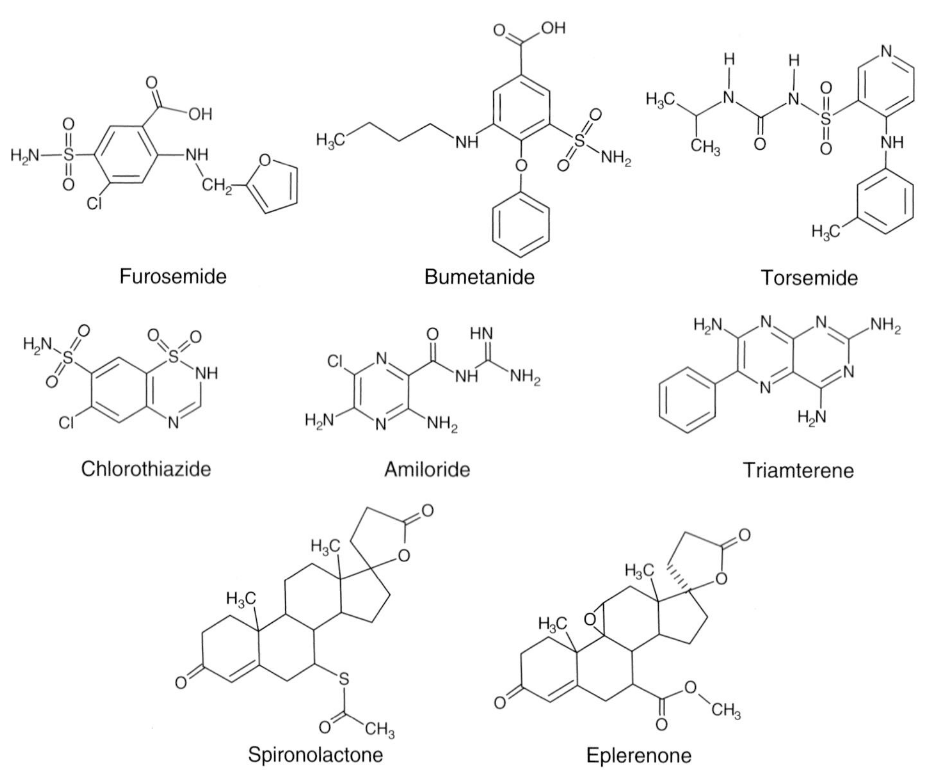

Figure 7 Structures of diuretic drugs.

including furosemide, bumetanide, and torsemide, reversibly inhibit the sodium-potassium-two chloride cotransporter at the luminal thick ascending limb of the loop of Henle, therefore inhibiting the reabsorption of sodium, potassium, and chloride ions. Loop diuretics produce relatively more urine formation and less loss of sodium and potassium than thiazides. Thiazide diuretics inhibit sodium and chloride reabsorption via the inhibition of sodium–chloride cotransporter in the distal tubule. This cotransporter is insensitive to loop diuretics. Spironolactone is a powerful potassium-sparing diuretic that inhibits the binding of aldosterone to mineralocorticoid receptors in many tissues, including epithelial cells of the distal convoluted tubule and collecting duct. This will increase sodium and water loss and spare potassium. Other potassium-sparing diuretics such as amiloride and triamterene act by inhibiting the sodium–hydrogen exchanger and indirectly decreasing potassium loss. They are relatively weak diuretics, which are often used in combination with thiazides and loop diuretics to increase sodium excretion without a major potassium loss.

Diuretics induce a loss of electrolytes and fluid, thereby stimulating several compensatory hemostatic mechanisms such as the RAS, which result in increased renal sodium retention by all nephron segments, a phenomenon known as diuretic resistance. If dietary salt intake is sufficiently high, a daily net negative sodium balance may not be achieved even with several daily doses of loop diuretics. Hence, salt intake must be restricted in patients with hypertension and heart failure to obtain a negative sodium balance.

The major side effect of loop and thiazide diuretics is the deficiency of the main electrolytes, particularly sodium and potassium. Hypokalemia may increase the risk of arrhythmia and cardiac arrest. Mild hypokalemia caused by these diuretics may result in leg cramps, polyuria, and muscle weakness. Most diuretics also decrease urate excretion and increase blood uric acid, causing gout in predisposed patients. Serum cholesterol levels may increase after diuretic therapy, and high doses of diuretics can impair glucose tolerance and precipitate diabetes mellitus, probably through an increase in insulin resistance.[63]

6.25.5.2 β-Adrenoceptor Blockers

In most elderly patients with essential hypertension, cardiac output is low, and systemic vascular resistance is elevated. Therefore, most antihypertensive agents, including diuretics, produce a decrease in vascular resistance while sparing systemic flow and cardiac output. Traditional β-adrenoceptor blockers (β_1-selective β-blockers) (**Figure 8**) are an exception as drugs such as atenolol and metoprolol lower arterial pressure by decreasing cardiac output and renin

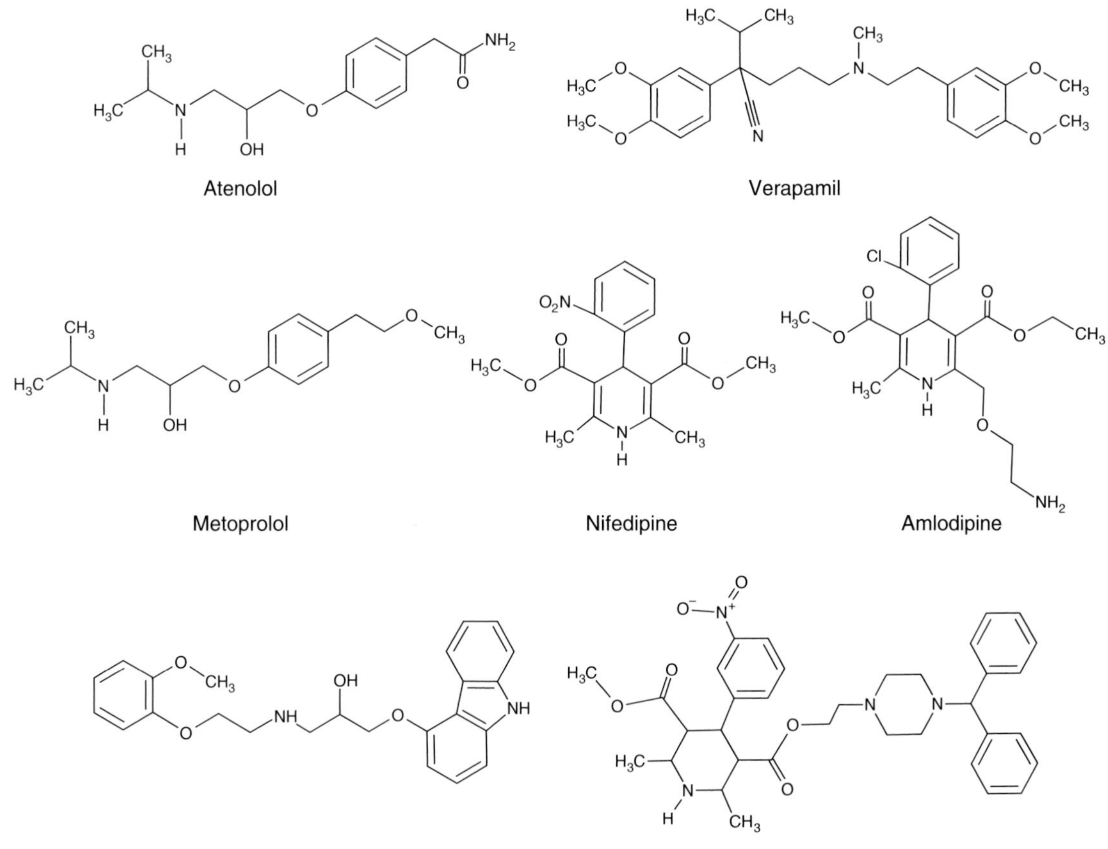

Atenolol

Verapamil

Metoprolol

Nifedipine

Amlodipine

Carvedilol

Manidipine

Figure 8 Structures of β-adrenoceptor and calcium channel blockers.

release while systemic vascular resistance remains unchanged. The use of β-adrenoceptor blockers reduces the risk of cardiovascular events such as myocardial infarction and sudden death in patients with coronary artery disease. β-Adrenoceptor blockers are effective in decreasing mortality rate in patients with reduced systolic function and heart failure.[64] Carvedilol, a β-adrenoceptor blocker with associated α_1-adrenoceptor blocking properties is an effective antihypertensive medication where the α_1-adrenergic blocking effect accounts for its vasodilatory effects in contrast to the β_1-selective agents.

Early stage renal disease in hypertension is characterized by a decrease in renal blood flow and a relatively well-preserved GFR. Microproteinuria parallels the progression of renal disease, particularly in hypertensive patients with diabetes mellitus. Traditional β-adrenoceptor blockers further diminish renal blood flow and may decrease GFR.[65] One of the main side effects of using β-adrenoceptor blockers is the disturbance of the lipid profile. Studies suggest that β-adrenoceptor blockers increase triglyceride levels and decrease HDL cholesterol levels. Those with intrinsic sympathomimetic activity have a lesser effect on triglycerides and HDL than those lacking intrinsic sympathetic activity.

6.25.5.3 Calcium Channel Blockers

Calcium channel blockers (**Figure 8**) decrease systemic blood pressure and increase or even preserve renal blood flow via the inhibition of calcium entry into vascular smooth muscle cells. Within the renal vasculature, calcium antagonists cause a preferential dilation of the afferent arteriole, with only modest action on the efferent arterioles. For example, the calcium channel blocker nifedipine causes a greater increase in GFR than renal plasma flow, resulting in an elevated filtration fraction. Furthermore, other calcium antagonists including verapamil increase filtration fraction, suggesting predominant effects on the afferent arteriole.[66] Because of the depressor action of the calcium antagonists, they can attenuate glomerular hypertension and provide renal protection. Unlike classical calcium antagonists, novel calcium channel blockers including manidipine are reported to dilate afferent and efferent arterioles.[67] For example, amlodipine reduces blood pressure and oxidative stress and improves endothelial dysfunction in AT-II-infused rats.[68] Calcium antagonists prevent the intrarenal glomerular and arteriolar histological lesion and physiological changes and also reduce blood pressure and renal injury in obesity models. For example, amlodipine reduces blood pressure and prevents renal fibrosis in fructose-fed rats.[69]

Calcium channel blockers can enhance NO activity. Amlodipine increases endothelial NO bioavailability in vitro by enhancing NO production and prolonging its half-life and can release NO to a similar extent as enalapril, although nifedipine does not. A synergistic effect on NO formation occurs in rats when amlodipine is combined with an ACE inhibitor.[70] Overall, calcium channel blockers decrease blood pressure and improve renal function by interfering with the vascular smooth muscle cell calcium and enhancing the endothelial release of NO.

6.25.5.4 Modulating the Renin–Angiotensin System

There is compelling evidence that modulating the RAS attenuates renal disease progression through blood pressure-independent mechanisms (**Figure 9**). For example, the ACE inhibitor enalapril slows the progression of end-stage renal failure compared with β-adrenoceptor blockers although blood pressure control is similar in both treatment groups. These data suggest that the protective effect of enalapril on renal function was not mediated through blood pressure and it could be attributed to the blocking of AT-II-induced increase in ROS, growth factors, inflammatory cytokines, and interstitial fibrosis. Therefore, interfering with the RAS may reduce proteinuria and provide a renoprotective effect in hypertension and diabetes.

6.25.5.5 Angiotensin Receptor Blockers

AT receptor blockers are effective antihypertensive agents where the antagonism of the AT-1 receptor and binding of angiotensin II to the AT-2 receptor probably underlies their effect. In contrast to ACE inhibitors, these drugs do not inhibit the breakdown of bradykinin. AT receptor blockers decrease blood pressure and prevent nephrosclerosis and renal damage in the stroke-prone spontaneous hypertensive rats via the blocking of AT-1 receptors.[71] Clinically, AT receptor blockers are widely used in the treatment of hypertension and hypertension-related end-organ damage (**Table 1**) and have also been recognized as regulators of glucose and lipid metabolism in adipose tissue. AT receptor blockers lower the risk for type 2 diabetes compared with other antihypertensive therapies. In addition, AT receptor blockade improves insulin sensitivity in humans and in animal models of insulin resistance. The underlying mechanisms of the insulin-sensitizing/antidiabetic effect of AT receptor blockers are not fully understood. AT receptor blockers also attenuate the

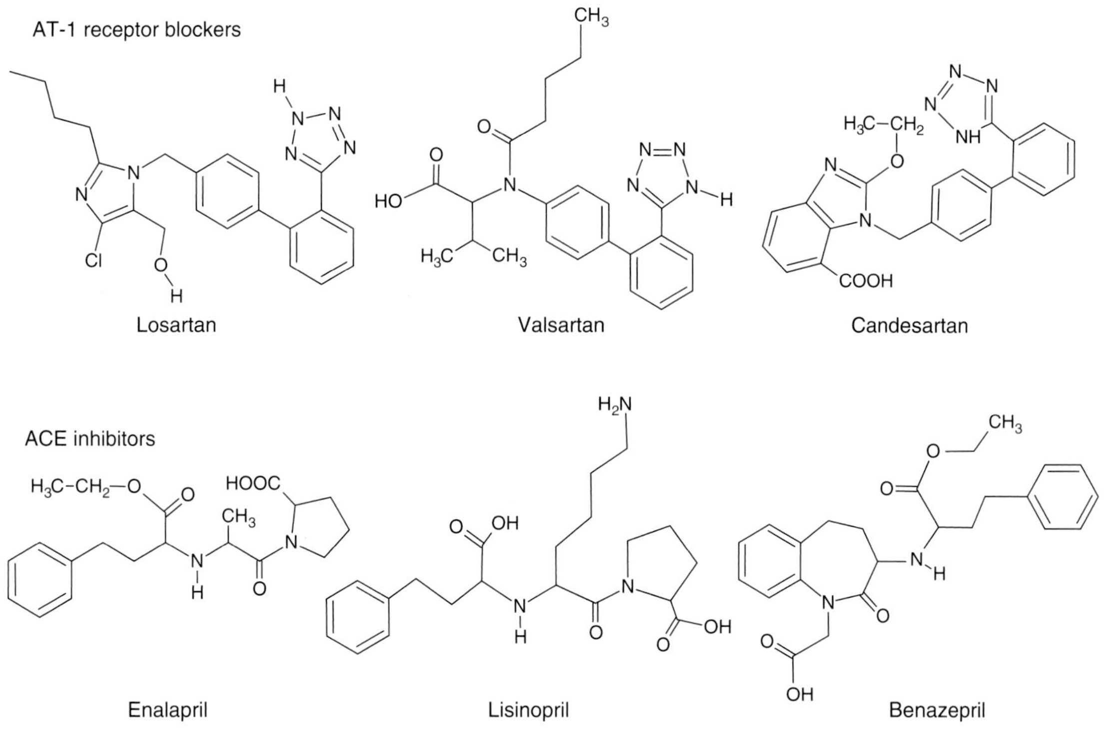

Figure 9 Structures of renin–angiotensin–aldosterone system blockers.

Table 1 Clinical trials of angiotensin receptor blockers in diabetic patients

Study	*Patients*	*Treatment*	*Study duration*	*Endpoints*	*Results*
Angiotensin II Receptor Antagonist Losartan (RENAAL)	Type 2 diabetic patient with nephropathy	Losartan versus placebo	3.4 years	Doubling serum creatinine, end-stage renal disease, death	15% risk reduction in endpoint parameters
Irbesartan Diabetic Nephropathy Trial (IDNT)	Hypertensive + type 2 diabetic patient with nephropathy	Irbesartan versus amlodipine versus placebo	2.6 years	Doubling serum creatinine, end-stage renal disease, death	24% risk reduction versus amlodipine and 19% risk reduction versus placebo in endpoint parameters
Irbesartan Microalbuminuria type-2 (IRMA-2)	Hypertensive + type 2 diabetic patient with microalbuminuria	Irbesartan verus placebo	2 years	Onset of diabetic nephropathy	39% risk reduction in endpoint parameters
Microalbuminuria Reduction with Valsartan (MARVAL)	Type 2 diabetic patient with microalbuminuria	Valsartan versus emlodipine	24 weeks	Percent change in baseline albumin excretion rate	44% risk reduction from baseline with valsartan and 8% risk reduction from baseline with amlodipine
Losartan Intervention for Endpoint Reduction (LIFE)	Old hypertensive patients with left ventricular hypertrophy	Valsartan verssus atenolol	4.8 years	Cardiovascular death due to stroke, diabetes, and	13% risk reduction of endpoint; 25% risk reduction of stroke and diabetes

development of atherosclerosis and reduce vascular inflammation and collagen deposition in diabetic apolipoprotein E-deficient mice. Blocking AT-1 receptors also reduces platelet-derived growth factor B (PDGF-B), MCP-1, and cell adhesion molecule expression in these mice suggesting that the AT-1 receptor plays a vital role in the development and acceleration of atherosclerosis in the setting of diabetes.[72]

6.25.5.6 ACE Inhibitors and Aldosterone Receptor Antagonism

Inhibition of ACE activity decreases the formation of AT-II and aldosterone and potentiates the vasodilatory effects of bradykinin. As a result, ACE inhibitors are used widely to treat hypertension. These agents reduce proteinuria and delay the progression of renal disease in patients with diabetic nephropathy or nondiabetic kidney disease. ACE inhibitors are effective in reducing renal damage in hypertensive animal models. For example, ACE inhibition prevents the development of glomerulosclerosis after subtotal nephrectomy in the remnant kidney rat model. The ACE inhibitor enalapril attenuates atherosclerosis and vascular inflammation induced by angiotensin II infusion in apolipoprotein E-deficient mice.[73] ACE inhibition also inhibits development of nephrosclerosis in young spontaneous hypertensive rats treated with N-nitro-L-arginine methyl ester (L-NAME). The increase in kinin levels by ACE inhibition provides an additional renoprotective effect in LNAME/spontaneous hypertensive rats. The development of atherosclerotic lesions and diabetes-induced overexpression of ACE, connective tissue growth factor, and vascular cell adhesion molecule-1 (VCAM-1) in the aorta of diabetic apolipoprotein E-deficient mice is prevented by ACE inhibition.[74] In obese Zucker rats, a model of the metabolic syndrome, ACE inhibition also reduces glomerular and tubulointerstitial lesions.[75]

Dual blockade of the RAS with ACE inhibitors and AT receptor blockers may provide a better therapy than either drug alone. In obese Zucker rats, combined administration of an AT blocker and an ACE inhibitor is more effective than therapy with either drug alone in controlling renal damage, although there is no difference in blood pressure.[75] Antagonism of the mineralocorticoid receptor with spironolactone, a competitive inhibitor of aldosterone, reduces blood pressure by enhancing renal sodium excretion and, consequently, may decrease the development of kidney disease. Animal experiments suggest that in addition to reducing renal injury from hypertension, spironolactone may also blunt the profibrotic effects of aldosterone. Aldosterone receptor antagonist shows a potential renoprotective effect by reducing proteinuria in diabetic nephropathy when used in combination with enalapril.[76] The aldosterone receptor antagonist, eplerenone, alone or in combination with enalapril, reduces albuminuria in hypertensive type 2 diabetic patients with microalbuminuria despite equivalent reductions in blood pressure.[77]

6.25.6 Unmet Medical Need

Renal disease may lead to dysfunction of the vascular endothelium that results in enhanced formation of vasoconstrictor peptides such as ET-1 and thromboxane, decreased formation of vasodilators such as EETs, NO, and prostacyclin, and increased formation of superoxide and inflammatory cytokines. These endothelial abnormalities impair pressure natriuresis and increase blood pressure. Although current antihypertensive drugs effectively reduce blood pressure by promoting the excretion of sodium and water and shift the pressure–natriuresis relationship back to normal, these drugs fail to correct all the pathophysiological changes that occur as a result of hypertension and endothelial dysfunction such a structural changes in renal vasculature, impaired renal autoregulation, tubular atrophy, interstitial fibrosis, and renal hypertrophy. Thus, future studies should focus on new drugs that can be used with current antihypertensive drugs to reduce blood pressure and correct the pathophysiological changes associated with renal dysfunction.

6.25.7 Emerging Approaches to Treat Renal Dysfunction

As mentioned before, treatment of renal dysfunction has focused on improving endothelial and renal function in the context of the treatment of hypertension and obesity. For example, increasing NO and EET levels and decreasing oxidative stress and ET-1 effects are emerging approaches to improve endothelial dysfunction and lower blood pressure in obesity and hypertension.

6.25.7.1 Antioxidants

Antioxidants act by reducing the superoxide that is elevated in hypertension and obesity. The reduction in superoxide levels may enhance NO bioavailability. For example, long-term treatment with the SOD mimetic tempol lowers arterial

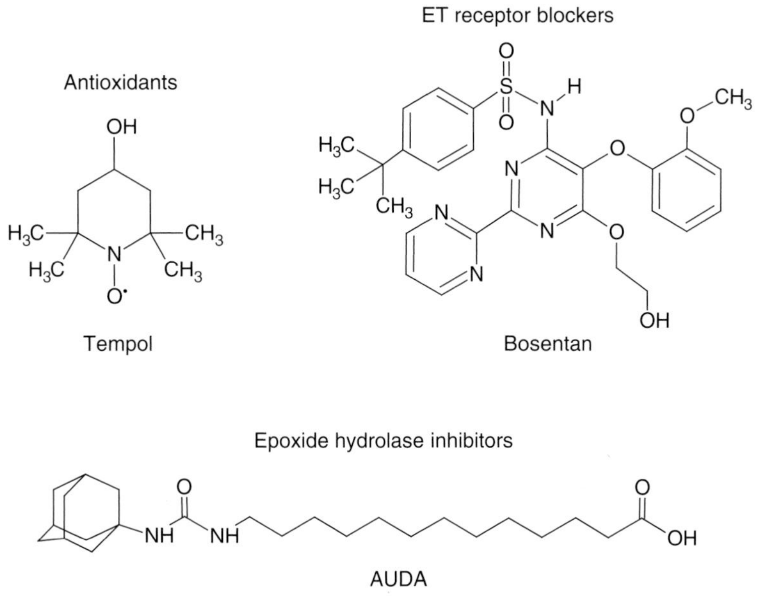

Figure 10 Structures of NCEs for novel targets to treat renal dysfunction.

pressure in several models of hypertension including chronic AT-II, DOCA-salt, and Dahl salt-sensitive rats and in salt-sensitive stroke-prone spontaneous hypertensive rats.[78]

In diabetic models, SOD mimetics may play a role in improving endothelial dysfunction. For example, tempol (**Figure 10**), improves insulin sensitivity and decreases oxidative stress in obese Zucker rats as a model of type 2 diabetes.[79] Tempol also reduces oxidative stress and improves endothelial dysfunction in streptozotocin-induced type 1 diabetic rats.[80] NADPH oxidase is the main source of superoxide production following AT-II stimulation and mice lacking the p47phox subunit of NADPH oxidase have an attenuated hypertensive response to chronic AT-II infusion.[81] Mice lacking the gp91phox subunit also have an attenuated hemodynamic response to acute AT-II.[82] Thus, inhibition of NADPH oxidase may be a novel approach to improve endothelial function and lower blood pressure in hypertension.

6.25.7.2 Endothelin Receptor Antagonists

The ET receptor antagonist, bosentan (**Figure 10**) is approved for the treatment of pulmonary hypertension in humans supporting the role of ET in the pathogenesis of this type of hypertension. ET_A receptor blockade reduces blood pressure and improves endothelial dysfunction in high-endothelin models of hypertension. For example, ET_A receptor blockade attenuated hypertension, oxidative stress, and NADPH oxidase activity during chronic aldosterone-induced hypertension.[83] ET_A receptor blockade also prolongs survival and reduces proteinuria, but has little effect on the progression of hypertension in salt-loaded stroke-prone spontaneously hypertensive rats.

ET may also have a role in obesity and its associated hypertension. Increased vascular production of ET-1 in hypertensive patients with increased body mass may be a potential mechanism for endothelial dysfunction. ET_A receptor-dependent vasoconstrictor tone increased in obese hypertensive patients and blockade of the ET_A receptor induced significant vasodilation in overweight and obese humans but not in lean hypertensive subjects. ET receptor antagonism also attenuates elevated blood pressure and reduces renal and cardiac damage in rats fed a high-fructose diet, suggesting that ET receptor blockers could be a potential approach for treatment of obesity-induced hypertension and renal dysfunction.[84]

6.25.7.3 Inhibitors of the Synthesis of 20-Hydroxyeicosatetraenoic Acid

As discussed previously, the potent vasoconstrictor metabolite 20-HETE plays a role in the pathogenesis of hypertension. While 20-HETE is a potent vasoconstrictor, increased renal production of 20-HETE provides natriuretic activity and lowers blood pressure. Induction of the renal formation of 20-HETE by fibrates lowers blood pressure and improves renal function in Dahl salt-sensitive rats, while chronic inhibition of 20-HETE formation results in salt-sensitive hypertension in normotensive rats.[32]

6.25.7.4 Epoxyeicosatrienoic Acid Mimetics and Soluble Epoxide Hydrolase Inhibitors

EETs are vasodilatory metabolites and are described as EDHF. Changes in the renal formation of EETs may also contribute to the development of AT-II hypertension with increased expression of soluble epoxide hydrolase (sEH), the enzyme responsible for EETs degradation in AT-II-infused rats.[33] Expression of epoxygenase enzymes is increased in the kidney of rats fed a high-salt diet,[36] while a deficiency in the renal formation of EETs may contribute to the development of AT-II-dependent forms of hypertension. Consistent with this, renal epoxygenase activity is reduced in transgenic rats overexpressing human renin and angiotensinogen genes.[85] These animals develop severe hypertension and renal interstitial fibrosis and inflammation. Pharmacological blockade of sEH increases the urinary excretion of EETs and lowers blood pressure in both spontaneous hypertensive rats and AT-II hypertensive rats.[86,87] Blood pressure is also reduced in sEH knockout mice in which production of EETs is elevated.[88] Chronic administration of the orally active sEH inhibitor 12-(3-adamantan-1-yl-ureido)-dodecanoic acid (AUDA) (**Figure 10**), increased urinary EET excretion and reduced arterial pressure and renal damage in AT-II-mediated hypertension.[89] Collectively, these studies suggest a potential role of EETs in lowering blood pressure and improving endothelial dysfunction.

6.25.8 Summary

In summary, obesity, diabetes, and hypertension are the most common risk factors for the progression of end-stage renal disease. Due to the vast increase in the incidence of renal disease, future efforts should focus on reducing the prevalence of these risk factors and preventing the progression of renal dysfunction. Many factors seem to be involved in the progression of renal dysfunction such as the sympathetic nervous system, RAS, ET, inflammatory cytokines, ROS, impaired NO signaling, and cytochrome P450 metabolites. Although current antihypertensive drugs slow the progression of renal dysfunction associated with diabetes, obesity, and hypertension, future studies need to focus more on the potential new drugs such as antioxidants, ET receptor antagonists, sEH inhibitors, EET mimetics, and 20-HETE inhibitors that may play a role in preventing renal dysfunction, and/or improving the effects of the traditional antihypertensive drugs.

References

1. Bommer, J. *Nephrol. Dial. Transplant.* **2002**, *17*, 8–12.
2. Bakris, G. L. *Am. J. Med.* **2003**, *115*, 49S–54S.
3. Norris, K.; Vaughn, C. *Exp. Rev. Cardiovasc. Ther.* **2003**, *1*, 51–63.
4. Arroyo, P.; Loria, A.; Fernandez, V.; Flegal, K. M.; Kuri-Morales, P.; Olaiz, G.; Tapia-Conyer, R. *Obes. Res.* **2000**, *8*, 179–185.
5. Hill, J. O.; Wyatt, H. R.; Reed, G. W.; Peters, J. C. *Science* **2003**, *299*, 853–855.
6. Reaven, G. M. *Diabetes* **1988**, *37*, 1595–1607.
7. Guerre-Millo, M. *J. Endocrinol. Invest.* **2002**, *25*, 855–861.
8. Zhang, Y.; Proenca, R.; Maffei, M.; Barone, M.; Leopold, L.; Friedman, J. M. *Nature* **1994**, *372*, 425–432.
9. Drazen, D. L.; Woods, S. C. *Curr. Opin. Clin. Nutr. Metab. Care* **2003**, *6*, 621–629.
10. Diez, J. J.; Iglesias, P. *Eur. J. Endocrinol.* **2003**, *148*, 293–300.
11. Hall, J. E. *Am. J. Hypertens.* **1997**, *10*, 49S–55S.
12. Gahagan, S.; Silverstein, J. *Pediatrics* **2003**, *112*, e328.
13. Atkinson, M. A.; Eisenbarth, G. S. *Lancet* **2001**, *358*, 221–229.
14. Stuveling, E. M.; Bakker, S. J.; Hillege, H. L.; Burgerhof, J. G.; de Jong, P. E.; Gans, R. O.; de Zeeuw, D. *Hypertension* **2004**, *43*, 791–796.
15. Bakris, G. L. *Am. J. Hypertens.* **2005**, *18*, 112S–119S.
16. Jafar, T. H.; Stark, P. C.; Schmid, C. H.; Landa, M.; Maschio, G.; de Jong, P. E.; de Zeeuw, D.; Shahinfar, S.; Toto, R.; Levey, A. S. *Ann. Intern. Med.* **2003**, *139*, 244–252.
17. Kopyt, N. P. *J. Am. Osteopath. Assoc.* **2005**, *105*, 207–215.
18. Yamada, Y.; Miyajima, E.; Tochikubo, O.; Matsukawa, T.; Ishii, M. *Hypertension* **1989**, *13*, 870–877.
19. Julius, S. *Curr. Opin. Nephrol. Hypertens.* **1992**, *1*, 299–305.
20. Anderson, E. A.; Sinkey, C. A.; Lawton, W. J.; Mark, A. L. *Hypertension* **1989**, *14*, 177–183.
21. Wofford, M. R.; Anderson, D. C., Jr.; Brown, C. A.; Jones, D. W.; Miller, M. E.; Hall, J. E. *Am. J. Hypertens.* **2001**, *14*, 694–698.
22. Carlyle, M.; Jones, O. B.; Kuo, J. J.; Hall, J. E. *Hypertension* **2002**, *39*, 496–501.
23. Rahmouni, K.; Morgan, D. A.; Morgan, G. M.; Liu, X.; Sigmund, C. D.; Mark, A. L.; Haynes, W. G. *J. Clin. Invest.* **2004**, *114*, 652–658.
24. Luscher, T. F.; Barton, M. *Clin. Cardiol.* **1997**, *20*, II-3–II-10.
25. Ignarro, L. J. *Biosci. Rep.* **1999**, *19*, 51–71.
26. Hirata, Y.; Hayakawa, H.; Kakoki, M.; Tojo, A.; Suzuki, E.; Kimura, K.; Goto, A.; Kikuchi, K.; Nagano, T.; Hirobe, M. et al. *Hypertension* **1996**, *27*, 672–678.
27. Fulton, D.; Gratton, J. P.; Sessa, W. C. *J. Pharmacol. Exp. Ther.* **2001**, *299*, 818–824.
28. Himmelfarb, J.; Stenvinkel, P.; Ikizler, T. A.; Hakim, R. M. *Kidney Int.* **2002**, *62*, 1524–1538.
29. Agarwal, R. *Clin. Nephrol.* **2004**, *61*, 377–383.
30. Papanicolaou, D. A.; Petrides, J. S.; Tsigos, C.; Bina, S.; Kalogeras, K. T.; Wilder, R.; Gold, P. W.; Deuster, P. A.; Chrousos, G. P. *Am. J. Physiol.* **1996**, *271*, E601–E605.

31. Zinman, B.; Hanley, A. J.; Harris, S. B.; Kwan, J.; Fantus, I. G. *J. Clin. Endocrinol. Metab.* **1999**, *84*, 272–278.
32. Roman, R. J. *Physiol. Rev.* **2002**, *82*, 131–185.
33. Imig, J. D. *Am. J. Physiol. Renal Physiol.* **2005**, *289*, F496–F503.
34. Archer, S. L.; Gragasin, F. S.; Wu, X.; Wang, S.; McMurtry, S.; Kim, D. H.; Platonov, M.; Koshal, A.; Hashimoto, K.; Campbell, W. B. et al. *Circulation* **2003**, *107*, 769–776.
35. Imig, J. D.; Falck, J. R.; Wei, S.; Capdevila, J. H. *J. Vasc. Res.* **2001**, *38*, 247–255.
36. Zhao, X.; Pollock, D. M.; Inscho, E. W.; Zeldin, D. C.; Imig, J. D. *Hypertension* **2003**, *41*, 709–714.
37. Pollock, D. M. *Curr. Opin. Nephrol. Hypertens.* **2000**, *9*, 157–164.
38. Schneider, J. G.; Tilly, N.; Hierl, T.; Sommer, U.; Hamann, A.; Dugi, K.; Leidig-Bruckner, G.; Kasperk, C. *Am. J. Hypertens.* **2002**, *15*, 967–972.
39. Elmarakby, A. A.; Loomis, E. D.; Pollock, J. S.; Pollock, D. M. *Hypertension* **2005**, *45*, 283–287.
40. Wedgwood, S.; Dettman, R. W.; Black, S. M. *Am. J. Physiol. Lung Cell. Mol. Physiol.* **2001**, *281*, L1058–L1067.
41. Elmarakby, A. A.; Dabbs Loomis, E.; Pollock, J. S.; Pollock, D. M. *J. Cardiovasc. Pharmacol.* **2004**, *44*, S7–S10.
42. Ammarguellat, F.; Larouche, I.; Schiffrin, E. L. *Circulation* **2001**, *103*, 319–324.
43. Muller, D. N.; Luft, F. C. *Basic Res. Cardiol.* **1998**, *93*, 7–14.
44. Mogensen, C. E. *J. Hum. Hypertens.* **2002**, *16*, S52–S58.
45. Goodfriend, T. L.; Calhoun, D. A. *Hypertension* **2004**, *43*, 518–524.
46. De Paula, R. B.; da Silva, A. A.; Hall, J. E. *Hypertension* **2004**, *43*, 41–47.
47. Sasser, J. M.; Pollock, J. S.; Pollock, D. M. *Am. J. Physiol. Regul. Integr. Comp. Physiol.* **2002**, *283*, R243–R248.
48. Ortiz, M. C.; Sanabria, E.; Manriquez, M. C.; Romero, J. C.; Juncos, L. A. *Hypertension* **2001**, *37*, 505–510.
49. Rajagopalan, S.; Kurz, S.; Munzel, T.; Tarpey, M.; Freeman, B. A.; Griendling, K. K.; Harrison, D. G. *J. Clin. Invest.* **1996**, *97*, 1916–1923.
50. Wolf, G.; Wenzel, U. O. *Hypertension* **2004**, *43*, 693–698.
51. Tamaki, K.; Okuda, S.; Ando, T.; Iwamoto, T.; Nakayama, M.; Fujishima, M. *Kidney Int.* **1994**, *45*, 525–536.
52. Cowley, A. W., Jr.; Mattson, D. L.; Lu, S.; Roman, R. J. *Hypertension* **1995**, *25*, 663–673.
53. Yasunari, K.; Maeda, K.; Nakamura, M.; Watanabe, T.; Yoshikawa, J.; Asada, A. *Am. J. Med.* **2004**, *116*, 460–465.
54. Palmer, B. F. *N. Engl. J. Med.* **2002**, *347*, 1256–1261.
55. Wright, J. T., Jr.; Bakris, G.; Greene, T.; Agodoa, L. Y.; Appel, L. J.; Charleston, J.; Cheek, D.; Douglas-Baltimore, J. G.; Gassman, J.; Glassock, R. *JAMA* **2002**, *288*, 2421–2431.
56. Lindholm, L. H.; Ibsen, H.; Borch-Johnsen, K.; Olsen, M. H.; Wachtell, K.; Dahlof, B.; Devereux, R. B.; Beevers, G.; de Faire, U.; Fyhrquist, F. et al. *J. Hypertens.* **2002**, *20*, 1879–1886.
57. Lewis, E. J.; Hunsicker, L. G.; Clarke, W. R.; Berl, T.; Pohl, M. A.; Lewis, J. B.; Ritz, E.; Atkins, R. C.; Rohde, R.; Raz, I. *N. Engl. J. Med.* **2001**, *345*, 851–860.
58. Russo, D.; Minutolo, R.; Pisani, A.; Esposito, R.; Signoriello, G.; Andreucci, M.; Balletta, M. M. *Am. J. Kidney Dis.* **2001**, *38*, 18–25.
59. Ruilope, L. M.; Aldigier, J. C.; Ponticelli, C.; Oddou-Stock, P.; Botteri, F.; Mann, J. F. *J. Hypertens.* **2000**, *18*, 89–95.
60. Palloshi, A.; Fragasso, G.; Piatti, P.; Monti, L. D.; Setola, E.; Valsecchi, G.; Galluccio, E.; Chierchia, S. L.; Margonato, A. *Am. J. Cardiol.* **2004**, *93*, 933–935.
61. Zhou, M. S.; Jaimes, E. A.; Raij, L. *Hypertension* **2004**, *44*, 186–190.
62. Hoagland, K. M.; Maier, K. G.; Roman, R. J. *Hypertension* **2003**, *41*, 697–702.
63. Hunter, S. J.; Harper, R.; Ennis, C. N.; Crothers, E.; Sheridan, B.; Johnston, G. D.; Atkinson, A. B.; Bell, P. M. *J. Hypertens.* **1998**, *16*, 103–109.
64. Poole-Wilson, P. A.; Cleland, J. G.; Di Lenarda, A.; Hanrath, P.; Komajda, M.; Metra, M.; W, J. R.; Swedberg, K.; Torp-Pedersen, C. *Eur. J. Heart Fail.* **2002**, *4*, 321–329.
65. Bauer, J. H. *South Med. J.* **1994**, *87*, 1043–1053.
66. Abe, Y.; Komori, T.; Miura, K.; Takada, T.; Imanishi, M.; Okahara, T.; Yamamoto, K. *J. Cardiovasc. Pharmacol.* **1983**, *5*, 254–259.
67. Ozawa, Y.; Hayashi, K.; Nagahama, T.; Fujiwara, K.; Wakino, S.; Saruta, T. *J. Cardiovasc. Pharmacol.* **1999**, *33*, 243–247.
68. Zhou, M. S.; Jaimes, E. A.; Raij, L. *Am. J. Hypertens.* **2004**, *17*, 167–171.
69. Bernobich, E.; Cosenzi, A.; Campa, C.; Zennaro, C.; Sasso, F.; Paoletti, S.; Bellini, G. *J. Cardiovasc. Pharmacol.* **2004**, *44*, 401–406.
70. Zhang, X.; Hintze, T. H. *Circulation* **1998**, *97*, 576–580.
71. Matsumoto, K.; Morishita, R.; Moriguchi, A.; Tomita, N.; Yo, Y.; Nishii, T.; Nakamura, T.; Higaki, J.; Ogihara, T. *Hypertension* **1999**, *34*, 279–284.
72. Candido, R.; Allen, T. J.; Lassila, M.; Cao, Z.; Thallas, V.; Cooper, M. E.; Jandeleit-Dahm, K. A. *Circulation* **2004**, *109*, 1536–1542.
73. Da Cunha, V.; Tham, D. M.; Martin-McNulty, B.; Deng, G.; Ho, J. J.; Wilson, D. W.; Rutledge, J. C.; Vergona, R.; Sullivan, M. E.; Wang, Y. X. *Atherosclerosis* **2005**, *178*, 9–17.
74. Candido, R.; Jandeleit-Dahm, K. A.; Cao, Z.; Nesteroff, S. P.; Burns, W. C.; Twigg, S. M.; Dilley, R. J.; Cooper, M. E.; Allen, T. J. *Circulation* **2002**, *106*, 246–253.
75. Toblli, J. E.; DeRosa, G.; Cao, G.; Piorno, P.; Pagano, P. *Kidney Int.* **2004**, *65*, 2343–2359.
76. Chrysostomou, A.; Becker, G. *N. Engl. J. Med.* **2001**, *345*, 925–926.
77. Brewster, U. C.; Perazella, M. A. *Am. J. Med.* **2004**, *116*, 263–272.
78. Park, J. B.; Touyz, R. M.; Chen, X.; Schiffrin, E. L. *Am. J. Hypertens.* **2002**, *15*, 78–84.
79. Banday, A. A.; Marwaha, A.; Tallam, L. S.; Lokhandwala, M. F. *Diabetes* **2005**, *54*, 2219–2226.
80. Nassar, T.; Kadery, B.; Lotan, C.; Da'as, N.; Kleinman, Y.; Haj-Yehia, A. *Eur. J. Pharmacol.* **2002**, *436*, 111–118.
81. Landmesser, U.; Cai, H.; Dikalov, S.; McCann, L.; Hwang, J.; Jo, H.; Holland, S. M.; Harrison, D. G. *Hypertension* **2002**, *40*, 511–515.
82. Haque, M. Z.; Majid, D. S. *Hypertension* **2004**, *43*, 335–340.
83. Pu, Q.; Neves, M. F.; Virdis, A.; Touyz, R. M.; Schiffrin, E. L. *Hypertension* **2003**, *42*, 49–55.
84. Cosenzi, A.; Bernobich, E.; Bonavita, M.; Bertola, G.; Trevisan, R.; Bellini, G. *J. Cardiovasc. Pharmacol.* **2002**, *39*, 488–495.
85. Kaergel, E.; Muller, D. N.; Honeck, H.; Theuer, J.; Shagdarsuren, E.; Mullally, A.; Luft, F. C.; Schunck, W. H. *Hypertension* **2002**, *40*, 273–279.
86. Yu, Z.; Xu, F.; Huse, L. M.; Morisseau, C.; Draper, A. J.; Newman, J. W.; Parker, C.; Graham, L.; Engler, M. M.; Hammock, B. D. et al. *Circ. Res.* **2000**, *87*, 992–998.
87. Imig, J. D.; Zhao, X.; Capdevila, J. H.; Morisseau, C.; Hammock, B. D. *Hypertension* **2002**, *39*, 690–694.
88. Sinal, C. J.; Miyata, M.; Tohkin, M.; Nagata, K.; Bend, J. R.; Gonzalez, F. J. *J. Biol. Chem.* **2000**, *275*, 40504–40510.
89. Imig, J. D.; Zhao, X.; Zaharis, C. Z.; Olearczyk, J. J.; Pollock, D. M.; Newman, J. W.; Kim, I. H.; Watanabe, T.; Hammock, B. D. *Hypertension* **2005**, *46*, 975–981.

Biographies

Ahmed Elmarakby is a postdoctoral fellow in the Vascular Biology Center at the Medical College of Georgia. He earned his BSc degree in pharmacy from Mansoura University, Egypt in 1993 and he also received an MS degree in pharmacology from the same University in 1998. Dr Elmarakby joined the graduate program at the Medical College of Georgia in 2000 where he was awarded his PhD degree in pharmacology in 2004. After he received his PhD degree, he started his postdoctoral training in the Vascular Biology Center at the Medical College of Georgia. He has won many fellowships for his research in vascular and renal physiology such as American Heart Association Pre-Doctoral Fellowship and National Kidney Foundation Fellowship. He is currently holding an American Heart Association Post-Doctoral Fellowship Award until 2007. Dr Elmarakby has authored many manuscripts related to hypertension and renal damage and his research area mainly focuses on the role of superoxide, endothelin, inflammatory cytokines, and epoxygenase metabolites in hypertension and renal diseases.

David M Pollock is a professor in the Vascular Biology Center at the Medical College of Georgia. He earned his BSc degree in biology from the University of Evansville in 1978 and a PhD in physiology in 1983 from the University of Cincinnati. Dr Pollock received his postdoctoral training at the University of North Carolina at Chapel Hill. Early in his career he worked as a research investigator in the Drug Discovery Division of Abbott Laboratories for nearly seven years before moving to the Medical College of Georgia in 1995. Dr Pollock has over 100 peer-reviewed publications from his research on the renal mechanisms of hypertension with a focus on the role of endothelin, nitric oxide, and reactive oxygen species in the control of salt balance and vascular function. His research is currently funded by several grants from the National Institutes of Health and an Established Investigator Award from the American Heart Association.

John D Imig is a professor in the Vascular Biology Center at the Medical College of Georgia. He earned his BA degree in biology from Blackburn College in 1985 and a PhD in physiology and biophysics in 1990 from the University of Louisville. Dr Imig moved to the Medical College of Wisconsin for postdoctoral training and accepted a faculty position in the Department of Physiology at Tulane University in 1993. He moved to his current position at the Medical College of Georgia in 2001. He has won numerous awards for his research in vascular and renal physiology and currently holds an American Heart Association Established Investigator Award. The National Institutes of Health and the American Heart Association have continuously funded Dr Imig's research since 1992. Dr Imig has authored numerous manuscripts related to the antihypertensive and renal and cardiovascular protective properties of angiotensin receptor antagonists, epoxygenase metabolites, and epoxide hydrolase inhibitors.

Comprehensive Medicinal Chemistry II
ISBN (set): 0-08-044513-6

ISBN (Volume 6) 0-08-044519-5; pp. 575–595

6.26 Gastrointestinal Overview

S Evangelista, Menarini Ricerche SpA, Firenze, Italy

6.26.1 Introduction

The gastrointestinal (GI) tract in essence is a long tube starting at the mouth and ending at the anal orifice (or, as viewed by gastroenterologists, starting with breakfast and ending at dinner). In the proximal part of the GI tract, food is taken in, and mastication prepares the food for digestion. When mechanical and chemical/enzymatic processes have finished this operation, nutrients can be absorbed by the intestinal mucosa in the form of relatively simple compounds. These then pass into the bloodstream and the lymphatic system, and subsequently become available for cellular needs.

Another function of the GI tract is excretion (e.g., elimination of waste deriving from the diet and products of the body metabolism) in order to avoid toxic elements being present inside the body.

Optimal environmental conditions for the treatment of various food components are produced by the secretory activity of the coating epithelium, the intramural glands, and extramural exocrine glands, such as the salivary glands, the liver, and the pancreas.

Between the inner part (the mucosal surface devoted to absorption) and the outer part (the serosal surface) of the GI tract lie the muscular structures. Some muscles have a circular arrangement (circular muscle), and are able to narrow the lumen. Other muscles are arranged lengthwise (longitudinal muscles), and their constriction causes shortening of the intestine. In some locations, circular muscle predominates and forms valves, the sphincters, whose constriction determines the closure of a passage or a natural opening. The circular and longitudinal muscles constrict and release in coordinated waves (peristaltic waves), thus mixing the contents of the intestine and allowing the progression of food from the mouth to the anus.

Pathologies of the GI tract are due to impairment of one or more of these simple functions (secretion, absorption, motility, etc.), and are divided into organic (e.g., inflammatory bowel disease (IBD) (*see* 6.30 Emesis/Prokinetic Agents), peptic ulcer, and gastroesophageal reflux disease (GERD) (*see* 6.28 Inflammatory Bowel Disease; 6.31 Cardiovascular Overview) or functional (e.g., irritable bowel syndrome (IBS) (*see* 6.29 Irritable Bowel Syndrome) and dyspepsia) types, depending on whether the diagnosis is based on the symptoms only or achieved with the aid of precise clinical tests.

6.26.2 Historical Overview

The therapy for GI diseases changed markedly with the introduction of histamine (H_2) receptor antagonists in the 1970s, which reduced surgical intervention and ameliorated problems with patient compliance.

With Schwarz's statement of 'no acid – no ulcer' in 1910,[1] the treatment of reflux esophagitis and gastroduodenal ulcers required new approaches to provide effective control of gastric acid secretion (*see* 6.27 Gastric and Mucosal Ulceration). In the late 1960s, antiacids were available, but these only partially neutralized the gastric acid that had been secreted, and therefore symptom control and ulcer healing rates were unsatisfactory. The persistence and the intuition of Black and Duncan led to the discovery of H_2 receptor antagonists.[2]

The first H_2 receptor antagonist developed was burimamide, followed by metiamide. With minimal structural alterations, Brimblecombe *et al.*[3] developed cimetidine, the first $1 billion a year blockbuster drug (*see* 8.03 Medicinal Chemistry as a Scientific Discipline in Industry and Academia: Personal Reflections). The modification of the chemical structure of cimetidine, with elimination of the imidazole ring, led to ranitidine, the pharmacokinetic characteristics of which provided improved patient compliance. These drugs provided effective inhibition of production and release of gastric acid via a pharmacologically proven mechanism, and became the gold standard therapy for peptic ulcers during the 1980s, leading to a marked improvement in the quality of life for a large number of patients, and substantial reductions in the societal costs of the treatment and incapacitation associated with gastric ulceration.

However, H_2 receptor antagonists do not block parietal cell stimulation by agonists other than histamine (e.g., vagal acetylcholine interacting with parietal cell muscarinic M_3 receptors, or gastrin released from G cells), develop rapid tolerance during therapy due to the elevation of cAMP levels in parietal cells, and are more effective in inhibiting nocturnal than daytime acid secretion.[4]

These shortcomings were overcome with the discovery and development of proton pump inhibitors (PPIs) during the late 1980s. The stimulation of several receptors on parietal cells converge on the final pathway to increase acid secretion via the H^+/K^+-ATPase pump.[5] By selectively blocking this pump, a novel class of efficient antisecretory agents was developed. Their superior antisecretory potency, long-lasting efficacy, and pharmacokinetic characteristics have established these compounds as drugs of choice for the therapy of peptic ulcers. Similarly to the H_2 receptor antagonists, lansoprazole, pantoprazole, and rabeprazole were subsequently developed by modifying the chemical structure of omeprazole, the first PPI discovered (*see* 8.17 Omeprazole).[5] In contrast to H_2 receptor antagonists, all PPIs possess a common structural element, the benzidamizole ring. The last compound of this class to be launched on the market was esomeprazole, the active isomer of omeprazole, which in 2004 was fourth best selling drug worldwide.

Another important discovery that changed peptic ulcer therapy occurred also during the 1980s, with the first culture of *Helicobacter pylori.* In April 1982, Warren and Marshall at the Royal Perth Hospital in Australia, after a long period of negative attempts, succeeded isolating this bacterium, at that time called *Campylobacter pyloridis*.[6] The close association of the bacterium with gastritis and peptic ulcer was subsequently confirmed by Marshall and co-workers in 114 infected patients.[6,7] *H. pylori*, a Gram-negative spiral organism localized on the luminal surface of epithelial cells, is the main cause of chronic gastritis. *H. pylori* infection is usually acquired in childhood and early adulthood, and up to 50% of the world population is infected, many without symptoms. In countries with low standards of hygiene, the infection rate is high, due to fecal–oral and oral–oral transmission of the bacterium. The infection usually remains in the stomach for decades, finally leading to atrophy and intestinal metaplasia. In a few patients this metaplasia may lead to dysplasia and gastric carcinoma.

H. pylori infects only gastric-type mucosal tissue, but when such tissue occurs in the duodenum (gastric metaplasia), the bacterium can cause duodenitis, and, with the aid of acid and pepsin, induce duodenal ulcers. In this tissue, *H. pylori* duodenitis is necessary but not sufficient for ulcer formation. Nonmalignant gastric ulcers and B cell gastric lymphoma are also associated with *H. pylori* infection. Gastric cancer is the second most common cause of cancer mortality worldwide, and a recent meta-analysis has shown that patients infected with *H. pylori* were nearly six times more likely to develop adenocarcinoma than uninfected controls,[8] while eradication of *H. pylori* improved gastric atrophy and intestinal metaplasia, thought to be premalignant changes.[9]

Owing to the availability of inexpensive and noninvasive tests such as the urea breath and the stool antigen tests, it is nowadays very easy to diagnose the presence of *H. pylori* in the stomach, and then to eradicate it by the use of at least two antibiotics (amoxicillin, clarithromycin and/or metronidazole) along with an antisecretory agent such as a PPI or an H_2 receptor antagonist.

6.26.3 Overall Market Size and Anticipated Growth

The efficiency of eradication of *H. pylori* infection (around 95–98%) and the use of antibiotics have completely changed the approach to peptic ulcer therapy, significantly reducing the use of antisecretory compounds. Nonetheless, as shown in **Table 1**, antiulcer drugs represent the second most frequently sold drug class.

Antisecretory compounds are used for the long-term treatment of GERD, a chronic disorder of the esophagus that results a decrease of the low esophageal sphincter (LES) barrier tone, leading to acid reflux, epithelial erosion, ulceration, and, sometimes, hyperplasia accompanied by inflammation. The introduction of endoscopes and ambulatory devices for the monitoring of esophageal pH has led to an improved diagnosis of GERD and its complications. Prokinetics and antacids are also used currently, but these therapies, although effective, do not influence the underlying causes of the disease. GERD is not sensitive to the eradication of *H. pylori*, and is highly prevalent in the population (8–10% in USA).[10] New approaches to GERD include endoscopic suturing devices to tighten the LES, endoscopic submucosal implantation of gelatinous microspheres in the lower esophagus, and radiofrequency energy delivery to the LES, but larger controlled trials with long-term follow-up are needed to establish their effective therapeutic role.

In view of the above, the interest of gastroenterologists and researchers has recently moved to diseases such as IBS, IBD, and colorectal cancer, where there is a large unmet medical need.

IBS, a dysregulation of the brain–gut axis resulting in altered gut motility and visceral sensitivity (*see* 6.29 Irritable Bowel Syndrome),[11] is a common disorder, affecting 9–23% of the total population, with a largely idiopathic causality,

Table 1 Leading therapy classes by global pharmaceuticals sales in 2004[a]

Position	*Audited world therapy class*	*2004 sales ($ billions)*	*Global sales (%)*	*Growth (year over year) (%)*
1	Cholesterol and triglyceride reducers	30.2	5.8	11.7
2	Antiulcers	25.5	4.9	1.4
3	Cytostatics	23.8	4.6	16.9
4	Antidepressants	20.3	3.9	1.3
5	Antipsychotics	14.1	2.7	12.1

[a] IMS Health data.

Table 2 Sales of the leading therapy classes for the treatment of IBS or IBD in 2004[a]

Audited world therapy class	*2004 sales ($ millions)*	*Growth (year over year) (%)*
Antispasmodics and anticholinergics	672	4
GI sensorimotor modulators	316	125
Intestinal antiinflammatory agents	1533	9

[a] IMS Health data.

resulting in abdominal pain, abdominal distention, and disturbed defecation, and is associated with significant disability and healthcare costs. IBS accounts for up to 50% of referrals to gastroenterology clinics; symptoms vary over the years, often overlapping with other functional disorders such as dyspepsia. Ten percent of the patients turn to their physicians, and the illness has a large economic impact on the healthcare system overall, on absenteeism, and on lost productivity. The annual direct cost for IBS is around $41 billion in the eight most industrialized countries,[12] although most individuals with IBS do not consult a physician, and the disease is therefore underestimated. IBS sufferers report substantially lower quality of life scores as compared with healthy individuals.[11]

A meta-analysis evaluating 213 studies related to the efficacy of pharmacological agents for IBS treatment that included smooth muscle relaxants, bulking agents, prokinetic agents, psychotropic agents, and loperamide[13] showed that only 70 of the 213 studies met the inclusion criteria for methodology, but the strongest evidence for efficacy was shown for smooth muscle relaxants in patients with abdominal pain as the predominant symptom. In contrast, the meta-analysis found that the efficacy of bulking agents has not been established, loperamide is only effective for diarrhea, and the evidence for the use of psychotropic agents is inconclusive. Among the smooth muscle relaxants, calcium channel inhibitors selective for the GI tract (e.g., pinaverium and otilonium bromide) are effective for pain relief; they are marketed in Europe, but not in the USA, where the available spasmolytics are *n*-butyl scopolammonium bromide or hyoscine butylbromide, and dicyclomine, whose efficacy in IBS has yet to be demonstrated.[14] For these reasons, the market for this class of pharmaceuticals is not as large as expected (**Table 2**; antispasmodics and anticholinergics).

New therapeutic opportunities for IBS are emerging from serotonin (5HT) receptor antagonists or agonists that have been described as potential modulators of the visceral sensitivity and gut motility.[15] The selective $5HT_3$ antagonist alosetron was effective in diarrhoic women, a subset of the IBS population, but was associated with several cases of ischemic colitis and death, and was initially withdrawn and then reapproved by the US Food and Drugs Administration with warning restrictions. A similar compound, cilansetron, with efficacy in diarrhea-predominant IBS patients of both sexes, is under development.[16] Tegaserod, a $5HT_4$ receptor partial agonist, is effective and safe in treating patients with IBS with constipation, and, in the absence of other competitors, is expanding in the world market of IBS (**Table 2**; gastrointestinal sensorimotor modulators). Apart from the limitation of the IBS patient population that can be treated (only one-third of the total) by tegaserod, some doubts regarding its efficacy exist, the therapeutic advantage over placebo being between 5% and 20% in different trials. Furthermore, tegaserod is not very efficient in reducing pain and discomfort.[17] New safer and more effective drugs are now awaited to treat this chronic and important disease.

Ulcerative colitis (UC) and Crohn's disease (CD) are idiopathic inflammatory disorders of unknown etiology, and are collectively termed IBD (*see* 6.28 Inflammatory Bowel Disease). A typical presentation is that of prolonged diarrhea often associated with rectal bleeding in UC, and abdominal pain as a key symptom in CD. The mucosa of the colon is inflamed, with erythema, edema, ulceration, and bleeding in UC, or with focal irregular deeper ulceration in CD. Although endoscopic

Table 3 Sales of the leading therapy classes for the treatment of nonulcer dyspepsia and emesis in 2004[a]

Audited world therapy class	*2004 sales ($ millions)*	*Growth (year over year) (%)*
Gastroprokinetics	1080	3
Antiemetics and antinauseants	3300	12

[a] IMS Health data.

biopsies of the actively inflamed colon are quite similar in these two diseases, the inflammatory infiltrate in CD extends down through all layers of the bowel wall, while in UC it is limited to the epithelium and lamina propria. Also, the localization may be different, with CD affecting any segment of the intestine, and UC only the rectum and colon. The incidence of these diseases is not high, but the inflammation needs to be treated continuously to prevent disease recurrence. Although the pathogenesis and appearance of CD and UC have some differences, the same medical therapy is used today for both.

Current therapies for IBD include aminosalicylates (sulfasalazine, mesalamine, and balsalazide) given by suppositories, enema, or oral formulations that deliver the active principle locally to the inflamed colonic mucosa. As noted in **Table 2** (intestinal antiinflammatory agents), the need to take these drugs for the life of the IBD patient to prevent the recurrence of symptoms makes the market an important one. Immunosuppressive agents (azathioprine, methotrexate, 6-mercaptopurine, and cyclosporin A) and corticosteroids (even though they are not effective in maintaining remission) are also used, but have a limited therapeutic index. Approximately one-third of IBD patients do not respond to these conventional medical treatments. This factor, and the observation that aminosalicylates have been used since the 1960s[18] with less than satisfactory results, has stimulated research on the pathophysiology of the disease, as reflected by the number of articles published on the topic – PubMed reported 1047 publications for IBD in 1994, 1536 in 2000, and 1927 in 2004.

Among potential new treatments are inflimax, a human–mouse chimeric monoclonal IgG_1 antibody directed against tumor necrosis factor α (TNF-α), which was recently recommended for the treatment of severe CD patients[19]; natalizumab, a monoclonal antibody to integrin α_4; and adalimumab, a fully human monoclonal antibody to TNF-α. Variations in the *NOD2* gene present on chromosome 16 are strongly associated with susceptibility to CD, and may be the basis for future gene therapy (*see* 1.06 Gene Therapy).[20]

Another functional disease affecting a large part of the population is nonulcer dyspepsia, defined as impairment of the motility of the upper GI tract. Delayed emptying of liquids and solids is also a component of other GI disorders such as GERD, and comorbidity of nonulcer dyspepsia occurs with GERD and IBS.[21] Nausea, vomiting, heartburn, postprandial discomfort, and acid reflux are the most common symptoms of nonulcer dyspepsia. Gastric motility is under the stimulatory control of cholinergic neurons and the modulatory influence of the enteric nervous system, where dopamine and 5HT play key roles.[22] Thus, antagonists of dopamine (D_2) and $5HT_3$ receptors and $5HT_4$ receptor agonists, including substituted benzamides (metoclopramide and cisapride) and benzamidazoles (domperidone), exert their prokinetic effects by the modulation of cholinergic transmission (*see* 6.30 Emesis/Prokinetic Agents). Further, the GI peptide motilin and the agonist of its receptor, an antibiotic erythromycin, are involved in gastric emptying and postprandial gastric contraction. In July 2000, cisapride, the most successful prokinetic worldwide at that time, was withdrawn because of concerns of adverse events, including QT prolongation, which may have led to fatal heart arrhythmias. These compounds have also been used as antiemetics, as both 5HT and dopamine are involved in gastric motor reflexes and the emetic signaling pathways involved in the vomiting reflex.[23] These agents are also used to treat chemotherapy-induced emesis. The $5HT_3/D_2$ antagonist metoclopramide and $5HT_3$ receptor antagonists (ondansentron, granisetron, tropisetron, and dolasetron) are used for severe cases. The phenothiazine D_2 antagonists (chlorpromazine, perphenazine, prochlorperazine, promethazine, thiethylperazine, and triflupromazine), benzimidazole derivatives (domperidone), butyrophenones (haloperidol and droperidol), the $5HT_3/D_2$ antagonist trimethobenzamide, corticosteroids (dexamethasone and methylprednisolone), and cannabinoids (dronabinol and nabilone) are used against mild to moderate chemotherapy-induced emesis.

The market for gastroprokinetics is growing slowly due to a lack of effective new drugs while that of antiemetics (**Table 3**) increased in 2004 by 12%, with more than half of the total being ascribable to ondansetron.

References

1. Schwarz, K. *Beitr. Klin. Chir.* **1910**, *67*, 96–128.
2. Black, J. W.; Duncan, W. A.; Durant, C. J.; Ganellin, C. R.; Parsons, E. M. *Nature* **1972**, *236*, 879–888.

3. Brimblecombe, R. W.; Duncan, W. A.; Durant, C. J.; Ganellin, C. R.; Parsons, E. M. *J. Int. Med. Res.* **1975**, *3*, 86–92.
4. Aihara, T.; Nakamura, E.; Amagase, K.; Tomita, K.; Fujishita, T.; Furutani, K.; Okabe, S. *Pharmacol. Ther.* **2003**, *98*, 109–127.
5. Fellenius, E.; Berglidh, T.; Sachs, G.; Olbe, L.; Elander, B.; Syostrand, S. E.; Wallmark, B. *Nature* **1981**, *290*, 159–161.
6. Marshall, B. J.; Warren, J. R. *Lancet* **1984**, *i*, 1311–1314.
7. Marshall, B. J.; McGechie, D. B.; Rogers, P. A.; Glancy, R. J. *Med. J. Austr.* **1985**, *142*, 439–444.
8. *Helicobacter* and Cancer Collaborative Group. *Gut* **2001**, *49*, 347–353.
9. Correa, P.; Fontham, E. T.; Bravo, J. C.; Bravo, L. E.; Ruiz, B.; Zarama, G.; Realpe, J. L.; Malcom, G. T.; Li, D.; Johnson, W. D. et al. *J. Natl. Cancer Inst.* **2000**, *92*, 1881–1888.
10. Storr, M.; Meining, A.; Allescher, H. D. *Exp. Opin. Pharmacother.* **2001**, *2*, 1099–1108.
11. Drossman, D. A.; Camilleri, M.; Mayer, E. A.; Whitehead, W. E. *Gastroenterology* **2002**, *123*, 2108–2131.
12. Camilleri, M.; Williams, D. E. *Pharmacoeconomics* **2000**, *17*, 331–338.
13. Jailwala, J.; Imperiale, T. F.; Kroenke, K. *Ann. Int. Med.* **2000**, *133*, 136–147.
14. Evangelista, S. *Curr. Pharm. Des.* **2004**, *10*, 3561–3568.
15. Farthing, M. J. G. *Br. Med. J.* **2005**, *330*, 429–430.
16. Chey, W. D.; Cash, B. D. *Expert Opinion Invest. Drugs* **2005**, *14*, 185–193.
17. Evans, B. W.; Clarck, W. K.; Moore, D. J.; Whorwell, P. J. *Cochrane Database Syst. Rev.* **2004**, *1*, CD003960.
18. Misiewicz, J. J.; Lennard-Jones, J. E.; Connel, A. M. *Lancet* **1965**, *i*, 185–188.
19. NICE. *Technology Appraisal Guidance*, No 40; National Institute for Clinical Excellence: London, UK, 2002.
20. Markowitz, J.; Hyams, J.; Mack, D.; Otley, A.; Rosh, J.; Pfefferkorn, M; Tolia, V.; Oliva-Hamker, M.; Mezoff, A.; Treem, W. et al. *Am. J. Gastroenterol.* **2004**, *99*, S254.
21. Whitehead, W. E.; Palsson, O.; Jones, K. R. *Gastroenterology* **2002**, *122*, 1140–1156.
22. Simren, M.; Tack, J. *Gastroenterol. Clin. North Am.* **2003**, *32*, 577–579.
23. Mahesh, R.; Perumal, R. V.; Pandi, P. V. *Pharmazie* **2005**, *60*, 83–96.

Biography

Stefano Evangelista PhD has a long history of research in gastrointestinal pharmacology at both academic and private institutes of research. He has successfully contributed to the characterization of the role of several neuropeptides in gastrointestinal function, and participated to the development of tachykinin antagonists. He is a member of a group developing basic research in the field of capsaicin-sensitive neurons, with pioneering input in the recognition of the function of neuropeptides such as calcitonin gene-related peptide, substance P, and neurokinin A in the gastrointestinal tract. Working in the pharmaceutical industry, first in the discovery section and then in development, he has made major contributions to the furthering of knowledge on the mechanism of action of drugs, from molecular pharmacology to the clinic. He is the author of 120 papers, and between 1993 and the present he has been a referee for 18 international journals. He is currently working in preclinical development for Menarini Ricerche, and has developed numerous collaborations with academic groups with significant input in exploratory pharmacology to find new and relevant targets for the treatment of irritable bowel syndrome.

Comprehensive Medicinal Chemistry II
ISBN (set): 0-08-044513-6

ISBN (Volume 6) 0-08-044519-5; pp. 597–601

6.27 Gastric and Mucosal Ulceration

G Sachs and J M Shin, University of California at Los Angeles, CA, USA, and Veterans' Administration Greater Los Angeles Healthcare System, Los Angeles, CA, USA

6.27.1 Disease State

There are two major types of acid-dependent diseases of the upper gastrointestinal (GI) tract, peptic ulcer disease (PUD) in the stomach and duodenum and gastroesophageal reflux disease (GERD) in the esophagus. Nocturnal acid secretion can also be related to hoarseness or laryngitis, and even symptoms of asthma. Although these have quite different etiologies, the main therapeutic approach has always been control of gastric acid secretion.

PUD, for most of human history, was the prevalent ailment of this region of the GI tract, and had very significant associated mortality due to perforation and hemorrhage. In the twentieth century, control of acid secretion, first surgically and then medically, became possible, reducing serious outcomes of this disease significantly.[1] In the mid-1980s an association with infection by *Helicobacter pylori* was suggested.[2] It is now universally accepted that, in addition to acid, infection by this organism is necessary for the development of PUD. Although PUD is now much less prevalent in developed countries, perhaps due to a decline in infection by *H. pylori*, it is highly significant in underdeveloped regions of the world, where the infection rate remains at approximately 90%. The mode of transmission of the bacteria remains unknown, but most suspicion has focused on the water supply and its treatment. The modality of treatment of PUD or its symptoms now requires acid suppression as well as confirmed eradication of *H. pylori*, if active infection is established. Eradication requires acid suppression in combination with at least two antibiotics twice a day for at least 7 days.[3]

GERD, on the other hand, is increasing in prevalence in the developed countries of the world, and may also predispose to, first, Barrett's esophagus and, then, esophageal cancer.[4,5] So, not only is there an emphasis on healing any erosions present in the lower esophagus, but there is also concern that treatment should extend to full symptomatic relief. When histamine (H_2) receptor antagonists were introduced, the first being cimetidine, these drugs enabled 8 week healing of peptic ulcers, but had a lesser effect on GERD. With the introduction of proton pump inhibitors (PPIs), such as omeprazole, GERD also became amenable to effective treatment in terms of healing, in that 8 weeks of treatment is usually sufficient. However, complete symptom relief is much more difficult: although healing occurs at a luminal esophageal pH of greater than 4.0,[6,7] this pH is still sufficiently high to activate the acid-sensitive ion channels (ASICs) in the nerves of the epithelium, which have an activation threshold of approximately pH 5.0, resulting in pain. Given the mechanism of action of the PPIs, as discussed below, this is particularly true for patients with night-time GERD. Night-time reflux also results in laryngitis and even asthma,[8] requiring highly effective acid suppression for treatment. Although treatment of pain with COX-2 inhibitors was thought to be a means to avoid the common gastric symptoms and lesions associated with treatment with COX-1 inhibitors, the nonsteroidal anti-inflammatory drugs (NSAIDs), the cardiac safety of the former class of drugs has been questioned, as have the data indicating an absence of gastric symptoms.[9] The combination of a PPI with a COX-1 inhibitor appears to reduce the incidence of gastric side-effects.[10,11]

6.27.2 Disease Basis

Since the beginning of the twentieth century, the concept of 'no acid–no ulcer' led to various measures to reduce acid secretion in the GI tract. These measures were, in the absence of drugs targeted to the stomach, inevitably surgical. Total or partial gastrectomy was introduced by Billroth, and, with the recognition that vagal stimulation was responsible in large measure for the stimulation of acid secretion, this somewhat dramatic procedure was followed by vagotomy and highly selective vagotomy.[12,13] These surgical measures were relatively effective for PUD, albeit accompanied by the risks associated with open abdominal surgery. With the recognition of GERD as a major affliction and also that abnormal reflux was responsible, measures such as fundoplication were introduced, and still are used today, but usually by laparoscopic approaches. In PUD, modern medications, such as the H_2 receptor antagonists or PPIs, greatly reduced the need for surgical intervention, as is the case also for treatment of GERD. The recognition of *H. pylori* as being required for most PUD resulted in eradication regiments that, for the first time, were able to cure, not only treat, this disorder. Nevertheless, either because of lack of appropriate diagnosis or timely access to medical care, or because of the excessive use of NSAIDs, there is still often hemorrhage or even perforation of the stomach or duodenum that is treated on an emergency basis, either surgically or by endoscopic cautery of the bleeding vessel. Hence, although advances in treatment have been truly remarkable, acid-related GI disease is still a major phenomenon throughout the world. With modern diagnostic approaches, symptoms are not always associated with visible findings, and nonulcer dyspepsia (NUD) remains a controversial topic, with the role of acid, *H. pylori*, and afferent neural hypersensitivity being considered causal. Up to 50% of NUD can be improved by acid suppression.

6.27.3 Disease Models

Until the development of targeted drugs, most research was carried on the pylorus ligated rat, and the number and size of ulcers determined with administration of any chemical available. This strategy was totally unsuccessful, and was replaced, during the development of H_2 receptor antagonists, by the measurement of acid secretion by the perfused rat stomach.[14] Another model that proved useful in the measurement of the inhibition of acid secretion was the gastric fistula dog.[15] With the arrival of the rabbit gastric gland preparation,[16] for the first time a model became available for determining the effect of drugs on a mammalian model of acid secretion. When the H^+/K^+-ATPase became a target, isolated enzyme preparations proved useful.[17]

6.27.4 Clinical Trials

The modern design of double-blind assessment for ulcer or erosion healing as monitored by endoscopy is straightforward and reliable. However, an issue still to be addressed is symptom relief. A detailed questionnaire is being implemented to try to better quantify the analysis of symptom relief data.

6.27.5 Current Treatment

6.27.5.1 Antacids

The earliest drugs used to combat gastric acid were antacids such as chalk ($CaCO_3$) or baking soda ($NaHCO_3$), used by the ancient Greeks. The problem with these drugs was that they only elevated the luminal gastric pH, and did not affect the pH of the gastric epithelium, and, given the capacity of the stomach to secrete 1.5 l of 1 N HCl per day, had to be taken very frequently. With the dawn of modern medical practice in the nineteenth century, use of these agents was combined with bed rest and a bland diet. These measures were largely ineffective when more serious consequences such as perforation or hemorrhage occurred.[18] Nevertheless, antacids continue to be used for symptomatic relief, and public awareness has significantly increased in the western world as to the need for medical treatment for frequent upper GI tract symptoms.

6.27.5.2 Anticholinergic Agents

Long before the mechanisms of activation of acid secretion by the stomach were recognized, extract of belladonna was found to ameliorate peptic ulcer symptoms. However, the active ingredient, atropine, has a generalized effect on all the muscarinic receptors in the body, resulting in side effects, such as blurred vision, dry mouth, and urinary tract dysfunction that resulted in abstinence from the drug. Even though atropine was later replaced by more selective M_1 receptor antagonists (e.g., pirenzepine), the general muscarinic side effects, although reduced compared with atropine remained.[19]

6.27.5.3 H_2 Receptor Antagonists

Histamine (**Figure 1**) was discovered by Dale, but a pupil of Pavlov, Piopielski, first described the stimulation of gastric acid secretion. There was much argument whether histamine or gastrin was the major direct stimulant of the parietal cell. In the 1950s, a series of histamine antagonists was synthesized, where the imidazole ring of histamine was modified to other aromatic structures by Bovet and co-workers.[20] These antagonists blocked the peripheral actions of histamine, such as vasodilation and mucus secretion, but had relatively little effect on acid secretion. It was therefore suggested that the histamine receptor in the stomach was a different subtype, a putative H_2 receptor, as contrasted to the H_1 receptor that was blocked by the compounds synthesized by Bovet. In 1963, Black began work on the synthesis and evaluation of compounds able to selectively block the H_2 receptor. By modifying the side chain rather than the imidazole ring of histamine. The first H_2 receptor antagonist, burimamide, was synthesized in 1970, rapidly followed by metiamide and cimetidine (**Figure 1**).[1,14] The last was the first H_2 receptor antagonist in 1977, marketed as Tagamet. Ranitidine, famotidine, and nizatidine were second-generation H_2 receptor antagonists where the imidazole ring was no longer retained (**Figure 1**). These all contained a protonatable nitrogen atom at the end of the side chain, and presumably bound to the receptor as cations.

The H_2 receptor antagonists effectively abolished gastrin-stimulated acid secretion, but were less effective against cholinergically stimulated acid secretion.[14] This resolved the issue as to whether histamine acting at H_2 receptors or gastrin was the primary stimulant of acid secretion. It is now accepted that the action of gastrin on acid secretion is due to stimulation of histamine release from the ECL cell, and is not due to a direct action of this hormone on the parietal cell.[21] The efficacy of H_2 receptor antagonists relies on their plasma half-life, as these are reversible inhibitors. Further, they are all inverse agonists, able to inhibit the receptor even in the absence of histamine. There were attempts to develop noncompetitive H_2 receptor antagonists (e.g., loxitidine),[22] but unexpected toxicity terminated their development.

6.27.5.4 Proton Pump Inhibitors

PPIs were introduced in 1989, the culmination of serendipity and drug design. It is probably correct to state that without the serendipity, the PPIs would never have been designed. Hassle AB in Sweden had for some years a program for the discovery of anti-ulcer drugs using the classic rat ulcer model and measurement of acid secretion. Pyridinyl-2-ethylamide was found to inhibit acid secretion. Modification to pyridinyl-2-ethylthioamide maintained inhibition of acid secretion, but the mechanism was unknown, and remains so. With the discovery of cimetidine, there was speculation that pyridinyl-2-ethylamide might be an H_2 receptor antagonist. 2-(Pyridinylmethylthio)benzimidazole was made, and then modified to generate 2-(pyridinylmethylsulfinyl) benzimidazole (timoprazole; **Figure 2**) to

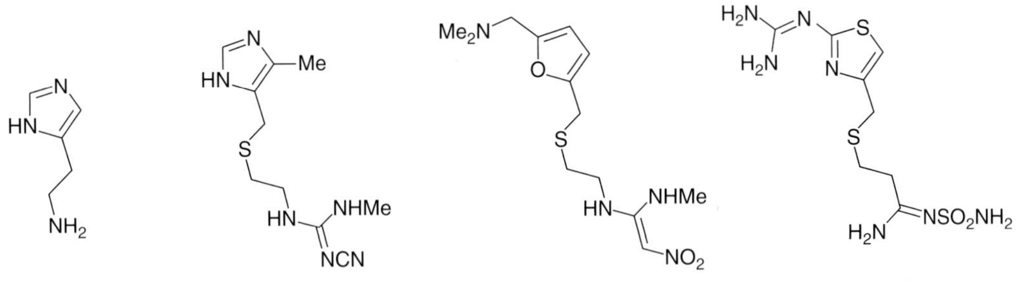

Histamine Cimetidine Ranitidine Famotidine

Figure 1 The structure of histamine, cimetidine, ranitidine, and famotidine, showing the retention of the imidazole structure in cimetidine, replaced by a furan ring in ranitidine and a thiazole ring in famotidine.

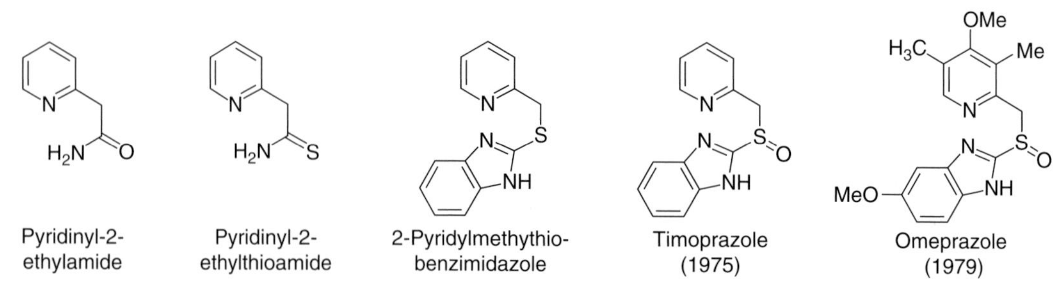

Figure 2 Compounds leading to the synthesis of omeprazole, the first PPI.

Omeprazole

Lansoprazole

Pantoprazole

Rabeprazole

Esomeprazole

Tenatoprazole

Ilaprazole

Figure 3 Various PPIs either on the market or in development. Esomeprazole is the (*S*) enantiomer of omeprazole, and all the other compounds are racemates. Different substitutions on the benzimidazole portion alter the pK_a of the benzimidazole nitrogen atom, resulting in different acid liability profiles.

improve stability. By 1975, Hassle was focusing more on acid secretion, and had available to them, in addition to rats and dogs, the rabbit gastric gland model. They then discovered that timoprazole, in contrast to H_2 receptor antagonists, inhibited acid secretion in vivo and in vitro, irrespective of the stimulus. Clearly this compound was not an H_2 receptor antagonist. Timoprazole was toxic to rats, causing thyroid enlargement and thymic degeneration. Modification of timoprazole to picoprazole resulted in decreased toxicity.

As a polyclonal antibody generated against the enzyme H^+/K^+-ATPase reacted by radial immunodiffusion with proteins of the thyroid and thymus, it appeared likely that these compounds inhibited the gastric H^+/K^+-ATPase. When tested against the ATPase in the absence of acid transport, no effect was found. However, as these compounds were acid-labile, further testing showed that timoprazole and picoprazole inhibited both enzyme activity and acid transport in vesicles under acid-transporting conditions. Moreover, there was a lag phase of inhibition when monitoring acid transport,[23,24] indicating that the mechanism of action was due to both the weak base properties of the compounds and their being acid-activated prodrugs. With this knowledge, omeprazole was synthesized and launched in 1989 (**Figure 2**). Subsequently, compounds with the same core structure were also introduced, and two further compounds with a slightly different core structure are under development (**Figure 3**).

The mechanism of acid activation is quite unique (**Figure 4**), and has only recently been described.[25] Protonation of the pyridine ring with a pK_a between 4.0 and 5.0 accounts for the accumulation of the prodrug in the acidic space of the active parietal cell canaliculus (pH ≈ 0.8). Protonation of the benzimidazole nitrogen atom activates the C-2 carbon, allowing reaction with the fraction of the pyridine that is unprotonated, to form a planar tetracyclic cation that is converted to the sulfenic acid and, by dehydration, to the sulfonamide.[26] In the presence of the acid-transporting enzyme, the PPIs react with one or more cysteine residues accessible from the luminal surface of the GI tract as the reactive species is a relatively impermeant cation. Omeprazole reacts primarily with Cys813 on the luminal face of the enzyme, and lansoprazole with Cys813 and Cys321, both present in a luminal vestibule, whereas pantoprazole and tenatoprazole react with Cyse813 and Cys822, the latter of which is present in the membrane domain of the enzyme.[27]

Advances in our understanding of the tertiary structure of the catalytic subunit of the H^+/K^+- ATPase based on homology modeling of different crystalline forms of the SERCA Ca^{2+}-ATPase[28,29] have allowed the generation of a specific structure of the PPI-bound form of the enzyme. During its transport cycle, the enzyme converts from the basal E_1 state to the phosphorylated state with a hydronium ion bound in the ion-binding site in the membrane, and then converts to the E_2P form, which then binds K^+ with dephosphorylation. This is the form that binds the active form of the PPI. Different conformations of the catalytic subunit of the pump are shown in **Figure 5**, with the binding sites of pantoprazole on the bottom left.

Knowledge of the mechanism of the PPIs as acid-activated prodrugs allows the prediction of many of their properties. First, since they are covalent inhibitors, their duration of action outlasts their presence in the blood. Since they require acid activation, they are most effective when the parietal cells are stimulated, hence are usually given 30–60 min after meals. Since they generally have a short plasma half-life (60–90 min), and not all acid pumps are active at any one time, their effective is cumulative, reaching a steady state on once-a-day dosing by the third day, inhibiting about 70% of all pumps. Restoration of pump activity depends on de novo synthesis of the pump, which has a half-life of about 54 h, and also on the partial reversal of inhibition by glutathione reduction of the disulfide bond at Cys813.[30,31] If there is significant acid secretion at night resulting in night-time GERD, it is difficult to control this with currently available PPIs. Their effect on GERD healing is also superior to their effect on symptom relief. They also heal peptic ulcers and esophageal erosions to greater than 90% within 8 weeks.

6.27.5.4 Eradication of *Helicobacter pylori*

H. pylori is a Gram-negative neutralophile (e.g., grows best at pH 7.0 and does not grow below pH 5.0) that is unique in having acclimated to the gastric environment to the extent that it is able to infect and colonize the human stomach.[32] Other pathogenic neutralophiles can survive in gastric acidity by acid-tolerant or -resistant mechanisms, but they cannot colonize the stomach.[33] Acid acclimation by *H. pylori* depends on the expression of several genes, in particular, the genes of the urease cluster.[34] These consist of a promoter, the structural subunit genes, *ureA* and *ureB*, followed by a second promoter and then *ureI*, *ureE*, *ureF*, *ureG*, and *ureH*. The last four genes are required for Ni^{2+} insertion into the UreA/UreB apoenzyme complex, to form active urease. UreI is a proton-gated urea channel that allows urea entry to the intrabacterial urease, enhancing the production of NH_3 and CO_2 up to 40-fold.[35] The NH_3 is able to neutralize entering acid in both the periplasm and cytoplasm of the organism, and CO_2 is converted to the buffer HCO_3^- in both the cytoplasm and periplasm.[36] *H. pylori* is unique in being able to control the pH of its periplasm to ~6.1, the effective pK_a of the HCO_3^-/CO_2 couple, at a medium pH as low as ~2.5 in the presence of urea. This property allows growth in specific regions of the stomach. Gastric juice contains 1–3 mM urea, and this is sufficient for colonization by the organism. Consistent with the important role of the urease system, deletion mutants of UreI or the carbonic anhydrases or urease negative mutants are not able to colonize animal models such as the mouse or the gerbil.[37,38] The urease/carbonic anhydrase mechanism for acid acclimation is shown in **Figure 6**.

The ability to acclimate to gastric acidity implies that *H. pylori* seeks a favorable milieu in which to multiply, but in other locations may just resist acid and not divide. Hence, the population of *H. pylori* in the human stomach consists of bacteria in both stationary and log phases.

H. pylori infection always results in gastritis, and can progress to gastric or duodenal ulceration, and increase the risk of gastric cancer up to 20-fold.[3] Although some believe that the organism is a harmless or even a beneficial commensal,[39] the vast majority of clinical data dispute this opinion. Eradication of *H. pylori*, if present, is the appropriate therapeutic choice.

Treatment with antibiotics (e.g., amoxicillin, clarithromycin, or metronidazole alone or in combination) cannot eradicate the organism, presumably because many of the bacteria are in the stationary phase and thus not affected by such growth-dependent antibiotics. An essential adjunct of therapy is the reduction of acid secretion by any PPI or

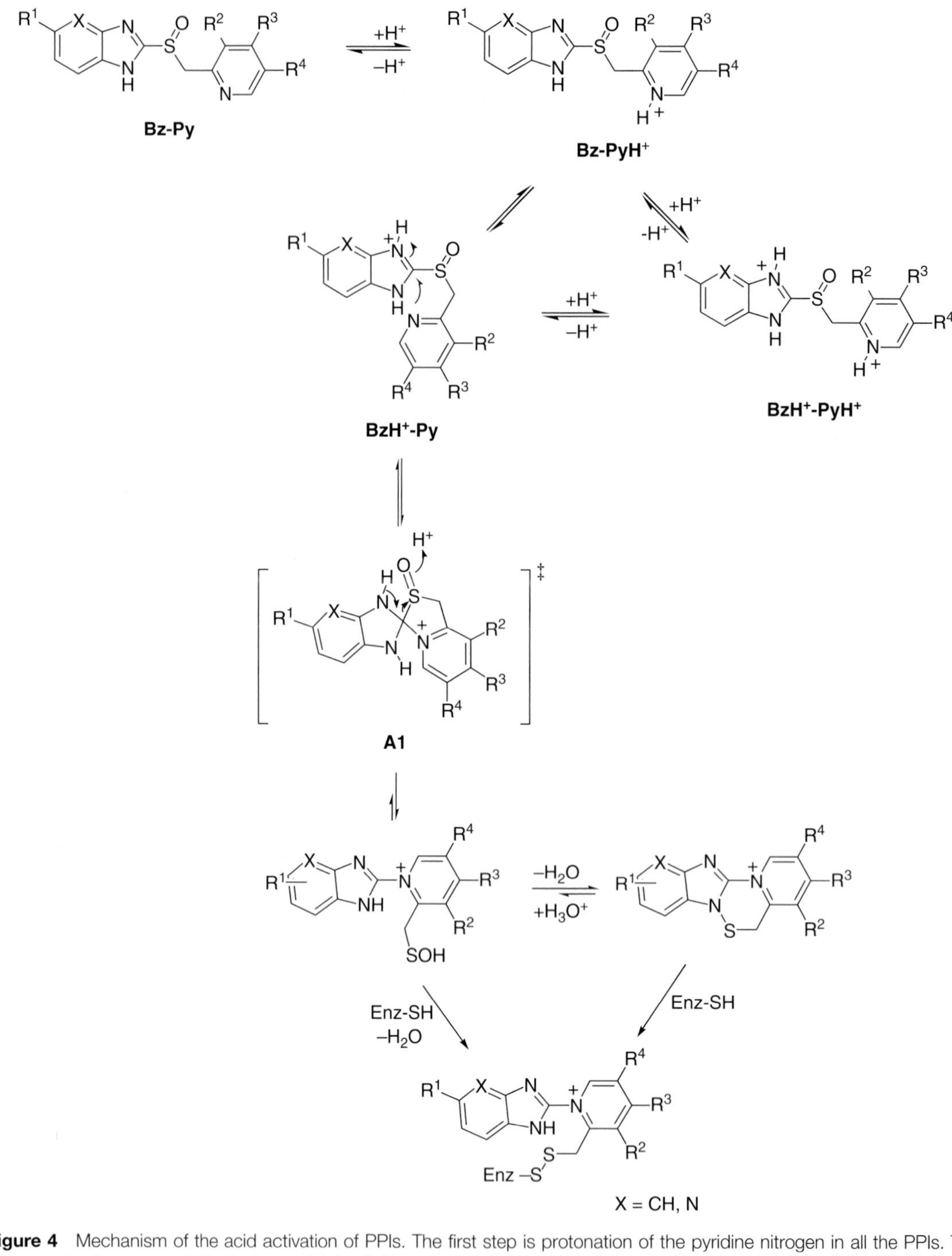

Figure 4 Mechanism of the acid activation of PPIs. The first step is protonation of the pyridine nitrogen in all the PPIs. This accounts for the accumulation in the parietal cell. The second protonation is the activation step. In all the PPIs, the activating protonation occurs at the nitrogen atom vicinal to the C-2 position of the imidazole ring. (In the particular case of tenatoprazole where X is N instead of C as in all other PPIs, the second protonation is spread throughout the imidazo-pyridine.) This second protonation results in a fraction of the species being present with an activated C-2 position and an unprotonated pyridine, which can then proceed to form the sulfenic acid via a transition state (A1). In solution, this proceeds to form the sulfenamide by dehydration. In the presence of thiols, the sulfenic acid reacts to form the disulfide, hence accelerating the reaction. Either the sulfenic acid or the sulfonamide can react with the luminally accessible cysteine residues of the gastric H^+/K^+-ATPase, but it is likely, given the selectivity of cysteine labeling by pantoprazole and tenatoprazole (slowly activated compounds compared with omeprazole, lansoprazole, and tenatoprazole) that is the sulfenic acid formed in the vestibule of the enzyme that is responsible for the inhibition of activity.

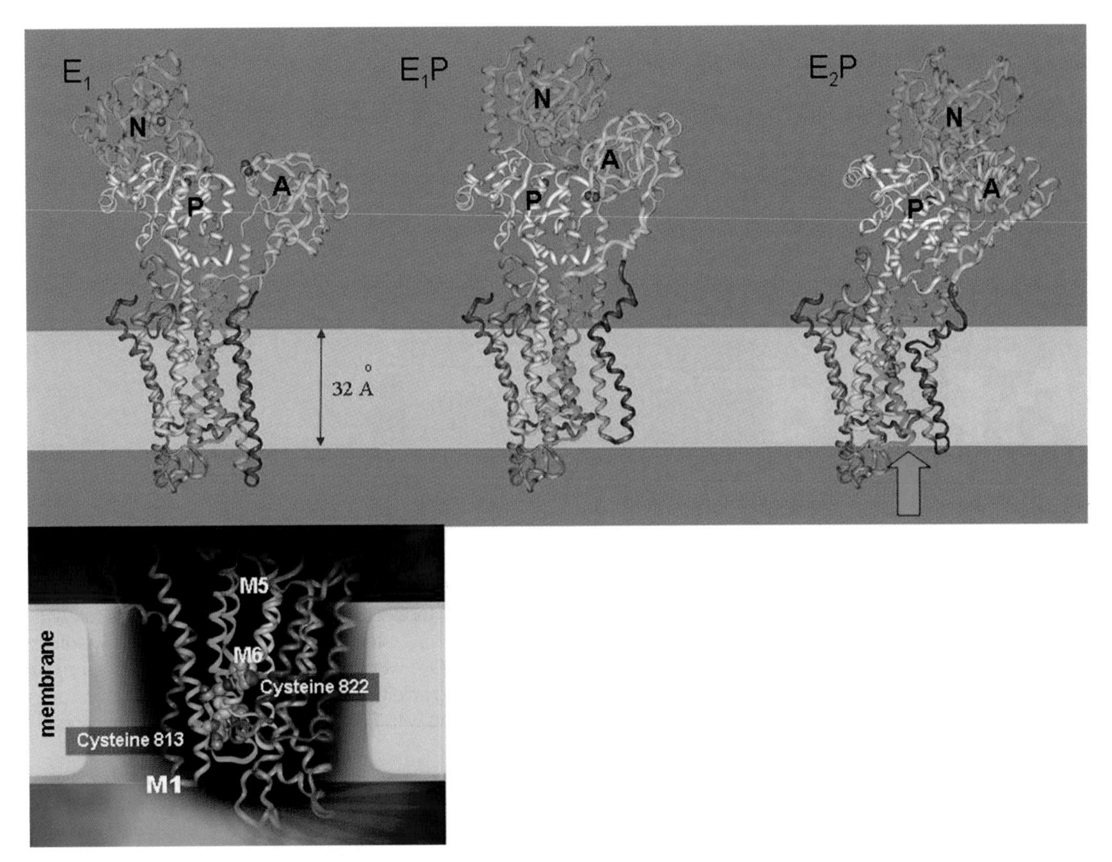

Figure 5 (a) An illustration of some of the conformations of the gastric H^+/K^+-ATPase, showing the movement of the cytoplasmic domains and also movement of the transmembrane segments that catalyze the outward transport of protons. The arrow shows where access by PPI occurs. E_1 is the resting conformation, with ion-binding sites facing the cytoplasm; E_1P is the conformation after phosphorylation of the pump before export of the proton (or hydronium); and E_2P is the phosphorylated conformation after release of the proton but before binding of K^+, which returns the pump to the E_1 form to complete the cycle. (b) Location of the two cysteine residues covalently bound by pantoprazole or tenatoprazole.

ranitidine bismuth subcitrate, along with two antibiotics.[40,41] This regimen provides eradication rates of about 80% in the field, and is increasingly confounded by either clarithromycin or metronidazole resistance. The reduction of acidity presumably increases the proportion of the organisms in the log phase, making them sensitive to amoxicillin and clarithromycin. More current therapy involves triple therapy, a combination of a PPI and, usually, amoxicillin and clarithromycin or metronidazole taken twice a day for at least 10 days. Analysis of infection is performed either by biopsy or measuring the $^{13}CO_2$ released from labeled ingested urea by a breath analyzer. Given the acid activation of UreI, it is important to ensure acidity of the gastric lumen for maximal sensitivity of this test.[42]

6.27.6 Unmet Medical Needs

Even though the introduction of PPIs as a therapy for acid-related disease combined with *H. pylori* eradication has drastically altered clinical outcomes, there are still areas where these drugs are less than optimal. For example, the short half-life of the PPIs and the need for stimulated acid secretion for maximal efficacy results in a continued fall in intragastric pH in the night. If the pH falls to below 3.0 and there is reflux, nocturnal GERD will result. Newer PPIs such as tenatoprazole or ilaprazole, because of their longer half-life, might reduce the incidence of nocturnal GERD; treatment with a PPI in the morning and a H_2 receptor antagonist at night might also be beneficial.[43] The complexity of *H. pylori* eradication treatment results in a loss of compliance, and is also difficult to administer in developing countries, where the incidence of infection can reach as high as 90%. A specific and simple monotherapy would be truly beneficial. A question that arises is also the treatment of choice for severe reflux disease in the young. Should they be placed on life-long PPIs or undergo laparoscopic fundoplication? There are advocates of both approaches.

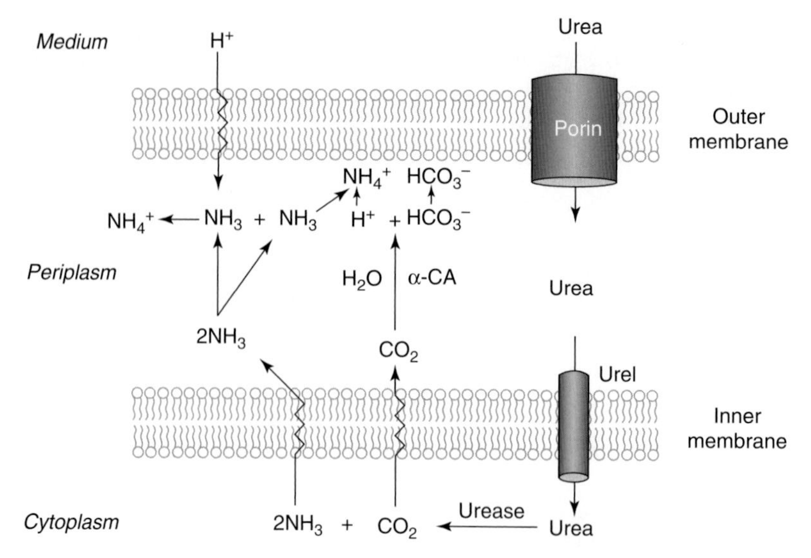

Figure 6 A model of two of the acid acclimation mechanism in *H. pylori*. The outer membrane contains porins permeable to urea and protons. With acidification, UreI opens and urea moves into the cytoplasm, increasing intrabacterial urease activity. This produces $2NH_3$ and CO_2, gases that readily exit the cell into the periplasm. Protons entering the cytoplasm are neutralized by NH_3 forming NH_4^+ while cytoplasmic carbonic anhydrase generates HCO_3^-, which is a stronger buffer at neutral pH than NH_3. Similarly, the NH_3 that effluxes into the periplasm can neutralize entering acidity and the CO_2 due to periplasmic carbonic anhydrase activity producing H^+ and HCO_3^-; a second NH_4^+ is formed along with HCO_3^-, the latter providing buffering in the range of pH 6.1.

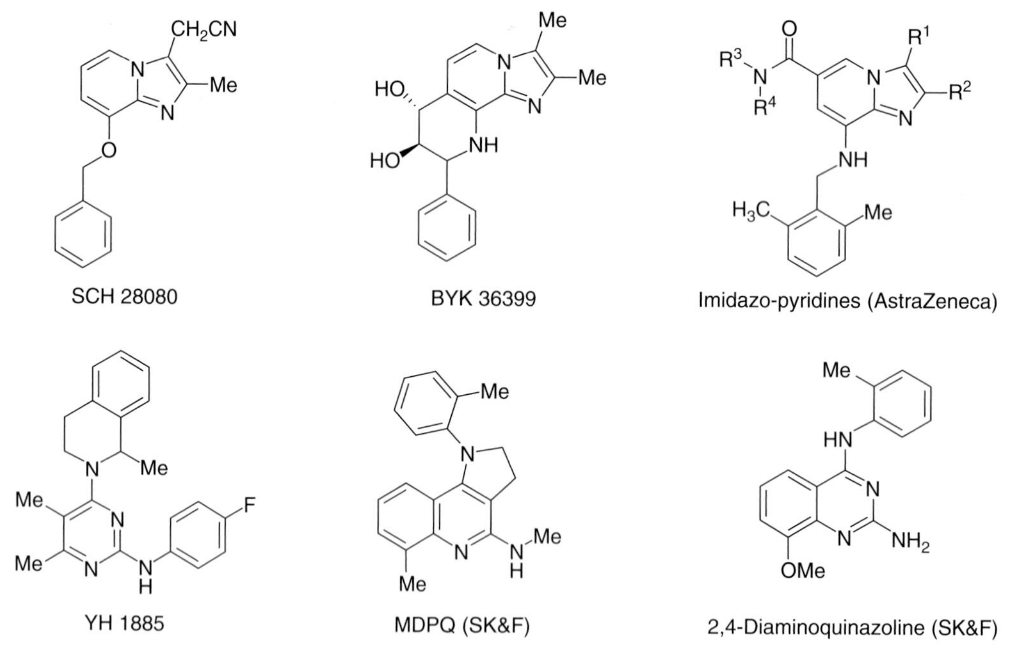

Figure 7 Some K^+ competitive inhibitors of the gastric H^+/K^+-ATPase. SCH 28080 was the first imidazopyridine shown to competitively inhibit the ATPase, and fused-ring analogs or derivatives of SCH 28080 have been tested in humans, showing superior pH control. The quinolines or quinazolines have not reached clinical trial, although they appear effective.

6.27.7 New Research Areas

Another type of inhibitor selective for the gastric H^+/K^+-ATPase, as compared to the other homologous ATPases such as the Na^+/K^+-, Ca^{2+}-, and nongastric H^+/K^+-ATPases, are the K^+ competitive inhibitors that bind to the outside face of the pump and prevent pump recycling.[44] Their mechanism of action results in a more rapid and more complete inhibition of acid secretion, promising better symptom relief. Their effectiveness depends entirely on the plasma half-life, and, with a half-life of about 6 h, dosing in the morning and the evening might provide more rapid healing and almost immediate symptom relief.[45] However, since their selectivity is entirely structure-dependent and does not have the acid activation benefit of the PPIs, the safety of these products (**Figure 7**), if introduced to the market, may be a concern.

In terms of monotherapy for *H. pylori*, there are some relatively obvious targets. For example, carbonic anhydrase inhibitors such as acetazolamide deserve clinical trials for eradication, especially given reports that treatment with acetazolamide resulted in a cure of PUD.[46] Inhibition of the unique proton-gated urea channel UreI would also result in eradication, but discovery of a compound with such an action requires high-throughput screening of a chemical library. For this to occur, the pharmaceutical industry needs to recognize the importance of prevention of gastric cancer worldwide rather than just focusing on the decline of PUD in the western world.

Acknowledgments

Supported in part by the US Veterans' Association and NIH grant numbers DK46917, 53462, 41301, and 58333.

References

1. Brimblecombe, R. W.; Duncan, W. A.; Durant, G. J.; Emmett, J. C.; Ganellin, C. R.; Leslie, G. B.; Parsons, M. E. *Gastroenterology* **1978**, *74*, 339–347.
2. Marshall, B. J.; Warren, J. R. *Lancet* **1984**, *i*, 1311–1315.
3. Suerbaum, S.; Michetti, P. *N. Engl. J. Med.* **2002**, *347*, 1175–1186.
4. Wong, A.; Fitzgerald, R. C. *Clin. Gastroenterol. Hepatol.* **2005**, *3*, 1–10.
5. Sharma, P. *Am. J. Med.* **2004**, *117*, 79S–85S.
6. Bell, N. J.; Hunt, R. H. *Aliment. Pharmacol. Ther.* **1996**, *10*, 897–904.
7. Bell, N.; Karol, M. D.; Sachs, G.; Greski-Rose, P.; Jennings, D. E.; Hunt, R. H. *Aliment. Pharmacol. Ther.* **2001**, *15*, 105–113.
8. Fass, R.; Achem, S. R.; Harding, S.; Mittal, R. K.; Quigley, E. *Aliment. Pharmacol. Ther.* **2004**, *20*, 26–38.
9. Fitzgerald, G. A. *N Engl. J. Med.* **2004**, *351*, 1709–1711.
10. Thompson, A. J.; Yeomans, N. D. *Curr. Pharm. Des.* **2003**, *9*, 2221–2228.
11. Wolfe, F.; Anderson, J.; Burke, T. A.; Arguelles, L. M.; Pettitt, D. *J. Rheumatol.* **2002**, *29*, 467–473.
12. Modlin, I. M.; Darr, U. *Yale J. Biol. Med.* **1994**, *67*, 63–80.
13. Walters, W.; Lynn, T. E.; Mobley, J. E. *Gastroenterology* **1957**, *33*, 685–690; discussion 691–682.
14. Black, J. W.; Duncan, W. A. M.; Durant, C. J.; Ganellin, C. R.; Parsons, M. E. *Nature* **1972**, *236*, 385–390.
15. Hirschowitz, B. I. *Fed. Proc.* **1968**, *27*, 1318–1321.
16. Berglindh, T.; Obrink, K. J. *Acta. Physiol. Scand.* **1976**, *96*, 150–159.
17. Sachs, G.; Chang, H. H.; Rabon, E.; Schackman, R.; Lewin, M.; Saccomani, G. *J. Biol. Chem.* **1976**, *251*, 7690–7698.
18. Modlin, I. M.; Sachs, G. *Acid Related Diseases. Biology and Treatment*, 2nd ed.; Lippincott Williams & Wilkins: Philadelphia, PA, 2004.
19. Wilkes, J. M.; Kajimura, M.; Scott, D. R.; Hersey, S. J.; Sachs, G. *J. Membr. Biol.* **1991**, *122*, 97–110.
20. Bovet, D.; Gertner, S. B.; Virno, M. *J. Pharmacol. Exp. Ther.* **1956**, *118*, 63–76.
21. Athmann, C.; Zeng, N.; Scott, D. R.; Sachs, G. *Am. J. Physiol. Gastrointest. Liver Physiol.* **2000**, *279*, 1048–1058.
22. Boyd, E. J.; Wormsley, K. G. *Eur. J. Clin. Pharmacol.* **1984**, *26*, 443–447.
23. Fellenius, E.; Berglindh, T.; Sachs, G.; Olbe, L.; Elander, B.; Sjostrand, S. E.; Wallmark, B. *Nature* **1981**, *290*, 159–161.
24. Wallmark, B.; Sachs, G.; Mardh, S.; Fellenius, E. *Biochim. Biophys. Acta* **1983**, *728*, 31–38.
25. Shin, J. M.; Cho, Y. M.; Sachs, G. *J. Am. Chem. Soc.* **2004**, *126*, 7800–7811.
26. Brandstrom, A.; Bergman, N.-A.; Lindberg, P.; Grundevik, I.; Johansson, S.; Tekenbergs-Hjelte, L.; Ohlson, K. *Acta Chem. Scand.* **1989**, *43*, 549–568.
27. Besancon, M.; Simon, A.; Sachs, G.; Shin, J. M. *J. Biol. Chem.* **1997**, *272*, 22438–22446.
28. Munson, K.; Garcia, R.; Sachs, G. *Biochemistry* **2005**, *44*, 5267–5284.
29. Toyoshima, C.; Nomura, H.; Sugita, Y. *Ann. NY Acad. Sci.* **2003**, *986*, 1–8.
30. Shin, J. M.; Sachs, G. *Gastroenterology* **2002**, *123*, 1588–1597.
31. Gedda, K.; Scott, D.; Besancon, M.; Lorentzon, P.; Sachs, G. *Gastroenterology* **1995**, *109*, 1134–1141.
32. Sachs, G.; Weeks, D. L.; Melchers, K.; Scott, D. R. *Annu. Rev. Physiol.* **2003**, *65*, 349–369.
33. Audia, J. P.; Foster, J. W. *J Mol. Microbiol. Biotechnol.* **2003**, *5*, 17–28.
34. Cussac, V.; Ferrero, R. L.; Labigne, A. *J. Bacteriol.* **1992**, *174*, 2466–2473.
35. Scott, D. R.; Marcus, E. A.; Weeks, D. L.; Sachs, G. *Gastroenterology* **2002**, *123*, 187–195.
36. Marcus, E. A.; Moshfegh, A. P.; Sachs, G.; Scott, D. R. *J. Bacteriol.* **2005**, *187*, 729–738.
37. De Reuse, H.; Mendz, G.; Ball, G.; LaBigne, A.; Thiberg, J.-M.; Bury-Mone, S. ASM 105th General Meeting, B-142, 2005.
38. Mollenhauer-Rektorschek, M.; Hanauer, G.; Sachs, G.; Melchers, K. *Res. Microbiol.* **2002**, *153*, 659–666.
39. Blaser, M. J. *Sci. Am.* **2005**, *292*, 38–45.
40. Gisbert, J. P.; Gonzalez, L.; Calvet, X. *Helicobacter* **2005**, *10*, 157–171.
41. Leodolter, A.; Naumann, M.; Malfertheiner, P. *Dig. Dis.* **2004**, *22*, 313–319.
42. Pantoflickova, D.; Scott, D. R.; Sachs, G.; Dorta, G.; Blum, A. L. *Gut* **2003**, *52*, 933–937.
43. Tutuian, R.; Castell, D. O. *MedGenMed* **2004**, *6*, 11.
44. Wallmark, B.; Briving, C.; Fryklund, J.; Munson, K.; Jackson, R.; Mendlein, J.; Rabon, E.; Sachs, G. *J. Biol. Chem.* **1987**, *262*, 2077–2084.
45. Andersson, K.; Carlsson, E. *Pharmacol. Ther.* **2005**, *108*, 294–307.
46. Puscas, I. *Ann. NY Acad. Sci.* **1984**, *429*, 587–591.

Biographies

George Sachs is a professor of the Department of Medicine and Physiology at the University of California at Los Angeles and Wilshire Chair in Medicine at University of California at Los Angeles. He has been Director of the Membrane Biology Laboratory and has a concurrent appointment with the Veterans Administration Greater Los Angeles Healthcare System in West Los Angeles since 1982. He received his BSc degree in 1957, his MB and ChB in 1960, and his DSc in 1980, all from the University of Edinburgh, Scotland. He received an honorary MD from the Medical Faculty of Gothenburg University, Sweden in 1996.

Jai Moo Shin is an associate researcher of the Department of Physiology at the University of California at Los Angeles. He joined the Membrane Biology Laboratory in 1990. He received his PhD degree in 1986 from Korea Advanced Institute of Science and Technology in the Republic of Korea, and has worked at the University of California at Los Angeles since 1990.

Comprehensive Medicinal Chemistry II
ISBN (set): 0-08-044513-6

ISBN (Volume 6) 0-08-044519-5; pp. 603–612

6.28 Inflammatory Bowel Disease

N Pullen and J D Gale, Pfizer Global Research and Development, Sandwich, UK

6.28.1 Disease State

Crohn's disease (CD; OMIM 266600) and ulcerative colitis (UC; OMIM 191390), the constellation of idiopathic conditions that constitute inflammatory bowel disease (IBD), are characterized by chronic and relapsing episodes of diarrhea, dense leukocytic mucosal infiltration, and extracellular matrix degradation (**Table 1**).[1] CD can occur anywhere along the length of the gastrointestinal (GI) tract, from the mouth to the anus, and is characterized histologically by transmural inflammation and epithelial ulceration and is often associated with formation of fistulae, fissures, and stenosis of segments of the entire tract. CD often appears multifocal, areas of active disease being segmented by histologically normal sections of gut. In contrast, UC only affects the colon and rectum. The inflammation, initiating distally and advancing proximally and giving rise to abscesses and ulcerations, is invariably

Table 1 Characteristics of human inflammatory bowel diseases

General definition

IBD is a chronic relapsing idiopathic inflammation of the GI tract. The two main clinical forms of IBD are Crohn's disease (CD) and ulcerative colitis (UC).

Epidemiology

Disease prevalence is highest in industrialized countries, in the range of 10–200 cases per 100 000 individuals in North America and Europe.

Areas of involvement

CD: most commonly the terminal ileum, cecum, colon, and perianal area, but any part of the GI tract can be affected. Disease is characterized by segments of normal bowel between affected regions, often referred to as 'skip lesions.'

UC: involves the rectum and extends proximally in a continuous fashion, and always remains restricted to the colon. Is sometimes restricted to the rectum as 'ulcerative proctitis.'

Histology and characterization of inflammation

CD: a transmural inflammation (affecting all layers of the bowel wall), dense infiltration of lymphocytes, and macrophages; granulomas in up to 60% of the patients; fissuring ulceration and submucosal fibrosis. The inflammation is characterized by a Th1 type response, with high levels of IL12, IFN-γ, and TNF-α.

UC: an inflammation affecting the superficial (mucosal) layers of the bowel wall, infiltration of lymphocytes and granulocytes, and loss of goblet cells, accompanied by ulcerations and crypt abscesses. The inflammation is characterized by a more Th2 type response, with high levels of IL5 and IL13 (but not IL4), and high levels of autoantibodies, such as antineutrophil cytoplasmic antibodies (pANCA), indicative of B cell activation.

Clinical features and complications

Both CD and UC present with diarrhea and, in more severe cases, weight loss, and can show extraintestinal inflammatory manifestations in joints, eyes, skin, mouth, and liver. Chronic IBD leads to an increased risk for colon carcinoma, especially in UC.

CD: abdominal pain is usually present and associated with rectal bleeding in $\sim$30% of cases. Narrowing of the gut lumen can lead to strictures and bowel obstruction, abscess formation, and fistulization to skin and internal organs.

UC: severe diarrhea often at night, blood loss, and, in more severe cases, anorexia and weight loss. Progressive loss of peristaltic function leading to rigid colonic tube; in severe cases, this can lead to 'toxic megacolon' and perforations.

Extraintestinal manifestations

IBD patients with HLA-B27 are more likely to develop skin, eye, and joint disease and up to 25% of patients with IBD will have an extraintestinal disease process including the following.

- Ankylosing spondylitis: a degenerative inflammatory disease affecting the joints of the spine causing calcification, ossification, and spinal rigidity.
- Erythema nodosum: red bruise-like dermal patches caused by panniculitis that normally results from hypersensitivity, but can be idiopathic.
- Pyoderma gangrenosum: typified by dermal ulceration and blisters, usually occurring on the legs.
- Uveitis: inflammation of the vascular tunic (uvea) of the eye, often also involving the sclera, retina, and cornea.
- Episcleritis: an inflammatory condition of the connective tissue between the conjunctiva and sclera.
- Sacroiliitis: inflammation of the sacroiliac joint in the lower back.
- Primary sclerosing cholangitis: inflammation, scarring, and narrowing of the bile ducts, causing liver damage.

continuous and confined to the mucosa and submucosa. A significant proportion of CD and UC patients either are or become refractory to standard pharmacotherapy during disease progression.

Refractory CD significantly impairs quality of life (QoL) resulting in debilitating complications and bowel resections. The latter tend to be repeated and the procedure is not considered curative. In UC, total proctocolectomy with ileal pouch–anal anastomosis has become the preferred surgical procedure for severe or refractory UC patients. Although most patients report a good functional outcome and an improvement in QoL after surgery, pouchitis remains the long-term complication. In both instances, there is still a high unmet medical need for therapies with superior efficacy and side effect profiles.

6.28.2 Diagnosis

The prevalence of IBD in developed countries is about 0.1% and the disease places a major burden on public healthcare resources. While considerable evidence suggests that CD and UC are distinct conditions, their similar clinicopathological phenotypes can make diagnosis difficult, especially in the significant proportion of patients that present with 'indeterminate' disease. The diagnosis of CD or UC is based on a combination of examinations[1]:

- Endoscopic sigmoidoscopy: examining the lining of the lower third of the large intestine; colonoscopy, examining the lining of the entire colon and potentially the ileum.
- Endoscopic retrograde cholangiopancreatography: examining the bile ducts in the liver and the pancreatic duct.
- Endoscopic ultrasound: using ultrasound to diagnose perianal fistulas.
- Capsule endoscopy: the state-of-the-art technique with which patients swallow a small capsule with a camera inside to produce images of sections of the small intestine that are beyond the reach of conventional techniques.
- X-ray (including computed tomography (CT) scan and leukocyte scintigraphy), as well as a battery of blood and histology tests.

For patients with CD, the intestinal lining will take the characteristic 'cobblestone' appearance of aphthous and linear ulcers. The CD intestine may also be characterized by strictures and fistulas, in contrast to UC. However, the first-order diagnosis is to determine whether there is an underlying inflammatory process occurring, or whether symptoms are related to some other condition. Flexible sigmoidoscopy or colonoscopy may be reasonably well-tolerated methods of diagnosis in the adult population, but this is less acceptable for pediatric patients. Complete blood count tests often detect high white blood cell counts, which could indicate intestinal inflammation or infection and low blood counts (anemia) that might reveal intestinal bleeding. A hallmark of active UC is the presence of a polymorphonuclear cell infiltrate into epithelial crypts and the lamina propria, an effect that is accompanied by an exudation of inflammatory cells into the colonic lumen. Although elevation is less consistent in UC, elevation in plasma C-reactive protein may also be a sensitive measure of ongoing inflammation. Several fecal assays exist, in particular those for the neutrophil-derived proteins calprotectin and lactoferrin, as sensitive measures that can be used to contribute to diagnosis and assess disease activity. Certain serological markers are also used routinely to reach a disease diagnosis, e.g., the prevalence of perinuclear antineutrophil antibodies in serum from UC patients (in the range of 50–80%), or the presence of anti-*Saccharomyces cerevisiae* antibodies in patients with CD.

6.28.3 Disease Basis: Genetic and Environmental Triggers

The crucial role that activated T cells, macrophages, and monocytes play in the initiation and maintenance of the inflammatory process and tissue damage in IBD has been widely reported.[2] However, the pathogenesis of CD and UC is still incompletely understood. A consistent emerging theme is that IBD results from an abnormal innate mucosal immune response to gut microflora in a genetically susceptible host.

The intestinal tract is unique in the body in regard to its constant exposure to insults from potentially pathogenic microbes and an abundant commensal bacterial microbiota. As sentinel, it performs a delicate balancing act between immune surveillance and waging war on opportunistic infections. In IBD, this balance is disrupted in favor of uncontrolled and excessive recruitment of activated lymphocytes, macrophages, monocytes, and neutrophils to the intestinal lamina propria. Once resident, these cells overproduce among others the pro-inflammatory cytokines tumor necrosis factor-alpha (TNF-α) and the interleukins IL1β, IL6, IL8, and IL18 and a proteolytic soup of matrix metalloproteinases (MMP-1 and MMP-3 especially) which together culminate in the prothrombotic destruction of the gut mucosa.

Genetic linkage analyses have generated some of the most compelling evidence relating a host microfloral trigger and an imbalance in innate and adaptive immunity with a causal link to IBD.[3] In 2001, mutations in the *NOD2/CARD12* gene were mapped to one of the five genetic susceptibility loci linked to CD, IBD1 (OMIM 605956) on chromosome 16q12.[4,5] NOD2 is an intracellular receptor for the bacteria-derived peptidoglycan, muramyl dipeptide (MDP), expressed on macrophages and dendritic cells, and is especially enriched in intestinal Paneth cells. CD patients harboring NOD mutations are especially predisposed to ileal lesions, the major location of Paneth cells. Upon binding to MDP, NOD2 oligomerizes and binds the scaffolding kinase RIP2/RICK. RIP2 then oligomerizes to transmit NOD2's signal directly to the IKK complex and the activation of nuclear factor kappa B (NFκB) (**Figure 1**). Individuals with one of the three major *NOD2* alleles have a two- to fourfold increased risk of developing CD; homozygous and compound heterozygous carriers have an up to 40-fold increased risk. Mutated *NOD2* mutations can be found in 20–30% of CD patients, but despite this strong disease association, mutated *NOD2* alleles are neither necessary nor sufficient for the development of CD, as they also occur in healthy individuals. While mutant *NOD2* is a predisposing factor, it is still not clear whether the effects of these mutations are 'gain-of-function' or 'loss-of-function' on NFκB activation. However, mice harboring the common CD susceptibility allele, *3020insC*, are more sensitive to dextran sodium sulfate (DSS)-induced colitis, and produce elevated amounts of IL1β, IL6, and other products of the NFκB pathway.

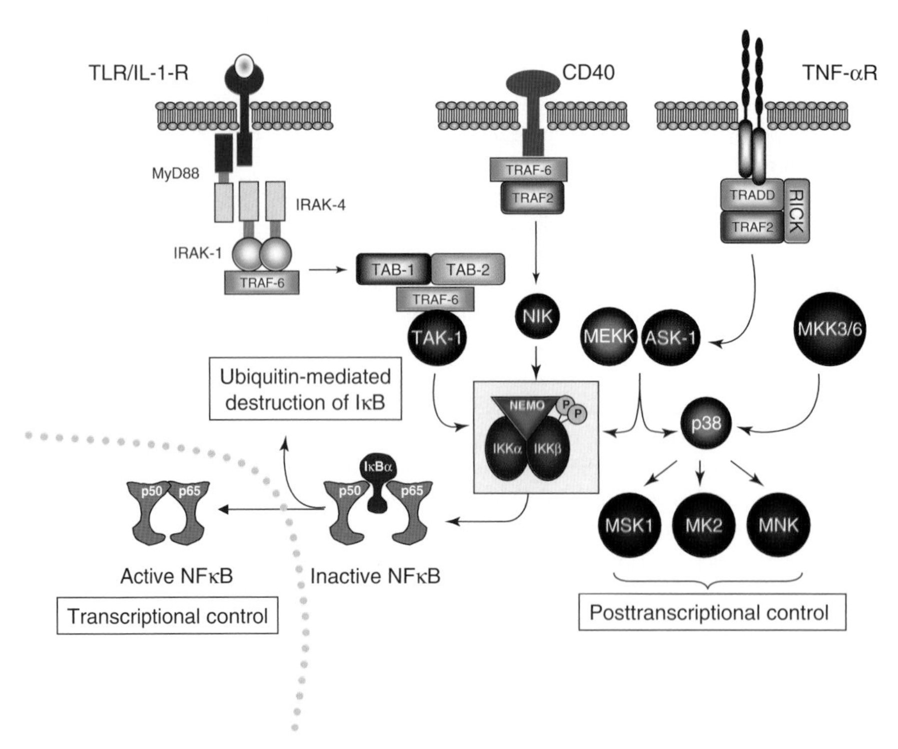

Figure 1 The activation of the nuclear factor kappa B (NFκB) pathway. NFκB heterodimers (p50/p65) are sequestered in the cytoplasm by IκB inhibitory proteins (IκBα). Stimulation by stress-inducing agents, or exposure to inflammatory cytokines (e.g., TNF-α), mitogens, or a diverse array of bacterial and viral pathogens leads to the activation of signaling cascades converging on the IκB kinase (IKK) complex (highlighted box). Phosphorylation of IκB by activated IKK is regulated by multiple pathways (TAK-1, NIK-1, ASK1, etc.) and is a signal for its ubiquitination and proteasome-dependent degradation. Free NFκB dimers translocate to the nucleus where they bind to κB elements, activating the transcription of a variety of genes involved in the control of cell proliferation, survival as well as inflammatory and immune response (transcriptional control). Other components of the signaling cascade (e.g., p38 pathway) also exert control at the level of translation or mRNA stability (posttranscriptional control).

The interplay with the innate immune system extends further into the family of toll-like receptors (TLRs).[6,7] The TLRs are a first-line, pattern-recognition defense, recognizing a diverse array of pathogen-associated molecular patterns, including lipopolysaccharide (LPS), bacterial lipoproteins and lipoteichoic acids, flagellin, double-stranded RNA (dsRNA), single-stranded RNA (ssRNA), and unmethylated CpG DNA of bacteria and viruses. In the gut setting, TLRs have to recognize and respond to both commensal bacteria and pathogenic bacteria in a manner that prevents infection, but does not damage the mucosal tissue. Ligand engagement through TLRs initiates an NFκB-mediated inflammatory response, which is characterized by the recruitment of leukocytes to the site of infection and polarization of Th1 pro-inflammatory responses, through dendritic cell production of IL12. Cell migration and tissue extravasation of cells from the peripheral blood involves a tightly controlled series of events, elicited by chemotactic factors (e.g., IL8, MCP-1, RANTES, MIP-1β). The relevance of this system to IBD, alongside what is being revealed by the function of NOD2, can be illustrated by a number of findings:

- TLR4 is the front-line defense receptor against Gram-negative bacteria and a major LPS transducer. C3H/HeJ mice carry a spontaneous and dominant-negative mutation in TLR4 that results in both LPS unresponsiveness, and a tendency to develop a spontaneous colitis (**Table 2**).[8] Supporting a clinical correlation, an increased association of the TLR4 Asp299Gly polymorphism, which impairs LPS signaling, in patients with UC and CD, compared with normal individuals,[9] has been reported. In patients with UC and CD, TLR4 expression also appears to be substantially upregulated.[10] That the C5a complement-derived anaphylatoxin is a negative regulator of TLR4 activity and C5a antagonists can attenuate block colitis induced by 2,4,6-trinitrobenzene sulfonic acid (TNBS) gives further support to the important role that TLR4 plays in the control of innate immunity and the host inflammatory response.[11,12]
- TLR5 recognizes bacterial flagellins, a major target of the T and B cell response in mouse models of IBD as well as in a subset of patients with CD.[13]

Table 2 Preclinical transgenic experimental disease models

Model		*Selected references*
Spontaneous models		
C3H/HeJBir	Inflammation is restricted to ileocecal lesions and right-hand side of colon	8
SAMP1/Yit mouse	Multifocal ileitis, enteritis, and cecitis	149
Cotton-top tamarin	Spontaneous model of UC, when kept in captivity	93,150
Augmented transgenic models		
NOD2^{2939iC}	Enhanced inflammatory response to DSS	151
STAT4	Inflammation in response to TNP-KLH, exaggerated IL12 signaling and IFN-γ production	152
$CD4^{+}CD45Rb^{hi}$	Adoptive transfer model into SCID or $Rag2^{-/-}$ recipients	153
MyD88	Enhanced inflammatory response to DSS, polarized Th2 response	14
ITF$^{-/-}$	Impaired mucosal repair to DSS	154
TIR8	Enhanced inflammatory response to DSS	15
Spontaneous transgenic models		
IL10$^{-/-}$	Depletion of regulatory T cells and activation of Th1 cells	155
TGFβ$^{-/-}$	Inactivation of regulatory T cell function	156
IL2$^{-/-}$	In surviving animals, colonic inflammation between 6 and 15 weeks of age	157
$TNF\alpha^{\Delta ARE}$	Elevates TNF-α by deletion of the 3′ AU-rich regulatory element.	158
Mdr1a$^{-/-}$	Intestinal efflux pump, mice develop colitis between 12 and 14 weeks of age	17
Gαi2$^{-/-}$	Colitis accompanied by marked increase in Th1 cytokines	159
$Gpx^{-/-}$	Profound colitis and ileitis by deletion of glutathione peroxidase 1 and 2	160
TCRα$^{-/-}$	A Th2-like colitis develops in mice by about 16 weeks of age	161
Spatial STAT3$^{-/-}$	Mice die by 6 weeks of age due to profound CD-like inflammation due to defects in oral tolerance	162
HLA-B27 rat	Inflammation affects stomach, ileum, and the entire colon	20

- TLRs and the related IL1 receptors share a conserved homophilic domain, termed the Toll/IL1 receptor (TIR) recognition domain, which recruits a common adaptor molecule, MyD88, following ligand binding. This leads to the formation of a complex with tumor necrosis factor receptor associated factor-6 (TRAF6) and the IL1 receptor associated protein kinase-1 (IRAK-1) and IRAK-4. IRAK-1 and TRAF-6 then dissociate from this complex and associate with transforming growth factor β (TGFβ)-activated kinase (TAK-1) and the TAK-1 binding proteins, TBP-1 and -2. This in turn activates TAK-1, leading to the phosphorylation and activation of IκB kinase complex and culminating in the degradation of IκB and activation of NFκB (**Figure 1**). Both MyD88 and TIR8 (an intracellular decoy receptor that traps TRAF-6/IRAK-1) knockout mice show increased susceptibility to DSS-induced colitis.[14,15]

A further example of how the TLR system plays a role in augmenting the breakdown in tolerance has come from preclinical models of autoimmune diabetes, in which transgenic mice expressing the lymphocytic choriomeningitis virus glycoprotein (LCMV-GP), under the control of the rat insulin promoter, were infected with LCMV-GP peptide.[16] Infection caused the generation of large numbers of autoreactive cytotoxic $CD8^{+}$ T cells, but did not induce autoimmune diabetes in the absence of treatment with certain TLR ligands. The relevance to IBD is compelling and underscores the significance of the preclinical phenotype observed in the *NOD2* mutant and *MyD88*$^{-/-}$ mice.

Are there other factors that create an abnormal immune response and precipitate chronic intestinal inflammation? The loss of local tolerance may also be due to the loss of regulatory cells (e.g., Th3 or Tr1), as is typified by the phenotypes of the IL10 and TGFβ1 knockout mice. Alternatively, a breakdown in barrier function, which would permit the access of inflammatory bacterial products to the local immune system, might overwhelm normal regulation. One example of how an apparent abnormality in mucosal permeability might contribute to the development of IBD has emerged from studies on the *MDR1a* locus (OMIM 171050). The *Mdr1a*$^{-/-}$ transgenic mouse was created, in part, to understand the contribution of the P-glycoprotein efflux transporter, expressed on intestinal epithelium, on drug absorption. However, these mice also display mucosal barrier dysfunction, characterized by increased basal colonic ion transport, altered tight junction function, and increased bacterial translocation, and consequently develop a severe spontaneous inflammatory colitis.[17] The *MDR1a* gene is expressed on chromosome 7q in human, a known IBD susceptibility locus. It is a highly polymorphic gene and genetic polymorphisms, which lower protein expression, have also been identified in CD and UC patients.[18,19] Whether the Mdr1a efflux pump performs a housekeeping function, by actively reducing mucosal exposure to commensal pathogenic antigens or xenobiotics, or is responsible from some other mechanism in maintaining mucosal barrier function, remains to be determined. Likewise, the identification of other pathways responsible for the potential breakdown in mucosal tolerance in IBD remains to be identified.

In contrast to the *NOD2/CARD15* gene susceptibility locus for CD, no specific gene defect has been linked to UC. A region of the major histocompatibility complex (MHC) locus on chromosome 6p that contains the human leukocyte antigen (HLA) class I and II histocompatability locus, however, has been implicated in increased UC susceptibility. In particular, the association between HLA-B27 and the development of extraintestinal conditions (**Table 2**) such as ankylosing spondylitis in patients with UC is reproduced in a transgenic rat model.[20] CD patients with *HLA-A2, DR1*, and *DQw5* are more likely to develop extraintestinal disease. Furthermore, in some populations, the *DRB1*1502* allele is positively associated with UC and the *DRB1*0103* allele is associated with both severe inflammation and an increased probability of colectomy.

Although the detection of disease-associated variants has greatly advanced our understanding of the primary events that lead to the development of IBD, in a subgroup of patients with CD especially, the implications of these findings for diagnostic algorithms and therapeutic modalities are less clear.

6.28.4 Experimental Disease Models

The number of experimental IBD models is large and continues to expand (**Table 2**).[2,21,22] Simplistically, they all manifest an intense intestinal inflammation, a consequence of an aberrant, chronic immune response triggered by enteric microflora or disruption of the mucosal barrier. Disruption of the mucosal barrier, either genetically, as evidenced by the phenotype of the *Mdr1a*$^{-/-}$ mice, or artificially by exposure of the GI tract to DSS or TNBS generates a fairly reproducible colitic phenotype resembling UC. However, animals treated in pathogen-free environments or coadministered antibiotics have a far lower disease severity and incidence.

The *Mdr1a*$^{-/-}$ mouse model is one of the few genetic disease models that develop a colitic phenotype in the absence of immune dysfunction. This is in contrast to virtually all other genetically targeted mice, which develop colitis as a result of either impaired immune function, a cytokine imbalance, or colitis that can be induced by reconstituting naive ($CD4^{+}CD45Rb^{hi}$) T cells into severe combined immunodeficiency (SCID) mice. Most of the experimental models of IBD in mice have inflammation only in the colon and resemble UC. The one striking exception to this is the SAMP1/Yit mouse and the derivative SAMP1/YitFc strain. These mice spontaneously develop a Crohn's-like transmural ileitis as early as 10 weeks of age, accompanied by prominent muscular hypertrophy, fibrosis, and activation of mesenteric lymph node lymphocytes, which produce high levels of interferon γ (IFN-γ). Furthermore, a subgroup of SAMP1/YitFc mice (~5%) also develops perianal fistulating disease.

The common underlying characteristic of the transgenic models of colitis is a chronic uncontrolled, mostly T cell-mediated inflammatory response. In many incidences, these models have helped inform the potential of new IBD treatment approaches and modalities, whether investigated clinically,[1,23] such as anti-TNF-α antagonists (Enbrel, CDP-870, Remicade/infliximab, and Humira/adalimumab), anti-IL12 mAbs,[24] rIL10,[25] intracellular adhesion molecule (ICAM) antisense oligonucelotides (alicaforsen/ISIS 2302), or in preclinical models.

6.28.5 Clinical Trial Issues

Investigating the efficacy of potential new therapies in the treatment of IBD through clinical trials is not straightforward. The heterogeneity of the condition, in particular CD, and the large number of potential endpoints that can be employed in studies are just two of the complexities that need to be considered in clinical trial design. Furthermore, clinical trials are predominantly of two types, those studying efficacy to treat active disease and those designed to study the efficacy in maintaining disease remission, and are likely to have different objectives and challenges. There is only limited regulatory guidance in this area.

6.28.5.1 Patient Selection

When selecting subjects for inclusion in an IBD clinical trial, many aspects of their disease need be considered. Firstly, the differentiation between CD and UC patients needs to be made. As already described, this is not necessarily completely straightforward, as no one feature is either pathognomonic or appears exclusively in one of these two conditions. However, in the majority of patients, a combination of clinical, laboratory, histopathological, and radiological observations can establish a reliable diagnosis. Still, in approximately 20% of patients, this cannot be made.

Other important factors relating to patient selection are severity and extent of disease. As already discussed, CD can affect any part of the GI tract and consideration must be given to whether patients with disease limited to a specific segment of the bowel (e.g., the colon) or to only a single site will be included. In addition, the presence or absence of complicated disease, including strictures and fistulae, also need to be considered, as penetrating complicated disease tends to be most resistant to medical treatment. In UC, the distal and continuous nature of the inflammation makes disease extent easier to define. The current consensus appears to be that in UC, patients with inflammation extending beyond the rectum (i.e., not limited to a proctitis) are suitable for inclusion. The chronic, yet relapsing, nature of both conditions adds an extra complexity. The therapeutic approach under investigation can also sometimes influence patient selection, as, for example, investigational drugs delivered by enema are likely to have reduced efficacy in all but the most distal instances of the disease.

Disease severity is an important consideration in patient selection. In addition to influencing the magnitude of the response to the investigational agent, disease severity can impact on issues such as likely size of the placebo response and selection of an appropriate comparator agent. Specifically, regression toward the mean may give rise to false positive results in patients with severe disease. Conversely, patients with mild disease may be more likely to spontaneously enter disease remission as part of the natural history of their disease, a factor that becomes more problematic in studies of longer duration. These factors, as well as others, contribute to the 20–50% placebo response observed in clinical trials in IBD. The selection of appropriate comparators is also influenced by disease severity. In severe disease, the use of a placebo is almost certainly unethical, unless no proven therapy exists, necessitating comparison with current standard of care. This can be avoided by studying patients in whom therapy has failed, such as steroid-refractory disease. In mild and moderate disease, the use of placebo is commonplace and can be justified, in particular by the downward pressure such a design has on sample size, thus reducing the number of IBD patients exposed to investigational agents.

6.28.5.2 Assessing Disease Activity

The selection of endpoints is of fundamental importance in the design of a clinical trial in IBD. A daunting array of options face the clinical investigator at the point of designing the clinical trial, ranging from QoL indices, measures of clinical disease activity, and endoscopic scores, to more detailed measures of systemic or mucosal inflammation. In addition, there are increasing opportunities to employ biomarkers, which can give valuable information concerning the pharmacological activity of the molecule under investigation.

The most frequently used primary outcomes are either disease remission or clinical improvement, both of these parameters being defined prospectively as either a fall beneath a threshold, or a reduction of a preset magnitude, on a clinical disease measure. Many such clinical disease indices exist, both for CD and UC, but none are without issues or can be considered unequivocally superior to the others. The limitations in some of these indices may limit their usefulness to the selection of patients for inclusion in clinical trials, rather than in demonstrating clinical utility for investigational drugs. In addition, the definition of remission or of the magnitude of reduction that constitutes a clinically significant improvement may be somewhat arbitrary.

In UC, the most commonly applied index is that of Truelove and Witts,[26] used initially to assess the efficacy of the corticosteroids. Although widely used, this index has a number of weaknesses, particularly the difficulty in defining improvement within its categorical framework. A number of indices based on modifications to this original index have also been suggested, the most widely known being the Powell–Tuck index,[27] which includes an endoscopic score as well as assessment of a greater range of symptoms and generating a quantitative index.

Scoring systems derived more recently vary mainly in their complexity. The desire for simplified indices led to the development of the Mayo scoring system,[28] which produces a quantitative index from four domains: bowel movements, fecal blood, and endoscopic score, together with a physician assessment, the latter of which is subjective by definition. In addition, the domains may be more closely associated with distal disease and may underestimate more proximal disease. Conversely, the complex index of Seo *et al.*[29] produces a quantitative index from a complex mathematical algorithm that combines symptom measures (fecal blood and bowel movement) with biochemical ones (hemoglobin, albumin, and erythrocyte sedimentation rate (ESR)). The simple clinical colitis index[30] has been constructed from the five clinical domains that correlate most closely with the Powell–Tuck index, although it also correlates closely with Seo index. It appears to have utility as a screening tool, but its usefulness as an outcome index in clinical trials is still unclear.

In CD, the use of the Crohn's disease activity index (CDAI) predominates. This index is constructed from eight domains that are given a relative weighting, and involves the collection of patient data through the use of a diary. Disease severity bands have been constructed within a scale of 0–750. Disease remission has been attributed to a score of less than 150 and clinical improvement defined as a reduction of greater than 70 points (or more recently 100 points), although the clinical significance of such a reduction in patients with severe disease is questionable. Used widely in clinical trials, criticisms still abound on the use of CDAI. These include the near reliance on subjective symptoms, the complexity of the index, and the need for a 1-week patient diary. Studies have also demonstrated substantial interobserver variation in scoring. The poor representation of parameters relating to perianal disease makes improvements in this troublesome presentation difficult to assess using CDAI and so a specific perianal disease activity index has been devised.

Several additional indices have been proposed, aimed at either simplifying the CDAI or reducing its subjectivity. The best known of these are the Harvey–Bradshaw and Oxford indices,[31,32] both of which remove the relative weightings of the parameters, while the Harvey–Bradshaw removes the need for the diary. Both correlate well with the CDAI.

The limitations of clinical scoring systems of disease activity strongly suggest that the demonstration of clinical efficacy should not rely entirely on such indices. The use of endoscopy can provide valuable support and to that end endoscopic indices of severity have been developed. In UC, the Baron score was first described in the mid-1960s, but is used with few modifications to the present day. For CD, the Crohn's disease endoscopic index of severity (CDEIS) has been developed but, although used in therapeutic trials, has to date not been used widely outside of Europe. CDEIS is complicated and time-consuming and so attempts have been made to simplify its use, most recently in producing the simple endoscopic score for Crohn's disease (SES-CD).[33] The SES-CD is simple to use and is highly correlated with the CDEIS, C-reactive protein, and clinical parameters.

Visualizing the mucosa of the small intestine in CD remains an obstacle to routine endoscopic evaluation in clinical trials in this patient population. The potential of wireless capsule endoscopy to overcome this hurdle remains to be fully assessed.

Further supportive data of efficacy can be provided from mucosal biopsies. The degree of inflammatory cell infiltration can be assessed microscopically and these data have been used to support the beneficial effect of a number of agents in development for the treatment of IBD. While relatively straightforward in UC patients, obtaining mucosal biopsy samples from CD patients with small bowel involvement is challenging.

6.28.5.3 Quality of Life

IBD patients have significantly lower health-related QoL across a number of domains when compared with healthy controls. Two IBD-specific questionnaires have been developed to allow the assessment of health-related QoL in this patient group: the inflammatory bowel disease questionnaire (IBDQ) and the rating form of inflammatory bowel disease patient concerns (RFIPCs). The IBDQ has been used in many clinical trials and is considered to be robust in measuring therapeutic efficacy. A short form of this questionnaire has been developed, but not yet tested in a clinical trial setting.

6.28.5.4 Use of Biomarkers

There are many ways in which biomarkers can be usefully employed during the clinical development of new treatments for IBD.[34] They can be used to select patients for study or to increase the homogeneity of the cohort. Clinical indices are a crude instrument for patient selection; a patient cohort defined clinically as having mild disease may show great variability in levels of intestinal inflammation. A fecal biomarker of intestinal inflammation, such as the neutrophil marker calprotectin, may help concentrate the cohort with regard to those patients with active intestinal inflammation. Biomarkers can also be used to define an acute response to drug treatment. For example, treatment of patients with UC or CD with either the anti-TNF-α antibodies CDP-571 or infliximab, or the anti-α_4 integrin antibody natalizumab, has been shown to reduce circulating C-reactive protein (CRP) levels, sometimes rapidly.[35] Of course, the clinical significance of these acute changes needs to be carefully understood and is not a replacement for improvements in disease activity assessments.

Biomarkers can also be used to confirm the mechanistic activity of an investigational drug and even identify pharmacologically active doses for further study. For example, acute administration of natalizumab has been shown to evoke a rise in circulating levels of T lymphocytes in both CD and UC patients.[36,37] These data confirm that natalizumab can inhibit the homing of these cells to the inflamed GI tract, increasing the levels in the peripheral circulation. Such mechanistic biomarkers do not need to be limited to analysis of blood. Biopsy material can provide a useful matrix in which to demonstrate the activity of an investigational drug using biomarkers of activity. For instance, treatment with infliximab has been shown to downregulate markers of mucosal inflammation from biopsies of CD patients, when studied ex vivo.[38]

The level of certain soluble biomarkers may also be useful in predicting therapeutic response. A number of clinical trials, with agents that target TNF-α, have demonstrated that the most significant improvements in clinical disease activity appear to be in those patients with elevations in CRP. These observations were first made in clinical trials with infliximab,[39] but have since been replicated with CDP-571.[35] Based on these data, threshold values for CRP have been proposed as predictors of therapeutic response.

6.28.5.5 Other Considerations

The use of concurrent medications is a significant challenge when conducting clinical trials in IBD. The question of what medications to allow and what to exclude is encountered almost universally, as it is extremely unlikely that any patient eligible for entry onto a clinical trial will not already be taking some medication. For example, the concomitant use of immunosuppressants can reduce the apparent immunogenicity of a biological therapy, thereby potentially overstating its safety profile, or in the context of a 'placebo' group, the magnitude of the 'placebo' effect may be enhanced. Still, there is no answer that is universally acceptable concerning concurrent medications and some issues will be specific to certain types of trial. In literature, allowable concurrent mediations have ranged from 5-aminosalicyclic acid (5-ASAs) only, to any medication as long as it is stable in terms of dose regimen at the time of entry into the trial and not increased during the study.

6.28.6 Current Treatment

Despite the introduction of anti-TNF-α antagonist therapies, such as Remicade/infliximab for steroid refractory and fistulating CD, the treatment of IBD has not substantially changed in the last decades (**Table 3**). Front-line therapy in the management of patients with mild-to-moderate UC and CD, and as a maintenance therapy to prevent disease relapse in UC, are the 5-ASAs (e.g., mesalazine, sulfasalazine, and olsalazine). The 5-ASAs appear to exert their anti-inflammatory activity by inhibiting activation of NFκB. For disease in the distal bowel, multiple delayed and sustained release formulations, as well as prodrugs such as olsalazine, have been designed to release the majority of an oral dose directly in the distal ileum/colon, thus preventing topical exposure in the proximal small intestine. For rectal disease, rectal foams are also administered. The most recently introduced prodrug, balsalazide (**1**), contains azo bonds that are cleaved by colonic bacterial azo-reductases and release the active 5-ASA mesalazine locally.

Although mild disease can be treated successfully with 5-ASAs, many patients eventually require corticosteroids to control symptoms. Oral glucocorticoids (e.g., prednisolone) are some of the most effective therapies for inducing clinical remission in patients with active CD and UC. However, the adverse effects (including obesity, osteoporosis, hypertension, and adrenal suppression) of chronic corticosteroid treatment are both extremely undesirable and invariably culminate in the development of clinically challenging, steroid-refractory disease. Budesonide (**2**) is a highly effective second-generation oral corticosteroid, but was engineered with metabolic vulnerability, to enable a topical mode of action coupled with extensive first-pass metabolism and low systemic exposure. Despite this potential improvement in safety burden for this class of agents, approximately 30% of patients will also be unable to discontinue steroid therapy without disease exacerbation, and approximately 20% will become steroid resistant or will not respond to steroids.

Table 3 Standard treatment options for active inflammatory bowel disease

Active disease	*CD treatment*	*Active disease*	*UC treatment*
Mild–moderate	Oral aminosalicylate	Mild–moderate	
	Metronidazole	Distal	Oral or topical aminosalicylate
	Oral corticosteroids		Topical corticosteroids
	Azathioprine or 6-mercaptopurine (6-MP)	Extensive colitis	Oral aminosalicylate
Severe	Oral or intravenous corticosteroids	Moderate–severe	
	Intravenous Remicade/infliximab	Distal	Oral or topical aminosalicylate
			Oral or topical corticosteroids
		Extensive colitis	Oral corticosteroids
Fistulating	Metronidazole or antibiotic therapy	Severe or fulminant	Intravenous corticosteroids or cyclosporin
	Intravenous Remicade/infliximab		
	Azathioprine or 6-MP		
Maintenance	Oral aminosalicylate	Maintenance	
	Metronidazole	Distal	Topical or oral aminosalicylate
	Methotrexate		Azathioprine or 6-MP
	Azathioprine or 6-MP		
		Extensive colitis	Oral aminosalicylate
			Azathioprine or 6-MP

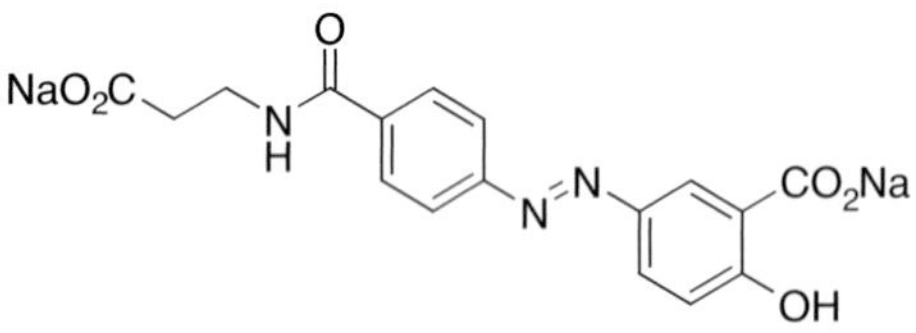

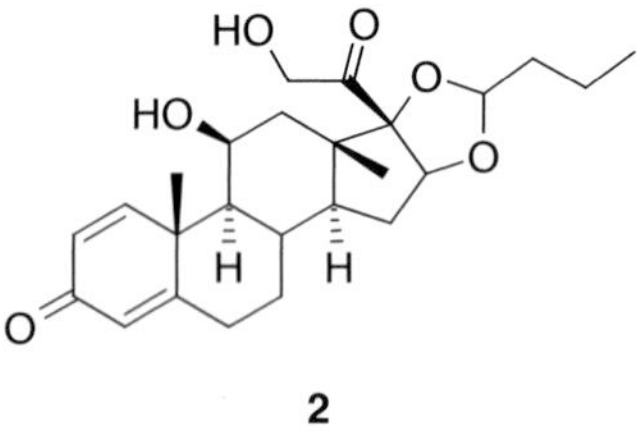

Outside of the approval and use of 5-ASAs, corticosteroids, and immunosuppressants, emerging treatments for IBD have become increasingly dominated by biological approaches. For instance, the use of anti-TNF-α antagonists such as Remicade/infliximab and Enbrel has revolutionized the treatment of clinically challenging steroid-refractory and fisulating CD; similar data are still awaited for utility in UC. Notwithstanding the rapid clinical efficacy observed with this class of agents, approximately 30% of CD patients do not respond to anti-TNF-α drugs and there are concerns over their long-term safety profile (opportunistic infections, increased carcinoma risk, immunogenicity).

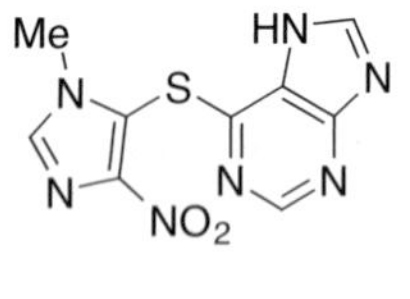

The chronic use of immunosuppressants, e.g., azathioprine (**3**), cyclosporin (**4**), methotrexate (**5**), and occasionally mycophenolate mofetil (**6**) and tacrolimus (**7**), also requires constant monitoring both for levels of drug and for side effects.[40] Thiopurines, such as 6-mercaptopurine and azathioprine, are metabolized to 6-thioguanine, blocking the nucleotide binding site and activity of G-protein Rac1 and inducing T cell apoptosis. While azathioprine is highly effective at inducing and maintaining disease remission, side effects include allergic reactions, pancreatitis, myelosuppression, nausea, infections, hepatoxicity, and malignancy. Bone marrow suppression is related to levels of 6-thioguanine, and therefore monitoring of complete blood counts at regular intervals is suggested. Of the other most commonly used immunosuppressants, methotrexate also induces myelosuppression and additionally it is associated with hepatoxicity, an increased risk of Epstein–Barr virus (EBV)-lymphoma, and opportunistic infections. Finally, cyclosporin is generally reserved for patients with severe UC that are refractory to steroids, but its chronic use is associated with renal insufficiency, hypertension, nausea, and opportunistic infections. As maintenance therapy for severe IBD, none of these agents is without considerable liabilities.

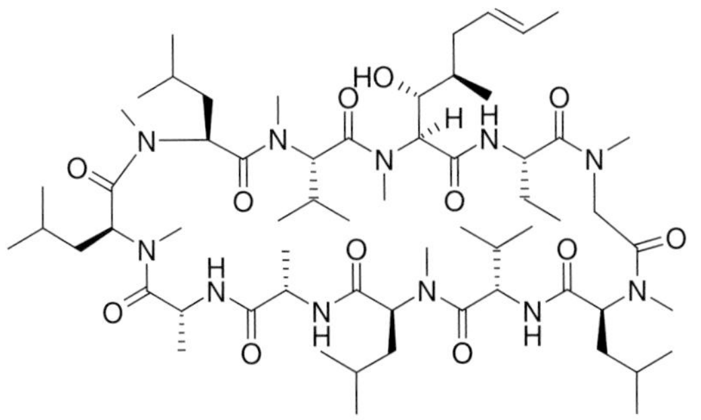

4 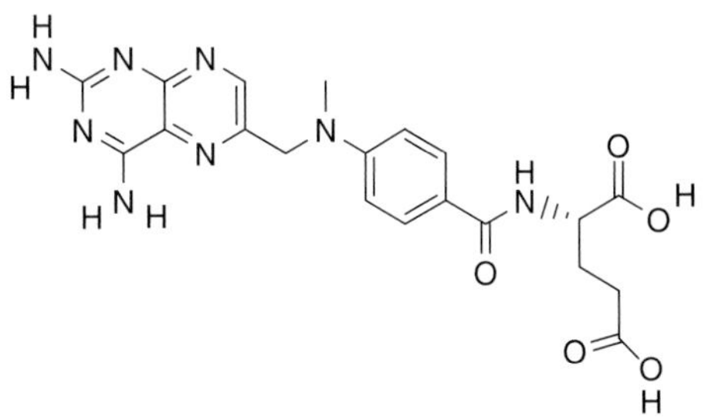 **5**

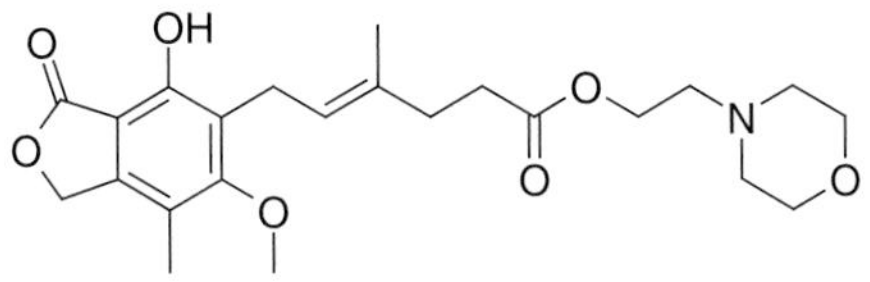

6 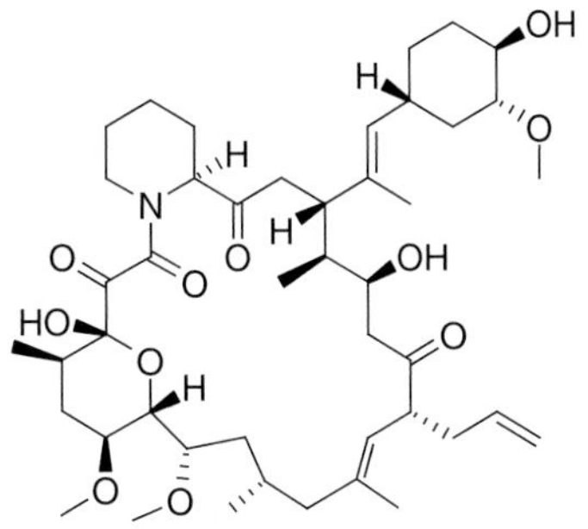 **7**

6.28.7 Unmet Medical Need

IBD affects approximately 1 million people in the USA, with an estimated indirect cost from loss of work alone in excess of $4 billion per year. The therapeutic goal in the management of IBD is to treat the active symptomatic disease of all patients and to restore mucosal integrity and function. The treatment of mild-to-moderate disease is adequately managed with current 5-ASA and topical corticosteroid therapies (**Table 3**). However, invariably the disease progresses and becomes more difficult to control. Furthermore, a significant proportion of CD and UC patients become refractory to standard treatment and this necessitates the use of powerful immunosuppressant drugs, such as azathioprine, methotrexate, or 6-mercaptopurine, or resective surgery. Patients are highly motivated to seek out alternative therapies with the promise of efficacy and which avoid the inevitability of surgery. The disease management of pediatric IBD remains a significant unmet challenge.

There is an increasing clinical use and reliance on anti-TNF-α agents for the treatment of steroid refractory, or fistulating CD. The effectiveness of these agents in severe CD has revolutionized the treatment of this patient group, despite their cost and notable side effects. For the induction and maintenance of remission infliximab is given as an intravenous infusion. It is a chimeric mouse/human immunoglobulin G1 (IgG1) anti-TNF-α antibody with high levels of immunogenicity, which is both a cause of erythema and reactions during infusion, and reduces the pharmacokinetic half-life, increasing the risk of systemic infections. While other anti-TNF-α approaches are in clinical development (Onercept, CDP-870, ISIS 104838, etc.), none is likely to avoid the core issue relating to increased risks of opportunistic infection and malignancy. Additionally, approximately 30% of CD patients will not respond to infliximab and currently the agent has not been approved for use in patients with UC, although clinical trials are ongoing. Embracing the precedence set by novel biological approaches to disease management, other agents such as Leukine/sargramostim (recombinant human granulocyte macrophage colony stimulating factor (GM-CSF)), daclizumab (humanized anti-IL2 receptor antibody), fontolizumab (humanized anti-interferon gamma (IFN-γ) antibody), CNTO 1275 and ABT-874 (anti-IL12 monoclonal antibodies (mAbs)), MLN-02 (humanized anti-$\alpha 4\beta 7$ antibody), and MDX-1100 (fully human anti-IP-10 antibody) have entered clinical trials in order to address the shortfall in safe and effective agents that have the widest utility within the IBD community. The long duration of action of biologics, compared with small-molecule approaches, is both their potential attraction and downfall. None so far is devoid of immunogenicity issues, even the fully human anti-TNF-α monoclonal antibody Humira/adalimumab. The discovery of novel small-molecule agents, which can fill the void in efficacy, is a rich and active area of research.

6.28.8 New Research Areas

As an inflammatory disease, IBD shares an association with other diseases such as rheumatoid arthritis, psoriasis, or chronic obstructive pulmonary disease (COPD). As a consequence, every potential protein and pathway that has been associated with or assessed in a preclinical model or clinical setting has invariably also been implicated in the disease biology of IBD or its partners. The disease pathophysiology of IBD has many unique features, but to the casual observer, the therapeutic possibilities seem endless. This section reviews how our better understanding of the disease is leading to improvements in existing therapeutic approaches, as well as identifying new ones.

6.28.8.1 Re-Establishing Mucosal Tolerance: Probiotics, Prebiotics, Worms, and Toll-Like Receptor Modulators

The coincidental rise in IBD in the developing world with the decline of helminthic parasitic infection is one aspect of a hygiene hypothesis that suggests that in a sanitary world the mucosal immune system has become functionally less important. An idle GI immune system may not necessarily be desirable. Indeed, reports indicating that administration of the helminths *Trichuris suis* or *Heligmosomoides polygyrus* is effective at reversing the inflammation and inducing clinical remission[41] directly support this notion.

The luminal microfloral environment profoundly influences the extent and severity of inflammation observed in many of the described preclinical models of colitis.[42] Alternatively, administration of beneficial bacterial species (probiotics), poorly absorbed dietary oligosaccharides (prebiotics), or combined agents (symbiotics) can restore the balance of beneficial *Lactobacillus* and *Bifidobacterium* species over the pathogenic flora contributing to disease. For instance, when either IL10$^{-/-}$ mice are treated with antibiotics or *Lactobacillus* GG,[43,44] or DSS-induced colitic mice are treated with a preparation of *Bifidobacterium longum*,[45] the extent of the inflammation is dramatically reduced. The most common chronic complication affecting UC patients following ileal pouch–anal anastamosis is the development of pouchitis, a further idiopathic inflammatory condition of the neorectal ileal mucosa. Antibiotics, in particular metronidazole and ciprofloxacin, are often used as first-line therapy, but increasingly probiotics are being considered. For instance the mixture VSL#3 (6 g day^{-1}), containing a number of strains of *Lactobacillus, Bifidobacterium*, and *Streptococcus*, induces remission.[46] While these observations are encouraging, and probiotics can contribute to sustaining clinical remission, a single combination of components has not been identified that induces remission and has clear benefit in the treatment of CD and UC.

As noted previously, the TLR system discriminates between many different microbial signatures and its role in controlling both normal gut immune homeostasis and pathogenesis is becoming dissected. Mucosal TLR4 expression is upregulated in UC patients and both the genetic epidemiology and the phenotype of the TLR4$^{-/-}$ mouse in response to DSS[47] would suggest that TLR4 antagonists might be one approach to the treatment of IBD, which builds on the experience with probiotics. E5564/Eritoran (8)[48] is a synthetic TLR4 antagonist currently in clinical assessment for sepsis, but an assessment either preclinically or clinically in IBD models is awaited. In contrast, CRX-526, a close structural homolog to E5564, prevents the development of colonic inflammation in both DSS-induced colitis and the *Mdr1a*$^{-/-}$ models, as well as suppressing TNF-α release from monocytes in vitro and in vivo.[49] Similarly, TLR9 agonists (unmethylated CpG DNA of bacteria and viruses, or immunostimulatory oligodeoxynucleotides, such as CPG 7909 [5′-TCG-TCG-TTT-TGT-CGT-TTT-GTC-GTT-3′][50]) prevent or ameliorate the severity of colonic inflammation in either the IL10$^{-/-}$ or DSS-induced models of colitis.[51,52] The mechanism by which TLR9 ligands are able to achieve this seems to be in part the ability to trigger an anti-inflammatory IFN-α/β response.[6] Initial clinical studies have also recently confirmed the beneficial effects of type I interferons in the treatment of UC,[53] supporting the notion that UC is a Th2 polarized disease. Our understanding of both the functional roles of specific TLR family members and the discovery of selective tools to investigate their function in disease is still in its infancy. However, the compelling connection between the components of innate and adaptive immunity (TLR, NOD2, etc.), the effect of probiotics, and the rise in IBD in a sanitized world, suggests that TLR-based therapeutics may have a significant contribution in future disease management.

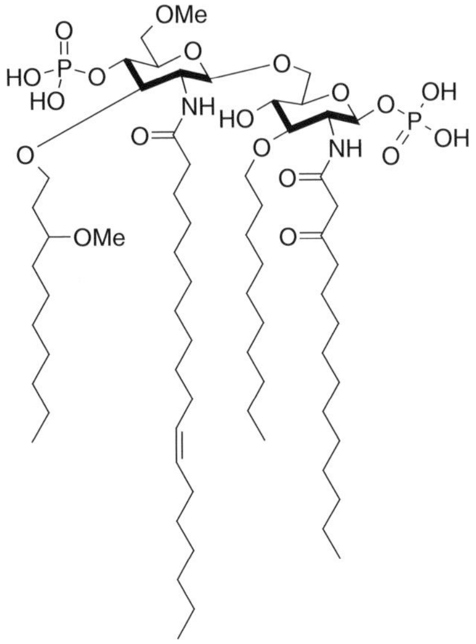

8 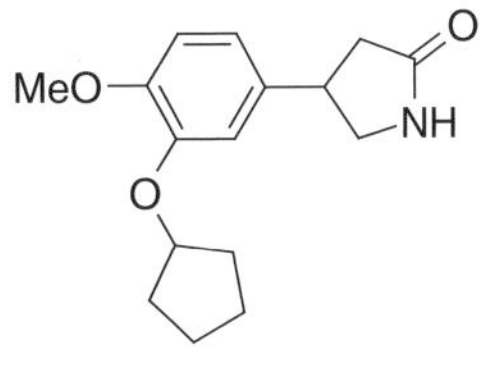 **9** 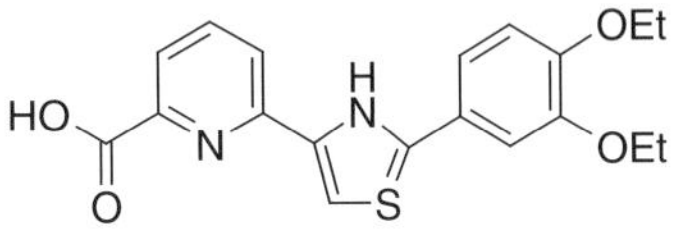 **10**

6.28.8.2 Anti-Inflammatory Approaches

NFκB is a transcription factor that is rapidly activated in response to pro-inflammatory stimuli, infections, and physical and chemical stressors (**Figure 1**). NFκB participates in the induction of numerous immunoregulatory genes, whose products include pro-inflammatory cytokines, growth factors, chemokines, adhesion molecules, and enzymes that produce secondary inflammatory mediators. Activated nuclear NFκB is frequently detected at sites of inflammation and infection. Multiple pathways control the activation and translation of NFκB-dependent reporter genes: the IκB kinase (IKK) complex, p38 kinase, TGFβ-activated kinase (TAK-1), etc. (**Figure 1**). For some time, aberrant activation of the NFκB transcriptional pathway has been considered a major mechanism in the disease pathogenesis of IBD. When antisense constructs to silence the p65 NFκB subunit are introduced into the murine IL10 knockout model of colitis, the disease severity greatly improves.[54] Evidence from a gut ischemic–reperfusion model[55] also indicates that the NFκB pathway maintains mucosal barrier function, protection from reactive oxygen species, and apoptosis induced by inflammatory mediators. The modulation of the pathways that converge on NFκB is an important therapeutic avenue for attenuating inflammation.

6.28.8.2.1 Phosphodiesterase-4 (PDE4) inhibitors

Cellular levels of cyclic AMP (cAMP) regulate T cell activation and endothelial adhesion, the synthesis and release of pro-inflammatory cytokines (IFN-γ, TNF-α, IL8, etc.) from macrophages and monocytes, and the generation of nitrous oxide and superoxide from neutrophils. PDE4A, PDE4B, and PDE4D are the predominant cAMP-hydrolyzing PDEs in most inflammatory cells, and their inhibition is associated with broad anti-inflammatory effects. The contribution of PDE7 activity to this process is now no longer considered significant.[56] In the rat DSS model of colitis, the PDE4 inhibitors rolipram (**9**) ($10\,mg\,kg^{-1}$, four times per day, intraperitoneally) and OPC-6535/Tetomilast (**10**) ($1\,mg\,kg^{-1}$, four times per day, orally) significantly reduce the extent of mucosal erosion, bleeding, and inflammation.[57,58]

There is tremendous structural diversity reported for PDE4 inhibitors,[59] but, as a class, their therapeutic utility and clinical development had been historically dogged by dose-limiting nausea and emetic side effects. It is now believed that tolerance to PDE4 inhibitors is dependent on two components, the relative receptor subtype selectivity for PDE4B (anti-inflammatory) over PDE4D (emesis), and the plasma C_{max} exposure. Cilomilast/SB-207499 (**11**), developed by GSK, and roflumilast/BY-217 (**12**) are both in phase III clinical trials for COPD. Cilomilast is more selective for PDE4D than the other subtypes, whereas roflumilast, and its major active metabolite N-oxide, are both more potent and do not discriminate between subtypes. Roflumilast is significantly better tolerated than cilomilast. Together, these data point to the potential clinical value of PDE4B selective inhibitors in inflammatory diseases, although neither agent has been assessed in patients with IBD. In contrast, tetomilast (25 or 50 mg per day orally for 8 weeks) has been assessed clinically in a phase II study with 186 UC patients.[60] Clinical improvement (defined as a three-point reduction in disease activity index) was achieved by 55% and 48% of patients taking 25 and 50 mg tetomilast, respectively, compared with 40% of patients taking placebo. A significantly higher proportion of patients taking tetomilast had reduced rectal bleeding and entered disease remission than those on placebo. The selectivity of tetomilast for the PDE4 isotypes is unknown, but drug-related side effects (nausea) accounted for 8% withdrawals in the 50 mg group. The future medicinal chemistry challenge for IBD is to prise apart efficacy from emesis. This might be achieved by a number of strategies, either by identifying pharmacophores that distinguish the close sequence and structural homologies of PDE4B and PDE4D, or by the design of less central nervous system (CNS) penetrant inhibitors, or PDE4 inhibitors that act through a topical mode of action and have lower systemic exposure.

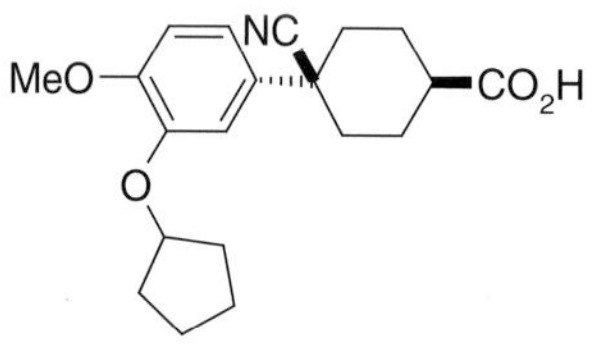

11 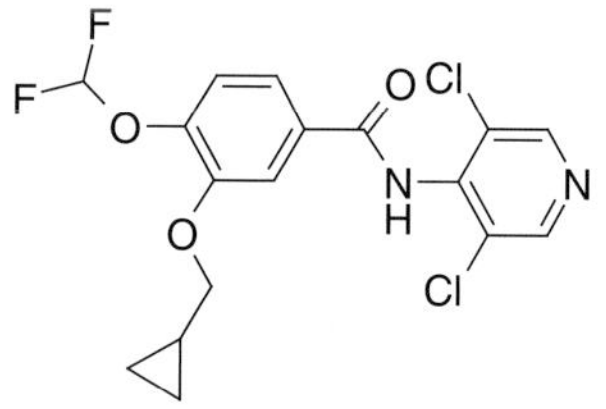 **12**

6.28.8.2.2 IκB kinase inhibitors

NFκB exists in a latent state in the cytoplasm and requires a signaling pathway for activation. Several distinct NFκB-activating pathways have emerged, and all rely on sequential kinase activation culminating in the activation of the classical IKK complex. Once activated, the IKK complex phosphorylates the NFκB inhibitor IκBα, subsequently targeting its polyubiquitination and degradation (**Figure 1**). Once activated, NFκB translocates to the nucleus to regulate the expression of multiple target genes.

The IKK complex consists of two kinases, IKKα and IKKβ, and a scaffold protein, NEMO/IKKγ. IKKα and IKKβ share a high degree of structural homology, but IKKβ has the higher kinase activity and is generally believed to be the critical component in the activation of NFκB. The deletion of IKKα of IKKβ in knockout mice causes pleiotropic and significant defects in organogenesis and immune function. IKKβ$^{-/-}$ mice, for instance, fail to mount an effective antibody response to T cell dependent and independent antigens, but also die during embryonic development with massive liver degeneration. Furthermore, individuals affected by the X-linked condition anhidrotic dysplasia with immunodeficiency have significant developmental defects due to impaired NFκB activation as a result of mutations in NEMO/IKKγ.[61]

A large number of commonly used anti-inflammatory agents such as aspirin, salicylates, and suldinac are reported to be competitive IKK inhibitors of the ATP binding site. However, more potent and selective agents are emerging from high-throughput screening (HTS) exercises. One of the first novel IKK inhibitors described was the quinoxaline BMS-345541 (**13**) (IKKβ $IC_{50}=300$ nM), which is approximately 10-fold more selective for IKKβ than IKKα and appears to inhibit IKK activity allosterically. In a collagen-induced arthritis model, BMS-345541 (100 mg kg^{-1}) blocked the induction of IL1β mRNA and reduced joint destruction.[62]

A series of 4-piperidin-3-ylpyridines and thiophencarboxamides, typified by **14** and **15**, respectively, emerged from lead optimization from an IKKβ HTS.[63–65] In contrast to BMS-345541, **14** ($K_i=2$ nM) and **15** ($IC_{50}=18$ nM[66]), are competitive ATP inhibitors of IKKβ and are selective over IKKα and other closely related kinases. Compound **14** has respectable absorption, distribution, metabolism, and excretion (ADME) properties in mice and rats and blocks

LPS-induced TNF-α production and septic shock ($30\,mg\,kg^{-1}$, orally) as well as allergen-induced inflammatory cytokine production, eosinophil and neutrophil recruitment, and airway inflammation.[67] A similar assessment of **15** in a murine collagen-induced arthritis model ($10\,mg\,kg^{-1}$, twice per day, intraperitoneally) has also demonstrated efficacy equivalent to etanercept. Significantly, no toxicological changes in histopathology have been reported in liver, heart, lung, and bone marrow of mice treated chronically with BMS-345541[62] or **14**.[67] These data contrast those obtained from IKK knockout mice, since, in addition to its role in protection from infection, basal NFκB activation is required to protect cells from apoptosis.[55] However, the early safety observations with BMS-345541 and **14** are encouraging and suggest that IKK inhibitors may have reduced potential for side effects, in contrast to those observed with glucocorticoids. Studies with selective IKK inhibitors are still in their infancy, and the tool agents described here have yet to be evaluated in preclinical colitis models.

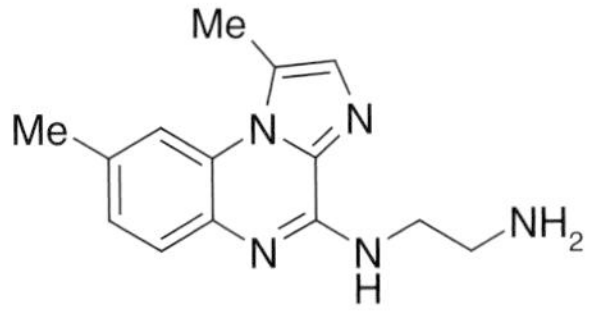

13 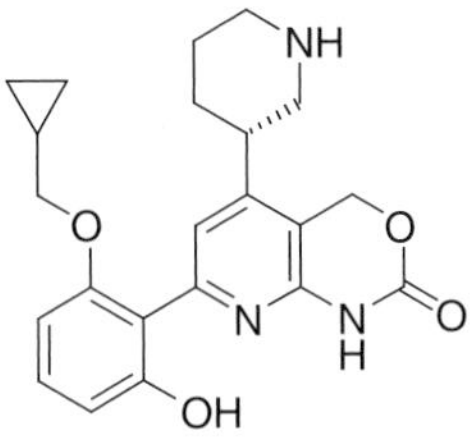 **14** 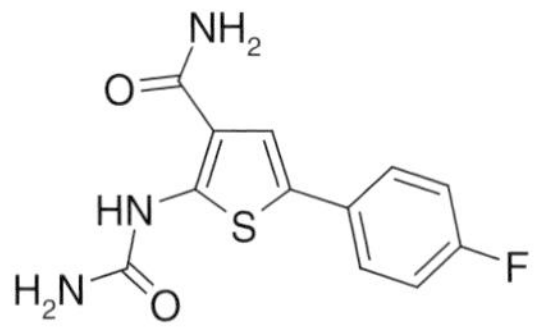 **15**

6.28.8.2.3 p38 kinase inhibitors

The p38 family of protein kinases lie downstream of the IL1β receptor and LPS/TLR to regulate, in part, the transcription/translation of pro-inflammatory cytokines and other NFκB-dependent genes, including IL1β, IL6, and TNF-α (**Figure 1**). Since the discovery of the founding member, the pyridylimidazole SB-203580 (**16**), the development of potent and selective p38 kinase inhibitors has been an area of active research and clinical assessment.[68] The dual p38/JNK tetra-guanylhydrazone inhibitor CNI-1493 (**17**) ($25\,mg\,kg^{-1}\,day^{-1}$, intravenously for 12 days) has been assessed in patients with severe CD,[69] inducing clinical remission in five out of 12 patients at 8 weeks and endoscopic improvement in all but one patient. More potent and selective inhibitors have been developed from either the pyridylimidazole pharmacophore, such as SB-242235 (**18**), L-790070 (**19**), and a series of recently discovered bisamides, exemplified by **20** (p38α $IC_{50} = 28\,nM$) which is effective ($30\,mg\,kg^{-1}$, orally) at reducing TNF-α production in Balb/c mice challenged with staphylococcal enterotoxin B.[70] The most clinically advanced agents, Doramapimod/BIRB-796 (**21**)[71] and VX-745 (**22**), have both demonstrated proof of concept in phase II trails for rheumatoid arthritis. BIRB-796 is an exceptionally high-affinity p38α kinase inhibitor ($K_d = 0.1\,nM$), for which clinical trials in patients with CD are under way. Vertex and Kissei were developing VX-745, the pyrimido pyridazinone that selectively inhibits the α and β isoforms of p38 ($IC_{50} = 10\,nM$ and 220 nM, respectively), but not p38γ ($IC_{50} > 20\,\mu M$), for rheumatoid arthritis. Although VX-745 demonstrated excellent efficacy in inflammatory rheumatoid arthritis models (and by analogy activity in colitic models), VX-745 was also associated with neurological toxicity. Genetic deletion of p38α in mice results in embryonic lethality, and it is unclear whether the toxicity observed for VX-745 is mechanism or pharmacophore mediated. Further preclinical work is required to determine whether systemic or topical p38 inhibitors, which target the gut mucosa specifically and akin to the mode of action of budesonide over other glucocorticoids, have utility in the treatment of IBD.

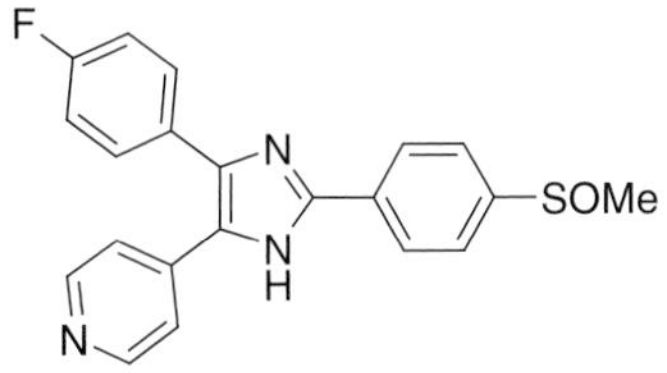

16 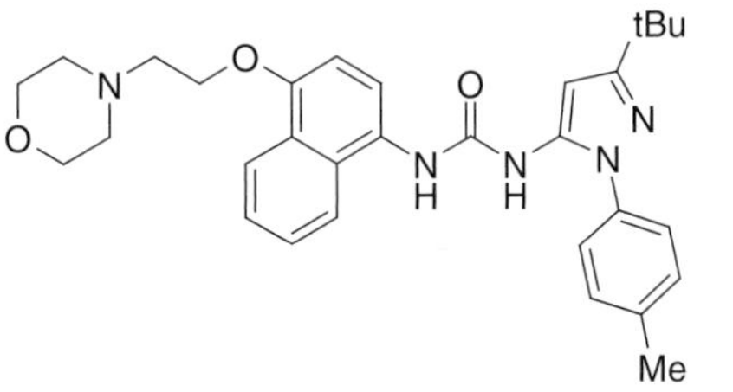 **17**

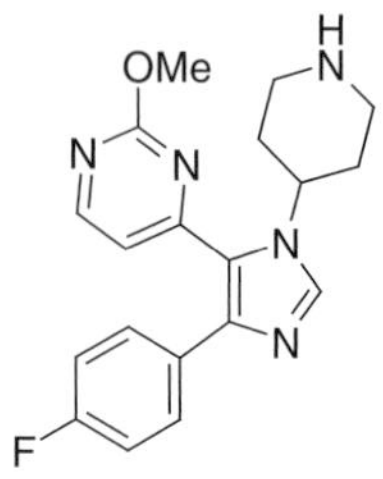

18 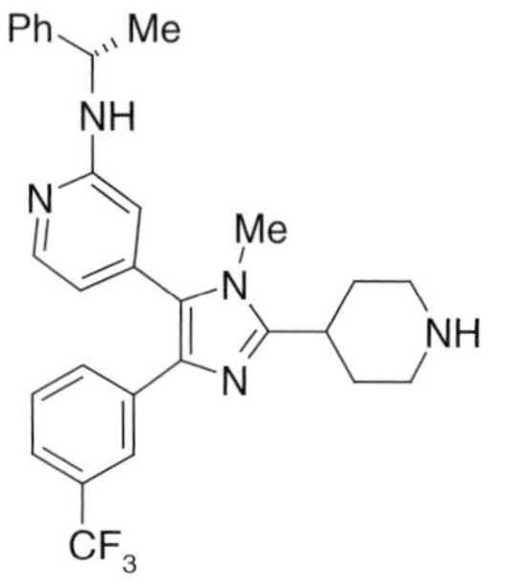 19 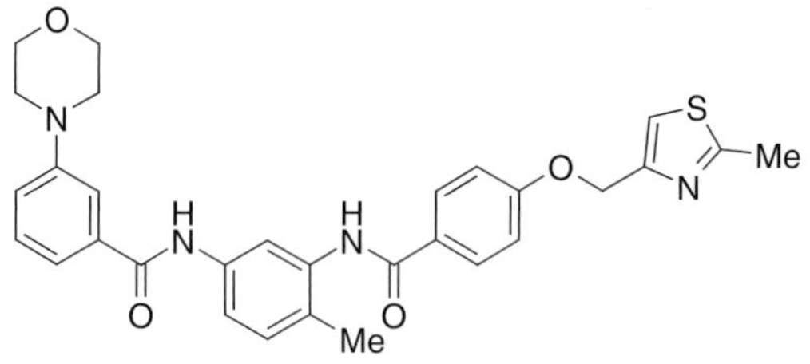 20

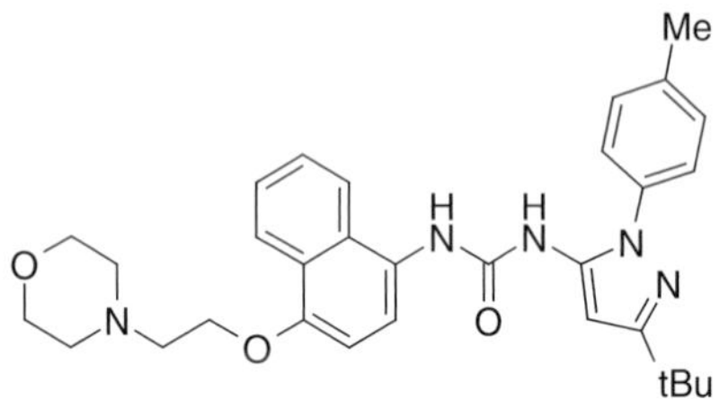

21 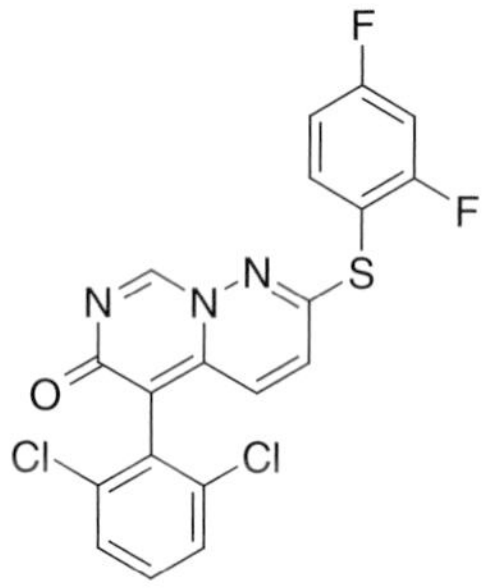 22

6.28.8.2.4 Interleukin-1β converting enzyme (ICE) inhibitors

The overproduction of active IL1β and IL18, observed in patients with IBD, is dependent on processing by the ICE/caspase-1 protease. Chronic administration of DSS to $ICE^{-/-}$ mice fails to induce colitis, an effect accompanied by reduced cell activation in the draining mesenteric lymph nodes and a significant reduction in colonic IL18, IFN-γ, and IL1β.[72] There are relatively few reported examples of selective ICE inhibitors. However, pralnacasan/VX-740 (**23**), designed on the peptide scaffold AcYVAD-CHO, is an ethyl-hemiacetal orally bioavailable prodrug ICE inhibitor currently in phase II development for rheumatoid arthritis. Once absorbed (reportedly, $F = 43\%$), it is rapidly hydrolyzed to the active but reversible aldehyde protease inhibitor. In the acute DSS-induced mouse model of colitis, pralnacasan administration (50 mg kg^{-1} twice daily intraperitoneally) resulted in significant reduction in disease activity.[72] Similarly, VX-765 (**24**) is also under early clinical assessment for inflammatory conditions,[73] but clearly further development work on this class of agents is warranted on the basis of the preclinical and clinical data reported so far.

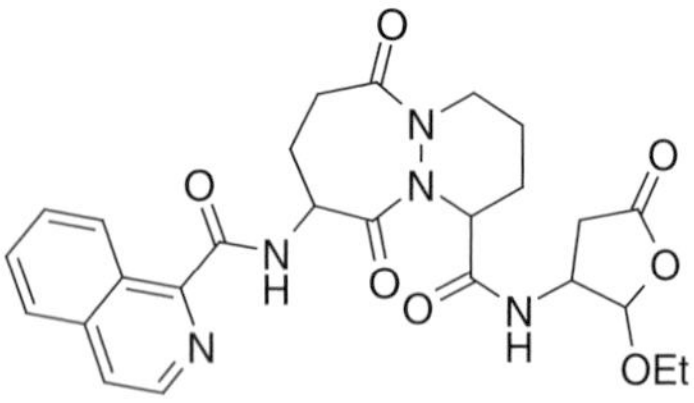

23 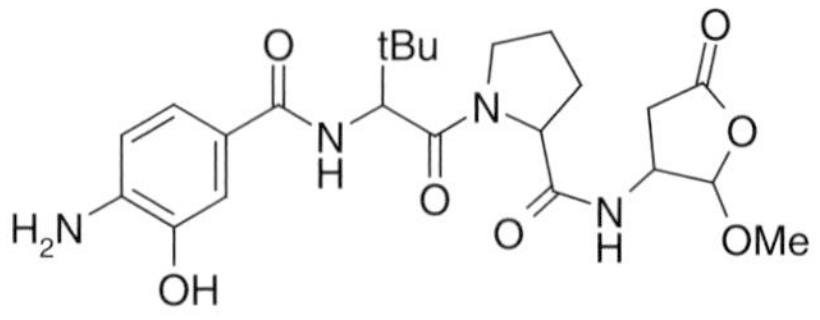 24

6.28.8.2.5 Poly(ADP-ribose) polymerase inhibitors

Poly(ADP-ribose) polymerase (PARP-1 and PARP-2) plays an essential role in the cellular response to genotoxic stress, facilitating DNA repair by binding to DNA breaks and attracting DNA repair proteins to the site of damage. Under conditions of oxidative stress and inflammation, PARP becomes overactivated and consumes NAD^+ and ATP, culminating in cell dysfunction and necrosis. The potential contribution of the PARP family to IBD is emphasized by a number of observations. When administered to the IL10$^{-/-}$ mouse, ISIS 110251 (25 mg kg^{-1} subcutaneously), a modified 2′-*O*-(2-methoxy) ethyl antisense oligonucleotide inhibitor of PARP-2 (CTT-TTG-CTT-TGT-TGA-GGT-CA) normalized colonic epithelial barrier and transport function, reduced pro-inflammatory cytokine secretion and inducible nitrous oxide synthase (iNOS) activity, and attenuated inflammation.[74] In contrast, PARP-1$^{-/-}$ mice exhibit a reduced tendency to inflammation and colon damage in the TNBS-induced model of colitis.[75] GP-1650 (**25**) ($K_i = 60$ nM), one of the founding members and most extensively studied of the isoquinoline series of nicotinamide mimetic PARP-1 inhibitors, significantly reduced the inflammatory response to rectal 4,6-dinitrobenzene sulfonic (DNBS) acid (40 mg kg^{-1} day^{-1}, intraperitoneally).[76] Several other classes of PARP inhibitors have emerged with improved potency and solubility, among them AG14361 (**26**) ($K_i < 5$ nM) that is effective at sensitizing cancer cells to cytotoxicity induced by chemotherapeutics,[77] but few studies have investigated the wider utility of these chemotypes in inflammatory models.

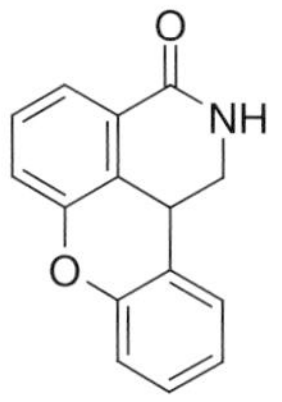

25 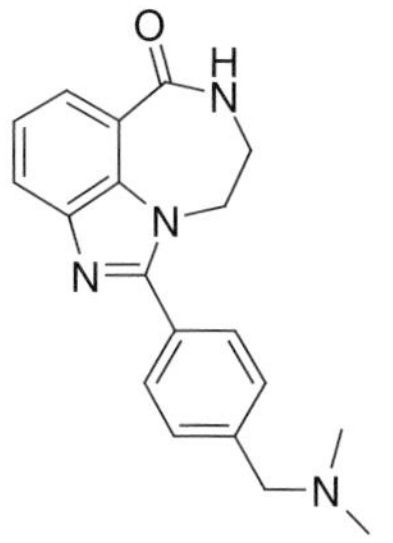 **26**

This may reflect a wider genotoxic concern, that in the development of novel treatment paradigms for oncology, disabling a genomic surveillance system may be an acceptable approach to acute therapy. In the context of chronic inflammatory diseases, this may be less desirable, but clearly further preclinical safety and efficacy data are awaited to place the importance of this approach alongside others for the treatment of IBD.

6.28.8.2.6 Cannabinoid receptor agonists/fatty acid amide hydrolase (FAAH) inhibitors

While it is well accepted that cigarette smoking protects against the development of UC, historically marijuana has also been used as adjunctive therapy to CD, to inhibit GI motility and secretion and to treat diarrhea, as well as to elicit anti-inflammatory and analgesic activity. Naturally occurring endocannabinoids are produced by the cleavage of membrane fatty acids, in particular arachidonic acids, that have varying specificities for the two cannabinoid 7 transmembrane (7TM) receptors, CB1 and CB2. Arachidonoyl ethanolamide (also referred to as anandamide) is produced and released by activated immune cells; its degradation is regulated by FAAH.

The tissue distribution of CB1 receptors accounts for the well-known psychotropic and peripheral effects of the cannabinoids. CB1 is abundantly expressed in the CNS, but it is also expressed in the human ileum and colon, and expression increases at sites of inflammation. CB1 and CB2 are both expressed on immune cells, where the effects of Δ^9-tetrahydrocannabinol (THC) (**27**), the psychoactive component in marijuana, are to reduce a range of pro-inflammatory cytokines including TNF-α, IL1β, IL6, IL12, and GM-CSF.

HU-210 (**28**), a nonselective cannabinoid agonist and THC derivative, reduces colonic inflammation, decreases the inflammatory cell influx, and reverses the electrophysiological signs of smooth muscle excitability, which mimic the intestinal dysmotility that is typical of the clinical condition, when administered to DNBS-induced colitic mice.[78] In the same study, CB1$^{-/-}$ mice or wild-type mice treated with the CB1 antagonist SR141716A/Acomplia developed a more severe inflammatory response to DNBS or DSS, while FAAH$^{-/-}$ animals were significantly protected. Collectively, these data indicate that the endocannabinoid system, through ligation of CB1 receptors, performs a protective role in the GI system during inflammation. However, anandamide has affects on other receptors and it cannot be excluded that decreased inflammation in FAAH$^{-/-}$ mice and the anti-inflammatory actions of HU-210 are due to CB2 receptor or TRPV1 calcium channel activation (*see* Section 6.28.8.4.2).

Treating a chronic inflammatory disease by stimulating CB1 receptors may be somewhat undesirable given the psychosocial, behavioral, and addictive properties of marijuana, unless their efficacy can be peripherally restricted. However, recent evidence has suggested that enhancing endocannabinoid activity by inhibition of FAAH activity may not have the same potential for abuse.[79]

FAAH is a membrane-bound enzyme, and uniquely within the serine hydrolase family, it bears an unusual Ser-Ser-Lys triad catalytic mechanism. As a class, URB597 (**29**, $IC_{50} = 4$ nM) is one of the best-characterized FAAH inhibitors,[79] but LY-2077855 (**30**, $IC_{50} = 41$ nM) is also reported to inhibit anandamide uptake in RBL-2H3 cells and inhibits FAAH activity,[80] and a series of exceptionally potent and selective 2-pyridyloxazole derivatives, typified by **31**, ($IC_{50} = 4.7$ nM) have also been described.[81] An assessment in against hyperalgesic and inflammatory endpoints in colitic models for this class of molecules is awaited.

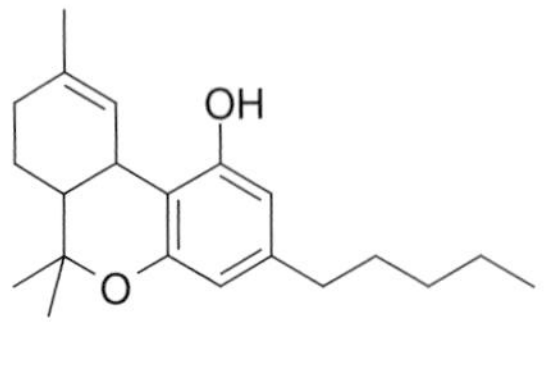

27 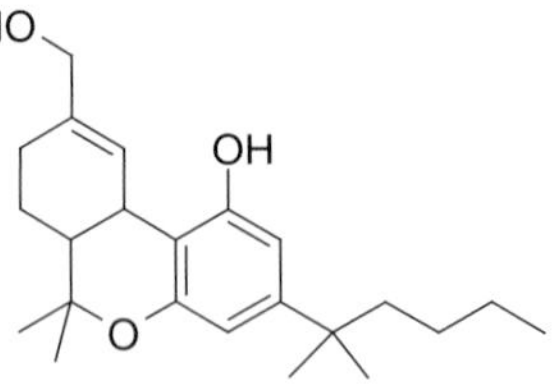 **28**

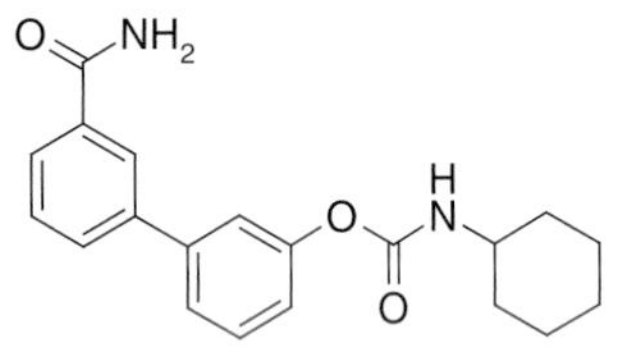

29 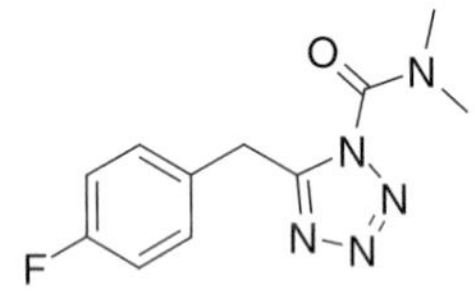 **30** 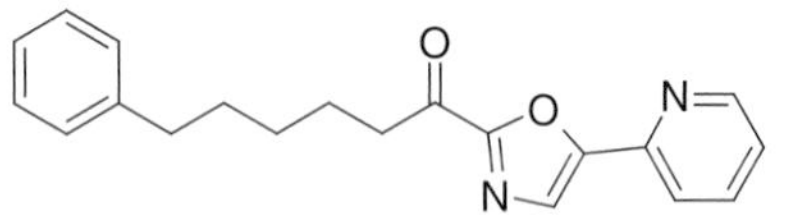 **31**

6.28.8.2.7 Peroxisome proliferator-activated receptor (PPAR)/retinoid X receptor (RXR) agonists

Recent evidence has suggested that the cyclopentenone prostaglandin 15-deoxy-$\Delta^{12,14}$ PGJ_2 functions as an early anti-inflammatory signal, by directly regulating the activation of IKKβ[82] and subsequently NFκB (*see* Section 6.28.8.2.2). 15-Deoxy-$\Delta^{12,14}$ PGJ_2 is a somewhat selective eicosanoid ligand for PPAR-γ, the nuclear receptor that plays a central role in adipocyte differentiation and insulin sensitivity, and is an important therapeutic target for diabetes. To activate transcription, PPAR-γ requires heterodimerization with RXR. The RXR–PPARγ heterodimers are permissive to activation by both PPAR-γ and RXR ligands, and several of the biological effects of PPAR-γ activation can be reproduced by specific RXR agonists or rexinoids.

PPAR-γ is activated by fatty acid derivatives and the class of insulin-sensitizing thiazolidinediones, such as troglitazone (**32**), rosiglitazone (**33**), and pioglitazone (**34**). In addition to its role in glucose homeostasis, PPAR-γ is expressed at high levels in colonic epithelium. Delivery of a replication-deficient adenovirus overexpressing PPAR-γ reduces TNBS-induced colitis,[83] and in contrast either PPAR-γ or $RXR\alpha^{+/-}$ mice display a significantly enhanced susceptibility to TNBS-induced colitis compared with their wild-type littermates.[84] When administered to either DSS- or TNBS-treated mice, troglitazone ($100\,mg\,kg^{-1}\,day^{-1}$, orally) or rosiglitazone ($20\,mg\,kg^{-1}\,day^{-1}$, orally) are effective in reducing histological lesions and intestinal bleeding.[84,85] A synergistic improvement in the extent of colonic inflammation induced by TNBS could be achieved by coadministering the RXRα agonist LG-101305,[84] a synergy that has been observed in a number of preclinical inflammatory, diabetes, and oncology models.

The significance of these preclinical observations is emphasized by further clinical findings. Firstly, PPAR-γ expression levels in patients with UC appear to be impaired compared to normals or patients with CD.[86] In a 12-week open label study with 15 UC patients with mild-to-moderate disease, rosiglitazone (4 mg twice per day orally) was able to induce clinical remission in 27% of patients.[87] Perhaps the most compelling evidence, however, of the role for PPAR-γ agonists has emerged from studies which have dissected the mechanism of action of the 5-ASA class of anti-inflammatory agents that are used extensively to treat IBD.[88] Desreumaux *et al.* demonstrated that $PPAR\text{-}\gamma^{+/-}$ mice treated with TNBS are refractory to the effects of 5-ASA and that 5-ASA is able to displace rosiglitazone from PPAR-γ with a value of K_i that is similar to the clinical exposure levels which are needed for efficacy data that collectively point to PPAR-γ being the possible molecular target of the 5-ASA drugs.[84] This observation could lead to the development of second-generation PPAR-γ ligands, as the clinical utility of the 5-ASA class of drugs is limited by their efficacy and side effect profile.

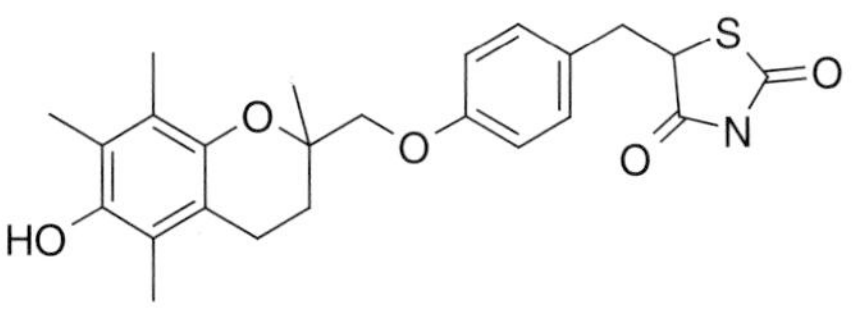

32 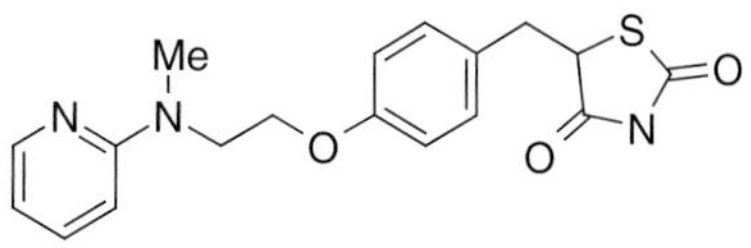 33

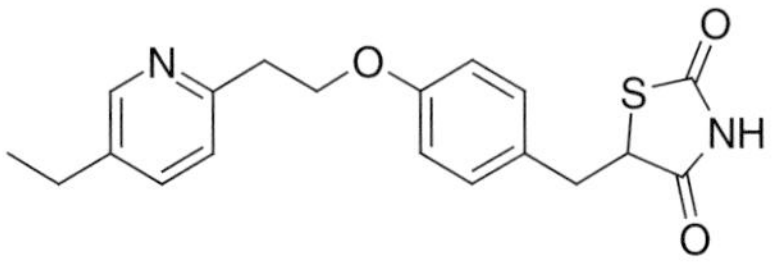

34 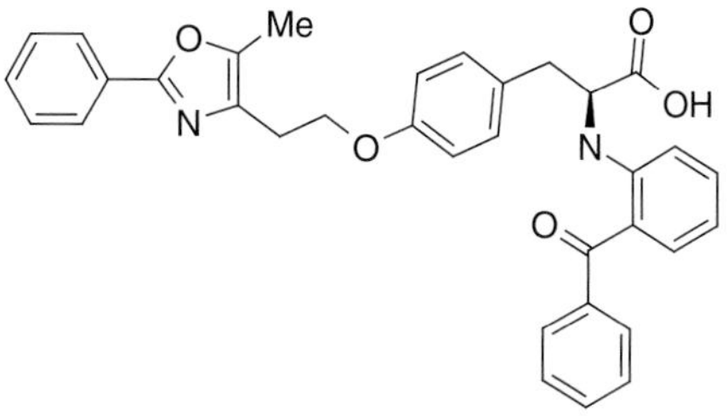 35

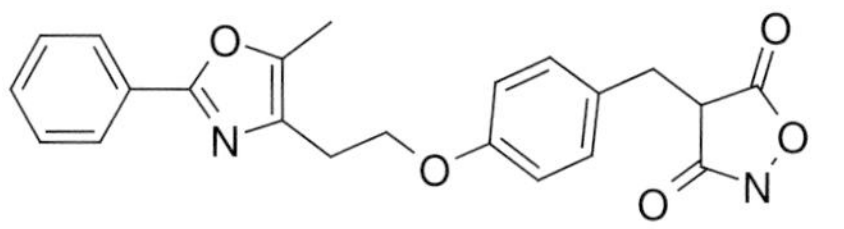

36 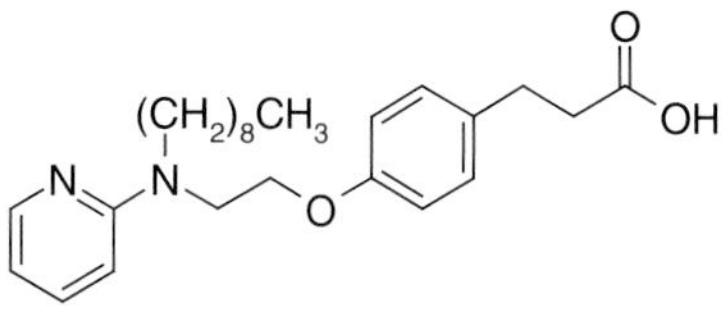 37

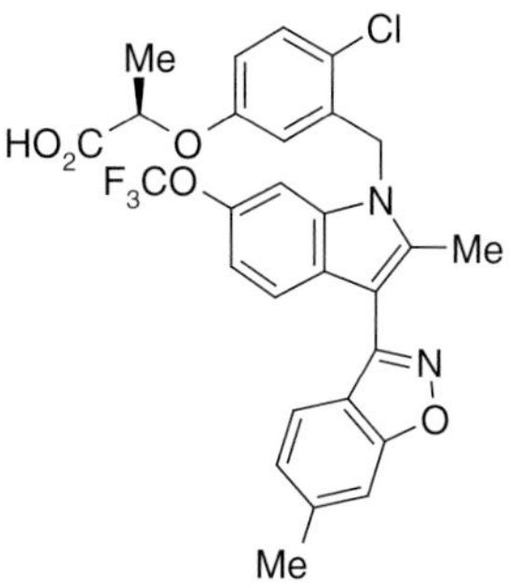

38 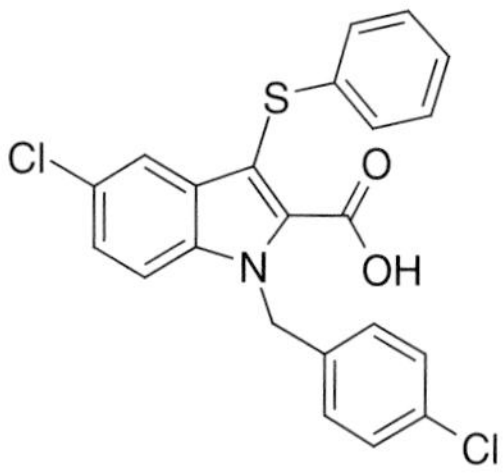 39

A serious issue with troglitazone, which led to its withdrawal, is hepatotoxicity. Troglitazone both activates the pregnane X receptor, to increase CYP3A4 levels, and is also a CYP3A4 substrate and generates a reactive quinone metabolite. Rosiglitazone is also reported to be associated with liver, cardiovascular, and hematological toxicities as well as edema and weight gain. It is not clear whether any of these clinical adverse effects are mediated by the thiazolidinedione moiety, but many series of nonthiazolidenedione-containing and selective PPAR-γ agonists have been described, such as the full agonists GW262570/Farglitazar (**35**) (previously in phase III trials), JTT-501/Reglitazar (**36**) previously in phase II/III trials), and **37**,[89] as well as the partial agonists **38** ($EC_{50} = 3$ nM, 24%)[90] and **39** ($EC_{50} = 250$ nM, 40%).[91] Further work is needed to establish the potential benefits of nonthiazolidenedione-containing PPAR-γ agonists as well as the extent of agonism that can achieve efficacy in inflammation over effects on glucose and lipid homeostasis.

6.28.8.3 Targeting Lymphocyte Extravasation

The excessive recruitment of activated leukocytes is one of the core components in the inflammatory disease pathogenesis. Several therapeutic approaches have emerged in recent years that specifically target this process. The potential advantage of this approach compared with anti-inflammatory mechanisms is that, conceptually, by specifically restricting lymphocyte trafficking, the immune system can still function normally and is not immunosuppressed.

6.28.8.3.1 Integrin receptor antagonists

Mature lymphocytes continuously circulate between peripheral blood, lymphoid tissues, and organs. This process (homing) is needed both to develop an effective immune system and to maintain immune surveillance. The recirculation is not a random event, but a process that is controlled by a repertoire of integrins, glycoproteins, and chemokine receptors, expressed on leukocytes, and cell adhesion molecules, selectins, chemokines, and chemoattractants, expressed or secreted from specialized endothelium.

The very late antigen-4 (VLA-4), also known as integrin α4β1, is expressed on monocytes, T and B lymphocytes, basophils, and eosinophils, and is involved in the massive recruitment of granulocytes in different pathological conditions, including IBD, multiple sclerosis, and asthma. VLA-4 interacts with its endogenous ligand VCAM-1 during chronic inflammation, and blockade of VLA-4 /VCAM-1 interaction is a potential target for immunosuppression. The α4β7 integrin, expressed on populations of lymphocytes, is a unique GI tract homing signal and specifically interacts with MAdCAM expressed on specialized gut endothelium.[92,93] At sites of inflammation, the expression of MAdCAM increases and this leads to an elevation in the recruitment of α4β7 lymphocytes. Blocking anti-MAdCAM or anti-α4β7 antibodies is effective in preventing rolling and adhesion of lymphocytes to the gut endothelium and reduce inflammation in animal models of colitis.[93,94] These findings have led to the development of the humanized monoclonal antibodies natalizumab (anti-α4) and MLN02 (anti-α4β7), both of which have demonstrated clinical efficacy in patients with UC and CD (*see* 6.09 Neuromuscular/Autoimmune Disorders). By inhibiting the recruitment of key inflammatory cells these agents have the potential to improve IBD symptoms and rebalance oral tolerance without inducing systemic immune suppression. Two classes of VLA-4/α4β7 antagonists have so far been reported and reviewed[95]: β-amino acid derivatives containing a diaryl urea moiety (BIO-1211 (**40**) α4β1 IC_{50} = 4 nM) and acylphenylalanine derivatives (examples including **41**, α4β1 IC_{50} = 0.5 nM; α4β7 IC_{50} = 4.4 nM[96], TR-14035 (**42**), α4β1 IC_{50} = 7 nM; α4β7 IC_{50} = 87 nM[97], **43**, α4β1 IC_{50} = 0.34 nM; α4β7 IC_{50} = 40.3 nM, F = 25%).[98] Both series are Arg-Gly-Asp (RGD) mimetics and require a carboxylic acid to interact with a metal ion in the MIDAS domain on the integrin β-chain. With a few exceptions,[98,99] however, these compounds tend to be of high molecular weight, have high metabolic clearance, are highly protein bound, and poorly bioavailable.

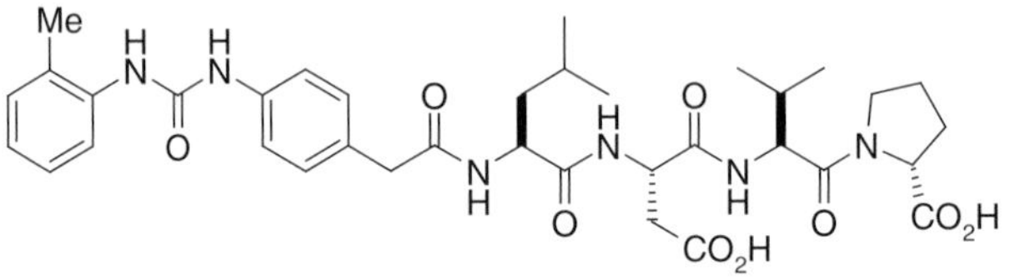

40

TR-14035 (10 mg kg^{-1}, intravenously) has been shown to block the binding of α4β7^{+} cells to MAdCAM in Peyer's patches by intravital microscopy,[100] but like many of the compounds described from these series, TR-14035 is rapidly metabolized and poorly bioavailable. In contrast, the *N*-benzylpyroglutamyl 4-substituted-L-phenylalanine derivative, R411 (RO0270608, Roche, IC_{50} = 0.37 nM), is currently under phase II clinical evaluation for asthma. The absolute oral bioavailability reported for R411 is approximately 27% and the terminal half-life is approximately 7.5 h, far higher than perhaps would have been anticipated from a high-molecular-weight acid.[101] Although the structure–activity relationship (SAR) for the acylphenylalanine series is fairly well developed, with many literature examples of potent dual α4β1/α4β7 and selective α4β1 inhibitors, few molecules have been engineered that sufficiently discriminate α4β7 from α4β1. One exception,[97] the trifluoromethansulfonamide (**44**), is nearly 400-fold selective for α4β7 relative to α4β1 (IC_{50} = 0.5 nM and 192 nM, respectively). Given the recent marketing withdrawal of natalizumab, due to fatal cases of progressive multifocal leukoencephalopathy in patients taking natalizumab and β-interferon,[102] perhaps this also highlights the desire to achieve selectivity over VLA-4 and focus more effort on α4β7-selective inhibitors.

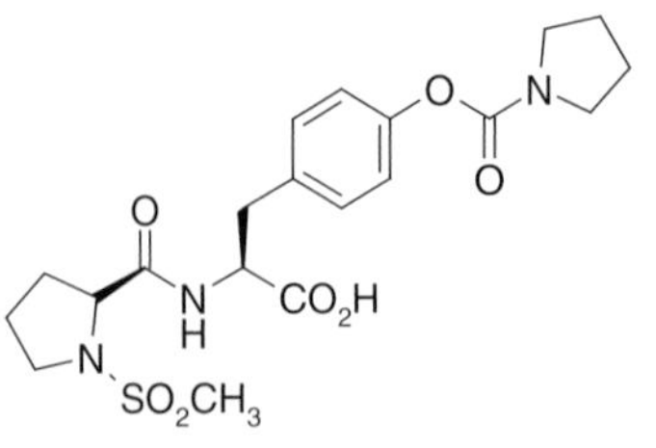

41

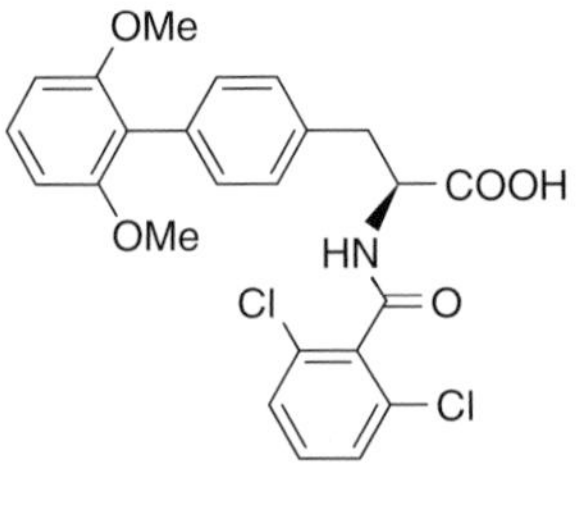

42

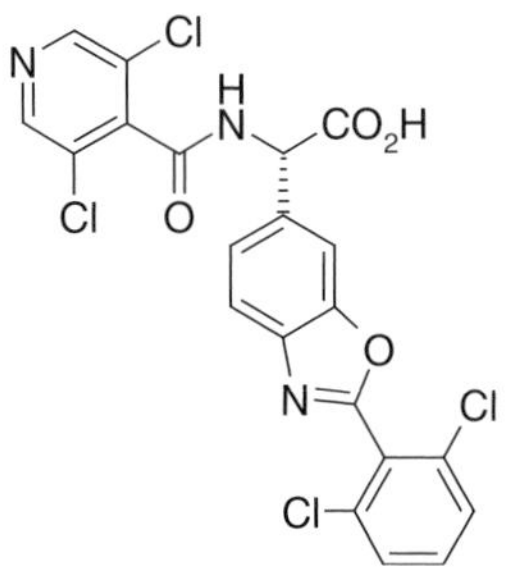

43

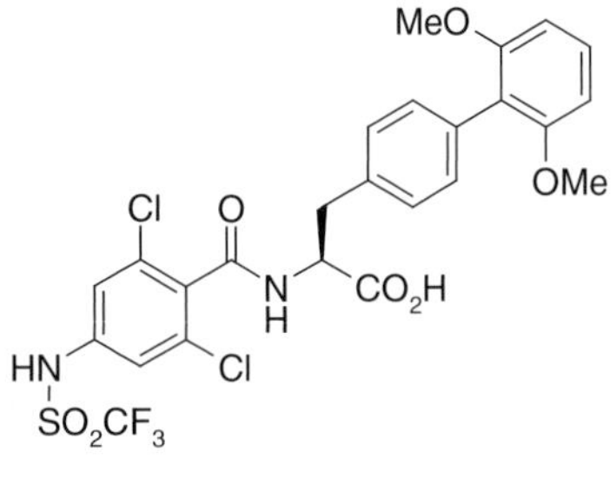

44

Similarly the contribution of the LFA-1/ICAM interaction to the disease pathophysiology has not been overlooked. Clinical trials with a humanized anti-LFA-1 antibody (efalizumab) in psoriasis have underscored the importance of this axis in the recruitment of activated T cells to cutaneous sites. In mouse models of colitis, either deletion of the ICAM gene or downregulation of expression by antisense oligonucleotides[103] reduces the extent of inflammation. Two series of synthetic allosteric antagonists have emerged from the finding that the hydroxymethylglutaryl coenzyme A (HMG-CoA) reductase inhibitor lovastatin binds to the I-domain allosteric sites of the CD11a subunit of LFA-1, locking its conformation in a low-affinity state and preventing its binding to ICAM-1.[104] The α1 allosteric antagonists, which bind and stabilize the closed conformation of the CD11a I domain, include BIRT-377 (**45**) ($K_d = 25.8$ nM) and LFA-703 (**46**) as well as compounds such as **47**[105] ($IC_{50} = 85$ nM) and **48**[106] ($IC_{50} = 4$ nM). There is significant structural diversity in the literature and further studies are warranted to determine the value of this axis in the modulation of leukocyte extravasation under conditions of inflammation.

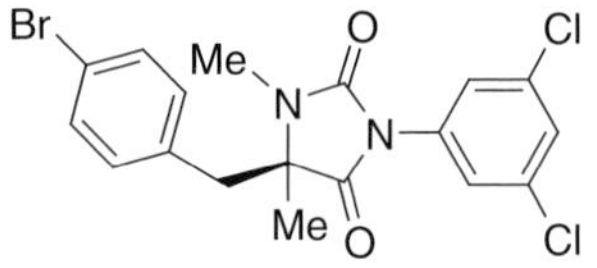

45

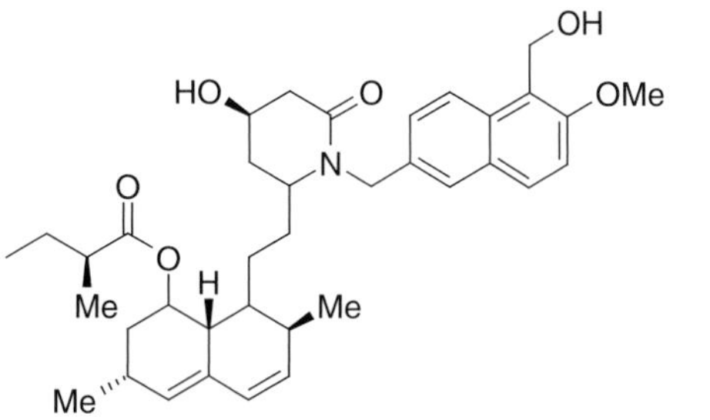

46

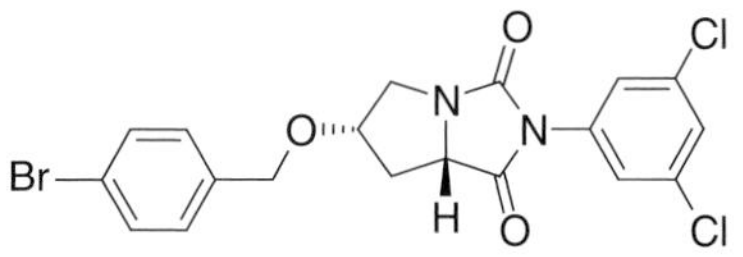

47

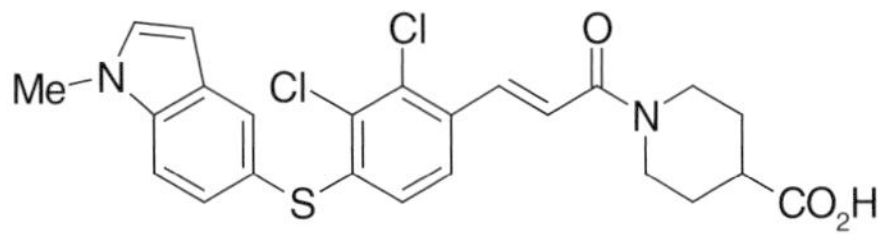

48

6.28.8.3.2 Chemokine receptor antagonists

Chemokines belong to a family of chemotactic cytokines that direct the migration of immune cells toward sites of inflammation. They are divided into two major (CXC and CC) and two minor (C and CX3C) groups dependent on the number and spacing of the first two conserved cysteine residues. They mediate their biological effects by binding to cell surface receptors, which belong to the 7TM G protein coupled-receptor superfamily.[107] As an additional means for modulating the ingress of activated lymphocytes to the GI tract in patients with inflammation, the development of specific chemokine receptor antagonists has been increasing in pace.

In IBD there is evidence of contribution of multiple chemokine receptors in leukocyte recruitment to the gut. The reported increased recruitment of $CCR2^{+}CD4^{+}$ and $CCR5^{+}CD4^{+}$ T cells in biopsy samples from IBD patients, for instance, is consistent with the observations of increased mucosal MIP-1α/β, RANTES, and MCP-1 expression.

Furthermore, CCR5 and CCR2 knockout mice are resistant to the effects of DSS-induced colitis.[108] CCR1, CCR2, and CCR5 have overlapping chemokine specificity, coligating and transducing signals alone from MIP-1α/β, RANTES, and MCP-1–4. To a greater or lesser extent all these chemokines have been implicated in the disease state. Remarkably few experimental studies have assessed the contribution of CCR1 directly in the pathophysiology of IBD, yet there is a quite large and diverse array of potent and selective nonpeptide CCR1 antagonists described. The 2-methylpiperazine BX-471 (**49**) ($IC_{50} = 2$ nM), the quinoxaline-2-carboxylic acid derivative CP-481715 (**50**) ($K_D = 0.2$ nM; $IC_{50} = 74$ nM), and a series of pyridylbenzoxepines, typified by **51** ($K_i = 2.3$ nM) for instance displace MIP-1β, RANTES, and MCP-3 with high-affinity functional potency.[109–111] BX-471 has anti-inflammatory activity in experimental models of MS and renal fibrosis. There is reported clinical development of a humanized anti-CCR2 mAb, MLN-1202, for RA and MS: however, apart from several series of cyclopentanes (of which **52** is a typical example; reported $IC_{50} < 1\ \mu M$[112,113]); the chemical landscape for CCR2 is less well defined. The finding that CCR5 is a major co-receptor for gp120-dependent HIV-1 entry, and that individuals carrying the Δ32 polymorphism are protected from infection, has resulted in the rapid development of specific CCR5 antagonists.[114] UK-427857/Maraviroc (**53**) is one of the most clinically progressed agents and prevents gp120 binding and viral replication ($IC_{50} = 43$ and 0.2 nM, respectively). Several other series of potent antagonists which also block viral replication, typified by **54** (Sch-D) ($IC_{50} = 0.45$ nM), **55** ($IC_{50} = 0.3$ nM), and **56** (GW-873140) ($IC_{50} = 1$ nM), have also been reported. Although CCR5 knockout mice appear to be protected from the effects of DSS-induced colitis, the Δ32 polymorphism appears equally in the IBD and healthy population,[115] suggesting that antagonism of CCR5 may not be sufficient to attenuate the disease process and that further multiple pathways contribute to the clinical condition.

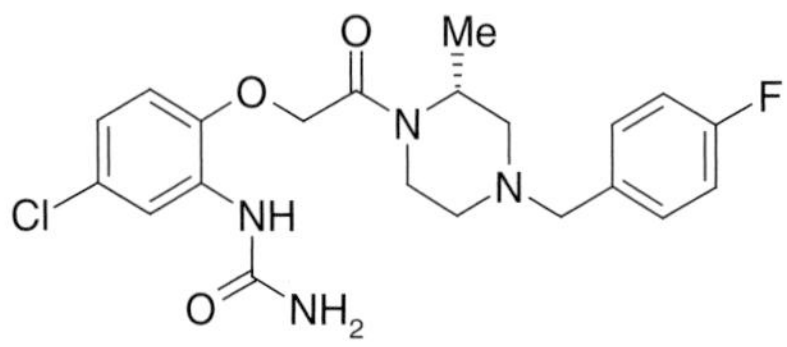

49 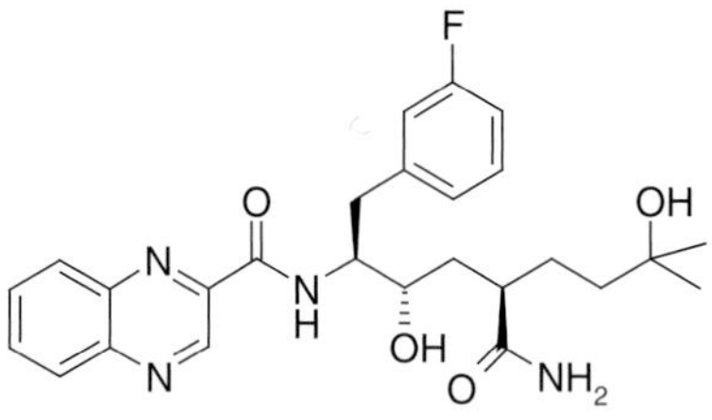 **50**

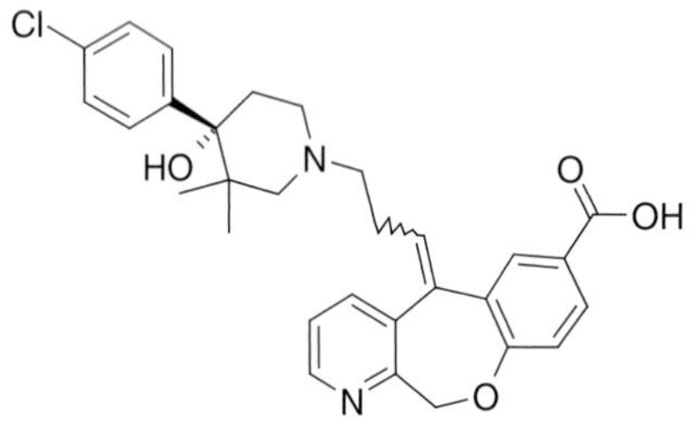

51 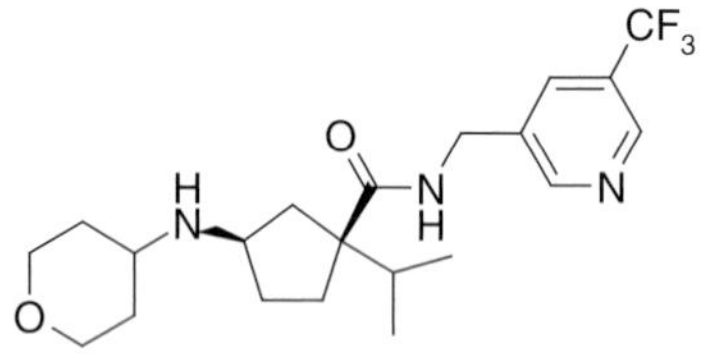 **52**

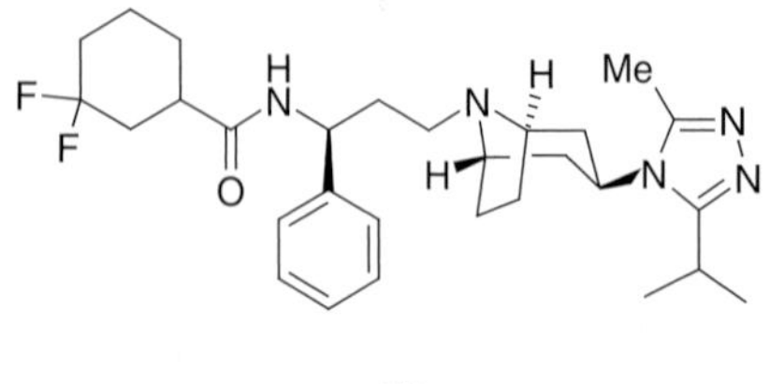

53 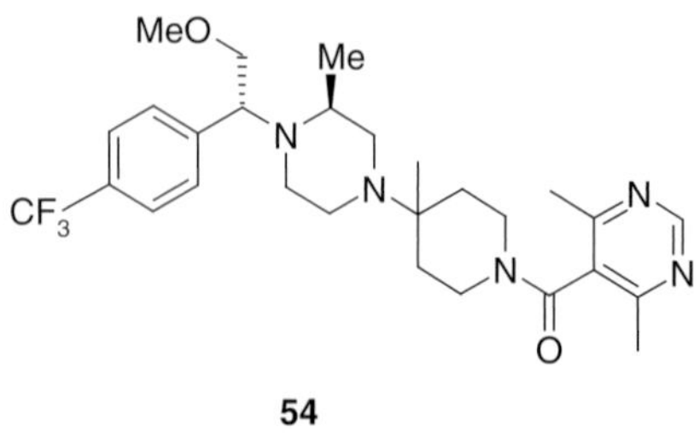 **54**

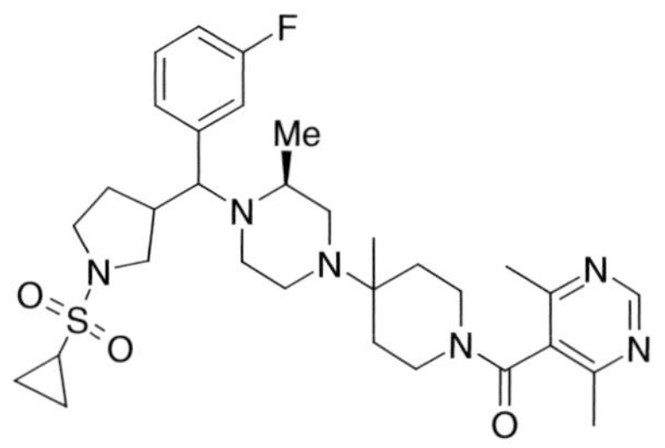

55 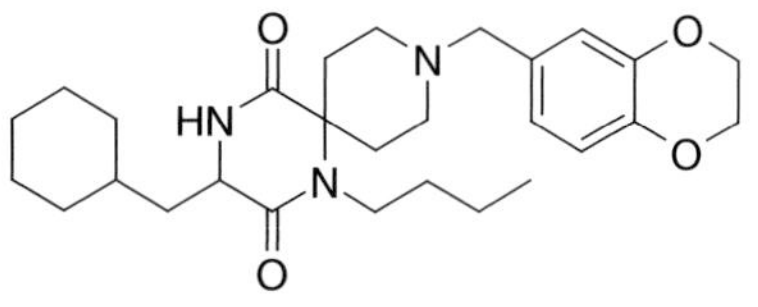 **56**

Several classes of dual-specificity antagonists have also been described. For instance, TAK-779 (**57**) is an inhibitor of mouse CXCR3 and CCR5 ($IC_{50}=369$ and 236 nM, respectively; the human CCR5 $IC_{50}=1.4$ nM) and displays significant inhibition at the human and mouse CCR2 receptor as well ($IC_{50}=27$ and 24 nM, respectively).[116,117] When dosed to DSS-treated mice, TAK-779 (50 mg kg^{-1}, intradermal) inhibited the recruitment of $CD11b^{+}$ leukocytes to the colon and delayed as well as suppressed the extent on the colonic injury.[118]

IFN-γ-inducible protein-10 (IP-10) and CXCR3 are highly expressed at sites of colitis. CXCR3 is expressed on a fraction of circulating blood T cells, B cells, and NK cells. T cells positive for CXCR3 are mostly $CD45RO^{+}$ memory cells, which express high levels of β1 integrins, and the $CXCR3^{+}/CD4^{+}$ T cell subset is enriched for Th1 cells. As with other chemokine receptors, T cell activation enhances CXCR3 expression and chemotactic responsiveness. Blockade of IP-10 with an antibody reduces inflammation in $IL10^{-/-}$ mice[119]; a fully human anti-IP10 antibody (MDX-1100) is currently undergoing early clinical evaluation in patients with CD.

One class of quinolines, typified by NBI-74330 (**58**), blocks the binding of IP-10 and interferon-inducible T cell alpha chemoattractant (ITAC) ($K_i=1.5$ and 3.2 nM, respectively) and CXCR3-dependent Ca^{2+} mobilization ($IC_{50}=$ 7–18 nM).[120] NBI-74330 was taken from a series of recent disclosures,[121,122] in which a lead clinical agent T-487 is putatively exemplified, but whose structure is not yet reported. Despite the encouraging developments in identification of CCR2, CCR5, and CXCR3 antagonists, none so far has been rigorously assessed in experimental models of colitis or clinically in the IBD population. CCR9 antagonists, however, are being progressed clinically for the condition.

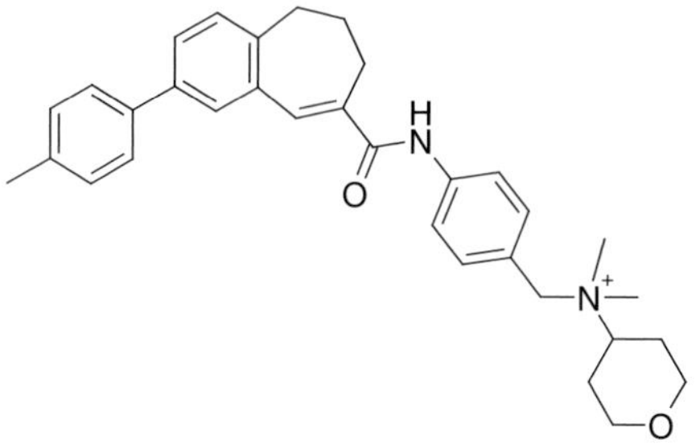

57 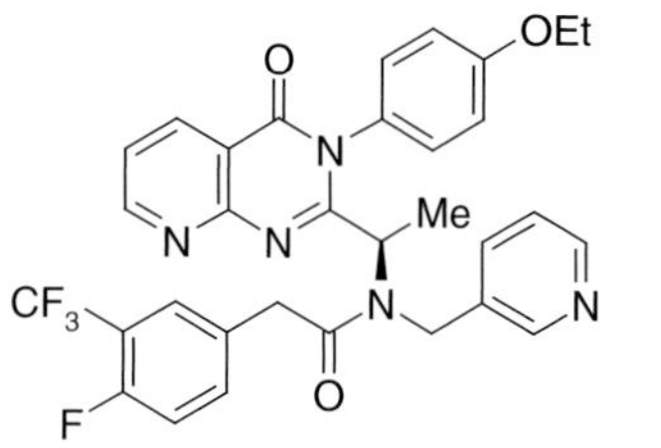 **58**

The chemokine receptor CCR9 is coexpressed on gut-homing α4β7 lymphocytes and additionally appears to contribute to both T cell activation and the production of pro-inflammatory cytokines, such as IFN-γ.[123–125] Either depletion of CCR9 or its ligand, CCL25/TECK, impairs the ability of these cells to home properly and attenuates the ileitis observed in the SAMP/Yit mouse model.[126,127] CCX282/Traficet is currently being evaluated in phase II clinical trials in CD. The structure of the CCX282 is undisclosed, but a series of arylsulphonamides have been reported,[128,129] of which **59** is also reported to reduce the onset of disease symptoms in the $Mdr1a^{-/-}$ mouse model of colitis (50 mg kg^{-1}, twice daily subcutaneously).[129] The CCR9/TECK interaction appears to confer some regional specialization to the recruitment and regulation of activation of gut-homing lymphocytes. Firstly, TECK expression appears to be restricted to small intestinal epithelial cells specifically, large bowel epithelium appears to be devoid of TECK expression, even at sites of inflammation.[125] A far lower proportion of colonic T cells are $CCR9^{+}$ (20%) than those in the small bowel, a pattern of restriction which suggests that CCR9 antagonists may be an effective modality in the treatment of small bowel inflammatory CD and celiac disease.

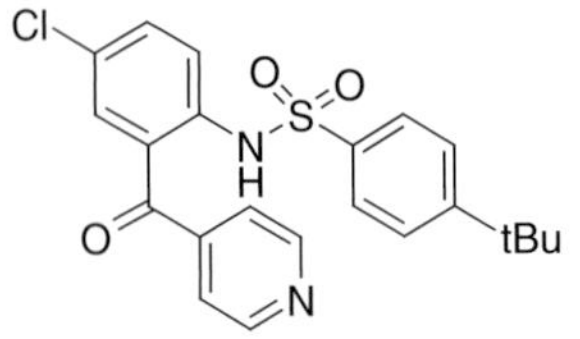

59

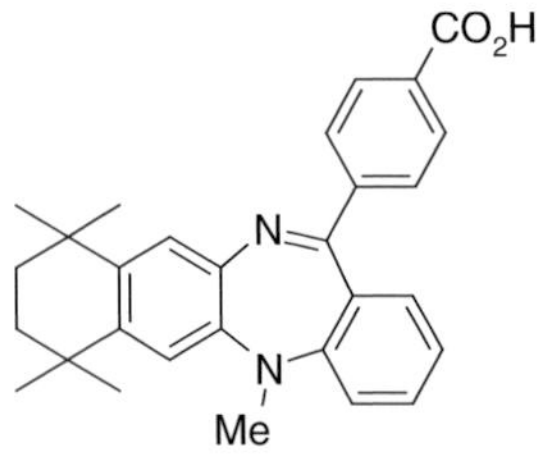

60

Although CD45Rb^{hi} cells from CCR9$^{-/-}$ mice fail to induce colitis when adoptively transferred into SCID mice, and CCR9 antagonists appear to be effective in treating the colitic phenotype in the Mdr1a$^{-/-}$ model, clinical studies aimed at investigating CCR9 antagonists in patients with UC are eagerly awaited to determine whether this approach has wider utility in the treatment of IBD and its extraintestinal manifestations.

Finally, gut tropism of $\alpha4\beta7$/CCR9^{+} lymphocytes is imprinted during maturation and activation in secondary lymphoid tissue.[127] Their expression can be induced directly by GALT-derived dendritic cells following antigen stimulation, an effect which is suppressed by the retinoic acid receptor-β (RARβ) antagonist LE135 (**60**).[130] Furthermore, as populations of splenic and Peyer's patch $\alpha4\beta7^{+}$ CD4^{+} T cells are dramatically reduced in retinoic acid-deficient mice, these findings further point to a possible involvement in the RAR/RXR pathway in imprinting gut-homing specificity.

6.28.8.3.3 Sphingosine 1-phosphate receptor modulators

In contrast to approaches that suppress lymphocyte recruitment to the intestinal lamina propria, an alternative strategy has emerged from studies of the novel immunosuppressant FTY720 (**61**).[131] Currently in phase III clinical studies for transplant rejection, FTY720 is a founding member of the class of sphingosine 1-phosphate (S1P) receptor agonists. Chronic agonism at S1P receptors expressed on thymocytes and lymphocytes causes desensitization and receptor internalization. As circulating S1P is an obligatory egress signal for lymphocytes leaving lymphoid organs, in the absence of S1P1/Edg1 receptor the cells are retained, inducing a peripheral lymphopenia. As such FTY720, and others of its class, do not appear to act as classical immunosuppressants, as B and T cell function per se is unaffected. In the IL10$^{-/-}$ models of spontaneous colitis, FTY720 (orally for 4 weeks) reduced the severity of inflammation.[132] FTY720, however, is a nonselective S1P receptor agonist and also activates the inward rectifying K^{+} channel and this may, in part, contribute to the negative chronotropic effects FTY720 has on isolated perfused guinea pig heart and the bradycardia reported clinically. In preclinical models, KRP-203 (**62**) has shown a reduced tendency to induce bradycardia, while inducing lymphopenia and maintaining efficacy on acute organ rejection.[133] As a divergence from these amino-propanolol derivatives, Merck has reported on a series of oxadiazoles, typified by **63**, which possess >100-fold selectivity for S1P1/Edg1 over the cardiac S1P3/Edg3 receptor.[134]

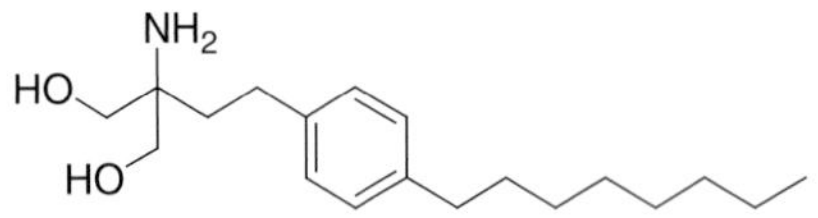

61

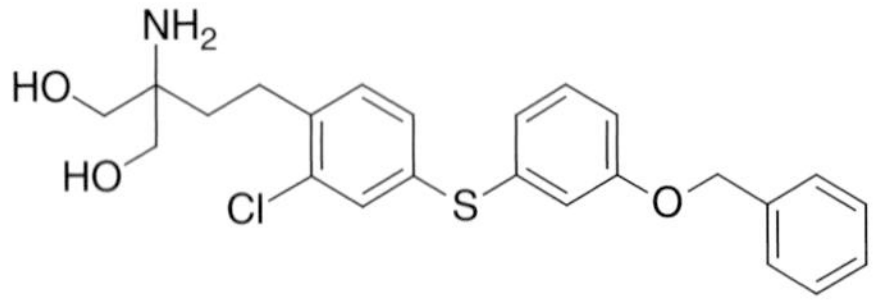

62

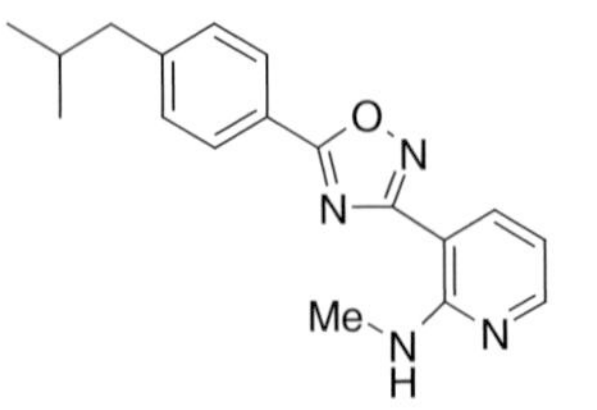

63

6.28.8.4 Miscellaneous Mechanisms

6.28.8.4.1 Macrophage migration inhibitory factor antagonists

Levels of the pro-inflammatory cytokine macrophage migration inhibitory factor (MIF) are elevated in IBD patients. Transgenic animals which overexpress MIF are more susceptible to DSS-induced colitis,[135] and either antibody blockade of MIF or genetic depletion in mice significantly ameliorates the DSS-induced inflammatory response in mice.[136] MIF-deficient macrophages are hyporesponsive to LPS, show reduced activation of NFκB and the production of TNF-α, an effect which appears to be in part due to a downregulation of TLR4.

MIF possesses the unique ability to catalyze the tautomerization of D,L-dopachrome methyl esters into their corresponding indole derivatives. Compounds that block the active site either inhibit MIF activity directly or may force MIF into a conformation which prevents its binding to CD74.[137] Despite the large structural diversity of MIF tautomerase inhibitors,[138] few compounds appear to have been studied in animal models of disease. One ISO-1, compound **64**,[139] is a reportedly selective MIF tautomerase inhibitor ($IC_{50} \sim 7\,\mu M$), which when administered to SDZ-treated mice (1 mg per mouse per day intraperitoneally), markedly reduced clinical and histopathological features of diabetes (hyperglycemia and insulitis), to an extent that was comparable with that of an anti-MIF antibody.[140] A number of MIF inhibitor candidates (AVP-13546, AVP-13748) are being clinically evaluated. When given orally to DSS-treated mice, AVP-13748 was effective at reducing the inflammation.[141] The structure of AVP-13748 is currently not known but may be one of a series of quinoline 3-carbonitriles as recently claimed.[142]

6.28.8.4.2 TRPV1 antagonists

One of the symptoms in IBD is abdominal pain. Additional neurogenic mechanisms have been also implicated in the induction of IBD. For instance, substance P receptors exhibit dramatic increases in binding in the colons of both CD and UC patients. TRPV1 is a nonselective cation channel, expressed on nerve terminals of intrinsic and extrinsic afferent neurons innervating the GI tract. TRPV1 plays a key role in the detection of noxious painful stimuli such as acid and heat, and $TRPV1^{-/-}$ mice have impaired inflammatory thermal hyperalgesia. Immunoreactive TRPV1 fibers increase in the submucosal plexus of diseased IBD tissue. On DSS-induced inflammation in Balb/c mice, the TRPV1 antagonist capsazepine (**65**) ($2.5\,mg\,kg^{-1}$, twice daily intraperitoneally), significantly reduced the severity of inflammation and epithelial damage.[143] Capsazepine was the first competitive TRPV1 antagonist identified, but since this discovery, several other series of compounds have been identified. Compounds such as A-425619 (**66**), which is 25-to 50-fold more potent at blocking TRPV1 activation ($IC_{50} = 5\,nM$) than capsazepine,[144] and GR-705498 (**67**) ($pK_B \sim 7.6$),[145,146] which are more druglike, have been described and warrant further preclinical evaluation in IBD models.

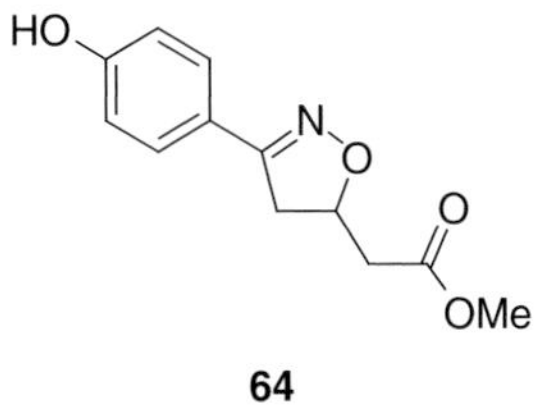

64 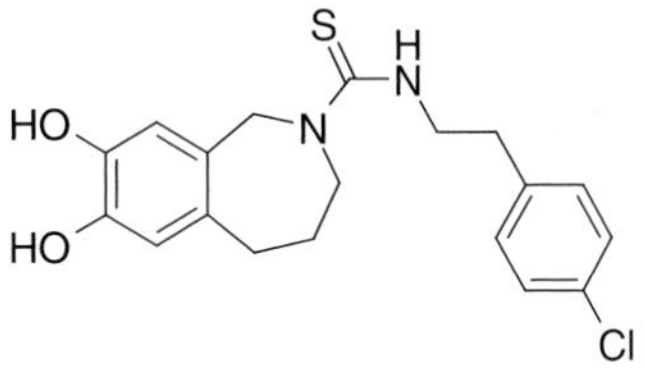 **65**

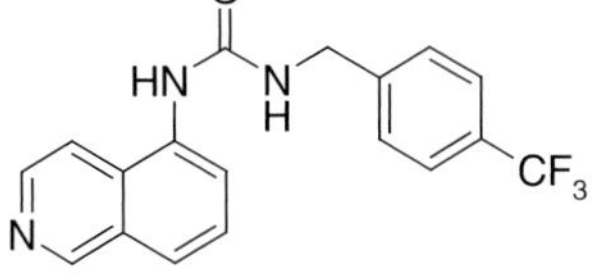

66 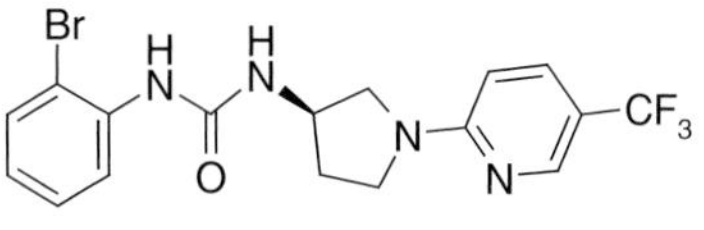 **67**

The observation that blockade of TRPV1 channels appears to be effective at reducing colitic inflammation opens up a number of other avenues, including the modulation of afferent activity by tetrodotoxin-resistant Na channel blockers, calcitonin gene-related peptide receptor antagonists, substance P and neurokinin receptor antagonists, and neuropeptide Y receptor modulators, many of which have already been implicated in nociception and neurogenic inflammation, as well as attenuating DSS-induced colitis,[147,148] but for which the limit of this review cannot cover.

References

1. Podolsky, D. K. *N. Engl. J. Med.* **2002**, *347*, 417–429.
2. Bouma, G.; Strober, W. *Nat. Rev. Immunol.* **2003**, *3*, 521–533.
3. Russell, R. K.; Nimmo, E. R.; Satsangi, J. *Curr. Opin. Genet. Devel.* **2004**, *14*, 264–270.
4. Hugot, J. P.; Chamaillard, M.; Zouali, H.; Lesage, S.; Cezard, J. P.; Belaiche, J.; Almer, S.; Tysk, C.; O'Morain, C. A.; Gassull, M. et al. *Nature* **2001**, *411*, 599–603.
5. Ogura, Y.; Bonen, D. K.; Inohara, N.; Nicolae, D. L.; Chen, F. F.; Ramos, R.; Britton, H.; Moran, T.; Karaliuskas, R.; Duerr, R. H. et al. *Nature* **2001**, *411*, 603–606.
6. Iwasaki, A.; Medzhitov, R. *Nat. Immunol.* **2004**, *5*, 987–995.
7. Abreu, M. T. *Curr. Opin. Gastroenterol.* **2003**, *19*, 559–564.
8. Sundberg, J. P.; Elson, C. O.; Bedigian, H.; Birkenmeier, E. H. *Gastroenterology* **1994**, *107*, 1726–1735.
9. Franchimont, D.; Vermeire, S.; El Housni, H.; Pierik, M.; Van Steen, K.; Gustot, T.; Quertinmont, E.; Abramowicz, M.; Van Gossum, A.; Deviere, J. et al. *Gut* **2004**, *53*, 987–992.
10. Cario, E.; Podolsky, D. K. *Infect. Immunol.* **2000**, *68*, 7010–7017.
11. Hawlisch, H.; Belkaid, Y.; Baelder, R.; Hildeman, D.; Gerard, C.; Kohl, J. *Immunity* **2005**, *22*, 415–426.
12. Woodruff, T. M.; Arumugam, T. V.; Shiels, I. A.; Reid, R. C.; Fairlie, D. P.; Taylor, S. M. *J. Immunol.* **2003**, *171*, 5514–5520.
13. Lodes, M. J.; Cong, Y.; Elson, C. O.; Mohamath, R.; Landers, C. J.; Targan, S. R.; Fort, M.; Hershberg, R. M. *J. Clin. Invest.* **2004**, *113*, 1296–1306.
14. Araki, A.; Kanai, T.; Ishikura, T.; Makita, S.; Uraushihara, K.; Iiyama, R.; Totsuka, T.; Takeda, K.; Akira, S.; Watanabe, M. *J. Gastroenterol.* **2005**, *40*, 16–23.
15. Garlanda, C.; Riva, F.; Polentarutti, N.; Buracchi, C.; Sironi, M.; De Bortoli, M.; Muzio, M.; Bergottini, R.; Scanziani, E.; Vecchi, A. et al. *Proc. Natl. Acad. Sci. USA* **2004**, *101*, 3522–3526.
16. Lang, K. S.; Recher, M.; Junt, T.; Navarini, A. A.; Harris, N. L.; Freigang, S.; Odermatt, B.; Conrad, C.; Ittner, L. M.; Bauer, S. *Nat. Med.* **2005**, *11*, 138–145.
17. Panwala, C. M.; Jones, J. C.; Viney, J. L. *J. Immunol.* **1998**, *161*, 5733–5744.
18. Schwab, M.; Schaeffeler, E.; Marx, C.; Fromm, M. F.; Kaskas, B.; Metzler, J.; Stange, E.; Herfarth, H.; Schoelmerich, J.; Gregor, M. et al. *Gastroenterology* **2003**, *124*, 26–33.
19. Brant, S. R.; Panhuysen, C. I.; Nicolae, D.; Reddy, D. M.; Bonen, D. K.; Karaliukas, R.; Zhang, L.; Swanson, E.; Datta, L. W.; Moran, T. et al. *Am. J. Hum. Genet.* **2003**, *73*, 1282–1292.
20. Hammer, R. E.; Maika, S. D.; Richardson, J. A.; Tang, J. P.; Taurog, J. D. *Cell* **1990**, *63*, 1099–1112.
21. Powrie, F.; Uhlig, H. *Novartis Found. Symp.* **2004**, *263*, 164–174, discussion 174–178, 211–218.
22. Jurjus, A. R.; Khoury, N. N.; Reimund, J.-M. *J. Pharmacol. Toxicol. Methods* **2004**, *50*, 81–92.
23. Sandborn, W. J.; Faubion, W. A. *Gut* **2004**, *53*, 1366–1373.
24. Mannon, P. J.; Fuss, I. J.; Mayer, L.; Elson, C. O.; Sandborn, W. J.; Present, D.; Dolin, B.; Goodman, N.; Groden, C.; Hornung, R. L. et al. *N. Engl. J. Med.* **2004**, *351*, 2069–2079.
25. Steidler, L.; Hans, W.; Schotte, L.; Neirynck, S.; Obermeier, F.; Falk, W.; Fiers, W.; Remaut, E. *Science* **2000**, *289*, 1352–1355.
26. Truelove, S. C.; Witts, L. J. *Br. Med. J.* **1955**, *4947*, 1041–1048.
27. Powell-Tuck, J.; Bown, R. L.; Lennard-Jones, J. E. *Scand. J. Gastroenterol.* **1978**, *13*, 833–837.
28. Schroeder, K. W.; Tremaine, W. J.; Ilstrup, D. M. *N. Engl. J. Med.* **1987**, *317*, 1625–1629.
29. Seo, M.; Okada, M.; Yao, T.; Okabe, N.; Maeda, K.; Oh, K. *Am. J. Gastroenterol.* **1995**, *90*, 1759–1763.
30. Walmsley, R. S.; Ayres, R. C.; Pounder, R. E.; Allan, R. N. *Gut* **1998**, *43*, 29–32.
31. Harvey, R. F.; Bradshaw, M. J. *Lancet* **1980**, *1*, 1134–1135.
32. Myren, J.; Bouchier, I. A.; Watkinson, G.; Softley, A.; Clamp, S. E.; de Dombal, F. T. *Scand. J. Gastroenterol.* **1984**, *95*, 1–27.
33. Daperno, M.; D'Haens, G.; Van Assche, G.; Baert, F.; Bulois, P.; Maunoury, V.; Sostegni, R.; Rocca, R.; Pera, A.; Gevers, A. et al. *Gastrointest. Endosc.* **2004**, *60*, 505–512.
34. Beaven, S. W.; Abreu, M. T. *Curr. Opin. Gastroenterol.* **2004**, *20*, 318–327.
35. Sandborn, W. J.; Feagan, B. G.; Radford-Smith, G.; Kovacs, A.; Enns, R.; Innes, A.; Patel, J. *Gut* **2004**, *53*, 1485–1493.
36. Gordon, F. H.; Lai, C. W.; Hamilton, M. I.; Allison, M. C.; Srivastava, E. D.; Fouweather, M. G.; Donoghue, S.; Greenlees, C.; Subhani, J.; Amlot, P. L. et al. *Gastroenterology* **2001**, *121*, 268–274.
37. Gordon, F. H.; Hamilton, M. I.; Donoghue, S.; Greenlees, C.; Palmer, T.; Rowley-Jones, D.; Dhillon, A. P.; Amlot, P. L.; Pounder, R. E. *Aliment. Pharmacol. Ther.* **2002**, *16*, 699–705.
38. Baert, F. J.; D'Haens, G. R.; Peeters, M.; Hiele, M. I.; Schaible, T. F.; Shealy, D.; Geboes, K.; Rutgeerts, P. J. *Gastroenterology* **1999**, *116*, 22–28.
39. Louis, E.; Vermeire, S.; Rutgeerts, P.; De Vos, M.; Van Gossum, A.; Pescatore, P.; Fiasse, R.; Pelckmans, P.; Reynaert, H.; D'Haens, G. et al. *Scand. J. Gastroenterol.* **2002**, *37*, 818–824.
40. Siegel, C. A.; Sands, B. E. *Aliment. Pharmacol. Ther.* **2005**, *22*, 1–16.
41. Summers, R. W.; Elliott, D. E.; Urban, J. F.; Thompson, R. A.; Weinstock, J. V. *Gastroenterology* **2005**, *128*, 825–832.
42. Cummings, J. H.; Kong, S. C. *Novartis Found. Symp.* **2004**, *263*, 99–111, discussion 111–114, 211–218.
43. Hoentjen, F.; Harmsen, H. J.; Braat, H.; Torrice, C. D.; Mann, B. A.; Sartor, R. B.; Dieleman, L. A. *Gut* **2003**, *52*, 1721–1727.
44. Madsen, K. L.; Doyle, J. S.; Jewell, L. D.; Tavernini, M. M.; Fedorak, R. N. *Gastroenterology* **1999**, *116*, 1107–1114.
45. Osman, N.; Adawi, D.; Ahrne, S.; Jeppsson, B.; Molin, G. *Dig. Dis. Sci.* **2004**, *49*, 320–327.
46. Gionchetti, P.; Rizzello, F.; Helwig, U.; Venturi, A.; Lammers, K. M.; Brigidi, P.; Vitali, B.; Poggioli, G.; Miglioli, M.; Campieri, M. *Gastroenterology* **2003**, *124*, 1202–1209.
47. Fukata, M.; Michelsen, K. S.; Eri, R.; Thomas, L. S.; Hu, B.; Lukasek, K.; Nast, C. C.; Lechago, J.; Xu, R.; Naiki, Y. et al. *Gastrointest. Liver. Physiol.* **2005**, *288*, G1055–G1065.
48. Hawkins, L. D.; Christ, W. J.; Rossignol, D. P. *Curr. Topics Med. Chem.* **2004**, *4*, 1147–1171.
49. Fort, M. M.; Mozaffarian, A.; Stover, A. G.; Correia Ida, S.; Johnson, D. A.; Crane, R. T.; Ulevitch, R. J.; Persing, D. H.; Bielefeldt-Ohmann, H.; Probst, P. et al. *J. Immunol.* **2005**, *174*, 6416–6423.
50. Krieg, A. M. *Curr. Oncol. Rep.* **2004**, *6*, 88–95.

51. Rachmilewitz, D.; Karmeli, F.; Takabayashi, K.; Hayashi, T.; Leider-Trejo, L.; Lee, J.; Leoni, L. M.; Raz, E. *Gastroenterology* **2002**, *122*, 1428–1441.
52. Rachmilewitz, D.; Katakura, K.; Karmeli, F.; Hayashi, T.; Reinus, C.; Rudensky, B.; Akira, S.; Takeda, K.; Lee, J.; Takabayashi, K. et al. *Gastroenterology* **2004**, *126*, 520–528.
53. Nikolaus, S.; Rutgeerts, P.; Fedorak, R.; Steinhart, A. H.; Wild, G. E.; Theuer, D.; Mohrle, J.; Schreiber, S. *Gut* **2003**, *52*, 1286–1290.
54. Neurath, M. F.; Pettersson, S.; Meyer zum Buschenfelde, K. H.; Strober, W. *Nat. Med.* **1996**, *2*, 998–1004.
55. Chen, L. W.; Egan, L.; Li, Z. W.; Greten, F. R.; Kagnoff, M. F.; Karin, M. *Nat. Med.* **2003**, *9*, 575–581.
56. Yang, G.; McIntyre, K. W.; Townsend, R. M.; Shen, H. H.; Pitts, W. J.; Dodd, J. H.; Nadler, S. G.; McKinnon, M.; Watson, A. J. *J. Immunol.* **2003**, *171*, 6414–6420.
57. Nagamoto, H.; Maeda, T.; Harutta, J.-P.; Miyakoda, G.; Mori, T.; Tominaga, M. *Gastroenterology* **2004**, *126*, M1045, Abst.
58. Hartmann, G.; Bidlingmaier, C.; Siegmund, B.; Albrich, S.; Schulze, J.; Tschoep, K.; Eigler, A.; Lehr, H. A.; Endres, S. *J. Pharmacol. Exp. Ther.* **2000**, *292*, 22–30.
59. Burnouf, C.; Pruniaux, M. P. *Curr. Pharmaceut. Des.* **2002**, *8*, 1255–1296.
60. Hanauer, S. B.; Miner, P. B.; Keshavarzian, A.; Isaacs, K.; Goff, J.; Harris, M. S. *Gastroenterology* **2004**, *126*, 814, Abst.
61. Doffinger, R.; Smahi, A.; Bessia, C.; Geissmann, F.; Feinberg, J.; Durandy, A.; Bodemer, C.; Kenwrick, S.; Dupuis-Girod, S.; Blanche, S. et al. *Nat. Genet.* **2001**, *27*, 277–285.
62. McIntyre, K. W.; Shuster, D. J.; Gillooly, K. M.; Dambach, D. M.; Pattoli, M. A.; Lu, P.; Zhou, X. D.; Qiu, Y.; Zusi, F. C.; Burke, J. R. *Arthritis Rheum.* **2003**, *48*, 2652–2659.
63. Murata, Y.; Shimada, M.; Kadono, H.; Sakakibara, S.; Yoshino, T.; Masuda, T.; Shimazaki, M.; Shintani, T.; Fuchikami, K.; Bacon, K. B. et al. *Bioorg. Med. Chem. Lett.* **2004**, *14*, 4013–4017.
64. Murata, T.; Shimada, M.; Sakakibara, S.; Yoshino, T.; Masuda, T.; Shintani, T.; Sato, H.; Koriyama, Y.; Fukushima, K.; Nunami, N. *Bioorg. Med. Chem. Lett.* **2004**, *14*, 4019–4022.
65. Baxter, A.; Brough, S.; Cooper, A.; Floettmann, E.; Foster, S.; Harding, C.; Kettle, J.; McInally, T.; Martin, C.; Mobbs, M. *Bioorg. Med. Chem. Lett.* **2004**, *14*, 2817–2822.
66. Podolin, P. L.; Callahan, J. F.; Bolognese, B. J.; Li, H. Y.; Carlson, K.; Gregg Davis, T.; Mellor, G. W.; Evans, C.; Roshak, A. K. *J. Pharmacol. Exp. Ther.* **2005**, *312*, 373–381.
67. Ziegelbauer, K.; Gantner, F.; Lukacs, N. W.; Berlin, A.; Fuchikami, K.; Niki, T.; Sakai, K.; Inbe, H.; Takeshita, K.; Ishimori, M. et al. *Br. J. Pharmacol.* **2005**, *145*, 178–192.
68. Pargellis, C.; Tong, L.; Churchill, L.; Cirillo, P. F.; Gilmore, T.; Graham, A. G.; Grob, P. M.; Hickey, E. R.; Moss, N.; Pav, S. et al. *Nat. Struct. Biol.* **2002**, *9*, 268–272.
69. Hommes, D.; van den Blink, B.; Plasse, T.; Bartelsman, J.; Xu, C.; Macpherson, B.; Tytgat, G.; Peppelenbosch, M.; Van Deventer, S. *Gastroenterology* **2002**, *122*, 7–14.
70. Brown, D. S.; Belfield, A. J.; Brown, G. R.; Campbell, D.; Foubister, A.; Masters, D. J.; Pike, K. G.; Snelson, W. L.; Wells, S. L. *Bioorg. Med. Chem. Lett.* **2004**, *14*, 5383–5387.
71. Regan, J.; Capolino, A.; Cirillo, P. F.; Gilmore, T.; Graham, A. G.; Hickey, E.; Kroe, R. R.; Madwed, J.; Moriak, M.; Nelson, R. et al. *J. Med. Chem.* **2003**, *46*, 4676–4686.
72. Loher, F.; Bauer, C.; Landauer, N.; Schmall, K.; Siegmund, B.; Lehr, H. A.; Dauer, M.; Schoenharting, M.; Endres, S.; Eigler, A. *J. Pharmacol. Exp. Ther.* **2004**, *308*, 583–590.
73. Wannamaker, M., Davies, R. Prodrug of an ICE Inhibitor. PCT Patent WO01/96003, Nov 29, 2001.
74. Popoff, I.; Jijon, H.; Monia, B.; Tavernini, M.; Ma, M.; McKay, R.; Madsen, K. *J. Pharmacol. Exp. Therap.* **2002**, *303*, 1145–1154.
75. Zingarelli, B.; Hake, P. W.; Burroughs, T. J.; Piraino, G.; O'Connor, M.; Denenberg, A. *Immunology* **2004**, *113*, 509–517.
76. Mazzon, E.; Dugo, L.; Li, J. H.; Di Paola, R.; Genovese, T.; Caputi, A. P.; Zhang, J.; Cuzzocrea, S. *Biochem. Pharmacol.* **2002**, *64*, 327–337.
77. Southan, G. J.; Szabo, C. *Curr. Med. Chem.* **2003**, *10*, 321–340.
78. Massa, F.; Marsicano, G.; Hermann, H.; Cannich, A.; Monory, K.; Cravatt, B. F.; Ferri, G. L.; Sibaev, A.; Storr, M.; Lutz, B. *J. Clin. Invest.* **2004**, *113*, 1202–1209.
79. Kathuria, S.; Gaetani, S.; Fegley, D.; Valino, F.; Duranti, A.; Tontini, A.; Mor, M.; Tarzia, G.; La Rana, G.; Calignano, A. et al. *Nat. Med.* **2003**, *9*, 76–81.
80. Porter, A. C.; Li, J.; Love, P. L.; Shannon, H. E.; Li, D. L.; Phebus, L. A.; Gleason, S. D.; Witkin, J. M.; Simmons, J. M. A.; Iyengar, S. et al. *Soc. Neurosci. Abstr.* **2004**, *34*, Abs 273.11.
81. Leung, D.; Du, W.; Hardouin, C.; Cheng, H.; Hwang, I.; Cravatt, B. F.; Boger, D. L. *Bioorg. Med. Chem. Lett.* **2005**, *15*, 1423–1428.
82. Rossi, A.; Kapahi, P.; Natoli, G.; Takahashi, T.; Chen, Y.; Karin, M.; Santoro, M. G. *Nature* **2000**, *403*, 103–108.
83. Katayama, K.; Wada, K.; Nakajima, A.; Mizuguchi, H.; Hayakawa, T.; Nakagawa, S.; Kadowaki, T.; Nagai, R.; Kamisaki, Y.; Blumberg, R. et al. *Gastroenterology* **2003**, *124*, 1315–1324.
84. Desreumaux, P.; Dubuquoy, L.; Nutten, S.; Peuchmaur, M.; Englaro, W.; Schoonjans, K.; Derijard, B.; Desvergne, B.; Wahli, W.; Chambon, P. et al. *J. Exp. Med.* **2001**, *193*, 827–838.
85. Su, C. G.; Wen, X.; Bailey, S. T.; Jiang, W.; Rangwala, S. M.; Keilbaugh, S. A.; Flanigan, A.; Murthy, S.; Lazar, M. A.; Wu, G. D. *J. Clin. Invest.* **1999**, *104*, 383–389.
86. Dubuquoy, L.; Jansson, E. A.; Deeb, S.; Rakotobe, S.; Karoui, M.; Colombel, J. F.; Auwerx, J.; Pettersson, S.; Desreumaux, P. *Gastroenterology* **2003**, *124*, 1265–1276.
87. Lewis, J. D.; Lichtenstein, G. R.; Stein, R. B.; Deren, J. J.; Judge, T. A.; Fogt, F.; Furth, E. E.; Demissie, E. J.; Hurd, L. B.; Su, C. G. et al. *Am. J. Gastroenterol.* **2001**, *96*, 3323–3328.
88. Rousseaux, C.; Lefebvre, B.; Dubuquoy, L.; Lefebvre, P.; Romano, O.; Auwerx, J.; Metzger, D.; Wahli, W.; Desvergne, B.; Naccari, G. C. et al. *J. Exp. Med.* **2005**, *201*, 1205–1215.
89. Usui, S.; Suzuki, T.; Hattori, Y.; Etoh, K.; Fujieda, H.; Nishizuka, M.; Imagawa, M.; Nakagawa, H.; Kohda, K.; Miyata, N. *Bioorg. Med. Chem. Lett.* **2005**, *15*, 1547–1551.
90. Liu, K.; Black, R. M.; Acton, J. J., III; Mosley, R.; Debenham, S.; Abola, R.; Yang, M.; Tschirret-Guth, R.; Colwell, L.; Liu, C. et al. *Bioorg. Med. Chem. Lett.* **2005**, *15*, 2437–2440.
91. Berger, J. P.; Petro, A. E.; Macnaul, K. L.; Kelly, L. J.; Zhang, B. B.; Richards, K.; Elbrecht, A.; Johnson, B. A.; Zhou, G.; Doebber, T. W. et al. *Mol. Endocrinol.* **2003**, *17*, 662–676.

92. Berlin, C.; Berg, E. L.; Briskin, M. J.; Andrew, D. P.; Kilshaw, P. J.; Holzmann, B.; Weissman, I. L.; Hamann, A.; Butcher, E. C. *Cell* **1993**, *74*, 185–195.
93. Hesterberg, P. E.; Winsor-Hines, D.; Briskin, M. J.; Soler-Ferran, D.; Merrill, C.; Mackay, C. R.; Newman, W.; Ringler, D. J. *Gastroenterology* **1996**, *111*, 1373–1380.
94. Hokari, R.; Kato, S.; Matsuzaki, K.; Iwai, A.; Kawaguchi, A.; Nagao, S.; Miyahara, T.; Itoh, K.; Sekizuka, E.; Nagata, H. et al. *Clin. Exp. Immunol.* **2001**, *126*, 259–265.
95. Tilley, J. W.; Chen, L.; Sidduri, A.; Fotouhi, N. *Curr. Topics Med. Chem.* **2004**, *4*, 1509–1523.
96. Chang, L. L.; Truong, Q.; Mumford, R. A.; Egger, L. A.; Kidambi, U.; Lyons, K.; McCauley, E.; Van Riper, G.; Vincent, S.; Schmidt, J. A. *Bioorg. Med. Chem. Lett.* **2002**, *12*, 159–163.
97. Sircar, I.; Gudmundsson, K. S.; Martin, R.; Liang, J.; Nomura, S.; Jayakumar, H.; Teegarden, B. R.; Nowlin, D. M.; Cardarelli, P. M.; Mah, J. R. *Bioorg. Med. Chem. Lett.* **2002**, *10*, 2051–2066.
98. Lin, L. S.; Lanza, T. J., Jr.; Castonguay, L. A.; Kamenecka, T.; McCauley, E.; Van Riper, G.; Egger, L. A.; Mumford, R. A.; Tong, X.; MacCoss, M. et al. *Bioorg. Med. Chem. Lett.* **2004**, *14*, 2331–2334.
99. Hoshina, Y.; Ikegami, S.; Okuyama, A.; Fukui, H.; Inoguchi, K.; Maruyama, T.; Fujimoto, K.; Matsumura, Y.; Aoyama, A.; Harada, T. *Bioorg. Med. Chem. Lett.* **2005**, *15*, 217–220.
100. Egger, L. A.; Kidambi, U.; Cao, J.; Van Riper, G.; McCauley, E.; Mumford, R. A.; Amo, S.; Lingham, R.; Lanza, T.; Lin, L. S. et al. *J. Pharmacol. Exp. Ther.* **2002**, *302*, 153–162.
101. Hijazi, Y.; Welker, H.; Dorr, A. E.; Tang, J. P.; Blain, R.; Renzetti, L. M.; Abbas, R. *J. Clin. Pharmacol.* **2004**, *44*, 1368–1378.
102. Berger, J. R.; Koralnik, I. J. *N. Engl. J. Med.* **2005**, *353*, 414–416.
103. Bowen-Yacyshyn, M. B.; Bennett, C. F.; Nation, N.; Rayner, D.; Yacyshyn, B. R. *J. Pharmacol. Exp. Ther.* **2002**, *302*, 908–917.
104. Shimaoka, M.; Springer, T. A. *Nat. Rev. Drug. Disc.* **2003**, *2*, 703–716.
105. Potin, D.; Launay, M.; Nicolai, E.; Fabreguette, M.; Malabre, P.; Caussade, F.; Besse, D.; Skala, S.; Stetsko, D. K.; Todderud, G. et al. *Bioorg. Med. Chem. Lett.* **2005**, *15*, 1161–1164.
106. Winn, M.; Reilly, E. B.; Liu, G.; Huth, J. R.; Jae, H. S.; Freeman, J.; Pei, Z.; Xin, Z.; Lynch, J.; Kester, J. et al. *J. Med. Chem.* **2001**, *44*, 4393–4403.
107. Moser, B.; Wolf, M.; Walz, A.; Loetscher, P. *Trends Immunol.* **2004**, *25*, 75–84.
108. Andres, P. G.; Beck, P. L.; Mizoguchi, E.; Mizoguchi, A.; Bhan, A. K.; Dawson, T.; Kuziel, W. A.; Maeda, N.; MacDermott, R. P.; Podolsky, D. K. et al. *J. Immunol.* **2000**, *164*, 6303–6312.
109. Carson, K. G., Harriman, G. C. B. CCR1 Antagonists and Methods of Use Therefor. US Patent 0,106,639, Jun 3, 2004.
110. Liang, M.; Mallari, C.; Rosser, M.; Ng, H. P.; May, K.; Monahan, S.; Bauman, J. G.; Islam, I.; Ghannam, A.; Buckman, B. et al. *J. Biol. Chem.* **2000**, *275*, 19000–19008.
111. Gladue, R. P.; Tylaska, L. A.; Brissette, W. H.; Lira, P. D.; Kath, J. C.; Poss, C. S.; Brown, M. F.; Paradis, T. J.; Conklyn, M. J.; Ogborne, K. T. et al. *J. Biol. Chem.* **2003**, *278*, 40473–40480.
112. Ge, M.; Goble, S. D.; Pasternack, A.; Yang, L. 7 and 8 Membered Heterocyclic Cyclopentyl Benzylamide Modulators of Chemokine Receptor Activity. PCT Patent WO2005/010154, 2005.
113. Butora, G.; Yang, L. L.; Goble, S. D. Tetrahydropyran Heterocyclic Cyclopentyl Heteroaryl Modulators of Chemokine Receptor Activity. PCT Patent WO2005/014537, 2004.
114. Castagna, A.; Biswas, P.; Beretta, A.; Lazzarin, A. *Drugs* **2005**, *65*, 879–904.
115. Rector, A.; Vermeire, S.; Thoelen, I.; Keyaerts, E.; Struyf, F.; Vlietinck, R.; Rutgeerts, P.; Van Ranst, M. *Hum. Genet.* **2001**, *108*, 190–193.
116. Arakaki, R.; Tamamura, H.; Premanathan, M.; Kanbara, K.; Ramanan, S.; Mochizuki, K.; Baba, M.; Fujii, N.; Nakashima, H. *J. Virol.* **1999**, *73*, 1719–1723.
117. Gao, P.; Zhou, X.-Y.; Yashiro-Ohtani, Y.; Yang, Y.-F.; Sugimoto, N.; Ono, S.; Nakanishi, T.; Obika, S.; Imanishi, T.; Egawa, T. et al. *J. Leukocyte Biol.* **2003**, *73*, 273–280.
118. Tokuyama, H.; Ueha, S.; Kurachi, M.; Matsushima, K.; Moriyasu, F.; Blumberg, R. S.; Kakimi, K. *Int. Immunol.* **2005**, *17*, 1023–1034.
119. Singh, U. P.; Singh, S.; Taub, D. D.; Lillard, J. W., Jr. *J. Immunol.* **2003**, *171*, 1401–1406.
120. Heise, C. E.; Pahuja, A.; Hudson, S. C.; Mistry, M. S.; Putnam, A. L.; Gross, M. M.; Gottlieb, P. A.; Wade, W. S.; Kiankarimi, M.; Schwarz, D. et al. *J. Pharmacol. Exp. Ther.* **2005**, *313*, 1263–1271.
121. Medina, J. C.; Johnson, M. G.; Li, A.-R.; Huang, A. X.; Zhu, L.; Marcus, A. P. CXCR3 Antagonists. PCT Patent WO2002/083143, 2002.
122. Collins, T. L.; Johnson, M. G.; Ma, J.; Medina, J. C.; Miao, S.; Schneider, M.; Tonn, G. R. CXCR3 Antagonists. PCT Patent WO2004/075863, 2004.
123. Papadakis, K. A.; Landers, C.; Prehn, J.; Kouroumalis, E. A.; Moreno, S. T.; Gutierrez-Ramos, J. C.; Hodge, M. R.; Targan, S. R. *J. Immunol.* **2003**, *171*, 159–165.
124. Zabel, B. A.; Agace, W. W.; Campbell, J. J.; Heath, H. M.; Parent, D.; Roberts, A. I.; Ebert, E. C.; Kassam, N.; Qin, S.; Zovko, M. et al. *J. Exp. Med.* **1999**, *190*, 1241–1256.
125. Papadakis, K. A.; Prehn, J.; Moreno, S. T.; Cheng, L.; Kouroumalis, E. A.; Deem, R.; Breaverman, T.; Ponath, P. D.; Andrew, D. P.; Green, P. H. et al. *Gastroenterology* **2001**, *121*, 246–254.
126. Svensson, M.; Marsal, J.; Ericsson, A.; Carramolino, L.; Broden, T.; Marquez, G.; Agace, W. W. *J. Clin. Invest.* **2002**, *110*, 1113–1121.
127. Mora, J. R.; Bono, M. R.; Manjunath, N.; Weninger, W.; Cavanagh, L. L.; Rosemblatt, M.; Von Andrian, U. H. *Nature* **2003**, *424*, 88–93.
128. Fleming, P.; Harriman, G. C. B.; Shi, Z.; Chen, S. CCR9 Inhibitors and Methods of Use Therof. PCT Patent WO2003/099773, 2004.
129. Ugashe, S.; Zheng, W.; Wright, J. J.; Pennel, A. Aryl Sulphonamides. PCT Patent WO2004/046092, 2004.
130. Iwata, M.; Hirakiyama, A.; Eshima, Y.; Kagechika, H.; Kato, C.; Song, S. Y. *Immunity* **2004**, *21*, 527–538.
131. Dumont, F. *IDrugs* **2005**, *8*, 236–253.
132. Mizushima, T.; Ito, T.; Kishi, D.; Kai, Y.; Tamagawa, H.; Nezu, R.; Kiyono, H.; Matsuda, H. *Inflamm. Bowel Dis.* **2004**, *10*, 182–192.
133. Shimizu, H.; Takahashi, M.; Kaneko, T.; Murakami, T.; Hakamata, Y.; Kudou, S.; Kishi, T.; Fukuchi, K.; Iwanami, S.; Kuriyama, K. et al. *Circulation* **2005**, *111*, 222–229.
134. Colandrea, V. J.; Doherty, G. A.; Hale, J. J.; Lynch, C.; Mills, S. G.; Neway, W. E.; Toth, L. 3-(2-Amino-1-Azacyclyl)-5-Aryl-1,2,4-Oxadiazoles as S1P Receptor Agonists. PCT Patent WO2004/103279, 2004.
135. Ohkawara, T.; Miyashita, K.; Nishihira, J.; Mitsuyama, K.; Takeda, H.; Kato, M.; Kondo, N.; Yamasaki, Y.; Sata, M.; Yoshiki, T. et al. *Clin. Exp. Immunol.* **2005**, *140*, 241–248.

136. de Jong, Y. P.; Abadia-Molina, A. C.; Satoskar, A. R.; Clarke, K.; Rietdijk, S. T.; Faubion, W. A.; Mizoguchi, E.; Metz, C. N.; Alsahli, M.; ten Hove, T. et al. *Nat. Immunol.* **2001**, *2*, 1061–1066.
137. Leng, L.; Metz, C. N.; Fang, Y.; Xu, J.; Donnelly, S.; Baugh, J.; Delohery, T.; Chen, Y.; Mitchell, R. A.; Bucala, R. *J. Exp. Med.* **2003**, *197*, 1467–1476.
138. Orita, M.; Yamamoto, S.; Katayama, N.; Fujita, S. *Curr. Pharm. Des.* **2002**, *8*, 1297–1317.
139. Lubetsky, J. B.; Dios, A.; Han, J.; Aljabari, B.; Ruzsicska, B.; Mitchell, R.; Lolis, E.; Al-Abed, Y. *J. Biol. Chem.* **2002**, *277*, 24976–24982.
140. Cvetkovic, I.; Al-Abed, Y.; Miljkovic, D.; Maksimovic-Ivanic, D.; Roth, J.; Bacher, M.; Lan, H. Y.; Nicoletti, F.; Stosic-Grujicic, S. *Endocrinology* **2005**, *146*, 2942–2951.
141. Ying, W.; Li, X.; Baclig, R.; Scholz, W.; Kumar, S.; Sircar, J. American Physiological Society; Translational Research Conference: Immunological and Pathophysiological Mechanisms in Inflammatory Bowel Disease Abstract 4.39. Sept. 8–11, 2004; Snowmass, Colorado.
142. Sircar, J.; Kumar, S.; Ying, W. Inhibitors of Macrophage Migration Inhibitory Factor and Methods for Identifying the Same. PCT Patent WO2004/074218, 2004.
143. Kimball, E. S.; Wallace, N. H.; Schneider, C. R.; D'Andrea, M. R.; Hornby, P. J. *Neurogastroenterol. Motil.* **2004**, *16*, 811–818.
144. El Kouhen, R.; Surowy, C. S.; Bianchi, B.; Neelands, T.; McDonald, H.; Niforatos, W.; Gomtsyan, A.; Lee, C.-H.; Honore, P.; Sullivan, J. et al. *J. Pharmacol. Exp. Ther.* **2005**, *414*, 400–409.
145. Rami, H. K. In *Anglo-Swedish Medicinal Chemistry, 2nd Symposium*, Are Fjallby, Sweden, Mar 13–16, 2005.
146. Rami, H. K.; Thompson, M.; Wyman, P.; Jerman, J. C.; Egerton, J.; Brough, S.; Stevens, A. J.; Randall, A. D.; Smart, D.; Gunthorpe, M. J. et al. *Bioorg. Med. Chem. Lett.* **2004**, *14*, 3631–3634.
147. Hassani, H.; Lucas, G.; Rozell, B.; Ernfors, P. *Am. J. Physiol. Gastrointest. Liver Physiol.* **2005**, *288*, G550–G556.
148. Stucchi, A. F.; Shofer, S.; Leeman, S.; Materne, O.; Beer, E.; McClung, J.; Shebani, K.; Moore, F.; O'Brien, M.; Becker, J. M. *Am. J. Physiol. Gastrointest. Liver. Physiol.* **2000**, *279*, G1298–G1306.
149. Matsumoto, S.; Okabe, Y.; Setoyama, H.; Takayama, K.; Ohtsuka, J.; Funahashi, H.; Imaoka, A.; Okada, Y.; Umesaki, Y. *Gut* **1998**, *43*, 71–78.
150. Podolsky, D. K.; Lobb, R.; King, N.; Benjamin, C. D.; Pepinsky, B.; Sehgal, P.; deBeaumont, M. *J. Clin. Invest.* **1993**, *92*, 372–380.
151. Maeda, S.; Hsu, L. C.; Liu, H.; Bankston, L. A.; Iimura, M.; Kagnoff, M. F.; Eckmann, L.; Karin, M. *Science* **2005**, *307*, 734–738.
152. Wirtz, S.; Finotto, S.; Kanzler, S.; Lohse, A. W.; Blessing, M.; Lehr, H. A.; Galle, P. R.; Neurath, M. F. *J. Immunol.* **1999**, *162*, 1884–1888.
153. Powrie, F.; Leach, M. W.; Mauze, S.; Caddle, L. B.; Coffman, R. L. *Int. Immunol.* **1993**, *5*, 1461–1471.
154. Mashimo, H.; Wu, D. C.; Podolsky, D. K.; Fishman, M. C. *Science* **1996**, *274*, 262–265.
155. Kuhn, R.; Lohler, J.; Rennick, D.; Rajewsky, K.; Muller, W. *Cell* **1993**, *75*, 263–274.
156. Shull, M. M.; Ormsby, I.; Kier, A. B.; Pawlowski, S.; Diebold, R. J.; Yin, M.; Allen, R.; Sidman, C.; Proetzel, G.; Calvin, D. et al. *Nature* **1992**, *359*, 693–699.
157. McDonald, S. A.; Palmen, M. J.; Van Rees, E. P.; MacDonald, T. T. *Immunology* **1997**, *91*, 73–80.
158. Kontoyiannis, D.; Pasparakis, M.; Pizarro, T. T.; Cominelli, F.; Kollias, G. *Immunity* **1999**, *10*, 387–398.
159. Hornquist, C. E.; Lu, X.; Rogers-Fani, P. M.; Rudolph, U.; Shappell, S.; Birnbaumer, L.; Harriman, G. R. *Immunology* **1997**, *158*, 1068–1077.
160. Esworthy, R. S.; Aranda, R.; Martin, M. G.; Doroshow, J. H.; Binder, S. W.; Chu, F. F. *Am. J. Physiol. Gastrointest. Liver. Physiol.* **2001**, *281*, G848–G855.
161. Mizoguchi, A.; Mizoguchi, E.; Chiba, C.; Spiekermann, G. M.; Tonegawa, S.; Nagler-Anderson, C.; Bhan, A. K. *J. Exp. Med.* **1996**, *183*, 847–856.
162. Welte, T.; Zhang, S. S.; Wang, T.; Zhang, Z.; Hesslein, D. G.; Yin, Z.; Kano, A.; Iwamoto, Y.; Li, E.; Craft, J. E. et al. *Proc. Natl. Acad. Sci. USA* **2003**, *100*, 1879–1884.

Biographies

Nick Pullen graduated in biochemistry and chemistry and completed his PhD in G protein-coupled receptor signaling from the University of Southampton. From there he moved to the Novartis-funded Friedrich Miescher Institute in Basel, to pursue postdoctoral studies on signaling mechanisms involved in diabetes and oncology. He joined the Pfizer Gastrointestinal therapeutic area in 1999.

Jeremy D Gale has 17 years pharmaceutical industry experience in Discovery and Exploratory Development. He graduated in pharmacology and completed his PhD in pharmacology and neuroscience from the University of London. Jem commenced his career with Glaxo in 1988 to work in gastrointestinal pharmacology, ultimately leading their Discovery biology programs in reflux disease, irritable bowel syndrome, and emesis. He joined Pfizer in 1995 to lead Discovery Biology's Gastrointestinal research team and moved into Exploratory Clinical Development in 2002.

Comprehensive Medicinal Chemistry II
ISBN (set): 0-08-044513-6

ISBN (Volume 6) 0-08-044519-5; pp. 613–642

6.29 Irritable Bowel Syndrome

G A Hicks, Novartis Pharmaceuticals Corporation, East Hanover, NJ, USA

6.29.1 Disease State

6.29.1.1 Epidemiology and Socioeconomic Burden

Irritable bowel syndrome (IBS), characterized by abdominal pain or discomfort associated with altered bowel habits, is one of the most prevalent disorders of the gastrointestinal (GI) tract. Estimates of the prevalence of IBS in Western countries typically range from 10% to 15%,[1–5] although the prevalence in the United States has been estimated to be as high as 20%.[5,6] Differences in these estimates probably result from the use of different diagnostic criteria and sampling methods, rather than actual variations in national prevalence.[1,4,5] Lower prevalence rates generally have been reported in non-Western and developing countries,[7–13] but it is interesting to note that the rapid economic development of many Asian countries has been accompanied by an increasing prevalence of IBS.[6,14]

Gender appears to be a risk factor for this disorder, but also contributes to care-seeking behavior. In Western populations, approximately twice as many women as men report symptoms consistent with IBS.[4,15,16] Among persons who seek medical care for their symptoms, the ratio of women to men is 3:1 in the primary care setting and 4–5:1 in the tertiary care setting.[15,17,18] However, in India and Sri Lanka, a greater proportion of men than women seek care for their IBS symptoms.[6,16–18]

Although not life-threatening, IBS is a severe disorder that produces quantifiable suffering. IBS symptoms have a markedly negative impact on the quality of life, personal relationships, social lives, daily activities, and work productivity of affected persons.[19,20] In one health-related quality-of-life (HRQoL) study, IBS sufferers, especially those with IBS with constipation (IBS-C), were found to have poorer HRQoL, based on SF-36 (Medical Outcome Study Short Form 36) scores, than individuals with asthma (women), migraine, or gastroesophageal reflux disease.[21]

A survey of US residents with medically diagnosed and nonmedically diagnosed IBS found that diet and food choice, going out for a meal, undertaking long trips, and going on vacation were significantly more problematic than for controls.[4] Almost one quarter of affected individuals had missed social engagements.[4] Eleven percent had missed work, nearly one quarter worked fewer hours, and 67% felt less productive at work (presenteeism) due to their symptoms.[4] The average number of sick days leading to absenteeism was 6.4 in persons with IBS versus 3.0 in controls.[4]

IBS symptoms also have a profound impact on healthcare utilization, managed care organizations, and employer costs. In the US, estimated annual direct costs of IBS treatment, excluding prescription and over-the-counter (OTC) drug costs, are between $1.7–10 billion.[22] In addition, the estimated indirect costs of IBS, which are largely borne by employers, are at least $20 billion.[23,24]

In the US, IBS is the sixth leading diagnosis for GI disorders made during outpatient clinic visits.[25] Not surprisingly, it is responsible for 12% of diagnoses made in primary care practices[26] and 28% of diagnoses made by gastroenterologists.[27] In 1998 alone, IBS led to 3.65 million physician visits.[24] The increased use of healthcare resources by patients with IBS results in substantial managed care costs. In one managed care organization, over a 1-year period, total costs for IBS patients were 51% higher (based on an adjusted mean difference of $1340.55) for patients with IBS ($3729) versus patients without IBS ($2608).[28] In an employer-based survey ($n = 1776$), employees with IBS reported a 21.1% mean reduction in overall productivity – equivalent to the loss of 1 day in a 40-hour work week – compared with to a 6.1% reduction experienced by employees without IBS.[29] Results of a study conducted at a Fortune, 100 company showed that employees with IBS cost the employer 1.5 times more on average than matched employees without IBS ($6364 versus $4245; $P < 0.001$); missed workdays and disability resulting from IBS accounted for 37% of the employer's total IBS-related costs.[30]

6.29.1.2 Physical Description of Symptoms

The multiple symptom complex of IBS encompasses lower-GI, upper-GI, and extraintestinal manifestations (**Table 1**).[5,31–34] In line with the Rome II diagnostic criteria, the cardinal symptom of IBS is abdominal pain/discomfort associated with three of three features: (1) relieved with defecation; and/or (2) onset is associated with a change in stool frequency; and/or (3) onset is associated with a change in stool form.[5,31] The pain can be generalized or localized, usually occurs in the lower abdomen, and is generally relieved after a bowel movement.[35,36] Patients with IBS are subcategorized further based on supportive symptoms. The presence of hard/lumpy stools, straining, bloating, and a feeling of incomplete evacuation are consistent with IBS-C, whereas the presence of loose/watery stools and urgency is consistent with IBS with diarrhea (IBS-D). Some patients with IBS alternate between these two categories: this subgroup is referred to as alternating IBS (IBS-A). At any time, most patients generally report one or the other as their main bowel habit. Patients with IBS may also experience numerous upper-GI (e.g., indigestion, nausea, dyspepsia)[32,33] and extraintestinal manifestations (e.g., urinary frequency, sexual dysfunction, poor sleep).[33–35] There is usually a positive correlation between number of extraintestinal symptoms and severity of IBS.[32,35]

Table 1 Select symptoms of IBS

Cardinal symptoms of IBS[5,31]

Abdominal pain/discomfort associated with altered bowel habits (stool frequency or stool form, including constipation, diarrhea, or both in alternation)

Lower GI symptoms

- Hard or lumpy stools[a]
- Straining[a]
- Feeling of incomplete evacuation[a]
- Bloating
- Presence of mucus in stool
- Loose watery stools[b]
- Urgency[b]

Upper GI symptoms[32,33]

- Indigestion
- Nausea
- Early satiety
- Loss of appetite
- Heartburn
- Vomiting

Extraintestinal manifestations[33,34]

- Back pain
- Headache
- Fatigue
- Poor sleep
- Decreased sex drive
- Shortness of breath/wheezing
- Muscle aches or soreness
- Dyspareunia
- Palpitations or heart pounding
- Dizziness
- Stiffness

[a] Supportive of IBS-C.
[b] Supportive of IBS-D.

6.29.2 Disease Basis

6.29.2.1 Basic Neurophysiology of the Gastrointestinal Tract

6.29.2.1.1 Neuronal control of gastrointestinal function

Consisting of as many neurons as are found in the spinal cord ($\sim$100 million),[37,38] the enteric nervous system (ENS), composed of the myenteric and submucosal plexuses,[39] can micromanage peristaltic and secretory reflexes of the intestines in an essentially independent manner (i.e., in the absence of control from the brain, spinal cord, dorsal root ganglia, or cranial nerve ganglia)[40]; it thus often is referred to as the second brain.[41–44]

To organize and manage the behavior patterns of the intestines, the ENS must possess the ability to sense the conditions prevailing in the enteric lumen (e.g., pressure, pH, nutrient status). In the absence of nerve fibers in the GI lumen, sensory signaling occurs transepithelially via enteroendocrine cells. An example of these is the enterochromaffin (EC) cells, the primary storage sites for serotonin (or 5-hydroxytryptamine (5HT)). Approximately 95% of the 5HT produced by the body is found in the bowel.[40] The release of 5HT from EC cells is a pivotal first step in the initiation of gut motility and secretion.[40]

Although able to perform its basic functions in isolation, neural regulation of the ENS is composed of both intrinsic and extrinsic components.[43,45–47] Intrinsically, gut activity is monitored by intrinsic primary afferent neurons (IPANs) located in the submucosal and myenteric plexuses.[45,47–49] IPANs are the first neurons to receive the signal from transmitters such as 5HT released from enteroendocrine cells and enable the ENS to mediate reflex responses independently of the central nervous system (CNS) influence. The CNS uses peripheral sensors to monitor the environment and transmits the information to the brain via dorsal root and cranial nerve ganglion cells. Similarly, IPANs are the first neuronal component in the intrinsic sensory system of the bowel responding to luminal stimuli. However, a key difference between CNS primary afferent nerves and IPANs is that the latter do not directly sense the luminal content. IPANs work by transmitting information from the enteroendocrine cell sensors (e.g., EC cells) to the motor neurons of the submucosal and myenteric plexuses. Submucosal IPANs, which secrete calcitonin gene-related peptide (CGRP) and acetylcholine (ACh), appear to be critical for mucosal-related actions, including peristaltic and secretory reflexes. Unlike the case with dorsal root and cranial nerve sensory ganglion neurons, IPANs are innervated, allowing them both to act as interneurons and initiate gut-related reflexes.[40]

Extrinsic innervation of the gut is composed of the two anatomically integrated branches of the autonomic nervous system (sympathetic and parasympathetic), with sensory afferent nerves traveling along the efferent fibers in the same nerve bundles.[41,49] The spinally directed afferent neurons travel primarily with the sympathetic pathway, composed of splanchnic (spinal) primary afferent neurons, but also in the pelvic nerve alongside parasympathetic fibers innervating the colon. These nerves are involved in the transmission of sensory signals that are perceived (e.g., pain, bloating, discomfort), and the overall action of the sympathetic efferent limb is to reduce GI activity via stimulation of contraction of sphincteric muscle and relaxation of nonsphincteric muscle. The vagal afferent fibers travel in parallel with the vagal efferent fibers in the vagus nerve. The vagal afferents relay primarily nonperceived (chemosensory, motor) gut-related physiological activity to the brain (although nausea is also mediated via signals beginning in the vagus nerve), and parasympathetic efferents exert an excitatory influence on enteric neurons (stimulate GI activity). Overall, the parasympathetic nervous system provides stimulatory activation of nonsphincteric muscle.

6.29.2.1.2 Key roles of 5-hydroxytryptamine in gastrointestinal function

Serotonin is a major player in the overall functioning of the bowel, with 95% of the serotonin produced by the body found in the GI mucosa, from which it is released as the pivotal first step in the initiation of gut motility and secretion.[40] Because evidence exists for a disruption of serotonin signaling in the GI tract as a pathophysiological factor in IBS and other GI diseases (discussed next) and because the major therapeutic agents currently used in the treatment of patients with IBS act upon the serotonin system, it is important that we consider this system in some detail in this review of the IBS disease area.

Of numerous neurotransmitters involved in communication along the brain–gut axis, 5HT is a common link in several key processes of the GI tract, including GI motility, intestinal secretion, and perception of pain (**Table 2**).[45,50–52]

Motor and secretory reflexes (e.g., peristalsis) are initiated via intrinsic ENS neurons, whereas bowel-related sensations (e.g., bloating, pain) are initiated via extrinsic afferent nerves.[5,45,53–57]

6.29.2.1.2.1 5-Hydroxytryptamine and gastrointestinal motility and secretion

5HT is involved in initiating and maintaining a cascade of coordinated events that comprise the peristaltic reflex (**Figure 1**),[51,52,58,59] which is triggered by distension of the gut lumen and/or mechanical disturbance of intestinal villi by a food bolus. These stimuli result in release of 5HT from the EC cells. There has been debate in the literature around the subtype of receptor responsible for the initiation of the ensuing enteric reflexes (*see* Section 6.29.2.1.2.3),

Table 2 Key mediators of gut function[50–52]

Motility	*Secretion*	*Visceral sensation*
Serotonin	Serotonin	Serotonin
Acetylcholine	Acetylcholine	Tachykinins
Nitric oxide	Vasoactive intestinal peptide	Calcitonin generelated peptide
Substance P		Neurokinin A
Vasoactive intential peptide		Enkephalins
Cholecystokinin		

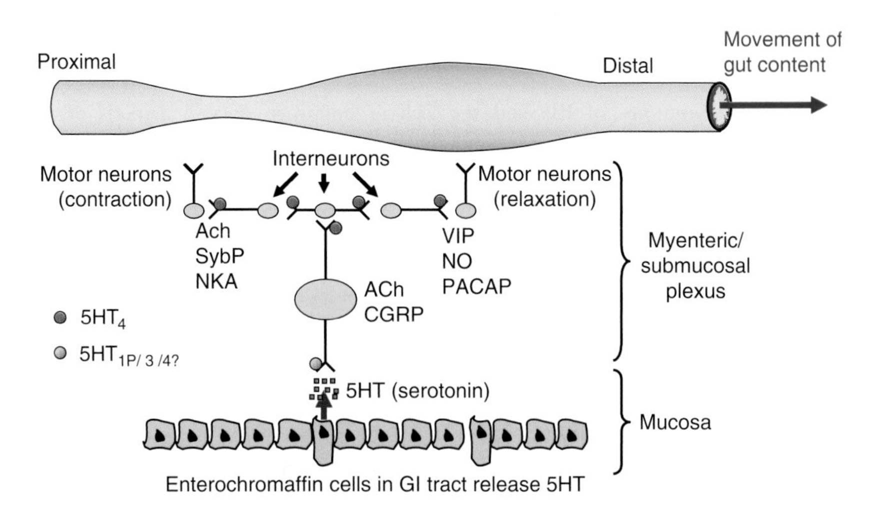

Figure 1 The peristaltic reflex.[51,52,58,59] (Adapted from Grider, J. R.; Foxx-Orenstein, A. E.; Jin, J. G. *Gastroenterology* **1998**, *115,* 370–380 © 1998. With permission from the American Gastroenterological Association. A version of this figure was originally published in Baker, D. E. *Am. J. Health Syst. Pharm.* **2005**, *62*, 700–711 © 2005, American Society of Health-System Pharmacists, Inc. All rights reserved. Adapted with permission.)

and in humans, the answer to this question remains unclear. However, from studies in animals, the current hypothesis is that 5HT released from EC cells binds to $5HT_{1P}$[60] and $5HT_3$[61] receptors on mucosal terminals of IPANs, initiating a nerve signal that triggers IPANs to release the excitatory neurotransmitters CGRP and ACh on to postsynaptic neurons within the ganglia. This signal then travels along proximal and distal interneurons to create proximal contraction and distal relaxation of circular smooth muscle, resulting in propulsive waves that propel the bolus forward. $5HT_4$ receptors are located presynaptically on terminals of IPANs and interneurons,[62–64] and in this position their activation results in enhancement, or amplification, of peristaltic neurotransmission via increased release of additional neurotransmitters (e.g., ACh and tachykinins).[45,52,65,66] Thus, initiation of peristalsis involves activation of $5HT_{1P}$ and $5HT_3$ receptors, whereas $5HT_4$ amplifies the response. Here, it might be useful to think of a radio, in which $5HT_{1P}$ and $5HT_3$ receptors are the aerial antennae receiving the signal and $5HT_4$ receptors turn up the volume.[45,65–67] 5HT is also involved in maintaining stool water content. For example, activation of $5HT_4$ receptors results in increased secretion of chloride ions, enhancing the net movement of water into the intestine.[68,69]

6.29.2.1.2.2 5-Hydroxytryptamine and visceral sensation

With greater than 90% of vagal fibers being afferent nerves (i.e., communicating information from the gut to the brain), sensory perception is a critical aspect of the bowel.[40] 5HT acts primarily via $5HT_3$ receptors to activate vagal and spinal afferent nerves.[70,71] However, recent data demonstrate an additional role for $5HT_4$ in the inhibition of spinal afferents, although whether this is a direct effect is uncertain.[72–74] Overall, 5HT is among the key neurotransmitters involved in modulating pain transmission signals from the bowel to the CNS[45] and therefore has an important role to play in the sensory symptoms of patients with IBS.[45,54]

6.29.2.1.2.3 5-Hydroxytryptamine receptors

To date, 14 subtypes of 5HT receptors have been identified; those considered the most relevant to motor, secretory, and sensory activities of the lower GI tract include $5HT_{1P}$, $5HT_3$, and $5HT_4$, but several others have also been implicated, including $5HT_1$, $5HT_2$, and $5HT_7$. Because effective therapeutic agents have been developed for both $5HT_3$ and $5HT_4$ subtypes, these are described in Section 6.29.5. However, an interesting recent development around $5HT_{1P}$ receptor deserves mention here. Although a distinct $5HT_{1P}$ receptor has not yet been cloned, its numerous actions on the submucosal plexus of the ENS have been identified and are referred to as $5HT_{1P}$ activity. Most recent evidence suggests that the molecular entity that underlies this $5HT_{1P}$ activity is a dimer, formed from the association of $5HT_{1B}$ and dopamine D_2 receptors.[75] As discussed, $5HT_{1P}$ receptor activation is responsible for initiating peristaltic and secretory reflexes by activating the submucosal IPANs.[40]

6.29.2.2 Clinical Observations with Potential Links to Irritable Bowel Syndrome Symptoms

Given the heterogeneity of symptoms of IBS and the lack of definitive organic markers of this disorder, the development of a unifying hypothesis explaining its underlying pathophysiology has been challenging. Nonetheless, substantial progress has been made. Whereas three decades ago the etiology of IBS focused on altered GI motility, research advances (particularly during the last 15–20 years) have led to a greater understanding of the additional critical role played by efficient communications along the brain–gut axis, particularly with respect to sensory symptoms of the disease (abdominal pain, bloating, urgency, incomplete evacuation). The putative role of immune-system alterations, genetic links, and environmental factors has also been studied. Although this research has brought us closer than ever before to understanding IBS as a real condition with a pathophysiological basis, there is a long way to go before the underlying mechanisms of this complex disorder are fully defined.

6.29.2.2.1 Gastrointestinal dysmotility

The GI tract neuromuscular system necessarily is highly coordinated, such that preprogrammed patterns of neuronally driven contractile activity, along with local and extrinsic modulating inputs based upon environmental factors, provide for the effective digestion of food and elimination of waste. Simply put, gut contents need to spend certain amounts of time in specific parts of the GI tract, where they may need to be mixed and distributed along its surface area to allow the addition of digestive enzymes and absorption of nutrients before subsequently being expelled. This is a highly regulated system that requires precise orchestration. Dysmotility describes a state in which the GI tract fails to perform the necessary concerto of contractile activity, such that GI transit is impaired and symptoms result. The orchestra analogy is perhaps useful here because there can be multiple reasons the system breaks down: the extrinsic control or conductor (the CNS via the efferent autonomic inputs), the local controllers or first violin (the intrinsic sensory neuron that sets the pace of reflex activity in the ganglia), or the musicians (the nerves, muscles, interstitial cells, inflammatory cells, neurotransmitters, etc.) who actually perform the work in a very professional and attentive manner, such that the output is concordant and comfortable.

Local contractile activity needs to follow specific patterns so that it is propulsive when it is time to move gut contents along the lumen. As discussed, this propulsive contractile activity is called peristalsis, and this pattern consists of contraction of the muscle behind the contents and relaxation ahead of them so that the contents are pushed along from behind, like squeezing the toothpaste along and out of the tube. An example of dysmotility is a breakdown of this peristaltic reflex, such that the contractile pattern becomes uncoordinated and nonpropulsive. This type of dysmotility can therefore result in impaired or slow intestinal transit, resulting in constipation. Other patterns of activity also occur, and dysmotility in the upper-GI tract, e.g., leading to reduced gastric emptying or accommodation, is associated with symptoms of dyspepsia and reflux disease. It is beyond the scope of this chapter to describe them all, but some have been linked to IBS symptoms and deserve mention. The migrating motor complex describes the cyclical motor activity that normally occurs approximately once every 60–90 min in the stomach and small bowel in the unfed state. Abnormalities in contractile activity of the colon and small bowel have been suggested to be present in between 25% and 75% of patients with IBS.[76] Patients with IBS-C have fewer than normal fast colonic and propagated contractions, as well as fewer high-amplitude propagated contractions (resulting in slowing of whole-gut transit). Conversely, patients with IBS-D may have a greater than normal number of fast colonic contractions and propagated contractions (resulting in accelerated motility).[76] However, no single motility abnormality has been demonstrated consistently in patients with IBS, and these abnormalities are not always associated with IBS symptoms. Furthermore, the specificity of the qualitative intestinal motor changes is low (i.e., similar changes can occur in non-IBS conditions, as well as in healthy individuals).[77] Therefore, although the presence of motility abnormalities is clear, they are not considered a diagnostic marker for IBS.[55,77]

An interesting area of study in recent years has been the dynamics of intestinal transit of gas.[78,79] Gas, either swallowed (or not expelled via a belch) during a meal or produced locally by gut flora, also needs to be moved along the gut lumen and expelled as flatus. When gas transit is impaired, symptoms of bloating and abdominal distension occur, and this is common in patients with IBS-C. While studies have not demonstrated abnormalities in the amount or content of intestinal gas in these patients, it has been shown that they may expel gas more slowly than normal through the GI tract and/or experience reduced tolerance to gas sensation (i.e., report more abdominal discomfort than healthy subjects).[80] Studies of patients with IBS have also shown that ingestion of dietary lipids results in delayed intestinal gas transit and symptoms of abdominal bloating.[81,82] Therapy with promotility agents, which provide relief of bloating in patients with IBS-C, has been demonstrated to accelerate transit of gas from the small to the large intestine,[83] whereas the opposite is true for fiber supplements, which are linked to a worsening of bloating in these patients.[84] The link between gas transit, sensory symptoms, and the efficacy of therapeutic agents will be an interesting area of future research.

6.29.2.2.2 Altered brain–gut axis communication

Normal GI function (including the processes of intestinal motility, secretion, and sensation) depends on efficient communications along bidirectional parallel circuits (known as the brain–gut axis) that integrate intestinal motor and sensory activities occurring in the ENS and extrinsic sensory nerves (spinal and vagal) with activities in the autonomic nervous system and CNS. The presence of dysregulated interactions at any level along this axis can lead to the cardinal IBS symptoms of altered GI motility, altered intestinal secretion, and enhanced visceral sensation.[47,52] Put simply, although the gut can contract and secrete in complete isolation, the extrinsic innervation provided by vagal and spinal sensory afferent and autonomic efferent nerves is essential to the coordination of overall digestive function and sensory perception, when required.

6.29.2.2.3 Visceral sensitivity and hypersensitivity

As discussed, although the ENS can operate essentially independently of CNS input, the two nervous systems usually work together to ensure proper GI function. It is therefore necessary for the CNS to be aware of events occurring in the bowel, particularly the enteric lumen.[40] Such CNS awareness may be unconscious information concerning content of the bowel (chemosensory, mechanosensory, pH, etc.) or the conscious perception of sensations such as the presence of stool, the urge to defecate, bloating, or pain. Although we have all experienced noxious sensations from our bowel, such as nausea, or bloating, or pain, alterations in CNS integration and processing of signals along the brain–gut axis may lead to heightened (and inappropriate) awareness of bowel events, such that normal activity is perceived as noxious. Such inappropriate sensations resulting from visceral hypersensitivity are thought to play a key role in IBS symptoms.[85–89] Thus, visceral hypersensitivity describes a phenomenon by which patients experience either an exaggerated response to normal stimuli or increased sensitivity to painful stimuli in the bowel. It is a common manifestation of IBS and may underlie the most distressing symptoms of the disorder.[40,45,87] Various balloon distension studies (designed to detect patients' discomfort and pain thresholds) have demonstrated that patients with IBS sense pain at lower levels of balloon inflation (in the rectum, as well as other intestinal areas) compared with healthy controls or patients with other GI disorders, confirming the presence of differences in visceral perception between IBS and non-IBS patient populations.[55,87,90,91] Interestingly, although this hypersensitivity appears not to be restricted to specific areas of the GI tract, patients with IBS have not been shown to experience generalized hypersensitivity to painful somatic stimuli.[55]

Brain-imaging studies have demonstrated differences in central mechanisms for pain modulation between patients with IBS and healthy controls. Patients with IBS may differ from patients without IBS in the way signals from the gut are received and/or processed by the CNS. For example, using positron emission tomography, one study demonstrated that the anterior cingulate cortex (ACC) was activated upon colorectal distension in healthy volunteers, but not in patients with IBS.[92] In another study using positron emission tomography, patients with IBS were shown to have increased activation of the dorsal ACC (involved in perception of emotional stimulus, such as fear, anxiety) and decreased activation of the periaquiductal gray region (involved in inhibition of endogenous pain). This combination of enhanced activation of perception to visceral stimulus and potential deficiency in cortical activation of endogenous pain inhibitory mechanisms may account for the visceral hypersensitivity commonly experienced by patients with IBS.[93]

In a study using functional magnetic resonance imaging, upon painful visceral stimulation, the ACC was activated in both patients with IBS and controls; however, those with IBS experienced greater pain sensitivity and heightened perception of visceral afferent signals along the brain–gut axis. These signals corresponded with subjective pain symptoms.[94] Based on results of brain imaging studies in patients with IBS for whom aversive visceral stimuli did not lead to increased activation of the insular cortex (which acts as the viscerosensory cortex), it has been suggested that the visceral afferent input received by the brain from the colon is not increased, but rather the management of the peripheral signal is enhanced.[95]

6.29.2.2.4 5-Hydroxytryptamine signaling abnormalities: role of altered release and/or reuptake

The importance of effective 5HT signaling in the ENS and the brain–gut axis to normal gut function was discussed. Perhaps unsurprisingly, recent evidence suggests that abnormalities in serotonin signaling play a critical role in IBS pathophysiology.[45,76,96,97] 5HT signaling refers to numerous components, including the synthesis, storage, release, receptor activation, reuptake, and degradation of 5HT. Research advances have demonstrated that patients with IBS may have alterations in several of these elements.[97–103]

Following a meal, 5HT is released into the GI tract wall. The distance separating EC cells from the nerves upon which they act is large; therefore, delivery of sufficient quantities of 5HT to neural receptors necessitates the secretion of very large amounts (both constitutively and especially upon stimulation). This mechanism leads to an overflow of 5HT into the portal circulation and intestinal lumen. Because extracellular enzymes do not catabolize 5HT, termination of signaling requires an efficient removal mechanism from the extracellular space to prevent overstimulation (potentially leading to diarrhea), eventual receptor desensitization (leading to constipation), or frank toxicity.[40] This is primarily accomplished through a specific serotonin reuptake transporter (SERT). This transporter is present in the plasma membranes of serotonergic neurons in the ENS and the brain, but, most importantly for our purposes, also in epithelial cells of the GI mucosa.[40,52]

Some researchers have used postprandial levels of 5HT as a surrogate marker for its release in the GI tract.[100,102,103] Although there are recognized issues with such techniques, direct measurement of release from the bowel wall in human subjects is clearly problematic. Others have tried to circumvent these problems by using in vitro techniques to study release from tissue biopsy specimens,[97] but this approach has its own inherent drawbacks, not the least of which is that the studies cannot be correlated with meal ingestion, symptoms, and so on, and the obvious loss of all extrinsic influences present in the in vivo state. However, these issues being accepted, the evidence supports a role for a dysfunction in serotonin release and/or reuptake in patients with IBS.

The easiest of these data to understand are those demonstrating altered postprandial serotonin levels in patients with IBS with differing phenotypes of the disease. Put simply, there appear to be exaggerated postprandial plasma levels of serotonin in patients whose major GI symptom is diarrhea[101,103] and reduced postprandial levels in those whose primary symptom is constipation.[100] The extrapolation to a conclusion that IBS-D is due to exaggerated release of 5HT and IBS-C is due to a blunting of this release is a tempting one, but the definitive experiments are still required.

A role for changes in SERT activity in patients with GI dysfunction has been demonstrated in both animal models and human studies. In a genetically manipulated mouse model, deletion of the SERT gene led to symptom manifestations of IBS-A: i.e., diarrhea, presumably caused by excessive 5HT signaling, and constipation, perhaps caused by desensitized 5HT receptors or hyperexcitability of enteric nerves, such that the coordination of reflexes breaks down.[104] Inflammation also can regulate SERT expression. In a study using a guinea pig model, during the first 7 days, induction of ileitis led to reduced SERT expression, an increased number of EC cells, and increased 5HT availability, with levels normalized by day 14.[105] Studies evaluating patients with IBS-C, IBS-D, and ulcerative colitis demonstrated reduced SERT gene and protein expression in the GI mucosa compared with healthy controls, supporting the concept that a diminution in SERT activity can lead to altered GI function.[97] This study also demonstrated molecular differences in other 5HT signaling components in patients with IBS compared with controls. Although numbers of EC cells and mucosal 5HT release did not differ, reductions in levels of tryptophan hydroxylase 1 (the rate-limiting enzyme in 5HT synthesis), SERT immunoreactivity, and 5HT content were shown.[97] Although the precise nature of the serotonin signaling dysfunction is unclear at this time, it is clear that the system is plastic, regulated, and plays a role in both physiology and pathophysiology of the bowel.

6.29.2.2.5 Role of stress/psychosocial factors

The presence of acute stress leads to release of stress-related hormones (e.g., corticotropin-releasing factor (CRF), associated with the hypothalamic–pituitary axis), which may lead to altered effects on GI motility, sensation, and inflammation. Studies using rat models have shown that acute stress affects the content of histamine in mast cells in the gut via release of interleukin-1 and CRF.[106] A recent study has demonstrated that infusion of CRF in patients with IBS leads to an exaggerated colonic motility response (likely as a result of CRF-induced activation of mast cells and release of mediators).[107] Patients who seek healthcare for IBS symptoms have been shown to have a high rate of psychiatric comorbidity (54–90%; e.g., depression, anxiety, hostility, somatization).[34] However, others have shown that only a small percentage of patients with IBS in tertiary referral centers have a high degree of distress; the majority do not demonstrate clinically evident psychosocial dysfunction.[108] It is important to note here that current evidence suggests that, although psychological factors do not cause IBS, they may affect how its symptoms are experienced (e.g., regularity, severity, or perception of severity), how they cope (whether they visit a physician), and the ultimate clinical outcome; this concept is known as the biopsychosocial model of IBS.[77,109]

6.29.2.2.6 Previous infection

Postinfectious IBS (PI-IBS) refers to the development of IBS symptoms (predominantly abdominal pain, diarrhea, and urgency) after an enteric infection, particularly with *Salmonella*, *Shigella*, or *Campylobacter* species. In prospective studies, recovery from bacterial gastroenteritis was followed by the development of IBS symptoms in 7–32% of patients.[76,110] Numerous risk factors for PI-IBS have been identified, including female sex, younger age, acute infectious gastroenteritis that lasts a long time and is severe, concomitant occurrence of psychological disturbances (e.g., anxiety, depression), presence of increased numbers of EC cells, and use of antibiotics to treat the acute bacterial gastroenteritis.[76,77,98,110] Studies have shown the presence of low-grade inflammation in the colonic mucosa of patients in whom symptoms have persisted after resolution of acute infection, suggesting difficulty in downregulation of intestinal inflammation.[110] It is also important to note that as an initial barrier to the external environment, there is a continual state of low-level inflammation in the GI mucosa. However, compared with healthy controls, patients with PI-IBS have demonstrated specific colonic changes that may explain the presence of altered bowel function and enhanced perception of symptoms, including increased mucosal EC cells (leading to excess production of 5HT?), increased permeability of the gut, and increased concentrations of mast cells and T lymphocytes in the lamina propria of the gut mucosa.[76,77,99]

6.29.2.2.7 Immune-system alterations/gut flora/food allergy

An interaction between the ENS and the gut immune system may partially explain the presence of IBS symptoms in a subset of patients. The colonic and ileal mucosa in certain patients with IBS has elevated numbers of inflammatory cells, including activated inflammatory cells (e.g., T lymphocytes, mast cells, macrophages) located within close proximity to colonic mucosal nerve endings, potentially resulting in the release of inflammation-related mediators (e.g., histamine, interleukins, nitric oxide) that can affect motor/secretory actions in the ENS.[98,111–113] These mediators can also affect gut-related sensory innervation, thereby enhancing visceral sensation and resultant abdominal pain/discomfort.[77] A change in intestinal microflora is another putative mechanism leading to IBS symptoms.[110,114] In patients with IBS, bacterial fermentation of foods may be increased.[115] In one study, the presence of small-intestinal bacterial overgrowth, a condition that most commonly affects patients predisposed to motility or structural abnormalities, was found in patients with IBS.[116,117] The possible role of gut microflora in IBS pathophysiology therefore deserves further study.

Although a concrete link between IBS symptoms and food intolerance has not been established, some patients link their IBS symptoms to intolerance to particular foods and report symptom relief as a result of eliminating specific foods from their diet. However, true detection of food intolerance is difficult because of the unclear cause and the nonspecific nature of symptoms. Similarly, use of an exclusion diet is time-consuming and generally difficult to implement. In the past, the majority of tests for food intolerance have focused on testing for immediate-type reactions triggered by the presence of IgE-mediated antibody response, which may occur rarely in patients with IBS. Recent research has evaluated the possibility that food-related symptoms in patients with IBS might be caused by immunoglobulin G (IgG) antibodies, which are associated with a more delayed onset following antigen exposures.[118,119] In a recently published randomized clinical trial, patients with IBS receiving an exclusion diet (based on food to which they demonstrated IgG antibodies) experienced significantly greater improvements in symptoms than patients for whom these foods were not excluded; reintroduction of the respective foods resulted in a return of symptoms.[118]

6.29.2.2.8 Genetic links

Although genetic factors may play a role in IBS development, the presence of environmental factors likely is an important contributor to the specific ways the condition is expressed, as well as patients' coping strategies, including healthcare-seeking behavior. Data demonstrate that IBS runs in families,[120] and studies show that the prevalence of IBS in monozygotic twins is twice as high as that in dizygotic twins.[121,122] Additionally, specific genotypes and polymorphisms in some genes have been associated with the disease (including interleukin-10, SERT, the α-adrenoceptor, and the G-protein GNb3).[77,101,110,123–125] Whereas all these are interesting findings, the data do not support an unequivocal association with IBS at this time.

6.29.3 Experimental Disease Models

The ideal future therapeutic advances for IBS are likely to be based on the treatment of the multiple components of this disorder, including both abdominal pain and bloating and multiple symptoms arising from dysmotility. Unfortunately, there is currently no animal model of IBS; thus, the experimental testing of new treatments is most challenging. Because IBS is a multifactorial disorder, it is unlikely that a single model (produced by a specific single insult or procedure) will ever be developed that mimics in full the diversity of the disorder in humans. Furthermore, the problem is compounded by the role of species differences in the physiology, pathophysiology, and pharmacology of the systems involved. For these reasons, select animal models have been developed to understand specific components of IBS or to test symptomatic treatment approaches independently of pathophysiology.

From the perspective of a researcher trying to understand IBS and its causes, the primary aim must be to find the condition or, more likely, the set of conditions that leads to the closest overall animal phenotype to the human IBS state. This area recently was reviewed extensively by two such researchers who have played significant roles in advancing our understanding of IBS pathophysiology.[126] Mayer and Collins[120] describe two broad categories of approaches taken to date are based upon either peripheral or central initial insults, and to these, they add a third growing area of genetic manipulation. They also consider the relationship of these models to symptoms in humans, the need for chronicity in its phenotype, the role of species and sex, and, perhaps most importantly, their predictive value with respect to therapeutic outcome. Researchers accept that it is not necessarily the case that we will ever be in a position in which we have a single model that mimics even one subset of IBS, but some attempts have led to interesting phenotypes that have encouraged continued effort. Notable examples of these are as follows:

1. The SERT knockout mouse, which produces a heavier and wetter stool than its wild-type littermates and has a colonic transit phenotype that is either increased (cf. diarrhea) or decreased (cf. constipation).[104]
2. The maternal separation model, which has its basis in mimicking stressful early life events and thus is based primarily upon a CNS disturbance manifesting itself in the periphery. The model has a phenotype of enhanced visceral sensitivity and altered colonic response to acute stress.[127]
3. The *Trichanella spiralis* PI-IBS model, which models the specific subset of PI-IBS and leads to intestinal dysmotility and colorectal hypersensitivity that outlasts the infection.[128,129]
4. The neonatal irritation model, which describes a long-lasting colorectal hypersensitivity and altered colonic function after either chemical or mechanical rectal irritation in neonatal life. The model has a most impressive chronicity, with the phenotype lasting several months.[130]
5. The postinflammatory-IBS model (trinitrobenzosulfonic acid (TNBS)), which shows a long-lasting (up to 60 days) colorectal hypersensitivity after intracolonic TNBS treatment in rats.[131]

All these models have advanced our understanding of the mechanisms that can lead to chronic visceral hypersensitivity, colonic dysmotility, and, perhaps, the resulting phenotypes of hyperalgesia and altered bowel function in patients with IBS. However, for the purposes of the drug discoverer and developer, the value of such models needs to be balanced with some other needs. Although active research within the industry is focused upon disorder pathophysiology and the hunt for an all-encompassing IBS model, in the absence of the perfect model, there is also a need to use technologies already available to evaluate new therapeutic approaches. This requirement necessitates the use of higher-throughput models that are primarily aimed at modeling or providing information around a single end point, rather than an entire disorder. Some of these techniques are better described as assays, rather than models, and this is an important distinction for the industrial scientist to bear in mind when considering the data or how to obtain it.

If we consider the scale of opportunity from the bottom up, in the IBS arena, we might perform the following actions:

1. Start with a simple assay of the effect of a tool compound upon normal GI transit, or visceral pain, regardless of pathophysiology. A positive effect here may provide the impetus for the development of new ligands at the target and for further studies in models.
2. Next, test whether the efficacy is reproduced in simple models of visceral hyperalgesia or dysmotility, for example, those employing acute inflammation or acute stress as insults.
3. Finally, end with a study of one or more of the more involved models of the disorder, if any are available.

However, an argument exists in the IBS arena for placing more emphasis on point 2 than point 1 and for not stopping in the event of a negative outcome of the initial simple assays. This occurs because the effect of a target may become apparent only following some kind of insult to produce the model phenotype. Furthermore, this may be the optimal approach for treatment because an agent that only affected the model and not the assay may provide more of a normalizing influence (e.g., an antihyperalgesic or anticonstipation activity), rather than a direct inhibitory or stimulatory one (e.g., analgesic or direct prokinetic/laxative effect). The benefit in the clinic would be that normal processes should be left unaffected, and only those that had been disturbed would be attacked. However, the drawback here is the potential for false-positive results in preclinical development that do not translate to a benefit in clinical trials several years down the line, perhaps due to the mechanisms involved in the model not matching those at work in the disorder itself.

6.29.4 Clinical Trial Issues

The methodologies of many of the clinical trials that have assessed potential drug treatments for patients with IBS have been flawed,[132,133] and several factors complicate IBS study design. Because IBS is a multiple-symptom complex, the mix of patients enrolled in trials is often heterogeneous, potentially complicating the interpretation of results.[134,135] It is also likely that the undetected presence of pelvic floor dysfunction negatively biases the results of IBS-C studies.[134]

Another challenge is the existence of a large placebo response rate in IBS trials.[135] An analysis of 27 randomized controlled trials evaluating the efficacy of various therapeutic options for patients with IBS found that the median placebo response was 47% (range, 0–84%).[132] A more recent analysis of 45 studies with a well-defined global response outcome found a mean placebo response of 40.2% (range, 16–71%).[136] Several factors may contribute to these high placebo rates. IBS is a chronic episodic disorder that waxes and wanes over long periods, and symptomatic cycles may alternate with symptom-free intervals of 1–3 months.[133] Symptoms therefore may lessen or resolve spontaneously during the course of a clinical trial. In addition, the positive interactive environment of a clinical trial may help relieve patients' stress and anxiety about their condition and give them better insight into the nature of IBS, resulting in symptomatic improvement.[132]

Surprisingly, one metaanalysis found that the placebo response declined significantly as the number of office visits increased, and that each visit was associated with a 4.4% reduction in placebo response rate.[136] Noting that this finding is counterintuitive, the investigators hypothesized that increased interaction with an investigator who is not the patient's primary doctor may have a negative effect, a greater number of office visits may be associated with inadequate blinding, and extra visits provide patients who are obtaining insufficient relief with study medication with increased opportunities for expressing their disappointment.[136] This analysis also found a significantly lower placebo response rate in clinical trials that used the stringent Rome I or II diagnostic criteria as an entry criterion compared with studies that used the more permissive Manning or other unspecified criteria.[136] Use of the Rome criteria may lead to a lower placebo response because it results in a more homogeneous population with a confirmed diagnosis of IBS.[136]

IBS is a disorder of GI motility and sensory disturbance, and assessment of efficacy is complicated by the absence of physiological markers and the variability of symptoms.[137,138] A major difficulty in trial design has been defining endpoints that will represent appropriate measures of efficacy.[132,133,139,140] Many authorities favor the use of a global assessment as the primary efficacy measure in IBS studies, but there is widespread disagreement about the type of global assessment that should be used.[132,138–141]

6.29.5 Current Treatment of Irritable Bowel Syndrome

The goal of IBS treatment is to provide rapid, sustained, global relief of the multiple symptoms of IBS with a single, effective, well-tolerated agent. However, because of the complexity and overlap of the neural circuitry of the gut and CNS and potential occurrence of multiple pathophysiological disturbances, it has proved difficult to identify a single optimal therapeutic target. The choice of therapy has traditionally been based on the primary bowel symptom. Because of the multiplicity of symptoms associated with IBS, patients often need to use a variety of agents to achieve relief. Traditional treatment approaches rely on a combination of dietary changes, bulking agents, laxatives, antispasmodics, antidiarrheal agents, and antidepressants. These therapies, in general, target individual symptoms and therefore do not address the multiple-symptom complex.[142] Clear-cut evidence for their use in patients with IBS is lacking. An evidence-based review of IBS therapies concluded that, although bulking agents, antidiarrheals, and tricyclic antidepressants (TCAs) relieved constipation, diarrhea, and abdominal pain, respectively, these agents did not provide global relief from multiple symptoms of this disorder.[5] The single-symptom approach often leaves many patients dissatisfied, leading to the use of multiple agents, frequent switching of medications, repeated doctor visits, and increased medical costs.[143] Furthermore, some treatments may have adverse effects that mimic IBS symptoms.

The use of serotonergic agents in patients with IBS is based on the critical role played by 5HT in maintaining normal motor, sensory, and secretory functions of the gut. Through their pharmacological action on serotonin receptors, serotonergic modulators can target multiple IBS symptoms. Two agents in this class are currently approved for clinical use, and others are in late-stage trials. Tegaserod **1**, a partial agonist of $5HT_4$, has been demonstrated to be efficacious against the symptoms of abdominal pain, bloating, and constipation in patients with IBS-C. Alosetron **2** is a $5HT_3$ antagonist that has proved effective in relieving abdominal pain, discomfort, and fecal urgency in patients with IBS-D.

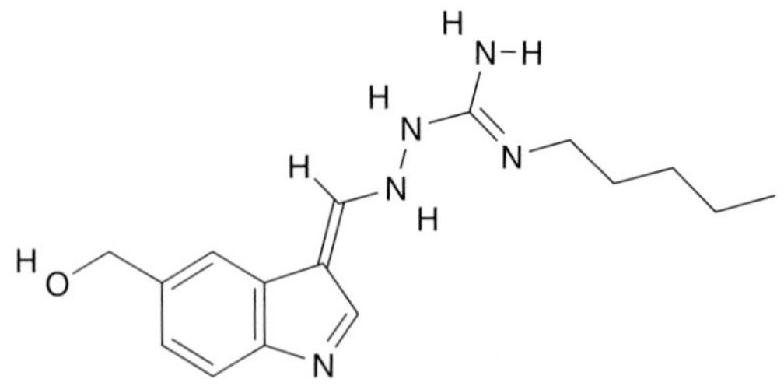

Tegaserod **1**

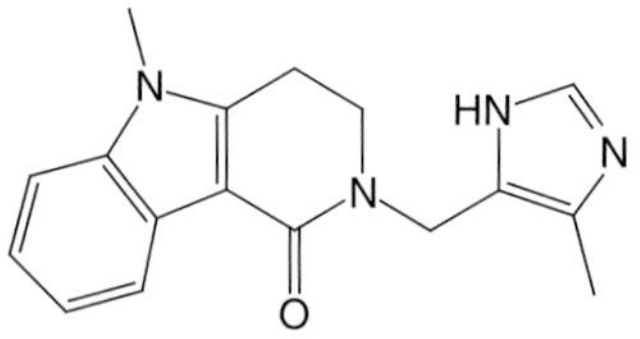

Alosetron **2**

The lack of a standardized treatment algorithm for patients with IBS poses a challenge with regard to optimal treatment. In general, the agent of choice depends on the severity of symptoms, response to traditional agents, and degree to which IBS symptoms are affecting daily life. The clinical pharmacology and potential place in therapy of single-symptom and multiple-symptom treatment options are discussed next and summarized in **Table 3**.[5,59,134,143–152]

6.29.5.1 Drugs Targeting the Multiple-Symptom Complex of Irritable Bowel Syndrome

6.29.5.1.1 Serotonergic agonists and antagonists

As discussed, 5HT is a common and key element linking the pathophysiological abnormalities observed in patients with IBS: altered GI motility, altered intestinal secretion, and visceral hypersensitivity. Although numerous serotonin receptor subtypes have been identified in the GI tract, the $5HT_3$ and $5HT_4$ subtypes have been studied most widely, primarily due to the availability of agents for clinical use. The critical role played by serotonergic mechanisms in GI function and dysfunction led to the clinical development and approval of two medications for the treatment of IBS: tegaserod, a $5HT_4$ agonist, and alosetron, a $5HT_3$ antagonist.

6.29.5.1.1.1 Tegaserod

$5HT_4$ receptors are unique in that they provide both excitatory and inhibitory influences on the GI tract with relevance to the treatment of disorders involving the multiple symptoms associated with dysmotility and visceral hypersensitivity. Their activation results in facilitation of peristalsis and secretion through the release of neurotransmitters involved in the ENS[75] and inhibition of visceral sensitivity via inhibition of extrinsic afferent nerves that transmit sensory information from the gut to the spinal cord.[40,54] $5HT_4$ agonists thus act via multiple mechanisms to treat the multiple symptoms associated with IBS.

Tegaserod **1** is a partial $5HT_4$ agonist, the first in a novel class of drugs, the aminoguanidole indoles.[60,153–155] Tegaserod is approved by the Food and Drug Administration (FDA) for the short-term (up to 12 weeks) treatment of women with IBS-C.[5] It also recently has been approved by the FDA for the treatment of women and men younger than 65 years with chronic idiopathic constipation.[156] The promotility and secretory actions of tegaserod have been confirmed in clinical pharmacology studies, which demonstrate that tegaserod accelerates overall GI transit, promotes gastric emptying, enhances the peristaltic reflex, increases fasting gastric compliance, and enhances fluid secretion in healthy subjects.[59] Similarly, in patients with IBS-C, tegaserod increases orocecal and proximal colon transit.[157]

The precise mechanisms by which $5HT_4$ agonists act as promotility agents are well defined (*see* Section 6.29.2). $5HT_4$ receptors are located presynaptically on terminals of IPANs and interneurons, and, in this position, their activation results in an enhancement, or amplification, of peristaltic neurotransmission via increased release of additional neurotransmitters (e.g., ACh and CGRP).[45,52,62–66]

A similar enhancement of reflexes in secretomotor circuits of the ENS is thought to underlie the prosecretory effect of tegaserod.[63,64] However, the demonstration of increased chloride ion and fluid secretion from isolated rat colonocytes suggests an additional action based upon the epithelium itself.[68]

The mechanisms by which tegaserod relieves abdominal pain and discomfort in patients with IBS are not as well defined. Peripheral $5HT_4$-mediated mechanisms appear to mediate visceral sensation and perception, and preclinical studies are beginning to reveal the visceroanalgesic properties of tegaserod. A feline model showed a direct action of tegaserod on the extrinsic spinal afferents innervating the colon.[72,73] These findings were confirmed by using an in vitro isolated rat colorectal inferior splanchnic nerve preparation.[74] Perceptions of fullness, urge to defecate,

Table 3 Select characteristics of current treatments for patients with IBS[5,59,134,143–152,156,186]

Drug class/agents	*Mechanism of action*	*Place in therapy for IBS*	*Potential adverse effects/ limitations*
Bulking agents			
Natural (derivatives of agar, psyllium, kelp, plant gums, ispaghula husk) Synthetic (methylcellulose, carboxymethylcellulose)	• Long-chain carbohydrates swell in intestinal fluid • Lead to softer bulkier stools • Promote peristalsis and eases intestinal transit	Initial treatment of patients with IBS-C	• May aggravate abdominal pain or bloating. Not recommended for patients with these symptoms or diarrhea • May hinder absorption of concurrently administered drugs
Laxatives			
Stimulant laxatives (bisacodyl, senna, danthron, ricinoleic acid)	Stimulate motility via nonspecific local irritation or a selective action on the intramural nerve plexus of intestinal smooth muscle	Initial therapy in patients with IBS-C	Potential adverse effects include severe cramping, fluid loss, malabsorption, and hypokalemia
Osmotic laxatives (magnesium salts, lactulose, sorbitol, polyethylene glycol)	• Increase intraluminal water content through an osmotic effect • Cause distension of the bowel and stimulate colonic peristalsis	Treatment of symptoms of IBS-C not improved with fiber	Abdominal cramping, flatulence, diarrhea
Stool softeners (docusate)	Facilitate uptake of intestinal fluids by fecal contents and softening of stools	Value in IBS-C not yet established	Abdominal cramping and diarrhea
Antidiarrheal agents			
Opioid agonists (loperamide, diphenoxylate-atropine, codeine)	• Stimulate μ-opiate receptors in enteric nervous system • Inhibit peristalsis • Increase fluid absorption secondary to prolonged intestinal transit • Increase basal and sphincter pressure	Loperamide is used in the treatment of patients with IBS-D	• Diphenoxylate-atropine and codeine cross the blood–brain barrier • Codeine may cause dependence
Tricyclic antidepressants (TCAs) (amitriptyline, desipramine, clomipramine, doxepin, trimipramine)	• Block reuptake of norepinephrine and serotonin and exhibit anticholinergic activity • Modulate sensorimotor activity and GI transit • Reduce pain	• Treatment of patients with moderate to severe IBS when pain is prominent or other therapies have failed • May be more effective in patients with IBS-D • Treatment of psychiatric comorbidity	• Potential adverse effects include fatigue, somnolence, dry mouth, urinary retention • May aggravate constipation; hence, use in patients with IBS-C or IBS-A is not recommended • High rate of discontinuation because of adverse effects
Selective serotonin reuptake inhibitors (SSRIs) (fluvoxamine, luvoxamine, sertraline, paroxetine)	Selective blockade of serotonin reuptake	• Alternative treatment in patients with IBS and abdominal pain after failure of TCAs or with intolerability to TCAs • May be preferred in patients with IBS-C • Superior to TCAs in treating coexisting panic, anxiety, or obsessive disorder • Treatment of older patients	• Fewer adverse effects and better safety than TCAs • Potential adverse effects include insomnia, night sweats, weight loss, and sexual dysfunction • More likely than TCAs to cause diarrhea

continued

Table 3 Continued

Drug class/agents	*Mechanism of action*	*Place in therapy for IBS*	*Potential adverse effects/ limitations*
Benzodiazepines	• Antianxiety effect • Mitigate exacerbation of IBS symptoms by psychological distress	Treatment of patients with IBS and anxiety disorders (weak treatment effects)	• Potential for physical dependence • Interactions with other drugs • Should be used on a limited basis
Serotonergic agents			
$5HT_4$ agonist (tegaserod)	• Accelerate GI transit • Reduce visceral sensitivity and hypersensitivity • Enhance peristalsis • Increase gastric emptying and compliance • Enhance fluid secretion	Treatment of women with IBS-C	• Headache and mild transient diarrhea • Rare case of ischemic colitis (transient and without long-term sequela) reported during postmarketing surveillance
$5HT_3$ antagonist (alosetron)	• Slow orocecal and colonic transit • Enhance intestinal absorption	Treatment of women with severe IBS-D refractory to conventional therapy (treatment under a restricted-use program)	• Constipation in 22–39% of patients • Ischemic colitis has occurred • Contraindicated in patients with IBS-C

discomfort, and pain are thought to be mediated by extrinsic mechanoreceptors.[158] Thus, it would be expected that modulation of the firing rate in afferent pathways by tegaserod would concomitantly result in changes in the intensity of pain sensations. In studies of its antinociceptive properties in rats, tegaserod has demonstrated both visceral analgesic and visceral antihyperalgesic activity in the postinflammatory (30-day post-TNBS) model[131,159,160] and in a neonatal colonic irritation model (induced by repeated injections of acetic acid during the neonatal period).[161,162] In the latter studies of rats with hypersensitivity, tegaserod potently inhibited the abdominal reflex and also attenuated the number of c-fos-positive neurons in limbic structures in a dose-dependent manner. It is likely, based upon the afferent nerve studies described previously, that the analgesic activity of tegaserod occurs through a decrease in signals ascending to the CNS from the periphery, rather than via a direct action in the CNS.

Data from the clinical use of tegaserod demonstrate its effect on the pain and bloating associated with IBS-C (discussed below), and recent studies of humans have added some further insight to the underlying mechanism for this action. A visceral analgesic action of tegaserod in humans, in both healthy volunteers and patients with IBS, has been demonstrated by using the RIII reflex technique in combination with rectal distention. The RIII reflex is a model of a nociceptive flexion reflex that provides an objective measure of antinociception in humans, is modulated by concomitant rectal (or gastric) distention,[163,164] and is altered in patients with IBS.[165] That the somatic reflex is altered in patients with IBS adds weight to the hypothesis that IBS involves widespread sensory dysfunction, likely including CNS circuits, rather than dysfunction centered solely in the bowel. Antinociceptive effects of tegaserod described initially in healthy volunteers by using this technique[166] recently have been reproduced in patients with IBS.[167] Recent studies of esophageal sensitivity in patients with functional heartburn have added further evidence for the visceral analgesic effect of tegaserod in a disease population with some similarity to the IBS state.

The efficacy of tegaserod was evaluated in several large, multicenter, randomized, double-blind, placebo-controlled trials of women with IBS-C. In these trials, tegaserod, 6 mg twice daily, was significantly more effective than placebo in providing rapid and sustained improvement in patients' overall well-being (assessed by using the Subjective Global Assessment of relief, a measure of the impact of treatment on overall well-being, abdominal pain/discomfort, and bowel function), as well as in relieving the multiple symptoms associated with IBS-C, including abdominal pain/discomfort, bloating, and constipation.[5,156,169] Patients using tegaserod also experienced greater satisfaction with their bowel habits compared with placebo-treated patients.[5]

Overall, tegaserod was well tolerated; headache and mild transient diarrhea were the adverse effects reported more commonly in patients administered tegaserod than in patients administered placebo (9% versus 4% and 15% versus

12%, respectively). Most episodes of diarrhea were mild, occurred during the first week of treatment, and resolved within several days without treatment discontinuation.[156] Studies have also demonstrated that long-term use of tegaserod (12 months), as well as continuous treatment versus re-treatment, is not associated with an increased incidence of adverse effects.[170,171]

Rare cases of serious consequences of diarrhea (e.g., those requiring hospitalization) have been reported during clinical trials and postmarketing surveillance with tegaserod; 4 of 100 000 (0.04%) have experienced clinically significant diarrhea.[156] The current Prescribing Information for this agent recently has been revised to reflect this fact.[156] In addition, although no reports of colonic ischemia (a vascular condition that results from reduced colonic blood flow)[172] have been reported during clinical trials with tegaserod, this condition occurred in a small number of patients during postmarketing surveillance.[173] Although pharmacological data from preclinical studies with tegaserod found no vascular mechanism that could result in mesenteric or colonic ischemia,[174] ischemic colitis currently appears as a precaution in the Prescribing Information.[156]

Unlike the case with cisapride, a substituted piperidinyl benzamide associated with serious cardiotoxicity (i.e., prolongation of the QT interval, resulting in torsades de pointes),[175,176] tegaserod, structurally distinct from cisapride, does not block the HERG channel that underlies this effect of cisapride.[177]

6.29.5.1.1.2 Alosetron

Alosetron, **2** a $5HT_3$ antagonist,[178–180] is FDA approved for the treatment of patients with severe IBS-D.

$5HT_3$ receptors have been described in myenteric neurons of guinea pigs and rodents,[61,181] where they may be involved in initiation of the peristaltic reflex. By preventing serotonin from binding to $5HT_3$, alosetron may attenuate symptoms of IBS by decelerating orocecal and colonic transit, enhancing intestinal absorption, and improving rectal compliance.[59] In line with this, blockade of this receptor is associated with constipation (discussed below). As well as producing excitation of motor and secretory systems, $5HT_3$ are associated with excitation of extrinsic sensory nerves. Thus, stimulation of $5HT_3$ promotes intestinal motility, secretion, and sensation.

Results from animal studies suggest that $5HT_3$ mediate noxious visceral sensory signals via spinal and vagal afferent neurons.[70,71] In a rat model with a pseudoaffective depressor endpoint, alosetron inhibited the reflex response to colorectal distension and significantly reduced the numbers of c-fos-expressing neurons in the lumbosacral spinal cord. These results suggested the involvement of $5HT_3$ in the transmission of visceral nociceptive signals and modulation of this transmission by alosetron in the rat. The site of action could be at the level of the enteric or primary sensory neuron or on spinal or supraspinal nociceptive neuronal circuits. A subsequent in vitro study demonstrated an excitatory effect of $5HT_3$ activation on colonic lumbar splanchnic afferent fibers,[70] suggesting that alosetron may modulate sensory perception, in part by a direct action on spinal afferent nerves.

Four randomized controlled trials compared alosetron, 1 mg twice daily, with placebo in patients with IBS-D (and a minority of patients with IBS-A).[182–185] In each trial, alosetron was superior to placebo in improving abdominal discomfort, stool frequency, stool consistency, and fecal urgency. In the single trial that evaluated global IBS symptoms, significantly more patients with severe IBS-D (the currently approved population) treated with alosetron reported an improvement compared with controls.[185]

Constipation was the most common adverse effect reported in these trials, occurring in 22–39% of alosetron-treated patients (versus 3–14% of controls), but led to discontinuation of the medication in only 10% of patients. In most cases, constipation was mild and resolved spontaneously. However, serious complications, including ischemic colitis and constipation requiring hospitalization, led to voluntary withdrawal of the drug in 2000.[134] In IBS clinical trials, serious constipation-associated complications occurred in approximately 1 of 1000 women receiving either alosetron or placebo. In women receiving alosetron, the cumulative incidence of ischemic colitis over a 3-month period was 2 per 1000 patients, and for a 6-month period was 3 per 1000 patients.[186]

Alosetron was reintroduced in June 2002 for female patients with severe IBS-D who are refractory to conventional IBS therapy, under a mandatory restricted-use program requiring both prescribing physicians and patients to follow specific protocols strictly and accept responsibility for proper drug use. Additionally, the starting dose has been reduced to 0.5 mg twice daily (from 1 mg twice daily). Alosetron is contraindicated in patients with IBS-C, and patients are advised to discontinue its use if constipation develops. Alosetron also must be used cautiously in patients with hepatic insufficiency because of its extensive metabolism by the liver.[143]

Alosetron was well tolerated in a recently published long-term (48-week) trial enrolling the current approved population: women with severe chronic IBS-D.[187] The most commonly reported adverse effect in this trial was constipation, which occurred in 23% of patients administered alosetron versus 5% of those administered placebo. No cases of ischemic colitis were reported during the trial.[187]

6.29.5.2 Drugs Targeting Single Symptoms

6.29.5.2.1 Dietary aids and bulking agents

Scientific evidence to support a role for dietary exclusions or modifications in patients with IBS remains conflicting, and no conclusive data have linked specific dietary components to the pathogenesis of IBS. However, bloating, diarrhea, and abdominal cramping have been shown to improve in some patients when tobacco, caffeine, dairy products, such gas-producing foods as legumes and cruciferous vegetables, and greasy or spicy foods are eliminated from the diet.[146]

Bulk-forming products, because of their safety, low cost, and physiological mechanism of action, have become a mainstay of therapy in patients with IBS and are used routinely as the initial treatment.[146] These agents are natural or semisynthetic hydrophilic polysaccharide and cellulose derivatives of agar, psyllium seed, kelp, and plant gums. They work by dissolving or swelling in intestinal fluid, leading to softer, wetter, bulkier stools, which promote peristalsis and facilitate colonic transit. Adequate intake of water is necessary in conjunction with fiber because the additional fluid is thought to expand and soften the fiber. Fiber leads to an improvement in stool frequency in patients with IBS-C and is recommended as first-line therapy in this group of patients.[146] However, symptoms of abdominal pain or bloating are not relieved by fiber supplementation; rather, these symptoms appear to be aggravated with fiber intake; hence, fiber is not recommended in patients with these symptoms or diarrhea as the predominant symptom.[5]

6.29.5.2.2 Laxatives

6.29.5.2.2.1 Stimulant laxatives

Stimulant laxatives include bisacodyl, senna, danthron, and ricinoleic acid. The drugs of choice in this group are the senna compounds. These agents enhance motility through either nonspecific local mucosal irritation or a more selective action on the intramural nerve plexus of intestinal smooth muscle.[147] They may also inhibit absorption of water or electrolytes from the large intestine, leading to increased volume and pressure that stimulates colonic motility.

Although used commonly by patients with IBS who have severe constipation, no clinical trials have evaluated their efficacy in this patient population. Stimulants may be used as initial therapy in patients with constipation, but for no longer than 1 week. The major hazards of stimulants are severe cramping, electrolyte and fluid loss, malabsorption, and hypokalemia. Their use is contraindicated in patients with undiagnosed rectal bleeding or signs of intestinal obstruction.[147]

6.29.5.2.2.2 Osmotic laxatives

Osmotic laxatives are subcategorized as saline (magnesium and sulfate ions, e.g., magnesium hydroxide, magnesium sulfate), poorly absorbed sugars (lactulose and sorbitol), and polyethylene glycol (metabolically inert). These agents work by drawing water into the intestine through an osmotic effect. This leads to distension and stimulation of colonic peristalsis.[147]

Osmotic laxatives are occasionally recommended as first-line treatment for patients with IBS-C whose symptoms have not improved with fiber supplementation. However, their use can result in such GI adverse effects as abdominal cramps and flatulence and aggravate diarrhea. Flatulence, cramping, and bloating are largely related to the fermentation of osmotic laxatives by colonic bacteria. Because polyethylene glycol is metabolically inert (i.e., resistant to bacterial fermentation), it is less likely to result in these adverse GI effects.[146]

6.29.5.2.3 Stool softeners

Stool softeners and emollients, such as docusate, are anionic surfactants that increase the wetting efficiency of intestinal fluid and facilitate softening of fecal contents. However, there is no evidence that they have any value in treating constipation in patients with IBS. Docusate works slowly and potentially can cause diarrhea and mild abdominal cramping.[5,147]

6.29.5.2.4 Antidiarrheal agents

A number of agents currently available improve symptoms of diarrhea in patients with IBS. Loperamide, diphenoxylate-atropine, and codeine are μ-opioid agonists that stimulate opiate receptors in the ENS and inhibit peristalsis. By prolonging intestinal transit time, these agents also increase fluid absorption indirectly and thus lead to firmer stools.[146]

In addition to slowing peristalsis, loperamide also increases basal anal sphincter pressure and thus improves symptoms of fecal incontinence in patients with diarrhea and urgency. Compared with placebo, loperamide significantly improved diarrhea, urgency, and borborygmi, but not abdominal pain, in patients with IBS-D.[146,149] Loperamide does not have CNS opioid activity at standard doses and has a more rapid onset and more prolonged duration of action than codeine or diphenoxylate. Loperamide thus has an important place in the treatment of the diarrhea component of IBS and is often used on an on-demand basis for this symptom.

No placebo-controlled clinical trials have evaluated diphenoxylate-atropine in patients with IBS-D. Both codeine and diphenoxylate cross the blood–brain barrier and are therefore best avoided.[146]

6.29.5.2.5 Antispasmodic agents

Antispasmodic drugs are a widely used class of compounds for the control of abdominal cramping, fecal urgency, and postprandial lower-abdominal discomfort associated with diarrhea that are common symptoms in patients with IBS. Anticholinergic agents (dicyclomine, hyoscyamine, cimetroprium bromide, belladonna, propantheline), calcium channel blockers (pinaverium bromide, otilonium bromide, mebeverine), and opioid agonists (trimebutine) all display antispasmodic activity by reducing the excessive contractility of smooth-muscle cells.[146,148] Although widely used to treat abdominal pain and bloating, most of these agents have not undergone rigorous scientific testing. A recent metaanalysis found that smooth muscle antispasmodics were superior to placebo in relieving abdominal pain in patients with IBS.[188] However, the randomized trials on which the metaanalysis was based all had limitations; hence, the results are considered equivocal.

The anticholinergic antispasmodic agents dicyclomine and hyoscyamine are the only agents in this class approved for treatment of patients with IBS in the US. Hyoscyamine is an antimuscarinic agent available in oral, sublingual, and liquid forms. The sublingual form acts rapidly, can be used as needed for abdominal pains and spasms, and is effective in attenuating a heightened gastrocolonic response.[146] Dicyclomine also has antimuscarinic activity and is prescribed before meals to reduce postprandial cramping. Because both agents are associated with bothersome anticholinergic-related adverse effects (e.g., dry mouth, dry eyes, urinary hesitancy, constipation, and drowsiness), they are not recommended for long-term use.

Formulations that combine an anticholinergic and antispasmodic agent with a barbiturate or benzodiazepine sedative are also available to treat patients with symptoms that remain refractory to anticholinergic agents alone. By reducing anxiety, which increases intestinal motility, the sedative modulates the CNS component of intestinal contractions, whereas the intestinal motor response is controlled by the antispasmodic agent. Donnatal consists of hyoscyamine, atropine, scopolamine, and phenobarbital; and Librax consists of the anticholinergic agent clinidium combined with the benzodiazepine chlordiazepoxide. Adverse effects associated with these agents are anticholinergic in nature.[146,148]

6.29.5.2.6 Antidepressants

Several classes of antidepressants are prescribed to patients with IBS, including TCAs, selective serotonin reuptake inhibitors (SSRIs), and, less frequently, such novel antidepressants as venlafaxine and mirtazapine. Antidepressants have neuromodulatory and analgesic properties that benefit patients independently of their psychotropic effects. The rationale for their use in patients with IBS includes treatment of psychiatric comorbidity, modulation of visceral sensitivity and motility, and reduction of central pain perception to afferent signals from the gut.[150] The TCAs and SSRIs primarily influence GI sensory and motor function through modulation of the 5HT and norepinephrine pathways. These agents act by blocking the uptake of 5HT and/or norepinephrine at presynaptic nerve endings. Because of their differential effects on gut function, TCAs and SSRIs may allow more directed treatment of IBS subgroups.[151] In addition to their primary GI effects, antidepressants may offer added benefits when used in conjunction with psychological treatment in the management of IBS.

6.29.5.2.6.1 Tricyclic antidepressants

The uptake of both 5HT and norepinephrine is blocked by TCAs. Because of this nonselective action, TCAs increase levels of both 5HT and norepinephrine in the gut, thus modulating sensorimotor activity and GI transit. Increased levels of norepinephrine can inhibit smooth muscle contractions and slow transit, thus potentially worsening symptoms of constipation.

The success of TCAs in treating neuropathic pain syndromes has led to some use of antidepressants in treating visceral pain, such as in patients with IBS. The following observations suggest that the neuromodulatory and analgesic effects of TCAs are unrelated to their antidepressant actions: low doses of TCAs are effective, improvements are noted within 2 weeks, and patients without depression also derive a benefit.[145] Despite widespread use in clinical practice, evidence supporting the use of TCAs for patients with IBS is mixed, mainly because of methodological deficiencies in several of the studies. A number of randomized controlled trials have demonstrated an improvement in IBS symptoms in patients administered low-dose TCAs, including amitriptyline, desipramine, clomipramine, doxepin, and trimipramine. A metaanalysis of randomized controlled trials concluded that TCAs were superior to placebo in reducing pain-related symptoms.[144] In its technical review, the American College of Gastroenterology (ACG) Functional Gastrointestinal Disorders (FGID) Task Force concluded that although evidence did not support the effectiveness of TCAs for improvement of global IBS symptoms, these agents may be useful for decreasing abdominal pain.[5]

TCAs have more established benefit than SSRIs in treating patients with IBS and hence remain the preferred antidepressant agents.[151] TCAs are recommended for patients with moderate to severe IBS in whom pain is a prominent symptom and for patients whose symptoms have not improved with other therapies.[148] The utility of TCAs is limited by their adverse effects, which contribute to a dropout rate as high as 40%.[134] Adverse effects often are dose related and include constipation, fatigue, somnolence, dry mouth, and urinary retention.[145] Although these effects are more prominent at higher doses, patients with IBS may manifest them at lower doses. Caution must be exercised in use of TCAs in patients with IBS-C or IBS-A.

6.29.5.2.6.2 Selective serotonin reuptake inhibitors

SSRIs preferentially inhibit 5HT reuptake. Increased levels of 5HT promote gut motor activity, enhance peristalsis, and accelerate orocecal transit; hence, SSRIs may be most beneficial in treating patients with IBS-C.[151] These agents lack many of the adverse effects of TCAs and may be particularly useful in patients with IBS with coexisting psychiatric disorders, such as anxiety and phobias. Also, their lack of disabling adverse effects makes them an attractive choice in older patients. No single SSRI has been identified as a preferred agent in the treatment of patients with visceral pain syndromes.[151]

Common SSRI-associated adverse effects include insomnia, diarrhea, night sweats, weight loss, and sexual dysfunction. In contrast to TCAs (that are constipating), SSRIs are more likely to cause diarrhea and may be a more appropriate choice for patients with IBS-C.[151]

6.29.6 Unmet Medical Needs

One of the chief issues facing patients with IBS is the inadequacy of many of the existing drugs used to treat this condition.[19] Most traditional treatment options target single symptoms.[19] Thus, in an attempt to achieve symptom relief, many patients with IBS try multiple traditional medications or switch medications.[19] One survey found that only 14% of patients with IBS were completely satisfied with their traditional IBS therapy.[19] Very small percentages of these patients reported that medical therapy was completely effective in relieving their symptoms of constipation (19%), diarrhea (18%), abdominal pain (15%), and bloating (10%).[19]

In addition to their limited efficacy, many of the traditional prescription and OTC medications used for patients with IBS can aggravate IBS symptoms.[19] The extent of the problem with traditional medications was indicated by an on-line survey of 668 subjects with physician-diagnosed IBS, 504 (75%) of whom had IBS-C.[189] Participants reported an average of 3.3 ± 2.7 adverse effects from their IBS medications.[189] The likelihood of experiencing a severe adverse event was greater with prescription drugs (antidepressants, 16%; antispasmodics, 11%; laxatives, 9%) than with OTC agents (laxatives, 8%; fiber supplements, 6%).[189] For 74% of participants, adverse events were the main reason for discontinuing treatment. Respondents also reported missing work or school and refrained from taking part in social or athletic activities due to the adverse effects of IBS medications.[189]

The inadequacy of many currently available drug treatments for patients with IBS is suggested by the results of a study of 135 patients who had sought care from a gastroenterologist and had a mean interval since IBS diagnosis of 13.7 ± 11.7 years.[190] When asked about their symptoms during the previous 7 days, 4% said they had no symptoms, 27% reported mild symptoms, 53% reported moderate symptoms, and 16% reported severe symptoms. Including both absenteeism and presenteeism, overall work productivity losses were 24.7%, 37.3%, and 41.7% for patients with no or mild, moderate, and severe symptoms, respectively. Assuming a 40 h work week, the overall estimated work productivity loss due to IBS was the equivalent of 14 h per week. Daily activity impairment increased significantly as symptom severity increased, and activity impairment levels of 25.0%, 45.3%, and 59.0% were reported by patients with no or mild, moderate, and severe symptoms, respectively ($P = 0.0001$).[190]

6.29.7 New Research Areas

Increased understanding of the basic processes underlying the motor and sensory dysfunction in patients with IBS is expected to yield new treatment targets. Several new classes of medications targeting GI sensorimotor function, gastrocolonic reflex, and visceral hypersensitivity are currently being investigated; select agents are discussed next and summarized in **Table 4**.[191–208]

Table 4 Selected new research areas in IBS[191–208]

Drug class	*Mechanism of action*	*Potential use/comments*
Newer serotonergic agents		
$5HT_3$ antagonist (cilansetron)	Same mechanism of action as alosetron	• Was in development for treatment of patients with IBS-D • Constipation is a side effect • Rare cases of ischemic colitis reported
$5HT_4$ partial agonist/$5HT_3$ antagonist (renzapride)	• Displays prokinetic activity in animal models • May reduce visceral sensation	In development for the treatment of patients with IBS
Neurokinin (NK) receptor modulators		
NK1 antagonists	• Inhibit peristalsis and attenuate abdominal contractions • Reduce visceral hypersensitivity	In development for treatment of patients with IBS
NK2 antagonists	• Reduce motor response to colorectal distension	In development for treatment of patients with IBS
NK3 antagonists	• Reduce visceral sensitivity • Reduce hypermotility	In development for treatment of patients with IBS
Opioid antagonists		
μ-Receptor antagonist (alvimopan)	• Peripherally acting opioid antagonist (does not cross blood–brain barrier) • Blocks inhibitory effect of opiate analgesics in GI motility and secretion	In development for treatment of patients with postoperative ileus and opiate bowel dysfunction
κ-Receptor agonist (asimadoline)	• Visceral analgesic • Does not cross blood–brain barrier • Decreases satiation and postprandial fullness	Treatment of visceral hypersensitivity in patients with IBS
Corticotropin-releasing factor (CRF) antagonists (astressin, antalarmin)	• Astressin relieves colorectal hyperalgia produced by CRF in animals • Antalarmin decreases visceromotor response to colonic distension in anxiety-prone rats	In development for the treatment of patients with IBS
α_2-Adrenoceptor agonists (clonidine)	• Increase visceral compliance without altering motility • Reduce colonic pain in response to distension	Possibly beneficial in patients with IBS-D Use is limited by adverse effects such as somnolence and hypotension
Cholecystokinin (CCK) receptor antagonists (dexloxiglumide)	• Block CCK- and lipid-induced delays in gastric emptying • Reduction of global IBS symptoms in patients with IBS-C in a preliminary study was not confirmed in larger trials	• Development for patients with IBS-C halted • May have efficacy in a different patient population
Chloride-channel openers (lubiprostone)	• Activate specific ClC2 chloride channel in gut-lining cells and increase intestinal fluid secretion • Soften stool and promote bowel movement due to increased volume	In development for treatment of patients with chronic idiopathic constipation, IBS-C, and postoperative ileus Major adverse effect is nausea; diarrhea and headache also reported
Cannabinoid receptor agonists and antagonists	• Antagonists enhance motor activity in colonic smooth muscle • Agonists (CB1 and CB2) may reduce pain and motility • Antiinflammatory (CB2)	IBS CNS adverse effects an issue for CB1

6.29.7.1 Cilansetron ($5HT_3$ Antagonist)

Cilansetron **3** is a $5HT_3$ antagonist under development for the treatment of patients with IBS-D. It is presumed to have the same mechanism of action as alosetron – inhibition of colonic transit and visceral sensory mechanisms. In two phase III trials of patients with IBS-D (a 3-month dose-ranging US study and a 6-month international study), combined analysis showed that a significantly greater proportion of patients who were administered cilansetron (2 mg three times daily) reported adequate relief of abnormal bowel habits and abdominal pain than control patients.[209] Individual measures, such as relief of urgency, stool frequency, and stool consistency, also were better with cilansetron. Subset analyses showed that cilansetron is of therapeutic benefit in both male and female patients.[197,199] In the 6-month phase III global study, 59% of patients treated with cilansetron reported adequate relief of global IBS symptoms versus 45% of placebo-treated patients.[198] Constipation was the most common adverse event noted in the phase III studies. Based on a safety database of more than 4000 patients enrolled in phase II and III clinical trials with cilansetron, the incidence of suspected cases of ischemic colitis was 1.3 per 1000 patients in the first 3 months and 2 per 1000 patients during 12 months; all patients recovered without complications within 20 days. Following a nonapproval action by the FDA in April 2005, future options for this agent currently are under review.[199]

6.29.7.2 Renzapride

Renzapride **4** is a substituted benzamide compound that is a $5HT_4$ partial agonist and $5HT_3$ antagonist. The presumed action of renzapride is activation of $5HT_4$ in cholinergic neurons to stimulate gut contractility; however, these actions may be opposed by its antagonistic activity at $5HT_3$. However, in laboratory studies, renzapride displayed prokinetic effects, stimulating peristalsis in guinea pig ileum segments and colonic motility in dogs,[204] and human pharmacodynamic data show that renzapride accelerates colonic transit, suggesting that the $5HT_4$ action is dominant.[205] Visceral sensations may be relieved through both actions of the drug, as described.[204] Results of preliminary reports suggest a positive effect of renzapride in improving stool consistency and symptoms of abdominal pain/discomfort.[192,206] However, in a phase II study of patients with IBS-A, renzapride was not more effective than placebo in the number of days of relief of overall IBS symptoms.[193]

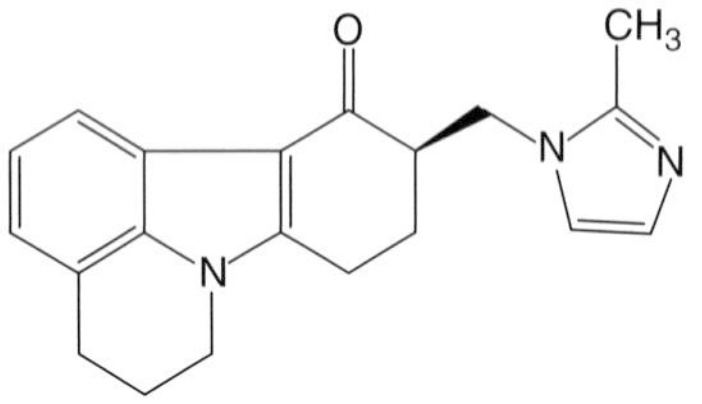

Cilansetron **3**

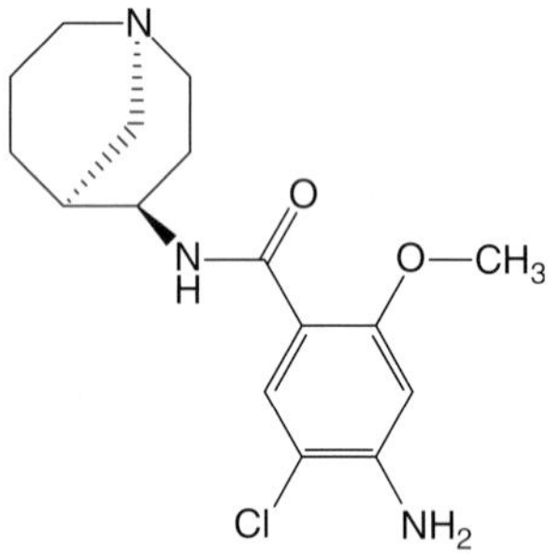

Renzapride **4**

6.29.7.3 Neurokinin Receptor Modulators

Neurokinin (NK) receptors NK1, NK2, and NK3 for tachykinin peptides are distributed widely in the CNS and on sensory enteric neurons. Targeting NK receptors to modulate sensorimotor function in the gut is therefore a potential treatment modality for patients with IBS.

NK1 are preferential receptors for substance P in myenteric neurons and afferent pathways and are expressed on neuronal and non-neuronal cells involved in gut motility. Blockade of NK1 in the gut inhibits peristalsis, and blockade of NK1 in the spinal cord attenuates abdominal contractions evoked by colorectal distension.[194] The NK1 antagonist aprepitant **5** relieves chemotherapy-induced emesis, an effect likely related to vagal afferent inhibition, but this agent has not been tested in patients with IBS. In a preliminary study of patients with IBS, treatment with another NK1 antagonist, ezlopitant **6**, reduced anger related to rectal balloon distension. In addition, blockade of substance P may reduce bowel inflammation and thus exert analgesic effects.[192]

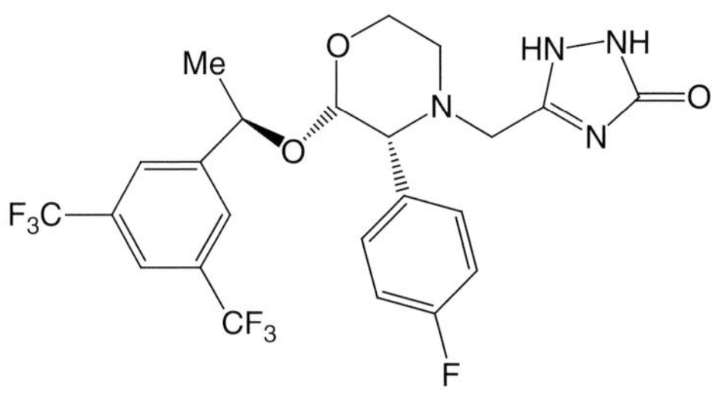

Aprepitant **5**

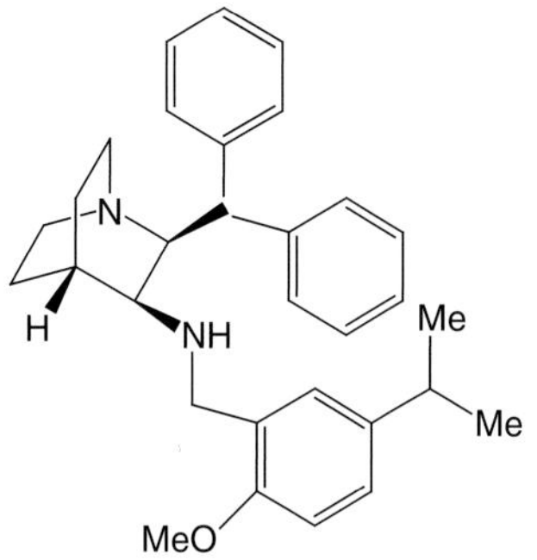

Ezlopitant **6**

NK2, which preferentially bind NKs A and B, are located predominantly on sensory neurons. However, these receptors are also involved in mediating GI motor activity. The NK2 antagonist nepadutant suppressed the stimulatory effects of NKA, but did not affect basal migrating motor complexes. In animal studies, saredutant, another NK2 antagonist, dose-dependently reduced agonist-induced fecal excretion and abdominal contractions in response to colorectal distension. However, no data are available on the effects of NK2 modulation in human subjects and patients with IBS.[192]

NK3 receptors are present within intrinsic neurons of the spinal cord and are also distributed to excitatory and inhibitory motor neurons and secretomotor neurons in the myenteric plexus and to secretomotor neurons in the submucosal plexus. These receptors appear to play a role in disrupted, but not normal, intestinal motility and also in visceral nociception.[202] Selective NK3 antagonists administered systemically reduced rat behavioral responses to colorectal distension.[196] A similar antinociceptive effect was noted with administration of talnetant, an antagonist that crosses the blood–brain barrier, and SB-235375 **7** , an antagonist that is restricted to the periphery, suggesting a peripheral site of action of NK3 antagonists.[194,202] Studies on the effects of NK3 antagonism in IBS in humans are underway.

6.29.7.4 Opioid Receptor Antagonists

Opioid receptor agonists have an inhibitory effect on enteric motility and secretion and hence have a constipating effect. The actions of opiates on GI motility are due to the suppression of excitability and neurotransmitter release from enteric neurons, which disrupts the normal coordinated contractions and relaxations necessary for propulsion of intraluminal content. Opiates also suppress the activity of intestinal secretomotor neurons, thus inhibiting intestinal secretions.[200]

μ-Opioid receptors are expressed widely on enteric neurons, as well as on nociceptive pathways conducting pain to the CNS. μ-Opioid receptor antagonists are currently most therapeutically promising in patients in whom long-term opioid therapy for relief of pain produces constipation and patients with postoperative ileus. Alvimopan **8** is a selective μ-opioid receptor antagonist that does not cross the blood–brain barrier. Postsurgical use of alvimopan speeds recovery of GI function and reduces the length of hospitalization. To date, studies on the effects of alvimopan in patients with IBS have not been performed.[204]

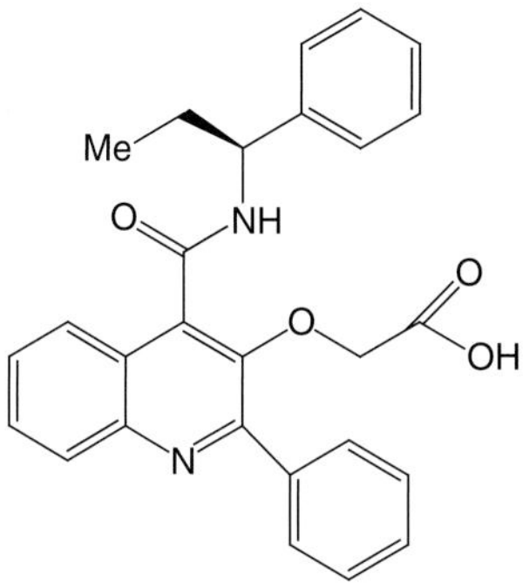

SB-235375 **7**

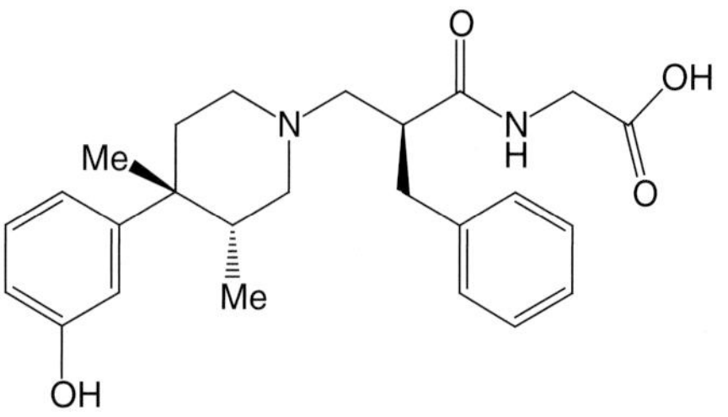

Alvimopan **8**

In contrast to available opioid agonists that relieve pain, but also result in constipation and central adverse effects, κ-opioid receptor agonists appear to be selective visceral analgesics. Asimadoline **9** is a κ-opioid receptor agonist that does not cross the blood–brain barrier. Pharmacodynamic studies of humans suggest that asimadoline can reduce colonic sensation at subnoxious levels of colonic distension and decrease satiation and postprandial fullness in humans without altering GI motor reflexes or transit.[191] Further support for investigation of asimadoline for treating visceral hypersensitivity in patients with IBS is derived from a preliminary study of patients with IBS-C in which asimadoline reduced hypersensitivity in response to colonic distension without affecting compliance or tone.[194]

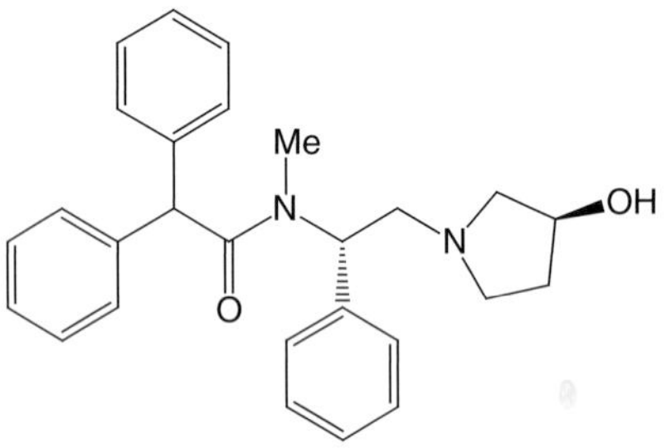

Asimadoline **9**

6.29.7.5 CRF-1 Antagonists

A key alteration in patients with IBS is increased responsiveness to stress, leading to excessive release of neurohormones and autonomic dysfunction.[126] CRF is a stress hormone that affects colonic sensorimotor function. CRF-1 receptors localized in the CNS and GI tract are involved in CRF-stimulated changes in GI function, including colonic motility, epithelial water transport, and permeability.[210] A recent demonstration of the role of this system in colorectal hyperalgesia has further increased interest in its potential as a therapeutic target.[211]

A selective CRF-1 antagonist, astressin, when injected intracerebrally, reduces visceral discomfort induced in animals by administration of CRF.[195] The CRF-1 antagonist antalarmin also decreases the visceromotor response to colonic distension in rats genetically predisposed to hyperanxiety coupled with a high basal level of CRF.[208] Thus, CRF-1 antagonism could be an effective treatment for the motility disturbance, abdominal pain, and comorbid CNS symptoms experienced by patients with IBS.

6.29.7.6 Clonidine

The sensorimotor disturbances in patients with IBS may be caused in part by adrenergic dysfunction. Functional polymorphisms in α_2-adrenoceptors may be associated with IBS-C.[192] The α_2-adrenoceptor agonist clonidine reduces colonic pain sensation in response to distension and reduces colonic motor activity. Clonidine also increases gut-wall compliance without altering motility, suggesting it may be beneficial in patients with urgency as the major symptom. In an initial study of patients with IBS-D, treatment with clonidine reduced bowel dysfunction and enhanced the proportion of patients achieving satisfactory relief of IBS, but did not alter GI transit.[207] However, adverse effects such as hypotension and somnolence may limit its use in patients with IBS. Agents in this class with greater GI selectivity would be advantageous.

6.29.7.7 Cholecystokinin Receptor Antagonists

An excessive gastrocolonic response has been associated with postprandial symptoms in patients with IBS. Cholecystokinin (CCK) is a neuropeptide released by duodenal and jejunal cells that modulates gastric relaxation in response to feeding. CCK may also contribute to abnormal colonic motility in patients with IBS. Endogenously released CCK acts at CCK-1 in gastric mucosa and the CNS to alter gastric motility.[192]

Dexloxiglumide is a potent CCK-1 antagonist that blocks CCK-induced delays in gastric emptying in rats and reverses lipid-induced delays in gastric emptying in healthy subjects.[204] This agent has been tested in patients with IBS-C. In a 12-week placebo-controlled trial, a significantly greater proportion of patients with IBS-C reported improvement in global symptoms after treatment with dexloxiglumide than with placebo. However, these preliminary results of a superiority of dexloxiglumide over placebo in improving global assessment scores were not confirmed in larger phase III trials. The development of this drug for patients with IBS-C has therefore been discontinued. Further studies of CCK-1 antagonists using different patient populations may be warranted.

6.29.7.8 Chloride Channel Openers

Chloride channels are proteins that control cell membrane transport of chloride ions and hence modulate intestinal fluid secretion. Chloride channel openers, such as lubiprostone **10**, are currently under development for the treatment of patients with constipation, IBS-C, and postoperative ileus. By activating the specific chloride channel ClC2 in cells lining the gut, lubiprostone increases intestinal fluid secretion, which softens the stool, promotes bowel movement, and thus indirectly decreases abdominal discomfort and bloating.[201] To date, two phase III trials of lubiprostone in patients with constipation showed that, relative to placebo, treatment with lubiprostone significantly improved symptoms of constipation, including stool frequency, stool consistency, and straining. The main adverse events were nausea, diarrhea, and headache. A phase III trial of lubiprostone in patients with IBS-C was initiated in mid-2005.[201]

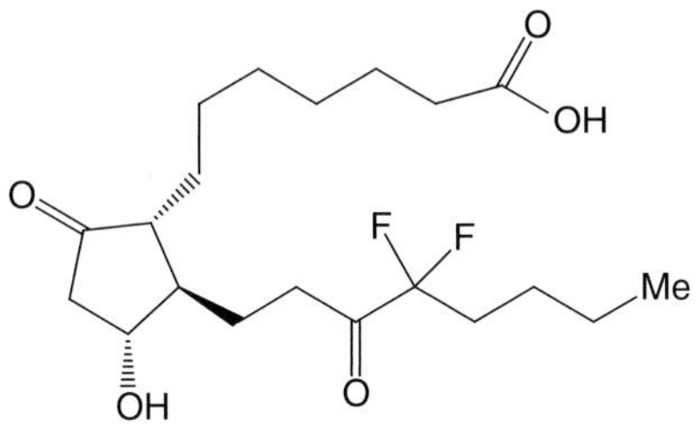

Lubiprostone **10**

6.29.7.9 Cannabinoid Receptor Modulators

Cannabis has been used for centuries to treat GI pain. Cannabinoids appear to have a neuromodulatory role in the GI system. CB1-cannabinoid receptors are present in the human ileum and colon, and their pharmacological activation leads to inhibition of excitatory cholinergic pathways involved in smooth-muscle contraction. Endogenous cannabinoids are physiological downregulators of colonic propulsion in mice; their actions on myenteric CB1-cannabinoid receptors leads to inhibition of colonic propulsion.[203] In keeping with this, in a study in mice, the competitive cannabinoid receptor antagonist rimonabant **11** enhanced motor activity in colonic smooth muscle. Modulation of peripheral cannabinoid receptors could thus provide relief from both sensory and motor components of IBS symptoms.[192]

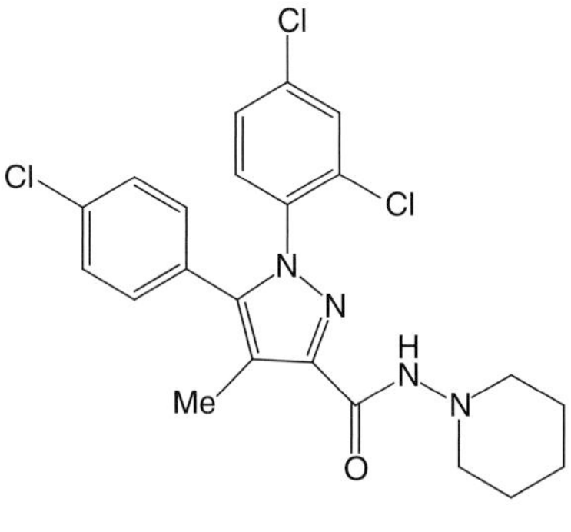

Rimonabant **11**

References

1. Mueller-Lissner, S. A.; Bollani, S.; Brummer, R. J.; Coremans, G.; Dapoigny, M.; Marshall, J. K.; Muris, J. W.; Oberndorff-Klein, W. A.; Pace, F.; Rodrigo, L. et al. *Digestion* **2001**, *64*, 200–204.
2. Hungin, P. A.; Tack, J.; Whorwell, P.; Mearin, F. Prevalence of Irritable Bowel Syndrome (IBS) Across Europe and Its Impact on Patients' Daily Lives: The Truth in IBS (T-IBS) Survey. Presented at 10th United European Gastroenterology Week, Geneva, Switzerland, October 19–23, 2002.
3. Dapoigny, M.; Bellanger, J.; Bonaz, B.; des Varannes, S. B.; Bueno, L.; Coffin, B.; Ducrotte, P.; Flourie, B.; Lemann, M.; Lepicard, A. et al. *Eur. J. Gastroenterol. Hepatol.* **2004**, *16*, 995–1001.
4. Hungin, A. P. S.; Chang, L.; Locke, G. R.; Dennis, E. H.; Barghout, V. *Aliment. Pharmacol. Ther.* **2005**, *21*, 1365–1375.

5. Brandt, L. J.; Bjorkman, D.; Fennerty, M. B.; Locke, G. R.; Olden, K.; Peterson, W.; Quigley, E.; Schoenfeld, P.; Schuster, M.; Talley, N. *Am. J. Gastroenterol.* **2002**, *97*, S7–S26.
6. Gwee, K.-A. *Neurogastroenterol. Motil.* **2005**, *17*, 317–324.
7. Hoseini-Asl, M. K.; Amra, B. *Indian J. Gastroenterol.* **2003**, *22*, 215–216.
8. Kwan, A. C.; Hu, W. H.; Chan, Y. K.; Yeung, Y. W.; Lai, T. S.; Yuen, H. *J. Gastroenterol. Hepatol.* **2002**, *17*, 1180–1186.
9. Lau, E. M.; Chan, F. K.; Ziea, E. T.; Chan, C. S.; Wu, J. C.; Sung, J. J. *Dig. Dis. Sci.* **2002**, *47*, 2621–2624.
10. Danivat, D.; Tankeyoon, M.; Sriratanaban, A. *BMJ (Clinical Research Ed.)* **1988**, *296*, 1710.
11. Jafri, N.; Raza, Z.; Jafri, S. W. *Am. J. Gastroenterol.* **2002**, *97*, S276A. Abstract.
12. Celebi, S.; Acik, Y. A.; Deveci, S. E.; Bahcecioglu, I. H.; Ayar, A.; Demir, A.; Durukan, P. *J. Gastroenterol. Hepatol.* **2004**, *19*, 738–743.
13. Kang, J. Y. *Aliment. Pharmacol. Ther.* **2005**, *21*, 663–676.
14. Gwee, K.-A.; Wee, S.; Wong, M.-L.; Png, D. J. C. *Am. J. Gastroenterol.* **2004**, *99*, 924–931.
15. Frissora, C. L.; Koch, K. L. *Curr. Gastroenterol. Rep.* **2005**, *7*, 257–263.
16. Chial, H. J.; Camilleri, M. *J. Gender Specific Med.* **2002**, *5*, 37–45.
17. Heitkemper, M.; Jarrett, M.; Bond, E. F.; Chang, L. *Biol. Res. Nurs.* **2003**, *5*, 56–65.
18. Chang, L.; Heitkemper, M. M. *Gastroenterology* **2002**, *123*, 1686–1701.
19. Hulisz, D. *J. Manag. Care Pharm.* **2004**, *10*, 299–309.
20. Hungin, A. P. S.; Tack, J.; Mearin, F.; Whorwell, P. J.; Dennis, E.; Barghout, V. *Am. J. Gastroenterol.* **2002**, *97*, S280–S281, Abstract.
21. Frank, L.; Kleinman, L.; Rentz, A.; Ciesla, G.; Kim, J. J.; Zacker, C. *Clin. Ther.* **2002**, *24*, 675–689.
22. Martin, R.; Barron, J. J.; Zacker, C. *Am. J. Manag. Care* **2001**, *7*, S268–S275.
23. Cash, B. D.; Sullivan, S.; Barghout, V. *Am. J. Manag. Care* **2005**, *11*, S7–S16.
24. American Gastroenterological Association. *The Burden of Gastrointestinal Diseases*; American Gastroenterological Association: Bethesda, MD, 2001; Vol. 01.
25. Russo, M. W.; Wei, J. T.; Thiny, M. T.; Gangarosa, L. M.; Brown, A.; Ringel, Y.; Shaheen, N. J.; Sandler, R. S. *Gastroenterology* **2004**, *126*, 1448–1453.
26. Drossman, D. A.; Whitehead, W. E.; Camilleri, M. *Gastroenterology* **1997**, *112*, 2120–2137.
27. Mitchell, C. M.; Drossman, D. A. *Gastroenterology* **1987**, *92*, 1282–1284.
28. Longstreth, G. F.; Wilson, A.; Knight, K.; Wong, J.; Chiou, C. F.; Barghout, V.; Frech, F.; Ofman, J. J. *Am. J. Gastroenterol.* **2003**, *98*, 600–607.
29. Dean, B. B.; Aguilar, D.; Barghout, V.; Kahler, K.; Frech, F.; Groves, D.; Ofman, J. J. *Am. J. Manag. Care* **2005**, *11*, S17–S26.
30. Leong, S. A.; Barghout, V.; Birnbaum, H. G.; Thibeault, C. E.; Ben-Hamadi, R.; Frech, F.; Ofman, J. J. *Arch. Intern. Med.* **2003**, *163*, 929–935.
31. Thompson, W. G.; Longstreth, G. F.; Drossman, D. A.; Heaton, K. W.; Irvine, E. J.; Muller-Lissner, S. A. *Gut* **1999**, *45*, II43–II47.
32. Talley, N. J.; Dennis, E. H.; Schettler-Duncan, V. A.; Lacy, B. E.; Olden, K. W.; Crowell, M. D. *Am. J. Gastroenterol.* **2003**, *98*, 2454–2459.
33. Frissora, C. L.; Koch, K. L. *Curr. Gastroenterol. Rep.* **2005**, *7*, 264–271.
34. Whitehead, W. E.; Palsson, O.; Jones, K. R. *Gastroenterology* **2002**, *122*, 1140–1156.
35. Lembo, T. J.; Fink, R. N. *J. Clin. Gastroenterol.* **2002**, *35*, S31–S36.
36. Ehrenpreis, E. D. *Fam. Pract. Recertification* **2004**, *26*, 53–60.
37. Ishiguchi, T.; Itoh, H.; Ichinose, M. *Dig. Endosc.* **2003**, *15*, 81–86.
38. Camilleri, M. *Rev. Gastroenterol. Disord.* **2001**, *1*, 2–17.
39. Schemann, M.; Neunlist, M. *Neurogastroenterol. Motil.* **2004**, *16*, 55–59.
40. Gershon, M. D. *J. Clin. Gastroenterol.* **2005**, *39*, S184–S193.
41. Goyal, R. K.; Hirano, I. N. *Engl. J. Med.* **1996**, *334*, 1106–1115.
42. Wood, J. D. *J. Clin. Gastroenterol.* **2002**, *35*, S11–S22.
43. Mulak, A.; Bonaz, B. *Med. Sci. Monit.* **2004**, *10*, RA52–RA62.
44. Gershon, M. D. *The Second Brain: A Groundbreaking New Understanding of Nervous Disorders of the Stomach And Intestine*; HarperCollins: New York, 1999, pp 1–320.
45. Crowell, M. D. *Am. J. Manag. Care* **2001**, *7*, S252–S260.
46. Ahn, J.; Ehrenpreis, E. D. *Expert. Opin. Pharmacother.* **2002**, *3*, 9–21.
47. Harris, M. L.; Aziz, Q. *Hosp. Med.* **2003**, *64*, 264–269.
48. Clerc, N.; Furness, J. B. *Neurogastroenterol. Motil.* **2004**, *16*, 24–27.
49. Wood, J. D.; Alpers, D. H.; Andrews, P. L. *Gut* **1999**, *45*, II6–II16.
50. Kim, D. Y.; Camilleri, M. *Am. J. Gastroenterol.* **2000**, *95*, 2698–2709.
51. Grider, J. R.; Foxx-Orenstein, A. E.; Jin, J. G. *Gastroenterology* **1998**, *115*, 370–380.
52. Gershon, M. D. *Rev. Gastroenterol. Disord.* **2003**, *3*, S25–S34.
53. De Ponti, F.; Tonini, M. *Drugs* **2001**, *61*, 317–332.
54. Talley, N. J. *Lancet* **2001**, *358*, 2061–2068.
55. Harris, L.; Chang, L. IBS: Improving Diagnosis, Serotonin Signaling, and Implications for Treatment. *Medscape Web Site* 2003. http://www.medscape.com/ (accessed July 2006).
56. Berthoud, H. R.; Blackshaw, L. A.; Brookes, S. J.; Grundy, D. *Neurogastroenterol. Motil.* **2004**, *16*, 28–33.
57. Grundy, D.; Scratcherd, T. Sensory Afferents from the Gastrointestinal Tract. In *Handbook of Physiology; Section 6: The Gastrointestinal System*; Schultz, S. G., Wood, J. D., Rauner, B. B., Eds.; Oxford University Press: Oxford, 1989, pp 593–620.
58. Cash, B. D.; Chey, W. D. *Aliment. Pharmacol. Ther.* **2005**, *22*, 1047–1060.
59. Baker, D. E. *Am. J. Health. Syst. Pharm.* **2005**, *62*, 700–711.
60. Pan, H.; Gershon, M. D. *J. Neurosci.* **2000**, *20*, 3295–3309.
61. Bertrand, P. P.; Kunze, W. A.; Furness, J. B.; Bornstein, J. C. *Neuroscience* **2000**, *101*, 459–469.
62. Pan, H.; Galligan, J. J. *Am. J. Physiol. Gastrointest. Liver Physiol.* **1994**, *266*, G230–G238.
63. Liu, M.; Geddis, M. S.; Wen, Y.; Setlik, W.; Gershon, M. D. *Am. J. Physiol. Gastrointest. Liver Physiol.* **2005**, *289*, G1148–G1163.
64. Hu, H.-Z.; Fang, X.; Liu, S.; Wang, G.-D.; Wang, X.-Y.; Gao, N.; Xia, Y.; Wood, J. D. *Neurogastroenterol. Motil.* **2005**, *17*, 614, Abstract.
65. Foxx-Orenstein, A. E.; Kuemmerle, J. F.; Grider, J. R. *Gastroenterology* **1996**, *111*, 1281–1290.
66. Grider, J. R.; Kuemmerle, J. F.; Jin, J. G. *Am. J. Physiol. Gastrointest. Liver Physiol.* **1996**, *270*, G778–G782.

67. Miwa, J.; Echizen, H.; Matsueda, K.; Umeda, N. *Digestion* **2001**, *63*, 188–194.
68. Stoner, M. C.; Arcuni, J.; Lee, J.; Kellum, J. M. *Gastroenterology* **1999**, *116*, A648, Abstract.
69. Borman, R. A.; Burleigh, D. E. *Br. J. Pharmacol.* **1993**, *110*, 927–928.
70. Hicks, G. A.; Coldwell, J. R.; Schindler, M.; Ward, P. A.; Jenkins, D.; Lynn, P. A.; Humphrey, P. P.; Blackshaw, L. A. J. *Physiol. Lond.* **2002**, *544*, 861–869.
71. Hillsley, K.; Grundy, D. *Neurosci. Lett.* **1998**, *255*, 63–66.
72. Mathis, C.; Schikowski, A.; Thewissen, M.; Ross, H.-G.; Pak, M. A.; Enck, P. *Neurogastroenterol. Motil.* **1998**, *10*, 459, Abstract.
73. Schikowski, A.; Thewissen, M.; Mathis, C.; Ross, H.-G.; Enck, P. *Neurogastroenterol. Motil.* **2002**, *14*, 221–227.
74. Wei, J. Y.; Wang, Y. H.; Mayer, E. *Gastroenterology* **2002**, *122*, A317–A318, Abstract.
75. Liu, M.; Gershon, M. D. *Neurogastroenterol. Motil.* **2005**, *17*, 614, Abstract.
76. Mach, T. *Med. Sci. Monit.* **2004**, *10*, RA125–RA131.
77. Barbara, G.; De Giorgio, R.; Stanghellini, V.; Cremon, C.; Salvioli, B.; Corinaldesi, R. *Aliment. Pharmacol. Ther.* **2004**, *20*, 1–9.
78. Hasler, W. L. *Gastroenterology* **2002**, *122*, 576–577.
79. Azpiroz, F.; Malagelada, J.-R. *Gastroenterology* **2005**, *129*, 1060–1078.
80. Serra, J.; Azpiroz, F.; Malagelada, J. R. *Gut* **2001**, *48*, 14–19.
81. Serra, J.; Salvioli, B.; Azpiroz, F.; Malagelada, J. R. *Gastroenterology* **2002**, *123*, 700–706.
82. Passos, M. C.; Serra, J.; Azpiroz, F.; Tremolaterra, F.; Malagelada, J.-R. *Gut* **2005**, *54*, 344–348.
83. Caldarella, M. P.; Serra, J.; Azpiroz, F.; Malagelada, J.-R. *Gastroenterology* **2002**, *122*, 1748–1755.
84. Gonlachanvit, S.; Coleski, R.; Owyang, C.; Hasler, W. *Gut* **2004**, *53*, 1577–1582.
85. Whitehead, W. E.; Engel, B. T.; Schuster, M. M. *Dig. Dis. Sci.* **1980**, *25*, 404–413.
86. Trimble, K. C.; Farouk, R.; Pryde, A.; Douglas, S.; Heading, R. C. *Dig. Dis. Sci.* **1995**, *40*, 1607–1613.
87. Whitehead, W. E.; Holtkotter, B.; Enck, P.; Hoelzl, R.; Holmes, K. D.; Anthony, J.; Shabsin, H. S.; Schuster, M. M. *Gastroenterology* **1990**, *98*, 1187–1192.
88. Mearin, F.; Cucala, M.; Azpiroz, F.; Malagelada, J.-R. *Gastroenterology* **1991**, *101*, 999–1006.
89. Mertz, H.; Naliboff, B.; Munakata, J.; Niazi, N.; Mayer, E. A. *Gastroenterology* **1995**, *109*, 40–52.
90. Chang, L.; Munakata, J.; Mayer, E. A.; Schmulson, M. J.; Johnson, T. D.; Bernstein, C. N.; Saba, L.; Naliboff, B.; Anton, P. A.; Matin, K. *Gut* **2000**, *47*, 497–505.
91. Olden, K. W. Diagnosis, pathophysiology, and treatment of irritable bowel syndrome. *Medscape Web Site* 2003. http://www.medscape.com/ (accessed July 2006).
92. Silverman, D. H.; Munakata, J. A.; Ennes, H.; Mandelkern, M. A.; Hoh, C. K.; Mayer, E. A. *Gastroenterology* **1997**, *112*, 64–72.
93. Naliboff, B. D.; Derbyshire, S. W. G.; Munakata, J.; Berman, S.; Mandelkern, M.; Chang, L.; Mayer, E. A. *Psychosom. Med.* **2001**, *63*, 365–375.
94. Mertz, H.; Morgan, V.; Tanner, G.; Pickens, D.; Price, R.; Shyr, Y.; Kessler, R. *Gastroenterology* **2000**, *118*, 842–848.
95. Schwetz, I.; Bradesi, S.; Mayer, E. A. *Minerva Med.* **2004**, *95*, 419–426.
96. Crowell, M. D.; Shetzline, M. A.; Moses, P. L.; Mawe, G. M.; Talley, N. J. *Curr. Opin. Invest. Drugs* **2004**, *5*, 55–60.
97. Coates, M. D.; Mahoney, C. R.; Linden, D. R.; Sampson, J. E.; Chen, J.; Blaszyk, H.; Crowell, M. D.; Sharkey, K. A.; Gershon, M. D.; Mawe, G. M. et al. *Gastroenterology* **2004**, *126*, 1657–1664.
98. Dunlop, S. P.; Jenkins, D.; Neal, K. R.; Spiller, R. C. *Gastroenterology* **2003**, *125*, 1651–1659.
99. Dunlop, S. *Am. J. Gastroenterol.* **2003**, *98*, 1578–1583.
100. Dunlop, S. P.; Coleman, N. S.; Blackshaw, E.; Perkins, A. C.; Singh, G.; Marsden, C. A.; Spiller, R. C. *Clin. Gastroenterol. Hepatol.* **2005**, *3*, 349–357.
101. Camilleri, M. *Gut* **2004**, *53*, 1396–1399.
102. Houghton, L. A.; Atkinson, W.; Whitaker, R. P.; Whorwell, P. J.; Rimmer, M. J. *Gut* **2003**, *52*, 663–670.
103. Atkinson, W.; Lockhart, S.; Keevil, B.; Whorwell, P.; Houghton, L. *Gastroenterology* **2004**, *126*, A93, Abstract.
104. Chen, J. J.; Li, Z.; Pan, H.; Murphy, D. L.; Tamir, H.; Koepsell, H.; Gershon, M. D. *J. Neurosci.* **2001**, *21*, 6348–6361.
105. O'Hara, J. R.; Ho, W.; Mawe, G. M.; Sharkey, K. A. *Gastroenterology* **2003**, *124*, A341, Abstract.
106. Eutamene, H.; Theodorou, V.; Fioramonti, J.; Bueno, L. *J. Physiol. Lond.* **2003**, *553*, 959–966.
107. Sagami, Y.; Shimada, Y.; Tayama, J.; Nomura, T.; Satake, M.; Endo, Y.; Shoji, T.; Karahashi, K.; Hongo, M.; Fukudo, S. *Gut* **2004**, *53*, 958–964.
108. Crowell, M. D.; Schettler-Duncan, A.; Dennis, E. H.; Lacy, B. E. *Am. J. Gastroenterol.* **2002**, *97*, S281–S282, Abstract.
109. Drossman, D. A. *J. Clin. Gastroenterol.* **2005**, *39*, S251–S256.
110. Barbara, G.; De Giorgio, R.; Stanghellini, V.; Cremon, C.; Corinaldesi, R. *Gut* **2002**, *51*, i41–i44.
111. Spiller, R. C.; Jenkins, D.; Thornley, J. P.; Hebden, J. M.; Wright, T.; Skinner, M.; Neal, K. R. *Gut* **2000**, *47*, 804–811.
112. Chadwick, V. S.; Chen, W.; Shu, D.; Paulus, B.; Bethwaite, P.; Tie, A.; Wilson, I. *Gastroenterology* **2002**, *122*, 1778–1783.
113. Barbara, G.; Stanghellini, V.; De Giorgio, R.; Cremon, C.; Cottrell, G. S.; Santini, D.; Pasquinelli, G.; Morselli-Labate, A. M.; Grady, E. F.; Bunnett, N. W. et al. *Gastroenterology* **2004**, *126*, 693–702.
114. O'Leary, C.; Quigley, E. M. *Am. J. Gastroenterol.* **2003**, *98*, 720–722.
115. King, T. S.; Elia, M.; Hunter, J. O. *Lancet* **1998**, *352*, 1187–1189.
116. Pimentel, E. R.; Castro, L. G.; Cuce, L. C.; Sampaio, S. A. *J. Dermatol. Surg. Oncol.* **1989**, *15*, 72–77.
117. Kurtovic, J.; Riordan, S. M. *Am. J. Gastroenterol.* **2004**, *99*, 961–962.
118. Atkinson, W.; Sheldon, T. A.; Shaath, N.; Whorwell, P. J. *Gut* **2004**, *53*, 1459–1464.
119. Isolauri, E.; Rautava, S.; Kalliomaki, M. *Gut* **2004**, *53*, 1391–1393.
120. Locke, G. R., III; Zinsmeister, A. R.; Talley, N. J.; Fett, S. L.; Melton, L. J., III *Mayo Clin. Proc.* **2000**, *75*, 907–912.
121. Levy, R. L.; Jones, K. R.; Whitehead, W. E.; Feld, S. I.; Talley, N. J.; Corey, L. A. *Gastroenterology* **2001**, *121*, 799–804.
122. Morris-Yates, A.; Talley, N. J.; Boyce, P. M.; Nandurkar, S.; Andrews, G. *Am. J. Gastroenterol.* **1998**, *93*, 1311–1317.
123. Camilleri, M.; Atanasova, E.; Carlson, P. J.; Ahmad, U.; Kim, H. J.; Viramontes, B. E.; McKinzie, S.; Urrutia, R. *Gastroenterology* **2002**, *123*, 425–432.
124. Kim, H. J.; Camilleri, M.; Carlson, P. J.; Cremonini, F.; Ferber, I.; Stephens, D.; McKinzie, S.; Zinsmeister, A. R.; Urrutia, R. *Gut* **2004**, *53*, 829–837.
125. Holtmann, G.; Siffert, W.; Haag, S.; Mueller, N.; Langkafel, M.; Senf, W.; Zotz, R.; Talley, N. J. *Gastroenterology* **2004**, *126*, 971–979.

126. Mayer, E. A.; Collins, S. M. *Gastroenterology* **2002**, *122*, 2032–2048.
127. Coutinho, S. V.; Plotsky, P. M.; Sablad, M.; Miller, J. C.; Zhou, H.; Bayati, A. I.; McRoberts, J. A.; Mayer, E. A. *Am. J. Physiol. Gastrointest. Liver Physiol.* **2002**, *282*, G307–G316.
128. Barbara, G.; Vallance, B. A.; Collins, S. M. *Gastroenterology* **1997**, *113*, 1224–1232.
129. De Giorgio, R.; Barbara, G.; Blennerhassett, P.; Wang, L.; Stanghellini, V.; Corinaldesi, R.; Collins, S. M.; Tougas, G. *Gut* **2001**, *49*, 822–827.
130. Al-Chaer, E. D.; Kawasaki, M.; Pasricha, P. J. *Gastroenterology* **2000**, *119*, 1276–1285.
131. Greenwood-Van meerveld, B.; Venkova, K.; Hicks, G.; Dennis, E.; Crowell, M. D. *Neurogastroenterol. Motil.* **2006**, *18*, 76–86.
132. Spiller, R. C. *Am. J. Med.* **1999**, *107*, 91S–97S.
133. Hawkey, C. J. *Am. J. Med.* **1999**, *107*, 98S–102S.
134. Talley, N. J. *Br. J. Clin. Pharmacol.* **2003**, *56*, 362–369.
135. Bergmann, J. F. *Am. J. Med.* **1999**, *107*, 59S–64S.
136. Patel, S. M.; Stason, W. B.; Legedza, A.; Ock, S. M.; Kaptchuk, T. J.; Conboy, L.; Canenguez, K.; Park, J. K.; Kelly, E.; Jacobson, E. et al. *Neurogastroenterol. Motil.* **2005**, *17*, 332–340.
137. Kamm, M. A. *Aliment. Pharmacol. Ther.* **2002**, *16*, 343–351.
138. Dunger-Baldauf, C., Nyhlin, H., Ruegg, P., Wagner, A. Subject's Global Assessment of Satisfactory Relief as a Measure to Assess Treatment Effect in Clinical Trials in Irritable Bowel Syndrome (IBS). Presented at the 28th Annual Meeting of the American College of Gastroenterology, Baltimore, Maryland, October 13–15, 2003.
139. Corazziari, E.; Bytzer, P.; Delvaux, M.; Holtmann, G.; Malagelada, J. R.; Morris, J.; Muller-Lissner, S.; Spiller, R. C.; Tack, J.; Whorwell, P. J. *Aliment. Pharmacol. Ther.* **2003**, *18*, 569–580.
140. Veldhuyzen Van Zanten, S. J.; Talley, N. J.; Bytzer, P.; Klein, K. B.; Whorwell, P. J.; Zinsmeister, A. R. *Gut* **1999**, *45*, II69–II77.
141. Mangel, A. W. *Aliment. Pharmacol. Ther.* **2004**, *19*, 141–142.
142. Fennerty, M. B. *Rev. Gastroenterol. Disord.* **2003**, *3*, S18–S24.
143. Johanson, J. F. *Neurogastroenterol. Motil.* **2004**, *16*, 701–711.
144. Jackson, J. L.; O'Malley, P. G.; Tomkins, G.; Balden, E.; Santoro, J.; Kroenke, K. *Am. J. Med.* **2000**, *108*, 65–72.
145. Mertz, H. *J. Clin. Gastroenterol.* **2005**, *39*, S247–S250.
146. Rosemore, J. G.; Lacy, B. E. *J. Clin. Gastroenterol.* **2002**, *35*, S37–S44.
147. Curry, C. E.; Butler, D. M. Constipation. In *Handbook of Nonprescription Drugs: An Interactive Approach to Self-Care*, 14th ed; Berardi, R. R., DeSimone, E. M., Newton, G. D., Oszko, M. A., Popovich, N. G., Rollins, C. J., Shimp, L. A., Tietze, K. J., Eds.; APhA Publications: Washington, DC, 2004, pp 367–403.
148. Mertz, H. R. *N. Engl. J. Med.* **2003**, *349*, 2136–2146.
149. Jailwala, J.; Imperiale, T. F.; Kroenke, K. *Ann. Intern. Med.* **2000**, *133*, 136–147.
150. Drossman, D. A.; Camilleri, M.; Mayer, E. A.; Whitehead, W. E. *Gastroenterology* **2002**, *123*, 2108–2131.
151. Crowell, M. D.; Jones, M. P.; Harris, L. A.; Dineen, T. N.; Schettler, V. A.; Olden, K. W. *Curr. Opin. Invest. Drugs* **2004**, *5*, 736–742.
152. Berrada, D.; Canenguez, K.; Lembo, T. *Curr. Gastroenterol. Rep.* **2003**, *5*, 337–342.
153. Buchheit, K. H.; Gamse, R.; Giger, R.; Hoyer, D.; Klein, F.; Kloppner, E.; Pfannkuche, H. J.; Mattes, H. J. *Med. Chem.* **1995**, *38*, 2326–2330.
154. Buchheit, K. H.; Gamse, R.; Giger, R.; Hoyer, D.; Klein, F.; Kloppner, E.; Pfannkuche, H. J.; Mattes, H. J. *Med. Chem.* **1995**, *38*, 2331–2338.
155. Hoyer, D.; Fehlmann, D.; Langenegger, D.; Kummer, J.; Giger, R.; Mattes, H.; Probst, A.; Pfannkuche, H. J. *Ann. NY Acad. Sci.* **1998**, *861*, 267–268.
156. Zelnorm (tegaserod maleate) tablets. Novartis Pharmaceuticals Corporation: East Hanover, NJ, 2004.
157. Prather, C. M.; Camilleri, M.; Zinsmeister, A. R.; McKinzie, S.; Thomforde, G. *Gastroenterology* **2000**, *118*, 463–468.
158. Janig, W.; Koltzenburg, M. *J. Neurophysiol.* **1991**, *65*, 1067–1077.
159. Coelho, A.-M.; Rovira, P.; Fioramonti, J.; Bueno, L. *Gastroenterology* **2000**, *118*, A835, Abstract.
160. Greenwood-Van meerveld, B.; Zumo, D.; Crowell, M. *Gastroenterology* **2003**, *124*, A1, Abstract.
161. Jiao, H. M.; Xie, P. Y. *World J. Gastroenterol.* **2004**, *10*, 2836–2841.
162. Liang, L. X.; Zhang, Q.; Qian, W.; Hou, X. H. *Chin. J. Dig. Dis.* **2005**, *6*, 21–25.
163. Bouhassira, D.; Sabate, J.-M.; Coffin, B.; Le Bars, D.; Willer, J.-C.; Jian, R. *Am. J. Physiol. Gastrointest. Liver Physiol.* **1998**, *275*, G410–G417.
164. Sabate, J.-M.; Coffin, B.; Jian, R.; Le Bars, D.; Bouhassira, D. *Am. J. Physiol. Gastrointest. Liver Physiol.* **2000**, *279*, G692–G699.
165. Coffin, B.; Bouhassira, D.; Sabate, J. M.; Barbe, L.; Jian, R. *Gut* **2004**, *53*, 1465–1470.
166. Coffin, B.; Farmachidi, J. P.; Rueegg, P.; Bastie, A.; Bouhassira, D. *Aliment. Pharmacol. Ther.* **2003**, *17*, 577–585.
167. Sabate, J.; Bouhassira, D.; Poupardin, C.; Loria, Y.; Wagner, A.; Coffin, B. *Gastroenterology* **2005**, *128*, A468, Abstract.
169. Novick, J.; Miner, P.; Krause, R.; Glebas, K.; Bliesath, H.; Ligozio, G.; Ruegg, P.; Lefkowitz, M. *Aliment. Pharmacol. Ther.* **2002**, *16*, 1877–1888.
170. Tougas, G.; Snape, W. J., Jr.; Otten, M. H.; Earnest, D. L.; Langaker, K.-E.; Pruitt, R. E.; Pecher, E.; Nault, B.; Rojavin, M. A. *Aliment. Pharmacol. Ther.* **2002**, *16*, 1701–1708.
171. The Latin America Investigation Group of Tegaserod; Cohen Munoz, V. *Gastroenterology* **2003**, *124*, A571. Abstract.
172. Higgins, P. D. R.; Davis, K. J.; Laine, L. *Aliment. Pharmacol. Ther.* **2004**, *19*, 729–738.
173. Chang, B.; Chey, W. Incidence of Ischemic Colitis and Serious Complications of Constipation among Patients using Alosetron: Systematic Review of Clinical Trials and Post-Marketing Surveillance Data. *Am. J. Gastroenterol.* in press.
174. Joelsson, B. E.; Shetzline, M. A.; Cunningham, S. *N. Engl. J. Med.* **2004**, *351*, 1363–1364, Reply.
175. Janssen Pharmaceutica stops marketing cisapride in the US. FDA talk paper. T00-14. 3-23-2000. Food and Drug Administration, US Department of Health and Human Services: Rockville, MD, Nov 16, 2004.
176. Drolet, B.; Khalifa, M.; Daleau, P.; Hamelin, B. A.; Turgeon, J. *Circulation* **1998**, *97*, 204–210.
177. Morganroth, J.; Ruegg, P. C.; Dunger-Baldauf, C.; Appel-Dingemanse, S.; Bliesath, H.; Lefkowitz, M. *Am. J. Gastroenterol.* **2002**, *97*, 2321–2327.
178. Clayton, N. M.; Sargent, R.; Butler, A.; Gale, J.; Maxwell, M. P.; Hunt, A. A.; Barrett, V. J.; Cambridge, D.; Bountra, C.; Humphrey, P. P. *Neurogastroenterol. Motil.* **1999**, *11*, 207–217.
179. Humphrey, P. P.; Bountra, C.; Clayton, N.; Kozlowski, K. *Aliment. Pharmacol. Ther.* **1999**, *13*, 31–38.

180. Kilpatrick, G. J.; Hagan, R. M.; Oxford, A. W.; North, P. C.; Tyers, M. B. *Drugs Future* **1992**, *17*, 660–664.
181. Mazzia, C.; Hicks, G. A.; Clerc, N. *Neuroscience* **2003**, *116*, 1033–1041.
182. Camilleri, M.; Mayer, E. A.; Drossman, D. A.; Heath, A.; Dukes, G. E.; McSorley, D.; Kong, S.; Mangel, A. W.; Northcutt, A. R. *Aliment. Pharmacol. Ther.* **1999**, *13*, 1149–1159.
183. De Angelis, C. D.; Drazen, J. M.; Frizelle, F. A.; Haug, C.; Hoey, J.; Kotzin, S.; Laine, C.; Marusic, A.; Overbeke, A. J. P. M.; Schroeder, T. V. et al. *N. Engl. J. Med.* **2004**, *351*, 1250–1251.
184. Camilleri, M.; Northcutt, A. R.; Kong, S.; Dukes, G. E.; McSorley, D.; Mangel, A. W. *Lancet* **2000**, *355*, 1035–1040.
185. Lembo, T.; Wright, R. A.; Bagby, B.; Decker, C.; Gordon, S.; Jhingran, P.; Carter, E. *Am. J. Gastroenterol.* **2001**, *96*, 2662–2670.
186. Lotronex (alosetron hydrochloride) tablets. GlaxoSmithKline: Research Triangle Park, NC, 2005.
187. Chey, W. D.; Chey, W. Y.; Heath, A. T.; Dukes, G. E.; Carter, E. G.; Northcutt, A. R.; Ameen, V. Z. *Am. J. Gastroenterol.* **2004**, *99*, 2195–2203.
188. Poynard, T.; Regimbeau, C.; Benhamou, Y. *Aliment. Pharmacol. Ther.* **2001**, *15*, 355–361.
189. Lembo, A. *J. Clin. Gastroenterol.* **2004**, *38*, 776–781.
190. Reilly, M. C.; Bracco, A.; Geissler, K.; Johanson, J.; Kahler, K. H. *Am. J. Gastroenterol.* **2004**, *99*, S232–S233, Abstract.
191. Delgado-Aros, S.; Chial, H. J.; Camilleri, M.; Szarka, L. A.; Weber, F. T.; Jacob, J.; Ferber, I.; McKinzie, S.; Burton, D. D.; Zinsmeister, A. R. *Am. J. Physiol. Gastrointest. Liver Physiol.* **2003**, *284*, G558–G566.
192. Cremonini, F.; Talley, N. J. *Nat. Clin. Pract. Gastroenterol. Hepatol.* **2005**, *2*, 82–88.
193. Henderson, J. C.; Palmer, R. M. J.; Meyers, N. L.; Spiller, R. C. *Gastroenterology* **2004**, *126*, A644, Abstract.
194. Camilleri, M. *Br. J. Pharmacol.* **2004**, *141*, 1237–1248.
195. Martinez, V.; Rivier, J.; Wang, L.; Tache, Y. *J. Pharmacol. Exp. Ther.* **1997**, *280*, 754–760.
196. Julia, V.; Su, X.; Bueno, L.; Gebhart, G. F. *Gastroenterology* **1999**, *116*, 1124–1131.
197. Coremans, G.; Clouse, R. E.; Carter, F.; Krause, G.; Caras, S.; Steinborn, C. *Gastroenterology* **2004**, *126*, A643, Abstract.
198. Bradette, M.; Moennikes, H.; Carter, F.; Krause, G.; Caras, S.; Steinborn, C. *Gastroenterology* **2004**, *126*, A42, Abstract.
199. Adis Data Information BV. Cilansetron: KC 9946 (review). *Drugs Res. Dev.* **2005**, *6*, 169–173.
200. Wood, J. D.; Galligan, J. J. *Neurogastroenterol. Motil.* **2004**, *16*, 17–28.
201. Lubiprostone: RU 0211, SPI 0211. *Drugs Res. Dev.* **2005**, *6*, 245–248.
202. Sanger, G. J. *Br. J. Pharmacol.* **2004**, *141*, 1303–1312.
203. Pinto, L.; Izzo, A. A.; Cascio, M. G.; Bisogno, T.; Hospodar-Scott, K.; Brown, D. R.; Mascolo, N.; Di Marzo, V.; Capasso, F. *Gastroenterology* **2002**, *123*, 227–234.
204. Galligan, J. J.; Vanner, S. *Neurogastroenterol. Motil.* **2005**, *17*, 643–653.
205. Camilleri, M.; McKinzie, S.; Fox, J.; Foxx-Orenstein, A.; Burton, D.; Thomforde, G.; Baxter, K.; Zinsmeister, A. R. *Clin. Gastroenterol. Hepatol.* **2004**, *2*, 895–904.
206. Meyers, N. L.; George, A.; Palmer, R. M. J. *Gastroenterology* **2004**, *4*, A640, Abstract.
207. Camilleri, M.; Kim, D.-Y.; McKinzie, S.; Kim, H. J.; Thomforde, G. M.; Burton, D. D.; Low, P. A.; Zinsmeister, A. R. *Clin. Gastroenterol. Hepatol.* **2003**, *1*, 111–121.
208. Greenwood-Van meerveld, B.; Johnson, A. C.; Cochrane, S.; Schulkin, J.; Myers, D. A. *Neurogastroenterol. Motil.* **2005**, *17*, 415–422.
209. Clouse, R. E.; Caras, S.; Cataldi, F.; Carter, F.; Krause, G.; Steinborn, C. *Cilansetron is Efficacious for Relief of Urgency in Patients with Irritable Bowel Syndrome with Diarrhea Predominance (IBS-D)*. Presented at the First European Society for Pediatric Gastroenterology, Hepatology, and Nutrition (ESPGHAN) Capri meeting, Naples, Italy, May 27–29, 2004.
210. Tache, Y.; Monnikes, H.; Bonaz, B.; Rivier, J. *Ann. NY Acad. Sci.* **1993**, *697*, 233–243.
211. Tache, Y. *Gut* **2004**, *53*, 919–921.

Biography

Gareth A Hicks has worked in the neuroscience field for over 15 years and in gastrointestinal research for the last eight of these. Following undergraduate studies in the Department of Pharmacology, University College London, Gareth trained as an electrophysiologist during doctoral and postdoctoral work in the Department of Pharmacology,

University of Cambridge, and the Vollum Institute in Portland, Oregon. After joining GlaxoWellcome (now GSK) in the UK in 1996, his research focus shifted to the gastrointestinal system, and irritable bowel syndrome in particular. Between that time and 2004, Gareth had various research and management roles, culminating in his appointment to the position of head of gastrointestinal research. From 2004 to the present, he has worked at Novartis Pharmaceuticals Corporation (New Jersey, USA) in the Medical Affairs Department of a multidisciplinary therapeutic area that covers gastrointestinal, urological, bone, and arthritic disease clinical development. Gareth's primary interest is understanding the pathophysiology of diseases, and the mechanisms of action of drugs used to treat them.

Comprehensive Medicinal Chemistry II
ISBN (set): 0-08-044513-6

ISBN (Volume 6) 0-08-044519-5; pp. 643–670

6.30 Emesis/Prokinetic Agents

J D Gale, Pfizer Global Research and Development, Sandwich, UK
I Mori, Pfizer Global Research and Development, Nagoya, Japan

6.30.1 Introduction

It may seem unusual to be reviewing antiemetic and prokinetic agents in the same chapter; however, as one explores these two areas of therapy, one encounters similar classes of agents being used to treat both indications (e.g., dopamine D_2 receptor antagonists). Alternatively, the same receptor system can sometimes be modulated as a therapeutic approach to both indications, with receptor agonists perhaps being utilized in one indication and receptor antagonists in the other (e.g., $5HT_3$ receptors). Antiemetic agents and prokinetic drugs are two classes of medicine that are not used to treat specific diseases, but rather are used to treat a wide range of conditions, including those that are either idiopathic or iatrogenic.

6.30.1.1 Disease State

The major use of antiemetic agents is in the treatment of nausea and vomiting that is evoked by treatment of malignancy with either chemo- or radiotherapy. The severity of this phenomenon is greatly influenced by the cocktail of chemotherapeutic agents or the dose of radiation that is used in treatment. In addition, with radiotherapy, the part of the body that is irradiated greatly influences the outcome, such that radiotherapy of the head and neck or the abdomen is significantly more likely to provoke emesis than in other regions of the body. This of course has a neuroanatomical/biochemical basis, which will be discussed later. The introduction of modern, powerful antiemetic agents has radically improved the patient's experience of chemotherapy and has enabled the use of more aggressive treatment regimes. Other forms of emesis that are also treated pharmacologically include motion sickness, emesis resulting from vestibular disturbances, such as Meniere's disease, postoperative nausea and vomiting (PONV), and emesis associated with pregnancy.

As mentioned previously, prokinetic agents are also used to treat a variety of medical conditions. Historically, the most common use of prokinetic agents has been in the treatment of gastroesophageal reflux disease (GERD), the most common symptom of which is heartburn. GERD is a highly prevalent condition affecting perhaps 10 million individuals in the USA on a weekly basis. The prevalence of GERD across the developed world appears to be broadly uniform and appears to be increasing with the increasing trends around age and obesity in these populations. In the developing world, the prevalence of GERD is much lower, although it does appear to increase with the adoption of Western diets and lifestyles. Other conditions for which treatment with prokinetic agents has been adopted or proposed include chronic constipation, postoperative ileus, constipation-predominant irritable bowel syndrome (C-IBS), diabetic gastroparesis, and functional dyspepsia.

6.30.2 Disease Basis

6.30.2.1 Emesis

Stimuli that can evoke an emetic response are many and varied. As noted, these include but are not limited to: pharmacologically active therapeutic agents such as platinum-containing cytostatic drugs, DNA alkylating agents such as cyclophosphamide, and drugs used to treat Parkinson's disease (including L-DOPA, apomorphine (**1**), and bromocriptine; *see* 6.08 Neurodegeneration); ionizing radiation; diseases affecting the vestibular apparatus of the middle ear; and environmental factors such as motion and pregnancy. While the range of emetic stimuli is diverse, the pathways of the central nervous system (CNS) through which the emetic response is mediated are surprisingly discrete and largely well described (**Figure 1**). What is less surprising is that the inputs to these CNS centers are more varied and comprise a number of different neuronal pathways, both peripheral and central, as well as circulatory and systemic routes.

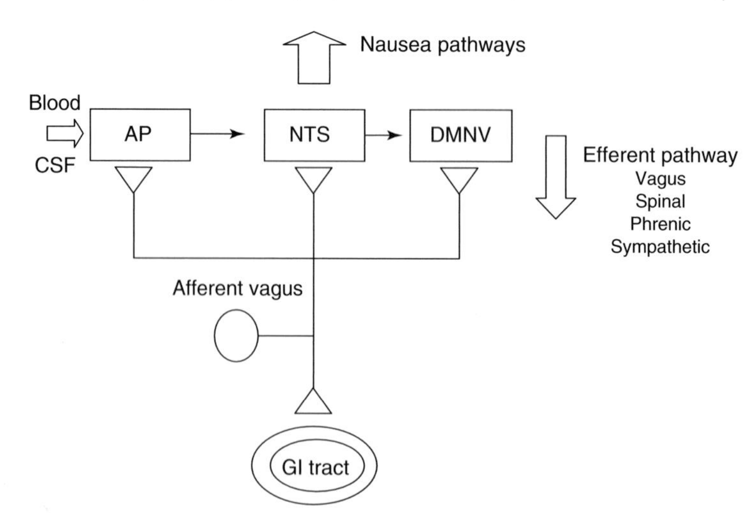

Figure 1 Schematic representation of the emetic reflex. AP, area postrema; CSF, cerebrospinal fluid; DMNV, dorsal medullary nucleus of the vagus nerve; NTS, nucleus of the solitary tract.

Historically, the central pathways pivotal to the emetic pathways were referred to by vague neuroanatomic terms including chemoreceptor trigger zone and vomiting center. The application of more advanced techniques for neuronal tracing has described the network of brain nuclei that comprise the emetic reflex. The key centers are located in the dorsomedullary region of the brainstem and include the area postrema (AP), the nucleus of the solitary tract (NTS), and the dorsal medullary nucleus of the vagus nerve (DMNV). The AP is situated on the floor of the fourth cerebral ventricle, is in direct contact with the cerebrospinal fluid, and lies outside the blood–brain barrier. This structure therefore has a unique opportunity to sample the systemic environment and through its close apposition to underlying structures, such as the NTS, relays this information directly to deeper CNS nuclei. The NTS and the AP both receive sensory inputs from the periphery via the afferent vagus nerve, inputs from the proximal parts of the gastrointestinal (GI) tract being particularly important in the emetic reflex. Sensory information can then be projected to higher cortical and hypothalamic centers or integrated and processed via the DMNV as the descending limb of the emetic reflex. These descending pathways are important in many of the mechanical elements of vomiting, including contraction of the diaphragm and intercostal muscles, retroperistalsis in the proximal intestine, relaxation of the stomach, and ejection of gastric contents. In contrast, the neurophysiology of nausea is less well defined. Animal correlates of this phenomenon are challenging to establish and most meaningful data come from the clinic. To date, evidence suggests the involvement of ascending pathways from the brainstem to the hypothalamus and to regions of the cortex such as the inferior frontal gyrus, but much detail remains to be elucidated.

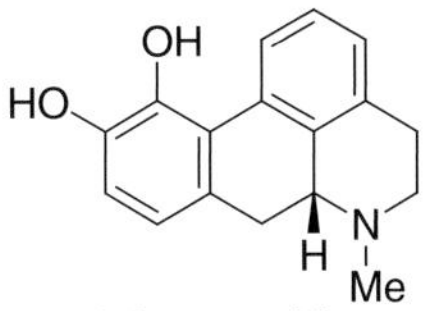

OH
HO
N
H
Me

1 Apomorphine

6.30.2.2 Gastrointestinal Motility Disorders and Prokinetics

Prokinetic agents are often employed to reverse an inhibition of physiological GI motility that has occurred as the result of disease or iatrogenic effects, such as in diabetic gastroparesis or following the use of opioid analgesia. However, prokinetic agents can also be employed in the absence of such deficits but where an increase of baseline motility can be demonstrated to generate benefit, such as in GERD. In several disease states patient symptoms result from a reduction in motility in the GI tract. For example, in type 1 diabetes, a retardation in gastric emptying, which appears to result from a neuropathy affecting the myenteric plexus in the proximal GI tract, can be demonstrated in up to 50% of patients. In a subset of these patients, this results in a significant symptom burden, most commonly manifest as nausea, vomiting, and early satiety, and in these patients the use of prokinetic agents can improve symptoms. Similarly, reduced motility in the distal GI tract, from either iatrogenic or idiopathic causes, can result in constipation, which can become chronic and present a significant problem to patients. In both circumstances, prokinetic agents have been shown to be beneficial to patients by improving their bowel habit and other symptoms. Prokinetic agents also appear to have a significant impact in the treatment of postoperative ileus. During surgery a combination of opioid analgesia and anesthetic use, incision of the peritoneum, and handling of the intestines results in a temporary cessation of GI motility, most probably through a combination of pharmacologic, physiologic, and inflammatory events. This can generate significant postoperative problems with symptoms such as bloating inhibiting resumption of feeding and delaying patient discharge from hospital. A number of less common conditions have been described in which GI motility is profoundly inhibited, such as systemic sclerosis and intestinal pseudo-obstruction. The value of prokinetic agents in these conditions is more controversial but these data serve to illustrate that one needs to consider the pathology that underlines the patient's condition before one can select a prokinetic agent with an appropriate mechanism of action to treat them. For example, $5HT_4$ receptor agonists increase GI motility through the stimulation of cholinergic neurotransmission. In disease states where there is loss of acetylcholine-containing neurons, the prokinetic activity of agents such as cisapride will be attenuated.

Historically, the most common usage of prokinetic agents was in the treatment of GERD. Before the withdrawal of the prokinetic agent cisapride (**2**) in 2000, over half of all uses were for the treatment of GERD. However, in only a small proportion of GERD patients has a delay in gastric emptying ever been demonstrated and a relationship between degree of delay and severity of reflux has not been found. Nonetheless, it makes empirical sense that if one can pharmacologically increase the rate of emptying of gastric contents, then the opportunity for retrograde reflux of gastric material will be reduced. Importantly, in GERD patients, a substantial amount of reflux is associated with transient relaxations of the lower esophageal sphincter (LES), normal physiological events most probably triggered by distension of the gastric fundus by intragastric gas. Pressure on the fundus triggers ascending intermural reflexes that result in relaxation of the LES. The benefit from prokinetic agents may result from the reduction in either pressure on the fundus or in the adaptive relaxation of the fundus in response to ingestion of food, which results from reductions in gastric retention.

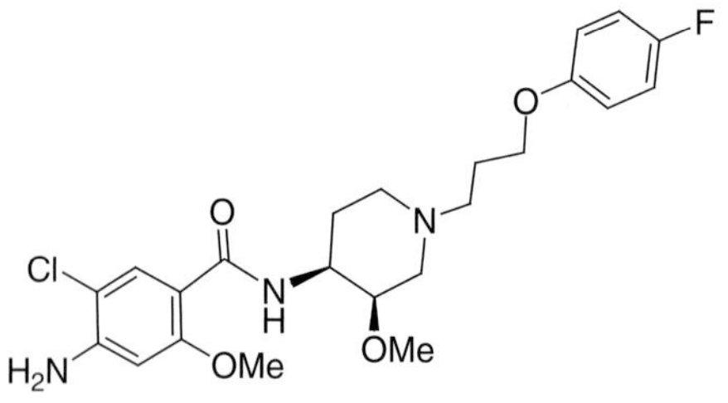

2 Cisapride (racemate)

6.30.3 Experimental Disease Models

6.30.3.1 Models of Emesis

A challenge encountered in the search for new antiemetic agents comes with the selection of a suitable experimental model. The most commonly used laboratory animal species do not possess an emetic reflex (e.g., mice, rats, guinea pigs, and rabbits) and will not vomit to any human emetogen. The most commonly used animal species in emesis research are the ferret and the dog, which have been studied widely. In addition models of emesis have been established in the cat, the monkey, the pigeon, and more recently in an insectivore, the house musk shrew.

Experimentally, the ferret has been shown to vomit to all human emetogens studied, including cisplatin, cyclophosphamide, x-irradiation, ipecacuanha, and hyperosmolar saline. Unlike human, the ferret will demonstrate a prolonged and substantial period of retching ahead of a vomiting episode, unlike the human, where only one or two retches will normally precede vomiting. This acute model of vomiting has been adapted to study the phenomenon of delayed emesis. In humans, nausea and vomiting can persist long after the original emetic stimulus has been given and compared to the initial, acute response, this has been problematic to control. In the ferret, lower doses of chemotherapeutic agents have been employed to provoke an emetic response that is prolonged and emesis has been observed for up to 24 h after administration of the emetogen. The usefulness of the ferret in the discovery and development of significant new antiemetic agents has been demonstrated with the discovery of the $5HT_3$ receptor antagonists, such as ondansetron (**3**) and granisetron (**4**). The dog has the longest pedigree in emesis research, having been used for many of the neurophysiologic studies that defined the emetic pathway in the 1950s and 1960s. The dog has been shown to respond to many of the same emetogens, often being highly sensitive to them, and to be susceptible to the blocking effects of antiemetics. However, since that time the dog has not been used as extensively, most probably for ethical questions raised by the terminal nature of many of the studies with radiation and chemotherapeutic agents. Other species have been used less frequently. The cat has been studied in motion-induced emesis, but in common with higher mammals, complex three-dimensional movement is required to evoke an emetic response. It is against this background that one of the smallest animals capable of demonstrating an emetic response was identified, the house musk shrew (*Suncus murinus*). In addition to demonstrating retching and vomiting to all emetogenic stimuli studied, this animal is sensitive to motion-induced emesis. In this case, the motion stimulus can be a simple two-dimensional one, which is simpler to apply than that required for other species. When selecting an animal model, one needs to understand the translational value of each model in terms of sensitivity and response to emetic stimuli, the pharmacology of the receptor under investigation in that species, and the behavior of the investigational drugs under study in the species selected.

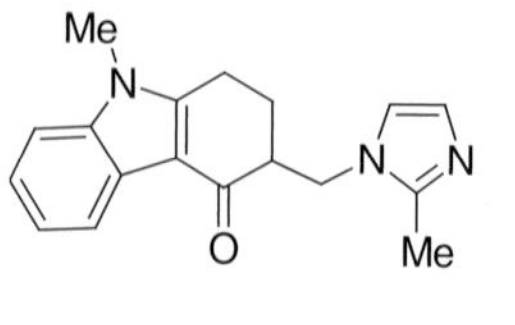

3 Ondansetron

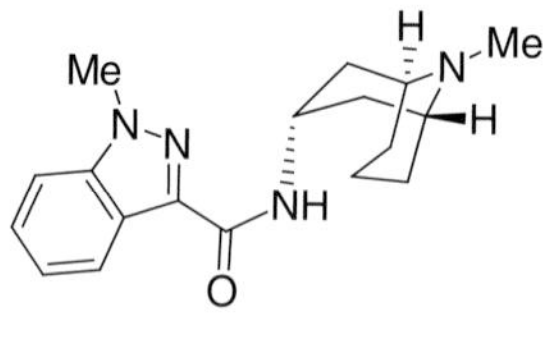

4 Granisetron

6.30.3.2 Models of Gastrointestinal Motility

Many animal models of GI motility have been described in the literature and have been used to illustrate the prokinetic activity of many of the pharmacological classes reviewed below. Nonmammalian models of GI motility suitable for the evaluation of new chemical entities (NCEs) effects are scarce. However recent studies of intestinal motor activity in the zebrafish (*Danio rerio*) and the nematode worm (*Caenorhabditis elegans*) raise the possibility of studying intestinal physiology in species that are highly manipulable at the genetic level and may be utilized in high-throughput experimental systems. The most commonly used species in motility research are the rat and the dog, although murine models have been developed that facilitate the study of GI motility in transgenic animals. When selecting an animal species in which to study the effects of NCEs, the relevance of the species to the human should be considered. For example, the anatomy of the rat stomach is quite different from that of the human, in having a defined fore-stomach that is quite different in structure, and probably function, from the human gastric fundus. The rat also lacks a structural equivalent to the human rectum and thus lacks the storage and control of continence offered by the human large bowel. Similarly the intestine, and in particular the colon, of the dog is considerably shorter than in the human and is in keeping with the largely carnivorous habits of this species. In this species the upper GI tract is considered a good model for the human whereas the lower GI tract may be less of a good model.

The effect of NCEs on gastric emptying has been widely studied and several experimental models have been established in a range of species. The most basic models involve the oral gavage of test meals and subsequent recovery of the gastric contents following euthanization of the animals. This type of model is only realistically applicable to rodents and is limited because it does not allow repeat observations in the same animals. This basic model has been adapted to allow recovery of a test meal, sometimes containing nonabsorbable markers, from artificial fistulae into the stomach or the duodenum. This model has been established in dogs but has not been widely adopted. The ideal model for use in higher animal species requires minimally invasive procedures and ideally is based on similar clinical experimental paradigms. This type of model offers the opportunity for establishing models with the greatest potential for translating across species from animals to human. Such models have been established for measuring gastric emptying.[1] In one such model, acetaminophen is included in the test meal and the appearance of this marker is measured in the systemic circulation. This model relies on the absorption of paracetamol/acetaminophen in the proximal regions of the duodenum as an indirect marker of gastric emptying and this model has been shown to be sensitive to drug effects. Another model of gastric emptying that is a direct correlate of a human model has been developed. In this model radioisotopes of indium and technetium are used to separately label liquids and solids and enable simultaneous monitoring of their egress from the stomach using gamma scintigraphy. One model that has been developed for use in animals, following its successful use in clinical research, relies on the metabolism of octanoate. In this method, rats or dogs are fed meals containing octanoate labeled with ^{13}C.[2,3] Gastric emptying is calculated from the area under the curve of the cumulative excretion of $^{13}CO_2$, collected from the expired air. This model has been shown to be sensitive to the effects of prokinetic agents. A challenge for gastric emptying models is that the baseline rate of emptying, especially of liquids, can be extremely rapid, making it extremely difficult to demonstrate a prokinetic effect of an investigational drug. An option for overcoming this is to slow the baseline rate of emptying, which can be achieved either pharmacologically or physiologically. Pharmacological slowing of gastric emptying, can be achieved by the administration of many endogenous GI hormones, such as cholecystokinin (CCK) or neuropeptide Y/peptide YY (NPY/PYY) or by pharmacologic agents such as morphine or clonidine. Such models have been well established in the literature and have proven useful in demonstrating prokinetic activity. The rate of gastric emptying can also be profoundly influenced by the composition of a test meal; for example, high-fat meals are emptied more slowly than ones containing lower levels of fat. However, meal composition can influence the effect of drugs under test as, for example, with $5HT_3$ receptor antagonists. These agents have little or no effect on rates of gastric emptying in most models but appear to accelerate gastric emptying in models using high-fat test meals. Fat retards gastric emptying through a mechanism involving the release of 5HT and the activation of $5HT_3$ receptors on vagal afferent neurons. Thus $5HT_3$ receptor antagonists can inhibit the retardation of gastric emptying induced by fat and thus appear as prokinetic agents, which is unlikely to be the case in any clinically meaningful sense.

In studying the effects of NCEs on intestinal motor function, investigators have utilized many approaches to monitor either motor activity directly or the transit of material along sections of the GI tract. Direct measurement of motor function, irrespective of the region of the GI tract under investigation, relies on the placement of strain gages on the intestinal serosa to measure contractile activity. Alternatively, measurement of patterns of electrical activity in the wall of the intestine have been made via the placement of electromyographic electrodes. Historically these techniques have required the physical connection of these strain gauges or electrodes to external recording equipment, which has necessitated at least partial restriction of the animals' movements and behavior. The development of telemetric methodologies allows long-term recordings of intestinal motility in free-living animals. The study of intestinal transit can be achieved either by identifying waves of orthograde and retrograde propulsive activity from motility recording at multiple sites along the intestine or by monitoring the passage of material along the intestine, either directly or indirectly. Both methodologies have strengths and weaknesses. The use of motility recordings to infer the passage of material is laborious and is prone to misinterpretation as not all contractile activity is associated with propulsion but rather can be involved in mixing intestinal contents. However, this type of record has the advantage of being a direct measurement of intestinal motor activity. The measurement of intestinal transit has been achieved using radioactive tracers, in which the 'center of gravity' of a radioactive meal is determined, or nonabsorbable marker such as charcoal, in which the leading edge of the charcoal is visualized. Both these methodologies are terminal procedures, as they require the animals to be euthanized for the data to be extracted. Noninvasive methodologies are clearly preferred and have been developed. The oral administration of the nonabsorbable carbohydrate lactulose, results in an increase in hydrogen in the exhaled breath. This occurs only when lactulose is metabolized by colonic bacteria and so can be used as an indirect measurement of orocecal transit. Fluoro- or radiometric methods have been developed to enable the measurement of large bowel transit. In these methodologies the passage of either a radioactive tracer or nonabsorbable radioopaque markers are monitored using a gamma camera or x-ray apparatus.

6.30.4 Clinical Trial Issues

6.30.4.1 Prokinetic Agents

Many technologies have been developed to enable the study of GI motility in the clinical setting, suitable for use in both healthy volunteers and patients. The 'gold standard' approach to the measurement of gastric emptying has involved the use of radioisotopes such as indium and technetium to label liquid and solid meal components followed by their detection using gamma scintigraphy. An alternative to this has been developed which negates the need for expensive gamma cameras and radioactivity. This uses ^{13}C-labeled octanoate or acetate which is given orally, with the rate of gastric emptying being determined through the emergence of expired $^{13}CO_2$. Strategies have been pursued to develop further nonnuclear techniques with which to determine gastric emptying. This has been driven by a desire to reduce the exposure of subjects to radiation and so facilitate repeated measurement of gastric emptying in the same individuals, so reducing variability. Such techniques have employed ultrasound[4] and magnetic resonance imaging (MRI)[5] and while these have been used to demonstrate the prokinetic effects of experimental compounds, these technologies have not yet been adopted widely.

However, MRI in particular offers the opportunity to study the regional distribution of test meals within the stomach and also the stimulation of regional motility by experimental drugs, for example in the gastric antrum. The measurement of intestinal motility in the clinic is more problematic. Rather than measuring motility directly, technologies have been developed to study the effects of drugs on the movement of material along the intestines, the so-called intestinal transit. Once again, the gold standard technique is to use a combination of radionuclides and scintigraphic recording and this can be applied to both the small and large bowel. Nonisotopic markers have been used much in the same way as in the stomach, with measurements of breath hydrogen following ingestion of a nonabsorbable carbohydrate, such as lactulose, being a useful index of orocecal transit time. The between- and within-subject variability of these techniques, together with the effects of gut function of the transit markers, needs to be fully understood. The continued reliance of radiographic measurements limits the number of investigations that can be carried out in an individual and so impacts the design of clinical trials of potential new prokinetics.

6.30.4.2 Antiemetic Agents

The study of potential novel antiemetic agents in the clinic presents particular challenges. While the major utility of such agents in clinical medicine is to treat emesis resulting from treatment with chemo- and radiotherapy, for ethical reasons such emetogens are not suitable for studying in healthy volunteer populations. Consequently, models of emesis have been developed in healthy volunteers that use ipecacuanha or motion as the emetic stimulus. These strategies have been highly successful in identifying novel agents and clinically effective doses, as illustrated during the development of the $5HT_3$ receptor antagonists. Once NCEs progress to patients, then further considerations apply. Given that effective therapies exist, it is inappropriate to withhold treatment and so the opportunity to determine the efficacy of a novel antiemetic as a stand-alone treatment may be limited, if for example in early patient studies the level of control of emesis appears lower than that of current standard of care. Under these circumstances, experimental agents may need to be given in combination with established agents.

6.30.5 Current Treatment

6.30.5.1 Prokinetic Agents

6.30.5.1.1 $5HT_4$ receptor agonists

$5HT_4$ receptors are located on the enteric nervous system within the GI tract of humans and a number of animal species. Activation of these receptors with agonists evokes the release of excitatory and inhibitory neurotransmitters, with the net result of increasing motility and orthograde peristalsis in the GI tract. The first $5HT_4$ receptor agonist prokinetic, metoclopramide (**5**), was adopted for clinical use before the target receptor had been characterized and the molecular mechanism of action of metoclopramide was disputed. Metoclopramide has moderate affinity and potency at the $5HT_4$ receptor in vitro and has prokinetic activity in models of gastric emptying. For example, in a rat model metoclopramide increased gastric emptying dose-dependently following oral dosing.[6] The mechanism of action of metoclopramide to increase gastric emptying was proposed as the enhancement of acetylcholine release, but this was the subject of controversy. Furthermore, the precise molecular target for metoclopramide remained in doubt, which arose from the complex pharmacology of this molecule. In addition to its activity at $5HT_4$ receptors, metoclopramide has significant antagonist activity at both $5HT_3$ and dopamine D_2 receptors. All of these activities have been proposed as underpinning the prokinetic activity of metoclopramide. Examination of compounds with $5HT_3$ receptor antagonist activity in in vitro and in vivo models of intestinal motility including: gastric emptying in the rat and gastric contraction in the dog and the classic in vivo model of $5HT_3$ receptor blockade, the Bezold–Jarisch reflex in the anesthetized rat, showed that compounds containing benzamide moieties (e.g., metoclopramide, zacopride, and cisapride) were highly active in all models of GI motor function tested, whereas those without this structural motif only increased rat gastric emptying.[7] Comparison of the rank orders of potency in the Bezold–Jarisch reflex and the motility models implicated a receptor other than $5HT_3$ in the promotility effects of the benzamide compounds, with metoclopramide and cisapride subsequently being shown to increase GI motor activity and propulsion via $5HT_4$ receptor activation.[8] Agonism of $5HT_4$ receptors increases lower esophageal sphincter tone,[9] increases esophageal peristalsis,[10] and promotes gastric emptying[11] in humans all of which might reduce exposure of the lower esophagus to gastric refluxate and so improve symptoms in GERD. In GERD patients, cisapride is superior to placebo in promoting mucosal healing and reducing heartburn.[12] Cisapride treatment resulted in healing in 50–70% of patients, underlining the potential of this drug class as stand-alone agents.[13]

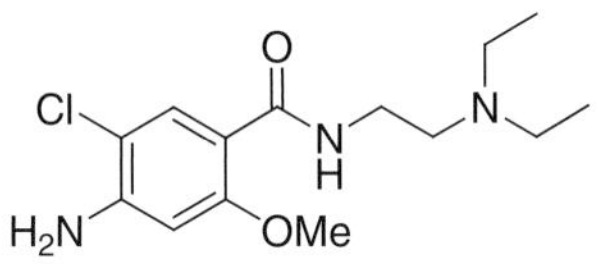

5 Metoclopramide

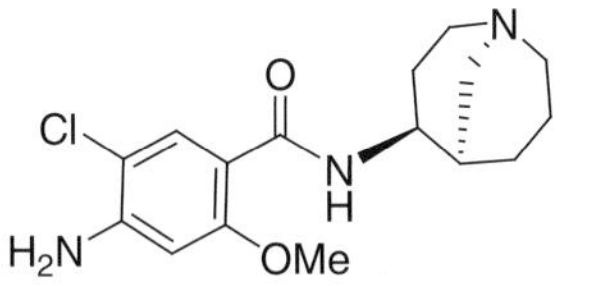

6 Renzapride (racemate)

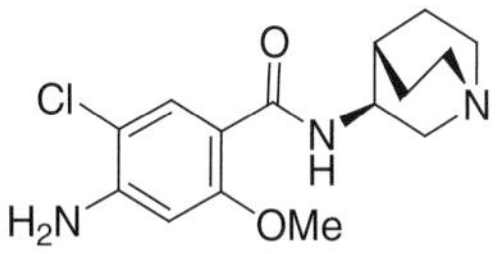

7 (*R*)-zacopride

Numerous $5HT_4$ receptor agonists have been identified and have prokinetic activity in animal models, although few have been tested clinically. Renzapride (**6**) and zacopride (**7**) are contemporaries of cisapride and both share its benzamide structural motif. Both are potent agonists at the $5HT_4$ receptor, but have significant $5HT_3$ antagonist activity. Both compounds accelerate gastric emptying in animal models. Renzapride reversed drug-induced gastroparesis in dogs[14] and zacopride partially reversed the delay in gastric emptying in a rodent model of diabetic gastroparesis.[15] In healthy volunteers, renzapride increases lower oesophageal sphincter pressure[16] and gastric emptying in patients with diabetic gastroparesis.[17] In conscious dogs, renzapride stimulated motility along the entire length of the colon.[18] In patients with C-IBS, renzapride accelerated colonic transit, apparently through an effect on the ascending colon, but was without effect on gastric emptying.[19] Renzapride is currently under clinical development for the treatment of C-IBS. Mosapride (**8**) is another $5HT_4$ receptor partial agonist in clinical use in Asia that has an interesting profile in animal models of gastric and intestinal motility. In conscious dogs, mosapride stimulates contractility in the gastric antrum, but has no effect on the colon at the doses tested.[20] In a mouse model of morphine-induced delay of gastric emptying, mosapride effectively reverses drug-induced gastroparesis.[21] Mosapride and other $5HT_4$ receptor agonists relax the esophagus of rats in vivo, as studied using the novel technique of digital sonomicrometry.[22] However, the effects of mosapride on the colon are controversial, having only been demonstrated in the guinea pig.[23] In humans, mosapride has been shown to increase gastric emptying in healthy volunteers[24] and in patients with diabetic gastroparesis.[25] In patients with GERD, mosapride has been shown to decrease reflux episodes and to reduce esophageal acid exposure.[26] The effects of mosapride on the human colon have not been investigated;

however, in a pilot study in patients with Parkinson's disease, mosapride treatment had some utility in the relief of constipation.[27]

Tegaserod (**9**) is another $5HT_4$ partial agonist approved for use in the treatment of C-IBS and chronic constipation. It stimulates colonic propulsion both in vitro[28] and in vivo[29] and also increases gastric emptying in animal models, including a mouse model of diabetic gastroparesis.[30] Tegaserod increases gastric emptying and intestinal transit in both healthy subjects[31] and in IBS patients.[32] In a model of colonic transit in healthy volunteers, tegaserod reversed the delay in colonic transit induced by dietary modification.[33]

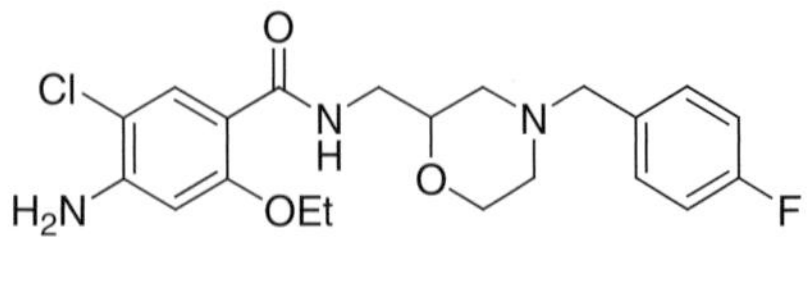

8 Mosapride (racemate)

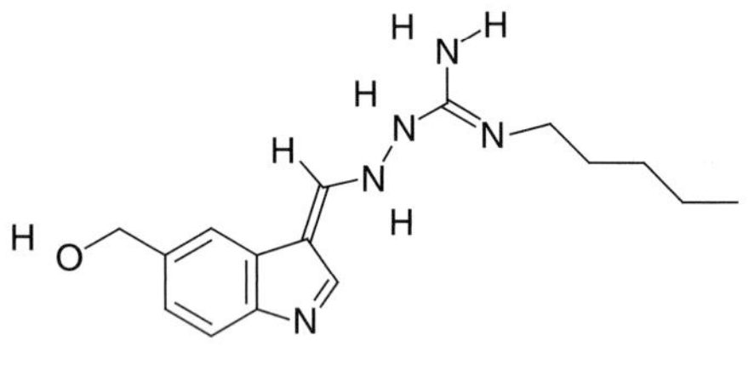

9 Tegaserod

Prucalopride (**10**) is a $5HT_4$ agonist whose activity in the lower GI tract has attracted attention. In conscious dogs, prucalopride stimulates motor activity in the proximal colon, although distal colon activity is inhibited[34] and in healthy human volunteers, prucalopride decreases colonic transit time.[35] This prokinetic effect has also been seen in patients with constipation, where prucalopride increased GI transit in the small and large intestine as well as increasing gastric emptying.[36] Prucalopride also provides symptomatic improvement in patients with constipation.[37] SK-951 (**11**) is a benzofuran derivative like prucalopride and is a potent agonist at the $5HT_4$ receptor, although the compound also has high affinity at $5HT_3$ receptors. SK-951 increases gastric emptying in rats and dogs, and in a canine model of postvagotomy gastroparesis.[38] In a further study, the ability of SK-951 to increase delayed gastric emptying in a canine model of diabetic gastroparesis has been demonstrated.[39]

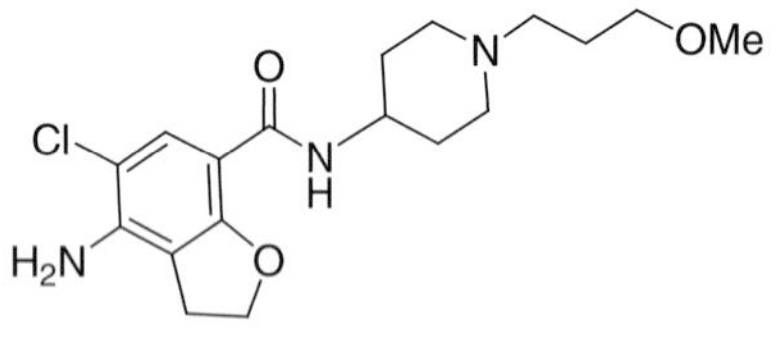

10 Prucalopride

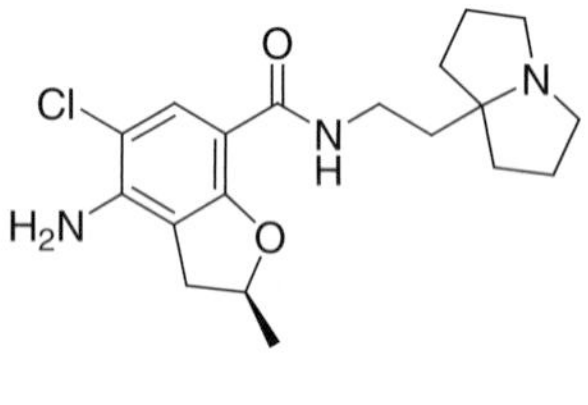

11 SK-951

Recently, several compounds containing the benzamide structural motifs have been described. These are novel $5HT_4$ receptor agonists but often combine this functionality with additional pharmacological activity. Such agents include KDR-5169 (**12**), TKS-159 (**13**), and YM-53389 (**14**). KDR-5169 combines $5HT_4$ agonism with dopamine D_2 receptor antagonism. It increases GI transit in three different rat models of gastroparesis[40] and in conscious dogs, not only stimulates fasting motility in the small intestine but also stimulates postprandial small intestinal motility that had been inhibited by previous administration of L-DOPA. YM-53389 is a $5HT_4$ receptor agonist that is highly selective for $5HT_4$ receptors compared to $5HT_3$ receptors. In mice, YM-53389 selectively increases colonic propulsion as whole gut transit time was increased by this NCE without any effect on upper gut motor function.[41] TKS-159 appears to be a selective and specific $5HT_4$ agonist that increases gastric motor activity in the whole stomach and in the Heidenhain pouch in conscious dog.[42] Novel $5HT_4$ agonists that do not contain the benzamide motif have also been described. ML-10302 (**15**) has been shown to stimulate motor activity in the small and large intestine of conscious dogs.[43] BIMU1 (**16**) represents another structural class of $5HT_4$ agonist, the benzimidazolones. This agent has agonist activity at $5HT_4$ receptors and antagonist activity at $5HT_3$ receptors. BIMU1 stimulates motility in the Heidenhain pouch dog and also increases the gastric emptying of liquids in a fistula dog model.[44] SL-65.0155 (**17**) has been confirmed pharmacologically as a $5HT_4$ receptor agonist, although to date its effects on GI motility have not yet been described.[45]

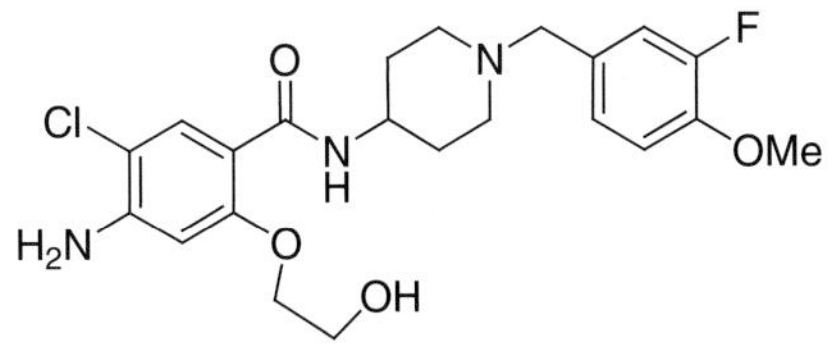

12 KDR-5169 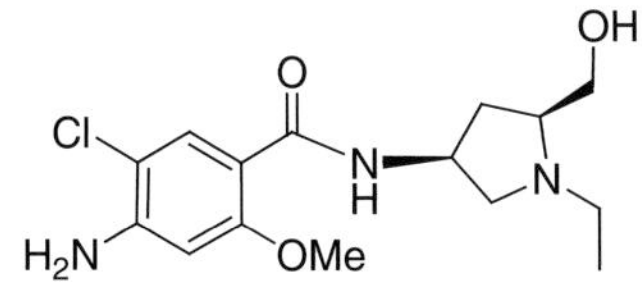 **13** TKS-159 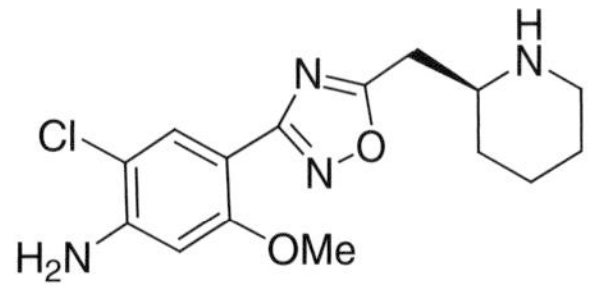 **14** YM-53389

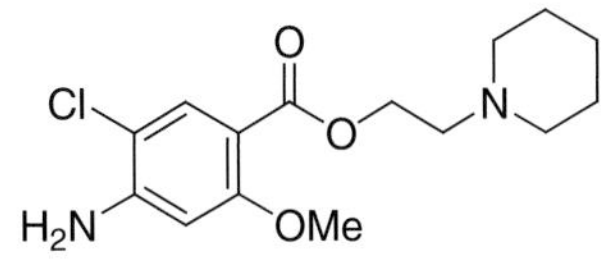

15 ML-10302 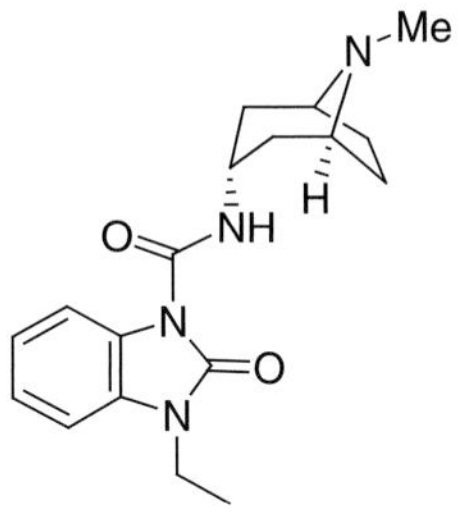 **16** BIMU1 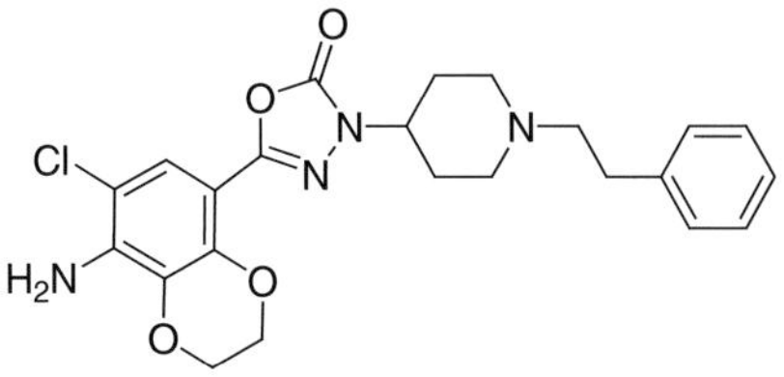 **17** SL-65.0155

6.30.5.1.2 Dopamine (DA) D_2 receptor antagonists

Unquestionably, metoclopramide has antagonist activity at dopamine (DA) D_2 receptors; however, its lack of receptor selectivity makes it difficult to categorize its prokinetic effects as resulting solely, or even predominantly, from DA receptor antagonism. The only selective DA receptor antagonist that has been established in clinical practice as a prokinetic agent is domperidone (**18**). It is differentiated from other DA receptor antagonists, developed for the treatment of schizophrenia, in only poorly crossing the blood–brain barrier, making the dystonic reactions that associated with metoclopramide use a rare occurrence. The precise mechanism through which domperidone increases GI motility remains unclear. Experiments in isolated gastric tissues suggest a potential explanation in that domperidone inhibits an ongoing inhibition of acetylcholine (ACh)-mediated excitatory enteric neurotransmission mediated via DA release.[46] In rat, domperidone is effective at stimulating gastric motor function and gastric emptying that has first been inhibited with a DA receptor agonist.[38,47] In conscious dogs, domperidone stimulates gastric motility[48] increasing gastric emptying, although this effect appears limited to the emptying of liquids.[49] The first human studies of the effects of domperidone on gastric emptying pre-date modern measurement techniques. Nonetheless, these pioneer studies clearly demonstrated a prokinetic effect of domperidone either when given alone or when given to reverse the gastroparesis evoked by DA receptor agonists like apomorphine. More recent studies have shown that domperidone accelerates gastric emptying in subjects with gastroparesis,[50] although data concerning the effects on basal gastric emptying rates in healthy volunteers are scarce. Following demonstration of the prokinetic effects of domperidone in patients with diabetic gastropathy,[51] this agent has become established as a treatment for patients with diabetic gastroparesis, where treatment can increase slowed gastric emptying and reduce symptoms.

The effects of domperidone on intestinal motility and transit have also been studied. In rats, DA antagonists, including domperidone, increase small intestinal transit.[52] Similarly, domperidone accelerates small intestinal transit when previously slowed with morphine.[21] The effects of domperidone on human GI motor function is more complex. In healthy volunteers, domperidone appears to prolong orocecal transit time,[53] while earlier studies suggested a modest acceleration in intestinal transit.[54] Another DA receptor antagonist, levosulpiride (**19**), increases gastric emptying of both liquid and mixed meals in patients with functional dyspepsia[55] and diabetic gastroparesis.[56] Some degree of symptomatic improvement can be obtained in dyspeptic patients with gastroparesis following treatment with levosulpiride.[57]

Itopride is a DA D_2 receptor antagonist activity with AChase activity, approved for the treatment of gastric motor disorders in Japan. In conscious dogs, itopride (and its hydrochloride salt, HSR-803, **20**) enhances gastric contractility induced by ACh, reverses the inhibition of gastric motility evoked by DA, and as a single agent, enhances postprandial gastric motility.[48,58] HSR-803 also stimulates gastric emptying and reverses drug-induced gastroparesis in dogs and rats,[59] increases small intestinal motility in mice, and reverses drug-induced ileus. Itopride also increases motility all along the dog intestine.[60] Itopride reduces esophageal acid exposure in patients with GERD, an observation consistent with a prokinetic effect.[61]

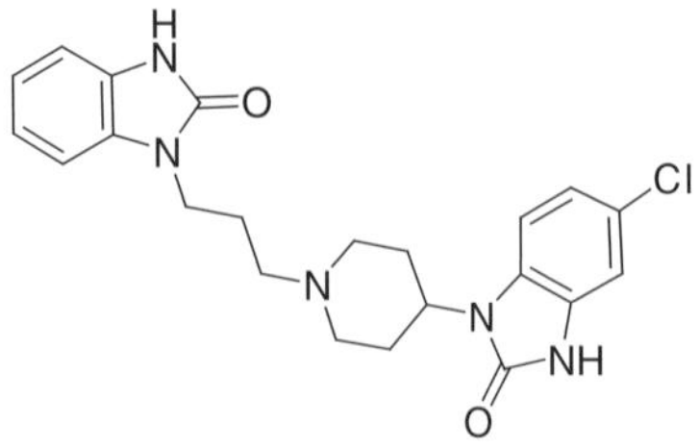

18 Domperidone

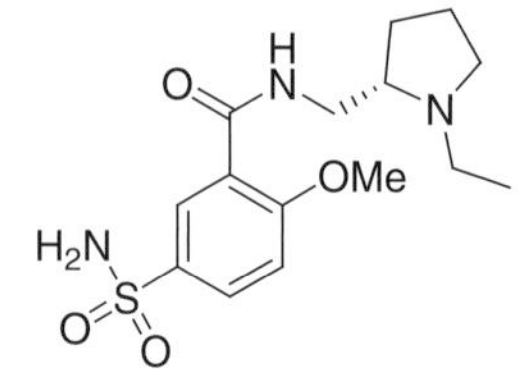

19 Levosulpiride

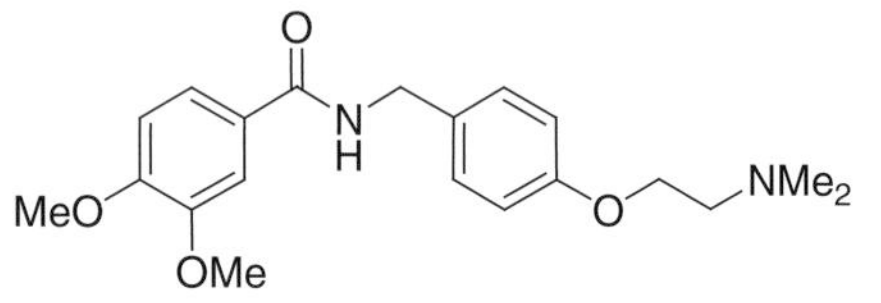

20 Itopride (HSR-803)

6.30.5.1.3 Modulators of cholinergic neurotransmission

Cholinergic mechanisms are critical to excitatory enteric neurotransmission. It is therefore perhaps unsurprising that modulation of this system, either through receptor agonism, antagonism, or through the inhibition of metabolism of endogenous neurotransmitters, has profound effects of GI motor function.

Both direct receptor agonism with bethanechol (**21**) or indirect receptor agonism through the inhibition of cholinesterase activity with neostigmine (**22**) have substantial historical precedent as strategies to increase GI motility. However, their claims as true prokinetic agents are less straightforward. Given the extensive preclinical data supporting the motility-enhancing and prokinetic effects of these agents, an historical perspective is beyond the scope of this chapter. However, studies of gastric emptying in healthy volunteers demonstrated that bethanechol increases gastric contractile activity, particularly in the antrum.[62] However, this effect was not accompanied by increases in gastric emptying. This lack of a true prokinetic effect in the stomach has been observed previously, where bethanechol even appeared to increase gastric residence time.[63] In the large bowel of healthy volunteers, bethanechol stimulated phasic contractile activity but failed to increase colonic transit.[64] Thus it appears that bethanechol, while able to stimulate contractile activity throughout the GI tract, is unable to stimulate propulsive motility.

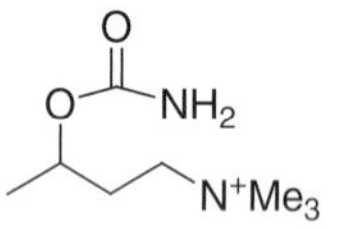

21 Bethanechol

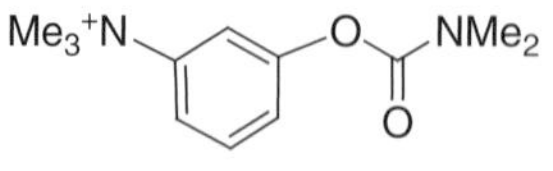

22 Neostigmine

Neostigmine stimulates GI motility throughout the length of the human GI tract. Neostigmine increases gastric emptying[65] and both colonic phasic and tonic activity in healthy volunteers.[64] Importantly, neostigmine also stimulates colonic transit. The ability to stimulate propulsive activity has been utilized in clinical medicine, where neostigmine has been used to benefit patients with conditions such as acute colonic pseudo-obstruction and postoperative ileus.

Z-338 (**23**) is a muscarinic M_1/M_2 receptor antagonist that increases ACh release from nerve terminals in the stomach, stimulating upper gut motility.[66] Z-338 appears to increase gastric motor activity in dogs and rats, but a significant effect on gastric emptying is only observed when this parameter is first inhibited pharmacologically.[67] In dyspeptic patients, Z-338 appears to have little effect on gastric emptying.[68]

KW-5092 (**24**) enhances the release of enteric ACh and may also have cholinesterase inhibitory activity. In animals, KW-5092 stimulates motility along the length of the GI tract, without evoking the behavioral and cardiovascular side effects seen with neostigmine.[69,70]

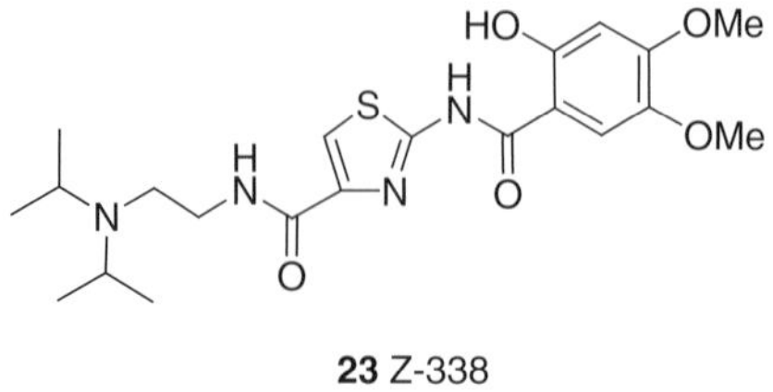

23 Z-338

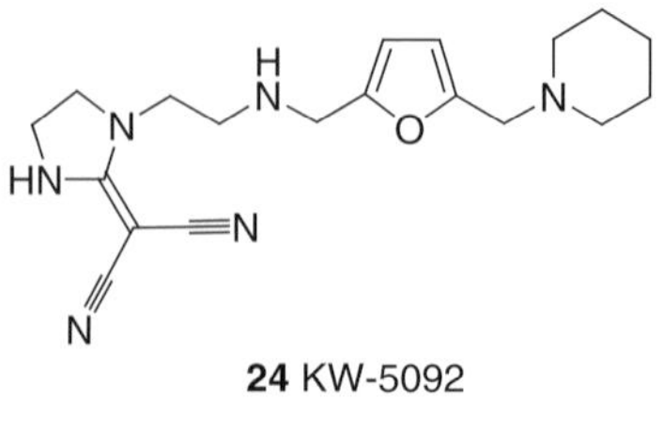

24 KW-5092

6.30.5.1.4 Motilin receptor agonists

Before recognition of its ability to increase GI motility, the antibiotic erythromycin was widely known for provoking intestinal side effects in patients receiving treatment for infections. In healthy human volunteers, erythromycin dose-dependently increased the gastric emptying of liquids.[71] The utility of erythromycin in the treatment of gastroparesis and pseudo-obstruction has been studied, but limitations in study design make conclusions difficult to draw. The search for clinically useful motilin receptor agonist prokinetics initially focused on the erythromycin analogs, EM-523 (**25**) and EM-574 (**26**), that stimulate gastric motility and gastric emptying in dogs.[72,73] EM-574 is a potent stimulant of gastric emptying in healthy human volunteers.[74] The clinical development of motilin receptor agonists has so far been unsuccessful, with ABT-229 (**27**), a synthetic macrolide devoid of antibiotic properties, being the most studied. ABT-229 has a powerful effect on gastric emptying in the dog, shortening the lag phase and increasing the rate of emptying of a solid meal.[75] The effect on motility in the human upper gut is controversial, with some studies showing an effect to increase gastric emptying,[76] and others no effect.[77] ABT-229 has been studied in diabetic gastroparesis and functional dyspepsia where it failed to improve symptoms, and in GERD patients where ABT-229 actually worsened reflux symptoms.[78–80] Attempts to reconcile these data have focused on the agonist-specific tachyphylaxis that is seen in this receptor system, with rapid desensitization occurring with many motilin receptor agonists. ABT-229 appears to be particularly active in this regard. In diabetic gastroparesis, another motilin agonist, KC-11458 (**28**), failed to increase the gastric emptying of either solids or liquids.[81] Another motilin receptor agonist, mitemcinal (**29**, GM-611) has useful prokinetic properties. In patients with diabetic or idiopathic gastroparesis, mitemcinal has been shown to enhance delayed gastric emptying.

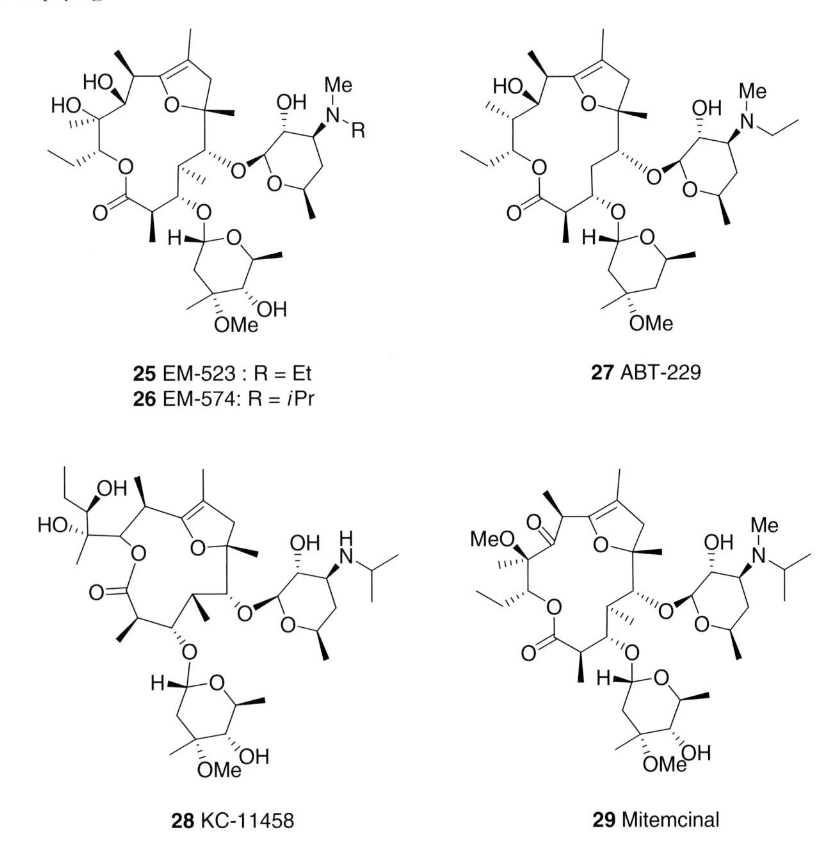

25 EM-523 : R = Et
26 EM-574: R = *i*Pr

27 ABT-229

28 KC-11458

29 Mitemcinal

6.30.5.2 Antiemetic Agents

6.30.5.2.1 $5HT_3$ receptor antagonists

The introduction of $5HT_3$ receptor antagonists into clinical practice as antiemetic agents revolutionized the treatment of patients undergoing therapy for malignancy. Once again, metoclopramide was at the vanguard, with high doses of this agent being an effective antiemetic in dog[82] and human.[83] The antiemetic efficacy of selective $5HT_3$ receptor antagonists was first demonstrated in the ferret.[84] In these studies, tropisetron and bemesetron profoundly inhibited the emetic response evoked by the chemotherapeutic agent cisplatin. Additional $5HT_3$ receptor antagonists ondansetron and granisetron prevented emesis evoked by high-dose cisplatin or cyclophosphamide or x-irradiation.[85,86] These antagonists were not only able to prevent the onset of emesis, but were also able to abolish established emesis, raising the possibility of using these agents as interventional therapies once emesis was established or following the failure of agents administered earlier.

The precise locus for the antiemetic activity of the $5HT_3$ receptor antagonists remains unclear. There is debate over whether the most important site of receptor blockade lies within the central or the peripheral nervous system, or indeed whether receptor blockade at both sites is important. What is clear is that many chemotherapeutic agents evoke a significant release of 5HT from small intestinal enterochromaffin cells. 5HT so released stimulates $5HT_3$ receptors located on vagal afferent nerve fibers that project into dorsomedullary structures in the brainstem. These sensory fibers also appear to have $5HT_3$ receptors on their central terminals, which modulate the release of neurotransmitters in this important modulatory region, and may represent an important site of drug activity.

Ondansetron, granisetron, and tropisetron are effective antiemetic agents in patients undergoing treatment with a wide range of chemotherapeutic agents[87,88] and with radiotherapy.[89] Additional $5HT_3$ receptor antagonists include dolasetron (**30**), ramosetron (**31**), and palonosetron (**32**). These agents vary in their affinity for $5HT_3$ receptors and in their routes of metabolism, and hence while they have broadly similar antiemetic profiles, individual responses to these agents may differ.

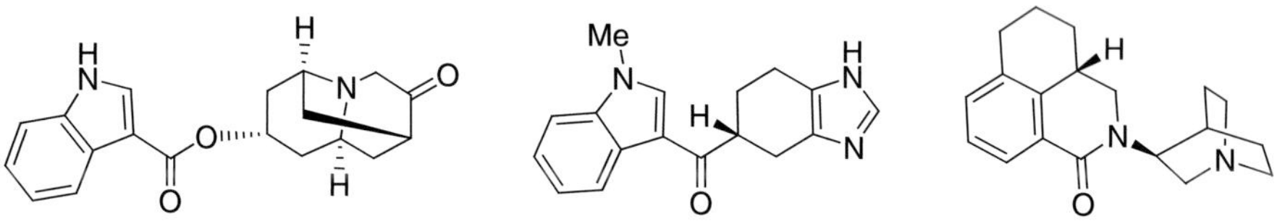

30 Dolasetron **31** Ramosetron **32** Palonosetron

6.30.5.2.2 NK_1 receptor antagonists

Substance P evokes neural excitation in the area postrema of the dorsal brainstem. The key role of this nucleus in the emetic reflex and its close proximity to the nucleus of the solitary tract, itself of fundamental importance in emetic pathways, raised the possibility that substance P or other tachykinins were important mediators of the emetic reflex. Administration of substance P evoked emesis in dogs and in the ferret; the prototypical tachykinin (NK_1) receptor antagonist, CP-99,994 (**33**) inhibited the emetic response to cisplatin. NK_1 receptor antagonists have a broad-spectrum antiemetic profile in a range of animals including dog, ferret, cat, pig, and shrew, preventing emetic responses to chemotherapeutic agents (both acute and delayed emetic responses), x-radiation, volatile and gaseous anesthetic agents, ipecacuanha, morphine, ethanol, and motion. Addition of the NK_1 receptor antagonist, ezlopitant, to an antiemetic regime of granisetron and dexamethasone was superior to use of the $5HT_3$ receptor antagonist and steroid alone in cisplatin-induced emesis.[90] Similar data were obtained with L-754,030 (**35**), an NK_1 receptor antagonist that has been superceded by its prodrug, aprepitant. Aprepitant was as effective as ondansetron in controlling acute emesis to cisplatin, but was more effective than ondansetron in controlling delayed emesis.[91] In a small study aimed at treatment of established PONV following gynecological surgery, the NK_1 receptor antagonist vofopitant (**37**) was superior to placebo at controlling symptoms over a 24 h period.[92] In a study of PONV prophylaxis following abdominal hysterectomy, the NK_1 receptor antagonist CP-122,721 (**34**) was superior to treatment with a $5HT_3$ receptor antagonist.[93] In studies of motion-induced nausea, vofopitant failed to control nausea at doses shown to be effective at controlling PONV.[94]

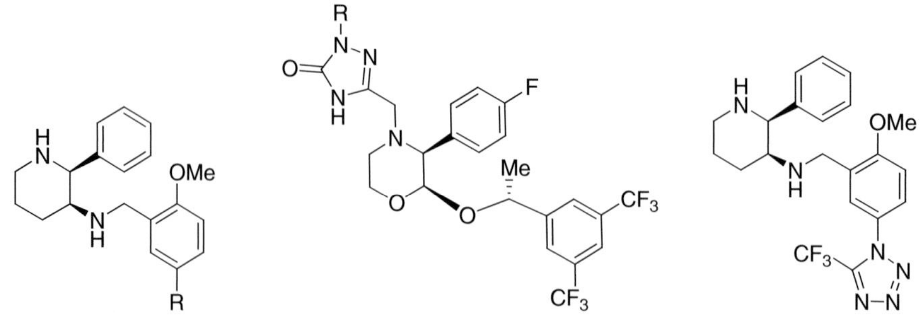

33 CP-99,994: R = H
34 CP-122,721: R = OCF_3

35 L-754,030/aprepitant: R = H
36 L-758,298: R = PO_3H_2

37 Vofopitant

6.30.5.2.3 Dopamine D_2/D_3 receptor antagonists

DA receptors are key mediators of the emetic reflex, with DA agonists, including apomorphine, evoking emesis in ferret and dog. The use of subtype-selective DA receptor agonists has implicated the D_2/D_3 receptors in the emetic response to dopamine agonists, a conclusion supported by the observation that apomorphine-induced emesis is blocked by domperidone and haloperidol (**38**). The effectiveness of DA receptor antagonists at inhibiting the emetic response to clinically relevant emetogens has also been studied. Radiation-induced emesis in the ferret was only weakly inhibited by domperidone, whereas emesis induced by cyclophosphamide or morphine was effectively inhibited by droperidol (**39**), which unlike domperidone, penetrates the blood–brain barrier.

DA receptors play a role in the emetic reflex in humans. Clinically experience with drugs that activate DA receptors, such as apomorphine, or increase DA (e.g., L-DOPA) induce nausea and vomiting. Clinical experience with DA antagonists in the control of nausea and vomiting has been mixed. Studies have shown that domperidone has only weak activity in controlling nausea and vomiting in response to powerful emetogenic agents like cisplatin, although more impressive results have been obtained with less emetogenic chemotherapeutic regimes. DA antagonists that cross the blood–brain barrier, e.g., droperidol, haloperidol, and prochlorperazine (**40**), are effective antiemetic agents, even controlling the emetic response to highly emetogenic therapy, although adverse events such as sedation are common.

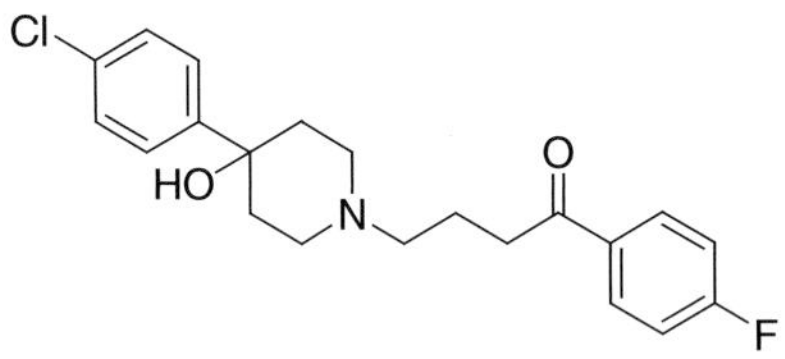

38 Haloperidol 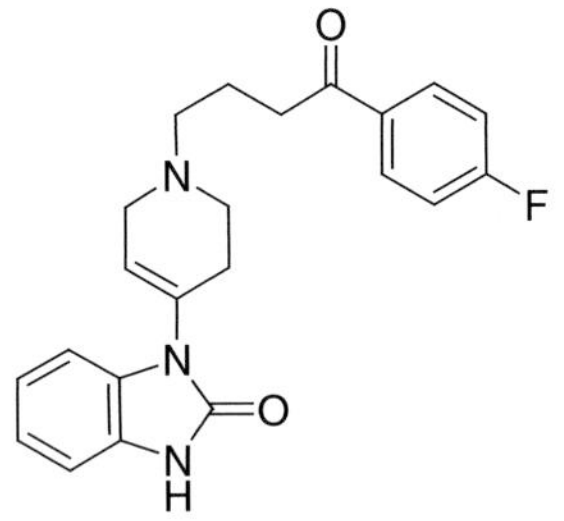 **39** Droperidol 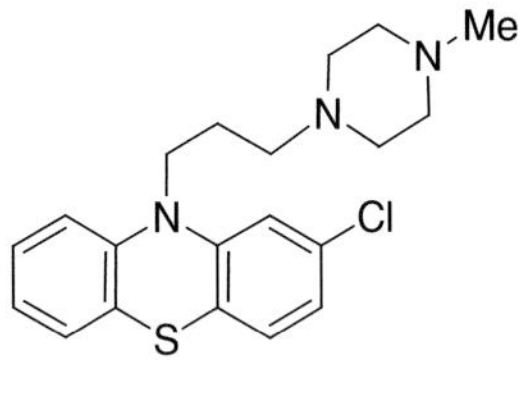 **40** Prochlorperazine

6.30.5.2.4 Modulators of cholinergic neurotransmission

There are limited data from animal experiments studying the activity of muscarinic receptor antagonists on the control of the emetic reflex. In the house musk shrew, scopolamine/hyoscine (**41**) was effective in inhibiting motion-induced vomiting. Muscarinic receptor antagonists, e.g., scopolamine, have long been used in clinical practice to control nausea and vomiting but have little or no value in treating or preventing emesis evoked by chemotherapeutic agents.

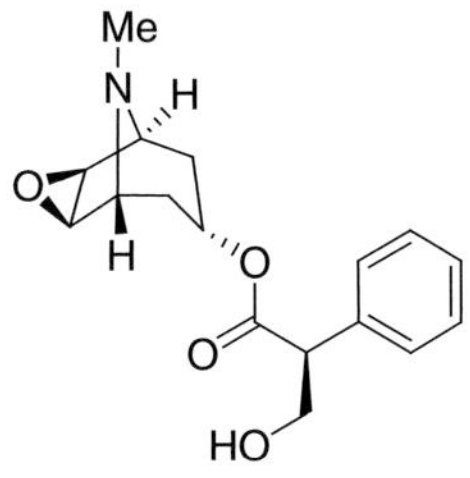

41 Scopolamine/hyoscine

6.30.5.2.5 Histamine H_1 antagonists

Histamine H_1 receptor antagonists are useful inhibitors of some forms of emesis, with agents such as thiethylperazine (**42**) being part of antiemetic drug cocktails used to control the emetic response to cancer chemotherapy. However, $5HT_3$ and NK_1 receptor antagonists have largely displaced H_1 receptor antagonists.

Circulating levels of histamine increase following gynecological surgical procedures and histamine receptor antagonists such as promethazine (**43**) and cyclizine (**44**) may be effective treatments for PONV when given prophylactically. H_1 receptor antagonists may also be useful as antiemetics in the postoperative setting. Morphine can evoke nausea and vomiting in patients and the rise in the use of patient-controlled analgesia has seen the incidence of this phenomenon increase. Coadministration of morphine with diphenhydramine (**45**) dramatically reduces morphine-induced nausea and vomiting.[95] Histamine H_1 antagonists are also clinically useful in the treatment of nausea and vomiting associated with vestibular disorders and motion sickness. Cinnarizine (**46**) is an effective agent at controlling sea-sickness.[96] Flunarizine (**47**) provides some benefit in preventing vomiting episodes in the pediatric condition, cyclic vomiting syndrome.[97]

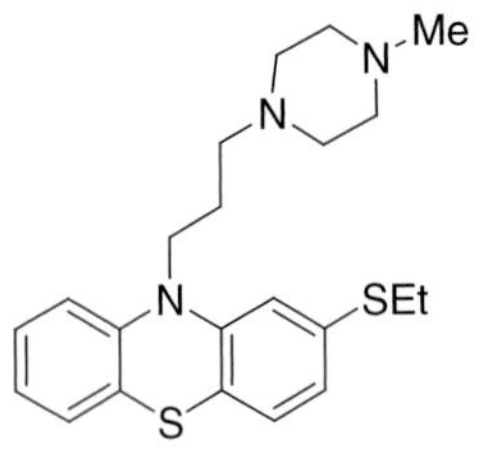

42 Thiethylperazine

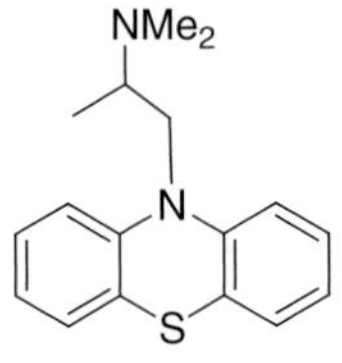

43 Promethazine

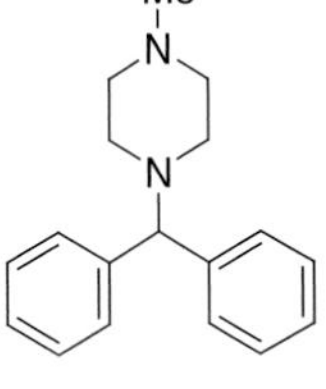

44 Cylizine

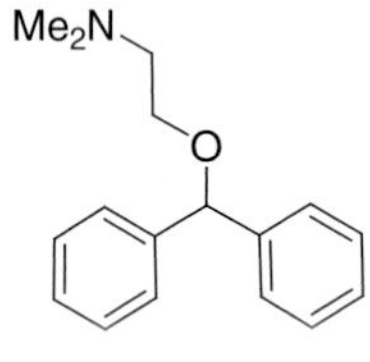

45 Diphenhydramine

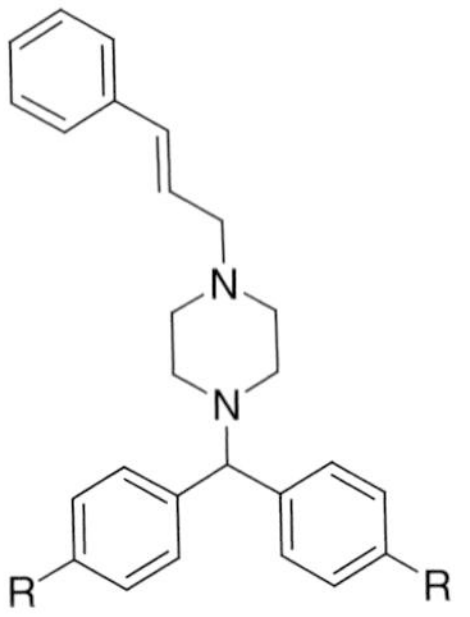

46 Cinnarizine: R = H
47 Flunarizine: R = F

6.30.5.2.6 Cannabinoid receptor agonists

Cannabinoid receptors are involved in the control of the emetic reflex. Receptor agonists, e.g., Δ^9-tetrahydrocannabinol (**48**, Δ^9THC), have a marked antiemetic effect in the ferret in response to emesis evoked with either hyperosmolar saline or cisplatin.[98] The cannabinoid receptor agonist WIN-55212 (**49**) and other subtype-selective antagonists has revealed a role for CB_1 receptors in the control of vomiting in this species, which also extends to morphine-induced emesis. In a novel animal model of anticipatory emesis in the house musk shrew, a phenomenon experienced in clinical practice, Δ^9THC and cannabidiol inhibit emesis, an effect not seen with ondansetron.[99] In the shrew, cannabinoid receptor agonists, including CP-55,940 (**50**), inhibit emesis evoked with cisplatin and lithium. In a pigeon model of emesis, the synthetic cannabinoid HU-211 (**51**) inhibited emesis evoked with cisplatin.

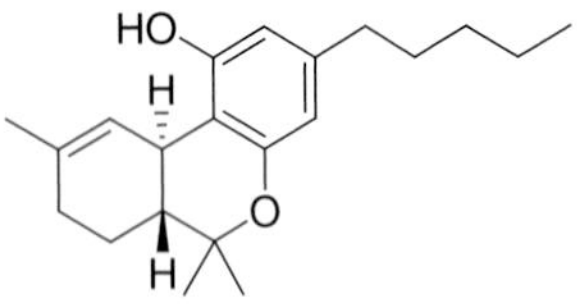

48 Δ^9-Tetrahydrocannabinol
dronabinol

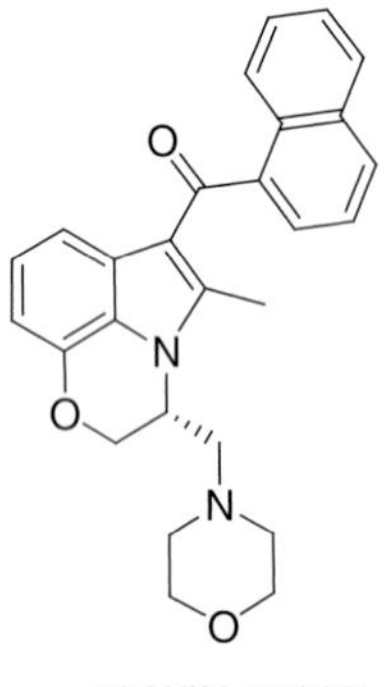

49 WIN-55212

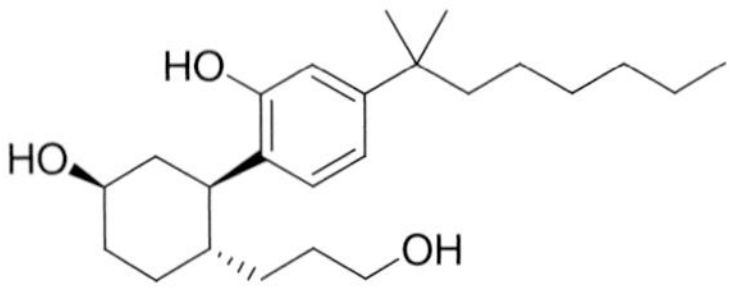

50 CP-55,940

Nabilone (**52**), dronabinol (Δ^9THC), and levonantradol (**53**) are used in clinical practice to control many forms of emesis, having inhibitory activity against nausea and vomiting evoked by cisplatin and other chemotherapeutic agents.[100] In addition the utility of this class of agent in the treatment of PONV has been suggested. Adverse events are common with this class of drug and include sedation, dizziness, and euphoria, and these limit their usefulness.

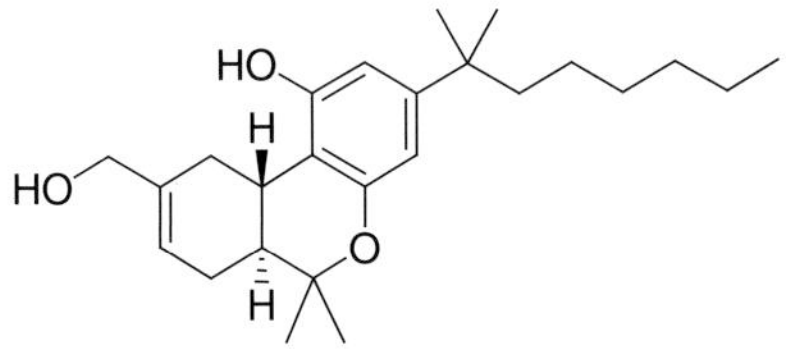

51 HU-211 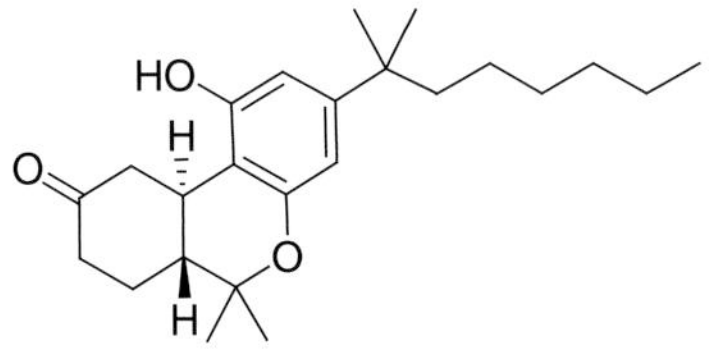 **52** Nabilone 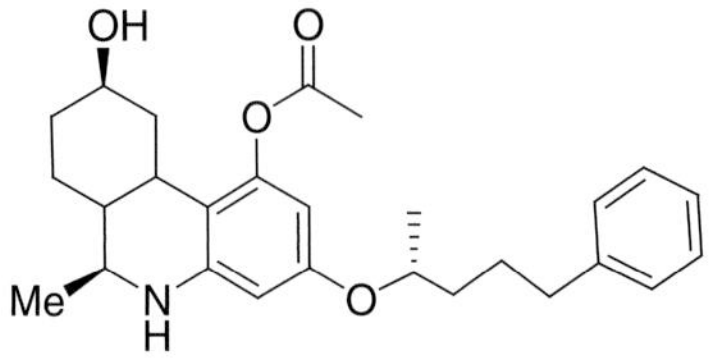 **53** Levonantradol

6.30.6 Unmet Medical Needs

The major unmet medical need in the field of prokinetic agents has resulted from the withdrawal of cisapride due to issues with QT prolongation. Cisapride had gained widespread use, primarily for the treatment of GERD, but also in the treatment of a wide range of GI disorders including ileus, functional dyspepsia, constipation, and IBS, thus leaving a gap in the pharmacopoeia. The need for a prokinetic agent continues, with a corresponding increased use of metoclopramide, although the adverse events associated with this agent, primarily dyskinesias, are considered unacceptable for most patients. Cholinergic modulators are not sufficiently tolerated for chronic use in outpatients and the efficacy of both motilin and DA receptor agonists have so far failed to convince the clinical community of their usefulness. The opportunity for a safe and effective prokinetic agents remains in the treatment of GERD, postoperative ileus, constipation, and potentially dyspepsia and IBS.

For emesis, the introduction of the NK_1 receptor antagonist aprepitant has improved on the prophylactic treatment and control of nausea and vomiting already offered by the $5HT_3$ receptor antagonists. Despite this advance, many patients on chemotherapy develop problematic nausea and vomiting, particularly in the period beyond the first 24 h postdose, a phenomenon known as delayed emesis.

6.30.7 New Research Areas

6.30.7.1 Prokinetic Agents

6.30.7.1.1 Growth hormone secretagog receptor agonists

Ghrelin is the endogenous ligand for the growth hormone secretagog receptor. In addition to its well-described role in regulating pituitary growth hormone release, this receptor is expressed on motor neurons in the enteric nervous system and there is increasing evidence that activation of these receptors can increase GI motility. Ghrelin administration increases gastric emptying in an animal model of gastroparesis,[101] an effect that has been replicated in healthy human subjects[102] and in patients with either idiopathic[103] or diabetic gastroparesis,[4] where administration of human recombinant ghrelin enhanced gastric emptying.

The synthetic ghrelin analog RC-1139 (structure not disclosed) is a potent prokinetic agent in rat, increasing gastric emptying and reversing postoperative gastroparesis.[104] Similarly, the ghrelin mimetic capromorelin (**54**) potently increases gastric emptying in mice.[105] In rats, another synthetic ghrelin agonist, CP-464,709 (**55**), has potent prokinetic activity, increasing gastric emptying of a semisolid mixed nutrient meal with efficacy at least equal to that of metoclopramide.[106] TZP-101, a macrocylic peptide analog (structure not disclosed), a novel, high-affinity ghrelin receptor agonist, dose-dependently increases gastric emptying in a rat model of gastric emptying of a liquid, nonnutrient meal.[107] In a rat model of postoperative ileus, TZP-101 normalizes the delay in gastric emptying and accelerates delayed small intestinal transit.[108] Additional novel ghrelin agonists, LY-444711 (**56**), L-692,429 (**57**), and ibutamoren (MK-677, **58**), have appeared in the literature, although their effects on GI motility have not yet been described.[109,110]

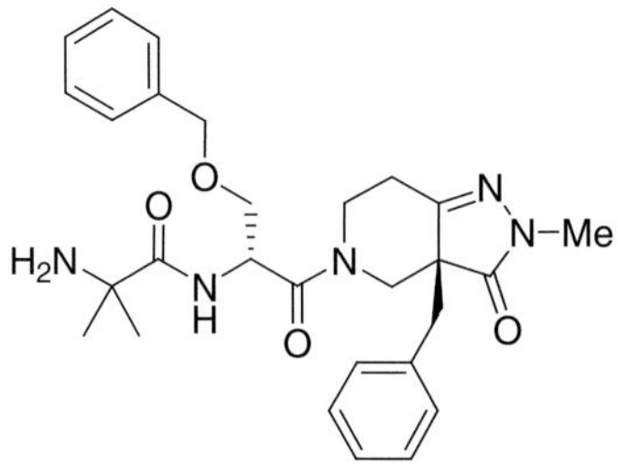

54 Capromorelin 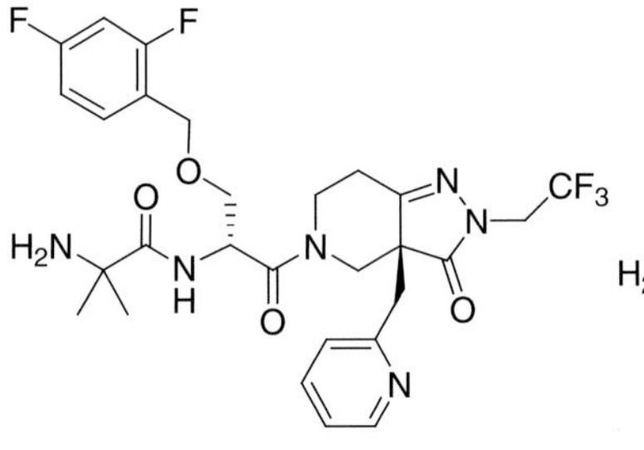 **55** CP-464,709 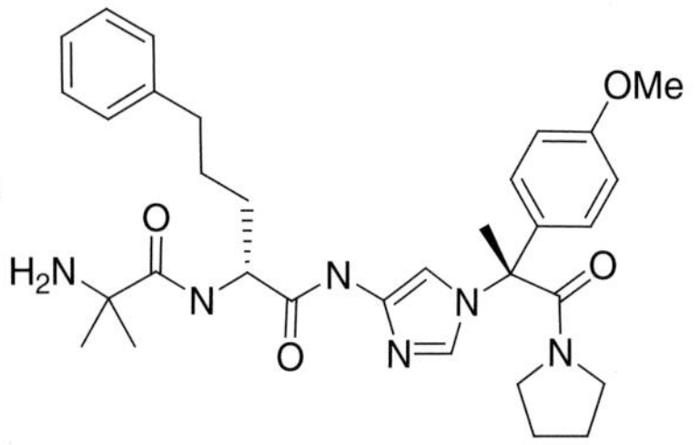 **56** LY-444711

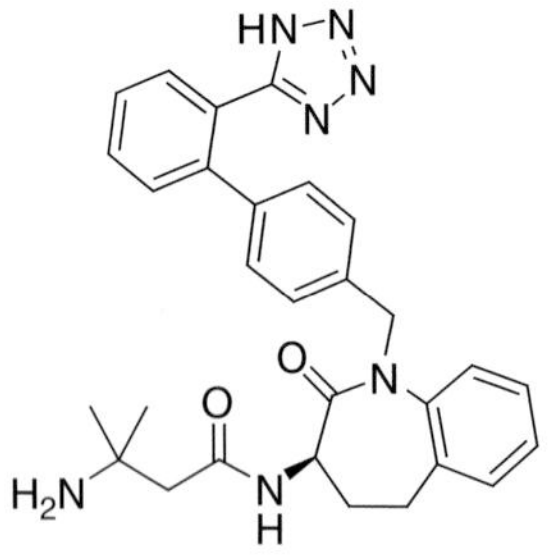

57 L-692,429

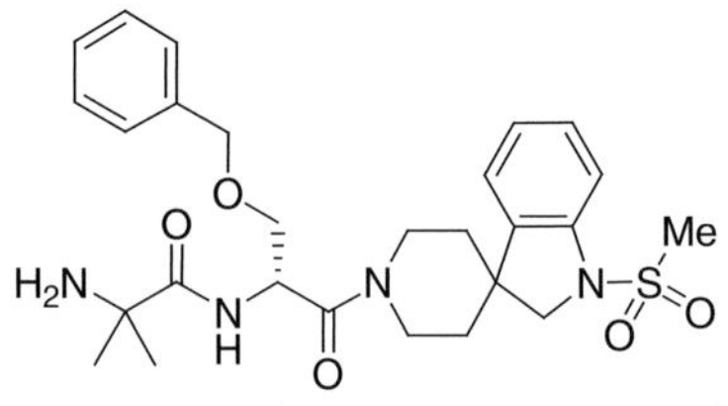

58 MK-677

6.30.7.1.2 $GABA_B$ receptor agonists

There has been considerable interest in the use of the γ-amino-butyric acid B ($GABA_B$) receptor agonist baclofen (**59**) as a potential prokinetic treatment for GERD. Baclofen reduces the occurrence of transient relaxations of the lower esophageal sphincter, reducing the reflux of acidic gastric contents.[111] Another $GABA_B$ receptor agonist, APFSiA (**60**), has similar activity.[112] In mice, baclofen administration has complex effects on gastric emptying, increasing the emptying of solids, but delaying the emptying of liquids.[113] The $GABA_B$ receptor agonist AZD-3355 (**61**)[114] has entered development for the treatment of GERD.

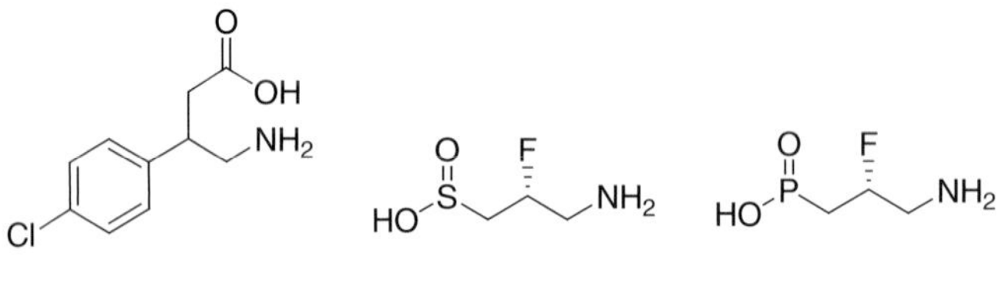

59 Baclofen

60 APFSiA

61 AZD-3355

6.30.7.1.3 $5HT_3$ receptor agonists

The $5HT_3$ receptor agonist, MKC-733 (**62**), has complex effects of human GI motility,[5] increasing small intestinal transit and anteroduodenal coordination and delaying the gastric emptying of liquids, probably as a result of a relaxation of the proximal stomach.

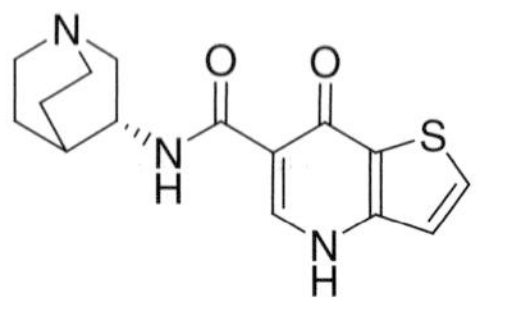

62 MKC-733 (5-HT_3 agonist)

6.30.7.1.4 CCK_1 receptor antagonists

CCK is a GI hormone with powerful effects on GI motility and secretion. CCK is released from endocrine cells within the GI mucosa and has a profound inhibitory effect on gastric emptying via CCK_1 receptor activation. This mucosal release of CCK is induced by many nutrients, in particular lipid. The effects of endogenously released or exogenous CCK can be blocked with selective CCK_1 receptor antagonists such as devazepide (**63**, MK-329) and A 70104 (**64**).[115] However, the potential of CCK_1 receptor antagonists to act as prokinetic agents in the clinic is less well described. In healthy volunteers, infusion of the CCK_1 receptor antagonist loxiglumide (**65**) accelerates the gastric emptying of liquids, although this effect is clearly dependent upon nutrient-induced release of CCK, such that a meal that does not release CCK is unaffected by loxiglumide.[116] A similar increase on gastric emptying was seen with an enantiomer of loxiglumide, dexloxiglumide, which increased gastric emptying in a study of IBS patients.[117] However, in a study of healthy volunteers, the gastric emptying of both solid and liquid components of a mixed nutrient meal was unaffected by administration of devazepide.[118] Thus the prokinetic effects of CCK_1 receptor antagonists has yet to be fully described and so the clinical utility of these agents in treated GI motility disorders is unknown.

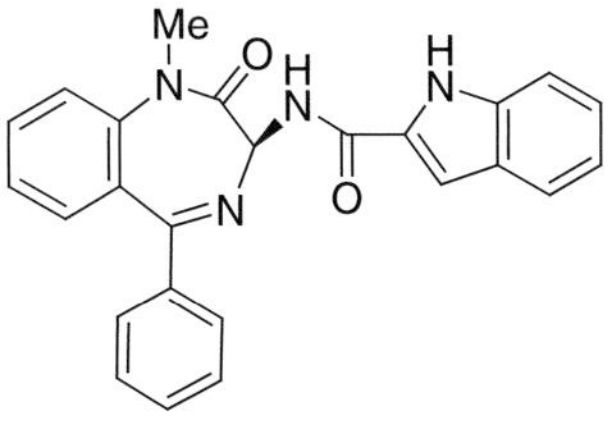

63 Devazepide (CCK_1 antagonist)

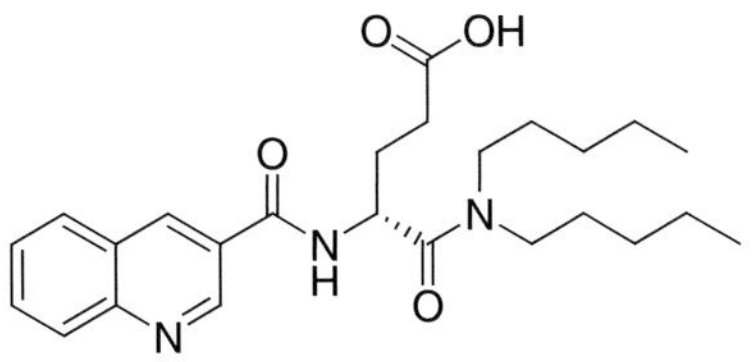

64 A 70104 (CCK_1 antagonist)

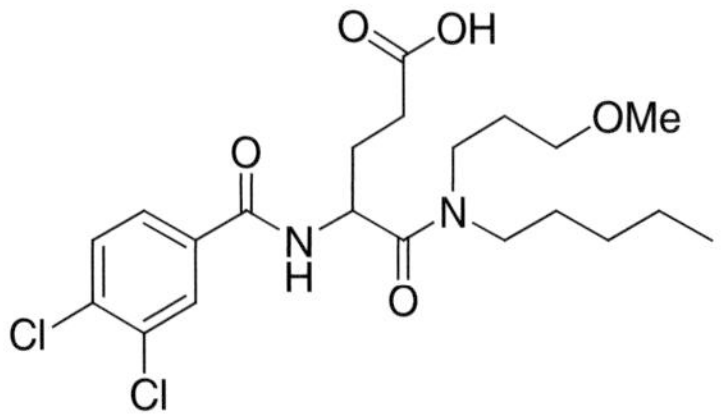

65 Loxiglumide (CCK_1 antagonist)

6.30.7.2 Antiemetic Agents

6.30.7.2.1 $GABA_B$ receptor agonists

The first indication for a role for $GABA_B$ receptors in the control of emesis came from a case study from a patient with muscular dystrophy whose intractable vomiting was controlled with baclofen, although the precise mechanism of action was unclear. Baclofen inhibits morphine-induced retching and vomiting in the ferret.[119] This is an important observation, as morphine-induced vomiting is only partially sensitive to inhibition by $5HT_3$ receptor antagonists. These observations have been replicated in the dog with baclofen and the $GABA_B$ receptor agonist, SKF-97541 (CGP 35024, **66**).[120]

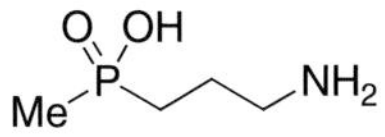

66 SKF-97541

In neurologically impaired children, baclofen reduced vomiting, but also significantly reduced acid reflux from the stomach.[121] The precise mechanism of action of baclofen in inhibiting emesis is unknown. GABA is an important inhibitory neurotransmitter in the human CNS, and $GABA_B$ receptors have been demonstrated in the brainstem, located in regions of importance in the emetic reflex. Furthermore, activation of the $GABA_B$ receptor both in vitro and in vivo inhibits the release of glutamate and substance P, substances of primary importance in afferent neurotransmission. Thus, baclofen could be inhibiting the emetic reflex at a number of central loci. Studies in patients with GERD and healthy volunteers have also shown that baclofen can reduce gastric reflux by reducing transient relaxations of the lower esophageal sphincter, which might also contribute to its antiemetic activity.[111]

6.30.7.2.2 D_3 receptor antagonists

The role of DA receptors in the emetic reflex has been focused on the D_2 receptor subtype. However, many of the DA antagonists studied have little selectivity for the D_2 receptor when compared to their affinity for the D_3 receptor and so the relative importance of each subtype with regard to efficacy or side effects is unclear. AS-9705 (**67**) is an antagonist for both the D_2 and the D_3 dopamine receptor subtype. In ferrets and dogs, AS-9705 inhibits emesis evoked by DA agonists, including the D_3 receptor-preferring agonist *R*(+)-7-OH-DPAT, and also has prokinetic activity increasing both normal and delayed gastric emptying in rats and increasing gastric motility in dogs.[122] The effect on gastric emptying could reflect the well-known prokinetic effects of D_2 receptor antagonists (see above) or may be an insight into an emerging role for the D_3 receptor. D_3 receptor agonists inhibit pyloric motility and decrease gastric emptying in rat.[123] If this system was tonically active, either in health or disease, then D_3 receptor antagonists might be expected to accelerate gastric emptying. Another D_2/D_3 receptor antagonist, *S*(–)-eticlopride (**68**), blocks emesis induced by morphine- and ipecacuanha-induced emesis in the ferret.[124] Due to its CNS penetration, *S*(–)-eticlopride evokes adverse events which limit its dosing.

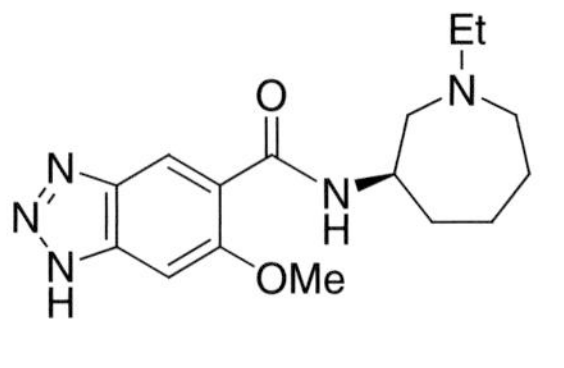

67 AS-9705

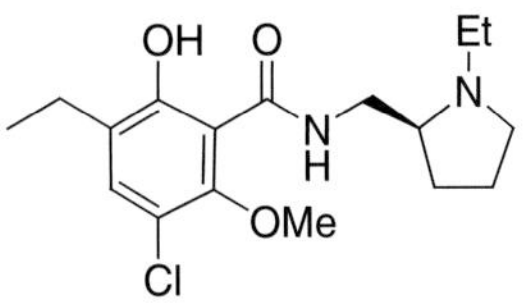

68 *S*(–)-eticlopride

6.30.8 Future Directions

In the prokinetic arena, the void left by the withdrawal from the market of cisapride has yet to be filled. There is still some confusion over whether the beneficial prokinetic effects of cisapride resulted wholly from activity at the $5HT_4$ receptor and only the successful development of a selective $5HT_4$ receptor agonist will allow that question to be fully answered. Since the withdrawal of cisapride, the treatment of GERD has been refined, with proton pump inhibitors becoming the standard of care. These drugs offer high levels of symptom relief in many patients and thus the clinical usefulness of the next generation of prokinetic drugs will need to be established within this context. The projected rise in the number of patients with diabetes suggests that diabetic gastroparesis could also substantially increase, offering another opportunity for prokinetic agents to bring benefit to patients (*see* 6.19 Diabetes/Syndrome X).

In the area of antiemetic therapy, the challenge is to develop agents with which to better control nausea. This will almost certainly depend upon the future development of reliable and predictive animal models in which to map the nausea pathways of the CNS and in which to ultimately understand the potential of novel antinausea medications.

References

1. Wyse, C. A.; McLellan, J.; Dickie, A. M.; Sutton, D. G.; Preston, T.; Yam, P. S. *J. Vet. Intern. Med.* **2003**, *17*, 609–621.
2. Wyse, C. A.; Preston, T.; Love, S.; Morrison, D. J.; Cooper, J. M.; Yam, P. S. *Am. J. Vet. Res.* **2001**, *62*, 1939–1944.
3. Schoonjans, R.; Van Vlem, B.; Van Heddeghem, N.; Vandamme, W.; Vanholder, R.; Lameire, N.; Lefebvre, R.; De Vos, M. *Neurogastroenterol. Motil.* **2002**, *14*, 287–293.
4. Murray, C. D.; Martin, N. M.; Patterson, M.; Taylor, S. A.; Ghatei, M. A.; Kamm, M. A.; Johnston, C.; Bloom, S. R.; Emmanuel, A. V. *Gut* **2005**, *54*, 1693–1698.
5. Coleman, N. S.; Marciani, L.; Blackshaw, E.; Wright, J.; Parker, M.; Yano, T.; Yamazaki, S.; Chan, P. Q.; Wilde, K.; Gowland, P. A. et al. *Aliment. Pharmacol. Ther.* **2003**, *18*, 1039–1048.
6. Decktor, D. L.; Pendleton, R. G.; Elnitsky, A. T.; Jenkins, A. M.; McDowell, A. P. *Eur. J. Pharmacol.* **1988**, *147*, 313–316.
7. Schiavone, A.; Volonte, M.; Micheletti, R. *Eur. J. Pharmacol.* **1990**, *187*, 323–329.
8. Gullikson, G. W.; Loeffler, R. F.; Virina, M. A. *J. Pharmacol. Exp. Ther.* **1991**, *258*, 103–110.
9. Pehlivanov, N.; Sarosiek, I.; Whitman, R.; Olyaee, M.; McCallum, R. *Aliment. Pharmacol. Ther.* **2002**, *16*, 743–747.
10. Wienbeck, M.; Li, Q. *Scand. J. Gastroenterol.* **1989**, *165*, 13–18.
11. Maddern, G. J.; Jamieson, G. G.; Myers, J. C.; Collins, P. J. *Gut* **1991**, *32*, 470–474.
12. Castell, D. O.; Sigmund, C., Jr.; Patterson, D.; Lambert, R.; Hasner, D.; Clyde, C.; Zeldis, J. B. *Am. J. Gastroenterol.* **1998**, *93*, 547–552.
13. Schutze, K.; Bigard, M. A.; Van Waes, L.; Hinojosa, J.; Bedogni, G.; Hentschel, E. *Aliment. Pharmacol. Ther.* **1997**, *11*, 497–503.
14. Gullikson, G. W.; Virina, M. A.; Loeffler, R.; Erwin, W. D. *Am. J. Physiol.* **1991**, *261*, G426–G432.
15. Yamano, M.; Kamato, T.; Miyata, K. *Arzneim.-Forsch.* **1997**, *47*, 1242–1246.
16. Robertson, C. S.; Ledingham, S. J.; Cooper, S. M.; Evans, D. F. *Br. J. Clin. Pharmacol.* **1989**, *28*, 323–327.
17. Mackie, A. D.; Ferrington, C.; Cowan, S.; Merrick, M. V.; Baird, J. D.; Palmer, K. R. *Aliment. Pharmacol. Ther.* **1991**, *5*, 135–142.
18. Nagakura, Y.; Kamato, T.; Nishida, A.; Ito, H.; Yamano, M.; Miyata, K. *Naunyn Schmiedebergs Arch. Pharmacol.* **1996**, *353*, 489–498.
19. Camilleri, M.; McKinzie, S.; Fox, J.; Foxx-Orenstein, A.; Burton, D.; Thomforde, G.; Baxter, K.; Zinsmeister, A. R. *Clin. Gastroenterol. Hepatol.* **2004**, *2*, 895–904.
20. Mine, Y.; Yoshikawa, T.; Oku, S.; Nagai, R.; Yoshida, N.; Hosoki, K. *J. Pharmacol. Exp. Ther.* **1997**, *283*, 1000–1008.
21. Suchitra, A. D.; Dkhar, S. A.; Shewade, D. G.; Shashindran, C. H. *World J. Gastroenterol.* **2003**, *9*, 779–783.
22. Armstrong, S. R.; McCullough, J. L.; Beattie, D. T. *J. Pharmacol. Toxicol. Methods* **2006**, *53*, 198–205.
23. Inui, A.; Yoshikawa, T.; Nagai, R.; Yoshida, N.; Ito, T. *Jpn. J. Pharmacol.* **2002**, *90*, 313–320.
24. Kanaizumi, T.; Nakano, H.; Matsui, Y.; Ishikawa, H.; Shimizu, R.; Park, S.; Kuriya, N. *Eur. J. Clin. Pharmacol.* **1991**, *41*, 335–337.
25. Asakawa, H.; Hayashi, I.; Fukui, T.; Tokunaga, K. *Diabetes Res. Clin. Pract.* **2003**, *61*, 175–182.
26. Ruth, M.; Hamelin, B.; Rohss, K.; Lundell, L. *Aliment. Pharmacol. Ther.* **1998**, *12*, 35–40.
27. Liu, Z.; Sakakibara, R.; Odaka, T.; Uchiyama, T.; Uchiyama, T.; Yamamoto, T.; Ito, T.; Asahina, M.; Yamaguchi, K.; Yamaguchi, T. et al. *Mov. Disord.* **2005**, *20*, 680–686.
28. Grider, J. R.; Foxx-Orenstein, A. E.; Jin, J. G. *Gastroenterology* **1998**, *115*, 370–380.
29. Nguyen, A.; Camilleri, M.; Kost, L. J.; Metzger, A.; Sarr, M. G.; Hanson, R. B.; Fett, S. L.; Zinsmeister, A. R. *J. Pharmacol. Exp. Ther.* **1997**, *280*, 1270–1276.
30. Crowell, M. D.; Mathis, C.; Schettler, V. A.; Yunus, T.; Lacy, B. E. *Neurogastroenterol. Motil.* **2005**, *17*, 738–743.
31. Degen, L.; Matzinger, D.; Merz, M.; Appel-Dingemanse, S.; Osborne, S.; Luchinger, S.; Bertold, R.; Maecke, H.; Beglinger, C. *Aliment. Pharmacol. Ther.* **2001**, *15*, 1745–1751.
32. Prather, C. M.; Camilleri, M.; Zinsmeister, A. R.; McKinzie, S.; Thomforde, G. *Gastroenterology* **2000**, *118*, 463–468.
33. Appel, S.; Kumle, A.; Meier, R. *Clin. Pharmacol. Ther.* **1997**, *62*, 546–555.
34. Briejer, M. R.; Prins, N. H.; Schuurkes, J. A. *Neurogastroenterol. Motil.* **2001**, *13*, 465–472.
35. Poen, A. C.; Felt-Bersma, R. J.; Van Dongen, P. A.; Meuwissen, S. G. *Aliment. Pharmacol. Ther.* **1999**, *13*, 1493–1497.
36. Bouras, E. P.; Camilleri, M.; Burton, D. D.; Thomforde, G.; McKinzie, S.; Zinsmeister, A. R. *Gastroenterology* **2001**, *120*, 354–360.
37. Coremans, G.; Kerstens, R.; De Pauw, M.; Stevens, M. *Digestion* **2003**, *67*, 82–89.
38. Takeda, M.; Tsukamoto, K.; Yamano, M.; Uesaka, H. *Jpn. J. Pharmacol.* **1999**, *81*, 292–297.
39. Takeda, M.; Mizutani, Y.; Yamano, M.; Tsukamoto, K.; Suzuki, T. *Pharmacology* **2001**, *62*, 23–28.
40. Tazawa, S.; Masuda, N.; Koizumi, T.; Kitazawa, M.; Nakane, T.; Miyata, H. *Eur. J. Pharmacol.* **2002**, *434*, 169–176.
41. Nagakura, Y.; Akuzawa, S.; Miyata, K.; Kamato, T.; Suzuki, T.; Ito, H.; Yamaguchi, T. *Pharmacol. Res.* **1999**, *39*, 375–382.
42. Haga, N.; Suzuki, H.; Shiba, Y.; Mochiki, E.; Mizumoto, A.; Itoh, Z. *Neurogastroenterol. Motil.* **1998**, *10*, 295–303.
43. De Ponti, F.; Crema, F.; Moro, E.; Nardelli, G.; Croci, T.; Frigo, G. M. *Neurogastroenterol. Motil.* **2001**, *13*, 543–553.

44. Rizzi, C. A.; Sagrada, A.; Schiavone, A.; Schiantarelli, P.; Cesana, R.; Schiavi, G. B.; Ladinsky, H.; Donetti, A. *Naunyn Schmiedebergs Arch. Pharmacol.* **1994**, *349*, 338–345.
45. Moser, P. C.; Bergis, O. E.; Jegham, S.; Lochead, A.; Duconseille, E.; Terranova, J. P.; Caille, D.; Berque-Bestel, I.; Lezoualc'h, F.; Fischmeister, R. et al. *J. Pharmacol. Exp. Ther.* **2002**, *302*, 731–741.
46. Takahashi, T.; Kurosawa, S.; Wiley, J. W.; Owyang, C. *Gastroenterology* **1991**, *101*, 703–710.
47. Nagahata, Y.; Azumi, Y.; Kawakita, N.; Wada, T.; Saitoh, Y. *Scand. J. Gastroenterol.* **1995**, *30*, 880–885.
48. Iwanaga, Y.; Miyashita, N.; Saito, T.; Morikawa, K.; Itoh, Z. *Jpn. J. Pharmacol.* **1996**, *71*, 129–137.
49. Gue, M.; Fioramonti, J.; Bueno, L. *Gastroenterol. Clin. Biol.* **1988**, *12*, 425–430.
50. Soykan, I.; Sarosiek, I.; McCallum, R. W. *Am. J. Gastroenterol.* **1997**, *92*, 976–980.
51. Horowitz, M.; Harding, P. E.; Chatterton, B. E.; Collins, P. J.; Shearman, D. J. *Dig. Dis. Sci.* **1985**, *30*, 1–9.
52. Zuccato, E.; Bertolo, C.; Salomoni, M.; Forgione, A.; Mussini, E. *Pharmacol. Res.* **1992**, *26*, 179–185.
53. Armbrecht, U.; Reul, W.; Stockbrugger, R. W. Z. *Gastroenterology* **1990**, *28*, 85–89.
54. Staniforth, D. H. *Eur. J. Clin. Pharmacol.* **1987**, *33*, 55–58.
55. Arienti, V.; Corazza, G. R.; Sorge, M.; Boriani, L.; Ugenti, F.; Biagi, F.; Maconi, G.; Sottili, S.; Van Thiel, D. H.; Gasbarrini, G. *Aliment. Pharmacol. Ther.* **1994**, *8*, 631–638.
56. Mansi, C.; Savarino, V.; Vigneri, S.; Perilli, D.; Melga, P.; Sciaba, L.; De Martini, D.; Mela, G. S. *Am. J. Gastroenterol.* **1995**, *90*, 1989–1993.
57. Mearin, F.; Rodrigo, L.; Perez-Mota, A.; Balboa, A.; Jimenez, I.; Sebastian, J. J.; Paton, C. *Clin. Gastroenterol. Hepatol.* **2004**, *2*, 301–308.
58. Iwanaga, Y.; Miyashita, N.; Morikawa, K.; Mizumoto, A.; Kondo, Y.; Itoh, Z. *Gastroenterology* **1990**, *99*, 401–408.
59. Iwanaga, Y.; Miyashita, N.; Mizutani, F.; Morikawa, K.; Kato, H.; Ito, Y.; Itoh, Z. *Jpn. J. Pharmacol.* **1991**, *56*, 261–269.
60. Tsubouchi, T.; Saito, T.; Mizutani, F.; Yamauchi, T.; Iwanaga, Y. *J. Pharmacol. Exp. Ther.* **2003**, *306*, 787–793.
61. Kim, Y. S.; Kim, T. H.; Choi, C. S.; Shon, Y. W.; Kim, S. W.; Seo, G. S.; Nah, Y. H.; Choi, M. G.; Choi, S. C. *World J. Gastroenterol.* **2005**, *11*, 4210–4214.
62. Parkman, H. P.; Trate, D. M.; Knight, L. C.; Brown, K. L.; Maurer, A. H.; Fisher, R. S. *Gut* **1999**, *45*, 346–354.
63. Kirby, M. G.; Dukes, G. E.; Heizer, W. D.; Bryson, J. C.; Powell, J. R. *Pharmacotherapy* **1989**, *9*, 226–231.
64. Law, N. M.; Bharucha, A. E.; Undale, A. S.; Zinsmeister, A. R. *Am. J. Physiol. Gastrointest. Liver Physiol.* **2001**, *281*, G1228–G1237.
65. van Wyk, M.; Sommers, D. K.; Meyer, E. C.; Moncrieff, J. *Methods Find. Exp. Clin. Pharmacol.* **1990**, *12*, 291–294.
66. Ueki, S.; Ogishima, M.; Higashino, R.; Yoneta, T.; Kurimoto, T.; Kaibara, M.; Taniyama, K. *Jpn. J. Pharmacol.* **2001**, *85*, P–186.
67. Ueki, S.; Matsunaga, Y.; Matsumura, T.; Hori, Y.; Yoneta, T.; Kurimoto, T.; Tamaki, H.; Itoh, Z. *Naunyn Schmeidebergs Arch. Pharmacol.* **1998**, *358*, P40.14.
68. Tack, J.; Masclee, A.; Headin, R.; Berstad, A.; Piessevaux, H.; Popiela, T.; Vandenberghe, A.; Kobayashi, S. *Gut* **2002**, *51*, A42.
69. Kishibayashi, N.; Tomaru, A.; Ichikawa, S.; Kitazawa, T.; Shuto, K.; Ishii, A.; Karasawa, A. *Jpn. J. Pharmacol.* **1994**, *65*, 131–142.
70. Kishibayashi, N.; Karasawa, A. *Jpn. J. Pharmacol.* **1995**, *67*, 45–50.
71. Boivin, M. A.; Carey, M. C.; Levy, H. *Pharmacotherapy* **2003**, *23*, 5–8.
72. Ohtawa, M.; Mizumoto, A.; Hayashi, N.; Yanagida, K.; Itoh, Z.; Omura, S. *Gastroenterology* **1993**, *104*, 1320–1327.
73. Tanaka, T.; Mizumoto, A.; Mochiki, E.; Suzuki, H.; Itoh, Z.; Omura, S. *J. Pharmacol. Exp. Ther.* **1998**, *287*, 712–719.
74. Choi, M. G.; Camilleri, M.; Burton, D. D.; Johnson, S.; Edmonds, A. *J. Pharmacol. Exp. Ther.* **1998**, *285*, 37–40.
75. Cowles, V. E.; Nellans, H. N.; Seifert, T. R.; Besecke, L. M.; Segreti, J. A.; Mohning, K. M.; Faghih, R.; Verlinden, M. H.; Wegner, C. D. *J. Pharmacol. Exp. Ther.* **2000**, *293*, 1106–11011.
76. Verhagen, M. A.; Samsom, M.; Maes, B.; Geypens, B. J.; Ghoos, Y. F.; Smout, A. J. *Aliment. Pharmacol. Ther.* **1997**, *11*, 1077–1086.
77. Netzer, P.; Schmitt, B.; Inauen, W. *Aliment. Pharmacol. Ther.* **2002**, *16*, 1481–1490.
78. Talley, N. J.; Verlinden, M.; Snape, W.; Beer, J. A.; Ducrotte, P.; Dettmer, A.; Brinkhoff, H.; Eaker, E.; Ohning, G.; Miner, P. B. et al. *Aliment. Pharmacol. Ther.* **2000**, *14*, 1653–1661.
79. Talley, N. J.; Verlinden, M.; Geenen, D. J.; Hogan, R. B.; Riff, D.; McCallum, R. W.; Mack, R. J. *Gut* **2001**, *49*, 395–401.
80. Chen, C. L.; Orr, W. C.; Verlinden, M. H.; Dettmer, A.; Brinkhoff, H.; Riff, D.; Schwartz, S.; Soloway, R. D.; Krause, R.; Lanza, F. et al. *Aliment. Pharmacol. Ther.* **2002**, *16*, 749–757.
81. Russo, A.; Stevens, J. E.; Giles, N.; Krause, G.; O'Donovan, D. G.; Horowitz, M.; Jones, K. L. *Aliment. Pharmacol. Ther.* **2004**, *20*, 333–338.
82. Gylys, J. A.; Doran, K. M.; Buyniski, J. P. *Res. Commun. Chem. Pathol. Pharmacol.* **1979**, *23*, 61–68.
83. Gralla, R. J.; Itri, L. M.; Pisko, S. E.; Squillante, A. E.; Kelsen, D. P.; Braun, D. W., Jr.; Bordin, L. A.; Braun, T. J.; Young, C. W. *N. Engl. J. Med.* **1981**, *305*, 905–909.
84. Costall, B.; Domeney, A. M.; Naylor, R. J.; Tattersall, F. D. *Neuropharmacology* **1986**, *25*, 959–961.
85. Stables, R.; Andrews, P. L.; Bailey, H. E.; Costall, B.; Gunning, S. J.; Hawthorn, J.; Naylor, R. J.; Tyers, M. B. *Cancer Treat. Rev.* **1987**, *14*, 333–336.
86. Bermudez, J.; Boyle, E. A.; Miner, W. D.; Sanger, G. J. *Br. J. Cancer* **1988**, *58*, 644–650.
87. Cunningham, D.; Hawthorn, J.; Pople, A.; Gazet, J. C.; Ford, H. T.; Challoner, T.; Coombes, R. C. *Lancet* **1987**, *1*, 1461–1463.
88. Cassidy, J.; Raina, V.; Lewis, C.; Adams, L.; Soukop, M.; Rapeport, W. G.; Zussman, B. D.; Rankin, E. M.; Kaye, S. B. *Br. J. Cancer* **1988**, *58*, 651–653.
89. Priestman, T. J. *Eur. J. Cancer Clin. Oncol.* **1989**, *25*, S29–S33.
90. Hesketh, P. J.; Gralla, R. J.; Webb, R. T.; Ueno, W.; DelPrete, S.; Bachinsky, M. E.; Dirlam, N. L.; Stack, C. B.; Silberman, S. L. *J. Clin. Oncol.* **1999**, *17*, 338–343.
91. Cocquyt, V.; Van Belle, S.; Reinhardt, R. R.; Decramer, M. L.; O'Brien, M.; Schellens, J. H.; Borms, M.; Verbeke, L.; Van Aelst, F.; De Smet, M. et al. *Eur. J. Cancer* **2001**, *37*, 835–842.
92. Diemunsch, P.; Schoeffler, P.; Bryssine, B.; Cheli-Muller, L. E.; Lees, J.; McQuade, B. A.; Spraggs, C. F. *Br. J. Anaesth.* **1999**, *82*, 274–276.
93. Gesztesi, Z.; Scuderi, P. E.; White, P. F.; Wright, W.; Wender, R. H.; D'Angelo, R.; Black, L. S.; Dalby, P. L.; MacLean, D. *Anesthesiology* **2000**, *93*, 931–937.
94. Reid, K.; Palmer, J. L.; Wright, R. J.; Clemes, S. A.; Troakes, C.; Somal, H. S.; House, F.; Stott, J. R. *Br. J. Clin. Pharmacol.* **2000**, *50*, 61–64.
95. Lin, T. F.; Yeh, Y. C.; Yen, Y. H.; Wang, Y. P.; Lin, C. J.; Sun, W. Z. *Br. J. Anaesth.* **2005**, *94*, 835–839.
96. Shupak, A.; Doweck, I.; Gordon, C. R.; Spitzer, O. *Clin. Pharmacol. Ther.* **1994**, *55*, 670–680.
97. Kothare, S. V. *Eur. J. Paediatr. Neurol.* **2005**, *9*, 23–26.
98. Van Sickle, M. D.; Oland, L. D.; Mackie, K.; Davison, J. S.; Sharkey, K. A. *Am. J. Physiol. Gastrointest. Liver Physiol.* **2003**, *285*, G566–G576.
99. Parker, L. A.; Kwiatkowska, M.; Mechoulam, R. *Physiol. Behav.* **2006**, *87*, 66–71.

100. Tramer, M. R.; Carroll, D.; Campbell, F. A.; Reynolds, D. J.; Moore, R. A.; McQuay, H. J. *Br. Med. J.* **2001**, *323*, 16–21.
101. Trudel, L.; Bouin, M.; Tomasetto, C.; Eberling, P.; St-Pierre, S.; Bannon, P.; L'Heureux, M. C.; Poitras, P. *Peptides* **2003**, *24*, 531–534.
102. Levin, F.; Edholm, T.; Degerblad, M.; Schmidt, P. T.; Gryback, P.; Jacobsson, H.; Hellstrom, P. M.; Maslund, E. *Gastroenterology* **2005**, *128*, A608.
103. Tack, J.; Depoortere, I.; Bisschops, R.; Verbeke, K.; Janssens, J.; Peeters, T. *Aliment. Pharmacol. Ther.* **2005**, *22*, 847–853.
104. Poitras, P.; Polvino, W. J.; Rocheleau, B. *Peptides* **2005**, *26*, 1598–1601.
105. Kitazawa, T.; De Smet, B.; Verbeke, K.; Depoortere, I.; Peeters, T. L. *Gut* **2005**, *54*, 1078–1084.
106. Edholm, T.; Naslund, E.; Hellstrom, P. M.; Gardner, C.; Darton, J.; O'Shaughnessy, C.; Witherington, J.; Sanger, G. J. *Gastroenterology* **2005**, *128*, A544.
107. Fraser, G. L.; Hoveyda, H. R. *Gastroenterology* **2005**, *128*, A608.
108. Venkova, K.; Johnson, A. C.; Fraser, G. L.; Hoveyda, H. R.; Greenwood-Van Meerveld, B. *Gastroenterology* **2005**, *128*, A611.
109. Lugar, C. W.; Clay, M. P.; Lindstrom, T. D.; Woodson, A. L.; Smiley, D.; Heiman, M. L.; Dodge, J. A. *Bioorgan. Med. Chem. Lett.* **2004**, *14*, 5873–5876.
110. Holst, B.; Brandt, E.; Bach, A.; Heding, A.; Schwartz, T. W. *Mol. Endocrinol.* **2005**, *19*, 2400–2411.
111. van Herwaarden, M. A.; Samsom, M.; Rydholm, H.; Smout, A. J. *Aliment. Pharmacol. Ther.* **2002**, *16*, 1655–1662.
112. Lehmann, A.; Holmberg, A. A.; Bhatt, U.; Bremner-Danielsen, M.; Branden, L.; Elg, S.; Elebring, T.; Fitzpatrick, K.; Geiss, W. B.; Guzzo, P. et al. *Br. J. Pharmacol.* **2005**, *146*, 89–97.
113. Symonds, E.; Butler, R.; Omari, T. *Eur. J. Pharmacol.* **2003**, *470*, 95–97.
114. Simonsson, R. In *Synthesis and Applications of Isotopically Labelled Compounds*, Proceedings of the International Symposium, Boston, MA, June 1–5, 2003; Dean, D. C., Filer, C. N., McCarthy, K. E., Eds.; John Wiley: Chichester, UK, 2004.
115. Reidelberger, R. D.; Kelsey, L.; Heimann, D.; Hulce, M. *Am. J. Physiol. Regul. Integr. Comp. Physiol.* **2003**, *284*, R66–R75.
116. Fried, M.; Erlacher, U.; Schwizer, W.; Lochner, C.; Koerfer, J.; Beglinger, C.; Jansen, J. B.; Lamers, C. B.; Harder, F.; Bischof-Delaloye, A. *Gastroenterology* **1991**, *101*, 503–511.
117. Cremonini, F.; Camilleri, M.; McKinzie, S.; Carlson, P.; Camilleri, C. E.; Burton, D.; Thomforde, G.; Urrutia, R.; Zinsmeister, A. R. *Am. J. Gastroenterol.* **2005**, *100*, 652–663.
118. Liddle, R. A.; Gertz, B. J.; Kanayama, S.; Beccaria, L.; Coker, L. D.; Turnbull, T. A.; Morita, E. T. *J. Clin. Invest.* **1989**, *84*, 1220–1225.
119. Suzuki, T.; Nurrochmad, A.; Ozaki, M.; Khotib, J.; Nakamura, A.; Imai, S.; Shibasaki, M.; Yajima, Y.; Narita, M. *Neuropharmacology* **2005**, *49*, 1121–1131.
120. Miner, W. D.; Casey, J. H.; Gale, J. D. *Br. J. Pharmacol.* **2005**, http://www.pa2online.org/Vol3Issue2abst033P (accessed June 2006).
121. Kawai, M.; Kawahara, H.; Hirayama, S.; Yoshimura, N.; Ida, S. *J. Pediatr. Gastroenterol. Nutr.* **2004**, *38*, 317–323.
122. Yoshikawa, T.; Mine, Y.; Morikage, K.; Yoshida, N. *Arzneim.-Forsch.* **2003**, *53*, 98–106.
123. Kashyap, P. C.; Micci, M.-A.; Pasricha, P. J. *Am. J. Gastroenterol.* **2005**, *100*, S94.
124. Lightbown, I. D.; Miner, W. D.; Gale, J. D. *Br. J. Pharmacol.* **2002**, *136*, 61P.

Biographies

Jeremy D Gale is a drug developer with 17 years experience in major R&D-based pharmaceutical companies working across the continuum of drug discovery and development. He graduated in pharmacology and then completed a PhD in pharmacology and neuroscience at the University of London. He joined Glaxo in 1988 and for almost 8 years worked in gastrointestinal pharmacology, leading the biology teams that formed part of the irritable bowel syndrome (IBS) and emesis research programes. Jeremy joined Pfizer in 1995 to lead the biology research team focused on gastroenterological diseases, developing expertise in IBS, inflammatory bowel disease (IBD), and gastroesophageal reflux disease (GERD). In 2002, he moved into Exploratory Clinical Development, becoming clinical leader for programs targeting GERD and IBD. In addition, Jeremy leads translational medicine activities and biomarker development for the gastroenterology and hepatology therapeutic area.

Ichiro Mori received an MA from Kyoto University (Prof H Nozaki), then his PhD from University of California, Berkeley in 1988 (Professors P A Bartlett and C H Heathcock) on development of stereoselective chemistry-related aldol reactions. In 1998, he spent 1 year at the Massachusetts Institute of Technology as a postdoctoral fellow with Professor R L Danheiser to investigate new synthetic methods of indols, and joined Ciba-Geigy Japan to develop inhibitors of histidine biosynthesis as herbicides. He then joined Glaxo Japan in 1997, working on the synthesis of kinase inhibitors. In 2003, he moved to Pfizer and is currently working on medicinal chemistry in the gastroenterology and hepatology therapeutic area.

Comprehensive Medicinal Chemistry II
ISBN (set): 0-08-044513-6

ISBN (Volume 6) 0-08-044519-5; pp. 671–691

6.31 Cardiovascular Overview

D A Taylor, East Carolina University, Greenville, NC, USA

6.31.1 Introduction

The cardiovascular system consists of the heart and the blood vessels that distribute blood to the entire body. The system is designed to provide oxygenated blood to tissues in order to maintain function. Much of this exchange occurs at the level of the capillaries and arterioles where surface contact between the blood cells carrying oxygen and the tissues is greatest. While the system is relatively simple in design, it is essential to the normal physiology of the body and is maintained at a level of activity that can supply energy under normal circumstances as well as under situations of stress where greater activity is required. Because of its importance to normal physiology, it should be obvious that abnormal functioning of the cardiovascular system can have an enormous deleterious impact on the homeostatic function. Disorders of the cardiovascular system are a primary contributor to the global morbidity and mortality that continues to increase in prevalence in spite of remarkable advances in the methods used to manage the disease. In fact, cardiovascular disease (CVD) has annually been the leading cause of mortality for over a century. For these reasons alone, the diseases that form the constellation of CVD are the focus of intensive research efforts by the biomedical research community. Considerable attention has been focused both on the underlying molecular basis of these diseases and the agents that can be used for the therapeutic management of CVD.

6.31.2 Economic Impact of Cardiovascular Diseases

The direct and indirect economic burden of CVDs in the USA has been estimated to be nearly \$400 billion annually.[1] Direct costs associated with CVD account for nearly \$250 billion while the indirect costs (lost productivity in the case of morbidity and lost future earnings in the case of mortality) account for slightly more than \$150 billion. Heart diseases (e.g., coronary heart disease (CHD), congestive heart failure (CHF), hypertension (*see* 6.32 Hypertension), cardiac arrhythmias, etc.) collectively account for nearly 65% of the total costs associated with CVD. Within the group of CVD, CHD represents the largest contributing disorder (56% of the group of heart diseases and 36% of the total economic burden) to both direct and indirect costs. Stroke accounts for nearly 14% of the total economic burden with the indirect costs associated with stroke (brain attack) being particularly burdensome.

6.31.3 Economic Opportunity of Cardiovascular Disease

CVD represents one of the most significant economic markets available to healthcare providers, including the pharmaceutical industry. Prescription agents used to treat CVD accounted for over $75 billion in prescriptions in 2004 and have shown steady growth of 2.5–3.0% annually for the past several years. The projected market is estimated to exceed $100 billion by 2008.[2,3] Therefore, safe and effective new agents will potentially reduce the direct and indirect costs associated with CVD. The probability of success in developing new cardiovascular agents is somewhat higher than other drug classes as the model systems employed to evaluate potential agents are more predictive of human efficacy.

The pharmaceutical industry devotes only slightly more than $80 million annually to develop new cardiovascular agents as opposed to nearly $200 million to market these agents.[4] However, the cost of developing a new chemical entity (NCE) in the cardiovascular area is unlikely to be recovered unless the NCE is a 'blockbuster.' This results in a close monitoring of the science that surrounds cardiovascular disease to be able to respond quickly and effectively to new concepts in disease management. Cardiovascular products represent nearly 17% of total spending on phase IV budgets with nearly 34% of the total product budget used to launch a new cardiovascular compound compared to 26% for an oncology product.[5]

6.31.4 Impact and Distribution of Cardiovascular Diseases

CVDs include a number of diseases which can be and are interrelated in very complex ways and include: CHD; acute coronary syndrome (myocardial infarction (MI), angina pectoris, or unstable angina); stroke; hypertension; congestive heart failure (CHF); arrhythmias, peripheral artery disease, rheumatic heart disease, valvular heart disease, and venous thromboembolism.[6] CVD is responsible for nearly one-third (approximately 16.7 million individuals) of the global mortality burden[7] and, for 2005, was estimated to have a direct and indirect economic burden of nearly $400 billion in the USA alone (see above).[1] In particular, CVD appears to be non-selective in the populations that are targeted with men, women, and children of all ethnic groups being nearly equally at risk for developing CVD but individuals older than 70 are the predominant age group who die from the effects of CVD.[7] Socioeconomic status does not preclude any group from CVD as it appears to target both developing (low and middle income countries) and developed countries.[8,9]

6.31.5 Prevalence of Cardiovascular Diseases

Globally, CVD represents the second greatest disease burden in men and third largest disease burden in women. The percentage of disability-adjusted life years (DALY) lost was nearly 7% for men compared to human immunodeficiency virus/acquired immune deficiency syndrome (HIV/AIDS) which represented nearly 7.5% of the DALY lost worldwide. For women, the primary contributors to DALY lost were unipolar depressive disorders and HIV/AIDS which represented 8.4% and 7.2%, respectively while CHD represented only 5.3%.[8,9] Cardiovascular disease accounts for nearly 10% of the DALY lost in developing countries and nearly 18% in industrialized countries. The DALY is an indicator of the total burden of the disease and reflects the healthy years of life lost to a specific disease.

In the USA, CVDs were the leading cause of death regardless of gender or ethnicity (**Table 2**)[1,8] and accounts for nearly 1 million deaths per year but afflicts over 70 million people (**Table 2**). CVDs were responsible for nearly 40% of all deaths in women while cancer ranked second with only about 20% of the total mortality (**Table 1**). Similarly, CVD accounted for 37% of all deaths in males with cancer being the second greatest cause of death (nearly 25%) with no apparent differences between white and non-Hispanic black men. Surprisingly, CVD appears to be a less significant contributor to mortality in Hispanic or Latino men and women (**Table 2**). However, this population seems to exhibit a higher prevalence mortality produced due to diabetes mellitus (**Table 2**). Also reflected in **Table 1**, there are a number of disorders that form the constellation of diseases considered in the spectrum of CVD. However, it is important to

Table 1 Prevalence of cardiovascular disorders in the USA

Cardiovascular disease	*Population (millions)*		
Coronary heart disease	Total 13.0	Males 7.1	Females 5.9
Acute myocardial infarction	Total 7.1	Males 4.1	Females 3.0
Angina pectoris	Total 6.4	Males 3.1	Females 3.3
Stroke	Total 5.4	Males 2.4	Females 3.0
Congestive heart failure	Total 4.9	Males 2.4	Female 2.5
High blood pressure	Total 65.0[a]	Males 29.4	Females 35.6
Congenital cardiovascular defects	Total 1.0	—	—
Arrhythmias	Total 2.2	—	—
Total	70.1 (Represents approximately 25% of total US population)		

[a] Many patients exhibit more than one form of cardiovascular disease and this is especially true for high blood pressure.

Table 2 Primary causes of death in the USA (data from 2002)

Gender and race	*Leading cause of death (% of population except total)*						
	Cardiovascular disease (Total[a])	*Cancer*	*Accidents*	*Chronic lower respiratory diseases*	*Diabetes mellitus*	*Alzheimer's disease*	*Other[b]*
White male	36.6	24.4	5.7	5.4	2.7	—	—
Black or African American male	33.4	22.2	5.9	—	—	—	9.6[c]
Hispanic or Latino male	27.1	18.6	11.7	—	4.2	—	4.0[d]
Males (Total)	433 825	288 768	69 257	60 713	34 301	—	—
White female	39.8	21.6	—	5.6	—	3.6	3.1[e]
Black or African American female	39.6	20.9	2.7	—	5.2	—	2.8[f]
Hispanic or Latino female	32.2	21.2	4.7	2.8	6.1	—	—
Females (Total)	493 623	268 503	—	64 103	38 948	41 877	—

Data compiled from *Heart Disease and Stroke Statistics – 2005 Update*, American Heart Association.
[a] The majority of 'Total cardiovascular disease' is derived from 'Diseases of the heart and stroke' which combined account for 91–93% of the total for all groups regardless of race or gender.
[b] 'Other diseases' includes a variety of disorders that are specific for a given gender and ethnic group.
[c] For Black or African American Males the 'Other' category consists of assault (homicide) 4.7% and HIV(AIDS) 3.7%.
[d] For Hispanic or Latino Males, the 'Other' category consists of assault (homicide).
[e] For White females, the 'Other' category consists of influenza and pneumonia.
[f] For Black or African-American Females, the 'Other' category consists of nephritis, nephrotic syndrome, and nephrosis.

recognize that several of these diseases are comorbid contributors to other forms of CVD. In fact, many of these diseases are so intermingled that it becomes difficult to identify one disease as being independent of others. Taken in that context, diabetes mellitus is considered one of the 'potential' major risk factors associated with the development of adverse cardiac or cardiovascular events. With the addition of diabetes mellitus to CVD, the potential for adverse cardiac or cardiovascular events to lead to mortality accounts for 40–45% of all deaths in white males and females and black females.[1] Excluding hypertension which is a comorbid factor to many other forms of CVD, coronary heart disease including myocardial infarction is the leading contributor to CVD in the USA followed by stroke and congestive heart failure and other cardiovascular diseases (**Table 2**). A similar ranking of cardiovascular diseases is seen globally with coronary heart disease and stroke being the leading causes of morbidity and mortality worldwide.[9]

CVD prevalence is expected to continue to rise in the population so that it will remain the leading cause of both death and disability worldwide with an estimated increase in the number of deaths to 20 million individuals by 2020 and to more than 24 million individuals by 2030.[9] These estimates are especially disconcerting when considering the fact that most CVD is preventable through lifestyle modification and drug treatment. Of even greater concern is the alarming worldwide increase in specific major risk factors for CVD that include: tobacco use, physical inactivity, nutrition, blood cholesterol, obesity and overweight, and diabetes.

6.31.6 Coronary Heart Disease

Coronary heart disease (CHD) is a set of syndromes that includes angina pectoris (comprising stable, unstable, and variant, or Prinzmetal's, angina), ischemic heart disease, and myocardial infarction (MI).[10] CHD is the most common cause of death worldwide accounting for over 7 million deaths per year (approximately 45% of the total number of deaths due to cardiovascular disease). Worldwide, CHD generally affects males to a greater extent than females.[9] However, in the USA, while MI has a greater prevalence among males, females have a greater propensity toward the development of angina.[1] In the USA, CHD and 'acute coronary syndrome' that includes acute MI and unstable angina will present in about 700 000 new patients annually while another 500 000 patients will experience a recurrent episode. In addition, an additional 175 000 individuals in the USA will experience a first silent heart attack in the next year while 3–4 million Americans will have an ischemic attack without recognizing it.[1,10] It is estimated that 565 000 Americans will experience a new MI annually while an additional 300 000 individuals will experience a recurrent event.

It is clear from these data that CHD is a significant contributor to morbidity and mortality and, therefore, the therapeutic management of these diseases is an important factor in efforts to reduce the global burden of cardiovascular disease. Angina pectoris represents a cardiovascular economic disorder where the demand exceeds the capacity of the coronary vasculature to supply oxygen. Chemical modification of this imbalance can be accomplished using a variety of agents that decrease the oxygen demand (e.g., β-adrenoceptor blockers), reduce the resistance against which the heart must work (e.g., antihypertensive medications), or improve oxygen delivery (e.g., coronary vasodilators, nitroglycerin, organic nitrates). Some of the agents perform both functions of reducing demand and increasing supply. Perhaps the greatest advances in the management of CHD have been made in the surgical arena. The use of new diagnostic tools has led to increased success in identifying and characterizing CHD early. Cardiac catheterization and coronary angiography can identify precise areas of artery occlusion that can then be managed using invasive techniques like percutaneous transluminal coronary angioplasty (PTCA), balloon angioplasty, laser angioplasty, or artherectomy to physically alter the vessel lumen and improve blood flow.[10] Newer additions to these techniques include the insertion of stents and/or drug-coated stents during PTCA to reduce the prospect of restenosis that occurs in 10–20% of patients after approximately 6 months. In addition, coronary artery bypass graft surgery (CABG) has markedly improved the success rate and decreased the morbidity associated with the CHD as patients are incapacitated for much shorter intervals. Over the past decade, this type of surgery has become one of the mainstays of the management of CHD. Newer diagnostic techniques such as radionuclide imaging, magnetic resonance imaging, digital cardiac angiography, computed tomography (including computerized axial tomography (CAT), electron beam computed tomography (EBCT), positron emission tomography (PET), and single photon emission computed tomography (SPECT)) have significantly improved the precise definition of the areas of impairment, directing physicians toward the most appropriate treatment regimen.

Acute severe episodes of coronary artery occlusion or thrombosis are called MI and require rapid treatment in order to reduce damage to the heart muscle produced by the loss of oxygen supply. As invasive cardiac techniques cannot be performed in the early minutes of an MI, the use of drugs that dissolve clots or reduce clotting has become the primary

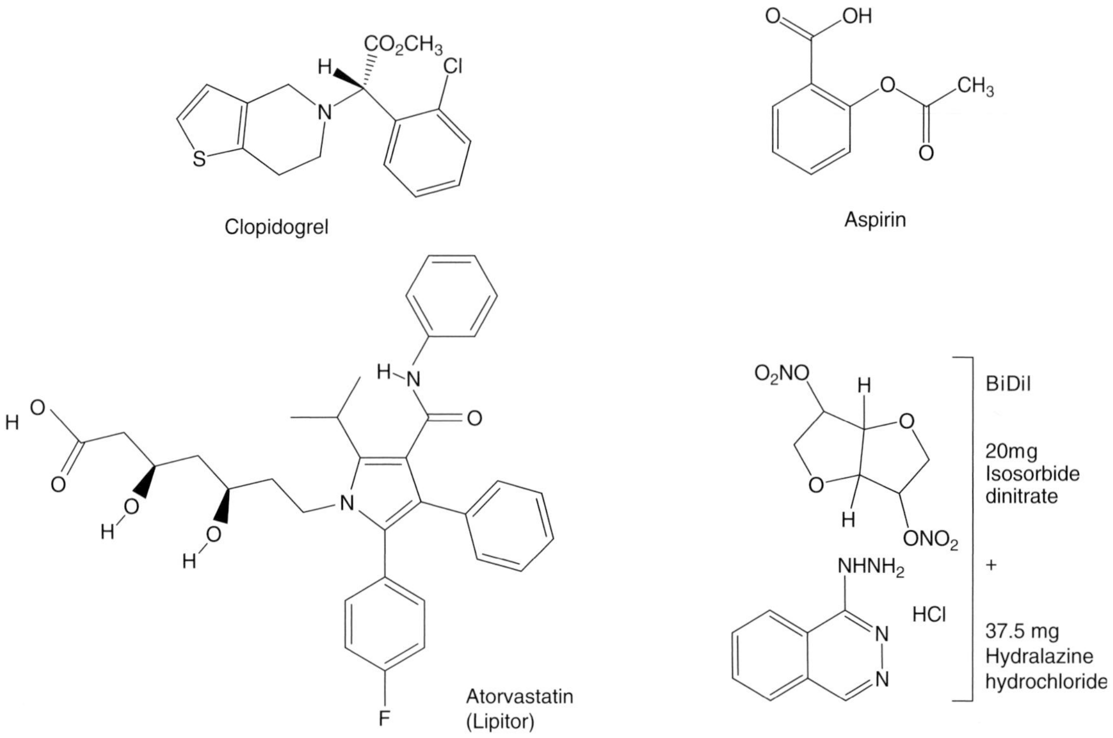

Figure 1 Cardiovascular drugs.

route of management until accurate diagnosis and treatment can be performed using techniques described above. Agents that are able to interfere with clot formation (e.g., the nonsteroidal anti-inflammatory drug (NSAID), aspirin, or the $P2Y_{12}$ antagonist clopidogrel) or to dissolve an existing thrombus (e.g., streptokinase) are often employed. Chemical structures of some of these agents are given in **Figure 1**.

6.31.7 Stroke

Stroke is the brain equivalent of a heart attack (brain attack) and can result from either the development of a thrombus (clot) in a cerebral blood vessel (ischemic stroke) or from a ruptured cerebral blood vessel that leads to a hemorrhage (hemorrhagic stroke). Ischemic strokes frequently occur at night or early in the morning and are often preceded by a transient ischemic attack – referred to as a 'mini-stroke.' In contrast, hemorrhagic strokes usually result from head trauma or the presence of other risk factors (e.g., hypertension). While ischemic strokes are most common (70–80% of all strokes), hemorrhagic strokes are the most devastating as the highest rates of fatality occur following this type of stroke.[10] Strokes represent the third most common cause of mortality in industrial countries accounting for about 5.5 million deaths worldwide. Nearly 15 million people annually suffer a stroke and about 5 million of these individuals are left permanently disabled.[9] Ischemic strokes result in mortality within 30 days of the event in about 10% of Americans while hemorrhagic strokes lead to mortality in over 35% over the same period of time. However, the mortality rate for white females is nearly twice that for white males. In contrast to CHD which seems to effect American populations equivalently, African Americans, American Indians, Alaskan Americans, and Hispanic Americans appear to possess a much higher incidence of stroke than whites regardless of gender.[1] African American males and females exhibit an incidence rate for stroke that is twice as high as that observed for white Americans. In addition, the risk of stroke increases dramatically with age as the chance of having a stroke more than doubles with each decade of life after age 55. Thus, while the incidence of stroke is higher in certain minority populations, the incidence of mortality following a stroke is significantly higher in white females than any other population group.

The therapeutic management of stroke (*see* 6.10 Stroke/Traumatic Brain and Spinal Cord Injuries; 6.34 Thrombolytics) primarily relies on the use of drugs that dissolve clots such as tissue plasminogen activator (TPA) which has significantly reduced the morbidity and recovery rate for patients who experience a stroke. However, as with

MI, this agent must be delivered very quickly after the onset of symptoms in order to be effective. In addition to 'clot busters' like TPA, other forms of therapeutic management rely on treating some of the other underlying risk factors such as hypertension or cardiac disorders (e.g., atrial fibrillation or damaged heart valves) that promote blood clotting. Newer forms of management include cerebral angioplasty or carotid endarterectomy that may be used to improve cerebral blood flow. In spite of these new techniques and agents for treatment, the primary methodology employed to manage stroke patients is appropriate physical and mental therapy to overcome the disabilities that normally accompany the event.

6.31.8 Congestive Heart Failure

CHF occurs when the heart is unable to fulfill its primary mission of pumping blood. The severity of heart failure ranges from mild to moderate forms which are treatable, to severe acute heart failure which usually requires more dramatic types of management, including cardiac transplantation. In most instances, heart failure develops after other antecedent cardiovascular diseases including hypertension (nearly 75% of all CHF cases) and MI or other forms of coronary heart disease. In 2002, there were 550 000 new cases of CHF in the USA with more than 50 000 deaths attributed to the disorder.[1] The total number of individuals with CHF was nearly 5 million with an economic burden of nearly $28 million or 7% of the economic burden of all cardiovascular diseases. The incidence of CHF is greater in African American males and females and in diabetic women (especially those with elevated body mass index (BMI) or reduced renal function) but increases dramatically with age, regardless of gender or ethnicity.[1] Since the incidence of CHF has not declined for the past two decades, it is likely to continue to increase worldwide due to the absence of any decline in the major risk factors that contribute to the development of the disease. Patient survival has however steadily increased as newer forms of treatment are developed.

Since the underlying cause of CHF is the inability of the heart to pump effectively and efficiently, one class of agents employed historically has been the inotropic agents that increase myocardial contractility like the cardiac glycosides (e.g., digoxin). While these are still used, they have a low therapeutic index. Newer inotropic agents have been developed that target intracellular enzymes like phosphodiesterase (e.g., milrinone and inamrinone) or mimic peptides that are released from failing heart (e.g., the atrial natriuretic factor (ANF) agonist nesiritide). Another class of agents routinely used to treat CHF are the diuretics that reduce the fluid retention, a classic sign of CHF. The therapeutic management of CHF has undergone a major and, in some minds, radical paradigm shift over the past decade as a result of an increased understanding of the process of cardiac remodeling that occurs in a failing heart that has identified the beneficial effect of drugs that previously had been contraindicated in the management of the disease. Cardiac remodeling can be significantly reversed through the use of angiotensin-converting enzyme (ACE) inhibitors, angiotensin II receptor blockers (ARB) and β-adrenoceptor blocking agents. A fixed combination of isosorbide dinitrate and hydralazine (BiDil, **Figure 1**) represents a novel management scheme for the treatment of CHF in African Americans who show an improved response relative to Caucasians.[11,12] However, uptake in the marketplace for this drug treatment that shows promise based on ethnic origin has been modest. The recognition that hypertension was an antecedent disorder in such a large percentage of patients presenting with CHF also led to the use of antihypertensive agents (e.g., vasodilators) to reduce the resistance against which the heart must work. In those cases where an identifiable cause of CHF can be identified (e.g., abnormal heart valves), surgical management has proven to be quite effective. Finally, in cases where the heart failure is too severe to be managed pharmacologically, heart transplants and mechanical assist devices have been employed.

6.31.9 Hypertension

Hypertension is the cardiovascular disorder that afflicts the largest number of individuals (**Table 1**).[1] Blood pressure in an individual is determined by two factors: the force with which the heart expels blood and the resistance to flow that blood vessels offer. High blood pressure exists in two different forms that are defined as 'essential hypertension' in which no definitive cause of the disease can be identified and 'secondary hypertension' in which a specific organic basis for the disease can be identified. For example, secondary hypertension has been associated with renovascular disease, primary aldosteronism, Cushing's syndrome, and pheochromacytoma as well as other diseases. A number of factors contribute to the development of essential hypertension that are risk factors associated with the development of cardiovascular disease in general (see below). Unfortunately for the management of the disease, nearly 90% of patients diagnosed with high blood pressure are in the essential hypertension category. Thus, defined bases for the disease are only identified in a very small percentage of the population. It is estimated that 30–40% of the world population has high blood pressure; and, that beyond age 50, this will exceed 50%.[9] High blood pressure is thought to contribute to nearly 50% of all CVDs and to about 5 million premature deaths per year. According to the World Health Organization (WHO), high blood pressure is responsible for causing 62% of strokes and 49% of heart attacks worldwide.[7] The WHO estimates that 600 million individuals with high blood pressure are at risk of developing heart attacks, strokes, or cardiac failure.[13] This places hypertension in the unenviable position of not only being a significant cardiovascular disorder in its own right but also being a major risk factor in many other cardiovascular diseases.

The American Heart Association (AHA) estimates that 65 million individuals in the USA have high blood pressure (defined as a systolic blood pressure of greater than 140 mmHg, diastolic blood pressure of greater than 90 mmHg, currently taking antihypertensive medication or having been told twice by a physician or other health professional that high blood pressure is present). It is the definition of high blood pressure and hypertension that appears to be the center of controversy surrounding this disease and the ultimate therapeutic management of hypertension. The Joint National Committee on Prevention, Detection, Evaluation, and Treatment of High Blood Pressure (JNC) has generally provided the guidelines for high blood pressure. The 7th report of the JNC[14,15] superceded the 6th report published in 1997 which established three stages of hypertension based upon diastolic and systolic blood pressures. The latest guidelines from the JNC established a new category called 'prehypertension' defined by a systolic pressure of 120–139 mmHg or a diastolic pressure of 80–89 mmHg. The three stages of hypertension described by JNC-6 were replaced by two stages in which Stage 1 was defined as a systolic pressure of 140–159 mmHg and/or a diastolic pressure of 90–99 mmHg and Stage 2 was defined by pressures of $\geq$160 and/or $\geq$100. As with the JNC-6 classification, prehypertension and each of the stages were associated with specific treatment guidelines based upon the urgency for treatment. More recently, the American Society of Hypertension has provided a new definition/classification of hypertension based on the complexity of the disorder rather than just the systolic and diastolic blood pressure values. This new system defines normal blood pressure as less than 120/80 mmHg with no risk factors for CVD and no identifiable early markers of CVD. The three stages correspond to the JNC-7 stages including prehypertension as follows: Stage 1 blood pressure 120/80–139/89 mmHg, multiple cardiovascular risk factors and early disease markers, and signs of functional or structural changes in the heart and small arteries but no end organ damage; Stage 2 blood pressure is $\geq$140/90 mmHg with widespread disease markers and early evidence of end-organ damage as well as risk factors; Stage 3 blood pressure is sustained $\geq$140/90 mmHg even when treated and the presence of end-organ damage and existing cardiovascular events.[16]

Within the USA, as in many other countries, nearly 1 in 3 adults has been diagnosed with established high blood pressure while an additional 25% of Americans over age 18 fall into the new category of prehypertension.[14,15] Even more significant is the fact that among those patients diagnosed with hypertension, nearly 1 in 3 individuals were not aware of the existence of the disease while 34% were adequately controlled using medications. In contrast, 25% of the individuals with high blood pressure were not adequately controlled with medication and 11% received no medication.[1] Hypertension impacts upon specific populations to a much greater extent that some other cardiovascular diseases. Between 1999 and 2002, the prevalence of hypertension for the total population evaluated in the National Health and Nutrition Examination Survey (NHANES) analysis was 29% while the age-adjusted prevalence in non-Hispanic blacks was 41%[17] which according to data obtained by the AHA is the highest in the world.[1] Furthermore, the disease appears to develop at an earlier age in blacks. The mortality associated with hypertension was significantly greater in females compared to males (59% versus 41%, respectively) and black women appear to exhibit a much greater prevalence of the disease than white females (45% versus 31%, respectively).

Agents to manage hypertension have been available for over half a century but, over the past 20 years, the development of newer drugs with improved therapeutic profiles has modified the manner in which the disorder is treated. The changing definitions and classifications of hypertension have contributed to confusion regarding the appropriate management of the disease. Clearly lifestyle modifications are one of the most important methods recommended for managing the disease; and JNC-7 currently recommends lifestyle modification for all classes of hypertension including even patients with normal blood pressure. These include alterations in diet and nutrition to reduce sodium intake and increase consumption of fruits, vegetables, and fat-free products. Individuals are also encouraged to control alcohol consumption and to increase physical activity to improve cardiovascular tone and regulate weight. As the severity of hypertension increases, medications may be added to assist in regulating the tone of the cardiovascular system. The agents that are frequently and, generally, successfully employed include diuretics (especially the thiazides), ACE inhibitors (captopril and enalapril), angiotensin receptor blocking agents (ARBs like losartan), calcium channel blockers (CCBs) (e.g., nifedipine), and β-blockers (e.g., propranolol). Structures of some of the representative agents are provided in **Figure 1**. Under conditions of greater severity or in the absence of adequate control of the high blood pressure with dietary and lifestyle modifications as well as these first line medications, combinations of first line agents (e.g., diuretic plus ACE inhibitor, ARB, CCB, or β-blocker) are used and, if unsuccessful, sympathetic nerve inhibitors that act within the CNS (e.g., α-methyldopa) or at peripheral nerves (e.g., guanethidine) are used. In resistant cases or in cases of hypertensive emergency (e.g., hypertensive crisis), vasodilator agents that act directly on the smooth muscle of the arteries and veins are employed. Individual agents in this class display a variety of different mechanisms of action that range from producing nitric oxide to activating potassium channels. The integral role that hypertension plays in so many other cardiovascular diseases will obviously create the need for continued efforts to develop newer and more effective agents.

6.31.10 Other Cardiovascular Diseases

There are a number of other cardiovascular diseases that contribute to total CVD but to a lesser extent. These include arrhythmias, peripheral arterial disease, atherosclerosis, rheumatic fever, and rheumatic heart disease. Among these diseases, peripheral arterial disease (PAD) and arrhythmias pose the greatest threats.

6.31.10.1 Peripheral Artery Disease

PAD is a disease in which blood vessels outside the heart and the brain are narrowed either due to build up of fatty deposits (atherosclerosis) or due to some other cause. PAD affects 8–12 million Americans with estimates of increasing prevalence to the point of affecting up to 16 million Americans age 65 or older by 2050.[1] This disorder usually causes cramping in the legs during activity that is termed intermittent claudication. A similar prevalence is observed in Europe suggesting that PAD may be a significant contributor to CVD as time progresses. Individuals with PAD have a four- to fivefold higher risk of mortality due to a cardiovascular event. The pharmacological management of PAD involves the use of antithrombotic agents that either prevents clotting through inhibition of coagulation (anticoagulants) or through prevention of platelet aggregation (antiplatelet agents). Surgical approaches to the management of this disease are also becoming popular. This particular area has seen the development of many new agents and should see the development of many more in the future.

6.31.10.2 Arrhythmias

Arrhythmias (*see* 6.33 Antiarrythmics) are disorders of cardiac rhythm that lead to nearly 40 000 deaths annually.[1] Arrhythmias also are observed following CHF or MI. Arrhythmias can occur at various levels of the heart including the atria, the atrioventricular node, and the ventricles. Arrhythmias of the ventricles are believed to be the primary factor in sudden cardiac death from coronary artery disease. These disorders are difficult to manage either pharmacologically or via invasive techniques such as ablation. Pharmacological agents used to manage cardiac arrhythmias are often classified according to their primary mechanism of action into four major categories. While they range in mechanism of action from sodium channel (Class 1A (procaineamide), 1B (lidocaine), and 1C (propafenone)), potassium channel (Class 3 (amiodarone)), and calcium channel (Class 4 (verapamil)) blockers to β-adrenoceptor blockers (Class 2 (metoprolol)) they all share the common property of being stabilizers of cardiac function and, therefore, are cardiac depressants. Agents in each class have serious side effects associated with their use that makes the decision to employ antiarrhythmic agents very delicate. This particular area of pharmacological management of CVD provides an opportunity to develop novel and useful agents as the current drugs employed to manage arrhythmias are not very satisfactory.

6.31.11 Risk Factors

Over 300 factors have been described that increase risk for development of stroke and CHD. The major risk factors (**Table 3**) have been defined by their specific appearance in many populations, by clear evidence that the specific factor exerts an independent impact on CVD and by the demonstration that treatment of that factor improves outcomes. A primary risk factor is hypertension that is present in many populations and singularly increases susceptibility to CAD, MI, heart failure, and stroke. Since the management of most cardiovascular diseases includes some form of lifestyle or dietary modification, the risk factors have been divided into two major categories: modifiable risk factors and nonmodifiable risk factors (**Table 3**). In industrialized and developing countries, the increased risk of cardiovascular disease development has been associated with a relatively limited number of modifiable risk factors that include: tobacco use, alcohol use, high blood pressure, abnormal blood lipids, and obesity (**Table 3**).

Table 3 Risk factors for cardiovascular disease

Modifiable risk factors	*Nonmodifiable risk factors*
High blood pressure	Age
Abnormal blood lipids	Hereditary factors
Tobacco use	Gender
Physical inactivity	Ethnicity
Obesity	Low socioeconomic status
Unhealthy diets	
Diabetes mellitus	
Metabolic syndrome	
Excessive alcohol consumption	
Medications	

6.31.11.1 Tobacco Use

The current estimated 1.3 billion smokers worldwide are projected to increase to 1.7 billion by 2025. Tobacco use is the single most preventable cause of death in the USA and costs an estimated $155 billion per year in direct medical and lost productivity costs. While most would believe that mortality associated with smoking would be caused by lung cancer, there is clear evidence now that the majority of smoking-related deaths are due to complications associated with CVD. Smoking and tobacco use increases the risk of coronary artery disease and stroke by 100% and of peripheral artery disease by more than 300%.[9] In fact, current cigarette smoking is one of the most powerful independent predictors of sudden cardiac death.[10] Smoking promotes CVD because it increases clotting, damages the endothelium, raises low-density lipoprotein (LDL) levels and lowers high-density lipoprotein (HDL) levels, and increases cholesterol plaques. The knowledge that smoking and tobacco use negatively impact health led to a reduction in tobacco use among high-school seniors between 1980 and 2002 (a decrease of 13% with substantial decreases in the female and African-American populations).[1] However, these positive data are offset by data obtained in 2003 that estimated nearly 30% of male students and 25% of female students in grades 9–12 reported current tobacco use. These same trends are seen globally with increased tobacco use in teenagers especially among young females who appear to exhibit greater risk for cardiovascular complications. Since the risk for CVD increases dramatically in individuals who began smoking prior to age 16, there is an increased level of concern that the number of teenagers using tobacco may ultimately increase the prevalence of cardiovascular disease at even earlier time periods. There is growing concern about the risk of developing CVD in children under age 18 who are exposed to tobacco through secondhand smoke. Since few studies have addressed the issue specifically, very little is known regarding the influence of secondhand smoke exposure on the prevalence of CVD. Estimates of teenage exposure to secondhand smoke suggest that nearly 25% of the population may be exposed with a range from 11% to 34%.[1] The potential impact of tobacco use on cardiovascular health certainly justifies the AHA position that individuals not take up smoking and those currently smoking quit, since the risk of CVD begins to decline immediately upon smoking cessation.

6.31.11.2 Abnormal Blood Cholesterol and Other Lipids

Cholesterol is necessary for a variety of bodily functions including the formation of plasma membranes and as a building block for hormones. Cholesterol is obtained through ingestion and synthesis in the body and is transported throughout the body by LDL and HDL. Elevation in plasma total cholesterol and LDL levels are clearly associated with increased risk of CVD, including CHD and stroke. High blood cholesterol levels are estimated to cause 4.4 million deaths annually that amount to over 18% of strokes and 56% of CHD.[7] The current guidelines for plasma cholesterol and other lipids indicate that patients with total cholesterol of $\geq 240\,\mathrm{mg\,dL^{-1}}$ are twice as likely to experience heart disease as a person whose total cholesterol is less than $200\,\mathrm{mg\,dL^{-1}}$. Similarly, patients with LDL cholesterol levels of $\geq 190\,\mathrm{mg\,dL^{-1}}$ are considered as 'high risk' patients for CVD. In patients with diabetes or existing CHD, the LDL cholesterol level that places an individual at risk decreases to 130–159 mg dL^{-1}, while individuals with no existing diabetes or CHD and no other risk factors should strive for an LDL cholesterol level less than 160 mg dL^{-1}. In contrast, elevated HDL cholesterol is apparently protective against CVD as HDL tends to carry excess cholesterol back to the liver for processing. Patients with HDL cholesterol levels of $<$40 mg dL^{-1} are at risk for the development of CVD. Therefore, individuals with high HDL and low LDL and total cholesterol levels assume the lowest risk for developing CVD.

The therapeutic management of cholesterol levels is accomplished using a number of approaches including dietary and lifestyle modifications. In addition, several classes of pharmacological agents that act at different sites of regulation in the process of absorption, synthesis and sequestration of cholesterol are available and include nicotinic acid, bile acid sequestrants and fibric acid derivatives, the statins (3-hydroxy-3-methylglutaryl coenzyme A reductase (HMG-CoA reductase) inhibitors) and newer agents ezetimibe (a substance that inhibits cholesterol uptake). The statins (a representative structure is provided in **Figure 1**) are very effective in lowering plasma LDL cholesterol and elevating HDL cholesterol. Aggressive regulation of plasma lipid levels reduces the risk of MI, stroke, and CHD by more than 20%.[17,18] Since the drug doses used in this aggressive approach to the management of plasma lipids are close to the levels associated with adverse effects, it is likely that newer agents or combinations of agents will be employed in the future.

Another important marker of cardiovascular risk is C-reactive protein which is a circulating protein produced by the liver that is elevated as part of the inflammatory process. In addition to being used as a marker of inflammation due to bacterial infections versus viral infections, the high sensitivity (hs) C-reactive protein has been promoted as a strong predictor of increased risk of negative cardiovascular events in women or for development of initial cardiovascular events like myocardial infarction. [20,21] These data were strong enough to prompt the AHA and CDC to recommend that hs-C-reactive protein levels be monitored as a marker for CHD.[22] However, recent studies suggest that using this inflammatory marker for CVD may not be as powerful as originally believed.[23] Whether hs-C-reactive protein level measurement is a powerful or moderately powerful predictor of risk for CVD, the data have clearly shown an important contribution of the inflammation process to several cardiovascular disorders of significance in humans. Therefore, the development of agents that act primarily as inhibitors of one or more of the various components of the inflammatory cascade could provide a significant cellular and molecular target for the future.

6.31.11.3 Diabetes Mellitus

Diabetes is a disorder in which the body does not make enough (type 1) or responds inappropriately to (type 2) insulin which converts glucose to energy (*see* 6.19 Diabetes/Syndrome X). The most common form of diabetes is type 2 in which insulin production is normal but tissues do not respond appropriately to the hormone. Worldwide there are about 150 million people with type 2 diabetes with that number is expected to double by 2025.[1,19] In the USA, the prevalence of diabetes has risen by 61% since 1990.[1] American Indians and Alaskan Native adults have an especially high prevalence of diabetes with over 15% of the population expressing the disease compared to 7% of the total population.[10] Similar high rates of prevalence are found in Mexican-Americans and non-Hispanic blacks with female diabetics appearing to be more susceptible to stroke.[1] Risk factors that are frequently associated with diabetes include obesity and physical inactivity and hypertension. Therefore, it is recommended that diabetic patients maintain blood pressure of less than 130/85 mm Hg.[9] The presence of diabetes is also associated with an increased risk of death due to MI, stroke, and/or CHD. Therefore, it is important to effectively manage the disease in order to reduce the risk of developing CVDs that are often the cause of mortality.

The therapeutic management of diabetes depends on the type, with type 1 requiring the administration of insulin since natural insulin is not produced. In contrast, type 2 diabetes can be managed with a number of agents that either promote the release of insulin from the pancreas, increase sensitivity to insulin in the target cells, or a combination of both.

6.31.11.4 Obesity, Overweight, and Metabolic Syndrome

Obesity has been described as one of the modern epidemics. The numbers that support this argument are real and staggering. Since 1991, the prevalence of obesity has increased 75% and has been observed in every ethnic group.[1] The direct and indirect costs of obesity-related diseases has been estimated to be over $100 billion annually. Perhaps the most alarming statistic relative to obesity is the magnitude of the increase in children. In the past 30 years, the prevalence of overweight and obese children has increased fourfold and currently has extended to preschool children where over 10% of children 2–5 years old are overweight.[1] Factors responsible for this alarming increase in obesity include alterations in diet and nutrition and physical inactivity, both of which are also risk factors for CVD. Obesity not only increases the risk for the development of CVD but also for the development of type 2 diabetes. Thus, the risk factors for CVD can actually synergize with each other and produce greater changes in morbidity and mortality. The concept of synergy in combined risk factors is further reinforced when the metabolic syndrome is considered This disorder actually combines obesity, diabetes, and high blood pressure in such a way that an individual exhibiting two of the three of these disorders would be considered as part of the metabolic syndrome population. Additional discussion of the metabolic syndrome and the management of this complex disorder is available in this volume (*see* 6.17 Obesity/ Metabolic Syndrome Overview).

While the obesity epidemic occurs worldwide, it is heavily concentrated in North America[9] and especially in the USA where it is estimated that over 60% of adults are overweight or obese as defined by a BMI of greater than 25 (representing approximately 135 million individuals). BMI is a measure of weight in relation to height and any value between 25 and 29.9 $kg\,m^{-2}$ is considered *overweight* while a value >30 is considered *obese*.[24] The contribution of obesity to CVD, diabetes, and the metabolic syndrome suggests that in the absence of some sort of regulation of obesity, there will be substantial increases in the number of cardiovascular deaths ultimately attributable in some form or another to obesity. While all of the risk factors are important and significant, the enormous increase in overweight and obese individuals is a high priority as there are few mechanisms for the management of the disorder that do not involve some sort of personal motivation. Drug regimens are inadequate in the absence of lifestyle changes. This area of drug development is clearly one that presents very few options with high efficacy and presents a significant opportunity for the development of truly life-changing 'blockbuster' drugs.

References

1. American Heart Association *Heart Disease and Stroke Statistics – 2005 Update*. http://www.americanheart.org/ (accessed April 2006).
2. *Research and Markets*, 2005. http://www.researchandmarkets.com/ (accessed April 2006).
3. *PR Leap*, 2005. http://www.prleap.com/ (accessed April 2006).
4. *Medical Patent News*, 2005. http://www.news-medical.net/ (accessed April 2006).
5. *Cutting Edge Information*, 2004. http://www.pharmacardio.com/ (accessed April 2006).
6. Fuster, V.; Alexander, R. W.; O'Rouke, R. A., Eds. *Hurst's The Heart*, 11th ed.; Roberts, R., King, S. B. III, Nash, I. S., Prystowsky, E. N., Assoc. Eds.; McGraw-Hill: New York, 2004.
7. World Health Organization *World Health Organization*, 2002. http://www.who.int/ (accessed April 2006).
8. NHANES 1999-2000 *Data Files Data, Docs, Codebooks, SAS Code*. CDC National Center for Health Statistics, 2005. http://www.cdc.gov/nchs/ (accessed April 2006).
9. *Atlas of Heart Disease and Stroke;* World Health Organization: Geneva, 2004. http://www.who.int/ (accessed April 2006).
10. American Heart Association. *Heart and Stroke Facts*, 2005. http://www.americanheart.org/ (accessed April 2006).
11. Taylor, A. L.; Ziesche, S.; Yancy, C.; Carson, P.; D'Agostino, R., Jr.; Ferdinand, K.; Taylor, M.; Adams, K.; Sabolinski, M.; Worcel, M. et al. *N. Engl. J. Med.* **2004**, *351*, 2049–2057.
12. Bloche, M. G. *N. Engl. J. Med.* **2004**, *342*, 2035–2037.
13. *Cardiovascular Disease: Prevention and Control*. World Health Organization: Geneva, 2004. http://www.who.int/ (accessed April 2006).
14. Chobanian, A. V.; Bakris, G. L.; Black, H. R.; Cushman, W. C.; Green, L. A.; Izzo, J. L., Jr.; Jones, D. W.; Materson, B. J.; Oparil, S.; Wright, J. T., Jr. et al. *JAMA* **2003**, *289*, 2560–2572.
15. Fields, L. E.; Burt, V. L.; Cutler, J. A.; Hughes, J.; Roccella, E. J.; Sorlie, P. *Hypertension* **2004**, *44*, 398–404.
16. Brookes, L. *Medscape Cardiology* **2005**, *9*, http://www.medscape.com/viewarticle/501272 (accessed April 2006).
17. Centers for Disease Control and Prevention (CDC). *MMWR Morb Mortal Wkly Rep* 2005, *54*, 7–9. http://www.cdc.gov/mmwr/ (accessed April 2006).
18. Koren, M. J.; Hunninghake, D. B. *J. Am. Coll. Cardiol.* **2004**, *44*, 1772–1779.
19. *Integrated Management of Cardiovascular Risk*. World Health Organization: Geneva, 2002. http://whqlibdoc.who.int/ (accessed April 2006).
20. Ridker, P. M.; Rifai, N.; Rose, L.; Buring, J. E.; Cook, N. R. *N. Engl. J. Med.* **2002**, *347*, 1557–1565.
21. Ridker, P. M.; Hennekens, C. H.; Buring, J. E.; Rifai, N. *N. Engl. J. Med.* **2000**, *342*, 836–843.
22. Pearson, T. A.; Mensah, G. A.; Alexander, R. W.; Anderson, J. L.; Cannon, R. O., III.; Criqui, M.; Fadl, Y. Y.; Fortmann, S. P.; Hong, Y.; Myers, G. L. et al. *Circulation* **2003**, *107*, 499–511.
23. Danesh, J.; Wheeler, J. G.; Hirschfield, G. M.; Eda, S.; Eiriksdottir, G.; Rumley, A.; Lowe, G. D. O.; Pepys, M. B.; Gudnason, V. *N. Engl. J. Med.* **2004**, *350*, 1387–1397.
24. Moore, S.A. *Overweight, Obesity: Prevention, Management and Treatment*; American Health Consultants: Atlanta, GA, 2001, pp 1–13.

Biography

David A Taylor received a BS degree in biology from Alderson-Broaddus College and began studies in pharmacology under Drs William W Fleming and David P Westfall at West Virginia University in 1970. After completing the PhD degree in the spring of 1974, he began a postdoctoral fellowship at the National Institutes of Health (NIH) through the Pharmacological Research Associate Training (PRAT) program where he studied under Drs Floyd E Bloom and Barry J Hoffer in the Laboratory of Neuropharmacology at the National Institute of Mental Health. After postdoctoral training, he took positions as an Assistant Professor in the Department of Pharmacology at the University of Colorado Medical School in Denver, CO, and as a Senior Laboratory Scientist at Merck, Sharp and Dohme Research Laboratories in West Point, PA, until returning to West Virginia University as an Associate Professor (1983–1988) and Professor (1988–2000) of Pharmacology and Toxicology. In January 2001, he assumed the position of Professor and Chair of the Department of Pharmacology (renamed the Department of Pharmacology and Toxicology in 2003) in the Brody School of Medicine at East Carolina University. Dr Taylor has maintained a very broad interest in research that focuses on the ability of cells and tissues to modify the manner in which they respond to drugs and neurotransmitters which may play an important role in the development of tolerance to drugs and in the process of addiction.

Comprehensive Medicinal Chemistry II
ISBN (set): 0-08-044513-6

ISBN (Volume 6) 0-08-044519-5; pp. 693–704

6.32 Hypertension

A Scriabine, Yale University School of Medicine, New Haven, CT, USA

6.32.1 Introduction

6.32.1.1 Definition, Classification, and Diagnosis of Hypertension

Hypertension is a disorder of the cardiovascular system characterized by elevated arterial blood pressure. The blood pressure in the arteries is dependent on the energy of cardiac contractions, elasticity, and contractile state of arterial walls, as well as on the volume and viscosity of the blood. It fluctuates with every heartbeat. The maximal pressure occurs near the end of the stroke output and is termed systolic. The minimal pressure occurs late in ventricular diastole and is termed diastolic.

Individuals with a systolic blood pressure at or under 120 mmHg and diastolic pressure at or under 80 mmHg are considered normotensive. The term 'prehypertensive' is used for individuals with a systolic pressure under 140, but above 120 mmHg and diastolic pressure under 90, but above 80 mmHg. Patients with 140–159 mmHg systolic and over 90 mmHg diastolic have phase I hypertension, while patients with systolic pressure over 160 mmHg and diastolic pressure over 100 mmHg have phase II hypertension.[1]

It has been estimated that up to 50 million, or over 20%, Americans are likely to be hypertensive. Only 70% of hypertensive patients are aware of their disease, 59% are treated, and only 34% are controlled.[1–3] The prevalence of hypertension in Europe has been estimated as 13%,[4] but in reality may not be any different from that in the US, since definitions of hypertension and populations studied were different. In the UK 12% of adults over 16 years of age have blood pressure values above 160/95 mmHg, and a further 21% have values above 140/90 mmHg.[5] In Spain 68% of elderly people are hypertensive and systolic blood pressure is controlled in only 32% of them.[6]

Most hypertensive patients have the so-called *essential* or *idiopathic* hypertension, a polygenic multifactorial disorder that involves the interaction of several genes with environmental factors.[7] Genetic abnormalities have been demonstrated in only a small percentage of patients with essential hypertension.

Risk factors for hypertension include ethnicity, obesity, psychosocial stress, limited physical activity, and diet. Hypertension itself is the major risk factor for atherosclerosis and for mortality from cardiovascular disease. The incidence of cardiovascular events increases starting with a systolic pressure of 115 and diastolic of 75 mmHg. There is a substantial risk of cardiovascular disease with a systolic pressure of 130–139 mmHg, even though these levels are not considered hypertensive.

Hypertension is the primary cause of stroke and a risk factor for coronary heart disease, myocardial infarction, sudden cardiac death, cardiac failure, and renal insufficiency. Patients with extremely high and sustained blood pressure levels over 210 mmHg systolic are likely to have malignant hypertension, a microvascular occlusive disease that affects the kidneys, brain, retina, and other organs. Malignant hypertension is fatal if untreated.

In the US, the Joint National Committee on Prevention, Detection, Evaluation, and Treatment of High Blood Pressure (JNC) establishes and reviews guidelines for the prevention, diagnosis, and management of hypertension. According to the last published guidelines of JNC (JNC VII),[1] the primary target for therapy should be systolic rather than diastolic blood pressure. All patients with a systolic pressure over 160 mmHg should receive drug therapy. To achieve goal blood pressure levels with minimal side effects, two drugs acting by different mechanisms are usually prescribed. The decision to start therapy should be based on the blood pressure level as well as on the evidence of organ damage. The presence of organ (e.g., kidney) damage calls for an immediate initiation of therapy. In patients without organ damage initial home measurements or 24-h monitoring of arterial pressure prior to the start of therapy are recommended. This recommendation is based on the existence of the so-called 'white-coat' hypertension – blood pressure elevation in the physician's office or clinic only. 'White-coat' hypertension is considered relatively benign; it carries only a low risk of cardiovascular mortality.

In the diagnosis of hypertension it is important to determine whether hypertension is primary or secondary to another disease. Only 5–10% of hypertensive patients have secondary hypertension, but it should be suspected if onset is sudden, particularly in childhood or in patients older than 50 years of age, if it is severe, resistant to therapy, and/or accompanied by unusual symptoms. Conditions that may cause secondary hypertension include aortic coarctation, eclampsia or preeclampsia, brain tumors, lead or mercury poisoning, illicit or prescribed drugs, renal disease, adrenal tumors, primary aldosteronism, Cushing's disease, and obstructive sleep apnea. Elimination or specific therapy of the cause of secondary hypertension should be attempted as the initial therapeutic approach.

Hypertension is usually systemic, but it may be limited to certain organs. The term 'pulmonary hypertension' is used to describe a selective elevation of pressure in pulmonary arteries. Portal hypertension, a persistent elevation of pressure in portal veins, is usually a secondary hypertension that often occurs as a consequence of liver cirrhosis.

6.32.1.2 Pathophysiology, Biochemistry, and Genetics of Hypertension

The pathophysiology of essential hypertension has been extensively studied over the last 50 years. Peripheral vascular resistance is usually increased in hypertensive individuals. Normally, the autonomic nervous system, kidneys, adrenal cortex, local hormones, and cytokines regulate vascular resistance. Failure of the normal regulation of vascular resistance leads to hypertension. The failure can theoretically occur in any part of the regulatory system. Initially, overactivity of the sympathetic nervous system plays a major role in the development and maintenance of hypertension. The excessive activation of the renin–angiotensin system (RAS) or enhanced sensitivity to its primary effector, Ang II, contributes to the development and maintenance of hypertension. The fact that inhibitors of the formation of Ang II or its antagonists at the receptor level are highly useful antihypertensive drugs supports the likely involvement of RAS in the pathogenesis of hypertension. Since calcium ions are required for the contraction of vascular smooth muscle (VSM), excessive permeability of VSM cells to Ca^{2+} or altered sodium–calcium exchange may also be involved. Activation of Ang II receptors leads to enhanced entry of calcium ions into VSM cells, so that the calcium and Ang II hypotheses are not mutually exclusive.

Various biochemical abnormalities in VSM cells have been proposed to play a role in the development of essential hypertension, including: (1) increased ratio of cyclic guanosine monophosphate/cyclic adenosine monophosphate (AMP); (2) decreased basal adenylyl cyclase; and (3) altered activity of cyclic AMP-dependent protein kinase. It is not clear whether any of these changes are consistently present in hypertensive individuals and whether they are causative or secondary to another abnormality in the biochemistry of VSM.

Over the last decade genetics of essential hypertension has been investigated.[8] The estimates for the extent of genetic contributions to the pathogenesis of essential hypertension range from 30% to 50%. The studies have failed, however, to identify one single gene responsible for essential hypertension. It appears that multiple candidate and susceptibility genes may contribute to the disease. Candidate genes have been identified in the RAS. Genetic linkages between ACE and essential hypertension have been suggested.[9] Polymorphisms of the Ang II receptor (AT_1) gene have been related to differential responses to antihypertensive drugs and severe forms of essential hypertension were found to be associated with a specific defect in this gene.[10] Mutations of subunits of the epithelial sodium-channel gene have been found in Liddle's syndrome that is known to be associated with increased renal reabsorption of sodium and hypertension.[11] Linkages between adrenergic receptor genes and hypertension have been reported. β_2-Adrenergic receptor gene, known to affect blood flow and arterial pressure, has been implicated in the genetics of essential hypertension. It has been suggested that essential hypertension is caused by a combination of small quantitative changes in the expression of many susceptibility genes with the environmental factors.[7,12]

The mutations of susceptibility genes may also be responsible for the excessive sensitivity to salt and higher risk for hypertension in black Americans. Hypertension appears to be primarily responsible for the higher mortality of black Americans from heart disease, renal failure, and stroke (*see* 6.10 Stroke/Traumatic Brain and Spinal Cord Injuries; 6.25 Renal Dysfunction in Hypertension and Obesity; 6.33 Antiarrythmics).[13]

6.32.2 Experimental Disease Models

In the development of antihypertensive drugs, animal models with surgically or chemically induced or spontaneous hypertension have been extensively used. The most commonly used model is the spontaneously hypertensive rat (SHR), initially bred by Okamoto and Aoki in 1963.

6.32.2.1 Surgically Induced Hypertension

One of the major discoveries in hypertension research in the twentieth century was that of Goldblatt and his associates in 1937. They occluded the main renal artery of one kidney in six dogs and found that renin levels in the peripheral blood of these animals substantially increased. Over the next 3–4 days the mean systemic blood pressure in these animals also increased by an average of 104–139 mmHg. The Goldblatt technique was later used in rats and more recently in mice and led to the establishment of models with transient or permanent renal artery occlusion with or without unilateral nephrectomy. These models are referred to as two kidneys/one clip (2K1C) or one kidney/one clip (1K1C) hypertension models. The initial phase of hypertension in either of these models is associated with a rapid increase in plasma renin levels and a subsequent increase in Ang II formation. In the 2K1C model, the RAS remains activated for at least 9 weeks after occlusion of the renal artery. After a few weeks the 1K1C model becomes a model of low-renin, volume-dependent hypertension, since volume retention by a single kidney tends to inhibit renin secretion.[14]

Eventually, however, the 2K1C rat model tends to develop volume-dependent hypertension. In the 2K1C mouse model, Ang II produces and maintains hypertension by activating the AT_{1A} receptor subtype, while nitric oxide (NO) seems to counteract the hypertensive effects of Ang II.[15]

6.32.2.2 Mineralocoticoid Hypertension

In 1942 Selye described experimental mineralocorticoid hypertension in rats produced by desoxycorticosterone acetate (DOCA) and salt. It is a low-renin salt-dependent model, which has been widely used in the evaluation of antihypertensive drugs. Recently this model has been shown to be useful in the evaluation of endothelin antagonists. The endothelin system was found to be involved in the pathogenesis of low-renin hypertension[16] and endothelin antagonists lower arterial pressure in DOCA-salt hypertensive rats. Iglarz *et al.*[17] reported that rosiglitazone, an activator of peroxisome proliferator-activated receptor γ (PPARγ), prevents the development of hypertension in DOCA-salt rats.

6.32.2.3 Spontaneous Hypertension

The first strain of rats with inheritable hypertension was bred in New Zealand by Smirk and Hall and is known as New Zealand hypertensive rats. Most of the currently available SHRs are derived from the strain developed in Japan by Okamoto and Aoki. The substrains of Okamoto–Aoki rats have undergone changes and differ in various aspects from the original strain. Okamoto–Aoki SHRs have been cross-bred with strains with other inheritable abnormalities, including hyperlipemia and obesity. The stroke-prone SHRs (SPSHRs) die from a type of stroke that resembles a rare form of human stroke, known as cerebral autosomal dominant arteriopathy with subcortical infarcts. If placed on a high-salt diet they do not survive longer than 8 weeks, but their life expectancy can be substantially prolonged by different types of drugs, including heparin or CCAs. Rossi *et al.*[18] found a markedly increased endothelin receptor (ET_A) density in the walls of cerebral arterioles of SPSHRs. This and other findings suggest involvement of endothelin in the pathogenesis of certain types of stroke.

The importance of salt intake in the pathogenesis of hypertension has been explored using Dahl salt-sensitive (DSS) and Dahl salt-resistant (DSR) rats. It has been reported that optimal dietary potassium chloride (2.6%) can prevent hypertension and protect cerebral and renal vasculature in DSS rats fed a 1% sodium chloride-containing diet for 8 months.[19] Lighthall *et al.*[20] detected three new salt-sensitive genes in the kidneys of DSS rats. Two of these genes encode enzymes required for the biosynthesis of L-arginine and the third encodes a small acid-soluble protein that is possibly involved in the transcription or translation of numerous genes.

The Milan hypertensive strain (MHS) of rats was developed by Bianchi *et al.* in Milan in the 1980s. Hypertension in these animals is transplantable with the kidneys. The strain is characterized by increased sodium reabsorption, hyperactivation of the Na^+-K^+ pump, and point mutation in the cytoskeletal membrane protein genes. More recently, MSH rats were used in the development of a novel antihypertensive drug, PST 2238, which has been shown to normalize the activity of the Na^+-K^+ pump.[21]

Many rat strains derived from SHRs and normotensive animals are used in genetic research to identify quantitative trait loci (QTLs) for blood pressure subphenotypes and gene mutations responsible not only for hypertension but also for its consequences: ventricular hypertrophy, kidney failure, and other cardiovascular abnormalities.[22] QTLs for blood pressure were identified on almost all rat chromosomes, with the exception of chromosomes 6, 11, 12, and 15. QTLs that have been present reproducibly in many experiments are viewed as candidate loci for human essential hypertension. Human genetic studies indicated that QTLs discovered in rats can be translated to human essential hypertension. This approach may lead to the identification of the major susceptibility genes for hypertension and its complications.

6.32.3 Treatment of Hypertension

The current JNC VII guidelines recommend that antihypertensive therapy should start with a thiazide diuretic or a β-adrenoceptor antagonist as the first choice and newer drugs should only be added in special circumstances.[1] Some investigators disagree, however, with the JNC recommendations and feel that drugs inhibiting RAS – ACE inhibitors (ACEIs) or angiotensin receptor blockers (ARBs) – should be the first-line drugs in the treatment of hypertension, since they provide benefits beyond blood pressure reduction. Extended-release formulations of calcium antagonists are also often used as first-line therapy.

6.32.3.1 Diuretics

The antihypertensive activity of diuretics was discovered in the clinic. Diuretics do not lower arterial pressure in acute experiments on normotensive animals. Only in chronic experiments on DSS or SH rats can the mild antihypertensive effect of diuretics be detected. The first diuretic found to lower arterial pressure in hypertensive patients was chlorothiazide, discovered by Beyer and his colleagues at Merck. Its discovery was soon followed by the introduction of hydrochlorothiazide, a closely related derivative with much higher relative potency that was discovered at Merck as well as at Ciba-Geigy. Many similar thiazides have been subsequently developed and marketed, but only a few are still available on the American market. The major side effect of thiazides is hypokalemia, due to the excessive excretion of K^+ ions. Other side effects include hyperuricemia and hyperglycemia. The search for diuretics with longer duration of action and fewer side effects led to the discovery of 'thiazide-like' diuretics, including metalozone, chlorthalidone, and indapamide (**Table 1**), which are chemically different from, but pharmacologically similar to, thiazides. Metolazone may produce diuresis in patients with a low glomerular filtration rate (below 20 mL min^{-1}), while thiazides tend to lose their effectiveness in such patients. Since the maximal obtainable saluretic effect of thiazides is low, diuretics with a higher ceiling effect – furosemide, bumetanide, torasemide and ethacrynic acid (**Table 1**) – have been introduced. These so-called loop diuretics have a much steeper saluretic dose–response curve than thiazides or 'thiazide-like' diuretics. Their antihypertensive effect is, however, not more pronounced or sustained than that of thiazides. Since water and/or electrolyte depletion is more likely to be produced by loop diuretics than by thiazides or thiazide-like diuretics, their use in the treatment of hypertension should be limited to patients who fail to respond to thiazides or patients with heart failure and/or severe edema who can benefit from the rapid salt depletion. The mechanism of action of diuretics within renal tubules is reasonably well understood. Thiazides and thiazide-like diuretics inhibit the Na^+-Cl^- symport in the distal convoluted tubules, blocking the reabsorption of these ions at the luminal site of

Table 1 Commonly used diuretics

Drugs	*Structure*	*Usual daily dose (mg)*	*Oral bioavailability (%)*	$t_{1/2}$ *(h)*
Thiazides				
Chlorothiazide	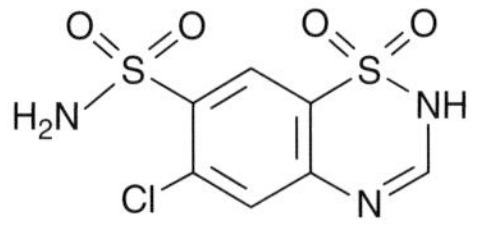	250–1000	9–56	1.5
Hydrochlorothiazide	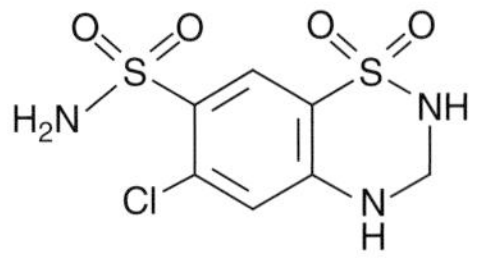	12.5–50	70	2.5
Thiazide-like diuretics				
Indapamide	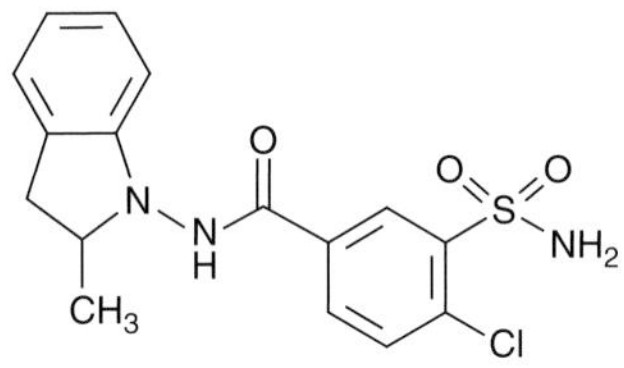	2.5–5.0	93	14
Chlorthalidone	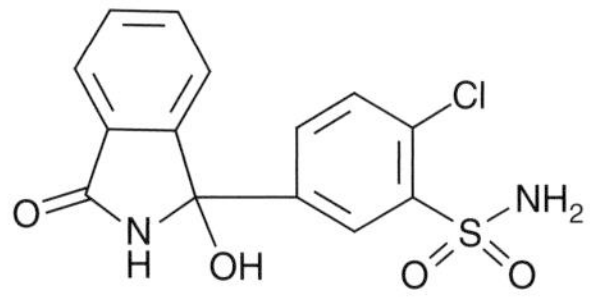	12.5–50	65	47
Metolazone	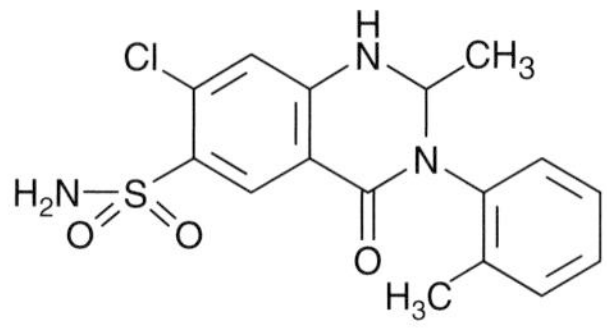	0.5–10 Dependent on formulation	65	50
K^+-sparing diuretics				
Triamterene	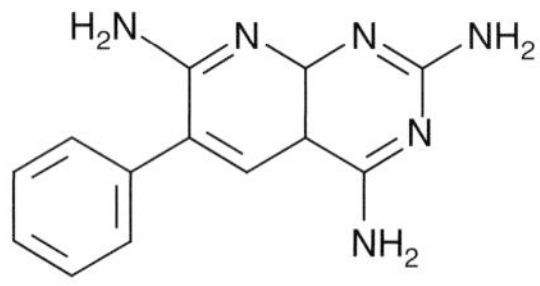	50–150	50	4.2
Amiloride	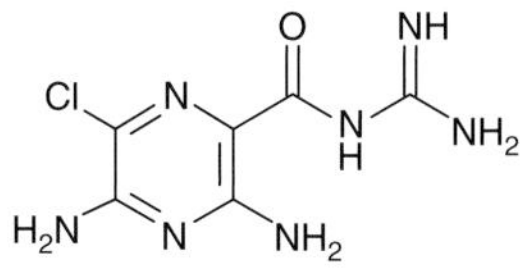	5–10	15–25	21
Spironolactone	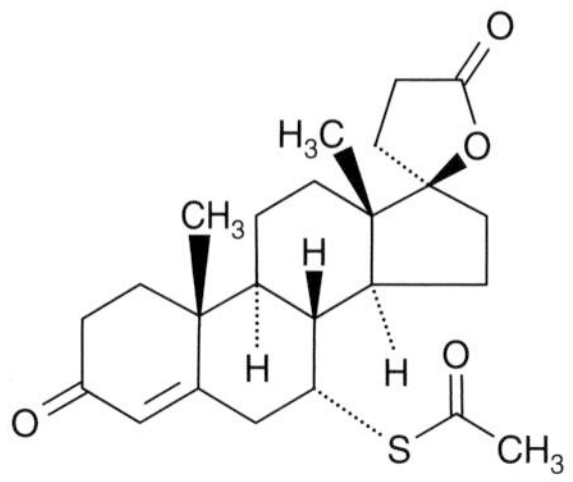	50–100	65	10–35

continued

Table 1 Continued

Drugs	*Structure*	*Usual daily dose (mg)*	*Oral bioavailability (%)*	$t_{1/2}$ *(h)*
Eplerenone	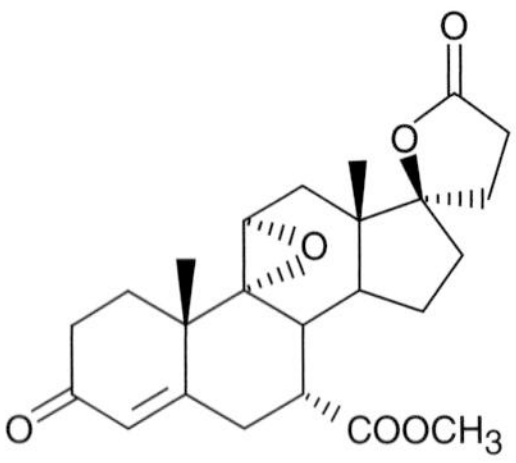	50–200	25–60	3–6
Loop diuretics				
Furosemide	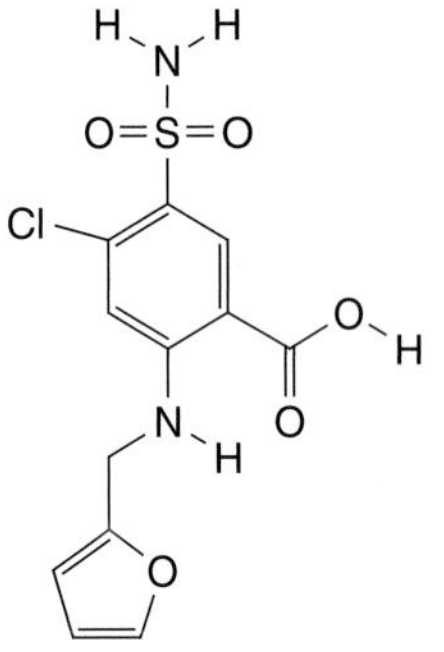	20–0	60	1.5
Bumetanide	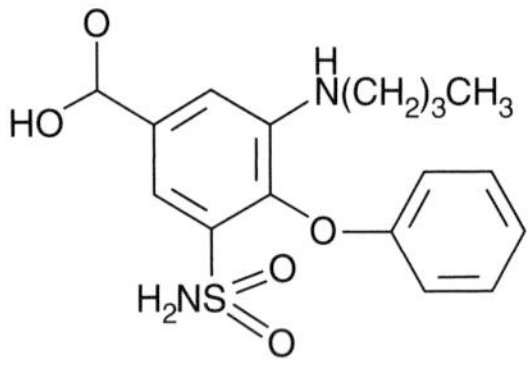	0.5–5.0	80	0.8
Torsemide	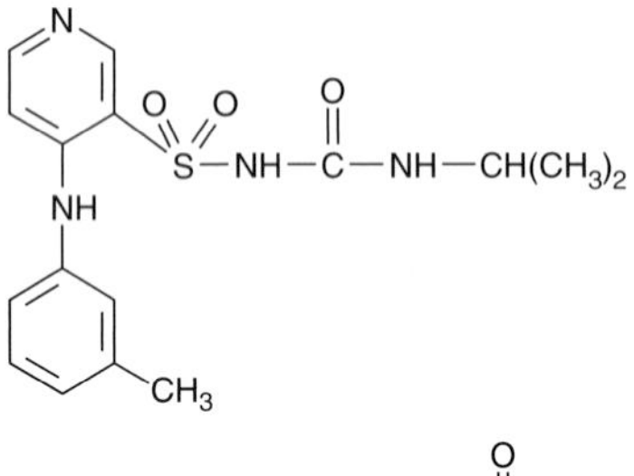	5.0–20	80	3.5
Ethacrynic acid	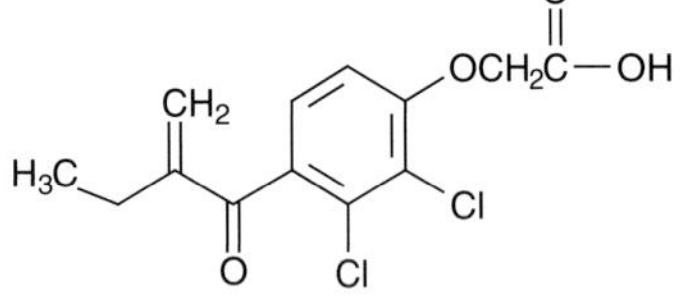	25–100	100	1.0

$t_{1/2}$, elimination half-life.

the tubular membrane, but increasing the excretion of K^+. Potassium-retaining diuretics, e.g., triamterene and amiloride (**Table 1**), inhibit epithelium sodium channels in the distal tubules and collecting duct, while furosemide and other loop diuretics inhibit Na^+K^+-$2Cl^-$ symport in the ascending limb of Henle's loop (**Figure 1**). Different sites of action within renal tubules justify the use of diuretic combinations to enhance the diuretic effect and to reduce side effects. Fixed combinations of hydrochlorothiazide and either triamterene or amiloride are widely used.

The mechanism of antihypertensive action of diuretics is less clear. It appears to depend on the reduction in body sodium, since diuretic-induced hypotension can be antagonized by salt infusion and in the anephric hypertensive patients diuretics do not lower arterial pressure. All diuretics tend to increase plasma renin activity, an effect that limits their antihypertensive efficacy, but justifies their combined use with the inhibitors of RAS.

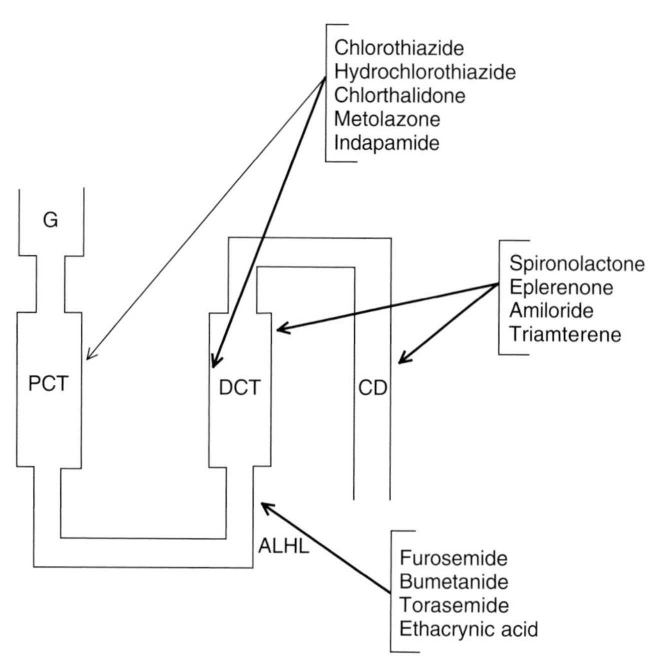

Figure 1 Renal sites of action of diuretics. G, glomerulus; PCT, proximal convoluted tubules; ALHL, ascending limb of Henle's loop; DCT, distal convoluted tubules; CD, collecting duct.

The aldosterone antagonists spironolactone and eplerenone (**Table 1**) have natriuretic, K^+-sparing, and antihypertensive effects. They inhibit the competitive binding of aldosterone to cytoplasmic mineralocorticoid receptors in the late distal tubules and collecting duct. Eplerenone is more selective for mineralocorticoid receptors and has lower affinity for androgen and progesterone receptors than spironolactone.[23] The antihypertensive activity of eplerenone was established in numerous clinical trials. Its efficacy was found to be independent of baseline aldosterone levels, age, race, or gender. In addition to its diuretic and antihypertensive effects, eplerenone has a cardioprotective effect: it prolongs survival in patients with heart failure secondary to myocardial infarction.[24] Daily doses, bioavailability, and elimination half-life of commonly used diuretics are listed in **Table 1**.

6.32.3.2 β-Adrenoceptor Antagonists (β-Blockers)

The first β-adrenoceptor antagonist, dichloroisoproterenol, was described by I. Slater in 1957. The first clinically useful β-adrenoceptor antagonists, pronethalol and propranolol (**Table 2**), were discovered and developed by the British pharmacologist Sir James Black. Their usefulness in the treatment of hypertension was first demonstrated in patients by Prichard. Pronethalol was found to be carcinogenic in mice,[25] while propranolol was marketed and is still used in the clinic. The discovery of subtypes of β-adrenoceptors, β_1 and β_2, facilitated the development of additional antagonists with various affinities for the two receptor subtypes. Propranolol is nonselective; it has similar affinity for β_1- and β_2-adrenoceptors. Blockade of β_2-adrenoceptors is likely to cause bronchial constriction, so that propranolol should not be used in patients with bronchial asthma. Propranolol enters the central nervous system and can produce vivid dreams as one of its side effects. Another nonselective β-adrenoceptor antagonist, nadolol, does not enter the brain and may, therefore, be preferable for some patients. Most β-adrenoceptor antagonists slow heart rate due to their affinity for cardiac β_1-adrenoceptors. Pindolol has mixed agonist–antagonist activity at cardiac β_1-adrenoceptors and at therapeutic doses does not appreciably change heart rate. β-Adrenoceptor antagonists also differ in their membrane-stabilizing properties. Some have other pharmacological effects in addition to the blockade of β-adrenoceptors. Sotalol has antiarrhythmic activity and carvedilol has α- and β-adrenoceptor blocking properties, as well as a cardioprotective effect. Carvedilol is recommended for heart failure patients with or without hypertension.[26] Another β_1-adrenoceptor antagonist, nebivolol, has vasodilator, antiproliferative, and antioxidant properties.[27] Pharmacokinetic properties can also determine indications for β-adrenoceptor antagonists. Esmolol has the shortest half-life among marketed β-adrenoceptor antagonists; it has rapid onset and ultrashort duration of action, and is used intravenously during surgery to control hypertension and arrhythmias.

The mechanism of antihypertensive action of β-adrenoceptor antagonists has been a subject of controversy for many years. The reduction of plasma renin activity and, consequently, plasma Ang II levels, appears to be, at least in part,

Table 2 β-Adrenoceptor antagonists

Drug	*Structure*	*Usual daily dose (mg)*	*Selectivity for receptors*	*ISA*	*MSA*	*Oral bioavailability (%)*	$t_{1/2}$ *(h)*	*Other effects*
Atenolol	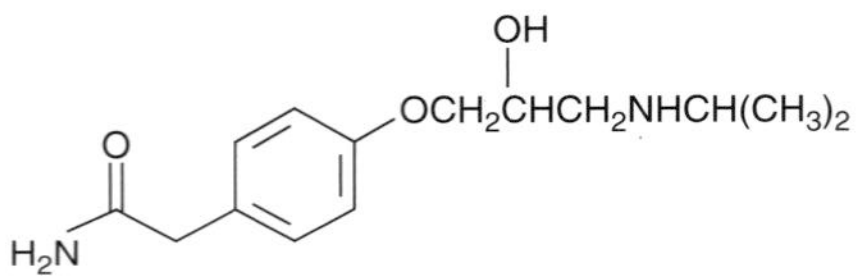 H_2N, O, OH, $OCH_2CHCH_2NHCH(CH_3)_2$	25–100	β_1	0	0	50–60	4–6	
Carvedilol	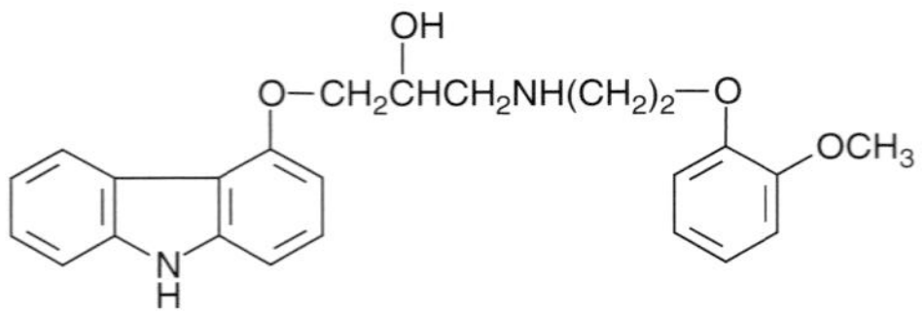 OH, $O-CH_2CHCH_2NH(CH_2)_2-O$, OCH_3, N H	3.125 q.d. to 25 b.i.d	0	0	+	20–25	4–7	α_1-Adrenoceptor antagonist Antioxidant Antiproliferative Cardioprotective Inhibits I_{Kr}, I_{to}, I_{Ca}
Esmolol	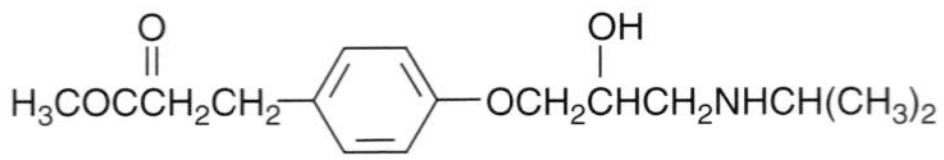 O, $H_3COCCH_2CH_2$, OH, $OCH_2CHCH_2NHCH(CH_3)_2$	$0.5\,mg\,kg^{-1}$ i.v prime, followed by $0.05\,mg\,kg^{-1}\,min^{-1}$ i.v. for 4 min	β_1	+	0	NA	2 min	Antiarrhythmic Antiischemic
Metoprolol	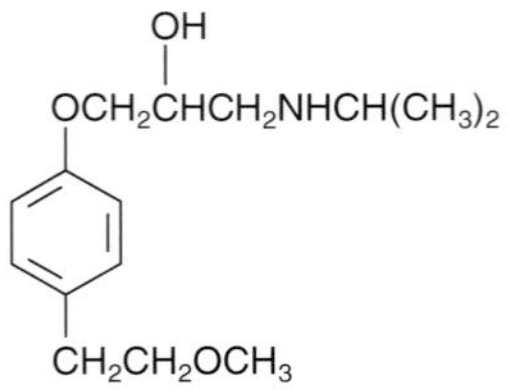 OH, $OCH_2CHCH_2NHCH(CH_3)_2$, $CH_2CH_2OCH_3$	150–300	β_1	0	0	20–60	3–4	Antiarrhythmic
Nadolol	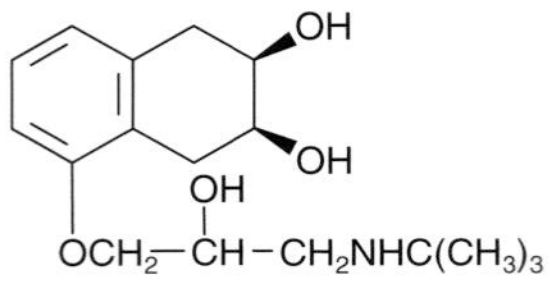 OH, OH, OH, $OCH_2-CH-CH_2NHC(CH_3)_3$	40–240	0	0	0	20–30	20–24	Renal vasodilator

Nebivolol	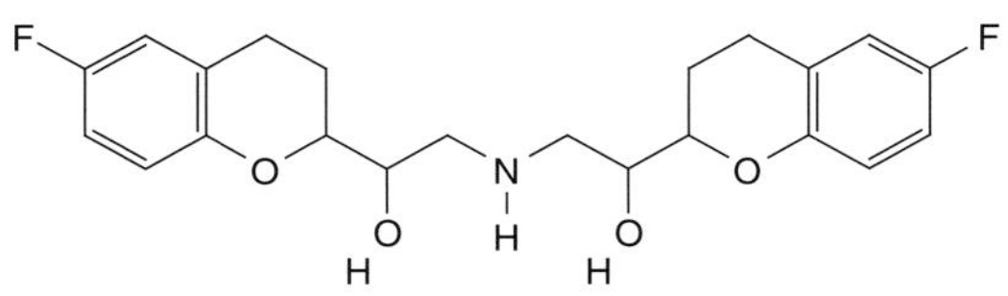	2.5–5.0	β_1	0	0	12	10–30	Cardioprotective Antioxidant Antiarrhythmic
Pindolol	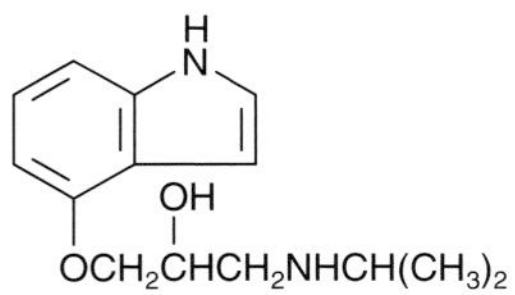 $OCH_2CH(OH)CH_2NHCH(CH_3)_2$	10–30	0	+	+	85–90	3–4	
Pronethalol	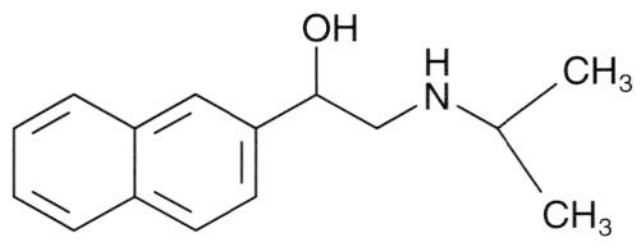	100–300 t.i.d.	0	0	+	?	?	Antiarrhythmic
Propranolol	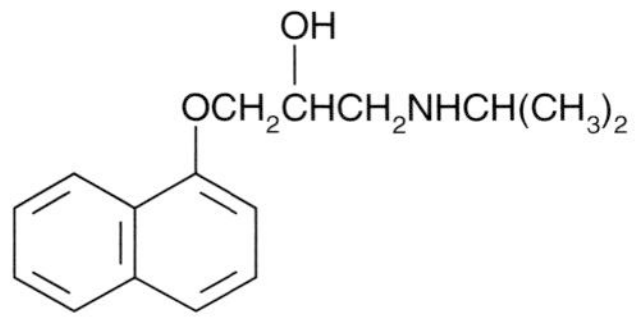 $OCH_2CH(OH)CH_2NHCH(CH_3)_2$	20–240	0	0	+	26	3.9	Local anesthetic
Sotalol	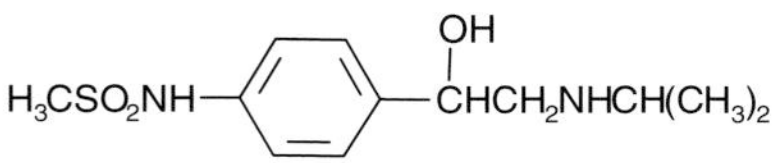 H_3CSO_2NH–C_6H_4–$CH(OH)CH_2NHCH(CH_3)_2$	320–640	0	0	0	100	10–15	Antiarrhythmic
Timolol	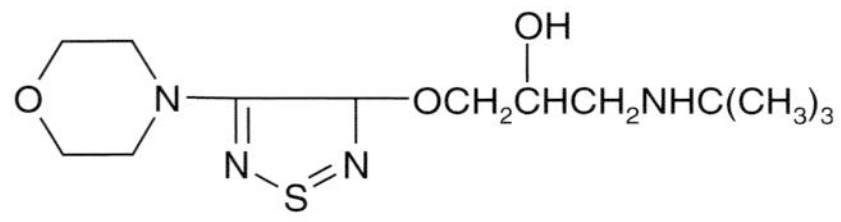 $OCH_2CH(OH)CH_2NHC(CH_3)_3$	30	0	0	+	55	2.5–5.5	Ocular hypotensive

ISA, intrinsic sympathomimetic activity; MSA, membrane-stabilizing action; +, present; 0, absent, NA, not applicable.

responsible for the antihypertensive efficacy of β-adrenoceptor antagonists and justifies their combined use with diuretics. **Table 2** gives the chemical structure and some pharmacological properties of the most commonly used β-adrenoceptor antagonists.

6.32.3.3 α_1-Adrenoceptor Antagonists

α-Adrenoceptors are subdivided into two major classes, α_1 and α_2, each of which has three subclasses. Antagonists of α_1-adrenoceptors, e.g., quinazolines (prazosin, doxazosin, and terazosin), are antihypertensive (**Figure 2**). They are selective for α_1-adrenoceptors, but not for any of the three subclasses (α_{1A}, α_{1B}, and α_{1D}) Unlike older α-adrenoceptor antagonists phenoxybenzamine and phentolamine, quinazolines do not increase heart rate. Prazosin was widely used in the treatment of hypertension for a quarter of a century. The onset of action of prazosin is rapid: the maximal antihypertensive effect is usually reached within 2 h of oral administration of the drug. An important advantage of prazosin and other quinazolines over thiazides or β-adrenoceptor antagonists is their favorable effect on blood lipids: they tend to decrease triglycerides and low-density lipoprotein, while increasing high-density lipoprotein cholesterol levels. Terazosin inhibits ex vivo platelet aggregation induced by epinephrine, collagen, or adenosine diphosphate.[28] A well-recognized side effect of quinazolines and particularly of prazosin is so-called 'first-dose phenomenon' – orthostatic hypotension with a transient loss of consciousness after the first dose of the drug.[29] This phenomenon can be avoided by a very gradual increase in the dose (starting with 0.5 mg of prazosin in the evening before retiring, maintaining it for 2 weeks, and then increasing to 1 and 2 weeks later to 2 mg).

α_1-Adrenoceptor antagonists are used primarily as second-line drugs, in combination with thiazides. Doxazosin differs from prazosin primarily in its pharmacokinetics; its half-life is 20 h. The onset of antihypertensive effect of doxazosin is more gradual than that of prazosin and it appears to be less likely to produce 'first-dose phenomenon.' Of concern is a recent report of increased heart failure risk in patients receiving doxazosin in comparison to those treated with chlorthalidone.[30] With the advance of ACEIs and Ang II antagonists, the use of quinazolines in the therapy of hypertension markedly declined.

Since α_1-adrenoceptors are present in the bladder musculature, α-adrenoceptor antagonists have been tested and found effective in improving urine flow in patients with benign prostatic hypertrophy (BPH) (*see* 6.24 Incontinence (Benign Prostatic Hyperplasia/Prostate Dysfunction)). Excellent results have been obtained in BPH with doxazosin, 4 mg day^{-1}, for periods up to 10 years.[31] Terazosin has also been reported to inhibit growth and to induce apoptosis of prostate cancer cells,[32] so that α_1-adrenoceptor antagonists are likely to have an additional therapeutic indication, not only in BPH, but also in the therapy of prostate cancer.

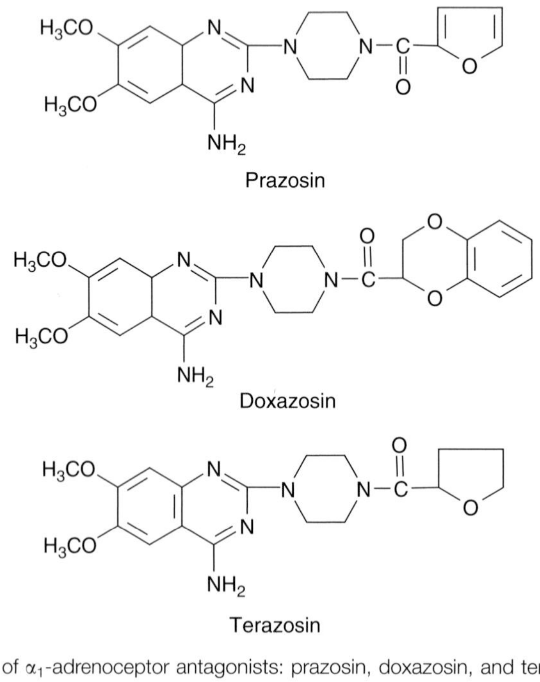

H3CO, H3CO, N, N, NH2, N, N–C, O, O

Prazosin

H3CO, H3CO, N, N, NH2, N, N–C, O, O, O

Doxazosin

H3CO, H3CO, N, N, NH2, N, N–C, O, O

Terazosin

Figure 2 Chemical structures of α_1-adrenoceptor antagonists: prazosin, doxazosin, and terazosin.

6.32.3.4 Ca^{2+} Channel Antagonists (Ca^{2+}Antagonists, Ca^{2+} Channel Blockers, CCAs)

Fleckenstein, late Professor of Physiology at the University of Freiburg, Germany, observed that drugs could mimic the effects of calcium withdrawal from the medium surrounding the cells. He termed such drugs calcium antagonists (**Figure 3** and **Tables 3** and **4**). The first compound reported by him to have such activity was verapamil, which is still used today in the treatment of hypertension. Subsequently, Fleckenstein found that nifedipine, discovered at Bayer AG in Germany,[33] is "the most potent and the most specific inhibitor of calcium entry into cells." In 1970 the first clinical studies with nifedipine were performed in patients with angina pectoris. The drug reduced the incidence of anginal attacks and was marketed for this purpose. Subsequently, nifedipine was found to reduce arterial pressure in patients with severe hypertension who did not previously respond to any other antihypertensive drugs available.

Sato and his associates at Tanabe Seiyaku Company in Japan discovered a chemically different CCA, diltiazem (**Figure 3**). All three first-generation CCAs, verapamil, nifedipine, and diltiazem, inhibit the same subunit (α_1) of L-type voltage-dependent calcium channels, but bind to different sites in the subunit. The commercial success of the first three compounds prompted many pharmaceutical companies to initiate their own research programs for CCAs. Most of the second-generation CCAs were, like nifedipine, dihydropyridines (DHPs) (**Tables 3** and **4**). These DHPs differed from nifedipine in potency, duration of action, lipophilicity, and relative specificity for vascular rather than cardiac L-type calcium channels. Amlodipine has the longest duration of action and was the most successful of the second-generation DHPs; it is still widely used in the treatment of hypertension. In the late 1990s once-a-day formulations of most marketed CCAs became available; they provided better compliance, improved safety, and

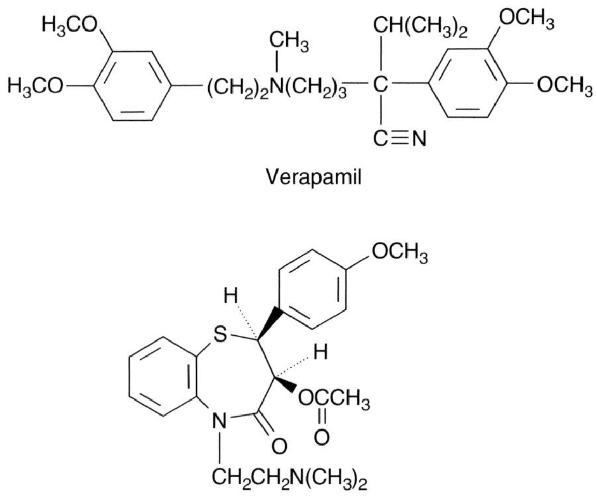

Verapamil

Diltiazem

Figure 3 Chemical structures of nondihydropyridine calcium-channel antagonists of first generation.

Table 3 Clinical doses, elimination half-life ($t_{1/2}$), and oral bioavailability of commonly used formulations of Ca^{2+}channel antagonists

Drug	*Clinical dose (mg)*	$t_{1/2}$ *(h)*	*Oral bioavailability (%)*
Amlodipine (Norvasc)	25–10 q.d.	31–37	52–88
Verapamil (Isoptin SR)	180 q.d.–240 b.i.d	2.8–7.4	20–35
Nifedipine (Adalat CC)	30–60 q.d.	7	84–89
Diltiazem (Cardizem SR)	120–360 q.d. or b.i.d.	6–9	40
Felodipine ER	2.5–10 q.d.	24	15
Nicardipine SR	60–120 q.d.	8	35
Nisoldipine (Sular CC)	20–40 q.d.	7–12	5–15

q.d., once a day; b.i.d., twice a day.

Table 4 Structure–activity relationship of some dihydropyridines

Drug	*R*[1]	*R*[2]	*R*[3]	*R*[4]	*R*[5]	*Advantages*
Nifedipine	CO_2CH_3	CO_2CH_3	H	NO_2	CH_3	First in class
Felodipine	CO_2CH_3	$CO_2C_2H_5$	Cl	Cl	CH_3	Vasoselective
Nitrendipine	CO_2CH_3	$CO_2C_2H_5$	NO_2	H	CH_3	Vasoselective Improves cognition?
Nisoldipine	CO_2CH_3	$CO_2CH_2CH(CH_3)_2$	NO_2	H	CH_3	Vasoselective
Nicardipine	CO_2CH_3	$CO_2(CH_2)_2{-}N(CH_3){-}CH_2Ph$	NO_2	H	CH_3	Water-soluble Neuroprotective?
Aranidipine	$CO_2CH_2COCH_3$	CO_2CH_3	H	NO_2	CH_3	Also blocks T channels Has active metabolite
Clevidipine	CO_2CH_3	$CO_2CH_2OCO(CH_2)_2CH_3$	Cl	Cl	CH_3	Ultrashort-acting Water-soluble
Lacidipine	$CO_2C_2H_5$	CO_2CH_5	H	$(CH)_2CO_2C(CH_3)_3$	CH_3	Vasoselective Antiathrosclerotic
Pranidipine	CO_2CH_3	$CO_2CH_2CH{=}CHC_6H_5$	NO_2	H	CH_3	Enhances NO Dilates veins Also blocks Na channels

Lercanidipine	CO_2CH_3	CH_3 $CO_2C(CH_3)_2CH_2N(CH_2)_2CH(Ph)_2$	NO_2	H	CH_3	Highly lipophilic Causes no reflex tachycardia
Efonidipine	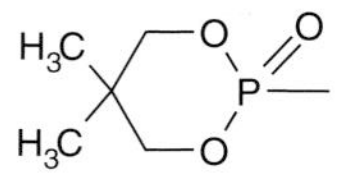 H_3C, H_3C, O, O, P, O	$CO_2(CH_2)_2NPhCH_2Ph$	NO_2	H	CH_3	Also blocks T-channels Diuretic Increases GFR
Benidipine	CO_2CH_3	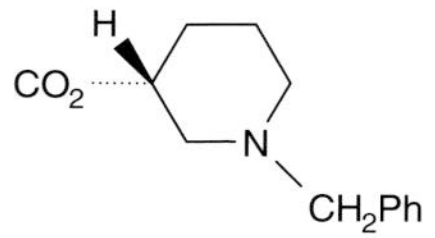 H, CO_2, N, CH_2Ph	NO_2	H	CH_3	Diuretic Increases GFR Antiapoptotic
Nimodipine	$CO_2CH(CH_3)_2$	$CO_2(CH_2)_2OCH_3$	NO_2	H	CH_3	Neuroprotective Reduces neurological deficits in SAH
Amlodipine	CO_2CH_3	$CO_2C_2H_5$	H	Cl	$CH_2O(CH_2)_2NH_2$	Increases GFR Slow onset and long duration of action

GFR, glomerular filtration rate; SAH, subarachnoid hemorrhage.

extended the patent life of older drugs. The safety of DHPs was of some concern, since Furberg *et al.* published a retrospective analysis of clinical studies with the rapid-release formulation of nifedipine and claimed that nifedipine may increase mortality of patients with heart disease.[34] The increased mortality, if true, may have been due to a rapid fall in arterial pressure, rather than a specific effect of nifedipine. Rapid fall in pressure should be avoided, particularly in patients with heart disease. In subsequent studies long-acting DHPs or extended-release formulations of first-generation compounds did not increase mortality in patients with diabetes or cardiovascular disease.[35] The World Health Association and International Society of Hypertension formed a subcommittee to evaluate the safety of CCAs. This subcommittee concluded that the available evidence does not prove any increased risk of cardiovascular events, cancer, or bleeding caused by CCAs.[36]

Many pharmacological properties of various CCAs are similar. There are, however, also substantial differences not only in their relative potency, but also in their pharmacokinetics, onset and duration of action, specificity for certain subunits of calcium channels, and effects unrelated to their action at Ca^{2+} channels. The clinical importance of differences in the properties of DHPs was recently emphasized in a review article by Meridith and Elliott.[37] Verapamil and diltiazem belong to different chemical classes and have direct effects on cardiac contractility and conduction. The extended-release formulations of DHPs differ from each other in their pharmacokinetic properties, but the elimination half-life of either of these formulations is shorter than that of amlodipine (**Table 3**). Not all pharmacological differences are, however, clinically relevant. By acute oral administration all DHPs have a high therapeutic index (ratio of LD_{50}:ED_{50}), so that differences in their relative potencies are not clinically important. Comparative clinical studies claim differences in efficacy or maximal obtainable therapeutic effect of a drug. These claims are misleading if the drugs are compared at only one commonly used dose, since this dose may represent an ED_{70} of one drug and ED_{40} of another. Higher efficacy can only be claimed if complete dose–response curves for both drugs are obtained and the maximal obtainable effect of one drug is significantly higher than that of the other. A claim for a lower incidence of side effects may also be misleading if the drugs are compared at nonequivalent doses.

Some of the third-generation CCAs developed in the 1990s were designed to incorporate additional pharmacological properties likely to provide beneficial effects, e.g., α-adrenoceptor blockade, NO release, diuresis, cardioprotection, lipid lowering, or antithrombotic activity. Some of these effects involve incorporation of an additional mechanism of action that is not related to the blockade of L-type Ca^{2+} channels. **Table 4** summarizes some structure–activity aspects of second- and third-generation DHPs. Vasoselectivity can apparently be achieved by replacing hydrogen in R_3 position with either Cl or NO_2 or by extending the side chain in R_4 position. The drugs are considered vasoselective if they inhibit Ca^{2+} entry into VSM cells at substantially lower concentrations than required for inhibition of Ca^{2+} entry into myocardial cells. Bulky substitution in the R_1 position may provide additional inhibitory effect at T-type channels, while inhibition of N-channels may be facilitated by the introduction of a longer side chain in R_2 position. Some of the additional pharmacological effects listed in **Table 4**, e.g., antiatherosclerotic effects, may represent a class effect, but have been described for only a few DHPs. Incorporation of additional activities is likely to change the therapeutic indications for individual drugs. Nimodipine is used in the therapy of subarachnoidal hemorrhage and benidipine can conceivably be used in the treatment of renal insufficiency.

Side effects of CCAs are predictable on the basis of their pharmacological properties. Headache, flushing, and peripheral edema are consequences of excessive vasodilatation. These effects are dose-dependent, usually transient, and seldom require discontinuation of the therapy. Ankle edema induced by CCAs is not due to sodium retention, but rather due to arteriolar vasodilatation without corresponding venous dilation. ACEIs alleviate this type of edema.

6.32.3.5 Angiotensin-Converting Enzyme Inhibitors

The discovery of ACEIs has been made possible by the finding of Ferreira *et al.* in Brazil that the venom of a Brazilian pit viper contains peptides that enhance the vasodilator effect of bradykinin by inhibiting its degradation. Erdös *et al.* found that the same enzyme that degrades bradykinin also converts Ang I to Ang II. Fereirra's peptides were expected not only to enhance bradykinin, but also to block the conversion of Ang I to Ang II. One of these, teprotide, was tested in animals and eventually humans, and found to inhibit ACE and to lower arterial pressure. Ondetti and Cushman at Squibb analyzed the structural requirements for the ACEI activity of teprotide, identified the active site of the enzyme, and designed a small molecule, a nonpeptide that inhibited ACE. Subsequent chemical optimization of this molecule led to the development of the first therapeutically useful ACEI, captopril (**Table 5**). The advantages of captopril and other ACEIs in the treatment of hypertension include their effectiveness in all types of hypertension, and the relative freedom from side effects. The major side effect of ACEIs is cough caused by bradykinin that accumulates due to inhibition of its degradation by the ACEIs. Another side effect is hyperkalemia; ACEIs should not be administered with K^+-retaining diuretics or with K^+ supplements.

Table 5 Structures, in vitro IC_{50}s, eilimination half-life ($t_{1/2}$), bioavailability, and clinical doses of commonly used ACEIs

Drugs	*Structure*	*In vitro ACE inhibition IC_{50} (nM)*	*$t_{1/2}$ (h)*	*Bioavailability (%)*	*Clinical doses (mg)*
Directly acting					
Captopril	H, CH_3, HS, O, N, H, CO_2H	7–13	2	60–70	6.25–150 b.i.d. or t.i.d.
Lisinopril	NH_2, OH, O, H, H, N, H, O, N, H, CO_2H	1.7	7	30	5–40 q.d. or b.i.d.
Prodrugs					
Benazepril	O, H, N, C, $COCH_2CH_3$, H, H, CH_2CH_2, N, O, CH_2CO_2H	2	22	37	10–40 q,d,
Fosinopril	H, N, CO_2H, H, O, $CH_2(CH_2)_2CH_2$, P, CH_2, C, O, O, C, $(CH_3)_2CH$, O, C, CH_2CH_3, H, O	2.2–20	12	32	10–60 q.d.
Moexipril	CH_3O, CH_3O, H, CO_2H, H, CH_3, H, $CO_2CH_2CH_3$, N, O, N, H	1	10	22	7.5–30 b.i.d.

continued

Table 5 Continued

Drugs	*Structure*	*In vitro ACE inhibition IC_{50} (nM)*	*$t_{1/2}$ (h)*	*Bioavailability (%)*	*Clinical doses (mg)*
Quinapril	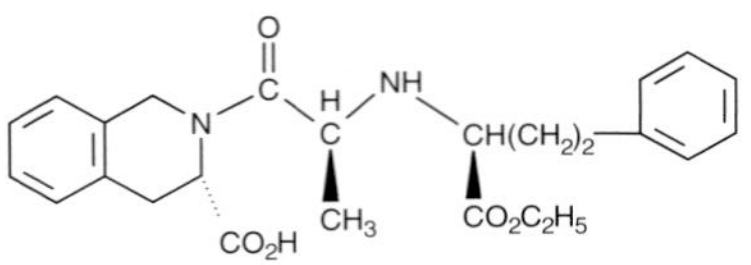 O, N, C, H, NH, C, $CH(CH_2)_2$, CO_2H, CH_3, $CO_2C_2H_5$	0.8–2.0	Biphasic 2 and 25	53–80	5–80 q.d.
Trandolapril	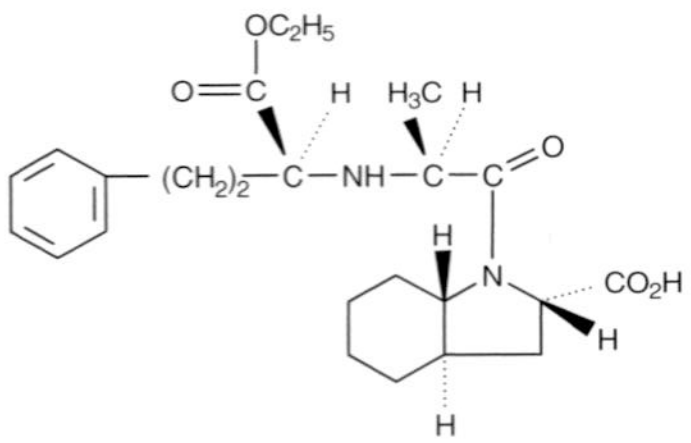 OC_2H_5, O=C, H, H_3C, H, $(CH_2)_2$–C–NH–C–C=O, H, N, CO_2H, H, H	0.9	0.7	40–60	1–4 q.d.
Enalapril	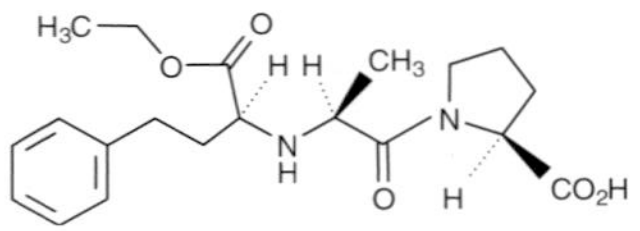 H_3C, O, O, H H, CH_3, N, N, H, O, H, CO_2H	2.6–9.0	11	40–60	5–40 q.d.
Perindopril	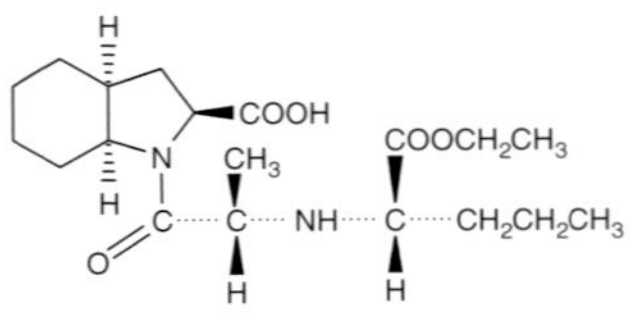 H, COOH, N, CH_3, $COOCH_2CH_3$, H, C, C, NH, C, $CH_2CH_2CH_3$, O, H, H	2.0	Biphasic 3–10 30–120	65–95	2–16 q.d.
Ramipril	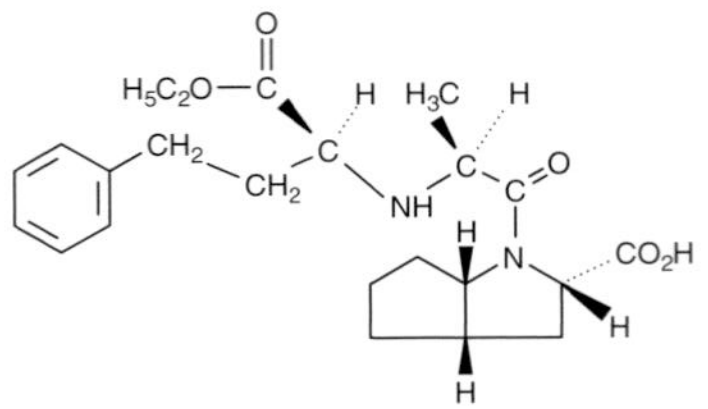 O, H_5C_2O–C, H, H_3C, H, CH_2, CH_2, C, NH, C, C=O, H, N, CO_2H, H, H	0.5–0.9	Triphasic 2–4 9–18 over 50	28–44	2.5–10 q.d.

$t_{1/2}$, elimination half-life.

Because of their vasodilator action, ACEIs reduce cardiac work and improve cardiac performance. They also protect heart, kidney, and possibly brain from hypoxic tissue damage. These effects are highly beneficial for heart failure patients. ACEIs are now indicated as adjunctive therapy in the management of heart failure in patients who are not responding adequately to diuretics and digitalis. Some ACEIs are also approved for the treatment of hemodynamically stable patients within 24 h of myocardial infarction to improve survival. The mechanism of the cardioprotective action of ACEIs is not well understood, but is considered to be the consequence of their major pharmacological action, reduction of formation of Ang II, which has also profibrotic and proapoptotic properties.

After the discovery of captopril several ACEIs were developed. With the exception of captopril and lisinopril, all are prodrug esters being hydrolyzed in the liver to the active metabolites that are identified by the suffix 'at,' e.g., the active metabolite of enalapril is enaprilat. Prodrugs, unlike their active metabolites, are well absorbed and are orally bioavailable. ACEIs differ in potency, tissue specificity, and pharmacokinetics. Their differences are mostly determined by their physicochemical properties, e.g., lipophilicity. Differences in their ability to cross cellular membranes determine tissue specificity, while duration of action often depends on the ability of an ACEI to dissociate from the enzyme. The ACE–ramipril complex dissociates much slower than the ACE–captopril complex. According to US Food and Drug Administration recommendations, antihypertensive drugs should have a trough-to-peak ratio of at least 50%. This means that, if they are to be administered once a day, they should retain at 24 h more than 50% of their activity at the time of their peak effect. At the recommended therapeutic doses enalapril, fosinopril, ramipril, and trandolapril have higher trough-to-peak ratios than 50% and higher than those of other ACEIs.[38]

The relative potency of ACEIs is first estimated in vitro by their affinity for the binding site of the enzyme in tissue (e.g., rabbit lungs) homogenates, expressed as an IC_{50} value in nanomolar. The IC_{50} value for one drug can vary substantially depending on the species and tissue used. Among the ACEIs listed in **Table 5**, ramipril has the greatest in vitro affinity for the enzyme. Its clinical doses are also low, but are higher than those of trandolapril, so that in vitro and in vivo relative potencies of ACEIs do not always correlate. ACEIs do not differ from each other in efficacy. Their maximal obtainable antihypertensive effects are identical. It is likely that any differences in efficacy claimed in the literature are based on a trial in which the drugs were compared at noncomparable dose levels.

The pharmacokinetic parameters account for the major differences among ACEIs. Drugs with longest elimination half-lives ($t_{1/2}$) are expected to have the longest duration of action. It is difficult, however, to establish such a correlation for ACEIs, in part due to biphasic or even triphasic declines in their blood levels, which are determined by differences in the rate of release of ACEIs from different organs and/or tissues.

All ACEIs lower blood pressure in renal hypertensive (2K1C) rats as well as in SHRs. Their relative potency in hypertensive rats is usually comparable to that in hypertensive patients. In normotensive animals ACEIs have little or no hypotensive activity. However, ACEIs can be assayed in normotensive animals on the basis of their ability to block pressor effects of Ang I, but not of Ang II.

6.32.3.6 Angiotensin II (Ang II) Antagonists or Blockers

The discovery of clinically useful nonpeptide antagonists of Ang II led to the recognition of the importance of this hormone in the control of circulation and to a new approach to the treatment of hypertension. Although at least four different types of Ang II receptors (AT_{1-4}) have been identified, pressor and other undesirable effects of Ang II are mediated by AT_1 receptors. Furukawa *et al.* from Takeda Chemical Industries initially described the antagonism of Ang II by *N*-benzylimidazole derivatives. Intensive structure–activity and optimization studies at DuPont Merck Pharmaceutical Company led to the discovery and development of the first marketed ARB, losartan (**Table 6**). Its discovery was followed by the development of a series of chemically related compounds, known as 'sartans.' Losartan and other 'sartans' block vasoconstrictor, aldosterone-releasing, and adrenergic-facilitating effects of Ang II. Losartan is converted in the liver to two carboxylic acid derivatives that are substantially more potent than the parent compound. Losartan and other ARBs lower arterial pressure in renal hypertensive (2K1C) and SHR. They also lower blood pressure in sodium-depleted normotensive animals. Unlike ACEIs, Ang II antagonists do not inhibit bradykinin breakdown and, therefore, do not produce cough. Also, selective AT_1 antagonists do not interfere with the activation of AT_2 by Ang II. This is likely to be beneficial, since AT_2 receptors are thought to oppose the actions of Ang II at AT_1 receptors. Another conceivable advantage of ARBs, compared to ACEIs, is the existence of non-ACE pathways for synthesis of Ang II, so that ACEIs cannot be expected to prevent the formation of Ang II completely.

Valsartan has no active metabolites, it is less potent, but its elimination half-life is longer than that of losartan. Irbesartan has better bioavailability and is more likely to enter the brain than either losartan or valsartan.[39]

Table 6 Chemical structures, elimination half-lives ($t_{1/2}$), and bioavailabilities of marketed angiotensin II antagonists (ARBs)

Drug	*Chemical structure*	*Elimination half-life, $t_{1/2}$ (h)*	*Bioavailability (%)*
Losartan	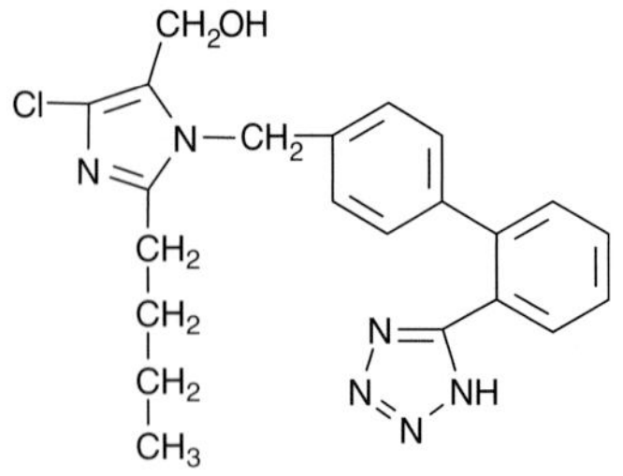	2–6	33
Valsartan	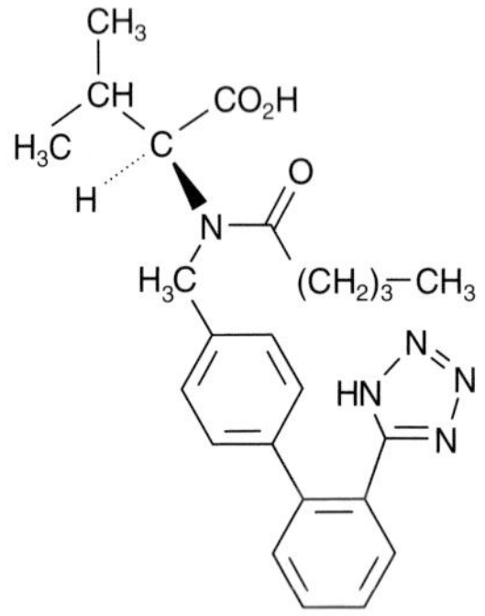	6	25
Irbesartan	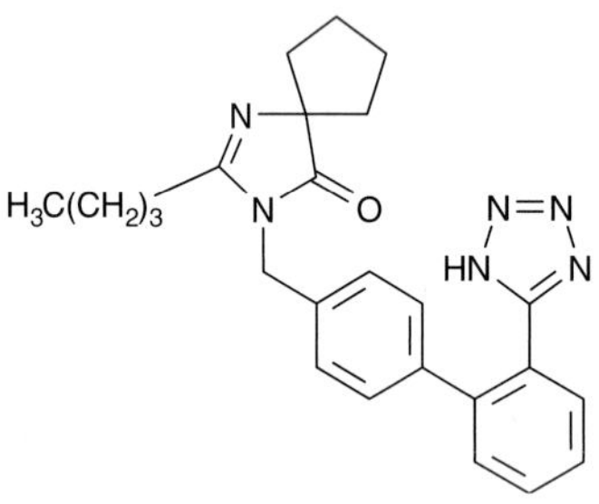	11–15	60–80
Telmisartan	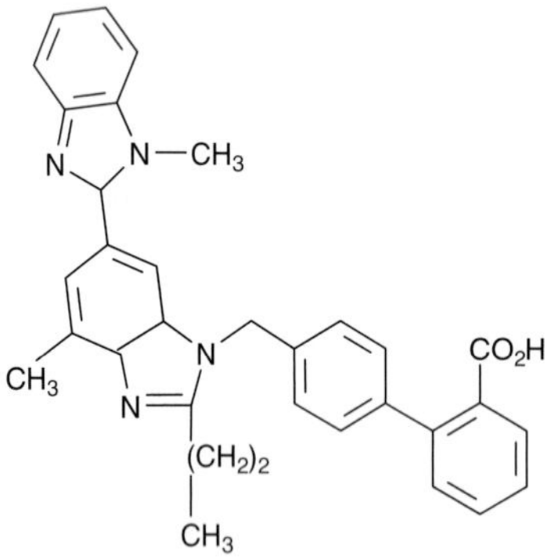	14–24	42–58
Eprosartan	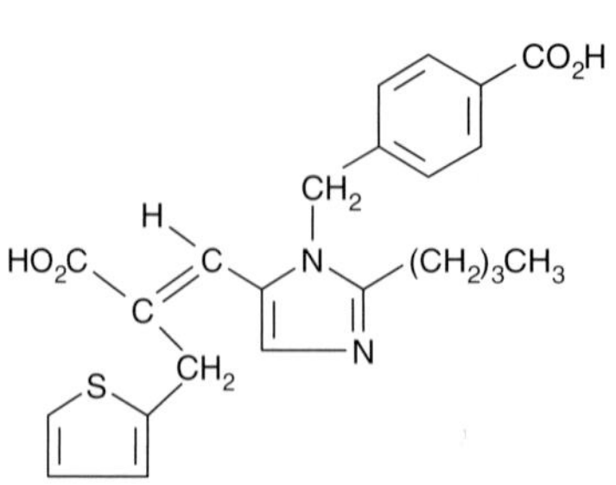	5–9	13

Table 6 Continued

Drug	*Chemical structure*	*Elimination half-life, $t_{1/2}$ (h)*	*Bioavailability (%)*
Candesartan cilexetil		4–9	15
Olmesartan		13–15	26

Telmisartan[40] offers distinct pharmacokinetic advantages over other ARBs. It is highly lipophilic, and has a high volume of distribution and long elimination half-life (24 h). Olmesartan has a shorter elimination half-life than telmisartan, but at the recommended clinical doses its duration of action is also 24 h. The relative potency of candesartan cilexetil, the prodrug of candesartan,[41] is higher, but its bioavailability and volume of distribution are lower than those of other ARBs (**Table 6**). In many clinical trials the side effects of ARBs were not distinguishable from those of placebo. However, like ACEIs they may induce hyperkalemia, particularly in patients with chronic renal failure and in those receiving potassium-sparing diuretics or potassium supplements. Also in volume-depleted patients ARBs may cause excessive hypotension and deterioration of renal function. Some clinical trials suggested differences in the antihypertensive efficacy of individual ARBs at the recommended clinical doses, but these differences are most likely due to the fact that ARBs have been compared at noncomparable doses. There is no convincing evidence that 'sartans' differ from each other in their maximal obtainable antihypertensive effects. ARBs appear to be as organ-protective as ACEIs in animal studies, but their primary indication is hypertension. Valsartan is approved for the treatment of heart failure, New York Heart Association grades II–IV, but only for patients who are intolerant to ACEIs. Losartan and irbesartan are also approved for the treatment of diabetic nephropathy.

6.32.3.7 Renin Inhibitors

The first and rate-limiting step in RAS, the conversion of angiotensinogen to Ang I, is catalyzed by renin. Several renin inhibitors have been synthesized, but low efficacy, poor oral bioavailability, high cost of synthesis, and the availability of highly successful ARBs prevented their development. One of the first clinically tested renin inhibitors was enalkiren (ABT-64662) (**Figure 4**). More recently developed, aliskiren[42] is currently in advanced clinical evaluation. At 75, 150, or 300 mg single doses, aliskiren effectively lowers ambulatory systolic pressure in hypertensive patients and is well tolerated. Its effects appear to be synergistic with valsartan. It still remains to be shown that aliskiren can affect vascular pathology and reduce morbidity and mortality in cardiovascular disease.

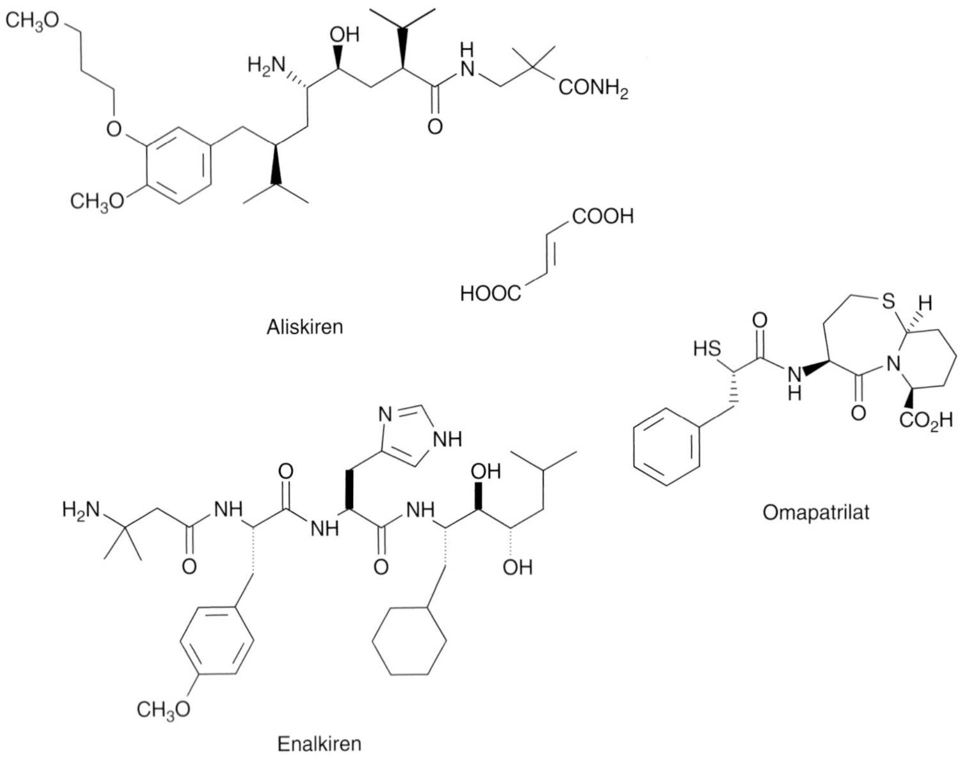

Aliskiren

Enalkiren

Omapatrilat

Figure 4 Chemical structures of aliskiren, enalkiren, and omapatrilat.

6.32.3.8 Vasopeptidase Inhibitors

Vasopeptidase inhibitors are drugs that are capable of simultaneously inhibiting ACE and the neutral endopeptidase (NEP, also known as EC 24.11). NEP is a cell surface metalloprotease that degrades various bioactive peptides, including big endothelin-1. Inhibition of NEP is anticipated to reduce the formation of endothelin-1 from big endothelin-1. None of the vasopeptidase inhibitors are currently marketed in the US. The first dual inhibitors, thiazepinones and oxazepinones, were synthesized at Bristol-Myers Squibb. In 1998 Trippodo *et al.* described the antihypertensive activity of the lead compound from this series, omapatrilat (see **Figure 4**).[43] Omapatrilat lowered arterial pressure in normal and high renin models of hypertension, prevented vascular remodeling, and provided long-term renoprotection to rats. In the initial clinical studies omapatrilat appeared to be highly promising, but a subsequent large clinical trial found that the rate of angioedema with omapatrilat was three times higher than that with enalapril.[44] Angioedema is considered to be a class effect and is likely to prevent further development of vasopeptidase inhibitors, unless drugs with substantially lower incidence of angioedema are discovered or the high-risk population for this side effect is better defined.

6.32.3.9 Combination Therapy

Very few hypertensive patients are currently being treated with one drug only, unless this drug is a fixed combination. The use of thiazide diuretics concurrently with β-adrenoceptor antagonists has been advocated ever since these drugs were discovered and is being currently recommended by JNC VII,[1] if a thiazide diuretic alone does not adequately lower arterial pressure. Almost all subsequently developed antihypertensive drugs have been used successfully with thiazide diuretics. There is currently substantially less opposition to the use of fixed combinations. The advantages of such combinations – better compliance, fewer side effects than with a higher dose of one of the components, and lower cost – appear to outweigh the disadvantages – less flexibility with dose adjustments and differences in the pharmacokinetic parameters of the two drugs. Fixed combinations of ACEIs, or ARBs with thiazide diuretics or of ACEIs with CCAs, are currently available and more combinations are likely to be marketed.

6.32.4 Unmet Clinical Needs

In the US only 59% of hypertensive patients are being treated. Forty-one percent of patients are either unaware of their hypertension, of the consequences of not treating it, or they are not insured and cannot afford the drugs. It is obvious that improvements in medical care delivery, patient education, and coverage of prescription drugs by Medicare and private insurance companies are urgently needed. There is also a need for new drugs for the treatment of drug-resistant, pulmonary, and portal hypertension.

6.32.4.1 Drug-Resistant Hypertension

Many cases of drug-resistant hypertension are in reality unrecognized failures of compliance. There are, however, patients who do not respond to any of the available drugs or their combinations. The most common cause of drug-resistant hypertension is atheroscleroric renal artery disease. These patients may respond to minoxidil, a highly effective vasodilator. Its side effects include tachycardia and hair growth, as well as sodium and water retention. Because of its side effects, it is rarely used clinically. The mechanism of its antihypertensive action involves activation of vascular K_{ATP} channels. There is a need for a safer K^+_{ATP} channel activator for the treatment of drug-resistant hypertension.

6.32.4.2 Pulmonary Arterial Hypertension

Pulmonary arterial hypertension is a debilitating and fatal lung disease. Patients are diagnosed as having pulmonary arterial hypertension if their mean pulmonary arterial pressure exceeds 25 mmHg at rest or 30 mmHg during exercise. Pulmonary arterial hypertension is subdivided into primary and secondary. Primary pulmonary hypertension (PPH) is idiopathic, while secondary pulmonary hypertension is caused by other diseases, e.g., pulmonary fibrosis, thromboembolism, or drugs. Some 6–12% of PPH patients have familial PPH, which is an autosomal dominant disease. In many families with PPH, mutations of the bone morphogenetic protein receptor type II (BMPR2) gene have been identified. Idiopathic pulmonary arterial hypertension has a particularly poor prognosis. Without lung transplantation the average survival time is 3 years after diagnosis.[45,46]

In the past the therapy of pulmonary arterial hypertension has been disappointing, although CCAs have been tried with some success. Anticoagulants were used routinely to prevent thromboembolic complications that were common in patients with pulmonary arterial hypertension and right heart failure. During the last decade new approaches to the therapy of pulmonary arterial hypertension have been developed. Prostacyclin (PGI_2, epoprostenol) was found to have a beneficial hemodynamic effect and to prolong life in patients with pulmonary arterial hypertension. It has to be administered by constant infusion through a central catheter. Treprostinil, a stable prostacyclin derivative, is currently used; it has to be administered by continuous subcutaneous infusion.

More recently, endothelin antagonists have been evaluated in the therapy of pulmonary arterial hypertension. There are at least two types of endothelin receptors: ET_A and ET_B. Activation of either receptor aggravates pulmonary arterial hypertension, but activation of ET_A receptors appears to be more detrimental. A nonselective endothelin antagonist, bosentan (**Figure 5**), is used in the therapy of pulmonary arterial hypertension. Its use is based on the evidence that endothelin is overexpressed in patients with pulmonary hypertension and that overproduction of endothelin may lead to pulmonary vascular remodeling. Bosentan is active orally and has been shown to produce short-term benefits in patients with pulmonary arterial hypertension: it improves exercise capacity and reduces pulmonary arterial pressure.[47] Its side effects include liver toxicity (elevation of hepatic aminotransferase in 7–14% of patients), teratogenicity, and drug interactions, since bosentan is a substrate for and inducer of CYP3A4 and CYP2C9 isoenzymes. Subsequent to the development of bosentan, other nonselective endothelin antagonists, tezosentan and enrasentan, have been developed. Two ET_A-selective endothelin antagonists, darusentan and sitaxsentan (**Figure 5**), are under clinical investigation. Other compounds reported to have beneficial effects in patients with pulmonary arterial hypertension include sildenafil, *N*-acetylcysteine, and adrenomedullin. Monocrotaline-induced pulmonary arterial hypertension in rats was found to be reversible by dichloroacetate. In spite of these new developments there is still a need for new drugs to treat pulmonary arterial hypertension, particularly if these drugs could reverse or arrest pulmonary fibrosis that may precede pulmonary arterial hypertension.

6.32.4.3 Portal Hypertension

Increased pressure in the portal vein is known as portal hypertension. Its pathogenesis is multifactorial, but liver cirrhosis is one of its most common causes. The increased vascular resistance in portal circulation due to liver fibrosis

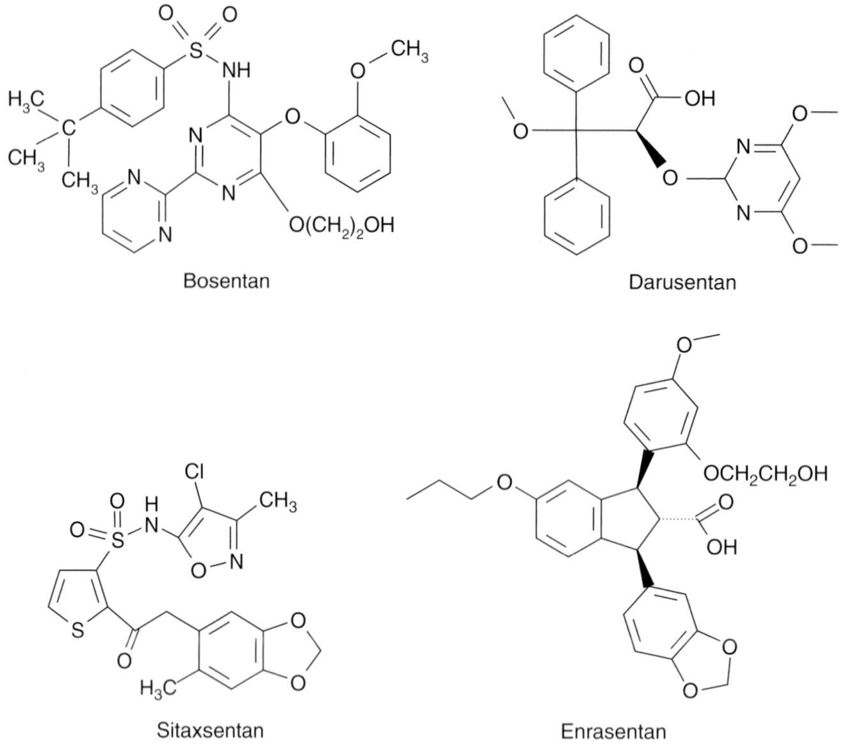

Bosentan

Darusentan

Sitaxsentan

Enrasentan

Figure 5 Chemical structures of endothelin antagonists.

associated with compression of portal and central venules is the likely initial event leading to portal hypertension. A serious consequence of portal hypertension is development of bleeding gastroesophageal varices. Nonselective β-adrenoceptor antagonists propranolol and carvedilol have successfully been used to prevent or reduce bleeding episodes.[48] They reduce portal venous pressure. Nitrates and more recently other NO donors are also effective. It has been proposed that in liver cirrhosis intrahepatic vascular tone increases due to an imbalance between increased sensitivity of hepatic blood vessels to endogenous vasoconstrictors and the reduced availability of NO. Pharmacotherapy of portal hypertension is in its infancy; it offers an opportunity for the development of new and more selective drugs.

6.32.4.4 Systolic Hypertension

Isolated systolic hypertension is common in patients over 65 years of age. Its prevalence in the elderly has been estimated to range from 34% to 65%. It is associated with the structural changes of arterial wall that include thickening of the intima, fibrosis of the media, and degeneration of the elastic fibers. These changes lead to stiffening of the arteries. Prevention of vascular changes should be an important goal in the therapy of isolated systolic hypertension. Since Ang II is one of the profibrotic factors, ARBs are likely to protect arteries from fibrotic degeneration. ARBs, ACEIs, CCAs as well as combinations of ACEIs and CCAs have been used in the therapy of isolated systolic hypertension. However, all these drugs reduce diastolic blood pressure as well. A large clinical trial designed to determine which drugs are more likely to reduce cardiovascular mortality and morbidity safely in patients with isolated systolic hypertension will be completed in 2008.[49]

6.32.5 Future Prospects

ARBs represent nearly optimal symptomatic therapy of hypertension. They are effective in the great majority of patients and have minimal side effects. They increase Ang II levels in blood, but, since AT_1 receptors are blocked, accumulation of Ang II appears not to lead to any harmful effects. Aliskiren or other renin inhibitors could conceivably be as effective as ARBs without increasing blood levels of Ang II. ACEIs are widely used. Their disadvantages include bradykinin-induced cough and inability to block Ang II synthesis completely. There are currently no other competitors for ARBs. It is anticipated that drug combinations will be more extensively used. Future research efforts are likely to be directed toward prevention of cardiovascular disease by mechanisms other than lowering of arterial pressure as well as to curative rather than symptomatic treatment of hypertension. It is conceivable that vascular changes can be prevented or reduced by antagonists of various cytokines better than by ACEIs or ARBs, leading to a greater decrease in morbidity and mortality from cardiovascular diseases.

Since genetic factors play an important role in the pathogenesis of essential hypertension, gene therapy of hypertension will be attempted in humans. Experimental studies in animals indicate its feasibility.[50] Genetic approaches to the treatment of hypertension do not necessarily require identification of the gene mutation responsible for hypertension in an individual or family. Sustained reduction of arterial pressure has been achieved in SHRs by retroviral delivery of ACE antisense,[51] and AT_1 receptor antisense cDNA prevented the development of hypertension in SHRs. Antisense oligonucleotides technology is considered a highly promising strategy to inhibit specific gene expression and consequently the progress of hypertension.[52] Modifications of the current gene delivery systems and use of efficacious therapeutic genes may lead to successful gene therapy of cardiovascular diseases. However, gene therapy of hypertension has many obstacles to overcome and may not become a reality in the near future.

References

1. Chobanian, A. V.; Bakris, G. L.; Black, H. R.; Cushman, W. C.; Green, L. A.; Izzo, J. L., Jr.; Jones, D. W.; Materson, B. J.; Oparil, S.; Wright, J. T., Jr. et al. *J. Hypertens.* **2003**, *42*, 1206–1252.
2. Lloyd-Jones, D. M.; Evans, J. C.; Larson, M. G.; Levy, D. *Hypertension* **2002**, *40*, 640–646.
3. Lloyd-Jones, D. M.; Evans, J. C.; Levy, D. *JAMA* **2005**, *294*, 466–472.
4. Valderrabano, F.; Gomez-Campdera, F.; Jones, E. H. *Kidney Int.* **1998**, *68*, S60–S66.
5. Lloyd, A.; Schmieder, C.; Marchant, N. *Pharmacoeconomics* **2003**, *21*, 33–41.
6. Banegas Banegas, J. R. *An. Real Acad. Nac. Med.* **2002**, *119*, 143–152.
7. Takahashi, N.; Smithies, O. *Trends Genet.* **2004**, *20*, 136–145.
8. Timberlake, D. S.; O'Connor, D. T.; Parmer, R. *Curr. Opin. Nephrol. Hypertens.* **2001**, *10*, 71–79.
9. Saeed, M. M.; Saboohi, K.; Osman, A. S.; Bokhari, A. M.; Frossard, P. M. *J. Hum. Hypertens.* **2003**, *17*, 719–723.
10. Baudin, B. *Pharmacogenomics* **2002**, *3*, 65–73.
11. Hannila-Handelberg, T.; Kontula, K.; Tikkanen, I.; Tikkanen, T.; Fyhrquist, F.; Helin, K.; Fodstad, H.; Piippo, K.; Miettinen, H. E.; Virtamo, J. et al. *BMC Med. Genet.* **2005**, *6*, 4.
12. Ruppert, V.; Maisch, B. *Herz* **2003**, *28*, 655–662.
13. Nesbitt, S. D. *Curr. Cardiol. Rep.* **2004**, *6*, 416–420.
14. Wiesel, P.; Mazzolai, L.; Nussberger, J.; Pedrazzini, T. *Hypertension* **1997**, *29*, 1025–1030.
15. Cervenka, L.; Horácek, V.; Vanecková, I.; Hubácek, J. A.; Oliverio, M. I.; Coffman, T. M.; Navar, L. G. *Hypertension* **2002**, *40*, 735–741.
16. van den Meiracker, A. H. *J. Hypertens.* **2002**, *20*, 587–589.
17. Iglarz, M.; Touyz, R. M.; Viel, E. C.; Paradis, P.; Amiri, F.; Diep, Q. N.; Schiffrin, E. L. *Hypertension* **2003**, *42*, 737–743.
18. Rossi, G. P.; Colonna, S.; Belloni, A. S.; Savoia, C.; Albertin, G.; Nussdorfer, G. G.; Hagiwara, H.; Rubattu, S.; Volpe, M. *J. Hypertens.* **2003**, *21*, 105–113.
19. Manger, W. M.; Simchon, S.; Stier, C. T., Jr.; Loscalzo, J.; Jan, K. M.; Jan, R.; Haddy, F. J. *Hypertension* **2003**, *21*, 2305–2313.
20. Lighthall, G. K.; Hamilton, B. P.; Hamlyn, J. M. *J. Hypertens.* **2004**, *22*, 1487–1494.
21. Ferrari, P.; Torielli, L.; Ferrandi, M.; Padoani, G.; Duzzi, L.; Florio, M.; Conti, F.; Melloni, P.; Vesci, L.; Corsico, N. et al. *J. Pharmacol. Exp. Ther.* **1998**, *285*, 83–94.
22. Dominiczak, A. F.; Negrin, D. C.; Clark, J. S.; Brosnan, M. J.; McBride, M. W.; Alexander, M. Y. *Hypertension* **2000**, *35*, 164.
23. Stier, C. T., Jr. *Cardiovasc. Drug Rev.* **2003**, *21*, 169–184.
24. Croom, K. F.; Perry, C. M. *Am. J. Cardiovasc. Drugs* **2005**, *5*, 51–69.
25. Paget, C. F. *Br. Med. J.* **1963**, *2*, 1266–1271.
26. Cheng, J.; Kamiya, K.; Kodama, I. *Cardiovasc. Drug Rev.* **2001**, *19*, 152–171.
27. Kuroedov, A.; Cosentino, F.; Lüscher, T. F. *Cardiovasc. Drug Rev.* **2004**, *22*, 155–168.
28. Hernández Hernández, R.; Angeli-Greaves, M.; Carvajal, A. R.; Guerrero Pajuelo, J.; Armas Padilla, M. C.; Armas-Hernández, M. J. *Am. J. Hypertens.* **1996**, *9*, 437–444.
29. Meredith, P. A. *Cardiology* **2001**, *96*, 1–19.
30. Barzilay, J. I.; Davis, B. R.; Bettencourt, J.; Margolis, K. L.; Goff, D. C., Jr.; Black, H.; Habib, G.; Ellsworth, A.; Force, R. W.; Wiegmann, T. et al. *J. Clin. Hypertens.* **2004**, *6*, 116–125.
31. Dutkiewicz, S. *Int. Urol. Nephrol.* **2004**, *36*, 169–173.
32. Xu, K.; Wang, X.; Ling, P. M.; Tsao, S. W.; Wong, Y. C. *Oncol. Rep.* **2003**, *10*, 1555–1560.
33. Vater, W.; Kroneberg, G.; Hoffmeister, F.; Kaller, H.; Meng, K.; Oberdorf, A.; Puls, W.; Schloβmann, K.; Stolpel, K. *Arzneimittelforschung* **1972**, *22*, 1–14.
34. Furberg, C. D.; Psaty, B. M.; Meyer, J. V. *Circulation* **1995**, *92*, 1326–1331.
35. Epstein, M. Safety of Calcium Antagonists as Antihypertensive Agents: An Update. In *Calcium Antagonists in Clinical Medicine*, 3rd ed; Epstein, M., Ed.; Hanley & Belfus: Philadelphia, PA, 2002, pp 807–832.
36. Ad Hoc Subcommittee of the Liaison Committee of the World Health Organization and the International Society of Hypertension. *J. Hypertens.* **1997**, *15*, 105–115.
37. Meredith, P. A.; Elliott, H. L. *J. Hypertens.* **2004**, *22*, 1641–1648.
38. Piepho, R. W. *Am. J. Health-System Pharmacy* **2000**, *57*, S3–S7.
39. Powell, J. R.; Reeves, R. A.; Marino, M. R.; Cazaubon, C.; Nusato, D. *Cardiovasc. Drug Rev.* **1998**, *16*, 169–194.

40. Wienen, W.; Entzeroth, M.; van Meel, J. C. A.; Stangier, J.; Busch, U.; Schmid, J.; Lehmann, H.; Matzek, K.; Kempthorne-Rawson, G. V.; Hauel, N. H. *Cardiovasc. Drug Rev.* **2000**, *18*, 127–156.
41. Gleiter, C. H.; Jägle, C.; Gresser, U.; Mörike, K. *Cardiovasc. Drug Rev.* **2004**, *22*, 263–284.
42. Stanton, A.; Jensen, C.; Nussberger, J.; O'Brien, E. *Hypertension* **2003**, *42*, 1137–1143.
43. Trippodo, N. C.; Robl, J. A.; Asaad, M. M.; Fox, M.; Panchal, B. C.; Schaeffer, T. R. *Am. J. Hypertens.* **1998**, *11*, 363–372.
44. Zanchi, A.; Maillard, M.; Burnier, M. *Curr. Hypertens. Rep.* **2003**, *5*, 346–352.
45. Lilienfeld, D. E.; Rubin, L. J. *Chest* **2000**, *117*, 796–800.
46. Rashid, A.; Lahrman, S.; Romano, P.; Frishman, W.; Dobkin, J.; Reichel, J. *Heart Dis.* **2000**, *2*, 422–430.
47. Cohen, H.; Chahine, C.; Hui, A.; Mukherji, R. *Am. J. Health-System Pharmacy* **2004**, *61*, 1107–1119.
48. Lin, H. C.; Yang, Y. Y.; Hou, M. C.; Huang, Y. T.; Lee, F. Y.; Lee, S. D. *Am. J. Gastroenterol.* **2004**, *99*, 1953–1958.
49. Jamerson, K. A.; Bakris, G. L.; Wun, C. C.; Dahlof, B.; Lefkowitz, M.; Manfreda, S.; Pitt, B.; Velazquez, E. J.; Weber, M. A. *Am. J. Hypertens.* **2004**, *17*, 793–801.
50. McKay, M. J.; Gaballa, M. A. *Cardiovasc. Drug Rev.* **2001**, *19*, 245–262.
51. Pachori, A. S.; Huentelman, M. J.; Francis, S. C.; Gelband, C. H.; Katovich, M. J.; Raizada, M. K. *Hypertension* **2001**, *37*, 357–364.
52. Tomita, N.; Morishita, R. *Curr. Pharm. Design* **2004**, *10*, 797–803.

Biography

Alexander Scriabine received his MS degree in pharmacology from Cornell University and an MD degree from the University of Mainz, in Germany. Dr Scriabine spent 36 years of his professional life in that pharmaceutical industry, rising from a bench pharmacologist to Research Manager at Pfizer and from Research Fellow to Executive Director at Merck, Sharp, and Dohme. For 14 years, Dr Scriabine was employed by Miles (now Bayer Corporation) in West Haven, CT. Dr Scriabine was also an Assistant Professor and subsequently an Associate Professor of Pharmacology at the University of Pennsylvania Medical School. He is currently the editor of *Cardiovascular Drug Reviews* and *CNS Drug Reviews*, a consultant to industry and lecturer at Yale University School of Medicine. He has published over 150 research articles, review articles, and edited 15 books. He has contributed chapters to many books, including *Cardiovascular Pharmacology*, *Pharmacology of Antihypertensive Drugs*, *Calcium Antagonists in Clinical Medicine*, *Pharmaceutical Innovation*, and *Neuroprotection in CNS Diseases*. He is best known for his contributions to the pharmacology of antihypertensive drugs, including α-, β-adrenoceptor and calcium channel antagonists and diuretics as well as neuroprotective drugs. He also participated in the discovery and the development of prazosin, polythiazide, timolol, nimodipine, nitrendipine, and many other drugs. He resides in Guilford, CT, USA.

Comprehensive Medicinal Chemistry II
ISBN (set): 0-08-044513-6

ISBN (Volume 6) 0-08-044519-5; pp. 705–728

6.33 Antiarrhythmics

M J A Walker, University of British Columbia, Vancouver, BC, Canada
P P S So, University of Toronto, Toronto, ON, Canada

6.33.1 Disease States

6.33.1.1 Types of Arrhythmias: Characteristics, Prevalence, Demographics, and Symptoms

Arrhythmias of the heart are any disorder of the orderly rhythmic beating initiated by the sinoatrial (SA) node (the normal pacemakers for the heart) such that the rate and/or rhythm is disturbed. Such disturbances of rate and rhythm occur as a result of electrophysiological perturbations of cardiac tissue that result in inappropriate changes in one, or more, of the inherent cardiac electrophysiological properties of automaticity, excitability, conduction, and refractoriness.

Arrhythmias are diagnosed primarily on the basis of electrocardiogram (ECG) findings although an initial diagnosis can be made on the basis of symptoms and heart rate. The generally accepted methods for classifying arrhythmias are based upon the anatomical site responsible for arrhythmia; the heart rate; the shape and nature of the ECG; as well as on other evidence.[1]

Simplistically, arrhythmias always have an inappropriate heart rate, and/or disorders of conduction. Increased heart rates are known as tachycardias while decreased heart rates are bradycardias. Conduction disturbances are either partial, or complete, and are often confined to particular cardiac anatomical sites. The morbidity and mortality associated with arrhythmias vary from the benign to the highly malignant, depending upon the type of arrhythmia. Some arrhythmias are essentially normal such as sinus arrhythmia where the heart rate increases and decreases with breathing. It is commonly seen in fit, healthy people. Others are abnormal, but not of any consequence, as with infrequent premature beats in atria or ventricles in mature adults. On the other hand, ventricular fibrillation is fatal (unless reverted) while atrial fibrillation carries a significant morbidity and mortality risk, especially in the elderly. The most common arrhythmias, based on their anatomical occurrence, are listed below.

6.33.1.1.1 Nodal arrhythmias

There are two anatomical and functional pacemaker nodes that control normal heart beating. These pacemakers can function independently of nerves, but their rates are modified by the autonomic nervous system in that they are accelerated by sympathetic nerve activity and slowed by parasympathetic nerves. The two nodes comprise the SA node, which normally dominates, and the atrioventricular (AV) node, which regulates impulse flow between atria and ventricles. This node assumes the role of the SA node if the latter fails.

Disturbances in the nodes are responsible for the following conditions.

6.33.1.1.1.1 Sick sinus syndrome

Sick sinus syndrome is an arrhythmia in which the rate in the SA node is erratic, or nonexistent, and cannot be relied upon for regular beating.

6.33.1.1.1.2 Sinoatrial nodal tachycardia and bradycardia

In SA nodal tachycardia and bradycardia there is an inappropriately fast (tachycardia) or slow (bradycardia) rate originating in the SA node. These rate disturbances are usually due to inappropriate activity in cardiac nerves (sympathetic for tachycardia, and parasympathetic for bradycardia). Tachycardia can also be due to circulating hormones such as epinephrine and thyroid hormone.

6.33.1.1.1.3 Atrioventricular nodal disturbances

Failure of AV node conduction can result in partial or complete AV block. Such conduction failure results in a ventricular rate that is slower than the atrial rate.

6.33.1.1.1.4 Atrioventricular nodal or paroxysmal supraventricular nodal tachycardias

These are tachycardias that occur with certain AV nodal dysfunctions. They give rise to paroxysmal attacks that may be maintained (for hours) or, in many cases, last only for a few minutes. The term paroxysmal is used to indicate an abrupt onset and abrupt spontaneous termination.

6.33.1.1.2 Atrial arrhythmias

The atria are particularly subject to arrhythmias when age and disease have caused structural damage. Developments in atrial defibrillation have been recently reviewed.[2]

6.33.1.1.2.1 Premature atrial contractions

Premature atrial contractions (PACs) are extra atrial beats that appear occasionally between normal SA nodal beats. They can consist of only an occasional extra beat or many such extra beats. PACs are normally not very troublesome to the patient, but they can indicate the presence of some degree of atrial pathology, and be a harbinger of the more serious atrial tachycardia or atrial fibrillation.

6.33.1.1.2.2 Atrial tachycardia

Atrial tachycardia (AT), also sometimes called atrial flutter, is an arrhythmia in which the atria beat very rapidly at 300 beats min^{-1} or more. Both the atrial and ventricular rate in atrial tachycardia is regular, and the atrial component of the ECG often has a characteristic sawtooth pattern in the intervals between QRS complexes (**Figure 1c**). This arrhythmia can be chronic or transient. The high beating rate of the atria cannot be conducted on a 1:1 basis through the AV node, and so the ventricles fail to follow each atrial beat. As a result, there is always some degree of AV block with the ratio of atrial to ventricular beats of 2 up to 4:1. AT is compatible with life and may produce few symptoms, although there is an associated increase in morbidity and mortality.[2]

6.33.1.1.2.3 Atrial fibrillation

Atrial fibrillation can be initiated when the recurring wave of excitation that constitutes atrial tachycardia breaks into multiple wavelets and gives rise to highly disorganized atrial fibrillation. The irregular and incoherent waves of contractions in atrial fibrillation that pass over the surface of the atria have been described as having the appearance of a 'can of worms.' The limited ability of the AV node to successfully transmit all atrial impulses through to the ventricle means that only the occasional wavelet from the fibrillating atria successfully passes through the AV node to activate the ventricles. The result is a fast ventricular rate that has no discernable regularity. The maximum rate at which ventricles can beat is limited to less than 200 beats min^{-1}. As a result, the ventricles can still fill, eject blood, and maintain cardiac output, but only to a limited extent. However, atrial fibrillation sometimes results in an inadequate cardiac output and this produces symptoms. The characteristics of atrial fibrillation on an ECG include a noisy, spiky baseline and the absence of atrial P waves (**Figure 1d**).

6.33.1.1.3 **Ventricular arrhythmias**

In an analogous manner to atria, ventricular arrhythmias include premature contractions, ventricular tachycardia (VT), and ventricular fibrillation (VF).

6.33.1.1.3.1 Ventricular tachycardia

Ventricular tachycardia occurs when the ventricles are not under the control of the SA node or the atria, but are beating rhythmically at a fast and inappropriate rate. If the AV node is removed by surgery or pathology, the ventricles will generally beat, but only at a much slower and dangerous rate (for example 30 beats min^{-1}) since it is not always sufficient to maintain an adequate cardiac output. However, in most VT the ventricular rate is much faster than the SA node rate. Providing the ventricular rate is not too high (an upper limit of around 170 beats min^{-1}), the ventricles will be able to fill and maintain an adequate cardiac output. At higher rates cardiac output, and hence blood pressure, can fall to catastrophically low levels. Furthermore, ventricular tachycardia can degenerate into ventricular fibrillation. This latter arrhythmia is fatal unless reverted within a few minutes. The ECG in ventricular tachycardia shows uniform repetitive cycles often at a constant wavelength (**Figure 1e**) and the usually narrow QRS complex of the ECG generally becomes much broader. This is because the QRS complex depends on the origin and direction of the wave of ventricular muscle depolarization sweeping repetitively across the ventricles. Usually the blood pressure during ventricular tachycardia is low, reflecting a low cardiac output.

6.33.1.1.3.2 *Torsades de pointes*

Torsades de pointes is a special form of VT. It most often occurs in the presence of a prolonged QT interval on the ECG, a slow heart rate, and/or disturbances in blood electrolyte balance. This can be an iatrogenic arrhythmia and due to drugs, or may be genetic in origin. It is a special form of VT in that it shows a cyclic pattern in which the wave heights of the QRS wave in the ECG vary in a progressive and repeating manner to produce an undulating pattern (**Figure 1f**), hence the French name, *torsades de pointes*. The best explanation for this pattern is a wave of depolarization (a rotor) moving around the heart. The arrhythmia is often self-terminating, but during its occurrence, syncope (fainting) can occur. Torsades can degenerate into ventricular fibrillation, and can be induced by drug therapy.

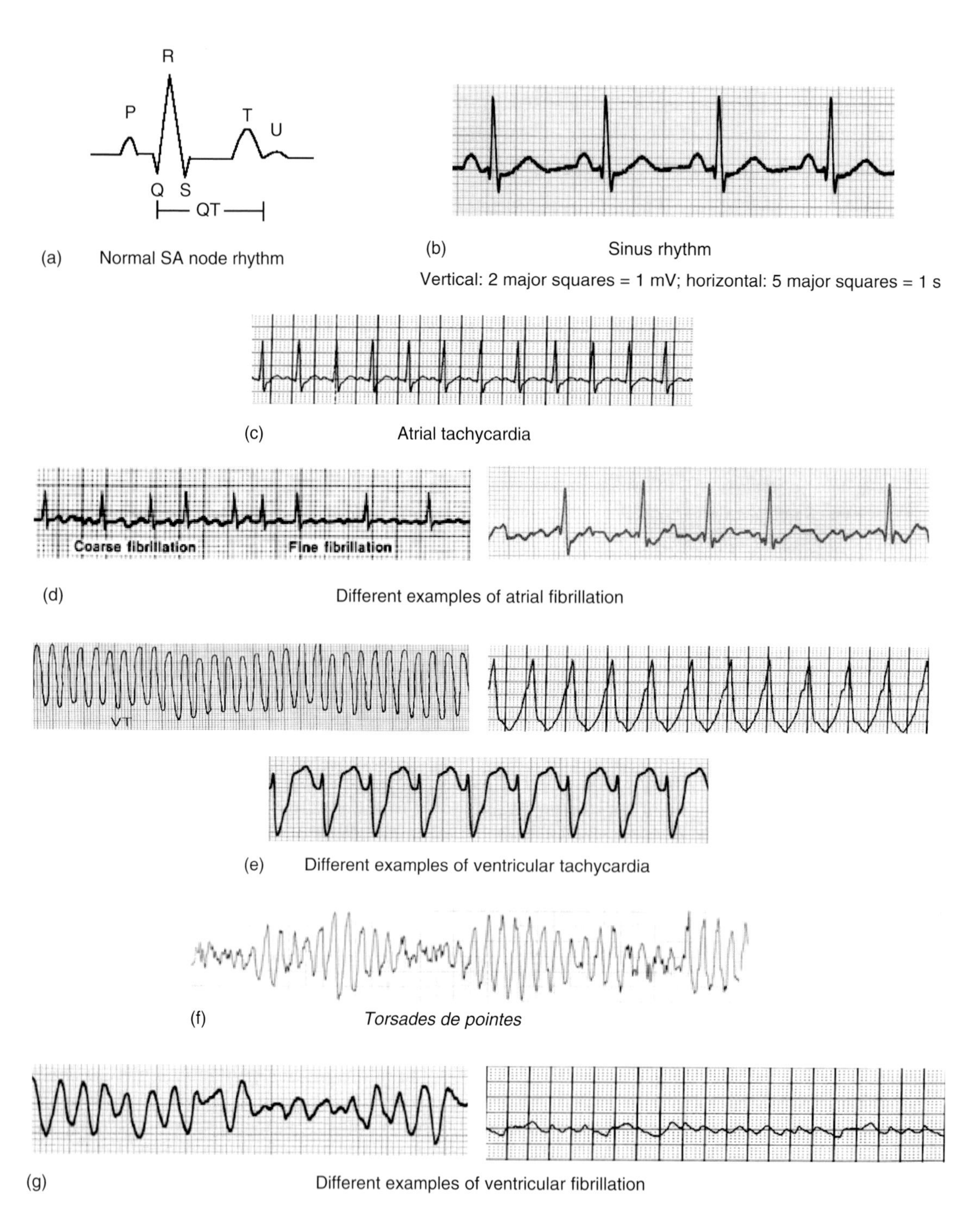

Figure 1 Typical ECG tracings for various atrial and ventricular arrhythmias. These tracings can be compared with a normal ECG shown at the top of the figure as (a) a stylized representation to show ECG intervals and as (b) an actual trace. The arrhythmias are: (c) atrial tachycardia, (d) atrial fibrillation, (e) ventricular tachycardia, (f) *Torsades de pointes*, and (g) ventricular fibrillation. In the traces, the vertical represents voltage and the horizontal, time.

6.33.1.1.3.3 Ventricular fibrillation

Ventricular fibrillation in humans is rapidly fatal unless it is reverted within a few minutes by DC cardioversion. Unconsciousness occurs within seconds of onset as a result of an instantaneous loss of cardiac output, and a resulting precipitous fall in blood pressure. The classic characteristic ECG pattern is of a noisy baseline lacking any discernible pattern (**Figure 1g**). However, since VT and torsades can degenerate into VF, intermediate ECG patterns between the three are seen. VF is the single largest cause of cardiac sudden death – a condition responsible for about 10% of all mortality.

Although not all arrhythmias result in significant morbidity and mortality, atrial tachyarrhythmias sometimes give rise to stroke and congestive heart failure (CHF) whereas ventricular tachyarrhythmias are the single biggest cause of death in many countries.

6.33.2 Disease Basis

The causes of arrhythmias can be considered at various levels of pathology:

1. At the underlying level of nonelectrophysiological causes such as cardiac pathologies that include: myocardial regional ischemia, myocardial scarring, excessive tension on myocardial chamber walls, inflammation, and/or excessive autonomic nervous system activity. These cause disturbances in ion channel function giving rise to arrhythmias, and/or to changes in conduction pathways.
2. At the level of release of arrhythmogens,[3] there are endogenous chemical substances that alter activity of ion channels and thereby precipitate arrhythmias. Arrhythmogens may act directly on myocytes, or indirectly, via nerves or blood vessels. Recognized arrhythmogens include catecholamines, H^+, K^+, eicosanoids, and other fatty acid products.
3. At the genetic level, where mutations are responsible for arrhythmias as a result of pathological changes in channel structure.[4] This is particularly so for *torsades* caused by a long QT syndrome due to mutations in potassium or sodium channels.
4. At the level of abnormal activation of channels that are usually closed. The classic example of this is the IK_{ATP} channel, normally closed but opened to provide outward potassium current when intracellular concentrations of ATP fall below critical levels. Cardiac tissue also contains ion channels that are sensitive to stretch and thus pathological stretch of cardiac tissue gives rise to arrhythmias (stretch-induced arrhythmias) via activation of this channel.
5. Iatrogenic causes such as inappropriate use of drugs, particularly antiarrhythmic drugs, other drugs that prolong the QT interval, or drugs and procedures that cause arrhythmogenic changes in ionic imbalances, e.g., hypo- or hyperkalemia.

All of the above have direct consequences on the electrophysiological behavior of cardiac muscle, and these changes are the fundamental cause of arrhythmias. Changes in cellular electrophysiological behavior change one or more of the basic cardiac electrophysiological functions of:

- Excitability – a measure of the ability of cardiac cells to be excited by invading electrical currents, so that they generate action potentials. Excitability is directly related to the number of sodium ion channels available for opening in atrial and ventricular tissue, and to calcium channels in nodal tissue.
- Pacemaking or automaticity – the ability, normally found in SA and AV nodes, and in ventricular Purkinje tissue, to spontaneously generate action potentials. While automaticity is normally restricted to the above tissues, damaged atrial and ventricular tissue may exhibit pacing.
- Conduction velocity – the speed of conduction of action potentials through cardiac tissue. Conduction velocity is sodium current dependent in ventricles as well as atria, and calcium current dependent in SA and AV nodal tissue.
- Refractoriness – the temporal resistance of cardiac tissue to excitation such that a second action potential cannot be initiated on a first action potential. Refractoriness is governed by the duration of the action potential which regulates sodium channel dependent excitability as they revert from an inactivated to a resting state (see later).

6.33.2.1 Cardiac Ion Channels

All of the above functions are dependent upon the opening and closing of transmembrane ion channels for sodium, potassium, and calcium in cardiac tissue.[5] The opening and closing of these ion channels depends upon both the voltage across cell membrane, and time. Channels that are open will close spontaneously with time. Simplistically, voltage dependent channels can exist in at least three states. These are:

1. Resting and ready to be opened by an appropriate voltage change across the cell membrane in which they are sited.
2. Open and in a conducting state for their particular ion.
3. Inactive and in a nonconducting state from where they convert back to the resting state with time, or they can be locked in this state at particular membrane voltages.

The major ion channel families in cardiac tissue include those for sodium, calcium, and potassium ions, and possibly chloride and stretch channels. There is only one major type of sodium channel in cardiac tissue, one major type of calcium channel, and many types of potassium channels. The genes for most of these channels have been identified, as have their proteomes, as well as the intimate details of their anatomical and functional mechanisms.[5]

The presence and functioning of these channels varies with cardiac tissue type to provide the basic electrical behavior characteristic of different cardiac tissues that include SA and AV nodes, various types of atrial tissue, His/Purkinje tissue (arising at the ventricular end of the AV node and spread throughout the ventricles providing a high-speed conduction pathway to the ventricles), and ventricular tissue that anatomically and electrophysiologically exists in at least three types (endocardial, mid-myocardial, and epicardial). Each type of tissue has channels that, when open, provide:

1. Depolarizing currents (Na^+ and/or Ca^{2+} dependent currents) that occur in all tissues.
2. Pacing currents normally found in the SA and AV nodes, and His/Purkinje tissue.
3. Repolarizing currents (K^+ dependent currents) which vary widely with different species and tissues, but provide the mechanism whereby the action potential returns to a stable resting state in most tissue.
4. Miscellaneous other ionic currents that are of lesser importance and are 'gated' by either voltage, hormones or neurotransmitters.

Some of the characteristic action potentials (and their underlying ionic currents) found in the various types of heart tissue are shown in **Figure 2** that includes both pacing and nonpacing cells. In pacing cells the membrane potential is not stable and the constant decay is due to the pacemaker current. The currents that participate in the generation of such potentials are also illustrated.

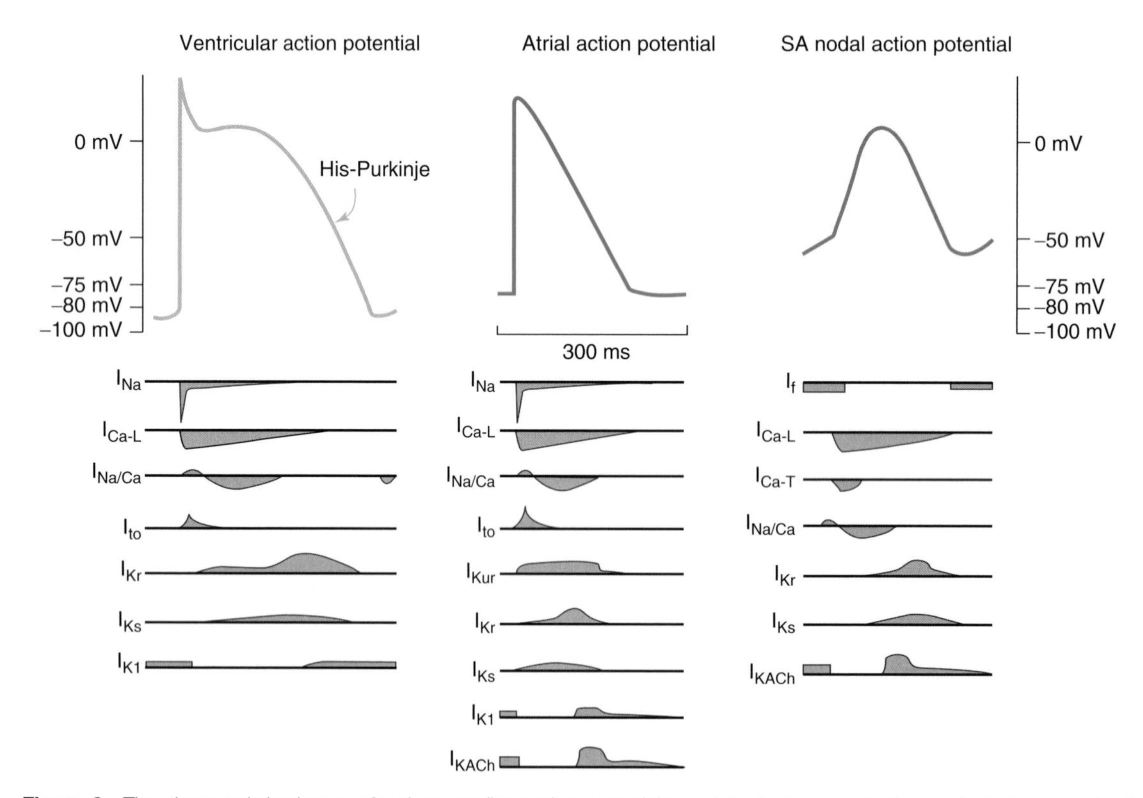

Figure 2 The characteristic shapes of various cardiac action potentials, and the ionic currents that underlie the genesis of such potentials. The figure shows representative action potentials seen in (left to right) ventricular, atrial, and nodal cardiac cells. An action potential for ventricular Purkinje cells is superimposed on the ventricular potential. In the case of the action potentials, calibration is in millivolts (mV) on the vertical axis, and time in milliseconds (ms) on the horizontal axis. The lower part of the figure shows the different currents responsible for the generation of the different action potentials. The time axis for these currents is the same as for potential, whereas that for currents is not to scale. A downward current indicates an inward current and vice versa.

Most of the important channels in the heart provide voltage-gated currents. A single gene encodes for alpha subunits that can provide current, although relatively normal behavior requires coexpression with beta subunits. However, the full range of normal behavior of currents requires the products of multiple genes.

Ion channels appear to have evolved from a primitive ion channel structure similar to the potassium inward-rectifier channel. Although many mutations have been found they rarely have consequences. In most channel types a loop between the 5th and 6th α helices (S5 and S6) enters the membrane to form a P loop. The pore region for all potassium channels has a signature motif conferring selectivity around an inverted cone, whereas for sodium and calcium channels, selectivity is conferred by four amino acids, one from each P loop. These are glutamate in calcium channels. Voltage-sensing regions are found in voltage-dependent channels. These regions contain regularly spaced arginine or lysine at every third amino acid in the transmembrane portions of the channels.

Ideally, intimate molecular knowledge of the different channels makes it possible to tailor small molecules to bind specifically to particular binding sites in specific channels in a manner analogous to targeting active sites in enzymes. Such targeting should make it possible in the future to achieve any desired level of specificity for ion channel modulating actions (activation or blocking) on the different types and subtypes of cardiac ion channels, but currently this cannot be done. Many of the current channel blockers lack specificity for particular channels. For example, verapamil blocks sodium as well as calcium channels; some sodium channel blockers block potassium channels while many potassium channel blockers fail to differentiate between the subtypes of potassium channels. The IK_r potassium channel is particularly promiscuous in that it is blocked by a multitude of different class and chemical types of drugs, although many of them appear to bind somewhere on the signature motif on its inverted cone. The more we understand the molecular structure of the voltage dependent cardiac ion channels, the greater is our expectation of creating ion channel modulators with very specific binding, resulting in selective actions against arrhythmias.

The process of inactivation can be viewed as being of two types:

1. N-type, or ball and chain, where an amino acid 'ball' swings into the intracellular mouth of pore of the previously opened channel.
2. C-type, which involves allosteric changes in the channel to terminate ion permeation.

Such models may be overly simplistic in that there are probably more than two types of inactivation with varying time courses.

Many naturally occurring toxins block channels by binding at extracellular sites and thereby disrupting ion movement by allosterically preventing opening, whereas synthetic small molecules may bind to relatively specific sites in channels where access to the sites are determined by channel state (e.g., S6 segment lining the pore) for sodium, potassium, and some calcium channels.

6.33.2.1.1 Sodium channels

Sodium channels as a family are encoded by nine genes for highly homologous α subunits with two ancillary beta subunits (β1 and β2 subunit) that are encoded by four genes. Together variations on this theme result in the different voltage and time dependent sodium channels found in various tissues.[6] The gene for the cardiac form of the sodium channel – *SCN5A* (2016 amino acid) – is located on chromosome 3p21.[7] The function of this channel is activation in atrial, His/Purkinje, and ventricular tissue. The S4 regions function as voltage sensors which lead to voltage dependent opening of the channel, and this action initiates time dependent coupled inactivation, although it is also possible for the channel to change directly from open to inactivated state. Tetrodotoxin, at nanomolar concentrations, characteristically blocks voltage and time dependent neuronal sodium currents although the cardiac form requires higher concentrations (1–10 μM). This difference is due to cysteine in position 373 in the heart form of the channel versus an aromatic residue in other forms.[7]

Mutations of *SCN5A*, including the LQT3 variant, are responsible for some of the genetic forms of long QT in the ECG.[6,7] These mutations cause a failure of fast inactivation. Mutations occurring at a variety of sites also produce the same ECG pathology. Other mutations cause idiopathic ventricular fibrillation, conduction block, and ventricular fibrillation associated with a J point aberration of the ECG (Brugada syndrome).

6.33.2.1.2 Calcium channels

Calcium channels exist in various forms (T, L, N, P, and Q) depending upon tissue type.[8] The dominant channel in the heart is L (large)-type calcium channel, with minor presence of T-type. Ten genes encode the main Cavα1 subunit of the voltage gated calcium channel with three major subfamilies: Cav1α1, Cav2α1, and Cav3α1. Neurons contain N, P/Q, and R type channels. The L-type channel is β-adrenoceptor-dependent since beta-adrenoceptor activation, via a cAMP dependent protein kinase, increases the L-type current and shifts activation toward the pacing range.

The L-type channel is highly voltage dependent and contains Cav1α1 subunits. It activates relatively slowly, but deactivates faster, shows weak voltage dependent inactivation, marked calcium dependent inactivation, and sensitivity to the dihydropyridine class of calcium channel blocking drugs. Conduction through the SA and AV nodes is critically dependent upon L-type calcium activation, as is the coupling of excitation to contraction in cardiac cells. The T-type channel is characterized by low voltage dependent characteristics. It has pronounced voltage dependent inactivation, is insensitive to dihydropyridines, and absent in ventricular tissue, but present in pacing cells and some atrial cells. It could contribute to pacemaking during the slow pacing depolarization seen in SA nodes.

Other types of channels include calcium activated currents such as $I_{Cl(Ca)}$ which is a Ca^{2+} activated Cl^- current, and the Ca^{2+} activated nonselective cation current $I_{NS(Ca)}$. The Na^+/Ca^{2+} exchanger produces an $I_{Na/Ca}$ current which is dependent upon membrane potential, and Ca^{2+} and Na^+ concentrations. $I_{Na/Ca}$ is important in all cardiac cells, both as a Ca^{2+} transporter, and in pacing and arrhythmogenic current generation.

6.33.2.1.3 Potassium channels

Potassium channels are responsible for generating repolarization currents and maintaining resting membrane potentials.[9] The resting potential is negative since only potassium channels are open at rest and potassium concentrations are high ($[K]_I = 140\,mM$) inside cells versus the outside ($[K]_o = 3.5\,mM$). All K^+ currents carry outward currents which serve to repolarize or stabilize resting potentials. These currents are divided into voltage activated or ligand activated currents and nongated currents: the voltage activated currents are I_{to}, IK_{ur}, IK_r, IK_s, and IK_{ss}; the ligand activated currents IK_{ATP}, IK_{ACh}, IK_{Na}, and IK_{AA}; and the nongated (background) currents IK_1.

The genes for most K^+ channels are known. Their molecular structure is analogous to Na^+ and Ca^{2+} channels. There is one domain with six transmembrane segments in which segment four (S4) acts as a voltage sensor. Voltage activated channels have N- and C-type inactivation where the C-type is sensitive to drugs and $[K]_o$. The ligand activated and background channels have only two transmembrane segments while the pore of the channel is located in the S5-S6 link.[9]

6.33.2.1.3.1 The main voltage activated K currents: I_{to}, IK_{ur}, IK_r, IK_s, and IK_{ss}

The transient outward K^+ current (I_{to}) occurs in most human cardiac tissue and is the major repolarizing current in mice and rat atrial and ventricular cells. It is also present in brain and pancreas. It is blocked by some antiarrhythmic drugs and regulated by glucocorticoids

The ultra rapid delayed rectifier (IK_{ur}-Kv1.5) is selectively present in human and dog atrium, and in all chambers of rat and mouse hearts. The gene for the rapid component of the delayed rectifier channel (IK_r) includes the human ether-a-go-go (hERG) gene that encodes the α subunit. It is most abundant in the heart, but also occurs in the hippocampus. Its amino acid sequence is highly conserved across mammals. As is discussed in detail later, the IK_r channel is blocked by many different drugs of different chemical types and classes, resulting in a potential for causing *torsades de pointes*. The first major example of this occurred with the long-acting H_1 antihistamines astemizole and terfenadine, which were unexpectedly observed to cause death, apparently due to sudden cardiac death, and the occurrence of torsades. The occurrence of torsades is related directly to prolongation of the QT interval of the ECG due to blockade of IK_r. The slow component of the delayed rectifier (IK_s) is also coexpressed in some species. Guinea pigs, humans, and dogs have both forms, and cats and rabbits primarily IK_r. IK_s is revealed by blockade of IK_r current. IK_s is a slowly activating outward current found in the heart and inner ear. Activation of this current may account for the action potential narrowing due to β-adrenoceptor activation, and acts to counter the increase in I_{Ca} seen with such activation. Mutation induced dysfunction of this channel may explain the increased arrhythmogenic actions of catecholamines seen in some arrhythmia phenotypes.

6.33.2.1.3.2 Ligand activated K currents include IK_{ATP}, IK_{ACh}, IK_{Na}, and IK_{AA}

The ATP dependent K^+ current (IK_{ATP}) consists of coexpressed Kir6.x with an attached sulfonylurea (antidiabetic drug) receptor.[9] The cardiac channel is a heteromultimeric form of Kir6.2 plus the 2A form of the sulfonylurea receptor found in both cell plasma membrane and mitochondrial membranes. The channel is activated by low levels of ATP or by nicorandil/pinacidil, drugs that lower channel affinity for ATP. Its channel kinetics are very complex and the sensitivity to low ATP is increased by acid, ADP, and lactate. To close the channel a critical concentration of ATP is required in the immediate vicinity of the channel. It is this channel that regulates the release of insulin in the β cells of the islets of Langerhans.

6.33.2.1.3.3 The K^+ channel activated by the cholinergic transmitter acetylcholine (ACh) acting on muscarinic (M_2 subtype) cholinoceptors

This channel (IK_{ACh}) consists of Kir3.1 and 3.4 (GIRK1 and GIRK4) in heart.[9] Acetylcholine activates muscarinic (M_2) receptors and the Gβγ pathway. The channel is predominantly expressed in nodes and atrial tissues, i.e., those exposed to ACh released from parasympathetic nerves that slow heart rate and conduction through the AV node. The channel is very selective for potassium ions and is sensitive to $[K]_o$. IK_{ACh} plays a role in atrial fibrillation since vagal stimulation leads to its activation and reduction of atrial refractoriness, and therefore increased liability to arrhythmias.

6.33.2.1.3.4 The sodium activated K^+ channel

This channel (IK_{Na}) is a Na^+-gated K^+ channel activated by high Na^+. It has two open states, but sometimes the channel is silent for minutes. It has a high selective conductance for K^+ ions (200pS) that is blocked by Rb^+ or Cs^+.

6.33.2.1.3.5 The fatty acid and amphiphile activated K^+ channel

This channel (IK_{AA}) is activated by arachidonic and other unsaturated fatty acids (in rats) and has two possible pores.

6.33.2.1.3.6 The inward rectifier K^+ channel

This channel (IK_1) is part of the Kir superfamily of potassium channels that subserve many physiological functions. The Kir2.x family consists of two transmembrane segments with a pore sequence.[9] In the heart, the IK_1 channel is important for maintaining resting negative membrane potential in all cardiac cells (except nodes), and participates in terminal repolarization. The highest density of channels is in Purkinje and ventricular cells with a lower density in atria (which has a corresponding lower resting membrane potential). They are almost absent in nodal cells. The inward rectification property of IK_1 results from a block of outward current by intracellular substances (Mg^{2+} and polyamines). It is selective for K^+ and regulated by $[K]_o$.

6.33.2.1.3.7 Other K^+ currents

These include the 'funny' (f) or hyperpolarizing (h) pacemaker current (I_f, I_h). I_h is a member of the HCN family with subtypes 1 and 4 occurring in the heart.[10] The channel has poor selectivity for Na^+ or K^+ ions, and a reversal potential of -10 to -20 mV. HCN has the typical six membrane spanning unit with a S4 voltage sensor. It is regulated by cyclic nucleotide signaling, down via mucarinic receptors and up via β-adrenoceptors. The mechanism underlying its hyperpolarization-induced activation is unknown.

6.33.2.2 Mechanistic Macro Models of Arrhythmias

Changes in the normal functioning of the above ion channels, in space and/or time as a result of arrhythmogenic stimuli, result in the initiation and maintenance of the two major macro mechanisms of arrhythmogenesis, namely ectopic foci and re-entry. Since regular rhythmic beating requires appropriate pacing, excitability, conduction, and refractoriness, disturbances of any of these four functions can give rise to arrhythmias via either, or both, of the two mechanisms.[11]

6.33.2.2.1 Ectopic foci

Ectopic foci (errors in pacing) occur when pacing occurs at sites that do not normally act as cardiac pacemakers. Most cardiac tissues do not exhibit pacing if entirely healthy, but previously stable tissue can be induced to pace as a result of damage and resulting stimuli that reduce the membrane potential, or its stability. It is not known whether such ectopic pacing involves revealing pacemaker currents that may be inherent in all tissue. The more obvious mechanism involved is early or late after-depolarizations due to membrane instability following intracellular calcium overload.

6.33.2.2.2 Re-entry

Re-entry (errors in conduction, and/or refractoriness) circuits (**Figure 3**)[12] have their origin in regions of depressed excitability and conduction. Over certain ranges, such changes can result in a unidirectional block of conduction such that by conducting in one direction, but not in the other, re-entry circuits (rotors) are set up (**Figure 3**). Both macro and micro re-entry circuits are possible. In AV nodal tissue, macro re-entry is responsible for supraventricular tachycardia which can involve Ca^{2+} dependent conduction, or Na^+ dependent aberrant conduction pathways. Re-entry circuits (rotors) can fragment into multiple circuits, i.e., fibrillation. *Torsades de pointes* is an arrhythmia that is probably due to single re-entry rotor wandering around the ventricles.

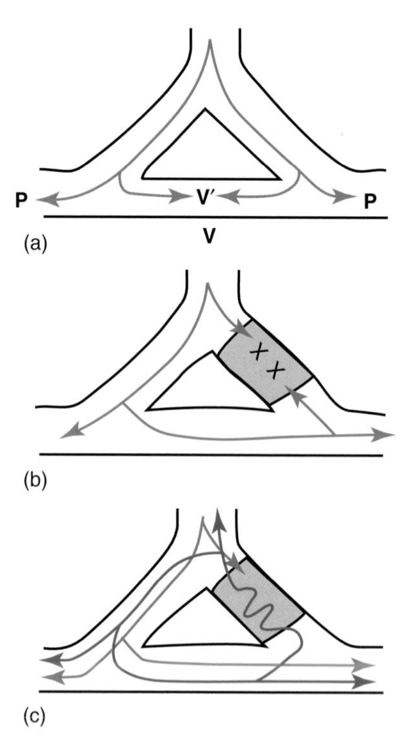

Figure 3 Functional model of re-entrant arrhythmias. (a) Normal conduction pathways for an impulse arriving from the AV node and down the Purkinje (**P**) tissue into ventricular (**V**) tissue. The impulses from the two pathways meet, collide, and mutually extinguish each other in ventricular tissue at **V′**. Panel (b) Shows where one of the Purkinje branches is damaged (shaded area) such that both antegrade (normal) conduction and retrograde (reverse) conduction are blocked at **X**. This condition is known as bidirectional block, and does not generate arrhythmias. (c) Damaged area in one of the Purkinje branches that blocks conduction in the antegrade direction but, because of special conditions, allows very slow retrograde conduction through the damaged area (wavy line). This is known as unidirectional block. The retrograde conduction passes into the normal tissue beyond the block, and since the normal tissue is no longer refractory by the time the retrograde impulse arrives, it is excited and an extra beat created. The result of this type of block is the generation of a sustained re-entry cycle of activation (ventricular tachycardia) circulating through the unidirectionally blocked area. The necessary conditions for re-entry are a critical degree of block in the damaged tissue, as well as appropriate conduction and refractoriness in all other parts of the circuit. Re-entry can be terminated by: (1) reducing excitability in the damaged area to convert unidirectional to bidirectional block; (2) changing conduction and/or refractoriness to abolish the re-entry circuit; the physical length of a re-entry circuit is the product of conduction velocity and refractory period. Therefore the limiting physical boundary for a re-entry circuit is the sum of (conduction velocity times refractory period in normal tissue) plus (conduction velocity times refractory period in the damaged tissue). Appropriate changes in the two properties in either tissue can make the occurrence of the re-entry circuit physically impossible.

How do the above models explain human arrhythmias? The answer to this question depends upon the type of arrhythmia (**Table 1**):

1. SA and AV nodal tachycardias are primarily due to excess neurotransmitter or hormone (norepinephrine or thyroxine), which increases pacing currents. Bradycardia is primarily due to ACh released from parasympathetic nerves. However, inherent defects in nodes or damage to nodes can cause alterations of rate. Fibrosis can damage such nodes.
2. Atrial arrhythmias are initiated by excess cardiac sympathetic nerve activity, scars, and fibrotic areas in the atria, or excessive stretch on atrial fibers. Such pathology can cause PACs from ectopic foci, or re-entry circuits in left, or right, or both atria (the two atria are not identical in this respect). A single maintained re-entry circuit will cause a tachycardia whereas multiple re-entries gives rise to atrial fibrillation. The site of origin of some atrial arrhythmias may be where large blood vessels enter the atria.
3. Conduction defects in the AV node will cause AV node block. Micro re-entry circuits can occur in the AV node and give rise to supraventricular nodal tachycardias.
4. The pathological mechanisms that give rise to atrial tachycardia and fibrillation also give rise to the same type of arrhythmias in ventricles. Thus, damage to Purkinje fibers will produce intraventricular conduction block, while in

Table 1 Possible mechanisms involved in the genesis of particular arrhythmias[a]

Arrhythmia[b]	*Cause*	*Mechanism*
SA and AV nodal tachycardia	Excess norepinephrine from sympathetic nerve, or excess of other hormones, such as thyroxine	Increase in pacing currents; micro re-entry circuits in the AV node can cause supraventricular nodal tachycardia
Bradycardia	Excess acetylcholine from parasympathetic nerve, damaged nodes due to nodal fibrosis	Decrease in pacing currents or a 'sick' SA node
Atrial arrhythmia	Excess sympathetic nervous activity, scars, and fibrosis or stretch in atrial tissue	Ectopic foci and re-entry; single re-entry; leads to atrial tachycardia and multiple re-entries lead to atrial fibrillation; source is often near or in the veins entering atria
AV nodal block	Damaged AV node/Purkinje tissues	Failure to conduct electrical signals from atria across to the ventricles
Ventricular arrhythmia	Excess sympathetic nervous activity, scars, and fibrosis or stretch in ventricular tissue	Ectopic foci and re-entry; Single re-entry leads to ventricular tachycardia and multiple re-entries lead to ventricular fibrillation

[a] Since ionic mechanisms are often of mixed origin, and cannot be readily identified, they are not considered in this table, only in the text.
[b] Arrhythmias are categorized according to anatomical site.

the ventricles, single re-entry circuits result in ventricular tachycardia, and multiple circuits in ventricular fibrillation. *Torsades de pointes* is thought to arise as a result of after-depolarizations due to calcium overload, while induction of a re-entry circuit due to inappropriate heterogeneity of repolarization provides substrate for its maintenance.

6.33.3 Experimental Disease Models

There are a number of levels and experimental approaches which are used in attempts to understand arrhythmic mechanisms and antiarrhythmic drug effects, and for the prediction of drug effects on the arrhythmias seen in humans. These can be considered according to a hierarchy of approaches:

- using computer modeling of cardiac electrophysiology, and/or arrhythmias, plus computer modeling of molecular targets (ion channels), and their interactions with compounds;
- studying ionic currents, and their characteristics, in single ion channels (expressed or wild-type) using patch clamp techniques;
- measuring whole cell currents, and their characteristics, using expressed or wild-type channels in special cells, or in isolated cardiac cells;
- measuring intracellular action potentials in cardiac tissue in vitro or in vivo;
- determining electrophysiological properties such as automaticity, excitability, conduction velocity, and refractoriness using electrical stimulation techniques in vitro in isolated cardiac tissue and/or isolated hearts, or in vivo on intact hearts;
- inducing arrhythmias in isolated or intact hearts, using various techniques such as chemical arrhythmogens or surgery to induce arrhythmic damage. The latter is used to stretch atria and ventricles, or create myocardial ischemia and infarction. This can be combined with electrical stimulation of the heart directly or indirectly through cardiac nerves.

No single one of these techniques definitively provides an unequivocal answer as to whether an experimental drug has antiarrhythmic actions in humans. Instead they have to be used in concert to build up a coherent picture of the clinical potential of a particular new chemical entity. The question of which technique to use depends upon the goal to be achieved. If, for example, one has decided to create a compound that acts potently and specifically on a particular ion channel it may be sufficient to concentrate on that channel. Of course the specificity for this chosen channel has to be compared with actions on all cardiac channels. This can be a daunting endeavor although it is being made easier as more channels become available in functional high-throughput assays.[13] The generally promiscuous nature of binding of ion channel blockers to channels, in terms of channel type and channel sites to which they bind, makes simple binding studies inadequate. Predictive in silico modeling is being used more often in terms of modeling putative drugs, binding sites, and pharmacological actions.[14] At the moment these approaches may be more in the nature of ad hoc rationalizations rather than predictions.

There are various approaches to modeling arrhythmias and to screen for antiarrhythmic actions. These screens range from precisely defined in vitro situations to ill-defined in vivo ones, as illustrated below:

1. Use of particular channels expressed in human embryonic kidneys cells or in high-throughput screens.
2. Use of the less well-defined models from patch clamp studies on wild-type ionic currents to intracellular potential studies in cardiac tissue preparations.
3. Use of isolated hearts or whole animals to screen for potential antiarrhythmic drug actions by determining their electrophysiological effects on the ECG or on responses to electrical stimulation.
4. Use of various arrhythmia models in isolated hearts or whole animals.

The in vivo models include induction of arrhythmias by use of electrical stimulation, arrhythmogenic chemicals, or pathological stimuli. For example, atrial arrhythmias can be induced by drugs and vagal stimulation, whereas heart attacks can be mimicked with regional myocardial ischemia, and/or infarction, by temporary or permanent occlusion of a coronary artery.

6.33.4 Clinical Trial Issues

Currently, clinical trials are required to assess effects of both acute and chronic drug treatments for the termination, prophylaxis, or prevention of the first presentation, or repeated episodes, of an arrhythmia. The ultimate measure of a drug's effectiveness is, of course, reductions in morbidity and mortality. It has to be remembered that in the past we fell into the trap of treating the ECG, but not the patient's arrhythmias. Thus we concentrated on ECG surrogates rather than true clinical outcomes. There are no surrogates for the actual arrhythmias, and so we ultimately always have to determine whether a new drug produces real beneficial clinical outcomes by improving quality of life and saving lives. One clear example of this is a recent trial to determine whether, in atrial fibrillation, it is better to re-establish sinus rhythm or simply use a drug to control ventricular rate – the so-called 'rate versus rhythm' controversy.[15] It seems that there is no clear advantage of rhythm control over rate control. Such a finding reflects the reality of a situation in which there are complex interplays between disease, patient, and drug. The importance of death as an unambiguous endpoint cannot be minimized, especially when a drug's disease target is, in the final analysis, a reduction in cardiac sudden death. The logistics and expense of using death as an endpoint is a further reason for the current lack of enthusiasm for antiarrhythmic drug discovery programs.

The easiest type of clinical trial for an antiarrhythmic drug is where it is given intravenously to terminate an arrhythmia. The endpoint is unequivocal, especially when the arrhythmia has been present for a period of hours, or longer. Since the drug is given intravenously, its presence at a potentially effective concentration ensures the most reliable type of clinical trial for an antiarrhythmic drug. This represents one end of the spectrum, as opposed to the other end where the endpoint of death is used.

A major problem with arrhythmias is their nature. Some arrhythmias are transient, others are chronic, and sometimes they subtly change their nature over time. Thus, recent-onset atrial fibrillation is electrophysiologically different from chronic atrial fibrillation as a result of remodeling. It appears that atrial fibrillation begets atrial fibrillation[16] and thereby ensures its continuation albeit in a slightly different form.

Given such problems with antiarrhythmic clinical trials, there is a clear need for arrhythmia surrogates, analogous to the manner in which blood pressure changes over a limited period can be used as a surrogate for the morbid and lethal adverse effects of sustained hypertension. Unfortunately, even for hypertension, this most obvious surrogate is inadequate, and can give a false picture. There are, of course, a number of potential surrogates for antiarrhythmic trials, but their validity has to be questioned. For example, the electrophysiological actions of an ion channel blocker can be shown explicitly in man by relatively noninvasive electrophysiological studies. For example, an IK_r blocker can prolong the QT interval and change refractoriness thereby revealing its electrophysiological actions in normal myocardium, but whether such actions necessarily result in antiarrhythmic effects still has to be tested in humans with arrhythmias.

Table 2 outlines the type of trial that could be used for particular arrhythmias and the difficulties encountered with such trials.

Table 2 The various factors encountered in clinical trials of antiarrhythmic drugs

Arrhythmia	*Endpoint*	*Patient population*	*Outcome measure*
A1: Acute atrial tachycardia and/or fibrillation	Termination of arrhythmia	Those with recent (hours) onset	Termination of arrhythmia in individuals
A2: Prophylaxis against recurrence of A1	Prevention of recurrence of the arrhythmia	Those converted from A1	Frequency of reoccurrence in treated versus control groups
B1: Chronic atrial tachycardia and/or fibrillation	Termination of arrhythmia	Those with long-standing arrhythmias (weeks to months) – a very heterogeneous group	Termination of arrhythmia
B2: Prophylaxis against recurrence of B1.	Prevention of recurrence of the arrhythmia	As in B1	Frequency of reoccurrence in treated versus control groups
C: Prophylaxis of ventricular tachycardia/fibrillation	Prevention of these fatal arrhythmias	Those who have had this arrhythmia or those judged to be at risk of the arrhythmia with or without an implantable cardioverter defibrillator (ICD)	Occurrence rate, or mortality; frequency of activation of ICD

6.33.5 Current Treatments

Current antiarrhythmic therapy is either with drugs or uses electrical devices to detect and treat arrhythmias by electroshock or pacing. Electrical techniques can either be applied externally to the chest, or can be implanted within the body together with batteries to power all its components. Electrical devices continue to improve in terms of utility, reduced cost, ease of implantation, and effectiveness. As a result we are in a period where such devices are increasingly used, due in no small part to their advantages over drug therapy in many situations. Drugs generally have inadequate efficacy, are too toxic, or have too many side effects, and this is unfortunate since it will never be practical to treat every arrhythmia electrically.

The clinical utility and prescribing trends for current antiarrhythmic drugs show small regional differences, but in essence they are reasonably consistent throughout the world. For instance, in Japan there is a tendency to treat premature ventricular contractions (PVCs) that would not be treated in the USA or Europe. Changes in antiarrhythmic drug use worldwide reflect a falling drug use, and the increasing use of electrical devices.

As indicated above, the major cause of death from arrhythmias is ventricular tachycardias, particularly ventricular fibrillation, and this remains a problem despite the increasing use of electrical devices. There is a lower morbidity and mortality associated with atrial fibrillation, but this arrhythmia also still remains a problem. In the past, treatment of arrhythmias often reflected the treatment of an ECG pattern, rather than the patient, but it is increasingly recognized that it is pointless to treat the ECG when the drugs used for this purpose cause more morbidity and mortality than when the arrhythmia is left untreated.[17] The reason for this conundrum is that all antiarrhythmic drugs, in addition to their own peculiar toxicity, are also arrhythmogenic when given in the wrong situation, and/or at the wrong dose. Thus, the overexpression of their main pharmacological actions (channel block) causes arrhythmias (i.e., many antiarrhythmic drugs are proarrhythmic). In addition, such drugs also have their own toxicity unrelated to their antiarrhythmic mechanism.[18]

The success of electrical devices has raised the question of whether antiarrhythmic drugs are still needed, or have drugs been superseded? The answer to this question has to be in the negative since, for ventricular fibrillation and atrial arrhythmias, there is a specific need for better drugs, as well as a need for drugs that are more efficacious and less toxic than those currently available.

The question becomes, can we create new and better antiarrhythmics? The unlimited potential of science means the answer has to be yes, although the routes to achieving this are not immediately clear. There has been a failure to answer clear questions, such as, do antiarrhythmic drugs have to be arrhythmia-specific? If the answer is yes, how are such drugs developed? Some of the answers to such questions occupy the remainder of this chapter.

6.33.5.1 Overview of Current Antiarrhythmic Drugs

All current drugs are synthetic compounds, except for the oldest, quinidine, a botanical alkaloid, and adenosine, an important endogenous chemical present in all living tissue where it acts locally to mediate many physiological processes. The use of adenosine is limited to its intravenous administration to terminate an episode of paroxysmal supraventricular nodal tachycardia where its short half-life limits the potential for adverse side effects.

6.33.5.2 Classification Systems for Antiarrhythmic Drugs

There are two accepted classification systems for antiarrhythmic drugs. The first, and most widely used, was created by Vaughan Williams and Singh,[19] and modified by Harrison and others. The second is known as the Sicilian Gambit.[20]

1. The Vaughan Williams–Singh classification has a number of advantages in that all drugs in a particular class tend to have the same clinical utility and similar, if not identical, actions on particular ion channels. The classification system has been found to be useful both experimentally and clinically with predictive utility, particularly in terms of clinical effectiveness and potential toxicity.
2. The Sicilian Gambit attempts to be more precise in its classification, but it ends up with almost every single drug being a special case which undermines any real attempt at classification. It does, however, fully describe the pharmacological spectrum of action of each antiarrhythmic drug.

In the following, drugs are considered on the basis of their classification with each class considered in terms of: (a) mechanism(s) and site of action; (b) structure–activity relationships (SARs) and basic pharmacophores; and (c) effectiveness, toxicity, and side effects. Emphasis is placed on comparisons within classifications with respect to particular drugs, their current limitations, and possible future directions in terms of making better drugs in the class.

6.33.5.2.1 Antiarrhythmic classes 1–5: actions, effectiveness, toxicity

The Vaughan Williams–Singh classification originally divided drugs into classes on the basis of their actions on action potential shape in atria, or on β-adrenoceptors. Modifications to the classification system became possible with an increased understanding of the ionic currents responsible for the action potentials found in the heart. In addition, clinical experience allowed for the clinical actions of the different classes of antiarrhythmic drugs on the ECG, and other clinical electrophysiological characteristics in man, to be accommodated within the classification system. The classes include (1) sodium channel blockers, (2) β-adrenoceptor blockers, (3) potassium channel blockers, (4) calcium channel blockers, and (5) selective bradycardic drugs.

6.33.5.2.1.1 Class 1 antiarrhythmic drugs are all sodium channel blockers

All Class 1 drugs block cardiac sodium channels. However, on the basis of the frequency-dependent nature of the block, and the presence or absence of concomitant potassium channel blocking actions resulting in action potential prolongation, Class 1 is divided into subclasses: 1a, 1b, and 1c.

6.33.5.2.1.1.1 Class 1a: quinidine, procainamide, disopyramide

- Mechanism: Class 1a drugs block sodium channels with moderate frequency dependence and also block potassium channels (IK_r, I_{to}). Some also block muscarinic receptors. Their ion channel blocking actions result in ECG changes including a widened QRS complex and QT duration on the ECG, as well as delayed AV conduction (PR prolongation). Blockade of muscarinic receptors, by blocking the actions of parasympathetic nerves on the heart, can cause an apparent paradoxical increase in heart rate and AV conduction.
- Effectiveness: Class 1a drugs are currently used to treat life-threatening sustained VT, but their use is associated with dangerous arrhythmias, especially in the presence of cardiac damage. Therefore, they are probably too dangerous for regular use in ventricular arrhythmias, but can be used together with an implantable cardioverter defibrillator (ICD). Quinidine and disopyramide (**Figure 4**) are also used to treat atrial flutter or fibrillation, while procainamide (**Figure 4**) is used to treat AF, but this does not have FDA approval for this use.
- Toxicity: Class 1a drugs all have subclass-wide adverse effects and toxicity profiles. They include ventricular arrhythmias and muscarinic antagonism, as well as nausea, vomiting, and diarrhea (in up to 30% of patients). Individual drugs can have their own unique toxicity such as the cinchona syndrome with quinidine, and an allergic systemic lupus-like syndrome with procainamide.

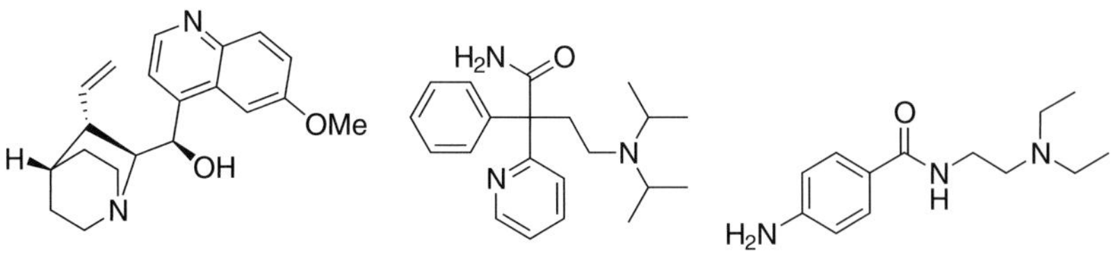

1 Quinidine **2** Disopyramide **3** Procainamide

Figure 4 Examples of class 1a antiarrhythmics.

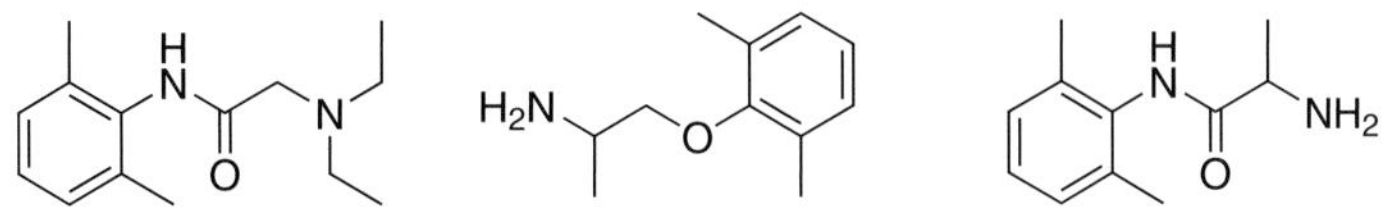

4 Lidocaine **5** Mexiletine **6** Tocainide

Figure 5 Examples of class 1b antiarrhythmics.

6.33.5.2.1.1.2 Class 1b: lidocaine, mexiletine, tocainide

- Mechanism: Class 1b drugs are selective sodium channel blockers with marked positive frequency dependence. As a result of this frequency dependence, they have few effects on the ECG at normal heart rates since they only block sodium channels when heart rate is high, or in ischemic cardiac tissue. All drugs in this subclass have low molecular weights, and high lipid solubility. The latter, combined with high frequency dependence, results in central nervous system (CNS) penetration and resulting CNS toxicity.
- Effectiveness: Class 1b drugs like lidocaine (**Figure 5**) have been used intravenously for the acute termination of ventricular arrhythmias, such as life-threatening VT, since they are mostly ineffective against atrial arrhythmias. Unlike lidocaine which undergoes rapid first-pass metabolism, mexiletine and tocainide (**Figure 5**) are orally effective. While there is evidence that such drugs reduce the incidence of VF in patients undergoing myocardial infarction, there is little evidence that they reduce mortality. Thus, they reduce the incidence of VF, but cause death as a result of asystole (absence of heart beat) resulting in no change in mortality.
- Toxicity: The toxicity of class 1b drugs includes loss of SA and AV node pacing resulting in a lack of a heart beat (asystole). CNS symptoms are common, and include initial paresthesias with subsequent convulsions. This is a result of their easy penetration (low molecular weight and high lipid solubility) into CNS tissue and nerves plus the very high frequency of CNS neuronal firing.

6.33.5.2.1.1.3 Class 1c: flecainide, encainide, propafenone, moricizine

- Mechanism: Class 1c drugs block sodium channels, but they have very limited frequency dependence and limited other pharmacological actions.
- Effectiveness: Class 1c drugs like flecainide and propafenone (**Figure 6**) are used to terminate paroxysmal AV nodal tachycardia, AV re-entrant tachycardia, and paroxysmal atrial flutter and fibrillation. They, and moricizine (**Figure 6**), are also indicated for the management of sustained VT, but only if it is judged to be life-threatening. However, they can be dangerous in ventricular arrhythmias, especially when there is cardiac damage, as was shown so graphically in the Cardiac Arrhythmia Suppression Trial (CAST),[17] resulting in the voluntary withdrawal of encainide (**Figure 6**) from the market.
- Toxicity: The toxicity of class 1c drugs includes arrhythmias and cardiovascular depression.

6.33.5.2.1.2 Class 2: β-adrenoceptor blockers: propranolol, atenolol, metoprolol, and many others

- Mechanism: There are many β-adrenoceptor blockers that vary slightly pharmacologically with some having limited specificity for cardiac β-adrenoceptors. Propranolol is the prototype while atenolol and metoprolol have some selectivity for the β_1-adrenoceptor (**Figure 7**). Some β_1-adrenoceptor blockers are also partial agonists.

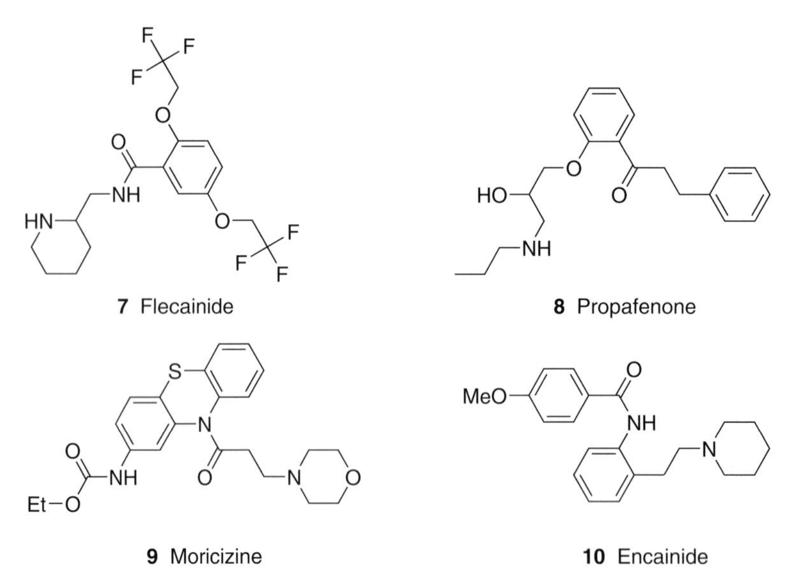

Figure 6 Examples of class 1c antiarrhythmics.

11 Propranolol

12 Atenolol

13 Metoprolol

Figure 7 Examples of class 2 antiarrhythmics with the prototypical β-adrenoceptor blocker propranolol, and two other blockers with some selectivity for the β_1-adrenoceptor (cardiac form).

An alternative description for partial agonism is intrinsic sympathetic activity (ISA) although the former is the preferred pharmacological term. This is the best term for an antagonist that actually stimulates β-adrenoceptor when they are not being stimulated by their neurotransmitter norepinephrine, or by epinephrine from the adrenal medulla, but block when such stimulation occurs. Thus β-blockers that are partial agonists only provide β-adrenoceptor blockade when activity of the sympathetic system on the heart is high.

β-Blockers also vary in their pharmacokinetic characteristics in terms of metabolism, half-life, and excretion. Since the catecholamines, norepinephrine from sympathetic nerves and epinephrine from the adrenal medulla, are well-identified arrhythmogens, blocking their actions provides prophylactic protection against catecholamine-induced arrhythmias. Indirect antiarrhythmic effects of β-adrenoceptor blockers in acute myocardial infarction may also involve elevation of serum potassium concentrations by β-adrenoceptor blockade since moderate elevations of serum potassium are antiarrhythmic.

- Effectiveness: β-Adrenoceptor blockers are also used routinely in postmyocardial infarction patients who can tolerate them. They reduce mortality by about 15%, although such protection may involve various mechanisms including antiarrhythmic actions, prevention of reinfarction, and beneficial effects in heart failure. They are also used to treat atrial and ventricular arrhythmias associated with sympathetic nervous system activation, or excess thyroid hormone secretion (more rarely), as well as slowing ventricular rate in atrial fibrillation.
- Toxicity: The adverse effects of β-adrenoceptor blockers are mostly due to β-adrenoceptor blockade. These include asthma due to blockade of the relaxant actions of catecholamines on bronchial smooth muscle, intermittent claudication (muscle pain associated with exercise), excessive bradycardia, and AV node blockade.

6.33.5.2.1.3 Class 3: Potassium channel blockers

There are many cardiac potassium channels including the transient outward (I_{to}), the ultrarapid, rapid, and slow components of delayed rectifier (IK_{ur}, IK_r, and IK_s, respectively) and inward rectifier (IK_1). Except for IK_r and IK_s

blockers, most potassium channel blockers, such as the I_{to} blocker tedisamil (**Figure 8**), are only partially specific for individual subtypes of potassium channels. Currently, only selective IK_r blockers and a mixed channel blocker like amiodarone are used clinically.

6.33.5.2.1.3.1 Selective IK_r blockers: dofetilide, ibutilide, D-sotalol The prototypical drug is D-sotalol, the adrenoceptor-inactive enantiomer of the β-adrenoceptor blocker, DL-sotalol (**Figure 9**). D-Sotalol, a methanesulfonanilide, prolongs ventricular cardiac action potentials by blocking IK_r. Methanesulfonanilide derivatives include ibutilide, dofetilide, E4031, and almokalant (**Figure 9**).

- Mechanism: Methanesulfonanilide derivatives block the open channel configuration. They appear to bind within the transmembrane pore at a fairly well-defined site. All the derivatives are very potent (acting at subnanomolar concentrations) and are specific for the IK_r current, especially dofetilide derivatives.
- Effectiveness: The IK_r blocker sotalol is used to treat atrial flutter and fibrillation only if the patient is highly symptomatic, since its use is associated with dangerous arrhythmias. Sotalol is also used to terminate life-threatening sustained VT. Dofetilide is used to maintain sinus rhythm in patients with atrial flutter and fibrillation who have been converted to sinus rhythm. On the other hand, rapid intravenous infusion of ibutilide is used for the conversion of atrial arrhythmias to sinus rhythm. However, the usefulness of these IK_r blockers is limited by a reverse frequency-dependent profile that results in excessive drug effects at slow heart rates. This not only increases their propensity to cause drug-induced proarrhythmias, but results in a loss of effectiveness at fast heart rates, the opposite of what is needed for tachyarrhythmias.

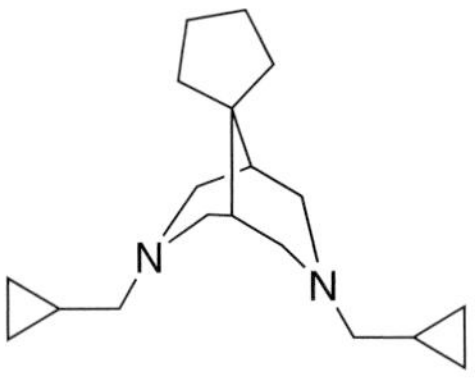

14 Tedisamil

Figure 8 Structure of tedisamil (I_{to} blocker).

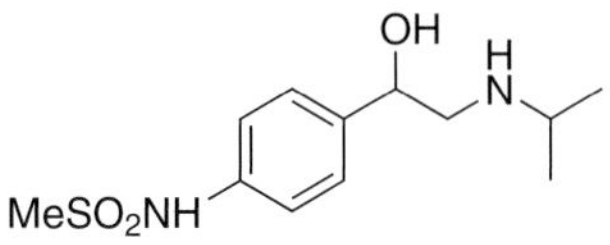

15 Sotalol

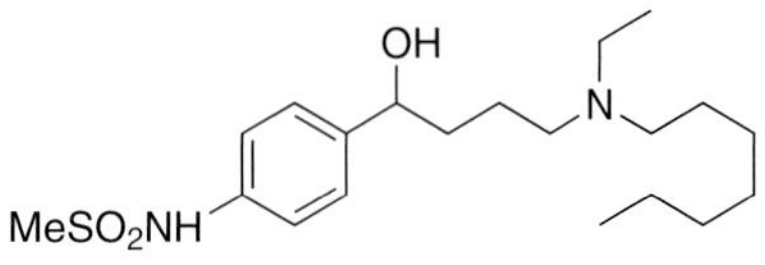

16 Ibutilide

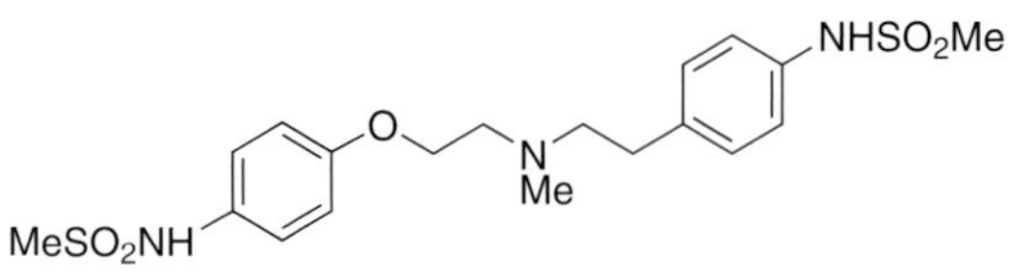

17 Dofetilide

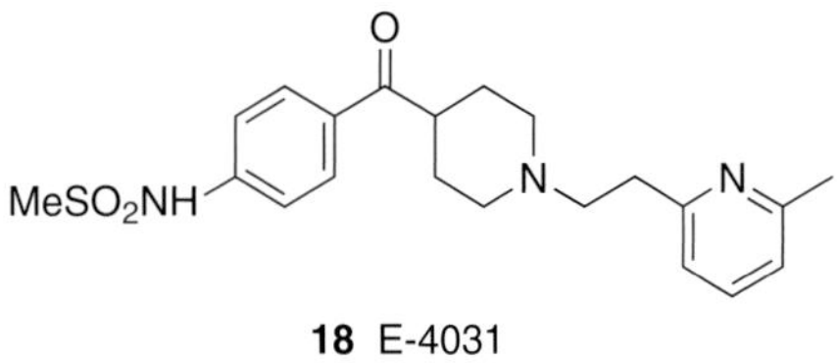

18 E-4031

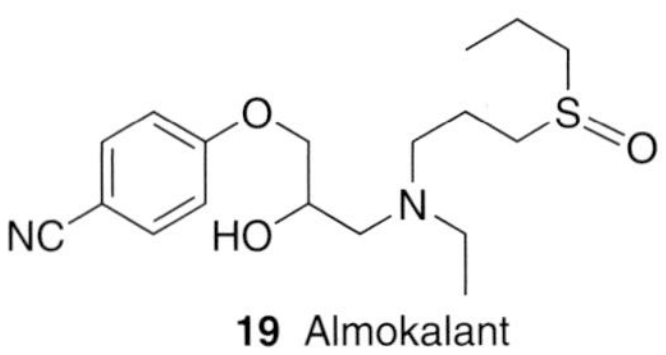

19 Almokalant

Figure 9 Examples of class 3 antiarrhythmics (methanesulfonanilide IK_r blockers).

- Toxicity: These drugs all induce *torsades*, the polymorphic ventricular tachycardia related to QT prolongation. Females have longer QT intervals than males, and are more prone to *torsades*. The mechanism responsible for *torsades* may be after-depolarizations acting as ectopic foci, or heterogeneity in repolarization causing re-entry. Since the therapeutic value of IK_r blockers is limited by proarrhythmic tendencies, they are used only for acute treatment of atrial tachyarrhythmias.

IK_r blockade also occurs with other drugs that are not methanesulfonanilde drugs. Such drugs include quinidine (**Figure 4**), cisapride, sulfamethoxazole, terfenadine, erythromycin, astemizole, ketoconazole, bepridil, and even antimalarials such as halofantrine (**Figure 10**). Currently such drugs are being withdrawn if an alternative is available that is equally effective, but does not block IK_r. Furthermore, current regulations do not allow the introduction of new drugs that block IK_r unless there are no alternatives, or they provide overwhelming benefit and save lives.

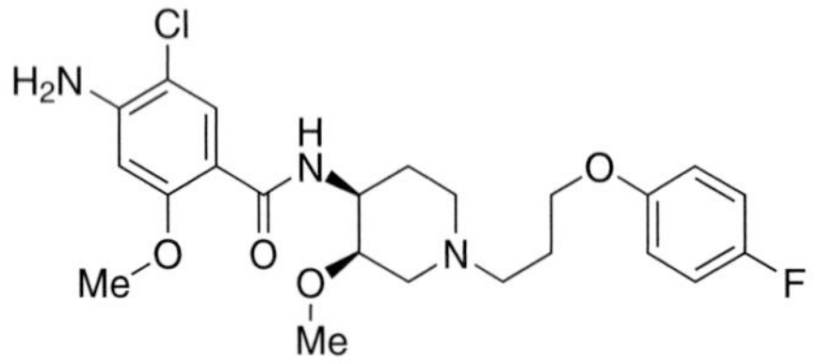

20 Cisapride

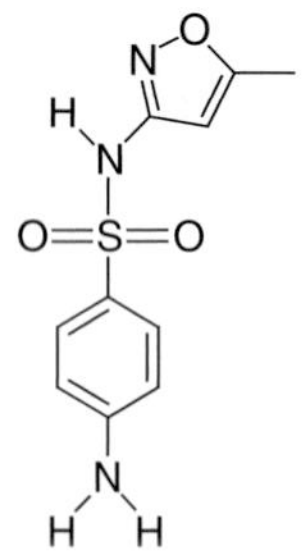

21 Sulfamethoxazole

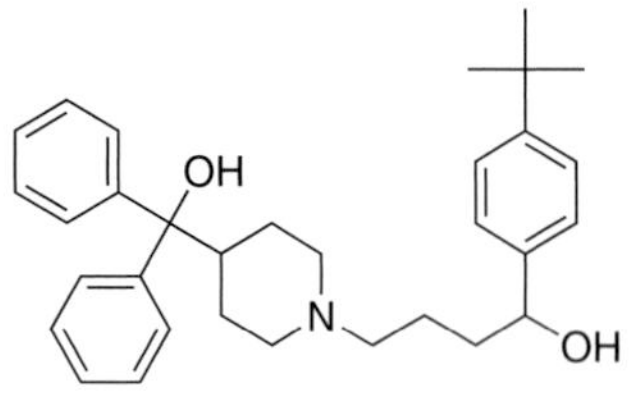

22 Terfenadine

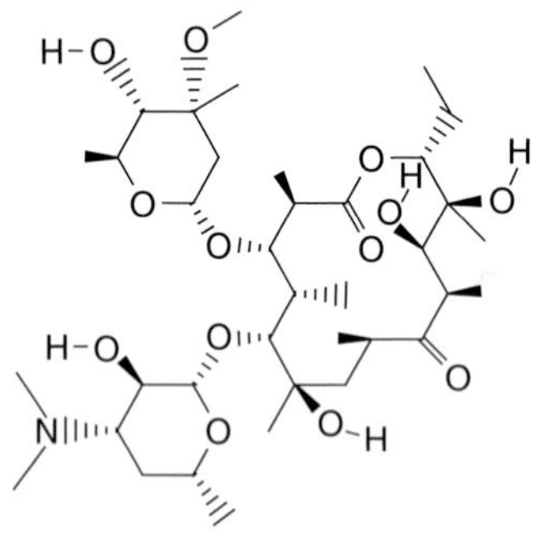

23 Erythromycin

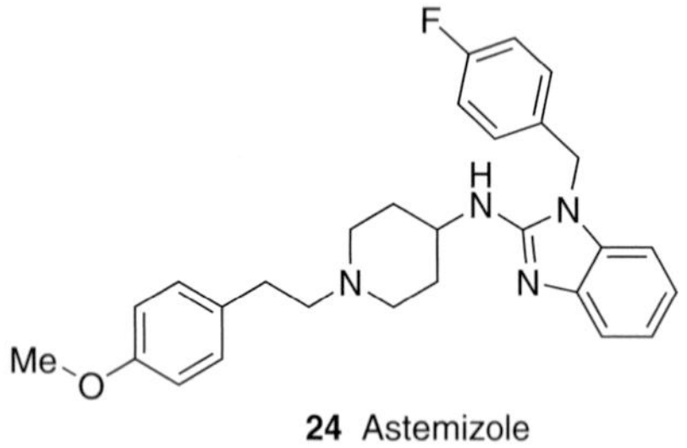

24 Astemizole

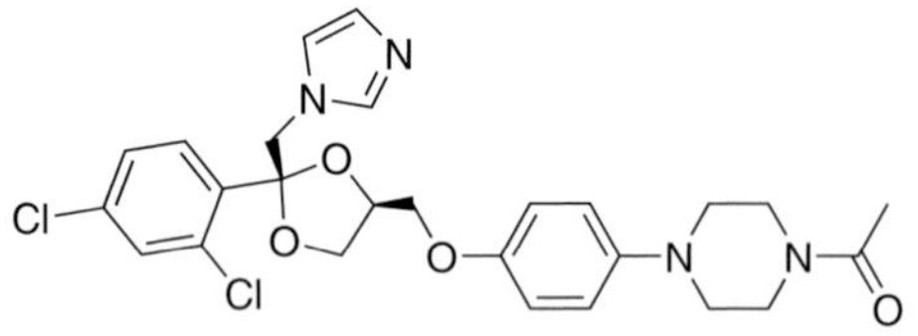

25 Ketoconazole

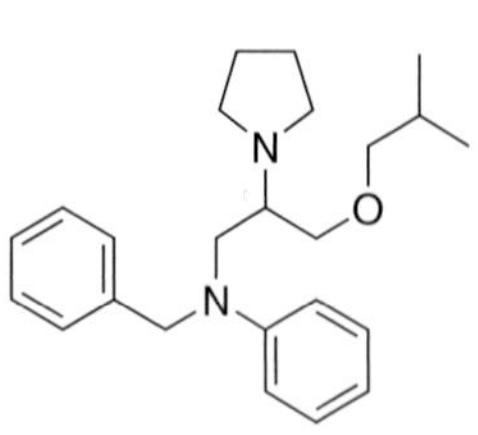

26 Bepridil

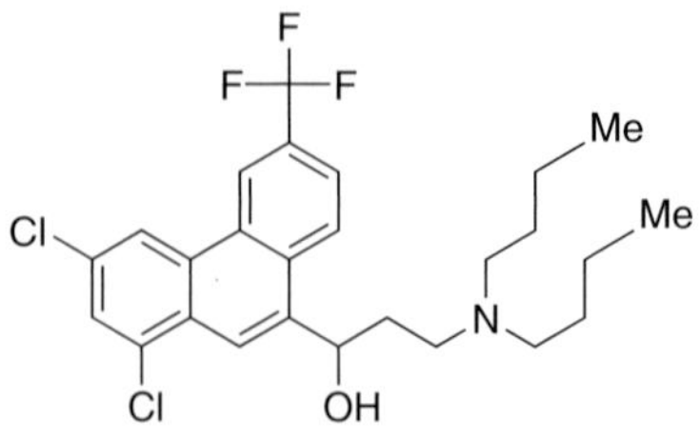

27 Halofantrine

Figure 10 Examples of non-methanesulfonanilide IK_r blockers.

6.33.5.2.1.4 Class 4: Calcium channel blockers

Verapamil and diltiazem (**Figure 11**) are not specific for the calcium channels in cardiac tissue. In fact they are more potent on vasculature tissue.

- Effectiveness: Although verapamil has been replaced by adenosine in the treatment of paroxysmal supraventricular nodal tachycardia, it is still useful in controlling ventricular rate in chronic atrial flutter and fibrillation. Verapamil has been used to slow ventricular rate in atrial fibrillation.
- Toxicity: Since the calcium current is important in mediating cardiac contraction and vasoconstriction, calcium channel blockers are associated with negative effects on cardiac contractility and with hypotension.

6.33.5.2.1.5 Class 5: pace maker current (I_f) blockers: alinidine, falipamil, zatebradine, ZD7288

These drugs, which currently have almost no clinical utility, are classified as Class 5 antiarrhythmics with the prototypical drug being alinidine (**Figure 12**). Others are zatebradine, falipamil, and ZD7288. They inhibit the pacemaker currents in the SA node and thereby induce bradycardia. However, little is known about their clinical utility.

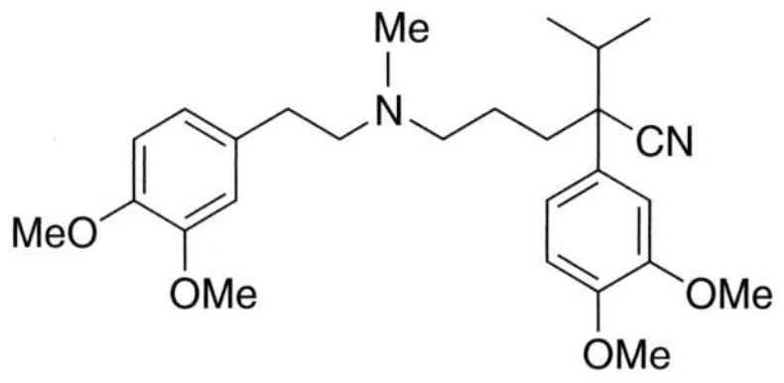

28 Verapamil

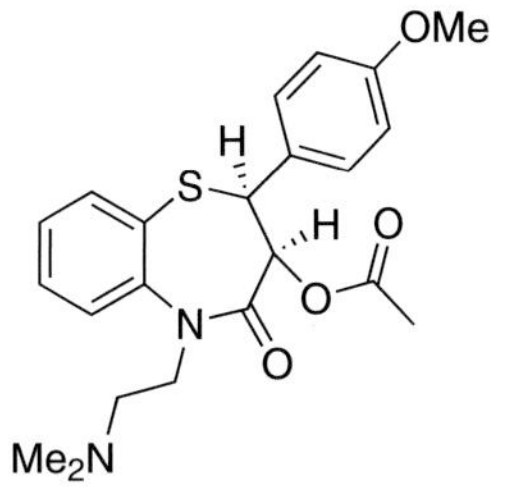

29 Diltiazem

Figure 11 Examples of class 4 antiarrhythmics (ICa blockers).

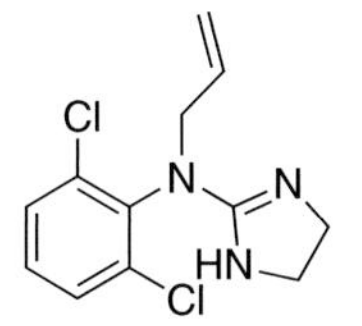

30 Alinidine

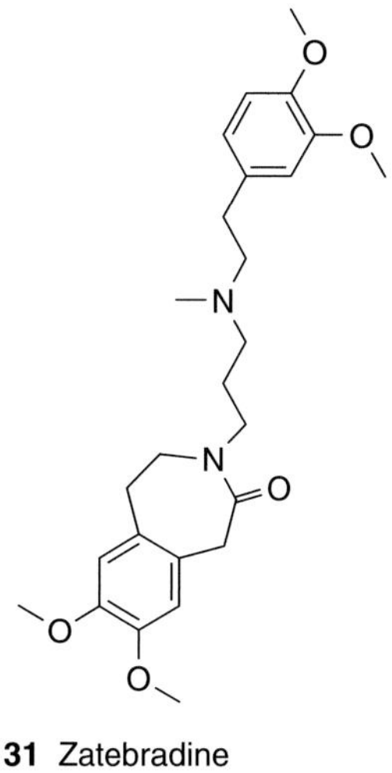

31 Zatebradine

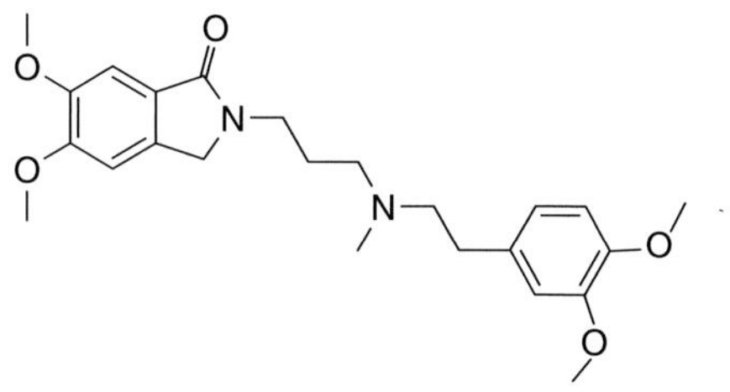

32 Falipamil

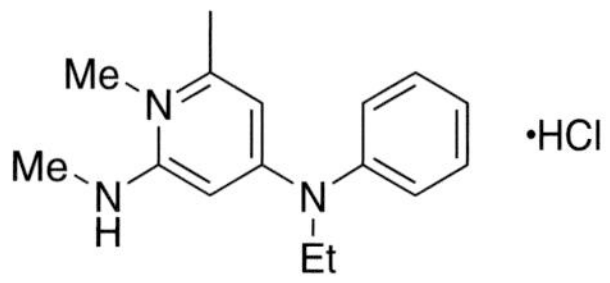

33 ZD7288

Figure 12 Examples of class 5 antiarrhythmics (I_f blockers).

6.33.5.2.2 Muscarinic receptor blockers

- Mechanism: Antagonists of the cardiac form of the muscarinic receptor (M_2) are useful in the abolition of an inappropriate bradycardia or excessively slow AV node conduction due to excessive activity in the cardiac parasympathetic nerves.
- Effectiveness: The nonselective muscarinic receptor blocker, atropine (**Figure 13**), is sometimes used in acute situations (e.g., bradycardia in acute myocardial infarction), in other emergency settings, and during anesthesia when excessive bradycardia is present. The cardiac form of the muscarinic receptor is the M_2 subtype. The only compounds with some selectivity for this subtype are gallamine and AF-DX 116 (**Figure 13**). Gallamine was originally introduced as a blocker at the skeletal neuromuscular junction (neuromuscular blocker), but it has some selectivity for M_2 receptors. AF-DX 116 is more specific for the M_2 receptor, but despite this, the usual muscarinic blocker used is typically atropine.
- Toxicity: Blockade of muscarinic receptors in other tissues account for side effects such as dry mouth, visual disturbances, and difficulties with micturition.

6.33.5.2.3 Adenosine

- Mechanism: The effects of adenosine (**Figure 14**) on atria and nodes are almost identical to those of ACh. Activation of the A1 subtype of adenosine receptors (P1) is linked to the potassium channels activated by acetylcholine (IK_{ACh}). Activation slows the rate of rise of pacemaking potentials and reduces nodal calcium currents, thereby reducing AV nodal conduction.
- Effectiveness: Adenosine's success rate in terminating paroxysmal supraventricular nodal tachycardia is as high as 95%. Thus it has replaced the less efficacious verapamil. Many episodes of paroxysmal supraventricular nodal tachycardia can be terminated by vagotonic maneuvers. These include the Valsalva maneuver (forced expiration against a closed glottis and the nostrils pinched shut), activation of the diving reflex (face in cold water), and carotid massage to activate the baroreceptor reflex.
- Toxicity: Although adenosine is a potent vasodilator, the half-life of adenosine is very short ($\sim$10 s) making its side effects tolerable, e.g., short-lived hypotension and occasional bronchoconstriction episode.

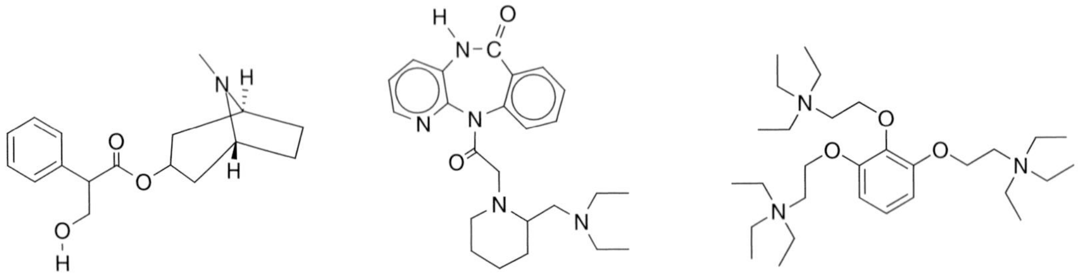

34 Atropine **35** AF-DX 116 **36** Gallamine

Figure 13 The prototypical muscarinic receptor blocker atropine and two other blockers with selectivity for M_2 receptor (cardiac form).

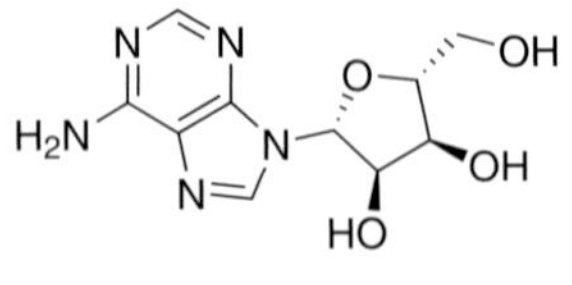

37 Adenosine

Figure 14 Structure of adenosine.

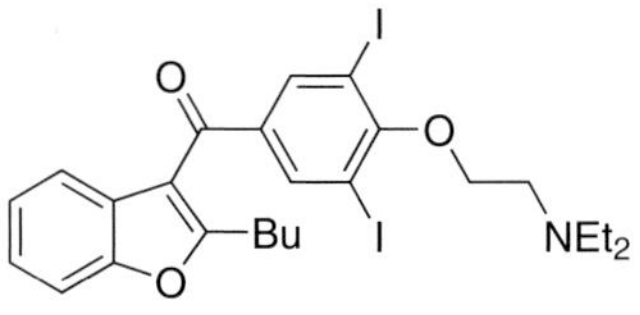

O

I

O

NEt$_2$

Bu I

O

38 Amiodarone

Figure 15 Structure of amiodarone.

6.33.5.2.4 Amiodarone

Amiodarone (**Figure 15**) is the only antiarrhythmic that, to a limited degree, reduces the incidence of cardiac arrhythmic sudden death, although with less efficacy than implanted defibrillators.[21]

- Mechanism: The mechanism of action of amiodarone is complex. Although originally classed as one of the prototype class 3 antiarrhythmic drugs that widen action potentials, it has since been found to block other channels, and even β-adrenoceptors. The importance of these latter effects is unknown.
- Effectiveness: Amiodarone reduces mortality due to ventricular tachyarrhythmias, and has value in the prophylaxis of atrial arrhythmias. It can be given both intravenously and orally.
- Toxicity: Amiodarone produces toxicity in the lungs, increases skin sensitivity to ultraviolet radiation, and induces corneal opacities in the eye. The toxicity in the lungs can be fatal. The drug has very unusual pharmacokinetics with a long half-life (days to weeks), but if given carefully, serious major toxicity can be avoided.

6.33.5.3 Overview of Current Electrical Treatments for Arrhythmias

As indicated previously, electrical therapy is becoming increasingly important as a method for treating arrhythmias. Some techniques that are being increasingly used include[22]:

1. Pacing – highly effective for controlling heart rate when there is AV block, and generally for control of atrial or ventricular rate where necessary.
2. DC conversion to sinus rhythm – this is highly effective with typical conversion rates of 89% for AF, 97% for AT, 67% for paroxysmal supraventricular tachycardia (PSVT), and 95% for VT and/or VF.
3. Ablation therapy – used where the site of the genesis of arrhythmias is known and is ablated by techniques of guided cryoablation, or radiofrequency ablation using intracardiac catheters, and x-ray guidance. By such means, anatomic or functional re-entry circuits can be abolished. It can be effective in up to 97% of cases of supraventricular tachycardia, and in 88–99% in cases of aberrant AV pathways such as occur in Wolf–Parkinson–White syndrome. It can also be effective in VT, but it has varying effectiveness for AF.
4. Implantable cardioverter defibrillators (ICDs or implanted defibrillators) – of great importance in the electrical control of arrhythmias since they are capable of both detecting an arrhythmia, and delivering an electroshock to revert the arrhythmia. Implantable ICDs reduce mortality in cases where ventricular tachycardia and/or fibrillation occur, and appear increasingly useful in atrial fibrillation.

6.33.5.4 Other Therapies for Arrhythmias

Other therapies for arrhythmias include cardiac surgery for the correction of conditions causing arrhythmias. The Maze technique involves surgical incisions across the atria or around the pulmonary veins to prevent atrial tachyarrhythmias. During anesthesia and surgery, as well as in emergency situations, imbalances in autonomic nerve activity, blood gases, and electrolyte can all cause arrhythmias, as can hypo- and hyperkalemia in outpatients. The correction of such arrhythmogenic abnormalities, by drugs or electrolytes, can rapidly correct these arrhythmias.

6.33.6 Unmet Medical Needs

The limitations to current therapy arise from many reasons, including incomplete knowledge of normal and arrhythmic electrogenesis. Models of electrogenesis (normal or arrhythmic) are still not truly predictive and we have still to identify specific arrhythmogens (except for catecholamines, potassium, and H^+) where blockade of their production could be reasonably expected to prevent arrhythmias.

While molecularly selective blockade can be produced with some drugs, such blockade is only antiarrhythmic under restricted conditions. Excessive ion channel blockade, regardless of the channel blocked, is arrhythmogenic.

In general, current antiarrhythmic drugs lack effectiveness, thereby emphasizing a need for new therapies for arrhythmias, whether these involve drugs or electrical devices. There are efficacy and toxicity limitations with most of the currently available drugs, despite new and improved knowledge of arrhythmias, and new concepts regarding strategies for better therapy. Consequently, there are many areas where treatment is far from ideal. A prime example is a need for prophylactic drug treatment in the prevention of cardiac sudden death due to ventricular fibrillation. Other areas of need include the following conditions.

6.33.6.1 Atrial Arrhythmias

Better antiarrhythmic drugs are needed for the acute termination of atrial tachycardia and fibrillation, whether of recent onset or long term. This remains a goal even if it is accepted that control of ventricular rate in atrial fibrillation is as good a therapeutic goal as reversion to sinus rhythm. However, to some extent the processes responsible for these arrhythmias are irreversible and the arrhythmia will eventually return. Additionally, reversion to sinus rhythm of long-standing atrial fibrillation carries the risk of emboli being dislodged from the atria, as a result of the sudden vigorous and coordinated contraction of atria upon resumption of sinus rhythm.

Despite the availability of electrical conversion for atrial arrhythmias, there is a need for:

1. Quick and simple conversion of supraventricular arrhythmias with drugs given intravenously or orally, together with a need for better prophylaxis against the recurrence of supraventricular arrhythmias.
2. Better antiarrhythmic drugs for the control of ventricular rate during atrial fibrillation. The current drugs (β-adrenoceptor blockers, calcium channel blockers, cardiac glycosides) are far from ideal. For instance, a cardiac selective L-type calcium channel blocker might have advantages in that it would be less likely to lower blood pressure.

6.33.6.2 Ventricular Arrhythmias

While it is apparent that electrical defibrillation is an effective treatment for ventricular tachycardia and fibrillation, this does not obviate the need for better drugs to prevent their occurrence. Thus there is a need for:

1. Prophylactic drugs for prevention of ventricular arrhythmias.
2. Background antiarrhythmic drug cover in those with implanted electrical devices. Even the latest devices are stressful to many patients when the defibrillation shock is delivered, and so there is a need to reduce their frequency at which they have to be activated.

6.33.7 New Research Areas

The following sections discuss the various approaches to finding new antiarrhythmic drugs, whether as variations on existing drugs, or as totally new antiarrhythmic drugs. These are discussed within the background of current limitations to antiarrhythmic drug therapy.

6.33.7.1 New Antiarrhythmics (Classes 1–5)

There is still muted interest in developing more effective and safer drugs to prevent and revert atrial and ventricular arrhythmias. The identification of new ion channels, enhanced knowledge regarding existing channels, new information on pathology, and the identification of putative new arrhythmogens provide the basis for ongoing research.

6.33.7.1.1 Class 1

Programs are ongoing to identify new class 1 antiarrhythmics, although there is little residual interest in this field because of the assumption that the class 1 mechanism is inherently flawed. This view is probably erroneous since all facets of cardiac sodium channel blockade have yet to be investigated. If there are critical differences in the nature and behavior of such channels in the initiation or maintenance arrhythmias, it is possible that specialized class 1 effects could provide better antiarrhythmic drugs.

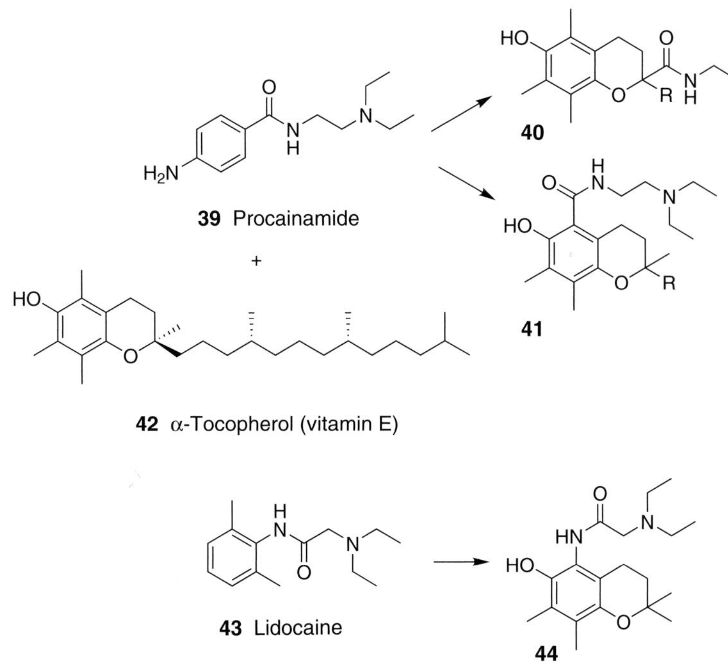

39 Procainamide

+

42 α-Tocopherol (vitamin E)

40

41

43 Lidocaine

44

Figure 16 Novel hybrid vitamin E/class 1 antiarrhythmics.

6.33.7.1.1.1 Novel hybrid vitamin E/class I antiarrhythmics

Antioxidants such as vitamin E protect against injury due to myocardial ischemia and subsequent reperfusion, although their highly hydrophobic properties reduce their availability to cardiac myocytes. One interesting approach has been to eliminate some of the hydrophobic moieties of vitamin E, and combine the resulting compound with the known class 1 antiarrhythmics such as lidocaine and procainamide to create a bifunctional class of antiarrhythmic antioxidants.[23] These compounds inhibit lipid peroxidation with antiarrhythmic effects against ischemia-reperfusion experimentally. **Figure 16** shows examples of these new hybrids where procainamide and lidocaine are covalently bound to α-tocopherol with the expectation that the saturated chain will confer selectivity for antiarrhythmic action.

6.33.7.1.2 Class 2

Since the introduction of propranolol in the early 1960s, many other β-adrenoceptor blockers have been introduced. It is not readily apparent that any have particular advantages over other drugs as antiarrhythmics, although they vary widely in their pharmacological and pharmacokinetic profiles. Differences in their pharmacological profiles are relatively minor, even for cardiac selectivity. There is no unequivocal evidence that such limited β-adrenoceptor selectivity provides better antiarrhythmic protection, neither does partial agonism. The same is probably true of pharmacokinetic differences, although it is reasonable to suppose that a drug providing the same level of blockade over 24 h would be better.

Apparently there have been few attempts to maximize the β-adrenoceptor selectivity of β-adrenoceptor blockers, specifically with respect to antiarrhythmic actions. The structural differences between propranolol and the cardiac selective blockers atenolol and metoprolol are shown in **Figure 7**.

In view of the above, it is not surprising that little effort has been expended on better class 2 antiarrhythmics. Possibly a markedly selective β-adrenoceptor blocker with a longer half-life would confer better protection against arrhythmias, but it would be very expensive to test this proposition clinically, since the endpoint for the necessary clinical trials would be mortality.

6.33.7.1.3 Class 3

6.33.7.1.3.1 IK_r blockers with faster recovery kinetics from channel binding

Most IK_r blockers have undesirable reverse rate dependence that is opposite to what is probably required clinically, since antiarrhythmic actions should be greater at the high rates seen during arrhythmias. Thus, compounds with

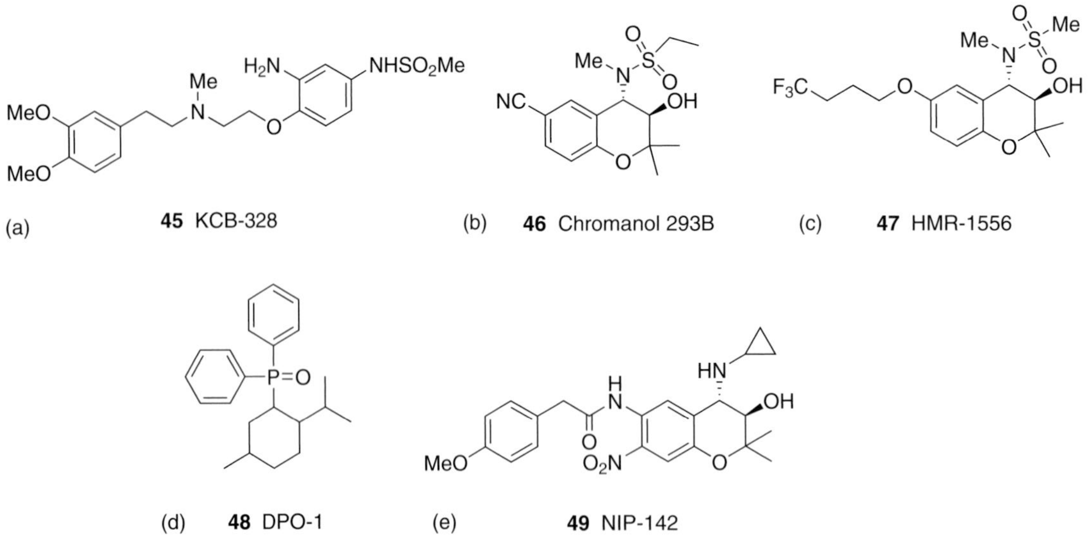

(a) **45** KCB-328 (b) **46** Chromanol 293B (c) **47** HMR-1556

(d) **48** DPO-1 (e) **49** NIP-142

Figure 17 (a) A recent example of an IK_r blocker with faster recovery kinetics from channel binding. Structure of IK_S blockers (b) chromanol 293B, (c) HMR 1556, (d) IK_{ur} blocker DPO-1, and (e) IK_{ur} and IK_{Ach}: NIP-142.

different recovery kinetics from IK_r block have been explored to improve their rate-dependent profile. KCB-328 (**Figure 17**)[24] has faster recovery kinetics than dofetilide. It unbinds from IK_r channels more readily, with little reverse rate dependence.

6.33.7.1.3.2 Selective IK_S blockers

IK_s is directly activated by increased sympathetic nervous activity. Moreover, the slow deactivation kinetics of IK_s favors its accumulation at higher heart rates. Thus its blockade and the accompanying action potential duration prolongation may result in increased effectiveness against catecholamine-induced tachyarrhythmias. Although IK_s blockers like chromanol 293B (**46**) and HMR 1556 (**47**) have not been tested clinically, they are antiarrhythmic in various animal models.[25] HMR 1556 has ototoxic side effects such as hearing impairment in cats, a finding in keeping with the fact that IK_s is found in the ear. Nonfunctioning mutations of the channel are associated with both arrhythmias related to long QT and hearing deficits.

6.33.7.1.3.3 IK_{ur} blockers

The occurrence of *torsades* limits the undoubted antiarrhythmic effect of action potential widening. Therefore, it is important to discover compounds that have this action, but do not induce torsades. The nature and complexity of the potassium channels that contribute to action potential duration emphasize a need to target each of these subtypes of channels individually, and then systematically study the effect of block when using combinations of specific subtype blockers. An example is the role of IK_{ur} in controlling action potential duration in atria, versus ventricles. IK_{ur} is found primarily in atria, and so it is suggested that IK_{ur} blockers would have selective class 3 actions in atria, without adverse QT prolonging effects in ventricles. This supposes that IK_{ur} regulates atrial repolarization and as a result increases atrial refractoriness thereby terminating, or preventing, atrial arrhythmias. For example, a novel IK_{ur} blocker, DPO-1 (**48**) preferentially increased atrial refractoriness without prolonging the ventricular refractory period, or QT interval, in nonhuman primates.[26] Thus, there is a predictable trend for the continuous development of atrial selective drugs that should potentially be safer in that they should have limited ventricular effects, and could be used prophylactically for atrial arrhythmias.

With chronic atrial tachyarrhythmias, there is a change in the density and importance of IK_{ur}, such that electrogenesis in chronically arrhythmic atria is different from that in normal atria. Thus, an IK_{ur} blocker may have differential effects in preventing the reoccurrence of atrial arrhythmias, or terminating acute atrial fibrillation.

6.33.7.1.3.3 IK_{ACh} blockers

The potential of IK_{ACh} blockade was aided by the discovery of tertiapin, a bee venom peptide, shown to block IK_{ACh} at nanomolar concentrations. Tertiapin prolongs guinea pig atrial, but not ventricular, action potentials.[27] The

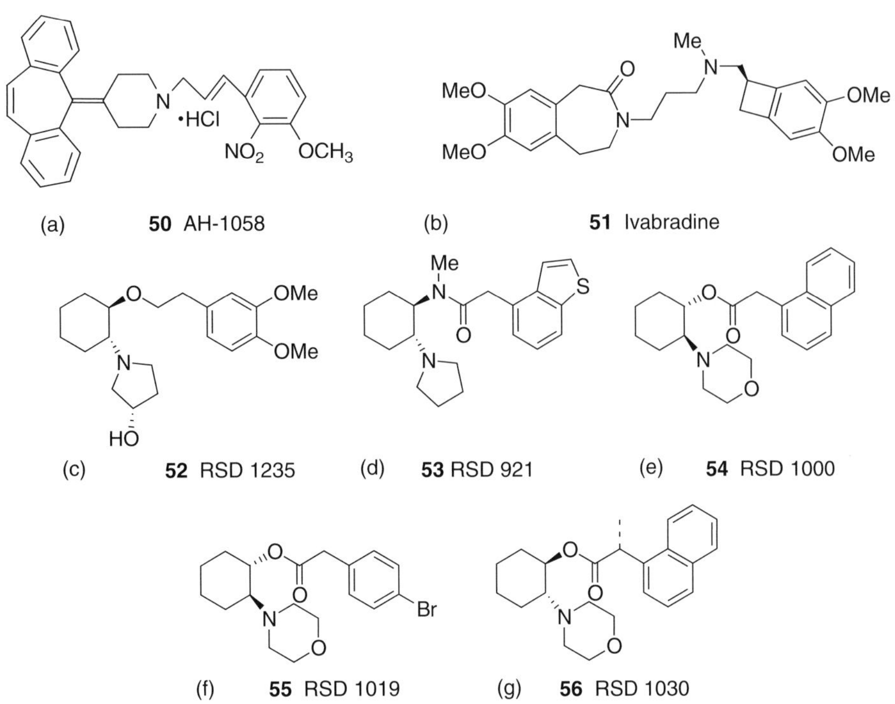

Figure 18 Structure of the (a) cardioselective calcium channel blocker AH-1058, (b) the pacemaker current (I_f) blocker ivabradine and (c–g) RSD1235 and its congeners.

combination of IK_{ur} and IK_{ACh} blockade is exemplified by NIP-142 (**49**) which terminates atrial fibrillation and flutter in dogs, and has electrophysiological actions confined to atria. The rationale for IK_{ACh} block arises from the fact that ACh from cardiac parasympathetic nerves acts on atrial tissue to increase (hyperpolarize) the resting potential and reduce action potential duration, changes that induce arrhythmias. Thus, if IK_{ACh} is blocked, a source of arrhythmias is removed. ACh muscarinic blockers would accomplish the same end result in the heart, but they are not cardiac selective, and therefore produce unacceptable side effects.

6.33.7.1.4 Class 4

6.33.7.1.4.1 Cardioselective Ca^{2+} channel blocker AH-1058

Unfortunately, the prototypical antiarrhythmic L-type Ca^{2+} channel blocker, verapamil (**28**), is not selective for cardiac versus vascular tissue. Early work suggested that if a calcium channel blocker had selective actions in ischemic cardiac tissue, it would be selective in its actions against ischemia-induced ventricular arrhythmias. There have been few efforts to follow this path although a new cyproheptadine derivative, AH-1058 (**50**), was developed to block L-type Ca^{2+} channels with better cardiac selectivity and limited vasodilator actions. AH-1058 (**Figure 18**) suppresses ventricular arrhythmias in dogs, but its negative effects on cardiac contractions coupled with limited hypotension may limit its utility in patients with cardiac contractile dysfunction, as in heart failure.[28]

6.33.7.1.5 Class 5

The class 5 designation was created for specific bradycardic drugs that slowed the SA node rate. Alinidine (**30**) (**Figure 12**) was the prototype for such compounds with newer ones including the novel pacemaker current (I_f) inhibitor ivabradine (**51**) which blocks I_f at high potency.[10] It is an open-channel blocker with a favorable 'positive use-dependence' that should potentially make ivabradine more effective in reducing heart rate during SA node tachycardia.

6.33.7.2 Antiarrhythmic Drugs Combining Actions of Classes 1–5

No selective channel blocker has yet been shown to unequivocally reduce mortality associated with atrial or ventricular arrhythmias. Amiodarone (**38**) (**Figure 15**), a mixed channel blocker that has class 1–4 actions, is the only ion channel blocking antiarrhythmic drug shown clinically to prevent cardiac sudden death. Thus, compounds with appropriate mixed channel blocking effects may be a useful route to new antiarrhythmic drugs. There is experimental evidence that when two different channel blockers are given together, the resulting antiarrhythmic effect is greater than either given alone. As a result, several compounds with combined ion channel blocking actions are under study, obvious examples being derivatives of amiodarone that lack its toxicity.[29]

RSD1235 (**52**) is a compound currently in phase III clinical trials for intravenous use in the termination of acute onset atrial fibrillation. This compound is a mixed channel blocker with combined Class 1b and 3 effects. Blockade of IK_{ur} provides an atrial selective widening of action potential and this is combined with class 1b actions and a degree of ischemia dependency. This combination of electropharmacological properties probably accounts for the ability of RSD1235 to terminate AF while having minimal effects on the ventricle. The positive rate dependence probably confers selectivity for tachyarrhythmias and widens the therapeutic window in the treatment of AF. RSD1235 showed no drug-induced arrhythmias or serious adverse events during phase I and phase II clinical trials.[30] In a phase III trial, RSD1235 terminated AF without inducing *torsades de pointes* (see later).

6.33.7.2.1 Combined class 1b and 3 actions

GYKI-16638 (**57**) (**Figure 19**) is one of a new series of *N*-(phenylalkyl)-*N*-phenoxyalkylamines, which combine the use dependent blockade of INa (Class 1b), with IK_r and IK_1 blockade. GYKI-16638 prolongs action potential duration in human ventricular muscle with little reverse rate dependence, unlike the typical selective IK_r blockers.[31] Such a neutral rate dependent effect, similar to amiodarone, is anticipated to lessen drug-induced proarrhythmia. Although GYKI-16638 is not a congener of amiodarone, it has certain structural similarities and might be expected to share some of the electrophysiological characteristics of amiodarone.

6.33.7.2.2 Combined class 3 and 4 actions

BRL-32872 (**58**) has blocking actions on both IK_r and ICa with respective IC_{50} values of 0.028 and 2.8 μM. At low concentrations, with IK_r blockade present, BRL-32872 prolongs action potential duration as expected of an IK_r blocker, but at higher concentrations the ICa blockade becomes significant and counteracts the action potential prolonging effects. Moreover, BRL-32872 does not prolong action potential with reverse rate dependency, suggesting that ICa inhibition attenuates the undesirable reverse rate dependence of IK_r block.[32]

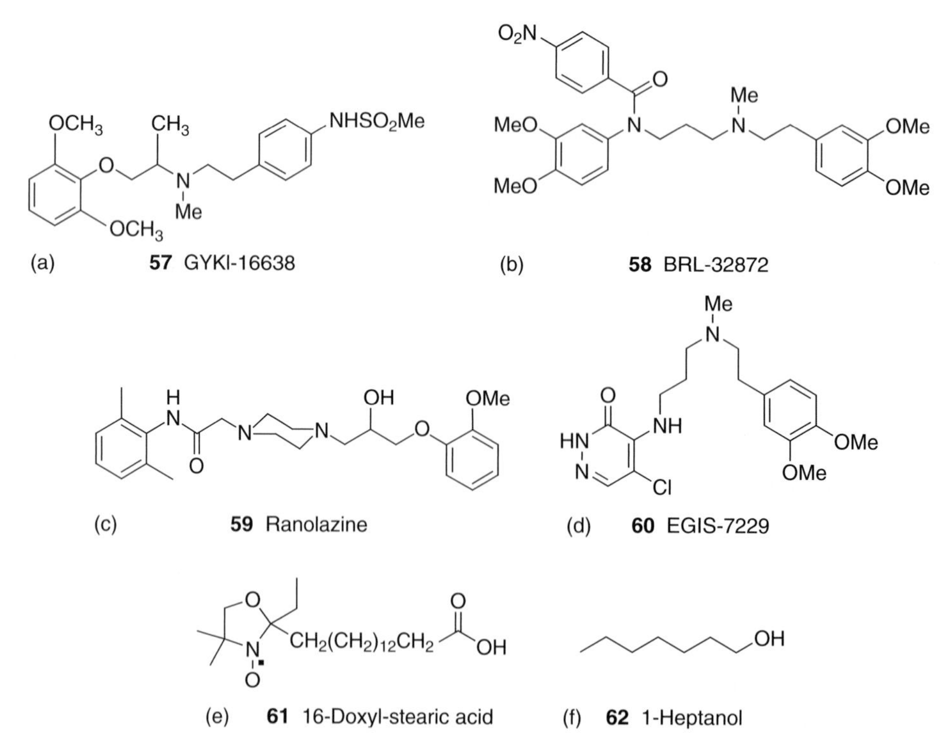

Figure 19 Structure of (a) GYKI-16638: combined class 1b and 3 actions, (b) BRL-32872: combined class 3 and 4 actions, (c–d) ranolazine and EGIS-7229: combined class 1, 3, and 4 actions, and (e–f) gap junction blockers: 16-doxyl-stearic acid and 1-heptanol.

BRL-32872 inhibits ischemia induced ventricular arrhythmias in minipigs, and it also reduces electrically induced VT in dogs. Moreover, it did not induce arrhythmias in rabbits, unlike selective IK_r blockers.

6.33.7.2.3 Combined class 1, 3, and 4 actions

The antianginal agent ranolazine (**59**) is an inhibitor of fatty acid oxidation having direct electrophysiological actions due to combined channel blockade of Na^+, K^+, and Ca^{2+} channels. It prolongs cardiac action potential duration without any rate dependency or induction of torsades. Interestingly, it suppresses drug-induced proarrhythmias induced by drugs that decrease IK_r or increase the late sodium current.[33]

EGIS-7229 (**60**) is another novel compound with combined class 1, 3, and 4 actions. Low concentration of EGIS-7229 inhibits IK_r and prolongs action potentials, but at higher concentrations it shortens action potentials with use-dependent inhibition of Na^+ channels.[34] Although the effects of EGIS-7229 are complex, it inhibits early after-depolarization, unlike sotatol which increases early after-depolarizations (EADs) in rabbit papillary muscles. Therefore, EGIS-7229, like BRL-32872, may have self-limiting action potential prolonging effects.

6.33.7.3 Drugs that Block Other Ion Channels, Pumps, or Transporters

In addition to the classic ion channels for sodium, potassium, and calcium, there are other ion channels that are thought to play a role in arrhythmias. There are stretch receptors that are candidates for explaining how stretch might induce arrhythmias. A common observation, both in vivo and in vitro, is that stretch of cardiac tissue can induce various types of arrhythmias.

Current flow between cardiac cells is vital for conduction of the impulse through all types of cardiac tissue. This is primarily by low-resistance pathways providing for easy electronic spread of potential from one cell to another. During ischemia and other conditions, changes in these pathways lead to the impediment of the spread of currents between cells, resulting in poor conduction, a condition where re-entry circuits become more likely.

In addition to the critical role played by voltage and ligand dependent ion channels, the maintenance of appropriate cardiac electrophysiology depends upon the existence of appropriate ionic gradients across cardiac cell membranes. The maintenance of such ionic gradients depends upon pumps and transporters for ions, principally sodium, potassium, calcium, and hydrogen. Since disorders of these mechanisms are common in conditions that result in arrhythmias, it is no surprise that they are potential targets for antiarrhythmic drugs.

6.33.7.3.1 Gap junction activators

Conduction though gap junctions is reduced during ischemia and contributes to re-entry arrhythmias. For example, expression of the gap junction protein connexin 43 (Cx43) is downregulated in guinea pigs with congestive heart failure and these animals are liable to arrhythmias. The reduction in gap junction conductance due to ischemia is related to intracellular acidosis. However, compounds that block gap junctions (16-doxyl-stearic acid (**61**) and 1-heptanol (**62**)) reduce the ventricular defibrillation threshold in isolated rabbit hearts.[35] Therefore, the role of gap junctions in arrhythmias is not totally clear, and there is uncertainty as to whether uncoupling, or maintaining, gap junctions is antiarrhythmic.

The short polypeptides, ZP123 (**63**) and AAP-10 (**64**) (**Figure 20**), prevent the closure of gap junctions induced by prolonged acidosis in guinea pig hearts. As a result, they reduce the heterogeneity of repolarization induced by acidosis. AAP-10 is a natural amino acid hexapeptide unlike the longer ZP123 which contains D amino acids. Not surprisingly AAP-10 has the limitation of a short half-life as compared with ZP123.[36]

6.33.7.3.2 Stretch activated channel blockers

The stretching that occurs in damaged atria or following overload of atria is a well-known cause of arrhythmias. Therefore, stretch-induced channels are considered a source of arrhythmogenic currents, and as a result a target for new antiarrhythmic drugs. Compounds such as DCPIB (**65**) can be expected to have antiarrhythmic actions, since it selectively inhibits the swelling-induced chloride current ($I_{Cl,swell}$) and prevents swell-induced shortening of atrial action potential in guinea pigs.[37] Unfortunately, there have been no studies into the effects of such blockers in arrhythmias.

6.33.7.3.3 Pumps, exchangers, and transporters

Ion exchangers or transporters are a fairly obvious target for antiarrhythmic drug discovery since they are integral to ionic homeostasis in cells, and generate electrogenic pump currents that supplement and complement the classical currents due to the opening of ion channels. Overexpression, or inappropriate activity, of such mechanisms is postulated to be responsible for certain types of arrhythmias.

N-Ac-(D)-Tyr-(D)-Pro-(D)-Hyp-Gly-(D)-Ala-Gly-NH_2

(a) **63** ZP123

Gly-Ala-Gly-Hyp-Pro-Tyr-NH_2

(b) **64** AAP-10

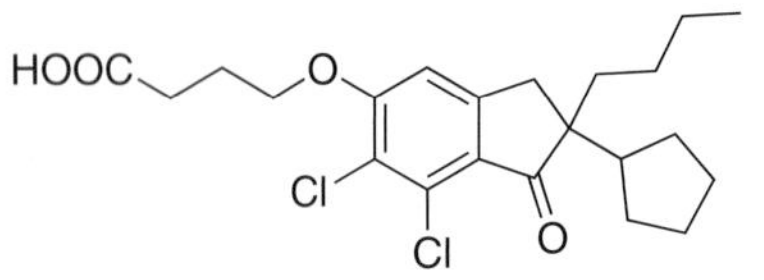

(c) **65** DCPIB

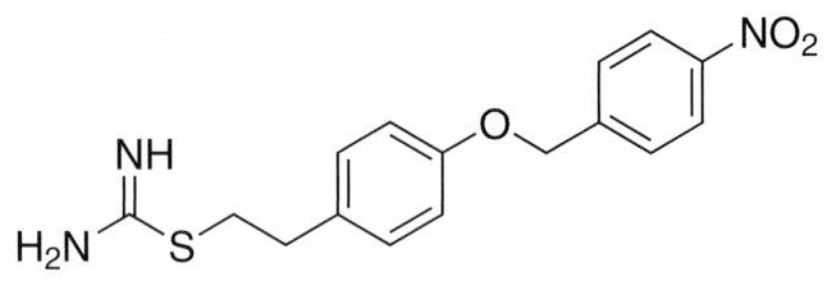

(d) **66** KB-R7943

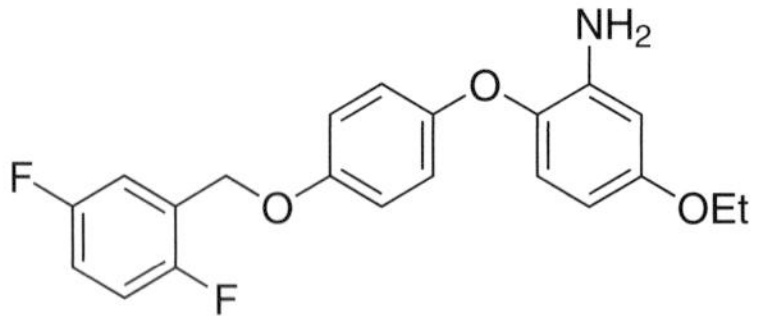

(e) **67** SEA0400

Figure 20 Polypeptide structure of gap junction activators: (a) AAP-10 and (b) ZP123. A stretch-activated channel ($I_{Cl,swell}$) blocker: (c) DCPIB. Sodium–calcium exchanger blockers: (d) KB-R7943 and (e) SEA0400.

6.33.7.3.3.1 Sodium–calcium exchanger blockers

The sodium–calcium exchanger is a particularly interesting target due to its role in arrhythmic situations where there is calcium overload, after potentials and other arrhythmic changes. The function of the sodium–calcium exchanger is perturbed by ischemia, heart failure, and by other pathological conditions of the heart, including the excessive action potential prolongation seen with IK_r blockers. Experimental studies have shown that sodium–calcium exchanger blockers, such as KB-R7943 (**66**) and SEA0400 (**67**), prevent arrhythmias due to ischemia and reperfusion.[38]

6.33.7.3.3.2 Sodium–hydrogen exchanger blockers

There have been many studies with the sodium–hydrogen exchanger (NHE) and the effects of drugs upon its activity. This exchanger is essential in cardiac cells since it moderates intracellular acidosis by exchanging one extracellular sodium ion for one intracellular hydrogen ion. In animals with infarction and ventricular hypertrophy, the level and activity of the exchanger are elevated. This increased activity may also lead to increases in intracellular sodium, which in turn increase intracellular Ca^{2+}, via the sodium–calcium exchanger, and result in Ca^{2+} overload. In rabbits, treatment with cariporide, an NHE inhibitor, attenuates the development of cardiac hypertrophy and failure, ionic remodeling, and arrhythmias.

Cariporide (**68**) was amongst the first blockers of the sodium–hydrogen exchanger. Cariporide and HOE-694 (**69**) are analogs of the diuretic amiloride (**70**) (**Figure 21**). Others include compounds such as eniporide (**71**) and BIIB-513 (**72**) (**Figure 21**), with the latter shown to be cardiac protective in a canine model of myocardial ischemia.[39]

6.33.7.3.3.3 Fatty acid and fish oil

There has been interest over the years regarding arrhythmogens released from endogenous fatty acids, and attempts have been made to manipulate the endogenous fatty acid composition by dietary means leading to sporadic experimental and clinical studies into the potential antiarrhythmic effects of diet. In a randomized control trial,[40] fish oil supplementation in patients with an ICD was associated with more episodes of recurrent VT/VF, thereby suggesting that, rather than reduce arrhythmias, fish oil might be proarrhythmic in some patients.

The mechanism by which long-chain ω-3 polyunsaturated fatty acids (PUFAs), like eicosapentaenoic acid (EPA, **73**) and docosahexaenoic acid (DHA, **74**) (**Figure 22**), affect arrhythmias is not known with any certainty, but Na^+ and K^+ channel blockade/modification is probably responsible for its potential class 1 and 3 effects.[41,42]

6.33.7.4 Rational Approaches to New Antiarrhythmic Drugs

It is possible to try to develop a rationale of how to discover new antiarrhythmic drugs. The most important factor to consider is the arrhythmia to be targeted. Insufficient attention has been paid to this factor since it was assumed too

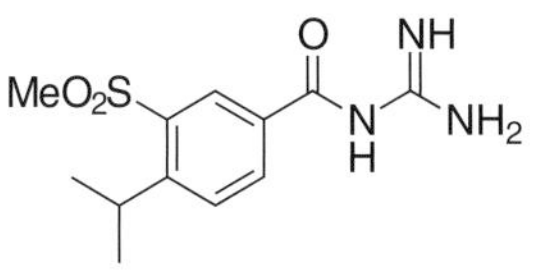

68 Cariporide

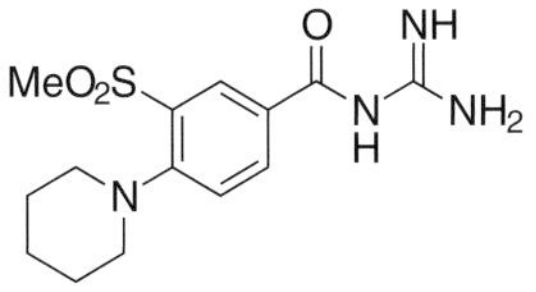

69 HOE-694

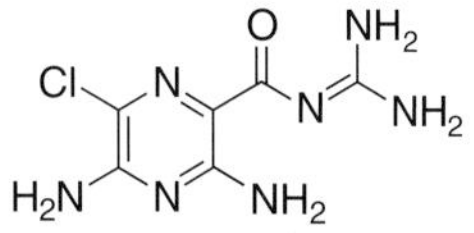

70 Amiloride

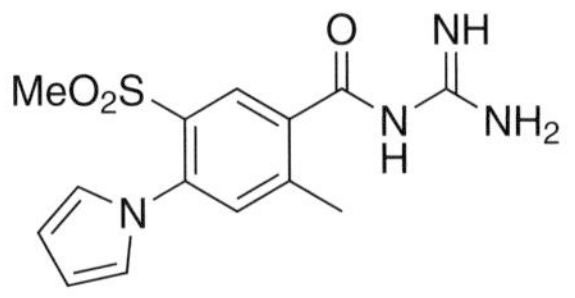

71 Eniporide

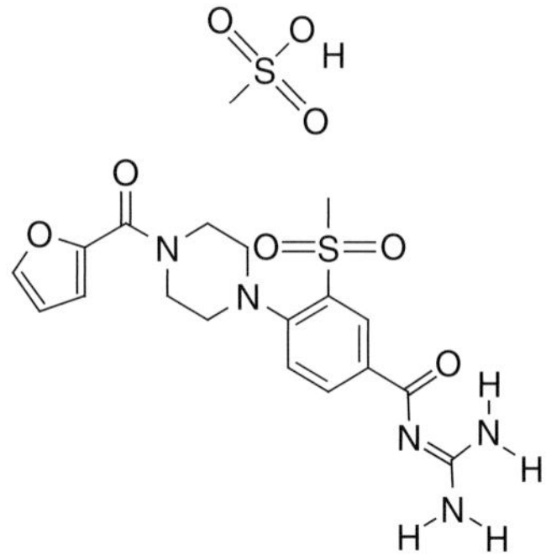

72 BIIB-513

Figure 21 Structure of several sodium–hydrogen exchanger blockers with cariporide as prototype.

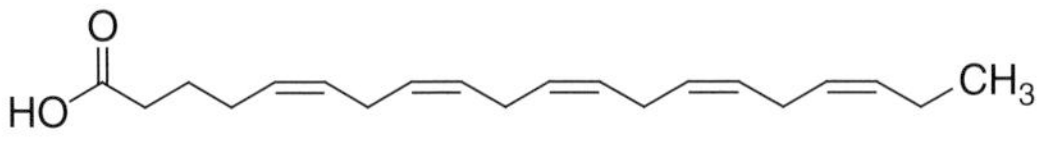

73 Eicosapentaenoic acid

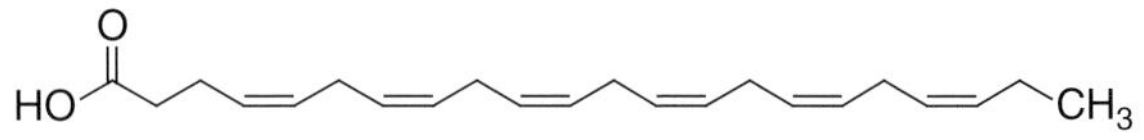

74 docosahexaenoic acid

Figure 22 Examples of ω-3 polyunsaturated fatty acids.

readily that all arrhythmias are similar. This is not so. With this in mind, various approaches can be considered such as:

1. Prevention of the generation, release, or action of arrhythmogens. Obvious examples of such an approach are class 2 antiarrhythmics which prevent the arrhythmogenic effects of catecholamines. Unfortunately, none of the other identified arrhythmogens have proved to be important enough to warrant targeting them with a full-scale research program. Similarly, a combination of arrhythmogens may also not be that important.[3]
2. Blockade of particular channels whose activation generates arrhythmias. Examples include IK_{ATP} and stretch channels. However, evidence for any particular channel being a sine qua non for arrhythmias is not overwhelming. It is the altered behavior of normal ion channels due to diverse pathology that appears to cause many arrhythmias.
3. Correction of pathologically disturbed normal channels to either revert behavior to normal; block them; or prevent their participation in the genesis/maintenance of arrhythmias.
4. Correction of the mutated channels responsible for the generation/maintenance of arrhythmias. The classic examples are those responsible for the prolonged QT syndrome, and associated torsades. It may prove possible in the future to use gene therapy, but currently drugs are needed to selectively alter the behavior of such aberrant channels, or prevent the consequence of their activation, e.g., intracellular calcium loading.
5. Selective blockade or activation of particular ion channels (e.g., INa, ICa, IK_r, IK_{ur}, I_f) might yet prove to be a useful antiarrhythmic strategy, for example, blockade of IK_{ur} for atrial selective drugs.
6. Combination blockers of various ion channels. Amiodarone probably owes its antiarrhythmic actions to simultaneous block of a number of different type of channels. The important question of which channels to block, and what degree of blockade, is currently difficult to answer.
7. Selective ion blocking actions confined to cardiac tissue that is subject to the pathological conditions causing an arrhythmia. The most obvious example is pathology due to myocardial ischemia and infarction. Similarly, damage due to infarction, or to stretch of cardiac tissue, provides another particular target. In the case of ischemia, its special conditions might be used to activate drugs so that they are only active in ischemic tissue.

As indicated above, all of these approaches have been studied, but with varying degrees of success. However they still provide a source for future directions in the search for new antiarrhythmic drugs.

6.33.7.5 Future Directions

In order to discover new antiarrhythmic drugs it is important to identify suitable arrhythmia targets, as well as molecular targets. It is also important to realize that both the induction, and maintenance, of arrhythmias is complex, may not be identical, and depends on the interplay of a number of different ion currents. Thus, simple reductionist models for arrhythmias and their treatment that are based upon analogies with diseases such as microbial infection and antibacterial drugs do not work well, since for arrhythmias there is often no one single agent responsible for the induction and/or the maintenance of arrhythmias.

A better disease analogy for arrhythmias might be epilepsy where multiple processes participate in the induction and maintenance of an epileptic attack. Despite this, it is possible to provide prophylaxis against epilepsy with a variety of drugs acting by different mechanisms.

It is important that certain conceptual steps are taken before embarking on an antiarrhythmic drug discovery program. Thus, in any antiarrhythmic drug program it is important to:

1. Clearly identify the target arrhythmia, and its anatomical location. A target arrhythmia should be as homogeneous as possible on the assumption that there is one arrhythmic mechanism responsible for a particular arrhythmia. Arguing by analogy from one type of arrhythmia to another may not be relevant.
2. Identify conditions and mechanisms unique to the target arrhythmia, its induction, and maintenance. Such knowledge provides possible avenues by which specificity of action can be achieved.
3. Choose a specific mechanistic objective for the drug so as to confer selectivity for the target arrhythmia (e.g., special characteristics such as blockade of a particular ion channel, or an arrhythmogen, tissue selectivity, frequency dependence).
4. Identify special factors that might confer therapeutic utility, e.g., special pharmacokinetic factors, or a special pharmacological profile. An example is the use of intravenous adenosine for termination of supraventricular nodal tachycardia. In this case the mechanism of action, and very short duration, confer special attributes that allow this drug to be used, but only against a specific arrhythmia.
5. Explicitly state the approach to be used.

The following section exemplifies the manner in which a new antiarrhythmic, RSD1235, was discovered. In this example the logic, procedures, and serendipitous findings are identified and discussed.

6.33.7.5.1 An example of an approach to discovering a new antiarrhythmic drug

The RSD1235 program began with an initial goal of discovering an antiarrhythmic drug selective for the fatal arrhythmias induced by acute myocardial infarction during the ischemic phase. This was considered a worthy target despite acknowledged difficulties and previous failure, since ventricular fibrillation is the single largest cause of mortality in Western Europe and North America. In accord with the approaches and criteria discussed above, five factors were considered:

1. The target arrhythmia was ventricular tachycardia and fibrillation caused by acute myocardial infarction in the ischemic phase of myocardial infarction. The ischemic phase is recognized to be when the most lethal of all ventricular arrhythmias occur.
2. The mechanism targeted was chosen on the basis that ischemia produces profound changes in excitability, conduction, and refractoriness that involve Na^+, $Ca^{2,+}$, and K^+ ion channels. Such changes directly precipitate in, and maintain, arrhythmias. It was assumed that abolition of the activity of such ion channels would render ischemic tissue quiescent, thereby preventing its participation in arrhythmias. Quiescent cardiac tissue cannot directly participate in arrhythmias, although injury currents might do so.
3. The condition targeted was ischemia since ischemic tissue differs from normal in numerous ways, and it is the site of the initiation and maintenance of arrhythmias. Most importantly it is acidic and has a high potassium concentration.
4. The specific pharmacological target(s) were the ion channels in ischemic tissue so as to prevent their participation in ischemic arrhythmogenesis. The major target channel was that for sodium, although potassium and calcium channels were also considered.
5. The specific approach was to create a blocker of sodium ion channels (plus other ion channels if necessary) that was selectively active in, or activated by, conditions found in ischemic tissue.

Ischemia-induced arrhythmias are most common early (minutes) after the onset of ischemia and are due to disturbed excitability, conduction velocity, and refractoriness resulting in multiple and complex re-entry circuits.[43] The antiarrhythmic requirement was for a mixed ion channel blocker, selective for ischemic tissue. In particular sodium channel block would reduce excitability and conduction thereby converting uni- to bidirectional conduction block. At the same time concurrent potassium channel block would increase refractoriness. While such mechanisms operate for conventional class 1 and 3 antiarrhythmics, their lack of ischemic tissue selectivity means that their beneficial effects are outweighed by proarrhythmias. Thus ischemic selectivity was vital.

Attempts have been made to target arrhythmia pathology but the history of targeting drugs for ion channels in normal cardiac tissue is replete with failures. Previous strategies tried to take advantage of the characteristic high rate of arrhythmias by concentrating on frequency dependent drugs, but such drugs have limited efficacy against ischemia-induced arrhythmias. One putative strategy is restoration of the disordered electrophysiology in ischemic tissue. However, this appears to be of limited value since ischemic tissue is doomed to die, unless the tissue is reperfused. Such an approach is feasible if only to keep ischemic tissue alive and functioning normally before reperfusion by surgery or drugs.

A strategy of rendering ischemic tissue electrically quiescent appears counterintuitive until it is recognized that the most lethal phase for ischemic arrhythmias is when the ischemic tissue is still electrically active, and that the occurrence of arrhythmias falls as the tissue becomes electrically quiescent before dying. If this is so, why not hasten and abbreviate the duration of the process of electrical quiescence so as to reduce the chance of arrhythmias? There are potential problems with this approach since there is evidence that sodium channel blockers acting upon ischemic tissue can initiate re-entry arrhythmias.[44] This mechanism was probably one of the possible mechanisms for the increased mortality seen in the CAST study.[17] However, this potential problem could be lessened by concomitant potassium channel blockade.

It has repeatedly been shown that selective sodium and potassium channel blockers have limited antiarrhythmic efficacy, and all have proarrhythmic activity. Mixed channel blockers may not be that much better although the multichannel blocker amiodarone is an antiarrhythmic drug that unequivocally saves lives.

How can an antiarrhythmic action be made selective for ischemic tissue, or for ischemia-induced arrhythmias? One approach is frequency dependence, as has been discussed previously, while the other takes advantage of the conditions in ischemic tissue. An antiarrhythmic drug activated by ischemic conditions would have ischemia selectivity. A Na^+/K^+ channel blocker acting selectively in ischemic tissue, by a combination of increased refractoriness, reduced conduction, and suppression of excitability, would render ischemic tissue quiet and thereby incapable of participating in ischemic arrhythmogenesis. However, since ischemic tissue still has a low membrane potential, it can be a source of injury currents that can act as arrhythmia generators.[45] However, ischemic arrhythmias are rare at a time when ischemic tissue is electrically quiet, but still partially polarized.

With this rationale, how can the conditions found in ischemia be utilized to provide selectively? There are examples of drug selectivity being achieved by using tissue-selective enzymes to release active drug from its prodrug. However, there are no known enzymes activated by ischemia that could be used for such a purpose. Another classic approach to site-specific activation is protonation, as with omeprazole which is activated by stomach acid.

Many sodium channel blocking antiarrhythmic drugs are tertiary amines that are active in their charged form within the cell. If a drug acts on the outer surface of the cell in its charged form, then the acidity in ischemic tissue could be used to elevate concentrations of the active species of such a drug.[46,47] In an analogous manner elevated extracellular potassium might make channels more vulnerable to some types of ion channel blocking drugs. Elevated K^+ also potentiates sodium channel blockers that bind to the inactivated sodium channel. Thus, a compound whose potency for sodium and potassium blockade is increased by raised potassium and hydrogen ion concentrations might have selectivity for ischemic tissue.

Such thinking led to a specific strategic search for such compounds as discussed in Walker and Guppy.[48] The following is a summary of a program that required an SAR approach to obtaining ischemia-selective compounds. It was centered on a lead compound with sodium-blocking actions that could potentially be increased by acidity. The choice of lead compound was based upon exploratory work with tetrodotoxin, sparteine analogs, and benzeneacetamide derivatives as well as calcium channel blockers. One of the latter compounds, RSD921, proved structurally suitable as a sodium channel blocker which also blocked the IK_{ur} channel.[49] However, its actions were not potentiated by acid or raised by extracellular potassium and it did not provide the required antiarrhythmic protection. However, since it was suitable as a pharmacophore a systematic SAR approach was made to obtain better compounds as outlined in **Figure 23**.

There were four sites on RSD921 that could be systematically studied by chemical substitutions or additions. Thus many changes could be made to A but its size and substituents were not of major importance. The pK_a of the nitrogen in the ring at B was very important. The best selectivity for antiarrhythmic activity was seen in those compounds where

The groundwork
Systematic exploration of the pharmacological properties (using various classes of drugs) that confer antiarrhythmic protection in the setting of myocardial ischemia.

Statement of requirements
Ion channel blockade (principally sodium?) potentiated in ischemic conditions (principally pH~6.5)

Potential pharmacophores
I. Benzeneacetamides (opiates) **II.** Sparteine analogs **III.** Tetrodotoxin
IV. Calcium channel blockers (verapamil – phenylalkylamine)

Choice of pharmacophores
I was chosen based upon the pharmacological profile, ease of chemical manipulation of the molecule, and potentially favorable pharmacokinetics. RSD 921 was the lead compound.

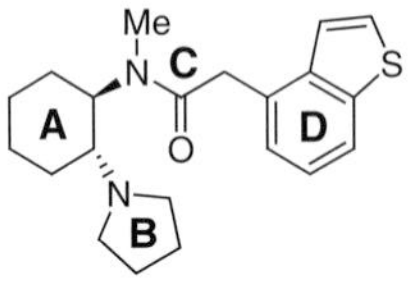

Me N C S A O D N B

RSD 921

Systematic changes were made to the four areas A–D

A	B	C	D
Important	Important	Amide/ester/ether	Lipophilic aryl group
Size important	Size important	Ether best	Size variable
Many substituents possible	Many substituents possible	Many substituents possible	Many substituents possible

Two partially successful compounds **FINALLY**

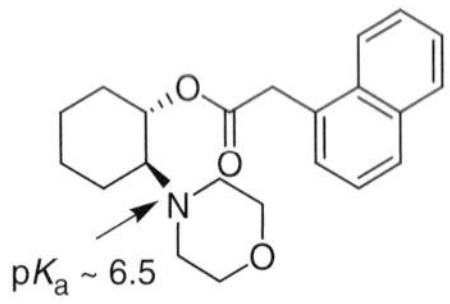

pK_a ~ 6.5

RSD 1000 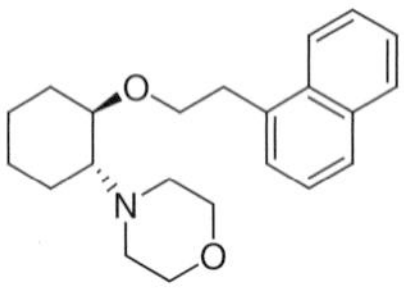 RSD 1070 HO RSD 1235 OMe OMe

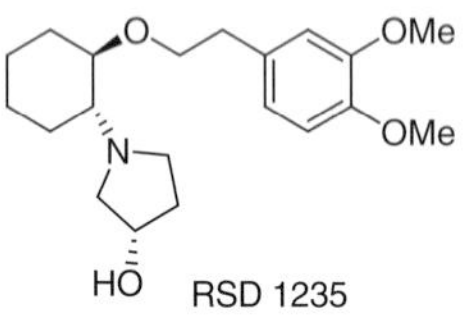

Figure 23 The path to RSD1235.

the pK_a was ~6.5 (i.e., the pH in ischemic myocardial tissue). With regard to site C an ester at this position was useful but, not unexpectedly, gave only a short duration of action. However an ether substitution was suitable and provided useful compounds. Further modifications provided compounds with better pharmacokinetic profiles and lesser toxicity. Thus, using classic SAR it proved possible to make compounds whose potency was increased at acid pH values (≈6.5). Patch clamp studies suggested an extracellular site of action as well as positive frequency (inactivation) dependency. When tested in vivo against ischemia induced arrhythmias in anesthetized rats, the best compounds selectively protected against the arrhythmias induced by ischemia versus those induced by electrical stimulation in normal cardiac tissue. Further studies showed that they also blocked potassium channels.

As indicated above a basic part of the approach was to make compounds with a pK_a of around 6.5, close to the pH found in ischemic tissue at the time when arrhythmias occur. This involved identifying the ring nitrogen that was essential for channel blockade and ascertaining that the pK_a of this nitrogen could be manipulated by substitution on, or next to, the ring. The second amide nitrogen in RSD921 played no role since it could be replaced by an ester-, or ether-oxygen without loss of activity. In those compounds which provided 100% protection against ischemia induced arrhythmias the pK_a of the ring nitrogen was around 6.5.[50]

Various studies have shown that in normal cardiac tissue, a moderately elevated $[K^+]_o$ (4–7 mM) is antiarrhythmic, but how important is such a mechanism in ischemic tissue? The effects of increased $[K^+]_o$ were therefore studied and raised potassium was found to add to the effects of acidity. Thus RSD1000 (**Figure 10**) was more than 10 times more potent than reference class 1 antiarrhythmics against ischemia induced arrhythmias.[46] Related morpholinocyclohexyls, including RSD1019 (**Figure 10**) and RSD1030, showed a similar increased efficacy with a combination of high $[K^+]_o$ and low pH, and all compounds had favorable therapeutic ratios versus the standard antiarrhythmic reference drugs.

The result of these studies were compounds that were reasonably potent, effective, and selective against ischemia induced arrhythmias in rats and pigs while having a satisfactory therapeutic ratios (compared with other antiarrhythmics) in terms of cardiovascular and CNS toxicity.

Compounds in the above series were also recognized as being potentially useful in the treatment of atrial arrhythmias in that, besides blocking sodium channels with a frequency dependent action, they also blocked IK_{to} and IK_{ur} potassium channels. Such actions resulted in atrial selectivity. In atrial fibrillation, blood flow through the fibrillating atria is compromised, and thus a condition of partial ischemia prevails. All these characteristics resulted in compounds in which further manipulations improved pharmacokinetic properties and finally resulted in RSD1235. This compound has been found effective in phase III clinical trials in terminating acute-onset atrial fibrillation when injected intravenously.

References

1. Olgin, J. E.; Zipes, D. P. In *Heart Disease: A Textbook of Cardiovascular Medicine*, 6th ed.; Braunwald, E., Zipes, D. P., Libby, P., Eds.; W. B. Saunders: Philadelphia, PA, 2001, pp 815–889.
2. Iqbal, M. B.; K Taneja, A.; Lip, G. Y.; Flather, M. *Br. Med. J.* **2005**, *330*, 238–243.
3. Clements-Jewery, H.; Curtis, M. J. In *Cardiac Drug Development Guide*; Pugsley, M. K., Ed.; Humana Press: Totowa, NJ, 2003, pp 203–225.
4. Wilde, A. A.; Bezzina, C. R. *Heart* **2005**, *91*, 1352–1358.
5. Roden, D. M.; Balser, J. R.; George, A. L., Jr.; Anderson, M. E. *Annu. Rev. Physiol.* **2002**, *64*, 431–475.
6. George, A. L., Jr. *J. Clin. Invest.* **2005**, *115*, 1990–1999.
7. Roden, D. M.; George, A. L., Jr. *Annu. Rev. Med.* **1996**, *47*, 135–148.
8. Catterall, W. A. *Annu. Rev. Cell Dev. Biol.* **2000**, *16*, 521–555.
9. Tamargo, J.; Caballero, R.; Gomez, R.; Valenzuela, C.; Delpon, E. *Cardiovasc. Res.* **2004**, *62*, 9–33.
10. DiFrancesco, D. *Curr. Med. Res. Opin.* **2005**, *21*, 1115–1122.
11. Zipes, D. P. *J. Cardiovasc. Electrophysiol.* **2003**, *14*, 902–912.
12. Antzelevitch, C. *Curr. Opin. Cardiol.* **2001**, *16*, 1–7.
13. Gonzalez, J. E.; Oades, K.; Leychkis, Y.; Harootunian, A.; Negulescu, P. A. *Drug Disc. Today* **1999**, *4*, 431–439.
14. Norinder, U. *SAR QSAR Environ. Res.* **2005**, *16*, 1–11.
15. Falk, R. H. *Circulation* **2005**, *111*, 3141–3150.
16. Wijffels, M. C.; Kirchhof, C. J.; Dorland, R.; Allessie, M. A. *Circulation* **1995**, *92*, 1954–1968.
17. The Cardiac Arrhythmia Suppression Trial (CAST) Investigators. *N. Engl. J. Med.* **1989**, *321*, 406–412.
18. Roden, D. M. In *Goodman & Gilman's: The Pharmacological Basis of Therapeutics*, 11th ed.; Brunton, L. L., Lazo, J. S., Parker, K. L., Eds.; McGraw-Hill: New York, 2005, pp 899–932.
19. Vaughan Williams, E. M. *J. Clin. Pharmacol.* **1992**, *32*, 964–977.
20. The Sicilian Gambit. *Circulation* **1991**, *84*, 1831–1851.
21. Bardy, G. H.; Lee, K. L.; Mark, D. B.; Poole, J. E.; Packer, D. L.; Boineau, R.; Domanski, M.; Troutman, C.; Anderson, J.; Johnson, G. et al. *N. Engl. J. Med.* **2005**, *352*, 225–237.
22. Luderitz, B. *J. Interv. Card. Electrophysiol.* **2003**, *9*, 75–83.
23. Koufaki, M.; Calogeropoulou, T.; Rekka, E.; Chryselis, M.; Papazafiri, P.; Gaitanaki, C.; Makriyannis, A. *Bioorg. Med. Chem.* **2003**, *11*, 5209–5219.
24. Lee, K.; Lee, J. Y.; Kim, H. Y.; Kwon, L. S.; Shin, H. S.; Tanabe, S.; Kozono, T.; Park, S. D.; Chung, Y. S. *J. Cardiovasc. Pharmacol.* **1998**, *31*, 609–617.
25. Bauer, A.; Koch, M.; Kraft, P.; Becker, R.; Kelemen, K.; Voss, F.; Senges, J. C.; Gerlach, U.; Katus, H. A.; Schoels, W. *Basic Res. Cardiol.* **2005**, *100*, 270–278.
26. Regan, C. P.; Wallace, A. A.; Cresswell, H. K.; Atkins, C. L.; Lynch, J. J. *J. Pharmacol. Exp. Ther.* **2006**, *316*, 727–732.
27. Matsuda, T.; Takeda, K.; Ito, M.; Yamagishi, R.; Tamura, M.; Nakamura, H.; Tsuruoka, N.; Saito, T.; Masumiya, H.; Suzuki, T. et al. *J. Pharmacol. Sci.* **2005**, *98*, 33–40.
28. Takahara, A.; Sugiyama, A.; Dohmoto, H.; Yoshimoto, R.; Hashimoto, K. *Eur. J. Pharmacol.* **2000**, *398*, 107–112.
29. Carlsson, B.; Singh, B. N.; Temciuc, M.; Nilsson, S.; Li, Y. L.; Mellin, C.; Malm, J. *J. Med. Chem.* **2002**, *45*, 623–630.
30. Roy, D.; Rowe, B. H.; Stiell, I. G.; Coutu, B.; Ip, J. H.; Phaneuf, D.; Lee, J.; Vidaillet, H.; Dickinson, G.; Grant, S. et al. *J. Am. Coll. Cardiol.* **2004**, *44*, 2355–2361.
31. Opincariu, M.; Varro, A.; Iost, N.; Virag, L.; Hala, O.; Szolnoki, J.; Szecsi, J.; Bogats, G.; Szenohradszky, P.; Matyus, P. et al. *Curr. Med. Chem.* **2002**, *9*, 41–46.
32. Bril, A.; Forest, M. C.; Cheval, B.; Faivre, J. F. *Cardiovasc. Res.* **1998**, *37*, 13–40.
33. Wu, L.; Shryock, J. C.; Song, Y.; Li, Y.; Antzelevitch, C.; Belardinelli, L. *J. Pharmacol. Exp. Ther.* **2004**, *310*, 599–605.
34. Pankucsi, C.; Banyasz, T.; Magyar, J.; Gyonos, I.; Kovacs, A.; Varro, A.; Szenasi, G.; Nanasi, P. P. *Naunyn Schmiedebergs Arch. Pharmacol.* **1997**, *355*, 398–405.
35. Qi, X.; Varma, P.; Newman, D.; Dorian, P. *Circulation* **2001**, *104*, 1544–1549.
36. Xing, D.; Kjolbye, A. L.; Nielsen, M. S.; Petersen, S.; Harlow, K. W.; Holstein-Rathlou, N. H.; Martins, J. B. *J. Cardiovasc. Electrophysiol.* **2003**, *14*, 510–520.
37. Decher, N.; Lang, H. J.; Nilius, B.; Bruggemann, A.; Busch, A. E.; Steinmeyer, K. *Br. J. Pharmacol.* **2001**, *134*, 1467–1479.
38. Hobai, I. A.; O'Rourke, B. *Exp. Opin. Investig. Drugs* **2004**, *13*, 653–664.

39. Gumina, R. J.; Buerger, E.; Eickmeier, C.; Moore, J.; Daemmgen, J.; Gross, G. J. *Circulation* **1999**, *100*, 2519–2526.
40. Raitt, M. H.; Connor, W. E.; Morris, C.; Kron, J.; Halperin, B.; Chugh, S. S.; McClelland, J.; Cook, J.; MacMurdy, K.; Swenson, R. et al. *JAMA* **2005**, *293*, 2884–2891.
41. Xiao, Y. F.; Kang, J. X.; Morgan, J. P.; Leaf, A. *Proc. Natl. Acad. Sci. USA* **1995**, *92*, 11000–11004.
42. Honore, E.; Barhanin, J.; Attali, B.; Lesage, F.; Lazdunski, M. *Proc. Natl. Acad. Sci. USA* **1994**, *91*, 1937–1941.
43. Ramaswamy, K.; Hamdan, M. H. *Crit. Care Med.* **2000**, *28*, N151–NN15.
44. Hondeghem, L. M. *Circulation* **1987**, *75*, 514–520.
45. Carmeliet, E. *Physiol. Rev.* **1999**, *79*, 917–1017.
46. Yong, S. L.; Xu, R.; McLarnon, J. G.; Zolotoy, A. B.; Beatch, G. N.; Walker, M. J. A. *J. Pharmacol. Exp. Ther.* **1999**, *289*, 236–244.
47. Bain, A. L.; Barrett, T. D.; Beatch, G. N.; Fedida, D.; Hayes, E. S.; Plouvier, B.; Pugsley, M. K.; Walker, M. J. A.; Walker, M. L.; Wall, R. A. et al. *Drug Dev. Res.* **1997**, *42*, 198–210.
48. Walker, M. J. A.; Guppy, L. J. In *Cardiac Drug Development Guide*; Pugsley, M. K., Ed.; Humana Press: Totowa, NJ, 2003, pp 175–201.
49. Ribeiro, W.; Ifa, D. R.; Corso, G.; Salmon, J.; Moraes, L. A.; Eberlin, M. N.; De Nucci, G. *J. Mass Spectrom.* **2001**, *36*, 1133–1139.
50. Barrett, T. D.; Hayes, E. S.; Yong, S. L.; Zolotoy, A. B.; Abraham, S.; Walker, M. J. A. *Eur. J. Pharmacol.* **2000**, *398*, 365–374.

Biographies

Michael J A Walker is an emeritus professor at the University of British Columbia and has spent over 30 years in the Department of Pharmacology and Therapeutics there. Michael Walker has been working for over 25 years in the area of antiarrhythmic research and has published widely in this, and other areas. His research interests, centered on the discovery of new antiarrhythmic drugs, led to the founding of a biotech company whose research goal was the discovery of new antiarrhythmic drugs, particularly for prevention of the ventricular fibrillation that occurs in the acute phase of a heart attack. That company, now known as Cardiome Pharma Corp., is currently finishing a series of phase III clinical trials with one of their discoveries, RSD1235, in preparation for a NDA. The drug's first use is expected to be as an intravenous drug for the termination of atrial fibrillation of recent onset.

Petsy P S So obtained a BSc Honours degree in the Department of Pharmacology and Therapeutics at the University of British Columbia, and is currently a PhD student working under the supervision of Dr Paul Dorian at the University of Toronto. Petsy is currently working on the antiarrhythmic actions of potassium channel blockers and looks forward to future antiarrhythmic drug research.

Comprehensive Medicinal Chemistry II
ISBN (set): 0-08-044513-6

ISBN (Volume 6) 0-08-044519-5; pp. 729–762

6.34 Thrombolytics

A E El-Gengaihy, S I Abdelhadi, J F Kirmani, and A I Qureshi, University of Medicine and Dentistry of New Jersey, Newark, NJ, USA

6.34.1 Introduction

Thrombus formation is a vital part of the hemostatic mechanism. Under normal circumstances formation of the thrombus is limited to the site of vessel injury and is regulated by different mechanisms to prevent further progression. Under pathological conditions, a thrombus can propagate into otherwise normal vessels, where it can lead to obstruction of blood flow and impair normal hemodynamic functions. Heart disease, ischemic stroke, pulmonary embolism, peripheral vascular disease, and many others all share a common pathophysiological process of thrombus formation.[1] The applications and use of thrombolytic therapy with reperfusion of occluded vessels has a major impact on the outcome of the above diseases. Thrombolysis is a standard treatment option for acute ST-segment elevation myocardial infarction (STEMI), stroke, pulmonary embolism, and other thromboembolic diseases. In this chapter the mechanisms of thrombus formation are reviewed; the various types, mechanism of action, and the use of different thrombolytic agents as well as major cardiovascular clinical trials are highlighted.

6.34.2 Pathophysiology of Thrombus Formation

Thrombosis is a pathologic event that results in the obstruction of coronary, cerebral, or peripheral blood flow.[2] A thrombus is formed by the two major components of the coagulation system: platelets and coagulation factors. Normally, the endothelial cell lining of the vessels maintain an antithrombotic status by several mechanisms that prevent platelet adhesion to the vessel wall, including: maintaining a transmural negative electrical charge; releasing prostacyclin (PGI_2), which prevents platelet aggregation; and activating the fibrinolytic pathway and protein C, which degrades the coagulation factors.[3]

In case of vascular injury, platelets and plasma clotting factors become exposed to the subendothelial collagen and the endothelial basement membrane, which releases adenosine diphosphate (ADP), a potent platelet aggregator, and tissue factor, which launches the clotting cascade (**Figures 1** and **2**). Platelets adhere to the vessel wall and undergo conformational changes, which enhances more platelet aggregation. Moreover, glycoprotein (GP) IIb/IIIa receptors are upregulated and subsequently fibrinogen and von Willebrand's factor bind to activated platelets via GP IIb/IIIa receptors forming the platelet plug.[4]

Activated platelets acquire an enhanced capacity to catalyze the interaction between activated coagulation factors. These factors are generally circulating in the blood in an inactive form (zymogens). The clotting cascade consists of two separate initial pathways (intrinsic and extrinsic) that converge into a final common pathway, which ends by activation of factor X to Xa and conversion of fibrinogen to fibrin (**Figure 2**). The extrinsic pathway is important for the initiation of fibrin formation while the intrinsic pathway is involved in fibrin growth and maintenance.[4]

The extrinsic pathway is triggered by tissue factor, also called thromboplastin (a lipid-rich protein material released after vascular injury). Thromboplastin combines with activated factor VIIa to convert factor IX to activated factor IXa in the intrinsic pathway and factor X to activated factor Xa in the extrinsic pathway. Factor VII is activated by IXa or XIIa of the contact system. This process takes 15 s.

The intrinsic pathway is initiated by activation of factor XII by the high-molecular-weight kininogen (the precursor of vasoactive peptide, bradykinin) and kallikrein in the presence of collagen. Factor XIIa converts factor XI to XIa, which in turn activates factor IX to factor IXa, and subsequently factor X to Xa. Activated factor Xa then binds with factor Va, Ca^{2+}, and phospholipids (PLs) from the platelet membrane to form Xa–Va–PL complex (prothrombinase), which converts prothrombin to thrombin. Thrombin finally cleaves fibrinogen to release isolated fibrin monomers, which polymerize with each other to form complex fibrin. Factor XIII, which is also activated by thrombin, stabilizes fibrin complexes and hence thrombus formation.

There is tight control over the coagulation system; once factor Xa is formed, it binds to tissue factor pathway inhibitor (TFPI), which inhibits further activation of factor Xa and factor Va. Other control mechanisms include

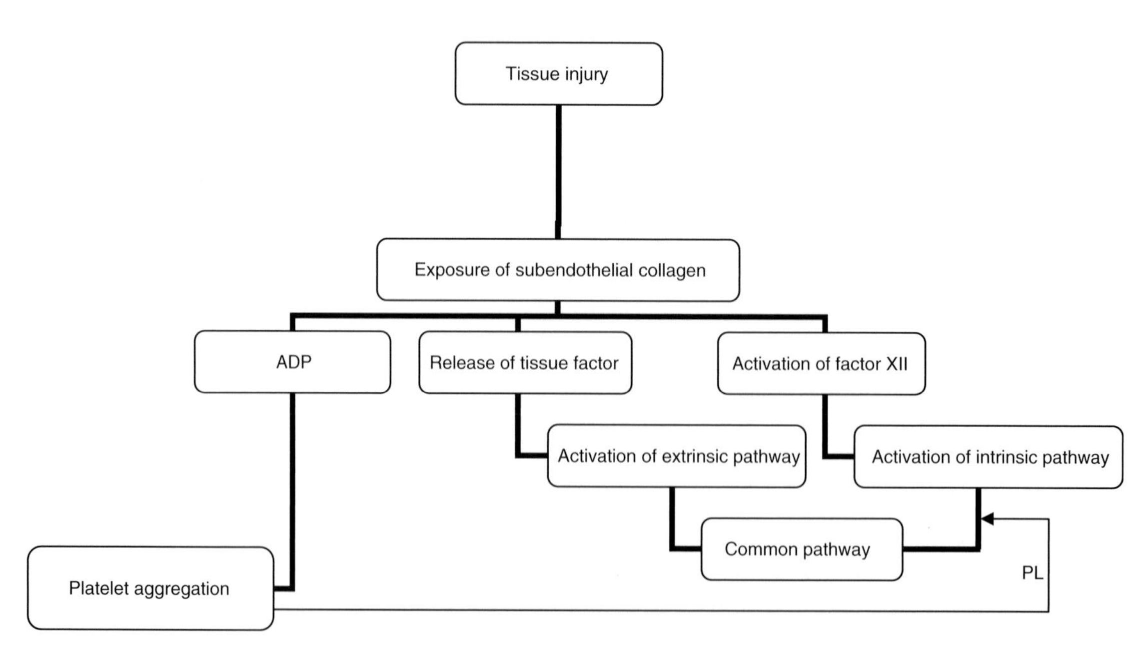

Figure 1 Triggering of coagulation cascade

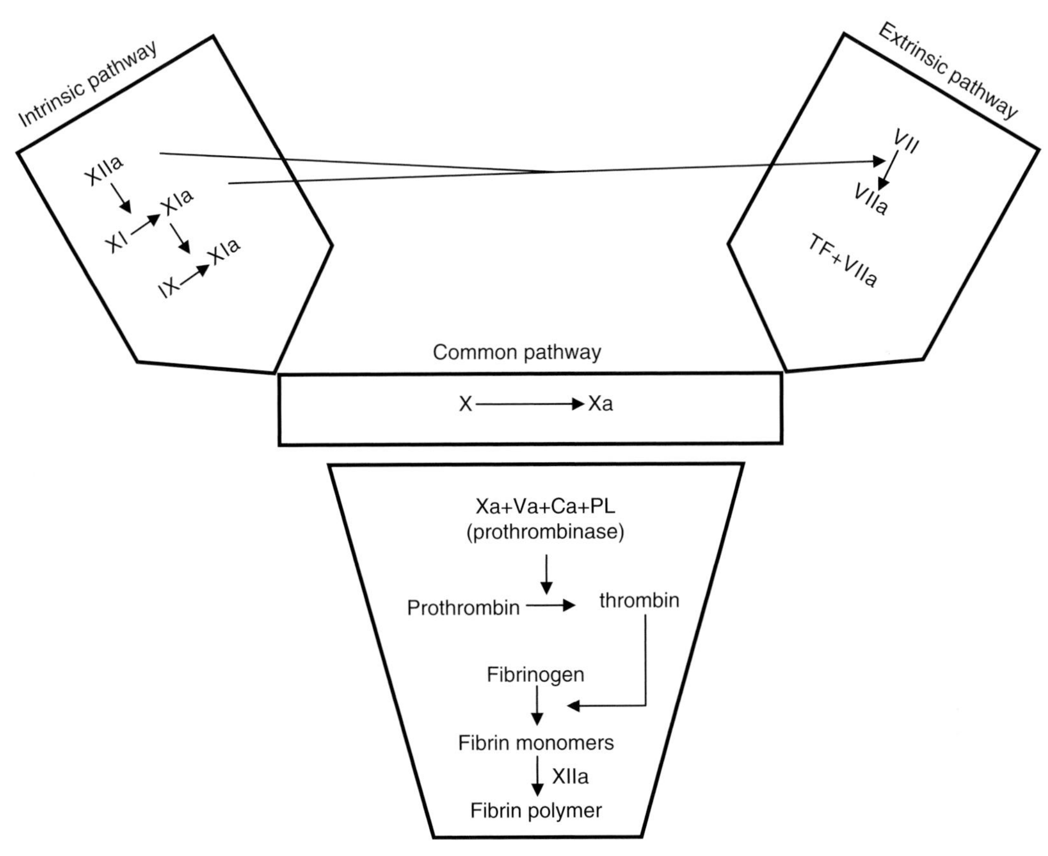

Figure 2 Diagram showing the coagulation cascade. PL, phospholipids.

activation of the fibrinolytic pathway and the presence of antithrombin III, which prevents clotting factor activation, protein C, which inactivates factor Va, and protein S, which acts as a cofactor for protein C where they form a complex to inactivate factor VIIIa.[3]

6.34.2.1 Factors Affecting Thrombus Formation

According to Virchow's triad,[5] there are three possible contributors to the formation of an abnormal clot (thrombus): vessel wall injury or inflammation, changes in the intrinsic properties of blood, and decrease in blood flow velocity (**Figure 1**).[5–9] Atherosclerotic plaques, which are found in most major arteries, are the main substrate for thrombus formation.[10] The atherosclerotic process can start even before birth[11] with approximately 65% of children between 12 and 14 years of age having intimal alteration.[12–14] The lipid core of the atherosclerotic lesion is rich in tissue factor, which initiates the clotting cascade upon plaque rupture.[15] Other factors that affect thrombus formation include the degree of plaque disruption and the content of tissue factor in the plaque.[16] Stenotic arteries and blood velocity also affect the platelet disposition and thrombus formation as they change the shear rate of flowing blood.[17] Certain systemic risk factors are also associated with thrombus formation, for example, lipoprotein(a) has a similar structure to plasminogen, which may impair thrombolysis.[18] Increased blood thrombogenicity is also associated with increased low-density lipoprotein (LDL).[19] Poorly controlled diabetes mellitus results in glycosylation of collagen and protein, increasing the levels of plasma fibrinogen. Furthermore, smoking has been found to increase tissue factor levels in thrombotic plaques.[10]

6.34.2.2 Mechanism of Thrombolysis

Under normal circumstances there is no plasminogen activation occurring in plasma, the action of the fibrinolytic system is confined to fibrin. When fibrin is formed, a small amount of plasminogen activator and plasminogen adsorb to it and plasmin is created in situ to catalyze the degradation of fibrin (**Figure 3**).[20] Naturally occurring plasminogen

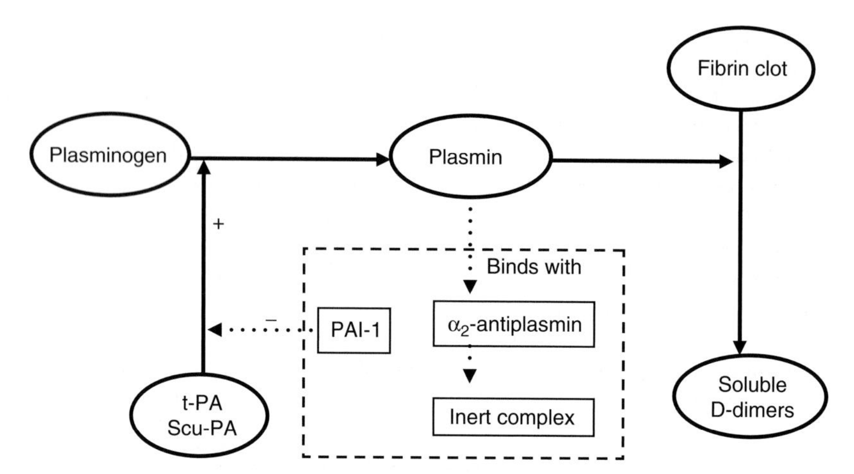

Figure 3 Schematic diagram showing the process of fibrinolysis and its control. D-dimers, breakdown product of fibrin.

activators are either tissue-type plasminogen activator (t-PA) or single chain urokinase-type plasminogen activator (scu-PA), both of which are synthesized by the endothelial wall.

The fibrinolytic system is tightly controlled and is normally restricted in the thrombus (**Figure 3**). Endothelial cells secrete plasminogen activator inhibitor-1 (PAI-1), which inactivates t-PA in the plasma but not that in the thrombus; in addition, the plasma t-PA has a short half-life (5–8 min). Also, α_2-antiplasmin, a physiological bound serine protease inhibitor, forms an inert complex with the circulating plasmin but does not act on fibrin-bound plasmin in the thrombus.[4] Any disturbance in this system can cause either excessive thrombus formation, if there is insufficient activation of the thrombolytic system or if there are excessive inhibitors,[21] or bleeding tendency, if there is overstimulation of the thrombolytic system and deficiency in the release of inhibitors.[20]

6.34.3 Animal Models and Preclinical Studies

The novel antithrombotic agents were discovered, validated, and characterized through animal studies. Animal models provide a great deal of information regarding the mechanism, doses, and treatment options, as well as interaction among different thrombolytic agents. As there are many animal models only a few selected ones are discussed here.

Infused intravenous (i.v.)/intracoronary t-PA in the left anterior descending (LAD) coronary artery in an animal model induced a thrombus within 1–2 h. Coronary reperfusion, intermediary metabolism, and nutritional myocardial blood flow were restored within 10 min without inducing a systemic fibrinolytic state.[22] Another study compared the thrombolytic effect of recombinant t-PA (rt-PA) with that of urokinase in dogs undergoing coronary occlusion for 1 h. Both agents were infused at the same rate. rt-PA elicited a faster reperfusion with less distal coronary embolization and systemic fibrinolysis as compared to urokinase.[23]

Dose and rate of infusion of rt-PA directly correlate to the degree of reperfusion of the coronary vessels. This was shown in open-chested, anesthetized dogs with induced thrombus in the LAD. Two hours after induction of the thrombus rt-PA was infused intravenously at different rates and doses. Lower rates of infusion and lower doses took longer to achieve coronary reperfusion when compared with higher doses, e.g., at $5\,\mu g\,kg^{-1}\,min^{-1}$ time-to-reperfusion was greater than 40 min, while at $25\,\mu g\,kg^{-1}\,min^{-1}$ lysis occurred within 13 min. Epicardial electrographic measurements showed a significant reduction in ST elevation in all reperfused hearts.[24]

Another important factor is the time of administration of the thrombolytic agent. Administration of rt-PA after 30–80 min in open-chested baboons showed a mean duration of reperfusion at 77 min and a decrease in infarct size by 38%. However, myocardial blood flow in the perfusion area of the LAD was only 70% of normal after 4 h in spite of perfect angiographic refilling.[25] From the above mentioned studies, it is concluded that t-PA i.v. may recanalize thrombosed coronary vessels without inducing a systemic fibrinolytic state. Timely reperfusion results in infarct sparing and restoration of nutritional blood flow.

Another comparative study was done using a canine model of coronary thrombosis; the aim of the study was to compare tissue plasminogen activator and urokinase. Urokinase was given in two doses, 15 000 (UK15) and 30 000 U kg^{-1} (UK30), and rt-PA was given in two doses, 0.25 (rtPA.25) and 0.75 mg kg^{-1} (rtPA.75). rtPA.75 showed a higher rate and extent of coronary thrombolysis; rtPA.25 was superior to UK15 but achieved the same results as UK30. Again, this study indicates that rt-PA is superior to urokinase.[26]

Platelet-rich coronary clot is resistant to thrombolytic therapy In order to investigate and overcome such a problem, investigators developed an animal model (dog) with platelet-rich coronary clot and investigated the effect of thrombolytic therapy combined with antiplatelet GP IIb/IIIa antibody. The use of a reduced dose thrombolytic therapy with antiplatelet GP IIa/IIIa can overcome the resistant clot and restore perfusion. This study concluded that platelet-rich clot can be treated pharmacologically without the need for mechanical thrombolysis.[27]

A canine model of combined coronary arterial and femoral venous thrombosis was induced to compare intravenous bolus injection versus infusion of recombinant unglycosylated full-length single chain urokinase-type plasminogen activator (rscu-PA, saruplase). Coronary artery thrombosis and femoral vein clots were produced in five randomized blinded groups, each consisting of five dogs. rscu-PA was infused over 60 min at 1 $mg\,kg^{-1}$ and 0.5 $mg\,kg^{-1}$ (group I and III, respectively), and was given as a bolus at 1 $mg\,kg^{-1}$ and 0.5 $mg\,kg^{-1}$ (group II and IV, respectively). Group V was the control group. Four out of five dogs showed coronary recanalization in groups I and IV, group I showed the highest percentage of clot lysis, while both groups I and II showed no coronary reocclusion Globally, this study proved that in both coronary and venous thrombolysis, intravenous bolus injection of rscu-PA is equipotent to an infusion over 60 min.[28]

Recent animal studies are now targeting inhibitors of PAI-1 or PAI-039. As noted above, PAI-1 functions to suppress t-PA and urokinase-type plasminogen activator (u-PA), so its inhibition should help in thrombolysis. Dogs were given PAI-039 orally and then subjected to coronary occlusion. Dogs treated with PAI-039 had more time for coronary reocclusion and thrombus weight was reduced compared to controls. Also the incidence of spontaneous reperfusion of the coronary artery was greater in dogs that received PAI-039 as compared to controls. Accordingly PAI-1 may be used as adjunctive therapy to lower the dose of thrombolytics and to increase their efficiency in clot lysis with fewer side effects such as cerebral hemorrhage and bleeding.[29]

6.34.4 Current Thrombolytics

Streptokinase entered clinical use in the mid-1940s.[30] The first use of a fibrinolytic drug in the treatment of acute myocardial infarction was reported in 1958, when intravenous infusion of streptokinase was used.[31] Shortly after that, in 1960, streptokinase and human plasmid were injected into the aortic root of a man following myocardial infarction.[32] However, the revolution of thrombolytic therapy started in 1976 in Russia, with direct intracoronary injection of fibrinolysin.[33] Several trials followed and it was soon recognized that early restoration of the blood flow preserved left ventricular function and yielded significant mortality benefit. However, the use of intracoronary therapy was limited, because it was time consuming and needed specially equipped hospitals and well-trained personnel.

t-PA and high-dose intravenous streptokinase became established as a life-saving treatment for acute myocardial infarction (MI),[34] and were approved for use by the US Food and Drug Administration (FDA) in the mid-1980s.[35] Trials to establish the efficacy of streptokinase for acute ischemic stroke were stopped because of a high rate of early death due to intracerebral hemorrhage. In 1996, the FDA approved intravenous thrombolysis with rt-PA as the 'first-ever' effective treatment for ischemic stroke during the first 3 h of onset of symptoms.[36] Reteplase was the first of the third-generation thrombolytics to be approved for use in acute MI.[35]

6.34.4.1 Thrombolytic Agents used in Trials and Clinical Practice

Thrombolytic drugs lyze preexisting thrombus in both arteries and veins and re-establish tissue perfusion.[4] They exert their action through conversion of plasminogen to plasmin, which then degrades fibrin, a major structural component of the thrombus.[37] The action of thrombolytic drugs is achieved by either potentiating endogenous fibrinolytic pathways or mimicking natural thrombolytic molecules.[1] Currently available thrombolytic agents are derived from bacterial products or manufactured using recombinant DNA technology. They differ in their efficiency, fibrin selectivity, and side-effect profile. Even for the same thrombolytic agent, different doses, different administration regimens, and concomitant use of adjunctive agents can modify its patency rates. However, these differences are only marginal.[1] Many limitations exist among the available generations of thrombolytic agents, raising the need for continued research for better agents. The indications of fibrinolytic therapy are summarized in **Table 1** and contraindications in **Table 2**.

6.34.4.1.1 First-generation thrombolytic agents

There are three agents belonging to this group: streptokinase, urokinase, and anisoylated purified streptokinase activator complex (APSAC). They are not fibrin specific or site specific, and they act anywhere in the blood, converting circulating plasminogen to plasmin. Eventually, this causes depletion of body plasminogen, disturbing the equilibrium between circulating plasminogen and plasminogen in the thrombus (plasminogen steal) and reducing clot lysis.

Table 1 Indications of fibrinolytic therapy (ACC guidelines)

Class I
1. In the absence of contraindications, fibrinolytic therapy should be administered to STEMI patients with symptom onset within the prior 12 h and ST elevation greater than 0.1 mV in at least 2 contiguous precordial leads or at least 2 adjacent limb leads *(Level of Evidence: A)*
2. In the absence of contraindications, fibrinolytic therapy should be administered to STEMI patients with symptom onset within the prior 12 h and new or presumably new LBBB *(Level of Evidence: A)*
Class IIa
1. In the absence of contraindications, it is reasonable to administer fibrinolytic therapy to STEMI patients with symptom onset within the prior 12 h and 12-lead ECG findings consistent with a true posterior MI *(Level of Evidence: C)*
2. In the absence of contraindications, it is reasonable to administer fibrinolytic therapy to patients with symptoms of STEMI beginning within the prior 12–24 h who have continuing ischemic symptoms and ST elevation greater than 0.1 mV in at least 2 contiguous precordial leads or at least 2 adjacent limb leads *(Level of Evidence: B)*
Class III
1. Fibrinolytic therapy should not be administered to asymptomatic patients whose initial symptoms of STEMI began more than 24 h earlier *(Level of Evidence: C)*
2. Fibrinolytic therapy should not be administered to patients whose 12-lead ECG shows only ST-segment depression except if a true posterior MI is suspected *(Level of Evidence: A)*

Table 2 Contraindications to fibrinolytic therapy (ACC guidelines)

Absolute contraindications
1. Any prior ICH
2. Known structural cerebral vascular lesion (e.g., AVM/aneurysm)
3. Known malignant intracranial neoplasm
4. Ischemic stroke within 3 months EXCEPT acute ischemic stroke within 3 h
5. Suspected aortic dissection
6. Active bleeding or bleeding diathesis (excluding menses)
7. Significant closed-headi or facial trauma within 3 months
Relative contraindications
1. History of chronic, severe, poorly controlled hypertension
2. Severe uncontrolled hypertension on presentation (SBP greater than 180 mmHg or DBP greater that 110 mmHg)
3. History of prior ischemic stroke greater than 3 months, dementia, or known intracranial pathology not covered in contraindications
4. Traumatic or prolonged (greater than 10 min) CPR or major surgery (within the last 3 weeks)
5. Recent (within 2–4 weeks) internal bleeding
6. Noncompressible vascular punctures
7. For streptokinase/anistreplase: prior exposure (more than 5 days ago) or prior allergic reaction to these agents
8. Pregnancy
9. Active peptic ulcer disease
10. Current use of anticoagulants: the higher the INR, the higher the risk of bleeding

CPR, cardiopulmonary resuscitation; AVM, arteriovenous malformation; SBP, systolic blood pressure; DBP, diastolic blood pressure; INR, international normalized ratio.

6.34.4.1.1.1 Streptokinase

Streptokinase is approved for use in myocardial infarction, pulmonary embolism, deep venous thrombosis, arteriovenous-cannula occlusions, and peripheral arterial occlusions.[30] Streptokinase is an enzyme with a molecular weight of 47 000 Da[35] derived from the culture filtrate of Lancefield group C[5] β-hemolytic streptococci.[1,35] Streptokinase lyses thrombus by potentiating the body's own fibrinolytic pathways.[1] It acts indirectly[2,37] through binding with free circulating plasminogen to form a 1:1 complex, resulting in a conformational change and exposure of an active site that can convert additional plasminogen into plasmin, the main thrombolytic enzyme in the body.[4] The resultant plasmin is of two types: fibrin-bound and unbound. Fibrin-bound plasmin causes direct fibrinolysis of the thrombus,[2] while unbound circulating plasmin leads to systemic fibrinolysis and a hypocoagulable state due to the depletion of fibrinogen, plasminogen, and factors V and VIII.[1,35,37] In addition, streptokinase increases levels of activated protein C, which enhances clot lysis.[35] The greatest benefit of streptokinase appears to be achieved by early i.v. administration.[4]

Streptokinase has no metabolites. The complex is inactivated, in part, by antistreptococcal antibodies, resulting from prior infection,[38] and eliminated through the liver.[35] It has two half-lives: a fast one (11–13 min) due to inhibition by the circulating antibodies; and a slow one (23–29 min) due to loss of the enzyme activity.[4] The unbound fraction, which constitutes about 15%, has a serum half-life of 80 min.[30] On the other hand, fibrin degradation products can be detected in the serum for up to 24 h, which means that the patient remains somewhat 'anticoagulated' even without the use of heparin.[35,37]

6.34.4.1.1.2 Urokinase

Urokinase is the most familiar thrombolytic agent among interventional radiologists and it is often used for peripheral vascular thrombosis.[37] It is approved for pulmonary embolism and for lysis of coronary thrombi, but not for mortality reduction in acute myocardial infarction.[4,35]

Urokinase is a trypsin-like[1] enzyme that is produced endogenously by renal parenchymal cells[30] and found in urine. Approximately 1500 L of urine are needed to yield enough urokinase to treat a single patient.[30] There are two forms of urokinase, which differ in molecular weight but have similar clinical effect.[39] Commercially available urokinase, the low-molecular-weight form (32 400 Da), is produced from cultured human neonatal kidney cells. It is a two-polypeptide chain serine protease, containing 411 amino acid residues[4,35]: an A chain of 2000 Da is linked by a sulfhydryl bond to a B chain of 30 400 Da. Recombinant techniques are used to produce urokinase in *Escherichia coli*.[30] Unlike streptokinase, urokinase directly cleaves plasminogen to produce plasmin.[1] Although this process is slightly increased in the presence of fibrin, urokinase produces circulating unbound plasmin, which means that it not only degrades fibrin clots but also fibrinogen and other plasma proteins leading to systemic fibrinolysis.[40]

In plasma, urokinase has a half-life of approximately 15 min[30] (7–20 min).[4,38] It is rapidly metabolized and cleared by the liver[38] with small fractions being excreted in bile and urine. Plasma levels can be elevated two- to fourfold[1] and clearance is reduced in patients with hepatic impairment. Moreover, due to its short half-life, re-thrombosis may occur within 15–30 min of therapy cessation. Heparin is commonly used during and after urokinase adminstration to minimize this risk.[40]

6.34.4.1.1.3 Anisoylated purified streptokinase activator complex (APSAC, anistreplase)

APSAC was the third thrombolytic agent to be developed. It has a molecular weight of 131 000 Da[35] and consists of streptokinase in a noncovalent 1:1 complex with plasminogen.[4] It does not require free circulating plasminogen to be effective. APSAC is catalytically inert because of the acylation of the catalytic site of plasminogen (having the catalytic site temporarily blocked by a *p*-anisoyl group), which protects the catalytic center of the complex from premature neutralization.[35] However, the affinity of plasminogen binding to fibrin is maintained.[4] APSAC acts as an indirect plasminogen activator[35] and is nonfibrin selective, and so activates both circulating and clot-bound plasminogen, but is most active within the thrombus.[2]

Inside the circulation, anistreplase undergoes spontaneous deacylation to form the active complex of plasminogen–streptokinase. This conversion occurs with a half-life of 90–100 min,[2,37] which lengthens its thrombolytic effect after i.v. injection.[4] The active complex is metabolized in the liver[35] and has the longest half-life among all thrombolytic agents, ranging from 90 min[4] to 100 min.[35]

6.34.4.1.2 **Second-generation thrombolytic agents**

These agents include t-PA and scu-PA (prourokinase). The second-generation agents are more fibrin selective and although they were developed to avoid systemic thrombolytic state, they can still cause a mild to moderate decrease in the levels of circulating fibrinogen and plasminogen.[35]

6.34.4.1.2.1 Tissue-type plasminogen activator (alteplase)

Alteplase was the first recombinant t-PA to be produced. It is the most familiar fibrinolytic agent in emergency departments and the most often used agent for treatment of coronary artery thrombosis, pulmonary embolism, and acute stroke.[30] Alteplase is expensive, costing approximately 8–10 times more than streptokinase per dose.[35,37] In vivo, t-PA is synthesized by the vascular endothelial cells and is considered the physiologic thrombolytic agent that is responsible for most of the body's natural efforts to prevent thrombus propagation. Based on the results of Stroke Trials sponsored by the National Institute of Neurological Disorders and Stroke (NINDS), intravenous thrombolysis with rt-PA was approved as the 'first-ever' effective treatment for ischemic stroke during the first 3 h of symptom onset.[36]

rt-PA is a sterile purified glycoprotein molecule consisting of 527 amino acids[35] that is structurally identical to endogenous t-PA.[1,41] It has a molecular weight of 70 000 Da[35] and is produced by recombinant technology[1,4,35] from a human melanoma cell line.[35] The molecule contains five domains: finger, epidermal growth factor, kringle 1, kringle 2, and serum protease.[2] There are two different forms of t-PA based on the number of chains: t-PA (the two-chain form duteplase and alteplase by recombinant technology) and rt-PA (one chain form).[2,35] It consists predominantly of the single-chain form (rt-PA), but upon exposure to fibrin, rt-PA is converted to the two-chain dimer.[4]

t-PA is a naturally occurring enzyme (serine protease).[42] It is the principal physiological activator of plasminogen in the blood and has a high binding affinity for fibrin[4] at the site of a thrombus, directly[2,35,37] activating clot-bound plasminogen only.[42] While this might seem an advantage, this selectivity is not absolute; circulating plasminogen may also be activated by large thrombolytic doses or lengthy treatment.[4] Moreover, its action is fibrin-enhanced; that is, in the absence of fibrin, t-PA is a weak plasminogen activator.[42] It is rapidly cleared from plasma with an initial half-life of approximately 5 min (4–10 min)[5,35]; however, its effect at the clot persists for over an hour[35] (72 min), but the concentration of circulating t-PA would be expected to return to endogenous circulating levels of 510 ng mL^{-1} within 30 min.[42] Having a short plasma half-life necessitates its administration as a bolus injection, followed by a short continuous infusion.[2] Heparin is usually coadministered to avoid reocclusion.[2,35] The plasma clearance is 380–570 mL min^{-1},[35] and is primarily mediated by the liver.[2,35,43]

t-PA is not associated with hypotension[35] and is not antigenic; it can be readministered as necessary[1] and may be considered for use in patients who have high antibody titer against streptokinase.[35] The activity of t-PA is enhanced in the presence of fibrin, resulting in thrombus-specific fibrinolysis.[1] In practice, however, t-PA causes a milder form of systemic fibrinogenolysis[4,35] than streptokinase at equi-effective doses, but the incidence of bleeding is similar with both agents.[4] Moreover, t-PA is considered more efficacious than streptokinase in establishing coronary reperfusion[4] and has a slight mortality advantage over streptokinase due to its accelerated administration because of the shorter half-life. Unfortunately, this occurs at the cost of a marginal increase in stroke rate.[1]

The risk of intracranial hemorrhage is higher with t-PA than streptokinase (0.7% versus 0.5%).[35] Having a shorter plasma half-life, the rate of re-thrombosis after t-PA is greater than streptokinase,[4] which raises the need for continuous infusion in order to achieve its greatest efficacy,[1] and since t-PA is more expensive and toxic than streptokinase, the latter is the agent of choice for coronary thrombolysis.[35]

6.34.4.1.2.2 SPB (scu-PA, prourokinase)

Prourokinase is a new fibrinolytic agent that is currently undergoing clinical trials for a variety of indications. It is a single chain urokinase[35] and has been produced both in glycosylated (ABT-74187) and nonglycosylated (saruplase) forms.[2,35] It is a relatively inactive precursor that must be converted to urokinase before it becomes active in vivo. However, it displays selectivity for clots by binding to fibrin before activation.[37] The mechanism of action of the nonglycosylated form is unclear, but it is known to be nonfibrin specific.[35] Its advantage over other plasminogen activators is that it is inactive in plasma and so does not consume circulating inhibitors. Also, it is somehow clot specific, where it needs fibrin to be converted by an unknown mechanism into active urokinase.[30] It is usually administered as a bolus followed by intravenous infusion, but single-bolus regimens are now being developed.[35]

6.34.4.1.3 **Third-generation thrombolytic agents**

These groups of agents have been developed through modifications of the basic t-PA structure. They are either: conjugates of plasminogen activators with monoclonal antibodies against fibrin, platelets, or thrombomodulin; mutants, variants, and hybrids of t-PA and prourokinase (amediplase); or new molecules of animal (vampire bat) or bacterial (*Staphylococcus aureus*) origin.[35] These molecular variations have yielded agents with better pharmacological properties than t-PA, a longer half-life, resistance to plasma protease inhibitors, and more selective fibrin binding.[4,35] Several of these agents are being developed including reteplase (r-PA, retevase), lanoteplase (nPA), tenecteplase (TNKase), pamiteplase (YM866; Solinase), staphylokinase, and novel modified tissue plasminogen activator (E6010).

6.34.4.1.3.1 Reteplase (rt-PA, r-PA, retevase)

Reteplase is the first third-generation thrombolytic agent to be approved for use in acute MI to improve postinfarct ventricular function, lessen the incidence of congestive heart failure, and reduce mortality.[35] It is a synthetic nonglycosylated deletion mutein of t-PA containing 355 of the 527 amino acids of the native tissue plasminogen activator; it lacks the finger, epidermal growth factor, and kringle 1 domains[2,44] as well as carbohydrate side chains.[2] This results in a prolonged half-life and less fibrin specificity than t-PA.[37,44] It has a molecular weight of 39 500 Da and is produced in *E. coli* by recombinant DNA technology.[1] The gene for a fragment of t-PA is inserted into *E. coli*, and the protein is then extracted from the bacteria and processed to convert it into an active thrombolytic.[30,37]

Patients receiving reteplase have faster clot resolution than those receiving t-PA, owing to the fact that reteplase binds less tightly to fibrin, allowing for more free diffusion through the clot rather than only binding to the surface as t-PA does. In a controlled trial, 64% of patients who received a double bolus of reteplase showed a decrease in fibrinogen levels to below 100 mg dL^{-1} within 2 h. However, the mean fibrinogen levels returned to baseline within 48 h.[45] Moreover, at high concentrations, it does not compete with plasminogen for fibrin-binding sites, allowing plasminogen at the site of the clot to be transformed into clot-dissolving plasmin.[30]

The above-mentioned structural modifications result in a fivefold decrease in fibrin binding and an extended half-life (11–19 min).[35,37,44] The longer half-life allows for administration of reteplase by double-bolus infusion rather than a prolonged infusion.[35,37,44] It undergoes renal (and some hepatic) clearance[30] at a rate of 250–450 mL min^{-1}.

6.34.4.1.3.2 Lanoteplase (nPA, novel plasminogen activator)

Lanoteplase is a deletion mutant of t-PA lacking the finger, epidermal growth factor, and one amino acid substitution in the kringle 1 domain, leading to deletion of a glycosylation site.[2,35]

6.34.4.1.3.3 Tenecteplase (TNK-t-PA, TNKase)

Tenecteplase is a tissue plasminogen activator with a molecular weight of 70 000 Da,[35] produced by genetic engineering.[37] To create tenecteplase, a 527 amino acid glycoprotein molecule, the human gene for t-PA was modified using 3 amino acid substitutions:[37] a substitution of threonine 103 with asparagine and asparagine 117 with glutamine within the kringle 1 domain, and a tetra-alanine substitution at amino acids 296–299 in the protease domain.[2,35] These mutations resulted in a decrease of plasma clearance, prolonged half-life, higher degree of fibrin specificity, and increased resistance to PAI-1 as compared to t-PA.[2,35,37,46] The FDA has approved the use of tenecteplase in acute MI.[37]

Tenecteplase is a modified form of human t-PA that binds more avidly to fibrin,[4] and directly converts plasminogen to plasmin.[35] This process relatively increases in the presence of fibrin, giving tenecteplase the advantage of being more specific with minimal systemic effect. However, this specificity is not absolute; a 4–15% decrease in circulating fibrinogen and 11–24% decrease in plasminogen has been reported following administration. Moreover, its clinical significance with regard to safety or efficacy has not been established.

A single bolus of tenecteplase administered to patients with acute myocardial infarction exhibits biphasic disposition from the plasma. The initial half-life is 20 min (15–24 min),[35] about four times that of t-PA,[4] and is considered to be the longest elimination half-life among t-PA derivatives.[4,35,37] The terminal phase half-life of tenecteplase is 90–130 min. The initial volume of distribution is weight related and approximates plasma volume. The main route of elimination is liver, at a clearance rate of 99–119 mL min^{-1}.[37]

6.34.4.1.3.4 Staphylokinase

Staphylokinase was known to possess profibrinolytic properties more than four decades ago.[35] It is produced by certain strains of *S. aureus*. It acts on the surface of the clot to form a plasmin–staphylokinase complex,[2] which has high fibrin specificity, only activating plasminogen trapped in the thrombus.[35] After administration, staphylokinase-related antigen disappears from plasma in a biphasic manner, with an initial half-life of 6.3 min and a terminal half-life of 37 min.[35]

As compared to t-PA, studies suggest that staphylokinase may have less procoagulant effects. Furthermore, it is highly antigenic; patients develop neutralizing antibodies in about 1–2 weeks and the titer remains elevated for several months after therapy cessation.[35] This limits the use of a second dose until safer, more effective new variants that have less immunogenicity are developed.[2]

6.34.4.1.4 **Fourth-generation thrombolytic agents**

6.34.4.1.4.1 Recombinant desmodus salivary plasminogen activator-1 (r DSPA-1, desmoteplase)

A naturally occurring enzyme in the saliva of the blood-feeding vampire bat (*Desmodus rotundus*) is genetically related to t-PA.[47] It consists of four different proteases – *D. rotundus* salivary plasminogen activators (DSPAs). DSPA-1 is the full-length variant with a greater than 72% sequence homology to human t-PA.[48] Unlike t-PA, it exists as single-chain molecules[48] and it is critically dependent on fibrin.[47,49,50] DSPA-1 targets and destroys fibrin;[47] it is more fibrin dependent and fibrin specific than t-PA.[47] Its catalytic efficiency is enhanced 13 000-fold[47] in the presence of fibrin, while that of t-PA increases only by 72-fold.[49]

It has high fibrin specificity and selectivity, and a longer half-life compared to other thrombolytic agents.[48,51] Compared to t-PA, it is non-neurotoxic, causes less fibrinogenolysis,[50] less antiplasmin consumption, and results in faster and more sustained reperfusion as demonstrated by animal studies.[52] Furthermore, it can be given to acute ischemic stroke patients within 3–9 h of onset of symptoms.[51] DSPA is safe and results in improved perfusion and low mortality rates without associated symptomatic intracerebral hemorrhage.[51]

6.34.5 Major Cardiovascular Clinical Trials

Numerous clinical trials with thrombolytic therapy have been performed to determine their clinical efficacy in cardiovascular settings, namely acute myocardial infarction. Only trials of major significance for each thrombolytic agent will be discussed in this chapter.

6.34.5.1 Streptokinase

The first large-scale thrombolytic trial was the GISSI-1 (Gruppo Italiano per lo Studio della Sopravvivenza nell'Infarto Miocardico), which evaluated the efficacy of a thrombolytic treatment with streptokinase on in-hospital mortality of patients with acute myocardial infarction (AMI). The GISSI demonstrated that overall in-hospital mortality was reduced in those who received streptokinase (10.7%) compared to controls (13%). The degree of benefit, which was sustained up to 1 year after the AMI episode, was related to the time between onset of symptoms and streptokinase treatment; the sooner thrombolytics were administered the greater the reduction in mortality. When thrombolytics were administered more than 6 h after AMI no difference was appreciated.[53]

Similar benefits were noted in the ISIS-2 trial (Second International Study of Infarct Survival), in which patients presenting to hospitals within 24 h (mean of 5 h) of onset of suspected AMI were randomly assigned to either: (1) 1 h i.v. infusion of streptokinase; (2) 1 month of 160 mg day^{-1} of enteric coated aspirin (with the first tablet crushed for a rapid antiplatelet effect); (3) both treatments; or (4) neither treatment Streptokinase alone and aspirin alone each produced a significant reduction in 5-week vascular mortality. A combination of streptokinase and aspirin was significantly better than either agent alone and displayed a synergistic effect in the reduction of mortality from 13.2% (placebo) to 8.0% (streptokinase + aspirin).[54] As in the GISSI trial, the ISIS-2 demonstrated that early therapy (within 6–24 h) is essential if mortality benefit and long-term benefit is to be achieved.[55]

6.34.5.2 Alteplase

Two initial trials, the GISSI-2 and ISIS-3 (Third International Study of Infarct Survival), compared the efficacy of altepase with that of streptokinase. In the GISSI-2 and its International Study Group extension no significant difference emerged regarding mortality in acute myocardial infarction between alteplase and streptokinase (9.6% versus 9.2%) and whether heparin was or was not administered (9.3% versus 9.4%). Also, no significant differences were observed regarding major cardiac complications. However, a small, albeit significant, increase in the incidence of stroke was found among patients treated with alteplase compared to those treated with streptokinase (1.3% versus 0.9%).[56,57] Long-term follow-up also showed no significant differences between these two thrombolytics in these main clinical findings.[58]

The ISIS-3 trial, another randomized trial, compared streptokinase and t-PA, but also included APSAC. In this study, there were no appreciable differences in mortality rates among thrombolytic agents. Moreover, rates for major in-hospital clinical events, including cardiogenic shock, heart failure requiring treatment, ventricular fibrillation, and cardiac rupture, were also similar. There was a small but significant deficit of in-hospital re-infarctions in the t-PA group. Also of note were higher rates of allergy and hypotension requiring treatment in those that received the bacterially derived proteins streptokinase and APSAC.

Altogether, these trials found no difference in efficacy between alteplase and streptokinase given with or without subcutaneous heparin.[56,59] Such data influenced the design of further studies, including GUSTO-1 (Global Utilization of Streptokinase and Tissue Plasminogen Activator for Occluded Coronary Arteries), which established accelerated rt-PA combined with i.v. heparin as the optimal thrombolytic strategy for patients with AMI, and found that successful reperfusion and myocardial salvage produced significant mortality benefits (1% absolute reduction in 30-day mortality) in the alteplase + heparin group when compared to the streptokinase + heparin group. This benefit persisted at 1-year follow-up.[60] As with all thrombolytic therapy, bleeding and stroke are of greatest concern. Combinations of streptokinase plus i.v. heparin and streptokinase plus t-PA were associated with a higher incidence of bleeding than the use of single agents.

The Continuous Infusion Versus Double-Bolus Administration of Alteplase (COBALT) study tested therapeutic equivalence between two alteplase dosing strategies in patients with acute myocardial infarction: a double bolus of alteplase versus accelerated infusion. The trial was stopped on January 5, 1996 because of higher rates of death, stroke, and cardiogenic shock in the double-bolus group than in the accelerated infusion group.[61]

A meta-analysis from the Fibrinolytic Therapy Trialists' (FTT) collaborative group sought to determine the indications and contraindications for fibrinolytic therapy and found a reduction in mortality in fibrinolytic-treated patients (9.6%) compared to controls (11.5%). Mortality was reduced among patients presenting with ST segment elevation (21%) or left bundle branch block (LBBB) (25%). Not surprisingly, absolute mortality reductions were greatest in patients treated earlier after symptom onset,[62] and FTT was beneficial in reducing mortality among various types of patients (all ages except those greater than 75 years, patients with or without prior MI, and diabetic and nondiabetic patients), but not in those without ST elevation or LBBB.[62]

6.34.5.3 Reteplase

The Recombinant Plasminogen Activator Angiographic Phase II International Dose-Finding Study (RAPID-1) was an angiographic study that compared reteplase (r-PA) with alteplase (t-PA).[45] Patients were randomized to: (1) t-Pa 100 mg i.v. over 3 h; (2) r-PA as a 15 MU single bolus; (3) r-PA as a 10 MU bolus followed by 5 MU 30 min later; or (4) r-PA as a 10 MU bolus followed by another 10 MU 30 min later. Coronary angiography was then performed at 30, 60, and 90 min after treatment and then again at discharge. Reteplase given as a double bolus (10 + 10 MU) achieved more rapid, complete, and sustained thrombolysis of the infarct-related artery than standard dose t-PA, without an apparent increased risk of complications.

The RAPID-II trial compared the double bolus of r-PA (10 + 10 MU) to a front-loaded regimen of alteplase (100 mg over 90 min). Again, reteplase proved to be superior to alteplase in achieving recanalization of coronary artery, better blood flow, and fewer acute (within 6 h) coronary interventions. However, there was no significant difference between groups for 35-day mortality, bleeding requiring transfusion, or hemorrhagic stroke.[63]

In both RAPID trials, mortality and other outcomes were more favorable in those who received reteplase. These hypotheses were further tested in the larger GUSTO-3 trial, which compared the efficacy and safety of alteplase and reteplase.[64] No difference was observed between the two thrombolytics regarding mortality at 30 days and incidence of stroke. Also, no difference in mortality was appreciated between the two thrombolytics at 1 year (11.2% versus 11.1%). This lack of significance, according to the GUSTO investigators, could reflect insufficient study size to detect such a difference.[64]

Before being compared to alteplase, double-dose reteplase (10 MU + 10 MU at 30 min) was compared to standard dose streptokinase (1.5 MU over 1) in the INJECT (International Joint Efficacy Comparison of Thrombolyics) trial to determine whether the effect of both thrombolytics on survival was equivalent.[65] At 35 days, mortality, recurrent myocardial infarction, in hospital stroke, and major bleeding events for the reteplase group and the streptokinase group were equivalent. Reteplase, however, proved to have significant benefits compared to streptokinase, including fewer cases of atrial fibrillation, asystole, heart failure, and hypotension. Therefore, the INJECT trial indicated that reteplase was therapeutically comparable to streptokinase in terms of safety and efficacy.

In 2001, the GUSTO-5 trial compared reperfusion therapy for acute myocardial infarction with standard dose reteplase versus half-dose reteplase (two boli of 5 U, 30 min apart) plus full-dose abciximab within the first 6 h of STEMI. The investigators concluded that for patients with STEMI, the combined reteplase and abciximab was noninferior (although not superior) to standard reteplase for decreasing the 30-day mortality.

6.34.5.4 Tenecteplase (TNK-t-PA)

Tenecteplase, a genetically engineered t-PA, was assessed for pharmokinetics, safety, and efficacy in humans in the Thrombolysis in Myocardial Infarction 10A (TIMI 10A) trial.[66] It was administered over 5–10 s as a single bolus of 5–50 mg and had a half-life of 17 min, nearly five times longer than t-PA.

Subsequently, the TIMI 10B trial, a phase II dosing trial,[67] was carried out to identify a specific bolus dosing regimen of TNK-t-PA that would achieve similar rates of TIMI-3 flow on 90-min angiography as t-PA, with a similar safety profile. Patients were randomized after 12 h of symptom onset and given either a bolus dose of 30 or 50 mg of TNK or a front-loaded regimen of t-PA. The 50 mg arm was changed to 40 mg due to a high incidence of intracranial hemorrhage. The results showed 50 and 40 mg of TNK had a similar rate of TIMI-3 flow when compared to t-PA. No differences in mortality and reinfarction were observed.

The Assessment of the Safety of a New Thrombolytic: TNK-t-PA (ASSENT-1) trial, a phase II dose-ranging trial, was conducted in conjunction with the TIMI 10B trial to test the clinical safety of three doses of tenecteplase-tissue plasminogen activator (TNK-t-PA) in ST-elevation myocardial infarction.[68] Doses of 30, 40, and 50 mg of TNK-t-PA i.v. bolus were selected. As discussed above, the 50 mg arm was discontinued. Overall, the safety profile of TNK-t-PA was similar to that of alteplase and no difference in incidence of total stroke, intracranial hemorrhage, severe bleeding, or death was observed among treatment groups.

Findings from the TIMI 10B and ASSENT-1 trials served as the basis for comparing TNK versus accelerated t-PA in the ASSENT-2 trial. In the ASSENT-2 trial patients were randomized to weight-adjusted single doses of TNK-t-PA or accelerated t-PA.[69] No difference in mortality was noted at 30 days nor was there a difference in the rate of stroke and intracerebral hemorrhage. Thus, TNK-t-PA was equivalent to t-PA in terms of its 30-day mortality benefit. Mortality at 1 year remained unchanged between the two agents.[70] However, for those treated with TNK-t-PA 4 h after the onset of symptoms, mortality at 30 days and 1 year was lower compared to t-PA.

In 2001, tenecteplase's efficacy was assessed with a GP IIb/IIIa inhibitor in the ASSENT-3 trial, a randomized trial that compared the efficacy and safety of tenecteplase plus enoxaparin or abciximab with that of tenecteplase plus weight-adjusted unfractionated heparin (UFH) in patients with AMI.[71] The tenecteplase plus enoxaparin or abciximab regimens reduced the frequency of ischemic complications in AMI patients. The combination of tenecteplase with enoxaparin was more efficacious than tenecteplase with heparin, and there was no increase in the risk of bleeding or intracranial hemorrhage, even in patients over the age of 75. In contrast, while efficacy was improved with the combination of tenecteplase plus abciximab, this was offset by a doubling in the rate of major hemorrhage and a higher event rate in patients over the age of 75 and in diabetic patients. Therefore, it was concluded that tenecteplase plus enoxaparin was a viable alternative regimen to tenecteplase plus UFH for the treatment of ST-elevation AMI.

6.34.5.5 Lanoteplase (nPA)

The Intravenous nPA for Treatment of Infarcting Myocardium Early (InTIME)-1 study compared a single bolus dose of lanoteplase (15–120 kU kg^{-1}) with an accelerated regimen of alteplase in patients and found that coronary patency at 90 min (83% versus 71%) and frequency of TIMI grade III flow were greater with the highest doses of lanoteplase compared to alteplase.[71]

The InTIME-1 study was followed by InTIME-2, which assessed mortality at 30 days in patients with acute myocardial infarction randomized to either nPA or t-PA.[72] The 30-day and 6-month mortality was similar between the two groups. Intracranial hemorrhage was more frequent in those treated with nPA. Overall bleeding complications were equal in the two treatment groups, however, mild complications occurred more frequently in nPA-treated individuals (19.6% versus 14.7%).

6.34.5.6 Saruplase

Saruplase has been evaluated in several trials in patients with acute MI or unstable angina. The Comparison Trial of Saruplase and Streptokinase (COMPASS), the largest of saruplase trials, attempted to demonstrate the equivalence of saruplase to streptokinase in the treatment of patients with ST-segment elevation myocardial infarction (STEMI). Patients were randomly assigned to receive heparin and saruplase (20 mg i.v. bolus followed by 60 mg i.v. over the next 60 min) or streptokinase without heparin. Saruplase was associated with a reduction in mortality compared to streptokinase. Patients treated with saruplase were also less likely to develop hypotension, but more likely to develop hemorrhagic stroke than those treated with streptokinase (0.9% versus 0.3%). The overall bleeding rates between the two treatment groups were similar (10.4% versus 10.9%). These findings suggested that saruplase is as effective as streptokinase in the treatment of acute STEMI, with a similar safety profile.[73]

6.34.6 Medical Management with Thrombolysis

Numerous studies have been performed and enough evidence accumulated showing an unequivocal benefit, in terms of mortality and morbidity, in regards to thrombolytic treatment of patients presenting with AMI. As discussed above, the FTT collaborative group found patients who presented within a 'therapeutic window' (less than 1 h) of symptom onset and found to have either LBBB or STEMI benefited most from thrombolytic therapy, most benefit being seen in those treated soonest after onset of symptoms. Patients presenting after 12 h should not receive fibrinolytics, unless there is evidence of ongoing ischemia, as no significant benefit has been shown. Unless clearly contraindicated (see **Table 2**), patients with infarction, particularly, STEMI (ST elevation greater than 0.1 mV in 2 contiguous leads) or new LBBB, should receive prompt treatment with fibrinolytic therapy and aspirin without delay. Patients with LBBB or anterior ST elevation are at greater inherent risk from MI and achieve greater benefit with fibrinolytic therapy. If possible, fibrinolysis should be started within 90 min of the patient calling for medical treatment ('call-to-needle' time) or within 30 min of arrival at the hospital ('door-to-needle' time).[74] As per the American College of Cardiology (ACC) guidelines, attainment of additional ECG leads (right sided and/or posterior) or an echocardiogram may help clarify the location and extent of infarction and anticipated risk of complications, but it is important that acquisition of such ancillary information does not interfere with the strategy of providing timely reperfusion in patients with STEMI.

6.34.7 New Approaches and Adjuvant Therapy

6.34.7.1 Aspirin

Aspirin (**Figure 4**) is a nonselective cyclooxygenase inhibitor that has antithrombotic effects mediated by inhibition of blood platelets. Inhibition of cyclooxygenase blocks production of thromboxane A2, which activates platelets leading to aggregation, an early step in thrombosis. Aspirin is more effective in preventing arterial thrombosis (myocardial infarction, stroke) than venous thrombosis (deep venous thrombosis, pulmonary embolism).

6.34.7.2 Dipyridamole

Dipyridamole (**Figure 4**) is an inhibitor of nucleoside transport (hENTs) acting to block platelet aggretion by increasing adeonsine levels. Dipyridamole can also reduce plasma von Willebrand factor levels and serum C-reactive protein.[75]

6.34.7.3 Warfarin, Heparin, and Low-Molecular-Weight Heparins

Warfarin (**Figure 4**) inhibits vitamin K, which is essential for effective production of clotting factors II, VII, IX, X, and anticoagulant proteins C and S. Heparin inhibits thrombin (factor IIa), and factors Xa and IXa. Heparin is used to treat unstable angina and to prevent and treat venous thromboembolism. Low-molecular-weight heparins (LMWHs) are fragments of the heparin molecule that inhibit clotting factor Xa more than factor IIa. Dalteparin, enoxaparin, and tinzaparin are three such LMWHs.

6.34.7.4 Glycoprotein IIb/IIIa Receptor Antagonists

Activation of platelets by several agonists results in the expression of specific functional receptors for fibrinogen on the platelet surface, referred to as GP IIb/IIIa receptors. The GP IIb/IIIa receptor is a member of the integrin family of receptors.[76] When platelets are activated by a variety of stimuli including thrombin, collagen, ADP, and epinephrine, the GP IIb/IIIa receptor changes its conformation to be receptive to one end of a fibrinogen dimer. Occupancy of a GP IIb/IIIa receptor by the other end of the dimer provides the basis for platelet aggregation. Thus, the GP IIb/IIIa receptor is considered the final common pathway of platelet aggregation.[77]

Abciximab (c7E3) is a chimeric mouse – human monoclonal antibody directed against the GP IIb/IIIa receptor. Its mechanism of action appears to be steric hindrance of the receptor as opposed to direct binding to the RGD binding site of the receptor. Abciximab also inhibits the vitronectin (a_v b_3)receptor, which mediates both platelet coagulation in addition to endothelial and vascular smooth muscle cell proliferation. The significance of vitronectin-receptor blockade is unknown. It produces a direct antithrombotic effect by inhibiting the binding of fibrinogen to the receptor and consequently inhibiting platelet aggregation.[78] Because abciximab is a more potent inhibitor of platelet function than aspirin, it potentially produces a greater degree of thrombolysis and prevents rethrombosis in short-term management of patients with STEMI.[78,79]

Eptifibatide (**Figure 4**) is another GP IIb/IIIa antagonist. It dose-dependently inhibits platelet aggregation and in a baboon model refractory to aspirin and heparin, inhibited aggregation, preventing acute thrombosis with only a modest prolongation (two- to threefold) of the bleeding time.

Tirofiban (**Figure 4**) is a nonpeptide inhibitor of the GP IIb/IIIa receptor. It reduces ischemic events at 48 h following infusion when compared to standard heparin therapy and is extensively used in percutaneous transluminal coronary angioplasty (PTCA) and acute coronary syndromes (ACS).

Platelet GP IIb/IIIa receptor antagonists have been shown to be effective and safe in reducing death and cardiac ischemic events among patients presenting with acute coronary syndromes without ST-segment elevation.[80]

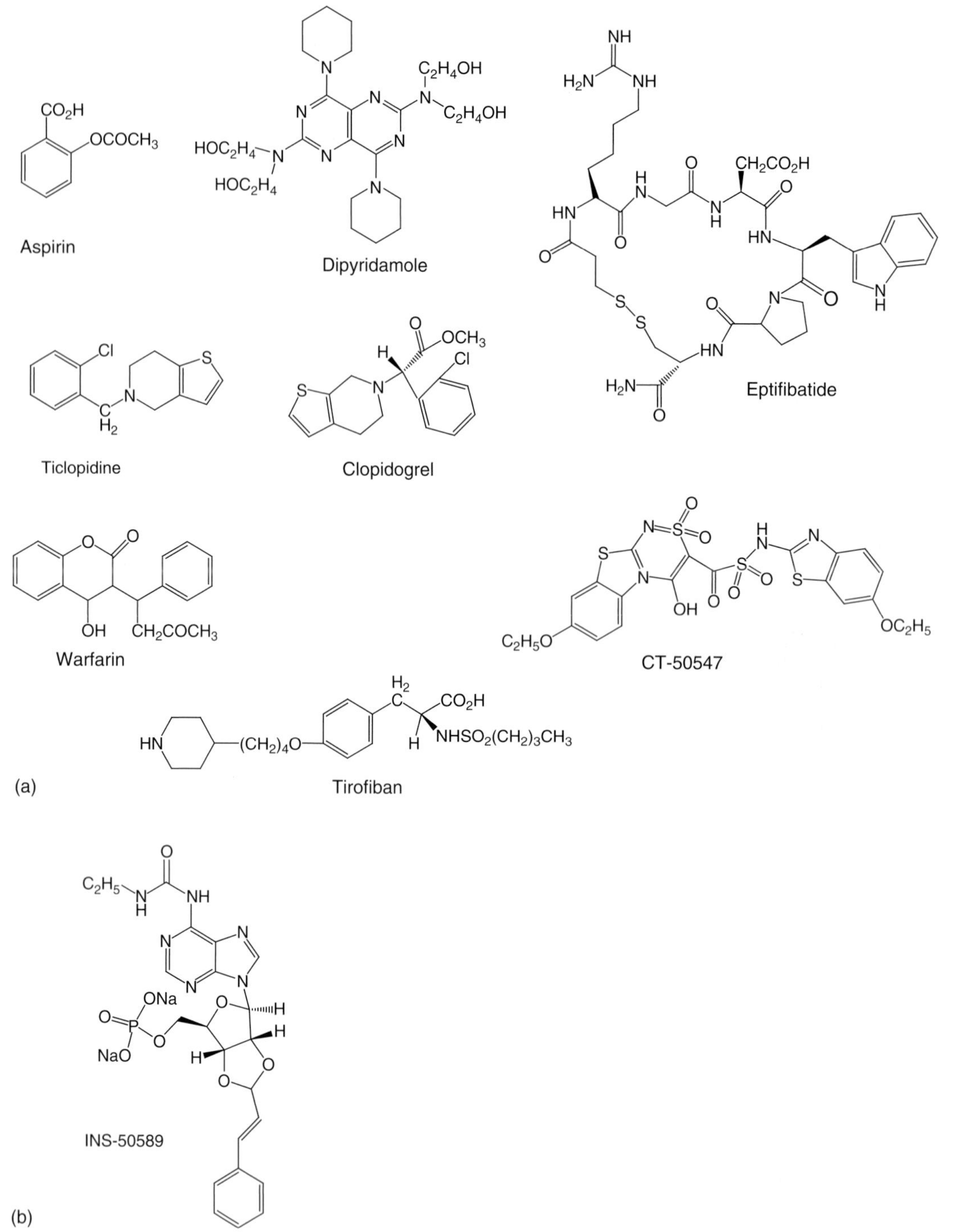

Aspirin

Dipyridamole

Eptifibatide

Ticlopidine

Clopidogrel

Warfarin

CT-50547

(a)

Tirofiban

INS-50589

(b)

Figure 4 Thrombolytics.

Angiographic observations suggest that abciximab facilitates dissolution of intracoronary thrombi without a significant risk of distal embolization.[81] Moreover, abciximab administered in conjunction with low-dose thrombolytic agents, e.g., reteplase, produces a sustained fibrinolytic effect and enhanced early reperfusion in patients with AMI with a favorable hemorrhagic profile as compared to full-dose thrombolytic therapy.[82] However, larger mortality trials need to be performed to provide definitive evidence as to whether such combination therapies improve mortality without increasing the risk of intracranial hemorrhage to an unacceptable level.[37]

6.34.7.5 $P2Y_{12}$ Antagonists

Ticlopidine and clopidogrel (**Figure 4**) are prototypic antagonists of the platelet $P2Y_{12}$ receptor that is involved in the inhibition of platelet function by selectively blocking ADP-induced platelet aggregation. These compounds were in clinical use long before the $P2Y_{12}$ receptor was identified and cloned. In patients with unstable angina, ticlopidine reduced endpoints that included nonfatal and fatal myocardial infarction and any cause of cardiovascular death in events at 6 months (13.6% placebo, 7.3% ticlopidine) with a noted trend toward mortality benefit as well. Clopidogrel produces platelet inhibition in a shorter period than ticlopidine, has fewer severe adverse effects, requires only once per day dosing, and is cheaper than a twice per day regimen of ticlopidine. Newer $P2Y_{12}$ receptor antagonists include CT-50547 and INS-50589 (**Figure 4**) both of which are in the preclinical stage.

6.34.8 Future Directions

Older agents, though effective, still have limitations such as risks of intracranial hemorrhage and 5–15% reocclusion rate. Ongoing efforts to develop newer thrombolytics that overcome the limitations seen with older agents are underway. Ideally, new thrombolytic agents should: (1) be fibrin specific; (2) be directed to newly formed fibrin without affecting normal hemostasis; (3) be nonantigenic; (4) be cost effective; and (5) have rapid onset. One such agent is BB-10153, an engineered variant of human plasminogen that is modified to be activated to plasmin by thrombin (a.k.a thrombin-activatable plasminogen). It was designed to act as a prodrug, persisting in the blood and activating only plasmin in fresh or forming thrombi. In essence, it should only act on clot-bound thrombin. Consequently, thrombus dissolution may be achieved without systemic destruction of hemostatic proteins, thus potentially reducing the risk of haemorrhage. The plasma half-life of BB-10153 was found to be 3–4 h and it also had no effect on plasma α_2-antiplasmin or fibrinogen levels, coagulation assays, or bleeding time. The long half-life and thrombus-selective thrombolytic activity of BB-10153 are promising and might allow it to overcome the bleeding and reocclusion shortfalls in the performance of current thrombolytics.[83,84]

Essentially, development of newer agents and an approach using a combination of pharmacological and mechanical strategies would increase the rates of recannalization, enhance the reversal of acute coronary syndromes, and improve patient outcomes.

References

1. Blann, A. D.; Landray, M. J.; Lip, G. Y. *Br. Med. J.* **2002**, *325*, 762–765.
2. Bizjak, E. D.; Mauro, V. F. *Ann. Pharmacother.* **1998**, *32*, 769–784.
3. Davoren, J. B. Blood Disorders. In *Pathophysiology of Disease: An Introduction to Clinical Medicine*; McPhee, S. J., Lingappa, V. R., Ganong, W. F., Eds.; The McGraw-Hill Companies, Inc.: New York, 2003, pp 113–142.
4. Fedan, J. S. Anticoagulant, Antiplatelet, and Fibrinolytic (Thrombolytic) Drugs. In *Modern Pharmacology with Clinical Applications*; Craig, C. R., Stitzel, R. E., Eds.; Lippincott Williams & Wilkins: Philadelphia, PA, 2004, pp 256–267.
5. Virchow, R. *A M von Meidinger Sohn.* **1856**, 520–525.
6. Fuster, V.; Badimon, L.; Badimon, J. J.; Chesebro, J. H. *N. Engl. J. Med.* **1992**, *326*, 242–250.
7. Fuster, V.; Badimon, L.; Badimon, J. J.; Chesebro, J. H. *N. Engl. J. Med.* **1992**, *326*, 310–318.
8. Fuster, V.; Fayad, Z. A.; Badimon, J. J. *Lancet* **1999**, *353*, SII5–SII9.
9. Fuster, V.; Gotto, A. M.; Libby, P.; Loscalzo, J.; McGill, H. C. *J. Am. Coll. Cardiol.* **1996**, *27*, 964–976.
10. Rauch, U.; Osende, J. I.; Fuster, V.; Badimon, J. J.; Fayad, Z.; Chesebro, J. H. *Ann. Intern. Med.* **2001**, *134*, 224–238.
11. Davies, M. J. *Thromb. Res.* **1996**, *82*, 1–32.
12. Stary, H. C. *Arteriosclerosis* **1989**, *9*, I19–I32.
13. Stary, H. C.; Blankenhorn, D. H.; Chandler, A. B.; Glagov, S.; Insull, W., Jr.; Richardson, M.; Rosenfeld, M. E.; Schaffer, S. A.; Schwartz, C. J.; Wagner, W. D. et al. *Circulation* **1992**, *85*, 391–405.
14. Wissler, R. W. *Ann. NY Acad. Sci.* **1991**, *623*, 26–39.
15. Toschi, V.; Gallo, R.; Lettino, M.; Fallon, J. T.; Gertz, S. D.; Fernandez-Ortiz, A.; Chesebro, J. H.; Badimon, L.; Nemerson, Y.; Fuster, V. et al. *Circulation* **1997**, *95*, 594–599.
16. Mallat, Z.; Hugel, B.; Ohan, J.; Leseche, G.; Freyssinet, J. M.; Tedgui, A. *Circulation* **1999**, *26*, 348–353.
17. Turitto, V. T.; Hall, C. L. *Thromb. Res.* **1998**, *92*, S25–S31.
18. Loscalzo, J. *Arteriosclerosis* **1990**, *10*, 672–679.
19. Brook, J. G.; Aviram, M. *Semin. Thromb. Hemost.* **1988**, *14*, 258–265.
20. Collen, D.; Lijnen, H. R. *Crit. Rev. Oncol. Hematol.* **1986**, *4*, 249–301.
21. Christ, G.; Hufnagl, P.; Kaun, C.; Mundigler, G.; Laufer, G.; Huber, K.; Wojta, J.; Binder, B. R. *Arterioscler. Thromb. Vasc. Biol.* **1997**, *17*, 723–730.
22. Bergmann, S. R.; Fox, K. A.; Ter-Pogossian, M. M.; Sobel, B. E.; Collen, D. *Science* **1983**, *220*, 1181–1183.
23. Van de Werf, F.; Bergmann, S. R.; Fox, K. A.; de Geest, H.; Hoyng, C. F.; Sobel, B. E.; Collen, D. *Circulation* **1984**, *69*, 605–610.
24. Gold, H. K.; Fallon, J. T.; Yasuda, T.; Leinbach, R. C.; Khaw, B. A.; Newell, J. B.; Guerrero, J. L.; Vislosky, F. M.; Hoyng, C. F.; Grossbard, E. et al. *Circulation* **1984**, *70*, 700–707.
25. Flameng, W.; Van de Werf, F.; Vanhaecke, J.; Verstraete, M.; Collen, D. *J. Clin. Invest.* **1985**, *75*, 84–90.

26. Gu, S.; Ducas, J.; Patton, J. N.; Greenberg, D.; Prewitt, R. M. *Chest* **1992**, *101*, 1684–1690.
27. Yasuda, T.; Gold, H. K.; Leinbach, R. C.; Saito, T.; Guerrero, J. L.; Jang, I. K.; Holt, R.; Fallon, J. T.; Collen, D. *J. Am. Coll. Cardiol.* **1990**, *16*, 1728–1735.
28. Rapold, H. J.; Wu, Z. M.; Stassen, T.; Van de Werf, F.; Collen, D. *Blood* **1990**, *76*, 1558–1563.
29. Hennan, J. K.; Elokdah, H.; Leal, M.; Ji, A.; Friedrichs, G. S.; Morgan, G. A.; Swillo, R. E.; Antrilli, T. M.; Hreha, A.; Crandall, D. L. *J. Pharmacol. Exp. Ther.* **2005**, *314*, 710–716.
30. Feied, C.; Handler, J. A. Thromboytic Therapy. *Emergency Medicine*, September 7, 2004
31. Fletcher, A. P.; Alkjaersig, N.; Smyrniotis, F. E.; Sherry, S. *Trans. Assoc. Am. Physicians* **1958**, *71*, 287–296.
32. Boucek, R. J.; Murphy, W. P., Jr. *Am. J. Cardiol.* **1960**, *6*, 525–533.
33. Chazov, E. I.; Matveeva, L. S.; Mazaev, A. V.; Sargin, K. E.; Sadovskaia, G. V.; Ruda, M. I. *Ter Arkh* **1976**, *48*, 8–19.
34. Gruppo Italiano per lo Studio della Streptochinasi nell'Infarto Miocardico (gissi). *Lancet* **1986**, *1*, 397–402.
35. Khan, I. A.; Gowda, R. M. *Int. J. Cardiol.* **2003**, *91*, 115–127.
36. Alexandrov, A. V.; Masdeu, J. C.; Devous, M. D., Sr.; Black, S. E.; Grotta, J. C. *Stroke* **1997**, *28*, 1830–1834.
37. Ohman, E. M.; Harrington, R. A.; Cannon, C. P.; Agnelli, G.; Cairns, J. A.; Kennedy, J. W. *Chest* **2001**, *119*, 253S–277S.
38. Majerus, P. W.; Tollefsen, D. M. Anticoagulant, Thrombolytic, and Antiplatelet Drugs. In *Goodman & Gilman's the Pharmacological Basis of Therapeutics*; McGraw-Hill: New York, 2001, pp 1519–1538.
39. Gulba, D. C.; Bode, C.; Runge, M. S.; Huber, K. *Ann. Hematol.* **1996**, *73*, S9–S27.
40. Maizel, A. S.; Bookstein, J. J. *Cardiovasc. Intervent. Radiol.* **1986**, *9*, 236–244.
41. Wagstaff, A. J.; Gillis, J. C.; Goa, K. L. *Drugs* **1995**, *50*, 289–316.
42. Collen, D.; Lijnen, H. R. Fibrinolysis and the Control of Hemostasis. In *The Molecular Basis of Blood Diseases*; Stamatoyannopoulos, G., Nienhui, A. W., Majerus, P. W., Varmus, H., Eds.; Saunders: Philadelphia, PA, 1994, pp 662–688.
43. Tanswell, P.; Tebbe, U.; Neuhaus, K. L.; Glasle-Schwarz, L.; Wojcik, J.; Seifried, E. *J. Am. Coll. Cardiol.* **1992**, *19*, 1071–1075.
44. Wooster, M. B.; Luzier, A. B. *Ann. Pharmacother.* **1999**, *33*, 318–324.
45. Smalling, R. W.; Bode, C.; Kalbfleisch, J.; Sen, S.; Limbourg, P.; Forycki, F.; Habib, G.; Feldman, R.; Hohnloser, S.; Seals, A. *Circulation* **1995**, *91*, 2725–2732.
46. Davydov, L.; Cheng, J. W. *Clin. Ther.* **2001**, *23*, 982–997, discussion 981.
47. Liberatore, G. T.; Samson, A.; Bladin, C.; Schleuning, W. D.; Medcalf, R. L. *Stroke* **2003**, *34*, 537–543.
48. Kratzschmar, J.; Haendler, B.; Langer, G.; Boidol, W.; Bringmann, P.; Alagon, A.; Donner, P.; Schleuning, W. D. *Gene* **1991**, *105*, 229–237.
49. Bringmann, P.; Gruber, D.; Liese, A.; Toschi, L.; Kratzchmar, J.; Schleuning, W. D.; Donner, P. *J. Biol. Chem.* **1995**, *270*, 25596–25603.
50. Toschi, L.; Bringmann, P.; Petri, T.; Donner, P.; Schleuning, W. D. *Eur. J. Biochem.* **1998**, *252*, 108–112.
51. Hacke, W.; Albers, G.; Al-Rawi, Y.; Bogousslavsky, J.; Davalos, A.; Eliasziw, M.; Fischer, M.; Furlan, A.; Kaste, M.; Lees, K. R. et al. *Stroke* **2005**, *36*, 66–73.
52. Mellott, M. J.; Stabilito, I. I.; Holahan, M. A.; Cuca, G. C.; Wang, S.; Li, P.; Barrett, J. S.; Lynch, J. J.; Gardell, S. J. *Arterioscler Thromb.* **1992**, *12*, 212–221.
53. Franzosi, M. G.; Santoro, E.; De Vita, C.; Geraci, E.; Lotto, A.; Maggioni, A. P.; Mauri, F.; Rovelli, F.; Santoro, L.; Tavazzi, L. et al. *Circulation* **1998**, *98*, 2659–2665.
54. Isis-2 (Second International Study of Infarct Survival) Collaborative Group. *Lancet* **1988**, *2*, 349–360.
55. Baigent, C.; Collins, R.; Appleby, P.; Parish, S.; Sleight, P.; Peto, R. *Br. Med. J.* **1998**, *316*, 1337–1343.
56. Gruppo Italiano per lo Studio della Sopravvivenza nell'Infarto Miocardico. *Lancet* **1990**, *336*, 65–71.
57. The International Study Group. *Lancet* **1990**, *336*, 71–75.
58. Gissi-2 and International Study Group: Gruppo Italiano per lo Studio della Sopravvivenza nell'Infarto. *Eur. Heart J.* **1992**, *13*, 1692–1697.
59. Isis-3 (Third International Study of Infarct Survival) Collaborative Group. *Lancet* **1992**, *339*, 753–770.
60. Califf, R. M.; White, H. D.; Van de Werf, F.; Sadowski, Z.; Armstrong, P. W.; Vahanian, A.; Simoons, M. L.; Simes, R. J.; Lee, K. L.; Topol, E. J. *Circulation* **1996**, *94*, 1233–1238.
61. The Continuous Infusion versus Double-Bolus Administration of Alteplase (Cobalt) Investigators. *N. Engl. J. Med.* **1997**, *337*, 1124–1130.
62. Fibrinolytic Therapy Trialists' (ftt) Collaborative Group. *Lancet* **1994**, *343*, 311–322.
63. Bode, C.; Smalling, R. W.; Berg, G.; Burnett, C.; Lorch, G.; Kalbfleisch, J. M.; Chernoff, R.; Christie, L. G.; Feldman, R. L.; Seals, A. A. et al. *Circulation* **1996**, *94*, 891–898.
64. The Global Use of Strategies to Open Occluded Coronary Arteries (Gusto iii) Investigators. *N. Engl. J. Med.* **1997**, *337*, 1118–1123.
65. International Joint Efficacy Comparison of Thrombolytics. *Lancet* **1995**, *346*, 329–336.
66. Cannon, G. P.; McCabe, C. H.; Gibson, M.; The TIMI 10 Investigators. *Proc. Natl. Acad. Sci. USA* **1994**, *95*, 351.
67. Cannon, C. P.; Gibson, C. M.; McCabe, C. H.; Adgey, A. A.; Schweiger, M. J.; Sequeira, R. F.; Grollier, G.; Giugliano, R. P.; Frey, M.; Mueller, H. S. et al. *Circulation* **1998**, *98*, 2805–2814.
68. Van de Werf, F.; Cannon, C. P.; Luyten, A.; Houbracken, K.; McCabe, C. H.; Berioli, S.; Bluhmki, E.; Sarelin, H.; Wang-Clow, F.; Fox, N. L. et al. *Am. Heart J.* **1999**, *137*, 786–791.
69. Assessment of the Safety and Efficacy of a new Thrombolytic Investigators. *Lancet* **1999**, *354*, 716–722.
70. The 22nd Congress of the European Society of Cardiology, Amsterdam, September 2000.
71. The Assent-3 Randomised Trial in Acute Myocardial Infarction. *Lancet* **2001**, *358*, 605–613.
72. den Heijer, P.; Vermeer, F.; Ambrosioni, E.; Sadowski, Z.; Lopez-Sendon, J. L.; von Essen, R.; Beaufils, P.; Thadani, U.; Adgey, J.; Pierard, L. et al. *Circulation* **1998**, *98*, 2117–2125.
73. Tebbe, U.; Michels, R.; Adgey, J.; Boland, J.; Caspi, A.; Charbonnier, B.; Windeler, J.; Barth, H.; Groves, R.; Hopkins, G. R. et al. *J. Am. Coll. Cardiol.* **1998**, *31*, 487–493.
74. Van de Werf, F.; Ardissino, D.; Betriu, A.; Cokkinos, D. V.; Falk, E.; Fox, K. A.; Julian, D.; Lengyel, M.; Neumann, F. J.; Ruzyllo, W. et al. *Eur. Heart J.* **2003**, *24*, 28–66.
75. Zhao, L.; Gray, L.; Leonardi-Bee, J.; Weaver, C. S.; Heptinstall, S.; Bath, P. M. *Platelets* **2006**, *17*, 100–104.
76. Coller, B. S.; Folts, J. D.; Smith, S. R.; Scudder, L. E.; Jordan, R. *Circulation* **1989**, *80*, 1766–1774.
77. Lefkovits, J.; Plow, E. F.; Topol, E. J. *N. Engl. J. Med.* **1995**, *332*, 1553–1559.
78. Coller, B. S. *Thromb. Haemost.* **1997**, *78*, 730–735.
79. Davies, C. H.; Ormerod, O. J. *Lancet* **1998**, *351*, 1191–1196.

80. Kong, D. F.; Califf, R. M.; Miller, D. P.; Moliterno, D. J.; White, H. D.; Harrington, R. A.; Tcheng, J. E.; Lincoff, A. M.; Hasselblad, V.; Topol, E. J. *Circulation* **1998**, *98*, 2829–2835.
81. Antman, E. M.; Giugliano, R. P.; Gibson, C. M.; McCabe, C. H.; Coussement, P.; Kleiman, N. S.; Vahanian, A.; Adgey, A. A.; Menown, I.; Rupprecht, H. J. et al. *Circulation* **1999**, *99*, 2720–2732.
82. Califf, R. M. *Am. Heart J.* **2000**, *139*, S33–S37.
83. Comer, M. B.; Cackett, K. S.; Gladwell, S.; Wood, L. M.; Dawson, K. M. *J. Thromb. Haemost.* **2005**, *3*, 146–153.
84. Curtis, L. D.; Brown, A.; Comer, M. B.; Senior, J. M.; Warrington, S.; Dawson, K. M. P. *J. Thromb. Haemost.* **2005**, *3*, 1180–1186.

Biographies

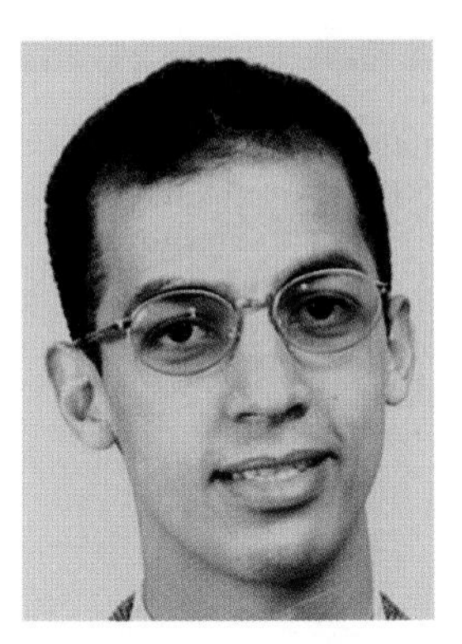

Ahmed E El-Gengaihy, MD, MSc is currently a Chief Neurology Resident at the University of Medicine and Dentistry of New Jersey (UMDNJ). He has completed his Neuropsychiatry residency at Ain Shams University, Cairo, Egypt. He subsequently joined Baylor College of Medicine for 1 year as a research fellow followed by his current residency training at UMDNJ. His research interests include neuroimaging of cerebrovascular diseases, neurocritical care, and basic and clinical aspects of interventional cerebrovascular diseases.

Samir I Abdelhadi, DO, is currently a second year resident of the Internal Medicine Program at the University of Medicine and Dentistry of New Jersey (UMDNJ). He hopes to obtain a Cardiology fellowship at the end of his residency training and is currently doing research within the Cardiology program at UMDNJ.

Jawad F Kirmani, MD, is currently an assistant professor with the stroke interventional team at the University of Medicine and Dentistry of New Jersey (UMDNJ). He completed his Neurology residency at the Ohio State University. He subsequently joined the State University of New York at Buffalo for a 2-year fellowship in neurocritical care and cerebrovascular diseases followed by two further years of training in interventional neurology at UMDNJ. His research interests include neuroimaging of cerebrovascular diseases, neurocritical care, neurocardiology, and basic and clinical aspects of interventional cerebrovascular diseases.

Adnan I Qureshi is Professor of Neurology and Neurosciences and Director of the Cerebrovascular Program at the University of Medicine and Dentistry of New Jersey (UMDNJ). Dr Qureshi is recognized as an international leader in stroke research and acute stroke management by physicians from multiple disciplines. He has written over 160 scientific publications in prestigious journals including the *New England Journal of Medicine*, *Lancet*, *Archives of Internal Medicine*, *Critical Care Medicine*, *Neurology*, *Stroke*, and *Circulation*. In addition, he has made over 500 presentations in various national and international meetings. He has served as an invited speaker at numerous national and international forums. He has also been invited as visiting professor to universities in the United States and abroad. He serves on editorial boards for several peer-review journals and guideline committees for American Heart Association and American Society of Neuroimaging. He is the present chair of the interventional section of the American Academy of Neurology and a Fellow of the American Heart Association. He is also the chair of the research committee of the American Society of Interventional and Therapeutic Neuroradiology. He was one of the founders of the subspecialty of interventional neurology.

Dr Qureshi's work has led to the development of new concepts for therapies and research pertaining to prevention of stroke. The work has importance for the general population and has appeared in numerous public information sources such as ABC, NBC, American Heart Association News, American Medical Association News, Web MD, Men's Health, and USA Today. His work has impacted upon the management of stroke by other physicians and incorporation into practice guidelines. Dr Qureshi has received several prestigious awards at the national level including the AMA Foundation Leadership award, the AAN-AB Baker Teacher Recognition award, NAPH Safety Net Community and Patient Service Award, Excellence in Care Award, the AHA/NCQA Achievement of Recognition for Delivery of Quality Stroke Care, and AMA Community Service award. He has been selected by his peers as one of *The Best Doctors in America* and was selected by the Consumer Research Council of America as one of *America's Top Physicians*.

Most recently, he laid the foundation of the Zeenat Qureshi Stroke Research Center at University of Medicine and Dentistry of New Jersey. Since its inauguration, the center has led the way in cutting edge research in epidemiology, clinical trials, and basic research pertaining to cerebrovascular diseases. Several investigators in the center receive funding through National Institutes of Health, American Heart Association, and Armed Forces Research Initiatives. Dr Qureshi has mentored several fellows and medical students. He has trained several clinical fellows in cerebrovascular diseases and endovascular procedures and is presently the program director for ACGME accredited vascular neurology program at UMDNJ. Several of his students and fellows have won national awards for research performed under his direct supervision. He continues to serve as course director for various educational seminars at a regional and national level. He is married to Aasma Qureshi who is presently completing her fellowship in Gastroenterology at Emory University in Atlanta.

Comprehensive Medicinal Chemistry II
ISBN (set): 0-08-044513-6
ISBN (Volume 6) 0-08-044519-5; pp. 763–781

INDEX FOR VOLUME 6

Notes

Abbreviations

SAR – structure–activity relationships

Cross-reference terms in italics are general cross-references, or refer to subentry terms within the main entry (the main entry is not repeated to save space). Readers are also advised to refer to the end of each article for additional cross-references – not all of these cross-references have been included in the index cross-references.

The index is arranged in set-out style with a maximum of three levels of heading. Major discussion of a subject is indicated by bold page numbers. Page numbers suffixed by T and F refer to Tables and Figures respectively. *vs.* indicates a comparison.

This index is in letter-by-letter order, whereby hyphens and spaces within index headings are ignored in the alphabetization. Prefixes and terms in parentheses are excluded from the initial alphabetization.

Any method, model or other subject, associated with the name of the developer (e.g. name's model) does NOT imply that Elsevier, nor the indexers, have assumed the right to name models/methods after the authors of the papers in which they are described. This is merely a succinct phrase to refer to a model/method developed/described by the relevant author, so that the subentry could be alphabetized under the most pertinent name.

A

B

C

D

E

F

G

H

M

N

O

Q

R

S

T

W

X

Y

Z